AutoCAD
and Its Applications

COMPREHENSIVE

2010

by

Terence M. Shumaker
Faculty Emeritus
Former Chairperson
Drafting Technology
Autodesk Premier Training Center
Clackamas Community College, Oregon City, Oregon

David A. Madsen
President, Madsen Designs Inc.
Faculty Emeritus, Former Department Chair Drafting Technology
Autodesk Premier Training Center
Clackamas Community College, Oregon City, Oregon
Director Emeritus, American Design Drafting Association

David P. Madsen
Vice President, Madsen Designs Inc.
Computer-Aided Design and Drafting Consultant and Educator
Autodesk Developer Network Member
American Design Drafting Association Member

Jeffrey A. Laurich
Instructor, Mechanical Design Technology
Fox Valley Technical College
Appleton, WI

Publisher
The Goodheart-Willcox Company, Inc.
Tinley Park, Illinois
www.g-w.com

The Goodheart-Willcox Company, Inc. Brand Disclaimer: Brand names, company names, and illustrations for products and services included in this text are provided for educational purposes only and do not represent or imply endorsement or recommendation by the author or the publisher.

The Goodheart-Willcox Company, Inc. Safety Notice: The reader is expressly advised to carefully read, understand, and apply all safety precautions and warnings described in this book or that might also be indicated in undertaking the activities and exercises described herein to minimize risk of personal injury or injury to others. Common sense and good judgment should also be exercised and applied to help avoid all potential hazards. The reader should always refer to the appropriate manufacturer's technical information, directions, and recommendations; then proceed with care to follow specific equipment operating instructions. The reader should understand these notices and cautions are not exhaustive.

The publisher makes no warranty or representation whatsoever, either expressed or implied, including but not limited to equipment, procedures, and applications described or referred to herein, their quality, performance, merchantability, or fitness for a particular purpose. The publisher assumes no responsibility for any changes, errors, or omissions in this book. The publisher specifically disclaims any liability whatsoever, including any direct, indirect, incidental, consequential, special, or exemplary damages resulting, in whole or in part, from the reader's use or reliance upon the information, instructions, procedures, warnings, cautions, applications, or other matter contained in this book. The publisher assumes no responsibility for the activities of the reader.

Cover Source: © Hervé Hughes/Hemis/Corbis

Library of Congress Cataloging-in-Publication Data

Shumaker, Terence M.
  AutoCAD and its applications. Comprehensive 2010 / by Terence M.
  Shumaker... [et al.]. -- 17th ed.
    p. cm.
  Includes bibliographical references and index.
  ISBN: 978-1-60525-163-9
  1. Computer graphics. 2. AutoCAD. I. Shumaker, Terence M.
  T385.S4616445 2010
  620'.00420285536--dc22
                          2009004710

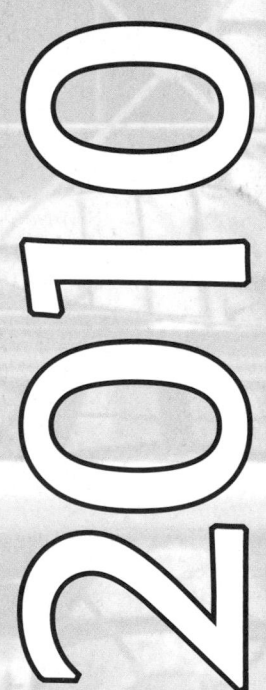

AutoCAD
and Its Applications

BASICS

by

Terence M. Shumaker
Faculty Emeritus
Former Chairperson
Drafting Technology
Autodesk Premier Training Center
Clackamas Community College, Oregon City, Oregon

David A. Madsen
President, Madsen Designs Inc.
Faculty Emeritus, Former Department Chairperson Drafting Technology
Autodesk Premier Training Center
Clackamas Community College, Oregon City, Oregon
Director Emeritus, American Design Drafting Association

David P. Madsen
Vice President, Madsen Designs Inc.
Computer-Aided Design and Drafting Consultant and Educator
Autodesk Developer Network Member
American Design Drafting Association Member

2010

Publisher
The Goodheart-Willcox Company, Inc.
Tinley Park, Illinois
www.g-w.com

The Goodheart-Willcox Company, Inc. Brand Disclaimer: Brand names, company names, and illustrations for products and services included in this text are provided for educational purposes only and do not represent or imply endorsement or recommendation by the author or the publisher.

The Goodheart-Willcox Company, Inc. Safety Notice: The reader is expressly advised to carefully read, understand, and apply all safety precautions and warnings described in this book or that might also be indicated in undertaking the activities and exercises described herein to minimize risk of personal injury or injury to others. Common sense and good judgment should also be exercised and applied to help avoid all potential hazards. The reader should always refer to the appropriate manufacturer's technical information, directions, and recommendations; then proceed with care to follow specific equipment operating instructions. The reader should understand these notices and cautions are not exhaustive.

The publisher makes no warranty or representation whatsoever, either expressed or implied, including but not limited to equipment, procedures, and applications described or referred to herein, their quality, performance, merchantability, or fitness for a particular purpose. The publisher assumes no responsibility for any changes, errors, or omissions in this book. The publisher specifically disclaims any liability whatsoever, including any direct, indirect, incidental, consequential, special, or exemplary damages resulting, in whole or in part, from the reader's use or reliance upon the information, instructions, procedures, warnings, cautions, applications, or other matter contained in this book. The publisher assumes no responsibility for the activities of the reader.

Cover Source: Dag Sundberg/GettyImages

Library of Congress Cataloging-in-Publication Data

Shumaker, Terence M.
    AutoCAD and Its Applications: BASICS 2010 / by Terence M.
Shumaker, David A. Madsen, David P. Madsen. – 17th ed.
        p. cm.

    Includes bibliographical references and index.
    ISBN 978-1-60525-161-5
    1. Computer graphics. 2. AutoCAD. I. Madsen, David A.
II. Madsen, David P.

T385.S461466 2010
    620'.00420285536--dc22   2009004709

Introduction

AutoCAD and Its Applications—Basics is a textbook providing complete instruction in mastering fundamental AutoCAD® 2010 tools and drawing techniques. Typical applications of AutoCAD are presented with basic drafting and design concepts. The topics are covered in an easy-to-understand sequence and progress in a way that allows you to become comfortable with the tools as your knowledge builds from one chapter to the next. In addition, *AutoCAD and Its Applications—Basics* offers the following features:

- Step-by-step use of AutoCAD tools.
- In-depth explanations of how and why tools function as they do.
- Extensive use of font changes to specify certain meanings.
- Examples and descriptions of industry practices and standards.
- Screen captures of AutoCAD features and functions.
- Professional tips explaining how to use AutoCAD effectively and efficiently.
- More than 250 exercises to reinforce the chapter topics. These exercises also build on previously learned material.
- Chapter tests for review of tools and key AutoCAD concepts.
- A large selection of drafting problems supplementing each chapter. Problems are presented as industrial drawings, engineering sketches, or architectural, civil, electrical, or other related industry drawings.

With *AutoCAD and Its Applications—Basics*, you learn AutoCAD tools and become acquainted with information in other areas:

- Preliminary planning and sketches.
- Drawing geometric shapes and constructions.
- Parametric drawing techniques.
- Special editing operations that increase productivity.
- Placing text and tables according to accepted industry practices.
- Making multiview drawings (orthographic projection).
- Dimensioning techniques and practices, based on accepted standards.
- Drawing section views and designing graphic patterns.
- Creating shapes and symbols.
- Creating and managing symbol libraries.
- Plotting and printing drawings.

Learning Objectives identify key items you will learn in the chapter.

Command Entry Graphics show ribbon, Application Menu, and keyboard entry options. Tool options are also shown where applicable.

Professional Tips increase your productivity in using AutoCAD tools and techniques.

Illustrations, including AutoCAD "screen shots," line art illustrations, and illustrated tables, make learning easy.

CHAPTER 18

Linear and Angular Dimensioning

Learning Objectives

After completing this chapter, you will be able to do the following:

✓ Add linear dimensions to a drawing.
✓ Add angular dimensions to a drawing.
✓ Draw datum and chain dimensions.
✓ Add dimensions for multiple items using the **QDIM** tool.

A drawing often requires a variety of dimensions to describe the size and shape of features and objects. Linear and angular dimensions are two of the most common. This chapter covers the process of adding linear and angular dimensions to a drawing using several dimensioning tools. You will also learn how to add a break symbol to a dimension line and use the **QDIM** tool.

Placing Linear Dimensions

Linear dimensions usually measure straight distances, such as distances between horizontal, vertical, or slanted surfaces. The **DIMLINEAR** tool allows you to place linear dimensions.

Dimension tools reference the current dimension style and the points or objects you select to create a single dimension object. When you use the **DIMLINEAR** tool, for example, you create a dimension object that includes all related dimension style characteristics, dimension and extension lines, arrowheads, and a dimension value associated with the distance between selected points.

Once you access the **DIMLINEAR** tool, pick a point to locate origin of the first extension line, and then pick a point to locate the origin of the second extension line. See **Figure 18-1.** Use object snap modes and other drawing aids to pick the exact points where extension lines begin. Once you establish the extension line origins, you can select from several options that appear at the Specify dimension line location or [Mtext/Text/Angle/Horizontal/Vertical/Rotated] prompt. To apply the default option and create a linear dimension, move the dimension line to the desired location and pick. See **Figure 18-2.**

Ribbon
Home
> Annotation
Annotate
> Dimensions

Linear

Type
DIMLINEAR
DLI

DIMLINEAR

475

Controlling Draw Order

Drawings often include overlapping objects. The overlap is difficult to see when all objects use a thin lineweight and when lineweight display is off. Controlling display order is better illustrated with an object that has width, such as the donuts shown in **Figure 6-8.** In this example, the drawing order of the donuts, and all other objects, to place the items above or below selected objects and to the front or back of all objects. You can

Use the **DRAWORDER** tool to change the order of objects in a drawing. You can also set draw order by picking an object to select it, right-clicking, and choosing **Draw Order. Figure 6-9** describes the options for changing draw order.

Ribbon
> Modify

Bring to Front

Type
DRAWORDER
DR

DRAWORDER

NOTE

You may need to use the **DRAWORDER** tool on several objects until the objects display correctly. Objects move to the front of the drawing when they are modified.

PROFESSIONAL TIP

Use **DRAWORDER** to help display and select objects that are hidden by other objects.

Figure 6-8.
The order of objects can be changed to place any object under or above other objects.

Figure 6-9.
Four options are available for rearranging the order of objects in a drawing.

Option	Function
Bring to Front	Places the selected objects at the front of the drawing.
Send to Back	Places the selected objects at the back of the drawing.
Bring Above Objects	Moves the selected objects above the reference object.
Send Under Objects	Moves the selected objects below the reference object.

183

Chapter 6 View Tools

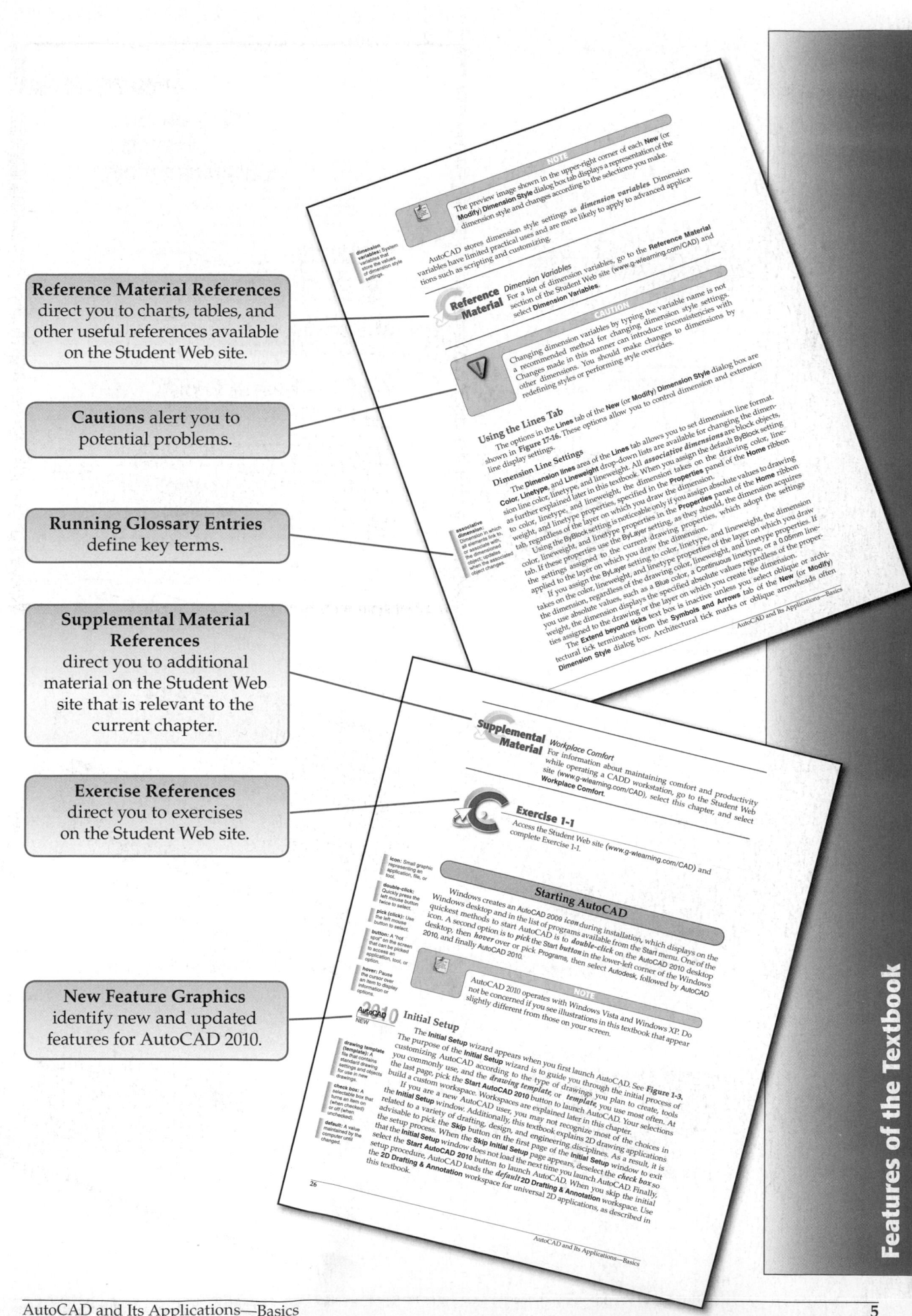

Reference Material References direct you to charts, tables, and other useful references available on the Student Web site.

Cautions alert you to potential problems.

Running Glossary Entries define key terms.

Supplemental Material References direct you to additional material on the Student Web site that is relevant to the current chapter.

Exercise References direct you to exercises on the Student Web site.

New Feature Graphics identify new and updated features for AutoCAD 2010.

Features of the Textbook

Notes explain important aspects of a topic.

Express Tool References direct you to Express Tool material on the Student Web site.

Template Development References direct you to Template Development material on the Student Web site.

Chapter Tests reinforce the knowledge gained by reading the chapter and completing the exercises.

Figure 10-15.
The **Existing** option of the **SCALETEXT** tool scales text objects using their individual justification settings.

BL Justification	BL Justification
MC Justification	MC Justification
TR Justification	TR Justification
Original Text	**Text Scaled Using Existing Base Point Option**

points using the **Existing** option. Notice how the text scales in relation to its own justification setting.

After you specify the justification to use as the base point, AutoCAD prompts for the scaling type. The default **Specify new model height** option allows you to type a new value for the text height of non-annotative objects. If the selected text is annotative, the value you enter is ignored. Use the **Paper height** option to type a new paper text height value for the text height of annotative objects. If the selected text is non-annotative, the value you enter is ignored.

The **Match object** option allows you to pick an existing text object. The height of the selected text object adopts the text height from the text object you pick. Use the **Scale factor** option to scale text objects that have different heights relative to their current heights. For example, using a scale factor of 2 scales all of the selected text objects to twice their current size.

> **NOTE**
> You should only use the **SCALETEXT** tool to scale non-annotative text.

JUSTIFYTEXT
Ribbon
Annotate
> Text
Justify
Type
JUSTIFYTEXT

Changing Text Justification

Use the **JUSTIFYTEXT** tool to change the justification point without moving the text. Pick the text for which you want to change justification, and enter the new justification option.

Exercise 10-6
Access the Student Web site (www.g-wlearning.com/CAD) and complete Exercise 10-6.

Express Tools
Chapter 10
The **Express Tools** ribbon tab includes additional tools for improved functionality and productivity during the drawing processes. The following Express Tools represent the most useful text express tools. For information about these tools, go to the Student Web site (www.g-wlearning.com/CAD), select this chapter, and select **Using Text Express Tools**.

Text Fit	Arc-Aligned Text
Text Mask	Enclose Text with Object
Unmask Text	Change Text Case
Convert Text to Mtext	

290   AutoCAD and Its Applications—Basics

Template Development
Chapter 7
For detailed instructions on setting object snaps and polar tracking to save time and increase efficiency, go to the Student Web site (www.g-wlearning.com/CAD), select this chapter, and select **Template Development**.

Chapter Test

Answer the following questions. Write your answers on a separate sheet of paper or go to the Student Web site (www.g-wlearning.com/CAD) and complete the electronic chapter test.

1. Define the term object snap.
2. What is an AutoSnap tooltip?
3. Name the following AutoSnap markers:

A.
B.
C.
D.
E.
F.
G.
H.
I.
J.
K.
L.

4. How do you set running object snaps?
5. Define the term *running object snap.*
6. How do you access the **Drafting Settings** dialog box to change object snap settings?
7. If you are using running object snaps and want to make several point specifications without the aid of object snap, but want to continue the same running object snaps after making the desired point selections, what is the easiest way to turn off the running object snaps temporarily?
8. If you are using running object snaps and you want to make a single point selection without the object snap override.
9. Describe the effects of the running object snaps, what do you do?
10. How do you activate the **Object Snap** shortcut menu?
11. Where are the four quadrant points on a circle?
12. What is the situation when the tooltip reads Extended Intersection?
13. What does it mean when the tooltip reads Deferred Perpendicular?

213

Chapter 7 Object Snaps and AutoTrack

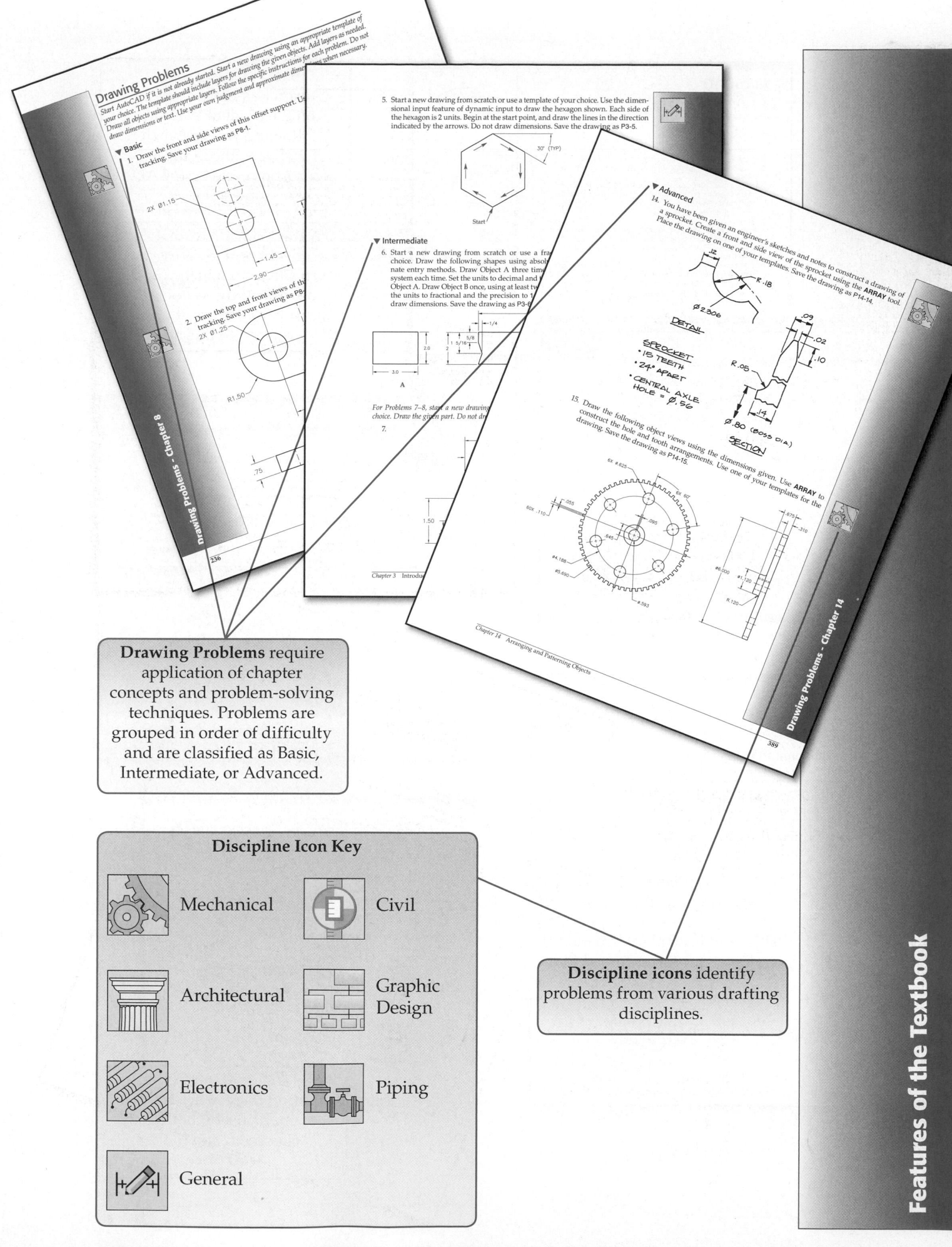

Drawing Problems require application of chapter concepts and problem-solving techniques. Problems are grouped in order of difficulty and are classified as Basic, Intermediate, or Advanced.

Discipline Icon Key

Mechanical

Civil

Architectural

Graphic Design

Electronics

Piping

General

Discipline icons identify problems from various drafting disciplines.

Features of the Textbook

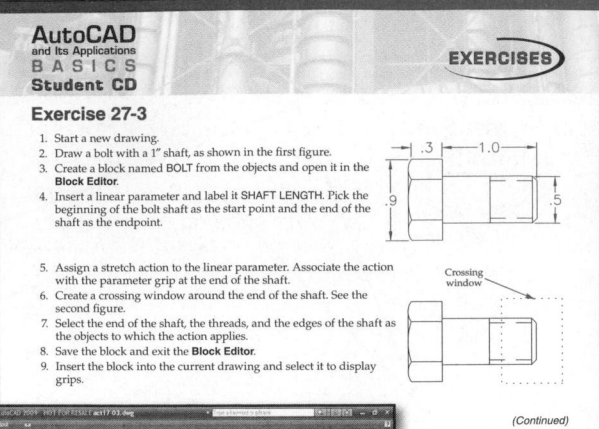

Exercise 27-3

1. Start a new drawing.
2. Draw a bolt with a 1" shaft, as shown in the first figure.
3. Create a block named BOLT from the objects and open it in the **Block Editor**.
4. Insert a linear parameter and label it SHAFT LENGTH. Pick the beginning of the bolt shaft as the start point and the end of the shaft as the endpoint.

5. Assign a stretch action to the linear parameter. Associate the action with the parameter grip at the end of the shaft.
6. Create a crossing window around the end of the shaft. See the second figure.
7. Select the end of the shaft, the threads, and the edges of the shaft as the objects to which the action applies.
8. Save the block and exit the **Block Editor**.
9. Insert the block into the current drawing and select it to display grips.

(Continued)

Exercises. Chapter exercises are provided on the Student Web site, allowing you to switch between the exercise directions and AutoCAD on-screen.

AutoCAD Software. Pick this button to access a Web site from which you can download the AutoCAD Electrical software at no cost for use with this book.

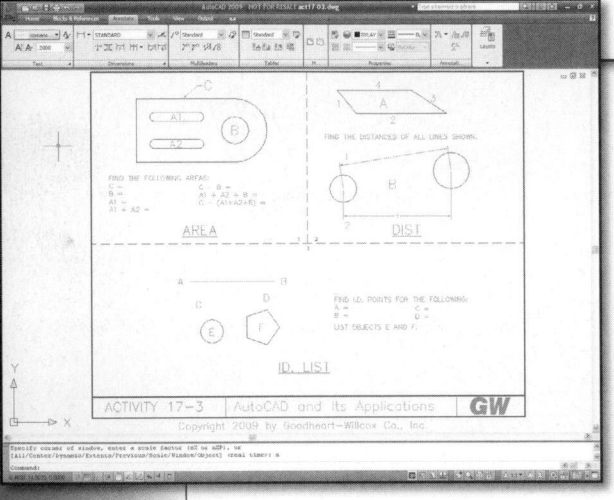

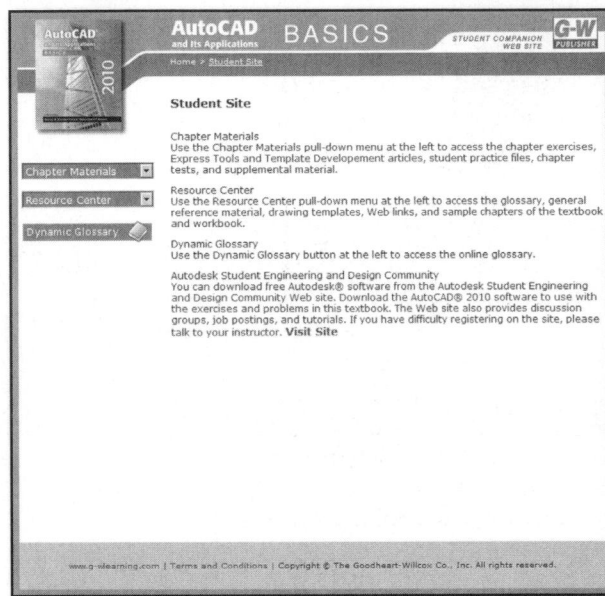

Additional Plotting Options

The **Plot and Publish** tab of the **Options** dialog box contains general plotting settings. Some options are described when applicable throughout the textbook. However, several additional options are also available. See **Figure S29-1**. These settings are seldom changed.

The **Default plot settings for new drawings** area defines the plot settings that are used by default when you access the plot dialog box. The default setting is **Use as default output device**. The device can be selected from the drop-down list. The **Use last successful plot settings** option retains the previous plot settings. Picking the **Add or Configure Plotters** button displays the **Plotters** window.

Figure S29-1.
Items in the shaded areas ra...
them in case you need to ad...

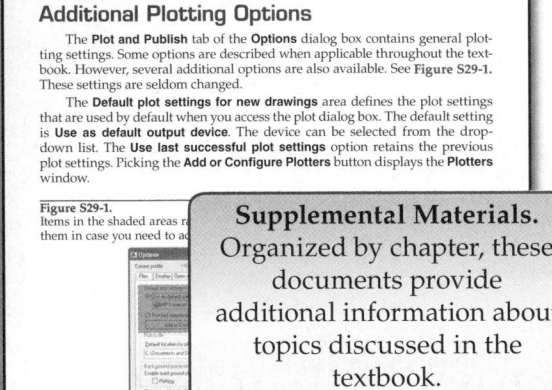

Supplemental Materials. Organized by chapter, these documents provide additional information about topics discussed in the textbook.

Copyright by Goodheart-Willcox Co., Inc.   Additional Plotting Options, page 1

Drafting Symbols

Symbols provide a "common language" for drafters all over the world. However, symbols can be meaningful only if they are created according to the relevant standards or conventions. This document describes and illustrates common dimensioning, GD&T, architectural, piping, and electrical symbols.

Standard Dimensioning Symbols

The size of dimensioning symbols varies with text size, but it should be consistent with the height of the text. In the following illustration, h = text height.

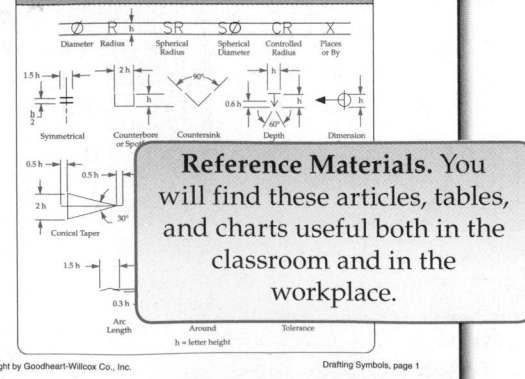

Reference Materials. You will find these articles, tables, and charts useful both in the classroom and in the workplace.

Copyright by Goodheart-Willcox Co., Inc.   Drafting Symbols, page 1

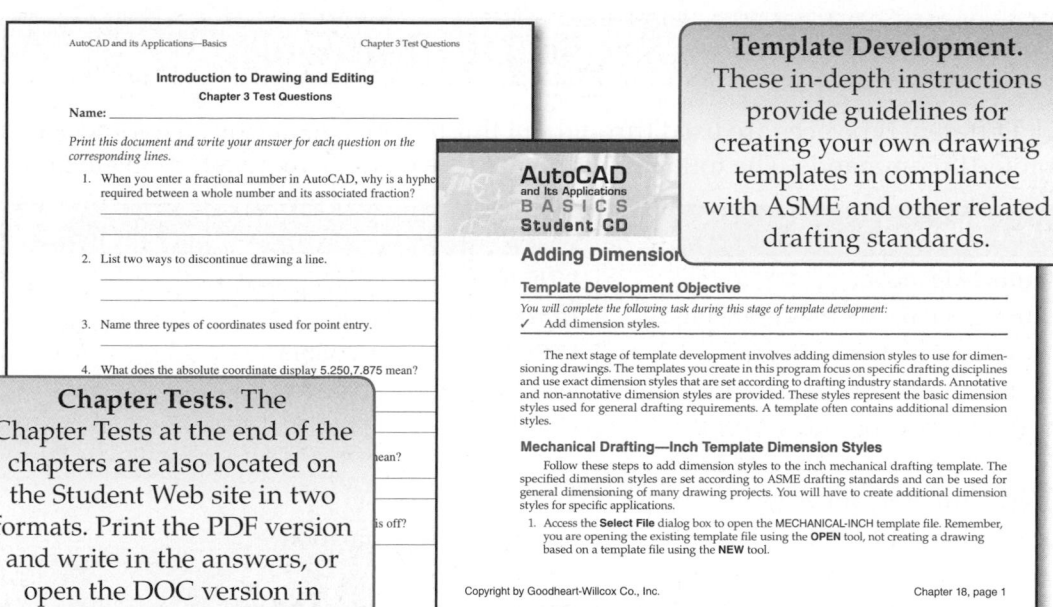

AutoCAD and its Applications—Basics Chapter 3 Test Questions

Introduction to Drawing and Editing

Chapter 3 Test Questions

Name: ______________________________

Print this document and write your answer for each question on the corresponding lines.

1. When you enter a fractional number in AutoCAD, why is a hyphe required between a whole number and its associated fraction?
2. List two ways to discontinue drawing a line.
3. Name three types of coordinates used for point entry.
4. What does the absolute coordinate display 5.250,7.875 mean?

AutoCAD and Its Applications BASICS Student CD

Adding Dimensio

Template Development Objective

You will complete the following task during this stage of template development:

✓ Add dimension styles.

The next stage of template development involves adding dimension styles to use for dimensioning drawings. The templates you create in this program focus on specific drafting disciplines and use exact dimension styles that are set according to drafting industry standards. Annotative and non-annotative dimension styles are provided. These styles represent the basic dimension styles used for general drafting requirements. A template often contains additional dimension styles.

Mechanical Drafting—Inch Template Dimension Styles

Follow these steps to add dimension styles to the inch mechanical drafting template. The specified dimension styles are set according to ASME drafting standards and can be used for general dimensioning of many drawing projects. You will have to create additional dimension styles for specific applications.

1. Access the **Select File** dialog box to open the MECHANICAL-INCH template file. Remember, you are opening the existing template file using the **OPEN** tool, not creating a drawing based on a template file using the **NEW** tool.

Copyright by Goodheart-Willcox Co., Inc. Chapter 18, page 1

Template Development. These in-depth instructions provide guidelines for creating your own drawing templates in compliance with ASME and other related drafting standards.

Chapter Tests. The Chapter Tests at the end of the chapters are also located on the Student Web site in two formats. Print the PDF version and write in the answers, or open the DOC version in Microsoft Word and enter your answers electronically.

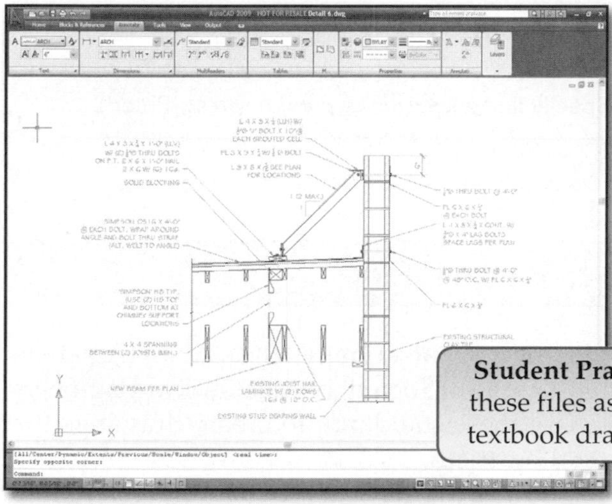

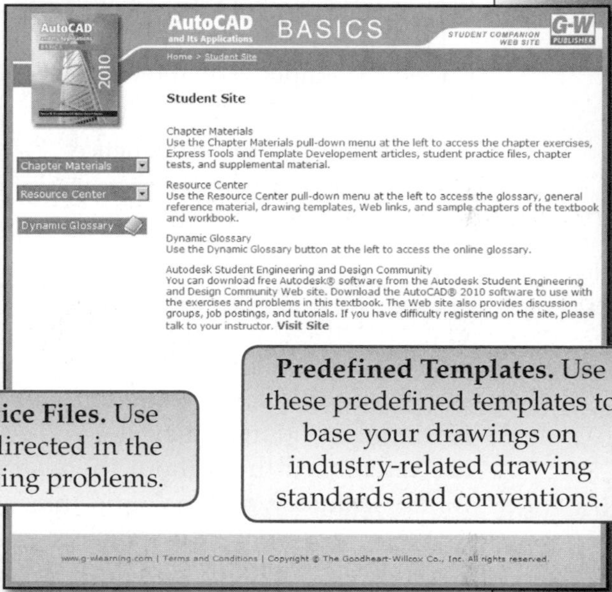

Student Practice Files. Use these files as directed in the textbook drawing problems.

Predefined Templates. Use these predefined templates to base your drawings on industry-related drawing standards and conventions.

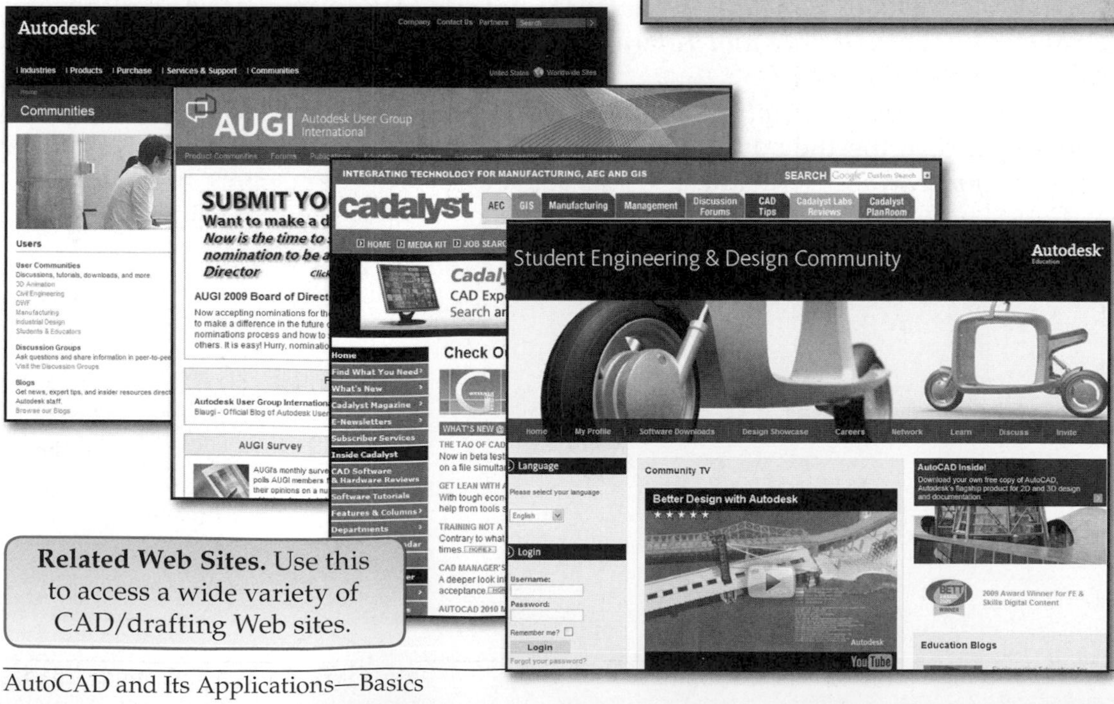

Related Web Sites. Use this to access a wide variety of CAD/drafting Web sites.

Fonts Used in This Textbook

Different typefaces are used throughout this textbook to define terms and identify AutoCAD commands. The following typeface conventions are used in this textbook:

Text Element	Example
AutoCAD tools	**LINE** tool
AutoCAD menu browser menus	**Draw** > **Arc** > **3 Points**
AutoCAD system variables	**LTSCALE** system variable
AutoCAD toolbars and buttons	**Quick Access** toolbar, **Undo** button
AutoCAD dialog boxes	**Insert Table** dialog box
Keyboard entry (in text)	Type LINE
Keyboard keys	[Ctrl]+[1] key combination
File names, folders, and paths	C:\Program Files\AutoCAD 2010\mydrawing.dwg
Microsoft Windows features	Start menu, Programs folder
Prompt sequence	Command:
Keyboard input at prompt sequence	Command: **L** *or* **LINE**⏎
Comment at a prompt sequence	Specify first point: (*pick a point or press* [Enter])

Other Text References

For additional information, standards from organizations such as ANSI (American National Standards Institute) and ASME (American Society of Mechanical Engineers) are referenced throughout the textbook. Use these standards to create drawings that follow industry, national, and international practices.

Also for your convenience, other Goodheart-Willcox textbooks are referenced. Referenced textbooks include *AutoCAD and Its Applications—Advanced* and *Geometric Dimensioning and Tolerancing*. These textbooks can be ordered directly from Goodheart-Willcox.

AutoCAD and Its Applications—Basics covers basic AutoCAD applications. For a textbook covering the advanced AutoCAD applications, please refer to *AutoCAD and Its Applications—Advanced*.

Contents in Brief

About the Authors

Terence M. Shumaker is Faculty Emeritus, the former Chairperson of the Drafting Technology Department, and former Director of the Autodesk Premier Training Center at Clackamas Community College. Terence taught at the community college level for over 25 years. He has professional experience in surveying, civil drafting, industrial piping, and technical illustration. He is the author of Goodheart-Willcox's *Process Pipe Drafting* and coauthor of the *AutoCAD and Its Applications* series.

David A. Madsen is the president of Madsen Designs Inc. (www.madsendesigns.com). David is Faculty Emeritus, the former Chairperson of Drafting Technology and the Autodesk Premier Training Center at Clackamas Community College and former member of the American Design and Drafting Association (ADDA) Board of Directors. David was honored by the ADDA with Director Emeritus status at the annual conference in 2005. David was an instructor and a department chair at Clackamas Community College for nearly 30 years. In addition to community college experience, David was a Drafting Technology instructor at Centennial High School in Gresham, Oregon. David also has extensive experience in mechanical drafting, architectural design and drafting, and construction practices. He is the author of Goodheart-Willcox's *Geometric Dimensioning and Tolerancing* and coauthor of the *AutoCAD and Its Applications* series (Release 10 through 2008 editions), *Architectural Drafting Using AutoCAD*, *AutoCAD Architecture and Its Applications*, *Architectural Desktop and its Applications*, *Architectural AutoCAD*, and *AutoCAD Essentials*.

David P. Madsen holds a Master of Science degree in Educational Policy, Foundations, and Administrative Studies with a specialization in Postsecondary, Adult, and Continuing Education; a Bachelor of Science degree in Technology Education; and an Associate of Science degree in General Studies and Drafting Technology. Dave has been involved in providing Drafting and Computer-Aided Design and Drafting instruction to adult learners since 1999. Dave has extensive and varied experience in the drafting, design, and engineering fields. He has worked in the drafting industry for over ten years and has created everything from mechanical and electronic to architectural and civil drawings.

Acknowledgments

Technical Assistance and Contribution of Materials

Margo Bilson of Willamette Industries, Inc.
Fitzgerald, Hagan, & Hackathorn
Bruce L. Wilcox, Johnson and Wales University School of Technology

Contribution of Photographs or Other Technical Information

Arthur Baker	International Source for Ergonomics
Autodesk	Jim Webster
CADalyst magazine	Kunz Associates
CADENCE magazine	Myonetics, Inc.
Chris Lindner	Norwest Engineering
EPCM Services, Ltd.	Schuchart & Associates, Inc.
Harris Group, Inc.	Willamette Industries, Inc.

Trademarks

Autodesk, the Autodesk logo, 3ds max, Autodesk VIZ, AutoCAD, DesignCenter, AutoCAD Learning Assistance, AutoSnap, and AutoTrack are either registered trademarks or trademarks of Autodesk, Inc., in the U.S.A. and/or other countries.

Microsoft, Windows, and Windows NT are registered trademarks of Microsoft Corporation in the United States and/or other countries.

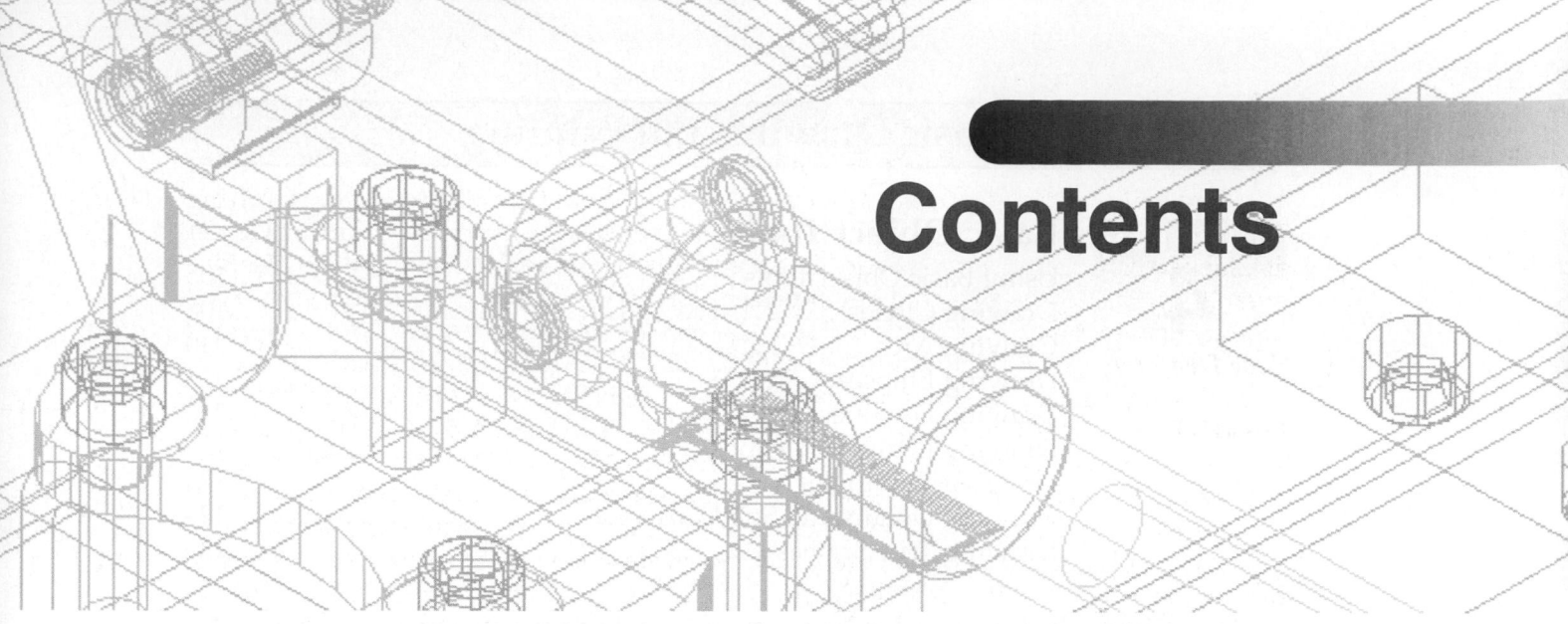

Contents

Introduction to AutoCAD

Basic Drawing and Printing

Creating Text and Tables

Editing Drawings

Dimensioning and Tolerancing

Additional AutoCAD Applications

Using Layouts

Student Web Site Content

Using the Web Site

Express Tools

Exercises

Template Development

Chapter Tests

Reference Materials

Supplemental Materials

Student Practice Files

Predefined Templates

Download Student AutoCAD

Related Web Links

CHAPTER 1

Introduction to AutoCAD

Learning Objectives

After completing this chapter, you will be able to do the following:

✓ Define computer-aided design and drafting.
✓ Describe typical AutoCAD applications.
✓ Explain the value of planning your work and system management.
✓ Describe the purpose and importance of drawing standards.
✓ Demonstrate how to start and exit AutoCAD.
✓ Describe the AutoCAD interface.
✓ Use a variety of methods to select AutoCAD tools.
✓ Use the features found in the **AutoCAD Help** window.

Computer-aided design and drafting (CADD) is the process of using a computer with software to design and produce models and drawings according to specific industry and company standards. The terms *computer-aided design (CAD)* and *computer-aided drafting (CAD)* refer to specific aspects of the CADD process. This chapter introduces the AutoCAD CADD system. In this chapter, you will learn how to begin working with AutoCAD and how to control the AutoCAD environment.

> **computer-aided design and drafting (CADD):** The process of using a computer with software to design and produce models and drawings.

AutoCAD Applications

AutoCAD *tools* and *options* are available for drawing objects of any size or shape. Use AutoCAD to prepare two-dimensional (2D) drawings, three-dimensional (3D) models, and animations. AutoCAD is a universal CADD program that applies to any drafting, design, or engineering discipline. For example, you can use AutoCAD to design and document mechanical parts and assemblies, architectural buildings, civil and structural engineering projects, electronics, and technical illustration. Using the AutoCAD software and this textbook, you will learn how to construct, lay out, dimension, and annotate 2D drawings. *AutoCAD and Its Applications—Advanced* provides detailed instruction on 3D modeling and 3D rendering.

> **tool (command):** An instruction issued to the computer to complete a specific task. For example, the **LINE** tool is used to draw lines.

> **option:** A choice associated with a tool, or an alternative function of a tool.

2D Drawings

2D drawings display object length and width, or width and height, in a flat (two-dimensional) form. A 2D drawing typically includes dimensions and annotations that fully describe features on the drawing. This practice results in a document used to manufacture or construct a product. 2D drawings are common in all drafting, design, and engineering fields. **Figure 1-1** shows an example of an architectural building floor plan created using AutoCAD.

3D Models

2D drawings are useful for documenting engineering and design requirements. However, 3D models are often more appropriate for design, visualization, analysis, and testing. AutoCAD provides tools and options for developing *wireframe*, *surface*, and *solid* models. 3D models are virtual representations of actual products. Add color, lighting, and texture to display a model in a realistic format. See **Figure 1-2A.** Use view tools to rotate and adjust a model to view it from any direction. See **Figure 1-2B.** Apply animation to a model, such as a *walkthrough* of a model home, to show product design or function.

wireframe model: A 3D model consisting of lines and curves connecting at the corners of an object to form edges; contains no surface properties or solid mass.

surface model: A 3D model consisting of volumeless surfaces, such as planes and curved faces that represent the exterior of an object.

solid model: A 3D model defined by object surfaces and volume; includes physical properties, such as mass and density, that can be analyzed.

walkthrough: A computer simulation that follows a path through or around a 3D model.

Reference Material

Glossaries

For detailed glossaries of CADD, AutoCAD, and computer terms, go to the **Reference Material** section of the Student Web site (www.g-wlearning.com/CAD) and select **Glossary of Computer Terms** or **Glossary of CADD Terms**.

Figure 1-1
AutoCAD provides tools and options for accurately creating 2D drawings such as the architectural floor plan shown here.

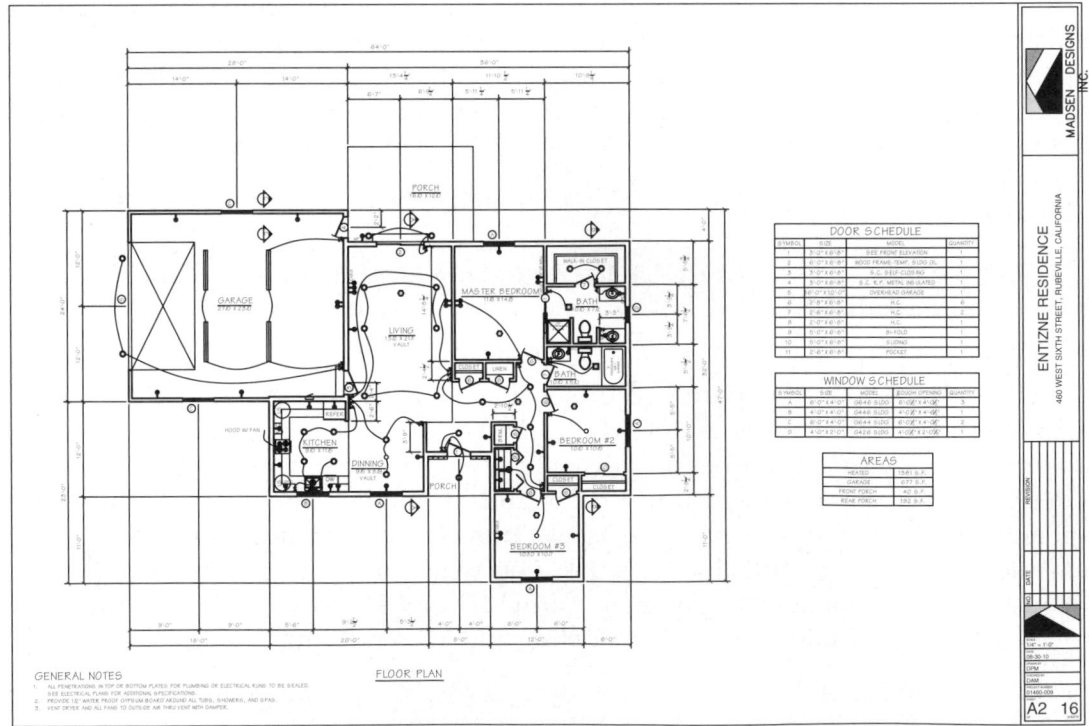

Figure 1-2.
A—A 3D wireframe model (left) with realistic colors and textures added (right). B—A model can be rotated, zoomed in and out, and viewed from any location in 3D space.

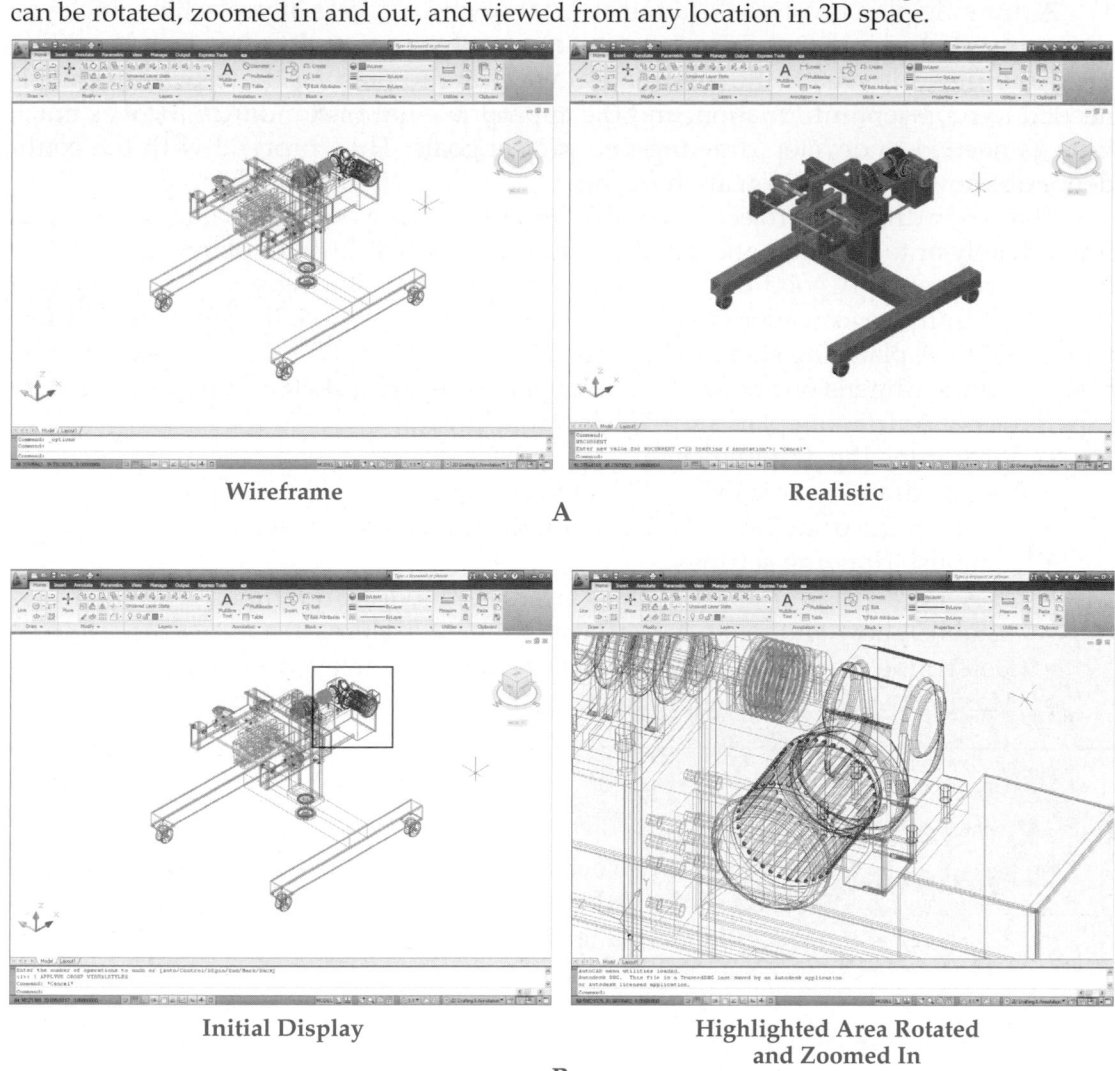

Wireframe   Realistic

A

Initial Display   Highlighted Area Rotated
                         and Zoomed In

B

Before You Begin

CADD offers greater speed, power, accuracy, and flexibility than traditional manual, or board, drafting. However, designing and drafting effectively with a computer requires a skilled CADD operator. Proficient AutoCAD users possess detailed knowledge of AutoCAD software operation. Therefore, it is important to be familiar with AutoCAD tools and to know how they work and when they are best suited for a specific task. There is no substitute for knowing AutoCAD tools. You must also have experience with design and drafting systems and conventions, and apply these rules when using AutoCAD.

As you begin your CADD training, develop effective methods for managing your work. First, plan your drawing sessions thoroughly to organize your thoughts. Second, learn and use industry and classroom, or office, standards. Third, save your work often. If you follow these three procedures, you will find it easier to use CADD tools and methods, and your experience with AutoCAD will be more productive and enjoyable.

Planning Your Work

A drawing plan involves thinking about the entire process or project in which you are involved. A plan determines how you approach a project. It focuses on the drawings you intend to create, the information you want to present, the types of symbols needed to represent information, and the appropriate use of standards. Take as much time as needed to develop drawing and project goals. Then proceed with the confidence of knowing where you are heading.

Plan your drawing projects carefully. You may want your applications to happen immediately or to be automatic, but if you hurry and do little or no planning, you may become frustrated. A good drawing plan can save time. During your early stages of AutoCAD training, consider creating a planning sheet, especially for your first few assignments. A planning sheet should document the tools you use, the selections you make, and the dimensions needed. You may also prepare a sketch as part of the planning process. A drawing plan and sketch can help you:

- Determine the drawing layout.
- Set the drawing area by laying out views and required free space.
- Confirm the drawing units, based on the dimensions provided.
- Establish drawing settings.
- Preset drawing variables, such as layers, text styles, and dimension styles.
- Establish how and when to perform various activities.
- Determine the best use of AutoCAD, resulting in an even workload.
- Maximize your use of equipment.

Reference Material

Planning Sheet

For a sample project planning sheet, go to the **Reference Material** section of the Student Web site (www.g-wlearning.com/CAD) and select **Planning Sheet**.

Using Drawing Standards

standards: Guidelines containing operating procedures, drawing techniques, and record keeping methods.

Most industries, schools, and companies have established *standards*. It is important that standards exist and are understood and used by all CADD personnel. Drawing standards can include:

- Methods of file storage (location and name)
- File naming conventions
- File backup methods and times
- Drawing templates with predefined settings
- Layout characteristics
- Borders and title blocks
- Drawing symbols
- Dimensioning styles and techniques
- Text styles
- Table styles
- Layer settings
- Plot styles

The standards you follow may vary in content. The most important aspect of standards is that people use them. When you follow drawing standards, drawings are consistent, you become more productive, and the classroom or office functions more efficiently.

The mechanical drafting standards used in this textbook are based on the American Society of Mechanical Engineers (ASME) and American National Standards Institute (ANSI) ASME Y series. Other drafting standards used in this textbook are based on appropriate discipline-specific standards, including those stated in the United States National CAD Standard.

Reference Material — *Drawing Standards*

For more information about drawing standards, go to the **Reference Material** section of the Student Web site (www. g-wlearning.com/CAD) and select **Drawing Standards**.

NOTE

You may consider other drafting standards when planning a drawing session and preparing drawings. *DIN* refers to the German standard *Deutsches Institut Für Normung*, established by the German Institute for Standardization. *Gb* refers to *Guo Biao* (Chinese) standards, *ISO* is the International Organization for Standardization, and *JIS* is the Japanese Industry Standard.

DIN: Deutsches Institut Für Normung.

Gb: Guo Biao (Chinese) standard.

ISO: International Organization for Standardization.

JIS: Japanese Industry Standard.

Saving Your Work

Drawings are lost due to software error, hardware malfunction, power failure, or accidents. Prepare for such an event by saving your work frequently. Develop the habit of saving your work at least every ten to fifteen minutes. The automatic save option, described in Chapter 2, can be set to save drawings automatically at predetermined intervals. However, you should also save your work manually at frequent intervals.

Working Procedures Checklist

As you begin learning AutoCAD, you will realize that several skills are required to become proficient. The following checklist provides you with some hints to help you become comfortable with AutoCAD. These hints also allow you to work quickly and efficiently.

- ✓ Carefully plan your work.
- ✓ Frequently check object and drawing settings, such as layers, styles, and properties, to see which object characteristics and drawing options are in effect.
- ✓ Read the prompts, tooltips, and *alerts* displayed by AutoCAD.
- ✓ Consistently check for the correct options, instructions, or keyboard entry of data.
- ✓ *Right-click* to access shortcut menus and review available options.
- ✓ Think ahead and know your next move.
- ✓ Learn new tools and options that can increase your speed and efficiency.
- ✓ Save your work every ten to fifteen minutes.
- ✓ Learn to use available resources, such as this textbook, to help solve problems and answer questions. You should also become familiar with the AutoCAD help system.

alert: A pop-up that indicates a potential problem or required action.

right-click: Use the right mouse button to select.

Workplace Comfort

For information about maintaining comfort and productivity while operating a CADD workstation, go to the Student Web site (www.g-wlearning.com/CAD), select this chapter, and select **Workplace Comfort**.

Exercise 1-1

Access the Student Web site (www.g-wlearning.com/CAD) and complete Exercise 1-1.

Starting AutoCAD

icon: Small graphic representing an application, file, or tool.

double-click: Quickly press the left mouse button twice to select.

pick (click): Use the left mouse button to select.

button: A "hot spot" on the screen that can be picked to access an application, tool, or option.

hover: Pause the cursor over an item to display information or options.

Windows creates an AutoCAD 2009 *icon* during installation, which displays on the Windows desktop and in the list of programs available from the Start menu. One of the quickest methods to start AutoCAD is to *double-click* on the AutoCAD 2010 desktop icon. A second option is to *pick* the Start *button* in the lower-left corner of the Windows desktop, then *hover* over or pick Programs, then select Autodesk, followed by AutoCAD 2010, and finally AutoCAD 2010.

NOTE

AutoCAD 2010 operates with Windows Vista and Windows XP. Do not be concerned if you see illustrations in this textbook that appear slightly different from those on your screen.

AutoCAD
2010
NEW

Initial Setup

drawing template (template): A file that contains standard drawing settings and objects for use in new drawings.

check box: A selectable box that turns an item on (when checked) or off (when unchecked).

default: A value maintained by the computer until changed.

The **Initial Setup** wizard appears when you first launch AutoCAD. See **Figure 1-3**. The purpose of the **Initial Setup** wizard is to guide you through the initial process of customizing AutoCAD according to the type of drawings you plan to create, tools you commonly use, and the *drawing template*, or *template*, you use most often. At the last page, pick the **Start AutoCAD 2010** button to launch AutoCAD. Your selections build a custom workspace. Workspaces are explained later in this chapter.

If you are a new AutoCAD user, you may not recognize most of the choices in the **Initial Setup** window. Additionally, this textbook explains 2D drawing applications related to a variety of drafting, design, and engineering disciplines. As a result, it is advisable to pick the **Skip** button on the first page of the **Initial Setup** window to exit the setup process. When the **Skip Initial Setup** page appears, deselect the *check box* so that the **Initial Setup** window does not load the next time you launch AutoCAD. Finally, select the **Start AutoCAD 2010** button to launch AutoCAD. When you skip the initial setup procedure, AutoCAD loads the *default* **2D Drafting & Annotation** workspace. Use the **2D Drafting & Annotation** workspace for universal 2D applications, as described in this textbook.

Figure 1-3.
Use the **Initial Setup** window to begin the process of customizing AutoCAD according to your discipline, interface, and template preferences.

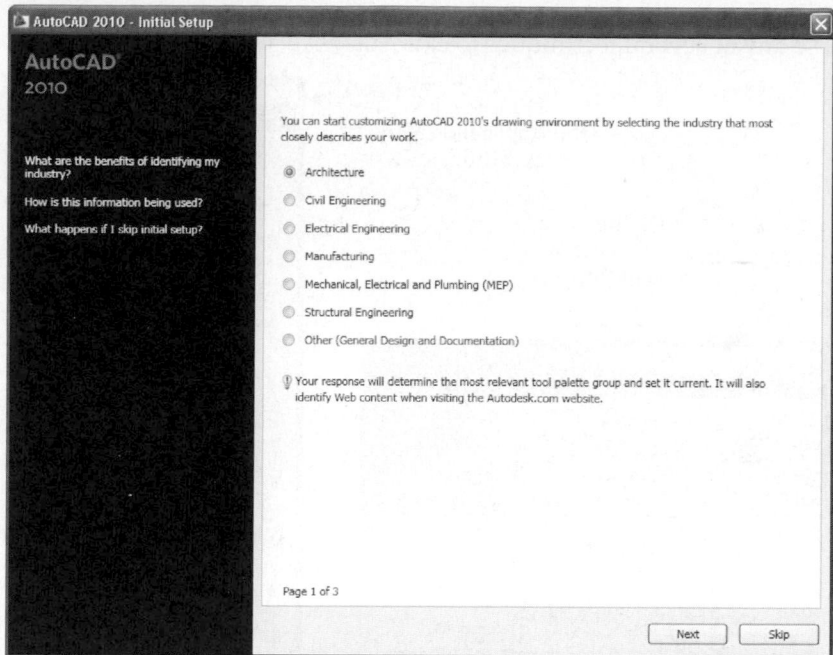

To access the **Initial Setup** wizard, pick the **Initial Setup...** button on the **User Preferences** tab of the **Options** dialog box. To display the **Options** dialog box, pick the **Options** button at the bottom of the **Application Menu**. The **Options** dialog box is described later in this chapter.

Exiting AutoCAD

Use the **EXIT** tool to end an AutoCAD session. To exit, pick the program **Close** button, located in the upper-right corner of the AutoCAD window; double-click the **Application Menu** button, found in the upper-left corner of the AutoCAD window; select the **Exit AutoCAD** button in the **Application Menu**; or with a file open, type EXIT or QUIT and press [Enter]. See **Figure 1-4.**

If you attempt to exit before saving your work, AutoCAD prompts you to save or discard changes.

Exercise 1-2

Access the Student Web site (www.g-wlearning.com/CAD) and complete Exercise 1-2.

Figure 1-4.
Use any of several techniques to exit AutoCAD when you finish a drawing session.

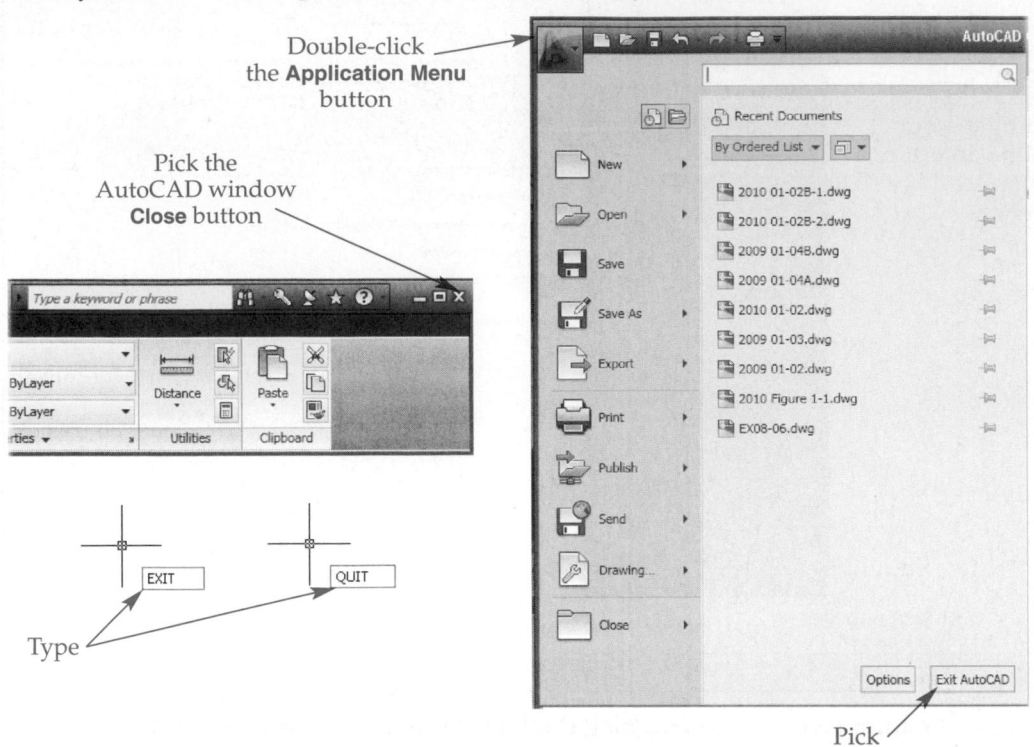

The AutoCAD Interface

Interface items include devices to input data, such as the keyboard and mouse, and devices to receive computer outputs, such as the monitor. AutoCAD uses a Windows-style *graphical user interface (GUI)* with an **Application Menu**, ribbon, dialog boxes, and AutoCAD-specific items. See **Figure 1-5.** You will explore the unique AutoCAD interface in this chapter and throughout this textbook. Learn the format, appearance, and proper use of interface items to help quickly master AutoCAD.

> **NOTE**
>
> As you learn AutoCAD, you may want to customize the graphical user interface according to common tasks and specific applications. Customize AutoCAD manually or begin customization using the **Initial Setup** window. *AutoCAD and Its Applications—Advanced* explains customizing the user interface.

Workspaces

The **2D Drafting & Annotation** *workspace,* shown in **Figure 1-5,** is active by default when you skip the initial setup procedure. The **2D Drafting & Annotation** workspace displays interface features above and below a large *drawing window* (also called the *graphics window*) and contains tools and options most often used for 2D drawing. To activate an alternative workspace, pick the **Workspace Switching** button on the status bar and select a different workspace. See **Figure 1-6.** You can also use the WSCURRENT *system variable.*

Figure 1-5.
The default AutoCAD window with the **2D Drafting & Annotation** workspace active.

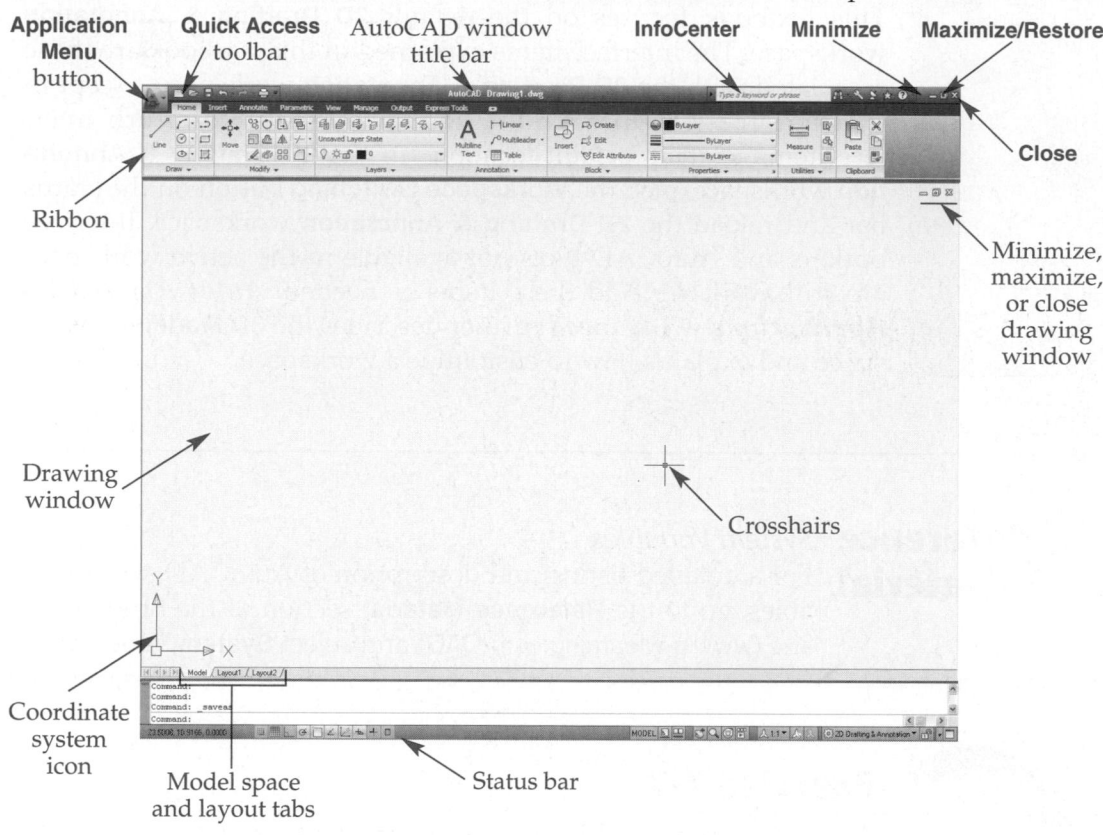

Figure 1-6.
Use the **Workspace Switching** button on the status bar to change to a different workspace, create a new workspace, or customize the user interface.

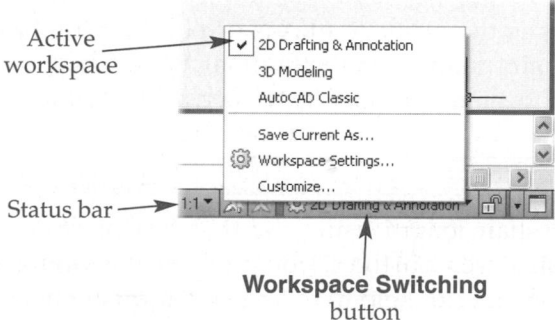

The **3D Modeling** workspace provides tools and options primarily used for 3D modeling applications. The **AutoCAD Classic** workspace displays the traditional AutoCAD menu bar, toolbars, and the **Tool Palettes** window with tools and options used for both 2D and 3D designs. The **Initial Setup Workspace**, if present, includes the default settings specified in the **Initial Setup** window. This workspace is available even if you skip the initial setup process. A new **Initial Setup Workspace** forms each time you use the **Initial Setup** function.

NEW

This textbook focuses on the default **2D Drafting & Annotation** workspace. The interface items explained in this textbook are those associated with the **2D Drafting & Annotation** workspace, except in specific situations that require additional items. To return interface items to their default locations in the **2D Drafting & Annotation** workspace, pick the **Workspace Switching** button on the status bar and reload the **2D Drafting & Annotation** workspace. Interface options and AutoCAD tools not available in the active workspace are still available. Add these items as needed. *AutoCAD and Its Applications—Advanced* further describes the **3D Modeling** workspace and explains how to customize a workspace.

Reference Material

System Variables
For a detailed listing and description of AutoCAD system variables, go to the **Reference Material** section of the Student Web site (www.g-wlearning.com/CAD) and select **System Variables**.

Exercise 1-3

Access the Student Web site (www.g-wlearning.com/CAD) and complete Exercise 1-3.

Crosshairs and Cursor

The AutoCAD crosshairs is the primary means of pointing to objects or locations within a drawing. The crosshairs changes to the familiar Windows cursor when moved outside of the drawing area or over an interface item, such as the status bar.

Control crosshair length using the *text box* or *slider* found in the **Crosshair size** area on the **Display** tab of the **Options** dialog box. Longer crosshairs can help to reference alignment between objects.

text box: A box in which you type a name, number, or single line of information.

slider: A movable bar that increases or decreases a value when you slide the bar.

tooltip: A pop-up that provides information about the item over which you are hovering.

Tooltips

A *tooltip* displays when you hover over most interface items. See **Figure 1-7.** The content presented in a tooltip varies depending on the item. Many tooltips expand as you continue to hover over a tool. The initial tooltip might only display the tool name, a brief description of the tool, and the command name. As you continue to hover, an explanation on how to use the tool and other information may appear.

Figure 1-7.
Examples of tooltips
displayed as you
hover the cursor
over an item.

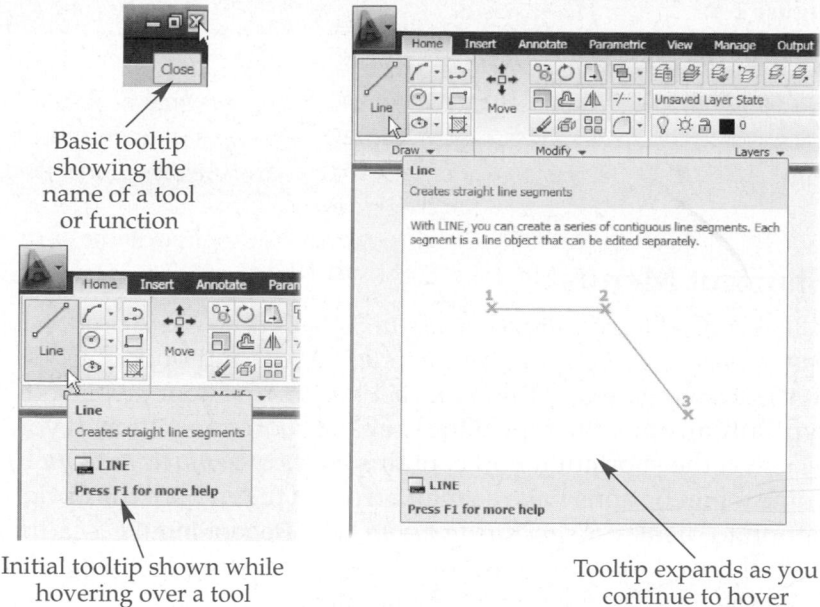

Basic tooltip
showing the
name of a tool
or function

Initial tooltip shown while
hovering over a tool

Tooltip expands as you
continue to hover

Controlling Windows

Control the AutoCAD and drawing windows using the same methods used to control other windows within the Windows operating system. To minimize, maximize, or close the AutoCAD window or individual drawing windows, pick the appropriate icon in the upper-right corner. You can also adjust the AutoCAD window by right-clicking on the title bar and choosing from the standard window control menu. Window sizing operations are also the same as those for other windows within the Windows operating system.

Floating and Docking

Several interface items, including the AutoCAD and drawing windows, can *float* or be *docked*. Floating features appear within a border. Some items, such as the drawing window, have a title bar at the top or side. You can move and resize floating windows in the same manner as other windows. However, drawing windows will only move and resize within the AutoCAD window. Different options are available depending on the particular interface item and the float or docked status of the item. Typically, the close and minimize or maximize options are available. Some floating items, such as sticky panels, include *grab bars*.

Locking

To prevent certain interface items from moving accidentally, lock the features in either a floating or a docked state. To access locking options, pick the **Toolbar/Window Positions** button on the status bar or select the **View** tab on the ribbon and then the **Window Locking** *flyout* from the **Windows** panel. **Figure 1-8** displays the **Lock Location** menu options.

Select an option to lock the interface items that reside in that group as floating or docked. To unlock a group, select the option again. To quickly lock or unlock all interface items, select **Locked** or **Unlocked** from the **All** cascading menu. Move a locked feature without unlocking it by holding down the [Ctrl] key while moving the feature.

float: Describes interface items that can be freely resized or moved about the screen.

docked: Describes interface items that are locked into position on an edge of the AutoCAD window (top, bottom, left, or right).

grab bars: Two thin bars at the top or left edge of a docked or floating feature; used to move the feature.

flyout: Set of related buttons that appears when you pick the arrow next to certain tool buttons.

Figure 1-8.
Some or all interface items can be locked in position.

Lock specific types of interface items

Lock all interface items

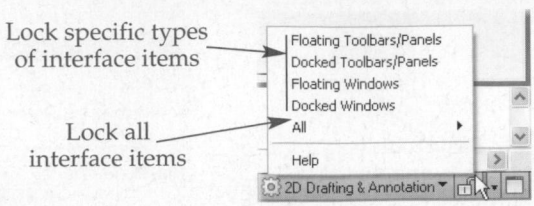

Shortcut Menus

AutoCAD uses *shortcut menus*, also known as *cursor menus, right-click menus,* or *pop-up menus,* to simplify and accelerate tool and option access. When you right-click in the drawing area with no tool active, the first item displayed in the shortcut menu is typically an option to repeat the previous tool or operation. If you right-click while a tool is active, the shortcut menu contains *context-sensitive menu options.* See **Figure 1-9.** Some menu options have a small arrow to the right of the option name. Hover over the option to display a *cascading menu.* The **Recent Input** cascading menu shows a list of recently used tools, options, or values, depending on the specific shortcut menu. Pick from the list to reuse a function or value.

Exercise 1-4

Access the Student Web site (www.g-wlearning.com/CAD) and complete Exercise 1-4.

AutoCAD 2010 NEW

Application Menu

The **Application Menu** provides access to application- and file-related tools and settings through a system of menus and menu options. The **Application Menu** displays when you pick the **Application Menu** button, located in the upper-left corner of the AutoCAD window. See **Figure 1-10.**

Figure 1-9.
Shortcut menus provide instant access to tools and options related to the current drawing or editing operation.

Pick to view and select the most recent tools

Cascading menu of recent tools

Tool-specific options

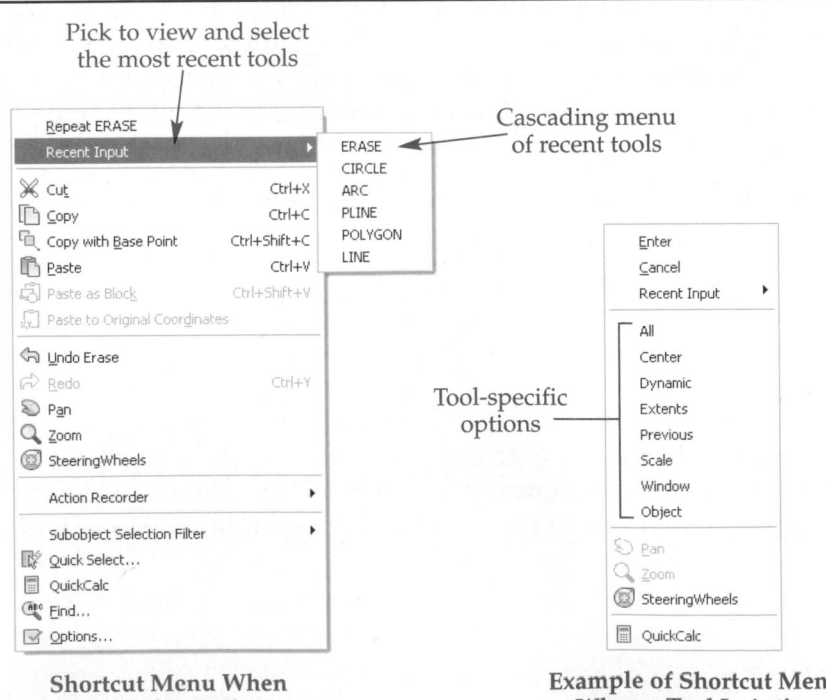

Shortcut Menu When No Tool Is Active

Example of Shortcut Menu When a Tool Is Active

Figure 1-10.
Use the **Application Menu** to access common application and file management tools and settings, search for commands, and view open and recently used documents.

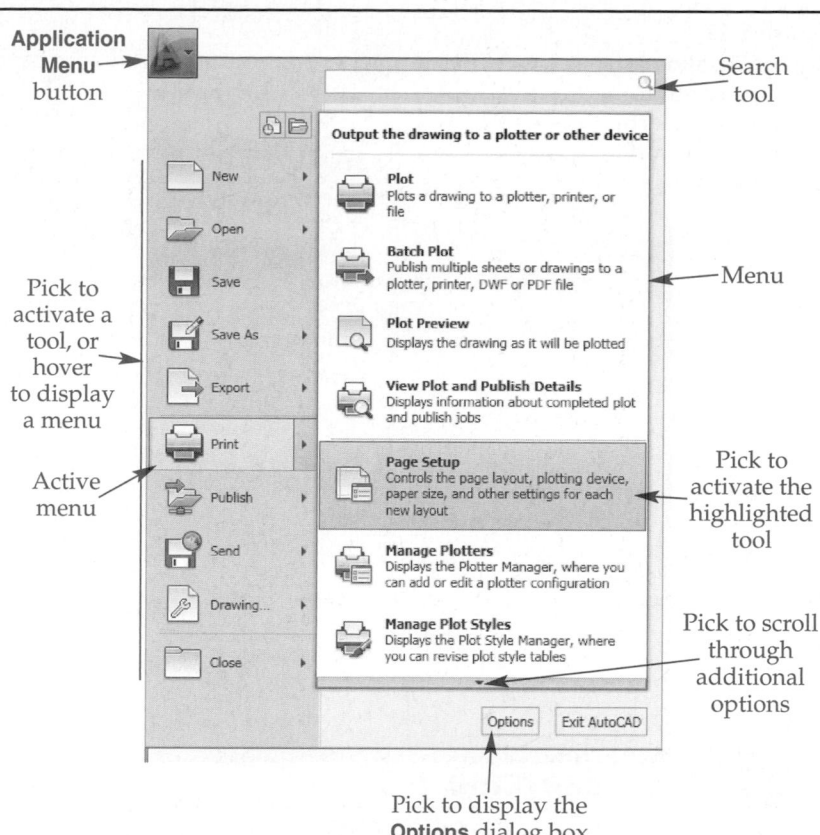

Items on the left side of the **Application Menu** function as buttons to activate common application tools, and except for the **Save** button, also display menus. For example, press the **New** button to begin a new file using the **QNEW** tool. To display a menu, hover the cursor over the menu name, or pick the arrow on the right side of the button. Long menus include small arrows at the top and bottom for scrolling through selections. Some options have a small arrow to the right of the item name that, when selected or hovered over, expands to provide a submenu. Pick the desired option to activate the tool.

Accessing Tools and Options

A tool or option accessible from the **Application Menu** appears as a graphic in the margin of this textbook. This graphic represents the process of picking the **Application Menu** button, then selecting a menu button, or hovering over a menu and picking a menu option or a submenu option. The example shown in this margin illustrates accessing the **PAGESETUP** tool from the **Application Menu**, as shown in **Figure 1-10**.

Searching for Commands

The **Application Menu** contains a search tool used to locate and access any AutoCAD command listed in the Customize User Interface (CUI) file. Type the name of the command you want to access in the **Search** text box. Commands that match the letters you enter appear as you type. Typing additional letters narrows the search, with the best-matched command listed first. **Figure 1-11** shows using the **Search** text box to locate the **SAVE** tool for saving a file. Pick a command from the list to start the command.

NOTE

The **Recent Documents** and **Open Documents** features of the **Application Menu** provide access to recently and currently open files. Chapter 2 describes these functions.

Figure 1-11.
Use the **Application Menu** to search for a command. Pick the command from the list to activate.

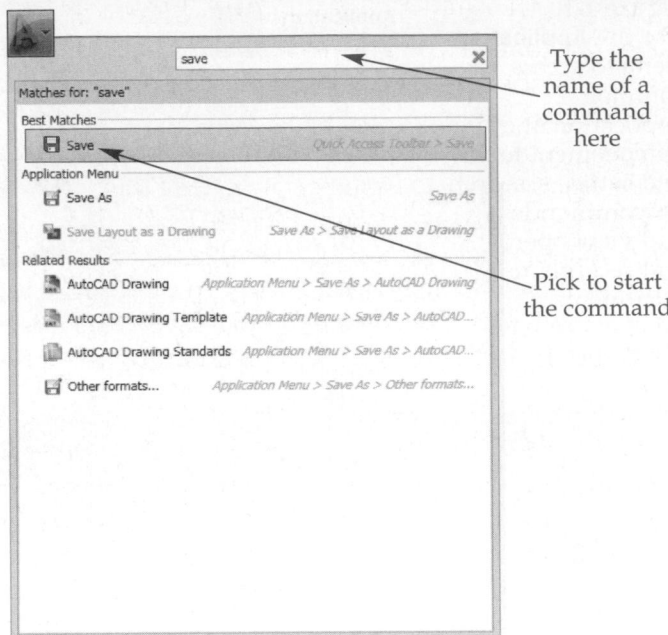

Type the name of a command here

Pick to start the command

Exercise 1-5

Access the Student Web site (www.g-wlearning.com/CAD) and complete Exercise 1-5.

Quick Access Toolbar

Toolbars contain *tool buttons*. Each tool button includes an icon that represents an AutoCAD tool or option. As you move the cursor over a tool button, the button highlights and may display a border. Use the tooltip to become familiar with the tool icons. Select a tool button to activate the associated tool.

The default **Quick Access** toolbar is located on the title bar in the upper-left corner of the AutoCAD window, to the right of the **Application Menu** button. See **Figure 1-12.** The **Quick Access** toolbar provides fast, convenient access to some of the most commonly used tools. Activating most tools from the **Quick Access** toolbar requires only a single pick. Most other interface items require two or more picks to activate a tool.

toolbars: Interface items that contain tool buttons or drop-down lists.

tool buttons: Interface items used to start tools.

Figure 1-12.
Use the **Quick Access** toolbar to access commonly used tools. Pick a tool button to activate the corresponding tool.

Pick to display a flyout

Pick to display options for customizing the **Quick Access** toolbar

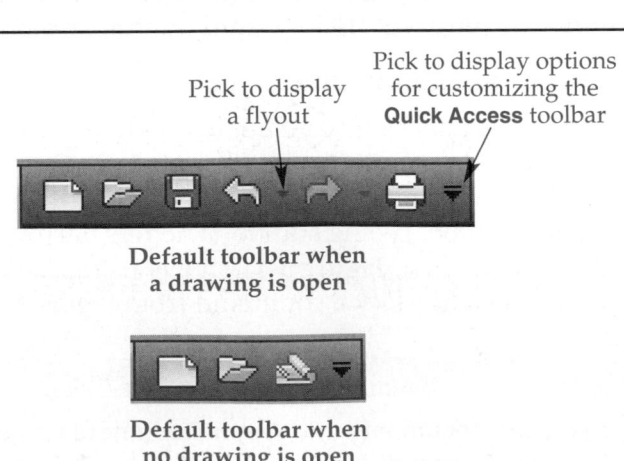

Default toolbar when a drawing is open

Default toolbar when no drawing is open

AutoCAD and Its Applications—Basics

When a drawing is open and the default **2D Drafting & Annotation** workspace is active, the toolbar contains **New, Open, Save, Undo, Redo,** and **Plot** buttons. When a drawing is not open, the **New, Open,** and **Sheet Set Manager** tool buttons display. The **Quick Access** toolbar is fully customizable by adding, removing, and relocating tool buttons. To make basic adjustments, pick the **Customize Quick Access Toolbar** flyout on the right side of the toolbar. *AutoCAD and Its Applications—Advanced* further explains customizing the user interface.

A tool or option accessible from the **Quick Access** toolbar appears as a graphic in the margin of this textbook. This graphic represents the process of picking a **Quick Access** toolbar button. The example shown in this margin illustrates accessing the **REDO** tool from the **Quick Access** toolbar to redo a previously undone operation.

REDO

NOTE

Several toolbars appear in the **AutoCAD Classic** workspace. These toolbars are usually application- or task-specific. The **Application Menu, Quick Access** toolbar, and ribbon replace classic toolbars in all other workspaces. Refer to *AutoCAD and Its Applications—Advanced* for information on customizing a workspace to display classic toolbars.

Exercise 1-6

Access the Student Web site (www.g-wlearning.com/CAD) and complete Exercise 1-6.

Ribbon

The ribbon, shown in **Figure 1-13,** is the primary means of accessing tools and options. The ribbon provides a convenient location from which to select tools and options that traditionally would require access by extensive typing, multiple toolbars, or several menus. The ribbon allows you to spend less time looking for tools and options, while reducing clutter in the AutoCAD window and increasing valuable drawing window space.

The ribbon appears by default in all workspaces except the **AutoCAD Classic** workspace. Use the *tabs* along the top of the ribbon to access individual collections of related *ribbon panels*, or *panels*. Each panel houses groups of similar tools. For example, the

tab: A small stub at the top or side of a page, window, dialog box, or palette, allowing access to other portions of the item.

ribbon panels (panels): Palette divisions that group tools.

Figure 1-13.
The ribbon is the most often used palette and is docked at the top of the drawing window. Palettes provide access to tools, options, properties, and settings.

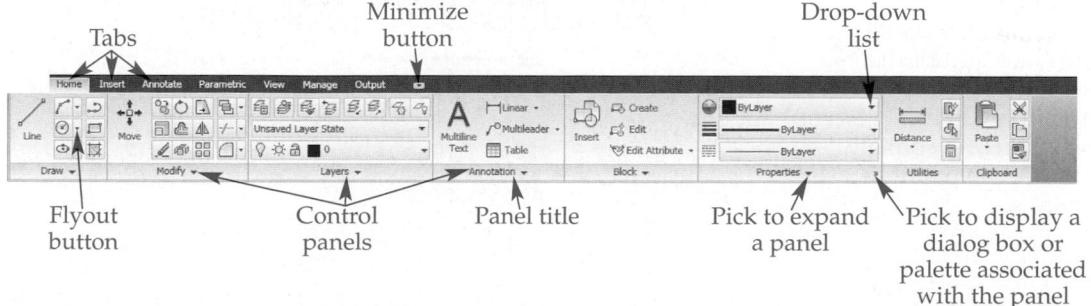

Tabs   Minimize button   Drop-down list

Flyout button   Control panels   Panel title   Pick to expand a panel   Pick to display a dialog box or palette associated with the panel

---

Annotate tab includes several panels, each with specific tools for creating, modifying, and formatting annotations, such as text. The tabs and panels shown when the **2D Drafting & Annotation** workspace is active provide access to 2D drawing tools.

A tool or option accessible from the ribbon appears in a graphic located in the margin of this textbook, like the example shown in this margin. The graphic identifies the tab and panel where the tool is located. You may need to expand the panel or pick a flyout to locate the tool. This example shows how to access the **LINE** tool using the ribbon.

Ribbon Panels

The large tool button in a panel signifies the most often used panel tool. In addition to tool buttons, panels can contain flyouts, *drop-down lists*, and other items. Some panels have a triangle, or arrow, next to the panel name. If you see this arrow, pick the bottom, or title, of the panel to display additional, related tools and functions. See **Figure 1-14**. To show the expanded list on-screen at all times, select the push pin button.

drop-down list: A list of options that appears when you pick a button that contains a down arrow.

> **NOTE**
>
> When you pick an option from a ribbon flyout, the option becomes the new default and appears in the ribbon. This makes it easier to select the same option the next time you use the tool.

Some panels include a small arrow in the lower-right corner or the panel. Pick this arrow to access a dialog box or palette closely associated with the panel function. For example, pick the arrow in the lower-right corner of the **Home** tab, **Properties** panel, as shown in **Figure 1-14**, to display the **Properties** palette. The **Properties** palette is one of the most often used tools for adjusting object properties.

Adjustments

Right-click on a portion of the ribbon that is not occupied by a panel to access a shortcut menu with a variety of ribbon display options. **Figure 1-15** provides a brief description of basic ribbon shortcut menu functions. Minimize options can also be activated by repeatedly pressing the **Minimize** button to the right of the tabs. You can also control the display of tabs and panels by right-clicking on a panel and selecting from the appropriate cascading submenu.

By default, the ribbon is docked horizontally below the AutoCAD window title bar. Use the **Undock** shortcut menu option described in **Figure 1-15** to change the ribbon to a floating state, as shown in **Figure 1-16**. Right-click on the title bar or pick the **Properties** button to select from a list of undocked ribbon control options. The **Auto-hide** option allows the ribbon to minimize when the cursor is away from the ribbon, conserving significant drawing space.

Figure 1-14.
An expanded panel provides additional, related tools and functions. In this case, the **Draw** panel has been expanded.

Pick to pin the expanded list to the screen

Pick to display the **Properties** palette

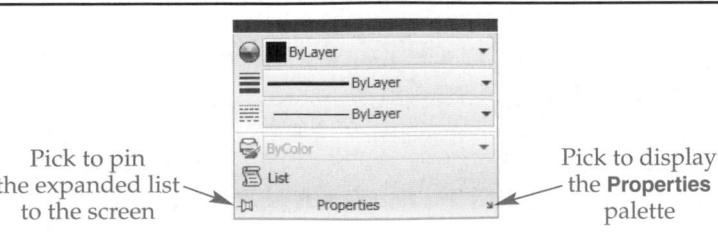

Figure 1-15.
Right-click options for displaying and organizing ribbon elements.

Selection	Result
Minimize > **Minimize to Tabs**	Shows only tabs. Pick a tab to show all panels in the tab.
Minimize > **Minimize to Panel Titles**	Displays tabs and panel titles. Pick a panel title to display the panel.
Minimize > **Show Full Ribbon**	Shows the default full ribbon.
Show Tabs	Allows you to choose which tabs to display.
Show Panels	Allows you to select which panels to display.
Show Panel Titles	Uncheck to hide panel titles.
Undock	Changes the ribbon to a floating state.
Close	Closes the ribbon. Use the **RIBBON** tool to redisplay the ribbon

Figure 1-16.
Floating palettes remain on-screen as you work in the drawing area.

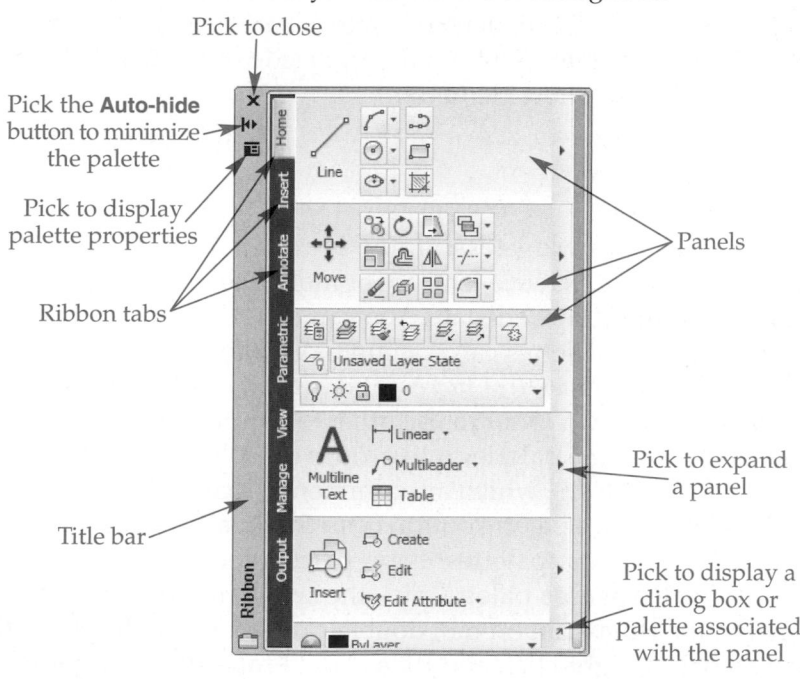

PROFESSIONAL TIP

Resize the floating ribbon using the resizing arrows that appear when you move the cursor over the ribbon edge. Then pick the **Auto-hide** button to take full advantage of the ribbon while displaying the largest possible drawing area.

To reposition a ribbon tab, hold down the left mouse button on a tab, and drag the tab right or left if the ribbon is in a horizontal orientation, or up or down if the ribbon is in a vertical orientation. Release the button when the tab is in the desired location. To reposition a panel within a tab, hold down the left mouse button on a panel title and drag the panel to the desired location.

Figure 1-17.
A sticky panel created by dragging the **Draw** panel from the **Home** tab and dropping it into the drawing window.

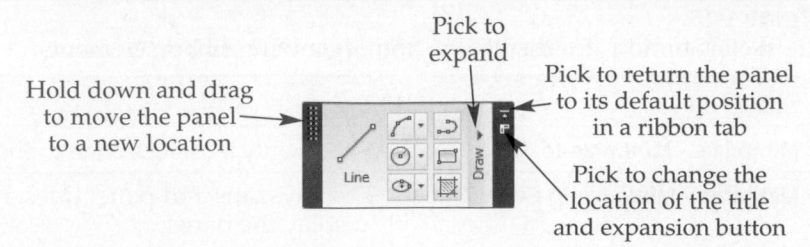

You can also drag a panel from a tab and drop it in the drawing window to create a *sticky panel*. See **Figure 1-17.** A sticky panel is much like a toolbar and conveniently remains on-screen when you select a different ribbon tab. Hover over a sticky panel to reveal grab bars for moving the panel and buttons for returning the panel to its appropriate ribbon tab and adjusting the title orientation. You can also drag and drop a sticky panel back into the appropriate tab.

sticky panel: A ribbon panel moved out of a tab and made to float in the drawing window.

> **NOTE**
> The **Application Menu**, **Quick Access** toolbar, and ribbon replace the traditional menu bar in workspaces other than the **AutoCAD Classic** workspace. To display the menu bar, pick the **Customize Quick Access Toolbar** flyout on the right side of the **Quick Access** toolbar and choose **Show Menu Bar**.

Palettes

Palettes, also known as *modeless dialog boxes*, control many AutoCAD functions. Palettes can look like extensive toolbars or more like dialog boxes, depending on the function and floating or docked state. You can consider the ribbon a palette used to access tools and options. Palettes can contain tool buttons, flyouts, drop-down lists, and many other features, such as *list boxes*, and *scroll bars*. Unlike a dialog box, you do not need to close a palette in order to use other tools and work on the drawing. Like the ribbon, panels divide some palettes into groups of tools. Large palettes are divided into separate pages or windows, which are commonly accessed using tabs.

To display a palette, pick a palette button from the **Palettes** panel in the **View** ribbon tab. You can also display most palettes using palette-specific access techniques. For example, to access the **Properties** palette, pick the arrow in the lower-left corner of the **Properties** panel in the **Home** ribbon tab; double-click on most objects in the drawing window; select an object, right-click and then select **Properties**; or type **PROPERTIES**.

When you display a palette for the first time, it is often in a floating state, although some palettes are dockable. Most of the palette control features available on the floating ribbon, such as docking and sizing, apply to other palettes as well. Deselect the **Allow Docking** palette property or menu option if you do not want to have the ability to dock a palette. The **Properties** button or shortcut menu on some palettes include other functions, such as the **Transparency...** option, which makes the palette transparent, allowing drawing geometry behind the palette to be viewed. See **Figure 1-18.**

palette (modeless dialog box): Special type of window containing tool buttons and other features found in dialog boxes. Palettes can remain open while other tools are in use.

list box: A boxed area that contains a list of items or options from which to select.

scroll bar: A bar tipped with buttons used to scroll through a list of options or information.

> **NOTE**
> Palettes play a major role in the operation of AutoCAD. Specific palettes are described when applicable throughout this textbook or in *AutoCAD and Its Applications—Advanced*.

Figure 1-18.
To make a palette transparent, pick the **Properties** button or right-click in the title bar and select **Transparency...**.

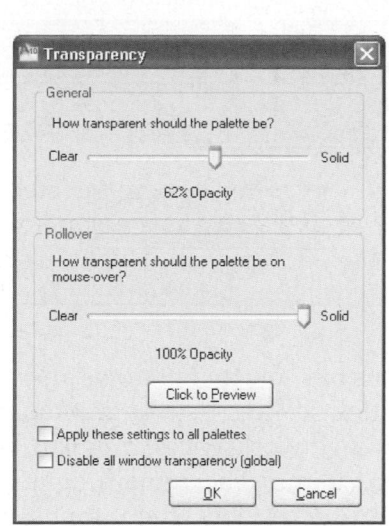

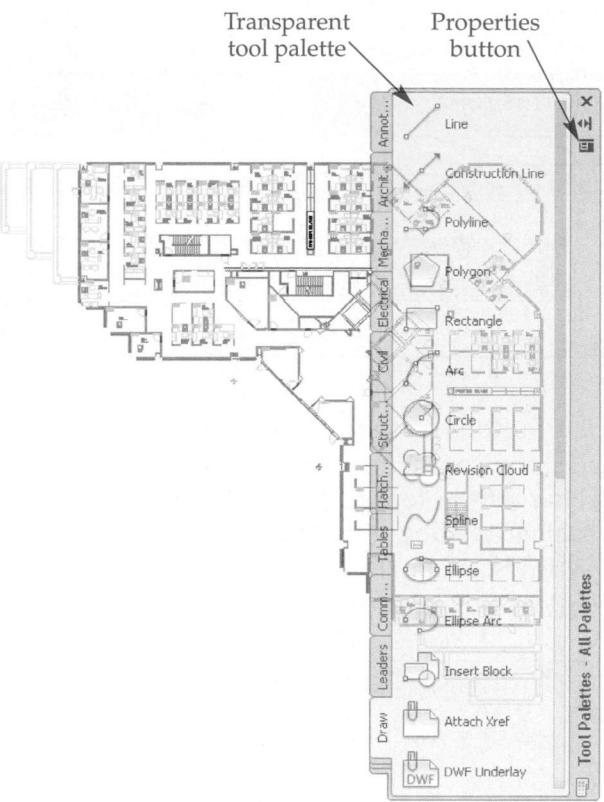

Transparent tool palette

Properties button

Exercise 1-7

Access the Student Web site (www.g-wlearning.com/CAD) and complete Exercise 1-7.

Status Bars

AutoCAD provides two types of status bars. The application status bar applies to all open drawings. A drawing status bar, when activated, appears above the *command line* and is specific to each drawing. Status bars are the quickest and most effective way to manage certain drawing settings.

command line: Area at the bottom of the screen where commands (tool names) and options may be typed.

Application Status Bar

The application status bar is located along the bottom of the AutoCAD window. See **Figure 1-19.** The application status bar is divided into areas that display and control a variety of drawing aids and tools. The coordinate display field, located on the left side of the application status bar, shows the XYZ coordinates of the crosshairs, identifying its location in drawing space. *Status toggle buttons* are located next to the coordinate display field.

status toggle buttons: Buttons that toggle drawing aids and tools on and off.

NOTE

Status toggle buttons appear as icons by default. To change the display from icons to names, right-click on any button and deselect **Use Icons**. This option applies only to the status toggle buttons.

Figure 1-19.
Picking buttons on the application status bar is the quickest and most effective way to manage certain drawing settings.

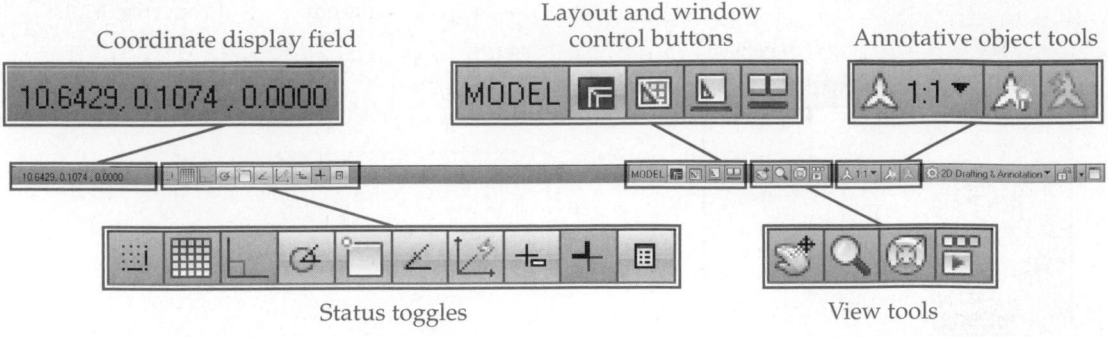

The buttons on the right side of the application status bar control windows and the drawing environment, activate tools, and adjust annotation scaling. Use the **Workspace Switching** button found in this area to change and manage workspaces. You can use the **Toolbar/Window Positions** button to lock interface items. The remaining tools and settings available on the application status bar are described when applicable throughout this textbook.

> **NOTE**
>
> Right-click on the application status bar, away from the coordinate display field or a button, to access a shortcut menu with options for modifying the application status bar display. Uncheck an item on the list to hide the item from the status bar. *AutoCAD and Its Applications—Advanced* further describes how to customize the status bar.

Drawing Status Bar

A **Drawing Status Bar** option is available from the application status bar shortcut menu and the **Status Bar** flyout in the **Windows** panel of the **View** ribbon tab. When this option is selected, a separate drawing status bar appears in the drawing window. The **Annotation Scale**, **Annotation Visibility**, and **AutoScale** tools move from the application status bar to the drawing status bar. See **Figure 1-20**. The settings are unique to each open file.

Figure 1-20.
The drawing status bar, when displayed, is specific to the current drawing. Each open drawing has its own drawing status bar.

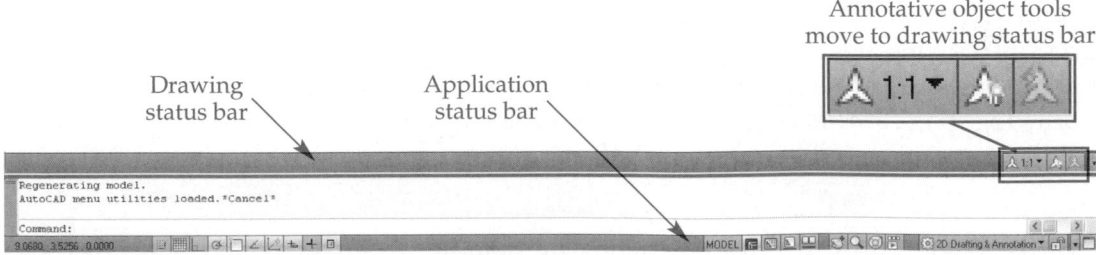

Right-click on the coordinate display field or a button in the application or drawing status bar to view a shortcut menu specific to the item. Picking options from a status bar shortcut menu is often the most efficient method of controlling drawing settings.

Exercise 1-8

Access the Student Web site (www.g-wlearning.com/CAD), and complete Exercise 1-8.

Dialog Boxes

You will see many *dialog boxes* during a drawing session, including those used to create, save, and open files. Dialog boxes contain many of the same features found in other interface items, including icons, text, buttons, and flyouts. **Figure 1-21** shows the dialog box that appears when you pick **Insert** from the **Block** panel of the **Insert** ribbon tab. This dialog box displays many common dialog box elements.

dialog box: A window-like part of the user interface that contains various kinds of information and settings.

A dialog box appears when you pick any menu selection or button displaying an ellipsis (…).

Use the cursor to set variables and select items in a dialog box. In addition, many dialog boxes include images, *preview boxes*, or other methods to help you to select appropriate options. When you pick a button in a dialog box that includes an ellipsis (…), another dialog box appears. You must make a selection from the second dialog box before returning to the original dialog box. A button with an arrow icon requires you to select in the drawing area.

Figure 1-21.
A dialog box appears when you pick an item that is followed by an ellipsis. The dialog box shown here displays when you select the **INSERT** tool.

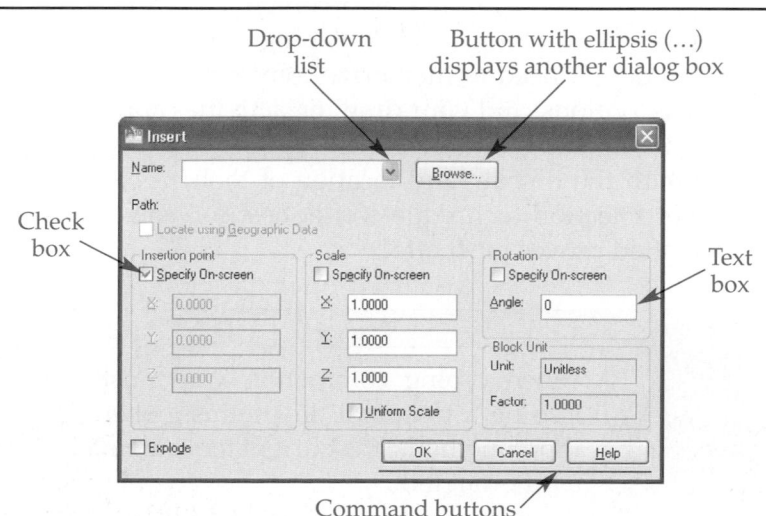

Drop-down list

Button with ellipsis (…) displays another dialog box

Check box

Text box

Command buttons

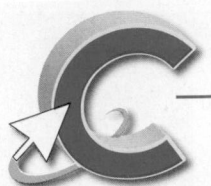

Exercise 1-9

Access the Student Web site (www.g-wlearning.com/CAD) and complete Exercise 1-9.

System Options

OPTIONS

Application Menu
Options
Options

Type
OPTIONS
OP

AutoCAD system options are contained in the **Options** dialog box. System options apply to the entire program and are not specific to a file. Many system options help configure the work environment. This textbook references the **Options** dialog box when applicable.

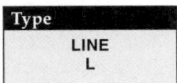

NOTE

The **Options** dialog box can also be accessed by right-clicking when no tool is active and selecting **Options....**

Selecting Tools

dynamic input: Area near the crosshairs where commands may be typed and context-oriented information is provided.

command aliases: Abbreviated command names entered at the keyboard.

Type
LINE
L

Tools are available by direct selection from the ribbon, shortcut menus, the **Application Menu**, the **Quick Access** toolbar, palettes, and the status bar. An alternative is to type the command that activates a tool using *dynamic input* or the command line. To activate a tool by typing, type the single-word command name or *command alias* and press [Enter] or the space bar, or right-click. You can use uppercase, lowercase, or a combination of uppercase and lowercase letters. You can only issue one command at a time.

You can activate all tools and options by typing commands. All tool names and aliases, along with other access techniques available in the **2D Drafting & Annotation** workspace, appear in a graphic in the margin of this textbook. The example displayed in the margin shows the command name (**LINE**) and alias (**L**) you can use to access the **LINE** tool.

Accessing tools using one or a combination of the methods explained earlier in this chapter offers advantages over typing commands at the keyboard. A major benefit is that you do not need to memorize command names or aliases. Another advantage is that tools, options, and your drawing activities appear on-screen as you work, using visual icons, tooltips, and prompts. As you work with AutoCAD, you will become familiar with the display and location of tools. As an AutoCAD drafter, you decide which tool selection technique works best for you. A combination of tool selection methods often proves most effective.

PROFESSIONAL TIP

When typing commands, you must exit the current tool before issuing a new tool. In contrast, when you use the ribbon or other input methods, the current tool automatically cancels when you pick a different tool.

Reference Material *Command Aliases*

For a detailed list of command aliases, go to the **Reference Material** section of the Student Web site (www.g-wlearning.com/CAD) and select **Command Aliases**.

Dynamic Input

Dynamic input allows you to keep your focus at the point where you are drawing. When dynamic input is on, a temporary area for tool input and information appears in the drawing window, below and to the right of the crosshairs by default. See **Figure 1-22.** When a tool is in progress, regardless of how you access the tool, the next action needed to proceed appears, along with additional tool options, an input area, and additional information.

Depending on the tool in progress, different information and options are available in the dynamic input area. For example, in **Figure 1-23**, the **RECTANGLE** tool has been issued. The first part of the dynamic input area is the tooltip, which reads Specify first corner point or. In this case, to draw a rectangle, you need to pick in the drawing area, enter *coordinates* to specify the first corner of the rectangle, or access available options as suggested by the "or" portion of the prompt.

coordinates: Numerical values used to locate a point in the drawing area.

Figure 1-22.
Use dynamic input to type or select tools and values from a temporary input area next to the crosshairs.

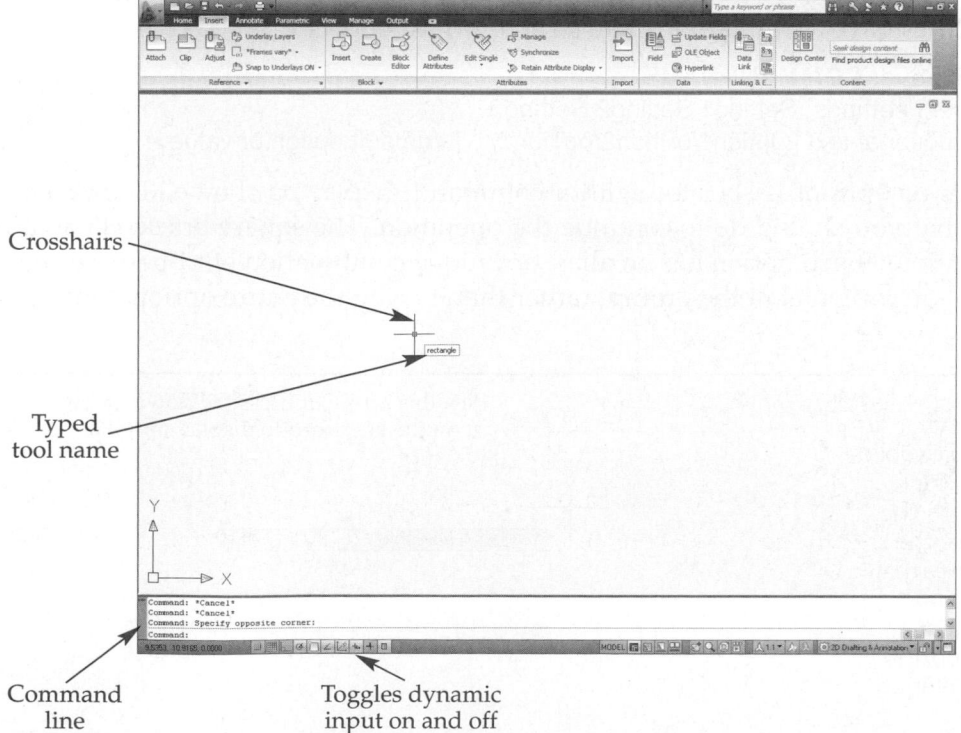

Crosshairs

Typed tool name

Command line

Toggles dynamic input on and off

Figure 1-23.
The dynamic input fields after the **RECTANGLE** tool has been started.

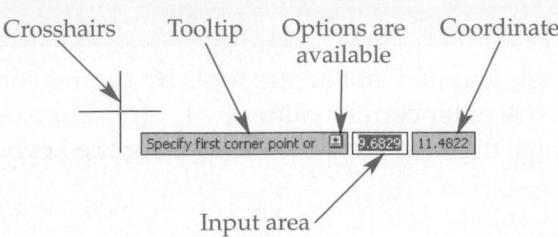

Crosshairs   Tooltip   Options are available   Coordinate

Specify first corner point or [ ] 9.6829 | 11.4822

Input area

Pressing the down arrow key displays the options available for the current tool. See **Figure 1-24.** Select an option using the cursor, or press the down arrow again to cycle through the available options, as indicated by a bullet next to the option. To select a bulleted option, press [Enter]. You can also select an option by right-clicking and selecting an option from the shortcut menu. The information displayed in the dynamic input area changes while you work with a tool, depending on the actions you choose. **Figure 1-25** shows the dynamic input display when the **LINE** tool is active.

NOTE

Dynamic input can be toggled on and off by picking the **Dynamic Input** button on the status bar or pressing the [F12] key. You can issue commands without dynamic input on.

Command Line

The command line, shown in **Figure 1-22,** provides the same function as dynamic input, but allows you to enter tools and context-specific information in a traditional window format. By default, the command line is docked at the bottom of the AutoCAD window, above the status bar. It acts like a palette and displays the Command: prompt and reflects any commands you issue. The command line also displays prompts that supply information or that request input.

When you issue a command, AutoCAD either performs the specified operation or displays prompts for additional information needed. The commands that activate AutoCAD tools have a standard format, structured as follows:

Command: **COMMANDNAME**↵
Current settings: Setting1 Setting2 Setting3
Instructional text [Option1/oPtion2/opTion3/...] <default option or value>:

Settings or options associated with a command display as shown. The prompt indicates what you should do to continue the operation. The square brackets contain available options. Each option has an alias, or unique combination of uppercase characters, that you can enter at the prompt rather than typing the entire option name. If a

Figure 1-24.
Press the down arrow key to expose tool options. Pick an option with the cursor, or use the up and down arrow keys to position the highlight at the desired option and press [Enter] to select.

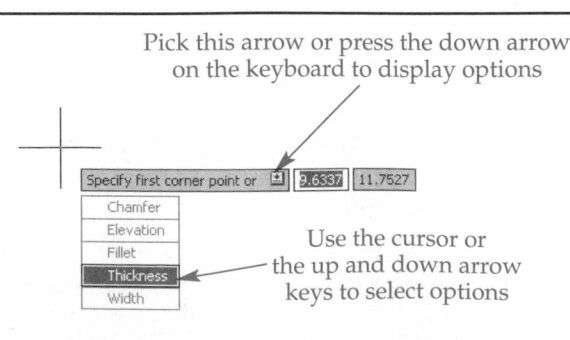

Pick this arrow or press the down arrow on the keyboard to display options

Specify first corner point or [ ] 9.6337 | 11.7527

Chamfer
Elevation
Fillet
Thickness
Width

Use the cursor or the up and down arrow keys to select options

AutoCAD and Its Applications—Basics

Figure 1-25.
Dynamic input fields change while a tool is in use. In this example, crosshair coordinates appear first. After you select the first endpoint, the distance and angle of the crosshairs relative to the first endpoint display.

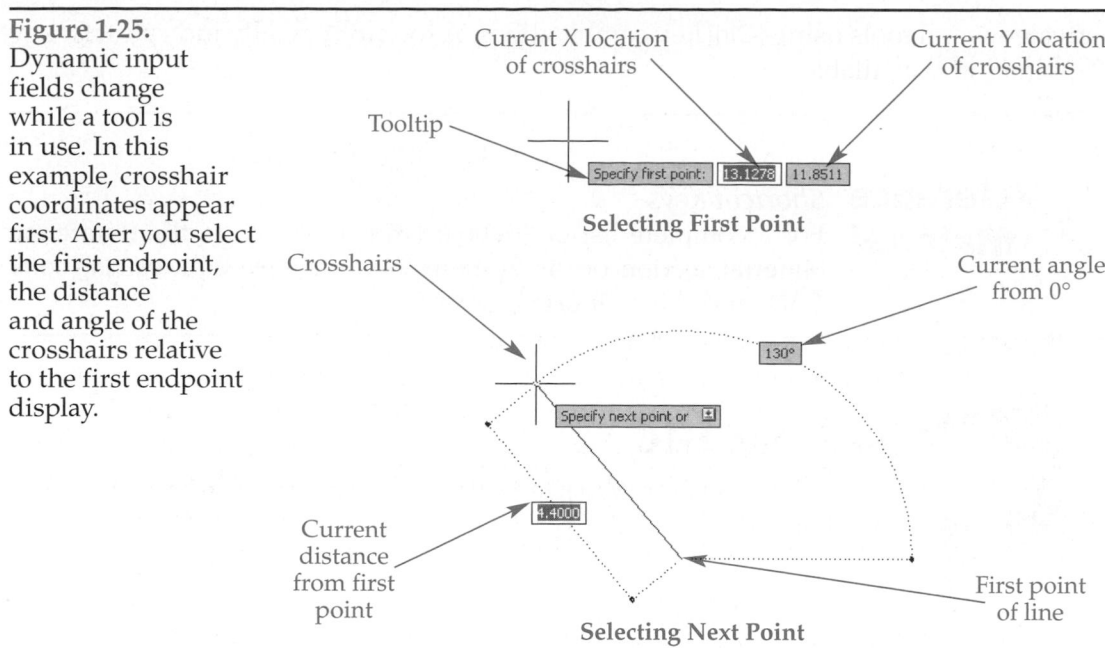

Current X location of crosshairs

Current Y location of crosshairs

Tooltip

Specify first point: 13.1278 11.8511

Selecting First Point

Crosshairs

Current angle from 0°

130°

Specify next point or

Current distance from first point

4.4000

First point of line

Selecting Next Point

default option is displayed in the angle brackets (<>), press [Enter] to accept the option rather than typing the value again.

Each default AutoCAD workspace includes the command line. The command line can float or be docked, resized, and locked. The floating command line contains the **Auto-hide** and **Properties** buttons found on palettes. Depending on your working preference, use the command line at the same time as dynamic input, or disable the command line if you use only dynamic input. To hide the command line, pick the **Close** button on the command line title bar, right-click on the command line and pick **Close**, type COMMANDLINEHIDE, or press [Ctrl]+[9].

PROFESSIONAL TIP

While learning AutoCAD, pay close attention to the prompts displayed in the dynamic input area and at the command line.

Keyboard Keys

Many keys on the keyboard, known as *shortcut keys* or *keyboard shortcuts*, allow you to perform AutoCAD functions quickly. Become familiar with these keys to improve your AutoCAD performance. Whenever it is necessary to cancel a tool or dialog box, press the *escape key* [Esc] in the upper-left corner of the keyboard. Some tool sequences require that you press [Esc] twice to cancel the operation.

Use the up and down arrow keys to select previously used tools. When no tool is active, press the up arrow to display the previously used tool. If dynamic input is active, previously used tools appear near the crosshairs by default. To display previously used tools at the command line, you must pick the command line before pressing the up arrow, or turn off dynamic input. If you continue to press the up arrow, AutoCAD continues to backtrack through the tools you have used. Press [Enter] to activate a displayed tool.

Function keys provide instant access to tools. They can also be programmed to perform a series of commands. Control and shift key combinations require that you press and hold the [Ctrl] or [Shift] key and then press a second character. You can

shortcut key (keyboard shortcut): Single key or key combination used to quickly issue a command or select an option.

escape key: Keyboard key used to cancel a tool or exit a dialog box.

function keys: The keys labeled [F1] through [F12] along the top of the keyboard.

activate several tools using [Ctrl] key combinations. A tooltip typically indicates if a key combination is available.

Reference Material

Shortcut Keys

For a complete list of keyboard shortcuts, go to the **Reference Material** section of the Student Web site (www.g-wlearning.com/CAD) and select **Shortcut Keys**.

Exercise 1-10

Access the Student Web site (www.g-wlearning.com/CAD) and complete Exercise 1-10.

Getting Help

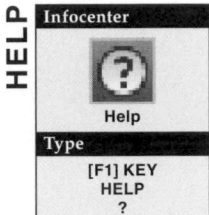

HELP

Infocenter

Help

Type

[F1] KEY
HELP
?

If you need help with a specific tool, option, or AutoCAD feature, use this textbook as a guide, or use the help system contained in the **AutoCAD Help** window. The graphic shown in the margin identifies several ways to access the **AutoCAD Help** window. You can also access the **AutoCAD Help** window from the **InfoCenter**, described later in this chapter, or by selecting **Help** from a shortcut menu.

The **AutoCAD Help** window consists of two frames. See **Figure 1-26**. The left frame has three tabs for locating help topics. The right frame displays the selected help topics. The **Contents** tab in the left frame displays a list of book icons and topic names. The book icons represent the organizational structure of books of topics within the AutoCAD documentation. Topics contain the actual help information; the icon used to represent a topic is a sheet of paper with a question mark. To open a book or a help topic, double-click on its name or icon.

NOTE

If you are unfamiliar with how to use a Windows help system, spend time now exploring all the topics under **AutoCAD Help** in the **Contents** tab of the **AutoCAD Help** window.

Although the **Contents** tab of the **AutoCAD Help** window is useful for displaying all the topics in an expanded table of contents manner, it may not be very useful when you are searching for a specific item. In this case, you should refer to the help file index. This is the function of the **Index** tab. Use the **Search** tab to search the help documents for specific words or phrases.

In addition to the two frames, six buttons reside at the top of the **AutoCAD Help** window. The **Hide/Show** button controls the visibility of the left frame. Use the **Back** button to view the previously displayed help topic. Use the **Forward** button to go forward to help pages you viewed before pressing the **Back** button. The **Home** button takes you to the AutoCAD Help page. The **Print** button prints the currently displayed help topic. The **Options** button presents a menu with a variety of items used to control other aspects of the **AutoCAD Help** window.

Figure 1-26.
Get help using the **AutoCAD Help** window or the **InfoCenter.**

InfoCenter

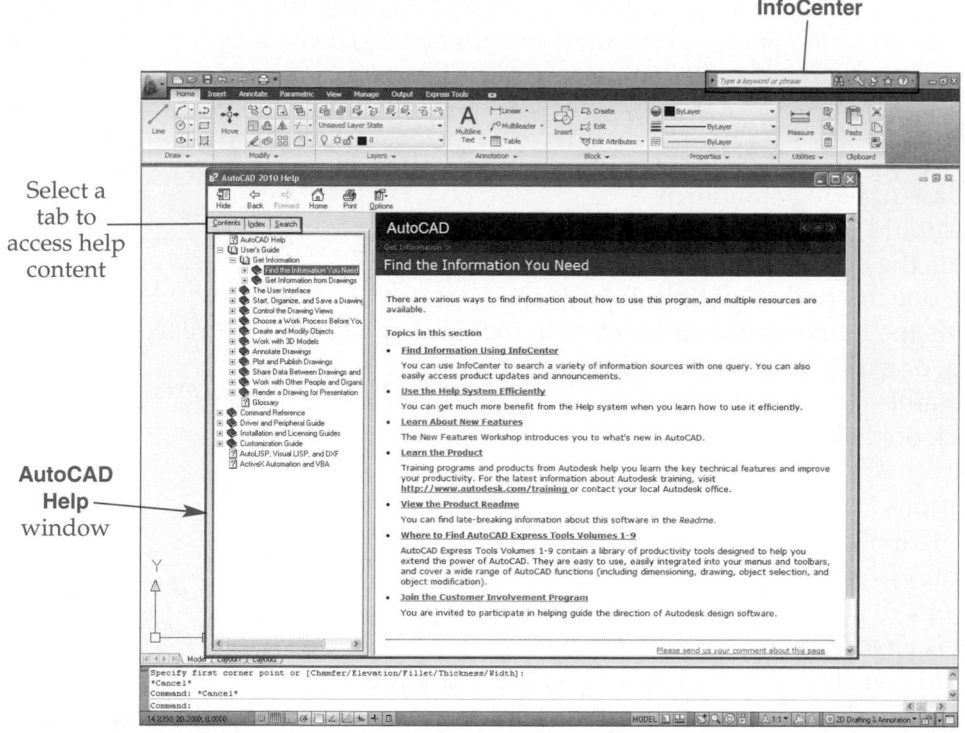

Select a tab to access help content

AutoCAD Help window

PROFESSIONAL TIP

If you press the [F1] key while you are in the process of using a tool, help information associated with the active tool displays. This *context-oriented help* saves valuable time, since you do not need to scan through the help contents or perform a search to find the information.

context-oriented help: Help information for the active tool.

Using the InfoCenter

The **InfoCenter**, located on the right side of the title bar as shown in **Figure 1-26**, allows you to search for help topics without first displaying the **AutoCAD Help** window. It also provides buttons for access to the **Subscription Center, Communication Center,** and the **Favorites** list. Type a question in the text box to search for topics. Then select the appropriate topic from the list to display it in the **AutoCAD Help** window. To add a topic to the **Favorites** list, pick the star next to the topic. Pick the **Subscription Center** button to access information associated with your license and subscription eligibility, options, and services. Pick the **Communication Center** button to access content on a variety of help topics. Pick the **Favorites** button to access any help topics you have stored.

Exercise 1-11

Access the Student Web site (www.g-wlearning.com/CAD) and complete Exercise 1-11.

Chapter Test

Answer the following questions. Write your answers on a separate sheet of paper or go to the Student Web site (www.g-wlearning.com/CAD), and complete the electronic chapter test.

1. Describe at least one application for AutoCAD software.
2. Briefly explain what is involved in planning a drawing.
3. What are drawing standards?
4. Why should you save your work every ten to fifteen minutes?
5. What is the quickest method for starting AutoCAD?
6. Name one method of exiting AutoCAD.
7. What is the name for the interface that includes on-screen features?
8. Define or explain the following terms:
   A. Default
   B. Pick or click
   C. Hover
   D. Button
   E. Function key
   F. Option
   G. Tool
9. What is a workspace?
10. How do you change from one workspace to another?
11. What is the difference between a docked interface item and a floating interface item?
12. How do you select the locking options to lock the interface items in either their floating and docked state?
13. What is a flyout?
14. How do you access a shortcut menu?
15. What does it mean when a shortcut menu is described as context-sensitive?
16. Explain the basic function of the **Application Menu**.
17. Describe the **Application Menu** search tool and briefly explain how to use it.
18. Briefly describe an advantage of using the ribbon.
19. What is the function of tabs in the ribbon?
20. What is another name for a palette?
21. Describe the function of the application status bar.
22. What is the meaning of the … (ellipsis) in a menu option or button?
23. What are the two primary methods for accessing AutoCAD tools? List interface items associated with each.
24. Briefly describe the function of dynamic input.
25. Briefly explain the function of the [Esc] key.
26. How do you access previously used tools when dynamic input is on?
27. Name the function keys that execute the following tasks. (Refer to the Shortcut Keys document in the **Reference Material** section on the Student Web Site.)
   A. Snap mode (toggle)
   B. Grid mode (toggle)
   C. Ortho mode (toggle)
28. Describe two ways to access the **AutoCAD Help** window.
29. Describe the purpose of the book icons in the **Contents** tab of the **AutoCAD Help** window.
30. What is context-oriented help, and how is it accessed?

Problems

Start AutoCAD if it is not already started. Follow the specific instructions for each problem.

▼ Basic

1. Perform the following tasks:
    A. Open the **AutoCAD Help** window.
    B. In the **Contents** tab, expand the **User's Guide** book.
    C. Expand the **Get Information** book.
    D. Expand the **Find the Information You Need** book.
    E. Pick and read each topic in the right pane.
    F. Close the **AutoCAD Help** window, and then close AutoCAD.

2. Launch AutoCAD using the Start button on the Windows task bar.
    A. Move the cursor over the buttons in the status bar and read the tooltip for each.
    B. Slowly move the cursor over each of the ribbon panels and read the tooltips.
    C. Pick the **Application Menu** to display it. Hover over the **File** menu, then use the right arrow key to move to the **File** options. Then use the down arrow key to move through all the menu options.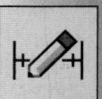
    D. Press the [Esc] key to dismiss the menu.
    E. Close AutoCAD.

▼ Intermediate

3. Interview your drafting instructor or supervisor and try to determine what type of drawing standards exist at your school or company. Write them down and keep them with you as you learn AutoCAD. Make notes as you progress through this textbook on how you use these standards. Also, note how the standards could be changed to match the capabilities of AutoCAD.

4. Research your drawing department standards. If you do not have a copy of the standards, acquire one. If AutoCAD standards have been created, make notes as to how you can use these in your projects. If no standards exist in your department or company, make notes about how you can help develop standards. Write a report on why your school or company should create CAD standards and how they would be used. Describe who should be responsible for specific tasks. Recommend procedures, techniques, and forms, if necessary. Develop this report as you progress through your AutoCAD instruction and as you read this textbook.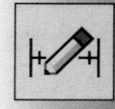

5. Develop a drawing planning sheet for use in your school or company. List items you think are important for planning a CAD drawing. Make changes to this sheet as you learn more about AutoCAD.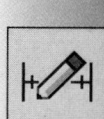

6. Create a freehand sketch of the default AutoCAD window with the **2D Drafting & Annotation** workspace active. Label each of the screen areas. To the side of the sketch, write a short description of the function of each screen area.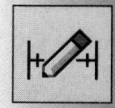

7. Create a freehand sketch showing three examples of tooltips displayed as you hover the cursor over an item. To the side of the sketch, write a short description of each example's function.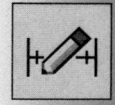

8. Using the **Application Menu** search tool, type the letter C and review the information provided in the **Application Menu**. Then add the letter L. How does the information change? Continue typing O, S, and E to complete the **CLOSE** command. Write a short paragraph explaining how you might use this search tool to find a command if you are unsure how the command is spelled or where it is located.

▼ Advanced

9. Research and write a report of 250 words or less covering the U.S. National CAD Standard.

10. Research and write a report of 250 words or less covering workplace ethics, especially as related to CAD applications and CAD-related software.

11. Research and write a report of 150 words or less covering an ergonomically designed CAD workstation. Include a sketch of what you might consider a high-quality design for a workstation and label its characteristics.

Learning Objectives

After completing this chapter, you will be able to do the following:

✓ Start a new drawing.
✓ Save your work.
✓ Close files.
✓ Open saved files.
✓ Work with multiple open documents.
✓ Create drawing templates.
✓ Determine and specify drawing units and limits.

In this chapter, you will learn how to start new drawings, save drawings, open existing drawings, and prepare drawing templates. This chapter also explains some of the basic drawing aids used in AutoCAD. You will find the drawing settings described in this chapter very useful as you begin working with drawing and drawing template files.

Starting a New Drawing

There are two primary AutoCAD file types: *drawing files*, which have a .dwg extension, and *drawing template files*, also known as *templates*, which have a .dwt extension. New drawings typically begin from templates that include standard drawing settings and objects. All template settings and contents are included in a new drawing. To help avoid confusion as you learn AutoCAD, remember that a new drawing file references a drawing template file, but the drawing file is where you draw.

Depending on your initial setup options or any customization you have done, when you launch AutoCAD, a new drawing file appears. The drawing references a default template. If you skip the initial setup process as advised in this textbook, the drawing you initially see references the acad.dwt template provided by AutoCAD. The drawing is appropriate for initial drawing applications and uses basic decimal U.S. Customary (inch) unit settings. You are ready to create a drawing, save, and close the file. To start another new drawing, you can use a template or start from scratch.

drawing files:
Files that contain the actual drawing geometry and information.

drawing template files (templates):
Files referenced to develop new drawings; templates contain standard drawing settings and objects.

Choosing a Template

QNEW

Quick Access

New

Application Menu

New
New
> Drawing

Type

QNEW
[Ctrl]+[N]

The **QNEW** tool is the primary means of starting a new drawing. By default, the **Select template** dialog box appears when you access the **QNEW** tool. See **Figure 2-1**. The **Select template** dialog box lists the templates found in the specified drawing template folder. The default template folder shown in **Figure 2-1** includes a variety of templates supplied with AutoCAD.

NOTE

All file navigation dialog boxes, including the **Select template** dialog box and those used to save, close, and open files, support auto-complete. Type in the **File Name** text box to view a list of files matching the characters you enter.

The tutorial templates for manufacturing and architecture include a border and title block. All other templates are blank, but include drawing settings specific to the requirements of a certain industry or drawing. For general 2D drawing applications, use the acad.dwt template, which includes basic U.S. Customary (inch) unit settings, or the acadiso.dwt template, which includes basic metric unit settings according to ISO standards. To use a template to begin a new drawing, double-click on the file name, right-click on the file and pick **Select**, or select the file and pick the **Open** button.

NOTE

You can also access the **QNEW** tool from the **Quick View Drawings** tool described later in this chapter.

PROFESSIONAL TIP

Use the **Options** dialog box to change the default drawing template folder displayed in the **Select template** dialog box. In the **Files** tab, expand Template Settings, and then expand Drawing Template File Location. Pick the **Browse...** button to select a folder.

Figure 2-1.
Use the **Select template** dialog box to begin a new drawing.

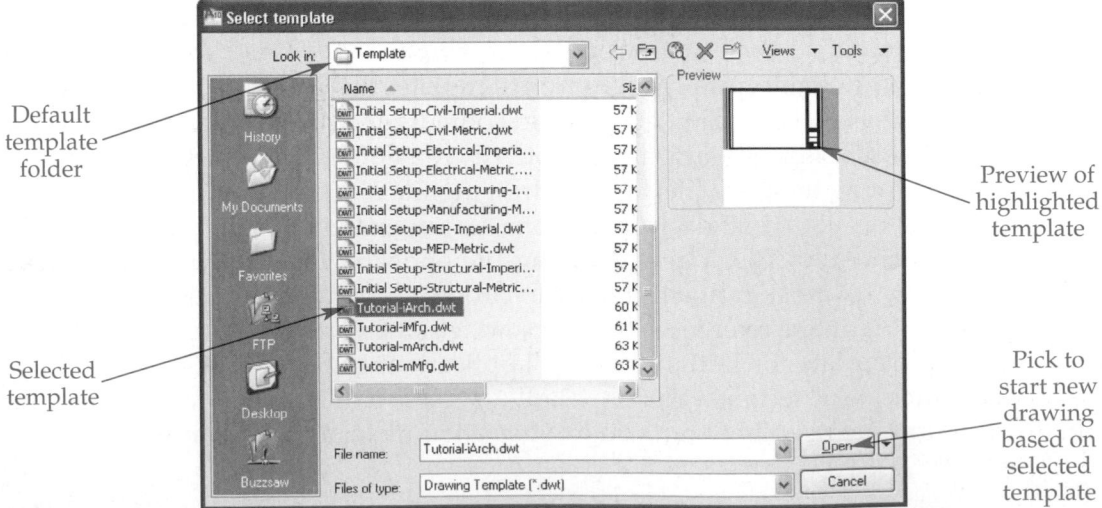

Starting from Scratch

For a new AutoCAD user, an effective way to begin a new drawing is to start "from scratch" using a blank drawing file without a border, title block, modified layouts, or customized drawing settings. Start from scratch when you are just beginning to learn AutoCAD, when you plan to create your own template, or when the start or end of a project is unknown. To start a drawing from scratch, pick the flyout next to the **Open** button in the **Select template** dialog box. See **Figure 2-2.** Then, to begin a drawing using basic inch unit settings, pick **Open with no Template-Imperial**, or to begin a drawing using basic metric unit settings, pick **Open with no Template-Metric**.

Setting a Quick Start Template

AutoCAD provides a quick start feature to begin a drawing using the **QNEW** tool and a specific template, skipping the **Select template** dialog box. Set this function in the **Options** dialog box. Pick the **Files** tab, expand the Template Settings option, and then expand the Default Template File Name for QNEW function. See **Figure 2-3.** None displays by default, which causes the **Select template** dialog box to appear. Pick the **Browse...** button to select a specific template to launch each time you use the **QNEW** tool.

PROFESSIONAL TIP

Use the **NEW** tool to override the quick start template and display the **Select template** dialog box. This allows you to use the **QNEW** tool to begin a drawing with the most frequently used template, but access other templates when necessary.

Type
NEW

Saving Your Work

You should save your drawing or template immediately after you begin work. Then, save every 10 to 15 minutes while working. Saving every 10 to 15 minutes results in less lost work if a software error, hardware malfunction, or power failure occurs. Several AutoCAD tools allow you to save your work. In addition, any tool or option ending the AutoCAD session provides an alert asking if you want to save changes to the drawing or template. This gives you a final option to save or not save changes.

Naming Drawings

Set up a system that allows you to determine the content of a drawing by the drawing name. Drawing names often identify a product by name and number—for example, VICE-101, FLOORPLAN, or 6DT1009. Use a standard drawing naming system that contains a clear and concise reference to the project, part number, process, sheet number, and revision level, depending on the product.

Figure 2-2.
Pick the arrow next to the **Open** button and select an **Open with no Template** option to start a drawing from scratch.

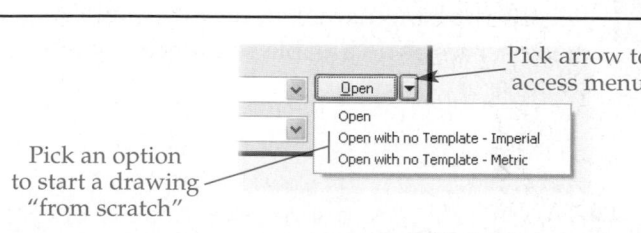

Pick arrow to access menu

Pick an option to start a drawing "from scratch"

Figure 2-3.
Specifying a template for the **QNEW** tool.

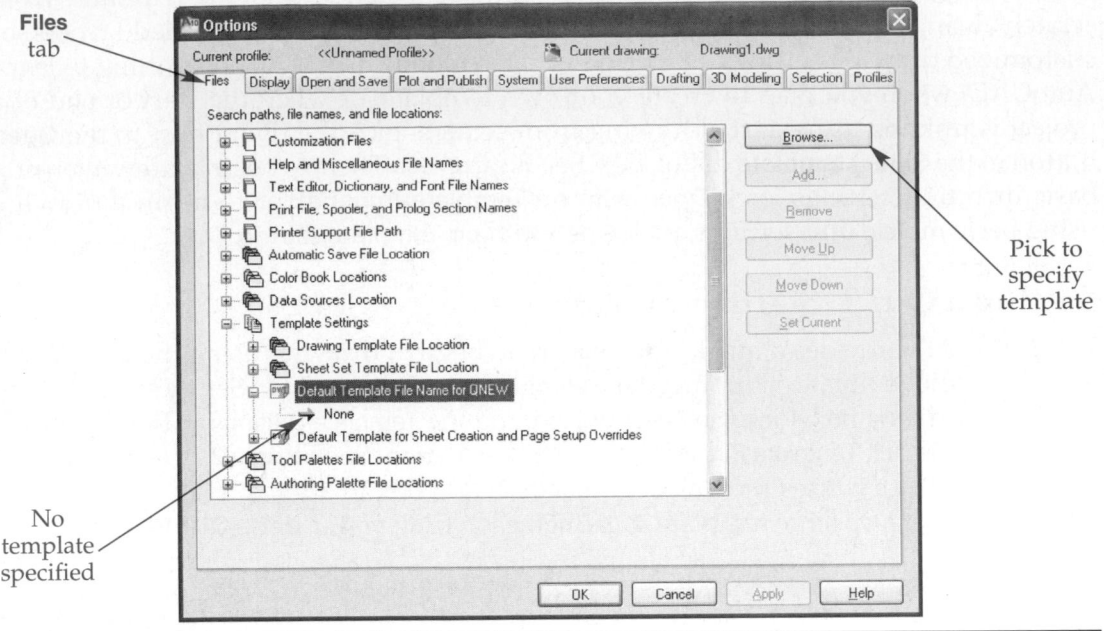

Although it is possible to give a drawing file an extended name, such as Details for Top Half of Compressor Housing for ACME, Inc., Part Number 4011A, Revision Level C, a standardized naming system that uses a shorter name, such as ACME.4011A.C, provides all of the necessary information. If you need to add information for easier recognition, add the content to the base name. Using the previous example, you might name the file ACME 4011A.C Compressor Housing.Top.Casting Details.

Some rules and restrictions apply to naming drawings. You can use most alphabetic and numeric characters and spaces, as well as most punctuation symbols. Characters that cannot be used include the quotation mark ("), asterisk (*), question mark (?), forward slash (/), and backward slash (\). You can use a maximum of 256 characters. You do not have to include the file extension, such as .dwg or .dwt, with the file name. File names are not case sensitive. For example, you can name a drawing PROBLEM 2-1, but Windows interprets Problem 2-1 as the same file name.

NOTE

The U.S. National CAD standard includes a comprehensive file naming structure for architectural and construction-related drawings. The system can be adapted for other disciplines.

PROFESSIONAL TIP

Every school or company should have a drawing naming system. Record drawing names in a part numbering or drawing name log. A log serves as a valuable reference if you forget what the drawings contain.

Using the QSAVE Tool

The **QSAVE** tool is the most frequently used tool for saving a drawing. If the current file has not yet been named, the **QSAVE** tool displays the **Save Drawing As** dialog box. See **Figure 2-4.** The **Save Drawing As** dialog box is a standard file selection dialog box. To save a file, first choose the type of file to save from the **Files of type:** drop-down list. Select the AutoCAD 2010 Drawing (*.dwg) option for most applications. Use the AutoCAD Drawing Template (*.dwt) to save a template file. Choose the appropriate older AutoCAD version from the list to convert an AutoCAD 2010 file to an older AutoCAD format. This allows someone using an older version of AutoCAD to view your file. You also have the option of saving to a *drawing exchange file (DXF)* or *drawing standards file (DWS)* format.

Next, select the folder in which to store the file from the **Save in:** drop-down list. To move upward from the current folder, pick the **Up one level** button. To create a new folder in the current location, pick the **Create New Folder** button and type the folder name.

The name Drawing1 appears in the **File name:** text box, if the file is the first file started since the launch of AutoCAD. Change the name to the desired file name. You do not need to include the extension. Once you specify the correct location and file name, pick the **Save** button to save the file. You can also press the [Enter] key to activate the **Save** button. If you already saved the current file, accessing the **QSAVE** tool updates, or resaves, the file based on the current file state. In this situation, **QSAVE** issues no prompts and displays no dialog box.

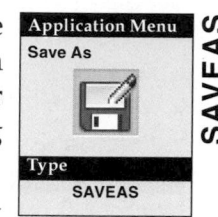

drawing exchange file (DXF): A file format often used by other CADD systems.

drawing standards file (DWS): A file used to check the standards of another file using AutoCAD standards-checking tools.

Using the SAVEAS Tool

Use the **SAVEAS** tool to save a copy of a file using a different name, or to save a file in an alternative format, such as a previous AutoCAD release format. You can also use the **SAVEAS** tool when you *open* a drawing template file to use as a basis for another drawing. This leaves the template unchanged and ready to use for starting other drawings.

The **SAVEAS** tool always displays the **Save Drawing As** dialog box. The name and location of a previously saved file appears. Confirm that the **Files of type:** drop-down list displays the desired file type and that the **Save in:** drop-down list displays the correct drive and folder. Type the new file name in the **File name:** text box and pick the **Save** button.

Figure 2-4.
The **Save Drawing As** dialog box.

Select folder where drawing will be saved

Move up one level from current folder

Create a new folder

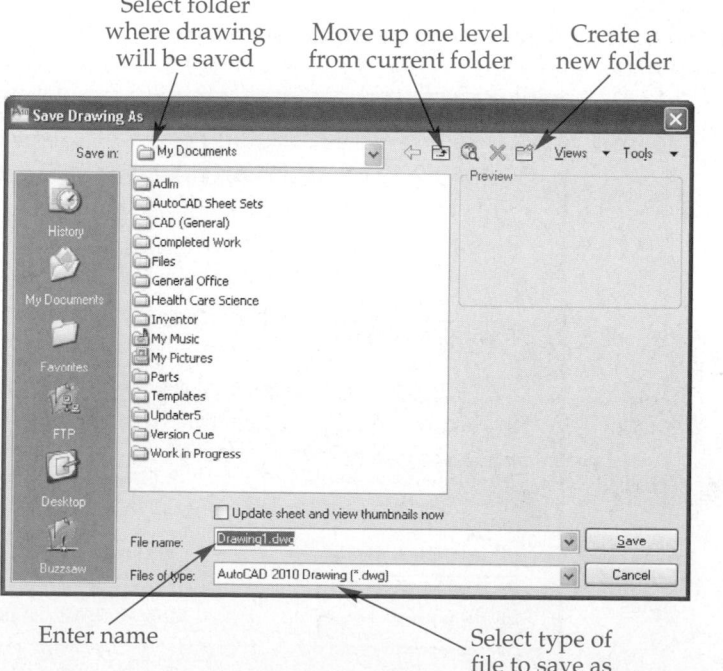

Enter name

Select type of file to save as

Automatic Saves

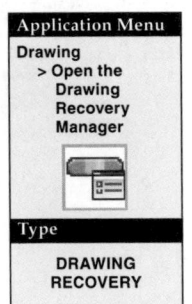

automatic save: A save procedure that occurs at specified intervals without your input.

AutoCAD provides an *automatic save* tool that automatically creates a temporary backup file during a work session. Settings in the **File Safety Precautions** area of the **Open and Save** tab in the **Options** dialog box control automatic saves. See **Figure 2-5**. Automatic save is on by default. Type the number of minutes between saves in the **Minutes between saves** text box. The default setting automatically saves every 10 minutes. By default, AutoCAD names automatically saved files *FileName_n_n_nnnn*. sv$ in the C:\Documents and Settings*user*\Local Settings\Temp folder.

The automatic save timer starts as soon as you make a change to the file and resets when you save the file. The file automatically saves when you start the first tool after reaching the automatic save time. Keep this in mind if you let the computer remain idle, as an automatic save does not execute until you return and use a tool. Be sure to save your file manually if you plan to be away from your computer for an extended period.

The automatic save feature is intended for use in case AutoCAD shuts down unexpectedly. Therefore, when you close a file, the automatic save file associated with that file automatically deletes. If AutoCAD shuts down unexpectedly, the automatic save file is available for use. By default, the next time you open AutoCAD after a system failure, the **Drawing Recovery Manager** displays, containing a node for every file to display all

Application Menu

Drawing
> Open the
Drawing
Recovery
Manager

Type

DRAWING
RECOVERY

Figure 2-5.
Use the **Open and Save** tab in the **Options** dialog box to set up the automatic save feature and backup files.

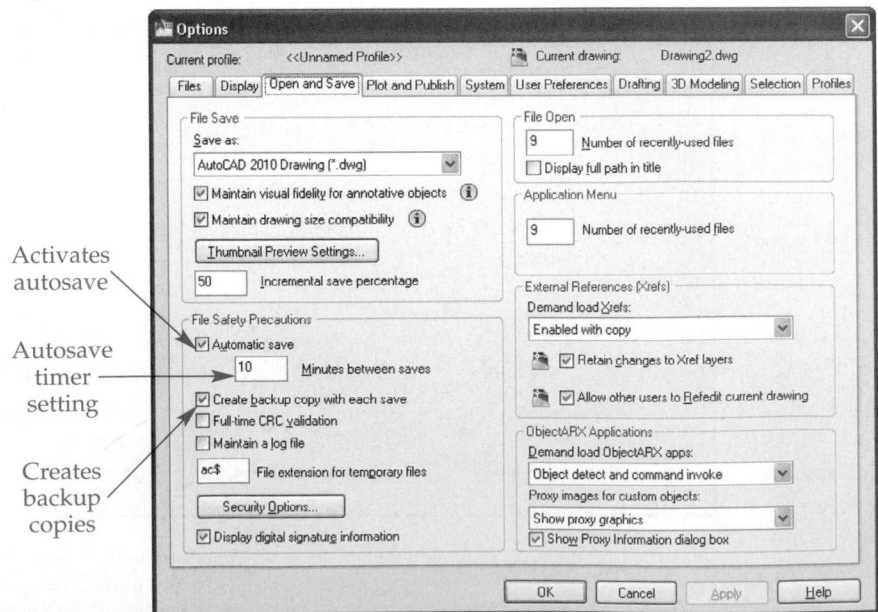

Activates autosave

Autosave timer setting

Creates backup copies

of the available versions of the file: the original file, the recovered file saved at the time of the system failure, the automatic save file, and the .bak file. Pick each version in the **Drawing Recovery Manager** to view it, determine which version you want to save, and then save that file. You can save the recovered file over the original file name.

> **NOTE**
>
> The **Automatic Save File Location** listing in the **Files** tab of the **Options** dialog box determines the folder where the automatic save files are stored.

Using Backup Files

By default, an AutoCAD backup file, with the extension .bak, automatically saves in the same folder where the drawing or template is located. When you save a drawing or template, the drawing or template file updates, and the old drawing or template file overwrites the backup file. Therefore, the backup file is always one save behind the drawing or template file.

The backup feature is on by default and is controlled using the **Create backup copy with each save** check box in the **Open and Save** tab of the **Options** dialog box. Refer again to **Figure 2-5.** If AutoCAD shuts down unexpectedly, you may be able to recover a file from the backup version using the **Drawing Recovery Manager.** You can also rename a backup and try to open it as a drawing or template. Use Windows Explorer to rename the file, changing the file extension from .bak to .dwg to restore a drawing, or from .bak to .dwt to restore a template.

Windows Explorer
For more information about using Windows Explorer, go to the Student Web site (www.g-wlearning.com/CAD), select this chapter, and select **Windows Explorer**.

Recovering a Damaged File
For information about recovering a damaged file, go to the Student Web site (www.g-wlearning.com/CAD) , select this chapter, and select **Recovering a Damaged File**.

Closing Files

Use the **CLOSE** tool to exit the current file without ending the AutoCAD session. One of the quickest methods of closing a file is to pick the **Close** button from the drawing window title bar. If you close a file before saving your work, AutoCAD prompts you to save or discard changes. Pick the **Yes** button to save the file. Pick the **No** button to discard any changes made to the file since the previous save. Pick the **Cancel** button if you decide not to close the drawing and want to return to the drawing area.

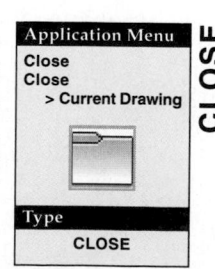

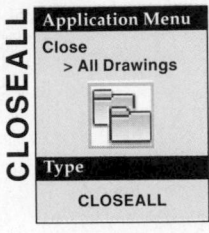

CLOSEALL

Application Menu
Close
> All Drawings

Type
CLOSEALL

As you will learn, AutoCAD allows you to have multiple files open at the same time. Use the **CLOSEALL** tool to close all open files. You are prompted to save each file in which you made changes.

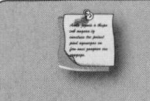

NOTE

You can also close files using the **Quick View Drawings** tool, described later in this chapter.

Exercise 2-1

Access the Student Web site (www.g-wlearning.com/CAD) and complete Exercise 2-1.

Opening Saved Files

Once you save and close a file, you will eventually want to reopen the file. To open a file, use the **OPEN** tool, select recently opened files from the **Application Menu**, or open a file from Windows Explorer.

Using the OPEN Tool

OPEN

Quick Access
Open...

Application Menu
Open
Open
> Drawing

Type
OPEN
[Ctrl]+[O]

When you access the **OPEN** tool, the **Select File** dialog box appears, containing a list of folders and files. See **Figure 2-6.** Double-click on a file folder to open it, and then double-click on the desired file to open the file.

Figure 2-6.
Use the **Select File** dialog box to open an AutoCAD file. In this illustration, the AutoCAD 2010\ Sample folder is open. The Architectural – Annotation Scaling and Multileaders drawing is selected and appears in the File name: text box and in the preview area.

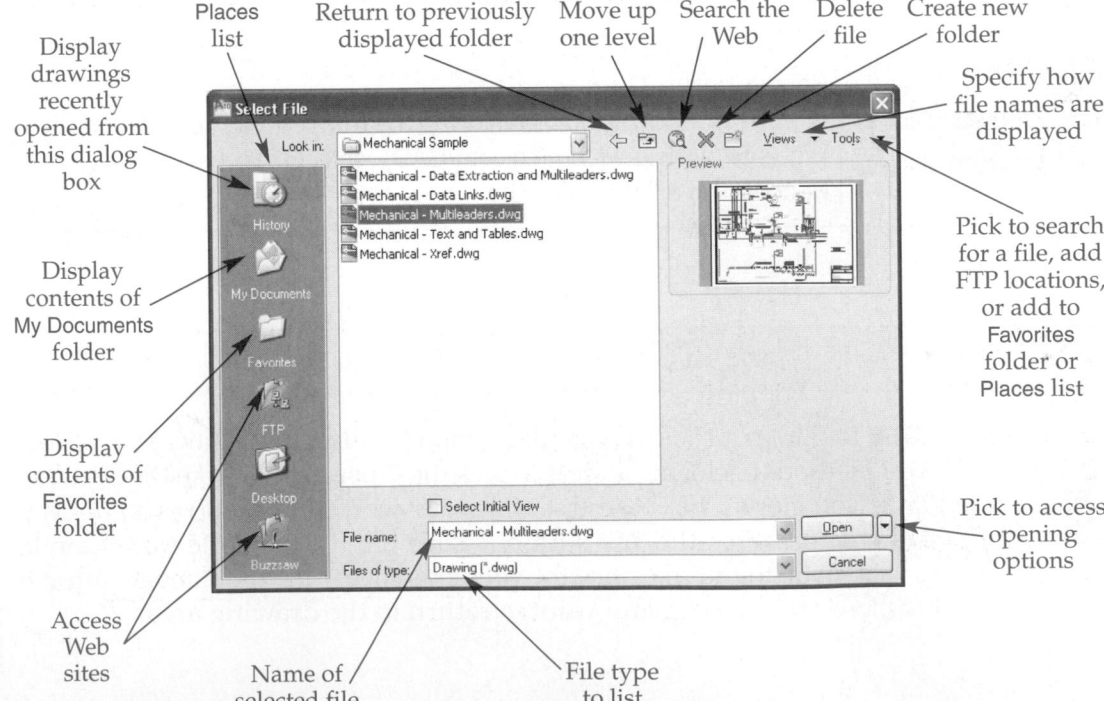

An image of the selected file appears in the **Preview** area. This provides an easy way for you to view the contents of a file without loading it into AutoCAD. After picking a file name to highlight, you can quickly highlight other files and scan through previews using the keyboard arrow keys. Use the up and down arrow keys to move vertically between files, and use the left and right arrow keys to move horizontally. The **Select File** dialog box includes a list on the left side that provides instant access to certain folders. See **Figure 2-7**.

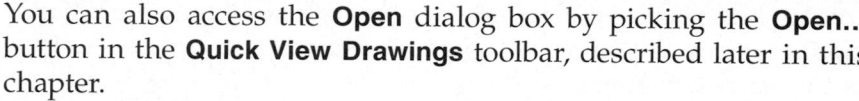 

You can also access the **Open** dialog box by picking the **Open...** button in the **Quick View Drawings** toolbar, described later in this chapter.

Figure 2-7.
Additional features in the **Select File** dialog box.

Button	Description
History	Lists drawing files recently opened from the **Select File** dialog box.
My Documents	Displays the files and folders contained in the My Documents folder for the current user.
Favorites	Displays files and folders located in the Favorites folder for the current user. To add the folder displayed in the **Look in:** box to the favorites list, select **Add to Favorites** from the **Tools** flyout.
FTP	Displays available FTP (file transfer protocol) sites. To add or modify the listed FTP sites, select **Add/Modify FTP Locations** from the **Tools** flyout.
Desktop	Lists the files, folders, and drives located on the computer desktop.
Buzzsaw	Displays projects on the Buzzsaw Web site. Buzzsaw.com is designed for the building industry. After setting up a project hosting account, users can access drawings from a given construction project on the Web site. This allows the various companies involved in the project to have instant access to the drawing files.

Exercise 2-2

Access the Student Web site (www.g-wlearning.com/CAD) and complete Exercise 2-2.

Finding Files

Pick **Find...** from the **Tools** flyout at the top of the **Select File** dialog box to search for files. The **Find** dialog box, shown in **Figure 2-8,** appears. If you know the file name, type it in the **Named:** text box. If you do not know the name, you can use wildcard characters, such as *, to narrow the search.

Choose the type of file to search for from the **Type:** drop-down list. You can search for DWG, DWS, DXF, or DWT files using the **Find** dialog box. Use Windows Explorer or Windows Search to search for other file types.

To complete a search more quickly, avoid searching the entire hard drive. If you know the folder in which the file is located, specify the folder in the **Look in:** text box. Pick the **Browse** button to select a folder from the **Browse for Folder** dialog box. Check the **Include subfolders** check box to search the subfolders within the selected folder. The **Date Modified** tab provides options to search for files modified within a certain time. This option is useful if you want to list all drawings modified within a specific week or month.

NOTE

Right-click on a folder or file to display a shortcut menu of options. Pick somewhere off the menu to close the menu.

PROFESSIONAL TIP

Certain file management capabilities are available in file dialog boxes, similar to those in Windows Explorer. For example, to rename an existing file or folder, pick it once and pause for a moment, then pick the name again. This places the name in a text box for editing. Type the new name and press [Enter].

Figure 2-8.
Use the **Find** dialog box to locate drawing files.

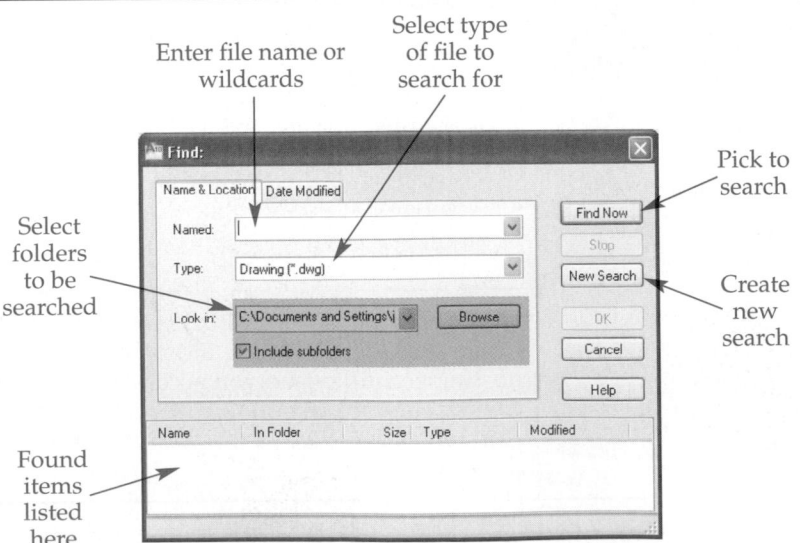

AutoCAD and Its Applications—Basics

Exercise 2-3

Access the Student Web site (www.g-wlearning.com/CAD) and complete Exercise 2-3.

Opening Recent Drawings

Pick the **Recent Documents** button in the **Application Menu** to display a list of recently opened documents. See **Figure 2-9.** This provides convenient access to files that may be related to your current project. By default, the name and location of up to 9 recent files appears.

Use the **Ordered List** drop-down list to organize recent files. Select **By Ordered List**, **By Access Date**, **By Size**, or **By Type** according to how you want files arranged. Select from the display options flyout to specify how recent files appear. Choose the appropriate option to display files as icons, small images, medium images, or large images.

When you hover over a file in the list of recent documents, a tooltip with file information and a preview of the file appears. Pick a file from the list to open the file. Pick the pushpin icon to the right of the file to keep the file on the recent documents list. Unpinned files are eventually removed from the recent documents list as you open other files.

Figure 2-9.
Use the **Recent Documents** menu on the **Application Menu** to locate and open a recent file quickly.

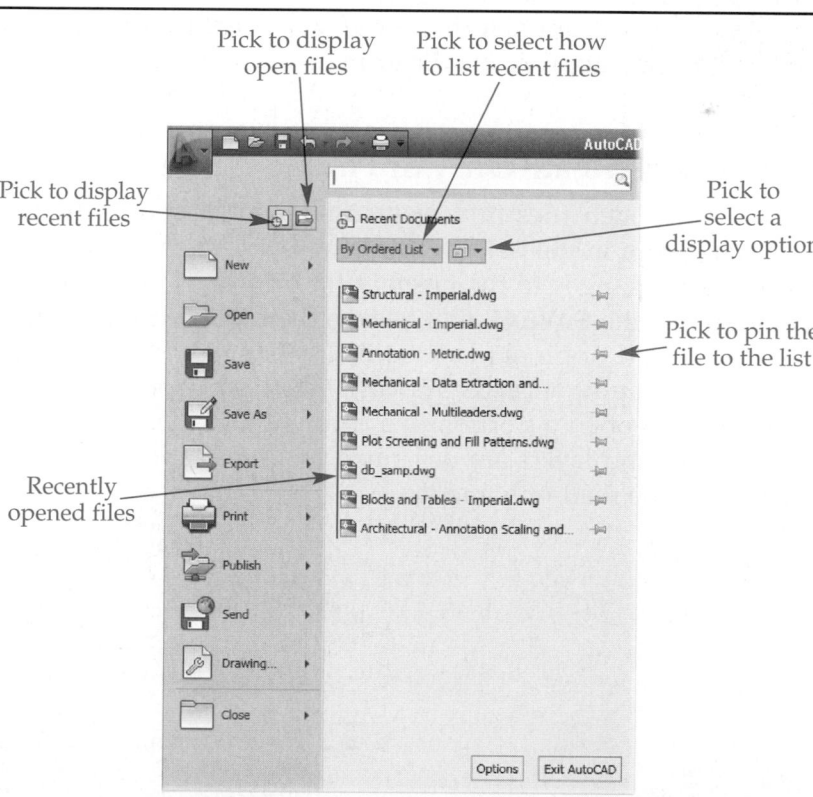

Pick to display open files

Pick to select how to list recent files

Pick to display recent files

Pick to select a display option

Pick to pin the file to the list

Recently opened files

Using Windows Explorer

Double-click on an AutoCAD file from Windows Explorer to open the file. If AutoCAD is not already running, it starts and the file opens. You can also drag-and-drop a file onto the AutoCAD **Command** line. If AutoCAD is not running, you can drag-and-drop the file onto the AutoCAD 2010 icon on your desktop. AutoCAD then starts and opens the drawing file.

Opening Old Files

In AutoCAD 2010, you can open AutoCAD files created in AutoCAD Release 12 or later. When you open and work on a file from a previous release, AutoCAD automatically updates the file to the AutoCAD 2010 file format when you save. A file saved in AutoCAD 2010 displays in the **Preview** image tile in the **Select File** dialog box.

Opening as Read-Only or Partial Open

You can open files in various modes by selecting the appropriate option from the **Open** flyout in the **Select File** dialog box. When you open a file as *read-only*, you cannot save changes to the original file. However, you can make changes to the file and then use the **SAVEAS** tool to save the modified file using a different name. This ensures that the original file remains unchanged.

When opening a large drawing you may choose to issue a *partial open*. This allows you to open a portion of a drawing by selecting specific views and layers to open. Views and layers are described in later chapters. You can also partially open a drawing in the read-only mode.

read-only: Describes a drawing file opened for viewing only. You can make changes to the drawing, but you cannot save changes without using the SAVEAS tool.

partial open: Opening a portion of a file by specifying only the views and layers you need to see.

Working with Multiple Documents

Most drafting projects are composed of a number of files; each file presents or organizes a different aspect of the project. For example, an architectural drafting project might include a site plan, a floor plan, electrical and plumbing plans, and assorted detail drawings. Another example is a mechanical assembly composed of several unique parts. The required drawings in this example might include an overall assembly drawing, plus individual detail drawings of each component.

Drawings and other files associated with a project are usually closely related to each other. By opening two or more files at the same time, you can easily reference information contained in existing files while working in a new file. AutoCAD allows you to copy all or part of the contents from one file directly into another using a drag-and-drop or similar operation.

Controlling Windows

Each file you start or open in AutoCAD appears in its own drawing window. The file name displays on the drawing window title bar if the drawing window is floating, or on the AutoCAD window title bar if you minimize the drawing window. The drawing windows and the AutoCAD window have the same relationship that program windows have with the Windows desktop. When you maximize a drawing window, it fills the available area in the AutoCAD window. Minimizing a drawing window displays it as a reduced-size title bar along the bottom of the drawing area. Pick the title bar of a minimized or floating drawing window to activate the file. You cannot move drawing windows outside the AutoCAD window. **Figure 2-10** illustrates drawing windows in a floating state and minimized.

Using the Quick View Drawings Tool

The **Quick View Drawings** tool is one of many AutoCAD tools that you can use to switch between open drawings. The visual display of this tool allows you to see and control open drawing files without actually changing drawing windows. The quickest way to access the **Quick View Drawings** tool is to pick the **Quick View Drawings** button on the status bar.

Figure 2-10.
Display drawing windows as floating windows when more than one file is open to move quickly from drawing to drawing. Minimized drawing windows display as reduced-size title bars. Pick the title bar to display a window control menu.

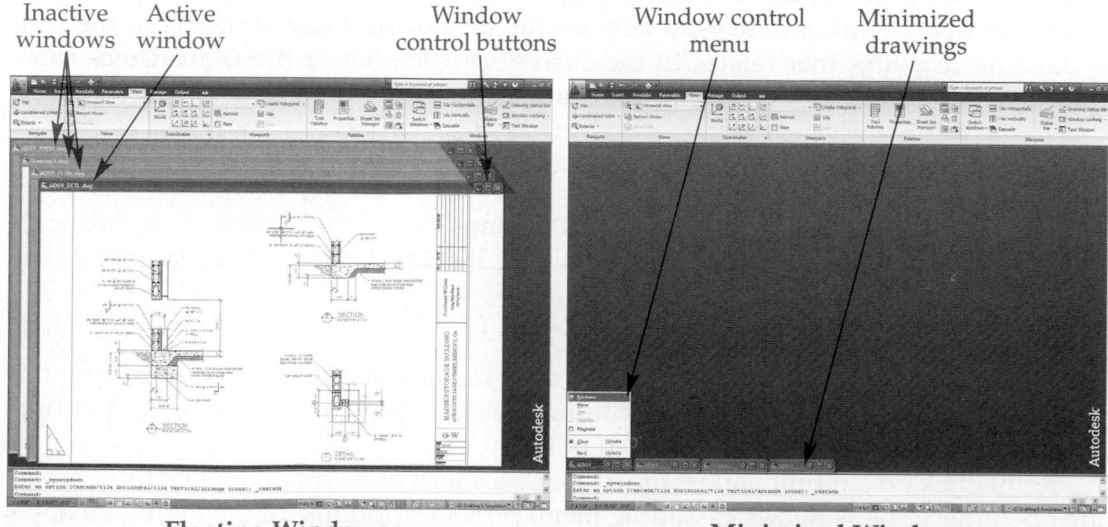

| Inactive windows | Active window | | Window control buttons | Window control menu | Minimized drawings |

Floating Windows **Minimized Windows**

Figure 2-11.
The **Quick View Drawings** tool offers an effective visual method for changing between open drawings.

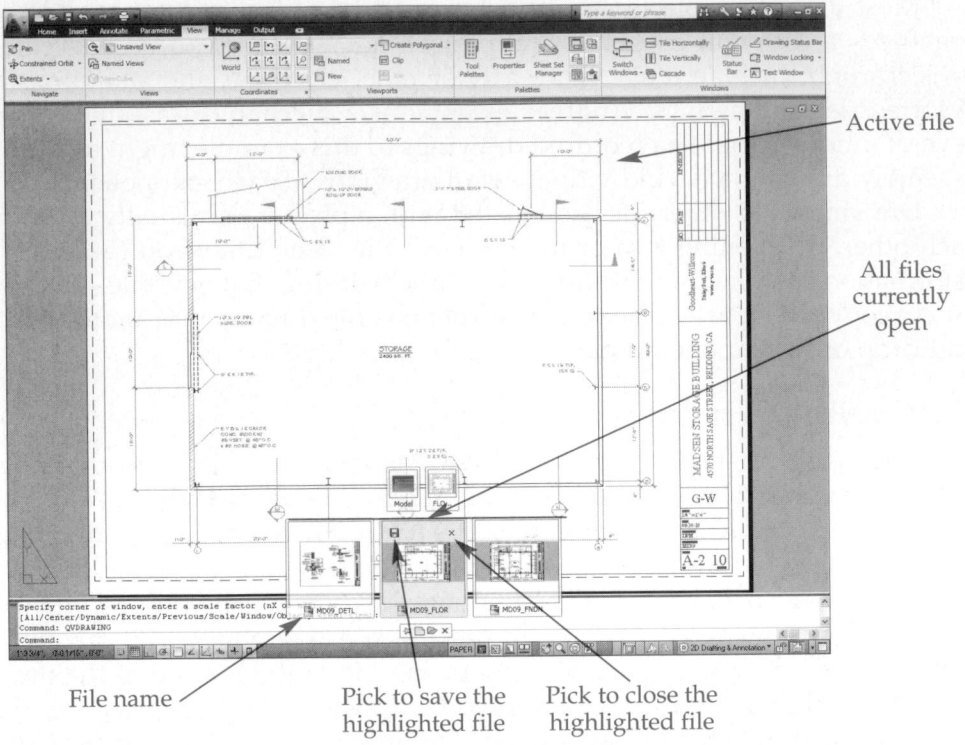

Active file

All files currently open

File name

Pick to save the highlighted file

Pick to close the highlighted file

The **Quick View Drawings** tool appears in the lower center of the AutoCAD window. See **Figure 2-11.** A thumbnail image and file name appear for each open file. Files are arranged in the order in which they were opened, with the file that was opened first on the left side of the row. The current file highlights when you initially access the **Quick View Drawings** tool. Pick the thumbnail of a different file to make the drawing window current, or move the cursor over a thumbnail to show additional options for controlling the drawing window.

The **Quick View Drawings** tool includes a small toolbar below the file thumbnail images. See **Figure 2-12.** By default, the **Quick View Drawings** tool hides when you pick a thumbnail to switch files. To keep the tool on-screen after you select a thumbnail, pick the **Pin Quick View Drawings** button on the left side of the toolbar. The **New...** button activates the **QNEW** tool and the **Open...** button activates the **OPEN** tool. The **New...** and **Open...** buttons are especially useful for starting a new drawing or opening an existing drawing that relates to the current project. Select the **Close Quick View Drawings** button to close the **Quick View Drawings** tool.

NOTE

If you pin the **Quick View Drawings** tool to the screen, close the tool and then access the tool again, the **Quick View Drawings** tool will still be in the pinned state.

You can also quickly display a specific drawing layout without first activating the drawing window. See **Figure 2-13.** Layouts are used to prepare a drawing for plotting and are fully described later in this textbook.

Right-click on a thumbnail image to access a shortcut menu of options for controlling open files. The **Windows** cascading menu provides options for arranging all open

Figure 2-12.
The **Quick View Drawings** toolbar provides access to basic functions directly from the **Quick View Drawings** feature.

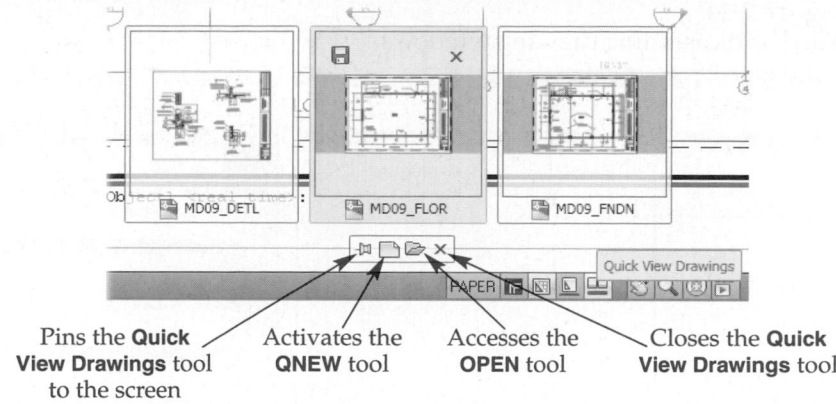

Pins the **Quick View Drawings** tool to the screen

Activates the **QNEW** tool

Accesses the **OPEN** tool

Closes the **Quick View Drawings** tool

Figure 2-13.
A—The initial display when you hover over a file thumbnail. B—Hover over the model space and layout thumbnails to enlarge.

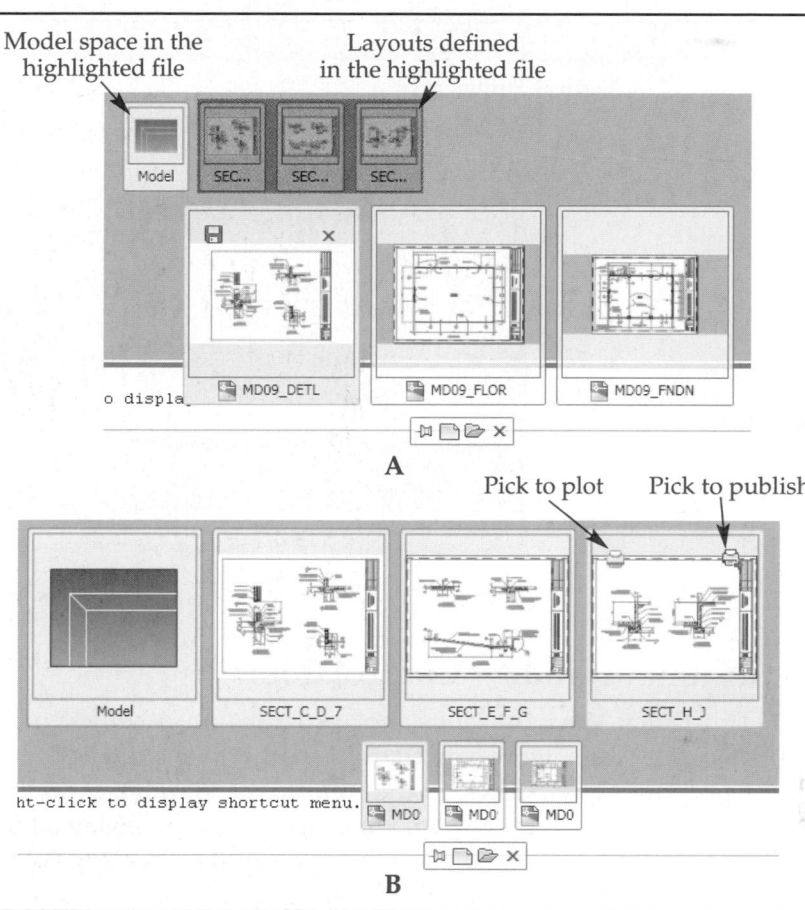

Model space in the highlighted file

Layouts defined in the highlighted file

A

Pick to plot   Pick to publish

B

files. Choose the **Arrange Icons** option to arrange minimized drawings neatly along the bottom of the drawing window area. Select the **Tile Vertically** option to tile unminimized drawing windows in a vertical arrangement, with the active window placed on the left. Pick the **Tile Horizontally** option to tile unminimized drawing windows in a horizontal arrangement, with the active drawing window placed on top. Pick the **Cascade** option to arrange the drawing windows in a cascading style. The effects of tiling the drawing windows vary depending on the number of open windows and whether they tile horizontally or vertically. See **Figure 2-14.**

The shortcut menu that appears when you right-click on a file's thumbnail image includes additional options. Pick **Copy File as a Link** to copy the entire file to the Clipboard for pasting into a drawing or document. Select **Close All** to close all open documents. Pick **Close other files** to close all files except the active file. Choose **Save All** to save all open documents. Select **Close** to close the active file.

Figure 2-14.
Tiled and cascading drawing windows.

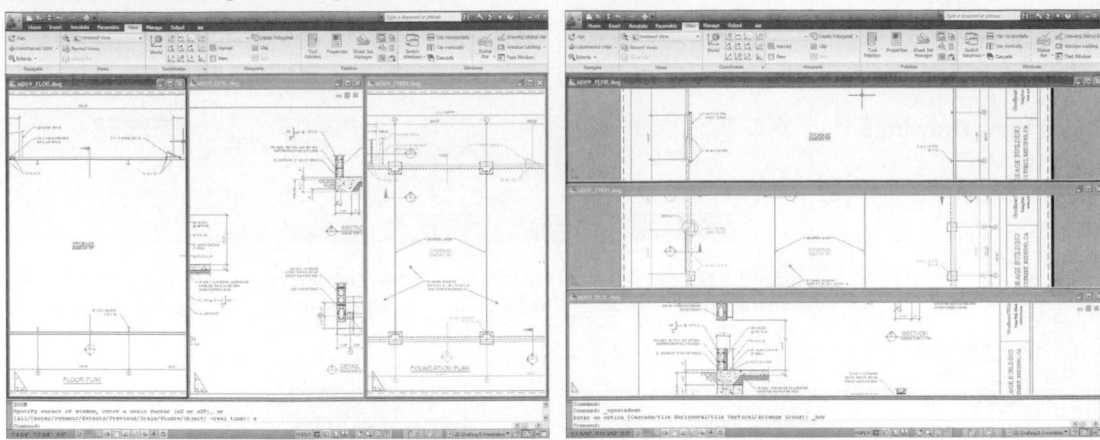

Vertical Tiling   Horizontal Tiling

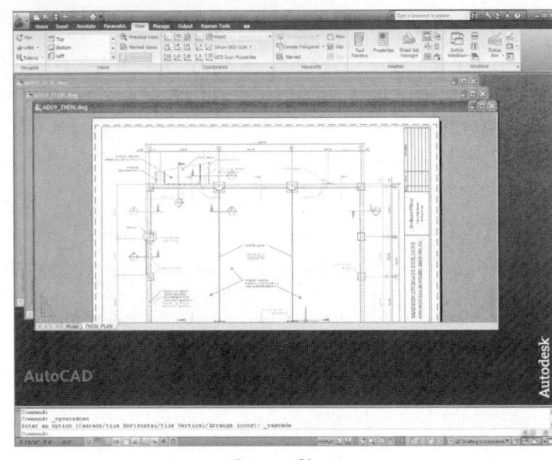

Cascading

Additional Window Control Tools

The **Quick View Drawings** tool provides an excellent method for working with multiple drawings. However, there are other ways to change the active drawing and manage drawing windows. In the ribbon, the **View** tab contains a **Windows** panel with several buttons. Pick the **Switch Windows** flyout to display a list of all open files. Select a file from the flyout to make it active. The **Tile Horizontally**, **Tile Vertically**, **Cascade**, and **Arrange Icons** buttons control the arrangement of open drawings. These are the same features found in the **Quick View Drawings** tool shortcut menu. The **Windows** panel also houses the status bar and window locking functions described in Chapter 1.

Pick the **Open Documents** button in the **Application Menu** to display currently open AutoCAD files. Files list alphanumerically in the order in which they were opened. You can display open files as icons, small images, medium images, or large images by picking the appropriate option from the display options flyout. To activate a different drawing window, pick the file from the list.

PROFESSIONAL TIP

Another technique for switching between open drawings is to press [Ctrl]+[F6]. This is a very effective way to cycle through open drawings quickly.

Typically, you can change the active drawing window as desired. However, you cannot activate a different window while a dialog box is open, or in certain other cases. You must either complete the operation or cancel the dialog box before switching is possible.

Exercise 2-4

Access the Student Web site (www.g-wlearning.com/CAD) and complete Exercise 2-4.

Creating and Using Templates

Drawing templates allow you to use an existing drawing as a starting point for a new drawing. Templates are incredible productivity boosters, and help ensure that everyone in a department, class, school, or company uses the same drawing standards. When you use a well-developed template, drawing settings are set automatically or are available each time you begin a new drawing. Templates usually include the following:

✓ Units and angle values
✓ Drawing limits
✓ Grid, snap, and other drawing aid settings
✓ Standard layouts with a border and title block
✓ Text styles
✓ Table styles
✓ Dimension styles
✓ Layer definitions and linetypes
✓ Plot styles
✓ Commonly used symbols and blocks
✓ General notes and other annotations

Reference Material

Drawing Sheets

For tables describing *sheet* characteristics, including *sheet size*, drawing scale, and drawing limits, go to the Reference Material section of the Student Web site (www.g-wlearning.com/CAD) and select **Drawing Sheets**. Additional information regarding sheet parameters and selection is described later in this textbook.

sheet: The paper used to lay out and plot drawings.

sheet size: Size of the paper used to lay out and plot drawings.

Creating Templates

If none of the drawing templates supplied with AutoCAD meet your needs, you can create and save your own custom templates. AutoCAD allows you to save *any* drawing as a template. Develop a template whenever several drawing applications require the same setup procedure. The template then allows you to apply the same setup to any number of future drawings. Using templates offers several advantages, including increased productivity by decreasing setup requirements. You can modify and then save an existing AutoCAD template as a new custom template, or you can construct a template from scratch. As you learn more about working with AutoCAD, you will find many settings to include in your templates.

When you have everything needed in the template, the template is ready for a final save and use. Use the **SAVEAS** tool and **Save Drawing As** dialog box to save a template. To save the drawing as a template, pick AutoCAD Drawing Template (*.dwt) from the **Files of type:** drop-down list. By default, the file list box shows the drawing templates currently found in the Template folder. See **Figure 2-15.**

You can store custom templates in any appropriate location, but if you place them in the Template folder, they automatically appear in the **Select template** dialog box by default. After specifying the name and location for the new template file, pick the **Save** button to save the template. The **Template Options** dialog box now appears. See **Figure 2-16.** Type a description of the template file in the **Description** area. A brief description usually works best. In the **Measurement** drop-down list, specify English or Metric units, and then pick the **OK** button.

The template name should relate to the template, such as Mechanical Template for mechanical part and assembly drawings. The template might be named for the drawing application, such as Architectural floor plans, or it might be as simple as Template 1. Write the template name and contents in a reference manual. This provides future reference for you and other users.

> ### NOTE
>
> Pick **AutoCAD Drawing Template** from the **Save As** menu of the **Application Menu** to preset the template file type in the **Save Drawing As** dialog box.

Figure 2-15.
Saving a template in the AutoCAD Template folder.

Folder where template is saved

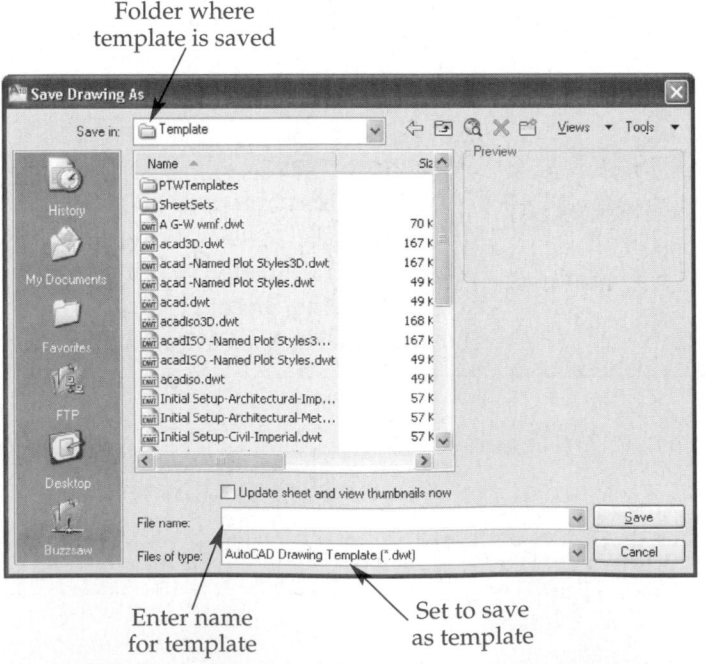

Enter name for template

Set to save as template

Figure 2-16.
Type a description of
the new template in
the **Template Options**
dialog box.

Enter description
for template

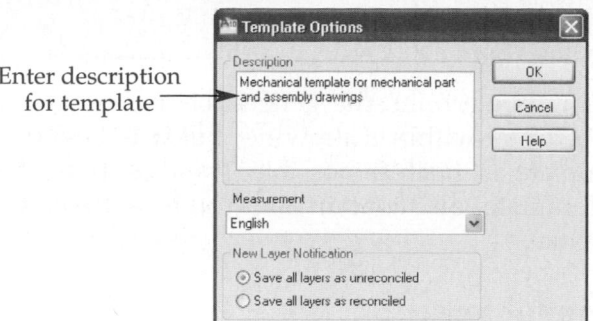

As you refine your setup procedure, you can open and revise template files. Use the **SAVEAS** tool to save a new template from an existing template.

Template Development

Template Development
Chapter 2

The Template Development section of the Student Web site contains several predefined templates that you can use to create drawings in accordance with correct mechanical, architectural, and civil drafting standards. **Figure 2-17** describes each of the available templates. The mechanical drafting templates are based on the American Society of Mechanical Engineers (ASME) and American National Standards Institute (ANSI) ASME Y series of drafting standards. The architectural and civil drafting templates are based on appropriate architectural and civil drafting standards, including standards specified in the United States National CAD Standard.

In addition to the complete, ready-to-use templates in the Template Development section on the Student Web site, the Template Development sections at the end of several chapters refer you to the Student Web site for important template creation topics and procedures. Use the Template Development feature of this textbook to learn gradually how to prepare templates in accordance with correct mechanical, architectural, and civil drafting standards.

Figure 2-17.
The predefined drawing templates available on the Student Web site.

Template File	Discipline	Units	Layout Sheet Sizes
MECHANICAL-INCH.dwt	Mechanical	Decimal inches	A-size, B-size, C-size, D-size
MECHANICAL-METRIC.dwt	Mechanical	Metric	A4-size, A3-size, A2-size, A1-size
ARCHITECTURAL-US.dwt	Architectural	Feet and inches	Architectural C-size, Architectural D-size
ARCHITECTURAL-METRIC.dwt	Architectural	Metric	Architectural A2-size, Architectural A1-size
CIVIL-US.dwt	Civil	Decimal inches	C-size, D-size
CIVIL-METRIC.dwt	Civil	Metric	A2-size, A1-size

Basic Drawing Settings

Drawing settings determine the general characteristics of a drawing. You can change drawing settings within a drawing, but it is best to adhere to the settings defined in the template as much as possible. The most basic drawing settings include units and limits. Templates also contain many other settings, as explained when applicable in this textbook.

Working in Model Space

Each drawing or template file includes two environments in which you can work. *Model space* is where you design and draft the *model* of a product. In mechanical drafting, for example, use model space to draw part and assembly views. In architectural drafting, use model space to draw building plans, elevations, and sections.

Once you complete the drawing or model in model space, you switch to *paper (layout) space*, where you prepare a *layout*. A layout represents the sheet of paper used to organize and scale, or lay out, and plot or export a drawing or model. Layouts typically include a border, title block, and general notes. A single drawing can have multiple layouts.

This textbook fully explains paper space when appropriate. For now, all of your drawing and drawing setup should take place in model space. Model space is active by default when you start a drawing from scratch and when you use many of the available templates.

A variety of on-screen characteristics indicate whether you are working in model space or paper space. You know you are in model space when you see the model space coordinate system icon, the active **Model** tab, the **MODEL** button on the status bar, and no representation of a sheet. See **Figure 2-18.** You know you are in a layout when you see the paper space coordinate system icon, an active layout tab, the **PAPER** button on

model: Any drawing composed of various objects, such as lines, circles, and text, and usually created at full size. However, this term is usually reserved for 3D drawings.

model space: The environment in AutoCAD where drawings and designs are created.

paper (layout) space: The environment in AutoCAD where layouts are created for plotting and display purposes.

layout: An arrangement in paper space of items drawn in model space.

Figure 2-18.
Model space is the environment in which drawings and designs are created.

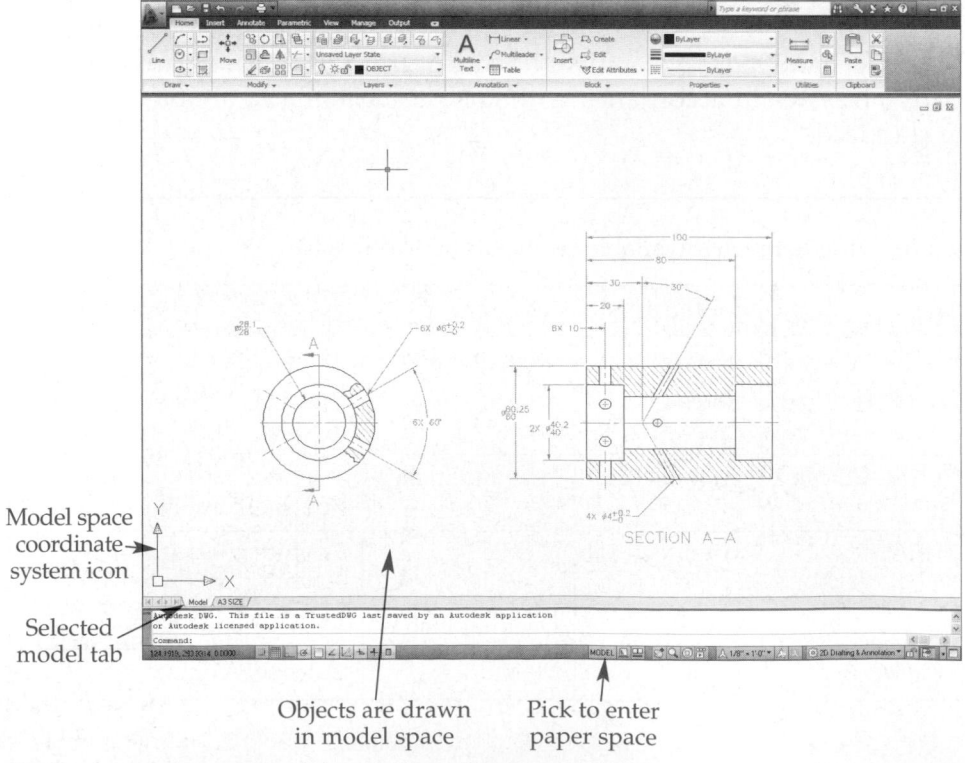

Model space coordinate system icon

Selected model tab

Objects are drawn in model space

Pick to enter paper space

Figure 2-19.
Paper space is the environment in which drawings and designs are laid out on paper for plotting.

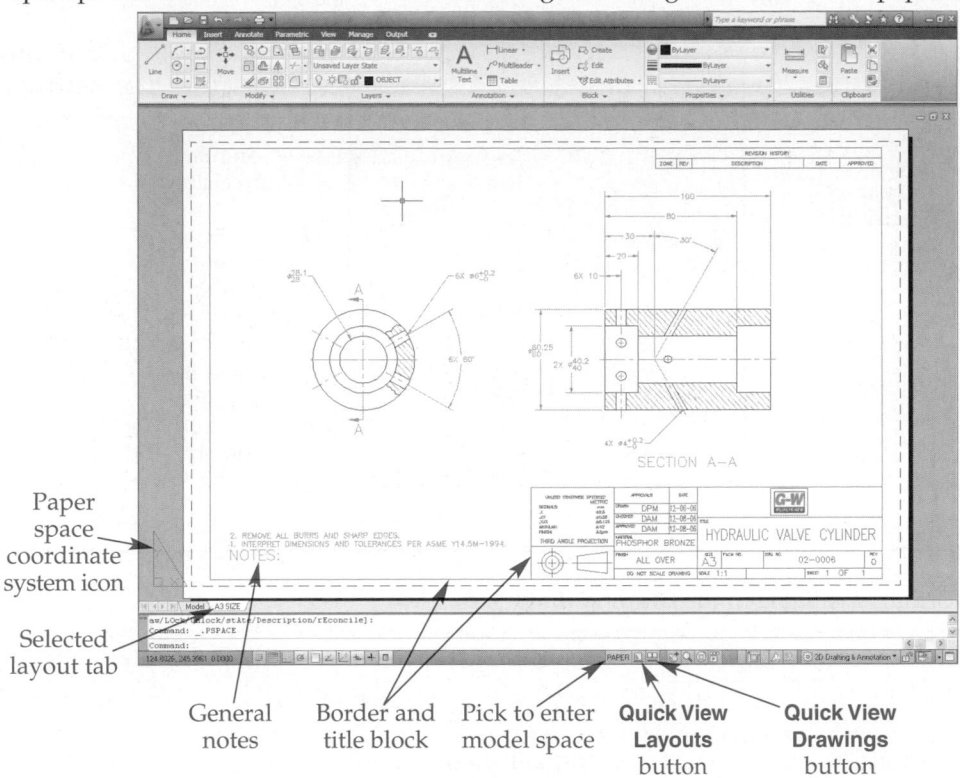

Paper
space
coordinate
system icon

Selected
layout tab

General
notes

Border and
title block

Pick to enter
model space

**Quick View
Layouts**
button

**Quick View
Drawings**
button

the status bar, and a representation of a sheet. See **Figure 2-19.** If you find that you are not in model space, pick the **Model** tab, the **PAPER** button on the status bar, or use **Quick View Layouts** or **Quick View Drawings**. **Quick View Layouts** is fully described later in this textbook.

CAUTION

This textbook assumes that all drawing takes place in model space, until it is appropriate to use a layout. For now, if you find you are not in model space, activate model space in the current drawing and make model space active in all of your custom templates.

Setting Drawing Units

Drawing units define the linear and angular measurements used while drawing and the precision to which these measurements display. In AutoCAD, 1 unit could be 1″, 1 mm, 1 m, or 1 mile. Most AutoCAD users generally think of 1 unit to be either 1″ or 1 mm. Use the **Drawing Units** dialog box to set linear and angular units. See **Figure 2-20.** Specify linear unit characteristics in the **Length** area. Use the **Type:** drop-down list to set the linear units format and use the **Precision:** drop-down list to specify the precision of linear units. **Figure 2-21** describes linear unit formats.

Set the angular unit format and precision in the **Type:** and **Precision:** drop-down lists in the **Angle** area of the **Drawing Units** dialog box. Select the **Clockwise** check box to change the direction for angular measurements to clockwise from the default setting of counterclockwise.

Pick the **Direction...** button to access the **Direction Control** dialog box. See **Figure 2-22.** Pick the **East**, **North**, **West**, or **South** *radio button* to set the compass

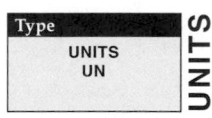

UNITS

Type
UNITS
UN

drawing units: The standard units for linear and angular measurements and the precision of the measurements.

radio button: A selection that activates a single item in a group of options.

Figure 2-20.
Use the **Drawing Units** dialog box to set linear and angular unit values.

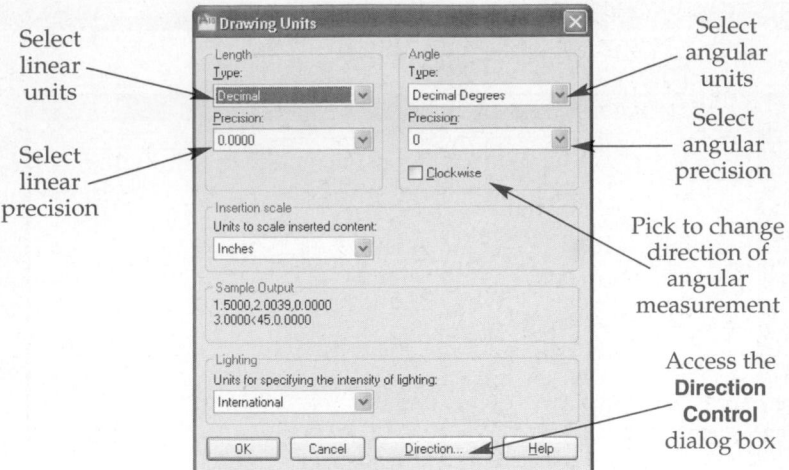

Select linear units

Select linear precision

Select angular units

Select angular precision

Pick to change direction of angular measurement

Access the **Direction Control** dialog box

Figure 2-21.
Linear unit formats available in the **Drawing Units** dialog box.

Type	Typical Applications	Characteristics	Example
Decimal	Mechanical—inch or metric, architectural, structural, civil—metric	• Decimal inches or millimeters. • Conforms to the ASME Y14.5M dimensioning and tolerancing standard. • Four decimal place default precision.	14.1655
Engineering	Civil—feet and inch	• Feet and decimal inches. • Four decimal place default precision.	1'-2.1655"
Architectural	Architectural, structural—feet and inch	• Feet, inches, and fractional inches. • 1/16" default precision.	1'-2 3/16"
Fractional	Mechanical—fractional	• Fractional parts of any common unit of measure. • 1/16" default precision.	14 3/16
Scientific	Chemical engineering, astronomy	• E+01 means the base number is multiplied by 10 to the first power. • Used when very large or small values are required. • Four decimal place default precision.	1.4166E+01

orientation. The **Other** radio button activates the **Angle:** text box and the **Pick an angle** button. Enter an angle for zero direction in the **Angle:** text box. The **Pick an angle** button allows you to pick two points on the screen to establish the angle zero direction.

CAUTION

Use the default direction of 0° East at all times, unless you have a specific need to change the compass direction angle, such as when measuring direction using azimuths (0° North). This textbook uses the 0° East direction, and this default should be set in order to complete most exercises and problems correctly.

AutoCAD and Its Applications—Basics

Figure 2-22.
The **Direction Control** dialog box.

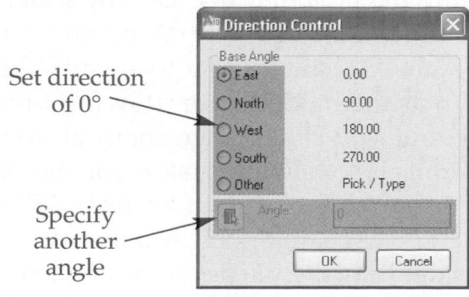

Set direction of 0°

Specify another angle

Figure 2-23.
Angular unit formats available in the **Drawing Units** dialog box.

Type	Applications and Characteristics	Example
Decimal Degrees	• Mechanical, architectural, and structural drafting applications. • Degrees and decimal parts of a degree. • Initial default setting.	45°
Deg/Min/Sec	• Civil and sometimes mechanical, architectural, and structural drafting applications. • Degrees, minutes, and seconds. • 1 degree = 60 minutes; 1 minute = 60 seconds.	45°0'0"
Grads	• Grad is the abbreviation for *gradient*. • One-quarter of a circle has 100 grads; a full circle has 400 grads.	50.000g
Radians	• A radian is an angular unit of measurement in which 2π radians = 360° and π radians = 180°. Pi (π) is approximately equal to 3.1416. • A 90° angle has $\pi/2$ radians and an arc length of $\pi/2$. • Changing the precision displays the radian value rounded to the specified decimal place.	0.785r
Surveyor	• Civil drafting applications. • Degrees, minutes, and seconds. • Uses bearings. A bearing is the direction of a line with respect to one of the quadrants of a compass. Bearings are measured clockwise or counterclockwise (depending on the quadrant), beginning from either north or south. • An angle measuring 55°45'22" from north toward west is expressed as N55°45'22"W. • Set precision to degrees, degrees/minutes, degrees/minutes/seconds, or decimal display accuracy of the seconds part of the measurement.	N45°E

Figure 2-23 describes angular unit formats. After selecting the linear and angular units and precision, pick the **OK** button to exit the **Drawing Units** dialog box.

Exercise 2-5

Access the Student Web site (www.g-wlearning.com/CAD) and complete Exercise 2-5.

Adjusting Drawing Limits

You prepare an AutoCAD drawing at actual size, or full scale, regardless of the type of drawing, the units used, or the size of the final layout on paper. Use model space to draw full-scale objects. AutoCAD allows you to specify the size of a virtual model space drawing area, known as the model space drawing limits, or *limits*. You typically set limits in a template, but you can change them at any time during the drawing process. The concept of limits is somewhat misleading, because the AutoCAD drawing area is infinite in size. For example, if you set limits to 17″ × 11″, you can still create objects that extend past the 17″ × 11″ area, such as a line that is 1200′ long. As a result, you can choose not to consider limits while developing a template or creating a drawing. Conversely, as you learn AutoCAD, you may decide that setting appropriate drawing limits is helpful, especially when you are drawing large objects. Regardless of whether you choose to acknowledge limits, you should be familiar with the concept, and recognize that some AutoCAD tools, such as the **ZOOM** and **PLOT** tools discussed in later chapters, provide options for using limits.

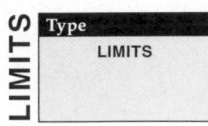

limits: The size of the virtual drawing area in model space.

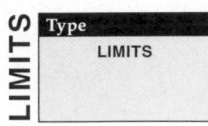

Use the **LIMITS** tool to set model space drawing limits. The first prompt asks you to specify the coordinates for the lower-left corner of the drawing limits. For now, when setting limits, the lower-left corner is 0,0. Press [Enter] to accept the default 0,0 value or enter a new value and press [Enter]. The next prompt asks you to specify the coordinates for the upper-right corner of the virtual drawing area. For example, type 17,11 and then press [Enter] to set limits of 17″ x 11″. The first value is the horizontal measurement, and the second value is the vertical measurement of the limits. A comma separates the values.

In general, you should set limits larger than the objects you plan to draw. You can determine limits accurately by identifying the drawing scale, converting the scale to a scale factor, and then multiplying the scale factor by the size of sheet on which you plan to plot the drawing. For now, calculate the approximate total length and width of all the objects you plan to draw, adding extra space for dimensions and notes. For example, if you are drawing a 48′ × 24′ building floor plan, allow 10′ on each side for dimensions and notes to make a total virtual drawing area of 68′ × 44′.

> **NOTE**
>
> The **LIMITS** tool provides a limits-checking feature that, when turned on, restricts your ability to draw outside of the drawing limits. Turn on limits checking by entering or selecting the **ON** option of the **LIMITS** tool. Turn off limits checking using the **OFF** option.

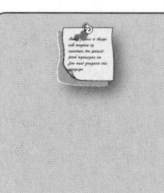

Exercise 2-6

Access the Student Web site (www.g-wlearning.com/CAD) and complete Exercise 2-6.

Template Development
Chapter 2

For detailed instructions to begin the development of drawing templates, go to the Student Web site (www.g-wlearning.com/CAD), select this chapter, and select **Template Development**.

Chapter Test

Answer the following questions. Write your answers on a separate sheet of paper or go to the Student Web site (www.g-wlearning.com/CAD) and complete the electronic chapter test.

1. What is a drawing template?
2. What is the name of the dialog box that opens by default when you pick the **New** button on the **Quick Access** toolbar?
3. Briefly explain how to start a drawing from scratch.
4. How often should work be saved?
5. Explain the benefits of using a standard system for naming drawing files.
6. Name the tool that allows you to save your work quickly without displaying a dialog box.
7. What tool allows you to save a drawing file in an older AutoCAD format?
8. How do you set AutoCAD to save your work automatically at designated intervals?
9. Identify the tool you would use if you wanted to exit a drawing file, but remain in the AutoCAD session.
10. What is the quickest way to close an AutoCAD drawing file?
11. How can you close all open drawing windows at the same time?
12. Which **Application Menu** function allows you to select the name of a recently opened drawing file and open it?
13. How can you set the number of recently opened files listed in the **Application Menu**?
14. What does the term *read-only* mean?
15. Describe the advantages of using the **Quick View Drawings** tool to work with multiple open drawings.
16. How do you keep the **Quick View Drawings** tool on-screen after you pick a thumbnail image?
17. How do you quickly cycle through all the currently open drawings in sequence?
18. What is sheet size?
19. How can you convert a drawing file into a drawing template?
20. Name three settings you can specify in the **Drawing Units** dialog box.

Drawing Problems

Start AutoCAD if it is not already started. Follow the specific instructions for each problem.

▼ Basic

1. Start a new drawing using the **acad-Named Plot Styles** template supplied by AutoCAD. Save the new drawing as a file named P2-1.dwg.

2. Start a new drawing using the **acadiso** template supplied by AutoCAD. Save the new drawing as a file named P2-2.dwg.

3. Start a new drawing using the **Tutorial-iArch** template supplied by AutoCAD. Save the new drawing as a file named P2-3.dwg.

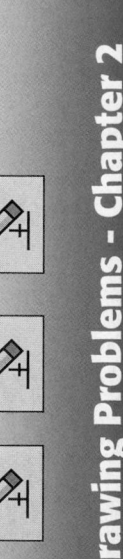

▼ Intermediate

Problems 4 through 7 can be done if the AutoCAD 2010\Sample file folder is loaded. All drawings listed are found in that folder.

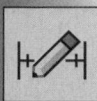

4. Locate and preview the Lineweights drawing. Open the drawing and describe it in your own words.

5. Locate and preview the TrueType drawing. Open the drawing and describe it in your own words.

6. Locate and preview the Tablet drawing. Open the drawing and describe it in your own words.

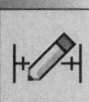

7. Locate and preview the 3D House drawing. Open the drawing and describe it in your own words.

▼ Advanced

The following problems can be saved as templates for future use.

8. Create a template with decimal units with 0.0 precision, decimal degrees with 0.0 precision, default angle measure and orientation, and a limits setting of 0,0 × 17,11. Save the template as P2-8.dwt. Enter an appropriate description for the template.

9. Create a template with metric units with 0.0 precision, decimal degrees with 0.0 precision, default angle measure and orientation, and a limits of 0,0 × 22,17. Save the template as P2-9.dwt. Enter an appropriate description for the template.

10. Create a template with architectural units with 0'-0" precision, decimal degrees with 0 precision, and a limits of 0,0 × 1632,1056. Save the template as P2-10.dwt. Enter an appropriate description for the template.

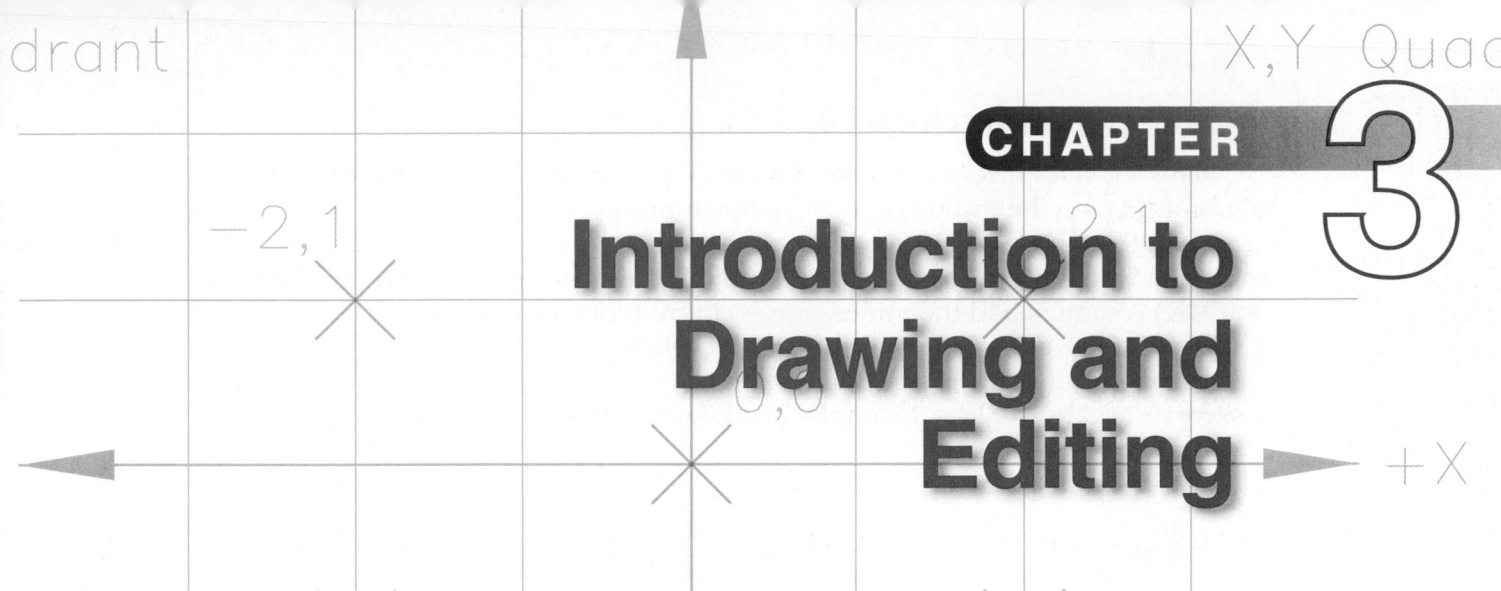

Learning Objectives

After completing this chapter, you will be able to do the following:

✓ Use appropriate values when responding to prompts.
✓ Describe the Cartesian coordinate system.
✓ Determine and specify drawing snap and grid.
✓ Draw given objects using the **LINE** tool.
✓ Describe and use several point entry methods.
✓ Demonstrate an ability to use dynamic input and the command line.
✓ Use direct distance entry with polar tracking and **Ortho** mode.
✓ Revise objects using the **ERASE** tool.
✓ Create selection sets using various selection options.
✓ Use the **UNDO**, **U**, **REDO**, and **OOPS** tools appropriately.

This chapter introduces a variety of fundamental drawing and editing concepts and processes, including point entry and object selection methods. You will learn to pick points and draw objects using the **LINE** tool. You will learn to make selections and changes using the **ERASE** tool. You can learn much about AutoCAD using the basic tools and operations described in this chapter.

Responding to Prompts

Most drawing and editing tools require that you respond to a prompt. A prompt "asks" you to perform a specific task. For example, when you are drawing a line, a prompt asks you to specify line endpoints. When you erase an object, a prompt asks you to select objects to erase. Many prompts provide options that you can select instead of responding to the immediate request. For example, after picking the first line endpoint, you are promoted to select the next point, or you can choose the **Undo** option to remove the previous selection instead of picking another point.

Responding with Numbers

Many tools require you to enter specific numeric data, such as the endpoint location of a line, or the radius of a circle. Some prompts require you to enter a whole number. Other entries require whole numbers that are positive or negative. AutoCAD understands that a number is positive without the plus sign (+) in front of the value. However, you must add the minus sign (–) in front of a negative number.

Much of the data you enter may not be whole numbers. In these cases, you can use any real number expressed as decimal, as a fraction, or in scientific notation using positive or negative values. Examples of acceptable real number entries include:

    4.250
    –6.375
    1/2
    1-3/4
    2.5E+4 *(25,000)*
    2.5E–4 *(0.00025)*

For fractions, the numerator and denominator must be whole numbers greater than zero. For example, 1/2, 3/4, and 2/3 are all acceptable fraction entries. Fractional numbers greater than one must have a hyphen between the whole number and the fraction. For example, type 2-3/4 for two and three-quarters. The hyphen (-) separator is needed because a space acts just like pressing [Enter] and automatically ends the input. The numerator can be larger than the denominator, as in 3/2, but *only* if you do not include a whole number with the fraction. For example, 1-3/2 is not a valid input for a fraction.

When you enter coordinates or measurements, the values used depend on the units of measurement. For decimal or fractional length units, AutoCAD interprets an entry of 2.500 or 2-1/2 to be 2.500 or 2-1/2 *units*, not a specific unit of measure. You are responsible for applying the appropriate units. Most AutoCAD users generally think of 1 unit to be either 1″ or 1 mm.

When using decimal or fractional units, you do not need to add a suffix after the numeral. Inch marks (″) are unnecessary and time-consuming to add, and adding mm for millimeters, for example, is not recognized.

AutoCAD assumes inch measurements when you use architectural and engineering length units. In this case, any value greater than 1″ expresses in inches, feet, or feet and inches. The values can be whole numbers, decimals, or fractions. For measurements in feet, the foot symbol (′) must follow the number, as in 24′. If a value is in feet and inches, there is no space and it is unnecessary to add a hyphen (-) between the feet and inch value. For example, 24′6 is the proper input for the value 24′-6″. If the inch part of the value contains a fraction, the inch and fractional part of an inch are separated by a hyphen, such as 24′6-1/2. Never mix feet with inch values greater than one foot. For example, 24′18 is an invalid entry. In this case, you should type 25′6.

NOTE

AutoCAD accepts the inch (″) and foot (′) symbols only when you specify architectural or engineering drawing units.

Ending and Canceling Tools

Some AutoCAD tools, such as the **LINE** tool, remain active until stopped. For example, you can continue to pick points to create new line segments until you end the **LINE** tool. You can usually end a tool by pressing [Enter] or the space bar, or by right-clicking and selecting **Enter**. Press [Esc] to cancel an active tool or abort data entry. It may be necessary to press [Esc] twice to cancel certain tools completely.

If you press the wrong key or misspell a word while using a tool or answering a prompt, use [Backspace] to correct the error. This works only if you notice your mistake *before* you accept the value. If you enter an inappropriate value or option, AutoCAD usually responds with an error message. Access the **AutoCAD Text Window** to view lengthy error messages, or review entries. Return to the graphics screen using the same method you used to access the text window, or pick any visible portion of the graphics screen.

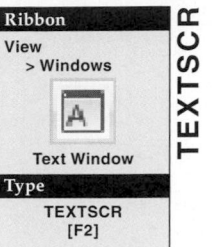

Ribbon
View
> Windows

Text Window

Type
TEXTSCR
[F2]

TEXTSCR

PROFESSIONAL TIP

You can cancel the active tool and access a new tool at the same time by picking a ribbon button or an **Application Menu** option.

Introduction to Drawing

You create most geometric shapes using drawing tools, such as the **LINE** tool. The most common method of locating and sizing objects while using drawing tools is through *point entry*. Specifying the two endpoints of a line segment is an example of point entry. To enter points, you use the *Cartesian (rectangular) coordinate system*. XYZ coordinate values describe the locations of points. These values, called *rectangular coordinates*, locate any point in 3D space.

In 2D drafting, the *origin* divides the coordinate system into four quadrants on the XY plane. The most basic point entry relates to the origin, where X = 0 and Y = 0, or 0,0. See **Figure 3-1**. The origin is usually at the lower-left corner of the drawing. This setup places all points in the upper-right quadrant of the XY plane, where both X and Y coordinate values are positive. See **Figure 3-2**. To locate a point in 3D space, a third dimension rises up from the surface of the XY plane and is given a Z value.

To describe a coordinate location, the X value is listed first, the Y value is second, and the Z value is third. A comma separates each value. For example, the coordinate location of 3,1,6 represents a point that is three units from the origin in the X direction, one unit from the origin in the Y direction, and six units from the origin in the Z direction. *AutoCAD and Its Applications—Advanced* describes how to use the Z axis to construct 3D models.

point entry: Identifying a point location in the AutoCAD coordinate system.

Cartesian (rectangular) coordinate system: A system in which points are located in space according to distances from three intersecting axes.

rectangular coordinates: A set of numerical values that identify the location of a point on the X, Y, and Z axes of the Cartesian coordinate system.

origin: The intersection point of the X, Y, and Z axes.

Figure 3-1.
The 2D Cartesian coordinate system consists of X and Y axes. The origin is located at the intersection of the axes.

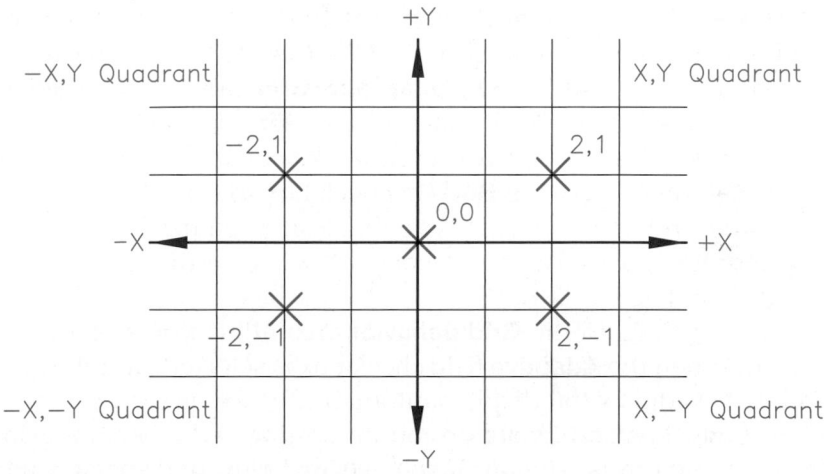

Figure 3-2.
By default, the upper-right quadrant of the Cartesian coordinate system fills the screen.

Increasing Y coordinate value

Increasing X coordinate value

Default origin

Exercise 3-1

Access the Student Web site (www.g-wlearning.com/CAD) and complete Exercise 3-1.

Introduction to Drawing Aids

AutoCAD includes many drawing aids that increase accuracy and productivity. The coordinate display, grid, and grid snap modes are examples of basic drawing aids. You will learn other useful drawing aids in later chapters. As an AutoCAD user, you will learn to apply whichever tools and options work best and quickest for a specific drawing task.

Using Grid Mode

GRID
Type
GRID
[Ctrl]+[G]
F7

DSETTINGS
Type
DSETTINGS
DS
SE

grid: A pattern of dots that appears on-screen to aid in the drawing process.

Turn on **Grid** mode to display a *grid* on the screen. See **Figure 3-3**. The grid is for reference only, for use as a visual aid to drawing layout. The quickest way to toggle **Grid** mode on and off is to pick the **Grid Display** button on the status bar.

Use the options on the **Snap and Grid** tab of the **Drafting Settings** dialog box to adjust the spacing between grid dots. See **Figure 3-4.** A quick way to access the **Drafting Settings** dialog box is to right-click on any of the status bar toggle buttons and select **Settings…**. Use the **Grid On** check box to turn the grid on or off. Grid spacing is set in the **Grid spacing** area. Type the desired values in the **Grid X spacing:** and **Grid Y spacing:** text boxes. Factors to consider when setting grid spacing are described later in this chapter.

The options in the **Grid behavior** area allow you to set how the grid appears on-screen. When the **Adaptive Grid** check box is selected, and the grid spacing is too dense, AutoCAD adjusts the display automatically so the grid can be shown on-screen. The **Display grid beyond Limits** option determines whether the grid appears only within the drawing limits. The **Allow subdivision below grid spacing** and **Follow Dynamic UCS** options apply to 3D applications.

Figure 3-3.
Dots represent the grid spacing when **Grid** mode is active.

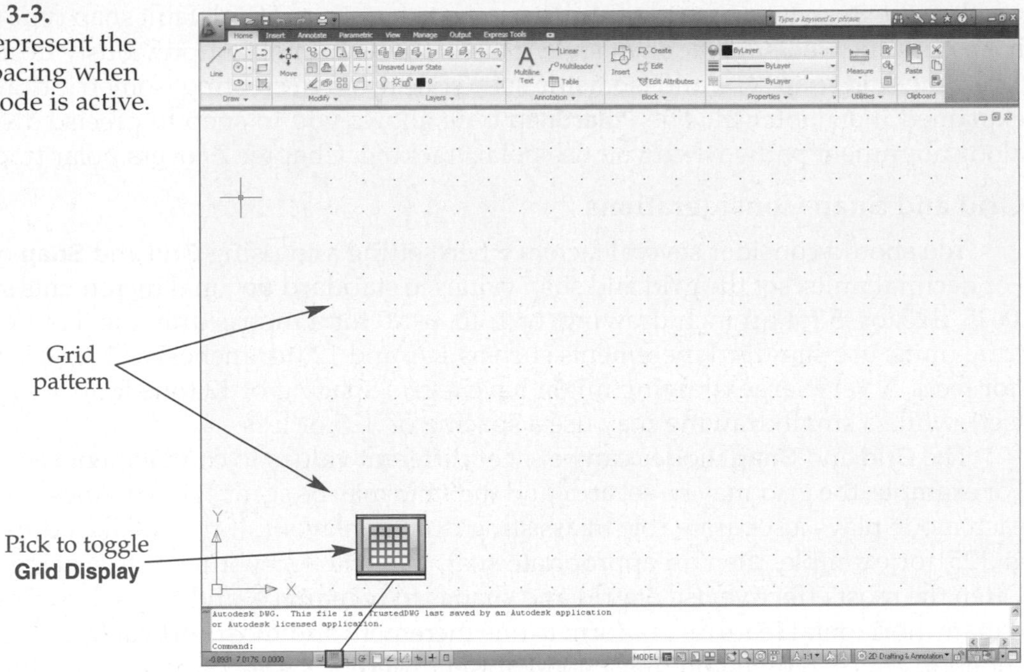

Grid pattern

Pick to toggle **Grid Display**

Figure 3-4.
Use the **Snap and Grid** tab of the **Drafting Settings** dialog box to specify grid and snap grid settings.

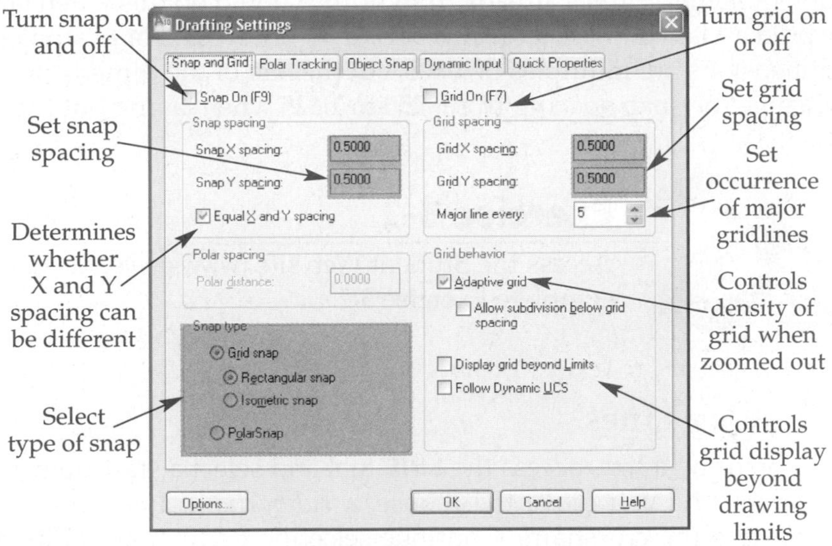

Turn snap on and off

Set snap spacing

Determines whether X and Y spacing can be different

Select type of snap

Turn grid on or off

Set grid spacing

Set occurrence of major gridlines

Controls density of grid when zoomed out

Controls grid display beyond drawing limits

Using Snap Mode

Turn on **Snap** mode to activate the *snap grid*, also known as *snap resolution* or *snap*. One of the quickest ways to toggle snap on and off is to pick the **Snap Mode** button on the status bar. By default, snap is off, and when you move the mouse, the crosshairs moves freely on the screen. Turn snap on to move the crosshairs in specific increments. Using the snap grid is different from using the on-screen grid. Snap controls the movement of the crosshairs, while the grid is only a visual guide. The grid and snap settings can, however, be used together.

The **Snap and Grid** tab of the **Drafting Settings** dialog box includes options for setting snaps. Refer again to **Figure 3-4.** Use the **Snap On** check box to turn snaps on or off. Snap increment is set in the **Snap spacing** area. Type the desired values in the **Snap X spacing:** and **Snap Y spacing:** text boxes.

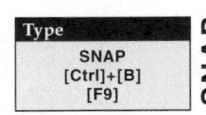

Type
SNAP
[Ctrl]+[B]
[F9]

SNAP

snap grid (snap resolution, snap): Invisible grid that allows the crosshairs to move only in exact increments.

Use the **Snap type** area to control how snaps function. The default snap type is **Grid snap** with the **Rectangular snap** style. This is the snap method previously described. Select the **Grid snap** type and **Isometric snap** style to aid in creating isometric drawings, explained in Chapter 24. The **PolarSnap** type allows you to snap to precise distances along alignment paths when you use polar tracking. Chapter 7 covers polar tracking.

Grid and Snap Considerations

You should consider several factors when setting and using **Grid** and **Snap** modes. For decimal units, set the grid and snap values to standard decimal increments such as .0625, .125, or .5 for an inch drawing, or 1, 10, or 20 for a metric drawing. For architectural units, use standard increments such as 1, 6, and 12 (for inches) or 1, 2, 4, 5, and 10 (for feet). A very large drawing might have a grid spacing of 12 (one foot), or 120 (ten feet), while a small drawing may use a spacing of .125 or less.

The **Grid** and **Snap** modes can be set at different values to complement each other. For example, the grid may be set at .5, and the snap may be set at .25. With these settings, each mode plays a separate role in assisting drawing layout. If the smallest dimension is .125, for example, then an appropriate snap value is .125 with a grid spacing of .25. Often the most effective use of grid and snap is to set equal X and Y spacing. However, if many horizontal features conform to one increment and most vertical features correspond to another, then you may choose to set different X and Y values.

You can change the snap and grid values at any time without changing the location of points or lines already drawn. You should do this when larger or smaller values would assist you with a certain part of the drawing. For example, suppose a few of the dimensions are multiples of .0625, but the rest of the dimensions are multiples of .250. Change the snap spacing from .250 to .0625 when laying out the smaller dimensions.

Exercise 3-2

Access the Student Web site (www.g-wlearning.com/CAD) and complete Exercise 3-2.

Drawing Lines

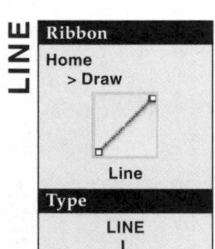

LINE

Ribbon
Home
> Draw

Line

Type
LINE
L

To draw a line, access the **LINE** tool and select a start point, which is the first line endpoint. As you move the mouse, a *rubberband line* appears connecting the first point and the crosshairs. Continue selecting additional points to connect a series of lines. Then press [Enter] or the space bar, or right-click and select **Enter** to end the **LINE** tool.

Undoing the Previously Drawn Line

If you make an error while still using the **LINE** tool, right-click and select **Undo**, pick the **Undo** dynamic input option, or type U and press [Enter]. This removes the previously drawn line and allows you to continue from the previous endpoint. You can use the **Undo** option repeatedly to delete line segments until the entire line is gone. See **Figure 3-5.**

rubberband line: A stretch line that extends from the crosshairs with certain drawing tools to show where an object will be drawn.

polygon: Closed plane figure with at least three sides. Triangles and rectangles are examples of polygons.

Using the Close Option

To aid in drawing a *polygon* using the **LINE** tool, after you draw two or more line segments, use the **Close** option to connect the endpoint of the last line segment to the start point of the first line segment. To use this option, right-click and select **Close**, pick the **Close** dynamic input option, or type C or CLOSE and press [Enter]. In **Figure 3-6**, the last line is drawn using the **Close** option.

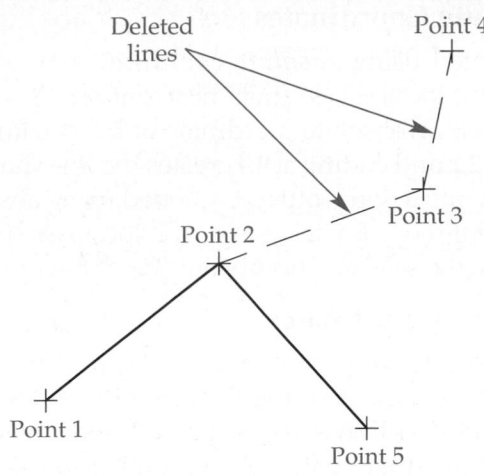

Figure 3-5.
Using the **Undo** option of the active **LINE** tool. Dashed lines are used here to represent the undone lines.

Deleted lines

Point 4

Point 3

Point 2

Point 1

Point 5

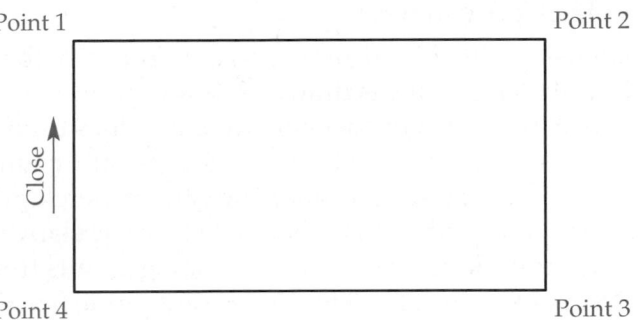

Figure 3-6.
Using the **Close** option to complete a box.

Point 1

Point 2

Close

Point 4

Point 3

Exercise 3-3

Access the Student Web site (www.g-wlearning.com/CAD) and complete Exercise 3-3.

Point Entry Methods

The most basic method of point entry is to pick a point using the crosshairs. Picking a random point in space is typically not accurate. One method of creating an accurate drawing is to use drawing aids. For example, with **Snap** mode on, you can select a specific point corresponding to the snap increment. Another option is to use a point entry method that locates an exact point on the rectangular coordinate system. This typically requires specialized coordinate input.

Exercise 3-4

Access the Student Web site (www.g-wlearning.com/CAD) and complete Exercise 3-4.

Using Absolute Coordinates

absolute coordinates: Coordinate distances measured from the origin.

Points located using *absolute coordinates* are measured from the origin (0,0). For example, a point located two units horizontally (X = 2) and two units vertically (Y = 2) from the origin is an absolute coordinate of 2,2. A comma separates the values. Drawing a line starting at 2,2 and ending at 4,4 creates the line shown in **Figure 3-7**. The first point you pick when drawing a line is often positioned using absolute coordinates. Remember, when using the absolute coordinate system, you locate each point from 0,0. If you enter negative X and Y values, the selection occurs outside of the upper-right XY plane quadrant.

Using Relative Coordinates

relative coordinates: Coordinates specified from, or relative to, the previous position, rather than from the origin.

When using *relative coordinates*, you may want to think of the previous point as the "temporary origin." Use the @ symbol to enter relative coordinates. For example, if the first point of a line is located at 2,2 and the second point is positioned using a relative @2,2 coordinate entry, the second point is located 4,4 from the origin. Refer again to **Figure 3-7**.

Using Polar Coordinates

polar coordinates: Coordinates based on the distance from a fixed point at a given angle.

When using *polar coordinate* entry, you specify the length of the line followed by the angle at which the line is drawn. A less than (<) symbol separates the distance and angle. **Figure 3-8** shows the default angular values used for polar coordinate entry. By default, 0° is to the right, or east, and angles measure counterclockwise. When preceded by the @ symbol, a polar coordinate point locates relative to the previous point. If the @ symbol is not included, the coordinate locates relative to the origin. For example, to draw a line 2 units long at a 45° angle, starting 2 units from 0,0 at a 45° angle, type 2<45 for the first point, and @2<45 for the second point. See **Figure 3-9**.

Figure 3-7.
Locating points using absolute and relative coordinates.

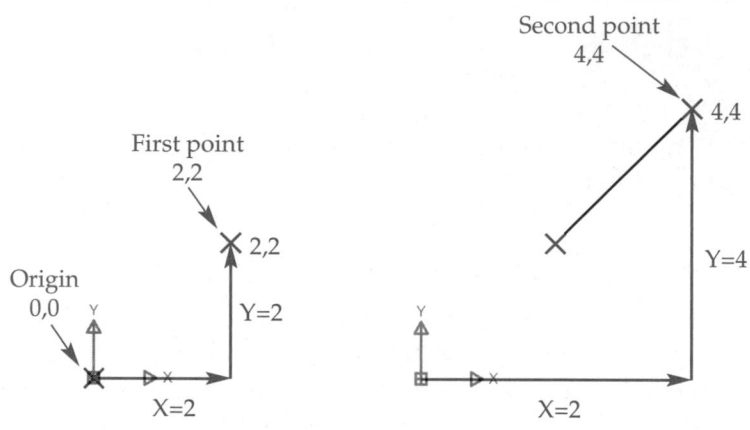

Absolute Coordinate Entry

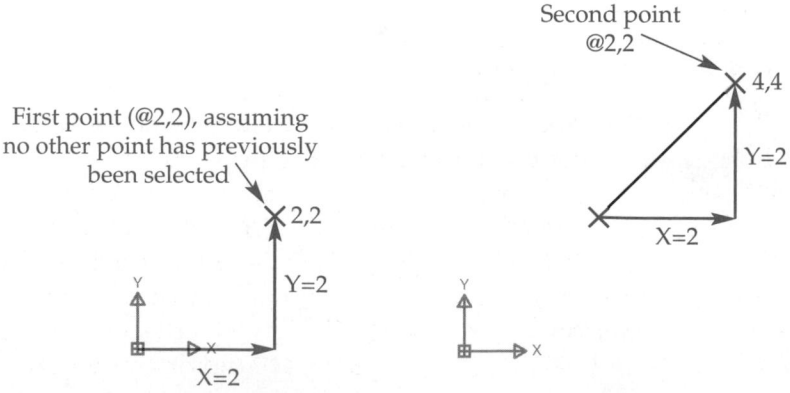

Relative Coordinate Entry

AutoCAD and Its Applications—Basics

Figure 3-8.
Angles used when
entering polar
coordinates.

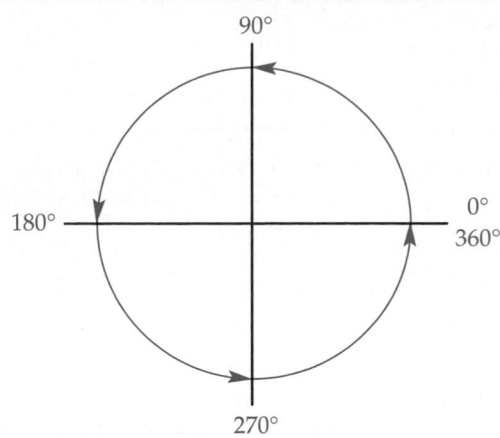

Figure 3-9.
Locating points
using polar
coordinates.

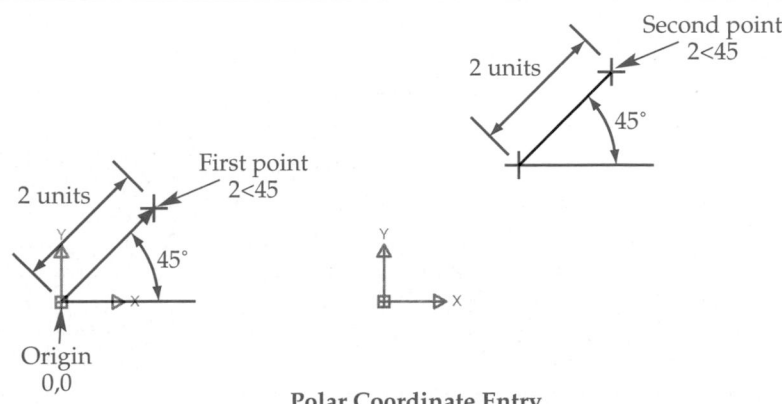

Polar Coordinate Entry

The Coordinate Display

The area on the left side of the status bar shows the coordinate display field. The drawing units setting determines the format and precision of the display. Pick the coordinate display in the status bar to toggle the display on and off. When coordinate display is on, the coordinates constantly change as the crosshairs moves. When coordinate display is off, the coordinates are "grayed out," but still update to identify the location of the last point selected.

When a tool such as **LINE** is active, you can choose the coordinate display mode. When the **Relative** mode is set and a tool is active, the coordinates of the crosshairs position display as polar coordinates relative to the previously picked point. The coordinates update each time you pick a new point. When you select the **Absolute** mode and a tool is active, the coordinate of the crosshairs location is set relative to the origin. The **Geographic** mode is available if you specify the geographic drawing location.

Using Dynamic Input

Dynamic input is on by default and is one of the most effective tools for entering coordinates. Dynamic input provides the same function as the command line, but it allows you to keep your focus at the point where you are drawing. Also, additional coordinate entry techniques are available with dynamic input. Common point entry methods function differently when dynamic input is active, depending on settings.

When you start the **LINE** tool, dynamic input prompts you to specify the first point. The X coordinate input field is active, and the Y coordinate input field appears. See **Figure 3-10.** The X and Y coordinates of the first point can be typed using *pointer input* with absolute, relative, or polar coordinate entry.

pointer input: The process of entering points using dynamic input.

Figure 3-10.
After you start the **LINE** tool, dynamic input displays these items. When you type the @ symbol to use relative coordinates, the symbol displays in the tooltip. When you use polar coordinates, the less than symbol (<) displays in the tooltip before the angle.

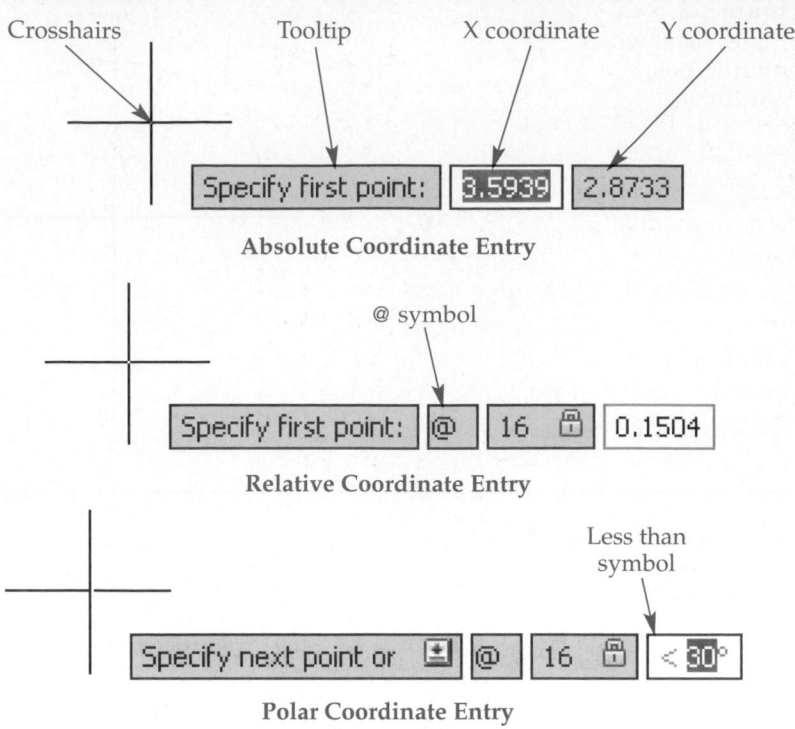

Absolute Coordinate Entry

Relative Coordinate Entry

Polar Coordinate Entry

Absolute coordinate entry is the default when you select the first point. Type the X value, and then press [Tab] or enter a comma to lock in the X value and move to the Y coordinate input field. Now type the Y value and right-click or press [Enter] to select the point. Polar coordinate entry is the other likely option for specifying the first point. Type the less than symbol (<) after entering the length of the line in the X coordinate input field. Then enter the angle of the line in the Y coordinate input field. Dynamic input fields automatically change to anticipate the next entry.

Once you enter the start point of the line, dynamic input provides a *dimensional input* feature that allows you to enter the length of a line and the angle at which the line is drawn, similar to polar coordinate entry. To use dimensional input, you must first establish the start point. Then, by default, distance and angle input fields appear. See **Figure 3-11.** Enter the length of the line in the active distance input field and press [Tab] to lock in the distance and move to the angle input field. Type the angle of the line and right-click or press [Enter] to select the point. The angular values used for

dimensional input: A method of entering points that is similar to polar coordinate entry, but uses dynamic input.

Figure 3-11.
Use dimensional input to define the length and angle of a line.

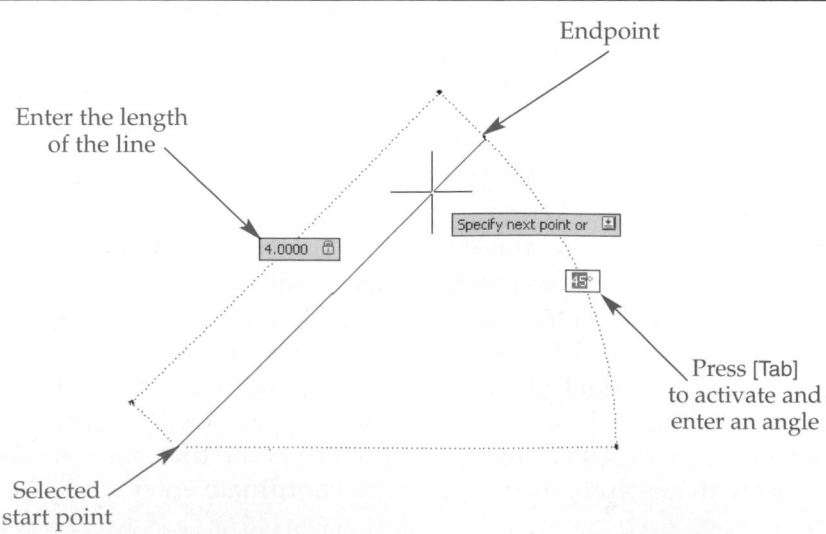

dimensional input are the same as those for polar coordinate entry. Refer again to **Figure 3-8.**

You can also use pointer input to pick additional points once you select the start point of the line. Relative coordinates are active by default, which means you do not need to type @ before entering the X and Y values. To select the second point using a relative coordinate entry, type the X value in the distance input field, then type a comma to lock in the X value and move to the Y coordinate input field. Type the Y value and right-click or press [Enter] to select the point.

Dimensional input is on by default, but you can temporarily turn it off by typing the # symbol before entering values. This makes dynamic input default to polar format, which means you can enter the length of the line in the active X coordinate input field, and then press [Tab] to lock in the length and move to the Y coordinate input field. Enter the angle of the line and right-click or press [Enter] to select the point. In order to use an absolute coordinate entry with the default settings, type the # symbol, enter the X coordinate in the active field, type a comma, type the Y value, and right-click or [Enter] to select the point.

PROFESSIONAL TIP

Use [Tab] to cycle through dynamic input fields. You can make changes to values before accepting the coordinates.

Supplemental Material *Dynamic Input Settings*

For more information about adjusting dynamic input options, go to the Student Web site (www.g-wlearning.com/CAD), select this chapter, and select **Dynamic Input Settings**.

Using the Command Line

Dynamic input is a very effective tool for locating points because of its on-screen display and ease of use. You can also use the command line for point entry, but you must closely adhere to point entry methods. Neither dimensional input nor quick input settings are available with the command line. You can use dynamic input at the same time as the command line, or you can disable dynamic input to use only the command line for tool input and information. Another option is to hide the command line to free additional drawing space and focus on using dynamic input.

Absolute, relative, and polar point entry methods accomplish the same tasks whether they are entered using dynamic input or typed at the command line. The following content provides examples of point entry using the command line. You can apply the same examples to dynamic input. Even if you choose not to use the command line, review these examples to help better understand point entry techniques. You must disable dynamic input in order for these exact command sequences to work properly.

Figure 3-12.
Drawing a shape using the **LINE** tool and absolute coordinates.

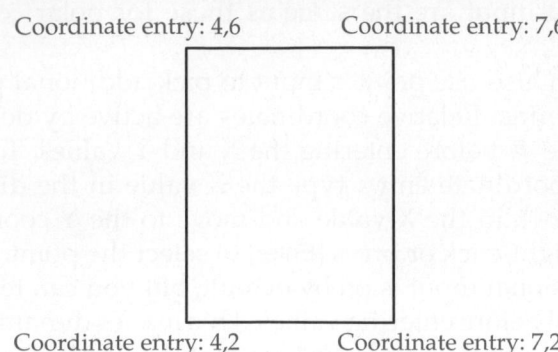

Coordinate entry: 4,6   Coordinate entry: 7,6

Coordinate entry: 4,2   Coordinate entry: 7,2

Absolute Coordinate Entry

Follow these commands and absolute coordinate entries at the command line as you refer to **Figure 3-12.**

> Command: **L** *or* **LINE**↵
> Specify first point: **4,2**↵
> Specify next point or [Undo]: **7,2**↵
> Specify next point or [Undo]: **7,6**↵
> Specify next point or [Close/Undo]: **4,6**↵
> Specify next point or [Close/Undo]: **4,2**↵
> Specify next point or [Close/Undo]: ↵
> Command:

Exercise 3-5

Access the Student Web site (www.g-wlearning.com/CAD) and complete Exercise 3-5.

Relative Coordinate Entry

Follow these commands and relative coordinate entry methods as you refer to **Figure 3-13.**

> Command: **L** *or* **LINE**↵
> Specify first point: **2,2**↵
> Specify next point or [Undo]: **@6,0**↵
> Specify next point or [Undo]: **@2,2**↵
> Specify next point or [Close/Undo]: **@0,3**↵
> Specify next point or [Close/Undo]: **@-2,2**↵
> Specify next point or [Close/Undo]: **@-6,0**↵
> Specify next point or [Close/Undo]: **@0,-7**↵
> Specify next point or [Close/Undo]: ↵
> Command:

Exercise 3-6

Access the Student Web site (www.g-wlearning.com/CAD) and complete Exercise 3-6.

Figure 3-13.
Drawing a shape using the **LINE** tool and relative coordinates. Notice that negative (–) values are used and the coordinates are entered counterclockwise from the first point, in this case, 2,2.

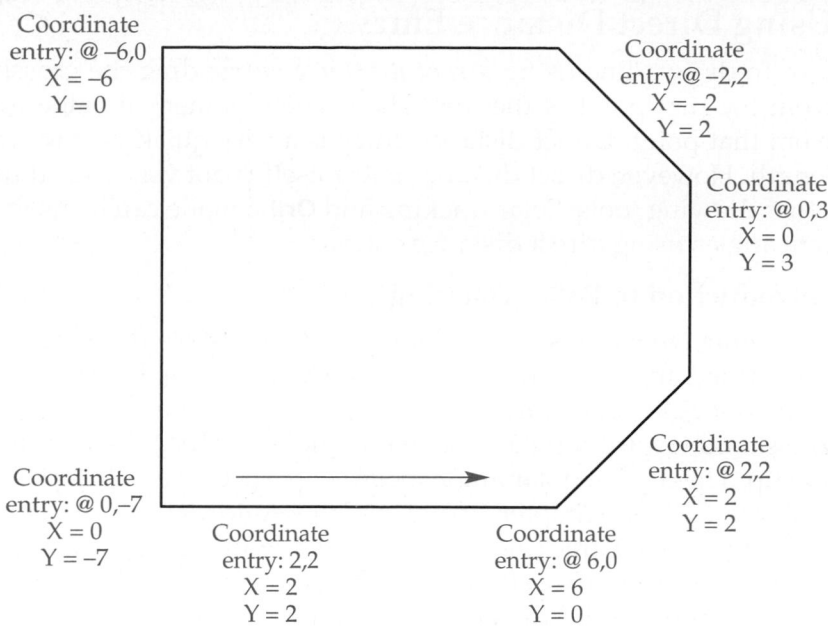

Coordinate entry: @ –6,0
X = –6
Y = 0

Coordinate entry:@ –2,2
X = –2
Y = 2

Coordinate entry: @ 0,3
X = 0
Y = 3

Coordinate entry: @ 2,2
X = 2
Y = 2

Coordinate entry: @ 0,–7
X = 0
Y = –7

Coordinate entry: 2,2
X = 2
Y = 2

Coordinate entry: @ 6,0
X = 6
Y = 0

Polar Coordinate Entry

Follow these commands and polar coordinate entry methods as you refer to **Figure 3-14.**

```
Command: L or LINE↵
Specify first point: 2,6↵
Specify next point or [Undo]: @2.5<0↵
Specify next point or [Undo]: @3<135↵
Specify next point or [Close/Undo]: 2,6↵
Specify next point or [Close/Undo]: ↵
Command: ↵
LINE Specify first point: 6,6↵
Specify next point or [Undo]: @4<0↵
Specify next point or [Undo]: @2<90↵
Specify next point or [Close/Undo]: @4<180↵
Specify next point or [Close/Undo]: @2<270↵
Specify next point or [Close/Undo]: ↵
Command:
```

Exercise 3-7

Access the Student Web site (www.g-wlearning.com/CAD) and complete Exercise 3-7.

Figure 3-14.
Using polar coordinates to draw a shape.

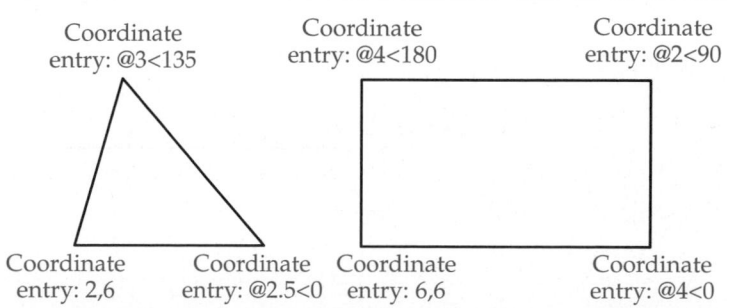

Coordinate entry: @3<135

Coordinate entry: @4<180

Coordinate entry: @2<90

Coordinate entry: 2,6

Coordinate entry: @2.5<0

Coordinate entry: 6,6

Coordinate entry: @4<0

Using Direct Distance Entry

direct distance entry: Entering points by dragging the crosshairs for direction and typing a number for distance.

To draw a line using *direct distance entry*, drag the crosshairs in any direction from the first point of the line. Then type a numerical value indicating the distance from that point. Direct distance entry is a very quick way to draw lines at a specific length. However, direct distance entry itself is not very useful unless you incorporate other drawing tools. Polar tracking and **Ortho** mode can be used to draw lines at accurate angles using direct distance entry.

Introduction to Polar Tracking

polar tracking: A drawing aid that causes the drawing crosshairs to "snap" to predefined angle increments.

Polar tracking is on by default and causes the drawing crosshairs to "snap" to predefined angle increments. Pick the **Polar Tracking** button on the status bar or press [F10] to toggle polar tracking on and off. When polar tracking is on, as you move the crosshairs toward a polar tracking angle, AutoCAD displays an alignment path and tooltip. The default polar angle increments are 0°, 90°, 180°, or 270°. Polar tracking is an AutoTrack mode. Chapter 7 fully explains AutoTrack.

To use polar tracking in combination with direct distance entry, first access the **LINE** tool and specify a start point. Then move the crosshairs in alignment with a polar tracking angle. Type the length of the line and press [Enter] or right-click and select **Enter**. See **Figure 3-15.** Chapter 7 fully explains polar tracking.

Drawing in Ortho Mode

ortho: From *orthogonal*, which means "at right angles."

Ortho mode constrains points selected while drawing and editing to be only horizontal or vertical from the previous point entry. See **Figure 3-16.** Pick the **Ortho Mode** button on the status bar to toggle **Ortho** mode on and off. If **Ortho** mode is turned off, you can temporarily turn it on when drawing an object by holding down [Shift].

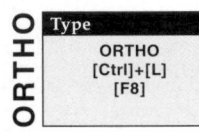

ORTHO

Type
ORTHO
[Ctrl]+[L]
[F8]

To use **Ortho** in combination with direct distance entry, access the **LINE** tool and specify a start point. Move the crosshairs to display a horizontal or vertical rubberband in the direction you want to draw. Then type the length of the line and press [Enter] or right-click and select **Enter**. See **Figure 3-17.**

Figure 3-15.
Using polar tracking to draw lines at predefined angle increments.

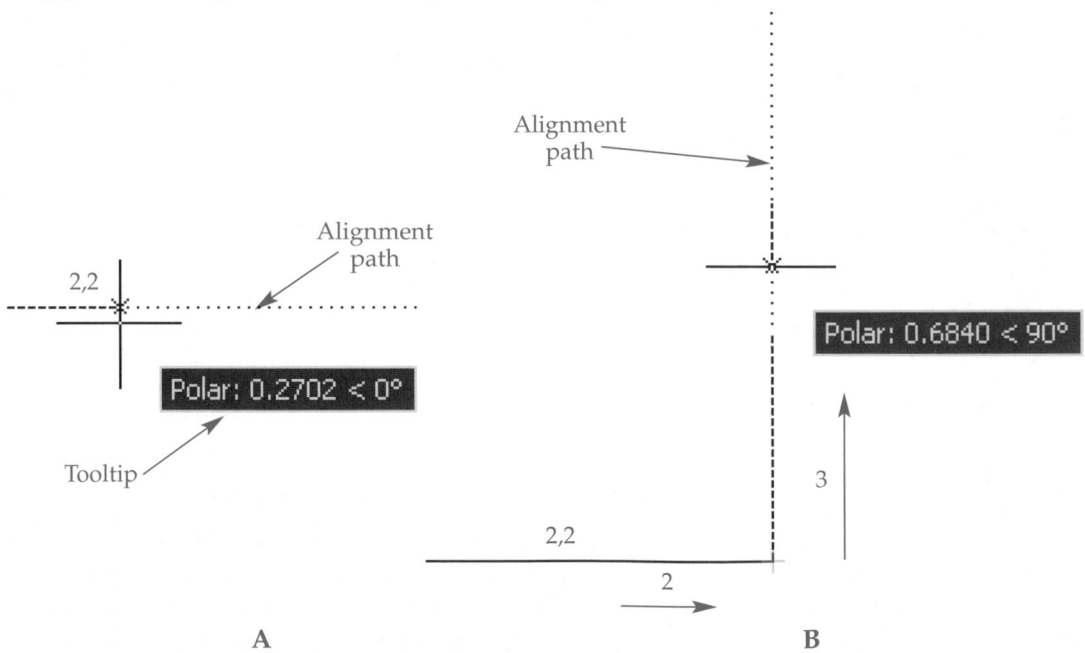

AutoCAD and Its Applications—Basics

Figure 3-16.
You can only draw horizontal and vertical lines when **Ortho** mode is on. Notice the location of the crosshairs when **Ortho** mode is on and off.

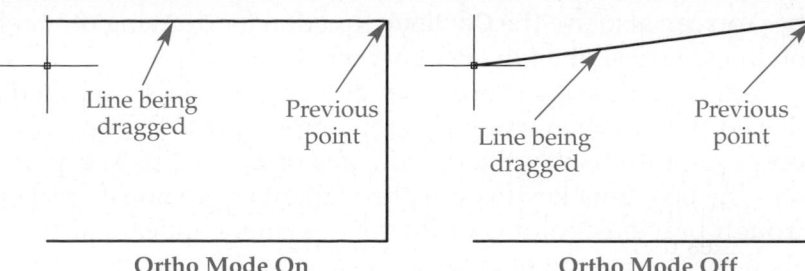

Figure 3-17.
Using direct distance entry to draw lines a designated distance from a current point. With **Ortho** mode on, move the cursor in the desired direction and type the distance.

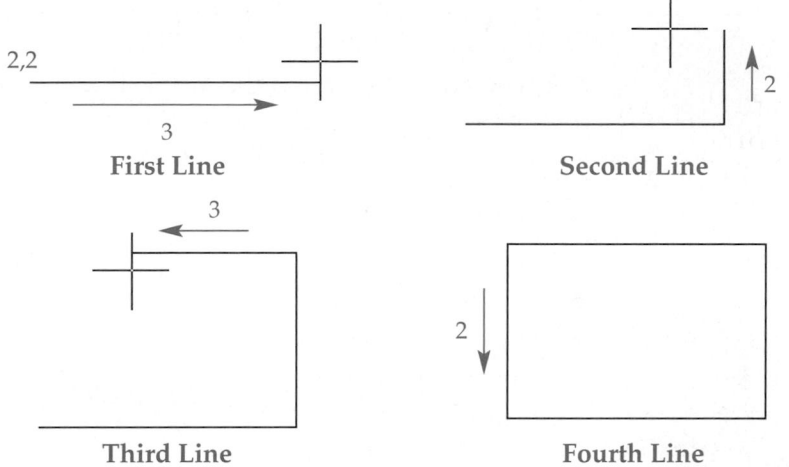

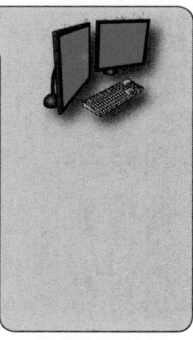

PROFESSIONAL TIP

Practice using different point entry techniques and decide which method works best for certain situations. Keep in mind that you can mix methods to help enhance your drawing speed. For example, absolute coordinates may work best to locate an initial point. Polar coordinates may work better to locate features in a circular pattern or in an angular relationship. Practice with direct distance entry using polar tracking and **Ortho** mode to see the advantages and disadvantages of each.

Exercise 3-8

Access the Student Web site (www.g-wlearning.com/CAD) and complete Exercise 3-8.

Using Previously Picked Points

Selecting previously picked points is a common need while drawing, especially if, for example, you draw a line, exit the **LINE** tool, and then decide to go back and connect a new line to the end of a previously selected point. The quickest way to reselect the last point entered is to first activate a tool, such as **LINE**. When you see the Specify first point: prompt, right-click or press [Enter] or the space bar. This action automatically connects the first endpoint of the new line segment to the endpoint of the previous line. You can also use the **Continue** function for drawing other objects such as arcs and polylines, as described in later chapters.

You can access the coordinates of many previously selected points, not just the last selected point. When using dynamic input, press the up arrow key at a point selection prompt to display the coordinates of the last picked point. You can continue to press the up arrow key to cycle through other previously picked points. As you scroll through previous point coordinates, a symbol appears at the location of each point. When the point symbol appears at the coordinates you want to pick, press [Enter] or right-click and choose **Enter**. Previous coordinates also display at the command line when a tool is active and you press the up arrow key.

PROFESSIONAL TIP

Another option to access the last selected point at a point selection prompt is to type @ and press [Enter]. This option is useful when you are connecting other shapes, such as circles, to the endpoint of a line.

Exercise 3-9

Access the Student Web site (www.g-wlearning.com/CAD) and complete Exercise 3-9.

Introduction to Editing

editing: Procedure used to modify an existing object.

selection set: A group of one or more drawing objects, typically defined to perform an editing operation.

Many *editing* tools exist to help increase productivity. This chapter introduces the basic editing tools **ERASE**, **OOPS**, **UNDO**, **U**, and **REDO**. To edit a drawing, you usually select one or more objects to create a *selection set*. The edit affects all objects in the selection set. The following information introduces several selection set methods using the **ERASE** tool. Keep in mind, however, that these techniques apply to most editing tools whenever the Select objects: prompt appears.

Erasing Objects

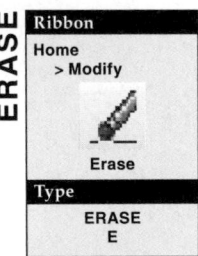

ERASE

Ribbon
Home
> Modify
Erase

Type
ERASE
E

Use the **ERASE** tool to remove unwanted drawing content. When you access the **ERASE** tool, the Select objects: prompt appears and an object selection target, or *pick box*, replaces the screen crosshairs. Move the pick box over the item you want to erase and pick that item. The object highlights and the Select objects: prompt redisplays, allowing you to select additional objects to erase. When you finish selecting objects, erase the selected objects by right-clicking, pressing [Enter], or pressing the space bar. See **Figure 3-18.**

pick box: Small box that replaces the crosshairs when objects are to be selected.

Figure 3-18.
Using the **ERASE** tool to erase a single object.

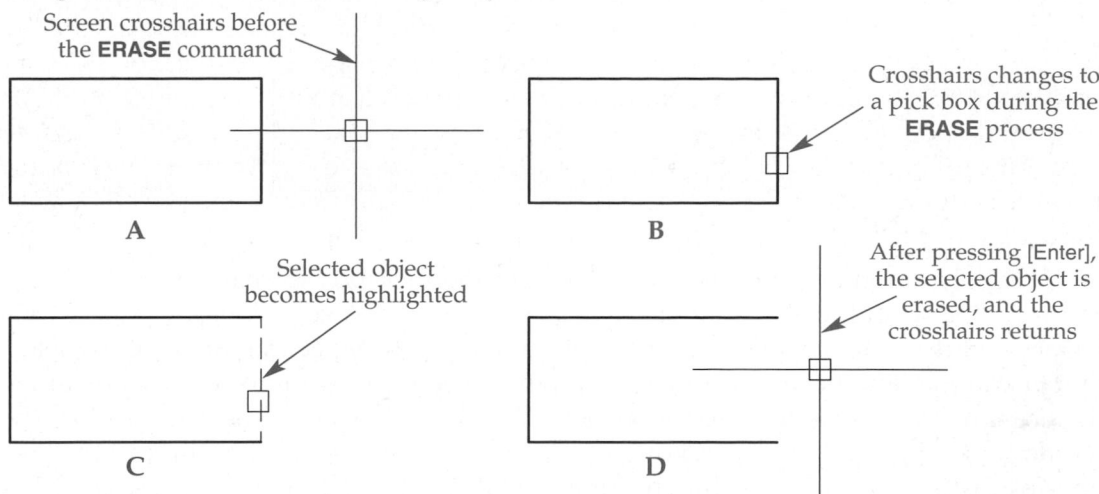

Screen crosshairs before
the **ERASE** command

A

Crosshairs changes to
a pick box during the
ERASE process

B

Selected object
becomes highlighted

C

After pressing [Enter],
the selected object is
erased, and the
crosshairs returns

D

NOTE

By default, when you move the crosshairs or pick box over an object and pause for a moment, the object changes to a thicker lineweight and becomes dashed. When you move the crosshairs or pick box off the object, the object display returns to normal. This allows you to preview the object before you select it. When many objects are in a small area, this feature helps you select the correct object the first time and often eliminates the need to cycle through stacked objects.

Exercise 3-10

Access the Student Web site (www.g-wlearning.com/CAD) and complete Exercise 3-10.

Window and Crossing Selection

Use window and crossing selection to select multiple objects, reducing the need to pick individual objects with the pick box. Window selection allows you to draw a box, or "window," around an object or group of objects to select for editing. Everything entirely within the window selects at the same time. Portions of objects that project outside the window remain unselected.

Crossing selection is similar to window selection, but with crossing selection, objects contained within the box *and crossing the box* are selected. The crossing selection box displays a dotted outline with a light green background to distinguish it from the window selection box, which displays a solid outline and light blue background.

The quickest and most effective way to use window or crossing selection is through a feature known as *automatic windowing*, or *implied windowing*, which is on by default. To apply automatic window selection, use the pick box to select a point clearly above or below and to the left of the objects to erase. A selection box replaces the pick box, and the Specify opposite corner: prompt appears. Move the corner of the selection box to the right and up or down so the box completely covers the objects to erase. Pick to locate the second corner. See **Figure 3-19.** All objects that lie completely within the box highlight. Right-click or press [Enter] or the space bar to complete the **ERASE** tool.

automatic windowing (implied windowing): Selection method that allows you to select multiple objects at one time without entering a selection option.

Figure 3-19.
Using automatic windowing to select all objects completely inside a window selection box.

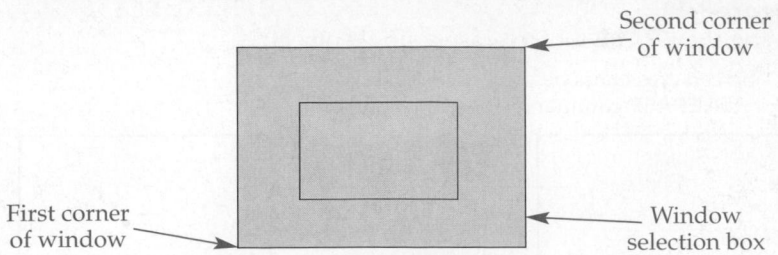

To apply automatic crossing selection, use the pick box to select a point clearly above or below and to the right of the object(s) to erase. A selection box replaces the pick box, and the Specify opposite corner: prompt appears. Move the corner of the selection box to the left and up or down, across the objects you want to select. Remember, the crossing box does not need to enclose the entire object to erase it, as does the window box. **Figure 3-20** shows how to use crossing selection to erase three of the four lines of a rectangle that was drawn using the **LINE** tool. Right-click or press [Enter] or the space bar to complete the **ERASE** tool.

NOTE

You can also type W or WINDOW at the Select objects: prompt to use manual window selection, or type C or CROSSING to use manual crossing selection. When you use manual window or crossing selection, the selection box remains in the window or crossing format whether the first pick is left or right of objects, and whether you move the cursor to the left or right.

Exercise 3-11

Access the Student Web site (www.g-wlearning.com/CAD) and complete Exercise 3-11.

Window Polygon and Crossing Polygon Selection

The window and crossing selection methods use a rectangular selection box, which may not allow you to select needed objects easily. An alternative is to form a window or crossing selection polygon. To use window polygon selection, type WPOLYGON or WP at the Select objects: prompt. Then pick points to draw a polygon

Figure 3-20.
Using crossing selection to select all objects inside and touching the crossing selection box.

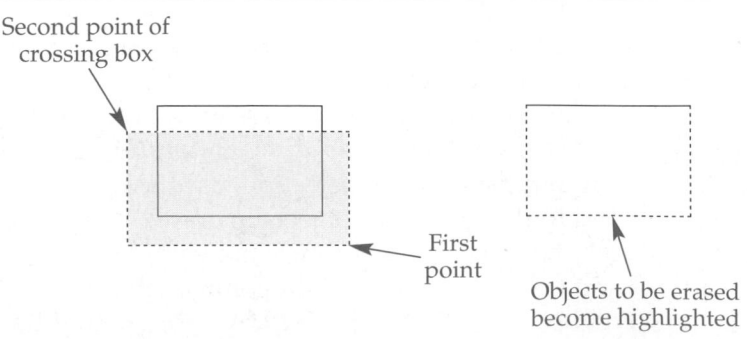

AutoCAD and Its Applications—Basics

Figure 3-21.
A—Using window
polygon selection
to erase objects.
B—Using crossing
polygon selection
to erase objects.
Everything enclosed
within and crossing
the polygon selects.

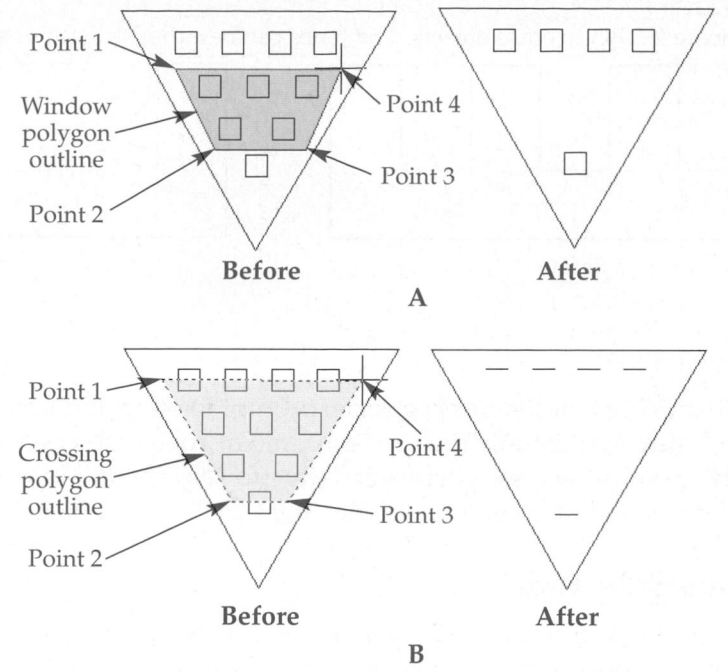

enclosing the objects you want to select. As you pick corners, the polygon drags into place. See **Figure 3-21A.**

Crossing polygon selection is similar to window polygon selection, but with crossing polygon selection, you create a crossing selection. To form a crossing selection polygon, type CPOLYGON or CP at the Select objects: prompt. Then pick points to draw a polygon. See **Figure 3-21B.**

PROFESSIONAL TIP

In window or crossing polygon selection, AutoCAD does not allow you to select a point that causes the lines of the selection polygon to intersect each other. Pick locations that do not result in an intersection. Use the **Undo** option if you need to go back and relocate a previous pick point.

Fence Selection

Fence selection is another method used to select several objects at the same time. When using fence selection, you place a fence, or connected lines, through the objects you want to select. Only the objects passing through the fence are included in the selection set. To use fence selection, type FENCE or F at the Select objects: prompt. Then pick points to draw a fence through the objects to select. The fence can be straight or staggered, as shown in **Figure 3-22.**

Exercise 3-12

Access the Student Web site (www.g-wlearning.com/CAD) and complete Exercise 3-12.

Figure 3-22.
Using fence selection to erase objects. The fence can be either straight or staggered.

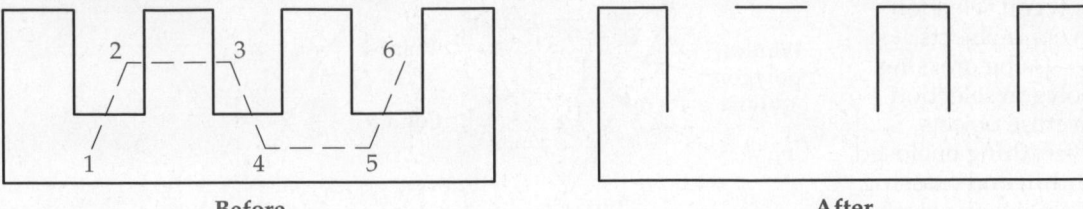

Before   After

Last Selection

Type LAST or L at the Select objects: prompt to select the last object drawn. The last selection feature selects only the last item drawn. You must repeatedly access a tool, such as **ERASE**, and use last selection every time to select individual items in reverse order. This is extremely slow compared to selecting the objects using other methods.

Previous Selection

Type Previous or P at the Select objects: prompt to reselect all the objects selected during the previous selection set. You usually use previous selection when you need to carry out more than one sequential editing operation on a specific group of objects. In this case, use previous selection to reselect the objects you just edited.

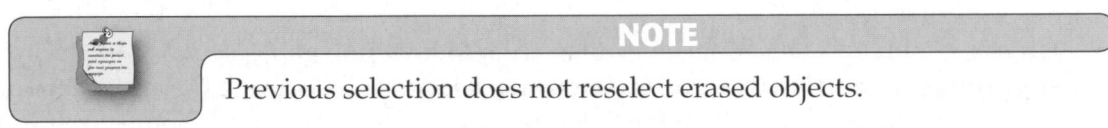

> **NOTE**
>
> Previous selection does not reselect erased objects.

Selecting All Objects

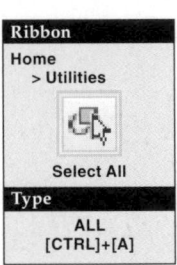

Ribbon
Home
> Utilities

Select All

Type
ALL
[CTRL]+[A]

Use the **Select All** tool to quickly select every object in the drawing. Everything in the drawing that is not on a frozen layer is selected, even objects that are outside of the current drawing window display. Chapter 5 describes layers.

Cycling through Stacked Objects

stacked objects:
Objects that overlap in the drawing. When you pick with the mouse, the topmost object selects by default.

cycle: Repeatedly select a series of stacked objects until the desired object highlights.

One way to deal with *stacked objects* is to let AutoCAD *cycle* through the overlapping objects. Cycling works best when several objects cross at the same place or are very close together. To cycle through stacked objects, first access a tool, such as **ERASE**. Next, with the Select objects: prompt shown, move the pick box over the intersection of the stacked objects. Hold down [Shift] and repeatedly press the space bar to cycle through the stacked objects. When the desired object highlights, release [Shift] and pick (left-click) to select the item. See **Figure 3-23**.

Changing the Selection Set

The quickest way to remove one or more objects from the current selection set is to hold down [Shift] and reselect the objects. This is possible only for individual picks and automatic windows. For automatic windowing, hold down [Shift] and pick the first corner. Then release [Shift] and pick the second corner. Select objects as usual to add them back to the selection set.

Another option for removing objects from a selection set is to type REMOVE or R at the Select objects: prompt. This enters the **Remove** option and changes the Select objects: prompt to Remove objects:, allowing you to pick the objects to remove from the selection set. To switch back to the selection mode, type ADD or A at the Remove objects:

Figure 3-23.
Cycling through a series of stacked circles until the desired object highlights.

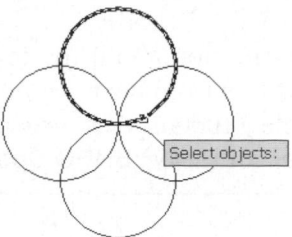
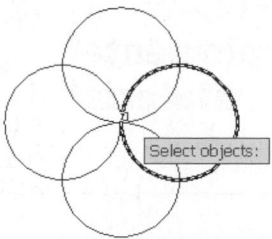

Original group of objects
with top item highlighted

Hold down shift key and press the
space bar to cycle through objects

prompt. This enters the **Add** option and restores the Select objects: prompt, allowing you to select additional objects.

PROFESSIONAL TIP

Removing items from a selection set is especially effective when used in combination with the **Select All** selection option. This allows you to keep very specific objects while erasing everything else.

Exercise 3-13

Access the Student Web site (www.g-wlearning.com/CAD) and complete Exercise 3-13.

Supplemental Material

Selection Display Options
For detailed information about adjusting selection display options, go to the Student Web site (www.g-wlearning.com/CAD), select this chapter, and select **Selection Display Options**.

Using the UNDO Tool

The **UNDO** tool offers several options that allow you to undo a single operation or a number of operations at once. The **UNDO** tool is different from the **Undo** option of certain tools, such as the **LINE** tool. This quickest way to use the **UNDO** tool is to pick the **Undo** button on the **Quick Access** toolbar. Select the button as many times as needed to undo multiple operations. An alternative is to pick the flyout and select all of the tools to undo from the list.

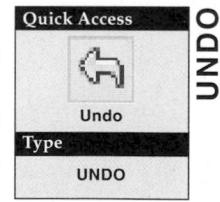

Quick Access

Undo

Type

UNDO

UNDO

Using the U Tool

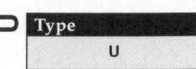

The **U** tool undoes the effect of the previously entered tool. You can reissue the **U** tool to continue undoing tool actions, but you can only undo one tool at a time. The actions are undone in the order in which they were used.

NOTE

You can also activate the **U** tool by right-clicking in the drawing window and selecting **Undo** *current*.

Redoing the Undone

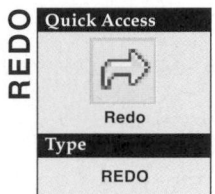

Use the **REDO** tool to reverse the action of the **UNDO** and **U** tools. The **REDO** tool works only *immediately* after undoing something. The **REDO** tool does not bring back line segments removed using the **Undo** option of the **LINE** tool. The quickest way to use the **REDO** tool is to pick the **Redo** button on the **Quick Access** toolbar. Select the button as many times as needed to redo multiple undone operations. An alternative is to pick the flyout and select one or more undone operations from the list to redo.

Exercise 3-14

Access the Student Web site (www.g-wlearning.com/CAD) and complete Exercise 3-14.

Using the OOPS Tool

The **OOPS** tool brings back the last object you *erased*. Unlike the **UNDO** and **U** tools, **OOPS** only returns the objects erased in the most recent procedure. It has no effect on other modifications. If you erase several objects in the same tool sequence, all of the objects return to the screen.

Chapter Test

Answer the following questions. Write your answers on a separate sheet of paper or go to the Student Web site (www.g-wlearning.com/CAD) and complete the electronic chapter test.

1. When you enter a fractional number in AutoCAD, why is a hyphen required between a whole number and its associated fraction?
2. Briefly describe the Cartesian coordinate system.
3. Name the tool used to place a pattern of dots on the screen.
4. Name two ways to access the **Drafting Settings** dialog box.
5. How do you activate the **Snap** mode?
6. How do you set a grid spacing of .25?
7. List two ways to discontinue drawing a line.
8. Give the tools and entries to draw a line from Point A to Point B to Point C and back to Point A. Return to the Command: prompt.
   A. Command: _____
   B. Specify first point: _____
   C. Specify next point or [Undo]: _____
   D. Specify next point or [Undo]: _____
   E. Specify next point or [Close/Undo]: _____
9. Name three types of coordinates used for point entry.
10. What does the absolute coordinate display 5.250,7.875 mean?
11. What does the polar coordinate display @2.750<90 mean?
12. How can you turn on the coordinate display field if it is off?
13. What two general methods of point entry are available when dynamic input is active?
14. Explain, in general terms, how direct distance entry works.
15. What are the default angle increments for polar tracking?
16. How can you turn on the **Ortho** mode?
17. Explain how you can continue drawing another line segment from a previously drawn line.
18. When you access the **ERASE** tool, what replaces the screen crosshairs?
19. How does the appearance of window and crossing selection boxes differ?
20. List five ways to select an object to erase.
21. Define *stacked objects*.
22. What is the difference between pressing the **Undo** button on the **Quick Access** toolbar and entering the **UNDO** tool?
23. How many tool sequences can you undo at one time with the **U** tool?
24. Name the tool used to bring back an object that was previously removed using the **UNDO** tool.
25. Name the tool used to bring back the last object(s) erased before starting another tool.

Drawing Problems

Start AutoCAD if it is not already started. Follow the specific instructions for each problem. Do not draw dimensions or text.

▼ Basic

1. Start a new drawing from scratch or use a template of your choice. Use the **LINE** tool and draw the following objects on the left side of the screen. Accurately draw the specified objects with grid and snap turned off.
   - Right triangle.
   - Isosceles triangle.
   - Rectangle.
   - Square.
   Save the drawing as P3-1.

2. Start a new drawing from scratch or use a template of your choice. Draw the same objects specified in Problem 1 on the right side of the screen. This time, make sure the snap grid is turned on. Observe the difference between having snap on for this problem and off for the previous problem. Save the drawing as P3-2.

3. Start a new drawing from scratch or use a decimal unit template of your choice. Draw an object by connecting the following point coordinates. Use dynamic input to enter the coordinates. Save your drawing as P3-3.

Point	Coordinates	Point	Coordinates
1	2,2	8	@-1.5,0
2	@1.5,0	9	@0,1.25
3	@.75<90	10	@-1.25,1.25
4	@.1.5<0	11	@2<180
5	@0,-.75	12	@-1.25,-1.25
6	@3,0	13	@2.25<270
7	@1<90		

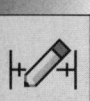

4. Start a new drawing from scratch or use a template of your choice. Use direct distance entry and polar tracking to draw the outline shown. Each grid square is one unit. Do not draw the grid lines. Save the drawing as P3-4.

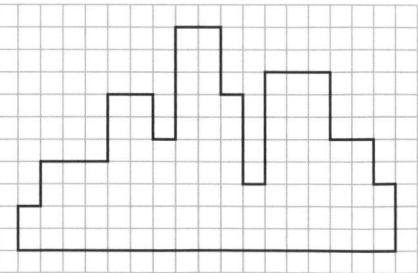

5. Start a new drawing from scratch or use a template of your choice. Use the dimensional input feature of dynamic input to draw the hexagon shown. Each side of the hexagon is 2 units. Begin at the start point, and draw the lines in the direction indicated by the arrows. Do not draw dimensions. Save the drawing as P3-5.

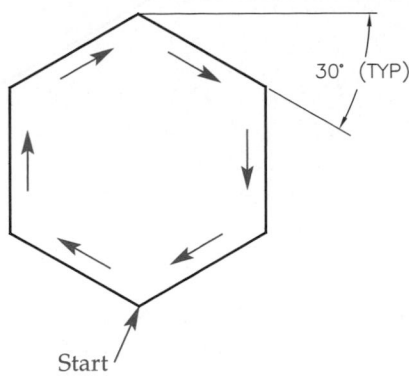

▼ Intermediate

6. Start a new drawing from scratch or use a fractional unit template of your choice. Draw the following shapes using absolute, relative, and polar coordinate entry methods. Draw Object A three times, using a different point entry system each time. Set the units to decimal and the precision to 0.0 when drawing Object A. Draw Object B once, using at least two methods of coordinate entry. Set the units to fractional and the precision to 1/16 when drawing Object B. Do not draw dimensions. Save the drawing as P3-6.

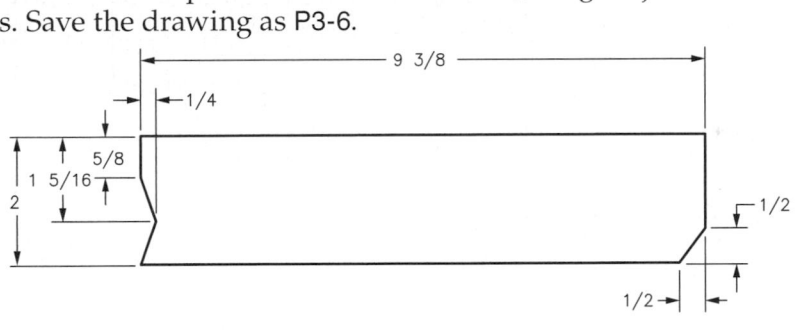

For Problems 7–8, start a new drawing from scratch or use a decimal unit template of your choice. Draw the given part. Do not draw dimensions. Save the drawings as P3-7 and P3-8.

7.

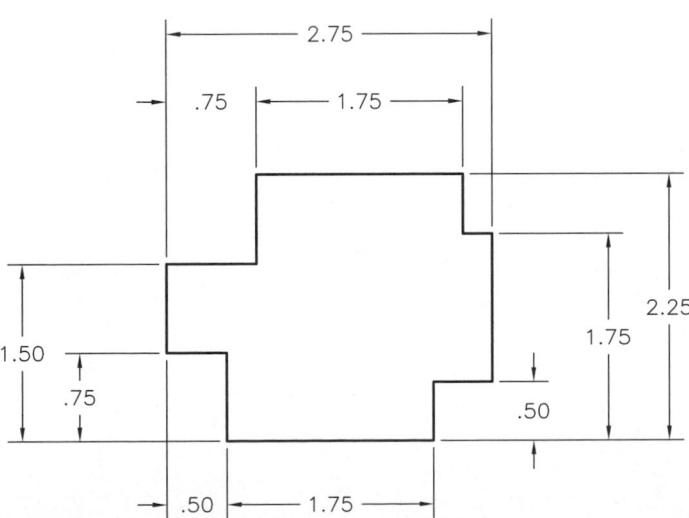

Drawing Problems - Chapter 3

8.

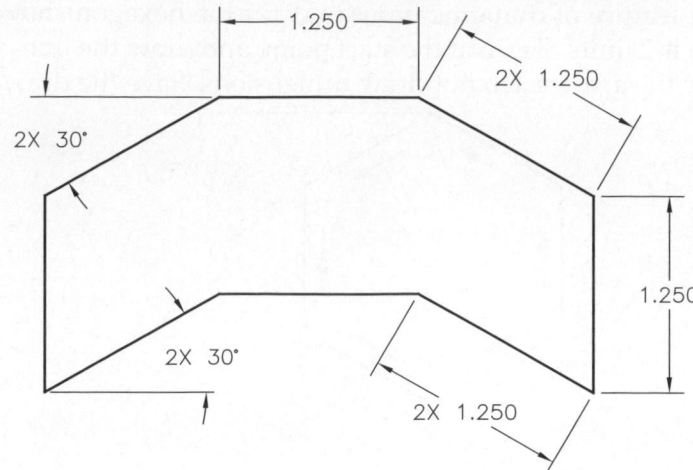

9. Start a new drawing from scratch or use a decimal unit template of your choice. Draw the objects shown in A and B. Begin at the start point and then discontinue the **LINE** tool at the point shown. Complete each object using the **Continue** option. Do not draw dimensions. Save the drawing as P3-9.

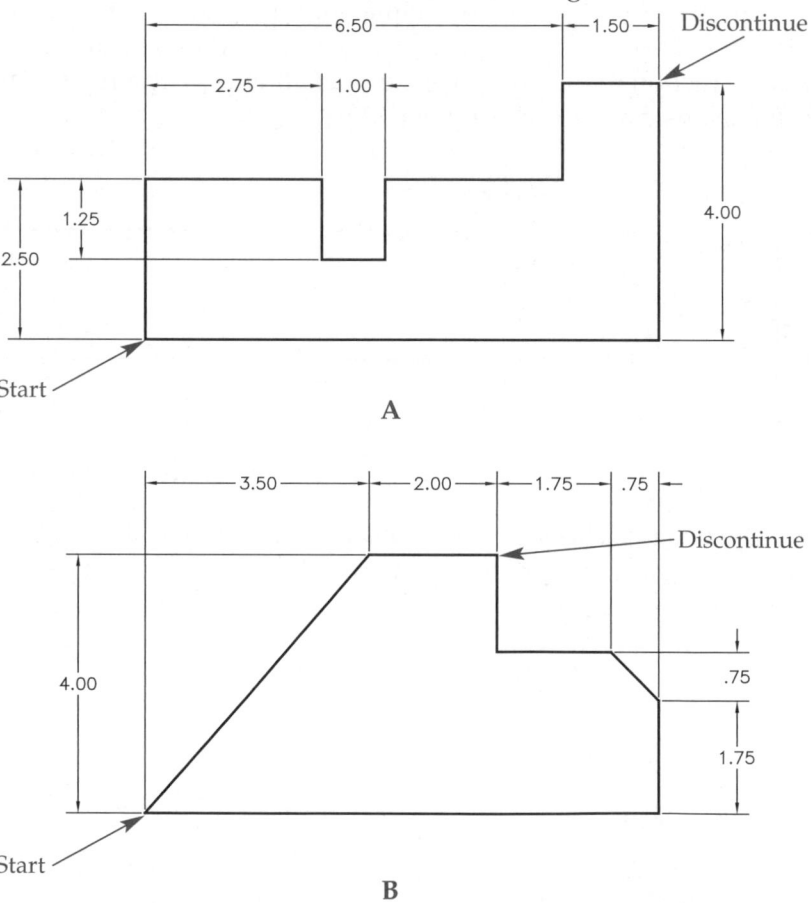

10. Sketch the X and Y axes on a sheet of paper. Label the origin, the positive values for X = 1 through X = 10, and the positive values for Y = 1 through Y = 10. Then sketch the object described by the following coordinate points:

2,2
8,2
8,7
7,7
7,3
6,3
6,6
4,6
4,3
3,3
3,7
2,7
2,2

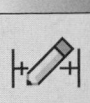

▼ Advanced

11. Sketch the X and Y axes of the Cartesian coordinate system as you did for the previous problem. Then sketch an object outline of your choice within the axes. List, in order, the rectangular coordinates of the points a drafter would need to specify in AutoCAD to recreate the object in your sketch.

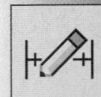

For Problems 12–14, start a new drawing from scratch or use a decimal unit template of your choice. Draw the given part. Do not draw dimensions. Save the drawings as P3-12, P3-13, *and* P3-14.

12.

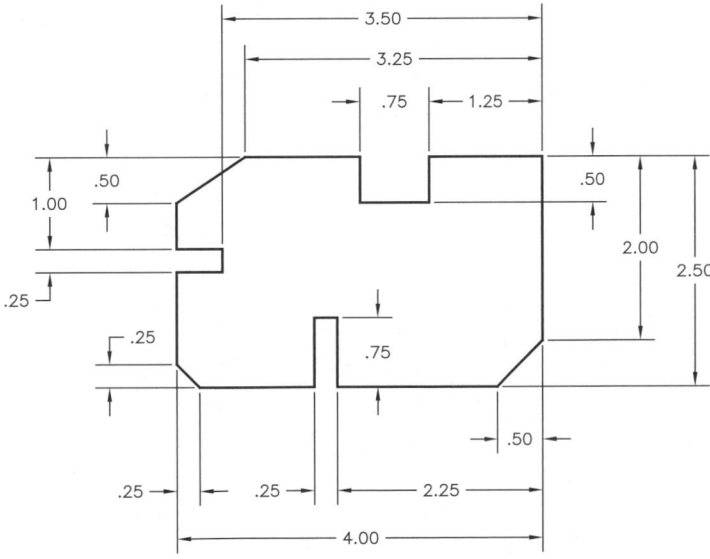

13.

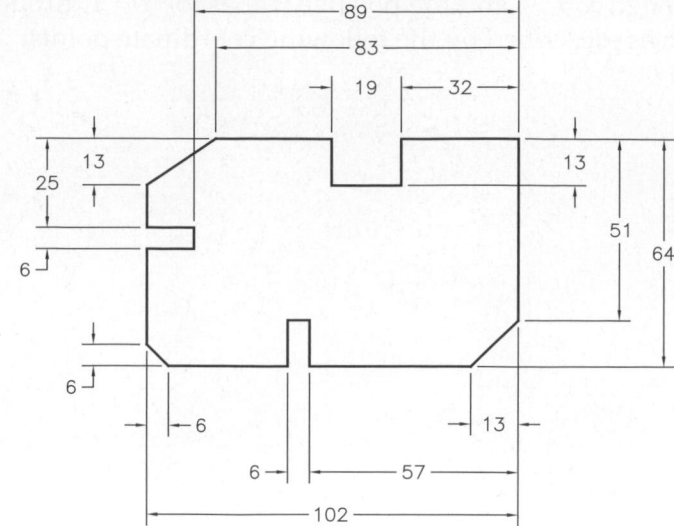

14.

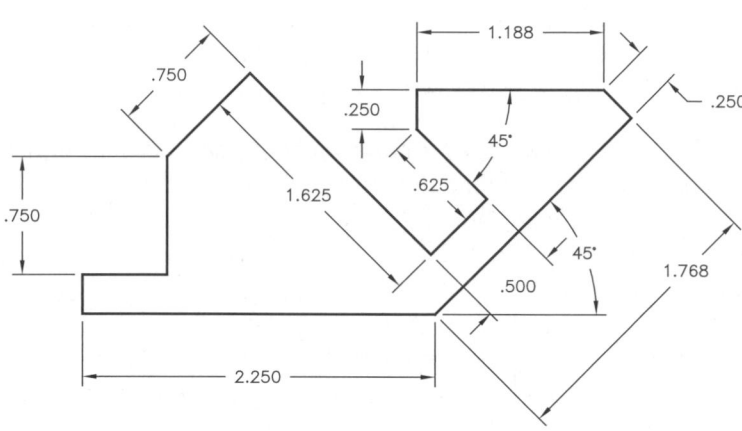

AutoCAD and Its Applications—Basics

Basic Object Tools

Learning Objectives

After completing this chapter, you will be able to do the following:

✓ Draw circles using the **CIRCLE** tool options.
✓ Draw arcs using the **ARC** tool options.
✓ Use the **ELLIPSE** tool to draw ellipses and elliptical arcs.
✓ Use the **PLINE** tool to draw polylines.
✓ Draw polygons using the **POLYGON** tool.
✓ Draw rectangles using the **RECTANGLE** tool options.
✓ Draw donuts and filled circles using the **DONUT** tool.
✓ Draw true spline curves using the **SPLINE** tool.

This chapter describes how to use a variety of tools to draw basic shapes. This chapter presents the ribbon as the primary means of accessing basic object tool options. When you select an object tool option from the ribbon, specific prompts associated with the option appear. When you issue a tool using dynamic input or the command line, you must enter specific options when prompted.

Using Basic Object Tools

When you draw basic shapes, AutoCAD prompts you to select coordinates, just as when you draw a line. For example, the Specify center point: prompt appears when you draw a circle using the **Center, Radius** or **Center, Diameter** method. Your response to this prompt is similar to your response to the Select first point: prompt that appears when you use the **LINE** tool. Use any appropriate point entry method to pick a location.

Some basic object tools give you the option of entering specific values to define the size and shape of the object. For example, the Specify radius of circle: prompt appears after you locate the center point of a circle using the **Center, Radius** option of the **CIRCLE** tool. Usually, the most effective response to this type of prompt is to type a value, such as the circle radius in this example. An alternative is to use an appropriate point entry technique. Refer to Chapter 3 to review proper numerical responses and point entry methods.

Figure 4-1.
Dragging a circle to a desired size. The circle attached to the crosshairs stretches like a rubberband until you pick a point to define the radius.

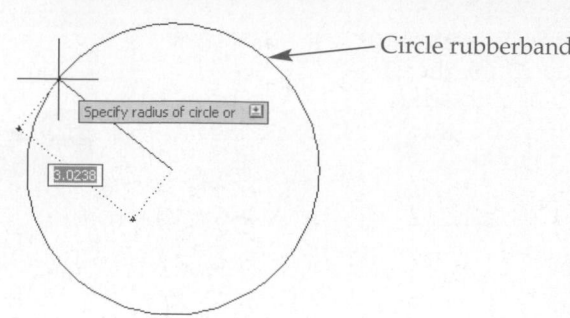

Circle rubberband

Specify radius of circle or

Similar to the **LINE** tool, many object tools include a rubberband shape that connects the first point selection to the crosshairs. The image aids in sizing and locating the object. For example, when you draw a circle using the **Center, Radius** option, a circle image appears on-screen after you pick the center point. The image gets larger or smaller as you move the pointer. When you pick the radius, the actual circle object replaces the rubberband image. See **Figure 4-1.**

Drawing Circles

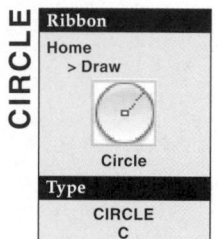

CIRCLE

Ribbon
Home
> Draw

Circle

Type
CIRCLE
C

circle: A closed curve with a constant radius around a center point; size is usually dimensioned according to the diameter.

The **CIRCLE** tool provides several methods for drawing *circles*. Choose the appropriate option based on how you want to locate the circle, and the information you know or want to use to construct the circle. The ribbon is an effective way to access circle tool options. See **Figure 4-2.**

Drawing a Circle by Radius

Use the **Center, Radius** option of the **CIRCLE** tool to specify the center point and the radius of a circle. After selecting the **Center, Radius** option, define the center point. Then type a value for the radius and press [Enter] or the space bar, or right-click and pick **Enter**. You can also define the radius using point entry. See **Figure 4-3.**

Figure 4-2.
The **Circle** options can be accessed from the **Draw** panel of the **Home** ribbon tab.

Pick the **Circle** flyout to display the **Circle** options

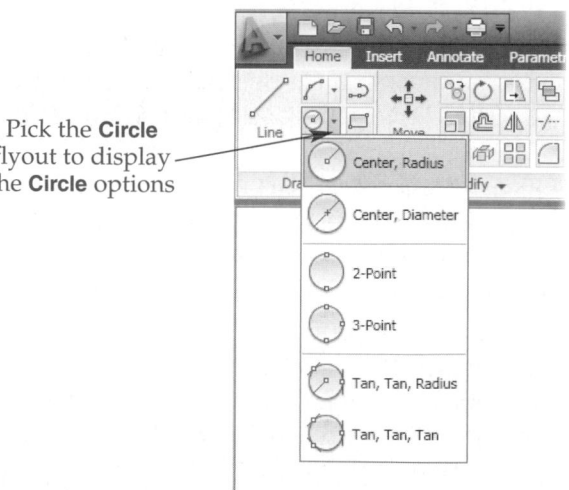

Center, Radius

Center, Diameter

2-Point

3-Point

Tan, Tan, Radius

Tan, Tan, Tan

Figure 4-3.
Drawing a circle by
specifying the center
point and radius.

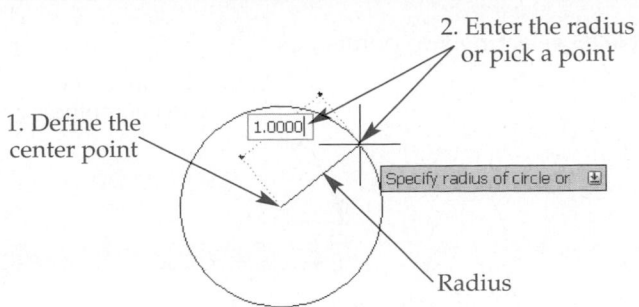

The radius value you enter is stored as the new default radius
setting, allowing you to quickly draw another circle with the same
radius.

Drawing a Circle by Diameter

You can also draw a circle by specifying the center point and the diameter. The
Center, Diameter option is convenient because most circular holes, shafts, and features
are sized according to diameter. After selecting the **Center, Diameter** option, define the
center point. Then type a value for the diameter and press [Enter] or the space bar, or
right-click and pick **Enter**. You can also define the diameter using point entry. In the
Center, Diameter option, the crosshairs measures the diameter, but the rubberband
circle passes midway between the center and the crosshairs. See **Figure 4-4.**

NOTE

If you use the **Center, Radius** option to draw a circle after using the
Diameter option, AutoCAD changes the default to a radius measurement based on the previous diameter.

Drawing a Two-Point Circle

A two-point circle is drawn by picking two points on opposite sides of the circle to
define its diameter. The **2-Point** option is useful if the diameter of the circle is known,
but the center is difficult to find. One example of this is locating a circle between
two lines. The process of drawing a two-point circle is very similar to drawing a line.
After selecting the **2 Points** option, enter or select a point for the first endpoint of the
circle's diameter. Then enter or select the second endpoint of the circle's diameter. See
Figure 4-5.

Figure 4-4.
When you use the
Center, Diameter
option, AutoCAD
calculates the circle's
position as you
move the crosshairs.

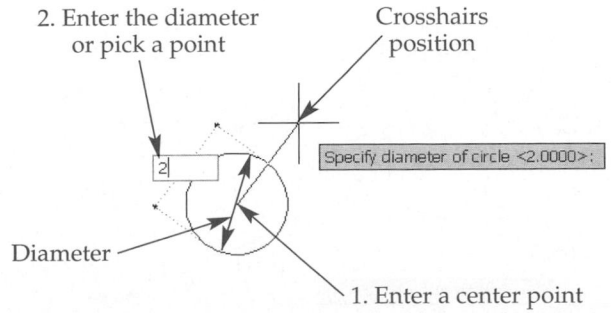

Figure 4-5.
Drawing a circle by selecting two points.

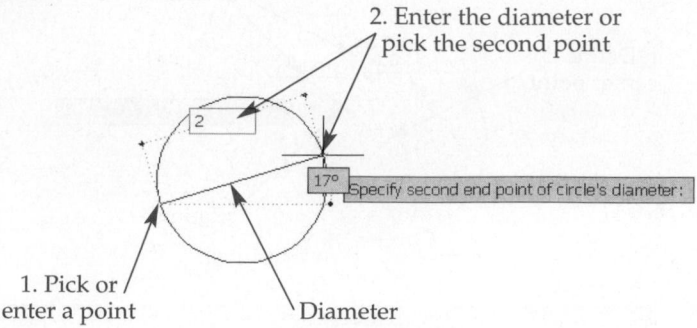

2. Enter the diameter or pick the second point

2

17° Specify second end point of circle's diameter:

1. Pick or enter a point

Diameter

Drawing a Three-Point Circle

The **3-Point** option is the best method to use if you know three points on the circumference of a circle. After selecting the **3-Point** option, enter or select three points in any order to define the circumference of the circle. See **Figure 4-6.**

Drawing a Circle Tangent to Two Objects

tangent: A line, circle, or arc that meets another circle or arc at only one point.

point of tangency: The point shared by tangent objects.

object snap: A tool that snaps to exact points, such as endpoints or midpoints, when you pick a point near these locations.

The **Tan, Tan, Radius** option creates a circle of a specified radius *tangent* to two objects. You can draw a circle tangent to given lines, circles, or arcs. The circle automatically positions at the *point of tangency*. The **Tan, Tan, Radius** option uses an *object snap* known as **Deferred Tangent** to assist you in picking a point exactly tangent to other objects. Object snaps are covered in detail in Chapter 7. After accessing the **Tan, Tan, Radius** option, hover the crosshairs over the first line, arc, or circle to which the new circle will be tangent. Pick when the deferred tangent symbol appears. Repeat the process to select the second line, arc, or circle to which the new circle will be tangent. Then enter or select the radius of the circle. See **Figure 4-7.**

NOTE

If the radius you enter while using the **Tan, Tan, Radius** option is too small, AutoCAD displays the message Circle does not exist.

Drawing a Circle Tangent to Three Objects

The **Tan, Tan, Tan** option allows you to draw a circle tangent to three existing objects. This option creates a three-point circle using the three points of tangency. Like the **Tan, Tan, Radius** option, the **Tan, Tan, Tan** option uses the **Deferred Tangent** object snap to assist you in picking a point exactly tangent to other objects. After accessing the **Tan, Tan, Tan** option, pick three lines, arcs, or circles to which the new circle will be

Figure 4-6.
Drawing a circle by picking three points that lie on the circle.

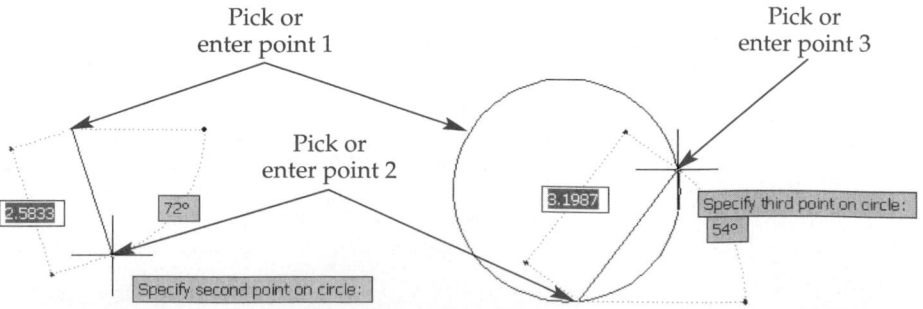

Pick or enter point 1

Pick or enter point 3

Pick or enter point 2

2.5833

72°

3.1987

Specify third point on circle:

54°

Specify second point on circle:

Figure 4-7.
Two examples of
drawing circles
tangent to two given
objects using the
Tan, Tan, Radius
option.

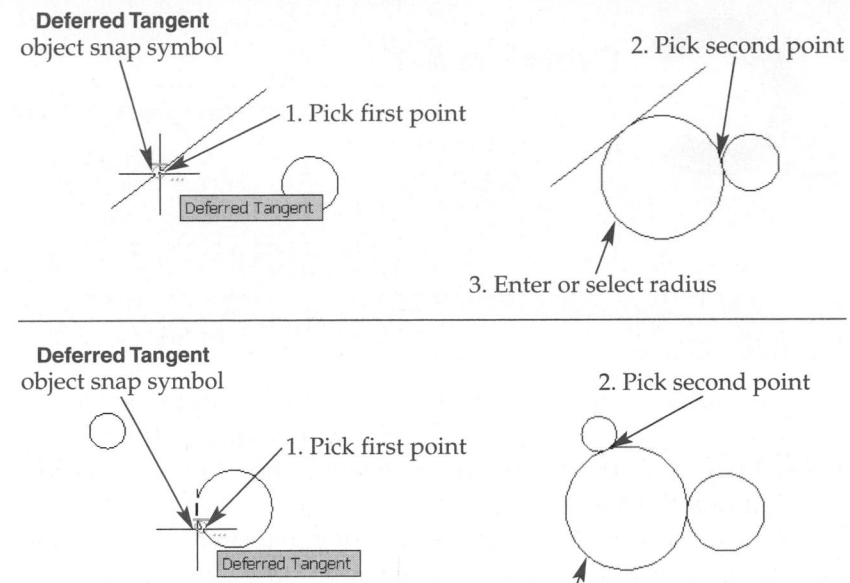

tangent. You must pick each item when the deferred tangent symbol appears, but the order in which you pick the items is not critical. See **Figure 4-8.**

> **NOTE**
>
> Unlike the **Tan, Tan, Radius** option, the **Tan, Tan, Tan** option does not automatically recover when you pick a point where no tangent exists. In such a case, you must manually reactivate the **Tangent** object snap to make additional picks. Chapter 7 describes the **Tangent** object snap. For now, type TAN and press [Enter] at the point selection prompt to return the pick box so you can pick again.

Figure 4-8.
Two examples of
drawing circles
tangent to three
given objects.

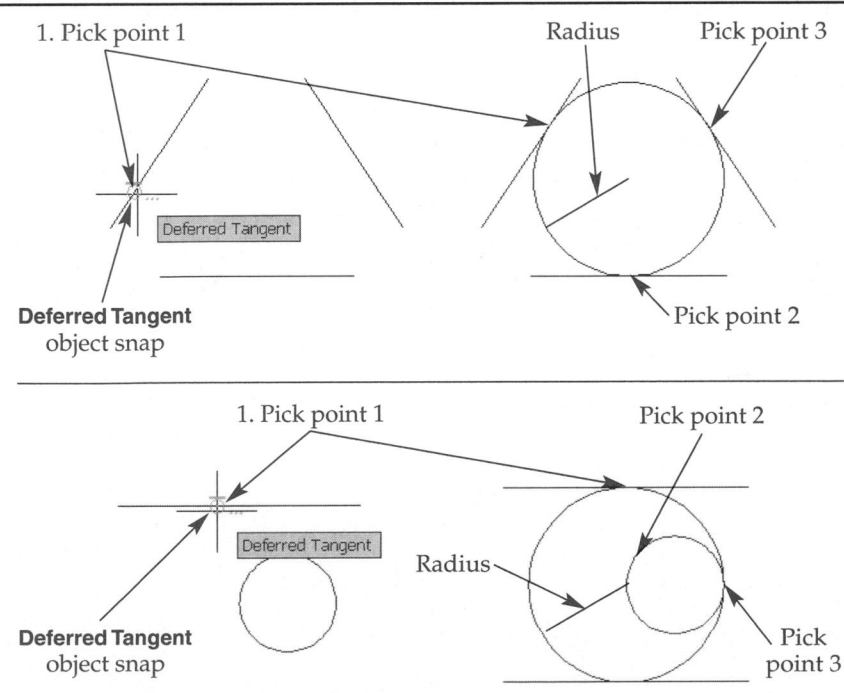

Exercise 4-1

Access the Student Web site (www.g-wlearning.com/CAD) and complete Exercise 4-1.

Drawing Arcs

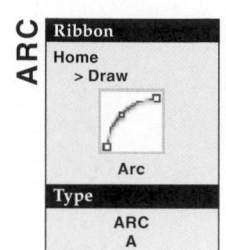

Ribbon

Home
> Draw

Arc

Type

ARC
A

arc: Any portion of a circle; usually dimensioned according to the radius.

included angle: The angle formed between the center, start point, and endpoint of an arc.

chord length: The linear distance between two points on a circle or arc.

The **ARC** tool offers a number of different options for drawing *arcs*. Select the appropriate option based on how you want to locate the arc, and the information you know or want to use to construct the arc. The ribbon is an effective way to access arc tool options. See **Figure 4-9**.

The various **ARC** tool options allow you to create an arc for any situation requiring an arc. **Figure 4-10** describes most methods. The step-by-step examples shown in **Figure 4-10** depict a single common application of each option. Arc placement is set according to your selections and the values you enter. Some arc options prompt for the *included angle*, and others prompt for the *chord length*. Locating points in a clockwise or counterclockwise pattern affects the result when using most arc options. The values you specify, including the use of positive or negative numbers, also affects the result.

> **NOTE**
>
> The **3-Point** option is default when you enter the **ARC** tool at the keyboard.

Figure 4-9.
Selecting an **Arc** option from the **Draw** panel of the **Home** ribbon tab.

Pick to display **Arc** options

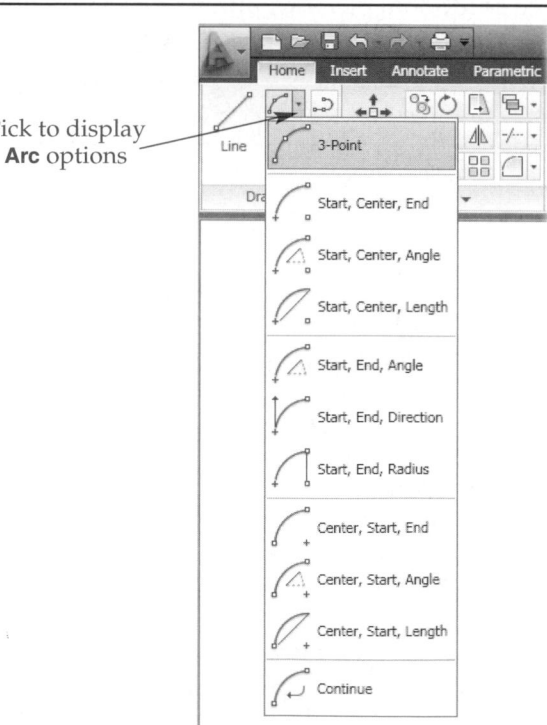

Figure 4-10.
Select the appropriate arc creation method based on how you want to locate the arc and the information you know or want to use to construct the arc.

Option	Direction	Steps
3-Point	Clockwise or counterclockwise	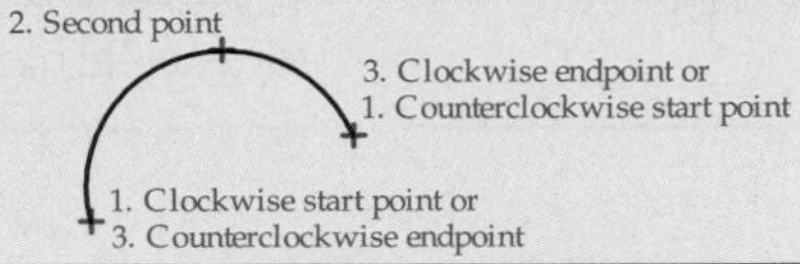 2. Second point; 3. Clockwise endpoint or 1. Counterclockwise start point; 1. Clockwise start point or 3. Counterclockwise endpoint
Start, Center, End	Counterclockwise	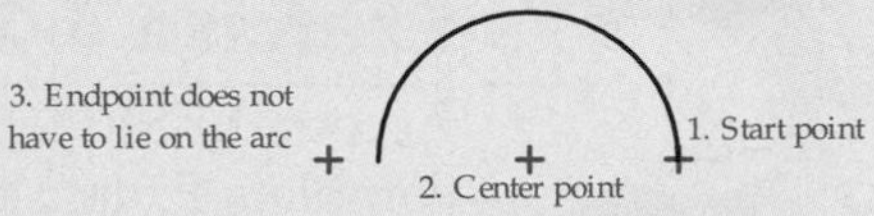 3. Endpoint does not have to lie on the arc; 1. Start point; 2. Center point
Start, Center, Angle	Positive angle = Clockwise; Negative angle = Counterclockwise	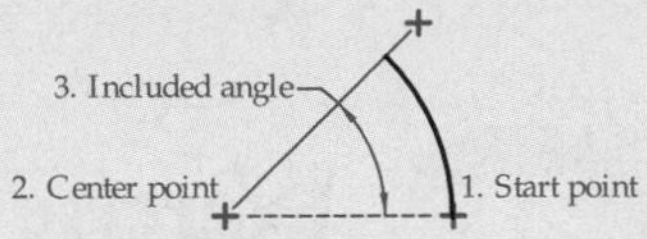 3. Included angle; 2. Center point; 1. Start point
Start, Center, Length	Counterclockwise	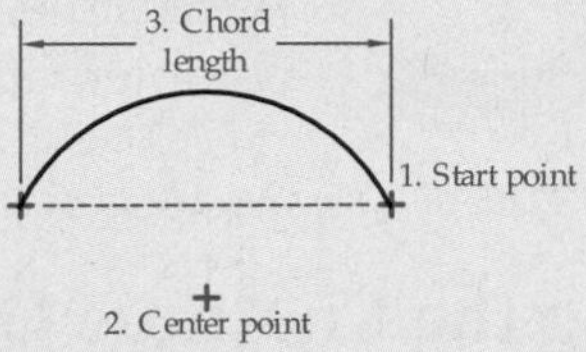 3. Chord length; 1. Start point; 2. Center point
Start, End, Angle	Positive angle = Clockwise; Negative angle = Counterclockwise	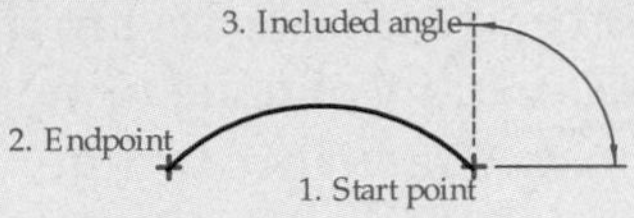 3. Included angle; 2. Endpoint; 1. Start point
Start, End, Direction	Tangent to specified direction	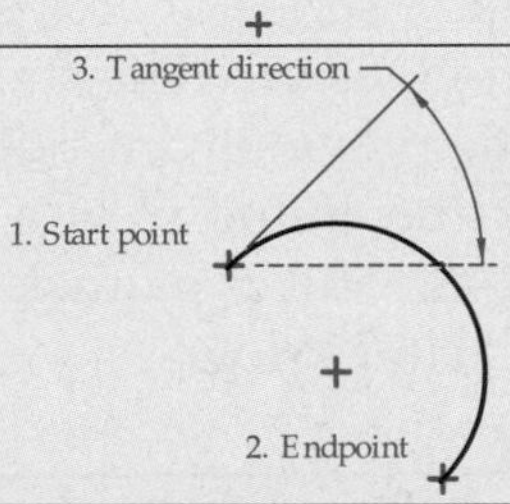 3. Tangent direction; 1. Start point; 2. Endpoint
Start, End, Radius	Counterclockwise	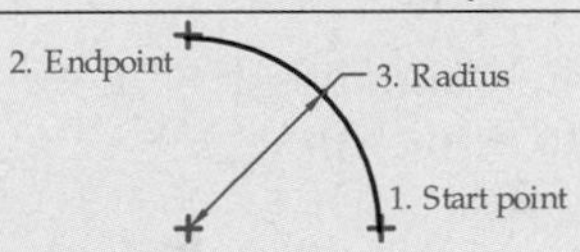 2. Endpoint; 3. Radius; 1. Start point
Center, Start, End	Counterclockwise	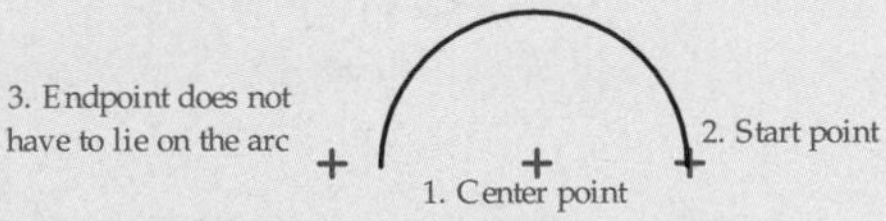 3. Endpoint does not have to lie on the arc; 2. Start point; 1. Center point
Center, Start, Angle	Positive angle = Clockwise; Negative angle = Counterclockwise	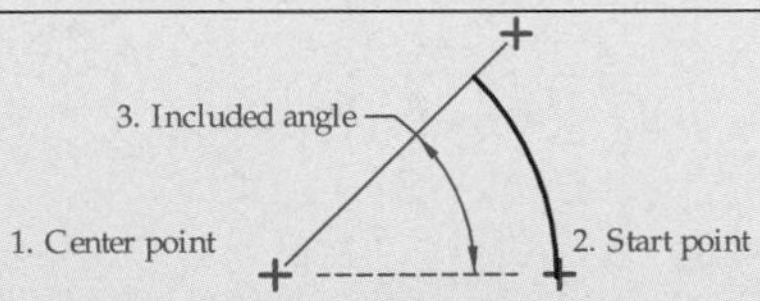 3. Included angle; 1. Center point; 2. Start point
Center, Start, Length	Counterclockwise	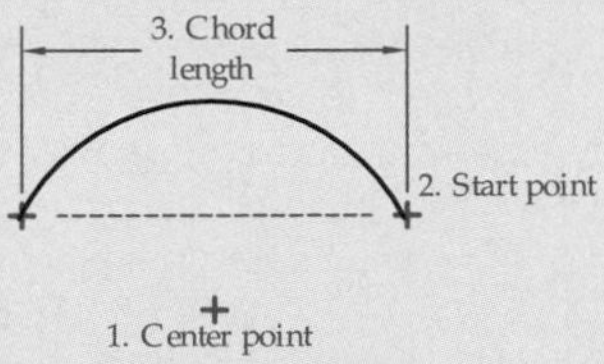 3. Chord length; 2. Start point; 1. Center point

Reference Material *Chord Length Table*
For a chord length table and other reference tables, go to the **Reference Material** section of the Student Web site (www.g-wlearning.com/CAD) and select **Standard Tables**.

Exercise 4-2

Access the Student Web site (www.g-wlearning.com/CAD) and complete Exercise 4-2.

Exercise 4-3

Access the Student Web site (www.g-wlearning.com/CAD) and complete Exercise 4-3.

Using the Continue Option

You can continue an arc from the endpoint of a previously drawn arc or line using the **ARC** tool's **Continue** option. The arc automatically attaches to the endpoint of the previously drawn arc or line, and the Specify endpoint of arc: prompt appears. Pick the endpoint to create the arc.

When a series of arcs are drawn using the **Continue** option, each arc is tangent to the previous arc. The start point and direction occur from the endpoint and direction of the previous arc. See **Figure 4-11.** When you use the **Continue** option to begin an arc at the endpoint of a previously drawn line, the arc is tangent to the line. See **Figure 4-12.** This is a quick way to draw an arc tangent to a line for a variety of applications, such as slots.

Figure 4–11.
Using the **Continue** option to draw three tangent arcs.

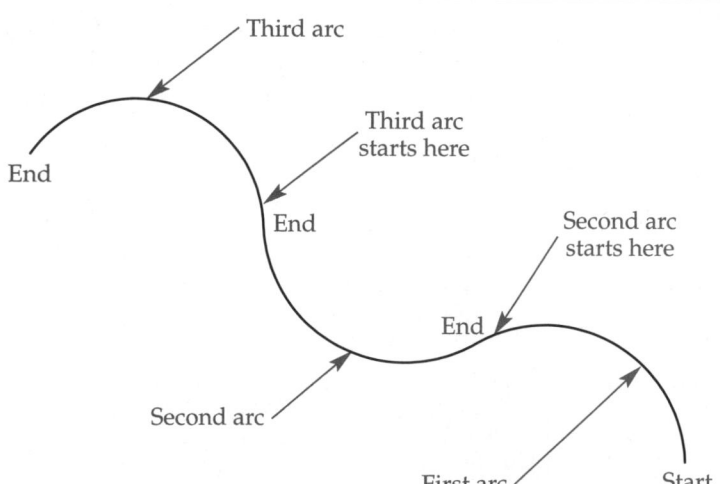

Figure 4–12.
An arc continuing from the previous line. Point 2 is the start of the arc, and Point 3 is the end of the arc. The arc and line are tangent at Point 2.

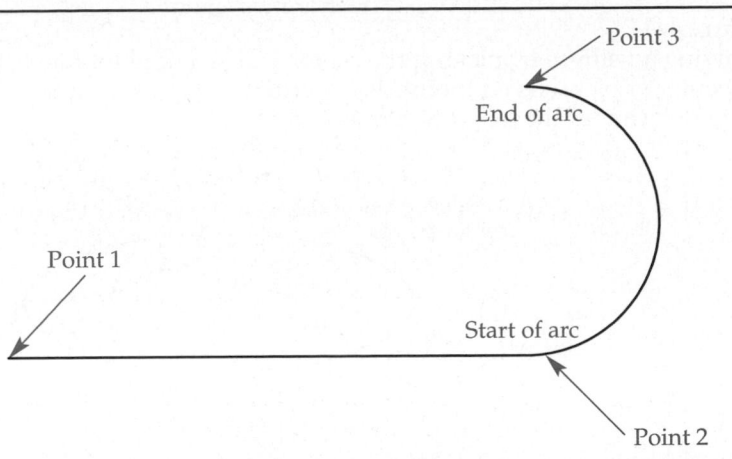

Point 3

End of arc

Point 1

Start of arc

Point 2

NOTE

You can also access the **Continue** option by beginning the **ARC** tool and then pressing [Enter] or the space bar, or by right-clicking and selecting **Enter** when prompted to specify the start point of the arc.

Exercise 4-4

Access the Student Web site (www.g-wlearning.com/CAD) and complete Exercise 4-4.

Drawing Ellipses

When you view a circle at an angle, it appears as an *ellipse* and contains both a *major axis* and a *minor axis*. For example, a 30° ellipse is a circle rotated 30° from the line of sight. **Figure 4-13** shows the parts of an ellipse.

The **ELLIPSE** tool provides several methods for drawing elliptical shapes. Choose the appropriate option based on how you want to locate the ellipse, the information you know or want to use to construct the ellipse, and whether the ellipse is whole or partial.

Using the Center Option

Use the **Center** option to construct an ellipse by specifying the center point and one endpoint for each of the two axes. Select the center point of the ellipse and then select the endpoint of one of the axes. Specify the endpoint of the other axis to complete the ellipse. See **Figure 4-14**.

ellipse: An oval shape that contains two centers of equal radius.

major axis: The longer of the two axes in an ellipse.

minor axis: The shorter of the two axes in an ellipse.

Ribbon
Home
> Draw

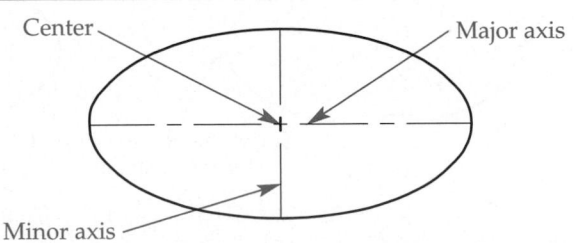

Ellipse

Type
ELLIPSE
EL

ELLIPSE

Figure 4–13.
The parts of an ellipse.

Center

Major axis

Minor axis

Figure 4–14.
Drawing an ellipse by picking the center and an endpoint for each axis. The order in which you enter or pick axis endpoints is not critical. The distance from each endpoint to the center point determines the major and minor axes.

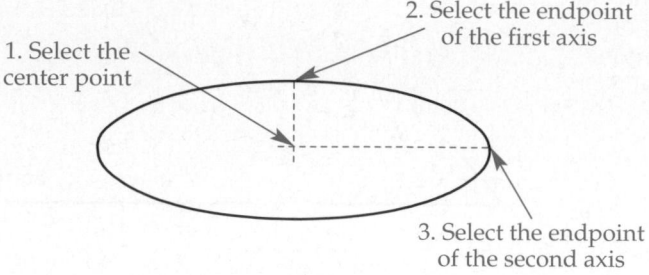

Using the Axis, End Option

The **Axis, End** option establishes the first axis and one endpoint of the second axis. Select the endpoint of one of the axes and then select the other endpoint of the same axis. Enter a distance from the midpoint of the first axis to the end of the second axis to complete the ellipse. The first axis can be the major or minor axis, depending on what you enter for the second axis. See **Figure 4-15**.

Using the Rotation Option

Use the **Rotation** option to create an ellipse by specifying the angle at which a circle is rotated from the line of sight. Begin by constructing an ellipse as usual, but be sure to select the major axis with the first axis endpoint. Then, when the Specify distance to other axis or: prompt appears, select the **Rotation** option instead of picking the second axis endpoint. Finally, enter the angle at which the circle is rotated from the line of sight to produce the ellipse.

For example, if you respond by entering 30 for a 30° rotation, an ellipse forms based on a circle rotated 30° from the line of sight. A 0 response draws an ellipse with the minor axis equal to the major axis, which is a circle. AutoCAD rejects any rotation angle between 89.99994° and 90.00006° or between 269.99994° and 270.00006°. **Figure 4-16** shows the relationship among several ellipses having the same major axis length, but different rotation angles.

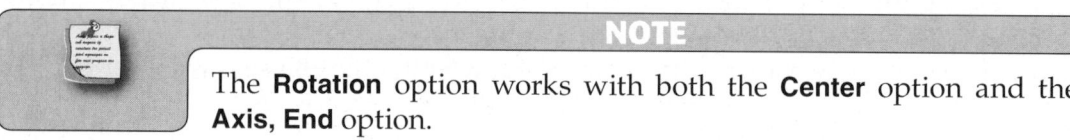

> **NOTE**
>
> The **Rotation** option works with both the **Center** option and the **Axis, End** option.

Figure 4–15.
Constructing the same ellipse by choosing different axis endpoints.

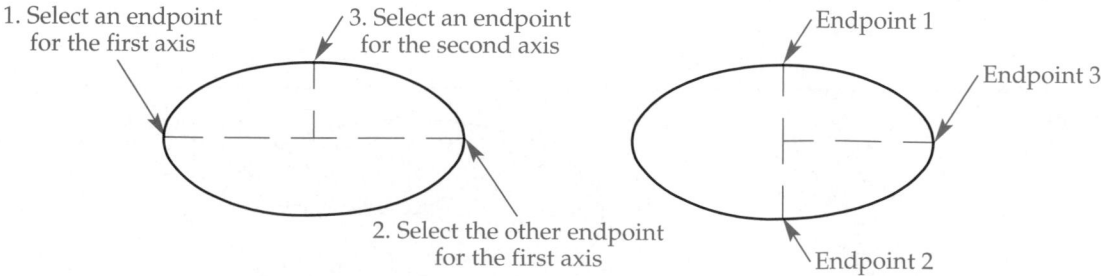

Figure 4–16.
Ellipse rotation angles.

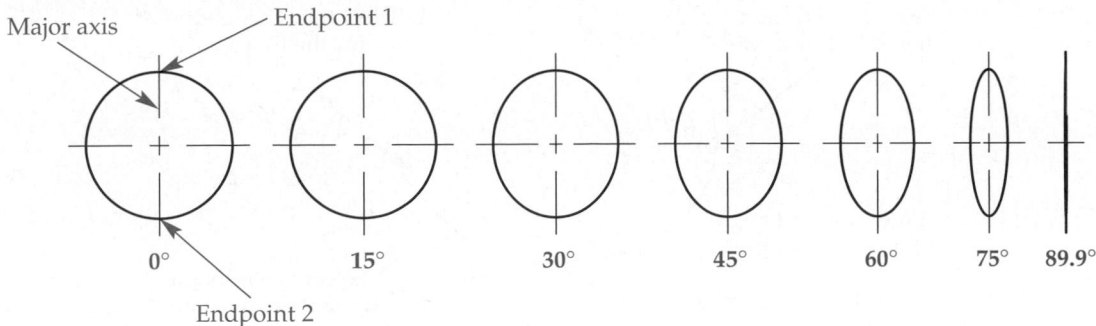

Major axis — Endpoint 1

0°   15°   30°   45°   60°   75°   89.9°

Endpoint 2

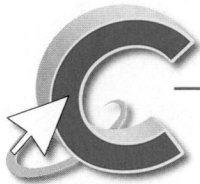

Exercise 4-5

Access the Student Web site (www.g-wlearning.com/CAD) and complete Exercise 4-5.

Drawing Elliptical Arcs

Use the **Arc** option of the **ELLIPSE** tool to draw elliptical arcs. Creating an elliptical arc is just like drawing an ellipse, but with two additional steps that define the start point and endpoint of the elliptical arc. Several options are available for defining the size and shape of an elliptical arc.

The default elliptical arc uses axis endpoints to define the ellipse, and then start and end angles to produce the elliptical arc. After selecting the **Elliptical Arc** option, specify the endpoint of one of the ellipse axes. Then select the other endpoint of the same axis. Enter a distance from the midpoint of the first axis to the end of the second axis to form an ellipse. Finally, select the start and end angles for the elliptical arc.

The start and end angles are the angular relationships between the ellipse's center and the arc's endpoints. The angle of the first axis establishes the angle of the elliptical arc. For example, a 0° start angle begins the arc at the first endpoint of the first axis. A 45° start angle begins the arc 45° counterclockwise from the first endpoint of the first axis. End angles are also counterclockwise from the start point. **Figure 4-17** shows an elliptical arc drawn using a 0° start angle and a 90° end angle and displays sample arcs with different start and end angles.

Additional elliptical arc options are also available. **Figure 4-18** briefly describes each option. Use the **Center** option when appropriate instead of the default axis endpoint method. The **Parameter**, **Included angle**, and **Rotation** options are available during the process of creating axis endpoint or center elliptical arcs.

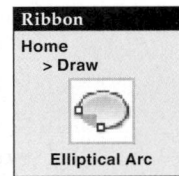

Ribbon
Home
> Draw

Elliptical Arc

Exercise 4-6

Access the Student Web site (www.g-wlearning.com/CAD) and complete Exercise 4-6.

Figure 4–17.
Drawing elliptical arcs. Note the three examples at the bottom created using three different angles.

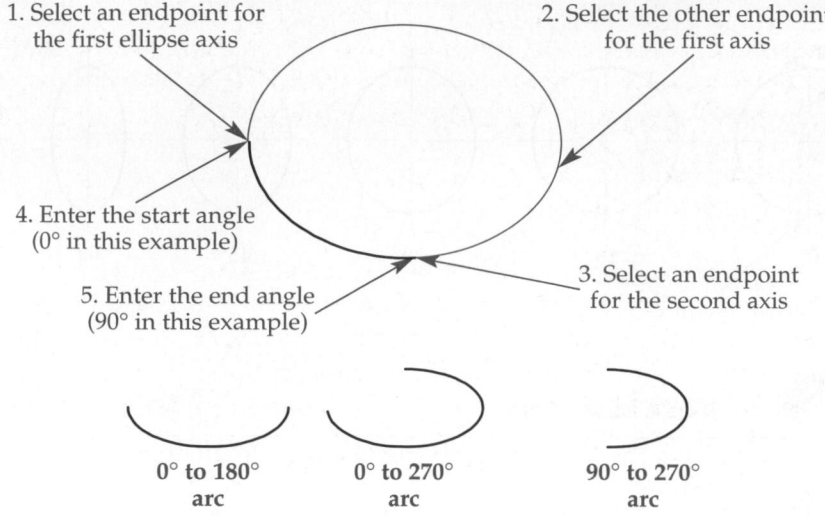

1. Select an endpoint for the first ellipse axis

2. Select the other endpoint for the first axis

4. Enter the start angle (0° in this example)

5. Enter the end angle (90° in this example)

3. Select an endpoint for the second axis

0° to 180°
arc

0° to 270°
arc

90° to 270°
arc

Figure 4–18.
Additional options available for drawing elliptical arcs.

Option	Application	Process
Center	Lets you establish the center of the elliptical arc. **Rotation**, **Parameter**, and **Included angle** options are available.	1. Select the ellipse center point. 2. Select the endpoint of one of the ellipse axes. 3. Pick the endpoint of the other axis to form the ellipse. 4. Enter the start angle for the elliptical arc. 5. Select the end angle.
Parameter	Use instead of picking the start angle of the elliptical arc. AutoCAD uses a different means of vector calculation to create the elliptical arc.	1. Specify the start parameter point. 2. Specify the end parameter point.
Included angle	Establishes an included angle beginning at the start angle.	1. Specify the included angle.
Rotation	Allows you to rotate the elliptical arc about the first axis by specifying a rotation angle. **Parameter** and **Included angle** options are available.	1. Specify the rotation around the major axis. 2. Specify the start angle for the elliptical arc. 3. Specify the end angle.

Drawing Polylines

PLINE

Ribbon
Home
> Draw

[icon] Polyline

Type
PLINE
PL

polyline: A series of lines and arcs that constitute a single object.

Use the **PLINE** tool to draw *polylines*. When you use the default polyline settings, the process of drawing polyline segments is identical to the process of creating line segments using the **LINE** tool. Access the **PLINE** tool and use point entry to locate polyline endpoints. When you finish specifying points, press [Enter], the space bar, or [Esc], or right-click and select **Enter**. The difference between a polyline and a line is that once it is drawn, all segments of a continuous polyline act as a single object. The **PLINE** tool also provides more flexibility than the **LINE** tool, allowing you to draw a single object composed of straight lines and arcs of varying thickness.

Undo and **Close** options are available for drawing polylines, and they function the same as those for drawing lines using the **LINE** tool. Use the **Undo** option to remove the last segment of a polyline without leaving the **PLINE** tool. This removes the previously drawn segment and allows you to continue from the previous endpoint. You can use the **Undo** option repeatedly to delete polyline segments until the entire object is gone. Use the **Close** option to connect the endpoint of the last polyline segment to the start point of the first polyline segment.

Setting the Polyline Width

The default polyline settings create a polyline with a constant width of 0. A polyline drawn using a constant width of 0 is similar to a standard line and accepts the lineweight applied to the layer on which the polyline is drawn. Chapter 5 explains layers. Adjust the polyline width to create thick and tapered polyline objects.

To change the width of a polyline segment, access the **PLINE** tool, select the first point, and then choose the **Width** option. AutoCAD prompts you to specify the starting width of the line, followed by the ending width of the line. Enter the same starting and ending width value to draw a polyline with constant width. The rubberband line from the first point reflects the width settings. **Figure 4-19A** shows a 4″ polyline with starting and ending widths of .25″. Notice that the starting and ending points of the line are located at the center of the line segment's width.

To create a tapered line segment, enter different values for the starting and ending widths. In the example shown in **Figure 4-19B,** starting width is .25 units, and the ending width is .5 units. One special use of a tapered polyline is the creation of arrowheads. To draw an arrowhead, use the **Width** option and specify 0 as the starting width, and then use an appropriate ending width.

NOTE

Using a starting or ending width value other than 0 overrides the lineweight applied to the layer on which the polyline is drawn.

Using the Halfwidth Option

The **Halfwidth** option allows you to specify the width of the polyline from the center to one side, as opposed to the total width of the polyline defined using the **Width** option. Access the **PLINE** tool, pick the first polyline endpoint, and then choose the **Halfwidth** option. Specify starting and ending values at the appropriate prompts.

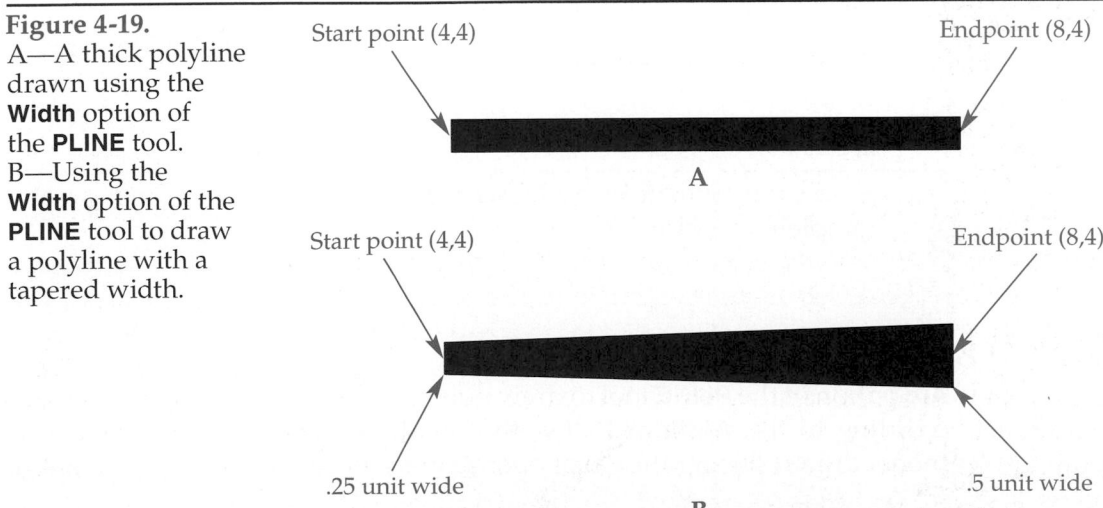

Figure 4-19.
A—A thick polyline drawn using the **Width** option of the **PLINE** tool.
B—Using the **Width** option of the **PLINE** tool to draw a polyline with a tapered width.

Start point (4,4)   Endpoint (8,4)

A

Start point (4,4)   Endpoint (8,4)

.25 unit wide   .5 unit wide

B

Figure 4-20.
Specifying the width of a polyline with the **Halfwidth** option. A starting value of .25 produces a polyline width of .5 unit, and an ending value of .5 produces a polyline width of 1 unit.

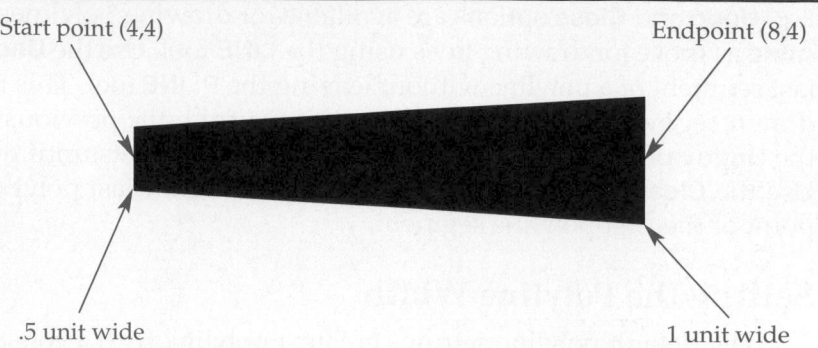

Start point (4,4)
Endpoint (8,4)
.5 unit wide
1 unit wide

Notice that the polyline in **Figure 4-20** is twice as wide as the polyline in **Figure 4-19B**, even though the same values are entered.

> **NOTE**
>
> All polyline objects with width, including polylines, polygons drawn using the **POLYGON** tool, rectangles drawn using the **RECTANGLE** tool, and donuts can appear filled or empty. The **Apply solid fill** setting in the **Display performance** area of the **Display** tab in the **Options** dialog box controls the appearance. Polyline objects are filled by default. You can also type FILL or FILLMODE and use the **On** or **Off** option. The fill display for previously drawn polyline objects updates when the drawing regenerates. You can regenerate the drawing manually by typing REGEN.

Using the Length Option

The **Length** option allows you to draw a polyline parallel to the previous polyline or line. After drawing a polyline, reissue the **PLINE** tool and pick a start point. Choose the **Length** option and enter the desired length. The second polyline draws parallel to the previous polyline using specified length.

Exercise 4-7

Access the Student Web site (www.g-wlearning.com/CAD) and complete Exercise 4-7.

Exercise 4-8

Access the Student Web site (www.g-wlearning.com/CAD) and complete Exercise 4-8.

Drawing Polyline Arcs

Use the **Arc** option of the **PLINE** tool to draw polyline arcs. Polyline arcs can include width, set according to the **Width** or **Halfwidth** option, and can continue from or to polyline segments drawn during the same operation to form a single object. Polyline

Figure 4-21.
An example of a polyline arc with different starting and ending widths.

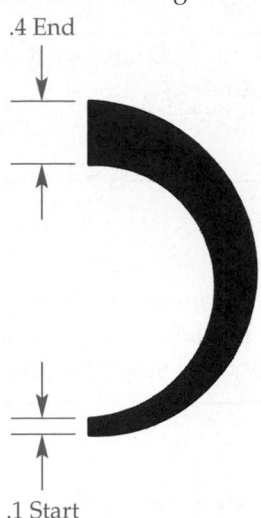

.4 End

.1 Start

arc width can range from 0 to the radius of the arc. You can set different starting and ending arc widths. See **Figure 4-21.** You can enter the **Width** or **Halfwidth** and **Arc** options in either order. Use the **Line** option to return the **PLINE** tool back to straight-line segment mode.

In addition to the **Close**, **Undo**, **Width**, and **Halfwidth** options, the **PLINE Arc** option includes functions for controlling the size and location of polyline arcs. Many of the **PLINE Arc** options allow you to create polyline arcs using the same methods available for drawing arcs using the **ARC** tool. Select the appropriate option and follow the prompts to create the polyline arc. Review the **ARC** tool options to help recognize the function of similar **PLINE Arc** options. **Figure 4-22** provides a brief description of each additional **PLINE Arc** option.

Figure 4–22.
Additional options available for drawing polyline arcs.

Option	Application	Options for Completion
Angle	Specify the polyline arc size according to an included angle.	1. Specify an endpoint. 2. Use the **Center** option to select the center point. 3. Use the **Radius** option to enter the radius.
Center	Specify the location of the polyline arc center point, instead of allowing AutoCAD to calculate the location automatically.	1. Specify an endpoint. 2. Use the **Angle** option to specify the included angle. 3. Use the **Length** option to specify the chord length.
Direction	Alter the polyline arc bearing, or tangent direction, instead of allowing the polyline arc to form tangent to the last object drawn.	1. Specify an endpoint.
Radius	Specify the polyline arc radius.	1. Specify an endpoint. 2. Use the **Angle** option to specify the included angle.
Second point	Draw a three-point polyline arc.	1. Pick the second point, followed by the endpoint.

Exercise 4-9

Access the Student Web site (www.g-wlearning.com/CAD) and complete Exercise 4-9.

Supplemental Material

Multilines

For information about drawing and editing multiline objects, go to the Student Web site (www.g-wlearning.com/CAD), select this chapter, and select **Multilines**.

Drawing Regular Polygons

POLYGON

Ribbon
Home
> Draw

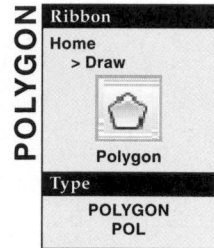

Polygon

Type
POLYGON
POL

regular polygon: A closed geometric figure with three or more equal sides and equal angles.

inscribed polygon: A polygon that is drawn inside an imaginary circle so that its corners touch the circle.

circumscribed polygon: A polygon that is drawn outside of an imaginary circle so that the sides of the polygon are tangent to the circle.

You can use the **POLYGON** tool to draw any *regular polygon* with up to 1024 sides. Polygons drawn using the **POLYGON** tool are single polyline objects. The first prompt asks for the number of sides. For example, to draw an octagon, which is a regular polygon with eight sides, enter 8. Next, decide how to describe the size and location of the polygon. The default setting involves choosing the center and radius of an imaginary circle. To use this method, after entering the number of polygon sides, specify a location for the polygon center point. A prompt then asks if you want to form an *inscribed polygon* or a *circumscribed polygon*. Select the appropriate option and specify the radius to create the polygon. See **Figure 4-23.**

NOTE

The number of polygon sides you enter, the **Inscribed in circle** or **Circumscribed about circle** option you select, and the radius you enter are stored as the new default settings, allowing you to quickly draw another polygon with the same characteristics.

Figure 4–23. Polygons can be inscribed in a circle (left) or circumscribed around a circle (right).

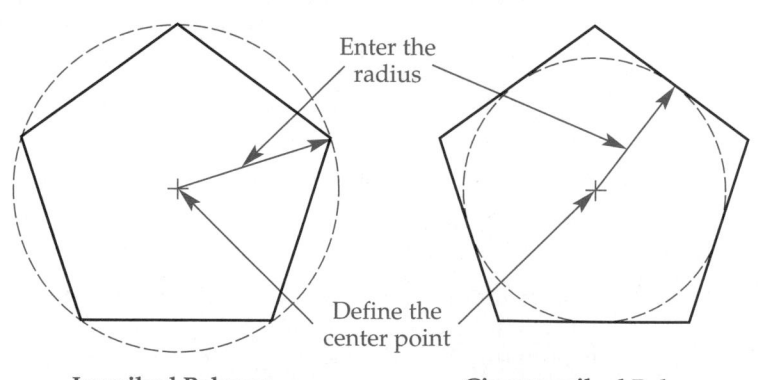

Enter the radius

Define the center point

Inscribed Polygon   Circumscribed Polygon

Regular polygons, such as the *hexagons* commonly drawn to represent bolt heads and nuts on mechanical drawings, are normally dimensioned across the flats. Use the **Circumscribed about circle** option to draw a polygon dimensioned across the flats. The radius you enter is equal to one-half the distance across the flats. Use the **Inscribed in circle** option to dimension a polygon across the corners, and to confine the polygon within a circular area.

Using the Edge Option

Use the **Edge** option to construct a polygon if you do not know the center point location or radius of the imaginary circle, but you do know the size and location of a polygon edge. After you access the **POLYGON** tool and enter the number of sides, choose the **Edge** option at the Specify center of polygon or [Edge]: prompt. Specify a point for the first endpoint of one of the polygon's sides. Then select the second endpoint of the polygon side. See **Figure 4-24.**

Exercise 4-10

Access the Student Web site (www.g-wlearning.com/CAD) and complete Exercise 4-10.

Figure 4–24.
Use the **Edge** option of the **POLYGON** tool to construct a polygon according to the location and size of an edge.

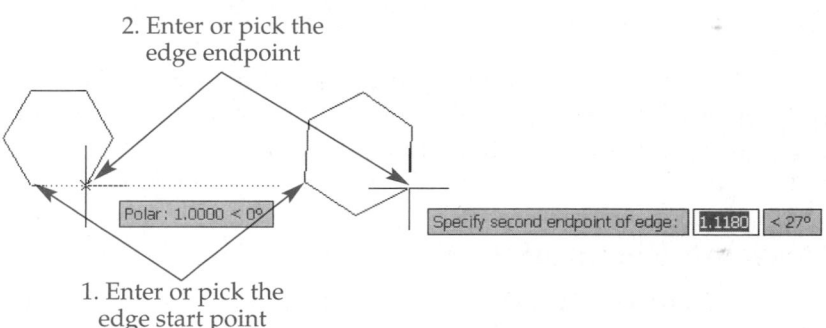

2. Enter or pick the edge endpoint

Polar: 1.0000 < 0°

Specify second endpoint of edge: 1.1180 < 27°

1. Enter or pick the edge start point

Drawing Rectangles

Use the **RECTANGLE** tool to draw rectangles easily. Rectangles drawn using the **RECTANGLE** tool are single polyline objects. To draw a rectangle using default settings, enter or pick one corner and then the opposite diagonal corner. See **Figure 4-25.** By default, the **RECTANGLE** tool draws a rectangle at a 0° angle with sharp corners.

Drawing Chamfered Rectangles

Use the **Chamfer** option to include *chamfered* corners during rectangle construction. See **Figure 4-26A.** When prompted, enter the first chamfer distance, followed by the second chamfer distance. Entering a value of 0 at the first or second chamfer distance prompt creates a rectangle with sharp corners. After setting the chamfer distances, you can either draw the rectangle or select another option. However, using the **Fillet**

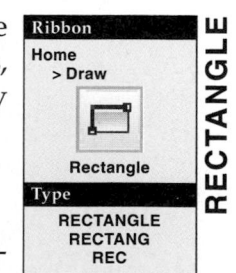

Figure 4–25.
Using the **RECTANGLE** tool.

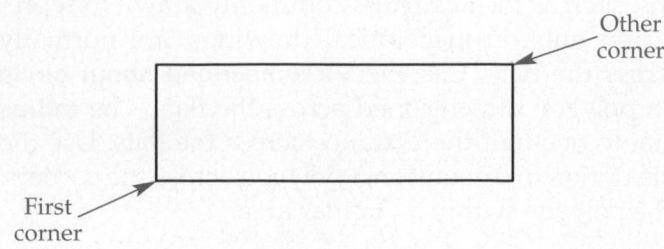

option overrides the **Chamfer** option. The rectangle you draw must be large enough to accommodate the specified chamfer distances. Otherwise, you will see sharp corners. New rectangles continue to produce chamfers until you reset the chamfer distances to 0 or use the **Fillet** option to create rounded corners.

Drawing Rounded Rectangles

fillet: A rounded interior corner.

round: A rounded exterior corner.

Use the **Fillet** option to include rounded corners during rectangle construction. See Figure 4-26B. AutoCAD uses the term *fillet* to describe both *fillets* and *rounds*. The **Fillet** option requires you to specify the round radius. When prompted, enter the round radius. After setting the round radius, you can either draw the rectangle or select another option. However, using the **Chamfer** option overrides the **Fillet** option. The rectangle you draw must be large enough to accommodate the specified round radius. Otherwise, you will see sharp corners. New rectangles continue to produce rounds until you reset the round radius to 0 or use the **Chamfer** option to create chamfered corners.

> **NOTE**
>
> This chapter introduces adding chamfers and rounds while creating rectangles. Chapter 12 covers adding chamfers and rounds using the **FILLET** tool in more detail.

Drawing Rectangles with Line Width

The **Width** option allows you to adjust rectangle line width, or "boldness." This should not to be confused with lineweight, described in Chapter 5. After you choose the **Width** option, a prompt asks you to enter the line width. For example, to create a rectangle with lines that are .5 wide, enter .5. After setting the rectangle width, you

Figure 4–26.
You can add chamfers or rounds to rectangles during construction.

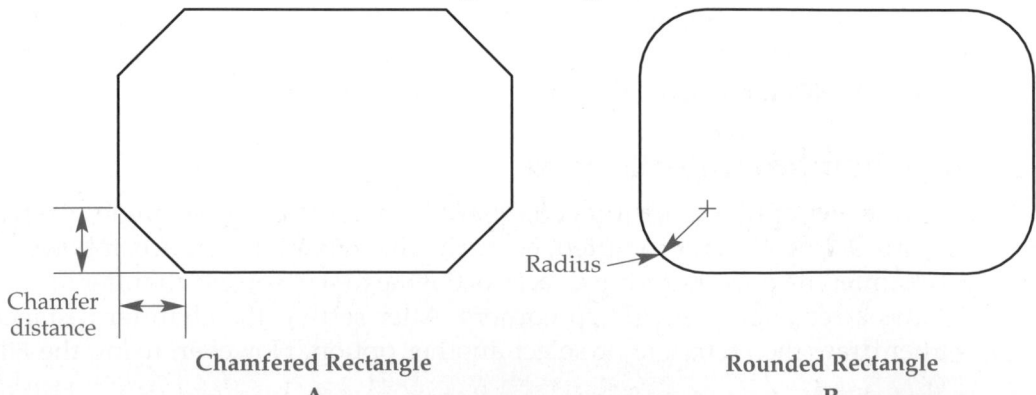

Chamfer distance

Radius

Chamfered Rectangle
A

Rounded Rectangle
B

can either draw the rectangle or set additional options. All new rectangles are drawn using the specified width. Reset the **Width** option to 0 to create new rectangles using a standard "0-width" line.

Specifying Rectangle Area

Use the **Area** option to draw a rectangle when you know the area of the rectangle and the length of one of its sides. This option is available after you pick the first corner point. Enter the **Area** option, and then specify the total area for the rectangle using a value that corresponds to the current units. For example, enter 45 to draw a rectangle with an area of 45 in^2. Next, choose the **Length** option if you know the length of a side, or the X value, or choose the **Width** option if you know the width of a side, or Y value. When prompted, enter the length or width to complete the rectangle. AutoCAD calculates the unspecified dimension automatically and draws the rectangle.

Specifying Rectangle Dimensions

An alternative to using point entry to specify the opposite corner of a rectangle is to enter the length and width of a rectangle using the **Dimensions** option. The option is available after you pick the first corner of the rectangle. Enter the **Dimensions** option and specify the length of a side, which is the X value. Next, enter the width of a side, which is the Y value. After you specify the length and width, AutoCAD asks for the other corner point. If you want to change the dimensions, select the **Dimensions** option again. If the dimensions are correct, specify the other corner point to complete the rectangle. The second corner point determines which of four possible rectangles you draw. See **Figure 4-27**.

Drawing a Rotated Rectangle

Use the **Rotation** option, available after you pick the first corner of the rectangle, to draw a rectangle at an angle other than 0°. A prompt asks you to specify the rotation angle. Enter or select an angle to rotate the rectangle. Then pick the opposite corner of the rectangle. An alternative is to choose the **Pick points** option when the Specify rotation angle or [Pick points]: prompt appears. If you select the **Pick points** option, the prompt asks you to select two points to define the angle. A new rotation value becomes the default angle for using the **Rotation** option.

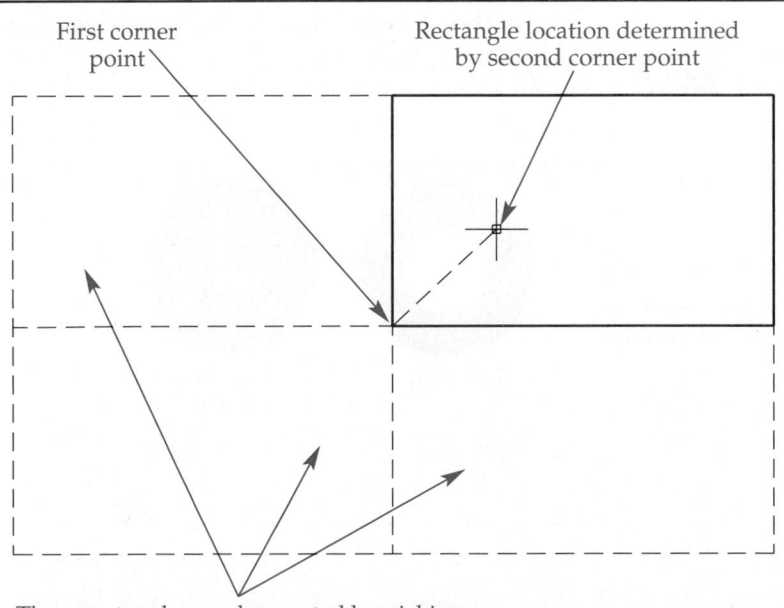

Figure 4–27.
The orientation of the rectangle relative to the first corner point is determined by the second corner point.

First corner point

Rectangle location determined by second corner point

These rectangles can be created by picking a different second corner point

NOTE

The **Elevation** and **Thickness** options of the **RECTANGLE** tool are appropriate for 3D applications and are explained in *AutoCAD and Its Applications—Advanced*.

PROFESSIONAL TIP

You can use a combination of rectangle settings to draw a single rectangle. For example, you can enter a width value, chamfer distances, and length and width dimensions to create a single rectangle.

Exercise 4-11

Access the Student Web site (www.g-wlearning.com/CAD) and complete Exercise 4-11.

Drawing Donuts and Filled Circles

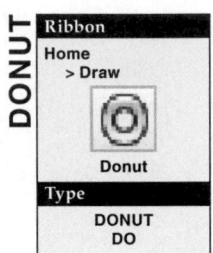

DONUT

| Ribbon |
| Home |
| > Draw |
| **Donut** |
| Type |
| DONUT |
| DO |

The **DONUT** tool allows you to draw a thick or filled circle. See **Figure 4-28.** A donut is a single polyline object. After activating the **DONUT** tool, enter the inside diameter and then the outside diameter of the donut. Enter a value of 0 for the inside diameter to create a completely filled donut, or solid circle.

The center point of the donut attaches to the crosshairs, and the Specify center of donut or <exit>: prompt appears. Pick a location to place the donut. The **DONUT** tool remains active until you right-click or press [Enter], the space bar, or [Esc]. This allows you to place multiple donuts of the same size using a single instance of the **DONUT** tool.

Exercise 4-12

Access the Student Web site (www.g-wlearning.com/CAD) and complete Exercise 4-12.

Figure 4–28.
The appearance of a donut depends on its inside and outside diameters and the current **FILL** mode.

Fill On

Fill On
I.D. = 0

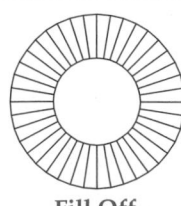

Fill Off

Fill Off
I.D. = 0

AutoCAD and Its Applications—Basics

Drawing True Splines

The **SPLINE** tool is used to create a special type of curve called a *nonuniform rational B-spline (NURBS) curve*, or *spline*. To create a spline, access the **SPLINE** tool and specify control points using any standard point entry method. For example, the spline shown in **Figure 4-29** uses absolute coordinate values of 2,2 for the first point, 4,4 for the second point, and 6,2 for the last point.

After you specify all spline control points, press [Enter] or the space bar, or right-click and select **Enter**, to end the point entry process. To complete the spline, enter values at the Specify start tangent: and Specify end tangent: prompts. Specifying the start and end tangents changes the direction in which the spline curve begins and ends. Right-click or press [Enter] or the space bar at the tangent prompts to accept the default direction, as calculated by AutoCAD.

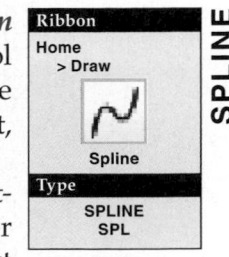

Ribbon
Home
> Draw

Spline

Type
SPLINE
SPL

nonuniform rational B-spline (NURBS) curve: A true (mathematically correct) spline.

> **NOTE**
>
> If you specify only two points for a spline curve and accept the AutoCAD default start and end tangents, an object that looks like a line is created, but the object is a spline.

Drawing Closed Splines

You can use the **Close** option of the **SPLINE** tool to draw a closed spline by connecting the last point to the first point. See **Figure 4-30**. After closing a spline, you are prompted to specify the tangent direction. Press [Enter] or the space bar or right-click and select **Enter** to accept the default calculated by AutoCAD.

Specifying the Spline Tangents

The previous **SPLINE** tool examples use the default AutoCAD tangent directions. You can set tangent directions by entering values at the prompts that appear after you pick spline points. The tangency is based on the tangent direction of the selected point. The results of specifying vertical and horizontal tangent directions are shown in **Figure 4-31**.

Exercise 4-13

Access the Student Web site (www.g-wlearning.com/CAD) and complete Exercise 4-13.

Figure 4-29.
A spline drawn with the **SPLINE** tool. AutoCAD's default start and end tangents were used for this spline.

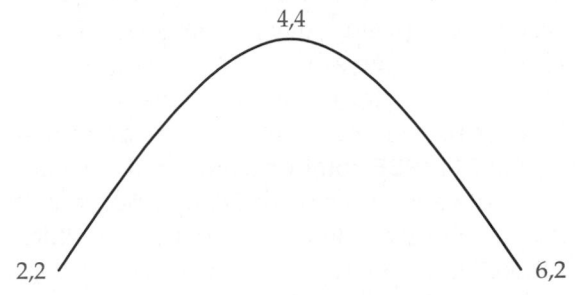

Figure 4-30.
Using the **Close** option of the **SPLINE** tool with AutoCAD default tangents to draw a closed spline. Compare this spline to the object shown in Figure 4-29.

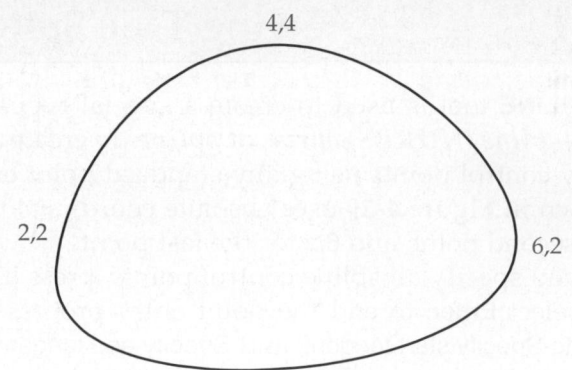

Figure 4-31.
These splines were drawn through the same points, but they have different start and end tangent directions. The arrows indicate the tangent directions.

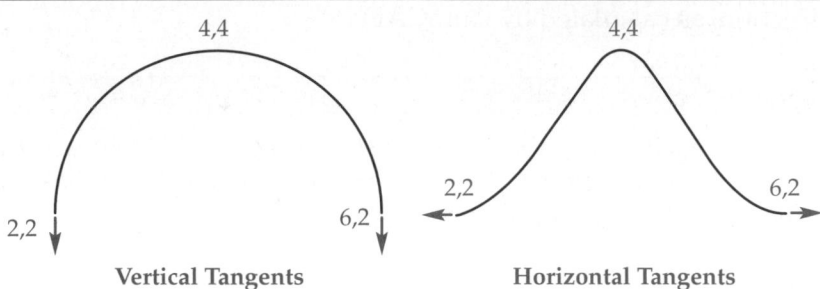

Vertical Tangents   Horizontal Tangents

Chapter Test

Answer the following questions. Write your answers on a separate sheet of paper or go to the Student Web site (www.g-wlearning.com/CAD) and complete the electronic chapter test.

1. Describe the rubberband display shown when you draw a **Center, Radius** circle. What is the purpose of the rubberband?
2. When you use the **CIRCLE** tool, what are the options for responding to the prompt Specify radius of circle?
3. Explain how to create a circle with a diameter of 2.5 units.
4. What option of the **CIRCLE** tool creates a circle of a specific radius that is tangent to two existing objects?
5. Define the term *point of tangency*.
6. Explain how to draw a circle tangent to three objects.
7. Briefly explain how to create a three-point arc.
8. Explain the procedure to draw an arc beginning with the center point and having a 60° included angle.
9. Define the term *included angle* as it applies to an arc.
10. What is the default option if you enter the **ARC** tool at the keyboard?
11. List three input options that you can use to draw an arc tangent to the endpoint of a previously drawn arc.
12. Name the two axes found on an ellipse.
13. Briefly describe the procedure to draw an ellipse using the **Axis, End** option.
14. What is the **ELLIPSE** rotation angle that causes you to draw a circle?
15. Identify two ways to access the **Arc** option for drawing elliptical arcs.
16. How do you draw a filled arrow using the **PLINE** tool?
17. Which **PLINE** tool option allows you to specify the width from the center to one side?
18. Explain how to turn the **FILL** mode off.
19. Briefly describe how to create a polyline parallel to a previously drawn polyline or line.

20. Explain how to draw a hexagon measuring 4″ (102 mm) across the flats.
21. Given the distance across the flats of a hexagon, would you use the **Inscribed** or **Circumscribed** option to draw the hexagon?
22. Name the ribbon tab and panel that contains the **RECTANGLE** tool.
23. Name at least three tools you could use to create a rectangle.
24. Name the tool option used to draw rectangles with rounded corners.
25. Name the tool option designed for drawing rectangles with a specific line thickness.
26. Give the easiest keyboard shortcut for the following tools:
    A. **CIRCLE**
    B. **ARC**
    C. **ELLIPSE**
    D. **POLYGON**
    E. **RECTANGLE**
    F. **DONUT**
27. Describe a method for drawing a solid circle.
28. Explain how to draw two donuts with an inside diameter of 6.25 and an outside diameter of 9.50.
29. Name the tool that can be used to create a true spline.
30. How do you accept the AutoCAD defaults for the start and end tangents of a spline?

Drawing Problems

Start AutoCAD if it is not already started. Start a new drawing from scratch or use an appropriate template of your choice. Follow the specific instructions for each problem. Do not draw centerlines. Do not draw dimensions or text. Use your own judgment and approximate dimensions when necessary.

▼ Basic

1. Use the **LINE** tool and the **CIRCLE** tool options to draw the objects below. Save the drawing as P4-1.

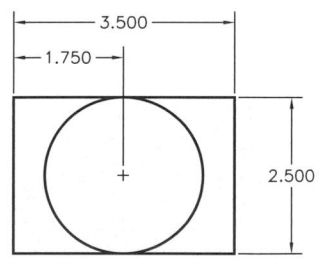
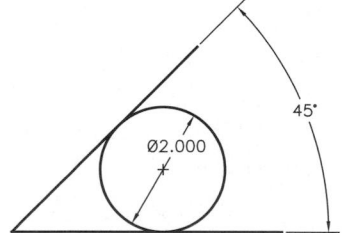
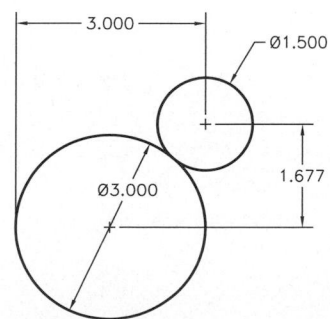
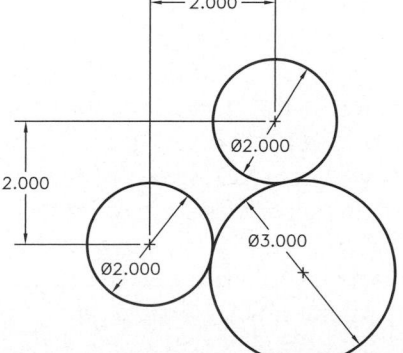

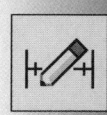

2. Use the **CIRCLE** and **ARC** tool options to draw the object below. Save the drawing as P4-2.

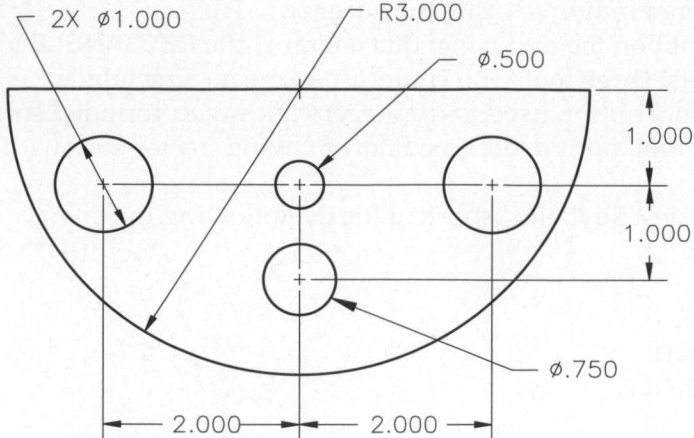

3. Draw the spacer below. Save the drawing as P4-3.

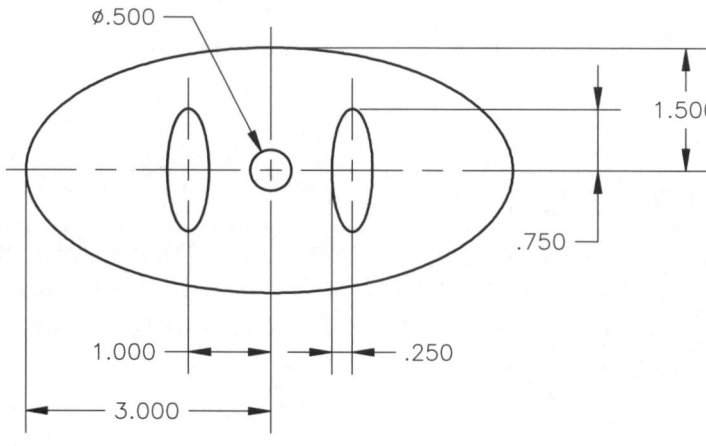

Drawing Problems - Chapter 4

4. Draw the following object. Save the drawing as P4-4.

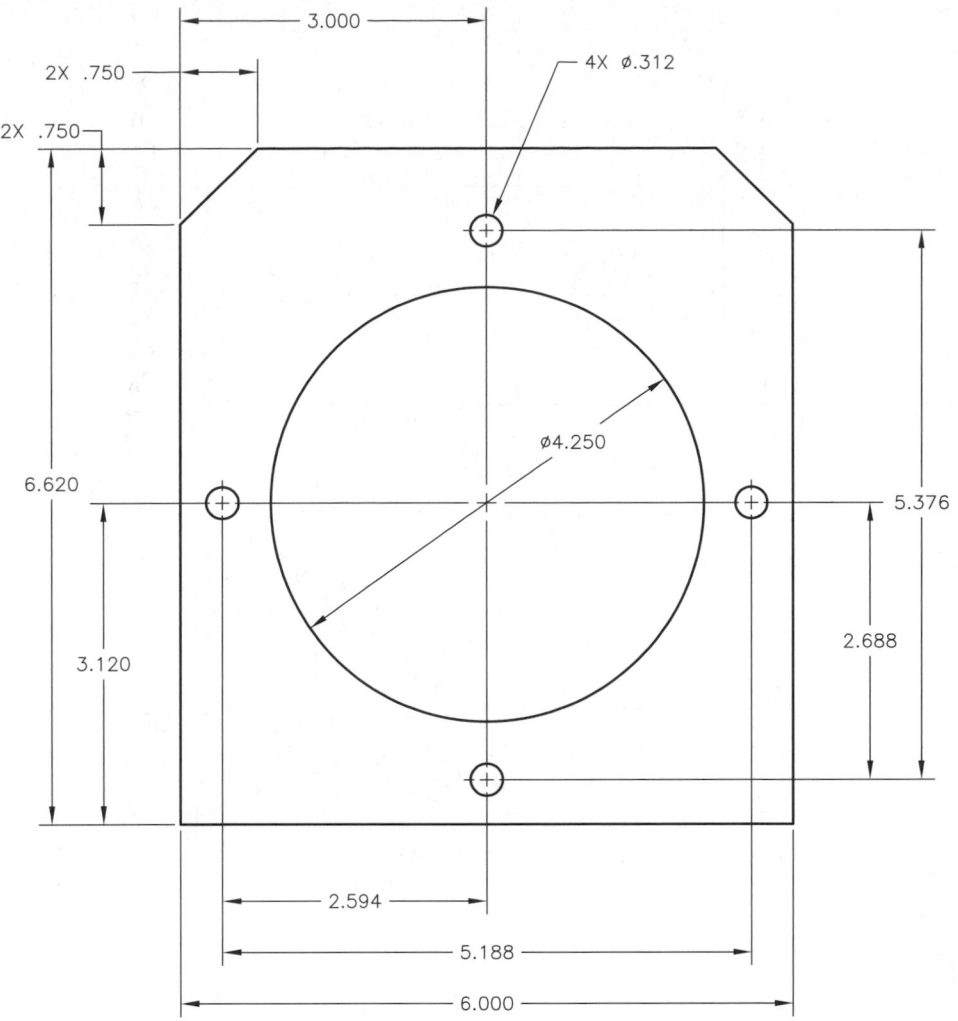

5. Draw the following object. Save the drawing as P4-5.

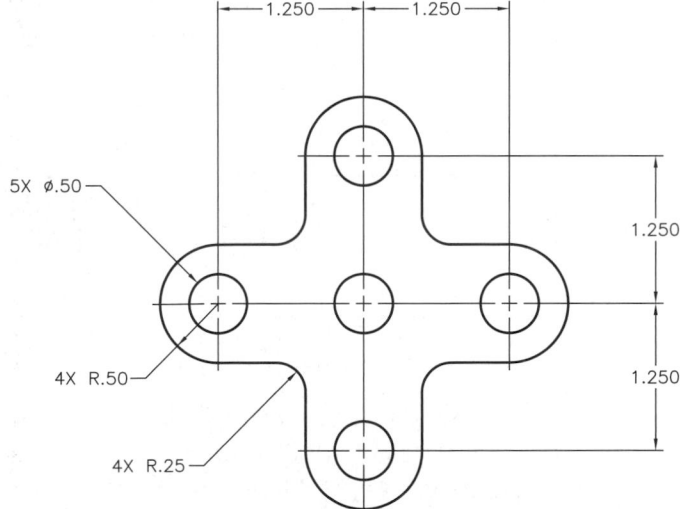

6. Draw the following object. Save the drawing as P4-6.

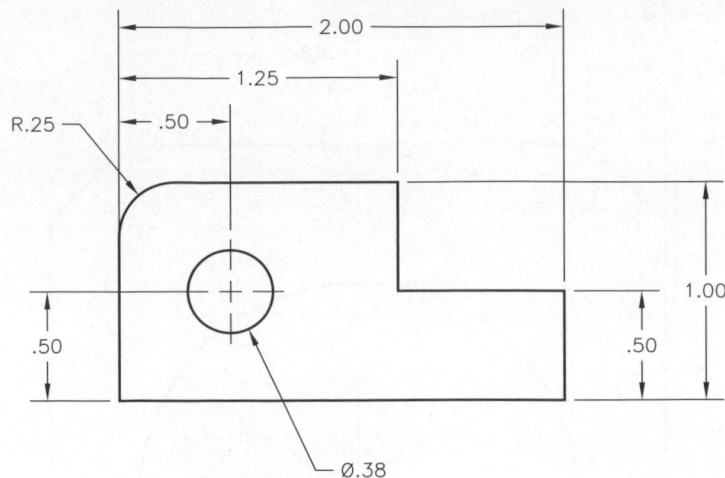

7. Draw the pipe spacer shown. Save the drawing as P4-7.

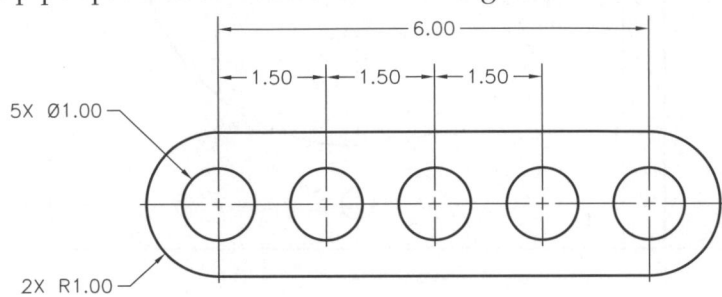

8. Use the **PLINE** tool to draw the following object with a .032 line width. Save the drawing as P4-8.

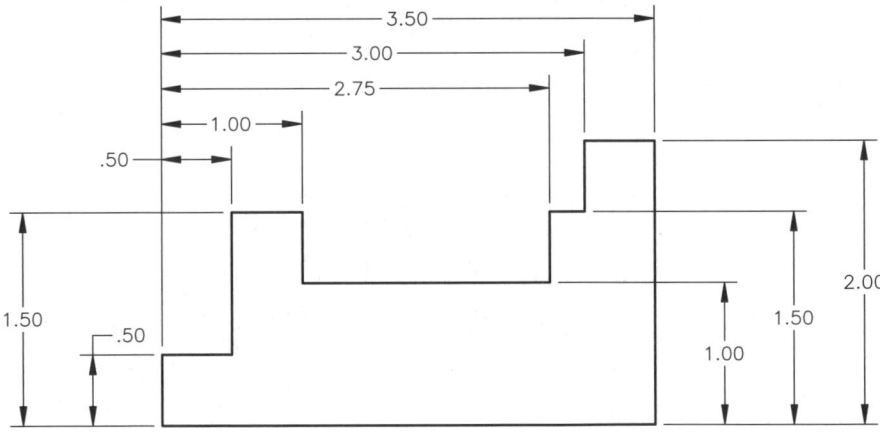

9. Use the **PLINE** tool to draw the following object with a .032 line width. Save the drawing as P4-9.

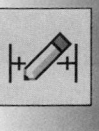

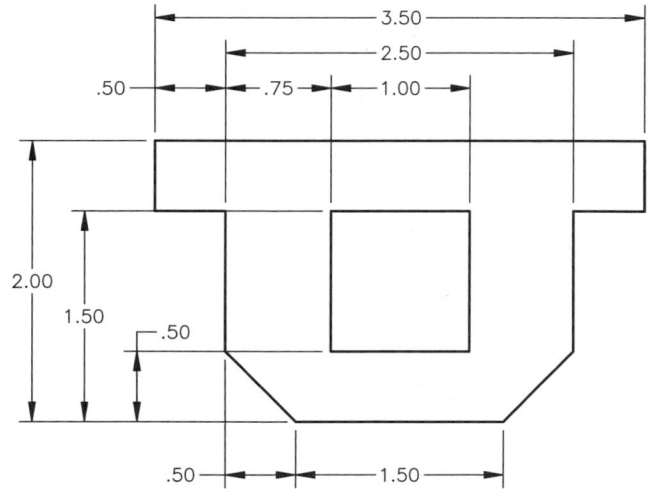

10. Use the **PLINE** tool to draw the following object with a .032 line width.
    A. Deactivate solid fills and use the **REGEN** tool, and reactivate solid fills and reissue the **REGEN** tool.
    B. Observe the difference with solid fills enabled.
    C. Save the drawing as P4-10.

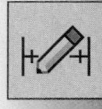

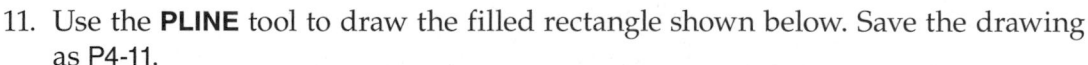

11. Use the **PLINE** tool to draw the filled rectangle shown below. Save the drawing as P4-11.

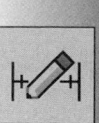

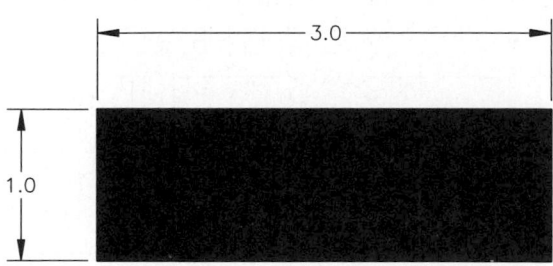

12. Draw the objects shown below. Save the drawing as P4-12.

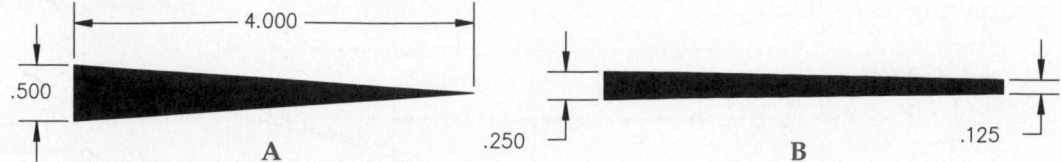

4.000

.500

.250

A

.125

B

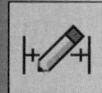

13. Draw the object shown below. Set decimal units, .25 grid spacing, .0625 snap spacing, and limits of 11,8.5. Save the drawing as P4-13.

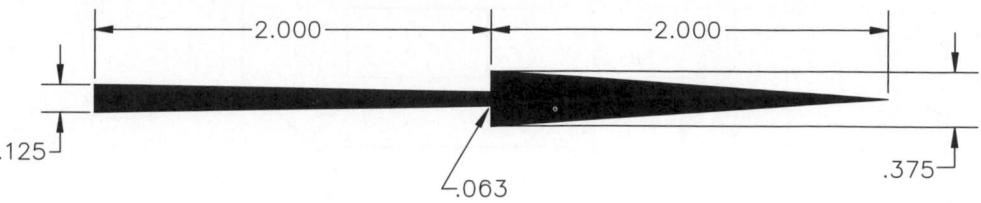

2.000

2.000

.125

.063

.375

▼ Intermediate

14. Draw the following object. Save the drawing as P4-14.

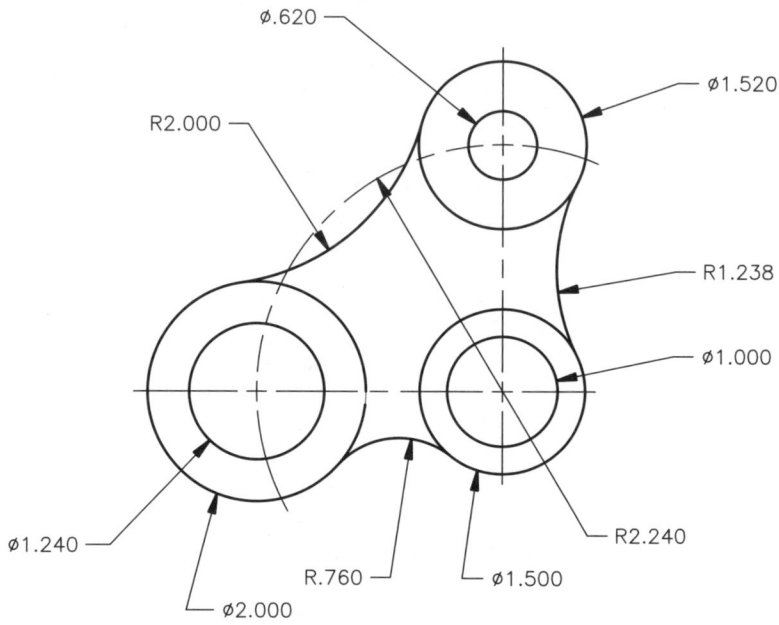

ø.620

ø1.520

R2.000

R1.238

ø1.000

R2.240

ø1.240

R.760

ø1.500

ø2.000

(Art courtesy of Bruce L. Wilcox)

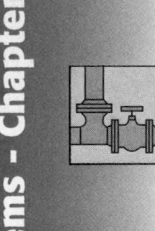

15. Draw the pressure cylinder shown below. Use the **Arc** option of the **ELLIPSE** tool to draw the cylinder ends. Save the drawing as P4-15.

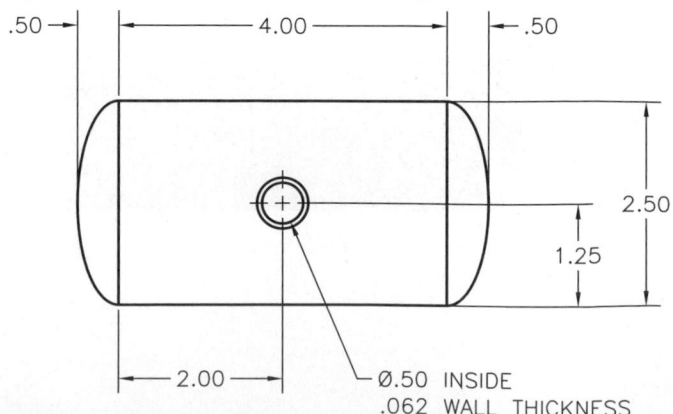

.50

4.00

.50

2.50

1.25

2.00

ø.50 INSIDE
.062 WALL THICKNESS

Drawing Problems - Chapter 4

16. Draw the gasket shown below. Save the drawing as P4-16.

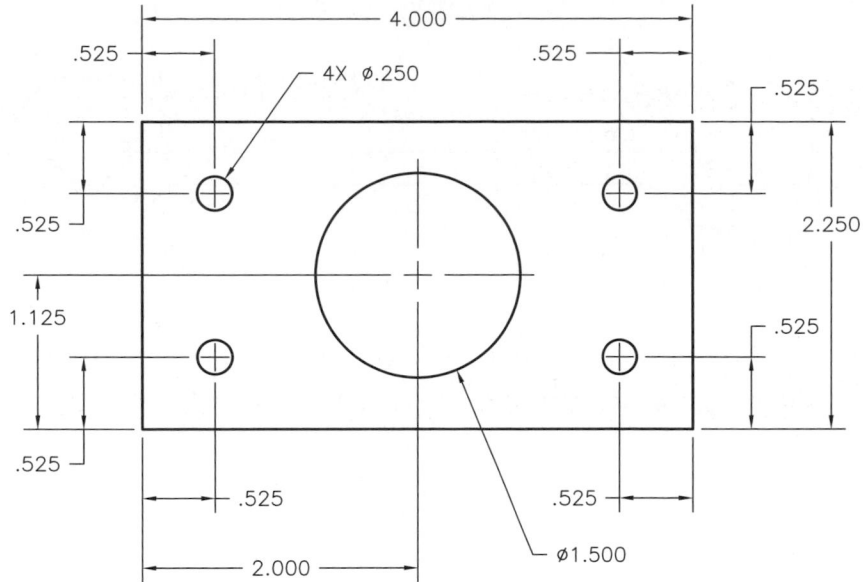

17. Draw the single polyline shown below. Use the **Arc**, **Width**, and **Close** options of the **PLINE** tool to complete the shape. Set the polyline width to 0, except at the points indicated. Save the drawing as P4-17.

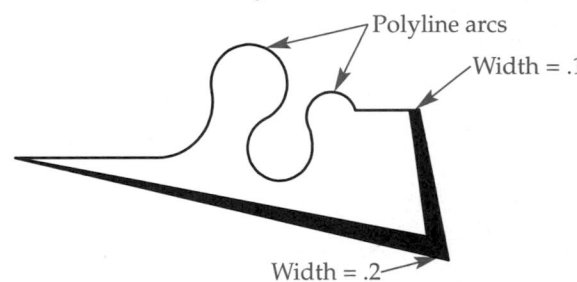

18. Draw the two curved arrows shown below using the **Arc** and **Width** options of the **PLINE** tool. The arrowheads should have a starting width of 1.4 and an ending width of 0. The body of each arrow should have a beginning width of .8 and an ending width of .4. Save the drawing as P4-18.

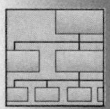

Drawing Problems - Chapter 4

19. Draw the following object. Save the drawing as P4-19.

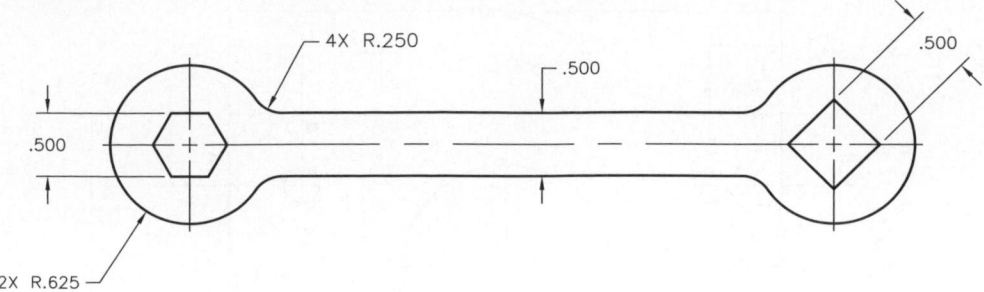

20. Draw the following object. Save the drawing as P4-20.

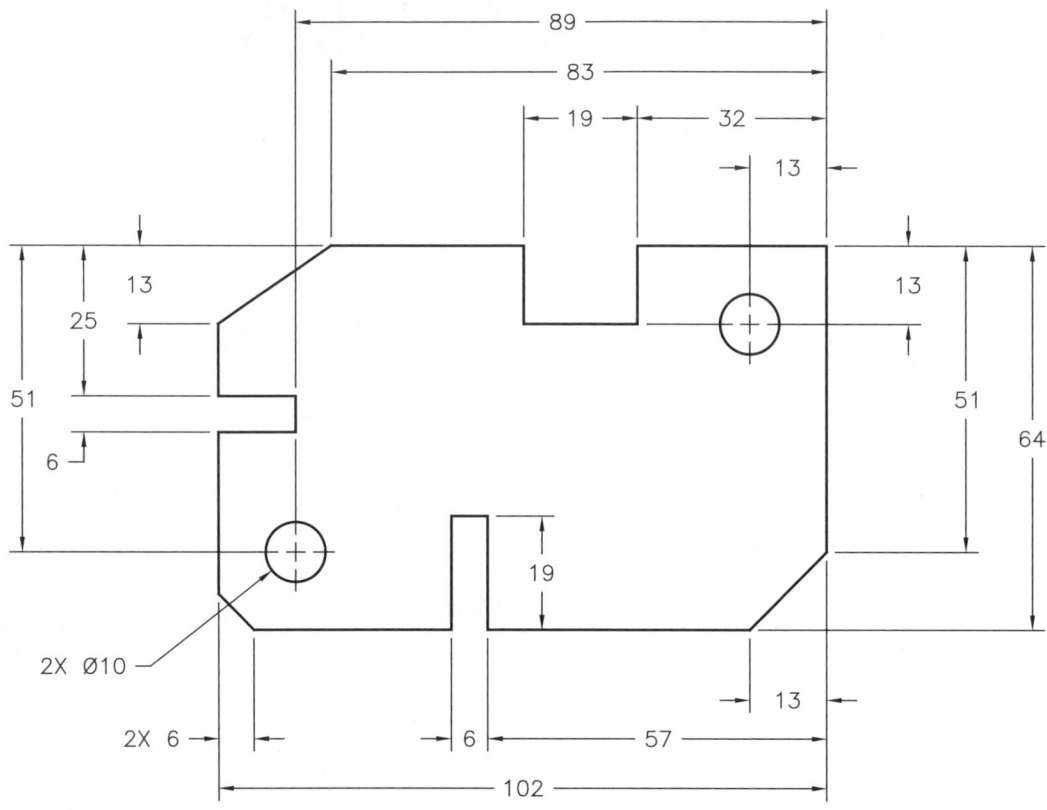

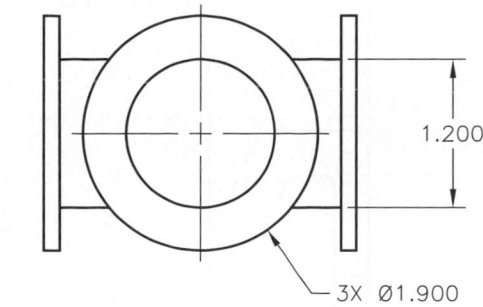

21. Draw the pipe fitting shown. Save the drawing as P4-21.

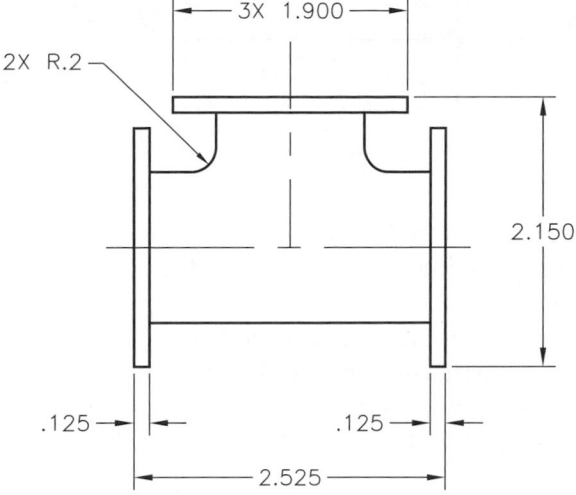

1.200

3X Ø1.900

3X 1.900

2X R.2

2.150

.125   .125

2.525

22. Draw the ellipse template shown. Save the drawing as P4-22.

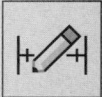

ELLIPSE TABLE		
KEY	MAJOR DIA	MINOR DIA
A	.9951	.5745
B	1.0717	.6187
C	1.1482	.6629
D	1.2247	.7071
E	1.3013	.7513
F	1.3778	.7955
G	1.4544	.8397
H	1.5309	.8839
I	1.6075	.9281
J	1.6840	.9723
K	1.7606	1.0165
L	1.8371	1.0607
M	1.9902	1.1490
N	2.1433	1.2374
O	2.2964	1.3258
P	2.4495	1.4142

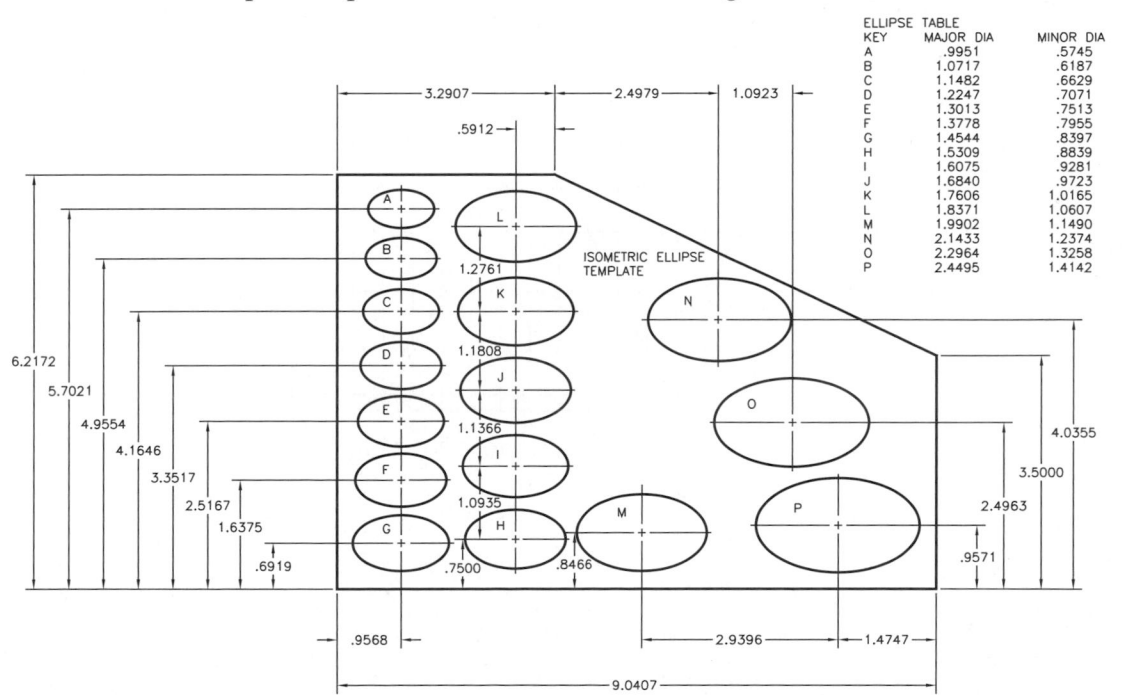

ISOMETRIC ELLIPSE TEMPLATE

3.2907   2.4979   1.0923

.5912

1.2761

1.1808

1.1366

1.0935

6.2172
5.7021
4.9554
4.1646
3.3517
2.5167
1.6375
.6919

.7500   .8466

.9568   2.9396   1.4747

9.0407

4.0355
3.5000
2.4963
.9571

23. Draw the gasket shown. Save the drawing as P4-23.

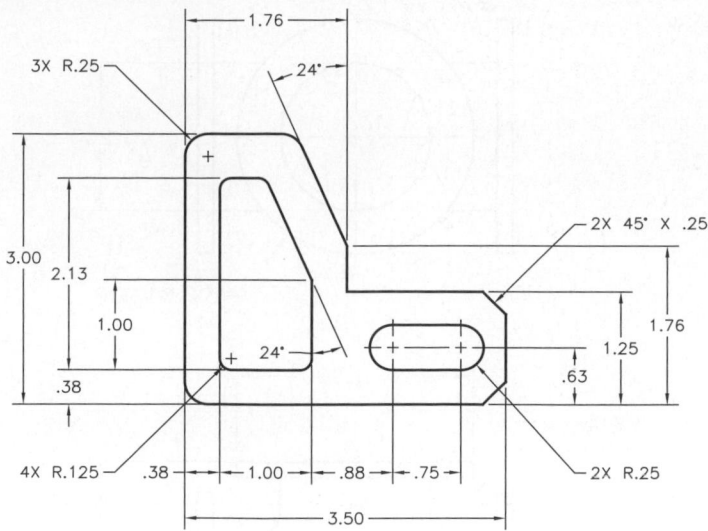

24. Use the **SPLINE** tool to draw the curve for the cam displacement diagram below. Use the following guidelines and the given drawing to complete this problem:
    A. The total rise equals 2.000.
    B. The total displacement can be any length.
    C. Divide the total displacement into 30° increments.
    D. Draw a half circle divided into 6 equal parts on one end.
    E. Draw a horizontal line from each division of the half circle to the other end of the diagram.
    F. Draw the displacement curve with the **SPLINE** tool by picking points where the horizontal and vertical lines cross.
    G. Label the displacement increments along the horizontal scale as shown. Save the drawing as P4-24.

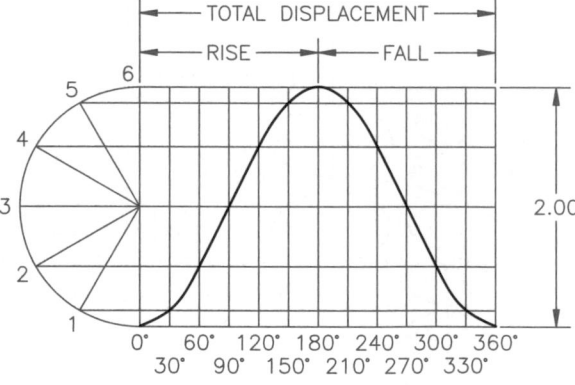

Drawing Problems - Chapter 4

▼ Advanced

25. Draw the object shown below. Save the drawing as P4-25.

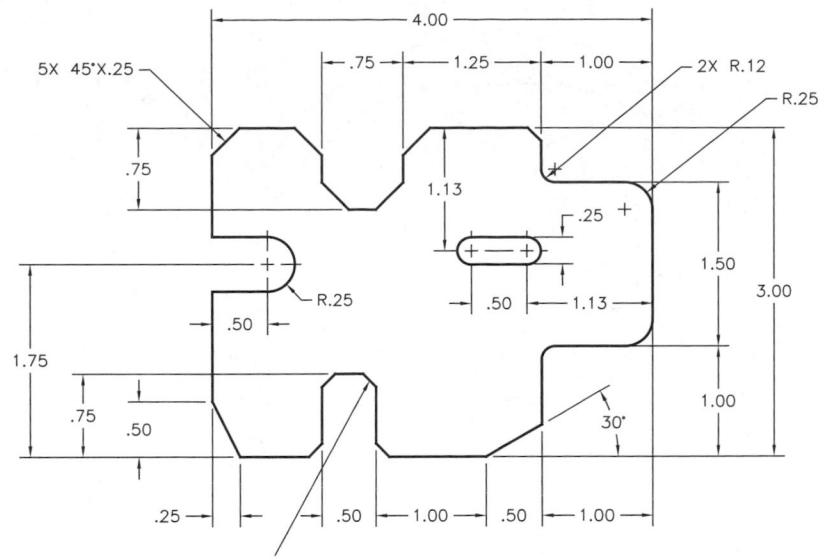

26. You have just been given the sketch of a new sports car design (shown below). You are asked to create a drawing from the sketch. Use the **LINE** tool and selected shape tools to draw the car. Do not be concerned with size and scale. Consider the tools and techniques used to draw the car, and try to minimize the number of objects. Save your drawing as P4-26.

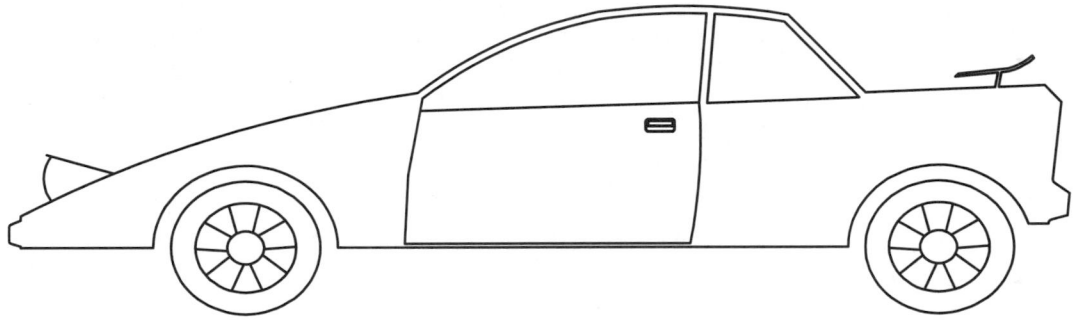

27. You have just been given the sketch of an innovative new truck design (shown below). You are asked to create a drawing from the sketch. Use the **LINE** tool and selected shape tools to draw a truck resembling the sketch. Do not be concerned with size and scale. Save your drawing as P4-27.

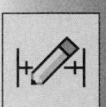

28. Draw the following object. Save the drawing as P4-28.

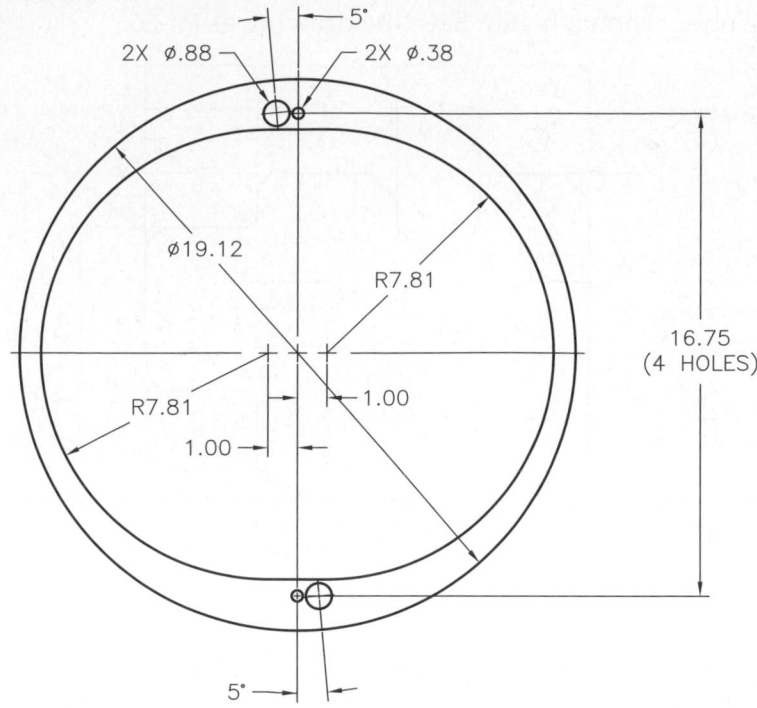

29. Draw this elevation using the **ARC**, **ELLIPSE**, **RECTANGLE**, and **DONUT** tools. Do not be concerned with size and scale. Save the drawing as P4-29.

AutoCAD and Its Applications—Basics

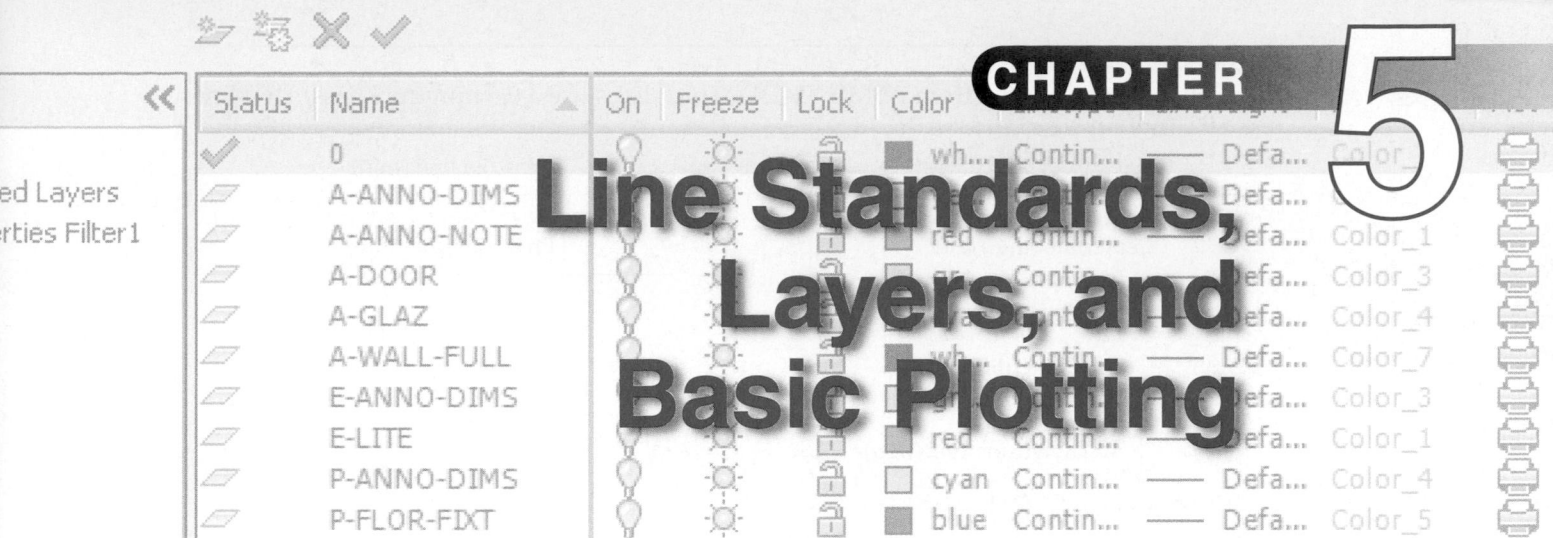

CHAPTER 5

Line Standards, Layers, and Basic Plotting

Learning Objectives

After completing this chapter, you will be able to do the following:

✓ Describe basic line conventions.
✓ Create and manage layers.
✓ Draw objects on separate layers.
✓ Use **DesignCenter** to copy layers and linetypes between drawings.
✓ Print and plot your drawings.

AutoCAD uses a layer system to organize linetypes and other object characteristics to conform to accepted standards and conventions. You can also use layer display options to help create different drawing sheets, views, and displays from a single drawing. This chapter introduces line conventions and the AutoCAD layer system. It also introduces printing and plotting so that you can begin printing and plotting your drawings. Printing and plotting are described in detail later in this textbook.

Line Standards

Drafting is a graphic language that uses lines, symbols, and words to describe products to be manufactured. *Line conventions* are designed to enhance the readability of drawings. This section introduces the line standards that you will apply later in this chapter when you begin loading linetypes and defining layers, as well as throughout your drafting career.

line conventions:
Standards related to line thickness and type.

Drawings include a variety of linetypes to classify drawing content. The ASME drafting standards recommend two line thicknesses to establish contrasting lines in a drawing. Lines are thick or thin. Thick lines are twice as thick as thin lines, with recommended widths of 0.6 mm and 0.3 mm, respectively. **Figure 5-1** shows recommended line width and type as defined in ASME Y14.2M, *Line Conventions and Lettering*. **Figure 5-2** describes the most often used linetypes. **Figure 5-3** shows an example of a drawing with several common linetypes. This textbook further explains and illustrates line standards when applicable.

Figure 5-1.
Line conventions adapted from ASME Y14.2M. Thick lines use a 0.6 mm thickness. Thin lines use a 0.3 mm line thickness.

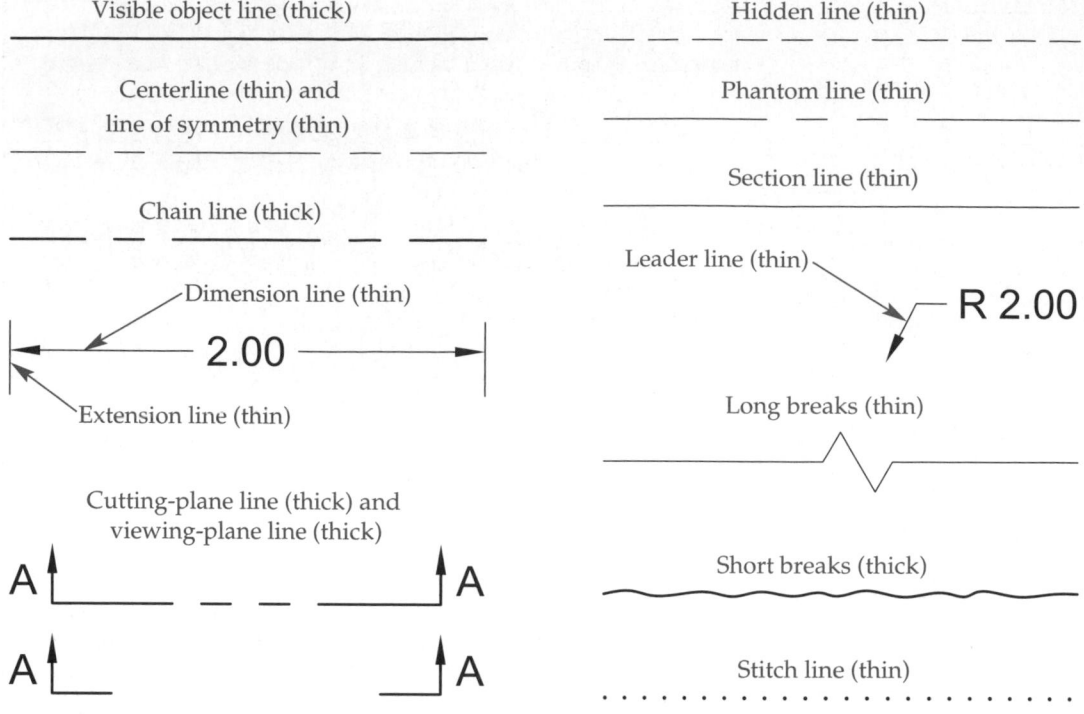

Figure 5-2.
Descriptions of common lines and line standards. Line characteristics and spacing are measured at full scale. Specifications vary according to drawing size.

Type	Purpose	Standards
Object lines (visible, outline)	Show the contour or outline of objects.	Thick, solid (Continuous).
Hidden lines	Represent features that are hidden in the current view.	Thin, dashed (HIDDEN or DASHED). Dashes are .125″ (3 mm) long and are spaced .06″ (1.5 mm) apart.
Centerlines	Locate the centers of circles and arcs, and show the axis of cylindrical or symmetrical shapes.	Thin (CENTER). Extend .125″ to .25″ (3 mm to 6 mm) past objects. Centerlines consist of one .125″ dash alternating with one .75″ to 1.5″ (19 mm to 38 mm) dash. A .06″ (1.5 mm) space separates the dashes. Small centerline dashes should cross only at the center of a circle.
Extension lines	Show the extent of a dimension.	Thin, solid (Continuous). Begin .06″ (1.5 mm) from an object and extend .125″ (3 mm) beyond the last dimension line. Can cross object lines, hidden lines, and centerlines, but should not cross dimension lines. Centerlines become extension lines when used to show the extent of a dimension.
Dimension lines	Show the distance being measured.	Thin, solid (Continuous). Broken near the center for placement of the dimension numeral in mechanical drafting. Unbroken in architectural and structural drawings, with dimension placed on top of the dimension line. Arrows terminate the ends of dimension lines, except in architectural drafting, where slashes (ticks) or dots are often used.

(Continued)

Figure 5-2.
(Continued)

Type	Purpose	Standards
Leader lines	Connect a specific note to a feature on a drawing.	Thin, solid (Continuous). Often terminate with an arrowhead at the feature. May be curved on architectural drawings. Straight leader lines often have a small shoulder at the note.
Cutting-plane lines	Identify the location and viewing direction of a section view.	Thick (PHANTOM or DASHED). Can be drawn in one of two ways.
Viewing-plane lines	Identify the location of a view	Thick (PHANTOM or DASHED). Can be drawn in one of two ways.
Section lines	In a section view, show where material has been cut away.	Thin, usually drawn in a pattern. Different linetypes can be used to indicate specific or different material.
Break lines	Show where a portion of an object has been removed for clarity or convenience.	Thin or thick depending on the symbol, solid (Continuous). Break representation is based on the object or material being broken.
Phantom lines	Identify repetitive details, show alternate positions of moving parts, and locate adjacent positions of related parts.	Thin, PHANTOM. Two .125″ (3 mm) dashes, alternating with one .75″ to 1.5″ (19 mm to 38 mm) dash. Spaces between dashes are .06″ (1.5 mm).
Chain lines	Indicate special features or unique treatment for a surface.	Thick (CENTER).

Figure 5-3.
An example of a mechanical drawing with several common types of lines.

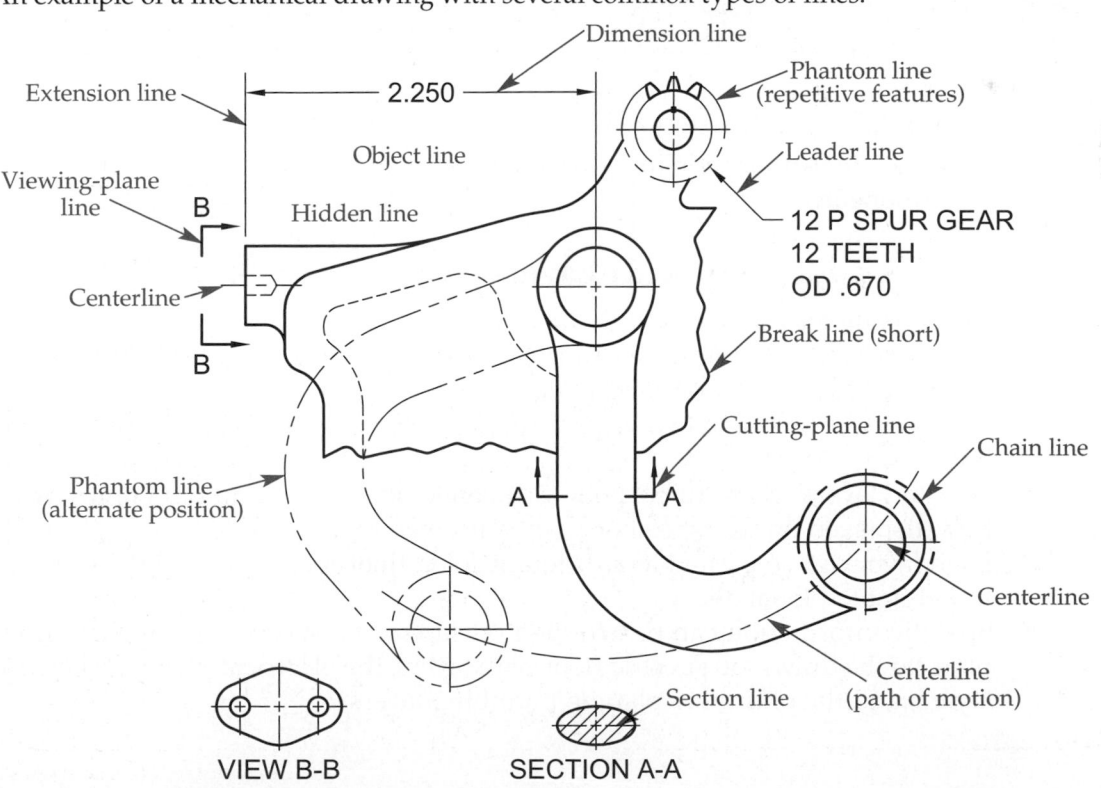

VIEW B-B   SECTION A-A

Figure 5-4.
A few of the most common U.S. National CAD Standard line thicknesses.

Thickness	Application
Thin, 0.25 mm	Dimensioning features, phantom lines, hidden lines, centerlines, long break lines, schedule grid lines, and background objects.
Medium, 0.35 mm	Object lines, text for dimension values, notes and schedules, terminator marks, door and window elevations, and schedule grid accent lines.
Wide, 0.5 mm	Major object lines at elevation edges, cutting-plane lines, short break lines, title text, minor title underlines, and border lines.
Extra wide, 0.7 mm	Major title underlines, schedule outlines, large titles, special emphasis object lines, elevation and section grade lines, property lines, sheet borders, and schedule borders.

The U.S. National CAD Standard (NCS) recommends a specific line thickness and characteristics for architectural and similar drawings. Thickness options range from extra fine, with a 0.13 mm thickness, to 4X wide, with a 2 mm thickness. You can use the range of NCS-recommended lineweights to provide accents to your drawings as needed. A common practice is to select a few of the lineweights that correlate best to specific applications. See **Figure 5-4.**

NOTE

Many AutoCAD tools simplify and automate the process of using and applying correct lines standards. You will learn applications and techniques for drawing specific types of lines in this book.

Introduction to Layers

overlay system:
A system of separating drawing components by layer.

layers:
Components of AutoCAD's overlay system that allow users to separate objects into logical groups for formatting and display purposes.

In AutoCAD, an *overlay system* separates different drawing components, referred to as *layers*. You can display all the layers together, or "overlaid," to reflect the entire design drawing. Display or hide individual layers as needed to show specific details or design components.

Increasing Productivity with Layers

Using layers increases productivity in several ways:
- ✓ Each layer can be assigned a different color, linetype, and lineweight to correspond to line conventions and to help improve clarity.
- ✓ Changes can be made to a layer promptly, affecting all objects drawn on the layer.
- ✓ Selected layers can be turned off or frozen to decrease the amount of information displayed on the screen or to speed screen regeneration.
- ✓ Each layer can be plotted in a different color, linetype, or lineweight, or it can be set not to plot at all.
- ✓ Specific information can be grouped on separate layers. For example, a floor plan can be drawn on specific floor plan layers, the electrical plan on electrical layers, and the plumbing plan on plumbing layers.

AutoCAD and Its Applications—Basics

✓ Several plot sheets can be created from the same drawing file by controlling layer visibility to separate or combine drawing information. For example, a floor plan and electrical plan can be reproduced together and sent to an electrical contractor for a bid. The floor plan and plumbing plan can be reproduced together and sent to a plumbing contractor.

Layers Used in Different Drafting Fields

Typically, the type of drawing you create determines the function of each layer. In mechanical drafting, you usually assign a specific layer to each different type of line or object. For example, draw object lines on an Object layer that is black in color, uses a solid (Continuous) linetype, and is 0.6 mm wide. Draw hidden lines on a green Hidden layer that uses a 0.3 mm hidden (HIDDEN or DASHED) linetype.

Architectural and civil drawings may have hundreds of layers, each used to produce a specific item. For example, draw full-height floor plan walls on a black A-WALL-FULL layer that uses a 0.5 mm solid (Continuous) linetype. Add plumbing fixtures to a floor plan on a blue P-FLOR-FIXT layer that uses a 0.35 mm solid (Continuous) linetype.

You can create layers for any type of drawing, including detail parts, assemblies, floor plans, foundation plans, partition layouts, plumbing systems, electrical systems, structural systems, roof drainage systems, reflected ceiling systems, HVAC systems, site plans, profiles, topographic maps, and details. Interior designers may use floor plan, interior partition, and furniture layers. In electronics drafting, you can draw each level of a multilevel circuit board on its own layer.

Creating and Using Layers

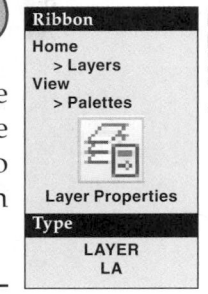

Ribbon
Home
  > Layers
View
  > Palettes

Layer Properties

Type
LAYER
LA

LAYER

The **LAYER** tool opens the **Layer Properties Manager** palette, which is used to create and delete layers and control layer properties. See **Figure 5-5.** Two panes divide the **Layer Properties Manager**. The list view pane on the right side uses a column format to list layers and provide layer property controls. Properties in each column appear as an

---

Figure 5-5.
The **Layer Properties Manager**. Layer 0 is the AutoCAD default layer.

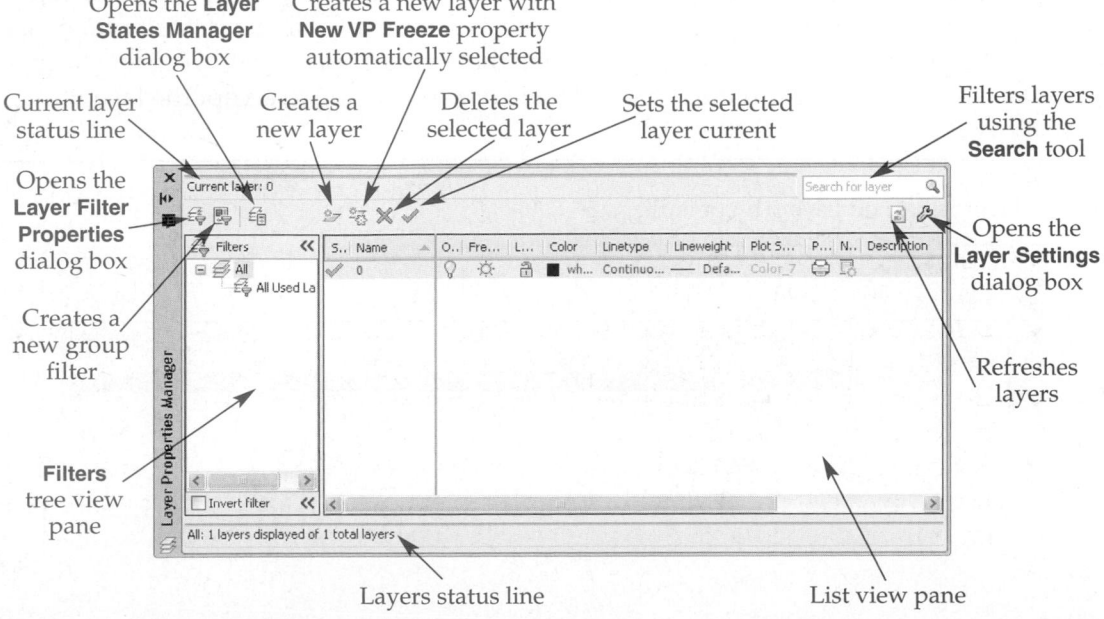

Figure 5-6.
Pick the icons in the **Layer Properties Manager** to change layer settings.

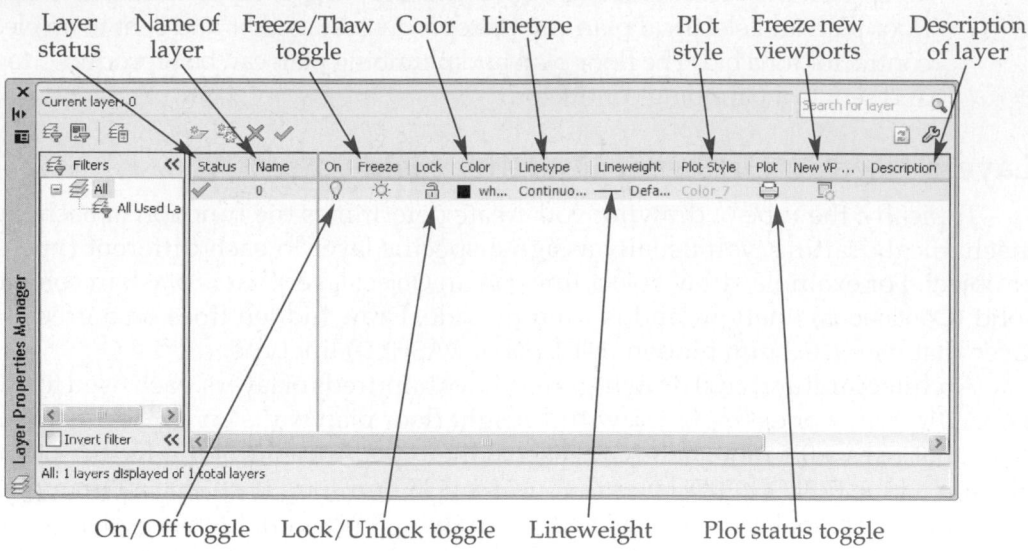

Defining New Layers

icon or as an icon and a name. See **Figure 5-6.** Pick a property to change layer settings. The tree view pane on the left side of the palette displays filters that you can use to limit the number of layers displayed in the list view pane.

Only one layer is required in an AutoCAD drawing. This default layer is named 0 and cannot be deleted, renamed, or purged from the drawing. The 0 layer is primarily reserved for drawing blocks, as described later in this textbook. Draw each object on a layer specific to the object. For example, draw object lines on an Object layer, draw floor plan walls on an A-WALL layer, and draw construction lines on a Construction or A-ANNO-NPLT layer.

Defining New Layers

The **Name** column in the list view shows the names of all the layers in the drawing. Add layers to a drawing to meet the needs of the current drawing project. To add a new layer, select an existing layer that contains properties similar to those you want to assign to the new layer. If this is the first new layer in a default template, only the 0 layer is available to reference. Then pick the **New Layer** button, right-click in the list view and select **New Layer**, or press [Enter] or [Alt] + [N]. A new layer appears, using a default name. See **Figure 5-7.** The layer name is highlighted when the listing appears, allowing you to type a new name. Pick away from the layer in the list or press [Enter] to accept the layer.

Figure 5-7.
A new layer is named Layer*n* by default.

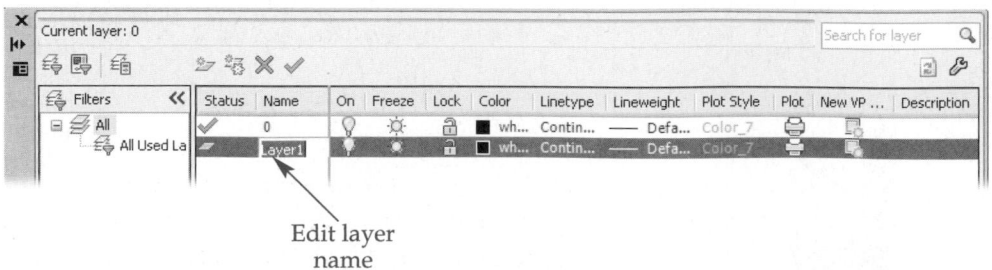

Edit layer name

AutoCAD and Its Applications—Basics

Layer Names

Layers should be given names to reflect drawing content. Layer names can have up to 255 characters and can include letters, numbers, and certain other characters, including spaces. Some examples of typical mechanical, architectural, and civil drafting layer names include:

Mechanical	Architectural	Civil
Object	A-WALL-FULL	C-BLDG
Hidden	A-GLAZ	C-WATR
Center	A-DOOR	C-TOPO
Dimension	E-LITE	C-PROP
Construction	P-FLOR-FIXT	C-NGAS
Section	S-FNDN	C-SSWR
Border	M-HVAC	C-ELEV

Layers names are usually set according to specific industry or company standards. However, for very simple drawings, layers might be named by linetype and color. For example, the layer name Continuous-White can have a continuous linetype drawn in white. The layer usage and color number, such as Object-7, may be used to indicate an object line with color 7. Another option is to assign the linetype a numerical value. For example, object lines can be 1, hidden lines can be 2, and centerlines can be 3. If you use this method, keep a written record of your numbering system for reference.

More complex layer names may be appropriate for some applications. The name might include the drawing number, color code, and layer content. The layer name Dwg100-2-Dimen, for example, could refer to drawing DWG100, color 2, for use when adding dimensions. The American Institute of Architects (AIA) *CAD Layer Guidelines*, associated with the NCS, specifies a layer naming system for architectural and related drawings. The system uses a highly detailed layer naming process that gives each layer a discipline designator and major group, and if necessary, one or two minor groups and a status field. The AIA system allows complete identification of drawing content.

Layer names automatically arrange alphanumerically as you create new layers. See **Figure 5-8.** Pick any column heading in the list view to sort layer names in ascending or descending order according to that column. The **Layer Properties Manager** is a palette, so new layers and changes made to existing layers automatically save and apply to the drawing. There is no need to "apply" changes or close the palette to see the effects of the changes in the drawing.

PROFESSIONAL TIP

If you need to create multiple layers, accelerate the process by pressing the comma key after typing each layer name to create another new layer.

Renaming Layers

To change an existing layer name using the **Layer Properties Manager**, pick the name in the **Name** column once to highlight it, pause for a moment, and then pick it again. Type the new name and press [Enter] or pick outside of the text box. You can also rename a layer by picking the name once to highlight it and then pressing [F2], or by right-clicking and selecting **Rename Layer**. You cannot rename layer 0 and layers associated with an external reference.

Figure 5-8.
Layer names are automatically placed in alphanumeric order when you create new layers or change layer names.

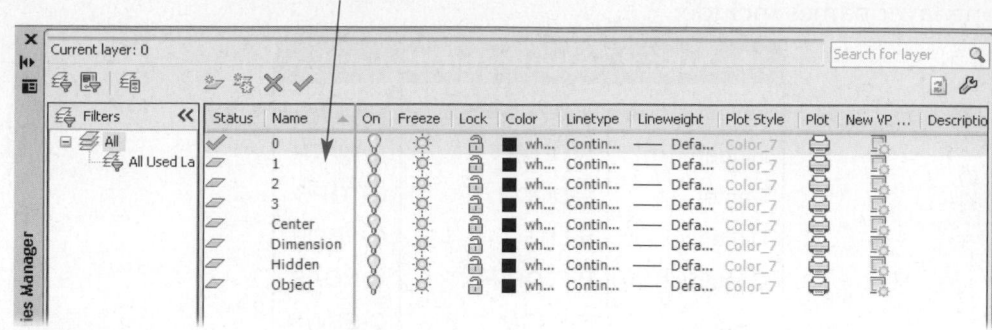

Layer names sort
automatically

Exercise 5-1

Access the Student Web site (www.g-wlearning.com/CAD) and complete Exercise 5-1.

Selecting Multiple Layers

Select multiple layers to speed the process of deleting or applying the same properties to several layers. To select multiple layers in the list view, use the same techniques you use to select files. You can highlight a single name by picking it, while picking another name deselects the previous name and highlights the new selection. You can use [Shift] to select two layers and all layers between the selections in the listing. Holding [Ctrl] while picking layer names highlights or deselects each selected name without affecting any other selections. You can also use a window selection to select all the layers that contact the window. The following additional selection options are available when you right-click in the list view:

- **Select All.** Selects all layers.
- **Clear All.** Deselects all layers.
- **Select All but Current.** Selects all layers except the current layer.
- **Invert Selection.** Deselects all selected layers and selects all deselected layers.

Exercise 5-2

Access the Student Web site (www.g-wlearning.com/CAD) and complete Exercise 5-2.

Layer Status

The icon in the **Status** column describes the status, or existing use of a layer. A green check mark indicates the *current layer*. The status line at the top of the **Layer Properties Manager** also identifies the current layer.

A white sheet of paper, or **Not In Use** icon, in the **Status** column indicates that the layer is not being used in any way by the drawing, the layer is not current, and no objects have been drawn on the layer. A blue sheet of paper, or **In Use** icon, in the

current layer:
The active layer. Whatever you draw is placed on the current layer.

Status column means that objects have been drawn with the layer, but the layer is not current. The **In Use** icon can also mean that the layer cannot be deleted or purged from the drawing, even if no objects are drawn on the layer.

Current

Not in Use

In Use

NOTE

If layers in use are not indicated in the **Layer Properties Manager**, pick the **Settings** button in the upper-right corner to display the **Layer Settings** dialog box and select the **Indicate layers in use** check box.

Setting the Current Layer

To set a different layer current using the **Layer Properties Manager,** double-click the layer name, pick the layer name in the layer list and select the **Set Current** button, or right-click on the layer and choose **Set Current.** You can also make a different layer current without using the **Layer Properties Manager** by selecting the layer you want to use from the **Layer Control** drop-down list of the **Home** ribbon tab. See **Figure 5-9.** This is often the most effective way to activate and manage layers while drawing. Pick the drop-down arrow and select a layer from the list to set current. Use the vertical scroll bar to move up and down through a long list.

NOTE

You can use the **Layer Properties Manager** or the **Layer Control** drop-down list to change the current layer or layer properties while a tool is active. For example, draw one line segment using the current layer, and then without exiting the **LINE** tool, make a different layer current to draw the next line segment on that layer.

Figure 5-9.
The **Layer Control** drop-down list allows you to change the current layer and change the properties of layers.

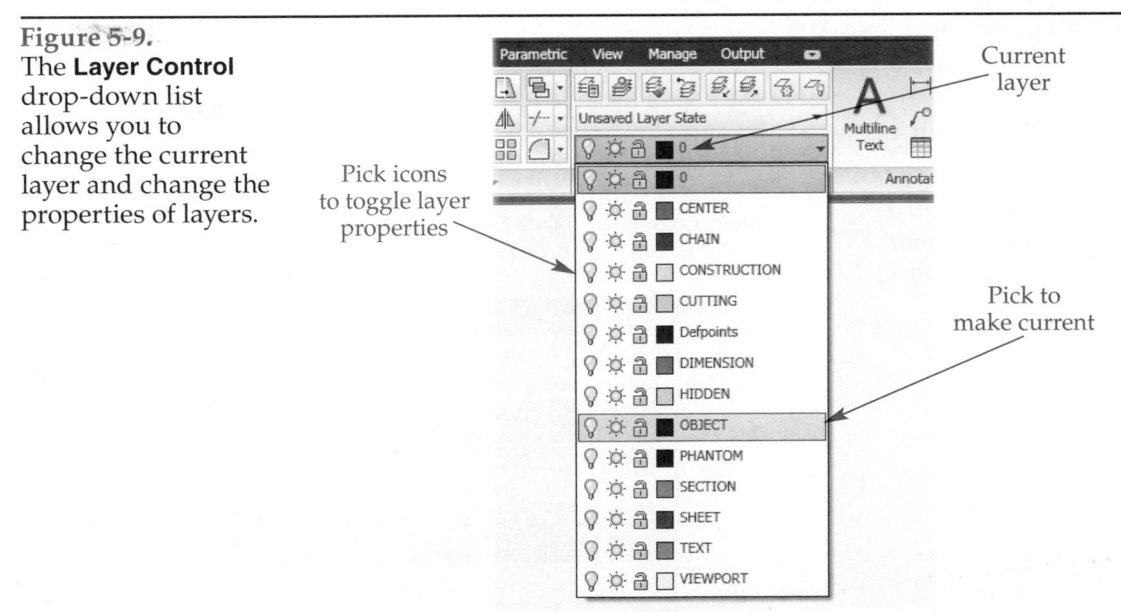

Current layer

Pick icons to toggle layer properties

Pick to make current

Exercise 5-3

Access the Student Web site (www.g-wlearning.com/CAD) and complete Exercise 5-3.

Setting Layer Color

You can assign a unique color to each layer to help differentiate drawing items on-screen. Layer colors can also affect the appearance of drawings plotted in color and can control object properties such as lineweight. Plotting using colors is not typical, but assigning color to layers is still very important for drawing clarity, organization, workability, and format. Layer colors should highlight the important features on the drawing and not cause eyestrain.

The **Color** column of the list view in the **Layer Properties Manager** indicates the color applied to each layer. Pick the color swatch to change the color of an existing layer using the **Select Color** dialog box. See **Figure 5-10**. This dialog box includes an **Index Color** tab, a **True Color** tab, and a **Color Books** tab from which you can select a color. Each tab uses a different method of obtaining colors.

The default **Index Color** tab includes 255 color swatches from which you can choose. This tab is commonly referred to as the AutoCAD Color Index (ACI) because colors are coded by name and number. The first seven colors in the ACI include both a numerical index number and a name: 1 = red, 2 = yellow, 3 = green, 4 = cyan, 5 = blue, 6 = magenta, and 7 = white.

To select a color, pick the appropriate color swatch or type the color name or ACI number in the **Color:** text box. The color white (number 7) shows up black with the default cream-colored drawing window background. The "white" name comes

Figure 5-10.
Use the **Select Color** dialog box to choose a layer color.

Index Color tab
255 colors

True Color tab
24-bit color

Color Books tab
Pantone colors

Selected color

Standard colors #1–9

Selected color index number

Previously selected color

New selected color

from the concept of using a black drawing window background, which was once the AutoCAD default display. If you change the drawing window background to a dark color such as black, color 7 shows up white on the screen.

As you move the cursor around the color swatches, the **Index color:** note updates to show you the number of the color over which the cursor is hovering. Beside the **Index color:** note is the **Red, Green, Blue:** (RGB) note. This indicates the RGB numbers used to mix the highlighted color. When you pick a color, the **Index color:** note appears in the **Color:** text box. A preview of the newly selected color and a sample of the previously assigned color appear in the lower right of the dialog box. An easy way to explore the ACI numbering system is to pick a color swatch and see what number appears in the **Color:** text box.

After selecting a color, pick the **OK** button to assign the specified color to the selected layer. The color appears as the color swatch for the layer. All objects drawn on this layer appear in the selected color, or ByLayer, by default.

NOTE

Your graphics card and monitor affect color display characteristics and sometimes the number of available colors. Most color systems usually support at least 256 colors.

PROFESSIONAL TIP

Use the color swatch in the **Layer Control** drop-down list to change the color assigned to a layer without accessing the **Layer Properties Manager**.

CAUTION

The **COLOR** tool provides access to the **Select Color** dialog box, which you can use to set an *absolute value* for color. If the absolute color value is set to red, for example, all objects appear red regardless of the color assigned to the layers on which objects are drawn. However, use layers and the **Layer Properties Manager** and set color as ByLayer to control object color for most applications.

absolute value: In property settings, a value set directly instead of being referenced by layer or a block. The current layer settings are ignored when an absolute value is set.

Exercise 5-4

Access the Student Web site (www.g-wlearning.com/CAD) and complete Exercise 5-4.

Setting Layer Linetype

Different line thicknesses and linetypes enhance the readability of drawings. You can apply standard line conventions to objects by assigning a linetype and thickness to each layer. AutoCAD provides standard linetypes that you can use to match ASME, NCS, or other applicable standards. You can also create your own custom linetypes. To achieve different line thicknesses, it is necessary to assign lineweights to layers.

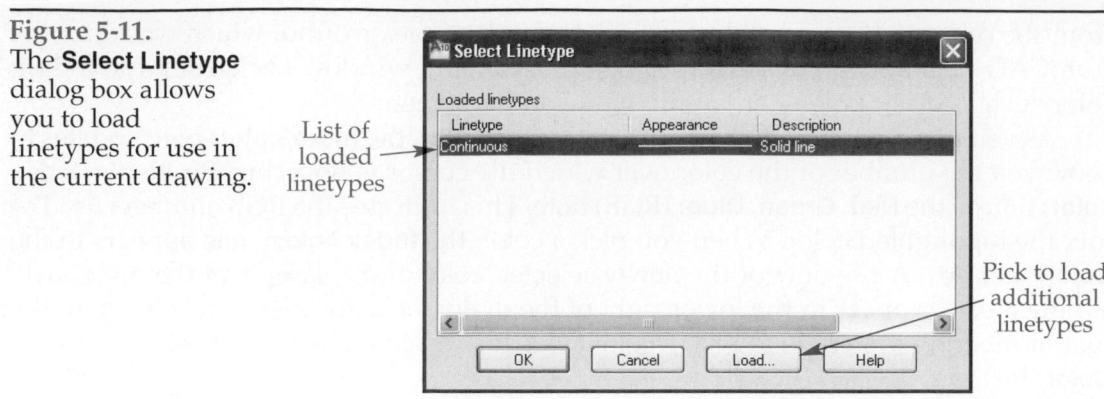

Figure 5-11.
The **Select Linetype** dialog box allows you to load linetypes for use in the current drawing.

List of loaded linetypes

Pick to load additional linetypes

The **Linetype** column of the list view in the **Layer Properties Manager** indicates the linetype applied to each layer. To change the linetype of an existing layer, pick the current linetype. This displays the **Select Linetype** dialog box, shown in **Figure 5-11.** Initially, the Continuous linetype is the only linetype listed in the **Loaded linetypes** list box. Use the Continuous linetype to draw solid lines with no breaks. AutoCAD maintains linetypes in external linetype definition files. Before you can apply a linetype other than Continuous to a layer, you must load the linetype into the **Select Linetype** dialog box.

Adding Linetypes

To add linetypes, pick the **Load...** button to display the **Load or Reload Linetypes** dialog box. See **Figure 5-12.** The acad.lin or acadiso.lin file is active, depending on the template you use to begin a new drawing. The acad.lin and acadiso.lin files are identical except that in the acadiso.lin file, the non-ISO linetypes are scaled up 25.4 times. The scale factor of 25.4 converts inches to millimeters for use in metric drawings. To switch to a different linetype definition file, pick the **File...** button in the **Load or Reload Linetypes** dialog box. Then use the **Select Linetype File** dialog box to select the desired file.

The **Available Linetypes** list displays the name and a description, which includes an image, of each linetype available from the specified linetype definition file. Use the scroll bars to view all available linetypes, and use the image in the **Description** column to aid in selecting the appropriate linetypes to load. Choose a single linetype, or select multiple linetypes using standard selection practices or the shortcut menu. Pick the **OK** button to return to the **Select Linetype** dialog box, where the linetypes you selected now appear. See **Figure 5-13.** In the **Select Linetype** dialog box, pick the desired linetype, and then pick **OK**. The HIDDEN linetype selected in **Figure 5-13** is now the linetype assigned to the layer named Hidden, as shown in **Figure 5-14.**

Figure 5-12.
The **Load or Reload Linetypes** dialog box displays linetypes available for loading.

Select file where linetype definitions are stored

Select linetypes to load into drawing

Figure 5-13.
Linetypes loaded using the **Load or Reload Linetypes** dialog box are added to the **Loaded linetypes** list box.

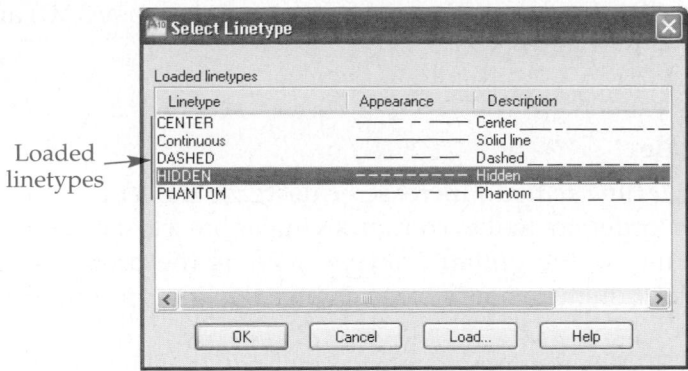

Loaded linetypes

Figure 5-14.
Objects drawn on the Hidden layer now display a HIDDEN linetype.

Linetype changed to HIDDEN

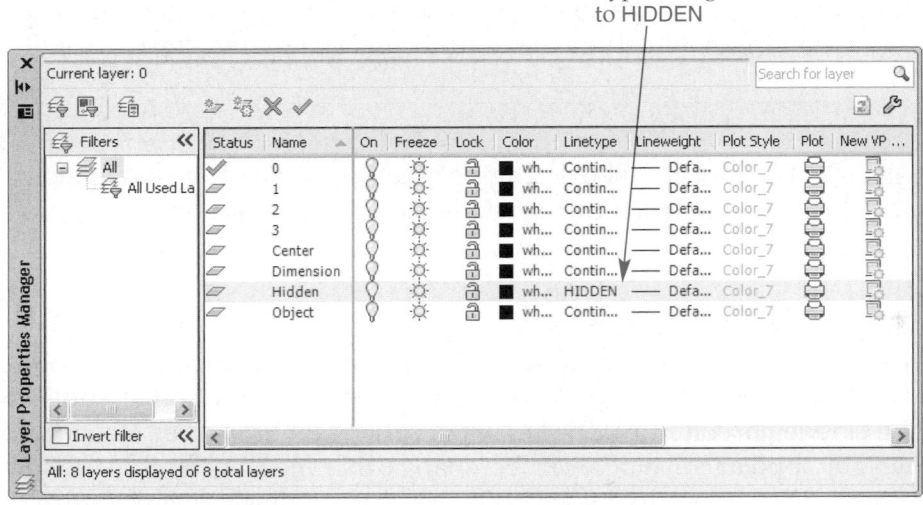

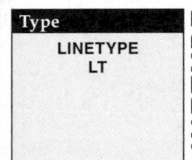

CAUTION

The current linetype can be an absolute linetype value. The **LINE-TYPE** tool provides access to the **Linetype Manager**, which you can use to control a variety of linetype characteristics, but you should set linetype as ByLayer for most applications. Changing linetype to a value other than ByLayer overrides layer linetype. Therefore, if the absolute linetype value is set to HIDDEN, for example, all objects are drawn using a HIDDEN linetype regardless of the linetype assigned to the layers on which objects are drawn. The **Linetype Manager** includes other options that are unnecessary for typical applications, or are more appropriately set using other techniques. Use layers and the **Layer Properties Manager** to control object linetype for most applications.

Type
LINETYPE
LT

LINETYPE

Exercise 5-5

Access the Student Web site (www.g-wlearning.com/CAD) and complete Exercise 5-5.

Setting Linetype Scale

linetype scale: The lengths of dashes and spaces in linetypes.

global linetype scale: A linetype scale applied to every linetype in the current drawing.

You can change *linetype scale* to increase or decrease the lengths of dashes and spaces in linetypes in order to make your drawing more closely match standard drafting practices. Changing the *global linetype scale* is the preferred method for adjusting linetype scale, though it is possible to change the linetype scale of individual objects.

You can use the **LTSCALE** system variable to make a global change to the linetype scale. The default global linetype scale factor is 1.0000. Any line with dashes initially assumes this factor. To change the linetype scale for the entire drawing, type LTSCALE and enter a new value. The drawing regenerates and the global linetype scale changes for all lines on the drawing. A value less than 1.0 makes the dashes and spaces smaller, and a value greater than 1.0 makes the dashes and spaces larger. See **Figure 5-15.** Experiment with different linetype scales until you achieve the desired results.

CAUTION

Be careful when changing linetype scales to avoid making your drawing look odd and not in accordance with drafting standards.

Setting Layer Lineweight

lineweight: The assigned width of lines for display and plotting.

Assign *lineweight* to a layer to manage the weight, or thickness, of objects. You can control the display of line thickness to match ASME, NCS, or other applicable standards. The **Lineweight** column of the list view in the **Layer Properties Manager** indicates the lineweight applied to each layer. To change the lineweight of an existing layer, pick the current lineweight to display the **Lineweight** dialog box. See **Figure 5-16.** The **Lineweight** dialog box displays fixed lineweights available in AutoCAD. Scroll through the **Lineweights:** list and select the lineweight you want to assign to the layer. Pick the **OK** button to apply the lineweight and return to the **Layer Properties Manager**.

Type
LINEWEIGHT
LWEIGHT
LW

LINEWEIGHT

The **LINEWEIGHT** tool provides access to the **Lineweight Settings** dialog box, shown in **Figure 5-17.** The **Units for Listing** area allows you to set the lineweight thickness to **Millimeters (mm)** or **Inches (in)**. The units apply only to values in the **Lineweight** and **Lineweight Settings** dialog boxes, allowing you to select lineweights based on a known unit of measurement.

Check the **Display Lineweight** box to turn lineweight on. Lineweight appears on-screen when lineweight display is turned on. Use the **Adjust Display Scale** slider to

Figure 5-15.
The CENTER linetype at different linetype scales.

Scale Factor	Line
0.5	— — — — — — — — — — — — — — —
1.0	— — — — — — — — — — — —
1.5	— — — — — — — — —

Figure 5-16.
Use the **Lineweight**
dialog box to assign
a lineweight to a
layer.

Select lineweight
from list

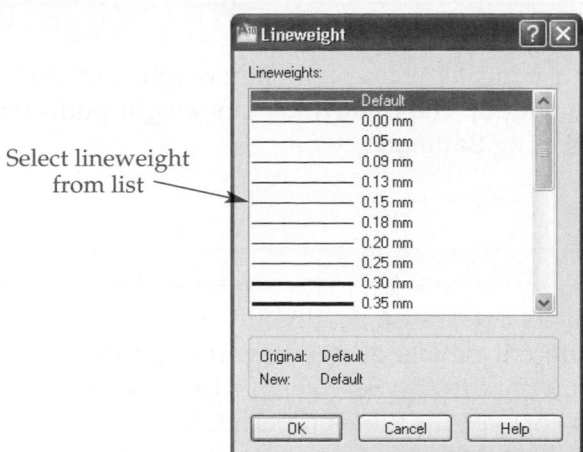

Figure 5-17.
The **Lineweight
Settings** dialog box.

Select
lineweight

Select units

Lineweight
display

Default
lineweight
setting

Display
scale

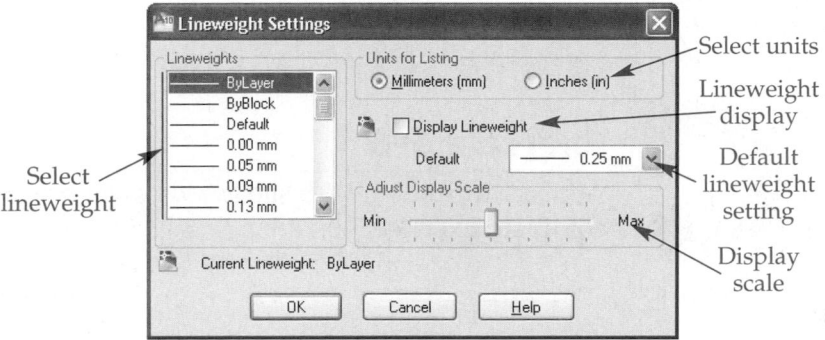

adjust the lineweight display scale to improve the appearance of different lineweights when lineweight display is on. When lineweight display is off, all objects display a 0, or one pixel, thickness regardless of the lineweight assigned to the object. You can also toggle screen lineweights by picking the **Show/Hide Lineweight** button on the status bar.

The value used when you assign the Default lineweight to a layer is set in the **Default** drop-down list. The Default lineweight is an application setting and applies to any drawing you open. The Default lineweight is not template-specific and remains set until you change the value. Do not assign layers the Default lineweight value if you anticipate using a different *default* lineweight for different drawing applications. Change each layer's lineweight individually. This rule maintains flexibility and consistency between drawings.

CAUTION

You can use the **Lineweights** area of the **Lineweight Settings** dialog box to set an absolute lineweight value. However, you should set lineweight as ByLayer for most applications. Changing lineweight to a value other than ByLayer overrides layer lineweight. Therefore, if the absolute linetype value is set to 0.30 mm, for example, all objects are drawn using a 0.30 mm weight regardless of the lineweight assigned to the layers on which objects are drawn. Use layers and the **Layer Properties Manager** to control object lineweight for most applications.

Exercise 5-6

Access the Student Web site (www.g-wlearning.com/CAD) and complete Exercise 5-6.

Plot   No Plot

Layer Plotting Properties

The **Plot Style** column of the **Layer Properties Manager** lists the plot style assigned to each layer. By default, the plot style setting is disabled. Plot styles are described later in this textbook.

The **Plot** column displays icons to show whether the layer plots. Select the default printer icon to turn off plotting for a particular layer. The **No Plot** icon appears when the layer is not available for plotting. The layer is still displayed and selectable, but it does not plot.

Adding a Layer Description

The **Description** column provides an area to type a short description for each layer. To add or change a description, pick the description once to highlight it, pause for a moment, and then pick it again. Type an appropriate description, and press [Enter] or pick outside of the **Description** text box. You can also define the layer description by right-clicking and selecting **Change Description**.

On   Off

Turning Layers On and Off

The **On** column shows whether a layer is on or off. The yellow light bulb, or **On** icon, means the layer is on. Objects on a layer that is turned on display on-screen and can be selected and plotted. If you pick the icon, the light bulb "turns off" (becomes gray), turning the layer off. Objects on a layer that is turned off do not display on-screen and are not plotted. Objects on an layer that is turned off can still be edited using advanced selection techniques, and they regenerate when a drawing regeneration occurs.

Freezing and Thawing Layers

Freeze   Thaw

The **Freeze** column shows whether a layer is thawed or frozen. Objects on a frozen layer do not display, plot, or regenerate when the drawing regenerates. You cannot edit objects on a frozen layer. Freeze layers to ensure that you do not accidentally modify objects they contain, and to increase system performance. The snowflake, or **Freeze**, icon displays when a layer is frozen. When a layer is thawed, objects on the layer appear on-screen, and they can be selected and regenerated. The sun, or **Thaw**, icon appears for thawed layers. Pick the **Freeze** or **Thaw** icon to toggle thawing and freezing.

Icons in the **New VP Freeze** column control freezing or thawing of layers when you create a new viewport. Additional layer functions also apply to layouts and viewports. Layouts and viewports are described later in this textbook.

Locking and Unlocking Layers

Lock   Unlock

The unlocked and locked padlock symbols (**Unlock** and **Lock** icons) located in the **Lock** column of the **Layer Properties Manager** control layer locking and unlocking. Objects on a locked layer remain visible, and you can use a locked layer to draw new objects. However, you cannot edit existing objects on a locked layer. Lock a layer whenever you want to see objects on-screen, but eliminate the possibility of selecting those objects.

Layers are unlocked by default. Pick an **Unlock** icon to lock the layer. When you rest the crosshairs over an object on a locked layer, the **Lock** icon appears next to the cursor. To lock all layers except specific layers, select the layers you want to remain unlocked, and then right-click on the selection and pick **Isolate selected layers**.

Chapter 5   Line Standards, Layers, and Basic Plotting

Figure 5-18.
Locked layers fade by default. You can increase or decrease fading and enable or disable locked layer fading as needed.

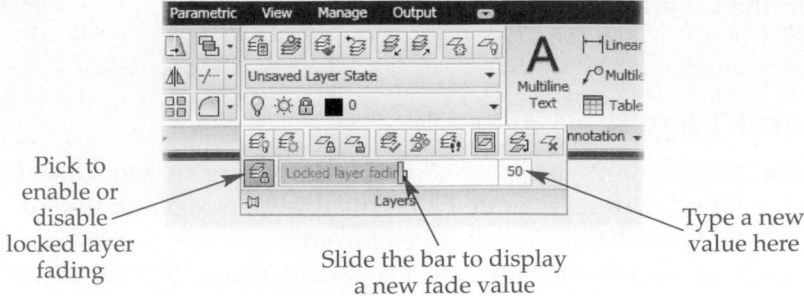

Pick to enable or disable locked layer fading

Slide the bar to display a new fade value

Type a new value here

Locked Layer Fading

By default, all locked layers fade, allowing unlocked layers to stand out on-screen. The quickest way to control locked layer fading is to use the options available in the expanded **Layers** panel of the **Home** ribbon tab. See **Figure 5-18**. Pick the **Locked layer fading** button to allow or disable locked layer fading. When you allow locked layer fading, use the **Locked Layer Fading** slider to increase or decrease fading, or type a new fading percentage. You can use a fade value between 0 and 90. A fade value of 0 fades the display of unisolated layers the least, while a fade value of 90 significantly fades unisolated layers. The default fade value is 50%. **Figure 5-19** shows the effect of locked layer fading on a drawing.

Deleting Layers

To delete a layer using the **Layer Properties Manager**, select the layer and pick the **Delete Layer** button, or right-click on the layer and choose **Delete Layer**. You cannot delete or purge the 0 layer, the current layer, layers containing objects, or layers associated with an external reference.

Figure 5-19.
All of the layers in this drawing are locked except A-WALL-FULL, which contains the walls. A—Locked layer fading is disabled. B—Locked layer fading is on and set to a fade value of 75.

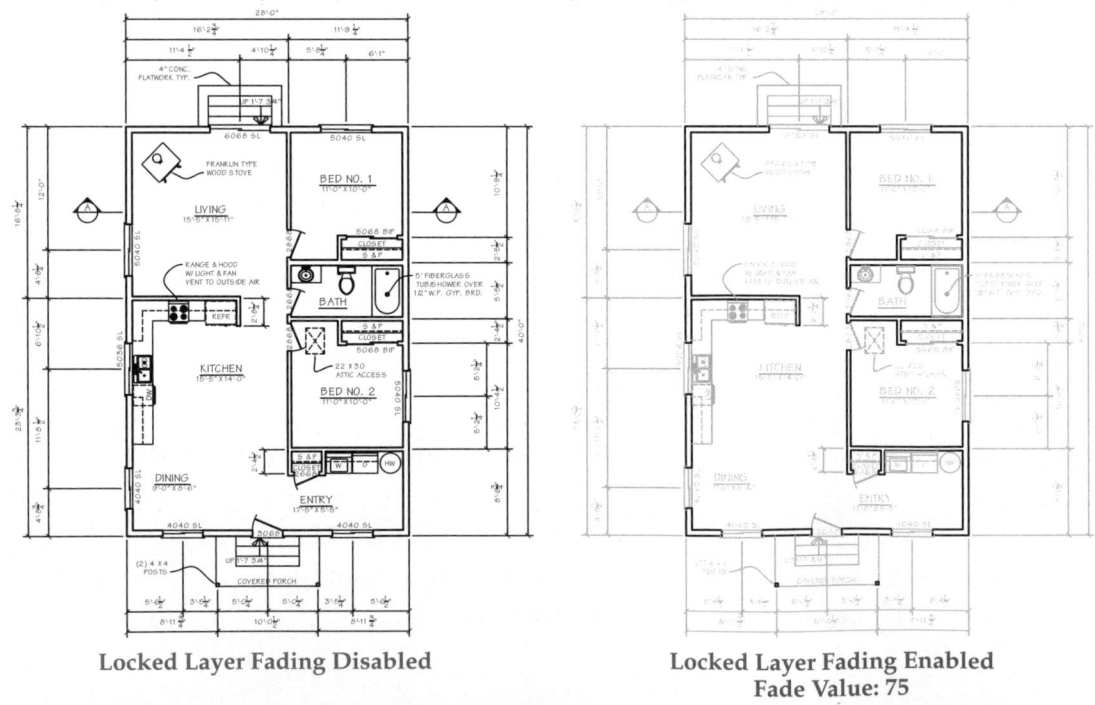

Locked Layer Fading Disabled

Locked Layer Fading Enabled
Fade Value: 75

A

B

Adjusting Property Columns

To resize a column in the **Layer Properties Manager**, move the cursor over the column edge to display the resize icon and drag the column to the desired width. You can maximize the width of an individual column to show the longest value in the column, or in the case of columns that list properties as icons, to display the full column heading. To maximize the width of a single property column, right-click on the column heading and select **Maximize column**. To maximize the width of all columns, right-click on any property column heading and select **Maximize all columns**.

You can optimize the width of columns to show the longest value in the column list for properties displayed as text, while reducing the width of columns that list properties as icons. To optimize the width of a single property column, right-click on the column heading and select **Optimize column**. To optimize the width of all columns, right-click on any property column heading and select **Optimize all columns**.

By default, a vertical bar appears on the right side of the **Name** column. Any column left of the bar is "frozen". Frozen columns remain in position when you move the scroll bar near the bottom of the **Layer Properties Manager**. Scroll columns to the right of the vertical bar using the horizontal scroll bar. To turn off the freeze function, right-click on any property column heading and select **Unfreeze column**. To freeze columns, right-click on a property column heading and select **Freeze column** to turn on the freeze function for every column left of the selected column.

You can hide columns in the **Layer Properties Manager** by right-clicking on any property column heading and deselecting the property column name from the menu. Another option is to right-click on any property column heading and select **Customize...** to display the **Customize Layer Columns** dialog box. You can use this dialog box to hide property columns by the associated check boxes. To move a column left or right in the **Layer Properties Manager**, pick a column name and select the **Move Up** or **Move Down** button. Reset the display of all property columns to default settings by right-clicking on any property column heading and selecting **Restore all columns to defaults**.

Supplemental Material

Additional Layer Tools

For information about additional layer tools, go to the Student Web site (www.g-wlearning.com/CAD), select this chapter, and select **Additional Layer Tools**.

Filtering Layers

A single drawing often includes a very large number of layers. Displaying all layers at the same time in the list view pane can make it more difficult to work with the layers. The filter tree view pane in the **Layer Properties Manager** manages *layer filters* that you can be apply to reduce the number of layers that appear in the list view. See **Figure 5-20**.

Select the **All** node of the filter tree view to display all layers in the drawing. Layer filters are listed in alphabetical order inside the **All** node. Pick the **All Used Layers** filter to hide all the layers that have no objects on them. When you insert external references, an Xref filter node appears, allowing you to filter the display of layers associated with external references. External references are described later in this textbook. You can create other filters as needed.

layer filters: Filters that screen out, or filter, layers you do not want to display in the list view pane of the **Layer Properties Manager**.

Figure 5-20.
Create and restore
layer filters using
the filter tree view of
the **Layer Properties
Manager.**

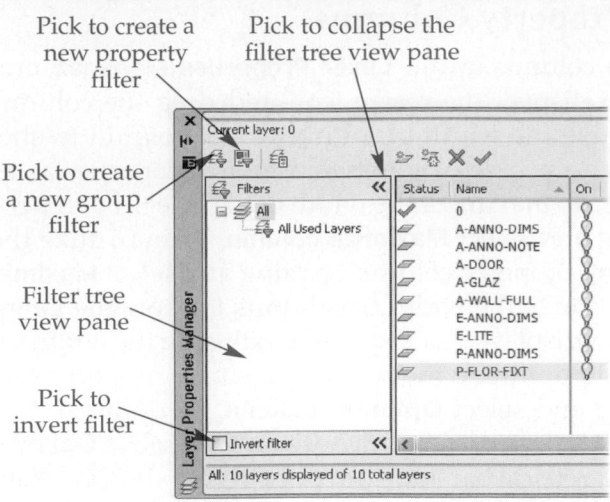

Pick to create a
new property
filter

Pick to collapse the
filter tree view pane

Pick to create
a new group
filter

Filter tree
view pane

Pick to
invert filter

> **NOTE**
>
> The filter tree view of the **Layer Properties Manager** can be collapsed by picking the **Collapse Layer filter tree** button. To display all filters and layers in the list view, right-click in the layer list area and select **Show Filters in Layer List**.

Creating a Property Filter

property filter: A filter that screens layers according to a specific layer property.

An example of using a *property filter* is filtering all layers that are turned on, or have a name beginning with the letter *A*, or both. The default **All Used Layers** filter is a property filter that filters layers according to layer status. To create a property filter, pick the **New Property Filter** button to display the **Layer Filter Properties** dialog box. See **Figure 5-21.** Enter a name for the new filter in the **Filter name:** text box. Pick the appropriate **Filter definition** area field box to define properties to filter.

The Status, On, Freeze, Lock, Plot, and New VP Freeze fields display a flyout. Select an option from the flyout to filter according to the selection. For example, pick the **On** icon from the On field to display only layers that are turned on. In the Name field, type a layer name or a partial layer name using the * wildcard character to filter according to layer name. For example, to see all layers that start with an *A*, type a*. You can filter using the Color, Linetype, Lineweight, or Plot Style fields by typing an appropriate value or selecting the ellipses (…) button to select from the corresponding dialog box.

After you define a property filter, another row appears in the **Filter definition:** area. This allows you to create a more advanced filter. **Figure 5-21** shows a filter named Floor Plan, in which two rows are used to filter out all the layers except the layers beginning with the letter A and the P-FLOR-FIXT layer. To save the filter, pick the **OK** button. The new filter now displays in the filter tree view area.

Creating a Group Filter

group filter: A filter created by adding layers to the filter definition.

An example of using a *group filter* is dragging and dropping all layers used to draw an electrical plan into a group filter. All of the layers are selectable through the group filter regardless of their individual properties. To create a group filter, select the **New Group Filter** button. A new group filter appears in the filter tree view. Select the **All** node at the top of the filter tree area to display all the layers in the drawing. Then, to add a layer to the group filter, drag a layer from the list and drop it onto the group filter name.

Figure 5-21.
Use the **Layer Filter Properties** dialog box to create and edit a property filter. Use multiple rows in the **Filter definition:** area to add layers to the filter set.

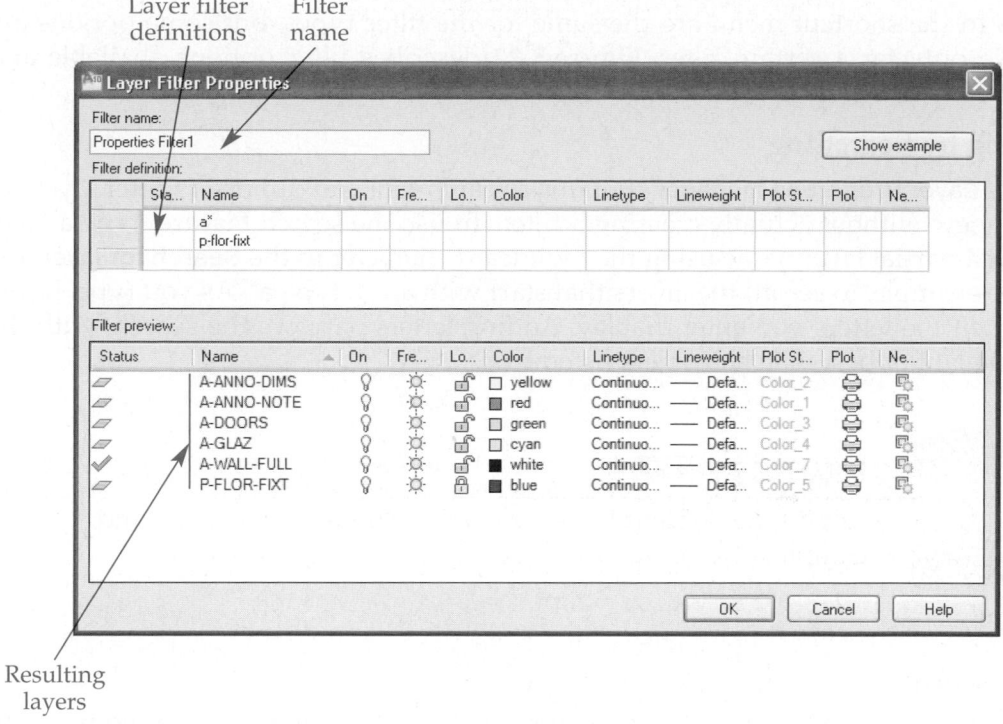

Layer filter definitions

Filter name

Resulting layers

You can also add layers to a group filter by selecting the group filter, right-clicking, and choosing **Add** from the **Select Layers** cascading submenu. This feature allows you to select objects on the layers you want to add to the group filter. After selecting the objects, right-click or press [Enter] to add the layers to the group filter.

Pick **Replace** from the **Select Layers** cascading submenu to select objects on the layers to replace all other layers in the group filter. After selecting the objects, right-click or press [Enter] to add the layers to the group filter. To remove a layer from a group filter, right-click on the layer in the layer list area of the **Layer Properties Manager** and choose **Remove From Group Filter**.

Activating a Layer Filter

Select the filter from the filter tree view of the **Layer Properties Manager** to activate the filter. When a layer filter is active, only those layers associated with the filter appear in the layer list area of the **Layer Properties Manager** and the **Layer Control** drop-down list in the **Layers** panel of the **Home** ribbon tab. To view all the layers again, pick the **All** node at the top of the filter tree area of the **Layer Properties Manager**.

NOTE

The lower **Layer Properties Manager** status bar provides a description of the active layer filter settings.

Inverting Layer Filters

To invert, or reverse, a layer filter using the **Layer Properties Manager**, pick the **Invert filter** check box located in the lower-left corner or right-click in the layer list area and select **Invert Layer Filter**. For example, selecting the **All Used Layers** filter shows only the layers that have objects on them. You can invert the **All Used Layers** filter to show all unused layers without creating an additional filter.

Additional Layer Filter Options

Other options associated with filters are accessible from a shortcut menu that appears when you right-click on a filter in the **Layer Properties Manager.** Most of the options in the shortcut menu are the same for the filter types, but some options are available only for a certain filter. **Figure 5-22** describes filter options available and applied only to the layers associated with the filter.

Filtering by Searching

The **Layer Properties Manager** contains a search tool you can use to filter layers in the list view without actually creating a filter. To use the search feature, type a layer name or a partial layer name using the * wildcard character in the **Search for layer** text box. For example, to see all the layers that start with an *A*, type a*. As you type, layers that match the letters you enter display. Adding letters narrows the search, with the most relevant or best-matched layers listed first.

Exercise 5-7

Access the Student Web site (www.g-wlearning.com/CAD) and complete Exercise 5-7.

Layer States

Once you save a *layer state*, you can readjust layer settings to meet your needs, with the option to restore a previously saved layer state at any time. For example, a basic architectural drawing might use the layers shown in **Figure 5-20**. Three different drawings can be plotted using this drawing file: a floor plan, a plumbing plan, and an electrical plan. The following chart shows the layer settings for each of the three drawings:

layer state: A saved setting, or state, of layer properties for all layers in the drawing.

Layer	Description	Floor Plan	Plumbing Plan	Electrical Plan
0		Off	Off	Off
A-ANNO-DIMS	Floor Plan Dimensions	On	Frozen	Frozen
A-ANNO-NOTE	Floor Plan Notes	On	Frozen	Frozen
A-DOOR	Doors	On	Frozen	Locked
A-GLAZ	Windows	On	Frozen	Locked
A-WALL-FULL	Full Height Walls	On	Locked	Locked
E-ANNO-DIMS	Electrical Plan Dimensions	Frozen	Frozen	On
E-LITE	Electrical Plan Lights	Frozen	Frozen	On
P-ANNO-DIMS	Plumbing Plan Dimensions	Frozen	On	Frozen
P-FLOR-FIXT	Plumbing Plan Fixtures	Locked	On	Locked

You can save each of the three groups of settings as an individual layer state. You can then restore a layer state to return the layer settings for a specific drawing. This is easier than changing the settings for each layer individually.

Figure 5-22.
Additional layer filter options available from the shortcut menu.

Option	Function
Visibility	Changes the **On/Off** and **Thawed/Frozen** states of the unfiltered layers.
Lock	Locks or unlocks the unfiltered layers.
Viewport	Freezes or thaws the unfiltered layers in the current layout viewport.
Isolate Group	Freezes all layers except those associated with the filter and the current layer.
New Properties Filter	Opens the **Layer Filter Properties** dialog box.
New Group Filter	Creates a new group filter.
Convert to Group Filter	Converts a property filter to a group filter.
Rename	Renames the selected filter.
Delete	Deletes the selected filter.
Properties	Edits a property filter.
Select Layers	Provides options to add layers or replace layers in an existing group filter.

Use the **Layer States Manager**, shown in **Figure 5-23**, to create a new layer state. Pick the **New...** button to display the **New Layer State to Save** dialog box. See **Figure 5-24**. Type a name for the layer state in the **New layer state name:** text box and enter a description. Pick the **OK** button to save the new layer state. Once you create a layer state, you can adjust layer properties as needed. **Figure 5-25** describes the areas, options, and buttons available in the **Layer States Manager**.

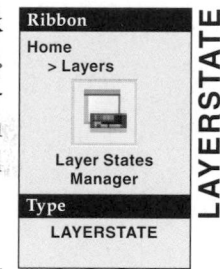

Ribbon
Home
> Layers

Layer States
Manager

Type
LAYERSTATE

LAYERSTATE

Figure 5-23.
The **Layer States Manager** allows you to save, restore, and manage layer settings.

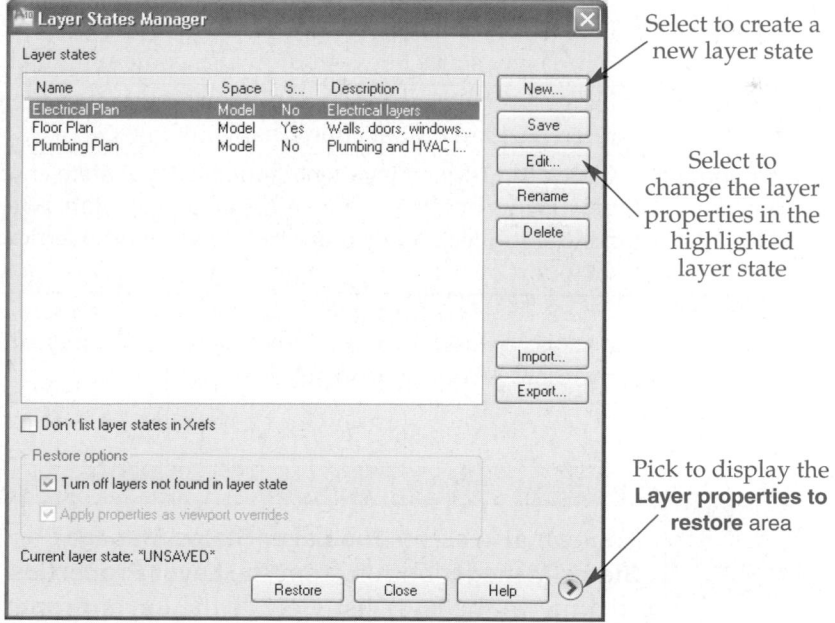

Select to create a new layer state

Select to change the layer properties in the highlighted layer state

Pick to display the **Layer properties to restore** area

Figure 5-24.
Creating a new layer state.

Enter the
layer state
name

Enter a
description
for the layer
state

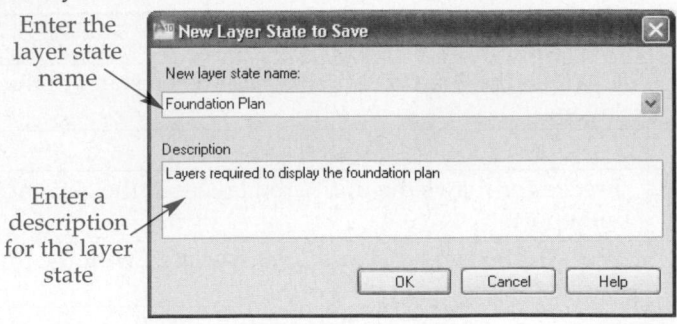

Figure 5-25.
Layer state options available in the **Layer States Manager**.

Item	Feature
Layer states	Displays saved layer states. The **Name** column provides the name of the layer state. The **Space** column indicates whether the layer state was saved in model space or paper space. The **Same as DWG** column indicates whether the layer state is the same as the current layer properties. The **Description** column lists the layer state description added when the layer state was saved.
Save	Pick to resave and override the selected layer state with the current layer properties.
Edit	Opens the **Edit Layer State** dialog box, where you can adjust the properties of each layer state without exiting the **Layer States Manager**.
Rename	Activates a text box that allows you to rename the current layer state.
Delete	Deletes the selected layer state.
Import	Opens the **Import layer state** dialog box, used to import an LAS file containing an existing layer state into the **Layer States Manager**.
Export	Opens the **Export layer state** dialog box, used to save a layer state as an LAS file. The file can be imported into other drawings, allowing you to share layer states between drawings containing identical layers.
Don't list layer states in Xrefs	Hides layer states associated with external reference drawings. External references are described in Chapter 30.
Restore options	Check the **Turn off layers not found in layer state** check box to turn off new layers or layers removed from a layer state when the layer state is restored. Check **Apply properties as viewport overrides** to apply layer viewport overrides when you are adjusting layer states within a layout.
Layer properties to restore	Check the layer properties that you want to restore when the layer state is restored. Pick the **Select All** button to pick all properties. Pick the **Clear All** button to deselect all properties.

NOTE

You can also access the **Layer States Manager** by picking the **Layer States Manager** button from the **Layer Properties Manager** or right-clicking in the layer list view of the **Layer Properties Manager** and selecting the **Restore Layer State** option. To save a layer state outside the **Layer States Manager**, pick the **New Layer State...** option from the **Layer States** drop-down list in the **Layers** panel on the **Home** ribbon tab.

AutoCAD and Its Applications—Basics

If you have a drawing that does not contain any layers other than 0, importing a layer state file (.las) adds the layers from the layer state to the drawing.

After you create a layer state, you can restore layer properties to the settings saved in the layer state at any time. To activate a layer state using the ribbon, select the layer state from the **Layer States** drop-down list in the **Layers** panel on the **Home** ribbon tab. You can also restore a layer state using the **Layer States Manager** by selecting the layer state from the list and picking the **Restore** button.

Supplemental Material

Layer Settings

For information about options available in the **Layer Settings** dialog box, go to the Student Web site (www.g-wlearning.com/CAD), select this chapter, and select **Layer Settings**.

Reusing Drawing Content

In nearly every drafting discipline, individual drawings created as part of a given project are likely to share a number of common elements. All the drawings within a specific drafting project generally have the same set of standards. *Drawing content*, such as layer names and properties, text size and font used for annotation, dimensioning methods and appearances, drafting symbols, drawing layouts, and even drawing details, is often duplicated in many different drawings. One of the most fundamental advantages of CADD systems is the ease with which you can share content between drawings. Once you define a commonly used drawing feature, you can reuse the item as needed in any number of drawing applications.

drawing content: All of the objects, settings, and other components that make up a drawing.

Drawing templates represent one way to reuse drawing content. Creating your own customized drawing template files provides an effective way to start each new drawing using standard settings. Drawing templates, however, provide only a starting point. During the course of a drawing project, you may need to add previously created content to the current drawing. Some drawing projects require you to revise an existing drawing rather than start a completely new drawing. For other projects, you may need to duplicate the standards used in a drawing a client has supplied.

Introduction to DesignCenter

AutoCAD provides a powerful drawing content manager called **DesignCenter**. The **DesignCenter** palette allows you to reuse drawing content defined in previous drawings using a drag-and-drop operation. **Figure 5-26** displays the main features of **DesignCenter**. **DesignCenter** can be used to manage several types of drawing content, including layers, linetypes, blocks, dimension styles, layouts, table styles, text styles, and externally referenced drawings. **DesignCenter** allows you to load content directly from any accessible drawing, without opening the drawing in AutoCAD.

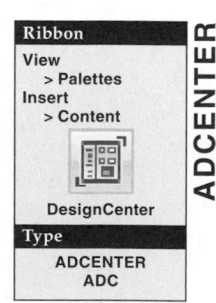

Ribbon
View
> Palettes
Insert
> Content

DesignCenter

Type
ADCENTER
ADC

ADCENTER

Figure 5-26.
Use **DesignCenter** to copy content from one drawing to another.

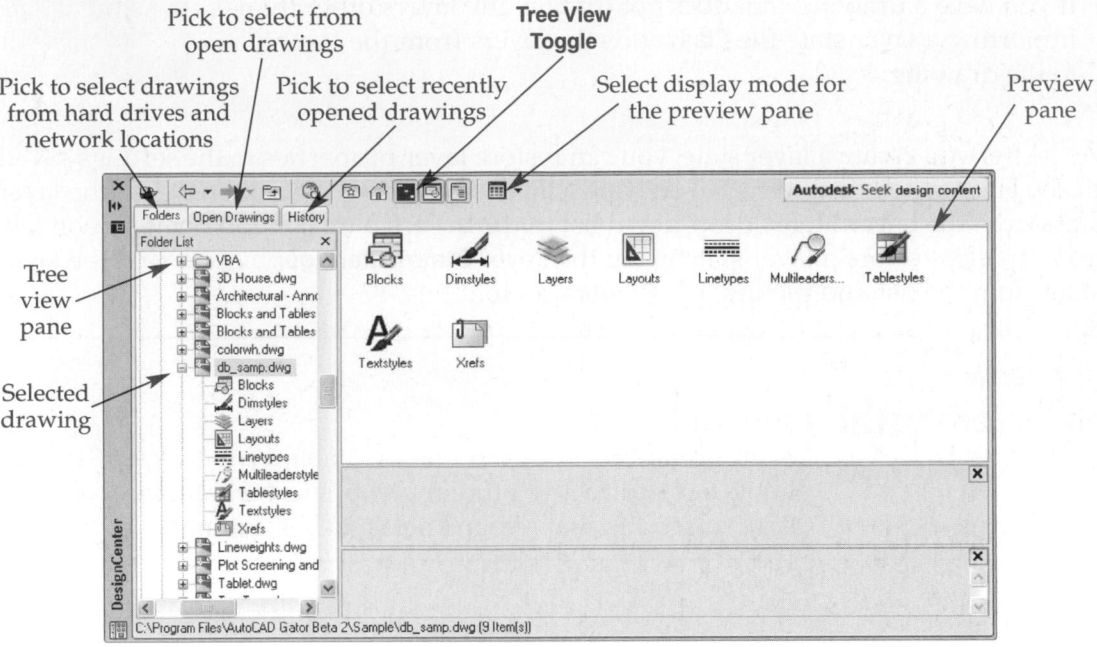

Copying Layers and Linetypes

To copy content using **DesignCenter**, first use the tree view pane to locate the drawing that includes the content you want to reuse. If the tree view is not already visible, toggle it on by picking the **Tree View Toggle** button in the **DesignCenter** toolbar. The three tabs on the **DesignCenter** toolbar control the tree view display. Select the **Folders** tab to display the folders and files found on the hard drive and network. Pick the **Open Drawings** tab to list only drawings that are currently open. Select the **History** tab to list recently opened drawings.

Pick the plus sign (+) next to a drawing icon to view the content categories for the drawing. Each category of drawing content includes a representative icon. Pick the **Layers** icon to load the preview pane with the layer content in the selected drawing. See **Figure 5-27.**

To use drag-and-drop to import layers into the current drawing, move the cursor over the desired icon in the preview pane in **DesignCenter**. Press and hold down the

Figure 5-27.
Displaying the layers found in a drawing using **DesignCenter**.

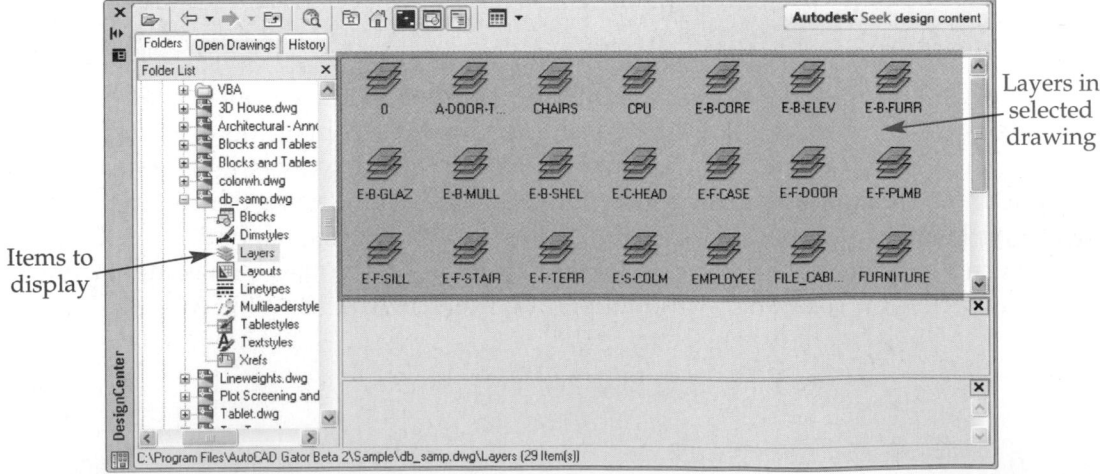

Figure 5-28.
To copy layers shown in **DesignCenter** into the current drawing, select the layers to be copied and then drag and drop them into the drawing area of the current drawing.

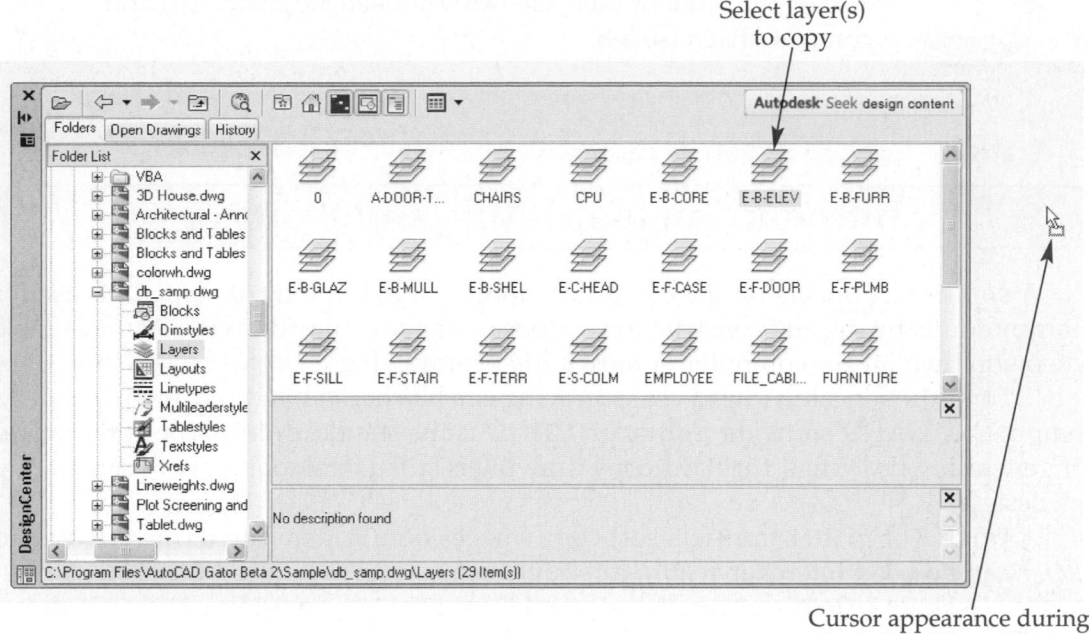

Select layer(s) to copy

Cursor appearance during drag-and-drop operation

pick button, and then drag the cursor to the open drawing. See **Figure 5-28.** Release the pick button to add the selected content to your current drawing file. You can also import layers into the current drawing by selecting the desired icons in the preview pane, right-clicking, and picking **Add Layer(s)**.

Use **DesignCenter** to copy linetypes from one file to another using the same procedure used to copy layers. In the tree view, select the drawing containing the linetypes you want to copy. Select the **Linetypes** icon to display the linetypes in the preview palette. Select the linetypes to copy, and then use drag-and-drop or the shortcut menu to add the linetypes to the current drawing.

NOTE

You cannot import drawing content if the content uses the same name as existing content. For example, if the name of a layer you try to reuse already exists in the destination drawing, the layer is ignored. The existing settings for the layer are preserved, and a message displays at the command line indicating that duplicate settings were ignored.

Supplemental Material

DesignCenter
For more information about the tools and features available for using **DesignCenter**, go to the Student Web site (www.g-wlearning.com/CAD), select this chapter, and select **DesignCenter**.

Exercise 5-8

Access the Student Web site (www.g-wlearning.com/CAD) and complete Exercise 5-8.

Introduction to Printing and Plotting

A *soft copy* appears on the computer monitor, making it inconvenient to use for many manufacturing and construction purposes. The soft-copy drawing is unavailable when you turn off the computer. A *hard copy* of a drawing is useful on the shop floor or at a construction site. A hard-copy drawing can be checked and redlined without a computer or CADD software. Although CADD is the standard throughout the world for generating drawings, the hard-copy drawing is still a vital tool for communicating the design.

A printer or plotter transfers soft-copy images onto paper. The terms *printer* and *plotter* can be used interchangeably, although *plotter* typically refers to a large-format printer. Desktop printers generally print 8 1/2″ × 11″ and sometimes 11″ × 17″ drawings. These are the printers common to computer workstations. Desktop printers print small drawings and reduced-size test prints. Large-format printers print larger drawings, such as C-size and D-size drawings. The most common types of both desktop and large-format printers are inkjet and laser printers. Pen plotters, which "draw" with actual ink pens, are still in use, but are less common.

Plotting in Model Space

You typically plot final drawings using a layout in paper space. A layout represents the sheet of paper used to organize and scale, or lay out, and plot or export a drawing or model. However, you can plot from model space as well as from a layout. The following content describes plotting from model space only.

Plotting from model space is common when a layout is unnecessary, to view how model space objects will appear on paper, and to make quick hard copies, such as when submitting basic assignments to your instructor or supervisor. The information in this chapter gives you only the basics, so you can make your first plot. This textbook fully explains creating and plotting layouts and additional printing and plotting information when appropriate.

Making a Plot

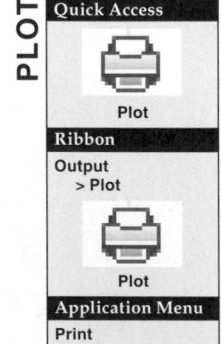

This section describes one of the many methods for creating a plot from model space. Refer to **Figure 5-29** as you read the following plotting procedure.

1. Access the **Plot** dialog box. If the column on the far right of the dialog box shown in **Figure 5-29** is not displayed, pick the **More Options** button (>) in the lower-right corner.
2. Check the plot device and paper size specifications in the **Printer/plotter** and **Paper size** areas.

Figure 5-29.
The **Plot** dialog box.

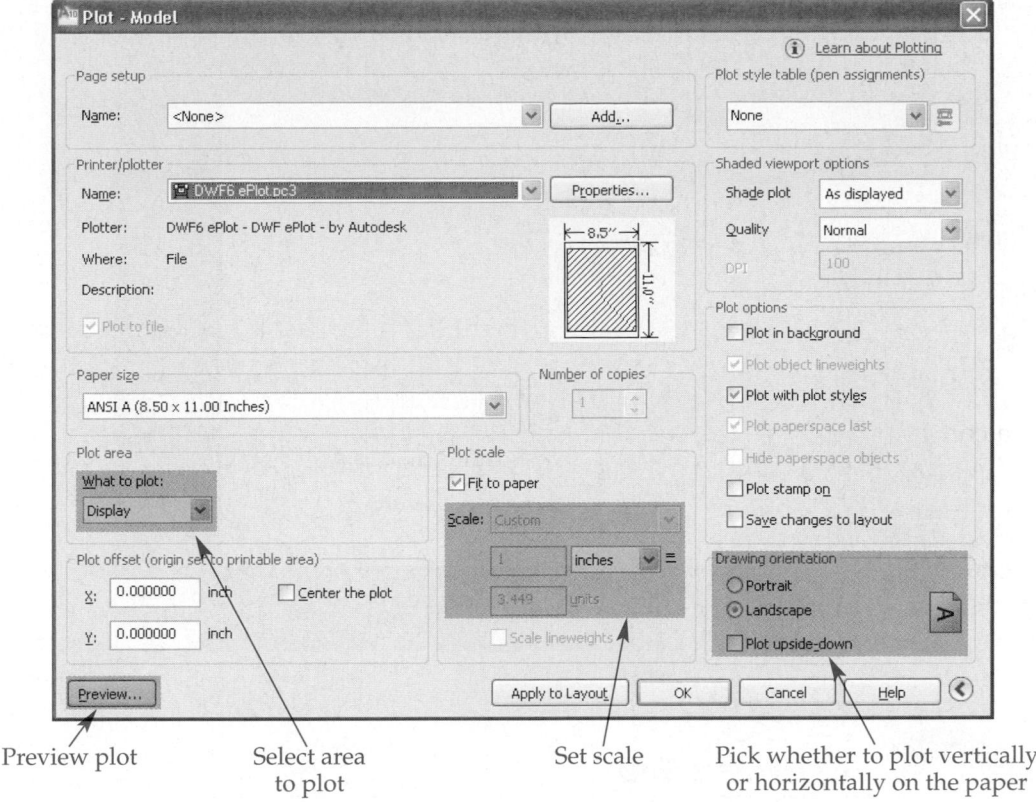

Preview plot   Select area to plot   Set scale   Pick whether to plot vertically or horizontally on the paper

3. Select what to plot in the **Plot area** section. The **Limits** option displays when you plot from model space. Select this option to plot everything inside the defined drawing limits. Pick the **Extents** option to plot the furthest extents of objects in the drawing. Select the **Display** option to plot the current screen display, exactly as it is shown. When you select the **Window** option, the **Page Setup** dialog box disappears temporarily so you can pick two opposite corners to define a window around the area to plot. Once you define the window, a **Window...** button appears in the **Plot area** section. Pick the button to redefine the opposite corners of a window around the portion of the drawing to plot.

4. Select an option in the **Drawing orientation** area. Choose **Portrait** to orient the drawing vertically (*portrait*) or **Landscape** to orient the drawing horizontally (*landscape*). The **Plot upside-down** option rotates the paper 180°.

5. Set the scale in the **Plot scale** area. Scale is measured as a ratio of either inches or millimeters to drawing units. Select a predefined scale from the **Scale:** drop-down list or enter values into the custom fields. Choose the **Fit to paper** check box to let AutoCAD automatically increase or decrease the plot area to fill the paper.

6. If desired, use the **Plot offset (origin set to printable area)** area to set additional left and bottom margins around the plot or to center the plot.

7. Pick the **Preview...** button to display the sheet as it will look when it is plotted. See **Figure 5-30.** The cursor appears as a magnifying glass with + and – symbols. Hold the left mouse button and move the cursor to increase or decrease the displayed image to view more or less detail. Press [Esc] to exit the preview.

8. Pick the **OK** button in the **Plot** dialog box to send the data to the plotting device.

portrait: A vertical paper orientation.

landscape: A horizontal paper orientation.

Figure 5-30.
A preview of the plot shows exactly how the drawing will appear on the paper.

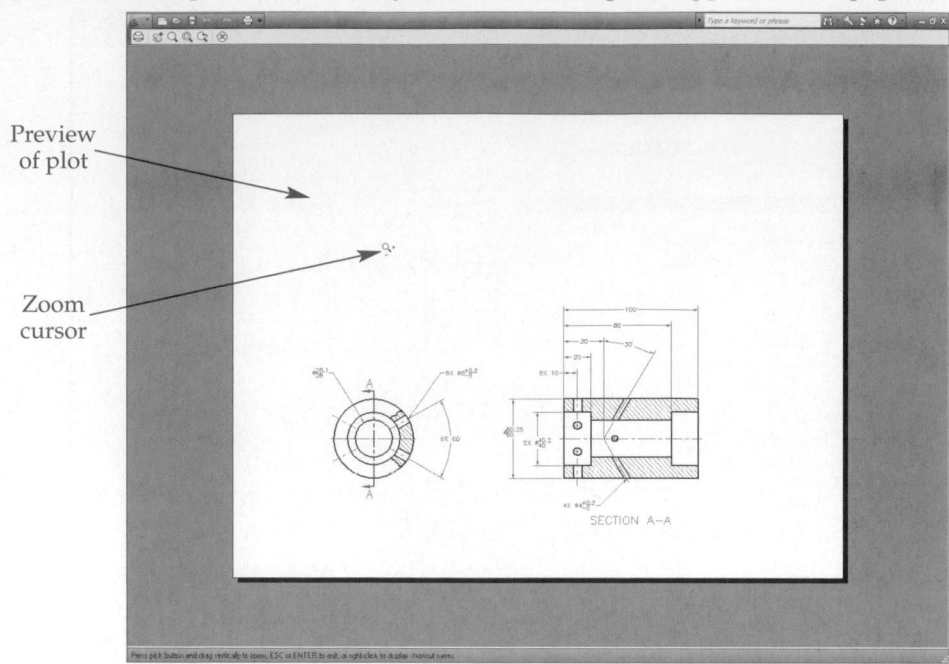

Preview
of plot

Zoom
cursor

Exercise 5-9

Access the Student Web site (www.g-wlearning.com/CAD) and
complete Exercise 5-9.

Template Development
Chapter 5

For detailed instructions on adding layers to each of your drawing
templates, go to the Student Web site (www.g-wlearning.com/CAD),
select this chapter, and select **Template Development**.

Chapter Test

Answer the following questions. Write your answers on a separate sheet of paper or go to the Student Web site (www.g-wlearning.com/CAD), and complete the electronic chapter test.

1. Identify the following linetypes:

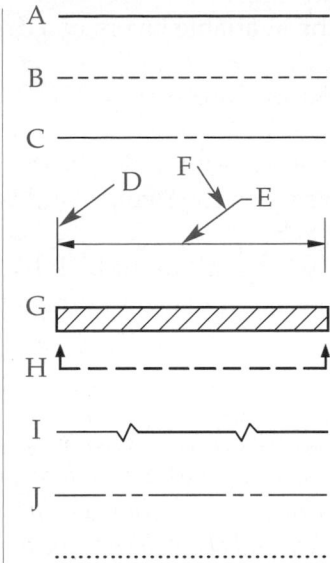

2. Identify two ways to access the **Layer Properties Manager**.
3. How can you tell if a layer is off, thawed, or unlocked by looking at the **Layer Properties Manager**?
4. Should you draw on layer 0? Explain.
5. How can several new layer names be entered consecutively without using the **New Layer** button in the **Layer Properties Manager**?
6. How do you make another layer current in the **Layer Properties Manager**?
7. How do you make another layer current using the ribbon?
8. How can you display the **Select Color** dialog box from the **Layer Properties Manager**?
9. List the seven standard color names and numbers.
10. How do you change a layer's linetype in the **Layer Properties Manager**?
11. What is the default linetype in AutoCAD?
12. What condition must exist before a linetype can be used in a layer?
13. Describe the basic procedure to change a layer's linetype to HIDDEN.
14. What is the function of the linetype scale?
15. Explain the effects of using a global linetype scale.
16. Why do you have to be careful when changing linetype scales?
17. What is the state of a layer *not* displayed on the screen and *not* calculated by the computer when the drawing is regenerated?
18. Explain the purpose of locking a layer.
19. Identify the following layer status icons:

A.   D.

B.   E.

C.   F.

20. Identify at least three layers that cannot be deleted from a drawing.
21. Describe the purpose of layer filters.
22. Name the two basic types of filters.
23. Which button in the **Layer Properties Manager** allows you to save layer settings so they can be restored later?
24. In the tree view area of **DesignCenter**, how do you view the content categories of one of the listed open drawings?
25. How do you display all the available layers in a drawing using the **DesignCenter** preview pane?
26. Briefly explain how drag-and-drop works.
27. Define *hard copy* and *soft copy*.
28. Identify four ways to access the **Plot** dialog box.
29. Describe the difference between the **Display** and **Window** options in the **Plot area** section of the **Plot** dialog box.
30. Explain how to examine what a plot will look like before you actually print the drawing. What is the major advantage of doing a plot preview?

Drawing Problems

Start AutoCAD if it is not already started. Start a new drawing using an appropriate template of your choice. The template should include layers for drawing the given objects. Add layers as needed. Draw all objects using appropriate layers. Follow the specific instructions for each problem. Do not draw dimensions or text. Use your own judgment and approximate dimensions when necessary.

▼ Basic

1. Draw the hex head bolt pattern shown below. Save the drawing as P5-1.

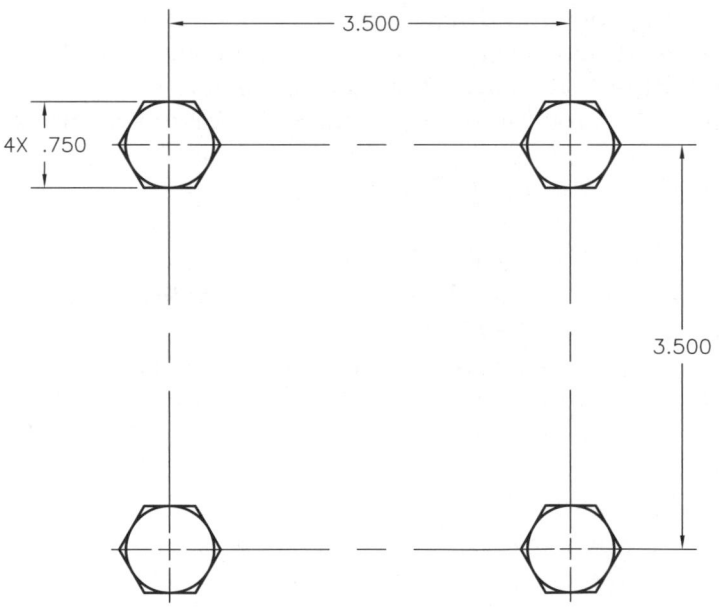

2. Create a 1/2" hex nut with 3/4" across the flats and a .422" root diameter as shown. Save the drawing as P5-2.

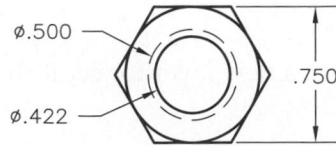

3. Draw the part shown below. Save the drawing as P5-3.

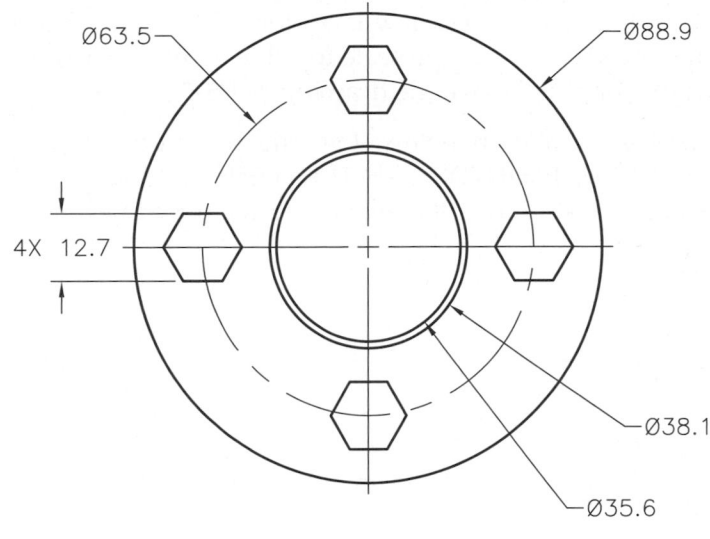

4. Draw the part shown below. Save the drawing as P5-4.

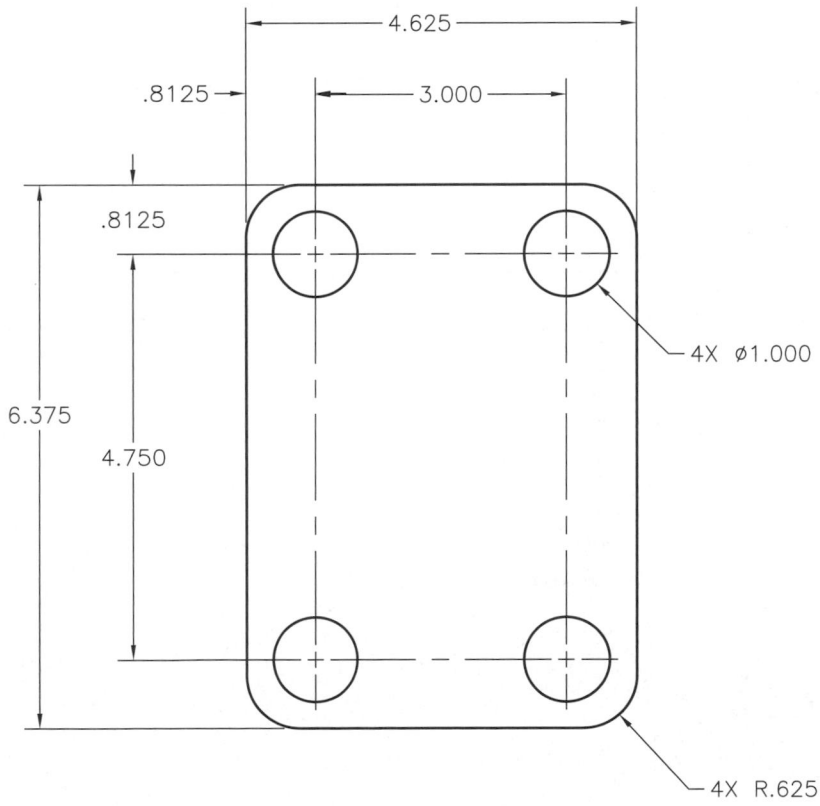

▼ Intermediate

5. Open P4-3, create or import a new layer for centerlines, and draw the centerlines. Save the drawing as P5-5.

6. Open P4-7, create or import a new layer for centerlines, and draw the centerlines. Save the drawing as P5-6.

▼ Advanced

7. Open P4-19, create or import a new layer for centerlines, and insert the centerlines. Change the global linetype scale to achieve an effect similar to the centerlines shown in Chapter 4. Save the drawing as P5-7.

8. Draw the plot plan shown below. Use the linetypes shown, which include Continuous, HIDDEN, PHANTOM, CENTER, FENCELINE2, and GAS_LINE. Make your drawing proportional to the example. Save the drawing as P5-8.

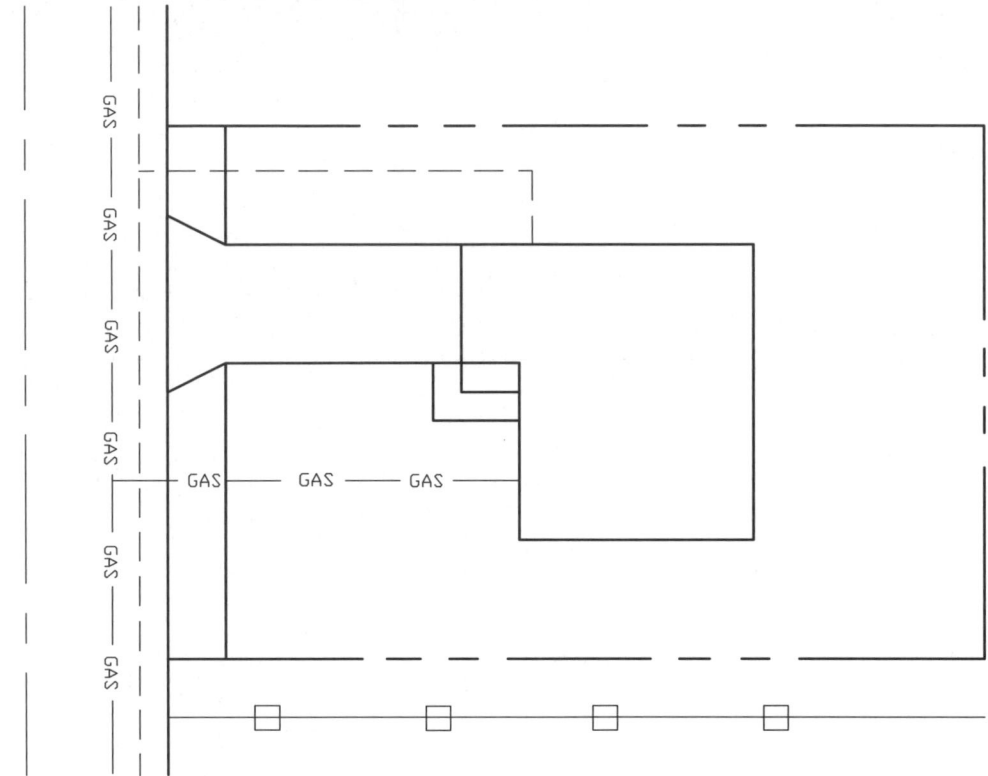

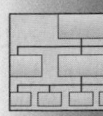

9. Draw the line chart shown below. Use the linetypes shown, which include Continuous, HIDDEN, PHANTOM, CENTER, FENCELINE1, and FENCELINE2. Make your drawing proportional to the given example. Save the drawing as P5-9.

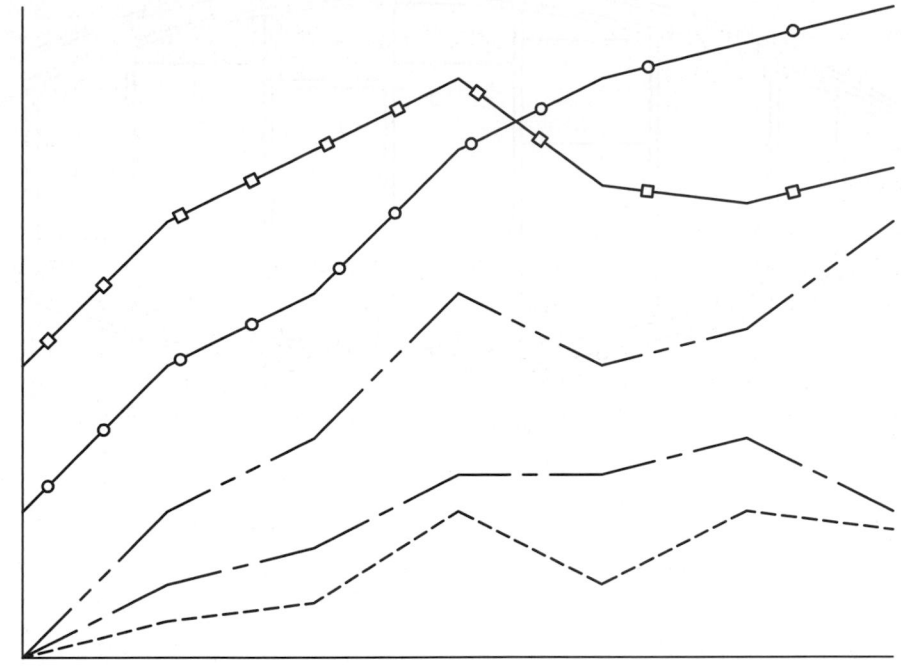

10. Create the controller integrated circuit diagram. Use a ruler or scale to keep the proportion as close as possible. Save the drawing as P5-10.

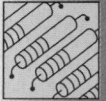

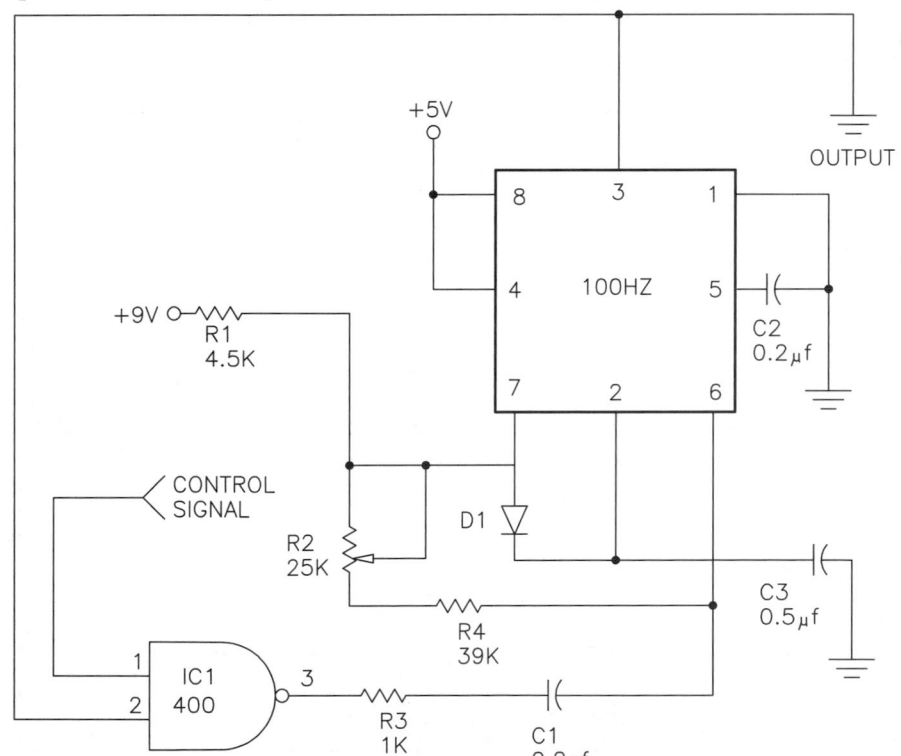

11. Draw a drift boat similar to the one shown below. Estimate dimensions. Save the drawing as P5-11.

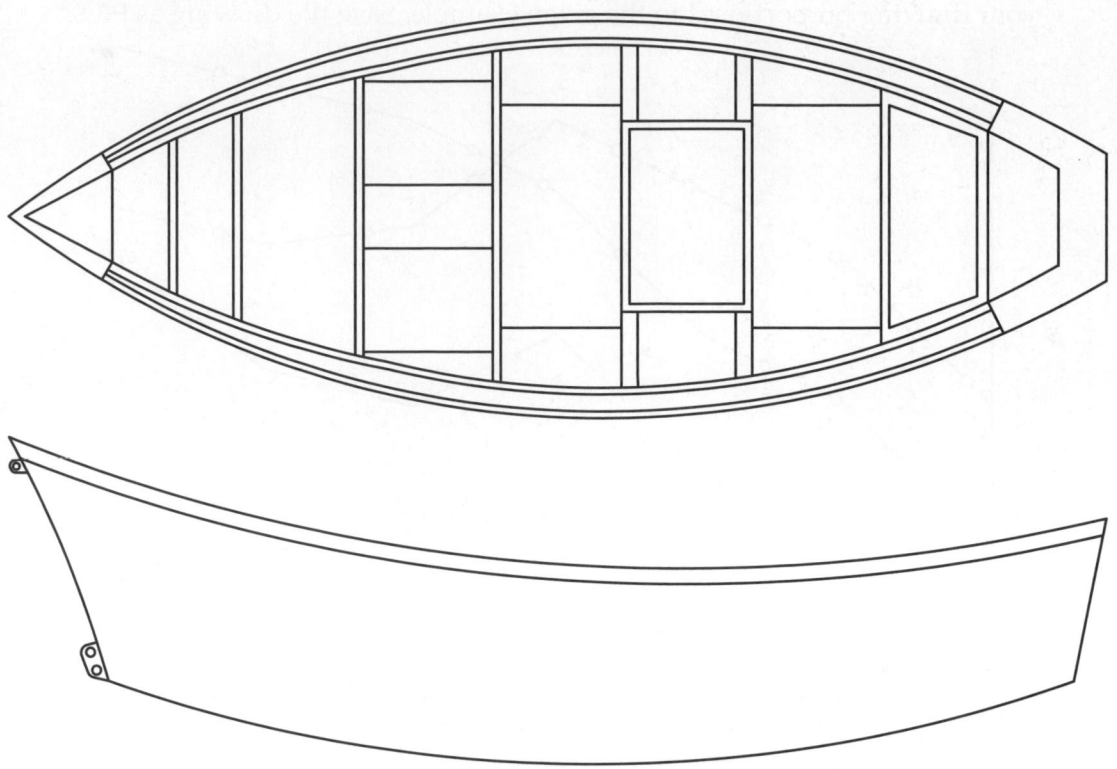

12. Draw the fishing boat shown. Save the drawing as P5-12.

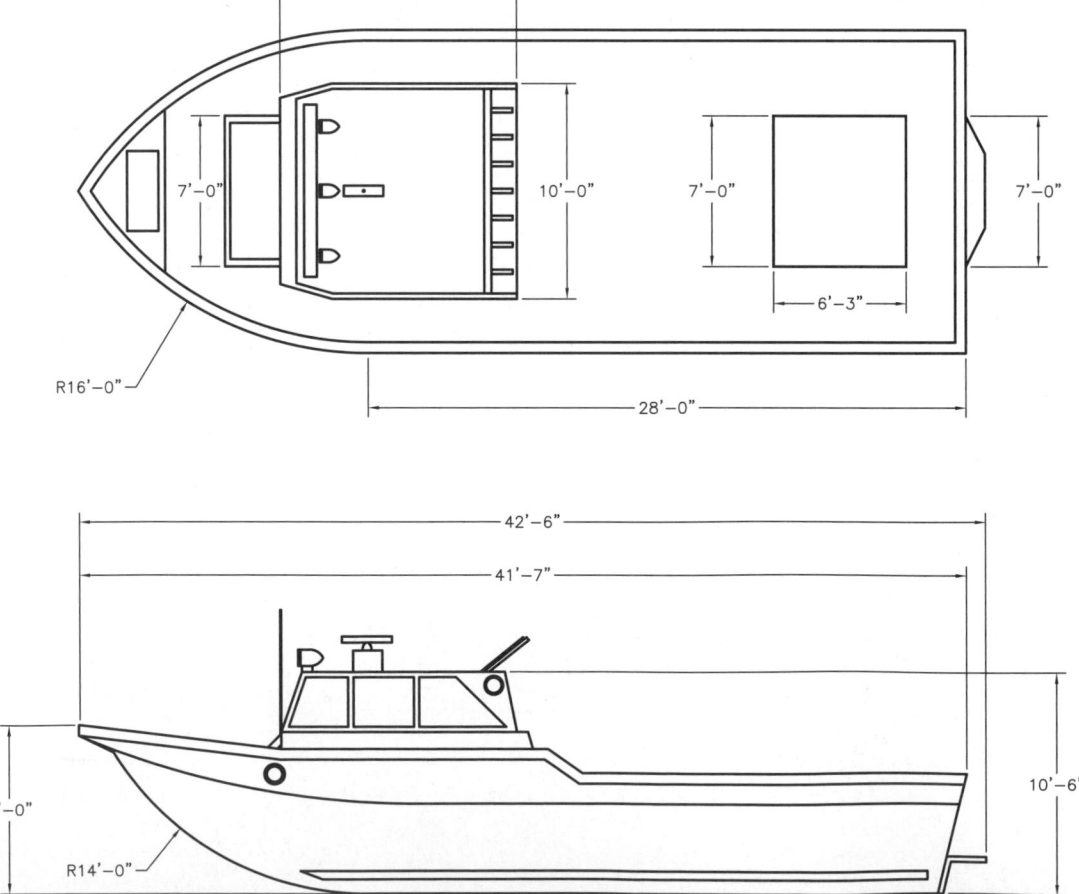

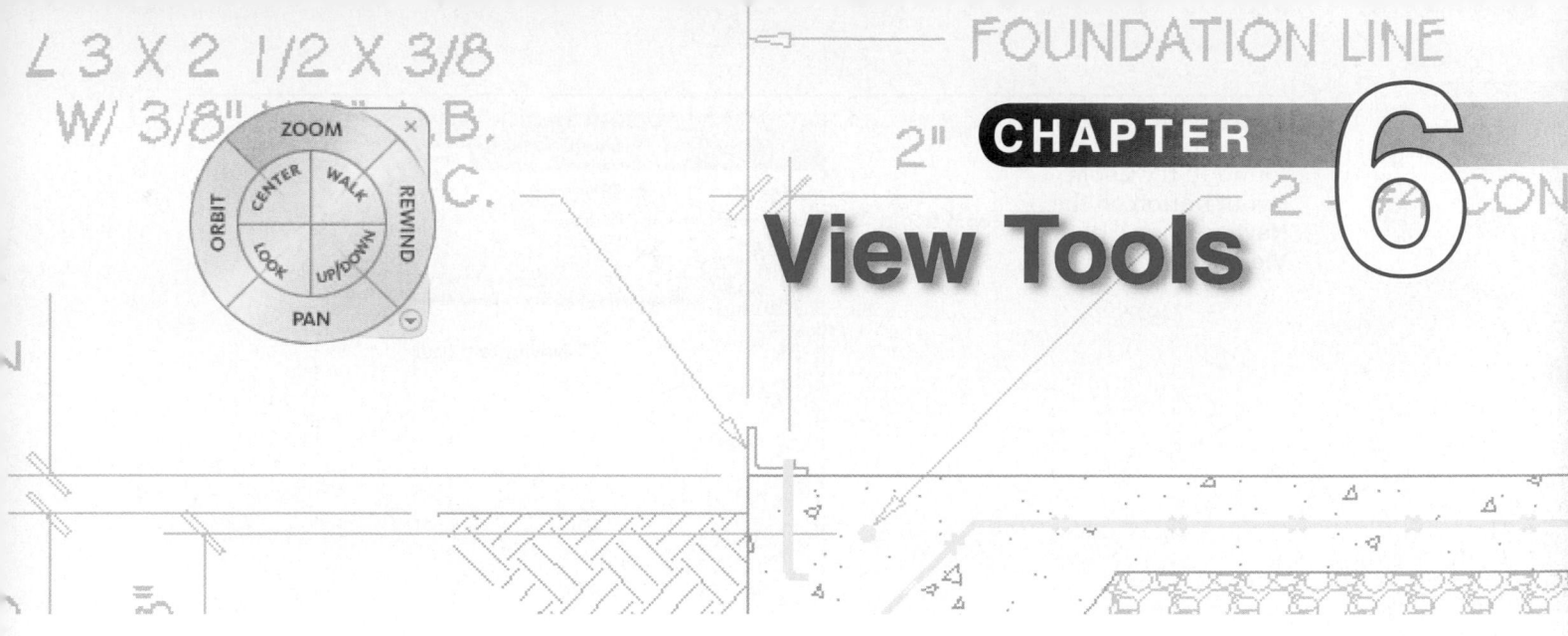

Learning Objectives

After completing this chapter, you will be able to do the following:

✓ Increase and decrease the displayed size of objects.
✓ Adjust the display window to view other portions of a drawing.
✓ Use SteeringWheels for 2D applications.
✓ Use transparent display tools and control display order.
✓ Create named views that can be recalled instantly.
✓ Create multiple viewports in the drawing window.
✓ Explain the difference between redrawing and regenerating the display.
✓ Use the **Clean Screen** tool.

View tools allow you to observe and work more efficiently with a specific portion of a drawing. As you create drawings that are more complex and draw large and small objects, you will realize the importance of adjusting the drawing display. This chapter describes a variety of view tools that you will use frequently during the drawing process.

zooming: Making objects appear bigger (zoom in) or smaller (zoom out) on the screen without affecting their actual sizes.

Zooming

The **ZOOM** tool provides several methods for *zooming*. Choose the appropriate zoom option based on the portion of the drawing you want to display and whether you want to *zoom in* or *zoom out*. This chapter focuses on the ribbon as the primary means of accessing **ZOOM** tool options. See **Figure 6-1.** When you select a zoom option from the ribbon, all prompts are specific to the selected option. When you use dynamic input or the command line to access zoom tools, you need to enter specific options when prompted.

zoom in: Change the display area to show a smaller part of the drawing at a higher magnification.

zoom out: Change the display area to show a larger part of the drawing at a lower magnification.

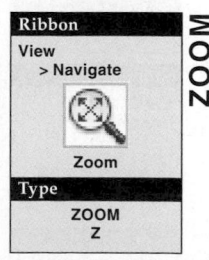

Ribbon
View
> Navigate
Zoom
Type
ZOOM
Z

ZOOM

NOTE

You can also activate zoom tools from various shortcut menus, or by picking the **Zoom** button on the status bar.

Figure 6-1.
ZOOM tool options found in the **Zoom** flyout button on the **Navigate** panel of the **View** ribbon tab.

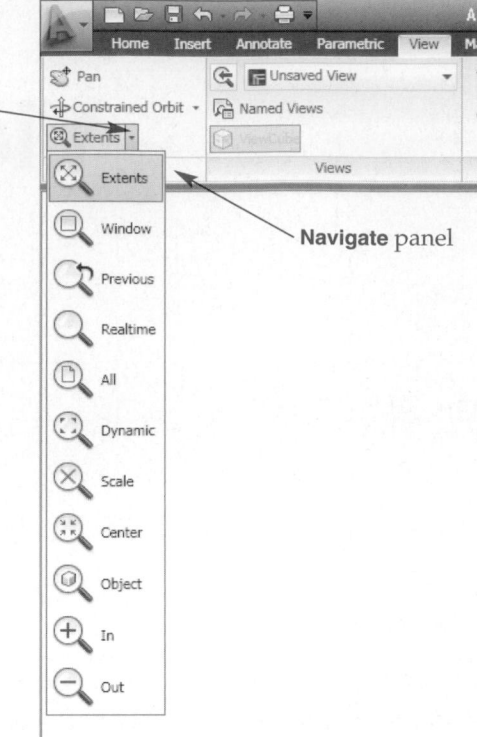

Realtime Zooming

realtime zoom:
A zoom that can be viewed as it is performed.

When you access the **Realtime** zoom option, known as *realtime zooming*, the zoom cursor appears as a magnifying glass with a plus and minus. Press and hold the left mouse button and move the cursor up to zoom in and down to zoom out. When you achieve the appropriate display, release the mouse button. Repeat the process to make further adjustments. Right-click while the zoom cursor is active to display a shortcut menu with several view options. This is a quick way to access alternative zoom options and related view options. **Figure 6-2** briefly describes each view option. This chapter further explains most of these options when applicable. When finished zooming, press [Esc], [Enter], or the space bar, or right-click and pick **Exit**.

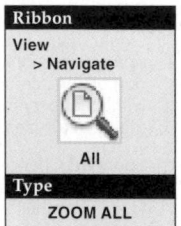

Ribbon
View
> Navigate

All

Type
ZOOM ALL

PROFESSIONAL TIP

AutoCAD supports most mice that have a scroll wheel between the two mouse buttons. Roll the wheel forward (away from you) to zoom in. Roll the wheel backward (toward you) to zoom out. This function also pans to the location of the crosshairs while zooming.

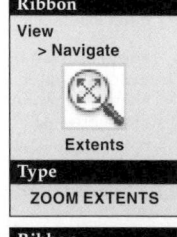

Ribbon
View
> Navigate

Extents

Type
ZOOM EXTENTS

Ribbon
View
> Navigate

Object

Type
ZOOM OBJECT

Additional Zoom Options

Depending on your drawing task, you may choose to use one or more of the additional zoom options instead of the **Realtime** option or the navigation wheels described later in this chapter. The **All** option zooms to the edges of the drawing limits. If objects are drawn beyond the limits, the **All** option zooms to the edges of your geometry. Always use this option after you change the drawing limits.

The **Extents** option zooms to the extents, or edges, of objects in a drawing. If you have a mouse with a scroll wheel, double-click the wheel to zoom to the drawing extents. The **Object** option allows you to select an object or set of objects. The selection is zoomed and centered to fill the display area.

Figure 6-2.
Options on the shortcut menu that are displayed when you right-click while the zoom cursor is active.

Selection	Cursor	Option	Function
Pan	🖐	Pan Realtime	Adjust the placement of the drawing on the screen.
Zoom	🔍+ −	Zoom Realtime	Toggle between **Pan** and **Zoom Realtime** to adjust the view.
3D Orbit	🔄	3D Orbit	Move around a 3D object. 3D navigation tools are described in *AutoCAD and Its Applications—Advanced*.
Zoom Window	▯	Zoom Window	Unlike the typical zoom window, described later in this chapter, this option requires you to press and hold the pick button while dragging the window box to the opposite corner, then release the pick button.
Zoom Original	(No cursor displayed)	(No tool name)	Restores the previous display before any realtime zooming or panning occurred; useful if the modified display is not appropriate.
Zoom Extents	(No cursor displayed)	Zoom Extents	Zoom to the extents of the drawing geometry.

The **Window** option allows you to pick opposite corners of a box. Objects in the box enlarge to fill the display. The **Window** option is the default if you pick a point on the screen after entering the **ZOOM** tool at the keyboard.

The **Scale** option allows you to zoom in or out according to a specific magnification scale factor. The **nX** option scales the display relative to the current display. The **nXP** option scales a drawing in model space relative to paper space, as described later in this textbook.

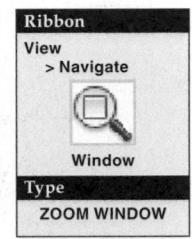

Ribbon
View
> Navigate

Window

Type
ZOOM WINDOW

NOTE

This textbook does not describe the **Previous**, **In**, **Out**, **Center**, and **Dynamic** options of the **ZOOM** tool. These options provide functions that you can achieve more easily using other view tools.

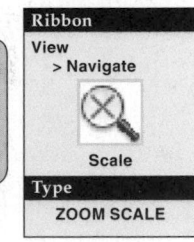

Ribbon
View
> Navigate

Scale

Type
ZOOM SCALE

Exercise 6-1

Access the Student Web site (www.g-wlearning.com/CAD) and complete Exercise 6-1.

PAN

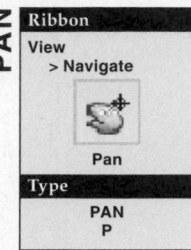

Ribbon

View
> Navigate

Pan

Type

PAN
P

panning: Changing the drawing display so that different portions of the drawing are visible on-screen.

realtime panning: A panning operation in which you can see the drawing move on the screen as you pan.

Panning

The **PAN** tool allows you to *pan*. Panning is usually required while drawing, especially while you are creating large objects, and while you are zoomed in on a part of the drawing to view fine detail. You often use the **PAN** and **ZOOM** tools together to change the display.

NOTE

You can also activate the **PAN** tool from various shortcut menus, or by picking the **Pan** button on the status bar.

When you first access the **PAN** tool, you are using the **Realtime** option, known as *realtime panning*. The **Realtime** option is the quickest and easiest method of adjusting the on-screen display. After starting the tool, press and hold the left mouse button and move the pan cursor in the direction to pan. A right-click displays the same shortcut menu available for realtime zooming. To exit realtime panning, press [Esc], [Enter], or the space bar, or right-click and pick **Exit**.

PROFESSIONAL TIP

If you have a mouse with a scroll wheel, press and hold the wheel button and move the mouse to perform a realtime pan.

NOTE

An alternative realtime panning method involves using drawing window scroll bars. To display the scroll bars, select the **Display scroll bars in drawing window** check box in the **Window Elements** area of the **Display** tab in the **Options** dialog box. Realtime panning is more efficient than using the scroll bars.

NOTE

By default, when you use the **U** tool after multiple zooming and panning operations, zooms and pans are grouped together, allowing you to return to the original view. To make each zoom and pan operation count individually, deselect the **Combine zoom and pan commands** check box in the **Undo/Redo** area of the **User Preferences** tab in the **Options** dialog box.

Exercise 6-2

Access the Student Web site (www.g-wlearning.com/CAD) and complete Exercise 6-2.

Introduction to SteeringWheels

AutoCAD SteeringWheels provide an alternative means of accessing and using certain view tools. Individual SteeringWheels are known as *navigation wheels*. Some navigation wheels and many of the tools available from navigation wheels are more appropriate for preparing 3D models. The **ZOOM**, **CENTER**, **PAN**, and **REWIND** tools are effective for 2D drafting applications. All other **SteeringWheels** tools and options are covered in *AutoCAD and Its Applications—Advanced*.

> **NOTE**
>
> You can also activate SteeringWheels from various shortcut menus, or by picking the **SteeringWheels** button on the status bar.

When you access SteeringWheels in model space, the **Full Navigation Wheel** appears by default, and the UCS icon changes to a 3D display. In layout space, the **2D Navigation Wheel** appears. See **Figure 6-3.** Navigation wheels display next to the cursor and are divided into *wedges*. Each wedge houses a navigation tool, similar to a tool button. Hover over a wedge to highlight the wedge. You can pick certain wedges to activate a tool. Other wedges require that you hold down the left mouse button to use the tool.

A navigation wheel remains on-screen until closed. This allows you to use multiple navigation tools. To close a navigation wheel, pick the **Close** button in the upper-right corner of the wheel, press [Esc], [Enter], or the space bar, or right-click and pick **Close Wheel**.

wedges: The parts of a wheel that contain navigation tools.

Zooming with the Navigation Wheel

The **ZOOM** navigation tool offers realtime zooming. Press and hold the left mouse button on the **ZOOM** wedge to display the pivot point icon and zoom navigation cursor. See **Figure 6-4.** The pivot point is the location where you press the **ZOOM** wedge. Move the zoom navigation cursor up to zoom in and down to zoom out. The pivot point icon also zooms in or out as a visual aid to zooming. When you achieve the appropriate display, release the left mouse button.

Using the Center Navigation Tool

The **CENTER** navigation tool centers the display screen at a picked point, without zooming. Press and hold the left mouse button on the **CENTER** wedge. The pivot point

Figure 6-3.
The **Full Navigation Wheel** appears by default when you access **SteeringWheels** in model space. The **2D Navigation Wheel** appears in a layout.

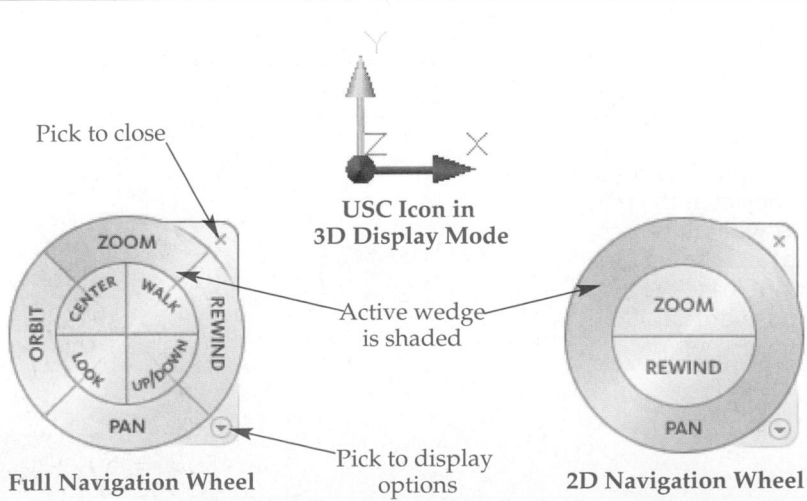

Pick to close

USC Icon in
3D Display Mode

Active wedge
is shaded

Pick to display
options

Full Navigation Wheel

2D Navigation Wheel

Figure 6-4.
To use the **ZOOM**
navigation tool,
move the cursor
up to zoom in and
down to zoom out.

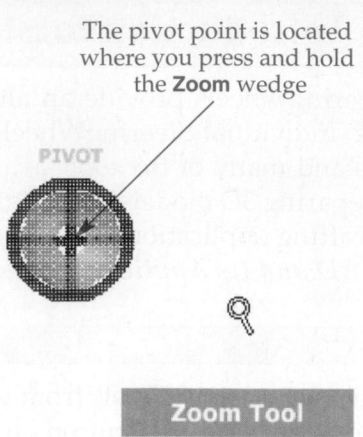

The pivot point is located
where you press and hold
the **Zoom** wedge

PIVOT

Zoom Tool

icon appears when you move the cursor over an object. Release the mouse button to pan so the location of the pivot point relocates to the center of the drawing window when you release the mouse button. See **Figure 6-5.**

Panning with the Navigation Wheel

The **PAN** navigation tool uses realtime panning to adjust the on-screen display. Press and hold the left mouse button on the **PAN** wedge to display the pan navigation cursor. Move the pan navigation cursor in the direction to pan. Release the left mouse button when you achieve the desired display.

Rewinding

The **REWIND** navigation tool allows you to observe the effects view tools have made on the drawing display, and return to a previous display. For example, if you use the navigation wheel to zoom in, then pan, then zoom out, you can rewind through each action and return to the original display, the zoomed-in view, the panned display, and then back to the current zoomed-out view. By default, you can rewind through view actions created using most view tools, not just those accessed from a navigation wheel.

Pick the **REWIND** tool once to return to the previous display. Thumbnail images appear in frames as the previous view restores. The orange-framed thumbnail surrounded by brackets indicates the restored display and its location in the sequence of events. See **Figure 6-6.** Repeatedly pick the **REWIND** button to cycle back through prior views. Another option is to press and hold the left mouse button on the **REWIND** wedge to display the framed view thumbnails. Then, while still holding the left mouse

Figure 6-5.
To use the **CENTER**
navigation tool,
move the cursor over
an object at the point
you want to center
in the drawing area
and release the left
mouse button.

Pivot icon appears only
when you move the
cursor over an object

This point pans to the
exact center of the
drawing window

PIVOT

Center Tool

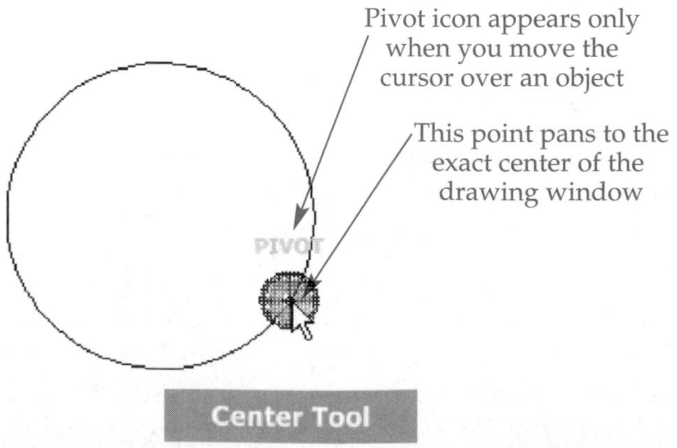

Figure 6-6.
Use the **REWIND** navigation tool to step back through and restore previous display configurations.

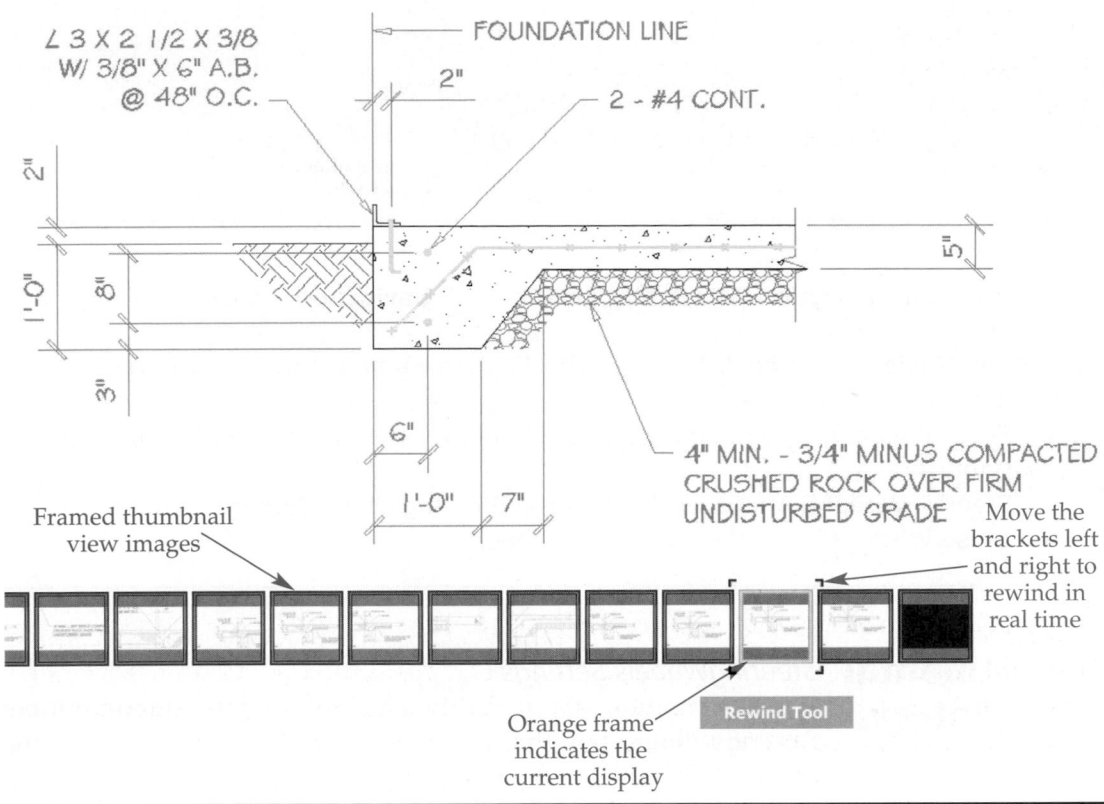

Framed thumbnail view images

4" MIN. - 3/4" MINUS COMPACTED CRUSHED ROCK OVER FIRM UNDISTURBED GRADE

Move the brackets left and right to rewind in real time

Orange frame indicates the current display

Rewind Tool

button, move the brackets left over the thumbnails to cycle through earlier views, and right to return to later views. Release the left mouse button when you achieve the desired display.

NOTE

By default, if you access and use a view tool, such as **ZOOM**, from a source other than the navigation wheel, such as the ribbon, a rewind icon appears in place of the thumbnail. A thumbnail displays as you move the brackets over the rewind icon.

Exercise 6-3

Access the Student Web site (www.g-wlearning.com/CAD) and complete Exercise 6-3.

SteeringWheel Options

A shortcut menu of options displays when you right-click while using a navigation wheel or when you pick the options button in the lower-right corner of a wheel. The options vary depending on the current work environment. Most of the options provide access to options and navigation wheels specifically for 3D modeling. The following options apply to 2D drafting applications:

Figure 6-7.
The **Full Navigation Wheel** in mini mode.

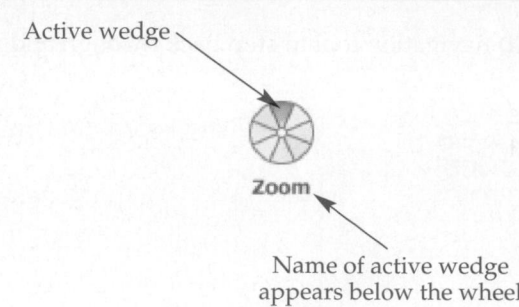

Active wedge

Zoom

Name of active wedge appears below the wheel

- **Mini Full Navigation Wheel.** Displays the **Full Navigation Wheel** in mini format. See **Figure 6-7.**
- **Full Navigation Wheel.** Displays the **Full Navigation Wheel** in the default big format.
- **Fit to Window.** Zooms and pans to show all objects centered in the drawing window.
- **SteeringWheel Settings….** Displays the **SteeringWheel Settings** dialog box.
- **Close Wheel.** Closes the navigation wheel.

Supplemental Material

SteeringWheels Settings

For information about options found in the **SteeringWheel Settings** dialog box that are specific to 2D drafting, go to the Student Web site (www.g-wlearning.com/CAD), select this chapter, and select **SteeringWheels Settings**.

Using Transparent Display Tools

transparently: When referring to tool access, a tool describes using while another tool is in progress.

Selecting a new tool usually cancels the tool in progress and then starts the new tool. However, you can use some tools *transparently*, temporarily interrupting the active tool. After the transparent tool has completed, the interrupted tool resumes. Therefore, it is not necessary to cancel the initial tool. You can use many display tools transparently, including **PAN**, **ZOOM**, and **SteeringWheels**.

An example of when transparent tools are useful is drawing a line when one end of the line is somewhere off the screen. One option is to cancel the **LINE** tool, zoom out to see more of the drawing, and select **LINE** again. A more efficient method is to use **PAN** or **ZOOM** transparently with the **LINE** tool. To do so, begin the **LINE** tool and pick the first point. At the Specify next point: prompt, access the **PAN** or **ZOOM** tool. You can use any access method, though right-clicking and selecting **Pan** or **Zoom**, or using the wheel mouse, is often quickest. Once you display the correct view, pick the second line endpoint. You can also activate tools transparently by typing an apostrophe (') before the tool name. For example, to enter the **ZOOM** tool transparently, type 'Z or 'ZOOM.

PROFESSIONAL TIP

You can use tools such as **Grid**, **Snap**, and **Ortho** transparently, but it is quicker to activate these modes with the appropriate button on the status bar or using a function key.

Controlling Draw Order

Drawings often include overlapping objects. The overlap is difficult to see when all objects use a thin lineweight and when lineweight display is off. Controlling display order is better illustrated with an object that has width, such as the donuts shown in **Figure 6-8.** In this example, the donuts were originally drawn after the other objects. You can change the drawing order of the donuts, and all other objects, to place the items above or below selected objects and to the front or back of all objects.

Use the **DRAWORDER** tool to change the order of objects in a drawing. You can also set draw order by picking an object to select it, right-clicking, and choosing **Draw Order. Figure 6-9** describes the options for changing draw order.

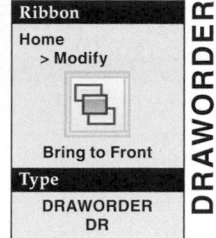

Ribbon

Home
> Modify

Bring to Front

Type

DRAWORDER
DR

DRAWORDER

NOTE

You may need to use the **DRAWORDER** tool on several objects until the objects display correctly. Objects move to the front of the drawing when they are modified.

PROFESSIONAL TIP

Use **DRAWORDER** to help display and select objects that are hidden by other objects.

Figure 6-8.
The order of objects can be changed to place any object under or above other objects.

Figure 6-9.
Four options are available for rearranging the order of objects in a drawing.

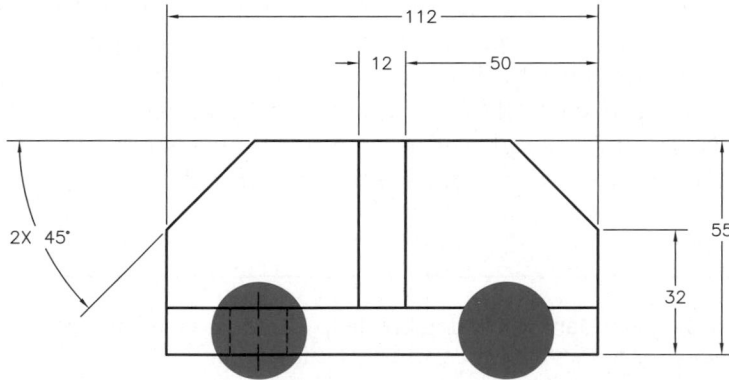

Option	Function
Bring to Front	Places the selected objects at the front of the drawing.
Send to Back	Places the selected objects at the back of the drawing.
Bring Above Objects	Moves the selected objects above the reference object.
Send Under Objects	Moves the selected objects below the reference object.

Exercise 6-4

Access the Student Web site (www.g-wlearning.com/CAD) and complete Exercise 6-4.

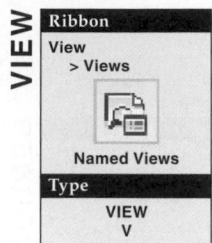
Creating Named Views

Use the **View Manager** to create and name specific views of the drawing. A view can be a portion of the drawing, such as the upper-left quadrant or a separate detail, or it can represent an enlarged area. When further drawing or editing operations are required, you can quickly and easily recall named views.

The left side of the **View Manager** contains a list of view types, or nodes. See **Figure 6-10.** You can expand each node, except **Current**, to reveal saved views. The **Current** node displays the properties of the current view. The **Model Views** node contains a list of saved model views. The **Layout Views** node contains a list of saved layout views. The **Preset Views** node lists all preset orthogonal and isometric views. Pick one of the view nodes to display information about the view type.

The right side of the **View Manager** contains buttons to control or modify the selected view or view type. These actions are also available in a shortcut menu when you right-click on the view or view type. Pick one of the view names to display information related to the current view in the middle area of the dialog box. Refer again to **Figure 6-10.** The first section, **General**, contains details such as the name of the view, layer settings saved with the view, and other specific view type settings. The **General** section is not visible when you select the **Current** node. The **Animation** section sets the properties for animated drawings and slide shows. The settings in the **View** section include camera position, target position, and perspective status. The **Clipping** section controls front plane and back plane location and the clipping status. Some of these items are covered in more detail in this chapter; others are reserved for later chapters or *AutoCAD and Its Applications—Advanced*, where the information is more relevant.

Figure 6-10.
The view nodes of the **View Manager** dialog box help organize saved and preset drawing views. Select a named view to see its properties and a preview image.

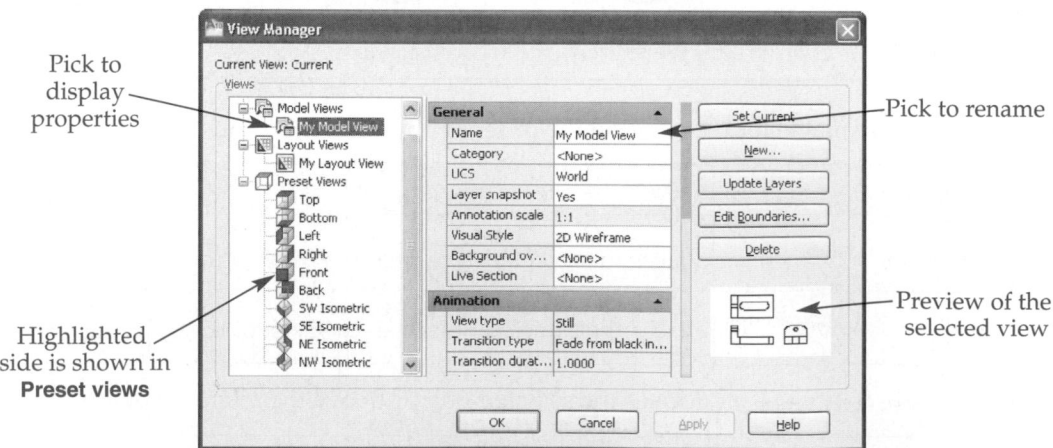

Pick to display properties

Highlighted side is shown in **Preset views**

Pick to rename

Preview of the selected view

New Views

To save the current display as a view, pick the **New...** button or right-click on a view node and select **New...** to access the **New View/Shot Properties** dialog box. See **Figure 6-11.** Type the view name in the **View name:** text box. If the named view is associated with a category in the **Sheet Set Manager**, you can select the category from the **View category** drop-down list. The **Sheet Set Manager** is described later in this textbook.

The **New View/Shot Properties** dialog box provides many options that are applicable to 3D modeling animations. To create a basic 2D view, select **Still** from the **View type** drop-down list, and focus on the settings in the **View Properties** tab. The **Current display** radio button is the default. Click **OK** to add the view name to the list. AutoCAD creates a view from the current display.

To use a window to define the view, pick the **Define window** radio button in the **New View** dialog box, and then pick the **Define view window** button. Pick two points to define a window. After you select the second corner, the **New View/Shot Properties** dialog box reappears. When you pick the **OK** button, the **View Manager** updates to reflect the new view.

Figure 6-11.
Use the **New View/Shot Properties** dialog box to save the current display as a view or define a window to create a view.

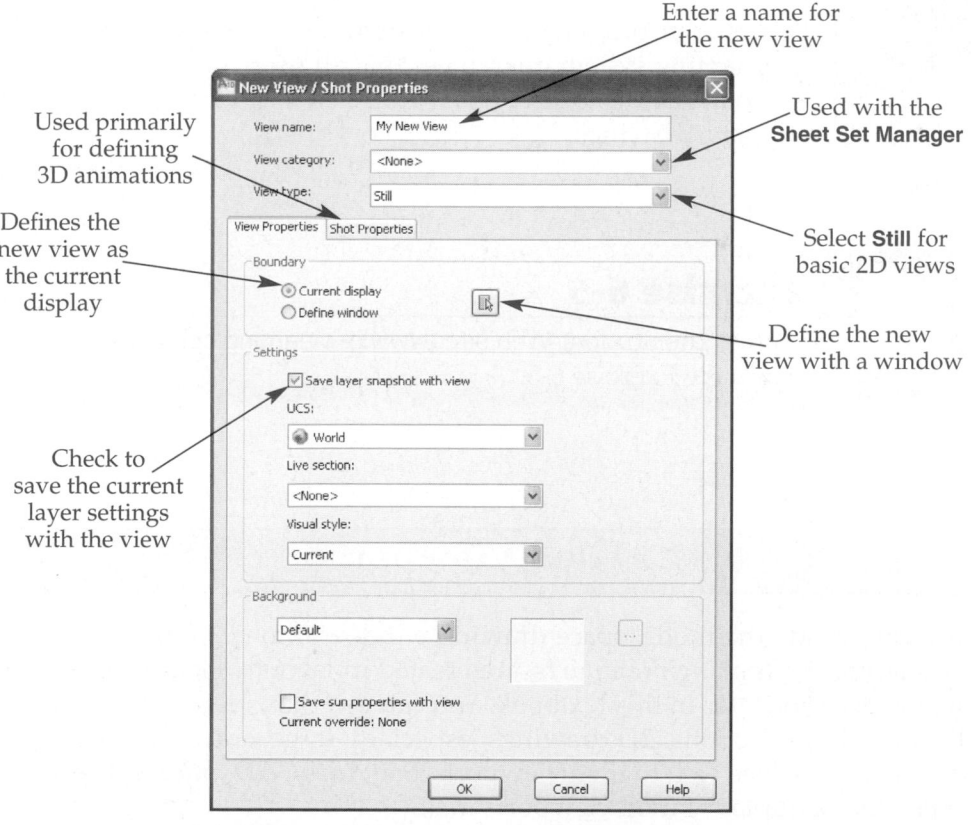

Select the **Save layer snapshot with view** check box to save the current layer settings when you save a new view. Saved layer settings are recalled each time the view is set current.

Activating Views

To display one of the listed views from inside the **View Manager**, pick the view name from the list in the **Views** area of the **View Manager** and pick the **Set Current** button. The name of the current view appears in the **Current View:** label above the **Views** area. Pick the **OK** button to display the selected view. To display a view without accessing the **View Manager**, select the name of the view from the list in the **Views** panel of the **View** ribbon tab.

Exercise 6-5

Access the Student Web site (www.g-wlearning.com/CAD) and complete Exercise 6-5.

Tiled Viewports

tiled viewports: Viewports created in model space.

viewport: The window or frame within which a drawing is visible.

floating viewports: Viewports created in paper space.

You can divide the model space drawing window into *tiled viewports*. Another type of *viewport, floating viewports*, are created in layouts. Floating viewports and layouts are describe later in this textbook. You can use tiled viewports in model space for 2D and 3D applications. 2D drawings, especially large drawings with significant detail, lend themselves well to tiled viewports. See *AutoCAD and Its Applications— Advanced* for examples of tiled viewports in 3D.

By default, the drawing window contains one viewport. Additional viewports divide the drawing window into separate tiles that butt against one another like floor tile. Tiled viewports cannot overlap. Multiple viewports contain different views of the same drawing, displayed at the same time. Only one viewport can be active at any given time. The active viewport has a bold outline around its edges. See **Figure 6-12.**

Creating Tiled Viewports

The **Viewports** dialog box provides one method for creating tiled viewports. **Figure 6-13** shows the **New Viewports** tab of the **Viewports** dialog box. The **Standard viewports:** list contains many preset viewport configurations. The configuration name identifies the number of viewports and the arrangement or location of the largest viewport. Select a configuration to see a preview of the tiled viewports in the **Preview** area on the right side of the **New Viewports** tab. Select *Active Model Configuration* to preview the current configuration. Pick the **OK** button to divide the drawing window into the selected viewport configuration.

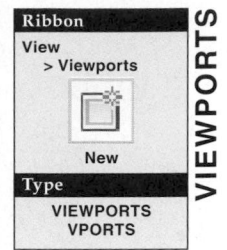

Ribbon
View
  > Viewports

New

Type
VIEWPORTS
VPORTS

VIEWPORTS

NOTE

You can activate the same preset viewport configurations available from the **New Viewports** tab of the **Viewport** dialog box using the **Set Viewports** drop-down list in the **Viewports** panel on the **View** ribbon tab. Prompts guide you through the process of creating the correct configuration. The arrangement you choose applies to the active viewport only.

The additional options in the **Viewports** dialog box are useful when two or more viewports already exist. The **Apply to:** drop-down list allows you to specify whether the viewport configuration applies to the entire drawing window or to the active viewport only. Select **Display** to apply the configuration to the entire drawing area.

Figure 6-12.
An example of three tiled viewports in model space. All of the viewports contain the same objects, but the display in each viewport can be unique.

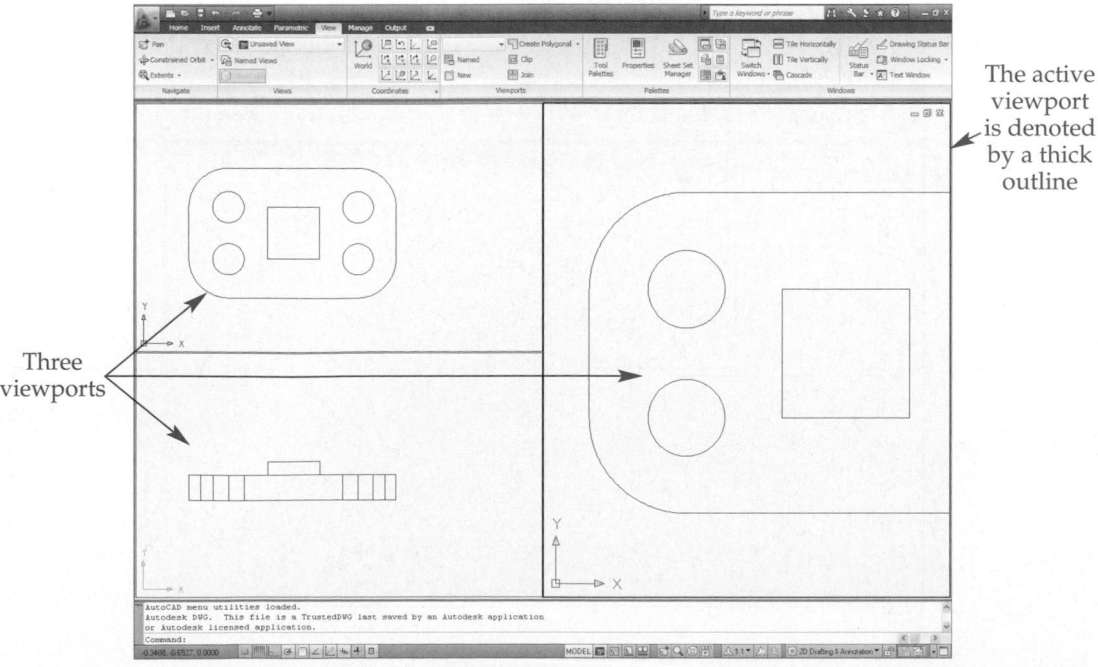

Three viewports

The active viewport is denoted by a thick outline

Figure 6-13.
Specify the number and arrangement of tiled viewports in the **New Viewports** tab of the
Viewports dialog box.

Enter a
name to
save

Previews
current
configuration

Preset
tiled viewport
configurations

Select if
configuration
is applied to
drawing area
or active
viewport

Preview
of selected
configuration

Pick 2D or 3D
views

Change
named view
of selected
viewport

Change the
visual style of
the selected
viewport

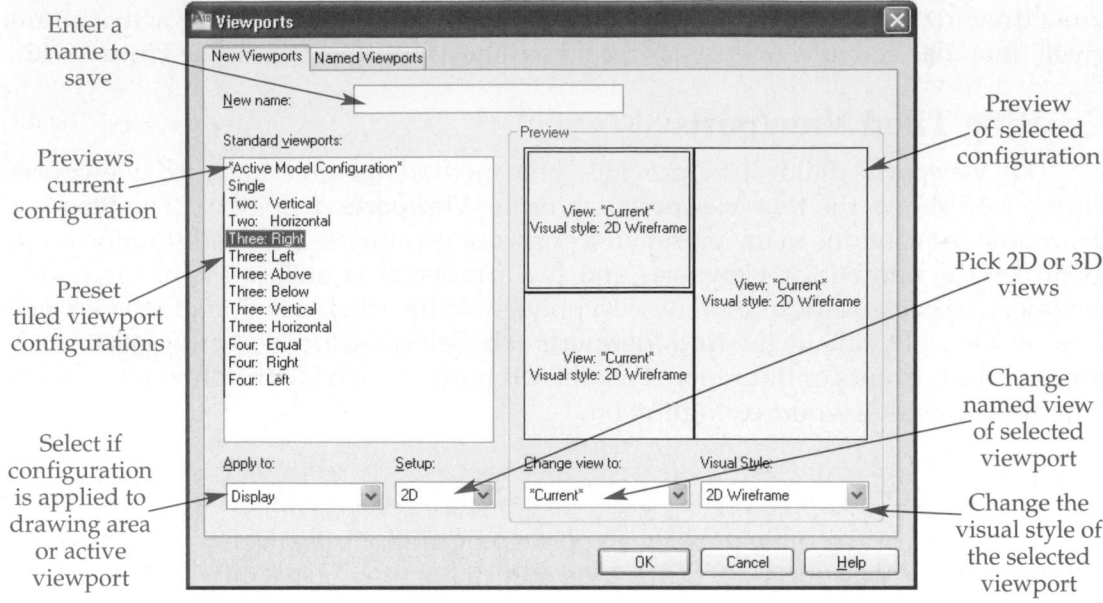

Select **Current Viewport** to apply the new configuration in the active viewport only. See
Figure 6-14.

The default setting in the **Setup:** drop-down list is **2D**. When you select the **2D**
option, all viewports show the top view of the drawing. If you choose the **3D** option,
the different viewports display various 3D views of the drawing. At least one viewport
is set up with an isometric view. The other viewports have different views, such as a
top view or side view. The viewport configuration displays in the **Preview** image. To

Figure 6-14.
You can subdivide a viewport by choosing **Current Viewport** in the **Apply to:** drop-down list.
Here, the top-left viewport is subdivided using the **Two: Vertical** preset configuration.

Configuration applied to
active viewport

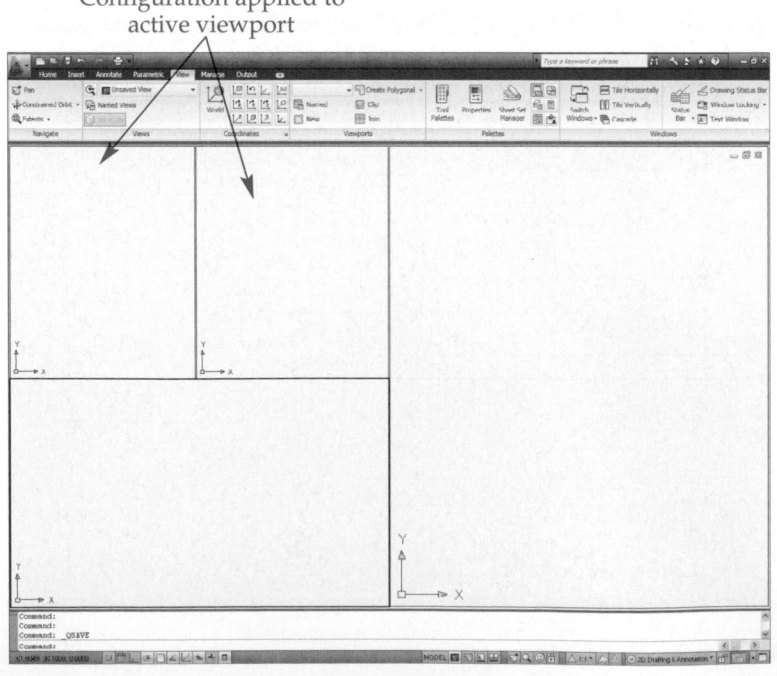

change a view in a viewport, pick the viewport in the **Preview** image and then select the new viewpoint from the **Change view to:** drop-down list.

If none of the preset configurations is appropriate, you can create and save a unique viewport configuration. After you create the custom viewport configuration, enter a descriptive name in the **New name:** text box. When you pick the **OK** button, the new named viewport configuration records and displays in the **Named Viewports** tab the next time you access the **Viewports** dialog box. See **Figure 6-15.** Select a different named viewport configuration and pick **OK** to apply it to the drawing area. Named viewport configurations can apply to the active viewport only.

> **NOTE**
>
>
> Apply the **1 Viewport** option to return the current viewport config-
> uration to the default single viewport.

Working in Tiled Viewports

After you select the viewport configuration and return to the drawing area, move the pointing device around and notice that only the active viewport contains cross-hairs. The cursor is an arrow in the other viewports. To make an inactive viewport active, move the cursor into the inactive viewport and pick.

As you draw in one viewport, the image appears in all viewports. Try drawing lines and other shapes and notice how the viewports are affected. Use a display tool, such as **ZOOM**, in the active viewport and notice the results. Only the active viewport reflects the use of the **ZOOM** tool.

Joining Tiled Viewports

Use the **Join Viewports** tool to join two viewports. When you select the **Join** tool, AutoCAD prompts you to select the dominant viewport. Select the viewport that has the view you want to display in the joined viewport, or press [Enter] to select the active viewport. Then select the viewport to join with the dominant viewport. AutoCAD "glues" the two viewports together and retains the dominant view.

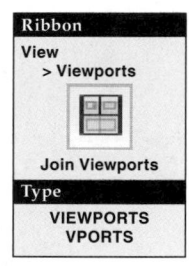

Ribbon
View
> Viewports

Join Viewports

Type
VIEWPORTS
VPORTS

Figure 6-15.
The **Named Viewports** tab displays custom viewports.

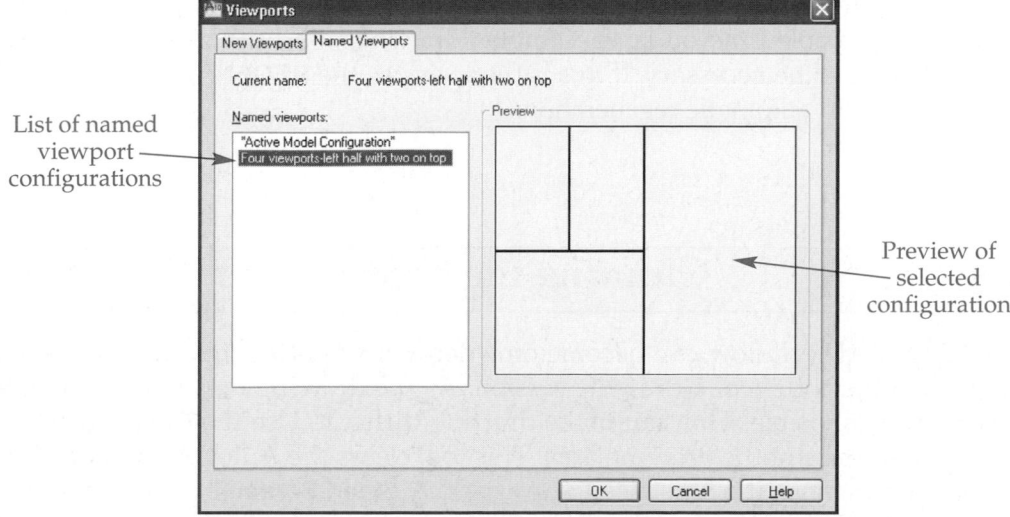

List of named viewport configurations

Preview of selected configuration

Exercise 6-6

Access the Student Web site (www.g-wlearning.com/CAD) and complete Exercise 6-6.

Redrawing and Regenerating the Screen

redrawing:
Refreshing the display of objects on the screen without recalculating the vectors.

regenerating:
Recalculating all objects based on the current zoom magnification and redisplaying them.

Redrawing refreshes the display of objects. *Regenerating* recalculates all object coordinates and displays them based on the current zoom magnification. For example, if curved objects appear as straight segments when you zoom in, you can regenerate the display to smooth the curves. Each viewport is a separate virtual screen. As a result, if you are using two or more viewports, you must decide whether to redraw or regenerate a single viewport or all the viewports.

Tool	Function
REDRAW	Redraws the display of the current viewport only.
REGEN	Regenerates the display in the current viewport only.
REDRAWALL	Redraws the display of the entire drawing.
REGENALL	Regenerates the display of the entire drawing.

PROFESSIONAL TIP

AutoCAD does an automatic regeneration when you use a tool that changes certain aspects of objects. This regeneration can take considerable time on large, complex drawings, and the regeneration may not be necessary. If this is the case, use the **REGENAUTO** tool to turn off automatic regenerations.

Cleaning the Screen

The AutoCAD window can become crowded with multiple interface items, such as palettes, in the course of a drawing session. As the drawing area gets smaller, less of the drawing is visible. This can make drafting difficult. Use the **Clean Screen** tool to maximize the size of the drawing area. This tool clears the AutoCAD window of all toolbars, palettes, and title bars. See **Figure 6-16.** A **Clean Screen** button is also available in the status bar. Accessing the **Clean Screen** tool toggles the clean screen display on and off.

Figure 6-16.
Using the **Clean Screen** tool. A—Initial display with the **Properties** palette and **Layer Properties Manager** displayed. B—Display after using the **Clean Screen** tool.

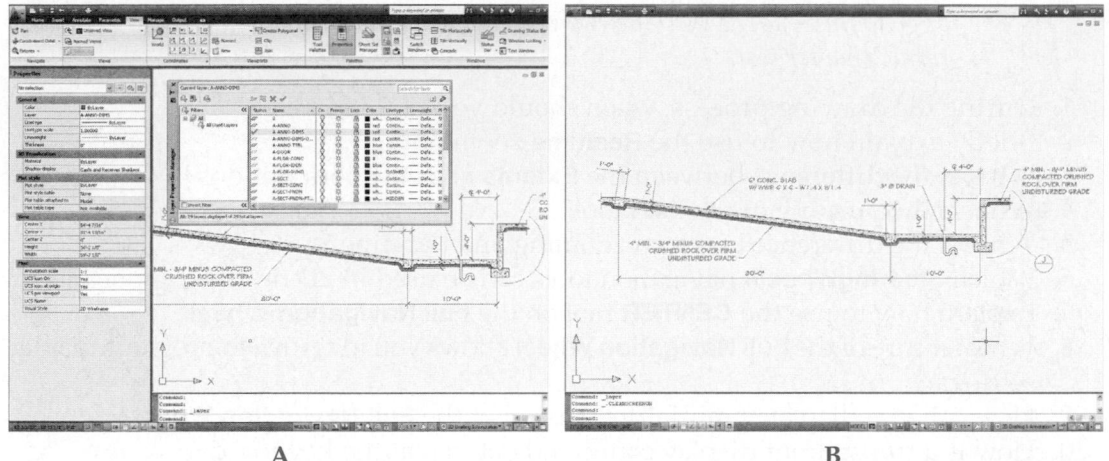

A   B

PROFESSIONAL TIP

The **Clean Screen** tool can be helpful when you have multiple drawings displayed. Only the active drawing appears when you use the **Clean Screen** tool. This allows you to work more efficiently within one of the drawings.

Supplemental Material

View Transitions and Resolution
For information about view transitions and view resolution, go to the Student Web site (www.g-wlearning.com/CAD), select this chapter, and select **View Transitions and Resolution**.

Template Development
Chapter 6

For detailed instructions on zooming to drawing limits in each template, go to the Student Web site (www.g-wlearning.com/CAD), select this chapter, and select **Template Development**.

Chapter Test

Answer the following questions. Write your answers on a separate sheet of paper or go to the Student Web site (www.g-wlearning.com/CAD) and complete the electronic chapter test.

1. During the drawing process, when should you use **ZOOM**?
2. Briefly explain how to use the **Realtime** zoom option.
3. What is the difference between the **Extents** and **All** zoom options?
4. What is the purpose of the **PAN** tool?
5. What is the difference between zooming and panning?
6. Which **SteeringWheels** navigation tools can be used in 2D drafting applications?
7. Explain how to use the **CENTER** tool on the **Full Navigation Wheel**.
8. What feature of the **Full Navigation Wheel** allows you to return to previous display settings?
9. How can you display a miniature version of the **Full Navigation Wheel**?
10. How is a transparent display command entered at the keyboard?
11. Name at least three display tools that can be used transparently.
12. Which tool changes the order in which objects are displayed in a drawing?
13. How can you obtain a list of existing views?
14. How do you create a named view of the current screen display?
15. How do you display an existing view?
16. What type of viewport is created in model space?
17. How can you specify whether a new viewport configuration applies to the entire drawing window or the active viewport?
18. Explain the procedures and conditions that need to exist for joining viewports.
19. What is the difference between the **REDRAW** and **REGEN** tools?
20. Which tool regenerates all of the viewports?

Drawing Problems

Start AutoCAD if it is not already started. Follow the specific instructions for each problem.

▼ Basic

1. Perform the following tasks:
    A. Draw a circle.
    B. Use realtime zooming to zoom in and out on the circle.
    C. Use realtime panning to pan the screen display.

2. Create a freehand sketch of the full-size **Full Navigation Wheel**. Label each of the wedges.

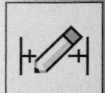

▼ Intermediate

3. Create a freehand sketch of the clean-screen AutoCAD window. Label each of the screen areas. To the side of the sketch, write a short description of each screen area's function.

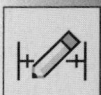

4. Open the drawing named 3D House.dwg found in the AutoCAD 2010\Sample folder.
    Perform the following display functions on the drawing:
    A. Zoom to the drawing extents.
    B. Create a view named Rendering.
    C. Replace the view with the Top view.
    D. Create a view named Plan using **Define Window** in the **New View** dialog box.
    E. Use realtime pan and realtime zoom to create a display of the dining room in the top-right area of the Plan view.
    F. Create a view of this display named Dining Room.
    G. Display the view named Rendering.
    H. Save the drawing as P6-4.

▼ Advanced

5. Use the **Select Template** dialog box to load the Tutorial-iMfg.dwt template or another mechanical template you have access to that contains a border and title block. Do the following:
    A. Zoom into the title block area. Create and save a view named Title.
    B. Zoom to the extents of the drawing and create and save a view named All.
    C. Determine the areas of the drawing that will contain notes, parts list, and revisions. Zoom into these areas and create views with appropriate names, such as Notes, Partlist, and Revisions.
    D. Divide the drawing area into commonly used multiview sections. Save the views with descriptive names such as Top, Front, Rightside, and Leftside.
    E. Restore the view named All.
    F. Save the drawing as P6-5.

6. Use the **Select Template** dialog box to load the Tutorial-iArch.dwt template or another architectural drafting template you have access to that contains a border and title block. Do the following:
   A. Zoom into the title block area. Create and save a view named Title.
   B. Zoom to the extents of the drawing and create and save a view named All.
   C. Determine the area of the drawing that will contain notes, schedules, or revisions. Zoom into these areas and create views with appropriate names such as Notes, Schedules, and Revisions.
   D. Restore the view named All.
   E. Save the drawing as P6-6.

Deferred Tangent
Specify next point or to ☐   4.0823 8.5220

Object Snaps and AutoTrack

Learning Objectives

After completing this chapter, you will be able to do the following:
- ✓ Set running object snap modes for continuous use.
- ✓ Use object snap overrides for single point selections.
- ✓ Select appropriate object snaps for various drawing tasks.
- ✓ Use AutoSnap™ features to speed up point specifications.
- ✓ Use AutoTrack™ to locate points relative to other points in a drawing.

This chapter explains how to use the powerful object snap and AutoTrack tools to draw accurate geometry. Object snap and AutoTrack tools are some of the most useful and efficient drawing aids in AutoCAD. Like coordinate entry, object snaps and AutoTrack allow you to produce accurate geometric constructions, but they do not constrain, or apply relationships between, objects. Chapter 22 explains how to use parametric tools to constrain objects.

Object Snap

Object snap increases your drafting performance and accuracy through the concept of *snapping*. Object snap *modes* identify the object snap point. The AutoSnap feature is on by default and displays snap mode information while you draw. AutoSnap uses visual signals that appear as *markers* displayed at the snap point. **Figure 7-1** shows two examples of visual AutoSnap cues. After a brief pause, a tooltip appears, indicating the object snap mode. **Figure 7-2** describes each object snap mode. Refer to **Figure 7-2** constantly as you learn to use object snaps.

object snap: A tool that locates exact points, such as endpoints or midpoints, on or in relation to existing objects.

snapping: Picking a point near the intended position to have the crosshairs "snap" exactly to the specific point.

markers: Visual cues to confirm object snap points.

NOTE

If you cannot see an AutoSnap marker because of the size of the current screen display, you can still confirm the point before picking by reading the tooltip, which indicates if a point acquires beyond the visible area.

Figure 7-1.
AutoSnap displays markers and related tooltips for object snap modes.

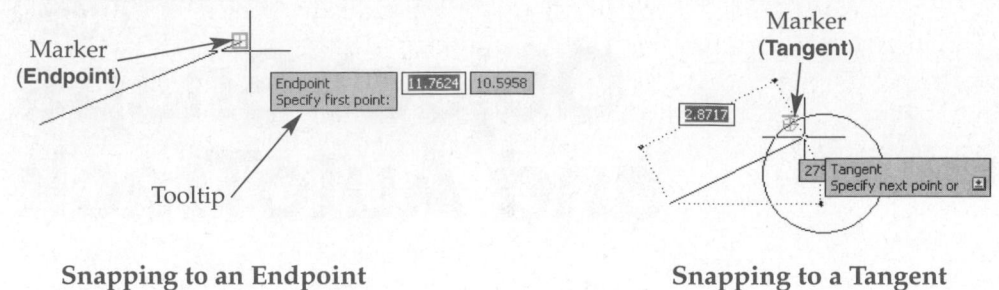

Snapping to an Endpoint   Snapping to a Tangent

Figure 7-2.
The object snap modes.

Mode	Marker	Description
Endpoint	□	Locates the nearest endpoint of a line, arc, polyline, elliptical arc, spline, ellipse, ray, solid, or multiline.
Midpoint	△	Finds the middle point of any object having two endpoints, such as a line, polyline, arc, elliptical arc, polyline arc, spline, xline, or multiline.
Center	○	Finds the center point of radial objects, including circles, arcs, ellipses, elliptical arcs, and radial solids.
Node	⊗	Locates a point object drawn with the **POINT**, **DIVIDE**, or **MEASURE** tool, or a dimension definition point.
Quadrant	◇	Locates the closest of the four quadrant points on circles, arcs, elliptical arcs, ellipses, and radial solids. (Some of these objects may not have all four quadrants.)
Intersection	✕	Locates the closest intersection of two objects.
Extension	+	Finds a point along the imaginary extension of an existing line, polyline, arc, polyline arc, elliptical arc, spline, ray, xline, solid, or multiline.
Insertion	⌐	Locates the insertion point of text objects and blocks.
Perpendicular	⌐	Finds a point that is perpendicular to an object from the previously picked point.
Tangent	○	Finds points of tangency between radial and linear objects.
Nearest	⊠	Locates the point on an object that is closest to the crosshairs.
Apparent Intersection	⊠	Locates the intersection between two objects that appear to intersect on-screen in the current view, but may not actually intersect in 3D space. Creating and editing 3D objects is described in *AutoCAD and Its Applications—Advanced*.
Parallel	⫽	Finds any point along an imaginary line parallel to an existing line or polyline.
None		Temporarily turns running object snap off during the current selection.

You can use object snaps for many drawing and editing applications. Practice with the different object snap modes to find the ones that work best for specific situations. With practice, object snap use becomes second nature, greatly increasing your productivity and accuracy.

Running Object Snaps

Running object snaps are on by default and are often the quickest and most effective way to use object snap. The **Endpoint**, **Center**, **Intersection**, and **Extension** running object snap modes are active by default. The quickest way to activate or deactivate running object snap modes is to right-click on the **Object Snap** or **Object Snap Tracking** button on the status bar and select the running object snaps to turn on or off.

You can also set running object snap modes using the **Object Snap** tab in the **Drafting Settings** dialog box. See **Figure 7-3.** To display the **Drafting Settings** dialog box, right-click on any of the status bar toggle buttons and selecting **Settings...**. Pick individual check boxes and use the **Select All** and **Clear All** buttons to help choose only those object snaps you want to run.

To use running object snaps, move the crosshairs near the location on an existing object where the object snap is to occur. When you see the appropriate marker, and if necessary, the tooltip, pick to locate the point at the exact position on the existing object. For example, with the **Endpoint** running object snap on, move the crosshairs toward the end of a curve to display the endpoint marker. Pick to locate the point at the exact endpoint of the existing object. See **Figure 7-4.**

Toggle running object snaps off and on by picking the **Object Snap** button on the status bar, pressing [F3], or using the **Object Snap On (F3)** check box on the **Object Snap** tab of the **Drafting Settings** dialog box. Turn off running object snaps to locate points without the aid, or to avoid possible confusion of object snap modes. The selected running object snap modes restore when you turn running object snaps back on.

running object snaps: Object snap modes that run in the background during all drawing and editing procedures.

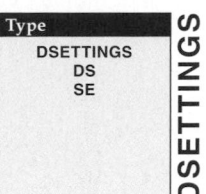

Type
DSETTINGS
DS
SE

DSETTINGS

Figure 7-3.
Running object snap modes can also be set in the **Drafting Settings** dialog box.

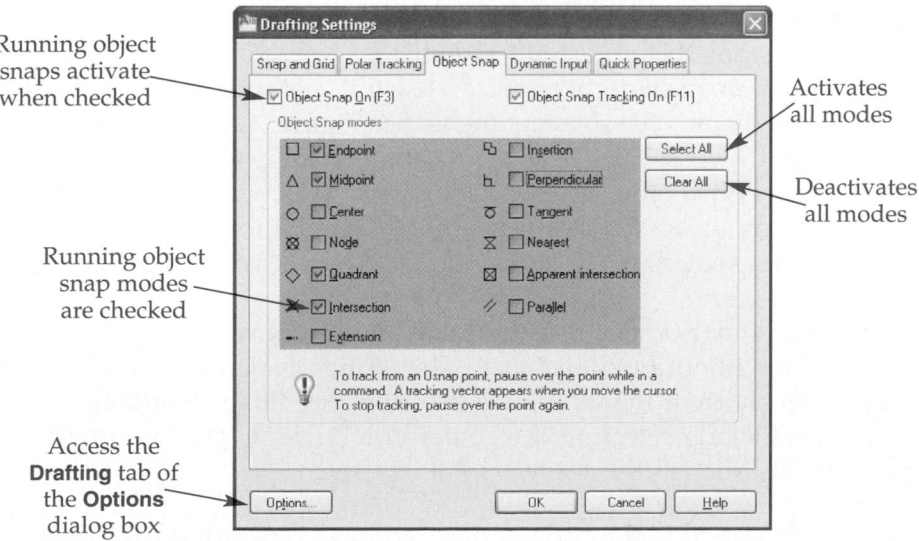

Running object snaps activate when checked

Activates all modes

Running object snap modes are checked

Deactivates all modes

Access the **Drafting** tab of the **Options** dialog box

Figure 7-4.
Using the **Endpoint** object snap. When using running object snaps, be sure the correct snap marker and tooltip appear before you pick.

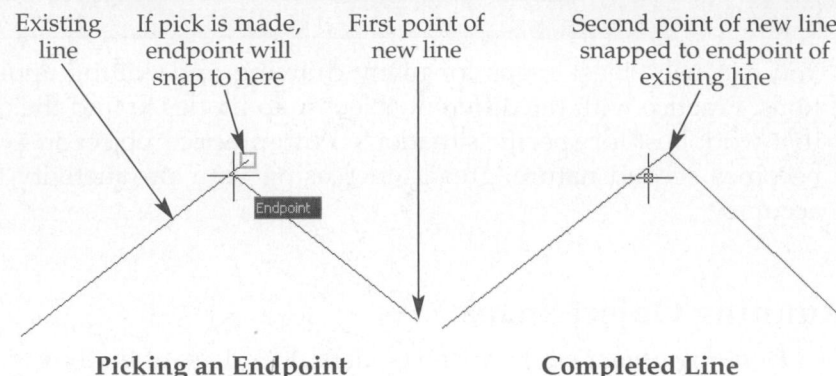

Existing line — If pick is made, endpoint will snap to here — First point of new line — Second point of new line snapped to endpoint of existing line

Picking an Endpoint   **Completed Line**

PROFESSIONAL TIP

Activate only the running object snap modes that you use most often. Too many running object snaps can make it difficult to snap to the appropriate location, especially on detailed drawings with several objects near each other. Use object snap overrides to access object snap modes that you use less often.

NOTE

By default, a keyboard point entry overrides running object snaps. Use the **Priority for Coordinate Data Entry** area on the **User Preferences** tab of the **Options** dialog box to adjust the default setting.

Object Snap Overrides

object snap override: A method of isolating a specific object snap mode while a tool is in use. The selected object snap temporarily overrides the running object snap modes.

Use an *object snap override* to select a specific point when running object snaps conflict with each other, or to use an object snap that is not running. Running objects snaps return after you make the object snap override selection. Most object snap modes are available as running object snaps or object snap overrides, although some modes are available only as object snap overrides.

After you access a tool, the preferred technique for activating an object snap override is to use the **Object Snap** shortcut menu. To use this method, press and hold [Shift] and then right-click to display the shortcut menu. See **Figure 7-5.** Select an object snap mode and then move the crosshairs near the location on an existing object where the object snap is to occur. When you see the marker, pick to locate the point at the exact position on the existing object. You can use this technique for accessing the **Object Snap** shortcut menu regardless of the active tool or whether you are picking the first point or an additional point.

When you locate a start point using some tools, an alternative for selecting an object snap is to right-click without holding [Shift]. This option functions the same except that the **Object Snap** shortcut menu is available from the **Snap Overrides** cascading submenu. Examples include selecting the center of a circle or placing an additional point using other tools, such as the second point of a line.

NOTE

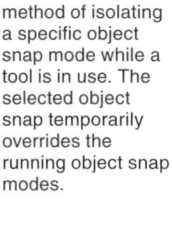

You can activate an object snap override by typing the first three letters of the name of the object snap. For example, enter END to activate the **Endpoint** object snap or CEN to activate the **Center** object snap.

Figure 7-5.
The **Object Snap**
shortcut menu
provides quick
access to object snap
overrides.

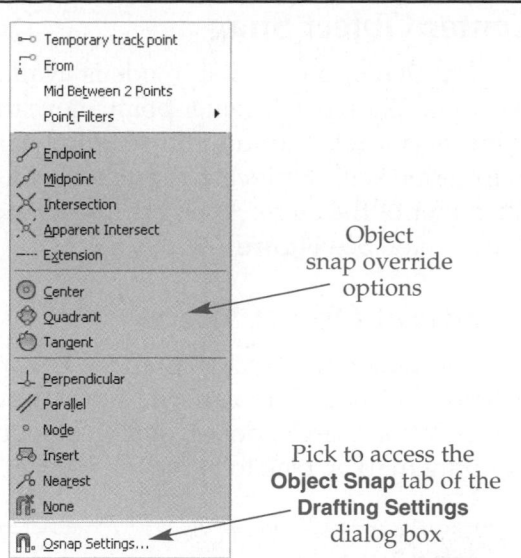

Temporary track point
From
Mid Between 2 Points
Point Filters

Endpoint
Midpoint
Intersection
Apparent Intersect
Extension

Center — Object snap override options
Quadrant
Tangent

Perpendicular
Parallel
Node
Insert — Pick to access the **Object Snap** tab of the **Drafting Settings** dialog box
Nearest
None

Osnap Settings...

PROFESSIONAL TIP

Remember that object snap modes are not tools, but are used with tools. An error message appears if you activate an object snap mode when no tool is active.

Endpoint Object Snap

The **Endpoint** object snap mode is available as a running object snap or object snap override. To snap to an endpoint, move the crosshairs near the endpoint of a line, arc, polyline, or spline. When the endpoint marker appears, pick to locate the point at the exact endpoint. Refer again to **Figure 7-4.**

Midpoint Object Snap

The **Midpoint** object snap mode is available as a running object snap or object snap override. To snap to a midpoint, move the crosshairs near the midpoint of a line, arc, or polyline. When the midpoint marker appears, pick to locate the point at the exact midpoint. See **Figure 7-6.**

Exercise 7-1

Access the Student Web site (www.g-wlearning.com/CAD) and complete Exercise 7-1.

Figure 7-6.
Using the **Midpoint** object snap.

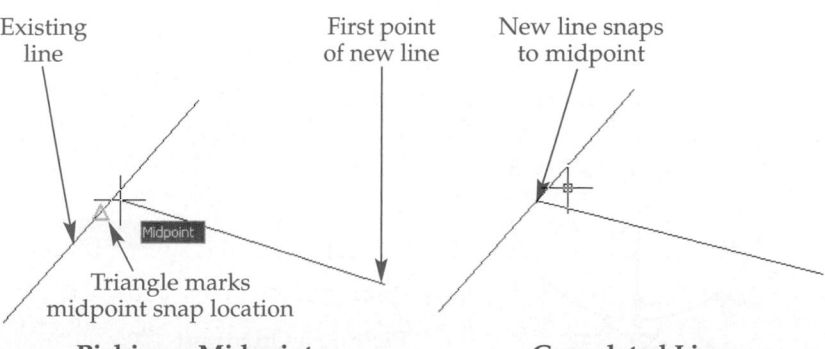

Existing line
First point of new line
New line snaps to midpoint

Triangle marks midpoint snap location

Picking a Midpoint   **Completed Line**

Center Object Snap

The **Center** object snap mode is available as a running object snap or object snap override. To snap to a center point, move the crosshairs near the *perimeter*, not the center point, of a circle, donut, ellipse, elliptical arc, polyline arc, or arc. The **Center** object snap mode will *not* locate the center of a large circle if the crosshairs is not near the perimeter of the circle. When the center marker appears, pick to locate the point at the exact center. See **Figure 7-7.**

Quadrant Object Snap

quadrant: Quarter section of a circle, donut, or ellipse.

The **Quadrant** object snap mode is available as a running object snap or object snap override. To snap to a *quadrant*, move the crosshairs near the appropriate 0°, 90°, 180°, or 270° point of a circle, donut, ellipse, elliptical arc, polyline arc, or arc. When you see the quadrant marker, pick to locate the point at the exact quadrant position. See **Figure 7-8.**

> **NOTE**
>
> Quadrant positions are unaffected by the current angle zero direction, but they always coincide with the angle of the X and Y axes. The quadrant points of circles, donuts, and arcs are at the right (0°), top (90°), left (180°), and bottom (270°), regardless of the rotation of the object. The quadrant points of ellipses and elliptical arcs, however, rotate with the objects.

Figure 7-7.
Using the **Center** object snap.

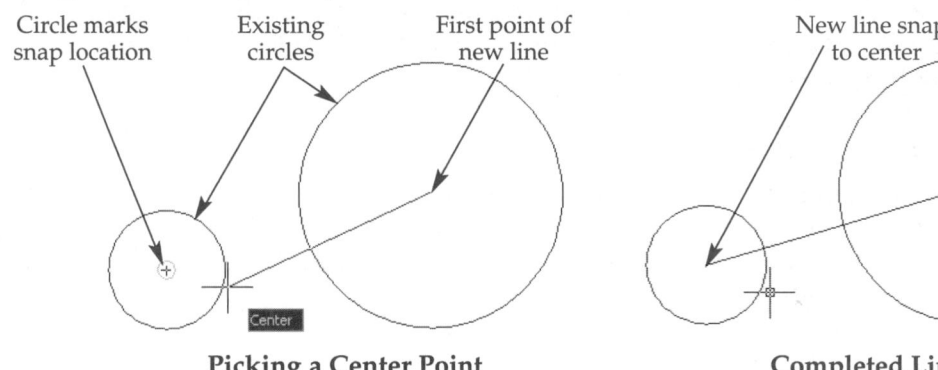

Picking a Center Point   **Completed Line**

Figure 7-8.
Using the **Quadrant** object snap.

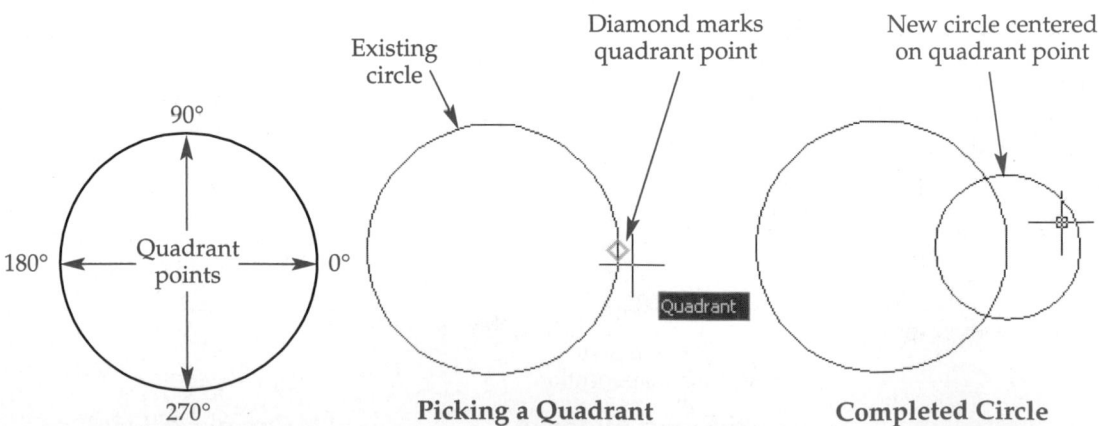

Picking a Quadrant   **Completed Circle**

AutoCAD and Its Applications—Basics

Exercise 7-2

Access the Student Web site (www.g-wlearning.com/CAD) and complete Exercise 7-2.

Intersection Object Snap

The **Intersection** object snap mode is available as a running object snap or object snap override. To snap to an intersection, move the crosshairs near the intersection of two or more objects. When you see the intersection marker, pick to locate the point at the exact intersection. See **Figure 7-9.**

Extension Object Snap

The **Extension** object snap mode is available as a running object snap or object snap override. The **Extension** object snap differs from most other object snaps because it uses *acquired points*, instead of direct point selection. To snap to an extension, move the crosshairs near the endpoint of a curve, but do not select. When an acquired point is found, a point symbol (+) marks the location. Move the crosshairs away from the acquired point to display an *extension path*. Pick a point along the extension path. **Figure 7-10** shows an example of using an **Extension** object snap twice to draw a line a specific distance away from two acquired points. In this example, type the .8 distance when you see the extension path. Dynamic input is not required to enter this value.

acquired point:
A point found by moving the crosshairs over a point on an existing object to reference for use when picking a new point.

extension path:
Dashed line or arc that extends from the acquired point to the current location of the crosshairs.

Exercise 7-3

Access the Student Web site (www.g-wlearning.com/CAD) and complete Exercise 7-3.

Figure 7-9.
Using the **Intersection** object snap.

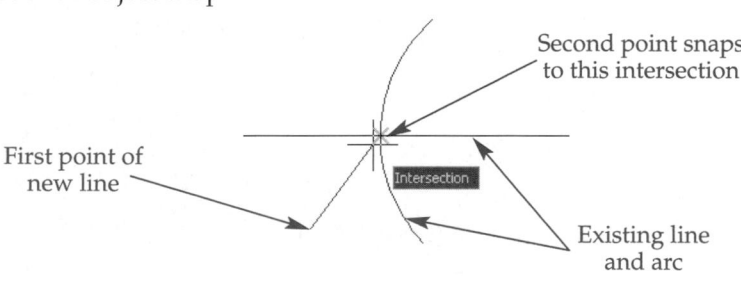

Second point snaps to this intersection

First point of new line

Intersection

Existing line and arc

Figure 7-10.
Using the **Extension** object snap to create a line .8 unit away from a rectangle.

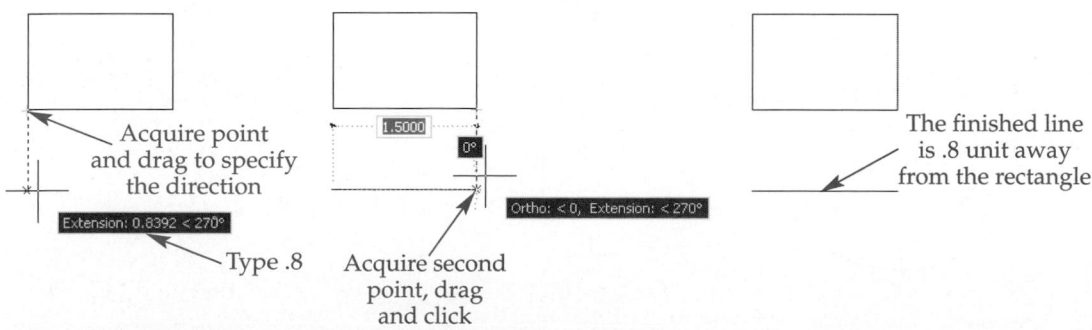

Acquire point and drag to specify the direction

Extension: 0.8392 < 270°

Type .8

1.5000

0°

Acquire second point, drag and click

Ortho: < 0, Extension: < 270°

The finished line is .8 unit away from the rectangle

Extended Intersection Object Snap

Even if objects do not intersect, you can snap to the location where the objects would intersect, if they were long enough. Make this selection using the **Extension** object snap or **Extended Intersection** object snap override. When using the **Extension** object snap mode, move the crosshairs near the endpoint of one curve to acquire the first point, and then move the crosshairs near the endpoint of another curve to acquire the second point. Now, move the crosshairs away from the acquired point, near the location of where the objects would intersect. When you see two extension paths and an intersection icon, pick to locate the point. See **Figure 7-11A.**

The **Extended Intersection** object snap override works by selecting objects one at a time using the **Intersection** object snap override. Once you activate the **Intersection** object snap override, move the cursor over one of the objects to display the intersection marker with an ellipsis (...), and pick the object. Then move the cursor over the other object to display the intersection marker at the extended intersection, and pick. See **Figure 7-11B.**

Figure 7-11.
Locating the center of a circle at the extended intersection of two objects. A—Using the **Extension** object snap. B— Using the **Extended Intersection** object snap.

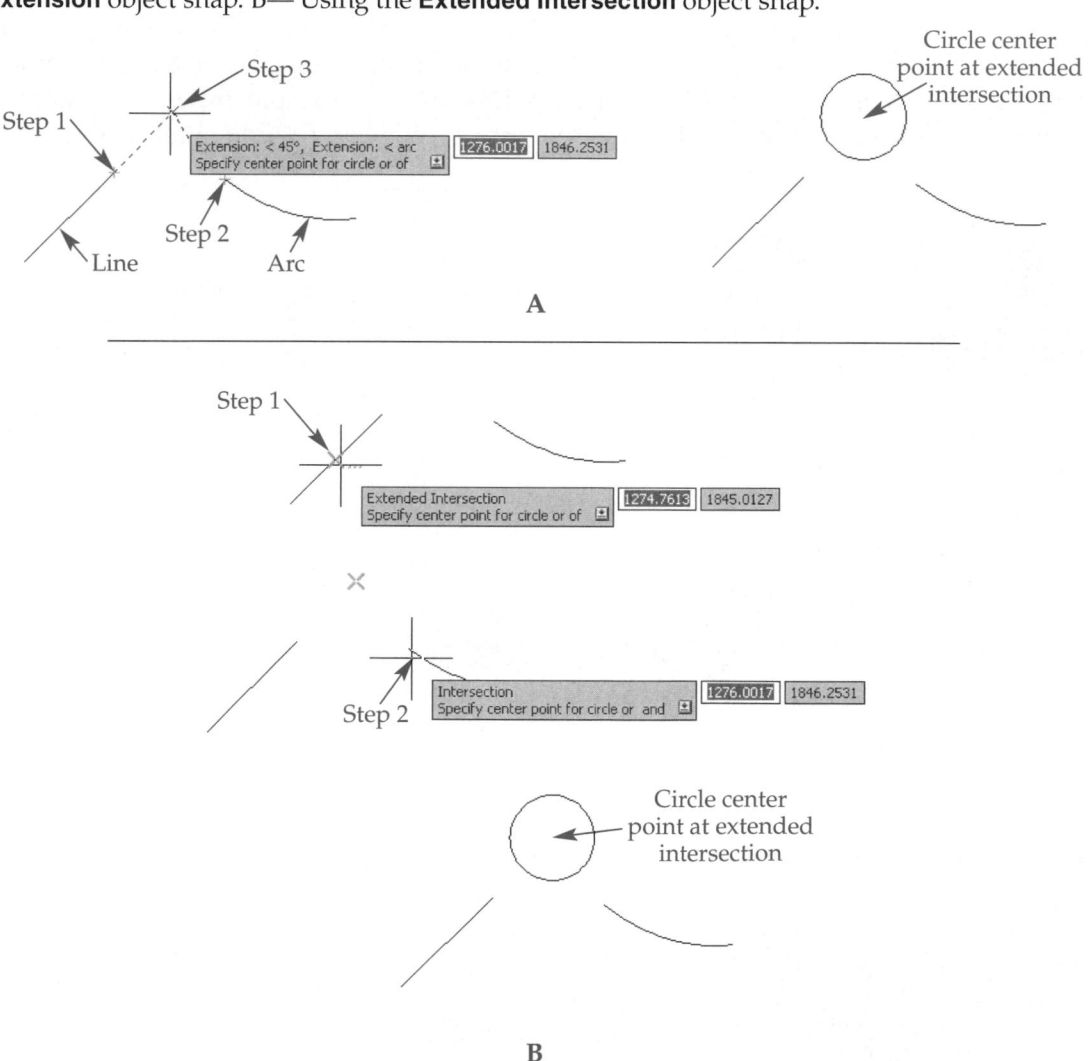

Exercise 7-4

Access the Student Web site (www.g-wlearning.com/CAD) and complete Exercise 7-4.

Perpendicular Object Snap

The **Perpendicular** object snap mode is available as a running object snap or object snap override. To snap to perpendicular, move the crosshairs near the point of perpendicularity on a line, arc, elliptical arc, ellipse, spline, xline, multiline, polyline, solid, trace, or circle, or the endpoint of a line, arc, polyline, or spline. When the perpendicular marker appears, pick to locate the point exactly perpendicular to the existing object. See **Figure 7-12.**

Figure 7-13 shows using the **Perpendicular** object snap mode to begin a line perpendicular to an existing object. The tooltip reads Deferred Perpendicular, and the perpendicular marker includes an ellipsis (...). The second endpoint determines the location of the line in a *deferred perpendicular* condition.

deferred perpendicular: A condition in which calculation of the perpendicular point is delayed until another point is picked.

> **NOTE**
>
> Perpendicularity measures from the point of intersection. Therefore, it is possible to draw a line perpendicular to a circle or arc.

Exercise 7-5

Access the Student Web site (www.g-wlearning.com/CAD) and complete Exercise 7-5.

Figure 7-12.
Drawing a line from a point perpendicular to an existing line.

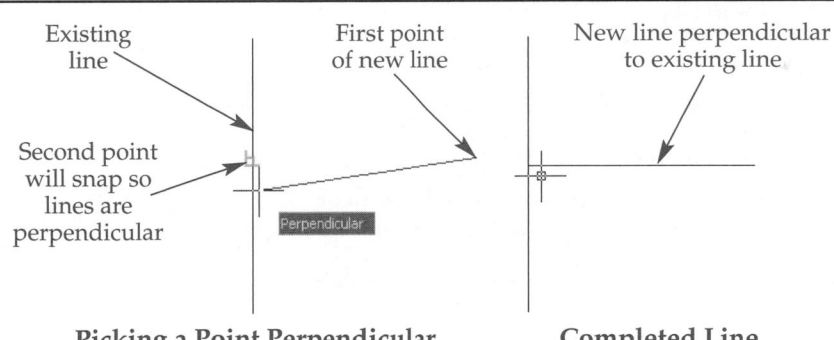

Figure 7-13.
Deferring the second point of a line to establish a perpendicular construction.

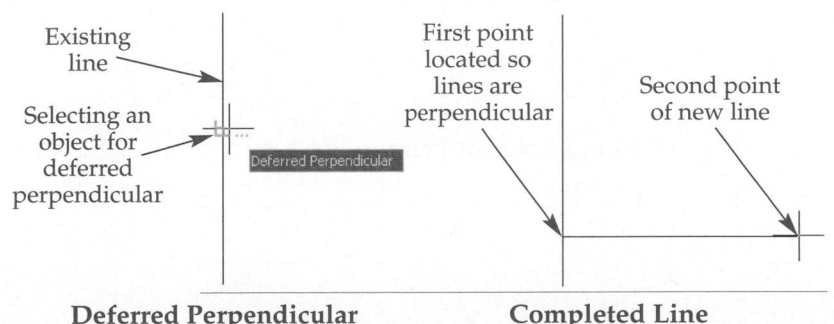

Tangent Object Snap

The **Tangent** object snap mode is available as a running object snap or object snap override. To snap to the point of tangency, move the crosshairs near an arc, circle, ellipse, elliptical arc, or spline. When the tangent marker appears, pick to locate the point at the exact point of tangency. See **Figure 7-14**.

When drawing an object tangent to two objects, you may need to pick multiple points to fix the point of tangency. For example, the point at which a line is tangent to a circle is found according to the locations of both ends of the line. Until both points are identified, the object snap specification is for *deferred tangency*. Once both endpoints are known, the tangency calculates, and the object is drawn in the correct location. See **Figure 7-15**.

Exercise 7-6

Access the Student Web site (www.g-wlearning.com/CAD) and complete Exercise 7-6.

Parallel Object Snap

The **Parallel** object snap mode is available as a running object snap or object snap override. To snap to a point parallel to a line or polyline, move the crosshairs near the existing object to display the parallel marker. Then, move the crosshairs away from and near parallel to the existing object. As you near a position parallel to the existing object, a *parallel alignment path* extends from the location of the crosshairs, and the parallel marker reappears to indicate acquired parallelism. Pick a point along the parallel alignment path. See **Figure 7-16**.

Exercise 7-7

Access the Student Web site (www.g-wlearning.com/CAD) and complete Exercise 7-7.

Figure 7-14.
Using the **Tangent** object snap.

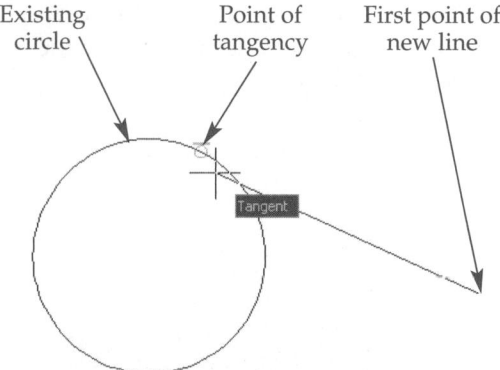

Existing circle   Point of tangency   First point of new line

Tangent

Picking a Tangent Point

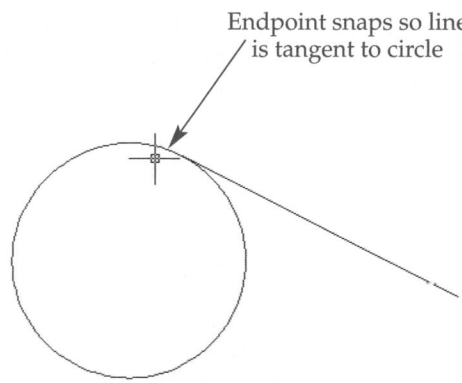

Endpoint snaps so line is tangent to circle

Completed Line

AutoCAD and Its Applications—Basics

Figure 7-15.
Drawing a line tangent to two circles.

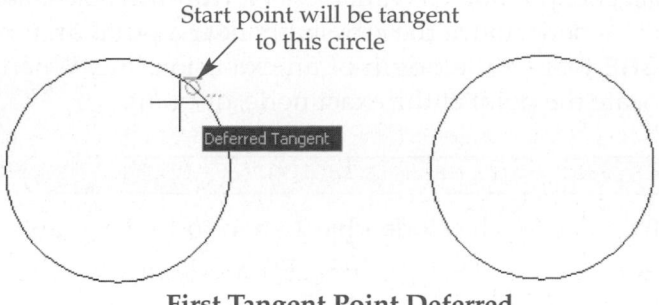

Start point will be tangent
to this circle

First Tangent Point Deferred

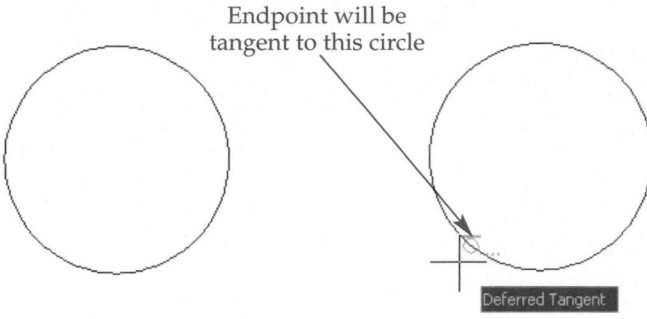

Endpoint will be
tangent to this circle

Picking Second Tangent Point

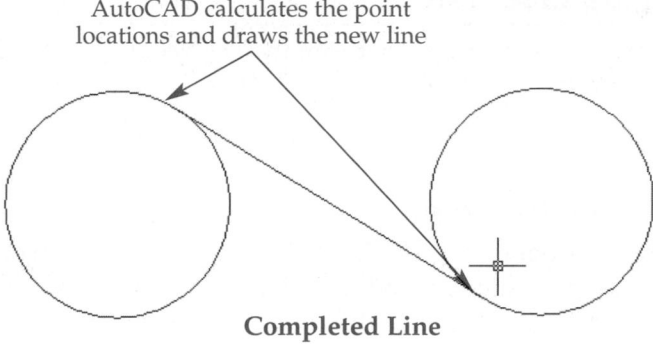

AutoCAD calculates the point
locations and draws the new line

Completed Line

Figure 7-16.
Using the **Parallel** object snap option to draw a line parallel to an existing line. A—Select the first endpoint for the new line, select the **Parallel** object snap, and then move the crosshairs near the existing line to acquire a point. B—Once the parallel point is acquired, move the crosshairs near the location of the parallel line to display an extension path.

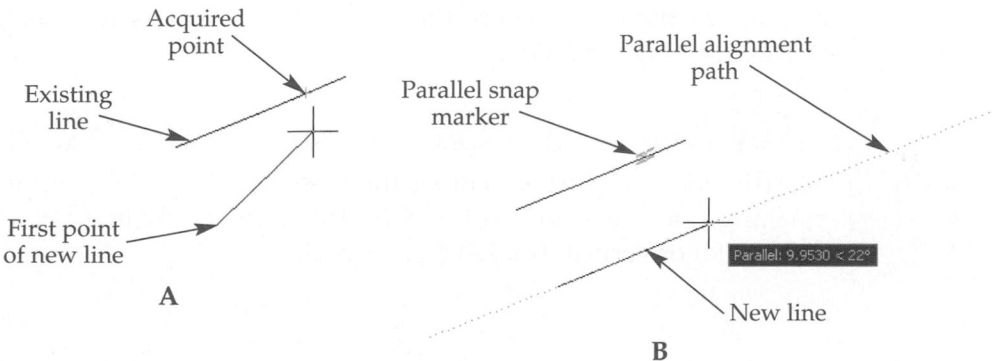

Acquired
point

Existing
line

First point
of new line

A

Parallel snap
marker

Parallel alignment
path

New line

B

Node Object Snap

The **Node** object snap mode is available as a running object snap or object snap override. To snap to a node, move the crosshairs near a point drawn using the **POINT**, **DIVIDE**, or **MEASURE** tool, or the origin of an extension line. When the node marker appears, pick to locate the point at the exact node, or point.

NOTE

In order for the **Node** object snap to find a point object, the point must be in a visible display mode. Chapter 8 explains point display mode controls.

Nearest Object Snap

The **Nearest** object snap mode is available as a running object snap or object snap override. Use the **Nearest** mode to specify a point that is directly on an object, but cannot be located with any of the other object snap modes, or when the location of the intersection is not critical. To snap to a nearest point, move the crosshairs near an existing object. When the nearest marker appears, pick to locate the point at a location on object closest to the crosshairs.

Exercise 7-8

Access the Student Web site (www.g-wlearning.com/CAD) and complete Exercise 7-8.

Temporary Track Point Snap

The **Temporary track point** snap mode is available only as an object snap override. It allows you to locate a point aligned with or relative to another point. For example, use the **Temporary track point** snap to place the center of a circle at the center of an existing rectangle. At the Specify center point for circle or [3P/2P/Ttr (tan tan radius)]: prompt, select the **Temporary tracking point** snap, and then use the **Midpoint** object snap to pick the midpoint of one of the vertical lines. This establishes the Y coordinate of the rectangle's center. See **Figure 7-17A.** When the Specify center point for circle or [3P/2P/Ttr (tan tan radius)]: prompt reappears, select the **Temporary tracking point** snap mode, and then use the **Midpoint** object snap mode to pick the midpoint of one of the horizontal lines. This establishes the X coordinate of the rectangle's center. See **Figure 7-17B.** Finally, pick to locate the center of the circle where the two tracking vectors intersect, and specify the circle radius. See **Figure 7-17C.**

NOTE

The direction in which you move the crosshairs from the temporary tracking point determines the X or Y alignment. Switch between horizontal or vertical tracking as needed.

Figure 7-17.
Using temporary tracking to locate the center of a rectangle. A—The midpoint of the left line is acquired. B—The midpoint of the bottom line is acquired. C—The center point of the circle is located at the intersection of the tracking vectors.

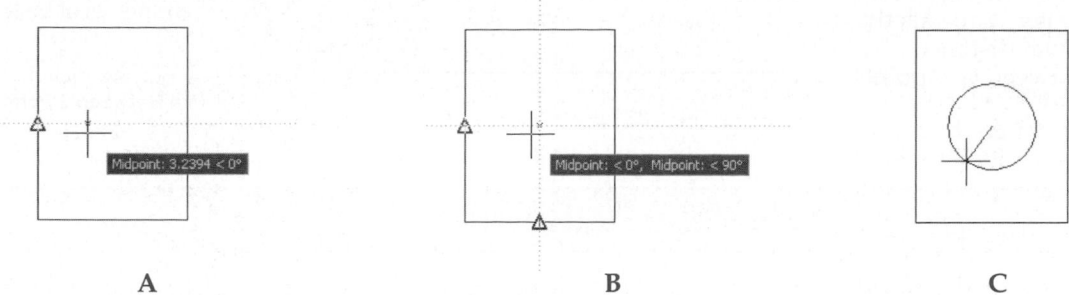

A   B   C

Snap From

The **From** snap mode is available only as an object snap override. It allows you to locate a point using coordinate entry from a specified reference base point. For example, use the **From** snap to place the center of a circle using a polar coordinate entry from the midpoint of an existing line. At the Specify center point for circle or [3P/2P/Ttr (tan tan radius)]: prompt, select the **From** snap mode, and then use the **Midpoint** object snap to pick the midpoint of the line. At the <Offset>: prompt, enter the polar coordinate @2<45 to establish the center of the circle 2 units and at a 45° angle from the midpoint of the line. Specify the radius of the circle to complete the operation. See **Figure 7-18.**

Mid Between 2 Points Snap

The **Mid Between 2 Points** snap feature is available only as an object snap override, and is very effective for locating a point exactly between two specified points. Use object snaps or coordinate point entry to pick reference points accurately. The example in **Figure 7-19** locates the center of a circle between two line endpoints.

Exercise 7-9

Access the Student Web site (www.g-wlearning.com/CAD) and complete Exercise 7-9.

Figure 7-18.
An example of using the **From** point selection mode to locate the center of a circle using a midpoint object snap and polar coordinate entry.

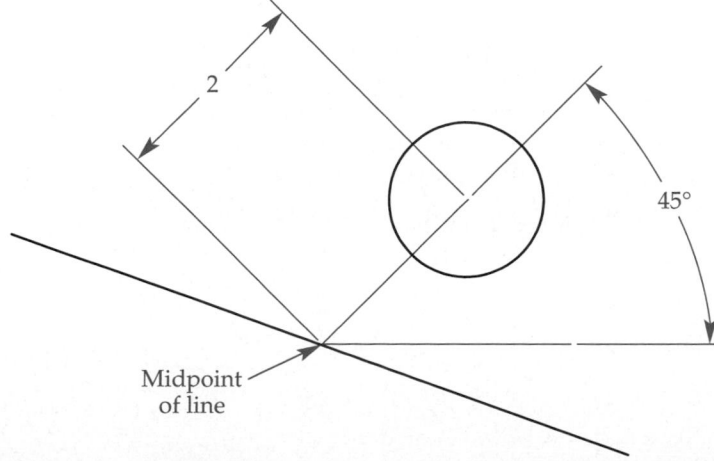

2

45°

Midpoint
of line

Figure 7-19.
Using the **Mid Between 2 Points** option to create a circle in which the center is an exactly equal distance between two points.

First pick of
Mid Between 2 Points

Center of circle at midpoint between endpoints of lines

Second pick of
Mid Between 2 Points

AutoTrack

AutoTrack offers an object snap tracking mode and a polar tracking mode. These tools are helpful for common drafting and design tasks, including basic geometric constructions. AutoTrack uses *alignment paths* and *tracking vectors* as drawing aids. You can use AutoTrack with any tool that requires a point selection.

alignment paths: Temporary lines and arcs that coincide with the position of existing objects.

tracking vectors: Temporary lines that display at specific angles, typically 0°, 90°, 180°, and 270°.

object snap tracking: Mode that provides horizontal and vertical alignment paths for locating points after a point is acquired with object snap.

Object Snap Tracking

Object snap tracking has two requirements: running object snaps must be active, and the crosshairs must pause over the intended selection long enough to acquire the point. Pick the **Object Snap Tracking** button on the status bar, press [F11], or use the **Object Snap Tracking On (F11)** check box in the **Object Snap** tab of the **Drafting Settings** dialog box to toggle object snap tracking on and off. Object snap tracking mode works with running object snaps. You must activate running object snaps and the appropriate running object snap modes in order for object snap tracking to function properly.

In **Figure 7-20**, object snap tracking is used with the **Perpendicular** and **Midpoint** running object snaps to draw a line 2 units long, perpendicular to the existing slanted

Figure 7-20.
Using object snap tracking to draw a line perpendicular to an existing line.

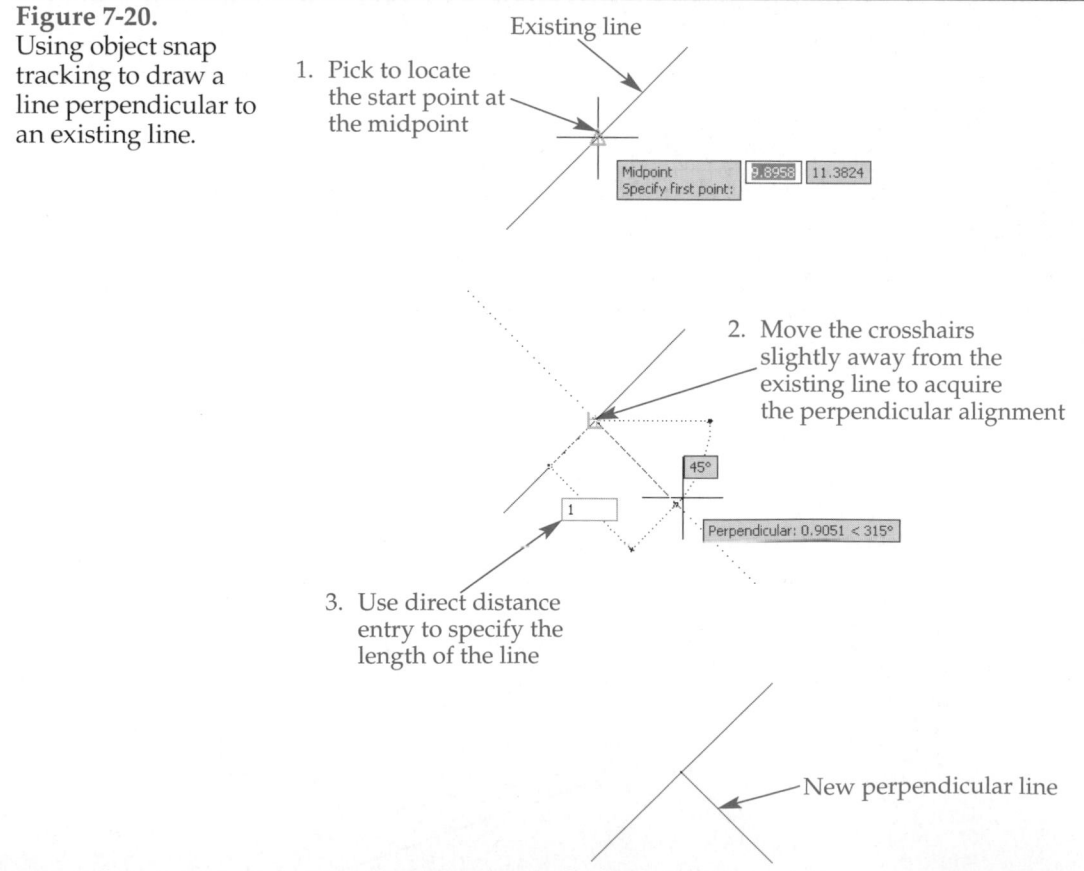

Existing line

1. Pick to locate the start point at the midpoint

Midpoint
Specify first point: 9.8958 11.3824

2. Move the crosshairs slightly away from the existing line to acquire the perpendicular alignment

45°

1

Perpendicular: 0.9051 < 315°

3. Use direct distance entry to specify the length of the line

New perpendicular line

AutoCAD and Its Applications—Basics

line. Running object snaps, the **Perpendicular** and **Midpoint** running object snap modes, and object snap tracking are active before using the **LINE** tool to draw the new line. Select the midpoint of the existing line to locate the start point of the new line. Then move the crosshairs slightly away from the existing line to display the perpendicular marker and the alignment path. Use direct distance entry along the alignment path to complete the line.

In **Figure 7-21**, object snap tracking is used with the **Midpoint** running object snap to position a circle directly above the midpoint of a horizontal line and to the right of the midpoint of an angled line. With running object snaps, the **Midpoint** running object snap mode, and object snap tracking active, use the **Circle, Radius** tool to draw the circle. Pause the crosshairs near the midpoint of the horizontal line to acquire the first point, and then pause the crosshairs near the midpoint of the angled line to acquire the second point. Move the crosshairs to the position as shown in the second step of **Figure 7-21** until two tracking vectors appear. Pick to locate the center of the circle, and complete the operation by entering a radius.

PROFESSIONAL TIP

Use object snap tracking whenever possible to complete tasks that require you to reference locations on existing objects. Often the combination of running object snaps and object snap tracking is the quickest way to construct geometry.

Exercise 7-10

Access the Student Web site (www.g-wlearning.com/CAD) and complete Exercise 7-10.

Figure 7-21.
Using object snap tracking to position a circle in line with the midpoints of two lines.

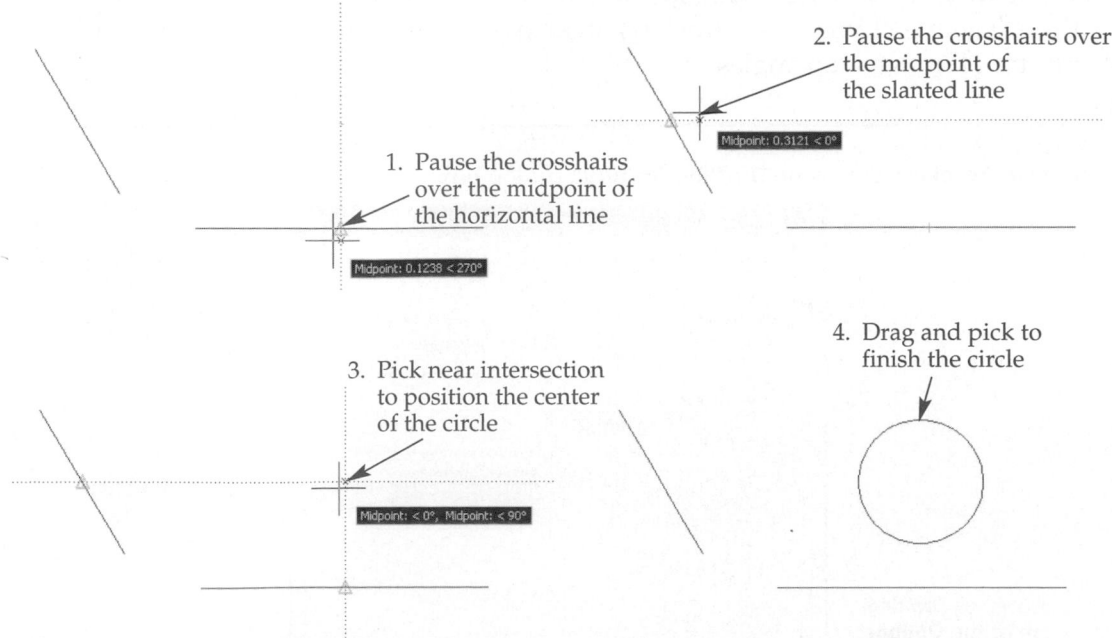

Polar Tracking

Chapter 3 introduces *polar tracking* as an accurate method of using direct distance entry. Pick the **Polar Tracking** button on the status bar, press [F10], or use the **Polar Tracking On (F10)** check box in the **Polar Tracking** tab of the **Drafting Settings** dialog box to toggle polar tracking on and off. When polar tracking is on, the crosshairs snaps to preset incremental angles when you locate a point relative to another point. For example, when you draw a line, polar tracking is not active for the first point selection, but it is available for the second and additional point selections. Polar tracking vectors appear as dotted lines whenever the crosshairs aligns with any of these preset angles.

To set incremental angles, use the **Polar Tracking** tab in the **Drafting Settings** dialog box. See **Figure 7-22**. To display the **Drafting Settings** dialog box, right-click on any of the status bar toggle buttons and select **Settings…**. The **Polar Angle Settings** area sets polar angle increments. Use the **Increment angle** drop-down list to select the angle increments at which polar tracking vectors occur. A variety of preset angles is available. The default increment is 90, which provides angle increments every 90°. The 30° setting shown in **Figure 7-22** provides polar tracking in 30° increments.

NOTE

The preset angle increments available in the **Polar Angle Settings** area of the **Drafting Settings** dialog box are also available in the shortcut menu displayed when you right-click on the **Polar Tracking** button on the status bar.

To add specific polar tracking angles, pick the **New** button in the **Polar Angle Settings** area, and type a new angle value in the text box that appears in the **Additional angles** window. Repeat the process to add other angles. Additional angles work with the increment angle setting when you use polar tracking. Only the specific additional angles you enter are recognized, not each increment of the angle. Use the **Delete** button to remove angles from the list. Make the additional angles inactive by unchecking the **Additional angles** check box.

The **Object Snap Tracking Settings** area sets the angles available with object snap tracking. If you select **Track orthogonally only**, only horizontal and vertical alignment paths are active. If you select **Track using all polar angle settings**, alignment paths are active for all polar snap angles.

Figure 7-22.
The **Polar Tracking** tab of the **Drafting Settings** dialog box.

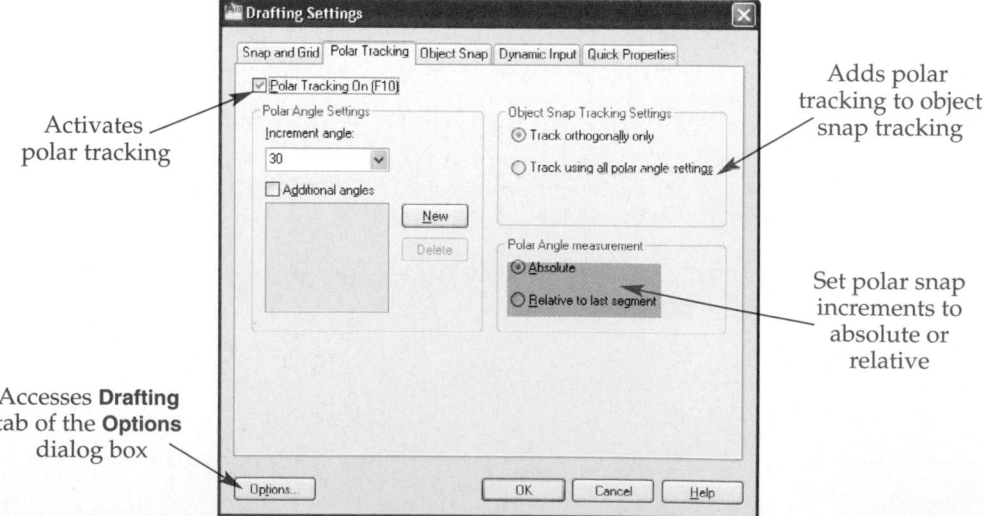

Activates polar tracking

Adds polar tracking to object snap tracking

Set polar snap increments to absolute or relative

Accesses **Drafting** tab of the **Options** dialog box

The **Polar Angle measurement** setting determines whether the polar snap increments are constant or relative to the previous segment. If you choose **Absolute**, polar snap angles measure from the base angle of 0° set for the drawing. If you pick **Relative to last segment**, each increment angle measures from a base angle established by the previously drawn segment.

Figure 7-23 shows how to draw a parallelogram using polar tracking and 30° angle increments. Access the **LINE** tool and select the first point. Then move the crosshairs to the right while the polar alignment path indicates <0°. Enter a direct distance value. Move the crosshairs to the 60° polar alignment path and enter a direct distance value. Move the crosshairs to the 180° polar alignment path and enter a value. Finally, use polar tracking with an angle of 240° and a specified line distance, or use the **Close** option to finish the parallelogram.

> **NOTE**
>
> You cannot use polar tracking and **Ortho** at the same time. AutoCAD automatically turns **Ortho** off when polar tracking is on, and it turns polar tracking off when **Ortho** is on.

Exercise 7-11

Access the Student Web site (www.g-wlearning.com/CAD) and complete Exercise 7-11.

Polar Tracking with Polar Snaps

You can also use polar tracking with polar snaps. For example, if you use polar tracking and polar snaps to draw the parallelogram in **Figure 7-23**, there is no need to type the length of the line, because you set the angle increment with polar tracking and a length increment with polar snaps. Establish angle and length increments using the **Snap and Grid** tab of the **Drafting Settings** dialog box. See **Figure 7-24.**

To activate polar snap, pick the **PolarSnap** radio button in the **Snap type & style** area of the dialog box. **PolarSnap** mode activates the **Polar spacing** area and deactivates the **Snap** area. Set the length of the polar snap increment in the **Polar distance:** text box. If the **Polar distance:** setting is 0, the polar snap distance is the orthogonal snap distance. **Figure 7-25** shows a parallelogram drawn with 30° angle increments and length increments of .75. The lengths of the parallelogram sides are 1.5 and .75.

Figure 7-23.
Using polar tracking with 30° angle increments to draw a parallelogram. A—After the first side is drawn, the alignment path and direct distance entry are used to create the second side. B—A horizontal alignment path is used for the third side. C—The parallelogram is completed with the Close option.

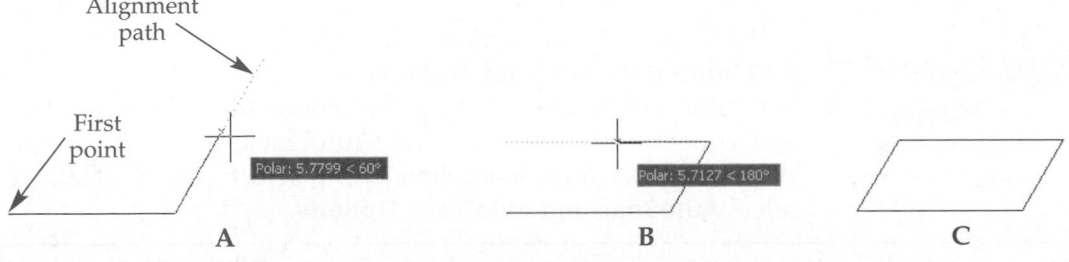

Figure 7-24.
The **Snap and Grid** tab of the **Drafting Settings** dialog box is used to set polar snap spacing.

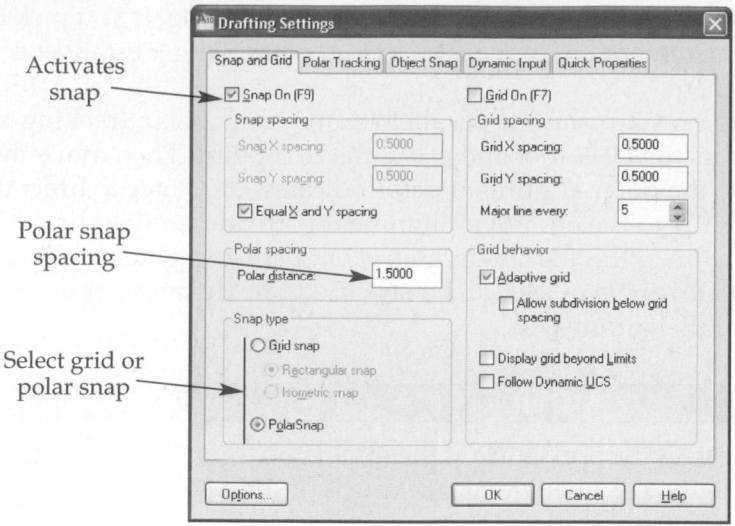

Activates snap

Polar snap spacing

Select grid or polar snap

Figure 7-25.
Drawing a parallelogram with polar snap.

A   B   C   D

Using Polar Tracking Overrides

It takes time to set up polar tracking and polar snap options, but it is worth the effort if you draw several objects that can take advantage of this feature. Use polar tracking overrides to perform polar tracking when you need to define only one point. Polar tracking overrides work for the specified angle whether polar tracking is on or off. To activate a polar tracking override, type a less than symbol (<) followed by the desired angle when AutoCAD asks you to specify a point. For example, after you access the **LINE** tool and pick a first point, enter <30 to set a 30° override. Then move the crosshairs in the desired 30° direction and enter a distance, such as 1.5 to draw a 1.5-unit line.

Exercise 7-12

Access the Student Web site (www.g-wlearning.com/CAD) and complete Exercise 7-12.

Supplemental Material
AutoSnap and AutoTrack Options
For information about options for controlling the appearance and function of AutoSnap and AutoTrack, go to the Student Web site (www.g-wlearning.com/CAD), select this chapter, and select **AutoSnap and AutoTrack Options**.

Template Development

Chapter 7

For detailed instructions on setting object snaps and polar tracking to save time and increase efficiency, go to the Student Web site (www.g-wlearning.com/CAD), select this chapter, and select **Template Development**.

Chapter Test

Answer the following questions. Write your answers on a separate sheet of paper or go to the Student Web site (www.g-wlearning.com/CAD) and complete the electronic chapter test.

1. Define the term *object snap*.
2. What is an AutoSnap tooltip?
3. Name the following AutoSnap markers:

   A. ⌐┘   E. ☐   I. ✕

   B. ◇   F. △   J. ┼

   C. ◯   G. ⌐   K. ⫽

   D. ⊗   H. ⌒   L. ⊠

4. How do you set running object snaps?
5. Define the term *running object snap*.
6. How do you access the **Drafting Settings** dialog box to change object snap settings?
7. If you are using running object snaps and want to make several point specifications without the aid of object snap, but want to continue the same running object snaps after making the desired point selections, what is the easiest way to turn off the running object snaps temporarily?
8. If you are using running object snaps and you want to make a single point selection without the effects of the running object snaps, what do you do?
9. Describe the object snap override.
10. How do you activate the **Object Snap** shortcut menu?
11. Where are the four quadrant points on a circle?
12. What is the situation when the tooltip reads Extended Intersection?
13. What does it mean when the tooltip reads Deferred Perpendicular?

14. Give the tool and entries needed to draw a line tangent to an existing circle and perpendicular to an existing line:
    A. Tool: _____
    B. Specify first point: _____
    C. to _____
    D. Specify next point or [Undo]: _____
    E. to _____
15. What is a deferred tangency?
16. What conditions must exist for the tooltip to read Tangent?
17. Which object snaps depend on acquired points to function?
18. What two display features does AutoTrack use to help you line up new objects with existing geometry?
19. What are the two requirements to use object snap tracking?
20. When are polar tracking vectors displayed as dotted lines?

Drawing Problems

Start AutoCAD if it is not already started. Start each new drawing using an appropriate template of your choice. Add layers as needed. Draw all objects on the appropriate layers. Follow the specific instructions for each problem. Do not draw dimensions or text. Use your own judgment and approximate dimensions when necessary.

▼ Basic

1. Draw the object below using object snap modes. Save the drawing as P7-1.

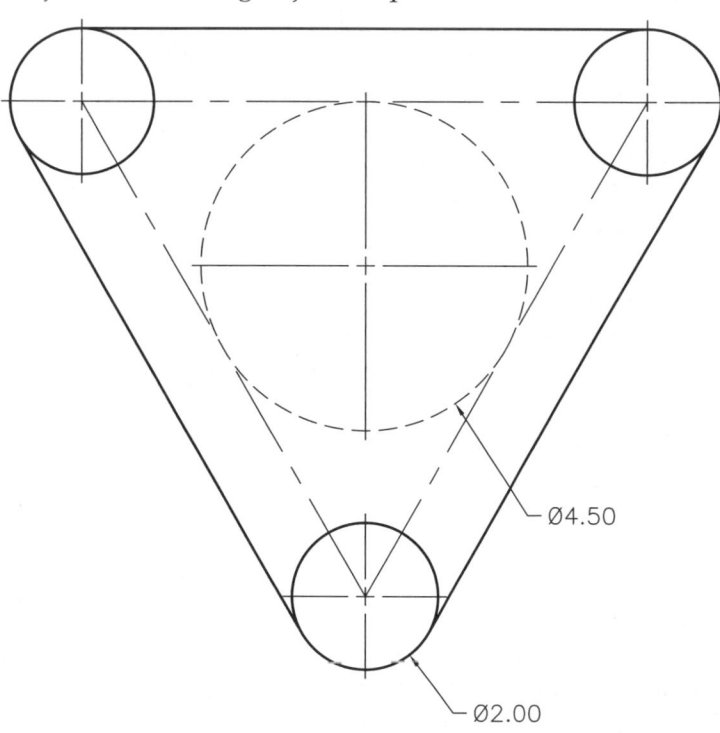

Ø4.50

Ø2.00

2. Draw the highlighted objects below, and then use the object snap modes indicated to draw the remaining objects. Save the drawing as P7-2.

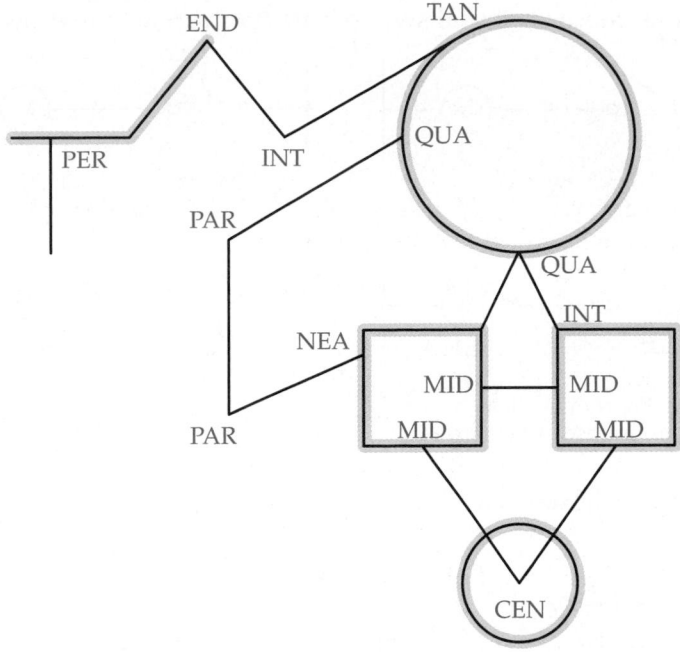

3. Draw the object below using the **Endpoint**, **Tangent**, **Perpendicular**, and **Quadrant** object snap modes. Save the drawing as P7-3.

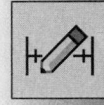

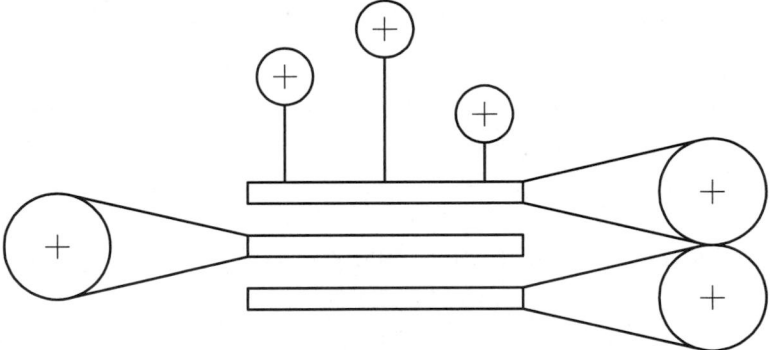

4. Draw the pipe separator shown below. Use object snaps and tracking to place the objects correctly. Save the drawing as P7-4.

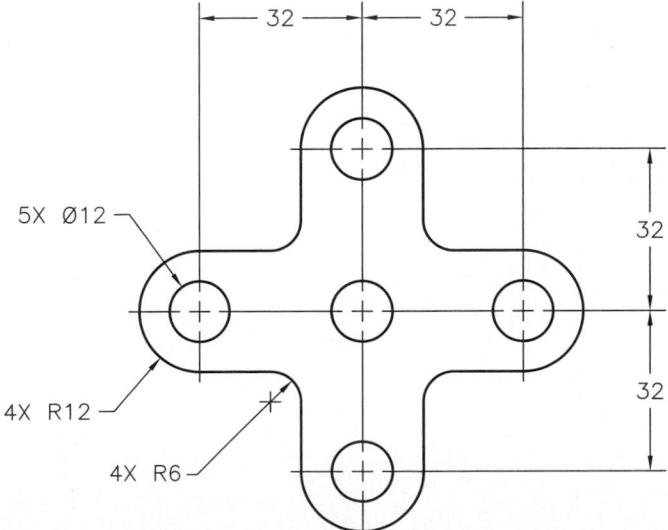

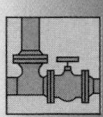

Drawing Problems - Chapter 7

5. Use the **Midpoint**, **Endpoint**, **Tangent**, **Perpendicular**, and **Quadrant** object snap modes to draw these electrical switch schematics. Save the drawing as P7-5.

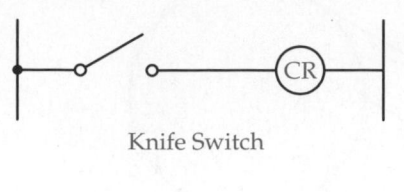

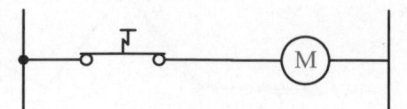

Knife Switch

Maintained Contact
Pushbutton

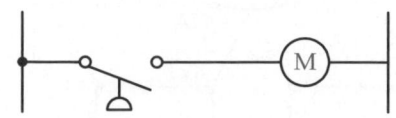

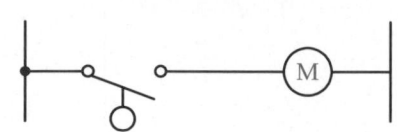

Pressure Switch
(Start on Rise in Pressure)

Level Switch
(Start on High Level)

6. Draw the elbow shown. Save the drawing as P7-6.

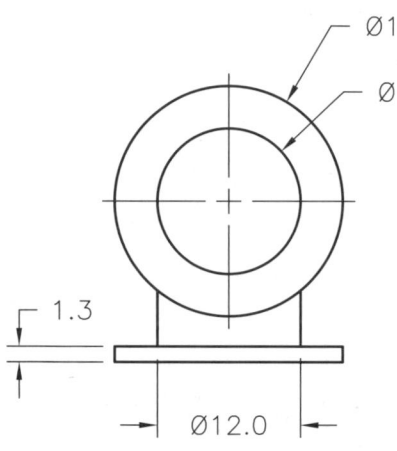

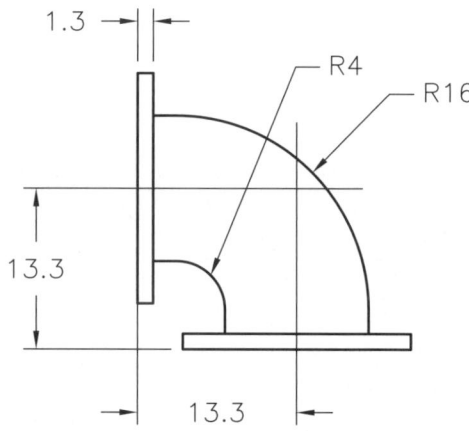

7. Draw the elbow shown. Save the drawing as P7-7.

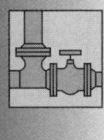

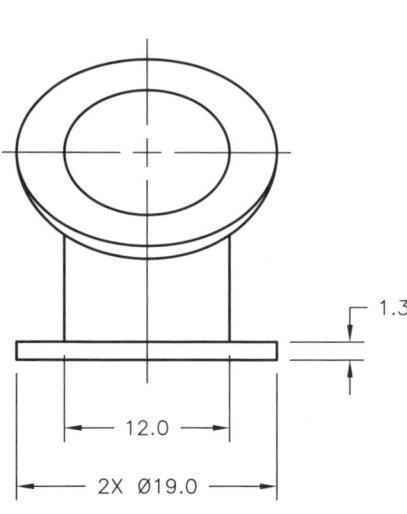

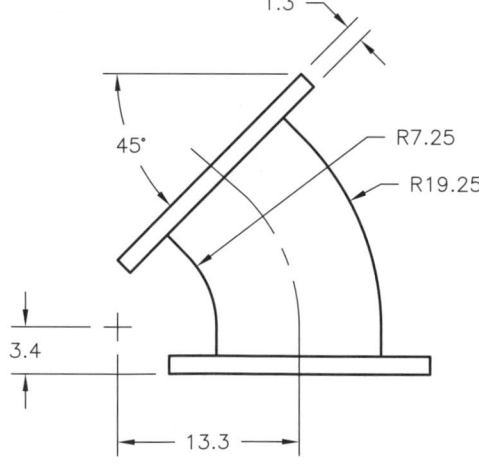

8. Draw the object shown. Save the drawing as **P7-8**.

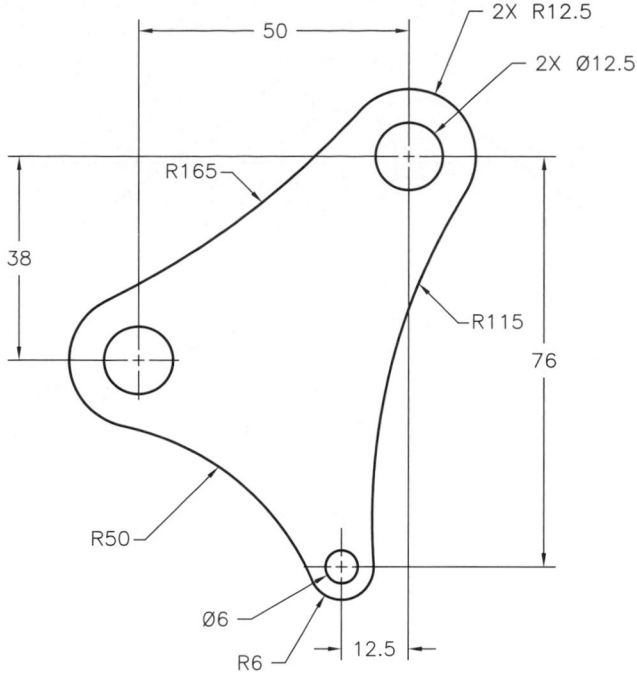

9. Draw the object shown. Save the drawing as **P7-9**.

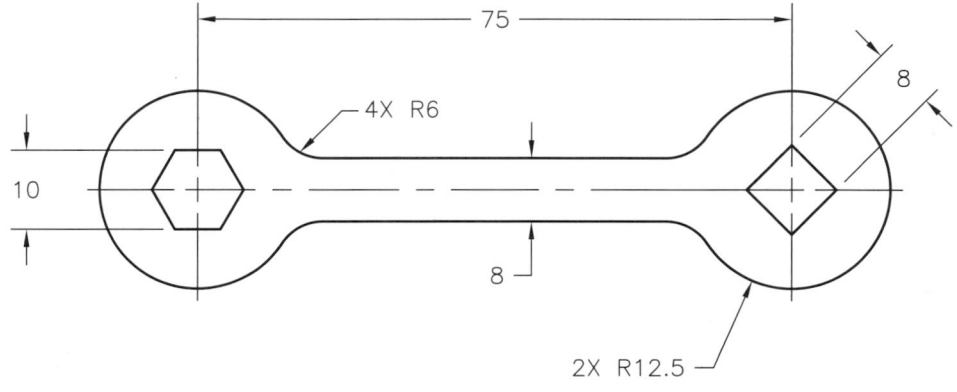

10. Draw the object shown. Save the drawing as **P7-10**.

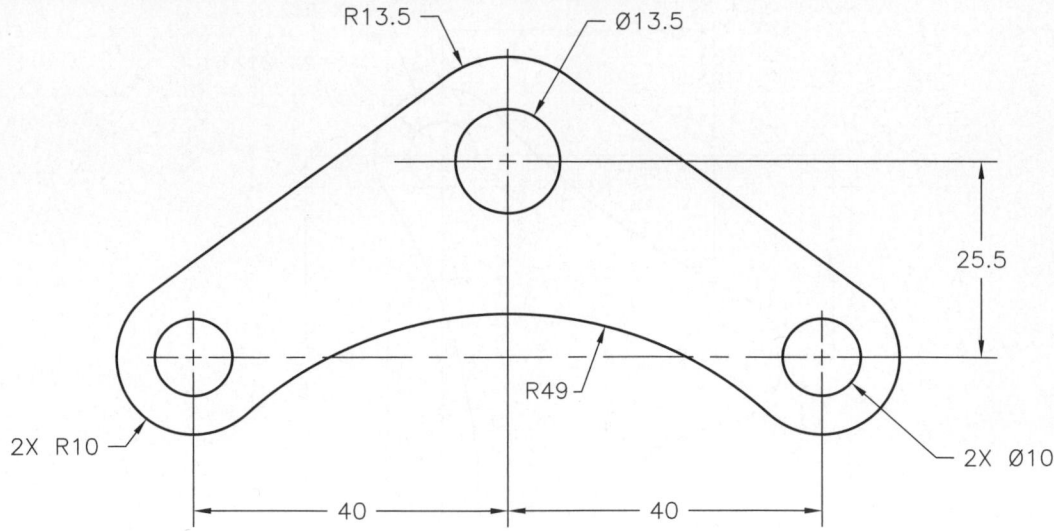

▼ Advanced

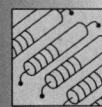

11. Use object snap modes to draw this elementary diagram. Save the drawing as P7-11.

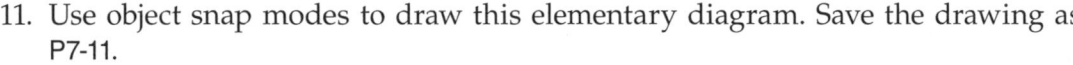

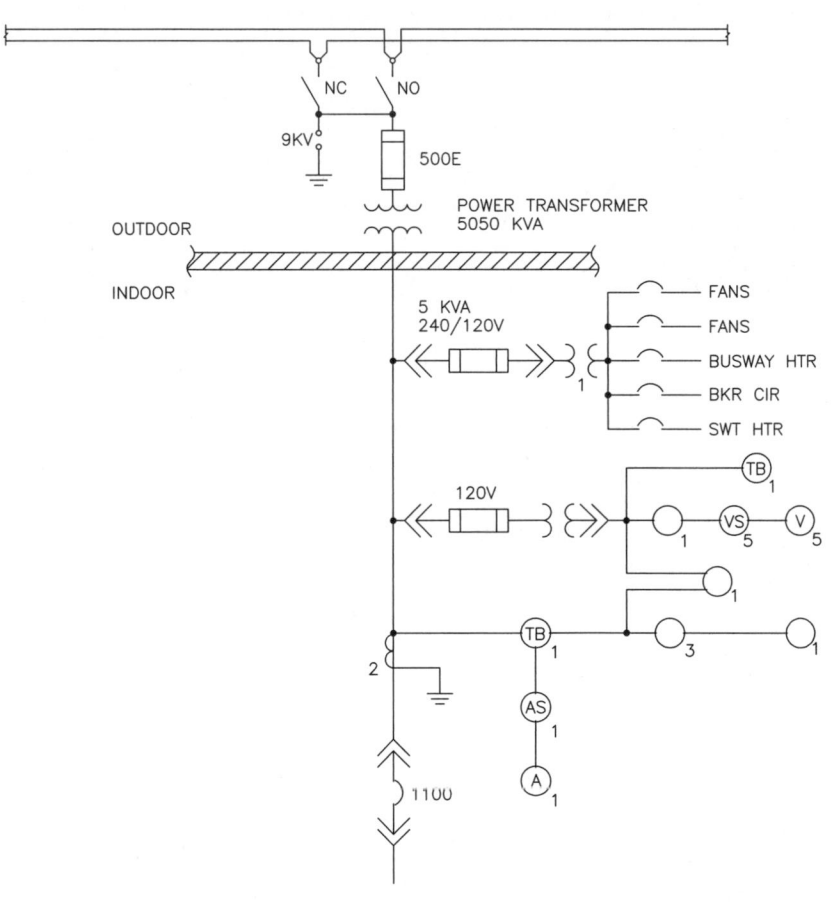

12. Draw the object shown. Save the drawing as P7-12.

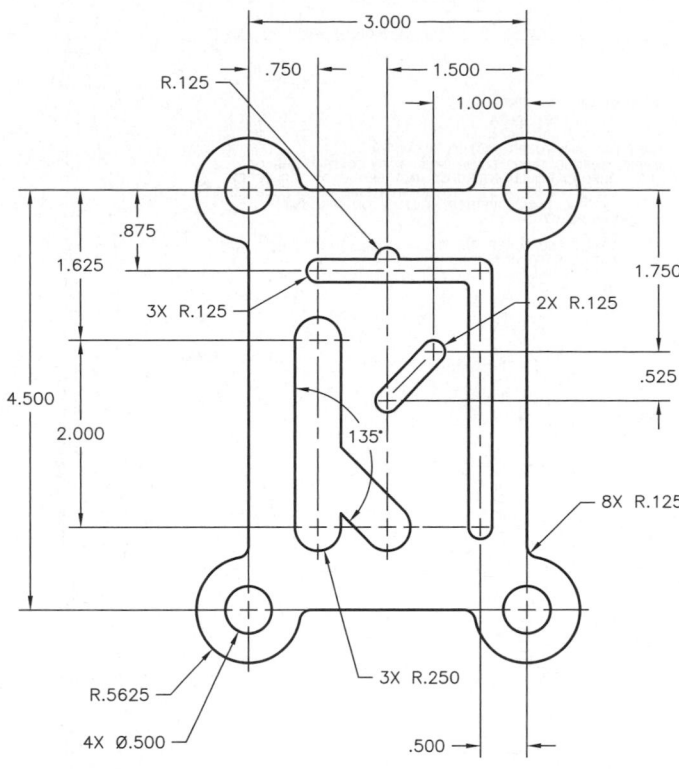

13. Design and draft a hammer similar to the one shown below using dimensions of your choice. Use the overall dimensions given as a basis for your design. Save the drawing as P7-13.

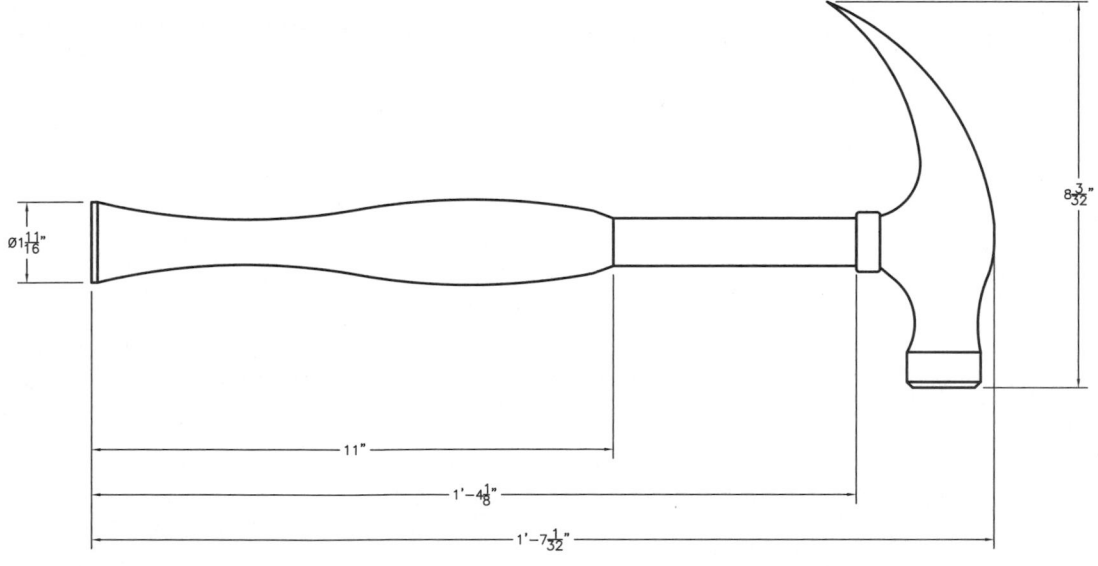

14. Draw the sailboat shown. Save the drawing as P7-14.

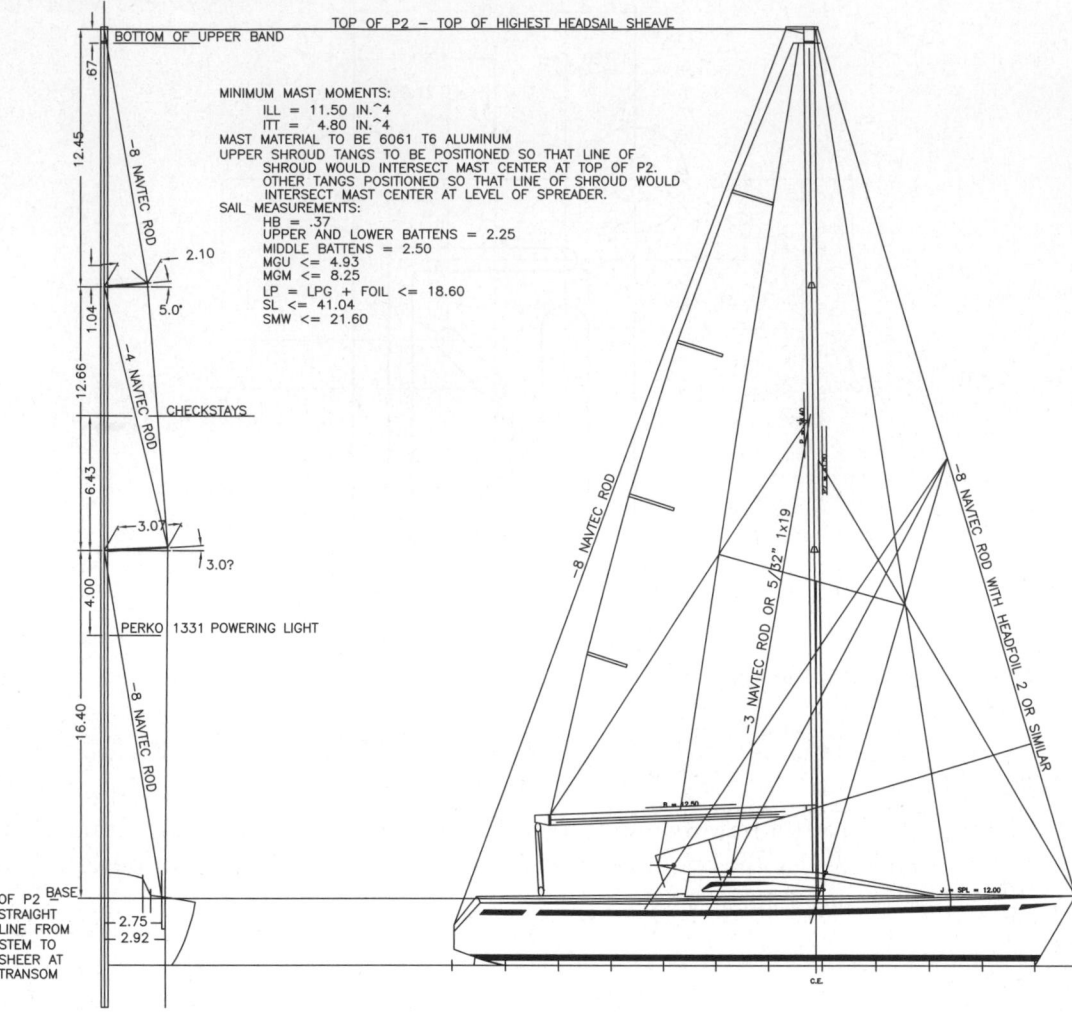

Construction Tools and Multiview Drawings

Learning Objectives

After completing this chapter, you will be able to do the following:

✓ Use the **OFFSET** tool to draw parallel objects.
✓ Place construction points.
✓ Mark points on objects at equal lengths using the **DIVIDE** tool.
✓ Mark points on objects at designated increments using the **MEASURE** tool.
✓ Create construction lines using the **XLINE** and **RAY** tools.
✓ Create orthographic multiview drawings.

This chapter explains how to create parallel offsets, divide objects, place point objects, and use construction lines. You can use these skills and the other geometric construction skills you have acquired to create multiview drawings. The tools described in this chapter allow you to produce accurate geometric constructions, but they do not apply relationships between objects. Chapter 22 explains how to use parametric tools to constrain objects.

Creating Parallel Offsets

The **OFFSET** tool is one of the most commonly used geometric construction tools. Offset lines and polylines for a variety of applications, such as constructing the thickness of architectural floor plan walls. Offset circles, arcs, and curves to form concentric objects. For example, offset a circle to form the wall thickness of a pipe.

Specifying the Offset Distance

Often the best way to use the **OFFSET** tool is to enter an offset value at the Specify offset distance or [Through/Erase/Layer] <current>: prompt. For example, to draw two concentric circles 1 unit apart, access the **OFFSET** tool, and specify an offset distance of 1. Pick the circle to offset, and then pick the side of the circle on which the offset occurs. See **Figure 8-1.** The **OFFSET** tool remains active, allowing you to pick another object to offset using the same offset distance. To exit the tool, press [Enter], [Esc], or the space bar, or choose the **Exit** option.

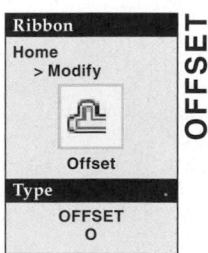

Ribbon
Home
> Modify

Offset

Type
OFFSET
O

OFFSET

Figure 8-1.
Drawing an offset circle using a designated distance.

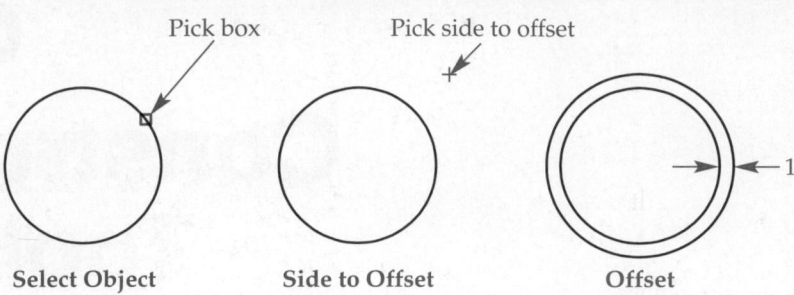

Pick box

Pick side to offset

Select Object   Side to Offset   Offset

Using the Through Option

Another option to specify the offset distance is to pick a point through which the offset occurs. After you access the **OFFSET** tool, activate the **Through** option at the Specify offset distance or [Through/Erase/Layer] <*current*>: prompt instead of picking an object to offset. Then pick the object to offset, and pick the point through which the offset occurs. See **Figure 8-2**. The **OFFSET** tool remains active, allowing you to pick another object to offset using the **Through** option. Exit the **OFFSET** tool when you are finished.

Erasing the Original Object

Use the **Erase** option of the **OFFSET** tool to erase the original, or source, object during the offset. Initiate the **OFFSET** tool, activate the **Erase** option, and choose **Yes** at the Erase source object after offsetting? prompt to erase the source object. The **Yes** option

Figure 8-2.
Drawing an offset through a given point.

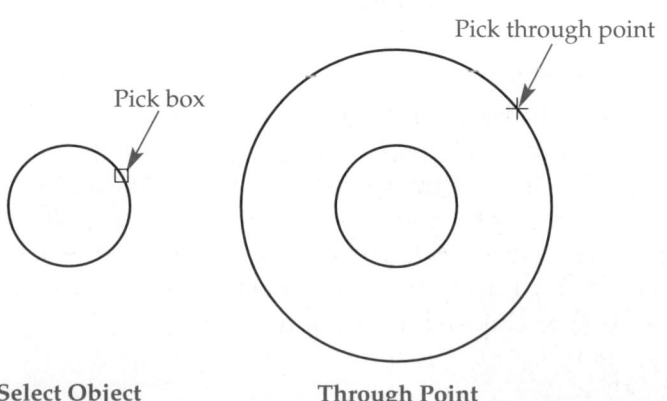

Pick box

Pick through point

Select Object   Through Point

remains set as default until you change the setting to **No**. Be sure to change the **Erase** setting back to **No** if you do not want the source offset object to erase the next time you use the **OFFSET** tool. Exit the **OFFSET** tool when you are finished.

Using the Layer Option

By default, offsets generate using the same properties, including layer, as the source object. Use the **Layer** option of the **OFFSET** tool to place the offset object on the current layer, regardless of the layer used to draw the source object. First, make the layer that you want to apply to the offset current. Then initiate the **OFFSET** tool, activate the **Layer** option, and choose **Current** at the Enter layer option for offset objects: prompt. The **Current** option remains set as default until you change the setting to **Source**. Be sure to change the **Layer** setting back to **Source** if you do not want the current layer applied to the offset the next time you use the **OFFSET** tool. Exit the **OFFSET** tool when you are finished.

Offsetting Multiple Times

After you select the object to offset, use the **Multiple** option to offset an object more than once, using the same distance between objects, without reselecting the object to offset. Initiate the **OFFSET** tool, specify the offset distance, and pick the source object. You can then select the **Multiple** option and begin picking to specify the offset direction. See **Figure 8-3**. Exit the **OFFSET** tool when finished.

NOTE

You can use the **Undo** option, when available, to undo the last offset without exiting the **OFFSET** tool.

Exercise 8-1

Access the Student Web site (www.g-wlearning.com/CAD) and complete Exercise 8-1.

Figure 8-3.
Use the **Multiple** option to create multiple offsets of the same distance, without picking the source object again.

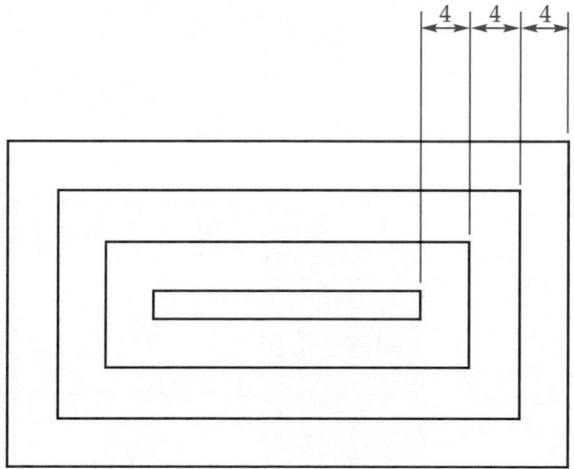

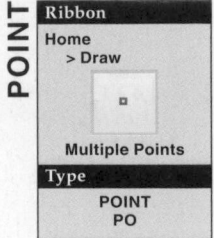

POINT

Ribbon
Home > Draw
Multiple Points
Type
POINT PO

Point objects are useful for identifying specific locations on a drawing and for marking positions on objects. You can draw points anywhere on the screen using the **POINT** tool. Use any appropriate method to specify the location of a point object. To place a single point object and then exit the **POINT** tool, enter **POINT** or **PO** at the keyboard. To draw multiple points without exiting the **POINT** tool, access the **Multiple Points** function from the ribbon. Press [Esc] to exit the tool.

Setting Point Style

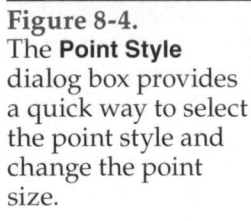

DDPTYPE

Ribbon
Home > Utilities
Point Style
Type
DDPTYPE

Point style and size are set using the **Point Style** dialog box. See **Figure 8-4**. By default, points appear as one-pixel dots and typically do not show up very well on-screen. Change the point style to make points more visible. The **Point Style** dialog box contains twenty different point styles. Pick the image of the desired style to make the point style current. All existing and new points change to the current style.

Setting Point Size

Set the point size by entering a value in the **Point Size:** text box of the **Point Style** dialog box. Pick the **Set Size Relative to Screen** button to change the point size in relation to different screen magnifications (zooming in or out). You may need to regenerate the display to view the relative sizes. Pick the **Set Size in Absolute Units** option button to make points appear the same size regardless of the screen magnification. See **Figure 8-5**.

Figure 8-4.
The **Point Style** dialog box provides a quick way to select the point style and change the point size.

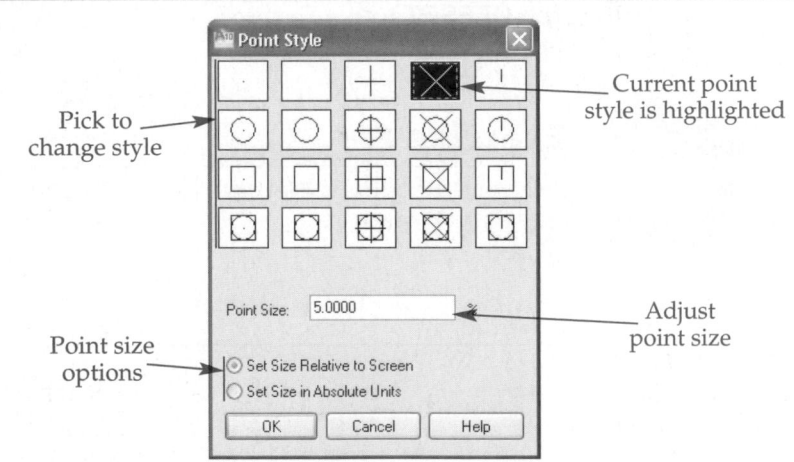

Pick to change style

Current point style is highlighted

Point size options

Adjust point size

Figure 8-5.
Points sized with the **Set Size Relative to Screen** setting change size as the drawing is zoomed. Points sized with the **Set Size in Absolute Units** setting remain a constant size.

Size Setting	Original Point Size	2X Zoom	.5 Zoom
Relative to Screen	⊠	⊠	⊠
Absolute Units	⊠	⊠	⊠

Exercise 8-2

Access the Student Web site (www.g-wlearning.com/CAD) and complete Exercise 8-2.

Marking an Object at Specified Increments

You can use the **DIVIDE** tool to place point objects or blocks at equally spaced locations on a line, circle, arc, or polyline. The **DIVIDE** tool *does not* break an object into an equal number of segments. Access the **DIVIDE** tool and select the object to mark. Enter the number of segments to mark with points and exit the tool. The point style determines the style of the marks placed on the object. **Figure 8-6** shows an example of using the **DIVIDE** tool to place points at seven equal increments.

The **Block** option of the **DIVIDE** tool allows you to place a *block* at each increment. Select the **Block** option at the Enter the number of segments or [Block]: prompt to insert a block. AutoCAD asks if the block should align with the object. Blocks are described in detail later in this textbook.

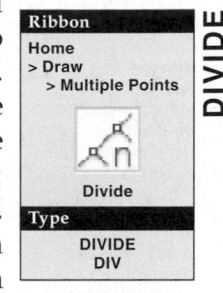

Ribbon
Home
> Draw
> Multiple Points

Divide

Type
DIVIDE
DIV

block: A previously drawn symbol or shape.

Marking an Object at Specified Distances

While the **DIVIDE** tool marks a line, circle, arc, or polyline according to a specified number of increments, the **MEASURE** tool places marks a specified distance apart. Access the **MEASURE** tool and select the object to mark. Measurement begins at the end closest to where you pick the object. Enter the distance between points to place points and exit the tool. All increments are equal to the specified segment length, except the last segment, which may be shorter. The point style determines the style of the marks placed on the object. The line shown in **Figure 8-7** was measured using a length of .75 unit. Use the **Block** option of the **MEASURE** tool to place a block at each measurement.

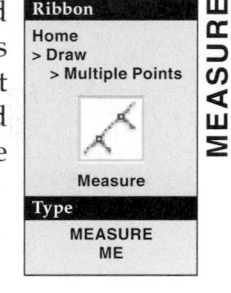

Ribbon
Home
> Draw
> Multiple Points

Measure

Type
MEASURE
ME

Exercise 8-3

Access the Student Web site (www.g-wlearning.com/CAD) and complete Exercise 8-3.

Figure 8-6.
Using the **DIVIDE** tool. Note that the default point style has been changed to ×.

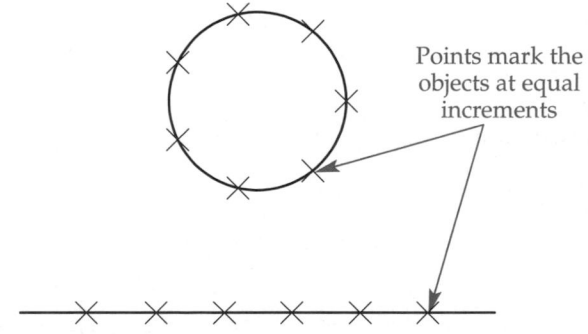

Points mark the objects at equal increments

Figure 8-7.
Using the **MEASURE** tool. Notice that the last segment may be shorter than the others, depending on the total length of the object.

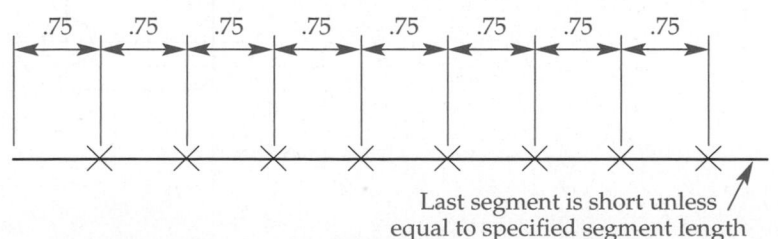

Last segment is short unless equal to specified segment length

Construction Lines and Rays

construction lines: Lines commonly used to lay out a drawing.

The tracking vectors and alignment paths available with object snap and AutoTrack tools are examples of *construction lines* generated by AutoCAD. These tools are very efficient for creating objects because vector and alignment lines appear only when needed. Often, however, drawings require construction lines that stay visible while you continue to create geometry. You can draw construction geometry using any drawing tool, such as **LINE**, **ARC**, or **CIRCLE**. However, the **XLINE** and **RAY** tools are specifically designed for adding construction lines to help lay out a drawing. See **Figure 8-8.** Use the **XLINE** tool to draw an AutoCAD construction line, or *xline.* Use the **RAY** tool to draw a *ray.*

xline: A line in AutoCAD that is infinite in both directions and is used to help build accurate geometry.

ray: An AutoCAD line object that is infinite in one direction only; considered semi-infinite.

PROFESSIONAL TIP

Create construction geometry on a separate construction layer, named CONST, CONSTRUCTION, or A-ANNO-NPLT for example. Turn off or freeze the construction layer when unneeded, or you can easily recognize and erase objects drawn on the construction layer if necessary.

Using the XLINE Tool

XLINE

Ribbon
Home
> Draw

Construction Line

Type
XLINE
XL

Use the **XLINE** tool to draw infinitely long construction lines, often called *xlines.* To draw a basic xline, specify the location of two points through which the xline passes. After you pick the *root point*, you can select as many points as needed to create additional xlines. When you place multiple xlines from the same root point, the root point acts as an axis point, through which all additional xlines pass. See **Figure 8-9.** To exit the tool, right-click or press [Enter], [Esc], or the space bar.

Using the Hor and Ver Options

root point: The first point specified to create a construction line.

Xline options are available as alternatives to selecting two points to create the xline. The **Hor** option draws a horizontal xline through a single specified point. The **Ver** option draws a vertical xline through a specified point. After the first xline is drawn, the **XLINE** tool remains active, allowing you to create multiple xlines using the same option. Place as many horizontal or vertical xlines as needed, and then exit the **XLINE** tool.

Figure 8-8.
An example of a drawing laid out using construction lines.

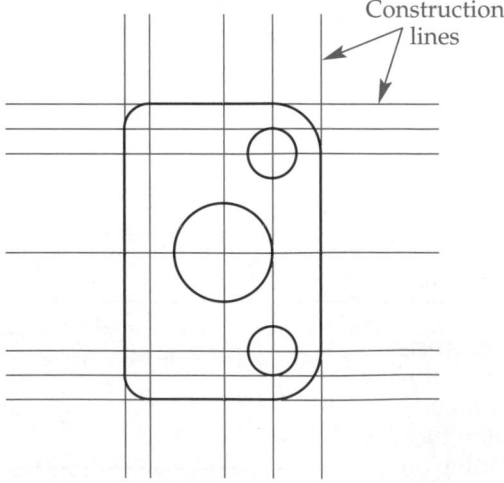

Construction lines

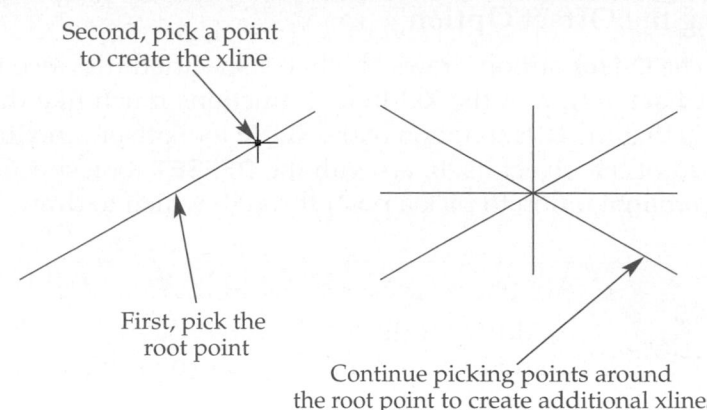

Figure 8-9.
Using the **XLINE** tool to draw an infinitely long construction line by picking two points through which the line passes.

Second, pick a point to create the xline

First, pick the root point

Continue picking points around the root point to create additional xlines

Using the Ang Option

Use the **Ang** option to draw an xline at a specified angle through a selected point. Access the **XLINE** tool and activate the **Ang** option. Enter an angle and then pick a point through which the xline passes. Another method is to select two points to define the angle. The **Reference** option of the **Ang** option allows you to reference the angle of an existing line object to use as the xline angle. This option is useful when you do not know the angle of the xline, but you know the angle between an existing object and the xline. See **Figure 8-10**.

Using the Bisect Option

The **Bisect** option draws an xline that bisects a specified angle, using the root point as the vertex. This is a convenient option for use in some geometric constructions. See **Figure 8-11**.

Figure 8-10.
Using **Reference** with the **Ang** option of the **XLINE** tool. A value of 90° is used in this example to make an xline perpendicular to the existing line.

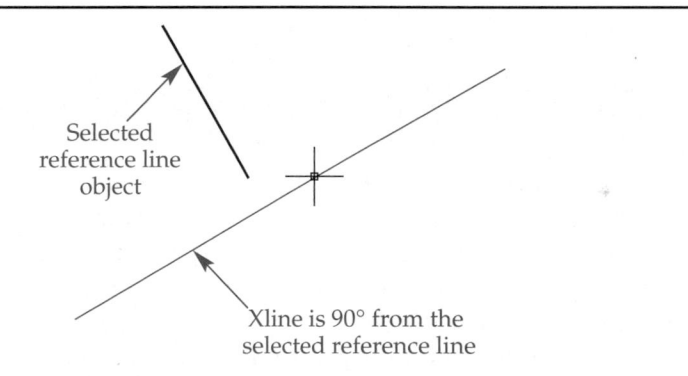

Selected reference line object

Xline is 90° from the selected reference line

Figure 8-11.
Using the **Bisect** option of the **XLINE** tool.

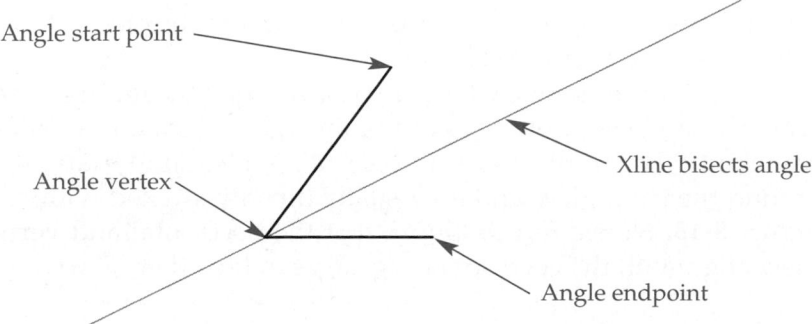

Angle start point

Angle vertex

Xline bisects angle

Angle endpoint

Using the Offset Option

The **Offset** option draws an xline a specified distance from a selected line object. The **Offset** option of the **XLINE** tool functions much like the **OFFSET** tool. The difference is that the **Offset** option of the **XLINE** tool offsets an xline from the selected object, instead of the object itself. As with the **OFFSET** tool, specify an offset distance or use the **Through** option to pick a point through which to draw the construction line.

> **NOTE**
>
> Although xlines are infinite, they do not change the drawing extents, and they have no effect on zooming operations.

Exercise 8-4

Access the Student Web site (www.g-wlearning.com/CAD) and complete Exercise 8-4.

Using the RAY Tool

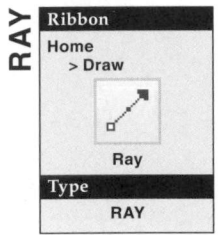

RAY

Ribbon
Home
> Draw
Ray
Type
RAY

The **RAY** tool allows you to specify the point of origin and a point through which the ray passes. In this respect, the **RAY** tool works much like the default option of the **XLINE** tool. The ray, however, is infinite only in the direction of the second pick point. Once you access the **RAY** tool, specify the root point of the ray. Then pick a second point to create the ray. Continue locating points to create additional rays from the same root point if necessary. To exit the tool, right-click or press [Enter], [Esc], or the space bar.

Multiview Drawings

Each drafting field has its own methods of presenting views of a product. Architectural drafting uses plan views, exterior elevations, and sections. In electronics drafting, a schematic diagram with electronic symbols shows circuit layout. In civil drafting, plan views and profiles show the topography of land. Mechanical drafting uses *multiview drawings*.

multiview drawings: Presentation of drawing views created through orthographic projection.

This textbook explains multiview drawings based on the ASME Y14.3M, *Multiview and Sectional View Drawings* standard. Multiview drawing views are created through *orthographic projection*. An imaginary *projection plane* is placed parallel to the object. Thus, the line of sight is perpendicular to the object. This results in 2D views of a 3D object. See **Figure 8-12.**

orthographic projection: Projecting object features onto an imaginary plane.

projection plane: The imaginary projection plane that is parallel to the object.

Six 2D orthographic views are possible: front, right side, left side, top, bottom, and rear. These views show all sides of an object, drawn in a standard arrangement for readability. The front view is the central, or most important, view. Other views occur around the front view and are usually directly aligned with or projected from it. See **Figure 8-13.** Notice in this figure that the horizontal and vertical edges in the front view align with the corresponding edges in the other views.

AutoCAD and Its Applications—Basics

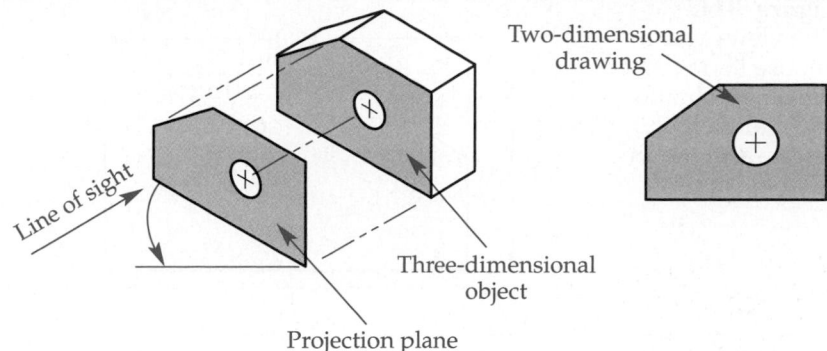

Figure 8-12. Obtaining a front view with orthographic projection.

Two-dimensional drawing

Line of sight

Three-dimensional object

Projection plane

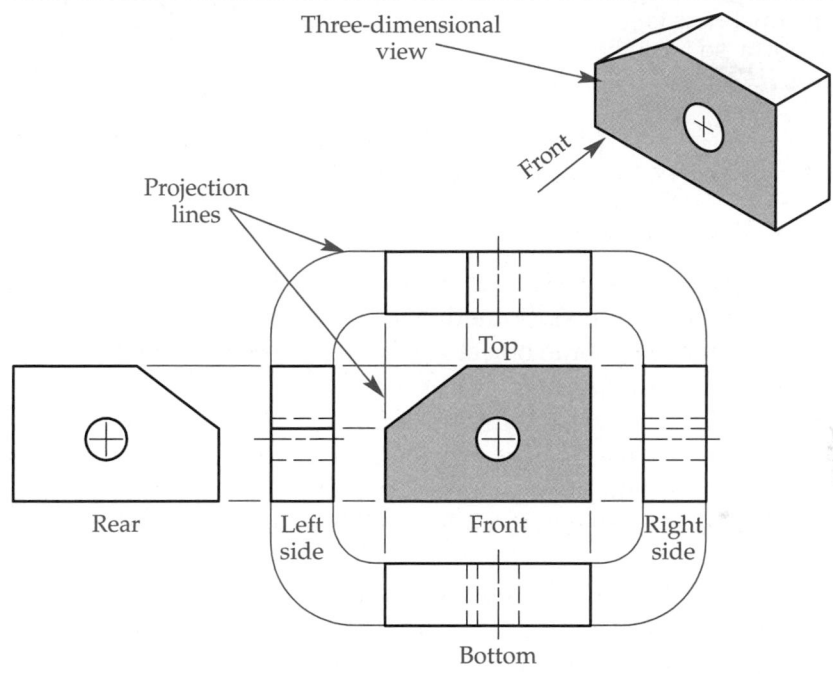

Figure 8-13. Arrangement of the six orthographic views.

Three-dimensional view

Front

Projection lines

Top

Rear

Left side

Front

Right side

Bottom

Selecting the Front View

The front view is usually the most descriptive view. Consider the following rules when selecting the front view:

✓ Most descriptive
✓ Most natural position
✓ Most stable position
✓ Provides the longest dimension
✓ Contains the least number of hidden features

Choosing Additional Views

Select additional views relative to the front view. Very few products require all six views. The required number of views depends on the complexity of the object. Use only enough views to completely describe the object. Drawing too many views is time-consuming and can clutter the drawing. The object shown in **Figure 8-14** needs only two views. These two views completely describe the width, height, depth, and features of the object. In some cases, a single view is enough to describe the object. You can often draw a thin part that has as uniform thickness, such as a gasket, with one view. **Figure 8-15** shows an example of a one-view drawing in which the part thickness is given as a note in the drawing or in the title block, eliminating the need for a second view.

Figure 8-14.
The views you choose to describe the object should show all height, width, and depth dimensions.

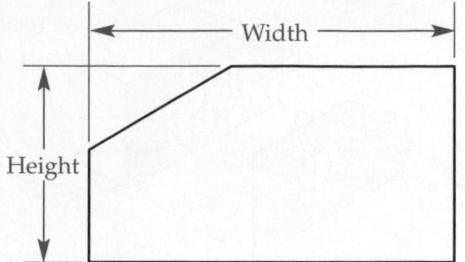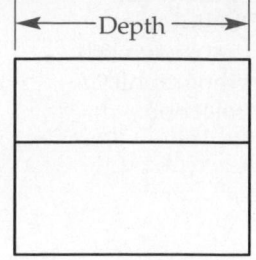

Figure 8-15.
A one-view drawing of a gasket. The thickness is uniform, so it can be given in a note.

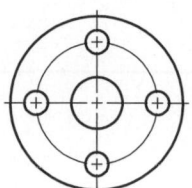

NOTE: THICKNESS 1.5mm

Auxiliary Views

Often you can completely describe an object using a combination of one or more of the six standard views. Sometimes, however, the orthographic views are not enough to properly identify some object surfaces. A *foreshortened* surface is typically described using an *auxiliary view*.

An auxiliary view is drawn by projecting lines perpendicular to a slanted surface. One projection line is often included on the drawing. It connects the auxiliary view to the view where the slanted surface appears as a line. The resulting auxiliary view shows the surface at its true size and shape. For most applications, a *partial auxiliary view* is enough. See **Figure 8-16.**

In some situations, there is not enough room on the drawing to project directly from the slanted surface. This requires that you locate the auxiliary view elsewhere. See **Figure 8-17.** A *viewing-plane line* is drawn next to the view where the slanted surface appears as a line. It terminates with bold arrowheads that point toward the slanted surface. A letter labels each end of the viewing-plane line. The letters relate the viewing-plane line with the proper auxiliary view. The auxiliary view includes a title, such as VIEW A-A, below the view to key the viewing plane with the view. When more

foreshortened: A surface at an angle to the line of sight. Foreshortened surfaces appear shorter than their true size and shape.

auxiliary view: View used to show a foreshortened surface at its true size and shape.

partial auxiliary view: An auxiliary view that shows only a single inclined surface of an object, rather than the entire object.

viewing-plane line: A thick dashed or phantom line identifying the viewing direction of a related view.

Figure 8-16.
Auxiliary views show the true size and shape of an inclined surface.

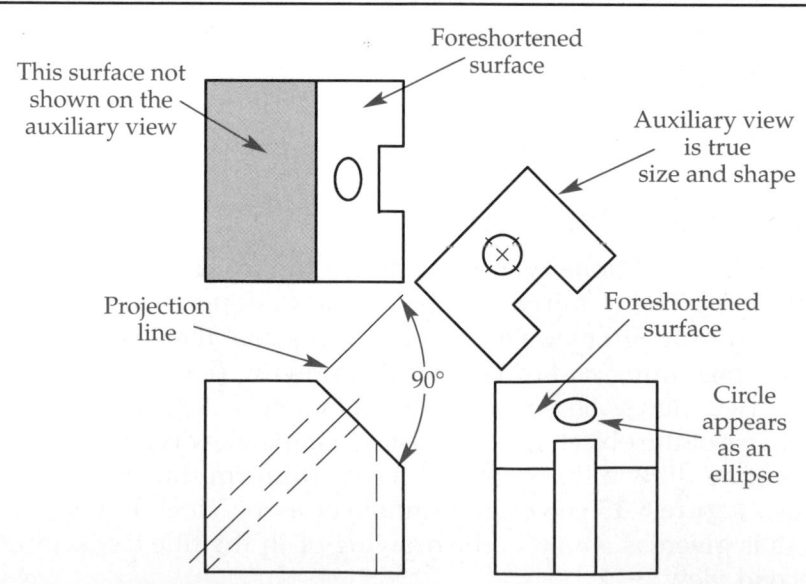

Figure 8-17.
When you add a viewing-plane line, you have the flexibility to move a view to a location where there is enough space to draw the view.

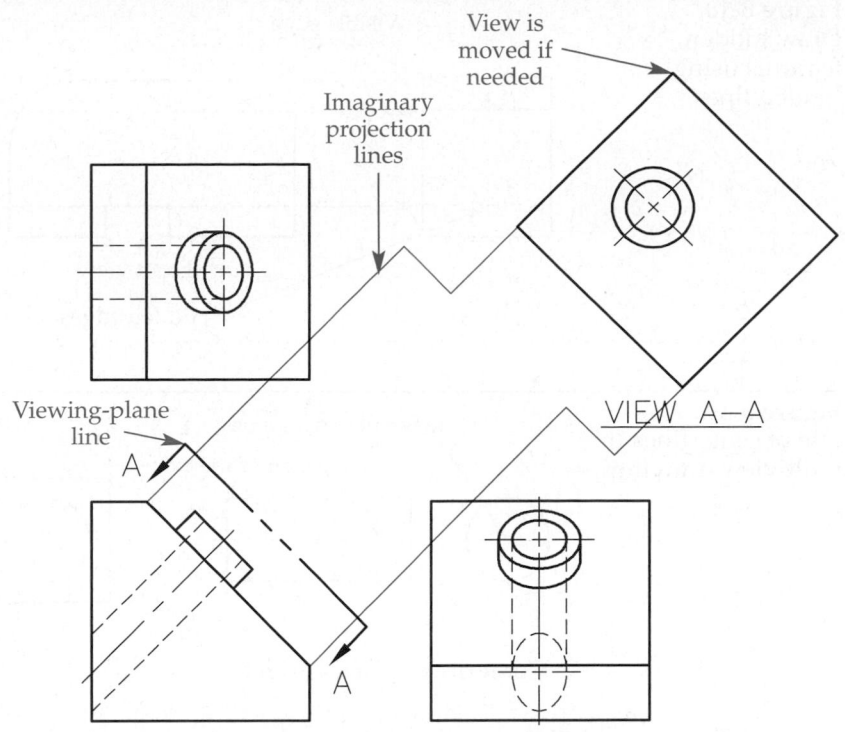

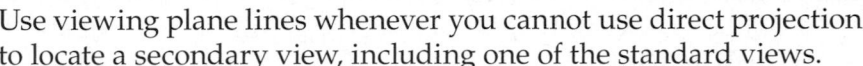

than one auxiliary view is drawn, labels continue with B-B through Z-Z, if necessary. The letters *I*, *O*, and *Q* are not used because they can be confused with numbers. An auxiliary view drawn away from the standard view retains the same angle as if it were projected directly.

> **NOTE**
>
> Use viewing plane lines whenever you cannot use direct projection to locate a secondary view, including one of the standard views.

Showing Hidden Features

Hidden features of an object are typically shown, even though they are not visible in the view at which you are looking. Visible edges appear as object lines. Hidden edges appear as hidden lines. Hidden lines are thin to provide contrast with thick object lines. See **Figure 8-18.**

Showing Symmetry and Circle Centers

Centerlines indicate the centerlines of symmetrical objects and the centers of circles. For example, in the circular view of a cylinder, centerlines cross to show the center of the cylinder. In the other view, a centerline identifies the axis. See **Figure 8-19.** The only place the small centerline dashes should cross is at the center of a circle, arc, ellipse, or other circular feature.

Figure 8-18.
Draw hidden features using hidden lines.

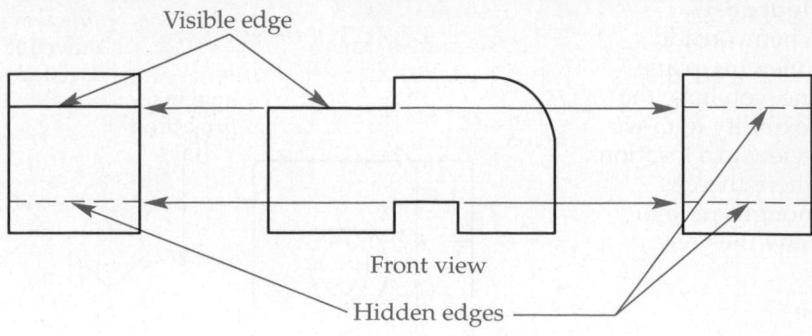

Visible edge

Front view

Hidden edges

Figure 8-19.
Use of centerlines in multiview drawings.

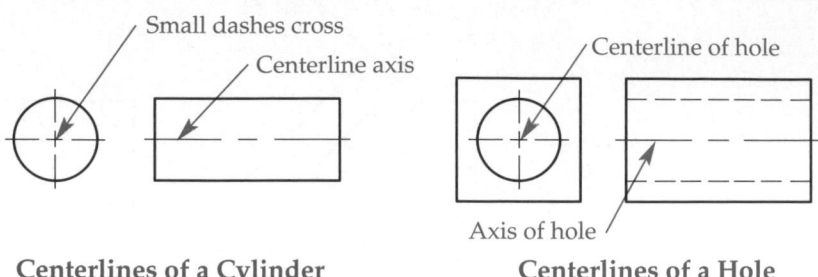

Small dashes cross

Centerline axis

Centerline of hole

Axis of hole

Centerlines of a Cylinder

Centerlines of a Hole

Constructing Multiview Drawings

You can construct multiview drawings using a variety of techniques, depending on the objects needed, personal working preference, and the information you know about the size and shape of items. Often you will use a combination of methods, including various coordinate point entries, object snaps, AutoTrack, and construction lines to produce drawings.

Drawing Orthographic Views

Figure 8-20 shows an example of how you can use object snap tracking to locate points for a new view by referencing points on existing views. In this example, a left-side view is constructed from an existing front view using object snap tracking and a running **Endpoint** object snap mode. Notice that the AutoTrack alignment path in **Figure 8-20A** provides a temporary construction line. Polar tracking vectors offer a similar type of temporary construction line. **Figure 8-20B** shows the completed front and left-side views.

Figure 8-21 shows an example of how you can use construction lines to form three views. In this example, vertical and horizontal xlines are offset to form an xline grid. Use the xline intersections to locate line and arc endpoints and the center point of the arc. Use the **Intersection** object snap mode to select the intersecting xlines. Notice that a single infinitely long xline can provide construction geometry for multiple views.

Exercise 8-5

Access the Student Web site (www.g-wlearning.com/CAD) and complete Exercise 8-5.

Figure 8-20.
An example of using object snap tracking to create an additional view. A—Referencing points from the front view to establish the first line of the left-side view. B—The completed front and left-side views.

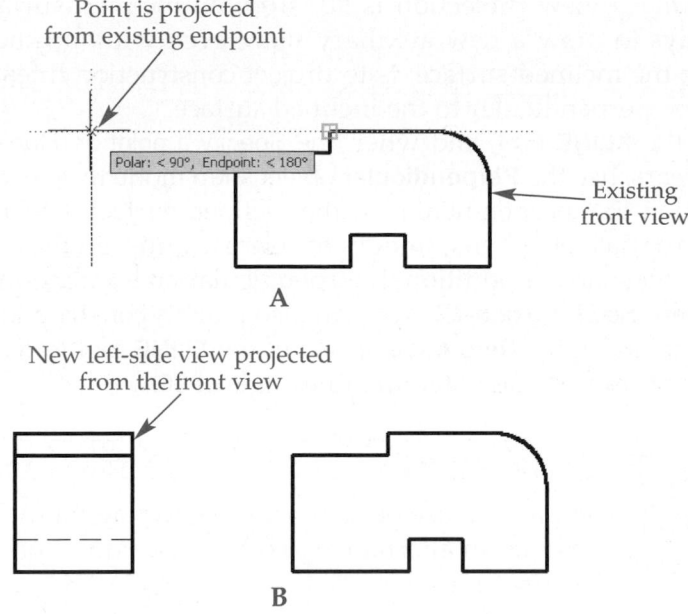

Figure 8-21.
Using a complete grid of construction lines to form a multiview drawing by "connecting the dots" at the construction line intersections. You can quickly draw the rectangular outlines of the right-side, left-side, and top views using the **RECTANGLE** tool.

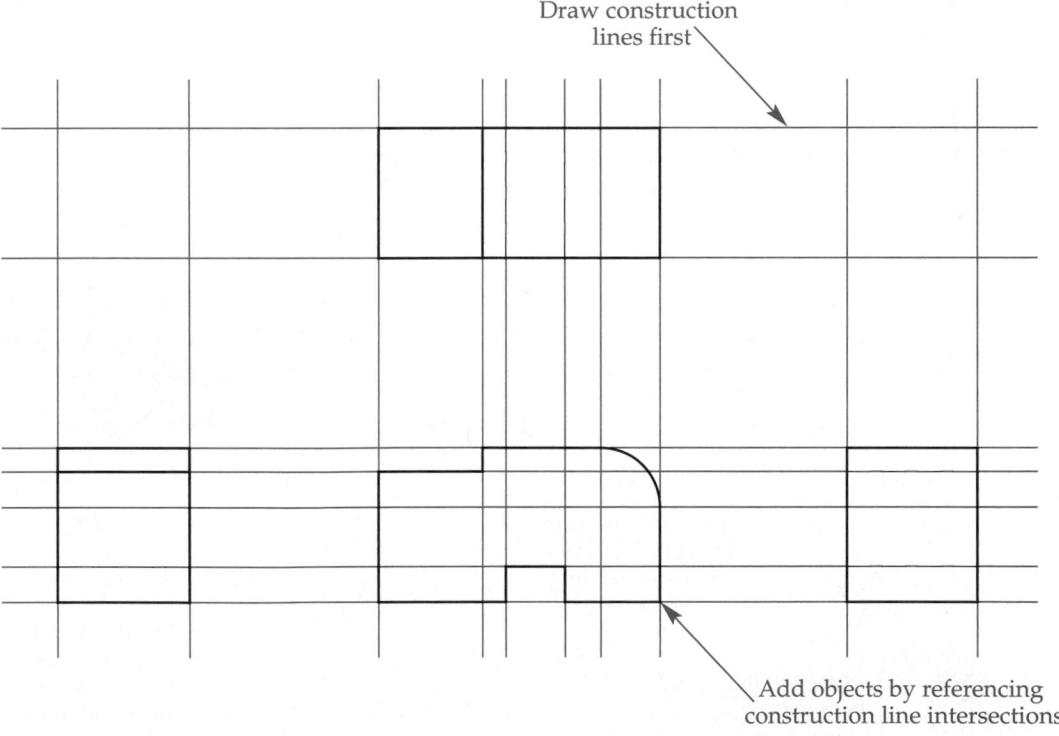

Drawing Auxiliary Views

Construct auxiliary views using the same tools and options you use to draw any of the six primary views. However, constructing auxiliary views presents unique challenges. Auxiliary view projection is 90° from an inclined surface. One of the most effective ways to draw a new auxiliary view, even without knowing or calculating the angle of the inclined surface, is to project construction lines from features on an existing view, perpendicular to the inclined surface.

Access the **XLINE** tool, and when the Specify a point or [Hor/Ver/Ang/Bisect/Offset]: prompt appears, use the **Perpendicular** object snap mode to select the inclined surface. A construction line perpendicular to the inclined surface attaches to the crosshairs. Use the appropriate object snap modes to select features on the existing view. You can then use object snaps or additional perpendicular construction lines to complete the auxiliary view. See **Figure 8-22.** You can also make a construction line perpendicular to a line object using the **Reference** option of the **XLINE** tool **Ang** option. Select the line object when prompted and enter an xline angle of 90.

NOTE

You can also use parametric tools, explained in Chapter 22, in addition or as an alternative to other techniques for constructing multi-view drawings.

Exercise 8-6

Access the Student Web site (www.g-wlearning.com/CAD) and complete Exercise 8-6.

Figure 8-22.
Using construction lines drawn perpendicular to the inclined surface on an existing view to construct an auxiliary view. The completed top and auxiliary views are for reference.

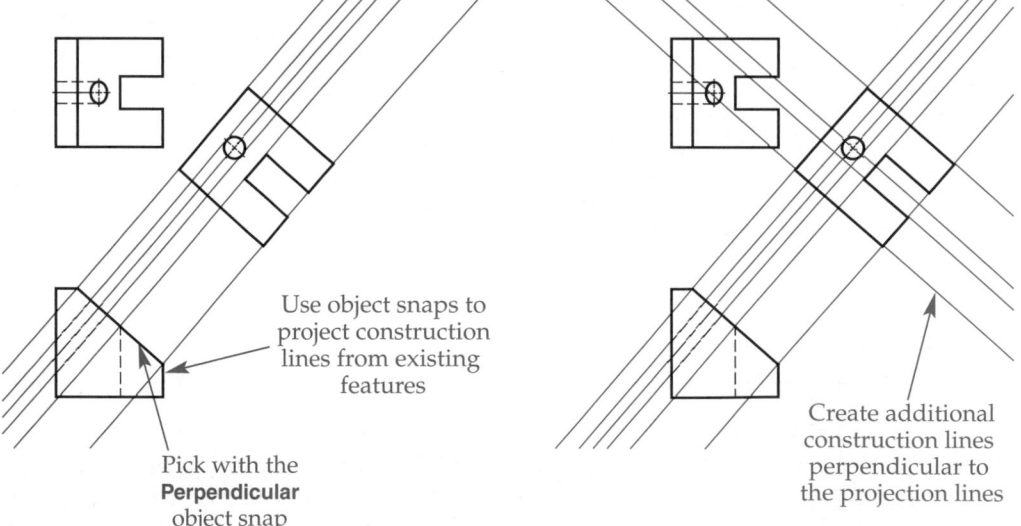

Use object snaps to project construction lines from existing features

Pick with the **Perpendicular** object snap

Create additional construction lines perpendicular to the projection lines

Template Development

Chapter 8

For detailed instructions on choosing a more appropriate point style to use for construction purposes, go to the Student Web site (www.g-wlearning.com/CAD), select this chapter, and select **Template Development**.

Chapter Test

Answer the following questions. Write your answers on a separate sheet of paper or go to the Student Web site (www.g-wlearning.com/CAD) and complete the electronic chapter test.

1. List two ways to establish an offset distance using the **OFFSET** tool.
2. Which option of the **OFFSET** tool allows you to remove the source offset object?
3. How do you draw a single point, and how do you draw multiple points?
4. How do you access the **Point Style** dialog box?
5. If you use the **DIVIDE** tool and nothing appears to happen, what should you do?
6. How do you change the point size in the **Point Style** dialog box?
7. What tool can you use to place point objects that divide a line into 24 equal parts?
8. What is the difference between the **DIVIDE** and **MEASURE** tools?
9. Why is it a good idea to put construction lines on their own layer?
10. Name the tool that allows you to draw infinite construction lines.
11. Name the option that allows you to bisect an angle with a construction line.
12. What is the difference between the construction lines drawn with the tool identified in Question 10 and rays drawn with the **RAY** tool?
13. What ASME drafting standard applies to multiview drawings?
14. Provide at least four guidelines for selecting the front view of an orthographic multiview drawing.
15. How do you determine how many views of an object are necessary in a multiview drawing?
16. When can you describe a part with only one view?
17. When does a drawing require an auxiliary view, and what does an auxiliary view show?
18. What is the angle of projection from the slanted surface into the auxiliary view?
19. List two methods of aligning the views in a multiview drawing.
20. Describe an effective method of constructing an auxiliary view even if you do not know the angle of the inclined surface.

Drawing Problems

Start AutoCAD if it is not already started. Start a new drawing using an appropriate template of your choice. The template should include layers for drawing the given objects. Add layers as needed. Draw all objects using appropriate layers. Follow the specific instructions for each problem. Do not draw dimensions or text. Use your own judgment and approximate dimensions when necessary.

▼ Basic

1. Draw the front and side views of this offset support. Use object snap modes and tracking. Save your drawing as **P8-1**.

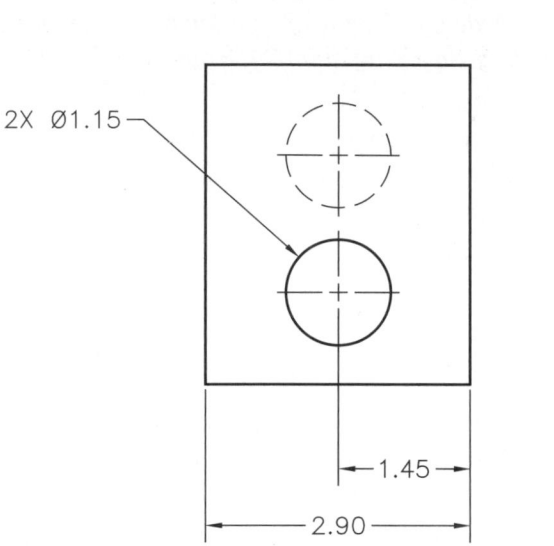
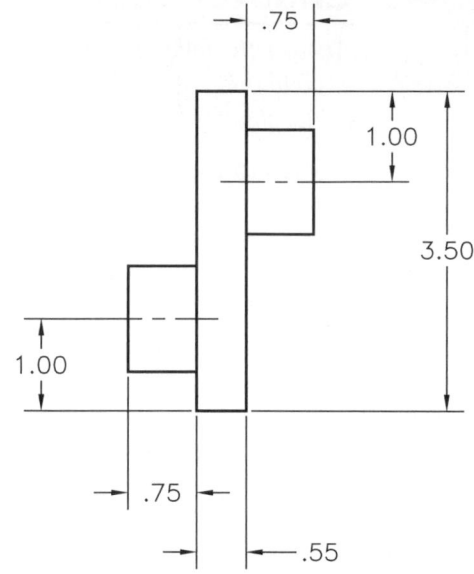

2. Draw the top and front views of this hitch bracket. Use object snap modes and tracking. Save your drawing as **P8-2**.

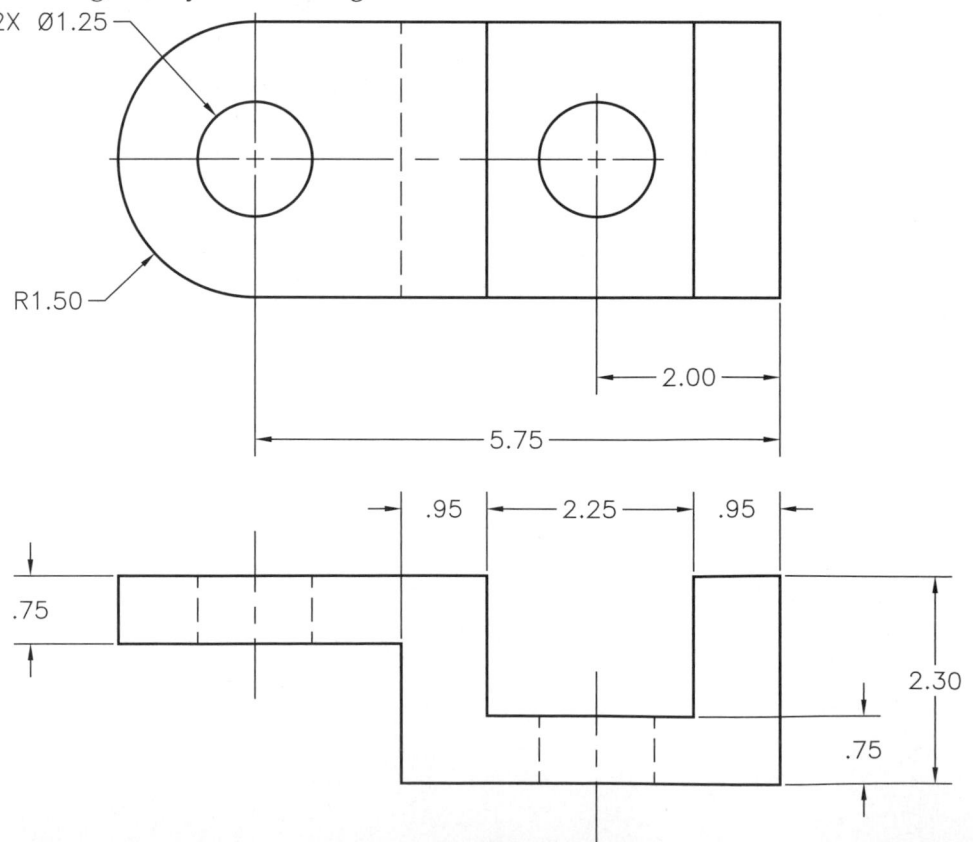

3. Draw this spring using the **OFFSET** tool for material thickness. Save the drawing as P8-3.

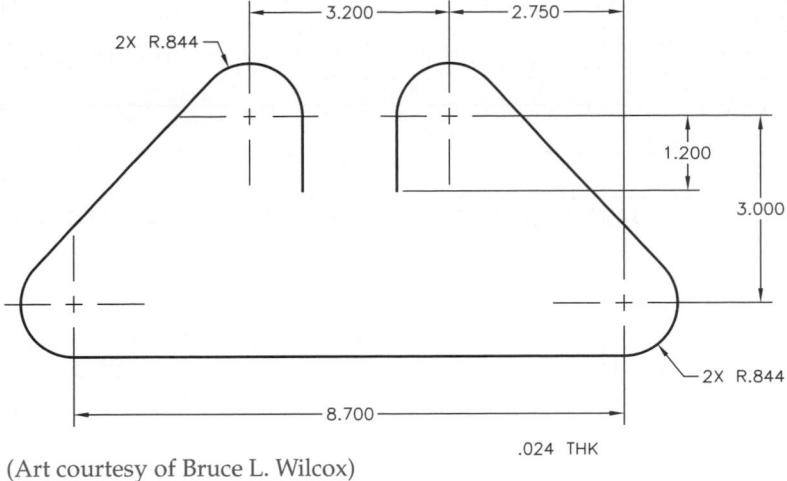

(Art courtesy of Bruce L. Wilcox)

4. Draw this sheet metal chassis. Use object snap tracking and polar tracking to your advantage. Save the drawing as P8-4.

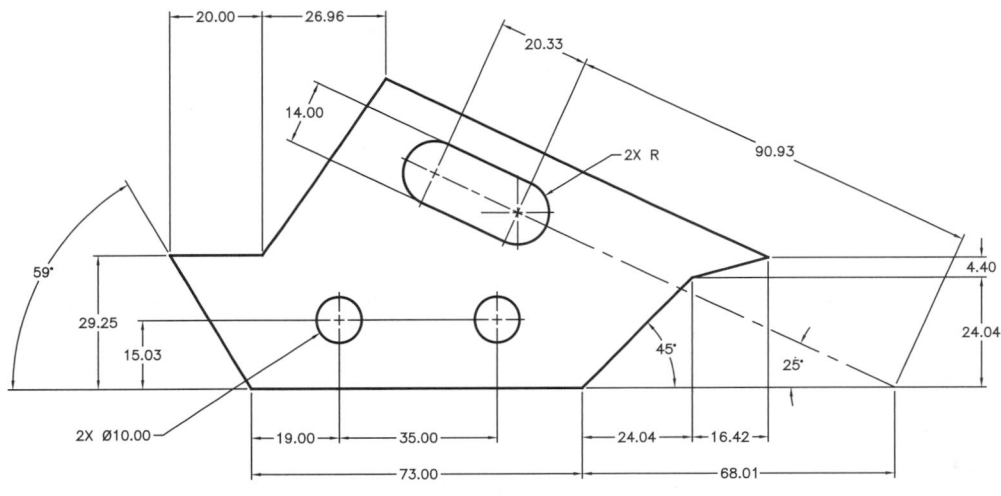

(Art courtesy of Bruce L. Wilcox)

▼ Intermediate

5. Use the **OFFSET** tool to draw the elevation of the desk shown. Center 1″ × 4″ rectangular drawer handles 2″ below the top of each drawer. The top of the legs begin 1″ from the edge of the bottom of the desk. Save the drawing as P8-5.

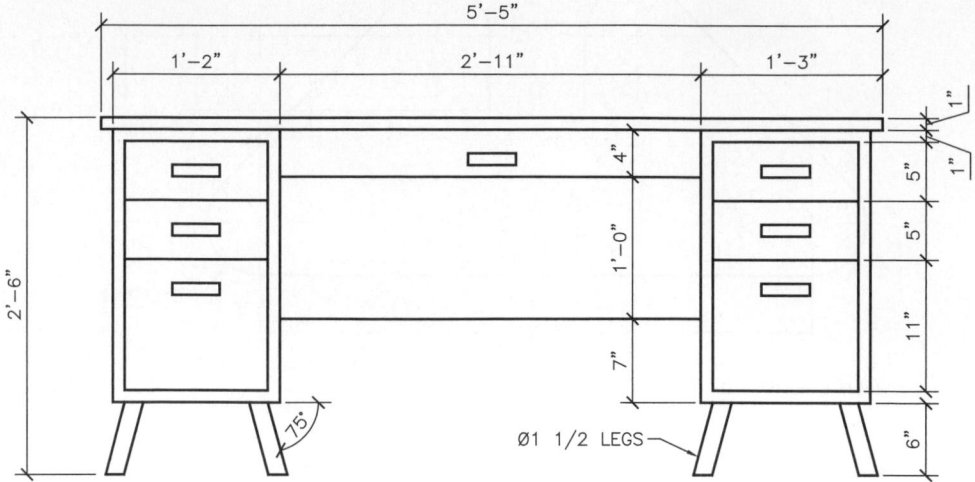

6. Draw this aluminum spacer. Use object snap modes and tracking. Save the drawing as P8-6.

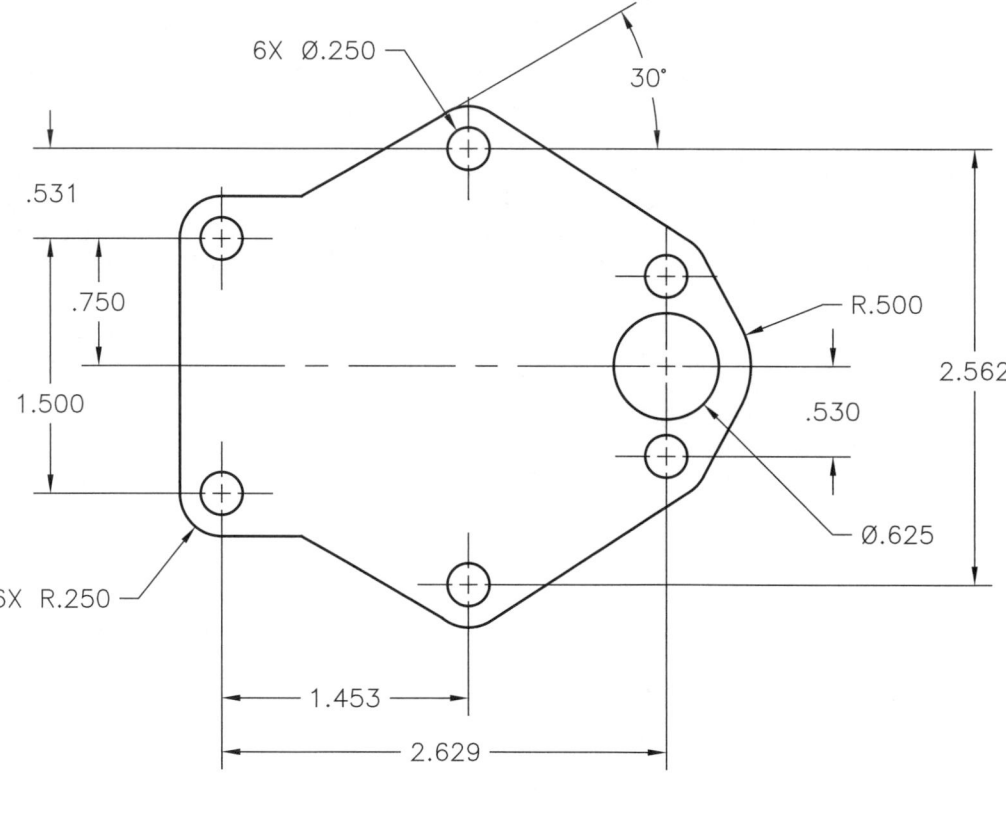

7. Draw this gasket. Save the drawing as **P8-7**.

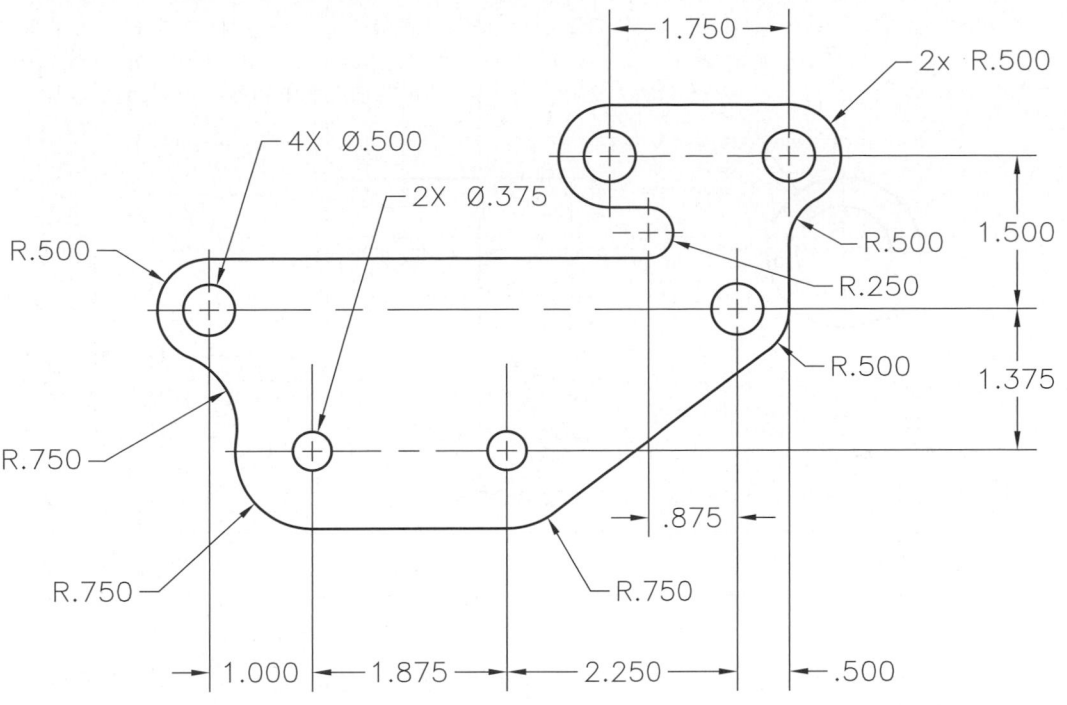

8. Draw this cup. Save the drawing as **P8-8**.

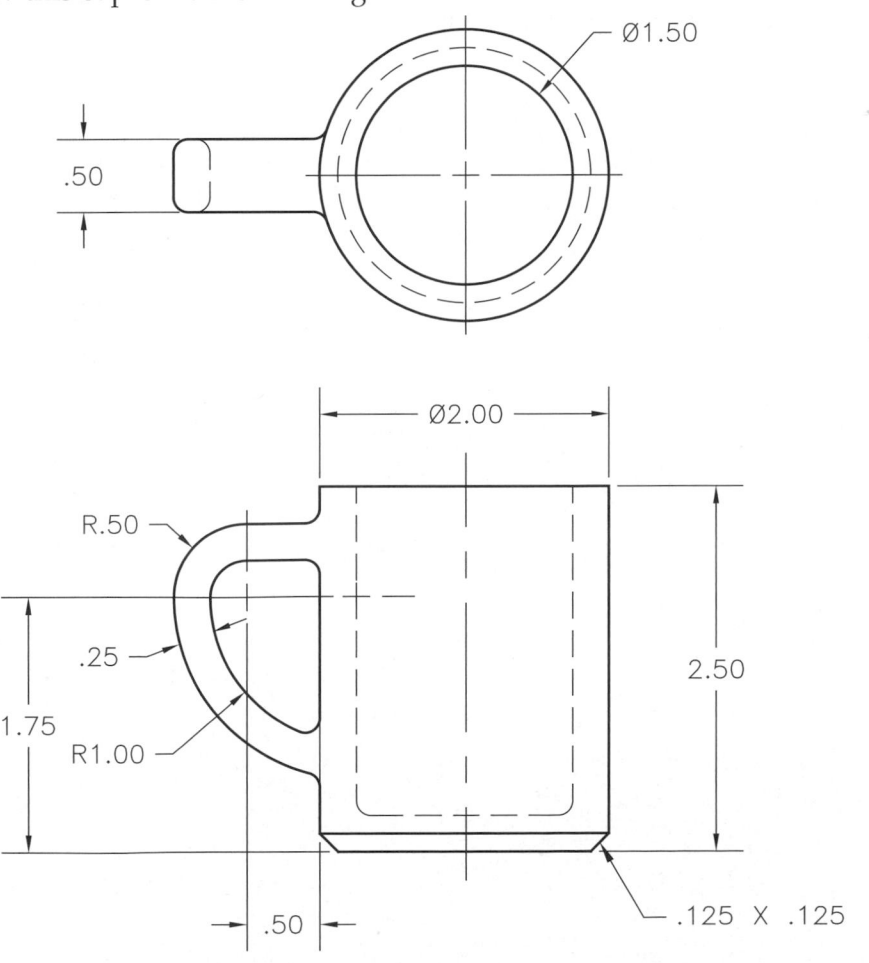

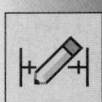

FILLETS AND ROUNDS R.10

9. Draw this bushing. Save the drawing as P8-9.

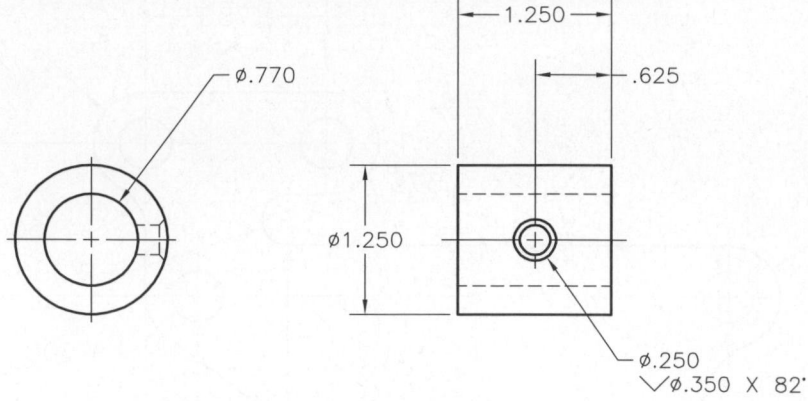

10. Draw this wrench. Save the drawing as P8-10.

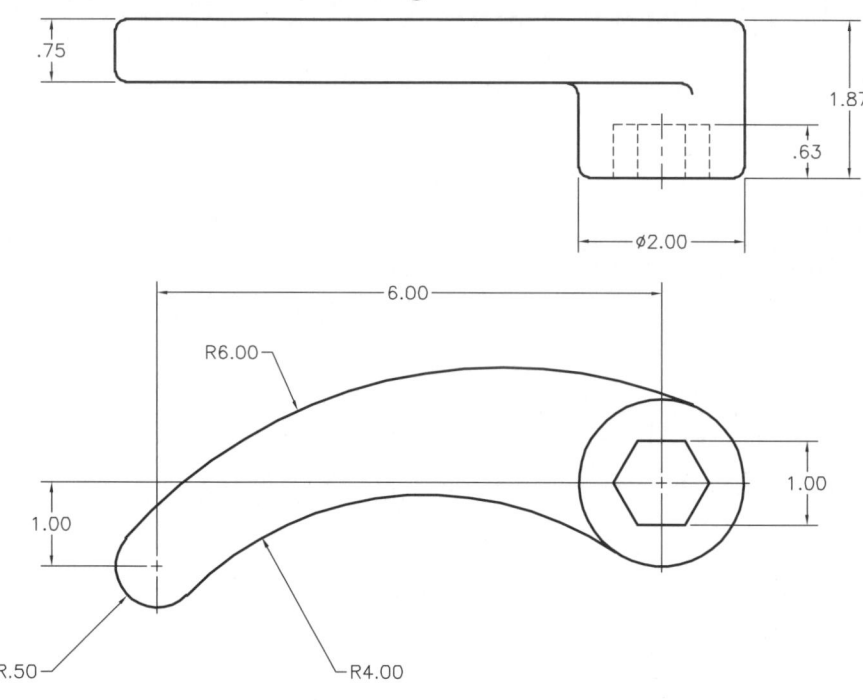

FILLETS AND ROUNDS = .125

11. Draw this support. Save the drawing as **P8-11**.

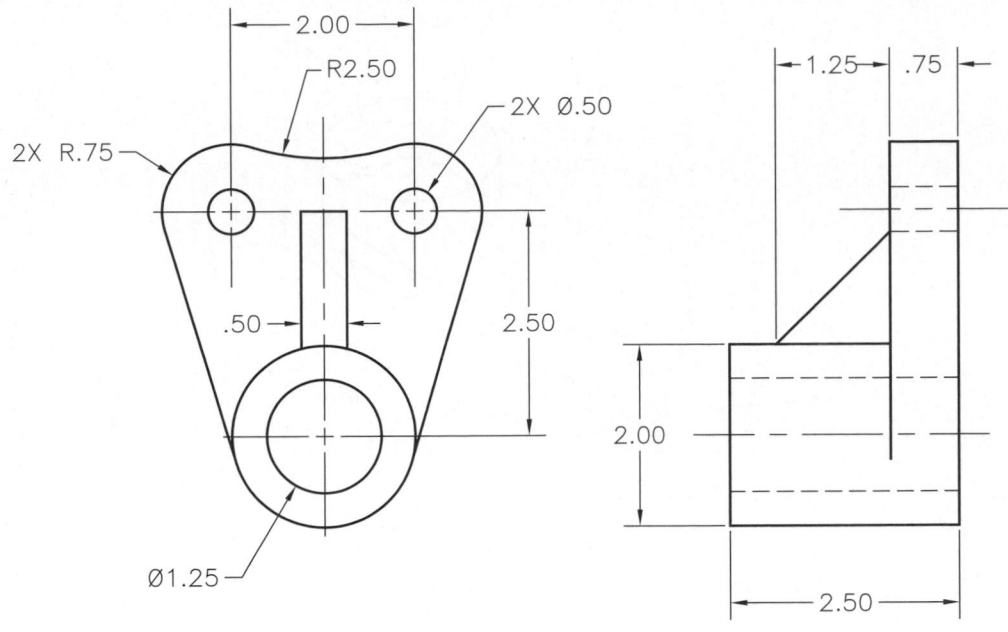

▼ Advanced

In Problems 12 through 16, draw the views needed to completely describe the objects. Use object snap modes, AutoTrack modes, xlines, rays, and offsets as needed. Save the drawings as P8-(problem number).

12.

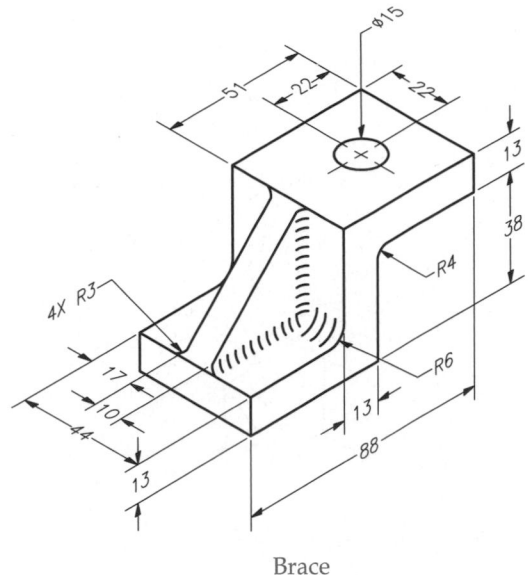

Brace

13.

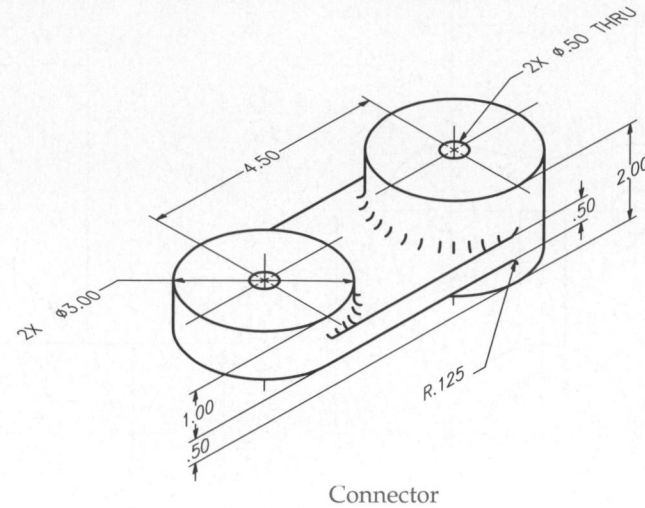

Connector

14.

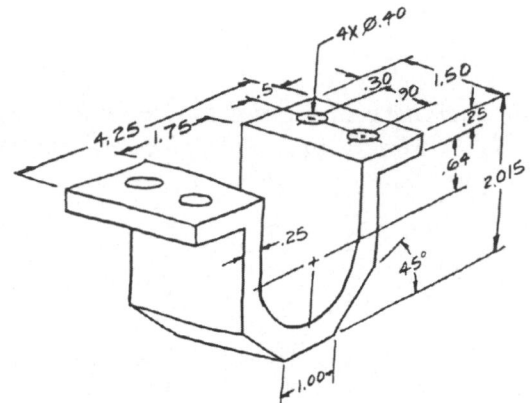

Journal Bracket (Engineer's Rough Sketch)

15.

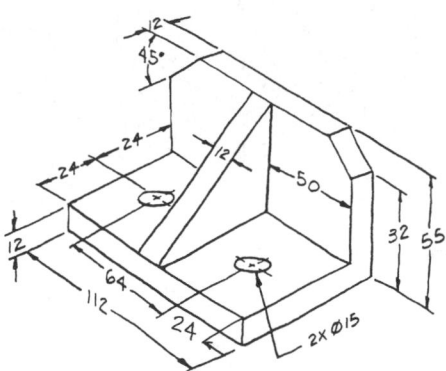

Angle Bracket (Engineer's Rough Sketch)
(Metric)

16.

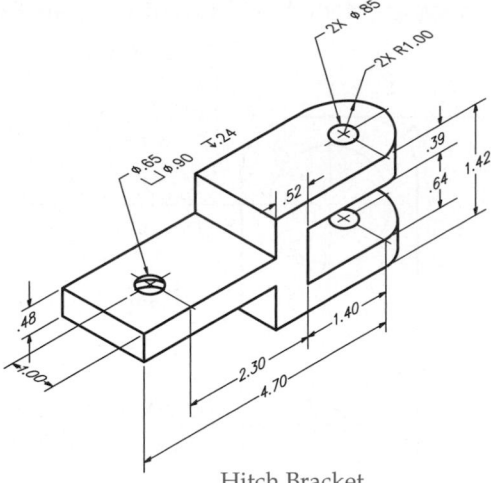

Hitch Bracket

17. Draw the views of this pillow block, including the auxiliary view. Save your drawing as P8-17.

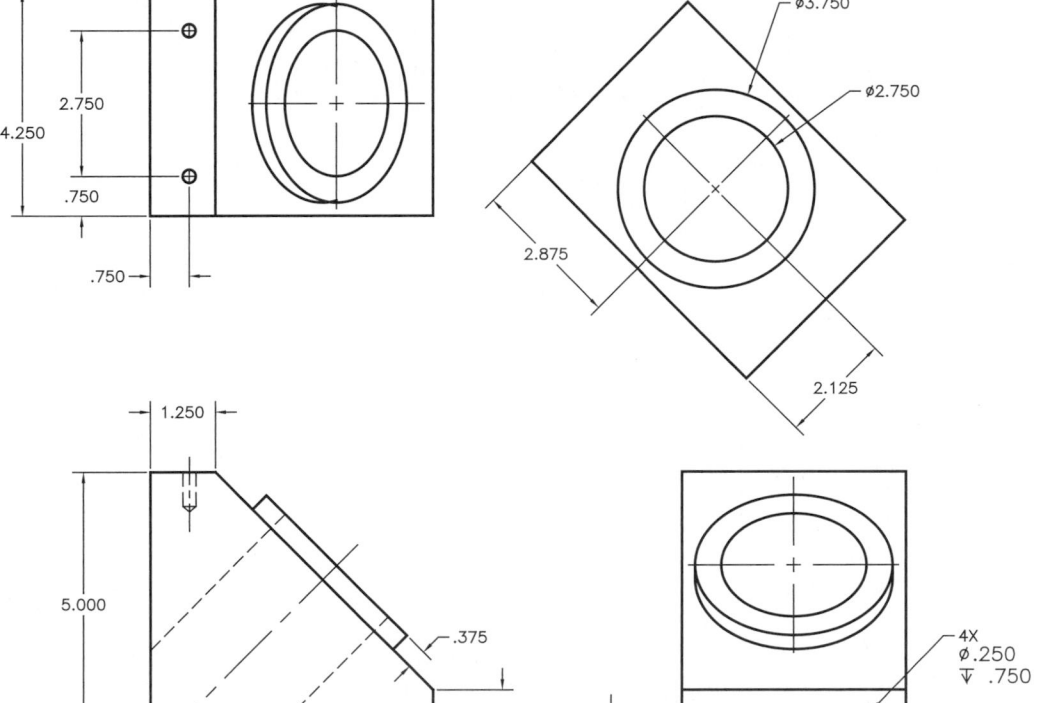

18. Draw the folded and flat pattern views of the sheet metal bracket shown. The part is made from 18-gauge steel. Save the drawing as P8-18.

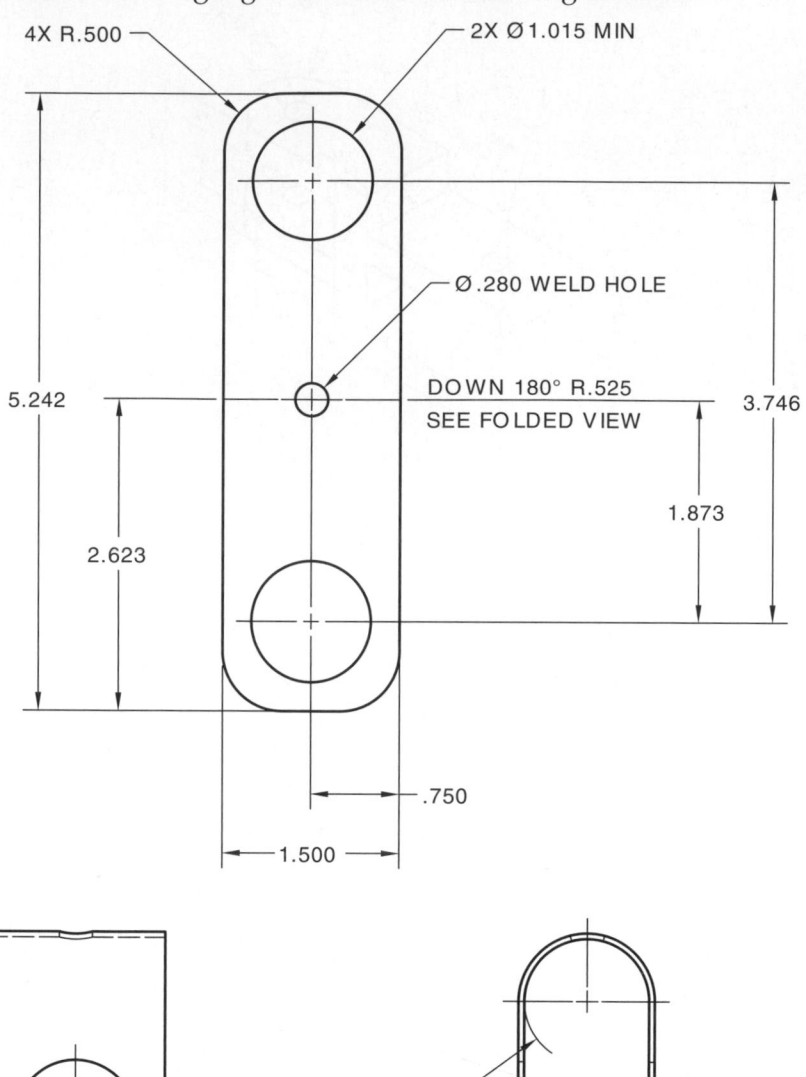

4X R.500

2X Ø1.015 MIN

Ø.280 WELD HOLE

DOWN 180° R.525
SEE FOLDED VIEW

5.242

3.746

2.623

1.873

.750

1.500

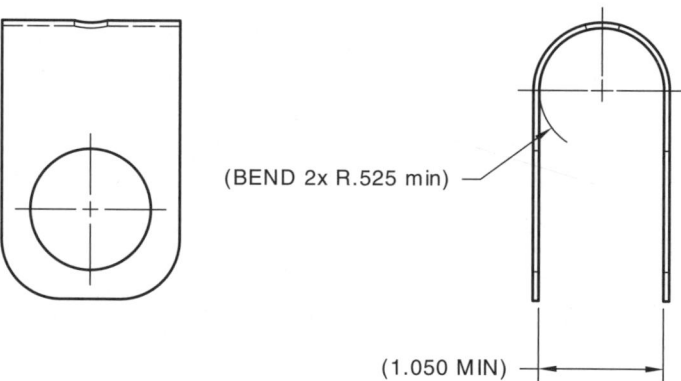

(BEND 2x R.525 min)

(1.050 MIN)

CHAPTER

9

Text Styles and Multiline Text

Learning Objectives

After completing this chapter, you will be able to do the following:

✓ Describe and use proper text standards.
✓ Calculate drawing scale and text height.
✓ Develop and use text styles.
✓ Use the **MTEXT** tool to create multiline text objects.

Letters, numbers, words, and notes, known as *annotation*, describe drawing information that is not shown using objects and symbols. CADD programs significantly reduce the tedious nature of adding *text* to a drawing. When drawn correctly, AutoCAD text is consistent and easy to read. This chapter introduces text standards and composition, and explains how to use the **MTEXT** tool to prepare a single text object that may consist of multiple lines of text, such as paragraphs or a list of general notes. Chapter 10 describes how to create single-line text using the **TEXT** tool, as well as other tools for working with text.

annotation: Letters, numbers, words, and notes used to describe information on a drawing.

text: Lettering on a CADD drawing.

Text Standards and Composition

Industry and company standards dictate how text appears on a drawing. **Figure 9-1** lists minimum letter heights based on the ASME Y14.2M, *Line Conventions and Lettering* standard. The U.S. National CAD Standard (NCS) and many companies, especially those who produce architectural and civil drawings, depart slightly from the ASME standard. The NCS specifies a minimum text height of 3/32″ (2.4 mm). Most text is drawn 1/8″ (3 mm) high, with titles and other unique text 1/4″ (3 mm) high. Regardless of the text height, all text should be consistent and easy to read.

Vertical text is standard on engineering drawings. Inclined text is used by some companies, especially those in structural drafting, and for specific drawing requirements, such as water features labeled on maps. **Figure 9-2** shows examples of vertical and inclined text. The recommended slant for inclined text is 68° from horizontal, though some drafters find 75° more appropriate. Text on a drawing is normally uppercase. Lowercase letters are sometimes used, most often on civil plans or maps. Typically, a drawing displays the same text format throughout. In some cases, a combination of text formats is used, such as text in architectural title blocks or on maps.

Figure 9-1.
Minimum letter heights based on the ASME Y14.2M, *Line Conventions and Lettering* standard.

Application	Height INCH	Height METRIC (mm)
Most text (dimension values, notes)	.12	3
Drawing title, drawing size, CAGE code, drawing number, revision letter	.24* .12**	6* 3**
Section and view letters	.24	6
Zone letters and numerals in borders	.24	6
Drawing block headings	.10	2.5

*D, E, F, H, J, K, A0, and A1 size sheets

**A, B, C, G, A2, A3, and A4 size sheets

Figure 9-2.
Vertical and inclined text.

ABC.. abc.. 123..
ABC.. abc.. 123..

Numbers in dimensions and notes are the same height as standard text. AutoCAD provides several methods for stacking text in fractions. When dimensions contain fractions, the fraction bar is usually placed horizontally between the numerator and denominator. However, many notes placed on drawings have fractions displayed with a diagonal fraction bar (/). In this case, use a dash or space between the whole number and the fraction. **Figure 9-3** shows examples of text for numbers and fractions in different unit formats.

AutoCAD text tools provide great control over text *composition*. You can lay out text horizontally, as is typical when adding notes, or draw text at any angle according to specific requirements. AutoCAD automatically spaces letters and lines of text. This helps maintain the identity of individual notes.

composition: The spacing, layout, and appearance of text.

NOTE

Several drawing variables can affect text height and format. Refer to appropriate industry, company, or school standards when adding text to your drawings.

Figure 9-3.
Examples of fractional text for different unit formats.

Decimal Inch	Fractional Inch	Millimeter
2.750 .25	$2\frac{3}{4}$ 2–3/4 2 3/4	2.5 3 0.7

Text presentation is important. Consider the following tips when adding text:

- Plan your drawing using rough sketches to allow room for text and notes.
- Arrange text to avoid crowding.
- Place related notes in groups to make the drawing easy to read.
- Place all general notes in a common location. Locate notes in the lower-left corner or above the title block when using ASME standards. Place notes in the upper-left corner when using military standards.
- Always use the spell checker.

Drawing Scale and Text Height

Ideally, you should determine drawing scale, scale factors, and text heights before you begin drawing. Incorporate these settings into your drawing template files, and make changes when necessary. The drawing scale factor is important because it determines how text appears on-screen and plots. To help understand the concept of drawing scale, look at the portion of a floor plan shown in **Figure 9-4.** You should draw everything in model space at full scale. This means that the bathtub, for example, is actually drawn 5' long. However, at this scale, text size becomes an issue, because full scale text that is 1/8" high is extremely small compared to the other full-scale objects. See **Figure 9-4A.** As a result, you must adjust the height of the text according to the drawing scale. See **Figure 9-4B.** You can calculate the scale factor manually and apply it to the text height, or you can allow AutoCAD to calculate it by using annotative text.

Figure 9-4.
An example of a portion of a floor plan drawn at full scale in model space. Text drawn at full scale, as shown in A, is very small compared to the large objects. You must scale text, as shown in B, in order to display and plot text correctly.

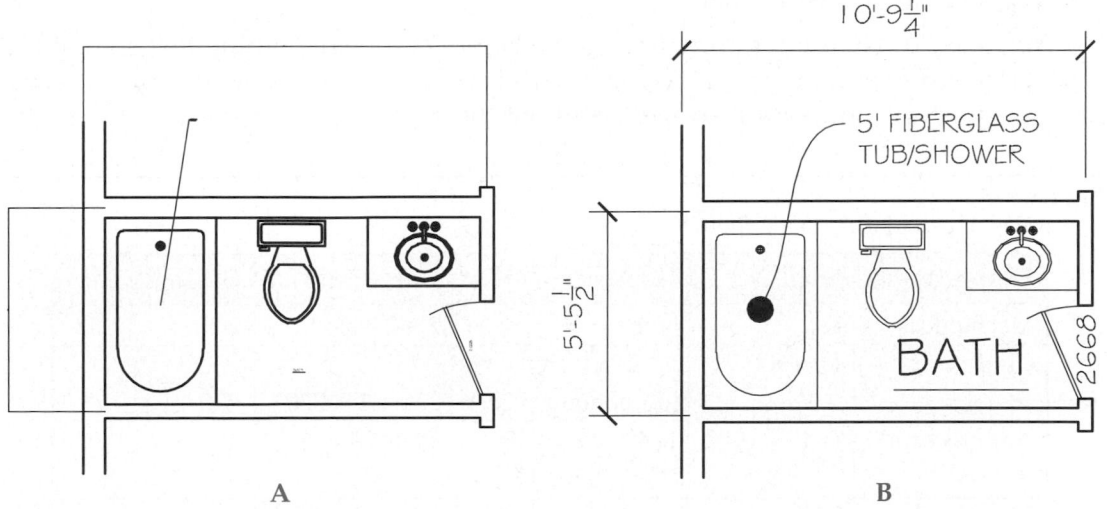

A   B

Scaling Text Manually

text height: The specified height of text, which may be different from the plotting size for text scaled manually.

scale factor: The reciprocal of the drawing scale.

paper text height: The plotted text height.

To adjust *text height* manually according to a specific drawing scale, you must calculate the drawing *scale factor*. **Figure 9-5** provides examples of calculating scale factor. Once you determine the scale factor, you then multiply the scale factor by the desired *paper text height* to get the model space text height.

For example, if you are working on a civil engineering drawing with a 1″ = 60′ scale, text drawn 1/8″ high is almost invisible. Remember, the drawing you are working on is 720 times larger than it is when plotted at the proper scale. Therefore, you must multiply the 720 scale factor by the desired text height. Text height multiplied by scale factor equals the model space scaled text height. So, in this example, multiply: 1/8″ (.125″) × 720 = 90″. The proper height of 1/8″ text in model space is 90″.

Annotative Text

annotative text: Text that is scaled by AutoCAD according to the specified annotation scale.

annotation scale: The drawing scale AutoCAD uses to calculate the height of annotative text.

AutoCAD scales *annotative text* according to the *annotation scale* you select, which eliminates the need for you to calculate the scale factor. Once you choose an annotation scale, AutoCAD determines the scale factor and applies it automatically to annotative text and all other annotative objects. For example, if you manually scale 1/8″ text for a drawing with a 1/4″ = 1′-0″ scale, or a scale factor of 48, you must draw the text using a text height of 6″ (1/8″ × 48 = 6″) in model space. When placing annotative text, using this example, you set an annotation scale of 1/4″ = 1′-0″. Then you draw the text using a paper text height of 1/8″ in model space. The 1/8″ high text scales to 6″ automatically because of the preset 1/4″ = 1′-0″ annotation scale.

Annotative text offers several advantages over manually scaled text, including the ability to control text scale based on scale, not scale factor. Annotative text is especially effective when the drawing scale changes or when a single sheet includes objects viewed at different scales.

> **PROFESSIONAL TIP**
>
> If you anticipate preparing scaled drawings, you should use annotative text and other annotative objects instead of traditional manual scaling. However, scale factor does influence non-annotative items and is still an important value to identify and use throughout the drawing process.

Setting the Annotation Scale

You should usually set annotation scale before you begin typing text so that the text height scales automatically. However, this is not always possible. It may be necessary to adjust the annotation scale throughout the drawing process, especially if you

Figure 9-5.
Examples of calculating scale factor.

Example	Scale	Conversion	Calculation	Scale Factor
Mechanical	1:2	None	2 ÷ 2 = 2	2
Civil	1″ = 60′	1″ = 720″ (60′ contains 720″)	720 ÷ 1 = 720	720
Architectural	1/4″ = 1′-0″	1/4″ (.25″) = 12″ (1′ contains 12″)	12 ÷ .25 = 48	48
Metric to Inch	1:1	1″ - 25.4 mm	25.4 ÷ 1 = 25.4	25.4
Metric to Inch	1:2	1″ = 25.4 mm × 2 (50.8)	50.8 ÷ 1 = 50.8	50.8

prepare multiple drawings with different scales on one sheet. This chapter approaches annotation scaling in model space only, using the process of selecting the appropriate annotation scale before typing text. To draw text using another scale, pick the new annotation scale and then type the text.

When you access a text tool and an annotative text style is current, the **Select Annotation Scale** dialog box appears. This is a very convenient way to set annotation scale before typing. Text styles are described later in this chapter. You can also select the annotation scale from the **Annotation Scale** flyout on the status bar. See **Figure 9-6.** The annotation scale is typically the same as the drawing scale.

NOTE

This textbook describes many additional annotative object tools. Some of these tools are more appropriate for working with layouts, as explained later in this textbook.

Editing Annotation Scales

If a certain scale is not available, or if you want to change existing scales, pick the **Annotation Scale** flyout on the status bar and choose **Custom...** to access the **Edit Scale List** dialog box. From this dialog box, you can move the highlighted scale up or down in the list by picking the **Move Up** or **Move Down** button. To remove the highlighted scale from the list, pick the **Delete** button.

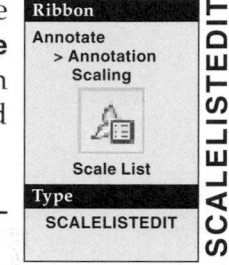

Ribbon
Annotate > Annotation Scaling
Scale List
Type
SCALELISTEDIT

SCALELISTEDIT

Figure 9-6.
Pick the **Annotation Scale** flyout on the status bar to activate an annotation scale.

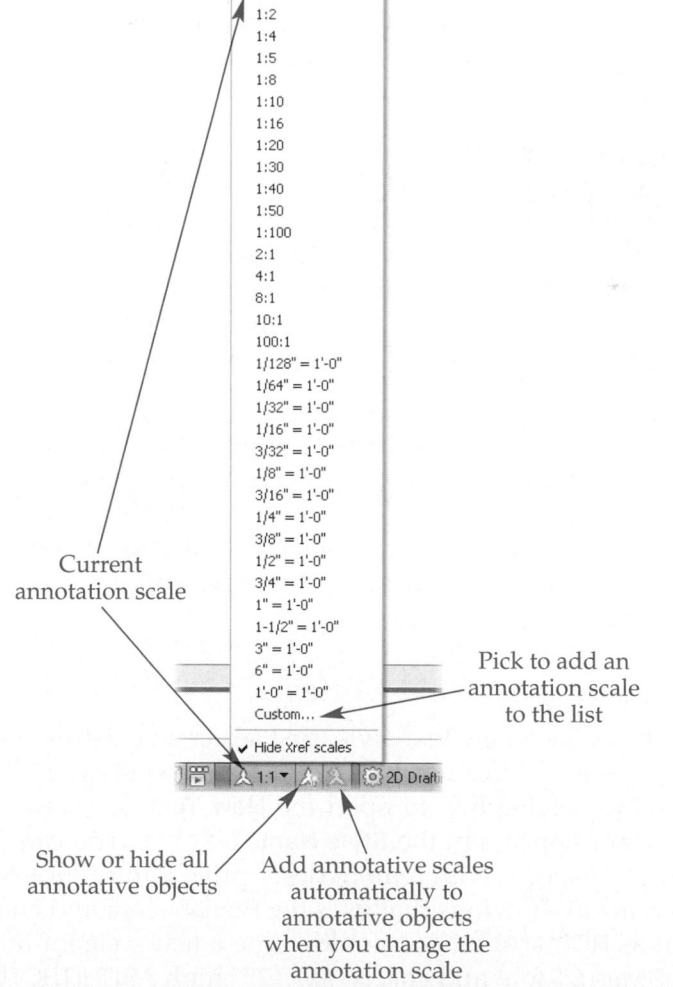

Current annotation scale

Pick to add an annotation scale to the list

Show or hide all annotative objects

Add annotative scales automatically to annotative objects when you change the annotation scale

Select the **Edit...** button to open the **Edit Scale** dialog box. Here you can change the name of the scale and adjust the scale by entering the paper and drawing units. For example, a scale of 1/4″ = 1′-0″ uses a paper units value of .25 or 1 and a drawing units value of 12 or 48.

To create a new annotation scale, pick the **Add...** button to display the **Add Scale** dialog box, which functions the same as the **Edit Scale** dialog box. Pick the **Reset** button to restore the default annotation scale. When the annotation scale is set current, you are ready to type annotative text that automatically displays at the correct text height according to the drawing scale.

Text Styles

Text styles control many text characteristics. You may have several text styles in a single drawing, depending on drawing requirements. Though you can adjust text format independently of a text style, you should create a text style for each unique application. For example, you may use a specific text style to draw most text objects, such as adding notes and dimensions, and a separate text style that uses different characteristics for adding text to a title block. You may also want to create an annotative and a non-annotative text style. Add text styles to drawing templates for repeated use.

Working with Text Styles

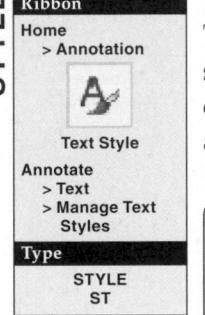

STYLE

Ribbon
Home
> Annotation

Text Style
Annotate
> Text
> Manage Text
Styles
Type
STYLE
ST

Create, modify, and delete text styles using the **Text Style** dialog box. See **Figure 9-7.** The **Styles** list box displays existing text styles. By default, Annotative and Standard text styles are available. The Annotative text style is preset to create annotative text, as indicated by the icon to the left of the style name. The Standard text style does not use the annotative function.

> **NOTE**
>
> Several panels on the ribbon contain small arrows in the lower-right corner that open appropriate dialog boxes. When a dialog box is accessible using one of these arrows, the arrow will be displayed in a margin graphic as shown next to this paragraph. In this case, picking the arrow opens the **Text Style** dialog box.

To make a text style current, double-click the style name; right-click the name and select **Current**; or pick the name and select the **Current** button. Below the **Styles** list box is a drop-down list that you can use to filter the number of text styles displayed in the **Text Style** dialog box. Pick the **All Styles** option to show all text styles in the file or pick the **Styles in use** option to show only the current style and styles used in the drawing.

Creating New Text Styles

To create a new text style, first select an existing text style from the **Styles** list box to use as a base for formatting the new text style. Then pick the **New...** button in the **Text Style** dialog box to open the **New Text Style** dialog box. See **Figure 9-8.** Notice that style1 appears in the **Style Name** text box. You can keep the default name, but you should replace it with a more descriptive name. For example, to create a text style for mechanical drawings that uses the Romans font and characters .12″ high, enter a name such as ROMANS-12. You could name a text style for architectural drawings that uses the Stylus BT font and characters .125″ high ARCHITECTURAL-125.

Figure 9-7.
Use the **Text Style** dialog box to create, rename, delete, and set the characteristics of a text style.

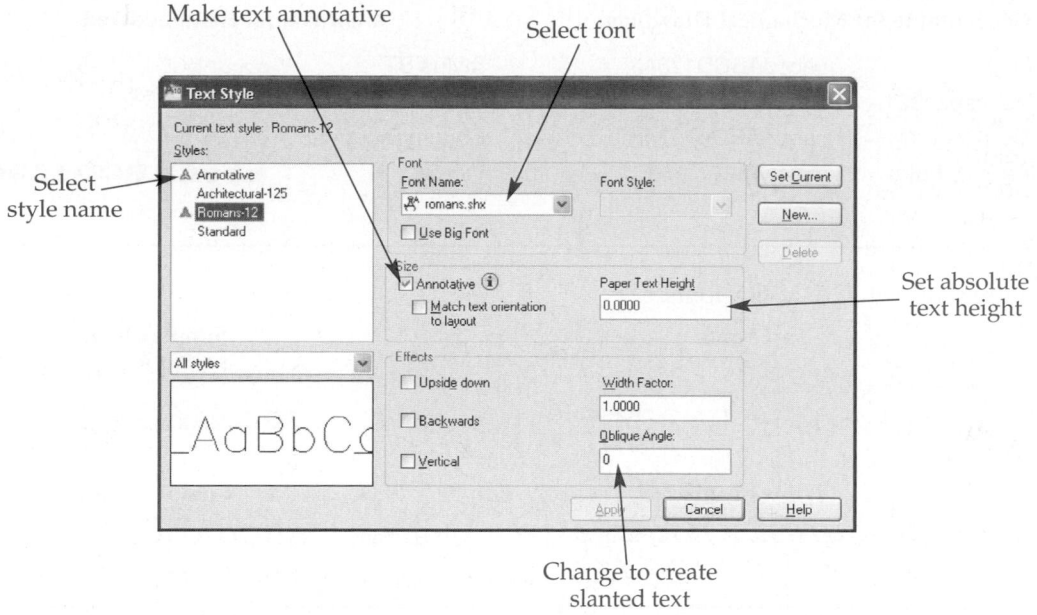

Figure 9-8.
Enter a descriptive name for the new text style in the **New Text Style** dialog box.

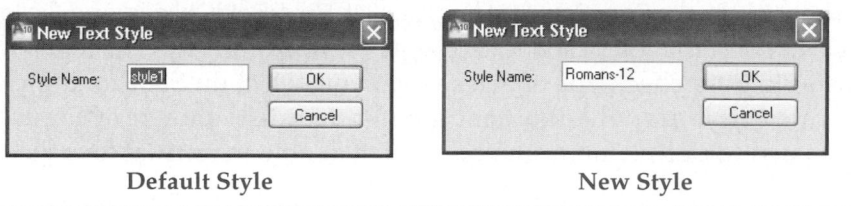

Text style names can have up to 255 characters, including uppercase and lowercase letters, numbers, dashes (–), underlines (_), and dollar signs ($). After entering the text style name, pick the **OK** button. The new text style appears in the **Styles** list box of the **Text Style** dialog box, and you are ready to adjust text style characteristics.

PROFESSIONAL TIP

It is a good idea to record the names and details about the text styles you create and keep this information in a log for future reference.

Setting the Font Style

The **Font** area of the **Text Style** dialog box is where you select a *font* and the style of the selected font. Use the **Font Name** drop-down list to access available fonts. Most fonts, including the default Arial font, are TrueType fonts identified by the TrueType icon. TrueType fonts are *scalable fonts* and have an outline. By default, TrueType fonts appear and plot filled. Examples of TrueType fonts are shown in **Figure 9-9.**

Fonts linked to AutoCAD shape files have .shx file extensions and display the AutoCAD compass icon. The Romans (roman simplex) font closely duplicates the single-stroke lettering that has long been the standard for most drafting. **Figure 9-10** shows examples of SHX fonts.

font: A letter face design.

scalable fonts: Fonts that can be displayed or printed at any size while retaining proportional letter thickness.

Figure 9-9.
A few of the many TrueType fonts available.

Styles Suitable for Mechanical Drawings		Architectural Hand Lettered	
Arial	abcdABCD12345	Stylus BT	abcdABCD I 2345
Monospac821	abcdABCD12345	CitiBlueprint	abcdABCD12345
Swis721 BT	abcdABCD12345	CountryBlueprint	abcdABCD12345
Technic Bold	ABCDABCDI2345	Vineta	**abcdABCD12345**

Figure 9-10.
Examples of AutoCAD SHX fonts.

	Fast Fonts			Simplex Fonts	
	abcdABCD12345	Txt	Romans	abcdABCD12345	
Monotxt	abcdABCD12345		Italic	*abcdABCD12345*	

	Triplex Fonts			Complex Fonts	
Romant	abcdABCD12345		Romanc	abcdABCD12345	
	abcdABCD12345	Italict	Italicc	*abcdABCD12345*	

The **Font Style** drop-down list is inactive unless the selected font includes options, such as bold or italic. The SHX fonts do not provide style options, but some of the TrueType fonts do. For example, the SansSerif font has Regular, Bold, BoldOblique, and Oblique options. Select a style or combination of styles to change the appearance of the font.

The **Use Big Font** check box enables when you select an SHX font. Pick the check box to activate *big fonts*. The **Big Font:** drop-down list, shown in **Figure 9-11,** is a supplement used to define many symbols not available in normal font files.

big fonts: Asian and other large-format fonts that have characters not present in other font files.

CAUTION

When a drawing contains a significant amount of text, especially TrueType font text, display changes may be slower and drawing regeneration time may increase.

Figure 9-11.
When you check the **Use Big Font** check box, the **Font Style:** drop-down list changes to display a list of available big fonts.

The **Big Font:** drop-down list displays the available nonstandard fonts

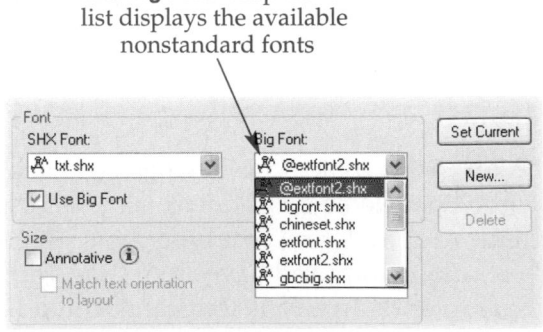

 **Reference Material** *AutoCAD Fonts*
For a sample of the many fonts available with AutoCAD, go to the **Reference Material** section on the Student Web site (www.g-wlearning.com/CAD) and select **AutoCAD Fonts**.

Text Style Height Options

The **Size** area of the **Text Style** dialog box contains options for defining text style height. Select the **Annotative** check box to set the text style as annotative and display the **Paper Text Height** text box. Deselect the **Annotative** check box to scale text manually and display the **Height** text box.

The default text height is 0. If you set a value other than 0, the text height is fixed for the text style and applies each time you use the text style. As a result, you can only use the specified text height to create single-line text, and you do not have the option of assigning a different text height to certain objects that include text, such as dimensions.

When you pick the **Annotative** check box, the **Match text orientation to layout** check box enables. Check this box to match the orientation of text in layout viewports with the layout orientation. Layouts are described later in this textbook.

PROFESSIONAL TIP

Use a text style text height of 0 to provide the greatest flexibility when drawing. A prompt asks you to specify a text height when you create single-line text, and you can specify a text height for dimension, multileader, and table styles. As an alternative, use a specific value to limit text height. In this case, the text style fully controls the height applied to single-line text, and dimension, multileader, and table styles. This restricts text objects to a certain height, which can increase productivity and accuracy, but requires that you create a text style for each unique height. Dimension, multileader, and table styles are explained later in this textbook.

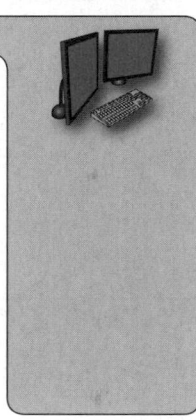

Adjusting Text Style Effects

The **Effects** area of the **Text Style** dialog box includes options to set the text format. It contains options for printing text upside-down, backwards, and vertically. See **Figure 9-12**. The Vertical check box is inactive for all TrueType fonts. A check in this box makes

Figure 9-12.
Special effects for text styles can be set in the **Effects** area of the **Text Style** dialog box.

UPSIDEDOWN   BACKWARDS   VERTICAL

Upside-Down Text   **Backwards Text**   **Vertical Text**

SHX font text vertical. Text on drawings is normally horizontal, but vertical text can be used for special effects and graphic designs. Vertical text usually works best with a 270° rotation angle.

Use the **Width Factor** text box to specify the text character width relative to its height. A width factor of 1 is default and is recommended for most drawing applications A width factor greater than 1 expands characters, and a factor less than 1 compresses characters. See **Figure 9-13.** You can set the width factor between 0.01 and 100.

The **Oblique Angle** text box allows you to set the angle at which text inclines. The 0 default draws characters vertically. A value greater than 0 slants characters to the right, and a negative value slants characters to the left. See **Figure 9-14.** Some fonts, such as the italic shape font, are already slanted.

PROFESSIONAL TIP

AutoCAD text slant measures from vertical. Use a 22° oblique angle to slant text according to the 68° horizontal incline standard.

NOTE

The **Preview** area of the **Text Style** dialog box displays an example of the selected font and font effects. This is a convenient way to see what the font looks like before using it in a new style.

Exercise 9-1

Access the Student Web site (www.g-wlearning.com/CAD) and complete Exercise 9-1.

Figure 9-13.
Examples of width factor settings for text.

Width Factor	Text
1	ABCDEFGHIJKLM
.5	ABCDEFGHIJKLMNOPQRSTUVWXY
1.5	ABCDEFGHI
2	ABCDEFG

Figure 9-14.
Examples of oblique angle settings for text.

Obliquing Angle	Text
0	ABCDEFGHIJKLM
15	*ABCDEFGHIJKLM*
−15	ABCDEFGHIJKLM

Changing, Renaming, and Deleting Text Styles

If you make changes to a text style, such as selecting a different font, all existing text objects drawn using the modified text style update to reflect the changes. Use a different text style with unique characteristics when appropriate. To rename a text style using the **Text Style** dialog box, slowly double-click the name or right-click on the name and select **Rename**.

To delete a text style using the **Text Style** dialog box, right-click on the name and select **Delete**, or pick the style and select the **Delete** button. You cannot delete a text style that is assigned to text objects. To delete a style that is in use, assign a different style to the text objects that reference the style you want to delete. You cannot delete or rename the Standard style.

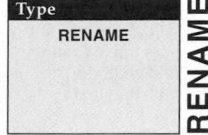

Setting a Text Style Current

You can set a text style current using the **Text Style** dialog box by double-clicking the style in the **Styles** list box, right-clicking on the name and selecting **Set current**, or picking the style and selecting the **Set current** button. To set a text style current without opening the **Text Style** dialog box, use the **Text Style** list located in the expanded **Annotation** panel of the **Home** ribbon tab or the **Text** panel of the **Annotate** ribbon tab. See **Figure 9-15**.

Figure 9-15.
The fastest way to set a style current is to use the drop-down list on the ribbon.

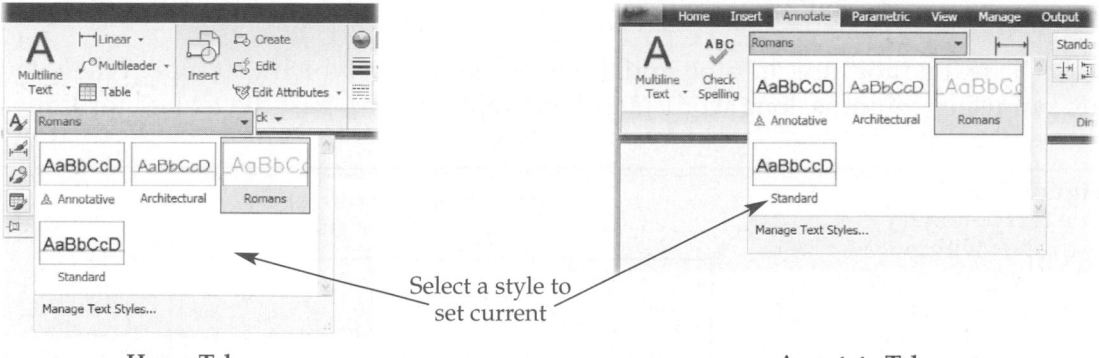

Select a style to
set current

Home Tab   **Annotate Tab**

Creating Multiline Text

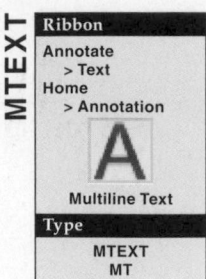

Ribbon

Annotate
> Text
Home
> Annotation

A

Multiline Text

Type

MTEXT
MT

text boundary: An imaginary box that sets the location and width for multiline text.

The **MTEXT** tool draws a single multiline text, or mtext, object that can include extensive paragraph formatting, lists, symbols, and columns. When you access the **MTEXT** tool, grayed-out letters appear next to the crosshairs to indicate the current text style and height settings. Pick the first corner of the *text boundary*. A prompt then asks you to specify the opposite corner or choose an available option. Use the options to preset specific mtext characteristics, including text height and style, text boundary justification, line spacing, rotation, width, and use of columns. You can control the same settings while typing, as explained in this chapter, or while editing mtext.

By default, mtext is set to use dynamic columns, which allow you to organize multiple text columns in a single mtext object. Although you can use dynamic columns to form a single paragraph, for typical text requirements without columns, disable columns. Before selecting the second corner of the text boundary, choose the **Columns** option and the **No columns** option. An arrow in the boundary shows the direction of text flow, where the boundary will expand as you type, if necessary. Pick the opposite corner of the text boundary to continue. See **Figure 9-16.** Columns are fully explained later in this chapter.

Using the Text Editor

text editor: The part of the multiline or single-line text system where text is typed.

Once you select the opposite corner of the text boundary, the **Text Editor** contextual ribbon tab and the *text editor* appear. **Figure 9-17** shows the display when you are not using columns. The controls and features for typing mtext are similar to those in software programs such as Microsoft® Word. The **Text Editor** ribbon tab provides tools for adjusting text typed into the text editor. You can access the shortcut menu shown in **Figure 9-18** by right-clicking anywhere outside of the ribbon. Many of the same options are available on the **Text Editor** ribbon tab.

> **NOTE**
>
> If you close the ribbon, the **Text Formatting** toolbar appears instead of the **Text Editor** ribbon tab. This textbook focuses on using the **Text Editor** ribbon tab to add mtext. The **Text Formatting** toolbar provides the same functions.

The text editor indicates the initial size of the area in which you type. When you are not using columns, long words and paragraphs extend past the text editor limits.

Figure 9-16.
The text boundary is the box within which you type. Consider the text boundary the extents of paragraph. Long words extend past the boundary in the direction the arrow indicates.

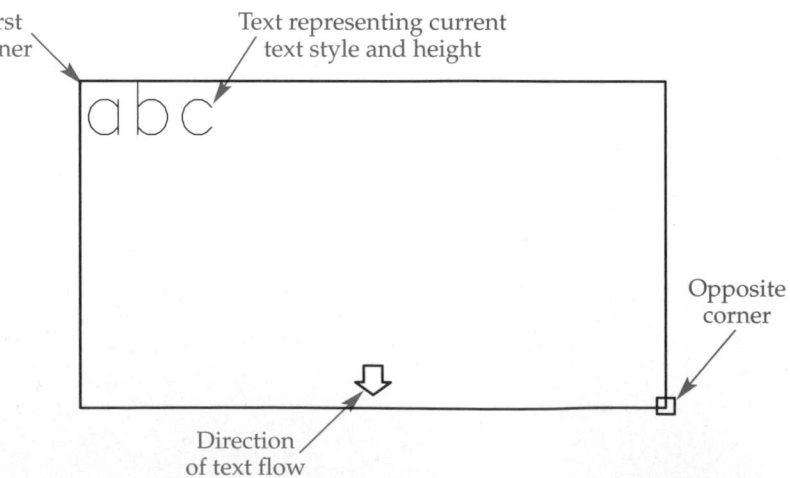

First corner

Text representing current text style and height

abc

Opposite corner

Direction of text flow

Figure 9-17.
The **Text Editor** ribbon tab provides many options for creating mtext.

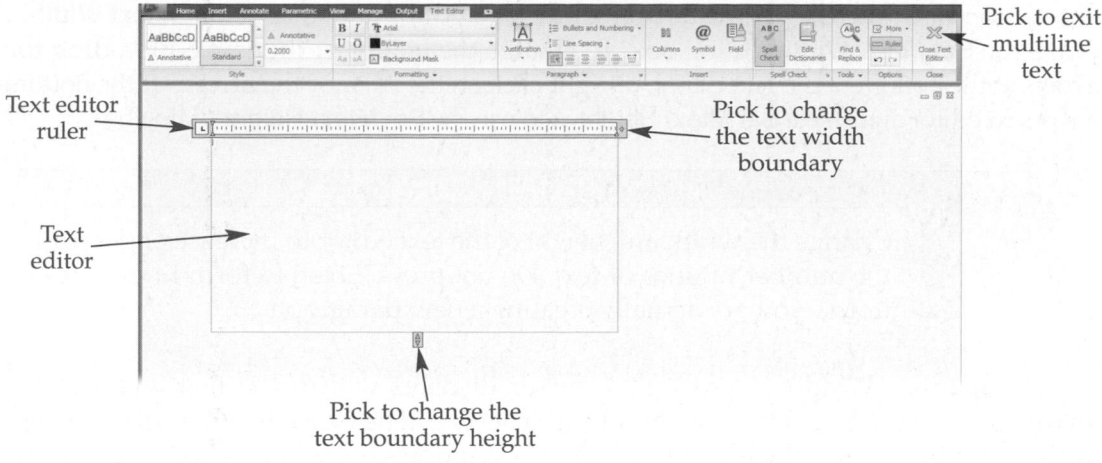

Pick to exit multiline text

Text editor ruler

Text editor

Pick to change the text width boundary

Pick to change the text boundary height

Figure 9-18.
Display the text editor shortcut menu by right-clicking anywhere except on the ribbon while the text editor is active.

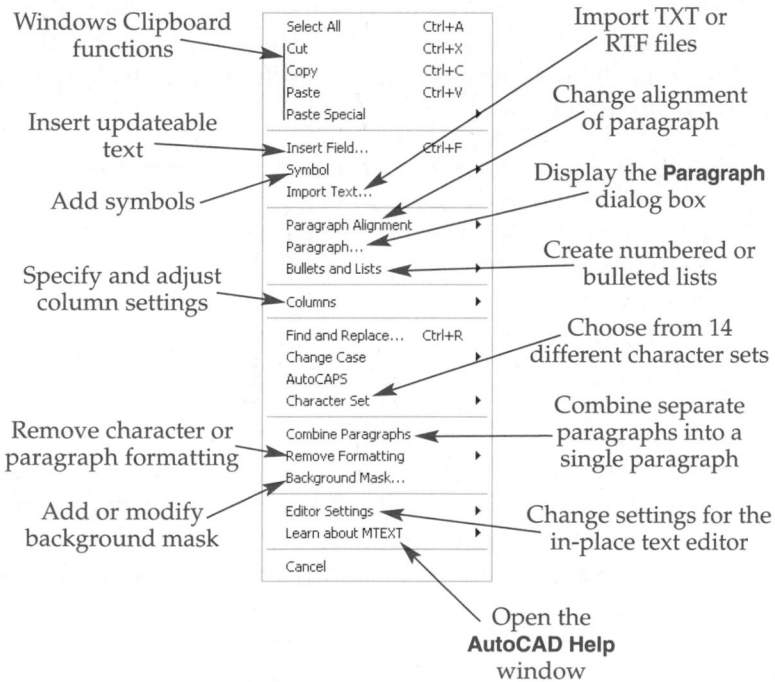

Windows Clipboard functions

Insert updateable text

Add symbols

Specify and adjust column settings

Remove character or paragraph formatting

Add or modify background mask

Import TXT or RTF files

Change alignment of paragraph

Display the **Paragraph** dialog box

Create numbered or bulleted lists

Choose from 14 different character sets

Combine separate paragraphs into a single paragraph

Change settings for the in-place text editor

Open the **AutoCAD Help** window

The text editor is transparent by default so that you can see how the text you type appears on-screen in relation to other objects. To make the text editor appear opaque, select the **Opaque Background** option available from the **Editor Settings** cascading submenu of the shortcut menu. The text editor includes a ruler where indent and tab stops and indent and tab markers are located.

NOTE

Pick the **Ruler** button in the **Options** panel of the **Text Editor** ribbon tab, or select the **Ruler** option from the **Editor Settings** cascading submenu of the shortcut menu to turn the ruler on or off. Use the **Undo** and **Redo** buttons, also found in the **Options** panel, to undo or redo text editor operations.

To change the width of the text editor, when you are not using columns, drag the arrows on the right-side end of the paragraph ruler. You can also right-click on the text editor ruler or the arrows at the bottom of the text editor and choose **Set Mtext Width...** to use the **Set Mtext Width** dialog box. To change the height of the text editor, drag the arrows at the bottom of the text editor, or right-click on the ruler or the arrows at the bottom of the text editor and select **Set Mtext Height...** to use the **Set Mtext Height** dialog box.

PROFESSIONAL TIP

Change the width and height of the text editor to increase or decrease the number of lines of text. Do not press [Enter] to form lines of text, unless you are actually creating a new paragraph.

The familiar Windows text editor cursor displays within the text editor at the current text height. Begin typing or editing as needed. The procedure for selecting and editing existing text is the same as in standard Windows text editors. To quickly select all text in the text editor, right-click inside the text editor and choose **Select All**. You can change the selected text highlight color by selecting the **Text Highlight Color...** option available from the **Editor Settings** cascading submenu of the shortcut menu.

When you finish typing text, exit the mtext system by picking the **Close Text Editor** button in the **Close** panel of the **Text Editor** ribbon tab or pick outside of the text editor. You can also press [Esc] or right-click and select **Cancel** to exit, but you are prompted to save changes. The easiest way to reopen the text editor to make changes to text content is to double-click on an mtext object.

NOTE

The text editor displays text horizontally, right-side up, and forward. Any special effects such as vertical, backwards, or upside-down take effect when you exit the text editor.

Reference Material *Shortcut Keys*

For a complete list of keyboard shortcuts for text editing, go to the **Reference Material** section of the Student Web site (www.g-wlearning.com/CAD) and select **Shortcut Keys**.

Exercise 9-2

Access the Student Web site (www.g-wlearning.com/CAD) and complete Exercise 9-2.

Stacking Text

When you enter a fraction in the **Text Editor**, the **AutoStack Properties** dialog box appears by default. See **Figure 9-19.** This dialog box allows you to activate AutoStacking, which causes the entered fraction to stack with a horizontal or diagonal fraction bar. You can also remove the leading space between a whole number and the fraction. This dialog box appears each time you enter a fraction. If you decide that you do not want this dialog box to pop up each time you create a fraction, pick the **Don't show this dialog again; always use these settings** check box.

Figure 9-19.
The **AutoStack Properties** dialog box.

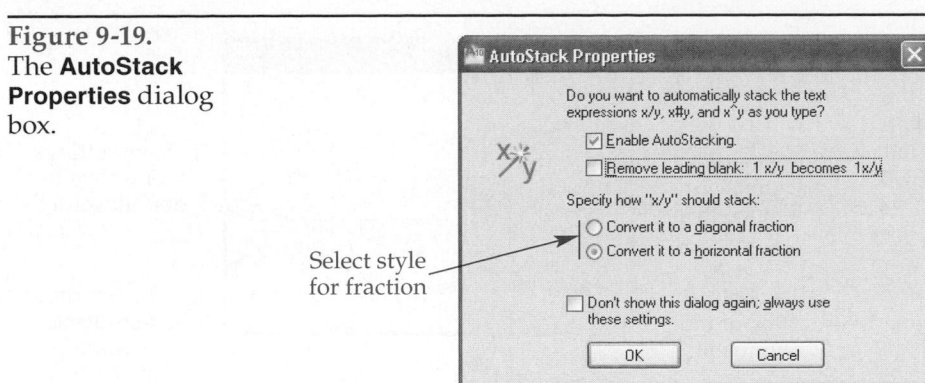

Select style for fraction

Manual Stacking

You can use the **Stack** feature to manually stack selected text vertically or diagonally. To use this feature to draw a vertically stacked fraction, place a forward slash between the top and bottom characters. Then select the text and pick the **Stack** option from the shortcut menu. To form a *tolerance stack*, use the caret (^) character between characters. To use a diagonal fraction bar, type a number sign (#) between characters. See **Figure 9-20.** To unstack text, select stacked text and pick the **Unstack** option from the shortcut menu. The stacked characters are placed on a single line with the appropriate character (^, #, or /) between the characters.

tolerance stack: Text that is stacked vertically without a fraction bar.

Stack Settings

The **Stack Properties** dialog box controls stack settings. To access the dialog box, select the stacked text and pick the **Stack Properties** option from the shortcut menu. **Figure 9-21** briefly describes the **Stack Properties** dialog box features.

Adding Symbols

You can insert a variety of common drafting symbols and other unique characters not found on a typical keyboard into the text editor. To insert a symbol, select the **Symbol** flyout on the **Insert** panel of the **Text Editor** ribbon tab, as shown in **Figure 9-22** or the **Symbol** cascading submenu available from the shortcut menu.

The **Symbol** menu allows the insertion of symbols at the text cursor location. The first two sections in the **Symbol** menu contain common symbols. The third section contains the **Non-breaking Space** option, which keeps two separate words together on one line. Pick a symbol to insert the symbol at the text cursor location and hide the **Symbol** menu. The **Other...** option opens the **Character Map** dialog box, shown in **Figure 9-23.** Use the following steps to insert a symbol from the **Character Map** dialog box:

1. Pick the font to display from the **Font:** drop-down list.
2. Pick the desired symbol and then pick the **Select** button. The symbol appears in the **Characters to copy:** box.
3. Pick the **Copy** button to copy the selected symbol or symbols to the Clipboard.

Figure 9-20.
Different types of stack characters. ASME standards recommend that the text height of stacked fraction numerals be the same as the height of other dimension numerals.

	Selected Text	Stacked Text
Vertical Fraction	1/2	$\frac{1}{2}$
Tolerance Stack	1^2	$\frac{1}{2}$
Diagonal Fraction	1#2	$\frac{1}{2}$

Figure 9-21.
The **Stack Properties** dialog box. Select 100% from the **Text size** drop-down list to conform to ASME standards.

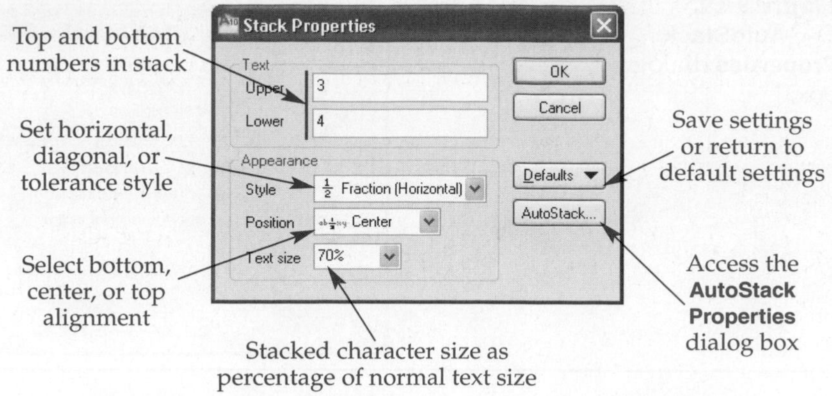

Top and bottom numbers in stack

Set horizontal, diagonal, or tolerance style

Select bottom, center, or top alignment

Save settings or return to default settings

Access the **AutoStack Properties** dialog box

Stacked character size as percentage of normal text size

Figure 9-22.
The **Symbol** menu options.

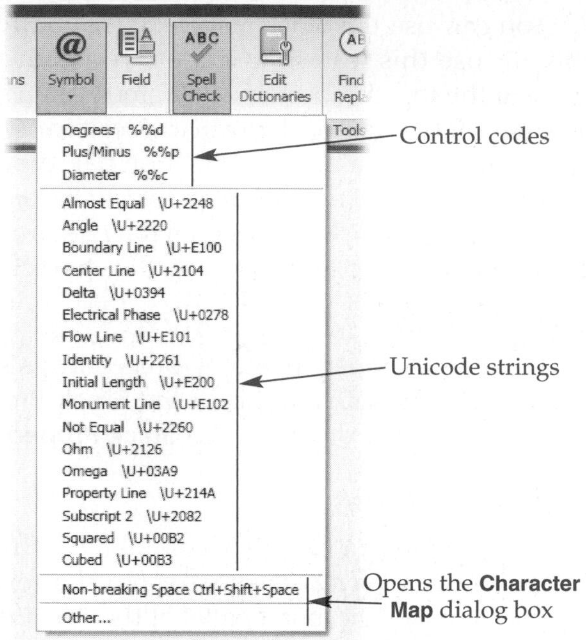

Control codes

Unicode strings

Opens the **Character Map** dialog box

Figure 9-23.
The **Character Map** dialog box.

Select font from drop-down list

Available symbols

Select to return to the **In-Place Text Editor**

Pick to select highlighted symbol

Pick to copy selected symbol(s) to Clipboard

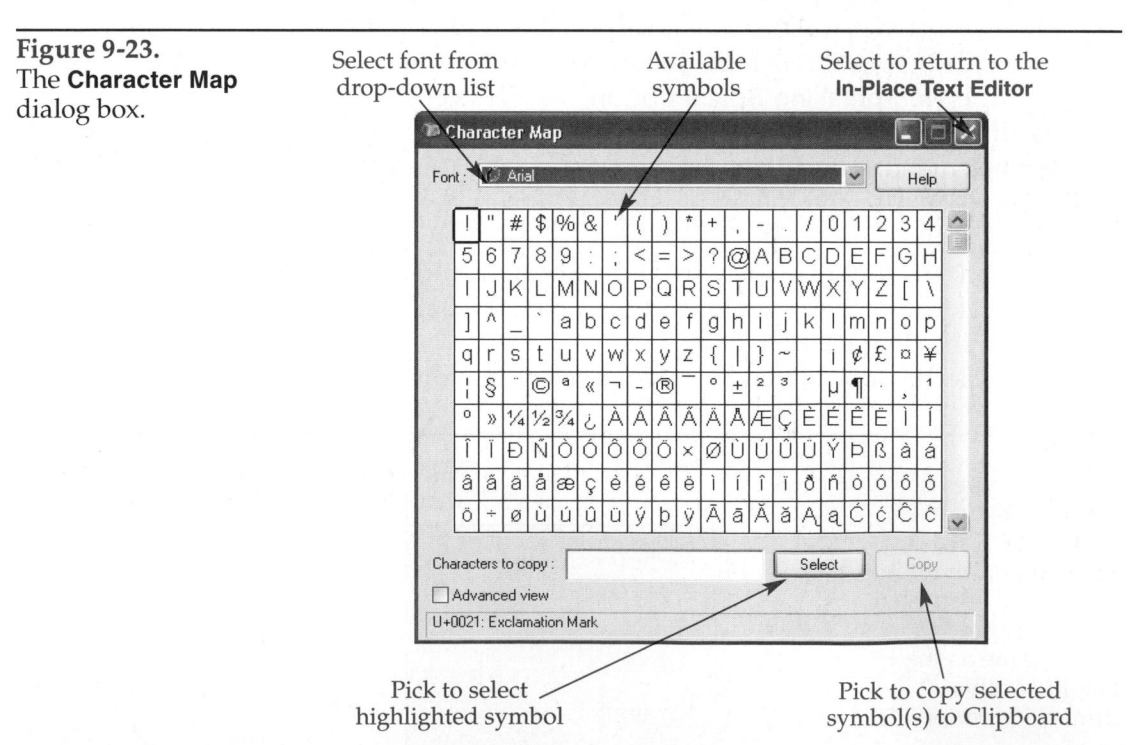

4. Close the **Character Map** dialog box.
5. In the text editor, place the cursor where you want the symbols to display.
6. Right-click and select **Paste** to paste the symbol at the cursor location.

Exercise 9-3

Access the Student Web site (www.g-wlearning.com/CAD) and complete Exercise 9-3.

Style Settings

The **Style** panel of the **Text Editor** ribbon tab includes options for changing the text style, annotative setting, and text height. Activate text style options before typing, or select existing text and make adjustments as necessary. A single mtext object can use a combination of text style settings. Remember, however, that making changes to some style settings overrides the settings specified in the text style, which is often not appropriate.

The text style flyout provides access to existing text styles. Use the scroll buttons to locate styles, or pick the expansion arrow to display a temporary window of styles. This allows you to use a text style other than the current text style while you are typing. Use the **Annotative** button to override the annotative setting of the current text style. Use the **Size** drop-down list to set the text height. If the current text style is annotative, or if you pick the **Annotative** button, the height you enter is the paper text height. If the current text style is not annotative, or if you deselect the **Annotative** button, the height you enter is the text height multiplied by the scale factor.

Character Formatting

The **Formatting** panel of the **Text Editor** ribbon tab includes options for adjusting character formatting. Activate character format options before typing, or select existing text and make adjustments as necessary. A single mtext object can use a combination of character formats. Remember, however, that making changes to character formatting overrides some of the settings specified in the text style and preset object properties, such as color. You should usually avoid this practice.

Pick the **Bold** button to make text bold. Select the **Italic** button to make text italic. The **Bold** and **Italic** settings work only with some TrueType fonts. Pick the **Underline** button to underline text, and select the **Overline** button to place a line over text. The **Font** drop-down list allows you to override the text font. The text color is set to ByLayer by default, but you can change the color by picking one of the colors in the **Color** drop-down list. Though you should usually define color as ByLayer, a single mtext object can use a combination of text colors.

Additional character formatting options are available from the expanded **Formatting** panel. The **Oblique Angle** text box overrides the angle at which text inclines. The value in the **Tracking** text box determines the amount of space between text characters. The default tracking value is 1, which results in normal spacing. Increase the value to add space between characters, or decrease the value to tighten the spacing between characters. You can enter any value between 0.75 and 4.0. See **Figure 9-24.** The value in the **Width Factor** text box overrides the text character width.

Figure 9-24.
The **Tracking** option for mtext determines the spacing between characters.

AutoCAD tracking
Normal Spacing

AutoCAD tracking
Tracking = 0.75

A u t o C A D   t r a c k i n g
Tracking = 2.0

Using a Background Mask

background mask: A mask that hides a portion of objects behind and around text so that the text is unobstructed.

Sometimes drawings require text to be placed over existing objects, such as graphic patterns, making the text hard to read. A *background mask* can solve this problem. Select the **Background Mask...** button to display the **Background Mask** dialog box. See **Figure 9-25.** To apply the mask settings to the current mtext object, check **Use background mask**. The **Border offset factor:** text box sets the amount of mask. This value, from 1 to 5, works with the text height value. If the border offset factor is set to 1, then the mask occurs directly within the boundary of the text. To offset the mask beyond the text boundary, use a value greater than 1. The formula is: border offset factor × text height = total masking distance from the bottom of the text. See **Figure 9-26.** The **Fill Color** area of the **Background Mask** dialog box allows you to apply color to the mask using the background color or a different color.

> **NOTE**
>
>
> The **Character Set** cascading submenu, available from the shortcut menu or the **More** flyout in the **Options** panel, displays a menu of code pages. A code page provides support for character sets used in different languages. Select a code page to apply it to the selected text.

Exercise 9-4

Access the Student Web site (www.g-wlearning.com/CAD) and complete Exercise 9-4.

Paragraph Formatting

justify: Align the margins or edges of text. For example, left-justified text is aligned along an imaginary left border.

The **Paragraph** panel of the **Text Editor** ribbon tab includes options for adjusting paragraph formatting. *Justify* the text boundary to control the arrangement and location of text within the text editor. You can also justify the text within the boundary independently of the text boundary justification. This provides great flexibility when

Figure 9-25.
The **Background Mask** dialog box controls text mask settings.

Determines how much of the background is masked

Sets mask color same as background color

Figure 9-26.
The border offset factor determines the size of the background mask. The text in the figure is 1/8″ with different border offset factors.

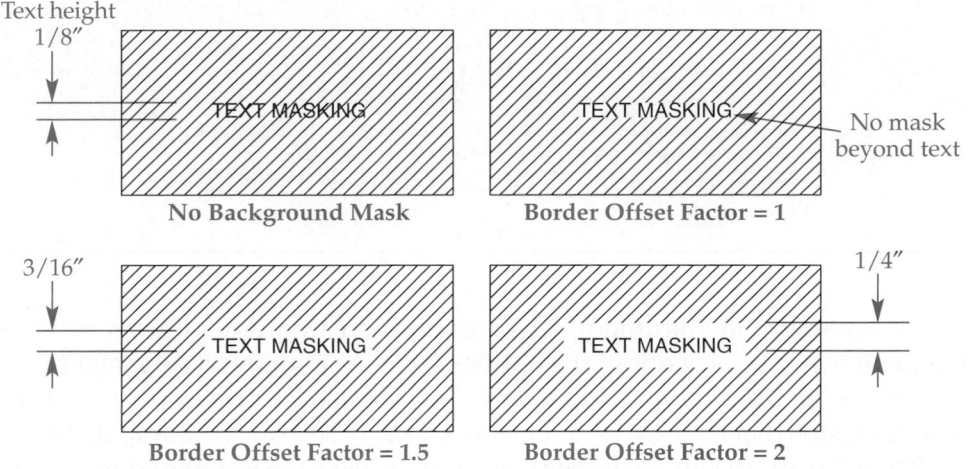

determining the location and arrangement of text. To justify the text boundary, select a justification option from the **Justification** flyout. Justification also determines the direction of text flow. **Figure 9-27** displays the options for justifying the mtext boundary vertically and horizontally.

    Paragraph alignment occurs inside the text boundary. For example, when you apply a **Middle Center** text box justification, then set the paragraph alignment to **Left**, the text inside the boundary aligns to the left edge of the text boundary, while the text boundary remains positioned according to the **Middle Center** justification. See **Figure 9-28.**

paragraph alignment: The alignment of multiline text inside the text boundary.

Figure 9-27.
Options for justifying the mtext boundary.

✗ This symbol represents the insertion point

▪ This symbol represents grips

— — This symbol represents the text boundary

Figure 9-28.
You can adjust paragraph alignment independently of text boundary justification.

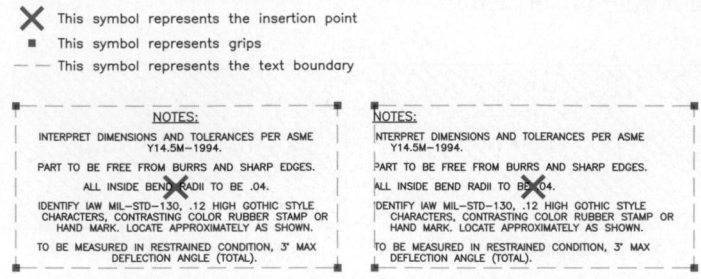

To adjust paragraph alignment, select one of the paragraph alignment buttons on the **Paragraph** panel, or pick one of the options from the **Paragraph Alignment** cascading submenu available from the shortcut menu. You can also control paragraph alignment using the **Paragraph** dialog box, shown in **Figure 9-29**. To display the **Paragraph** dialog box, pick the **Paragraph** button on the **Paragraph** panel or select the **Paragraph** option available from the shortcut menu. To set paragraph alignment in the **Paragraph** dialog box, pick the **Paragraph Alignment** check box, then choose the appropriate paragraph alignment radio button. You can choose from five paragraph alignment options, as shown in **Figure 9-30**.

Tabs, indents, paragraph spacing, and paragraph line spacing can also be set in the **Paragraph** dialog box. Use the **Tab** area to set custom tab stops. Pick the tab type radio button, enter a value for the tab in the text box, and pick the **Add** button to add it to the list and insert the tab on the ruler. **Figure 9-31** shows and briefly describes each tab option. Add as many custom tabs as necessary. You can also add custom tabs to the ruler by picking the tab button on the far left side of the ruler until the desired tab symbol appears. Then pick a location on the ruler to insert the tab.

The options in the **Left Indent** area set the indentation for the first line of a paragraph of text as well as the remaining portion of a paragraph. The **First line** indent is used each time you start a new paragraph. As text wraps to the next line, the **Hanging** indent is used. The options in the **Right Indent** area set the indentation for the right side of a paragraph. As text is typed, the right indent value, not the right edge of the text boundary, determines when the text wraps to the next line.

Figure 9-29.
The **Paragraph** dialog box.

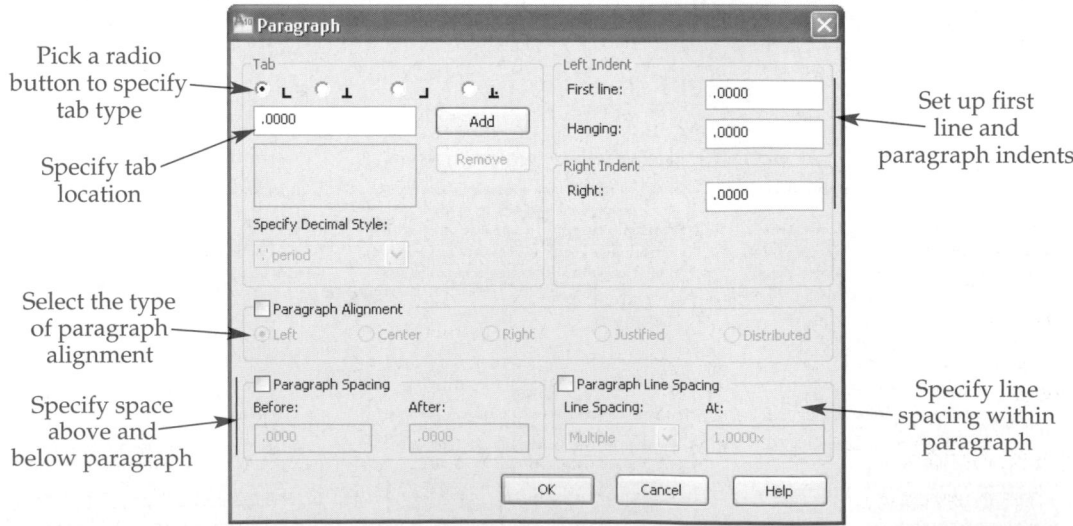

Figure 9-30.
Paragraph alignment options for mtext. In each of these examples, the text boundary justification is set to Top Left.

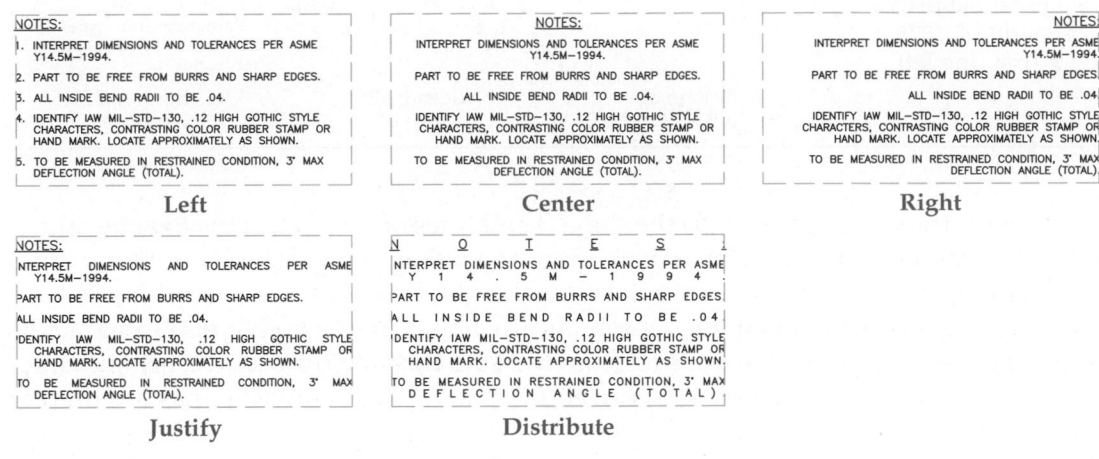

Left

Center

Right

Justify

Distribute

Figure 9-31.
Using custom tabs to position text in the text editor. When you press [Tab], the cursor moves to the tab position. The type of tab determines text behavior.

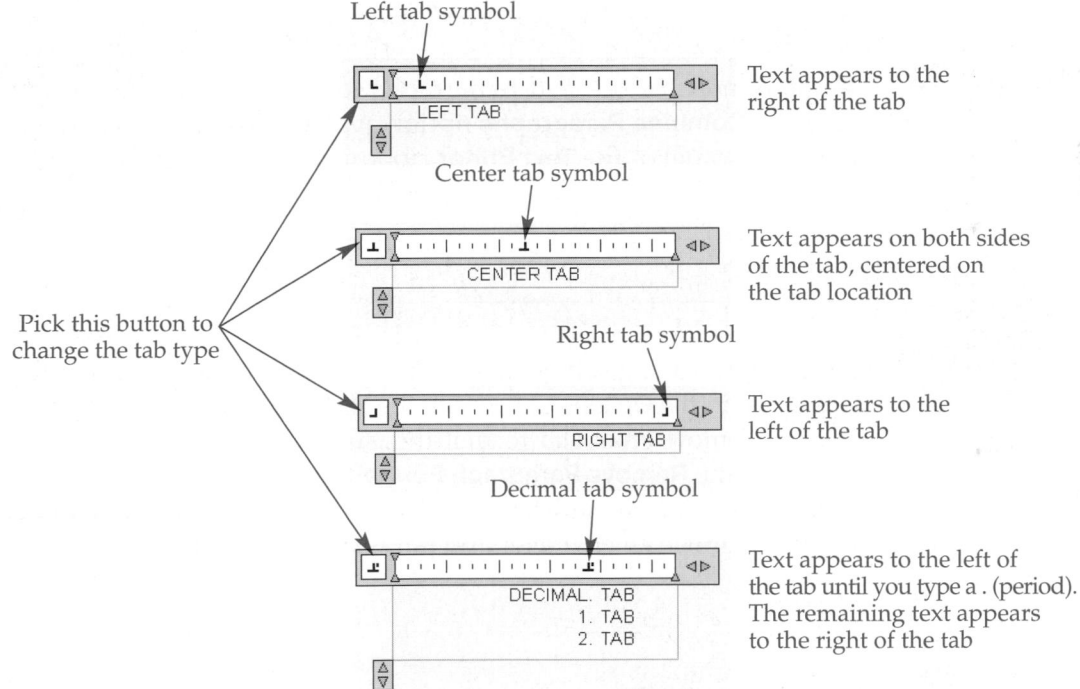

Left tab symbol

Text appears to the right of the tab

Center tab symbol

Text appears on both sides of the tab, centered on the tab location

Pick this button to change the tab type

Right tab symbol

Text appears to the left of the tab

Decimal tab symbol

Text appears to the left of the tab until you type a . (period). The remaining text appears to the right of the tab

The options in the **Paragraph Spacing** area define the amount of space before and after paragraphs. To set paragraph line spacing, pick the **Paragraph Spacing** check box. Then enter the spacing above a paragraph in the **Before** text box, and the spacing below a paragraph in the **After** text box. **Figure 9-32** shows examples of paragraph spacing settings.

The options in the **Paragraph Line Spacing** area adjust *line spacing*. Default line spacing for single lines of text is equal to 1.5625 times the text height. To adjust the line spacing, pick the **Paragraph Line Spacing** check box. Select the **Multiple** option from the **Line Spacing** drop-down list to enter a multiple of the text height in the **At** text box. For example, lines with a text height of .12″ are spaced .1875″ apart. To double-space lines, you could enter a value of **3.125x**, making the space between lines of text .375″.

line spacing: The vertical distance from the bottom of one line of text to the bottom of the next line.

Figure 9-32.
Examples of paragraph spacing. Each example uses a text height of .1875″ and a first line left indent of .5″.

Paragraph one typed with no paragpah spacing. Paragraph two typed with no paragraph spacing	Paragraph one typed with .25 before spacing and no after spacing. Paragraph two typed with .25 before spacing and no after spacing.	Paragraph one typed with .125 before spacing and .5 after spacing. Paragraph two typed with 125 before spacing and .5 after spacing.
Before: 0″ After: 0″	Before: .25″ After: 0″	Before: .125″ After: .5″

To force the line spacing to be the same for all lines of text, select the **Exactly** option from the **Line Spacing** drop-down list and enter a value in the **At** text box. If you enter an exact line spacing that is less than the text height, lines of text stack on top of each other. To add spaces between lines automatically based on the height of the characters in the line, choose the **At Least** option from the **Line Spacing** drop-down list and enter a value in the **At** text box. The result is an equal spacing even between lines of different height text.

You can also set line spacing using the **Line Spacing** flyout button on the **Paragraph** panel of the **Text Editor** ribbon tab. Select one of the available multiple options, pick the **More...** button to display the **Paragraph** dialog box, or choose the **Clear Line Spacing** option to apply an automatic spacing similar to the **At Least** function.

NOTE

Combine multiple selected paragraphs to form a single paragraph using the **Combine Paragraphs** option available from the expanded **Paragraph** panel of the **Text Editor** ribbon tab or the shortcut menu.

PROFESSIONAL TIP

Quickly remove formatting from selected text using the **Remove Formatting** cascading submenu available from the shortcut menu or the **More** flyout in the **Options** panel. Pick the **Remove Character Formatting** option to remove character formatting, such as bold, italic, or underline. Select the **Remove Paragraph Formatting** option to remove paragraph formatting, including lists. Pick the **Remove All Formatting** option to remove all character and paragraph formatting.

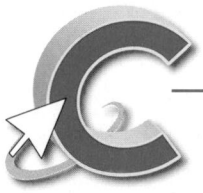

Exercise 9-5

Access the Student Web site (www.g-wlearning.com/CAD) and complete Exercise 9-5.

Creating Lists

Drawings often use lists to organize information. Lists provide a way to arrange related items in a logical order and help make lines of text more readable. General notes are usually in list format.

You can create lists as you enter text or apply list formatting to existing text. Numbering or lettering automatically adjusts when you add or remove listed items. You must select the **Allow Bullets and Lists** option in order to create a list. This option is active by default. Unchecking this option converts any list items in the text object to plain text characters and disables the other options in the menu.

Lists can also contain sublevel items designated with double numbers, letters, or bullets. Default tab settings apply unless you adjust paragraph options. List tools are available from the **Numbering** flyout on the **Paragraph** panel of the **Text Editor** ribbon tab or the **Bullets and Lists** cascading submenu of the shortcut menu. These tools allow you to create numbered, bulleted, and alphabetical lists.

Choose an option from the **Lettered** cascading submenu to create an alphabetical list. Pick **Uppercase** to use uppercase lettering or choose **Lowercase** to use lowercase lettering. The **Uppercase** option is the default. Pick the **Numbered** option to form a numbered list. To create a bulleted list in the default style, select the **Bulleted** option. This places the default solid circle bullet symbol at the beginning of each line of text.

Another method of creating lists is to use the **Allow Auto-list** option, which is active by default. When using the **Allow Auto-list** option, AutoCAD detects characters that frequently start a list and automatically assigns the first list item. For example, if a line of text begins with a number or letter and a period, AutoCAD assumes that you are starting a list and formats any additional lines of text to continue the list.

To create a numbered or lettered auto-list, you must include punctuation, such as a period, parenthesis, or colon, and press [Tab] after the number or letter that begins the first item. When you press [Enter] to start a new line of text, the new line uses the same formatting as the previous line, and the next consecutive number or letter appears. To end the list, press [Enter] twice. **Figure 9-33** shows an example of a numbered list.

When creating a bulleted auto-list, you can use typical keyboard characters, such as a hyphen [-], tilde [~], bracket [>], or asterisk [*], at the beginning of a line. Another option is to insert a symbol at the beginning of a line. Then, to form the list, press [Tab] and type the line of text. When you press [Enter] to start a new line of text, the new line uses the same formatting bullet symbol as the previous line. To end the list, press [Enter] twice. See **Figure 9-34**.

Figure 9-33.
Framing notes arranged in a numbered list.

FRAMING NOTES:
1. ALL FRAMING NOTES TO BE DFL #2 OR BETTER.
2. ALL HEATED WALLS @ HEATED LIVING AREA TO BE 2 X 6 @ 16" OC.
    FRAME ALL EXTERIOR NON-BEARING WALLS W/2 X 6 STUDS @ 24" OC.
3. USE 2 X 6 NAILER AT THE BOTTOM OF ALL 2-2 X 12 OR 4 X HEADERS
    @ EXTERIOR WALLS, BACK HEADER W/2" RIGID INSULATION.
4. BLOCK ALL WALLS OVER 10'-0" HIGH AT MID HEIGHT.

Figure 9-34.
In addition to the regular bullet symbol, other keyboard characters can be used for items in bulleted lists.

•   An elevation of the beam with end views or sections
•   Complete locational dimensions for holes, plates, and angles
•   Length dimensions

Bulleted List with Bullet Symbols

~ Connection specifications
~ Cutouts
~ Miscellaneous notes for the fabricator

Bulleted List with Tilde Characters

Picking the **Use Tab Delimiter Only** option limits unwanted list formatting by instructing AutoCAD to recognize only tabs to start a list. If the **Use Tab Delimiter Only** option is unchecked, list formatting occurs when a space or tab follows the initial list item character.

You can convert multiple lines of text to a list by selecting all of the lines of text and then picking a list formatting option. AutoCAD detects where you type [Enter] to start a new line of text and lists the lines in sequence. When you create a list in this manner, a tab automatically occurs after the number, letter, or symbol preceding the text. Set tabs and indents to adjust spacing and appearance.

Additional options are available for creating lists. Use the **Off** option to remove any list characters or bulleting from selected text. Pick the **Restart** option to renumber or re-letter selected items in a new sequence. The numbering or lettering restarts from the beginning, using 1 or A, for example. Choose the **Continue** option to add selected items to a list that exists above the currently selected item. The number of the selected item continues from the previous list. Items below the selected item also renumber.

Exercise 9-6

Access the Student Web site (www.g-wlearning.com/CAD) and complete Exercise 9-6.

Forming Columns

Sometimes it is necessary to break up text into multiple sections, or columns. This is especially true when you add lengthy general notes or when you must group information together. See **Figure 9-35.** Mtext columns are created in the text editor as a single object. This eliminates the need to create multiple text objects to form separate columns of text. You can create columns as you enter text or apply column formatting to existing text.

This chapter previously suggests turning columns off before accessing the text editor. This approach is appropriate for typical text requirements without columns, especially as you learn to create mtext. By default, however, mtext is set to form

Figure 9-35.
An example of drawing notes created as a single mtext object and divided into three columns.

dynamic columns, with the associated **Manual height** option. Column tools are also available while the text editor is active, from the **Columns** flyout on the **Insert** panel of the **Text Editor** ribbon tab or from the **Columns** cascading submenu of the shortcut menu. You can create *dynamic columns* or *static columns*.

dynamic columns:
Columns calculated automatically by AutoCAD according to the amount of text and the height and width of the columns.

static columns:
Columns in which you divide the text into a specified number of columns.

Forming Dynamic Columns

To form dynamic columns, choose an option from the **Dynamic Columns** cascading submenu. Pick the **Auto height** option to produce columns of equal height. **Figure 9-36** shows methods for adjusting dynamic columns using **Auto height**. Increasing column width or height reduces the number of columns, while decreasing column width or height produces more columns. Pick the **Manual height** option to produce columns you can adjust individually for height to produce distinct groups of information. Pick and drag the arrows at the bottom of each column to adjust column height. See **Figure 9-37.**

Forming Static Columns

To form static columns, choose the number of columns from the **Static Columns** cascading submenu. The display of text in static columns depends on how much text is in the text editor and the height and width of the columns. However, the selected number of columns does not change even if text does not fill or extends past a column. **Figure 9-38** shows methods for adjusting static columns. Increasing column width or height rearranges the text in the specified number of columns, but the number of static columns does not change based on column width or height.

NOTE

To create more than six static columns, pick the **More...** option to access the **Column Settings** dialog box and enter the number of columns in the **Column Number** text box.

Figure 9-36.
Controlling columns using the dynamic column **Auto Height** option. Notice how column text flows automatically from one column to the next.

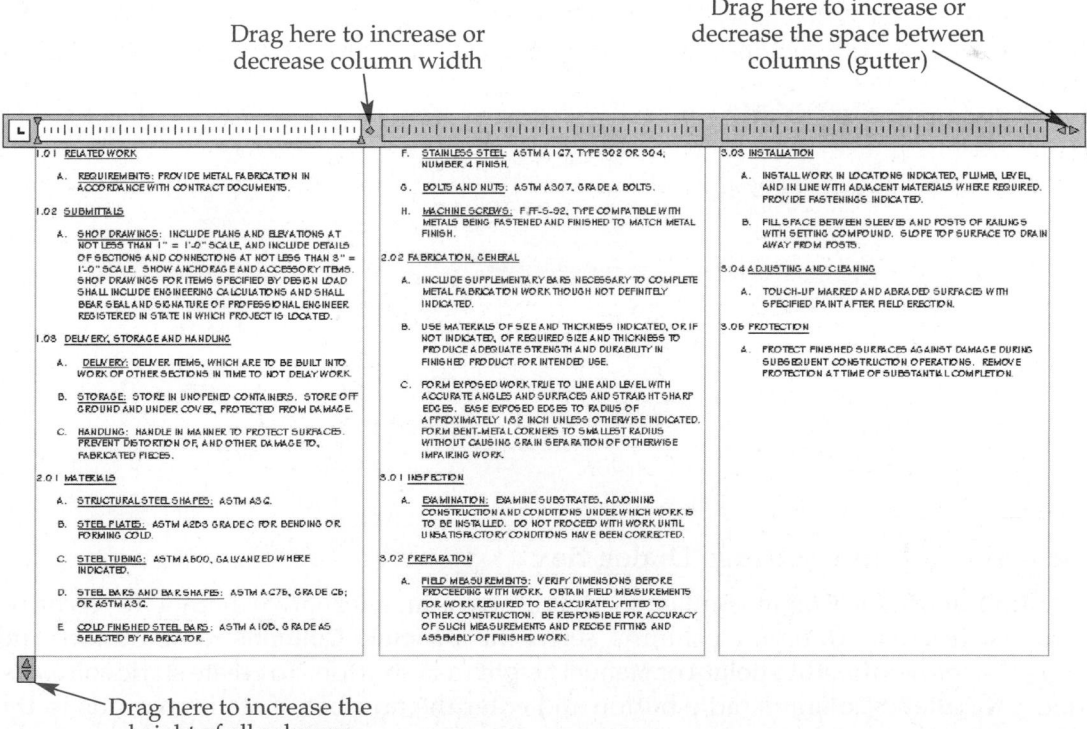

Drag here to increase or decrease column width

Drag here to increase or decrease the space between columns (gutter)

Drag here to increase the height of all columns

Figure 9-37.
Controlling the length of dynamic columns individually (manually).

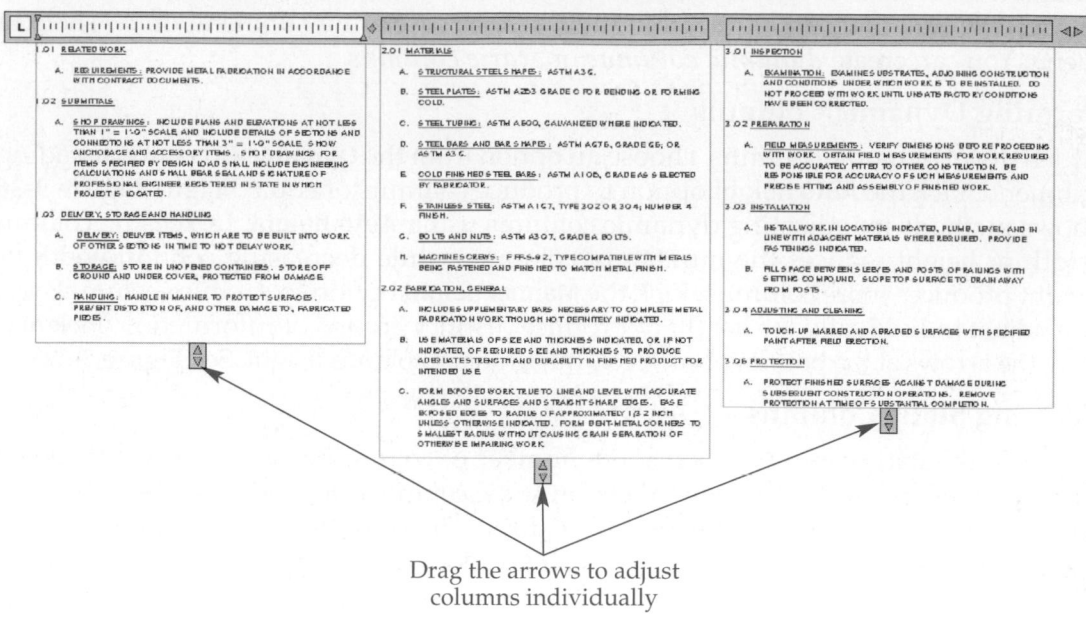

Drag the arrows to adjust
columns individually

Figure 9-38.
Controlling static columns.

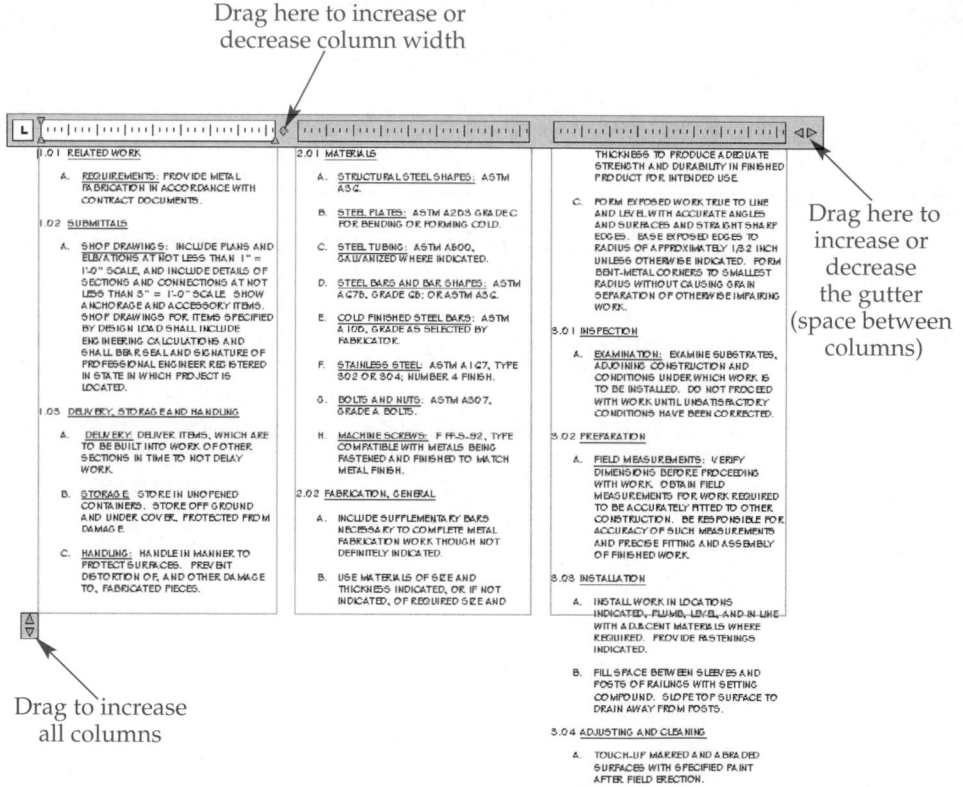

Drag here to increase or
decrease column width

Drag here to
increase or
decrease
the gutter
(space between
columns)

Drag to increase
all columns

Using the Column Settings Dialog Box

You can use the **Column Settings...** dialog box as an alternative method for creating columns. To create dynamic columns, select the **Dynamic Columns** radio button, and then select either the **Auto height** or **Manual height** radio button. To create static columns, choose the **Static Columns** radio button and enter the number of static columns in the **Column Number** text box.

Additional controls become available depending on the selected column type radio buttons. Enter the number of static columns in the **Column Number** text box. The **Height** text box allows you to enter the height for all static or dynamic columns. The **Width** area allows you to set column width and the *gutter*. Enter the column width in the **Column** text box and the gutter width in the **Gutter** text box. The **Total** text box is available only with static columns. It allows you to enter the total width of the text editor, which is the sum of the width of all columns and the gutter spacing between columns. To eliminate columns, pick the **No Columns** radio button.

gutter: The space between columns of text.

NOTE

If you choose to remove columns using the **No Columns** option, any column breaks added using the **Insert Column Break** function remain set. Backspace to remove column breaks.

Controlling Column Breaks

You can identify the line of text at which a new column begins using the **Insert Column Break** option. To apply this technique, you must first form a dynamic or static column. Then place the cursor at a location in the text editor where you want a new column to start, such as the start of a paragraph. Pick the **Insert Column Break** option to form the break. The text shifts to the next column at the location of the break. Continue applying column breaks as needed to separate sections of information.

Exercise 9-7

Access the Student Web site (www.g-wlearning.com/CAD) and complete Exercise 9-7.

Importing Text

The **Import Text** option available from the expanded **Tools** panel of the **Text Editor** ribbon tab, or from the shortcut menu, allows you to import text from an existing text file directly into the text editor. The text file can be either a standard ASCII text file (TXT) or a rich text format (RTF) file. The imported text becomes a part of the current mtext object. The **Select File** dialog box appears when you access the **Import Text...** option. Select the text file to be imported and pick the **Open** button. The text inserts at the current cursor location.

PROFESSIONAL TIP

Importing text is useful if someone has already created specification notes in a program other than AutoCAD and you want to place the same notes in drawings.

Template Development
Chapter 9

For detailed instructions on adding text styles to each drawing template, go to the Student Web site (www.g-wlearning.com/CAD), select this chapter, and select **Template Development**.

Chapter Test

Answer the following questions. Write your answers on a separate sheet of paper or go to the Student Web site (www.g-wlearning.com/CAD) and complete the electronic chapter test.

1. Which ASME standard contains guidelines for lettering?
2. What is text composition?
3. Determine the AutoCAD text height for text to be plotted .188″ high using a half (1:2) scale. Show your calculations.
4. Determine the AutoCAD text height for text to be plotted .188″ high using a scale of 1/4″ = 1′-0″. Show your calculations.
5. Explain the function of annotative text and give an example.
6. What is the relationship between the drawing scale and the annotation scale for annotative text?
7. Define *text style*.
8. Describe how to create a text style that has the name ROMANS-12_15, uses the romans.shx font, has a fixed height of .12, a text width of 1.25, and an oblique angle of 15.
9. Define *font*.
10. What are "big fonts"?
11. When setting text height in the **Text Style** dialog box, what value do you enter so text height can be altered each time the **TEXT** tool is used?
12. How would you specify text to display vertically on the screen?
13. What does a width factor of .5 do to text when compared to the default width factor of 1?
14. Explain how to make a text style current quickly.
15. Name the tool that lets you create multiline text objects.
16. How does the width of the mtext boundary affect what you type?
17. What happens if the mtext you are entering exceeds or is not as long as the boundary length that you initially establish?
18. In the text editor, how do you open the text editor shortcut menu?
19. What happens when you enter a fraction in the mtext editor, and what does this allow you to do?
20. How can you draw stacked fractions manually when using the **MTEXT** tool?
21. What happens when you pick the **Other...** option in the **Symbol** cascading menu of the text editor shortcut menu?
22. What is the purpose of tracking?
23. What text feature allows you to hide parts of objects behind and around text?
24. What is the difference between text boundary justification and paragraph alignment?
25. Define *line spacing*.
26. Explain the function of the **Allow Auto-list** option.
27. Explain how to convert multiple lines of text into a numbered list.
28. Briefly describe the difference between dynamic columns and static columns.
29. How can you insert a column break in static columns?
30. In what two formats can text be imported into the text editor?

AutoCAD and Its Applications—Basics

Drawing Problems

Start AutoCAD if it is not already started. Start a new drawing using an appropriate template of your choice. The template should include layers and text styles, when necessary, for drawing the given objects. Add layers and text styles as needed. Draw all objects using appropriate layers, and text styles, justification, and format. Follow the specific instructions for each problem. Use your own judgment and approximate dimensions when necessary.

▼ Basic

1. Use the **MTEXT** tool to type your name using a text style of your choosing and a text height of 1″. Print or plot your name as a name tag. Save the drawing as P9-1.

2. Use the **MTEXT** tool to type the definition of the following terms using a text style with the Romand font and a .12 text height. Save the drawing as P9-2.
   - scale factor
   - annotative text
   - annotation scale
   - text height
   - paper text height

3. Use the **MTEXT** tool to type the following text using a text style with the Stylus BT font and a .125 text height. The heading text height is .25. Save the drawing as P9-3.

   KEY NOTES
   1.   SLOPING SURFACE
   2.   DIAGONAL SUPPORT STRUT
   3.   VENT- PROVIDE NEW CANT FLASHING
   4.   BRICK CHIMNEY- REMOVE TO BELOW
        DECK SURFACE

▼ Intermediate

4. Use the **MTEXT** tool to type the following text using a text style with the Romans font and a .12 text height. The heading text height is .24. Check your spelling. Save the drawing as P9-4.

   NOTES:

   1. INTERPRET ALL DIMENSIONS AND TOLERANCES PER ANSI Y14.4M—1994.
   2. REMOVE ALL BURRS AND SHARP EDGES.
   CASTING NOTES UNLESS OTHERWISE SPECIFIED:
   1. .31 WALL THICKNESS
   2. R.12 FILLETS
   3. R.06 CORNERS
   4. 1.5°—3.0° DRAFT
   5. TOLERANCES
      ± 1° ANGULAR
      ± .03 TWO—PLACE DIMENSIONS
   6. PROVIDE .12 THK MACHINING STOCK ON ALL MACHINED SURFACES.

5. Use the **MTEXT** tool to type the following text using a text style with the Stylus BT font and a .125 text height. The heading text height is .188. After typing the text exactly as shown, edit the text with the following changes:
   A. Change the \ in item 7 to 1/2.
   B. Change the [in item 8 to 1.
   C. Change the 1/2 in item 8 to 3/4.
   D. Change the ^ in item 10 to a degree symbol.
   E. Check your spelling after making the changes.
   F. Save as drawing P9-5.

COMMON FRAMING NOTES:
1. ALL FRAMING LUMBER TO BE DFL #2 OR BETTER.
2. ALL HEATED WALLS @ HEATED LIVING AREAS TO BE 2 X 6 @ 24" OC.
3. ALL EXTERIOR HEADERS TO BE 2-2 X 12 UNLESS NOTED, W/ 2" RIGID INSULATION BACKING UNLESS NOTED.
4. ALL SHEAR PANELS TO BE 1/2" CDX PLY W/8d @ 4" OC @ EDGE, HDRS, & BLOCKING AND 8d @ 8" OC @ FIELD UNLESS NOTED.
5. ALL METAL CONNECTORS TO BE SIMPSON CO. OR EQUAL.
6. ALL TRUSSES TO BE 24" OC. SUBMIT TRUSS CALCS TO BUILDING DEPT. PRIOR TO ERECTION.
7. PLYWOOD ROOF SHEATHING TO BE \ STD GRADE 32/16 PLY LAID PERP TO RAFTERS. NAIL W/8d @ 6" OC @ EDGES AND 12" OC @ FIELD.
8. PROVIDE [1/2" STD GRADE T&G PLY FLOOR SHEATHING LAID PERP TO FLOOR JOISTS. NAIL W/10d @ 6" OC @ EDGES AND BLOCKING AND 12" OC @ FIELD.
9. BLOCK ALL WALLS OVER 10'-0" HIGH AT MID.
10. LET-IN BRACES TO BE 1 X 4 DIAG BRACES @ 45 ^ FOR ALL INTERIOR LOAD-BEARING WALLS.

6. Draw the general caulking notes shown below. Save your drawing as P9-6.

CAULKING NOTES:

CAULKING REQUIREMENTS BASED ON 1992 OREGON RESIDENTIAL ENERGY CODE

1. SEAL THE EXTERIOR SHEATHING AT CORNERS, JOINTS, DOORS, WINDOWS, AND FOUNDATION SILL WITH SILICONE CAULK.
2. CAULK THE FOLLOWING OPENINGS W/ EXPANDED FOAM, BACKER RODS, OR SIMILAR:
   • ANY SPACE BETWEEN WINDOW AND DOOR FRAMES
   • BETWEEN ALL EXTERIOR WALL SOLE PLATES AND PLY SHEATHING
   • ON TOP OF RIM JOIST PRIOR TO PLYWOOD FLOOR APPLICATION
   • WALL SHEATHING TO TOP PLATE
   • JOINTS BETWEEN WALL AND FOUNDATION
   • JOINTS BETWEEN WALL AND ROOF
   • JOINTS BETWEEN WALL PANELS
   • AROUND OPENINGS

AutoCAD and Its Applications—Basics

7. Draw the basic organizational chart shown below. Save your drawing as P9-7.

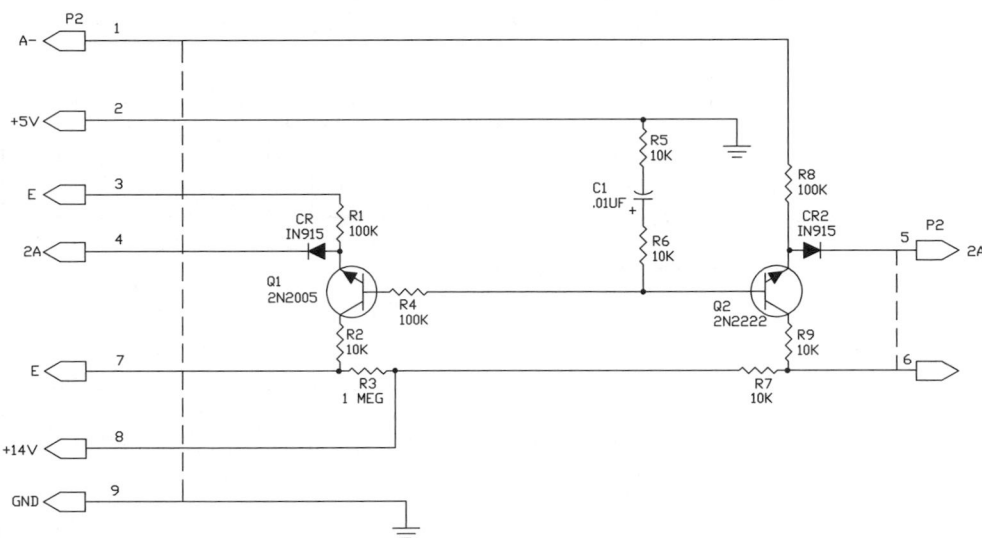

▼ Advanced

8. Draw the controller schematic shown below. Save your drawing as P9-8.

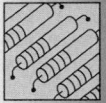

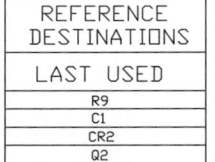

NOTES:

1. INTERPRET ELECTRICAL AND ELECTRONICS DIAGRAMS PER ANSI Y14.15.

2. UNLESS OTHERWISE SPECIFIED:

    RESISTANCE VALUES ARE IN OHMS.
    RESISTANCE TOLERANCE IS 5%.
    RESISTORS ARE 1/4 WATT.
    CAPACITANCE VALUES ARE IN MICROFARADS.
    CAPACITANCE TOLERANCE IS 10%.
    CAPACITOR VOLTAGE RATING IS 20V.
    INDUCTANCE VALUES ARE IN MICROHENRIES.

REFERENCE DESTINATIONS	
LAST USED	
R9	
C1	
CR2	
Q2	

9. Draw the electrical notes shown below. Save your drawing as P9-9.

ELECTRICAL NOTES:

1. ALL GARAGE AND EXTERIOR PLUGS AND LIGHT FIXTURES TO BE ON GFCI CIRCUIT.

2. ALL KITCHEN PLUGS AND LIGHT FIXTURES TO BE ON GFCI CIRCUIT.

3. PROVIDE A SEPARATE CIRCUIT FOR MICROWAVE OVEN.

4. PROVIDE A SEPARATE CIRCUIT FOR PERSONAL COMPUTER. VERIFY LOCATION WITH OWNER.

5. VERIFY ALL ELECTRICAL LOCATIONS W/ OWNER.

6. EXTERIOR SPOTLIGHTS TO BE ON PHOTOELECTRIC CELL W/ TIMER.

7. ALL RECESSED LIGHTS IN EXTERIOR CEILINGS TO BE INSULATION COVER RATED.

8. ELECTRICAL OUTLET PLATE GASKETS SHALL BE INSULATED ON RECEPTACLE, SWITCH, AND ANY OTHER BOXES IN EXTERIOR WALL.

9. PROVIDE THERMOSTATICALLY CONTROLLED FAN IN ATTIC WITH MANUAL OVERRIDE. VERIFY LOCATION WITH OWNER.

10. ALL FANS TO VENT TO OUTSIDE AIR. ALL FAN DUCTS TO HAVE AUTOMATIC DAMPERS.

11. HOT WATER TANKS TO BE INSULATED TO R-11 MINIMUM.

12. INSULATE ALL HOT WATER LINES TO R-4 MINIMUM. PROVIDE ALTERNATE BID TO INSULATE ALL PIPES FOR NOISE CONTROL.

13. PROVIDE 6 SQ. FT. OF VENT FOR COMBUSTION AIR TO OUTSIDE AIR FOR FIREPLACE CONNECTED DIRECTLY TO FIREBOX. PROVIDE FULLY CLOSABLE AIR INLET.

14. HEATING TO BE ELECTRIC HEAT PUMP. PROVIDE BID FOR SINGLE UNIT NEAR GARAGE OR FOR A UNIT EACH FLOOR (IN ATTIC).

15. INSULATE ALL HEATING DUCTS IN UNHEATED AREAS TO R-11. ALL HVAC DUCTS TO BE SEALED AT JOINTS AND CORNERS.

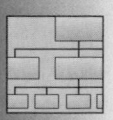

10. Draw the flowchart shown below. Save your drawing as P9-10.

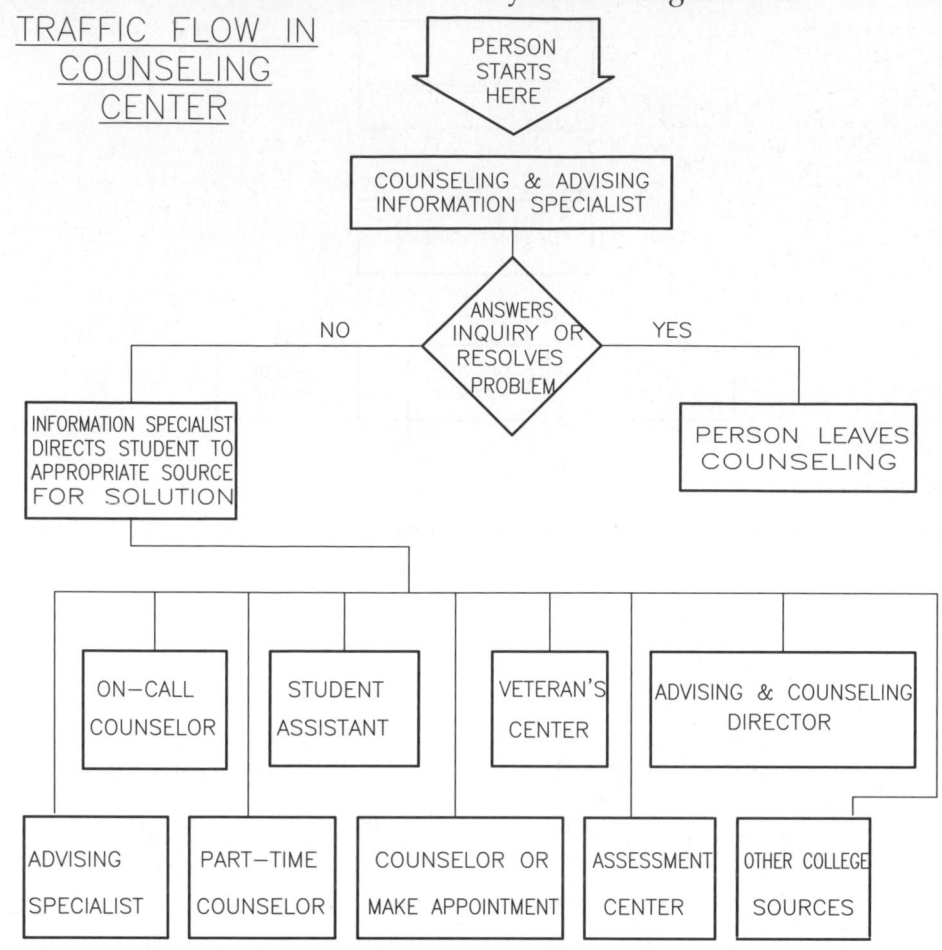

TRAFFIC FLOW IN
COUNSELING
CENTER

PERSON
STARTS
HERE

COUNSELING & ADVISING
INFORMATION SPECIALIST

ANSWERS
INQUIRY OR
RESOLVES
PROBLEM

NO   YES

INFORMATION SPECIALIST
DIRECTS STUDENT TO
APPROPRIATE SOURCE
FOR SOLUTION

PERSON LEAVES
COUNSELING

ON–CALL COUNSELOR

STUDENT ASSISTANT

VETERAN'S CENTER

ADVISING & COUNSELING DIRECTOR

ADVISING SPECIALIST

PART–TIME COUNSELOR

COUNSELOR OR MAKE APPOINTMENT

ASSESSMENT CENTER

OTHER COLLEGE SOURCES

11. Draw the electrical legend shown below. Save your drawing as P9-11.

ELECTRICAL LEGEND:

	I I0 VOLT DUPLEX CONVENIENCE OUTLET
GFCI	I I0 VOLT GROUND FAULT CIRCUIT INTERRUPT DUPLEX OUTLET
GFCI WP	I I0 VOLT WATERPROOF GFCI DUPLEX OUTLET
	I I0 VOLT SPLIT WIRED OUTLET
	220 VOLT OUTLET
	JUNCTION BOX
	CABLE TELEVISION OUTLET
	CLOCK OUTLET
	DOOR BELL
	SINGLE POLE SWITCH
	THREE-WAY SWITCH
	CEILING-MOUNTED LIGHT
	WALL-MOUNTED LIGHT
	FLUORESCENT LIGHT
	CIRCULAR RECESSED LIGHT
	SQUARE RECESSED LIGHT
	LIGHT, FAN COMBINATION
	LIGHT, FAN, HEAT COMBINATION
SD	CEILING-MOUNTED SMOKE DETECTOR
SD	WALL-MOUNTED SMOKE DETECTOR

Single-Line Text and Additional Text Tools

Learning Objectives

After completing this chapter, you will be able to do the following:
- ✓ Use the **TEXT** tool to create single-line text.
- ✓ Insert and use fields.
- ✓ Check your spelling.
- ✓ Edit existing text.
- ✓ Search for and replace text automatically.

This chapter describes how to use the **TEXT** tool to place single-line text objects. The **TEXT** tool is most useful for text items that require a single character, word, or line of text. You should typically use mtext to type paragraphs or if the text requires mixed fonts, sizes, colors, or other characteristics. This chapter also presents text editing functions and other valuable tools, including fields, spell checking, and methods for finding and replacing text.

Creating Single-Line Text

Access the **TEXT** tool to create a single-line text object. Pick a point to define the lower-left corner of the text, using default justification. Next, if the current text style uses a height of 0, enter the text height. If the current text style is annotative, enter the paper text height. If the current text style is not annotative, enter the text height multiplied by the scale factor. The next prompt asks for the text rotation angle. The default value is 0, which draws horizontal text. Other values rotate text in a counterclockwise direction. The text pivots about the start point, as shown in **Figure 10-1**.

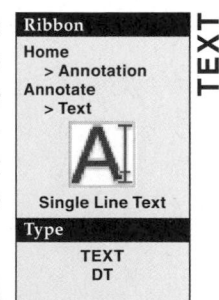

Ribbon
Home
> Annotation
Annotate
> Text
Single Line Text
Type
TEXT
DT

TEXT

NOTE

Changes you make to the default text angle orientation or direction affect the text rotation.

After you set the text height and rotation angle, a text editor and cursor equal in height to the text height appears on-screen at the start point. As you type, the text editor increases in size to display the characters. See **Figure 10-2**. Type additional lines

Figure 10-1.
Rotation angles for text. The plus sign indicates the start point.

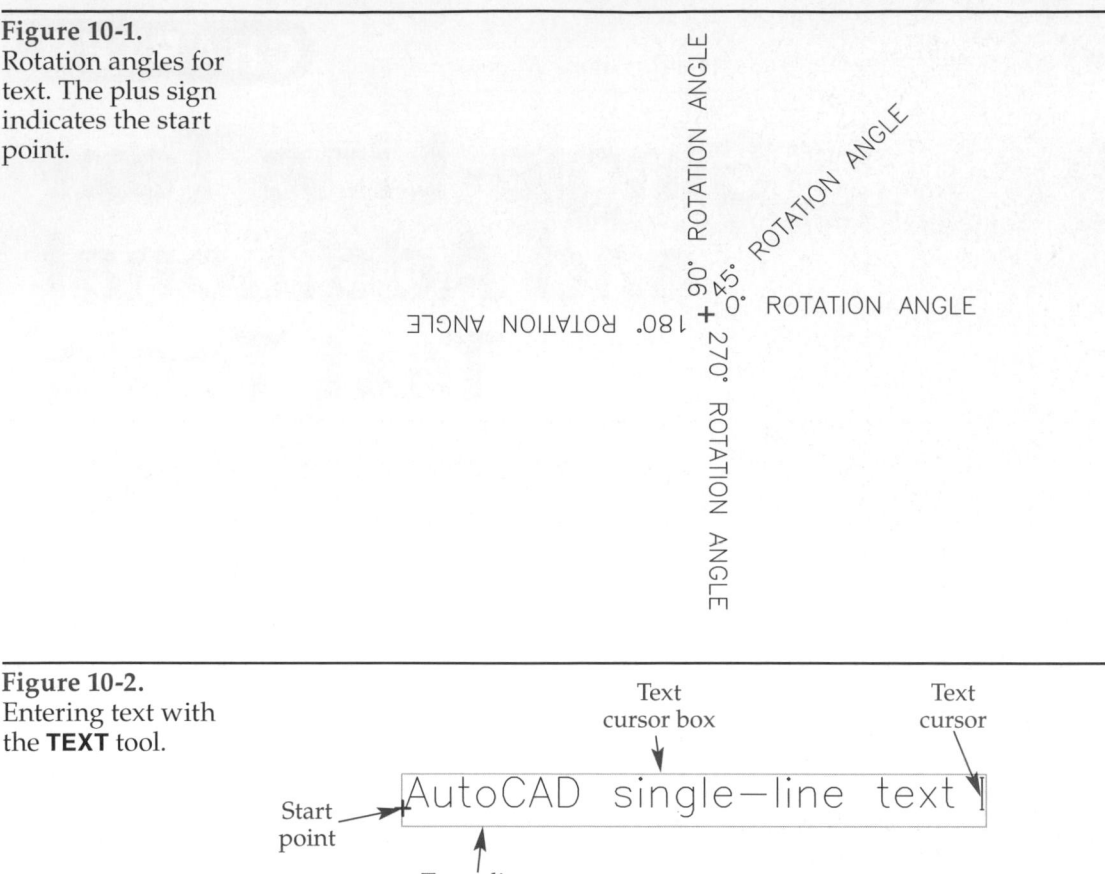

Figure 10-2.
Entering text with the **TEXT** tool.

AutoCAD single—line text

Start point

Text editor

Text cursor box

Text cursor

of text by pressing [Enter] at the end of each line. The text cursor automatically moves to a start point one line below the preceding line. Press [Enter] twice to exit the **TEXT** tool and keep what you have typed. You can cancel the tool at any time by pressing [Esc]. This action removes any incomplete lines of text.

While typing, you can right-click to display a shortcut menu of text options. These options function much like those for the **MTEXT** tool. Options for accessing help files and canceling the **TEXT** tool are also available from the shortcut menu.

NOTE

Set the text style you want to use current before accessing the **TEXT** tool. A **Style** option is available before you pick the text start point to set the text style, but it is difficult to use.

Text Justification

The **TEXT** tool offers a variety of justification options. Left justification is the default. To use a different justification option, choose the **Justify** option at the Specify start point of text [Justify/Style]: prompt before you pick the text start point.

The **Center** option allows you to select the center point for the baseline of the text. The **Middle** option allows you to center text horizontally and vertically at a given point. The **Right** option justifies text at the lower-right corner. You can also change the letter height and rotation when using these options. **Figure 10-3** compares the **Center**, **Middle**, and **Right** options. A number of text alignment options allow you to place text on a drawing in relation to the top, bottom, middle, left side, or right side of the text. **Figure 10-4** shows these alignment options.

AutoCAD and Its Applications—Basics

Figure 10-3.
The **Center**, **Middle**, and **Right** text justification options.

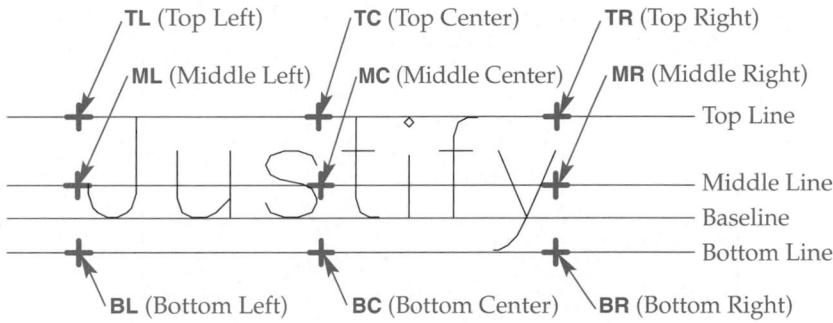

AUTOCAD CENTERED TEXT
Center Option

AUTOCAD MIDDLE TEXT
Middle Option

AUTOCAD RIGHT—JUSTIFIED TEXT
Right Option

Figure 10-4.
Using the **TL**, **TC**, **TR**, **ML**, **MC**, **MR**, **BL**, **BC**, and **BR** text alignment options. Notice the meaning of each abbreviation.

TL (Top Left) **TC** (Top Center) **TR** (Top Right)
ML (Middle Left) **MC** (Middle Center) **MR** (Middle Right)
Top Line
Middle Line
Baseline
Bottom Line
BL (Bottom Left) **BC** (Bottom Center) **BR** (Bottom Right)

When you use the **Align** justification option, AutoCAD automatically adjusts the text height to fit between the start point and endpoint. The height varies according to the distance between the points and the number of characters. The **Fit** option is similar to the **Align** option, except you can select the text height. AutoCAD adjusts character width to fit between the two given points, while keeping text height constant. **Figure 10-5** shows the effects of the **Align** and **Fit** options.

PROFESSIONAL TIP

The **Align** and **Fit** options of the **TEXT** tool are not recommended because the text height or width is inconsistent from one line of text to another. In addition, text height adjusted by the **Fit** option can cause one line of text to run into another.

Exercise 10-1

Access the Student Web site (www.g-wlearning.com/CAD) and complete Exercise 10-1.

Figure 10-5.
Examples of aligned and fit text. In aligned text, the text height is adjusted. In fit text, the text width is adjusted.

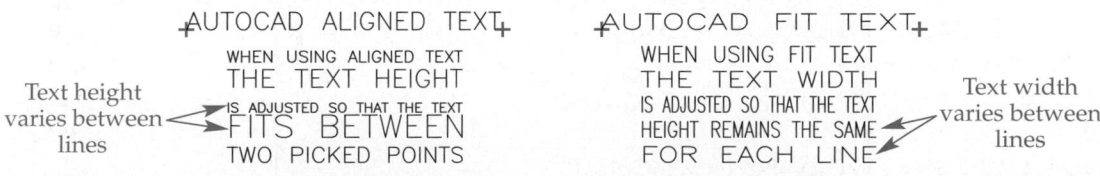

AUTOCAD ALIGNED TEXT
WHEN USING ALIGNED TEXT
THE TEXT HEIGHT
IS ADJUSTED SO THAT THE TEXT
FITS BETWEEN
TWO PICKED POINTS

Text height varies between lines

AUTOCAD FIT TEXT
WHEN USING FIT TEXT
THE TEXT WIDTH
IS ADJUSTED SO THAT THE TEXT
HEIGHT REMAINS THE SAME
FOR EACH LINE

Text width varies between lines

Align Option   **Fit Option**

Inserting Symbols

control code sequence: A key sequence beginning with %% that defines symbols in text created with the **TEXT** tool.

In order to insert a symbol with the **TEXT** tool, you must type a *control code sequence*. For example, to add the note ⌀2.75, type %%C2.75 in the text editor. In this example, %%C is the code used to add the diameter symbol. **Figure 10-6** shows many symbol codes and the symbols the codes create. Add a single percent sign normally. However, when a percent sign must precede another control code sequence, you can use %%% to force a single percent sign.

Figure 10-6.
Common control code sequences used to add symbols to single-line text.

Control Code or Unicode	Type of Symbol	Appearance
%%d	Degrees	°
%%p	Plus/Minus	±
%%c	Diameter	⌀
%%%	Percent	%
\U+2248	Almost equal	≈
\U+2220	Angle	∠
\U+E100	Boundary line	BL
\U+2104	Centerline	℄
\U+0394	Delta	△
\U+0278	Electrical phase	⏀
\U+E101	Flow line	FL
\U+2261	Identity	≡
\U+E200	Initial length	◯→
\U+E102	Monument line	ML
\U+2260	Not equal	≠
\U+2126	Ohm	Ω
\U+03A9	Omega	Ω
\U+214A	Property line	PL
\U+2082	Subscript 2	2
\U+00B2	Squared	2
\U+00B3	Cubed	3

Drawing Underscored or Overscored Text

Type %%O in front of the line of text to overscore text and %%U in front of the line of text to create underscored (underlined) text. To create the note UNDERSCORING TEXT, for example, type %%UUNDERSCORING TEXT. A line of text may require both underscoring and overscoring. To do this, use both control code sequences. For example, the control code sequence %%O%%ULINE OF TEXT produces LINE OF TEXT.

The %%O and %%U control codes are toggles that turn overscoring and underscoring on and off. Type %%U preceding a word or phrase to turn underscoring on. Type %%U after the desired word or phrase to turn underscoring off. Any text following the second %%U appears without underscoring. For example, enter DETAIL A HUB ASSEMBLY as %%UDETAIL A%%U HUB ASSEMBLY.

PROFESSIONAL TIP

Many drafters prefer to underline labels such as SECTION A-A or DETAIL B. Rather than draw line or polyline objects under the text, use the **Middle** or **Center** justification mode and underscoring. The view labels automatically underline and center under the views or details they identify.

Exercise 10-2

Access the Student Web site (www.g-wlearning.com/CAD) and complete Exercise 10-2.

Working with Fields

Fields display information related to a specific object, general drawing properties, or the current user or computer system. You can set field information to update automatically. This makes fields useful tools for displaying information that may change throughout the course of a project. For example, you could insert the **Date** field into a title block. The field updates automatically with the current date throughout the life of the drawing file.

field: A special type of text object that can display a specific property value, setting, or characteristic.

Inserting Fields

Use the **FIELD** tool and **Field** dialog box, shown in **Figure 10-7,** to add fields to mtext or text objects. To insert a field in an active mtext editor, pick the **Field** button from the **Insert** panel of the **Text Editor** ribbon tab, pick the **Insert Field** option available from the shortcut menu, or press [Ctrl]+[F]. To insert a field in an active single-line text editor, right-click and select **Insert Field...** or press [Ctrl]+[F].

The **Field** dialog box includes many preset fields. Notice that fields are grouped into categories. When you select a category from the **Field category** drop-down list, only the fields within the category appear in the **Field names** list box. This makes it much easier to locate the desired field. Pick the field category, and then pick the field to insert from the **Field names** list box. You can also select from a list of formats to determine the display of the field. The **Format** list varies, depending on the selected field.

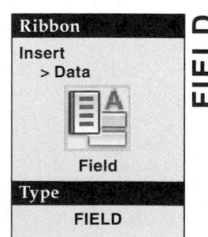

Ribbon

Insert
> Data

Field

Type

FIELD

FIELD

Figure 10-7.
Select fields using the **Field** dialog box.

Selected field

Current value for field

Select category to limit field list

All available fields listed

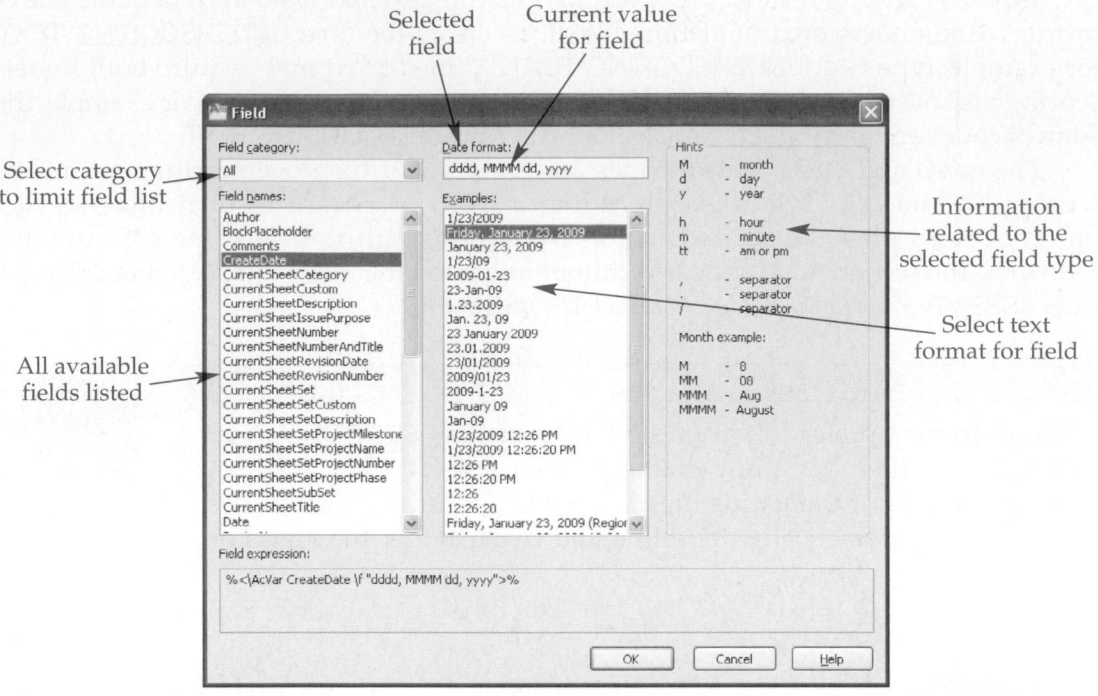

Information related to the selected field type

Select text format for field

After you select the field and format, pick the **OK** button to insert the field. The field assumes the current text style. By default, field text displays a gray background. See **Figure 10-8.** This keeps you aware that the text is actually a field, and the value may change. You can deactivate the background in the **Fields** area of the **User Preferences** tab of the **Options** dialog box. See **Figure 10-9.**

Updating Fields

After you insert a field into a drawing, the displayed value may change. For example, a field indicating the current date changes every day. A field displaying the file name changes if the file name changes. A field displaying the value of an object property changes if modifications to the object cause the property to change.

Automatic or manual field *updating* is possible. Automatic updating is set using the **Field Update Settings** dialog box. To access this dialog box, pick the **Field Update Settings...** button in the **Fields** area of the **User Preferences** tab of the **Options** dialog box. Whenever a selected event (such as saving or regenerating) occurs, all associated fields automatically update.

Update fields manually using the **Update Fields** tool. After selecting the tool, pick the fields to update. You can use the **All** selection option to update all fields in a single operation. You can also update a field within the text editor by right-clicking on the field and selecting **Update Field**.

Updating:
AutoCAD's procedure for changing text in a field based on the field's current value.

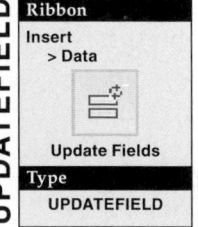

UPDATEFIELD

Ribbon
Insert
> Data

Update Fields

Type
UPDATEFIELD

Figure 10-8.
A date and time field inserted into multiline text. The gray background identifies the text as a field.

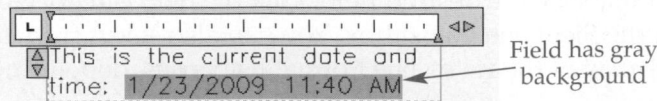

Field has gray background

Figure 10-9.
Control the background display for fields in the **User Preferences** tab of the **Options** dialog box.

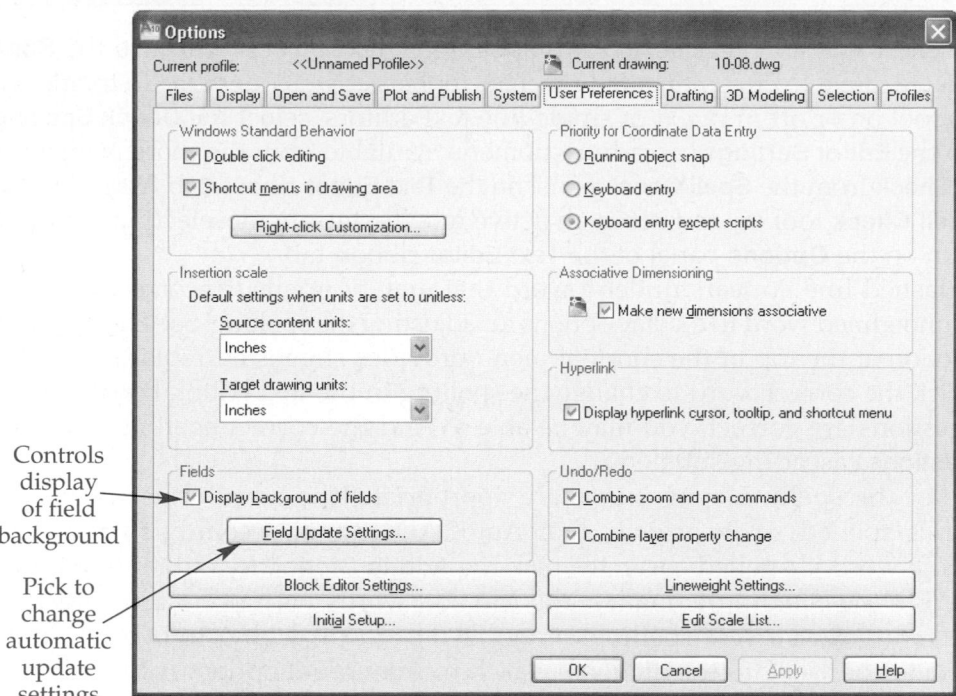

Controls display of field background →

Pick to change automatic update settings

Editing Fields

To edit a field, you must first select the text object containing the field for editing. You can do this quickly by double-clicking on the text object. Then double-click on the field to display the **Field** dialog box. You can also right-click in the field and pick **Edit Field…**. Use the **Field** dialog box to modify the field settings and pick **OK** to apply the changes.

You can also convert a field to standard text. When you convert a field, the currently displayed value becomes text, the association to the field is lost, and the value no longer updates. To convert a field to text, select the text for editing, right-click in the field, and pick **Convert Field To Text**.

NOTE

You can use fields with many AutoCAD tools, including inquiry tools, drawing properties, attributes, and sheet sets. Specific field applications are described where appropriate throughout this textbook.

Exercise 10-3

Access the Student Web site (www.g-wlearning.com/CAD) and complete Exercise 10-3.

Checking Your Spelling

The quickest way to check for correct spelling in text objects is to use the **Spell Check** tool available in a current text editor. This tool is active by default. To toggle the **Spell Check** tool on or off in mtext or single-line text editors, select the **Check Spelling** option from the **Editor Settings** cascading submenu available from the shortcut menu or select **Spell Check** from the **Spell Check** panel on the **Text Editor** ribbon tab. You can also turn the **Spell Check** tool on and off in an active mtext editor by deselecting the **Spell Check** button on the **Options** panel of the **Text Editor** ribbon tab.

A red dashed line appears under a word that may be spelled incorrectly. Right-click on the underlined word to display options for adjusting the spelling. See **Figure 10-10**. The first section at the top of the shortcut menu provides suggested replacements for the word. Pick the correct word to change the spelling in the text editor. If none of the initial suggestions are correct, you may be able to find the correct spelling from the **More Suggestions** cascading submenu.

If none of the spelling suggestions are appropriate, the word either is spelled correctly or is spelled so incorrectly that AutoCAD cannot recommend the right spelling. If the word is spelled correctly, pick the **Add to Dictionary** option to add the current word to the custom dictionary. You can add words with up to 63 characters. If you want to use the current spelling, recognized as incorrect by AutoCAD, but do not want to add the word to the dictionary, pick the **Ignore All** option. All words that match the currently found misspelled word in the active text editor are ignored and the underline is hidden. Common drafting words and abbreviations, such as the abbreviation for the word SCHEDULE (SCH) in **Figure 10-10** can be added to the dictionary or ignored.

Figure 10-10.
Quickly checking your spelling using the **Spell Check** tool. Spell checking in the multiline text editor is shown. The tool functions the same in the single-line text editor.

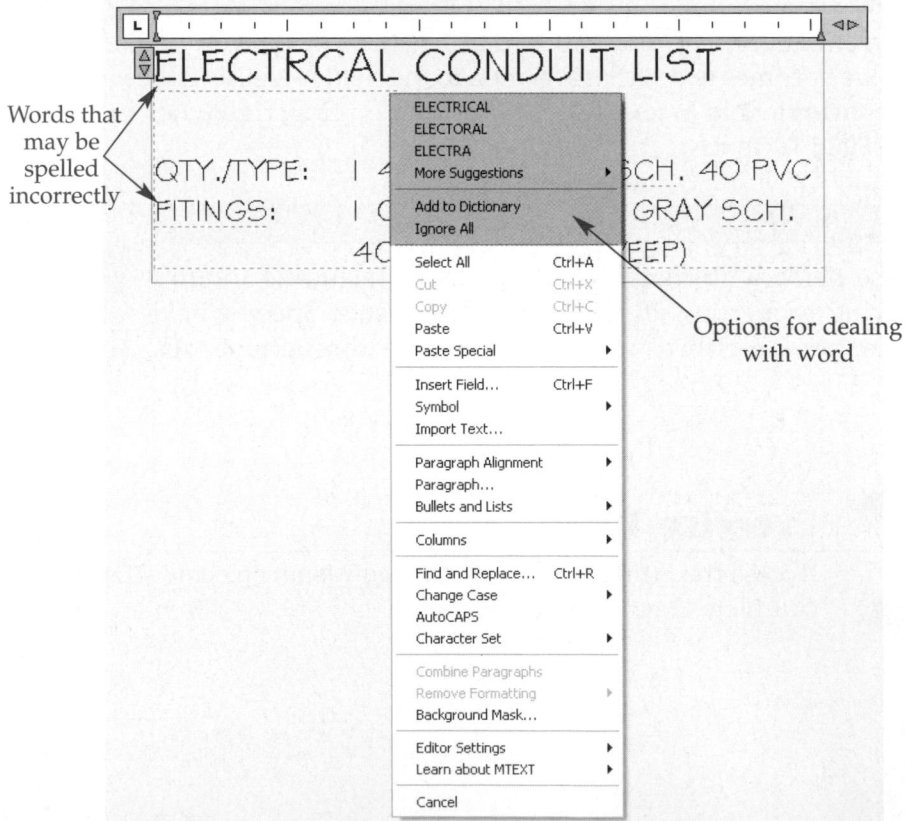

Using the SPELL Tool

An alternative method to check spelling is to use the **SPELL** tool to access the **Check Spelling** dialog box, shown in **Figure 10-11.** The **Check Spelling** dialog box checks spelling of all text objects, without activating a text editor. To check spelling, first identify the portion of the drawing you want to spell-check by selecting an option from the **Where to Check** drop-down list. Pick the **Entire drawing** option to check the spelling of all text objects in the drawing file, including model space and all layouts, or choose the **Current space/layout** to check spelling only of text objects in the active layout or in model space, if model space is active. You can also check the spelling of certain text objects by picking the **Select text objects** button, next to the **Where to Check** drop-down list, to enter the drawing window and select all the text objects for which you want to check the spelling. You do not need to choose the **Selected objects** option from the **Where to Check** drop-down list to check selected objects.

After you define what and where to check, pick the **Start** button to begin to check spelling. The first word that may be misspelled is highlighted in the drawing window and is active in the **Check Spelling** dialog box. **Figure 10-11** briefly describes the options found in the **Check Spelling** dialog box.

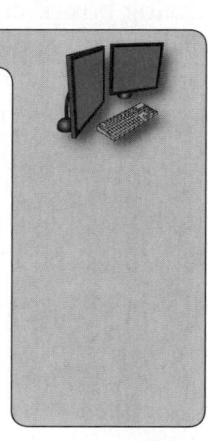

Ribbon
Annotate
> Text

Check Spelling

Type
SPELL
SP

SPELL

PROFESSIONAL TIP

Before you check spelling, you may want to adjust some of the spell-checking preferences provided in the **Check Spelling Settings** dialog box. Access this dialog box by picking the **Settings...** button in the **Check Spelling** dialog box, or select the **Check Spelling Settings...** option from the **Editor Settings** cascading submenu available from the text editor shortcut menu. If the mtext editor is already open, pick the small arrow in the lower-right corner of the **Spell Check** panel on the **Text Editor** ribbon tab. The settings apply to spelling checked using the **Spell Check** tool in an active text editor and using the **Check Spelling** dialog box.

Figure 10-11.
The **Check Spelling** dialog box.

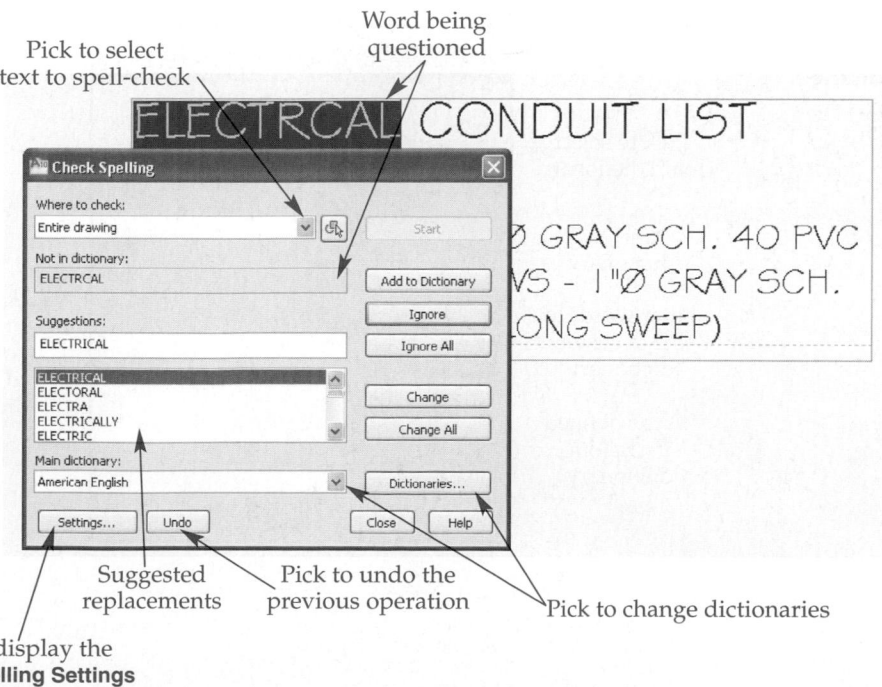

Pick to select text to spell-check

Word being questioned

Suggested replacements

Pick to undo the previous operation

Pick to change dictionaries

Pick to display the **Check Spelling Settings** dialog box

Ribbon

Text Editor
> Spell Check

Edit Dictionaries

Changing Dictionaries

AutoCAD provides a variety of spelling dictionaries, including dictionaries for several non-English languages. Pick the **Dictionaries...** button of the **Check Spelling** dialog box or select the **Dictionaries...** option from the **Editor Settings** cascading submenu available from the text editor shortcut menu to access the **Dictionaries** dialog box. See **Figure 10-12.**

You can use the **Main dictionary** list to select one of the many language dictionaries to use as the current main dictionary. The main dictionary is protected; you cannot add definitions to it. You can use the **Custom dictionary** list to select the active custom dictionary. The default custom dictionary is sample.cus. Type a word in the **Content** text box that you want to either add or delete from the custom dictionary. For example, ASME Y14.5M is custom text used in engineering drafting. Pick the **Add** button to accept the custom words in the text box, or pick the **Delete** button to remove the words from the custom dictionary. Custom dictionary entries may be up to 63 characters in length.

You can create and manage a custom dictionary by picking the **Manage Custom Dictionaries...** option from the drop-down list to access the **Manage Custom Dictionary** dialog box. Pick the **New** button to create a new custom dictionary by entering a new file name with a .cus extension. You can add and delete words and combine dictionaries using any standard text editor. If you use a word processor such as Microsoft® Word, be sure to save the file as *text only*, with no special text formatting or printer codes. Add a custom dictionary by picking the **Add** button, and choose the **Remove** button to delete a custom dictionary from the list. You can also add existing custom dictionaries by picking the **Import...** button from the **Custom dictionary** area.

PROFESSIONAL TIP

You can create custom dictionaries for various disciplines. For example, you might add common abbreviations and brand names for mechanical drawings to a mech.cus file. A separate file named arch.cus might contain common architectural abbreviations and frequently used brand names.

Figure 10-12.
The **Dictionaries** dialog box.

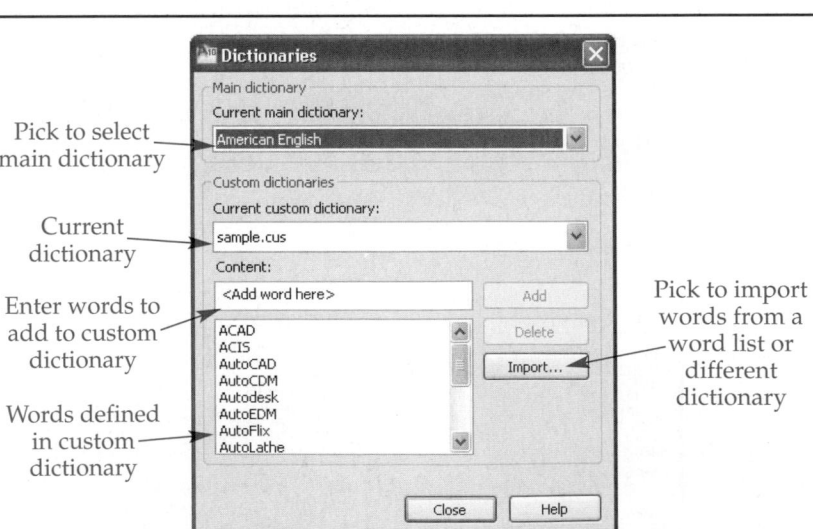

Pick to select main dictionary

Current dictionary

Enter words to add to custom dictionary

Words defined in custom dictionary

Pick to import words from a word list or different dictionary

AutoCAD and Its Applications—Basics

Revising Text

The easiest way to reopen the text editor to make changes to text content is to double-click an mtext or text object. Another technique to re-enter the text editor is to pick the text object to modify and then right-click and select **Mtext Edit...** to revise mtext, or **Edit...** to modify single-line text.

NOTE

You can also type **MTEDIT** to edit mtext or **DDEDIT** to edit either single-line text or mtext.

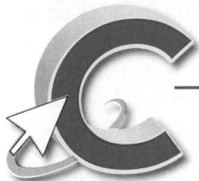

Exercise 10-4

Access the Student Web site (www.g-wlearning.com/CAD) and complete Exercise 10-4.

Exercise 10-5

Access the Student Web site (www.g-wlearning.com/CAD) and complete Exercise 10-5.

Changing Case

If you forget to type text using uppercase letters, you can quickly set all text to uppercase by selecting the text and picking the **UPPERCASE** option from the **Change Case** cascading submenu available from the mtext or text editor shortcut menu. While editing mtext, you can also select the **Make Uppercase** button from the **Formatting** panel of the **Text Editor** ribbon tab. Select the **lowercase** option, also found in the **Change Case** cascading submenu, to change selected text to lowercase characters. While editing mtext, you can also select the **Make Lowercase** button from the **Formatting** panel of the **Text Editor** ribbon tab.

The **AutoCAPS** option available from the mtext editor shortcut menu or the expanded **Tools** panel of the **Text Editor** ribbon tab turns [Caps Lock] on when you open an mtext editor. [Caps Lock] turns off when you exit the text editor so that text in other programs is not all uppercase.

Cutting, Copying, and Pasting Text

Clipboard functions allow you to copy, cut, and paste text from any text-based application, such as Microsoft® Word, into a text editor. You can quickly access clipboard functions from the shortcut menu when an mtext or text editor is active. Pasted text retains its properties. You can also copy or cut and paste text from the text editor into other text-based applications.

AutoCAD provides three additional paste options for pasting text into the mtext text editor. These options are available from the **Paste Special** cascading submenu of the text editor shortcut menu. Pick the **Paste without Character Formatting** option to paste text without applying preset character formatting such as bold, italic, or underline. Select the **Paste without Paragraph Formatting** option to paste text without applying current paragraph formatting, including lists. Pick the **Paste without Any Formatting** option to paste text without applying any current character and paragraph formatting.

Cutting or copying and pasting text is useful if someone has already created specification notes in a program other than AutoCAD and you want to place the same notes in drawings.

Finding and Replacing Text

AutoCAD provides tools for searching for a piece of text in your drawing and replacing it with an alternative piece of text. You can search for and replace text in an active mtext or single-line text editor, or without opening a text editor.

Using the Find and Replace Tool

You can activate the **Find and Replace** tool in the mtext and single-line text editors by selecting the **Find and Replace...** option available from the shortcut menu. Another way to turn on the tool in an active mtext editor is by selecting the **Find & Replace** button from the **Tools** panel of the **Text Editor** ribbon tab. The **Find and Replace** dialog box displays. See **Figure 10-13.**

Enter the text you are searching for in the **Find what:** text box. Enter the text to substitute in the **Replace with:** text box. Then pick the **Find Next** button to highlight the next instance of the search text. You can then pick the **Replace** or **Replace All** button to replace just the highlighted text or all words that match your search criteria. Check boxes control the characters and words recognized when finding and replacing text.

Using the FIND Tool

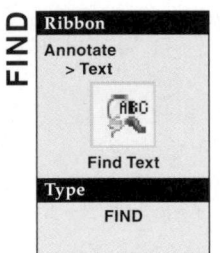

FIND

Ribbon
Annotate
> Text
Find Text
Type
FIND

Use the **FIND** tool to find text throughout the entire drawing and replace it with a different piece of text. Access the **FIND** tool when a text editor is not active to search the entire drawing. AutoCAD displays the **Find and Replace** dialog box when you access the **FIND** tool. See **Figure 10-14.** Another method for accessing the **Find and Replace** dialog box is to enter the text string you want to find in the **Find Text** text box in the **Text** panel of the **Annotation** ribbon tab, and press [Enter].

Figure 10-13.
Using the **Find and Replace** dialog box in an active text editor.

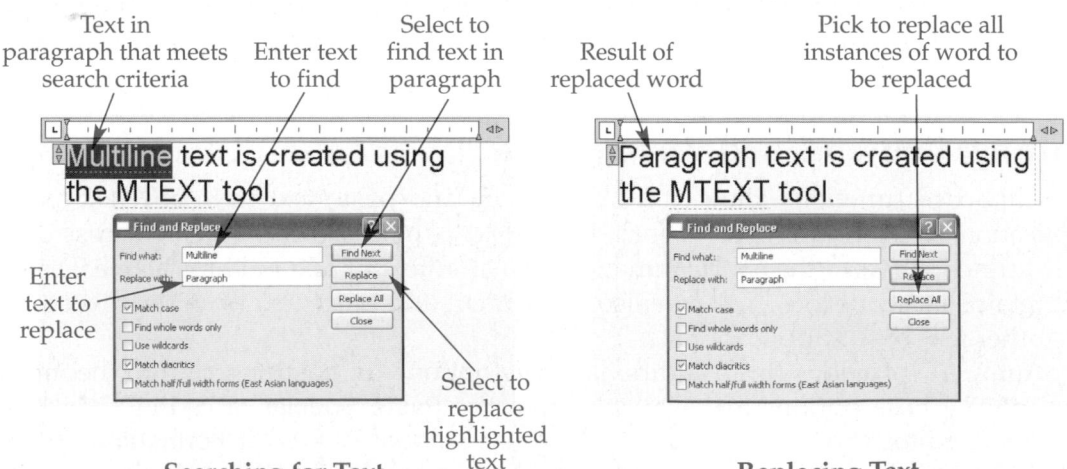

Searching for Text

Replacing Text

Figure 10-14.
Using the version of the **Find and Replace** dialog box that appears when you use the **Find** tool.

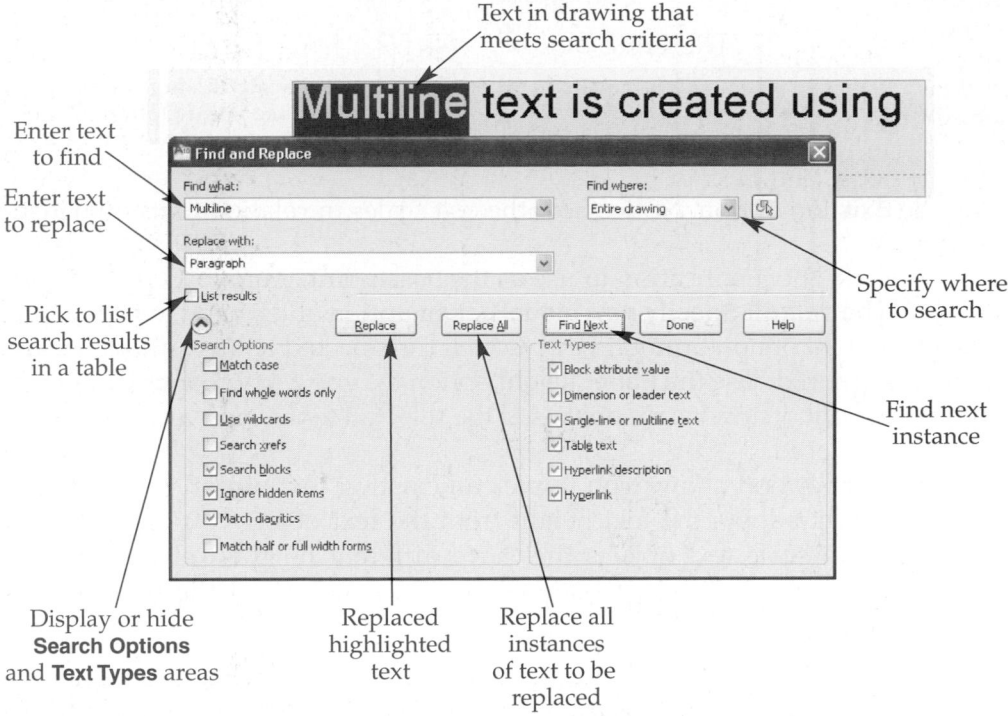

Text in drawing that meets search criteria

Enter text to find

Enter text to replace

Pick to list search results in a table

Specify where to search

Find next instance

Display or hide **Search Options** and **Text Types** areas

Replaced highlighted text

Replace all instances of text to be replaced

You can use the **Find and Replace** dialog box to locate and replace text in multiple text objects. As a result, to find and replace text, you must first identify the portion of the drawing you want to search by selecting an option from the **Find Where** drop-down list. These options are similar to those described for the **Find Where** drop-down list in the **Check Spelling** dialog box.

The **Find and Replace** dialog box is much like the dialog box of the same name that appears when you find and replace text within a text editor. However, this version allows you to display the search results in a table within the dialog box. It also provides more search options. Pick the **More Options** button to display several check boxes used to control the characters and words recognized when finding and replacing text.

NOTE

The find and replace strings are saved with the drawing file and can be reused.

Scaling Text

One option for changing the height of text objects is to use the **SCALETEXT** tool. This tool allows you to scale text objects in relation to their individual insertion points or in relation to a single base point. The **SCALETEXT** tool works with mtext and text objects. You can also select both types of text objects at the same time. After you access the **SCALETEXT** tool, select the text objects to scale. Then, at the Enter a base point option for scaling [Existing/Left/Center/Middle/Right/TL/TC/TR/ML/MC/MR/BL/BC/BR] <Existing>: prompt, specify the justification for the base point.

Figure 10-4 and **Figure 10-5** show all of the justification options except **Left** and **Existing.** The **Left** option scales text objects using their lower-left corner point as the base point. The **Existing** option scales text objects using their existing justification setting as the base point. **Figure 10-15** shows scaling text with different justification

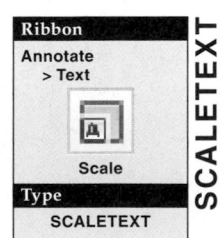

Ribbon
Annotate
> Text

Scale

Type
SCALETEXT

SCALETEXT

Figure 10-15.
The **Existing** option of the **SCALETEXT** tool scales text objects using their individual justification settings.

+BL Justification
MC Justification
TR Justification†
Original Text

+BL Justification
MC Justification
TR Justification†
**Text Scaled Using
Existing Base Point Option**

points using the **Existing** option. Notice how the text scales in relation to its own justification setting.

After you specify the justification to use as the base point, AutoCAD prompts for the scaling type. The default **Specify new model height** option allows you to type a new value for the text height of non-annotative objects. If the selected text is annotative, the value you enter is ignored. Use the **Paper height** option to type a new paper text height value for the text height of annotative objects. If the selected text is non-annotative, the value you enter is ignored.

The **Match object** option allows you to pick an existing text object. The height of the selected text object adopts the text height from the text object you pick. Use the **Scale factor** option to scale text objects that have different heights relative to their current heights. For example, using a scale factor of 2 scales all of the selected text objects to twice their current size.

> **NOTE**
>
> You should only use the **SCALETEXT** tool to scale non-annotative text.

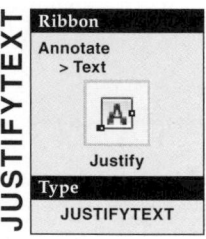

JUSTIFYTEXT

Ribbon
Annotate > Text
🅰
Justify
Type
JUSTIFYTEXT

Changing Text Justification

Use the **JUSTIFYTEXT** tool to change the justification point without moving the text. Pick the text for which you want to change justification, and enter the new justification option.

Exercise 10-6

Access the Student Web site (www.g-wlearning.com/CAD) and complete Exercise 10-6.

Express Tools
Chapter 10

The **Express Tools** ribbon tab includes additional tools for improved functionality and productivity during the drawing processes. The following Express Tools represent the most useful text express tools. For information about these tools, go to the Student Web site (www.g-wlearning.com/CAD), select this chapter, and select **Using Text Express Tools**.

Text Fit　　　　　　　　　**Arc-Aligned Text**
Text Mask　　　　　　　　**Enclose Text with Object**
Unmask Text　　　　　　　**Change Text Case**
Convert Text to Mtext

Chapter Test

Answer the following questions. Write your answers on a separate sheet of paper or go to the Student Web site (www.g-wlearning.com/CAD) and complete the electronic chapter test.

1. List two ways to access the **TEXT** tool.
2. Write the control code sequence required to draw the following symbols when using the **TEXT** tool:
   A. 30°
   B. 1.375 ± .005
   C. Ø24
   D. <u>NOT FOR CONSTRUCTION</u>
3. Briefly explain the function and purpose of fields.
4. What is different about the on-screen display of fields compared to that of text?
5. How can you access the **Field Update Settings** dialog box?
6. Explain how to convert a field to text.
7. What is the quickest way to check your spelling within a current text editor?
8. How do you change the **Current word** if you do not think the word that is displayed in the **Suggestions:** text box of the **Check Spelling** dialog box is the correct word, but one of the words in the list of suggestions is the correct word?
9. Identify three ways to access the AutoCAD spell checker.
10. How do you change the main dictionary for use in the **Check Spelling** dialog box?
11. What appears if you double-click on multiline text?
12. What is the purpose of the **AutoCAPS** option available from the mtext editor shortcut menu?
13. Briefly describe how to find and replace text when an mtext or single-line text editor is open.
14. Name the tool that allows you to find a piece of text and replace it with an alternative piece of text in a single instance or for every instance in your drawing.
15. When using the **SCALETEXT** tool, which base point option would you select to keep the text object's current justification point?

Drawing Problems

Start AutoCAD if it is not already started. Start a new drawing using an appropriate template of your choice. The template should include layers and text styles, when necessary, for drawing the given objects. Add layers and text styles as needed. Draw all objects using appropriate layers, and text styles, justification, and format. Follow the specific instructions for each problem. Use your own judgment and approximate dimensions when necessary.

▼ Basic

1. Use the **TEXT** tool to type the following information. Change the text style to represent each of the four fonts named. Use a .25 unit text height and 0° rotation angle. Save the drawing as P10-1.

   TXT–AUTOCAD'S DEFAULT TEXT FONT, WHICH IS AVAILABLE FOR USE WHEN YOU BEGIN A DRAWING.
   ROMANS–SMOOTHER THAN TXT FONT AND CLOSELY DUPLICATES THE SINGLE-STROKE LETTERING THAT HAS BEEN THE STANDARD FOR DRAFTING.
   ROMANC–A MULTISTROKE DECORATIVE FONT THAT IS GOOD FOR USE IN DRAWING TITLES.
   ITALICC–AN ORNAMENTAL FONT SLANTED TO THE RIGHT AND HAVING THE SAME LETTER DESIGN AS THE COMPLEX FONT.

Drawing Problems - Chapter 10

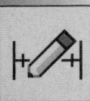

2. Use the list below to create text styles. Then use the **TEXT** tool to type the text, changing the text style to represent each style. Use a .25 unit text height. Save the drawing as P10-2.

> TXT–EXPAND THE WIDTH BY THREE.
> MONOTXT–SLANT TO THE LEFT –30°.
> ROMANS–SLANT TO THE RIGHT 30°.
> ROMAND–BACKWARDS.
> ROMANC–VERTICAL.
> ITALICC–UNDERSCORED AND OVERSCORED.
> ROMANS–USE 16d NAILS @ 10″ OC.
> ROMANT–⌀32 (812.8).

3. Open P5-10 and add text to the circuit diagram. Use a text style with the Romans font. Save the drawing as P10-3.

▼ Intermediate

4. Create text styles with a .375 height and the following fonts: Arial, BankGothic Lt BT, CityBlueprint, Stylus BT, Swis721 BdOul BT, Vineta BT, and Wingdings. Use the **TEXT** tool to type the complete alphabet and numbers 1–10 for the text styles. Also, type all symbols available on the keyboard and the diameter, degree, and plus/minus symbols. Save the drawing as P10-4.

5. Create the window schedule shown below. Create the text using a text style with the Stylus BT font. Draw the hexagonal symbols in the SYM column. Save the drawing as P10-5.

WINDOW SCHEDULE				
SYM.	SIZE	MODEL	ROUGH OPEN	QTY.
Ⓐ	12 x 60	JOB BUILT	VERIFY	2
Ⓑ	96 x 60	W4N5 CSM.	8'-0 3/4" x 5'-0 7/8"	1
Ⓒ	48 x 60	W2N5 CSM.	4'-0 3/4" x 5'-0 7/8"	2
Ⓓ	48 x 36	W2N3 CSM.	4'-0 3/4" x 3'-6 1/2"	2
Ⓔ	42 x 42	2N3 CSM.	3'- 6 1/2" x 3'-6 1/2"	2
Ⓕ	72 x 48	G64 SLDG.	6'-0 1/2" x 4'-0 1/2"	1
Ⓖ	60 x 42	G536 SLDG.	5'-0 1/2" x 3'-6 1/2"	4
Ⓗ	48 x 42	G436 SLDG.	4'-0 1/2" x 3'-6 1/2"	1
Ⓙ	48 x 24	A41 AWN.	4'-0 1/2" x 2'-0 7/8"	3

6. Create the door schedule shown below. Create the text using a text style with the Stylus BT font. Draw the circle symbols in the SYM column. Save the drawing as P10-6.

DOOR SCHEDULE			
SYM.	SIZE	TYPE	QTY.
①	36 x 80	S.C. R.P. METAL INSULATED	1
②	36 x 80	S.C. FLUSH METAL INSULATED	2
③	32 x 80	S.C. SELF CLOSING	2
④	32 x 80	HOLLOW CORE	5
⑤	30 x 80	HOLLOW CORE	5
⑥	30 x 80	POCKET SLDG.	2

7. Create the interior finish schedule shown below. Create the text using a text style with the Stylus BT font. Save the drawing as P10-7.

INTERIOR FINISH SCHEDULE											
ROOM	FLOOR					WALLS				CEILING		
	VINYL	CARPET	TILE	HARDWOOD	CONCRETE	PAINT	PAPER	TEXTURE	SPRAY	SMOOTH	BROCADE	PAINT
ENTRY					•							
FOYER		•				•			•			•
KITCHEN		•						•		•		
DINING			•			•			•		•	•
FAMILY		•				•					•	•
LIVING		•				•					•	•
MSTR. BATH			•			•				•		
BATH #2			•			•				•	•	
MSTR. BED		•				•					•	•
BED #2		•				•				•	•	
BED #3		•				•				•	•	
UTILITY	•					•				•	•	•

8. Create the block diagram shown below. Create the text using a text style with the Romans font. Save the drawing as P10-8.

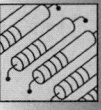
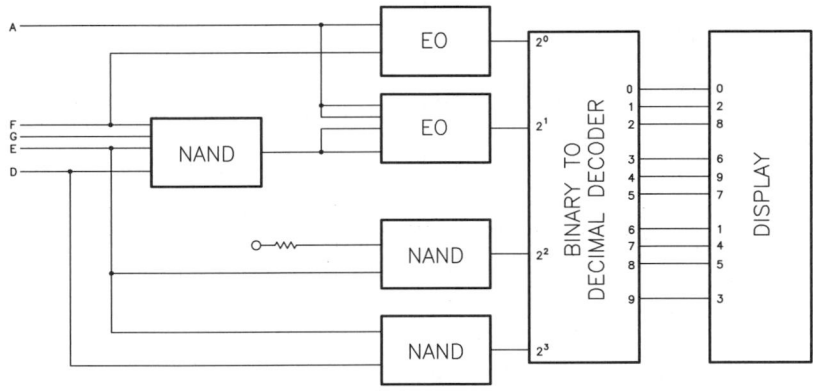

9. Create the block diagram shown below. Create the text using a text style with the Romans font. Use polylines to create the arrowheads. Save the drawing as P10-9.

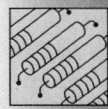
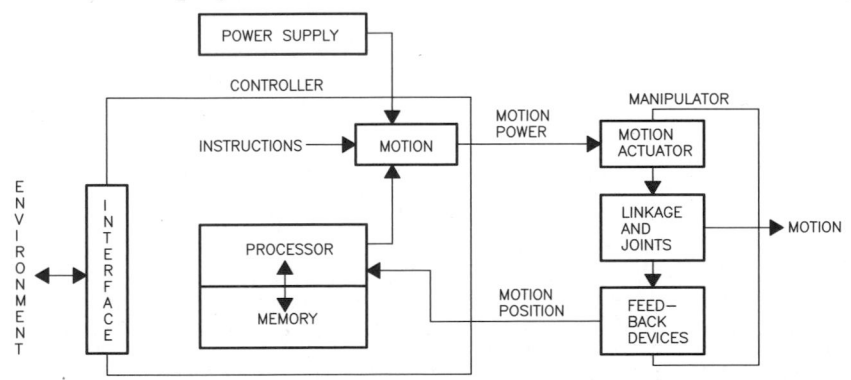

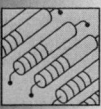

10. Draw the AND/OR schematic shown below. Save your drawing as P10-10.

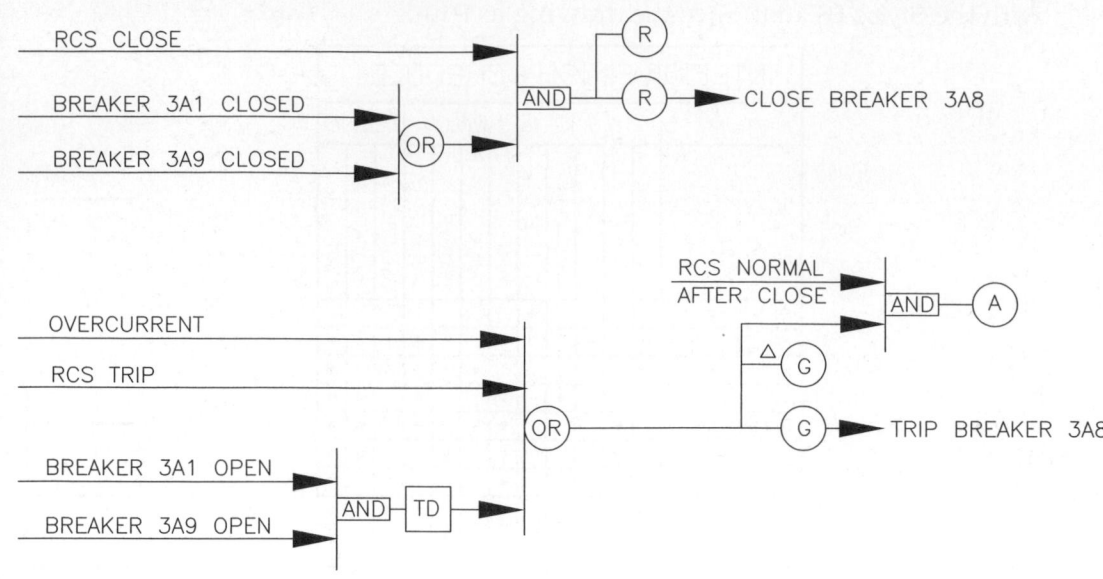

11. Draw the finish schedule shown below. Save your drawing as P10-11.

INTERIOR FINISH SCHEDULE										
ROOM	FLOOR				WALLS				CEIL	
	CARPET	VINYL	TILE	HARDWOOD	PAINT	PAPER	TEXTURE	SPRAY	SMOOTH	BROCADE	PAINT
FOYER			●		●		●			●	●
KITCHEN			●			●		●	●		●
DINING				●	●		●			●	●
FAMILY	●				●		●			●	●
LIVING	●				●		●			●	●
MASTER BED	●				●		●			●	●
MASTER BATH			●			●		●	●		●
BATH 2		●				●		●	●		●
BED 2	●				●		●			●	●
BED 3	●				●		●			●	●
UTILITY		●				●		●	●		●

▼ Advanced

12. Create the title block shown below.

R -	CHANGE		DATE	ECN

SPECIFICATIONS

HYSTER COMPANY

THIS PRINT CONTAINS CONFIDENTIAL INFORMATION WHICH IS THE PROPERTY OF HYSTER COMPANY. BY ACCEPTING THIS INFORMATION THE BORROWER AGREES THAT IT WILL NOT BE USED FOR ANY PURPOSE OTHER THAN THAT FOR WHICH IT IS LOANED.

UNLESS OTHERWISE SPECIFIED DIMENSIONS ARE IN
~~INCHES~~ MILLIMETERS AND TOLERANCES FOR:
_____ PLACE DIMS± _____ : _____ PLACE DIMS± _____
ANGLES ± _____ : WHOLE DIMS± _____

DR.		SCALE		DATE	
CK. MAT'L.		CK. DESIGN		REL. ON ECN	
NAME					

MODEL	DWG. FIRST USED	SIMILAR TO

DEPT.	PROJECT	LIST DIVISION

H | PART NO. | | R

13. Start a new drawing using the C-size mechanical drawing template located on the Student Web site. Draw a small parts list connected to the title block, similar to the one shown below. Save the drawing as P10-13.
    A. Enter PARTS LIST with a style containing a complex font.
    B. Enter the other information using text and the **TEXT** tool.

3	HOLDING PINS	12
2	SIDE COVERS	3
1	MAIN HOUSING	1
KEY	DESCRIPTION	QTY

PARTS LIST

UNLESS OTHERWISE SPECIFIED
ALL DIMENSIONS IN

INCHES

AND TOLERANCES FOR:
1 PLACE DIMS: ±.1
2 PLACE DIMS: ±.01
3 PLACE DIMS: ±.005
ANGULAR: ±30'
FRACTIONAL: ±.1/32
FINISH: 125? in.

JANE'S
DESIGN

DR:	SCALE:	DATE:	APPD:
JANE	FULL	XX—XX—XX	

MATERIAL:
MILD STEEL

NAME:
XXX—XXXX

FIRST USED ON:	SIMILAR TO:

B | PART NO: 123—321 | REV: 0

14. Create the architectural title block shown below. Use the same guidelines given for Problem 13. Save the drawing as P10-14.

15. Draw title blocks with borders for your electrical, piping, and general drawings. Use the same guidelines provided in Problem 13. The title block can be similar to the one displayed with Problem 12, but the area for mechanical drafting tolerances is not required. Research sample title blocks to come up with your design. Save the drawings as P10-15A, P10-15B, and P10-15C.

16. Draw the engineering change notice form shown below. Save your drawing as P10-16.

Engineering Change Notice

ECN NO.

Disposition of production stock:
A =Alter or rework U=Use in production
T=Transfer to service stock S=Scrap

Qty.	Drawing Size Part No.	R/N	Description	Change	Other Usage in Production	D/S
01						
02						
03						
04						
05						
06						
07						
08						
09						
10						
11						
12						
13						
14						
15						
16						
17						
18						

Reason:

Castings & forgings affected? ☐ Yes ☐ No	Design engineer:	Supervisor approval:	Release date:	Page

PARTS LIST

ITEM NO.	QTY.	PART NO.	DESCRIPTION
1	1	1DT1001	PBASE
2	1	1DT1002	PEDESTAL HOUSING
3	1	1DT1003	PEDESTAL
4	1	1DT1004	MAIN HOUSING

CHAPTER 11

Tables

Learning Objectives

After completing this chapter, you will be able to do the following:
- ✓ Create and modify table styles.
- ✓ Insert tables into a drawing.
- ✓ Edit tables.
- ✓ Insert formulas into table cells to perform calculations on numeric data.

This chapter describes how to create *tables*, which are common to a variety of drafting applications including bills of materials, door and window schedules, legends, and title block information. Review the tables and terminology shown in **Figure 11-1**. This will help you better understand tables and table information as you read this chapter.

table: An arrangement of rows and columns that organize data to make it easier to read.

Table Styles

Table styles control a variety of table characteristics. You may have several table styles in a single drawing, depending on drawing requirements. Though you can adjust table format independently of a table style, you should create a table style for each unique application. For example, you can have one table style for creating door and window schedules, and another table style with different characteristics for adding

table style: A saved collection of table settings, including direction, text appearance, and margin spacing.

Figure 11-1.
You can create tables in AutoCAD with the title and header rows at the top or the bottom.

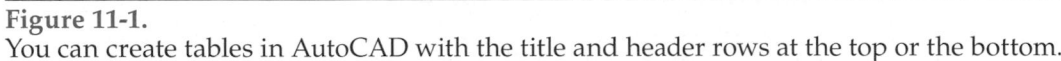
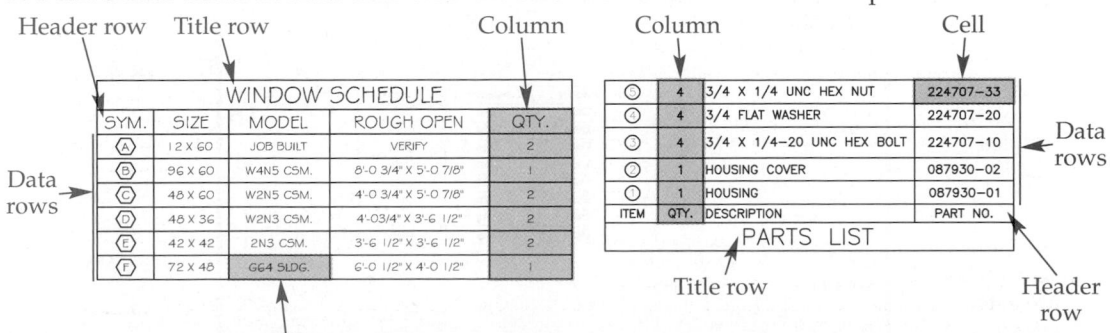

an interior finish schedule. In mechanical drafting, you might prepare a table style for parts lists and another table style for gear data tables. Add table styles to drawing templates for repeated use.

Working with Table Styles

TABLESTYLE

Ribbon
Home
> Annotation

Table Style
Annotate
> Tables

Table Style
Type
TABLESTYLE
TS

Create, modify, and delete table styles using the **Table Style** dialog box. See **Figure 11-2.** You can also open the **Table Style** dialog box from the **Insert Table** dialog box, described later in this chapter, by picking the **Launch the Table Style dialog** button.

The **Styles** list box displays existing table styles. The Standard table style is available and current by default. To make a table style current, double-click the style name; right-click the name and select **Set current**; or pick the name and select the **Set current** button. Below the **Styles** list box is a drop-down list that you can use to filter the number of table styles displayed in the **Table Style** dialog box. Pick the **All Styles** option to show all table styles in the file or pick the **Styles in use** option to show only the current style and styles used in the drawing.

Creating New Table Styles

To create a new table style, first select an existing table style from the **Styles** list box to use as a base for formatting the new table style. Then pick the **New...** button in the **Table Style** dialog box to open the **Create New Table Style** dialog box. See **Figure 11-3**. In the **New Style Name** text box, type a name for the new table style. You can base the new table on the formatting settings from a different table style by selecting the name of the table style from the **Start With** drop-down list.

The default new table style name is Copy of followed by the name of the selected existing style. You can keep the default name, but you should usually enter a more descriptive name, such as Parts List, Parts List No Heading, or Door Schedule. Table style names can have up to 255 characters, including uppercase or lowercase letters, numbers, dashes (–), underlines (_), and dollar signs ($). After entering the table style name, pick the **Continue** button to open the **New Table Style** dialog box and adjust table style settings. See **Figure 11-4.**

PROFESSIONAL TIP

It is a good idea to record the names and details about the table styles you create and keep this information in a log for future reference.

Figure 11-2.
The **Table Style** dialog box.

Preview of the selected style

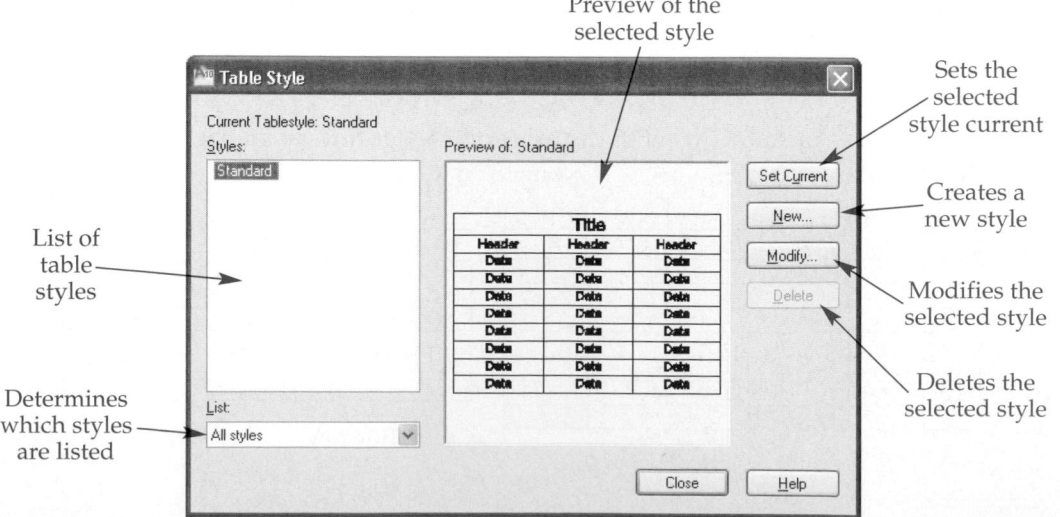

List of table styles

Determines which styles are listed

Sets the selected style current

Creates a new style

Modifies the selected style

Deletes the selected style

Figure 11-3.
In the **Create New Table Style** dialog box, specify the name of the new table style and the existing style that will be copied as a basis for the new style.

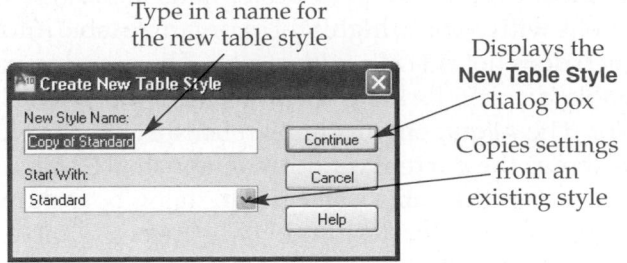

Type in a name for
the new table style

Displays the
New Table Style
dialog box

Copies settings
from an
existing style

Figure 11-4.
The formatting properties for a new style are specified in the **New Table Style** dialog box. The **Data** cell style is shown in the **Cell styles** area in this figure.

Pick to create a
starting table style

Pick to remove
the starting table
reference

Select a
cell style

Pick to create a
new cell style

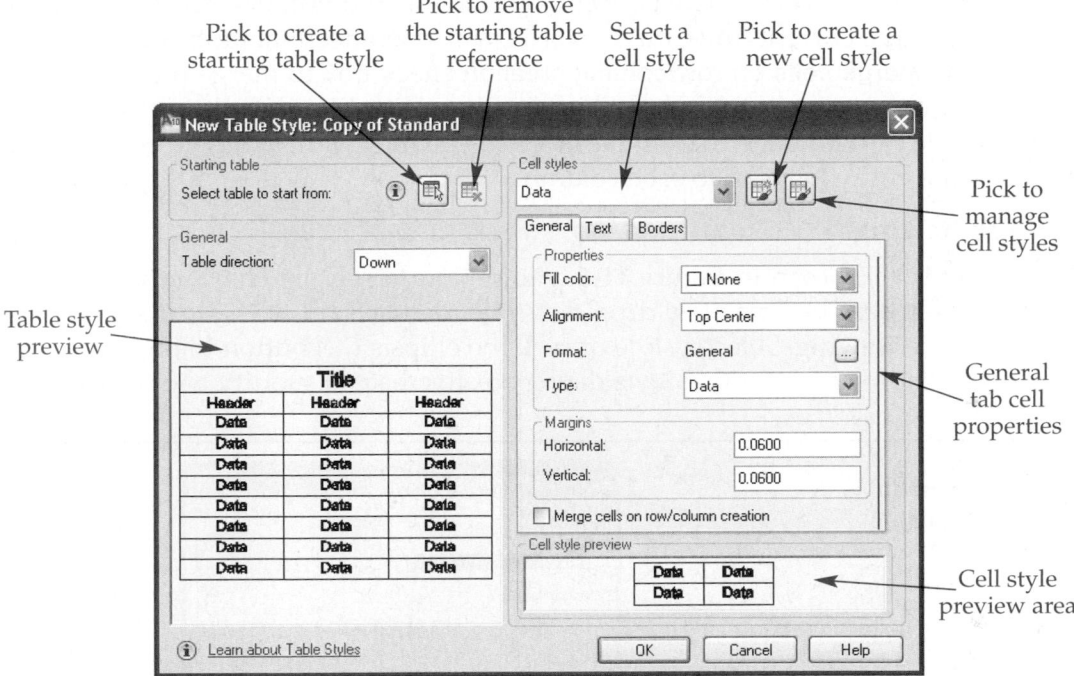

Pick to
manage
cell styles

Table style
preview

General
tab cell
properties

Cell style
preview area

Adjusting Table Direction

The **Table direction** setting in the **General** area of the **New Table Style** dialog box determines the placement of title and header rows and the order of data rows. Select **Down** from the drop-down list to place data rows below the title and header rows. Select **Up** to place data rows above the title and header rows. Refer again to **Figure 11-1.**

Cell Styles

Cell styles allow data cell rows, the column header row, and the title row to have their own formatting properties. Three default cell styles are available in the **Cell Styles** area of the **New Table Style** dialog box: **Data**, **Header**, and **Title**. Pick a cell style from the drop-down list to display the properties for the corresponding element. **Figure 11-4** shows the **Data** cell style selected. Cell formatting properties are set using the **General**, **Text**, and **Borders** tabs. The options in these tabs are the same for adjusting data, header, and title cell styles.

General Tab Settings

The **General** tab, shown in **Figure 11-4**, allows you to set general table characteristics. The **Fill color** drop-down provides options for adjusting the color used to fill cells. You can fill cells with color to highlight or organize table information. The default setting is None, which does not fill cells with a color. The drawing window color determines the on-screen table display. Pick a color from the drop-down list to fill the cells with the selected color. The **Alignment** drop-down list specifies text justification within the cell.

Format shows the current cell format, which is General by default. Pick the ellipsis (**...**) button to access the **Table Cell Format** dialog box, shown in **Figure 11-5**. The **Data Type** area lists options for formatting the selected table cell. Pick the appropriate format, such as **Text** or **Currency**, to view options for adjusting the format characteristics. Different options are available depending on the selected format.

Use the **Type** drop-down list to specify the cell data type. Pick **Data** to define a data cell type. Choose **Label** if the cell is a label type, such as a column heading or the table title. The **Margins** area provides a **Horizontal** and **Vertical** text box for controlling the horizontal and vertical space between cell content and borders. The default varies depending on the current units, but is .06 (1.5 mm) when decimal units are used.

Pick the **Merge cells on row/column creation** check box to merge the row of cells together to form a single cell. This check box is selected by default for the **Title** cell style. The title cell provides an example of when you may want to merge cells. The title applies to the entire table, or to each column.

Text Tab Settings

The **Text** tab, shown in **Figure 11-6**, allows you to set text characteristics for the selected cell style. The **Text style** drop-down list displays all of the text styles found in the current drawing. Select a style or pick the ellipsis (**...**) button to the right of the drop-down list to open the **Text Style** dialog box to create or modify a text style.

Figure 11-5.
Many different data types are available to format a table cell.

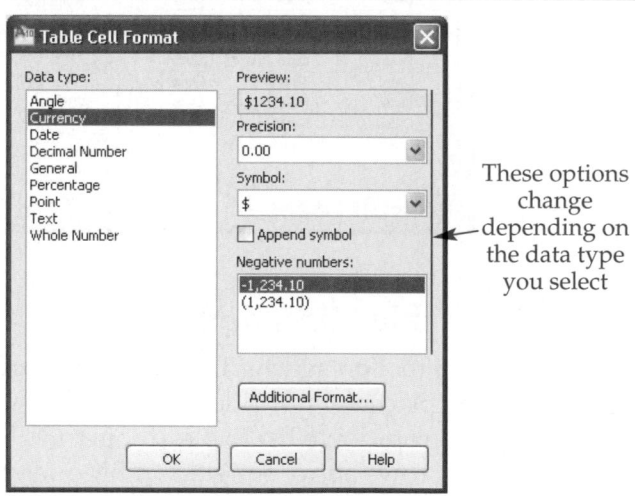

These options change depending on the data type you select

Figure 11-6.
The **Text** tab in the **New Table Style** dialog box allows you to set text properties.

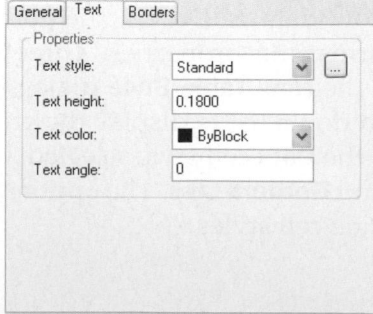

AutoCAD and Its Applications—Basics

Use the **Text height** text box to specify the height of the text. The default setting for data row and column header cells varies depending on the current units, but is .18 (4.5 mm) when decimal units are used. If you assign a text height other than 0 to the text style, the **Text height** text box is grayed out. Use the **Text color** drop-down list to set the color of the text. The **Text angle** text box controls the rotation angle of text within the table cell. **Figure 11-7** shows an example of a 90° text angle applied to the **Header** cell style.

Borders Tab Settings

The **Borders** tab, shown in **Figure 11-8,** allows you to control the border display and characteristics for the selected cell style. Use the **Lineweight** drop-down list to assign a unique lineweight to cell borders. Use the **Linetype** drop-down list to assign a unique linetype to cell borders. As when creating layers, you must load linetypes if they are not currently loaded in the file in order to apply them to the cell border. Use the **Color** drop-down list to set the color of the cell borders.

Pick the **Double line** check box to add another line around the default single line border style. When checked, the **Spacing** edit box becomes available, allowing you to enter the distance between the double lines. The default spacing varies depending on the current units, but is .045 (1.125 mm) when decimal units are used.

The **Border** buttons control how the **Lineweight**, **Linetype**, **Color**, and **Double line** border properties apply to the cell borders. From right to left, the options are **All Borders, Outside Borders, Inside Borders, Bottom Border, Left Border, Top Border, Right Border**, and **No Borders**. Once you set the desired border properties, select or deselect these buttons according to how you want cell borders to display. **Figure 11-9** shows an example of each border style applied to the data cell borders.

NOTE

Lineweights appear on-screen only if you display lineweights. Pick the **Show/Hide Lineweight** button on the status bar to display lineweights.

Figure 11-7.
In this table, a 90° text angle has been applied to the header cell style.

NUMBER	ROOM SCHEDULE					
	NAME	LENGTH	WIDTH	HEIGHT	AREA	
1	BEDROOM 1	11'-0"	10'-0"	9'-0"	110 SQ. FT.	
2	BEDROOM 2	10'-0"	11'-0"	9'-0"	110 SQ. FT.	
3	MASTER BEDROOM	12'-0"	14'-0"	9'-0"	168 SQ. FT.	
4	LIVING ROOM	12'-0"	16'-0"	9'-0"	192 SQ. FT.	
5	DINING ROOM	11'-0"	12'-0"	9'-0"	132 SQ. FT.	
6	KITCHEN	11'-0"	10'-0"	9'-0"	110 SQ. FT.	

Figure 11-8.
The **Borders** tab in the **New Table Style** dialog box allows you to set cell border properties.

General | Text | Borders

Properties
Lineweight: ——— ByBlock
Linetype: ——— ByBlock
Color: ■ ByBlock
☐ Double line
Spacing: 0.0450

Apply the selected properties to borders by clicking the buttons above.

Figure 11-9.
Several border options are available for table cells. The settings shown are for data row borders.

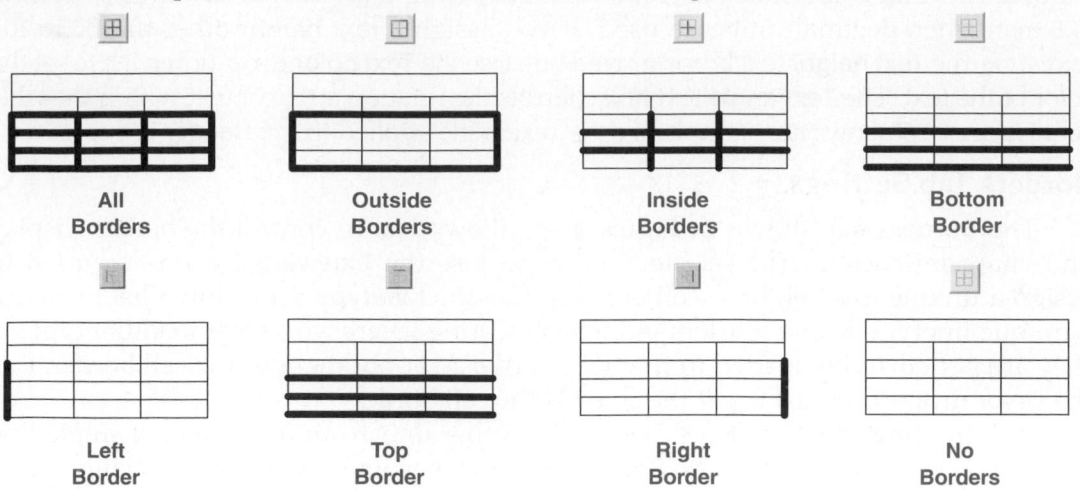

| All Borders | Outside Borders | Inside Borders | Bottom Border |

| Left Border | Top Border | Right Border | No Borders |

Creating Cell Styles

The default **Data**, **Header**, and **Title** cell styles are adequate for typical table applications. However, you can increase the flexibility and options for creating tables by developing additional cell styles. For example, you can create a cell style called Data Yellow that is the same as the **Data** cell style but fills cells with a yellow color. Then when you draw a table, you can choose from either the **Data** or the **Data Yellow** cell style, depending on the application.

To create a new cell style, first select an existing cell style from the **Cell Styles** area drop-down list to use as a base for formatting the new cell style. Then pick the **Create new cell style...** button from the **Cell Styles** area, or select **Create new cell style...** from the **Cell Styles** area drop-down list to display the **Create New Cell Style** dialog box. In the **New Style Name** text box, type a name for the new cell style. You can base the new cell style on the formatting settings of a different cell style by selecting the name of the cell style from the **Start With** drop-down list.

Use the **Manage Cell Styles** dialog box, shown in **Figure 11-10**, to create, rename, and delete cell styles. To access this dialog box, pick the **Manage Cell Style dialog...** button from the **Cell Styles** area, or select **Manage cell styles...** from the **Cell Styles** area drop-down list.

Figure 11-10.
Use the **Manage Cell Styles** dialog box to create new cell styles and to rename and delete existing cell styles.

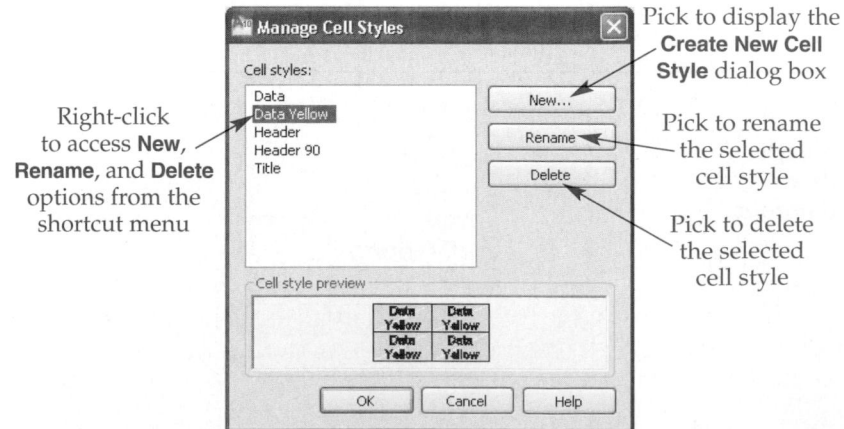

Right-click to access **New**, **Rename**, and **Delete** options from the shortcut menu

Pick to display the **Create New Cell Style** dialog box

Pick to rename the selected cell style

Pick to delete the selected cell style

The **Preview** and **Cell style preview** areas of the **New Table Style** dialog box allow you to see how the selected table style characteristics appear in a table. This provides a convenient way to observe changes made to a table style, without creating a table.

Exercise 11-1

Access the Student Web site (www.g-wlearning.com/CAD) and complete Exercise 11-1.

Developing a Starting Table Style

One technique for creating a table is to use a starting table style to base a new table on an existing table. You can consider a starting table style a table template that includes preset table style characteristics and specific rows, columns, and data entries. Using a starting table style is much like copying a complete table and editing the table as needed. A starting table style can save time if you often prepare similar tables. For example, use a starting table style if you already created a complete door schedule, and want to add a very similar door schedule to a new drawing project that contains most of the same doors.

Before you can create a starting table style, an existing table must be available in the drawing. A starting table style references the characteristics of a selected table, including the number of columns and rows and the table direction. Other table style characteristics, such as text style, are set according to the selected base table style. As a result, it is usually most appropriate to create a new starting table style using a base table style that is the same as that used to draw the reference table. For example, if you create a door schedule using a table style named Door Schedule, you should base the new starting table style on the Door Schedule table style.

To create a starting table style, pick the **Select table to start from** button in the **Starting table** area. Then pick a border line of the table you want to reference to form the starting table style. The preview displays the selected table and the table style settings of the base table style. See **Figure 11-11**. Modify the table direction and cell style options using the **General** and **Cell Styles** areas. Pick the **Remove Table** button to remove the table reference from the table style. Pick the **Start from Table style** insertion option, described later in this chapter, to add a table to a drawing using a starting table style.

Changing, Renaming, and Deleting Table Styles

To change the characteristics of an existing table style, select the table style and pick the **Modify** button to access the **Modify Table Style** dialog box, which is the same as the **New Table Style** dialog box. If you change the characteristics of an existing table style, all tables added using that style redraw with the new values.

To rename a table style using the **Table Style** dialog box, slowly double-click the name or right-click on the name and select **Rename**. To delete a table style, right-click on the name and choose **Delete**, or pick the style and select the **Delete** button. You cannot delete a table style that is assigned to a table. To delete a style that is in use, assign a different style to the tables that reference the style you want to delete. You cannot delete or rename the Standard style.

Figure 11-11.
Creating a starting table style that references an existing table.

Pick to create a
starting table style

Pick to remove the
starting table reference

Select existing
table

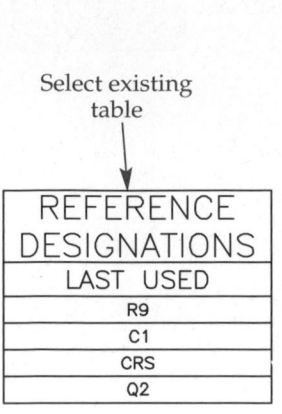

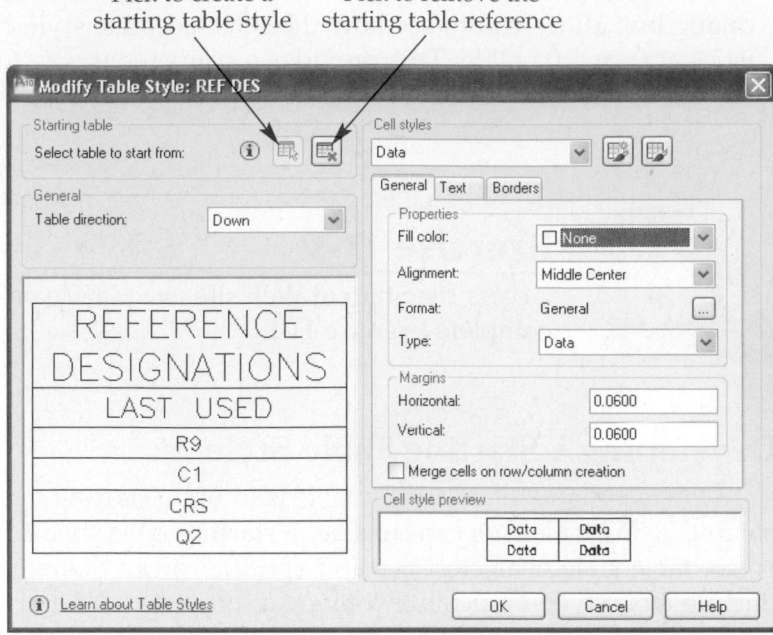

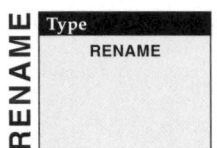

RENAME

Type
RENAME

Setting a Table Style Current

You can set a table style current using the **Table Style** dialog box by double-clicking the style in the **Styles** list box, right-clicking the style and selecting **Set current**, or picking the style and selecting the **Set current** button. To set a table style current without opening the **Table Style** dialog box, use the **Table Style** flyout located in the expanded **Annotate** panel of the **Home** ribbon tab, or the **Tables** panel of the **Annotate** ribbon tab. See **Figure 11-12.**

Figure 11-12.
The fastest way to set a style current is to use one of the drop-down lists on the ribbon.

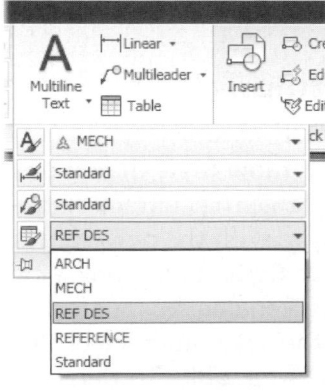
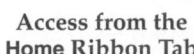
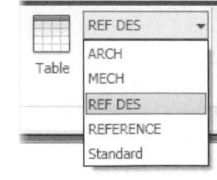

**Access from the
Home Ribbon Tab**

**Access from the
Annotate Ribbon Tab**

Inserting Tables

Ribbon
Home
> Annotation
Annotate
> Tables

Table

Type
TABLE
TB

TABLE

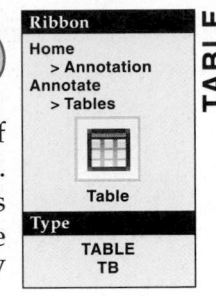

The **TABLE** tool allows you to insert an empty table by specifying the number of rows and columns. After you insert the table, you can type text into the table cells. You can also insert blocks and fields into table cells. The **TABLE** tool also provides other methods for inserting tables, such as beginning a table using a starting table style, forming a table from data in an existing Microsoft® Excel spreadsheet or CSV (comma-separated) file, and creating a table by referencing AutoCAD data. Accessing the **TABLE** tool opens the **Insert Table** dialog box, shown in **Figure 11-13**.

Placing an Empty Table

An empty table is the most basic type of table and requires you to specify all table characteristics and content. To place an empty table, first select a table style from the **Table Style** drop-down list, or pick the ellipsis (**...**) button to create or modify a style. The preview area shows a preview of a table with the current table style settings. The preview area does not adjust to the column and row settings, but it does show table style properties.

Next, pick the **Start from empty table** radio button in the **Insert options** area. The empty table insertion option is set in the **Insertion Behavior** area. Pick the **Specify insertion point** radio button to create a table using the values in the **Column & row settings** area, and then select a single point to place the table in the drawing.

Figure 11-13.
The **Insert Table** dialog box, shown with the **Start from empty table** insert option selected.

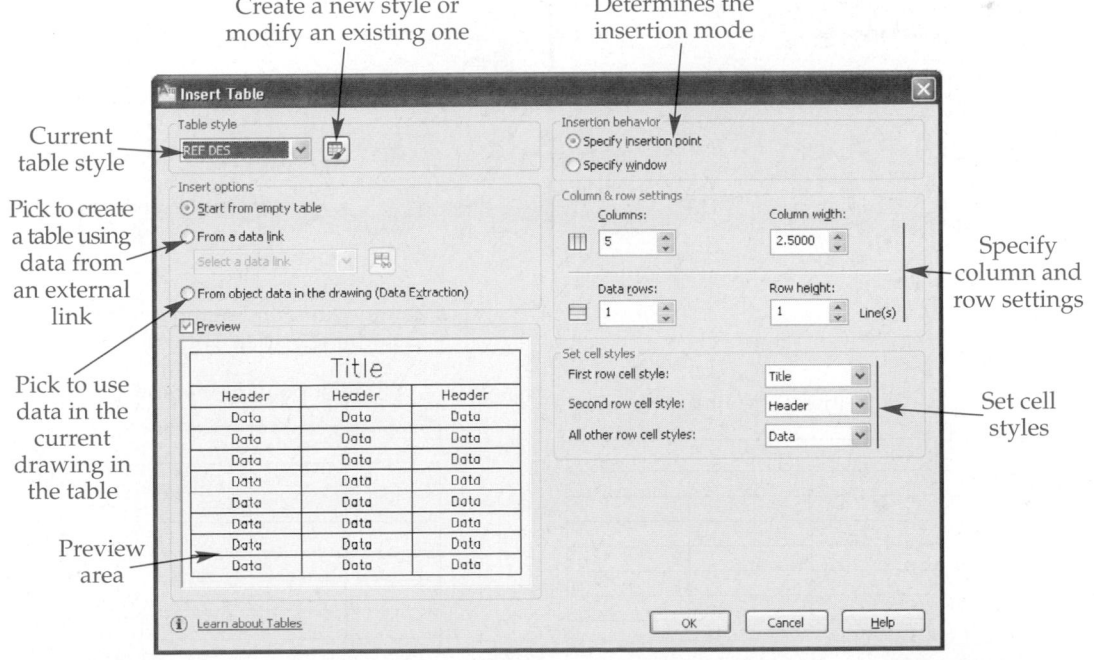

The number in the **Columns** text box determines the total number of table columns. The **Column width** value specifies the initial width of each column. You may want to enter a width larger than necessary in the **Column width** text box and then resize the columns later. The number in the **Data rows** text box determines the total number of data rows. The **Row height** value specifies the initial height of each row based on the number of lines typed and the table style margin settings. When you pick the **OK** button, AutoCAD prompts you to specify the table insertion point. **Figure 11-14A** shows a table created with three columns and five data rows using the **Insertion point** option.

NOTE

When you use the **Specify insertion point** option to place a table, the cursor attaches to the table based on the table style direction.

Pick the **Specify window** radio button to create a table that fits within a rectangular area. The radio buttons that appear in the **Column & row settings** area control which column and row settings are active. To set a fixed number of columns, choose the **Columns** radio button. The selected table width determines the width of the specified number of columns. The alternative is to set a fixed column width by selecting the **Column width** radio button. The selected table width determines the total number of columns.

To set a fixed number of rows, choose the **Data rows** radio button. The selected table height determines the height of the specified number of data rows. The alternative is to set a fixed row height by selecting the **Row height** radio button. The selected table height determines the total number of rows.

When you pick the **OK** button, AutoCAD prompts you to select the upper-left and lower-right table corners. The fixed **Column & row settings** values are used and the other settings adjust to fit the window. **Figure 11-14B** shows a table created with three columns and five data rows using the **Specify window** option.

Figure 11-14.
Two ways to insert a table. A—Specifying a single insertion point. B—Windowing an area with two pick points.

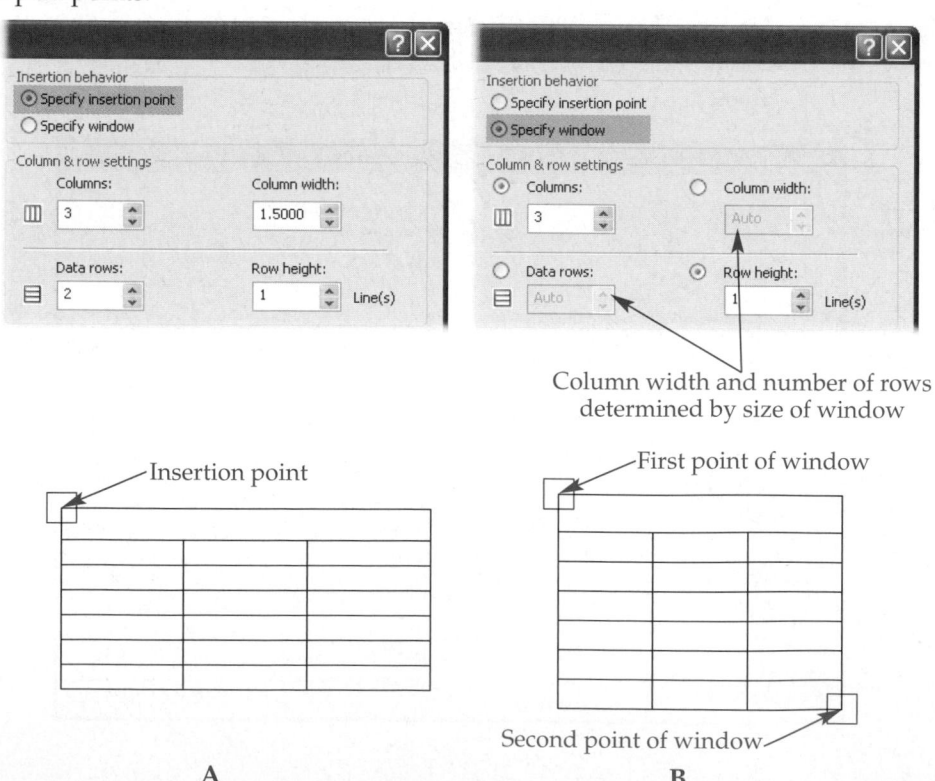

The number of data rows you set in the **Insert Table** dialog box does not include the title and header rows. If you specify 1 data row, for example, the table will have three rows because the top two rows are used for the table title and content headers, depending on cell style settings. The current table style text height and cell margin settings determine the default value for the row height. For example, enter a row height of 1 if you plan to have a single line of text in each cell.

NOTE

You can add, delete, and fully adjust rows and columns as needed. As a result, it is not critical that you enter the exact number and size of columns and rows before inserting the table.

Exercise 11-2

Access the Student Web site (www.g-wlearning.com/CAD) and complete Exercise 11-2.

Adding Cell Content

When you insert a table, the **Text Editor** ribbon tab appears by default, with the text editor cursor in the title cell ready for typing. See **Figure 11-15**. A dashed line around the border and a light gray background indicates the active cell. The *table indicator* identifies individual cells in the table. The identification system helps you to

table indicator: The grid of letters and numbers that appear around the table while the table is being edited to identify individual cells.

Figure 11-15.
The **Text Editor** ribbon tab is used to add and modify text table cell text. The active cell is indicated by a blinking cursor, a dashed border, and a light gray background.

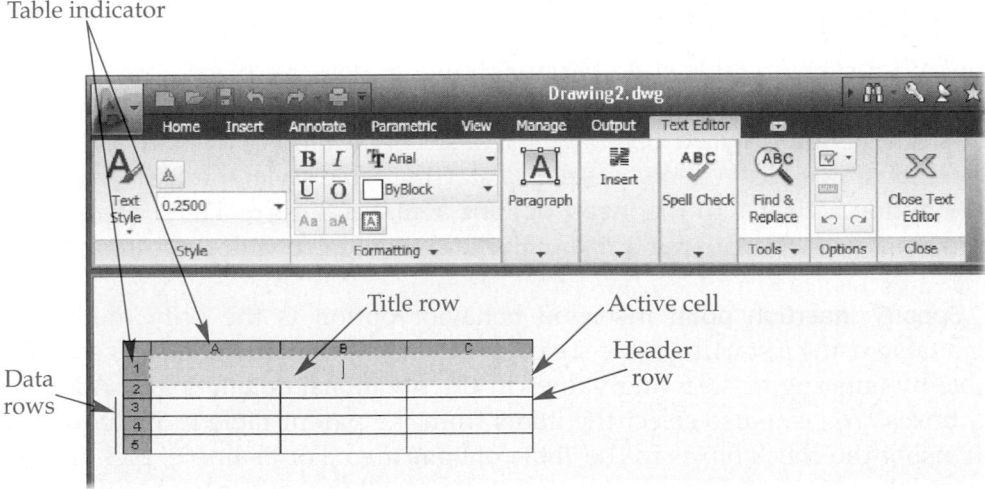

assign formulas to table cells for calculation purposes. Formulas are described later in this chapter. Before adding content to a cell, adjust the text settings in the **Text Editor** ribbon, if needed. Remember, however, that making changes to some text characteristics overrides the settings specified in the text style or table style, which is often not appropriate.

Hold [Alt] and press [Enter] to insert a return within the cell. When you finish entering text in the active cell, press [Tab] to move to the next cell. Hold [Shift] and press [Tab] to move the cursor backward (to the left or up) and make the previous cell active. Press [Enter] to make the cell directly below the current cell active, or exit the text editor if the cursor is at bottom cell. You can also use the arrow keys to navigate through table cells.

When you finish typing in the table, exit the text editor by picking the **Close Text Editor** button in the **Close** panel of the **Text Editor** ribbon tab, or by picking outside of the table. You can also press [Esc] or right-click and select **Cancel** to exit, but you are prompted to save changes. The easiest way to reopen the text editor to make changes to text in a cell is to double-click in the cell. **Figure 11-16** shows a completed table.

> **NOTE**
>
> The options and settings available in the **Text Editor** ribbon tab and shortcut menu function the same in table cells as they do when you edit mtext.

Exercise 11-3

Access the Student Web site (www.g-wlearning.com/CAD) and complete Exercise 11-3.

Using a Starting Table Style

If you add a starting table style to your drawing, you can place a new table by referencing the starting table style. To start from a table style, first select a starting table style from the **Table Style** drop-down list or select the ellipsis (...) button to create or modify a starting style. When you select a starting table style, the **Start from Table Style** radio button activates in the **Insert options** area. See **Figure 11-17.** The preview area shows a preview of the parent table with the current table style settings and table options.

The **Specify insertion point** insertion behavior option is the only method for inserting a table using a starting table style. However, you can add columns and rows to the table by entering or selecting values in the **Additional columns** and **Additional rows** text boxes. You can also select the items from the parent table to include in the new table using the check boxes in the **Table options** area. For example, pick the **Data cell text** check box to create a new table that contains all the text entries added to the data cells of the parent table.

Figure 11-16.
A completed parts list table.

1	PARTS LIST		
2	PART NUMBER	PART TYPE	QTY.
3	100−SCR−45	SCREW	18
4	202−BLT−32	BOLT	18
5	340−WSHR−06	WASHER	18

Figure 11-17.
You can use the **Insert Table** dialog box to create a new table using a starting table style.

Pick to create a new table using a starting table style

Start from Table Style becomes active

Preview of parent table

Determines table properties to copy from starting table style

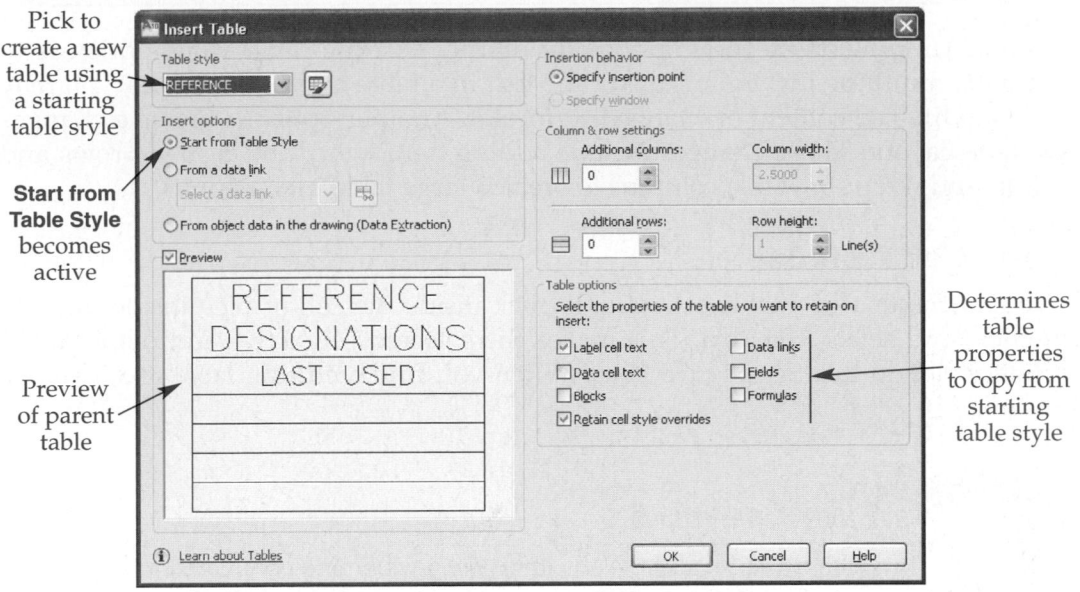

Pick the **OK** button and specify the insertion point of the table. The **Text Editor** ribbon tab appears with the text editor cursor in the title cell ready for adding new content or editing existing values. Exit the text editor when you finish typing text. **Figure 11-18** shows a table created by referencing an existing starting table style, with two additional rows.

Exercise 11-4

Access the Student Web site (www.g-wlearning.com/CAD) and complete Exercise 11-4.

Figure 11-18.
You can create a new table quickly by referencing a starting table style.

Existing table style used to form a starting table style

REFERENCE DESIGNATIONS
LAST USED
R9
C1
CRS
Q2

In the new table, all table options are retained and two rows are added

REFERENCE DESIGNATIONS
LAST USED
R9
C1
CRS
Q2
T4
R6

AutoCAD provides several options for editing existing tables. One option is to re-enter the multiline text editor to edit the text in a table cell. For example, you may need to modify cell content or change text format. Another option is to make changes to the table layout. These changes include adding, removing, and resizing rows and columns, and wrapping table columns to break a large table into sections.

Editing Cell Content

To edit the text in a table cell, double-click inside the cell or pick inside the cell, right-click, and select **Edit Text**. This makes the selected cell active and displays the **Text Editor** ribbon tab. See **Figure 11-19**. When you finish editing table text, exit the text editor.

Exercise 11-5

Access the Student Web site (www.g-wlearning.com/CAD) and complete Exercise 11-5.

Picking Inside a Cell to Edit Table Layout

You can access several table layout settings by picking (single-clicking) *inside* a cell to make the cell active and display the **Table Cell** ribbon tab. The highlighted cell includes *grips*. The **Table Cell** ribbon tab contains options for adjusting table and individual cell layout. The same tools and options found in the **Table Cell** ribbon tab, as well as additional options, are available from a shortcut menu. See **Figure 11-20**. To display the shortcut menu, select a cell and then right-click anywhere in the drawing window. The first section of the shortcut menu contains Windows Clipboard functions. These options affect the entire cell contents. Select **Recent Input** to display a list of previous entries.

> **grips:** Small boxes that appear at strategic points on an object, allowing you to edit the object directly.

Figure 11-19.
Double-click inside a cell to open the text editing function, allowing you to make changes to the text.

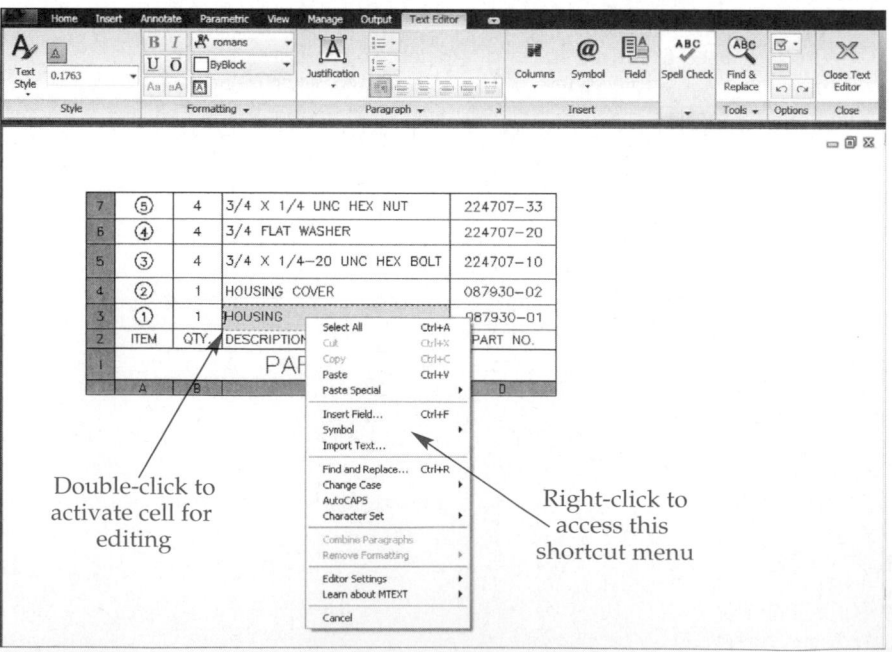

Double-click to activate cell for editing

Right-click to access this shortcut menu

Figure 11-20.
Several table layout modification options are available when you pick inside a cell.

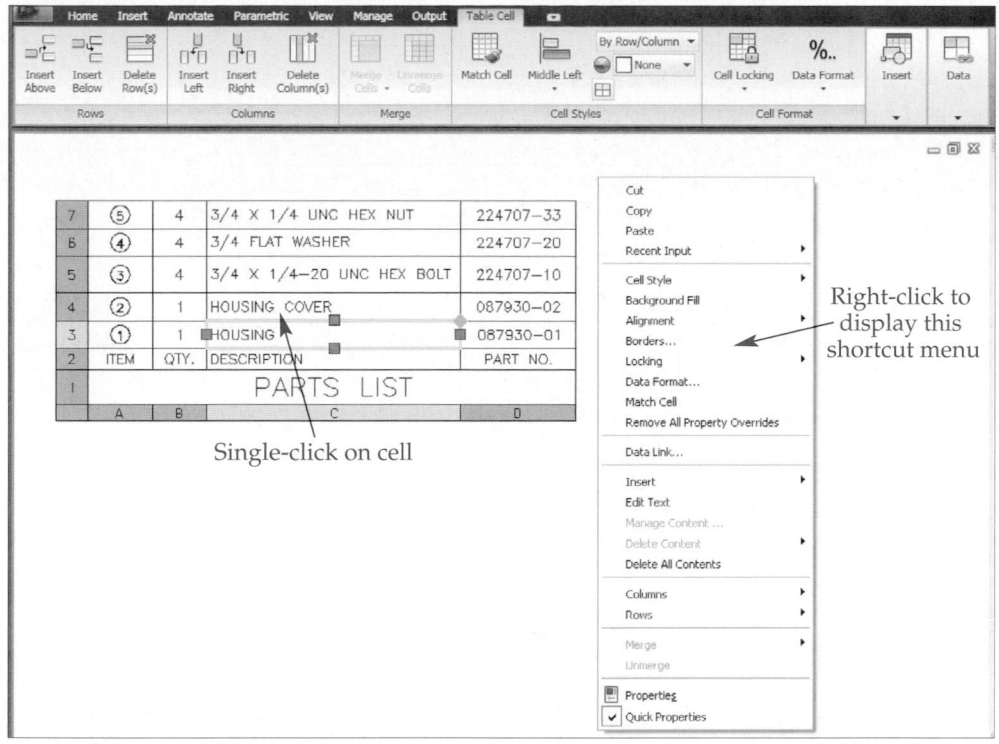

Single-click on cell

Right-click to display this shortcut menu

AutoCAD provides several ways to select multiple cells and apply changes to all the cells at once, as shown in **Figure 11-21.** One option is to pick in a cell and drag the window over the other cells. When you release the pick button, all of the cells touching the window become selected. You can also select multiple cells by picking a cell, holding down [Shift], and then picking another cell. This process selects the picked cells and all of the cells between. You can select entire columns or rows by picking the column or row number in the table indicator. To select the entire table, pick the corner of the table indicator.

NOTE

Make sure you pick completely inside of the cell. If you accidentally select one of the cell borders, the entire table becomes the selected object. Editing table layout by selecting a cell border is described later in this chapter.

Quickly Copying Cell Content

The most effective method for copying the content of one cell to multiple cells is to use the *auto-fill* function. To use auto-fill, pick inside the cell that contains the content you want to copy. The auto-fill grip is a diamond-shaped grip located in a corner of the cell. See **Figure 11-22.** Select the auto-fill grip, as shown in **Figure 11-23A,** and then right-click directly on the auto-fill grip and select an option to define how to adjust the cell content of auto-fill copies.

The **Fill Series** option fills cells with the content of the selected cell and applies format overrides. This option also automatically increases or decreases values of certain data types, such as dates, as the fill occurs. See **Figure 11-23B.** The **Fill Series Without Formatting** option fills cells with the content of the selected cell, but does not include format overrides.

auto-fill: A table function that automatically fills selected cells based on the contents of a specified cell.

Figure 11-21.
Selecting multiple cells in a table for editing. A—Using the pick-and-drag method.
B—Picking a range of cells using [Shift]. C—Selecting a row, column, or the entire table.

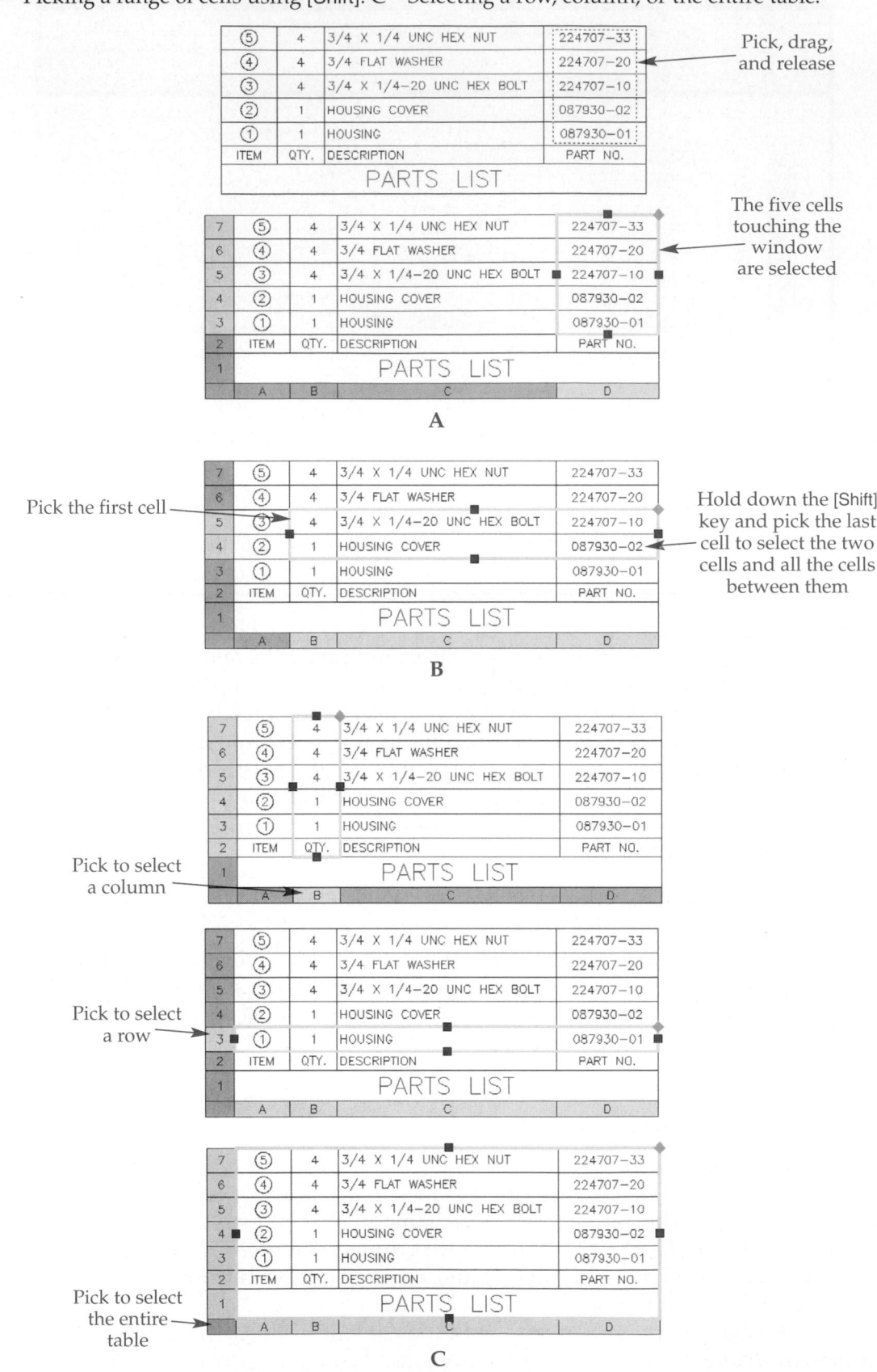

AutoCAD and Its Applications—Basics

Figure 11-22.
Using the auto-fill function to copy cell content to multiple cells.

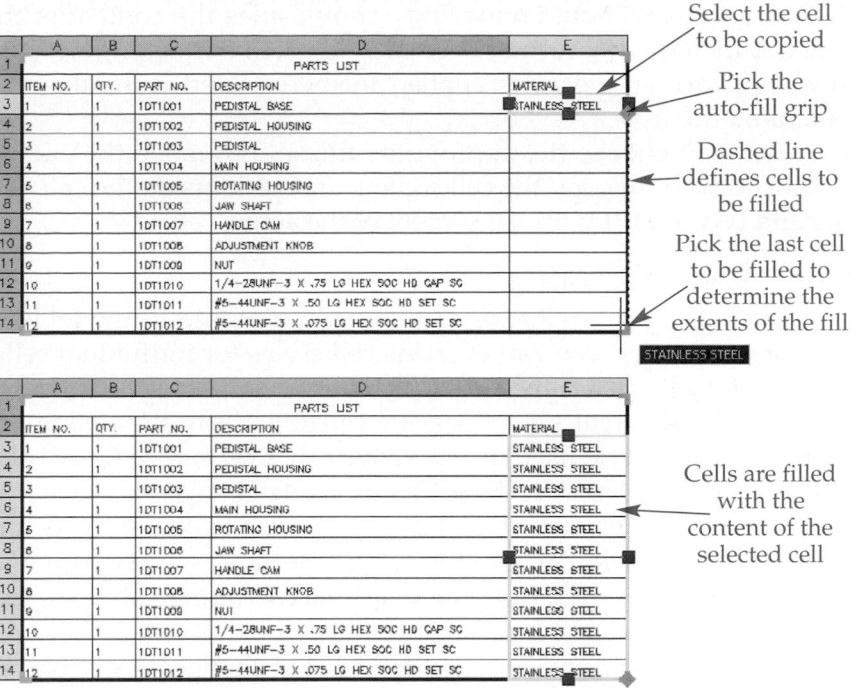

Select the cell to be copied

Pick the auto-fill grip

Dashed line defines cells to be filled

Pick the last cell to be filled to determine the extents of the fill

Cells are filled with the content of the selected cell

Figure 11-23.
Using the auto-fill options to control fill characteristics.

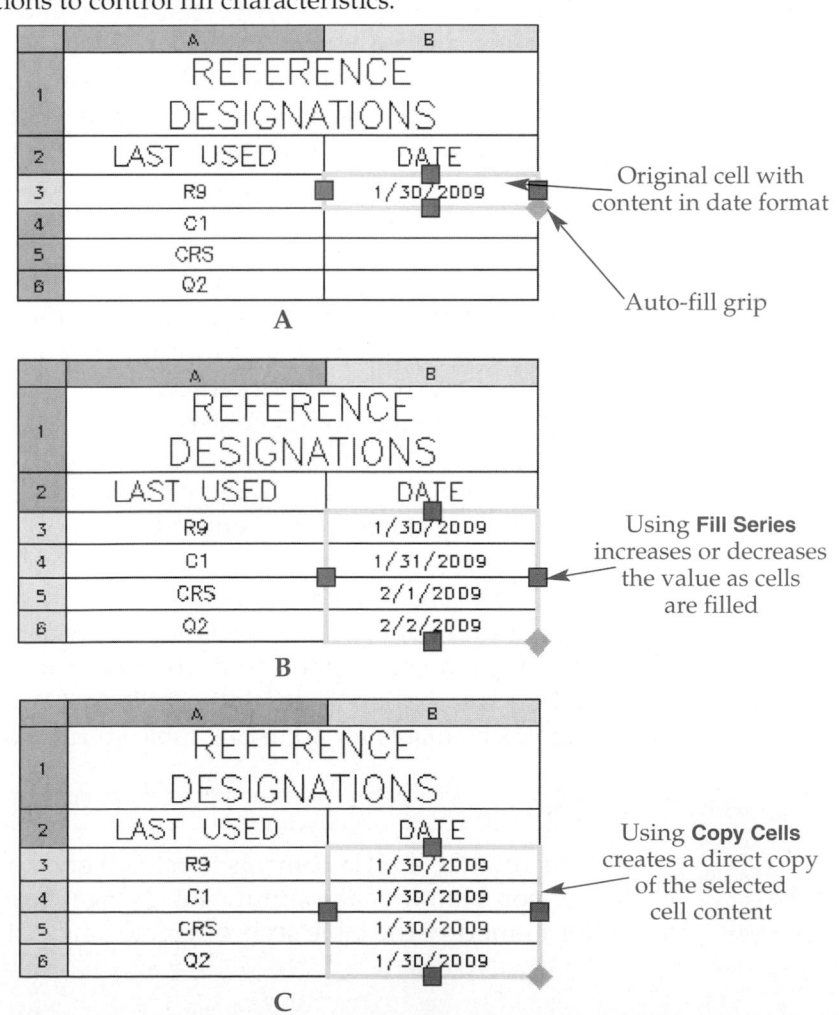

Original cell with content in date format

Auto-fill grip

A

Using **Fill Series** increases or decreases the value as cells are filled

B

Using **Copy Cells** creates a direct copy of the selected cell content

C

The **Copy Cells** option copies the content of the selected cell and applies format overrides, but creates a static cell copy that does not adjust data values. See **Figure 11-23C**. The **Copy Cells Without Formatting** option copies the content of the selected cell, but does not include any format overrides. The **Fill Formatting Only** option fills the cells only with format overrides applied to the selected cell, allowing you to enter cell content manually.

After you choose the appropriate fill option, move the cursor to the last cell to which you want to copy the cell content and pick inside the cell. All cells between the first and last cell fill with the content of the active cell.

Modifying Cell Styles

Most cell style properties are defined according to the table style used to create the table. However, you can override cell styles for individual cells or groups of cells as needed. Cell style options are available from the **Cell Styles** panel of the ribbon and from the shortcut menu. Select a cell style from the **Cell Style** drop-down list to override the cell style used for the active cell. **Create New Cell Style...** and **Manage Cell Style...** options are also available. If you made changes to the active cell, you can save the changes as a new cell style that you can use to adjust the display of other cells or apply to a table style. To save the cell properties as a new cell style, right-click, pick **Save as New Cell Style...** from the **Cell Style** cascading submenu, and enter a name for the style in the **Save as New Cell Style** dialog box.

Pick the **Background Fill** drop-down list to change the cell background color. Select a cell alignment option from the **Alignment** flyout to override the justification of content within the selected cell. Cell content is located in relation to the cell borders. Pick the **Cell Borders** button to open the **Cell Border Properties** dialog box. This dialog box allows you to override cell border display properties and contains the same options found in the **Border** tab of the **Table Style** dialog box.

Select the **Match Cell** button to format settings from one cell to another. First select the cell that has the settings you want to copy, and then pick the **Match Cell** button. AutoCAD prompts you to select a destination cell. Pick the cell to which you want to copy the settings. Select another cell or right-click to exit.

Adjusting Cell Format

Cell format options are available from the **Cell Format** panel of the ribbon and from the shortcut menu. The **Cell Locking** flyout lists options for locking cells. You can lock cells to protect them from unintended or inappropriate changes. The locked icon appears when you move the cursor over a locked cell. The **Unlocked** option unlocks the cell so that you can make changes to cell content and format. The **Content Locked** option locks only the content of the cell, allowing you to make changes to the cell format. The **Format Locked** option locks only the cell format, allowing you to make changes to the cell content. The **Content and Format Locked** option locks cell content and format so that you cannot make changes.

The table style you use to create the table defines the data format. However, you can override the format of individual cells or groups of cells as needed. Override the data format of the cells by picking an option from the **Data Format...** flyout, or choose the **Custom Table Cell Format...** option to open the **Table Cell Data Format** dialog box. This is the same dialog box available for adjusting table style data cell format.

PROFESSIONAL TIP

The current table style controls most cell style and format properties. If you plan to make significant changes to cell properties, it is better to modify the table style or create a new style.

Inserting Fields and Blocks

In addition to text, table cells can contain fields, blocks, and formulas. Formulas are described later in this chapter. Blocks are useful for creating a legend or for displaying images of parts in a parts list. The options for inserting a block in a table are briefly described here. Detailed information about blocks is provided later in this textbook. To insert fields, blocks, and formulas, use the tools available from the **Insert** panel of the **Table Cell** ribbon tab, or from the **Insert** cascading submenu of the shortcut menu.

Pick the **Field...** button to open the **Field** dialog box, which allows you to insert a field into a table cell. This is the same dialog box you use to insert fields into mtext and text objects. You can also insert fields into a cell using the mtext text editor.

Select the **Block...** button to insert a block into a table cell using the **Insert a Block in a Table Cell** dialog box. **Figure 11-24** describes the options available in this dialog box. A cell can contain text and blocks. Double-clicking a block in a cell opens the **Insert a Block in a Table Cell** dialog box. Double-clicking on text in a cell opens the **Text Formatting** function.

Adding and Resizing Columns and Rows

You can add, delete, and resize columns and rows after you create a table. The following options are available from the **Columns** panel on the ribbon or the **Columns** cascading submenu in the shortcut menu:

- **Insert Left.** Adds a new column to the left of the selected cell.
- **Insert Right.** Adds a new column to the right of the selected cell.
- **Delete.** Deletes the entire column (or set of columns) containing the selected cell(s).
- **Size Equally.** Automatically sizes the selected columns to the same width. This option is only available when you select cells belonging to multiple columns.

The following options are available from the **Rows** panel of the ribbon or the **Rows** cascading submenu in the shortcut menu:

Figure 11-24.
Options in the **Insert a Block in a Table Cell** dialog box.

Feature	Description
Name	Used to choose the block from a drop-down list of the blocks stored in the current drawing.
Browse	Displays the **Select Drawing File** dialog box, where a drawing file can be selected and inserted into the table cell as a block.
AutoFit	Scales the block automatically to fit inside the cell.
Scale	Sets the block insertion scale. For example, a value of 2 inserts the block at twice its original size. A value of .5 inserts the block at half its created size. The **Scale** option is not available if the **AutoFit** check box is checked.
Rotation angle	Rotates the block to the specified angle.
Overall cell alignment	Determines the justification of the block in the cell and overrides the current cell alignment setting.

- **Insert Above.** Adds a new row above the selected cell.
- **Insert Below.** Adds a new row below the selected cell.
- **Delete.** Deletes the row (or set of rows) containing the selected cell(s).
- **Size Equally.** Automatically sizes the selected rows to the same height. This option is only available when you select cells belonging to multiple rows.

NOTE

A quick way to insert a new row at the bottom of a table using a **Down** table direction is to position the cursor in the lower-right cell and press [Tab]. To insert a new row at the top of a table using an **Up** table direction, position the cursor in the upper-right cell and press [Tab].

PROFESSIONAL TIP

You can also adjust column and row size using grips. Grips are the boxes located in the middle of cell border lines. To resize a column or row, select a grip, move the mouse, and pick. Chapter 15 explains grips in detail.

Merging Cells

Merging allows you to combine adjacent cells. The default tile cell style is an example of a merged cell. Merge tools are available from the **Merge** panel of the ribbon and from the **Merge** cascading submenu in the shortcut menu. To merge cells, select the cells to merge, and then select the appropriate merge option from the **Merge cells** flyout. Select the **All** option to merge all cells into one cell. The **By Row** and **By Column** options allow you to merge cells in multiple rows or columns without removing the horizontal or vertical borders. Pick the **Unmerge Cells** button to separate merged cells back into individual cells.

NOTE

Selecting the **Delete All Contents** option deletes the contents in the selected cell. This is the same as picking a cell and pressing [Delete].

Exercise 11-6

Access the Student Web site (www.g-wlearning.com/CAD) and complete Exercise 11-6.

Picking a Cell Edge to Edit Table Layout

Additional methods for adjusting table layout are available when you pick the edge, or border, of a cell. This displays the table indicator grid, grips you can use to adjust row height and column width, and the table break function. After picking a cell border, right-click anywhere in the drawing window to display the shortcut menu shown in **Figure 11-25.**

Adjusting Table Style

Select a table style from the **Table Style** cascading submenu of the shortcut menu to apply a different table style to the selected table. Picking the **Set as Table in Current Table Style** option creates a starting table style based on the selected table and the current table style. This is an alternative technique for creating a starting table style without opening the **Table Style** dialog box. If the selected table was drawn using a starting table style, selecting the **Set as Table in Current Table Style** redefines the starting table. To save modifications made to the table as a new table style, pick the **Save as New Table Style...** option and enter a name for the style in the **Save as New Table Style** dialog box.

Resizing Columns and Rows

You can use the grips boxes that appear when you select a table cell border to adjust column and row size. Grips are the boxes and arrowheads located at the corners of columns and the table. To resize a column or row, select a grip box, move the cursor, and pick. You can use the arrowhead grips to increase or decrease row height and/or column width uniformly. Chapter 15 explains grips in detail.

The **Size Columns Equally** option sizes all columns to the same width. The total width of the table is divided evenly among the columns. The **Size Rows Equally** option sizes all rows to the same height. All rows increase in height to match the height of the tallest row in the table.

Using Table Breaks

Use the table break function to break a table into separate sections while maintaining a single table object. This is a common requirement when it is necessary to fit a long table in a specific area or on a certain size sheet. Pick a cell edge to access table breaking. The table breaking grip is located midway between the sides of the table at the top or the bottom of the table, depending on the table direction. See **Figure 11-26.**

To break a table, select the table breaking grip and then move the cursor into the table to display a preview of the table sections and a dashed line with crosshairs.

Figure 11-25.
Several additional table layout modification options are available when you pick any cell border.

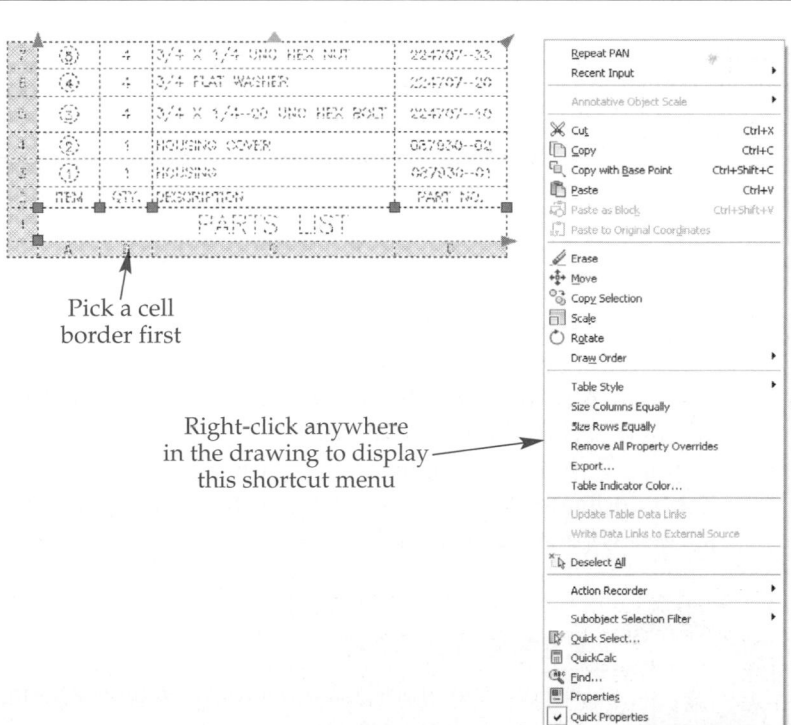

Pick a cell border first

Right-click anywhere in the drawing to display this shortcut menu

Figure 11-26.
The procedure for breaking, or wrapping, a table into sections.

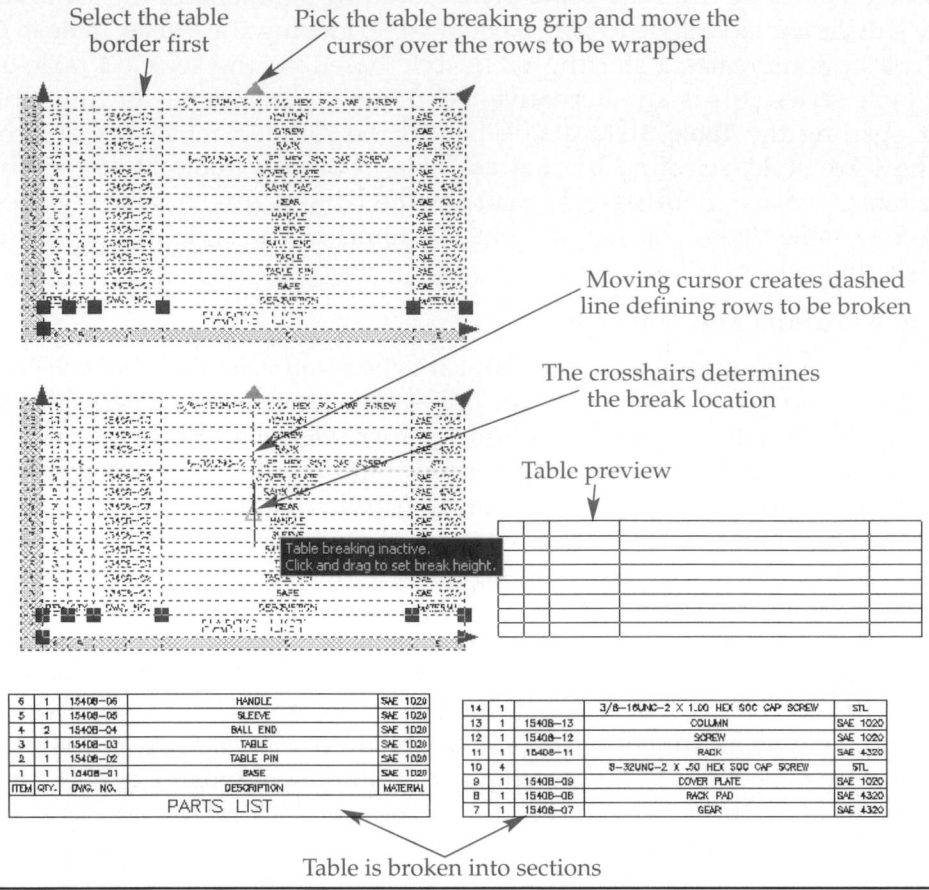

Select the table
border first

Pick the table breaking grip and move the
cursor over the rows to be wrapped

Moving cursor creates dashed
line defining rows to be broken

The crosshairs determines
the break location

Table preview

Table breaking inactive.
Click and drag to set break height.

Table is broken into sections

The crosshairs determines the location of the break. The closer to the table title and headers the crosshairs appears, the more sections you create, as shown in the table preview. When the preview of the table looks correct, pick the location to form the table breaks.

NOTE

After you add table breaks, several options become available from the **Properties** palette for adjusting the table sections. Chapter 15 covers using the **Properties** palette.

Additional Table Layout Options

The following additional table options are available from the shortcut menu:
- **Remove All Property Overrides.** Restores the table to its original properties, defined according to the selected table style.
- **Export.** Exports the table as a CSV file.
- **Table Indicator Color....** Allows you to change the color of the table indicator shown when you pick inside a cell.

Exercise 11-7

Access the Student Web site (www.g-wlearning.com/CAD) and complete Exercise 11-7.

Calculating Values in Tables

Performing calculations on table data using *formulas* is a common requirement. For example, in a parts list, you can add the number of parts and show the total. In a door or window schedule, you can calculate the total number of doors or windows. A room schedule often includes square footage calculations for various areas. Formulas calculate operations based on numeric data in table cells. AutoCAD allows you to write formulas for sums, averages, counts, and other mathematical functions.

formulas: Mathematical expressions that allow you to perform calculations within table cells.

The table indicator grid that appears when you edit a table cell provides a numbering system for the cells. Letters identify columns, and numbers identify rows. Together, the column letter and row number describe cells in formulas. For example, A3 identifies the cell located in Column A, Row 3. **Figure 11-27** illustrates the naming system. Cell C6 is highlighted in the table shown.

Creating Formulas

When you enter a formula in a cell, it evaluates values from other cells and displays the resulting value. Formulas are field objects. As with other types of fields, the expression and the resulting value display a gray background. The value changes if values in the corresponding expression change. This enables you to update data in a table cell automatically when you update the data in other cells.

Typically, you reference table cells with existing numeric values when writing formulas. For example, you can add the values of all cells in a single column and display the total at the bottom of the column, in a new cell. You can define a formula that evaluates a range of continuous cells or cells that do not share a common border.

In a formula, you enter common symbols used for mathematical functions as operators in the expression: + for addition, – for subtraction, * for multiplication, / for division, and ^ for exponentiation. Use parentheses to enclose expressions for table cell formulas. To perform an operation correctly, you must enter the proper syntax in the table cell. The syntax uses the following conventions:

Entry	Description
=	The equal sign is placed at the beginning of a formula. This tells AutoCAD that you want to perform a calculation.
(	An open parenthesis after the equal sign begins the expression.
)	A closing parenthesis ends the expression.
(expression)	Write the expression by typing the cells to evaluate and the desired operator symbols.

Figure 11-27.
Table cells are identified by column letter and row number. The table indicator grid provides a reference for identifying each cell individually.

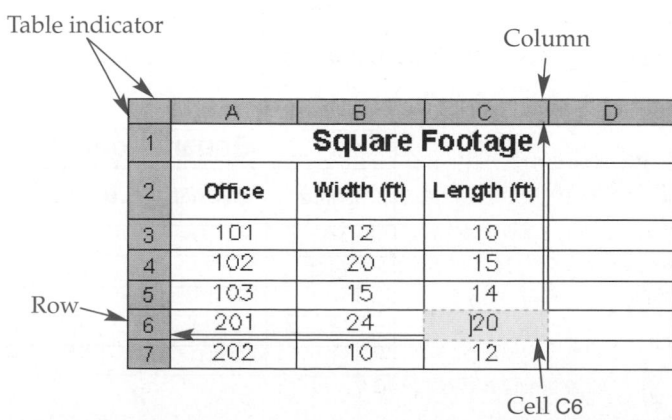

An example of a complete expression is =(C3+D4). This expression tells AutoCAD to add the value of cell C3 to the value of cell D4 and display the sum in the current cell. When identifying a cell in an expression, you must enter the letter before the number. For example, you cannot enter 3C to designate the cell C3. If you enter an incorrect expression or an expression evaluating cells without numeric data, AutoCAD displays the pound character (#) to indicate the error.

NOTE

Parentheses are unnecessary in some expressions, but other expressions will not calculate without them. It is good practice to use parentheses in all expressions.

You can use the text editor to input a formula. **Figure 11-28** shows an example of a multiplication formula. The expression =(B3*C3) is entered in cell D3. The resulting value appears after the text editor is closed.

You can also use grouped expressions in formulas by enclosing expression sets in parentheses. For example, the expression =(E1+F1)*E2 multiplies the sum of E1 and F1 by E2. Another example: =(E1+F1)*(E2+F2)/G6 multiplies the sum of E1 and F1 by the sum of E2 and F2 and divides the product by G6.

Creating Sum, Average, and Count Formulas

In addition to entering basic mathematical formulas in table cells manually, you can select from one of AutoCAD's formulas to calculate the sum, average, or count of a range of cells. Select a cell and choose an option from the **Formula** flyout on the **Table Cell** ribbon tab, or pick **Formula** from the **Insert** cascading submenu of the shortcut menu to display a cascading menu with formula options. See **Figure 11-29.**

The **Sum** option allows you to add the values of a range of cells by specifying a selection window on-screen. AutoCAD prompts you to pick the first corner of a window defining the cell range. The range you specify can include cells from several columns and rows. Pick inside the top or bottom cell that you want to include in the calculation. Then move the cursor and pick inside the lowest or highest cell, making sure that all of the desired cells are included in the window selection. See **Figure 11-30.** When you select the second point, the expression automatically appears in the cell. In the example

Figure 11-28.
Entering a multiplication formula. A—Type the expression in the table cell using the correct syntax. B—The resulting value displays after the expression is calculated.

	A	B	C	D
1	**Square Footage**			
2	Office	Width (ft)	Length (ft)	Sq Ft
3	101	12	10	=(B3*C3) ←Expression
4	102	20	15	
5	103	15	14	
6	201	24	20	
7	202	10	12	

A

Square Footage			
Office	Width (ft)	Length (ft)	Sq Ft
101	12	10	120 ← Result
102	20	15	
103	15	14	
201	24	20	
202	10	12	

B

AutoCAD and Its Applications—Basics

Figure 11-29.
The **Formula** flyout on the **Table Cell** ribbon tab allows you to insert a formula into a cell.

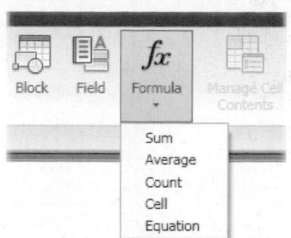

Figure 11-30.
Creating a sum formula in a table cell. A—Pick a cell to hold the formula and select a range of cells for the formula by windowing around the cells. B—After you pick the second point of the window, the formula displays in the cell.

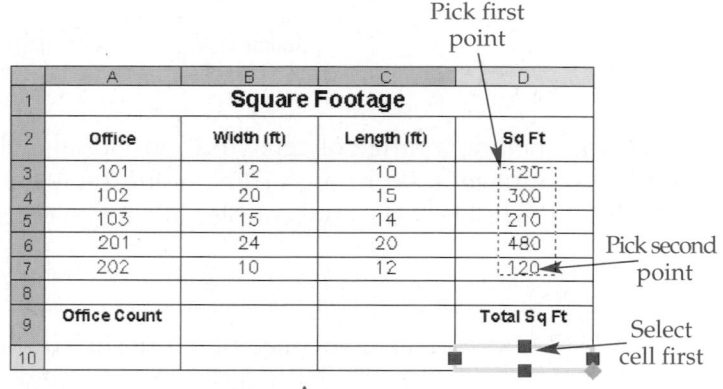

Pick first point

	A	B	C	D
1	Square Footage			
2	Office	Width (ft)	Length (ft)	Sq Ft
3	101	12	10	120
4	102	20	15	300
5	103	15	14	210
6	201	24	20	480
7	202	10	12	120
8				
9	Office Count			Total Sq Ft
10				

Pick second point

Select cell first

A

	A	B	C	D
1	Square Footage			
2	Office	Width (ft)	Length (ft)	Sq Ft
3	101	12	10	120
4	102	20	15	300
5	103	15	14	210
6	201	24	20	480
7	202	10	12	120
8				
9	Office Count			Total Sq Ft
10				=Sum(D3:D7)

Sum formula

B

shown, the square footage for each office is first calculated in the Sq Ft column on the right in **Figure 11-30A**. The values are then selected for a sum formula that calculates the total square footage of all of the offices, as shown in **Figure 11-30B**.

Notice in **Figure 11-30B** that the resulting expression is =Sum(D3:D7) This formula specifies that the selected cell is equal to the sum of cells D3 through D7. The colon (:) indicates the range of cells for the calculation.

The **Average** option creates a formula that calculates the average value of the cells you select. The average is the sum of the selected cells divided by the number of cells selected. The **Count** option creates a formula that counts the number of selected cells. Only cells that contain a value are included in the count.

You can type sum, average, and count formulas directly into a cell without using the **Formula** flyout button or the **Insert** cascading menu. If you calculate a value over a range of cells, use the colon symbol to designate the range. You can also write an expression that evaluates individual cells instead of a range. The cells do not have to share a common border. To write an expression using nonadjacent cells, use a comma to separate the cell names. For example, to average cells D1, D3, and D6, type =Average(D1,D3,D6). This formula calculates the average of the cell values for cells D1, D3, and D6.

Figure 11-31.
Sum, average, and count formulas and their resulting values.

	A	B	C	D
1			Square Footage	
2	Office	Width (ft)	Length (ft)	Sq Ft
3	101	12	10	120
4	102	20	15	300
5	103	15	14	210
6	201	24	20	480
7	202	10	12	120
8				
9	Office Count	Average Sq Ft Per Room		Total Sq Ft
10	5	246.000000		1230

=Count(A3:A7)   =Average(D3:D7)   =Sum(D3:D7)

You can include a range of cells and individual cells in the same expression. For example, to count cells A1 through B10 in addition to cells C4 and C6, enter =Count(A1: B10,C4,C6). **Figure 11-31** shows examples of sum, average, and count formulas.

> **NOTE**
>
> When using architectural units in a drawing, you can type the foot (′) and inch (″) symbols in table cells for use in values and formulas. When you use the foot symbol for a cell value, a formula in another cell automatically converts the resulting value to inches and feet.

Other Formula Options

The **Formula** flyout button and **Insert** cascading submenu contain additional options for writing table formulas. The **Cell** option allows you to select a table cell from a different table and insert its contents in the current cell. You can then use the cell value in a new formula. When you select the **Cell** option, AutoCAD prompts you to select the cell. The value of the selected cell appears in the current cell.

The **Equation** option allows you to enter an expression manually. Selecting this option places an equal sign (=) in the current cell. You can then type the expression.

> **NOTE**
>
> You can use the **Field** tool to insert and edit table cell formulas. Selecting **Formula** from the **Field names** list in the **Field** dialog box displays option buttons for creating sum, average, and count formulas. You can also select a cell value from a different table as a starting point. Select table cells on-screen to define the formula. You can use the **Formula** text box in the **Field** dialog box to add to or edit the formula. Unit format options are also available.

Exercise 11-8

Access the Student Web site (www.g-wlearning.com/CAD) and complete Exercise 11-8.

Supplemental Material

Linking a Table to Excel Data

For information about using existing data entered in a Microsoft® Excel spreadsheet or a CSV file to create an AutoCAD table, go to the Student Web site (www.g-wlearning.com/CAD), select this chapter, and select **Linking a Table to Excel Data**.

Supplemental Material

Extracting Table Data

For information about using existing AutoCAD text to create a table, go to the Student Web site (www.g-wlearning.com/CAD), select this chapter, and select **Extracting Table Data**.

Template Development
Chapter 11

For detailed instructions on adding table styles to each drawing template, go to the Student Web site (www.g-wlearning.com/CAD), select this chapter, and select **Template Development**.

Chapter Test

Answer the following questions. Write your answers on a separate sheet of paper or go to the Student Web site (www.g-wlearning.com/CAD) and complete the electronic chapter test.

1. What is the purpose of creating a table style?
2. Briefly describe the procedure for creating a table style based on an existing table style.
3. What does the **Alignment** setting in the **New Table Style** dialog box do?
4. Which setting would you adjust in the **New Table Style** dialog box to increase the spacing between the text and the top of the cell?
5. How can creating a new table using a starting table style save time?
6. How can you make a table style current without opening the **Table Style** dialog box?
7. List two ways to open the **Insert Table** dialog box.
8. Describe the two ways to insert a table and explain how the methods differ.
9. By default, what two types of rows are at the top of a table when it is inserted?
10. What ribbon tab opens when you insert a table?
11. If you are finished typing in one cell and want to move to the next cell in the same row, what two keyboard keys can you use?
12. List two ways to make a cell active for editing.
13. Explain how to insert a block into a table cell.
14. How can you insert a new row quickly at the bottom of a table?
15. How are table cells identified in formulas?
16. Write the table cell formula that adds the value of C3 plus the value of D4.
17. What is the function of the colon symbol (:) in the formula =Sum(D3:D7)?
18. What is the difference between a sum formula and a count formula?
19. Write the table cell formula that averages the values of cells D1, D3, and D6.
20. Explain how to write a formula that calculates a function for cells that do not share common borders.

Drawing Problems

Follow these instructions to complete the drawing problems for this chapter:

- *Start AutoCAD if it is not already started.*
- *Start a new drawing using an appropriate template of your choice. The template should include layers, text styles, and table styles when necessary, for drawing the given tables.*
- *Add layers, text styles, and table styles as needed. Draw all objects using appropriate layers, text styles, table styles, justification, and format.*
- *Follow the specific instructions for each problem. Use your own judgment and approximate dimensions when necessary.*
- *Make the measurements for rows and columns approximately the same as in the given table.*

▼ Basic

1. Create the parts list shown below. Save the drawing as P11-1.

	1	CAPS	1/2–12UNC HEX NUT	210014–29
	1	CAPS	1/2 FLAT WASHER	320014–33
	2	CAPS	7/16 EXTERNAL SNAP RING	632043–43
	2	CAPS	1/4–20UNC WING NUT	255010–41
	2	CAPS	3/4 X 1/4–20UNC BOLT	803010–11
KEY	QTY	NAME	DESCRIPTION	PART NO.
			PARTS LIST	

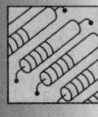

2. Create the table shown below. Save the drawing as P11-2.

REFERENCE DESIGNATIONS	
LAST USED	DATE
R9	1/30/2009
C1	1/30/2009
CRS	1/30/2009
Q2	1/30/2009

▼ Intermediate

3. Create the door schedule shown below. Save the drawing as P11-3.

DOOR SCHEDULE			
SYM.	SIZE	TYPE	QTY.
①	36x80	S.C. RP. METAL INSULATED	1
②	36x80	S.C. FLUSH METAL INSULATED	2
③	32x80	S.C. SELF CLOSING	2
④	32x80	HOLLOW CORE	5
⑤	30x80	HOLLOW CORE	5
⑥	30x80	POCKET SLDG.	2

4. Create the window schedule shown below. Save the drawing as P11-4.

SYM.	SIZE	MODEL	ROUGH OPEN	QTY.
Ⓐ	12x60	JOB BUILT	VERIFY	2
Ⓑ	96x60	W4N5 CSM.	8'-0 3/4" x 5'-0 7/8"	1
Ⓒ	48x60	W2N5 CSM.	4'-0 3/4" x 5'-0 7/8"	2
Ⓓ	48x36	W2N3 CSM.	4'-0 3/4" x 3'-6 1/2"	2
Ⓔ	42x42	2N3 CSM.	3'-6 1/2" x 3'-6 1/2"	2
Ⓕ	72x48	G64 SLDG.	6'-0 1/2" x 4'-0 1/2"	1
Ⓖ	60x42	G536 SLDG.	5'-0 1/2" x 3'-6 1/2"	4
Ⓗ	48x42	G436 SLDG.	4'-0 1/2" x 3'-6 1/2"	1
Ⓙ	48x24	A41 AWN.	4'-0 1/2" x 2'-0 7/8"	3

WINDOW SCHEDULE

5. Create the door schedule shown below. Save the drawing as P11-5.

DOOR SCHEDULE

SYMBOL	SIZE	MODEL	QUANTITY	SYMBOL	SIZE	MODEL	QUANTITY
1	3'-0" X 6'-8"	S.C. R.P. METAL INSULATED	1	11	4'-0" X 6'-8"	BI-FOLD	1
2	3'-0" X 6'-8"	S.C.-FLUSH-METAL INSULATED	2	12	2'-0" X 6'-0"	SHATTER PROOF	1
3	2'-8" X 6'-8"	S.C.-SELF CLOSING	2	13	6'-0" X 6'-8"	WOOD FRAME-TEMP. SLDG GL.	1
4	2'-8" X 6'-8"	H.C.	5	14	9'-0" X 7'-0"	OVERHEAD GARAGE	2
5	2'-6" X 6'-8"	H.C.	3				
6	2'-6" X 6'-8"	POCKET	2				
7	2'-4" X 6'-8"	POCKET	1				
9	5'-0" X 6'-0"	BI-PASS	2				
10	3'-0" X 6'-8"	BI-FOLD	1				

6. Open P11-5 and make the following changes:
   - Change the DOOR SCHEDULE so that SYMBOL items 11, 12, 13, and 14 continue directly below SYMBOL items 1 through 10, with only one DOOR SCHEDULE heading at the top.
   - Change the following abbreviations to full words:
   - S.C.R.P. – SOLID CORE RAISED PANEL
   - S.C. – SOLID CORE
   - H.C. – HOLLOW CORE
   Save the drawing as P11-6.

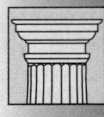

▼ Advanced

7. Create a parts list for a mechanical drawing with the content of your choice, or locate a drawing with a parts list and make a similar drawing. Save the drawing as P11-7.

8. Create a door and window schedule for an architectural drawing with the content of your choice, or locate a drawing with a door and window schedule and make a similar drawing. Save the drawing as P11-8.

9. Create a legend for a civil drawing with the content of your choice, or locate a drawing with a legend and make a similar drawing. Save the drawing as P11-9.

10. Create a parts list for a mechanical drawing with the content of your choice, or locate a drawing with a parts list and make a similar drawing. Save the drawing as P11-10.

11. Open P11-1 and make the following changes:
    - Change PARTS LIST to PURCHASE PARTS LIST.
    - Change Part Number 803010-11 as follows: Key: 7, Name: HEX HD, Description: $\frac{1}{4}$20UNC-2 X $\frac{3}{4}$ BOLT.
    - Change Part Number 255010-41 as follows: Key: 11, Name: WING NUT, Description: $\frac{1}{4}$20UNC.
    - Change Part Number 632043-43 as follows: Key: 15, Name: SNAP RING, Description: Ø $\frac{7}{16}$EXTERNAL.
    - Change Part Number 320014-33 as follows: Key: 19, Name: WASHER, Description: Ø $\frac{1}{2}$FLAT.
    - Change Part Number 210014-29 as follows: Key: 21, Name: NUT, Description: $\frac{1}{2}$-12UNC-2 HEX.

    Save the drawing as P11-11.

12. Create a table of your own design and use at least six of the applications of calculating values in tables described in this chapter. Save the drawing as P11-12.

13. Access the Student Web site content for this chapter and review Supplement 11A, "Linking a Table to Excel Data." Use this information to create a table by linking to Microsoft® Excel data. To do this, create your own Excel spreadsheet or find an existing Excel spreadsheet containing the desired data. Link a table to the Excel data. Save the drawing as P11-13.

14. Access the Student Web site content for this chapter and review Supplement 11B, "Extracting Table Data." Use the description to create a table the same as or similar to the given examples and then extract the table data into an AutoCAD drawing. Save the drawing as P11-14.

CHAPTER 12

Basic Object Editing Tools

Learning Objectives

After completing this chapter, you will be able to do the following:

✓ Use the **FILLET** tool to draw fillets, rounds, and other rounded corners.
✓ Place chamfers and angled corners with the **CHAMFER** tool.
✓ Separate objects using the **BREAK** tool and combine objects using the **JOIN** tool.
✓ Use the **TRIM** and **EXTEND** tools to edit objects.
✓ Modify objects using the **STRETCH** and **LENGTHEN** tools.
✓ Change the size of objects using the **SCALE** tool.
✓ Use the **EXPLODE** tool.

This chapter explains methods for changing objects using basic editing tools. You will learn to use various editing tools to increase drawing efficiency. The editing tools described in this chapter include many options. As you work through this chapter, experiment with each option to see which is the most effective in different situations.

Using the FILLET Tool

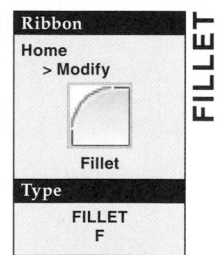

Ribbon
Home
> Modify

Fillet

Type
FILLET
F

FILLET

Drafting terminology refers to a rounded interior corner as a *fillet*, and a rounded exterior corner as a *round*. AutoCAD refers to all rounded corners as fillets. The **FILLET** tool draws a rounded corner between intersecting and nonintersecting lines, circles, arcs, and polylines. See **Figure 12-1**.

Setting Fillet Radius

After initiating the **FILLET** tool, use the **Radius** option to enter the fillet radius dimension. The fillet radius determines the size of a fillet and must be set before you select objects. Once you specify the radius, select the objects to fillet. The specified fillet radius is stored as the new default radius, allowing you to place additional fillets of the same size.

fillet: A rounded interior corner used to relieve stress or ease the contour of inside corners.

round: A rounded exterior corner used to remove sharp edges or ease the contour of exterior corners.

Figure 12-1.
Using the **FILLET** tool.

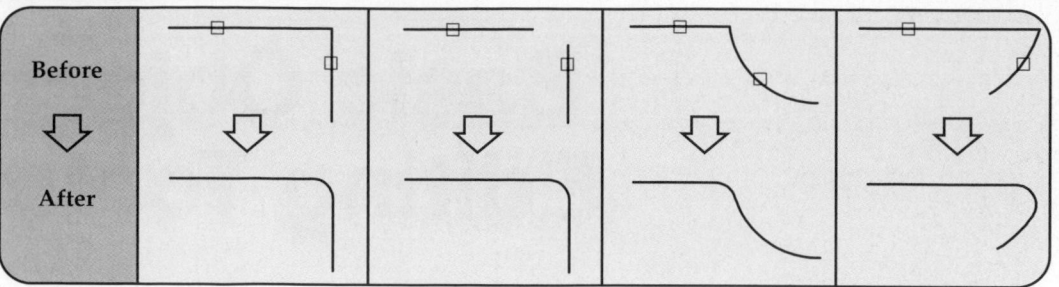

NOTE

Only corners large enough to accept the specified fillet radius are eligible for filleting. If the specified fillet radius is too large, AutoCAD displays a message, such as Distance is too large *Invalid*.

Exercise 12-1

Access the Student Web site (www.g-wlearning.com/CAD) and complete Exercise 12-1.

Forming Sharp Corners

Use a fillet radius of 0 to connect two objects at a sharp corner. You can also create a zero-radius fillet without setting the radius to 0 by holding down the [Shift] key when you pick the second object. This is a convenient way to connect objects at a corner, or to form a square corner if edges are perpendicular.

Filleting Parallel Lines

You can use the **FILLET** tool to draw a full radius between parallel lines. When you set the **Trim** option to **Trim**, a longer line trims to match the length of a shorter line. The radius of a fillet between parallel lines is always half the distance between the two lines, regardless of the radius setting. You can use this method to create a full radius, such as the end radii applied to a slot.

Rounding Polyline Corners

You can use the **Polyline** option to fillet all corners of a closed polyline. See **Figure 12-2**. Remember to set the appropriate radius before filleting. If you drew the polyline without using the **Close** option, the beginning corner does not fillet, as shown in **Figure 12-2**.

Fillet Trim Settings

The **Trim** option controls whether the **FILLET** tool trims object segments that extend beyond the fillet radius point of tangency. See **Figure 12-3**. Use the default **Trim** setting to trim objects. When you set the **Trim** option to **No Trim**, the fillet occurs, but filleted objects do not change.

Figure 12-2.
Using the **Polyline** option of the **FILLET** tool.

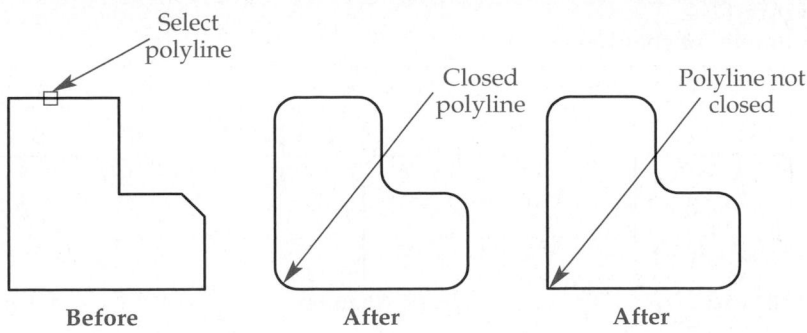

Select polyline

Closed polyline

Polyline not closed

Before

After

After

Figure 12-3.
Comparison of the **Trim** and **No trim** options of the **FILLET** tool.

Before Fillet	Fillet with Trim	Fillet with No Trim

PROFESSIONAL TIP

You can fillet objects even when the corners do not meet. If the **Trim** option is set to **Trim,** objects extend as required to generate the fillet and complete the corner. If the **Trim** option is set to **No Trim**, objects do not extend to complete the corner.

Making Multiple Fillets

Use the **Multiple** option to make several fillets without exiting the **FILLET** tool. The prompt for a first object repeats. To exist, press [Enter], the space bar, [Esc], or right-click and select **Enter**. When you use **Multiple** mode, use the **Undo** option to discard the previous fillet.

Exercise 12-2

Access the Student Web site (www.g-wlearning.com/CAD) and complete Exercise 12-2.

Ribbon
Home
> Modify

Chamfer
Type
CHAMFER
CHA

CHAMFER

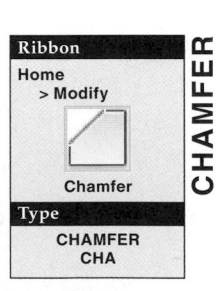

Using the CHAMFER Tool

Drafting terminology often refers to a *chamfer* as a small, angled surface used to relieve a sharp corner. The **CHAMFER** tool allows you to draw an angled corner between intersecting and nonintersecting lines, polylines, xlines, and rays. Chamfer size is determined based on the distance from the corner. A 45° chamfer is the same distance from the corner in each direction. See **Figure 12-4.** Typically, two distances or one distance and one angle identify the size of a chamfer. The defaults are zero units for the length and the angle. A value of .5 for both distances produces a 45° × .5 chamfer.

chamfer: In mechanical drafting, a small, angled surface used to relieve a sharp corner.

Figure 12-4.
Examples of chamfers.

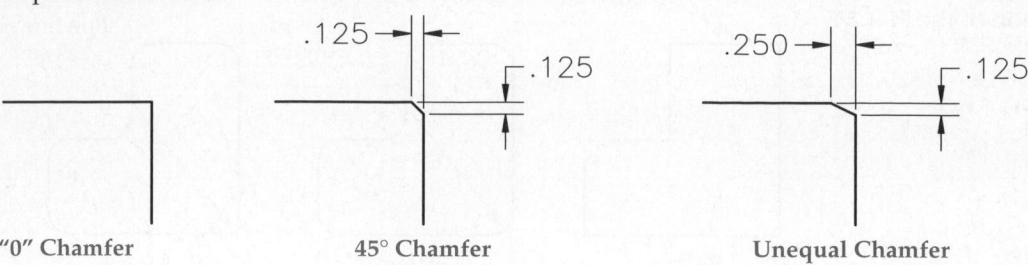

"0" Chamfer   45° Chamfer   Unequal Chamfer

Setting Chamfer Distances

After initiating the **CHAMFER** tool, use the **Distance** option to enter the chamfer distances. Chamfer distances determine the size of a chamfer from a corner and must be set before you select objects. Once you specify the distances, select the objects to chamfer. **Figure 12-5** shows several chamfering operations. The specified chamfer distances are stored as the new default distances, allowing you to place additional chamfers of the same size.

Setting the Chamfer Angle

Instead of setting two chamfer distances, you can use the **Angle** option to set the chamfer distance for one edge and set an angle to determine the chamfer to the second edge. See **Figure 12-6.** After entering the distance and angle, select the two objects to chamfer. The specified distance and angle remain active until changed, allowing you to place additional chamfers of the same size.

Figure 12-5.
Using the **CHAMFER** tool.

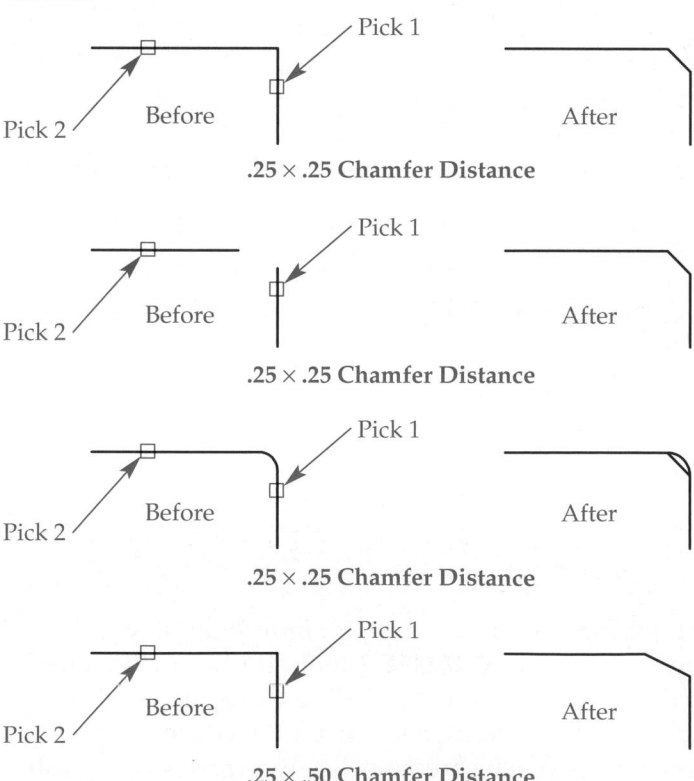

AutoCAD and Its Applications—Basics

Figure 12-6.
Using the **Angle**
option of the
CHAMFER tool with
the chamfer length
set at .5 and the
angle set at 45°.

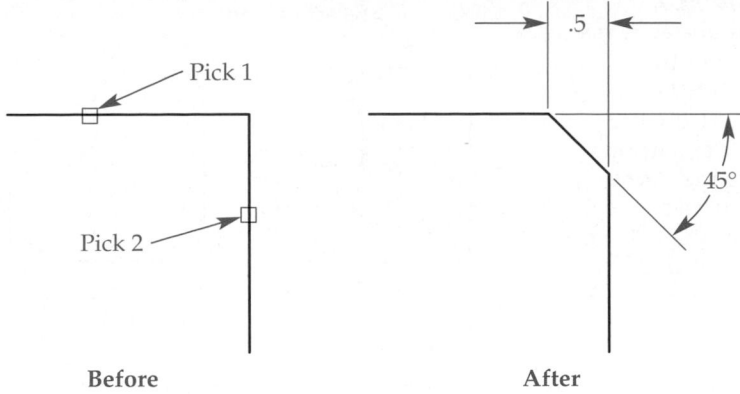

Before   After

Setting the Chamfer Method

AutoCAD maintains the specified chamfer distances, or a distance and an angle, until you change the values. You can set the values for each method without affecting the other. Use the **Method** option to toggle between drawing chamfers using the **Distance** and **Angle** options.

PROFESSIONAL TIP

You can use the **CHAMFER** tool to form sharp corners by specifying chamfer distances or an angle and distance of 0; or by holding [Shift] when you pick the second object.

NOTE

Only corners large enough to accept the specified chamfer size are eligible for chamfering. If the chamfer is too large, AutoCAD displays a message, such as Distance is too large *Invalid*.

Exercise 12-3

Access the Student Web site (www.g-wlearning.com/CAD) and complete Exercise 12-3.

Additional Chamfer Options

The **CHAMFER** tool includes the same **Polyline**, **Trim**, and **Multiple** options available with the **FILLET** tool, and similar rules apply when using these options with the **CHAMFER** tool. Use the **Polyline** option to chamfer all corners of a closed polyline, as shown in **Figure 12-7**. The **Trim** option controls whether the **CHAMFER** tool trims object segments that extend beyond the intersection, as shown in **Figure 12-8**. Use the **Multiple** option to make several chamfers without exiting the **CHAMFER** tool.

Exercise 12-4

Access the Student Web site (www.g-wlearning.com/CAD) and complete Exercise 12-4.

Figure 12-7.
Using the **Polyline** option of the **CHAMFER** tool. If you drew the polyline without using the **Close** option, the beginning corner does not chamfer.

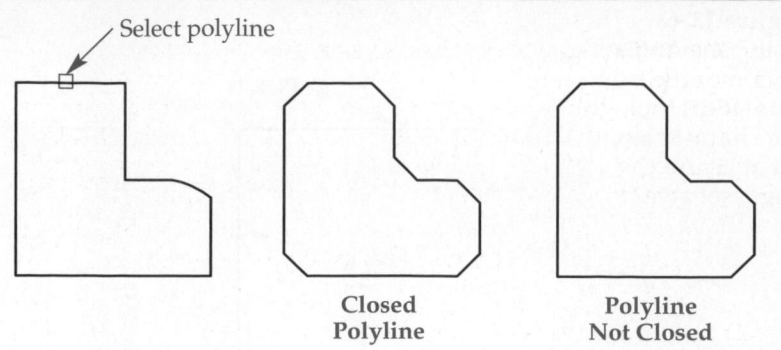

Select polyline

Closed
Polyline

Polyline
Not Closed

Figure 12-8.
Use the default **Trim** setting to trim objects. When you set the **Trim** option to **No Trim**, the chamfer occurs, but chamfered objects do not change.

Before Chamfer	Chamfer with Trim	Chamfer with No Trim

Using the BREAK Tool

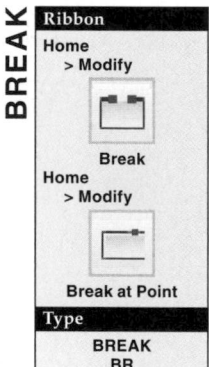

You can use the **BREAK** tool to separate a single object into two objects. A break can remove a portion of an object or split the object at a single point, depending on the selected points. The **BREAK** tool requires you to select the object to break and the first and second break points. By default, the point you pick when you select the object to break also locates the first break point. To select a different first break point, use the **First point** option at the Specify second break point or [First point]: prompt. The portion of the object between the two points deletes. See **Figure 12-9.**

If you select the same point for the first and second break points, the **BREAK** tool splits the object into two pieces without removing a portion. You can accomplish this by entering @ at the Specify second break point or [First point]: prompt. The @ symbol repeats the coordinates of the previously selected point. You can also pick the **Break at Point** button on the expanded **Modify** panel of the **Home** ribbon tab. **Figure 12-10** shows the process of breaking without removing a portion of the object.

Always work in a counterclockwise direction when breaking arcs or circles. Otherwise, you may break the portion of the arc or circle you want to keep. If you want to break off the end of a line or an arc, pick the first point on the object. Pick the second point slightly beyond the end to be cut off. See **Figure 12-11.** When you pick a second point not on the object, AutoCAD selects a point on the object nearest the point you pick.

Figure 12-9.
Using the **BREAK** tool to break an object. You can use the first pick to select both the object and the first break point.

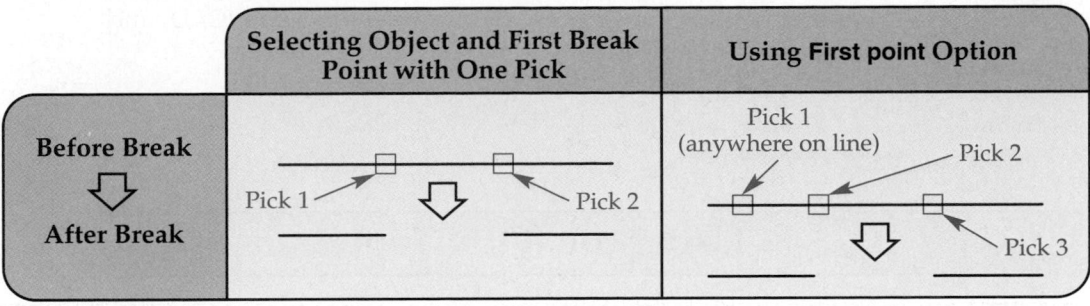

	Selecting Object and First Break Point with One Pick	Using **First point** Option
Before Break ⬇ **After Break**		

Figure 12-10.
Using the **BREAK** tool to break an object at a single point without removing any of the object. Select the same point as the first and second break points, or use the **Break at Point** button.

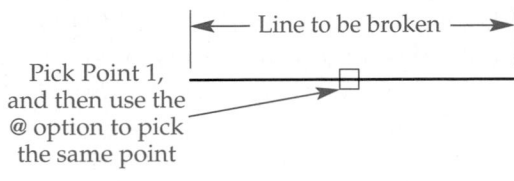

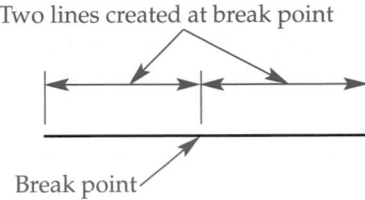

Figure 12-11.
Work counterclockwise when using the **BREAK** tool on circles and arcs.

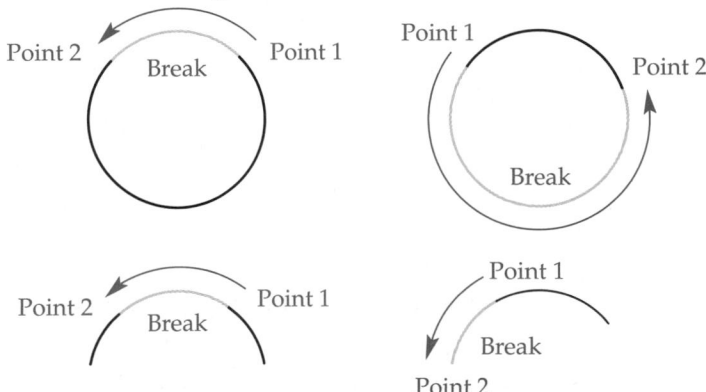

Exercise 12-5

Access the Student Web site (www.g-wlearning.com/CAD) and complete Exercise 12-5.

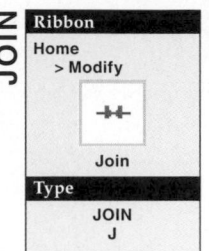

JOIN

Ribbon
Home
> Modify

Join

Type
JOIN
J

Using the JOIN Tool

Often multiple objects that should be one object form because of the drawing and editing process. These multiple objects make the drawing file size larger and the drawing more cumbersome. You can use the **JOIN** tool to join lines, polylines, splines, arcs, and elliptical arcs together to make one object. You can only join objects of the same type. For example, you can join a line to another line, but you cannot join a line to a polyline. In addition, joined objects must be in the same 2D plane.

Each object type has different rules for joining. Lines must be collinear, but they can touch, overlap, or have gaps between segments. See **Figure 12-12.** Polylines and splines must share a common endpoint and cannot have gaps between segments or overlap. **Figure 12-13** shows rules for joining polylines. The same rules apply for joining splines.

Arcs and elliptical arcs must share the same center point and circular path, but they can touch, overlap, or have gaps between segments. **Figure 12-14** shows joining two arcs separated by a gap. The same rules apply for joining elliptical arcs. Pick arcs or elliptical arcs in a clockwise direction to close the nearest clockwise gap, or pick in a counterclockwise direction to close the nearest counterclockwise gap. Depending on your selections, you may receive a prompt to convert arcs to a circle. If you do not want to form a circle, choose the **No** option and reselect the arcs in a counterclockwise direction.

Figure 12-12.
Lines must be collinear to join, but the lines can overlap, and there can be gaps between segments.

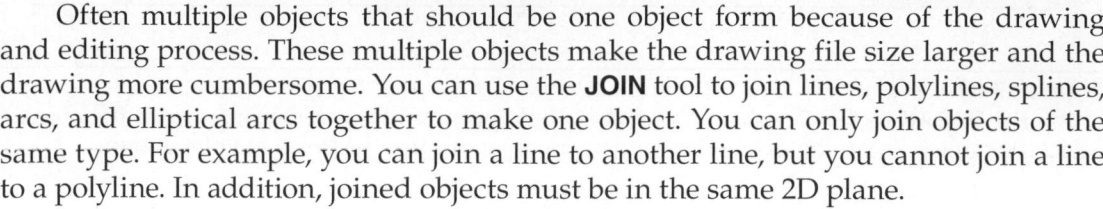

Figure 12-13.
You can join polylines only if they share an endpoint. The same rule applies to joining splines.

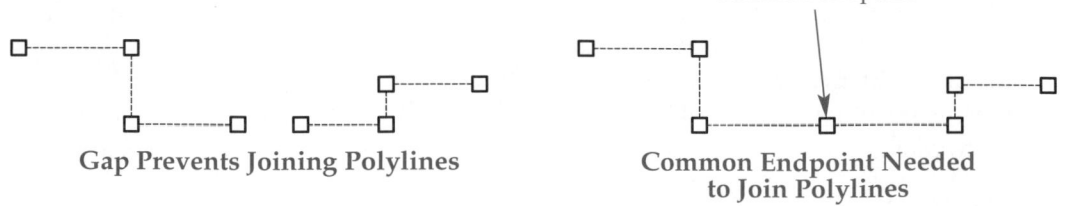

Figure 12-14.
Arcs can have a gap or be overlapping, as shown on the left, but they must share the same circular path. On the right, the two arcs are joined.

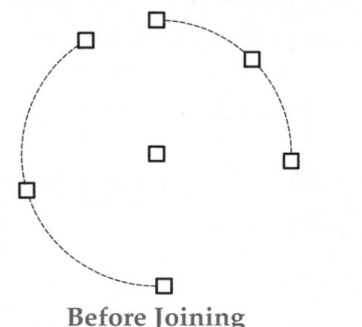
Before Joining

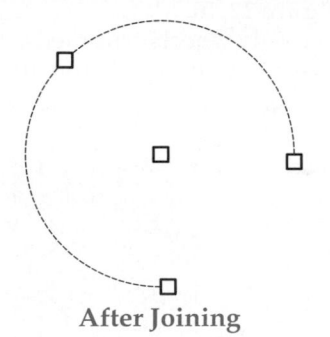
After Joining

> **NOTE**
>
> After selecting an arc or elliptical arc segment to join, you can use the **cLose** option to form a circle from the arc, or to form an ellipse from the elliptical arc. This option closes the two ends of the selected segment and does not join the segment with any other objects.

Trimming Objects

The **TRIM** tool cuts lines, polylines, circles, arcs, ellipses, splines, xlines, and rays that extend beyond a desired point of intersection. Once you access the **TRIM** tool, pick as many *cutting edges* as necessary and then right-click or press [Enter] or the space bar. Then pick the objects to trim to the cutting edges. To exit, right-click or press [Enter] or the space bar. See **Figure 12-15.**

The **TRIM** tool presents specific **Crossing** and **Fence** options that function the same as standard crossing and fence selection overrides, described in Chapter 3. **Figure 12-16** shows using the **Crossing** option to trim multiple objects. **Figure 12-17** shows an example of using the **Fence** option to trim multiple objects. However, automatic windowing with the crossing function is often the quickest and most effective method for trimming multiple objects. You can also use window or crossing polygons.

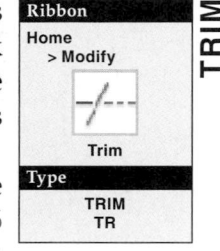

Ribbon
Home > Modify
Trim
Type
TRIM TR

cutting edge: An object such as a line, an arc, or text that defines the point (edge) at which the object you trim will be cut.

> **NOTE**
>
> To access the **EXTEND** tool while using the **TRIM** tool, after selecting the cutting edge(s), hold [Shift] and pick objects to extend to the cutting edge. The **EXTEND** tool is described later in this chapter.

Figure 12-15.
Using the **TRIM** tool. Note the cutting edges.

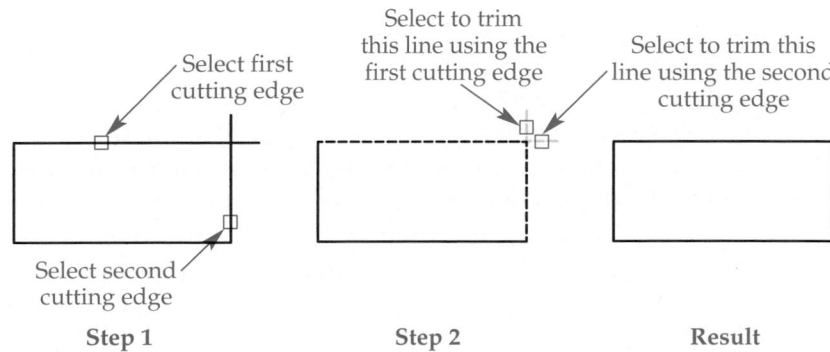

Figure 12-16.
The only objects trimmed with the **Crossing** option are those that cross the edges of the crossing window. Automatic windowing accomplishes the same task.

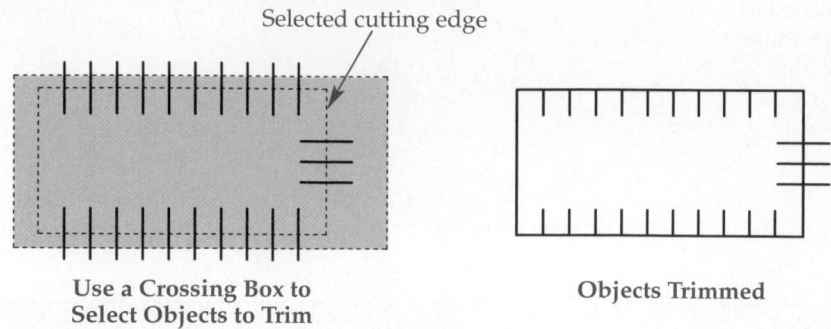

Use a Crossing Box to
Select Objects to Trim

Objects Trimmed

Figure 12-17.
The **Fence** option allows you to select around objects. In this case, the **RECTANGLE** tool was used to create the rectangle, so the cutting edge consists of the entire rectangle.

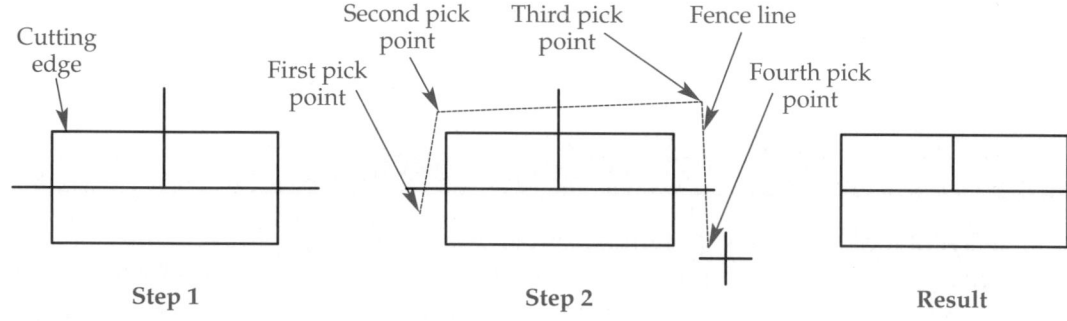

Step 1

Step 2

Result

Trimming without Selecting a Cutting Edge

To trim objects to the nearest intersection without selecting a cutting edge, access the **TRIM** tool and, at the first Select objects or <select all>: prompt, right-click or press [Enter] or the space bar instead of picking a cutting edge. Then pick the objects to trim. You can continue selecting objects to trim without restarting the **TRIM** tool. To exit, right-click or press [Enter], the space bar, or [Esc].

Trimming to an Implied Intersection

implied intersection: The point at which objects would meet if they were extended.

Trimming to an *implied intersection* is possible using the **Edge** option of the **TRIM** tool. Access the **TRIM** tool, pick the cutting edges, and then select the **Extend** option. The **No extend** mode is active by default, and as a result, you cannot trim objects that do not intersect. Choose the **Extend** mode to recognize implied intersections, and then pick objects to trim. See **Figure 12-18.** This does not change the selected cutting edges.

> **NOTE**
>
> The **TRIM** tool includes additional options. Use the **eRase** option to erase objects selected to trim. Use the **Undo** option to restore previously trimmed objects without leaving the tool. You must activate the **Undo** option immediately after performing an unwanted trim. The **Project** option applies to trimming 3D objects. *AutoCAD and Its Applications—Advanced* explains drawing and editing 3D objects.

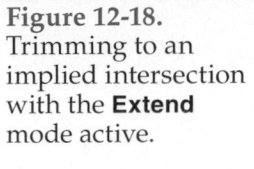

Figure 12-18.
Trimming to an implied intersection with the **Extend** mode active.

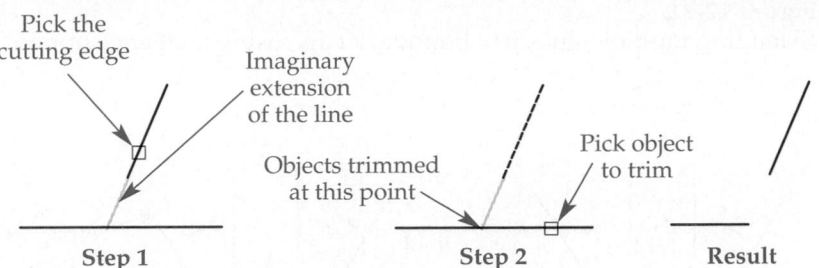

Pick the cutting edge

Imaginary extension of the line

Objects trimmed at this point

Pick object to trim

Step 1   Step 2   Result

Extending Objects

The **EXTEND** tool allows you to extend lines, elliptical arcs, rays, open polylines, and arcs to meet another object. **EXTEND** does not work on closed polylines because an unconnected endpoint does not exist. Once you access the **EXTEND** tool, pick as many *boundary edges* as necessary, and then right-click or press [Enter] or the space bar. Then pick the objects to extend to the boundary edges. To exit, right-click or press [Enter] or the space bar. See **Figure 12-19**.

Like the **TRIM** tool, the **EXTEND** tool presents specific **Crossing** and **Fence** options. **Figure 12-20** shows using the **Crossing** option to extend multiple objects. **Figure 12-21** shows an example of using the **Fence** option to extend multiple objects. However, automatic windowing with the crossing function is often the quickest and most effective method for extending multiple objects. You can also use window or crossing polygons.

The **EXTEND** tool includes the same **Edge**, **Undo**, and **Project** options available for the **TRIM** tool, and similar rules apply when using these options with the **EXTEND** tool. **Figure 12-22** illustrates how to combine the **EXTEND** and **TRIM** tools, without selecting

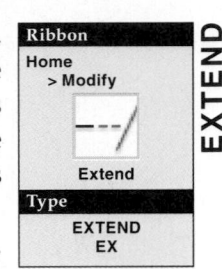

| Ribbon |
| Home |
| > Modify |
| Extend |
| Type |
| EXTEND |
| EX |

boundary edge: The edge to which objects such as lines, arcs, and polylines are extended.

Figure 12-19.
Using the **EXTEND** tool. Note the boundary edges.

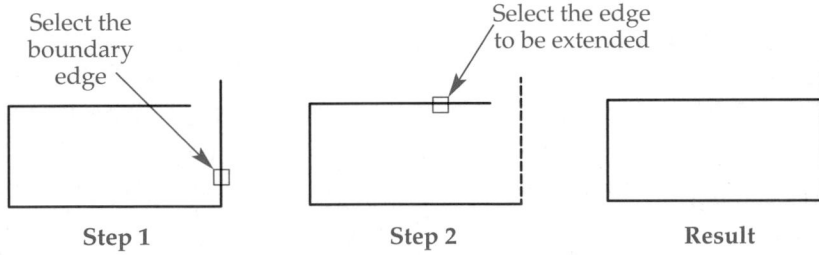

Select the boundary edge

Select the edge to be extended

Step 1   Step 2   Result

Figure 12-20.
Selecting objects to extend using the **Crossing** option. Automatic windowing accomplishes the same task.

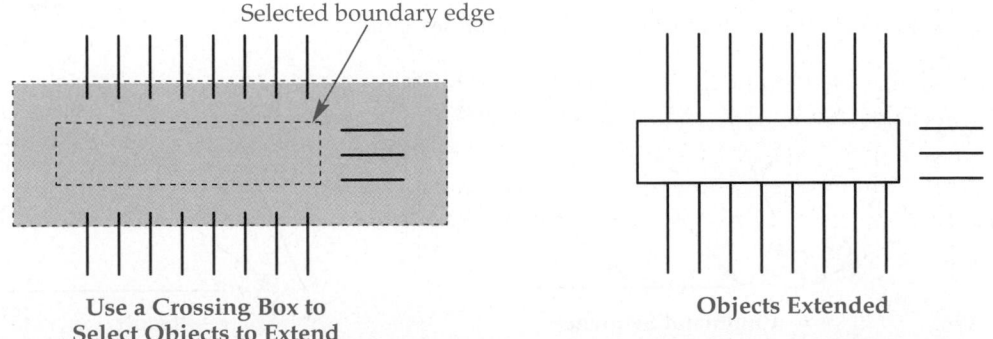

Selected boundary edge

Use a Crossing Box to Select Objects to Extend

Objects Extended

Figure 12-21.
Extending multiple lines to a boundary edge using the **Fence** option.

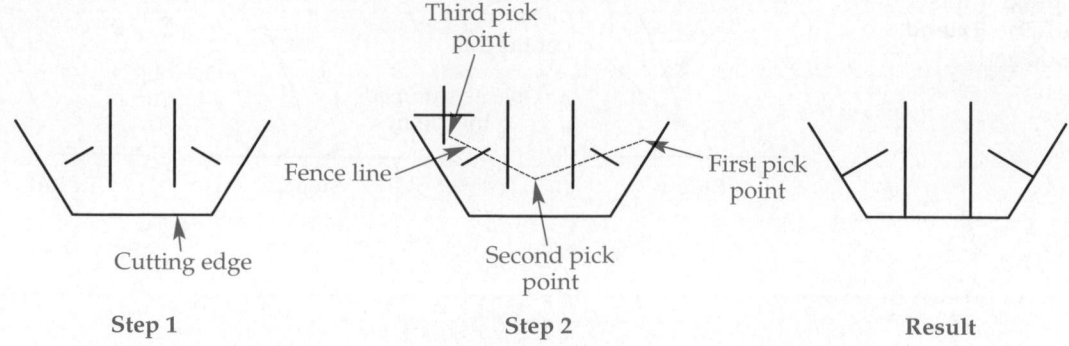

Step 1   Step 2   Result

Figure 12-22.
To extend objects to the nearest intersection without selecting a boundary edge, right-click or press [Enter] or the space bar instead of picking a boundary edge. Then pick object(s) to extend. Hold down the [Shift] key to toggle between extending and trimming.

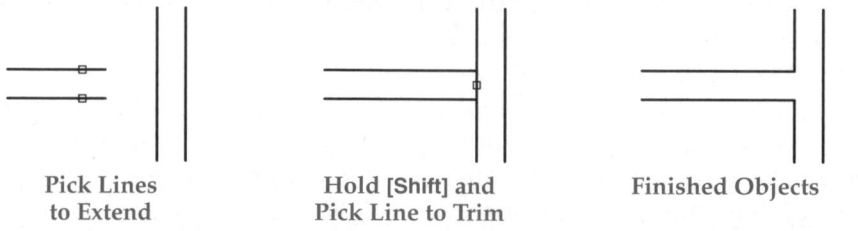

Pick Lines   Hold [Shift] and   Finished Objects
to Extend   Pick Line to Trim

a boundary edge, to insert a wall in a floor plan. You can apply this process to a variety of applications.

Use the **Extend** mode of the **Edge** option to extend to an implied intersection, as shown in **Figure 12-23.** Select the **Undo** option immediately after performing an unwanted extend to restore previous objects without leaving the tool. The **Project** option applies to extending 3D objects, as explained in *AutoCAD and Its Applications—Advanced.*

PROFESSIONAL TIP

Construction lines created using the **XLINE** and **RAY** tools are modified using standard editing tools. When you trim one infinite end of an xline, the object becomes a ray. When you trim both infinite ends of an xline, or the infinite end of a ray, the object becomes a line object. Therefore, in many cases, you can modify construction lines to become a portion of the actual drawing. This approach can save a significant amount of time for a variety of applications.

Figure 12-23.
Extending to an implied intersection with the **Extend** mode.

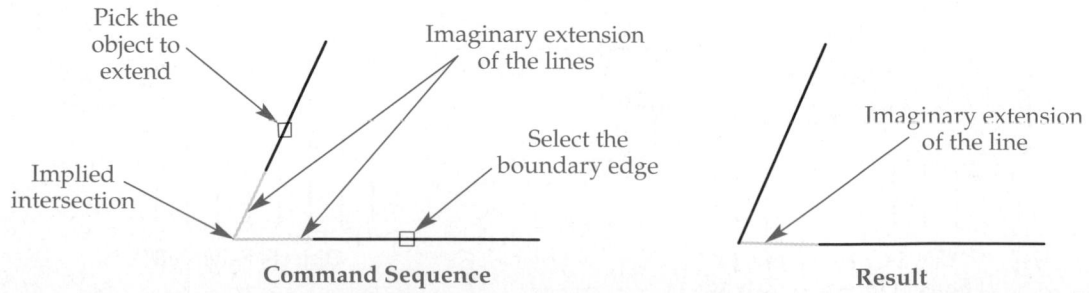

Command Sequence   Result

Exercise 12-6

Access the Student Web site (www.g-wlearning.com/CAD) and complete Exercise 12-6.

Stretching Objects

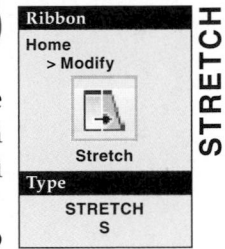

The **STRETCH** tool allows you to modify certain dimensions of an object while leaving other dimensions the same. In mechanical drafting, for example, you can stretch a screw body to create a longer or shorter screw. In architectural design, you can stretch room sizes to increase or decrease square footage.

Once you access the **STRETCH** tool, you must use a crossing box or polygon to select only the objects to be stretched. This is a very important requirement and is different from selection using other editing tools. See **Figure 12-24.** If you select using the pick box or a window, the **STRETCH** tool works like the **MOVE** tool, described in Chapter 14.

After selecting the objects to stretch, specify the *base point* from which the objects will stretch. Although the position of the base point is often not critical, you may want to select a point on an object, the corner of a view, or the center of a circle. As you move the crosshairs, the selection stretches or compresses. Pick a second point to complete the stretch.

base point: The initial reference point AutoCAD uses when stretching, moving, copying, and scaling objects.

PROFESSIONAL TIP

Use object snap modes to your best advantage while editing. For example, to stretch a rectangle to make it twice as long, use the **Endpoint** object snap to select the endpoint of a rectangle for the base point, and another **Endpoint** object snap to select the opposite endpoint of the rectangle.

Figure 12-24.
Using the **STRETCH** tool.

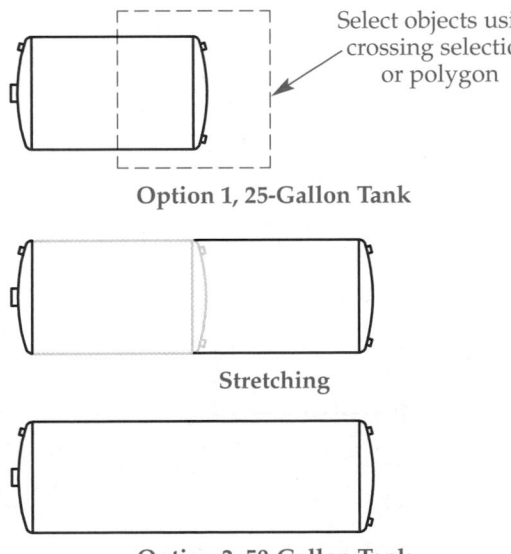

Select objects using crossing selection or polygon

Option 1, 25-Gallon Tank

Stretching

Option 2, 50-Gallon Tank

Using the Displacement Option

The **Displacement** option allows you to stretch objects relative to the origin, or 0,0,0 point. To stretch using a *displacement*, access the **STRETCH** tool and use a crossing box or polygon to select only the objects to stretch. Then select the **Displacement** option instead of defining the base point. At the Specify displacement <0,0,0>: prompt, enter an absolute coordinate to stretch the objects from the origin to the coordinate point. See **Figure 12-25.**

Using the First Point As Displacement

Another method for stretching an object is to use the first point as the displacement. This means the coordinates you use to select the base point automatically define the coordinates for the direction and distance for stretching the object. To apply this technique, access the **STRETCH** tool and use a crossing box or polygon to select only the objects to stretch. Then specify the base point, and instead of locating the second point, right-click or press [Enter] or the space bar to accept the <use first point as displacement> default. See **Figure 12-26.**

Figure 12-25.
Using the **Displacement** option of the **STRETCH** tool. A—An example of a 1 × 1 rectangle to stretch. B—Stretching the rectangle using a 1,0 displacement.

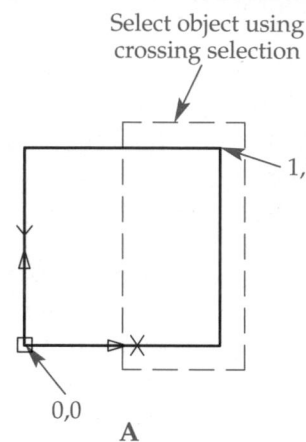
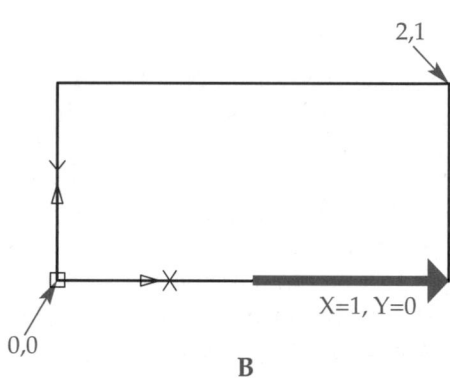

Figure 12-26.
A—An example of a 1 × 1 rectangle to stretch. B—Stretching using the selected base point (1,1) as the displacement.

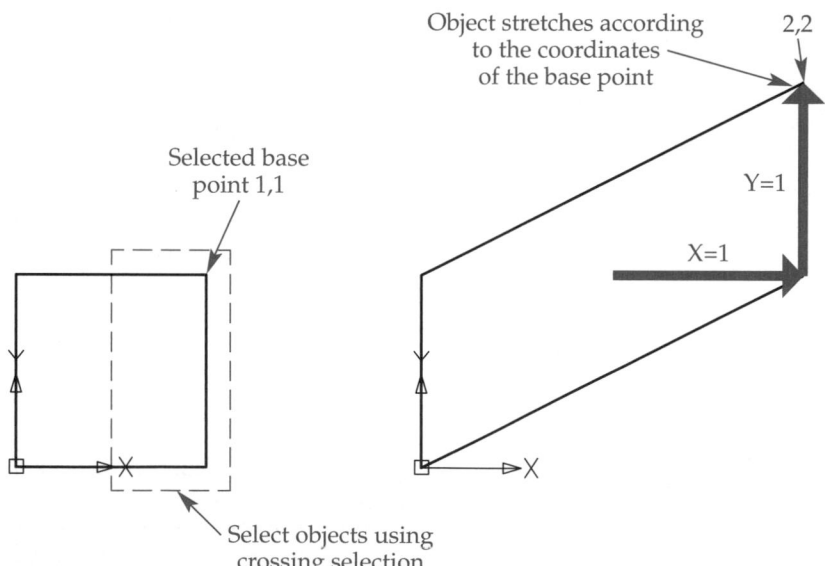

Exercise 12-7

Access the Student Web site (www.g-wlearning.com/CAD) and complete Exercise 12-7.

Using the LENGTHEN Tool

You can use the **LENGTHEN** tool to change the length of objects and the included angle of an arc. The **LENGTHEN** tool does not affect closed objects. For example, you can lengthen a line, a polyline, an arc, an elliptical arc, or a spline, but you cannot lengthen a closed ellipse, polygon, or circle. You can only lengthen one object at a time.

Once you access the **LENGTHEN** tool, select the object to change. AutoCAD gives you the current length if the object is linear or the included angle if the object is an arc. Choose one of the four options and follow the prompts. The **DElta** option allows you to specify a positive or negative change in length, measured from the endpoint of the selected object. The lengthening or shortening happens closest to the selection point and changes the length by the amount entered. See **Figure 12-27**. The **DElta** option has an **Angle** function that lets you change the included angle of an arc according to a specified angle. See **Figure 12-28**.

The **Percent** option allows you to change the length of an object or the angle of an arc by a specified percentage. The original length is 100 percent. Make the object shorter by specifying less than 100 percent or longer by specifying more than 100 percent. See **Figure 12-29**.

The **Total** option allows you to set the total length or angle of the object after the **LENGTHEN** operation. See **Figure 12-30.** The **DYnamic** option lets you drag the endpoint of the object to the desired length or angle using the crosshairs. See **Figure 12-31**. It is

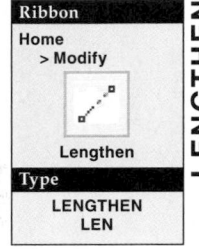

Ribbon
Home
> Modify

Lengthen

Type
LENGTHEN
LEN

LENGTHEN

Figure 12-27.
Using the **DElta** option of the **LENGTHEN** tool with values of .75 and –.75.

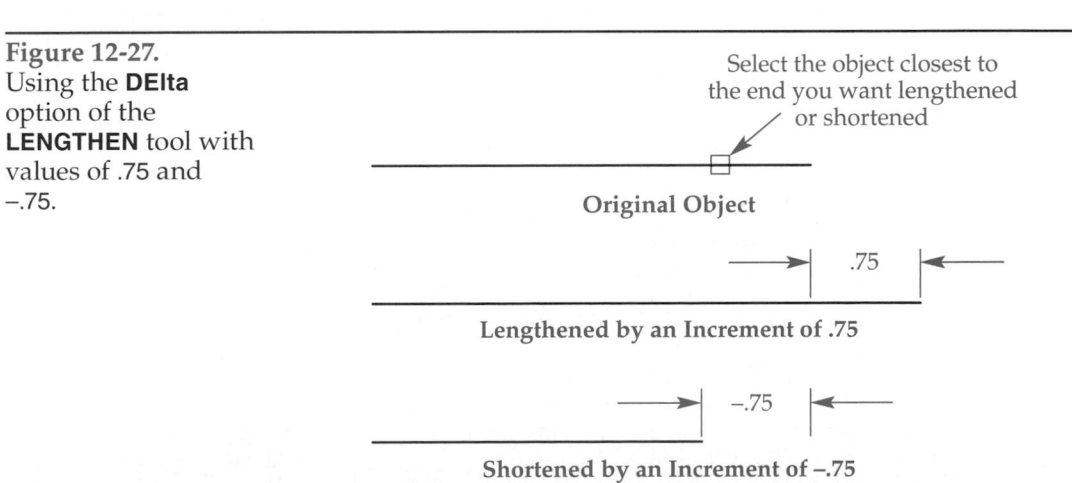

Select the object closest to the end you want lengthened or shortened

Original Object

.75

Lengthened by an Increment of .75

–.75

Shortened by an Increment of –.75

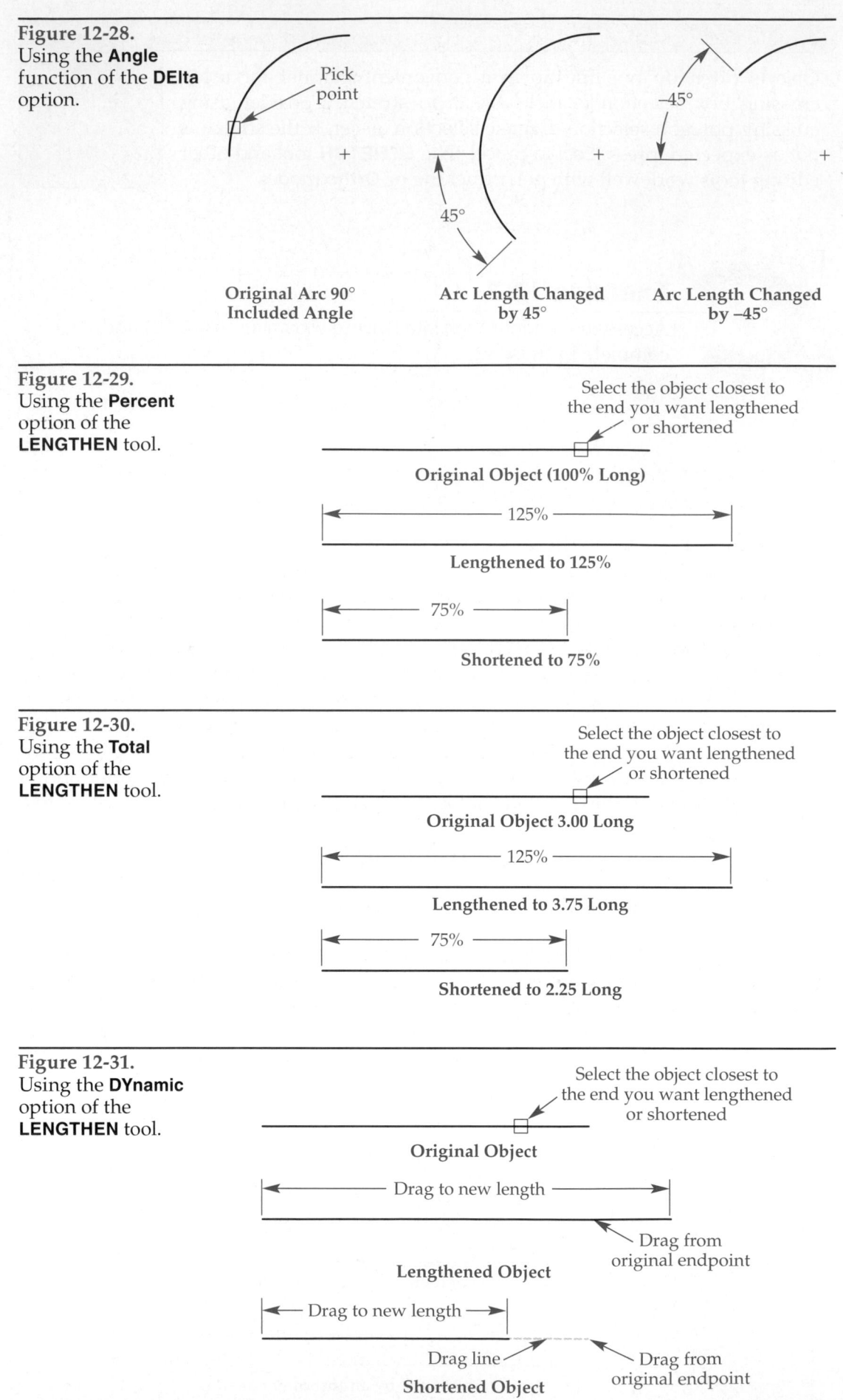

Figure 12-28.
Using the **Angle** function of the **DElta** option.

Pick point

45°

−45°

Original Arc 90° Included Angle

Arc Length Changed by 45°

Arc Length Changed by −45°

Figure 12-29.
Using the **Percent** option of the **LENGTHEN** tool.

Select the object closest to the end you want lengthened or shortened

Original Object (100% Long)

125%

Lengthened to 125%

75%

Shortened to 75%

Figure 12-30.
Using the **Total** option of the **LENGTHEN** tool.

Select the object closest to the end you want lengthened or shortened

Original Object 3.00 Long

125%

Lengthened to 3.75 Long

75%

Shortened to 2.25 Long

Figure 12-31.
Using the **DYnamic** option of the **LENGTHEN** tool.

Select the object closest to the end you want lengthened or shortened

Original Object

Drag to new length

Lengthened Object

Drag from original endpoint

Drag to new length

Drag line

Drag from original endpoint

Shortened Object

helpful to use dynamic input with polar tracking or **Ortho** mode, or have the grid and snap set to usable increments when using this option.

Exercise 12-8

Access the Student Web site (www.g-wlearning.com/CAD) and complete Exercise 12-8.

Using the SCALE Tool

The **SCALE** tool allows you to proportionately enlarge or reduce the size of objects. After you access the **SCALE** tool, pick a base point to define where the increase or decrease in size occurs. The selected objects move away from or toward the base point during scaling. The next step is to specify the scale factor. Enter a number to indicate the amount of enlargement or reduction. For example, to make the selection twice the current size, type 2 at the Specify scale factor or [Copy/Reference] <*current*>: prompt, as shown in **Figure 12-32. Figure 12-33** provides examples of scale factors.

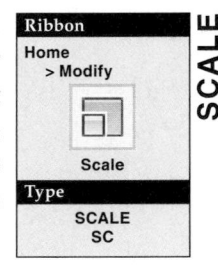

Ribbon
Home
> Modify

Scale

Type
SCALE
SC

SCALE

Using the Reference Option

You can use the **Reference** option instead of entering a scale factor by specifying a new size in relation to an existing dimension. For example, suppose you want to proportionately change the size of a part with an overall dimension of 2.50″ to an overall dimension of 3.00″. Access the **Reference** option and enter the current length, in this case 2.5, at the Specify reference length: prompt. Next, enter the length you want the object to be, in this example 3. See **Figure 12-34.**

Figure 12-32.
Using the **SCALE** tool. The base point does not move, but every other point in the object does.

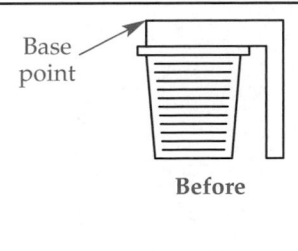

Base point

Before

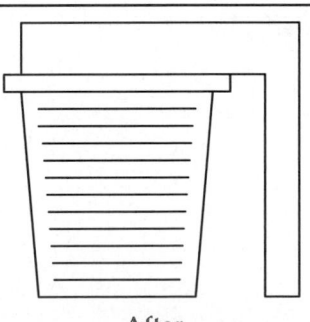

After

Figure 12-33.
Scale factors and the resulting sizes.

Scale Factor	Resulting Size
10	10 times bigger
5	5 times bigger
2	2 times bigger
1	Equal to existing size
.75	3/4 of original size
.50	1/2 of original size
.25	1/4 of original size

Figure 12-34.
Using the **Reference** option of the **SCALE** tool.

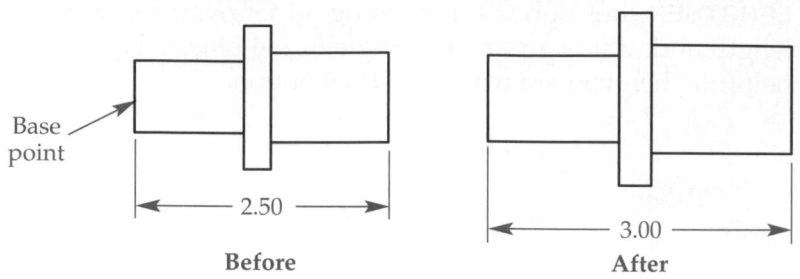

Base point

2.50

Before

3.00

After

Specify the current and reference lengths using specific values; or choose points, often on existing objects. Picking points is especially effective when you do not know the exact current and reference lengths.

Copying While Scaling

The **Copy** option of the **SCALE** tool copies and scales the selected object, leaving the original object unchanged. The copy is moved to a location you specify.

The **SCALE** tool changes all dimensions of an object proportionately. Use the **STRETCH** or **LENGTHEN** tool to change only the length, width, or height.

Exercise 12-9

Access the Student Web site (www.g-wlearning.com/CAD) and complete Exercise 12-9.

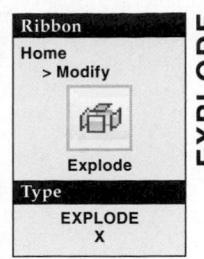
Exploding Objects

The **EXPLODE** tool allows you to change a single object that consists of multiple items into a series of individual objects. For example, you can explode a polyline object into individual line and arc segments that you can edit individually. Explode a multi-line text object to convert each line of text to a single-line text object. You can explode a variety of other objects, including multilines, regions, dimensions, leaders, and blocks. This textbook explains these objects when appropriate.

Access the **EXPLODE** tool, pick the object to explode, and right-click or press [Enter] or the space bar to cause the explosion. **Figure 12-35** shows an example of exploding a polyline object. In this example, the polyline becomes two collinear arcs with no polyline width or tangency information. Exploded polyline lines and arcs occur along the centerline of the original polyline.

Exercise 12-10

Access the Student Web site (www.g-wlearning.com/CAD) and complete Exercise 12-10.

Figure 12-35.
Exploding a polyline converts the object into individual lines and arcs and removes all polyline information.

Original Polyline   Exploded Polyline

Chapter Test

Answer the following questions. Write your answers on a separate sheet of paper or go to the Student Web site (www.g-wlearning.com/CAD) and complete the electronic chapter test.

1. How do you specify the size of a fillet?
2. Explain how to set the radius of a fillet to .50.
3. Which option of the **CHAMFER** tool would you use to specify a .125 × .125 chamfer?
4. What is the purpose of the **Method** option in the **CHAMFER** tool?
5. Describe the difference between the **Trim** and **No trim** options when using the **CHAMFER** and **FILLET** tools.
6. How can you split an object in two without removing a portion?
7. In what direction should you pick points to break a portion out of a circle or arc?
8. What tool can you use to combine two collinear lines into a single line object?
9. What two requirements must be met before two arcs can be joined?
10. Which tool performs the opposite function of the **EXTEND** tool?
11. Name the tool that trims an object to a cutting edge.
12. Name the tool associated with boundary edges.
13. Name the option in the **TRIM** and **EXTEND** tools that allows you to trim or extend to an implied intersection.

14. Which panel of the ribbon contains the **TRIM**, **EXTEND**, and **STRETCH** tools?
15. List two locations drafters normally choose as the base point when using the **STRETCH** tool.
16. Define the term *displacement*, as it relates to the **STRETCH** tool.
17. Identify the **LENGTHEN** tool option that corresponds to each of the following descriptions:
    A. Allows a positive or negative change in length from the endpoint.
    B. Changes a length or an arc angle by a percentage of the total.
    C. Sets the total length or angle to the value specified.
    D. Drags the endpoint of the object to the desired length or angle.
18. What tool would you use to reduce the size of an entire drawing by one-half?
19. Write the command aliases for the following tools:
    A. **CHAMFER**
    B. **FILLET**
    C. **BREAK**
    D. **TRIM**
    E. **EXTEND**
    F. **SCALE**
    G. **LENGTHEN**
20. Which tool removes all width characteristics and tangency information from a polyline?

Drawing Problems

Follow these instructions to complete the drawing problems for this chapter:

- *Start AutoCAD if it is not already started.*

- *Start a new drawing using an appropriate template of your choice. The template should include layers and text styles when necessary for drawing the given objects.*

- *Add layers and text styles as needed. Draw all objects using appropriate layers and text styles, justification, and format.*

- *Follow the specific instructions for each problem. Do not draw dimensions. Use your own judgment and approximate dimensions when necessary.*

▼ Basic

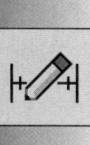

1. Draw Object A using the **LINE** and **ARC** tools. Make sure the corners overrun and the arc is centered on the lines, but does not touch the lines. Use the **TRIM**, **EXTEND**, and **STRETCH** tools to make Object B. Save the drawing as P12-1.

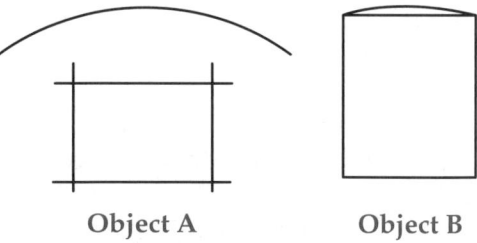

Object A   Object B

2. Open drawing P12-1 for further editing (Object A). Using the **STRETCH** tool, change the shape to create Object B. Save the drawing as P12-2.

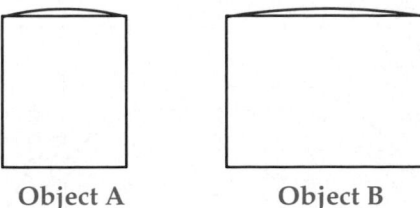

Object A   Object B

3. Refer to **Figure 12-24** in this chapter. Draw the object shown in Option 1. Stretch the object to twice its length, as shown in Option 2. Stretch the object again to one and a half times its length. *Hint:* use endpoint and midpoint object snap modes to stretch accurately. Save the drawing as P12-3.

4. Draw the object shown. Use the **CHAMFER** tool to create the inclined surface. Save the drawing as P12-4.

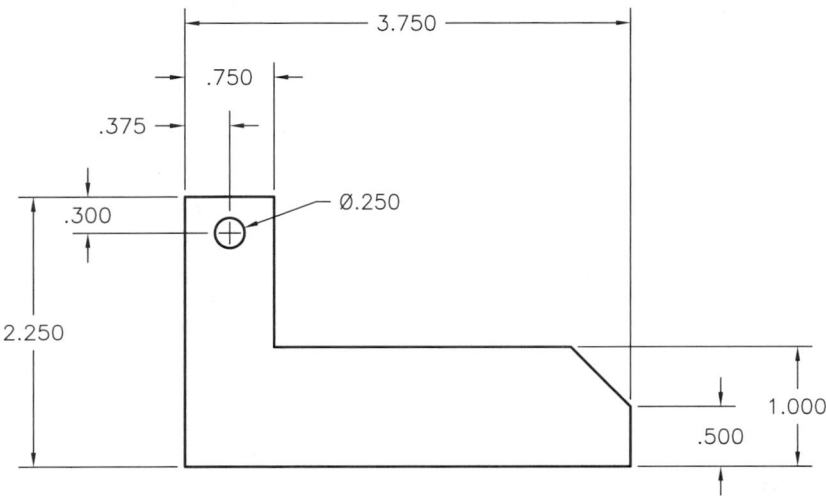

▼ Intermediate

5. Draw the object shown using the **ELLIPSE** and **LINE** tools. Use the **BREAK** or **TRIM** tool when drawing and editing the lower ellipse. Save the drawing as P12-5.

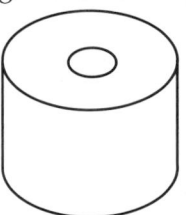

6. Open drawing P12-4 for further editing. Shorten the height of the object using the **STRETCH** tool, as shown below. Next, add to the object as indicated. Save the drawing as P12-6.

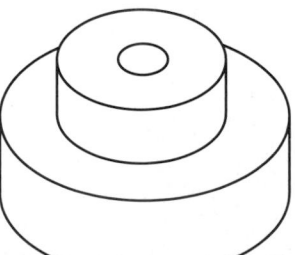

7. Use the **TRIM** and **OFFSET** tools to assist you in drawing this object. Do not draw centerlines or dimensions. Save the completed drawing as P12-7.

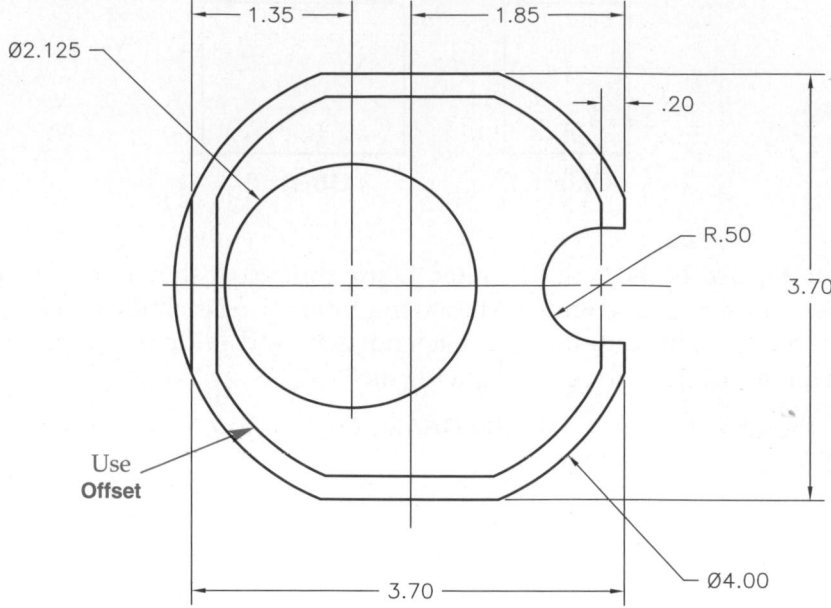

8. Draw the following plate. Use the **FILLET** tool where appropriate. Save the drawing as P12-8.

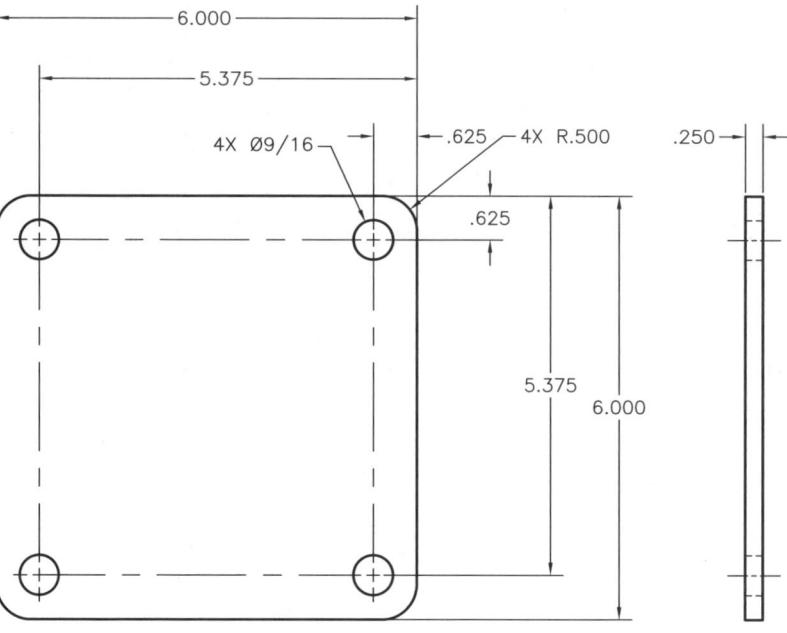

9. Draw the following toilet. Do not include dimensions. Use dimensions of your choice for objects that are not fully dimensioned. Save the drawing as P12-9.

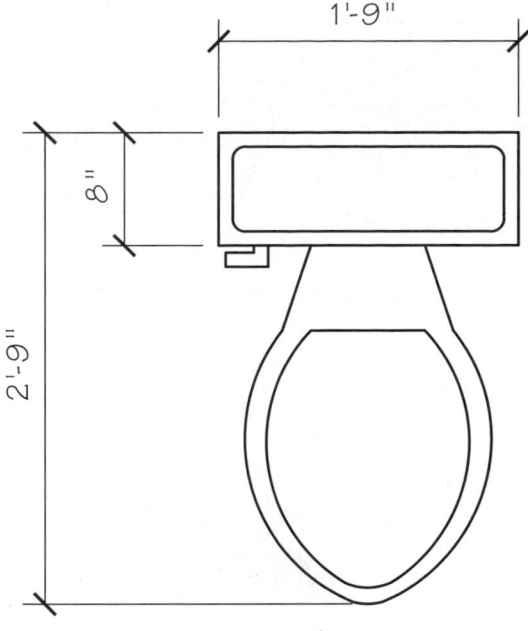

▼ Advanced

10. Draw the following object. Add rounds using the **FILLET** tool and chamfers using the **CHAMFER** tool. Use the trim mode setting to your advantage. Save the drawing as P12-10.

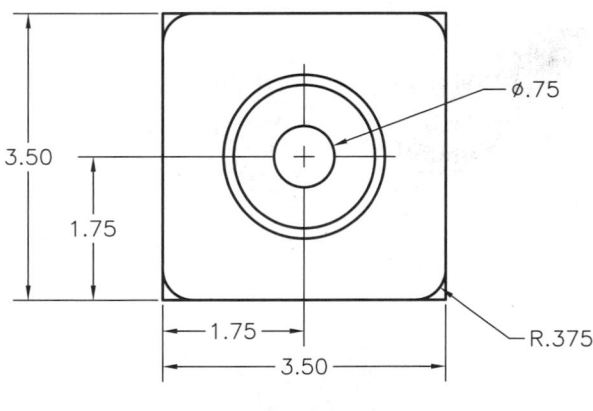

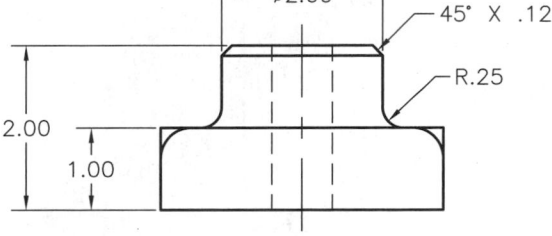

11. Draw the following bracket. Use the **FILLET** tool where appropriate. Save the drawing as P12-11.

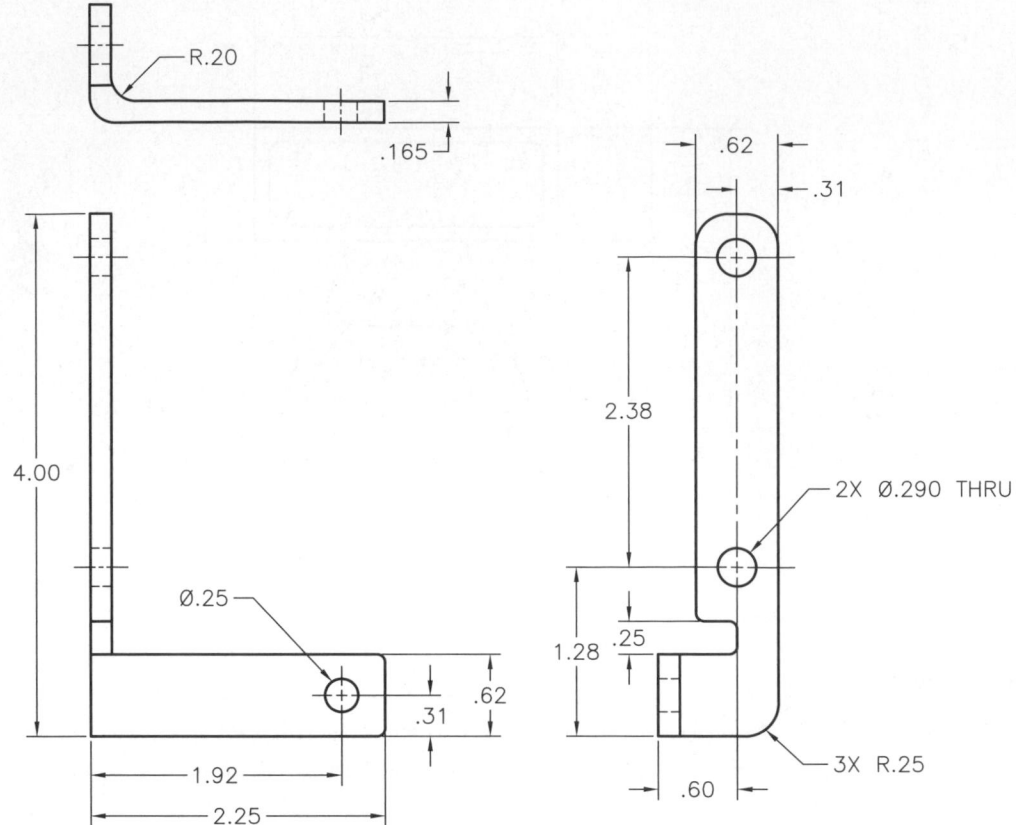

12. Draw the beam wrap detail shown. Save the drawing as P12-12.

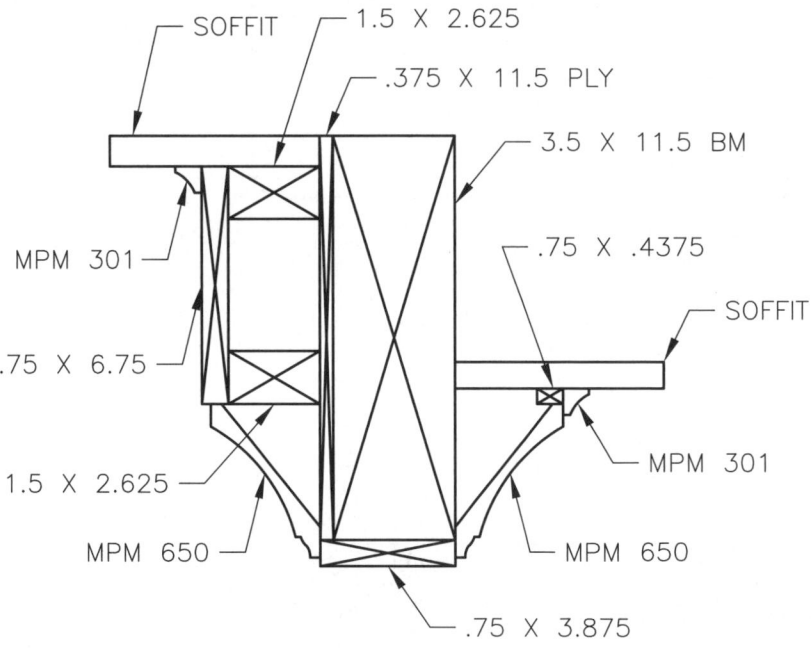

Drawing Problems - Chapter 12

Polyline and Spline Editing Tools

Learning Objectives

After completing this chapter, you will be able to do the following:
- ✓ Edit polylines with the **PEDIT** tool.
- ✓ Create polyline boundaries.
- ✓ Edit splines with the **SPLINEDIT** tool.
- ✓ Convert polylines and splines.

You can modify polyline and spline objects using standard editing tools such as **ERASE**, **STRETCH**, and **SCALE**. In addition to normal editing practices, you can modify polylines using the **PEDIT** tool and adjust splines using the **SPLINEDIT** tool. This chapter also describes how to create polyline boundaries and explores additional options for converting polylines and splines.

Using the PEDIT Tool

Access the **PEDIT** tool and select the polyline to edit, or activate the **Multiple** option to edit multiple polylines. To select a wide polyline, pick the edge of a polyline segment rather than the center. Choose from the list of options to activate the appropriate editing function.

You can also use the **PEDIT** tool to convert a line, arc, or spline into a polyline. Access the **PEDIT** tool and select the object to convert. A prompt asks you if you want to turn the object into a polyline. Select the **Yes** option to make the conversion. Once the object is converted, the **PEDIT** tool continues normally.

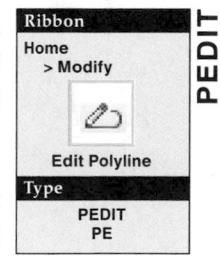

Ribbon
Home
> Modify

Edit Polyline

Type
PEDIT
PE

PEDIT

NOTE

You can also access the **PEDIT** tool by selecting a polyline, right-clicking, and choosing **Polyline Edit**.

Opening and Closing Polylines

The **Open** and **Close** options allow you to close an open polyline or open a closed polyline, as shown in **Figure 13-1.** The **Open** option is unavailable if you closed the polyline by drawing the final segment manually. Instead, the **Close** option appears. The **Open** option is also available if you used the **Close** option of the **PLINE** tool. If you select an open polyline, the **Close** option appears instead of the **Open** option. Enter the **Close** option to close the polyline.

Joining Polylines

Use the **Join** option to create a single polyline object from connected but ungrouped polylines or from a polyline connected to lines or arcs. The **Join** option works only if objects connect appropriately. Segments cannot cross and there cannot be any spaces or breaks between the segments. See **Figure 13-2.** You can include the original polyline in the selection set, but it is not necessary. See **Figure 13-3.** AutoCAD automatically converts selected lines and arcs to polylines.

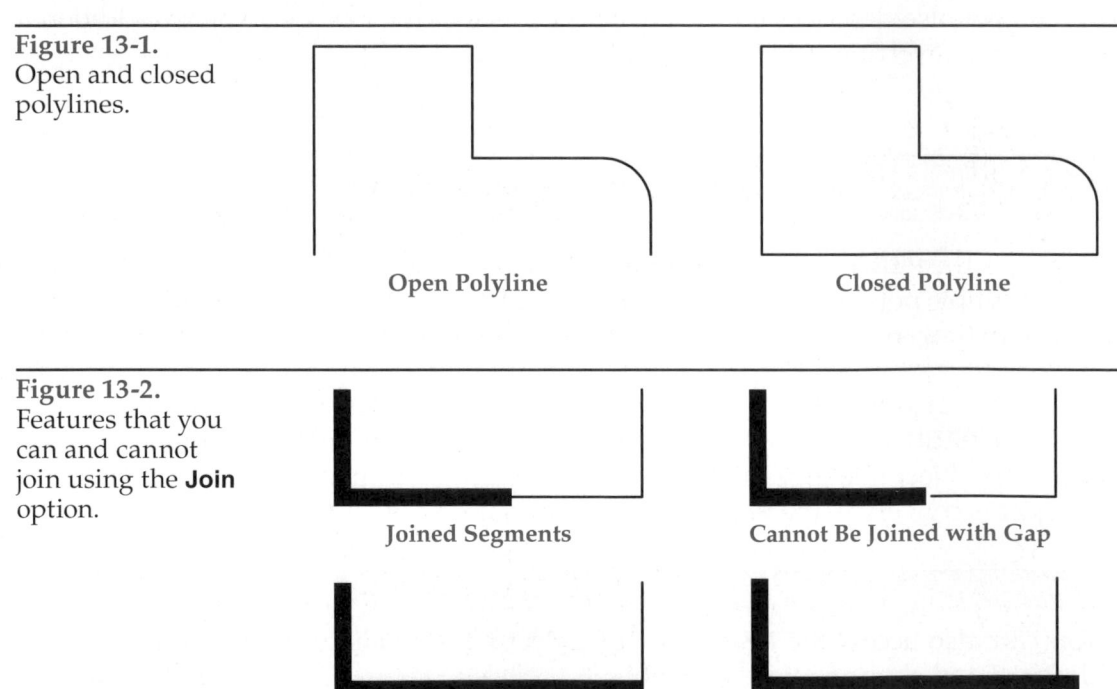

Figure 13-1.
Open and closed polylines.

Open Polyline   Closed Polyline

Figure 13-2.
Features that you can and cannot join using the **Join** option.

Joined Segments   Cannot Be Joined with Gap

Joined Segments   "T" Crossing Cannot Be Joined

Figure 13-3.
Joining a polyline
to other connected
lines and arcs.

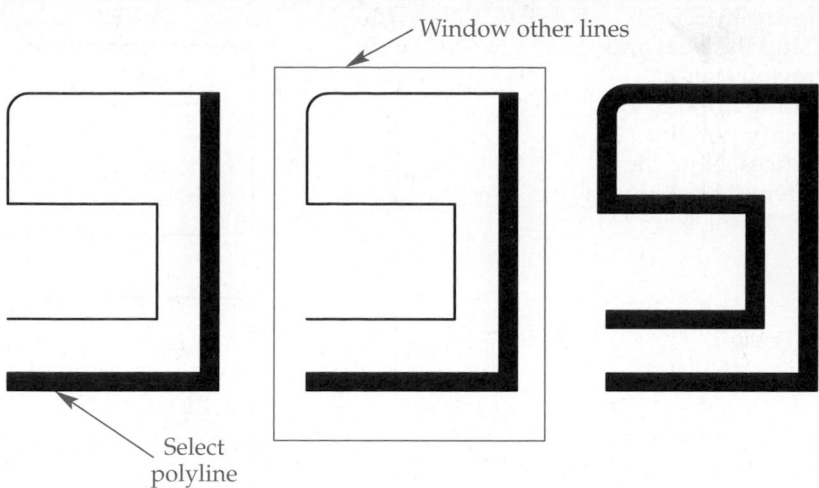

Window other lines

Select
polyline

Changing Polyline Width

The **Width** option allows you to assign a new width to a polyline or donut. The width of the original polyline can be constant, or it can vary, but *all* segments change to the constant width you specify. See **Figure 13-4.**

Exercise 13-1

Access the Student Web site (www.g-wlearning.com/CAD) and complete Exercise 13-1.

Editing a Polyline Vertex or Point of Tangency

The **Edit vertex** option allows you to edit a *polyline vertex* and a *point of tangency*. When you enter the **Edit vertex** option, an "X" marker appears on-screen at the first polyline vertex or point of tangency. The **Edit vertex** option contains several functions that affect only the point identified by the "X" marker.

In **Figure 13-5,** the marker is moved clockwise through the points using the **Next** option and counterclockwise using the **Previous** option. If you edit the vertices of a polyline and nothing appears to happen, use the **Regen** option to regenerate the polyline. Select the **eXit** option to return to the **PEDIT** prompt.

polyline vertex:
The point at which two straight polyline segments meet.

point of tangency:
The point at which a polyline arc meets another polyline arc or a straight polyline segment.

Making Breaks

You can use the **Break** function to break a polyline into separate polylines. Enter the **Edit vertex** option, move the "X" marker to the first vertex where you want to break the polyline, and activate the **Break** function. Move the "X" marker to the second vertex of the break, and enter the **Go** option to remove the portion of the polyline between

Figure 13-4.
Changing the width
of a polyline.

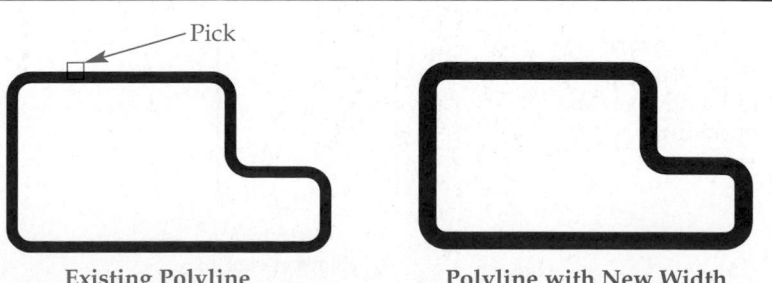

Pick

Existing Polyline   Polyline with New Width

Figure 13-5.
Using the **Next** and **Previous** vertex editing options to specify polyline vertices. Note the different positions of the "X" marker.

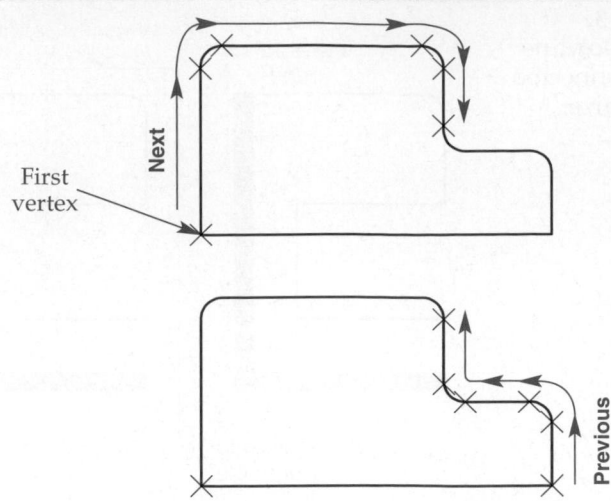

First vertex

the two points. See **Figure 13-6.** To break the polyline at a point, without removing a segment, choose **Go** without moving to a second vertex.

Inserting a New Vertex

The **Insert** function allows you to add a new vertex to a polyline. Enter the **Edit vertex** option, move the "X" marker to the appropriate point near the desired new vertex, and activate the **Insert** function. Specify the location of the new vertex on or away from an existing polyline segment. See **Figure 13-7.**

Moving a Vertex

The **Move** function allows you to move a polyline vertex to a new location. Enter the **Edit vertex** option, move the "X" marker to the vertex you want to move, and enter the **Move** function. Then, specify the new vertex location on or away from an existing polyline segment. See **Figure 13-8.**

Straightening Polyline Segments or Arcs

The **Straighten** function allows you to straighten polyline segments or arcs between two points. Enter the **Edit vertex** option, move the "X" marker to one end of

Figure 13-6.
Using the **Break** vertex editing option to break a polyline and remove a portion.

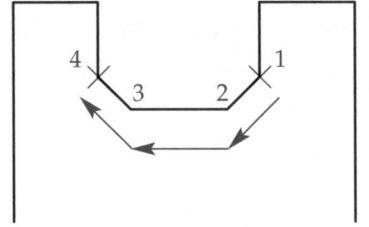

Break Points Specified

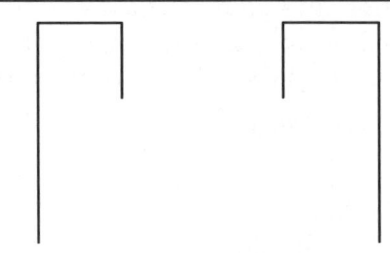

New Polylines

Figure 13-7.
Using the **Insert** vertex editing option to add a new vertex to a polyline.

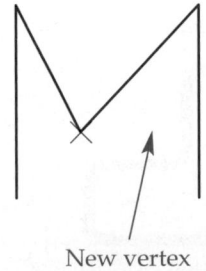

New vertex location

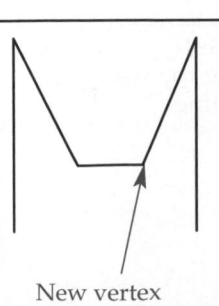

New vertex inserted

Figure 13-8.
Using the **Move** vertex editing option to move a polyline vertex to a new location.

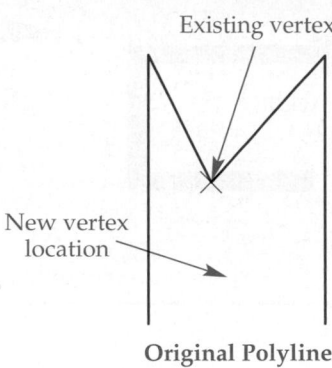
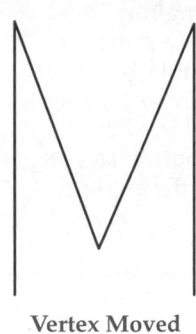

Existing vertex

New vertex location

Original Polyline   Vertex Moved

the polyline segment you want to straighten, and enter the **Straighten** function. Move the "X" marker to the other end of the segment you want to straighten, and enter the **Go** option to straighten the polyline between the two points. See **Figure 13-9.** To straighten between two consecutive vertices, choose **Go** without moving to a second vertex. This provides a quick way to straighten an arc, as shown in **Figure 13-9.**

Changing Polyline Segment Widths

You can use the **Width** function to change the starting and ending widths of a polyline segment. Enter the **Edit vertex** option and move the "X" marker to the first vertex where you want to change the polyline width. Activate the **Width** function and specify the starting and ending width of the polyline segment. See **Figure 13-10.** If nothing appears to happen to the segment, press [Enter] and use the **Regen** option of the **PEDIT** tool to regenerate the polyline.

Exercise 13-2

Access the Student Web site (www.g-wlearning.com/CAD) and complete Exercise 13-2.

Fitting a Curve to a Polyline

In some situations, you may need to convert a polyline into a series of smooth curves. A graph, for example, may show a series of plotted points as a smooth curve instead of straight segments. You can accomplish this process, known as *curve fitting*, using the **Fit** option and the **Tangent** function of the **Edit vertex** option. The **Fit** option

curve fitting: Converting a polyline into a series of smooth curves.

Figure 13-9.
The **Straighten** vertex editing option allows you to straighten polyline segments and arcs.

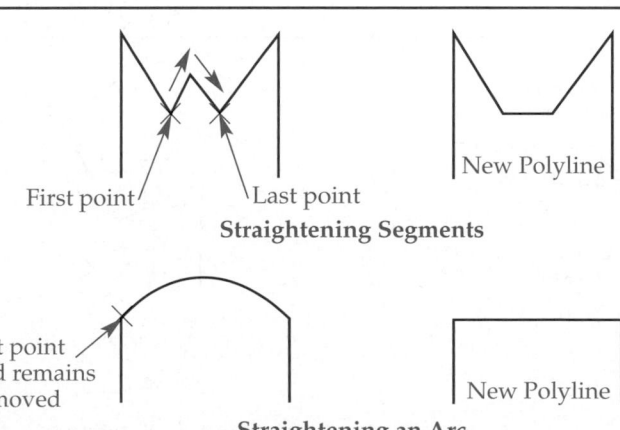

First point   Last point

New Polyline

Straightening Segments

First point marked remains unmoved

New Polyline

Straightening an Arc

Figure 13-10.
Changing the
width of a polyline
segment with the
Width vertex editing
option. Use the
Regen option to
display the change.

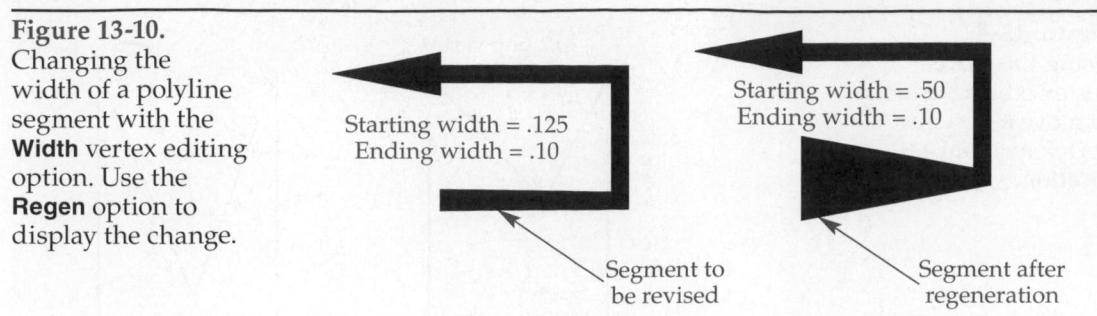

Starting width = .125
Ending width = .10

Starting width = .50
Ending width = .10

Segment to
be revised

Segment after
regeneration

fit curve: A curve
that passes through
all of its control
points.

creates a *fit curve* by constructing pairs of arcs that pass through control points. You
can specify the control points, or use the vertices of the polyline.

Prior to curve fitting, you have the option of assigning each vertex a tangent direction. AutoCAD then fits the curve based on the tangent directions you set. Specifying
tangent directions is a way to edit vertices when the **Fit** option of the **PEDIT** tool does
not produce the best results.

The **Tangent** function of the **Edit vertex** option allows you to edit tangent directions. After entering the **PEDIT** tool and the **Edit vertex** option, move the "X" marker to
the first vertex to change. Enter the **Tangent** option and specify a tangent direction in
degrees or pick a point in the expected direction. An arrow placed at the vertex then
indicates the selected direction.

Continue by moving the marker to each vertex to change, entering the **Tangent**
option for each vertex and selecting a tangent direction. Once all vertices to change
include a specified tangent direction, enter the **Fit** option to create the curve.

You can also enter the **PEDIT** tool, select a polyline, and enter the **Fit** option without
adjusting tangencies. **Figure 13-11** shows a polyline formed into a smooth curve using
the **Fit** option. If the resulting curve does not look correct, enter the **Edit vertex** option
and make changes as necessary.

Using the Spline Option

When you edit a polyline with the **Fit** option, the resulting curve passes through
each polyline vertex. The **Spline** option also smoothes the corners of a straight-segment
polyline. This option, however, creates a *spline curve* that passes through the first and
last control points or vertices only. The curve pulls toward the other vertices, but does
not necessarily pass through them. See **Figure 13-12**.

The **Spline** option creates a curve that approximates a true B-spline. You can
choose a *cubic* and *quadratic* calculation to create the curve. Like a cubic curve, a
quadratic curve passes through the first and last control points. The remainder of the

spline curve: A
curve that passes
through the first and
last control points
and is influenced
by the other control
points.

cubic curve:
A very smooth
curve created
by the **PEDIT**
Spline option with
SPLINETYPE set
at 6.

quadratic curve: A
curve created by the
PEDIT Spline option
with **SPLINETYPE**
set at 5.

Figure 13-11.
Using the **Fit** option
of the **PEDIT** tool to
turn a polyline into
a smooth curve.

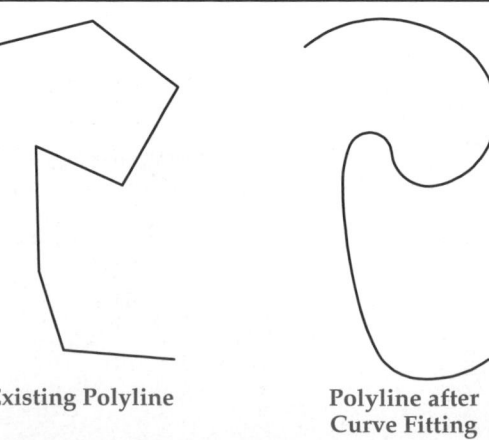

Existing Polyline

Polyline after
Curve Fitting

Figure 13-12.
A comparison of polylines edited with the **Fit** and **Spline** options of the **PEDIT** tool.

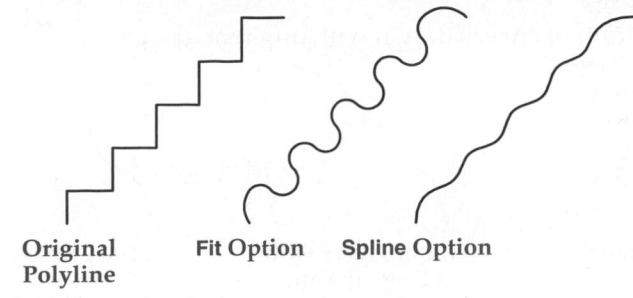

Original Polyline   Fit Option   Spline Option

Figure 13-13.
The **SPLINETYPE** system variable controls whether the **Spline** option uses a quadratic or cubic curve.

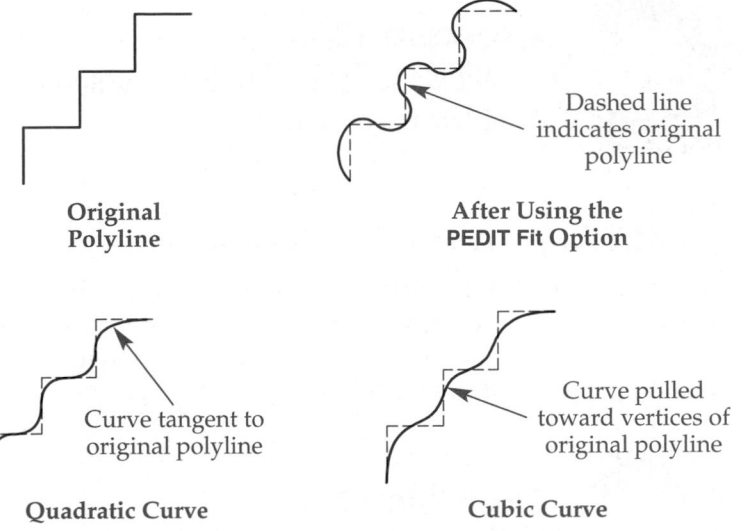

Original Polyline

After Using the PEDIT Fit Option

Dashed line indicates original polyline

Curve tangent to original polyline

Quadratic Curve

Curve pulled toward vertices of original polyline

Cubic Curve

curve is tangent to the polyline segments between the intermediate control points, as shown in **Figure 13-13.**

The **SPLINETYPE** system variable determines whether AutoCAD draws cubic or quadratic curves. The default setting is 6. At this setting, the **Spline** option of the **PEDIT** tool draws a cubic curve. Set the **SPLINETYPE** system variable to 5 to generate a quadratic curve. The only valid values for **SPLINETYPE** are 5 and 6.

You can set the number of line segments used to construct spline curves by entering a value in the **Segments in a polyline curve** text box in the **Display resolution** area of the **Display** tab of the **Options** dialog box. After changing the value, you must reissue the **Spline** option of the **PEDIT** tool and select the polyline to see the result. The default value is 8, which creates a smooth spline curve with moderate regeneration time. If you decrease the value, the resulting spline curve is less smooth. The resulting spline curve is smoother if you increase the value, but the regeneration time and drawing file size increase. See **Figure 13-14.**

NOTE

The **Fit** and **Spline** options of the **PEDIT** tool create approximations of a B-spline curve. Use the **SPLINE** tool to create a true B-spline curve.

Figure 13-14.
A comparison of curves drawn with different display resolution settings.

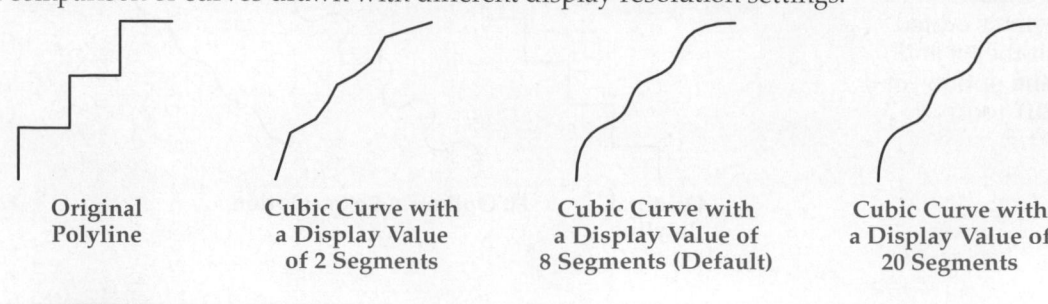

| Original Polyline | Cubic Curve with a Display Value of 2 Segments | Cubic Curve with a Display Value of 8 Segments (Default) | Cubic Curve with a Display Value of 20 Segments |

Exercise 13-3

Access the Student Web site (www.g-wlearning.com/CAD) and complete Exercise 13-3.

Straightening All Polyline Segments

The **Decurve** option returns a polyline edited with the **Fit** or **Spline** option to its original form. The information entered for tangent directions remains, however, for future reference. You can also use the **Decurve** option to straighten the curved segments of a polyline. See **Figure 13-15.**

Exercise 13-4

Access the Student Web site (www.g-wlearning.com/CAD) and complete Exercise 13-4.

Changing the Appearance of Polyline Linetypes

The **Ltype gen** (linetype generation) option determines how linetypes other than Continuous appear in relation to the vertices of a polyline. For example, if you use a Center linetype and disable the **Ltype gen** option, the polyline has a long dash at each vertex. When you activate the **Ltype gen** option, the polyline generates with a constant pattern in relation to the polyline as a whole. See **Figure 13-16.**

Figure 13-15.
The **Decurve** option allows you to straighten the curved segments of a polyline.

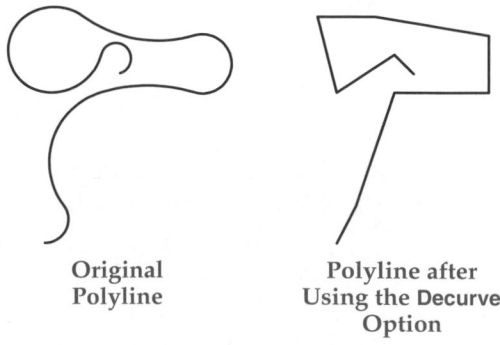

| Original Polyline | Polyline after Using the Decurve Option |

AutoCAD and Its Applications—Basics

Figure 13-16.
A comparison of polylines and splined polylines with the **Ltype gen** option of the **PEDIT** tool on and off.

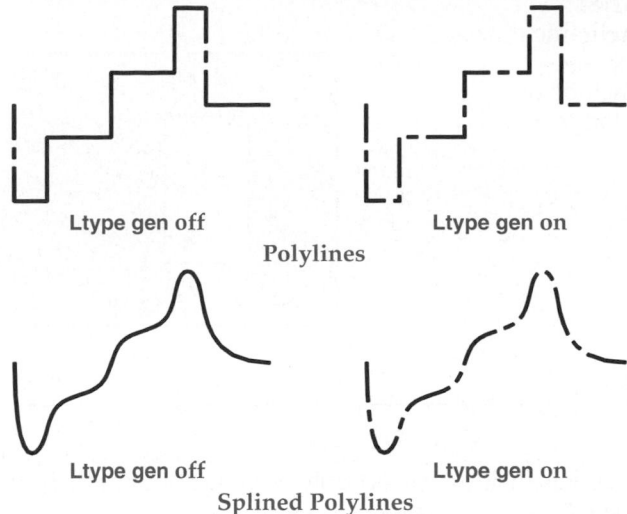

Ltype gen off   Ltype gen on

Polylines

Ltype gen off   Ltype gen on

Splined Polylines

Creating a Polyline Boundary

You can use the **BOUNDARY** tool to create a polyline boundary from line segments that form a closed area. The **Boundary Creation** dialog box appears when you access the **BOUNDARY** tool. See **Figure 13-17.**

Select the default **Polyline** option from the **Object type:** drop-down list to create a polyline around the specified area. Select **Region** from the **Object type:** drop-down list to create a *region* that you can use for area calculations, shading, extruding a solid model, and other purposes.

In the **Boundary set** drop-down list, the **Current viewport** setting is active. The **Current viewport** option defines the *boundary set* from everything visible in the current viewport, even if it is not in the current display. The **New** button allows you to define a different boundary set. The **Boundary Creation** dialog box closes and the Select objects: prompt appears allowing you to select the objects to use to create a boundary set. When you are finished, right-click or press [Enter] or the space bar. The **Boundary Creation** dialog box returns with **Existing set** active in the **Boundary set** drop-down list. This means the boundary set references the selected objects.

The **Island detection** setting specifies whether *islands* within the boundary apply as boundary objects. See **Figure 13-18.** Check **Island detection** to form separate boundaries from islands within a boundary.

When you select the **Pick Points** button, located in the upper-left corner, the **Boundary Creation** dialog box closes and the Pick internal point: prompt appears. If the

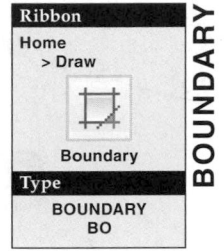

Ribbon
Home
> Draw

Boundary

Type
BOUNDARY
BO

BOUNDARY

region: A closed 2D area that can have physical properties such as centroids and products of inertia.

boundary set: The part of the drawing AutoCAD evaluates to define a boundary.

island: A closed area inside a boundary.

Figure 13-17.
The **Boundary Creation** dialog box.

Pick to create a polyline or region boundary

Check to include automatic island detection

Select the boundary set

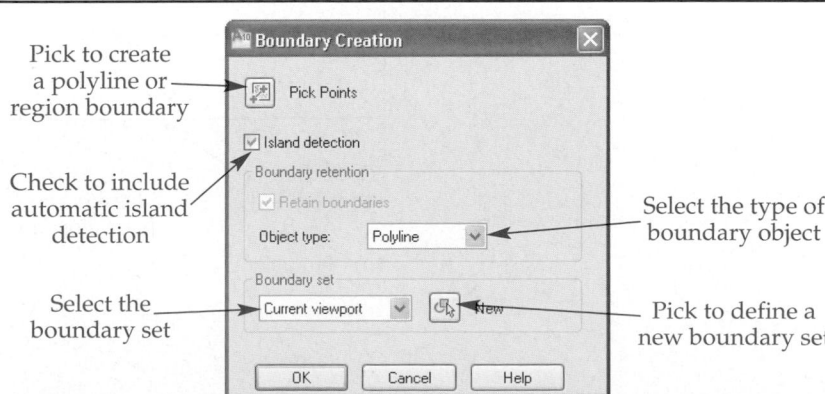

Select the type of boundary object

Pick to define a new boundary set

Figure 13-18.
When you define
a boundary set,
you can include or
exclude islands.

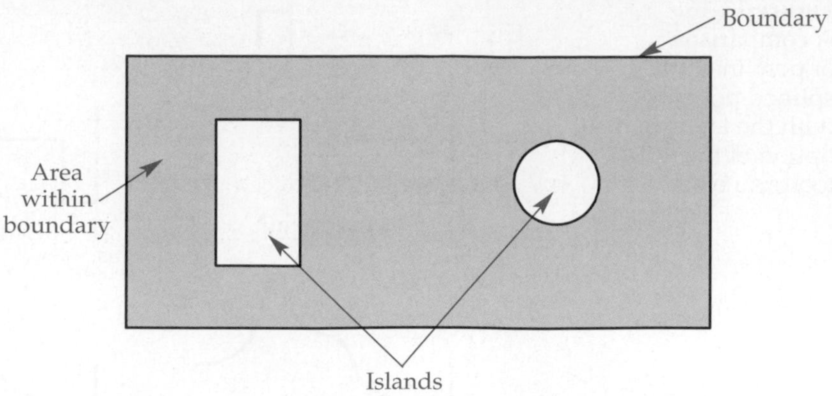

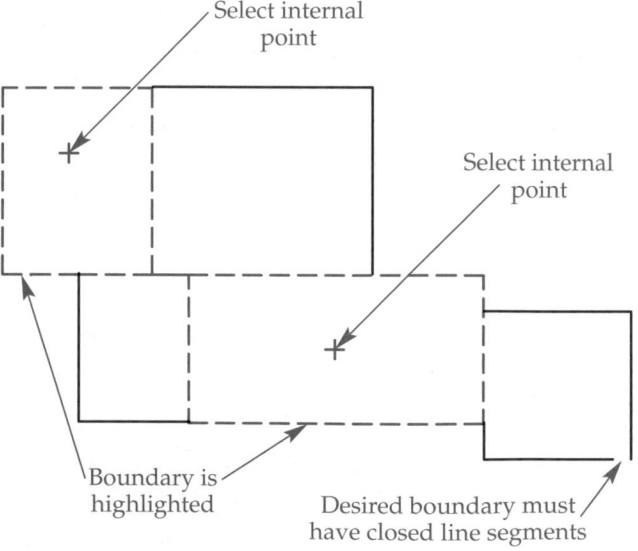

point you pick is inside a closed polygon, the boundary highlights, as shown in **Figure 13-19.** The **Boundary Definition Error** alert box appears if the point you pick is not within a closed polygon. Pick **OK** and try again.

Unlike an object created with the **Join** option of the **PEDIT** tool, a polyline boundary created with the **BOUNDARY** tool does not replace the original objects. The polyline traces over the defining objects with a polyline. The separate objects still exist underneath the newly created boundary. To avoid duplicate geometry, move the boundary to another location on the screen, erase the original objects, and then move the boundary back to its original position.

> **PROFESSIONAL TIP**
>
> You can simplify area calculations by using the **BOUNDARY** tool or by joining objects with the **Join** option of the **PEDIT** tool before issuing the **AREA** tool. To retain the original objects, explode the polyline after the area calculation if you used the **Join** option of the **PEDIT** tool. Erase the polyline boundary after the calculation if you used the **BOUNDARY** tool. Chapter 16 covers the **AREA** tool.

Figure 13-19.
When you select
a point inside a
closed polygon,
the boundary
highlights.

AutoCAD and Its Applications—Basics

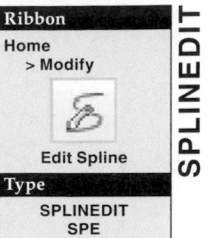

Using the SPLINEDIT Tool

Ribbon
Home
> Modify

Edit Spline
Type
SPLINEDIT
SPE

The **SPLINEDIT** tool allows you to edit spline objects. Access the **SPLINEDIT** tool and select the spline to edit. Grips identify spline control points, as shown in **Figure 13-20**. Choose from the list of options to activate the appropriate editing function.

NOTE

You can also access the **SPLINEDIT** tool by selecting a spline, right-clicking, and choosing **Spline**.

PROFESSIONAL TIP

Select the **Undo** option immediately after performing an unwanted edit to restore the previous spline without leaving the tool. Use the **Undo** option more than once to step back through each operation.

Editing Fit Data

The **Fit data** option allows you to edit spline *fit points*. The **Fit data** option has several functions. Use the **eXit** function to return to the **SPLINEDIT** option prompt. **Figure 13-21** provides examples of using fit data options.

fit points: Spline control points.

The **Add** option adds new fit points to a spline definition. You can locate a new fit point by picking a point or entering coordinates. Fit points appear as unselected grips. When you select a fit point, it highlights, along with the next spline fit point. You can then add a fit point between the two highlighted points. If you select the endpoint of the spline, only the endpoint highlights. If you select the start point, choose the **After** or **Before** option to insert the new fit point after or before the existing point. When you add a fit point, the spline curve updates according to the new point.

The **Add** option is in a running mode, which means you can continue to add points as needed. Press [Enter] or the space bar, or right-click and select **Enter** at a Specify new point <exit>: prompt to select other existing fit points and add points anywhere on the spline.

The **Delete** option deletes fit points as needed. However, at least two fit points must remain to define the spline. Like the **Add** option, the **Delete** option operates in a running mode, allowing as many deletions as needed. The spline curve updates to reflect changes made by deleting each point.

The **Move** option allows you to move fit points as necessary. The start point of the spline highlights. You can specify a different location by picking a new point. You can also specify other fit points to move. Pick the **Specify new location** option to move the

Figure 13-20.
The control points for a spline appear as grips.

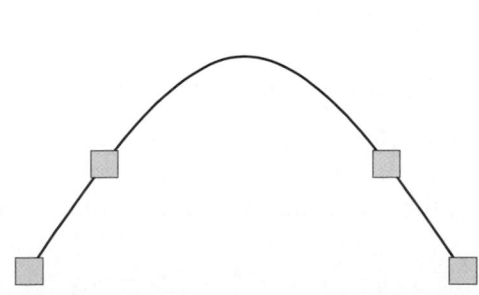

Figure 13-21.
Examples of using the **Fit data** options of the **SPLINEDIT** tool to edit a spline. Compare the original spline to each of the edited objects.

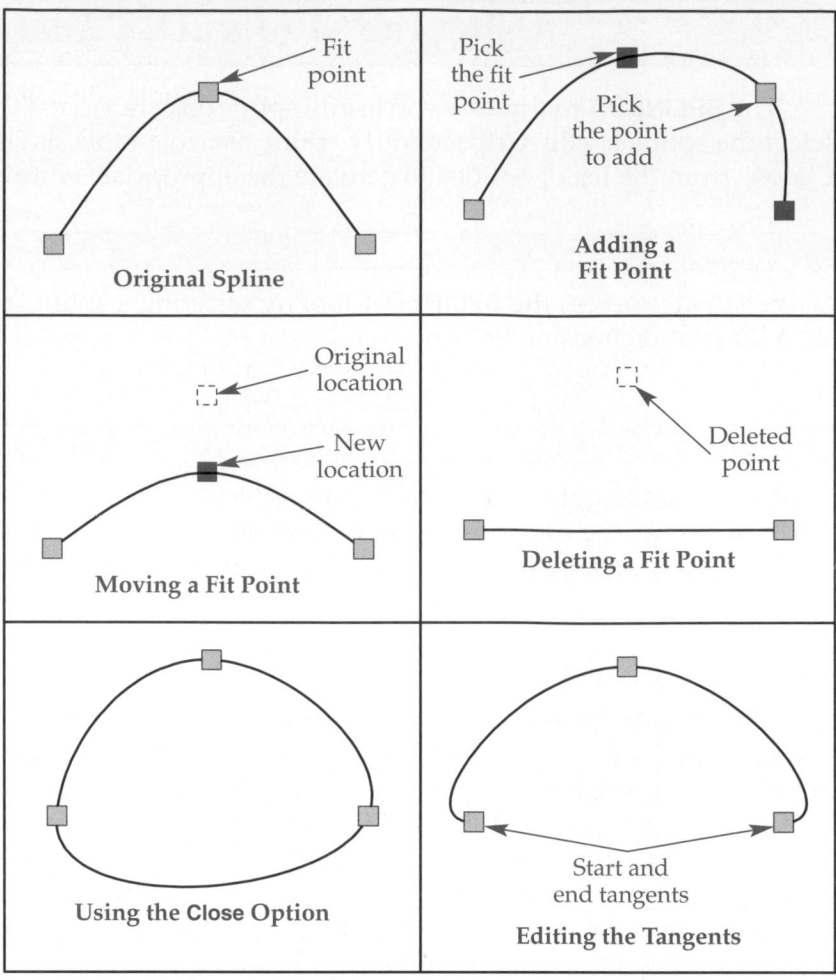

highlighted point to a new location. Select the **Next** option to highlight the next fit point, and then press [Enter]. Select the **Previous** option to highlight the previous fit point. Pick the **Select point** option to select a different fit point to move. Pick the **eXit** option to return to the **Fit data** option prompt.

The **Purge** option lets you remove fit point data from a spline. In very large drawings that contain many complex splines, purging fit point data reduces the file size by simplifying spline definitions. However, after you use this option, the resulting spline is not as easy to edit. After you purge a spline, the **Fit data** option is no longer available in the **SPLINEDIT** tool.

The **Tangents** option allows you to edit the start and end tangents for an open spline and the start tangent for a closed spline. The direction of the selected point determines the tangency. You can also use the **System default** option to set the tangency values to the AutoCAD defaults.

You can use the **toLerance** option to adjust fit tolerance values. The results are immediate, so you can adjust the fit tolerance as necessary to produce different results.

Opening or Closing a Spline

The **Open** and **Close** options are alternately displayed, depending on the status of the spline object. If the spline is open, the **Close** option displays. The **Open** option appears if the spline is closed. Use the available option to open or close the selected spline.

Moving a Vertex

The **Move vertex** option allows you to move the fit points of a spline. When you access this option, you can specify a new location for a selected fit point. The options displayed are identical to those used with the **Move** function of the **Fit data** option. You can pick a new location for the highlighted fit point, or you can enter another option. The following **Move vertex** options are available:

- **Specify new location.** Moves the currently highlighted point to a specified location.
- **Next.** Highlights the next fit point.
- **Previous.** Highlights the previous fit point.
- **Select point.** Picks a different fit point to move; provides an alternative to cycling through points with the **Next** or **Previous** functions.
- **eXit.** Returns to the **SPLINEDIT** prompt.

Exercise 13-5

Access the Student Web site (www.g-wlearning.com/CAD) and complete Exercise 13-5.

Smoothing or Reshaping a Spline Section

The **Refine** option allows you to fine-tune the spline curve and includes several functions. If necessary, use the **Add control point** option to add fit points to help smooth or reshape a section of the spline.

The **Elevate order** option causes more control points to appear on the curve for greater control and spline *order* refinement. For example, a cubic spline has an order of 4. In **Figure 13-22A,** the order of the spline is elevated from 4 to 6. You can use an order setting from 4 to 26, but you cannot adjust the value downward. For example, if the order is set to 24, the only remaining settings are 25 and 26.

order: In a spline, the degree of the spline polynomial + 1.

The **Weight** option changes the weight of a control point. When all control points have the same weight, they exert the same amount of pull on the spline. When you reduce a weight value for a control point, the spline does not pull as close to the point. When you increase a weight value, the control point exerts more pull on the spline.

Figure 13-22.
A—The effects of elevating the order of a spline. B—Increasing the weight of an individual control point.

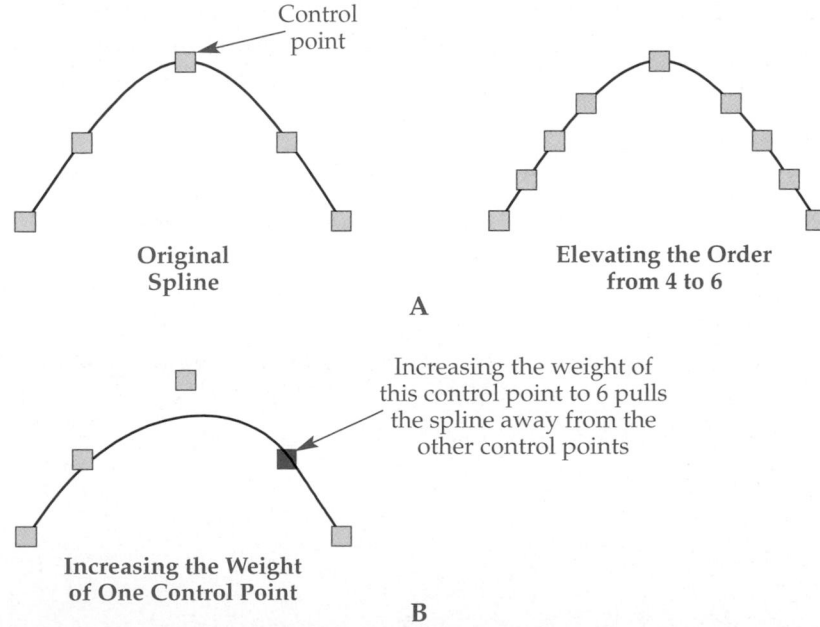

See **Figure 13-22B.** You can adjust the default setting of 1, but the weight setting must be positive. The control point selection functions of the **Weight** option are the same as those used with the **Move vertex** option of the **SPLINEDIT** tool. You can use the **Enter new weight** option to specify a new weight for the highlighted control point.

Exercise 13-6

Access the Student Web site (www.g-wlearning.com/CAD) and complete Exercise 13-6.

Converting Polylines and Splines

You can use the **Object** option of the **SPLINE** tool to convert a spline-fitted polyline object, created using the **PEDIT** tool, to a spline object. Once you access the **SPLINE** tool, activate the **Object** option instead of defining control points. Then pick a spline-fitted polyline object to convert the polyline to a spline.

The **Convert to Polyline** option of the **SPLINEDIT** tool allows you to convert a spline to a polyline. Select the spline to convert and enter a value at the Specify a precision <current>: prompt. The higher the specified precision, the more vertex points are added to the polyline, making the polyline smoother. See **Figure 13-23.**

NOTE

You can achieve different spline results by altering the specifications used with the **Fit tolerance** option. The setting specifies a tolerance within which the spline curve falls as it passes through the control points.

PROFESSIONAL TIP

A spline created by fitting a spline curve to a polyline is a linear approximation of a true spline and is not as accurate as a spline drawn using the **SPLINE** tool. An additional advantage of spline objects over smoothed polylines is that splines use less disk space.

Figure 13-23.
The effects of changing precision when converting a spline to a polyline.

| Original Spline | Converted with Default Precision of 10 | Converted with Precision of 1 |

Chapter Test

Answer the following questions. Write your answers on a separate sheet of paper or go to the Student Web site (www.g-wlearning.com/CAD) and complete the electronic chapter test.

1. Name the tool and option required to turn three connected lines into a single polyline.
2. When you enter the **Edit vertex** option of the **PEDIT** tool, where does AutoCAD place the "X" marker?
3. How do you move the "X" marker to edit a different polyline vertex?
4. Name the **Edit vertex** option of the **PEDIT** tool that relates to each definition below.
   A. Moves the "X" marker to the next position.
   B. Moves a polyline vertex to a new location.
   C. Breaks a polyline at a point or between two points.
   D. Generates the revised version of a polyline.
   E. Specifies a tangent direction.
   F. Adds a new polyline vertex.
   G. Returns to the **PEDIT** tool prompt.
5. Which **PEDIT** tool option and function allow you to change the starting and ending widths of a polyline?
6. Why might it appear that nothing happens when you change the starting and ending widths of a polyline?
7. Name the **PEDIT** tool option and function used for curve fitting.
8. Explain the difference between a fit curve and a spline curve.
9. Compare a quadratic curve, cubic curve, and fit curve.
10. Which **SPLINETYPE** system variable setting allows you to draw a quadratic curve?
11. Explain how you can adjust the way polyline linetypes are generated using the **PEDIT** tool.
12. Name the tool used to create a polyline boundary.
13. Name the tool that allows you to edit splines.
14. What is the purpose of the **Add** function of the **Fit data** option of the **SPLINEDIT** tool?
15. What is the minimum number of fit points for a spline?
16. Name the **SPLINEDIT** option that allows you to move the fit points in a spline.
17. Which function of the **SPLINEDIT Fit data** option allows you to reduce file size, but also makes the resulting spline harder to edit?
18. Identify the **Refine** option of the **SPLINEDIT** tool that lets you increase, but not decrease, the number of control points appearing on a spline curve.
19. Name the **Refine** function of the **SPLINEDIT** tool that controls the pull exerted by a control point on a spline.
20. Name the **SPLINE** tool option that allows you to turn a spline-fitted polyline into a true spline.

Drawing Problems

Follow these instructions to complete the drawing problems for this chapter:

- *Start AutoCAD if it is not already started.*
- *Start a new drawing using an appropriate template of your choice. The template should include layers and text styles necessary for drawing the given objects.*
- *Add layers and text styles as needed. Draw all objects using appropriate layers and text styles, justification, and format.*
- *Follow the specific instructions for each problem. Do not draw dimensions. Use your own judgment and approximate dimensions when necessary.*

▼ Basic

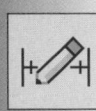

1. Use the **LINE** tool to draw two connected lines. Use the **PEDIT** tool to convert one of lines to a polyline, and then use the **Join** option to convert the line and polyline into a single polyline object. Use the **LINE** tool to draw a rectangle. Use the **PEDIT** tool to convert one of lines to a polyline, and then use the **Join** option to convert the three remaining lines and polyline into a single polyline object. Save the completed drawing as P13-1.

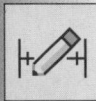

2. Use the **SPLINE** tool to draw a spline of your own design with at least four control points. Use the **Convert to Polyline** option of the **SPLINEDIT** tool to convert the spline to a polyline. Save the completed drawing as P13-2.

▼ Intermediate

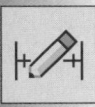

3. Open drawing P4-17 and save it as P13-3. In the P13-3 drawing file, use the **PEDIT** tool to change the object into a rectangle. Use the **Decurve** and **Width** options and the **Straighten**, **Insert**, and **Move** vertex editing options of the **PEDIT** tool. Save the completed drawing as P13-3.

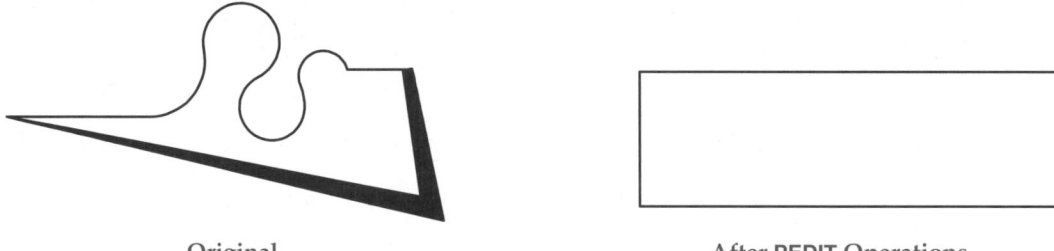

Original	After **PEDIT** Operations

4. Open drawing P4-18 and make the following changes. Save the drawing as P13-4.
   A. Combine the two polylines using the **Join** option of the **PEDIT** tool.
   B. Change the beginning width of the left arrow to 1.0 and the ending width to .2.
   C. Draw a polyline .062 wide, similar to Line A, as shown.

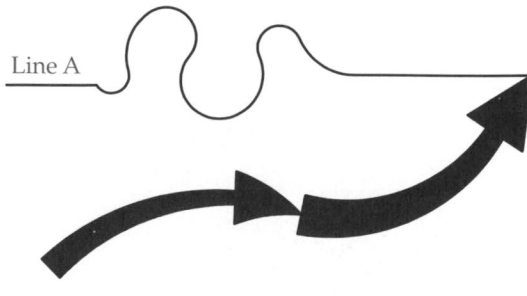

Line A

5. Draw four polylines .032 wide, using the following absolute coordinates for all four objects.

Point	Coordinates	Point	Coordinates	Point	Coordinates
1	1,1	5	3,3	9	5,5
2	2,1	6	4,3	10	6,5
3	2,2	7	4,4	11	6,6
4	3,2	8	5,4	12	7,6

Leave the first polyline as drawn. Use the **Fit** option of the **PEDIT** tool to smooth the second polyline. Use the **Spline** option of the **PEDIT** tool to turn the third polyline into a quadratic curve. Make the fourth polyline into a cubic curve. Use the **Decurve** option of the **PEDIT** tool to return one of the three edited polylines to its original form. Save the drawing as P13-5.

6. Use the **PLINE** tool to draw four copies of a patio plan similar to the one shown in Example A below. Draw the house walls 6″ wide. Leave the first plan as drawn. Use the **PEDIT** tool to create the designs shown. Use the **Fit** option for Example B, a quadratic spline for Example C, and a cubic spline for Example D. Change the **SPLINETYPE** system variable as required. Save the drawing as P13-6.

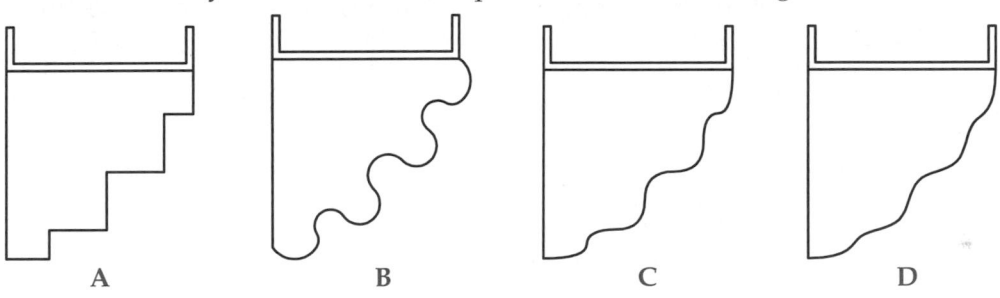

A   B   C   D

7. Open drawing P13-6 and create four new patio designs. This time, use grips to edit the polylines and create designs similar to Examples A, B, C, and D below. Save the drawing as P13-7.

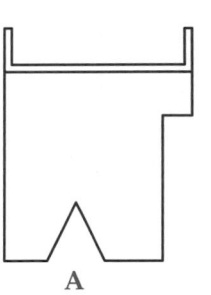 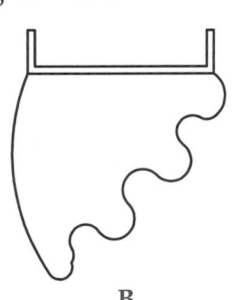 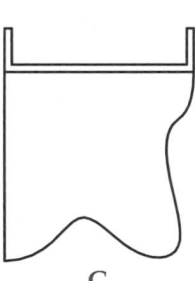 

A   B   C   D

Chapter 13   Polyline and Spline Editing Tools   367

Drawing Problems - Chapter 13

8. Draw a spline similar to the original spline shown below seven times in a layout similar to the layout shown. Perform the **SPLINEDIT** operations identified under each of the seven splines. Save the drawing as P13-8.

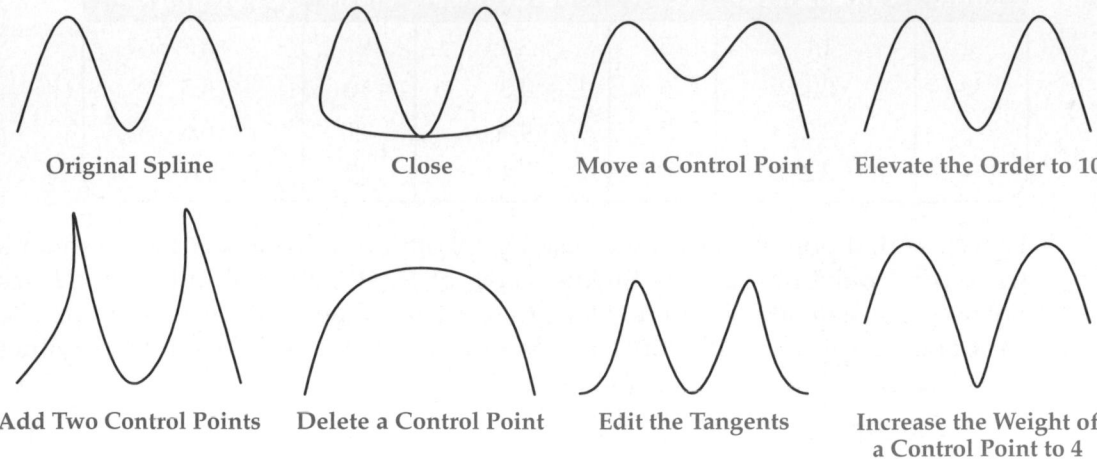

| Original Spline | Close | Move a Control Point | Elevate the Order to 10 |

| Add Two Control Points | Delete a Control Point | Edit the Tangents | Increase the Weight of a Control Point to 4 |

▼ Advanced

9. Draw the flow chart shown below. Use polylines to draw the connecting lines, arrows, and diamonds. Save the drawing as P13-9.

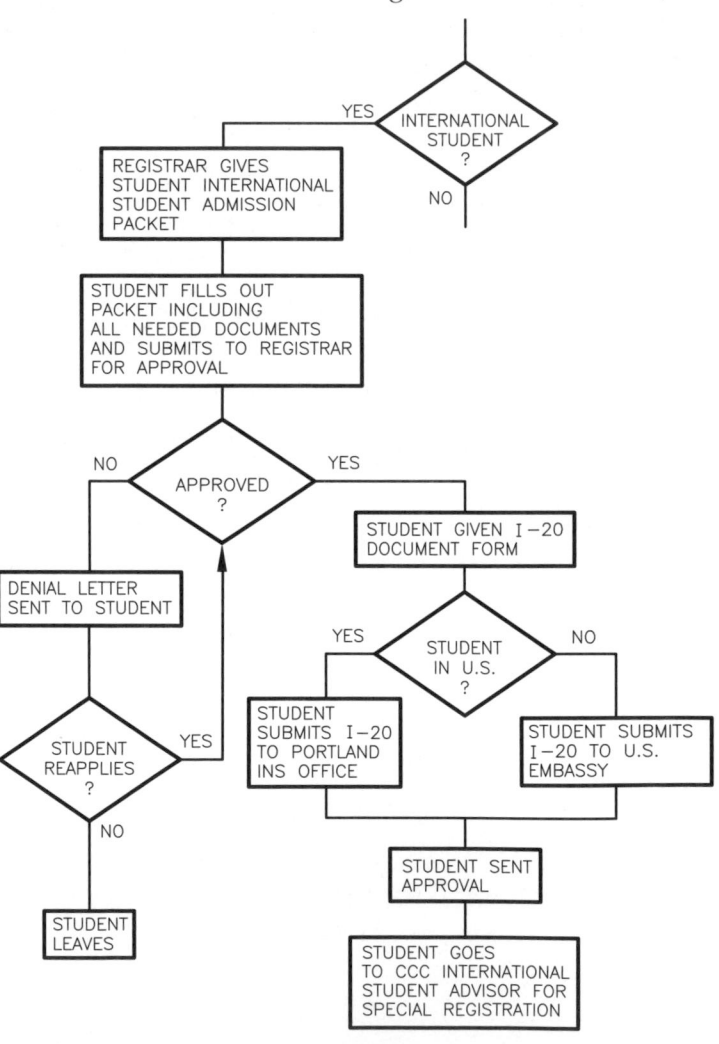

Drawing Problems - Chapter 13

10. Draw the flow chart shown below. Use polylines to draw the connecting lines, arrows, and diamonds. Save the drawing as P13-10.

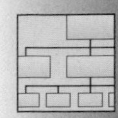

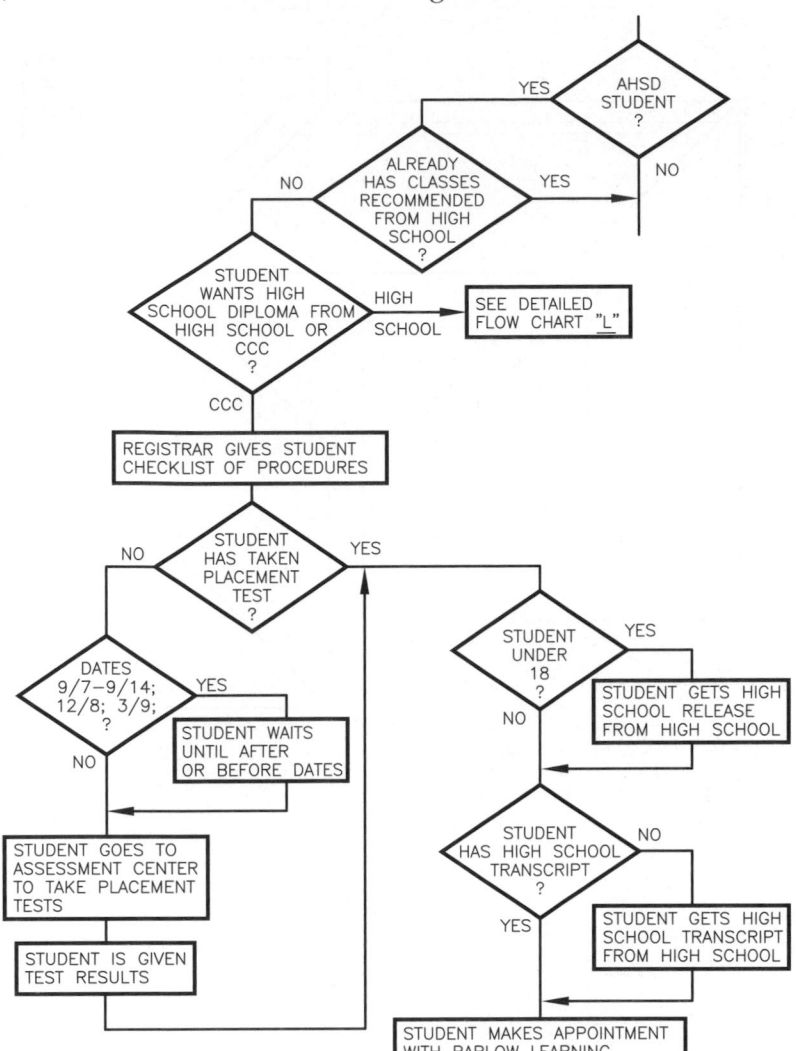

11. Draw the following roof plan. Save the drawing as P13-11.

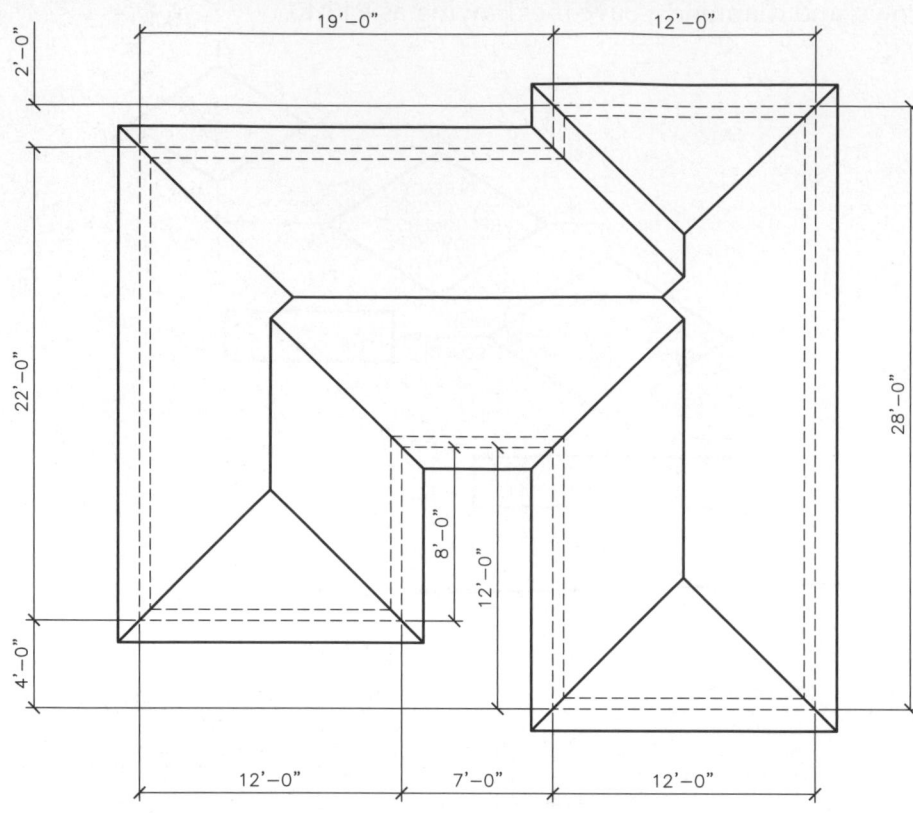

CHAPTER 14

Arranging and Patterning Objects

Learning Objectives

After completing this chapter, you will be able to do the following:

✓ Relocate objects using the **MOVE** tool.
✓ Change the angular positions of objects using the **ROTATE** tool.
✓ Use the **ALIGN** tool to simultaneously move and rotate objects.
✓ Make copies of objects using the **COPY** tool.
✓ Draw mirror images of objects using the **MIRROR** tool.
✓ Use the **REVERSE** tool.
✓ Create patterns of objects using the **ARRAY** tool.

This chapter explains methods for arranging and patterning objects using basic editing tools. You will learn to use various editing tools to increase drawing efficiency. The editing tools described in this chapter include many options. As you work through this chapter, experiment with each option to see which is the most effective in different situations.

Using the MOVE Tool

The **MOVE** tool provides an easy way for you to move a view or feature to a more appropriate location. The **MOVE** tool functions similar to the **STRETCH** tool, except that when objects move, they retain the same size and shape. Access the **MOVE** tool and select the objects to move. Proceed to the next prompt and specify the base point from which the objects will move. Though the position of the base point is often not critical, you may want to select a point on an object, the corner of a view, or the center of a circle. The selection moves as you move the crosshairs. Pick a second point to complete the move. See **Figure 14-1.**

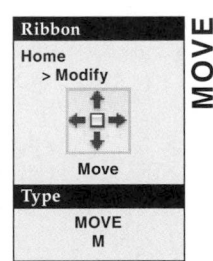

Ribbon
Home
> Modify
Move

Type
MOVE
M

MOVE

Using the Displacement Option

The **Displacement** option allows you to move objects relative to the origin, or 0,0,0 point. To move using a displacement, access the **MOVE** tool and select objects to move. Then select the **Displacement** option instead of defining the base point. At the Specify displacement <0,0,0>: prompt, enter an absolute coordinate to stretch the objects from the origin to the coordinate point. See **Figure 14-2.**

Figure 14-1.
Using the **MOVE** tool.

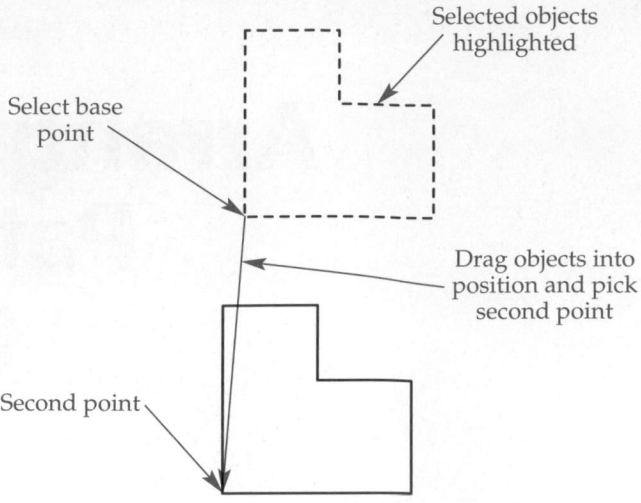

Selected objects highlighted

Select base point

Drag objects into position and pick second point

Second point

Figure 14-2.
Using the **Displacement** option of the **MOVE** tool to move objects. In this example, the origin is the base point and the absolute coordinate point 2,2 is the displacement.

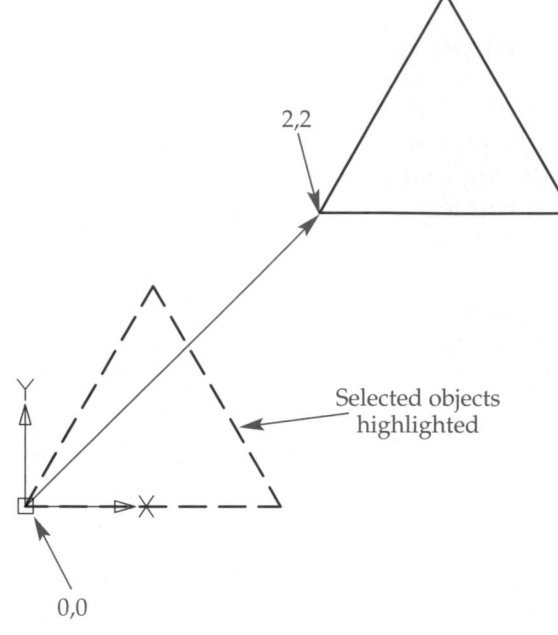

2,2

Selected objects highlighted

0,0

Using the First Point As Displacement

Another method for moving an object is to use the first point as the displacement. This means the coordinates you use to select the base point automatically define the coordinates for the direction and distance for moving the object. To apply this technique, access the **MOVE** tool and select objects to move. Then specify the base point, and instead of locating the second point, right-click or press [Enter] or the space bar to accept the <use first point as displacement> default. See **Figure 14-3.**

PROFESSIONAL TIP

Use object snap modes to your best advantage while editing. For example, to move an object to the center of a circle, use the **Center** object snap mode to select the center of the circle.

AutoCAD and Its Applications—Basics

Figure 14-3.
Moving a circle using the selected base point, 1,1 in this example, as the displacement.

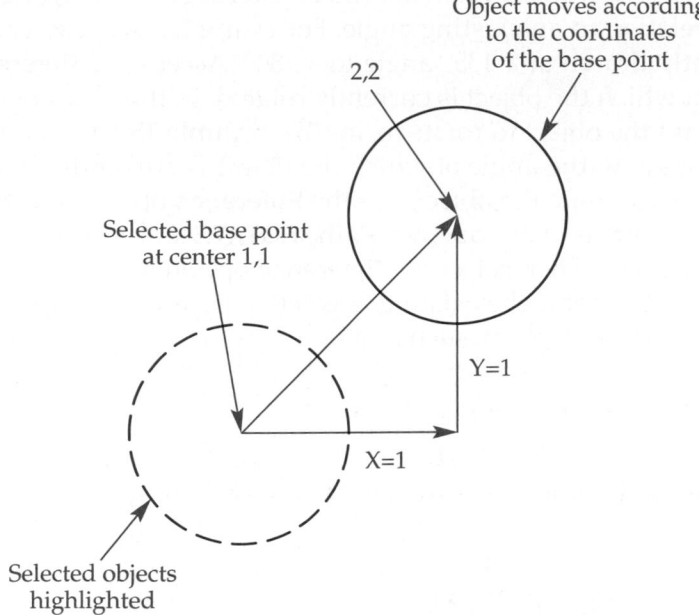

Object moves according
to the coordinates
of the base point

2,2

Selected base point
at center 1,1

Y=1

X=1

Selected objects
highlighted

Exercise 14-1

Access the Student Web site (www.g-wlearning.com/CAD) and
complete Exercise 14-1.

Rotating Objects

Design changes often require you to rotate an object, feature, or view. For example,
you may have to rotate the furniture in an office layout for an interior design. Use the
ROTATE tool to revise the layout to obtain the final design. Access the **ROTATE** tool
and select the objects to rotate. Proceed to the next prompt and specify the base point,
or axis of rotation, around which the objects rotate. Next, enter a rotation angle at the
Specify rotation angle or [Copy/Reference] <current>: prompt. A negative rotation angle
revolves the object clockwise. A positive rotation angle revolves the object counterclockwise. See **Figure 14-4.**

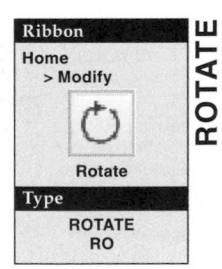

Ribbon
Home
> Modify

Rotate

Type
ROTATE
RO

ROTATE

Figure 14-4.
Rotation angles.

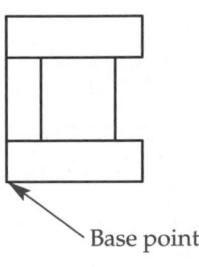

Base point

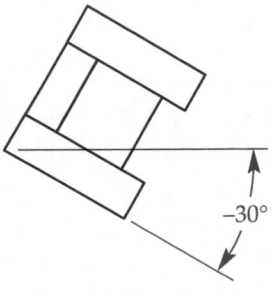

−30°

−30° Rotation

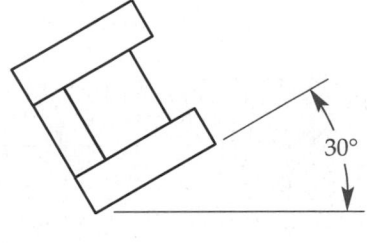

30°

30° Rotation

Using the Reference Option

You can use the **Reference** option instead of entering a rotation angle by specifying a new angle in relation to an existing angle. For example, suppose you want to rotate an object currently drawn at a 135° angle to a 180°. Access the **Reference** option and enter the angle at which the object is currently rotated, in this example 135. Next, enter the angle you want the object to rotate to, in this example 180. See **Figure 14-5A.**

If you do not know the angle at which the object is currently drawn or the angle at which you want to rotate the object, use the **Reference** option to rotate according to reference lines. To use this technique, access the **ROTATE** tool, select the objects to rotate, and pick the base point. Then select the **Reference** option and pick the two endpoints of a reference line that forms the existing angle. Finally, enter the new angle, as shown in **Figure 14-5B,** or select a point, such as a point on a correctly rotated object.

Creating a Copy While Rotating

The **Copy** option of the **ROTATE** tool copies and rotates the selected object. This option leaves the original object unchanged and rotates only the copy of the object.

Exercise 14-2

Access the Student Web site (www.g-wlearning.com/CAD) and complete Exercise 14-2.

Using the ALIGN Tool

Use the **ALIGN** tool to move and rotate an object with one operation. **ALIGN** is primarily a 3D tool, but it can be used for 2D drawings. After you access the **ALIGN** tool, select objects to align. Then, you must specify *source points* and *destination points*. Pick the first source point and then the first destination point. Next, pick the

ALIGN

| Ribbon |
| Home |
| > Modify |
| Align |
| Type |
| ALIGN |
| AL |

source points: Points to define a reference line relative to the object's original position for an **ALIGN** operation.

destination points: Points to define the location of the reference line of the object's new location in an **ALIGN** operation.

Figure 14-5.
Using the **Reference** option of the **ROTATE** tool. A—Entering reference angles. B—Selecting points on a reference line.

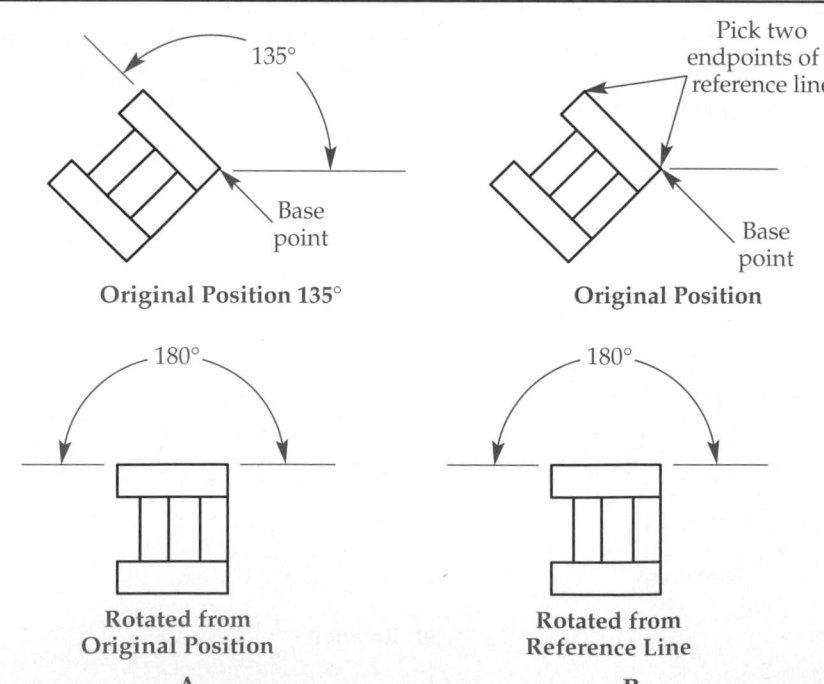

Original Position 135°

Original Position

Rotated from Original Position

Rotated from Reference Line

A

B

AutoCAD and Its Applications—Basics

Figure 14-6.
Using the **ALIGN** tool to move and rotate a kitchen cabinet layout against a wall.

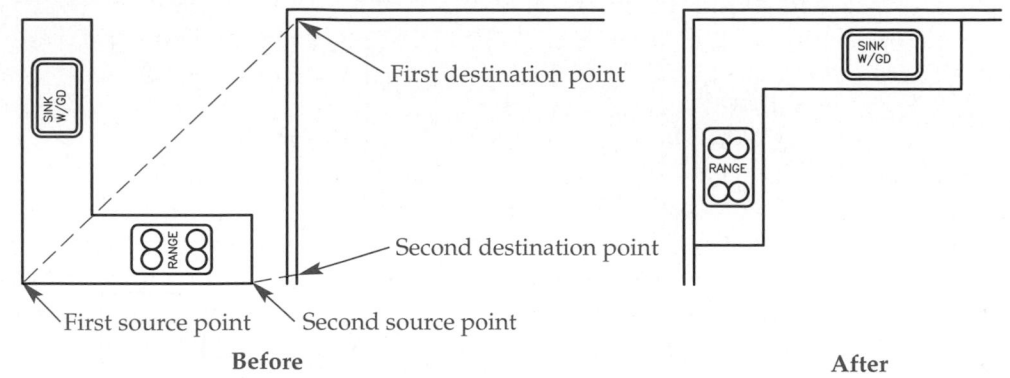

Before   After

second source point, followed by the second destination point. For 2D applications, you only need two source points and two destination points. Right-click or press [Enter] or the space bar when the prompt requests the third source and destination points. See **Figure 14-6.** The last prompt allows you to change the size of the selected object. Choose the **Yes** option to scale the object if the distance between the source points is different from the distance between the destination points. **Figure 14-7** illustrates using the **Scale** option of the **ALIGN** tool.

Exercise 14-3

Access the Student Web site (www.g-wlearning.com/CAD) and complete Exercise 14-3.

Using the COPY Tool

The **COPY** tool allows you to copy existing objects. The **COPY** tool functions similar to the **MOVE** tool, except that when you pick a second point, the original object remains in place and a copy is drawn. See **Figure 14-8.** Access the **COPY** tool, select objects to

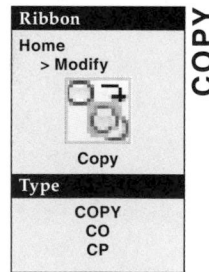

Ribbon
Home
> Modify
Copy

Type
COPY
CO
CP

COPY

Figure 14-7.
The **Scale** option of the **ALIGN** tool allows you to change the size of an object during the alignment.

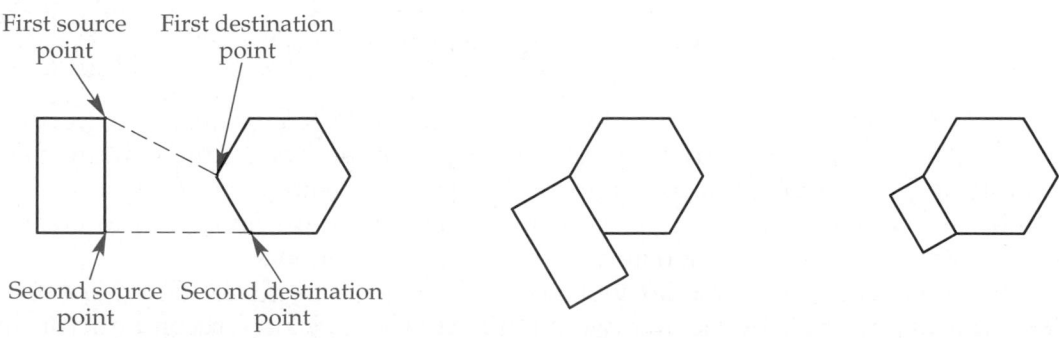

Original Objects   Rectangle   Rectangle Scaled
                                   Not Scaled

Figure 14-8.
Using the **COPY** tool.

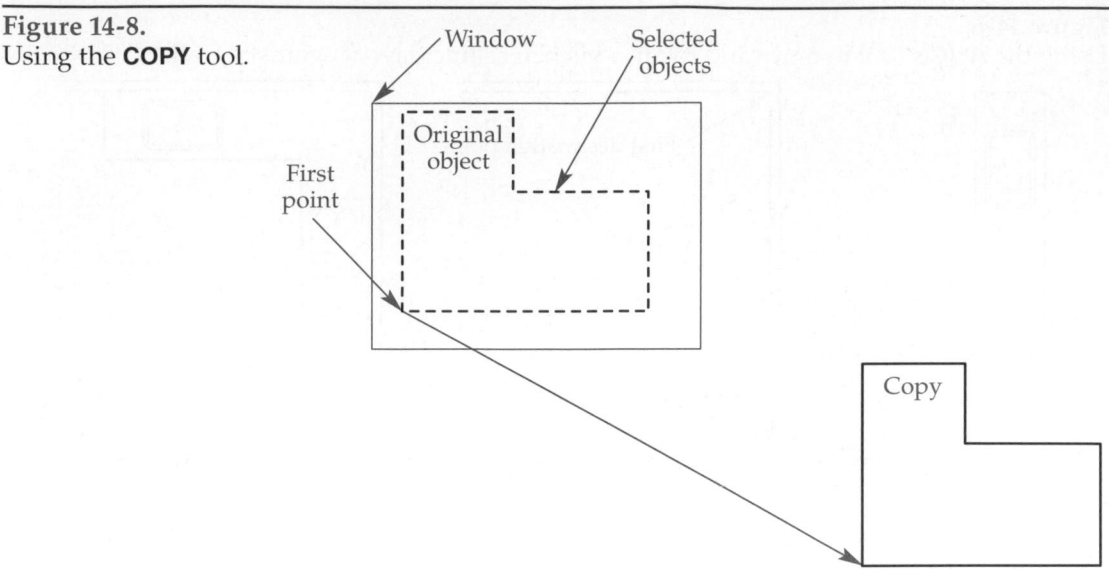

copy, specify a base point, and pick a location to place the first copy. By default, you can continue creating copies of the selected objects by specifying additional "second" points. Press [Enter] or the space bar or right-click and select **Enter** to exit.

As with the **MOVE** tool, you can specify a base point and a second point, specify a displacement using the **Displacement** option, or define the first point as displacement. These techniques function the same when making copies as when moving objects.

NOTE

By default, the **Multiple** copy mode is active, allowing you to create several copies of the same object using a single **COPY** operation. To make a single copy and exit the tool after placing the copy, use the **mOde** option and activate the **Single** function.

Exercise 14-4

Access the Student Web site (www.g-wlearning.com/CAD) and complete Exercise 14-4.

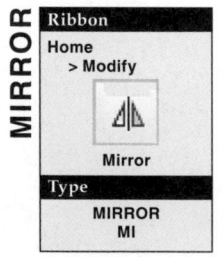

MIRROR

Ribbon
Home
> Modify

Mirror

Type
MIRROR
MI

Mirroring Objects

The **MIRROR** tool allows you to draw objects in a reflected, or mirrored, position. Mirroring is common to a variety of drafting applications. For example, in mechanical drafting you can mirror a part to form the opposite component of a symmetrical assembly. In architectural drafting, you can mirror an entire floor plan to create a duplex residence or to accommodate a different site orientation.

Once you access the **MIRROR** tool, select the objects to mirror. Then specify a *mirror line* at any angle by picking two points. After you pick the second mirror line point, you have the option to delete the original objects. The objects and any space between the objects and the mirror line are reflected. See **Figure 14-9.**

mirror line: The line of symmetry across which objects mirror.

Figure 14-9.
The **MIRROR** tool gives you the option to delete old objects. When you reflect an object about a mirror line, the space between the object and the mirror line also mirrors.

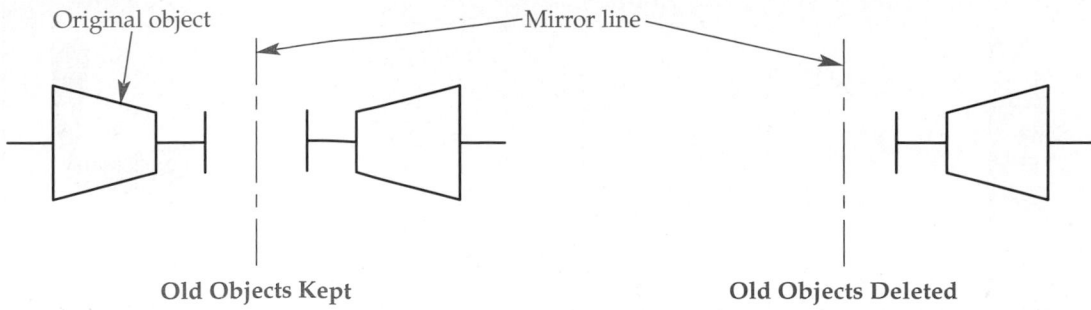

Original object

Mirror line

Old Objects Kept

Old Objects Deleted

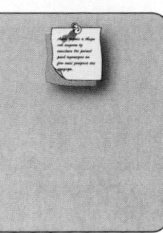

Exercise 14-5

Access the Student Web site (www.g-wlearning.com/CAD) and complete Exercise 14-5.

Figure 14-10.
The **MIRRTEXT** system variable options.

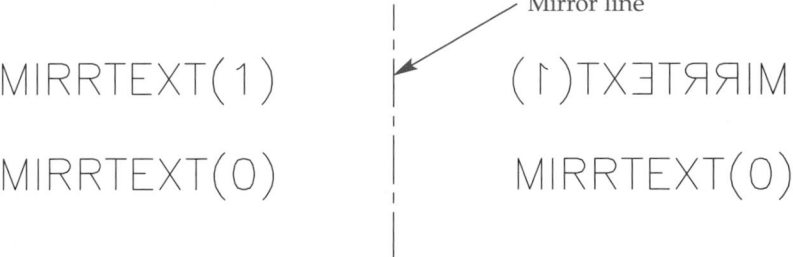

Mirror line

MIRRTEXT(1)

MIRRTEXT(0)

(1)TXƎTЯЯIM

MIRRTEXT(0)

Using the REVERSE Tool

AutoCAD 2010
NEW

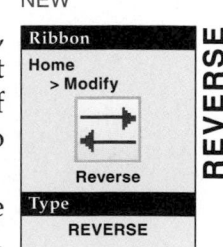
Ribbon
Home
> Modify
Reverse
Type
REVERSE
REVERSE

You can use the **REVERSE** tool to reverse the calculation of points along lines, polylines, splines, and helixes. This makes the previous start point the new endpoint and the previous endpoint the new start point. You can also use the **rEverse** option of the **PEDIT** tool to reverse polylines, and the **rEverse** option of the **SPLINEDIT** tool to reverse splines.

As shown in **Figure 14-11**, reversing is most apparent when you apply a linetype that includes text or specific objects, or when you draw polylines with varying width. Reversing also affects the various vertex options of the **PEDIT** tool and control point options of the **SPLINEDIT** tool. Typically, it is improper to reverse text included with linetypes, although you may find specific applications where this is an appropriate requirement.

Figure 14-11.
Examples of objects before and after using the **REVERSE** tool.

	Line	Spline	Polyline	Polyline
Before				
After				

Patterning with ARRAY

array: Multiple copies of an object arranged in a pattern.

rectangular array: A pattern made up of columns and rows of objects.

polar array: A circular pattern of objects.

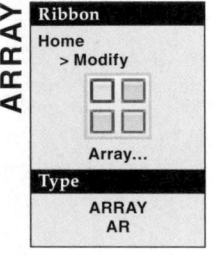

ARRAY

Ribbon
Home
> Modify

Array...

Type
ARRAY
AR

Designs often require a rectangular or circular pattern objects. In interior planning, for example, you might draw a rectangular pattern, or *array*, of office desks. Suppose your design calls for four rows, each having four desks. You can create the pattern by drawing one desk and copying it fifteen times, but this operation is time-consuming. A quicker method is to create a rectangular pattern known as a *rectangular array*. A circular pattern is a *polar array*. **Figure 14-12** shows some basic arrays.

Arranging Objects in a Rectangular Pattern

To create arrays, use the **ARRAY** tool and the corresponding **Array** dialog box, shown in **Figure 14-13**. The **Rectangular Array** radio button is selected by default, allowing you to form a rectangular pattern of rows and columns. You can create an

Figure 14-12.
Examples of arrays.

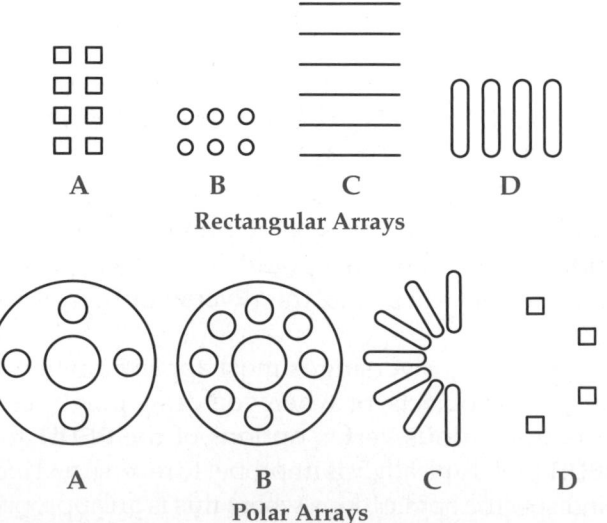

A   B   C   D

Rectangular Arrays

A   B   C   D

Polar Arrays

Figure 14-13.
The **Array** dialog box options for a rectangular array.

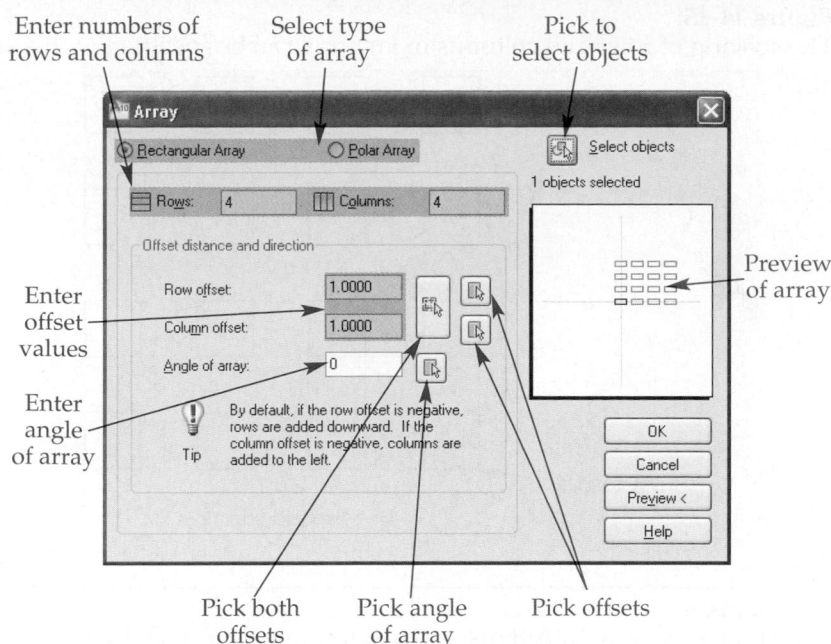

Enter numbers of rows and columns

Select type of array

Pick to select objects

Enter offset values

Enter angle of array

Pick both offsets

Pick angle of array

Pick offsets

Preview of array

array that has a single row, a single column, or multiple rows and columns. Pick the **Select objects** button to return to the drawing area and select the objects to array. Right-click or press [Enter] or the space bar to return to the **Array** dialog box.

Next, specify the array characteristics. For example, suppose you want to create a rectangular pattern of a .5-unit square having four rows, four columns, and a .5 spacing between squares. Enter 4 in the **Rows:** and **Columns:** text boxes and 1.0000 in the **Row offset:** and **Column offset:** text boxes. See **Figure 14-14.** Notice that the distances do not refer to the space between the objects, but the distances between the same point on each object.

You can also enter row and column distances by picking points. One option is to use the **Pick Row Offset** and **Pick Column Offset** buttons in the **Array** dialog box to select each distance separately. The second option is to select the **Pick Both Offsets** button to specify both distances in one pick, as illustrated in **Figure 14-15. Figure 14-16** shows the four directions in which an array can grow. The direction is based on the use of positive and negative distance values for row and column offsets.

Figure 14-14.
The original object (in dashed lines) and the rectangular array. Note how the distances between rows and columns are determined.

Row distance

Column distance

Figure 14-15.
The spacing of rows and columns in an array can be specified with a single point.

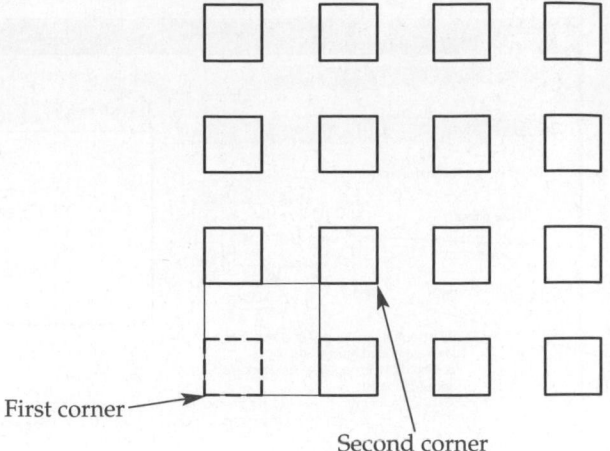

First corner

Second corner

Figure 14-16.
Positive and negative offset distances determine the direction in which an array will grow.

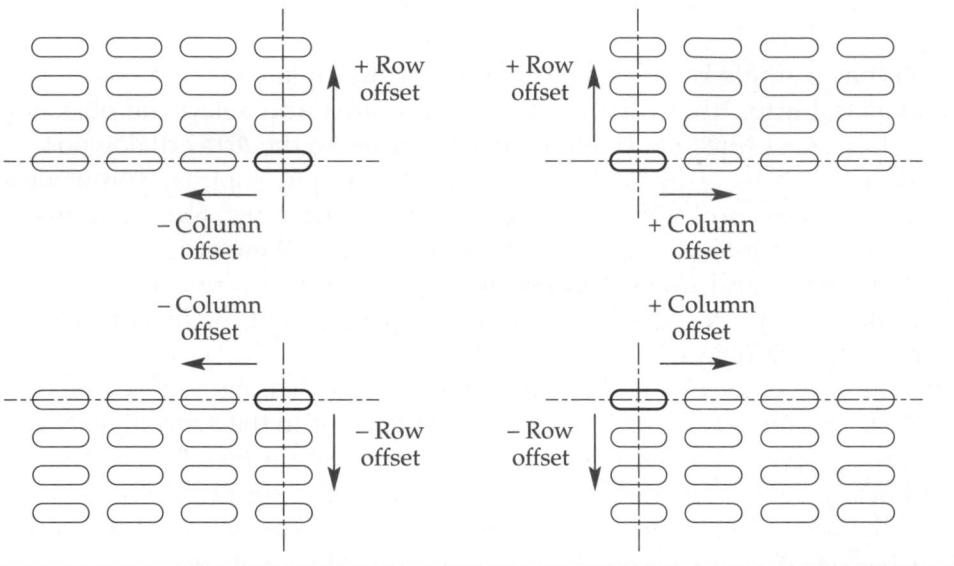

+ Row offset

− Column offset

+ Row offset

+ Column offset

− Column offset

+ Column offset

− Row offset

− Row offset

You can also create an angled rectangular array. Enter the angle in the **Angle of array:** text box or choose the **Pick Angle of Array** button to pick the angle on-screen. The column and row alignment rotate, not the objects. See **Figure 14-17.**

Arranging Objects around a Center Point

Pick the **Polar Array** radio button to create a circular pattern of objects around a center point. See **Figure 14-18.** Pick the **Select objects** button to return to the drawing area and select the objects to array. Right-click or press [Enter] or the space bar to return to the **Array** dialog box.

Next, specify the center point about which the objects in the array rotate. Enter the coordinates for the center point in the **X:** and **Y:** text boxes or select the **Pick Center Point** button to pick the center point on-screen. After locating the center point, you must specify the type of polar array to create using the **Method:** drop-down list. The selected method determines which settings in the dialog box are available. Three methods are available:

- Total number of items & Angle to fill
- Total number of items & Angle between items
- Angle to fill & Angle between items

Figure 14-17.
Rectangular arrays can be set at an angle using the **Angle of array:** setting.

0° Angle of Array   30° Angle of Array   45° Angle of Array

Figure 14-18.
The **Array** dialog box options for a polar array.

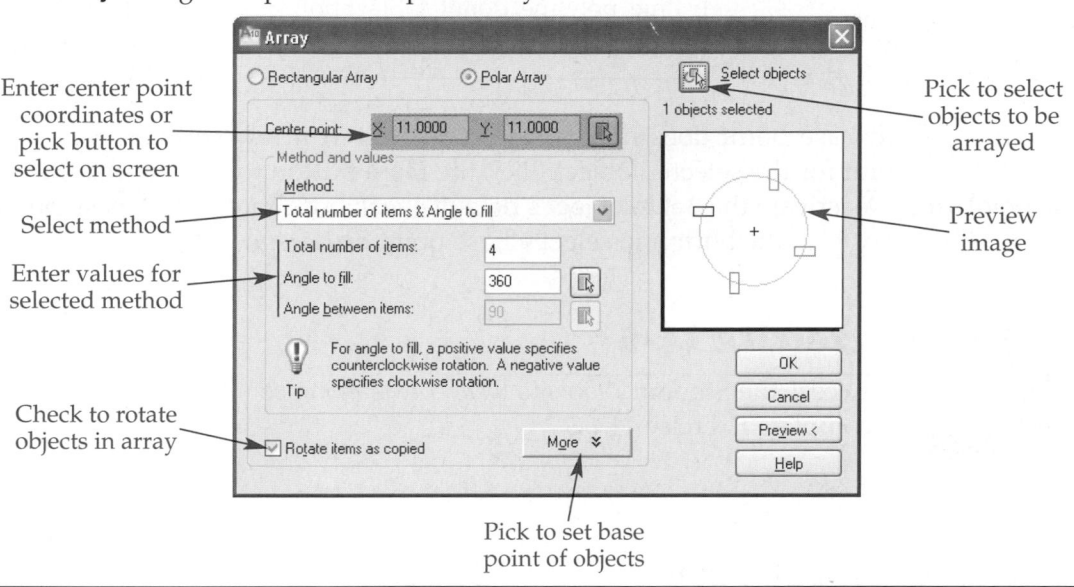

Enter center point coordinates or pick button to select on screen

Select method

Enter values for selected method

Check to rotate objects in array

Pick to select objects to be arrayed

Preview image

Pick to set base point of objects

The **Total number of items:** setting is the total number of objects in the array, including the original object. Use the Angle to fill & Angle between items method if you do not know the number of items to array. Enter a positive angle in the **Angle to fill:** text box to array the object in a counterclockwise direction, or enter a negative angle to array the object in a clockwise direction. Enter 360 to create a complete circular array. The **Angle between items:** setting specifies the angular distance between adjacent objects in the array. For example, to create a circular pattern of five items spaced 18° apart, enter 5 in the **Total number of items:** text box and 18 in the **Angle between items:** text box.

You can set objects to rotate as they array by checking **Rotate items as copied**. This keeps the same face of each object pointing toward the center point. If you do not rotate objects as they array, they remain in the same orientation as the original object. See **Figure 14-19.**

When you create a polar array, the base point of the object rotates and remains at a constant distance from the center point. The default base point varies for different types of objects, as shown in **Figure 14-20.**

Figure 14-19.
Rotating objects in a polar array. A—The square rotates as it arrays. B—The square does not rotate as it arrays.

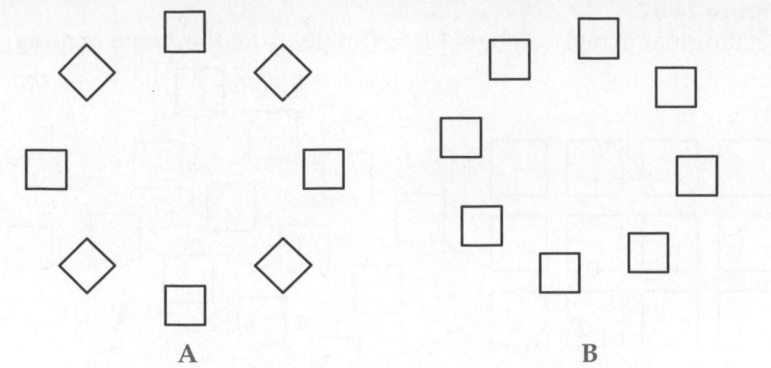

A   B

Figure 14-20.
Default base points for objects.

Object Type	Default Base Point
Arc, circle, ellipse	Center
Rectangle, polygon	First corner
Line, polyline, donut	Start point
Block, text	Insertion point

If the default base point does not produce the desired array, you can choose a different base point for the selected object. Pick the **More** button to display the **Object base point** area. Deactivate the **Set to object's default** check box. Enter a new base point in the text boxes or pick the button to select a base point on-screen.

Exercise 14-6

Access the Student Web site (www.g-wlearning.com/CAD) and complete Exercise 14-6.

Chapter Test

Answer the following questions. Write your answers on a separate sheet of paper or go to the Student Web site (www.g-wlearning.com/CAD) and complete the electronic chapter test.

1. List two locations drafters normally choose as the base point when using the **MOVE** tool.
2. How would you go about rotating an object 45° clockwise?
3. Briefly describe the two methods of using the **Reference** option of the **ROTATE** tool.
4. Name the tool that can be used to move and rotate an object simultaneously.
5. How many points must you select to align an object in a 2D drawing?
6. Which ribbon tab and panel contains the **MOVE** and **COPY** tools?
7. Explain the difference between the **MOVE** and **COPY** tools.
8. Briefly explain how to make several copies of the same object.
9. Which tool allows you to draw a reverse image of an existing object?
10. What is the purpose of the **REVERSE** tool?
11. What is the difference between polar and rectangular arrays?
12. What four values should you know before you create a rectangular array?
13. Suppose an object is 1.5″ (38 mm) wide and you want to create a rectangular array with .75″ (19 mm) spacing between objects. What should you specify for the distance between columns?
14. How do you specify a clockwise polar array rotation?
15. What values should you know before you create a polar array?

Drawing Problems

Follow these instructions to complete the drawing problems for this chapter:

- *Start AutoCAD if it is not already started.*
- *Start a new drawing using an appropriate template of your choice. The template should include layers and text styles necessary for drawing the given objects.*
- *Add layers and text styles as needed. Draw all objects using appropriate layers and text styles, justification, and format.*
- *Follow the specific instructions for each problem. Do not draw dimensions. Use your own judgment and approximate dimensions when necessary.*

▼ Basic

1. Open P12-1. Rotate the object 90 degrees to the right and mirror the object to the left. Use the vertical base of the object as the mirror line. Your final drawing should look like the example below. Save the drawing as P14-1.

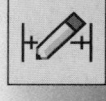

2. Open drawing P12-2. Make two copies of the object to the right of the original object. Scale the first copy 1.5 times size of the original object. Scale the second copy 2 times the size of the original object. Move the objects so they are approximately centered in your drawing area. Move the objects as needed to align the bases of all objects and provide an equal amount of space between the objects. Save the drawing as P14-2.

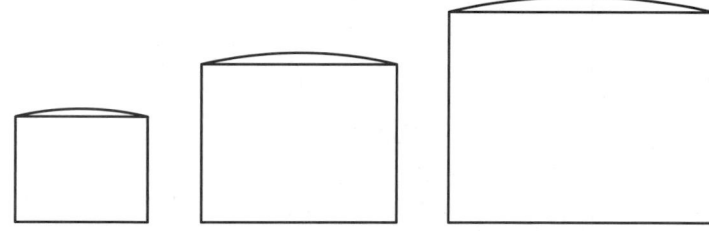

3. Draw Objects A, B, and C at the sizes shown below. Make a copy of Object A two units up. Make four copies of Object B three units up, center to center. Make three copies of Object C three units up, center to center. Save the drawing as P14-3.

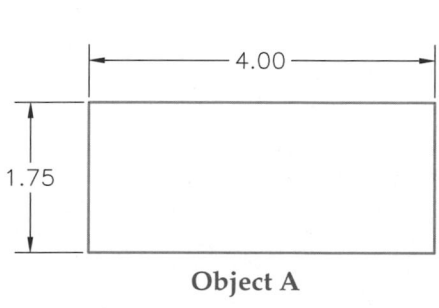

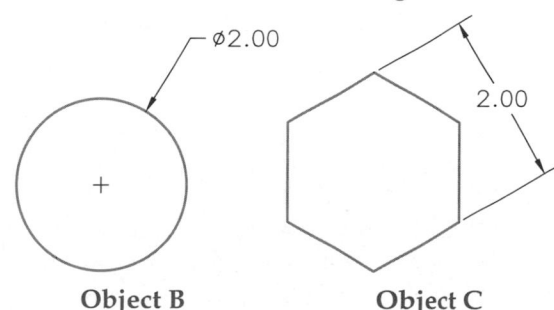

| Object A | Object B | Object C |

4. Open P12-4. Draw a mirror image as Object B. Then remove the original view and move the new view so that Point 2 is at the original Point 1 location. Save the drawing as P14-4.

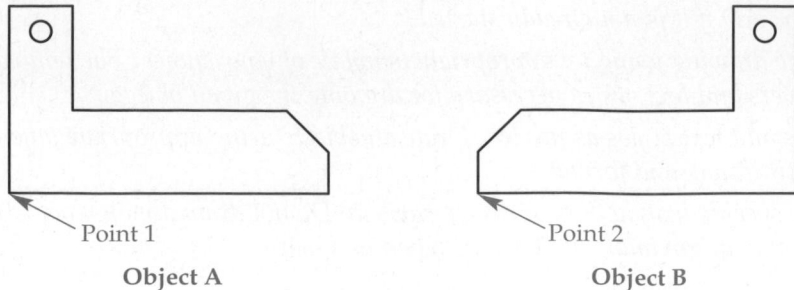

Point 1

Point 2

Object A

Object B

▼ Intermediate

5. Draw the object shown below. The object is symmetrical; therefore, draw only one-half. Mirror the other half into place. Use the **CHAMFER** and **FILLET** tools to your best advantage. All fillets and rounds are .125. Use the **JOIN** tool where necessary. Save the drawing as P14-5.

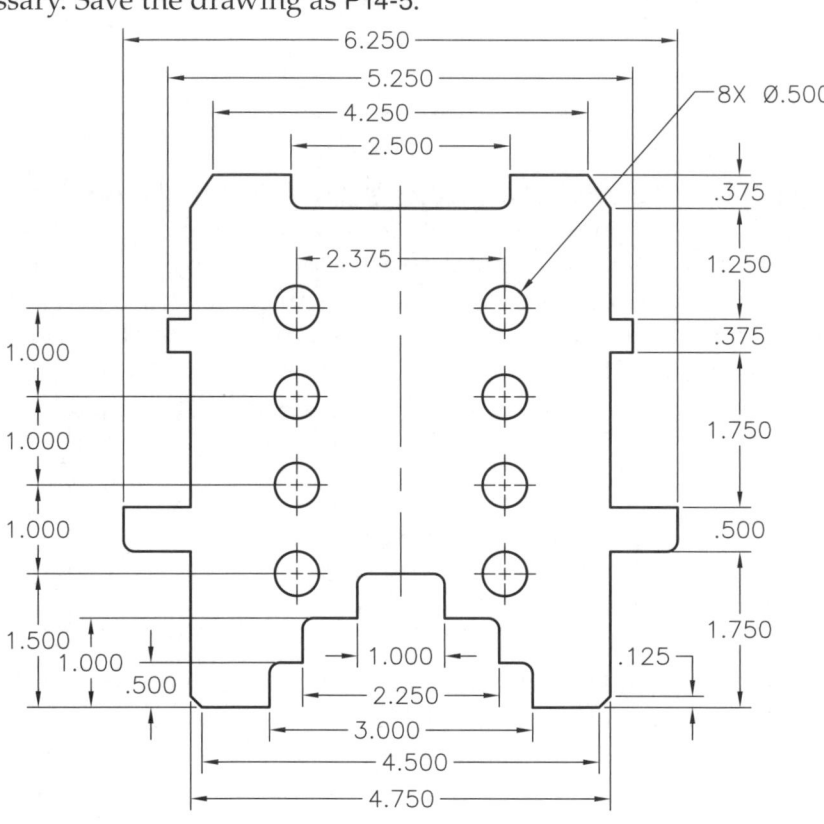

6. Draw the object shown below. Mirror the right half into place. Use the **CHAMFER** and **FILLET** tools to your best advantage. Save the drawing as P14-6.

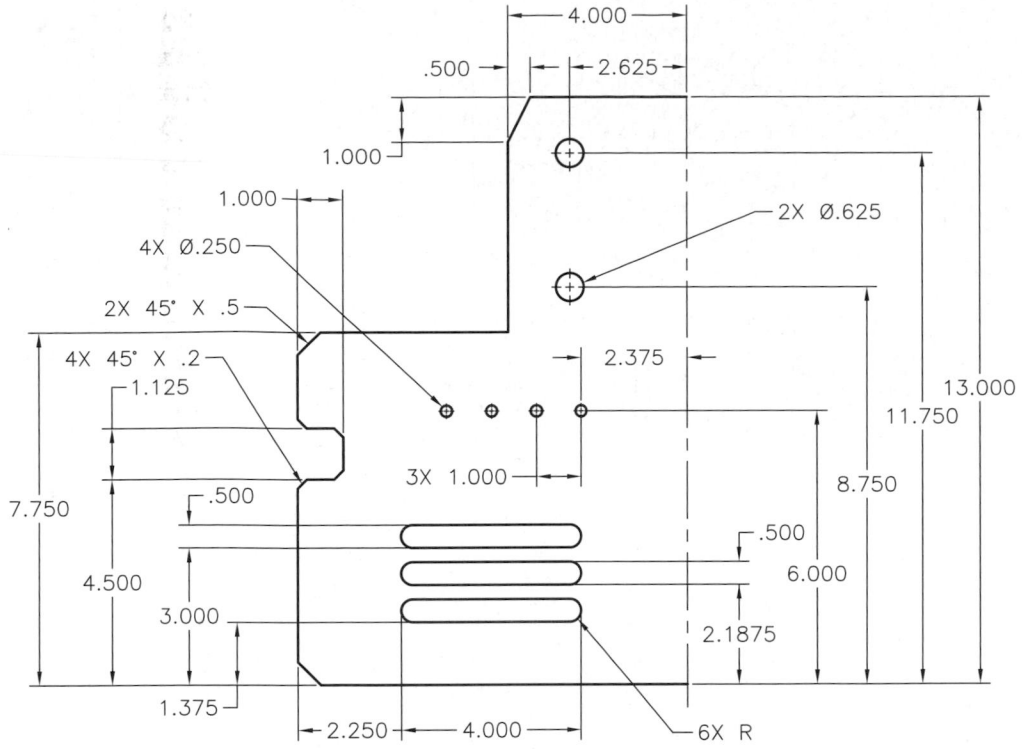

7. Redraw the objects shown below. Mirror the drawing, but make sure the text remains readable. Delete the original image during the mirroring process. Save the drawing as P14-7.

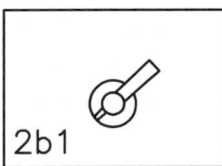 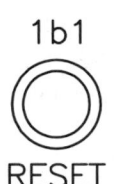

2b1

TRANSFER

5a2 4a1 8a1

LTS. HTRS. FANS

2b1

1b1

RESET

11b1

BYPASS

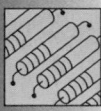

8. Draw this timer schematic. Save the drawing as P14-8.

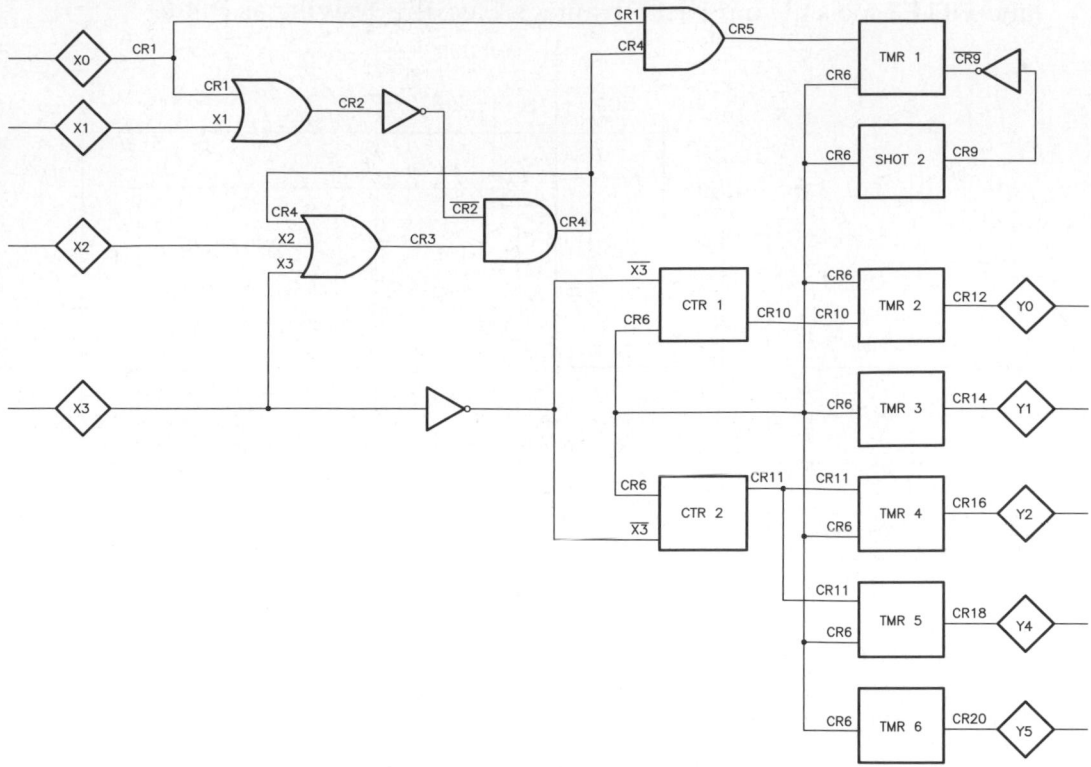

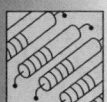

9. Use tracking and object snaps to draw the object shown below based on these instructions:
   A. Draw the outline of the object first, followed by the ten ⌀.500 holes (A).
   B. The holes labeled B are located vertically halfway between the centers of the holes labeled A. They have a diameter one-quarter the size of the holes labeled A.
   C. The holes labeled C are located vertically halfway between the holes labeled A and B. Their diameter is three-quarters of the diameter of the holes labeled B.
   D. The holes labeled D are located horizontally halfway between the centers of the holes labeled A. These holes have the same diameter as the holes labeled B.
   E. Draw the rectangles around the circles as shown.
   F. Do not draw dimensions, notes, or labels.
   G. Save the drawing as P14-9.

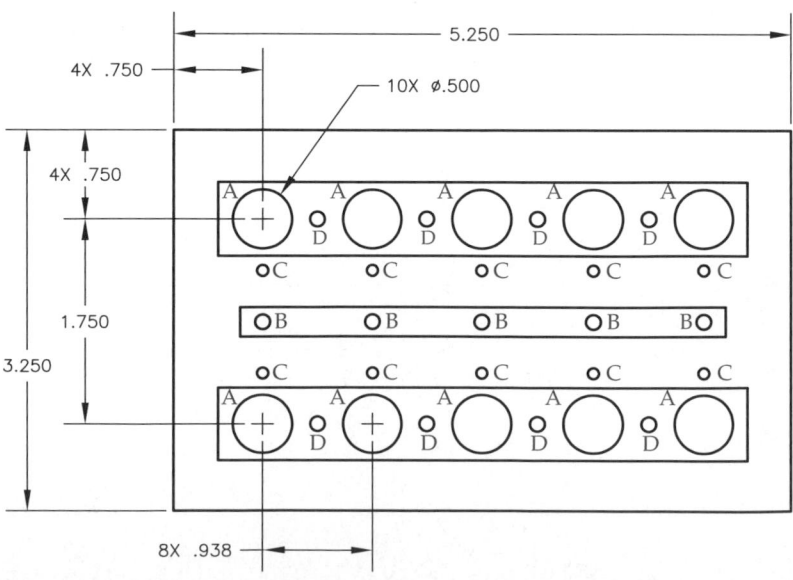

10. Draw the portion of the gasket shown on the left. Use the **MIRROR** tool to complete the gasket as shown on the right. Save the drawing as P14-10.

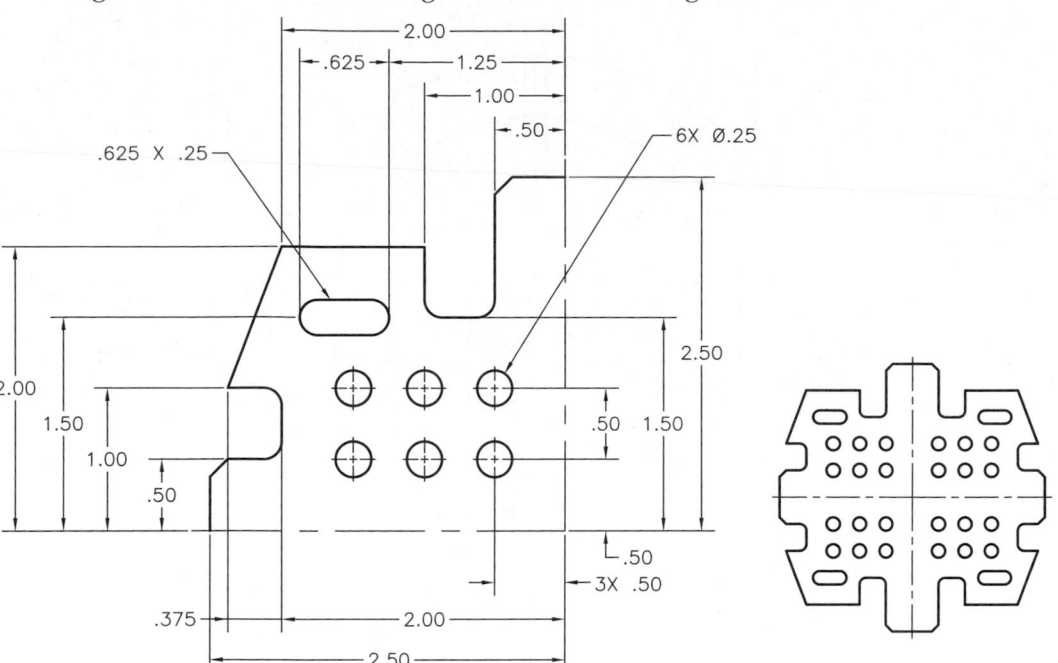

ALL FILLETS AND ROUNDS R.125.
CHAMFERS 45° X .125

11. Draw the padded bench. Use the **COPY** and **ARRAY** tools as needed. Save the drawing as P14-11.

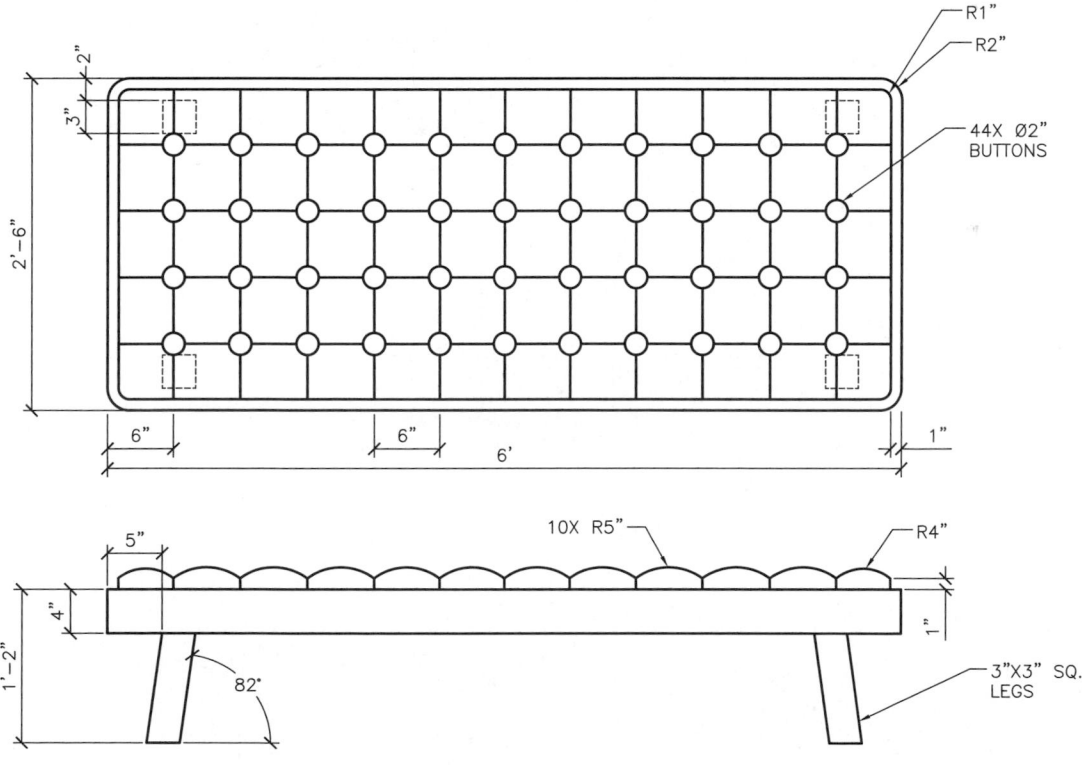

12. Draw the hand wheel shown below. Use the **ARRAY** tool to draw the spokes. Save the drawing as P14-12.

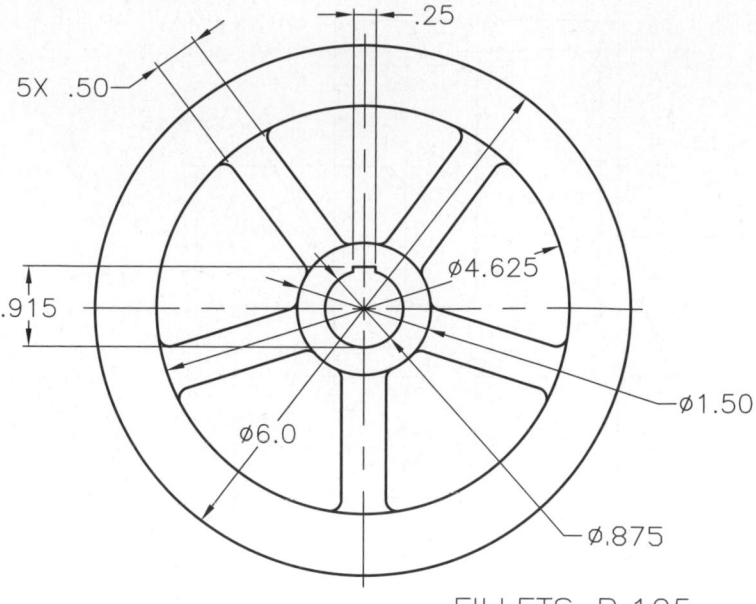

13. Draw the control diagram. Draw one branch (including text) and use the **COPY** tool to your advantage. Use text editing tools as needed. Save the drawing as P14-13. (Design and drawing by EC Company, Portland, Oregon)

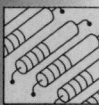

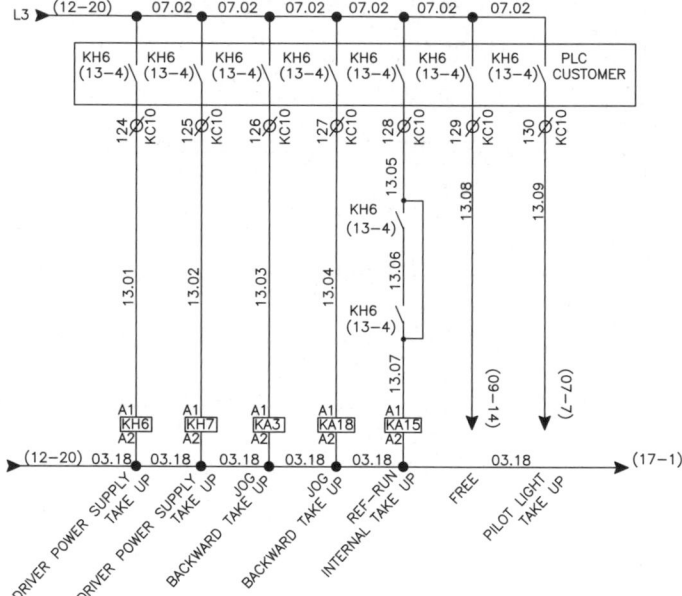

▼ Advanced

14. You have been given an engineer's sketches and notes to construct a drawing of a sprocket. Create a front and side view of the sprocket using the **ARRAY** tool. Place the drawing on one of your templates. Save the drawing as P14-14.

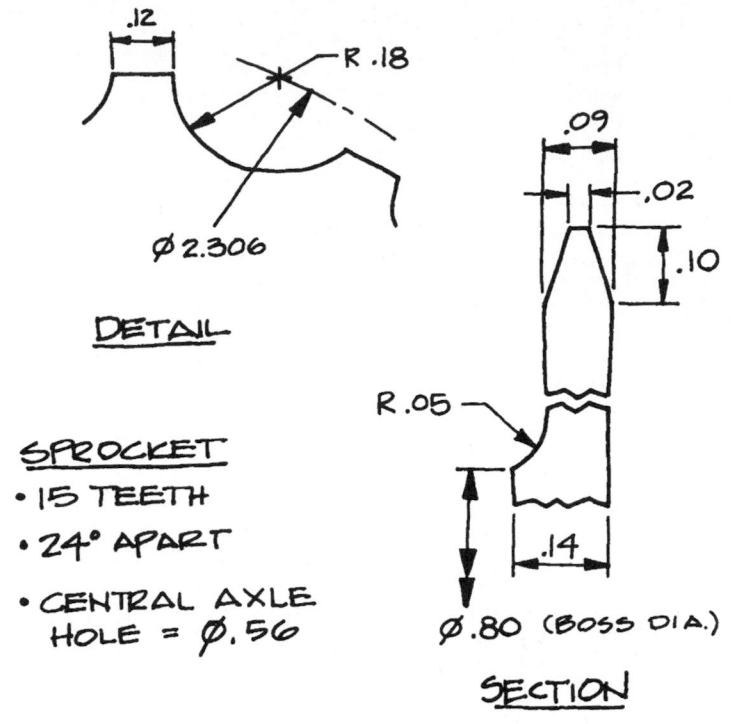

DETAIL

SPROCKET
- 15 TEETH
- 24° APART
- CENTRAL AXLE
  HOLE = ∅.56

SECTION

15. Draw the following object views using the dimensions given. Use **ARRAY** to construct the hole and tooth arrangements. Use one of your templates for the drawing. Save the drawing as P14-15.

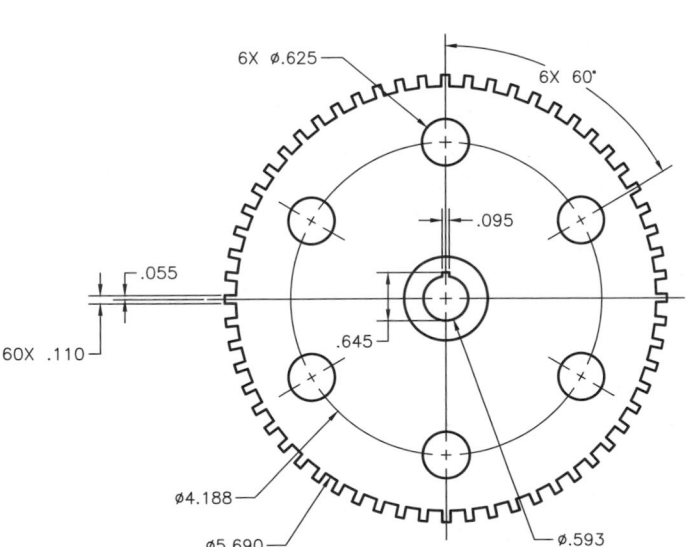

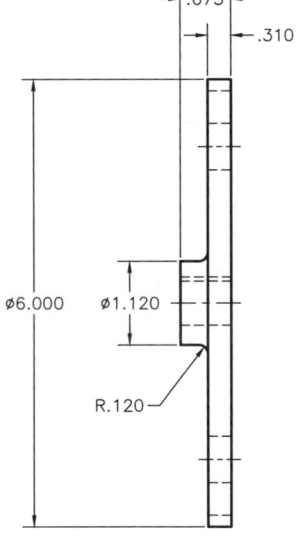

Chapter 14   Arranging and Patterning Objects

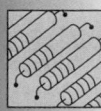

16. Draw this refrigeration system schematic. Save the drawing as P14-16.

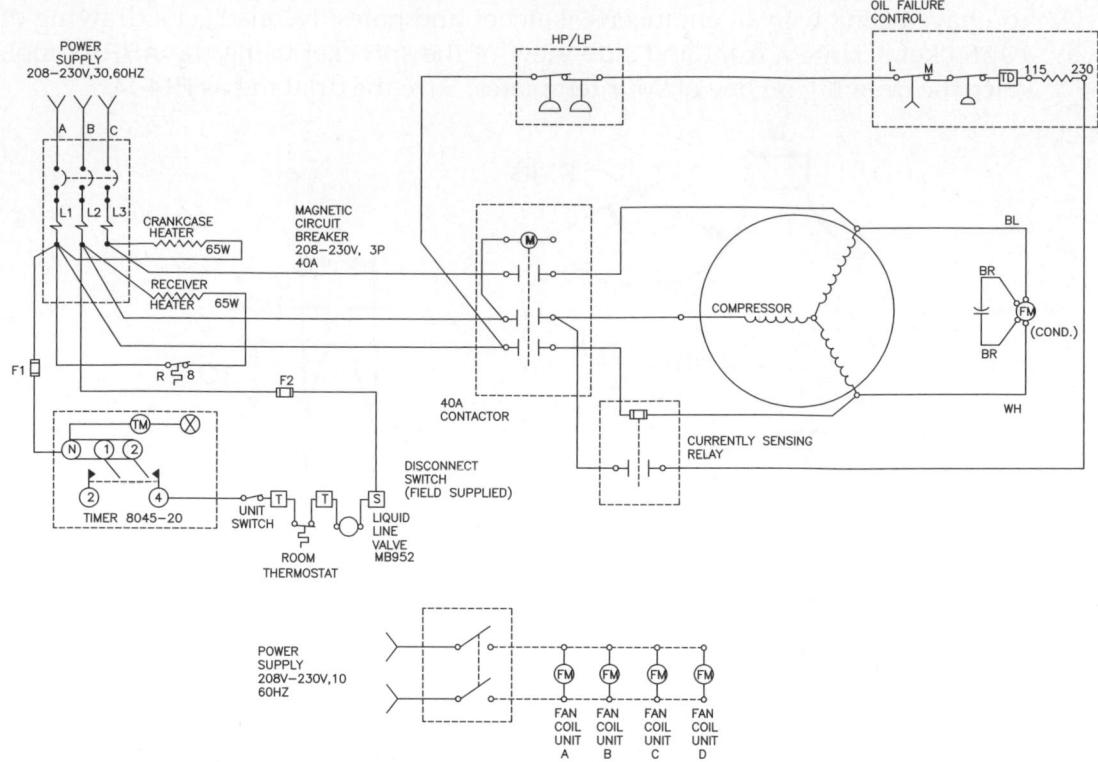

17. The following structural sketch shows a steel column arrangement on a concrete floor slab for a new building. The *I*-shaped symbols represent the steel columns. The columns are arranged in "bay lines" and "column lines." The column lines are numbered *1*, *2*, and *3*. The bay lines are labeled *A* through *G*. The width of a bay is 24'-0". Line balloons, or tags, identify the bay and column lines. Draw the arrangement, using **ARRAY** for the steel column symbols and the tags. The following guidelines will help you:
    A. Begin a new drawing using an architectural template.
    B. Select architectural units and set up the drawing to print on a 36 × 24 sheet size. Determine the scale required for the floor plan to fit on this sheet size and specify the drawing limits accordingly.
    C. Draw the steel column symbol to the dimensions given.
    D. Set the grid spacing at 2'-0" (24").
    E. Set the snap spacing at 12".
    F. Draw all other objects.
    G. Place text inside the balloon tags. Set the running object snap mode to **Center** and justify the text to **Middle**. Make the text height 6".
    H. Save the drawing as P14-17.

AutoCAD and Its Applications—Basics

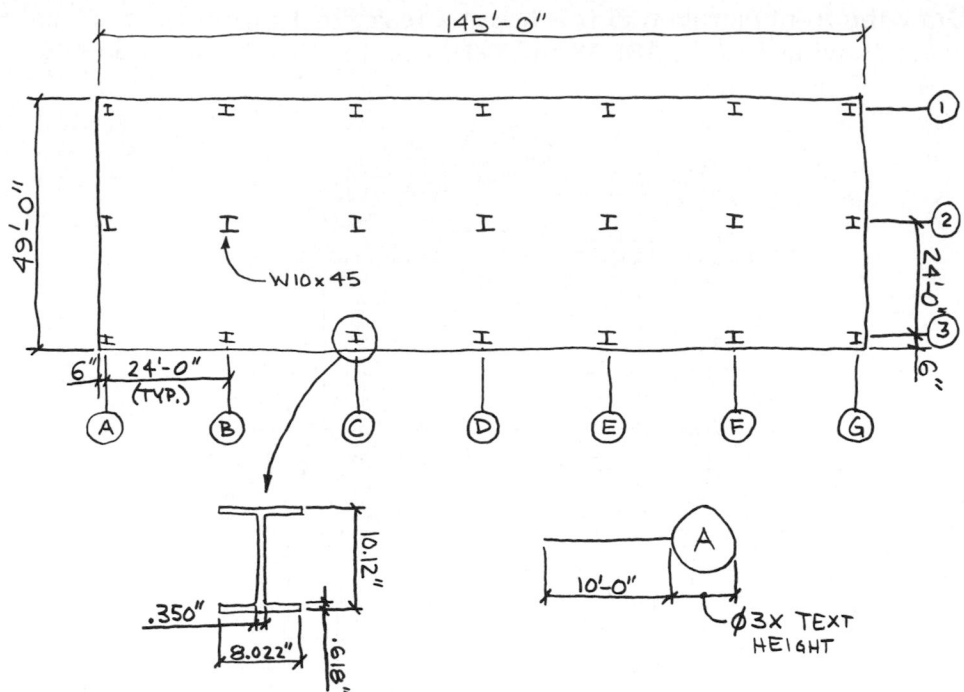

18. The sketch shown below is a proposed classroom layout of desks and chairs. One desk is shown with the layout of a chair, keyboard, monitor, and tower-mounted computer (drawn with dotted lines). All of the desk workstations should have the same configuration. The exact sizes and locations of the doors and windows are not important for this problem. Use the following guidelines to complete this problem:
   A. Begin a new drawing.
   B. Choose architectural units.
   C. Set up the drawing to print on a C-size sheet, and be sure to create the drawing in model space.
   D. Use the appropriate drawing and editing tools to complete this problem quickly and efficiently.
   E. Draw the desk and computer hardware to the dimensions given.
   F. Do not dimension the drawing.
   G. Save the drawing as P14-18.

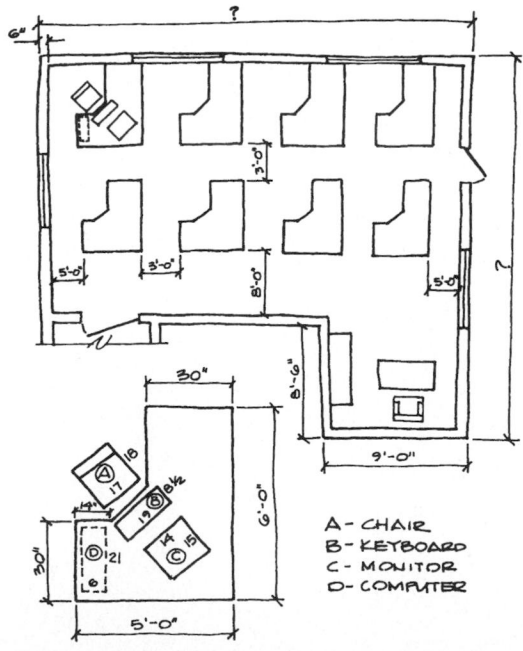

A - CHAIR
B - KEYBOARD
C - MONITOR
D - COMPUTER

19. Draw the front elevation of this house. Create the features proportional to the given drawing. Use the **ARRAY** and **TRIM** tools to place the siding and porch rails evenly. Save the drawing as P14-19.

20. Use **SPLINE** and other tools, such as **ELLIPSE**, **MIRROR**, and **OFFSET**, to design an architectural door knocker similar to the one shown. Use an appropriate text tool and font to place your initials in the center. Save the drawing as P14-20.

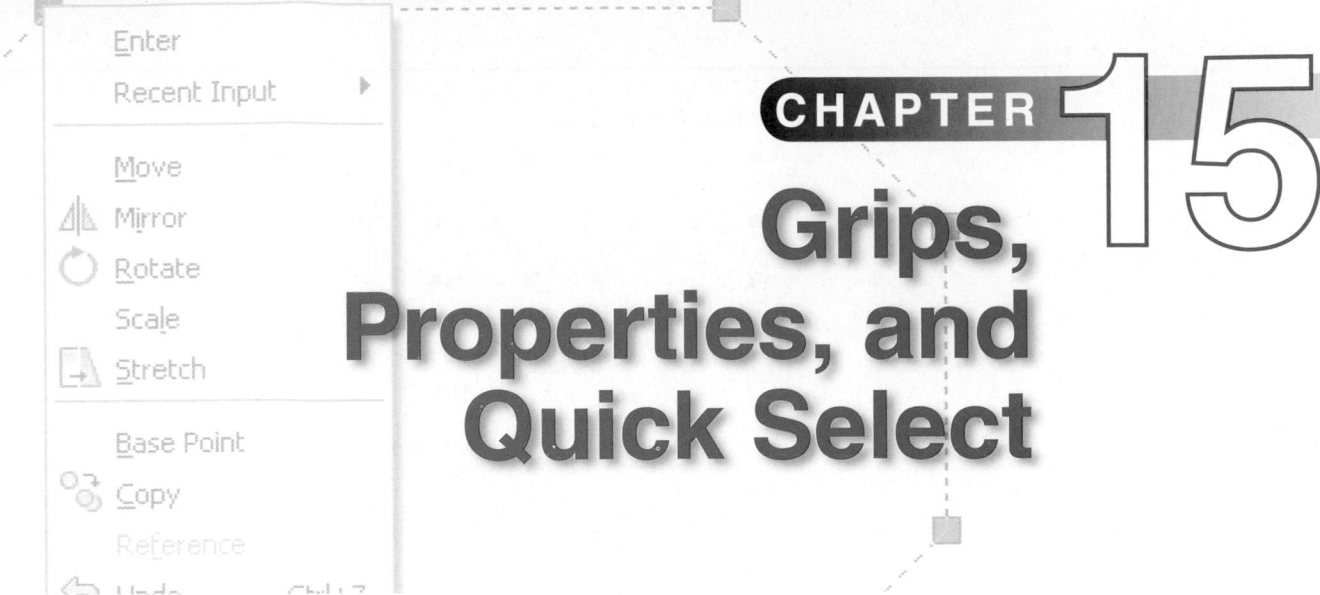

Grips, Properties, and Quick Select

Learning Objectives

After completing this chapter, you will be able to do the following:

✓ Use grips to stretch, move, rotate, scale, mirror, and copy objects.
✓ Adjust object properties using the **Quick Properties** panel and the **Properties** palette.
✓ Use the **MATCHPROP** tool to match object properties.
✓ Edit between drawings.
✓ Create selection sets using the **Quick Select** dialog box.

Typically, when using tools such as **ERASE, FILLET, MOVE,** and **COPY,** you access the tool first and then follow prompts that require you to select the object to edit. This chapter describes the process of editing objects and changing object properties by selecting an object first and then performing editing operations. This chapter also explains tools for selecting objects using selection set filters.

Using Grips

Grips appear on an object when you select the object while no tool is active. **Figure 15-1** shows the locations of grips on several different types of objects. When you select an object, the object highlights and grips appear in an *unselected grip* state. By default, unselected grips display as blue (Color 150) filled squares or arrows. Some objects include other, specialized grips, as described when applicable in this textbook. Move the crosshairs over an unselected grip to snap to the grip. Then pause to change the color of the grip to pink (Color 11). Hovering over an unselected grip and allowing it to change color helps you select the correct grip, especially when multiple grips are close together.

A *selected grip* appears as a red (Color 12) filled square or arrow by default. You can use a selected grip to perform several editing operations. If you select more than one object displaying unselected grips, what you do with the selected grips affects all of the selected objects. Objects having unselected and selected grips highlight and become part of the current selection set.

grips: Small boxes that appear at strategic points on an object when you select it, allowing you to edit the object.

unselected grips: Grips that have not yet been picked to perform an operation.

selected grip: A grip that has been picked to perform an operation.

Figure 15-1.
Grips appear at specific locations on objects.

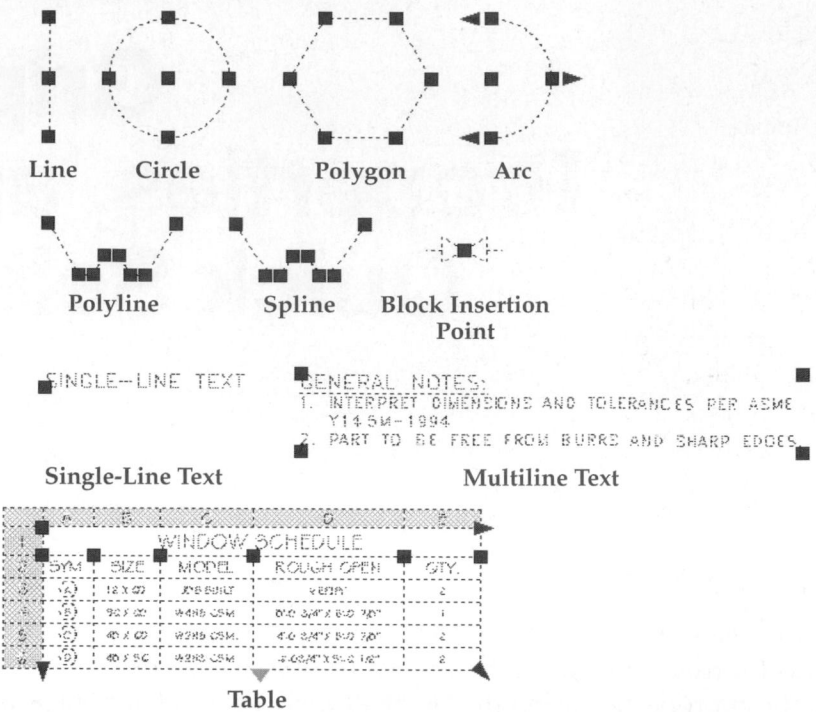

To remove highlighted objects from a selection, hold down the [Shift] key and pick the objects to remove. [Shift] also allows you to add or remove selected grips. To make a second grip selected, hold [Shift] down while selecting both the first and second grip. To add more grips to the selected grip set, continue to hold [Shift] down and pick additional grips. With [Shift] held down, picking a selected (red) grip returns it to the unselected (blue) state. **Figure 15-2** shows an example of modifying two circles at the same time using selected grips.

Return objects to the selection set by picking them again. Remove all selected grips from the selection set by pressing [Esc] to cancel. Press [Esc] again to deselect all objects and remove grips from the selection set. You can also right-click and pick **Deselect All** to remove all grips.

Figure 15-2.
You can modify multiple objects simultaneously by pressing [Shift] to select additional grips.

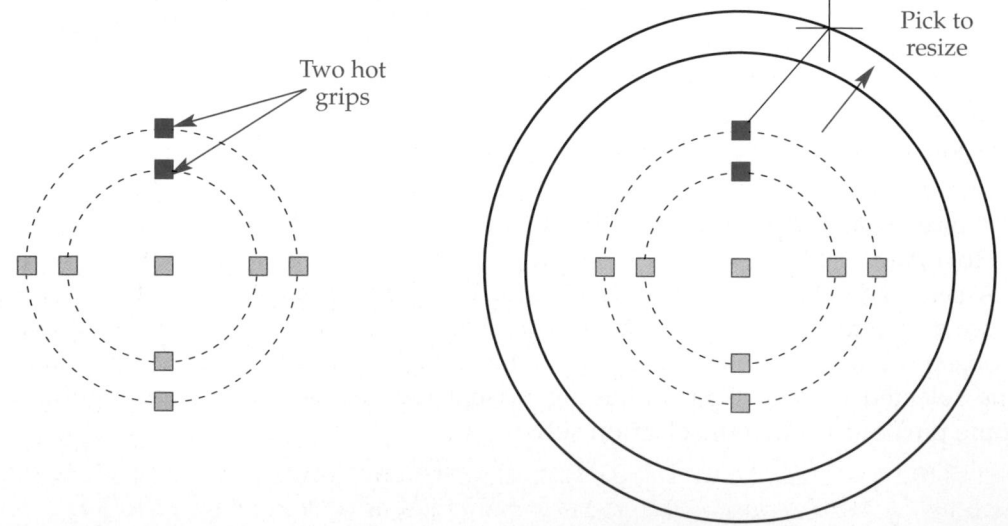

noun/verb selection: Performing tasks in AutoCAD by selecting the objects before entering a tool.

verb/noun selection: Performing tasks in AutoCAD by entering a tool before selecting objects.

Using Grip Tools

Grips provide access to the **STRETCH, MOVE, ROTATE, SCALE,** and **MIRROR** tools. In addition, the **Copy** option of the **MOVE** tool and sometimes, depending on the selected grip, the **STRETCH** tool imitate the **COPY** tool. Grip tools become available at the dynamic input cursor and the command line when you select a grip. Do not attempt to use conventional means of tool access, such as the ribbon. The first tool is **STRETCH**, as indicated by the ** STRETCH ** Specify stretch point or [Base point/Copy/Undo/eXit]: prompt. You can use the **STRETCH** tool at this prompt, or press [Enter] or the space bar or right-click and select **Enter** to cycle through additional tool options:

```
** STRETCH **
Specify stretch point or [Base point/Copy/Undo/eXit]: ↵
** MOVE **
Specify move point or [Base point/Copy/Undo/eXit]: ↵
** ROTATE **
Specify rotation angle or [Base point/Copy/Undo/Reference/eXit]: ↵
** SCALE **
Specify scale factor or [Base point/Copy/Undo/Reference/eXit]: ↵
** MIRROR **
Specify second point or [Base point/Copy/Undo/eXit]: ↵
** STRETCH **
Specify stretch point or [Base point/Copy/Undo/eXit]:
```

As an alternative to cycling through the tool options, you can right-click to access a grips shortcut menu, which is available only after you select a grip. The shortcut menu allows you to access the five grip editing tools without using the keyboard. A third option to activate a tool is to enter the first two characters of the desired tool. Type MO for **MOVE**, MI for **MIRROR**, RO for **ROTATE**, SC for **SCALE**, or ST for **STRETCH**.

Stretching Objects

The process of stretching using grips is similar to stretching using the traditional **STRETCH** tool. The main difference is that the selected grip acts as the stretch base point. Move the crosshairs to stretch the selected object. See **Figure 15-3.** If you pick the middle grip of a line or arc, the center grip of a circle, or the insertion point grip of a block, single-line text, multiline text, or table, the object moves instead of stretching.

Figure 15-4 shows two methods to stretch features of an object. Step 1 in **Figure 15-4A** stretches the first corner, and Step 2 stretches the second corner. You can combine the two operations by holding down [Shift] to select two grips, as shown in **Figure 15-4B.**

Figure 15-3.
Using the **STRETCH** grip tool. Note the selected grip in each case.

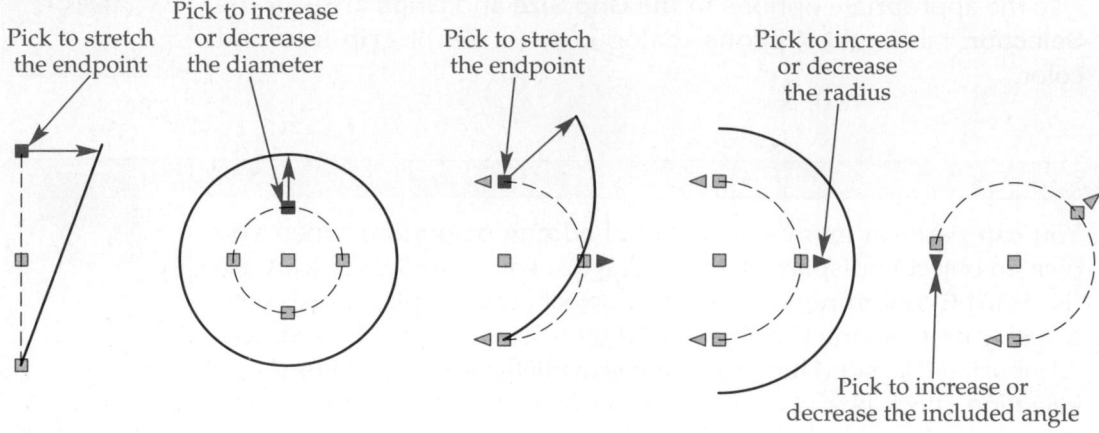

Pick to stretch the endpoint

Pick to increase or decrease the diameter

Pick to stretch the endpoint

Pick to increase or decrease the radius

Pick to increase or decrease the included angle

Figure 15-4.
Stretching an object. A—Select corners to stretch individually. B—Select several grips by holding down [Shift].

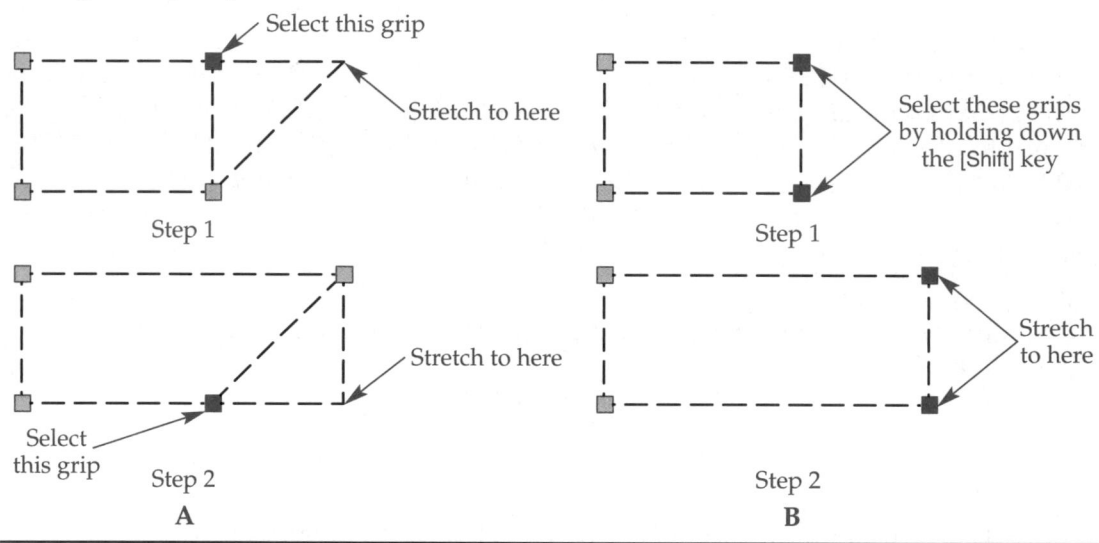

Select this grip

Stretch to here

Step 1

Select this grip

Step 2

A

Select these grips by holding down the [Shift] key

Step 1

Stretch to here

Step 2

B

Use the **Base point** option to specify a base point instead of using the selected grip as the base point. Activate the **Undo** option to undo the previous operation. Choose the **eXit** option or press [Esc] to exit without completing the stretch. When you finish stretching, or exit the tool, the selected grip is gone, but unselected grips remain. Press [Esc] to remove the unselected grips.

Dynamic input is especially effective with the **STRETCH** grip tool. **Figure 15-5** shows an example of how you can use dimensional input to quickly modify the size of a circle or offset a circle by a specific distance using the **STRETCH** grip tool. In this example, enter the new radius of the circle in the distance input field, or press [Tab] to enter an offset in the other distance input field. This is just one example of how you can use dynamic input with grips. Most objects include similar operations.

> **PROFESSIONAL TIP**
>
> You can use grid snaps, coordinate entry techniques, polar tracking, object snaps, and object snap tracking with any of the grip editing tools to improve accuracy. Remember to use these functions as you edit your drawings.

Figure 15-5.
Using the dimensional input feature of dynamic input with the **STRETCH** grip tool.

Enter a value to redefine the radius of the circle

Press [Tab] and enter an offset distance here to modify the size of the circle

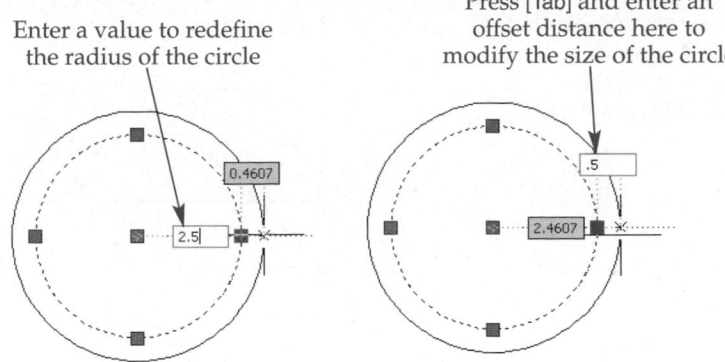

Exercise 15-1

Access the Student Web site (www.g-wlearning.com/CAD) and complete Exercise 15-1.

Moving Objects

To move an object with grips, select the object, pick a grip to use as the base point, and then activate the **MOVE** tool. Specify a new location for the base point to move the object. See **Figure 15-6**. The **Base point**, **Undo**, and **eXit** options are similar to those for the **STRETCH** tool.

Exercise 15-2

Access the Student Web site (www.g-wlearning.com/CAD) and complete Exercise 15-2.

Rotating Objects

To rotate an object using grips, select the object, pick a grip to use as the base point, and then activate the **ROTATE** tool. Specify a rotation angle from the base point to rotate the object. The **Base point**, **Undo**, and **eXit** options are similar to those for the **STRETCH** tool.

You can use the **Reference** option to specify a new angle in relation to an existing angle. The reference angle is the current angle of the object. If you know the value of

Figure 15-6.
When you specify the **MOVE** tool, the selected grip becomes the base point for the move.

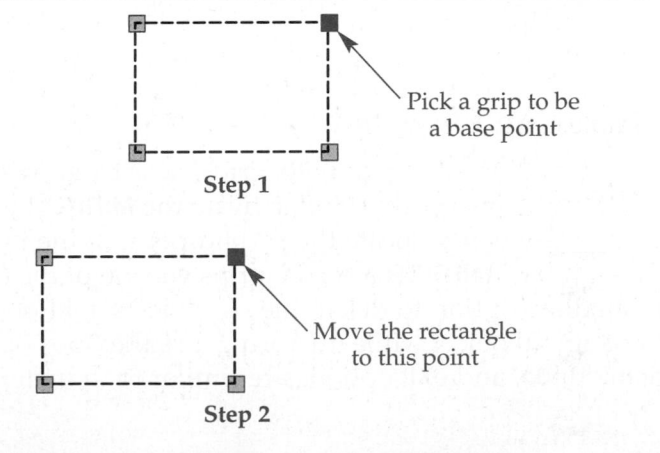

Pick a grip to be a base point

Step 1

Move the rectangle to this point

Step 2

Figure 15-7.
The rotation angle
and **Reference**
options of the
ROTATE grip tool.

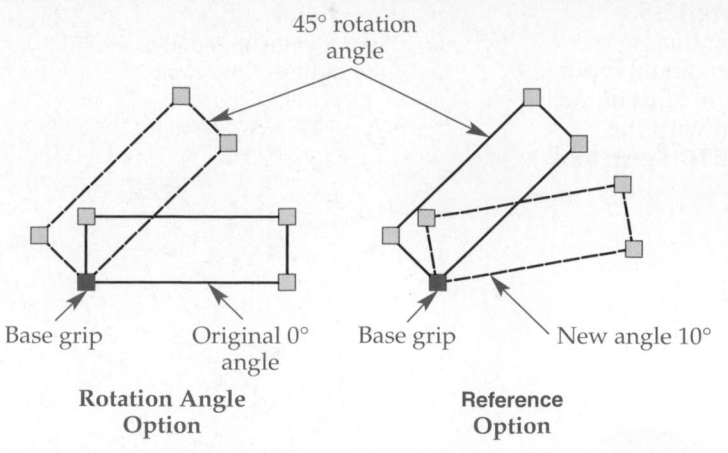

45° rotation
angle

Base grip   Original 0°   Base grip   New angle 10°
angle

Rotation Angle   **Reference**
Option   **Option**

the current angle, enter the value at the prompt. Otherwise, pick two points on the reference line to identify the existing angle. Enter a value for the new angle or pick a point. **Figure 15-7** shows the **ROTATE** options.

Exercise 15-3

Access the Student Web site (www.g-wlearning.com/CAD) and complete Exercise 15-3.

Scaling Objects

To scale an object with grips, select the object, pick a grip to use as the base point, and then activate the **SCALE** tool. Enter a scale factor or pick a point to increase or decrease the size of the object. The **Base point**, **Undo**, and **eXit** options are similar to those for the **STRETCH** tool.

You can use the **Reference** option to specify a new size in relation to an existing size. The reference size is the current length, width, or height of the object. If you know the current size, enter the value at the prompt. Otherwise, pick two points on the reference line to identify the existing size. Enter a value for the new size or pick a point. **Figure 15-8** shows the **ROTATE** options.

Exercise 15-4

Access the Student Web site (www.g-wlearning.com/CAD) and complete Exercise 15-4.

Mirroring Objects

To mirror an object with grips, select the object, pick a grip to use as the first point of the mirror line, and activate the **MIRROR** tool. Then pick another grip or any point on-screen to locate the second point of the mirror line. See **Figure 15-9**. Unlike the standard **MIRROR** tool, the grips version of the **MIRROR** tool does not give you the immediate option to delete the old objects. Old objects delete automatically. To keep the original object while mirroring, use the **Copy** option of the **MIRROR** tool. The **Base point**, **Undo**, and **eXit** options are similar to those for the **STRETCH** tool.

Figure 15-8.
When using the
SCALE tool with
grips, you can enter
a scale factor or
use the **Reference**
option.

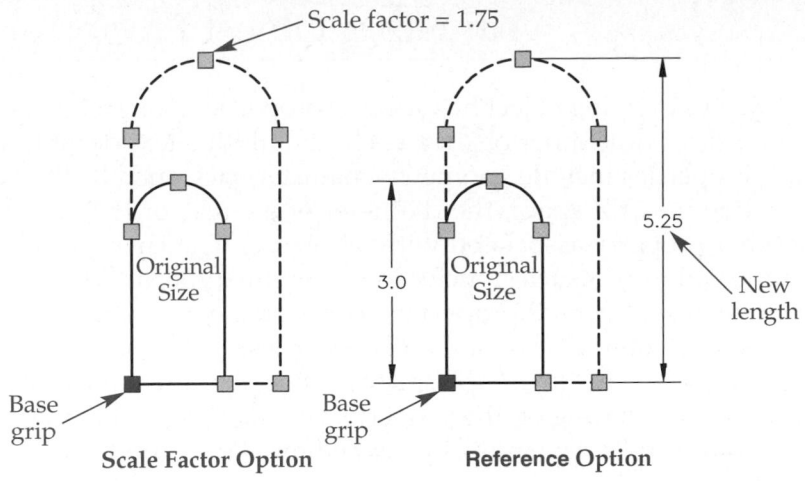

Scale factor = 1.75

Original Size

Original Size

Base grip

Base grip

5.25

3.0

New length

Scale Factor Option

Reference Option

Figure 15-9.
When you use
grips to access
the **MIRROR** tool,
the selected grip
becomes the first
point of the mirror
line, and the original
object automatically
deletes.

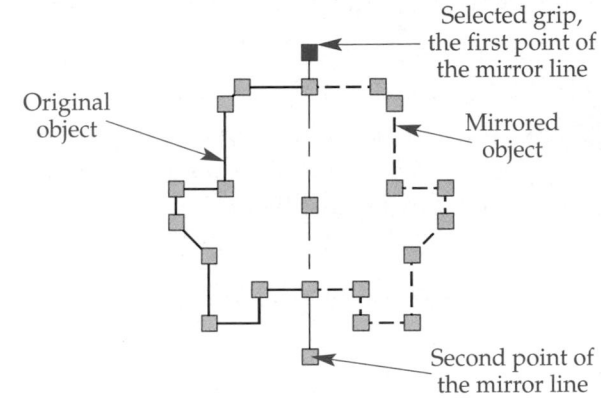

Selected grip, the first point of the mirror line

Original object

Mirrored object

Second point of the mirror line

Exercise 15-5

Access the Student Web site (www.g-wlearning.com/CAD) and complete Exercise 15-5.

Copying Objects

The **Copy** option is included in each of the grip editing tools. The effect of using the **Copy** option depends on the selected grip and tool. The original selected object remains unchanged, and the copy stretches when the **STRETCH** tool is active, rotates when the **ROTATE** tool is active, or scales when the **SCALE** tool is active. The **Copy** option of the **MOVE** tool is the true form of the **COPY** tool, allowing you to copy from any selected grip. The selected grip acts as the copy base point. Create as many copies of the selected object as needed, and then exit the tool.

Exercise 15-6

Access the Student Web site (www.g-wlearning.com/CAD) and complete Exercise 15-6.

Adjusting Object Properties

Every drawing object has specific properties. Some objects, such as lines, have few properties, while other objects, such as multiline text or tables, contain many properties. Properties include geometry characteristics, such as the location of the endpoint of a line in X,Y,Z space, the diameter of a circle, or the area of a rectangle. Layer is another property associated with all objects. The layer on which you draw an object defines other properties, including color, linetype, and lineweight. Most objects also include object-specific properties. For example, text objects have text properties, and tables have column, row, and cell properties.

You can edit object properties using modification tools or a variety of other methods, depending on the property. For example, you can adjust layer characteristics using layer tools. The multiline text editor allows you to adjust existing multiline text properties. Another technique to view and make changes to object properties is to use the **Quick Properties** panel or the **Properties** palette. These tools are especially effective for modifying a particular property or set of properties for multiple objects at once.

PROFESSIONAL TIP

You can view object, color, layer, and linetypes properties by hovering over an object. This is a quick way to reference basic object information. See **Figure 15-10.**

Using the Quick Properties Panel

The **Quick Properties** panel, shown in **Figure 15-11,** appears by default when you pick an object. The **Quick Properties** panel only floats, and by default, it appears above and to the right of the crosshairs. The drop-down list at the top of the **Quick Properties** panel indicates the type of object selected. Properties associated with the selected object display below the drop-down list in rows.

Figure 15-10.
Hover over an object to quickly view object, color, layer, and linetype properties.

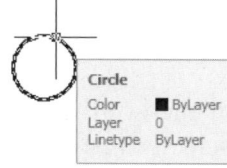

Figure 15-11.
You can use the **Quick Properties** panel to modify some object properties. A—The initial display of the **Quick Properties** panel for the **LINE** tool. B—The expanded list of properties.

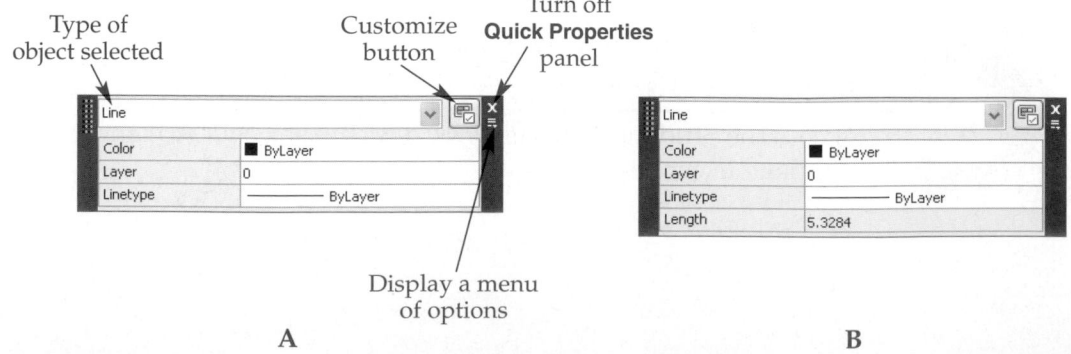

Figure 15-12.
The **Quick Properties** panel with three objects selected. You can edit the objects individually or select All (3) to edit them all together.

Total number of objects selected

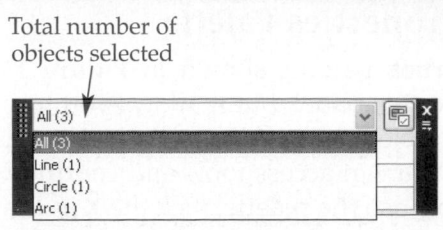

If you pick a circle, for example, rows of circle properties are listed. When you pick multiple objects, use the **Quick Properties** panel to modify all of the objects, or pick only one of the object types from the drop-down list to modify. See **Figure 15-12**. Select All (*n*) to change the properties of all selected objects. Only properties shared by all selected objects appear when you choose All (*n*). Select the appropriate object type to modify a single type of object.

The **Quick Properties** panel lists the most common properties associated with the selected objects. You should recognize most of the properties; they are the same values you use to create the objects. By default, three properties display, unless the selected objects contain fewer properties. If more properties are available, hover over a row or a **Quick Properties** panel side bar to expand the list.

To change a property, pick the property or its current value. The way you change a value depends on the property. A text box opens when you select certain properties, such as the **Radius** property of an arc or circle or the **Text height** property of a single-line or multiline text object. Enter a new value in the text box to change the property. Most text boxes display a calculator icon on the right side that opens the **QuickCalc** tool for calculating values. Chapter 16 covers using **QuickCalc**. Other properties, such as the **Layer** property, display a drop-down list of selections. A pick button is available for geometric properties, such as the **Center X** and **Center Y** properties of a circle. Select the pick button to specify a new coordinate location. When you choose an available **...** (ellipsis) button, a dialog box related to the property opens.

Press [Esc] or pick the **Close** button in the upper-right corner of the panel to hide the **Quick Properties** panel. Closing the **Quick Properties** panel does not disable the tool. If you choose not to use the **Quick Properties** panel, a quick way to disable or enable the panel is to pick the **Quick Properties** button on the status bar.

Exercise 15-7

Access the Student Web site (www.g-wlearning.com/CAD) and complete Exercise 15-7.

Quick Properties Panel Options
For information about adjusting **Quick Properties** panel options, go to the Student Web site (www.g-wlearning.com/CAD), select this chapter, and select **Quick Properties Panel Options**.

Using the Properties Palette

PROPERTIES

Ribbon
View
> Palettes

Properties
Home
> Properties

Properties

Type
PROPERTIES
PROPS
CH
MO
[Ctrl]+[1]

The **Properties** palette, shown in **Figure 15-13,** provides the same function as the **Quick Properties** panel, but it allows you to view and adjust all properties related to the selected objects. You can dock, lock, and resize the **Properties** palette in the drawing area. You can access tools and continue to work while the **Properties** palette is displayed. To close the palette, pick the **X** in the top-left corner, select **Close** from the options menu, or press [Ctrl]+[1].

> **NOTE**
>
> If you have already selected an object, you can access the **Properties** palette by right-clicking and selecting **Properties**. You can also double-click many objects to select the object and open the **Properties** palette automatically.

Categories divide the **Properties** palette. When no object is selected, the **General**, **3D Visualization**, **Plot style**, **View**, and **Misc** categories list the current settings for the drawing. Underneath each category are rows of object properties. For example, in **Figure 15-13,** the Color row in the **General** category displays the current color. The color property in this example is ByLayer.

The upper-right portion of the **Properties** palette contains three buttons. Pick the **Quick Select** button to access the **Quick Select** dialog box, where you can create object selection sets, as described later in this chapter. Picking the **Select Objects** button deselects the currently selected objects and changes the crosshairs to a pick box, allowing you to select other objects. The third button toggles the value of the **PICKADD** system variable, which determines whether you need to hold down [Shift] when adding objects to a selection set.

In order to modify an object using the **Properties** palette, you must open the palette and select an object. For example, if your drawing includes a circle and a line and you want to modify the circle, either double-click the circle or pick the circle and then open the **Properties** palette. Working with properties in the **Properties** palette is very similar working with properties in the **Quick Properties** panel. When you select objects, the drop-down list at the top of the **Properties** palette indicates the types of

Figure 15-13.
You can use the **Properties** palette to modify object properties.

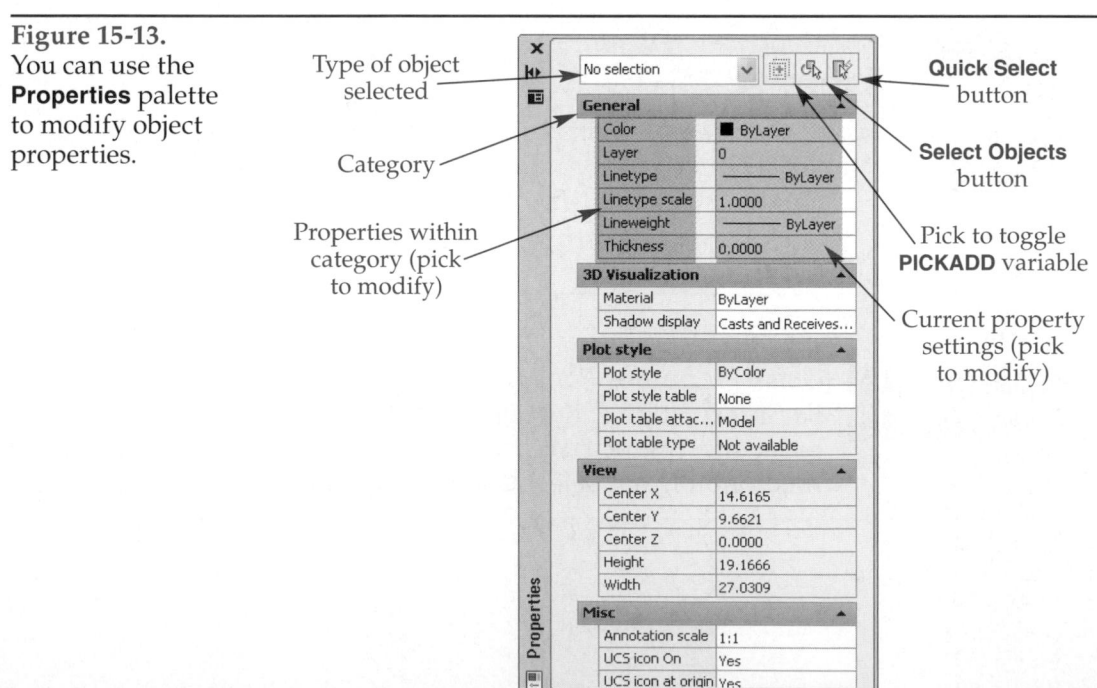

objects selected. The categories and property rows update to display properties associated with your selections.

If you pick a circle, for example, circle properties are listed. When you select multiple objects, use the **Properties** palette to modify all of the objects, or pick only one of the object types from the drop-down list to modify. See **Figure 15-14.** This drop-down list works exactly like the drop-down list in the **Quick Properties** panel.

You will recognize many of the properties listed in the **Properties** palette. However, the **Properties** palette lists all properties related to the object, including some you may not recognize. For example, the **3D Visualization** category and any properties related to the Z axis are for use in 3D applications. You should not adjust these properties or any other properties that you are not familiar with for basic drawing applications.

Change properties using the **Properties** palette the same way you change properties using the **Quick Properties** palette. Use the appropriate text box, drop-down list, or button to modify the value. After you make all the changes to the object, press [Esc] to clear the grips and remove the object from the **Properties** palette. The object now appears in the drawing window with the desired changes.

General Properties

All objects have a **General** category in the **Properties** palette. Refer again to **Figure 15-13.** The **General** category allows you to modify properties such as color, layer, linetype, linetype scale, plot style, lineweight, and thickness. The **Quick Properties** panel also lists certain general properties.

NOTE

You can change the layer of a selected object by choosing a layer from the **Layer Control** drop-down list in the **Layers** panel on the **Home** ribbon tab. You can change color, linetype, plot style, and lineweight by choosing from the appropriate drop-down list in the **Properties** panel on the **Home** ribbon tab.

CAUTION

Colors, linetypes, and lineweights should typically be set as ByLayer. Changing color, linetype, or lineweight to a value other than ByLayer overrides logical properties, making the property an *absolute value.* Therefore, if the color of an object is set to red, for example, it appears red regardless of the layer on which it is drawn.

For most applications, linetype scale should be set globally so the linetype scale of all objects is constant. Adjusting the linetype scale of individual objects can create nonstandard drawings and make it difficult to adjust linetype scale globally. For most applications, you should not override color, linetype, linetype scale, plot style, lineweight, or thickness.

absolute value: In property settings, a value set directly instead of being referenced by layer or a block. An absolute value ignores the current layer settings.

Figure 15-14.
The **Properties** palette with four objects selected. You can edit the objects individually or select All (4) to edit them all together.

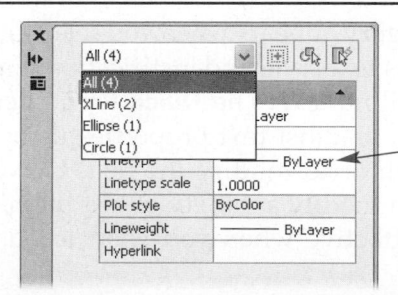

When All is selected only properties common to all selected objects appear

---

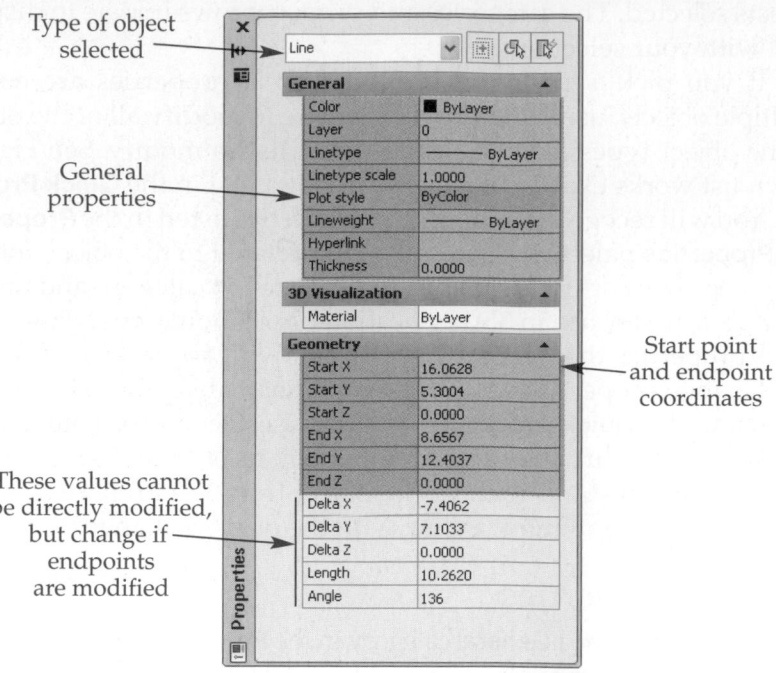

Figure 15-15.
The **Properties**
palette with a Line
object selected.
Line objects have
options in only three
property categories.

Type of object selected

General properties

Start point and endpoint coordinates

These values cannot be directly modified, but change if endpoints are modified

Geometry Properties

One of the most common categories in the **Properties** palette is **Geometry**. See **Figure 15-15.**
Although most objects have a **Geometry** category, the properties within the category
vary depending on the object. The **Quick Properties** panel also lists certain geometry
properties. Typically, three properties allow you to change the absolute coordinates for
the object by specifying the X, Y, and Z coordinates. When you select one of these properties, a pick button appears, allowing you to pick a point in the drawing for the new
location. In addition to choosing a point with the pick button, you can change the value of
the coordinate in a text box or use the calculator button to calculate a new location.

Figure 15-16 shows an example of the properties displayed when you select a
circle. The **Geometry** category displays the current location of the center of the circle
by showing three properties: **Center X**, **Center Y**, and **Center Z**. To choose a new center
location for the circle, select the appropriate property. Pick a new point or type or
calculate the coordinate values. You can also modify other circle size properties, such
as the radius, diameter, circumference, and area.

Exercise 15-8

Access the Student Web site (www.g-wlearning.com/CAD) and
complete Exercise 15-8.

Text Properties

The **Text** category appears when you select single-line or multiline text, and includes
properties such as text style and justification. **Figure 15-17** shows text properties associated with multiline text. The **Quick Properties** panel also lists certain text-specific
properties. You can adjust text properties using traditional techniques, such as reentering the text editor to modify the text. However, the **Properties** palette provides a
convenient way to modify a variety of text properties without reentering the text editor.
It is especially effective when you want to adjust a particular property for multiple

Figure 15-16.
The **Properties** palette with a Circle object selected for editing.

Type of object selected

Pick to modify location

Calculator button

Pick button

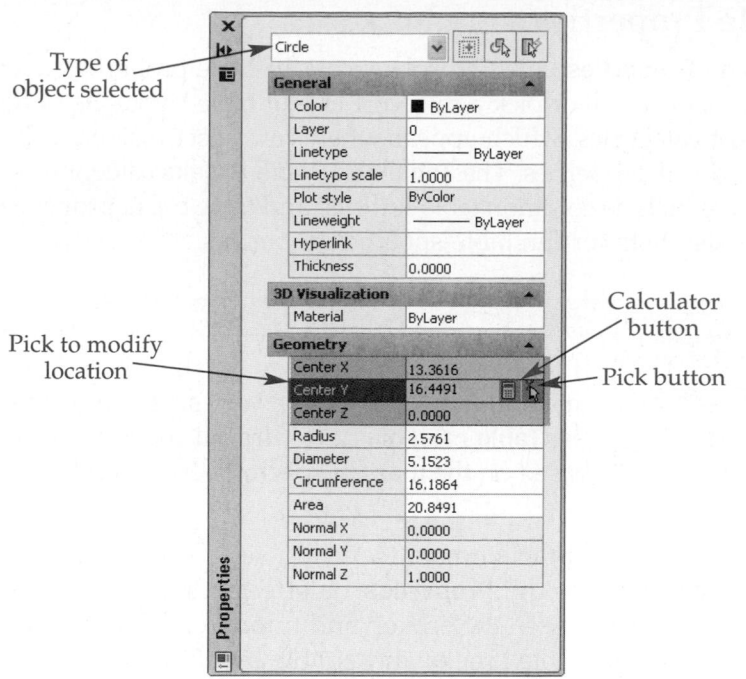

Figure 15-17.
The **Properties** palette shows the properties of the selected text. The properties of single-line text are slightly different from those of multiline text.

Text category

Properties specific to text objects

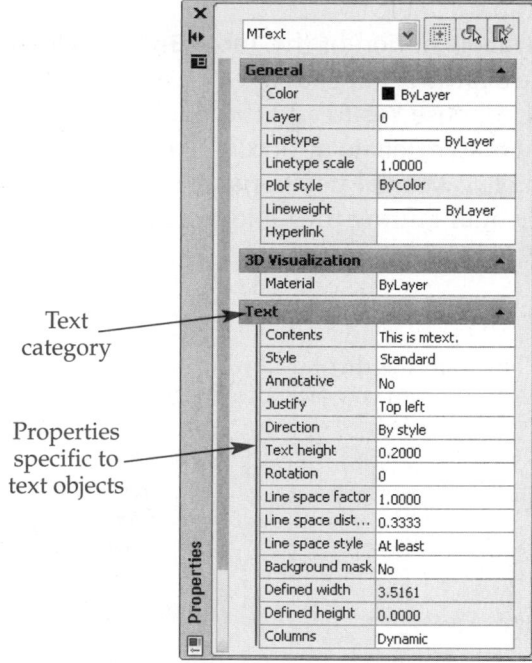

selected text objects. For example, you can change the annotative setting of all text in the drawing using the **Annotative** property row, or reset the height of multiple single-line or multiline text objects using the **Height** property row.

Exercise 15-9

Access the Student Web site (www.g-wlearning.com/CAD) and complete Exercise 15-9.

Table Properties

The **Properties** palette displays certain table properties depending on whether you select inside a cell or pick a cell edge to edit table layout. See **Figure 15-18**. The **Cell** and **Content** categories, which appear when you select inside a cell, allow you to adjust the selected cell properties. The **Table** and **Table Breaks** categories shown when you select a cell edge include common table settings and table break properties. The **Quick Properties** panel also lists certain table-specific properties.

PROFESSIONAL TIP

If the height of table rows is taller than desired, or if rows become unequal in height, enter a very small value in the **Table height** row of the **Table** category to return all rows to the smallest height possible based on the margin spacing between cell content and cell borders.

After you add table breaks, several useful options become available in the **Table Breaks** category of the **Properties** palette for adjusting table sections. The **Enabled** option toggles between the broken and unbroken table display. The **Yes** value displays when you create table breaks and enable breaking. Pick **No** to return the table to an unbroken display. The **Direction** option defines direction of broken table flow, or wrap. The default **Right** option wraps the table to the right. Select **Left** to wrap the table to the left or pick **Up** to wrap the table above.

The **Repeat top labels** option of the **Table Breaks** category repeats cells that use a **Label** cell type at the beginning of each section. Typically, the title cell and header cells use a **Label** cell type. Choose **Yes** to add the title and header cells to the wrapped table sections. The **Repeat bottom labels** option repeats cells that use a **Label** cell type at the end of each section. The **Manual positions** option allows you to move table sections independently while maintaining the table as a single object. When you select **No**, table sections move as a group.

Figure 15-18.
Table properties. A—The properties displayed when you select inside a cell to edit table layout. B—The properties displayed when you pick a cell edge to edit table layout.

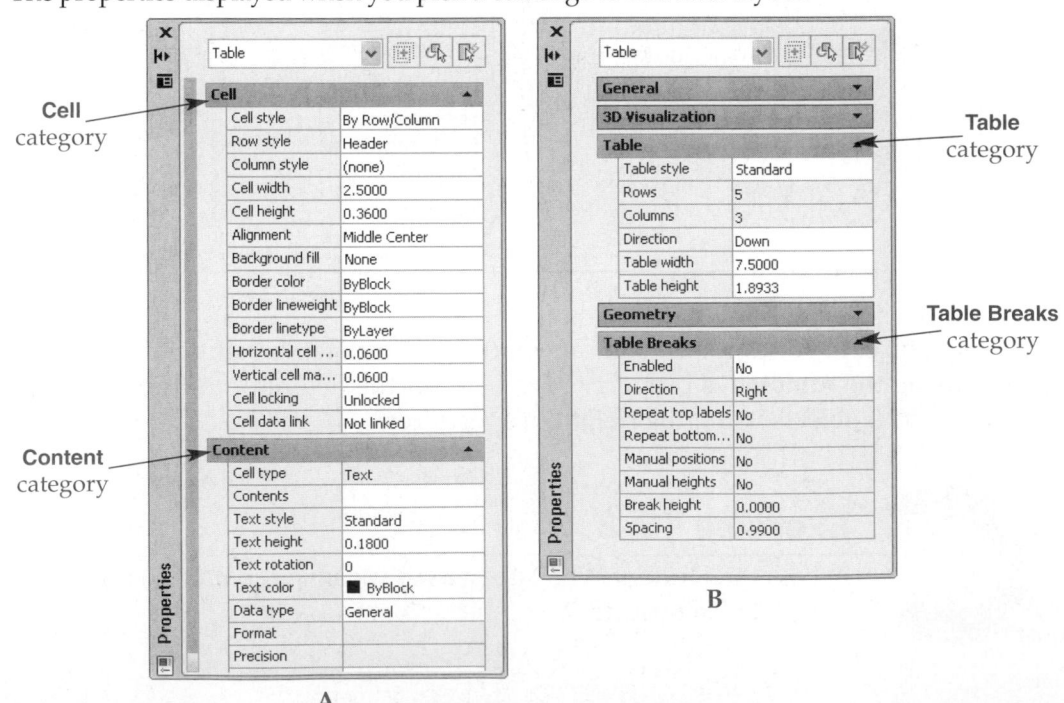

AutoCAD and Its Applications—Basics

The **Manual height** option of the **Table Breaks** category adds a table-breaking grip to each section. This allows you to adjust the number of rows in each section independently and add additional breaks. When you select **No**, the table-breaking grip appears at the original section only and controls the number of breaks. The **Break height** text box allows you to define the height of each table section. The selected height determines the number of sections. The **Spacing** text box allows you to define the spacing between table sections. A value of 0 places the sections together.

Exercise 15-10

Access the Student Web site (www.g-wlearning.com/CAD) and complete Exercise 15-10.

Matching Properties

The **MATCHPROP** tool allows you to copy, or "paint," properties quickly from one object to other objects. You can match properties in the same drawing or between drawings. When you first access the **MATCHPROP** tool, AutoCAD prompts you for the *source object*. After you select the source object, AutoCAD displays the properties it will paint. The next prompt allows you to pick the *destination object*.

To change the paint properties, select the **Settings** option before picking the destination objects. The **Property Settings** dialog box appears, showing the properties to paint. See **Figure 15-19.** The **Basic Properties** area lists the general properties of the source object. The **Special Properties** area contains check boxes that allow you to paint over a variety of additional properties related to object styles. Properties replace in the destination object if the corresponding **Property Settings** dialog box check boxes are active. For example, if you want to paint only the layer property and text style of one text object to another text object, uncheck all boxes except the **Layer** and **Text** property check boxes.

Ribbon
Home
> Clipboard

Match Properties

Type
MATCHPROP
PAINTER
MA

MATCHPROP

source object:
When matching properties, the object with the properties you want to copy to other objects.

destination object:
When matching properties, the object that receives the properties of the source object.

Figure 15-19.
The **Property Settings** dialog box for the **MATCHPROP** tool. Select the properties to paint onto a new object.

Properties to be painted to other objects

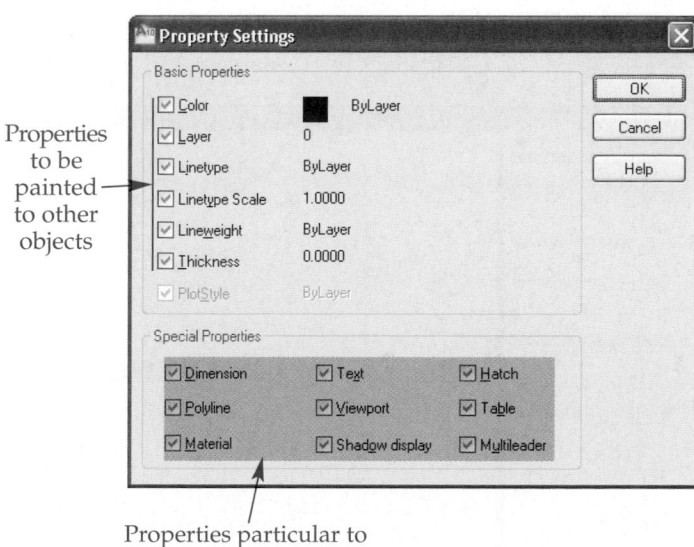

Properties particular to specific objects

Exercise 15-11

Access the Student Web site (www.g-wlearning.com/CAD) and complete Exercise 15-11.

<table><tr><td style="text-align:center">**Editing between Drawings**</td></tr></table>

You can edit in more than one drawing at a time and edit between open drawings. For example, you can copy objects from one drawing to another. You can also refer to a drawing to obtain information, such as a distance, while working in a different drawing.

copy and paste: A Windows function that allows an object to be copied from one location and pasted into another.

Figure 15-20 shows two drawings (the drawings created during Exercises 15-8 and 15-11) tiled horizontally. The Windows *copy and paste* function allows you to copy an object from one drawing to another. To use this feature, select the object you intend to copy. For example, if you want to copy the pentagon from drawing EX15-11 to drawing EX15-8, first select the pentagon. Then right-click to display the shortcut menu shown in **Figure 15-21**.

The shortcut menu has two options that allow you to copy to the Windows Clipboard. The **Copy** option copies selected objects from AutoCAD onto the Windows Clipboard to use in an AutoCAD drawing or another application. The **Copy with Base Point** option also copies the selected objects to the Clipboard, but it allows you to specify a base point to position the copied object for pasting. When you use this option, AutoCAD prompts you to select a base point. Select a logical base point, such as a corner or center point of the object.

After you select one of the copy options, make the second drawing active by picking the drawing or by pressing [Tab]+[Ctrl]. Right-click to display the shortcut menu shown in **Figure 15-22**. Notice that the copy options remain available, but three

---

Figure 15-20.
Tile multiple drawings to make editing between drawings easier.

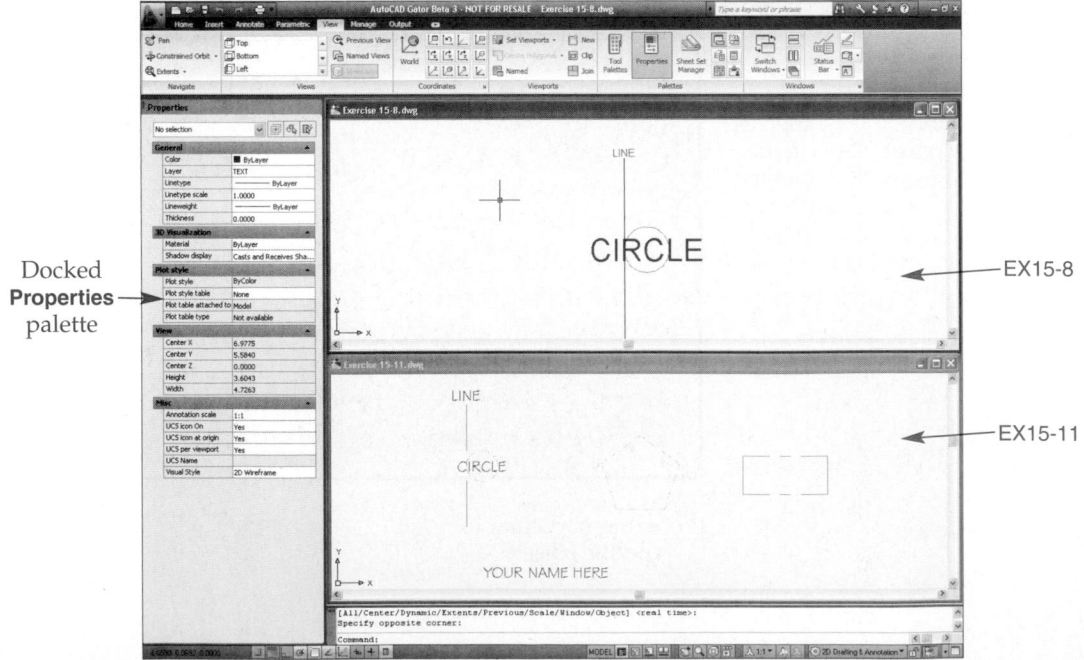

Docked **Properties** palette

EX15-8

EX15-11

AutoCAD and Its Applications—Basics

Figure 15-21.
Two copy options
appear on the
shortcut menu.

Copy options

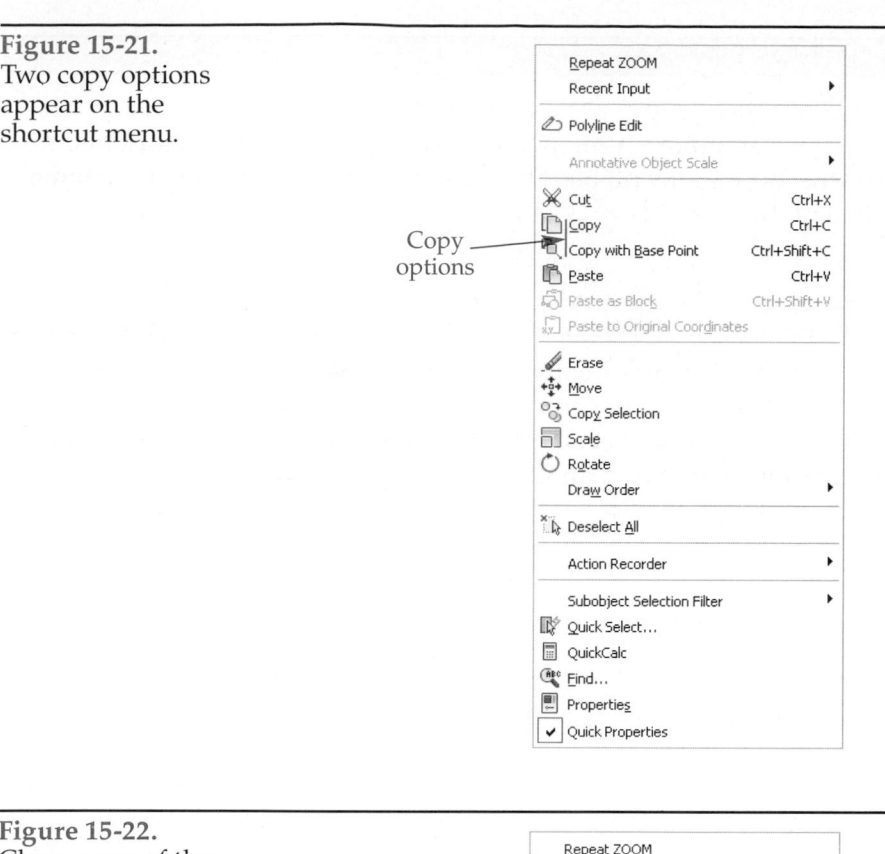

Figure 15-22.
Choose one of the
three paste options
from the shortcut
menu.

Paste options

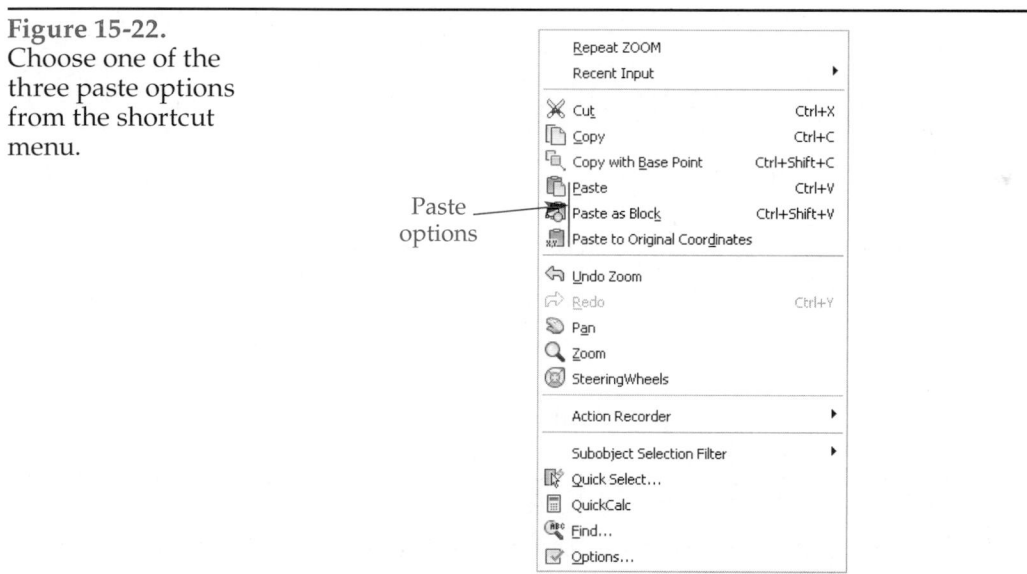

paste options are now available. These options are available only if there is something on the Clipboard. The **Paste** option pastes the information on the Clipboard into the current drawing. If you used the **Copy with Base Point** option, the objects to paste are attached to the crosshairs at the specified base point.

The **Paste as Block** option "joins" all objects on the Clipboard when they are pasted into the drawing. The pasted objects act like a block in that they are single objects grouped together to form one object. Blocks are covered later in this textbook. Use the **EXPLODE** tool to break up the block so that the objects act individually. The **Paste to Original Coordinates** option pastes the objects from the Clipboard to the same coordinates at which they were located in the original drawing.

Exercise 15-12

Access the Student Web site (www.g-wlearning.com/CAD) and complete Exercise 15-12.

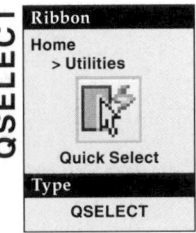

QSELECT

Ribbon
Home
> Utilities

Quick Select

Type
QSELECT

Using Quick Select

When creating complex drawings, you often need to perform the same editing operation on many objects. For example, suppose you designed a complex metal part with more than 40 holes to accept 1/8" screws. A design change occurs, and you are notified that 3/16" screws are to be used instead. Therefore, the hole size must also change. You could select and modify each circle individually, but it would be more efficient to create a selection set of all the circles and then modify them all at the same time. The **Quick Select** dialog box is the most common tool for creating selection sets by specifying object types and property values for selection. See **Figure 15-23**.

With the **Quick Select** dialog box, you can quickly create a selection set based on specified filtering criteria. One method is to specify an object type (such as text, line, or circle) to select throughout the drawing. Another option is to specify a property (such as a color or layer) that objects must possess in order to be selected. A third option is to

Figure 15-23.
Selection sets can be defined in the **Quick Select** dialog box.

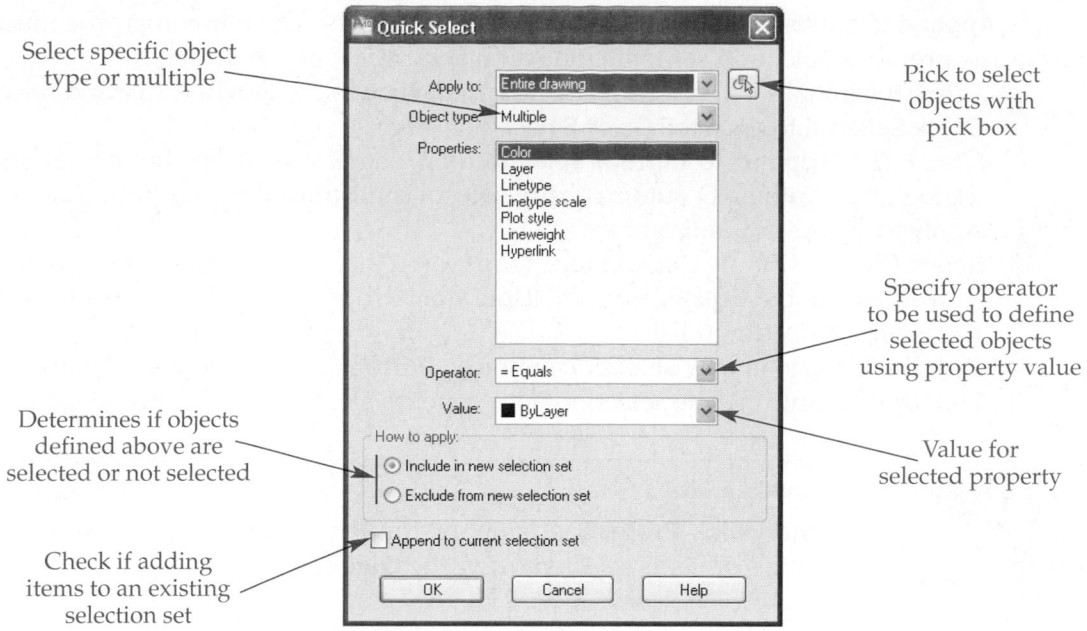

Select specific object type or multiple

Pick to select objects with pick box

Specify operator to be used to define selected objects using property value

Determines if objects defined above are selected or not selected

Value for selected property

Check if adding items to an existing selection set

pick the **Select objects** button and select objects on-screen. Once the selection criteria are defined using one of these techniques, you can use the radio buttons in the **How to apply:** area to include or exclude the selected objects.

Look at **Figure 15-24** as you read the following steps for using the **Quick Select** tool:

1. Open the **Quick Select** dialog box.
2. Select **Entire drawing** from the **Apply to:** drop-down list. (If you access the **Quick Select** dialog box after you select objects, a **Current selection** option allows you to create a subset of the existing set.)
3. Select **Multiple** from the **Object type:** drop-down list. This allows you to select any object type. The drop-down list contains all the object types in the drawing.
4. Select **Color** from the **Properties:** list. The items in the **Properties:** list vary depending on what you specify in the **Object type:** drop-down list.
5. Select **= Equals** from the **Operator:** drop-down list.
6. The **Value:** drop-down list contains values corresponding to the entry in the **Properties:** drop-down list. In this case, color values are listed. Select the color of the right-hand objects in **Figure 15-24A**.
7. Pick the **Include in new selection set** radio button in the **How to apply:** area.
8. Pick the **OK** button to select all objects with the color specified in the **Value:** drop-down list. See **Figure 15-24B**.

Figure 15-24.
Creating selection sets with the **Quick Select** dialog box. A—Objects in the drawing. B—Selection set containing objects with the display color specified. C—Circle object added to the initial selection set.

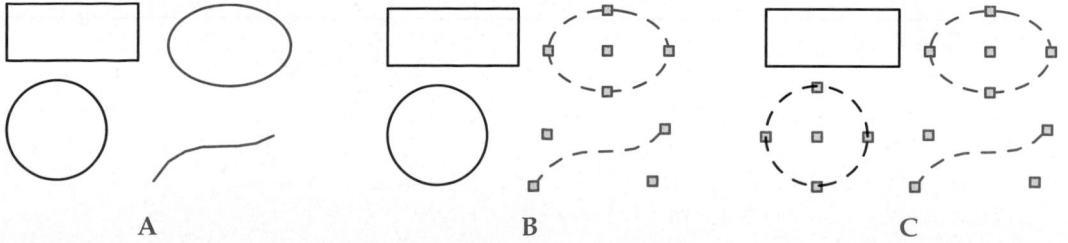

A   B   C

Once you select a set of objects, you can use the **Quick Select** dialog box to refine the selection set. Use the **Exclude from new selection set** option to exclude objects, or use the **Append to current selection set** option to add objects. The following procedure refines the previous selection set to include all black circles in the drawing.

1. While the initial set of objects is selected, right-click in the drawing area and select **Quick Select...** to open the **Quick Select** dialog box.
2. Check the **Append to current selection set** check box at the bottom of the dialog box. AutoCAD automatically selects the **Entire drawing** option in the **Apply to:** drop-down list.
3. Select **Circle** from the **Object type:** drop-down list, **Color** from the **Properties:** drop-down list, **= Equals** from the **Operator:** drop-down list, and **Black** from the **Value:** drop-down list.
4. Pick the **Include in new selection set** radio button in the **How to apply:** area.
5. Pick the **OK** button. The selection now appears as shown in **Figure 15-24C.**

PROFESSIONAL TIP

Use the **Quick Properties** panel or the **Properties** palette to adjust properties of items selected using the **Quick Select** dialog box.

Supplemental Material

Object Selection Filters

For detailed information about selecting multiple objects using the **Object Selection Filters** dialog box, go to the Student Web site (www.g-wlearning.com/CAD), select this chapter, and select **Object Selection Filters**.

Supplemental Material

Creating Object Groups

For information about creating object groups, go to the Student Web site (www.g-wlearning.com/CAD), select this chapter, and select **Creating Object Groups**.

Express Tools

Chapter 15

AutoCAD includes additional tools for improved functionality and productivity during the drawing process. The following Express Tools apply to creating selection sets. For information about these tools, go to the Student Web site (www.g-wlearning.com/CAD), select this chapter, and select **Using Selection Express Tools**.

Get Selection Set
Fast Select

Chapter Test

Answer the following questions. Write your answers on a separate sheet of paper or go to the Student Web site (www.g-wlearning.com/CAD) and complete the electronic chapter test.

1. Name the editing tools that can be accessed automatically using grips.
2. How can you select a grip tool other than the default **STRETCH**?
3. What is the purpose of the **Base Point** option in the grip tools?
4. Explain the function of the **Undo** option in the grip tools.
5. What happens when you choose the **eXit** option from the grips shortcut menu?
6. Which option of the **ROTATE** grip tool option would you use to rotate an object from an existing 60° angle to a new 25° angle?
7. What scale factor would you use to scale an object to become three-quarters of its original size?
8. Describe the options for editing object properties.
9. Where does the **Quick Properties** panel appear by default when an object is selected?
10. By default, how many properties are shown in the **Quick Properties** panel?
11. Describe three items that might be displayed when you pick a property from a **Quick Properties** panel or **Properties** palette row.
12. Identify at least two ways to access the **Properties** palette.
13. Explain how you would change the radius of a circle from 1.375 to 1.875 using the **Properties** palette.
14. How can you change the linetype of an object using the **Properties** palette?
15. For most applications, what value should you use for the color, linetype, and lineweight of objects?
16. What tool is used to change the properties of objects to match the properties of a different object?
17. Briefly explain how the Windows copy and paste function works to copy an object from one drawing to another.
18. Name the paste option that joins a group of objects as a block when they are pasted.
19. When you use the option described in Question 18, how do you separate the objects back into individual objects?
20. Identify three ways to open the **Quick Select** dialog box.

Drawing Problems

Follow these instructions to complete the drawing problems for this chapter:

- *Start AutoCAD if it is not already started.*
- *Start a new drawing using an appropriate template of your choice. The template should include layers, text styles, and table styles necessary for drawing the given objects.*
- *Add layers, text styles, and table styles as needed. Draw all objects using appropriate layers and styles, justification, and format.*
- *Use grips and the associated editing tools or other editing techniques described in this chapter.*
- *Follow the specific instructions for each problem. Do not draw dimensions. Use your own judgment and approximate dimensions when necessary.*

▼ Basic

1. Draw the objects labeled A. Then use the grips **STRETCH** tool to make them look like the objects labeled B. Save the drawing as P15-1.

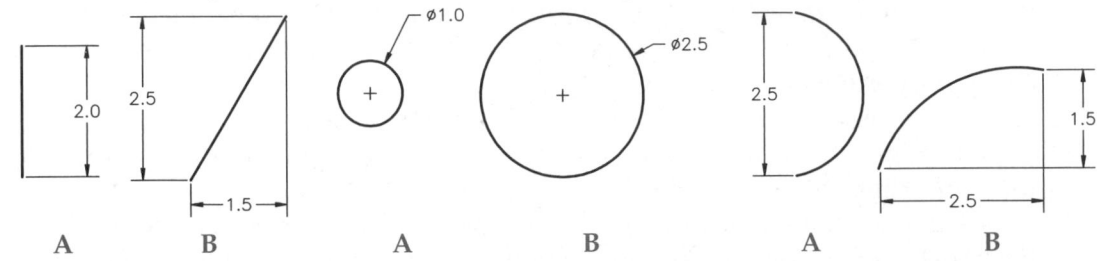

2. Draw the object labeled A. Using the **Copy** option of the grips **MOVE** tool, copy the object to the position labeled B. Edit Object A so it resembles Object C. Edit Object B so it looks like Object D. Save the drawing as P15-2.

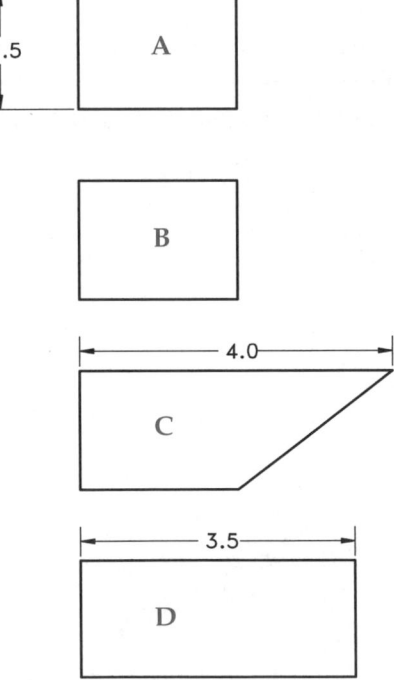

3. Draw the object labeled A. Copy the object, without rotating it, to a position below, as indicated by the dashed lines. Rotate the object 45°. Copy the rotated object labeled B to a position below, as indicated by the dashed lines. Use the **Reference** option to rotate the object labeled C to 25°, as shown. Save the drawing as P15-3.

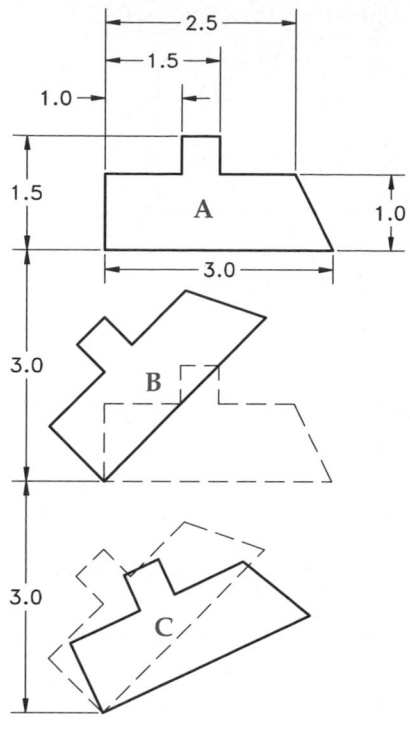

▼ Intermediate

4. Draw the individual objects (vertical line, horizontal line, circle, arc, and C shape) in A using the dimensions given. Use grips and the editing tools to create the object shown in B. Save the drawing as P15-4.

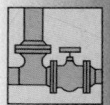

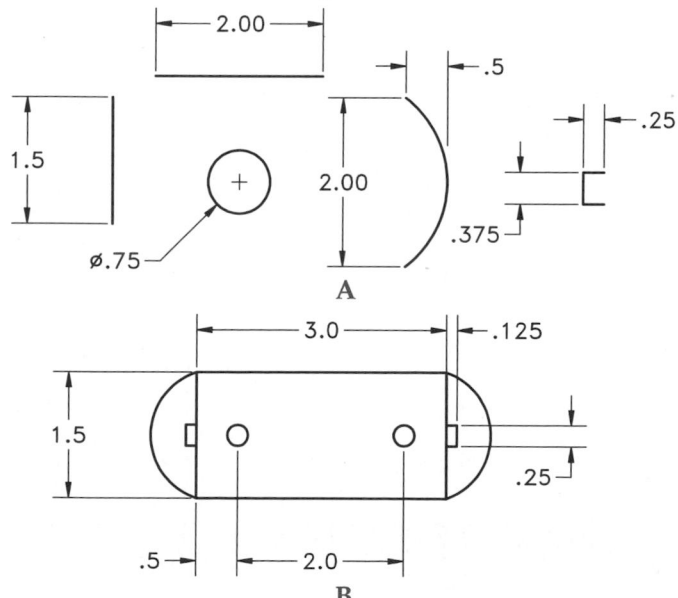

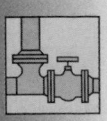

5. Use the completed drawing from Problem 15-4. Erase everything except the completed object and move it to a position similar to that shown in A. Copy the object two times to positions B and C. Use the **SCALE** grip tool to scale the object in position B to 50 percent of its original size. Use the **Reference** option of the **SCALE** tool to enlarge the object in position C from the existing 3.0 length to a 4.5 length, as shown in C. Save the drawing as P15-5.

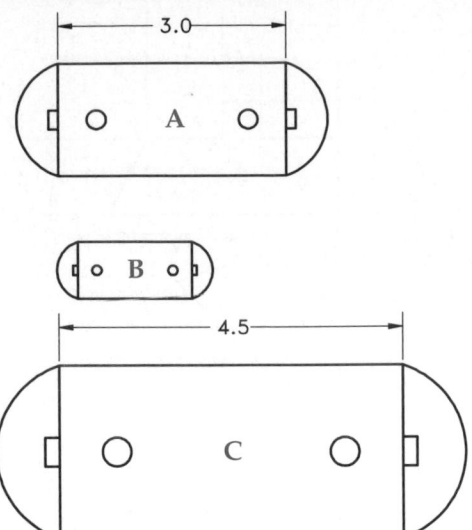

6. Draw the dimensioned partial object shown in A. Mirror the drawing to complete the four quadrants, as shown in B. Change the color of the horizontal and vertical parting lines to Red and the linetype to CENTER. Save the drawing as P15-6.

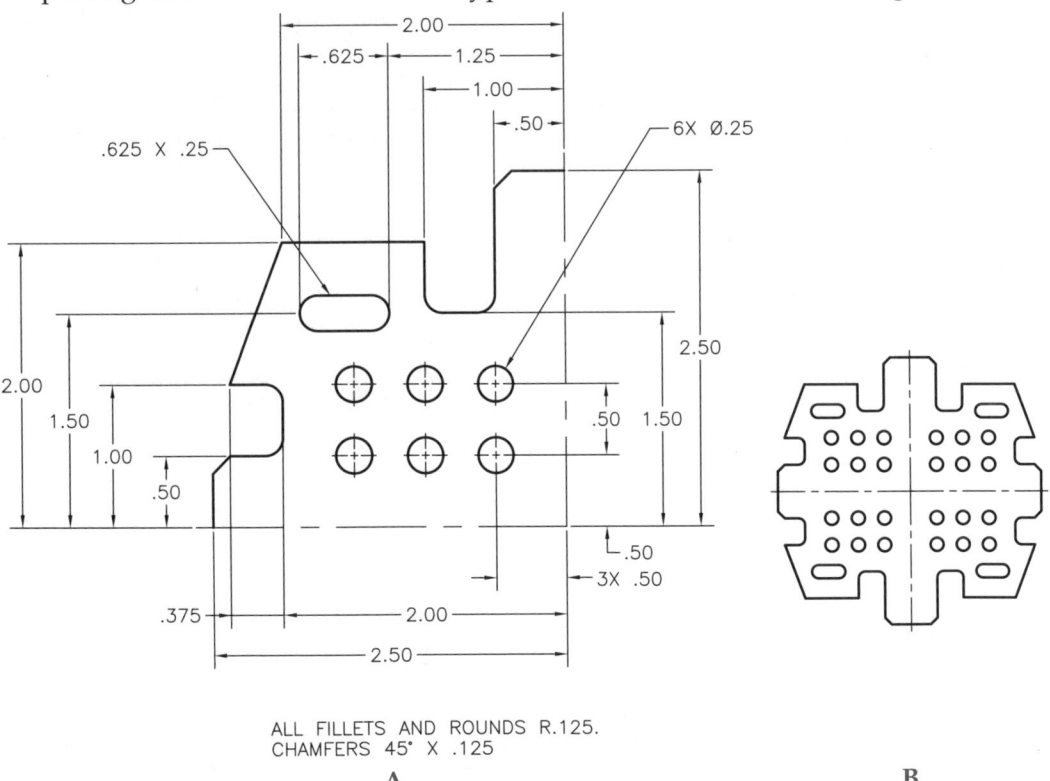

ALL FILLETS AND ROUNDS R.125.
CHAMFERS 45° X .125

A

B

7. Load the final drawing you created in Problem 15-6. Use the **Properties** palette to change the diameters of the circles from .25 to .125. Change the linetype of the slots to PHANTOM. Be sure the linetype scale allows the linetypes to be displayed. Save the drawing as P15-7.

8. Use the editing tools described in this chapter to assist you in drawing the following object. Draw the object within the boundaries of the given dimensions. All other dimensions are flexible. Save the drawing as P15-8.

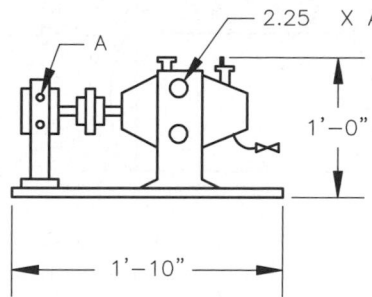

9. Draw the gasket half shown below. Mirror the drawing to complete the other half of the gasket. Save the drawing as P15-9.

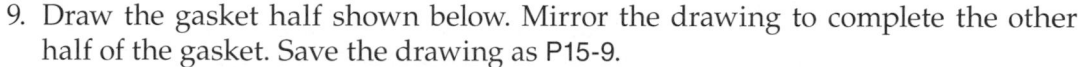

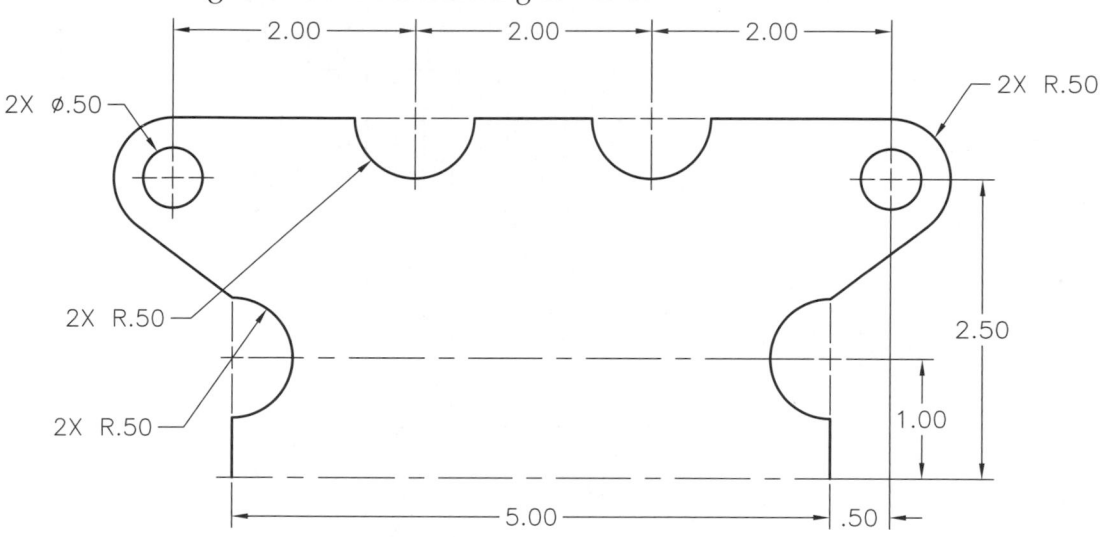

▼ Advanced

10. Draw the following object within the boundaries of the given dimensions. All other dimensions are flexible. After drawing the object, create a page for a vendor catalog, as follows:

   • All labels should be ROMAND text, centered directly below the view. Use a text height of .125".

   • Label the drawing ONE-GALLON TANK WITH HORIZONTAL VALVE.

   • Keep the valve the same scale as the original drawing in each copy.

   • Copy the original tank to a new location and scale it so it is 2 times its original size. Rotate the valve 45°. Label this tank TWO-GALLON TANK WITH 45° VALVE.

   • Copy the original tank to another location and scale it to 2.5 times the size of the original. Rotate the valve 90°. Label this tank TWO-AND-ONE-HALF GALLON TANK WITH 90° VALVE.

   • Copy the two-gallon tank to a new position and scale it so it is 2 times this size. Rotate the valve to 22°30'. Label this tank FOUR-GALLON TANK WITH 22°30' VALVE.

- Left-justify this note at the bottom of the page: Combinations of tank size and valve orientation are available upon request.
- Use the **Properties** palette to change all tank labels to ROMANC, .25" high.
- Change the note at the bottom of the sheet to ROMANS, centered on the sheet, using uppercase letters.
- Save the drawing as P15-10.

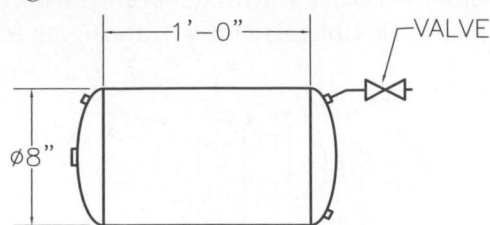

11. Create the interior finish schedule shown below using the **TABLE** tool. Make the measurements for the rows and columns approximately the same as in the given table. Use the **Properties** palette to assist you in constructing the schedule. Save the drawing as P15-11.

INTERIOR FINISH SCHEDULE

ROOM	FLOOR					WALLS			CEILING			
	VINYL	CARPET	TILE	HARDWOOD	CONCRETE	PAINT	PAPER	TEXTURE	SPRAY	SMOOTH	BROCADE	PAINT
ENTRY					•							
FOYER			•			•				•		•
KITCHEN			•					•		•		•
DINING				•		•				•	•	•
FAMILY		•				•				•	•	•
LIVING		•				•		•			•	•
MSTR. BATH			•			•				•		•
BATH #2			•			•			•	•		•
MSTR. BED		•				•		•			•	•
BED #2		•				•				•	•	•
BED #3		•				•				•	•	•
UTILITY	•					•				•	•	•

12. Draw the three views of a sports car, using **SPLINE** to create the curved shapes. Save the drawing as P15-12.

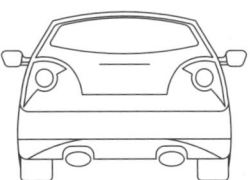 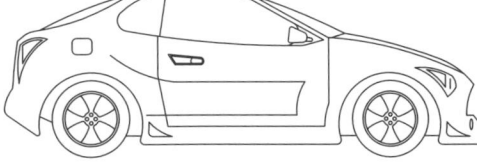 

CHAPTER 16

Obtaining Drawing Information

Learning Objectives

After completing this chapter, you will be able to do the following:

✓ Identify a point location and basic object dimensions.
✓ Find the distance between points.
✓ Measure radii, diameters, and angles.
✓ Calculate the area of objects.
✓ List data related to a single point, an object, a group of objects, or an entire drawing.
✓ Determine the drawing status.
✓ Determine the amount of time spent in a drawing session.
✓ Use fields to display object properties in a drawing.
✓ Perform basic and advanced calculations using the **QuickCalc** calculator.

This chapter describes tools that allow you to retrieve geometric values such as distances, angles, and areas. You will also learn how to access and use additional drawing data, such as object properties and overall drawing status. You can even check to see how much time you spend working on a drawing. This chapter also describes how to use the **QuickCalc** tool to calculate values while you work.

Taking Measurements

Taking measurements from your drawing is a common requirement of designing and drafting. In mechanical drafting, for example, you may have to confirm the size of a hole, or identify the angle between two surfaces. In architectural drafting, you often need to check the dimensions of a room, or calculate building square footage.

You can use grips to view basic geometry dimensions. To identify the location of a point that corresponds to an object grip, confirm that the coordinate display field in the status bar is on. Then pick the object to activate grips and hover over a grip. The exact coordinates of the point appear in the coordinate display field. Dynamic input does not have to be active to identify the coordinates of a grip point, but it must be active to view relevant dimensions between grips. Pick the object to activate grips and hover over a grip to display dimensions. The information that appears varies depending on the object type and the selected grip. See **Figure 16-1**.

Figure 16-1.
Examples of hovering over grips to display geometry dimensions.

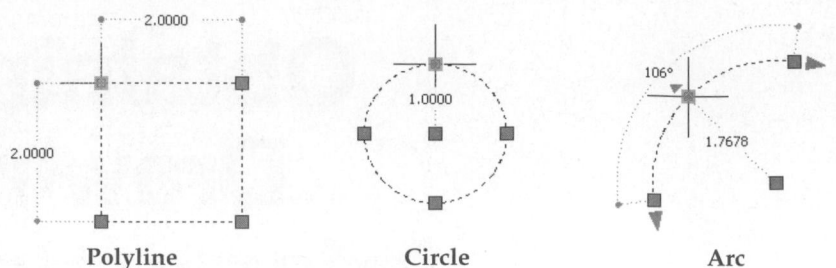

Polyline   Circle   Arc

Exercise 16-1

Access the Student Web site (www.g-wlearning.com/CAD) and complete Exercise 16-1.

NEW

Ribbon
Home
   > Utilities
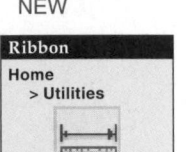
Measure flyout
Type
MEASUREGEOM

MEASUREGEOM

Using the MEASUREGEOM Tool

The **MEASUREGEOM** tool allows you to take a variety of common measurements, including distance, radius, angle, area, perimeter, and volume. The ribbon is an effective way to access **MEASUREGEOM** tool options. See **Figure 16-2.** When you access the **MEASUREGEOM** tool by typing, you must activate a measurement option before you begin. You will notice that the **MEASUREGEOM** tool remains active after you take measurements using the appropriate option. This allows you to continue measuring without reselecting the tool. Select the **eXit** option or press [Esc] to exit the tool.

Measuring Distance

Use the **Distance** option of the **MEASUREGEOM** tool to find the distance between points. Specify the first point followed by the second point. The linear distance between the points, angle *in* the XY plane, and delta X and Y values appear on-screen, as shown in **Figure 16-3,** and at the command line. In a 2D drawing, the angle *from* the XY plane and delta Z values are always 0, as indicated at the command line. As shown in **Figure 16-3,** the first point you specify defines the vertex of the angular dimension.

Once you specify the first point, you can choose the **Multiple points** function to measure the distance between multiple points. AutoCAD calculates the distance between each point and displays the value at the command line during selection.

Figure 16-2.
Use the flyout in the **Utilities** panel of the **Home** ribbon tab to access specific **MEASUREGEOM** tool options.

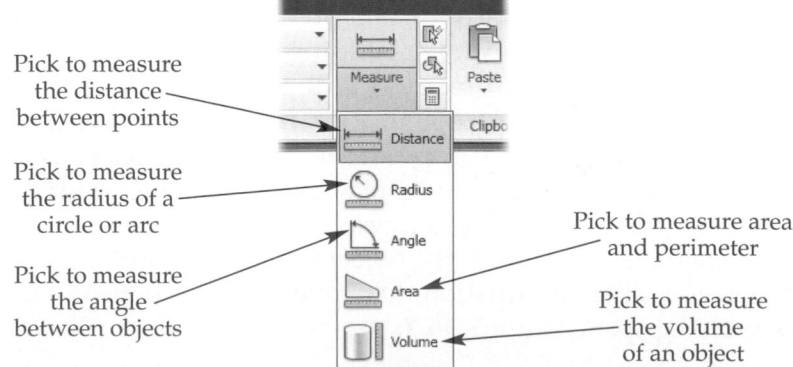

Pick to measure the distance between points

Pick to measure the radius of a circle or arc

Pick to measure the angle between objects

Pick to measure area and perimeter

Pick to measure the volume of an object

Figure 16-3.
The data provided by the **Distance** option. Notice that the first point defines the vertex of the angular value.

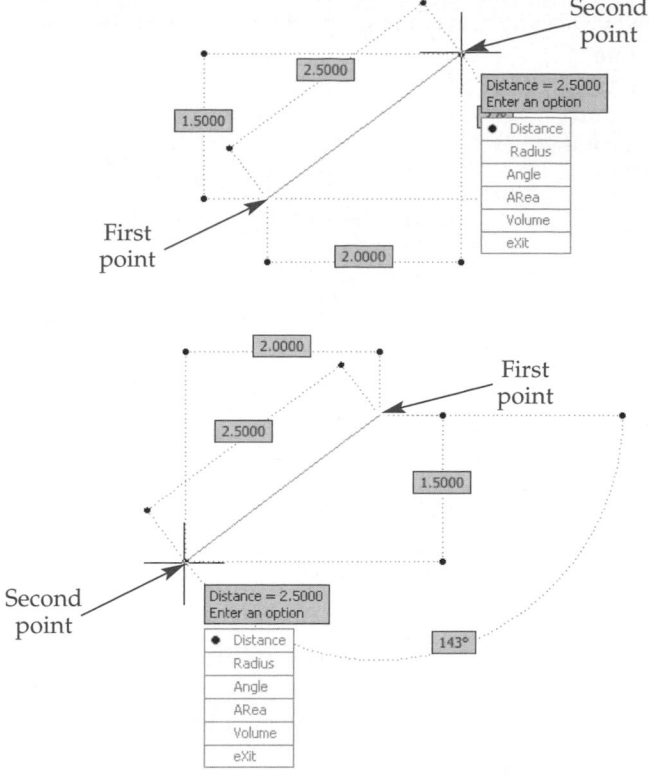

Several options are available for picking multiple points, as described in **Figure 16-4.** When you are finished, use the **Total** or **Close** option to display the distance between all points. **Figure 16-5** shows an example of using the **Multiple points** function to calculate the perimeter of a shape.

NOTE

Use coordinate entry, object snap modes, and other drawing aids to pick points when you use the **MEASUREGEOM** tool.

Figure 16-4.
Options available for the **Multiple points** function of the **Distance** option.

Option	Description
Arc	Measures the length of an arc; includes the same functions available for drawing arcs. Choose the **Line** function to return to measuring the distance between linear points.
Length	Measures the specified length of a line.
Undo	Cancels the effects of an unwanted selection, returning to the previous measurement point.
Total	Finishes multiple point selection and calculates the total distance between points.
Close	Connects the current point to the first point; finishes multiple point selection and calculates the total distance between points.

Figure 16-5.
An example of
using the **Multiple
points** function of
the **Distance** option
to calculate the total
distance between
several points along
lines and an arc.

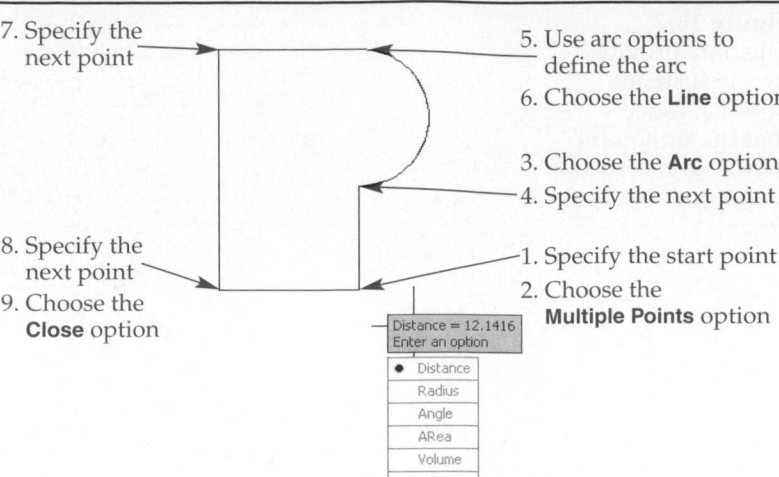

7. Specify the next point

5. Use arc options to define the arc
6. Choose the **Line** option

3. Choose the **Arc** option
4. Specify the next point

8. Specify the next point

9. Choose the **Close** option

1. Specify the start point
2. Choose the **Multiple Points** option

Distance = 12.1416
Enter an option
• Distance
Radius
Angle
ARea
Volume
eXit

PROFESSIONAL TIP

You should typically use the **Multiple points** function of the **Distance** option to calculate the total distance between points of an open shape. The **Area** option of the **MEASUREGEOM** tool, described later in this chapter, is often easier for calculating the perimeter of a closed shape.

Measuring Radius and Diameter

Specify the **Radius** option of the **MEASUREGEOM** tool and pick an arc or circle to find its radius and diameter. The dimensions appear on-screen, as shown in **Figure 16-6,** and at the command line.

Measuring Angle

Use the **Angle** option of the **MEASUREGEOM** tool to find the angle between lines or points. Measure the angle between lines by selecting the first line followed by the second line. The dimension appears on-screen, as shown in **Figure 16-7A,** and at the command line. Select an arc to measure the angle between arc endpoints. See **Figure 16-7B.**

NOTE

Measuring the angle by selecting a circle, followed by a point, is typically most suitable when you are measuring 3D objects such as a cylinder.

Figure 16-6.
The data provided by the **Radius** option. You can use the same option to measure an arc.

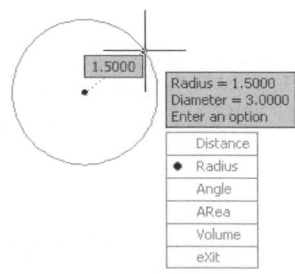

1.5000

Radius = 1.5000
Diameter = 3.0000
Enter an option
Distance
• Radius
Angle
ARea
Volume
eXit

Figure 16-7.
The data provided by the **Angle** option. A—Selecting two lines. B—Picking an arc.

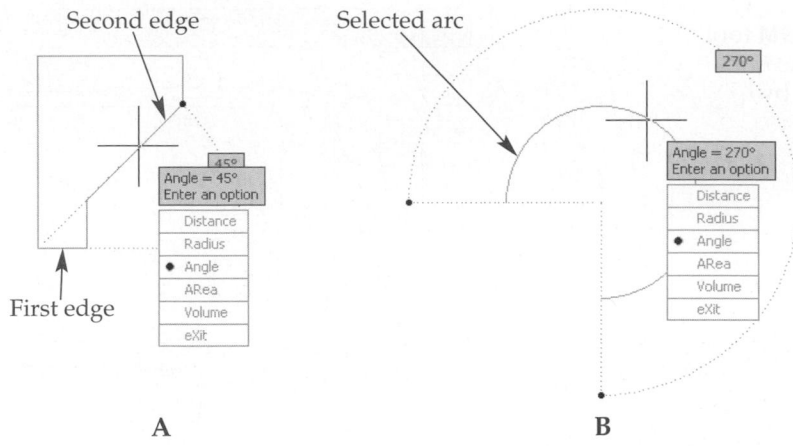

When selecting objects is not appropriate, measure the angle between points by choosing the **Specify vertex** option instead of picking objects. Then select the angle vertex, followed by the first angle endpoint, and finally the second angle endpoint. See **Figure 16-8.**

Exercise 16-2

Access the Student Web site (www.g-wlearning.com/CAD) and complete Exercise 16-2.

Measuring Area

Use the **Area** option of the **MEASUREGEOM** tool to find the area of an object or the area encompassed by selected points. Specify the first corner of the area to measure, followed by all other perimeter corners. Use the **Arc, Length,** or **Undo** functions as needed. These options work the same as they do for measuring distance using the **Multiple points** function of the **Distance** option. A default green (color 100) background fills the area to help you visualize the area. When you are finished specifying perimeter corners, use the **Total** or **Close** option to display the area encompassed by all the points. The area and perimeter appear on-screen, as shown in **Figure 16-9,** and at the command line.

Figure 16-8.
Examples of when it is more appropriate to specify a vertex to measure an angle.

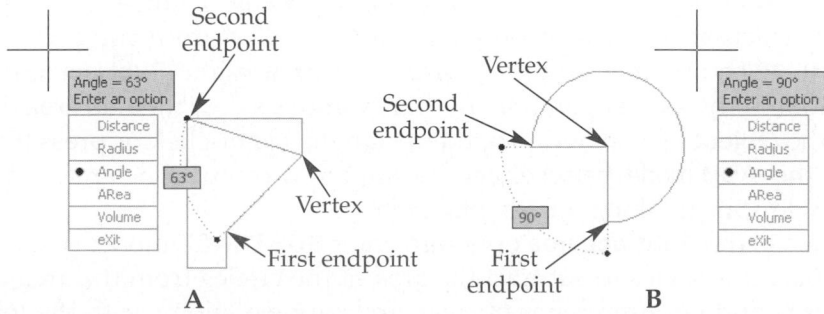

Figure 16-9.
Using the **Area**
option of the
MEASUREGEOM tool
to calculate the area
encompassed by
selected corners.

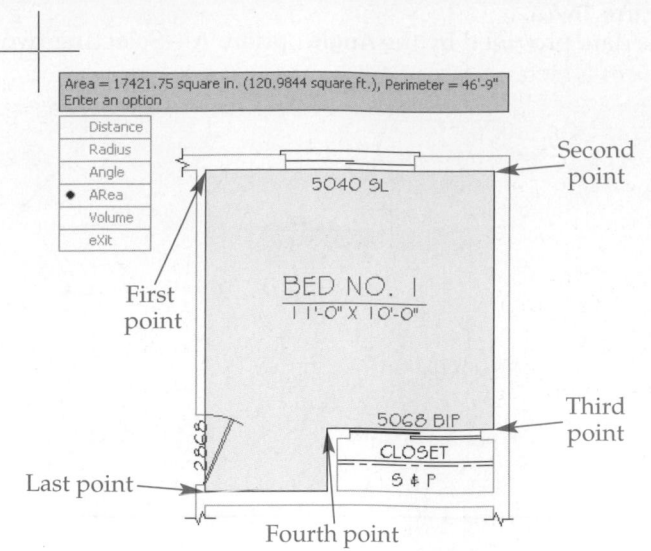

You can find the area of a polyline object, circle, or spline without picking corners by using the **Object** function of the **Area** option. Once you access the **Area** option, activate the **Object** function instead of picking vertices. Then select the object to display the area. AutoCAD displays the area of the object and a second value. The second value returned by the **Object** function varies, depending on the object type, as shown in the following table:

Object	Values Returned
Polyline	Area and length or perimeter
Circle	Area and circumference
Spline	Area and length or perimeter
Rectangle	Area and perimeter

NOTE

You can calculate the area of an open polyline or spline object, but only if the endpoints show an apparent closure.

The **Area** option includes functions that allow you to calculate the sum of multiple different areas during a single operation. Before selecting corners or an object, activate the **Add area** option and define the first area by picking corners or using the **Object** function. Continue adding areas as needed, or use the **Subtract area** option to remove areas from the selection set. As you add or remove areas, a running total of the area automatically calculates. The **Area** option remains in effect until you exit.

Figure 16-10 shows an example of using the **Add area** and **Subtract area** functions of the **Area** tool in the same operation. In this example, select the **Add area** option, and then select the **Object** option and pick the rectangle. Right-click or press [Enter] or the space bar at the (ADD mode) select objects: prompt to continue. The area and perimeter of the rectangle appear, along with a total area.

Then choose the **Subtract area** option to enter **SUBTRACT** mode. Select the **Object** option and pick the circles to subtract the area of the circles from the area of the rectangle. The area and circumference of each circle appear, along with the total area of the rectangle, minus the total area of the subtracted circles. Right-click or press [Enter]

Figure 16-10.
To calculate the area of a rectangle drawn with the **RECTANGLE** tool, first select the outer boundary of the rectangle using the **Add area** function of the **Area** option. Select the inner circle boundaries using the **Subtract area** option. The total calculation is shown.

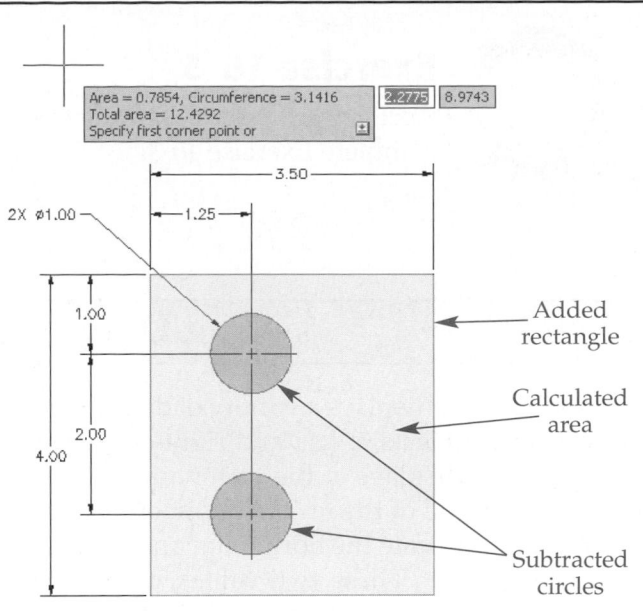

Area = 0.7854, Circumference = 3.1416   2.2775 8.9743
Total area = 12.4292
Specify first corner point or

3.50
2X ⌀1.00
1.25
1.00
2.00
4.00

Added rectangle

Calculated area

Subtracted circles

or the space bar at the (SUBTRACT mode) select objects: prompt to continue. Press [Enter] or the space bar, or right-click and choose **Enter** to display the total area and return to the **MEASUREGEOM** tool prompt.

PROFESSIONAL TIP

Calculating area, circumference, and perimeter values of shapes drawn with the **LINE** tool can be time-consuming, because you must specify each vertex. If you need to calculate areas, it is best to create lines and arcs with the **PLINE** tool. Use the **Object** function of the **Area** option to add or subtract objects.

NOTE

You can use the **Volume** option of the **MEASUREGEOM** tool to measure the volume of a basic drawing in 2D multiview format, but the option is most appropriate for measuring 3D objects. The **Region/Mass Properties** tool provides data related to the properties of a 2D region or 3D solid. *AutoCAD and Its Applications—Advanced* describes the **Volume** option of the **MEASUREGEOM** tool and the **Region/Mass Properties** tool.

NOTE

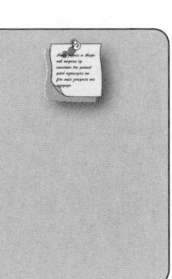

Traditional *inquiry* tools are available by typing command names. The **ID** tool allows you to display the coordinates of a single selected point. You can use the **DIST** tool to find the distance between points, and the **AREA** tool to calculate areas. You will find that these tools function much like the options available with the **MEASUREGEOM** tool, but they are not as interactive.

Exercise 16-3

Access the Student Web site (www.g-wlearning.com/CAD) and complete Exercise 16-3.

Listing Drawing Data

LIST

Ribbon

Home
> Properties

List

Type

LIST
LI
LS

The **LIST** tool displays a variety of data about any AutoCAD object. Access the **LIST** tool, select the objects to list, and right-click or press [Enter] or the space bar. The data for each object displays at the command line and in the text window. **Figure 16-11A** shows an example of the text window displayed when you list a line. The Delta X and Delta Y values indicate the horizontal and vertical distances between the *from point* and *to point* of the line. These two values, along with the length and angle, provide you with four measurements for a single line. See **Figure 16-11B**.

Figure 16-12 shows an example of the text window displayed with a selection set of multiple objects, including a circle, and a rectangle, and multiline text. When you select multiple objects and not all of the information fits in the window, AutoCAD prompts you to press [Enter] to display additional information.

Figure 16-11.
A—An example of the text window displayed when you use the **LIST** tool to list the properties of a line. B—The various data and measurements of a line provided by the **LIST** tool.

```
AutoCAD Text Window - 16-11.dwg
Edit
Command: LIST

Select objects: 1 found

Select objects:
                LINE      Layer: "0"
                          Space: Model space
                Handle = 86
          from point, X=   4.2766  Y=   4.5303  Z=   0.0000
            to point, X=   6.1992  Y=   5.0454  Z=   0.0000
    Length =   1.9904,   Angle in XY Plane =      15
            Delta X =   1.9226, Delta Y =    0.5152, Delta Z =   0.0000

Command:
```

A

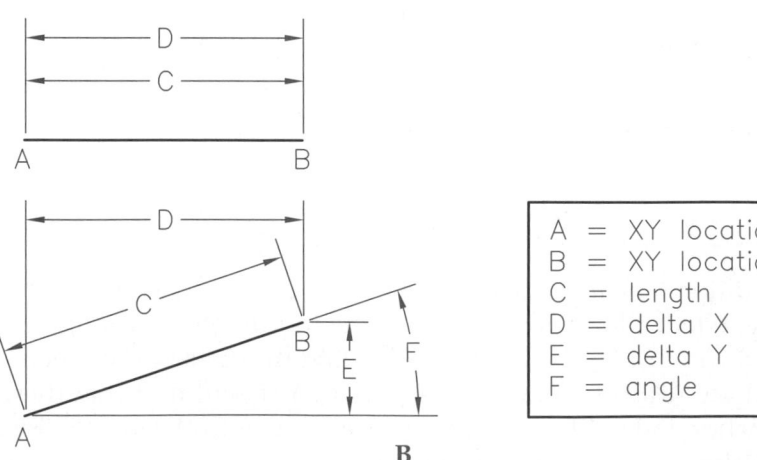

A = XY location
B = XY location
C = length
D = delta X
E = delta Y
F = angle

B

Figure 16-12.
Using the **LIST** tool
to list the properties
of multiline text,
a circle, and a
rectangle.

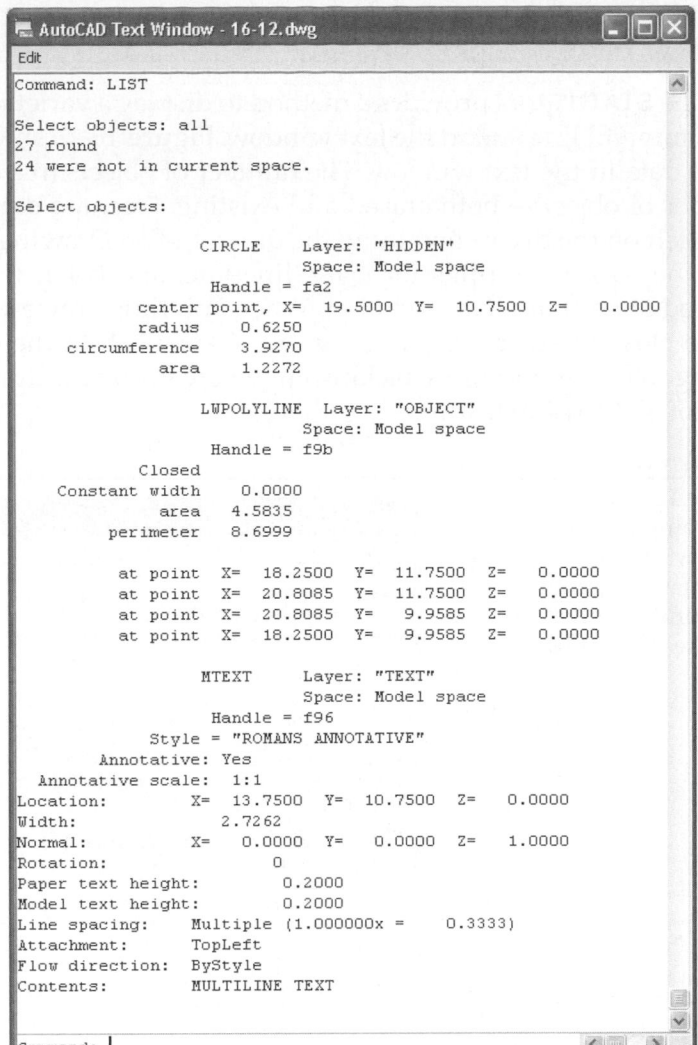

```
AutoCAD Text Window - 16-12.dwg                                    _ □ X
Edit
Command: LIST

Select objects: all
27 found
24 were not in current space.

Select objects:

                    CIRCLE     Layer: "HIDDEN"
                               Space: Model space
                Handle = fa2
        center point, X=  19.5000  Y=  10.7500  Z=    0.0000
          radius    0.6250
    circumference   3.9270
           area     1.2272

                    LWPOLYLINE  Layer: "OBJECT"
                                Space: Model space
                Handle = f9b
          Closed
   Constant width    0.0000
            area     4.5835
        perimeter    8.6999

        at point  X=  18.2500  Y=  11.7500  Z=   0.0000
        at point  X=  20.8085  Y=  11.7500  Z=   0.0000
        at point  X=  20.8085  Y=   9.9585  Z=   0.0000
        at point  X=  18.2500  Y=   9.9585  Z=   0.0000

                    MTEXT      Layer: "TEXT"
                               Space: Model space
                Handle = f96
            Style = "ROMANS ANNOTATIVE"
        Annotative: Yes
    Annotative scale:  1:1
Location:        X=  13.7500  Y=  10.7500  Z=    0.0000
Width:           2.7262
Normal:          X=   0.0000  Y=   0.0000  Z=    1.0000
Rotation:           0
Paper text height:       0.2000
Model text height:       0.2000
Line spacing:    Multiple (1.000000x =    0.3333)
Attachment:      TopLeft
Flow direction:  ByStyle
Contents:        MULTILINE TEXT

Command: |
```

NOTE

The **DBLIST** (database list) tool lists all data for every object in the
current drawing. To use this tool, enter DBLIST. The information
appears in the same format used by the **LIST** tool, although the text
window does not appear automatically.

PROFESSIONAL TIP

The **LIST** tool provides most information about an object, including the
area and perimeter of polylines. The **LIST** tool also reports object color
and linetype, unless both are BYLAYER.

Exercise 16-4

Access the Student Web site (www.g-wlearning.com/CAD) and
complete Exercise 16-4.

Reviewing the Drawing Status

Type
STATUS

The **STATUS** tool provides a method to display a variety of drawing information at the command line and in the text window. **Figure 16-13** shows an example of drawing status data in the text window. The number of objects in a drawing refers to the total number of objects—both erased and existing. Free dwg disk (C:) space: represents the space left on the drive containing the drawing file. Drawing aid settings appear, along with the current settings for layer, linetype, and color. Press [Enter] if necessary to proceed to additional information. When you finish reviewing the information, press [F2] to close the text window. You can also switch to the drawing window without closing the text window by picking anywhere inside the drawing window or using the Windows [Alt]+[Tab] feature.

Figure 16-13.
An example of drawing information in the text window when you use the **STATUS** tool.

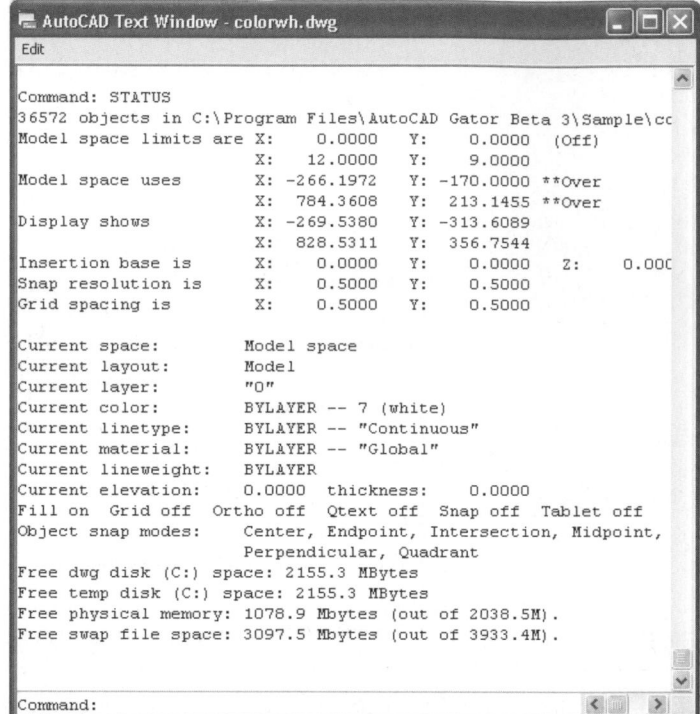

```
AutoCAD Text Window - colorwh.dwg
Edit

Command: STATUS
36572 objects in C:\Program Files\AutoCAD Gator Beta 3\Sample\cc
Model space limits are X:      0.0000   Y:      0.0000   (Off)
                       X:     12.0000   Y:      9.0000
Model space uses       X:   -266.1972   Y:   -170.0000 **Over
                       X:    784.3608   Y:    213.1455 **Over
Display shows          X:   -269.5380   Y:   -313.6089
                       X:    828.5311   Y:    356.7544
Insertion base is      X:      0.0000   Y:      0.0000   Z:      0.000
Snap resolution is     X:      0.5000   Y:      0.5000
Grid spacing is        X:      0.5000   Y:      0.5000

Current space:        Model space
Current layout:       Model
Current layer:        "0"
Current color:        BYLAYER -- 7 (white)
Current linetype:     BYLAYER -- "Continuous"
Current material:     BYLAYER -- "Global"
Current lineweight:   BYLAYER
Current elevation:      0.0000   thickness:      0.0000
Fill on  Grid off  Ortho off  Qtext off  Snap off  Tablet off
Object snap modes:      Center, Endpoint, Intersection, Midpoint,
                        Perpendicular, Quadrant
Free dwg disk (C:) space: 2155.3 MBytes
Free temp disk (C:) space: 2155.3 MBytes
Free physical memory: 1078.9 Mbytes (out of 2038.5M).
Free swap file space: 3097.5 Mbytes (out of 3933.4M).

Command:
```

Checking the Time

Type
TIME

The **TIME** tool allows you to display the current time and time related to the current drawing session. **Figure 16-14** shows an example of drawing time data in the text window. The drawing creation time starts when you begin a new drawing, not when you first save a new drawing. The **SAVE** tool affects the Last updated: time. However, all drawing session time erases when you exit AutoCAD and do not save the drawing.

You can time a specific drawing task using the **Reset** option of the **TIME** tool to reset the elapsed timer. The timer is on by default when you enter the drawing area. Use the **OFF** option to stop the timer. If the timer is off, use the **ON** option to start it again. Time information is static, which means the times you view may be old. Use the **Display** option to request an update.

Figure 16-14.
An example of
the text window
displayed when you
use the **TIME** tool.

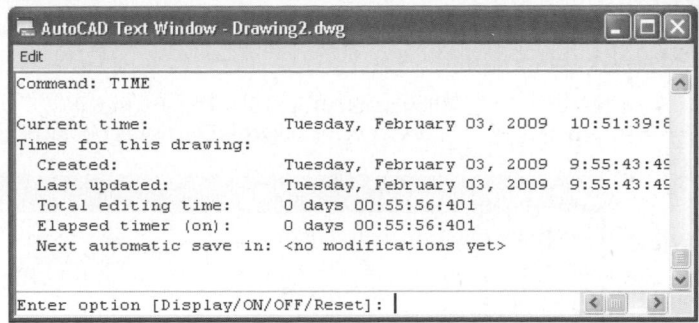

```
AutoCAD Text Window - Drawing2.dwg
Edit
Command: TIME

Current time:                 Tuesday, February 03, 2009   10:51:39:8
Times for this drawing:
  Created:                    Tuesday, February 03, 2009   9:55:43:49
  Last updated:               Tuesday, February 03, 2009   9:55:43:49
  Total editing time:         0 days 00:55:56:401
  Elapsed timer (on):         0 days 00:55:56:401
  Next automatic save in:  <no modifications yet>

Enter option [Display/ON/OFF/Reset]: |
```

NOTE

The Windows operating system maintains the date and time settings for the computer. You can change these settings in the Windows Control Panel. To access the Control Panel, pick Settings and then Control Panel from the Start menu.

Exercise 16-5

Access the Student Web site (www.g-wlearning.com/CAD) and complete Exercise 16-5.

Displaying Information with Fields

You can use *fields* to list a variety of object properties and drawing information. Each object type has different properties that you can display in a field. For example, you can use a field in an mtext or text object near a circle to list the area and circumference of the circle.

Use the **FIELD** tool and **Field** dialog box, shown in **Figure 16-15,** to add fields to mtext or text objects. To insert a field in an active mtext editor, pick the **Field** button from the **Insert** panel of the **Text Editor** ribbon tab, pick the **Insert Field** option available from the shortcut menu, or press [Ctrl]+[F]. To insert a field in an active single-line text editor, right-click and select **Insert Field...** or press [Ctrl]+[F].

In the **Field** dialog box, pick Objects from the **Field category:** drop-down list, and then pick Object in the **Field names:** list box. Pick the **Select object** button to return to the drawing window and pick an object. When you select an object, the **Field** dialog box reappears with the available properties listed. See **Figure 16-16.** Pick the property, select the format, and pick the **OK** button to insert the field. Once you create the field, whenever you modify the object, the value displayed in the field updates to reflect the new value. In addition to object property settings such as layer, linetype, lineweight, and plot style, you can include many dimensional properties in a field.

field: Text object that displays a property, setting, or value for an object, drawing, or computer system.

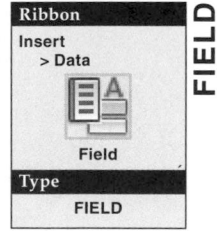

Ribbon
Insert
> Data
Field
Type
FIELD

FIELD

Exercise 16-6

Access the Student Web site (www.g-wlearning.com/CAD) and complete Exercise 16-6.

Figure 16-15.
Pick the Object field to add a property for a specific object to a field. Pick the **Select object** button to select the object.

Select category

Pick to list properties of a specific object

Pick to select object

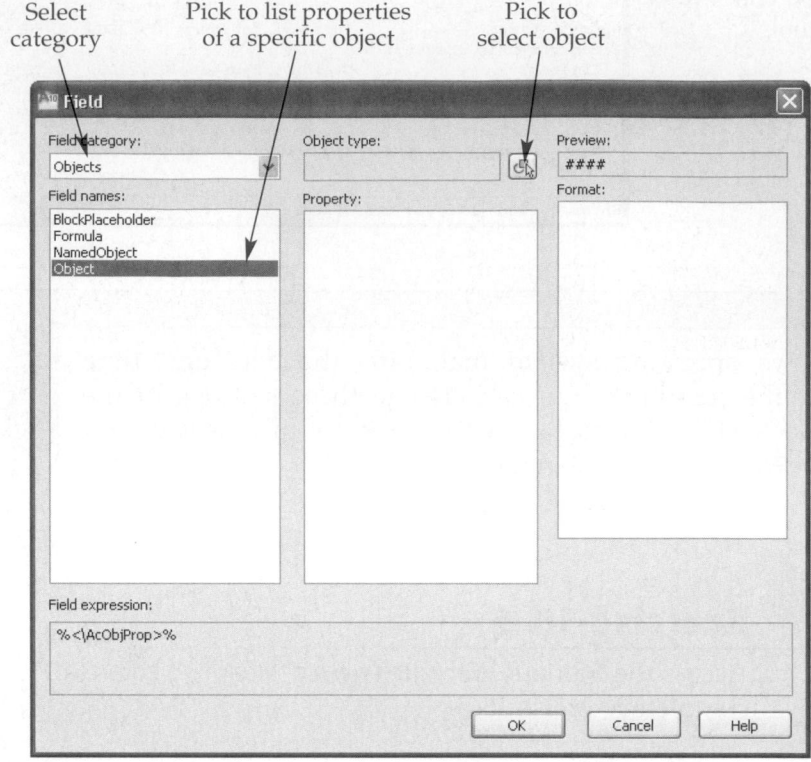

Field expression:
%<\AcObjProp>%

Figure 16-16.
After you pick the object, properties specific to the object type are listed. Select the property and format for the field.

Properties available to field

Value of selected property

Format options are property specific

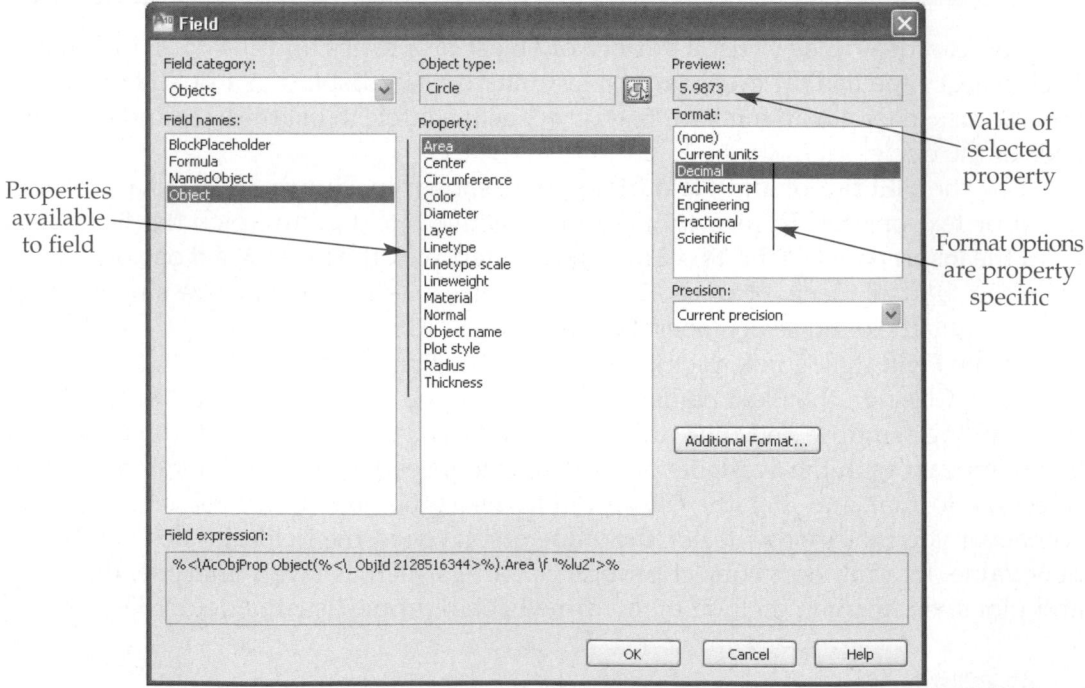

Field expression:
%<\AcObjProp Object(%<_ObjId 2128516344>%).Area \f "%lu2">%

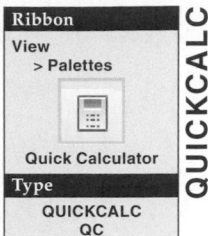

Ribbon
View
   > Palettes

Quick Calculator

Type
  QUICKCALC
    QC

QUICKCALC

Using QuickCalc

Most drafting projects require you to make calculations. For example, when working from a sketch with missing dimensions, you may need to calculate a distance or angle, or you may need to double-check dimensions. Often drafters make calculations using a handheld calculator. An alternative is to use **QuickCalc**, which is a palette containing a basic calculator, scientific calculator, units converter, and variables feature. See **Figure 16-17.** You can use **QuickCalc** as you would a handheld calculator, but you can also use it while a tool is active to paste calculations when a prompt asks for a specific value.

> **NOTE**
>
> You can also access the **QuickCalc** palette by right-clicking in the drawing window and selecting **QuickCalc**, or from specific areas such as the button that sometimes appears next to a text box in the **Properties** palette.

Entering Expressions

The basic math functions used in numeric expressions include addition, subtraction, multiplication, division, and exponential notation. You can enter grouped expressions by using parentheses to break up the expressions that calculate separately. For example, to calculate 6 + 2 and then multiply the sum by 4, enter (6+2)*4. The result will be wrong if you do not add the parentheses. The following table shows the symbols used for basic math operators.

Symbol	Function	Example
+	Addition	3+26
–	Subtraction	270–15.3
*	Multiplication	4*156
/	Division	256/16
^	Exponent	22.6^3
( )	Grouped expressions	2*(16+2^3)

Figure 16-17.
The **QuickCalc** palette allows you to perform a variety of calculations.

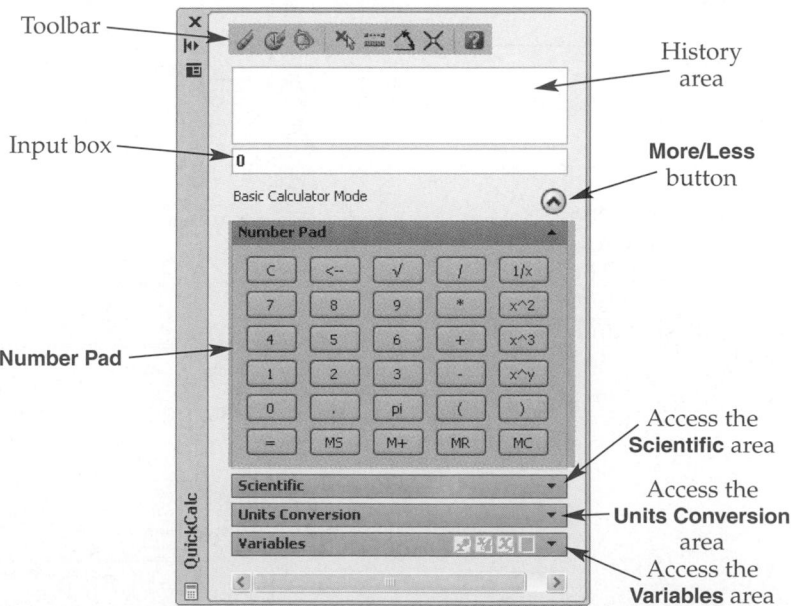

Toolbar

History area

Input box

More/Less button

Basic Calculator Mode

Number Pad

Access the **Scientific** area

Access the **Units Conversion** area

Access the **Variables** area

Figure 16-18.
The basic **Number Pad** area contains additional options that you cannot access using the keyboard.

To get the square root of a value

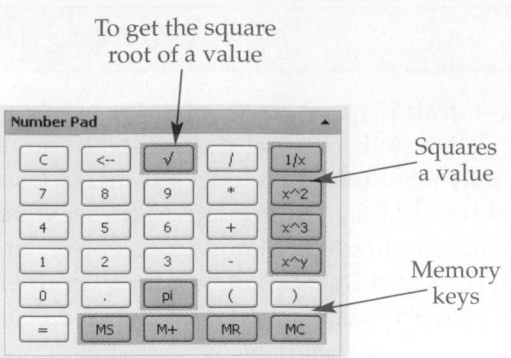

Squares a value

Memory keys

You can add expressions in the input box by picking buttons on the **Number Pad** or by pressing keyboard keys. After creating the expression, pick the equal (=) button on the number pad or press [Enter] to evaluate the expression. **Figure 16-18** shows options found on the basic **Number Pad** that are not available from the keyboard.

The result of an evaluated expression appears in the input box, and the expression moves to the history area. **Figure 16-19** displays the **QuickCalc** palette after calculating 96.27 + 23.58. When you are using only the input box of **QuickCalc**, pick the **More/Less** button below the input box to hide the additional sections, saving valuable drawing space. When the **QuickCalc** palette displays all areas, the button is an up arrow and its tooltip reads **Less**. Pick the button again to display hidden areas.

NOTE

If you move the cursor outside of the **QuickCalc** palette, the drawing area automatically becomes active. Pick anywhere inside the **Quick-Calc** palette to reactivate it.

PROFESSIONAL TIP

If you make a mistake in the input box, you do not need to clear the input box and start over again. Use the left and right arrow keys to move through the field. Right-click in the input box to display a shortcut menu that includes useful options for copying and pasting.

Figure 16-19.
A—Add an expression into the input box.
B—After creating the expression, press [Enter] to evaluate the expression. The history area stores the expression and result.

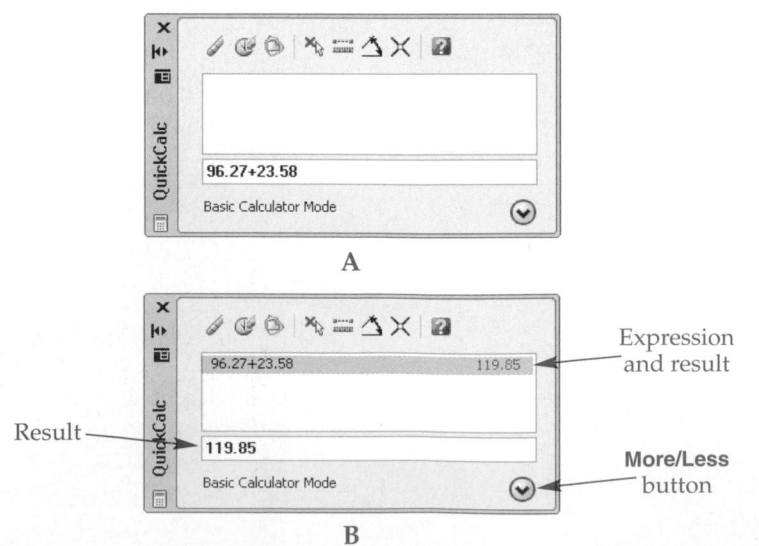

Result

Expression and result

More/Less button

Clearing the Input and History Areas

After picking the equal (=) button or pressing [Enter] to evaluate an expression, you can create a new expression without clearing the last result. AutoCAD automatically starts a new expression. You can clear the input box manually by placing the cursor in the input box and pressing [Backspace] or [Delete], or by picking the **Clear** button from the **QuickCalc** toolbar. To clear the history area, pick the **Clear History** button. See **Figure 16-20.**

CAUTION

If you enter an expression that **QuickCalc** cannot evaluate, AutoCAD displays an **Error in Expression** dialog box. Pick the **OK** button, correct the error, and try it again.

Exercise 16-7

Access the Student Web site (www.g-wlearning.com/CAD) and complete Exercise 16-7.

Scientific Calculations

The **Scientific** area of **QuickCalc** includes trigonometry, exponential, and some geometric functions. See **Figure 16-21.** To use one of the functions, add a value to the input box, pick the appropriate function button, and pick the equal (=) button or press [Enter]. When you pick a function button, the input box value appears in parentheses after the expression. For example, to get the *sine* of 14, clear the input box, type 14 in the input box and pick the **sin** button. The input box now reads sin(14). Pick the equal (=) button or press [Enter] to view the result.

NOTE

You can pick a function button first, but doing so puts a default value of 0 in parentheses. You can then place the cursor in the input box to type a different number in the parentheses if needed.

Figure 16-20.
You can use the buttons on the **QuickCalc** toolbar to clear the input box and history areas.

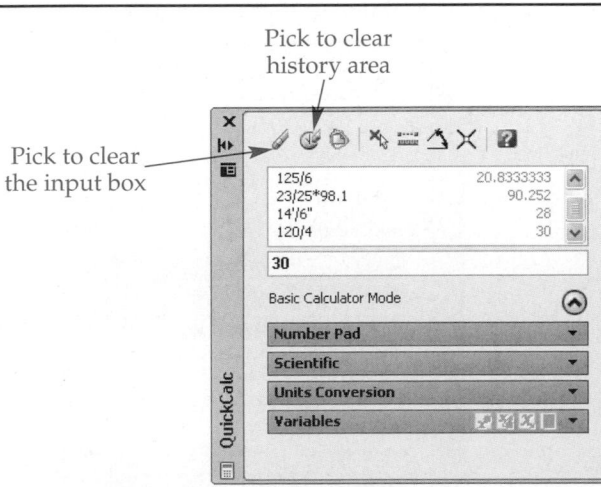

Pick to clear history area

Pick to clear the input box

Figure 16-21.
The scientific functions available in **QuickCalc**.

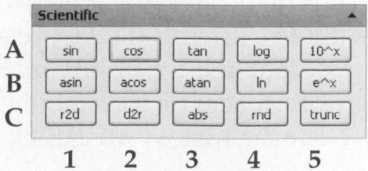

	1	2	3	4	5
A	Sine	Cosine	Tangent	Base–10 Log	Base–10 Exponent
B	Arcsine	Arccosine	Arctangent	Natural Log	Natural Exponent
C	Convert Radians to Degrees	Convert Degrees to Radians	Absolute Value	Round	Truncate

Converting Units

The **Units Conversion** area allows you to convert one unit type to another. The unit types available are **Length**, **Area**, **Volume**, and **Angular**. For example, to use the unit converter to convert 23 centimeters to inches, pick in the **Units type** field to display the drop-down list. See **Figure 16-22.** Pick the drop-down list button to display the different unit types and select Length. Activate the **Convert from** field and select Centimeters from the drop-down list. Activate the **Convert to** field and select Inches from the drop-down list. Type 23 in the **Value to convert** field and pick the equal (=) button or press [Enter]. The **Converted value** field displays the converted units.

To pass the converted value to the input box for use in an expression, pick the **Return Conversion to Calculator Input Area** button. See **Figure 16-23.** If the button is not visible, pick once on the converted units in the **Converted value** field.

Figure 16-22.
Picking the current unit type activates the field and displays the drop-down list button.

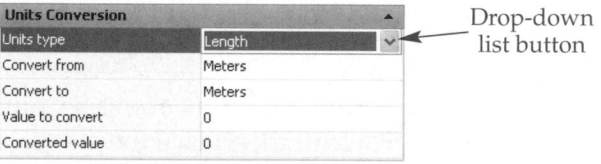

Figure 16-23.
After you convert a value, pass the value to the input box for use in an expression.

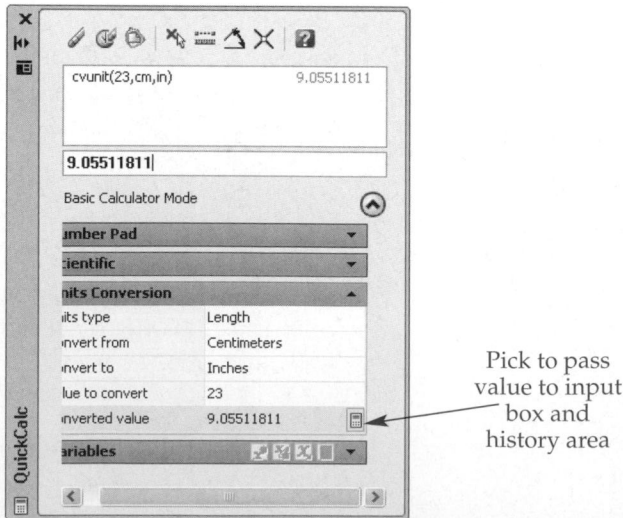

Exercise 16-8

Access the Student Web site (www.g-wlearning.com/CAD) and complete Exercise 16-8.

Using Variables

If you use an expression or value frequently, you can save it as a *variable*. Use the **Variables** area, shown in **Figure 16-24,** to create, edit, delete, and pass variables to the input box. The **Variables** area of **QuickCalc** includes two types of predefined variables: *constant* and *function*.

To create a new variable, select the **New Variable...** button to open the **Variable Definition** dialog box shown in **Figure 16-25.** Type a name for the variable in the **Name:** field. Select a group to contain the variable using the **Group with:** field. Type the value or the expression for the variable in the **Value or expression:** field. Give a description for the variable in the **Description** field. Pick the **OK** button to save the variable and display it in the **Variables** area.

To edit a variable, pick the variable to modify and select the **Edit Variable** button to reopen the **Variable Definition** dialog box. Pick the **Delete** button to delete the selected variable. Pick the **Return Variable to Input Area** button to pass the selected variable to the input box. You can also pass the variable to the input box by double-clicking the variable name.

You can also access variable tools by right-clicking in the **Variable** area to display a shortcut menu. Two additional options are available from the menu. Pick the **New Category** option to create a new category for saving variables. Select the **Rename** option to rename the selected variable. You can also rename a variable by selecting it, pausing, and then selecting it again to activate the name for editing.

variable: Text item that represents another value and can be accessed later as needed.

constant: Expression or value that stays the same.

function: Expression or value that asks for user input to get values that can be passed to the expression.

Figure 16-24.
The **Variables** area of **QuickCalc** allows you to store values and expressions for later use.

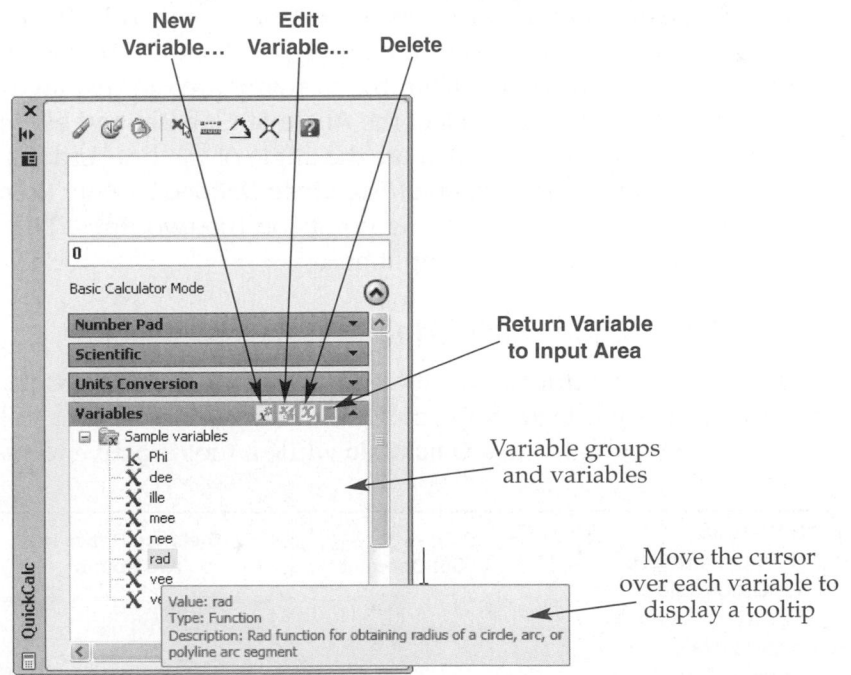

Figure 16-25.
Use the **Variable Definition** dialog box to define new variables.

Choose the type of variable

Group the new variable with the samples or create a new category

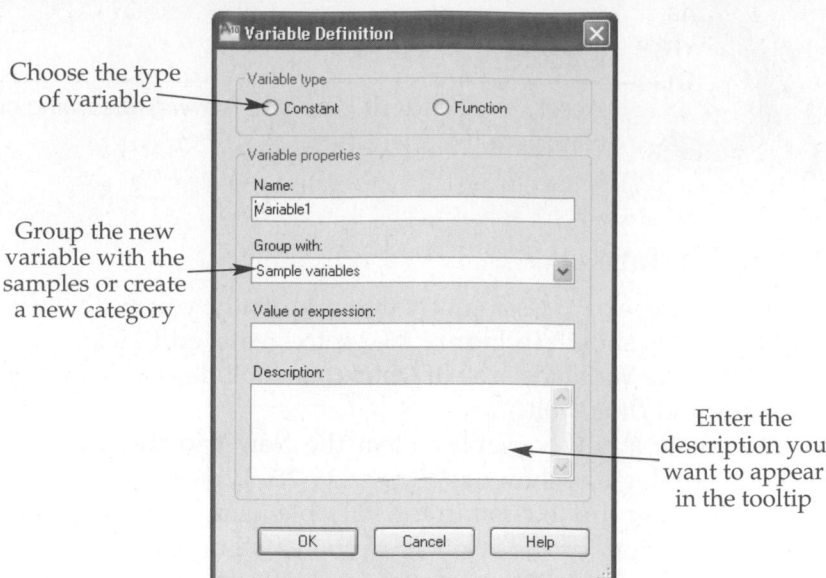

Enter the description you want to appear in the tooltip

Using Drawing Values

Tools available from the **QuickCalc** toolbar allow you to pass values from the drawing to the **QuickCalc** input box, and from the input box to the drawing. When you select any of the buttons shown in **Figure 16-26,** the **QuickCalc** palette temporarily hides so that you can select points from the drawing window.

Pick the **Get Coordinates** button to select a point from the drawing window and display the X,Y,Z coordinates in the input box. Pick the **Distance Between Two Points** button and select two points from the drawing area to display the distance between the points in the input box. Select the **Angle of Line Defined by Two Points** button and pick two points on a line to calculate the angle of the line and display the angle in the input box. Select the **Intersection of Two Lines Defined by Four Points** button to find the intersection of two lines by picking points on the two lines. The X,Y,Z coordinates of the intersection appears in the input box.

Using QuickCalc with Tools

The previous information focuses on using the **QuickCalc** palette to calculate unknown values while drafting, much like using a handheld calculator or the Windows Calculator. You can also use **QuickCalc** while a tool is active to *pass* a calculated value

Figure 16-26.
You can use buttons on the **QuickCalc** toolbar to pass values from **QuickCalc** to AutoCAD and retrieve values from AutoCAD to pass to **QuickCalc.**

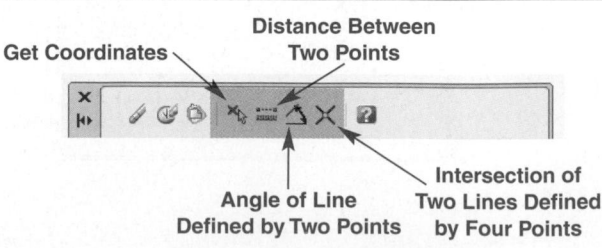

Get Coordinates

Distance Between Two Points

Angle of Line Defined by Two Points

Intersection of Two Lines Defined by Four Points

to the command line as a response to a prompt. There are a few alternatives for using **QuickCalc** while a tool is active.

If the **QuickCalc** palette is active when you access a tool, when the prompt requesting an unknown value appears, calculate the value using the **QuickCalc** palette and then press the **Paste value to command line** button to pass the value to the command line. For example, to draw a line a distance of 14'8" + 26'3" horizontally from a start point, activate the **QuickCalc** palette, access the **LINE** tool, and pick a start point. Then use polar tracking or **Ortho** mode to move the crosshairs to the right or left of the start point so the line is at a 0° angle. At the Specify next point or [Undo]: prompt, enter 14'8" + 26'3" in the **QuickCalc** palette input box and pick the equal (=) button or press [Enter]. The result in the input box is 40'11". Pick the **Paste value to command line** button to make 40'11" appear at the command line. Press [Enter] or the space bar or right-click and select **Enter** to draw the 40'11" line.

If the **QuickCalc** palette is not active while you are using a tool, you can still calculate and use a value. When the prompt requesting an unknown value appears, access **QuickCalc** by right-clicking and selecting **QuickCalc** or typing 'QC. A **QuickCalc** *window*, which is not the same as the **QuickCalc** palette, opens in command calculation mode. See **Figure 16-27.** Use the necessary tools to evaluate an expression. Then pick the **Apply** button to pass the value back to the command line, and close the **QuickCalc** window.

PROFESSIONAL TIP

The units you use in **QuickCalc** must match the drawing units. In the preceding example, which uses architectural units, the drawing units must also be architectural. If needed, use the **Drawing Units** dialog box to change the drawing units to Architectural.

Exercise 16-9

Access the Student Web site (www.g-wlearning.com/CAD) and complete Exercise 16-9.

Figure 16-27.
The **QuickCalc** window that appears when you access **QuickCalc** while a tool is active. The **Apply** and **Close** buttons are available at the bottom of the window.

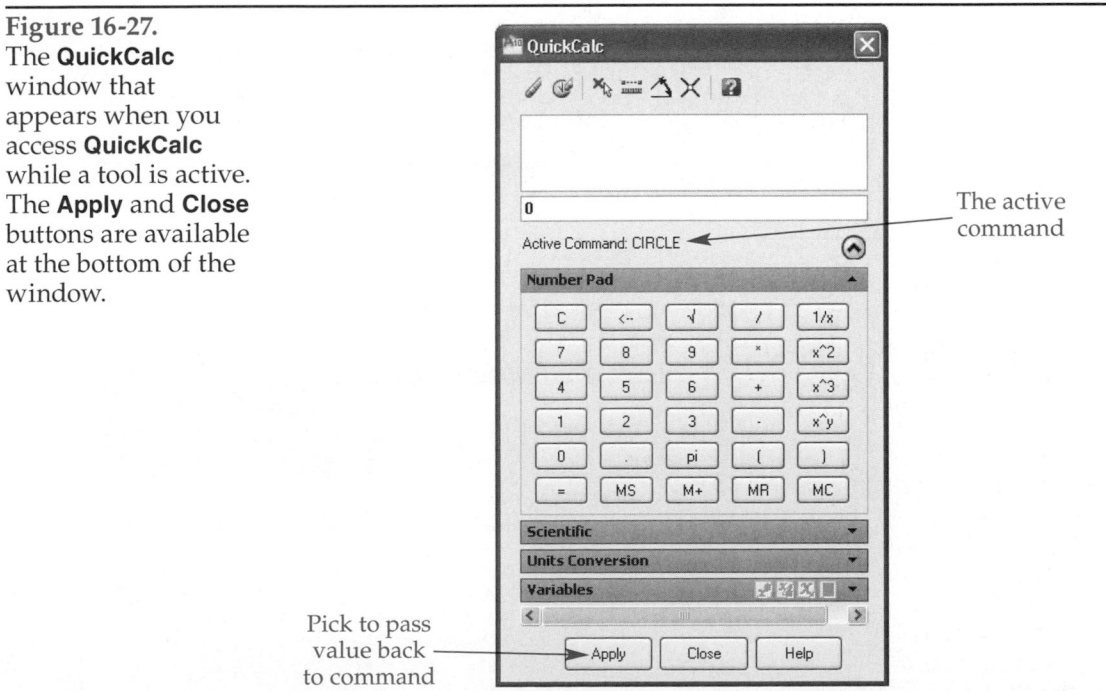

The active command

Pick to pass value back to command

Using QuickCalc with Object Properties

QuickCalc also allows you to calculate expressions for an object while using the **Properties** palette. Pick a field that contains a numeric value to display the calculator icon. **Figure 16-28** shows a selected circle and the active **Radius** field in the **Properties** palette. Pick the calculator icon to open the **QuickCalc** window, again not the same item as the **QuickCalc** palette, in property calculation mode. Use expressions and values in the same manner as when using **QuickCalc** at any other time. Once you evaluate the expression in the input box, pick the **Apply** button to pass the value to the property field in the **Properties** palette. The object automatically updates based on the new value.

Additional QuickCalc Options

The history area contains some settings and features that you can only access from a shortcut menu. This menu, shown in **Figure 16-29,** displays when you right-click anywhere in the history area. The following options are available:

- **Expression Font Color.** Changes the color of the expression font.
- **Value Font Color.** Changes the color of the value font.
- **Copy.** Copies the expression and value to the Windows clipboard.
- **Append Expression to Input Area.** Passes the expression to the input box.
- **Append Value to Input Area.** Passes the value to the input box.
- **Clear History.** Clears the history area.
- **Paste to Command Line.** Passes the value to the Command: prompt.

As with other palettes, picking the **Properties** button on the **QuickCalc** palette displays the **Properties** menu. The settings allow you to change the palette appearance, including its ability to dock, hide, or appear transparent.

Figure 16-28.
The calculator icon appears when you select a numeric field in the **Properties** palette.

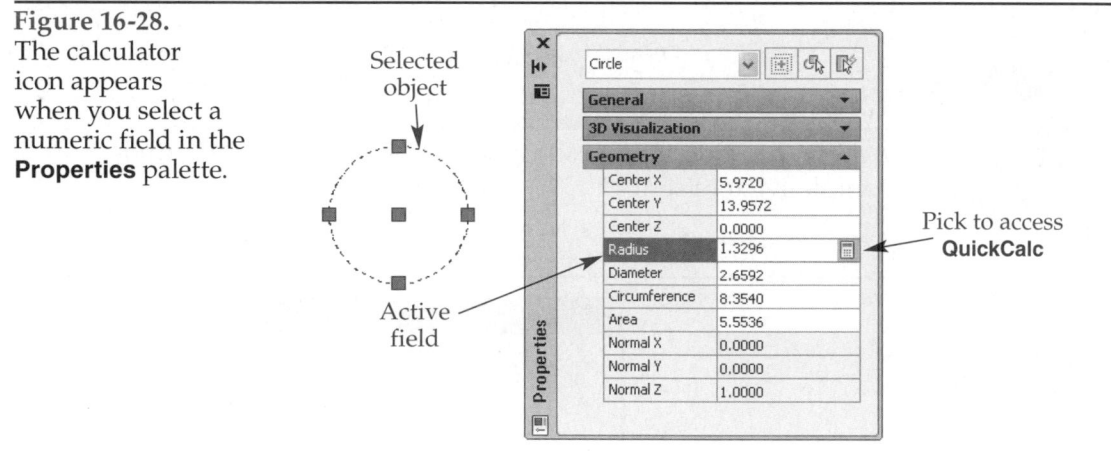

Figure 16-29.
The history area shortcut menu contains additional functions and settings.

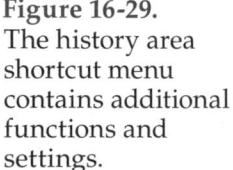
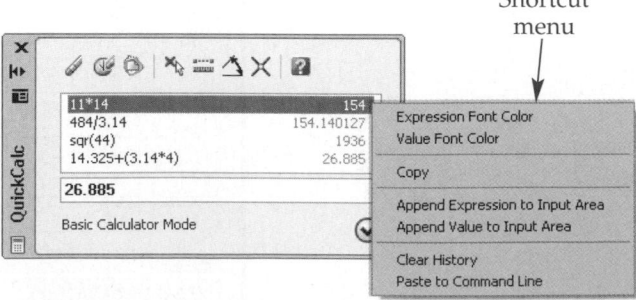

Chapter Test

Answer the following questions. Write your answers on a separate sheet of paper or go to the Student Web site (www.g-wlearning.com/CAD) and complete the electronic chapter test.

1. Explain how to use grips to identify the location of a point, and dimensions of an object.
2. What types of information does the **Distance** option of the **MEASUREGEOM** tool provide?
3. What information does the **Area** option of the **MEASUREGEOM** tool provide?
4. To add the areas of several objects when using the **Area** option of the **MEASUREGEOM** tool, when do you select the **Add area** function?
5. Explain how picking a polyline when using the **Area** option of the **MEASUREGEOM** tool is different from measuring the area of an object drawn with the **LINE** tool.
6. What is the purpose of the **LIST** tool?
7. Describe the meanings of *delta X* and *delta Y*.
8. What is the function of the **DBLIST** tool?
9. What tool, other than **MEASUREGEOM** and **AREA**, provides the area and perimeter of an object?
10. Which tool allows you to list drawing aid settings for the current drawing?
11. What information is provided by the **TIME** tool?
12. When does the drawing creation time start?
13. What term describes a text object that displays a set property, setting, or value for an object?
14. List three ways to open the **QuickCalc** palette.
15. Name the four sections of the **QuickCalc** palette.
16. Give the proper symbol to use for the following math functions:
    A. Addition
    B. Subtraction
    C. Multiplication
    D. Division
    E. Exponent
    F. Grouped expressions
17. Under which section of the **QuickCalc** palette can you find the square root function?
18. Under which section of the **QuickCalc** palette can you find the arccosine function?
19. When using one of the scientific functions, which should you do first: pick the scientific function button or type in the value to be used in the input box? Why?
20. Name the four types of units that can be converted using **QuickCalc**.
21. What term describes a text item that represents another value and can be accessed later as needed?
22. Which tool button is used to pass the value in the **QuickCalc** input box to the command line?
23. How can you start **QuickCalc** while a command is active?
24. When you are using **QuickCalc** while a tool is active, how do you pass the value to the command line?
25. When the **Properties** palette is open, what do you need to do first to see the calculator icon so that **QuickCalc** can be used?

Drawing Problems

Follow these instructions to complete the drawing problems for this chapter:

- *Start AutoCAD if it is not already started.*
- *Start a new drawing using an appropriate template of your choice. The template should include layers and text styles necessary for drawing the given objects.*
- *Add layers and text styles as needed. Draw all objects using appropriate layers and styles, justification, and format.*
- *Follow the specific instructions for each problem. Do not draw dimensions. Use your own judgment and approximate dimensions when necessary.*

▼ Basic

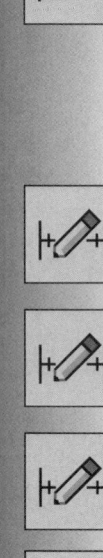

1. Use **QuickCalc** to calculate the result of the following equations.
   A. 27.375 + 15.875
   B. 16.0625 – 7.1250
   C. 5 × 17'-8"
   D. 48'-0" ÷ 16
   E. (12.625 + 3.063) + (18.250 – 4.375) – (2.625 – 1.188)
   F. 7.25^2

2. Convert 4.625" to millimeters.

3. Convert 26 mm to inches.

4. Convert 65 miles to kilometers.

5. Convert 5 gallons to liters.

6. Find the square root of 360.

7. Calculate 3.25 squared.

▼ Intermediate

8. Show the calculation and answer that would be used with the **LINE** tool to make an 8" line .006 in./in. longer in a pattern to allow for shrinkage in the final casting. Show only the expression and answer.

9. Solve for the deflection of a structural member. The formula is written as $PL^3/48EI$, where P = pounds of force, L = length of beam, E = modulus of elasticity, and I = moment of inertia. The values to be used are P = 4000 lbs, L = 240", and E = 1,000,000 lbs/in². The value for I is the result of the beam (width × height³)/12, where width = 6.75" and height = 13.5".

10. Calculate the coordinate located at 4,4,0 + 3<30.

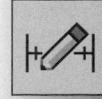

11. Calculate the coordinate located at (3 + 5,1 + 1.25,0) + (2.375,1.625,0).

12. Draw the object shown below using the dimensions given. Check the time when you start the drawing. Draw all the features using the **PLINE** and **CIRCLE** tools. Use the **Area** option of the **MEASUREGEOM** tool and the **Object**, **Add area**, and **Subtract area** functions to calculate the following measurements:
    A. The area and perimeter of Object A.
    B. The area and perimeter of area B. The slot ends are full radius.
    C. The area and circumference of one of the circles.
    D. The area of Object A, minus the area of Object B.
    E. The area of Object A, minus the areas of the other three features.
    Enter the **TIME** tool and note the editing time spent on your drawing. Save the drawing as P16-12.

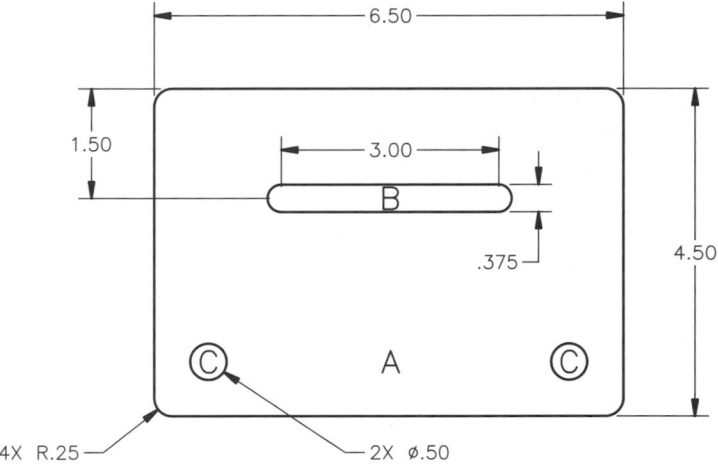

13. Draw the deck shown using the **PLINE** tool. Using the **POLYGON** tool, draw the hot tub. Use the following guidelines to complete this problem:
    A. Specify architectural units for your drawing. Use 1/2" fractions and decimal degrees. Leave the remaining settings for the drawing units at the default values.
    B. Set the limits to 100',80' and use the **All** option of the **ZOOM** tool.
    C. Set the grid spacing to 2' and the snap spacing to 1'.
    D. Calculate the measurements listed below.
        a. The area and perimeter of the deck.
        b. The area and perimeter of the hot tub.
        c. The area of the deck minus the area of the hot tub.
        d. The distance between Point C and Point D.
        e. The distance between Point E and Point C.
        f. The coordinates of Points C, D, and F.
    E. Enter the **DBLIST** tool and check the information listed for your drawing.
    F. Enter the **TIME** tool and note the total editing time spent on your drawing.
    G. Save the drawing as P16-13.

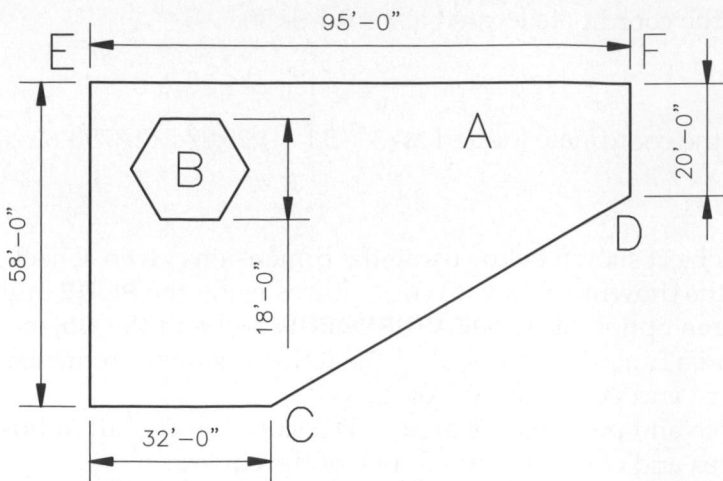

14. The drawing shown below is a side view of a pyramid. The pyramid has four sides. Create an auxiliary view showing the true size of a pyramid face. Save the drawing as **P16-14**. Using inquiry techniques, calculate the following:
    A. The area of one side.
    B. The perimeter of one side.
    C. The area of all four sides.
    D. The area of the base.
    E. The true length (distance) from the midpoint of the base on one side to the apex.

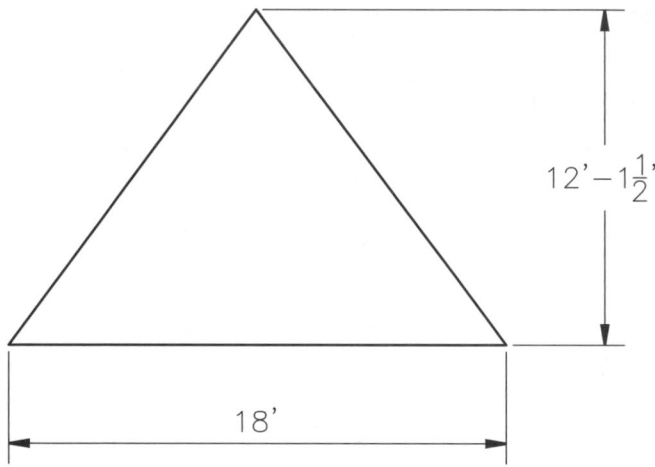

▼ Advanced

15. Given the following right triangle, make the required trigonometry calculations.
    A. Length of side c (hypotenuse).
    B. Sine of angle *A*.
    C. Sine of angle *B*.
    D. Cosine of angle *A*.
    E. Tangent of angle *A*.
    F. Tangent of angle *B*.

AutoCAD and Its Applications—Basics

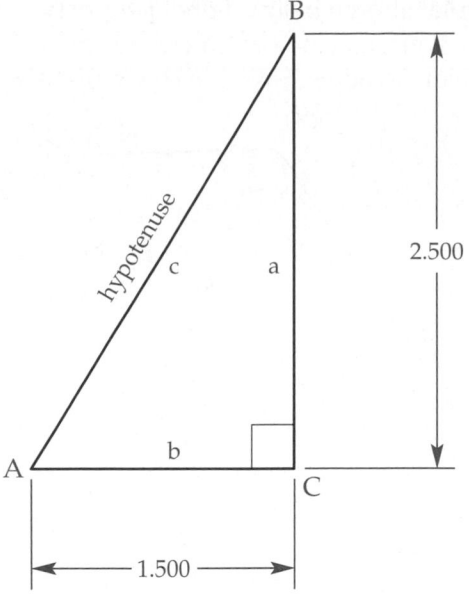

16. The drawing below is a view of the gable end of a house. Draw the house using the dimensions given, and draw the windows as single lines only (the location of the windows is not important). The spacing between each of the second-floor windows is 3". The width of this end of the house is 16'-6". The length of the roof is 40'. You may want to use the **PLINE** tool to assist in creating specific shapes in this drawing, except as noted above. Save the drawing as P16-16. Calculate the following:

A. The total area of the roof.
B. The diagonal distance from one corner of the roof to the other.
C. The area of the first-floor window.
D. The total area of all second-floor windows, including the 3" spaces between each of them.
E. Siding will cover the house. What is the total area of siding for this end?

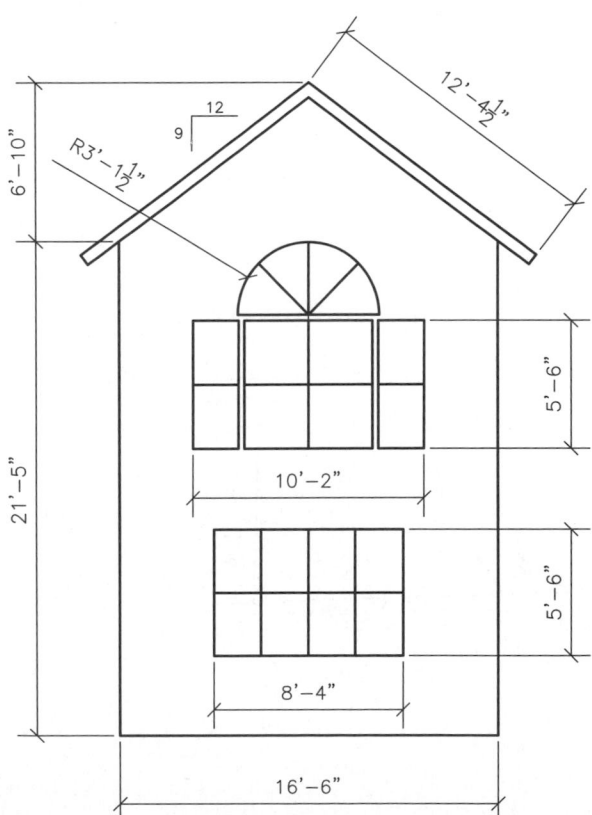

17. Draw the property plat shown below. Label property line bearings and distances only if required by your instructor. Calculate the area of the property plat in square feet and convert to acres. Save the drawing as P16-17.

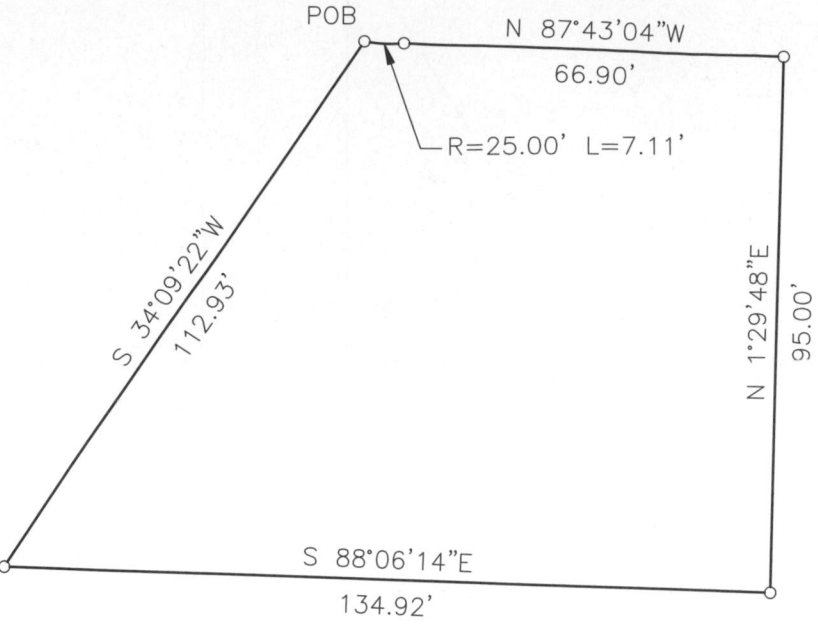

18. Draw the subdivision plat shown below. Label the drawing as shown. Calculate the acreage of each lot and record each value as a label inside the corresponding lot (for example, .249 AC). Save the drawing as P16-18.

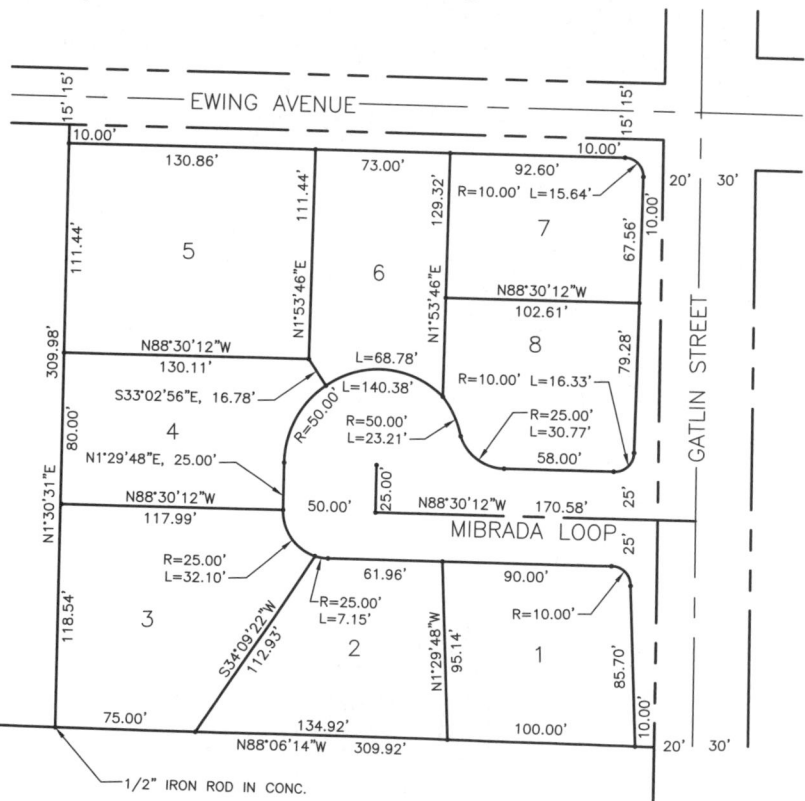

AutoCAD and Its Applications—Basics

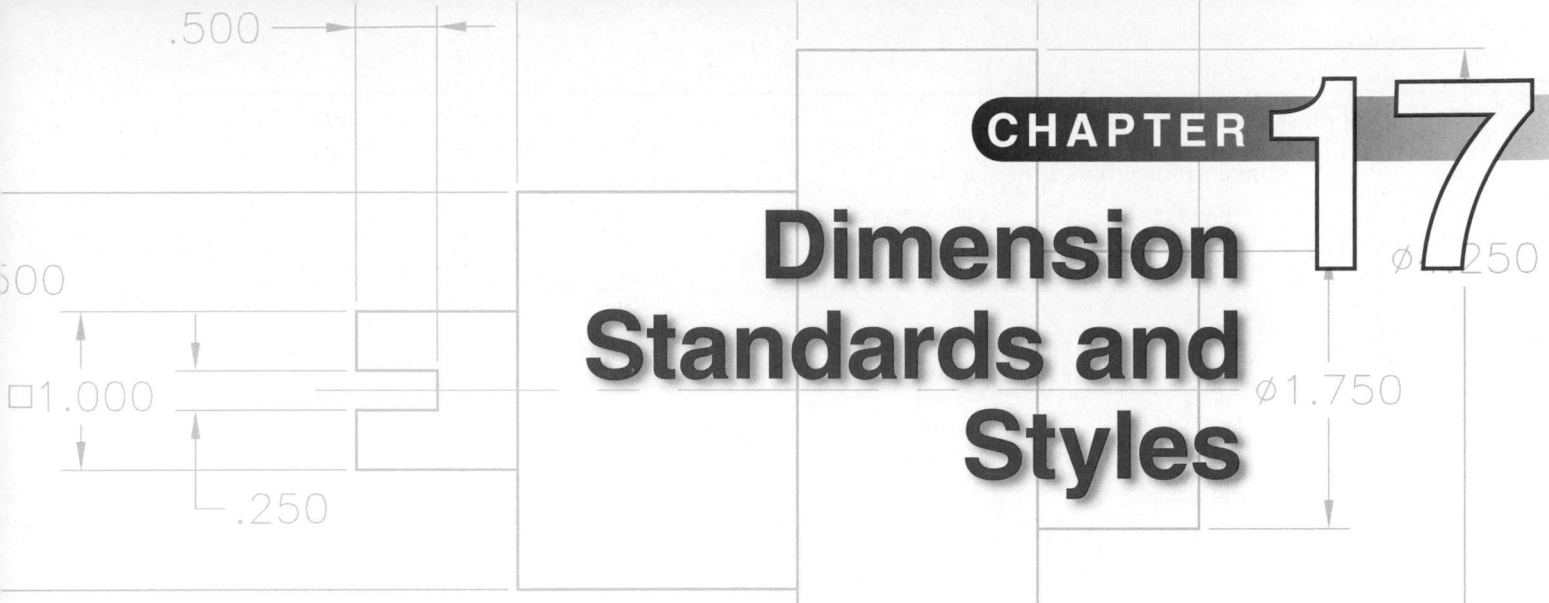

CHAPTER 17

Dimension Standards and Styles

Learning Objectives

After completing this chapter, you will be able to do the following:

✓ Describe common dimension standards and practices.
✓ Create dimension styles.
✓ Manage dimension styles.
✓ Set a dimension style current.

A *dimension* can consist of numerical values, lines, symbols, and notes. **Figure 17-1** shows typical dimension elements and dimensioning applications. Use dimension tools to dimension the size and location of features and objects. Dimension styles control the appearance of dimension elements. Dimensional constraint tools, explained in Chapter 22, allow you to use dimensions to control object size and location.

dimension: A description of the size, shape, or location of features on an object or structure.

Dimension Standards and Practices

Dimensions communicate drawing information. Each drafting field uses different dimensioning practices. Dimensioning practices often depend on product requirements, manufacturing or construction accuracy, standards, and tradition. It is important for you to place dimensions in accordance with industry and company standards and produce drawings that are as clear and easy to read as possible. Dimension standards help to ensure that product manufacturing or construction is accurate.

The standard emphasized in this textbook is ASME Y14.5M-1994, *Dimensioning and Tolerancing*, published by the American Society of Mechanical Engineers (ASME). The *M* in Y14.5M indicates metric numeric values for dimensions. This textbook describes the correct application of inch and metric dimensioning, as well as additional, discipline-specific dimensioning standards and applications.

Unidirectional Dimensioning

Unidirectional dimensioning is common in mechanical drafting. The term *unidirectional* means "in one direction." This type of dimensioning allows you to read all dimensions from the bottom of the sheet. Unidirectional dimensions normally have arrowheads on the ends of dimension lines. The dimension value is usually centered in a break near the center of the dimension line. See **Figure 17-2.**

unidirectional dimensioning: A dimensioning system in which all dimension values display horizontally on the drawing.

Figure 17-1.
Dimensions describe the size and location of objects and features. Follow accepted conventions when dimensioning.

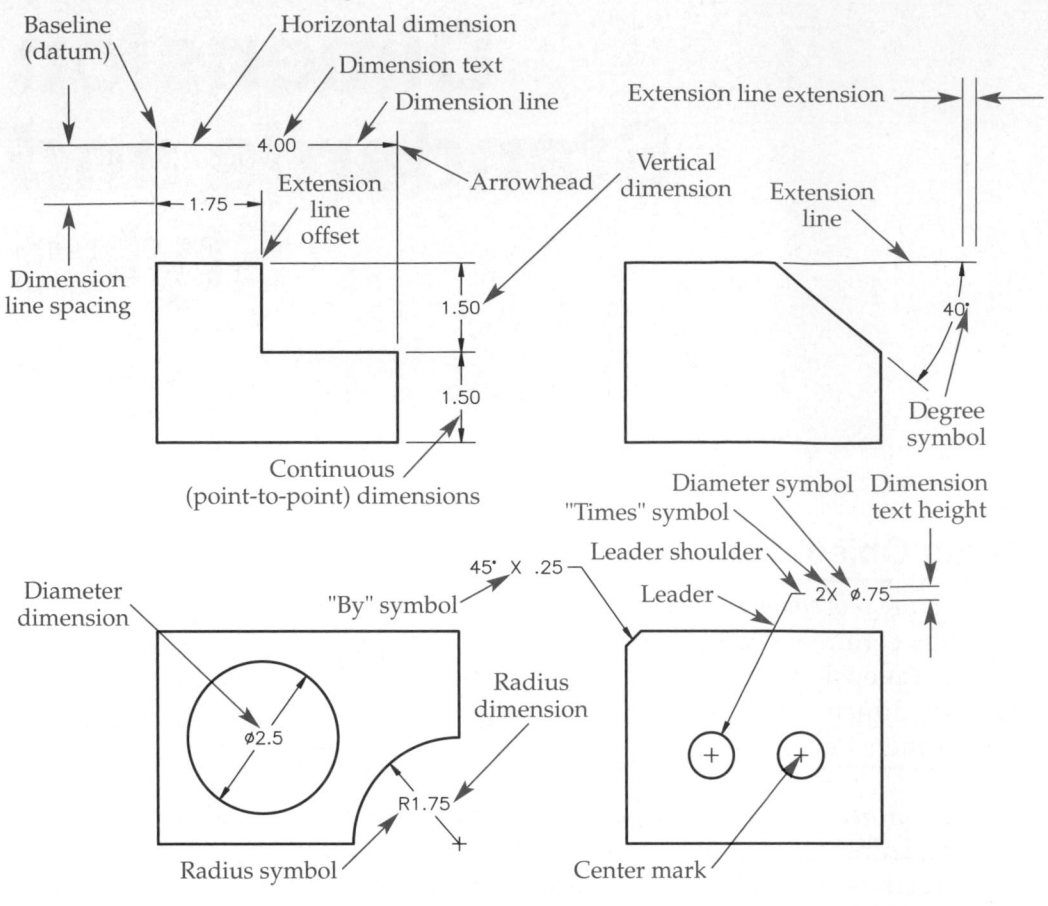

Figure 17-2.
When unidirectional dimensions are used, all dimension numbers and notes are horizontal on the drawing.

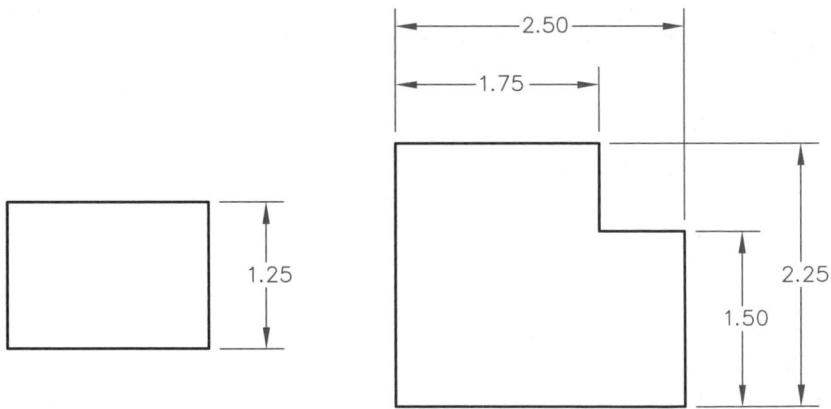

Aligned Dimensioning

Aligned dimensions are common on architectural and structural drawings. Text for dimensions reads at the same angle as the dimension line. Text applied to horizontal dimensions reads horizontally, and text applied to vertical dimensions rotates 90° to read from the right side of the sheet. Notes usually read horizontally. Tick marks, dots, or arrowheads may terminate aligned dimension lines. In architectural drafting, you generally place the dimension number above the dimension line and use tick mark terminators. See **Figure 17-3.**

Figure 17-3.
An example of aligned dimensioning in architectural drafting. Notice the tick marks used instead of arrowheads and the placement of the dimensions above the dimension line.

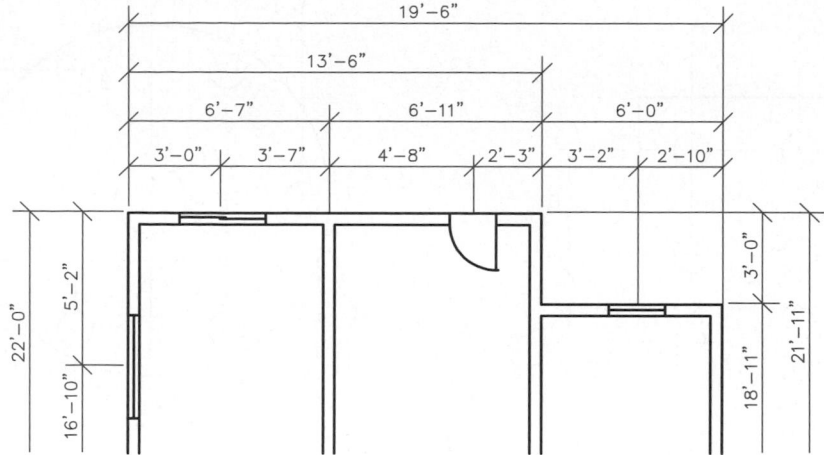

Size and Location Dimensions

Size dimensions provide the size of physical geometric *features*. See **Figure 17-4.** *Location dimensions* locate features on an object. See **Figure 17-5.** Dimension circular features, such as holes and arcs, to their centers in the view in which they appear

size dimensions: Dimensions that provide the size of physical features.

feature: Any physical portion of a part or object, such as a surface, hole, window, or door.

location dimensions: Dimensions used to locate features on an object without specifying the size of the feature.

Figure 17-4.
Size dimensions.

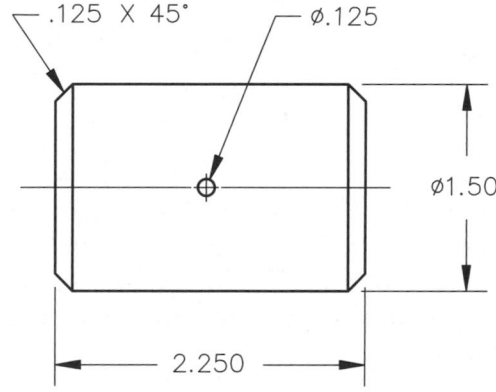

Figure 17-5.
Using location dimensions to locate circular and rectangular features.

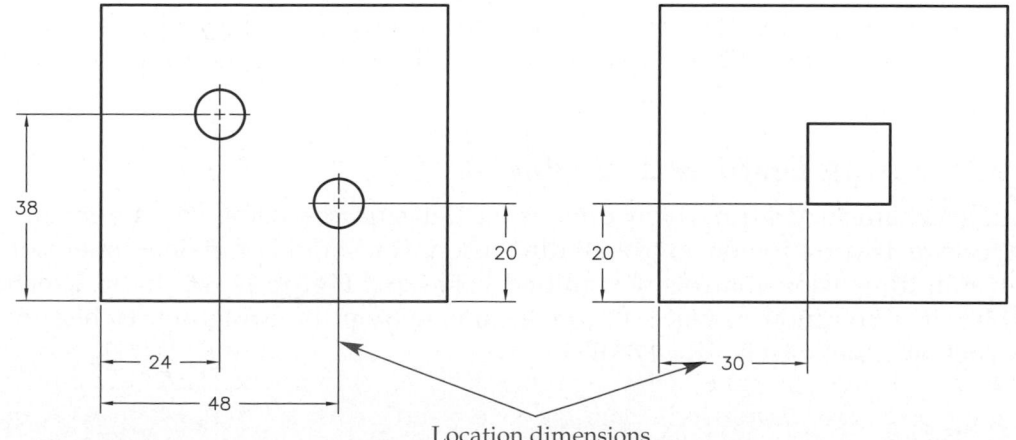

Location dimensions

Figure 17-6.
A—Rectangular coordinate location dimensions. B—Polar coordinate location dimensions.

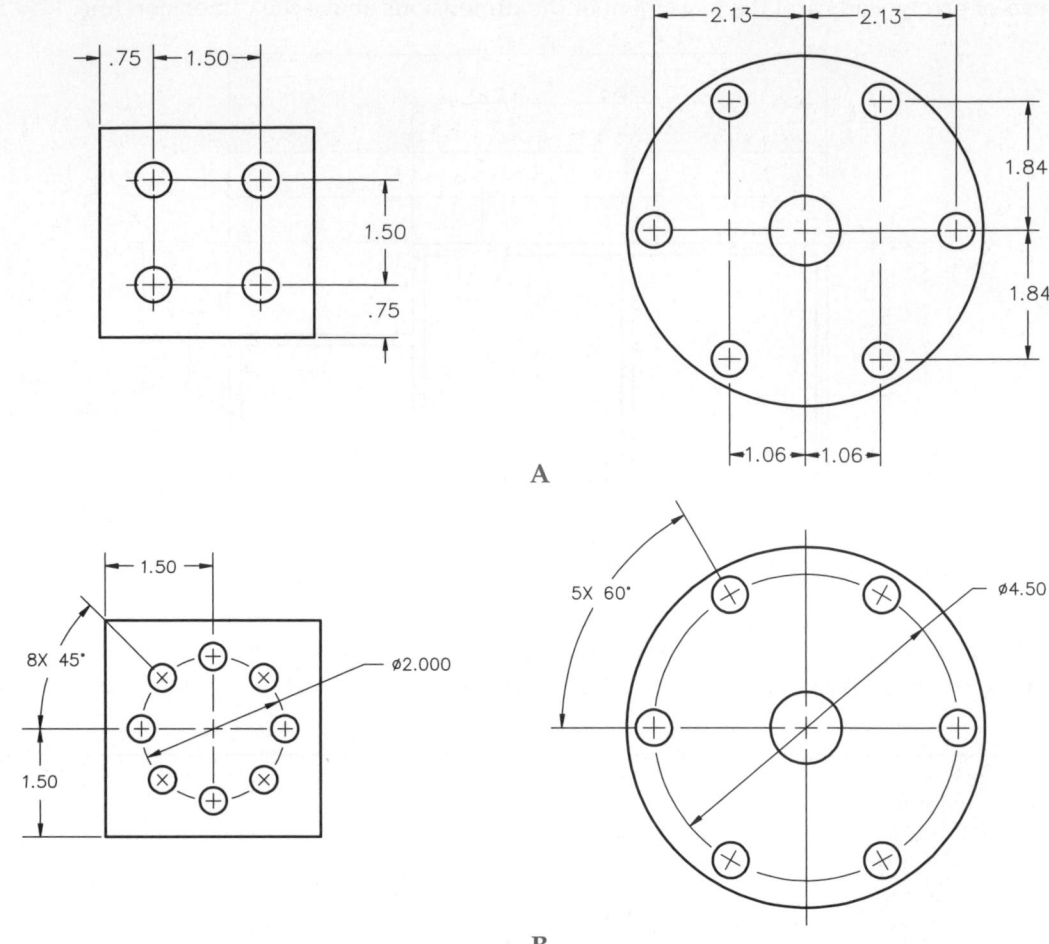

A

B

circular. Dimension rectangular features to their edges. An example of location dimensions used in architectural drafting is the dimensioning of windows and doors, usually to their centers, on a floor plan. The *rectangular coordinate system* and the *polar coordinate system* are the two basic systems for creating location dimensions. See **Figure 17-6**.

Notes

Specific notes and *general notes* are another way to describe feature size, location, or additional information. See **Figure 17-7**. Specific notes attach to the dimensioned feature using a leader line. Place general notes in the lower-left corner, upper-left corner, or above or next to the title block, depending on sheet size and industry, company, or school practice.

Dimensioning Features and Objects

In mechanical drafting, you dimension flat surfaces using measurements for each feature. If you provide an overall dimension, you should omit one dimension, as the overall dimension controls the omitted value. See **Figure 17-8A**. In architectural drafting, it is common to place all dimensions without omitting any to help make construction easier. See **Figure 17-8B**.

rectangular coordinate system: A system for locating dimensions from surfaces, centerlines, or center planes using linear dimensions.

polar coordinate system: A coordinate system in which angular dimensions locate features from surfaces, centerlines, or center planes.

specific notes: Notes that relate to individual or specific features on the drawing.

general notes: Notes that apply to the entire drawing.

Figure 17-7.
A—An example of a specific note. B—An example of general notes.

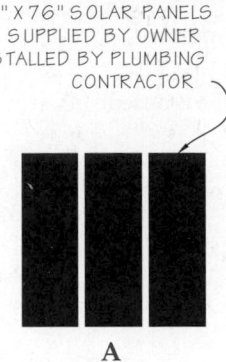

(3) 24" X 76" SOLAR PANELS
SUPPLIED BY OWNER
INSTALLED BY PLUMBING
CONTRACTOR

A

GENERAL NOTES:

1. PROVIDE SCREENED VENTS @ EA. 3RD. JOIST SPACE @ ALL ATTIC EAVES.
2. PROVIDE SCREENED ROOF VENTS @ 10'-0" O.C. (1/300 VENT TO ATTIC SPACE).
3. USE 1/2" CCX PLY. @ ALL EXPOSED EAVES.
4. USE 300# COMPOSITION SHINGLES OVER 15# FELT.

B

Figure 17-8.
A—Dimensioning flat surfaces. B—Dimensioning architectural features.

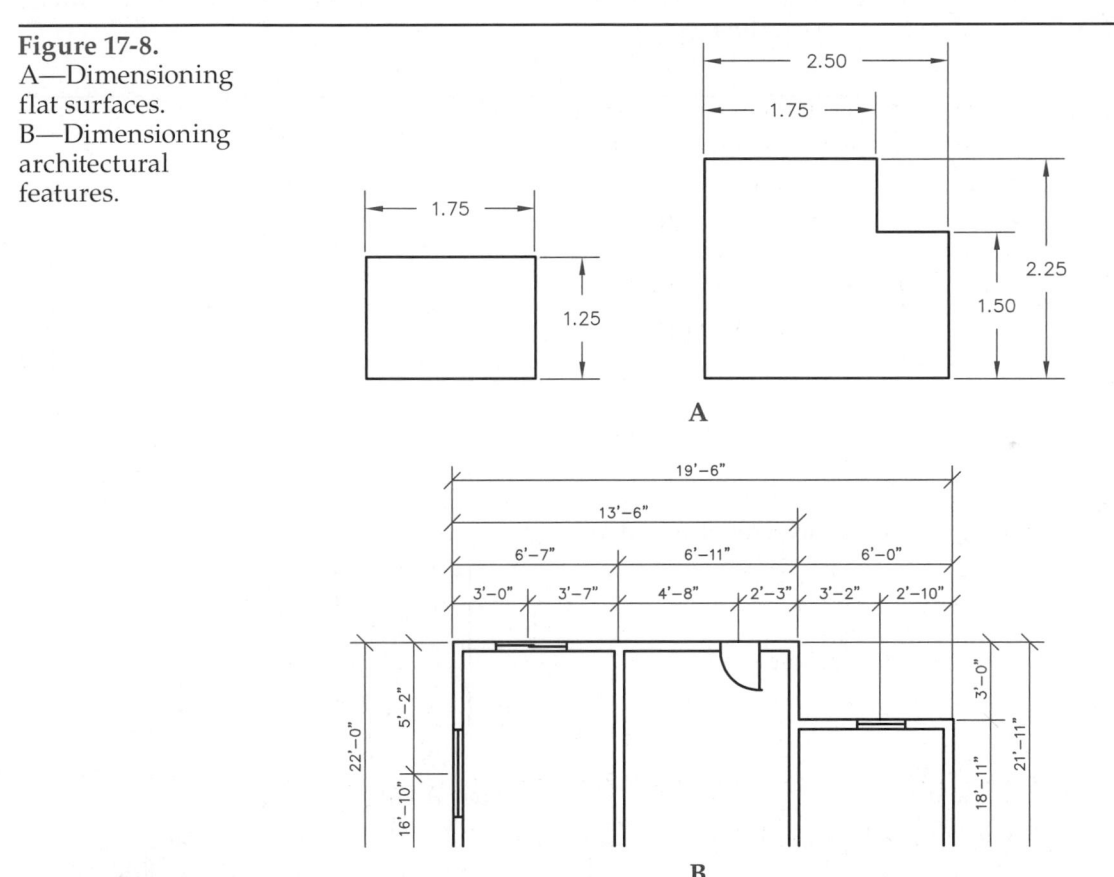

Dimensioning Cylindrical Shapes

You typically dimension the diameter and the length of a cylindrical shape in the view in which the cylinder appears rectangular. See **Figure 17-9.** The diameter symbol next to the dimension indicates that the part is a cylinder. This allows you to omit the view in which the cylinder appears as a circle.

Figure 17-9.
Dimensioning cylindrical shapes.

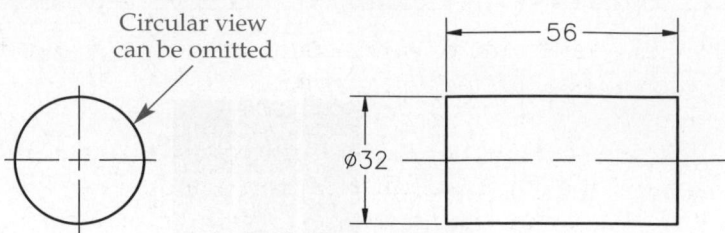

Dimensioning Square and Rectangular Features

You usually dimension square and rectangular features in the views that show the length and height. If appropriate, add a square symbol preceding the dimension for a square feature, to eliminate the need for an additional view. See **Figure 17-10.**

Dimensioning Cones and Regular Polygons

One method to dimension a conical shape is to dimension the length and the diameters at both ends. An alternative is to dimension the taper angle and the length. You usually dimension regular polygons that have an even number of sides by giving the distance across the flats and the length. **Figure 17-11** shows examples of dimensioned cones and regular polygons.

Figure 17-10.
Dimensioning square and rectangular features.

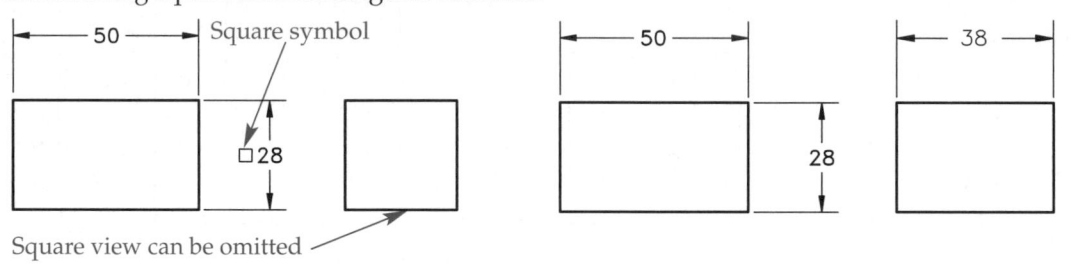

Figure 17-11.
Dimensioning cones and hexagonal cylinders.

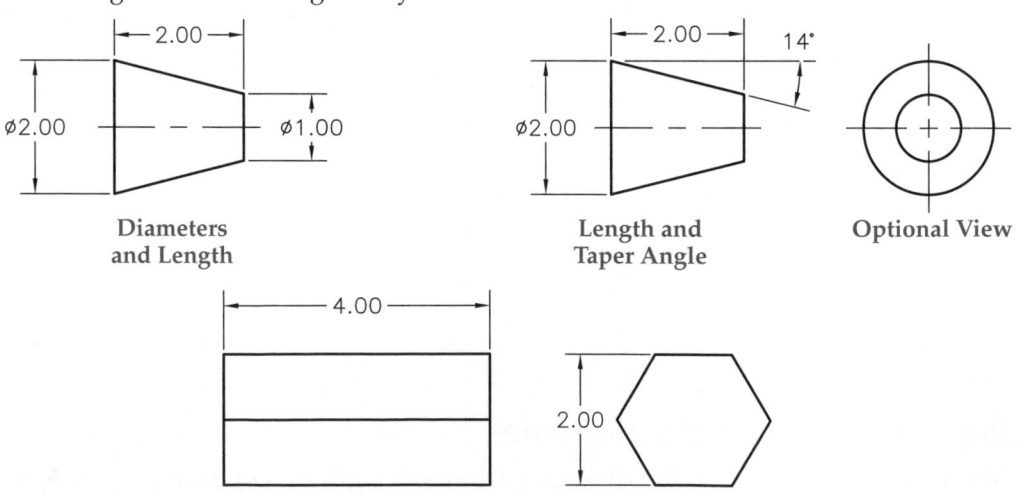

Drawing Scale and Dimensions

Ideally, you should determine drawing scale, scale factors, and dimension size characteristics before you begin drawing. Incorporate these settings into your drawing template files, and make changes when necessary. The drawing scale factor is important because it determines how dimensions appear on-screen and plot.

To help understand the concept of drawing scale, look at the portion of a floor plan shown in **Figure 17-12.** You should draw everything in model space at full scale. This means that the bathtub, for example, is actually drawn 5′ long. However, at this scale, dimension appearance becomes an issue. Full-scale dimension characteristics, such as 1/8″ high dimension text, are extremely small compared to the other full-scale objects. See **Figure 17-12A.** To see the dimensions clearly, you must adjust the size of dimension characteristics according to the drawing scale. See **Figure 17-12B.** You can calculate the scale factor manually and apply it to dimensions, or you can allow AutoCAD to calculate the scale factor using annotative dimensions.

Scaling Dimensions Manually

To adjust the size of dimension elements manually according to a specific drawing scale, you must first calculate the drawing scale factor. Once you determine the scale factor, you then multiply the scale factor by the desired plotted dimension size to get the model space dimension size. You can apply this calculation to all dimension elements by entering the scale factor in the **Fit** tab of the **New** (or **Modify**) **Dimension Style** dialog box, described later in this chapter. Refer to Chapter 9 for information on determining the drawing scale factor.

Annotative Dimensions

AutoCAD scales annotative dimensions according to the annotation scale you select, which eliminates the need for you to calculate the scale factor. Once you choose an annotation scale, AutoCAD determines the scale factor and applies it to annotative dimensions and all other annotative objects. For example, if you scale dimensions manually at a drawing scale of 1/4″ = 1′-0″, or a scale factor of 48, you must enter 48 in the **Fit** tab of the **New** (or **Modify**) **Dimension Style** dialog box. When you place annotative dimensions, using this example, you set an annotation scale of 1/4″ = 1′-0″. Then, when you add annotative dimensions, AutoCAD scales them automatically according to the 1/4″ = 1′-0″ annotation scale.

Figure 17-12.
An example of a portion of a floor plan drawn at full scale in model space. If you draw dimensions at full scale, as shown in A, the dimensions are very small compared to the large objects. You must scale the dimensions, as shown in B, in order to see and plot them correctly.

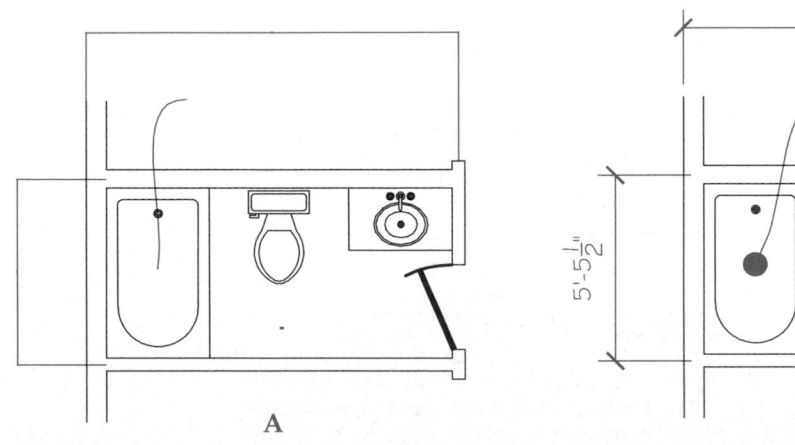

A

B

Annotative dimensions offer several advantages over manually scaled dimensions, including the ability to control dimension appearance based on scale, not scale factor. Annotative dimensions are especially effective when the drawing scale changes or when a single sheet includes objects viewed at different scales.

Setting Annotation Scale

You should usually set annotation scale before you begin adding dimensions so that dimension characteristics scale automatically. However, this is not always possible. It may be necessary to adjust the annotation scale throughout the drawing process, especially if you prepare multiple drawings with different scales on one sheet. This textbook approaches annotation scaling in model space only, using the process of selecting the appropriate annotation scale before placing dimensions. To draw dimensions at another scale, pick the new annotation scale and then place the dimensions.

When you access a dimension tool and an annotative dimension style is current, the **Select Annotation Scale** dialog box appears. This is a very convenient way to set annotation scale before adding dimensions. Dimension styles are described later in this chapter. You can also select the annotation scale from the **Annotation Scale** flyout located on the status bar. See **Figure 17-13.** Remember that the annotation scale is typically the same as the drawing scale.

Figure 17-13.
The status bar provides access to annotation scale options.

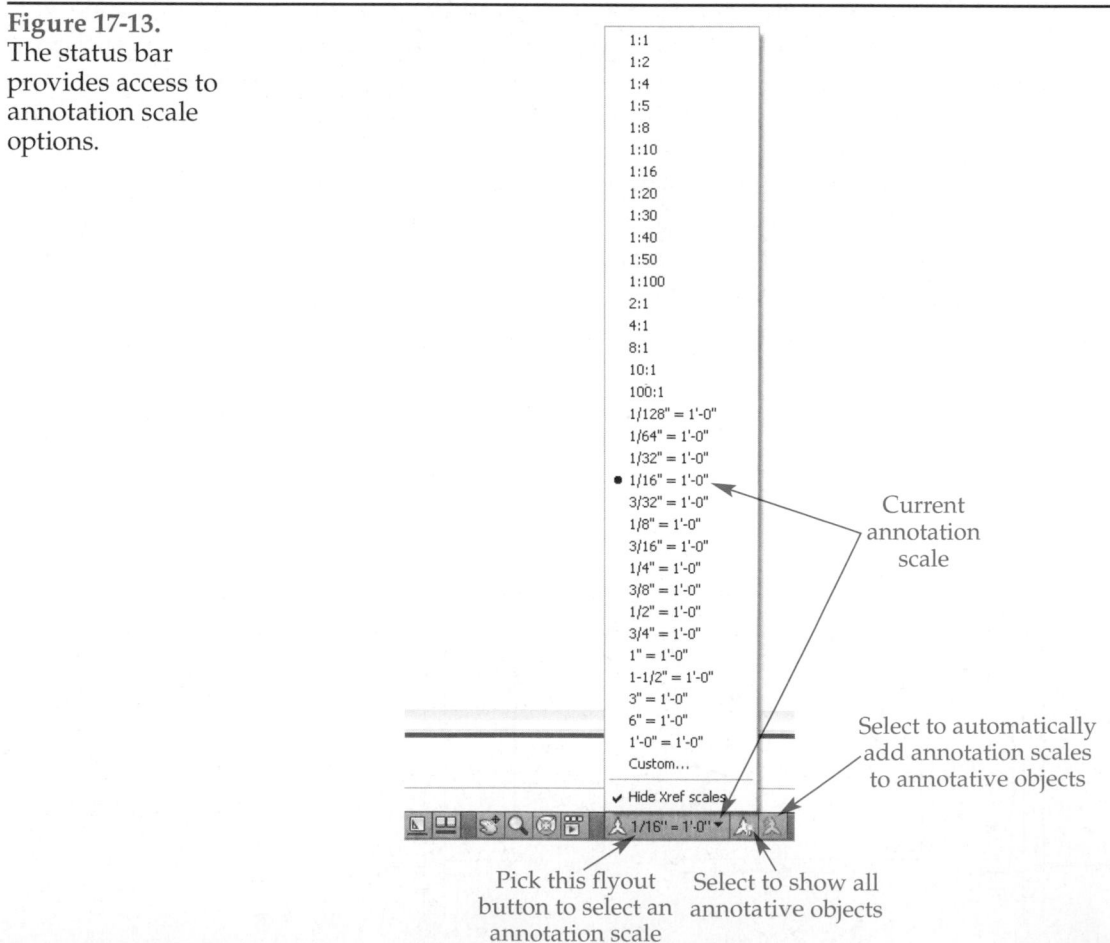

Editing Annotation Scales

If a certain scale is not available, or to change existing scales, pick the **Annotation Scale** flyout in the status bar and choose the **Custom...** option to access the **Edit Scale List** dialog box. From this dialog box, you can move the highlighted scale up or down in the list by picking the **Move Up** or **Move Down** button. To remove the highlighted scale from the list, pick the **Delete** button.

Select the **Edit...** button to open the **Edit Scale** dialog box. Here you can change the name of the scale and adjust the scale by entering the paper and drawing units. For example, a scale of 1/4″ = 1′-0″ uses a paper units value of .25 or 1 and a drawing units value of 12 or 48.

To create a new annotation scale, pick the **Add...** button to display the **Add Scale** dialog box, which functions the same as the **Edit Scale** dialog box. Pick the **Reset** button to restore the default annotation scale. When the correct annotation scale is set current, you are ready to place dimensions that automatically appear at the correct size according to the drawing scale.

> **NOTE**
>
> This textbook describes many additional annotative object tools. Some of these tools are more appropriate for working with layouts, as described later in this textbook.

Ribbon
Annotate
> Annotation
Scaling

Scale List

Type
SCALELISTEDIT

SCALELISTEDIT

Dimension Styles

Dimension styles control many dimension appearance characteristics. You might think of a dimension style as a grouping of dimensioning standards. Dimension styles usually apply to a specific type of drafting field or dimensioning application and correspond to appropriate drafting standards. For example, a dimension style used for mechanical drafting may use unidirectional dimensions, the Romans text font placed in a break in the dimension line, and dimension lines terminated with arrowheads. Refer again to **Figure 17-2**. A dimension style for architectural drafting may use aligned dimensions, the Stylus BT text font placed above the dimension line, and dimension lines terminated by tick marks, as shown in **Figure 17-3.**

dimension style: A saved configuration of dimension appearance settings.

Some drawings only require a single dimension style. However, you may need multiple dimension styles, depending on the variety of dimensions you apply and different dimension characteristics. You should generally create a dimension style for each unique dimensioning requirement. You can also override dimension appearance settings for individual dimensions. Add dimension styles to drawing templates for repeated use.

Working with Dimension Styles

Create, modify, and delete dimension styles using the **Dimension Style Manager** dialog box. See **Figure 17-14.** The **Styles:** list box displays existing dimension styles. Standard and Annotative dimension styles are available by default. The Annotative dimension style is preset to create annotative dimensions, as indicated by the icon to the left of the style name. The Standard dimension style does not use the annotative function. To make a dimension style current, double-click the style name, right-click on the name and select **Set current**, or pick the name and selecting the **Current** button.

Ribbon
Home
> Annotation

Dimension Style
Annotate
> Dimensions

Dimension Style
Type
DIMSTYLE
DIMSTY
DDIM
DST
D

DIMSTYLE

Figure 17-14.
The **Dimension Style Manager** dialog box. The non-annotative dimension style Standard and the annotative dimension style Annotative are available by default.

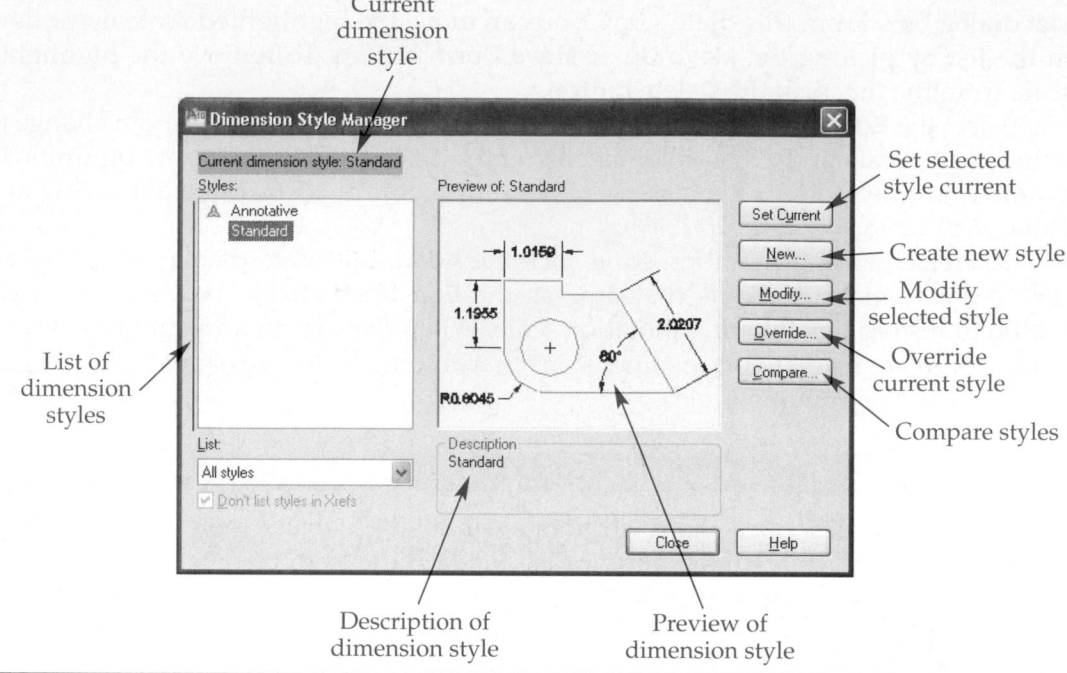

Current dimension style

Set selected style current

Create new style

Modify selected style

Override current style

Compare styles

List of dimension styles

Description of dimension style

Preview of dimension style

The **List:** drop-down list allows you to control whether all styles or only the styles in use appear in the **Styles:** list box. If the current drawing contains external references drawings (xrefs), you can use the **Don't list styles in Xrefs** box to eliminate xref-dependent dimension styles from the **Styles:** list box. This is often valuable because you cannot set xref dimension styles current or use them to create new dimensions. External references are described later in this textbook.

The **Description** area and **Preview of:** image provide information about the selected dimension style. If you change any of the default dimension settings without first creating a new dimension style, the changes are automatically stored as a dimension style override.

NOTE

The **Preview of:** image displays a representation of the dimension style and changes according to the selections you make.

Creating New Dimension Styles

To create a new dimension style, first select an existing dimension style from the **Styles:** list box to use as a base for formatting the new dimension style. Then pick the **New...** button in the **Dimension Style Manager** to open the **Create New Dimension Style** dialog box. See **Figure 17-15.**

Enter a descriptive name for the new dimension style, such as Architectural or Mechanical, in the **New Style Name** text box. If necessary, select a different style from the **Start With** drop-down list from which to base the new style. Pick the **Annotative** check box to make the dimension style annotative. You can also make the dimension style annotative by selecting the **Annotative** check box in the **Fit** tab of the **New** (or **Modify**) **Dimension Style** dialog box, described later in this chapter.

Figure 17-15.
The **Create New Dimension Style** dialog box.

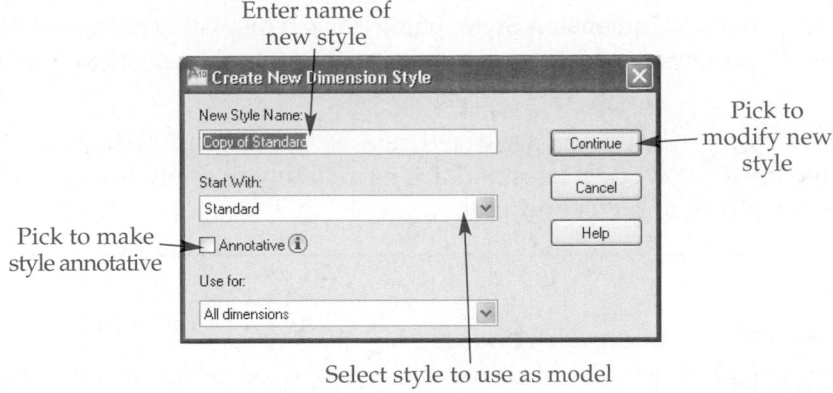

Enter name of new style

Pick to modify new style

Pick to make style annotative

Select style to use as model

The Use for drop-down list specifies the type of dimensions to which the new style applies. Use the All dimensions option to create a new dimension style for all types of dimensions. If you select the Linear dimensions, Angular dimensions, Radius dimensions, Diameter dimensions, Ordinate dimensions, or Leaders and Tolerances option, you create a sub-style of the dimension style specified in the Start With: text box.

Pick the **Continue** button to access the **New Dimension Style** dialog box, shown in **Figure 17-16,** and adjust dimension style characteristics. The **Lines**, **Symbols and Arrows**, **Text**, **Fit**, **Primary Units**, **Alternate Units**, and **Tolerances** tabs display groups of settings for specifying dimension appearance. The next sections of this chapter describe each tab. After completing the style definition, pick the **OK** button to return to the **Dimension Style Manager** dialog box.

Figure 17-16.
The **Lines** tab of the **New** (or **Modify**) **Dimension Style** dialog box.

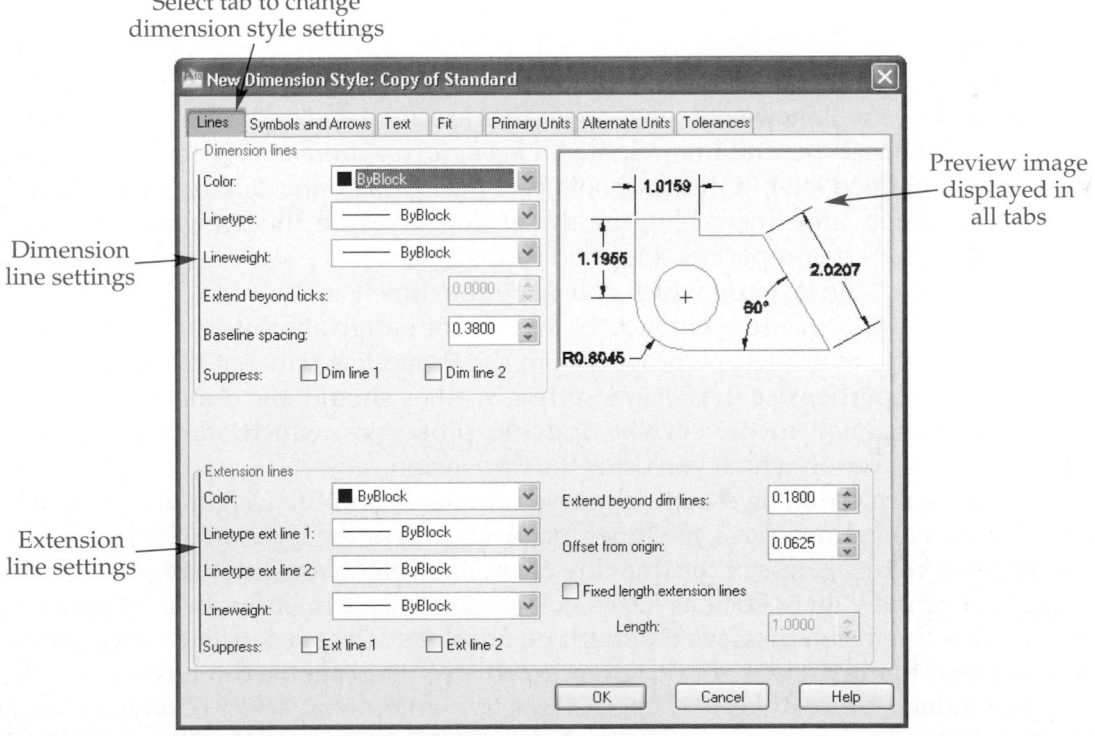

Select tab to change dimension style settings

Dimension line settings

Extension line settings

Preview image displayed in all tabs

The preview image shown in the upper-right corner of each **New** (or **Modify**) **Dimension Style** dialog box tab displays a representation of the dimension style and changes according to the selections you make.

AutoCAD stores dimension style settings as *dimension variables*. Dimension variables have limited practical uses and are more likely to apply to advanced applications such as scripting and customizing.

Reference Material *Dimension Variables*

For a list of dimension variables, go to the **Reference Material** section of the Student Web site (www.g-wlearning.com/CAD) and select **Dimension Variables**.

Changing dimension variables by typing the variable name is not a recommended method for changing dimension style settings. Changes made in this manner can introduce inconsistencies with other dimensions. You should make changes to dimensions by redefining styles or performing style overrides.

Using the Lines Tab

The options in the **Lines** tab of the **New** (or **Modify**) **Dimension Style** dialog box are shown in **Figure 17-16.** These options allow you to control dimension and extension line display settings.

Dimension Line Settings

The **Dimension lines** area of the **Lines** tab allows you to set dimension line format. **Color**, **Linetype**, and **Lineweight** drop-down lists are available for changing the dimension line color, linetype, and lineweight. All *associative dimensions* are block objects, as further explained later in this textbook. When you assign the default ByBlock setting to color, linetype, and lineweight, the dimension takes on the drawing color, lineweight, and linetype properties, specified in the **Properties** panel of the **Home** ribbon tab, regardless of the layer on which you draw the dimension.

Using the ByBlock setting is noticeable only if you assign absolute values to drawing color, lineweight, and linetype properties in the **Properties** panel of the **Home** ribbon tab. If these properties use the ByLayer setting, as they should, the dimension acquires the settings assigned to the current drawing properties, which adopt the settings applied to the layer on which you draw the dimension.

If you assign the ByLayer setting to color, linetype, and lineweight, the dimension takes on the color, lineweight, and linetype properties of the layer on which you draw the dimension, regardless of the drawing color, lineweight, and linetype properties. If you use absolute values, such as a Blue color, a Continuous linetype, or a 0.05mm lineweight, the dimension displays the specified absolute values regardless of the properties assigned to the drawing or the layer on which you create the dimension.

The **Extend beyond ticks** text box is inactive unless you select oblique or architectural tick terminators from the **Symbols and Arrows** tab of the **New** (or **Modify**) **Dimension Style** dialog box. Architectural tick marks or oblique arrowheads often

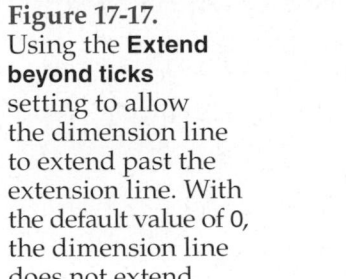

Figure 17-17.
Using the **Extend beyond ticks** setting to allow the dimension line to extend past the extension line. With the default value of 0, the dimension line does not extend.

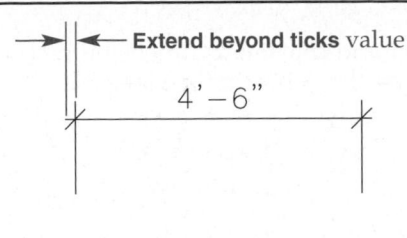

terminate dimension lines on architectural drawings. In this style of dimensioning, dimension lines often extend past extension lines, as shown in **Figure 17-17.** The 0.00 default draws dimension lines that do not extend past extension lines.

The **Baseline spacing** text box allows you to change the spacing between the dimension lines of baseline dimensions created with the **DIMBASELINE** tool. The default spacing is too close for most drawings, as shown in **Figure 17-18.** The ASME minimum dimension line spacing for baseline dimensioning is .375 (10mm). A value of .5 (12mm) or .75 (19mm) is usually more appropriate.

The **Suppress** feature has two toggles that prevent the display of the first, second, or both dimension lines and their arrowheads. The **Dim line 1** and **Dim line 2** check boxes refer to the first and second points picked when you create a dimension. Both dimension lines appear by default. **Figure 17-19** shows the results of using dimension line suppression options.

Extension Line Settings

The **Extension lines** area of the **Lines** tab allows you to set extension line format. **Color, Linetype ext line 1, Linetype ext line 2,** and **Lineweight** drop-down lists are available for changing the extension line color, linetype, and lineweight from the default ByBlock setting, if necessary. You can use the **Linetype ext line 1** and **Linetype ext line 2** drop-down lists to specify the linetype applied to each extension line. Lines 1 and 2 correspond to the first and second points you pick when drawing a dimension.

The **Extend beyond dim lines** option allows you to set the distance the extension line runs past the dimension line. See **Figure 17-20.** An extension line extension of .125 (3mm) is recommended by ASME standards. The **Offset from origin** option specifies the distance between the object and the beginning of the extension line. Most applications require this small offset. ASME standards recommend an extension line offset distance of .063 (1.5mm). When an extension line meets a centerline, however, use a setting of 0.0 to prevent a gap.

Figure 17-18.
The **Baseline spacing** setting controls the spacing between dimension lines.

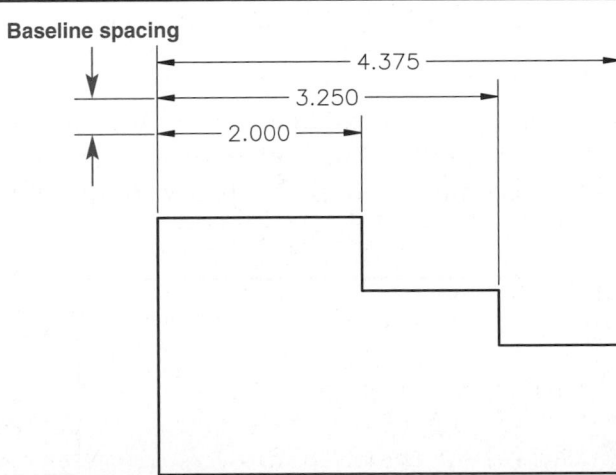

Figure 17-19.
Using the **Dim line 1** and **Dim line 2** dimensioning settings. "Off" is equivalent to an unchecked **Suppress** check box in the **Lines** tab.

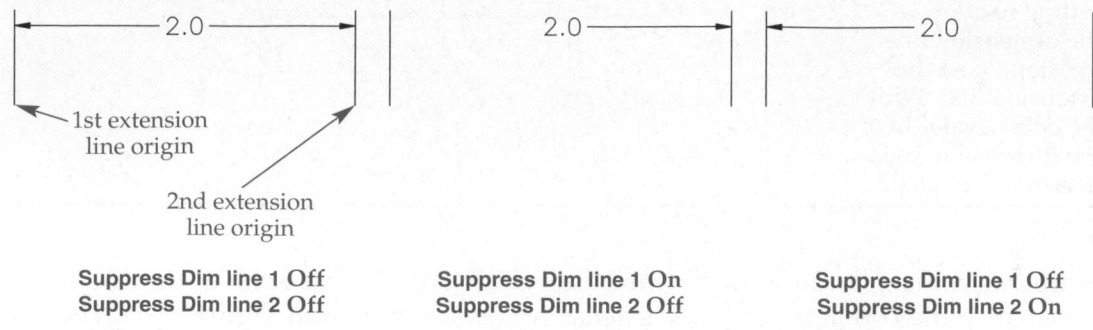

Suppress Dim line 1 Off   Suppress Dim line 1 On   Suppress Dim line 1 Off
Suppress Dim line 2 Off   Suppress Dim line 2 Off   Suppress Dim line 2 On

Figure 17-20.
The extension line extension and extension line offset settings.

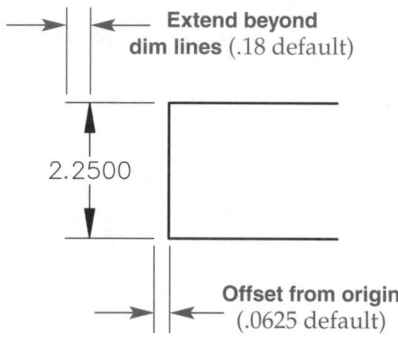

The **Fixed length extension lines** check box sets a given length for extension lines. When this box is checked, the **Length** text box becomes active. The value in the **Length** text box sets a restricted length for extension lines, measured from the dimension line toward the extension line origin.

Extension lines display by default. Use the **Ext line 1** and **Ext line 2** check boxes to suppress extension lines. Though extension line suppression is typically applied to individual dimensions, not a dimension style, you might suppress an extension line, for example, if it coincides with an object line. See **Figure 17-21.**

Using the Symbols and Arrows Tab

The options in the **Symbols and Arrows** tab of the **New** (or **Modify**) **Dimension Style** dialog box are shown in **Figure 17-22.** These options allow you to control the appearance of arrowheads, center marks, and other symbol components of dimensions.

Arrowhead Settings

Use the appropriate drop-down list in **Arrowheads** area to select the arrowhead to use for the first, second, and leader arrowheads. The default arrowhead is **Closed filled**, which is recommended by ASME standards, although **Closed blank**, **Closed**, or

Figure 17-21.
Suppressing extension lines.

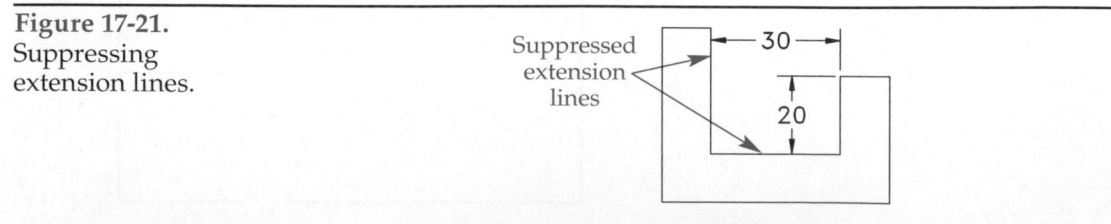

Figure 17-22.
The **Symbols and Arrows** tab of the **New** (or **Modify**) **Dimension Style** dialog box.

Select tab to specify arrow style

Arrowhead properties

Center mark properties

Dimension break size

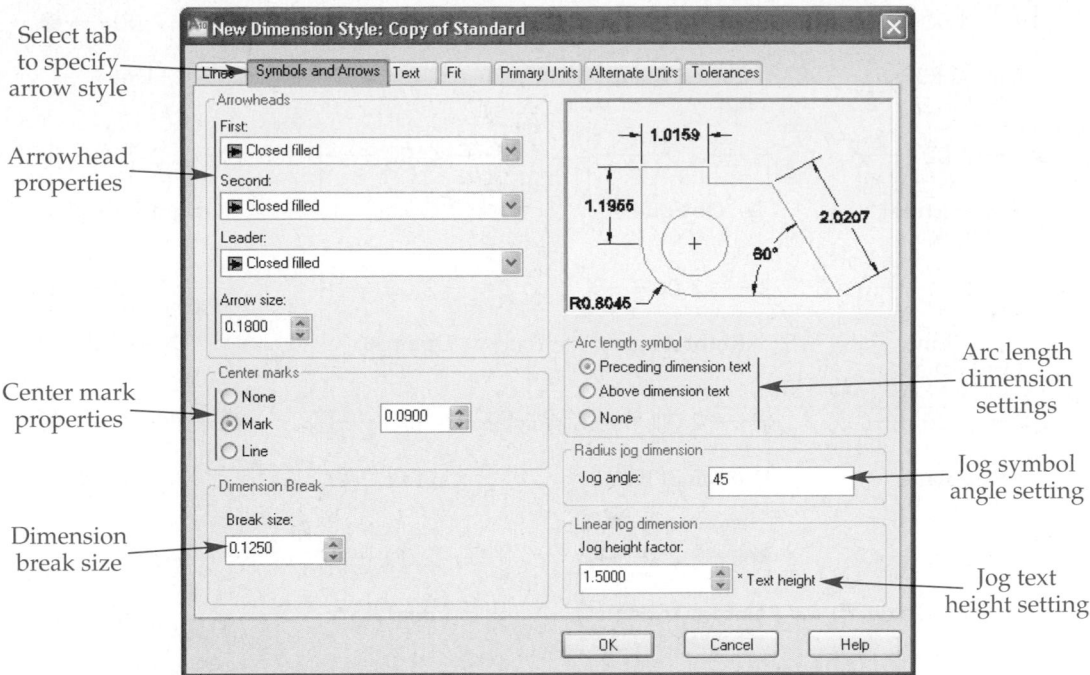

Arc length dimension settings

Jog symbol angle setting

Jog text height setting

Open arrowheads are sometimes used. A leader pointing to a surface uses a small dot. **Figure 17-23** shows arrowhead styles. If you pick a new arrowhead in the **First:** drop-down list, AutoCAD automatically makes the same selection for the **Second:** drop-down list. When you select the **Oblique** or **Architectural tick** arrowhead, the **Extend beyond ticks:** text box in the **Lines** tab activates.

Notice that **Figure 17-23** does not contain an example of a user arrow. This option allows you to access an arrowhead of your own design. For this to work, you must first design an arrowhead that fits inside a 1 unit square (unit block) with a dimension line "tail" of 1 unit in length, and save the arrowhead as a block. Blocks are described later in this textbook. The **Select Custom Arrow Block** dialog box appears when you pick **User Arrow...** from an **Arrowheads** drop-down list. Type the name of the custom arrow block in the **Select from Drawing Blocks:** text box or pick a block from the drop-down list and then pick **OK** to apply the arrow to the style.

The **Arrow size:** text box allows you to change arrowhead size. A .125″ (3 mm) arrowhead size is common on mechanical drawings. See **Figure 17-24**.

Center Mark Settings

The **Center marks** area allows you to select the way center marks appear in circles and arcs when you use circular feature dimensioning tools. Fillets and rounds generally have no center marks. The **None** option provides for no center marks to occur in circles and arcs. The **Mark** option places center marks without centerlines. The **Line** option places center marks and centerlines. Use the **Size:** text box in the **Center marks** area to change the size of the center mark and centerline. The size defines half the length of a centerline dash and the distance that the centerline extends past the object. A value of .0625″ (1.5 mm) is appropriate for the centerline dash half-length, but does not provide for the preferred .125″ (3 mm) extension past the object. **Figure 17-25** shows the results of specifying center marks and centerlines.

Figure 17-23.
Examples of dimensions drawn using the options found in the **Arrowheads** drop-down lists.

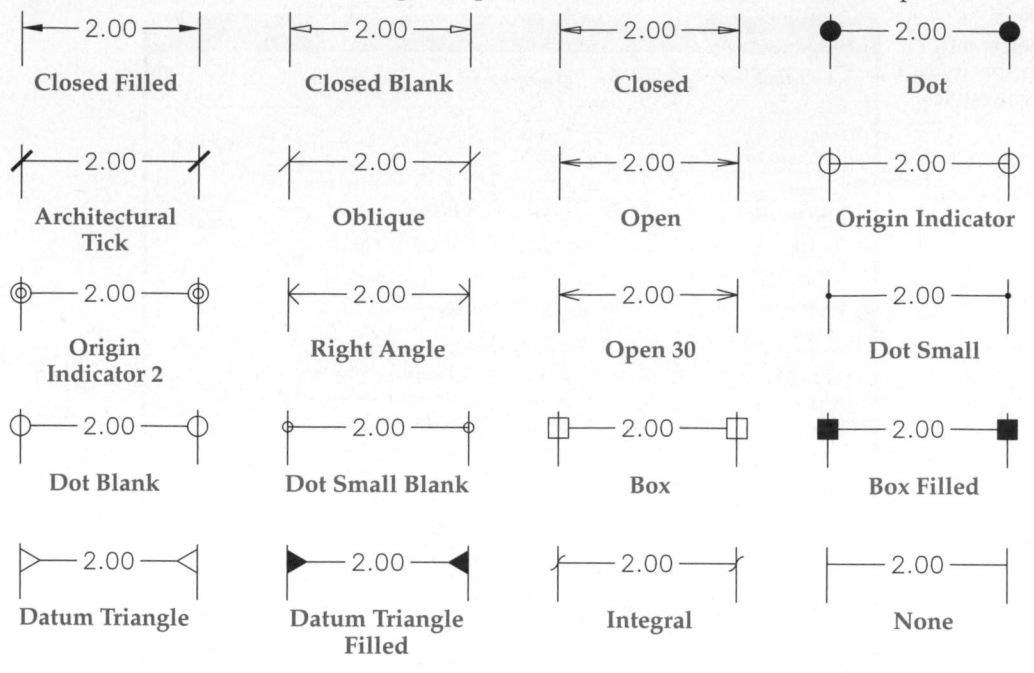

Figure 17-24.
ASME standards specify an arrowhead size of .125". The **Closed filled, Closed blank,** and **Closed** arrowhead styles adhere to the standard 3:1 ratio of length to width.

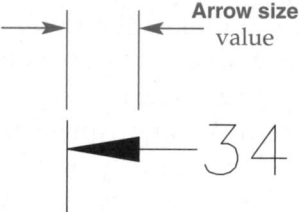

Figure 17-25.
Arcs and circles displayed with center marks and centerlines.

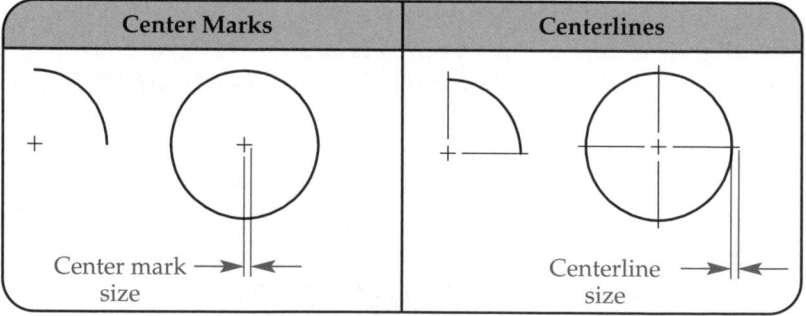

Adjusting Break Size

The **Dimension Break** area controls the amount of extension line removed when you use the **DIMBREAK** tool. Specify a value in the **Break size:** text box to set the total break length. **Figure 17-26** shows an example of a 3 mm extension line break. The default size is .125" (3 mm). ASME standards do not recommend breaking extension lines.

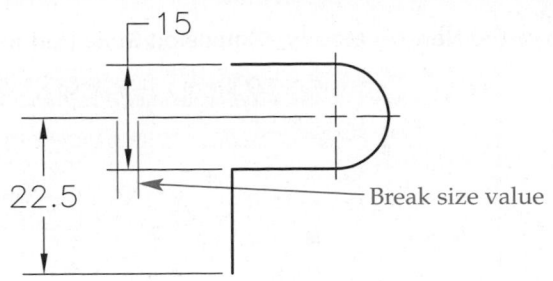

Figure 17-26.
Use the **Break size** setting to specify the length of the break created using the **DIMBREAK** tool.

15

22.5

Break size value

Adding an Arc Length Symbol

The **Arc Length Symbol** area controls the placement of the arc length symbol when you use the **DIMARC** tool. The default **Preceding dimension text** option places the symbol in front of the dimension value. Select the **Above dimension text** radio button to place the arc length symbol over the length value. See **Figure 17-27.** Pick the **None** radio button to suppress the symbol so that it does not show.

Adjusting Jog Angle

The **Jog angle** setting in the **Radius jog dimension** area controls the appearance of the break line applied to the jog symbol when you use the **DIMJOGGED** tool. This value sets the incline formed by the line connecting the extension line and dimension line. The default angle is 45°.

Setting Jog Height

The **Jog height factor** setting in the **Linear jog dimension** area controls the size of the break symbol created using the **DIMJOGLINE** tool. This value sets the height of the break symbol based on a multiple of the text height. For example, the default value of 1.5 creates a break symbol that is .18″ tall if the text height is .12″. The default angle is 45°.

NOTE

Chapter 18 describes the **DIMJOGLINE** tool in more detail, and Chapter 19 covers the **DIMJOGGED** , **DIMARC**, and **DIMBREAK** tools.

Exercise 17-1

Access the Student Web site (www.g-wlearning.com/CAD) and complete Exercise 17-1.

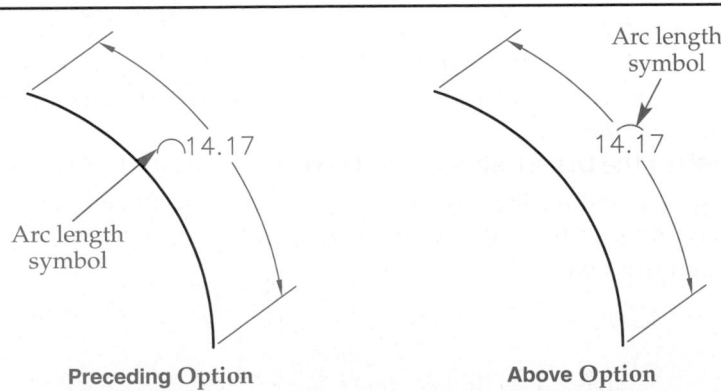

Figure 17-27.
You can place the arc length symbol in front of or above the arc dimension text.

Arc length
symbol

⌒14.17

Arc length
symbol

Preceding Option

Arc length
symbol

14.17

Above Option

Select tab
to set up
dimension
text

Set
appearance
of the text

Set location
of text
relative to
dimension line

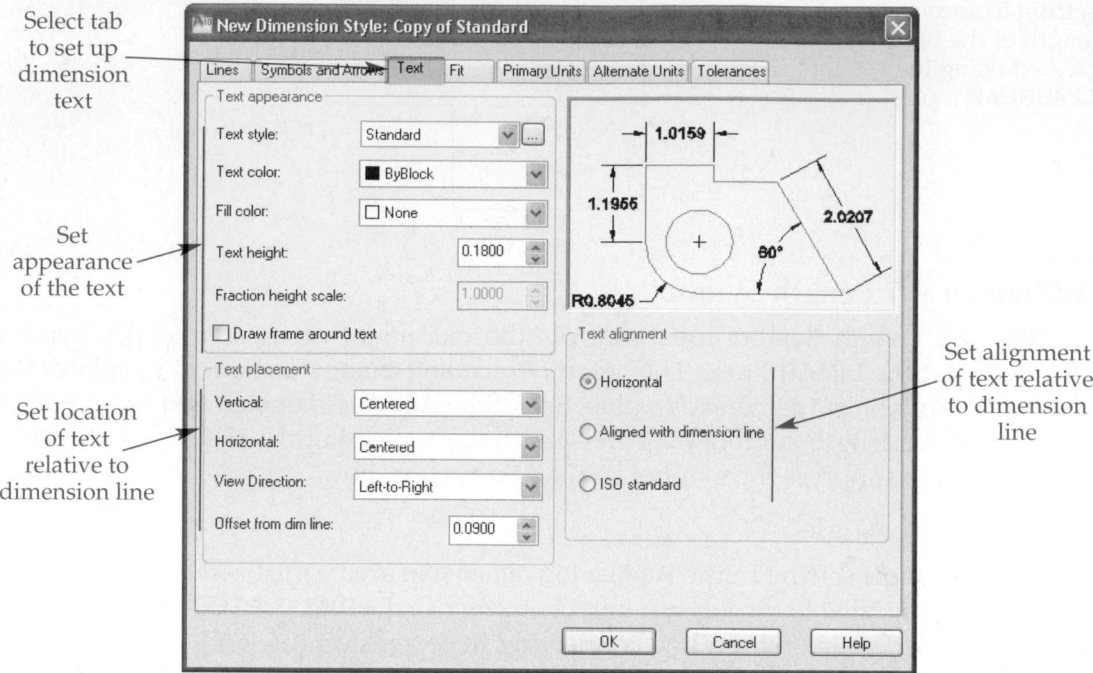

Set alignment
of text relative
to dimension
line

Using the Text Tab

The **Text** tab of the **New** (or **Modify**) **Dimension Style** dialog box is shown in Figure 17-28. Use this tab to control the display of dimension values.

Text Appearance Settings

Use the **Text appearance** area to set the dimension text style, color, height, and frame. A text style must be loaded in the current drawing before it is available for use in dimension text. Pick the desired text style from the **Text style** drop-down list. To create or modify an existing text style, pick the ellipsis (…) button next to the drop-down list to launch the **Text Style** dialog box. Use the **Text color** drop-down list to specify the appropriate text color, which should be ByBlock for typical applications.

Use the **Text height** text box to specify the dimension text height. Dimension text height is commonly the same as the text height used for most other drawing text, except for titles, which are often larger. The default dimension text height of .18″ (2.5 mm) is an acceptable standard. Many companies use a text height of .125″ (3 mm). The ASME standard recommends a .12″ (3 mm) text height. The text height for titles and labels is usually .24 (6 mm).

The **Fraction height scale** setting controls the height of fractions for architectural and fractional unit dimensions. The value in the **Fraction height scale** box is multiplied by the text height value to determine the height of the fraction. A value of 1.0 creates fractions that are the same text height as regular (nonfractional) text, which is the normally accepted standard. A value less than 1.0 makes the fraction smaller than the regular text height.

Select the **Draw frame around text** check box to draw a rectangle around the dimension text. A rectangle is most often used to describe a *basic dimension*. The setting for the **Offset from dim line** value, explained later in this section, determines the distance between the text and the frame.

basic dimension:
A theoretically perfect dimension used to describe the exact size, profile, orientation, and location of a feature.

Figure 17-29.
Dimension text justification options. A—Vertical justification options, with the horizontal Centered justification. B—Horizontal justification options, with the vertical Centered justification.

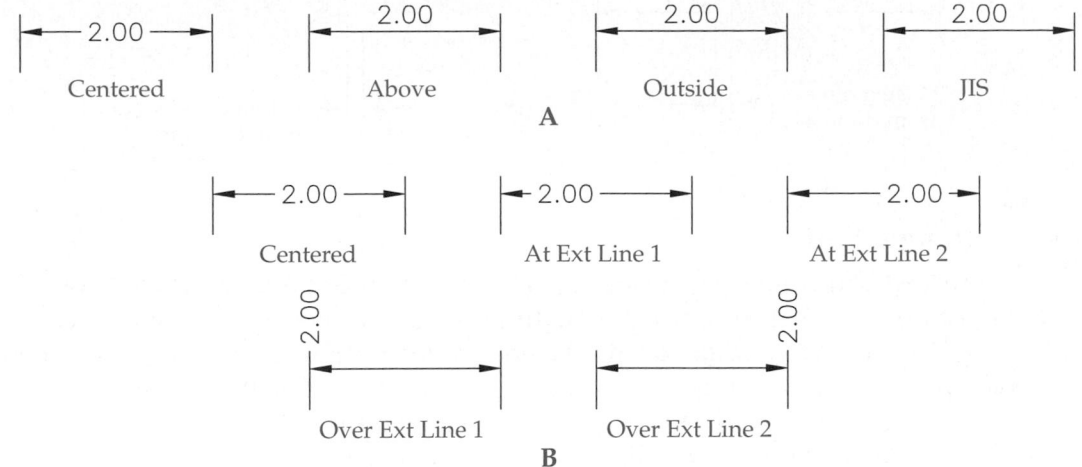

Text Placement Settings

The **Text placement** area controls text placement relative to the dimension line. See **Figure 17-29.** The **Vertical:** drop-down list provides vertical justification options. Use the default **Centered** option to place dimension text centered in a gap provided in the dimension line. This is the most common dimensioning practice in mechanical drafting and many other fields.

Select the **Above** option to place the dimension text horizontally and above horizontal dimension lines. For vertical and angled dimension lines, the text appears in a gap provided in the dimension line. This option is generally used for architectural drafting and building construction. Architectural drafting commonly uses aligned dimensioning, in which the dimension text aligns with the dimension lines and all text reads from either the bottom or the right side of the sheet.

Pick the **Outside** option to place the dimension text outside the dimension line and either above or below a horizontal dimension line or to the right or left of a vertical dimension line. The direction you move the cursor determines the above/below and left/right placement. Choose the **JIS** option to align the text according to the Japanese Industrial Standard (JIS).

The **Horizontal:** drop-down list provides options for controlling the horizontal placement of dimension text. Pick the default **Centered** option to place dimension text centered between the extension lines. Select the **At Ext Line 1** option to locate the text next to the extension line placed first, or choose **At Ext Line 2** to locate the text next to the extension line placed second. Pick **Over Ext Line 1** to place the text aligned with and over the first extension line, or select **Over Ext Line 2** to place the text aligned with and over the second extension line. Placing text aligned with and over an extension line is not common practice.

The **View Direction:** drop-down list determines how you read dimension text. Use the default **Left-to-Right** option to read text from left-to-right or bottom-to-top, depending on the text placement and alignment. Choose the **Right-to-Left** option to flip dimension text. Text may appear inverted and reads from right-to-left or top-to-bottom, depending on the text placement and alignment. Changing text view direction to right-to-left is not common practice.

The **Offset from dim line:** text box sets the gap between the dimension line and dimension text, the distance between the leader shoulder and text, and the space between text and the rectangle drawn around it. The gap should be set to half the text height for most applications. **Figure 17-30** shows the gap in linear and leader dimensions.

Figure 17-30.
The gap (offset) used for text in a linear dimension and a leader dimension.

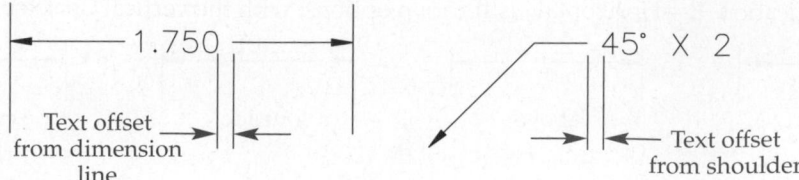

Text offset from dimension line

Text offset from shoulder

Text Alignment Settings

Use the **Text alignment** area to specify unidirectional or aligned dimensions. The **Horizontal** option draws the unidirectional dimensions commonly used for mechanical drafting applications. The **Aligned with dimension line** option creates aligned dimensions, typically used for architectural drafting applications. The **ISO Standard** option creates aligned dimensions when the text falls between the extension lines and horizontal dimensions when the text falls outside the extension lines.

Using the Fit Tab

The **Fit** tab of the **New** (or **Modify**) **Dimension Style** dialog box is shown in **Figure 17-31.** The settings on this tab allow you to establish dimension *fit format*.

fit format: The arrangement of dimension text and arrowheads on a drawing.

Fit Options

The **Fit options** area controls how text, dimension lines, and arrows behave when there is not enough room between extension lines to accommodate all of the items. The amount of space between the extension lines and the size of the dimension value, offset, and arrowheads, influence fit performance. All fit options place text and dimension lines with arrowheads inside the extension lines if space is available. All except the **Always keep text between ext lines** option place arrowheads, dimension lines, and text outside of the extension lines when space is limited.

Figure 17-31.
The **Fit** tab of the **New** (or **Modify**) **Dimension Style** dialog box.

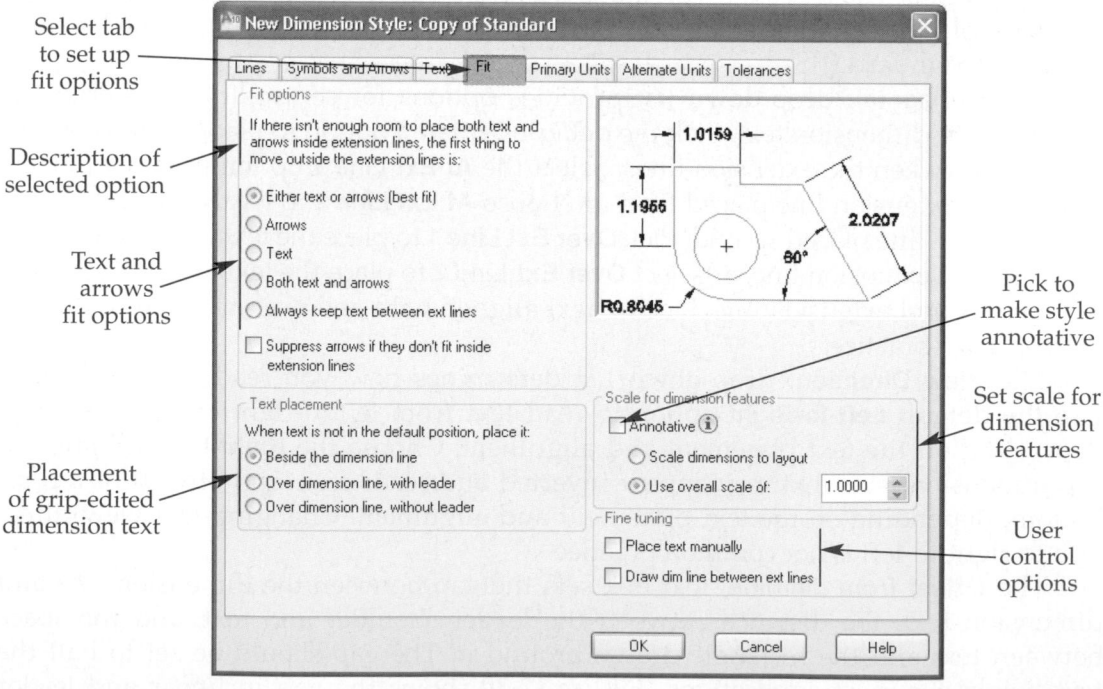

Select tab to set up fit options

Description of selected option

Text and arrows fit options

Placement of grip-edited dimension text

Pick to make style annotative

Set scale for dimension features

User control options

AutoCAD and Its Applications—Basics

Choose the default **Either text or arrows (best fit)** radio button to move either the dimension value or the arrows outside of the extension lines first. Pick the **Arrows** radio button to attempt to place arrowheads outside of the extension lines first, followed by text. Pick the **Text** radio button to attempt to place text outside of the extension lines first, followed by arrowheads. Choose the **Both text and arrows** radio button to move both text and arrowheads outside of the extension lines.

Select the **Always keep text between ext lines** radio button to place the dimension value between the extension lines. This option typically causes interference between the dimension value and extension lines when there is limited space between extension lines. Pick the **Suppress arrows if they don't fit inside extension lines** radio button to remove the arrowheads if they do not fit inside the extension lines. Use this option with caution, because it can create dimensions that violate standards.

Text Placement Settings

Sometimes it becomes necessary to move the dimension value from its default position. You can use grips to move the value independently of the dimension. The options in the **Text placement** area specify how these grip-editing situations function.

Select the **Beside the dimension line** radio button to restrict dimension text movement. You can grip-move the text with the dimension line, but only within the same plane as the dimension line. If you pick the **Over dimension line, with leader** radio button, you can grip-move the dimension text in any direction away from the dimension line. A leader line forms connecting the text to the dimension line. Choose the **Over dimension line, without leader** radio button to have the ability to move the dimension text in any direction away from the dimension line without a connecting leader.

PROFESSIONAL TIP

To return the dimension text to its default position, select the dimension, right-click and select **Home text** from the **Dim Text position** cascading submenu.

Text Scale Options

Use the **Scale for dimension features** area to set the dimension scale factor. Select the **Annotative** check box to create an annotative dimension style. The **Annotative** check box is already set when you modify the default Annotative dimension style or pick the **Annotative** check box in the **Create New Dimension Style** dialog box.

You can select the **Scale dimensions to layout** radio button to dimension in a floating viewport in a paper space layout. You must add dimensions to the model in a floating viewport in order for this option to function. Scaling dimensions to the layout allows the overall scale to adjust according to the active floating viewport by setting the overall scale equal to the viewport scale factor.

Pick the **Use overall scale of** radio button to scale a drawing manually, and enter the drawing scale factor to be applied to all dimension settings. The scale factor is multiplied by the desired plotted dimension size to get the model space dimension size. For example, if the height of dimension text is set to .12 and the value for the overall scale is set to 2 for a half scale drawing, then the dimension text measures .24 units ($2 \times .12 = .24$). If you then plot the drawing at a plot scale of 1:2 (half), the size of the dimension text on the paper measures .12 units.

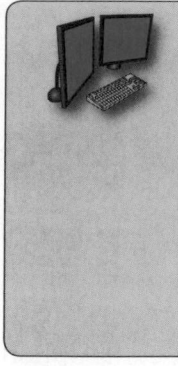

You can draw dimensions in either model space or layout space. Model space dimension scale is set by the drawing scale factor to achieve the correct dimension appearance. Associative paper space dimensions automatically adjust to model modifications and do not require scaling. In addition, if you dimension in paper space, you can dimension the model differently in two viewports. However, paper space dimensions are not visible when you work in model space, so you must be careful not to move a model space object into a paper space dimension. Avoid using nonassociative paper space dimensions.

Fine Tuning Settings

The **Fine tuning** area provides flexibility in controlling the placement of dimension text. Select the **Place text manually** check box to have the ability to place text where you want it, such as to the side within the extension lines or outside of the extension lines. However, this feature can make equally offsetting dimension lines somewhat more cumbersome, and it is not necessary for standard dimensioning practices.

The **Draw dim line between ext lines** option forces AutoCAD to place the dimension line inside the extension lines, even when the text and arrowheads are outside. The default application is to place the dimension line and arrowheads outside the extension lines. See **Figure 17-32**. Though some companies prefer the appearance, forcing the dimension line inside the extension lines is not an ASME standard.

Exercise 17-2

Access the Student Web site (www.g-wlearning.com/CAD) and complete Exercise 17-2.

Using the Primary Units Tab

The **Primary Units** tab of the **New** (or **Modify**) **Dimension Style** dialog box is shown in **Figure 17-33**. This tab controls linear and angular dimension units.

Linear Dimension Settings

The **Linear dimensions** area allows you to specify settings for primary linear dimensions. The options from the **Unit format** drop-down list are the same as those in the **Length** area of the **Drawing Units** dialog box. Typically, primary linear dimension unit format is the same as the corresponding drawing units.

The **Precision** drop-down list allows you to specify the precision applied to dimensions, which may be the same as the related drawing units precision. A variety of dimension precisions are often found on the same drawing. When you are using decimal units, precision determines how many zeros follow the decimal place. Precision settings in mechanical drafting depend on the accuracy required to manufacture specific features. Some features require greater precision, generally due to fits

Figure 17-32.
The effect of the **Draw dim line between ext lines** option in the **Fine tuning** area of the **Fit** tab.

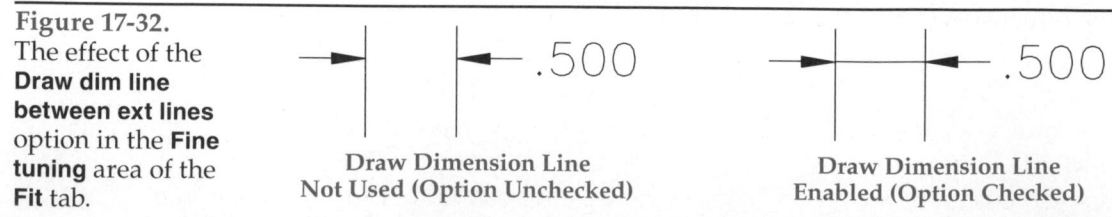

Draw Dimension Line
Not Used (Option Unchecked)

Draw Dimension Line
Enabled (Option Checked)

Figure 17-33.
The **Primary Units** tab of the **New** (or **Modify**) **Dimension Style** dialog box.

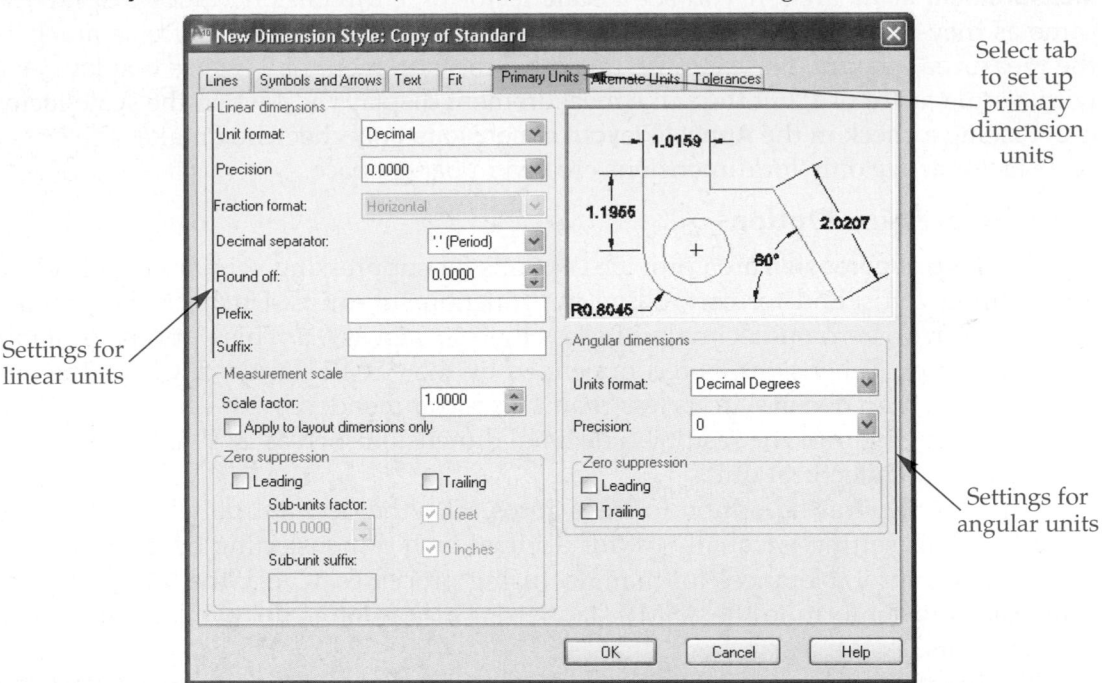

Settings for linear units

Select tab to set up primary dimension units

Settings for angular units

between mating parts. For example, a precision setting of 0.00 represents less exactness than a setting of 0.0000. Chapter 20 further explains this concept. Common precisions in mechanical drafting include 0.0000, 0.000, and 0.00. When you specify fractional units, precision values identify the smallest desired fractional denominator. Precisions of 1/16 to 1/64 are common, but you can choose other options, ranging from 1/256 to 1/2; 0 displays no fractional values.

Use the **Decimal separator** drop-down list to specify commas, periods, or spaces as separators for decimal numbers. The '**.**' **(Period)** option is default and is appropriate for typical applications. The **Decimal separator** option is not available if the unit format is **Architectural** or **Fractional**. The **Fraction format** drop-down list is available if the unit format is **Architectural** or **Fractional**. The options for controlling the display of fractions are **Diagonal**, **Horizontal**, and **Not Stacked**.

The **Round off** text box specifies the accuracy of rounding for dimension numbers. The default is zero, which means that no rounding takes place and all dimensions specify the value exactly as measured. No rounding is appropriate for most applications. If you enter a value of .1, all dimensions are rounded to the closest .1 unit. For example, an actual measurement of 1.188 is rounded to 1.2.

Add a *prefix* to a dimension by entering a value in the **Prefix** text box. A typical application for a prefix is SR3.5, where SR means "spherical radius." When a prefix is used on a diameter or radius dimension, the prefix replaces the ∅ or R symbol. Add a *suffix* to a dimension by entering a value in the **Suffix** text box. A typical application for a suffix is 3.5 MAX, where MAX is the abbreviation for "maximum." The abbreviation in could be used when one or more inch dimensions are placed on a metric-dimensioned drawing. Conversely, a suffix of mm could be used on one or more millimeter dimensions placed on an inch drawing.

prefixes: Special notes or applications placed in front of the dimension text.

suffixes: Special notes or applications placed after the dimension text.

PROFESSIONAL TIP

A prefix or suffix is normally a special specification, used in only a few cases on a drawing. As a result, often it is easiest to enter a prefix or suffix using the **MText** or **Text** option of the related dimensioning tool.

Set the scale factor of linear dimensions in the **Scale factor:** text box of the **Measurement scale** area. If you set a scale factor of 1, dimension values display the same as they measure. If the scale factor is 2, dimension values are twice as much as the measured amount. For example, an actual measurement of 2 inches displays as 2 with a scale factor of 1, but the same measurement displays as 4 when the scale factor is 2. Placing a check in the **Apply to layout dimensions only** check box makes the linear scale factor active only for dimensions created in paper space.

Zero Suppression Options

The **Zero suppression** area provides options for suppressing primary unit leading and trailing zeros, and for controlling the function sub-units. Uncheck the **Leading** option to leave a zero on decimal units less than 1, such as 0.5. This option is suitable to create metric dimensions as recommended by the ASME standard. Check the box to remove the 0 on decimal units less than 1, as recommended by the ASME standard for inch dimensioning. The result is a decimal dimension such as .5. This option is not available for architectural units.

Uncheck the **Trailing** option to leave zeros after the decimal point based on the precision. This setting is usually off for decimal inch dimensioning because trailing zeros often control tolerances for manufacturing processes. Check the box for metric dimensions to conform to the ASME standard. This option is not available for architectural units.

The **0 feet** check box is enabled for architectural and engineering units. Check the box to remove the zero in dimensions given in feet and inches when there are zero feet. For example, when this box is checked, a dimension reads 11″. When **0 feet** is unchecked, however, the same dimension reads 0′-11″.

The **0 inches** check box is also enabled for architectural and engineering units. Check the box to remove the zero when the inch portion of dimensions displayed in feet and inches is less than one inch, such as 12′-7/8″. If this box is unchecked, the same dimension reads 12′-0 7/8″. In addition, this option removes the zero from a dimension with no inch value; for example, 12′ appears instead of 12′-0″.

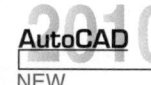

sub-units: Unit formats that are smaller than the primary unit format. For example, centimeters can be defined as a sub-unit of meters.

The **Sub-units factor** and **Sub-unit suffix** text boxes become enabled when you use decimal units and select the **Leading** check box. Most drawings use a single format for all dimension values. For example, all dimensions on a decimal inch drawing measure in inches, or decimals of an inch. *Sub-units* allow you to apply a different unit format to dimensions that are smaller than the primary unit format, without using decimals. For example, if you use meters to dimension most objects on a metric civil engineering drawing, you can use a **Sub-units factor** value of 100 (100 cm/m) and a **Sub-unit suffix** of cm to dimension objects smaller than one meter using centimeters, instead of decimals of a meter. Now when you dimension an object that is 0.5 meters, the dimension reads 500 cm.

> **NOTE**
>
> For drawings that do not require sub-units, but do suppress leading zeros, specify no sub-unit suffix. As long as you do not add a suffix, there is no need to change the sub-unit factor, though a factor of 0 also disables sub-units.

Angular Dimension Settings

The **Angular dimensions** area allows you to specify settings for primary angular dimensions. The options from the **Units format** drop-down list are the same as those in the **Angle** area of the **Drawing Units** dialog box. Typically, the primary angular dimension unit format is the same as the corresponding drawing units. Use the **Precision** drop-down list to set the appropriate angular dimension value precision. The **Zero**

suppression area has check boxes for suppressing angular dimension **Leading** and **Trailing** zeros. Zero suppression for angular units is usually the same as applied to linear dimensions.

Using the Alternate Units Tab

The **Alternate Units** tab of the **New** (or **Modify**) **Dimension Style** dialog box, shown in **Figure 17-34**, allows you to set *alternate units*, or *dual dimensioning units*. Dual dimensioning practices are no longer a recommended ASME standard. ASME recommends that drawings be dimensioned using inch or metric units only. However, many other applications do use alternate units.

To use alternate units, select the **Display alternate units** check box to enable the settings. The **Alternate Units** tab includes most of the same settings found in the **Primary Units** tab. The **Multiplier for alt units** setting is multiplied by the primary unit to establish the value for the alternate unit. The default 25.4 allows you to use millimeters as alternate units on an inch unit drawing. The **Placement** area controls the location of the alternate-unit dimension. You can choose to place the alternate-unit dimension after or below the primary value.

alternate units (dual dimensioning units): Dimensions in which measurements in one system, such as inches, are followed by bracketed measurements in another system, such as millimeters.

NOTE

Chapter 20 describes the **Tolerances** tab found in the **New** (or **Modify**) **Dimension Style** dialog box.

Exercise 17-3

Access the Student Web site (www.g-wlearning.com/CAD) and complete Exercise 17-3.

Figure 17-34.
The **Alternate Units** tab of the **New** (or **Modify**) **Dimension Style** dialog box.

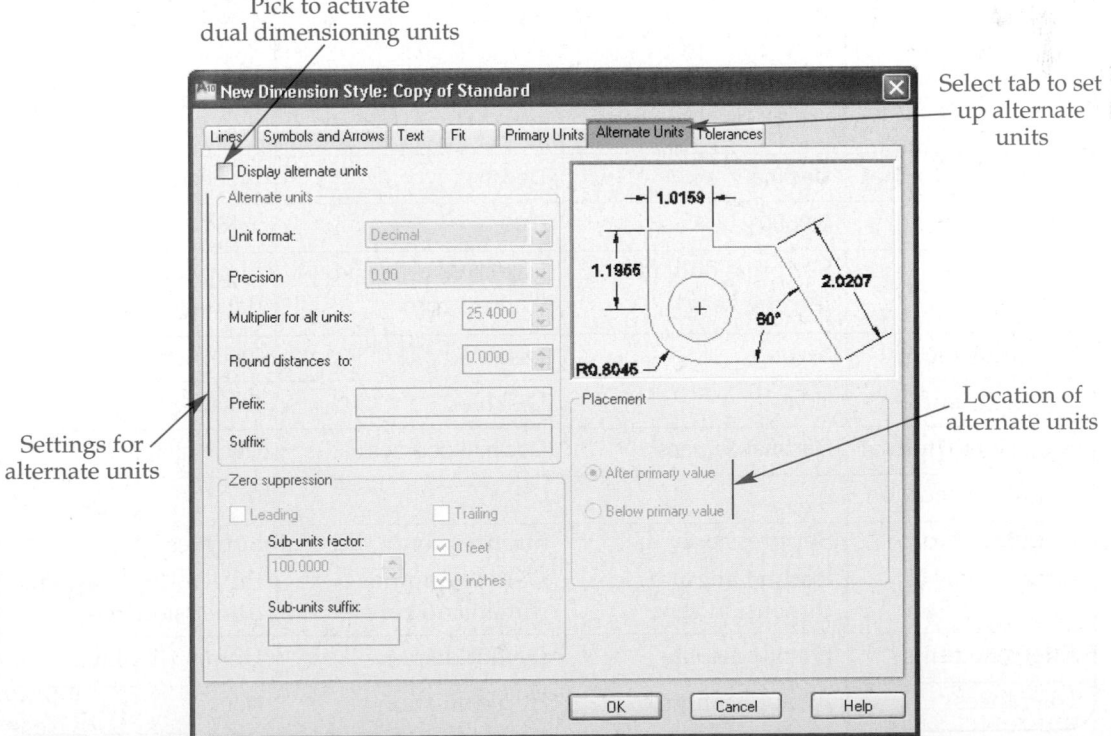

Making Your Own Dimension Styles

Creating and using dimension styles is an important element of drafting with AutoCAD. Carefully evaluate the characteristics of the dimensions you will add to drawings. Check school, company, and national standards to verify the accuracy of the dimension settings you plan to use. When you are ready, use the **Dimension Style Manager** dialog box to establish appropriate dimension styles. **Figure 17-35** provides possible settings for three common dimension styles. Use the AutoCAD default values for settings not listed.

Figure 17-35.
This chart shows dimension settings for typical mechanical and architectural drawings.

Setting	Mechanical—Inch	Mechanical—Metric (mm)	Architectural—U.S. Customary
Baseline spacing	.5	12	1/2″
Extend beyond dimension lines	.125	3	1/8″
Offset from origin	.063	1.5	1/16″ or 3/32″
Arrowhead options	Closed filled, Closed, or Open	Closed filled, Closed, or Open	Architectural tick, Dot, Closed filled, Oblique, or Right angle
Arrow size	.125	3	1/8″
Center marks	Line	Line	Mark
Center mark size	.0625	1.5	1/16″
Text style	Romans	Romans	Stylus BT
Text height	.12	3	1/8″
Vertical and horizontal text placement	Centered	Centered	Vertical: Above Horizontal: Centered
View direction	Left-to-right	Left-to-right	Left-to-right
Offset from dimension line	.063	1.5	1/16″
Text alignment	Horizontal	Horizontal	Aligned with dimension line
Linear unit format	Decimal	Decimal	Architectural
Linear precision	0.0000	0.00	1/16″
Linear zero suppression	Suppress only the leading zero	Suppress only the trailing zero	Suppress only the 0 feet zero
Sub-units factor	0	Disabled	Disabled
Sub-units suffix	None	Disabled	Disabled
Angular unit format	Decimal degrees	Decimal degrees	Decimal degrees
Angular precision	0	0	0
Angular zero suppression	Suppress only the leading angular dimension zero	Suppress only the trailing angular dimension zero	Suppress only the leading angular dimension zero
Alternate units	Do not display	Do not display	Do not display
Tolerances	By application	By application	None

To save valuable drafting time, add dimension styles to your template drawings.

Exercise 17-4

Access the Student Web site (www.g-wlearning.com/CAD) and complete Exercise 17-4.

Changing Dimension Styles

Use the **Dimension Style Manager** to change the characteristics of an existing dimension style. Pick the **Modify** button to open the **Modify Dimension Style** dialog box, which allows you to make changes to the selected style. When you make changes to a dimension style, such as selecting a different text style or linear precision, all existing dimensions drawn using the modified dimension style update to reflect the changes. Use a different dimension style with unique characteristics when appropriate.

To *override* a dimension style, pick the **Override** button in the **Dimension Style Manager** to open the **Override Current Style** dialog box. An example of an override is including a text prefix for a few of the dimensions in a drawing. The **Override** button is only available for the current style. Once you create an override, it is current and appears as a branch, called the *child*, of the *parent* style. The override settings are lost when any other style, including the parent, is set current.

override: A temporary change to the current style settings; the process of changing a current style temporarily.

child: A style override.

parent: The dimension style from which a style override is created.

Sometimes it is useful to view the details of two styles to determine their differences. Select the **Compare...** button in the **Dimension Style Manager** to display the **Compare Dimension Styles** dialog box. Here you can compare two styles by selecting the name of one style from the **Compare:** drop-down list and the name of the other in the **With:** drop-down list. The differences between the selected styles display in the dialog box.

NOTE
The **New Dimension Style**, **Modify Dimension Style**, and **Override Current Style** dialog boxes have the same tabs.

Renaming and Deleting Dimension Styles

To rename a dimension style using the **Dimension Style Manager**, slowly double-click on the name or right-click the name and select **Rename**. To delete a dimension style using the **Dimension Style Manager**, right-click the name and select **Delete**. You cannot delete a dimension style that is assigned to dimensions. To delete a style that is in use, assign a different style to the dimensions that reference the style to be deleted.

NOTE
You can also rename styles using the **Rename** dialog box. Select **Dimension styles** in the **Named Objects** list to rename a dimension style.

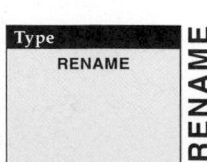
Type
RENAME
RENAME

Setting a Dimension Style Current

You can set a dimension style current using the **Dimension Style Manager** by double-clicking the style in the **Styles** list box, right-clicking on the name and selecting **Set current**, or picking the style and selecting the **Set current** button. To set a text style current without opening the **Dimension Style Manager** dialog box, use the **Dimension Style** flyout located in the expanded **Annotation** panel of the **Home** ribbon tab and on the **Dimensions** panel of the **Annotate** ribbon tab.

PROFESSIONAL TIP

You can import dimension styles from existing drawings using **DesignCenter**. See Chapter 5 for more information about using **DesignCenter** to import file content.

Exercise 17-5

Access the Student Web site (www.g-wlearning.com/CAD) and complete Exercise 17-5.

Template Development
Chapter 9

For instructions on adding dimension styles to each drawing template, go to the Student Web site (www.g-wlearning.com/CAD), select this chapter, and select **Template Development**.

Chapter Test

Answer the following questions. Write your answers on a separate sheet of paper or go to the Student Web site (www.g-wlearning.com/CAD) and complete the electronic chapter test.

1. List at least three factors that influence company dimensioning practices.
2. What does the *M* mean in the title of the standard ASME Y14.5M-1994?
3. Name two basic coordinate systems that are used to create location dimensions.
4. Define the term *general notes.*
5. Briefly explain the difference between placing specific and general notes on a drawing.
6. Explain how to dimension a cylinder using only one view.
7. Describe two ways to dimension a cone.
8. When is the best time to determine the drawing scale and scale factors for a drawing?
9. Explain how to add a scale to the **Annotation Scale** flyout in the status bar.
10. Define *dimension style.*
11. Name the dialog box that is used to create dimension styles.
12. Identify two ways to access the dialog box identified in Question 11.
13. Name the dialog box tab used to control the appearance of dimension lines and extension lines.

14. Name at least four arrowhead types that are available in the **Symbols and Arrows** tab for common use on architectural drawings.
15. Name the dialog box tab used to control the settings that display the dimension text.
16. What has to happen before a text style can be accessed for use in dimension text?
17. What is the ASME recommended height for dimension numbers and notes on drawings?
18. Name the dialog box tab used to control settings that adjust the location of dimension lines, dimension text, arrowheads, and leader lines.
19. How can you delete a dimension style from a drawing?
20. How do you set a dimension style current?

Drawing Problems

Start AutoCAD if it is not already started. Start a new drawing for each problem using an appropriate template of your choice. Follow the specific instructions for each problem.

▼ Basic

1. Start a new drawing using one of your templates and create a RomanS text style using the romans font. Create the Mechanical (Inch) dimension style shown in **Figure 17-35.** Use the default AutoCAD settings for the dimension style settings not listed. Save the drawing as P17-1.

2. Start a new drawing using one of your templates and create a RomanS text style using the romans font. Create the Mechanical (Metric) dimension style shown in **Figure 17-35.** Use the default AutoCAD settings for the dimension style settings not listed. Save the drawing as P17-2.

3. Start a new drawing using one of your templates and create a Stylus BT text style using the Stylus BT font. Create the Architectural dimension style shown in **Figure 17-35.** Use the default AutoCAD settings for the dimension style settings not listed. Save the drawing as P17-3.

4. Write a short report explaining the difference between unidirectional and aligned dimensioning. Use a word processor and include sketches giving examples of each method.

5. Write a short report explaining the difference between size and location dimensions. Use a word processor and include sketches giving examples of each method.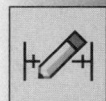

6. Write a short report describing the basic difference between dimensioning for mechanical drafting (drafting for manufacturing) and architectural drafting. Use a word processor and include sketches giving examples of each method.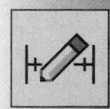

7. Make sketches showing the standard practice for dimensioning a cylindrical object, a square object, and a conical object.

8. Make sketches showing the standard practice for dimensioning angles. Make one sketch showing coordinate dimensioning and another showing angular dimensioning.

▼ Intermediate

9. Find a copy of the ASME Y14.5M, *Dimensioning and Tolerancing* standard and write a report of approximately 350 words explaining the importance and basic content of this standard.

10. Interview your drafting instructor or supervisor and determine what dimension standards exist at your school or company. Write them down and keep them with you as you learn AutoCAD. Make notes as you progress through this textbook on how you use these standards. Also, note how the standards could be changed to better match the capabilities of AutoCAD.

▼ Advanced

11. Create a freehand sketch of **Figure 17-1.** Label each of the dimension items. To the side of the sketch, write a short description of each item.

12. Research civil drafting and create a template establishing the dimension styles for a civil drawing.

13. Visit at least three local manufacturing companies where design drafting work is done as part of their business. Write a report with sketched examples identifying the standards used at each company.

14. Find a local manufacturing company where design drafting work is done as part of their business. Write a report with sketched examples identifying the standards used at the company.

15. Find and visit two local companies, one architectural and one civil, where design drafting work is done as part of their business. Write a report with sketched examples identifying the standards used at each company.

Drawing Problems - Chapter 17

CHAPTER 18

Linear and Angular Dimensioning

Learning Objectives

After completing this chapter, you will be able to do the following:
✓ Add linear dimensions to a drawing.
✓ Add angular dimensions to a drawing.
✓ Draw datum and chain dimensions.
✓ Add dimensions for multiple items using the **QDIM** tool.

A drawing often requires a variety of dimensions to describe the size and shape of features and objects. Linear and angular dimensions are two of the most common. This chapter covers the process of adding linear and angular dimensions to a drawing using several dimensioning tools. You will also learn how to add a break symbol to a dimension line and use the **QDIM** tool.

Placing Linear Dimensions

Linear dimensions usually measure straight distances, such as distances between horizontal, vertical, or slanted surfaces. The **DIMLINEAR** tool allows you to place linear dimensions.

Dimension tools reference the current dimension style and the points or objects you select to create a single dimension object. When you use the **DIMLINEAR** tool, for example, you create a dimension object that includes all related dimension style characteristics, dimension and extension lines, arrowheads, and a dimension value associated with the distance between selected points.

Once you access the **DIMLINEAR** tool, pick a point to locate origin of the first extension line, and then pick a point to locate the origin of the second extension line. See **Figure 18-1.** Use object snap modes and other drawing aids to pick the exact points where extension lines begin. Once you establish the extension line origins, you can select from several options that appear at the Specify dimension line location or [Mtext/Text/Angle/Horizontal/Vertical/Rotated] prompt. To apply the default option and create a linear dimension, move the dimension line to the desired location and pick. See **Figure 18-2.**

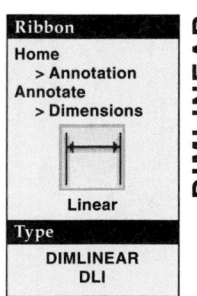

Ribbon
Home
> Annotation
Annotate
> Dimensions

Linear

Type
DIMLINEAR
DLI

DIMLINEAR

Figure 18-1.
Establishing extension line origins. The **Endpoint** and **Intersection** object snap modes are useful for locating origins accurately.

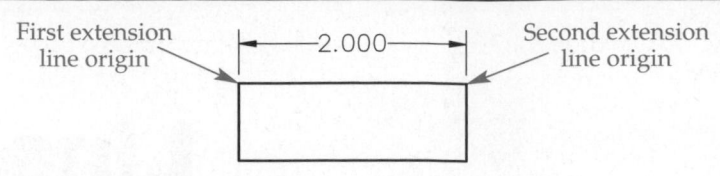

First extension line origin

2.000

Second extension line origin

Figure 18-2.
Establishing the location of a dimension line.

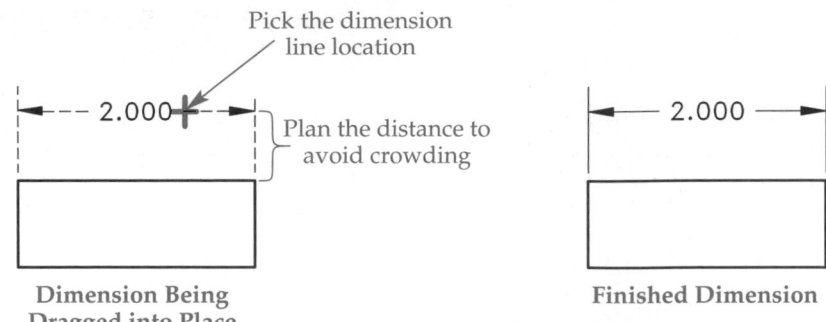

Pick the dimension line location

2.000

Plan the distance to avoid crowding

2.000

Dimension Being Dragged into Place

Finished Dimension

NOTE

When you dimension objects with AutoCAD, the objects automatically measure exactly as you have drawn them. This makes it important for you to draw objects and features accurately and to select the origins of the extension lines accurately.

PROFESSIONAL TIP

Use preliminary plan sheets and sketches to help you determine proper dimension line location and distances between dimension lines to avoid crowding.

Exercise 18-1

Access the Student Web site (www.g-wlearning.com/CAD) and complete Exercise 18-1.

Selecting an Object to Dimension

An alternative method for locating extension line origins involves picking a single line, circle, or arc to dimension. You can use this option whenever you see the Specify first extension line origin or <select object>: prompt. Press [Enter] or the space bar or right-click and then pick the object to dimension. When you select a line or arc, extension lines begin from endpoints. When you pick a circle, extension lines begin from the closest quadrant and its opposite quadrant. See **Figure 18-3.**

Figure 18-3.
AutoCAD can determine the extension line origins automatically when you select a line, arc, or circle.

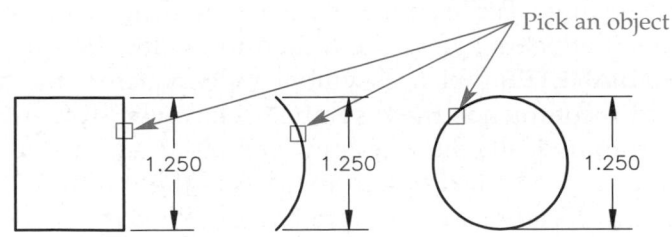

Adjusting Dimension Text

The value attached to the dimension corresponds to the distance between extension lines. Use the **Mtext** option to access the multiline text editor to adjust the dimension value. See **Figure 18-4.** The highlighted value represents the current dimension value. Add to or modify the dimension text and then close the text editor. The tool continues, allowing you to pick the dimension line location.

The **Text** option allows you to use the single-line text editor to change dimension text, even though the final dimension value is an mtext object. The current dimension value appears in brackets. Add to or modify the value as necessary, and then press [Enter] to exit the option. The tool continues, allowing you to pick the dimension line location.

NOTE

Dimension values are horizontal or aligned with the dimension line, according to the current dimension style format. The **Angle** option has limited applications, but allows you to rotate the dimension text. Enter the desired angle at the Specify angle of dimension text: prompt to use this option.

Figure 18-4.
When you use the **Mtext** option, the **Text Editor** ribbon tab appears, and AutoCAD's calculated dimension value appears in a text box for editing.

Multiline
Text tab
of the
ribbon
appears

Represents
the
dimension
calculated
by AutoCAD

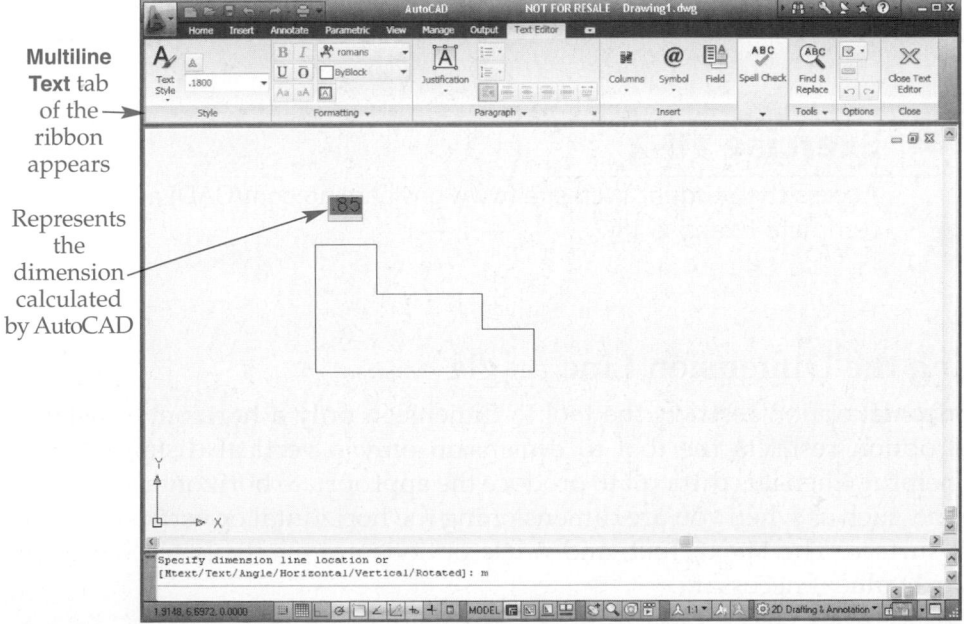

Including Symbols with Dimension Text

With some dimension tools, AutoCAD automatically places appropriate symbols with the dimension value. For example, when you dimension an arc using the **DIMRADIUS** tool, an R appears before the dimension value. When you dimension a circle using the **DIMDIAMETER** tool, a ∅ symbol appears before the dimension value. The ASME standard recommends these symbols. However, dimension tools such as **DIMLINEAR** do not automatically place certain symbols or add necessary characters.

One option is to use the **Mtext** option to activate the multiline text editor. Place the cursor at the correct location and use options from the **Symbol** flyout or type characters to add information. For example, pick the **Diameter** symbol from the **Symbol** flyout to add ∅. Another example is enclosing a reference dimension in parentheses, as recommended by the ASME standard. To create a reference dimension, type open and close parentheses around the highlighted value.

You can also add content using the **Text** option and the single-line text editor. Place the cursor at the correct location and use control codes or characters to add content. For example, type %%C to display ∅ or type parentheses around the value to create a reference dimension.

Still another way to place symbols with dimension text is to create a dimension style that references a text style using the gdt.shx font. A text style with the gdt.shx font allows you to place common dimension symbols, including geometric dimensioning and tolerancing (GD&T) symbols, using the lowercase letter keys.

PROFESSIONAL TIP

Although you can add a prefix and suffix to a dimension style, usually it is more appropriate to adjust the limited number of dimensions that require a prefix or suffix.

Reference Material

Drafting Symbols

For more information about common drafting symbols and the gdt.shx font, go to the **Reference Material** section of the Student Web site (www.g-wlearning.com/CAD) and select **Drafting Symbols** in the list.

Exercise 18-2

Access the Student Web site (www.g-wlearning.com/CAD) and complete Exercise 18-2.

Controlling the Dimension Line Angle

The **Horizontal** option restricts the tool to dimension only a horizontal distance. The **Vertical** option restricts the tool to dimension only a vertical distance. These options are helpful when it is difficult to produce the appropriate horizontal or vertical dimension line, such as when you are dimensioning the horizontal or vertical distance of a slanted surface. The **Mtext**, **Text**, and **Angle** options are available to change the dimension text value if necessary.

The **Rotated** option allows you to specify a dimension line angle. A practical application is dimensioning to angled surfaces and auxiliary views. This technique is different from other dimensioning tools because you provide a dimension line angle. See **Figure 18-5.** At the Specify angle of dimension line <0>: prompt, enter a value or pick two points on the line to dimension.

Exercise 18-3

Access the Student Web site (www.g-wlearning.com/CAD) and complete Exercise 18-3.

Dimensioning Angled Surfaces and Auxiliary Views

When you dimension a surface drawn at an angle, such as an auxiliary view, it is often necessary to align the dimension line with the surface, with extension lines perpendicular to the surface. In order to dimension these features properly, use the **DIMALIGNED** tool or the **Rotated** option of the **DIMLINEAR** tool.

Figure 18-5.
Rotating a dimension for an angled view.

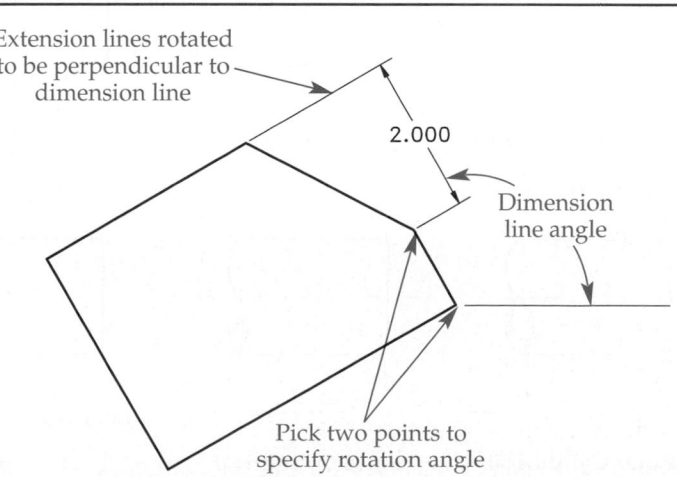

Extension lines rotated to be perpendicular to dimension line

2.000

Dimension line angle

Pick two points to specify rotation angle

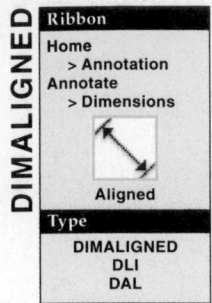

DIMALIGNED

Ribbon
Home
> Annotation
Annotate
> Dimensions

Aligned

Type
DIMALIGNED
DLI
DAL

Figure 18-6 shows the results of using the **DIMALIGNED** tool. Notice the difference between the aligned dimension in this figure and the rotated dimension in **Figure 18-5**. You can usually use the **DIMALIGNED** tool when the length of the extension lines is equal. The **Rotated** option of the **DIMLINEAR** tool is often necessary when extension lines are unequal.

Exercise 18-4

Access the Student Web site (www.g-wlearning.com/CAD) and complete Exercise 18-4.

Dimensioning Long Objects

When you create a drawing of a long part that has a constant shape, the view may not fit on the desired sheet size, or it may look strange compared to the rest of the drawing. To overcome this problem, use a *conventional break* (or *break*) to shorten the view. **Figure 18-7** shows examples of standard break lines. For many long parts, a conventional break is required to display views or increase the view scale without increasing the sheet size. Dimensions added to conventional breaks describe the actual length of the product in its unbroken form. The dimension line often includes a break symbol to indicate that the drawing view is broken and that the feature is longer than it appears. See **Figure 18-8**.

You can use the **DIMJOGLINE** tool to add a break symbol to dimension lines created using the **DIMLINEAR** or **DIMALIGNED** tools. Once you access the **DIMJOGLINE** tool, pick a linear or aligned dimension line. Then pick a location on the dimension line to place the break symbol, as shown in **Figure 18-8**. An alternative to selecting the

conventional break (break): Removal of a portion of a long, constant-shaped object that has been removed from the drawing to make the object fit better on the drawing sheet.

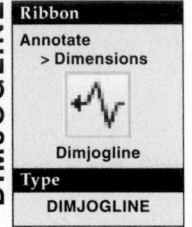

DIMJOGLINE

Ribbon
Annotate
> Dimensions

Dimjogline

Type
DIMJOGLINE

Figure 18-6.
The **DIMALIGNED** tool allows you to place dimension lines parallel to angled features.

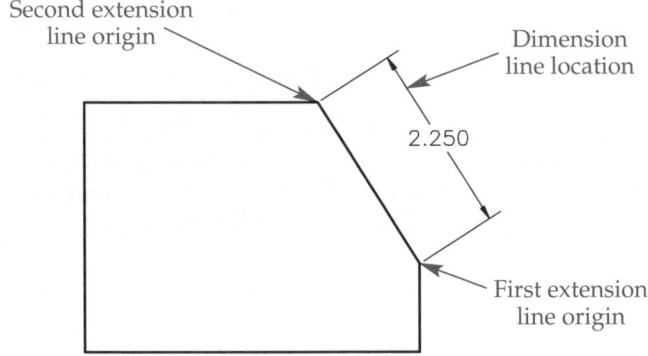

Figure 18-7.
Standard break lines.

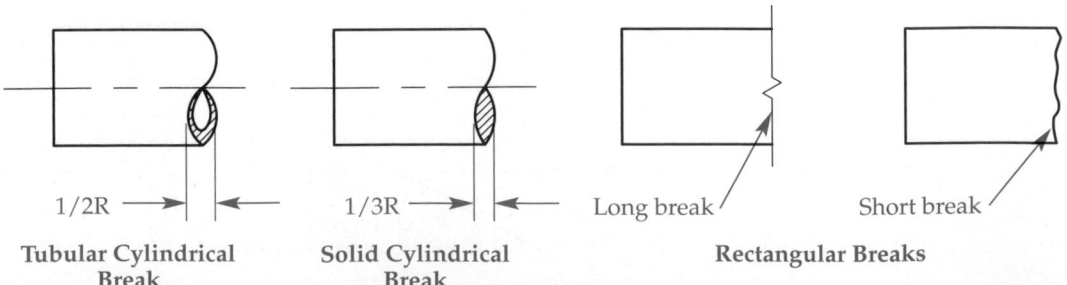

1/2R 1/3R Long break Short break

Tubular Cylindrical Break   **Solid Cylindrical Break**   **Rectangular Breaks**

AutoCAD and Its Applications—Basics

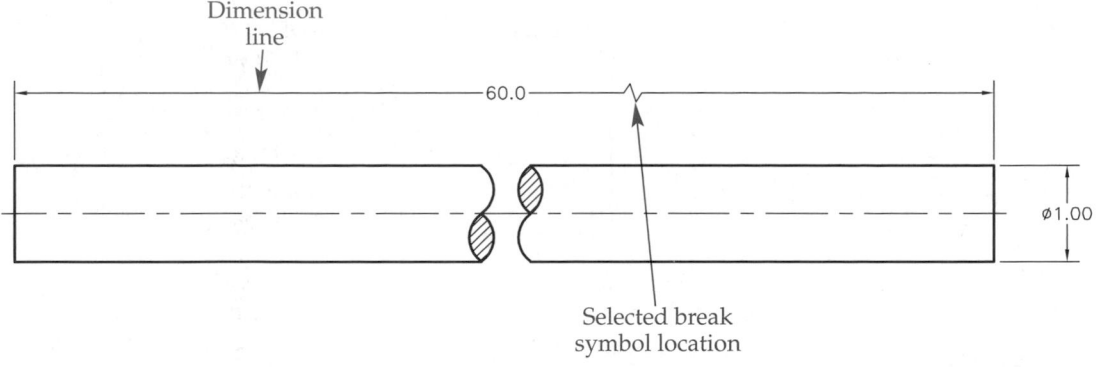

Dimension line

60.0

Selected break symbol location

Ø1.00

location of the break symbol is to press [Enter] to accept the default location. You can move the break later using grip editing or by reusing the **DIMJOGLINE** tool to select a different location. To remove the break symbol, access the **DIMJOGLINE** tool and select the **Remove** option.

NOTE

You can add only one break symbol to a dimension line.

Exercise 18-5

Access the Student Web site (www.g-wlearning.com/CAD) and complete Exercise 18-5.

Dimensioning Angles

Coordinate and angular dimensioning are both accepted methods for dimensioning angles. **Figure 18-9** shows an example of *coordinate dimensioning* using the **DIMLINEAR** tool.

Figure 18-10 shows an example of *angular dimensioning* using the **DIMANGULAR** tool. You can dimension the angle between any two nonparallel lines from the *vertex* of the angle. AutoCAD automatically draws extension lines if needed.

Once you access the **DIMANGULAR** tool, pick the first leg of the angle to dimension, and then pick the second leg of the angle. The last prompt asks you to pick the location of the dimension line arc. **Figure 18-11** shows examples of angular dimensions and the effect that limited space may have on dimension fit and placement. Fit characteristics apply to most dimensions.

coordinate dimensioning: A method of dimensioning angles in which dimensions locate the corner of the angle.

angular dimensioning: A method of dimensioning angles in which one corner of an angle is located with a dimension and the value of the angle is provided in degrees.

vertex: The point at which the two lines that form an angle meet.

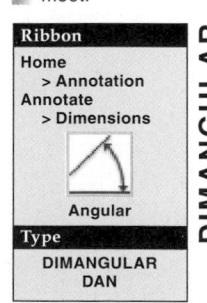

Ribbon
Home
  > Annotation
Annotate
  > Dimensions

Angular

Type
DIMANGULAR
DAN

DIMANGULAR

Figure 18-9.
Coordinate
dimensioning of
angles.

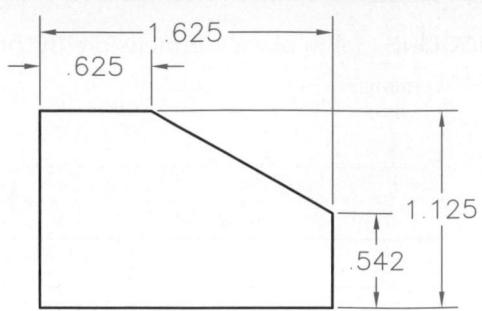

Figure 18-10.
Two examples of
drawing angular
dimensions.

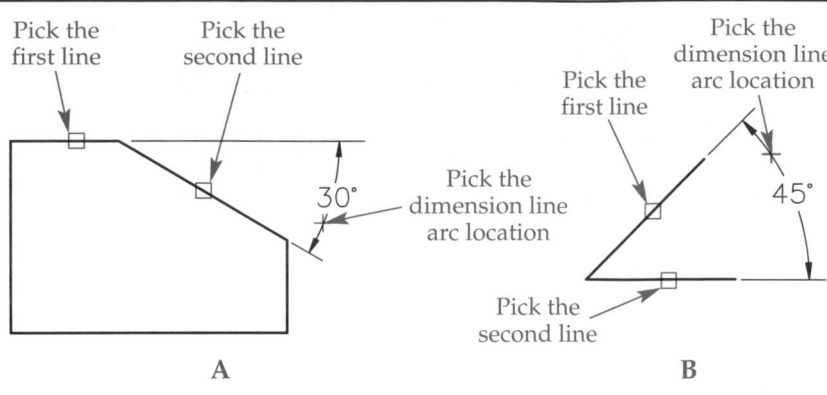

A

B

Figure 18-11.
The dimension
line arc location
determines where
the dimension line
arc, text, and arrows
display.

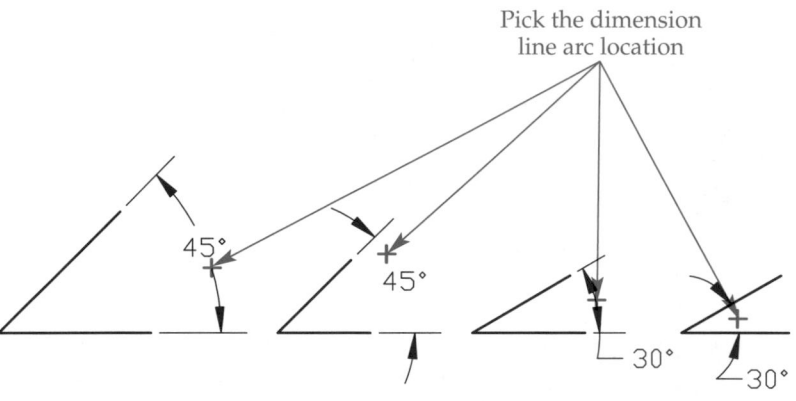

PROFESSIONAL TIP

You can create four different dimensions (two different angles) with an angular dimension. To preview these options before selecting the dimension line location, use the cursor to move the dimension around an imaginary circle. You can also use the **Quadrant** option to isolate a specific quadrant of the imaginary circle and force the dimension to produce the value found in the selected quadrant.

Dimensioning Angles on Arcs and Circles

You can use the **DIMANGULAR** tool to dimension the included angle of an arc or a portion of a circle. When you dimension an arc, the center point becomes the angle vertex, and the two arc endpoints establish the extension line origins. See **Figure 18-12.**

When you dimension a circle using **DIMANGULAR**, the center point becomes the angle vertex and two picked points specify the extension line origins. See **Figure 18-13.** The point you pick to select the circle locates the origin of the first extension line. You then select the second angle endpoint, which locates the origin of the second extension line.

Figure 18-12.
Placing angular dimensions on arcs.

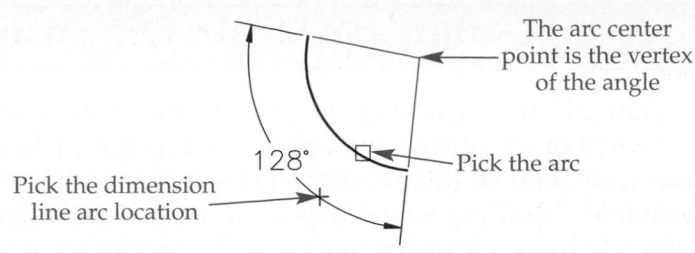

The arc center point is the vertex of the angle

128°

Pick the dimension line arc location

Pick the arc

Figure 18-13.
Placing angular dimensions on circles.

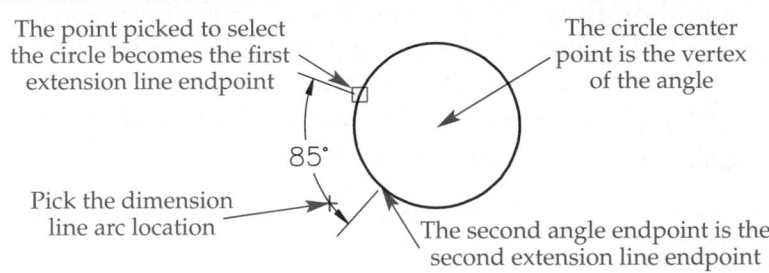

The point picked to select the circle becomes the first extension line endpoint

The circle center point is the vertex of the angle

85°

Pick the dimension line arc location

The second angle endpoint is the second extension line endpoint

PROFESSIONAL TIP

Using angular dimensioning for circles increases the number of possible solutions for a given dimensioning requirement, but the actual uses are limited. One application is dimensioning an angle from a quadrant point to a particular feature without first drawing a line to dimension. Another benefit of this option is the ability to specify angles that exceed 180°.

Angular Dimensioning through Three Points

You can also establish an angular dimension through three points. The points are the angle vertex and two angle line endpoints. See **Figure 18-14.** To apply this technique, press [Enter] or the space bar, or right-click after the first prompt. Then pick the vertex, followed by the two points. This method also dimensions angles over 180°.

Exercise 18-6

Access the Student Web site (www.g-wlearning.com/CAD) and complete Exercise 18-6.

Figure 18-14.
Placing angular dimensions using three points.

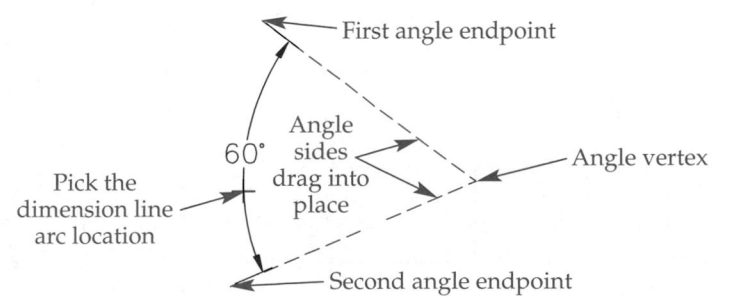

First angle endpoint

Angle sides drag into place

Angle vertex

60°

Pick the dimension line arc location

Second angle endpoint

Datum and Chain Dimensioning

datum dimensioning: A method of dimensioning in which several dimensions originate from a common surface, centerline, or center plane.

datum: Theoretically perfect surface, plane, point, or axis from which measurements can be taken.

chain dimensioning (point-to-point dimensioning): A method of dimensioning in which dimensions are placed in a line from one feature to the next.

Mechanical drafting often requires *datum dimensioning* because each dimension is independent of the others, and references a *datum*. This achieves more accuracy in manufacturing. **Figure 18-15** shows an object dimensioned with surface datums.

Mechanical drafting sometimes uses *chain dimensioning*, also called *point-to-point dimensioning*. However, this method provides less accuracy than datum dimensioning because each dimension is dependent on other dimensions in the chain. In mechanical drafting, it is common to leave one dimension out and provide an overall dimension that controls the missing value. See **Figure 18-16.** Architectural drafting uses chain dimensioning in most applications to reduce the need to calculate or find dimension values during construction. Architectural drafting practices usually show dimensions for all features, plus an overall dimension.

Figure 18-15.
Datum dimensioning.

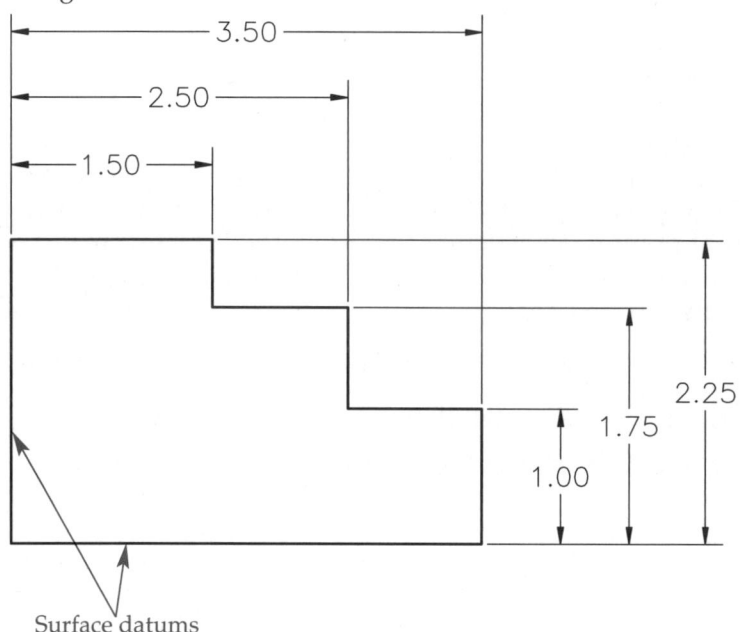

Figure 18-16.
Chain dimensioning. The example on the right is most common, but either technique is acceptable, depending on the design and dimensioning requirement.

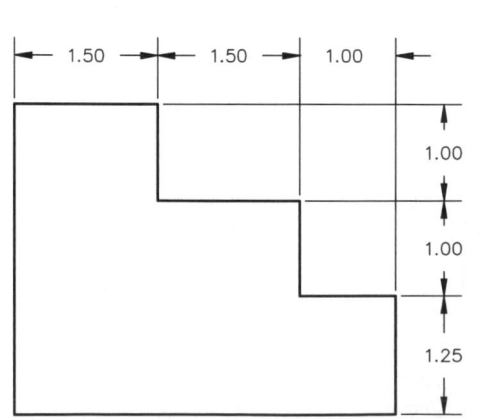
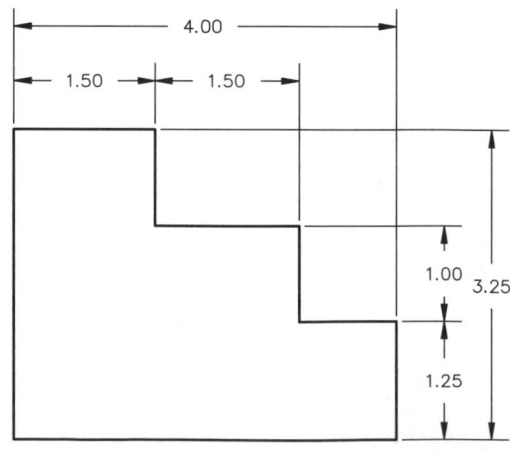

Placing Datum Dimensions

AutoCAD refers to datum dimensioning as *baseline dimensioning*. The **DIMBASELINE** tool controls datum dimensioning and allows you to select several points to define a series of datum dimensions. You can create baseline dimensions with linear, angular, and ordinate dimensions. Chapter 19 describes ordinate dimensions.

The **DIMBASELINE** tool continues from an existing dimension. Therefore, you must create the first dimension using the **DIMLINEAR** tool. The first point you select when drawing the linear dimension defines the datum. Then access the **DIMBASELINE** tool and pick the next second extension line origin. Continue picking extension line origins until you have dimensioned all of the features. Press [Enter] or the space bar or right-click twice, once at the Specify a second extension line origin or: prompt, and again at the Select base dimension: prompt, to create the dimensions and exit the tool. Notice that as you pick additional extension line origins, AutoCAD automatically places the dimension text; you do not specify a location. **Figure 18-17** shows an example of using the **DIMBASELINE** tool to pick two additional extension line origins to an existing linear dimension.

AutoCAD automatically selects the most recently drawn dimension as the base dimension unless you specify a different dimension. To add datum dimensions to an existing dimension other than the most recently drawn dimension, use the **Select** option by pressing [Enter] or the space bar, or by right-clicking and selecting **Enter,** at the first prompt. At the Select base dimension: prompt, pick the dimension to serve as the base. The extension line nearest the point where you select the dimension establishes the datum. Then select the new second extension line origins as previously described.

You can also draw baseline dimensions to angular features. First, draw an angular dimension. Then enter the **DIMBASELINE** tool. **Figure 18-18** shows angular baseline dimensions. As with linear dimensions, you can pick an existing angular dimension other than the dimension most recently drawn.

Placing Chain Dimensions

AutoCAD refers to chain dimensioning as *continued dimensioning*. The **DIMCONTINUE** tool controls chain dimensioning and allows you to select several points to define a series of chain dimensions. **Figure 18-19** shows chain dimensioning.

Use the **DIMBASELINE** and **DIMCONTINUE** tools in the same manner. When creating chain dimensions, you will see the same prompts and options you see when creating datum dimensions. Like datum dimensions, you can create continued dimensions with linear, angular, and ordinate dimensions.

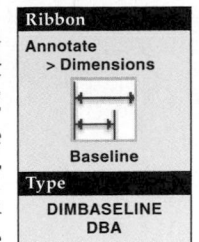

baseline dimensioning: The AutoCAD term for datum dimensioning.

Ribbon
Annotate
> Dimensions

Baseline

Type
DIMBASELINE
DBA

DIMBASELINE

continued dimensioning: The AutoCAD term for chain dimensioning.

Ribbon
Annotate
> Dimensions

Continue

Type
DIMCONTINUE
DCO

DIMCONTINUE

Figure 18-17.
Using the **DIMBASELINE** tool. AutoCAD automatically places the extension lines, dimension lines, arrowheads, and text.

First dimension line location

Dimensions are automatically drawn

4.375
3.250
2.000

First extension line origin

Second extension line origin

Next second extension line origin

Next second extension line origin

Figure 18-18.
Using the
DIMBASELINE
tool to add datum
dimensions to
angular features.

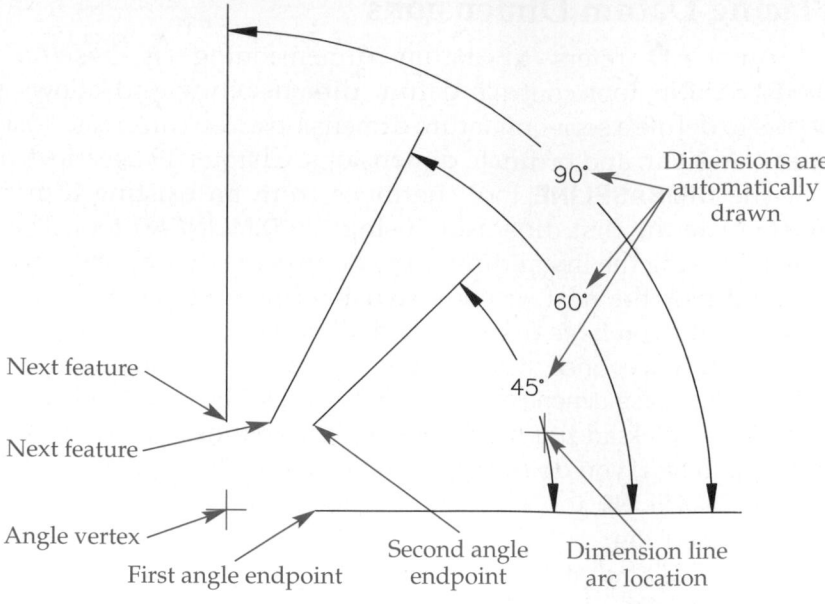

Figure 18-19.
Using the
DIMCONTINUE
tool to create chain
dimensions.

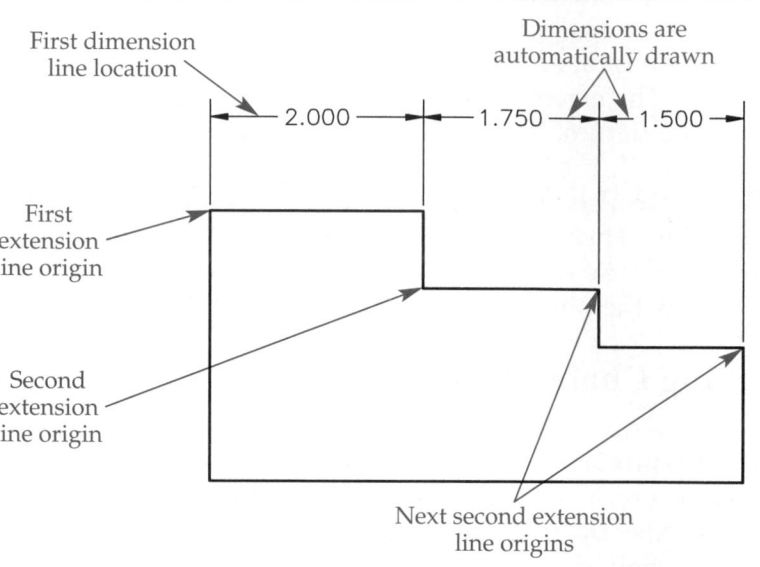

NOTE

Use the **Undo** option in the **DIMBASELINE** and **DIMCONTINUE** tools to undo previously drawn dimensions.

PROFESSIONAL TIP

You do not have to use **DIMBASELINE** or **DIMCONTINUE** immediately after you create the base or chain dimension. Use the **Select** option later during the drawing session to pick the datum or chain dimension.

AutoCAD and Its Applications—Basics

Exercise 18-7

Access the Student Web site (www.g-wlearning.com/CAD) and complete Exercise 18-7.

Using QDIM to Dimension

The **QDIM**, or quick dimension, tool makes chain and datum dimensioning easier by eliminating the need to define exact dimension points. Often, the points you select for dimensioning are the endpoints of lines or the center points of arcs. AutoCAD automates the process of point selection in the **QDIM** tool by finding those points for you.

The type of geometry you select affects the **QDIM** output. If you choose a single polyline, **QDIM** attempts to draw linear dimensions to every vertex. If you pick a single arc or circle, **QDIM** draws a radius or diameter dimension. If you select multiple objects, linear dimensions occur from the vertex of every line or polyline and to the center of every arc or circle. In each case, AutoCAD finds the points automatically.

Once you access the **QDIM** tool, pick several lines, polylines, arcs, and/or circles and press [Enter] or the space bar, or right-click. Then pick a position for the dimension lines to create the dimensions and exit the tool. **Figure 18-20** shows examples of using the **QDIM** tool to dimension different types of objects. To create the upper dimensions, select each object separately. To create the lower dimensions, select all of the objects at once.

The **Continuous** option of the **QDIM** tool creates chain dimensions. The **Baseline** option creates datum dimensions. The **Staggered** option creates staggered (noncontinuous) dimensions. The **Ordinate**, **Radius**, and **Diameter** options provide methods of adding ordinate, radius, and diameter dimensions. Chapter 19 describes these types of dimensions.

To dimension as shown in **Figure 18-20A,** access the **QDIM** tool and select the polyline. Then issue the **Baseline** option and select a position for the dimension line above the polyline to create the dimension and exit the tool. To create the dimensions at the bottom of **Figure 18-20**, use the **Continuous** option of the **QDIM** tool and select all three objects.

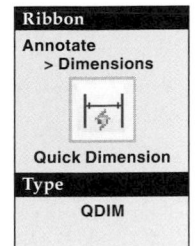

Ribbon
Annotate
> Dimensions

Quick Dimension

Type
QDIM

QDIM

Figure 18-20.
The **QDIM** tool can dimension multiple features or objects at the same time.

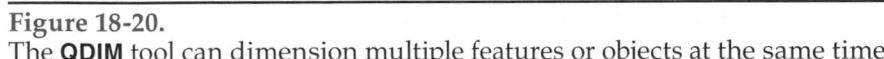
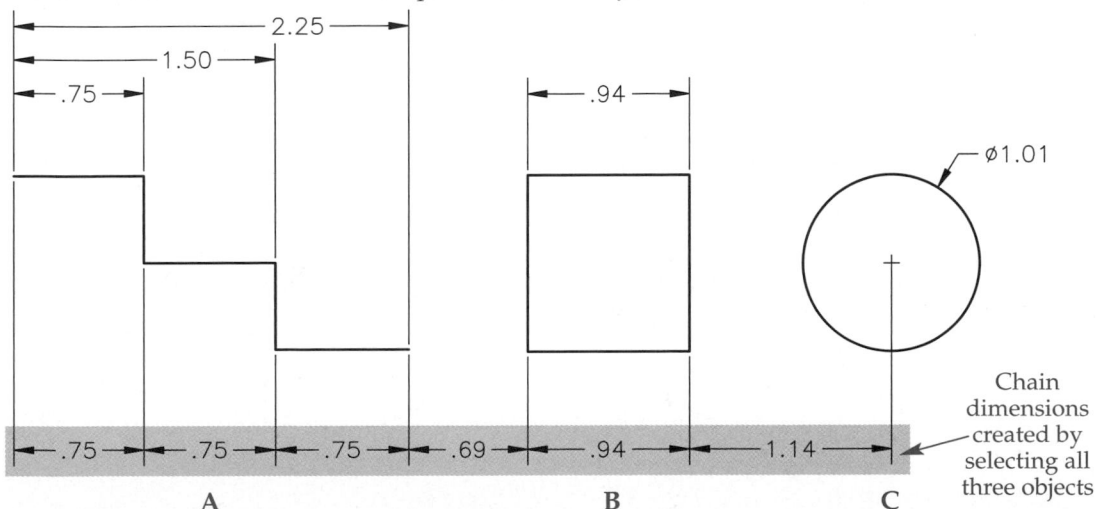

You can use the **datumPoint** option to change the datum point for datum or chain dimensions. The **Settings** option allows you to set the object snap mode for establishing the extension line origins to **Endpoint** or **Intersection**.

> **NOTE**
>
> You can also use the **QDIM** tool to edit any existing associative dimension. Chapter 21 describes editing dimensions and using the **Edit** option of the **QDIM** tool.

Chapter Test

Answer the following questions. Write your answers on a separate sheet of paper or go to the Student Web site (www.g-wlearning.com/CAD) and complete the electronic chapter test.

1. Name the two **DIMLINEAR** options that allow you to change dimension text.
2. Give two examples of symbols that automatically appear with some dimensions.
3. What is the purpose of AutoCAD's gdt.shx font?
4. Name the two dimensioning tools that provide linear dimensions for angled surfaces.
5. Which tool allows you to place a break symbol in a dimension line?
6. Name the tool used to dimension angles in degrees.
7. Describe a way to specify an angle in degrees if the angle is greater than 180°.
8. Which type of dimensioning is generally preferred for manufacturing because of its accuracy?
9. What is the AutoCAD term for datum dimensioning?
10. How do you place a datum dimension from the origin of the previously drawn dimension?
11. How do you place a datum dimension from the origin of a dimension drawn during a previous drawing session?
12. What is the conventional term for the type of dimensioning AutoCAD refers to as continuous dimensioning?
13. Which type of dimensions are created when you select multiple objects in the **QDIM** tool?
14. Which tool other than **DIMBASELINE** is used to create baseline dimensions?
15. Name at least three modes of dimensioning available through the **QDIM** tool.

Drawing Problems

- *Start AutoCAD if it is not already started.*

- *Start a new drawing using an appropriate template of your choice. The template should include layers, text styles, and dimension styles appropriate for drawing the given objects.*

- *Add layers, text styles, and dimension styles as needed. Draw all objects using appropriate layers, text styles, dimension styles, justification, and format.*

- *Follow the specific instructions for each problem. Use your own judgment and approximate dimensions when necessary.*

- *Apply dimensions accurately using ASME or appropriate industry standards. Use object snap modes to your best advantage.*

- *For mechanical drawings, place the following general notes 1/2" from the lower-left corner:*

    NOTES:
    1. INTERPRET DIMENSIONS AND TOLERANCES PER ASME Y14.5M-1994.
    2. REMOVE ALL BURRS AND SHARP EDGES.
    3. UNLESS OTHERWISE SPECIFIED, ALL DIMENSIONS ARE IN INCHES (or MILLIMETERS as applicable).

▼ Basic

1. Open a new drawing and save the file as P18-1. Open P3-6 and copy one instance of Object A and Object B to the P18-1 drawing. The P18-1 file should be active. Dimension the two objects as shown.

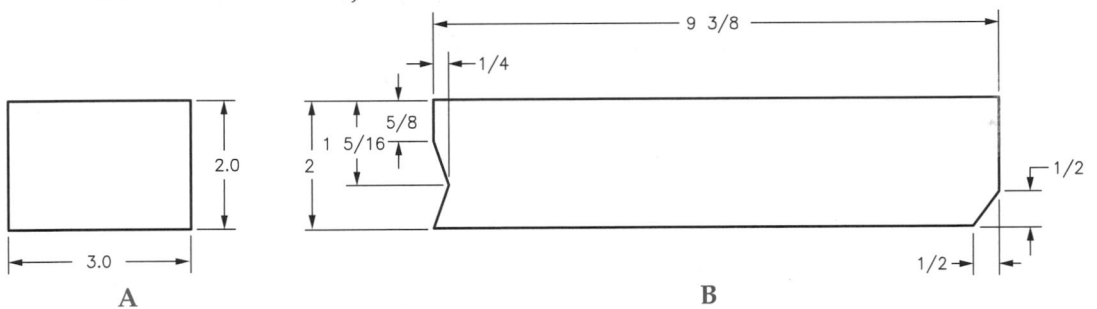

2. Open P3-4 and save the file as P18-2. The P18-2 file should be active. Dimension the drawing as shown using datum dimensions.

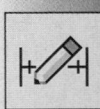

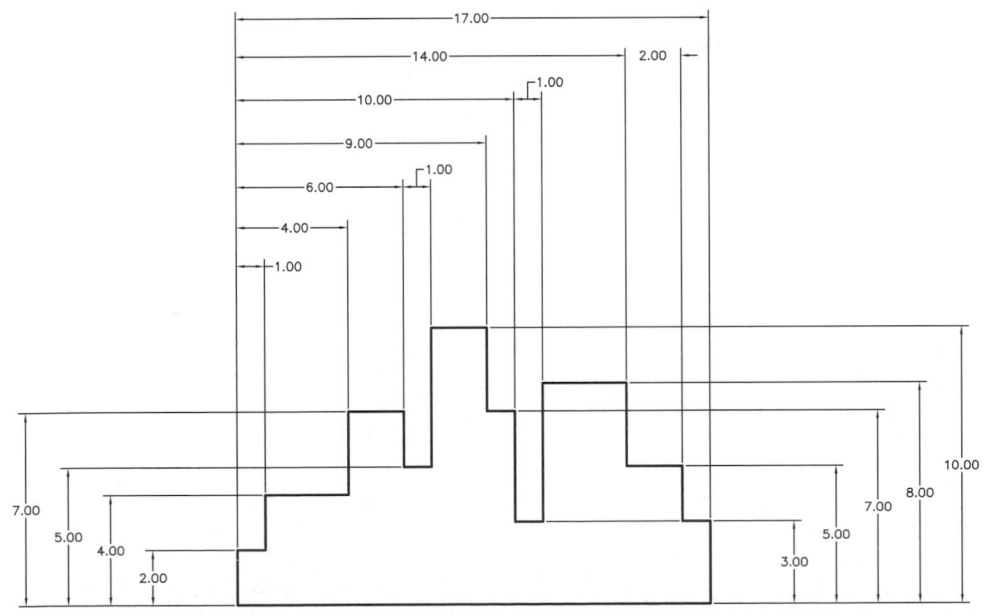

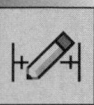

3. Open P3-9 and save the file as P18-3. The P18-3 file should be active. Dimension both objects as shown.

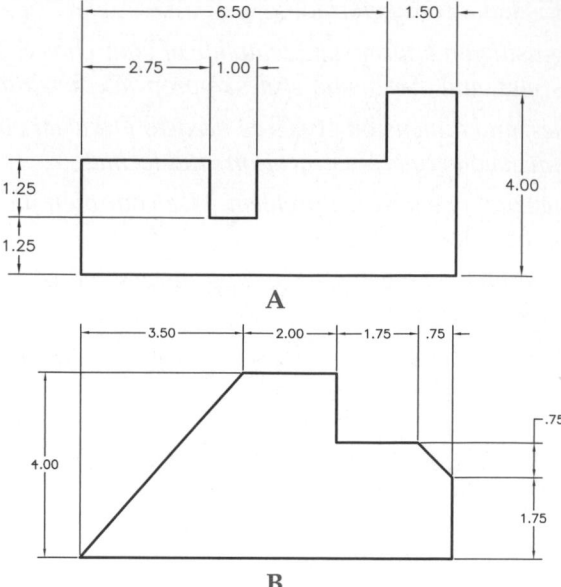

4. Open P3-10 and save the file as P18-4. The P18-4 file should be active. Dimension the drawing as shown.

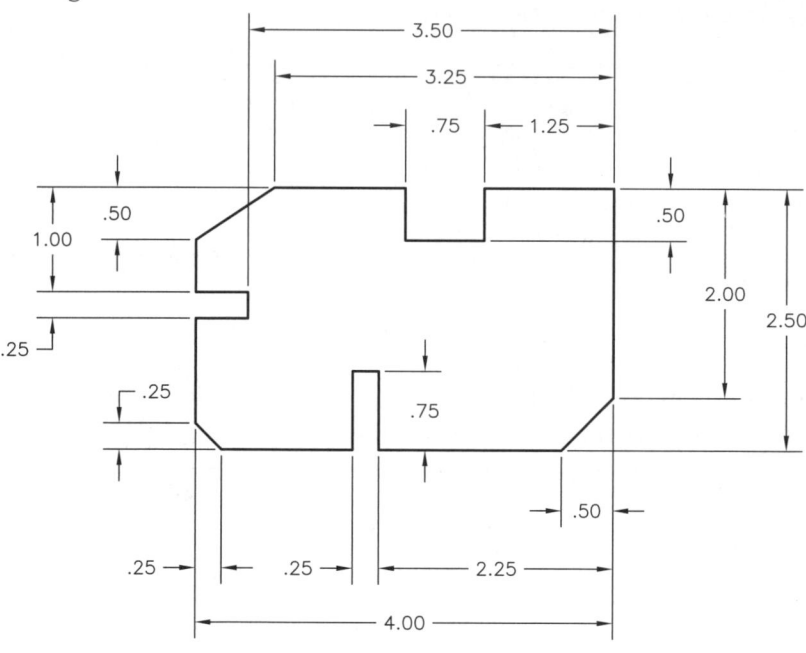

5. Open P3-11 and save the file as P18-5. The P18-5 file should be active. Dimension the drawing as shown. Note that this is a metric drawing.

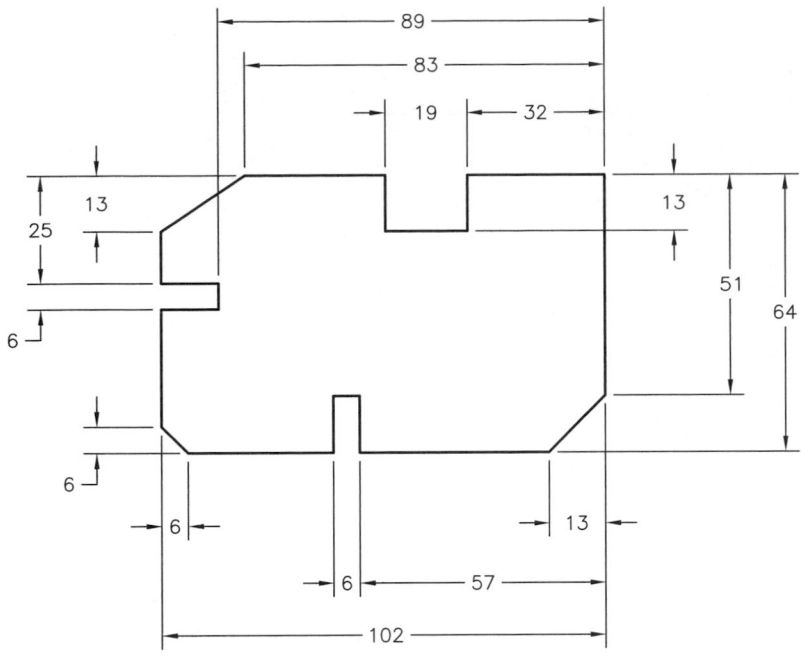

6. Open P3-7 and save the file as P18-6. The P18-6 file should be active. Dimension the drawing as shown.

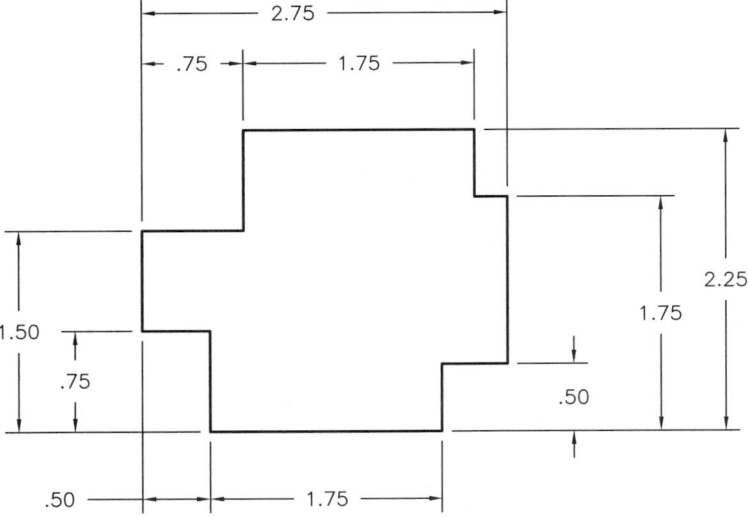

7. Open P3-8 and save the file as P18-7. The P18-7 file should be active. Dimension the drawing as shown.

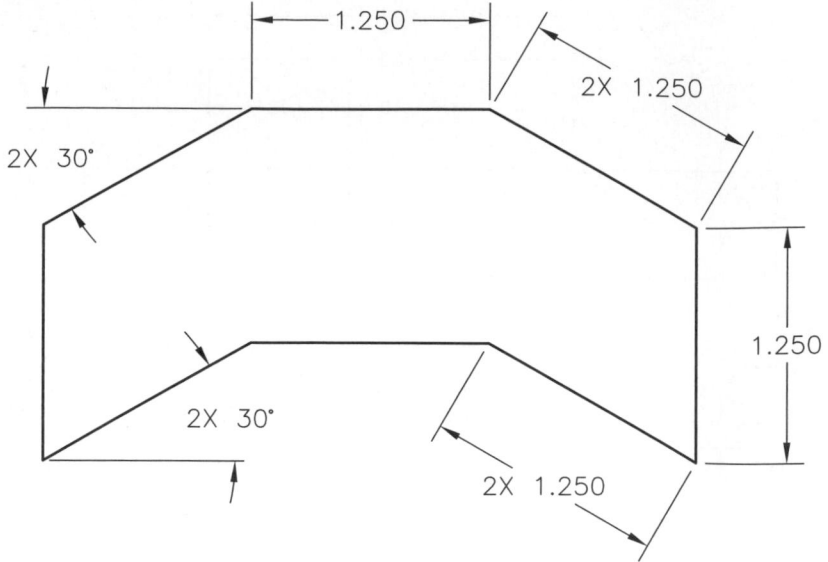

8. Write a report explaining the difference between datum and chain dimensioning. Use a word processor and include sketches giving examples of each method.

▼ Intermediate

9. Create the views of a shaft and dimension as shown. Save the drawing as P18-9.

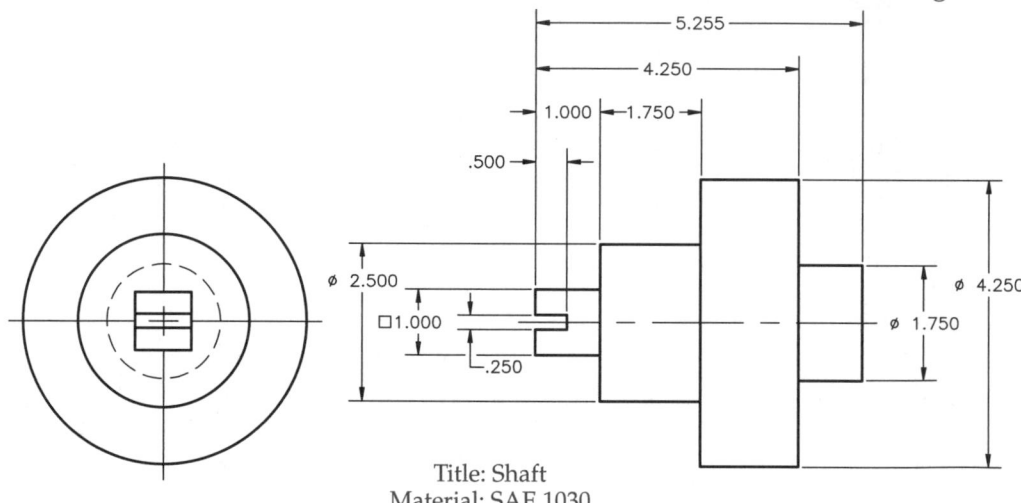

Title: Shaft
Material: SAE 1030

10. Open P3-12 and save the file as P18-10. The P18-10 file should be active. Dimension the drawing as shown.

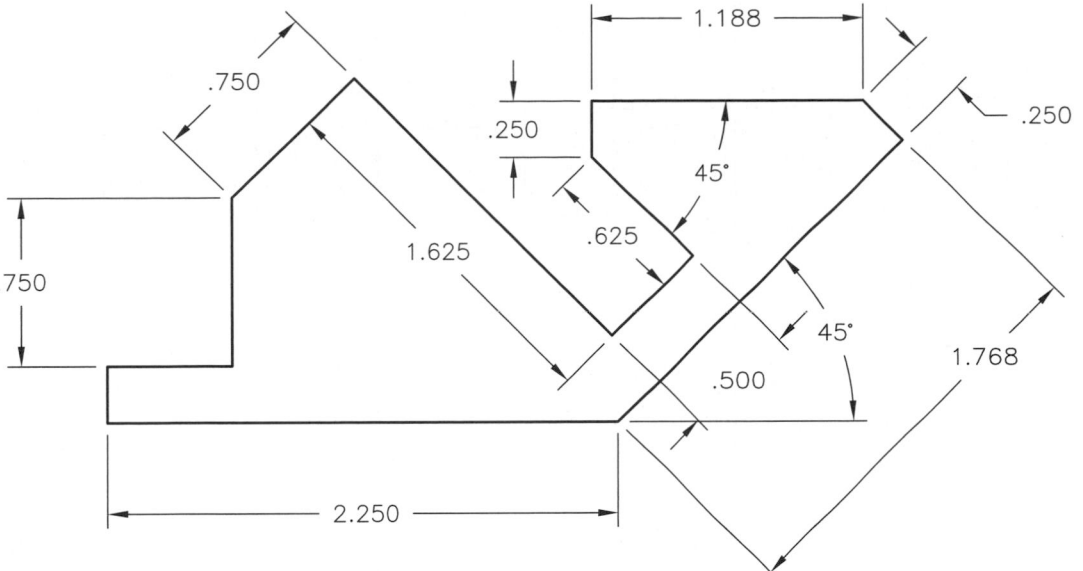

11. Create the partial floor plan shown below and dimension as shown. Save the drawing as P18-11.

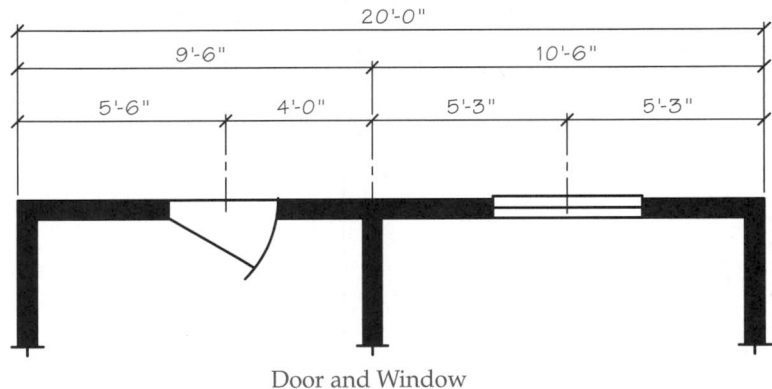

Door and Window

12. Create the partial floor plan and dimension as shown. Save the drawing as P18-12.

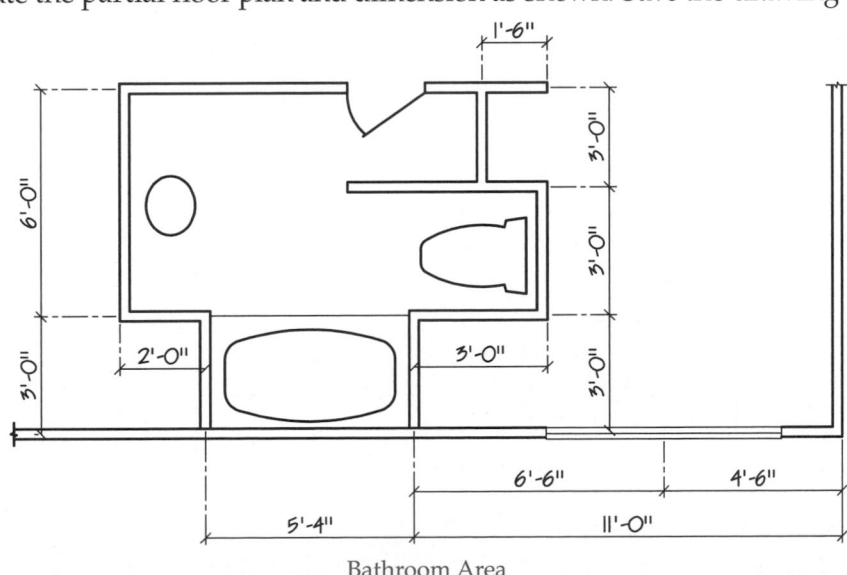

Bathroom Area

Drawing Problems - Chapter 18

13. Create the object and dimension as shown. Save the drawing as P18-13.

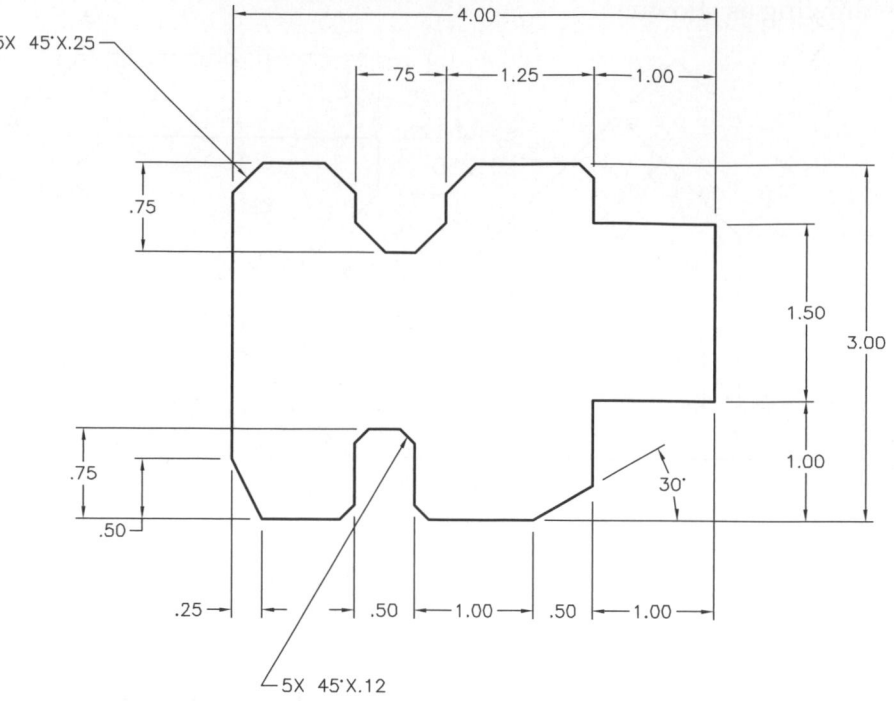

14. Create the object and dimension as shown. Save the drawing as P18-14.

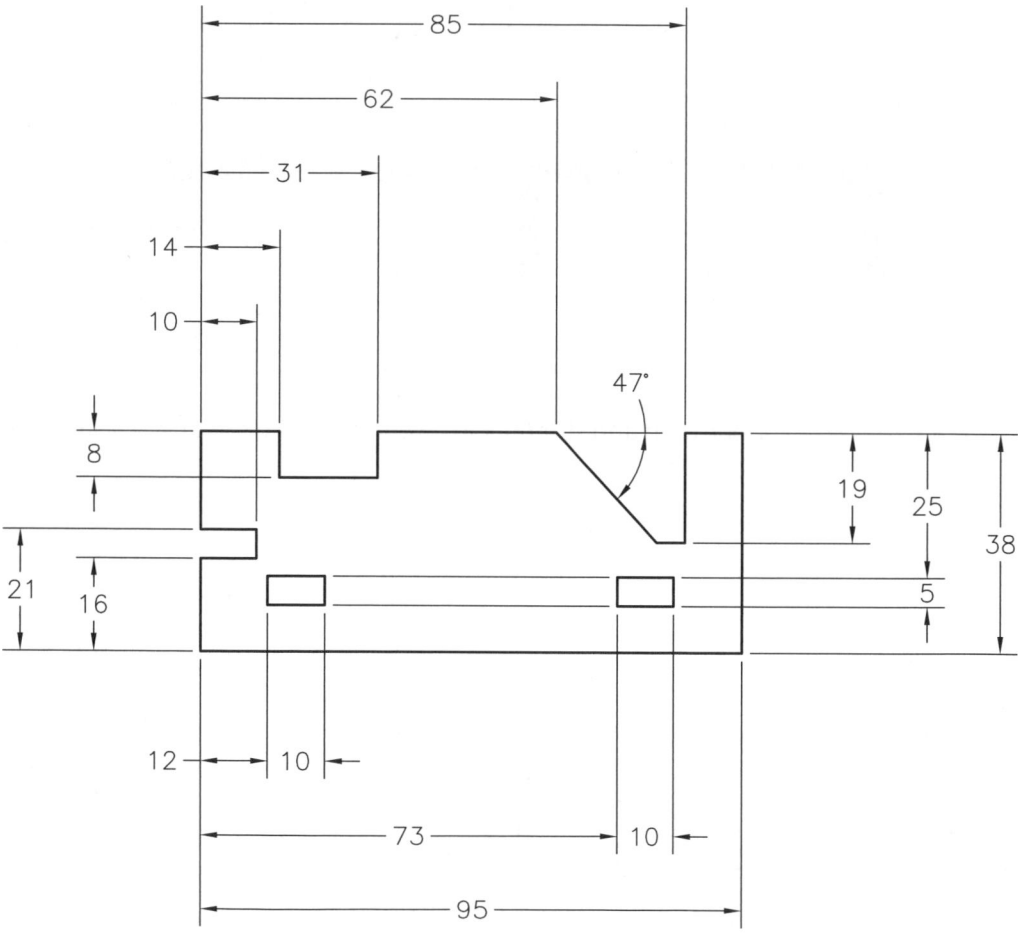

▼ Advanced

15. Open **P3-3** as shown below and save the file as **P18-15**. The P18-15 file should be active. Make one copy of the object at a new location to the right of the original object. Dimension the object on the left using datum dimensioning. Dimension the object on the right using chain dimensioning.

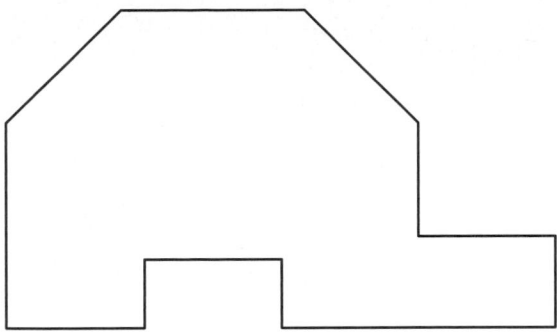

16. Open **P8-5** and save the file as **P18-16**. The P18-16 file should be active. Dimension the drawing as shown.

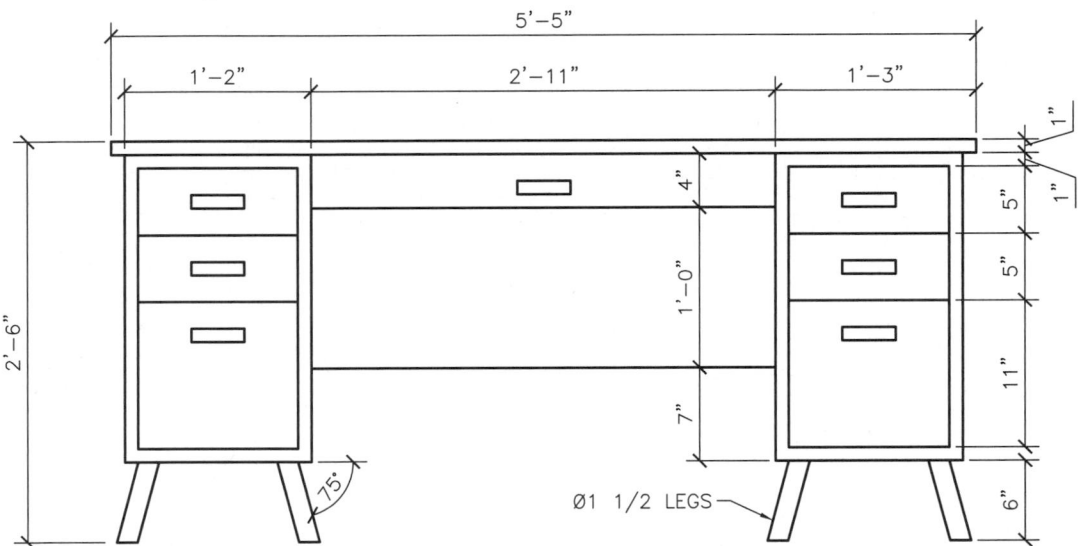

17. Draw this floor plan. Size the windows and doors to your own specifications. Dimension the drawing as shown. Save the drawing as P18-17.

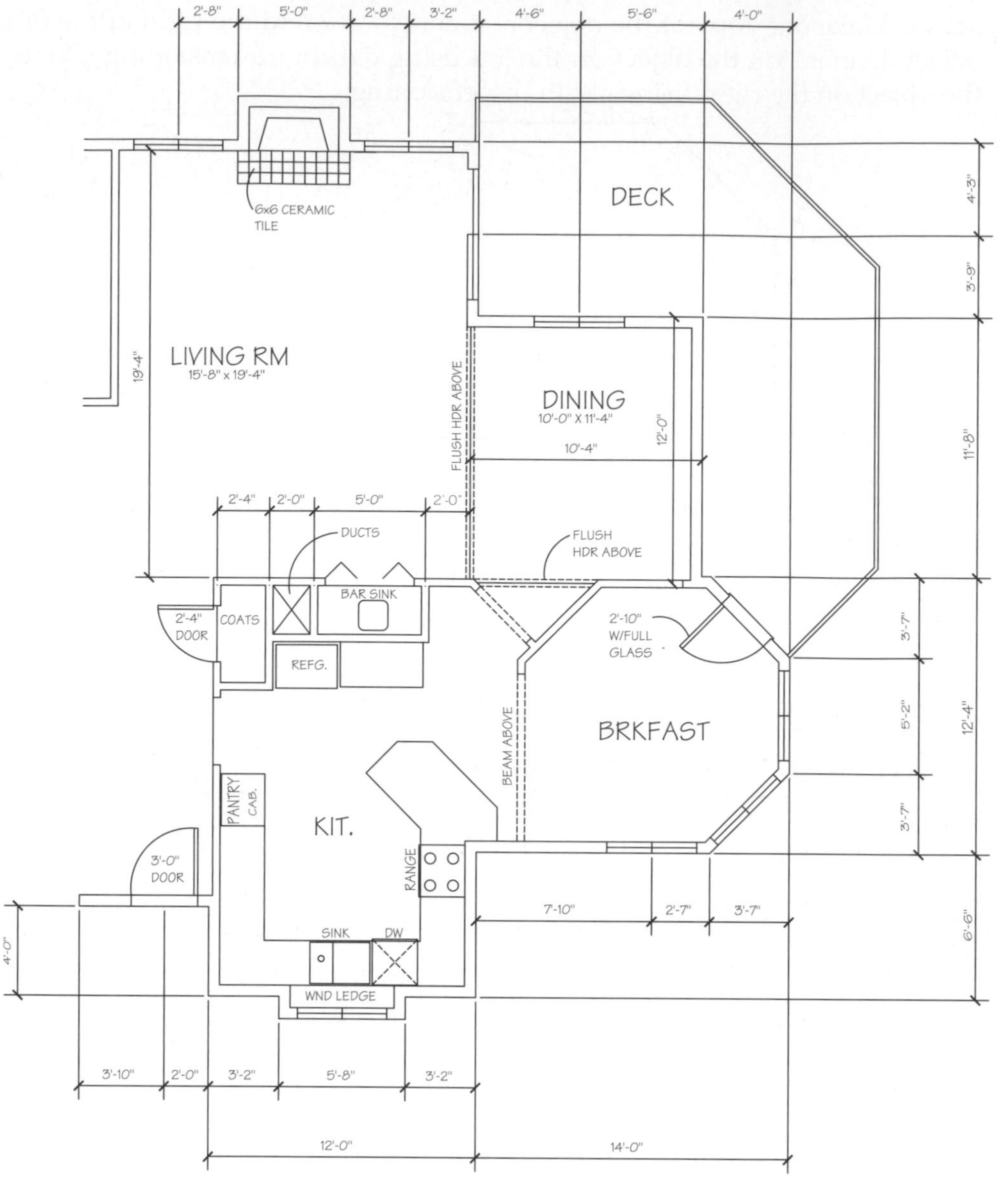

2'-8" | 5'-0" | 2'-8" | 3'-2" | 4'-6" | 5'-6" | 4'-0"

DECK

6x6 CERAMIC TILE

4'-3"

3'-9"

LIVING RM
15'-8" x 19'-4"

19'-4"

FLUSH HDR ABOVE

DINING
10'-0" X 11'-4"

10'-4"

12'-0"

11'-8"

2'-4" | 2'-0" | 5'-0" | 2'-0"

DUCTS

FLUSH HDR ABOVE

2'-4" DOOR

COATS

BAR SINK

REFG.

2'-10" W/FULL GLASS

3'-7"

BEAM ABOVE

BRKFAST

5'-2"

12'-4"

PANTRY CAB.

KIT.

RANGE

3'-0" DOOR

3'-7"

SINK

DW

7'-10" | 2'-7" | 3'-7"

6'-6"

4'-0"

WND LEDGE

3'-10" | 2'-0" | 3'-2" | 5'-8" | 3'-2"

12'-0"

14'-0"

Dimensioning Features and Alternate Practices

Learning Objectives

After completing this chapter, you will be able to do the following:

✓ Dimension circles and arcs.
✓ Create and use multileader styles.
✓ Draw leaders using the **MLEADER** tool.
✓ Apply alternate dimensioning practices.
✓ Dimension using the **DIMORDINATE** tool.
✓ Mark up a drawing using the **REVCLOUD** and **WIPEOUT** tools.

A drawing must describe the size and location of all features for manufacturing or construction. This chapter describes dimensioning practices for object features and introduces alternate mechanical drafting dimensioning practices. This chapter also introduces basic redlining techniques using the **Revision Cloud** and **Wipeout** tools.

Dimensioning Circles

Diameter usually describes the size of circles. The ASME standard for dimensioning arcs is to give the radius. However, AutoCAD allows you to dimension a circle or an arc with a diameter using the **DIMDIAMETER** tool. Access the **DIMDIAMETER** tool and select a circle or arc to display a leader line and a diameter dimension value attached to the crosshairs. Move the leader to the desired location and pick to place the dimension. See **Figure 19-1.**

Like other dimension tools, the **DIMDIAMETER** tool references the current dimension style and the object you select to create a single dimension object. The dimension you create includes all related dimension style characteristics, centerlines, leader, arrowheads, and a dimension value associated with the diameter. The leader points to the center of the circle or arc, as recommended by the ASME standard.

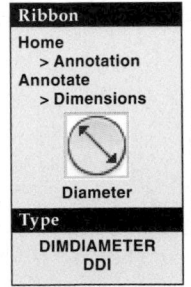

Ribbon
Home
 > Annotation
Annotate
 > Dimensions

Diameter
Type
DIMDIAMETER
DDI

DIMDIAMETER

Figure 19-1.
Using the **DIMDIAMETER** tool.

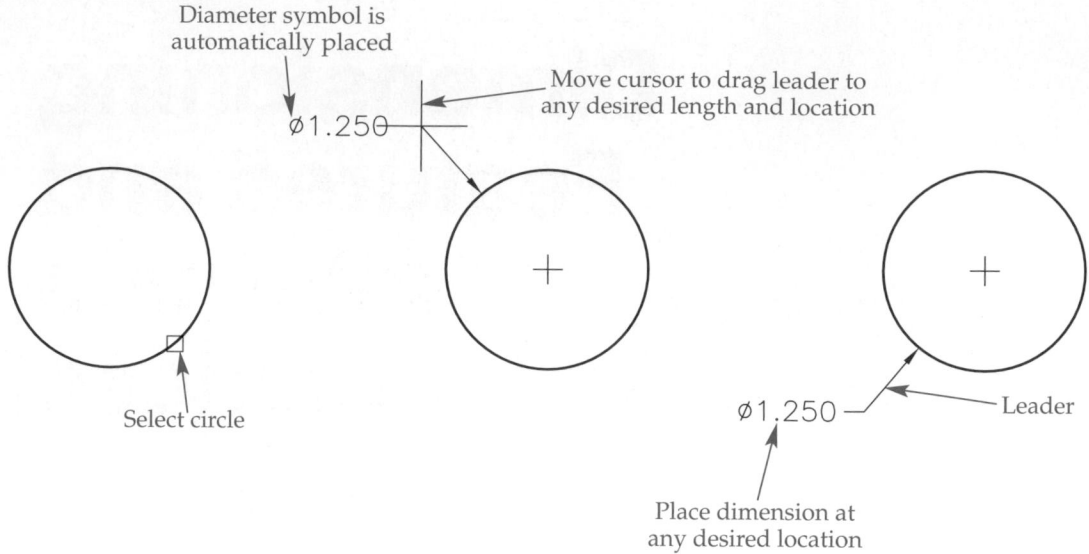

Diameter symbol is automatically placed

Ø1.250

Move cursor to drag leader to any desired length and location

Select circle

Ø1.250 — Leader

Place dimension at any desired location

Exercise 19-1

Access the Student Web site (www.g-wlearning.com/CAD) and complete Exercise 19-1.

Dimensioning Holes

Dimension holes in the view in which they appear as circles. Give location dimensions to the center and a leader showing the diameter. The **DIMDIAMETER** tool is effective for dimensioning holes. To note multiple holes of the same size, dimension the size of one hole using the **DIMDIAMETER** tool and the **Mtext** or **Text** option. Precede the diameter with the number of holes followed by X and then a space. The 2X Ø.50 dimension in **Figure 19-2** is an example of this practice.

Figure 19-2.
Dimensioning holes.

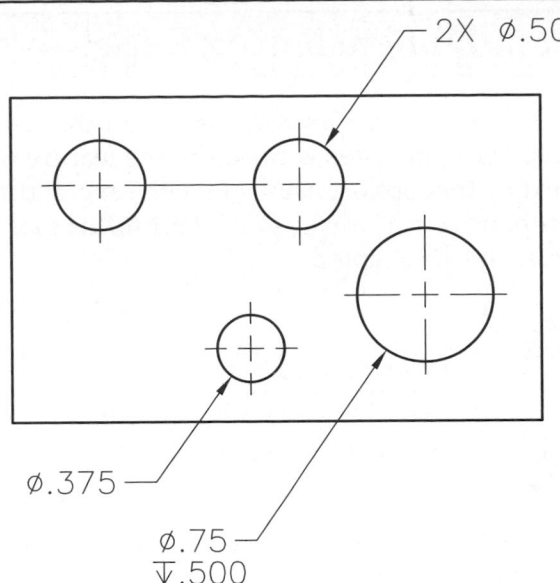

Dimensioning for Manufacturing Processes

Counterbore, *spotface*, and *countersink* manufacturing processes are examples of features dimensioned using symbols. Dimension these processes in the circular view, like holes, with a leader providing machining information in a note. See **Figure 19-3.** The **Mtext** option of the **DIMDIAMETER** tool provides a convenient method for dimensioning manufacturing processes. Many symbols are available from the gdt.shx font. You can also create custom symbols as blocks. Blocks are described later in this textbook.

counterbore: A larger-diameter hole machined at one end of a smaller hole that provides a place for the screw head.

spotface: A larger-diameter hole machined at one end of a smaller hole that provides a smooth, recessed surface for a washer; similar to a counterbore, but not as deep.

countersink: A cone-shaped recess at one end of a hole that provides a mating surface for a screw head of the same shape.

Reference Material

Drafting Symbols

For more information about common drafting symbols and the gdt.shx font, go to the **Reference Material** section of the Student Web site (www.g-wlearning.com/CAD) and select **Drafting Symbols** in the list.

Figure 19-3.
Dimension notes for machining processes. You can insert symbols as blocks or use lowercase letters in the gdt.shx font.

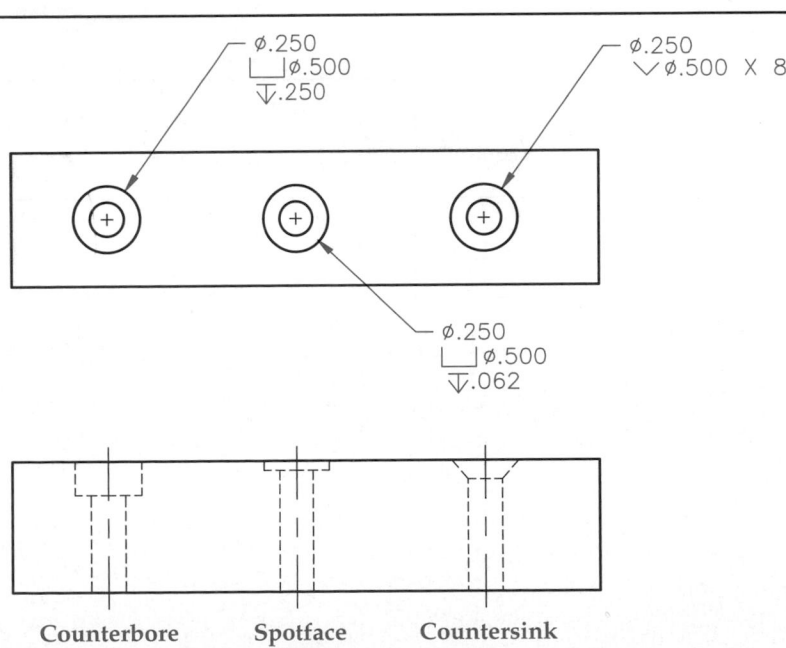

Dimensioning Repetitive Features

repetitive features:
Many features having the same shape and size.

For *repetitive features*, an X, a space, and the size dimension follow the number of repetitions. The dimension connects to the feature with a leader. **Figure 19-4** shows how the **Mtext** or **Text** option, available with several dimensioning tools, allows you to dimension repetitive features. Use the **MLEADER** tool to create the 8X note shown, as described later in this chapter.

Exercise 19-2

Access the Student Web site (www.g-wlearning.com/CAD) and complete Exercise 19-2.

Figure 19-4.
Dimensioning repetitive features (shown in color).

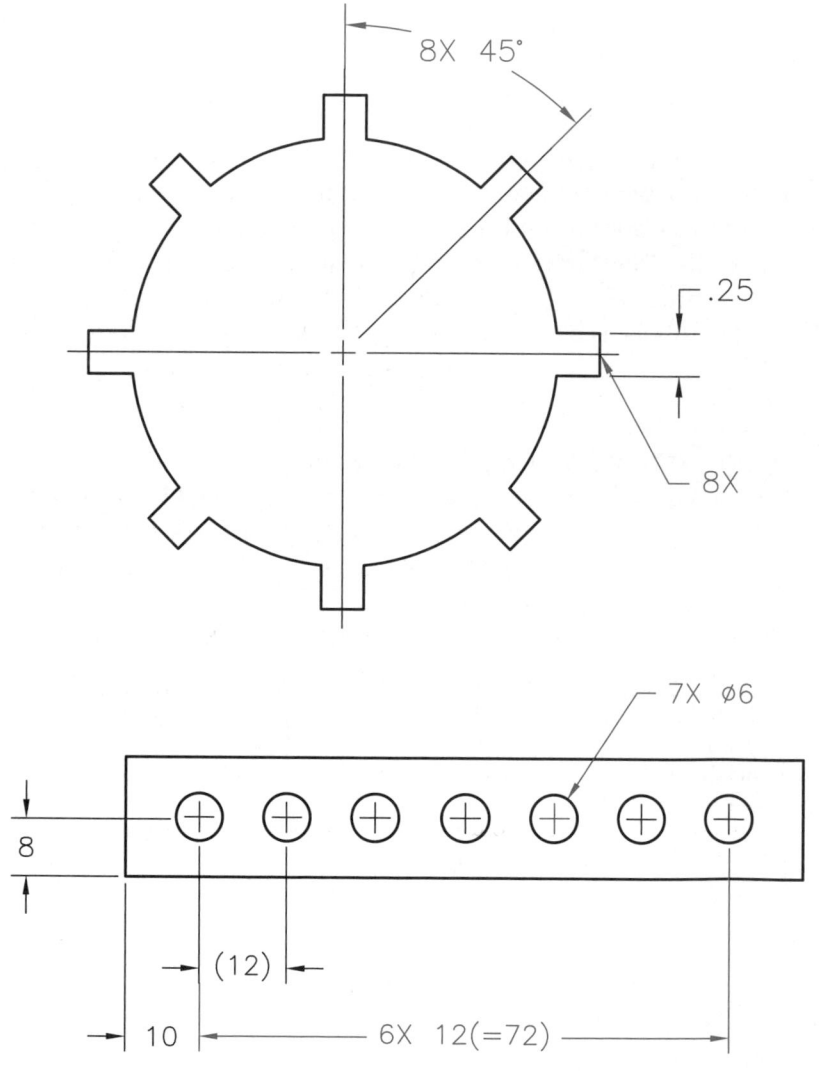

Dimensioning Arcs

The standard for dimensioning arcs is a radius dimension, which you can create using the **DIMRADIUS** tool. Access the **DIMRADIUS** tool and select an arc or circle to display a leader line and radius dimension value attached to the crosshairs. Move the leader to the desired location, and pick to place the dimension. See **Figure 19-5.** The resulting leader points to the center of the arc or circle, as recommended by the ASME standard. Centerlines or center marks appear, depending on the current dimensions style setting.

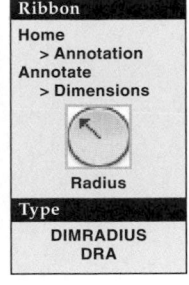

Ribbon
Home
> Annotation
Annotate
> Dimensions
Radius
Type
DIMRADIUS
DRA

DIMRADIUS

NOTE

The **DIMRADIUS** tool includes the **Mtext**, **Text**, and **Angle** options. Use the **Mtext** or **Text** option to add information to or change the dimension value. The **Angle** option changes the dimension text angle, although this practice is not common.

Dimensioning Arc Length

You can use the **DIMARC** tool to dimension the length of an arc. The length measures the distance along the arc segment. Access the **DIMARC** tool and select an arc or polyline arc segment to display the arc length symbol and dimension value attached to the crosshairs. Move the text to the desired location and pick. By default, the symbol occurs before the text. The ASME standard recommends placing the symbol over the text, as shown in **Figure 19-6.** The dimension style controls the symbol placement.

Before placing the arc length dimension, add information to or change the dimension value using the **Mtext** or **Text** option. The **Angle** option is available to change the text angle. Use the **Partial** option to dimension a portion of the arc length. Select a first point on the arc followed by a second point to dimension the length between the points. When the arc is greater than 90°, the **Leader** option is available. This option allows you to add a leader pointing to the arc you are dimensioning.

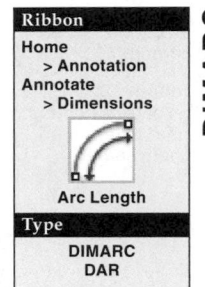

Ribbon
Home
> Annotation
Annotate
> Dimensions
Arc Length
Type
DIMARC
DAR

DIMARC

Dimensioning Large Circles and Arcs

When a circle or arc is so large that the center point cannot appear on the layout, use the **DIMJOGGED** tool to create the dimension. Access the **DIMJOGGED** tool and pick an arc or circle. Then pick a location for the origin of the center. This is the point

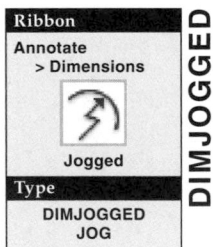

Ribbon
Annotate
> Dimensions
Jogged
Type
DIMJOGGED
JOG

DIMJOGGED

Figure 19-5.
Using the **DIMRADIUS** tool to dimension arcs.

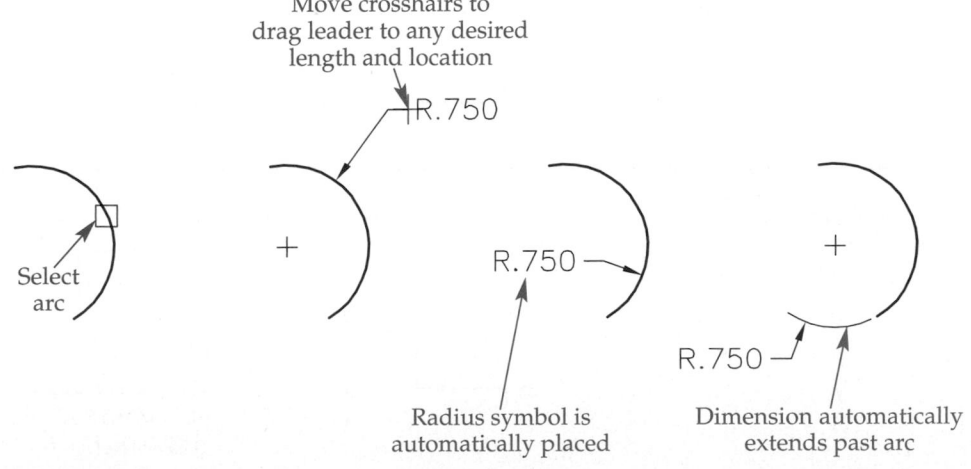

Figure 19-6.
Using the **DIMARC** tool to dimension the length of an arc.

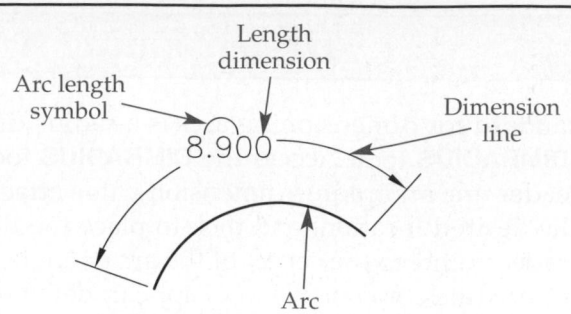

that represents, or overrides, the center of the arc or circle. The associated radius value does not change. Select a location for the dimension line and then pick a location for the break symbol. See **Figure 19-7.** You can move the components of the dimension by grip editing after you place the dimension.

Dimensioning Fillets and Rounds

fillets: Small inside arcs designed to strengthen inside corners.

rounds: Small arcs on outside corners used to relieve sharp corners.

You can dimension *fillets* and *rounds* individually as arcs, using the **DIMRADIUS** tool, or collectively in a general note. See **Figure 19-8.** On mechanical drawings, it is common to include a general note such as ALL FILLETS AND ROUNDS R.125 UNLESS OTHERWISE SPECIFIED on the drawing.

Exercise 19-3

Access the Student Web site (www.g-wlearning.com/CAD) and complete Exercise 19-3.

Figure 19-7.
Using the **DIMJOGGED** tool to place a radius dimension for a large arc.

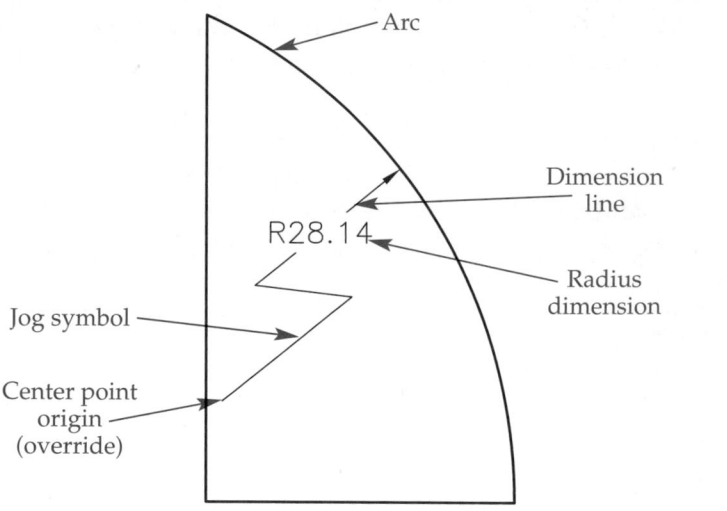

Figure 19-8.
Dimensioning fillets and rounds.

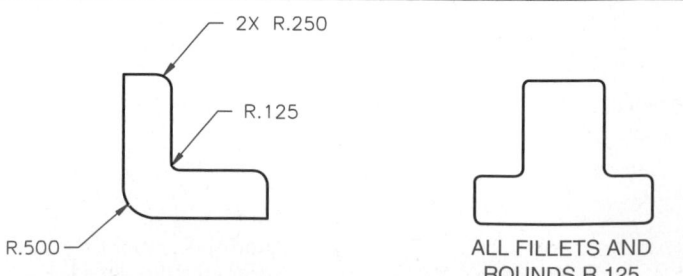

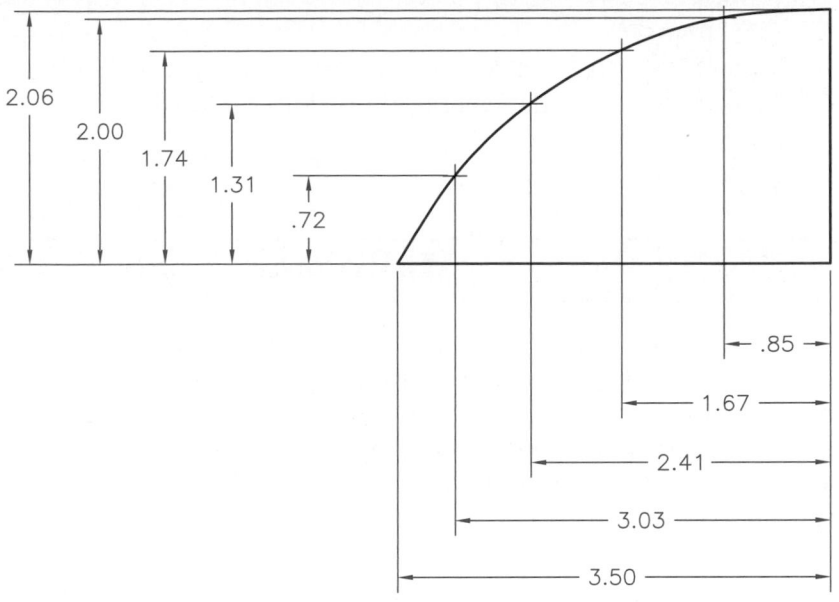

Figure 19-9.
Dimensioning curves that do not have a constant radius.

Dimensioning Curves

Dimension curves as arcs when possible. When an arc is not in the shape of a constant-radius arc, you should dimension to points along the curve using the **DIMLINEAR** tool. See **Figure 19-9.**

Adding Center Dashes and Centerlines

Depending on the current dimension style setting, when you use the **DIMDIAMETER** and **DIMRADIUS** tools, small circles or arcs automatically receive center dashes, and large circles or arcs display center dashes or centerlines. Use the **DIMCENTER** tool to add center dashes or centerlines to objects that are not dimensioned using the **DIMDIAMETER** or **DIMRADIUS** tools. Access the **DIMCENTER** tool and pick a circle or an arc to display center marks.

The **DIMCENTER** tool references the current dimension style and the size of the circle or arc to place center dashes, centerlines, or no symbol. The ASME standard refers to AutoCAD center marks as the center dashes of centerlines. Center marks are typically applied to circular objects that are too small to receive centerlines. Center marks are also common for rectangular coordinate dimensioning without dimension lines, regardless of circular object size, as described later in this chapter.

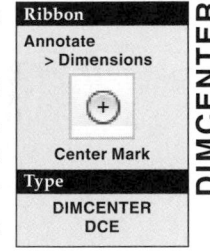

Ribbon
Annotate
> Dimensions

Center Mark

Type
DIMCENTER
DCE

DIMCENTER

Drawing Leader Lines

The **DIMDIAMETER** and **DIMRADIUS** tools automatically place *leader lines* when you dimension circles and arcs. AutoCAD multileaders created using the **MLEADER** tool allow you to begin and end a leader line at a specific location. Multileaders consist of single or multiple lines of *annotations*, including symbols, with the leader. Multileader styles control multileader characteristics, such as leader format, annotation style, and arrowhead size. You can create multiple-segment leaders and align and group separate leaders. Chapter 21 describes adding and removing multiple leader lines and aligning leaders.

leader line: A line that connects note text to a specific feature or location on a drawing.

annotation: Text or similar information on a drawing, such as a note or dimension.

Multileader Styles

multileader styles: Saved configurations for the appearance of leaders.

shoulder: A short horizontal line usually added to the end of straight leader lines.

Multileader styles allow you to control multileader settings as needed to achieve the desired leader appearance. This process is similar to using a dimension style. In mechanical drafting, properly drawn leaders have one straight segment extending from the feature to a horizontal *shoulder* that is 1/8"–1/4" (3 mm–6 mm) long. While most other fields also use straight leaders, AutoCAD provides the option of drawing curved leaders, which are common in architectural drafting. See **Figure 19-10.**

Working with Multileader Styles

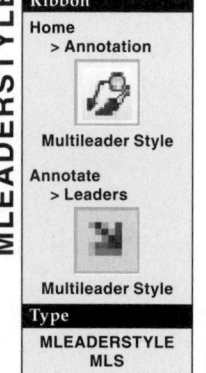

MLEADERSTYLE

Ribbon
Home
> Annotation

Multileader Style

Annotate
> Leaders

Multileader Style

Type
MLEADERSTYLE
MLS

Create, modify, and delete multileader styles using the **Multileader Style Manager** dialog box. See **Figure 19-11.** The **Styles:** list box displays existing multileader styles. Standard and Annotative multileader styles are available by default in some templates. The Annotative multileader style is preset to create annotative leaders, as indicated by the icon to the left of the style name. The Standard multileader style does not use the annotative function. The **List:** drop-down list allows you to control whether all styles or only the styles in use appear in the **Styles:** list box. To make a multileader style current, double-click the style name, right-click on the name and select **Set current**, or pick the name and select the **Current** button.

NOTE

The **Preview of:** image displays a representation of the multileader style and changes according to the selections you make.

Figure 19-10.
Multileader styles control the display of leaders created using the **MLEADER** tool.

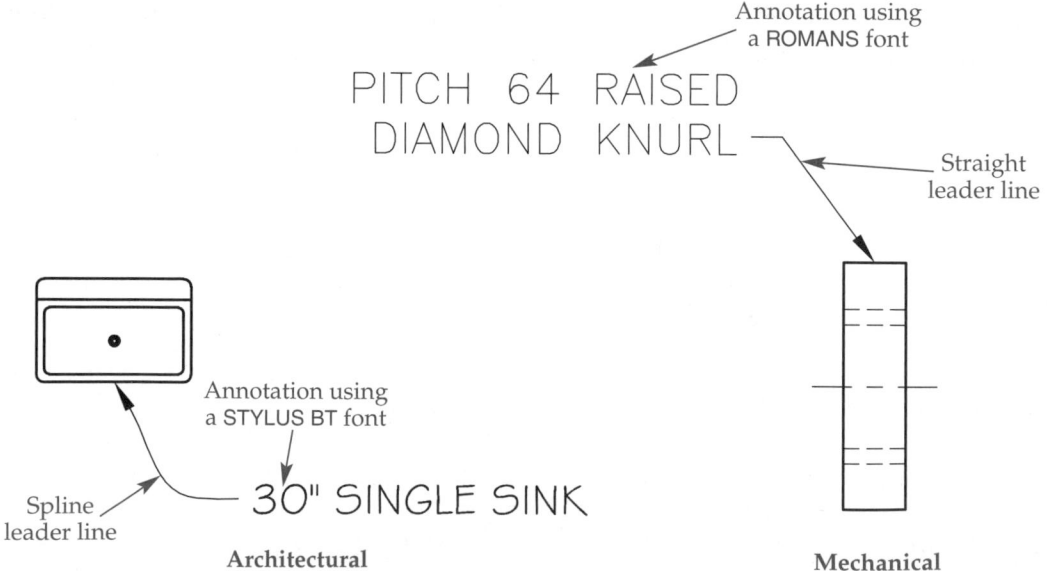

Figure 19-11.
The **Multileader Style Manager** dialog box. The Standard multileader style is the AutoCAD default.

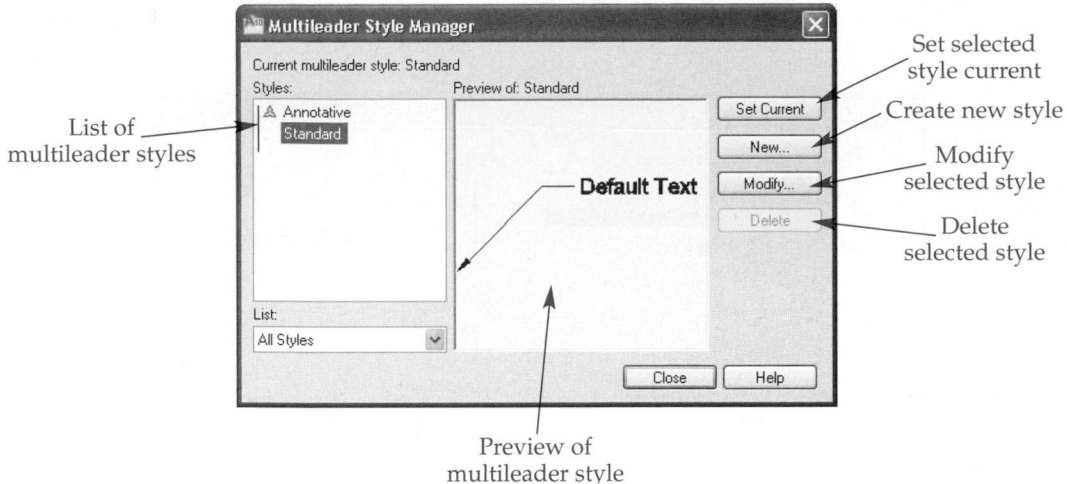

List of multileader styles

Set selected style current

Create new style

Modify selected style

Delete selected style

Preview of multileader style

Creating a New Multileader Style

To create a new multileader style, first select an existing multileader style from the **Styles:** list box to use as a base for formatting the new multileader style. Then pick the **New...** button in the **Multileader Style Manager** to open the **Create New Multileader Style** dialog box. See **Figure 19-12**.

Enter a descriptive name for the new multileader style, such as Architectural, Mechanical, Straight, or Spline, in the **New Style Name** text box. If necessary, select a different style from the **Start With** drop-down list on which to base the new style. Pick the **Annotative** check box to make the multileader style annotative.

Pick the **Continue** button to access the **Modify Multileader Style** dialog box, shown in **Figure 19-13**. The **Leader Format**, **Leader Structure**, and **Content** tabs display groups of settings for specifying leader appearance. The next sections of this chapter describe each tab. After completing the style definition, pick the **OK** button to return to the **Multileader Style Manager** dialog box.

> **NOTE**
>
> The preview image in the upper-right corner of each **Modify Multileader Style** dialog box tab displays a representation of the multileader style and changes according to the selections you make.

Figure 19-12.
The **Create New Multileader Style** dialog box.

Name new style

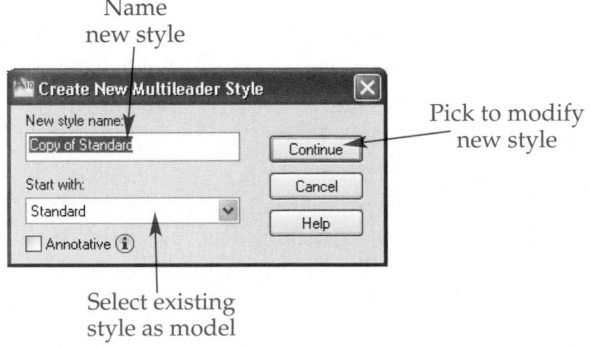

Pick to modify new style

Select existing style as model

Figure 19-13.
The **Leader Format** tab of the **Modify Multileader Style** dialog box.

Preview image is
displayed in all tabs

General
multileader
format
settings

Multileader
arrowhead
settings

Multileader
break size
setting

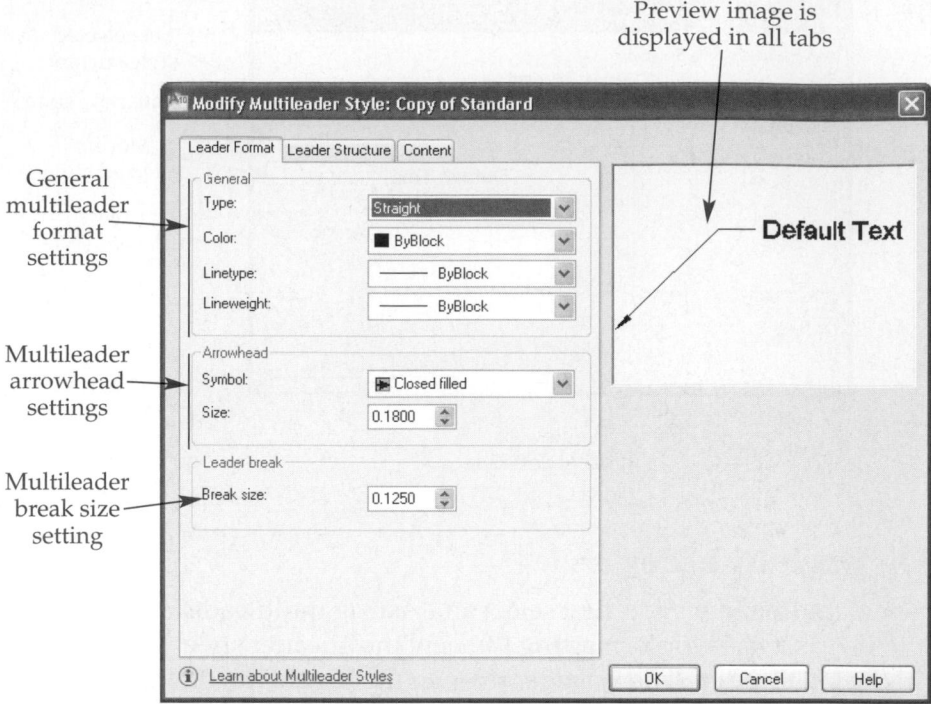

Leader Format Settings

The **Leader Format** tab of the **Modify Multileader Style** dialog box, shown in **Figure 19-13**, controls leader line settings. It allows you to set the appearance of the leader line and arrowhead.

General Leader Format Settings

The **General** area contains a **Type** drop-down list that that you can use to specify the leader line shape. The **Straight** option produces leaders with straight-line segments. The **Spline** option produces the curved leader lines common in architectural drafting. Pick the **None** option to create a multileader style that does not use a leader line. Use this option to create a leader that you can associate with other leaders using the **MLEADERALIGN** and **MLEADERCOLLECT** tools, described in Chapter 21.

Color, **Linetype**, and **Lineweight** drop-down lists are available for changing the multileader color, linetype, and lineweight. These options function the same as they do for adjusting dimension elements.

Arrowhead Settings

The **Arrowhead** area sets the leader arrowhead style and size. Select the arrowhead style from the **Symbol:** drop-down list. The arrowhead symbol options are the same as those for dimension style arrowheads. Set the arrowhead size using the **Size:** text box. A common arrowhead size is .125″ (3 mm). Leader arrowheads are typically the same size as dimension arrowheads.

Adjusting Break Size

The **Leader Break** area controls the amount of leader line removed by the **DIMBREAK** tool. Specify a value in the **Break size:** text box to set the total length of the break. The default size is .125″ (3 mm). ASME standards do not recommend breaking leader lines.

AutoCAD and Its Applications—Basics

Figure 19-14.
The **Leader Structure** tab of the **Modify Multileader Style** dialog box.

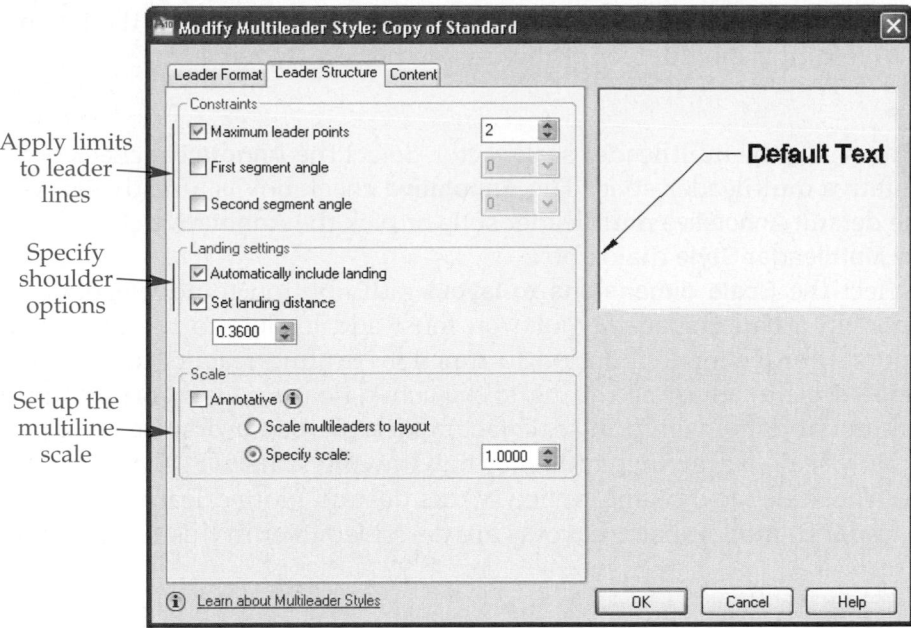

Apply limits to leader lines

Specify shoulder options

Set up the multiline scale

Leader Structure Settings

The **Leader Structure** tab of the **Modify Multileader Style** dialog box is shown in **Figure 19-14.** This tab contains settings that control leader construction and size.

Setting Constraints

The **Constraints** area restricts the number of points you can select to create a leader, as well as the leader line angle. Pick the **Maximum leader points** check box to set a maximum number of vertices on the leader line. The multileader automatically forms once you pick the maximum number of points. To use fewer than the maximum number of points, press [Enter] at the Specify next point: prompt. Deselect the **Maximum leader points** check box to allow an unlimited number of vertices.

You can use the **First segment angle** and **Second segment angle** check boxes to restrict the first two leader line segments to certain angles. Deselect the check boxes to draw leader lines at any angle. Select the appropriate check boxes and pick a value from the drop-down list to restrict the angle of the leader segment according to the selected value. The **Ortho** mode setting overrides the angle constraints, so it is advisable to turn **Ortho** mode off while you are placing leaders.

PROFESSIONAL TIP

The ASME standard for leaders recommends that leader lines have angles not less than 15° and not greater than 75° from horizontal. Use the **First segment angle** and **Second segment angle** settings to help maintain this standard.

Landing Settings

The **Landing settings** area controls the display and size of the *landing* and is only available with straight multileader styles. Select the **Automatically include landing** check box to display a shoulder automatically when you select the second leader line point. This is the preferred method for creating straight leader lines. Deselect the check box to create leaders without shoulders, or to pick a third point to manually draw

landing: The AutoCAD term for a leader shoulder.

the shoulder. When you check **Automatically include landing**, the **Set landing distance** check box enables. Pick the **Set landing distance** check box to define a specific shoulder length, typically 1/8″–1/4″ (3 mm–6 mm), in the text box. If you deselect the text box, a prompt asks you for the shoulder length when you place a leader.

Scale Options

The **Scale** area sets the multileader scale factor. Select the **Annotative** check box to create an annotative multileader style. The **Annotative** check box is already set when you modify the default Annotative multileader style or pick the **Annotative** check box in the **Create New Multileader Style** dialog box.

You can select the **Scale dimensions to layout** radio button to add leaders in a floating viewport in a paper space layout. You must add leaders to the model in a floating viewport in order for this option to function. Scaling leaders to the layout allows the overall scale to adjust according to the active floating viewport by setting the overall scale equal to the viewport scale factor. Pick the **Use overall scale of** radio button to manually scale a drawing, and enter the drawing scale factor applied to all leader settings. The scale factor is multiplied by the desired plotted leader size to get the size of the leader in model space. Layouts are described later in this textbook.

Content Settings

The **Content** tab of the **Modify Multileader Style** dialog box, shown in **Figure 19-15**, controls the display of text or a block with the leader line. Use the **Multileader type:** drop-down list to select the type of object to attach to the end of the leader line or shoulder. **Figure 19-16** shows an example of a leader drawn with each content option.

Attaching Mtext

Pick the **Mtext** option from the **Multileader type:** drop-down list, as shown in **Figure 19-15**, to attach a multiline text object to the leader. The **Text options** and **Leader connection** areas of the **Content** tab appear when you select the **Mtext** content type. The **Default text** option allows you to specify a value to attach leaders during leader

---

Figure 19-15.
The **Content** tab of the **Modify Multileader Style** dialog box with the **Mtext** multileader type selected.

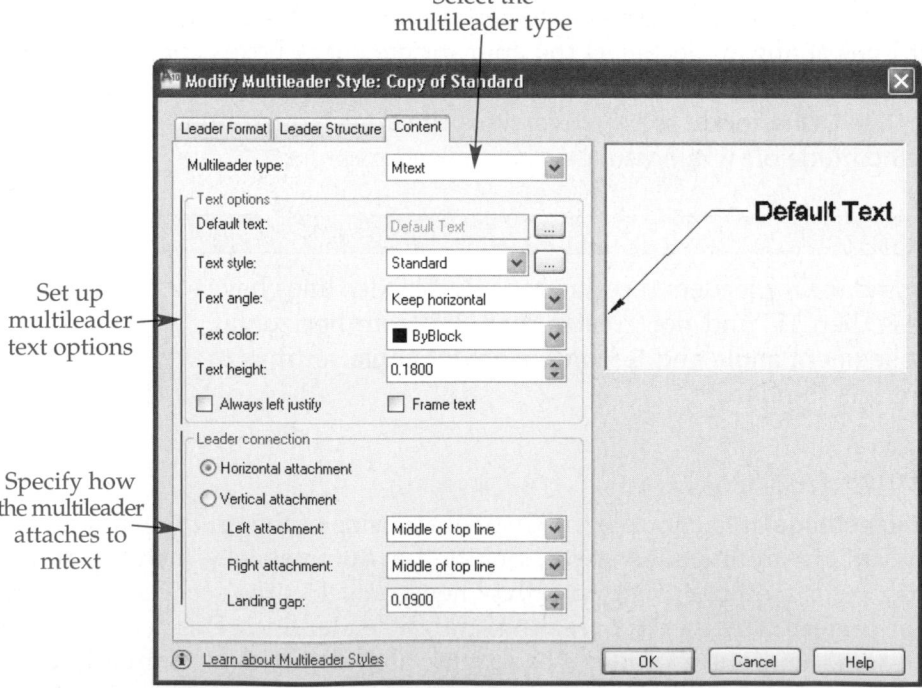

AutoCAD and Its Applications—Basics

Figure 19-16.
Examples of each multileader content type.

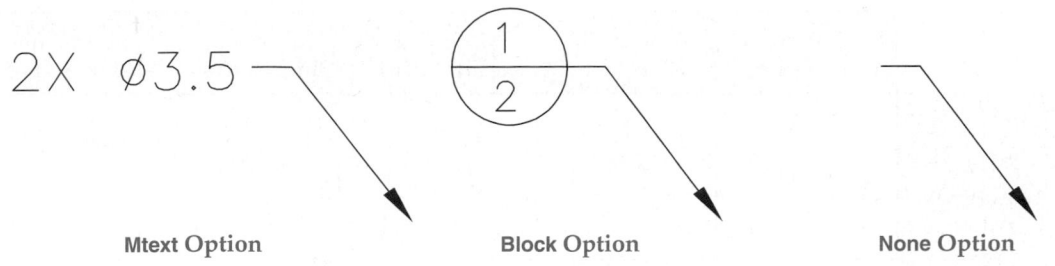

Mtext Option   Block Option   None Option

placement. This is useful when the same note or symbol is required throughout a drawing. Pick the ellipsis (...) button to return to the drawing window and use the multiline text editor to enter the default text value. Close the text editor to return to the **Modify Multileader Style** dialog box.

A text style must be loaded in the current drawing before it is available for use in leader text. Pick the desired text style from the **Text style** drop-down list. To create or modify an existing text style, pick the ellipsis (...) button next to the drop-down list to launch the **Text Style** dialog box.

Select an option from the **Text angle** drop-down list to control the angle at which text appears in reference to the angle of the leader line or shoulder. **Figure 19-17** shows the effects of applying each text angle option to the same leader. Use the **Text color** drop-down list to specify the text color, which should be ByBlock for typical applications. Use the **Text height** text box to specify the leader text height. Leader text height is commonly the same as the text height used for dimensions.

The **Always left justify** option forces leader text to left-justify, regardless of the leader line direction. Pick the **Frame text** check box to create a box around the multiline text box. The current multileader style settings control the default properties of the frame.

The **Leader connection** area contains options that determine how the mtext object positions relative to the endpoint of the leader line or shoulder. Most drawings require leaders that use the **Horizontal attachment** option. Use the **Left attachment:** drop-down list to define how multiple lines of text are positioned when the leader is on the left side

Figure 19-17.
Text angle options available for mtext.

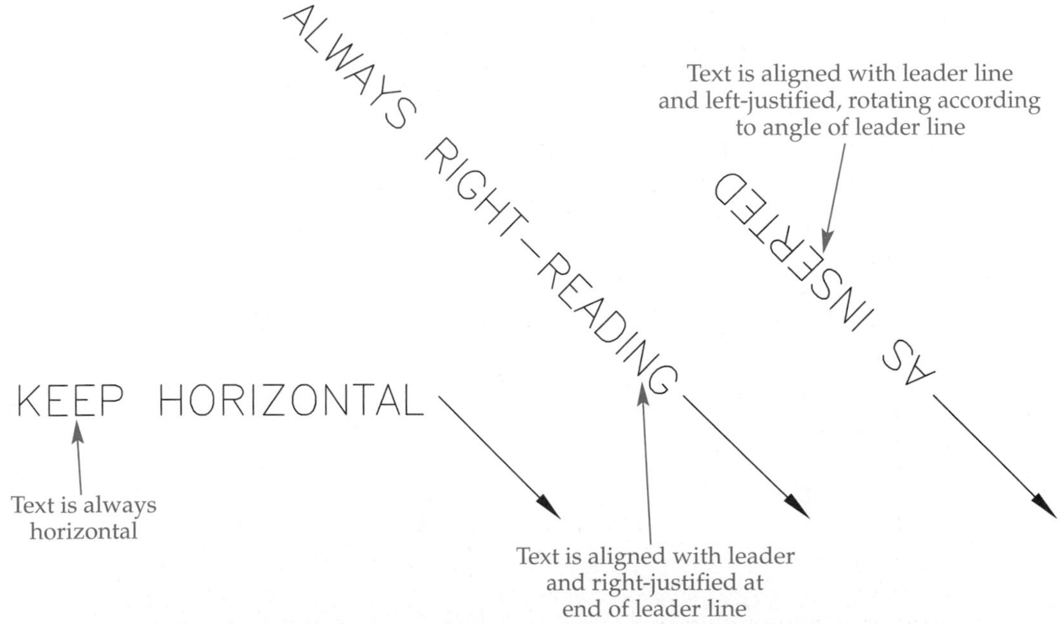

ALWAYS RIGHT-READING

AS INSERTED

Text is aligned with leader line
and left-justified, rotating according
to angle of leader line

KEEP HORIZONTAL

Text is always
horizontal

Text is aligned with leader
and right-justified at
end of leader line

Figure 19-18.
Horizontal alignment options. The shaded examples are the recommended ASME standards.

	Top of Top Line	Middle of Top Line	Middle of Multiline Text	Middle of Bottom Line	Bottom of Bottom Line
Text on Left Side	⌀.250 ⌴⌀.500 ▽.062	⌀.250 ⌴⌀.500 ▽.062	⌀.250 ⌴⌀.500 ▽.062	⌀.250 ⌴⌀.500 ▽.062	⌀.250 ⌴⌀.500 ▽.062
Text on Right Side	⌀.250 ⌴⌀.500 ▽.062	⌀.250 ⌴⌀.500 ▽.062	⌀.250 ⌴⌀.500 ▽.062	⌀.250 ⌴⌀.500 ▽.062	⌀.250 ⌴⌀.500 ▽.062

of the text. Use the **Right attachment:** drop-down list to define how multiple lines of text are positioned when the leader is on the right side of the text. **Figure 19-18** shows typical selections. The **Underline bottom line** option draws a line along the bottom of the multiline text box. The **Underline all text** option underlines each line of leader text. The **Leading gap** text box specifies the space between the leader line or shoulder and the text. The default is .09, but .063 (1.5 mm) is standard.

PROFESSIONAL TIP

Common drafting practice is to use the **Middle of bottom line** option for left attachment and the **Middle of top line** option for right attachment.

For some applications, you may find it necessary to select the **Vertical attachment** radio button, although this is not common. This option eliminates the possible use of a shoulder and connects the leader endpoint to the top center or bottom center of the text, depending on the leader line position. Use the **Top attachment:** drop-down list to define how text is positioned when the leader is above the text. Use the **Bottom attachment:** drop-down list to define how text is positioned when the leader is below the text. **Figure 19-19** shows each option.

Figure 19-19.
Vertical alignment options.

Center	Underline and Center	Center	Overline and Center
↑ A ↓	A ↓	↑ A	A

Figure 19-20.
The **Content** tab of the **Modify Multileader Style** dialog box with the **Block** multileader type selected.

Select the multileader type

Set the options for the block

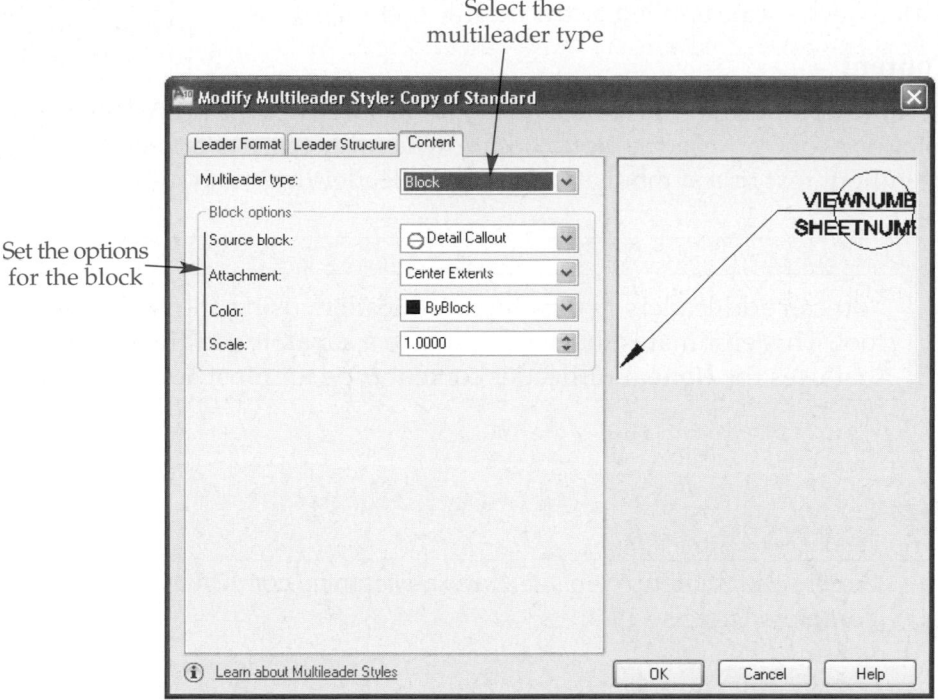

Attaching a Symbol

Pick the **Block** option from the **Multileader type:** drop-down list, as shown in **Figure 19-20,** to attach a *block* to the leader. Chapters 25 through 28 describe blocks in detail. Several blocks are available by default from the **Source block:** drop-down list. You also have the option of picking the **User Block...** option to select your own saved block. The **Select Custom Content Block** dialog box appears when you pick the **User Block...** option. Pick a block in the current drawing from the **Select from Drawing Blocks:** drop-down list and then pick the **OK** button.

Use the **Attachment:** drop-down list to specify how to attach the block to the leader. Pick the **Insertion point** option to attach the block to the leader according to the block insertion point, or base point. Choose the **Center extents** option to attach the block directly to the leader, aligned to the center of the block, even if the block insertion point is not on the block itself. See **Figure 19-21.**

block: A symbol that was previously created and saved for reuse.

Figure 19-21.
Adjusting multileader block attachment.

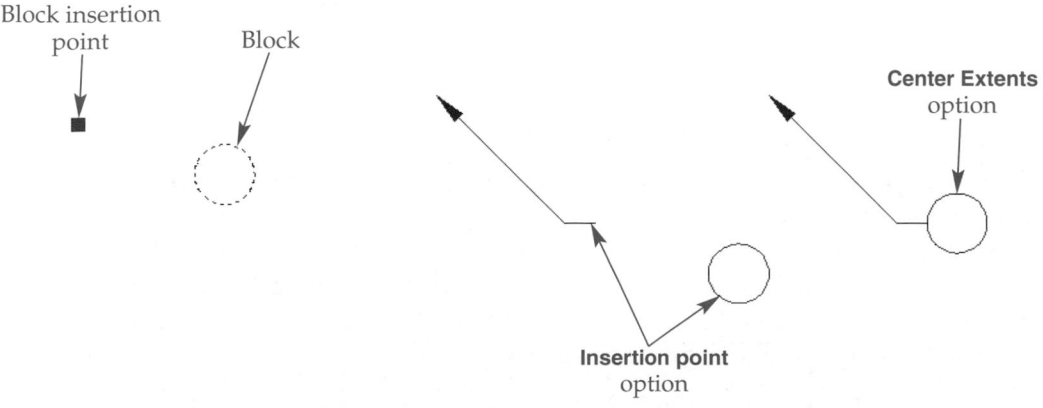

Use the **Color** drop-down list to specify the appropriate block color, which should be ByBlock for typical applications. Use the **Scale** text box to proportionately increase or decrease the block size. Scale does not affect the appearance of the leader line, arrowhead, or shoulder, or the scale applied to the multileader object.

Using No Content

Select the **None** option from the **Multileader type:** drop-down list to end the leader with no annotation. You can use the **None** option whenever there is a need to create only a leader, without text or a symbol attached to the leader line or shoulder.

> **NOTE**
>
> You can add leaders to existing multileaders using the **Add Leader** tool. This eliminates the need to create a separate multileader style that uses the **None** multileader content type for most applications.

Exercise 19-4

Access the Student Web site (www.g-wlearning.com/CAD) and complete Exercise 19-4.

Modifying Multileader Styles

Use the **Multileader Style Manager** to change the characteristics of an existing multileader style. Pick the **Modify** button to open the **Modify Multileader Style** dialog box, which allows you to make changes to the selected style. If you make changes to a multileader style, such as selecting a different text or arrowhead style, existing leaders you drew using the modified multileader style update to reflect the changes. Use a different multileader style with unique characteristics when appropriate.

Renaming and Deleting Multileader Styles

To rename a multileader style using the **Multileader Style Manager**, slowly double-click on the name or right-click the name and select **Rename**. To delete a multileader style using the **Multileader Style Manager**, right-click the name and select **Delete**. You cannot delete a multileader style that is assigned to leaders. To delete a style that is in use, assign a different style to the leaders that reference the style to be deleted.

> **NOTE**
>
> You can also rename styles using the **Rename** dialog box. Select **Multileader styles** in the **Named Objects** list to rename the style.

Setting a Multileader Style Current

You can set a multileader style current using the **Multileader Style Manager** by double-clicking the style in the **Styles** list box, right-clicking on the name and selecting **Set current**, or picking the style and selecting the **Set Current** button. To set a multileader style current without opening the **Multileader Style Manager**, use the **Multileader Style** drop-down list located in the expanded **Annotation** panel on the **Home** ribbon tab and on the **Leaders** panel on the **Annotate** ribbon tab.

Inserting Multileaders

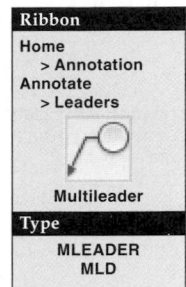

Once you develop a multileader style and make the style current, you are ready to insert leaders using the **MLEADER** tool. How you insert a leader depends on the current multileader style settings and the option you choose to construct the leader. In general, there are three methods for inserting a multileader, depending on what portion of the leader you locate first. Review the components of a leader, shown in **Figure 19-22,** before reading the options for creating a multileader.

The first option for inserting a leader is **Specify leader arrowhead location**. To use this option, first select the location at which you want the arrowhead to point. Then choose where the leader ends and the shoulder begins. If the **Mtext** option is active, enter leader text using the multiline text editor.

The second method uses the **leader Landing first** option. To use this technique, first select where the leader ends and the shoulder begins. Then choose the location where the arrowhead points. If the **Mtext** option is active, enter leader text using the multiline text editor.

The third method involves using the **Content first** option. To use this technique, first define the leader content. The **Mtext** option allows you to type text using the multi-line text editor. Then you can select the location where the arrowhead points.

Select **Options** to access a list of options that allow you to override the current multileader style characteristics. These options are the same as those found in the **Modify Multileader Style** dialog box.

Figure 19-22.
Examples of leaders created using the **MLEADER** tool. A—An architectural leader created using a spline leader line, the **Specify leader arrowhead location** option, and three leader points. B—A mechanical leader created using a straight leader line, the **leader Landing first** option, and two leader points.

Exercise 19-5

Access the Student Web site (www.g-wlearning.com/CAD) and complete Exercise 19-5.

Dimensioning Chamfers

chamfer: An angled surface used to relieve sharp corners.

Dimension *chamfers* of 45° either with a leader giving the angle and linear dimension, or with two linear dimensions. See **Figure 19-23.** Place the leader using the **MLEADER** tool. Chamfers other than 45° must include either the angle and a linear dimension or two linear dimensions. See **Figure 19-24.** Use the **DIMLINEAR** and **DIMANGULAR** tools for this purpose.

Exercise 19-6

Access the Student Web site (www.g-wlearning.com/CAD) and complete Exercise 19-6.

Thread Drawings and Notes

Figure 19-25 shows the elements of a screw thread. However, threads commonly appear on a drawing as a simplified representation in which a hidden line indicates thread depth. See **Figure 19-26.** Both external and internal threads use this method. An externally threaded part often includes a chamfer to help engage the mating thread.

Thread representations show the reader that a thread exists, but the thread note gives the exact specifications. The thread note typically connects to the thread with a leader. See **Figure 19-27.** The most common thread forms are the Unified and metric screw threads, but a variety of other thread forms are required for specific applications.

Figure 19-23.
Dimensioning 45°
chamfers.

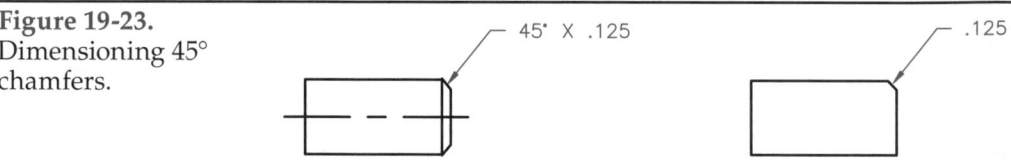

Figure 19-24.
Dimensioning
chamfers that are
not 45°.

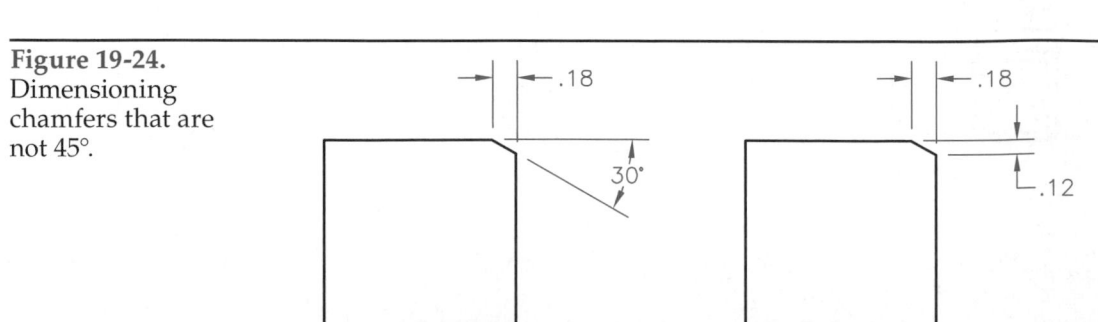

Figure 19-25.
Features of a screw thread.

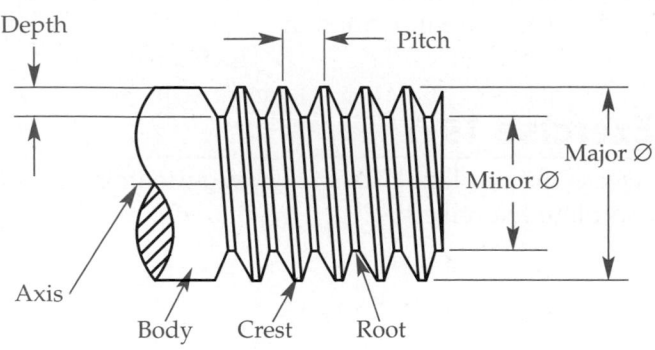

Figure 19-26.
Simplified thread representations.

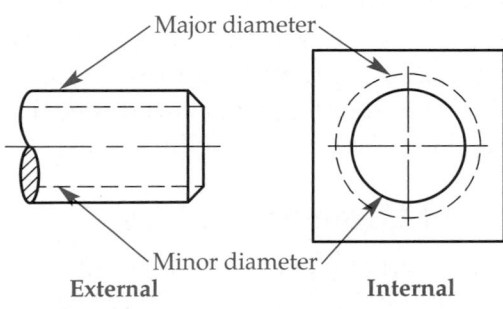

Figure 19-27.
Displaying the thread note with a leader.

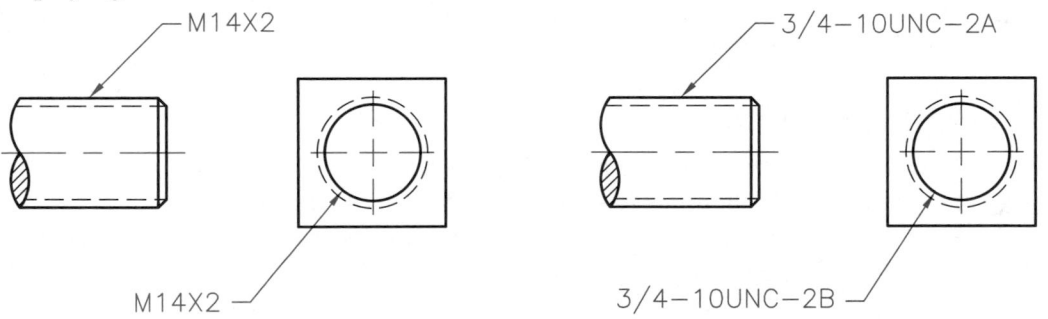

The following format specifies the thread note for Unified screw threads.

3/4-   10UNC-   2A
  (1)   (2)   (3)   (4)(5)

(1) Major diameter of thread, given as a fraction or number.
(2) Number of threads per inch.
(3) Thread series: UNC = Unified National Coarse, UNF = Unified National Fine.
(4) Class of fit: 1 = large tolerance, 2 = general-purpose tolerance, 3 = tight tolerance.
(5) Thread type: A = external thread. B = internal thread.

The following format specifies the thread note for metric threads.

M   14X2
(1)   (2) (3)

(1) M = metric thread.
(2) Major diameter in millimeters.
(3) Pitch in millimeters.

There are too many screw threads to describe in detail in this textbook. Refer to the *Machinery's Handbook*, published by Industrial Press Inc., or a comprehensive mechanical drafting text for more information.

Exercise 19-7

Access the Student Web site (www.g-wlearning.com/CAD) and complete Exercise 19-7.

Alternate Dimensioning Practices

Dimension lines are often omitted on drawings in industries in which computer-controlled machining processes are used, and unconventional dimensioning practices are sometimes required because of product features. Rectangular coordinate dimensioning without dimension lines, tabular, and chart dimensioning are three examples of dimensioning methods that omit dimension lines.

Dimensioning without Dimension Lines

rectangular coordinate dimensioning without dimension lines (arrowless dimensioning): A type of dimensioning that includes only extension lines and text aligned with the extension lines.

Rectangular coordinate dimensioning without dimension lines, traditionally known as *arrowless dimensioning*, is popular in mechanical drafting for specific applications. Common applications include precision sheet metal part drawings and electronics drafting, especially for chassis layout. Each dimension represents a measurement originating from a *datum*. Identification letters label holes or similar features. Often a table, keyed to the identification letters, indicates feature size or specifications. See **Figure 19-28**.

datum: The 0 dimension, baseline, or common point from which all measurements are made while dimensioning.

Tabular Dimensioning

tabular dimensioning: A form of rectangular coordinate dimensioning without dimension lines in which dimensions are shown in a table.

In *tabular dimensioning*, each feature receives a label with a letter or number that correlates to a table. The table gives the location of features from the X and Y axes. See **Figure 19-29**. The table also provides the depth of features from the Z axis and other specifications when appropriate.

Figure 19-28. Rectangular coordinate dimensioning without dimension lines, or arrowless dimensioning.

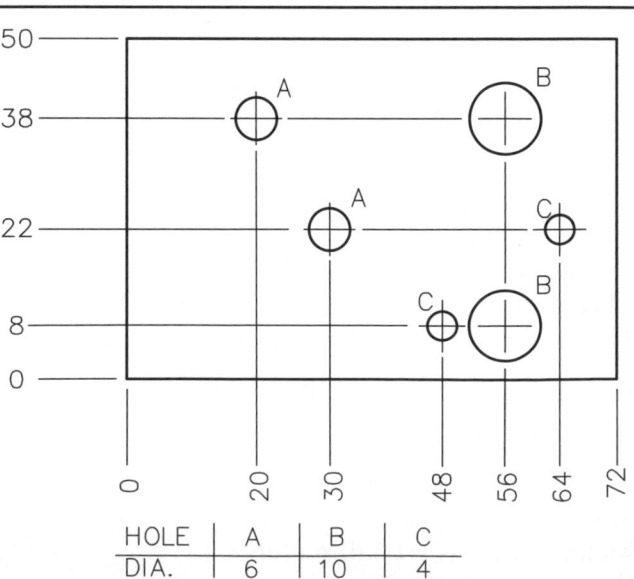

HOLE	A	B	C
DIA.	6	10	4

Figure 19-29.
Tabular dimensioning. (Doug Major)

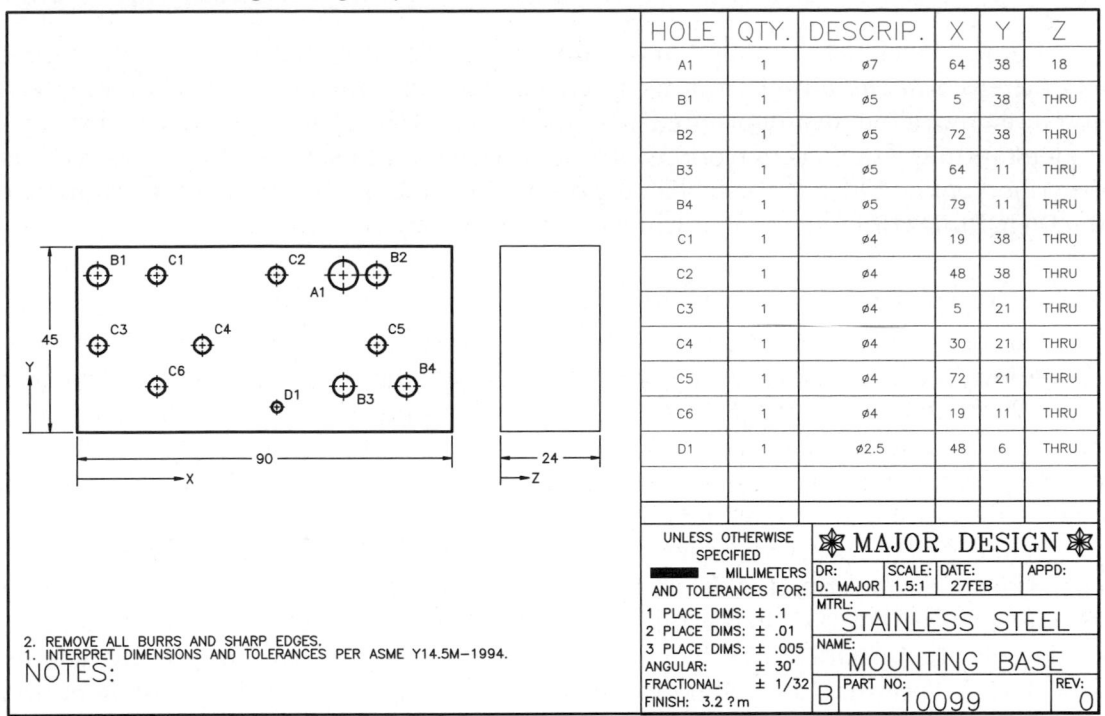

HOLE	QTY.	DESCRIP.	X	Y	Z
A1	1	⌀7	64	38	18
B1	1	⌀5	5	38	THRU
B2	1	⌀5	72	38	THRU
B3	1	⌀5	64	11	THRU
B4	1	⌀5	79	11	THRU
C1	1	⌀4	19	38	THRU
C2	1	⌀4	48	38	THRU
C3	1	⌀4	5	21	THRU
C4	1	⌀4	30	21	THRU
C5	1	⌀4	72	21	THRU
C6	1	⌀4	19	11	THRU
D1	1	⌀2.5	48	6	THRU

UNLESS OTHERWISE SPECIFIED
▬ – MILLIMETERS
AND TOLERANCES FOR:
1 PLACE DIMS: ± .1
2 PLACE DIMS: ± .01
3 PLACE DIMS: ± .005
ANGULAR: ± 30'
FRACTIONAL: ± 1/32
FINISH: 3.2 ? m

✾ MAJOR DESIGN ✾
DR: D. MAJOR | SCALE: 1.5:1 | DATE: 27FEB | APPD:
MTRL: STAINLESS STEEL
NAME: MOUNTING BASE
B | PART NO: 10099 | REV: 0

2. REMOVE ALL BURRS AND SHARP EDGES.
1. INTERPRET DIMENSIONS AND TOLERANCES PER ASME Y14.5M–1994.
NOTES:

Chart Dimensioning

Chart dimensioning may take the form of unidirectional, aligned, rectangular coordinate dimensioning without dimension lines, or tabular dimensioning. Chart dimensioning provides flexibility when dimensions change as requirements of the product change. See **Figure 19-30**.

chart dimensioning: A type of dimensioning in which the variable dimensions are shown with letters that correlate to a chart where the possible dimensions are given.

Figure 19-30.
Chart dimensioning.

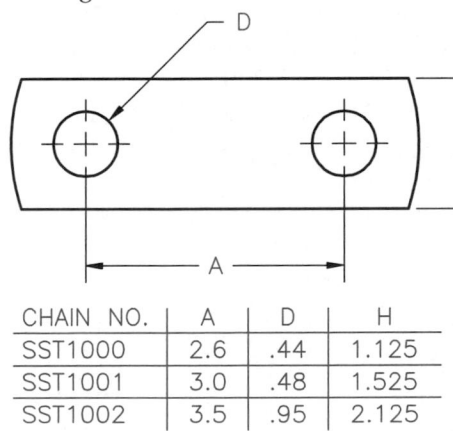

CHAIN NO.	A	D	H
SST1000	2.6	.44	1.125
SST1001	3.0	.48	1.525
SST1002	3.5	.95	2.125

NOTE:
OVERALL LENGTH IS 1.5 X A
END RADII ARE .9 X A

ordinate dimensioning: The AutoCAD term for rectangular coordinate dimensioning without dimension lines.

AutoCAD refers to rectangular coordinate dimensioning without dimension lines as *ordinate dimensioning*. In order to create ordinate dimension objects accurately, you must move the default origin (0,0,0 coordinate) to the object datum. This involves understanding AutoCAD's world coordinate system and user coordinate systems, as described below. Once you establish the datum by temporarily moving the origin, use the **DIMORDINATE** tool to place ordinate dimension objects.

Introduction to WCS and UCS

world coordinate system (WCS): AutoCAD's rectangular coordinate system. In 2D drafting, the WCS contains four quadrants, separated by the X and Y axes.

The origin (0,0,0 coordinate) of the *world coordinate system (WCS)* has been in the lower-left corner of the drawing window for the drawings you have created throughout this textbook. In most cases, this is appropriate. However, when you dimension without dimension lines, it is best to have the dimensions originate from a primary datum, which is often a corner of the object. Depending on how you draw the object, this point may or may not align with the WCS origin.

user coordinate system (UCS): A temporary override of the WCS in which the origin (0,0,0) is moved to a location specified by the user.

While the WCS is fixed, a *user coordinate system (UCS)* can move to any orientation. The UCS is described in detail in *AutoCAD and Its Applications—Advanced*. In general, a UCS allows you to set your own coordinate system and origin. Measurements made with the **DIMORDINATE** tool originate from the current UCS origin. By default, this is the 0,0 origin. One method for relocating the origin is to pick the **Origin** button from the **Coordinates** panel of the **View** ribbon tab. Then specify a new origin point, such as the corner of an object or another appropriate datum. See **Figure 19-31**.

When you finish drawing dimensions from a datum, you can leave the UCS origin at the datum or move it back to the WCS origin. To return to the WCS, pick the **World** button from the **Coordinates** panel of the **View** ribbon tab.

Using the DIMORDINATE Tool

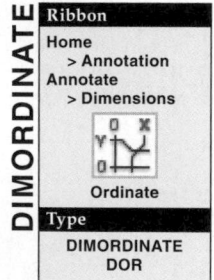

DIMORDINATE

Ribbon
Home
> Annotation
Annotate
> Dimensions

Ordinate

Type
DIMORDINATE
DOR

When you use the **DIMORDINATE** tool, AutoCAD automatically places an extension line and a dimension at the location you pick. The dimension is measured as an X or Y coordinate distance from the UCS origin.

Since you are working in the XY plane, you may want to set vertical and horizontal polar tracking or turn **Ortho** mode on before using the **DIMORDINATE** tool. Also, if the drawing includes circular features, use the **DIMCENTER** tool to place center dashes as shown in **Figure 19-31**. This conforms to ASME standards and provides something to pick when you dimension circular features.

Figure 19-31.
Before you use the **DIMORDINATE** tool, move the UCS origin to the appropriate datum location. Also, add center marks to circular features that you plan to dimension.

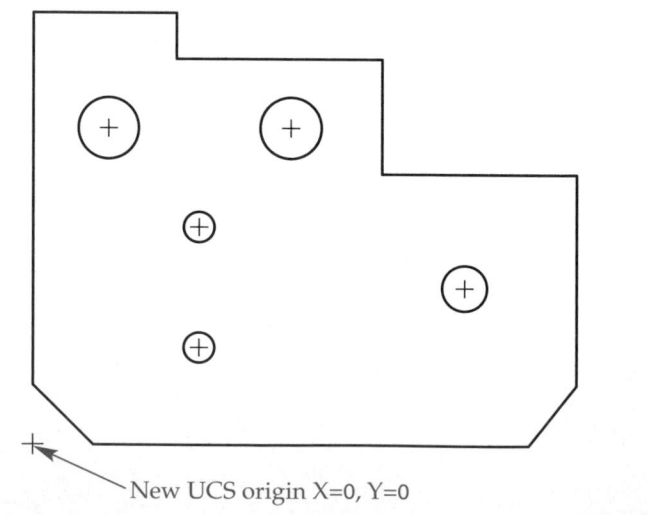

New UCS origin X=0, Y=0

Now you are ready to start placing ordinate dimensions. Access the **DIMORDINATE** tool. When the Specify feature location: prompt appears, pick a point to locate the origin of the extension line. If the feature is the corner of the object, pick an endpoint. If the feature is a circle, pick the *end* of the center dash, as shown in **Figure 19-32**, not the center of the object. This leaves the required space between the center mark and the extension line. Zoom in if needed and use object snap modes when necessary. The next prompt asks for the leader endpoint. This actually refers to the extension line endpoint, so pick the endpoint of the extension line.

If the X axis or Y axis distance between the feature and the extension line endpoint is large, the axis AutoCAD uses for the dimension by default may not be correct. When this happens, use the **Xdatum** or **Ydatum** option to specify the axis from which the dimension originates. The **Mtext**, **Text**, and **Angle** options are identical to the options available with other dimensioning tools. Pick the extension line endpoint to complete the process.

Figure 19-33A shows ordinate dimensions placed on the object. Notice that the dimension text aligns with the extension lines. Aligned dimensioning is standard with ordinate dimensioning. Finally, complete the drawing by adding any missing lines, such as centerlines or fold lines. Identify the holes with letters and create a correlated dimensioning table if appropriate. See **Figure 19-33B**.

PROFESSIONAL TIP

Most ordinate dimensioning tasks work best with polar tracking or **Ortho** mode on. However, when the extension line is too close to an adjacent dimension, it is best to stagger the extension line as shown in the following illustration. With **Ortho** mode off, the extension line automatically staggers when you pick the second extension line point, as shown here.

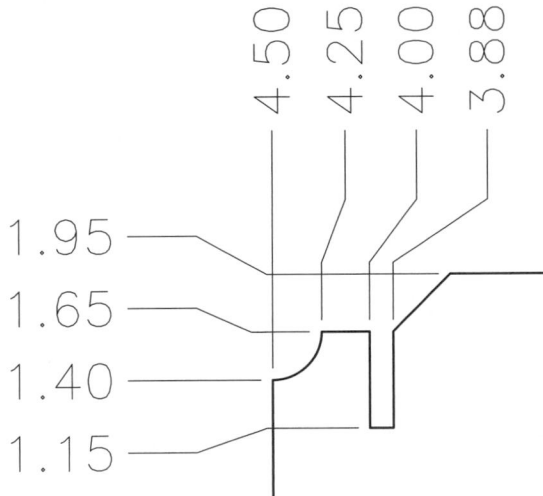

Figure 19-32.
Pick the endpoints of center marks to establish the correct offset, or develop a specific dimension style with an extension line origin offset of 0 for placing dimensions from center marks.

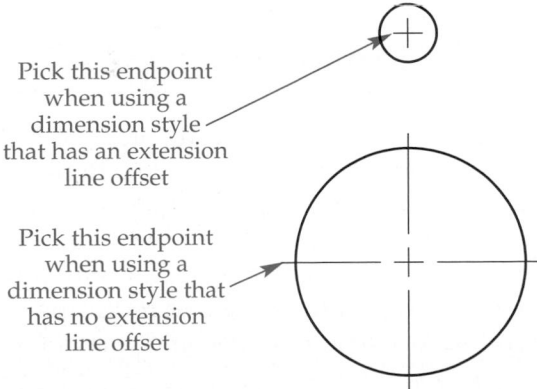

Pick this endpoint when using a dimension style that has an extension line offset

Pick this endpoint when using a dimension style that has no extension line offset

---

Figure 19-33.
A—Placing ordinate dimensions.
B—Completing the drawing.

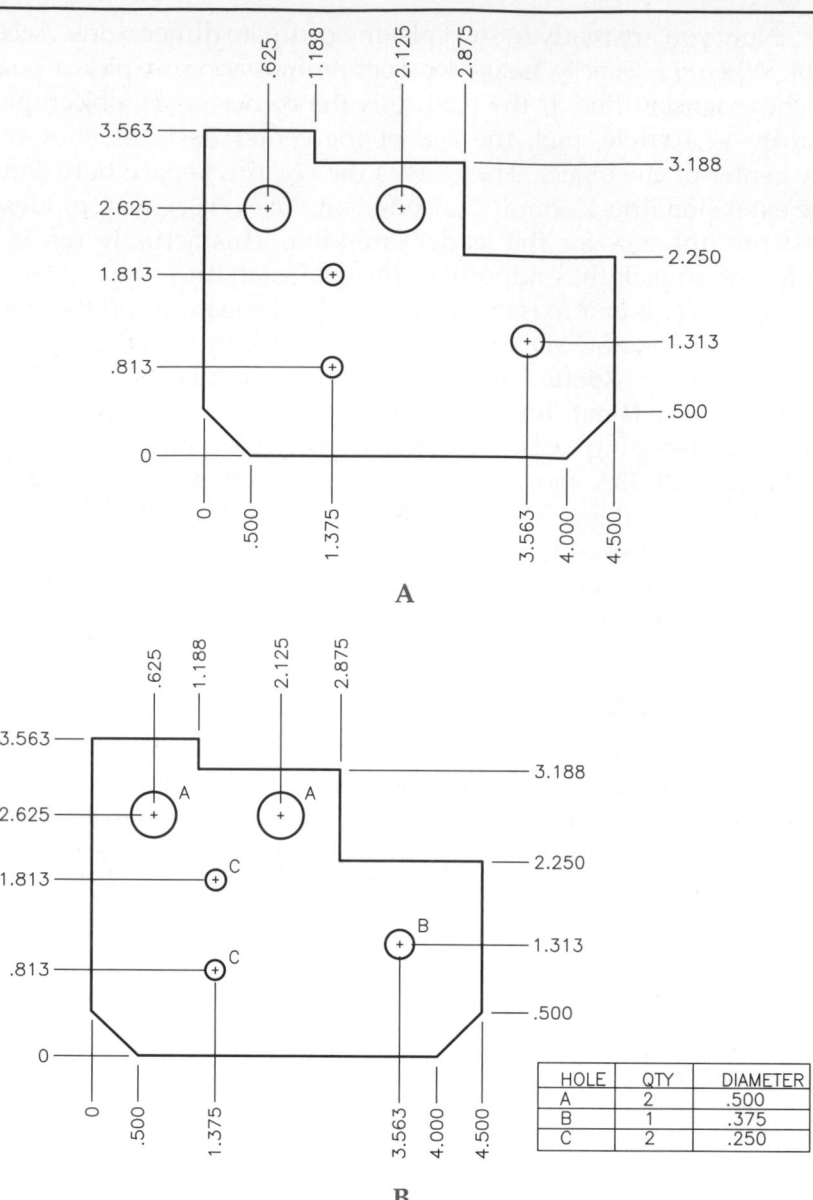

A

B

HOLE	QTY	DIAMETER
A	2	.500
B	1	.375
C	2	.250

Exercise 19-8

Access the Student Web site (www.g-wlearning.com/CAD) and complete Exercise 19-8.

Marking Up Drawings

marking up (redlining): The process of reviewing a drawing and marking required changes.

Marking up, or *redlining*, is not a dimensioning practice, but it is similar to dimensioning in that it explains information to the reader. Redlines are typically added directly to a final drawing by someone who reviews the drawing for accuracy and design changes. You might experience this process with your instructor or supervisor. Common mark-up techniques include redlining a plot with a red pen, using separate mark-up software to review exported drawings, or redlining directly in the drawing file. Redlines drawn in AutoCAD sometimes become part of the drawing to document revision history.

You can redline a drawing using any appropriate AutoCAD tools, typically using a separate layer. Redlining often includes basic objects, text, and leaders. In some cases, you may add redline dimensions and even an entire drawing or detail. The **REVCLOUD** and **WIPEOUT** tools are also common mark-up tools.

Creating Revision Clouds

A *revision cloud* is a polyline of sequential arcs forming a cloud-shaped object. **Figure 19-34** shows a revision cloud with a leader and note attached. The revision cloud points the drafter to a specific portion of the drawing that may require an edit.

Drawing a revision cloud using the **REVCLOUD** tool is somewhat different than drawing most other objects, because a single pick is all that is required. To begin drawing the revision cloud, pick a start point in the drawing, and then move the cross-hairs around the objects to enclose until you return close to the start point. AutoCAD closes the cloud automatically and exits the tool. Options are available before you pick the start point.

Defining Arc Length

Use the **Arc length** option to specify the size of revision cloud arcs. The value measures the length of an arc from the arc start point to the arc endpoint. AutoCAD prompts for the minimum arc length and then for the maximum arc length. Specifying different minimum and maximum values causes the revision cloud to have an uneven, hand-drawn appearance.

Converting Objects to Revision Clouds

Use the **Object** option to convert a circle, closed polyline, ellipse, polygon, or rectangle to a revision cloud. Pick the object to convert to a revision cloud. Enter the **No** option at the Reverse direction: prompt, or use the **Yes** option to reverse the direction of the cloud arcs.

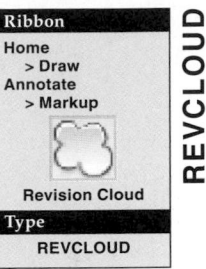

Ribbon
Home
> Draw
Annotate
> Markup

Revision Cloud

Type
REVCLOUD

REVCLOUD

revision cloud: A polyline of sequential arcs used to form a cloud shape around changes in a drawing.

Figure 19-34.
An example of a revision cloud identifying a modified area of a drawing. Notice the leader describing the change.

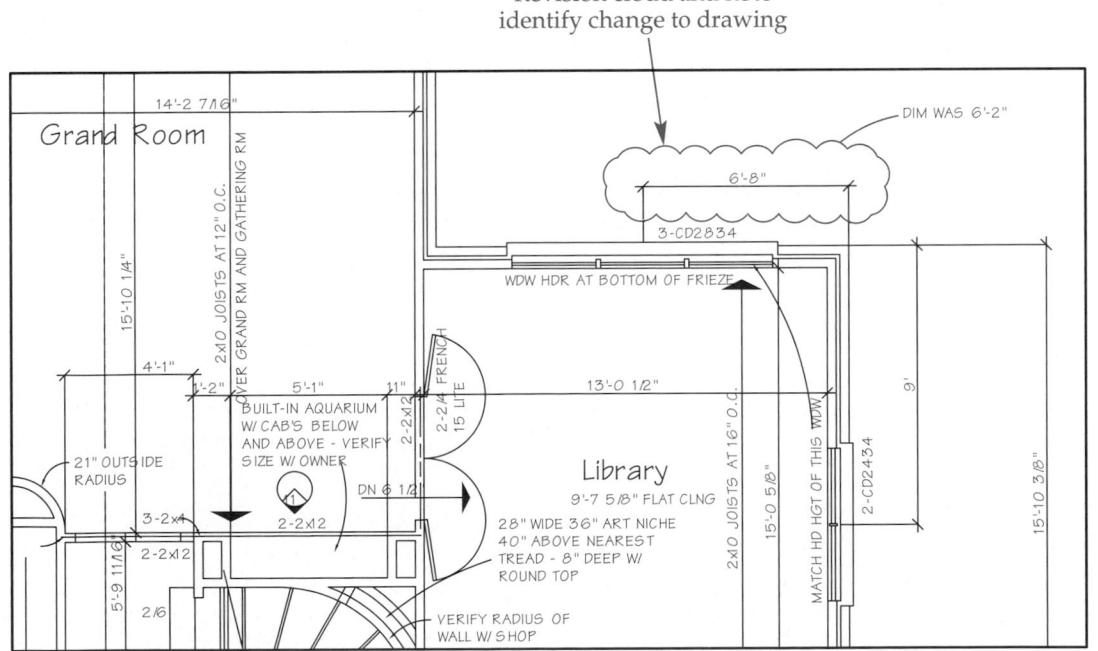

Changing Revision Cloud Style

The **Style** option offers two style choices: **Normal** and **Calligraphy**. The default style is **Normal**, which displays arcs with a consistent width. When you specify the **Calligraphy** style, the start and end widths of the individual arcs are different, creating a more stylized revision cloud. See **Figure 19-35**.

Exercise 19-9

Access the Student Web site (www.g-wlearning.com/CAD) and complete Exercise 19-9.

Using the WIPEOUT Tool

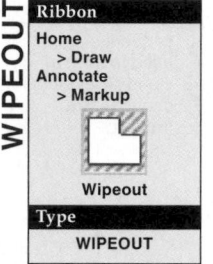

WIPEOUT

Ribbon
Home
> Draw
Annotate
> Markup

Wipeout

Type
WIPEOUT

The **WIPEOUT** tool allows you to clear a portion of the drawing without erasing objects. This tool is sometimes appropriate for applications similar to those for **REVCLOUD**, most often redlining. **Figure 19-36** shows an example of a wipeout used to lay out the location of a proposed building sight.

Specify the first corner of the wipeout, followed by all other perimeter corners. Use the **Undo** option as needed to reverse the effects of an incorrect selection. When you are finished selecting points, use the **Close** option, press [Enter] or the space bar, or right-click and choose **Enter** to create the wipeout.

An alternative to picking points is to use the **Polyline** option and select a closed polyline object to convert to a wipeout. Use the **Frames** option to turn the display of all wipeout boundaries on or off. You may need to regenerate the display to observe the effects of changing the frame setting. To reveal objects hidden by a wipeout, freeze or turn off the wipeout layer, use draw order tools, or erase the wipeout if it is no longer needed.

Template Development

Chapter 19

For detailed instructions on adding multileader styles to each drawing template, go to the Student Web site (www.g-wlearning.com/CAD), select this chapter, and select **Template Development**.

Figure 19-35.
You can create revision clouds in two different styles: the **Normal** style and the **Calligraphy** style.

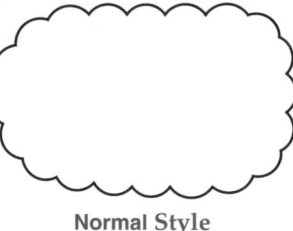

Normal Style

Calligraphy Style

Figure 19-36.
An example of using the **WIPEOUT** tool to clear a portion of a drawing. Objects below the wipeout still exist. Additional information is added to the wipeout in this example.

Pick points to create the wipeout

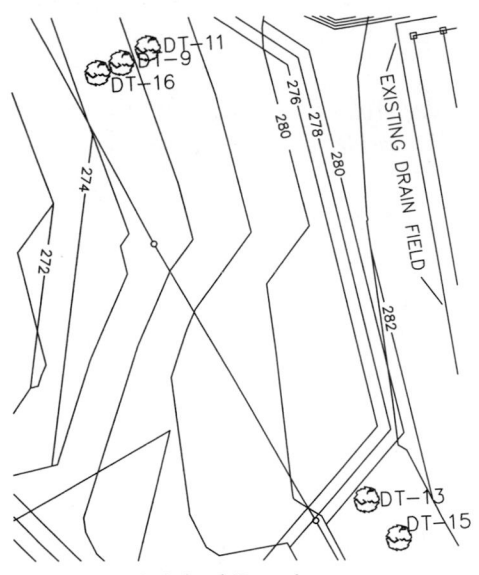

Original Drawing

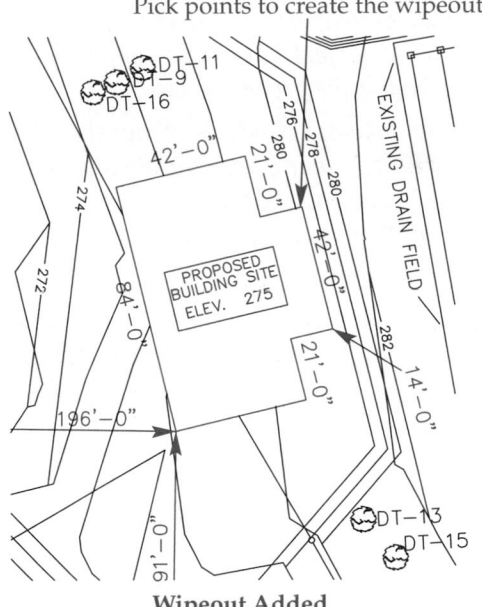

Wipeout Added

Chapter Test

Answer the following questions. Write your answers on a separate sheet of paper or go to the Student Web site (www.g-wlearning.com/CAD) and complete the electronic chapter test.

1. Which tool provides diameter dimensions for circles?
2. Which tool provides radius dimensions for arcs?
3. Explain how to add a center mark to a circle without using the **DIMDIAMETER** or **DIMRADIUS** tool.
4. What is the most common size for leader arrowheads?
5. What angle constraints should you use for leaders to maintain the ASME standard?
6. What is the usual length for the shoulder of a leader in mechanical drafting?
7. Describe two ways to dimension a 45° chamfer.
8. Identify the elements of this Unified screw thread note: 1/2-13UNC-2B.
   A. 1/2
   B. 13
   C. UNC
   D. 2
   E. B
9. Identify the elements of this metric screw thread note: M 14 X 2.
   A. M
   B. 14
   C. 2
10. Define *rectangular coordinate dimensioning without dimension lines.*
11. What term does AutoCAD use to refer to rectangular coordinate dimensioning without dimension lines?
12. Explain the importance of the user coordinate system (UCS) for drawing ordinate dimension objects.
13. What is the purpose of a revision cloud?
14. How do you close a revision cloud?
15. What is the purpose of the **WIPEOUT** tool?

Chapter 19 Dimensioning Features and Alternate Practices

Drawing Problems

- *Start AutoCAD if it is not already started.*
- *Start a new drawing using an appropriate template of your choice. The template should include layers, text styles, dimension styles, and multileader styles appropriate for drawing the given objects.*
- *Add layers, text styles, dimension styles, and multileader styles as needed. Draw all objects using appropriate layers, text styles, dimension styles, multileader styles, justification, and format.*
- *Follow the specific instructions for each problem. Use your own judgment and approximate dimensions when necessary.*
- *Apply dimensions accurately using ASME or appropriate industry standards. Use object snap modes to your best advantage.*
- *For mechanical drawings, place the following general notes 1/2″ from the lower-left corner:*

> NOTES:
>
> 1. INTERPRET DIMENSIONS AND TOLERANCES PER ASME Y14.5M-1994.
> 2. REMOVE ALL BURRS AND SHARP EDGES.
> 3. UNLESS OTHERWISE SPECIFIED, ALL DIMENSIONS ARE IN INCHES (or MILLIMETERS as applicable).

▼ Basic

1. Draw the view and add dimensions as shown. Save the drawing as P19-1.

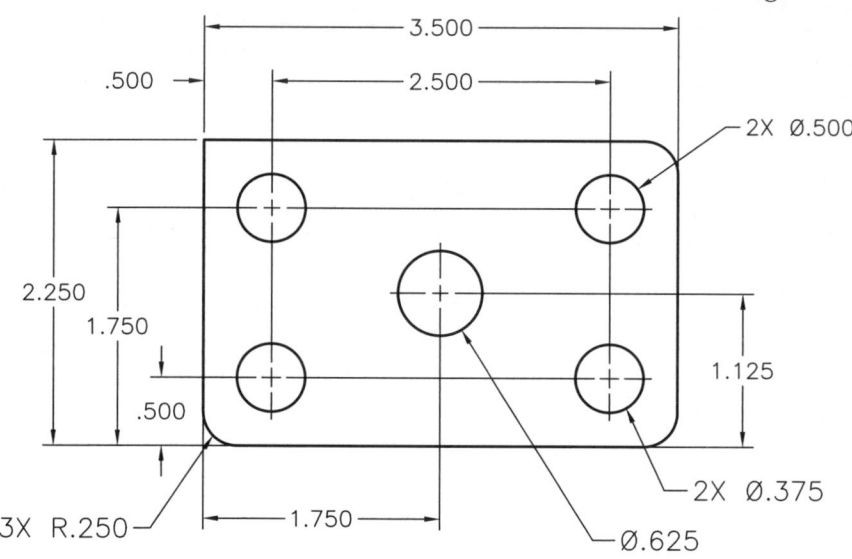

2. Draw the views and add dimensions as shown. Save the drawing as P19-2.

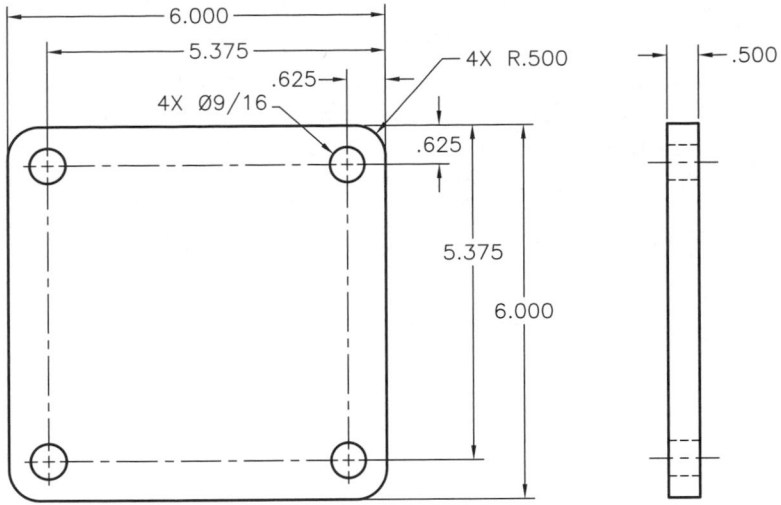

3. Draw the view and add dimensions as shown. Save the drawing as P19-3.

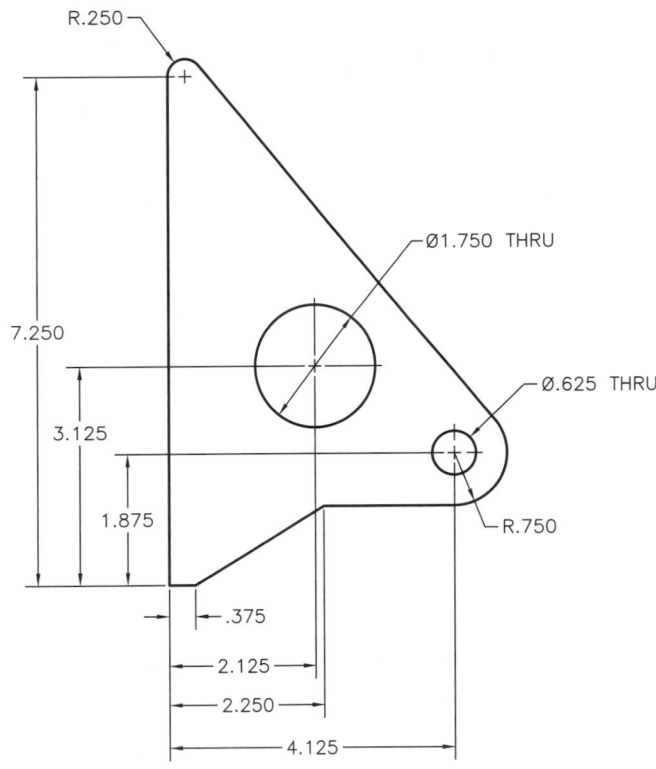

4. Draw the pin and add dimensions as shown. Save the drawing as P19-4.

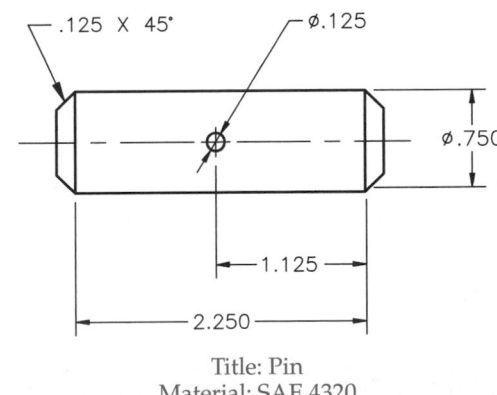

Title: Pin
Material: SAE 4320

Drawing Problems - Chapter 19

5. Draw the spline and add dimensions as shown. Save the drawing as P19-5.

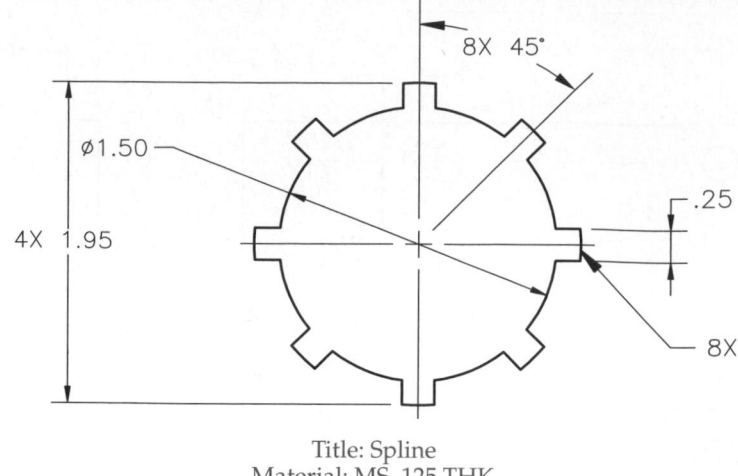

Title: Spline
Material: MS .125 THK

6. Draw the gasket and add dimensions as shown. Save the drawing as P19-6.

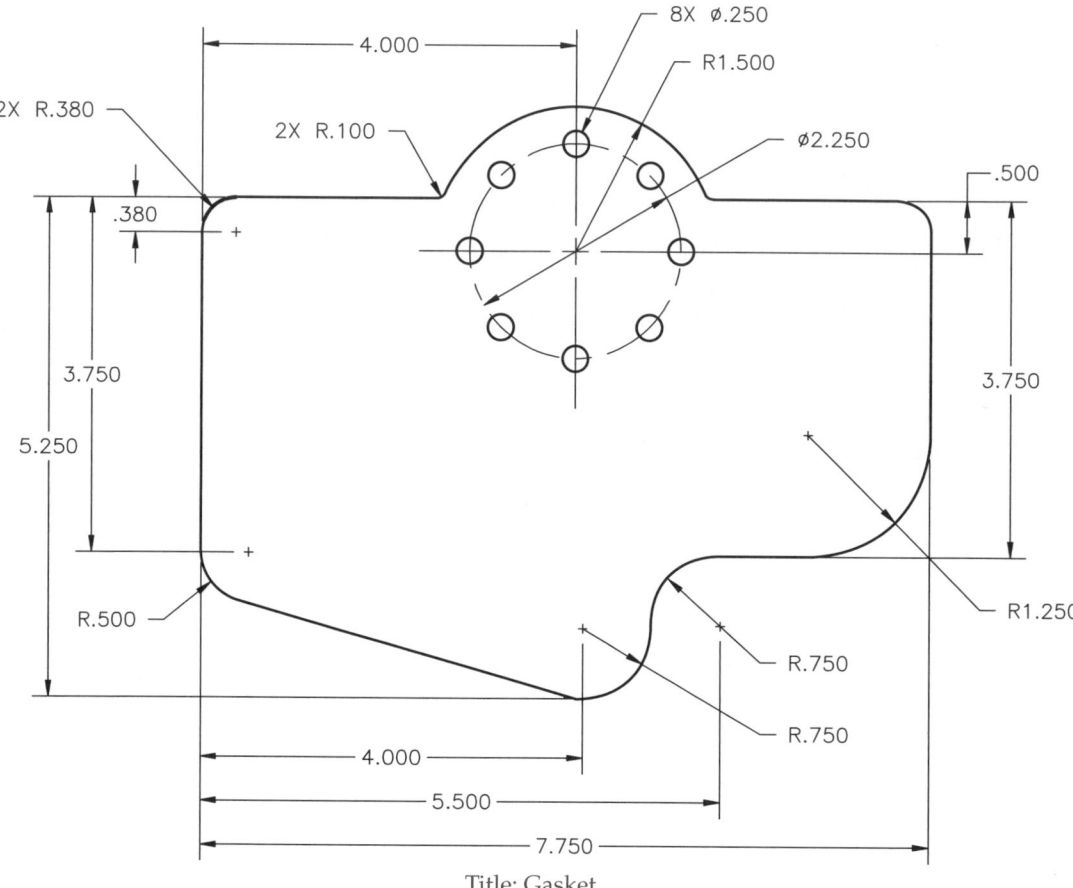

Title: Gasket
Material: 00 Phosphor Bronze

Drawing Problems - Chapter 19

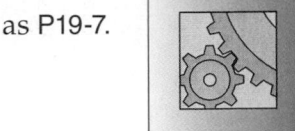

7. Draw the chain link and add dimensions as shown. Save the drawing as P19-7.

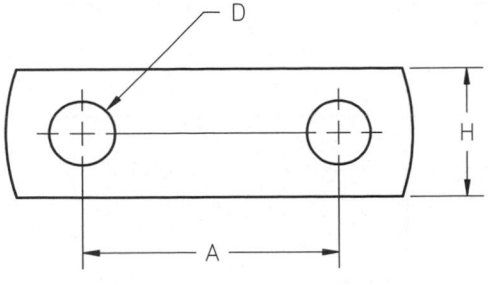

CHAIN NO.	A	D	H
SST1000	2.6	.44	1.125
SST1001	3.0	.48	1.525
SST1002	3.5	.95	2.125

Note:
Overall Length is 1.5XA
end radii are .9XA

Title: Chain Link
Material: Steel

8. Draw the view and add dimensions as shown. Save the drawing as P19-8.

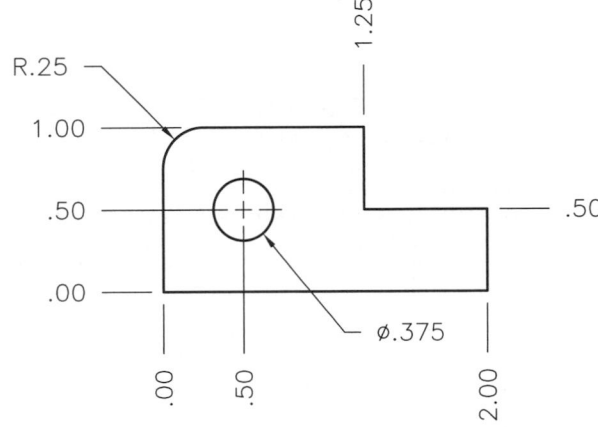

9. Draw the chassis spacer and add dimensions as shown. Save the drawing as P19-9.

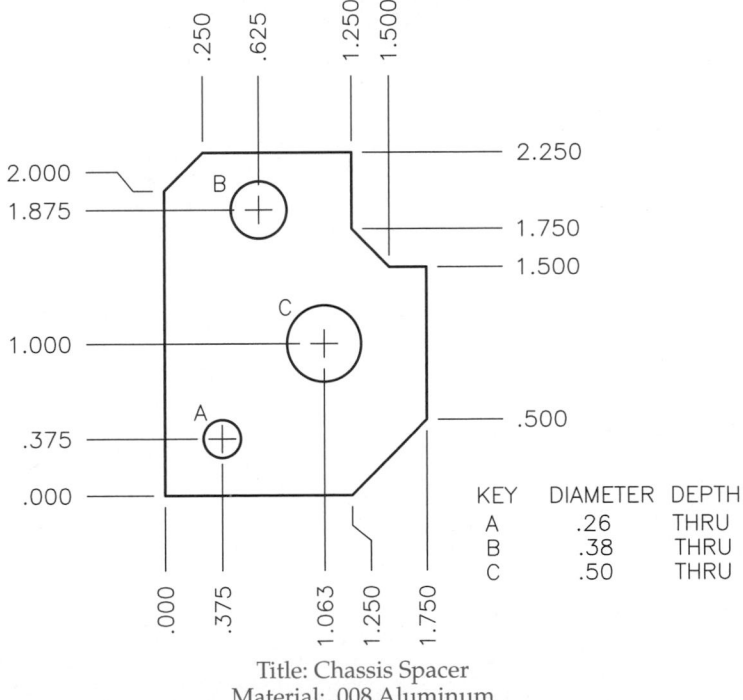

KEY	DIAMETER	DEPTH
A	.26	THRU
B	.38	THRU
C	.50	THRU

Title: Chassis Spacer
Material: .008 Aluminum

10. Draw the chassis and add dimensions as shown. Save the drawing as P19-10.

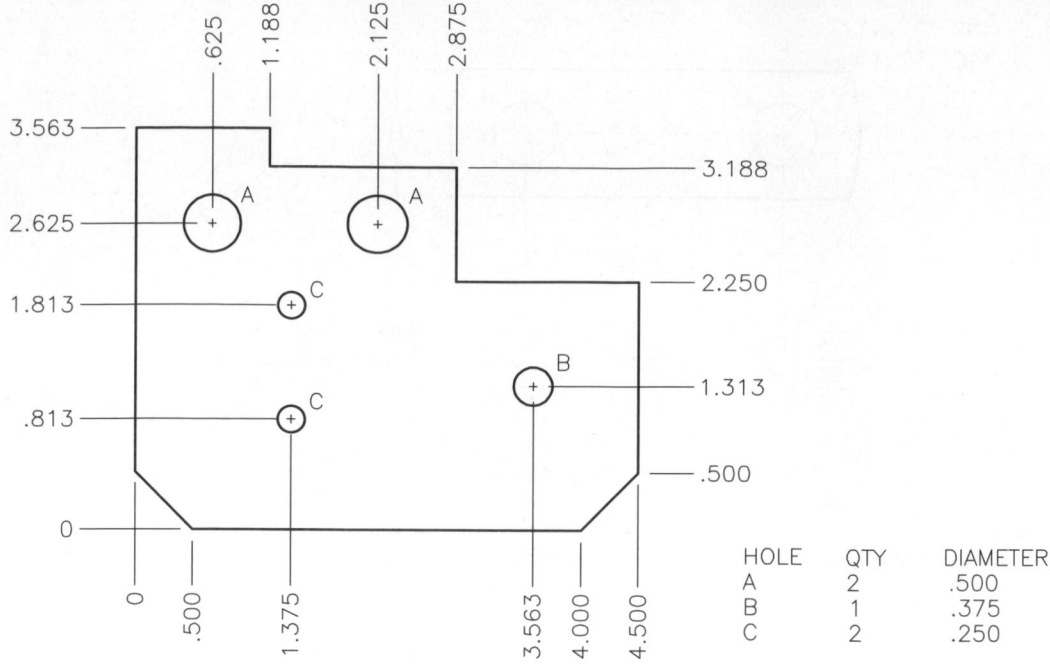

HOLE	QTY	DIAMETER
A	2	.500
B	1	.375
C	2	.250

Title: Chassis
Material: Aluminum .100 THK

11. Draw the view and add dimensions as shown. Save the drawing as P19-11.

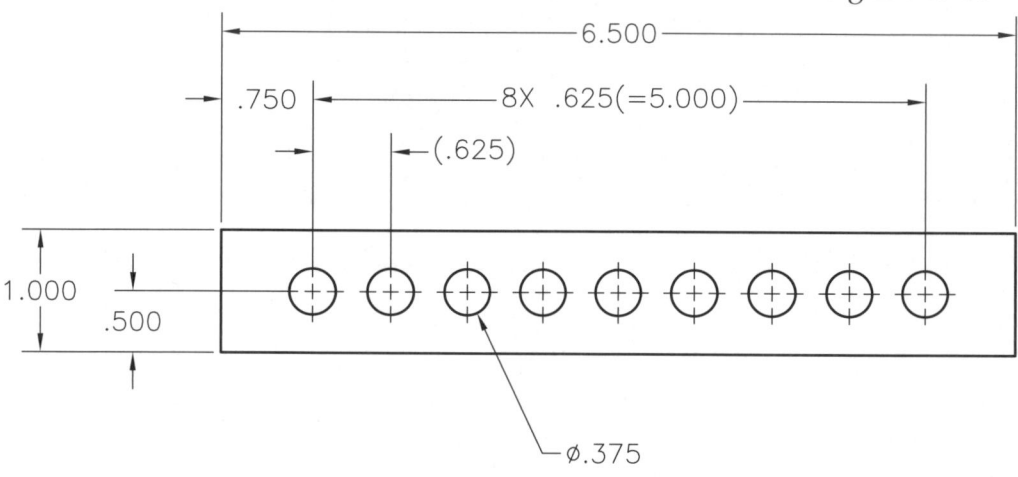

▼ Intermediate

12. Convert the given drawing to a drawing with the holes located using the **DIMORDINATE** tool based on the X and Y coordinates given in the table. Place a table above your title block with columns for Hole (identification), Quantity, Description, and Depth (Z axis). Save the drawing as P19-12.

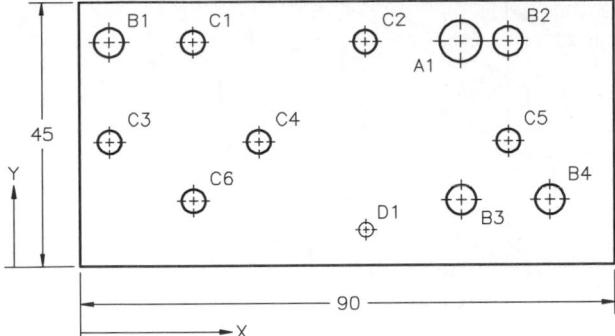 

HOLE	QTY	DESC	X	Y	Z
A1	1	⌀7	64	38	18
B1	1	⌀5	5	38	THRU
B2	1	⌀5	72	38	THRU
B3	1	⌀5	64	11	THRU
B4	1	⌀5	79	11	THRU
C1	1	⌀4	19	38	THRU
C2	1	⌀4	48	38	THRU
C3	1	⌀4	5	21	THRU
C4	1	⌀4	30	21	THRU
C5	1	⌀4	72	21	THRU
C6	1	⌀4	19	11	THRU
D1	1	⌀2.5	48	6	THRU

Title: Base
Material: Bronze

For Problems 13-15, use the isometric drawing provided to create a multiview orthographic drawing for the part. Include only the views necessary to fully describe the object. Add all required dimensions according to the ASME standards.

13. Save the drawing as P19-13.

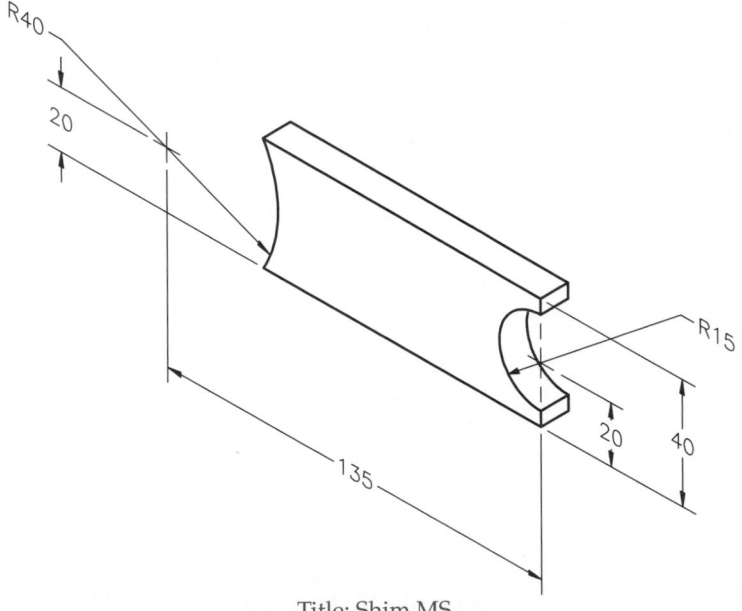 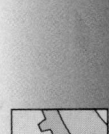

Title: Shim MS
Metric 10 THK

Chapter 19   Dimensioning Features and Alternate Practices

14. Half of the object is removed for clarity. The entire object should be drawn. Save the drawing as P19-14.

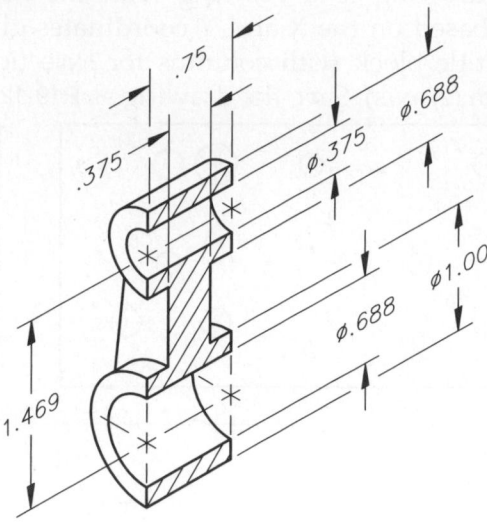

.75

.375

Ø.375 Ø.688

Ø.688

Ø1.00

1.469

FILLETS R.125

Title: Shaft Support
Material: Cast Iron (CI)

15. Half of the object is removed for clarity. The entire object should be drawn. Save the drawing as P19-15.

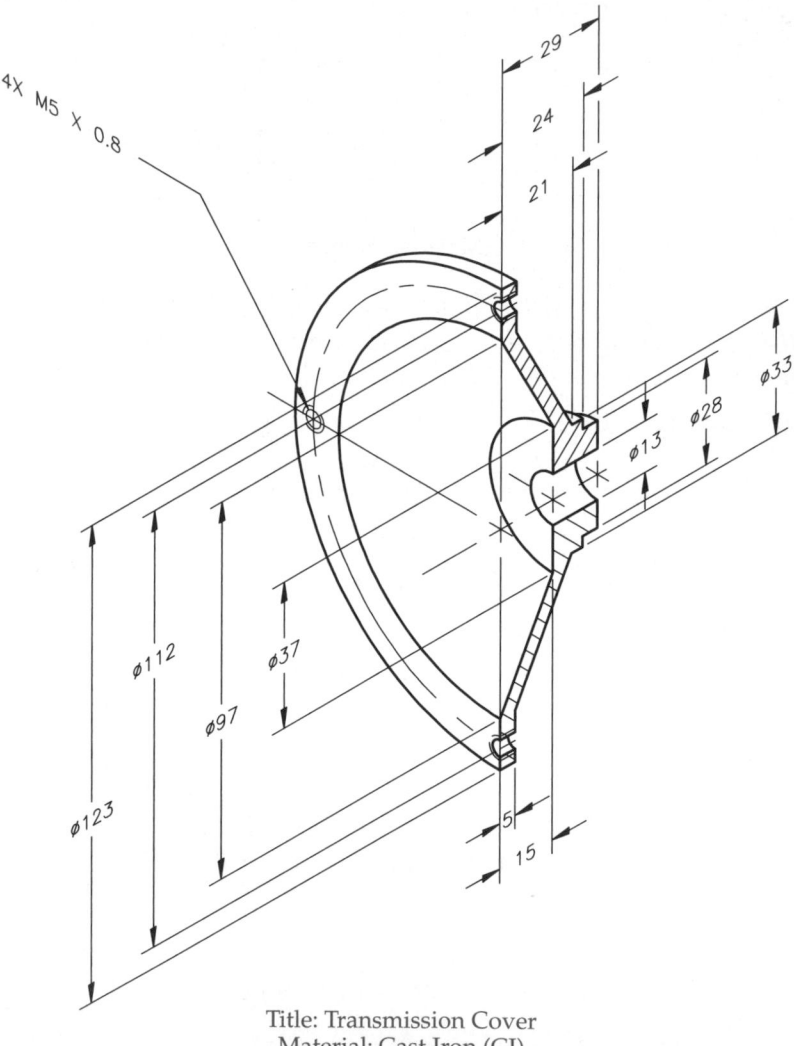

4X M5 X 0.8

29

24

21

Ø112

Ø97

Ø37

Ø123

Ø13

Ø28

Ø33

5

15

Title: Transmission Cover
Material: Cast Iron (CI)
Metric

16. Create the drawing as shown and save it as P19-16.

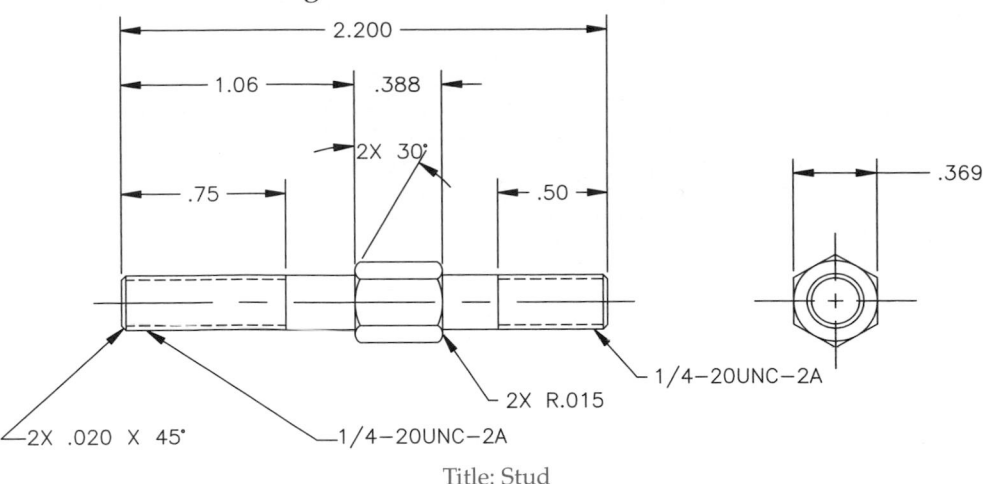

Title: Stud
Material: Stainless Steel

17. Create the drawing as shown and save it as P19-17.

HOLE LAYOUT			
KEY	SIZE	DEPTH	NO. REQD
A	⌀.250	THRU	6
B	⌀.125	THRU	4
C	⌀.375	THRU	4
D	R.125	THRU	2

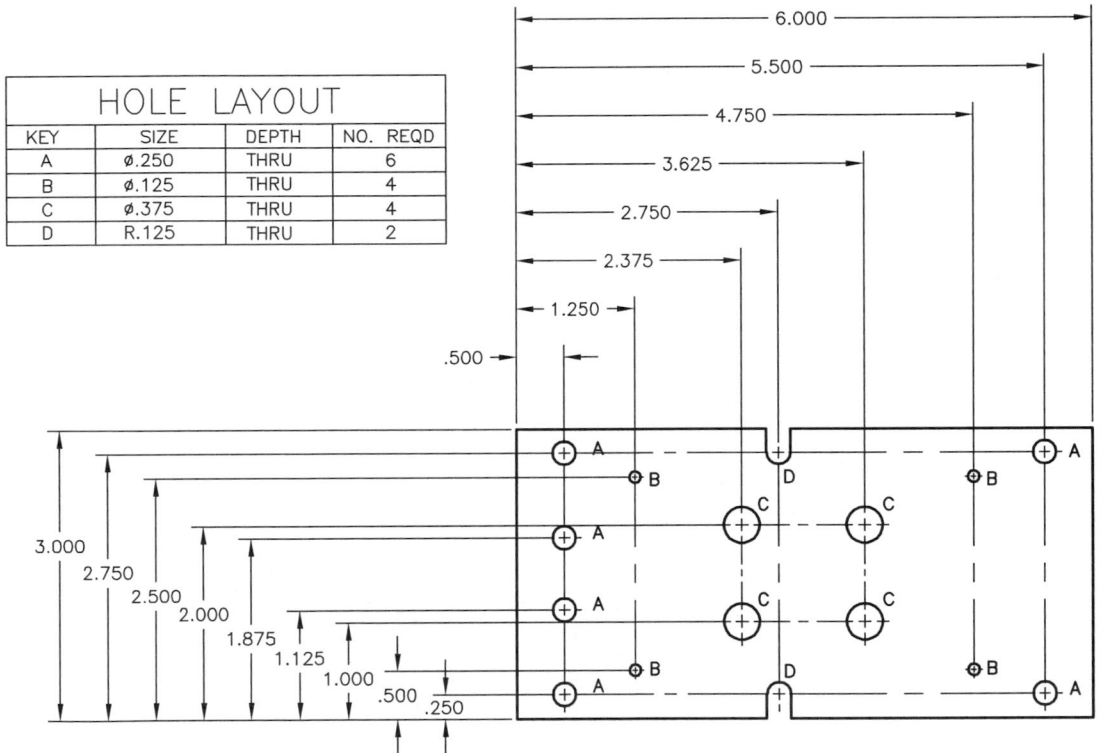

Title: Chassis Base (datum dimensioning)
Material: 12 gage Aluminum

18. Create the drawing as shown and save it as P19-18.

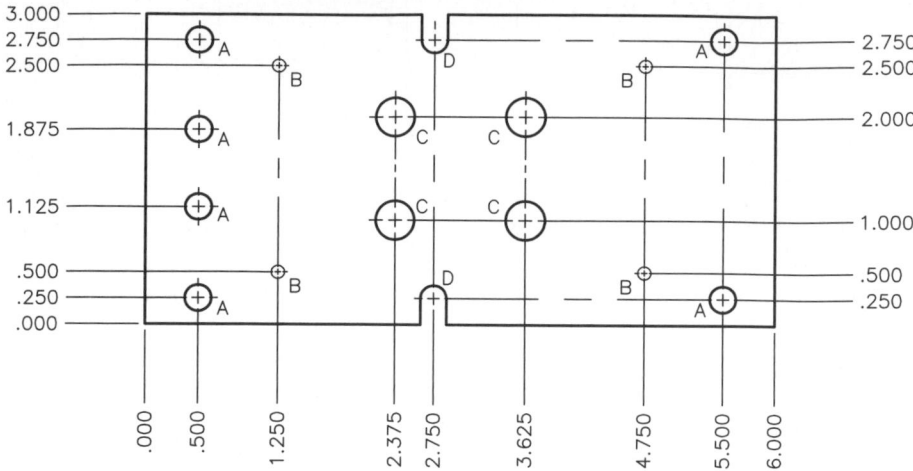

HOLE LAYOUT			
KEY	SIZE	DEPTH	NO. REQD
A	⌀.250	THRU	6
B	⌀.125	THRU	4
C	⌀.375	THRU	4
D	R.125	THRU	2

Title: Chassis Base (arrowless dimensioning)
Material: 12 gage Aluminum

19. Create the drawing as shown and save it as P19-19.

HOLE LAYOUT				
KEY	X	Y	SIZE	TOL
A1	.500	2.750	⌀.250	±.002
A2	.500	1.875	⌀.250	±.002
A3	.500	1.125	⌀.250	±.002
A4	.500	.250	⌀.250	±.002
A5	5.500	2.750	⌀.250	±.002
A6	5.500	.250	⌀.250	±.002
B1	1.250	2.500	⌀.125	±.001
B2	1.250	.500	⌀.125	±.001
B3	4.750	2.500	⌀.125	±.001
B4	4.750	.500	⌀.125	±.001
C1	2.375	2.000	⌀.375	±.005
C2	2.375	1.000	⌀.375	±.005
C3	3.625	2.000	⌀.375	±.005
C4	3.625	1.000	⌀.375	±.005
D1	2.750	2.750	R.125	±.002
D2	2.750	.250	R.125	±.002

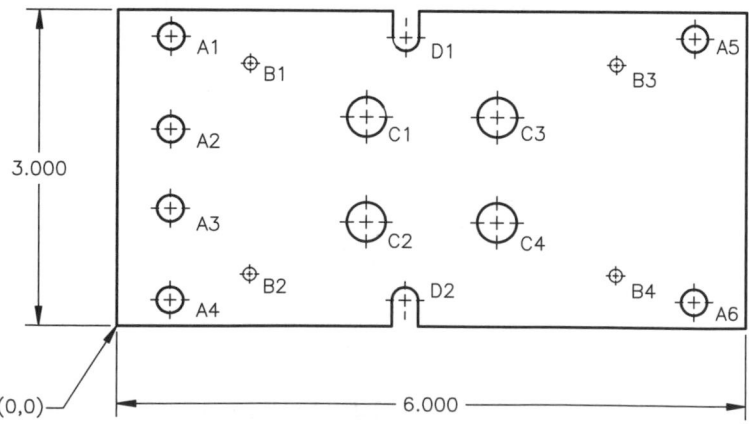

Title: Chassis Base (arrowless tabular dimensioning)
Material: 12 gage Aluminum

▼ Advanced

20. Create the drawing as shown and save it as P19-20.

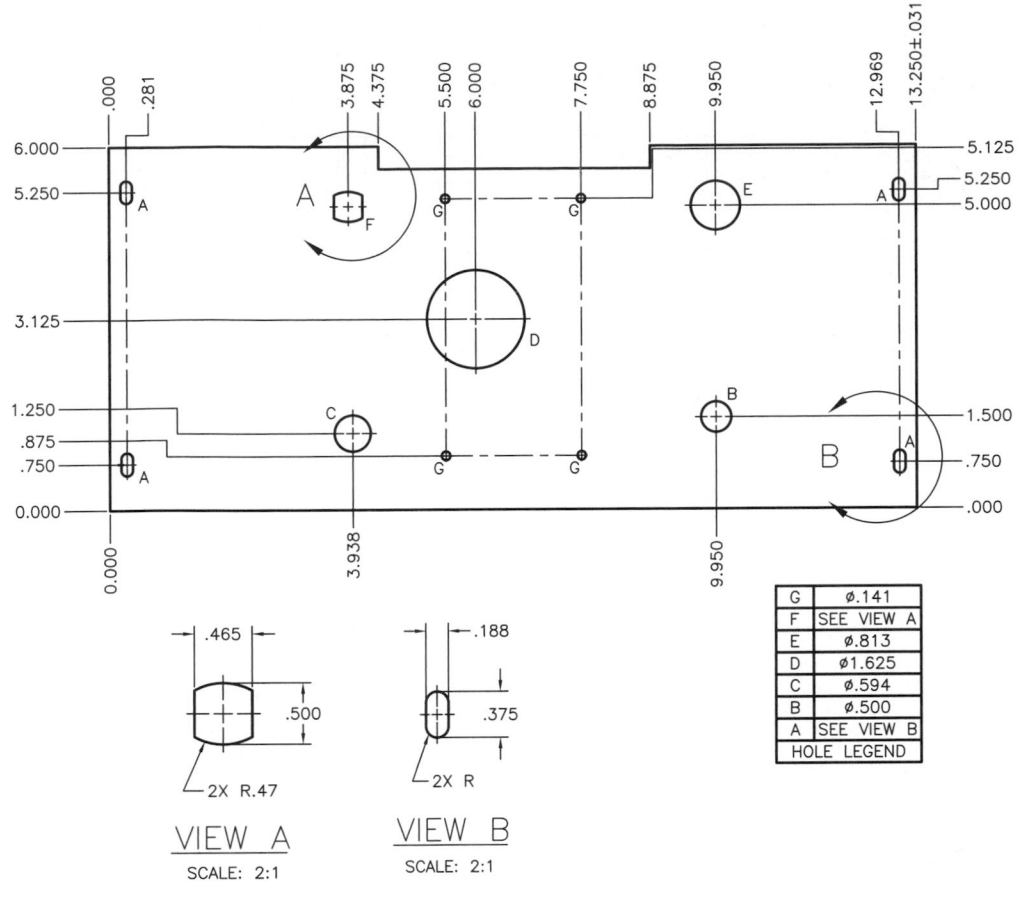

	HOLE LEGEND
G	⌀.141
F	SEE VIEW A
E	⌀.813
D	⌀1.625
C	⌀.594
B	⌀.500
A	SEE VIEW B

VIEW A
SCALE: 2:1

VIEW B
SCALE: 2:1

21. Create the drawing as shown and save it as P19-21.

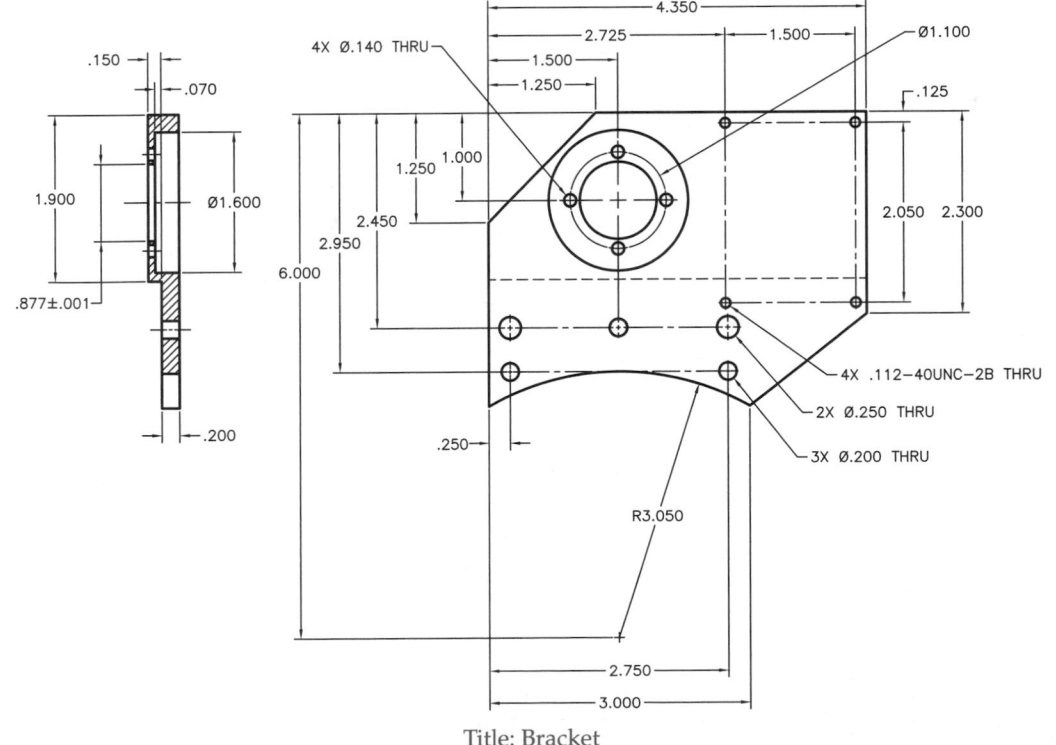

Title: Bracket
Material: SAE 1040

22. Create the drawing as shown and save it as P19-22.

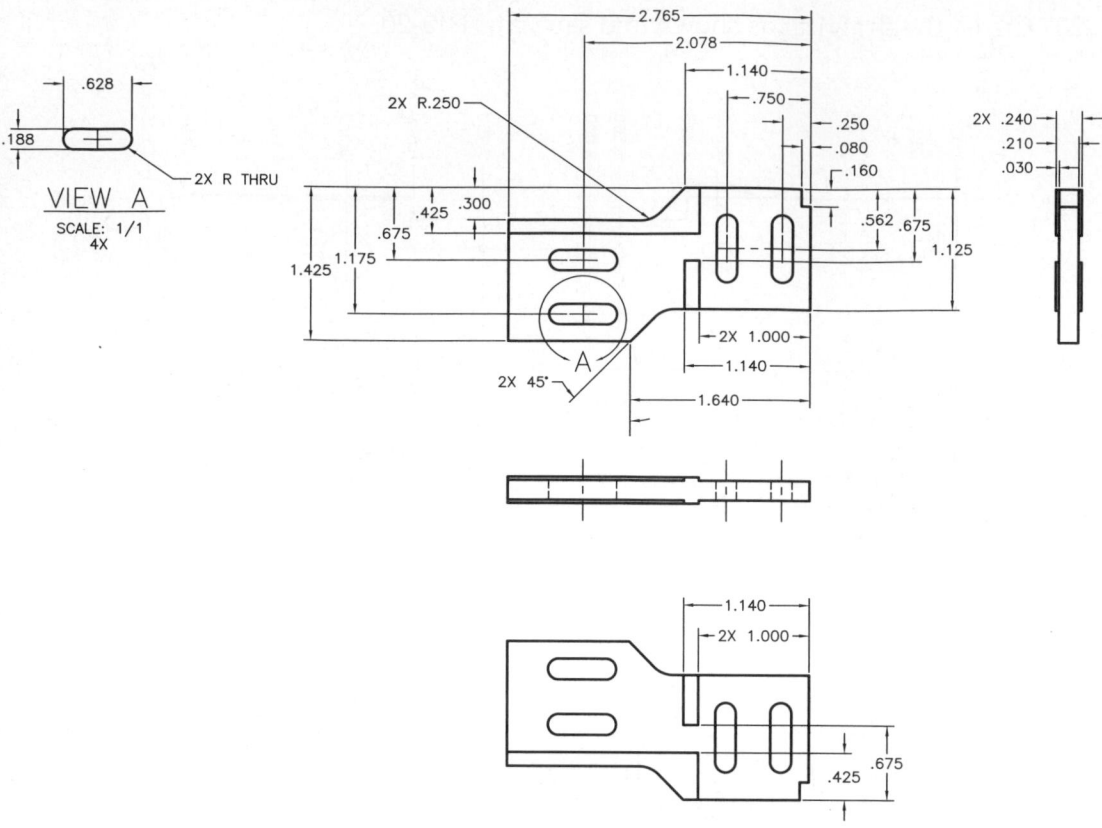

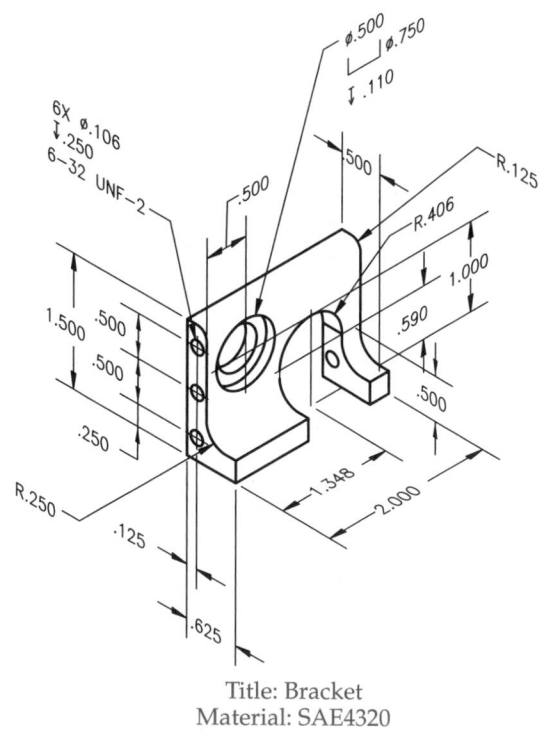

For Problems 23 and 24, use the isometric drawing provided to create a multiview orthographic drawing for the part. Include only the views necessary to fully describe the object. Add all required dimensions according to the ASME standards.

23. Create the drawing as shown and save it as P19-23.

Title: Bracket
Material: SAE4320

24. Carefully evaluate the given problem before beginning. Many given dimensions are provided to the inside surfaces of the bracket. This application is incorrect. Calculate the dimensions as needed to place datum dimensioning from the surfaces labeled A and B. Do not place the A and B on your final drawing. Save the drawing as P19-24.

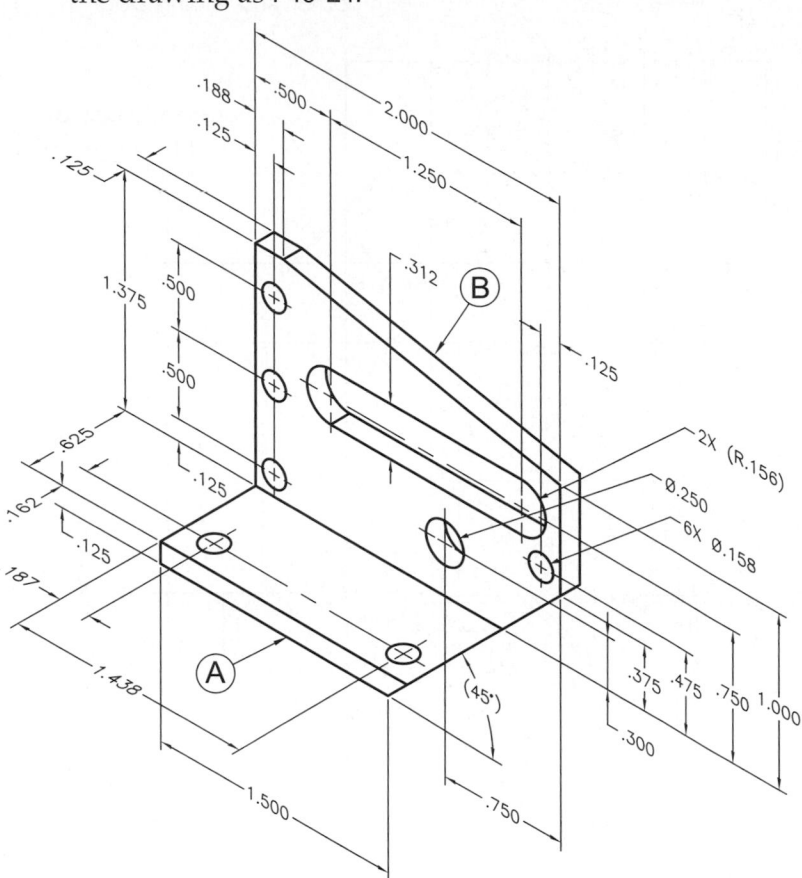

25. Open P18-17 and save the file as P19-25. The P19-25 file should be active. Add the client-requested redlines to the floor plan as shown.

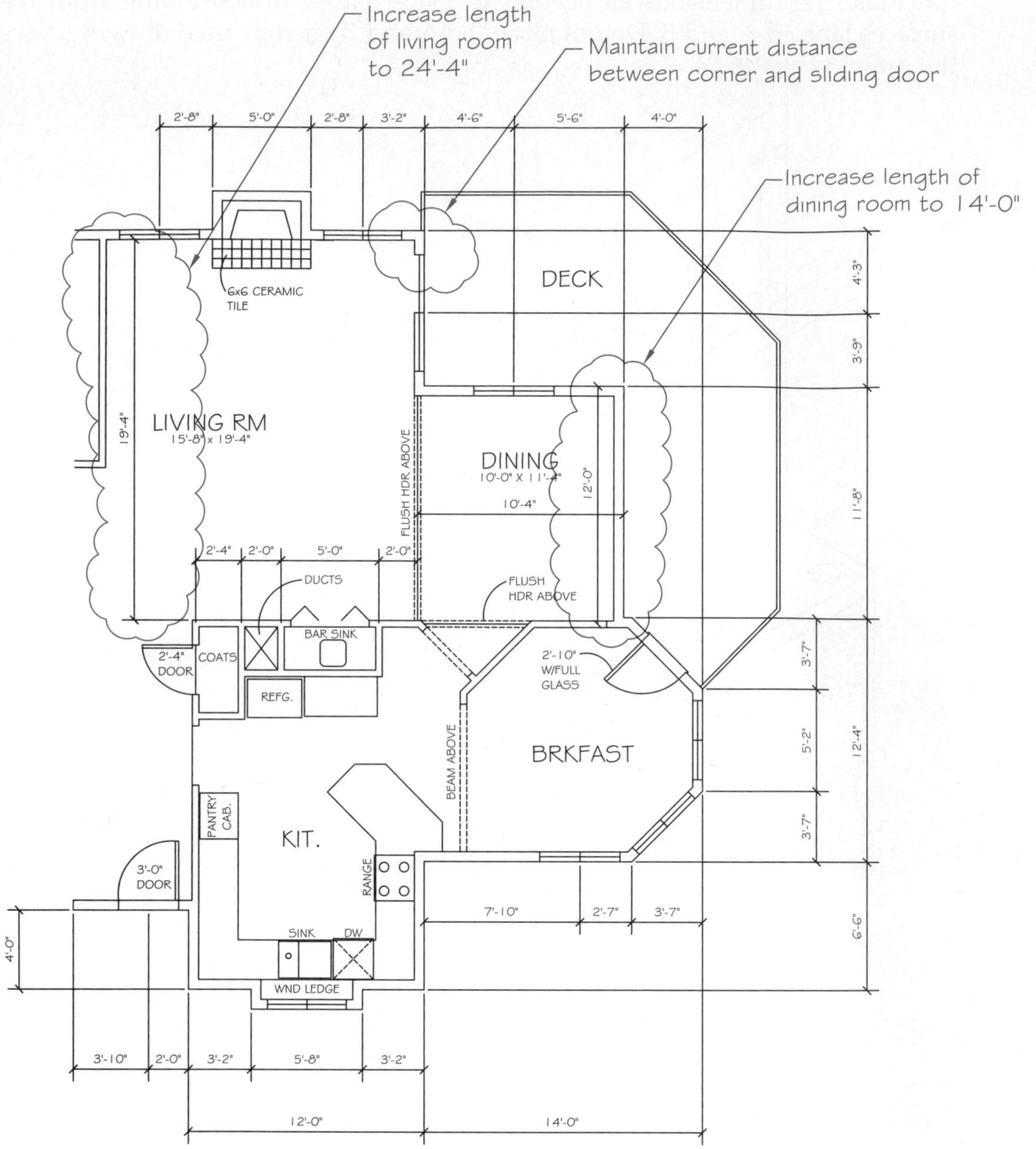

Increase length of living room to 24'-4"

Maintain current distance between corner and sliding door

Increase length of dining room to 14'-0"

2'-8" 5'-0" 2'-8" 3'-2" 4'-6" 5'-6" 4'-0"

DECK

4'-3"

3'-9"

6x6 CERAMIC TILE

19'-4"

LIVING RM
15'-8" x 19'-4"

FLUSH HDR ABOVE

DINING
10'-0" X 11'-4"
10'-4"

12'-0"

11'-8"

2'-4" 2'-0" 5'-0" 2'-0"

DUCTS

FLUSH HDR ABOVE

BAR SINK

2'-4"
DOOR COATS

3'-7"

REFG.

2'-10"
W/FULL
GLASS

PANTRY
CAB.

KIT.

BEAM ABOVE

BRKFAST

5'-2"

12'-4"

3'-0"
DOOR

RANGE

3'-7"

4'-0"

SINK DW

7'-10" 2'-7" 3'-7"

6'-6"

WND LEDGE

3'-10" 2'-0" 3'-2" 5'-8" 3'-2"

12'-0" 14'-0"

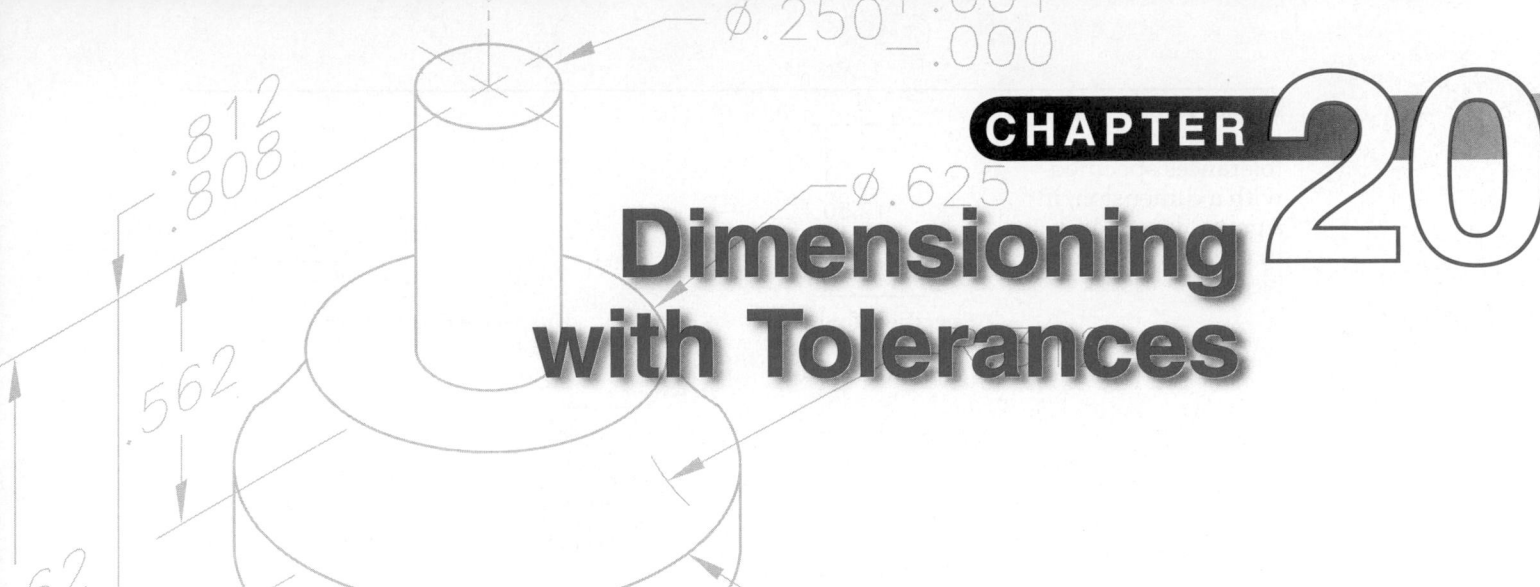

Dimensioning with Tolerances

Learning Objectives

After completing this chapter, you will be able to do the following:

✓ Define and use dimensioning and tolerancing terminology.
✓ Set the precision for dimensions and tolerances.
✓ Set up the primary units for use with inch or metric dimensions.
✓ Create and use dimension styles with various tolerance settings.
✓ Explain the purpose of geometric dimensioning and tolerancing (GD&T).

This chapter introduces tolerancing and explains how to prepare dimensions with tolerances for mechanical manufacturing drawings. This chapter also introduces drawing geometric dimensioning and tolerancing (GD&T) symbols and offers information on how you can learn more about GD&T.

Tolerancing Fundamentals

Methods of specifying *tolerance* include direct placement with a dimension, a general note, and specifications in the drawing title block. See **Figure 20-1**. The dimension stated as 12.50±.25 in **Figure 20-2A** is in a style known as *plus-minus dimensioning*. The tolerance of this dimension is the difference between the maximum and minimum *limits*. This tolerance style applies when the variance is the same in the positive and negative directions. In this case, the upper limit is 12.75 (12.50 +.25 = 12.75), and the lower limit is 12.25 (12.50 − .25 = 12.25). To find the tolerance, subtract the lower limit from the upper limit. The tolerance in this example is .50 (12.75 − 12.25 = .50). The *specified dimension* of the feature shown in **Figure 20-2** is 12.50.

Limits dimensioning, shown in **Figure 20-2B,** is another common method of showing and calculating tolerance. Many schools and companies prefer this method because it does not require calculating limits. Additional tolerance methods are also common, depending on the design requirement. **Figure 20-3** shows examples of equal and unequal *bilateral tolerances*. **Figure 20-4** shows an example of a *unilateral tolerance*.

tolerance: The total amount by which a specific dimension is permitted to vary.

plus-minus dimensioning: A tolerance style in which the positive and negative variance is equal and is preceded by a ± symbol.

limits: The largest and smallest numerical values the feature can have.

specified dimension: The part of the dimension from which the limits are calculated.

limits dimensioning: Method in which the upper and lower limits are given, instead of the specified dimension and tolerance.

bilateral tolerance: A tolerance style that permits variance in both the positive and negative directions from the specified dimension.

unilateral tolerance: A tolerance style that permits a variation in only one direction from the specified dimension.

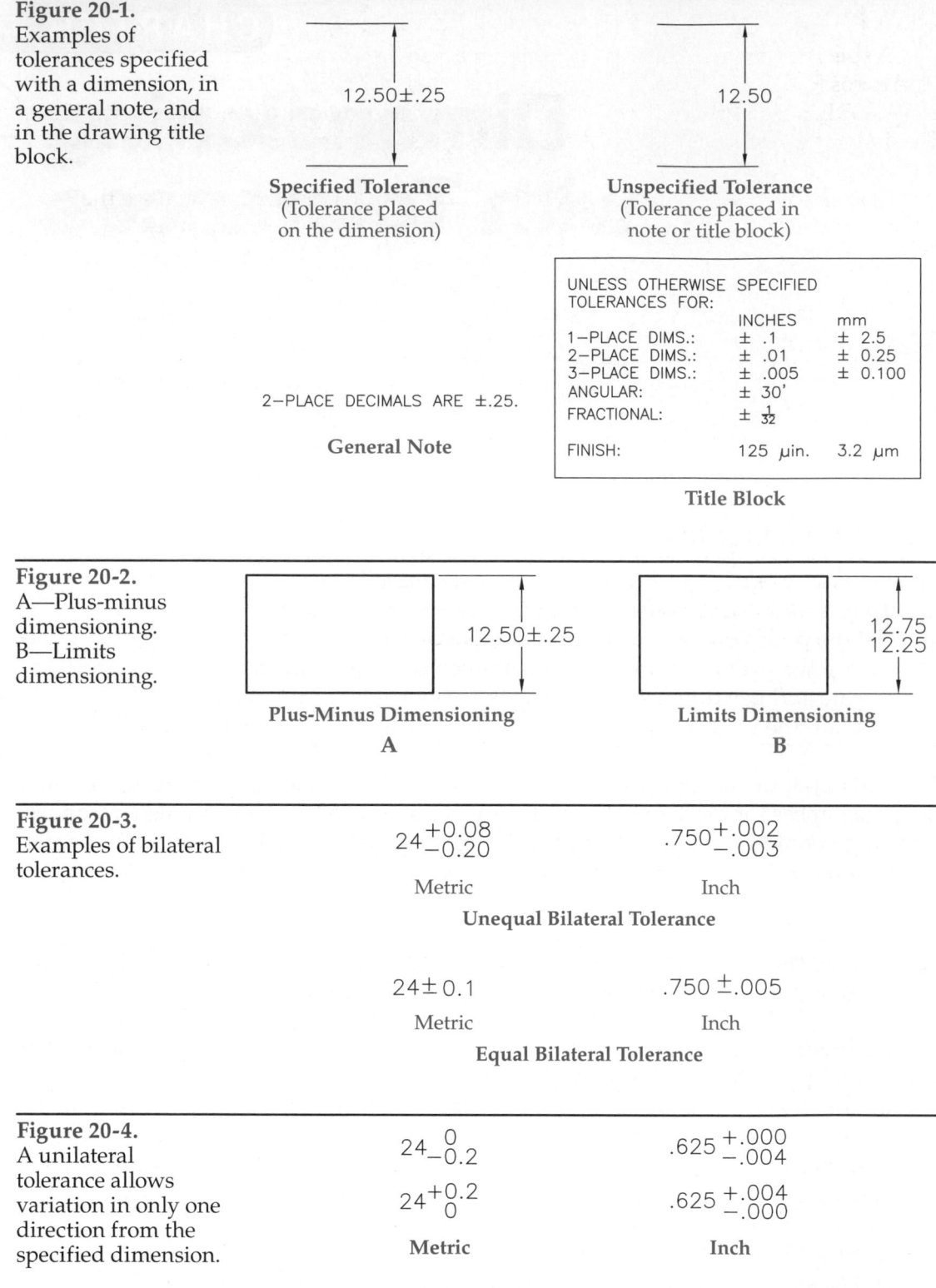

Figure 20-1.
Examples of tolerances specified with a dimension, in a general note, and in the drawing title block.

12.50±.25

Specified Tolerance
(Tolerance placed on the dimension)

12.50

Unspecified Tolerance
(Tolerance placed in note or title block)

2—PLACE DECIMALS ARE ±.25.

General Note

UNLESS OTHERWISE SPECIFIED TOLERANCES FOR:		
	INCHES	mm
1—PLACE DIMS.:	± .1	± 2.5
2—PLACE DIMS.:	± .01	± 0.25
3—PLACE DIMS.:	± .005	± 0.100
ANGULAR:	± 30'	
FRACTIONAL:	± $\frac{1}{32}$	
FINISH:	125 μin.	3.2 μm

Title Block

Figure 20-2.
A—Plus-minus dimensioning.
B—Limits dimensioning.

12.50±.25

Plus-Minus Dimensioning
A

12.75
12.25

Limits Dimensioning
B

Figure 20-3.
Examples of bilateral tolerances.

$24^{+0.08}_{-0.20}$

Metric

$.750^{+.002}_{-.003}$

Inch

Unequal Bilateral Tolerance

$24±0.1$

Metric

$.750±.005$

Inch

Equal Bilateral Tolerance

Figure 20-4.
A unilateral tolerance allows variation in only one direction from the specified dimension.

$24^{0}_{-0.2}$

$24^{+0.2}_{0}$

Metric

$.625^{+.000}_{-.004}$

$.625^{+.004}_{-.000}$

Inch

Assigning Decimal Places

The ASME Y14.5M *Dimensioning and Tolerancing* standard has separate recommendations for the display of decimal places in inch and metric dimensions. **Figure 20-3** and **Figure 20-4** show examples of decimal dimension values in inches and metric units.

Inch Dimensioning

A specified inch dimension uses the same number of decimal places as its tolerance. Add zeros to the right of the decimal point if needed. For example, the inch dimension .250±.005 has an additional zero added to the .25 to match the three-decimal tolerance. Similarly, the dimensions 2.000±.005 and 2.500±.005 have zeros added to match the tolerance.

Both of the values in a plus-minus tolerance for an inch dimension have the same number of decimal places. Add zeros to fill in where needed. For example:

$$\begin{matrix} +.005 \\ -.010 \end{matrix} \quad not \quad \begin{matrix} +.005 \\ -.01 \end{matrix}$$

Metric Dimensioning

Omit the decimal point and zeros from the specified dimension of a metric whole number. For example, the metric dimension 12 has no decimal point followed by a zero. This rule is true unless the drawing displays tolerance values. When a metric dimension includes a decimal portion, a zero does not follow the last digit to the right of the decimal point. For example, the metric dimension 12.5 has no zero to the right of the 5. This rule is true unless the drawing displays tolerance values.

Both values in a bilateral tolerance for a metric dimension have the same number of decimal places. Add zeros to fill in where needed. Typically, no additional zeros appear after the specified dimension to match the tolerance. For example, both 24±0.25 and 24.5±0.25 are correct. However, some companies prefer to add zeros after the specified dimension to match the tolerance, in which case 24.00±0.25 and 24.50±0.25 are both correct.

Setting Primary Units

The dimension style controls the appearance of dimensions, including dimension values and tolerance. The initial phase of dimensioning with tolerances involves setting the appropriate values for the primary units of the dimension style you plan to use. Use the **Primary Units** tab of the **New** (or **Modify**) **Dimension Style** dialog box, shown in **Figure 20-5**, to set the dimension units and precision.

The **Precision** drop-down list in the **Linear dimensions** area allows you to specify the number of zeros displayed after the decimal point of the specified dimension. The ASME standard recommends that the precision for the dimension and the tolerance be the same for inch dimensions, but it may be different for metric values, as previously described.

The **Zero suppression** settings control the display of zeros before and after the decimal point. For inch dimensions, the **Leading** options should be on, and the **Trailing** options should be off. For typical metric dimensions, without using sub-units, the **Leading** options should be off, and the **Trailing** options should be on.

Figure 20-5.
The **Primary Units** tab of the **New** (or **Modify**) **Dimension Style** dialog box sets the unit format and precision of linear dimensions.

Set the precision for specified dimensions

Set the zero suppression

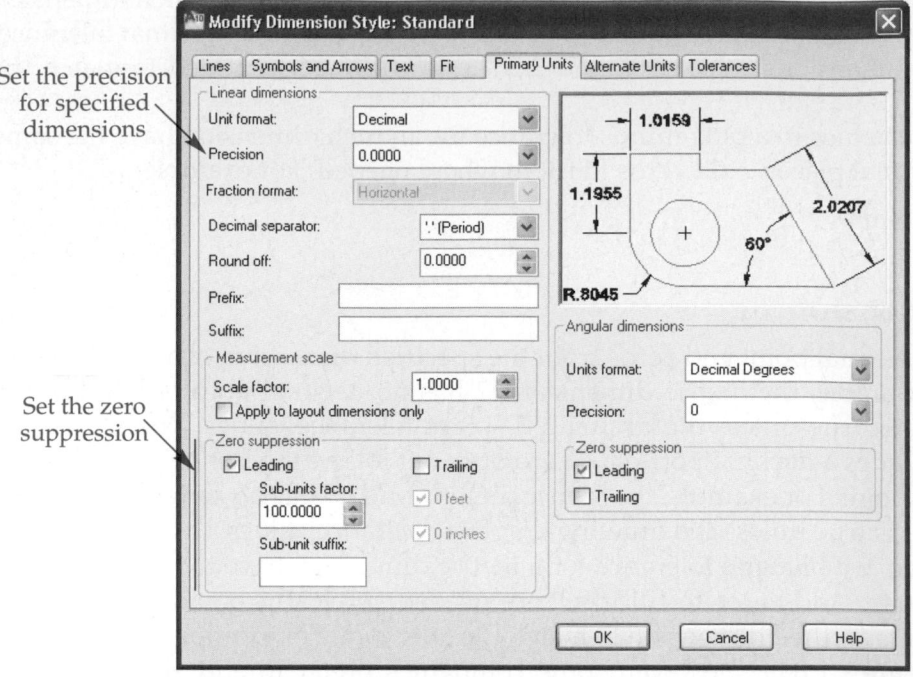

Setting Tolerance Methods

The **Tolerances** tab of the **New** (or **Modify**) **Dimension Style** dialog box, shown in **Figure 20-6**, allows you to apply a tolerance method to your drawing. The default option in the **Method:** drop-down list is None. This means dimensions do not include

Figure 20-6.
The **Tolerances** tab of the **New** (or **Modify**) **Dimension Style** dialog box contains formatting settings for tolerance dimensions.

Select a tolerance method

Set the precision for tolerance dimensions

Settings should match the **Zero suppression** linear dimension settings in the **Primary Units** tab

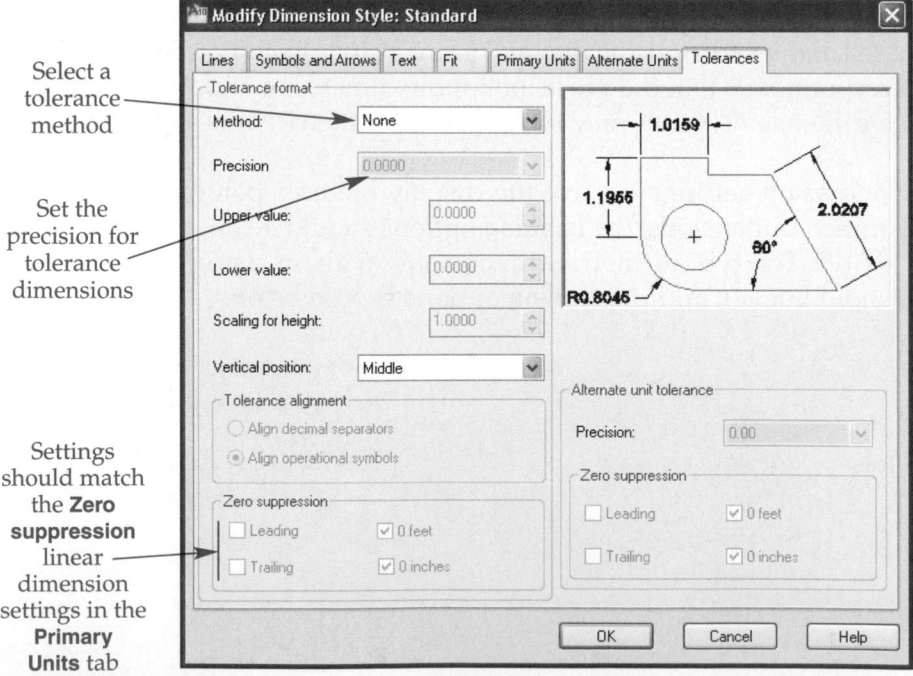

Figure 20-7.
Select a tolerance dimensioning method from the **Method:** drop-down list, in the **Tolerance format** area.

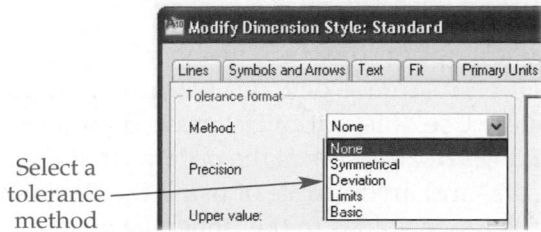

Select a tolerance method

a tolerance method, or specific tolerance. As a result, most of the options in the **Tolerances** tab are disabled. When you pick a tolerance method from the drop-down list, appropriate options enable, and the image in the tab reflects the selected method. **Figure 20-7** shows the drop-down list options.

> **NOTE**
>
> The following information describes tolerance methods and the settings unique to each. General settings, including tolerance precision, height, vertical position, alignment, and zero suppression, are explained later in this chapter.

Symmetrical Tolerance Method

Select the **Symmetrical** option from the **Method:** drop-down list to create a *symmetrical tolerance*. Use this option to draw dimensions that display an equal bilateral tolerance in the plus-minus format. **Figure 20-8** shows the options that enable symmetrical tolerances, the preview that appears, and an example of using the **Symmetrical** option. Enter a tolerance value in the **Upper value:** text box. Although it is disabled, you can see that the value in the **Lower value:** text box matches the value in the **Upper value:** text box.

symmetrical tolerance: AutoCAD's term for an equal bilateral tolerance.

Exercise 20-1

Access the Student Web site (www.g-wlearning.com/CAD) and complete Exercise 20-1.

Figure 20-8.
Setting the **Symmetrical** tolerance method option current, with an equal bilateral tolerance value of .005.

Specified tolerance method

Equal bilateral tolerance value

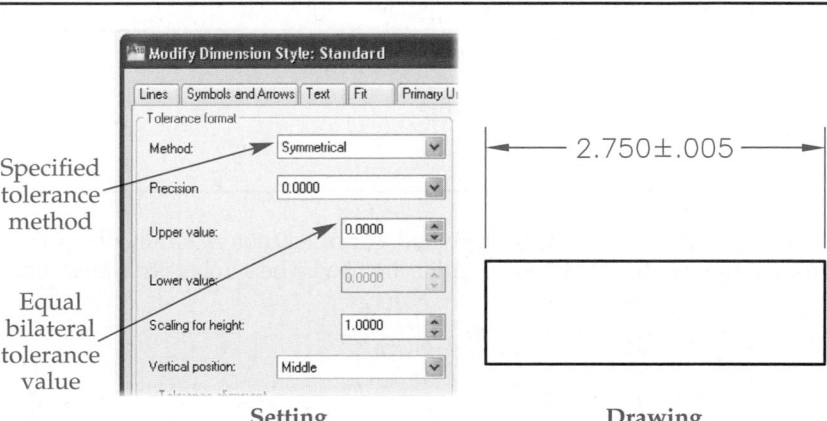

Setting   Drawing

Deviation Tolerance Method

**deviation
tolerance:**
AutoCAD's term for
an unequal bilateral
tolerance.

Pick the **Deviation** option from the **Method:** drop-down list to create a *deviation tolerance*. A deviation tolerance deviates (varies) from the specified dimension with two different values. Use this option to draw dimensions that display an unequal bilateral tolerance. **Figure 20-9** shows the options that enable deviation tolerances, the preview that appears, and an example of using the **Deviation** option. Enter the desired upper and lower tolerance values in the **Upper value:** and **Lower value:** text boxes.

You can also use the deviation option to draw a unilateral tolerance by entering 0 for either the **Upper value:** or **Lower value:** setting. If you are using inch units, AutoCAD includes the plus or minus sign before the zero tolerance. When you use metric units, AutoCAD omits the sign for the zero tolerance. See **Figure 20-10**.

Exercise 20-2

Access the Student Web site (www.g-wlearning.com/CAD) and complete Exercise 20-2.

Limits Tolerance Method

Select the **Limits** option from the **Method:** drop-down list to apply the limits tolerance method. In limits dimensioning, the tolerance limits are given, and no calculations from the specified dimension are required (unlike plus-minus dimensioning). **Figure 20-11** shows the options that enable the limits method, the preview that appears, and an example of using the **Limits** option. Use the **Upper value:** and **Lower value:** text

---

Figure 20-9.
Setting the **Deviation** tolerance method option current, with unequal bilateral tolerance values.

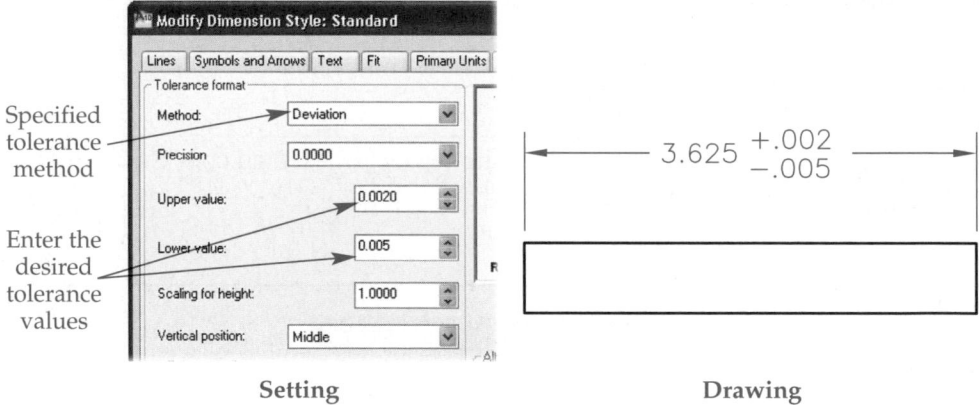

Setting   Drawing

Figure 20-10.
When a unilateral tolerance is specified, AutoCAD automatically places the plus or minus symbol in front of the zero tolerance, if inch units are used. The symbol is omitted with metric units.

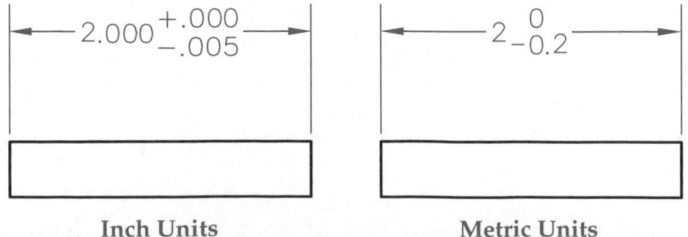

Inch Units   Metric Units

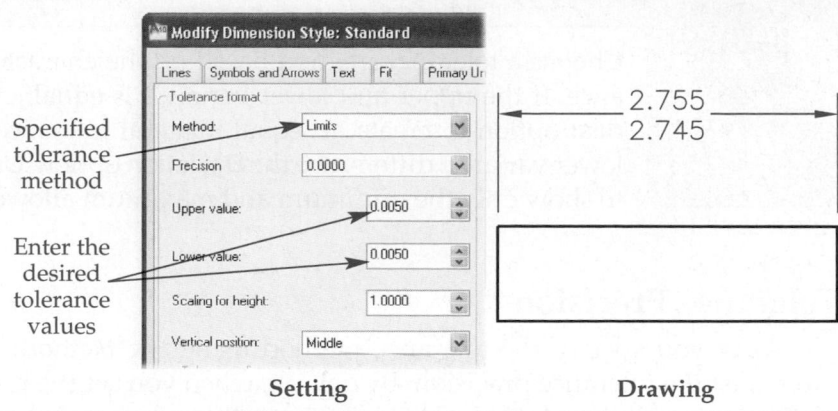

Figure 20-11.
Selecting the limits tolerance method and setting limit values.

Specified tolerance method

Enter the desired tolerance values

Setting

Drawing

boxes to enter the upper and lower tolerance values to add and subtract from the specified dimension. The upper and lower values can be equal or different.

Exercise 20-3

Access the Student Web site (www.g-wlearning.com/CAD) and complete Exercise 20-3.

Basic Tolerance Method

Pick the **Basic** option from the **Method:** drop-down list to draw *basic dimensions*. **Figure 20-12** shows the options that enable the basic method, the preview that appears, and an example of using the **Basic** option. Few options are enabled because a basic dimension has no tolerance. A rectangle placed around the dimension number distinguishes a basic dimension from other dimensions.

basic dimension: A theoretically perfect dimension used in geometric dimensioning and tolerancing.

> **NOTE**
>
> Picking the **Draw frame around text** check box in the **Text** tab of the **New** (or **Modify**) **Dimension Style** dialog box also activates the basic tolerance method.

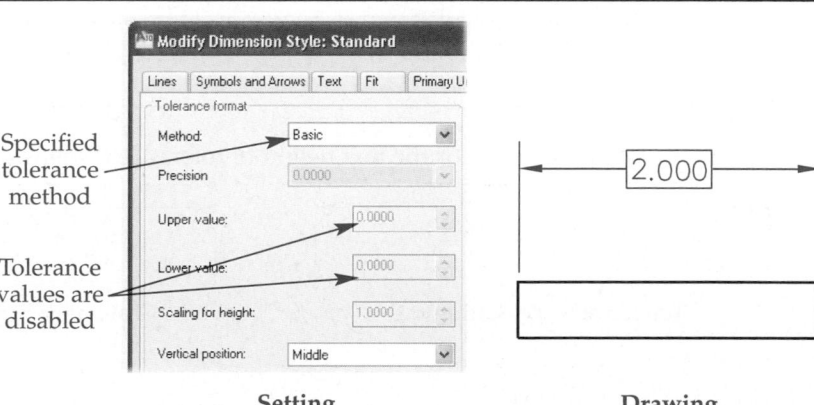

Figure 20-12.
Use the basic tolerance method for basic dimensioning. The dimension text for a basic dimension appears inside a rectangle.

Specified tolerance method

Tolerance values are disabled

Setting

Drawing

Tolerance Precision

After you specify the tolerance method using the **Method:** drop-down list, you can adjust the tolerance precision. By default, when you set the primary unit precision on the **Primary Units** tab, AutoCAD automatically makes the tolerance precision in the **Tolerances** tab the same unit precision. If the setting does not reflect the correct level of precision, change it using the **Precision** drop-down list in the **Tolerance format** area.

Tolerance Height

You can set the text height of the tolerance dimension in relation to the text height of the specified dimension using the **Scaling for height:** text box in the **Tolerance format** area. The default of 1.0000 makes the tolerance dimension text the same height as the specified dimension text. This is recommended in the ASME standard.

To make the tolerance dimension height three-quarters as high as the specified dimension height, type .75 in the **Scaling for height:** text box. Some companies prefer this practice to keep the tolerance part of the dimension from taking up additional space. **Figure 20-13** shows examples of tolerance dimensions with different text heights.

Vertical Position

Use the options in the **Vertical position:** drop-down list in the **Tolerance format** area to control the alignment, or justification, of deviation tolerance dimensions. The **Middle** option, which is the default, centers the tolerance with the specified dimension. This is the recommended ASME practice. The other justification options are **Top** and **Bottom**. **Figure 20-14** displays deviation tolerance dimensions with each of the justification options.

Tolerance Alignment

The options in the **Tolerance alignment** area become available for selection when you use a deviation or limits tolerance method. The setting controls the left and right tolerance justification. When using a deviation tolerance method, pick the **Align decimal**

Figure 20-13.
Using different scale settings for the text height of tolerance dimensions.

3.250 $^{+.005}_{-.002}$

Tolerance Scale Setting = 1

3.250 $^{+.005}_{-.002}$

Tolerance Scale Setting = .75

3.255
3.248

Tolerance Scale Setting = 1

3.255
3.248

Tolerance Scale Setting = .75

Figure 20-14.
Examples of the tolerance justification options for deviation tolerance dimensions.

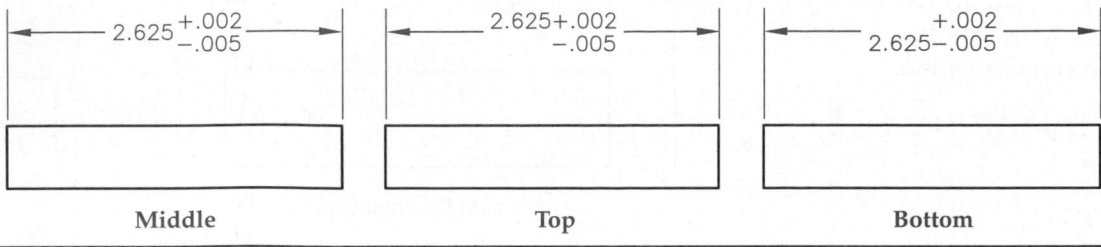

| Middle | Top | Bottom |

separators radio button to align the upper and lower tolerance value decimal points vertically. Select the **Align operational symbols** radio button to align the upper and lower tolerance plus and minus symbols vertically. See **Figure 20-15.** When using the limits tolerance method, pick the **Align decimal separators** radio button to align the upper and lower limit decimal points vertically. Select the **Align operational symbols** radio button to left-justify the upper and lower limits. See **Figure 20-16.**

Zero Suppression

You must select a tolerance method to enable the options in the **Zero suppression** area. The suppression settings for linear dimensions in the **Tolerances** tab should be the same as the **Zero suppression** tolerance format settings in the **Primary Units** tab. AutoCAD does not automatically match the tolerance setting to the primary units setting.

Select the **Leading** check box in the **Zero suppression** area of the **Tolerances** tab when you are drawing inch tolerance dimensions. Activate the same option for linear dimensions in the **Primary Units** tab. You can then draw inch tolerance dimensions without placing the zero before the decimal point, as recommended by ASME standards. These settings allow you to draw a tolerance dimension such as .625±.005.

Deselect the **Leading** check box in the **Zero suppression** area of the **Tolerances** tab when drawing metric tolerance dimensions. Deactivate the same option for linear dimensions in the **Primary Units** tab. This allows you to place a metric tolerance dimension with the zero before the decimal point, such as 12±0.2, as recommended by ASME standards.

Figure 20-15.
Changing tolerance alignment for use with a deviation tolerance method.

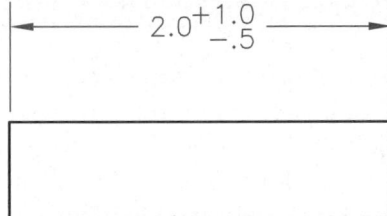

Aligned on Decimal Separator

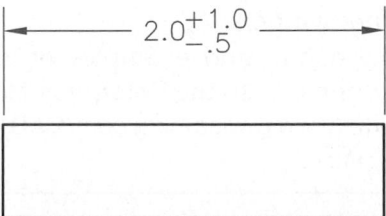

Aligned on Operational Symbols

Figure 20-16.
Changing tolerance alignment for use with a limits tolerance method.

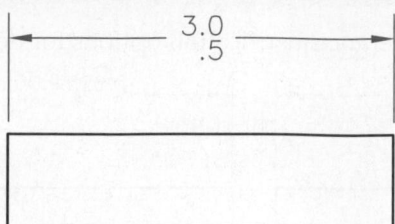

Aligned on Decimal Separator

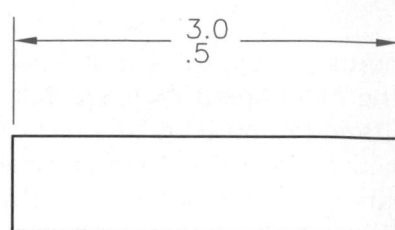

Aligned on Operational Symbols

NOTE

The options in the **Alternate unit tolerance** area become enabled when you pick the **Display alternate units** check box in the **Alternate Units** tab of the **New** (or **Modify**) **Dimension Style** dialog box. Use the **Alternate unit tolerance** area to set specific tolerances for alternate units.

Exercise 20-4

Access the Student Web site (www.g-wlearning.com/CAD) and complete Exercise 20-4.

Introduction to GD&T Symbols

geometric dimensioning and tolerancing (GD&T): The dimensioning and tolerancing of individual features of a part where the permissible variations relate to characteristics of form, profile, orientation, runout, or the relationship between features.

Geometric dimensioning and tolerancing (GD&T) is the dimensioning and tolerancing of individual features of a part where the permissible variations relate to characteristics of form, profile, orientation, runout, or the relationship between features. For complete coverage of GD&T, refer to *Geometric Dimensioning and Tolerancing* by David A. Madsen, published by Goodheart-Willcox Company, Inc.

Reference Material

Drafting Symbols
For the names and examples of GD&T symbols and symbol applications, go to the **Reference Material** section of the Student Web site (www.g-wlearning.com/CAD) and select **Drafting Symbols** in the list.

Supplemental Material   *GD&T with AutoCAD*

For information about creating GD&T symbols using AutoCAD, go to the Student Web site (www.g-wlearning.com/CAD), select this chapter, and select **GD&T with AutoCAD**.

Chapter Test

Answer the following questions. Write your answers on a separate sheet of paper or go to the Student Web site (www.g-wlearning.com/CAD) and complete the electronic chapter test.

1. Define the term *tolerance*.
2. What are the limits of the tolerance dimension 3.625±.005?
3. Give an example of an equal bilateral tolerance in inches and in metric units.
4. Give an example of an unequal bilateral tolerance in inches and in metric units.
5. Give an example of a unilateral tolerance in inches and in metric units.
6. What is the purpose of the **Symmetrical** tolerance method option?
7. What is the purpose of the **Deviation** tolerance method option?
8. What is the purpose of the **Limits** tolerance method option?
9. How do you set the number of zeros displayed after the decimal point for a tolerance dimension?
10. Explain the result of setting the **Scaling for height:** option to 1 in the **Tolerances** tab.
11. What setting would you use for the **Scaling for height:** option if you wanted the tolerance dimension height to be three-quarters of the specified dimension height?
12. Name the tolerance dimension justification option recommended by the ASME standards.
13. Which **Zero suppression** settings should you specify for linear and tolerance dimensions when you are using inch units?
14. Which **Zero suppression** settings should you specify for linear and tolerance dimensions when you are using metric units?
15. What is the purpose of geometric dimensioning and tolerancing?

Drawing Problems

- *Start AutoCAD if it is not already started.*

- *Start a new drawing using an appropriate template of your choice. The template should include layers, text styles, dimension styles, and multileader styles appropriate for drawing the given objects.*

- *Add layers, text styles, dimension styles, and multileader styles as needed. Draw all objects using appropriate layers, text styles, dimension styles, multileader styles, justification, and format.*

- *Follow the specific instructions for each problem. Use your own judgment and approximate dimensions when necessary.*

- *Apply dimensions accurately using ASME or appropriate industry standards. Use object snap modes to your best advantage.*

- *For mechanical drawings, place the following general notes 1/2" from the lower-left corner:*

      NOTES:
      1. INTERPRET DIMENSIONS AND TOLERANCES PER ASME Y14.5M-1994.
      2. REMOVE ALL BURRS AND SHARP EDGES.
      3. UNLESS OTHERWISE SPECIFIED, ALL DIMENSIONS ARE IN INCHES
         (or MILLIMETERS as applicable).

For Problems 1-7, use the isometric drawing provided to create a multiview orthographic drawing for the part. Include only the views necessary to fully describe the object. Add all required dimensions according to the ASME standards.

▼ Basic

1. Create the drawing as shown and save it as P20-1.

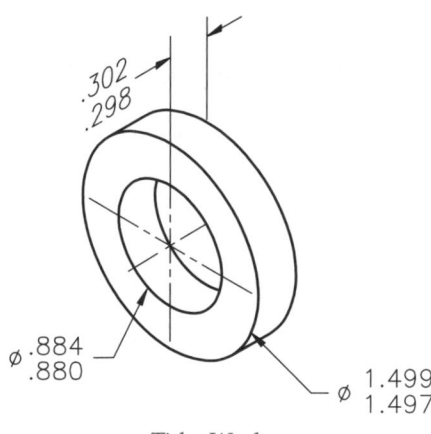

Title: Washer
Material: SAE 1020
Inch

2. Create the drawing as shown and save it as **P20-2**.

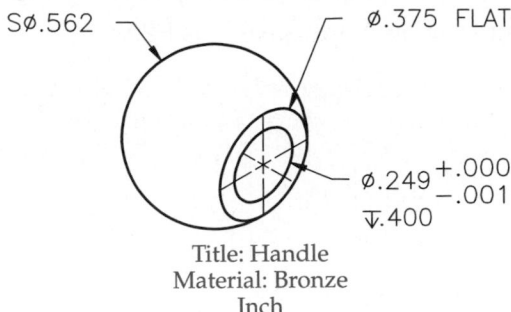

SØ.562

Ø.375 FLAT

Ø.249 $^{+.000}_{-.001}$

▽.400

Title: Handle
Material: Bronze
Inch

3. Create the drawing as shown and save it as **P20-3**.

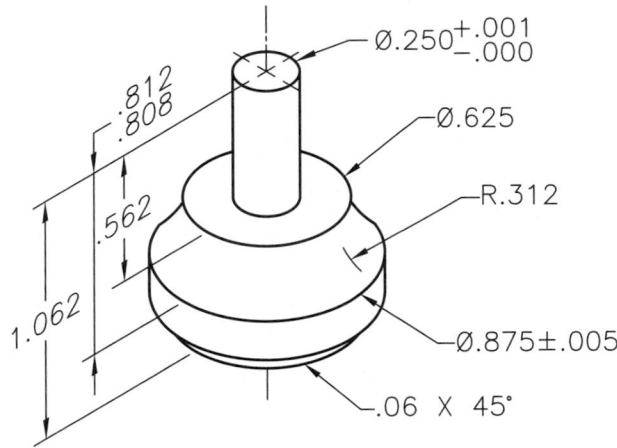

Ø.250 $^{+.001}_{-.000}$

.812
.808

Ø.625

.562

R.312

1.062

Ø.875±.005

.06 X 45°

ALL OTHER THREE PLACE DECIMALS ±.010

4. Create the drawing as shown and save it as **P20-4**.

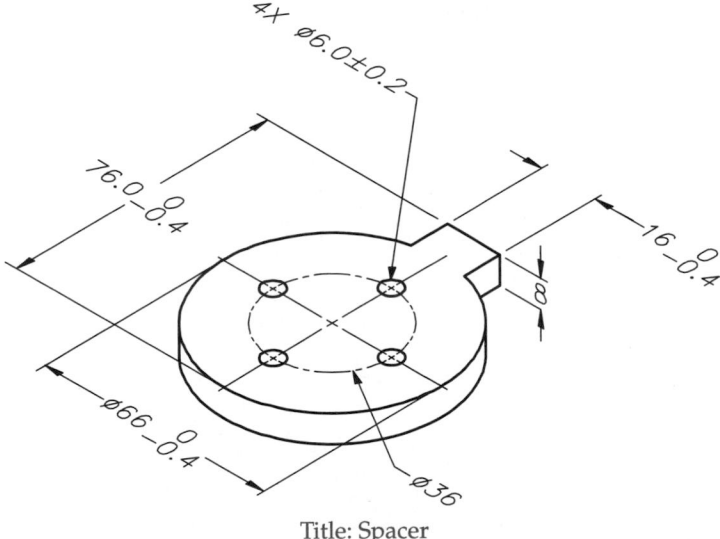

4X Ø6.0±0.2

76.0 $^{0}_{-0.4}$

16 $^{0}_{-0.4}$

8

Ø66 $^{0}_{-0.4}$

Ø36

Title: Spacer
Material: Cold Rolled Steel
Metric

▼ Intermediate

5. Create the drawing as shown and save it as P20-5.

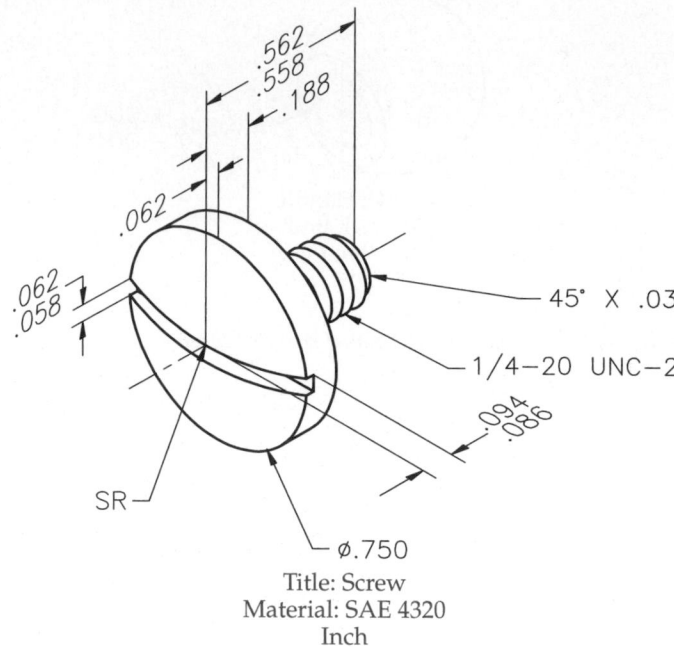

.562
.558
.188

.062

.062
.058

45° X .03

1/4—20 UNC—2

.094
.086

SR

⌀.750

Title: Screw
Material: SAE 4320
Inch

6. This object is shown as a section for clarity. Do not draw a section. Save the drawing as P20-6.

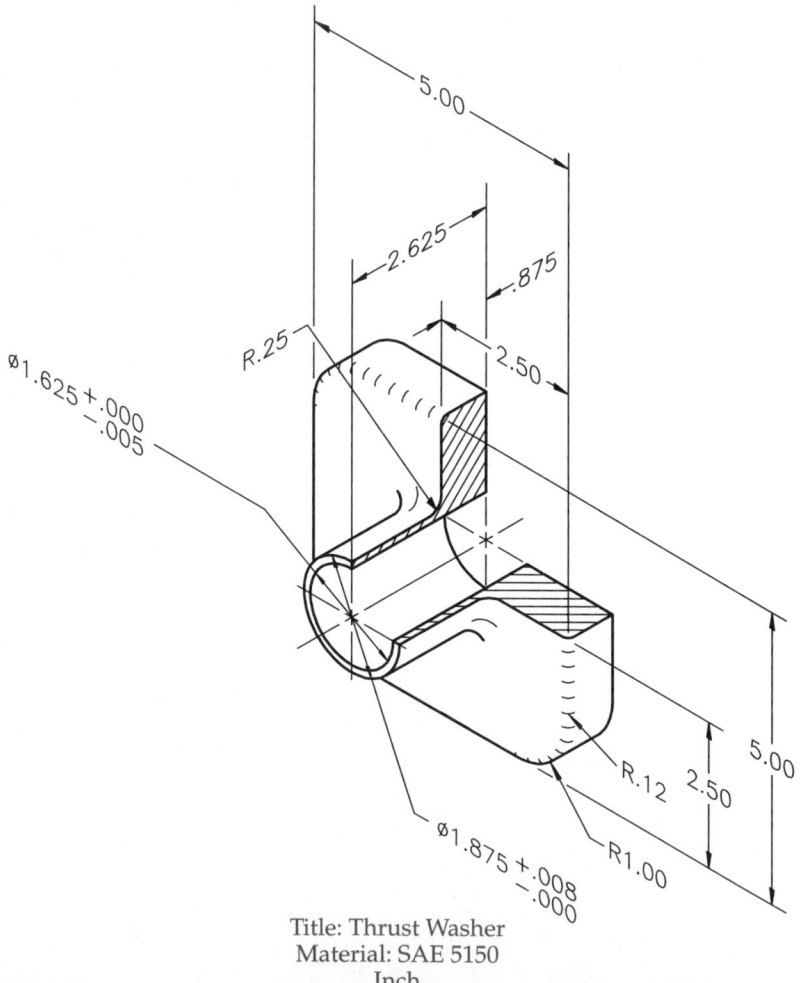

5.00

2.625

.875

2.50

R.25

⌀1.625 +.000
        −.005

R.12

2.50

5.00

⌀1.875 +.008
        −.000

R1.00

Title: Thrust Washer
Material: SAE 5150
Inch

▼ Advanced

7. Create the drawing as shown and save it as P20-7.

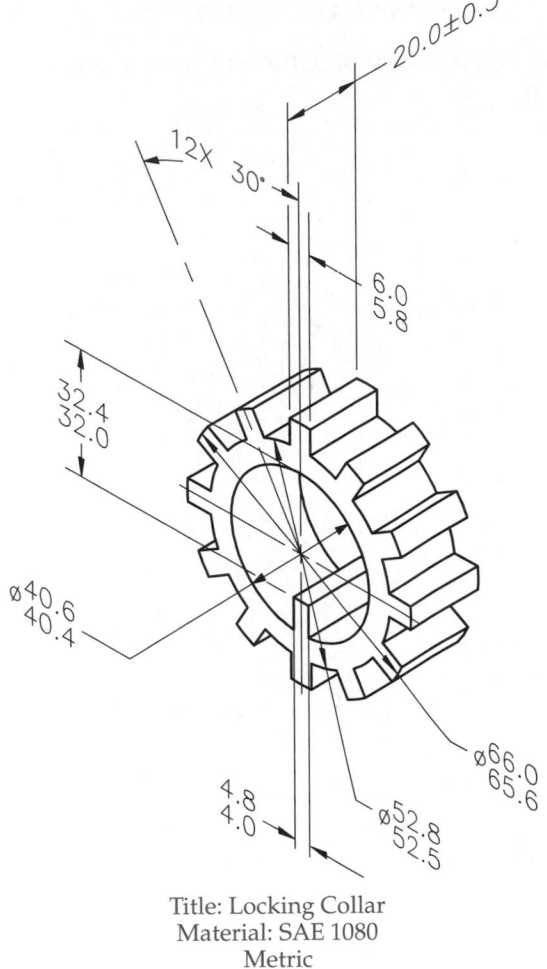

Title: Locking Collar
Material: SAE 1080
Metric

8. Draw the vise clamp and add dimensions as shown. Save the drawing as P20-8.

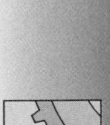

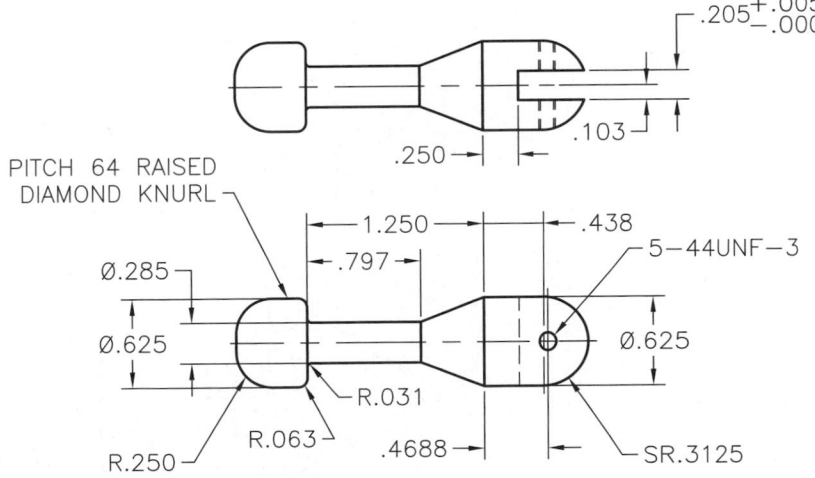

Chapter 20   Dimensioning with Tolerances

For Problems 9-15, use the isometric drawing provided to create a multiview orthographic drawing for the part. Include only the views necessary to fully describe the objects. Add all required dimensions according to the ASME standards. Use GD&T tools and practices as described in the GD&T with AutoCAD supplement available in the Supplemental Material for this chapter on the Student Web site.

9. Create the drawing as shown. Untoleranced dimensions are ±0.3. Save the drawing as P20-9.

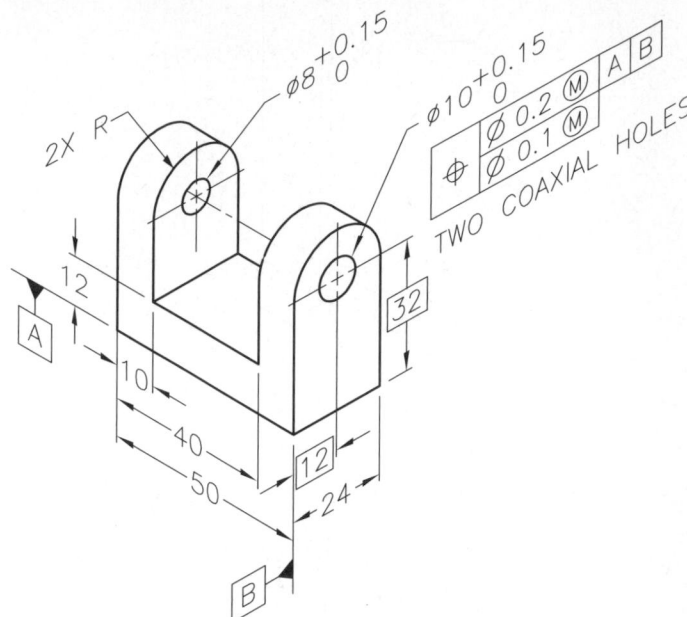

10. Open P20-6 and save the file as P20-10. The P20-10 file should be active. Add the geometric tolerancing applications shown. Untoleranced dimensions are ±.02 for two-place decimal precision and ±.005 for three-place decimal precision.

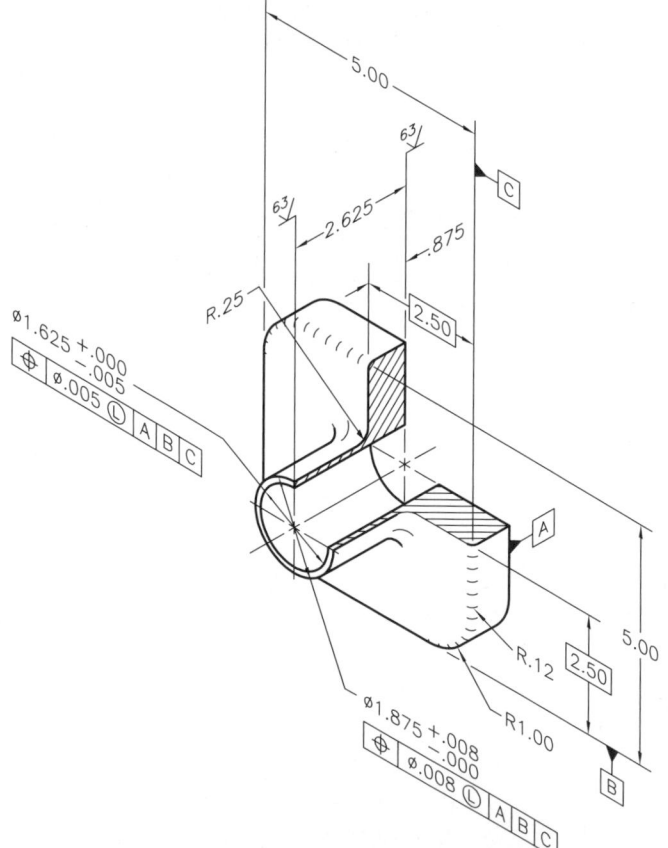

11. Open P20-4 and save the file as P20-11. The P20-11 file should be active. Add the geometric tolerancing applications shown. Untoleranced dimensions are ±0.5.

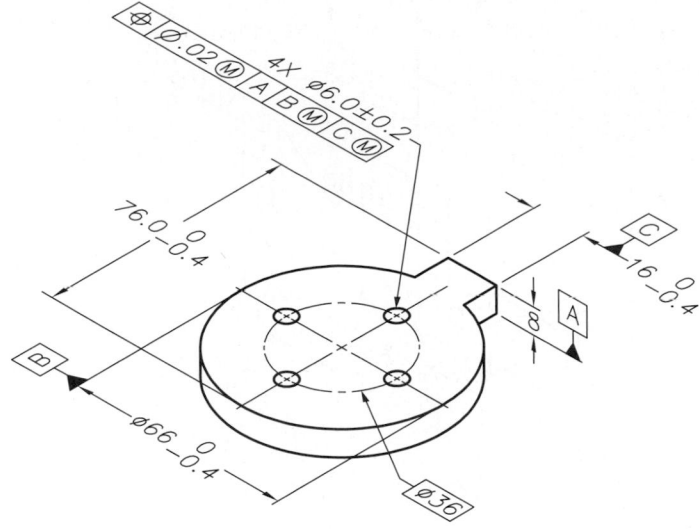

12. Open P20-7 and save the file as P20-12. The P20-12 file should be active. Add the geometric tolerancing applications shown.

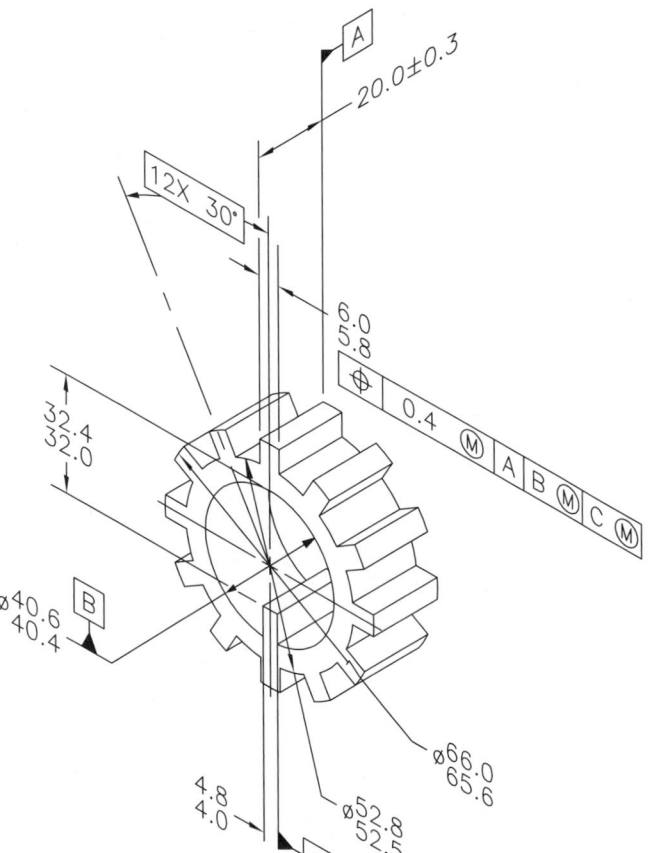

13. This problem is shown with a full section for clarity. You do not need to draw a section. Untoleranced dimensions are ±.010. Save the drawing as P20-13.

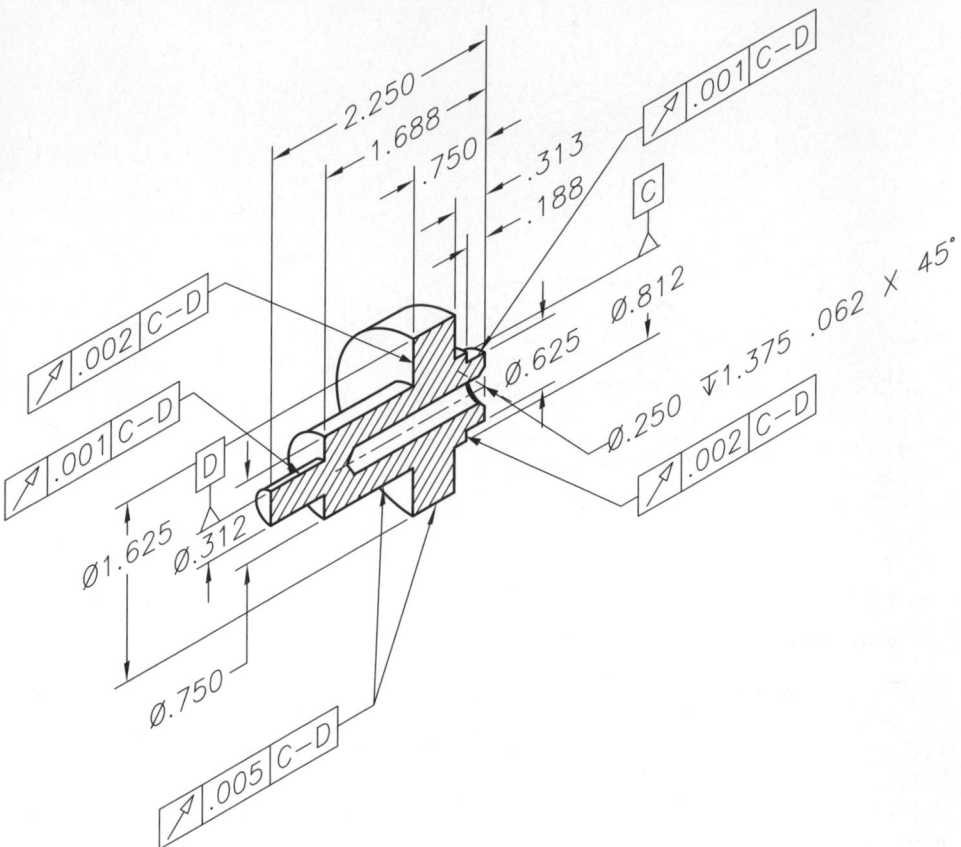

Drawing Problems - Chapter 20

14. Open P19-21 and save the file as P20-14. The P20-14 file should be active. Add the geometric tolerancing applications shown.

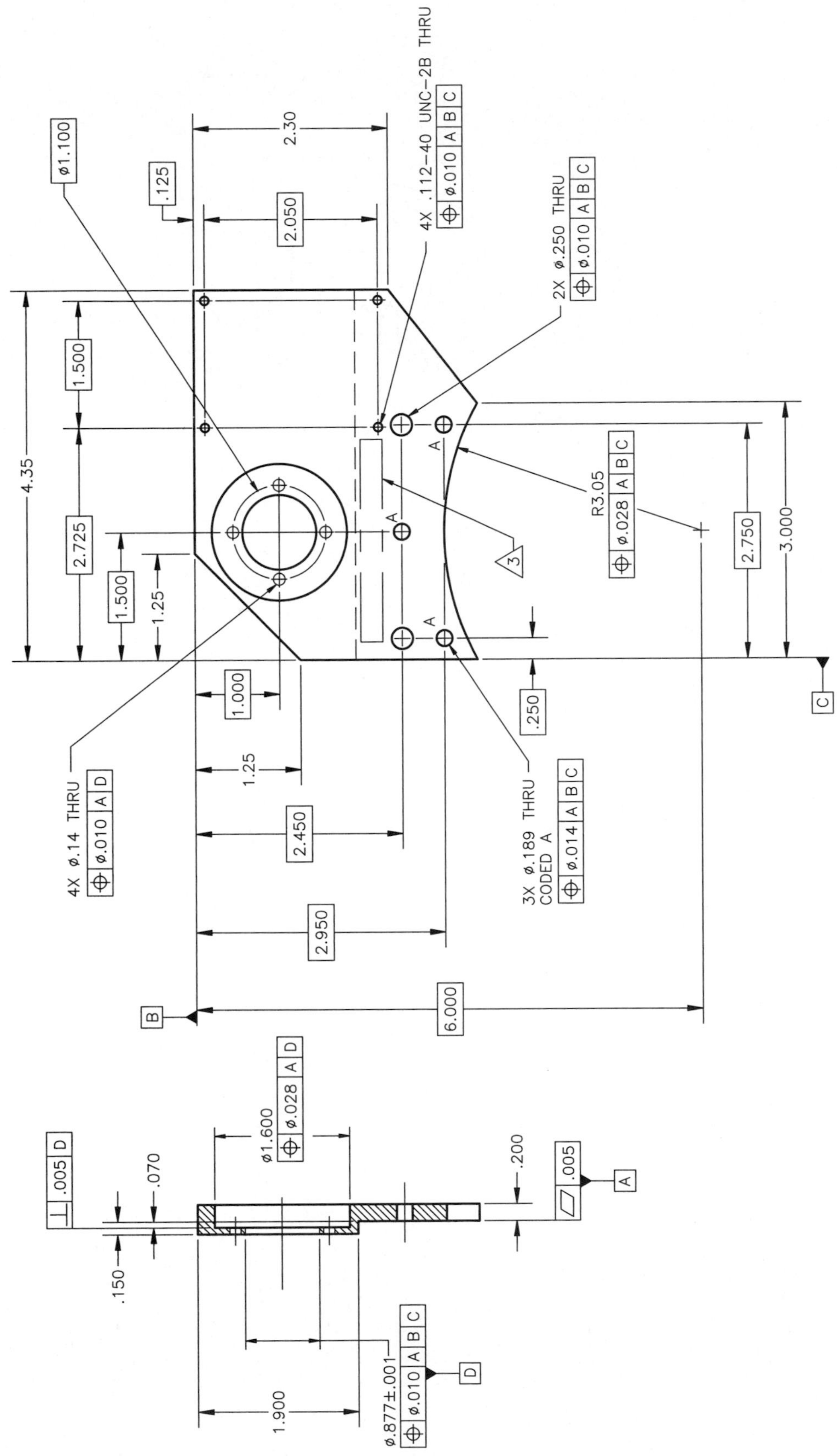

15. The problem is shown with a half section for clarity. You do not need to draw a section. Untoleranced dimensions are ±.010. Save the drawing as P20-15.

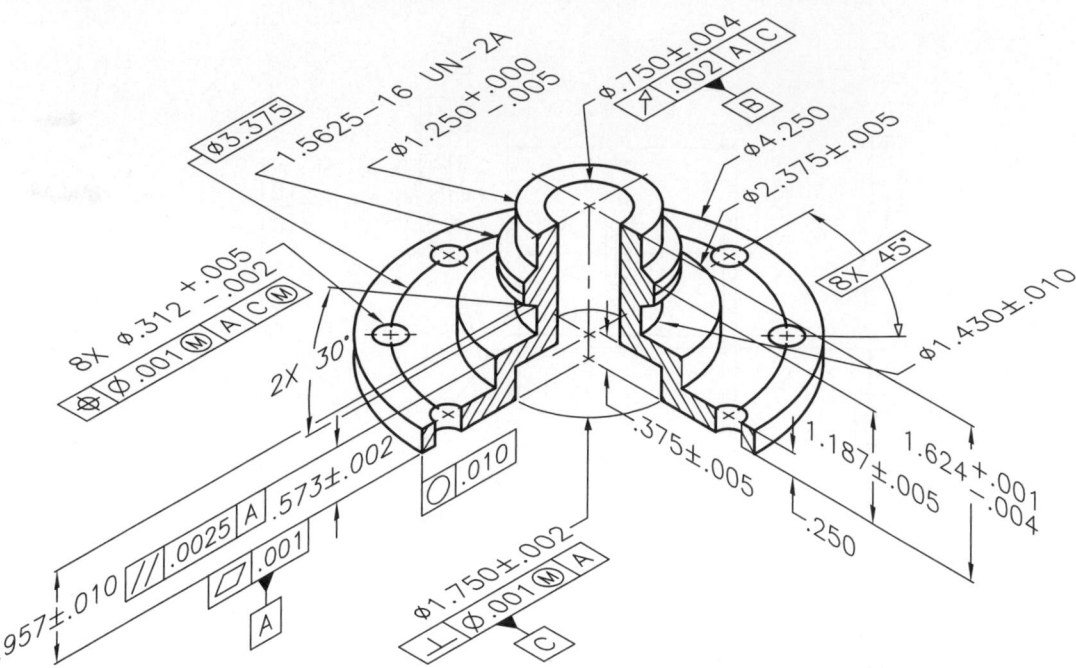

CHAPTER

Editing Dimensions

21

Learning Objectives

After completing this chapter, you will be able to do the following:
- ✓ Describe and control associative dimensions.
- ✓ Control the appearance of existing dimensions and dimension text.
- ✓ Update dimensions to reflect the current dimension style.
- ✓ Override dimension style settings and match dimension properties.
- ✓ Change dimension line spacing and alignment.
- ✓ Break dimension, extension, and leader lines.
- ✓ Create inspection dimensions.
- ✓ Edit existing multileaders.

You can use most typical object editing tools to modify and arrange dimensions. In addition to these modification techniques, there are specific considerations and tools for adjusting dimension objects. This chapter describes a variety of useful techniques for editing dimension placement, value, and appearance.

Associative Dimensioning

A dimension is a group of elements treated as a single object. When you select a dimension to edit, all dimension elements highlight. If you use the **ERASE** tool, for example, you pick the dimension as a single object and erase all elements at once. Additionally, dimensions reference objects or points. This allows you to use editing tools such as **STRETCH**, **MOVE**, **ROTATE**, and **SCALE** to make changes to the drawing that also affect dimensions. See **Figure 21-1**.

An *associative dimension* forms by default when you create a dimension using object selection, or by picking points using object snaps. For example, when you dimension the ∅1.0 circle in **Figure 21-1** using the **DIMDIAMETER** tool, and then change size of the circle to ∅2.00, the diameter dimension adapts to correctly reflect the size of the modified circle. Create associative dimensions when possible and practical by selecting objects or using appropriate object snaps. This allows dimensions to relate best to object size, and often makes revisions easier.

associative dimension: Dimension associated with an object. The dimension value updates automatically when the object changes.

557

Figure 21-1.
An example of typical drawing revisions. The dimensions adjust with the changes, and dimension values update to reflect the size and location of the modified geometry.

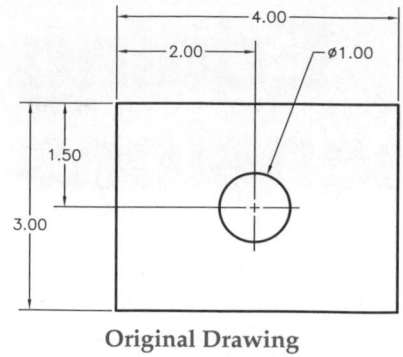

Original Drawing

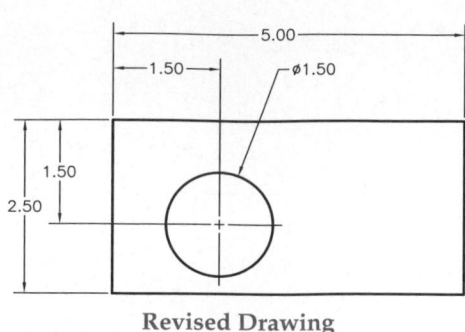

Revised Drawing

nonassociative dimension: A dimension linked to point locations, not an object; does not update when the object changes.

A *nonassociative dimension* forms when you create a dimension by selecting points without using object snaps. A nonassociative dimension is still a single object that updates when you make changes to the dimension, such as stretching the extension line origin. Nonassociative dimensions are appropriate for applications in which using associative dimensions would result in dimensioning difficulty or unacceptable standards. When using nonassociative dimensions, remember to edit the dimension with the object it dimensions, or adjust the dimension after the object changes.

NOTE

Refer to the **Associative** property in the **General** area of the **Properties** palette to determine whether a dimension is associative.

PROFESSIONAL TIP

Dimension tools allow you to dimension a drawing, but they do not control object size and location. Chapter 22 explains how to use dimensional constraint tools to control object size and location. If you anticipate creating a drawing with features that may require significant or constant change, you may want to use dimensional constraints instead of, or in addition to, traditional dimensioning tools.

Associating Dimensions with Objects

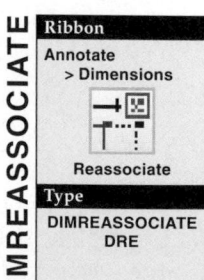

Ribbon
Annotate
> Dimensions

Reassociate

Type
DIMREASSOCIATE
DRE

DIMREASSOCIATE

Dimensions associate with objects by default when you use object selection or pick points using object snaps. To deactivate associative dimensioning for new objects, access the **Options** dialog box, and deselect the **Make new dimensions associative** check box, in the **Associative Dimensioning** area of the **User Preferences** tab.

Often the easiest way to convert a nonassociative dimension to an associative dimension is to select the dimension for grip editing and stretch the appropriate grip to the corresponding object snap point. You can also make the conversion using the **DIMREASSOCIATE** tool. Select the dimension to associate with an object. An X marker appears at a dimension origin, such as the origin of a linear dimension extension line or the center of a radial dimension. Select a point on an object to associate with the marker location. Repeat the process to locate the second object point for the first extension line, if required.

Use the **Next** option to advance to the next definition point. Use the **Select object** option to select an object to associate with the dimension. The extension line endpoints automatically associate with the object endpoints.

Type
DIMDISASSOCIATE

NOTE

To disassociate a dimension from an object, use grip editing to stretch an appropriate grip point away from the associated object, or use the **DIMDISASSOCIATE** tool.

Dimension Definition Points

Definition points, or *defpoints*, occur as a dimension element whenever you create a dimension. Use the **Node** object snap to snap to a definition point. If you select an object to edit and want to include dimensions in the edit, you must include the definition points in the selection set. AutoCAD automatically creates a Defpoints layer and places definition points on the layer. By default, the Defpoints layer does not plot. You can only plot definition points if you rename and then set the Defpoints layer to plot. Definition points display even if you turn off or freeze the Defpoints layer.

definition points (defpoints): The points used to specify the dimension location and the center point of the dimension text.

Exercise 21-1

Access the Student Web site (www.g-wlearning.com/CAD) and complete Exercise 21-1.

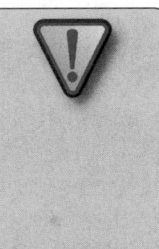
CAUTION

A dimension is a single object even though it consists of extension lines, a dimension line, arrowheads, and text. You may be tempted to explode the dimension using the **EXPLODE** tool to modify individual dimension elements. You should rarely, or never, explode dimensions. Exploded dimensions lose their layer assignments and their association to related features and dimension styles.

PROFESSIONAL TIP

You can edit individual dimension properties without exploding a dimension by using dimension shortcut menu options or the **Properties** palette to create a dimension style override.

Dimension Editing Tools

Attention to dimension style settings and careful placement of dimensions allow you to dimension a drawing according to specific drafting standards. As a drawing process evolves and design changes occur, you will find it necessary to make changes to dimensioned objects and dimensions. Dimension-specific editing tools and techniques are available to help you adjust dimensions as necessary.

Figure 21-2.
Select a dimension
and then right-
click to access this
shortcut menu.

Repeat DIMLINEAR	
Recent Input	▶
Dim Text position	▶
Precision	▶
Dim Style	▶
⟋ Flip Arrow	

Dimension
editing
options

Above dim line
Centered
Home text
Move text alone
Move with leader
Move with dim line

0
0.0
0.00
0.000
0.0000
0.00000
0.000000

Save as New Style...
Standard
Annotative
Other...

Annotative Object Scale	▶
✂ Cut	Ctrl+X
▢ Copy	Ctrl+C
▢ Copy with Base Point	Ctrl+Shift+C
▢ Paste	Ctrl+V
▢ Paste as Block	Ctrl+Shift+V
▢ Paste to Original Coordinates	
▱ Erase	
✛ Move	
▢ Copy Selection	
▢ Scale	
↻ Rotate	
Draw Order	▶
▢ Deselect All	
Action Recorder	▶
Subobject Selection Filter	▶
▢ Quick Select...	
▢ QuickCalc	
▢ Find...	
▢ Properties	
✔ Quick Properties	

Dimension Shortcut Menu Options

Select a dimension and then right-click to display the shortcut menu shown in **Figure 21-2.** The **Dim Text position** cascading menu provides options for adjusting the dimension value location. Pick **Above dim line** to move the dimension text above the dimension line. Select **Centered** to center the dimension text on the dimension line. Pick **Home text** to reposition the text at its original position. **Move text alone** allows you to move the text away from the dimension line. **Move with leader** allows you to move the text away from the dimension line with a leader attaching the text to the dimension line. **Move with dimension line** allows you to move the text, but maintain its alignment with the dimension line.

The options in the **Precision** cascading menu allow you to adjust the number of decimal places displayed in a dimension text value. This is often the easiest way to specify an alternative tolerance. The **Dim Style** cascading menu allows you to create a new dimension style based on the properties of the selected dimension. You can also apply a different dimension style to the dimension.

The **Flip Arrow** option allows you to flip the direction of a dimension arrowhead to the opposite side of the extension line or object that the arrow touches. For example, if the arrowheads and dimension value are crowded inside the extension lines, you can flip the arrowheads to the outside of the extension lines to make the dimension easier to read. If the selected dimension includes two arrowheads, only one of the arrowheads flips, allowing you to control the arrowheads independently. The arrowhead that flips is the one closest to the point you pick when you select the dimension (not the right-click point).

Assigning a Different Dimension Style

To assign a different dimension style to existing dimensions, you can use the options from the **Dim Style** cascading menu of the dimension shortcut menu. A second option is to pick the dimensions to change and select a different dimension style from the **Dimension Style** drop-down list on the **Home** and **Annotation** ribbon tabs. A third option is to select the dimensions to change and choose a new dimension style from the **Quick Properties** panel or the **Properties** palette.

Another technique for changing the dimension style of existing dimensions is to use the **Update** dimension tool. Before you access the **Update** dimension tool, be sure the current dimension style is the dimension style you want to assign to existing dimensions. Then access the **Update** dimension tool and pick the dimensions to change to the current style.

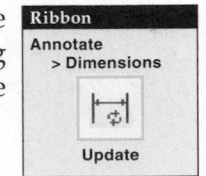

Ribbon
Annotate
> Dimensions
Update

Editing the Dimension Value

The **DDEDIT** tool allows you to add a prefix or suffix to the text or edit the dimension text format. For example, use the **DDEDIT** tool to dimension a linear diameter if you forget to use the **Mtext** or **Text** option of the **DIMLINEAR** tool to add a diameter symbol to the value. See **Figure 21-3.** Access the **DDEDIT** tool and select a dimension to enter the multiline text editor with the current dimension value highlighted. Add to or modify the dimension text and then close the text editor. The tool continues, allowing you to edit other text if necessary.

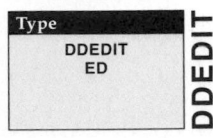

Type
DDEDIT
ED

DDEDIT

CAUTION

You can replace the highlighted text, which represents the dimension value, with numeric values. However, this action disassociates the dimension value with the object or points it dimensions. Therefore, leave the default value intact whenever possible.

Exercise 21-2

Access the Student Web site (www.g-wlearning.com/CAD) and complete Exercise 21-2.

Editing Dimension Text Placement

Proper dimensioning practice requires dimensions that are clear and easy to read. This sometimes involves moving the text of adjacent dimensions to separate the text elements. See **Figure 21-4.** You can use the dimension shortcut menu to adjust dimension text position, but often the quickest method is to use grips. Select the dimension, pick the dimension text grip, and stretch the text to the new location. AutoCAD automatically reestablishes the break in the dimension line when you pick the new location.

Using the DIMTEDIT Tool

The **DIMTEDIT** tool allows you to change the placement and orientation of existing dimension text. Access the **DIMTEDIT** tool and select the dimension to alter. Drag the dimension text to a new location and pick to move the text and automatically reestablish the break in the dimension line.

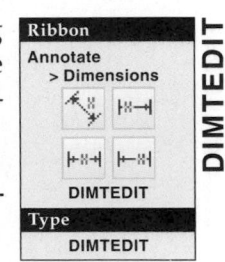

Ribbon
Annotate
> Dimensions
DIMTEDIT
Type
DIMTEDIT

DIMTEDIT

Figure 21-3.
Using the **DDEDIT** tool to add a diameter symbol to an existing dimension.

|← 3.250 →| |← ⌀3.250 →|

Original   Diameter Symbol Added

Figure 21-4.
Staggering
dimension text
for improved
readability.

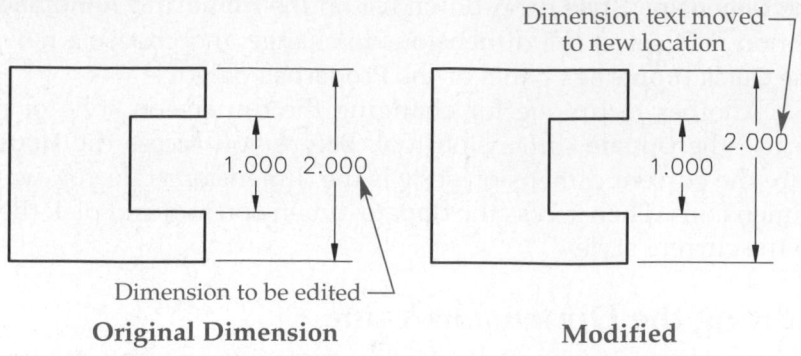

Dimension text moved
to new location

1.000 2.000

2.000

1.000

Dimension to be edited

Original Dimension   **Modified**

The **DIMTEDIT** tool also provides options for relocating dimension text to a specific location and for rotating the text. However, it is usually quicker to select the appropriate button from the expanded **Dimensions** panel of the **Annotation** ribbon tab or select a similar option from the dimension shortcut menu. Select **Text Angle (Angle)** to rotate the dimension text. **Left Justify (Left)** moves horizontal text to the left and vertical text down. **Center Justify (Center)** centers the dimension text on the dimension line. **Right Justify (Right)** moves horizontal text to the right and vertical text up. Select the **Home** option to relocate text back to its original position. **Figure 21-5** shows the result of using each **DIMTEDIT** tool option.

PROFESSIONAL TIP

You may want to activate the **Place text manually** check box in the **Fit** tab of the **New** (or **Modify**) **Dimension Style** dialog box to provide greater flexibility for the initial placement of dimensions.

Exercise 21-3

Access the Student Web site (www.g-wlearning.com/CAD) and complete Exercise 21-3.

Figure 21-5.
A comparison of
the **DIMTEDIT** tool
options.

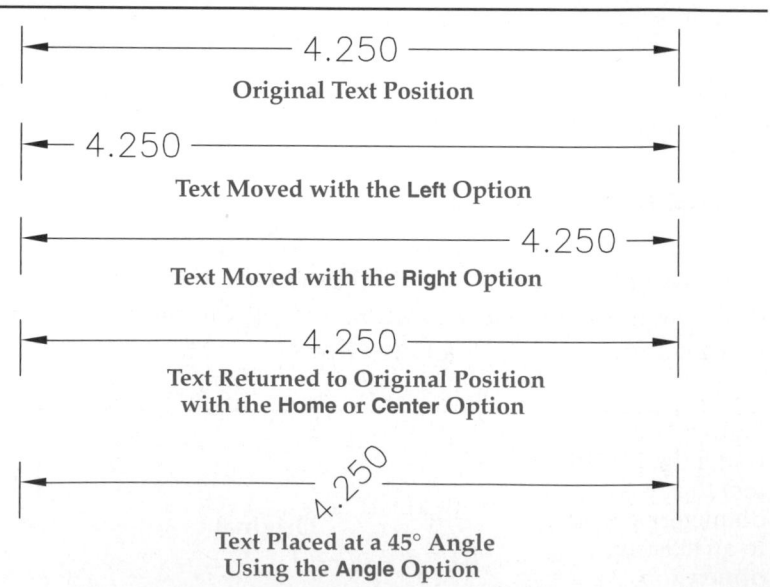

4.250
Original Text Position

4.250
Text Moved with the Left Option

4.250
Text Moved with the Right Option

4.250
**Text Returned to Original Position
with the Home or Center Option**

4.250
**Text Placed at a 45° Angle
Using the Angle Option**

AutoCAD and Its Applications—Basics

Using the DIMEDIT Tool

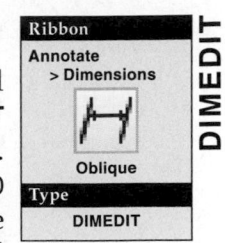
The **DIMEDIT** tool, not to be confused with the **DIMTEDIT** tool, provides **Home** and **Rotate** options that function the same as the **Home** and **Angle** options of the **DIMTEDIT** tool. The **New** option is similar to using the **DDEDIT** tool to edit dimension text values. When you activate the **New** option, the multiline text editor displays with a 0.0000 (depending on units and precision) value highlighted. The highlighted 0.0000 value represents the associated dimension value. Add to or modify the dimension text and then close the text editor.

Creating Oblique Extension Lines

The **Oblique** option is unique to the **DIMEDIT** tool and allows you to change the extension line angle without affecting the associated dimension value. In **Figure 21-6A,** for example, the curve on the left is dimensioned using linear dimensions. On the right, the extension lines of two of the linear dimensions have been made oblique to adjust the placement of dimensions when space is limited. **Figure 21-6B** shows an example of oblique extension lines used to orient extension lines properly with the angle of the stairs in a stair section. Notice that the associated values and orientation of the dimension lines in these examples do not change.

To create oblique extension lines, first dimension the object using the **DIMLINEAR** tool as appropriate, even if dimensions are crowded or overlap. Then access the **Oblique** option of the **DIMEDIT** tool. The quickest way to access the **Oblique** option is to pick the corresponding button from the expanded **Dimensions** panel of the **Annotation** ribbon

Figure 21-6.
Drawing dimensions with oblique extension lines.

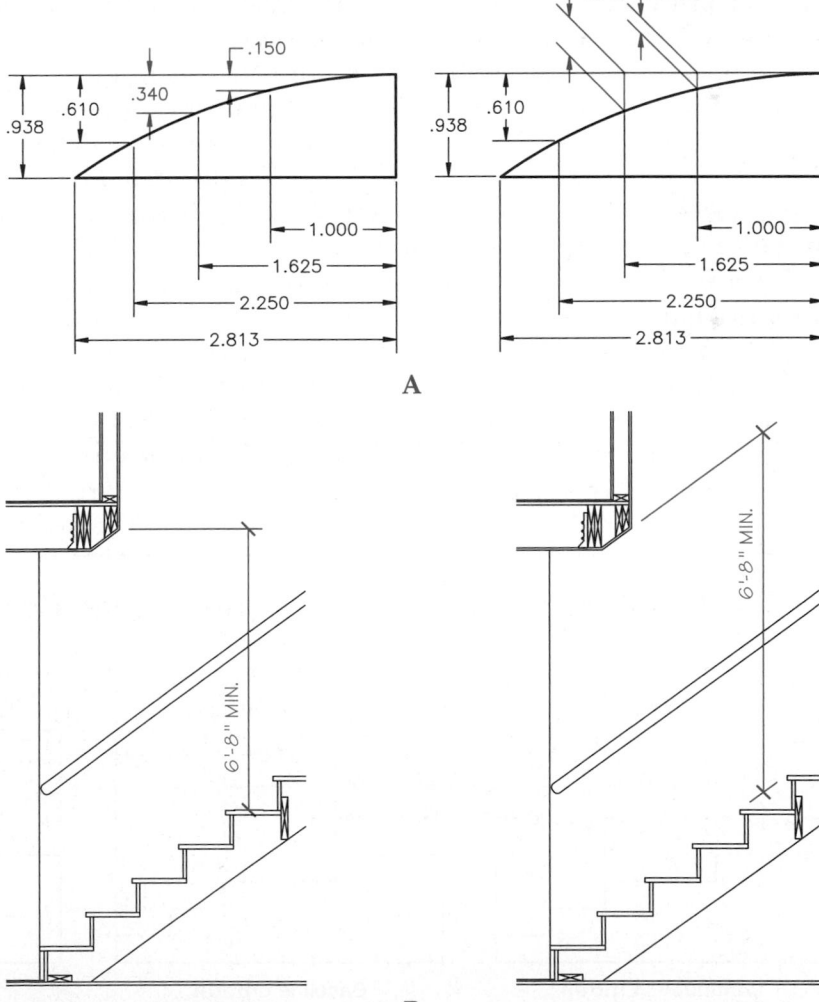

tab. Once you activate the **Oblique** option, pick the dimensions to redraw at an oblique angle. Next, specify the obliquing angle. Plan carefully to make sure you enter the correct obliquing angle. Obliquing angles originate from 0° East and revolve counter-clockwise. Enter a specific value or pick two points to define the obliquing angle.

> **NOTE**
>
> You can use the **Oblique** option of the **DIMEDIT** tool to dimension oblique and isometric drawings, as described in Chapter 24.

Exercise 21-4

Access the Student Web site (www.g-wlearning.com/CAD) and complete Exercise 21-4.

Editing Dimensions with the QDIM Tool

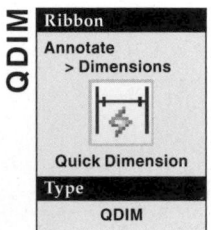

QDIM

Ribbon
Annotate
> Dimensions

Quick Dimension

Type
QDIM

The **QDIM** tool includes options for changing the arrangement of existing dimensions, adding a dimension, and removing a dimension. The **Continuous** option allows you to change a selected group of dimensions to chain dimensions. See **Figure 21-7A.** The **Baseline** option allows you to change a selected group of dimensions to datum dimensions. **Figure 21-7B** shows an example of using the **Baseline** option to change the dimensioning arrangement from chain to datum.

You can use the **Edit** option to add dimensions to, or remove dimensions from, a selected group of existing dimensions and automatically reorder the group. Use the **Add** function of the **Edit** option to add a dimension, or use the **Remove** function to remove a dimension. For example, to add the dimension shown in **Figure 21-7C,** access the **QDIM** tool, select all of the dimensions in the group to change, and right-click or press [Enter] or the space bar. Then activate the **Edit** option and the **Add** function. Pick the location or feature where the new dimension is to originate and right-click or press [Enter] or the space bar. Finally, pick a location for the baseline dimension arrangement.

The dimensions automatically realign after you pick a location for the arrangement. You do not have to pick all of the dimensions in the group. However, if you do not pick the entire group, you must carefully select the location. In addition, the spacing for the edited dimension and the dimensions in the group that you do not select may be inconsistent.

Figure 21-7.
You can use the **QDIM** tool to change existing dimension arrangements and add or remove dimensions.

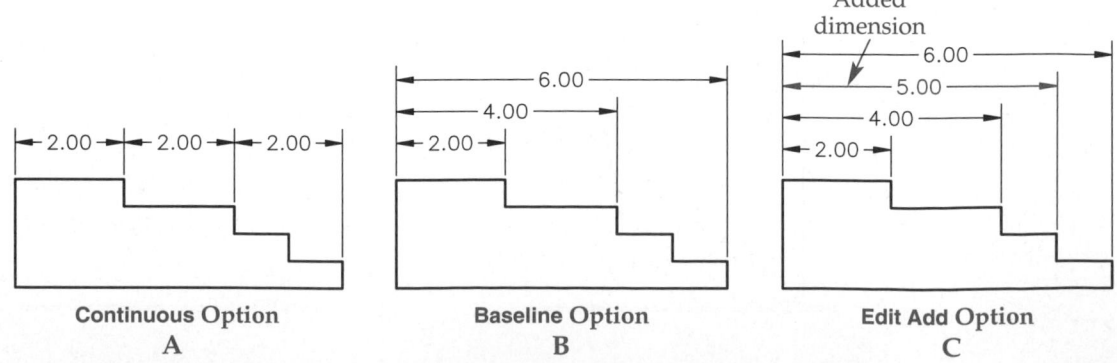

Continuous Option	Baseline Option	Edit Add Option
A	B	C

Exercise 21-5

Access the Student Web site (www.g-wlearning.com/CAD) and complete Exercise 21-5.

Overriding Existing Dimension Style Settings

Generally, you should set up one or more dimension styles to perform specific dimensioning tasks. However, in some situations, a few dimensions require specific settings that an existing dimension style cannot provide. These situations may be too few to merit creating a new style. For example, assume you have the value for **Offset from origin** set at .063, which conforms to ASME standards. However, three dimensions in your final drawing require a 0 **Offset from origin** setting. For these dimensions, you can perform a *dimension style override*.

dimension style override: A temporary alteration of settings for the dimension style that does not actually modify the style.

Dimension Style Overrides for Existing Dimensions

The **Properties** palette is the most effective tool to use for overriding the dimension style settings of existing dimensions. The **Properties** palette divides dimension properties into several categories. See **Figure 21-8**. To change a property, access the proper category, pick the property to highlight, and adjust the corresponding value. Most changes made using the **Properties** palette are overrides to the dimension style for the selected dimension. The changes do not alter the original dimension style and do not apply to new dimensions.

> **NOTE**
>
> The **Quick Properties** panel provides a limited number of dimension properties and style overrides.

Dimension Style Overrides for New Dimensions

Use the **Dimension Style Manager** to override the dimension style for dimensions you are about to create. An example of an override is including a text prefix for a few of the dimensions in a drawing. Select the dimension style to override from the **Styles** list and then pick the **Override** button to open the **Override Current Style** dialog box. The **Override** button is only available for the current style. The **Override Current Style** dialog box includes the same tabs as the **New Dimension Style** and **Modify Dimension Style** dialog boxes. Make the necessary changes and pick the **OK** button. Once you create an override, it is current and appears as a branch under the original style labeled **<style overrides>**. Close the **Dimension Style Manager** and draw the needed dimensions.

To clear style overrides, return to the **Dimension Style Manager** and set any other style current. The override settings are lost when any other style, including the parent, is set current. To incorporate the overrides into the overridden style, right-click on the **<style overrides>** name and select **Save to current style**. To save the changes to a new style, pick the **New...** button. Then select **<style overrides>** in the **Start With** drop-down list in the **Create New Dimension Style** dialog box. In the **New Dimension Style** dialog box, pick **OK** to save the overrides as a new style.

Figure 21-8.
The **Properties** palette allows you to edit dimension properties and create a dimension style override.

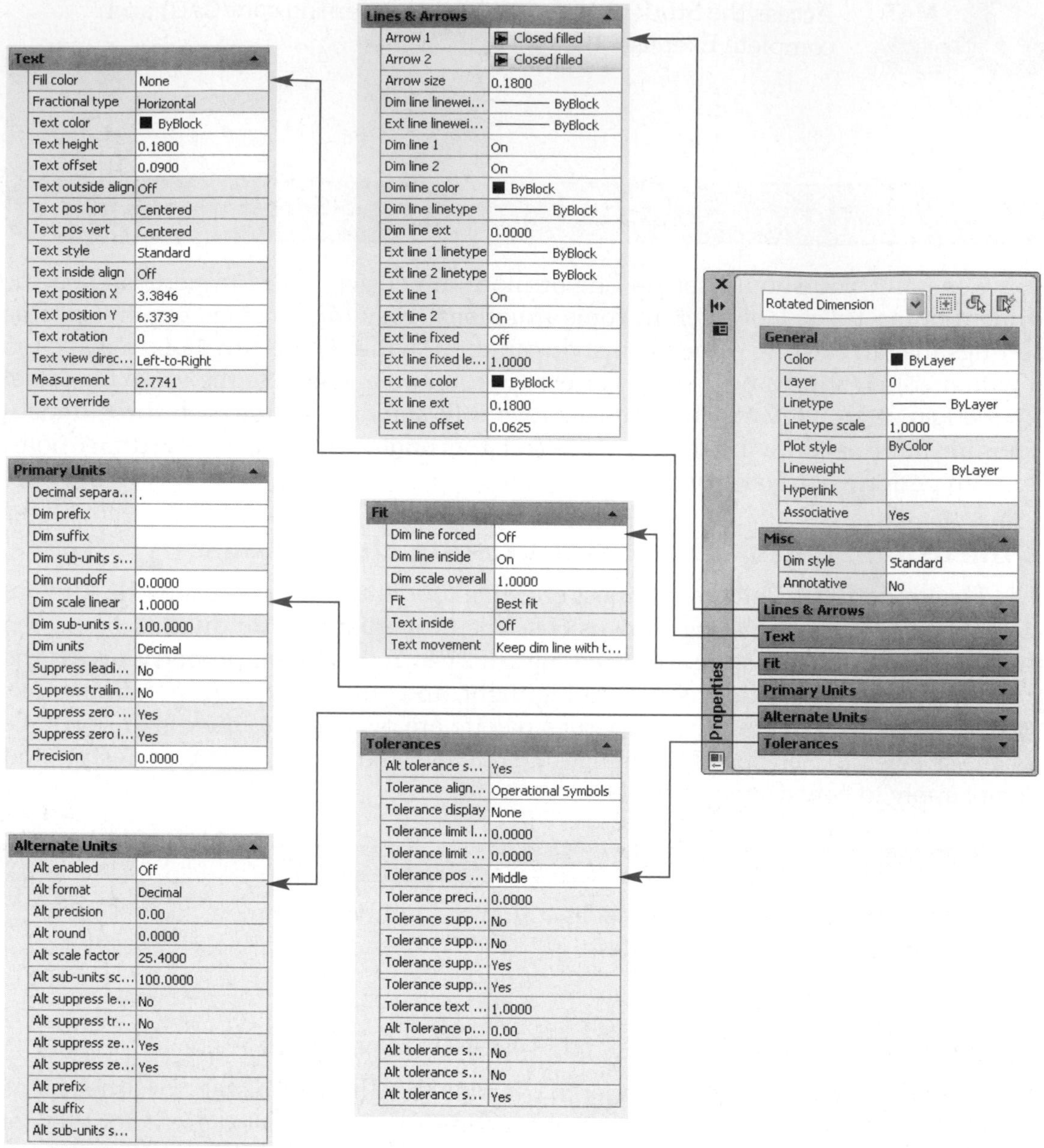

Text

Fill color	None
Fractional type	Horizontal
Text color	ByBlock
Text height	0.1800
Text offset	0.0900
Text outside align	Off
Text pos hor	Centered
Text pos vert	Centered
Text style	Standard
Text inside align	Off
Text position X	3.3846
Text position Y	6.3739
Text rotation	0
Text view direc...	Left-to-Right
Measurement	2.7741
Text override	

Lines & Arrows

Arrow 1	Closed filled
Arrow 2	Closed filled
Arrow size	0.1800
Dim line linewei...	ByBlock
Ext line linewei...	ByBlock
Dim line 1	On
Dim line 2	On
Dim line color	ByBlock
Dim line linetype	ByBlock
Dim line ext	0.0000
Ext line 1 linetype	ByBlock
Ext line 2 linetype	ByBlock
Ext line 1	On
Ext line 2	On
Ext line fixed	Off
Ext line fixed le...	1.0000
Ext line color	ByBlock
Ext line ext	0.1800
Ext line offset	0.0625

Primary Units

Decimal separa...	.
Dim prefix	
Dim suffix	
Dim sub-units s...	
Dim roundoff	0.0000
Dim scale linear	1.0000
Dim sub-units s...	100.0000
Dim units	Decimal
Suppress leadi...	No
Suppress trailin...	No
Suppress zero ...	Yes
Suppress zero i...	Yes
Precision	0.0000

Fit

Dim line forced	Off
Dim line inside	On
Dim scale overall	1.0000
Fit	Best fit
Text inside	Off
Text movement	Keep dim line with t...

Rotated Dimension

General

Color	ByLayer
Layer	0
Linetype	ByLayer
Linetype scale	1.0000
Plot style	ByColor
Lineweight	ByLayer
Hyperlink	
Associative	Yes

Misc

Dim style	Standard
Annotative	No

Lines & Arrows
Text
Fit
Primary Units
Alternate Units
Tolerances

Alternate Units

Alt enabled	Off
Alt format	Decimal
Alt precision	0.00
Alt round	0.0000
Alt scale factor	25.4000
Alt sub-units sc...	100.0000
Alt suppress le...	No
Alt suppress tr...	No
Alt suppress ze...	Yes
Alt suppress ze...	Yes
Alt prefix	
Alt suffix	
Alt sub-units s...	

Tolerances

Alt tolerance s...	Yes
Tolerance align...	Operational Symbols
Tolerance display	None
Tolerance limit l...	0.0000
Tolerance limit ...	0.0000
Tolerance pos ...	Middle
Tolerance preci...	0.0000
Tolerance supp...	No
Tolerance supp...	No
Tolerance supp...	Yes
Tolerance supp...	Yes
Tolerance text ...	1.0000
Alt Tolerance p...	0.00
Alt tolerance s...	No
Alt tolerance s...	No
Alt tolerance s...	Yes

PROFESSIONAL TIP

Carefully evaluate the dimensioning requirements in a drawing before performing a style override. It may be better to create a new style. For example, if a number of the dimensions in the current drawing require the same overrides, generating a new dimension style is a good idea. If only one or two dimensions need the same changes, performing an override may be more productive.

Exercise 21-6

Access the Student Web site (www.g-wlearning.com/CAD) and complete Exercise 21-6.

AutoCAD and Its Applications—Basics

Using the MATCHPROP Tool

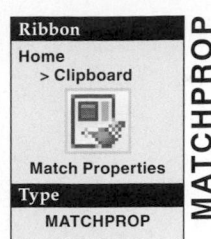

Ribbon
Home
> Clipboard

Match Properties
Type
MATCHPROP

MATCHPROP

The **MATCHPROP** tool provides an effective method of applying the properties of a selected dimension to existing dimensions in the same drawing or a different drawing. Access the **MATCHPROP** tool, pick the source dimension that has the desired properties, and then pick the destination dimensions to change. Press [Enter] to update all of the destination dimensions to reflect the properties of the source dimension.

For the **MATCHPROP** tool to work with dimensions, the **Dimension** setting must be active. You can check this after you select the source object. When the Current active settings: prompt appears, Dim should appear with the other settings. If Dim does not appear when you are prompted to select a destination object, use the **Settings** option to display the **Property Settings** dialog box. Activate the **Dimension** check box in the **Special Properties** area and pick **OK**. Then select the destination dimensions.

PROFESSIONAL TIP

The style of the source dimension applies to destination dimensions. If you override the dimension style of the source dimension, the "base" style applies, along with the dimension style override. Reapplying the "base" style removes the overrides.

Exercise 21-7

Access the Student Web site (www.g-wlearning.com/CAD) and complete Exercise 21-7.

Using the DIMSPACE Tool

The amount of space between a drawing view and the first dimension line, and the space between dimension lines, varies depending on the drawing and industry or company standard. ASME standards recommend a minimum spacing of .375″ (10 mm) from a drawing feature to the first dimension line and a minimum spacing of .25″ (6 mm) between dimension lines. A minimum spacing of 3/8″ is common for architectural drawings. These minimum recommendations are generally less than the spacing required by actual company or school standards.

Typically, the spacing between dimension lines is equal, and chain dimensions align. See **Figure 21-9.** As a result, it is important to determine the correct location and spacing of dimension lines. However, you can adjust dimension line spacing and alignment after you place dimensions. This is a common requirement when there is a need to increase or decrease the space between dimension lines, such as when the drawing scale changes, or when dimensions are spaced unequally or misaligned.

The **STRETCH** and **DIMTEDIT** tools or grips are common methods for adjusting the location and alignment of dimension lines. You must determine the exact location or amount of stretch applied to each dimension line before using these tools. An alternative is to use the **DIMSPACE** tool, which allows you to adjust the space equally between dimension lines or align dimension lines.

Figure 21-9.
Correct drafting practice requires equal space and alignment between dimension lines for readability.

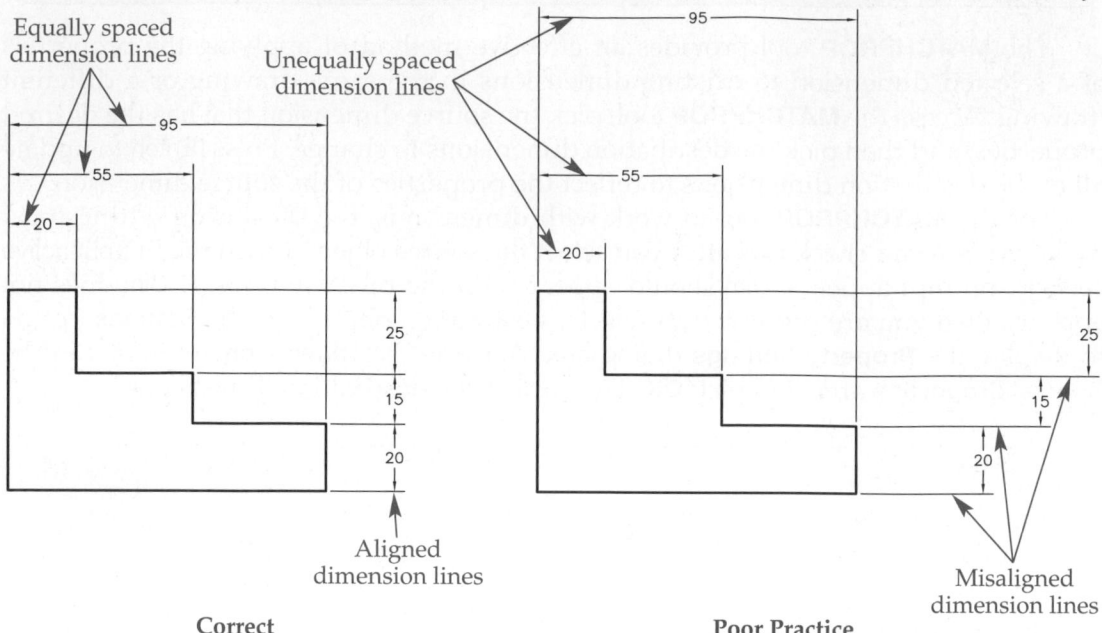

Correct   Poor Practice

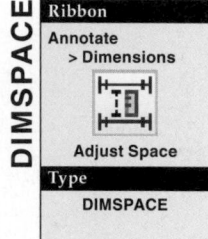
Access the **DIMSPACE** tool and select the *base dimension*, followed by each dimension to space. Right-click or press [Enter] or the space bar to display the Enter value or [Auto]: prompt. Enter a value to space the dimension lines equally. For example, enter .5 to space the selected dimension lines .5″ apart. Enter a value of 0 to align the dimensions. See **Figure 21-10.** Use the **Auto** option to space dimension lines using a value that is twice the height of the dimension text.

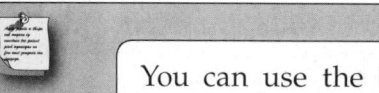

NOTE

You can use the **DIMSPACE** tool to space and align linear and angular dimensions.

Figure 21-10.
Using the **DIMSPACE** tool to space and align dimension lines correctly.

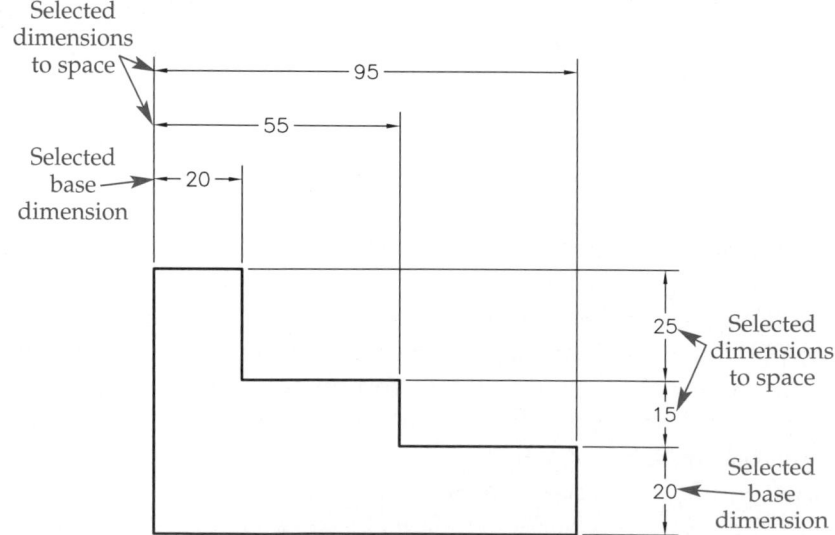

Exercise 21-8

Access the Student Web site (www.g-wlearning.com/CAD) and complete Exercise 21-8.

Using the DIMBREAK Tool

Drafting standards state that when dimension, extension, or leader lines cross a drawing feature or another dimension, neither line is broken at the intersection. See **Figure 21-11.** However, you can use the **DIMBREAK** tool to create breaks if desired.

Access the **DIMBREAK** tool and select the dimension to break. This is the dimension that contains the dimension, extension, or leader line that you want to break across an object. If you pick a single dimension to break, the Select object to break dimension or [Auto/Restore/Manual]: prompt appears.

The **Auto** option of the **DIMBREAK** tool, which is the default, breaks the dimension, extension, or leader line at the selected object. The **Dimension Break** setting of the current dimension style controls the break size. You can pick additional objects if necessary to break the dimension at additional locations. See **Figure 21-12.** Use the **Manual** option to define the size of the break by selecting two points along the dimension, extension, or leader line, instead of using the break size set in the current dimension style. Activate the **Restore** option to remove an existing break created using the **DIMBREAK** tool.

Another technique is to use the **Multiple** option to select more than one dimension. The **Break** and **Restore** options are available when you use the **Multiple** option to select multiple dimensions. Right-click or press [Enter] or the space bar after you select the dimensions to display the Enter and option [Break/Restore]: prompt. Select the **Break** option to break the selected dimension, extension, or leader lines everywhere they intersect another object. Use the **Restore** option to remove any existing breaks added to the selected dimensions using the **DIMBREAK** tool.

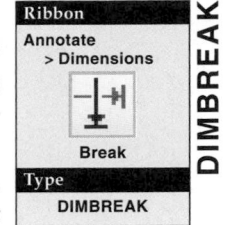

Ribbon
Annotate
> Dimensions

Break

Type
DIMBREAK

DIMBREAK

Figure 21-11.
Drafting standards state that when dimension, extension, or leader lines cross a drawing feature or another dimension, the line is not broken at the intersection.

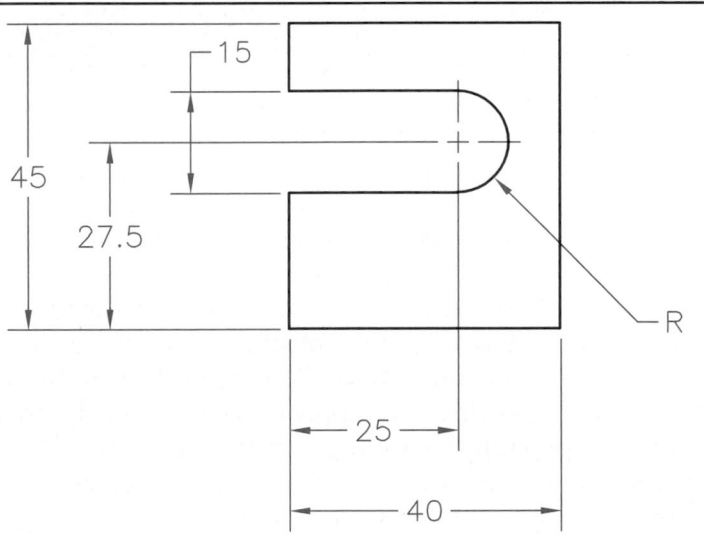

Figure 21-12.
Use the **DIMBREAK** tool to break dimension, extension, or leader lines when they cross an object. *Caution:* This example violates ASME standards and is for reference only. Extension and leader lines do not break over object lines, but some drafters prefer to break an extension line when it crosses a dimension line.

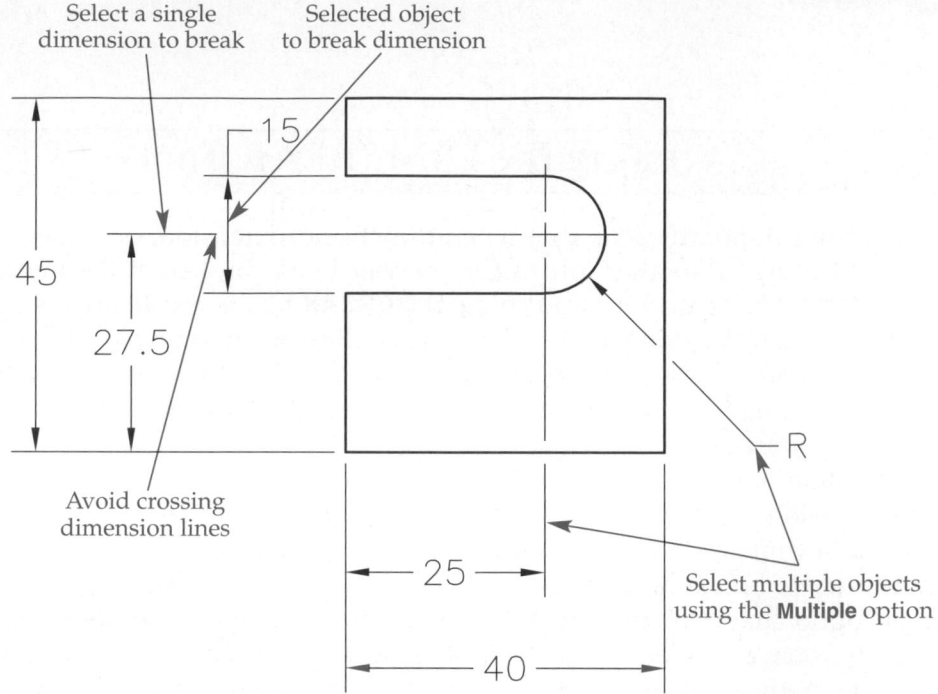

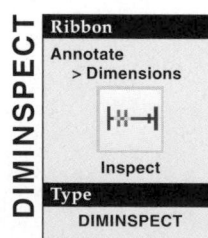

Exercise 21-9

Access the Student Web site (www.g-wlearning.com/CAD) and complete Exercise 21-9.

Creating Inspection Dimensions

Inspections and tests occur throughout the design and manufacturing of a product. These tests help ensure the correct size and location of product features. In some cases, size and location dimensions include information about how frequently a test on the dimension occurs for consistency and tolerance during the manufacturing process. See **Figure 21-13.** You can use the **DIMINSPECT** tool to add inspection information to most types of existing dimensions.

Access the **DIMINSPECT** tool to display the **Inspection Dimension** dialog box, shown in **Figure 21-14.** Pick the **Select dimensions** button and choose the dimensions to receive inspection information. You can select multiple dimensions, although the same inspection specifications apply to each. Define the shape of the inspection dimension frame by picking the appropriate radio button in the **Shape** area. The inspection dimension contains the inspection label, the dimension value, and the inspection rate. Select the **None** option to omit frames around values.

To include a label, pick the **Label** check box and type the label in the text box. The label is located on the left side of the inspection dimension and identifies the specific dimension. The inspection dimension shown in **Figure 21-13** is labeled A. The dimension frame houses the dimension value specified when you create the

AutoCAD and Its Applications—Basics

Figure 21-13.
An example of an inspection dimension added to a part drawing. This example shows an angular shape with a label, dimension, and inspection rate frame.

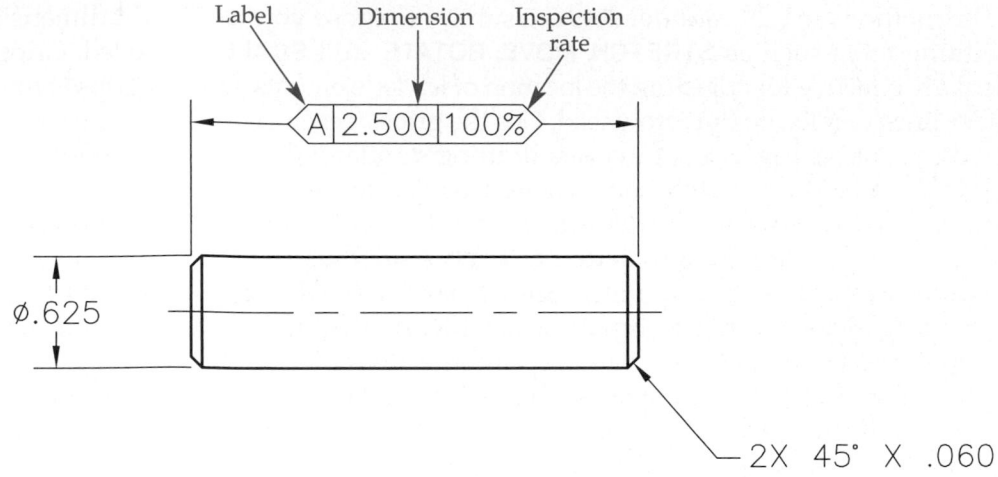

Figure 21-14.
The **Inspection Dimension** dialog box allows you to add inspection information to existing dimensions.

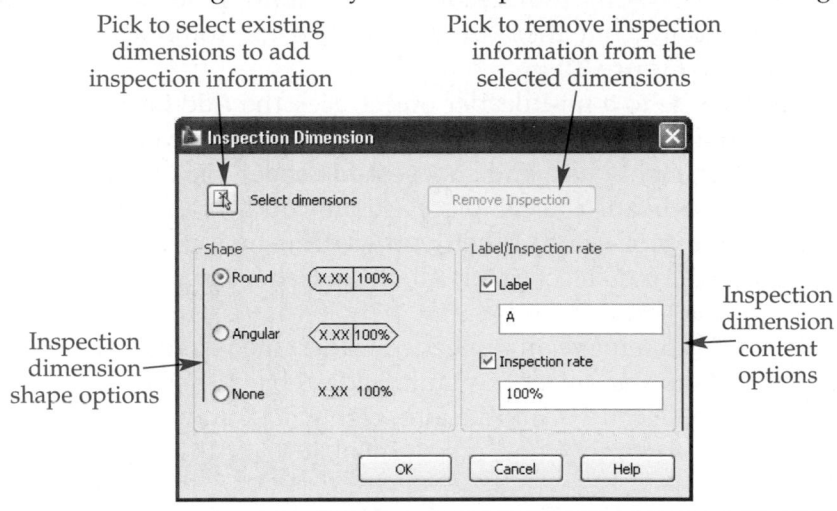

dimension. The length of the part shown in **Figure 21-13** is 2.500, as specified when using the **DIMLINEAR** tool. The **Inspection rate** check box is selected by default. Enter a value in the text box to indicate how often a test should be performed on the dimension. The inspection rate for the dimension shown in **Figure 21-13** is 100%. This rate can have different meanings depending on the application. In this example, the inspection rate of 100% means that the length of the part must be checked for tolerance every time the part is added to an assembly.

To remove an inspection dimension, access the **DIMINSPECT** tool, pick the **Select dimensions** button in the **Inspection Dimension** dialog box, and choose the dimensions that contain the inspection information to remove. Right-click or press [Enter] or the space bar to return to the **Inspection Dimension** dialog box, and pick the **Remove Inspection** button to return the dimension to its condition prior to adding the inspection content.

Exercise 21-10

Access the Student Web site (www.g-wlearning.com/CAD) and complete Exercise 21-10.

Editing Multileaders

The methods to edit multileaders are similar to those you use to edit dimensions. Use editing tools such as **STRETCH**, **MOVE**, **ROTATE**, and **SCALE** as needed. Grips are particularly effective for adjusting the location of leader elements. Use the grips at the end of leader lines to relocate the arrowhead. Use the grips at each end of a landing to stretch the landing, but be careful not to violate drafting standards. Use the grips positioned at the middle of a landing or with leader content to relocate content.

To make changes to multileader text, double-click on the text to re-enter the multiline text editor. Use the **Properties** palette or **Quick Properties** panel to override specific multileader properties. You can also use the **MATCHPROP** tool. In addition to these general multileader editing techniques, specific tools allow you to add and remove leader lines, and space, align, and group multileader objects. Select a multileader and right-click to display a shortcut menu with specific options for adjusting multileaders and assigning a different multileader style.

Adding and Removing Multiple Leader Lines

The **MLEADEREDIT** tool provides options for adding leader lines to, and removing leader lines from, an existing multileader object. Multiple leaders are not a recommended ASME standard, but they are appropriate for some applications, such as welding symbols. See **Figure 21-15.**

To add a leader line to a multileader object, pick the **Add Leader** button from the ribbon and select the multileader that will receive the new leader line. You can also select the multileader, right-click, and choose **Add Leader**. Pick a location for the additional leader line arrowhead. You can place as many additional leader lines as needed without accessing the tool again. When you are finished, press [Enter], [Esc] or the space bar or right-click and select **Enter.** All of the leader lines group to form a single object.

The quickest way to remove an unneeded leader line is to pick the **Remove Leader** button from the ribbon and select the multileader object that includes the leader to remove. You can also select the multileader, right-click, and choose **Remove Leader**. Select the leader lines to remove and press [Enter], [Esc] or the space bar or right-click and select **Enter.**

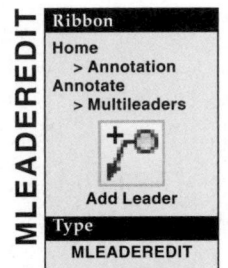

MLEADEREDIT

Ribbon
Home
> Annotation
Annotate
> Multileaders

Add Leader
Type
MLEADEREDIT

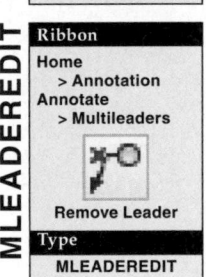

MLEADEREDIT

Ribbon
Home
> Annotation
Annotate
> Multileaders

Remove Leader
Type
MLEADEREDIT

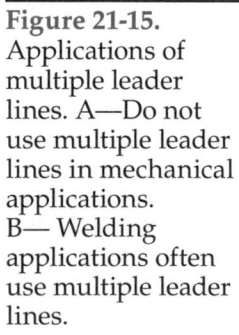

NOTE

If you type **MLEADEREDIT** to access the tool, you must activate the **Remove leaders** option to remove leader lines.

Figure 21-15. Applications of multiple leader lines. A—Do not use multiple leader lines in mechanical applications. B— Welding applications often use multiple leader lines.

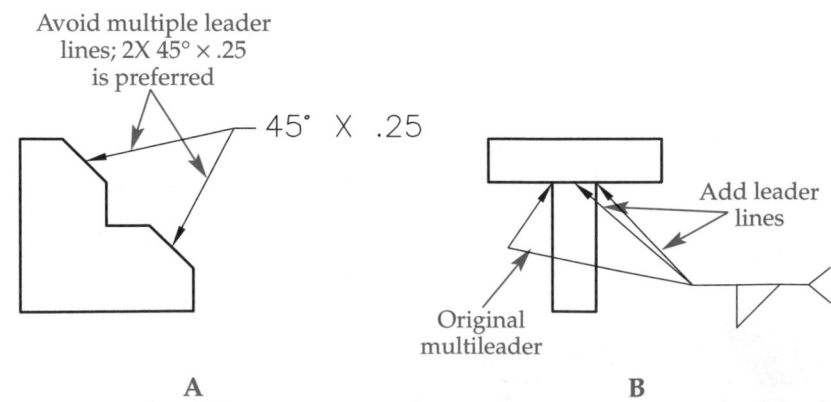

Avoid multiple leader lines; 2X 45° × .25 is preferred

45° X .25

Add leader lines

Original multileader

A   B

PROFESSIONAL TIP

To adjust the properties of a specific leader line in a group of leaders attached to the same content, hold down [Ctrl] and pick the leader to modify. Then access the **Properties** palette. Options specific to the selected leader appear, and all other properties are filtered out.

Exercise 21-11

Access the Student Web site (www.g-wlearning.com/CAD) and complete Exercise 21-11.

Aligning Multileaders

An advantage of using multileaders is the ability to space and align leaders in an easy-to-read pattern. It is good practice to determine the correct location and spacing of the leaders before placement, but you can also adjust leader spacing and alignment after you add leaders. This is a common requirement when there is a need to increase or decrease the space between leader lines, such as when the drawing scale changes, or when leaders are spaced unequally or misaligned. See **Figure 21-16**.

The **STRETCH** tool or grips are common methods for adjusting the location and alignment of leaders. You must determine the exact location of or amount of stretch applied to each leader before using these tools. An alternative is to use the **MLEADERALIGN** tool, which allows you to align and adjust the space between leaders.

Access the **MLEADERALIGN** tool and select the leaders to space and align. You can use the **MLEADERALIGN** tool to adjust the location of a single leader in reference to another leader, but for most applications, you should select several leaders. Select each

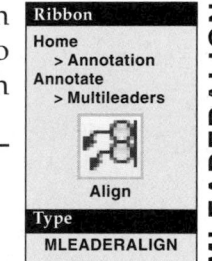

Figure 21-16.
Leaders that are equally spaced and aligned improve drawing readability.

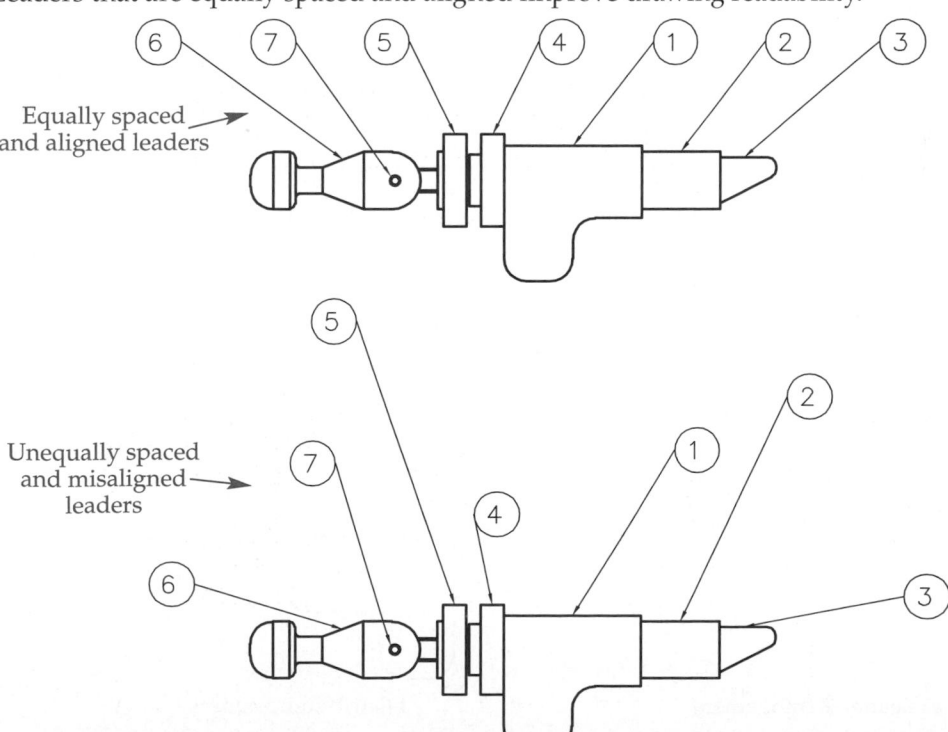

leader to space or align and right-click or press [Enter] or the space bar. When a prompt asks you to select the multileader to align to, you can activate **Options** to change the multileader alignment.

Using the Current Leader Spacing

Apply the **Use current spacing** option to align and space the selected leaders equally according to the distance between one of the selected leaders and the next closest leader. Select the multileader with which all other leaders are aligned and spaced. Then specify the direction of the leader arrangement by entering or picking a point. The space between leaders is maintained if possible, depending on the selected direction. See **Figure 21-17.**

Using the Distribute Option

Select the **Distribute** option to align and distribute the leaders, or place them at equally spaced locations between two points. The first point you pick identifies the location of one of the leaders and determines where distribution begins. The second point you pick identifies the location of the last leader. All other leaders distribute equally between the two points. Leaders align with the first point. See **Figure 21-18.**

Figure 21-17.
Using the **Use current spacing** option to align and equally space leaders.

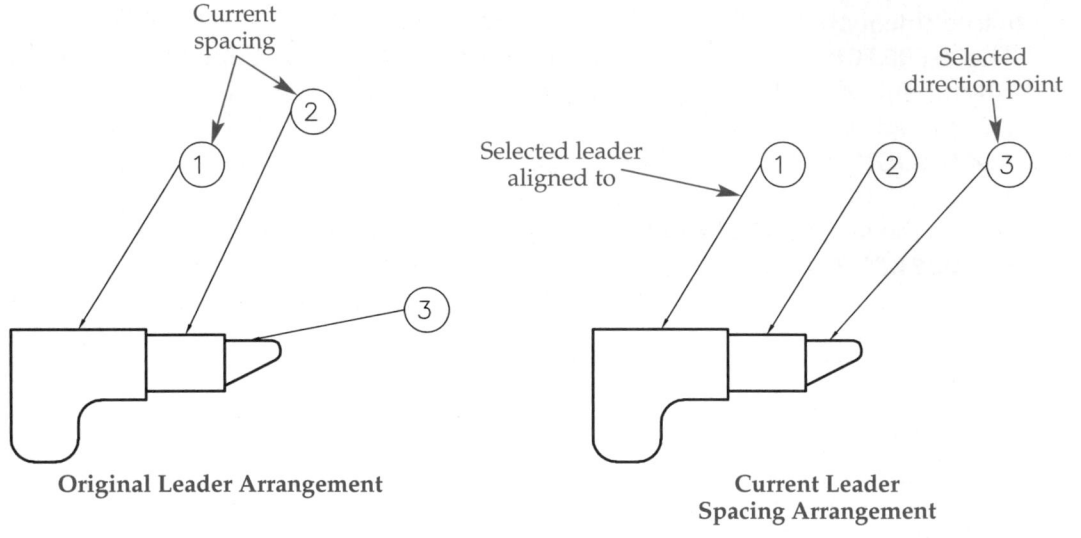

Figure 21-18.
Using the **Distribute** option to align and equally space leaders. In this example, horizontally aligned points are used.

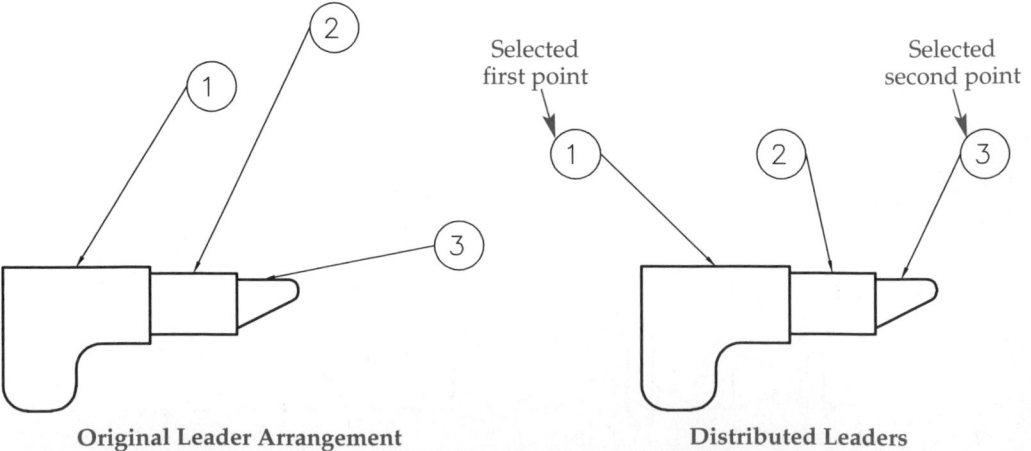

Making Leader Segments Parallel

Use the **make leader segments Parallel** option to make all the selected leader lines parallel to one of the selected leader lines. Select the existing leader to keep in the same location and at the same angle. All other leaders become parallel to this selection. The length of each leader line, except for the leader aligned to, increases or decreases in order to become parallel with the first leader. See **Figure 21-19.**

Specifying the Leader Spacing

Choose the **Specify spacing** option to align and equally space the selected leaders according to the distance, or clear space, between the extents of the content of each leader. Select the multileader with which all other leaders align and space. Finally, specify the direction of the leader arrangement by entering or picking a point. See **Figure 21-20.**

Exercise 21-12

Access the Student Web site (www.g-wlearning.com/CAD) and complete Exercise 21-12.

Figure 21-19.
Using the **make leader segments Parallel** option to make leader lines parallel to each other.

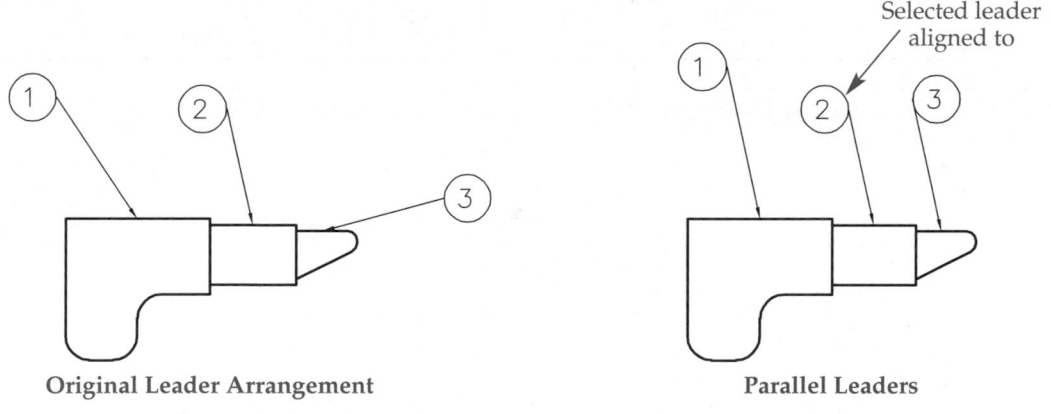

Original Leader Arrangement

Parallel Leaders

Figure 21-20.
Using the **Specify spacing** option to align and equally space leaders.

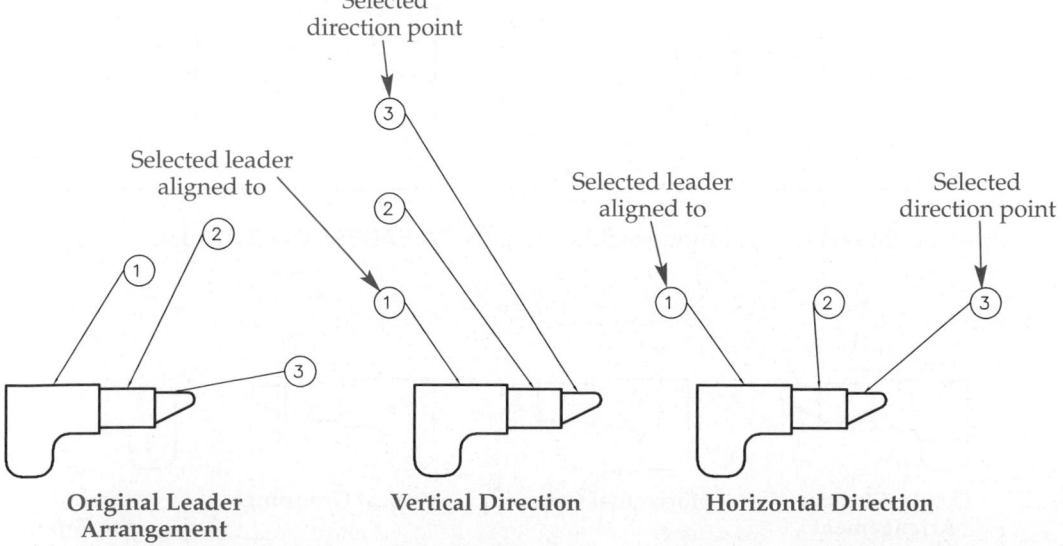

Original Leader Arrangement

Vertical Direction

Horizontal Direction

MLEADERCOLLECT

Grouping Multileaders

You can group separate multileaders created using a **Block** multileader content style using a single leader line. This practice is common when adding *balloons* to assembly drawings. *Grouped balloons* allow you to identify closely related clusters of assembly components, such as a bolt, washer, and nut. See **Figure 21-21.** Use the **MLEADERCOLLECT** tool to group multiple existing leaders together using a single leader line.

Access the **MLEADERCOLLECT** tool and select the leaders to group. The order in which you select the leaders determines how the leaders group. Select leaders in a sequential order, ending with the leader line to keep. The options illustrated in **Figure 21-22** become available after you select the leaders.

Select the **Horizontal** option to align the grouped content horizontally or the **Vertical** option to align the grouped content vertically. Pick a point to locate the grouped leader. Select the **Wrap** option to wrap the grouped content to additional lines as needed when the number of items exceeds a specified width or quantity. Enter the width at the Specify width prompt, or use the **Number** option to enter a quantity not to exceed before the grouped leaders wrap. Then pick a point to locate the grouped leader.

balloons: Circles that contain a number or letter to identify the part and correlate it to a parts list or bill of materials. Balloons connect to a part with a leader line.

grouped balloons: Balloons that share the same leader, which typically connects to the most obviously displayed component.

> **NOTE**
>
> You can only use the **MLEADERCOLLECT** tool to group symbols attached to leaders created using the **Block** content style.

Figure 21-21.
An example of grouped balloons identifying closely related parts. Often some parts or features are hidden.

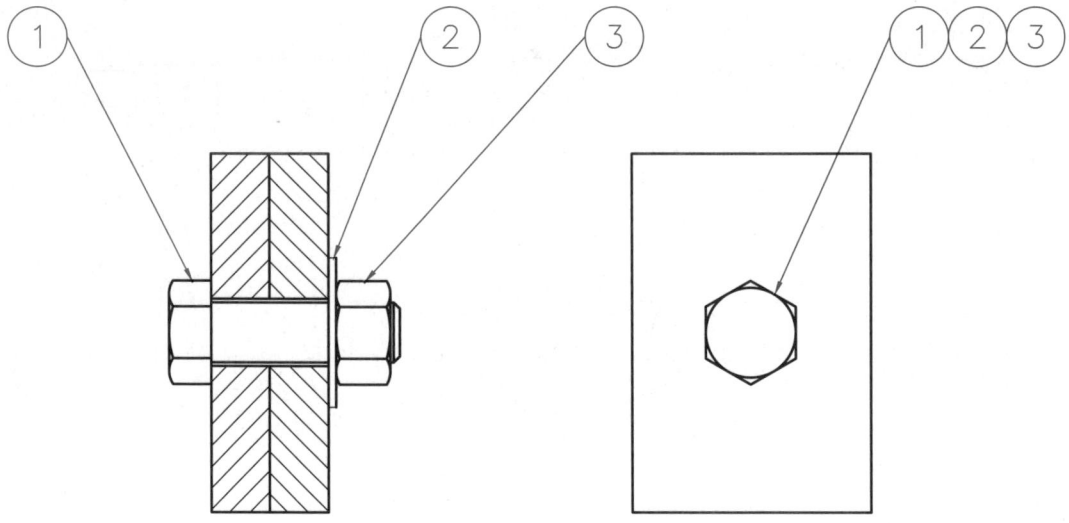

Figure 21-22.
Examples of options for grouping leaders using the **MLEADERCOLLECT** tool.

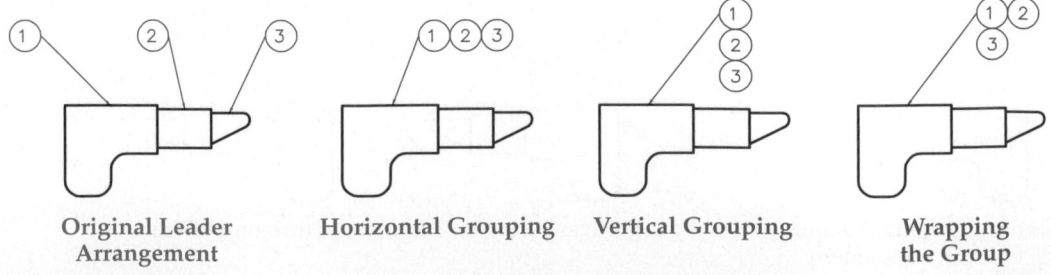

Original Leader Arrangement   Horizontal Grouping   Vertical Grouping   Wrapping the Group

Chapter Test

Answer the following questions. Write your answers on a separate sheet of paper or go to the Student Web site (www.g-wlearning.com/CAD) and complete the electronic chapter test.

1. Define *associative dimension*.
2. Why is it important to have associative dimensions for editing objects?
3. Which **Options** dialog box setting controls associative dimensioning?
4. Which tool allows you to convert nonassociative dimensions to associative dimensions?
5. Which tool allows you to convert associative dimensions to nonassociative dimensions?
6. What are definition points?
7. Which four tool options related to dimension editing appear in the shortcut menu accessed when you select a dimension?
8. Name three methods of changing the dimension style of a dimension.
9. How does the **Dimension Update** tool affect selected dimensions?
10. Explain how to add a diameter symbol to a dimension text value using the **DDEDIT** tool.
11. Name the tool that allows you to control the placement and orientation of an existing associative dimension text value.
12. Name two applications in which you might need to create oblique extension lines.
13. Which tool and option can you use to add a new baseline dimension to an existing set of baseline dimensions?
14. When you use the **Properties** palette to edit a dimension, what is the effect on the dimension style?
15. How do you access the **Property Settings** dialog box?
16. Which tool can you use to adjust the space equally between dimension lines or align dimension lines without having to determine the exact location or amount of stretch needed?
17. What two options are available when you use the **Multiple** option of the **DIMBREAK** tool?
18. What tool allows you to add information about how frequently the dimension should be tested for consistency and tolerance during the manufacturing of a product?
19. Name an application in which leaders with multiple leader lines are commonly used.
20. Identify the four options available to change the multileader alignment.

Drawing Problems

- *Start AutoCAD if it is not already started.*
- *Start a new drawing using an appropriate template of your choice. The template should include layers, text styles, dimension styles, and multileader styles appropriate for drawing the given objects.*
- *Add layers, text styles, dimension styles, and multileader styles as needed. Draw all objects using appropriate layers, text styles, dimension styles, multileader styles, justification, and format.*
- *Follow the specific instructions for each problem. Use your own judgment and approximate dimensions when necessary.*
- *Apply dimensions accurately using ASME or appropriate industry standards. Use object snap modes to your best advantage.*
- *For mechanical drawings, place the following general notes 1/2" from the lower-left corner:*

  NOTES:
  1. INTERPRET DIMENSIONS AND TOLERANCES PER ASME Y14.5M-1994.
  2. REMOVE ALL BURRS AND SHARP EDGES.
  3. UNLESS OTHERWISE SPECIFIED, ALL DIMENSIONS ARE IN INCHES (or MILLIMETERS as applicable).

▼ Basic

1. Open P18-9 and save the file as P21-1. The P21-1 file should be active. Edit the drawing as follows:
   A. Erase the front (circular) view.
   B. Stretch the vertical dimensions to provide more space between dimension lines. Be sure the space you create is the same between all vertical dimensions.
   C. Stagger the existing vertical dimension text numbers if they are not staggered as shown in the original problem.
   D. Erase the 1.750 horizontal dimension and then stretch the 5.255 and 4.250 dimensions to make room for a new datum dimension from the baseline to where the 1.750 dimension was located. This should result in a new baseline dimension that equals 2.750. Be sure all horizontal dimension lines are equally spaced.

2. Open P19-1 and save the file as P21-2. The P21-2 file should be active. Edit the drawing as follows:
   A. Stretch the total length from 3.500 to 4.000, leaving the holes the same distance from the edges.
   B. Fillet the upper-left corner. Modify the 3X R.250 dimension accordingly.

3. Open P18-12 and save the file as P21-13. The P21-3 file should be active. Edit the drawing as follows:
   A. Make the bathroom 8'-0" wide by stretching the walls and vanity that are currently 6'-0" wide to 8'-0". Do this without increasing the size of the water closet compartment. Provide two equally spaced oval sinks where there is currently one.

4. Open P19-18 and save the file as P21-4. The P21-4 file should be active. Edit the drawing as follows:
   A. Lengthen the part .250 on each side for a new overall dimension of 6.500.
   B. Change the width of the part from 3.000 to 3.500 by widening an equal amount on each side.

5. Open P19-16 and save the file as P21-5. The P21-5 file should be active. Edit the drawing as follows:
   A. Shorten the .75 thread on the left side to .50.
   B. Shorten the .388 hexagon length to .300.

6. Draw the shim as shown at A. Then edit the .150 and .340 values using oblique dimensions as shown at B. Save the drawing as P21-6.

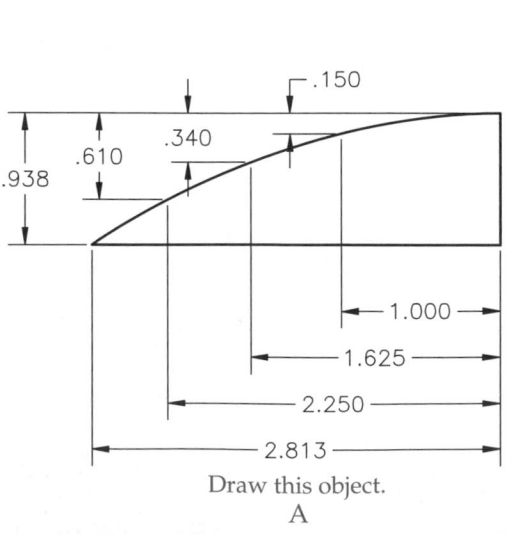

Draw this object.
A

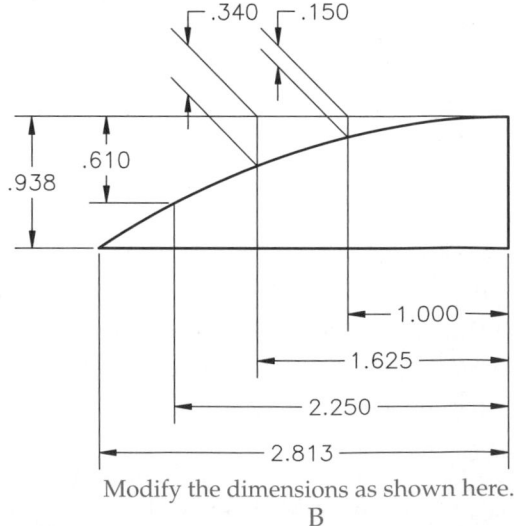

Modify the dimensions as shown here.
B

Title: Shim

7. Open P19-4 and save the file as P21-7. The P21-7 file should be active. Edit the drawing as follows:
   A. Use the existing drawing as the model and make four copies.
   B. Leave the original drawing as it is and edit the other four pins in the following manner, keeping the Ø.125 hole exactly in the center of each pin.
   C. Give one pin a total length of 1.500.
   D. Create the next pin with a total length of 2.000.
   E. Edit the third pin to a length of 2.500.
   F. Change the last pin to a length of 3.000.
   G. Organize the pins on your drawing in a vertical row ranging in length from the smallest to the largest. You may need to change the drawing limits.

8. Open P19-5 and save the file as P21-8. The P21-8 file should be active. Edit the drawing as follows:
   A. Modify the spline to have twelve projections, rather than eight.
   B. Change the angular dimension, linear dimension, and 8X dimension to reflect the modification.

9. Open P19-11 and save the file as P21-9. The P21-9 file should be active. Edit the drawing as follows:
   A. Stretch the total length from 6.500 to 7.750.
   B. Add two more holes that continue the equally spaced pattern of .625 apart.
   C. Change the 8X .625(=5.00) dimension to read 10X .625(=6.250).

▼ Advanced

10. Draw the stairs cross section using the dimensions and notes provided. Use oblique dimensions where necessary. Save the drawing as P21-10.

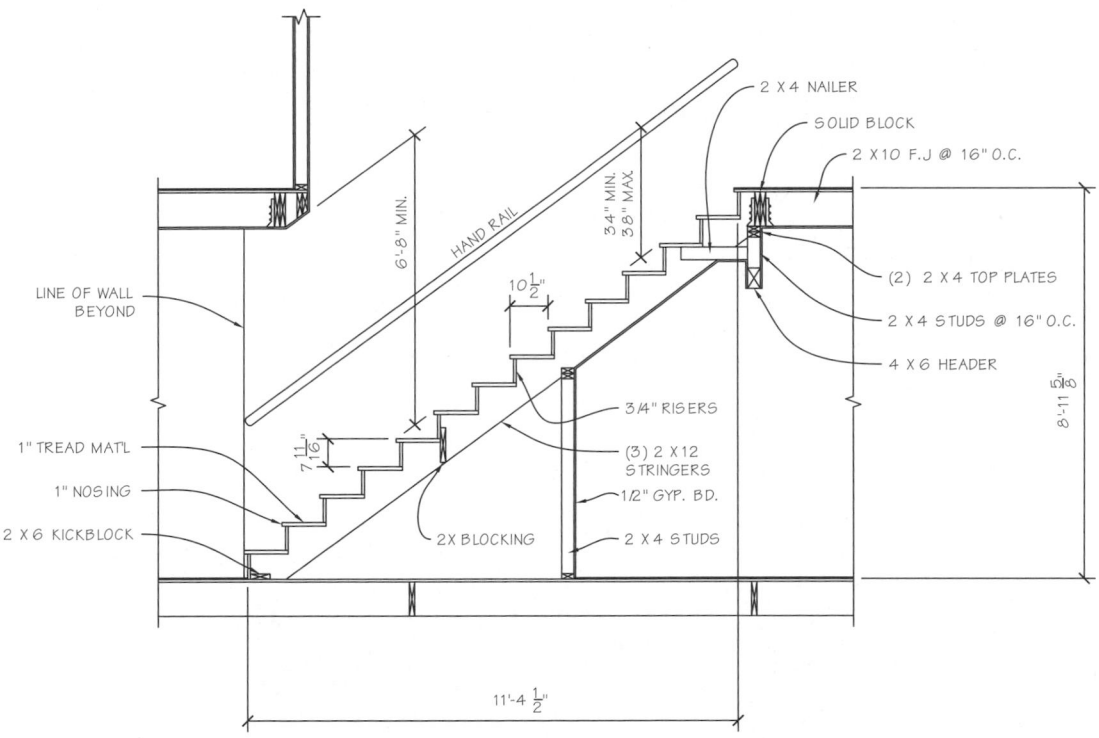

Drawing Problems - Chapter 21

11. Use a word processor to write a report of at least 250 words explaining the importance of associative dimensioning. Site at least three examples from actual industry applications. Show at least four drawings illustrating your report.

12. Open P18-17 and save the file as P21-12. The P21-12 file should be active. Make the client-requested revisions to the floor plan as shown. Make sure the dimensions reflect the changes.

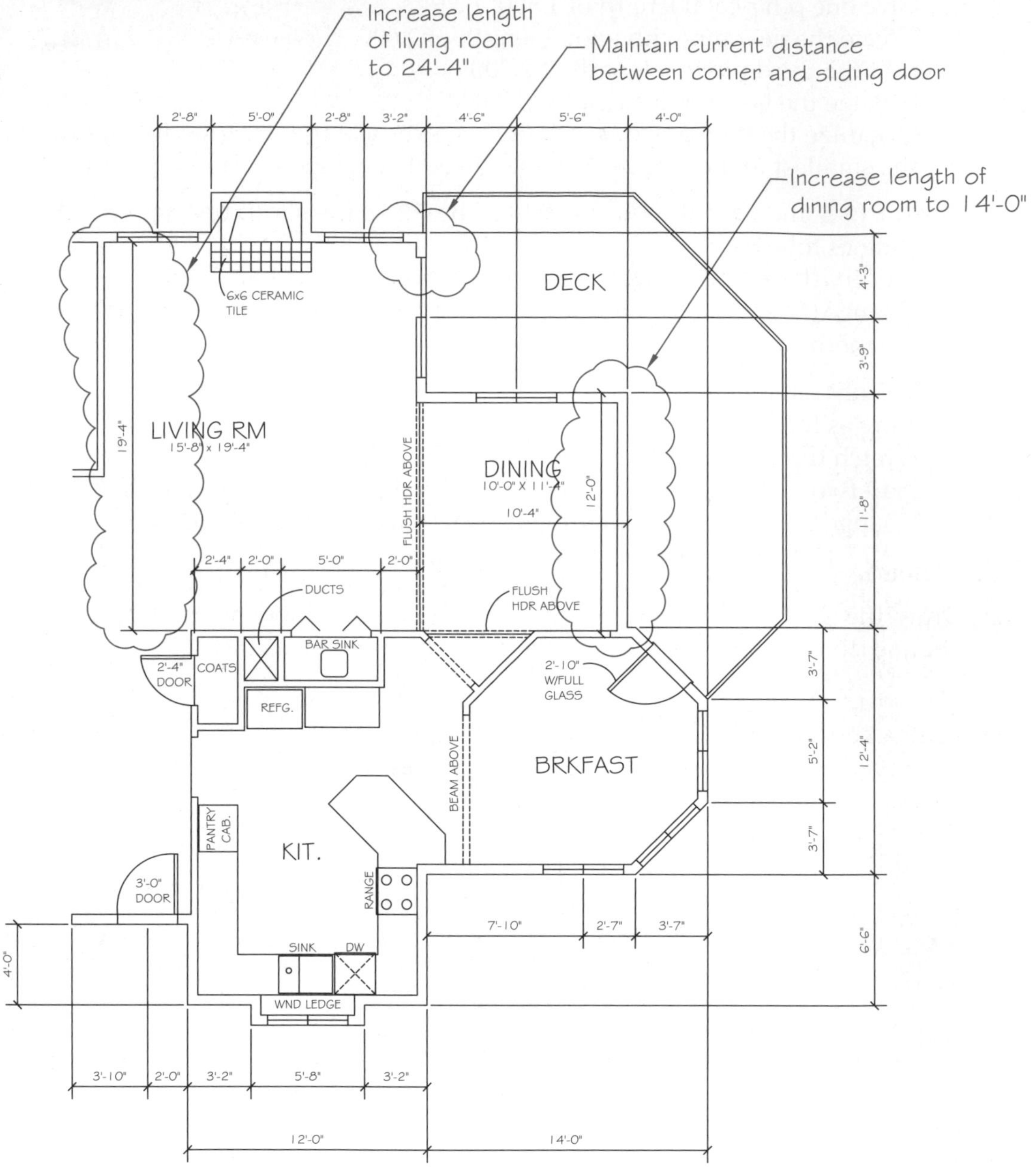

Increase length of living room to 24'-4"

Maintain current distance between corner and sliding door

Increase length of dining room to 14'-0"

2'-8" 5'-0" 2'-8" 3'-2" 4'-6" 5'-6" 4'-0"

6x6 CERAMIC TILE

DECK

4'-3"

3'-9"

19'-4"

LIVING RM
15'-8" x 19'-4"

FLUSH HDR ABOVE

DINING
10'-0" X 11'-4"

10'-4"

12'-0"

11'-8"

2'-4" 2'-0" 5'-0" 2'-0"

DUCTS

FLUSH HDR ABOVE

BAR SINK

FLUSH HDR ABOVE

2'-4" DOOR COATS

REFG.

2'-10" W/FULL GLASS

3'-7"

PANTRY CAB.

KIT.

BEAM ABOVE

RANGE

BRKFAST

5'-2"

12'-4"

3'-0" DOOR

3'-7"

SINK DW

7'-10" 2'-7" 3'-7"

6'-6"

WND LEDGE

4'-0"

3'-10" 2'-0" 3'-2" 5'-8" 3'-2"

12'-0" 14'-0"

Drawing Problems - Chapter 21

AutoCAD and Its Applications—Basics

Parametric Drafting

Learning Objectives

After completing this chapter, you will be able to do the following:

✓ Explain parametric drafting processes and applications.
✓ Create and edit parametric drawings.
✓ Add and manage geometric constraints.
✓ Add and manage dimensional constraints.
✓ Convert dimensional constraints.

Parametric drafting tools allow you to assign *parameters*, or *constraints*, to objects. The parametric concept, also known as *intelligence*, provides a way to associate objects and limit design changes. You cannot change a constraint so that it conflicts with other parametric geometry. A database stores and allows you to manage all parameters. You typically use parametric tools with standard drafting practices to create a more interactive drawing.

> **parametric drafting:** A form of drafting in which parameters and constraints drive object size and location to produce drawings with features that adapt to changes made to other features.

> **parameters (constraints):** Geometric characteristics and dimensions that control the size, shape, and position of drawing geometry.

Parametric Fundamentals

AutoCAD 2010
NEW

Parametric drafting allows you to control every aspect of a drawing during and after the design and documentation process. You should always use standard drafting practices and drawing aids to construct initial geometry. When used correctly, these techniques allow you to produce accurate drawings efficiently. You typically use parametric tools as a supplement or drawing aid. First, create a drawing using the same tools and techniques you use to create a non-parametric drawing. Then add parameters using *geometric constraints* and *dimensional constraints*.

Understanding Constraints

Well-defined constraints allow you to incorporate and preserve specific design intentions and increase revision efficiency. For example, if two holes through a part, drawn as circles, must always be the same size, use a geometric constraint to make the circles equal and add a dimensional constraint to size one of the circles. The size of both circles changes when you modify the dimensional constraint value. See **Figure 22-1.**

> **geometric constraints:** Geometric characteristics applied to restrict the size or location of geometry.

> **dimensional constraints:** Measurements that numerically control the size or location of geometry.

Figure 22-1.
An example of a
basic parametric
relationship. The
dimensional
constraint controls
the size of both
circles with
the aid of an
equal geometric
constraint.

Icon indicates
dimensional constraint

dia1=1.000 🔒

Dimensional constraint
added to one circle only

Equal geometric
constraints

=   =

Original Circles

Value controls
circle diameter

Circle diameter changes
parametrically

dia1=2.000 🔒

=   =

Revised Circles

You must add constraints to make an object parametric. Dimensional constraints create parameters that direct object size and location. A traditional associative dimension is associated with an object, but it does not control object size or location. **Figure 22-2** shows an example constraint levels, including *under-constrained*, *fully constrained*, and *over-constrained*. As you progress through the design process, you will often fully or near fully constrain the drawing to ensure that your design is accurate. However, if you attempt to over-constrain the drawing, a message appears allowing you to decide what to do. See **Figure 22-3**. AutoCAD does not allow you to over-constrain a drawing, as shown by the *reference dimension* in **Figure 22-2**.

Figure 22-4 shows an extreme example of constraining, for reference only. Study the figure to help understand how constraints work, and how applying constraints differs from and compliments traditional drafting. Typically, you should prepare initial objects as accurately as possible using the many tools and drawing aids available, and then add constraints.

Parametric Applications

Parametric tools aid the design and revision process, place limits on geometry to preserve design intent, and help form geometric constructions. You may want to consider using constraints to help maintain relationships between objects in a drawing, especially during the design process, when changes are often frequent. However, you must decide if the additional steps required to make a drawing parametric are appropriate and necessary for your applications.

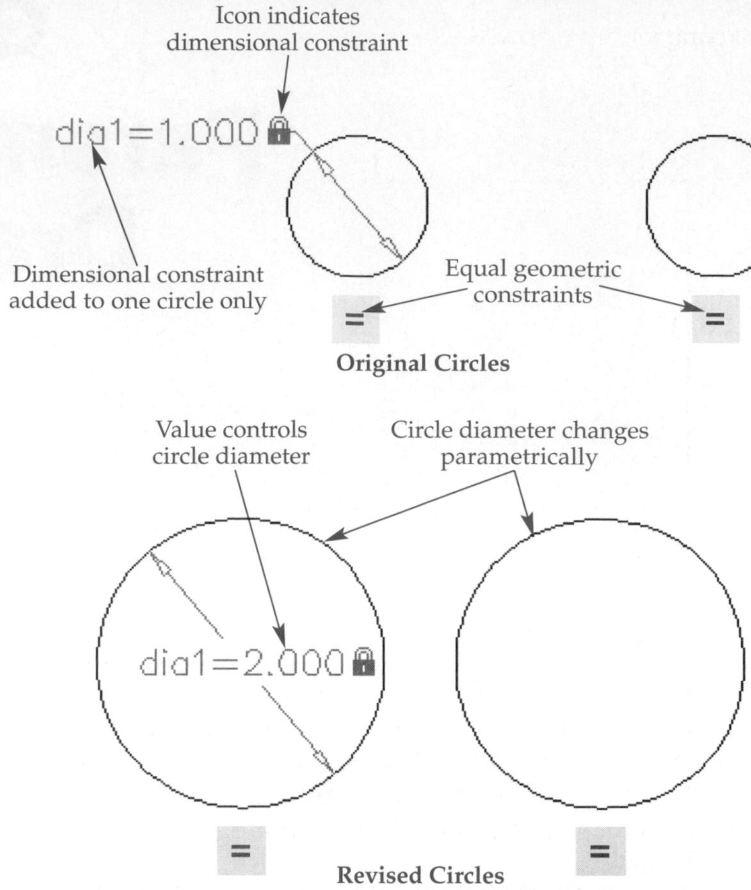

under-constrained: Describes a drawing that includes constraints, but not enough to size and locate all geometry.

fully constrained: Describes a drawing in which objects have no freedom of movement.

over-constrained: Describes a drawing that contains too many constraints.

reference dimension: A dimension used for reference purposes only. Parentheses enclose reference dimensions to differentiate them from other dimensions.

Figure 22-2.
Levels of parametric constraint.

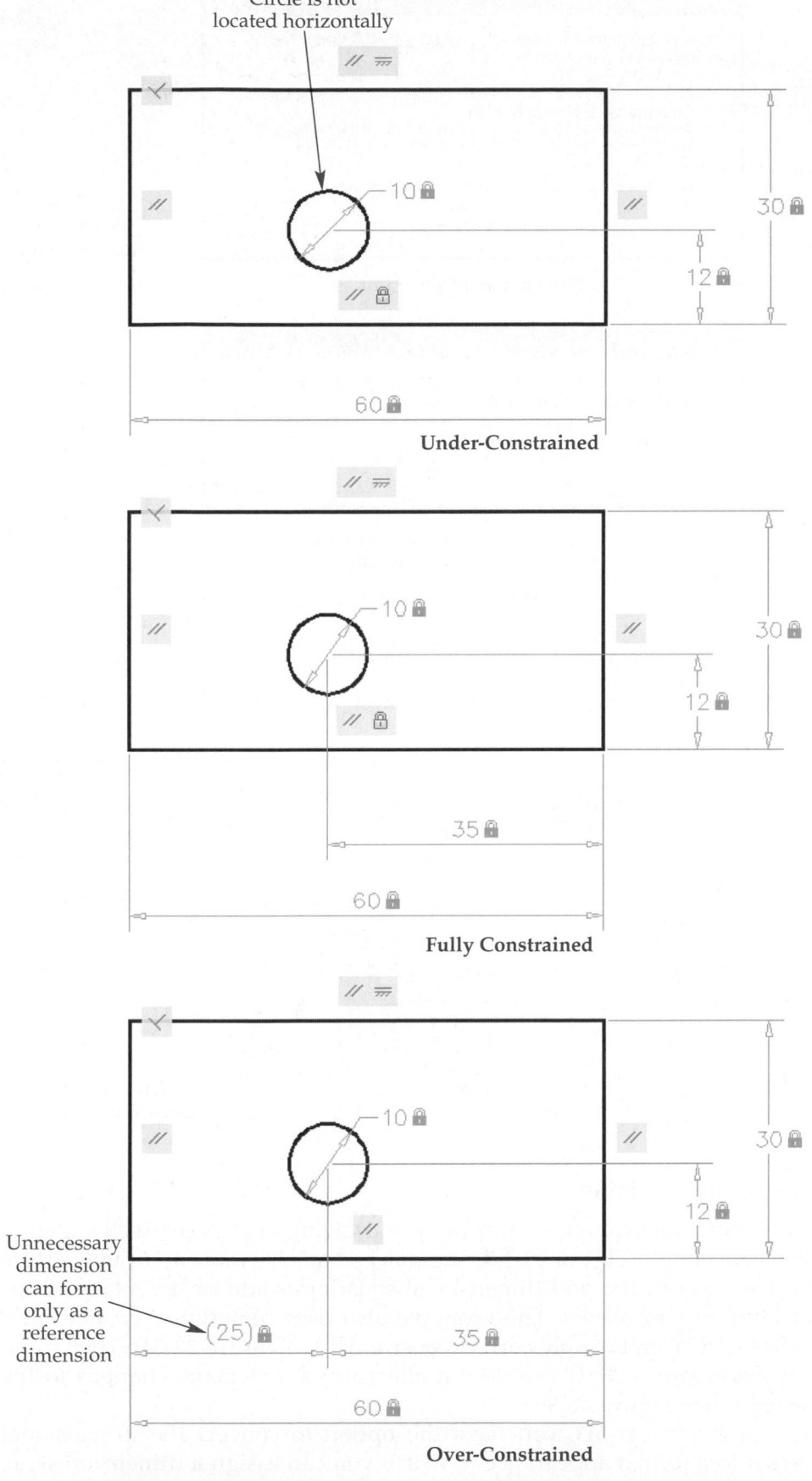

Circle is not located horizontally

10

30

12

60

Under-Constrained

10

30

12

35

60

Fully Constrained

10

30

12

Unnecessary dimension can form only as a reference dimension

(25)

35

60

Over-Constrained

Figure 22-3.
Error messages that appear when you attempt to over-constrain objects.

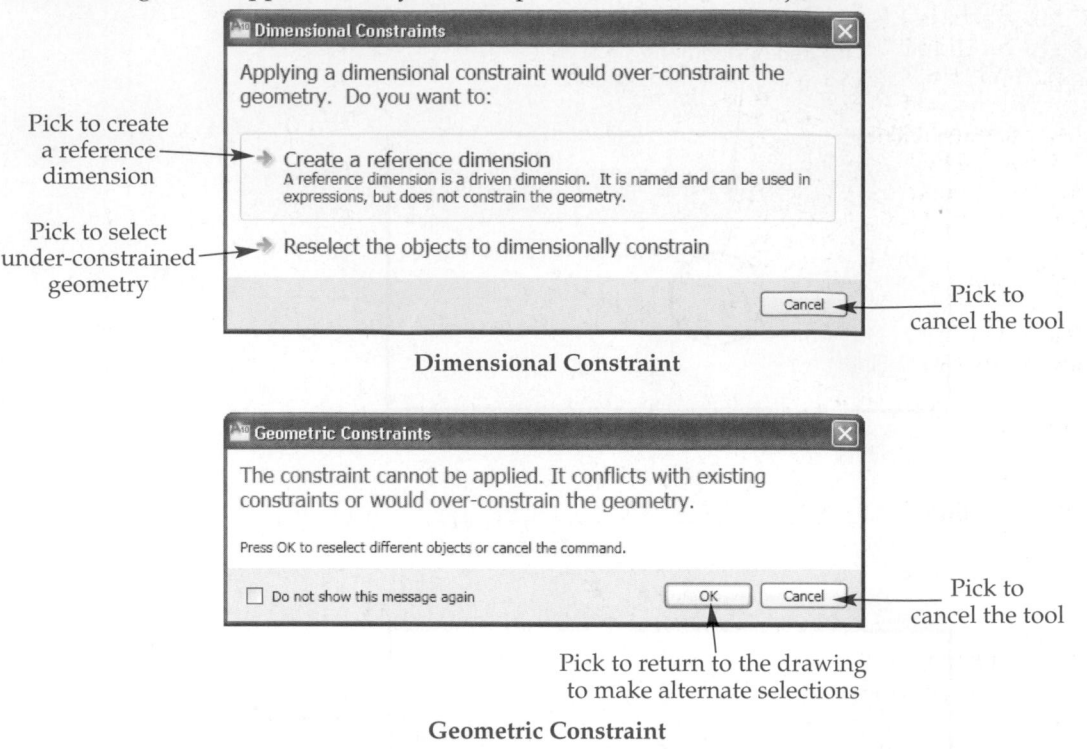

Pick to create a reference dimension

Pick to select under-constrained geometry

Pick to cancel the tool

Dimensional Constraint

Pick to cancel the tool

Pick to return to the drawing to make alternate selections

Geometric Constraint

Figure 22-4.
An extreme example of constraining a drawing to help you understand how you apply constraints.

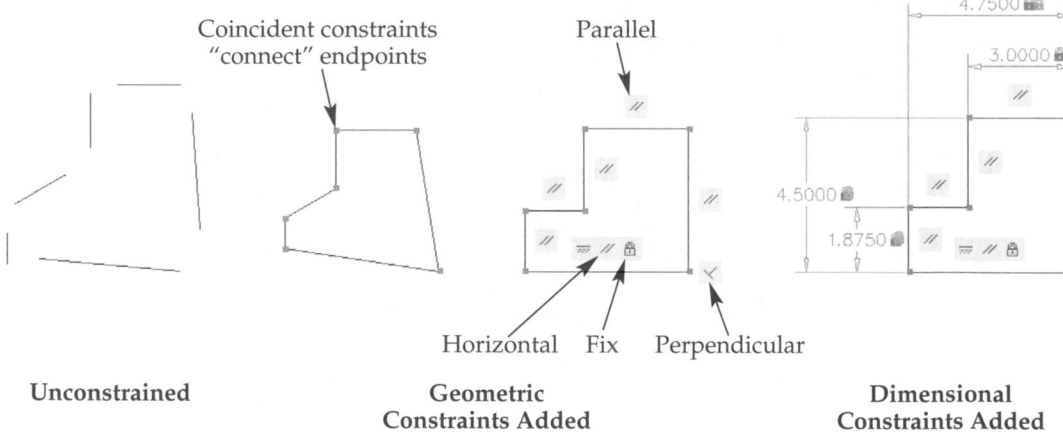

Coincident constraints "connect" endpoints

Parallel

Horizontal   Fix   Perpendicular

Unconstrained

Geometric Constraints Added

Dimensional Constraints Added

Product Design and Revision

Figure 22-5 shows an example of a front view drawing of a spacer, well suited to parametric construction. In **Figure 22-5A,** standard AutoCAD tools accurately create the geometry. Next, geometric and dimensional constraints add object relationships and size and location parameters. This example also uses centerlines, created on a separate construction layer, to apply correct constrains. See **Figure 22-5B.** Then, with parameters in place, you can explore design alternatives and make changes to the drawing efficiently. See **Figure 22-5C.**

As shown in **Figure 22-5D,** you have the option to convert the dimensional constraint format to a formal appearance to which you can assign a dimension style. You can still use converted dimensions to adjust geometry parametrically. In this

Figure 22-5.
The front view of this spacer is a good candidate for parametric associations. A—Accurate view geometry constructed using standard AutoCAD practices. B—Adding geometric and dimensional constraints to constrain the drawing. C—Changing the values of a few dimensional constraints to update the entire drawing. D—Reusing dimensional constraints to help prepare a formal drawing.

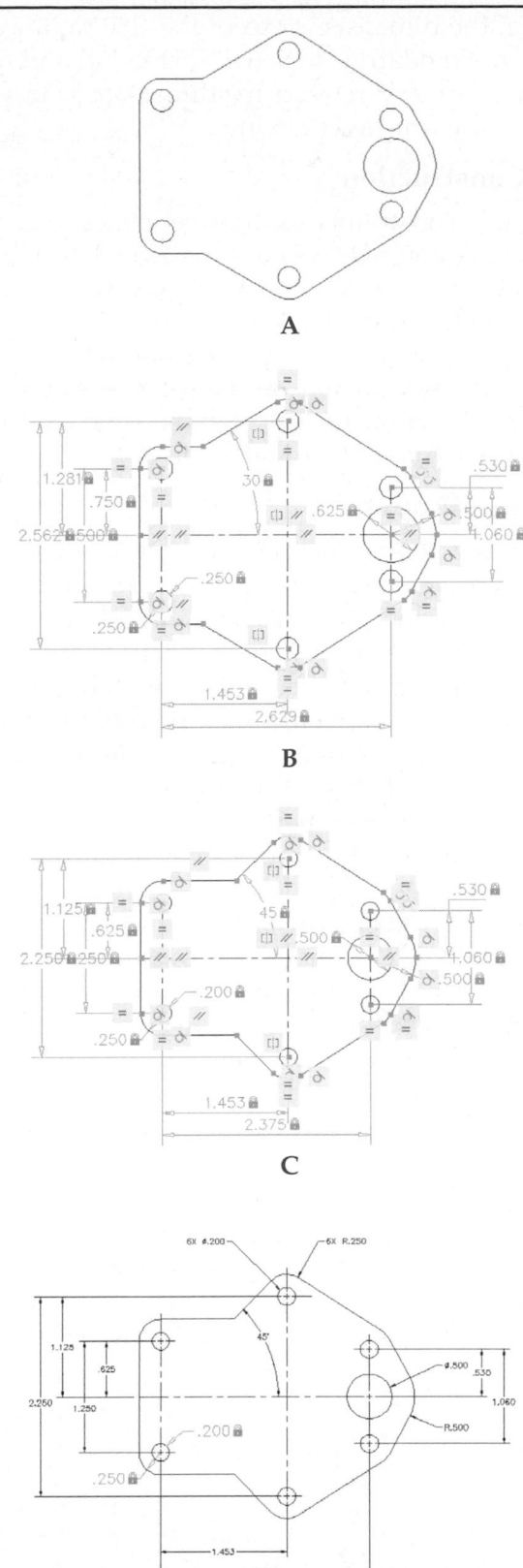

A

B

C

D

example, all of the dimensions except the .250 radius and diameter dimensions have been converted. Standard associative .250 radial and diameter dimensions allow you to add the 6X prefix and relocate the dimensions. You can then hide constraint information to view the finished drawing.

Geometric Construction

You can use constraints to form geometric constructions in specific situations when standard AutoCAD tools are inefficient or ineffective. For example, suppose you know that the angle of a line is 30°, and you know the line is tangent to a circle. However, you do not know the length of the line or the location of the line endpoints. One option is to position a 30° construction line, using the **Ang** option of the **XLINE** tool, anywhere in the drawing. See **Figure 22-6A.** Use a **Tangent** geometric constraint to form a tangent relationship between the xline and circle. See **Figure 22-6B.** You can then hide or delete the constraint if necessary.

Unsuitable Applications

You may find that parametric drafting is unsuitable or ineffective for some applications. For example, it may be unsuitable to add parameters to a drawing if the drawing is of a finalized product that will not require extensive revision, or if you can easily modify drawing geometry without associating objects.

In addition, if your drawing includes a large number of objects, you may find it cumbersome to add the constraints required to form an intelligent drawing. For instance, you can use constraints to form all desired relationships between objects in a floor plan. See **Figure 22-7.** In this example, dimensional constraints can specify wall thickness, position windows between walls, locate sinks on vanities, and form many other parametric relationships. You can then adjust dimensional constraints as needed to update the drawing.

If you effectively constrain *all* objects shown in **Figure 22-7,** you have the ability to change the 11′-10 1/2″ dimensional constraint to increase the width of the master bedroom, for example. The entire floor plan adjusts to the modified room size. Consider, however, what this process requires. You must constrain all wall endpoints; the points where doors and windows meet walls; the distance between walls and objects, such as cabinets, sinks, and water closets; and form all other geometric and dimensional constraints.

Figure 22-6.
A—A 30° xline placed near an associated circle. B—Using a **Tangent** geometric constraint to form a tangent construction. Notice the appropriate selection process. C—The final drawing.

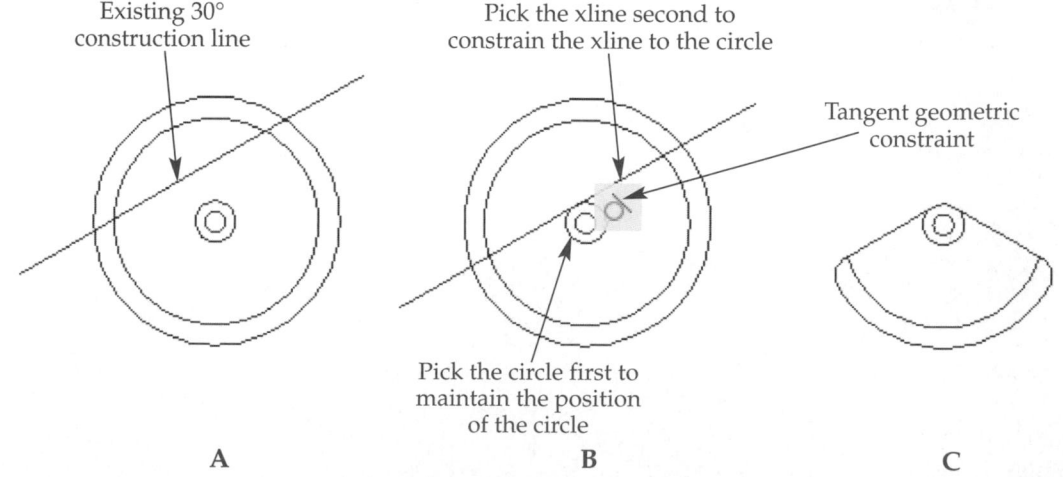

AutoCAD and Its Applications—Basics

Figure 22-7.
An architectural floor plan usually includes too many objects to effectively and efficiently constrain.

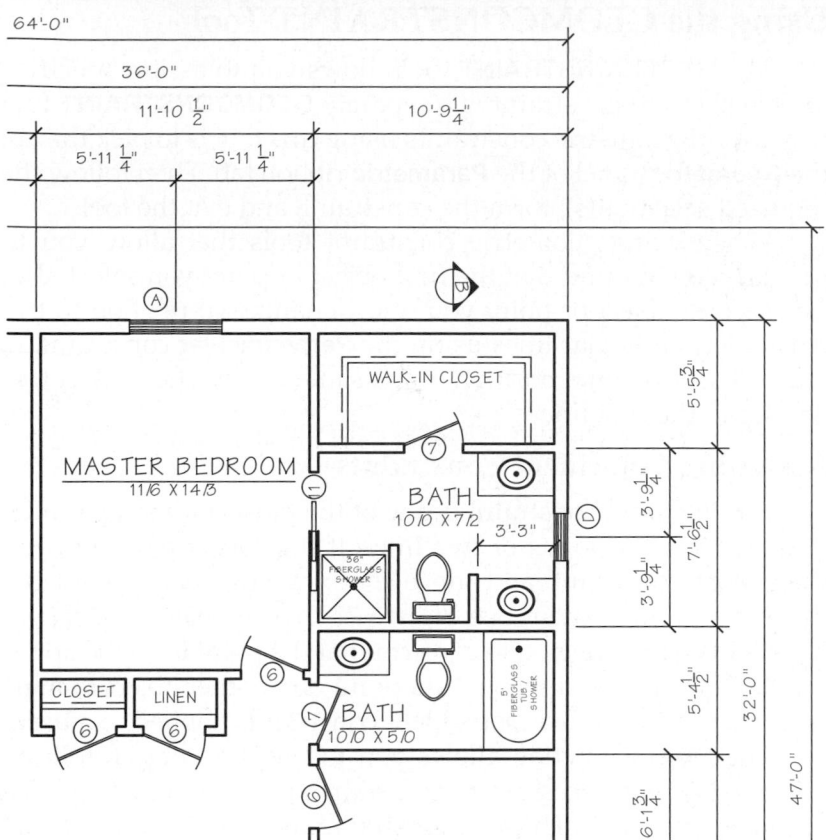

PROFESSIONAL TIP

Constraints can prove effective for multiview layout, in order to help maintain alignment between views.

Exercise 22-1

Access the Student Web site (www.g-wlearning.com/CAD) and complete Exercise 22-1.

Adding Geometric Constraints

Geometric constraint tools allow you to add the geometric relationships required to build a parametric drawing. You should typically add geometric constraints, or at least a portion of the necessary geometric constraints, before dimensional constraints to help preserve design intent. By default, a constraint-specific icon is visible to indicate the presence of a geometric constraint. Once you have placed geometric constraints, you can view, adjust, and remove them as needed.

Using the GEOMCONSTRAINT Tool

Type
GEOMCONSTRAINT
GCON

The **GEOMCONSTRAINT** tool allows you to assign specific geometric constraints to objects. Each constraint is a separate **GEOMCONSTRAINT** tool option. The quickest way to add geometric constraints using this tool is to pick the appropriate button from the **Geometric** panel of the **Parametric** ribbon tab. Then follow the prompts to make the required selection(s), form the constraint, and exit the tool.

When using geometric constraint tools that allow you to pick two objects or points, keep in mind that the first object or point you select always remains the same. The second object or point you select changes in relation to the first. For example, to create perpendicular lines using the **Perpendicular** constraint, first select the line that remains in the same position at the same angle. Then select the line to make perpendicular to the first line.

coincident:
A geometric
construction that
specifies two points
sharing the same
position.

Assigning Coincident Constraints

A *coincident* constraint is one of the most common parametric constructions. For example, the endpoints of two lines at the corner of a rectangle coincide. Access the **Coincident** constraint, and move the pick box near a point on an existing object to display a point marker. See **Figure 22-8.** The object associated with a specific point highlights, allowing you to confirm that the point is on the appropriate object. Pick the marked location, and then pick a point on another object to make the two points coincide. **Figure 22-9** shows coincident constraints applied to a basic multiview drawing.

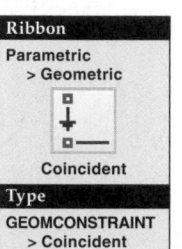

Ribbon
Parametric
> Geometric

Coincident
Type
GEOMCONSTRAINT
> Coincident

The **Object** function allows you to select an object, followed by a point. This is useful when it is necessary to constrain a point along a curve. The point does not have to contact the curve. The **Autoconstrain** function allows you to select multiple objects to form coincident constraints at every possible coincident intersection, in a single operation. Right-click or press [Enter] or the space bar to use the **Autoconstrain** function.

> **NOTE**
>
> Connected polyline segments already include a form of coincident constraint, although no icon displays.

Figure 22-8.
Examples of objects with selectable constraint points. Study these points, because they are the same points used to add other geometric constraints when point selection is necessary.

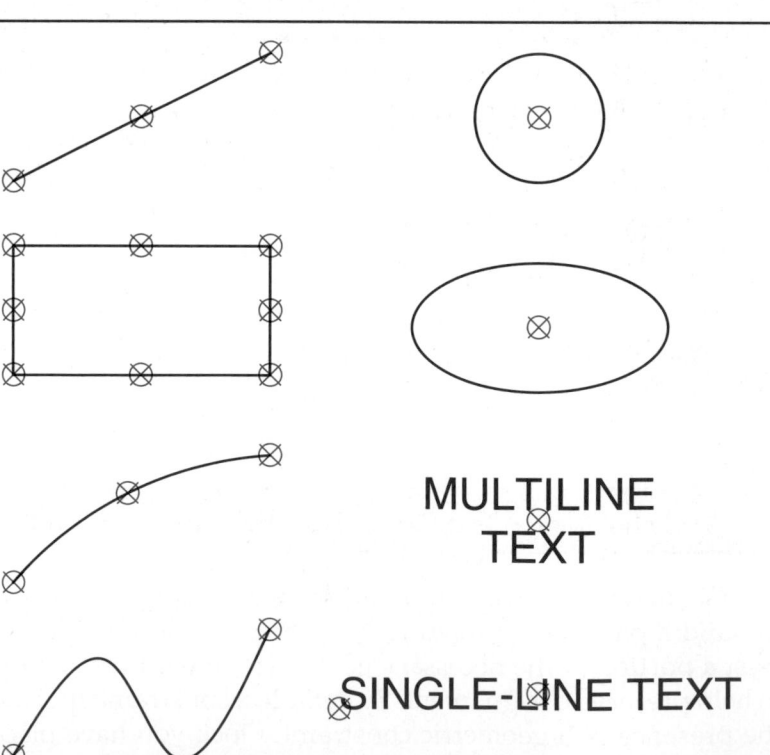

Figure 22-9.
Coincident constraints required for a basic multiview drawing. For clarity, labels do not indicate all required coincident constraints. This drawing is used as an example throughout this chapter.

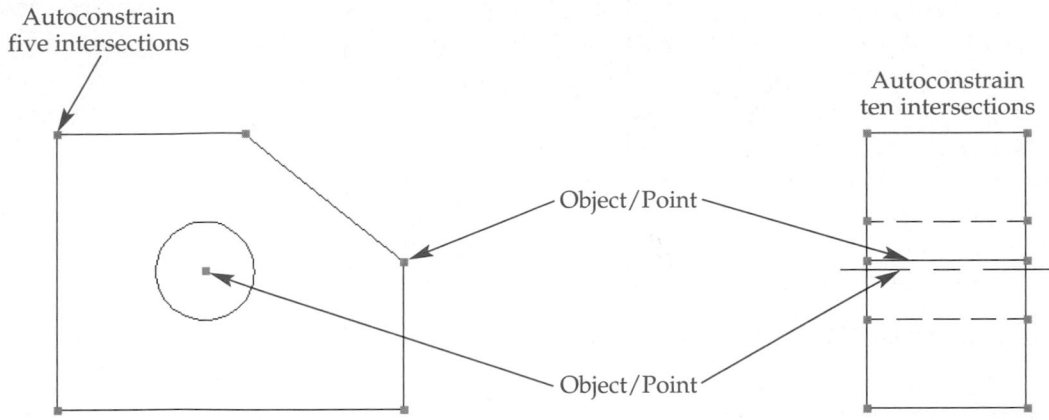

Autoconstrain five intersections

Autoconstrain ten intersections

Object/Point

Object/Point

PROFESSIONAL TIP

When you are selecting points, be sure to select the points corresponding to the surface to constrain.

Exercise 22-2

Access the Student Web site (www.g-wlearning.com/CAD) and complete Exercise 22-2.

Assigning Horizontal and Vertical Constraints

The **Horizontal** constraint aligns lines, polylines, or points along the X axis, or horizontally. The **Vertical** constraint aligns lines, polylines, or points along the Y axis, or vertically. These constraints are commonly used to define a horizontal or vertical surface datum or to align points. Access the **Horizontal** or **Vertical** constraint and select the object to constrain. Both options include a **2Points** function that allows you to pick two points to align horizontally or vertically. See **Figure 22-10.**

Assigning Parallel and Perpendicular Constraints

The **Parallel** constraint creates a *parallel* constraint between lines or polylines. The **Perpendicular** constraint creates a *perpendicular* constraint between lines or polylines. These constraints are used for a variety of applications. Access the **Parallel** or **Perpendicular** constraint, and select two lines or polylines, or a line and a polyline. See **Figure 22-11.**

Exercise 22-3

Access the Student Web site (www.g-wlearning.com/CAD) and complete Exercise 22-3.

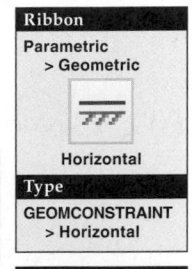

Ribbon
Parametric
> Geometric

Horizontal

Type
GEOMCONSTRAINT
> Horizontal

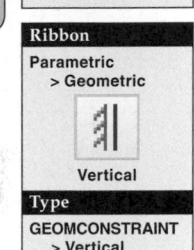

Ribbon
Parametric
> Geometric

Vertical

Type
GEOMCONSTRAINT
> Vertical

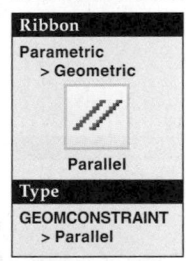

Ribbon
Parametric
> Geometric

Parallel

Type
GEOMCONSTRAINT
> Parallel

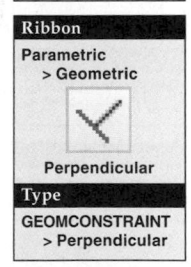

Ribbon
Parametric
> Geometric

Perpendicular

Type
GEOMCONSTRAINT
> Perpendicular

parallel: A geometric construction that specifies that objects such as lines will never intersect, no matter how long they become.

perpendicular: A geometric construction that defines a 90° angle between objects such as lines.

Figure 22-10.
Examples of vertically and horizontally constrained objects and points.

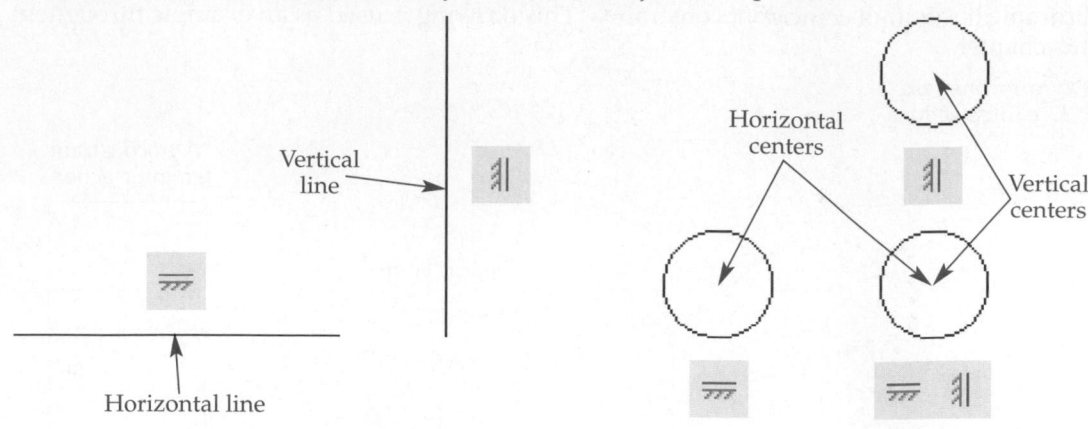

Figure 22-11.
Parallel and perpendicular constraints required for the example multiview drawing.

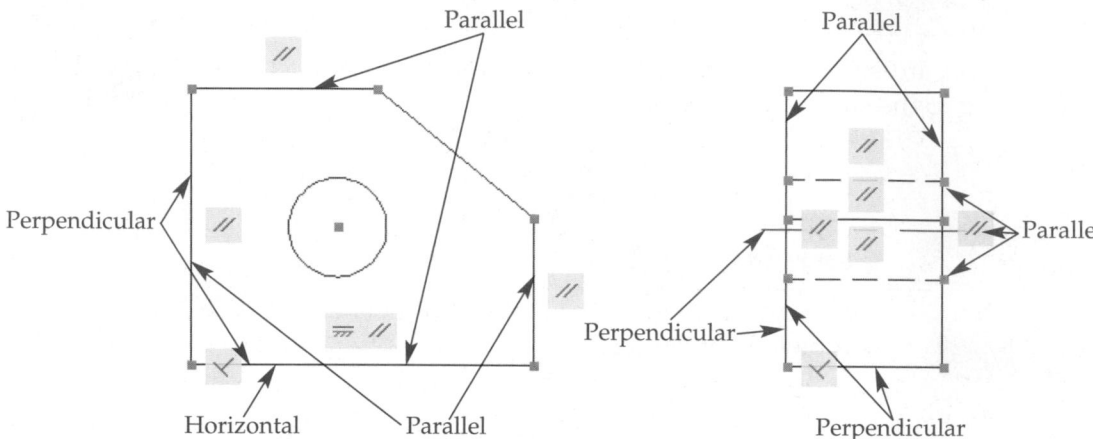

Assigning Collinear and Tangent Constraints

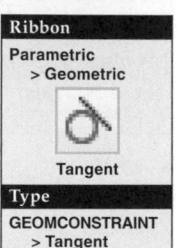
Ribbon
Parametric
> Geometric

Collinear

Type
GEOMCONSTRAINT
> Collinear

The **Collinear** constraint allows you to align two lines or polylines along the same line. A collinear constraint is common for applications such as aligning multiview surfaces. Access the **Collinear** constraint and select two lines or polylines, or activate the **Multiple** function to select multiple lines and/or polylines to align in a single operation. Right-click or press [Enter] or the space bar to complete a multiple constrain.

Ribbon
Parametric
> Geometric

Tangent

Type
GEOMCONSTRAINT
> Tangent

Use the **Tangent** constraint option to form a tangent constraint between a line or polyline and an arc, circle, or ellipse. You can also make two circular objects tangent. Access the **Tangent** constraint and select two appropriate objects. **Figure 22-12** shows examples of collinear and tangent constraints.

Exercise 22-4

Access the Student Web site (www.g-wlearning.com/CAD) and complete Exercise 22-4.

Figure 22-12.
A—Collinear and tangent constraints required for the example multiview drawing. B—
Example of common tangent constraint applications.

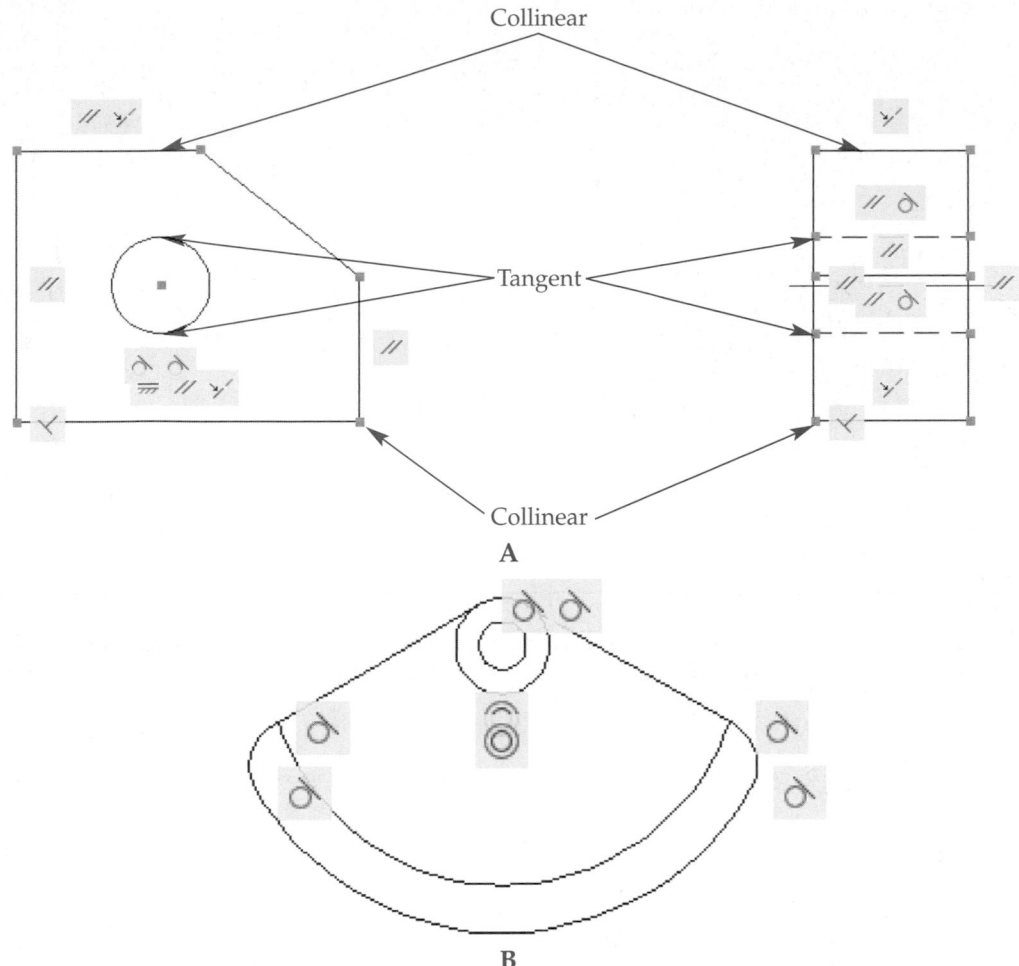

A

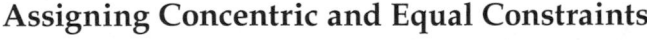

B

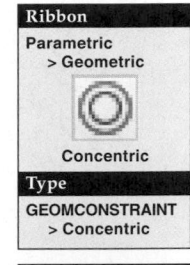

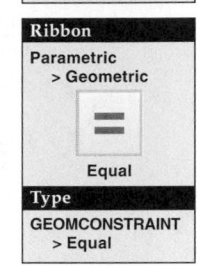

concentric: Arcs, circles, and/or ellipses sharing the same center point.

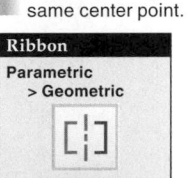

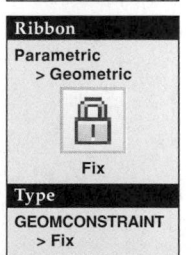

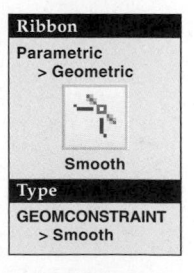

Assigning Concentric and Equal Constraints

To assign a *concentric* constraint, access the **Concentric** constraint and select any combination of arcs, circles, or ellipses. The **Equal** constraint allows you to size objects equally and, in some cases, locate objects. Access the **Equal** constraint and select two objects, or activate the **Multiple** function to select multiple objects to equalize in single operation. Right-click or press [Enter] or the space bar to complete a multiple constrain. **Figure 22-13** shows examples of concentric and equal constraints.

Assigning Symmetric, Fix, and Smooth Constraints

By default, the **Symmetric** constraint allows you to establish symmetry by selecting one object, followed by another, and finally, a line of symmetry. The **2Points** option allows you to pick points followed by the line of symmetry to constrain symmetrical points.

The **Fix** constraint secures a point or object to its current location in space to help preserve design intent. A single fix constraint is often required to fully constrain a drawing. Use the default method to fix a point, or activate the **Object** function to select an object to fix. **Figure 22-14** shows examples of symmetric and fix constraints. Use the **Smooth** constraint to create a curvature-continuous situation, or G2 curve, between a selected spline and a line, second spline, or arc connected to the spline endpoint.

Figure 22-13.
Examples of concentric and equal constraints. All circles are concentric to an arc. All small arcs are equal. All small circles are equal. Use the **Multiple** option to help make multiple objects equal in a single operation.

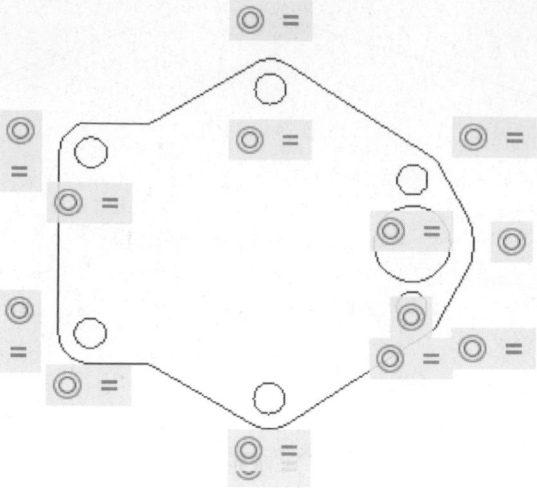

Figure 22-14.
An example of a symmetrical parametric drawing created by adding symmetric constraints to circles and arcs. A fix constraint secures the drawing in space.

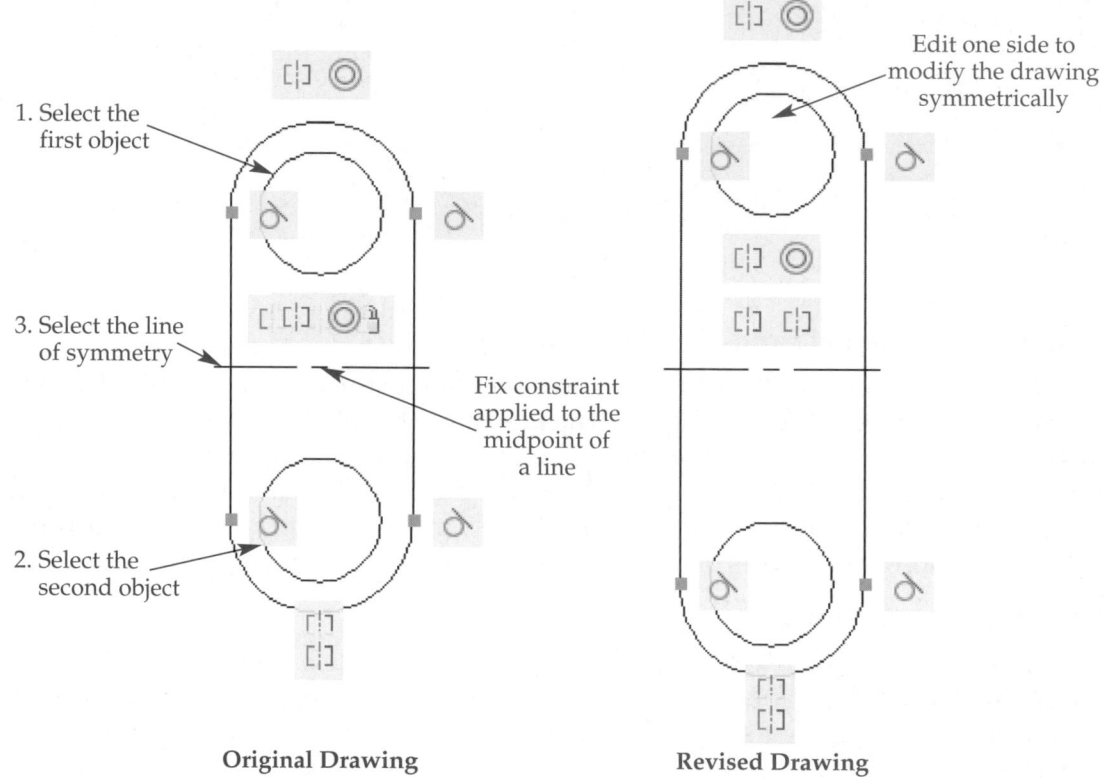

Original Drawing   Revised Drawing

Exercise 22-5

Access the Student Web site (www.g-wlearning.com/CAD) and complete Exercise 22-5.

Using the AUTOCONSTRAIN Tool

You can use the **AUTOCONSTRAIN** tool in an attempt to add all required geometric constraints in a single operation. Before using this tool, access the **Constraint Settings** dialog box and use the options in the **AutoConstrain** tab to specify which geometric constraints to apply. The constraint priority determines which constraints apply first. The higher the priority, the more likely and often the constraint will form if appropriate geometry is available. Select a constraint and use the **Move Up** and **Move Down** buttons to change its priority. Use the corresponding **Apply** check marks and the **Select All** and **Clear All** buttons if there are specific constraints you want to omit during the constraining procedure.

The check boxes determine whether tangent and perpendicular constraints can form if objects do not intersect. See **Figure 22-15**. The **Tolerances** area controls how specific constraints form based on the distance between and angle of objects. A distance less than or equal to the value specified in the **Distance** text box receives constraints. An angle less than or equal to the value specified in the **Angle** text box receives constraints. See **Figure 22-16**.

Once you specify the settings, access **AUTOCONSTRAIN** tool and select the objects to assign geometric constraints. The **Settings** option is available before selection to access the **AutoConstrain** tab of the **Constraint Settings** dialog box. A fix constraint does not occur.

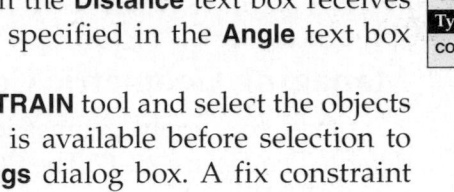

Figure 22-15.
Examples of constraints that will form when you select the **Tangent objects must share an intersection** the **Perpendicular objects must share an intersection** check boxes.

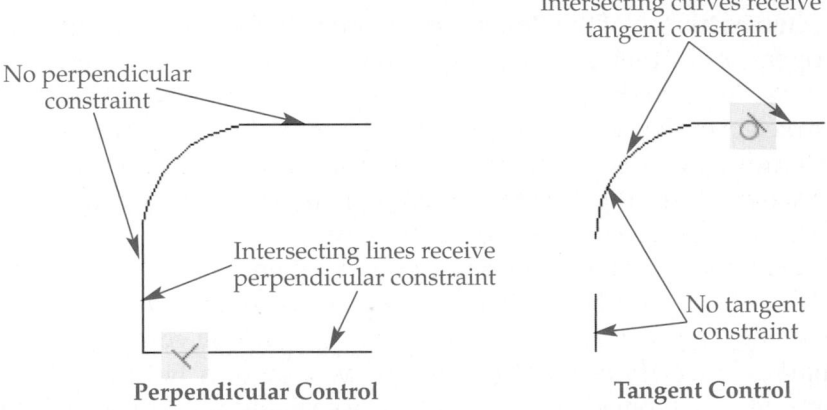

No perpendicular constraint

Intersecting curves receive tangent constraint

Intersecting lines receive perpendicular constraint

No tangent constraint

Perpendicular Control   **Tangent Control**

Figure 22-16.
Examples of constraints that form based on **Distance** and **Angle** tolerance values.

	Distance between Objects > Distance Tolerance	Distance between Objects ≥ Distance Tolerance	Angle > Angle tolerance	Angle ≤ Angle tolerance
Before				
After				

Managing Geometric Constraints

constraint bars:
Toolbars that allow you to view and remove geometric constraints.

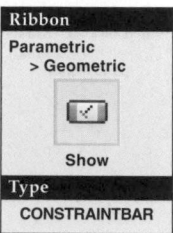

Ribbon
Parametric > Geometric
☑
Show
Type
CONSTRAINTBAR

Once you apply geometric constraints, geometric *constraint bars* appear by default. See **Figure 22-17.** The **CONSTRAINTBAR** tool includes options to show specific constraint bars, show all constraint bars, or hide all constraint bars. The quickest way to access these options is to pick the appropriate button from the **Geometric** panel of the **Parametric** ribbon tab. Select the **Show** button, and then pick objects to display hidden constraint bars. Choose the **Show All** button to display all constraint bars, or choose the **Hide All** button to hide all constraint bars. Hiding constraint bars does not remove geometric constraints.

Use the **Geometric** tab of the **Constraint Settings** dialog box to specify the geometric bars that appear and other geometric bar characteristics. Select the check boxes for individual constraints to display them in constraint bars. Use the **Select All** and **Clear All** buttons to select or deselect all constraint type check boxes. By default, all constraint types show. Limiting constraint bar visibility to specific constraint types often helps to locate and adjust constraints. Use the **Constraint bar transparency** slider or text box to increase or decease the constraint bar transparency.

A coincident constraint appears as a dot that, when hovered over, shows the coincident constraint bar. All other constraints appear as constraint bar icons. When you hover over or select a constraint bar icon, the corresponding constrained objects highlight and markers identify constrained points, as shown in **Figure 22-17.** This allows you to recognize which objects and points are associated with the constraint.

If a constraint bar blocks your view, drag it to a new location. To close a constraint bar, pick the **Hide Constraint Bar** button located to the right of the symbols, or right-click and choose **Hide.** The right-click shortcut menu also includes options for hiding all constraint bars, and for accessing the **Geometric** tab of the **Constraint Settings** dialog box. In some cases, you may need to delete existing constraints in order to apply

Figure 22-17.
Use geometric constraint bars to view and delete geometric constraints.

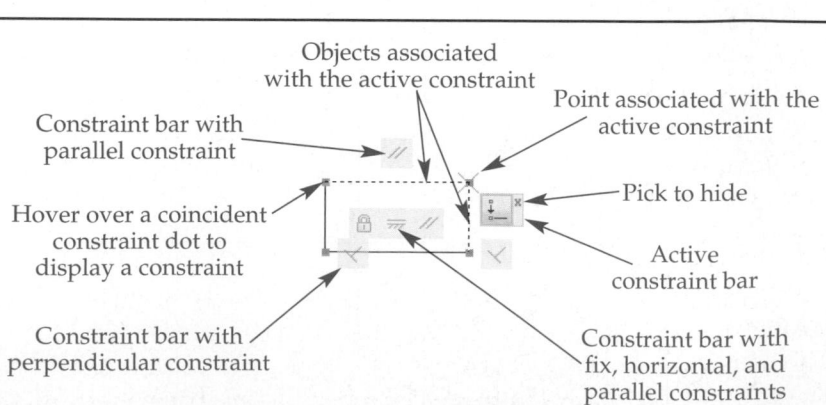

different constructions. To delete constraints, hover over an icon in the constraint bar and press [Delete], or right-click and select **Delete**.

NOTE

You can also right-click with no objects selected to access a **Parametric** cascading submenu that provides options for displaying and hiding constraints, and for accessing the **Constraint Settings** dialog box.

PROFESSIONAL TIP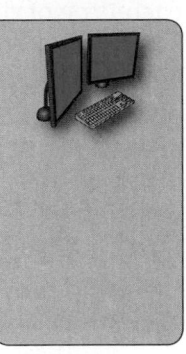

To confirm that constraints, especially geometric constraints, are present and appropriate, select an object to display grips and attempt to stretch a grip point. As an object becomes constrained, you should observe less freedom of movement. Stretching or attempting to stretch grip points is one of the fastest ways to assess design options and to analyze where a constraint is still required. You know your drawing is becoming constrained when you are no longer able to stretch geometry.

Exercise 22-6

Access the Student Web site (www.g-wlearning.com/CAD) and complete Exercise 22-6.

Adding Dimensional Constraints

NEW

Dimensional constraints establish size and location parameters. You must include dimensional constraints to create a truly parametric drawing. Dimensional constraints use a *dynamic format* by default and display a unique appearance. You cannot modify how dimensional constraints appear, but you can convert them to an *annotational format*. Once you have placed dimensional constraints, you can view, adjust, and remove them as needed.

Using the DIMCONSTRAINT Tool

The **DIMCONSTRAINT** tool allows you to assign linear, diameter, radius, and angular dimensional constraints. You can also use the tool to convert dimensional constraints and associative dimensions. Each dimensional constraint is a separate **DIMCONSTRAINT** tool option. The quickest way to add or convert dimensional constraints using this tool is to pick the appropriate button from the **Dimensional** panel of the **Parametric** ribbon tab.

Dimensional Constraint Procedures

To create a dimensional constraint, follow the prompts to make the required selections, pick a location for the dimension line, enter a value to form the constraint, and exit the tool. The process of selecting points or an object to create a dimensional constraint is the same as that for adding geometric constraints. When a dimensional

dynamic format: A dimensional constraint format specifically for controlling the size or location of geometry.

annotational format: A dimensional constraint format in which the constraints look like traditional dimensions, using a dimension style. Constraints displayed in this format can still control the size or location of geometry.

constraint tool requires you to pick two points or objects, the first point or object you select always remains the same. In some cases, you should consider the first point or object to be the datum. The second object or point you select changes in relation to the first. When selecting points, be sure to select the points corresponding to the surface you want to constrain.

A text editor appears after you select the location for the dimension line, allowing you to specify the dimension value. See **Figure 22-18A.** Each dimensional constraint is a parameter with a specific name, expression, and value. By default, linear dimensions receive d names, angular dimensions receive ang names, diameter dimensions receive dia names, and radial dimensions receive rad names. The name of the first of each type of dimension includes a 1, such as d1. The next dimension includes a 2, such as d2, and so on. You can use the text editor to enter a more descriptive name, such as Length, Width, or Diameter. Every parameter must have a unique name. Follow the name with the = symbol and then the dimension value. Press [Enter] or pick outside of the text editor to form the constraint. See **Figure 22-18B.**

You can specify the dimension value for a dimensional constraint in different ways. The most basic option is to type a value in the text editor. Dimensional constraint units reflect the current work environment and unit settings, including length, angle type, and precision. If the drawing is accurate, you should be able to accept the current value. Enter a different value if the drawing is inaccurate, or to change object size.

Another option is to enter an expression in the text editor. You can enter an expression if you do not know an exact value, much like using a calculator. Usually, however, expressions include parameters to associate dimensional constraints. This enables the drawing to adapt according to dimension changes. In the active text editor, include the parameter name in the expression. **Figure 22-19** shows a basic example of using an existing parameter in an expression to control the size of an object. In this example, the design requires that the height of the object always be half the value of the length.

NOTE

If you enter an existing parameter in the text editor without making a calculation, such as d1 = d2, the dimensional constraint references another dimensional constraint value. This is necessary for many applications. Use an equal geometric constraint when possible.

Applying Linear Dimensional Constraints

The **Linear** option allows you to place horizontal or vertical linear dimensional constraints. The **Horizontal** option sets the tool to constrain only a horizontal distance. The **Vertical** option sets the tool to constrain only a vertical distance. The **Horizontal** and **Vertical** options are helpful when it is difficult to produce the appropriate linear

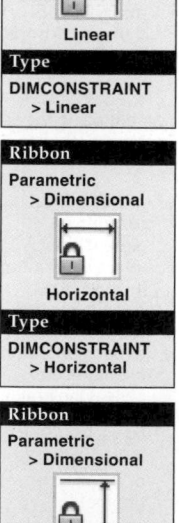

Ribbon
Parametric > Dimensional
Linear

Type
DIMCONSTRAINT > Linear

Ribbon
Parametric > Dimensional
Horizontal

Type
DIMCONSTRAINT > Horizontal

Ribbon
Parametric > Dimensional
Vertical

Type
DIMCONSTRAINT > Vertical

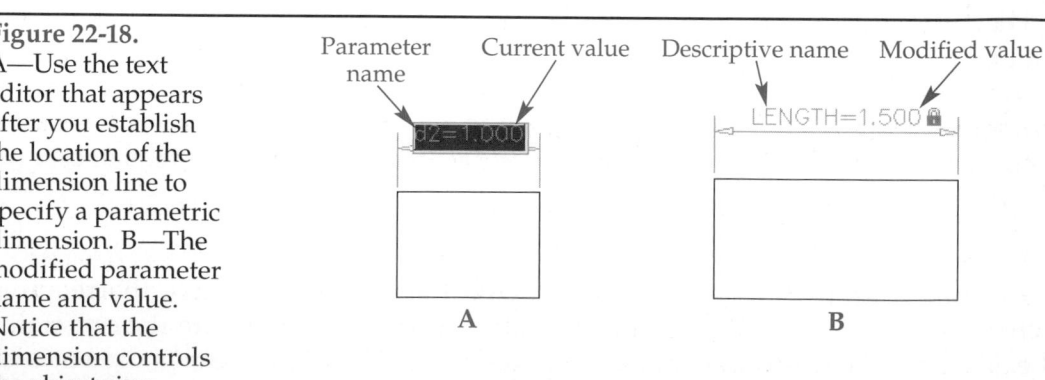

Figure 22-18.
A—Use the text editor that appears after you establish the location of the dimension line to specify a parametric dimension. B—The modified parameter name and value. Notice that the dimension controls the object size.

Figure 22-19.
A—An example of a modified parameter name and expression that references another parameter.
B—The dimension that includes a parameter references the parameter when changes occur.

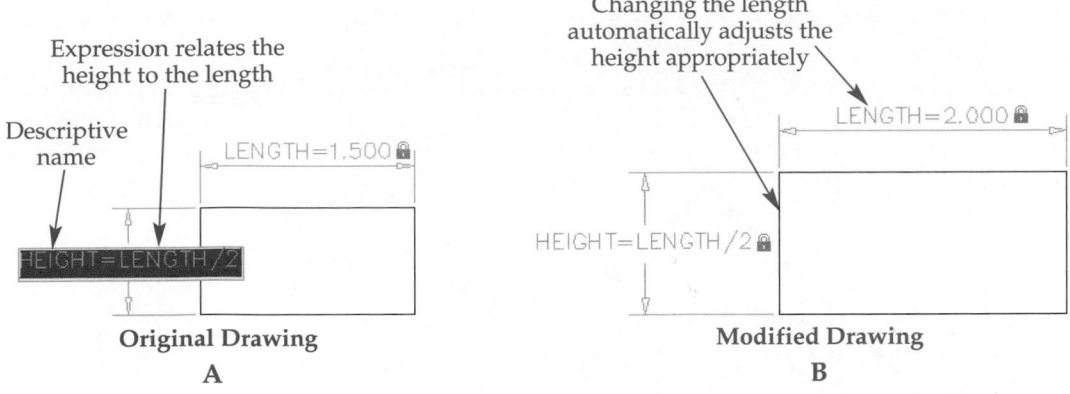

Original Drawing
A

Modified Drawing
B

dimensional constraint, such as when dimensioning the horizontal or vertical distance of an angled surface.

Access the appropriate linear dimensional constraint option and pick two points to specify the origin of the dimensional constraint. The **Object** function allows you to select a line, polyline, or arc to constrain, instead of two points. After you select points or an object, move the dimension line to an appropriate location and pick. Specify the dimension value and adjust the parameter name if desired. Press [Enter] or pick outside of the text editor to form the constraint. **Figure 22-18** and **Figure 22-19** show examples of linear dimensions.

Exercise 22-7

Access the Student Web site (www.g-wlearning.com/CAD) and complete Exercise 22-7.

Applying Aligned Dimensional Constraints

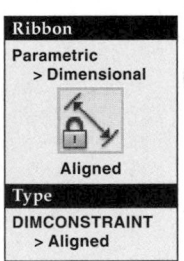

The **Aligned** option allows you to place a linear dimensional constraint with a dimension line aligned with, and extension lines perpendicular to, an angled surface. Access the **Aligned** option and pick two points to specify the origin of the dimensional constraint. The **Object** function allows you to select a line, polyline, or arc to constrain, instead of two points. Often when you apply aligned dimensions, such as when you are dimensioning an auxiliary view, it is necessary to pick a point and an aligned surface or two aligned surfaces. Use the **Point & Line** function to select a point and an alignment line. Use the **2Line** function to select two alignment lines.

Once you make the selections, move the dimension line to an appropriate location and pick. Specify the dimension value, and adjust the parameter name if desired. Press [Enter] or pick outside of the text editor to form the constraint. **Figure 22-20** shows examples of aligned dimensional constraints created using each method.

Applying Angular Dimensional Constraints

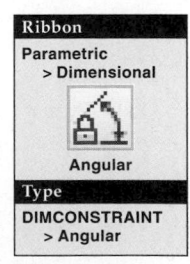

The **Angular** option allows you to place an angular dimension between two objects or three points. Access the **Angular** option, and by default, pick two lines, polylines, or arcs. The **3Point** function allows you to select the angle vertex, followed by two points to locate each side of the angle. After you make the selections, move the dimension

Figure 22-20.
An example of an auxiliary view with full dimensional constraints created using the **Aligned** option.

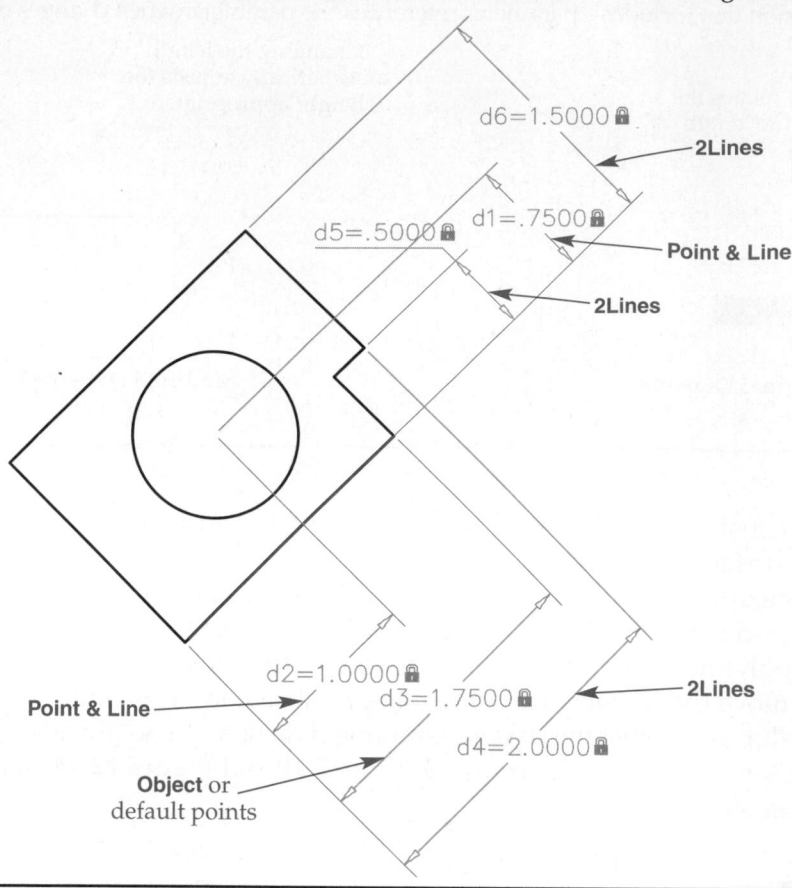

line to an appropriate location and pick. Specify the dimension value, and adjust the parameter name if desired. Press [Enter] or pick outside of the text editor to form the constraint. **Figure 22-21** shows examples of angular dimensional constraints created using each method.

Exercise 22-8

Access the Student Web site (www.g-wlearning.com/CAD) and complete Exercise 22-8.

Figure 22-21.
Forming angular dimensional constraints.

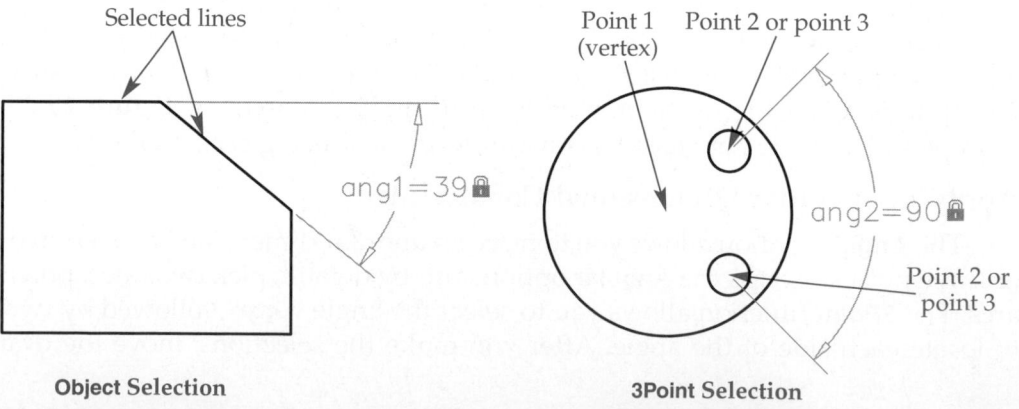

Adding Diameter and Radial Dimensional Constraints

Use the **Diameter** option to create a diameter dimensional constraint, and use the **Radial** option to form a radial dimensional constraint. You can select a circle or arc when using either tool. In formal drafting, circles receive diameter constraints and arcs receive radial constraints. See **Figure 22-22.** In some parametric applications, however, you may find it appropriate to constrain arcs using a diameter and circles using a radius.

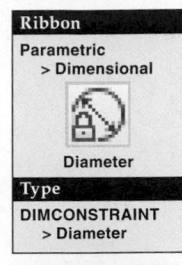

Ribbon
Parametric
> Dimensional

Diameter

Type
DIMCONSTRAINT
> Diameter

Creating Reference Dimensional Constraints

When you constrain a defined object, the drawing should become over-constrained. However, AutoCAD does not allow over-constraining to occur. You can either cancel the tool without accepting the dimension, or accept the dimension and allow it to become a reference dimension. See **Figure 22-23.** Reference dimensions are sometimes required to form specific parameters or expressions. You cannot edit a reference dimension to change the size of an object, but a reference dimension changes when you modify corresponding dimensions.

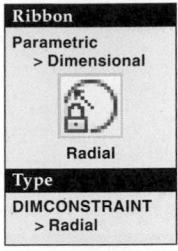

Ribbon
Parametric
> Dimensional

Radial

Type
DIMCONSTRAINT
> Radial

Figure 22-22.
Forming diameter and radial dimensional constraints. In this example, equal geometric constraints control the size of the undimensioned arcs.

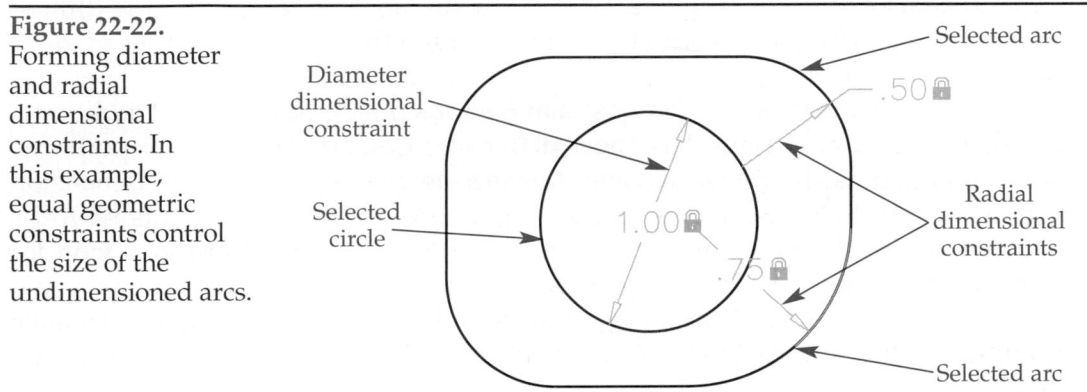

Figure 22-23.
You cannot directly modify reference dimensions, which are identified by parentheses.

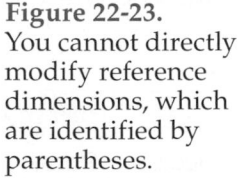

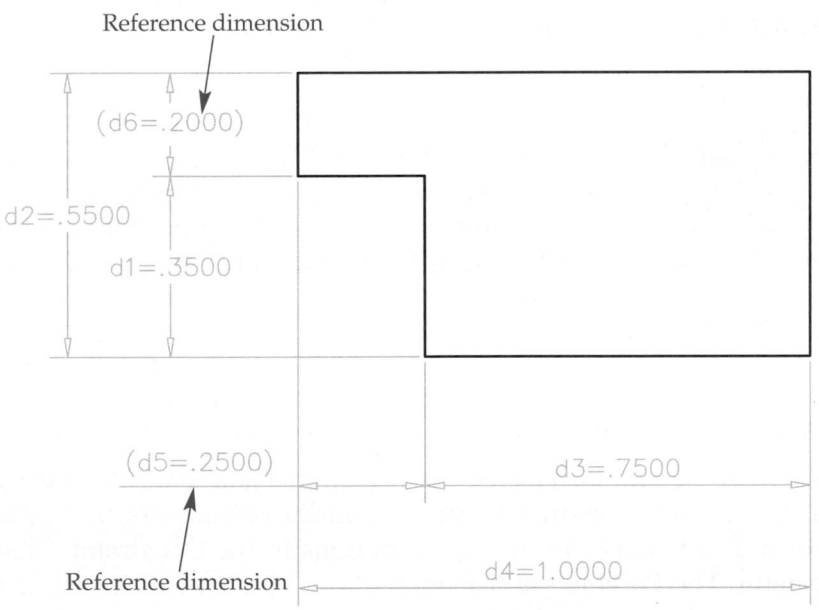

NOTE

Dimensional constraints define and constrain drawing geometry. Dimensional constraints use a unique AutoCAD style and do not comply with ASME standards. Do not be overly concerned about the placement or display characteristics of these dimensions, but if possible, apply dimensional constraints just as you would add dimensions to a drawing, using correct drafting practices. In addition, move and manipulate dimensional constraints so the drawing is as uncluttered as possible. Use grips to make basic adjustments to dimensional constraint position.

Managing Dimensional Constraints

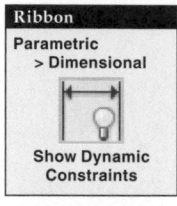

Ribbon
Parametric
> Dimensional

Show Dynamic
Constraints

By default, all dimensional constraints display in the dynamic format, include a lock icon, and include the parameter name and dimension value. The quickest way to hide all dimensional constraints is to deselect the **Show Dynamic Constraints** button from the **Dimensional** panel of the **Parametric** ribbon tab. Typically, you hide dimensional constraints to prepare a formal drawing or when they are no longer needed for the current design phase.

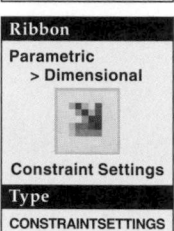

Ribbon
Parametric
> Dimensional

Constraint Settings

Type
CONSTRAINTSETTINGS

Use the **Dimensional** tab of the **Constraint Settings** dialog box to adjust additional dimensional constraint settings. The **Show All Dynamic Constraints** check box performs the same function as the **Show Dynamic Constraints** ribbon button. The **Dimension name format** drop-down list allows you to display dimensional constraints with the parameter name and value, parameter name, or value. Use the **Show lock icon for additional constraints** check box to toggle the lock icon on or off for new dimensional constraints. If you hide dimensional constraints and select the **Show hidden dynamic constraints of selected objects** check box, you can pick an object to show associated dimensional constraints temporarily.

Hiding dimensional constraints does not remove them. In some cases, you may need to delete existing constraints in order to apply different parameters. Use the **ERASE** tool to eliminate specific dimensional constraints.

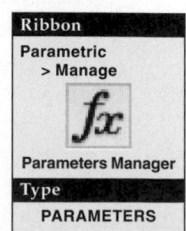

NOTE

You can also right-click with no objects selected to access a **Parametric** cascading submenu that provides options for displaying and hiding constraints, changing the dimension name format, and accessing the **Constraint Settings** dialog box. The options are also available when you pick a dimensional constraint and then right-click.

dimensional constraint parameters: Parameters added when you insert a dimensional constraint.

Working with Parameters

A *dimensional constraint parameter* automatically forms every time you add a dimensional constraint. You can adjust parameters by changing the dimensional constraint value or by using the options in the **Constraint** category of the **Properties** palette. The **Parameters Manager** allows you to manage all of the parameters in the drawing. See **Figure 22-24**.

Ribbon
Parametric
> Manage

fx

Parameters Manager

Type
PARAMETERS

You can use the **Filter** flyout to show only those parameters used in expressions. To change a parameter name, pick inside a text box in the **Name** column to activate it, type the new name, and press [Enter] or pick outside of the text box. Enter a new value or expression for the parameter in a text box in the **Expression** column. The value appears in the **Value** column display boxes for reference. Use the **Delete** button or

Figure 22-24.
The **Parameters Manager** is a good resource for reviewing and editing parameters and for creating user parameters.

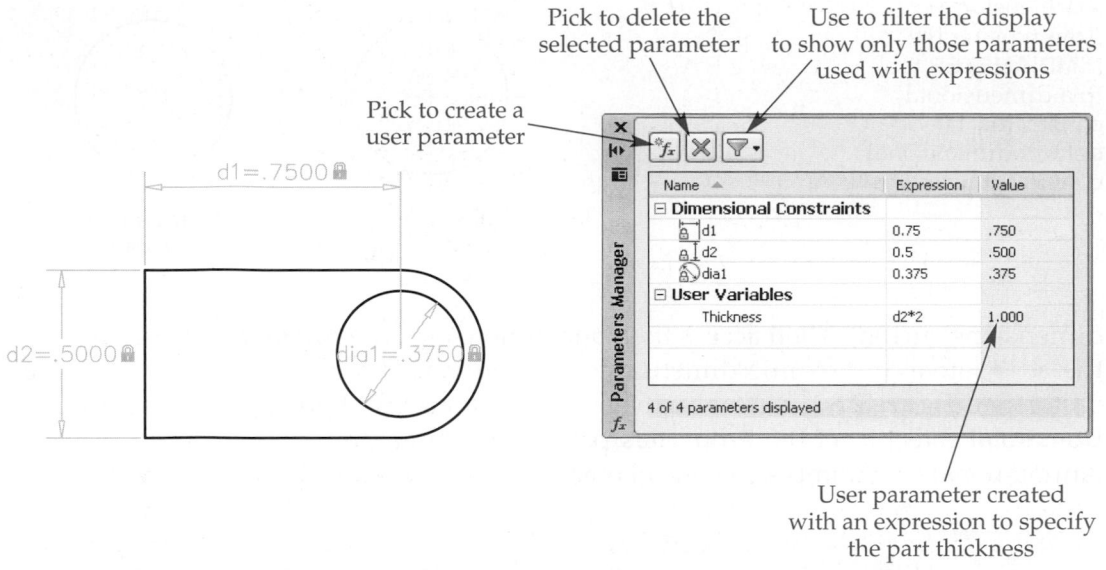

Pick to delete the selected parameter

Use to filter the display to show only those parameters used with expressions

Pick to create a user parameter

User parameter created with an expression to specify the part thickness

right-click on a parameter and select **Delete** to remove the parameter and the corresponding dimensional constraint.

You can also specify your own *user parameters* by picking the **User Parameters** button to display a **User Variables** node. User parameters function like dimensional constraints in the **Parameters Manager**. Create parameters in order to access specific parameters throughout the design process. For example, if you know the thickness of a part will always be twice a certain dimensional constraint, create a user parameter similar to the one in **Figure 22-24** to define the thickness. You can then use the custom parameter for reference and in expressions when you place additional dimensional constraints.

user parameters: Additional parameters you define.

Exercise 22-9

Access the Student Web site (www.g-wlearning.com/CAD) and complete Exercise 22-9.

Converting Dimensional Constraints

The **Convert** option of the **DIMCONSTRAINT** tool allows you to convert an associative dimension to a dimensional constraint. This allows you to prepare a parametric drawing using existing associative dimensions. Access the **Convert** option, pick the associative dimensions to convert, and press [Enter] or pick outside of the text editor. By default, new dimensional constraints and converted dimensions use the dynamic format. See **Figure 22-25.** Recall that this is the primary format for creating parametric geometry, but it is not necessarily associated with formal dimension practices. **Figure 22-26A** shows all of the dynamic dimensions and geometric constraints required to fully constrain the example multiview drawing.

You can use the **Form** option of the **DIMCONSTRAINT** tool to change the dimensional constraint format. As explained earlier, annotational dimensions reference the current dimension style. To convert dynamic dimensions to the annotational format, first make the dimension style and dimension layer to apply to the annotational

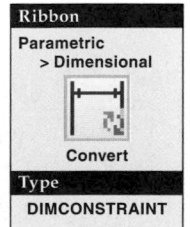

Ribbon

Parametric > Dimensional

Convert

Type

DIMCONSTRAINT

Figure 22-25.
An example
of converting
a dimension
drawn using the
DIMDIAMETER tool
to a dimensional
constraint. The
default dimensional
constraint format is
dynamic.

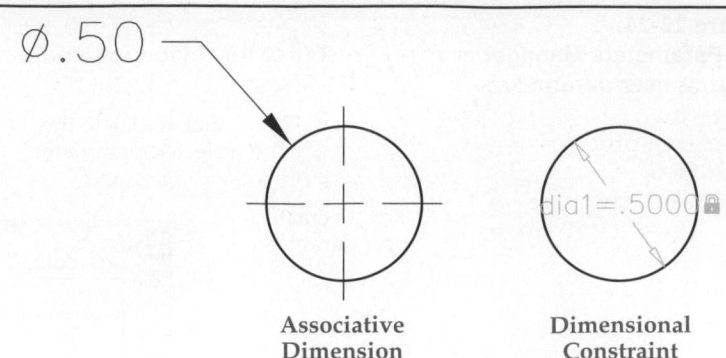

Associative
Dimension

Dimensional
Constraint

dimensions current. Then access the **Form** option, pick the **Annotational** format, select the associative or dynamic dimensions, and press [Enter] or pick outside of the text editor. See **Figure 22-26B.** You can also use the **Constraint Form** drop-down list in the **Constraint** category of the **Properties** palette to change dimensional constraint form. Annotational constraints still control object size and location.

NOTE

The specified **Annotational** or **Dynamic** form remains set and applies to new dimensional constraints until you change to the alternate setting.

Annotational dimensions, especially those converted from dynamic dimensions, may require that you make some format and organizational changes to prepare the final drawing. You may also have to add non-parametric dimensions. Make the following changes to create the final drawing shown in **Figure 22-26C.**

- Hide all geometric constraints and dynamic dimensions.
- Disable the lock icon display.
- Change the dimension name format to **Value**.
- Make basic dimension style overrides and dimension location adjustments if necessary.

Exercise 22-10

Access the Student Web site (www.g-wlearning.com/CAD) and complete Exercise 22-10.

Parametric Editing

You can adjust existing parametric drawings in a variety of ways. Drawing objects adds additional unconstrained geometry. You may need to delete or replace existing constraints in order to constrain new objects. Erasing objects and exploding polyline objects removes constraints. If you erase geometry associated with an expression, an alert appears asking if you want to convert the dimensional constraint to a user parameter, maintain the information, or remove the parameter with the dimensional constraint.

AutoCAD and Its Applications—Basics

Figure 22-26.
A typical
dimensional
constraint
conversion process.
A—A fully
constrained drawing
with dynamic
dimensional
constraints. Notice
that dimensions
apply to all items,
including centerline
extensions and
the distance
between views.
B—Converting
dynamic dimensions
to annotational
dimensions.
Convert only
those dimensions
required for formal
dimensioning.

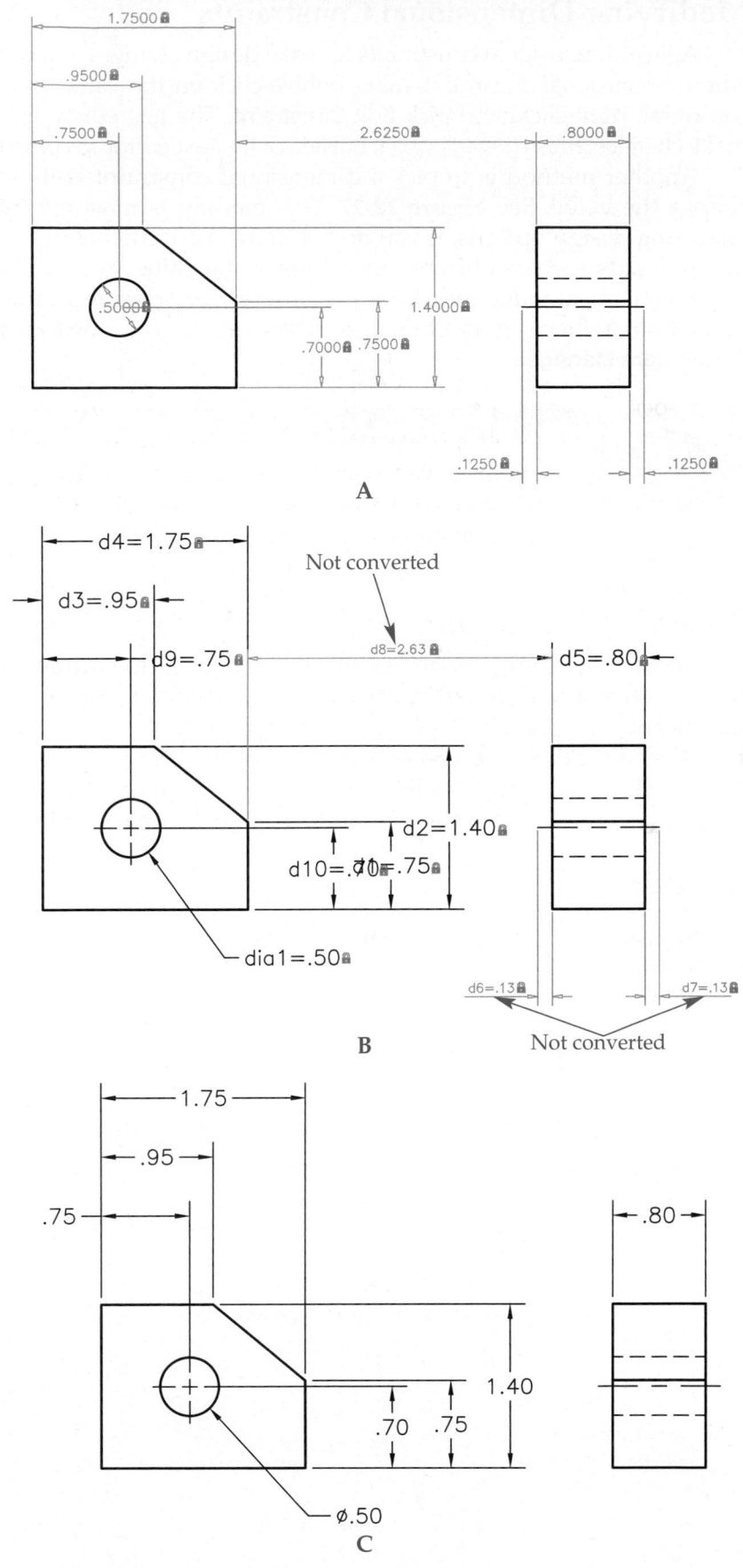

Modifying Dimensional Constraints

Adjust dimensional constraints to make design changes to a constrained drawing. To edit a dimensional constraint value, double-click on the value, or select the dimensional constraint, right-click, and pick **Edit Constraint**. The text editor appears, allowing you to make changes. Press [Enter] or pick outside of the text editor to complete the operation.

Another method is to pick a dimensional constraint and use parameter grips to change the value. See **Figure 22-27**. This method is most appropriate when you are analyzing design options, if you do not know a specific value, or when you are using drawing aids such as object snaps to adjust the value. You can also modify a dimensional constraint value by entering a different expression in the **Expression** text box in the **Constraint** category of the **Properties** palette, or in the **Expression** text box of the **Parameters Manager**.

> **NOTE**
>
> If your drawing includes enough geometric constraints, and the geometric constraints are accurate, changing dimensions should maintain all geometric relationships.

Removing Constraints

Constraints limit the ability to make changes to a drawing. For example, you cannot rotate a horizontally or vertically constrained line. If it is necessary to make significant changes to a drawing, you may have to relax or delete the constraints that limit the edit. Constraints relax using standard editing practices. When you are using a basic editing tool such as **ROTATE**, a message appears asking if you want to relax constraints. To relax constraints while grip stretching, moving, or scaling, you may

---

Figure 22-27.
Using a parameter grip to change the dimensional constraint value.

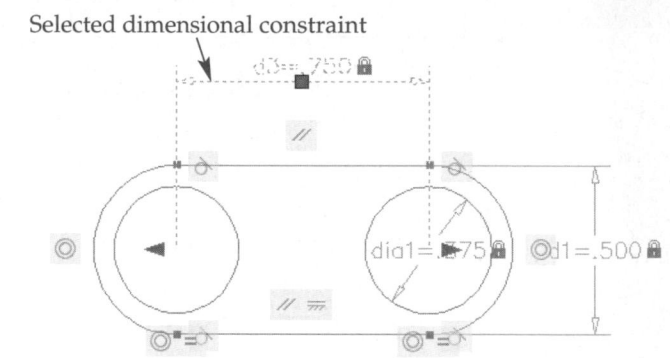

Original Drawing

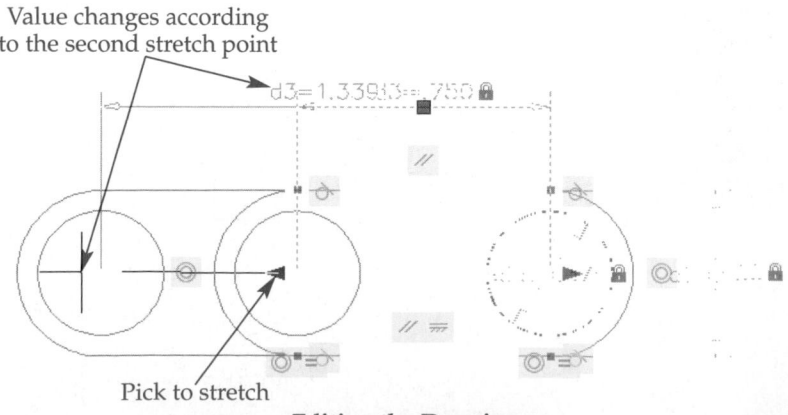

Editing the Drawing

need to press [Ctrl] to toggle relaxing on and off. See **Figure 22-28A.** Other edits automatically remove constraints. See **Figure 22-28B.** The only constraints, if any, that are removed are those required to achieve the edit.

In addition to the methods described for deleting individual constraints, a **DELCONSTRAINT** tool is available. Access the tool and select the geometric and dimensional constraints to remove. This tool is especially effective for removing a significant number of constraints. Use the **All** selection option to remove all constraints from the drawing.

Ribbon
Parametric > Manage
Delete Constraints
Type
DELCONSTRAINT

NOTE

Some editing techniques, such as mirroring or moving, may only require removal of fix constraints, especially when they are applied to an entire drawing.

PROFESSIONAL TIP

Occasionally, constraining geometry causes objects to twist out of shape, making it difficult to control the size and position of the drawing. Use the **UNDO** tool to return to the previous design. To help avoid this situation, consider the following suggestions:
- Construct objects at or close to their finished size using standard drafting practices.
- Add as many geometric constraints as appropriate before dimensioning.
- Dimension the largest objects first.
- Move objects to a more appropriate location, if necessary, and change object size before constraining.

Exercise 22-11

Access the Student Web site (www.g-wlearning.com/CAD) and complete Exercise 22-11.

Figure 22-28.
Examples of relaxing constraints. A—Access the MOVE tool and then press [Ctrl] to remove the concentric constraint. The process is similar for stretching and scaling. B—Rotating automatically removes, in this example, a vertical constraint.

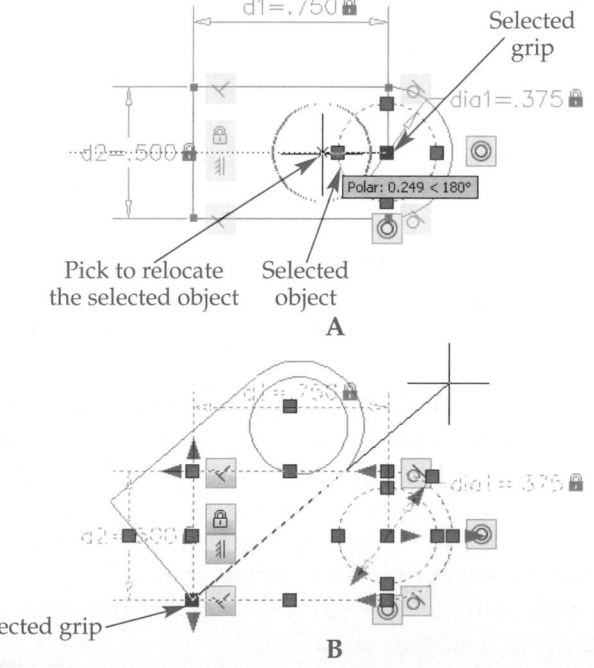

Chapter Test

Answer the following questions. Write your answers on a separate sheet of paper or go to the Student Web site (www.g-wlearning.com/CAD) and complete the electronic chapter test.

1. Give an example demonstrating how you can use constraints to form geometric constructions in specific situations when standard AutoCAD tools are inefficient or ineffective.
2. Describe two applications in which parametric drafting is unsuitable or ineffective.
3. Briefly describe what geometric constraint tools allow you to do, and identify what you see that indicates the presence of a geometric constraint.
4. Name the tool that allows you to assign specific geometric constraints to objects.
5. When you use geometric constraint tools that allow you to pick two objects or points, describe what happens to the first and second objects you select.
6. A coincident constraint is one of the most common parametric constructions. Give an example of objects that coincide.
7. Identify common uses for horizontal and vertical constraints.
8. Name the two basic object types that can form parallel or perpendicular constraints.
9. What does the **Collinear** constraint allow you to do?
10. Name the types of objects you can constrain with the **Tangent** constraint.
11. Explain the basic function of the **Equal** constraint.
12. Describe the default function of the **Symmetric** constraint.
13. Name the tool you can use if you want to attempt to add all required geometric constraints in a single operation.
14. Briefly describe how to specify the appearance and characteristics of geometric bars.
15. Compare the appearance of a coincident constraint with the display of other constraints.
16. Explain how you can determine which objects and points are associated with constraints.
17. What should you do if constraint bars block your view or if you want to hide constraint bars?
18. Name the tool that allows you to assign linear, diameter, radius, and angular dimensional constraints.
19. Describe the basic process used to create a dimensional constraint.
20. What is the most basic method to specify dimension values when you create a dimensional constraint?
21. Name the option that allows you to place horizontal or vertical linear dimensional constraints.
22. Name the option that allows you to place a linear dimensional constraint with a dimension line aligned with, and extension lines perpendicular to, an angled surface.
23. Which option allows you to place an angular dimension between two objects or between three points?
24. Explain the options AutoCAD provides when you try to over-constrain a defined object.
25. What happens every time you add a dimensional constraint?
26. How do you adjust parameters?
27. Name the tool and option that allow you to convert an associative dimension to a dimensional constraint, and give the advantage of using this option.
28. Explain how to edit a dimensional constraint value.
29. Briefly describe how to relax constraints.
30. Which tool provides an efficient method of removing a significant number of constraints in a single operation?

Drawing Problems

- *Start AutoCAD if it is not already started.*
- *Start a new drawing using an appropriate template of your choice. The template should include layers, text styles, dimension styles, and multileader styles appropriate for drawing the given objects.*
- *Add layers, text styles, dimension styles, and multileader styles as needed. Draw all objects using appropriate layers, text styles, dimension styles, multileader styles, justification, and format.*
- *Follow the specific instructions for each problem. Use your own judgment and approximate dimensions when necessary.*
- *Apply formal dimensions accurately using ASME or appropriate industry standards.*

Note: Dimensional constraints shown for reference are created using AutoCAD and may not comply with ASME standards.

▼ Basic

1. Use the **POLYGON** tool to draw the hexagon and fully constrain it as shown. All sides are equal. Edit the d1 parameter to change the distance across the flats to 4.000. Save the drawing as P22-1.

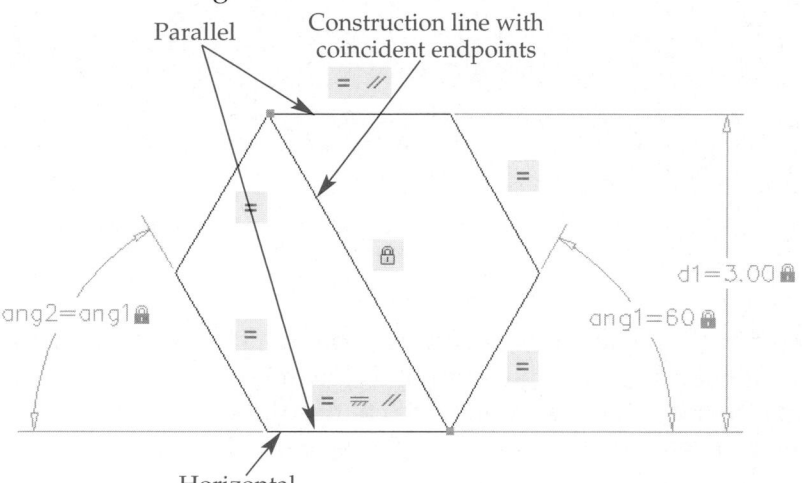
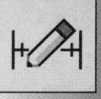

2. Use the **RECTANGLE** and **CIRCLE** tools to draw the view and fully constrain it as shown. The circle is tangent to the rectangle in two locations. Edit the d2 parameter to change the distance to 4.500. Edit the d3 parameter to change the distance to 2.000. Save the drawing as P22-2.

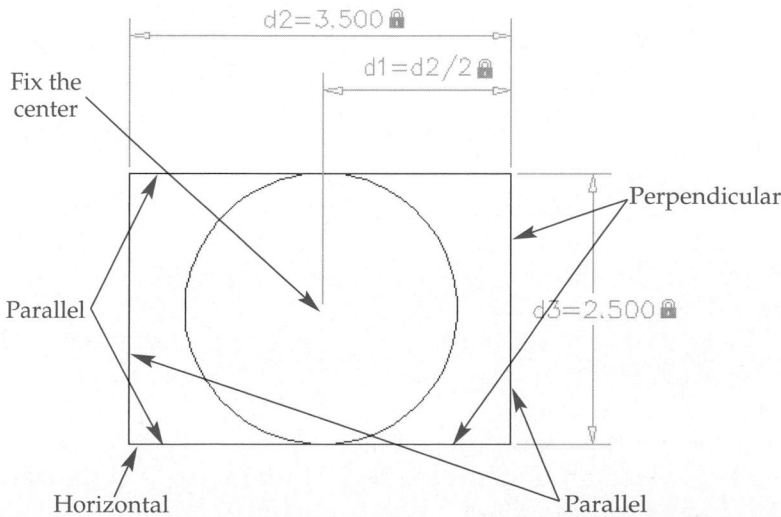
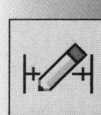

3. Draw and fully constrain the circles shown. Both of the smaller circles are tangent to the larger circle. Edit the dia1 parameter to change the diameter to 2.000. Save the drawing as P22-3.

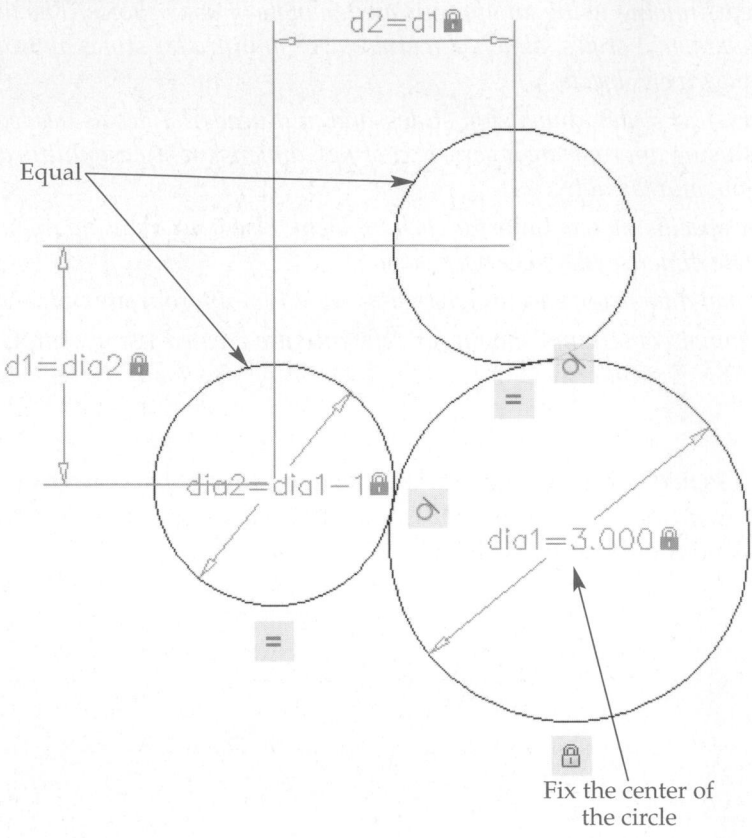

4. Draw and fully constrain the pipe spacer shown. Use the **AUTOCONSTRAIN** tool, with default settings, to apply most of the geometric constraints. Make all circles equal in size. Edit the d1 parameter to change the distance to 2.000. Edit the rad1 parameter to change the radius to 1.250. Save the drawing as P22-4.

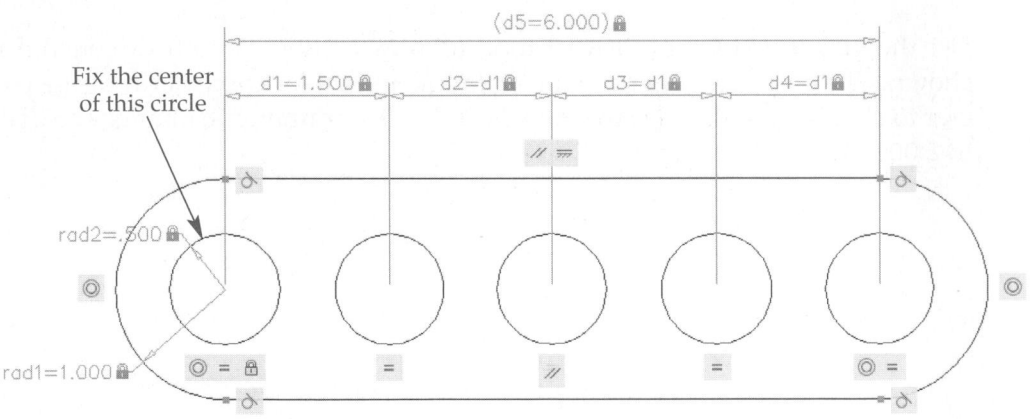

5. Draw and fully constrain the view shown. Follow the guidelines below.
   - Begin by using the **Fillet** option of the **RECTANGLE** tool to create the 4.6250 by 6.3750 rectangle with .6250 rounded corners. Notice that the circles are not concentric to the arcs.
   - Use the **RECTANGLE** tool to create the construction rectangle.
   - Apply geometric constraints in the following order: fix the center of the lower-left circle; use the **Autoconstrain** function of the **Coincident** option to apply all coincident constraints; make all circles equal; make all arcs equal.
   - Use the **AUTOCONSTRAIN** tool with default settings to apply the remaining geometric constraints.
   - Edit the d1 parameter to change the distance to 1.0000.
   - Edit the d5 parameter to change the distance to 5.1250.
   - Save the drawing as P22-5.

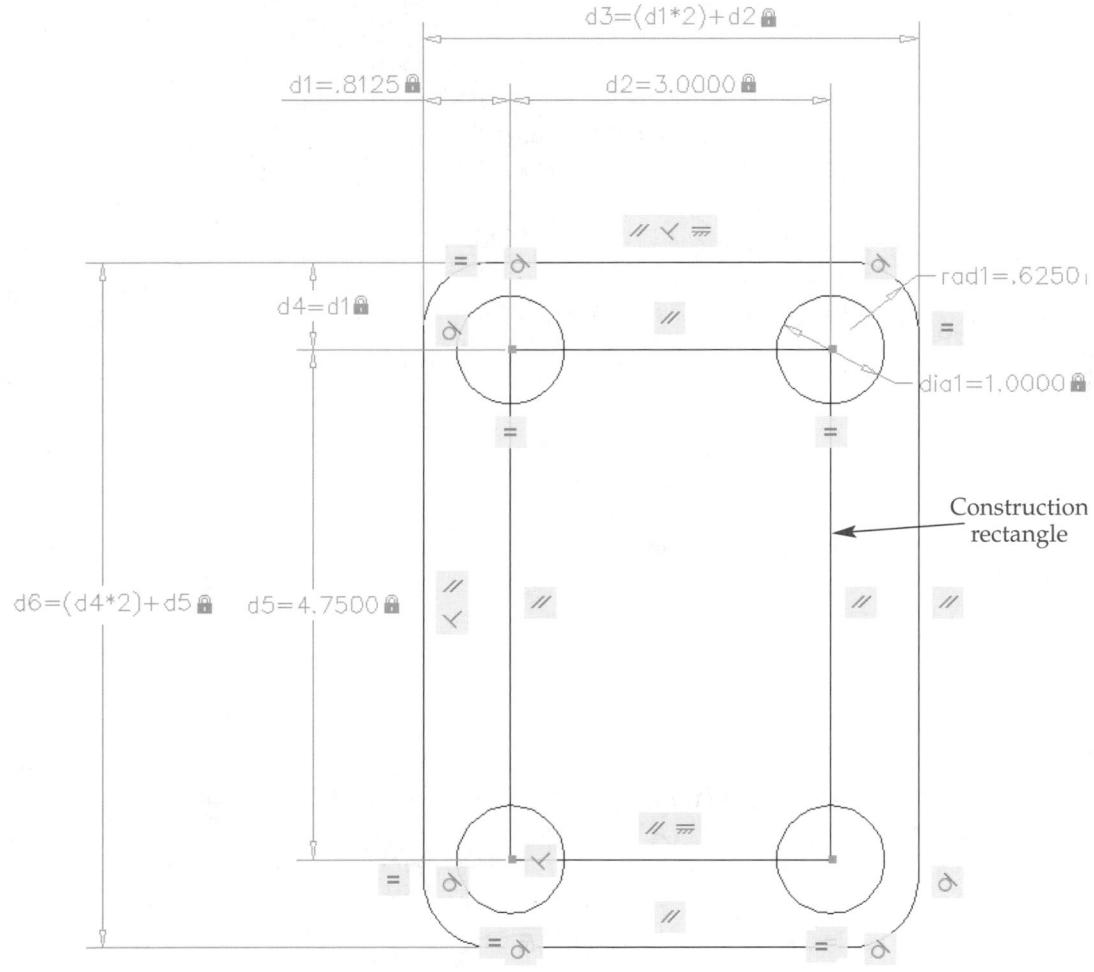

Construction rectangle

Chapter 22   Parametric Drafting

6. Draw and fully constrain the view shown. Apply geometric constraints in the following order: fix the center of the center circle; use the **Autoconstrain** function of the **Coincident** option to apply all coincident constraints; make all circles equal; make all small arcs equal; make all large arcs concentric to the appropriate circle; use the **AUTOCONSTRAIN** tool with default settings to apply the remaining geometric constraints. Edit the d1 parameter to change the distance to 1.750. Save the drawing as P22-6.

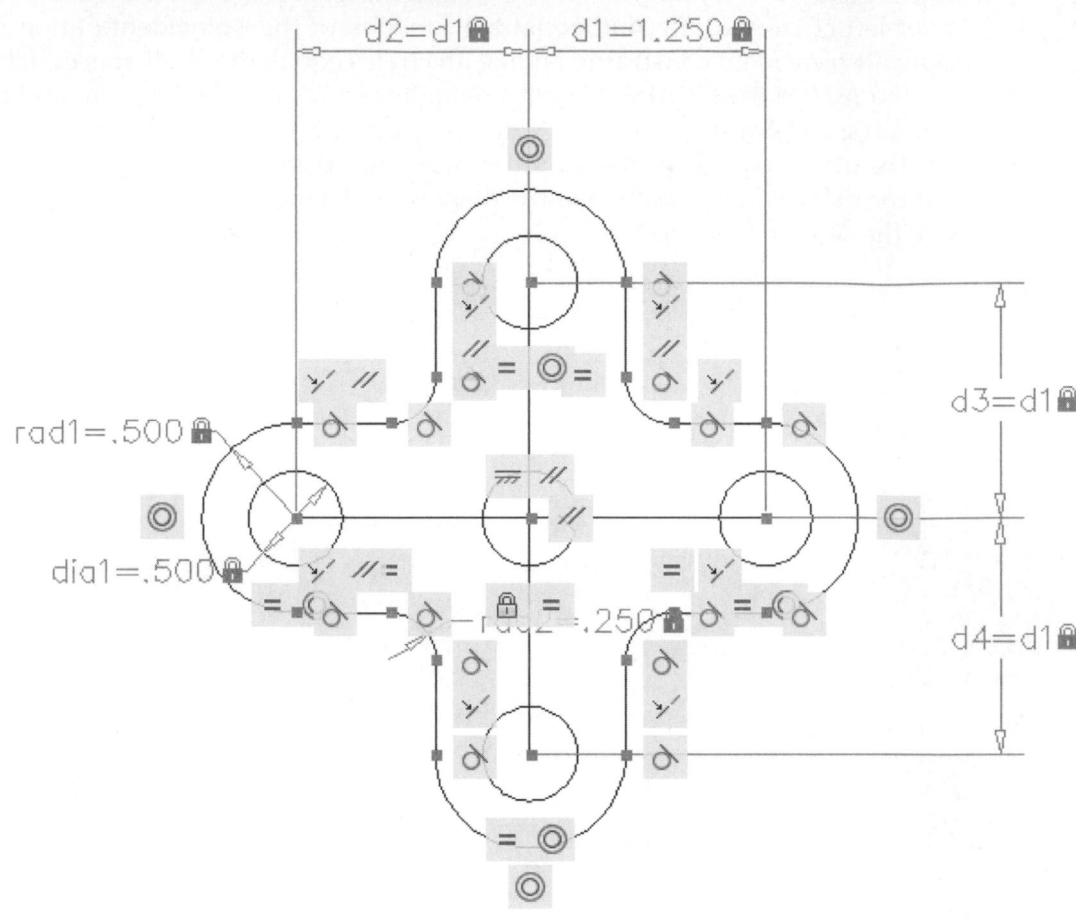

▼ Advanced

7. Draw and fully constrain the view shown. Save the drawing as P22-7.

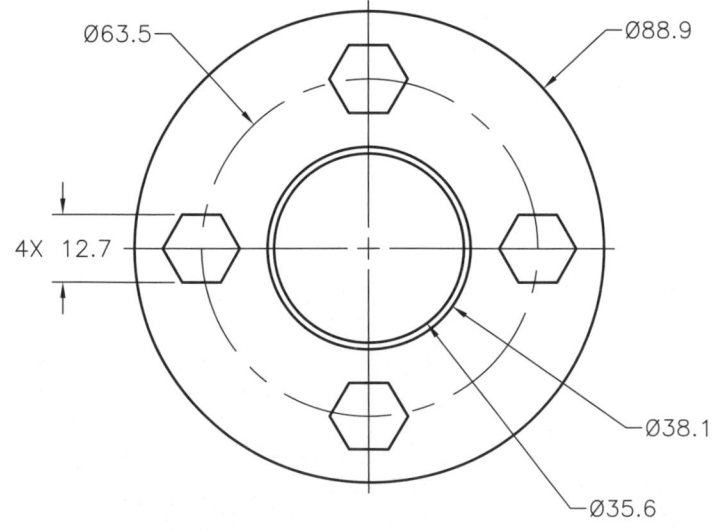

8. Use the information below to draw the non-parametric view shown in A. Then fully constrain the view as shown in B. Edit the view as shown in C. Finish by converting dimensional constraints to create the formal drawing shown in D. Save the drawing as P22-8.

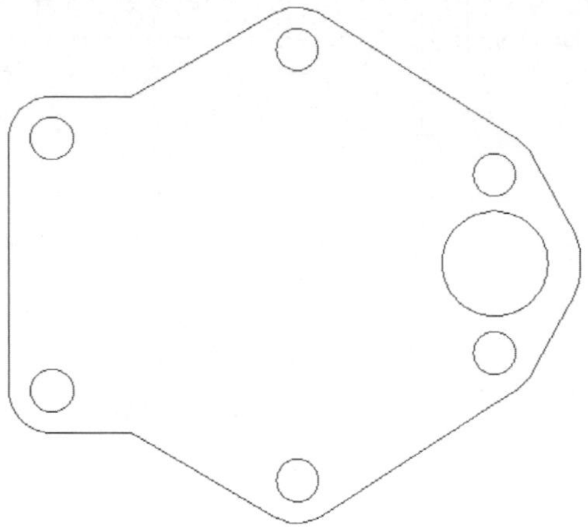

A

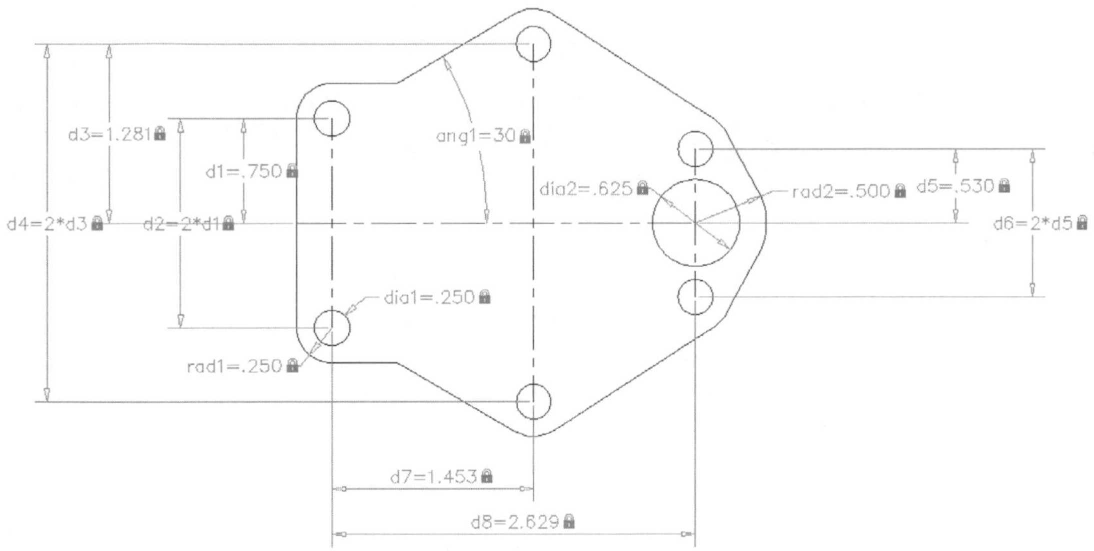

d3=1.281

d1=.750

ang1=30

dia2=.625

rad2=.500   d5=.530

d4=2*d3   d2=2*d1

d6=2*d5

dia1=.250

rad1=.250

d7=1.453

d8=2.629

B

(Continued)

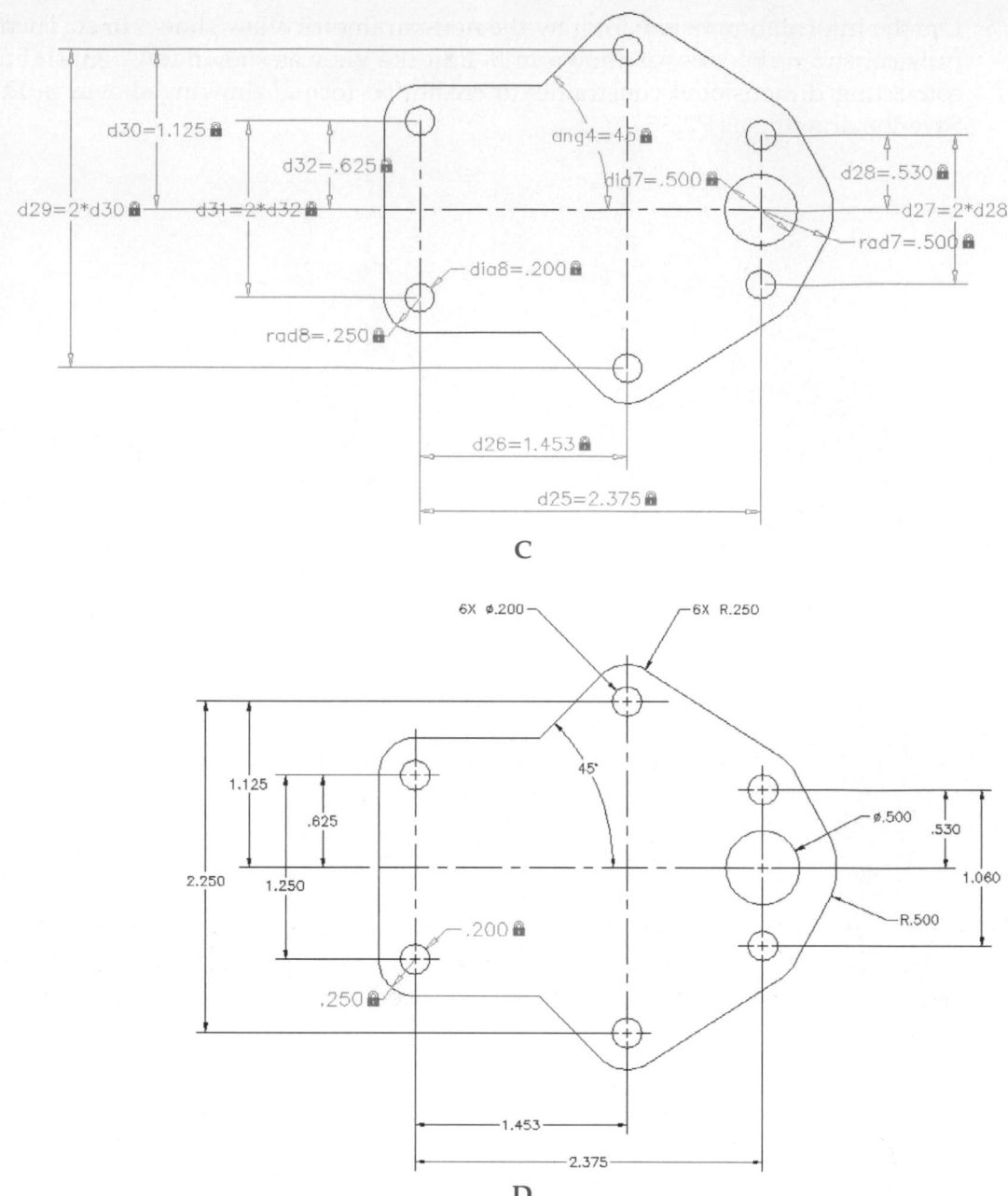

C

D

AutoCAD and Its Applications—Basics

9. Draw and fully constrain the view shown. Convert dynamic dimensional constraints to annotational dimensional constraints. Adjust the drawing and add associative dimensions as needed to create the formal drawing shown. Save the drawing as P22-9.

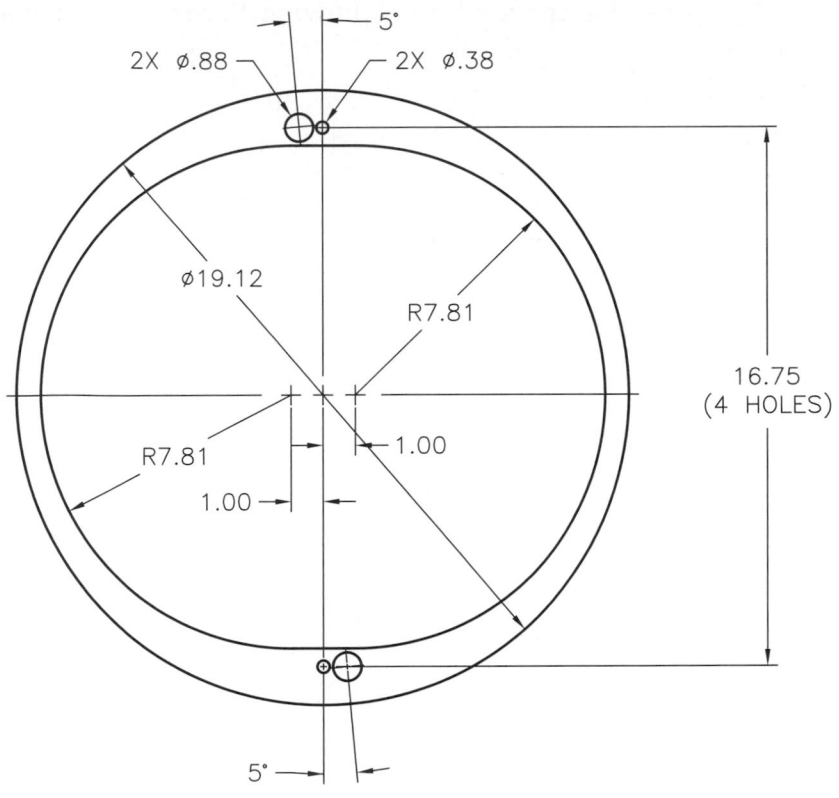

10. Draw and fully constrain the multiview drawing of the support. Convert dynamic dimensional constraints to annotational dimensional constraints. Adjust the drawing and add associative dimensions as needed to create the formal drawing shown. Save the drawing as P22-10.

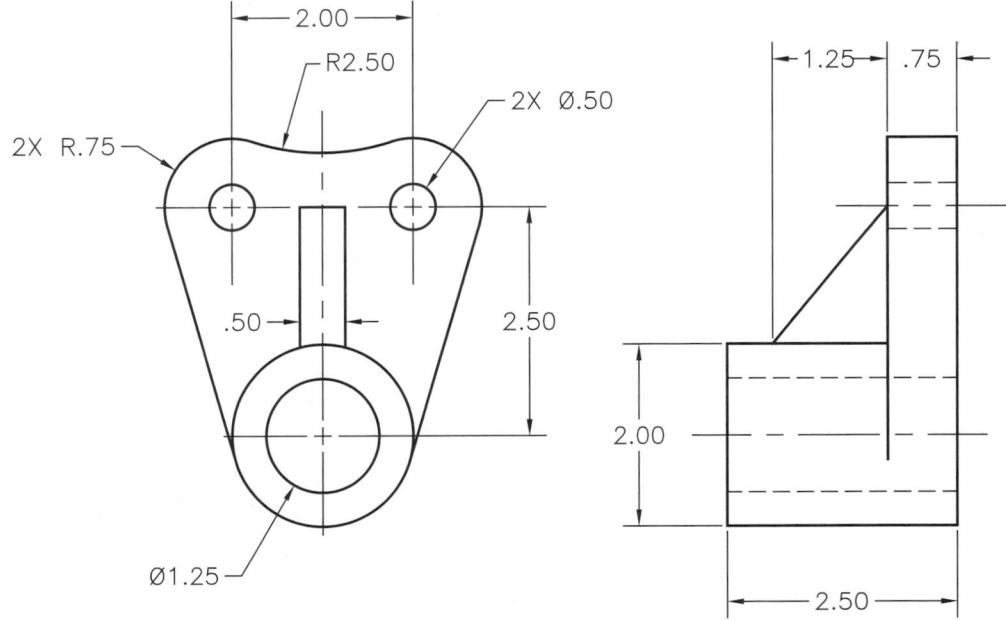

11. Use the isometric drawing provided to create a multiview orthographic drawing for the part. Include only the views necessary to fully describe the object. Draw and fully constrain the drawing. Convert dynamic dimensional constraints to annotational dimensional constraints. Adjust the drawing and add associative dimensions as needed to create a formal drawing. Save the drawing as P22-11.

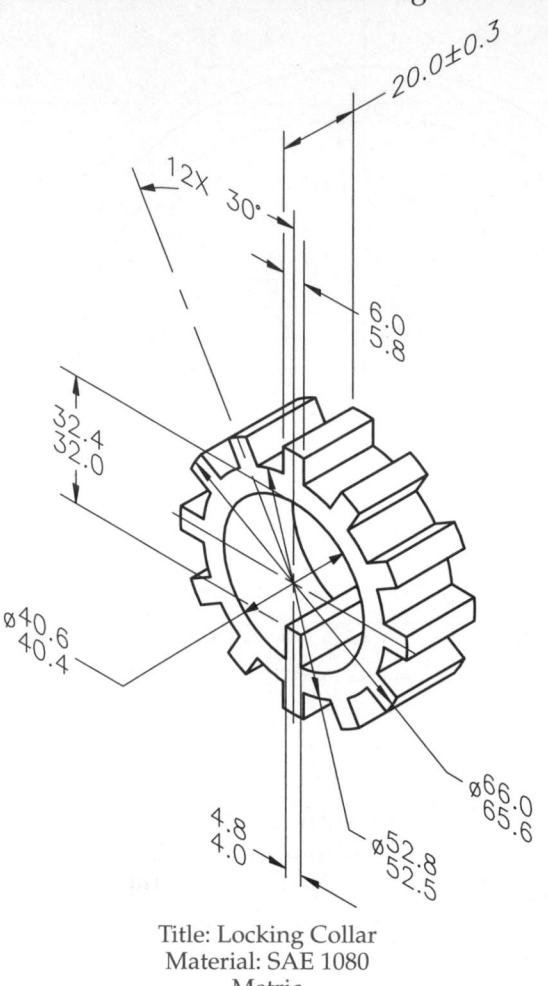

Title: Locking Collar
Material: SAE 1080
Metric

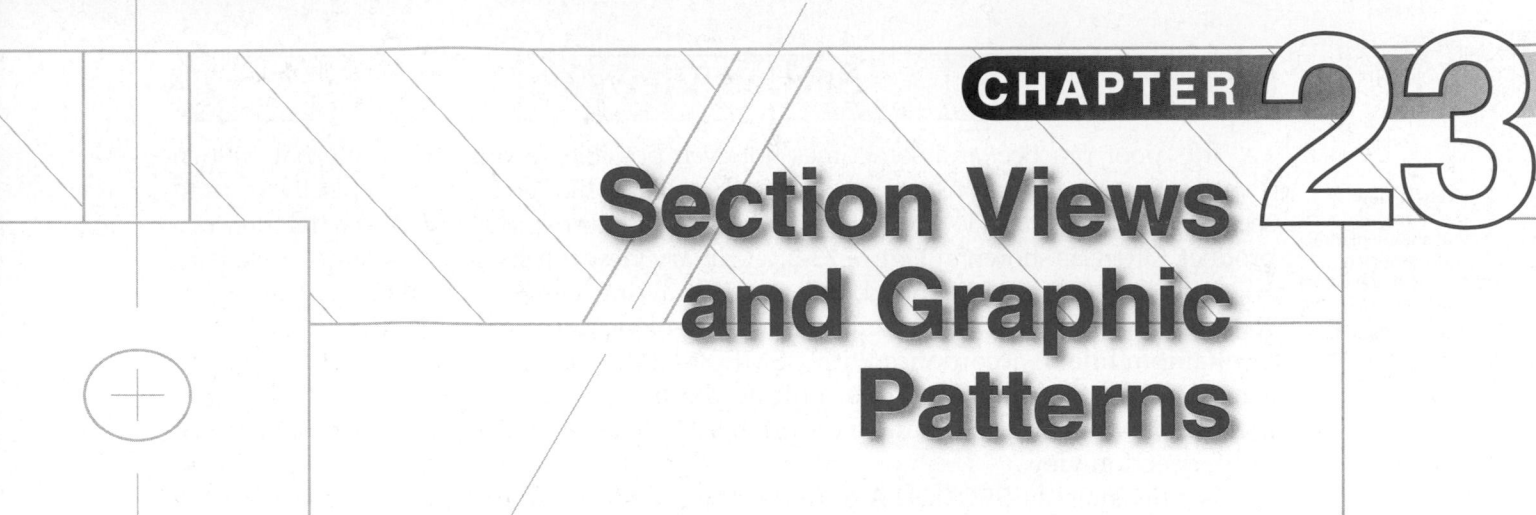

Section Views and Graphic Patterns

Learning Objectives

After completing this chapter, you will be able to do the following:

✓ Identify sectioning techniques.
✓ Add graphic patterns using the **HATCH** tool.
✓ Insert hatch patterns into drawings using **DesignCenter** and tool palettes.
✓ Edit existing hatch patterns with the **HATCHEDIT** tool and grips.

Many drawings require *graphic patterns* to describe specific information. For example, the front elevation of the house shown in **Figure 23-1** contains patterns of lines that create graphic representations of building materials. One of the most common graphic patterns is a group of section lines added to a section view. This chapter focuses on using the **HATCH** tool to draw graphic patterns quickly.

graphic pattern:
A patterned arrangement of objects or symbols.

Figure 23-1.
Graphic patterns describe repetitive drawing information, such as the siding, brick, roof, and concrete step materials added to the front elevation of a house. In this illustration, all of the items shown in color were drawn as graphic patterns.

Section Views

It is poor practice, and sometimes not even possible, to dimension internal, hidden features. *Section views,* or *sections,* allow you to clarify hidden features. Typically, you add section views to a multiview drawing to describe both the exterior and interior features of a product. Often, as shown in **Figure 23-2,** a primary view is a section or includes a section.

When a section is included, one of the drawing views contains a *cutting-plane line* to show the location of the cut. The cutting-plane line is drawn with a thick dashed or phantom line in accordance with ASME Y14.2M, *Line Conventions and Lettering.* The arrows on the cutting-plane line indicate the line of sight when you are looking at the section view. Cutting-plane lines often include labels with letters that relate to the proper section view.

A title, such as SECTION A-A, under the section view links the section with the labeled cutting-plane. When more than one section view is drawn, labels continue with B-B through Z-Z. The letters *I, O,* and *Q* are not used because they may be confused with numbers. Labeling section views is necessary for drawings with multiple sections. The label is sometimes omitted when only one section view is present and its location is obvious. *Section lines* help distinguish hidden features from exterior objects.

Other drafting fields, including architectural, structural, and civil, also use sectioning. Cross sections through buildings show the construction methods and materials. See **Figure 23-3.** The cutting-plane lines used in these fields are often composed of letter and number symbols. This helps coordinate the large number of sections found in a set of architectural drawings.

Section Lines

Section views include section line symbols to show where material has been cut to reveal hidden features. In most cases, 45° section lines are standard, unless another angle is required to satisfy the next two rules. Avoid section lines placed at angles greater than 75° or less than 15° from horizontal. Section lines should never be parallel or perpendicular to any other adjacent lines on the drawing. In addition, section lines should not cross object lines.

---

section view (section): A view that shows internal features as if a portion of the object has been cut away.

cutting-plane line: The line that cuts through the object to expose internal features.

section lines: Lines that show where material has been cut away.

Figure 23-2.
A two-view drawing with a full section view and a broken-out section.

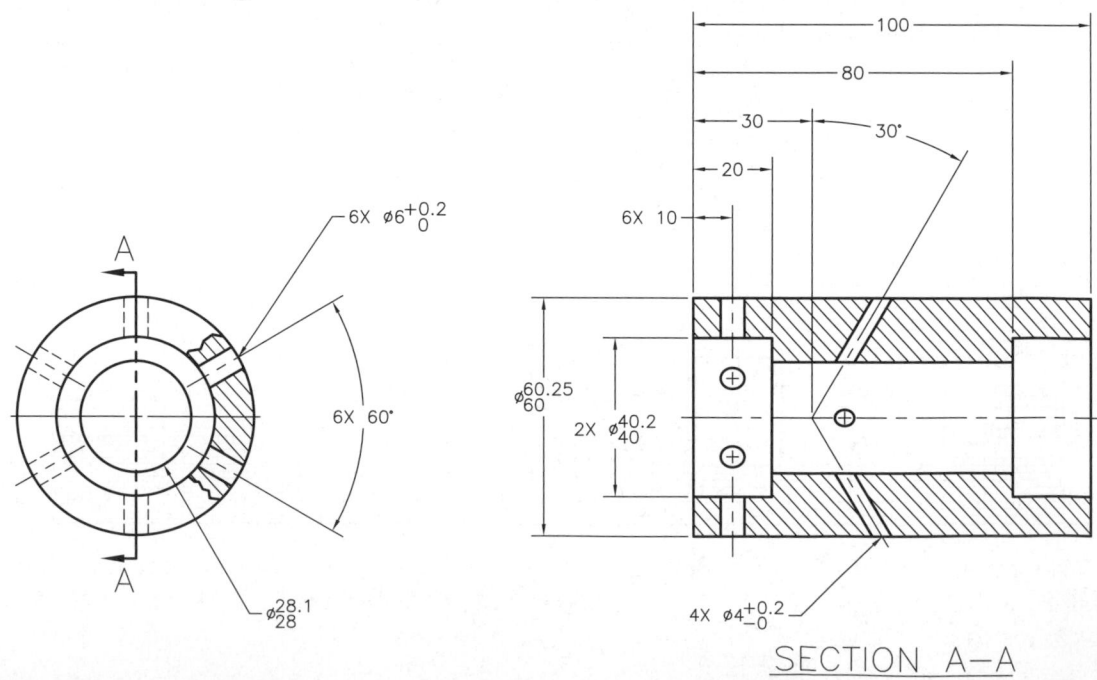

SECTION A-A

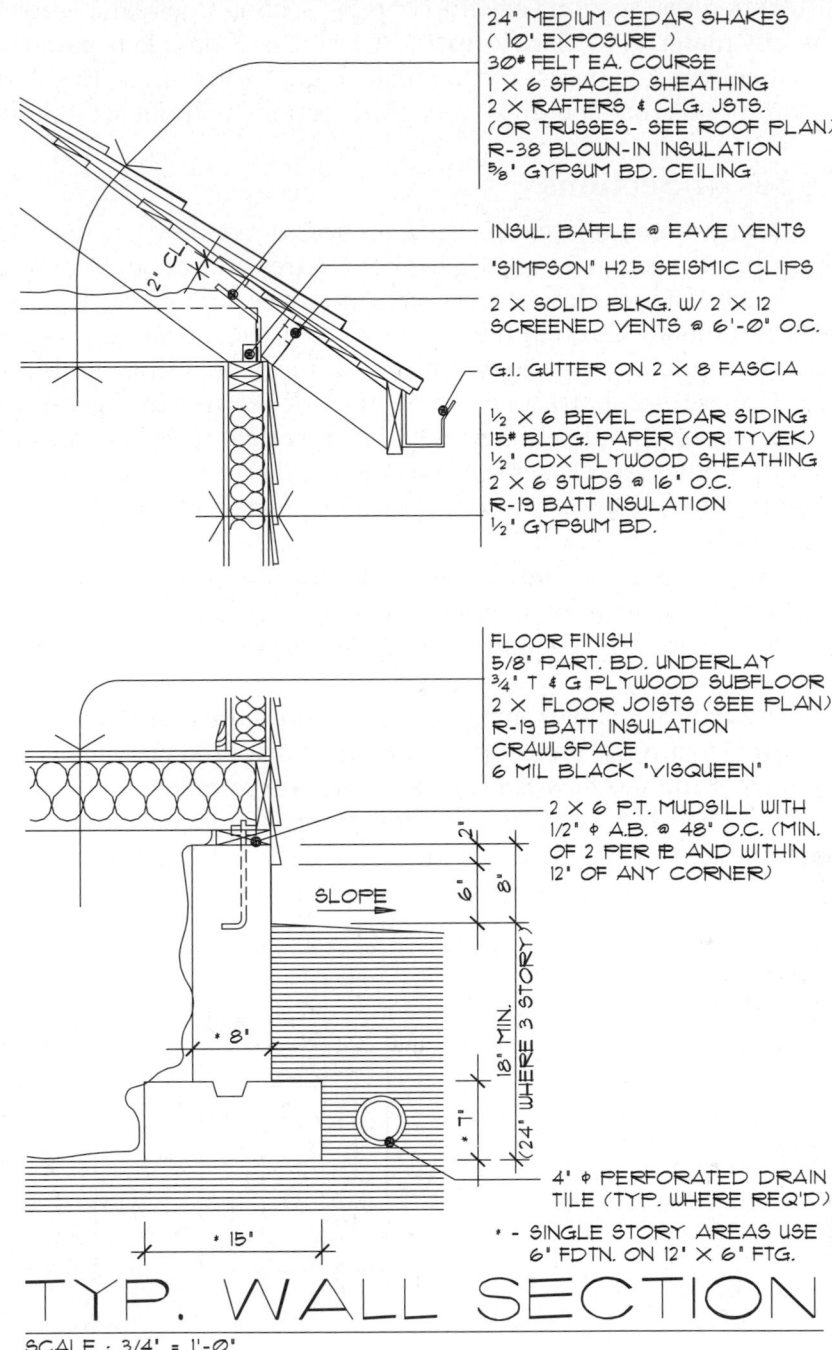

Figure 23-3.
An architectural section view. (Alan Mascord Design Associates)

24" MEDIUM CEDAR SHAKES
(10' EXPOSURE)
30# FELT EA. COURSE
1 X 6 SPACED SHEATHING
2 X RAFTERS & CLG. JSTS.
(OR TRUSSES- SEE ROOF PLAN)
R-38 BLOWN-IN INSULATION
⅝' GYPSUM BD. CEILING

INSUL. BAFFLE @ EAVE VENTS

"SIMPSON" H2.5 SEISMIC CLIPS

2 X SOLID BLKG. W/ 2 X 12
SCREENED VENTS @ 6'-0" O.C.

G.I. GUTTER ON 2 X 8 FASCIA

½ X 6 BEVEL CEDAR SIDING
15# BLDG. PAPER (OR TYVEK)
½' CDX PLYWOOD SHEATHING
2 X 6 STUDS @ 16' O.C.
R-19 BATT INSULATION
½' GYPSUM BD.

FLOOR FINISH
5/8' PART. BD. UNDERLAY
¾' T & G PLYWOOD SUBFLOOR
2 X FLOOR JOISTS (SEE PLAN)
R-19 BATT INSULATION
CRAWLSPACE
6 MIL BLACK 'VISQUEEN'

2 X 6 P.T. MUDSILL WITH
1/2' Φ A.B. @ 48' O.C. (MIN.
OF 2 PER ₽ AND WITHIN
12' OF ANY CORNER)

SLOPE

18' MIN.
(24' WHERE 3 STORY)

4' Φ PERFORATED DRAIN
TILE (TYP. WHERE REQ'D)

* - SINGLE STORY AREAS USE
6' FDTN. ON 12' X 6' FTG.

TYP. WALL SECTION

SCALE : 3/4' = 1'-0'

Section lines may be drawn using different patterns to represent specific types of material. Equally spaced section lines like those in **Figure 23-2** represent a general application. This is adequate in most situations. Additional patterns are not necessary if the title block or a note clearly indicates the type of material. However, different section line material symbols are needed when connected parts of different materials are sectioned.

AutoCAD provides standard section line symbols known as *hatch patterns*. The acad.pat file stores standard hatch pattern symbols. The ANSI31 pattern is a general section line symbol and is the default pattern in some templates. The ANSI31 pattern also represents cast iron in a section. The ANSI32 symbol identifies steel in a section. When you change to a different hatch pattern, the new pattern becomes the default in the current drawing until changed.

hatch patterns:
AutoCAD section line symbols and graphic patterns.

When you section very thin objects, section lines can completely blacken or shade the cut material to clarify features. Use the Solid hatch pattern for this application, or whenever it is necessary to create a solid filled area. The ASME Y14.2M standard recommends that you draw very thin sections without section lines or solid fills.

Types of Sections

Many types of sections can be used, depending on the application. Choose the appropriate section according to the features and detail to be sectioned. For example, an object that includes a significant amount of hidden detail may require a section that cuts completely through the object. In contrast, a drawing may require that you only remove a small portion to expose and dimension a single, minor interior feature.

Figure 23-2 shows an example of a *full section*. In this type of section, the cutting plane passes completely through the object along the center plane, as shown by the cutting-plane line. *Offset sections* are similar to full sections, except the cutting plane staggers. This allows you to cut through features that are not in a straight line. See **Figure 23-4.**

Figure 23-5 provides an example of an *aligned section*. The cutting plane cuts through the feature to section, and then *rotates* to align with the center plane before projecting onto the section view. Using an offset section for this application would distort the image.

Figure 23-6 shows an example of a *revolved section*. A revolved section may appear in place within the object, or a portion of the view may be broken away to make dimensioning easier. *Removed sections* serve much the same function as revolved sections. See **Figure 23-7.** A cutting-plane line identifies the location of the section. Multiple removed sections require labeled cutting-plane lines and related views. Drawing only the ends of the cutting-plane lines simplifies the views.

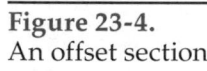

full sections: Sections in which half the object is removed.

offset sections: Sections that have a staggered cutting plane.

aligned sections: Sections used when a feature is out of alignment with the center plane.

revolved sections: Sections that clarify the contour of objects that have the same shape throughout their length.

removed sections: Section views that are similar to revolved sections, but are removed from the regular view.

Figure 23-4.
An offset section.

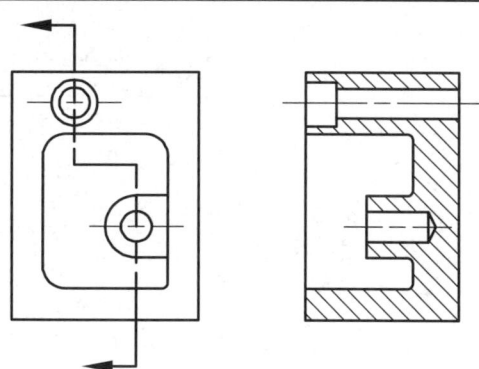

Figure 23-5.
An aligned section.

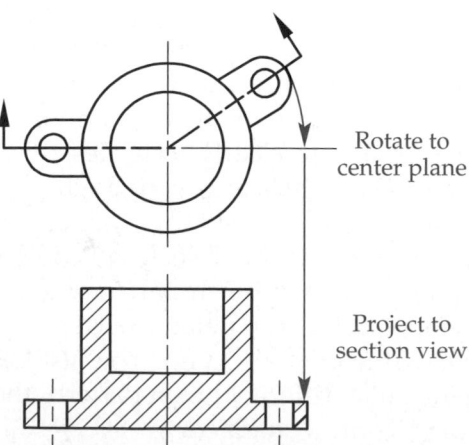

Rotate to center plane

Project to section view

AutoCAD and Its Applications—Basics

Figure 23-6.
A revolved section.

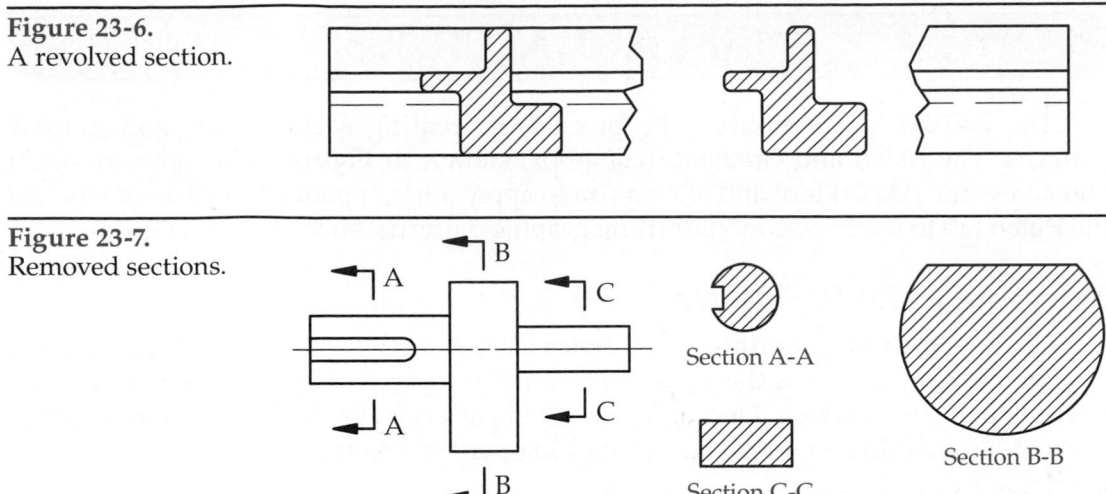

Figure 23-7.
Removed sections.

Figure 23-8 shows an example of a *half section*. The term *half* describes how half of the view appears in section, while the other half remains as an exterior view. Half sections are common when drawing symmetrical objects. A centerline separates the sectioned part of the view from the unsectioned part. You normally omit hidden lines from the unsectioned side. *Broken-out sections* clarify a specific hidden feature(s). See **Figure 23-9**. **Figure 23-2** also shows a broken-out section.

NOTE

Section lines are a basic application for hatch patterns. Many other drawings also require hatching, such as for material representation on an architectural elevation, shading on a technical illustration, and artistic patterns for graphic layouts.

Figure 23-8.
A half section.

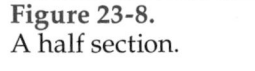

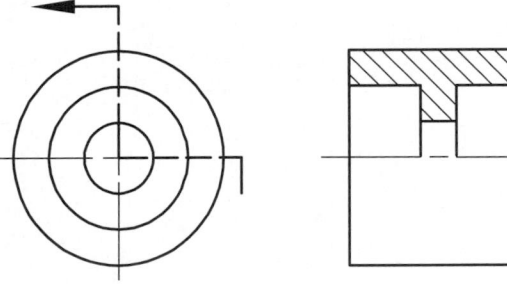

Figure 23-9.
A broken-out section.

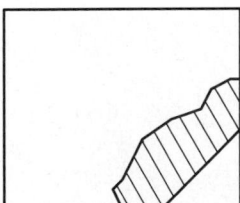

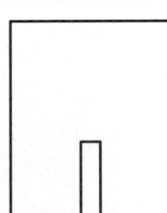

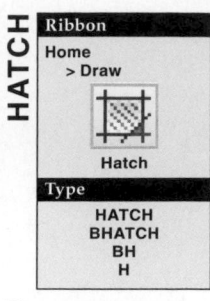

HATCH

Ribbon
Home
> Draw

Hatch

Type
HATCH
BHATCH
BH
H

boundary: The area filled by a hatch pattern.

The **HATCH** tool simplifies the process of creating section lines and graphic patterns. The **Hatch and Gradient** dialog box, shown in **Figure 23-10**, appears when you access the **HATCH** tool and allows you to apply a hatch pattern to a *boundary*. Use the **Hatch** tab to apply common drafting graphic patterns, such as section lines.

Selecting a Hatch Pattern

Use the **Type and pattern** area of the **Hatch** tab to select a hatch pattern. Hatch pattern categories are available in the **Type:** drop-down list. Hatch types include **Predefined**, **User defined**, and **Custom**. Once you select the pattern type, specific options become enabled for selecting a pattern and controlling pattern characteristics.

Predefined Hatch Patterns

The **Predefined** hatch type provides patterns stored in the acad.pat and acadiso.pat files. Use the **Pattern:** drop-down list to select a predefined hatch pattern by name. Alternatively, pick the ellipsis (**...**) button next to the **Pattern:** drop-down list or pick in the **Swatch:** preview box to display the **Hatch Pattern Palette** dialog box. See **Figure 23-11**. The **ANSI, ISO, Other Predefined**, and **Custom** tabs divide the hatch patterns into groups. The name and an image identify each pattern. Select the pattern from the appropriate tab and pick the **OK** button to return to the **Hatch and Gradient** dialog box. The selected pattern appears in the **Pattern:** text box and the **Swatch:** preview box.

Figure 23-10.
The **Hatch** tab of the **Hatch and Gradient** dialog box.

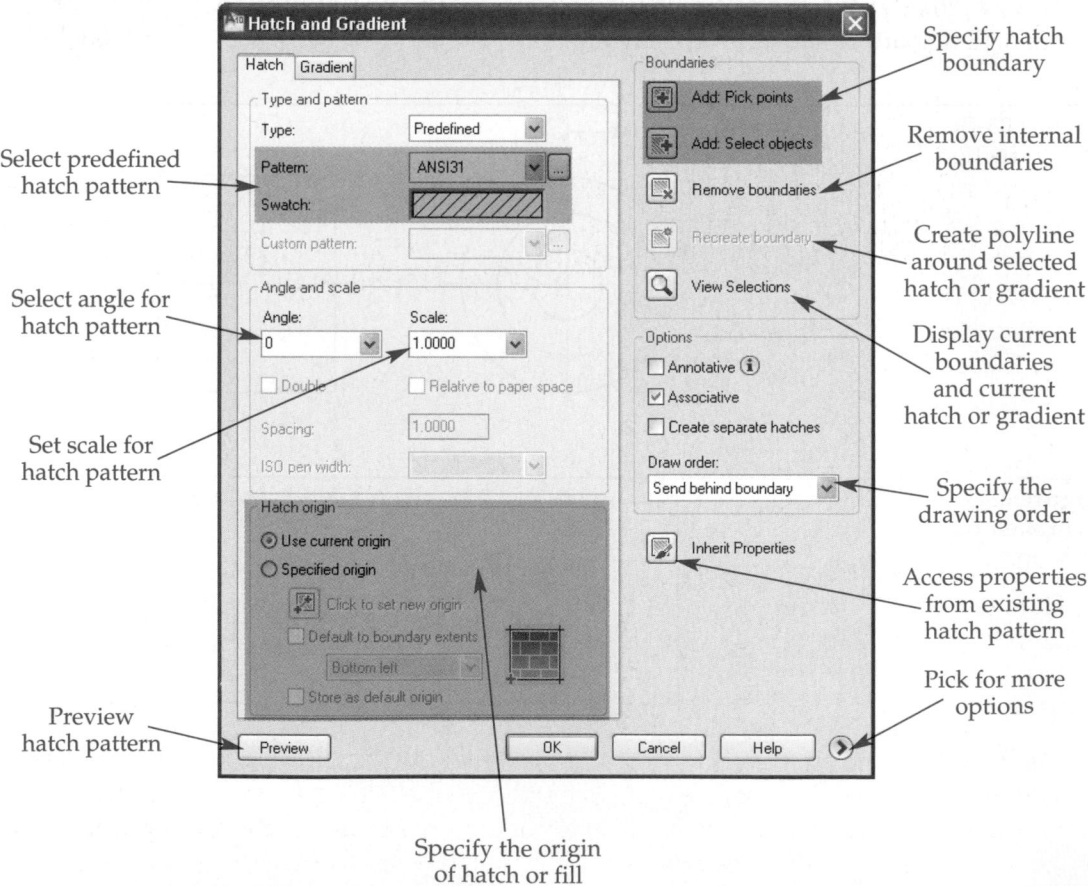

Select predefined hatch pattern

Select angle for hatch pattern

Set scale for hatch pattern

Preview hatch pattern

Specify hatch boundary

Remove internal boundaries

Create polyline around selected hatch or gradient

Display current boundaries and current hatch or gradient

Specify the drawing order

Access properties from existing hatch pattern

Pick for more options

Specify the origin of hatch or fill

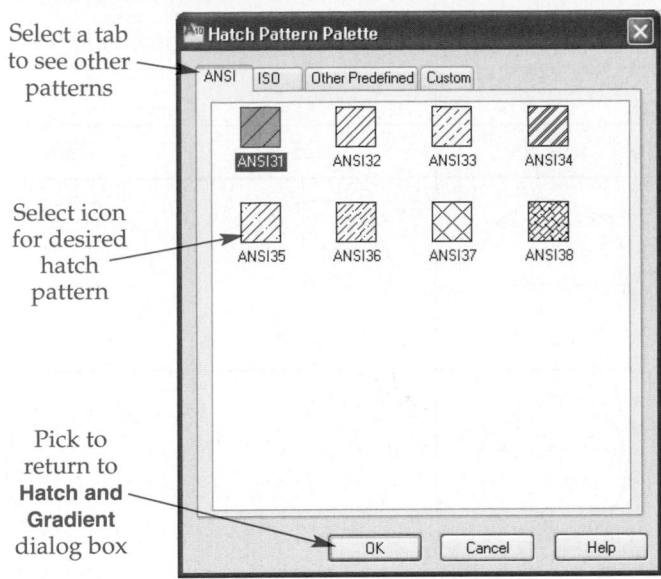

Figure 23-11.
The **Hatch Pattern Palette** dialog box is often the best place to select a predefined or custom hatch pattern.

Select a tab to see other patterns

Select icon for desired hatch pattern

Pick to return to **Hatch and Gradient** dialog box

Predefined hatch patterns use specific angle and scale characteristics. You can modify these settings for unique applications using the options in the **Angle and scale** area. Specify a value in the **Angle:** text box to rotate the pattern. Specify a value in the **Scale:** text box to change the pattern size. For example, by default, the ANSI31 hatch is a pattern of 45° lines spaced .125″ (3 mm) apart. If you change the angle to 15° and the scale to 2, a pattern of 60° (45+15=60) lines spaced .25″ (6 mm) apart forms.

NOTE

You can control the ISO pen width for predefined ISO patterns using the **ISO pen width:** drop-down list.

User-Defined Hatch Patterns

The **User defined** hatch type creates a pattern of equally spaced lines for basic hatching applications. The lines use the current linetype. The options in the **Angle and scale** area allow you to form a specific pattern of lines. Specify a value in the **Angle:** text box to rotate the pattern relative to the X axis. Use the **Spacing:** text box to specify the distance between lines in the pattern. Check **Double** to create a pattern of double lines. **Figure 23-12** shows examples of user-defined hatch patterns.

Custom Hatch Patterns

You create and save custom hatch patterns in PAT files. The **Custom** hatch type allows you to specify a pattern defined in any custom PAT file you add to the AutoCAD search path. Use the **Custom pattern:** drop-down list to select a custom pattern name, or pick the ellipsis (...) button or the **Swatch:** preview box to select the pattern from the **Custom** tab of the **Hatch Pattern Palette** dialog box. You can set the angle and scale of custom hatch patterns using the same techniques as for predefined hatch patterns.

Setting the Hatch Pattern Size

Adjust the size of predefined and custom hatch patterns using the **Scale:** text box. The default scale is 1. If the pattern appears too small or large, specify a different scale. See **Figure 23-13.** User-defined hatch pattern size is set using the **Spacing:** text box and defines the exact distance between lines. Refer again to **Figure 23-12.**

Figure 23-12.
Examples of user-defined hatch patterns with different hatch angles and spacing.

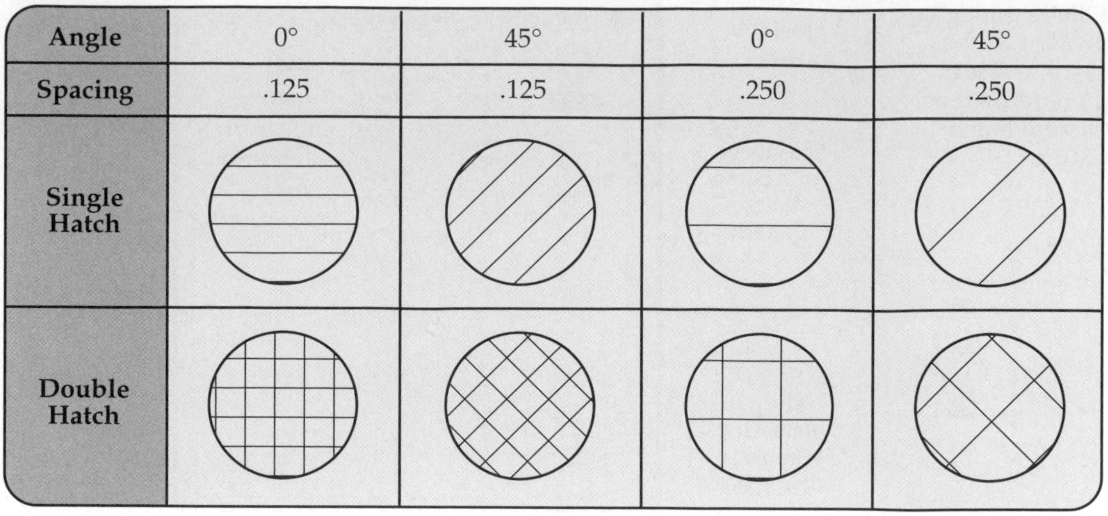

Angle	0°	45°	0°	45°
Spacing	.125	.125	.250	.250
Single Hatch				
Double Hatch				

Figure 23-13.
Hatch pattern scale factors.

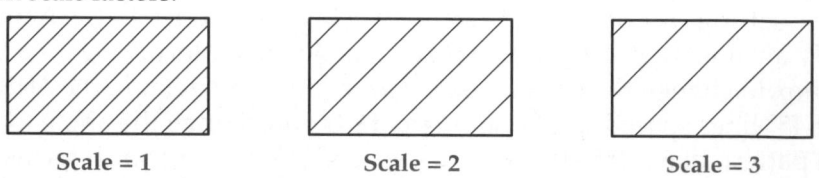

Scale = 1   Scale = 2   Scale = 3

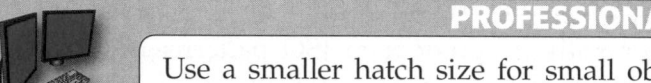

PROFESSIONAL TIP

Use a smaller hatch size for small objects and a larger hatch size for larger objects. This makes section lines look appropriate for the drawing scale. Often you must use your best judgment when selecting a hatch size.

The drawing scale is an important consideration when you select the hatch pattern scale or spacing. You must use an appropriate hatch size in order to make sure the hatch pattern appears on-screen and plots correctly. To understand the concept of hatch size, look at the section view shown in **Figure 23-14.** In this example, which uses the ANSI31 hatch pattern, the section line spacing should be the same distance apart regardless of drawing scale. The section lines on the full-scale (1:1) drawing display correctly. However, the section lines are too close on the half-scale (1:2) drawing, and they are too far apart on the double-scale (2:1) drawing. To overcome this issue, you must adjust the hatch pattern according to the drawing scale. You can calculate and manually apply the scale factor to the hatch pattern scale or spacing, or you can allow AutoCAD to calculate the scale factor by using annotative hatch patterns.

Scaling Hatch Patterns Manually

To adjust hatch size manually according to a specific drawing scale, you must calculate the drawing scale factor and then multiply the scale factor by the desired plotted hatch scale or spacing to get the model space hatch scale or spacing. Enter the adjusted scale of predefined or custom hatch patterns in the **Scale:** text box. Enter the adjusted spacing of a user-defined hatch pattern in the **Spacing:** text box. **Figure 23-15** shows examples of adjusting hatch scale according to drawing scale. Refer to Chapter 9 for information on determining the drawing scale factor.

Figure 23-14.
The hatch pattern may appear incorrect if the drawing scale changes.

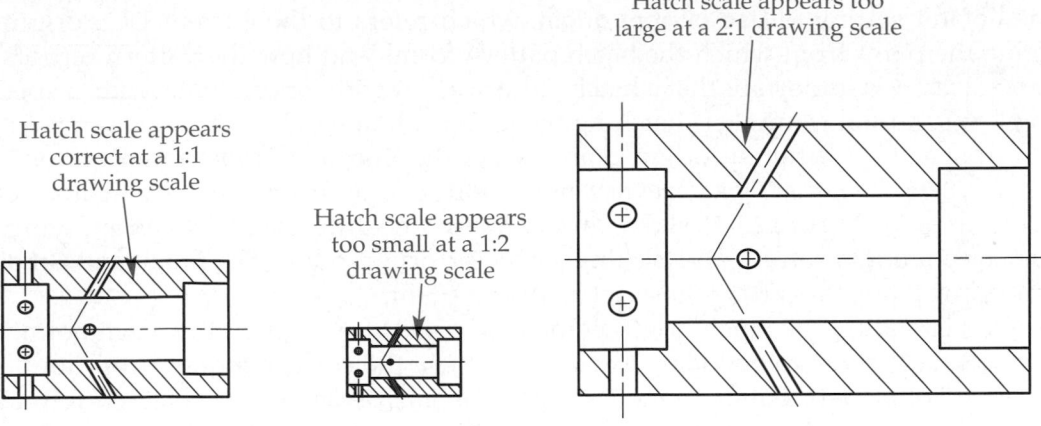

Hatch scale appears
correct at a 1:1
drawing scale

Hatch scale appears
too small at a 1:2
drawing scale

Hatch scale appears too
large at a 2:1 drawing scale

Figure 23-15.
The hatch pattern scale may require adjusting, depending on the drawing scale.

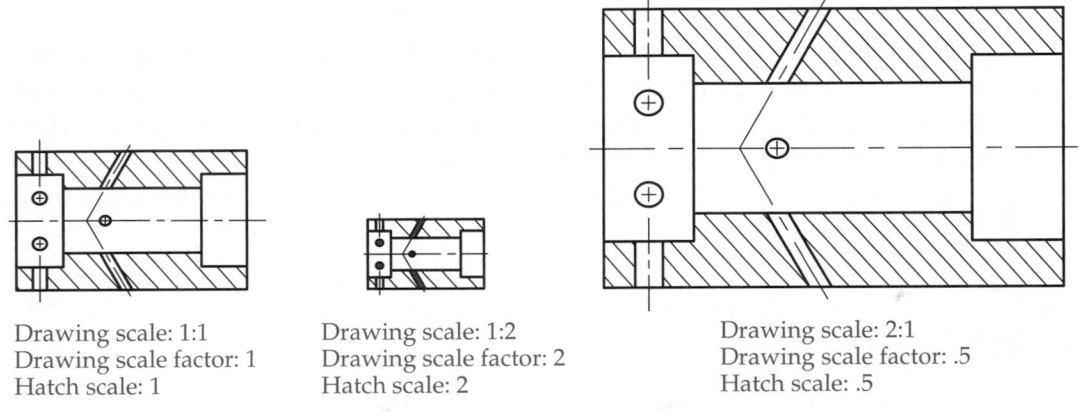

Drawing scale: 1:1
Drawing scale factor: 1
Hatch scale: 1

Drawing scale: 1:2
Drawing scale factor: 2
Hatch scale: 2

Drawing scale: 2:1
Drawing scale factor: .5
Hatch scale: .5

Annotative Hatch Patterns

Pick the **Annotative** check box in the **Options** area to make the hatch pattern annotative. AutoCAD scales annotative hatch patterns according to the annotation scale you select, which eliminates the need for you to calculate the scale factor. When you select an annotation scale from the **Annotation Scale** flyout button on the status bar, AutoCAD determines the scale factor and automatically applies it to annotative hatch patterns, or any annotative object. The result is a hatch pattern that displays the proper size regardless of the drawing scale, much like the example shown in **Figure 23-15**, but without requiring you to change the hatch scale or spacing between different drawing scales. For example, if you enter a value in the **Scale:** text box or **Spacing:** text box that is appropriate for an annotation scale of 1/4″ = 1′-0″, and then change the annotation scale to 1″ = 1′-0″, the appearance of the hatch pattern relative to the drawing scale does not change. It looks the same on the 1/4″ = 1′-0″ scale drawing as it does on the 1″ = 1′-0″ scale drawing.

Scaling Relative to Paper Space

The **Relative to paper space** check box in the **Angle and scale** area allows you to scale the hatch pattern relative to the scale of the active layout viewport. You must enter a floating layout viewport in order to select the **Relative to paper space** check box. The hatch scale automatically adjusts according to the viewport scale. For example, a floating viewport scale set to 4:1 uses a scale factor of .25 (1 ÷ 4 = .25). If you enter a hatch scale of 1, the hatch automatically appears at a scale of .25 (1 × .25 = .25).

Setting the Hatch Origin Point

The **Hatch origin** area includes options that control the position of hatch patterns. The default setting is **Use current origin**, which refers to the current UCS origin, to define the point from which the hatch pattern forms and how the pattern repeats. In some cases, it is important that a hatch pattern align with, or originate from, a specific point. A common example is hatching the representation of bricks.

To specify a different origin point, select the **Specified origin** radio button. See **Figure 23-16**. Pick the **Click to set new origin** button to return to the drawing and select an origin point. **Figure 23-17** shows an example of the difference between applying the **Use current origin** setting and picking a specific origin point. The **Hatch and Gradient** dialog box reappears after you select an origin point.

You can align the hatch origin point with a specific point on the hatch boundary by checking **Default to boundary extents**. Then use the corresponding drop-down list to select **Bottom left, Bottom right, Top right, Top left**, or **Center**. The hatch origin positions at the selected point on the boundary. For example, you could use the **Bottom right** option to create the pattern shown in **Figure 23-17B**. Check **Store as default origin** to save the custom origin point.

Specifying the Hatch Boundary

The **Add: Pick points** button often provides the easiest method of defining the hatch boundary. Pick the **Add: Pick points** button to return to the drawing and pick a point within the area to hatch. AutoCAD automatically defines and highlights the boundary around the selected point. You can select more than one internal point. Press [Enter] or the space bar, or right-click and select **Enter** to return to the **Hatch and Gradient** dialog box. See **Figure 23-18**.

Figure 23-16.
The **Specified origin** setting in the **Hatch origin** area of the **Hatch** tab activates the other hatch origin settings.

Pick to specify a different origin point

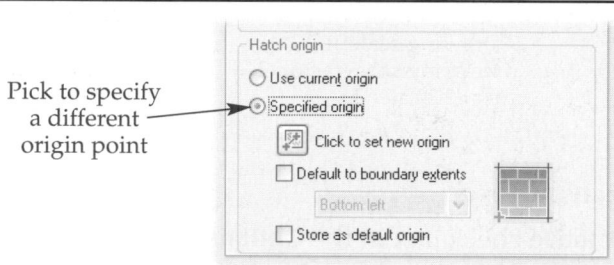

Figure 23-17.
A—The default **Use current origin** setting. B—The **Specified origin** option is selected and the **Click to set new origin** button is used to select the lower-left corner (endpoint) of the rectangle. Notice how the pattern, or in this example the first brick, starts exactly at the corner of the hatched area.

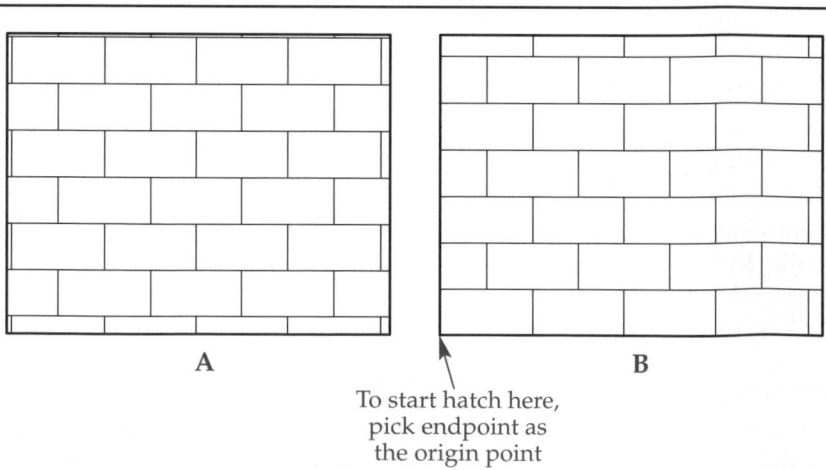

A

B

To start hatch here, pick endpoint as the origin point

Figure 23-18.
Picking a point to define the hatch boundary.

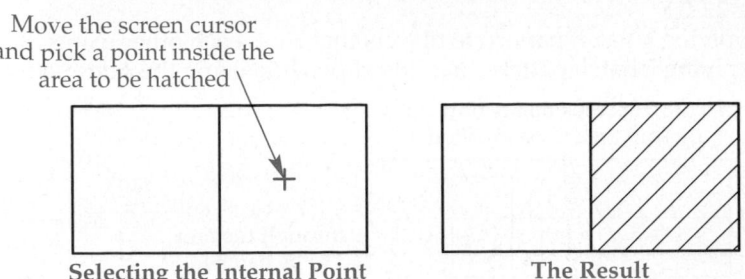

Move the screen cursor and pick a point inside the area to be hatched

Selecting the Internal Point   The Result

NOTE

If you pick the wrong area, use the **UNDO** tool at the Select internal point: prompt to undo the selection. You can also undo the hatch pattern after the pattern is drawn. However, to save time, you can preview the hatch before applying it.

Use the **Add: Select objects** button to define the hatch boundary by selecting objects, rather than picking inside an area. See **Figure 23-19.** Remember to select objects within the hatch boundary to exclude from the hatch pattern if necessary. See **Figure 23-20.** Press [Enter] or the space bar, or right-click and select **Enter** to return to the **Hatch and Gradient** dialog box. Selecting objects works especially well if other objects cross the area to hatch, forming several possible boundaries. **Figure 23-21** shows the difference between picking points and objects when hatching a bar graph.

The **Remove boundaries** button is available after you select a point or objects to define a boundary. If necessary, pick the **Remove boundaries** button to return to the drawing and select objects to remove from the hatch boundary. Press [Enter] or the space bar, or right-click and select **Enter** to return to the **Hatch and Gradient** dialog box.

Figure 23-19.
Selecting objects to hatch.

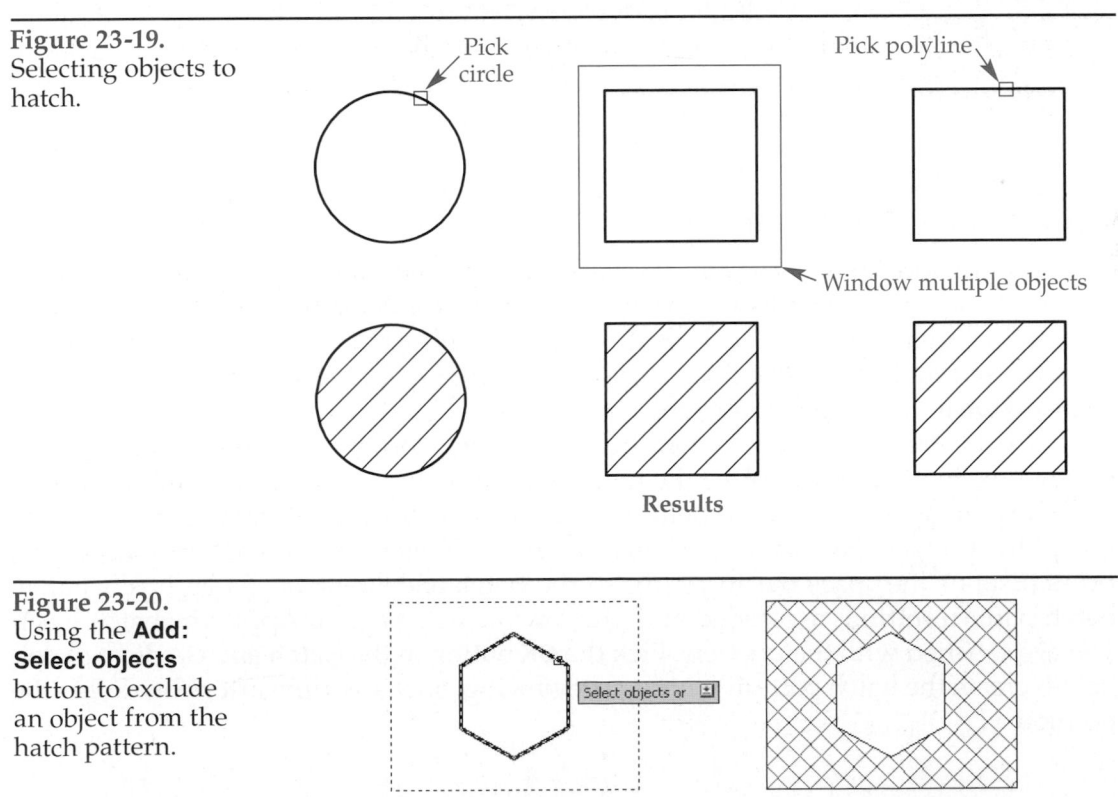

Pick circle

Pick polyline

Window multiple objects

Results

Figure 23-20.
Using the **Add: Select objects** button to exclude an object from the hatch pattern.

Select objects or

Selected Objects   Result

Figure 23-21.
A—Applying a hatch pattern to objects that cross each other using the **Add: Pick points** button.
B—Applying a hatch pattern to a closed polygon using the **Add: Select objects** button.

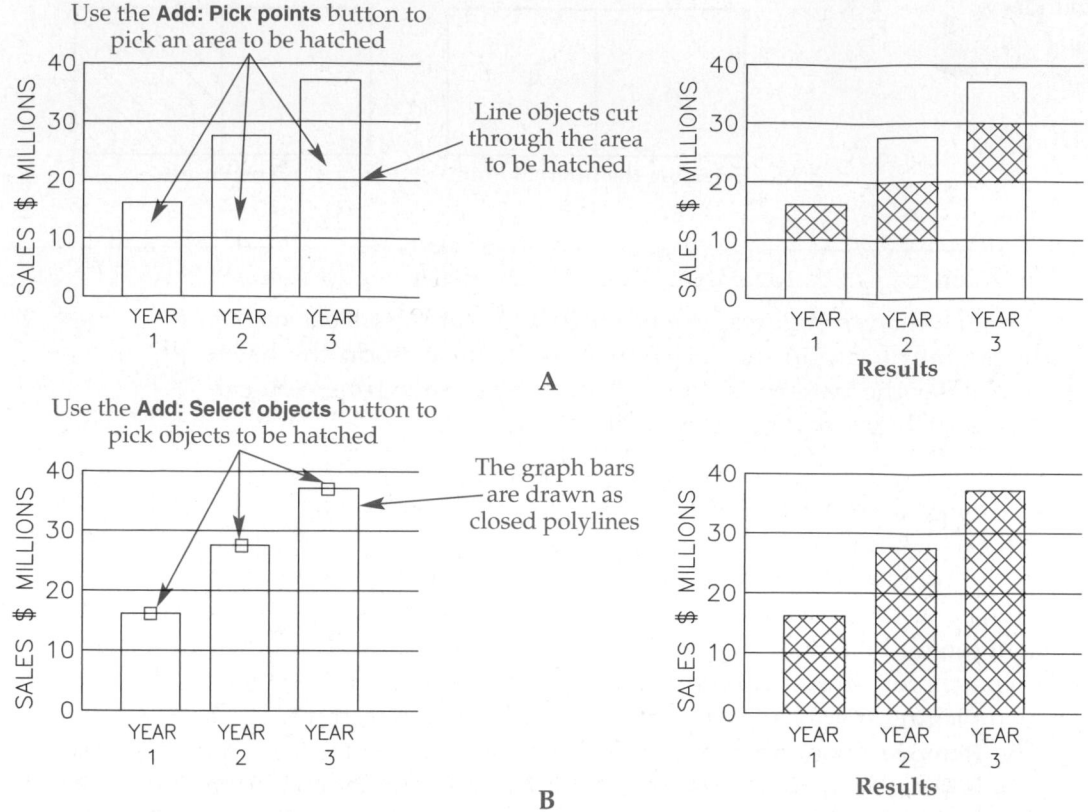

A

B

NOTE

Prompt options are available for toggling between point selection, object selection, or object removal while you are defining a boundary.

Previewing the Hatch

Use preview tools to be sure the hatch pattern and hatch boundary settings are correct before applying a hatch pattern. The **View Selections** button is available after you define a hatch boundary. Pick this button to return to the drawing window to view the selected hatch boundary. Press [Enter] or the space bar, or right-click to return to the **Hatch and Gradient** dialog box.

Pick the **Preview** button, located in the lower-left corner of the **Hatch and Gradient** dialog box, to temporarily place the hatch pattern on the drawing. This allows you to see if any changes are required before the hatch is drawn. Press [Enter] or right-click to accept the preview and create the hatch. To make changes after previewing the hatch, press [Esc] or the space bar to return to the **Hatch and Gradient** dialog box. Change hatch pattern settings as needed and preview the hatch again. Apply the hatch when you are satisfied with the preview. Pick the **OK** button in the **Hatch and Gradient** dialog box to create the hatch pattern without previewing, or at any time after you define the boundary.

Exercise 23-1

Access the Student Web site (www.g-wlearning.com/CAD) and complete Exercise 23-1.

Hatch Pattern Composition Options

The **HATCH** tool creates an *associative hatch pattern* by default. To create a *nonassociative hatch pattern*, deselect the **Associative** check box in the **Options** area of the **Hatch and Gradient** dialog box. An associative hatch is appropriate for most applications. If you stretch, scale, or otherwise edit the objects that define the boundary of an associative hatch, the pattern automatically adjusts to and fills the modified boundary. A nonassociative hatch pattern does not respond this way. Instead, nonassociative hatch boundary grips are available for changing the extents of the hatch, separate from the original boundary objects.

You can select multiple points and objects during a single hatch operation. By default, multiple boundaries form a single hatch object. Selecting and editing one of the hatch patterns selects and edits all patterns created during the same operation. If this is not the preferred result, check the **Create separate hatches** check box in the **Options** area before applying the hatch patterns. Individual hatch patterns will then form for each boundary.

> **associative hatch pattern:** Pattern that updates automatically when the associated objects are edited.

> **nonassociative hatch pattern:** A pattern that is independent of objects; it updates when the boundary changes, but not when changes are made to objects.

Controlling the Draw Order

The **Draw order** drop-down list in the **Options** area provides options for controlling the order of display when a hatch pattern overlaps other objects. The **Send behind boundary** option is default and makes the hatch pattern appear behind the boundary. Select the **Bring in front of boundary** option to make the hatch pattern appear on top of the boundary. Select the **Do not assign** option to have no automatic drawing order setting assigned to the hatch.

Use the **Send to back** option to send the hatch pattern behind all other objects in the drawing. Any objects that are in the hatching area appear as if they are on top of the hatch pattern. Use the **Bring to front** option to bring the hatch pattern in front of, or on top of, all other objects in the drawing. Any objects that are in the hatching area appear as if they are behind the hatch pattern.

NOTE

Use the **DRAWORDER** tool to change the draw order setting after creating a hatch pattern.

Hatching Objects with Islands

When defining a hatch boundary using the **Add: Pick points** button, you may need to adjust how AutoCAD treats *islands*, like those shown in **Figure 23-22**. By default, islands do not hatch, as shown in **Figure 23-22B**. One option to hatch islands is to use the **Remove boundaries** button after selecting the internal point. Then pick the islands to remove and press [Enter] or the space bar, or right-click and pick **Enter** to return to the dialog box and create the hatch. See **Figure 23-22C**.

Another method is to adjust island detection in the **Islands** area. Pick the **More Options** button in the lower-right corner of the **Hatch and Gradient** dialog box to expand the dialog box to show the **Islands** area and additional hatch settings. See **Figure 23-23**.

> **islands:** Boundaries inside another boundary.

Figure 23-22.
A—Original objects. B—Using the **Add: Pick points** button to hatch an internal area leaves islands unhatched. C—After picking an internal point, use the **Remove boundaries** button and pick the islands to hatch over the islands.

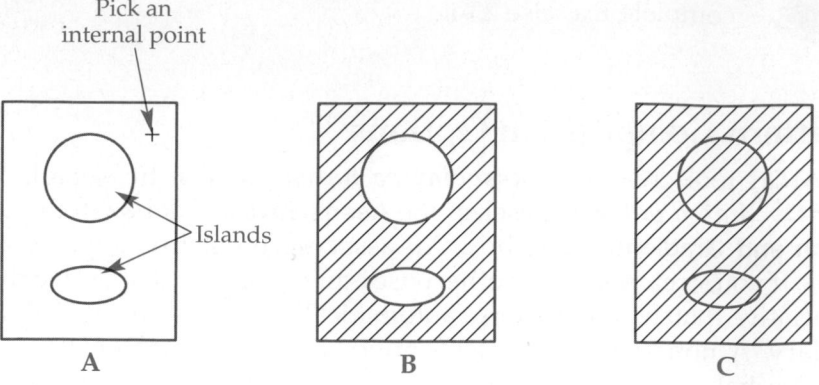

Figure 23-23.
The **Islands** area and additional hatch settings appear when you pick the **More Options** button. The **Boundary retention** and **Boundary set** areas of the **Hatch and Gradient** dialog box provide options to improve hatching efficiency.

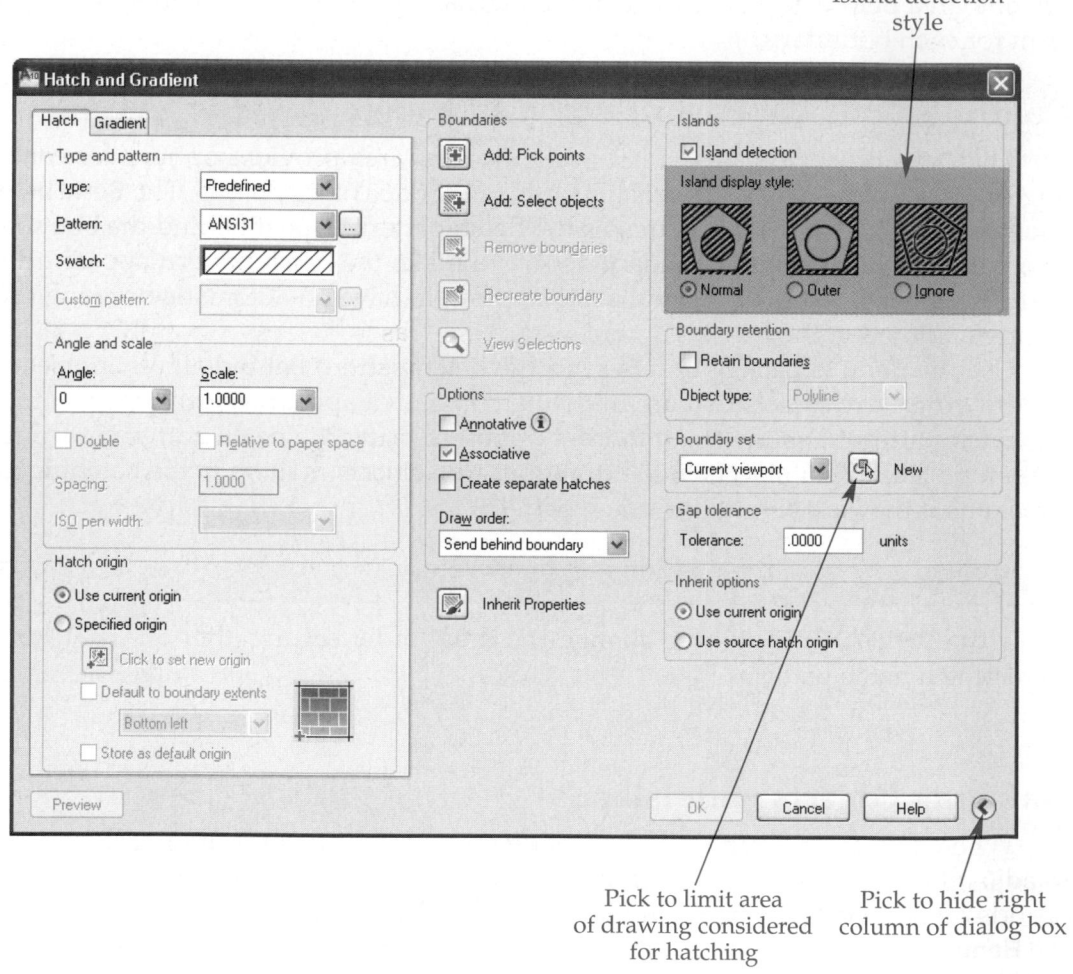

Select the **Island detection** check box to adjust the island display style. The island display style images illustrate the effect of each island detection option.

Pick the **Normal** radio button to hatch every other boundary, stepping inward from the outer boundary. If AutoCAD encounters an island, it turns off hatching until it encounters another island, and then reactivates hatching. Select the **Outer** radio button to hatch only the outermost area inward from the outer boundary. AutoCAD stops hatching when it encounters the first island. Choose the **Ignore** radio button to ignore all islands and hatch everything within the selected boundary.

Exercise 23-2

Access the Student Web site (www.g-wlearning.com/CAD) and complete Exercise 23-2.

Improving Boundary Hatching Speed

You can improve the hatching speed and resolve other hatching problems using options in the expanded **Hatch and Gradient** dialog box. By default, a hatch pattern forms according to objects in the drawing. The boundary is essentially temporary. This is appropriate for most applications. To form the boundary as a separate object that overlaps the original objects, pick the **Retain boundaries** check box in the **Boundary retention** area. See **Figure 23-24.** Select the Polyline option from the **Object type:** drop-down list to specify the boundary as a polyline object around the hatch area. Choose the Region option to specify the hatch boundary as a hatched *region*. Retaining a boundary forms additional boundary objects that you can use even if you remove or edit the original objects.

region: A closed two-dimensional area.

The **Boundary set** area specifies the amount of drawing area evaluated during the hatch operation. The default setting is Current viewport. To limit the evaluation area, possibly increasing hatch performance, pick the **New** button and use a window to select the area to evaluate. See **Figure 23-25.** After you define the new boundary set, the **Hatch and Gradient** dialog box returns, and the drop-down list in the **Boundary set** area displays Existing set. **Figure 23-26** shows a new hatch pattern applied to a boundary within the boundary set. You can make as many boundary sets as necessary. The last boundary set remains current until you create another.

Figure 23-24.
You can form a boundary as a polyline or a region. These options are only available if the **Retain boundaries** option is checked.

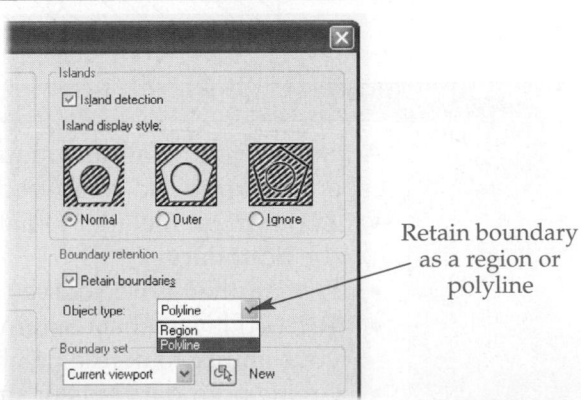

Retain boundary as a region or polyline

Figure 23-25.
The boundary set limits the area that AutoCAD evaluates during a hatching operation.

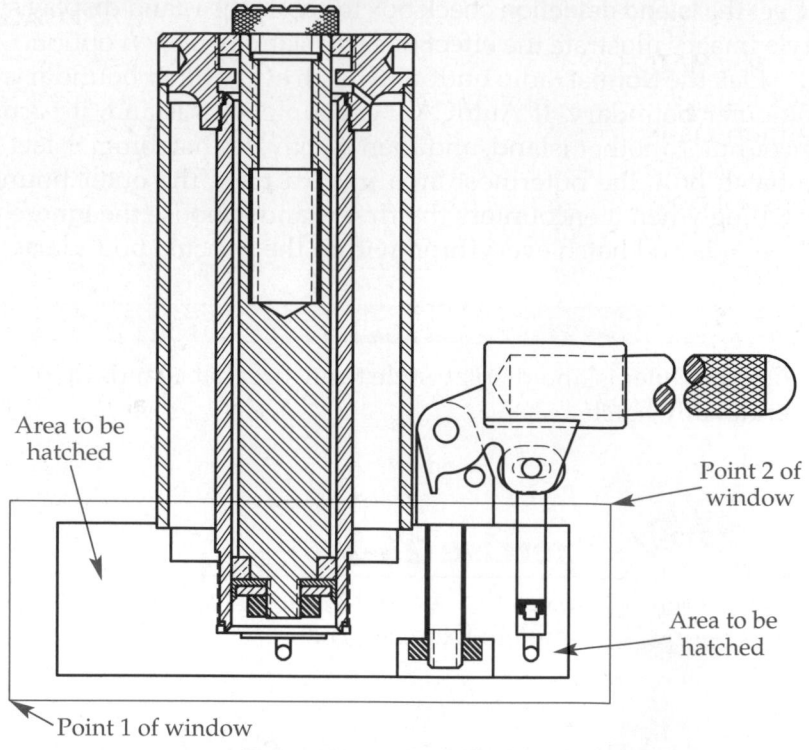

Area to be hatched

Point 2 of window

Area to be hatched

Point 1 of window

Figure 23-26.
Results of hatching the drawing in **Figure 23-25** after selecting a boundary set.

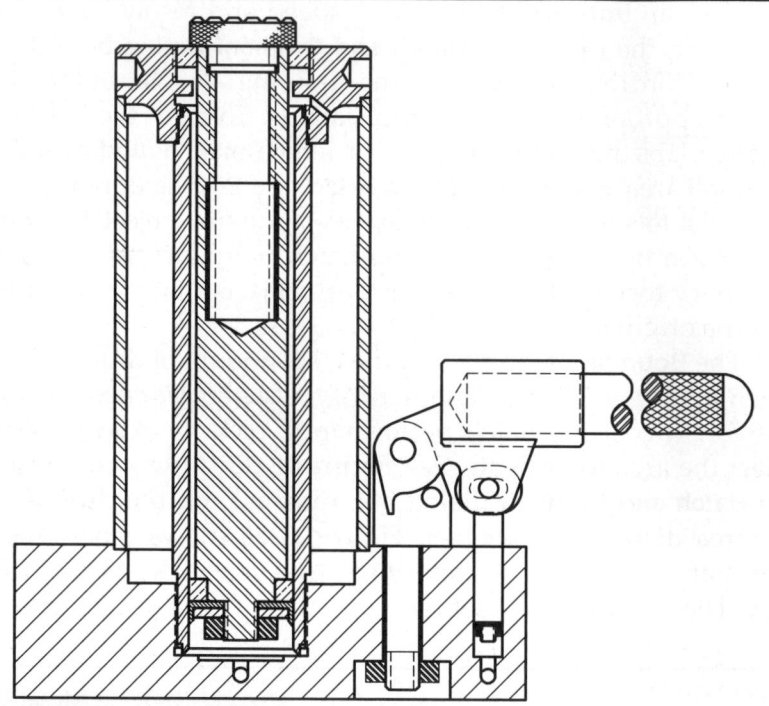

PROFESSIONAL TIP

Apply the following techniques to help save time when hatching, especially large and complex drawings:
- Zoom in on the area to hatch to make it easier for you to define the boundary.
- Preview the hatch before you apply it to make last-minute adjustments.
- Turn off layers that contain lines or text that might interfere with your ability to define hatch boundaries accurately.
- Create boundary sets of small areas within a complex drawing.

Hatching Unclosed Areas and Correcting Boundary Errors

The **HATCH** tool works well unless there is a gap in the hatch boundary or you pick a point outside a likely boundary. When you select a point where no boundary can form, an error message states that a valid boundary cannot be determined. Close the message and try again to specify the boundary. When you try to hatch an area that does not close because of a small gap, you will see the error message and circles shown in **Figure 23-27.** Close the message and eliminate the gap to create the hatch.

For most applications, it is best to identify and close a gap. However, you can hatch an unclosed boundary by setting a *gap tolerance* in the **Gap Tolerance** area. AutoCAD ignores any gaps in the boundary less than or equal to the value specified in the **Tolerance:** text box. Before generating the hatch, AutoCAD displays a message allowing you to hatch the unclosed area or return to the **Hatch and Gradient** dialog box.

gap tolerance: The amount of gap allowed between segments of a boundary to be hatched.

PROFESSIONAL TIP

When you create an associative hatch, it is often best to specify a single internal point per hatch. If you specify more than one internal point in the same operation, AutoCAD creates one hatch object from all points picked. This can cause unexpected results when you try to edit what appears to be a separate hatch object.

Creating Solid and Gradient Fills

The Solid predefined hatch pattern, available from the **Pattern:** drop-down list or **Other Predefined** tab of the **Hatch Pattern Palette** dialog box, is an effective way to fill a boundary with a solid. See **Figure 23-28.** To create a more advanced gradient fill, use the options in the **Gradient** tab of the **Hatch and Gradient** dialog box. See **Figure 23-29.**

Gradient fills are commonly used to simulate color-shaded objects. Gradients create the appearance of a lit surface with a gradual transition from an area of highlight to a filled area. Use two colors to simulate a transition from light to dark between

gradient fill: A shading transition between the tones of one color or two separate colors.

Figure 23-27.
Close a boundary to create a hatch pattern.

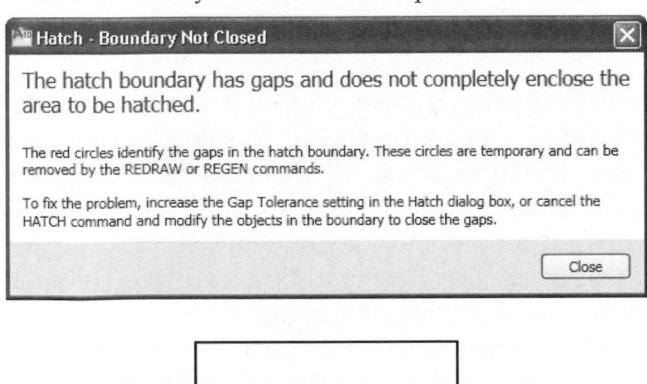

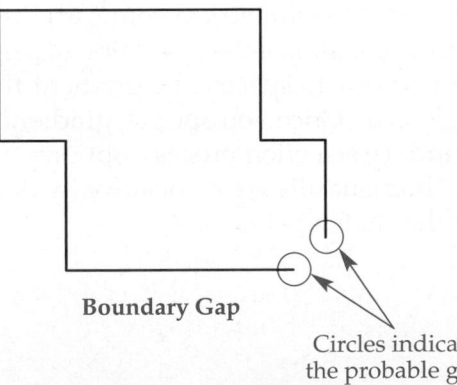

Boundary Gap

Circles indicate
the probable gap

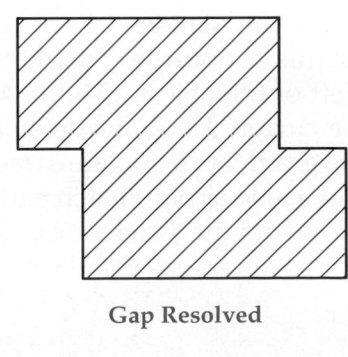

Gap Resolved

Figure 23-28.
Using the Solid hatch pattern to make a solid hatch object.

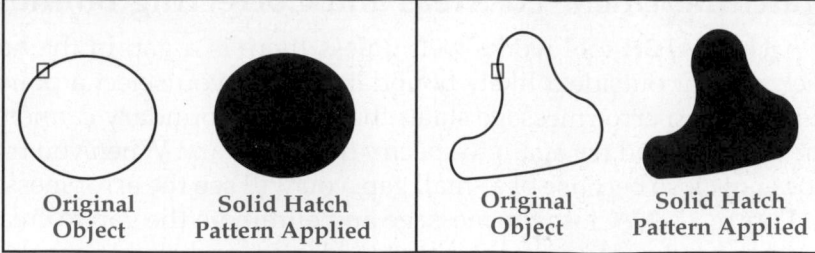

Original Object   Solid Hatch Pattern Applied   Original Object   Solid Hatch Pattern Applied

Figure 23-29.
The **Gradient** tab of the **Hatch and Gradient** dialog box contains options for creating gradient fills.

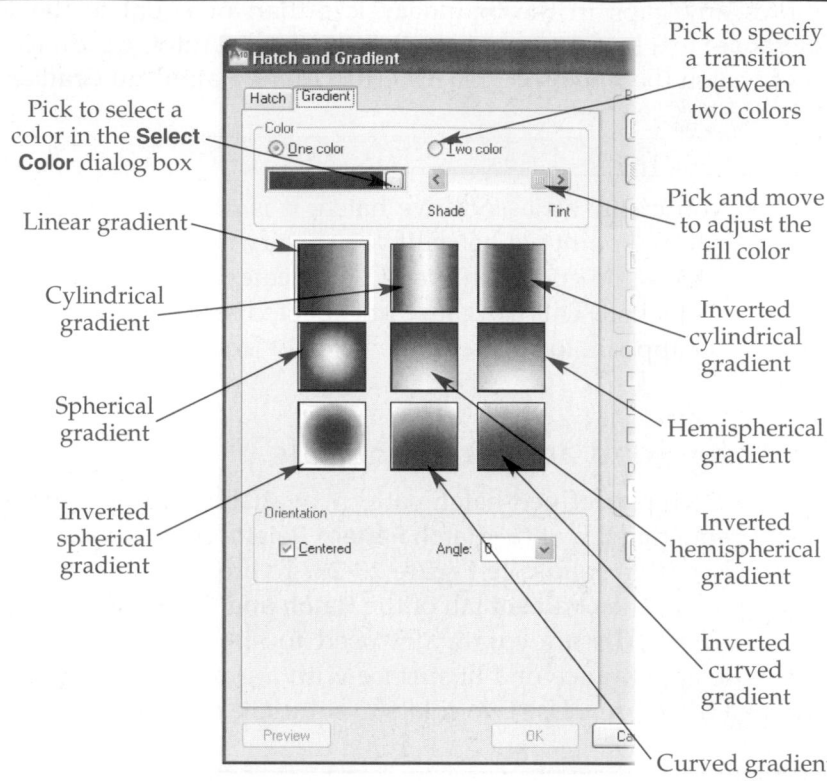

Pick to select a color in the **Select Color** dialog box

Pick to specify a transition between two colors

Pick and move to adjust the fill color

Linear gradient

Cylindrical gradient

Inverted cylindrical gradient

Spherical gradient

Hemispherical gradient

Inverted spherical gradient

Inverted hemispherical gradient

Inverted curved gradient

Curved gradient

the colors. Several different gradient fill patterns are available to create linear sweep, spherical, radial, or curved shading.

The **One color** radio button is the default and creates a fill that has a smooth transition between the darker shades and lighter tints of one color. To select a color, pick the ellipsis (**...**) button next to the color swatch to access the **Select Color** dialog box. When the **One color** option is active, the **Shade** and **Tint** slider appears. Use the slider

tint: A specific color mixed with white.

shade: A specific color mixed with gray or black.

to specify the *tint* or *shade* of a color used for a one-color gradient fill. Pick the **Two color** radio button to specify a fill using a smooth transition between two colors. A color swatch with an ellipsis (**...**) button is available for each color.

Pick the **Centered** check box to apply a symmetrical configuration. If you do not select this option, the gradient fill shifts to simulate the projection of a light source from the left of the object. Use the **Angle** text box to specify the gradient fill angle relative to the current UCS. The default angle is 0°. Once you specify gradient characteristics, create gradients using the same boundary selection process, options, and settings you would use to apply a hatch pattern. Gradient fills are associative by default and can be edited using the same methods as other hatch patterns.

The **SOLID** tool allows you to draw basic solid-filled shapes without creating a boundary. Access the **SOLID** tool and pick points in a specific sequence to form different shapes. Using the **HATCH** tool with the Solid predefined hatch pattern is typically a better method of creating solid fills.

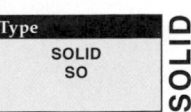

Type
SOLID
SO

SOLID

Exercise 23-3

Access the Student Web site (www.g-wlearning.com/CAD) and complete Exercise 23-3.

Reusing Existing Hatch Properties

You can specify hatch pattern characteristics by referencing an identical hatch pattern from the drawing. Pick the **Inherit Properties** button and select an existing hatch pattern. The crosshairs appears with a paintbrush icon. Pick a point inside a different area to define the boundary, and then press [Enter] or the space bar, or right-click and pick **Enter** to return to the **Hatch and Gradient** dialog box. The dialog box displays the settings of the selected pattern.

The **Inherit options** area in the **Hatch and Gradient** dialog box controls the hatch origin. The default **Use current origin** option uses the origin point setting specified in the **Hatch origin** area on the **Hatch** tab. Select the **Use source hatch origin** option to originate the new hatch pattern from the origin of the hatch selected with the **Inherit Properties** button.

Exercise 23-4

Access the Student Web site (www.g-wlearning.com/CAD) and complete Exercise 23-4.

Using DesignCenter to Insert Hatch Patterns

To pattern a boundary using **DesignCenter**, locate and select a PAT file to display the contents in the preview pane. See **Figure 23-30A.** Usually the quickest and most effective technique for transferring a hatch pattern from **DesignCenter** into the active drawing is to use a drag-and-drop operation. Pick the hatch pattern from **DesignCenter** and hold down the pick button. When you move the cursor into the active drawing, a hatch pattern symbol appears with the cursor. See **Figure 23-30B.** Place the cursor in the area to hatch and release the pick button to apply the hatch pattern. See **Figure 23-30C.**

An alternative to drag and drop is copy and paste. Right-click on a hatch pattern in **DesignCenter** and pick **Copy**. Move the cursor into the active drawing, right-click, and select **Paste**. The hatch pattern symbol appears with the crosshairs. Pick in the area to hatch to apply the hatch pattern. You can also use **DesignCenter** in combination with the **Hatch and Gradient** dialog box. Right-click on a hatch pattern in **DesignCenter** and select **BHATCH...** to access the **Hatch and Gradient** dialog box with the selected hatch pattern active.

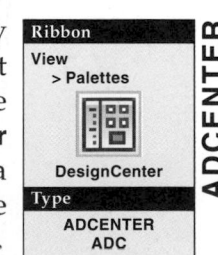

Ribbon
View
> Palettes

DesignCenter

Type
ADCENTER
ADC

ADCENTER

Figure 23-30.
A—Pick a PAT file in **DesignCenter** to display the available hatch patterns in the preview palette. B—The hatch pattern symbol appears under the cursor during the drag-and-drop and paste operations. C—Pick a point to apply the hatch pattern.

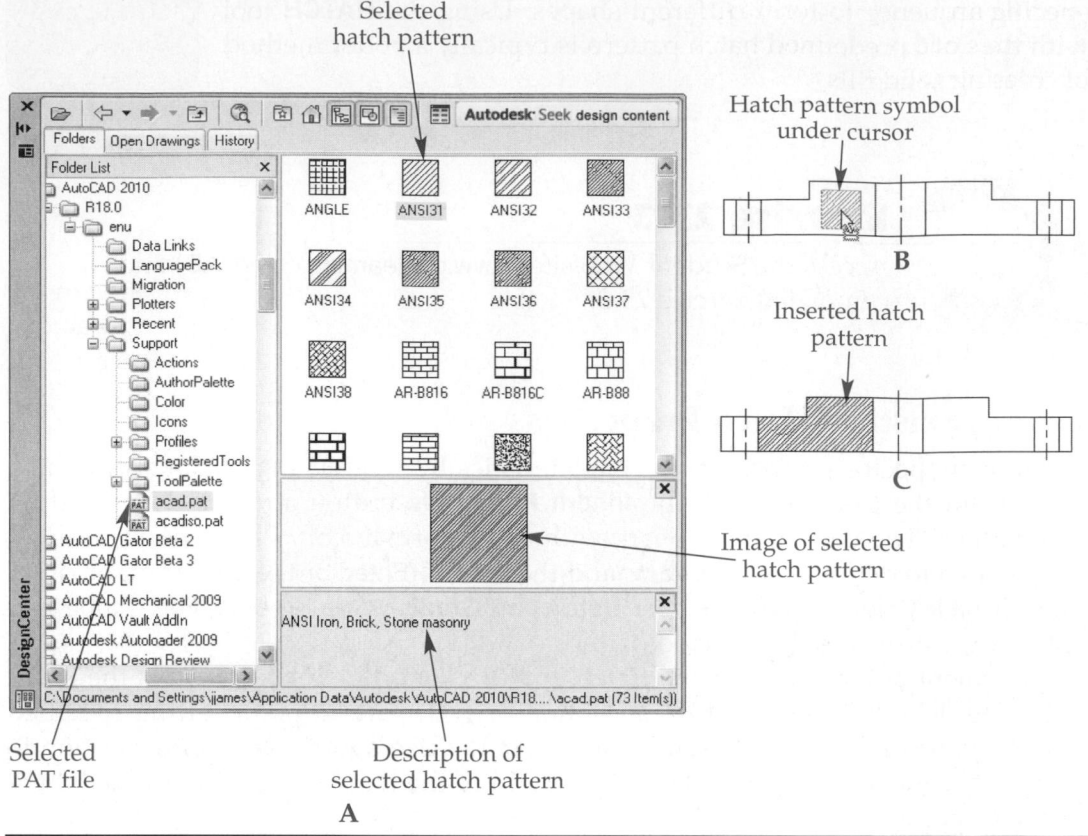

When you drag and drop or paste a hatch pattern into an area that is not a closed boundary, the same rules apply as when using the **HATCH** tool. When you insert or reference hatch patterns using **DesignCenter**, the angle, scale, and island detection settings match the settings of the previous hatch pattern. Edit the hatch pattern, as described later in this chapter, to change the settings after you insert the hatch pattern.

NOTE

AutoCAD includes two PAT files: acad.pat and acadiso.pat. These files are located in Program Files/AutoCAD 2010/UserDataCache/ Support. To verify the location of AutoCAD support files, access the **Files** tab in the **Options** dialog box and check the path listed under the Support File Search Path.

Exercise 23-5

Access the Student Web site (www.g-wlearning.com/CAD) and complete Exercise 23-5.

Using Tool Palettes to Insert Hatch Patterns

The **Tool Palettes** palette, shown in **Figure 23-31,** provides an alternative means of storing and inserting hatch patterns. *Tool palettes* can also store and activate many other types of drawing content and tools, such as blocks, images, tables, external reference files, drawing and editing tools, user-defined macros, script files, and AutoLISP routines. The **Command Tools Samples** tool palette contains examples of custom tools. For more information on AutoCAD customization and using tool palettes, refer to *AutoCAD and Its Applications—Advanced.*

Locating and Viewing Content

Each tool palette in the **Tool Palettes** palette has its own tab along the side of the window. To view the content in a tool palette, pick the related tab. If the **Tool Palettes** palette contains more palettes than can display on-screen, pick on the edge of the lowest tab to display a selection menu listing the palette tabs. Select the name of the tab to access the related tool palette. If not all of the content of a selected tool palette fits in the window, you can view the remainder using the scroll bar or the scroll hand. The scroll hand appears when you place the cursor in an empty area in the tool palette. Picking and dragging scrolls the tool palette up and down.

NOTE

By default, icons represent the tools in each tool palette. Several tool palette view options are available, including the ability to display a tool as an image of your choice. For more information on adjusting tool palette display, refer to *AutoCAD and Its Applications—Advanced.*

Inserting Hatch Patterns

To insert a hatch pattern from the **Tool Palettes** palette, access the tool palette in which the pattern resides. To drag and drop the pattern, hold down the pick button on the image. When you move the cursor into the active drawing, a hatch pattern symbol appears with the cursor. Place the cursor in the area to hatch and release the

---

Figure 23-31.
You can use the **Tool Palettes** palette to access and insert hatch patterns.

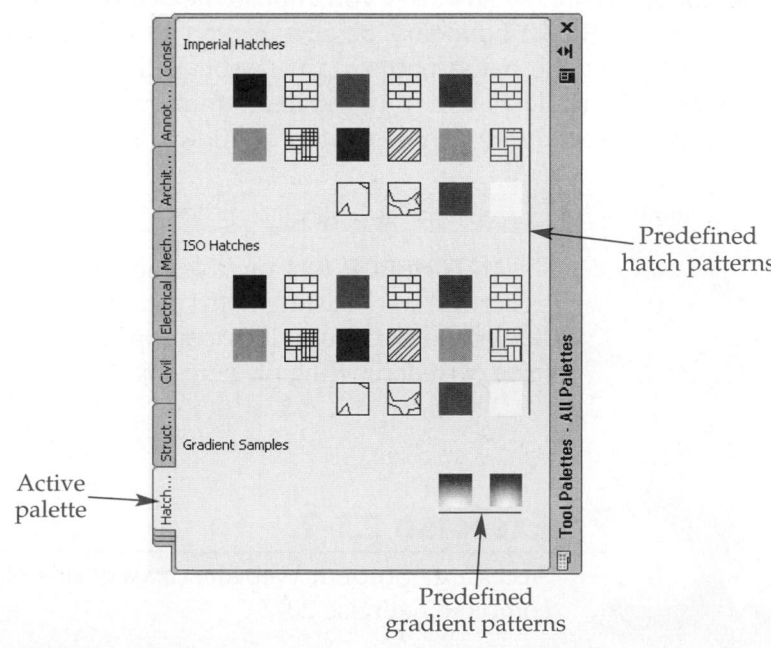

Active palette

Predefined hatch patterns

Predefined gradient patterns

---

pick button to apply the hatch pattern. An alternative to the drag-and-drop method is to pick once on the hatch image to attach the hatch pattern to the crosshairs, and then pick a boundary in the drawing to apply the hatch pattern. Edit the hatch pattern, as described later in this chapter, to change the settings after inserting the hatch pattern.

NOTE

You can add tool palettes to the **Tool Palettes** window and add tools to tool palettes. For more information on creating and modifying tool palettes, refer to *AutoCAD and Its Applications—Advanced*.

Exercise 23-6

Access the Student Web site (www.g-wlearning.com/CAD) and complete Exercise 23-6.

Editing Hatch Patterns

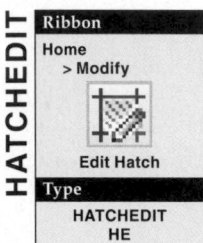

HATCHEDIT

Ribbon
Home
> Modify

Edit Hatch
Type
HATCHEDIT
HE

A hatch pattern is a single object that you can edit with tools such as **ERASE**, **COPY**, and **MOVE**, grips, or the **Properties** palette. You can also use the **HATCHEDIT** tool to adjust the characteristics of an existing hatch pattern. A quick way to access the **HATCHEDIT** tool is to double-click the hatch pattern to be edited. The **HATCHEDIT** tool opens the **Hatch Edit** dialog box shown in Figure **23-32.** The **Hatch Edit** dialog box is identical to the **Hatch and Gradient** dialog box except for the available **Recreate boundary** and **Select boundary objects** buttons in the **Boundaries** area.

You can use the **Recreate boundary** button to trace boundary objects over the original objects defining the boundary. However, a more practical application is to recreate the geometry if you have erased the original boundary object. See **Figure 23-33.** Pick the button and follow the prompts to create the boundary objects as a region or polyline. You can also specify whether to associate the hatch with the objects. The **Hatch Edit** dialog box reappears after you choose the desired options.

Pick the **Select boundary objects** button to exit the **Hatch and Gradient** dialog box and select the hatch pattern and associated boundary object or nonassociated boundary. This allows you to use grips to make changes to the size and shape of the associated object or the nonassociated boundary.

PROFESSIONAL TIP

The **MATCHPROP** tool provides an alternate method of inheriting the properties of an existing hatch pattern and applying the properties to a different hatch pattern. This tool applies existing hatch patterns to objects in the current drawing file or to objects in other open drawing files.

Exercise 23-7

Access the Student Web site (www.g-wlearning.com/CAD) and complete Exercise 23-7.

Figure 23-32.
The **Hatch Edit**
dialog box allows
you to edit hatch
patterns. Notice that
only the options
related to hatch
characteristics are
available.

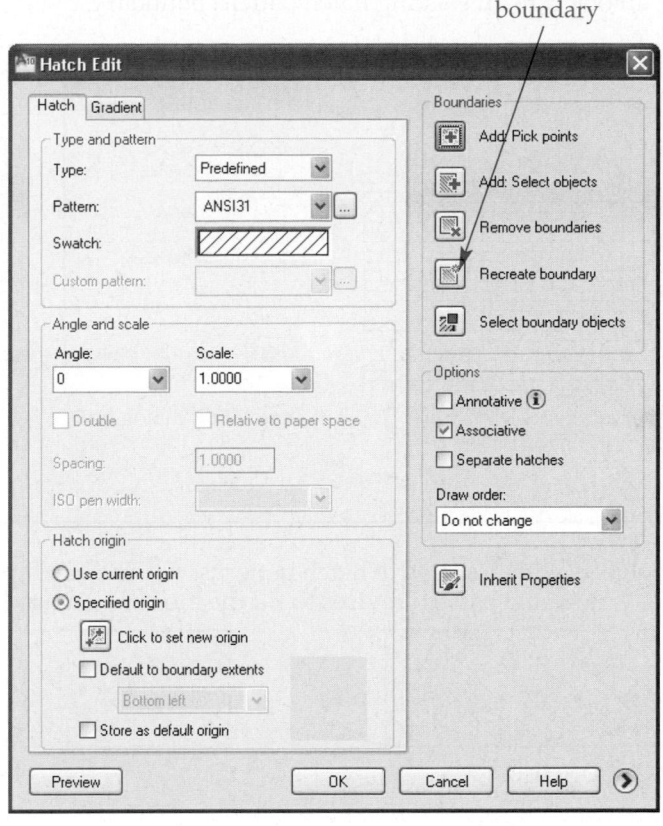

Figure 23-33.
An example of
using the **Recreate
boundary** button to
recreate a lost object
associated with the
hatch boundary.

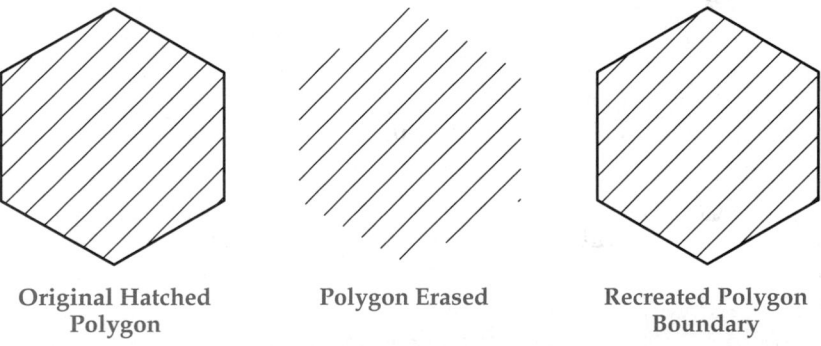

Original Hatched
Polygon

Polygon Erased

Recreated Polygon
Boundary

Adding and Removing Boundaries

Use the **Add: Pick points**, **Add: Select objects**, and **Remove boundaries** buttons in
the **Hatch Edit** dialog box to add boundaries to and remove them from existing associative and nonassociative hatch patterns. In **Figure 23-34**, for example, a rectangle is
drawn to create a window. To add the window as an island in the boundary, double-
click the hatch pattern to open the **Hatch Edit** dialog box. Pick the **Add: Select objects**
button to return to the drawing and pick the rectangle. Return to the **Hatch Edit** dialog
box and complete the operation.

Editing Associative Hatch Patterns

When you edit an object associated with a hatch pattern, the hatch pattern changes
to adapt to the edit. **Figure 23-35** shows examples of stretching an associated object
and removing an island from an associative boundary. As long as you edit the objects
associated with the boundary, the hatch pattern will update.

Figure 23-34.
Adding an object to an existing hatch pattern boundary.

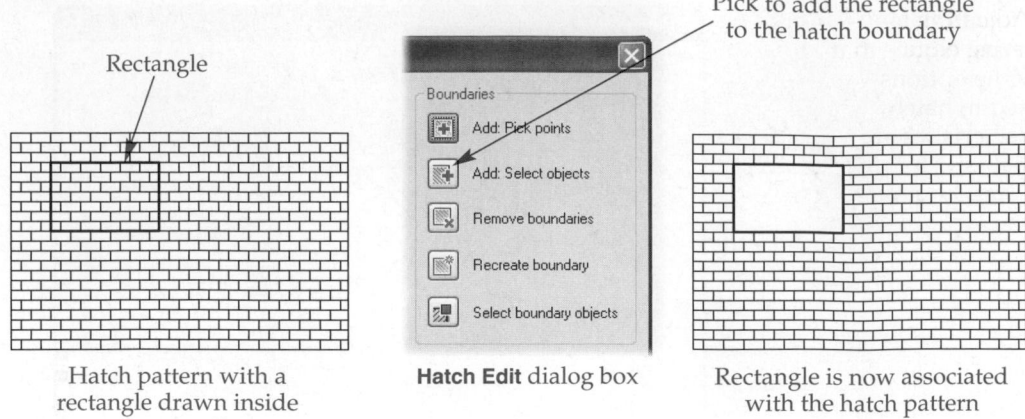

Rectangle

Pick to add the rectangle
to the hatch boundary

Boundaries
Add: Pick points
Add: Select objects
Remove boundaries
Recreate boundary
Select boundary objects

Hatch pattern with a
rectangle drawn inside

Hatch Edit dialog box

Rectangle is now associated
with the hatch pattern

Figure 23-35.
Editing objects with associative hatch patterns. A—The hatch pattern stretches with the object. B—The hatch pattern revises to fill the area of an erased island.

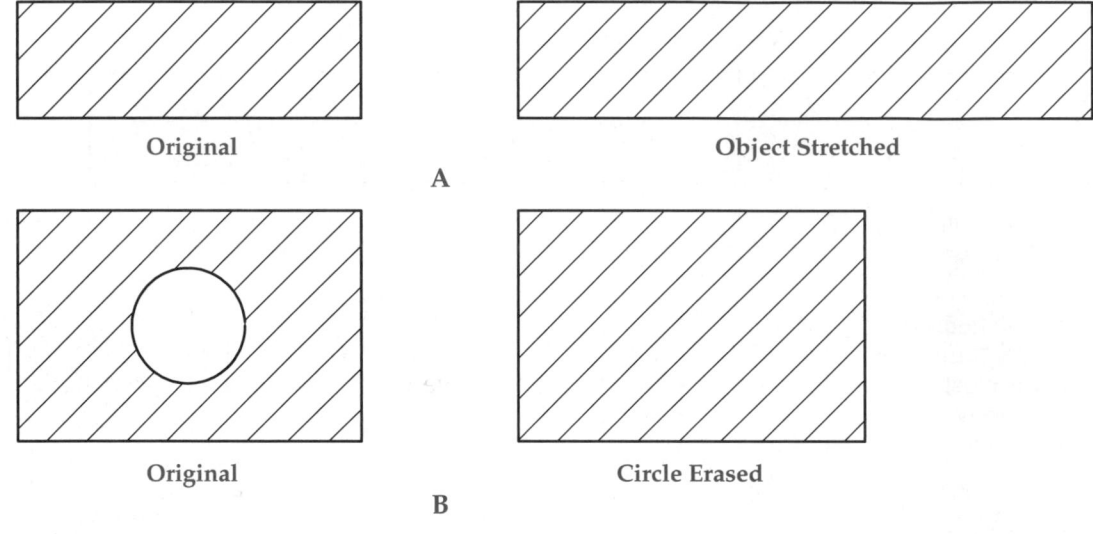

Original

Object Stretched

A

Original

Circle Erased

B

Exercise 23-8

Access the Student Web site (www.g-wlearning.com/CAD) and complete Exercise 23-8.

NEW

Editing Nonassociative Hatch Patterns

You can create a nonassociative hatch pattern using the **HATCH** or **HATCHEDIT** tool, or by moving an associative hatch pattern away from the associated objects. The objects you reference to create a nonassociative hatch pattern cannot control the size and shape of the hatch. However, you can edit nonassociative hatch patterns using many tools, such as **ROTATE**, **COPY**, and **MOVE**. In addition, special grips allow you to make changes to the boundary. See **Figure 23-36.** Standard grip editing tools and other editing options are available, depending on the selected grip and boundary geometry.

Figure 23-36.
Using the grips that appear when you select a nonassociative hatch pattern. A—Original nonassociative hatch. B—Moving an island out of the boundary and adjusting edge grips. C—Adjusting edge and point grips. D—Moving an island back into the boundary and adjusting edge and point grips.

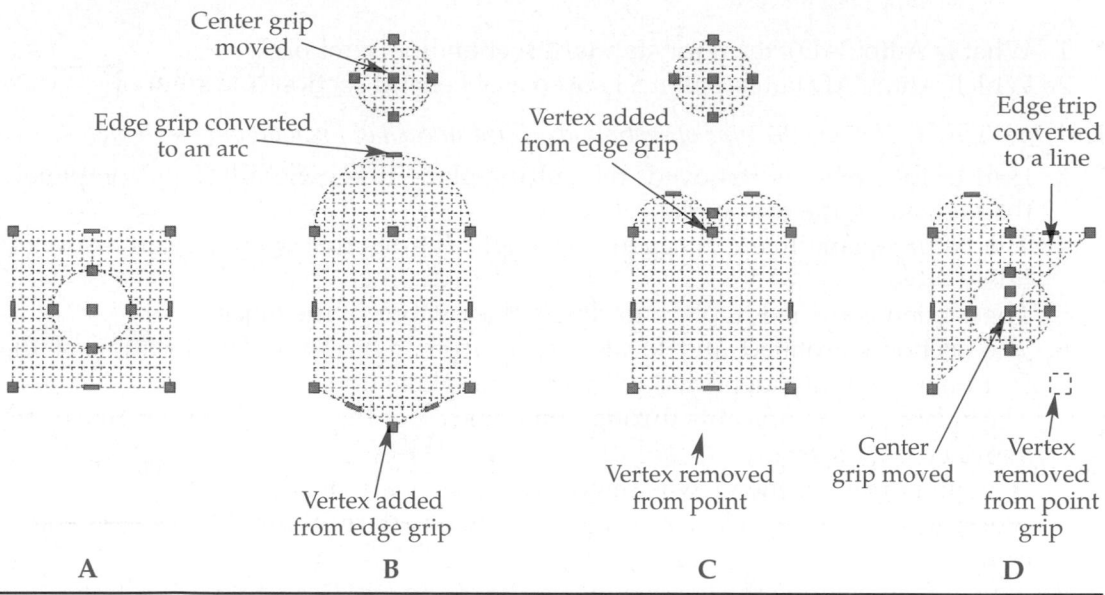

Center grip moved

Edge grip converted to an arc

Vertex added from edge grip

Vertex added from edge grip

Vertex removed from point

Edge trip converted to a line

Center grip moved

Vertex removed from point grip

A   B   C   D

Once you select a grip, press [Ctrl] to cycle though unique editing functions. You may be able to add a vertex to create a new line or arc, remove a vertex to eliminate a line or arc, or convert a line to an arc or an arc to a line. **Figure 23-36** shows the process of making several changes to a boundary using grip editing techniques.

PROFESSIONAL TIP

When working with associative and nonassociative hatch patterns, remember that associative hatch patterns are associated with objects. The objects define the hatch boundary. Nonassociative hatch patterns are not associated with objects, but they do show association with the hatch boundary.

Exercise 23-9

Access the Student Web site (www.g-wlearning.com/CAD) and complete Exercise 23-9.

Express Tools
Chapter 23

The *Express Tools* ribbon tab includes additional tools for improved functionality and productivity during the drawing processes. The following Express Tool is a hatch express tool. For information about this tool, go to the Student Web site (www.g-wlearning.com/CAD), select this chapter, and select *Inserting Hatch Patterns with the SUPERHATCH Tool*.
Super Hatch

Chapter Test

Answer the following questions. Write your answers on a separate sheet of paper or go to the Student Web site (www.g-wlearning.com/CAD) and complete the electronic chapter test.

1. What is AutoCAD's term for standard section line symbols?
2. Which AutoCAD hatch pattern is used as a general section line symbol?

For Questions 3–8, name the type of section identified in each of the following statements:

3. Half of the object is removed; the cutting-plane line generally cuts completely through along the center plane.
4. The cutting-plane line is staggered through features that do not lie in a straight line.
5. The section is turned in place to clarify the contour of the object.
6. The section is rotated and located away from the object. The location of the section is normally identified with a cutting-plane line.
7. The cutting-plane line cuts through one-quarter of the object; used primarily on symmetrical objects.
8. A small portion of the view is removed to clarify an internal feature.
9. Identify two ways to select a predefined hatch pattern in the **Hatch and Gradient** dialog box.
10. How do you change the hatch angle in the **Hatch and Gradient** dialog box?
11. Describe the fundamental difference between using the **Add: Pick points** and the **Add: Select objects** buttons in the **Hatch and Gradient** dialog box.
12. Define *associative hatch pattern*.
13. What is the result of stretching an object that is hatched with an associative hatch pattern?
14. If you use the **Add: Pick points** button inside the **Hatch and Gradient** dialog box to hatch an area, how do you hatch around an island inside the area to be hatched?
15. Explain the three island detection style options.
16. How do you limit AutoCAD hatch evaluation to a specific area of the drawing?
17. What is the purpose of the **Gap Tolerance** setting in the **Hatch and Gradient** dialog box?
18. What are gradient fill hatch patterns? How are they created with the **HATCH** tool?
19. Explain how to use an existing hatch pattern on a drawing as the pattern for your next hatch.
20. Explain how to use drag and drop to insert a hatch pattern from **DesignCenter** into an active drawing.
21. Name the two files that contain hatch patterns that can be copied from **DesignCenter**.
22. Explain two ways to use drag and drop for inserting a hatch pattern from a tool palette into the drawing.
23. Name the tool used to edit existing associative hatch patterns.
24. How does the **Hatch Edit** dialog box compare to the **Hatch and Gradient** dialog box?
25. What happens if you erase an island inside an associative hatch pattern?

Drawing Problems

Start AutoCAD if it is not already started. Start a new drawing using an appropriate template of your choice. The template should include layers, text styles, dimension styles, and multileader styles appropriate for drawing the given objects. Add layers, text styles, dimension styles, and multileader styles as needed. Follow the specific instructions for each problem. Use your own judgment and approximate dimensions when necessary. Draw all objects using appropriate layers, text styles, dimension styles, multileader styles, justification, and format.

▼ Basic

1. Draw the game board as shown. Save the drawing as P23-1.

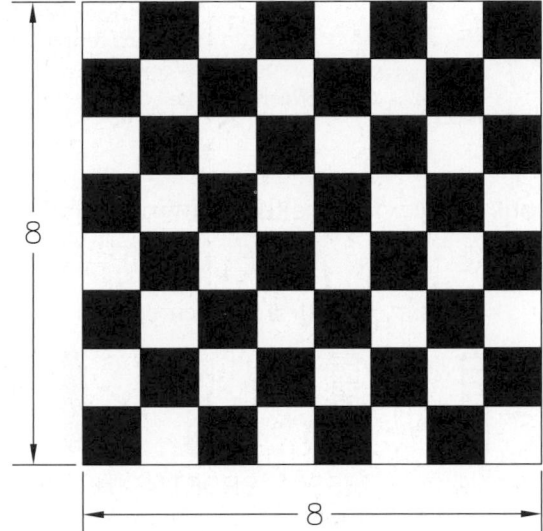

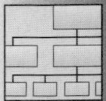

2. Draw the bar graph as shown. Save the drawing as P23-2.

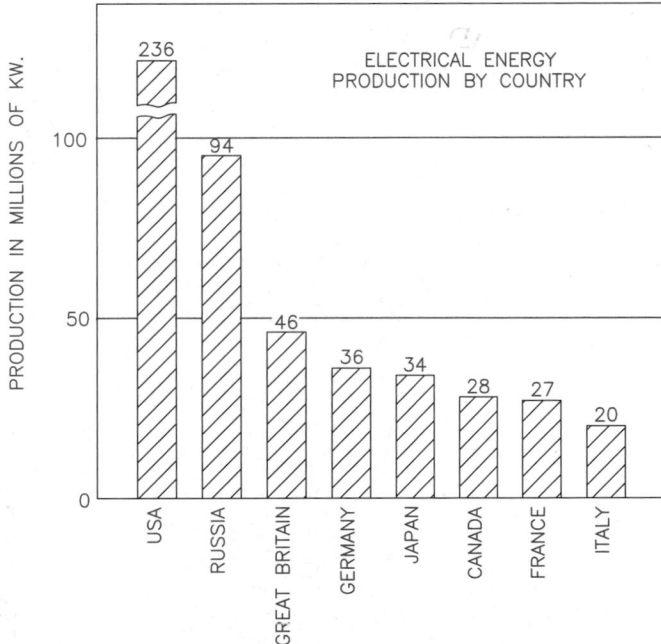

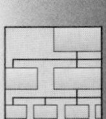

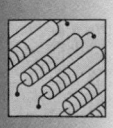

3. Draw the component layout as shown. Save the drawing as P23-3.

COMPONENT LAYOUT

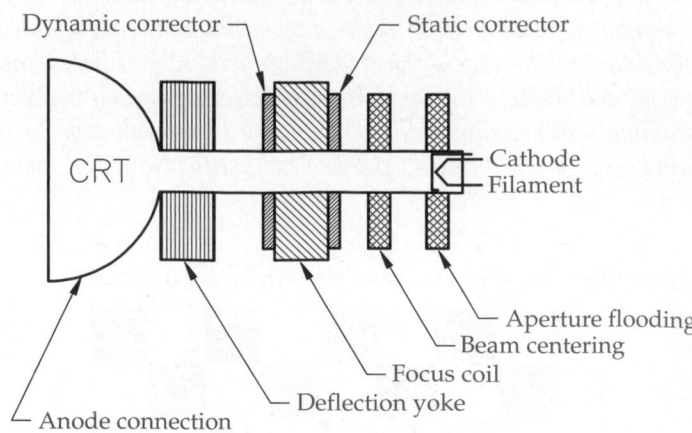

Dynamic corrector — — Static corrector

CRT

Cathode
Filament

Aperture flooding
Beam centering
Focus coil
Deflection yoke
Anode connection

4. Draw the bar graphs as shown. Save the drawing as P23-4.

SOLOMAN SHOE COMPANY

PERCENT OF TOTAL SALES EACH DIVISION

CASUAL   DRESS   SPORTS   BOOTS

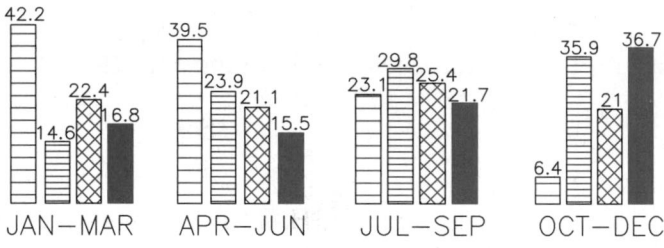

42.2   39.5   29.8   35.9   36.7
22.4   23.9   25.4   21
14.6 16.8   21.1   23.1   21.7   6.4
15.5

JAN—MAR   APR—JUN   JUL—SEP   OCT—DEC

5. Draw the pie chart as shown. Save the drawing as P23-5.

DIAL TECHNOLOGIES
EXPENSE BUDGET
FISCAL YEAR

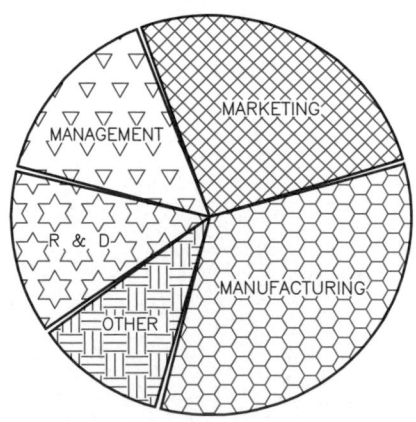

MARKETING

MANAGEMENT

R & D

MANUFACTURING

OTHER

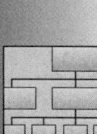

6. Draw the bar graph as shown. Save the drawing as P23-6.

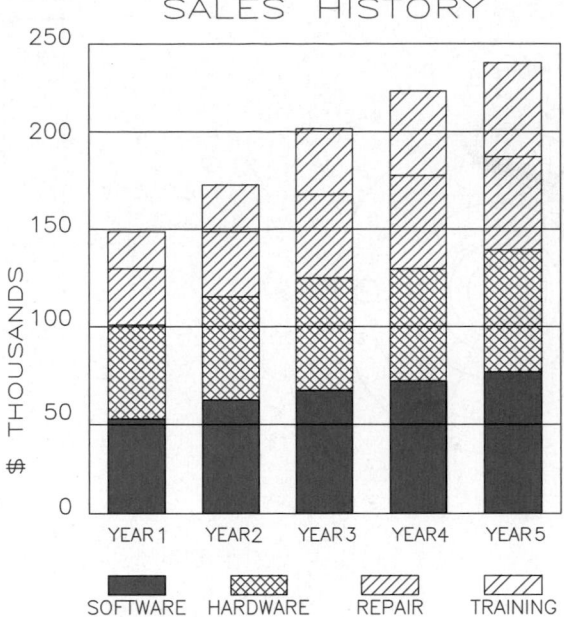

SALES HISTORY

For Problems 7–15, use the following additional guidelines:

- *Apply dimensions accurately using ASME or appropriate industry standards. Use object snap modes to your best advantage.*
- *For mechanical drawing problems in which no notes are specified, place the following notes 1/2" from the lower-left corner:*

  NOTES:
  1. INTERPRET DIMENSIONS AND TOLERANCES PER ASME Y14.5M-1994.
  2. REMOVE ALL BURRS AND SHARP EDGES.
  3. UNLESS OTHERWISE SPECIFIED, ALL DIMENSIONS ARE IN INCHES (or MILLIMETERS as applicable).

▼ Intermediate

7. Draw and dimension the views given, which include aligned and broken-out sections. Add the following notes: FINISH ALL OVER 1.63 mm UNLESS OTHERWISE SPECIFIED and ALL DIMENSIONS ARE IN MILLIMETERS. All arc and circle contours are tangent. Save the drawing as P23-7.

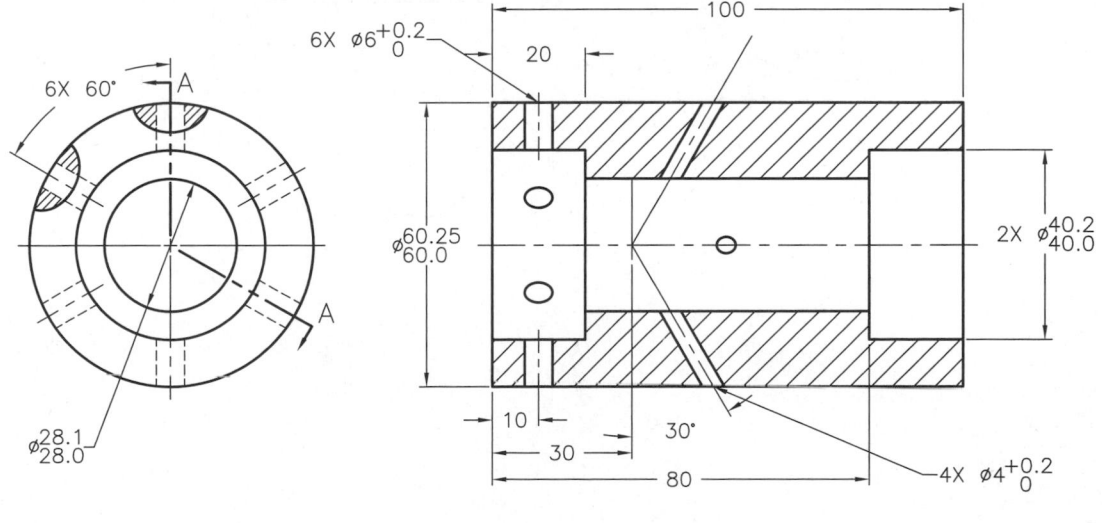

SECTION A–A

Name: Nozzle
Material: Phosphor Bronze

8. Draw and dimension the given views. Save the drawing as P23-8.

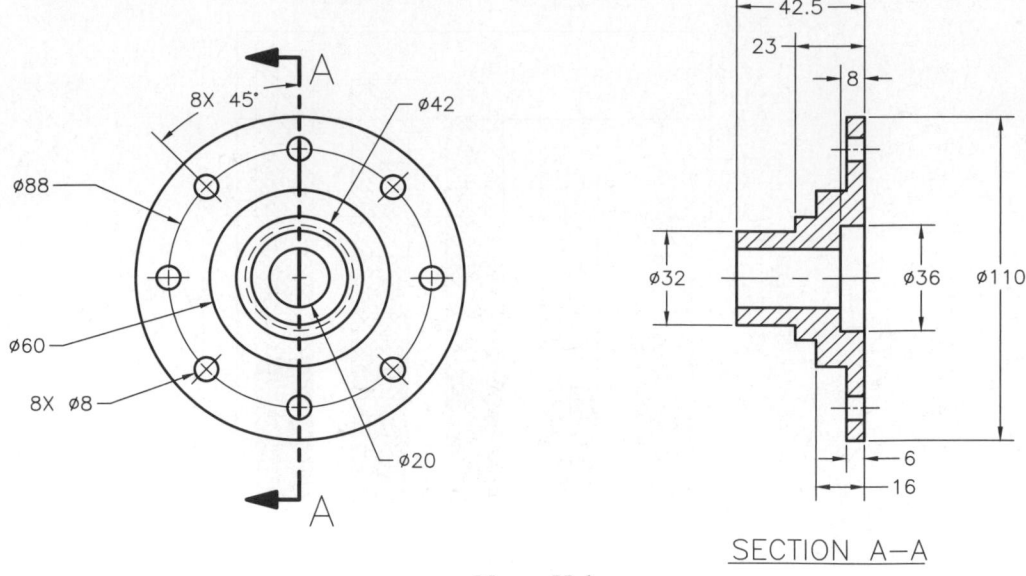

SECTION A–A

Name: Hub
Material: Cast Iron

9. Draw and dimension the given views, including the aligned section shown on the right. Add the following notes: FINISH ALL OVER 1.63 mm UNLESS OTHERWISE SPECIFIED and ALL DIMENSIONS ARE IN MILLIMETERS. Save the drawing as P23-9.

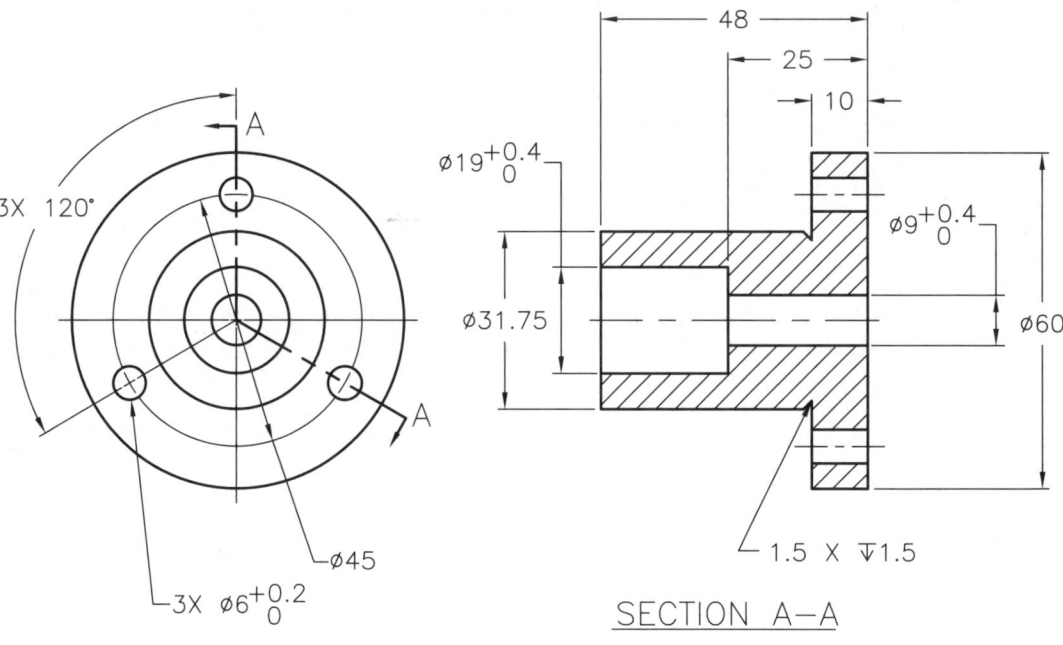

SECTION A–A

10. Draw and dimension the views of the chain guide as shown. Save the drawing as P23-10.

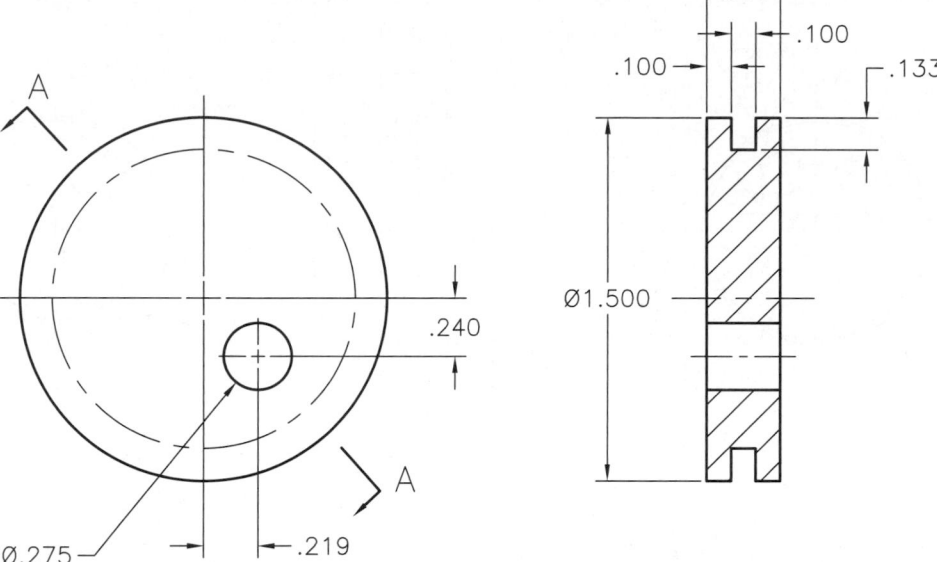

SECTION A–A

11. Draw and dimension the views of the sleeve and add the notes as shown. Save the drawing as P23-11.

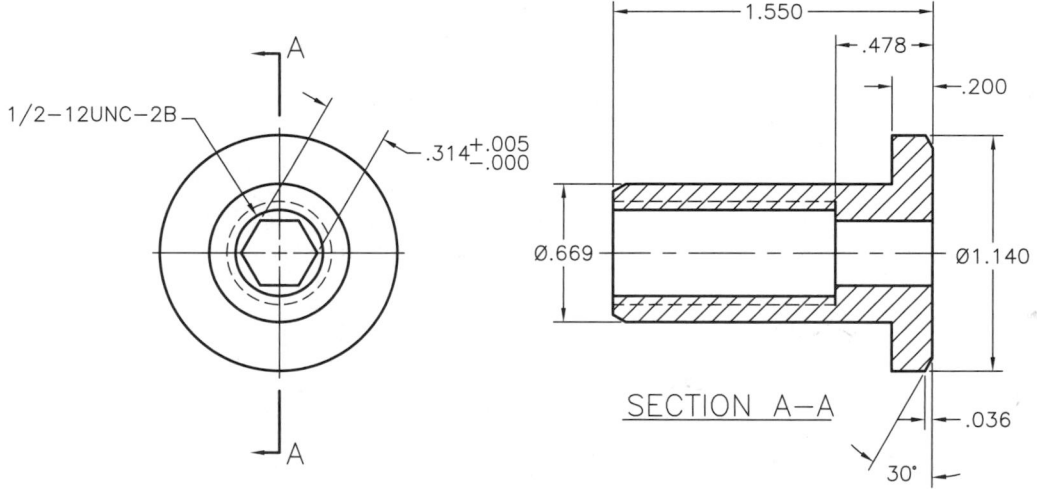

SECTION A–A

4. PAINT ACE GLOSS BLACK ALL OVER.
3. CASE HARDEN 45–50 ROCKWELL.
2. REMOVE ALL BURRS AND SHARP EDGES.
1. INTERPRET ALL DIMENSIONS AND
   TOLERANCES PER ASME Y14.5M–1994.

12. Draw and dimension the given views. Add the following notes: OIL QUENCH 40-45C, CASE HARDEN .020 DEEP, and 59-60 ROCKWELL C SCALE. Save the drawing as P23-12.

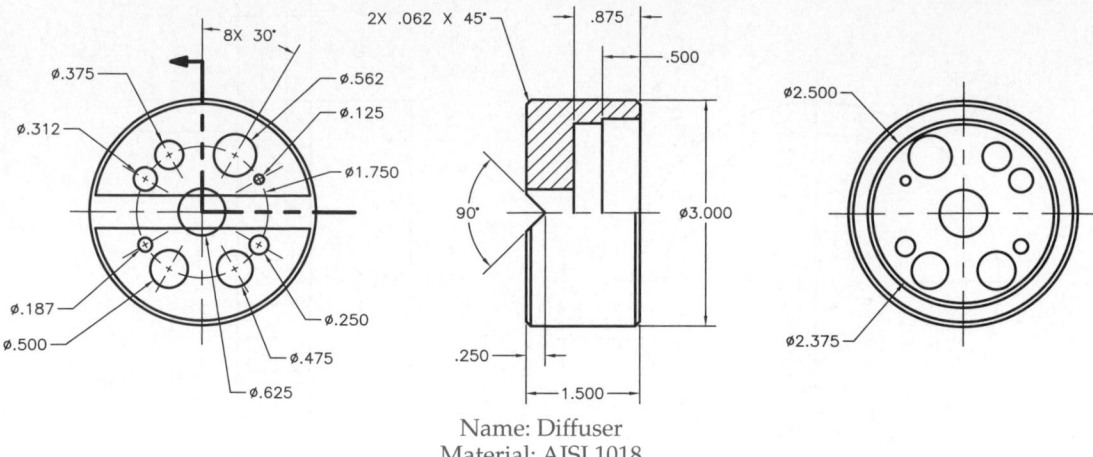

Name: Diffuser
Material: AISI 1018

13. Draw and dimension the views of the tow hook as shown. Save the drawing as P23-13.

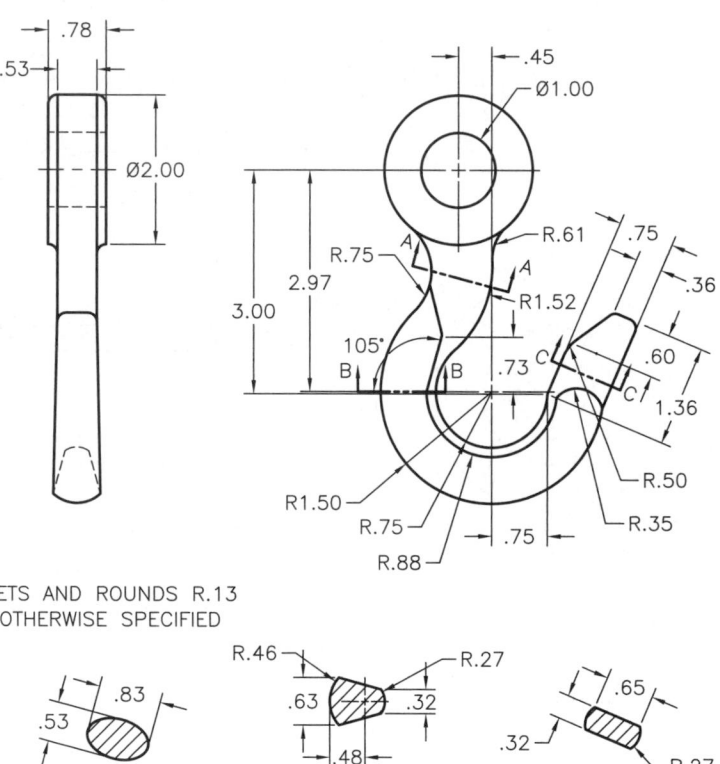

ALL FILLETS AND ROUNDS R.13
UNLESS OTHERWISE SPECIFIED

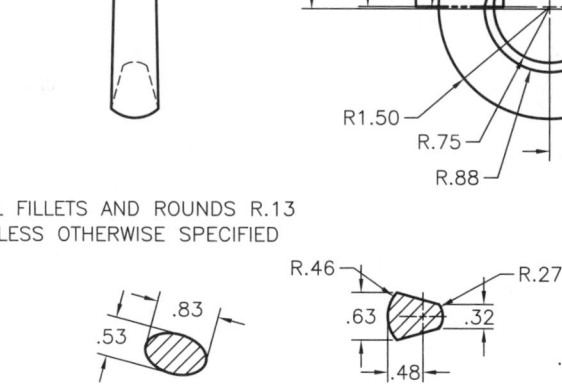

SECTION A–A   SECTION B–B   SECTION C–C

14. Draw the foundation detail as shown. Save the drawing as P23-14.

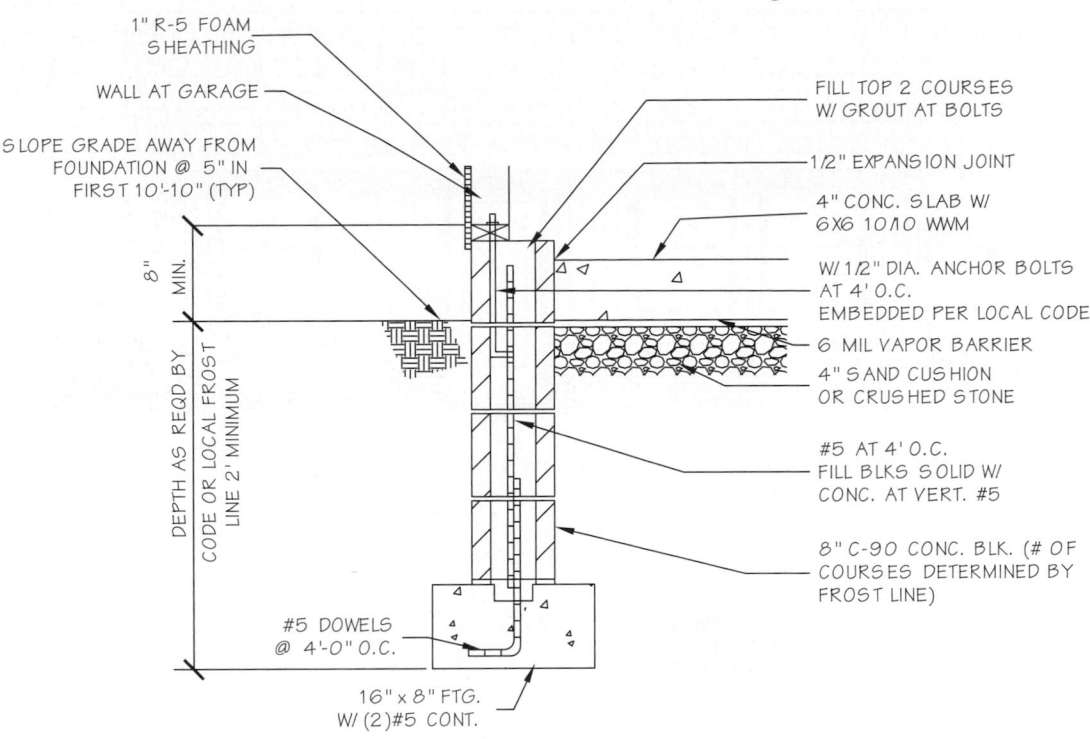

1" R-5 FOAM SHEATHING

WALL AT GARAGE

SLOPE GRADE AWAY FROM FOUNDATION @ 5" IN FIRST 10'-10" (TYP)

8" MIN.

DEPTH AS REQD BY CODE OR LOCAL FROST LINE 2' MINIMUM

#5 DOWELS @ 4'-0" O.C.

16" x 8" FTG. W/ (2)#5 CONT.

FILL TOP 2 COURSES W/ GROUT AT BOLTS

1/2" EXPANSION JOINT

4" CONC. SLAB W/ 6X6 10/10 WWM

W/ 1/2" DIA. ANCHOR BOLTS AT 4' O.C. EMBEDDED PER LOCAL CODE

6 MIL VAPOR BARRIER

4" SAND CUSHION OR CRUSHED STONE

#5 AT 4' O.C. FILL BLKS SOLID W/ CONC. AT VERT. #5

8" C-90 CONC. BLK. (# OF COURSES DETERMINED BY FROST LINE)

15. Draw the stair detail as shown. Save the drawing as P23-15.

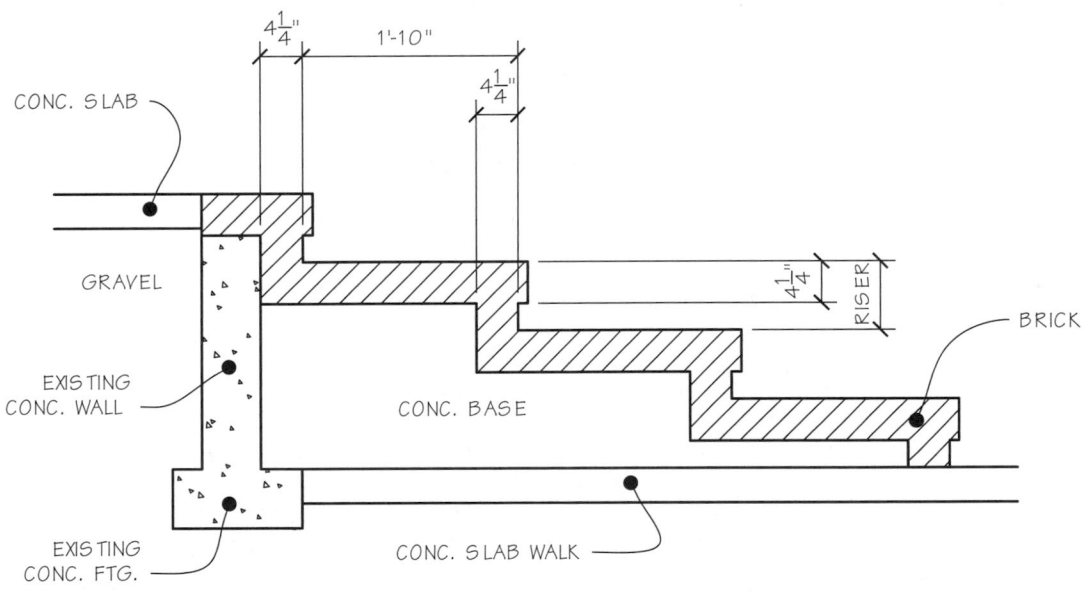

4 1/4" 1'-10" 4 1/4"

CONC. SLAB

GRAVEL

EXISTING CONC. WALL

EXISTING CONC. FTG.

CONC. BASE

CONC. SLAB WALK

4 1/4" RISER

BRICK

16. Draw the front elevation as shown. Save the drawing as P23-16.

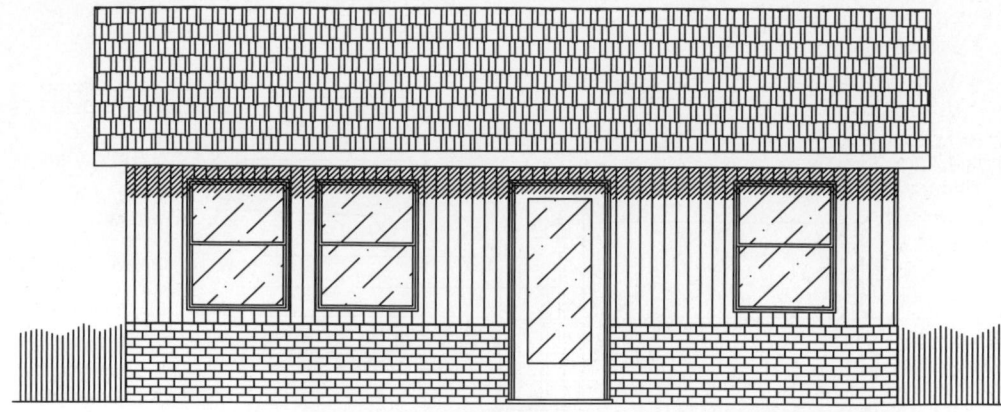

17. Draw the plan as shown. Save the drawing as P23-17.

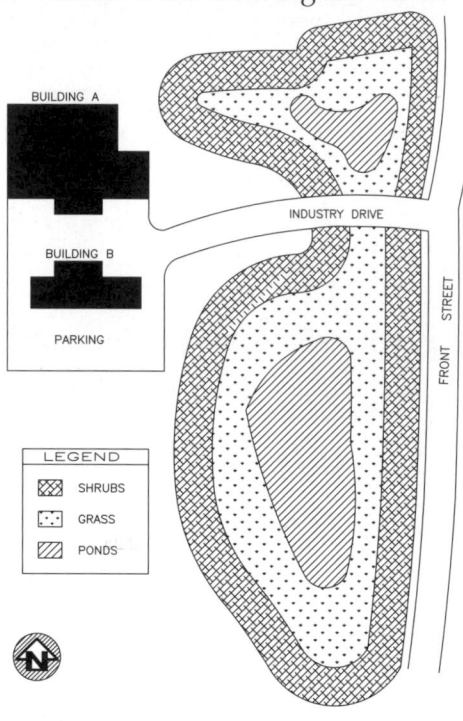

18. Draw the plan as shown. Save the drawing as P23-18.

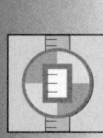

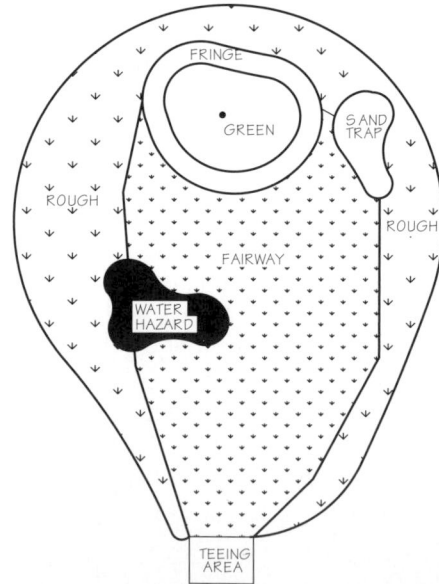

19. Draw the plan as shown. Save the drawing as P23-19.

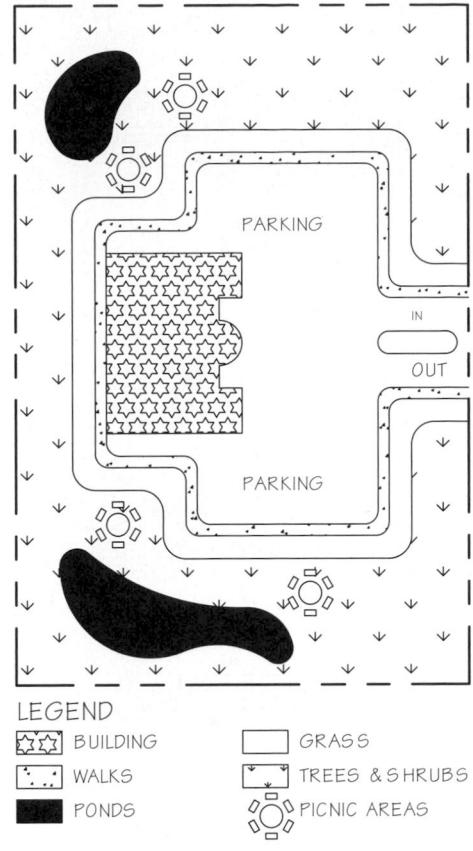

PARKING

IN

OUT

PARKING

LEGEND

BUILDING	GRASS
WALKS	TREES & SHRUBS
PONDS	PICNIC AREAS

20. Draw the profile as shown. Save the drawing as P23-20.

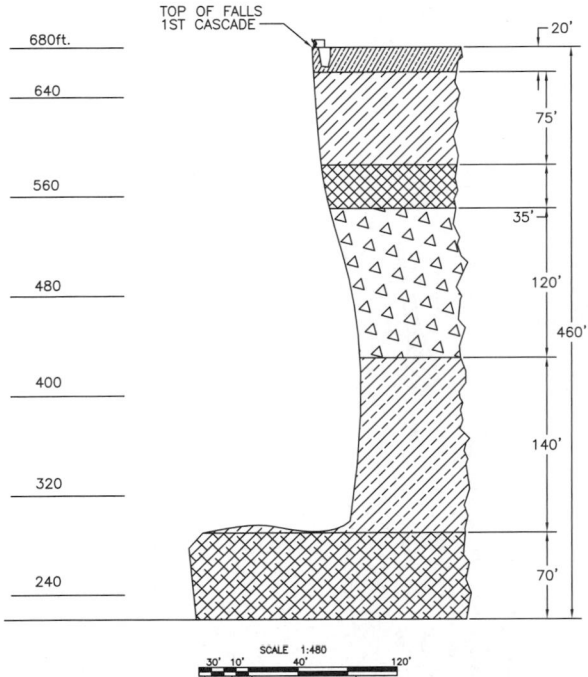

TOP OF FALLS
1ST CASCADE

680ft.
640
560
480
400
320
240

20'
75'
35'
120'
460'
140'
70'

PROFILE OF MULTNOMAH FALLS
&
GEOLOGIC INFORMATION

20' COLLONADE OF AN 80-FOOT THICK FLOW, NOTCHED BY MULTNOMAH CREEK.

75' PILLOW LAVA

35' A GLASSY FLOW, WITH WELL-FORMED ENTABLATURE AND COLLONADE.

120' CONSISTING OF TWO TIERS OF HACKLY-JOINTED BASALT, WITH NO COLLONADE.

140' ENTABLATURE WITH THIN COLUMNS, TOPPED BY A VESICULAR ZONE.

70' OF ENTABLATURE BENEATH THE LOWER FALLS.

BRIEF DESCRIPTION OF TERMS.
COLONNADE: THE LOWER PORTION OF A LAVA FLOW OF COLUMNAR-JOINTED BASALT.
ENTABLATURE: THE UPPER MASSIVE OF A LAVA FLOW OF HACKLY-JOINTED BASALT.

* INFORMATION TAKEN FROM:
"THE MAGNIFICENT GATEWAY"
AUTHOR: JOHN ELIOT ALLEN
PAGES: 89-91

SCALE 1:480
30' 10'   40'   120'
40' 20' 0'   80'

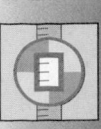

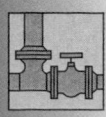

21. Draw the drop cleanout detail as shown. Save the drawing as P23-21.

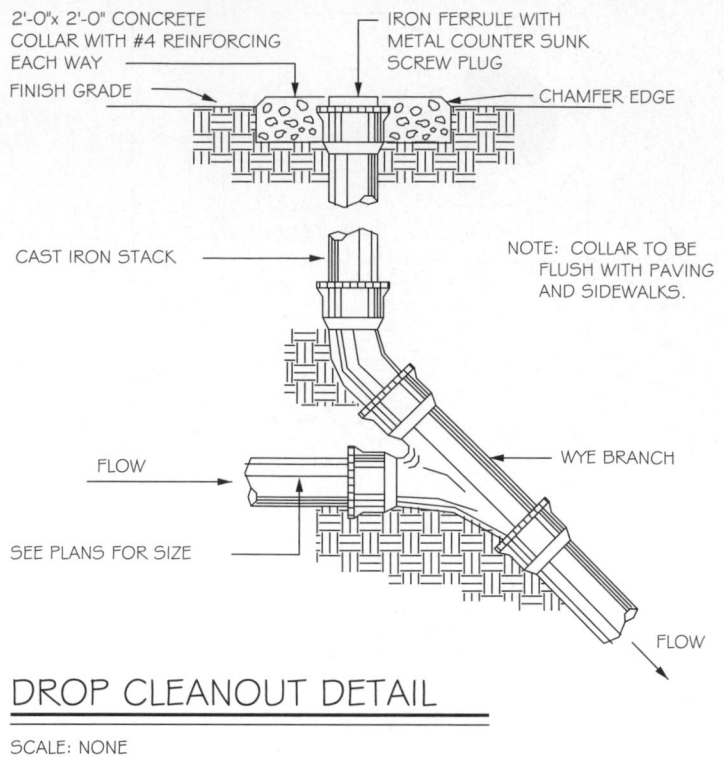

2'-0"x 2'-0" CONCRETE
COLLAR WITH #4 REINFORCING
EACH WAY

IRON FERRULE WITH
METAL COUNTER SUNK
SCREW PLUG

FINISH GRADE

CHAMFER EDGE

CAST IRON STACK

NOTE: COLLAR TO BE
FLUSH WITH PAVING
AND SIDEWALKS.

FLOW

WYE BRANCH

SEE PLANS FOR SIZE

FLOW

DROP CLEANOUT DETAIL

SCALE: NONE

22. Draw the map shown below, using **SPLINE** to create the curved shapes. Try to make the shapes as similar to those on the map as possible. Save the drawing as P23-22.

CATFISHING ANCHOR LOCATIONS

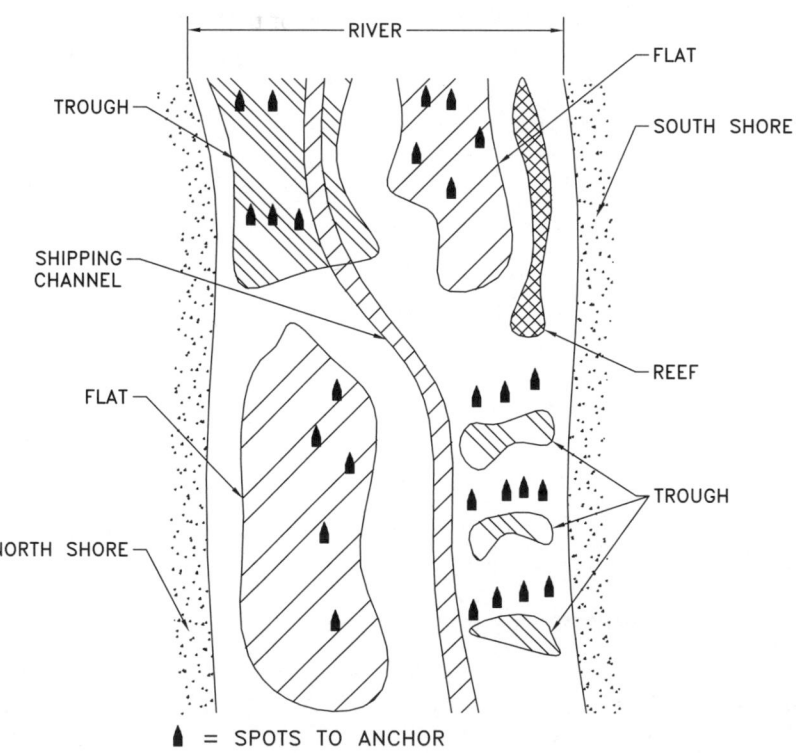

RIVER

FLAT

TROUGH

SOUTH SHORE

SHIPPING
CHANNEL

REEF

FLAT

TROUGH

NORTH SHORE

♠ = SPOTS TO ANCHOR

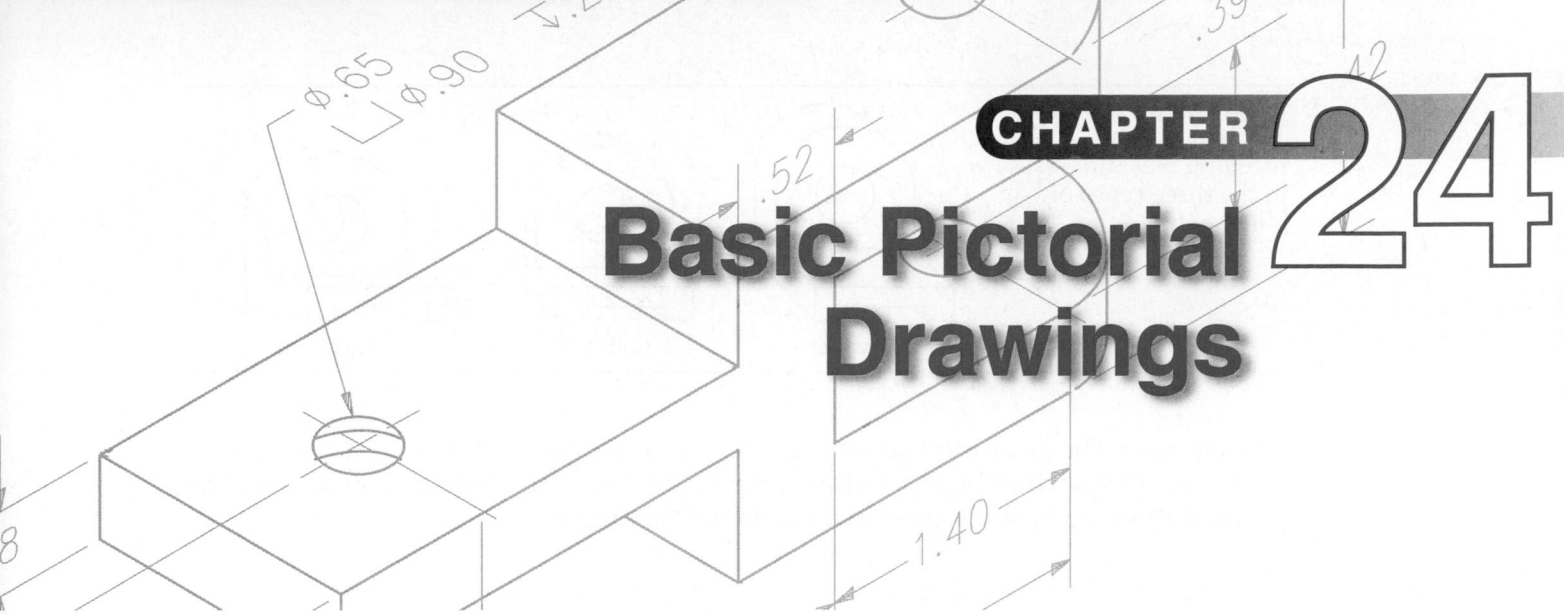

CHAPTER 24

Basic Pictorial Drawings

Learning Objectives

After completing this chapter, you will be able to do the following:

✓ Describe the three basic types of pictorial drawings.
✓ Construct accurate isometric drawings.
✓ Dimension isometric drawings.

Being able to visualize and draw three-dimensional shapes is a skill that every drafter, designer, and engineer should possess. This is especially important in 3D modeling. However, there is a distinct difference between drawing a view that *looks* three-dimensional and creating a *true* 3D model.

A 3D model can be rotated on the display screen and viewed from any angle. The computer calculates the points, lines, and surfaces of the objects in space. For information on 3D modeling, viewing, and visualization techniques, see *AutoCAD and Its Applications—Advanced*. This chapter is provided as background information on the classic drawing techniques used in pictorial drawing. The focus of this chapter is creating views that *look* three-dimensional using some special AutoCAD functions and two-dimensional coordinates and objects.

Pictorial Drawing Overview

The word *pictorial* means "like a picture." It refers to any form of 2D drawing that illustrates height, width, and depth. Several forms of pictorial drawings are used in industry today. The least realistic is oblique. However, this is the simplest type. The most realistic, but also the most complex, is perspective. The realism and complexity of isometric drawing falls midway between the two.

pictorial: A 2D drawing that shows height, width, and depth; similar to a picture.

Oblique Drawings

An *oblique drawing* shows objects with one or more parallel faces at their true shape and size. A scale is selected for the orthographic, or front, faces. Then an angle for the depth (receding axis) is chosen. The three types of oblique drawings are cavalier, cabinet, and general. See **Figure 24-1.** These types vary in the scale of the receding axis. The receding axis is drawn at full scale for a cavalier view and at half scale for a

oblique drawing: A drawing that shows objects with one or more parallel faces having true shape and size.

Figure 24-1.
The scale of the receding axis differs in the three types of oblique drawings.

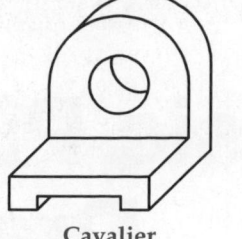 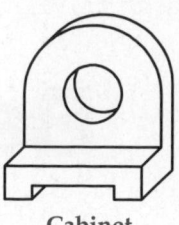 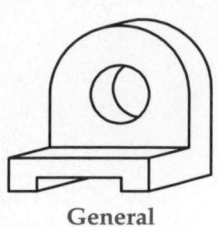

| Cavalier | Cabinet | General |

cabinet view. The receding axis on a general oblique view is normally drawn with a 3/4 scale. Oblique drawings were traditionally used by cabinetmakers, but this type of pictorial drawing is rarely used in the CAD environment.

Axonometric Drawings

The general term for drawings that display three dimensions on a two-dimensional surface such as a drawing sheet is *axonometric drawings*. The three types of axonometric drawings are isometric, dimetric, and trimetric.

Isometric drawings are more realistic than oblique drawings. The entire object appears as if it is tilted toward the viewer. The word *isometric* means "equal measure." This equal measure refers to the angle between the three axes (120°) after the object has been tilted. The tilt angle is 35°16′. This is shown in **Figure 24-2.** The 120° angle corresponds to an angle of 30° from horizontal. In the construction of isometric drawings, lines that are parallel in the orthogonal views must be parallel in the isometric view.

The most appealing aspect of isometric drawing is that all three axis lines can be measured using the same scale. This saves time, while still producing a pleasing pictorial of the object.

Dimetric and trimetric drawings are closely related to isometric drawing. These forms of pictorial drawing differ from isometric in the scales used to measure the three axes. *Dimetric* drawing uses two different scales and *trimetric* uses three scales. Using different scales is an attempt to create *foreshortening*. The relationship between isometric, dimetric, and trimetric drawings is illustrated in **Figure 24-3.**

axonometric drawings: Drawings in which a 3D object is rotated for display on a 2D drawing sheet so all three dimensions can be seen.

isometric drawings: Drawings in which the three axes are equally spaced at 120°.

dimetric: An axonometric drawing in which two different scales are used to measure the three axes.

trimetric: An axonometric drawing in which three different scales are used to measure the three axes.

foreshortening: Property of a drawing in which objects appear to recede in the distance.

Figure 24-2.
An object is tilted 35°16′ to achieve an isometric view having 120° between the three axes. Notice the highlighted face in each view.

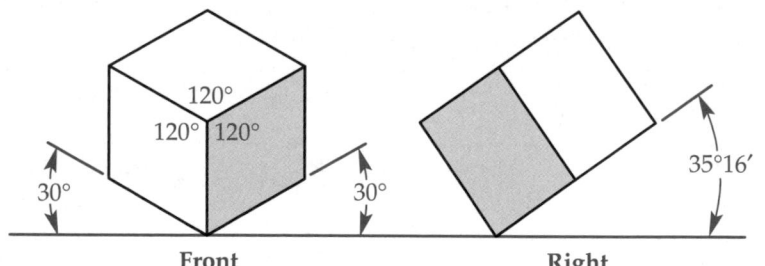

Figure 24-3.
Isometric, dimetric, and trimetric drawings differ in the scales used to draw the three axes. In the isometric shown here, all three sides are drawn at full scale, or 1. You can see how the dimetric and trimetric scales vary.

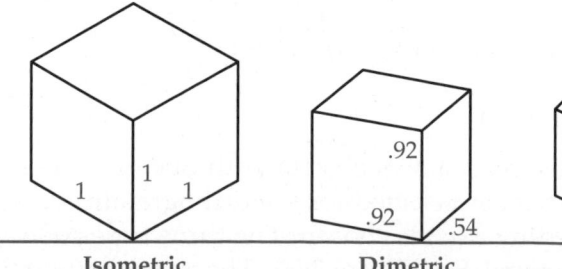

NOTE

When objects are created as 3D models, they can be quickly displayed in a variety of isometric views. See *AutoCAD and Its Applications—Advanced* for a complete discussion of 3D modeling techniques.

Perspective Drawings

The most realistic form of pictorial drawing is a *perspective drawing*. The eye naturally sees objects in perspective. Look down a long hall and notice that the wall and floor lines seem to converge in the distance. The point at which they converge is called the *vanishing point*. The most common types of perspective drawing are one-point and two-point perspectives. These forms of pictorial drawing are often used in architecture. They are also used in the automotive and aircraft industries. Examples of one-point and two-point perspectives are shown in **Figure 24-4**. A true perspective of a 3D model can be produced in AutoCAD using the **ViewCube**. See *AutoCAD and Its Applications—Advanced* for complete coverage of the **ViewCube** and 3D modeling.

perspective drawing: The most realistic form of pictorial drawing, in which receding objects meet at one or more vanishing points on the horizon.

vanishing point: The point at which objects seem to converge in the distance.

Figure 24-4.
An example of one-point perspective and two-point perspective.

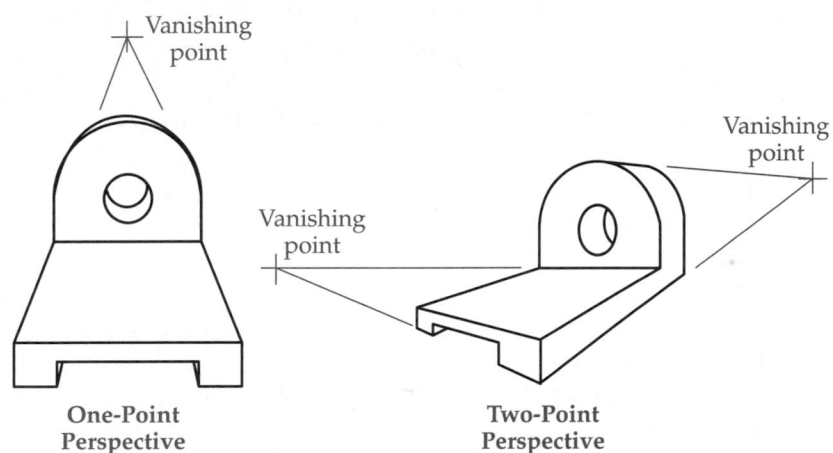

One-Point Perspective

Two-Point Perspective

Isometric Drawing

The method of pictorial drawing that is used most commonly in industry is isometric. Isometric drawings provide a single view showing three sides that can be measured using the same scale. An isometric view has no perspective and may appear somewhat distorted. Two of the isometric axes are drawn at 30° to horizontal, and the third is drawn at 90°. See **Figure 24-5**.

The three axes shown in **Figure 24-5** represent the width, height, and depth of the object. Lines that appear horizontal in an orthographic view are placed at a 30° angle. Lines that are vertical in an orthographic view are placed vertically. These lines are parallel to the axes. Any line parallel to an axis can be measured and is called an *isometric line*. Lines that are not parallel to the axes are called *nonisometric lines* and cannot be measured. Note the two nonisometric lines in **Figure 24-5**.

Circles appear as ellipses in an isometric drawing. Circular features shown on isometric objects must be oriented properly or they appear distorted. The correct orientation of isometric circles on the three principal planes is shown in **Figure 24-6**. The small diameter (minor axis) of the ellipse must always align with the axis of the

isometric line: Any line that is parallel to an axis in an isometric drawing.

nonisometric lines: Lines that are not parallel to the axes in an isometric drawing.

Figure 24-5.
Layout of the isometric axes.

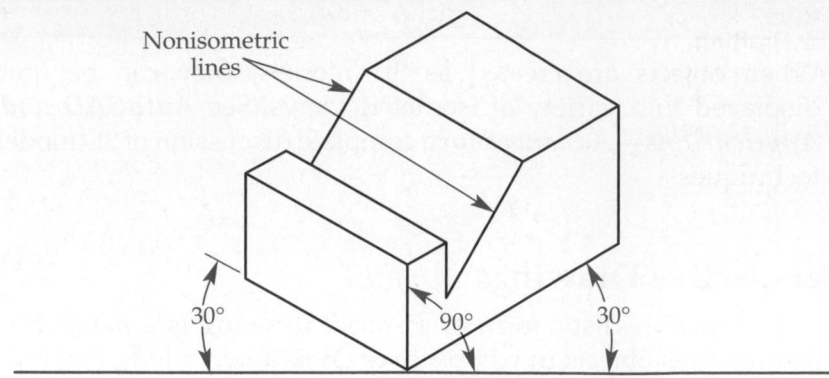

Nonisometric lines

30°   90°   30°

Figure 24-6.
Proper isometric circle (ellipse) orientation on isometric planes. The minor axis always aligns with the axis centerline.

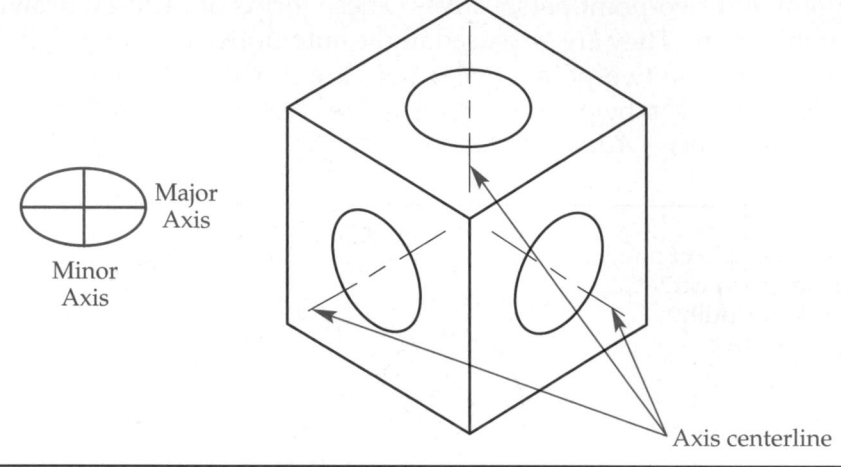

Major Axis

Minor Axis

Axis centerline

circular feature. Notice that the centerline axes of the holes in **Figure 24-6** are parallel to one of the isometric planes.

A basic rule to remember about isometric drawing is that lines that are parallel in an orthogonal view must also be parallel in the isometric view. AutoCAD's **ISOPLANE** tool makes that task, and the positioning of ellipses, easy.

PROFESSIONAL TIP

If you are ever in doubt about the proper orientation of an ellipse in an isometric drawing, remember that the minor axis of the ellipse must always be aligned with the centerline axis of the circular feature. This is shown clearly in **Figure 24-6.**

Settings for Isometric Drawing

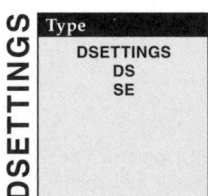

Type
DSETTINGS
DS
SE

DSETTINGS

You can quickly set isometric variables in the **Snap and Grid** tab of the **Drafting Settings** dialog box. See **Figure 24-7.** This dialog box can also be accessed by right-clicking the **Snap Mode** or **Grid Display** status bar button and then selecting **Settings...** from the shortcut menu.

To activate the isometric snap grid, pick the **Isometric snap** radio button in the **Snap type** area. Notice that the **Snap X spacing** and **Grid X spacing** text boxes are now grayed out. Since X spacing relates to horizontal measurements, it is not used in **Isometric snap** mode. You can only set the Y spacing for grid and snap in isometric. Check the **Snap On (F9)** and **Grid On (F7)** check boxes if you want **Snap** and **Grid** modes to be activated. Pick the **OK** button to display the grid dots on the screen in an isometric orientation, as shown in **Figure 24-8.**

Figure 24-7.
The **Drafting Settings** dialog box allows you to specify the settings needed for isometric drawing.

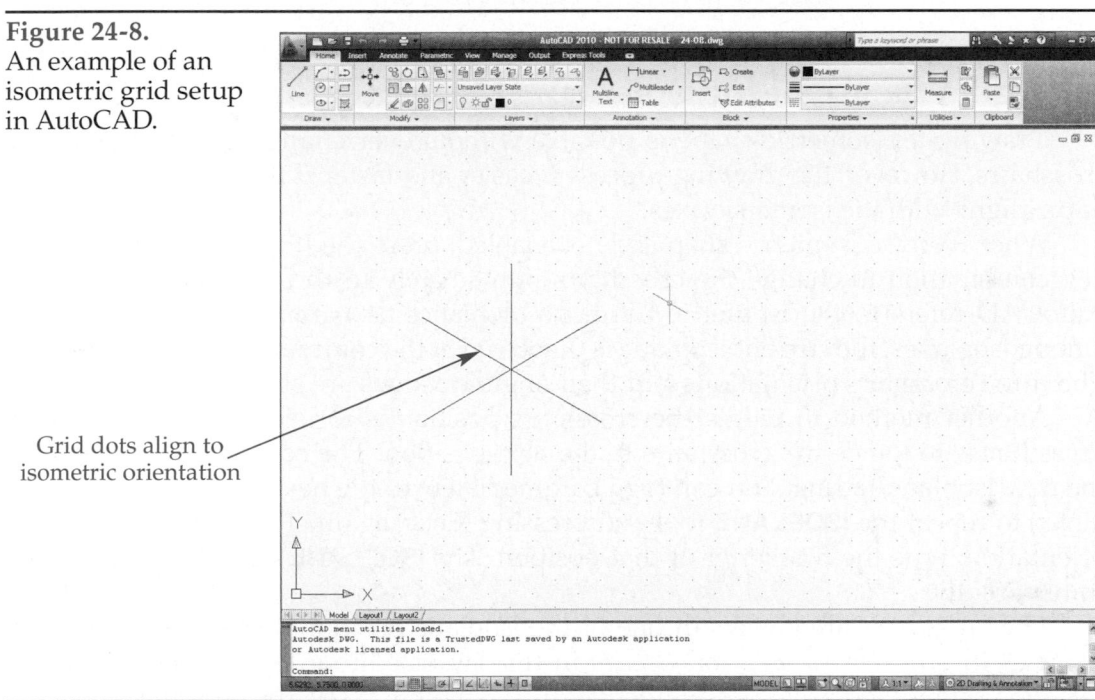

Pick to activate isometric snap grid

Figure 24-8.
An example of an isometric grid setup in AutoCAD.

Grid dots align to isometric orientation

Notice that the crosshairs appears angled. This aids you in drawing lines at the proper isometric angles. Try drawing a four-sided surface using the **LINE** tool. Draw the surface so it appears to be the left side of a box in an isometric layout. See **Figure 24-9.** To draw nonparallel surfaces, you can change the angle of the crosshairs to make your task easier, as discussed in the next section.

To turn off the **Isometric snap** mode, pick the **Rectangular snap** radio button in the **Snap type** area. The **Isometric snap** mode is turned off and you are returned to the drawing area when you pick the **OK** button.

> **NOTE**
>
> You can also set the **Isometric snap** mode by typing SNAP or SN, selecting the **Style** option, and then typing I to select **Isometric**.

Figure 24-9.
A four-sided object drawn with the **LINE** tool can be used as the left side of an isometric box.

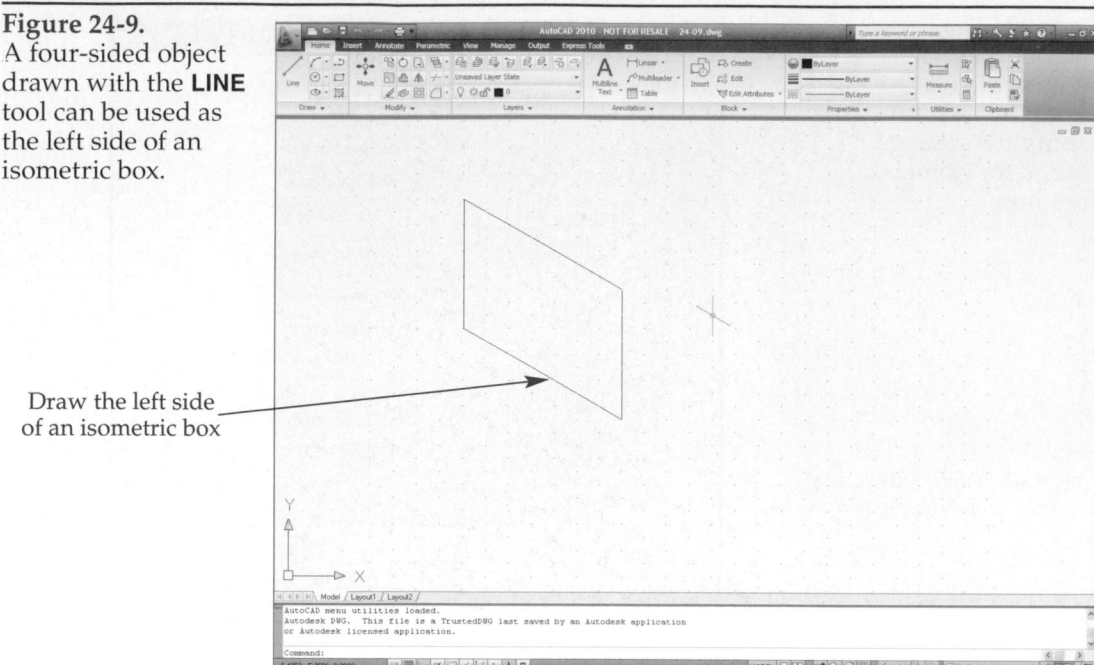

Draw the left side of an isometric box

Changing the Isometric Crosshairs Orientation

Drawing an isometric shape is possible without ever changing the angle of the crosshairs. However, the drawing process is easier and faster if the angle of the crosshairs aligns with the isometric axes.

Whenever the isometric snap style is enabled, press the [F5] key or the [Ctrl]+[E] key combination to change the crosshairs immediately to the next isometric plane. AutoCAD refers to the isometric positions or planes as *isoplanes*. As you change among isoplanes, the current isoplane is displayed at the command line as a reference. The three crosshairs orientations and their angular values are shown in **Figure 24-10.**

isoplanes: The three isometric positions or planes.

Another method to toggle the crosshairs position is to use the **ISOPLANE** tool. Press [Enter] to toggle the crosshairs to the next position. The command line displays the new isoplane setting. You can toggle immediately to the next position by pressing [Enter] to repeat the **ISOPLANE** tool and pressing [Enter] again. To specify the plane of orientation, type the first letter of that position. The **ISOPLANE** tool can also be used transparently.

The crosshairs are always in one of the isoplane positions when **Isometric snap** mode is in effect. An exception occurs in display or editing tools when a multiple selection set method, such as a window, is used. In these cases, the crosshairs changes to the normal vertical and horizontal position. When the display or editing tool is closed, the crosshairs reverts to its former isoplane orientation.

Figure 24-10.
You can toggle among the three isometric crosshairs positions using the [F5] function key, the [Ctrl]+[E] key combination, or the **ISOPLANE** tool.

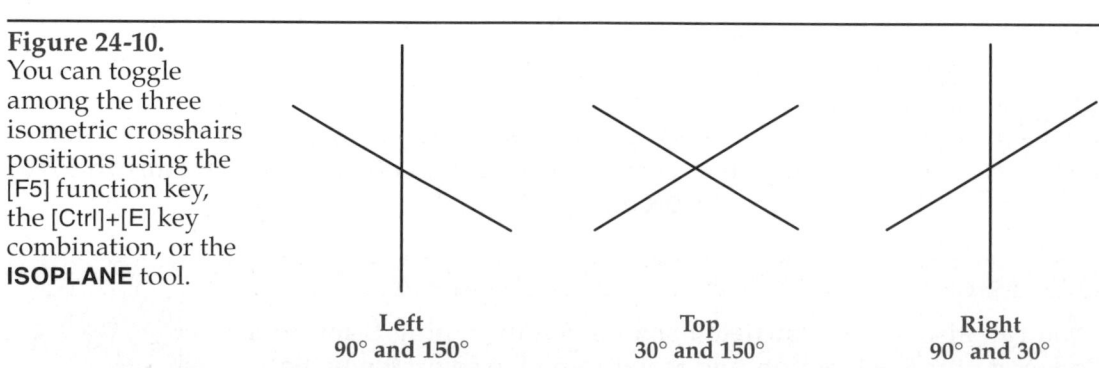

Left	Top	Right
90° and 150°	30° and 150°	90° and 30°

Exercise 24-1

Access the Student Web site (www.g-wlearning.com/CAD) and complete Exercise 24-1.

Isometric Ellipses

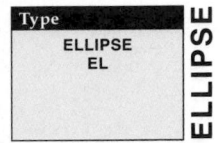

Type

ELLIPSE
EL

ELLIPSE

Placing an isometric ellipse on an object is easy with AutoCAD because of the **Isocircle** option of the **ELLIPSE** tool. An ellipse is positioned automatically on the current isoplane. To use the **ELLIPSE** tool, first make sure you are in **Isometric snap** mode. Access the **ELLIPSE** tool and select the **Isocircle** option. Then pick the center point and set the radius or diameter. Note that the **Isocircle** option only appears when you are in **Isometric snap** mode.

Always check the isoplane position before placing an ellipse (isocircle) on your drawing. You can dynamically view the three positions an ellipse can take. Access the **ELLIPSE** tool, enter the **Isocircle** option, pick a center point, and toggle the crosshairs orientation. See **Figure 24-11.** The ellipse rotates each time you toggle the crosshairs.

The isometric ellipse (isocircle) is a true ellipse. When an ellipse is selected, grips are displayed at the center and four quadrant points. See **Figure 24-12.** However, do not use grips to resize or otherwise adjust an isometric ellipse. As soon as you resize an isometric ellipse in this manner, its angular value is changed and it is no longer isometric. You can use the center grip to move the ellipse. Also, if you rotate an isometric ellipse while **Ortho** mode is on, it will not appear in a proper isometric plane. You *can* rotate an isometric ellipse from one isometric plane to another, but you must enter a value of 120°.

Figure 24-11.
The orientation of an isometric ellipse is determined by the current isometric plane.

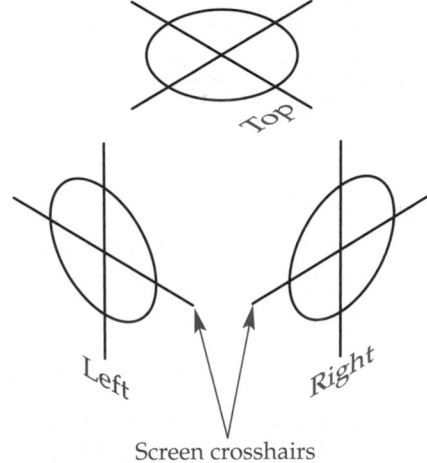

Figure 24-12.
An isometric ellipse has grips at its four quadrant points and center.

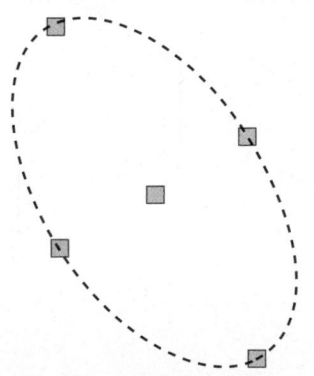

Exercise 24-2

Access the Student Web site (www.g-wlearning.com/CAD) and complete Exercise 24-2.

Constructing Isometric Arcs

The **ELLIPSE** tool can also be used to draw an isometric arc of any included angle. To construct an isometric arc, use the **Arc** option of the **ELLIPSE** tool while in **Isometric snap** mode. Access the **ELLIPSE** tool, type A to initiate the **Arc** option, select the **Isocircle** option, and then pick the center of the arc. Select values for the radius, start angle, and end angle.

A common application of isometric arcs is drawing fillets and rounds. Once a round is created in isometric, the edge (corner) of the object sits back from its original, unfilleted position. See **Figure 24-13A.** You can draw the object first and then trim away the excess after locating the fillets, or you can draw the isometric arcs and then the connecting lines. Either way, the center point of the ellipse is a critical feature and should be located first. The left-hand arc in **Figure 24-13A** was drawn first and copied to the back position. Use **Ortho** mode to help quickly draw 90° arcs.

The next step is to move the original edge to its new position, which is tangent to the isometric arcs. You can do this by snapping the endpoint of the line to the quadrant point of the arc. See **Figure 24-13B.** Notice the grips on the line and on the arc. The endpoint of the line is snapped to the quadrant grip on the arc. The final step is to trim away the excess lines and arc segment. The completed feature is shown in **Figure 24-13C.**

Rounded edges, when viewed straight on, cannot be shown as complete-edge lines that extend to the ends of the object. Instead, a good technique to use is a broken line in the original location of the edge. This is clearly shown on the right-hand edge in **Figure 25-13C.**

Exercise 24-3

Access the Student Web site (www.g-wlearning.com/CAD) and complete Exercise 24-3.

Figure 24-13.
Fillets and rounds can be drawn with the **Arc** option of the **ELLIPSE** tool. A broken line is used to represent an edge that is viewed straight on.

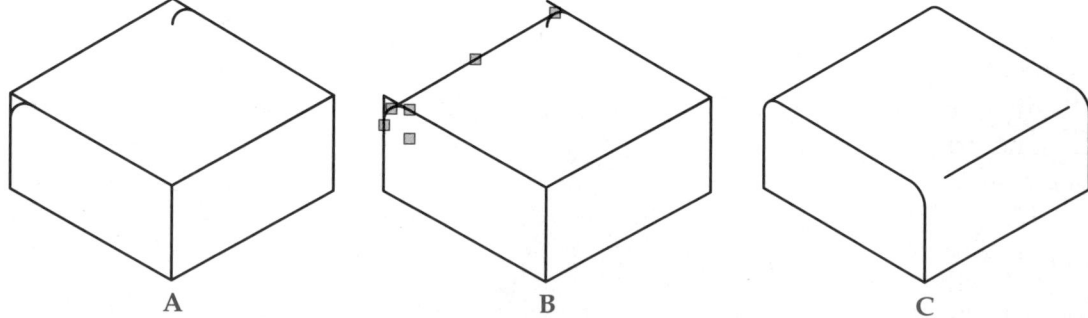

A   B   C

Creating Isometric Text Styles

Text placed in an isometric drawing should appear to be parallel to one of the isometric planes. It should align with the plane to which it applies. Text may be located on the object or positioned away from it as a note. Drafters and artists occasionally neglect this aspect of pictorial drawing, and it shows on the final product.

Properly placing text on an isometric drawing involves creating new text styles. **Figure 24-14** illustrates possible orientations of text on an isometric drawing. These examples were created using only two text styles. The text styles have an obliquing angle of either 30° or –30°. The labels in **Figure 24-14** refer to the following table. The angle in the figure indicates the rotation angle entered when using one of the text tools. For example, ISO-2 90 means that the ISO-2 style was used and the text was rotated 90°. This technique can be applied to any font.

Name	Font	Obliquing Angle
ISO-1	Romans	30°
ISO-2	Romans	–30°

Exercise 24-4

Access the Student Web site (www.g-wlearning.com/CAD) and complete Exercise 24-4.

Figure 24-14.
Isometric text applications. The text shown here indicates which style and angle were used.

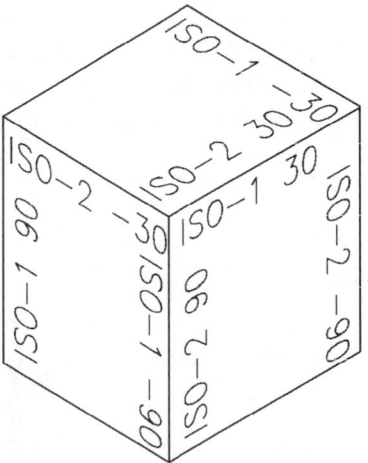

Isometric Dimensioning

An important aspect of dimensioning in isometric is to place dimension lines, text, and arrowheads on the proper plane. Remember these guidelines:
- Extension lines should always extend in the plane being dimensioned.
- The heel of the arrowhead should always be parallel to the extension line.
- Strokes of the text that would normally be vertical should always be parallel with the extension lines or dimension lines.

These techniques are shown on the dimensioned isometric part in **Figure 24-15**. AutoCAD does not automatically dimension isometric objects. You must first create isometric arrowheads and text styles. Then, manually draw the dimension lines and text as they should appear in each of the three isometric planes.

Figure 24-15.
A dimensioned
isometric part.
Note the text
and arrowhead
orientation in
relation to the
extension lines.

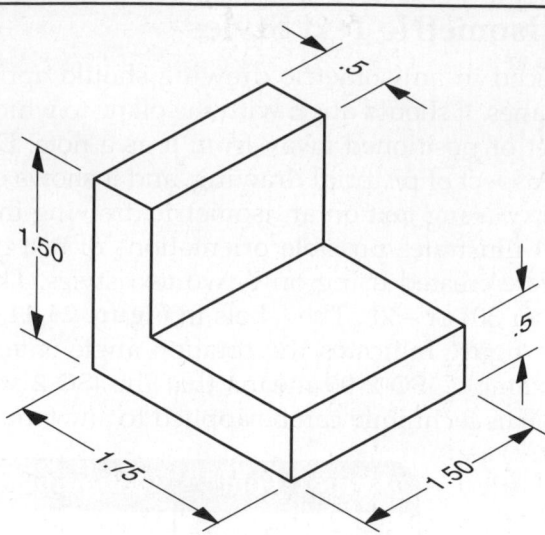

You have already learned how to create isometric text styles. You can set up appropriate text styles in an isometric template drawing if you draw isometrics often.

Isometric Arrowheads

You can draw isometric arrowheads and fill them in with a solid hatch pattern. A variable-width polyline cannot be used because the heel of the arrowhead will not be parallel to the extension lines. Examples of arrowheads for the three isometric planes are shown in **Figure 24-16.**

You do not need to draw every arrowhead individually. First, draw two isometric axes, as shown in **Figure 24-17A.** Then draw one arrowhead like the one shown in **Figure 24-17B.** Use the **MIRROR** tool to create additional arrowheads. As you create new arrowheads, move each one to its proper plane.

Figure 24-16.
Examples of
arrowheads in
each of the three
isometric planes.

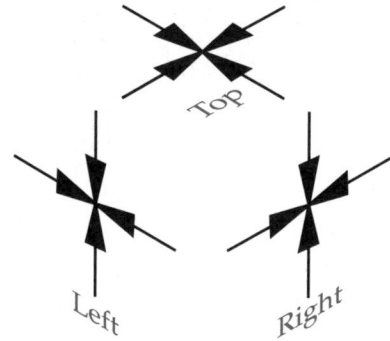

Figure 24-17.
Creating isometric
arrowheads.
A—Draw the two
isometric axes
for arrowhead
placement. B—Draw
the first arrowhead
on one of the axis
lines. Then mirror
the arrowhead to
create the others.

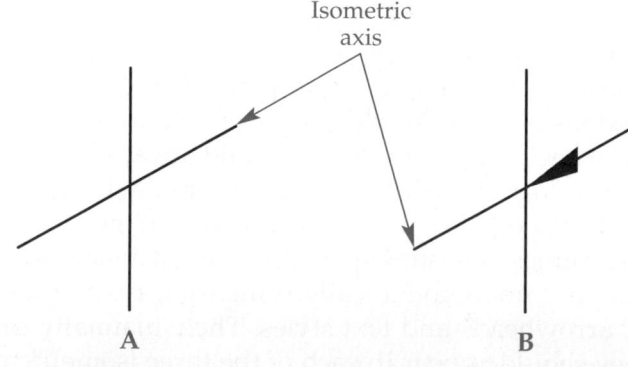

You can save each arrowhead as a block in your isometric template or prototype. Use block names that are easy to remember. Blocks are discussed in Chapters 25–28.

Oblique Dimensioning

AutoCAD has a way to dimension isometric and oblique lines semiautomatically. First, draw the dimensions using any of the linear dimensioning tools. The object in **Figure 24-18A** was dimensioned using the **DIMALIGNED** and **DIMLINEAR** tools. Then use the **Oblique** option of the **DIMEDIT** tool to rotate the extension lines. See **Figure 24-18B**.

To use the **Oblique** option, access the **DIMEDIT** tool and then type O for **Oblique**. When prompted, select the dimension and enter the obliquing angle. This technique creates suitable dimensions for an isometric drawing and is quicker than the method previously discussed. However, this method does not rotate the arrows to align the arrowhead heels with the extension lines. It also does not draw the dimension text aligned in the plane of the dimension. Therefore, this method does not produce technically correct dimensions.

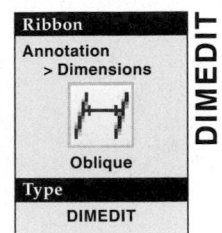

Figure 24-18.
Using the **Oblique** option of the **DIMEDIT** tool, you can create semiautomatic isometric dimensions by editing existing dimensions. A—Create the dimensions using the **DIMALIGNED** or **DIMLINEAR** tool. B—Adjust the positions of the dimensions using **DIMEDIT**. C—The obliquing angles used to achieve the results shown.

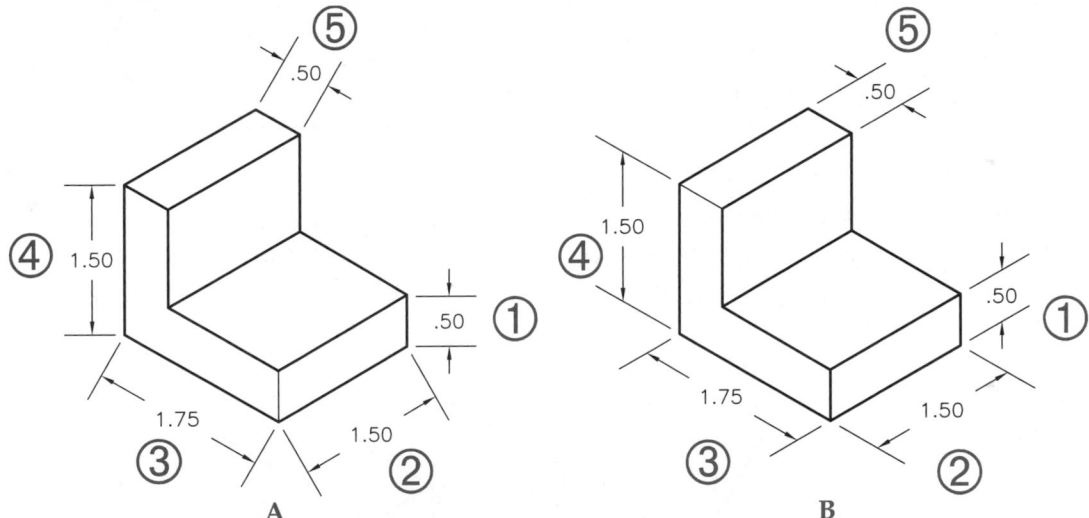

Dimension	Obliquing Angle
1	30°
2	−30°
3	30°
4	−30°
5	30°

C

Chapter Test

Answer the following questions. Write your answers on a separate sheet of paper or go to the Student Web site (www.g-wlearning.com/CAD) and complete the electronic chapter test.

1. Which is the simplest form of pictorial drawing?
2. How does isometric drawing differ from oblique drawing?
3. How do dimetric and trimetric drawings differ from isometric drawings?
4. Which is the most realistic form of pictorial drawing?
5. What must be set in the **Drafting Settings** dialog box to turn on **Isometric snap** mode and set a snap spacing of .2?
6. What function does the **ISOPLANE** tool perform?
7. Name the tool and option used to draw an isometric circle.
8. What factor determines the orientation of an isometric ellipse?
9. Where are grips located on a circle drawn in isometric?
10. Can grips be used to resize an isometric circle correctly? Explain your answer.
11. How are isometric arcs drawn?
12. Briefly explain how to create text that can be used on an isometric drawing.
13. How can you create isometric arrowheads?
14. What technique does AutoCAD provide for semiautomatically dimensioning isometric objects?
15. Why does the technique in Question 14 not produce technically correct dimensions?

Drawing Problems

▼ Basic

Create an isometric template drawing. Items that should be set in the template include grid spacing, snap spacing, ortho setting, and text size. Save the template as isoproto.dwt. Use the template to construct the isometric drawings in Problems 1–10. Measure the drawings to obtain the dimensions. Save the drawing problems as P24-(problem number).

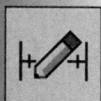

1.
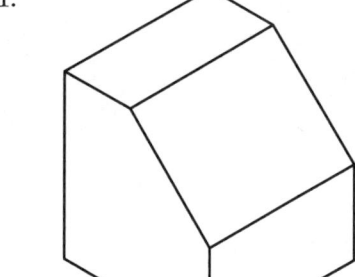

2.
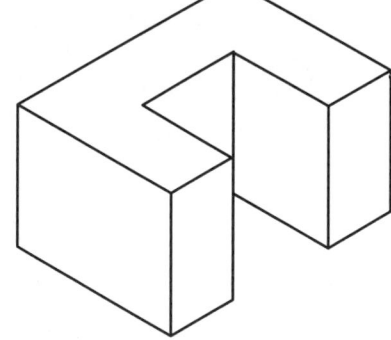

3.
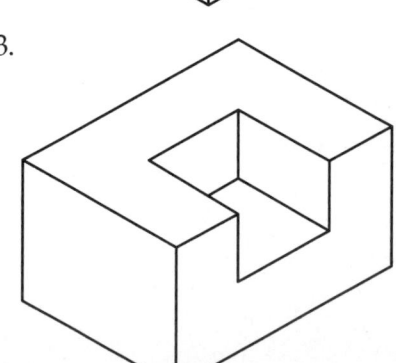

4.
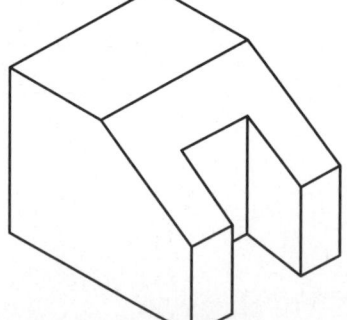

5.

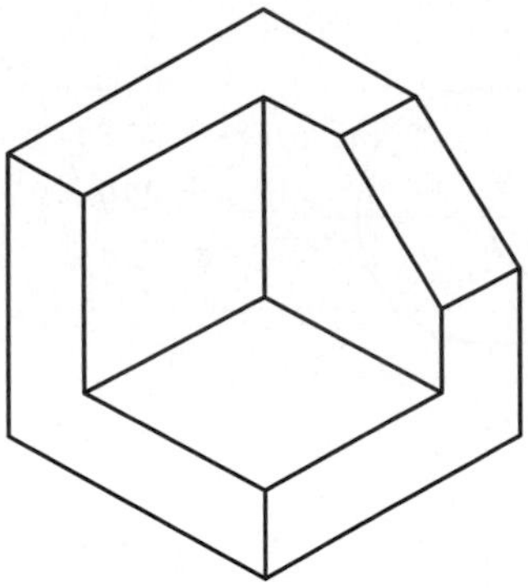

6.

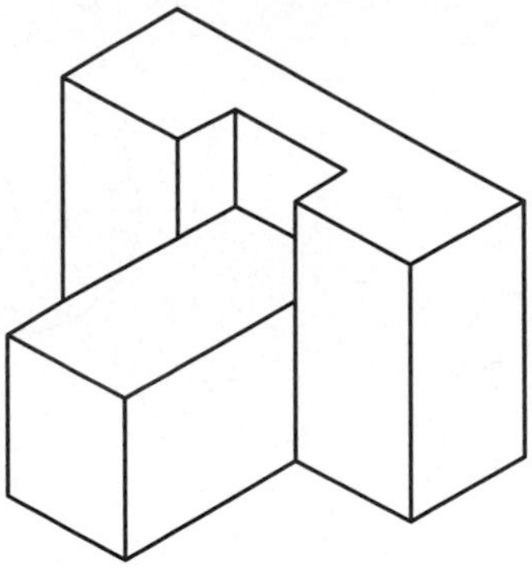

7.

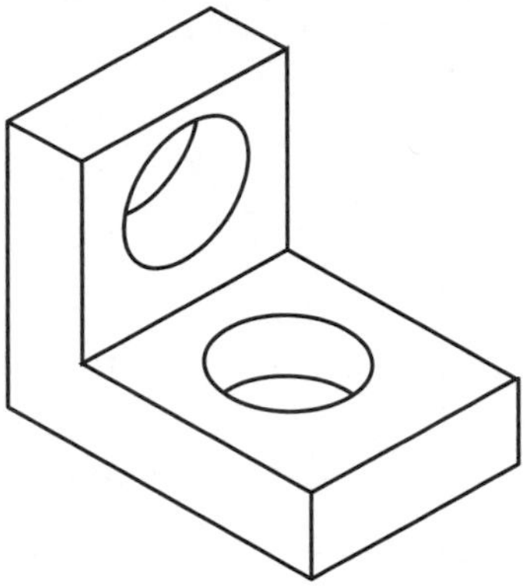

8.

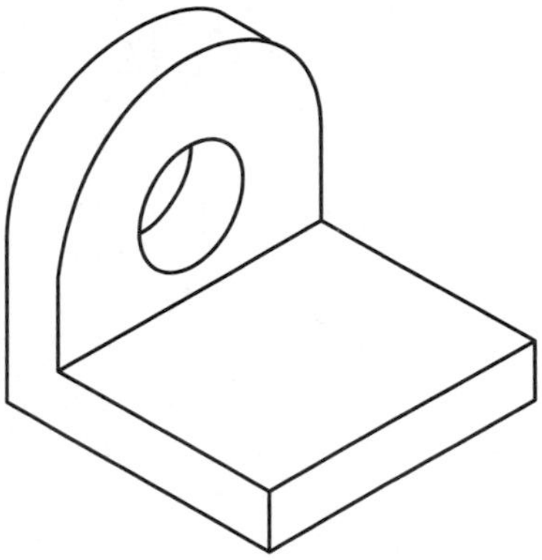

9.

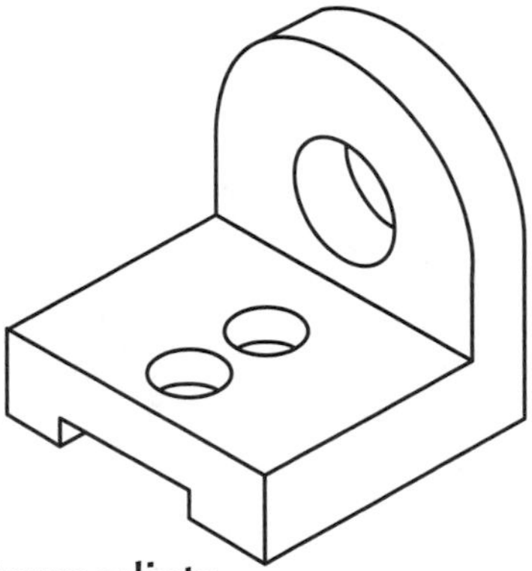

10.

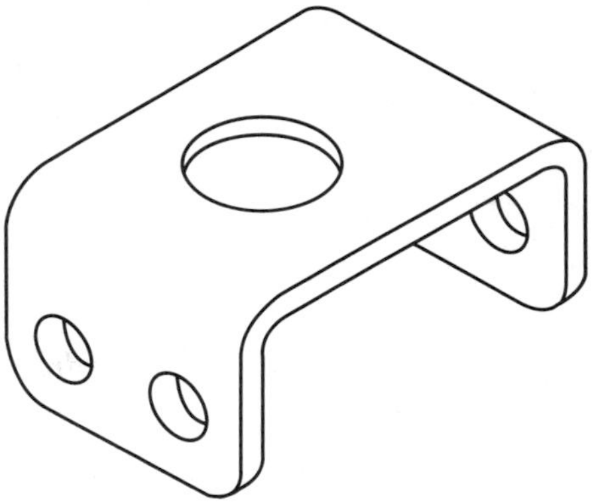

▼ Intermediate

For Problems 11–14, create isometric drawings using the views shown. Measure the drawings to obtain the dimensions.

11.

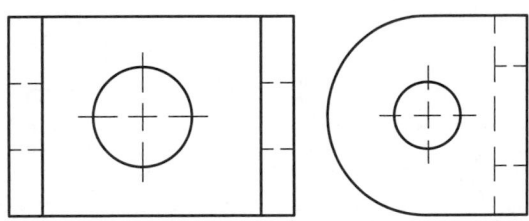

12.

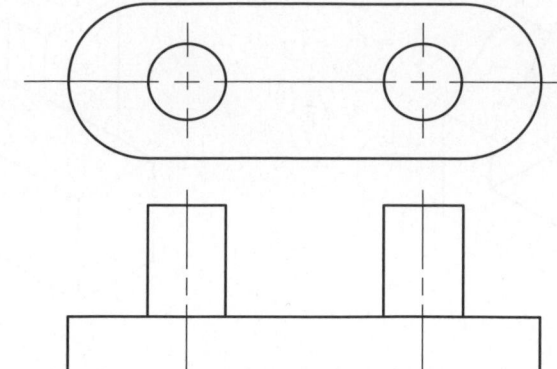

13.

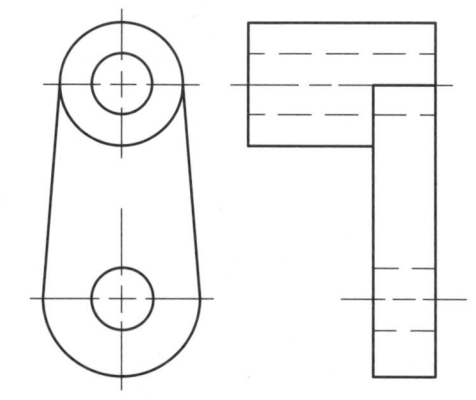

14.

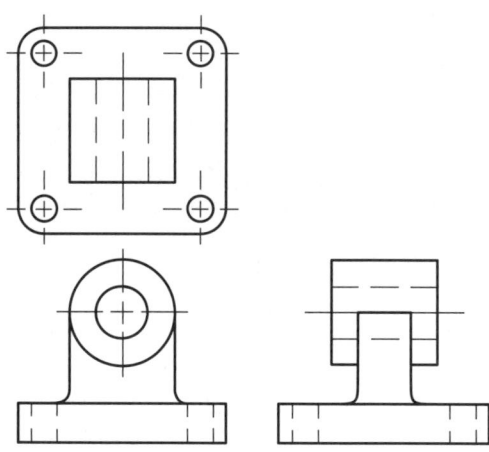

15. Construct a set of isometric arrowheads to use when dimensioning isometric drawings. Load your isometric template drawing. Create arrowheads for each of the three isometric planes. Name them with the first letter indicating the plane: T for top, L for left, and R for right. Also, number them clockwise from the top. See the example below for the right isometric plane. Do not include the labels in your drawing. Save the template again when you are finished.

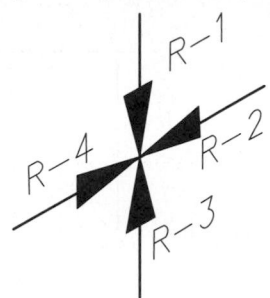

16. Create a set of isometric text styles like those shown in **Figure 24-14.** Load your template drawing and make a complete set in one font. Make additional sets in other fonts if you wish. Enter a text height of 0 so you can specify the height when placing the text. Save the template when finished.

17. Begin a new drawing using your isometric template. Select one of the following problems from this chapter and dimension it: Problem 5, 7, 8, or 9. When adding dimensions, be sure to use the proper arrowhead and text style for the plane on which you are working. Save the drawing as P24-17.

▼ Advanced

18. Create an isometric drawing of the switch plate shown below. Select a view that best displays the features of the object. Do not include dimensions. Save the drawing as P24-18.

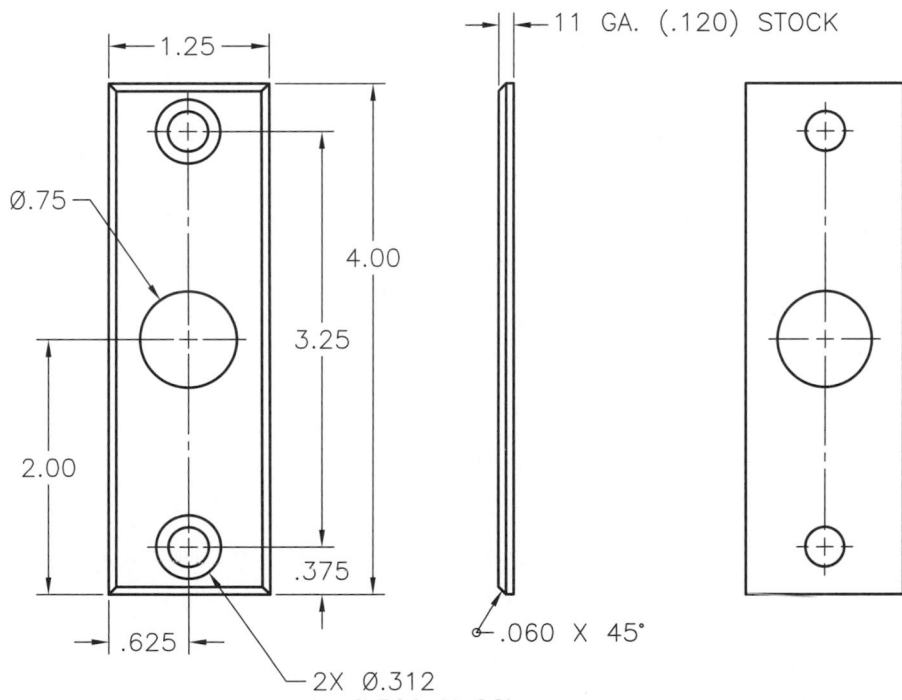

19. Create an isometric drawing of the retainer shown below. Select a view that best displays the features of the object. Do not include dimensions. Save the drawing as P24-19.

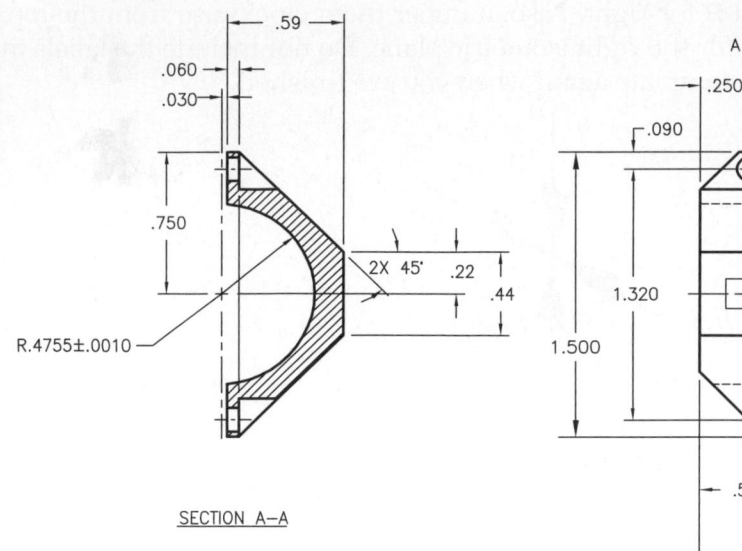

SECTION A—A

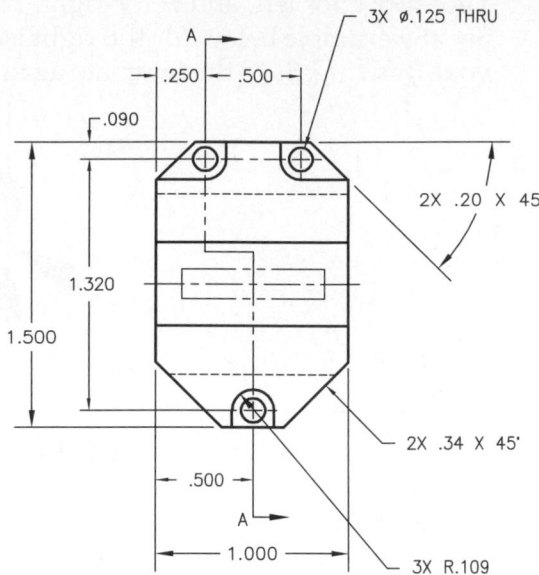

AutoCAD and Its Applications—Basics

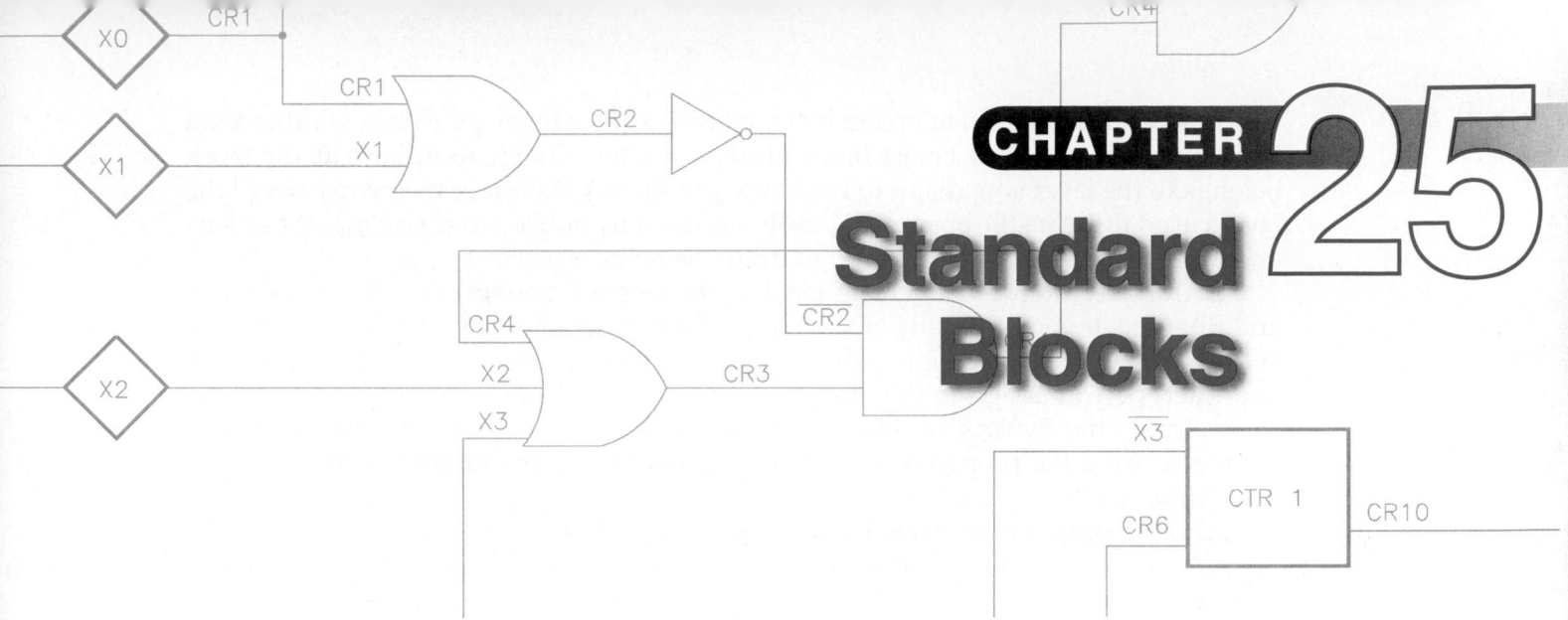

Standard Blocks

Learning Objectives

After completing this chapter, you will be able to do the following:

✓ Create and save blocks.
✓ Insert blocks into a drawing.
✓ Edit a block and update the block in a drawing.
✓ Create blocks as drawing files.
✓ Construct and use a symbol library of blocks.
✓ Purge unused items from a drawing.

The ability to create and use *blocks* is one of the greatest benefits of drawing with AutoCAD. The **BLOCK** tool stores a block within a drawing as a *block definition*. The **WBLOCK** tool saves a *wblock* as a separate drawing file. You can insert blocks as often as needed and share blocks between drawings. You also have the option to scale, rotate, and adjust blocks to meet specific drawing requirements.

block: A symbol saved and stored in a drawing for future use.

block definition: Information about a block that is stored within the drawing file.

wblock: A block definition saved as a separate drawing file.

Constructing Blocks

A block can consist of any shape, group of objects, symbol, annotation, view, or drawing. Review the drawing to identify any item you plan to use more than once. Screws, punches, subassemblies, plumbing fixtures, and appliances are examples of items you may want to convert to blocks. Draw the item once and then save the objects as a block for multiple use.

Selecting a Layer

Before you begin drawing block components, you should identify the appropriate layer on which to create the objects. It is critical that you understand how layers and object properties apply when using blocks. The 0 layer is the preferred layer on which to draw block objects. If you originally create block objects on layer 0, the block assumes, or inherits, the properties of the layer you assign to the block. Draw the objects for all blocks on layer 0 and then assign the appropriate layer to each block when you insert the block. If you draw block objects on a layer other than layer 0, place all the objects on layer 0 before creating the block.

A second method is to create block objects using one or more layers other than layer 0. If you originally create block objects on a layer other than layer 0, the block belongs to the layer you assign to the block, but the objects retain the properties of the layers used to create the objects. The difference is only noticeable if you place the block on a layer other than the layer used to draw the block objects.

A third technique is to create block objects using the ByBlock color, lineweight, and linetype. If you originally create block objects using ByBlock properties, the block belongs to the layer you assign to the block, but the objects take on the color, lineweight, and linetype you assign to the block, regardless of the layer on which you place the block. Using the ByBlock setting is only noticeable if you assign absolute values to the block using the properties in the **Properties** panel of the **Home** ribbon tab or the **Properties** palette.

Another option is to create block objects using an absolute color, lineweight, and linetype. If you originally create block objects using absolute values, such as a Blue color, a 0.05mm lineweight, and a Continuous linetype, the block belongs to the layer you assign to the block, but the objects display the specified absolute values regardless of the properties assigned to the drawing or the layer on which you place the block.

Drawing Block Components

insertion base point: The point on a block that defines where the block is positioned during insertion.

Draw a block as you would any other geometry. When you finish drawing the objects, determine the best location for the *insertion base point*. When you insert the block into a drawing, the insertion base point positions the block. **Figure 25-1** shows several examples of common blocks and a possible insertion base point for each.

Creating Blocks

Once you draw objects and identify an appropriate insertion base point, you are ready to save the objects as a block. Use the **BLOCK** tool and the corresponding **Block Definition** dialog box to create a block. See **Figure 25-2.**

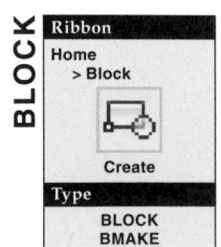

BLOCK

Ribbon
Home
> Block

Create

Type
BLOCK
BMAKE
B

Naming and Describing the Block

Enter a descriptive name for the block in the **Name:** text box. For example, name a vacuum pump PUMP or a 3′ × 6′-8″ door DOOR_3068. The block name cannot exceed 255 characters. It can include numbers, letters, and spaces, as well as the dollar sign ($), hyphen (-), and underscore (_). The drop-down list allows you to access an existing name to recreate a block, or to use as reference when naming a new block with a similar name.

Figure 25-1.
Common drafting symbols and their insertion points for placement on drawings. Colored dots indicate the insertion points for reference.

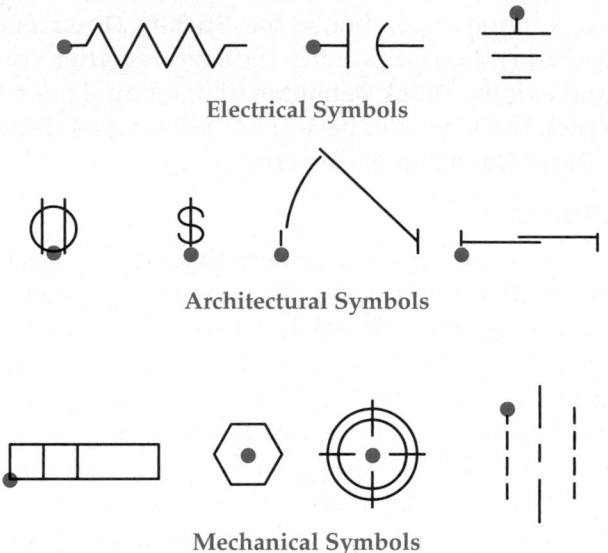

Electrical Symbols

Architectural Symbols

Mechanical Symbols

Figure 25-2.
Use the **Block Definition** dialog box to create a block.

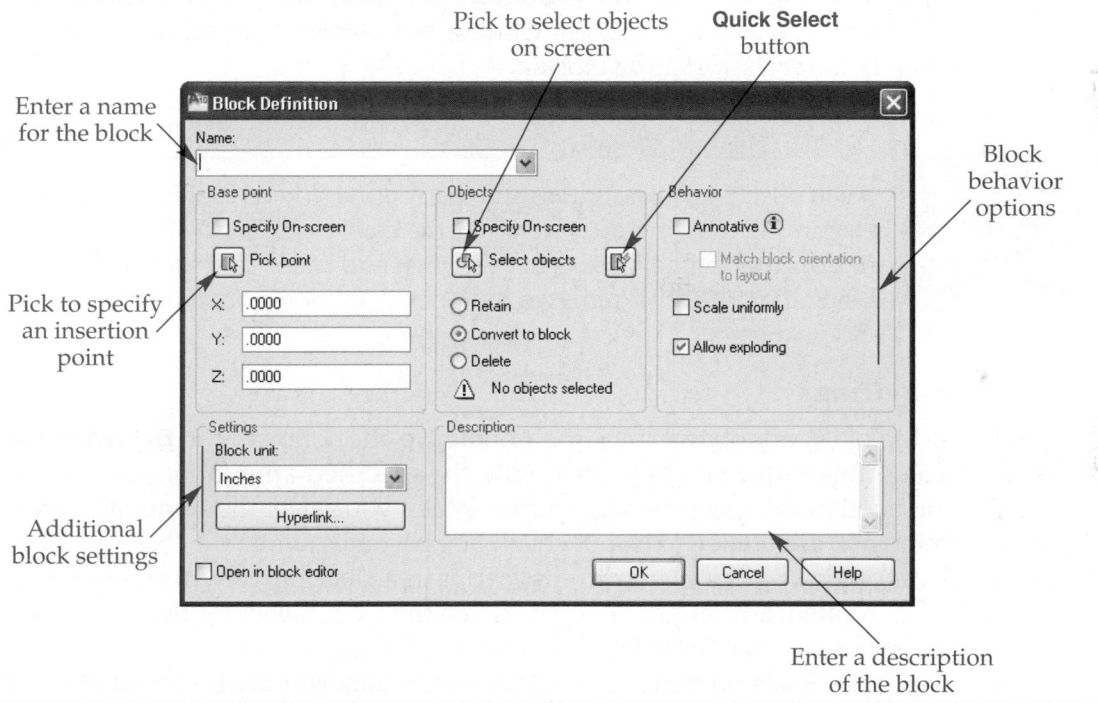

Pick to select objects on screen

Quick Select button

Enter a name for the block

Block behavior options

Pick to specify an insertion point

Additional block settings

Enter a description of the block

A block name is often descriptive enough to identify the block. However, you can enter a description of the block in the **Description:** text box to help identify the block. For example, the PUMP block might include the description This is a vacuum pump symbol, or the DOOR_3068 block might include the description This is 3′ wide by 6′-8″ tall interior single-swing door.

Defining the Block Insertion Base Point

The **Base point** area allows you to define the insertion base point. If you know the coordinates for the insertion base point, type values in the **X:**, **Y:**, and **Z:** text boxes. However, often the best way to specify the insertion base point is to use object snap to select a point on an object. Choose the **Pick point** button to return to the drawing and select an insertion base point. The **Block Definition** dialog box reappears after you select the insertion base point.

An alternative technique is to choose the **Specify On-screen** check box, which allows you to pick an insertion base point in the drawing after you pick the **OK** button to create the block and exit the **Block Definition** dialog box. This method can save time by allowing you to pick the insertion base point without using the **Pick point** button and re-entering the **Block Definition** dialog box.

Selecting Block Objects

The **Objects** area includes options for selecting objects for the block definition. Pick the **Select objects** button to return to the drawing and select the objects that will make up the block. Press [Enter] or the space bar or right-click to redisplay the **Block Definition** dialog box. The number of selected objects appears in the **Objects** area, and an image of the selection displays next to the **Name:** drop-down list. To create a selection set, use the **QuickSelect** button and **Quick Select** dialog box to define a filter.

An alternative method for selecting objects is to choose the **Specify On-screen** check box, which allows you to pick objects from the drawing after you pick the **OK** button to create the block and exit the **Block Definition** dialog box. This method can save time by allowing you to select objects without using the **Select objects** button and re-entering the **Block Definition** dialog box.

Pick the **Retain** radio button to keep the selected objects in the current drawing in their original, unblocked state. Select the **Convert to block** radio button to replace the selected objects with the block definition. Choose the **Delete** radio button to remove the selected objects after defining the block.

PROFESSIONAL TIP

If you select the **Delete** option and then decide to keep the original geometry in the drawing after defining the block, use the **OOPS** tool. This returns the original objects to the screen and keeps the block definition. Using the **UNDO** tool removes the block definition from the drawing.

Block Scale Settings

To make the block annotative, pick the **Annotative** check box in the **Behavior** area. AutoCAD scales annotative blocks according to the specified annotation scale, which eliminates the need to calculate the scale factor. When you select the **Annotative** check box, the **Match block orientation to layout** check box becomes available. Pick this check box to keep annotative blocks planar to the layout in a floating viewport, even if the drawing view is rotated, as it might be if you rotate the UCS. Selecting this option also prohibits you from using the **ROTATE** tool to rotate a block.

If you check the **Scale uniformly** check box in the **Behavior** area, you do not have the option of specifying different X and Y scale factors when you insert the block. Block scaling options are described later in this chapter.

Additional Block Definition Settings

If the **Allow exploding** check box in the **Behavior** area is checked, you have the option of exploding the block. If the box is not checked, you cannot explode the block, even after inserting it. Select a unit type from the **Block unit** drop-down list in the **Settings** area to specify the insertion units of the block. Pick the **Hyperlink...** button to access the **Insert Hyperlink** dialog box to insert a hyperlink in the block. If **Open in block editor** is checked, the new block immediately opens in the **Block Definition Editor** when you create the block and exit the **Block Definition** dialog box. The **Block Definition Editor** is described later in this chapter.

nesting: Creating a block that includes other blocks.

Exercise 25-1

Access the Student Web site (www.g-wlearning.com/CAD) and complete Exercise 25-1.

Inserting Blocks

Once you create a block, you have several options for inserting it into a drawing. Remember to make the layer you want to assign to the block current *before* inserting the block. You should also determine the proper size and rotation angle for the block before insertion. The term *block reference* describes an inserted block. *Dependent symbols* are any named objects, such as blocks and layers. AutoCAD automatically updates dependent symbols in a drawing the next time you open the drawing.

block reference: A specific instance of a block inserted into a drawing.

dependent symbols: Named objects in a drawing that has been inserted or referenced into another drawing.

Using the INSERT Tool

The **INSERT** tool is one of the most common methods for inserting blocks in a drawing. The **Insert** dialog box, shown in **Figure 25-3,** appears when you access the **INSERT** tool.

Selecting the Block to Insert

Pick the **Name:** drop-down list button to show the blocks defined in the current drawing and select the name of the block to insert. You can also type the name of the block in the **Name:** text box. Pick the **Browse...** button to display the **Select Drawing File** dialog box, used to locate and select a drawing or DXF file (wblock) for insertion as a block. This process is described later in this chapter.

Specifying the Block Insertion Point

The **Insertion point** area contains options for specifying where to insert the block. Select the **Specify On-screen** check box to specify a location in the drawing to insert the block when you pick the **OK** button. To insert the block using absolute coordinates, deselect the **Specify On-screen** check box and enter coordinates in the **X:**, **Y:**, and **Z:** text boxes.

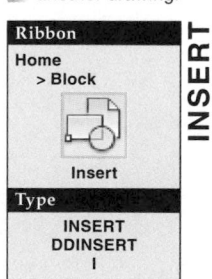

Ribbon
Home
> Block

Insert
Type
INSERT
DDINSERT
I

INSERT

Figure 25-3.
The **Insert** dialog box allows you to select and prepare a block for insertion. Select the block to insert from the drop-down list or enter the block name in the **Name:** text box.

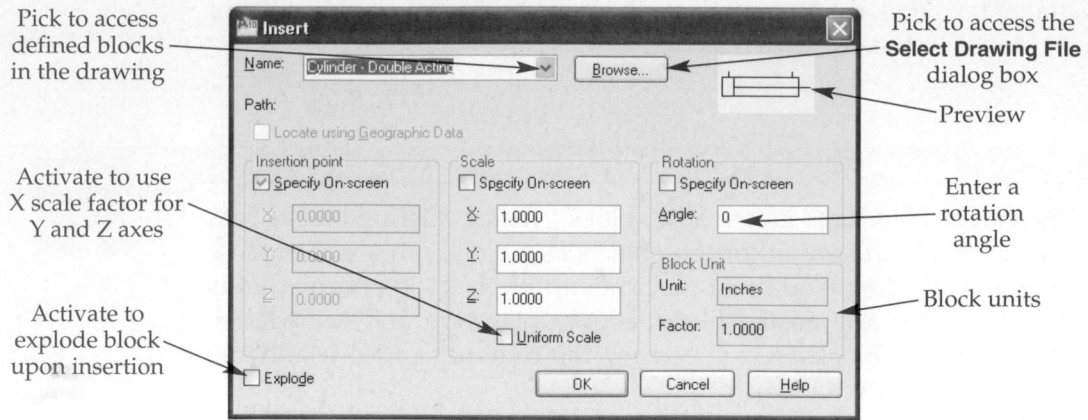

Pick to access defined blocks in the drawing

Pick to access the **Select Drawing File** dialog box

Preview

Activate to use X scale factor for Y and Z axes

Enter a rotation angle

Block units

Activate to explode block upon insertion

Scaling Blocks

The **Scale** area allows you to specify scale values for the block in relation to the X, Y, and Z axes. Deselect the **Specify On-screen** check box to enter scale values in the **X:**, **Y:**, and **Z:** text boxes. If you activate the **Uniform Scale** check box, you can specify a scale value for the X axis that also applies to the scale of the Y and Z axes. The **X** value is the only active axis value if you created the block with **Scale uniformly** checked in the **Block Definition** dialog box. Select the **Specify On-screen** check box to receive prompts for scaling the block during insertion.

It is possible to create a mirror image of a block by entering a negative scale factor values. For example, enter –1 for the X and Y scale factor to mirror the block to the opposite quadrant of the original orientation, but retain the original size. **Figure 25-4** shows different scale and mirroring techniques.

Blocks can be classified as real blocks, schematic blocks, or unit blocks, depending on how you scale the block during insertion. Examples of *real blocks* include a bolt, a bathtub, a pipe fitting, and the car shown in **Figure 25-5A.** Examples of *schematic blocks* include notes, detail bubbles, and section symbols. See **Figure 25-5B.** Schematic blocks typically include annotative blocks. When you insert an annotative schematic block, AutoCAD automatically determines the block scale based on the annotation scale. When you insert a non-annotative schematic block, you must specify the scale factor.

real block: A block originally drawn at a 1:1 scale and then inserted using 1 for both the X and Y scale factors.

schematic block: A block originally drawn at a 1:1 scale and then inserted using the drawing scale factor for both the X and Y scale values.

Figure 25-4.
Negative and positive scale factors have different effects when used to insert a block. Colored dots indicate the insertion points for reference.

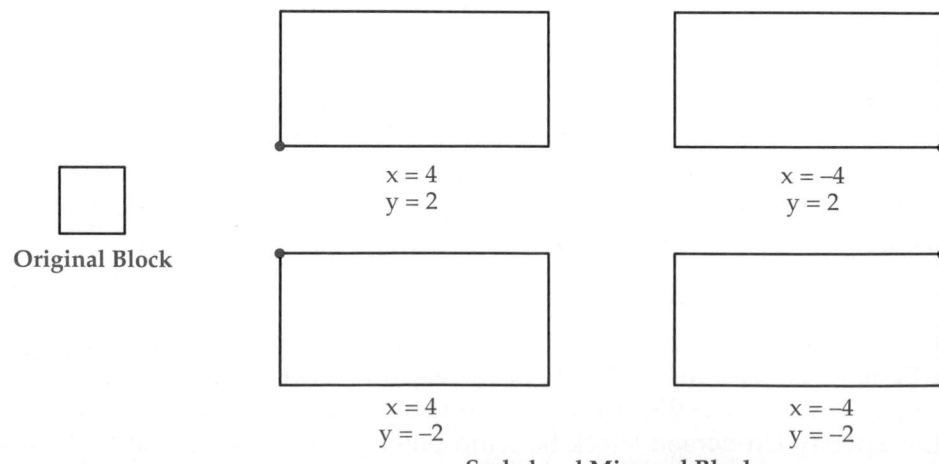

Original Block

x = 4
y = 2

x = –4
y = 2

x = 4
y = –2

x = –4
y = –2

Scaled and Mirrored Blocks

Figure 25-5.
A—Real blocks, such as this car, are drawn at full scale and inserted using a scale factor of 1 for both the X and Y axes. B—A schematic block is inserted using the scale factor of the drawing for the X and Y axes. C—A 2D unit block is often inserted at different scales for the X and Y axes.

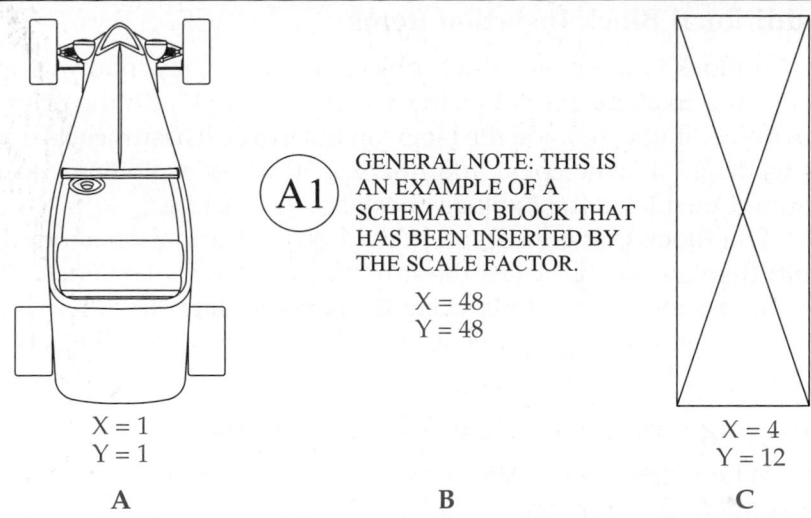

GENERAL NOTE: THIS IS AN EXAMPLE OF A SCHEMATIC BLOCK THAT HAS BEEN INSERTED BY THE SCALE FACTOR.

X = 48
Y = 48

X = 1
Y = 1

X = 4
Y = 12

A   B   C

> **NOTE**
>
> For most applications, you should insert annotative blocks at a scale of 1 in order for the annotation scale to apply correctly. Entering a scale other than 1 adjusts the scale of the block by multiplying the scale value by the annotative scale factor.

There are three different types of *unit blocks*. An example of a *1D unit block* is a 1-unit (1″, for example) blocked line object. A *2D unit block* is any blocked object that can fit inside a 1-unit × 1-unit (1″ × 1″, for example) square. A *3D unit block* is any blocked object that can fit inside a 1-unit × 1-unit × 1-unit (1″ × 1″ × 1″ for example) cube. To use a unit block, insert the block and determine the individual scale factors for each axis. For example, insert a 1″ 1D unit block line at a scale of 4 to create a 4″ line. When inserting a 2D unit block, assign different scale factors for the X and Y axes, such as 4 for the X axis and 12 for the Y axis, to create the 4″ × 12″ beam shown in **Figure 25-5C.** A 3D unit block allows you to adjust the scale of the X, Y, and Z axes.

unit block: A 1D, 2D, or 3D block drawn to fit in a 1-unit, 1-unit-square, or 1-unit-cubed area so that it can be scaled easily.

1D unit block: A 1-unit, one-dimensional object, such as a straight line segment, saved as a block.

2D unit block: A 2D object that fits into a 1-unit × 1-unit square, saved as a block.

3D unit block: A 3D object that fits into a 1-unit × 1-unit × 1-unit cube, saved as a block.

Rotating Blocks

The **Rotation** area allows you to insert the block at a specific angle. Deselect the **Specify On-screen** check box to enter a value in the **Angle:** text box. The default angle of 0° inserts the block as saved. Select the **Specify On-screen** check box to receive a prompt for rotating the block during insertion.

> **NOTE**
>
> You cannot rotate a block defined using the **Match block orientation to layout** option.

> **PROFESSIONAL TIP**
>
> You can rotate a block based on the current UCS. Be sure the proper UCS is active, and then insert the block using a rotation angle of 0°. If you decide to change the UCS later, any inserted blocks retain their original angle.

Additional Block Insertion Items

A block is saved as a single object, no matter how many objects the block includes. Select the **Explode** check box to explode the block into the original objects for editing purposes. If you explode the block on insertion, it assumes its original properties, such as its original layer, color, and linetype. If **Allow exploding** was unchecked when you defined the block, the **Explode** check box is inactive.

The **Block Unit** area displays read-only information about the selected block. The **Unit:** display box indicates the units for the block. The **Factor:** display box indicates the scale factor. The **Locate using Geographic Data** check box is active when the block and current drawing include *geographic data*. Pick the check box to position the block using geographic data.

geographic data: Information added to a drawing to describe specific locations and directions on Earth.

Working with Specify On-Screen Prompts

When you pick the **OK** button, prompts appear for any values defined as **Specify On-screen** in the **Insert** dialog box. If you specify the insertion point on-screen, the Specify insertion point or [Basepoint/Scale/X/Y/Z/Rotate/PScale/PX/PY/PZ/PRotate]: prompt appears. Enter or select a point to insert the block. The options allow you to specify a different base point; enter a value for the overall scale; enter independent scale factors for the X, Y, and Z axes; enter a rotation angle; and preview the scale of the X, Y, and Z axes or the rotation angle before entering actual values. If you use an available option, the new value overrides the related setting in the **Insert** dialog box.

If you specify the X scale factor on screen, the Enter X scale factor, specify opposite corner, or [Corner/XYZ] <1>: prompt appears. Pick a point or enter a value for the scale. You can also use the **Corner** option to scale the block. The Enter Y scale factor <use X scale factor>: prompt appears if you enter an X scale factor. Press [Enter] or the space bar, or right-click to accept the default scale value. Specify a value different from the X scale factor, or press [Enter] or the space bar, or right-click to accept the same scale specified for the X axis.

The X and Y scale factors allow you to stretch or compress the block to create modified versions of the block. See **Figure 25-6.** This is why it is a good idea to draw blocks to fit inside a one-unit square when appropriate. It makes the block easy to scale because you can enter the exact number of units for the X and Y dimensions. For example, if you want the block to be three units long and two units high, enter 3 when prompted to enter the X scale factor, and enter 2 when prompted to enter the Y scale factor.

The insertion base point specified when the block was created may not always be the best point when you actually insert the block. Instead of inserting and then moving the block, you can use the **Basepoint** option to specify a different base point before locating the block. Select the **Basepoint** option when prompted to specify the insertion point. The block temporarily appears on-screen, allowing you to choose an alternate insertion base point. The block reattaches to the crosshairs at the new point and a message appears indicating that the tool is resuming, allowing you to pick the insertion point in the drawing.

Figure 25-6.
A comparison of different X and Y scale factors used for inserting a 2D unit block.

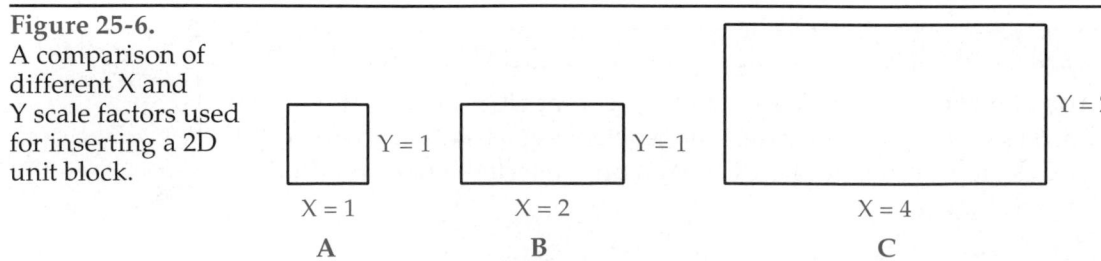

Exercise 25-2

Access the Student Web site (www.g-wlearning.com/CAD) and complete Exercise 25-2.

Inserting Multiple Arranged Copies of a Block

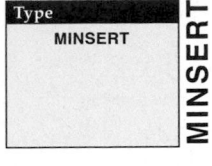

Type
MINSERT

MINSERT

The **MINSERT** tool combines the functions of the **INSERT** and **ARRAY** tools. **Figure 25-7** shows an example of an **MINSERT** tool application. To follow this example, set architectural units, draw a 4′ × 3′ rectangle, and save the rectangle as a block named DESK. Then access the **MINSERT** tool and enter DESK. Pick a point as the insertion point and then accept the X scale factor of 1, the Y scale factor of use X scale factor, and the rotation angle of 0. The arrangement is to be three rows and four columns. In order to make the horizontal spacing between desks 2′ and the vertical spacing 4′, you must consider the size of the desk when entering the distance between rows and columns. Enter 7′ (3′ desk depth + 4′ space between desks) at the Enter distance between rows of specify unit cell: prompt. Enter 6′ (4′ desk width × 2′ space between desks) at the Specify distance between columns: prompt.

The complete pattern takes on the characteristics of a block, except that you cannot explode the pattern. Therefore, you must use the **Properties** palette to modify the number of rows and columns, change the spacing between objects, or change other properties. If you rotate the initial block, all objects in the pattern rotate about their insertion points. If you rotate the patterned objects about the insertion point while using the **MINSERT** tool, all objects align on that point.

PROFESSIONAL TIP

As an alternative to the previous example, if you were working with different desk sizes, a 2D unit block may serve your purposes better than an exact size block. To create a 5′ × 3′-6″ (60″ × 42″) desk, for example, insert a one-unit-square block using either the **INSERT** or **MINSERT** tool, and enter 60 for the X scale factor and 42 for the Y scale factor.

Figure 25-7.
Creating an arrangement of desks using the **MINSERT** tool.

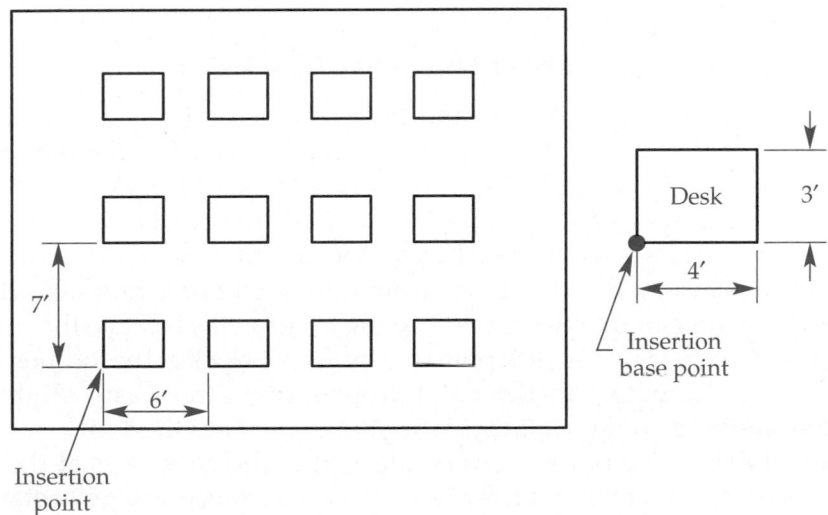

Exercise 25-3

Access the Student Web site (www.g-wlearning.com/CAD) and complete Exercise 25-3.

Inserting Entire Drawings

The **INSERT** tool also allows you to insert an entire drawing into the current drawing as a block. Access the **INSERT** tool and pick the **Browse...** button in the **Insert** dialog box. Use the **Select Drawing File** dialog box to select a drawing or DXF file to insert.

When you insert one drawing into another, the inserted drawing becomes a block reference and functions as a single object. The drawing is inserted on the current layer, but it does not inherit the color, linetype, or thickness properties of that layer. You can explode the inserted drawing back to its original objects if desired. Once exploded, the drawing objects revert to their original layers. An inserted drawing brings any existing block definitions and other drawing content, such as layers and dimension styles, into the current drawing.

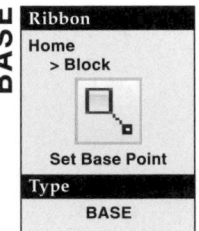

By default, every drawing has an insertion base point of 0,0,0 when you insert the drawing into another drawing. To change the insertion base point of the drawing, access the **BASE** tool and select a new insertion base point. Save the drawing before inserting it into another drawing.

When inserting a drawing as a block, you have the option of using the existing drawing to create a block with a different name. For example, to define a block named BOLT from an existing drawing named Fastener.dwg, access the **INSERT** tool and use the **Browse...** button to select the Fastener.dwg file. Use the **Name:** text box to change the name from Fastener to BOLT, and pick the **OK** button. You can then insert the file into the drawing or press [Esc] to exit the tool. A BOLT block definition is now available to use as desired.

Exercise 25-4

Access the Student Web site (www.g-wlearning.com/CAD) and complete Exercise 25-4.

Using DesignCenter to Insert Blocks

DesignCenter provides an effective way to insert blocks or entire drawings as blocks in the current drawing. To insert a block, locate the file containing the block to insert and select the **Blocks** branch in the tree view, or double-click on the **Blocks** icon in the content area. See **Figure 25-8.** Usually the quickest and most effective technique for transferring a block from **DesignCenter** into the active drawing is to use a drag-and-drop operation. Pick the block from the content area and hold down the pick button. Move the cursor into the active drawing to attach the block to the cursor at the block insertion point. Release the pick button to insert the block at the location of the cursor.

An alternative to drag-and-drop is copy and paste. Right-click on a block in **DesignCenter** and pick **Copy**. Move the cursor into the active drawing, right-click, and select **Paste**. The block appears attached to the crosshairs at the insertion base point. Specify a point to insert the block. You can also use **DesignCenter** in combination with the **Insert** dialog box. Right-click on a block in **DesignCenter** and select **Insert Block...** to access the **Insert** dialog box with the selected block active. This allows you to scale, rotate, or explode the block during insertion.

Figure 25-8.
Use the **DesignCenter** to insert blocks from files or drawings from folders. Several example blocks are available from the AutoCAD Sample folder, as shown.

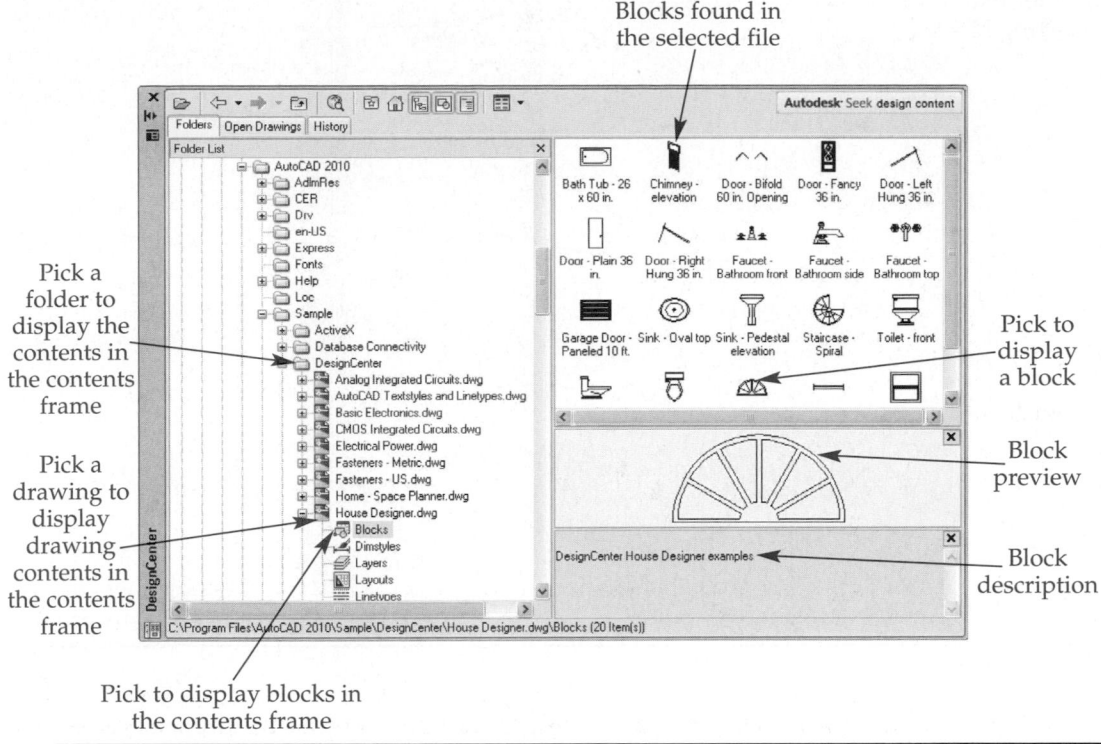

Blocks found in the selected file

Pick a folder to display the contents in the contents frame

Pick a drawing to display drawing contents in the contents frame

Pick to display a block

Block preview

Block description

Pick to display blocks in the contents frame

To insert a drawing or DXF file using **DesignCenter**, select the folder in the tree view that contains the file to display the contents in the content area. Drag and drop or copy and paste the file into the current drawing. You can also right-click a file icon in the content area and select **Insert as Block…**.

NOTE

Blocks are inserted from **DesignCenter** based on the type of block units you specified when you created the block. For example, if the original block was a 1 × 1 square and you specified the block units as feet when you created the block, then the block will insert as a 12″ × 12″ square.

Using Tool Palettes to Insert Blocks

The **Tool Palettes** palette, shown in **Figure 25-9**, provides another means of storing and inserting blocks. Blocks located in a tool palette are known as *block insertion tools.* Tool palettes can also store and activate many other types of drawing content and tools. For more information on AutoCAD customization and using tool palettes, refer to *AutoCAD and Its Applications—Advanced.*

block insertion tools: Blocks located on a tool palette.

To insert a block from the **Tool Palettes** palette, access the tool palette in which the block resides. Place the cursor over the block icon to display the name and description. Hold down the pick button on the block image. Move the cursor into the active drawing to attach the block to the cursor at the block insertion point. Release the pick button to insert the block at the location of the cursor.

An alternative to drag-and-drop is to pick once on the block image to attach the block to the crosshairs, and then pick a location for the block. This method offers an

Figure 25-9.
The **Tool Palettes** palette provides another means of inserting blocks.

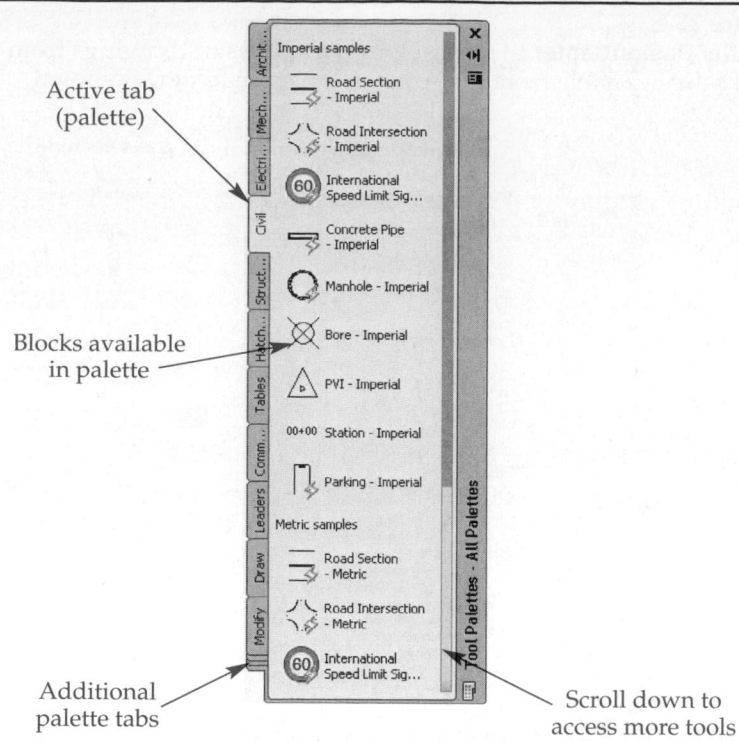

Active tab (palette)

Blocks available in palette

Additional palette tabs

Scroll down to access more tools

advantage over drag-and-drop by presenting options for adjusting the insertion base point, scale, and rotation. These options function the same as when you insert a block from the **Insert** dialog box.

Exercise 25-5

Access the Student Web site (www.g-wlearning.com/CAD) and complete Exercise 25-5.

Editing Blocks

The first form of block editing involves modifying a block that has already been inserted into a drawing using tools such as **MOVE**, **COPY**, **ROTATE**, or **MIRROR**. You can use grip editing by selecting the grip box that appears at the insertion base point of the block. You can also use the **Properties** palette and **Quick Properties** panel to make limited changes to inserted blocks. Remember, once you insert a block, it is treated as a single object.

The second type of block editing involves redefining the block by editing the block definition or changing the objects within the block. You can redefine a block using the **Block Editor** or by exploding and then recreating the block.

Changing Block Properties to ByLayer

If block component properties such as color and linetype were originally set to absolute values, and you want to change the properties to ByLayer, you can edit the block definition or use the **SETBYLAYER** tool to accomplish the same task without editing the block definition.

Access the **SETBYLAYER** tool and use the **Settings** option to display the **SetByLayer Settings** dialog box. Select the check boxes that correspond to the object properties to convert to ByLayer. Pick the **OK** button to exit the **SetByLayer Settings** dialog box.

Next, select the blocks with the properties to set to ByLayer and press [Enter] or the space bar, or right-click to display the Change ByBlock to ByLayer? prompt. Select the **Yes** option to change all object properties currently set to ByBlock to ByLayer. Pick the **No** option to change all object properties set to values other than ByBlock to ByLayer. The next prompt asks to include blocks in the conversion. If the selected object is a block, choose **Yes** to convert the properties of all references of the same block in the drawing to ByLayer. If you pick **No**, only the properties of the selected block are converted. All other references of the same block remain unchanged.

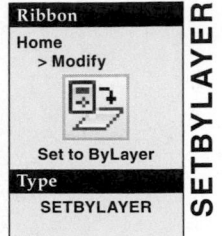

PROFESSIONAL TIP

To change the properties of several blocks, you can use the **Quick Select** tool to create a selection set of block reference objects.

Using the Block Editor

The **BEDIT** tool allows you to edit a block using the **Block Editor**. Access the **BEDIT** tool to display the **Edit Block Definition** dialog box, shown in **Figure 25-10.** To edit an existing block, select the name of the block from the list box. Pick the <Current Drawing> option to edit a block saved as the current drawing, such as a wblock. A preview and the description of the selected block appear. You can create a new block by typing a unique name in the **Block to create or edit** field. Pick the **OK** button to open the selected block in the **Block Editor,** as shown in **Figure 25-11.** If you typed a new block name, the drawing area is empty, allowing you to create a new block.

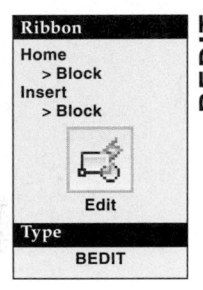

Modifying a Block

Use drawing and editing tools to modify or create the block. Specify the UCS origin, or 0,0,0 point, as the block insertion base point. The tools in the panels of the **Block Editor** ribbon tab are specific for modifying and creating block geometry. **Figure 25-12** describes some of the basic tools available in the **Block Editor** ribbon. The parametric tools allow you to constrain block geometry and form block tables. Many of the tools and options found on the **Block Editor** ribbon tab relate to dynamic blocks. This textbook explains dynamic blocks, block tables, and other block editing tools in later chapters.

Figure 25-10.
The **Edit Block Definition** dialog box.

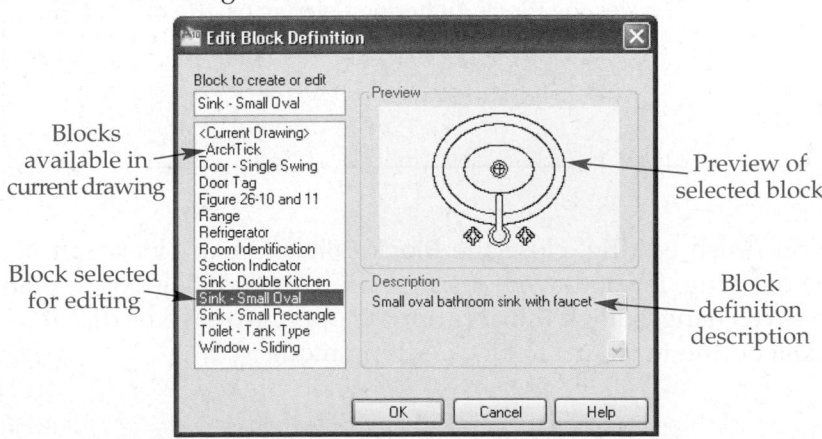

Figure 25-11.
The **Block Editor** ribbon tab and the **Block Authoring Palettes** palette are available in block editing mode. Only the block geometry appears in the **Block Editor**.

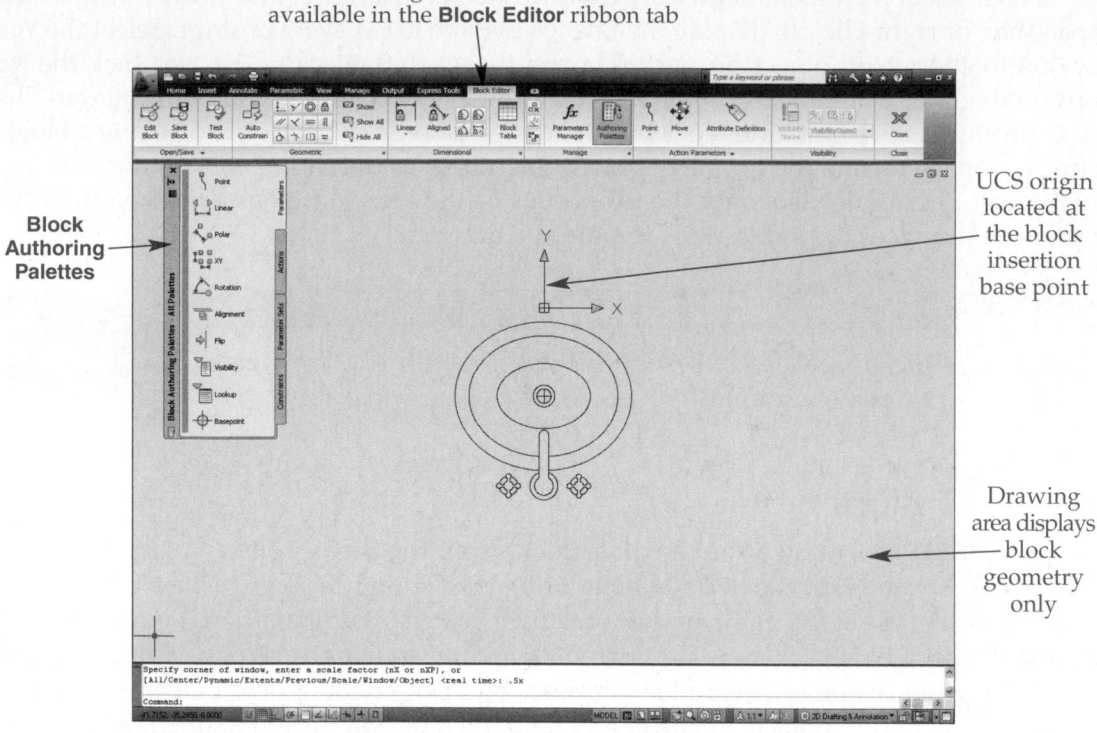

Block editing and construction tools available in the **Block Editor** ribbon tab

Block Authoring Palettes

UCS origin located at the block insertion base point

Drawing area displays block geometry only

Figure 25-12.
The **Block Editor** ribbon tab contains several tools and options specifically for editing and constructing blocks. This table describes the most basic functions.

Button	Description
	Saves changes to the block and updates the block definition.
	Opens the **Save Block As** dialog box, allowing you to save the block as a new block, using a different name.
	Opens the **Edit Block Definition** dialog box, which is the same dialog box displayed when you enter block editing mode. You can select a different block to edit or specify the name of a new block to create from scratch.
	Toggles the **Block Authoring Palettes** palette off and on.
	Closes the **Block Editor**.

When you finish editing, close the **Block Editor** to exit block editing mode and return to the drawing. If you have not saved changes, a dialog box appears asking if you want to save changes. Pick the appropriate option to save or discard changes, or pick the **Cancel** button to return to block editing mode.

> **NOTE**
>
> Double-click a block to display the **Edit Block Definition** dialog box with the block selected. You can open a block directly in the **Block Editor** by selecting a block and then right-clicking and choosing **Block Editor**. Another option is to open a block directly in the **Block Editor** when you create the block by selecting the **Open in block editor** check box in the **Block Definition** dialog box.

Adding a Block Description

To change the description assigned to a block when it was originally created, open the block in the **Block Editor** and then display the **Properties** palette with no objects selected. Make changes to the description using the **Description** property in the **Block** category. Pick the **Save Block Definition** button and the **Close Block Editor** button to return to the drawing.

> **NOTE**
>
> Blocks can also be edited "in-place" using the **REFEDIT** tool. Chapter 32 describes in-place editing using the **REFEDIT** tool, as it applies to external references. You can use the same techniques to edit blocks.

Exercise 25-6

Access the Student Web site (www.g-wlearning.com/CAD) and complete Exercise 25-6.

Exploding and Redefining a Block

You have the option of exploding a block during insertion using the **Insert** dialog box. This is useful when you want to edit the individual objects of the block. You can also use the **EXPLODE** tool after inserting the block to break it into the original objects. Access the **EXPLODE** tool, select the objects to explode and press [Enter] or the space bar or right-click to complete the operation.

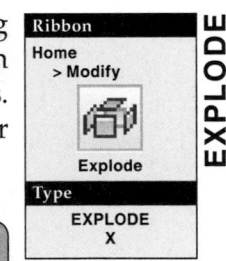

> **NOTE**
>
> You cannot explode a block that was created with **Allow exploding** unchecked in the **Block Definition** dialog box.

Follow this procedure to redefine an existing block using the **EXPLODE** and **BLOCK** tools:
1. Insert the block to redefine.
2. Make sure you know the exact location of the insertion point, because the point is lost during explosion.
3. Use the **EXPLODE** tool to explode the block.
4. Edit the components of the block as needed.
5. Recreate the block definition using the **BLOCK** tool.

6. Assign the block the same original name and, if appropriate, the same insertion point.
7. Select the objects to include in the block.
8. Pick the **OK** button in the **Block Definition** dialog box to save the block. When a message appears asking if you want to redefine the block, pick **Yes**.

A common mistake is to forget to use the **EXPLODE** tool before redefining the block. When you try to create the block again with the same name, an alert box indicates that the block references itself. This means you are trying to create a block that already exists. When you pick the **OK** button, the alert box disappears and the **Block Definition** dialog box redisplays. Press the **Cancel** button, explode the block, and try again to redefine the block.

> ### NOTE
> Once a block is modified, whether from changes made using the **BEDIT** tool or from redefinition using the **EXPLODE** and **BLOCK** tools, all instances of that block in the drawing update according to the changes.

Exercise 25-7

Access the Student Web site (www.g-wlearning.com/CAD) and complete Exercise 25-7.

Understanding the Circular Reference Error

When you try to redefine a block that already exists using the same name, a *circular reference error* occurs. AutoCAD informs you that the block references itself or that it has not been modified. The concept of a block referencing itself may be confusing unless you fully understand how AutoCAD works with blocks. A block can be composed of many objects, including other blocks. When you use the **BLOCK** tool to incorporate an existing block into a new block, AutoCAD makes a list of all the objects that compose the new block. This means AutoCAD refers to any existing block definitions added to the new block. A problem occurs if you select an instance, or reference, of the redefined block as a component object for the new definition. The new block refers to a block of the same name, or references itself. **Figure 25-13A** illustrates the process of correctly redefining a block named BOX to avoid a circular reference error. **Figure 25-13B** shows an incorrect redefinition resulting in a circular reference error.

circular reference error: An error that occurs when a block definition references itself.

Renaming Blocks

Use the **RENAME** tool to rename a block without editing the block definition. Access the **RENAME** tool to display the **Rename** dialog box shown in **Figure 25-14**. Select Blocks from the **Named Objects** list, and then pick the block to rename in the **Items** list. The current name appears in the **Old Name:** text box. Type the new block name in the **Rename To:** text box. Pick the **Rename To:** button to display the new name in the **Items** list. Pick the **OK** button to exit the **Rename** dialog box.

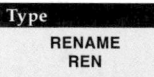

Type
RENAME
REN

Figure 25-13.
A—The correct procedure for redefining a block. B—Redefining a block without first exploding the block creates an invalid circular reference.

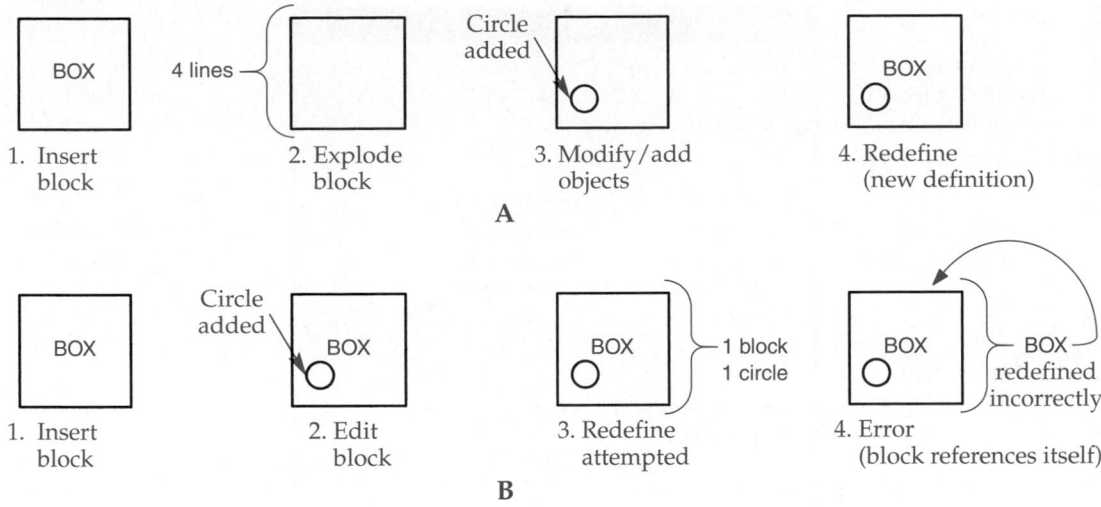

A

B

Figure 25-14.
The **Rename** dialog box allows you to change the name of blocks and other named objects.

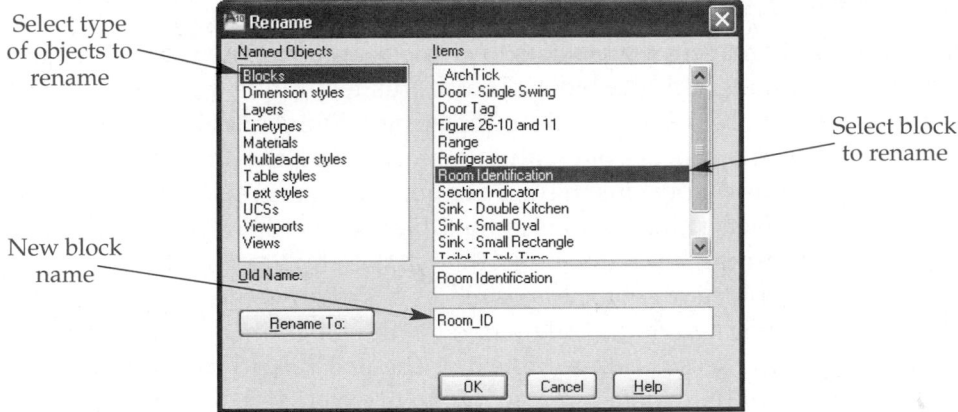

Updating Block Icons

A block icon forms when you build a block. The icon appears when you insert and edit blocks to help you recognize the block. Block icons require updating when an icon does not appear, as is often the case when you store a block in a drawing created with an older version of AutoCAD, or if the icon does not reflect changes made to the block. Use the **BLOCKICON** tool to create or update a block icon. Open the drawing in which the block is stored, access the **BLOCKICON** tool, enter the name of the block, and press [Enter].

Creating Blocks as Drawing Files

Blocks created with the **BLOCK** tool are stored in the drawing in which they are defined. A write block, created using the **WBLOCK** tool, saves the block as a separate drawing (DWG) file. You can also use the **WBLOCK** tool to create a global block from any object. It does not have to be a previously saved block. You can insert the resulting drawing file as a block into any drawing. Access the **WBLOCK** tool to display the **Write Block** dialog box shown in **Figure 25-15**.

Type
WBLOCK

Figure 25-15.
Using the **Write Block** dialog box to create a wblock from selected objects without first defining a block.

Pick to save selected objects as a wblock

Pick to select the insertion point

Pick to select the objects defining the wblock

File location and name

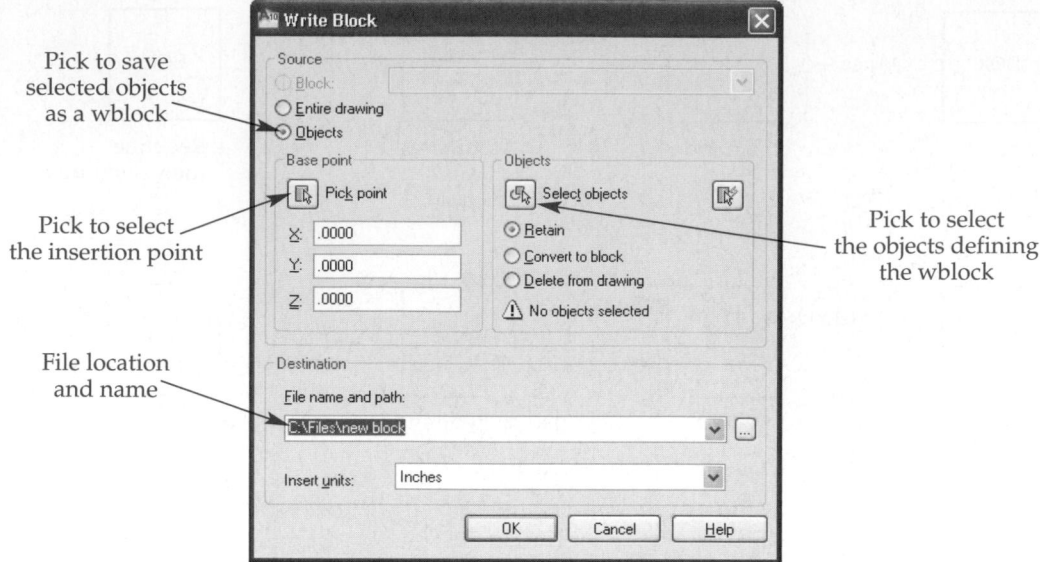

Creating a New Wblock

One method of creating a wblock is to create a drawing file from existing objects that you have not converted to a block. To use this technique, pick the **Objects** radio button in the **Source** area. The process of creating a wblock from existing non-block objects is similar to the process of creating a block using the **BLOCK** tool. Specify an insertion base point using options in the **Base point** area, and select the objects and the disposition of the objects using options in the **Objects** area. The **Base point** and **Objects** areas function the same as those found in the **Block Definition** dialog box.

In contrast to a block, a wblock is saved as a drawing file, *not* as a block in the current drawing. Enter a path and file name for the block in the **File name and path:** text box or pick the ellipsis (...) button next to the text box to display the **Browse for Drawing File** dialog box. Navigate to the folder where you want to save the file, confirm the name of the file in the **File name:** text box, and pick the **Save** button. The **Write Block** dialog box redisplays with the path and file name shown in the **File name and path:** text box. Finally, select the type of units that **DesignCenter** should use to insert the block in the **Insert units:** drop-down list. This is also located in the **Destination** area. Pick the **OK** button to finish. The objects are saved as a wblock in the specified folder. Now you can use the **INSERT** tool in any drawing to insert the block.

Saving an Existing Block As a Wblock

To create a wblock from an existing block, pick the **Block** radio button in the **Source** area. See **Figure 25-16.** Select the block to save as a wblock from the drop-down list. Use the options in the **Destination** area to locate the wblock, and pick the **OK** button to finish.

Storing a Drawing As a Wblock

To store an entire drawing as a wblock, pick the **Entire drawing** radio button in the **Source** area. Use the options in the **Destination** area to locate the wblock. In this case, the whole drawing is saved as if you were using the **SAVE** tool. However, all uninserted, or unused, blocks in the drawing are deleted. If the drawing contains any unused blocks, this method may reduce the size of a drawing considerably. Pick the **OK** button to finish.

Figure 25-16.
Using the **Write Block** dialog box to create a wblock from an existing block.

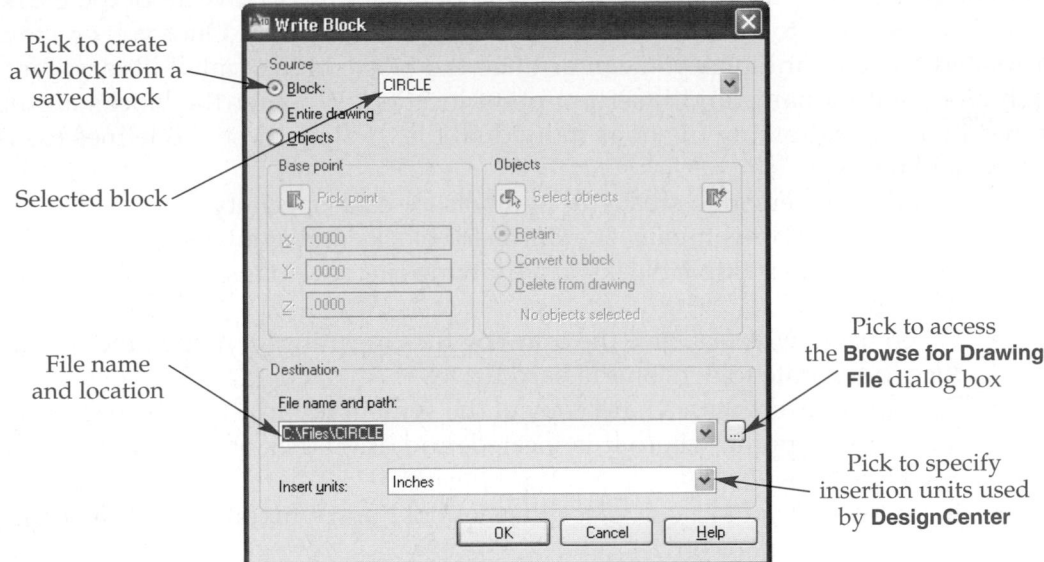

Pick to create a wblock from a saved block

Selected block

File name and location

Pick to access the **Browse for Drawing File** dialog box

Pick to specify insertion units used by **DesignCenter**

Exercise 25-8

Access the Student Web site (www.g-wlearning.com/CAD) and complete Exercise 25-8.

Revising an Inserted Drawing

If you insert a wblock into multiple drawings and then need to make changes to the wblock, use the **INSERT** tool to access the original drawing file with the **Select Drawing File** dialog box. Then activate the **Specify On-screen** check box in the **Insertion point** area and pick the **OK** button. When a message asks if you want to redefine the block, pick the **Yes** option. All of the wblock references automatically update. Press [Esc] to cancel the tool so that you do not insert a new block.

PROFESSIONAL TIP

If you work on projects in which inserted drawings require revisions, it is far more productive to use reference drawings instead of inserted drawing files. Chapter 32 explains reference drawings placed using the **XREF** tool. All referenced drawings automatically update when you open a drawing file that contains the externally referenced material.

Symbol Libraries

As you become proficient with AutoCAD, you may want to start constructing *symbol libraries*. Arranging a storage system for frequently used symbols significantly increases productivity. Establish how to store the symbols, as either blocks or drawing files, and identify a storage location and system.

symbol library: A collection of related blocks, shapes, views, symbols, or other content.

Creating a Symbol Library

The two basic options for creating a symbol library are to save all of the blocks in a single drawing or to save each block to a separate wblock file. Once you create a set of related block definitions, you can arrange the blocks in a symbol library. Identify each block with a name and insertion point location. Whether the blocks are being stored in a single drawing file or as individual files, follow these guidelines to create the symbol library:

- Assign one person to create the symbols for each specialty.
- Follow school or company standards for blocks and symbols.
- When saving multiple blocks in a drawing file, save one group of symbols per drawing file.
- When using wblocks, give the drawing files meaningful names and assign the files to separate folders on the hard drive.
- Provide all users with a hard copy of the symbol library showing each symbol, its insertion point, where it is located, and any other necessary information. See **Figure 25-17**.
- If a network is not in use, place the symbol library file(s) on each workstation in the classroom or office.
- Keep backup copies of all files in a secure place.
- When you revise symbols, update all files containing the edited symbols.
- Inform all users of any changes to saved symbols.

Storing Symbol Drawings

The local or network hard drive is one of the best places to store a symbol library. It is easy to access, quick, and more convenient to use than portable media. Removable media, such as a removable hard drive, USB flash drive, or a CD, are most appropriate

Figure 25-17.
A printed copy of electrical blocks stored in a symbol library. The "X" symbols indicate insertion points and are not part of the blocks.

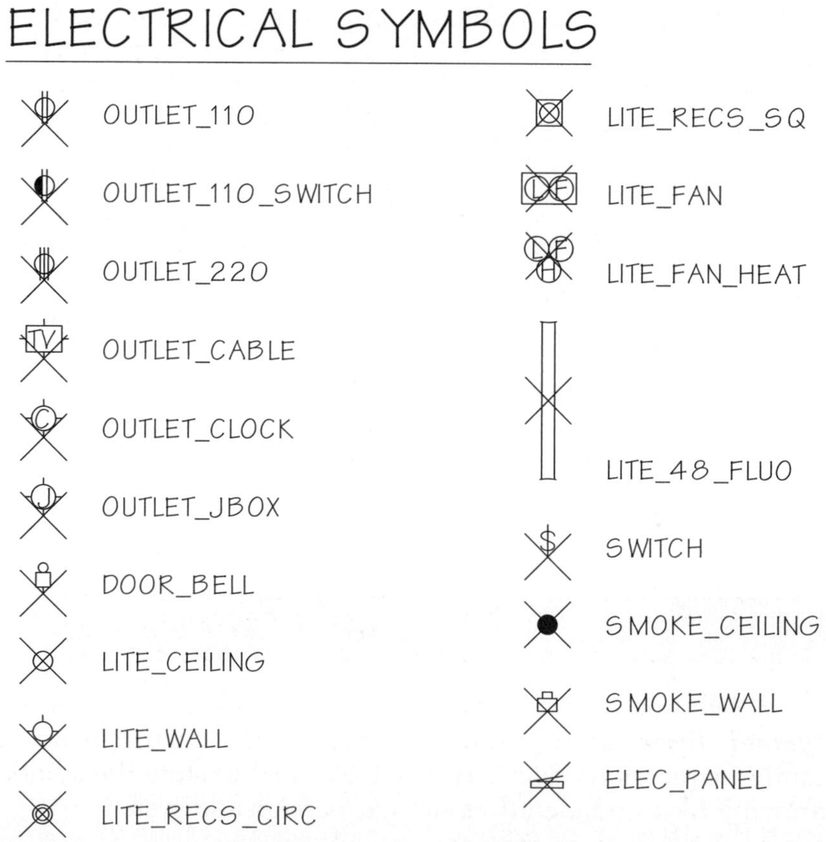

ELECTRICAL SYMBOLS

OUTLET_110
OUTLET_110_SWITCH
OUTLET_220
OUTLET_CABLE
OUTLET_CLOCK
OUTLET_JBOX
DOOR_BELL
LITE_CEILING
LITE_WALL
LITE_RECS_CIRC

LITE_RECS_SQ
LITE_FAN
LITE_FAN_HEAT
LITE_48_FLUO
SWITCH
SMOKE_CEILING
SMOKE_WALL
ELEC_PANEL

for backup purposes if a network drive with an automatic backup function is not available. In the absence of a network or modem, you can also use removable media to transport files from one workstation to another.

There are several methods of storing symbols on the hard drive. One option is to save symbols as wblocks in organized folders. You should store content outside of the AutoCAD folder to keep the system folder uncluttered and to allow you to differentiate your folders and files from AutoCAD system folders and files. A good method is to create a \Blocks folder for storing your blocks, as shown in **Figure 25-18**.

If you save multiple symbols within a single drawing, use **DesignCenter** or the **Tool Palettes** palette to insert the symbols as needed. When you use this system, it is often a good idea to use several drawing files to group similar symbols. For example, you may want to create different symbol libraries based on electronic, electrical, piping, mechanical, structural, architectural, landscaping, and mapping symbols. Limit the symbols in a drawing to a reasonable number so you can easily find the symbols. If there are too many blocks in a drawing, it may be difficult to locate the desired symbol.

You should arrange drawing files saved on the hard drive in a logical manner. All workstations in a non-networked classroom or office should have folders with the same names. Assign one person to update and copy symbol libraries to all workstations. Copy drawing files onto each workstation from a master CD. Keep the master and backup versions of the symbol libraries in separate locations.

Figure 25-18.
An efficient way to store blocks saved as drawing files is to set up a Blocks folder containing folders for each type of block on the hard drive.

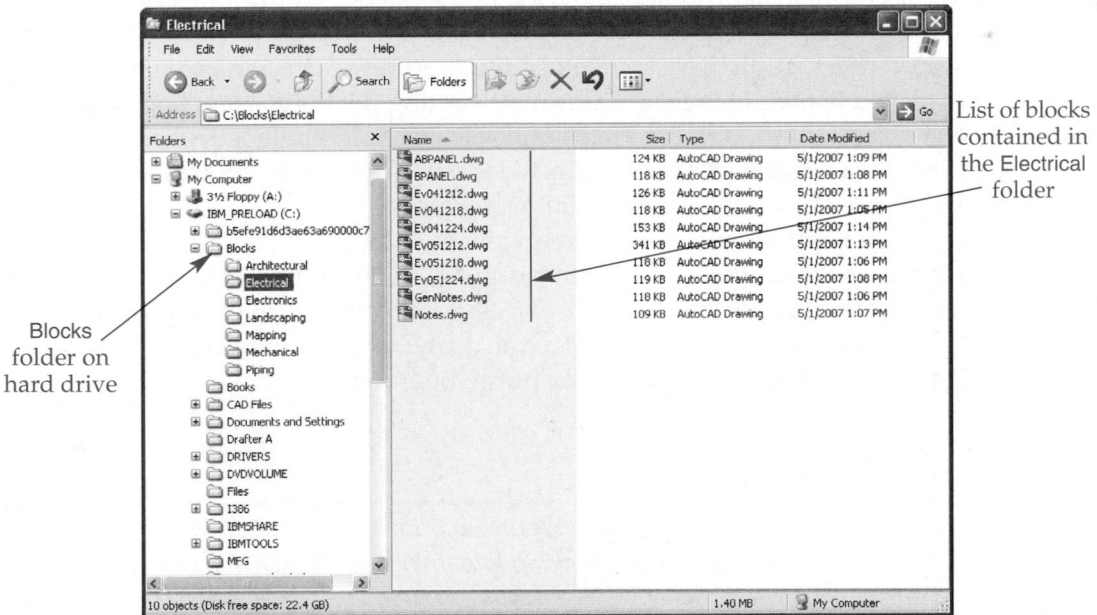

Purging Named Objects

named objects: Blocks, dimension styles, layers, linetypes, materials, multileader styles, plot styles, shapes, table styles, text styles, and visual styles that have specific names.

purge: Delete unused named objects from a drawing file.

A drawing often accumulates several *named objects* that are unused and may be unnecessary. Unused named objects increase the drawing file size and may make it difficult to locate and use items that are often required or referenced in the drawing. As a result, you may want to use the **PURGE** tool to delete or *purge* unused objects from the drawing. Access the **PURGE** tool to display the **Purge** dialog box, shown in **Figure 25-19**.

Type
PURGE
PU

Figure 25-19.
The **Purge** dialog box.

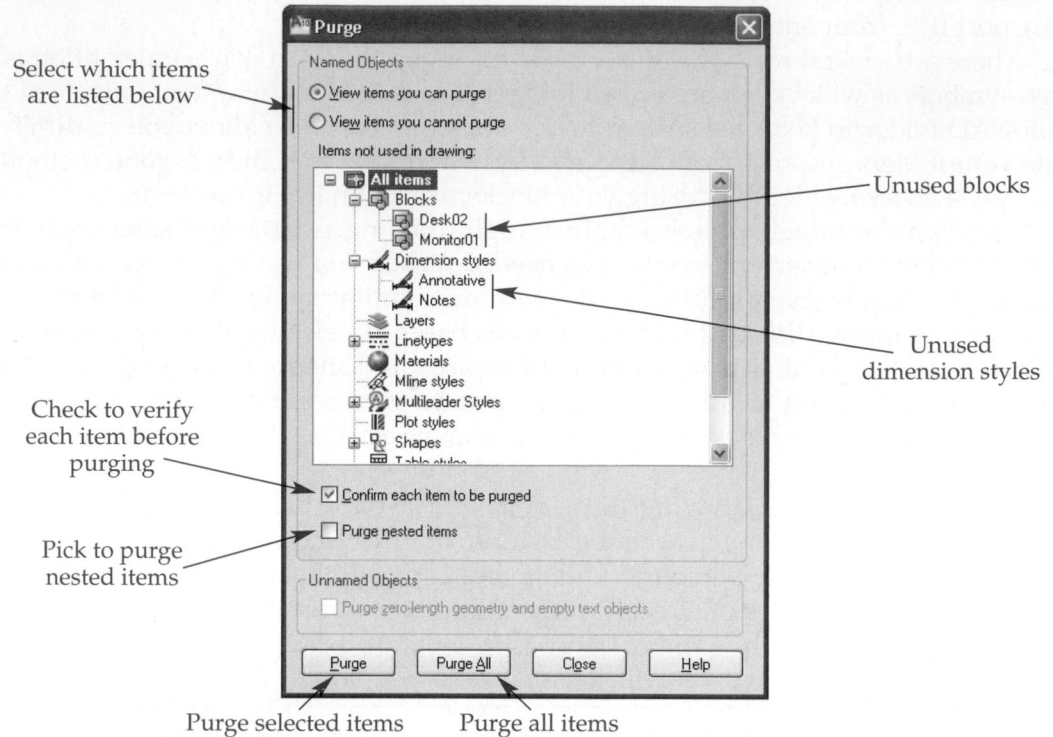

Select which items
are listed below

Unused blocks

Unused
dimension styles

Check to verify
each item before
purging

Pick to purge
nested items

Purge selected items   Purge all items

Select the appropriate radio button at the top of the dialog box to view content that you can purge or to view content that you cannot purge. Before purging, select the **Confirm each item to be purged** check box to have an opportunity to review each item before it is deleted. Check **Purge nested items** to purge nested items. Selecting the **Purge zero-length geometry and empty text objects** check box is an effective way to erase all zero-length objects, such as a line or arc drawn as a dot and text that only includes spaces. These objects are often mistakes or unintended results of the drawing and editing processes.

To purge only some items, use the tree view to locate and highlight the items to purge, and then pick the **Purge** button. To purge all unused items, pick the **Purge All** button. Purging may cause other named objects to become unreferenced. Thus, you may need to purge more than once to purge the drawing of all unused named objects. Messages appear to guide you through the purge operation.

Chapter Test

Answer the following questions. Write your answers on a separate sheet of paper or go to the Student Web site (www.g-wlearning.com/CAD) and complete the electronic chapter test.

1. Why would you draw blocks on layer 0?
2. What properties do blocks drawn on a layer other than layer 0 assume when they are inserted?
3. What is the maximum number of characters allowed in a block name?
4. Define the term *nesting* in relation to blocks.
5. What is a block reference?
6. How can you access a listing of all blocks in the current drawing?
7. Describe the effect of entering negative scale factors when inserting a block.
8. What type of block is a one-unit line object?
9. How do you preset block insertion variables using the **Insert** dialog box?
10. Name a limitation of an array pattern created with the **MINSERT** tool.
11. What is the purpose of the **BASE** tool?

12. Briefly explain how to insert a block into a drawing from **DesignCenter**.
13. What tool allows you to change a block's layer without editing the block definition?
14. Identify the tool that allows you to break an inserted block into its individual objects for editing purposes.
15. Suppose you have found that a block was incorrectly drawn. Unfortunately, you have already inserted the block 30 times. How can you edit all of the blocks quickly?
16. What is the primary difference between blocks created with the **BLOCK** and **WBLOCK** tools?
17. Explain the advantage of storing a drawing as a wblock if you anticipate the need to insert the drawing into other drawings.
18. Define *symbol library*.
19. Explain two ways to remove all unused blocks from a drawing.
20. What is the purpose of the **PURGE** tool?

Drawing Problems

Start AutoCAD if it is not already started. Start a new drawing using an appropriate template of your choice. The template should include layers, text styles, dimension styles, and multileader styles appropriate for drawing the given objects. Add layers, text styles, dimension styles, and multileader styles as needed. Draw all objects using appropriate layers, text styles, dimension styles, multileader styles, justification, and format. Follow the specific instructions for each problem. Use your own judgment and approximate dimensions when necessary.

▼ Basic

1. Open P14-17 and save as P25-1. The P25-1 file should be active. The sketch for this drawing is shown below. Erase all copies of the symbols, leaving the original objects intact. These include the steel column symbols and the bay and column line tags. Then do the following:
   A. Make blocks of the steel column symbol and the tag symbols.
   B. Use the **MINSERT** tool or the **ARRAY** tool to place the symbols in the drawing.
   C. Dimension the drawing as shown in the sketch.
   D. Resave the drawing.

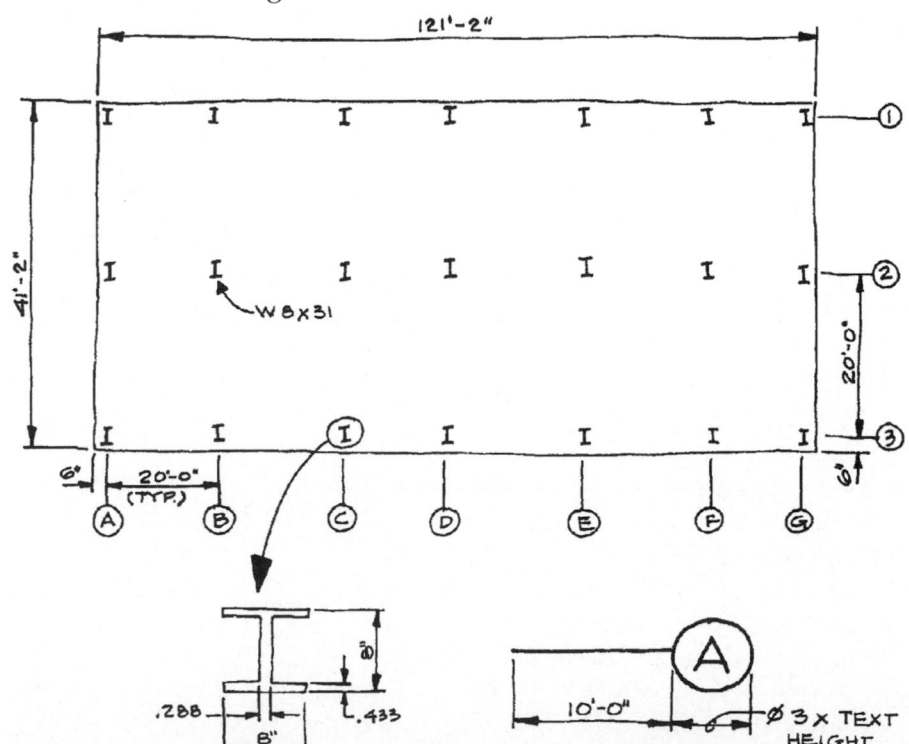

2. Open P14-18 and save as P25-2. The P25-2 file should be active. The sketch for this drawing is shown below. Erase all of the desk workstations except one. Then do the following:
   A. Create a block of the workstation.
   B. Insert the block into the drawing using the **MINSERT** tool.
   C. Dimension one of the workstations as shown in the sketch.
   D. Resave the drawing.

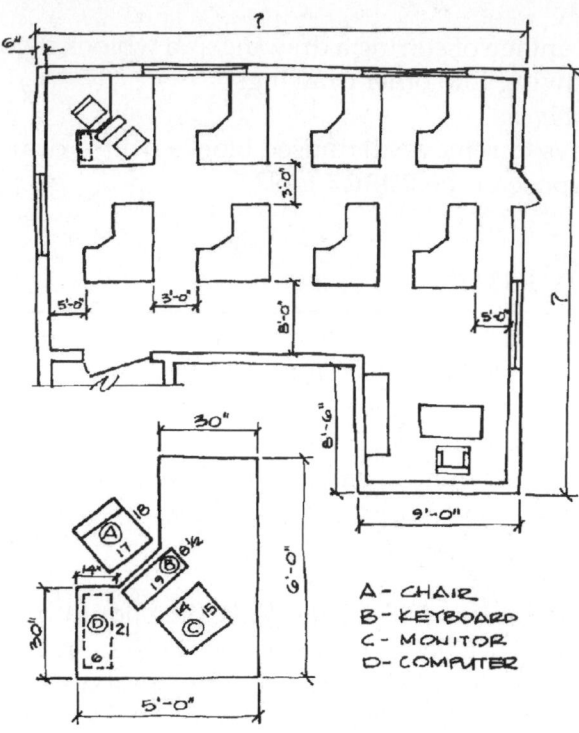

A - CHAIR
B - KEYBOARD
C - MONITOR
D - COMPUTER

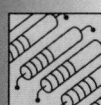

3. Complete this problem after completing Problem 25-7. Open P25-7 and save as P25-3. The P25-3 file should be active. Modify the NAND gates to become XNOR gates, as shown below, by modifying the block definition. Save the drawing as P25-3.

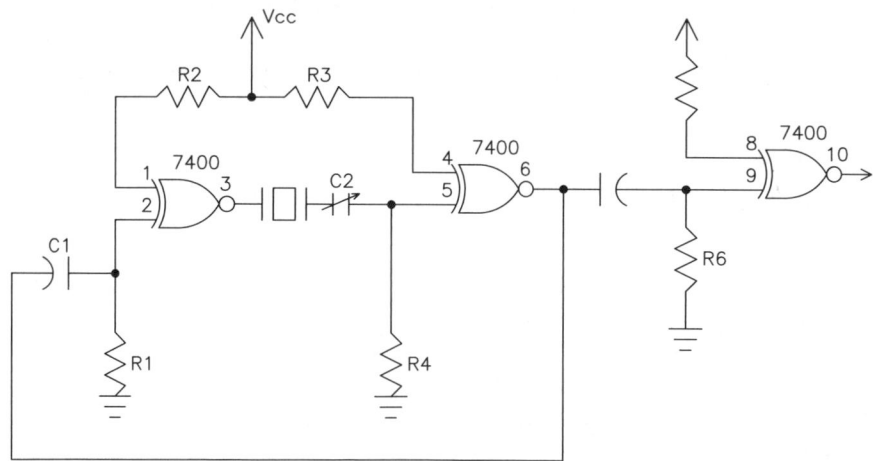

▼ Intermediate

Problems 4–7 represent a variety of diagrams created using symbols as blocks. Create each drawing as shown. (The drawings are not to scale.) Create the symbols first as blocks or wblocks and then save them in a symbol library using one of the methods described in this chapter.

4. Draw the integrated circuit schematic for a clock as shown. Save the drawing as P25-4.

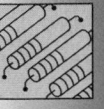

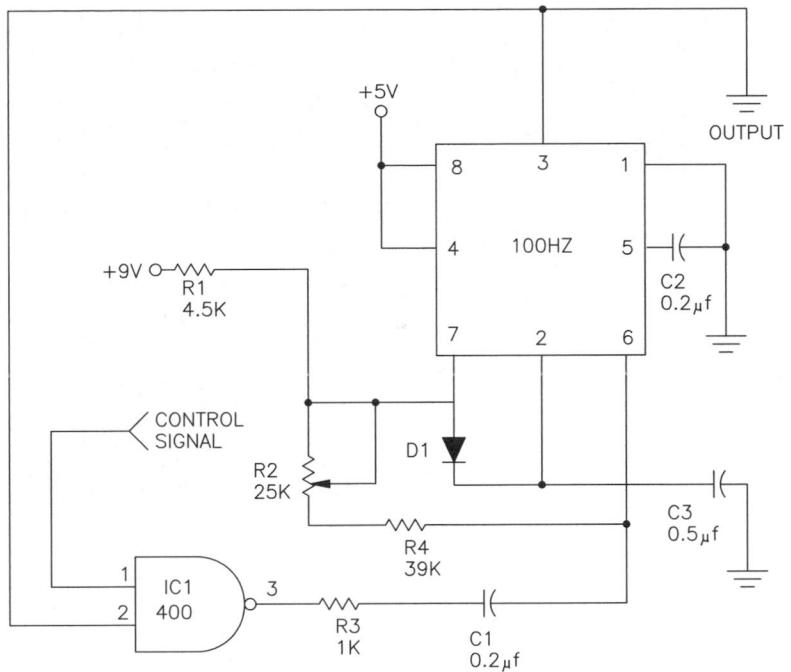

Integrated Circuit for Clock

5. Draw the piping flow diagram as shown. Save the drawing as P25-5.

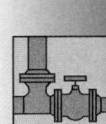

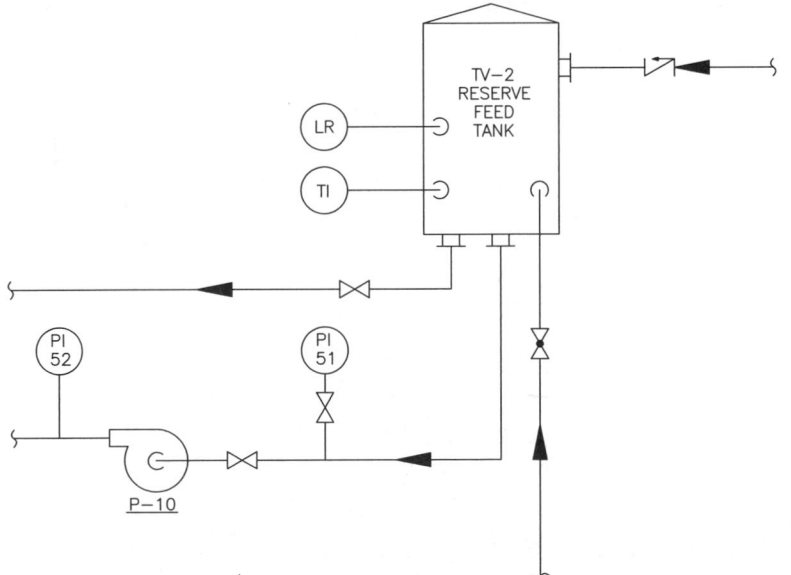

Piping Flow Diagram

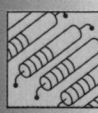

6. Draw the logic diagram of a marking system as shown. Save the drawing as P25-6.

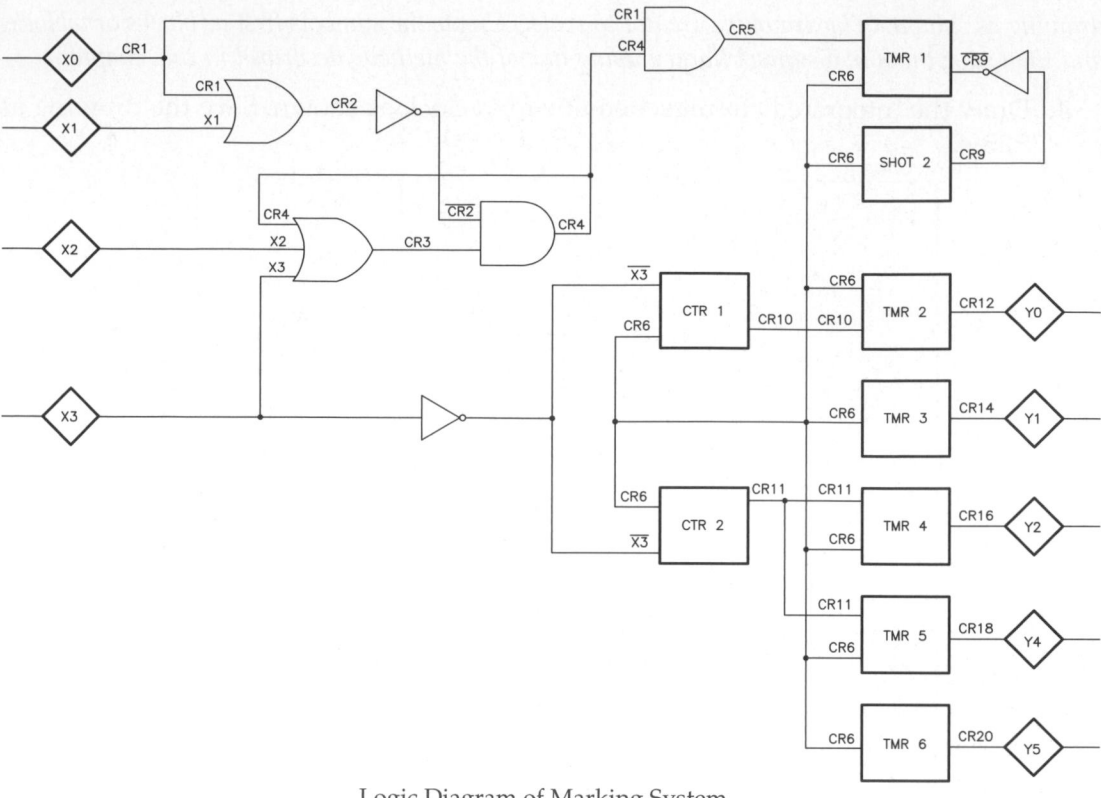

Logic Diagram of Marking System

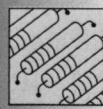

7. Draw the digital logic circuit shown. Create each type of component in the circuit as a block. Save the drawing as P25-7.

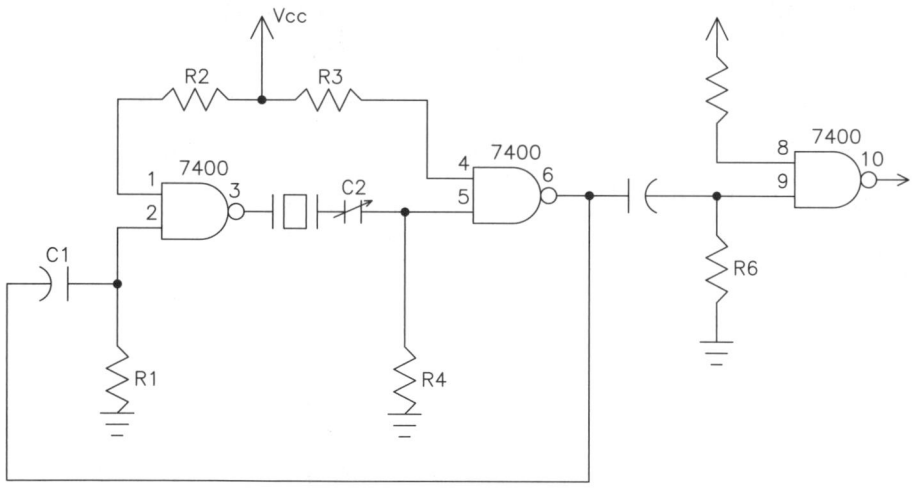

AutoCAD and Its Applications—Basics

▼ Advanced

Problems 8–12 are presented as engineering sketches. They are schematic drawings created using symbols and are not drawn to scale. Create the symbols first as blocks or wblocks and then save them in a symbol library using one of the methods described in this chapter.

8. The rough sketch shown below is a logic diagram of a portion of the internal components of a computer. Save the drawing as P25-8.

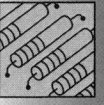

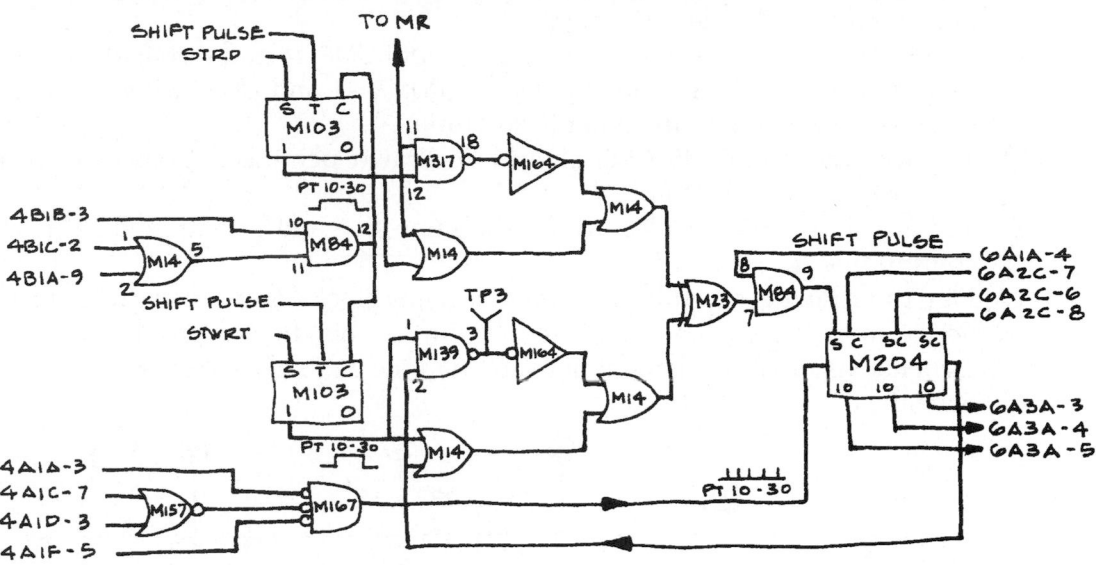

9. The rough sketch shown below is a piping flow diagram of a cooling water system. Look closely at this drawing before you begin. Draw the thick flow lines with polylines. Save the drawing as P25-9.

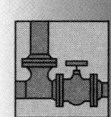

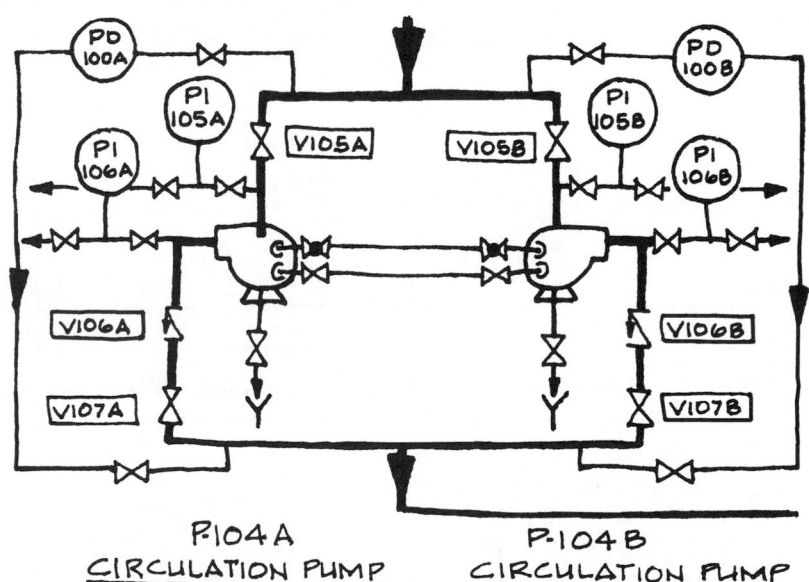

Chapter 25   Standard Blocks

Drawing Problems - Chapter 25

10. The rough sketch shown below is the general arrangement of a basement floor plan for a new building. The engineer has shown one example of each type of equipment. Use the following instructions to complete the drawing.

A. All text should be 1/8" high, except the text for the bay and column line tags, which should be 3/16" high. The diameter of the line balloons for the bay and column lines should be twice the diameter of the text height.

B. The column and bay steel symbols represent wide-flange structural shapes and should be 8" wide × 12" high.

C. The PUMP and CHILLER installations (except PUMP #4 and PUMP #5) should be drawn per the dimensions given for PUMP #1 and CHILLER #1. Use the dimensions shown for the other PUMP units.

D. TANK #2 and PUMP #5 (P-5) should be drawn per the dimensions given for TANK #1 and PUMP #4.

E. Tanks T-3, T-4, T-5, and T-6 are all the same size and are aligned 12' from column line A.

F. Plan this drawing carefully and create as many blocks as necessary to increase your productivity. Dimension the drawing as shown, and provide location dimensions for all equipment not shown in the engineer's sketch.

G. Save the drawing as P25-10.

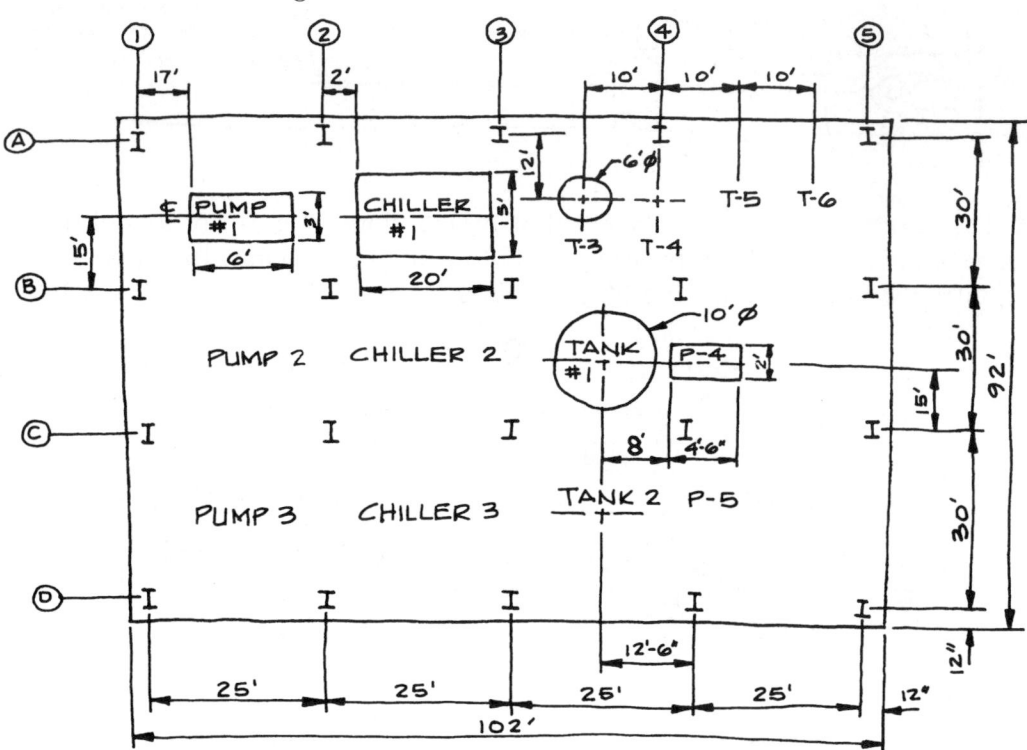

11. Open P25-10 and save as P25-11. The P25-11 file should be active. The engineer has provided you with a sketch of the necessary revisions to the drawing. It is up to you to alter the drawing as quickly and efficiently as possible. The dimensions shown on the sketch below *do not* need to be added to the drawing; they are provided for construction purposes only. Revise the drawing so all chillers and the four tanks reflect the changes. Save the drawing as P25-11.

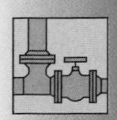

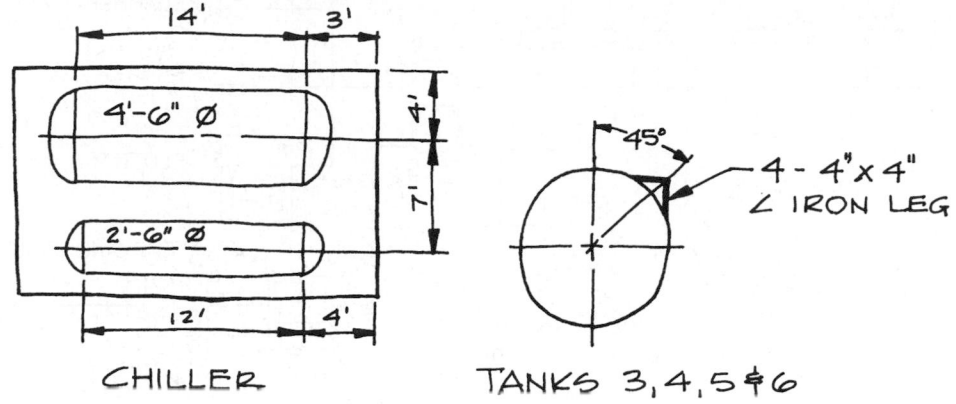

12. The rough sketch of a piping flow diagram shown below is part of an industrial effluent treatment system. Eliminate as many bends in the flow lines as possible. Place arrowheads at all flow line intersections and bends. The flow lines should not run through any valves or equipment. Use polylines for the thick flow lines. Save the drawing as P25-12.

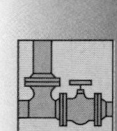

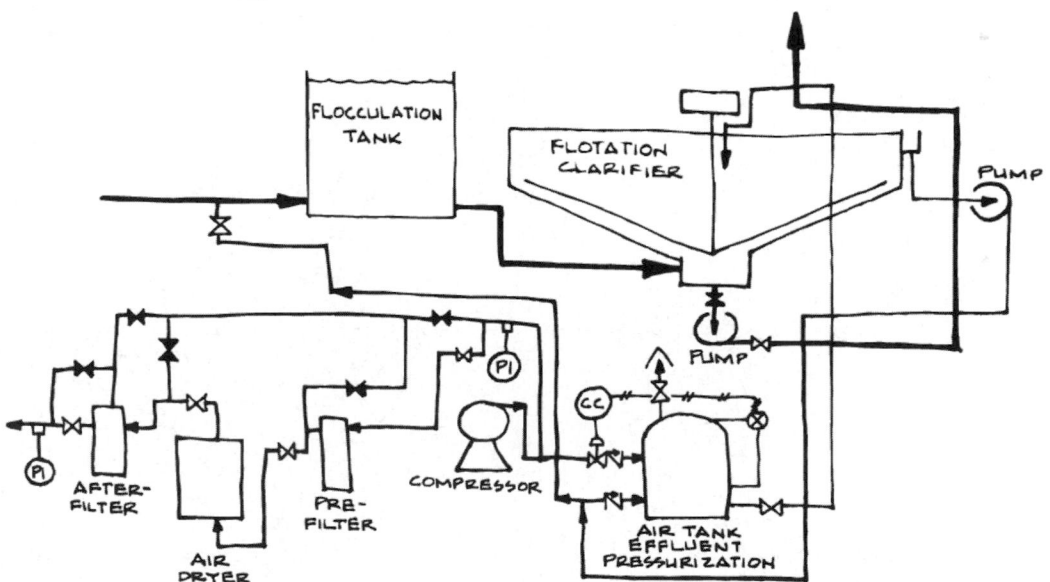

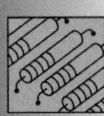

13. Create computer, plotter, and printer/copier blocks and then draw the network diagram. Save the drawing as P25-13.

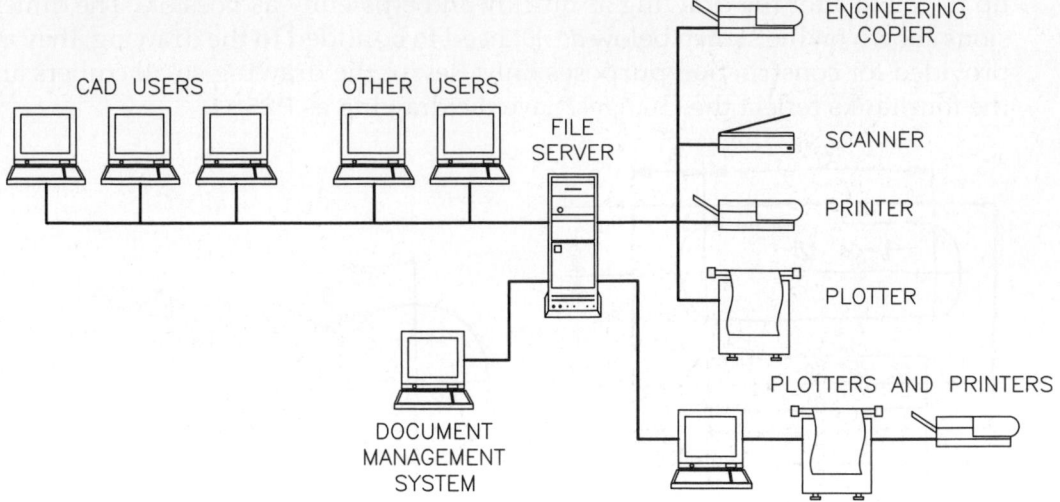

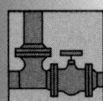

14. Draw the piping diagram shown, creating blocks for each type of fitting. Save the drawing as P25-14.

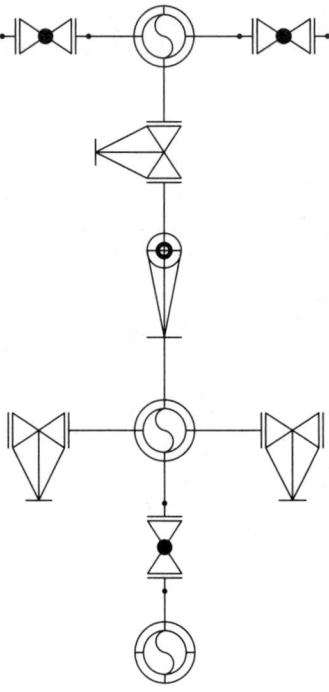

15. Create component blocks based on the dimensions shown. Then use the blocks to draw the schematic below. Save the drawing as P25-15.

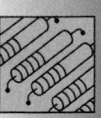

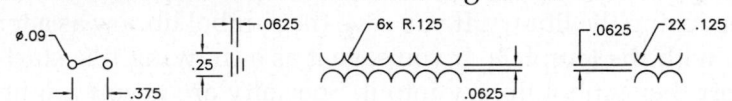

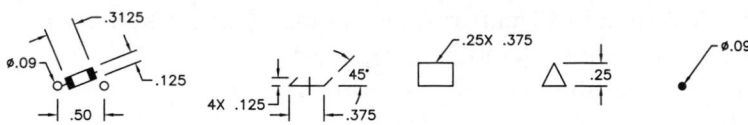

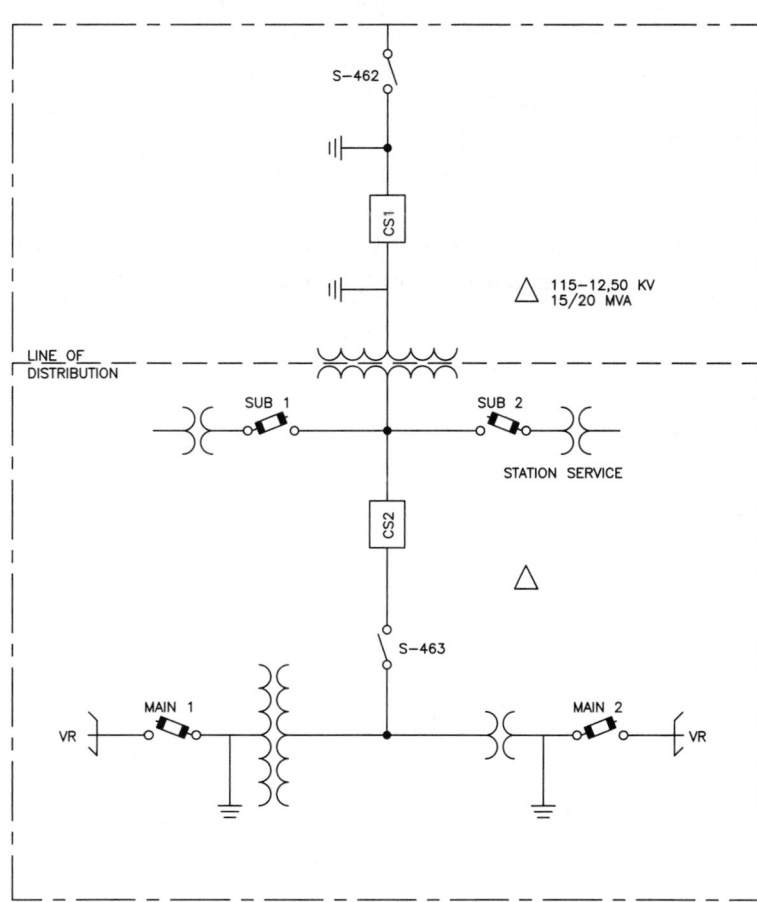

16. Create a symbol library for one of the drafting disciplines listed below and save it as a template or drawing file. Then, after checking with your instructor, draw a problem using the library. If you save the symbol library as a template, start the problem with the template. If you save it as a drawing file, start a new drawing and insert the symbol library into it. Specialty areas you might create symbols for include:

- Mechanical (machine features, fasteners, tolerance symbols)
- Architectural (doors, windows, fixtures)
- Structural (steel shapes, bolts, standard footings)
- Industrial piping (fittings, valves)
- Piping flow diagrams (tanks, valves, pumps)
- Electrical schematics (resistors, capacitors, switches)
- Electrical one-line (transformers, switches)
- Electronics (IC chips, test points, components)
- Logic diagrams (AND gates, NAND gates, buffers)
- Mapping, civil (survey markers, piping)
- Geometric tolerancing (feature control frames)

Save the drawing as P25-16 or choose an appropriate file name, such as ARCH-PRO or ELEC-PRO. Display the symbol library created in this problem and print a hard copy. Put the printed copy in your notebook as a reference.

Drawing Problems - Chapter 25

Mode

☑ Invisible

☐ Constant

☑ Verify

☐ Preset

☑ Lock position

Attribute

Prompt: Which manufacturer?

Default:

CHAPTER 26

Block Attributes

Learning Objectives

After completing this chapter, you will be able to do the following:
- ✓ Define attributes.
- ✓ Create and insert blocks that contain attributes.
- ✓ Edit attribute values and definitions in existing blocks.
- ✓ Create title blocks, revision blocks, and parts lists with attributes.
- ✓ Display attribute values in fields.

Attributes significantly enhance blocks that require text or numerical information. For example, a door identification block contains a letter or number that links the door to a door schedule. Adding an attribute to the door identification symbol allows you to include any letter or number with the symbol, without adding block definitions. You can also *extract* attribute data to automate drawing applications, such as preparing schedules, parts lists, and bills of materials.

attributes: Text-based data assigned to a specific object. Attributes turn a drawing into a graphical database.

extract: Gathering content from the drawing file database to display in the drawing or in an external document.

Defining Attributes

Attributes and geometry are often used together to create a block. See **Figure 26-1**. However, you can prepare stand-alone blocks that only include attributes. Create attributes along with other objects during the initial phase of block development. You can add as many attributes as needed to describe the symbol or product, such as the name, number, manufacturer, type, size, price, and weight of an item. To assign attributes, access the **ATTDEF** tool to display the **Attribute Definition** dialog box. See **Figure 26-2**.

Setting Attribute Modes

The **Mode** area allows you to set attribute modes. Symbols often require attributes to appear with the block. Select the **Invisible** check box to hide attributes, but still include attribute data in the drawing that you can reference and extract. The geranium symbol in **Figure 26-1** is an example of a block with attributes that you may want to hide. The other blocks show examples in which the attributes should appear. Blocks often include both visible and invisible attributes.

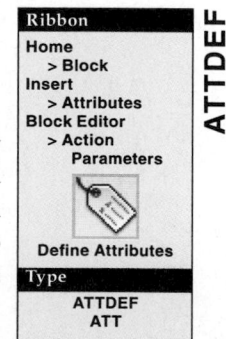

Ribbon
Home
> Block
Insert
> Attributes
Block Editor
> Action
   Parameters

Define Attributes
Type
ATTDEF
ATT

ATTDEF

Figure 26-1.
Examples of blocks with defined attributes.

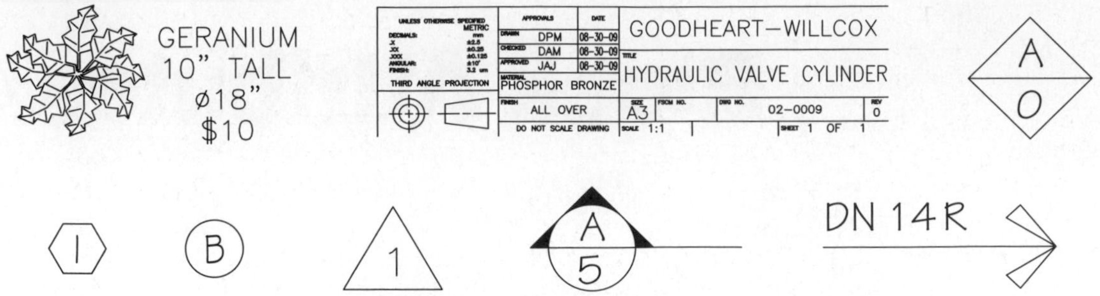

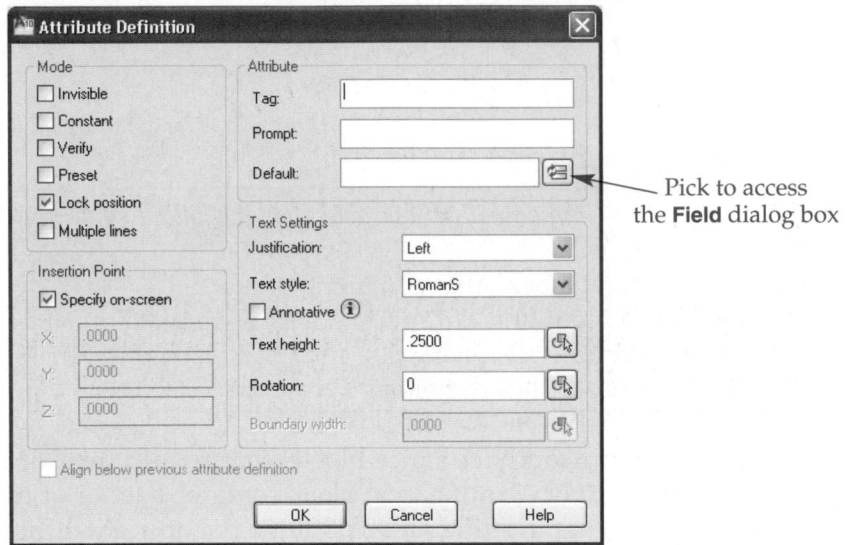

Figure 26-2.
Use the **Attribute Definition** dialog box to assign attributes to blocks.

Pick the **Constant** check box if the value of the attribute should always be the same. All insertions of the block display the same value for the attribute, without prompting for a new value when you insert the block. Deselect the **Constant** check box to use different attribute values for multiple insertions of the block. Pick the **Verify** check box to display a prompt that asks if attribute value is correct when you insert the block. Choose the **Preset** check box to have the attribute assume preset values during block insertion. This option disables the attribute prompt. Uncheck **Preset** to display the normal prompt.

Deselect the **Lock position** check box to have the ability to move the attribute independently of the block after insertion. In addition, you must deselect the **Lock position** check box to include the attribute with the action selection set when you assign an action to a dynamic block. If the box is checked, the attribute filters out when you assign the action to the dynamic block. Dynamic blocks are covered later in this textbook.

You can create single-line or multiple-line attributes. Pick the **Multiple lines** check box to activate options for creating a multiple-line attribute. Deselect the check box to create a single-line attribute.

Using the Attribute Area

The **Attribute** area provides text boxes for assigning a tag, prompt, and default value to the attribute. Attribute values can include up to 256 characters. If the first character in an entry is a space, start the string with a backslash (\). If the first character is a backslash, begin the entry with two backslashes (\\).

Use the **Tag** text box to enter the attribute name, or tag. For example, the tag for a size attribute for a valve block could be SIZE. You must enter a tag in order to create an attribute. The tag cannot include spaces. The attribute definition applies uppercase characters to the tag, even if you type lowercase characters in the text box.

Enter a statement in the **Prompt** text box that will display when you insert or edit the block. For example, if you specify SIZE as the attribute tag, you might specify What is the valve size? or Enter valve size: as the prompt. You have the option to leave the prompt blank. The **Prompt** text box is disabled when you select the **Constant** attribute mode.

The **Default** text box allows you to enter a default attribute value, or a description of an acceptable value for reference. For example, you might enter the most common size for the SIZE attribute, or a message regarding the type of information needed, such as 10 SPACES MAX or NUMBERS ONLY. If you deselect the **Multiple lines** attribute mode, enter the default value directly in the text box. When using the **Multiple lines** attribute mode, select the ellipsis (**...**) button to enter the drawing area and place multiline text. The **Text Editor** ribbon tab appears, along with the **Text Formatting** toolbar shown in **Figure 26-3.** Enter the default text, and then pick the **OK** button on the toolbar to return to the **Attribute Definition** dialog box. Use the **Insert field** button to include a field in the default value. You also have the option to leave the default value blank.

NOTE

The abbreviated **Text Formatting** toolbar shown in **Figure 26-3** appears by default. Set the **ATTIPE** system variable to 1 to display the complete **Text Formatting** toolbar. The **ATTIPE** system variable is set to 0 by default.

Adjusting Attribute Text Options

The **Text Options** area allows you to specify attribute text settings. Many of these options function like the text settings for single-line and multiline text. Use the **Justification** drop-down list to select a justification for the attribute text. The default option is Left. In single-line attributes, the text itself is justified. In the **Multiple lines** attribute mode, the text boundary is justified.

Use the **Text Style** drop-down list to select a text style for the attribute from the styles available in the current drawing. Pick the **Annotative** check box to make the attribute text height annotative. AutoCAD scales annotative attributes according to the specified annotation scale, which eliminates the need to calculate the scale factor.

Figure 26-3.
You can define multiple-line attributes directly on-screen. The abbreviated **Text Formatting** toolbar appears in addition to the **Text Editor** ribbon tab.

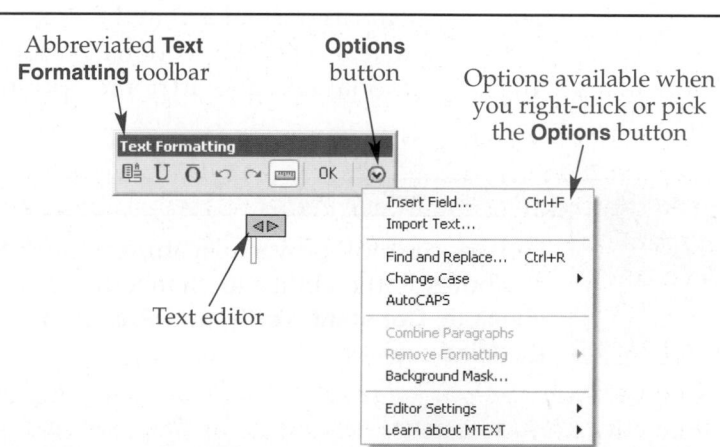

Abbreviated **Text Formatting** toolbar

Options button

Options available when you right-click or pick the **Options** button

Text editor

Specify the height of the attribute text in the **Height** text box, or pick the **Text Height** button next to the text box to pick two points in the drawing to set the text height. Identify the rotation angle for the attribute text in the **Rotation** text box, or pick the **Rotation** button next to the text box to pick two points in the drawing to set the text rotation. The **Boundary width** option is available only in the **Multiple lines** attribute mode. Enter a width for the multiple-line attribute boundary in the **Boundary width** text box, or pick the **Boundary width** button next to the text box to pick two points in the drawing to set a text boundary width.

Defining the Attribute Insertion Point

The **Insertion Point** area of the **Attribute Definition** dialog box provides options for defining how and where to position the attribute during insertion. Choose the **Specify On-screen** check box to pick an insertion point in the drawing after you pick the **OK** button to create the attribute and exit the **Attribute Definition** dialog box. This method can save time by allowing you to pick the insertion base point without using the **Pick point** button and then reentering the **Block Definition** dialog box. As an alternative, if you know the coordinates for the insertion point, deselect the **Specify On-screen** check box and type values in the **X:**, **Y:**, and **Z:** text boxes.

The **Align below previous attribute definition** check box becomes enabled if the drawing already contains at least one attribute. Check the box to place the new attribute directly below the most recently created attribute using the justification of that attribute. This is an effective technique for placing a group of different attributes in the same block. When this box is checked, the **Text Options** and **Insertion Point** areas are deactivated.

Placing the Attribute

After defining all elements of the attribute, pick **OK** to close the **Attribute Definition** dialog box. The attribute tag appears on-screen automatically if coordinates specify the insertion point, or if you are using the **Align below previous attribute definition** option. Otherwise, AutoCAD prompts you to select a location. If the attribute mode is set to **Invisible**, do not be concerned that the tag is visible; this is the only time the tag appears.

Editing Attribute Properties

The **Properties** palette provides expanded options for editing attributes. **Figure 26-4** shows the **Properties** palette with an attribute selected. You can change the color, linetype, or layer of the selected attribute in the **General** section. In the **Text** section, you can select **Tag**, **Prompt**, or **Value** to change the corresponding entries. If the value contains a field, it appears as normal text in the **Properties** palette. Modified field text automatically converts to text. The **Text** section also contains options to change the attribute text settings. Additional text and attribute options are available in the **Misc** section.

PROFESSIONAL TIP

Perhaps the most powerful feature of the **Properties** palette for editing attributes is the ability to change the original attribute modes. The **Invisible**, **Constant**, **Verify**, and **Preset** mode settings are available in the **Misc** category.

Figure 26-4.
Using the **Properties**
palette to modify
attributes.

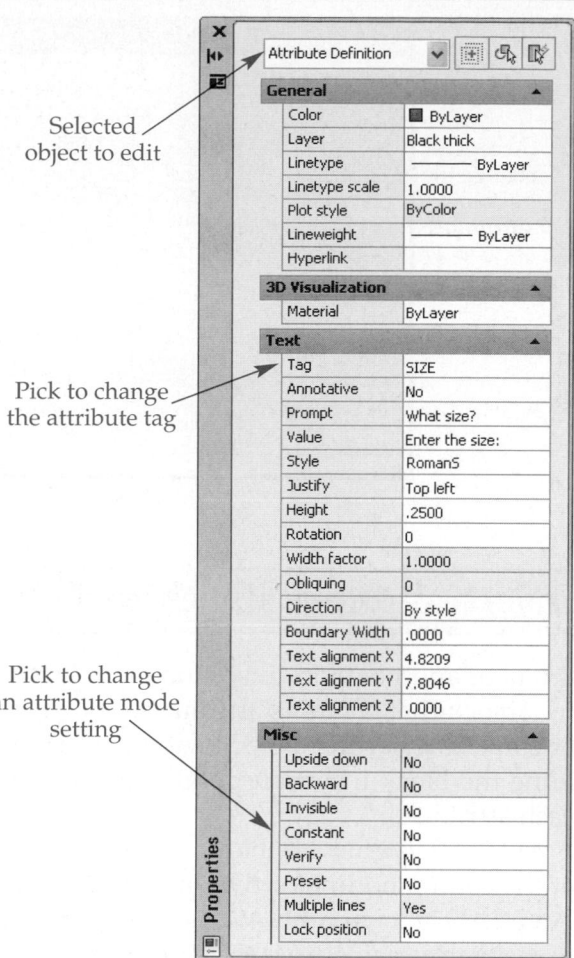

Selected
object to edit

Pick to change
the attribute tag

Pick to change
an attribute mode
setting

Creating Blocks with Attributes

Once you create attributes, use the **BLOCK** or **WBLOCK** tool to define a block with attributes. When creating the block, be sure to select all of the objects and attributes to include with the block. The order in which you select the attribute definitions is the order of prompts, or the order in which the attributes appear in the **Edit Attributes** dialog box. If you select the **Convert to Block** radio button in the **Block Definition** dialog box, the **Edit Attributes** dialog box appears when you create the block. See **Figure 26-5**. This dialog box allows you to adjust attribute values when you insert or edit the block.

PROFESSIONAL TIP

If you create attributes in the order in which you want to receive prompts and then use window or crossing selection to select the attributes, the attribute prompts are displayed in the *reverse* order of the desired prompting. To change the order, insert, explode, and then redefine the block, using window or crossing selection to pick the attributes. The attribute prompt order reverses again, placing the prompts in the desired order.

Figure 26-5.
The **Enter Attributes** dialog box allows you to enter attribute definitions when you insert or edit a block.

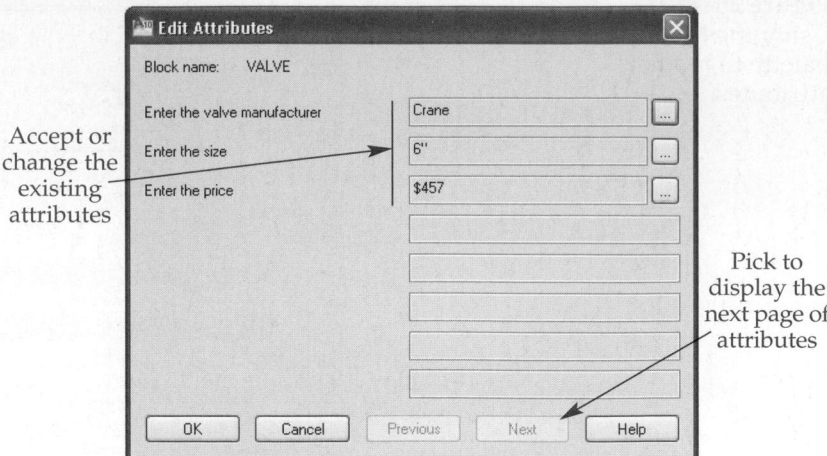

Accept or change the existing attributes

Pick to display the next page of attributes

Inserting Blocks with Attributes

Use the **INSERT** tool or another block insertion method, such as **DesignCenter** or a tool palette, to insert a block that contains attributes. The process of inserting a block with attributes is the same as inserting a block without attributes. The only difference is that after you define the block insertion point, scale, and rotation angle, prompts request values for each attribute.

By default, the **ATTDIA** system variable is set to 0, which displays single-line attribute prompts at the command line or dynamic input, and multiple-line attribute prompts using the AutoCAD text window. A better method of entering attribute values is to set the **ATTDIA** system variable to 1 before inserting blocks, to enable the **Edit Attributes** dialog box. The dialog box appears after you enter the insertion point, scale, and rotation angle, allowing you to answer each attribute prompt. Type single-line attribute values in the text boxes. To define multiple-line attributes, select the ellipsis (**...**) button next to the text boxes to enter values on-screen as multiline text. If a value includes a field, you can right-click on the field to edit it or convert it to text.

You can quickly move forward through the attributes and buttons in the **Edit Attributes** dialog box by pressing [Tab]. Press [Shift]+[Tab] to cycle through the attributes and buttons in reverse order. If the block includes more than eight attributes, pick the **Next** button at the bottom of the **Edit Attributes** dialog box to display the next page of attributes. When you finish entering values, pick the **OK** button to close the dialog box and create the block with all visible, defined attributes.

Exercise 26-1

Access the Student Web site (www.g-wlearning.com/CAD) and complete Exercise 26-1.

Attribute Prompt Suppression

Some drawings may use blocks with attributes that always retain their default values. In this case, there is no need to answer prompts for the attribute values when you insert the block. You can turn off the attribute prompts by setting the **ATTREQ** system variable to 0. After making this setting, try inserting the VALVE block created in Exercise 26-1. Notice that none of the attribute prompts appear. To display attribute prompts again, change the setting back to 1. The **ATTREQ** system variable setting is saved with the drawing.

Controlling Attribute Display

Attributes contain valuable drawing information. Some attributes only provide content to generate parts lists or bills of materials and to speed accounting. These types of attributes usually do not display on-screen or plot. Use the **ATTDISP** tool to control the display of attributes on-screen. The easiest way to activate an **ATTDISP** tool option is to pick the corresponding button from **Attributes** panel of the **Insert** ribbon tab.

Use the **Normal** (**Retain display**) option to display attributes exactly as created. This is the default setting. Use the **ON** (**Display all**) option to display *all* attributes. Apply the **OFF** (**Hide all**) option to suppress the display of all attributes, including visible attributes.

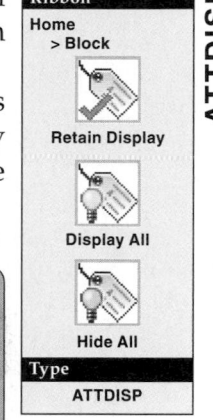

ATTDISP

Changing Attribute Values

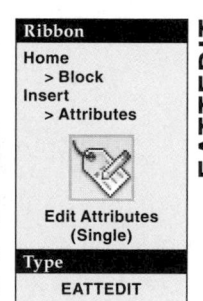

EATTEDIT

Once you create a block with attributes, tools are available for editing attribute values and settings. One option is modify the attributes of a single block using the **EATTEDIT** tool. Access the **EATTEDIT** tool and pick the block containing the attributes you want to modify to display the **Enhanced Attribute Editor**. See **Figure 26-6.**

The **Attribute** tab, shown in **Figure 26-6,** displays all attributes assigned to the selected block. Pick the attribute to modify and enter a new value in the **Value:** text box. If the attribute is a multiple-line attribute, the ellipsis (**...**) button is available for selection, allowing you to modify the text on-screen. Pick the **Apply** button after adjusting the value.

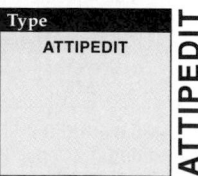

ATTIPEDIT

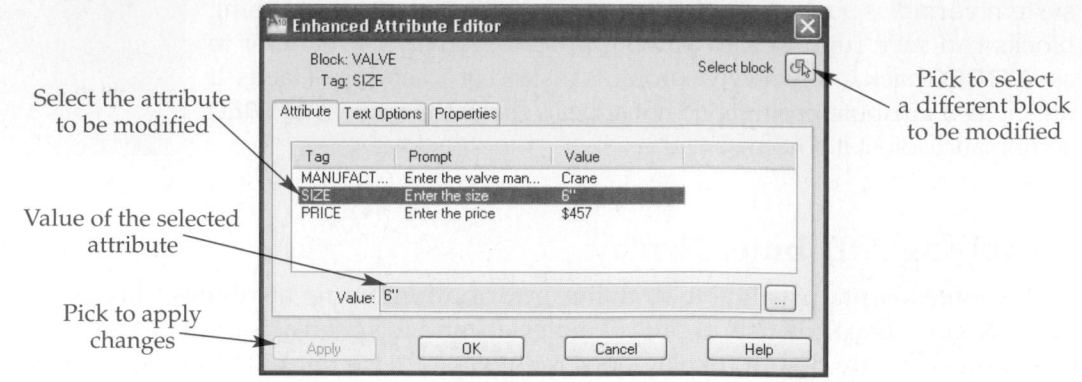

Select the attribute to be modified

Pick to select a different block to be modified

Value of the selected attribute

Pick to apply changes

To select a different block to modify, pick the **Select block** button in the dialog box. The dialog box hides to allow you to select a different block in the drawing. Then the dialog box reappears and displays the attributes for the selected block.

The **Text Options** tab, shown in **Figure 26-7A** allows you to modify the text properties of an attribute. The **Properties** tab, shown in **Figure 26-7B,** provides object property adjustments for an attribute. Each attribute in a block is a separate item. The settings you apply in the **Text Options** and **Properties** tabs affect the active attribute in the **Attribute** tab. Pick the **Apply** button to view changes made to attributes. Pick the **OK** button to close the dialog box.

Exercise 26-2

Access the Student Web site (www.g-wlearning.com/CAD) and complete Exercise 26-2.

Using the FIND Tool to Edit Attributes

One of the quickest ways to edit attributes is to use the **FIND** tool. With no tool active, right-click in the drawing area and select **Find…** to display the **Find and Replace** dialog box. You can search the entire drawing or a selected group of objects for an attribute.

Editing Attribute Values and Properties Globally

The **Enhanced Attribute Editor** allows you to edit attribute values by selecting blocks one at a time. You can use the **-ATTEDIT** tool to edit the attributes of several blocks. When you access the **-ATTEDIT** tool, a prompt asks if you want to edit attributes individually. Use the default Yes option to select specific blocks with attributes to edit. Use the No option to apply *global attribute editing*.

If you choose the Yes option, prompts appear to specify the block name, attribute tag, and attribute value. To edit attribute values selectively, respond to each prompt with the correct name or value, and then select one or more attributes. If you receive the message "0 found" after selecting attributes, you picked an incorrectly specified attribute. It is often quicker to press [Enter] at each of the three specification prompts and then pick the attribute to edit. Select an option and follow the prompts to edit the attribute(s).

-ATTEDIT

Ribbon
Home
  > Block
Insert
  > Attributes

Edit Attribute (Multiple)

Type
-ATTEDIT
-ATE

global attribute editing: Editing or changing all insertions, or instances, of the same block in a single operation.

Figure 26-7.
A—The **Text Options** tab provides options in addition to those set in the **Attribute Definition** dialog box. B—The **Properties** tab allows you to modify the properties of an attribute.

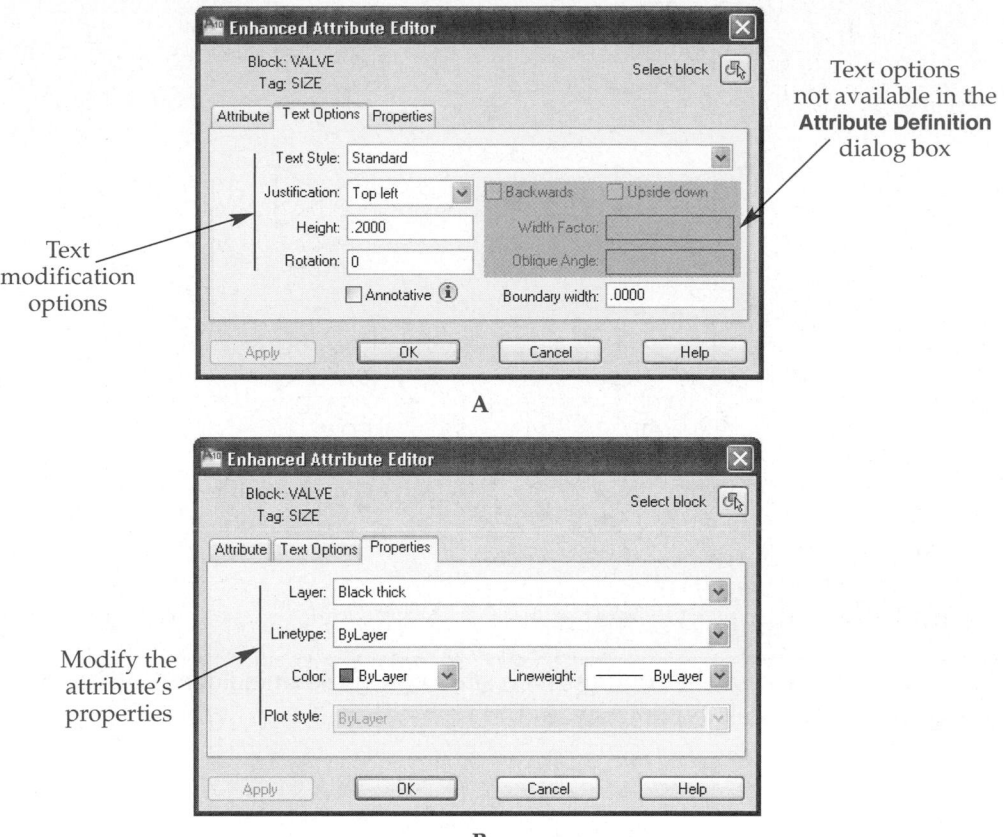

Text modification options

Text options not available in the **Attribute Definition** dialog box

A

Modify the attribute's properties

B

If you choose the No option, the Edit only attributes visible on screen? prompt appears. Select Yes to edit all visible attributes or No to edit all attributes, including those that are invisible. The same three prompts previously described for individual block editing now appear.

Figure 26-8A shows the VALVE block from Exercise 26-1 inserted three times with the manufacturer specified as CRANE. In this example, the manufacturer was supposed to be POWELL. To change the attribute for each insertion, enter the **-ATTEDIT** tool and specify global editing. Press [Enter] at each of the three specification prompts. When the Select attributes: prompt appears, pick CRANE on each of the VALVE blocks and press [Enter]. At the Enter string to change: prompt, enter CRANE, and at the Enter new string: prompt, enter POWELL. See the result in **Figure 26-8B**.

PROFESSIONAL TIP

Use care when assigning the **Constant** mode to attribute definitions. The **-ATTEDIT** tool displays 0 found if you attempt to edit a block attribute that has a **Constant** mode setting. Assign the **Constant** mode only to attributes you know will not change.

NOTE

You can also use the **-ATTEDIT** tool to edit individual attribute values and properties. However, it is more efficient to use the **Enhanced Attribute Editor** to change individual attributes.

Figure 26-8.
Using the global editing technique with the **-ATTEDIT** tool allows you to change the same attribute on several block insertions.

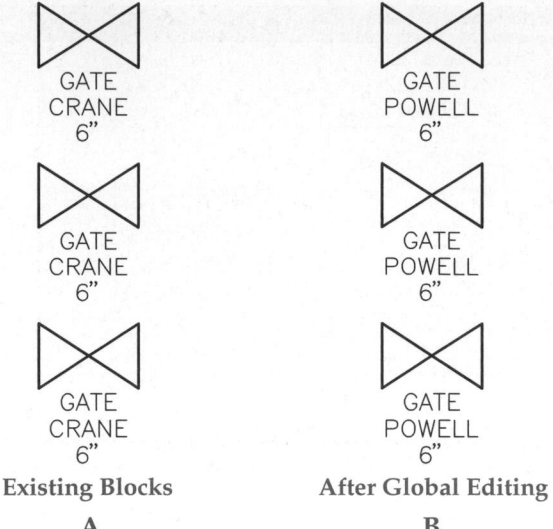

Existing Blocks
A

After Global Editing
B

Exercise 26-3

Access the Student Web site (www.g-wlearning.com/CAD) and complete Exercise 26-3.

Changing Attribute Definitions

Ribbon
Home
> Block
Insert
> Attributes

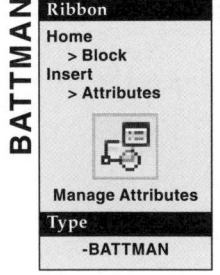

Manage Attributes
Type
-BATTMAN

Once you create a block with attributes, tools are available for modifying attribute definitions. One option is to modify attribute definitions using the **BATTMAN** tool, which displays the **Block Attribute Manager**. See **Figure 26-9.** To manage the attributes in a block, choose the block name from the **Block:** drop-down list or pick the **Select block** button to return to the drawing and pick a block.

The tag, prompt, default value, and modes for each attribute are listed by default. To select the attribute properties listed in the **Block Attribute Manager**, pick the **Settings...** button to open the **Block Attribute Settings** dialog box. See **Figure 26-10.** Check the

Figure 26-9.
Use the **Block Attribute Manager** to change attribute definitions, delete attributes, and change the order of attribute prompts.

Select the block
to modify

Pick to apply the
current attribute
definitions to
existing blocks

Pick to change the
attribute order

Attributes
in the selected block

Pick to edit the
attribute definition

Pick to delete the
attribute from
the block

Pick to set
**Block Attribute
Manager** settings

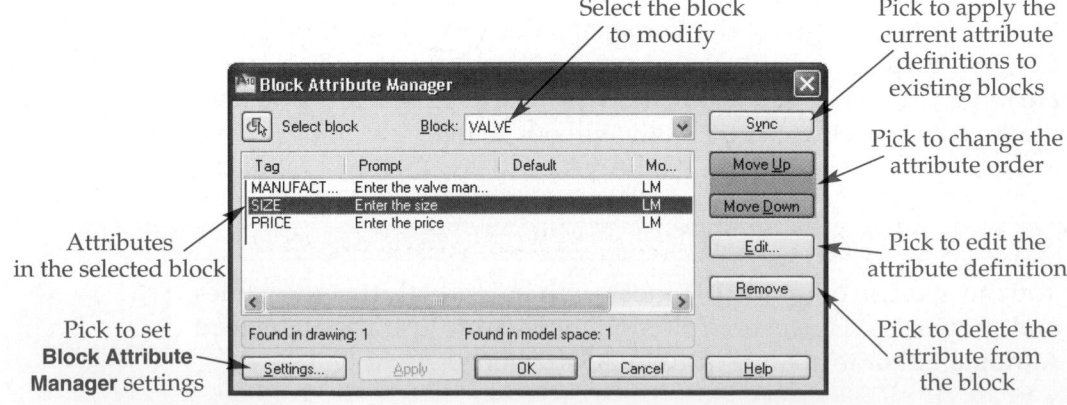

Figure 26-10.
The **Block Attribute Settings** dialog box controls the types of attributes displayed in the **Block Attribute Manager**.

Select the attribute properties to list in the **Block Attribute Manager**

Identifies duplicate tags

Updates existing blocks

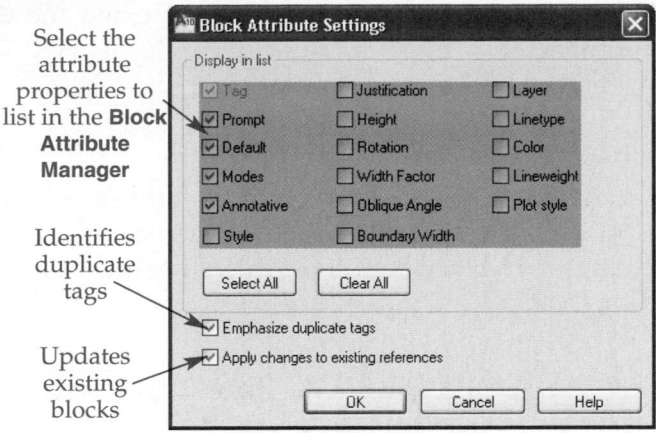

properties to list in the **Display in list** area. When you select the **Emphasize duplicate tags** check box, attributes with identical tags highlight in red. To apply the changes you make in the **Block Attribute Manager** to existing blocks, check **Apply changes to existing references**. Pick the **OK** button to return to the **Block Attribute Manager**.

The attribute list in the **Block Attribute Manager** reflects the order in which prompts appear when you insert a block. Use the **Move Up** and **Move Down** buttons to change the order of the selected attribute within the list, modifying the prompt order. To delete an attribute, pick the **Remove** button. To modify an attribute, select the attribute to edit and pick the **Edit...** button to display the **Edit Attribute** dialog box. See **Figure 26-11.** The **Attribute** tab allows you to modify the modes, tag, prompt, and default value. The **Text Options** and **Properties** tabs of the **Edit Attribute** dialog box are identical to the tabs found in the **Enhanced Attribute Editor**. If you check **Auto preview changes** at the bottom of the dialog box, changes to attributes display immediately in the drawing area.

After modifying the attribute definition in the **Edit Attribute** dialog box, pick the **OK** button to return to the **Block Attribute Manager**. Then pick the **OK** button to return to the drawing. When you modify attributes within a block, future insertions of the block reflect the changes. Existing blocks update only if you select the **Apply changes to existing references** check box in the **Settings** dialog box.

Figure 26-11.
Use the **Edit Attribute** dialog box to modify attribute definitions and properties.

Use these tabs to modify attribute properties

Select modes

Modify attribute definition

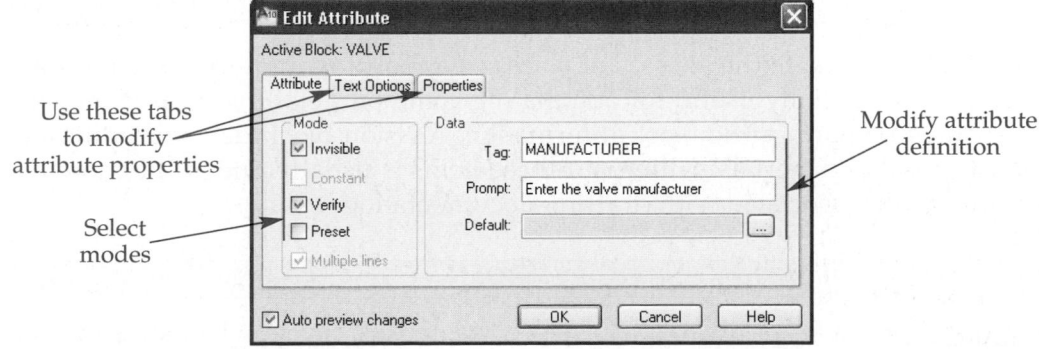

Redefining a Block and Its Attributes

To add attributes to, or revise the geometry of, a block, edit the block definition using the **BEDIT** or **REFEDIT** tools. The **REFEDIT** tool is covered in Chapter 32. Both tools allow you to make changes to a block definition, including attributes assigned to the block, without exploding the block.

NEW

Synchronizing Attributes

Redefining a block automatically updates the properties of all of the same blocks in the drawing, but not changes made to attributes. For example, if you add an object to a block, all existing blocks of the same name update to display the new object. However, if you add an attribute to a block, all existing blocks of the same name continue to display the original attributes, without the new attribute. Synchronize the blocks to update the attribute redefinition.

You can synchronize blocks in the **Block Attribute Manager** by picking the **Sync** button. This is convenient because of the ability to make changes to and remove attributes using the **Block Attribute Manager**. Use the **ATTSYNC** tool to synchronize attributes from outside the **Block Attribute Manager**. Access the **ATTSYNC** tool and use the default **Select** option to pick any of the blocks containing the attributes to synchronize. An alternative is to use the **Name** option to type the block name, or use the **?** option to list the names of all blocks in the drawing. Then choose the **Yes** option to synchronize attributes, or **No** to select a different block.

ATTSYNC

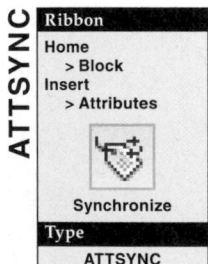

Ribbon
Home
> Block
Insert
> Attributes

Synchronize
Type
ATTSYNC

Automating Drafting Documentation

Attributes automate the process of placing symbols that require textual information. They are especially useful for automating common detailing or documentation tasks such as preparing title block information, revision block data, schedules, or a parts list or bill of materials. Filling out these items is usually one of the more time-consuming tasks associated with drafting documentation.

Creating Title Blocks

To create an automated title block, first use the correct layer(s), typically layer 0, to draw title block objects and add text that does not change, such as titles. Format the title block in accordance with industry or company standards. Include your company or school logo if appropriate. If you work in an industry that produces items for the federal government, also include the applicable *Federal Supply Code for Manufacturers (FSCM)*. **Figure 26-12** shows a title block drawn in accordance with the ASME Y14.1 *Decimal Inch Drawing Sheet Size and Format* standard.

Next, define attributes for each area of the title block. As you create attributes, determine the appropriate text height and justification for each definition. Common title block attributes include drawing title, drawing number, drafter, checker, dates, drawing scale, sheet size, material, finish, revision letter, and tolerance information. See **Figure 26-13**. Create approval attributes with a prompt such as ENTER INITIALS OR SEEK SIGNATURE, providing the flexibility to type initials or leave the cell blank for written initials. Apply the same practice to date attributes. Include any other information that may be specific to your organization or application. Assign default values to the attributes wherever possible. For example, if your organization consistently specifies the same overall tolerances for drawing dimensions, assign default values to the tolerance attributes.

> **Federal Supply Code for Manufacturers (FSCM):** A five-digit numerical code identifier applicable to any organization that produces items used by the federal government.

PROFESSIONAL TIP

The size of each area within the title block limits the number of characters displayed in a line of text. You may want to include a reminder about the maximum number of characters in the attribute prompt. For example, the prompt could read Enter drawing name (15 characters max). Each time you insert a block or drawing containing the attribute, the prompt displays the reminder.

Figure 26-12.
A title block must comply with applicable standards. This title block complies with the ASME Y14.1 standard, *Decimal Inch Drawing Sheet Size and Format*.

Figure 26-13.
Define attributes for each area of the title block. Attributes should define all information that might possibly change, including general tolerances.

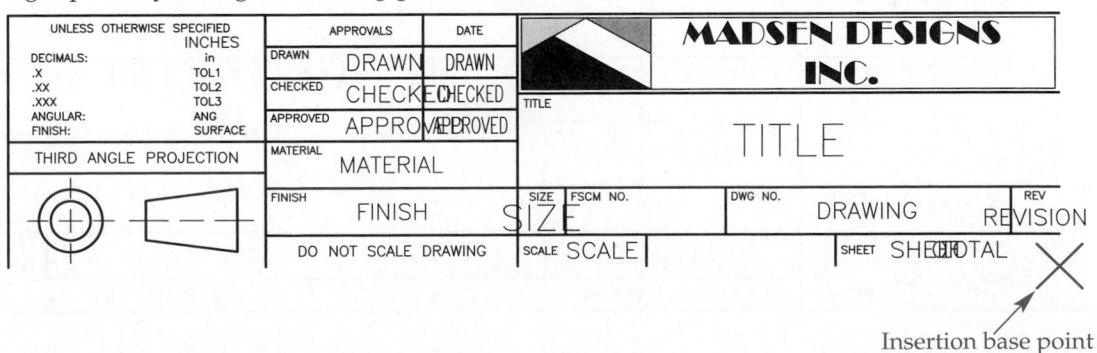

Insertion base point

After you define each attribute in the title block, you are ready to create the block. One option is to use the **BLOCK** tool to create a block of the title block within the current file. When specifying the insertion base point, pick a corner of the title block that is convenient to use each time you insert the block. The point indicated in **Figure 26-13** is an insertion base point for this particular title block. Use the **Delete** option in the **Block Definition** dialog box to remove the selected objects from the drawing. Another option is to use the **WBLOCK** tool to save the drawing as a file. Give the file a descriptive name, such as TITLE_B or FORMAT_B for a B-size title block. **Figure 26-14** shows the attribute block created in **Figure 26-13**, inserted and completely filled out using attributes.

> **NOTE**
>
> If you are creating a template, insert the block at the appropriate location and save the file as a drawing template. Edit the values in an existing title block using the **Enhanced Attribute Editor**.

Creating Revision Blocks

It is almost certain that a detail drawing will require revision at some time. Typical changes include design improvements and the correction of drafting errors. The first revision usually receives the revision letter *A*. If necessary, revision letters continue with *B* through *Y*, but the letters *I*, *O*, *Q*, *S*, *X*, and *Z* are not used because they might be confused with numbers.

revision block: A block that provides space for the revision letter, a description of the change, the date, and approvals.

zones: A system of letters and numbers used on large drawings to help direct the print reader's attention to the correction location on the drawing.

Drawing layout formats include an area with columns specifically designated to record all drawing changes. This area, commonly called the *revision block*, is normally located at the upper-right corner of the drawing sheet. A column for *zones* is included only if applicable.

The **TABLE** tool is an excellent tool for preparing a revision block. An alternative is to use blocks and attributes to document revisions. The process is similar to creating a title block, but a revision block requires two separate blocks. The first block consists of only lines and text and forms the title and heading rows. See **Figure 26-15A**. The second block includes attributes and is inserted whenever a revision is required. See **Figure 26-15B**.

Format the revision block according to industry or company standards, and use the correct layer, typically layer 0. As you create attributes, determine the appropriate text height and justification for each definition. Define attributes for the zone (if necessary), revision letter, description, date, and approval. Assign the APPROVED attribute a prompt such as ENTER INITIALS OR SEEK SIGNATURE, providing the flexibility to type initials or leave the cell blank for written initials. Apply the same practice to the date attribute.

Figure 26-14.
The title block after insertion of the attributes. Dates and approvals are added when the drawing is complete.

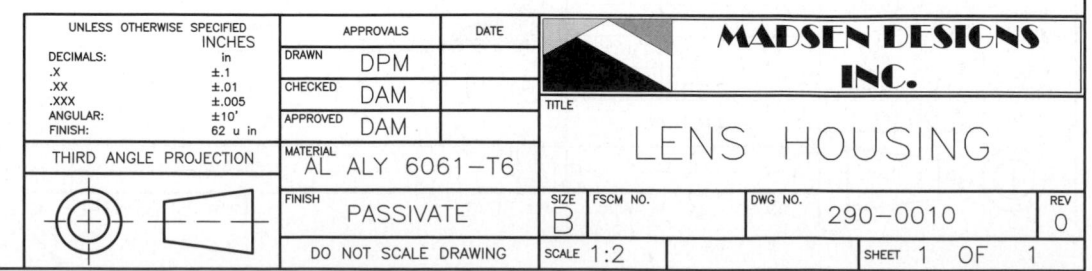

Figure 26-15.
You can create a revision block using two separate blocks. A—The first block forms the title and heading rows. B—The second block includes attributes and is added each time an engineering change is employed. The revision block shown complies with the ASME Y14.1 standard, *Decimal Inch Drawing Sheet Size and Format.*

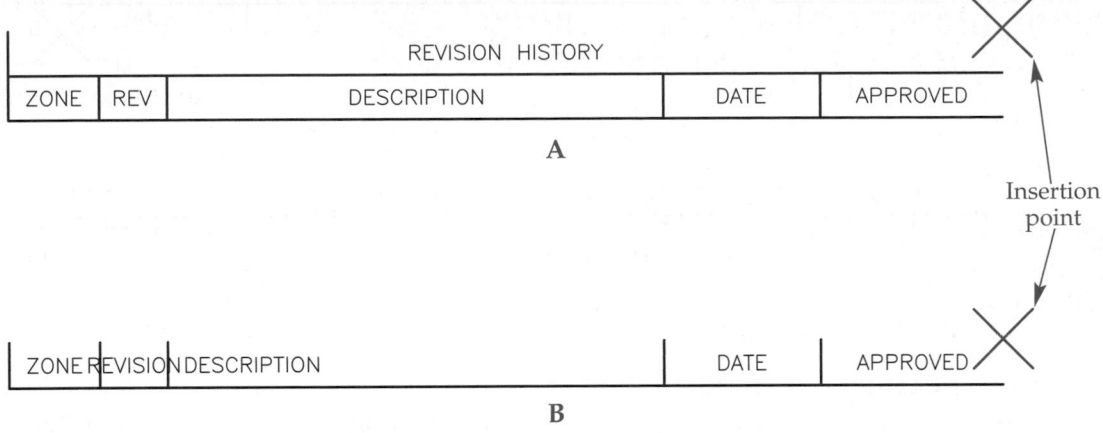

Use the **BLOCK** or **WBLOCK** tool to create the blocks. If you create wblocks, use descriptive file names such as REVBLK or REV. **Figure 26-16** shows an example of revision information added by inserting the two blocks created in **Figure 26-15** in the upper-left inside corner of the border.

Creating Parts Lists

Assembly drawings require a parts list, or bill of materials, that provides information about each component of the assembly or subassembly. This information includes the quantity, FSCM (when necessary), part number, description, and item number for each component. In some organizations, the parts list is a separate document, usually in an 8-1/2″ × 11″ format. At other companies, it is common practice to include the parts list on the face of the assembly drawing. A parts list on an assembly drawing usually appears directly above the title block, depending on industry and company standards.

The **TABLE** tool is an excellent tool for preparing a parts list. An alternative is to use blocks and attributes. The process is very similar to creating a revision block. The first block consists of only lines and text and forms the title (if used) and heading rows. See **Figure 26-17A.** The second block includes attributes and is inserted as many times as necessary to document each assembly component. See **Figure 26-17B.**

Format the parts list according to industry or company standards, and use the correct layer, typically layer 0. As you create attributes, select the appropriate text height and justification for each definition. Define attributes for the item number, quantity, FSCM (when necessary), part number, item description, and material specification.

Figure 26-16.
The completed revision block after inserting two blocks.

		REVISION HISTORY		
ZONE	REV	DESCRIPTION	DATE	APPROVED
C3	A	ADDED .125 CHAMFER	08−30−10	

Figure 26-17.
Creating a parts list using two separate blocks. A—The first block forms the title (if used) and heading rows. B—The second block includes attributes and is inserted as many times as necessary to define each assembly component. The revision block shown complies with the ASME Y14.1 standard, *Decimal Inch Drawing Sheet Size and Format.*

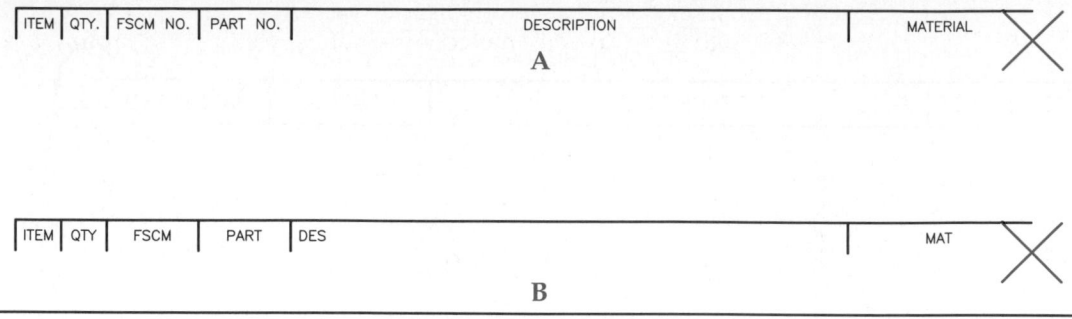

ITEM	QTY.	FSCM NO.	PART NO.	DESCRIPTION	MATERIAL

A

ITEM	QTY	FSCM	PART	DES	MAT

B

Use the **BLOCK** or **WBLOCK** tool to create the blocks. If you save wblocks, use descriptive file names, such as PL for parts list or BOM for bill of materials. **Figure 26-18** shows an example of the beginning of a parts list developed by inserting the blocks shown in **Figure 26-17.**

Figure 26-18.
The beginning of a parts list after inserting blocks and editing attribute values.

ITEM	QTY.	FSCM NO.	PART NO.	DESCRIPTION	MATERIAL
4	4		74–0080	SLEEVE	SAE 1020
3	12		85741	8–32UNC–2 X .50 HEX SOC CAP SCREW	SAE 4320
2	2		2569–01	RACK PAD	UHMW
1	1		52451	PLATE, MOUNTING	6061–T6 ALUM

UNLESS OTHERWISE SPECIFIED
INCHES
DECIMALS: in
.X ±.1
.XX ±.01
.XXX ±.005
ANGULAR: ±10'
FINISH: 62 u in

THIRD ANGLE PROJECTION

APPROVALS	DATE
DRAWN DPM	
CHECKED DAM	
APPROVED DAM	

MATERIAL VARIES

FINISH ALL OVER

DO NOT SCALE DRAWING

MADSEN DESIGNS INC.

TITLE

VRF MULTIPLIER

SIZE C FSCM NO.

DWG NO. 290010–A REV 0

SCALE 1:2 SHEET 25 OF 1

Using Fields to Reference Attributes

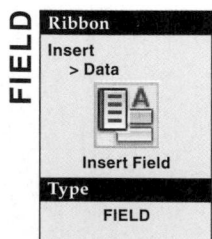

FIELD

Ribbon
Insert
> Data

Insert Field

Type
FIELD

Use fields to display the value of an attribute in a location away from the block. To display an attribute value in a field, access the **Field** dialog box from within the **MTEXT** or **TEXT** tool, from the ribbon, or by typing FIELD. In the **Field** dialog box, pick **Objects** from the **Field category:** drop-down list, and pick **Object** in the **Field names:** list box. Then pick the **Select object** button to return to the drawing window and select the block containing the attribute.

When you select the block, the **Field** dialog box reappears with the available properties (attributes) listed. Pick the desired attribute tag to display the corresponding value in the **Preview:** box. Select the format and pick **OK** to insert the field in the text object.

Chapter Test

Answer the following questions. Write your answers on a separate sheet of paper or go to the Student Web site (www.g-wlearning.com/CAD) and complete the electronic chapter test.

1. What is an attribute?
2. Explain the purpose of the **ATTDEF** tool.
3. Define the function of the following attribute modes:
   A. **Invisible**
   B. **Constant**
   C. **Verify**
   D. **Preset**
4. What is the purpose of the **Default** text box in the **Attribute Definition** dialog box?
5. How can you edit attributes before they are included within a block?
6. How can you change an existing attribute from visible to invisible?
7. If you select attributes using the **Window** or **Crossing** selection method to define a block, in what order will attribute prompts appear?
8. What purpose does the **ATTREQ** system variable serve?
9. List the three options for attribute display.
10. Explain how to change the value of an inserted attribute.
11. What does *global attribute editing* mean?
12. After you save a block with attributes, what method can you use to change the order of prompts when you insert the block?
13. What three detailing or documentation tasks can be automated using attributes?
14. What section of an assembly drawing provides information about each component of the assembly or subassembly?
15. How can you display the value of an attribute in a location away from the associated block?

Drawing Problems

Start AutoCAD if it is not already started. Start a new drawing using an appropriate template of your choice. The template should include layers, text styles, dimension styles, and multileader styles appropriate for drawing the given objects. Add layers, text styles, dimension styles, and multileader styles as needed. Draw all objects using appropriate layers, text styles, dimension styles, multileader styles, justification, and format. Follow the specific instructions for each problem. Use your own judgment and approximate dimensions when necessary.

▼ Basic

1. Use a word processor to list each attribute mode. Provide a brief description of each.

2. Draw the structural steel wide flange shape shown below using the dimensions given. Do not dimension the drawing. Create attributes for the drawing using the information given. Make a block of the drawing and name it W12 X 40. Insert the block once to test the attributes. Save the drawing as P26-2.

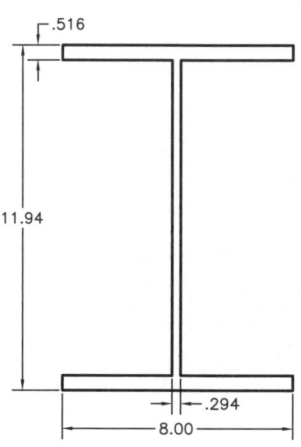

Attributes			
	Steel	W12 × 40	Visible
	Mfr.	Ryerson	Invisible
	Price	$.30/lb	Invisible
	Weight	40 lbs/ft	Invisible
	Length	10′	Invisible
	Code	03116WF	Invisible

▼ Intermediate

3. Open P26-2 and save it as P26-3. The P26-3 file should be active. Construct the floor plan shown using the dimensions given. Dimension the drawing. Insert the block **W12 X 40** six times as shown. The chart below the drawing provides the required attribute data. Enter the appropriate information for the attributes as prompted. Note that the steel columns labeled 3 and 6 require slightly different attribute data. You can speed the drawing process by using **ARRAY** or **COPY**. Resave the drawing.

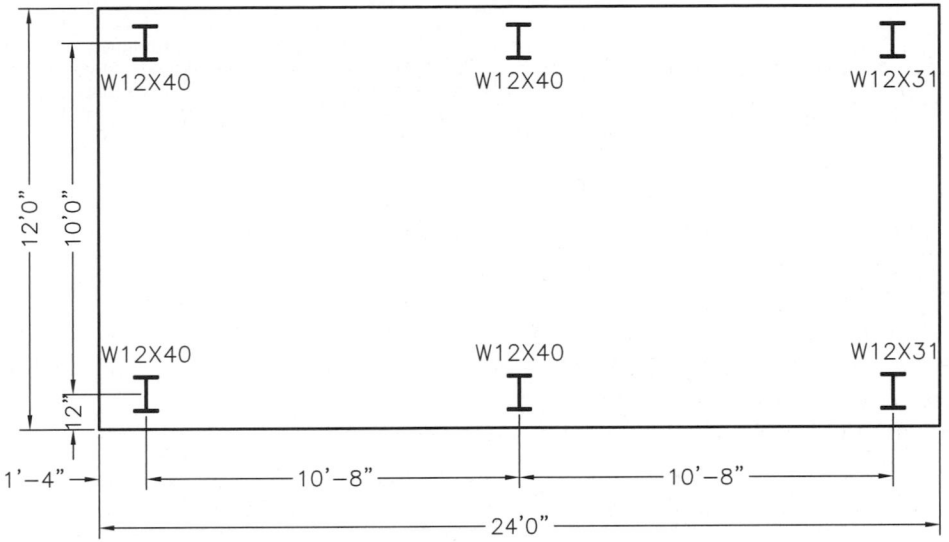

	Steel	Mfr.	Price	Weight	Length	Code
Blocks ①, ②, ④, & ⑤	W12 × 40	Ryerson	$.30/lb	40 lbs/ft	10'	03116WF
Blocks ③ & ⑥	W12 × 31	Ryerson	$.30/lb	31 lbs/ft	8.5'	03125WF

4. Open P26-2 and save it as P26-4. The P26-4 file should be active. Edit the W12 X 40 block in the newly saved drawing according to the following information. Resave the drawing.

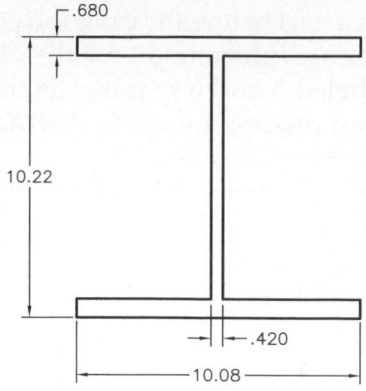

	Steel	W10 × 60	Visible
Attributes	Mfr.	Ryerson	Invisible
	Price	$.25/lb	Invisible
	Weight	60 lbs/ft	Invisible
	Length	10′	Invisible
	Code	02457WF	Invisible

▼ **Advanced**

5. Open P26-2 and save it as P26-4. The P26-4 file should be active. Create a tab-separated extraction file for the blocks in the drawing. Extract the following information for each block:
   - Block name
   - Steel
   - Manufacturer
   - Price
   - Weight
   - Length
   - Code

   Resave the drawing. Save the tab-separated extraction file as P26-4.

6. Open P26-2 and save it as P26-5. The P26-5 file should be active. Create a table from the block attribute data and insert it into the drawing. Resave the drawing.

7. Select a drawing from Chapter 25 and create a bill of materials for it using the **Data Extraction** wizard. Use the comma-separated format to display the file. Display the file in Windows Notepad. Save the drawing and the comma-separated extraction file as P26-6.

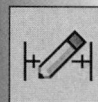

8. Create a drawing of the computer workstation layout in the classroom or office in which you are working. Provide attribute definitions for all of the items listed here.
   - Workstation ID number
   - Computer brand name
   - Model number
   - Processor chip
   - Amount of RAM
   - Hard disk capacity
   - Video graphics card brand and model
   - CD-ROM/DVD-ROM speed
   - Date purchased
   - Price
   - Vendor's phone number
   - Other data as you see fit

   Generate and extract a file for all of the computers in the drawing. Save the drawing and extracted file as P26-7.

Introduction to Dynamic Blocks

Learning Objectives

After completing this chapter, you will be able to do the following:

✓ Explain the function of dynamic blocks.

✓ Assign action parameters and actions to blocks.

✓ Modify parameters and actions.

A standard block typically represents a very specific item, such as a specific style of a 1″ long bolt. In this example, if the same style of bolt is available in three other lengths, you must create three additional standard blocks. An alternative is to create a single *dynamic block* that adjusts according to each unique bolt length. Creating and using dynamic blocks can increase productivity and reduce the size of symbol libraries, making them more manageable.

Dynamic Block Fundamentals

A dynamic block is a parametric symbol that you can adjust to change the symbol size, shape, and even geometry, without drawing additional blocks, and without affecting other instances of the block reference. **Figure 27-1** shows an example of a dynamic block of a single-swing door symbol. In this example, the dynamic properties of the block allow you to create many different single-swing door symbols according to specific parameters, such as door size, wall thickness, swing location, swing angle representation, wall angle, and exterior or interior usage.

The process of constructing and using dynamic blocks is identical to the process for standard blocks, except for the addition of *action parameters* and (usually) *actions* that control block geometry. Action parameters are commonly known as *parameters* in the context of dynamic blocks. A dynamic block can contain multiple parameters, and a single parameter can include multiple actions. Geometric constraints and *constraint parameters* are available to use as an alternative or in addition to parameters and actions. Many different tools and options exist for constructing dynamic blocks, depending on the purpose of the block.

Figure 27-2A shows an example of a bolt symbol created as a dynamic block and selected for grip editing. The bolt shaft objects include a linear parameter with a stretch action, as indicated by the *parameter grips*. The length of the bolt increases when you stretch the right-hand linear parameter grip to the right. See **Figure 27-2B.**

dynamic block: An editable block that can be assigned parameters, actions, and/or geometric constraints and constraint parameters.

action parameter (parameter): A specification for block construction that controls block characteristics such as the positions, distances, and angles of dynamic block geometry.

action: A definition that controls how dynamic block parameters behave.

constraint parameters: Dimensional constraints available for block construction to control the size or location of block geometry numerically.

parameter grips: Special grips that allow you to change the parameters of a dynamic block.

Figure 27-1.
A—A dynamic block of a single swing door symbol. B—The dynamic block allows you to create many unique door symbols without creating new blocks or affecting other instances of the same block.

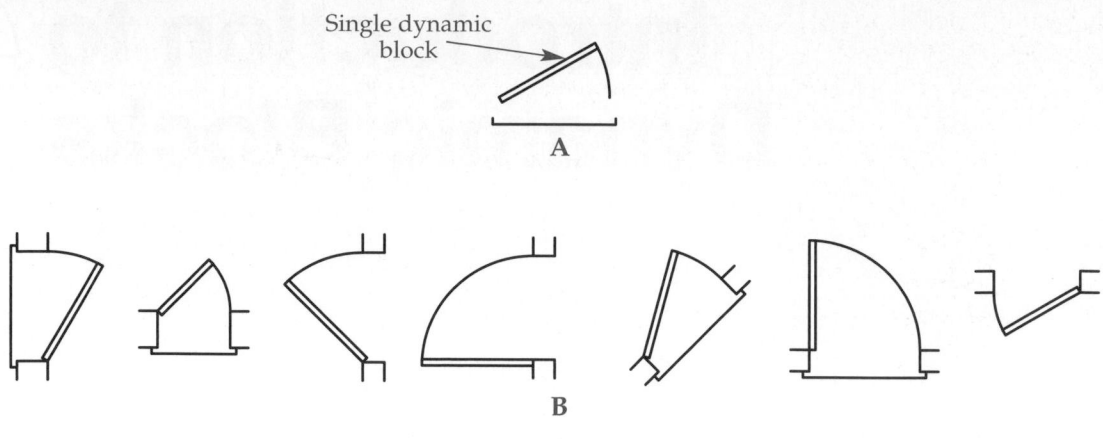

Figure 27-2.
A linear parameter with a stretch action assigned to the shaft objects in the block of a bolt. A—Selecting the block displays the linear grips. B—Selecting a linear grip and dragging it stretches the shaft of the bolt.

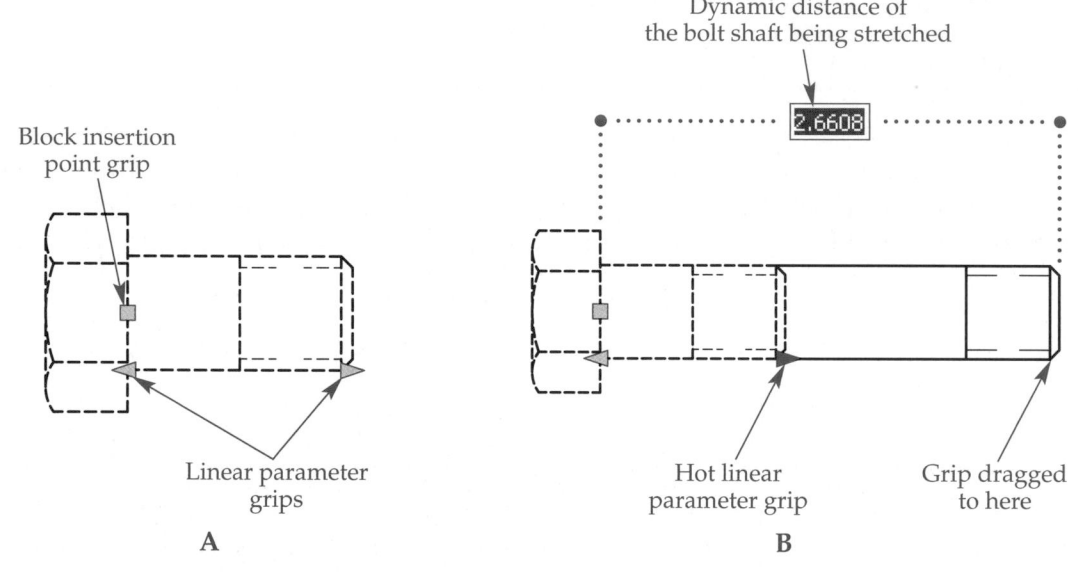

Assigning Dynamic Properties

Edit a block in the **Block Editor** to assign dynamic properties. Access the **BEDIT** tool to display the **Edit Block Definition** dialog box shown in **Figure 27-3**. To create a dynamic block from scratch from within the **Block Editor,** type a name for the new block in the **Block to create or edit** field. To edit a block saved as the current drawing, such as a wblock, pick the <Current Drawing> option. To add dynamic properties to an existing block, select the block name from the list box. A preview and the description of the selected block appear. Pick the **OK** button to open the selection in the **Block Editor**. See **Figure 27-4**.

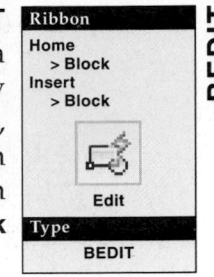

Ribbon
Home
> Block
Insert
> Block

Edit

Type
BEDIT

BEDIT

Figure 27-3.
The **Edit Block Definition** dialog box.

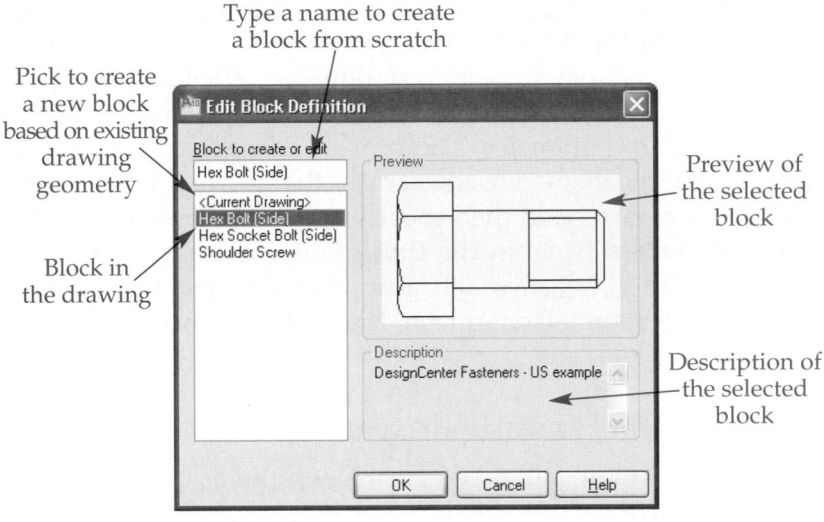

Type a name to create a block from scratch

Pick to create a new block based on existing drawing geometry

Block in the drawing

Preview of the selected block

Description of the selected block

Figure 27-4.
In block editing mode, the **Block Editor** ribbon tab and the **Block Authoring Palettes** are available.

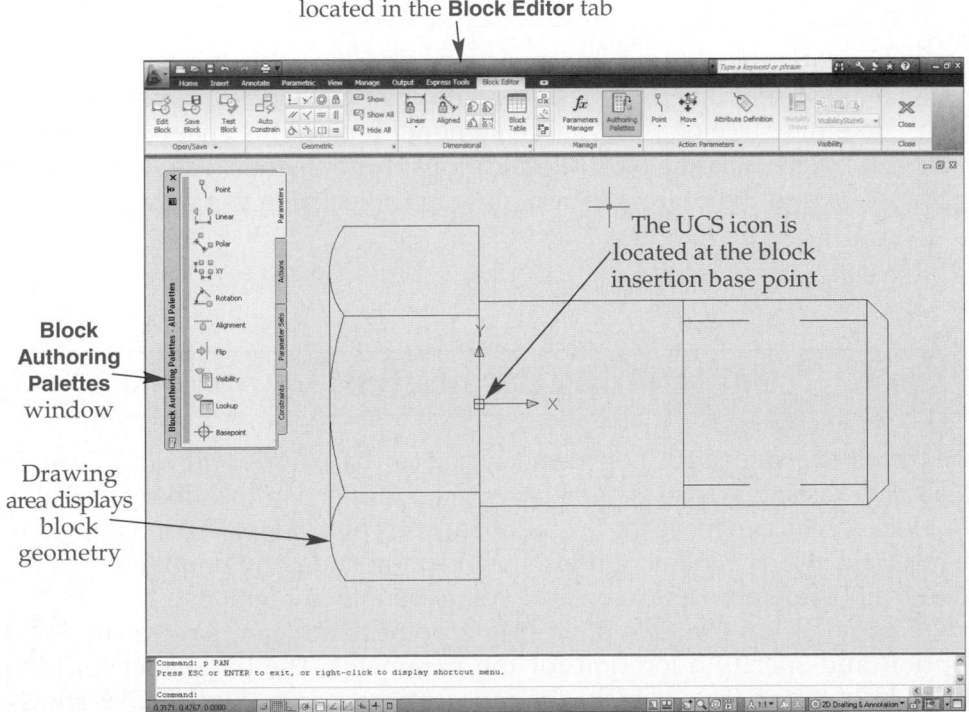

Block editing tools are located in the **Block Editor** tab

The UCS icon is located at the block insertion base point

Block Authoring Palettes window

Drawing area displays block geometry

Chapter 27 Introduction to Dynamic Blocks

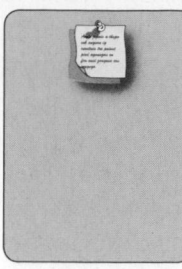
Double-click a block to display the **Edit Block Definition** dialog box with the block selected. You can open a block directly in the **Block Editor** by selecting a block, right-clicking, and choosing **Block Editor**. Another option is to open a block directly in the **Block Editor** during block creation by selecting the **Open in block editor** check box in the **Block Definition** dialog box.

The **Block Editor** ribbon tab and **Block Authoring Palettes** window provide easy access to tools and options for assigning dynamic block properties and creating attributes. The **Block Authoring Palettes** contain parameter, action, and constraint tools. Though you can type BPARAMETER or BACTION to activate the **BPARAMETER** or **BACTION** tool and then select a parameter or action as an option, it is easier to use the **Block Editor** ribbon tab or the **Block Authoring Palettes**.

You can also assign actions to certain parameters, such as point parameters, by double-clicking on the parameter and selecting an action option. You can assign only specific actions to a given parameter. The process of assigning an action is slightly different, depending on the method used to access the action. If you type BACTION, you must first select the parameter and then specify the action type. If you pick the action from the **Action Parameters** panel in the **Block Editor** ribbon or the **Block Authoring Palettes**, the specific action is active and a prompt asks you to pick the parameter. Finally, if you double-click on the parameter, the parameter becomes selected, but you must choose the action type.

Saving a Block with Dynamic Properties

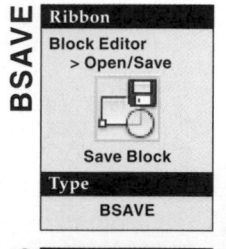

BSAVE

Ribbon
Block Editor
> Open/Save

Save Block

Type
BSAVE

Once you add one or more parameters to a block and assign actions to the parameters, you are ready to save and use the dynamic block. Use the **BSAVE** tool to save the block, or use the **BSAVEAS** tool to save the block using a different name. Remember that saving changes to a block updates all blocks of the same name in the drawing. Use the **BCLOSE** tool to exit the **Block Editor** when you are finished.

BSAVEAS

Ribbon
Block Editor
> Open/Save

Save Block As

Type
BSAVEAS

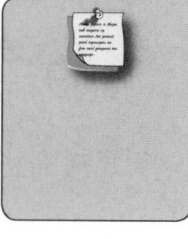
Dynamic blocks can become very complex with the addition of many dynamic properties. A single dynamic block can potentially take the place of a very large symbol library. This chapter focuses on basic dynamic block applications, fundamental use of parameters, and the process of assigning a single action to a parameter.

point parameter:
A parameter that defines an XY coordinate location in the drawing.

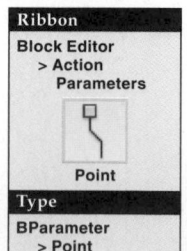

Ribbon
Block Editor
> Action
Parameters

Point

Type
BParameter
> Point

Using Point Parameters

A *point parameter* creates a position property and can be assigned move and stretch actions. For example, assign a point parameter with a move action to a door tag that is part of a door block so you can move the tag independently of the door. Point parameters also provide multiple insertion point options. For example, add point parameters to the ends of a weld symbol reference line to create two insertion point options.

Figure 27-5 provides an example of adding a point parameter. Access the **Point** parameter option and specify a location for the parameter. The parameter location determines the base point from which dynamic actions occur. **Figure 27-5** shows

Figure 27-5.
A point parameter
consists of the grip
location and a label.

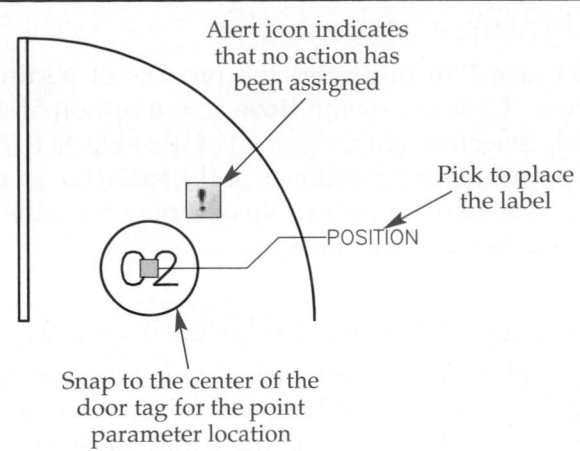

Alert icon indicates
that no action has
been assigned

Pick to place
the label

POSITION

Snap to the center of the
door tag for the point
parameter location

picking the center of the door tag circle to identify the base point of a move action. You can adjust the parameter location after initial placement if necessary. The yellow alert icon indicates that no action is assigned to the parameter.

Once you specify the parameter location, pick a location for the *parameter label*. All parameters include and require you to locate a parameter label. The label appears only in block editing mode. By default, the label for the first point parameter is Position. You can move the label as needed after initial placement.

Next, enter the number of grips to associate with the parameter. The default **1** option creates a single grip at the parameter location that allows you to use grip editing to carry out the assigned action. If you choose the **0** option, you can only use the **Properties** palette to adjust the block.

Parameter options are available before you specify the parameter location. Most of the options are also available from the **Properties** palette if you have already created the parameter. Use the **Label** option to enter a more descriptive label name. The **Name** option allows you to specify a name for the parameter that displays as the **Parameter type** in the **Properties** palette. The **Chain** option specifies whether a chain action can affect the parameter. Chain actions are described later in this chapter. The **Description** option allows you to type a description, such as the purpose of or application for the parameter. The description displays in the drawing area as a tooltip. The **Palette** option determines whether the label is displayed in the **Properties** palette when you select the block.

parameter label: A label that indicates the purpose of a parameter.

PROFESSIONAL TIP

Change the parameter label name to something more descriptive. Naming labels helps you organize parameters and recognize each parameter during editing. This is especially important when you are adding multiple parameters to a block. You may want to keep the default parameter type as part of the name. For example, change the name of the door symbol point parameter from Point to Point – Tag Center.

Exercise 27-1

Access the Student Web site (www.g-wlearning.com/CAD) and complete Exercise 27-1.

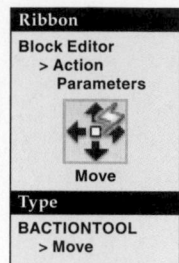

move action: An
action used to
move a block object
independently of
other objects in the
same block.

Assigning a Move Action

Figure 27-6 illustrates the process of adding a *move action* to the door block example. First, access the **Move** action option and pick the point parameter if it is not already selected. Then select all of the objects that make up the door tag and the associated parameter. Press [Enter] or the space bar or right-click to place the action. Test the block, as explained later in this chapter. Save the block, and exit the **Block Editor**. The dynamic block is now ready to use.

NOTE

Typically, when you select objects to include with an action, you should also select the associated parameter. If you do not select the parameter, the parameter grip is not included with the action and can be left behind when the action is applied.

PROFESSIONAL TIP

After you create an action, use the **Properties** palette to change the action name to something more descriptive, but keep the default action type with the name. For example, change the name of the door symbol move action from Move to Move - Tag.

Using a Move Action Dynamically

Figure 27-7A shows the door block reference, selected for editing. The point parameter grip displays as a light blue square in the center of the door tag. The insertion base point specified when the block was created appears as a standard unselected grip. Select the point parameter grip and move the door tag as shown in **Figure 27-7B**. Pick a point to specify a new location for the door tag. See **Figure 27-7C**.

NOTE

During block insertion, you can cycle through the positions of any parameters added to the dynamic block by pressing [Ctrl]. This is one method of selecting a different insertion point, corresponding to the position of a parameter, to use when inserting the block.

Figure 27-6.
Assigning a move action to a point parameter.

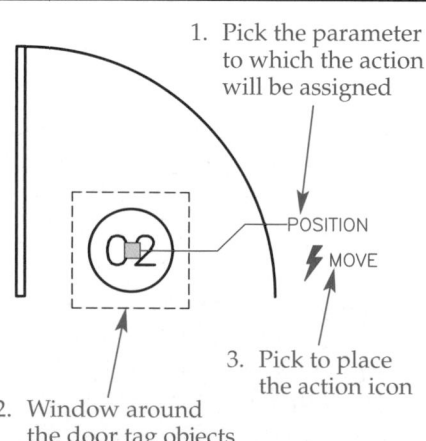

Figure 27-7.
Dynamically moving an action assigned to a point parameter. A—When the block is selected to display grips, the point parameter grip is shown as a light blue square. B—Select the point parameter grip and move it. C—The door tag is at a new location, but it is still part of the block.

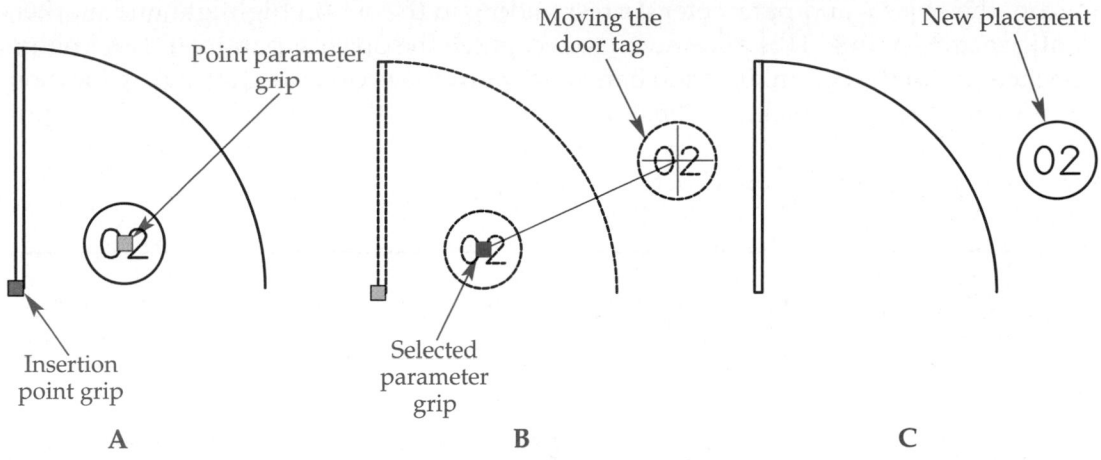

A   B   C

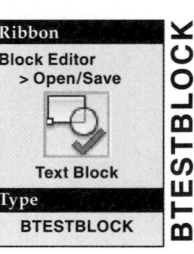

PROFESSIONAL TIP

During block insertion, use the **Properties** palette to adjust dynamic properties. If you do not include grips with a parameter, the options in the **Custom** category are the only way to adjust the block.

Testing and Managing Dynamic Properties

AutoCAD 2010 NEW

After you exit the **Block Editor**, you can insert a block and use grip editing or the **Properties** palette to confirm appropriate dynamic function. If the block does not respond as desired, however, you must re-enter the **Block Editor** to make change. A more convenient option is to access the **BTESTBLOCK** tool from inside the **Block Editor** to enter the **Test Block Window**. This window provides all of the standard AutoCAD tools and options, allowing you to test dynamic function without exiting block editing mode. Pick the **Close Test Block Window** button to re-enter the **Block Editor**. Any changes made while testing are discarded so that you can adjust the original block as needed. Block testing is especially important when a block includes multiple dynamic properties.

Ribbon
Block Editor > Open/Save

Text Block
Type
BTESTBLOCK

BTESTBLOCK

Managing Parameters

Use standard editing tools such as **MOVE** to make changes to existing parameter labels or grips. You can move parameter grips independently of the parameter label, which is often required if multiple grips are stacked or are near the same location. You can remove a parameter or parameter grips using the **ERASE** tool.

Grip editing is especially effective for managing parameters. When you select a parameter, grips appear at the parameter location and label. Use the **Properties** palette to adjust the properties of the elected parameter. The settings in the **Properties** palette change depending on the type of parameter you select. Limited property options are also available by selecting a parameter and right-clicking. Use the **Grip Display** cascading submenu to redefine the number of grips or relocate grips with the parameter location. Use the **Rename** option to change the name of the label.

Managing Actions

action bars:
Toolbars that allow you to view, remove, and adjust actions.

Action bars appear by default when you insert actions to identify and control the actions. See **Figure 27-8A**. Each action displays an icon. When you hover over or select an icon, the objects and parameter corresponding to the action highlight and markers identify action points. This allows you to recognize the objects, parameter, and points associated with the action. If action bars block your view, drag them to a new location. To hide an action bar, pick the **Close** button located to the right of the icons. Hiding action bars does not remove actions. **Figure 27-8B** briefly describes the options available when you right-click on an action icon.

Figure 27-8.
A—Action bars appear by default when you add actions. B—Options for managing actions when you right-click on an action icon.

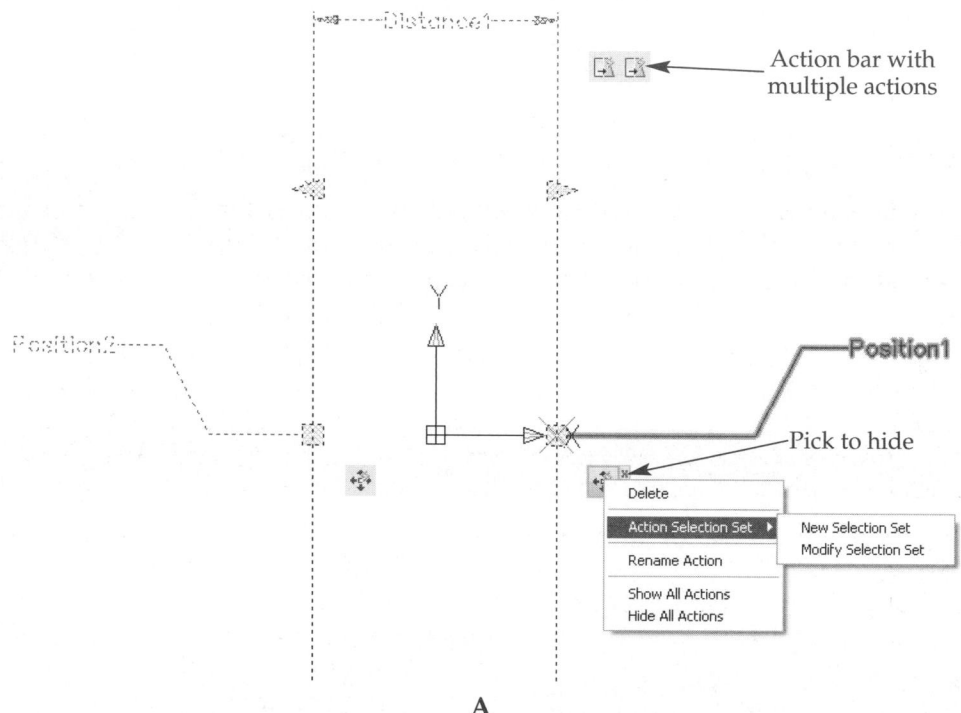

A

Option	Description
Delete	Deletes the action.
New Selection Set	Allows you to select objects to associate with the action; eliminates the original selection set.
Modify Selection Set	Adds objects to associate with the action.
Rename Action	Provides a text box for renaming the action.
Show All Actions	Displays all actions.
Hide All Actions	Hides all actions.

B

Exercise 27-2

Access the Student Web site (www.g-wlearning.com/CAD) and complete Exercise 27-2.

Using Linear Parameters

A *linear parameter* creates a distance property and can be assigned move, scale, stretch, and array actions. For example, assign a linear parameter with a stretch action to a bolt block so you can make the bolt shaft longer or shorter. Assign a second linear parameter and stretch action to the bolt head to control the bolt head diameter.

Figure 27-9 provides an example of adding a linear parameter. In this example, activate the **Linear** parameter option and use the **Label** function to name the linear parameter Shaft Length. Next, pick the start and endpoints of the linear parameter. The start and endpoints determine the locations from which dynamic actions occur. If you plan to assign a single action to the parameter, select the point associated with the action second. **Figure 27-9** shows picking the endpoint of the lower edge of the shaft, and then using polar tracking or the extension object snap to pick the point where the edge of the shaft would meet the end if extended. You must select points that are horizontal or vertical to each other to create a horizontal or vertical linear parameter.

Once you select the start and endpoints, pick a location for the parameter label. Next, enter the number of grips to associate with the parameter. The default **2** option creates grips at the start and endpoints, allowing you to use grip editing to carry out the action assigned to either point. Select the **1** option to assign a grip at the endpoint

linear parameter:
A parameter that creates a measurement reference between two points.

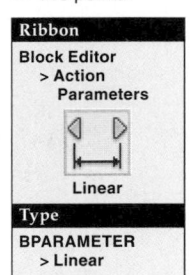

Ribbon
Block Editor
> **Action Parameters**

Linear

Type
BPARAMETER
> **Linear**

Figure 27-9.
Defining a linear parameter.

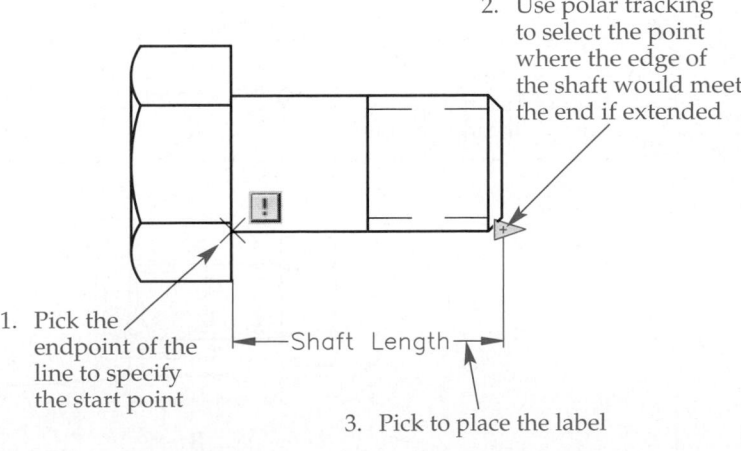

2. Use polar tracking to select the point where the edge of the shaft would meet the end if extended

1. Pick the endpoint of the line to specify the start point

Shaft Length

3. Pick to place the label

only, as shown in **Figure 27-9**. You will be able to grip-edit the block only if an action is associated with the endpoint. If you choose the **0** option, you can only use the **Properties** palette to adjust the block.

Name, **Label**, **Chain**, **Description**, **Base**, **Palette**, and **Value set** options are available before you specify points. Most of the options are also available from the **Properties** palette if you have already created the parameter. The **Base** option allows you to assign the start point or midpoint of the linear parameter as the action base point. The **Value set** option allows you to specify values for the action. Both options are described later in this chapter.

Assigning a Stretch Action

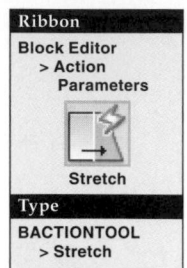

stretch action:
An action used to change the size and shape of block objects with a stretch operation.

Figure 27-10 illustrates the process of adding a *stretch action* to the bolt symbol example. Access the **Stretch** action option and pick the Shaft Length parameter if it is not already selected. Then specify a parameter point to associate with the action. Move the crosshairs near the appropriate parameter point to display the red snap marker, and pick to select. An alternative is to choose the **sTart point** option to pick the start point of the linear parameter, or the **Second point** option to select the endpoint. If you plan to use grip editing to control the block, and added a single grip, pick the point with the grip.

Next, create a window to define the stretch frame. This is the same technique you apply when using the **STRETCH** tool. See **Figure 27-10A**. Pick the objects to stretch, including the associated parameter. You do not need to use a crossing window, because the previous operation defines the stretch. However, crossing selection is often quicker. See **Figure 27-10B**. Press [Enter] or the space bar or right-click to place the action icon. See **Figure 27-10C**. Test and save the block, and exit the **Block Editor**. The dynamic block is now ready to use.

Using a Stretch Action Dynamically

Figure 27-11 shows the bolt block reference selected for editing. The linear parameter grip displays as a light blue arrow at the far end of the bolt shaft. The insertion base point specified when the block was created appears as a standard unselected grip. Select the parameter grip and stretch the shaft to the new length. Use dynamic input to view the stretch dimension, and enter an exact length value in the distance field. You can also use the **Properties** palette to define the distance.

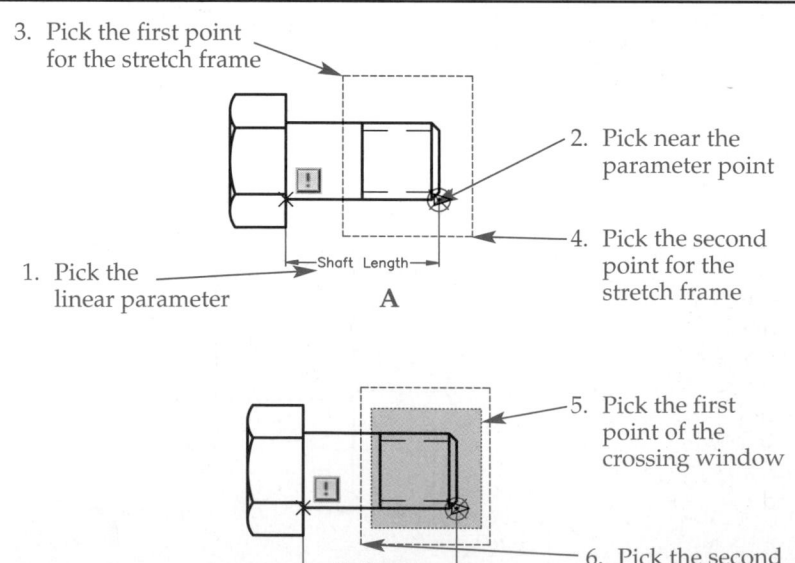

Figure 27-10.
Assigning a stretch action to a linear parameter. A—Specify the parameter and parameter grip, and create a window for the stretch frame. B—Use a crossing window to select the objects affected by the stretch action.

3. Pick the first point for the stretch frame

2. Pick near the parameter point

4. Pick the second point for the stretch frame

1. Pick the linear parameter

Shaft Length

A

5. Pick the first point of the crossing window

6. Pick the second point of the crossing window

Shaft Length

B

Figure 27-11.
Selecting the
inserted block in the
drawing displays
the parameter grips.

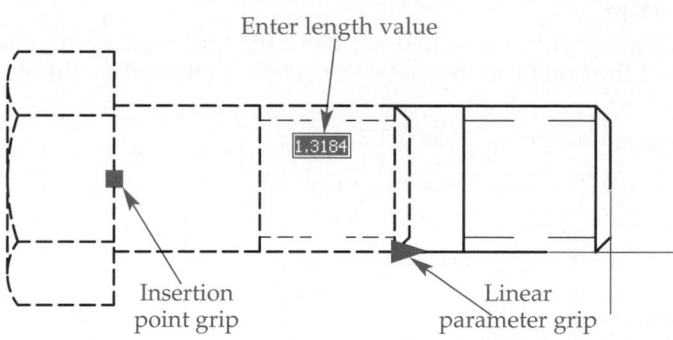

Enter length value

1.3184

Insertion
point grip

Linear
parameter grip

PROFESSIONAL TIP

The dynamic input distance field is a property of the linear parameter, allowing you to enter an exact distance. To get the best results when using a linear parameter, it is important that you locate the first and second points correctly.

Exercise 27-3

Access the Student Web site (www.g-wlearning.com/CAD) and complete Exercise 27-3.

Stretching Objects Symmetrically

The **Linear** parameter option includes a **Base** function that allows you to assign the start point or midpoint of the linear parameter as the action base point. Use the **Midpoint** setting to specify the midpoint as the action base point. This maintains symmetry when you adjust the block. You can set the **Base** preference before picking the first point or later using the **Properties** palette.

Figure 27-12 shows an example of a linear parameter with a stretch action assigned to the objects composing the bolt head. In this example, activate the **Linear** parameter option and use the **Base** option to choose the **Midpoint** setting. Next, use the **Label** option to change the label name to Head Diameter. Select the start and endpoints of the linear parameter to define the parameter and automatically calculate the midpoint. This example uses the upper-right and lower-left corners of the bolt head. After you select the start and endpoints, pick a location for the parameter label. Next, enter the number of grips to associate with the parameter. The **Figure 27-12** example uses the default **2** option to create grips at the start and endpoints.

Figure 27-12.
The base point of
a linear parameter
appears as an X.
Use the **Midpoint**
option to locate the
base point halfway
between the start
and endpoints.

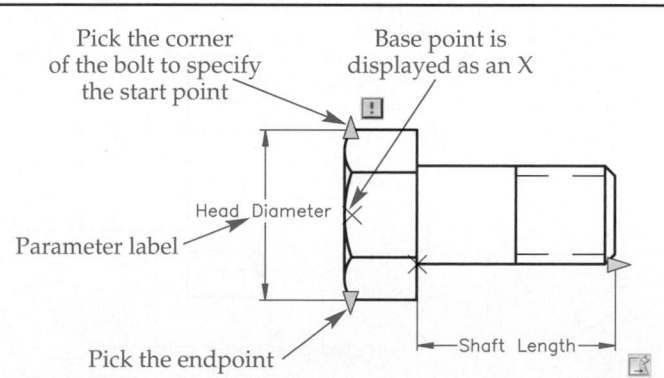

Pick the corner
of the bolt to specify
the start point

Base point is
displayed as an X

Head Diameter

Parameter label

Pick the endpoint

Shaft Length

Figure 27-13.
Assigning a stretch action to one side of the bolt head. A—Create a crossing window around the top of the bolt head. B—Select the objects affected by the stretch action.

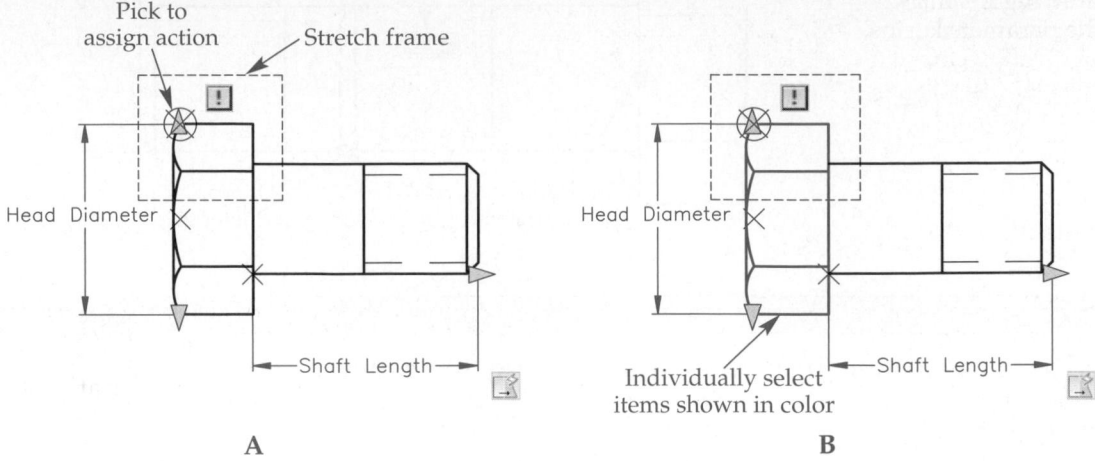

Figure 27-13 demonstrates the process of assigning a stretch action to one side of the bolt head. First, access the **Stretch** action option and pick the Head Diameter parameter if it is not already selected. Then pick the upper linear parameter point to associate with the action. Create a crossing window to define the stretch frame, as shown in **Figure 27-13A**. Then select the objects to stretch, including the associated parameter, as shown in **Figure 27-13B**. Press [Enter] or the space bar or right-click to place the action icon.

Repeat the previous sequence to assign a second stretch action to the opposite side of the bolt head. Test and save the block, and exit the **Block Editor**. The dynamic block is now ready to use. **Figure 27-14** illustrates using the lower grip point or dynamic input to stretch the bolt block reference. You can use either grip to stretch the bolt head. You can also use the **Properties** palette to define the distance.

Assigning a Scale Action

scale action: An action used to scale some of the objects within a block independently of the other objects.

Figure 27-15 shows a countertop and sink block. In this example, a *scale action* is assigned to a linear parameter to adjust the size of the sink while maintaining the dimensions of the countertop. Activate the **Linear** parameter option and use the **Base** option to choose the **Midpoint** setting. Next, use the **Label** option to change the label name to SINK LENGTH. Select the start and endpoints of the linear parameter to define the parameter and automatically calculate the midpoint. This example uses

Figure 27-14.
Dynamically stretching the bolt head. Notice that the head stretches symmetrically.

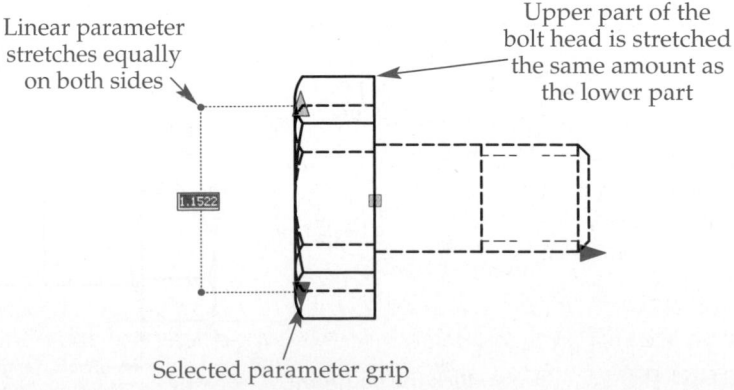

Figure 27-15.
A linear parameter assigned to the sink objects in a block of a sink and countertop. You must locate the parameter base point and independent base type at the center of the sink to scale the sink symmetrically.

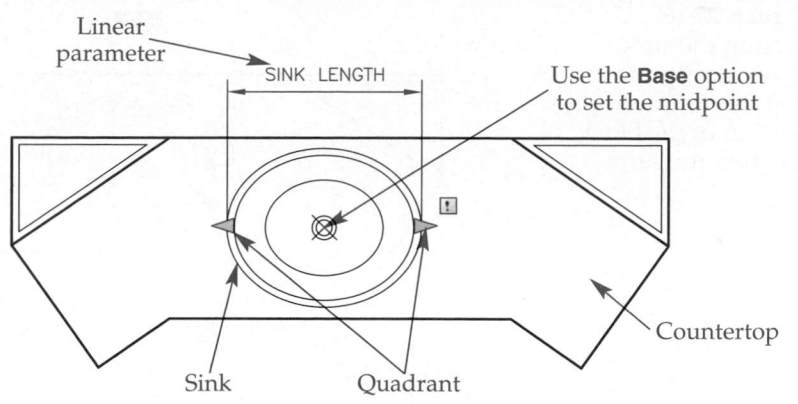

Linear parameter — SINK LENGTH

Use the **Base** option to set the midpoint

Countertop

Sink   Quadrant

two quadrants of the sink. After you select the start and endpoints, pick a location for the parameter label. Next, enter the number of grips to associate with the parameter. The **Figure 27-15** example uses the default **2** option to create grips at the start and endpoints.

Now assign a scale action to the parameter. First, access the **Scale** action option and pick the SINK LENGTH linear parameter if it is not already selected. Then select the objects to scale, including the associated parameter. Press [Enter] or the space bar or right-click to place the action.

When using a scale action, it is critical to scale objects relative to the correct base point. Access the **Properties** palette and display the properties of the scale action. The **Overrides** category includes options for adjusting the base point. The default **Base type** option is **Dependent**, which scales the objects relative to the base point of the associated parameter. Choose the **Independent** option to specify a different location. The **Base X** and **Base Y** values default to the parameter start point. Enter the coordinates relative to the block insertion base point, or use the pick button that appears when you select the **Base X** and **Base Y** values to choose points on-screen. For the sink example, it is important that the objects be scaled relative to the exact center of the sink to center the sink within the countertop as the scale changes. Test and save the block, and exit the **Block Editor**. The dynamic block is now ready to use.

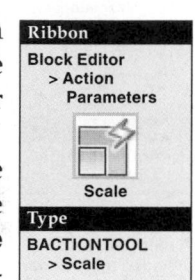

Ribbon

Block Editor
> Action
Parameters

Scale

Type

BACTIONTOOL
> Scale

PROFESSIONAL TIP

In the previous example, the **Independent** option allows you to set the center of the sink as the base point for the scale action. This is necessary because the base point of the linear parameter is not the specified midpoint. The parameter uses a midpoint base to scale the parameter, and the parameter grips, from the parameter midpoint. The **Independent** option of the scale action controls the point from which the geometry, not the parameter, is scaled.

Using a Scale Action Dynamically

Figure 27-16 shows using the right grip or dynamic input to scale the sink in the countertop and sink block reference. You can use either grip to scale the sink. You can also use the **Properties** palette to define the distance.

Exercise 27-4

Access the Student Web site (www.g-wlearning.com/CAD) and complete Exercise 27-4.

Figure 27-16.
Scaling the sink dynamically. Notice that the countertop portion of the block remains the same.

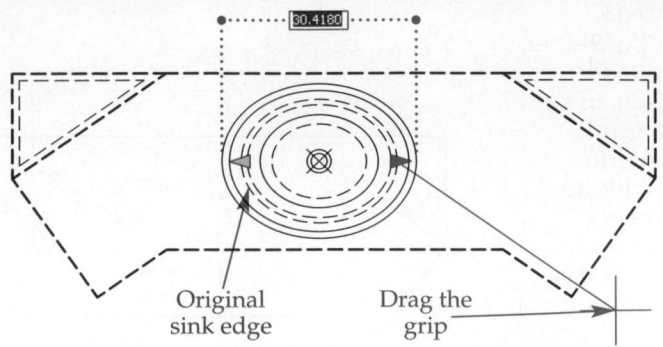

Original sink edge

Drag the grip

Using Polar Parameters

Ribbon

Block Editor > Action Parameters

Polar

Type

BPARAMETER > Polar

polar parameter: A parameter that includes a distance property and an angle property.

A *polar parameter* creates distance and angle properties and can be assigned move, scale, stretch, polar stretch, and array actions. For example, assign a polar parameter with a move action to the small circle in **Figure 27-17** to move the small circles a specified distance and angle without affecting the larger circle. To insert the polar parameter, access the **Polar** parameter option and specify the base point as the center of the large circle. Then pick the center of the small circle to specify the endpoint.

After you select the start and endpoints, pick a location for the parameter label. Next, enter the number of grips to associate with the parameter. Select the **1** option to assign a grip at the endpoint only, as shown in **Figure 27-17**. You will be able to grip-edit the block only if an action is associated with the endpoint.

NOTE

Name, **Label**, **Chain**, **Description**, **Palette**, and **Value set** options are available before you specify the parameter. Most of these options are also available from the **Properties** palette if you have already created the parameter.

Assigning a Move Action

Figure 27-18 illustrates the process of assigning a move action to the polar parameter. First, access the **Move** action option and pick the polar parameter if it is not already selected. Then select the parameter point in the center of the small circle to associate the point with the move action. Select the small circle and the associated parameter.

Figure 27-17.
Adding a polar parameter.

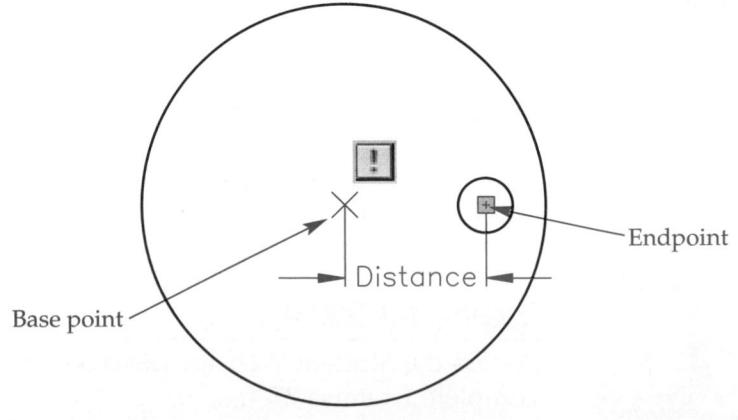

Base point

Distance

Endpoint

Figure 27-18.
Assigning a move
action to a polar
parameter.

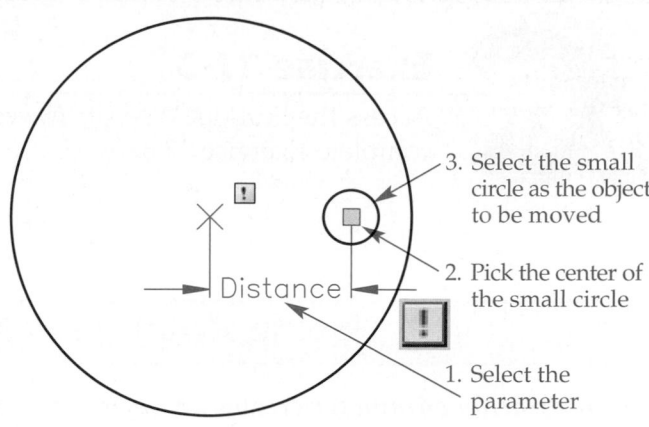

3. Select the small
circle as the object
to be moved

2. Pick the center of
the small circle

1. Select the
parameter

Distance

Press [Enter] or the space bar or right-click to place the action. Test and save the block, and exit the **Block Editor**. The dynamic block is now ready to use.

Using a Move Action Dynamically

Figure 27-19 shows using the grip point or dynamic input to change the location and angle of the small circle from the base point in a block reference. For example, to move the small circle three inches away from the center of the large circle at 45°, type @3<45 and press [Enter], or enter 3 in the distance field and 45 in the angle field and press [Enter]. You can also use the **Properties** palette to define the distance.

NOTE

Move, stretch, and polar stretch actions include **Multiplier** and **Angle Offset** options available in the **Properties** palette. Enter a value in the **Multiplier** text box to multiply by the parameter value when adjusting the block. For example, if you assign a distance multiplier of 2 to a move action and move an object 4 units, the object actually moves 8 units. Enter an angle in the **Angle Offset** text box to change the parameter grip angle. For example, if you assign an offset angle of 45 to a move action and move an object 10°, the object actually moves 55°.

Figure 27-19.
Moving an object
with a polar
parameter displays
the distance and the
angle from the base
point when dynamic
input is on.

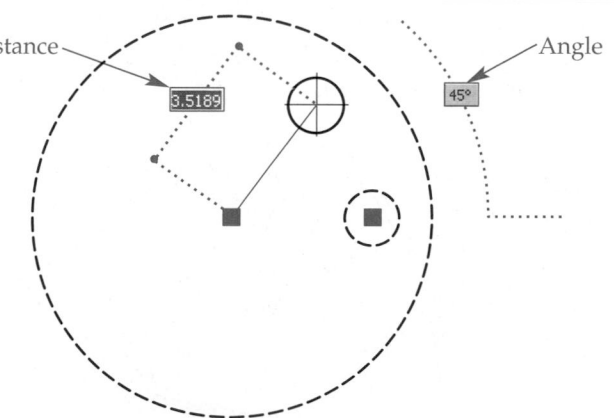

Distance

3.5189

Angle

45°

Exercise 27-5

Access the Student Web site (www.g-wlearning.com/CAD) and complete Exercise 27-5.

Using Rotation Parameters

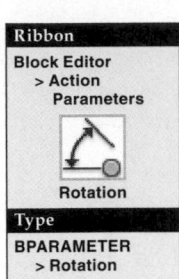
A *rotation parameter* creates an angle property to which you can assign a rotate action. For example, assign a rotation parameter with a rotate action to the needle in the speedometer block shown in **Figure 27-20** to rotate the needle around the circumference of the dial. To insert the rotation parameter, access the **Rotation** parameter option and pick the center of circular base of the needle as the rotation base point. Then pick a point, such as the needle endpoint shown, to specify the parameter radius. Set the default rotation angle from 0° east, or if rotation should originate from an angle other than 0°, use the **Base angle** option. **Figure 27-20** shows using the **Base angle** option to base the rotation at 0° and specify a default rotation angle of 200° to align the rotation with the 100 and 0 marks.

After you define the rotation parameter, pick a location for the parameter label. Next, enter the number of grips to associate with the parameter. The default **1** option creates a single grip at the parameter radius that allows you to use grip editing to carry out the rotate action.

> **NOTE**
>
> **Name**, **Label**, **Chain**, **Description**, **Palette**, and **Value set** options are available before you specify the parameter. Most of these options and the **Base angle** setting are also available from the **Properties** palette if you have already created the parameter.

Figure 27-20.
A rotation parameter with a rotate action allows you to rotate the needle in a speedometer block to indicate different speeds. The **Base angle** option allows you to set a base angle other than 0°.

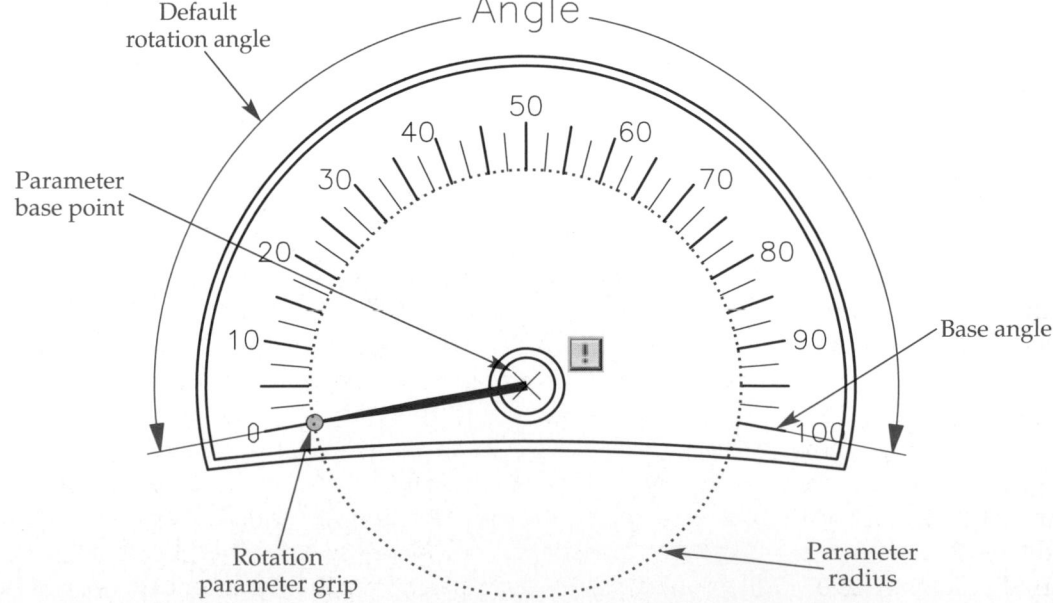

Assigning a Rotate Action

To assign a *rotate action* to the speedometer example, access the **Rotate Action** option and pick the rotation parameter if it is not already selected. Then select the objects that make up the needle and the rotation parameter. Press [Enter] or the space bar, or right-click to place the action. If necessary, access the **Properties** palette and adjust the **Base type** option. The default **Dependent** option sets the rotation point as the base point of the rotation parameter, which is appropriate for the speedometer example. Test and save the block, and exit the **Block Editor**. The dynamic block is now ready to use.

Ribbon

Block Editor
> Action
  Parameters

Rotate

Type

BACTIONTOOL
> Rotate

rotate action: An action used to rotate individual objects within a block without affecting the other objects in the block.

Using a Rotate Action Dynamically

Figure 27-21 shows using the rotation parameter grip or dynamic input to rotate the needle inside a reference of the speedometer block. Selecting a specific speed using an endpoint object snap is most appropriate for this example. You can also use the **Properties** palette to define the distance.

Exercise 27-6

Access the Student Web site (www.g-wlearning.com/CAD) and complete Exercise 27-6.

Figure 27-21.
Dynamically rotating the needle in a speedometer block using a rotate action assigned to a rotation parameter.

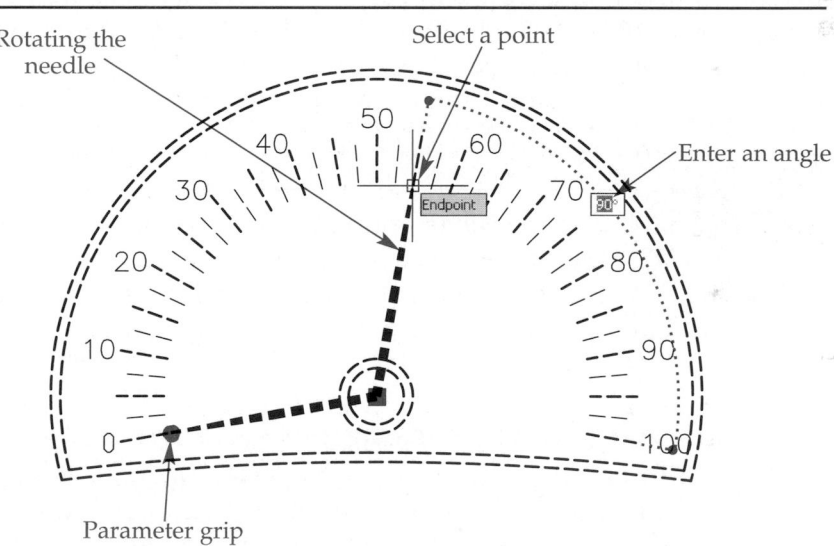

- Rotating the needle
- Select a point
- Enter an angle
- Endpoint
- Parameter grip

Using Alignment Parameters

alignment parameter: A parameter that aligns a block with another object in the drawing.

An *alignment parameter* creates an alignment property. When you move a block with an alignment parameter near another object, the block rotates to align with the object based on the angle and alignment line defined in the block. This parameter saves time by eliminating the need to rotate a block or assign a rotation parameter. An alignment parameter affects the entire block, and therefore requires no action.

Figure 27-22 provides an example of adding an alignment parameter to the block of a gate valve symbol to align the gate valve with pipes. Access the **Alignment** parameter option and pick the point in the center of the valve to locate the parameter grip and define the first point of the alignment line. Next, specify the alignment direction, or use the **Type** option to specify the alignment type. Alignment type does not

Ribbon

Block Editor
> Action
  Parameters

Alignment

Type

BPARAMETER
> Alignment

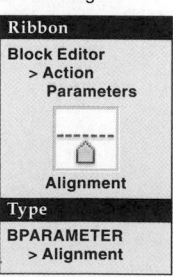

Figure 27-22.
Adding an alignment parameter to a gate valve block.

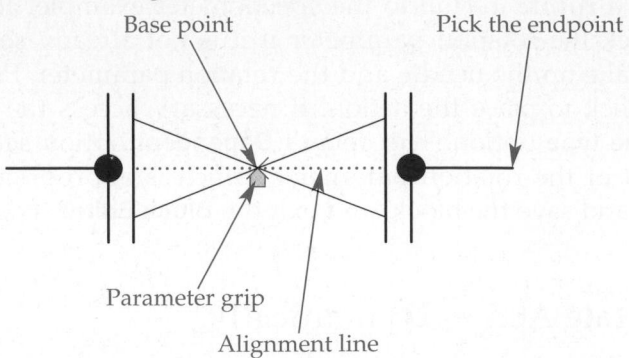

affect how the block aligns; it determines the direction of the alignment grip. Select the **Perpendicular** option to point the grip perpendicular to the alignment line, or choose the **Tangent** option to point the grip tangent to the alignment line. Set the **Tangent** option for the gate valve example.

After specifying the base point, and if necessary the alignment type, pick a second point to set the alignment direction. The angle between the first point and the second point defines the alignment line. The alignment line determines the default rotation angle. **Figure 27-22** shows selecting the endpoint of the value symbol. The alignment parameter grip is an arrow that points in the direction of alignment, perpendicular or tangent to the object to align. Test the block by drawing a line in the **Test Block Window** and attempting to align the block with the line. When you are finished, save the block and exit the **Block Editor**. The dynamic block is now ready to use.

> **NOTE**
>
> Use the **Name** option before you specify the parameter to rename the parameter. Alignment parameters do not include labels. You can also adjust the alignment **Type** from the **Properties** palette if you have already created the parameter.

Using an Alignment Parameter Dynamically

Figure 27-23 shows using the alignment parameter grip to align a reference of the gate valve with a pipeline. Select the block to display grips and then pick the parameter grip. Move the block near another object to align the block with the object. The rotation depends on the alignment path and type, and the angle of the other object.

Figure 27-23.
When you move the gate valve block near the angled line, the block aligns with the line.

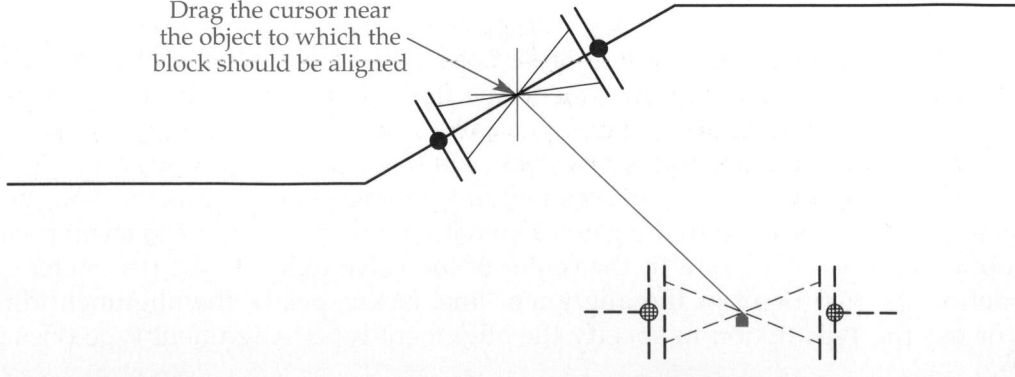

AutoCAD and Its Applications—Basics

Exercise 27-7

Access the Student Web site (www.g-wlearning.com/CAD) and complete Exercise 27-7.

Using Flip Parameters

A *flip parameter* creates a flip property to which you can assign a flip action. For example, assign a flip parameter with a flip action to a door symbol to provide the option to place the door on either side of a wall. Another example is using a flip parameter to control the side of a reference line where a weld symbol displays for arrow side or other side applications.

Figure 27-24 provides an example of adding a flip parameter. Access the **Flip Parameter** option and pick the base point, followed by the endpoint of the reflection line. See **Figure 27-24A**. Pick a location for the parameter label, and then enter the number of grips to associate with the parameter. The default **1** option creates a single flip grip that allows you to use grip editing to carry out the flip action.

Flipping a block mirrors the block over the reflection line. However, for the door symbol, with the line in the current position, as shown in **Figure 27-24A**, an incorrect flip will result when you flip the block to the other side of a wall. To mirror the block properly, you must locate the reflection line to account for the wall thickness. To place the door on a 4″ wall, for example, use the **MOVE** tool to move the reflection line 2″ lower than the door. The label and parameter grip also move. In addition, you may want to move the parameter grip horizontally to the middle of the door opening. This may help in placing and flipping the block. See **Figure 27-24B**.

> **flip parameter:**
> A parameter that mirrors selected objects within a block.

Ribbon
Block Editor
> Action
Parameters
⇨⁞
Flip

Type
BPARAMETER
> Flip

Figure 27-24.
A—Inserting a flip parameter. B—Moving the parameter so the block will correctly flip about the centerline of a wall.

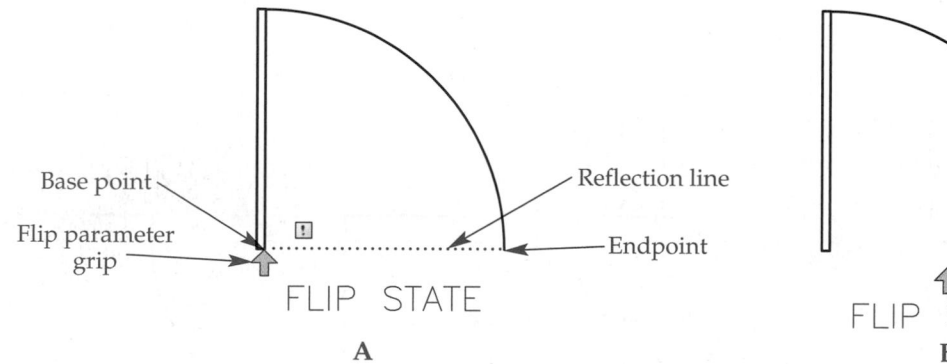

Assigning a Flip Action

Ribbon

Block Editor
> Action
Parameters

Flip

Type

BACTIONTOOL
> Flip

flip action: An action used to flip the entire block.

To assign a *flip action* to the door example, access the **Flip** action option and pick the flip parameter if not already selected. Then select the objects that make up the door and the flip parameter. Press [Enter] or the space bar or right-click to place the action. Test and save the block, and exit the **Block Editor**. The dynamic block is now ready to use.

Using a Flip Action Dynamically

Figure 27-25A shows a reference of the door block, selected for editing. Pick the flip parameter grip to flip the block to the other side of the reflection line, as shown in Figure 27-25B. Unlike other parameters and actions that require stretching, moving, or rotating, a single pick initiates a flip action.

Figure 27-25.
A—Select the block to display the flip parameter grip. B—Pick the flip parameter grip to flip the block about the reflection line. The entire block flips because all of the objects within the block are included in the selection set for the action.

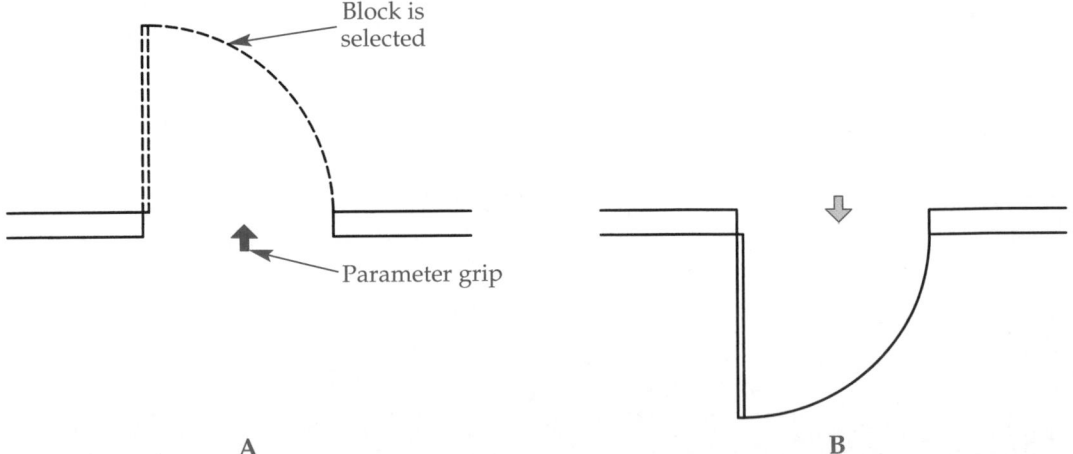

Block is
selected

Parameter grip

A

B

AutoCAD and Its Applications—Basics

Exercise 27-8

Access the Student Web site (www.g-wlearning.com/CAD) and complete Exercise 27-8.

Using XY Parameters

An *XY parameter* creates horizontal and vertical distance properties and can be assigned move, scale, stretch, and array actions. The XY parameter can include up to four parameter grips—one at each corner of a rectangle defined by the parameter. You can use the XY parameter for a variety of applications, depending on the assigned actions.

Figure 27-26 provides an example of inserting an XY parameter. Access the **XY** parameter option and pick the base point. The base point is the origin of the X and Y distances. Next, pick a point to specify the XY point, which is the *corner* opposite the base point. Finally, enter the number of grips to associate with the parameter. The default **2** option creates grips at the start and endpoints, allowing you to use grip editing to carry out the action assigned to either point. Select the 4 option, as shown in **Figure 27-26**, to assign a grip at each XY corner to maximize flexibility, or choose a smaller number to limit dynamic options. If you choose the **0** option, you can only use the **Properties** palette to adjust the block.

> **XY parameter:**
> A parameter that specifies distance properties in the X and Y directions.

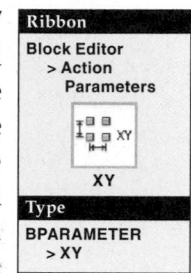

Ribbon
Block Editor
> Action
   Parameters

XY

Type
BPARAMETER
> XY

NOTE

Name, **Label**, **Chain**, **Description**, **Palette**, and **Value set** options are available before you specify the parameter. Most of these options and the **Base angle** setting are also available from the **Properties** palette if you have already created the parameter.

Assigning an Array Action

Figure 27-27 illustrates using an *array action* assigned to an XY parameter. This example shows dynamically arraying the block of an architectural glass block, allowing you to create an architectural feature of glass blocks without using a separate array operation. Access the **Array** action option and pick the XY parameter if it is not already selected. Then select the objects to include in the array, and press [Enter] or the space bar or right-click to accept the selection.

> **array action:**
> An action used to array objects within the block based on preset specifications.

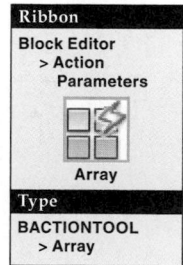

Ribbon
Block Editor
> Action
   Parameters

Array

Type
BACTIONTOOL
> Array

Figure 27-26.
Adding an XY parameter to a block of an architectural glass block. The XY parameter consists of X and Y distance properties and four grips.

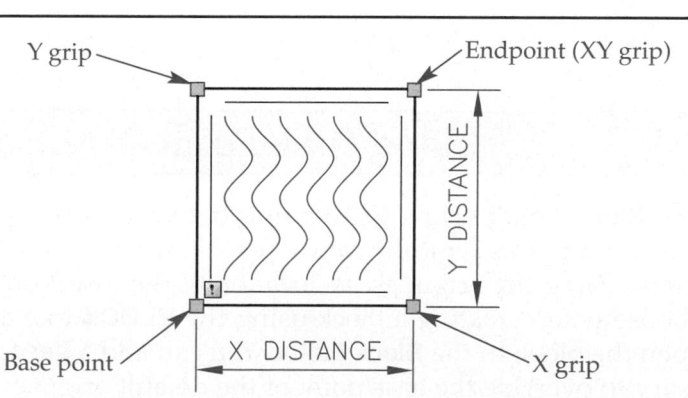

Figure 27-27.
Dynamically creating an array of architectural glass blocks using a block with an XY parameter and an array action. The pattern of rows and columns forms as you move the XY parameter. Notice the grout lines that form between the glass blocks because of proper action definition.

XY grip selected

Drag the XY grip

2.1057

Y distance

3.2128

X distance

At the Enter the distance between rows or specify unit cell: prompt, enter a value for the distance between rows or pick two points to set the row and column values. At the Enter the distance between columns: prompt, specify a value for the distance between columns. The second prompt does not appear if you select two points to define the row and column values. In the example of the glass block, be sure to allow for a grout joint when setting the row and column distance. Before assigning the action, you may want to draw a construction point offset from the block by the width of the grout joint. Then you can pick two points to define the row and column values. Be sure to erase the construction point before saving the block. Test and save the block, and exit the **Block Editor**. The dynamic block is now ready to use.

Using an Array Action Dynamically

Figure 27-27 illustrates using the upper-right grip or dynamic input to array a reference of the architectural glass block. You can use any available grip to apply the array, depending on where you want the array to occur. You can also use the **Properties** palette to define the array. Notice that proper action definition produces grout lines. The resulting array remains a single block.

Exercise 27-9

Access the Student Web site (www.g-wlearning.com/CAD) and complete Exercise 27-9.

Using Base Point Parameters

The **Block Editor** origin (0,0,0 point) determines the default location of the block insertion base point. Typically, you construct blocks in the **Block Editor** in reference to the origin, using the origin as the location of the insertion base point. The base point you choose when creating a block using the **BLOCK** tool attaches to the origin when you open the block in the **Block Editor**. You can add a *base point parameter* when it is necessary to override the base point of the default origin.

base point parameter: A parameter that defines an alternate base point for a block.

AutoCAD and Its Applications—Basics

Access the **Basepoint** parameter option and pick a point to place the base point parameter. The parameter displays as a circle with crosshairs. After you save the block, the location of the base point parameter becomes the new base point for the block. You cannot assign actions to a base point parameter, but you can include a base point parameter in the selection set for actions.

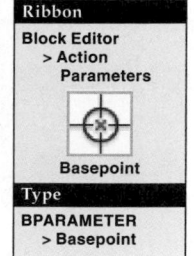

Ribbon

Block Editor
> Action
    Parameters

Basepoint

Type

BPARAMETER
> Basepoint

Using Parameter Value Sets

A *value set* helps to ensure that you select an appropriate value when editing a block, and can often increase the usefulness of a dynamic block. For example, if a window style is only available in widths of 36", 42", 48", 54", and 60", add a value set to a linear parameter with a stretch action to limit selection to these sizes. See **Figure 27-28**. You can use a value set with linear, polar, XY, and rotation parameters.

value set: A set of allowed values for a parameter.

To create a value set, select the **Value set** option available at the first prompt after you access a parameter option, and then pick a value set type. Choose the **List** option to create a list of possible sizes. Type all of the valid values for the parameter separated by commas. For the window block example, enter 36,42,48,54,60. Then press [Enter] or the space bar or right-click to return to the initial parameter prompt, and add the parameter as you normally would. After you insert the parameter, the valid values appear as tick marks.

Select the **Increment** option to specify an incremental value. Minimum and maximum values are also set to provide a limit for the increments. For the window block example, use the **Value set** option again to set 6" width increments. This time choose the **Increment** option and type 6 for the distance increment, 36 for the minimum distance, and 60 for the maximum distance. The initial parameter prompt returns after you enter the maximum distance.

After you add a parameter with a value set, you must assign an action to the parameter. For the window block in **Figure 27-28**, assign a stretch action to the linear parameter. This allows the window to stretch to the valid widths specified in the value set. Test and save the block, and exit the **Block Editor**. The dynamic block is now ready to use.

NOTE

You can also use the options in the **Value Set** category of the **Properties** palette to specify value sets during block definition.

Figure 27-28.
When you adjust a block that includes a value set, tick marks appear at locations corresponding to the values in the value set. You can only stretch the block to one of the tick marks.

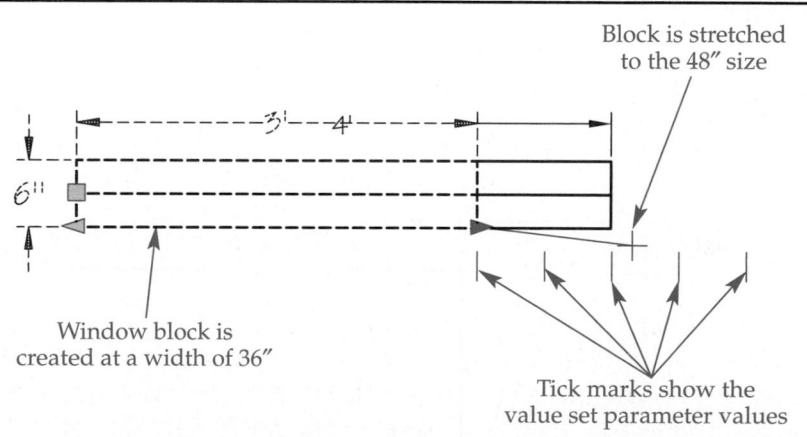

Block is stretched to the 48" size

Window block is created at a width of 36"

Tick marks show the value set parameter values

Using a Value Set with a Parameter

Figure 27-28 shows using a linear parameter grip with a stretch action to specify the width of a window block reference. Tick marks appear, indicating the positions of valid values. As you stretch the grip, the modified block snaps to the nearest tick mark. When using dynamic input, you can also enter a value in the input field. If you type a value that is not included in the value set, the nearest valid value applies. You can also use the **Value Set** category of the **Properties** palette to select a value.

Exercise 27-10

Access the Student Web site (www.g-wlearning.com/CAD) and complete Exercise 27-10.

Using a Chain Action

chain action: An action that triggers another action when a parameter is modified.

A *chain action* limits the number of edits that you have to perform by allowing one action to trigger other actions. For example, **Figure 27-29** shows using a chain action to stretch the block of a table and chairs, and array the chairs along the table at the same time. Point, linear, polar, XY, and rotation parameters can be part of a chain action.

Creating a Chain Action

To create a chain action, select the **Chain** option available at the first prompt after you access a parameter option to display the Evaluate associated actions when parameter is edited by another action? [Yes/No]: prompt. The default **No** setting does not create a chain action. Select the **Yes** option to create a chain action.

Figure 27-29.
A—A block of a table with six chairs. B—Using a chain action with a linear parameter, you can array the chairs automatically when the table is stretched.

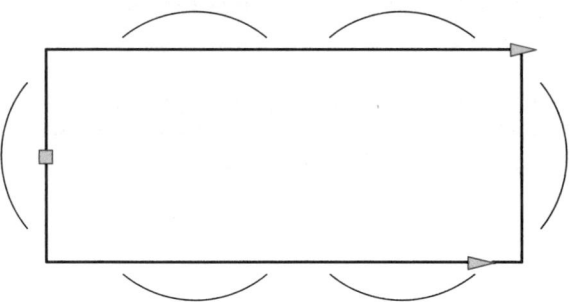

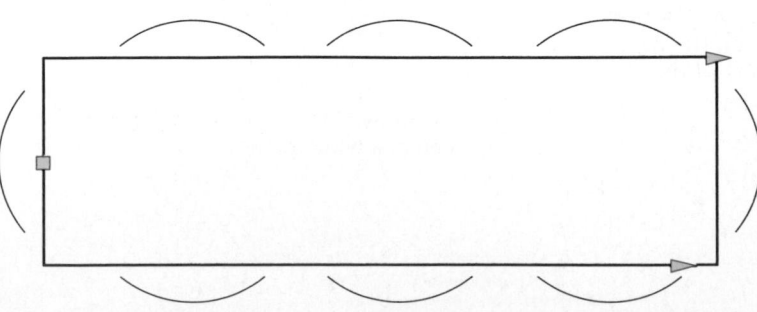

AutoCAD and Its Applications—Basics

Figure 27-30 shows the default arrangement of the table and chairs block example. For this example, access the **Linear** parameter option and use the **Label** option to change the label name to CHAIR ARRAY. Next, choose the **Chain** option and select **Yes**. To complete the parameter, select the start and endpoints shown in **Figure 27-30A**, and assign a single grip to the endpoint. Then assign an array action to the parameter, selecting the chairs on the top and bottom of the table as the objects to array. At the

Figure 27-30.
A—Inserting a linear parameter to use with an array action for the chairs. B—Inserting a linear parameter to stretch the table. C—Assigning a stretch action to the linear parameter. When you specify the crossing window, be sure the CHAIR ARRAY parameter grip is included in the frame.

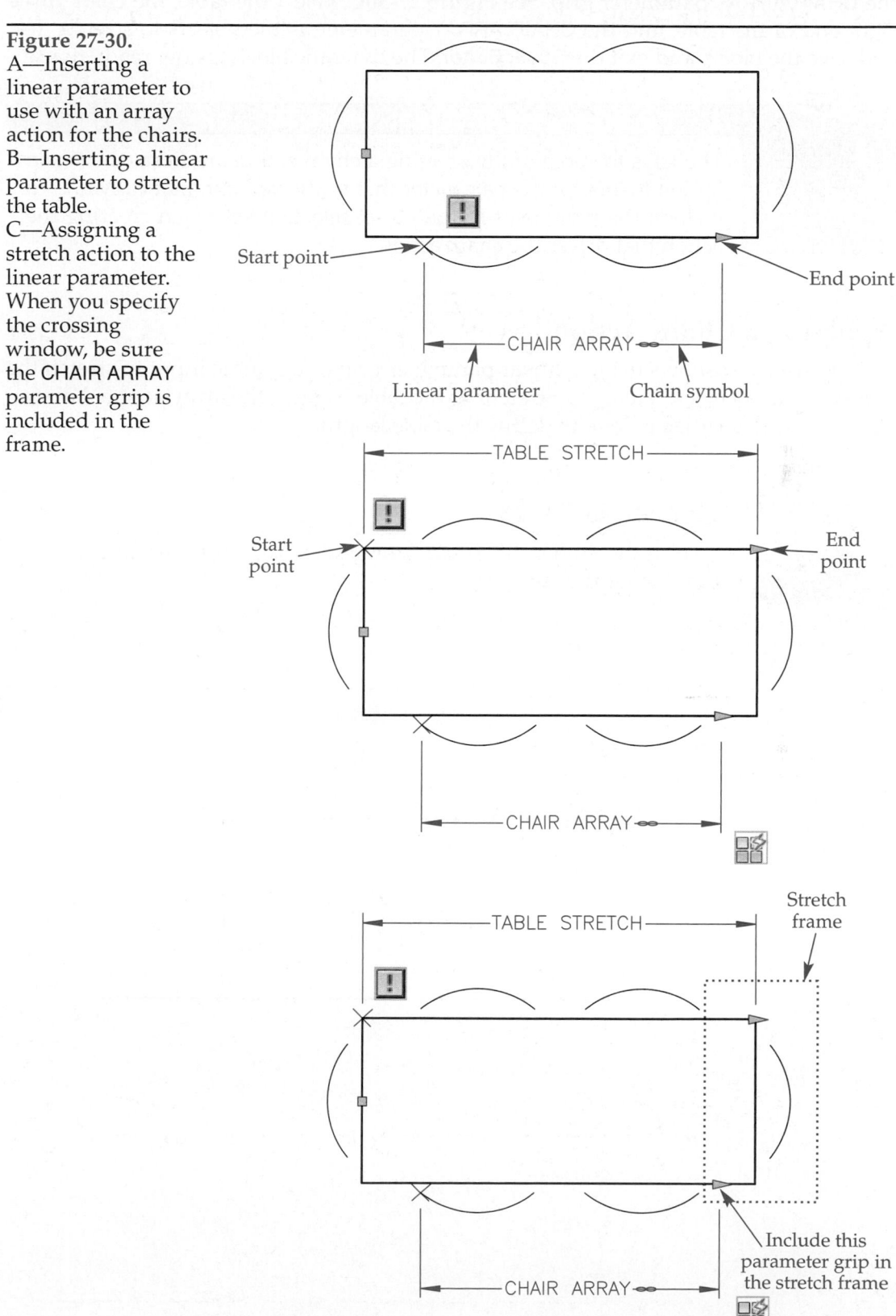

Enter the distance between rows or specify unit cell: prompt, use object snaps to snap to the endpoint of one of the chairs and then snap to the equivalent endpoint on the chair next to the first chair.

Add another single-grip linear parameter, labeled TABLE STRETCH, as shown in **Figure 27-30B.** Assign a stretch action to the parameter associated with the TABLE STRETCH parameter grip. Use a crossing window around the right end of the table and the CHAIR ARRAY parameter grip. See **Figure 27-30C.** Select the table, the chair at the right end of the table, and the CHAIR ARRAY parameter as the objects to stretch. Test and save the block, and exit the **Block Editor.** The dynamic block is now ready to use.

PROFESSIONAL TIP

The keys to successfully creating a chain action are to set the **Chain** option to **Yes** for the parameter that is affected automatically and to include the parameter in the object selection set when creating the action that drives the chain action.

Applying a Chain Action

Figure 27-31 shows using a linear parameter grip or dynamic input to stretch the table and chairs block reference. Stretching the table triggers the array action. You can also use the **Properties** palette to define the table length.

Exercise 27-11

Access the Student Web site (www.g-wlearning.com/CAD) and complete Exercise 27-11.

Figure 27-31.
As you stretch the parameter grip, the table stretches and the chairs are arrayed at the same time.

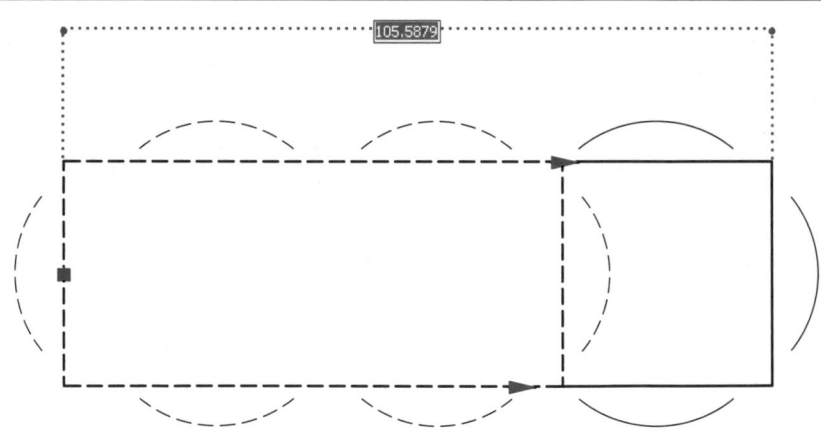

Chapter Test

Answer the following questions. Write your answers on a separate sheet of paper or go to the Student Web site (www.g-wlearning.com/CAD) and complete the electronic chapter test.

1. Define *dynamic block.*
2. What is the function of a dynamic block?
3. How are standard blocks and dynamic blocks the same? How are they different?
4. Identify the property that forms when you create a point parameter and list the actions you can assign.
5. What is the purpose of a move action?
6. When do action bars appear?
7. What information can you get by hovering over or selecting an action bar?
8. What is the basic function of a linear parameter?
9. Explain the function of a stretch action.
10. Briefly describe how to use a stretch action symmetrically.
11. What is the basic function of a scale action?
12. Describe the polar parameter type and list the actions that can be assigned to it.
13. Identify the property that forms when you create a rotation parameter and list the action you can assign.
14. Describe what happens when you move a block with an alignment parameter near another object in the drawing. How does this save drawing time?
15. Give at least one practical example of using a flip parameter.
16. Briefly describe the properties an XY parameter creates and list the actions that can be assigned.
17. Give an example of using an array action assigned to an XY parameter.
18. When would you add a base point parameter?
19. Describe the basic use of a value set.
20. Explain the basic function of a chain action.

Drawing Problems

Start AutoCAD if it is not already started. Start a new drawing using an appropriate template of your choice. The template should include layers, text styles, dimension styles, and multileader styles appropriate for drawing the given objects. Add layers, text styles, dimension styles, and multileader styles as needed. Draw all objects using appropriate layers, text styles, dimension styles, multileader styles, justification, and format. Follow the specific instructions for each problem. Use your own judgment and approximate dimensions when necessary.

▼ Basic

1. Open P25-1 and save it as P27-1. The P27-1 file should be active. Erase all copies of the steel column symbols except for the one in the lower-left corner. Insert an XY parameter into the steel column block and associate an array action with the parameter. Use the proper values for the array action to dynamically array the block to match the drawing. Use the one dynamic block to create the rest of the steel columns in the drawing. Resave the drawing.

2. Create a block named WIRE ROLL as shown below. Do not include the dimensions. Insert a linear parameter on the entire length of the roll. Use a value set with the following values: 36″, 42″, 48″, and 54″. Assign a stretch action to the parameter and associate the action with either parameter grip. Create a crossing window that will allow the length of the roll to stretch. Select all of the objects on one end and the length lines as the objects to stretch. Insert the WIRE ROLL block four times into a drawing and stretch each block to use a different value set length. Save the drawing as P27-2.

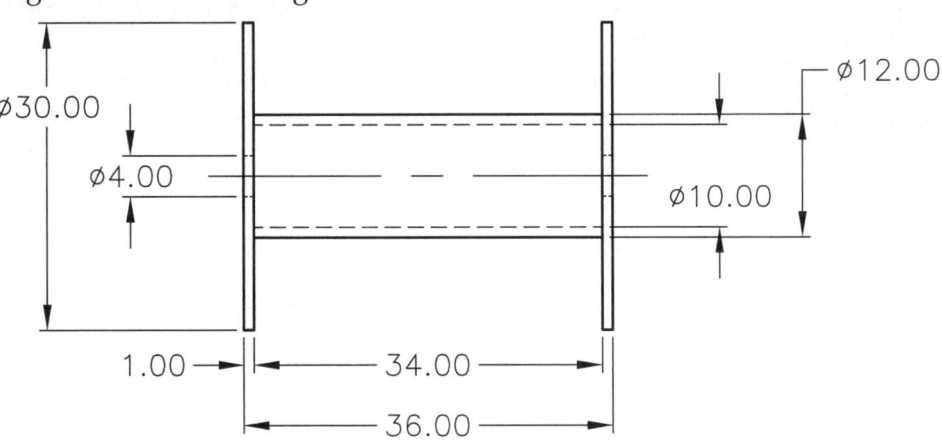

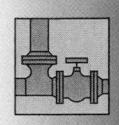

3. Create a block named **90D ELBOW** as shown below on the left. Do not include the dimensions. Insert two flip parameters and two flip actions. One of the flip parameter/action combinations is to flip the elbow horizontally. The second flip parameter/action combination is to flip the elbow vertically. Use the dynamic block to create the drawing shown below on the right. Save the drawing as P27-3.

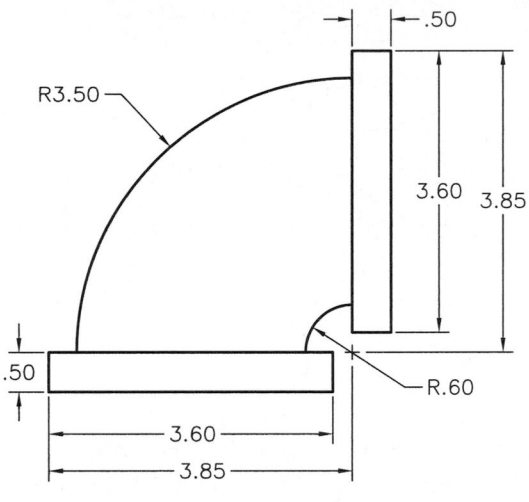

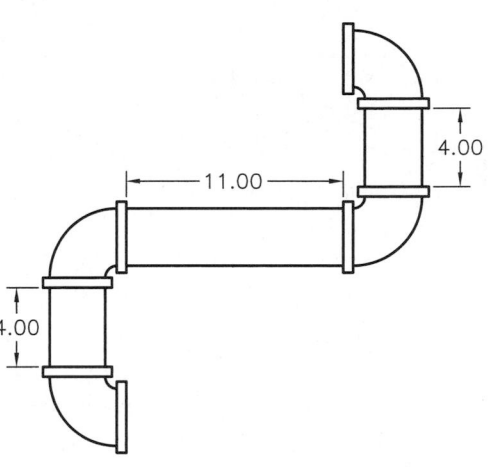

4. Create a block of the 48″ window shown below on the left. Do not include the dimensions. Insert an alignment parameter so the length of the window can align with a wall. Then draw the walls shown below on the right. Insert the window block as needed. Use the alignment parameter to align the window to the walls. Center the windows on wall segments unless dimensioned. Save the drawing as P27-4.

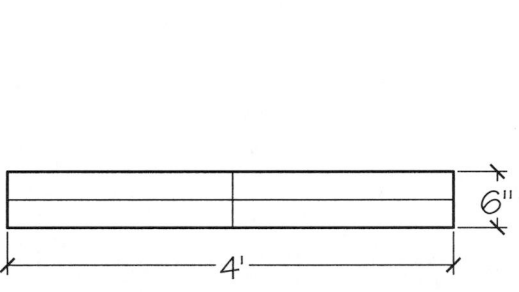

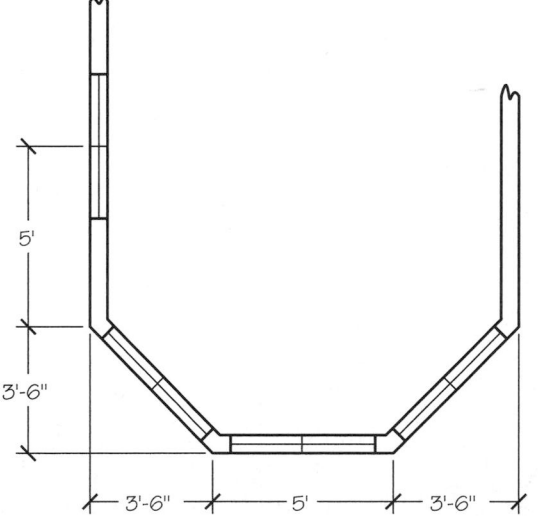

5. Create a block named **CONTROL VALVE** as shown below on the left. Include the label in the block. Insert a point parameter and assign a move action to it. Select the two lines of text as the objects to which the action applies. Insert the **CONTROL VALVE** block into the drawing three times. Use the point parameter to move the text to match the three positions shown below. Save the drawing as **P27-5**.

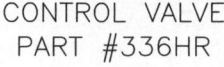

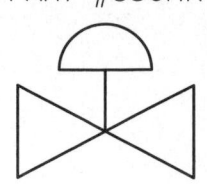

▼ Intermediate

6. Create a block named **FLANGE** as shown below. Do not include dimensions. Insert a rotation parameter specifying the center of the flange as the base point. Assign a rotate action to the parameter, selecting the six Ø.2 circles as the objects to which the action applies. Insert the **FLANGE** block into the drawing twice. Use the rotation parameter to create the two configurations shown below. Save the drawing as **P27-6**.

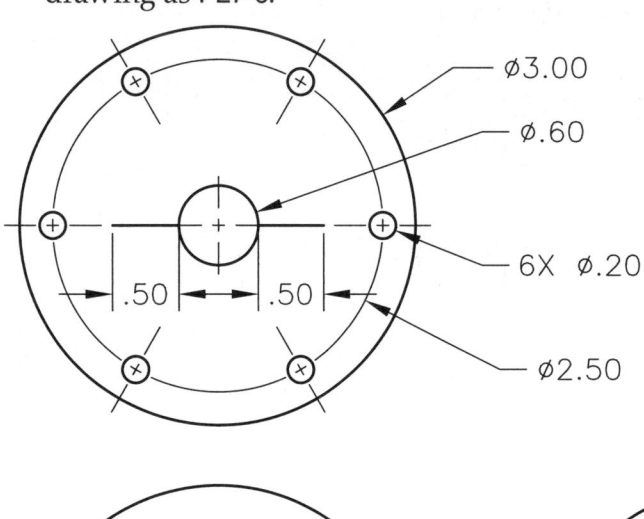

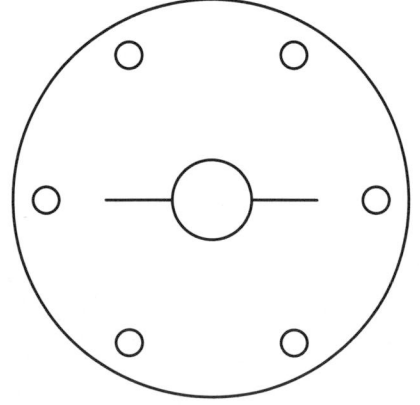
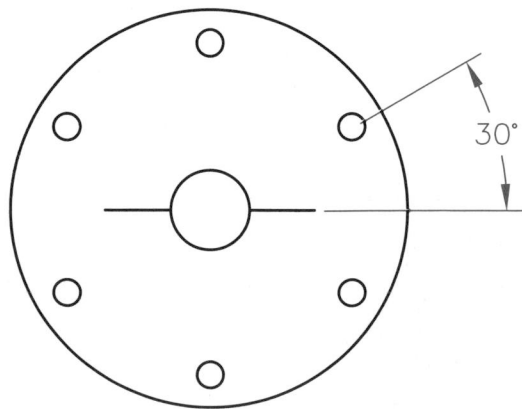

7. Open P27-6 and save it as P27-7. The P27-7 file should be active. Open the FLANGE block in the **Block Editor** and use the **Properties** palette to give the following settings to the rotation parameter:
   A. **Angle label**—BOLT HOLES
   B. **Angle description**—ROTATION OF BOLT HOLE PATTERN
   C. **Ang type**—INCREMENT
   D. **Ang increment**—30
   E. Save the changes and exit the **Block Editor**. Save the drawing.

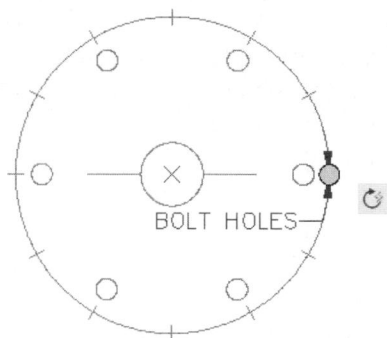

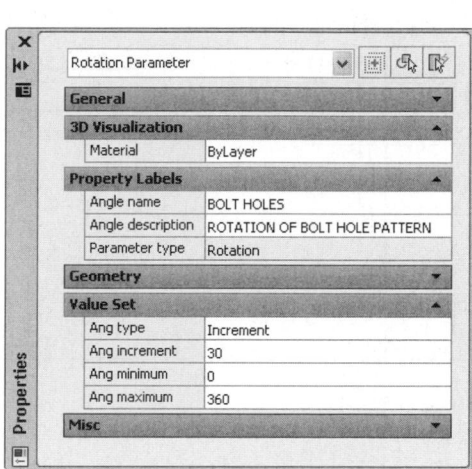

Chapter 27   Introduction to Dynamic Blocks

▼ Advanced

8. The drawing below shows a fan with an enlarged view of the motor. This fan can have one of three motors of different sizes. Create the fan as a dynamic block.
   A. Draw all the objects. Do not dimension the drawing or draw the enlarged view.
   B. Create a block named **FAN** consisting of the objects shown in the enlarged view.
   C. Open the block in the **Block Editor** and insert a linear parameter along the top of the motor (the 1.50″ dimension). Use a value set with the following values: 1.5, 1.75, and 2.
   D. Assign a scale action to the linear parameter. Select all of the objects that make up the motor as the objects to which the action applies. Use an independent base point type and specify the base point as the lower-left corner of the motor (the implied intersection).
   E. Save the block and exit the **Block Editor**.
   F. Insert the block three times into the drawing. Use the linear parameter grip to scale the motor to the three different sizes, as shown below on the right.
   G. Save the drawing as P27-8.

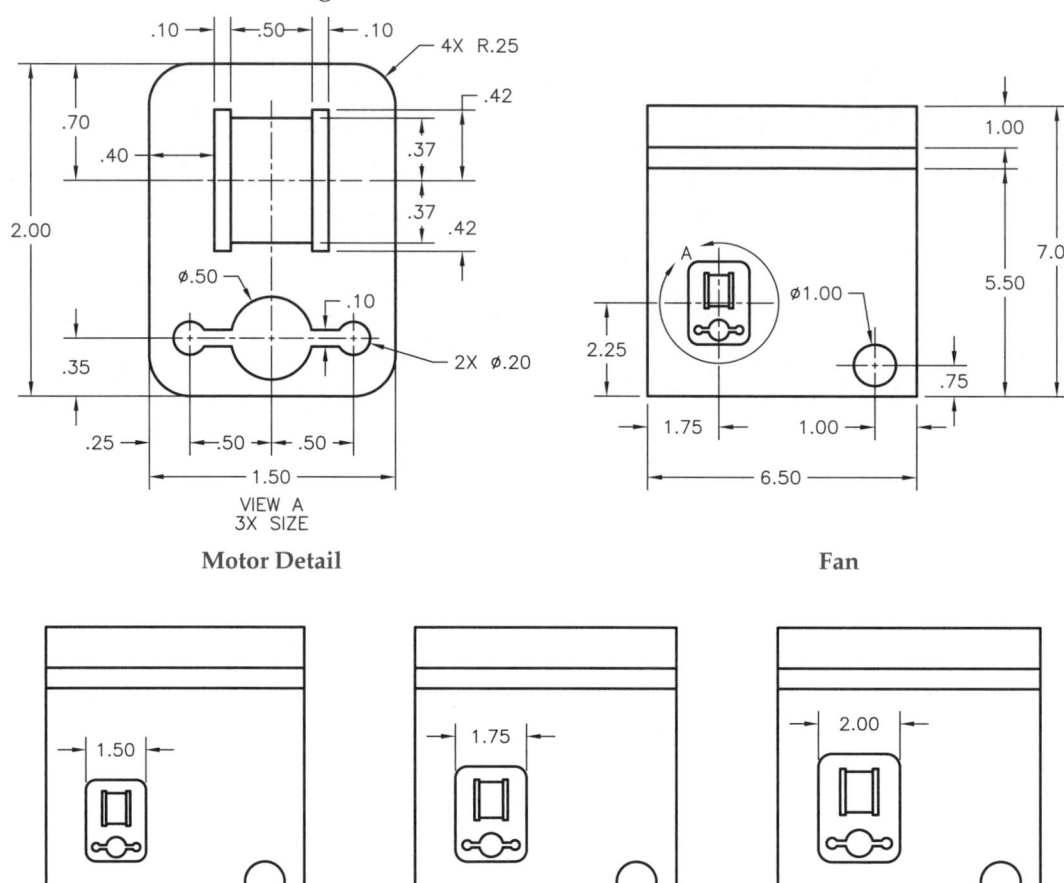

Motor Detail

Fan

ction name:

ookup1

Add Proper

CHAPTER 28

Additional Dynamic Block Tools

nput Properties

Middle Line	Start Line	End Line	Lookup
0	.5000	5000	0 ees
10	.3700	6300	10 Degrees
20	.2400	.7600	20 Degrees
<Unmatched>			Custom
			Allow reverse lookup

Learning Objectives

After completing this chapter, you will be able to do the following:
- ✓ Apply visibility and lookup parameters.
- ✓ Use parameter sets.
- ✓ Constrain block geometry.
- ✓ Use a block properties table.

This chapter describes adding visibility and lookup parameters to enhance the usefulness of blocks. You will also learn to apply geometric constraints and constraint parameters to blocks to use as an alternative or in addition to parameters and actions. Finally, this chapter explores the process of using a block properties table.

Using Visibility Parameters

A *visibility parameter* allows you to assign *visibility states* to objects within a block. Selecting a visibility state displays the only objects in the block that are associated with the visibility state. Visibility states expand on the ability to make a block into a symbol library by allowing you to hide or make visible specific objects and even completely different symbols. A block can include only one visibility parameter. Visibility parameters do not require an action.

Figure 28-1 provides a basic example of using a visibility parameter to create four different symbols from a single block. To create the block, you must draw all objects representing the different variations, as shown in **Figure 28-1A**. Then assign a visibility parameter and add visibility states that identify the objects that are visible with each variation. Insert the block and select a visibility state to display the associated objects. See **Figure 28-1B**.

To add a visibility parameter, access the **Visibility Parameter** option and pick a location for the parameter label. The parameter automatically includes a single grip. When you insert the block and select the grip, a shortcut menu appears listing visibility states. There is no prompt to select objects because the visibility parameter is associated with the entire block.

visibility parameter: A parameter that allows you to assign multiple different views to objects within a block.

visibility states: Views created by selecting block objects to display or hide.

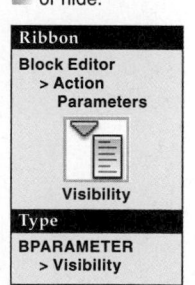

Ribbon
Block Editor
> Action
  Parameters

Visibility

Type
BPARAMETER
> Visibility

Figure 28-1.
A—All of the objects composing each unique valve are shown together. B—All four of these different valves are created from one block using a visibility parameter with different visibility states.

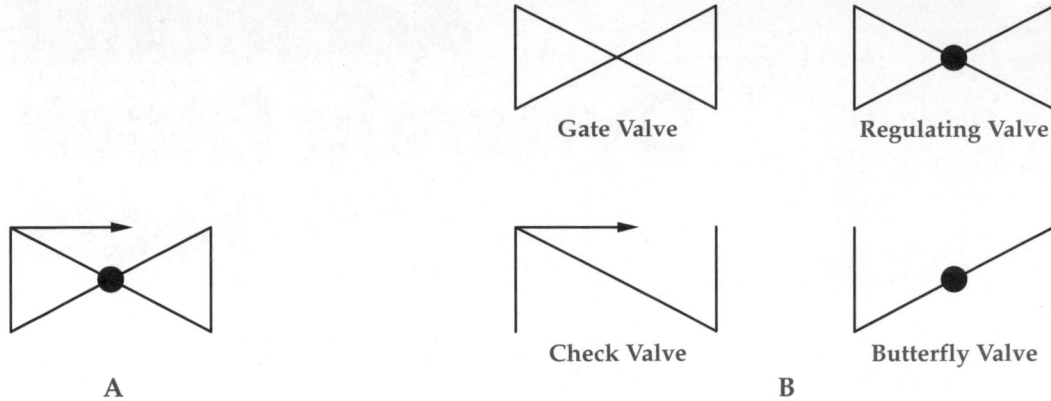

Gate Valve   Regulating Valve

Check Valve   Butterfly Valve

A   B

NOTE

Name, **Label**, **Description**, and **Palette** options are available before you specify the parameter. Most of the options are also available from the **Properties** palette if you have already created the parameter.

Creating Visibility States

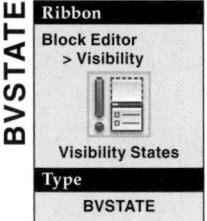
BVSTATE

Ribbon

Block Editor
> Visibility

Visibility States

Type

BVSTATE

The tools in the **Visibility** panel of the **Block Editor** ribbon tab enable when you add a visibility parameter. See **Figure 28-2.** To create a visibility state, access the **BVSTATE** tool to display the **Visibility States** dialog box. See **Figure 28-3A.** Pick the **New...** button to open the **New Visibility State** dialog box shown in **Figure 28-3B.** Type the name of the new visibility state in the **Visibility state name:** text box. For the valve block example shown in **Figure 28-1,** an appropriate name could be GATE VALVE, REGULATING VALVE, CHECK VALVE, or BUTTERFLY VALVE, depending on which valve the visibility state represents.

Pick the **Hide all existing objects in new state** radio button to make all of the objects in the block invisible when you create the new visibility state. This allows you to display only the objects that should be visible for the visibility state. Pick the **Show all existing objects in new state** radio button to make all of the objects in the block visible when you create the new visibility state. This allows you to hide objects that should be invisible for the state. Select the **Leave visibility of existing objects unchanged in new state** radio button to display the objects that are currently visible when you create the new visibility state.

Figure 28-2.
The visibility tools found in the **Visibility** panel of the **Block Editor** ribbon tab.

Visibility   Make   Make
Mode   Visible   Invisible

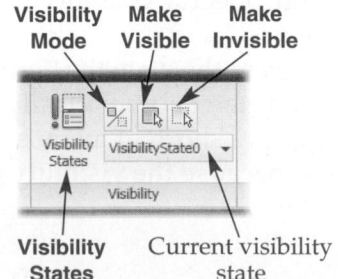

Visibility   Current visibility
States   state

Figure 28-3.
A—Manage
visibility states
using the **Visibility
States** dialog box.
B—Create new
visibility states
using the **New
Visibility State**
dialog box.

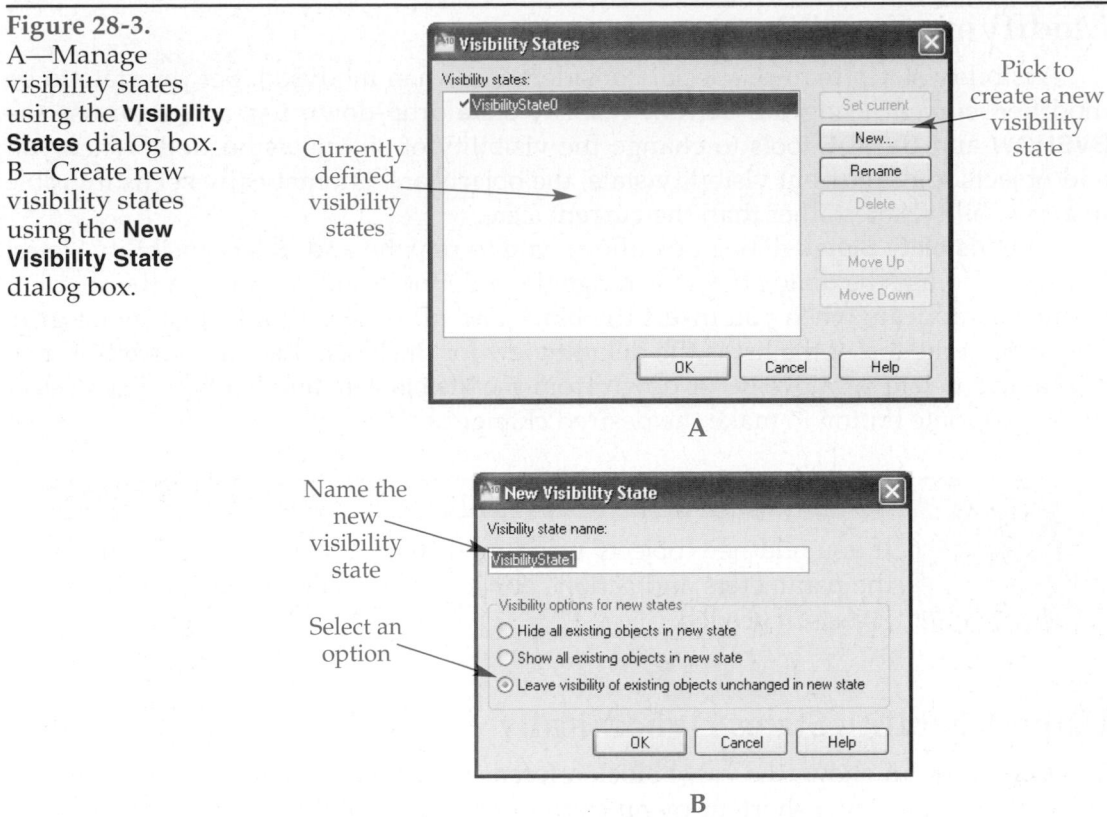

Currently
defined
visibility
states

Pick to
create a new
visibility
state

A

Name the
new
visibility
state

Select an
option

B

Pick the **OK** button to create the new visibility state. The new state adds to the list in the **Visibility States** dialog box and is current, as indicated by the check mark next to the name. Pick the **OK** button to return to block editing mode.

Next, use the **BVSHOW** and **BVHIDE** tools to display only the objects that should be visible in the current state. Pick the **Make Visible** button to select objects to make visible. Invisible objects temporarily display as semitransparent for selection. Pick the **Make Invisible** button to select objects to make invisible. For example, to make a visibility state to depict the gate valve shown in **Figure 28-4B** from the valve block shown in **Figure 28-4A,** use the **Make Invisible** tool to turn off the filled circle and the arrow. The changes are automatically saved to the visibility state. Use the **BVMODE** tool to toggle the visibility mode on and off. Turn on visibility mode to display invisible objects as semitransparent. Turn off visibility mode to display only visible objects.

Repeat the process to create additional visibility states for the block. The valve block example requires four visibility states. The **Current visibility state** drop-down list displays the current visibility state. Select a state from the drop-down list to make the state current. After you create all the visibility states, test and save the block, and exit the **Block Editor**. The dynamic block is now ready to use.

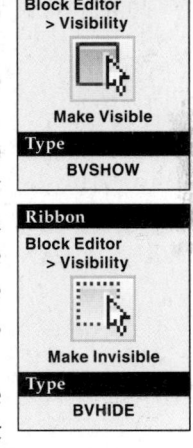

Ribbon
**Block Editor
> Visibility**

Make Visible

Type
BVSHOW

Ribbon
**Block Editor
> Visibility**

Make Invisible

Type
BVHIDE

Figure 28-4.
A—The VALVE block
with all objects
visible. B—The
VALVE block after
making the arrow
and filled circle
invisible to create
the GATE VALVE
visibility state.

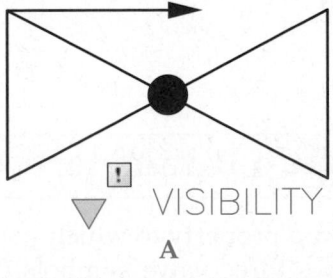

A

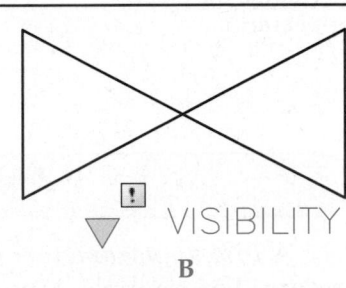

B

Modifying Visibility States

Visibility states require special consideration when modified. Set the state to be modified current using the **Current visibility state** drop-down list, and then use the **BVSHOW** and **BVHIDE** tools to change the visibility of objects as needed. When you add objects to the current visibility state, the objects are automatically set as invisible in all visibility states other than the current state.

The **Visibility States** dialog box allows you to rename and delete visibility states. You can also use the dialog box to arrange the order of visibility states in the shortcut menu that appears when you insert the block and pick the visibility parameter grip. The state at the top of the list is the default view for the block. Pick the visibility state to rename, delete, or move up or down from the **Visibility states:** list box. Then select the appropriate button to make the desired change.

> **PROFESSIONAL TIP**
>
> If you add new objects when modifying a state, be sure to update the parameters and actions applied to the block to include the new objects, if needed.

Using Visibility States Dynamically

Figure 28-5A shows the valve block reference selected for editing. Select the visibility grip to display a shortcut menu containing each visibility state. A check mark indicates the current visibility state. To switch to a different view of the block, select the name of the visibility state from the list. See **Figure 28-5B**. You can also use the **Properties** palette to select a visibility state.

Exercise 28-1

Access the Student Web site (www.g-wlearning.com/CAD) and complete Exercise 28-1.

Figure 28-5.
A—Pick the visibility parameter grip to display a shortcut menu with the available visibility states. The current state is checked.
B—Select a different visibility state from the shortcut menu to change the block appearance.

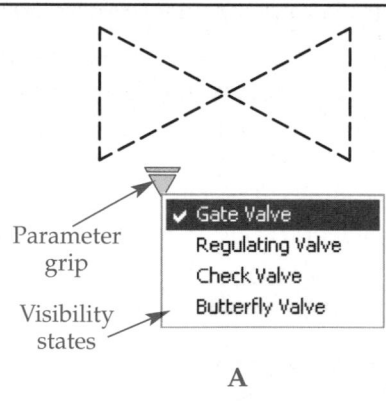

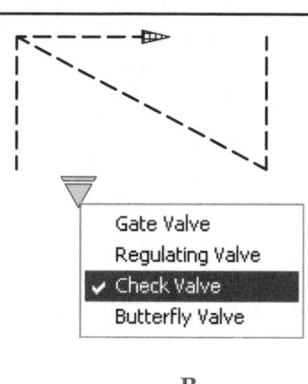

Parameter grip

Visibility states

A

B

Using Lookup Parameters

A *lookup parameter* creates a lookup property to which you can assign a *lookup action*. For example, **Figure 28-6** shows three valve symbols created from a single block by adjusting the rotation parameter of the middle line. The lookup action allows the middle line rotation to control the length of the start and end lines.

Figure 28-6.
A lookup parameter allows you to create these three views of the same block. Notice how the geometry changes.

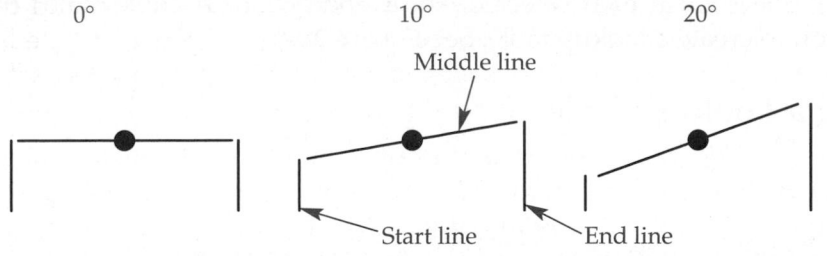

To create the valve block shown in **Figure 28-7,** first draw the geometry of the 0° symbol. Then add a linear parameter, labeled Start Line. Select the start point as the bottom of the start line, and the endpoint as the top of the start line. Assign a stretch action to the parameter that is associated with the top parameter grip. Draw the crossing window around the top of the start line and select the start line as the object to stretch.

Add another linear parameter, labeled End Line. Select the start point as the bottom of the end line and the endpoint as the top of the end line. Assign a stretch action to the parameter that is associated with the top parameter grip. Draw the crossing window around the top of the end line and select the end line as the object to stretch.

Next, add a rotation parameter, labeled Middle Line. Specify the base point as the center of the circle. Select the right endpoint of the middle line to set the radius, and specify the default rotation angle as 0. Assign a rotation action to the parameter. Pick the center of the circle as the rotation base point, and select the middle line as the object to rotate.

To add a lookup parameter, access the **Lookup Parameter** option and pick a location for the parameter label. Then enter the number of grips to associate with the parameter. The default **1** option creates a single lookup grip that allows you to use grip editing to carry out the lookup action. When you insert the block and select the grip, a shortcut menu appears listing rotation options. There is no prompt to select objects because a lookup parameter is associated with the entire block.

Ribbon
Block Editor
> Action
Parameters
Lookup
Type
BPARAMETER
> Lookup

NOTE

Name, Label, Description, and **Palette** options are available before you specify the parameter. Most of the options are also available from the **Properties** palette if you have already created the parameter.

Figure 28-7.
The example block with linear parameters and stretch actions assigned to the start and end lines and a rotation parameter and rotate action assigned to the middle line.

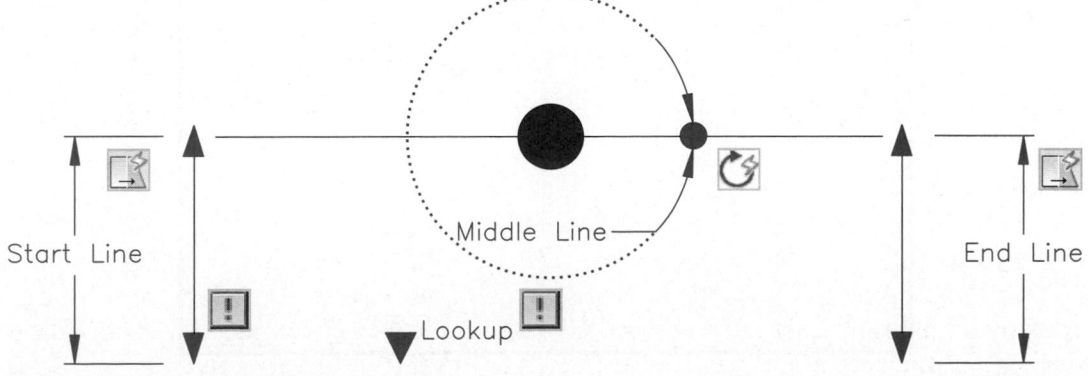

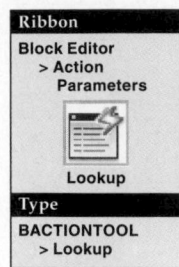

Ribbon

**Block Editor
> Action
Parameters**

Lookup

Type

**BACTIONTOOL
> Lookup**

lookup table: A table that groups the properties of parameters into custom-named lookup records.

Assigning a Lookup Action

To assign a lookup action, access the **Lookup Action** option and pick a lookup parameter if one is not already selected. The **Property Lookup Table** dialog box appears, allowing you to create a lookup table. See **Figure 28-8.**

Creating a Lookup Table

A *lookup table* groups the properties of parameters into custom-named lookup records. The **Action name:** display box indicates the name of the lookup action associated with the table. The table is initially blank. To add a parameter property, pick the **Add Properties...** button to open the **Add Parameter Properties** dialog box. See **Figure 28-9.**

All parameters in the block containing property values appear in the **Parameter properties:** list. Lookup, alignment, and base point parameters do not contain property values. Notice that the property name is the parameter label. The **Property type** area determines the type of property parameters shown in the list. By default, the **Add input properties** radio button is active, which displays available input property parameters. To display available lookup property parameters, select the **Add lookup properties** radio button.

To add parameter properties to the lookup table, select the properties in the **Parameter properties:** list and pick the **OK** button. A new column, named as the parameter property, forms for each parameter in the **Input Properties** area of the **Property Lookup Table** dialog box. See **Figure 28-10.** The **Input Properties** area allows you to specify a value for parameters added to the table. Type a value in each cell in the column. Add a custom name for each row, or record, in the **Lookup** column in the **Lookup Properties** area. This area displays the name that appears in the shortcut menu when you insert the block and select the lookup parameter grip.

For the valve block example, add the Middle Line, Start Line, and End Line parameter properties to the table. Then complete the lookup table as shown in **Figure 28-10.** Start with the Middle Line values. Press [Enter] after typing the value to add a new blank row and then type the remaining values in each cell. Use the [Enter], [Tab], arrow keys, or pick in a different cell to navigate through the table.

Figure 28-8.
The **Property Lookup Table** dialog box.

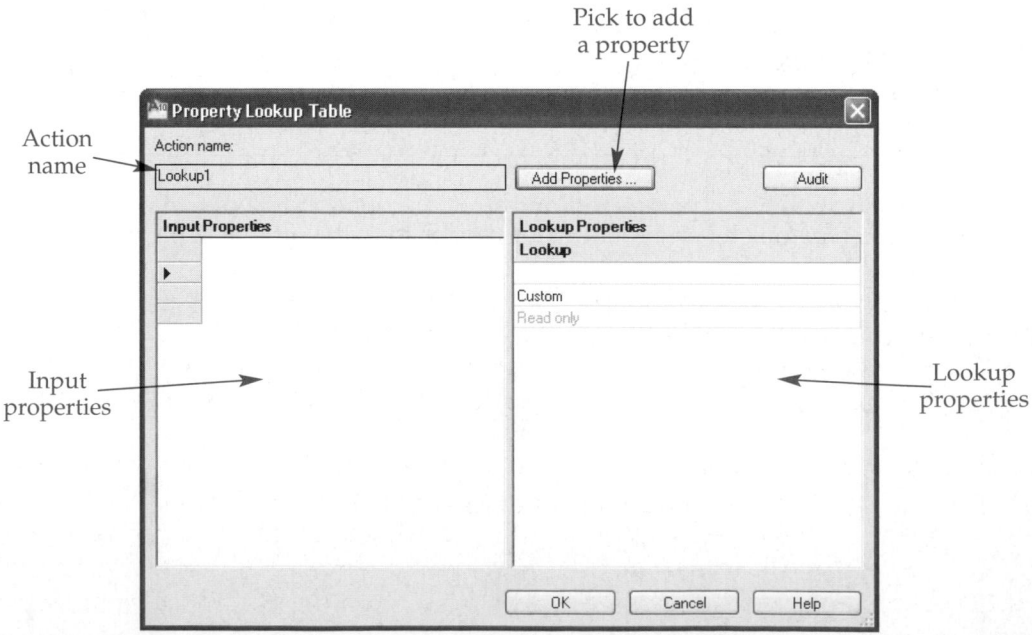

Figure 28-9.
Parameter properties are listed in the **Add Parameter Properties** dialog box.

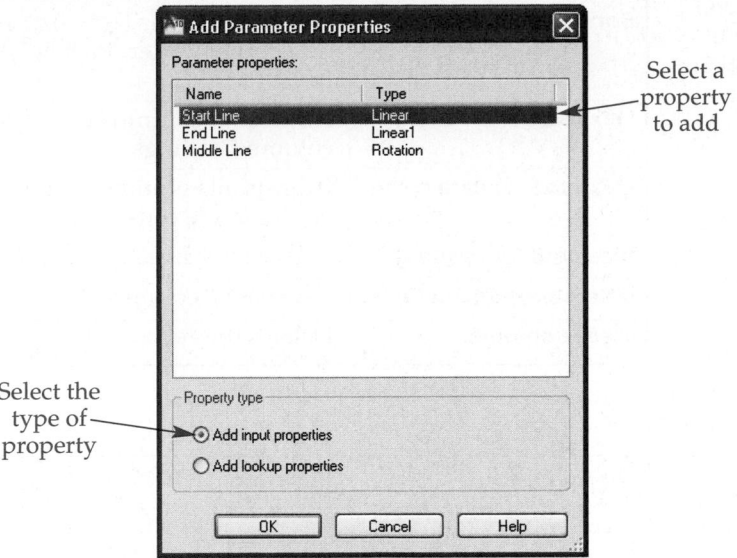

Select a property to add

Select the type of property

Figure 28-10.
A lookup table with multiple parameters and values added.

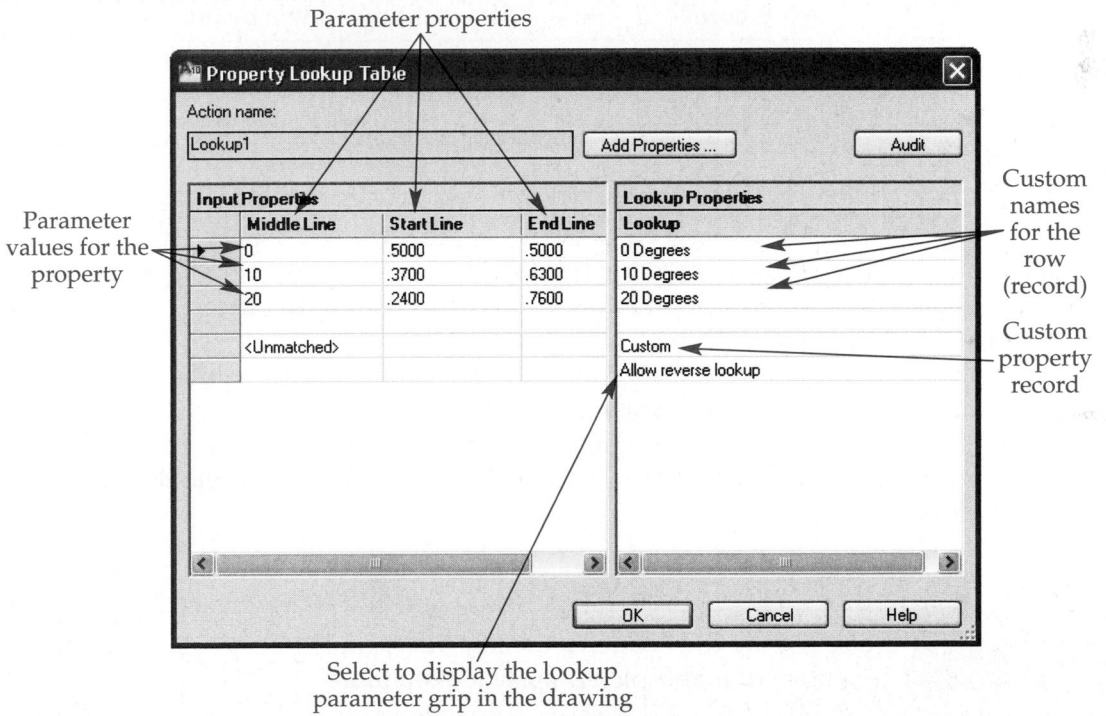

Parameter properties

Parameter values for the property

Custom names for the row (record)

Custom property record

Select to display the lookup parameter grip in the drawing

The row, or record, that contains the <Unmatched> value, named Custom in the **Lookup** column, applies when the current parameter values of the block do not match any of the records in the table. This allows you to adjust the block using parameter values other than those specified in the lookup table. You cannot add any values to the row, but you can change the name of **Custom**.

The Allow reverse lookup setting at the bottom of the **Lookup** column is available only if all of the names in the lookup table are unique. The option allows the lookup parameter grip to display when you select the block. Pick the grip to choose a specific lookup record. The Read only setting appears if you do not name a lookup property, or if two or more properties have the same name. Selecting **Read only** from the drop-down list disallows selecting a lookup record.

Figure 28-11.
A—Options
available when you
right-click on a
column. B—Options
available when you
right-click on a row.

Menu Option	Function
Sort	Sorts the records (rows) in ascending or descending order. Pick again to reverse the sort order.
Maximize all headings	Adjusts all columns to the width of the column headings.
Maximize all data cells	Adjusts all columns to the width of the values in the cells.
Size columns equally	Makes all columns equal in width.
Delete property column	Deletes the column.
Clear contents	Deletes the cell values.

A

Menu Option	Function
Insert row	Inserts a new row above the selected row.
Delete row	Deletes the record (row).
Clear contents	Deletes the cell values.
Move up	Moves the row up by one row.
Move down	Moves the row down by one row.
Range syntax examples	Displays the online documentation examples of how to enter values into a lookup table.

B

You can right-click on a column heading to access a menu with options for adjusting columns, or right-click on a row to access a menu with options for adjusting rows. **Figure 28-11** briefly describes each option.

After you add all required properties to the table and assign values to each, pick the **Audit** button in the **Property Lookup Table** dialog box to check each record in the table to make sure they are all unique. If no errors are found, pick the **OK** button to return to the **Block Editor**. Test and save the block, and exit the **Block Editor**. The dynamic block is now ready to use.

NOTE

To redisplay the **Property Lookup Table** dialog box, right-click on a lookup action and pick **Display lookup table**.

Using a Lookup Action Dynamically

Figure 28-12 shows a valve block reference selected for editing. The figure shows the **Property Lookup Table** dialog box for reference only. Since Allow reverse lookup is set in the lookup table, the lookup parameter grip appears along with the other parameter grips. Pick the lookup parameter grip to display a shortcut menu containing each lookup record. The entries in this shortcut menu match the entries in the **Lookup** column of the **Property Lookup Table** dialog box. A check mark indicates the current record. To switch to a different view of the block, select the name of the record from the list.

Figure 28-12.
The lookup parameter grip displays when you select the block. The list of available lookup records displays when you pick the lookup parameter grip. Notice the correlation between the available options and the lookup property names in the **Property Lookup Table** dialog box.

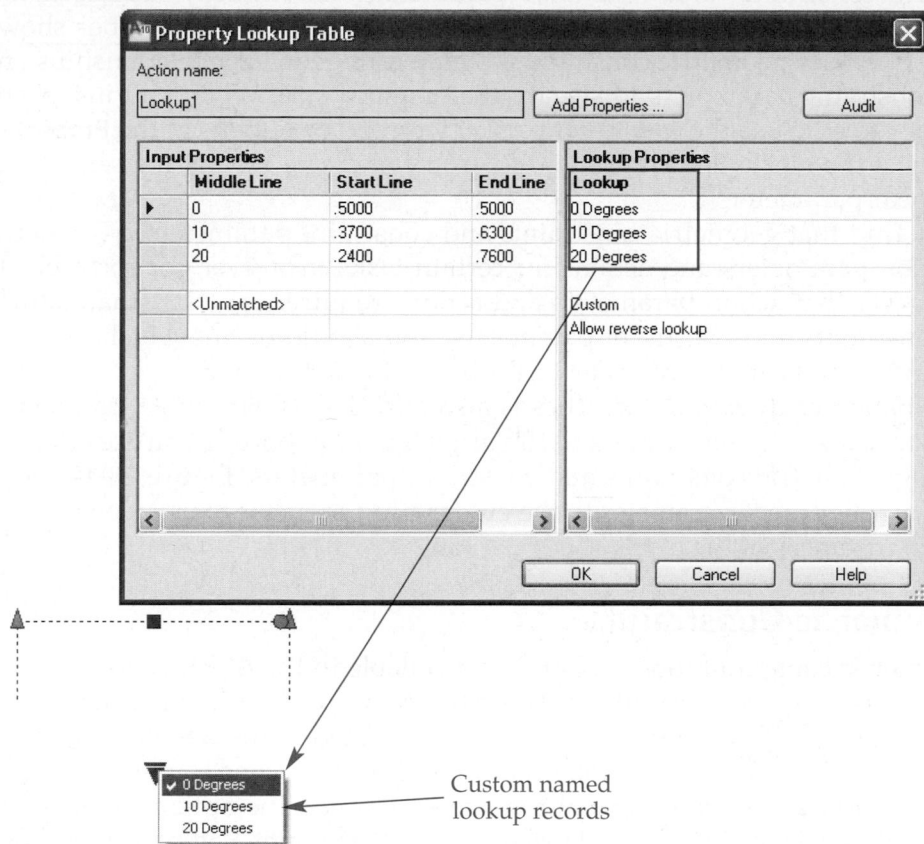

Custom named lookup records

You can change other parameters assigned to the block, such as the linear and rotation parameters of the example block, independently of the named records. When you change any of the parameters, the lookup parameter becomes Custom, because the current parameter values do not match one of the records in the lookup table.

Exercise 28-2

Access the Student Web site (www.g-wlearning.com/CAD) and complete Exercise 28-2.

Using Parameter Sets

The **Parameter Sets** tab of the **Block Authoring Palettes** window contains common parameter and actions grouped to enhance productivity. Follow the prompts to create the parameter and automatically associate an action with the parameter. The action forms without any selected objects, as is indicated by the yellow alert icon. If the parameter set contains an action that must include associated objects, as most do, double-click on the action icon and select objects. The prompts may differ depending on the type of action.

Geometric constraints and constraint parameters can directly replace action parameters and actions. For example, the block of the cut framing member shown in **Figure 28-13A** uses geometric constraints to maintain geometric relationships and two linear constraint parameters to specify the member size. When you insert and select the block to edit, use the constraint parameter grips or options in the **Properties** palette to adjust the block. See **Figure 28-13B.** You can create exactly the same block using two linear parameters.

You may find that geometric constraints and constraint parameters are easier to use than action parameters and actions for certain tasks. However, for some blocks you will discover that action parameters and actions require less effort than adding geometric constraints and constraint parameters. You should decide which dynamic block tools and options are appropriate for the blocks you create.

A combination of dynamic properties is also effective. For example, parameters and actions such as alignment, array, and flip offer dynamic controls that are often not possible using geometric constraints and constraint parameters. **Figure 28-14** shows how adding an alignment parameter to the cut framing member block allows you to size and align instances of the block.

Using Geometric Constraints

The geometric constraint tools and options available in the **Block Editor** are identical to those you use to geometrically constrain a parametric drawing. The tools in the **Geometric** panel of the **Parametric** ribbon tab are duplicated in the **Geometric** panel of the **Block Editor** ribbon tab for convenience. Use the geometric constraints in the **Block Editor** as you would in the drawing environment, including the options for relaxing and deleting constraints. The same shortcut menu, **Constraint Settings** dialog box, and **Properties** palette functions apply. Review Chapter 22 for information on adding geometric constraints.

Figure 28-13.
A—A cut framing member block made dynamic using geometric constraints and constraint parameters. B—Using the default 2x4 block to create a 4x4 symbol.

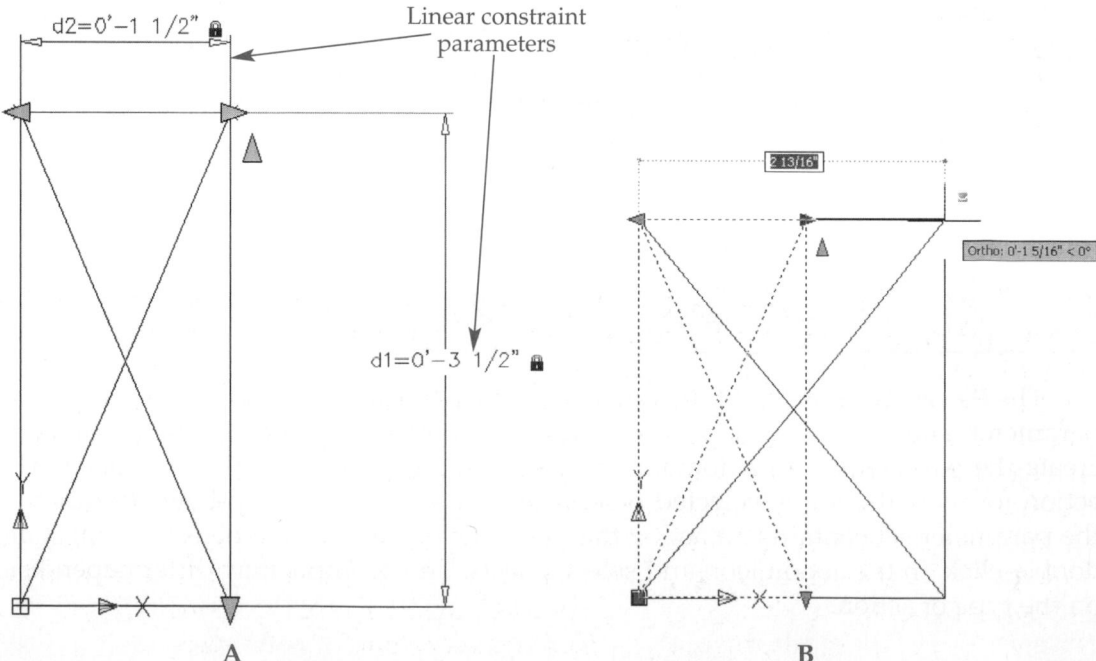

A   B

Figure 28-14.
An example of using geometric constraints and constraint parameters to adjust the cut framing member size. An alignment parameter aligns each member for unique applications.

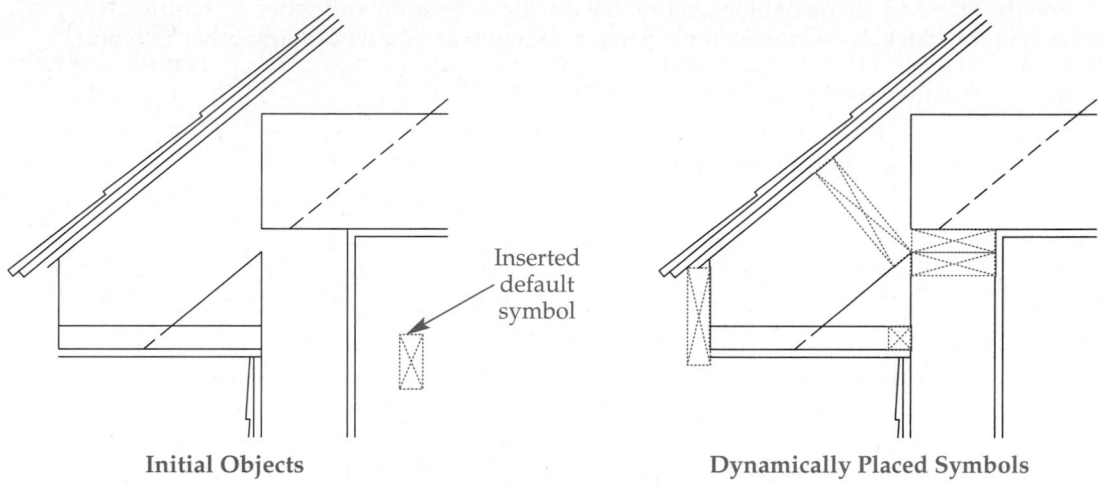

Inserted default symbol

Initial Objects   Dynamically Placed Symbols

You can assign constraints to block objects before you define the block or during block editing to create a dynamic block. See **Figure 28-15A.** Once you define and insert the block, only constraint parameters, action parameters, or actions influence geometric constraints. This allows you to use blocks as objects in parametric drawings. For example, you can insert and rotate the block, as shown in **Figure 28-15B,** even though the block definition includes a horizontal constraint. Use constraints in the drawing to locate blocks and establish geometric relationships between blocks and other objects. See **Figure 28-15C.**

> **NOTE**
>
> You can also use geometric constraints in the block environment to form geometric constructions in specific situations when standard AutoCAD tools are inefficient or ineffective.

Exercise 28-3

Access the Student Web site (www.g-wlearning.com/CAD) and complete Exercise 28-3.

Using Constraint Parameters

Constraint parameters replace dimensional constraints in the **Block Editor**. To help avoid confusion, remember that dimensional constraints constrain a parametric drawing, including block references, as shown in **Figure 28-15C.** Constraint parameters constrain the size and location of block components. By default, dimensional constraints are gray and constraint parameters are blue. You also have the option of converting dimensional constraints into constraint parameters.

You can often use constraint parameters instead of action parameters and actions. If you do not use action parameters, you must include constraint parameters to create a dynamic block. The constraint parameter tools and options available in the **Block Editor** function much like those you use to dimensionally constrain a parametric drawing. Review Chapter 22 for information on adding dimensional constraints.

constraint parameters: Dimensional constraints available for block construction to control the size or location of block geometry numerically.

Figure 28-15.
A—A wide flange block made dynamic using geometric constraints, including a horizontal constraint, and constraint parameters. B—You can rotate the block, because constraints define the size and shape of block geometry during definition and when dynamically adjusting the block. C—Constrain blocks in a drawing as you would any other geometry.

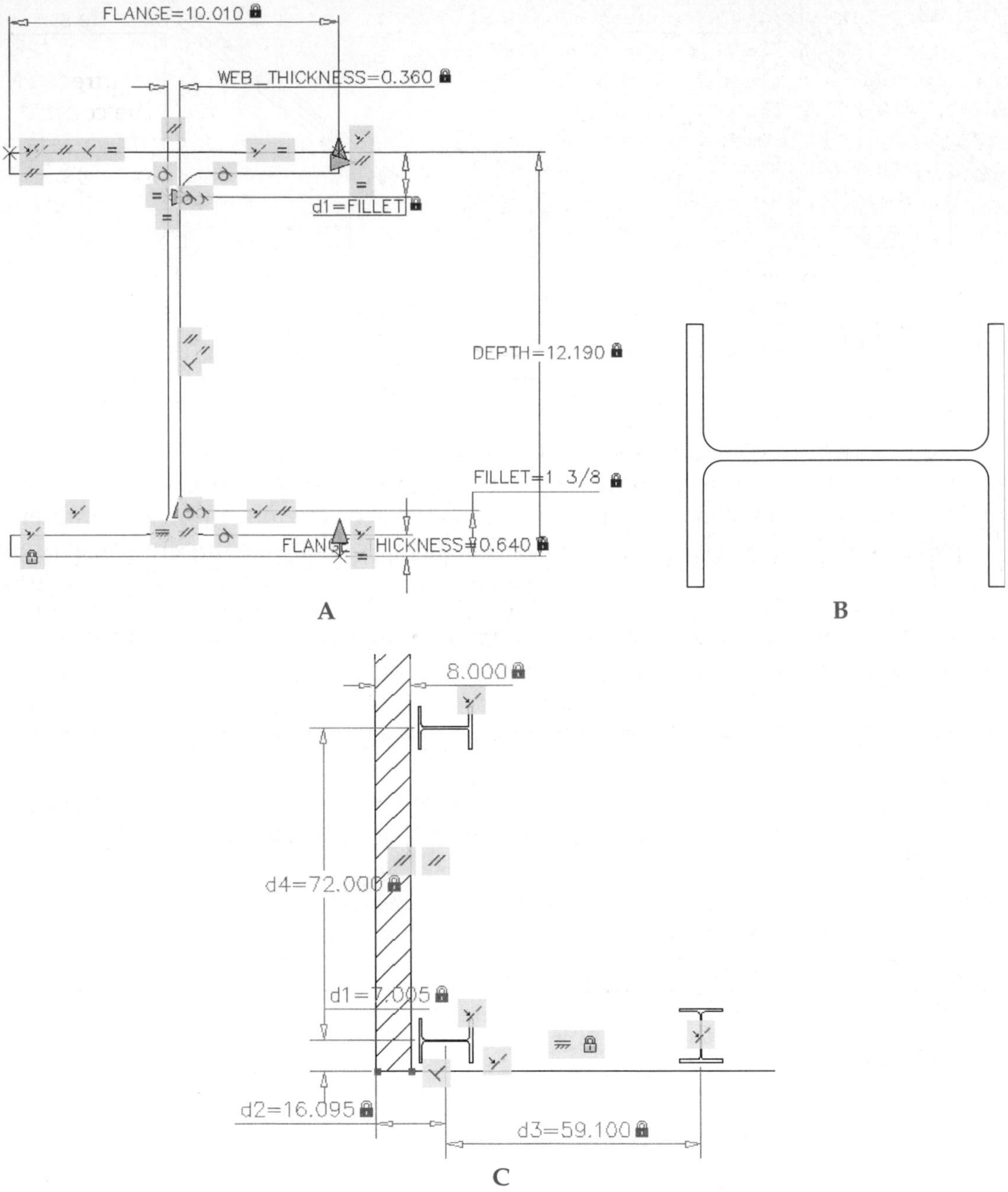

FLANGE=10.010

WEB_THICKNESS=0.360

d1=FILLET

DEPTH=12.190

FILLET=1 3/8

FLANGE_THICKNESS=0.640

A

B

8.000

d4=72.000

d1=7.005

d2=16.095

d3=59.100

C

Using the BCPARAMETER Tool

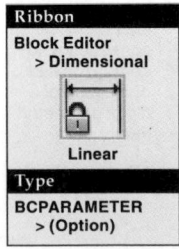

Ribbon

**Block Editor
> Dimensional**

Linear

Type

**BCPARAMETER
> (Option)**

The **BCPARAMETER** tool replaces the **DIMCONSTRAINT** tool in the **Block Editor**, and provides **Linear**, **Horizontal**, **Vertical**, **Aligned**, **Diameter**, and **Radius** options. The **Linear** option shows by default in the **Block Editor** ribbon tab. You can also use the **BCPARAMETER** tool to convert dimensional constraints to constraint parameters. Each constraint parameter is a separate **DIMCONSTRAINT** tool option. The quickest way to add or convert constraint parameters using this tool is to pick the appropriate button from the **Dimensional** panel of the **Block Editor** ribbon tab.

The process of adding constraint parameters is identical to adding dimensional constraints, except that constraint parameters can include grips. Constraint parameters are essentially a combination of dimensional constraints and action parameters. In the example shown in **Figure 28-16,** the constraint parameters given custom names are those that adjust for unique block references. As when creating a parametric drawing, the other constraint parameters are required to define the block and associate specific geometric relationships. Notice the expressions applied to these values.

To create a constraint parameter, follow the prompts to make the required selections, pick a location for the dimension line, and enter a value to form the constraint. When prompted, specify the number of grips. The radius constraint parameter allows you to add 0 or 1 grip. All other constraint parameters can include 0, 1, or 2 grips. If you plan to assign a single grip to a constraint parameter, select the point associated with the grip second. If you choose the **0** option, you can only use the **Properties** palette to adjust the block.

NOTE

If you attempt to over-constrain a block, a message appears indicating that adding the geometric constraint or constraint parameter is not allowed. You cannot create reference constraint parameters.

PROFESSIONAL TIP

As when adding dimensional constraints or action parameters, change the constraint parameter name to a custom, more descriptive name. Naming labels helps you organize parameters and recognize the parameter during editing. Custom parameters also appear in the **Custom** category of the **Properties** palette.

The **Convert** option of the **BCPARAMETER** tool allows you to convert a dimensional constraint to a constraint parameter. This allows you to prepare a dynamic block using existing dimensional constraints. Access the **Convert** option and pick the dimensional constraint to convert. The dimensional constraint becomes the corresponding constraint parameter and includes the default number of grips.

Ribbon
Block Editor > Dimensional
Convert
Type
BCPARAMETER > Convert

Figure 28-16.
Using constraint parameters to form a dynamic block of a spacer. A single grip is all that is required for these constraint parameters. Do not assign or rename constraint parameters that do not control geometry.

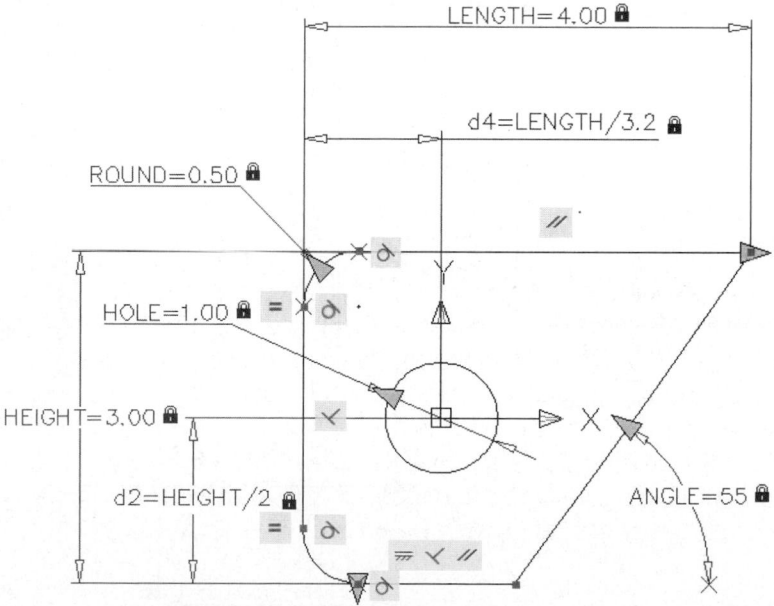

Controlling Constraint Parameters

Control and adjust constraint parameters using a combination of the same techniques you use to manage dimensional constraints and action parameters. Many of the options from shortcut menus, the **Constraint Settings** dialog box, and the **Properties** palette apply. Right-click with no objects selected to access options for displaying and hiding parametric constraints and for accessing the **Constraint Settings** dialog box. Select a constraint parameter and then right-click to display a shortcut menu with options for editing the constraint, changing the format of the name, and redefining the grips.

As with dimensional constraints and the action parameters, the **Properties** palette provides an effective way to control and enhance constraint parameters. You can also use the **Parameters Manager**. **Figure 28-17** shows a foundation detail block with linear constraint parameters. Notice the multiple options available in the **Properties** palette for adjusting the selected constraint parameter.

The options in the **Value set** category of the **Properties** palette allow you to assign value sets to a constraint parameter. Each constraint parameter in the **Figure 28-17** example uses an incremental value to help ensure that you select an appropriate value when adjusting a block reference. You can also create a list of possible sizes. The processes of creating a value set in the **Properties** palette and using value sets are identical for constraint parameters and action parameters.

Exercise 28-4

Access the Student Web site (www.g-wlearning.com/CAD) and complete Exercise 28-4.

Figure 28-17.
Adjust constraint parameters as you would dimensional constraints and action parameters. Use the **Properties** palette to add value sets.

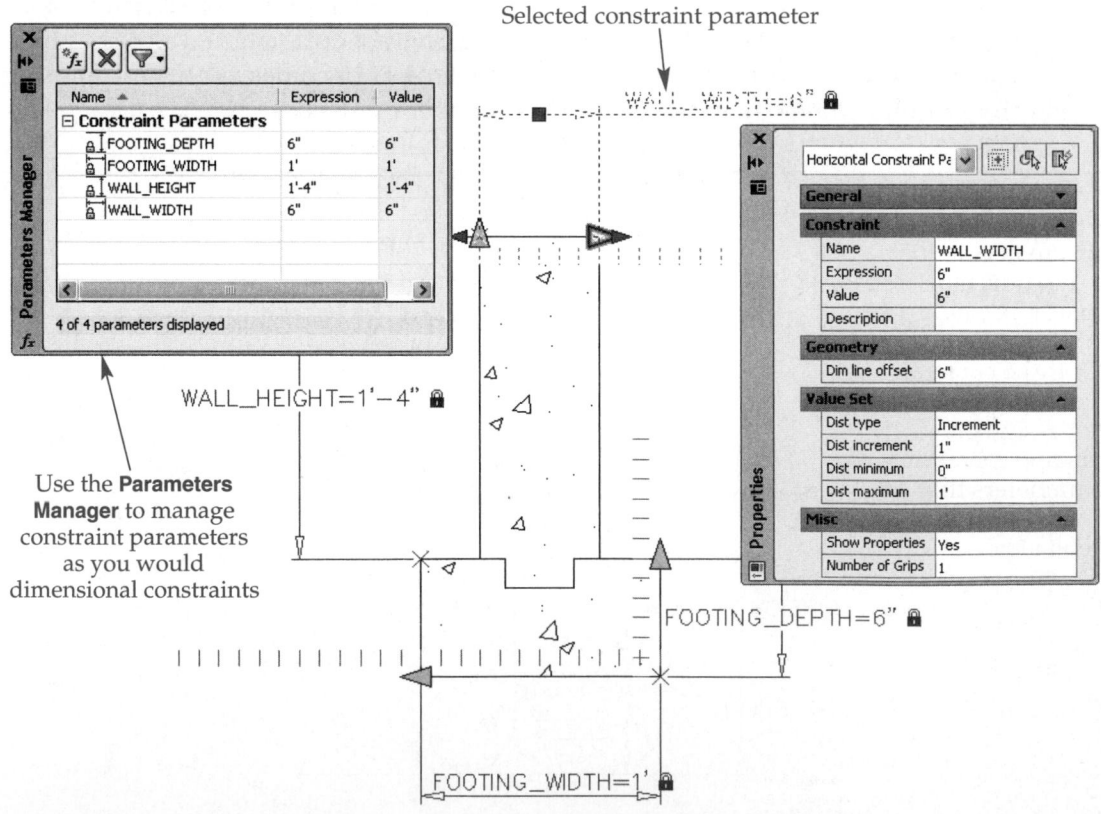

Additional Parametric Tools

The **Block Editor** offers additional options for adding constraints to blocks. Many of the tools, such as the **DELCONSTRAINT** tool, function the same in block editing mode as in drawing mode. However, the **Block Editor** does offer some unique parametric construction tools.

The **BCONSTRUCTION** tool allows you to create construction geometry to aid geometric construction and constraining. Construction geometry appears only in the block definition. See **Figure 28-18.** Access the **BCONSTRUCTION** tool and select the objects to convert to or revert from construction geometry. Press [Enter] or the space bar, or right-click and pick **Enter.** Next, choose the **Convert** option to convert non-construction objects to the construction format, or **Revert**, to return construction geometry to the standard format. You can also use the **Hide all** option to hide all existing construction geometry before selecting objects, or use the **Show all** option to display all construction geometry.

Use the **BCONSTATUSMODE** tool to toggle constraint status identification on and off. When you turn on constraint status mode, objects with no constraints appear white (black) by default, objects assigned some form of constraints are blue, and fully constrained geometry is magenta. If the block contains a constraint error, objects associated with the error are red. Using constraint status is helpful, especially if you want to constrain objects in a certain order or confirm that geometry has been fully constrained.

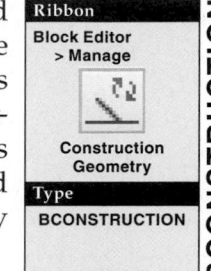

Ribbon
Block Editor > Manage
Construction Geometry
Type
BCONSTRUCTION

BCONSTRUCTION

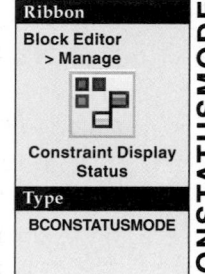

Ribbon
Block Editor > Manage
Constraint Display Status
Type
BCONSTATUSMODE

BCONSTATUSMODE

NOTE

Use the **BESETTINGS** tool to access the **Block Editor Settings** dialog box. There you can adjust parameter and parameter grip color and appearance, constraint status colors, and other **Block Editor** settings.

Using a Block Properties Table

AutoCAD 2010 NEW

A *block properties table* allows you to assign specific values to multiple block properties, and then select a unique group, or row, of properties to create block references. The concept is similar to using a lookup action parameter. A block properties table can include action parameters and/or constraint parameters. You can also add attributes to the table, which is often appropriate for naming each record, or row.

block properties table: A table of action parameters and/or constraint parameters that allows you to create multiple block properties and then select them to create block references.

Figure 28-18.
An example of a weld nut block that uses construction geometry to aid geometric construction and constraining.

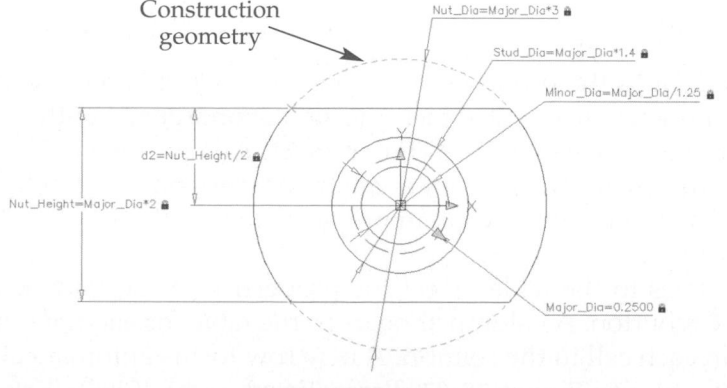
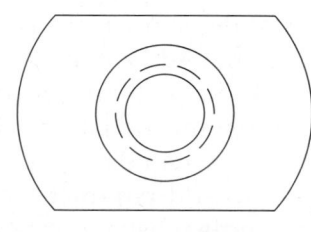

Construction geometry

Block Definition

Block Reference

Figure 28-19.
A heavy hex nut block definition ready to use to create a block properties table.

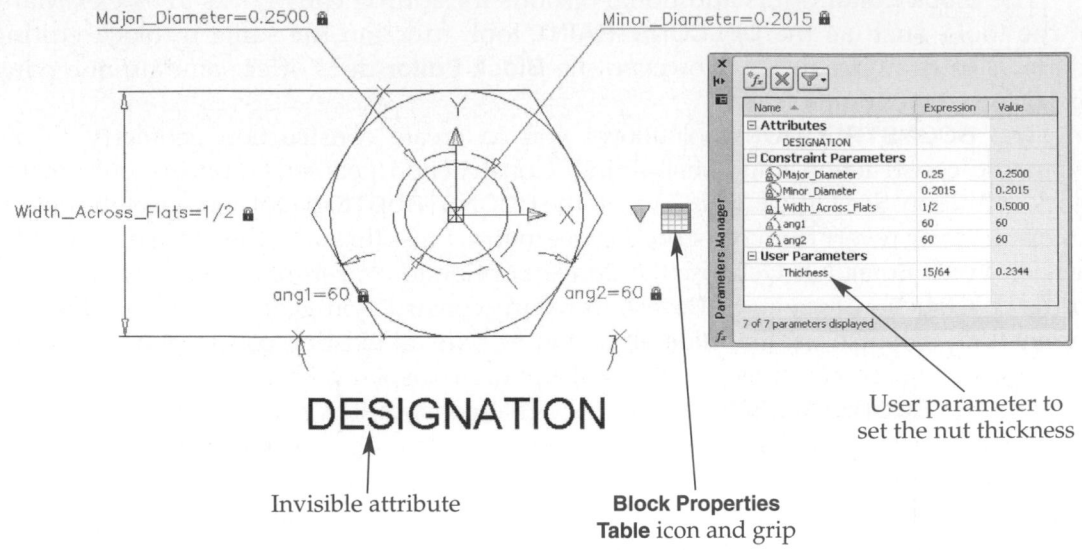

Figure 28-19.
A heavy hex nut block definition ready to use to create a block properties table.

Invisible attribute

Block Properties Table icon and grip

User parameter to set the nut thickness

Figure 28-19 shows the block of the front view of a heavy hex nut in the **Block Editor**. The block includes an appropriate level of constraints and includes constraint parameters to direct dynamic changes. The block also includes an invisible and preset attribute for defining the designation of each different nut and, as shown in the **Parameters Manager**, a user-defined parameter for the nut thickness.

PROFESSIONAL TIP

It is critical that you assign the **Preset** mode to attributes that you include in a block properties table. This allows the attribute value to adjust to the selected block record. The **Preset** mode requires no default value, and you will not receive a prompt to adjust the value.

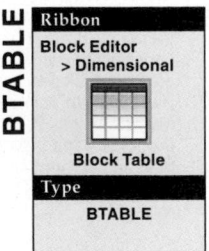

BTABLE

Ribbon
Block Editor > Dimensional

Block Table

Type
BTABLE

After you create parameters and attributes, access the **BTABLE** tool and select the parameter location. Next, enter the number of grips to associate with the parameter. The default **1** option creates a single grip that allows you to select a table record from the grip shortcut menu. If you choose the **0** option, you can only use the **Properties** palette to select a record. The **Palette** option, available before you specify the parameter location or from the **Properties** palette, determines whether the label displays in the **Properties** palette when you select the block reference. The **Block Properties Table** dialog box appears, allowing you to create a block properties table. See **Figure 28-20.**

Creating a Block Properties Table

A block properties table groups the properties of parameters into custom named records, or rows. To add parameter properties, pick the **Add Properties...** button to open the **Add Parameter Properties** dialog box. See **Figure 28-21.** All parameters in the block that contain property values appear in the **Parameter properties:** list. Lookup, alignment, and base point parameters do not contain property values. Notice that the property name is the parameter label.

To add parameter properties to the table, select the properties in the **Parameter properties:** list and pick the **OK** button. A column appears in the table for each parameter property. Type a value in each cell in the column. A new row forms automatically when you enter a value in a cell. See **Figure 28-22.** Press [Enter], [Tab], [Shift]+[Enter], arrow keys, or pick in a different cell to navigate through the table.

AutoCAD and Its Applications—Basics

Figure 28-20.
The **Block Properties Table** dialog box.

Pick to add
block properties

Pick to specify
a user property

Block
properties
appear here

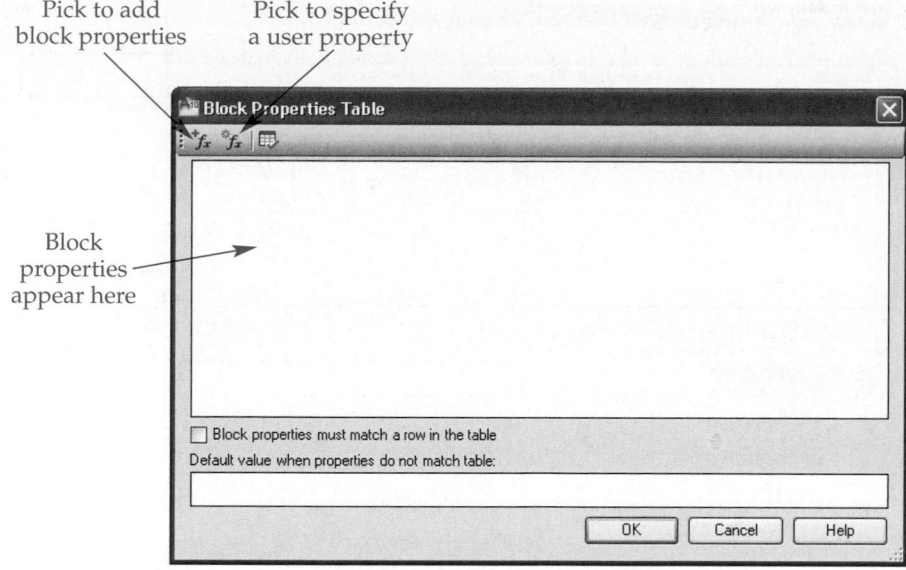

Figure 28-21.
Parameter properties are listed in the **Add Parameter Properties** dialog box.

Select the
properties
to include
in the table

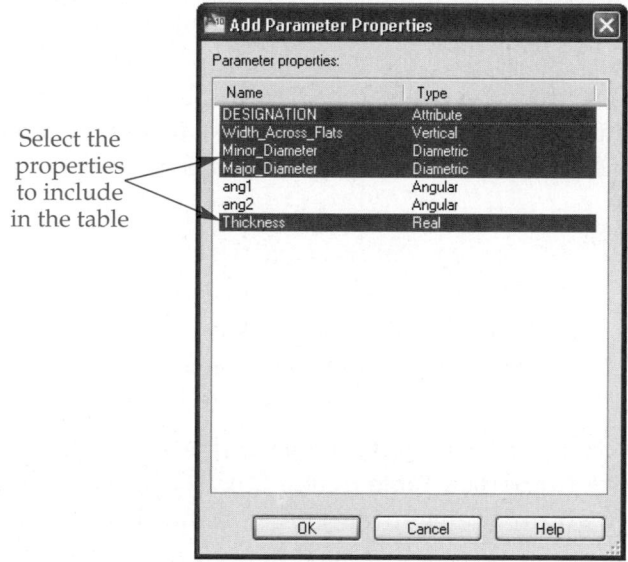

For the nut block example, complete the table as shown in **Figure 28-22**. The **DESIGNATION** column references the attribute property. The value you enter in this text box in each row specifies the row, or record, name. This value appears in the shortcut menu when you insert the block and select the block properties table parameter grip.

> **NOTE**
>
> Right-click on a column heading to access a menu with options for adjusting columns. Right-click on a row to access a menu with options for adjusting rows. The options function the same as those for adjusting lookup table columns and rows.

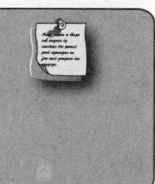

Figure 28-22.
A block properties table with multiple parameters and values added.

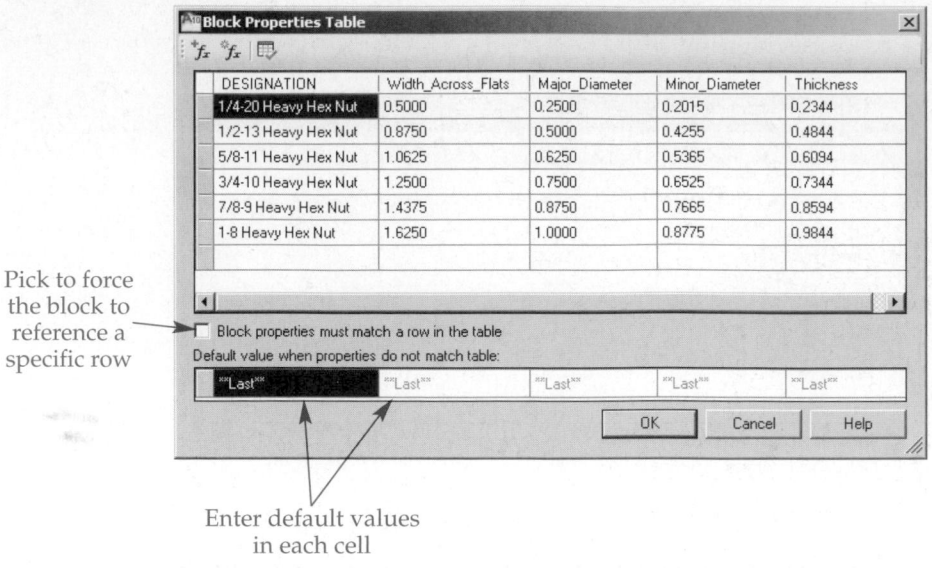

Pick to force
the block to
reference a
specific row

Enter default values
in each cell

By default, you can adjust the block using parameter values other than those speci-fied in the table. You may be able to enter a value, such as the value of an attribute prop-erty, in a text box found in the **Default property when values do not match table** area of the **Block Properties Table** dialog box. Use the ****Last**** option to use the value assigned to the previous block reference when you specify a value not found in the table. Often it is appropriate to choose the **Block properties must match a row in the table** check box to force the selection of a specific record, matching all values in a row.

NOTE

It is critical that all block definition values match the values speci-fied in the default block row in the block properties table, especially if you force the selection of a specific record.

After you add all required properties to the table and assign values to each, pick the **Audit** button in the **Block Properties Table** dialog box to check each record in the table to make sure they are all unique, and that there are no discrepancies between the block definition and the table values. If no errors are found, pick the **OK** button to return to the **Block Editor**. Test and save the block, and exit the **Block Editor**. The dynamic block is now ready to use.

NOTE

To redisplay the **Block Properties Table** dialog box, double-click on the parameter, or access the **BTABLE** tool.

Using a Block Properties Table Dynamically

Figure 28-23 shows the inserted nut block selected for editing. Since the block table parameter includes a grip, a grip appears that you can select to choose a specific block style. The entries in the grip shortcut menu match the rows in the **Block Properties Table** dialog box. A check mark indicates the current record. To switch to a different view of

Figure 28-23.
The lookup parameter grip displays when you select the block. The list of available lookup records displays when you pick the lookup parameter grip. Notice the correlation between the available options and the lookup property names in the **Property Lookup Table** dialog box.

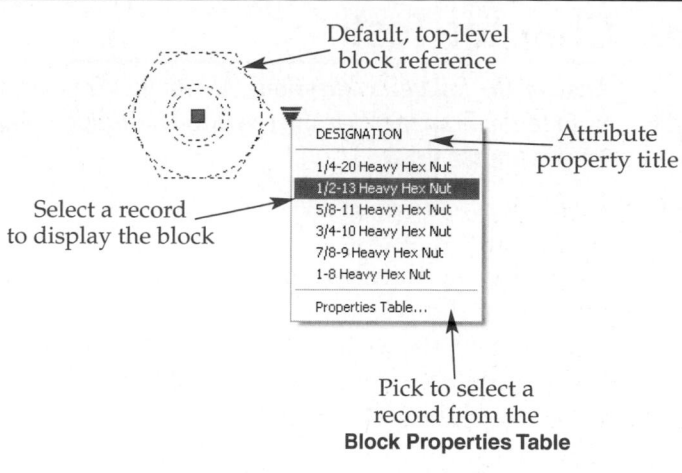

Default, top-level block reference

DESIGNATION — Attribute property title

Select a record to display the block

Pick to select a record from the **Block Properties Table**

the block, select the name of the record from the list. You can also pick the **Properties Table...** option to display the **Block Properties Table** in drawing mode. Double-click a row to activate. In this example, no other grips were assigned to blocks. This makes the table and the **Properties** palette the only two methods of selecting a block reference.

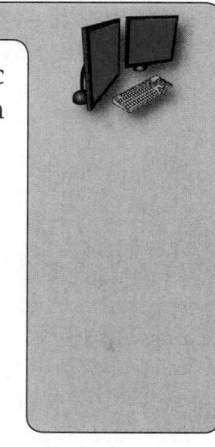

Exercise 28-5

Access the Student Web site (www.g-wlearning.com/CAD) and complete Exercise 28-5.

Chapter Test

Answer the following questions. Write your answers on a separate sheet of paper or go to the Student Web site (www.g-wlearning.com/CAD) and complete the electronic chapter test.

1. Define *visibility parameter*.
2. What are visibility states?
3. How do you display the shortcut menu that allows you to select from a block's existing visibility states?
4. When you display the shortcut menu when a block reference is selected for editing, what indicates the current visibility state?
5. Briefly describe a lookup parameter.
6. Explain the basic function of a lookup action.
7. Identify the basic function of a lookup table.
8. What is a parameter set?
9. What ribbon tab, in addition to the **Parametric** ribbon tab, contains geometric constraint tools?
10. When can you assign constraints to block objects to create a dynamic block?
11. What takes the place of dimensional constraints in the **Block Editor**?
12. What tool and option allow you to convert a dimensional constraint to a constraint parameter?
13. How can you create construction geometry to aid geometric construction in the **Block Editor**?
14. Explain how to toggle constraint status identification on and off in the **Block Editor**.
15. What does a block properties table allow you to do?

Drawing Problems

Start AutoCAD if it is not already started. Start a new drawing using an appropriate template of your choice. The template should include layers, text styles, dimension styles, and multileader styles appropriate for drawing the given objects. Add layers, text styles, dimension styles, and multileader styles as needed. Draw all objects using appropriate layers, text styles, dimension styles, multileader styles, justification, and format. Follow the specific instructions for each problem. Use your own judgment and approximate dimensions when necessary.

Note: Constraint parameters shown for reference are created using AutoCAD and may not comply with ASME standards.

▼ Basic

1. Use the **DesignCenter** tool to insert the HEAVY HEX NUT block you created in Exercise 28-5. Insert or copy the block to create six total symbols. Use the block table parameter to display each size nut as shown. Save the drawing as P22-1.

   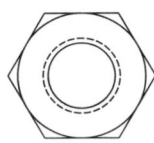 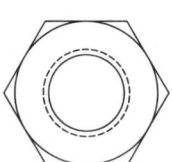 

2. Open P22-2 and save as P28-2. The P28-2 file should be active. Create a block from the objects named FIXTURE, select all objects, and pick the center of the circle as the insertion base point. Open the block in the **Block Editor** and convert the dimensional constraints to constraint parameters. Insert the FIXTURE block into the drawing three times to create the 4x4, 6x6, and 4x5 symbols as shown. Resave the drawing.

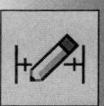

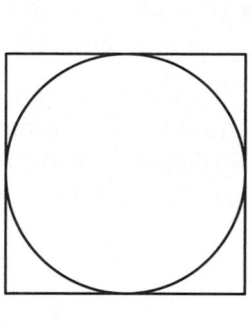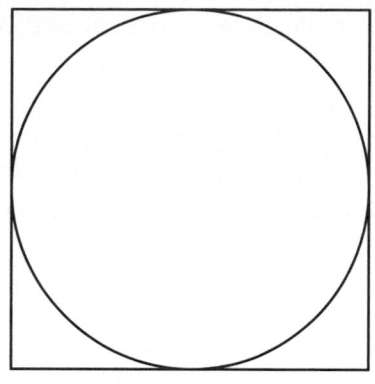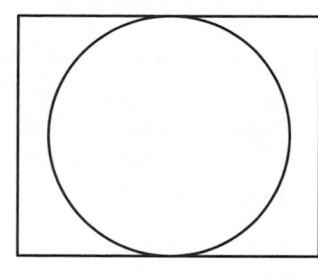

3. Open P22-5 and save as P28-3. The P28-3 file should be active. Create a block from the objects named PLATE, select all objects and pick the center of the plate as the insertion base point. Open the block in the **Block Editor** and convert the construction rectangle to construction geometry. Convert the dimensional constraints to constraint parameters. Insert the PLATE into the drawing and create the block shown. Do not add dimensions. Resave the drawing.

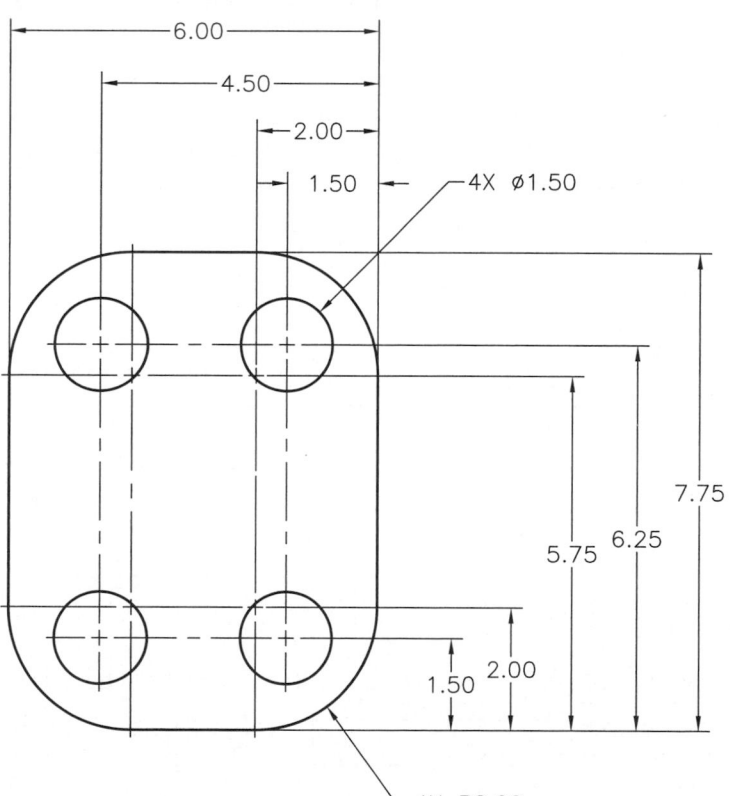

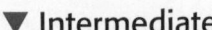

▼ Intermediate

4. Open P22-6 and save as P28-4. The P28-4 file should be active. Create a block named SELECTOR from the objects, select all objects, and pick the center of the center circle as the insertion base point. Open the block in the **Block Editor** and convert the construction lines to construction geometry. Convert the dimensional constraints to constraint parameters. Insert the SELECTOR block into the drawing three times and create three different symbols of your own design. Resave the drawing.

5. Create a single block that can be used to represent each of the three door blocks shown below. Name the block 30 INCH DOOR. Do not include labels. Create an appropriately named visibility state for each view: 90 OPEN, 60 OPEN, and 30 OPEN. Insert the 30 INCH DOOR block into the drawing three times. Set each block to a different visibility state. Save the drawing as P28-5.

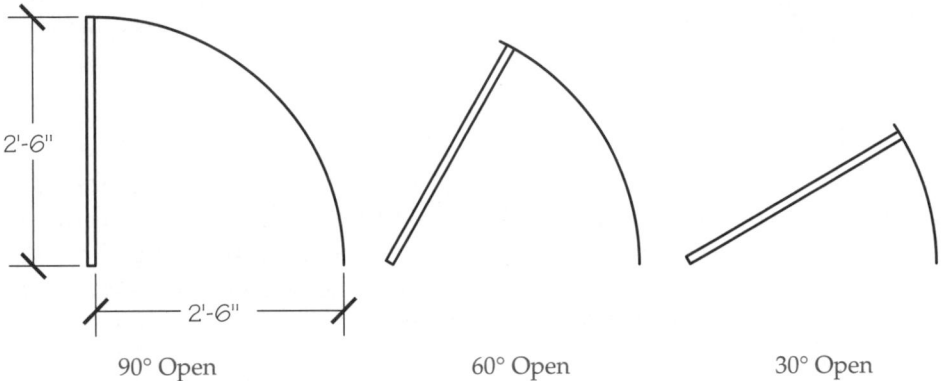

90° Open   60° Open   30° Open

6. Create a cut framing member block that can be used to represent each of the symbols shown below. Name the block FRAME. Add geometric constraints and constraint parameters as needed, and assign an alignment parameter. Use the block to create the portion of the detail shown. Save the drawing as P28-6.

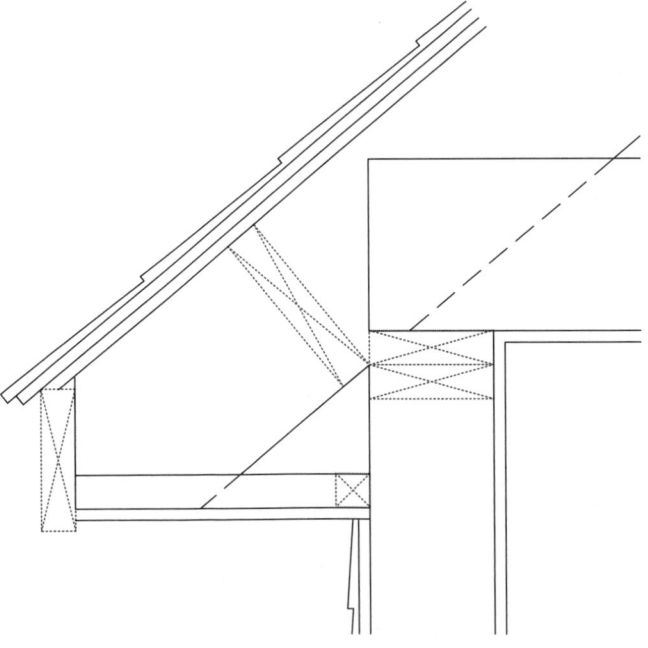

Drawing Problems - Chapter 28

▼ Advanced

7. Create a foundation footing and wall block that can be used to represent unique construction requirements. Name the block Foundation. Add geometric constraints and constraint parameters as shown. Include 1″ increment value sets for each constraint parameter. The block should stretch symmetrically as shown. Save the drawing as P28-7.

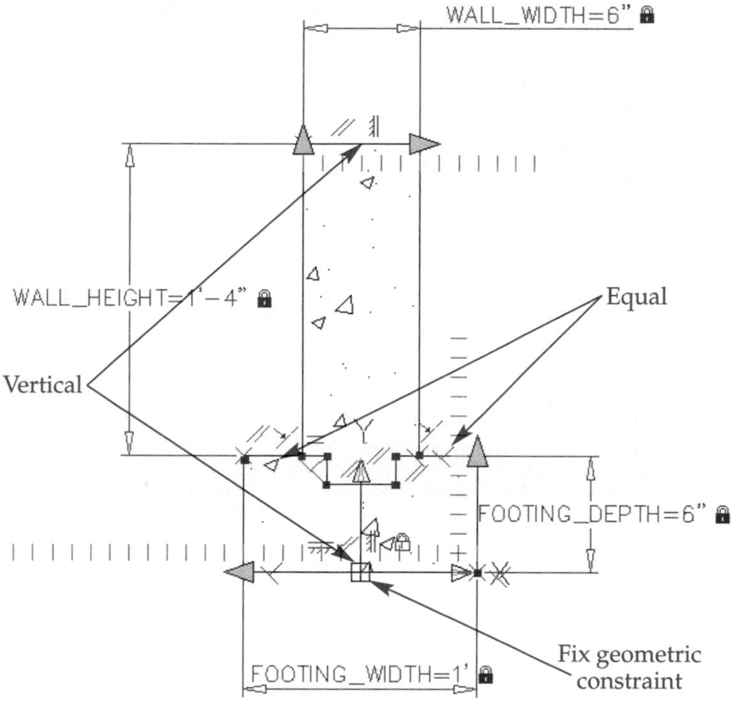

WALL_WIDTH=6″ 🔒

WALL_HEIGHT=1′–4″ 🔒

Equal

Vertical

FOOTING_DEPTH=6″ 🔒

Fix geometric constraint

FOOTING_WIDTH=1′ 🔒

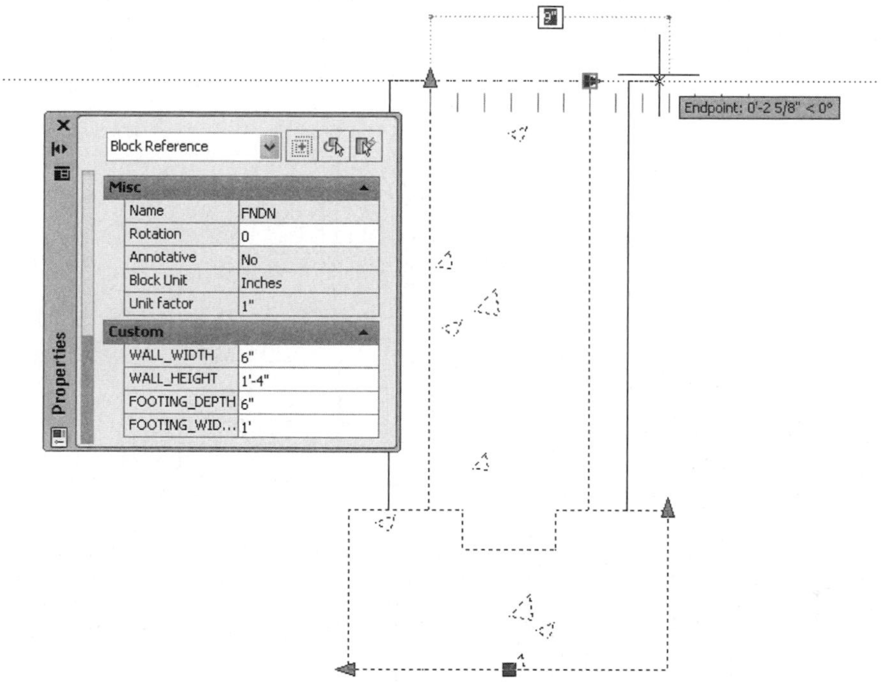

Block Reference	
Misc	
Name	FNDN
Rotation	0
Annotative	No
Block Unit	Inches
Unit factor	1″
Custom	
WALL_WIDTH	6″
WALL_HEIGHT	1′-4″
FOOTING_DEPTH	6″
FOOTING_WID...	1′

Endpoint: 0'-2 5/8" < 0°

8. The bolt shown in the drawing below is available in four different lengths. As the length increases, the size of the bolt head increases for added strength. Create a dynamic block that will allow the length of the shaft and the size of the bolt head to be changed in a single operation.

A. Draw the objects composing the bolt and create a block named BOLT. Do not include dimensions.

B. Insert a linear parameter along the length of the shaft from the bottom of the bolt head to the end of the shaft. Label it SHAFT LENGTH.

C. Assign a stretch action to the SHAFT LENGTH parameter. Associate the action with the parameter grip at the end of the shaft. Create a crossing window around the end of the shaft that includes the threads. Select the end of the shaft, threads, and edges of the shaft.

D. Insert a linear parameter along the depth of the bolt head (the .3″ dimension). Label it HEAD THICKNESS.

E. Assign a scale action to the HEAD THICKNESS parameter and select the objects that compose the bolt head. Use an independent base point type and specify the midpoint of the vertical line where the shaft meets the bolt head.

F. Insert a lookup parameter and assign a lookup action to it.

G. Add the SHAFT LENGTH and the HEAD THICKNESS parameters to the lookup table. Complete the table with the following properties:

Shaft Length	Head Thickness	Lookup
1	0.3	1″ Length
1.5	0.333	1.5″ Length
2	0.366	2″ Length
2.5	0.4	2.5″ Length

H. Set the table to allow reverse lookup, save the block, and exit the **Block Editor**.

I. Insert the block four times into the drawing. Specify a different lookup property for each block.

J. Save the drawing as P28-8.

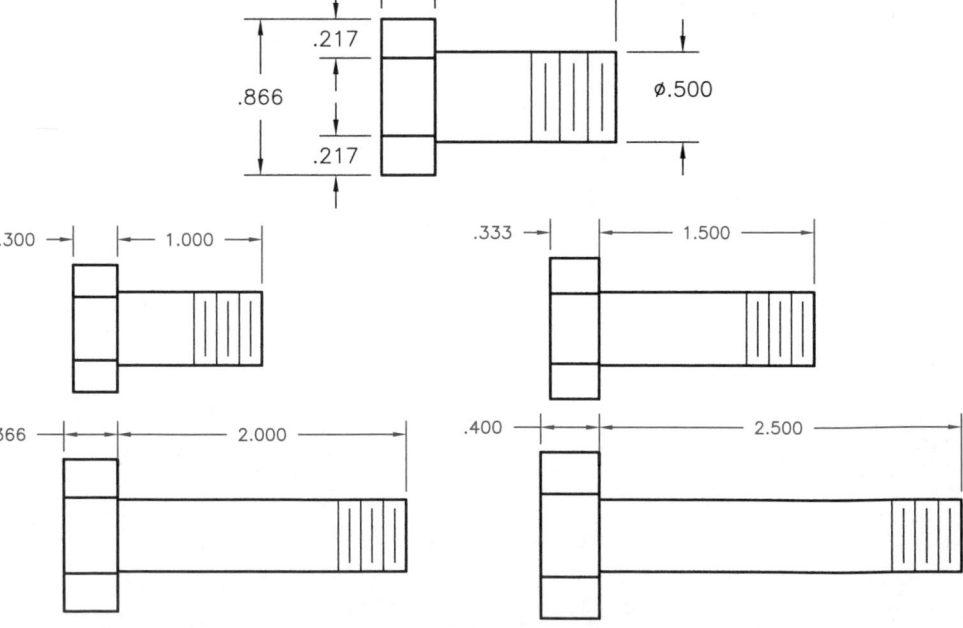

9. Repeat Problem 28-8, but this time use geometric constraints, constraint parameters, and a block properties table instead of action parameters. Save the drawing as P28-9.

Page Setup Manager

Current layout: Layout1

Page setups

Current page setup: <None>

Layout1

Set Current

CHAPTER 29

Layout Setup

Learning Objectives

After completing this chapter, you will be able to do the following:
- ✓ Describe the purpose for and proper use of layouts.
- ✓ Begin to prepare layouts for plotting.
- ✓ Manage layouts.
- ✓ Use the **Page Setup Manager** to define plot settings.
- ✓ Use plot styles and plot style tables.

You usually print or plot a drawing in model space to make a quick hard copy, often for check or reference purposes. Ordinarily, however, you create a drawing in model space and then lay out the drawing for plotting in paper space. Hard-copy plots are required for a variety of reasons. For example, it is typically easier for workers in a machine shop or a construction crew in the field to refer to a print than to use a computer to view the drawing file. Additionally, the process of preparing a drawing to plot uses many of the steps needed to publish or export a drawing.

Introduction to Layouts

The first step in making an AutoCAD drawing is to create a *model* in *model space*. See **Figure 29-1A.** Model space is usually active by default. You have been using model space throughout this textbook to create objects and dimensioned drawing views. Once you complete a model, use a *layout* in *paper space* to prepare the final drawing for plotting. See **Figure 29-1B.** A layout represents the sheet of paper used to lay out and plot a drawing. It may include the following items:
- Floating viewports
- Border
- Title block
- Revision block
- General notes
- Bill of materials, parts list, or schedules
- Page setup information

model: A 2D or 3D drawing composed of various objects, such as lines, circles, and text, usually created at full size.

model space: The environment in AutoCAD in which you create drawings and designs.

layout: A specific arrangement of views or drawings for plotting or printing on paper.

paper space: The environment in AutoCAD in which you create layouts.

Figure 29-1.
A—Create drawings and designs in model space. B—Use paper space to finalize and lay out drawings and designs on paper for plotting.

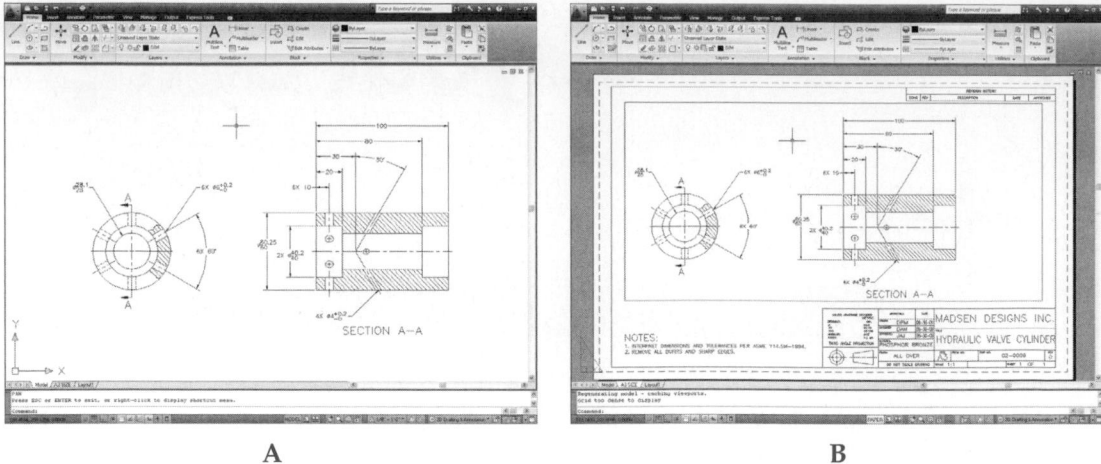

A   B

floating viewport:
A viewport added to a layout in paper space to display objects drawn in model space.

A major element of the layout system is the *floating viewport.* Consider a layout to be a virtual sheet of paper and a floating viewport as a hole cut into the paper to show objects drawn in model space. In **Figure 29-1B,** a single viewport exposes objects drawn in model space. You should usually draw the floating viewport on a layer that you can turn off or freeze so the viewport does not plot and is not displayed on-screen. Chapter 30 explains using floating viewports.

A single drawing can have multiple layouts, each representing a different paper space, or plot, definition. Each layout can include multiple floating viewports to provide additional or alternate drawing views, prepared at different scales if necessary. Layouts with floating viewports offer the ability to construct properly scaled drawings and use a single drawing file to prepare several unique final drawings and drawing views. For example, an architectural drawing file might include several details that are too large to place on a single sheet of paper. You can use multiple layouts, and if necessary differently scaled floating viewports, to prepare as many sheets as needed to plot all of the details found in the drawing.

Working with Layouts

Before preparing a layout for plotting, you should be familiar with tools and options for displaying and managing layouts. The layout and model tabs and model space and paper space tools in the status bar are available by default. These are the most effective tools for navigating to and from model space and layouts, and managing layouts. You can also return to model space from a layout by typing MODEL.

Using the Layout and Model Tabs

The model and layout tabs that appear directly below the drawing window by default are among the most useful options for accessing and preparing layouts. See **Figure 29-2A.** The model space tab is furthest to the left, followed by layout tabs arranged in the order created, from left to right. If the drawing includes so many layouts that tabs spread past the screen, use the forward and reverse buttons left of the tabs to access the appropriate layout or model space. Hover over an inactive tab to display a preview of the contents. Pick a layout tab to enter paper space with the selected layout current, or pick the **Model** tab to re-enter model space.

Figure 29-2.
A—Using the layout and model tabs to activate model space and paper space. B—Options available when you right-click on a tab.

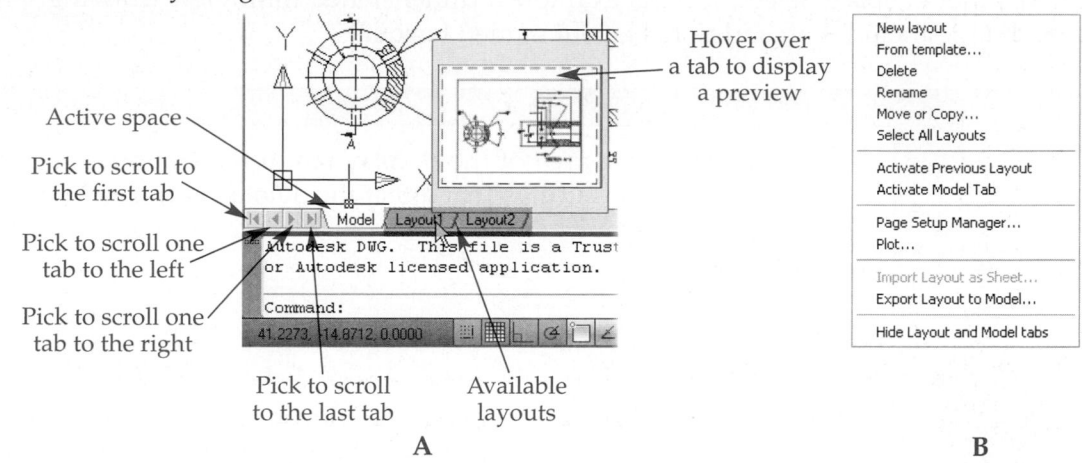

Active space

Pick to scroll to the first tab

Pick to scroll one tab to the left

Pick to scroll one tab to the right

Pick to scroll to the last tab

Available layouts

Hover over a tab to display a preview

New layout
From template...
Delete
Rename
Move or Copy...
Select All Layouts

Activate Previous Layout
Activate Model Tab

Page Setup Manager...
Plot...

Import Layout as Sheet...
Export Layout to Model...

Hide Layout and Model tabs

A

B

Right-click on a tab to access a shortcut menu of options to control layouts and an option to hide the layout and model tabs. See **Figure 29-2B.** Pick **Activate Model Tab** to enter model space. Select **Activate Previous Layout** to make the previously current layout current. Pick **Select All Layouts** to select all layouts in the drawing. This is a valuable option for selecting all layouts for editing purposes, such as deleting or publishing. In addition to these basic functions, the shortcut menu is the primary resource for adding layouts and moving, renaming, and deleting existing layouts.

Figure 29-3 shows an example of using the supplied acad.dwt drawing template and picking the **Layout1** tab to display the default **Layout1** layout. The layout uses default settings based on an 8.5″ × 11″ sheet of paper in a landscape (horizontal)

Figure 29-3.
Pick the **Layout1** tab to display **Layout1**, which is provided in the default acad.dwt template.

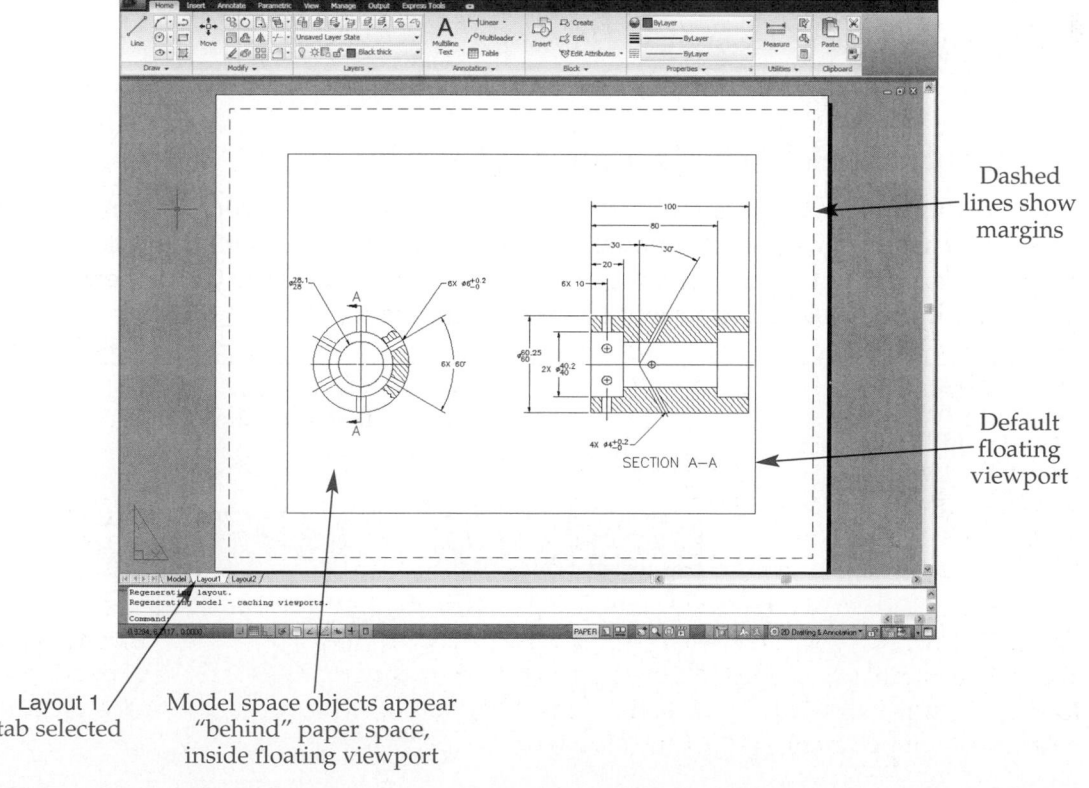

Dashed lines show margins

Default floating viewport

Layout 1 tab selected

Model space objects appear "behind" paper space, inside floating viewport margin: The extent of the printable area; objects drawn past the margin (dashed lines) do not print.

orientation. The white rectangle you see on the gray background is a representation of the sheet. Dashed lines mark the sheet *margin*. A large, rectangular floating viewport reveals model space objects, in this example a dimensioned multiview drawing. The acad.dwt file includes an additional layout named **Layout2**.

NOTE

The **Layout elements** area of the **Display** tab in the **Options** dialog box includes several settings that affect the display and function of layouts. Use the default settings until you are comfortable working with layouts.

PROFESSIONAL TIP

Look at the user coordinate system (UCS) icon to confirm whether you are in model space or paper space. When you enter a layout, the UCS icon changes from two arrows to a triangle that indicates the X and Y coordinate directions.

Using the Model and Paper Buttons

The status bar provides other convenient tools for managing layouts. See **Figure 29-4**. The **Quick View Layouts** and **Quick View Drawings** tools are described later in this chapter. Pick the **MODEL** button to exit model space and enter paper space. If the file contains multiple layouts, the top-level layout displays unless you previously accessed a different layout. While a layout is active, pick the **PAPER** button to use the **MSPACE** tool, which activates a floating viewport. It does not return you to model space. Select the **PAPER** button to deactivate a floating viewport.

Type
MSPACE

Exercise 29-1

Access the Student Web site (www.g-wlearning.com/CAD) and complete Exercise 29-1.

Using the Quick View Layouts Tool

Type
QVLAYOUT

The **Quick View Layouts** tool is very similar to using the layout and model tabs. However, the tool provides additional options and a visual format for displaying and adjusting layouts in the current file. The quickest way to access the **Quick View Layouts** tool is to pick the **Quick View Layouts** button on the status bar.

Figure 29-4.
The status bar provides additional tools for activating model and paper space and managing layouts.

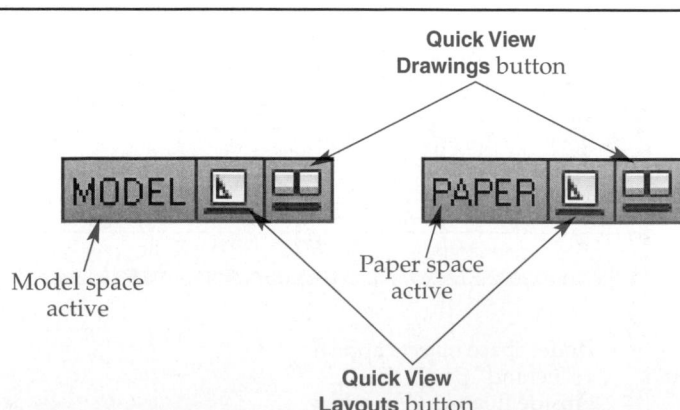

Figure 29-5.
The **Quick View Layouts** tool offers an effective visual method for changing between model space and paper space and provides options for managing layouts.

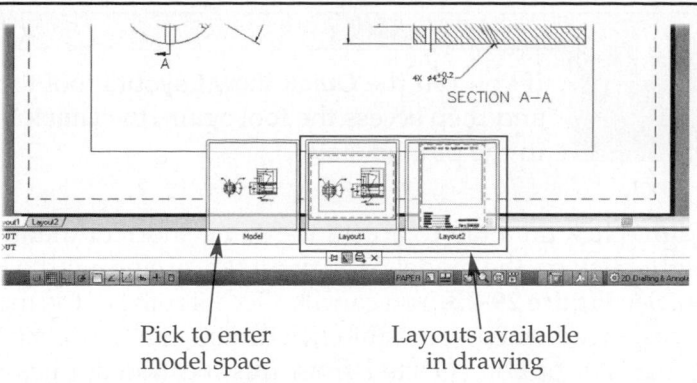

Pick to enter model space

Layouts available in drawing

When activated, the **Quick View Layouts** tool appears in the lower center of the AutoCAD window. See **Figure 29-5.** The **Model** thumbnail image is furthest to the left, followed by layout thumbnails in the order created, from left to right. If the drawing includes so many layouts that thumbnails spread past the screen, hover the cursor over the furthest right and left thumbnails to scroll though the options. Hover over a thumbnail to highlight the image and show additional options. See **Figure 29-6.** Pick a layout thumbnail to enter paper space with the selected layout current, or pick the **Model** thumbnail to re-enter model space.

NOTE

Icons represent model and layout thumbnails until you enter a layout for the first time (initialize the layout). The icon then changes to a thumbnail image of model space and eventually the layout.

The **Quick View Layouts** tool provides a small toolbar below the thumbnail images, as shown in **Figure 29-6.** By default, the **Quick View Layouts** tool disappears when you pick a thumbnail to switch layouts or enter model space. To keep the tool on-screen, pick the **Pin Quick View Layouts** button. Pick the **New Layout** button to create a new layout from scratch, as described later in this chapter. Pick the **Publish...** button to access the **Publish** dialog box, explained later in this textbook. Select the **Close** button to exit the **Quick View Layouts** tool.

Figure 29-6.
Hover over a thumbnail to display **Plot...** and **Publish...** buttons.

Display when you hover over a thumbnail

Pick to access the **Plot** dialog box

Pick to access the **Publish** dialog box

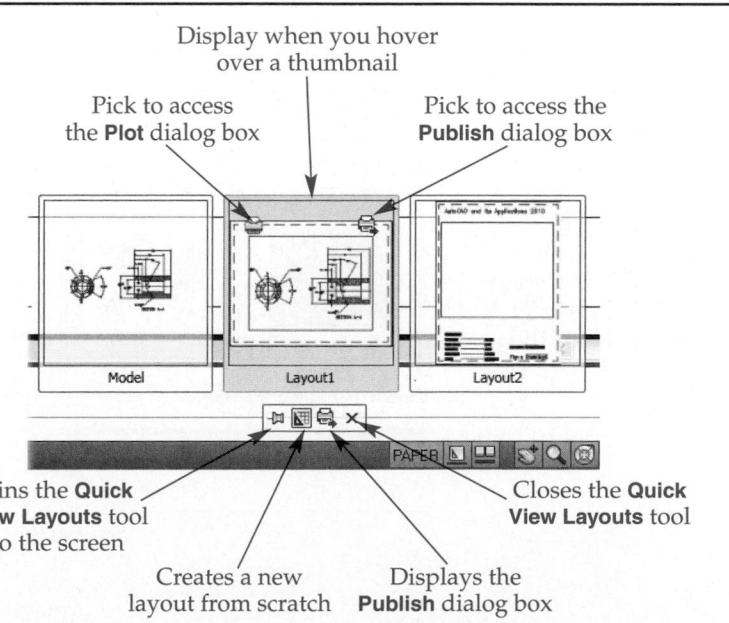

Pins the **Quick View Layouts** tool to the screen

Closes the **Quick View Layouts** tool

Creates a new layout from scratch

Displays the **Publish** dialog box

Right-click on a tab to access the same shortcut menu of options available when you right-click on the model or a layout tab, excluding the **Hide Layout and Model tabs** option. See **Figure 29-2B.** You can also access some of the menu options from a shortcut menu displayed when you right-click directly on the **Quick View Layouts** button on the status bar. An option selected from this location applies to the current file and the current layout.

Using the Quick View Drawings Tool

The **Quick View Drawings** tool provides the same features for working with layouts as the **Quick View Layouts** tool, but it allows you to manage the layouts in all open drawings. Use this tool to increase productivity when you are working between existing drawings. The quickest way to access the **Quick View Drawings** tool is to pick the **Quick View Drawings** button on the status bar. Refer to Chapter 2 for information on basic **Quick View Drawings** tool features, such as using the tool to work with multiple open documents.

Access the **Quick View Drawings** tool and hover the cursor over the drawing file to control. The model and layout thumbnail images appear above the highlighted drawing. Move the cursor over the model or a layout thumbnail to enlarge the display. See **Figure 29-7.** Pick a layout thumbnail to switch to the highlighted file and enter paper space with the selected layout active, or pick the **Model** thumbnail to switch to the highlighted file in model space. Right-click on a thumbnail to access the same shortcut menu displayed when you right-click on a thumbnail using the **Quick View Layouts** tool. Right-clicking on the thumbnail makes the associated file current.

Figure 29-7.
Use the **Quick View Drawings** tool to manage layouts located in other open files. Hover over a file thumbnail to display model and layout thumbnails for the file.

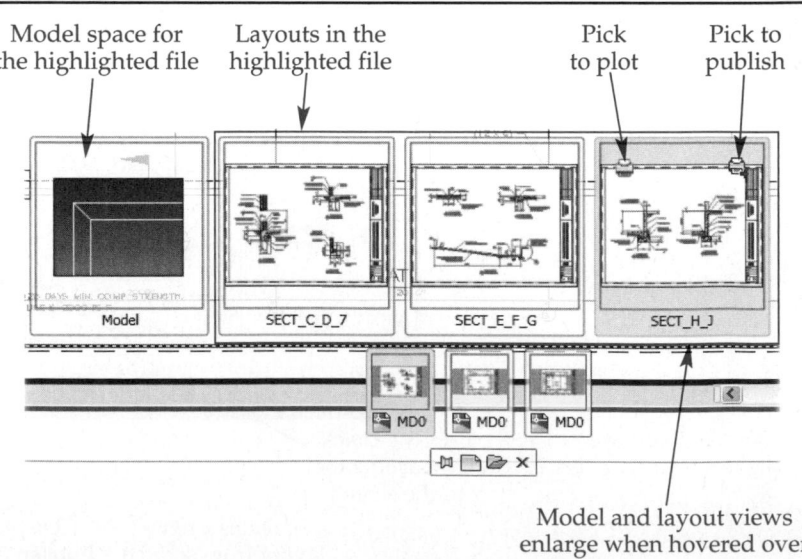

Model space for the highlighted file   Layouts in the highlighted file   Pick to plot   Pick to publish

Model and layout views enlarge when hovered over

Adding Layouts

To add a new layout to a drawing, create a new layout from scratch, use the **Create Layout** wizard, or reference an existing layout. Referencing an existing, preset layout is often the most effective approach. You can also insert a layout from a different DWG, DWT, or DXF file into the current file or create a copy of a layout from the current file.

Starting from Scratch

To create a new layout from scratch, right-click on the model or a layout tab, a **Quick View Layouts** or **Quick View Drawings** thumbnail image, or the **Quick View Layouts** button on the status bar, and pick **New Layout**. A new layout appears on the far right of the layout list. The settings applied to the new layout depend on the template used to create the original file. The name of the layout is set according to the names of other existing layouts. For example, when you add a new layout to a default drawing started from the acad.dwt template, a new layout named **Layout3** appears and includes an 8.5″ × 11″ sheet of paper, a landscape (horizontal) orientation, and a large floating viewport.

Using the Create Layout Wizard

Use the **Create Layout** wizard to build a layout from scratch using values and options you enter in the wizard. The pages of the wizard guide you through the process of developing the layout. They provide options for naming the new layout and selecting a printer, paper size, drawing units, paper orientation, title block, and viewport configuration.

Using a Template

To create a new layout from a layout stored in an existing DWG, DWT, or DXF file, right-click on the model or a layout tab, a **Quick View Layouts** or **Quick View Drawings** thumbnail image, or on the **Quick View Layouts** button on the status bar, and pick **From Template….** The **Select Template From File** dialog box appears. See **Figure 29-8A**. The Template folder in the path set by the AutoCAD Drawing Template File Location is the default. Select the file containing the layout to add to the current drawing and pick the **Open** button. The **Insert Layout(s)** dialog box appears, listing all layouts in the selected file. See **Figure 29-8B**. Highlight the layout or layouts to copy and pick the **OK** button.

Using DesignCenter

DesignCenter provides an effective way to add existing layouts to the current drawing. To view the layouts found in a drawing, select the **Layouts** branch in the tree view or double-click on the **Layouts** icon in the content area. See **Figure 29-9.** Select the layout(s) to copy from the content area and then drag and drop, or use the **Add Layout(s)** or **Copy** and **Paste** options from the shortcut menu to insert the layouts in the current drawing.

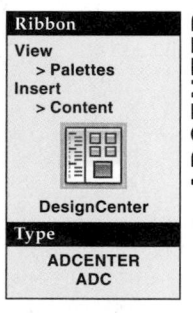

Copying and Moving Layouts

To create a copy of a layout, right-click on a layout tab, a **Quick View Layouts** or **Quick View Drawings** thumbnail image, or make the layout to copy current and right-click on the **Quick View Layouts** button on the status bar. Then pick **Move or Copy…** to display the **Move or Copy** dialog box. See **Figure 29-10.** To create a copy, select the **Create a copy** check box and pick the layout that will appear to the right of the new layout, or pick (move to end) to place the copy right of all other layouts. The default name of the new layout is the name of the current or selected layout plus a number in parentheses.

Move a layout using the **Move or Copy…** dialog box without selecting the **Create a copy** check box. When you add and rename layouts, the layouts do not automatically rearrange into a predetermined order. Organize layouts in an appropriate order to reduce confusion and aid in the publishing process, as described in Chapter 33.

Figure 29-8.
Adding a layout using an existing layout stored in a different drawing, drawing template, or DXF file. A—Select the file containing the layout. B—Highlight the layout to add to the current drawing.

Template folder opens by default

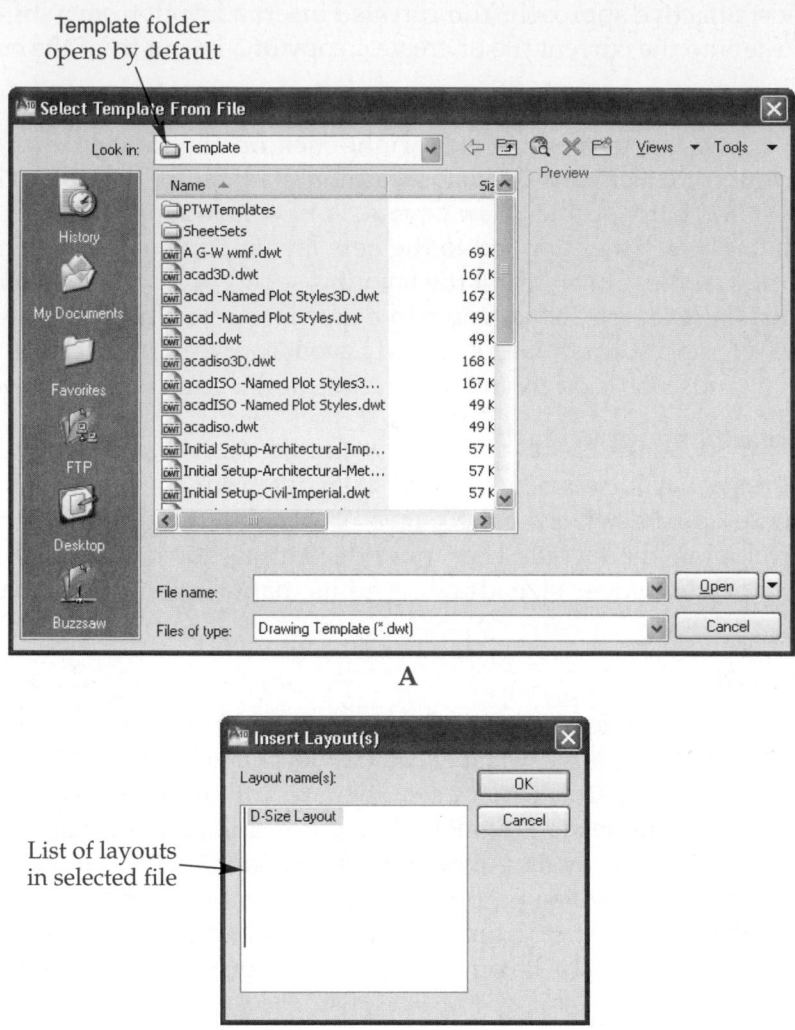

A

List of layouts in selected file

B

Figure 29-9.
Using **DesignCenter** to share layouts between drawings.

Select drawing

Copy layout to current drawing

Pick to display layouts defined in the drawing

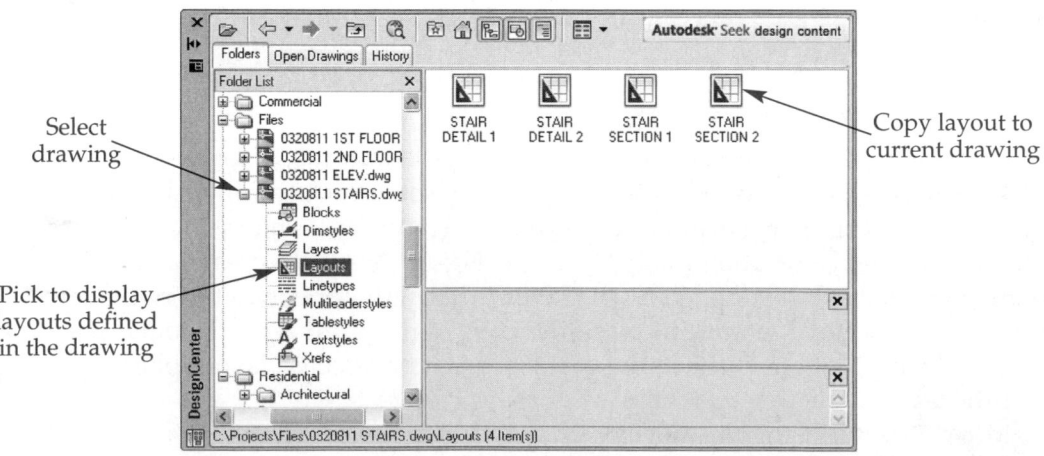

Figure 29-10.
The **Move or Copy** dialog box allows you to reorganize layouts and copy layouts within a drawing.

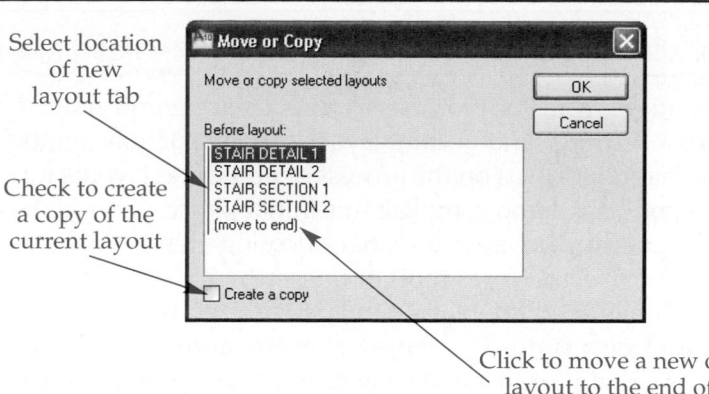

Select location of new layout tab

Check to create a copy of the current layout

Click to move a new or existing layout to the end of the list

Renaming Layouts

Layouts are easier to recognize and use when they have descriptive names. To rename a layout, right-click a layout tab or a **Quick View Layouts** or **Quick View Drawings** thumbnail image and pick **Rename**. You can also double-click slowly on the current name to activate it for editing. Once the layout name highlights, type a new name and press [Enter].

Deleting Layouts

To delete an unused layout from the drawing, right-click on the layout tab or the **Quick View Layouts** or **Quick View Drawings** thumbnail image and pick **Delete**. An alert message warns you that the layout will be deleted permanently. Pick the **OK** button to remove the layout.

Exporting a Layout to Model Space

The **EXPORTLAYOUT** tool allows you to save the layout display as a separate DWG file. This tool produces a "snapshot" of the current layout display that you can use for applications in which it is necessary to combine model space and paper space objects, such as when exporting a file as an image. (Model space and paper space do not export together as an image.)

The quickest way to access the **EXPORTLAYOUT** tool is to right-click on a layout tab or a **Quick View Layouts** or **Quick View Drawings** thumbnail image and pick **Export Layout to Model…**. The **Export Layout to Model Space** dialog box appears and functions much like the **Save As** dialog box. Pick a location for the file, use the default file name or enter a different name, and pick the **SAVE** button. Everything shown in the layout, including objects drawn in model space, are converted to model space and are saved as a new file.

CAUTION

The **EXPORTLAYOUT** tool eliminates the relationship between model space and paper space. Export a layout only when it is necessary to export model space and paper space together as a single unit.

Exercise 29-2

Access the Student Web site (www.g-wlearning.com/CAD) and complete Exercise 29-2.

Initial Layout Setup

Preparing a layout for plotting involves creating and modifying floating viewports, adjusting plot settings, and adding layout content such as symbols, a border, and a title block. This chapter focuses on the process of preparing layouts for plotting using the **Page Setup Manager**. When you complete this initial phase, you will be better prepared to add content to layouts and create and manage floating viewports, as described in Chapter 30.

A *page setup* establishes most of the settings that determine how a drawing plots. Plot settings include printer selection, paper size and orientation, plot area and offset, plot scale, and plot style. The **Page Setup Manager** and related **Page Setup** dialog box allow you to create and modify saved page setups that control how layouts appear on-screen and plot. This is where initial layout setup occurs. You then use the **Plot** dialog box to create the actual plot using the saved page setup. The **Page Setup** and **Plot** dialog boxes include most of the same settings.

PROFESSIONAL TIP

Layout setup usually involves several steps. A well-defined page setup decreases the amount of time required to prepare a drawing for plotting. Once a layout is set up, only a few steps are required to produce a plot. Add fully defined layouts to your drawing templates for convenient future use.

Working with Page Setups

Ribbon
Output
> Plot

Page Setup Manager

Type
PAGESETUP

Access the **PAGESETUP** tool to create page setups using the **Page Setup Manager**. See **Figure 29-11**. The **Page setups** area of the **Page Setup Manager** contains a list box that lists available page setups, as well as buttons to add and modify page setups. The **Selected page setup details** area provides information about the highlighted page setup.

NOTE

You can also access the **Page Setup Manager** by right-clicking on the model or a layout tab, a **Quick View Layouts** or **Quick View Drawings** thumbnail image, or on the **Quick View Layouts** button on the status bar, and selecting **Page Setup Manager...**.

When you access the **Page Setup Manager** in model space, *Model* appears in the **Page Setups** list box. When you access the **Page Setup Manager** in paper space, the name of the current layout appears in the **Page Setups** list box. When preparing a layout for plotting, check to be sure that you are in paper space and that the appropriate layout is current. Each layout can have a unique page setup. Asterisks (*) before and after the layout name indicate the page setup assigned to the current layout. You have the option of creating or using other page setups instead of the page setup associated with the layout.

Create a new page setup to use different plot characteristics without overriding plot settings or spending time making page setup changes. For example, you can create two page setups to plot to two different printers or plotters. Pick the **New...** button to create a new page setup using the **New Page Setup** dialog box. See **Figure 29-12**. Type a name for the new page setup and choose an option from the **Start with:** list box. Pick

Figure 29-11.
Use the **Page Setup Manager** to modify existing page setups and to create and import page setups.

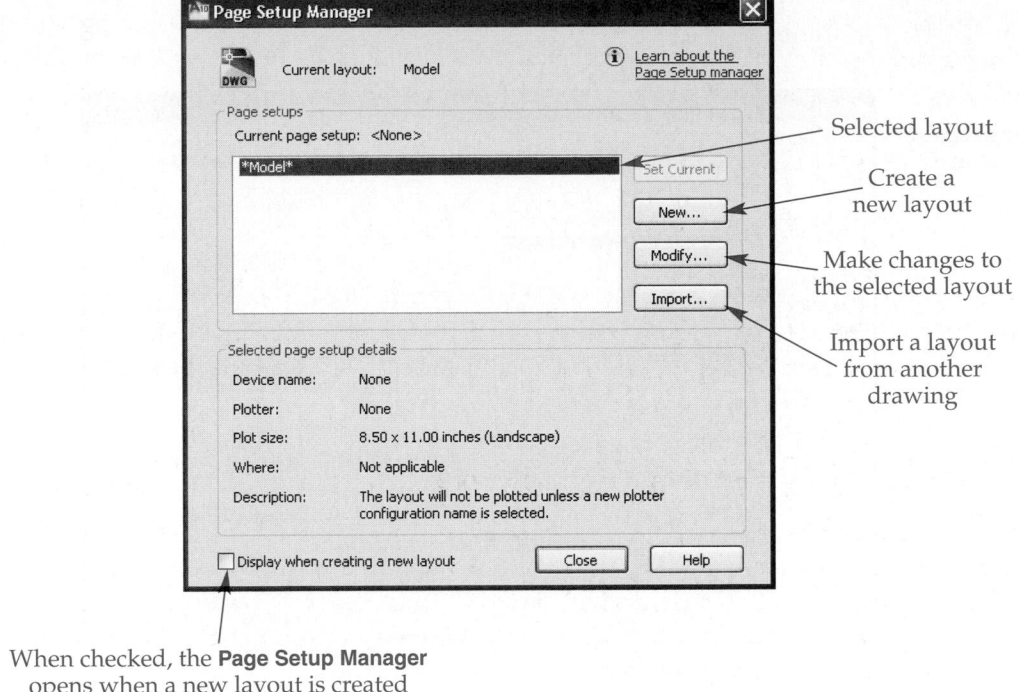

Selected layout

Create a new layout

Make changes to the selected layout

Import a layout from another drawing

When checked, the **Page Setup Manager** opens when a new layout is created

Figure 29-12.
The **New Page Setup** dialog box appears when you create a new page setup. Selecting **None** in the **Start with:** list box does not select a printer. **Default** selects the computer's default printer.

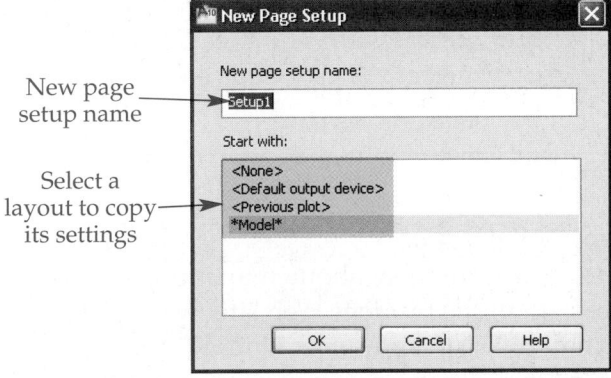

New page setup name

Select a layout to copy its settings

the **OK** button to create the page setup and display the **Page Setup** dialog box. Pick the **Import...** button to use existing page setups from a DWG, DWT, or DXF file.

To attach a different page setup to the current layout, select a page setup from the list in the **Page Setup Manager** and pick the **Set Current** button, or right-click on a page setup and choose **Set Current**. The layout will now plot according to the selected page setup. When you make a different page setup current, the selected page setup overrides the layout page setup. The page setup name appears in parentheses next to the layout name. To rename or delete an existing page setup, right-click on the page setup and pick **Rename** or **Delete**.

Select the **Modify...** button in the **Page Setup Manager** to change the settings of an existing page setup using the **Page Setup** dialog box. See **Figure 29-13.** The **Page Setup** dialog box defines page setup characteristics. The settings control layout appearance and plot function. Each area, as described in the following sections, controls a specific plot setting.

Figure 29-13.
The **Page Setup** dialog box allows you to adjust the plot settings for the selected page setup. This is where the initial phase of layout setup begins.

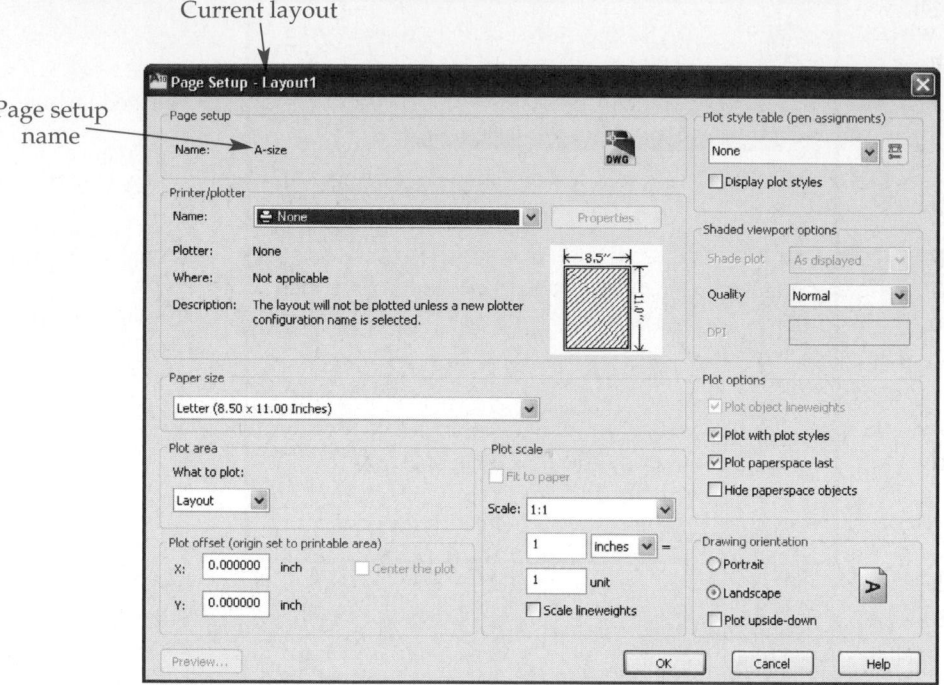

Selecting a Plot Device

The **Printer/plotter** area of the **Page Setup** dialog box, shown in **Figure 29-14,** allows you to select the appropriate *plot device* and adjust the plot device configuration if necessary. The default None setting indicates that no plot device has been specified. If a plot device has been *configured*, select the plot device you want from the **Name:** drop-down list.

 **Supplemental Material** *Plotter Configuration*
For more information about managing and configuring plot devices, go to the Student Web site (www.g-wlearning.com/CAD), select this chapter, and select **Plotter Configuration**.

Figure 29-14.
The **Printer/plotter** area of the **Page Setup** dialog box.

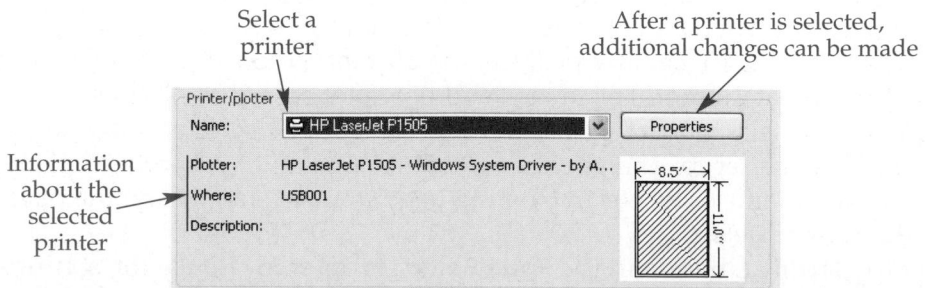

Figure 29-15.
The **Paper size** area
allows you to define
the sheet size applied
to the layout, which
corresponds to the
sheet size on which
you plan to plot.

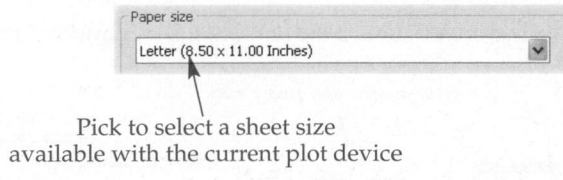

Pick to select a sheet size
available with the current plot device

Choosing a Sheet Size

The **Paper size** area of the **Page Setup** dialog box, shown in **Figure 29-15**, controls the *sheet size*. The sheet size determines the size of the virtual sheet of paper displayed in a layout, as well as the actual sheet size you plan to use when plotting. When determining sheet size, you should take into account the size of the drawing and additional space for dimensions, notes, border, clear space between the drawing and border, title block, revision block, zoning, and an area for general notes. Select the appropriate sheet size from the drop-down list in the **Paper size** area.

sheet size: The size of the paper you use to lay out and plot the final drawing.

Standard Sheet Sizes

The ASME Y14.1, *Decimal Inch Drawing Sheet Size and Format,* and ASME Y14.1M, *Metric Drawing Sheet Size and Format* documents specify the American Society of Mechanical Engineers (ASME) and American National Standards Institute (ANSI) standard sheet sizes and formats. ASME Y14.1 lists sheet size specifications in inches, as follows:

Size Designation	Size (in inches)	
	Vertical	Horizontal
A	8 1/2	11 (horizontal format)
	11	8 1/2 (vertical format)
B	11	17
C	17	22
D	22	34
E	34	44
F	28	40
Sizes G, H, J, and K are roll sizes.		

ASME Y14.1M lists sheet size specifications in metric units, as follows:

Size Designation	Size (in millimeters)	
	Vertical	Horizontal
A0	841	1189
A1	594	841
A2	420	594
A3	297	420
A4	210	297

Longer lengths are known as elongated and extra-elongated drawing sizes. These are available in multiples of the short side of the sheet size. **Figure 29-16** shows standard ASME/ANSI sheet sizes. To describe sheet size values verbally, generally state the vertical measurement and then the horizontal measurement. For example, describe a C-size sheet as 22 (horizontal) × 17 (vertical).

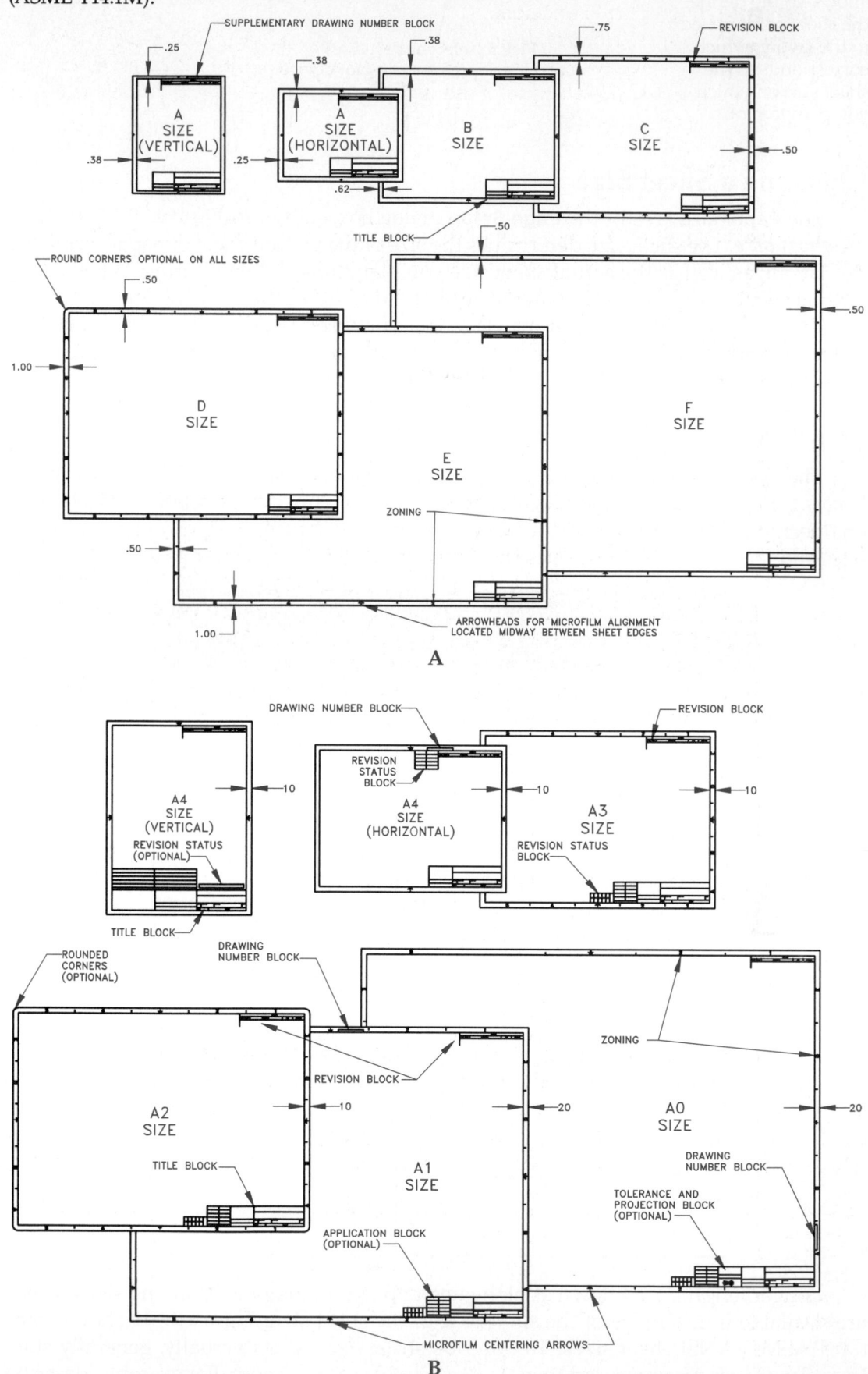

Drawing Sheets
For tables describing sheet characteristics, including sheet size, drawing scale, and drawing limits, go to the **Reference Material** section of the Student Web site (www.g-wlearning.com/CAD) and select **Drawing Sheets**.

Specifying the Drawing Orientation

The **Drawing orientation** area controls the plot rotation. See **Figure 29-17.** Landscape orientation is the most common engineering drawing orientation and is the default in most AutoCAD-supplied templates. Portrait orientation is the standard for most written documents printed on 8.5″ × 11″ paper.

Pick the **Plot upside-down** check box to produce variations of the standard landscape and portrait orientations. When you select an upside-down orientation, it may help to consider landscape format to be a rotation angle of 0° and portrait format to be a rotation angle of 90°. Therefore, an upside-down landscape format rotates the drawing 180°, and an upside-down portrait orientation rotates the drawing 270°. Use the preview image to help select the appropriate orientation.

PROFESSIONAL TIP

The way in which a sheet feeds into a printer or plotter can affect the sheet size and drawing orientation you select. Sheets of paper, especially large sheets, often feed into a plotter with the short side of the sheet entering first. This may require you to use a sheet size that orients the sheet in a portrait format, for example D-Size 22x34 instead of D-Size 34x22, while still using a landscape drawing orientation.

Exercise 29-3

Access the Student Web site (www.g-wlearning.com/CAD) and complete Exercise 29-3.

Choosing the Plot Area

The **Plot area** section, shown in **Figure 29-18,** allows you to choose the portion of the drawing to plot. Select an option from the **What to plot:** drop-down list. The **Layout** option is available when you plot a layout. When you select this option, everything inside the margins of the layout plots. Plotting the layout using the **Layout** option is the default and is the most common setting for plotting a layout. Use other plot area

---

Figure 29-17.
The **Drawing orientation** area contains options for adjusting the drawing angle of rotation.

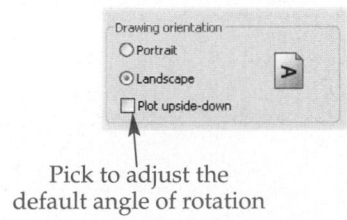

Pick to adjust the
default angle of rotation

Figure 29-18.
Use the **Plot area**
section to define
what portion of the
drawing to plot.

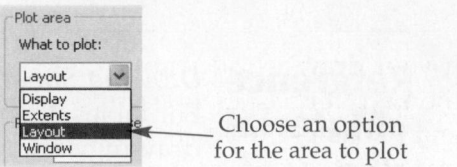

Choose an option
for the area to plot

options primarily for plotting in model space or when it is necessary to adjust the area plotted in paper space.

The **View** option is available when named views exist in the drawing, and if the model space or paper space environment associated with the current page setup includes a named view. When you pick the **View** option, an additional drop-down list appears in the **Plot area** section, allowing you to select a specific view to define as the plot area. Refer to Chapter 5 for information about other **Plot area** options.

Defining the Plot Offset

The **Plot offset** area, shown in **Figure 29-19,** controls how far the drawing is offset from the plot origin. You can specify the plot origin as the lower-left corner of the printable area or the lower-left corner of the sheet by selecting the appropriate radio button in the **Specify plot offset relative to** area in the **Plot and Publish** tab of the **Options** dialog box.

The **Plot offset (origin set to printable area)** title appears when you use the default **Printable area** option of the **Options** dialog box. The values you enter in the **X:** and **Y:** text boxes define the offset from the printable area. Use the default values of 0 to locate the plot origin at the lower-left corner of the printable area, which corresponds to the lower-left corner of the layout margin (dashed rectangle). To move the drawing away from the default printable area origin, enter positive or negative values in the text boxes. For example, to move the drawing one unit to the right and two units above the lower-left corner of the margin, enter 1 in the **X:** text box and 2 in the **Y:** text box.

Select the **Edge of paper** radio button in the **Options** dialog box to display the **Plot offset (origin set to layout border)** title. This causes the values you enter in the **X:** and **Y:** text boxes to define the offset from the edge of the sheet. Use the default values of 0 to locate the plot origin at the lower-left corner of the sheet. Change the values in the text boxes to move the drawing away from the sheet origin.

When you select any of the options from the **What to plot:** drop-down list in the **Plot area** section, except the **Layout** option, the **Center the plot** check box becomes enabled. Pick this check box to shift the plot origin automatically as needed to center the selected plot area in the printable area.

Figure 29-19.
The **Plot offset** area
controls how far
the drawing offsets
from the lower-
left corner of the
printable area or
layout border.

Enter offset from
lower-left corner

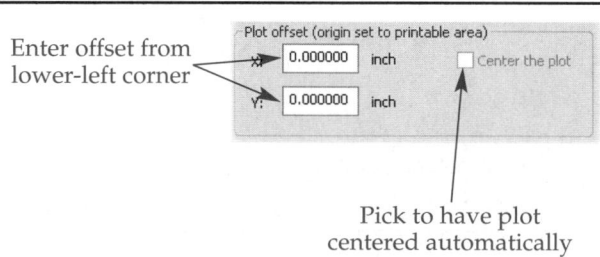

Pick to have plot
centered automatically

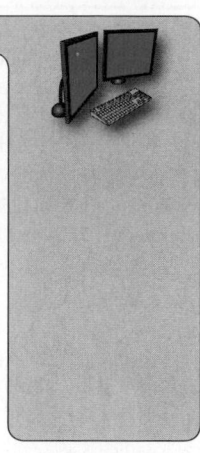

If your drawing does not center on the sheet when you plot using the **Layout** option, open the **Options** dialog box and select the **Plot and Publish** tab. Pick the **Edge of paper** radio button in the **Specify plot offset relative to** area. Then use values of 0 in the X and Y offset text boxes in the **Page Setup** dialog box to locate the plot origin at the lower-left corner of the sheet. Depending on the specific plot configuration, you may also need to pick the **Plot upside-down** check box in the **Drawing orientation** area of the **Page Setup** dialog box to locate the origin exactly at the lower-left corner of the sheet. Alternatively, resolve the issue by changing the plot offset to the X and Y values of the lower-left corner of the printable area.

Selecting a Plot Scale

You should always draw objects at their actual size, or full scale, in model space, regardless of the size of the objects. For example, if you draw a small machine part and the length of a line in the drawing is 2 mm, draw the line 2 mm long in model space. If you draw a building and the length of a line in the drawing is 80', draw the line 80' long in model space. For layout and printing purposes, you must then scale most drawings to fit properly on a sheet, according to a specific *drawing scale*.

When you scale a drawing, you increase or decrease the *displayed* size of model space objects. A properly scaled floating viewport in a layout allows for this process, as described in Chapter 30. When setting up a layout for plotting, remember that the layout is also at full scale. One difference between model space and paper space is that objects in model space can be very large or very small, while objects in paper space always correspond to sheet size. In order for objects on the layout and in model space to appear correct when plotted, you must plot a layout at full scale, or 1:1.

> **drawing scale:**
> The ratio between the actual size of objects in the drawing and the size at which the objects plot on a sheet of paper.

The **Plot scale** area of the **Page Setup** dialog box allows you to specify the plot scale. See **Figure 29-20.** The **Scale:** drop-down list provides several predefined decimal and architectural scales, as well as a **Custom** option. For most applications, when setting up a layout for plotting, the plot scale is set to 1:1. This ensures that the layout and scaled floating viewports plot correctly. If you choose a scale other than 1:1, the layout does not plot to scale.

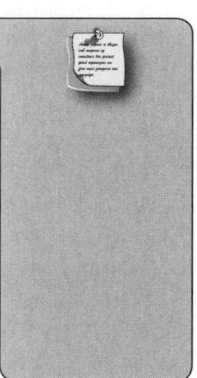

When preparing to plot in model space, you may choose to select a scale other than 1:1. For example, to plot an architectural floor plan, you might want to set the plot scale to 1/4" = 1'-0". If the desired scale is not available, enter values in the text boxes below the predefined scales drop-down list and select the correct unit of measure from the drop-down list. The **Custom** option is automatically displayed when you enter values. For example, 1 inch = 600 units is a custom scale entry used to plot at a scale of 1" = 50' (50' x 12" = 600). Refer to Chapter 9 for more information about drawing scale and scale factors.

Pick the **Fit to paper** check box to adjust the plot scale automatically to fit on the selected sheet. This is useful if you are not concerned about plotting to scale, such as when you are creating a "check copy" on a sheet that is too small to plot at the appropriate scale. Select the **Scale lineweights** check box to scale (increase or decrease the weight of) lines when the plot scale changes.

Figure 29-20.
Use the **Plot scale** area to adjust the scale at which the drawing plots. The plot scale is typically set to 1:1 to plot a layout even though the drawing scale may not be 1:1.

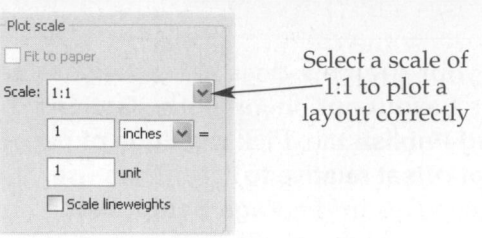

Select a scale of 1:1 to plot a layout correctly

PROFESSIONAL TIP

In the rare event that you need to plot an inch drawing on a metric sheet, use a custom scale of 1 inch = 25.4 units. Conversely, use a custom scale of 25.4 mm = 1 unit to plot a metric drawing on an inch sheet.

Exercise 29-4

Access the Student Web site (www.g-wlearning.com/CAD) and complete Exercise 29-4.

Using Plot Styles

Object properties control the appearance of objects on-screen. By default, what you see on-screen is what plots. For example, if you draw objects on a layer that uses a Red color, Continuous linetype, and 0.60 mm lineweight, the objects display and plot red, continuous, and thick, assuming you show lineweights on-screen and use a color plotter. If this situation is desirable, you are ready to continue with the page setup and plotting process.

plot styles: Properties, including color, linetype, lineweight, line end treatment, and fill style, that are applied to objects for plotting purposes only.

However, to define exactly how objects plot regardless of what displays on-screen, you must assign *plot styles* to objects. Use plot styles to maintain object properties in the drawing, but plot objects according to specific plotting properties. For example, you may want all of the objects in a drawing to plot as dark as possible, which requires that all objects use the color Black when plotted. In this example, plot styles allow you to plot all objects black without making them black on-screen. Plot styles also allow you to plot objects using shades of gray instead of color, or to plot objects lighter or darker than they display on-screen.

Plot Style Tables

plot style table: A configuration, saved as a separate file, that groups plot styles and provides complete control over plot style settings.

Plot styles are contained in *plot style tables*. You can choose to use either a *color-dependent plot style table* or a *named plot style table*. When you use a color-dependent plot style table, objects plot according to object color. Color-dependent plot style tables contain 256 preset plot styles—one for each AutoCAD Color Index (ACI) color. Each color-dependent plot style links to an index color. Plot style properties control how to treat objects of a certain color when they are plotted. For example, the plot style Color 1, which is Red, defines how all objects that are red on-screen plot. If you assign a Black plot color to plot style Color 1, all objects drawn using a Red color plot black, even though the objects are red on-screen.

color-dependent plot style table: A file that contains plot style settings used to assign plot values to object colors.

named plot style table: A file that contains plot style settings used to assign plot values to objects or layers.

When you use a named plot style table, drawing objects plot according to named plot style values, which you can assign to a layer or object. Any layer or object assigned a named plot style plots using the settings specified for that plot style. For example, create a layer named OBJECT that uses a Red color, and a plot style named BLACK that uses a Black color. Then assign the BLACK plot style to the OBJECT layer. Objects drawn on the OBJECT layer plot using the BLACK plot style and plot black in color, even though the objects are red on-screen.

Ideally, you should decide which plot style table is appropriate for your application before you begin drawing. The templates created in the Template Development feature of this textbook, for example, assume that drawings plot so that all objects appear dark, or black, with different object linetypes and lineweights. You can use a color-dependent plot style table or a named plot style table to create this effect. Default AutoCAD plot style behavior is set to use color-dependent plot style tables. For basic applications, it is usually best to use color-dependent plot style tables, because they are the default and do not require you to assign named plot styles to layers or objects.

Configuring Plot Style Table Type

You must choose a type of plot style table (color-dependent or named) to apply to new drawings *before* you start a new drawing file, unless you use a template that already is assigned a plot style table type. To configure the plot style type used by default when you create new drawings, pick the **Plot Style Table Settings...** button in the **Plot and Publish** tab of the **Options** dialog box. This opens the **Plot Style Table Settings** dialog box shown in **Figure 29-21.**

To use a named plot style table, pick the **Use named plot styles** radio button from the **Default plot style behavior for new drawings** area *before* you start a new drawing file. Select a specific plot style table to use as the default for new drawings from the **Default plot style table:** drop-down list in the **Current plot style table settings** area. When you select the **Use named plot styles** radio button, the **Default plot style for layer 0:** and **Default plot style for objects:** options become enabled.

NOTE

When you use a template to create a new drawing, the plot style settings defined in the template override the settings you specify in the **Plot Style Table Settings** dialog box. For example, if you configure the template to use named plot style tables, you can only select a named plot style table to apply to the plot, even if you select the **Use color dependent plot styles** radio button in the **Plot Style Table Settings** dialog box.

Figure 29-21.
Use the **Plot and Publish** tab of the **Options** dialog box to setup default plot style types and tables for new drawings.

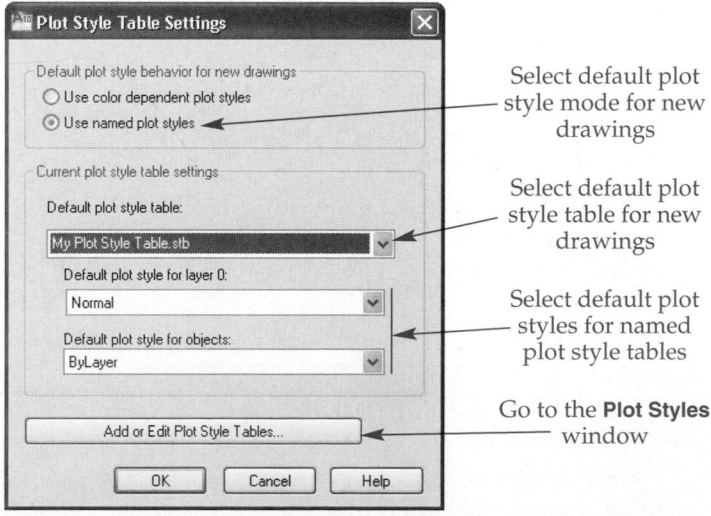

Select default plot style mode for new drawings

Select default plot style table for new drawings

Select default plot styles for named plot style tables

Go to the **Plot Styles** window

Pick the **Add or Edit Plot Style Tables...** button in the **Plot Style Table Settings** dialog box to open the **Plot Styles** window. See **Figure 29-22.** The **Plot Styles** window lists all available color-dependent and named plot style tables saved in the Plot Style Table Search Path, as defined in the expanded **Printer Support File Path** option in the **Files** tab of the **Options** dialog box. Color-dependent plot style table files (CTB) use the .ctb extension. Named plot style table files (STB) include an .stb extension. Double-click on **Add-A-Plot Style Table Wizard** to create a new plot style table, or double-click on an existing plot style table file to edit the file.

Supplemental Material

Creating and Editing Plot Style Tables

For detailed information about creating and editing plot style tables, go to the Student Web site (www.g-wlearning.com/CAD), select this chapter, and select **Creating and Editing Plot Style Tables**.

Selecting a Plot Style Table

Use the **Plot style table (pen assignments)** area of the **Page Setup** dialog box, shown in **Figure 29-23,** to activate and manage plot style tables. Select a plot style table to use from the drop-down list. The **None** option is default and can be used instead of selecting a specific color-dependent or named plot style table. Select **None** to plot exactly what appears on-screen, without using plot styles, assuming you use a color plotter to plot objects with color.

Only color-dependent or named plot style tables appear, depending on the type of plot style tables assigned to the current drawing. The most often used color-dependent plot style tables are:

- Monochrome.ctb—plots the drawing in monochrome (black and white).
- Grayscale.ctb—plots the drawing using shades of gray.
- Screening files—plot the drawing using faded, or screened, colors.

When using named plot style tables, it is common to select one of the following:

- Autodesk-Color.stb—provides access to named plot styles for plotting the drawing using solid and faded colors.

Figure 29-22.
The **Plot Styles** window lists all available plot style files and allows you to create and edit plot style tables.

Named plot style icon

Double-click to create a new plot style table

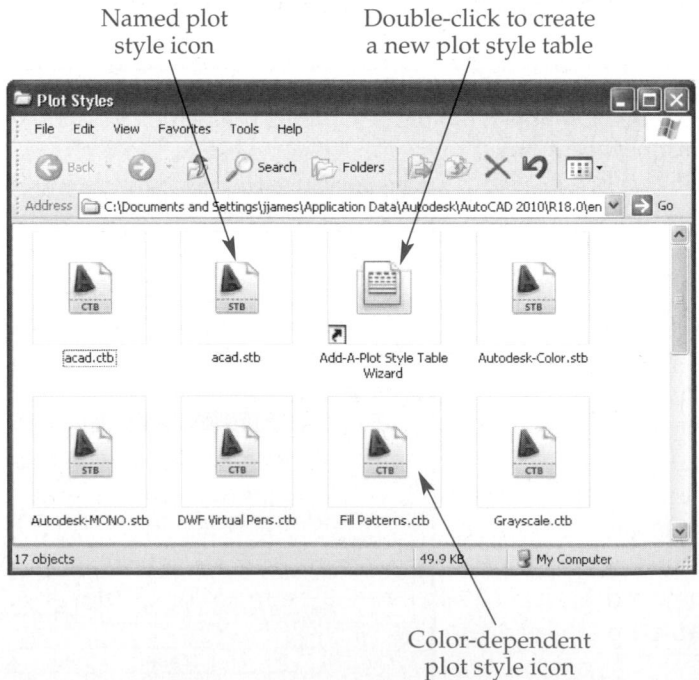

Color-dependent plot style icon

Figure 29-23.
Use the **Plot style table (pen assignments)** area to select, create, and edit plot style tables.

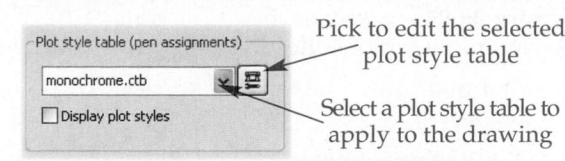

Pick to edit the selected plot style table

Select a plot style table to apply to the drawing

- Monochrome.stb—references a named plot style for plotting the drawing in monochrome.
- Autodesk-MONO.stb—provides access to named plot styles for plotting objects monochrome, in color, and in faded monochrome.

PROFESSIONAL TIP

You cannot select named plot style tables if you started the drawing with color-dependent plot style tables. Conversely, you cannot select color-dependent plot style tables if you started the drawing with named plot style tables. You can type CONVERTPSTYLES to access the **CONVERTPSTYLES** tool, which allows you to switch between table modes in a drawing. However, you should avoid converting plot style tables when possible. Instead, use the appropriate plot style type when beginning a new drawing.

Exercise 29-5

Access the Student Web site (www.g-wlearning.com/CAD) and complete Exercise 29-5.

Applying Plot Styles

After you select a plot style table from the drop-down list in the **Plot style table (pen assignments)** area of the **Page Setup** dialog box, the plot styles contained in the selected plot style table are ready to assign to objects in the drawing. When you select a color-dependent plot style table, plot styles are automatically applied to objects in the drawing according to the color of the objects. No additional steps are required to apply color-dependent plot styles to objects.

When you select a named plot style table, the named plot styles contained in the table are ready to assign to layers or individual objects. Any layer or object assigned a named plot style plots using the settings specified for that style. Named plot style tables contain as many plot styles as have been created. For example, the monochrome.stb plot style table contains the default Normal plot style and a style named Style 1. When you apply the Normal plot style to layers or objects, objects plot exactly as they appear on-screen. Style 1 assigns the color Black to all layers or objects that use Style 1 for plotting. In order to plot all objects in the drawing black, even if the objects display different colors on-screen, you must apply Style 1 to all layers.

Assign named plot styles to layers in the **Layer Properties Manager** palette. See **Figure 29-24.** Identify the layer to assign a named plot style and select the default plot style, such as **Normal,** from the **Plot Style** column. The **Select Plot Style** dialog box appears. See **Figure 29-25.** Select a named style to assign it to the highlighted layer. Any object drawn on a layer assigned to a named plot style plots using the settings in the named plot style.

Figure 29-24.
Assign named plot styles to layers using the **Layer Properties Manager**.

Pick plot style name for layer
to select a different plot style

Figure 29-25.
Use the **Select Plot Style** dialog box to select a plot style to assign to a layer.

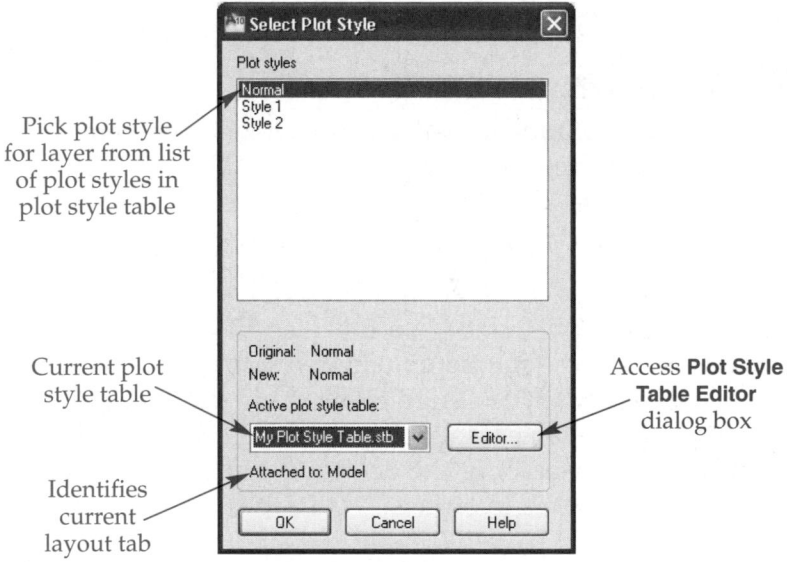

Pick plot style
for layer from list
of plot styles in
plot style table

Current plot
style table

Identifies
current
layout tab

Access **Plot Style
Table Editor**
dialog box

You can also assign named plot styles to individual objects. If you assign a plot style to an object drawn on a layer given a named plot style, the plot style assigned to the object overrides the plot style settings given to the layer. Assign plot styles to objects using the **Properties** palette or the **Plot Style Control** drop-down list in the **Properties** panel on the **Home** ribbon tab. See **Figure 29-26.**

> **NOTE**
>
> If the current drawing is set to use a named plot style table, the default plot style for all layers is Normal. The default plot style for all objects is ByLayer. Objects plotted with these settings keep their original properties.

Figure 29-26.
You can assign a plot
style to individual
objects using
A— the **Properties**
palette or B—the
Plot Style Control
drop-down list in
the ribbon.

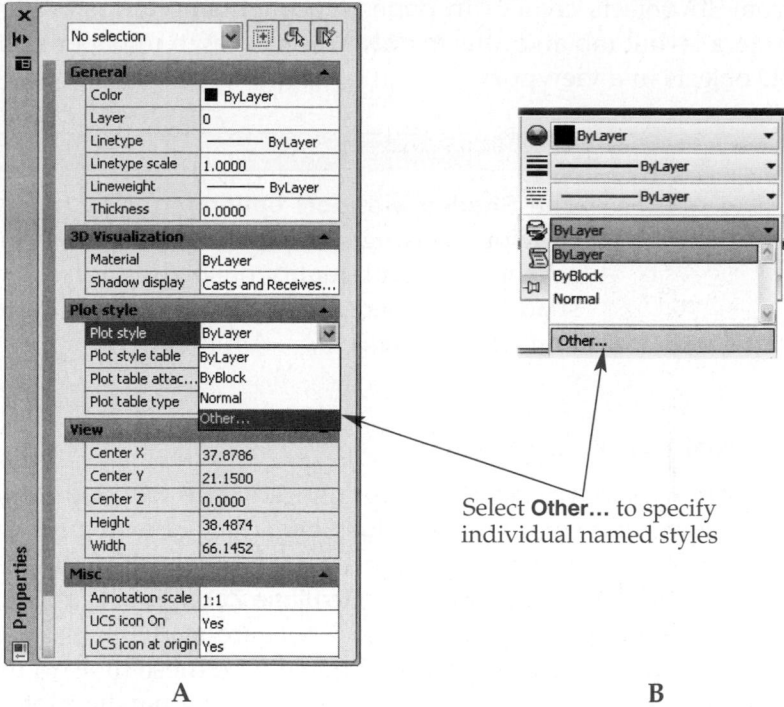

Select **Other...** to specify
individual named styles

A   B

Exercise 29-6

Access the Student Web site (www.g-wlearning.com/CAD) and
complete Exercise 29-6.

Viewing Plot Style Effects On-Screen

To display plot style effects on-screen, pick the **Display plot styles** check box in the
Plot style table area in the **Page Setup** dialog box. Objects on-screen appear as they will
plot. Typically, it is not appropriate to work with objects displayed as they will plot,
especially if you print in monochrome. In this example, the color assigned to a layer
appears black, which defeats the purpose of assigning unique colors to layers. A better
practice is to use the preview feature of the **Page Setup** or **Plot** dialog box, as described
later in this chapter, to preview the effects of plot styles before plotting.

Other Plotting Options

The **Plot options** area of the **Page Setup** dialog box contains additional plot options
that affect how specific items plot. If you plan to plot objects using a plot style table, be
sure to select the **Plot with plot styles** check box. The main purpose of this check box is
to toggle the use of plot styles on and off. This allows you to create a plot quickly with
or without using plot styles.

When you deselect the **Plot with plot styles** check box, the **Plot object lineweights**
check box is enabled and selected by default. When this box is checked, all objects with
a lineweight greater than 0 plot using the assigned lineweight. Deselect the check box
to plot all objects using a 0, or thin, lineweight.

Pick the **Plot paperspace last** check box to plot paper space objects after model
space objects. This option ensures that objects in paper space that overlap objects in
model space plot on top of model space objects. The **Plot paperspace last** check box is
disabled when you plot in model space, because model space does not include paper
space objects. Select the **Hide paperspace objects** check box to remove hidden lines

from 3D objects created in paper space. This option is only available when you plot from a layout tab and affects only objects drawn in paper space. It does not affect any 3D objects in a viewport.

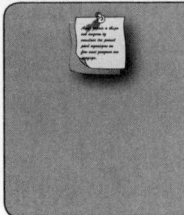

NOTE

The **Shaded viewport options** area of the **Page Setup** dialog box provides settings that control viewport shading. These options set the type and quality of shading for plotting 3D models from a shaded or rendered viewport. 3D models are explained in *AutoCAD and Its Applications—Advanced*.

Completing Page Setup

After you have selected all appropriate page setup options, pick the **Preview** button in the lower-left corner of the **Page Setup** dialog box to preview the effects of the page setup. What you see on-screen is the exact plot appearance, assuming you use a color plotter to make color prints. The **Realtime Zoom** tool is activated automatically in the preview window. Additional view tools are available from the toolbar near the top of the window or from a shortcut menu. Use these tools to help confirm that the plot settings are correct. When you finish previewing the plot, press [Esc] or [Enter], or right-click and select **Exit** to return to the **Page Setup** dialog box. Pick the **OK** button to exit the **Page Setup** dialog box, and pick **Close** button to exit the **Page Setup Manager**.

Exercise 29-7

Access the Student Web site (www.g-wlearning.com/CAD) and complete Exercise 29-7.

Chapter Test

Answer the following questions. Write your answers on a separate sheet of paper or go to the Student Web site (www.g-wlearning.com/CAD) and complete the electronic chapter test.

1. What is a model?
2. Define *model space*.
3. How is a layout used?
4. What is paper space?
5. What is the purpose of a floating viewport?
6. What is the purpose of the dashed rectangle that appears on the default layout?
7. If you pick the **MODEL** button to enter paper space and the file contains multiple layouts, none of which has previously been accessed, which layout displays by default? Which layout displays if a different layout has already been opened?
8. Briefly describe the function of the **Quick View Layouts** tool.
9. What does the **Quick View Drawings** tool allow you to do?
10. What is a page setup?
11. Briefly describe the basic function of the **Page Setup Manager** and the related **Page Setup** dialog box.
12. Briefly describe the function of the **Plot** dialog box and explain when it is used in relation to the **Page Setup Manager** and **Page Setup** dialog box.

13. Explain the importance of a well-defined page setup.
14. How can you identify the page setup that is tied to the current layout?
15. What is a plot device?
16. Define *sheet size*.
17. What factors should you consider when you select a sheet size for a drawing?
18. Which ANSI standards specify sheet sizes?
19. What is the plot offset of a layout?
20. Define *drawing scale*.
21. What are plot styles?
22. Briefly explain the purpose of a plot style table.
23. How are plot styles assigned in a color-dependent plot style?
24. Briefly explain the function of a named plot style table.
25. How can you be certain that your page setup options will produce the desired plot?

Drawing Problems

Start AutoCAD if it is not already started. Follow the specific instructions for each problem.

▼ Basic

1. Use a word processor to list five items commonly found in a layout. Provide a brief description of each item.

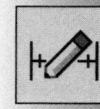

2. Start a new drawing using the acad.dwt template. Create a plot style table named Black35mm.cbt that will plot all colors in the AutoCAD drawing in black ink on the paper, with a lineweight of 0.35 mm. (*Hint:* To make the same change to a property of all the plot styles, select the first plot style in the list, in this case Color 1, then scroll to the end of the list, hold down the [Shift] key, and select the last plot style in the list, in this case Color 255). Save the drawing as P29-2.

3. Start a new drawing using the acad -Named Plot Styles.dwt template. Create a plot style table named BlackShades.stb. Create the following plot styles:
   - Black100% with color set to black, all other properties set to their default values.
   - Black50% with color set to black, screening set to 50, all other properties set to their default values.
   - Black25% with color set to black, screening set to 25, all other properties set to their default values.
   Save the drawing as P29-3.

4. Open P8-14 and save as P29-4. The P29-4 file should be active. Delete the default **Layout2**. Create a new B-size sheet layout by following these steps:
   A. Rename the default **Layout1** to **B-SIZE**.
   B. Select the **B-SIZE** layout and access the **Page Setup Manager**.
   C. Modify the **B-SIZE** page setup according to the following settings:
      • **Printer/Plotter:** Select a printer or plotter that can plot a B-size sheet
      • **Paper size:** Select the appropriate B-size sheet (varies with printer or plotter)
      • **Plot area:** Layout
      • **Plot offset:** 0,0
      • **Plot scale:** 1:1 (1 inch = 1 unit)
      • **Plot style table:** monochrome.ctb
      • **Plot with plot styles**
      • **Plot paperspace last**
      • Do not check **Hide paperspace objects**
      • **Drawing orientation:** Select the appropriate orientation (varies with printer or plotter)
      Resave P29-4.

5. Open P8-15 and save as P29-5. The P29-5 file should be active. Delete the default **Layout2**. Create a new A2-size sheet layout according to the following steps:
   A. Rename the default **Layout1** to **A2-SIZE**.
   B. Select the **A2-SIZE** layout and access the **Page Setup Manager**.
   C. Modify the **A2-SIZE** page setup according to the following settings:
      • **Printer/Plotter:** Select a printer or plotter that can plot an A2-size sheet
      • **Paper size:** Select the appropriate A2-size sheet (varies with printer or plotter)
      • **Plot area:** Layout
      • **Plot offset:** 0,0
      • **Plot scale:** 1:1 (1 mm = 1 unit)
      • **Plot style table:** monochrome.ctb
      • **Plot with plot styles**
      • **Plot paperspace last**
      • Do not check **Hide paperspace objects**
      • **Drawing orientation:** Select the appropriate orientation (varies with printer or plotter)
      Resave P29-5.

6. Open P8-16 and save as P29-6. The P29-6 file should be active. Delete the default **Layout2**. Create a new B-size sheet layout according to the following steps:
   A. Rename the default **Layout1** to **B-SIZE**.
   B. Select the **B-SIZE** layout and access the **Page Setup Manager**.
   C. Modify the **B-SIZE** page setup according to the following settings:
      • **Printer/Plotter:** Select a printer or plotter that can plot a B-size sheet
      • **Paper size:** Select the appropriate B-size sheet (varies with printer or plotter)
      • **Plot area:** Layout
      • **Plot offset:** 0,0
      • **Plot scale:** 1:1 (1 inch = 1 unit)
      • **Plot style table:** monochrome.ctb
      • **Plot with plot styles**
      • **Plot paperspace last**
      • Do not check **Hide paperspace objects**
      • **Drawing orientation:** Select the appropriate orientation (varies with printer or plotter)
      Resave P29-6.

▼ Advanced

7. Draw the wood beam details using the dimensions and notes provided. Establish the missing information using your own specifications, or determine the correct size of items not dimensioned. Prepare a single layout for plotting on a C-size sheet, using the monochrome.ctb plot style. Delete all other layouts. Save the drawing as P29-7.

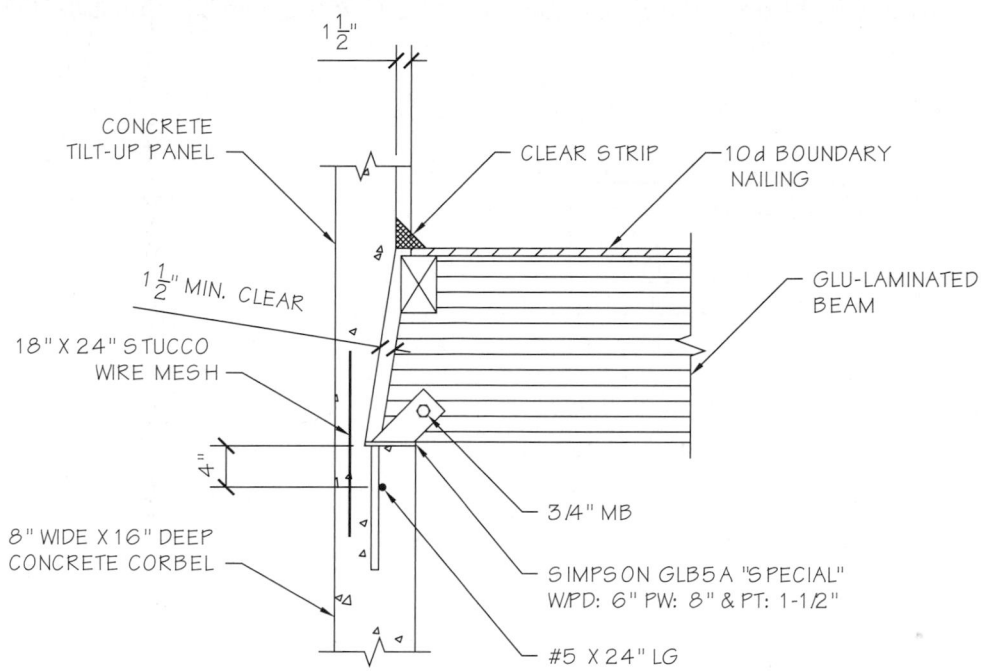

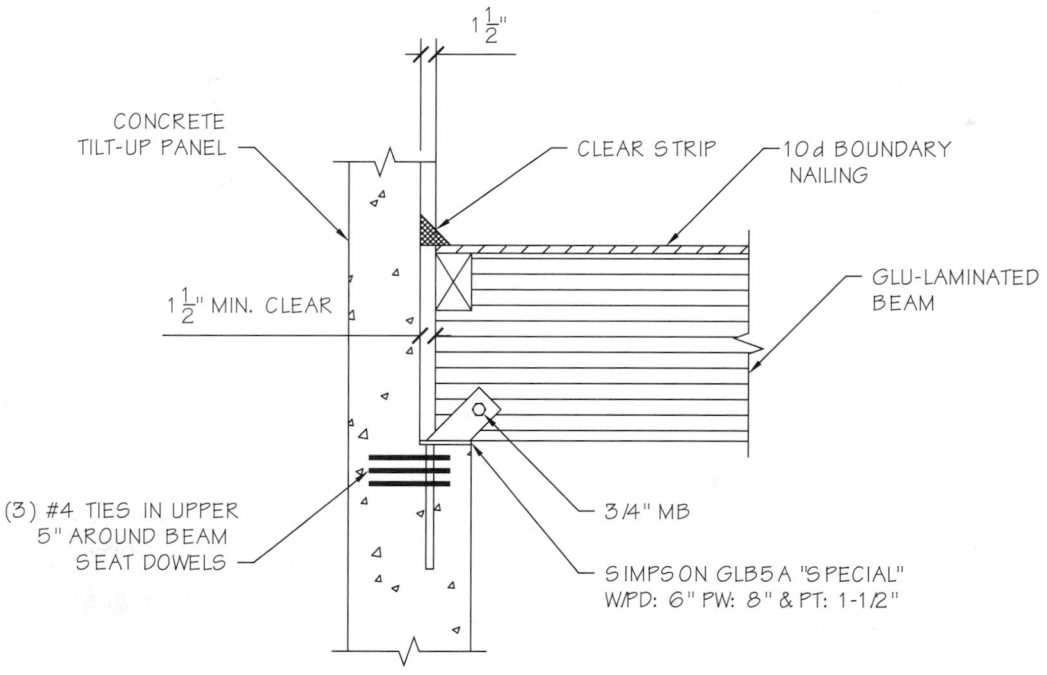

8. Use a word processor to write a report of approximately 250 words explaining the difference between model space and paper space and describing the importance of using layouts. Cite at least three examples from actual industry applications of using layouts to prepare a multisheet drawing. Use at least four drawings to illustrate your report.

9. Create the multiview drawing using the dimensions provided. Prepare a single layout for plotting on a C-size sheet, using the monochrom.ctb plot style. Delete all other layouts. Save the drawing as P29-9.

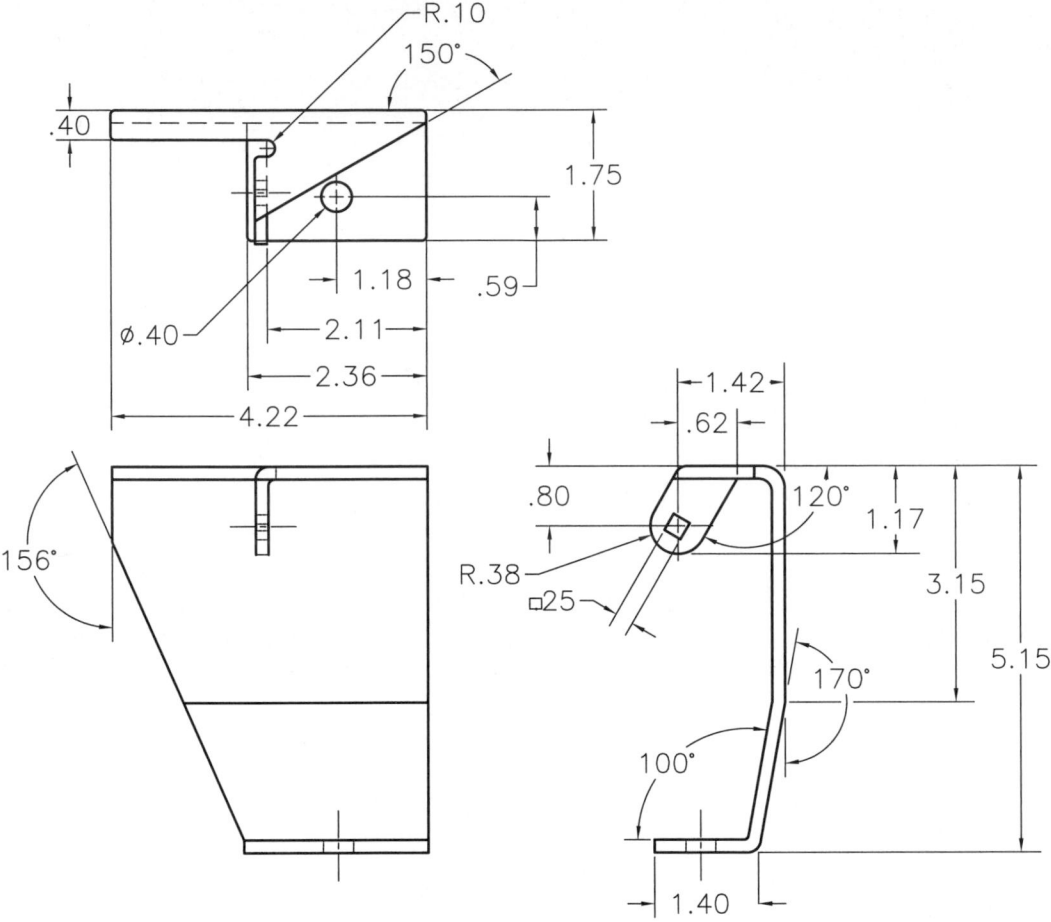

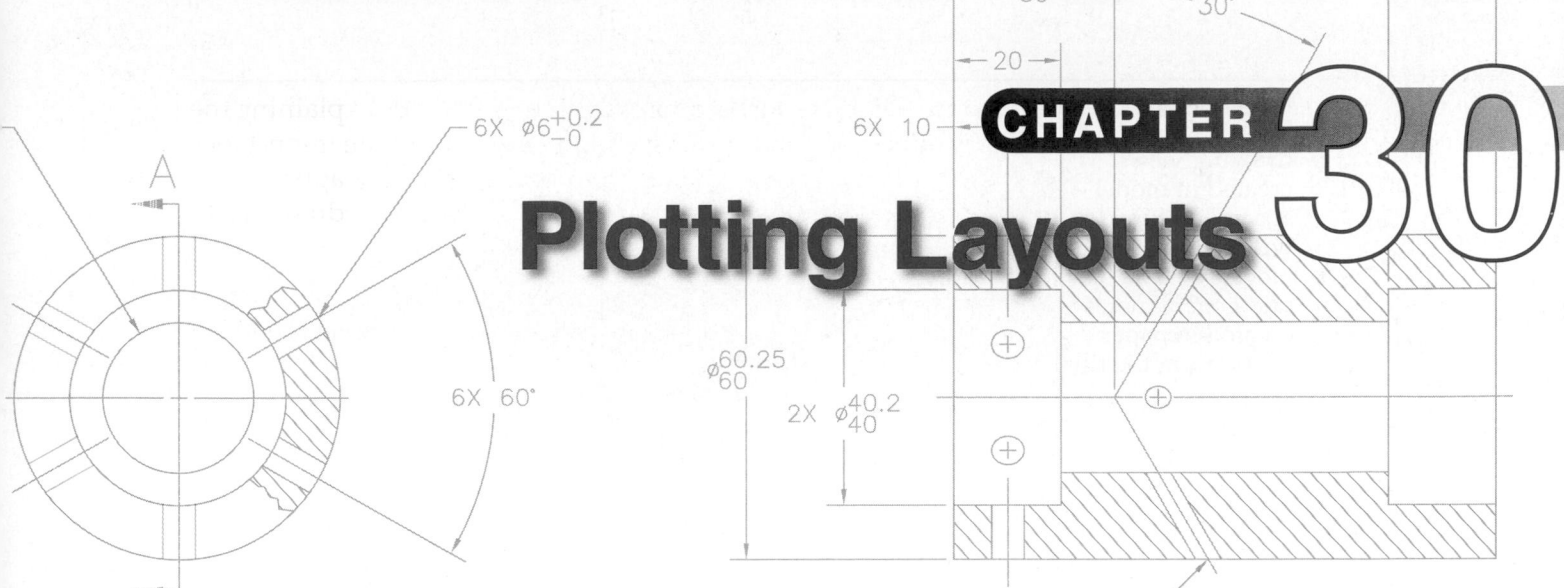

CHAPTER

Plotting Layouts

Learning Objectives

After completing this chapter, you will be able to do the following:
- ✓ Add layout content.
- ✓ Use floating viewports.
- ✓ Create properly scaled final drawings.
- ✓ Preview and plot layouts.

This chapter explores the additional steps to complete layout setup. You will learn to add content to a layout and place and use floating viewports. You will also use the **PLOT** tool to preview the plot and send the layout to a printer or plotter.

Layout Content

Model space provides an environment to create drawing views and add dimensions and annotations directly to views. Layouts provide an effective method to display model space content and add items such as a border, title block, revision block, general notes, and a bill of materials, parts list, or schedules. Layouts provide flexibility for laying out, scaling, and preparing a final drawing. Consider the objects placed in model space to be *drawing* content and the objects you add to a layout to be *sheet* content. A complete drawing forms when you bring sheet and drawing content together. See **Figure 30-1**.

Drawing in Paper Space

A layout is a representation of a flat piece of paper. As a result, paper space is a 2D drawing environment. Most 2D drawing and editing tools and options described throughout this textbook function the same in paper space as in model space. Some tools, however, are specific to or are most often used in either model space or paper space. For example, the **Full Navigation Wheel** appears by default when you access the **SteeringWheel** view tool in model space. The **2D Navigation Wheel** appears in paper space and is specific to 2D drafting.

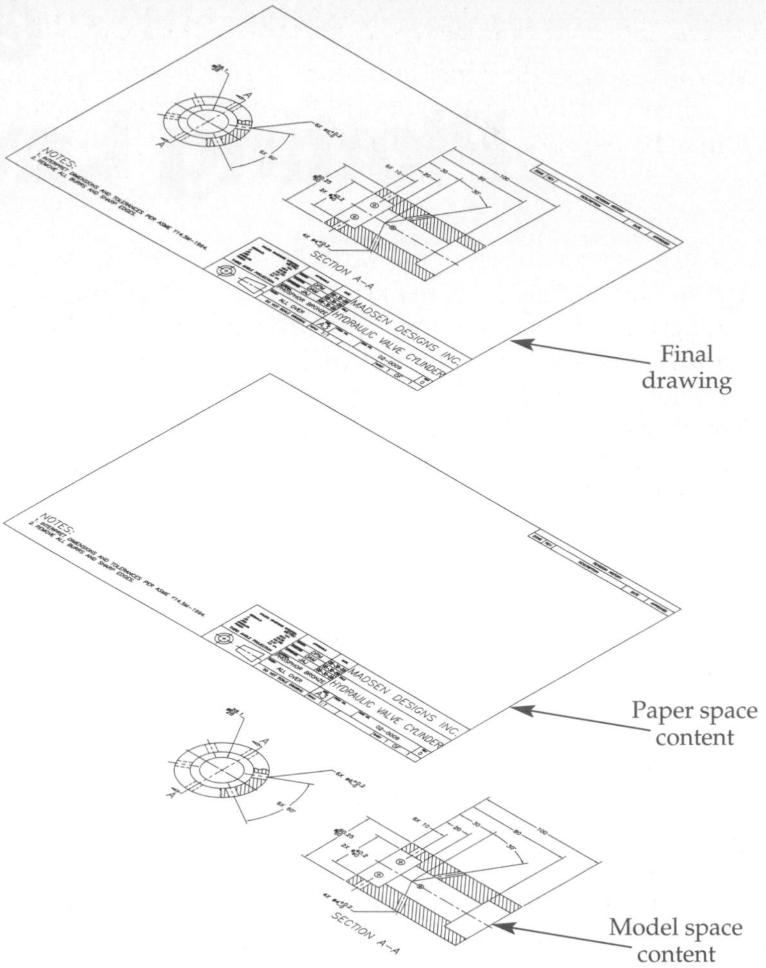

Figure 30-1.
Dimensioned
drawing views
created in model
space, combined
with a border, title
block, revision block,
and general notes
created in paper
space, form the final
drawing.

Final
drawing

Paper space
content

Model space
content

> **NOTE**
>
> Although paper space is a 2D environment, you can display 3D models created in model space in paper space floating viewports.

As in model space, you typically draw geometry on a layout at full scale. One difference between model space and paper space is that objects in model space may be very large or very small, while paper space objects always correspond to the sheet size. Draw all layout content using the actual size you want the objects to appear on the plotted sheet. Add layout content such as general notes, view titles, and similar annotations as multiline or single-line text or as blocks with attributes. Place a bill of materials, parts list, schedule, or similar tabular information using the **TABLE** tool or blocks with attributes. You typically create most other items, such as a border, title block, and revision block as blocks, often with attributes, and then insert these items into the layout.

When you draw in paper space, use layers appropriate for the layout and the objects added to the layout. You may want to create a single layer named SHEET, for example, on which you draw all layout content. Another option is to use layers specific to layout items, such as a BORDER layer for the border and a TITLE or A-ANNO-TTBL layer for the title block. A layer named Viewport, VPORT, or A-ANNO-NPLT is typically assigned to floating viewports so the viewport boundary can be turned off, frozen, or set to "no plot" before plotting.

NOTE

The default floating viewport boundary assumes the layer that is current when you first access a layout. If you use the default viewport, it may be necessary to change the layer on which it is drawn.

The Layout Origin

Drawing and editing in paper space is most often done on the sheet, which is the white rectangle you see on the gray background. However, it is possible, and necessary in some applications, to create objects off the sheet. When drawing and editing on a layout, remember that the origin (0,0) is controlled by the X and Y values you enter in the **Plot offset** area of the **Page Setup** dialog box. For example, if you draw a line with a start point of (1,1), the line begins 1 unit to the left and 1 unit up from the plot origin. The default origin position is at the lower-left corner of the printable area. See **Figure 30-2.** Refer to Chapter 29 for more information on plot origin and the **Plot offset** area of the **Page Setup** dialog box.

Layout content often references the edge of the sheet. For example, you might position a border 1/2″ inside of the sheet edge. In this situation, it is usually best to define the layout origin as the lower-left corner of the sheet. The best option is to select the **Edge of paper** radio button in the **Specify plot offset relative to** area in the **Plot and Publish** tab of the **Options** dialog box. In the **Plot offset** area of the **Page Setup** dialog box, use values of 0.000 in the X and Y offset text boxes to locate the plot origin at the lower-left corner of the sheet. This is an excellent way to center the drawing on the sheet.

NOTE

Depending on the specific plot configuration, you may need to pick the **Plot upside-down** check box in the **Drawing orientation** area of the **Page Setup** dialog box in addition to using the **Edge of paper** offset option. Draw an object to see if (0,0) is actually located exactly at the lower-left corner of the sheet. If not, pick the **Plot upside-down** check box to solve the problem.

PROFESSIONAL TIP

Defining the plot offset relative to the edge of the paper has the benefit of maintaining the plot offset location from the sheet edge even when you use a different plotter or plot configuration, as long as you use the same sheet size.

A second option to relate content to the lower-left corner of the sheet is to select the **Printable area** radio button in the **Specify plot offset relative to** area in the **Plot and Publish** tab of the **Options** dialog box. Then identify the values of the lower-right corner of

Figure 30-2.
The default location of the plot origin is often not appropriate, because all point entry is in reference to the lower-left corner of the printable area. Change the location of the plot origin to define the lower-left corner of the sheet as the (0,0) point.

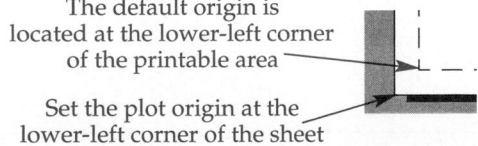

The default origin is located at the lower-left corner of the printable area

Set the plot origin at the lower-left corner of the sheet

the printable area. This information is available in the **Device and Document Settings** tab of the plotter's **Configuration Editor**. Next, in the **Plot offset** area of the **Page Setup** dialog box, change the X and Y plot offset values to the values of the lower-left corner of the printable area. The values you enter must be negative. The origin is offset from the printable area, which means if you use a different plotter or plot configuration, the printable area may change, causing the offset to shift.

Exercise 30-1

Access the Student Web site (www.g-wlearning.com/CAD) and complete Exercise 30-1.

Using Floating Viewports

The primary advantage of using floating viewports in a paper space layout is the ability to prepare scaled drawings without increasing or decreasing the actual size of drawing views or sheet content. You can also create multiple viewports on a single layout to show uniquely scaled or alternate drawing views. For example, a single sheet might contain a floor plan drawn at a 1/4″ = 1′-0″ scale, an eave detail drawn at a 3/4″ = 1′-0″ scale, and a foundation detail drawn at a 3/4″ = 1′-0″ scale. See **Figure 30-3**.

A floating viewport boundary is the portion of the viewport that you see. Everything inside the viewport is "showing through" from model space. Use tools such as **MOVE**, **ERASE**, **STRETCH**, and **COPY** to modify the viewport boundary in

Figure 30-3.
An example of a layout, ready to plot, that includes three floating viewports used to display drawing views at different scales. The layer on which the viewport is drawn is off or frozen for plotting.

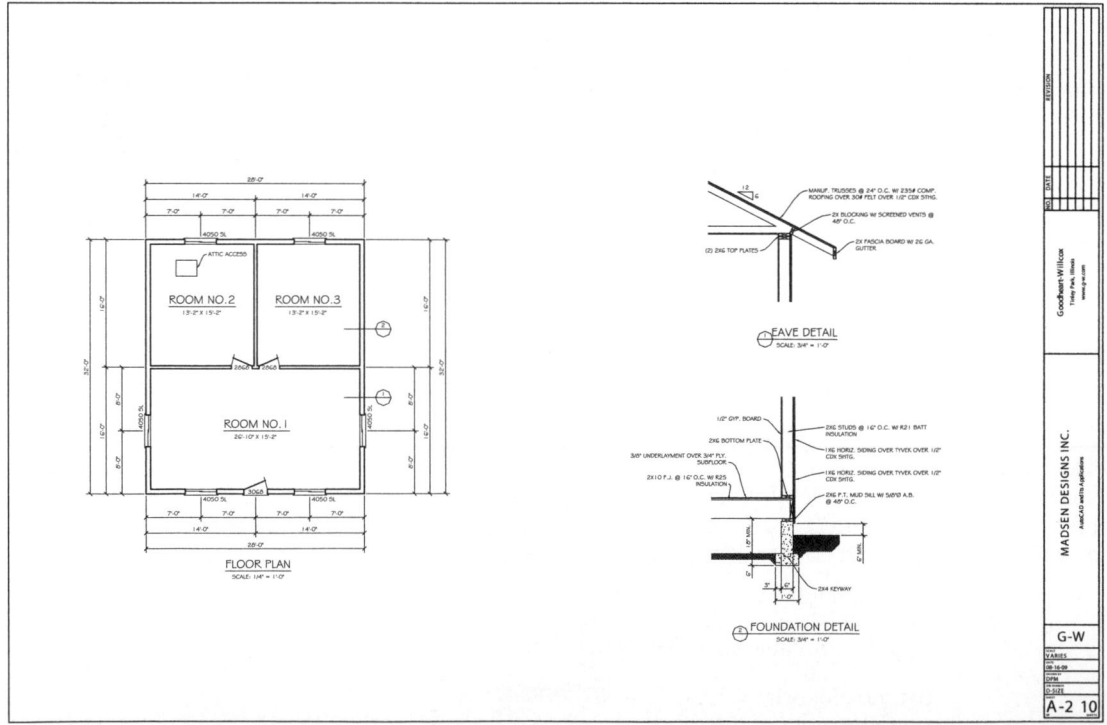

AutoCAD and Its Applications—Basics

paper space. Display tools such as **VIEW**, **PAN**, and **ZOOM** modify the display of model space objects in the floating viewport. Additional options are available for adjusting how objects appear, according to the layers on which you draw objects. This allows you to define how the drawing appears within the viewport. Be sure a layout tab is current as you work through the following sections describing floating viewports.

PROFESSIONAL TIP

When you first select a layout and enter paper space, a rectangular floating viewport appears around objects in model space. As part of layout setup, you may want to use the **ERASE** tool to erase the default viewport and draw a new viewport (or several viewports). Erasing the default viewport, or erasing everything in the layout, does not erase objects in model space.

Creating Floating Viewports

You can use a variety of techniques to create new floating viewports in paper space. The **Viewports** dialog box provides options for creating one to four new viewports according to a specific viewport configuration. The **MVIEW** tool is a text-based method for creating new viewports. The **MVIEW** tool provides the same options found in the **Viewports** dialog box, plus additional viewport definition options. You can also access many of the methods for creating new viewports directly from the **Viewports** panel on the **View** ribbon tab.

Using the Viewports Dialog Box

The **Viewports** dialog box, shown in **Figure 30-4**, looks and functions the same in paper space as in model space, except for a few differences. One difference is that the **Viewport spacing:** text box replaces the **Apply to:** drop-down list in the **New Viewports** tab in paper space. This text box is available only when you select a standard viewport configuration that places two or more viewports, as shown in **Figure 30-4.** Enter a value to define the space between multiple viewports.

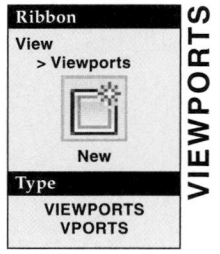

Ribbon

View
> Viewports

New

Type
VIEWPORTS
VPORTS

VIEWPORTS

Figure 30-4.
The **Viewports** dialog box with the **New Viewports** tab selected. The **Viewport Spacing:** setting is available when paper space is active.

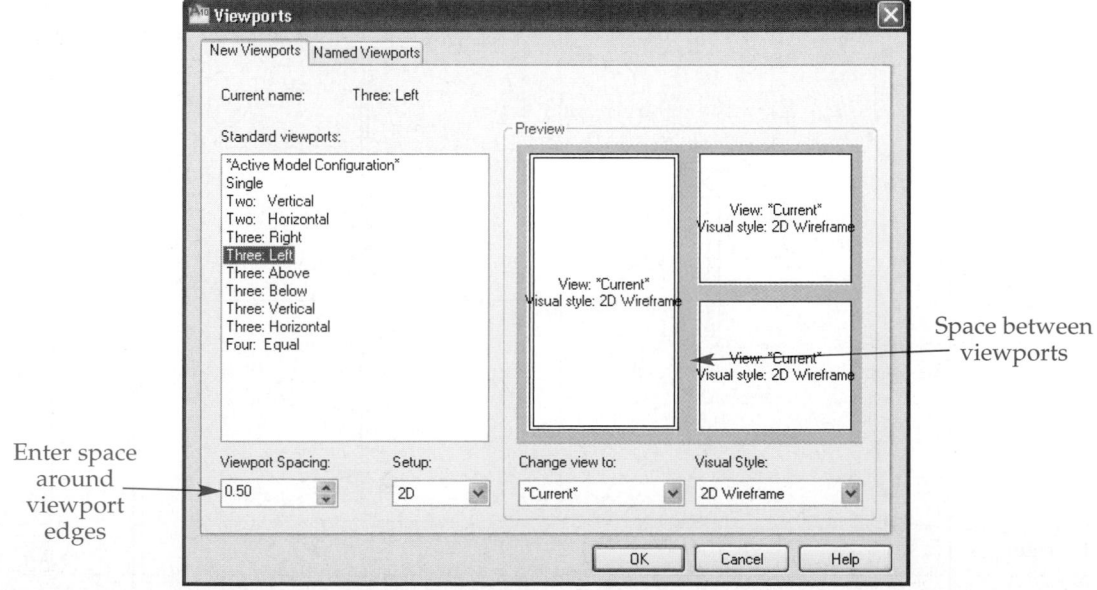

Another difference is that when you pick the **OK** button to create floating view-ports, the viewports do not automatically appear, as in model space. Instead, you must specify a first and second corner to define the area occupied by the viewport configuration. See **Figure 30-5.** If you use the **Fit** option, AutoCAD fits the viewport(s) into the printable area without requiring you to pick points. Refer to Chapter 6 for more information about the **Viewports** dialog box.

NOTE

You can also create new floating viewports by selecting a specific configuration from the Application Menu. Select **View > Viewports** and then select the **1 Viewport**, **2 Viewports**, **3 Viewports**, or **4 Viewports** option and follow the prompts to create the appropriate configuration.

Exercise 30-2

Access the Student Web site (www.g-wlearning.com/CAD) and complete Exercise 30-2.

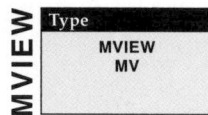

MVIEW

Type
MVIEW
MV

Using the MVIEW Tool

Once you access the **MVIEW** tool, you can create a single viewport by selecting opposite corners of the viewport, press [Enter] or the space bar, or right-click and choose **Enter** to activate the **Fit** option. The **Fit** option creates a viewport that fills the printable area. The **2**, **3**, and **4** options provide preset viewport configurations similar to those available from the **Viewports** dialog box. The **Fit** function is available when you create multiple viewports. The **Shadeplot** option includes the same options available from the **Visual Style** drop-down list in the **Viewports** dialog box, which is explained in *AutoCAD and Its Applications—Advanced.*

Figure 30-5.
Enter or select two points on the layout to specify the area filled by the viewport configuration. Notice the viewport spacing that forms as specified in the **Viewports** dialog box.

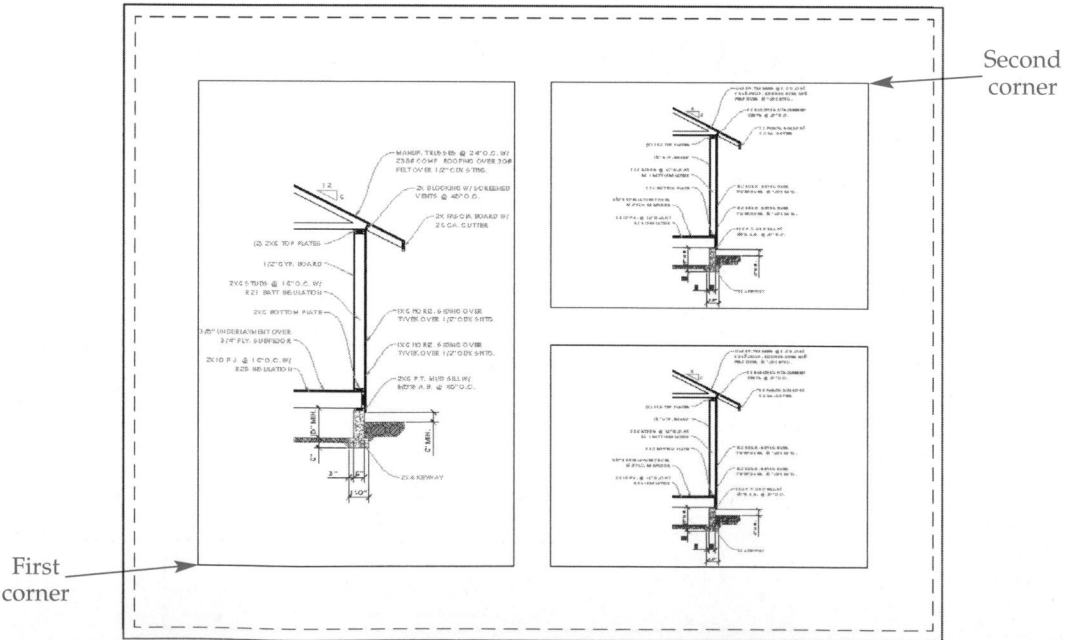

Choose the **Restore** option to convert a saved viewport configuration into individual floating viewports. This option is typically used to convert tiled viewports into floating viewports. For example, if model space displays two tiled viewports, use the **Restore** option to create two floating viewports. See **Figure 30-6.** After you access the **Restore** option, enter the viewport configuration name. In the previous example, you would select the **Active** option to use the two tiled viewports from model space. This is the active model space viewport configuration. Next, select opposite corners of the viewport, press [Enter] or the space bar, or right-click and choose **Enter** to activate the **Fit** option. The same model space tiled viewport configuration now displays in paper space as floating viewports.

Forming Polygonal Floating Viewports

The most common shape for a floating viewport is rectangular, which is suitable for many applications. As an alternative, use the **Polygonal** option of the **MVIEW** tool to form a polygonal floating viewport boundary. This option is easily accessed from the ribbon. Construct a polygonal viewport using the same techniques as when drawing a closed polyline object. The viewport can be any closed shape composed of lines and arcs. **Figure 30-7** shows an example of a polygonal floating viewport used to define the maximum drawing view area 1/2″ in from the border and title block.

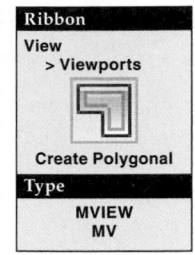

Ribbon

View
> Viewports

Create Polygonal

Type

MVIEW
MV

Figure 30-6.
A—A **Two: Horizontal** tiled viewport configuration in model space. B—The model space tiled viewports converted to floating viewports using the **Restore** option of the **MVIEW** tool.

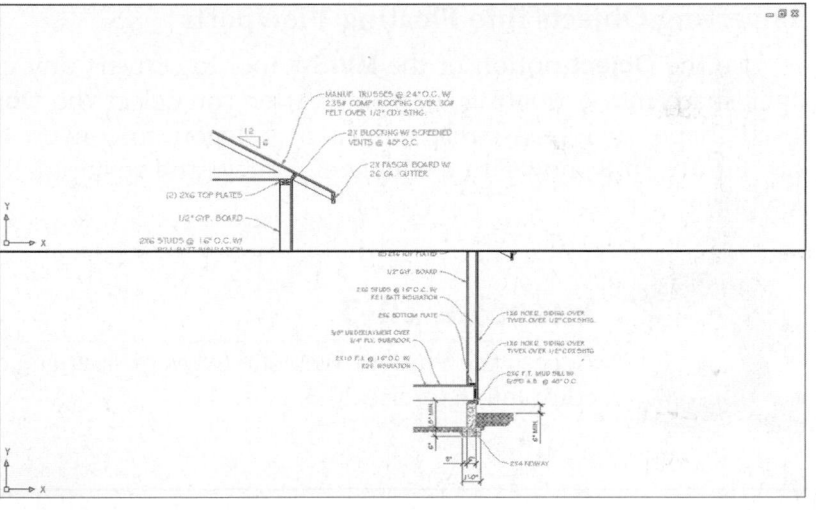

A

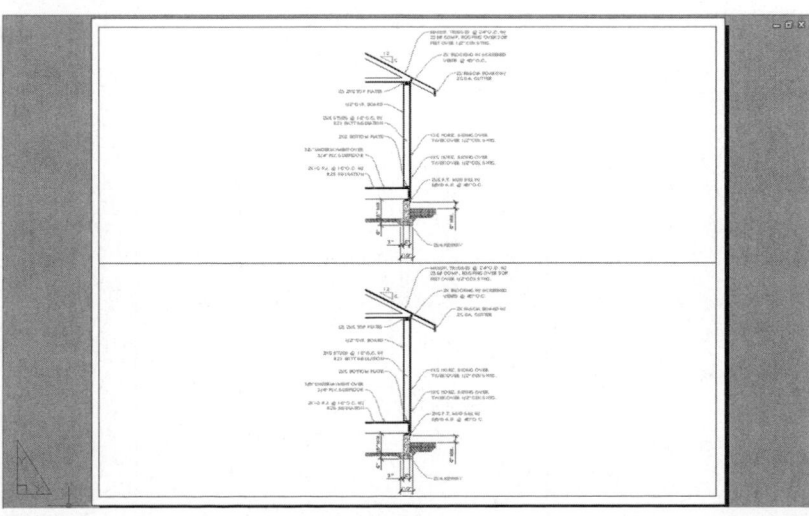

B

Figure 30-7.
An example of a polygonal floating viewport. The layer on which the viewport is drawn is turned off or frozen for plotting.

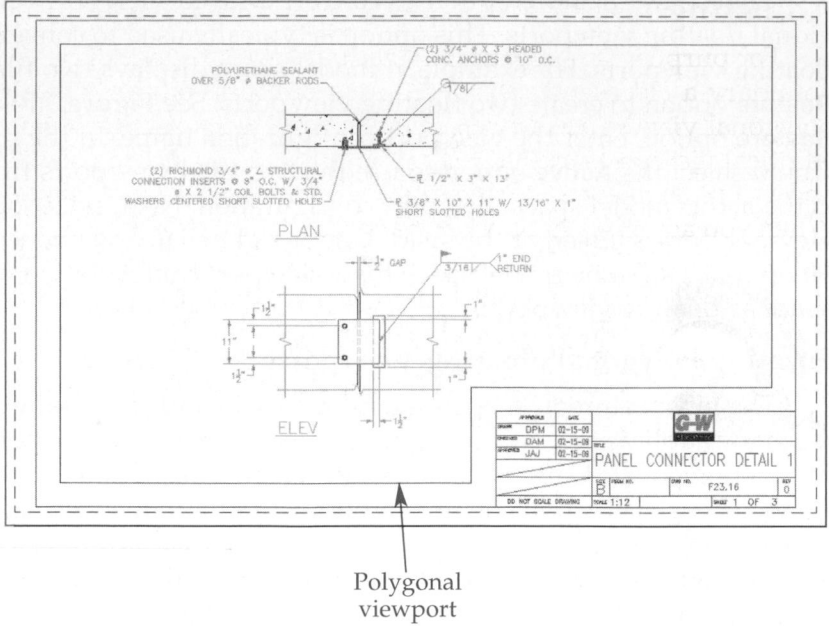

Polygonal viewport

Converting Objects into Floating Viewports

Use the **Object** option of the **MVIEW** tool to convert any closed object drawn in paper space into a floating viewport. After you select the **Object** option, select any closed shape, such as a circle, ellipse, or polygon, to convert the object into a viewport. **Figure 30-8** shows an example of a circle and rectangle converted into floating viewports.

Exercise 30-3

Access the Student Web site (www.g-wlearning.com/CAD) and complete Exercise 30-3.

Figure 30-8.
You can convert any closed object into a floating viewport. The layer on which these particular viewports are drawn remains on and thawed for plotting.

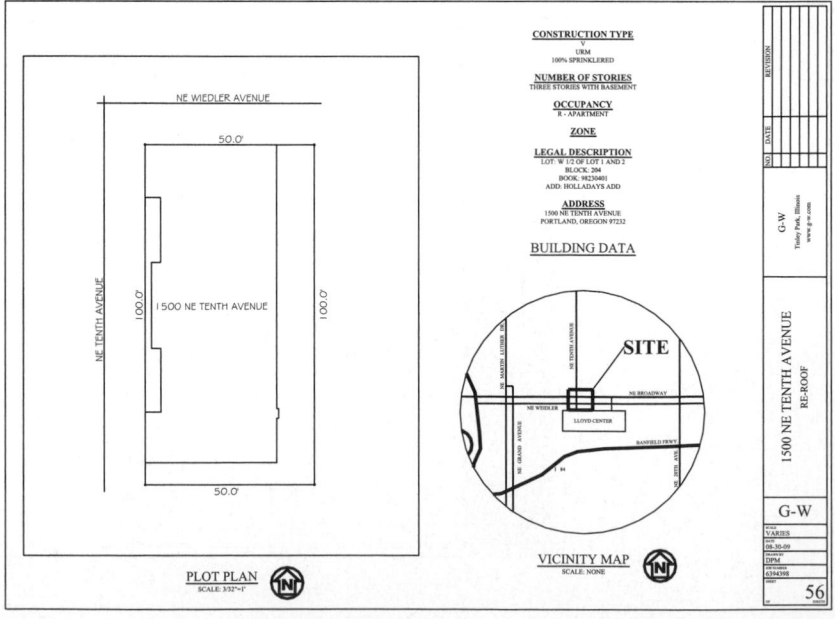

Adjusting the Floating Viewport Boundary

For purposes of adjusting a floating viewport boundary, you should consider the boundary a closed object. For example, treat rectangular viewports like rectangles, polygonal viewports like closed polyline objects, circular viewports like circles, and elliptical viewports like ellipses. Use grips or tools such as **MOVE**, **ERASE**, **STRETCH**, and **COPY** as needed to modify the size, shape, and location of floating viewports. When you adjust a floating viewport, the "hole" cut through the sheet changes.

Exercise 30-4

Access the Student Web site (www.g-wlearning.com/CAD) and complete Exercise 30-4.

Clipping Viewports

The **VPCLIP** tool allows you to redefine the boundary of an existing viewport. To access the tool from a shortcut menu, select a viewport and then right-click and pick **Viewport Clip**. You can clip a floating viewport to an existing closed object that you draw before accessing the **VPCLIP** tool, or you can clip the viewport to a polygonal shape that you create while using the tool.

After you access the **VPCLIP** tool, select the viewport to clip. Then select an existing closed shape, such as a circle, ellipse, or polygon, to recreate the viewport in the shape of the selected object. See **Figure 30-9**. An alternative, once you select the existing closed shape, is to use the **Polygonal** option to redefine the viewport to a polygonal shape. This option functions the same as the **Polygonal** option of the **MVIEW** tool, except the existing viewport transforms into the new shape.

AutoCAD recognizes a clipped floating viewport as clipped. The **VPCLIP** tool offers a **Delete** option when you select a clipped viewport. Use the **Delete** option to remove the clipped definition and convert the shape into a rectangle sized to fit the extents of the original clipping object or polygonal shape.

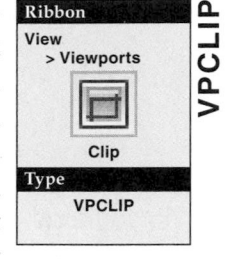

Ribbon

View
> Viewports

Clip

Type

VPCLIP

VPCLIP

Figure 30-9.
An example of clipping a viewport to an existing rectangle. The original viewport is removed and the rectangle converts into a viewport.

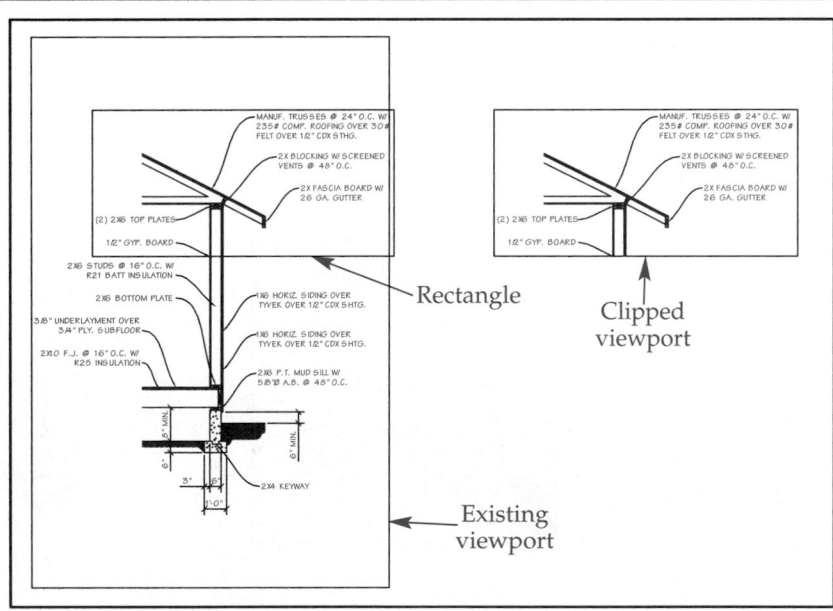

AutoCAD
NEW

Rotating Model Space Content

The **VPROTATEASSOC** system variable setting determines what happens to model space content when you rotate a floating viewport. By default, the variable uses a value of 1. As a result, when you rotate a viewport, objects shown in model space rotate to align with the viewport. See **Figure 30-10A.** The orientation of objects in model space does not change. In order to maintain the original alignment of model space content in a floating viewport, as shown in **Figure 30-10B,** access the **VPROTATEASSOC** system variable *before* rotating the viewport, and enter a value of 0.

Figure 30-10.
A—The default **VPROTATEASSOC** setting of 1 rotates model space content on a floating viewport to align with viewport rotation. B—Change the value to 0 to maintain the original model space orientation when you rotate a viewport.

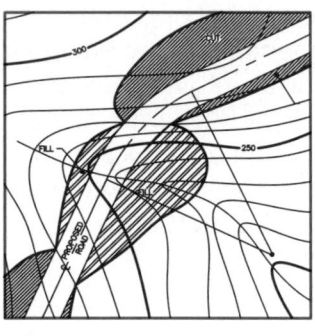
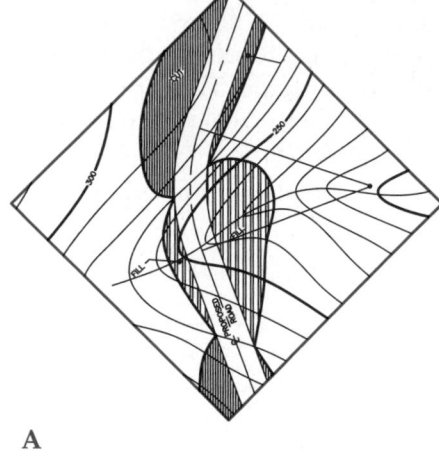

A

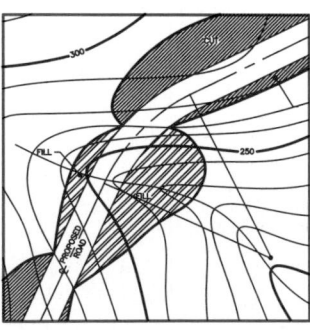
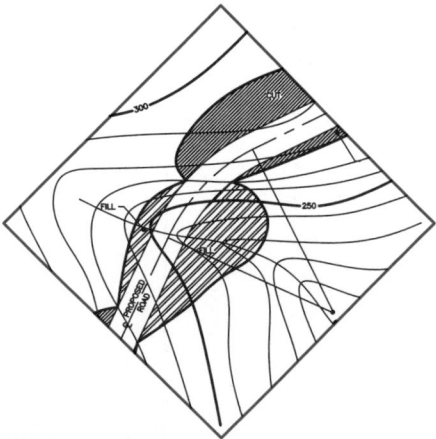

B

Other tools and options, such as **UCS** and **MVSETUP**, include options for rotating items shown through a viewport. The **VPROTATEASSOC** system variable automates the process. The entire display of model space rotates with the rotation of the viewport. Drawing characteristics, such as unidirectional dimensions, do not update according to the rotation.

Activating and Deactivating Floating Viewports

Activate a floating viewport to work with model space objects while in paper space. This allows you to adjust the display of the model space drawing shown in the viewport. Repeat the process of activating and adjusting a viewport for every floating viewport in the layout to achieve the final drawing.

To activate a floating viewport, double-click inside the viewport area, press the **PAPER** button on the status bar, or type MSPACE or MS. If the layout contains a single viewport, the viewport appears highlighted, indicating that it is current. On the layout, the UCS icon disappears, and the model space UCS icon displays in the corner of each layout viewport. You are now working directly in model space, through the paper space viewport. The active and highlighted viewport is the viewport you double-click on or the newest viewport depending on how you access the **MSPACE** tool. See **Figure 30-11**. To make a different viewport active, pick once inside the viewport.

After you adjust the display of all floating viewports, you must re-enter paper space to plot and continue working with the layout. To activate paper space, double-click outside the viewport area, press the **MODEL** button on the status bar, or type PSPACE or PS. The layout space UCS icon reappears, and the model space UCS icon disappears from the corners of the viewports.

Figure 30-11.
The active viewport appears highlighted and allows you to work in model space while AutoCAD displays paper space.

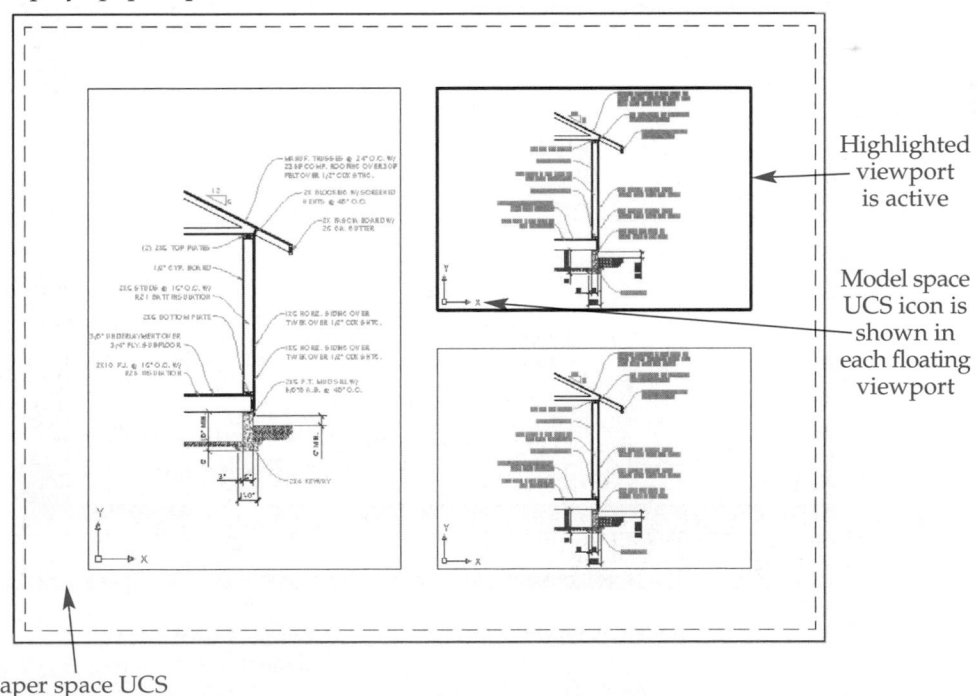

Highlighted viewport is active

Model space UCS icon is shown in each floating viewport

Paper space UCS icon is removed

Scaling a Floating Viewport

The scale you assign to a floating viewport is the same as the drawing scale. The quickest way to set viewport scale is to activate a viewport or pick a viewport boundary in paper space, *without* activating the viewport. Then select the appropriate scale from the **Viewport Scale** flyout on the status bar. See **Figure 30-12.** An alternative is to pick the viewport to scale in paper space and access the **Properties** palette. Then choose a viewport scale from the **Standard scale** drop-down list.

If a certain scale is not available from the **Viewport Scale** or **Standard scale** list, or if you want to change existing scales, pick the **Annotation Scale** flyout on the status bar and choose **Custom...** to access the **Edit Scale List** dialog box. From this dialog box, you can move the highlighted scale up or down in the list by picking the **Move Up** or **Move Down** button. To remove the highlighted scale from the list, pick the **Delete** button.

Select the **Edit...** button to open the **Edit Scale** dialog box. Here you can change the name of the scale and adjust the scale by entering the paper and drawing units. For example, a scale of 1/4″ = 1′-0″ uses a paper units value of .25 or 1 and a drawing units value of 12 or 48.

To create a new annotation scale, pick the **Add...** button to display the **Add Scale** dialog box, which functions the same as the **Edit Scale** dialog box. Pick the **Reset** button to restore the default annotation scale. When the annotation scale is set current, you are ready to type annotative text that automatically displays at the correct text height according to the drawing scale.

NOTE

Setting viewport scale is a zoom function that increases or decreases the *displayed* size of the drawing in the viewport. You can also use the **XP** option of the **ZOOM** tool to specify the scale of the active viewport.

Figure 30-12.
Using the **Viewport Scale** flyout button on the status bar to set the drawing scale.

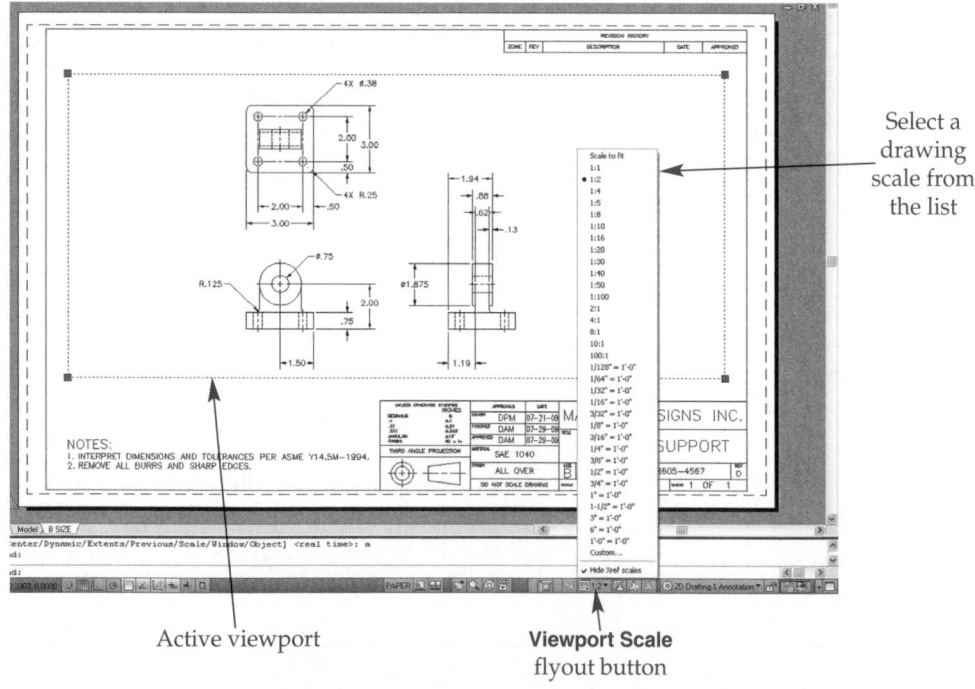

Select a drawing scale from the list

Active viewport

Viewport Scale flyout button

CAUTION

If you use an option of the **ZOOM** tool other than a specific **XP** value to adjust the drawing inside an active floating viewport, the drawing loses the correct scale. Once the viewport scale is set, do not zoom in or out. Lock the viewport, as described later in this chapter, to help ensure that the drawing remains properly scaled.

Scaling Annotations

You should always draw objects at their actual size, or full scale, in model space, regardless of the size of the objects. However, this method requires special consideration for annotations, hatches, and similar items added to objects in model space. You can adjust the appearance of these items manually, but most often, it is best to use annotative objects to automate the process. Scaled viewports and annotative objects function together to scale drawings properly and increase multiview drawing flexibility. Chapter 31 explains annotative objects.

Controlling Linetype Scale

The **LTSCALE** system variable allows you to make a global change to the linetype scale to increase or decrease the lengths of the dashes and spaces found in some linetypes. You must often modify the **LTSCALE** value to make linetypes match standard drafting practices. However, depending on the size of objects in model space and the specified floating viewport scale, an **LTSCALE** value in model space may not be appropriate for paper space.

For example, an **LTSCALE** value of .5 is appropriate for a U.S. customary mechanical drawing plotted at full scale. In this example, apply an **LTSCALE** value of .5 to model space and paper space because both environments function at full scale. If you scale the drawing to 2:1, a linetype scale of .25 (scale factor of 1/2 × **LTSCALE** value of .5 = .25) is needed in model space and paper space in order for lines to appear correct in both environments. By default, AutoCAD calculates the appropriate linetype scale display in model space and paper space according to the **LTSCALE** setting.

The **CELTSCALE**, **PSLTSCALE**, and **MSLTSCALE** system variables control how the **LTSCALE** system variable applies, or does not apply, to linetypes in model space and paper space. The **CELTSCALE**, **PSLTSCALE**, and **MSLTSCALE** system variables are set to 1 by default, and should be set to 1 in order for the **LTSCALE** value to apply correctly in model space and paper space. All linetypes will then appear with the same lengths of dashes and dots regardless of the floating viewport scale, and no matter whether you are in paper space or model space.

Using the previous example, lines will appear correctly in model space and at a scale of 1:1 and 2:1 in paper space. However, when you scale a floating viewport or change the annotation scale in model space, you must remember to use the **REGEN** tool to regenerate the display. Otherwise, the linetype scale will not update according to the new scale. The **MSLTSCALE** system variable is associated with the selected annotation scale and is further described in Chapter 31.

Adjusting the View

When you first create a floating viewport, AutoCAD performs a **ZOOM Extents** to display everything in model space through the viewport. The **Scale to fit viewport scale** option accomplishes the same task. When you scale a viewport, AutoCAD adjusts the view from the center of the viewport. This is often the appropriate display. However, if you change the size or shape of the viewport, if a centered view is not appropriate, or if you want to display a specific portion of the drawing, you must adjust the view. Use the **PAN** tool in an active viewport to redefine the location of the view.

Viewport edges can "cut off" a scaled model space drawing. This may be acceptable to display a portion of a view. However, to display the entire view, you can increase the size of the viewport boundary or select a different scale to reduce the displayed size of the view to fit the viewport. If it is not appropriate to increase the size of the viewport or decrease the scale, use a larger sheet size. **Figure 30-13** shows a drawing with two viewports. One shows everything in model space, the other "cuts off" model space objects and displays objects at a higher zoom level to create a detail.

Locking and Unlocking Floating Viewports

Once you adjust the drawing in the viewport to reflect the proper scale and view, lock the viewport so the scale or view orientation does not accidentally change. This allows you to use display tools such as **ZOOM** and **PAN** to aid in working with objects in model space without changing the scale or position of the view.

The quickest way to lock or unlock a viewport is to select a viewport in paper space and right-click. From the **Display Locked** cascading submenu, select **Yes** to lock the viewport or select **No** to unlock the viewport. A second option is to select a viewport in paper space and access the **Properties** palette. From the **Display Locked** drop-down list, select **Yes** to lock the viewport or select **No** to unlock the viewport. You can also use the **Lock** option of the **MVIEW** tool. Follow the prompts to select the viewport(s) to lock or unlock.

Exercise 30-5

Access the Student Web site (www.g-wlearning.com/CAD) and complete Exercise 30-5.

Controlling Layer Display

Layers function the same in paper space as in model space. The **On**, **Freeze**, **Color**, **Linetype**, **Lineweight**, **Plot Style**, and **Plot** settings described throughout this textbook are *global layer settings*. **On**, **Freeze**, and **Plot** are global layer states. **Color**, **Linetype**, **Lineweight**, and **Plot Style** are global layer properties. Changing a global layer setting

global layer settings: Layer settings applied to both model space and paper space.

Figure 30-13.
This drawing shows examples of when it is appropriate to show all model space objects, and when it is necessary to display only a portion of model space. Use the **PAN** tool to adjust the position of model space objects in floating viewports.

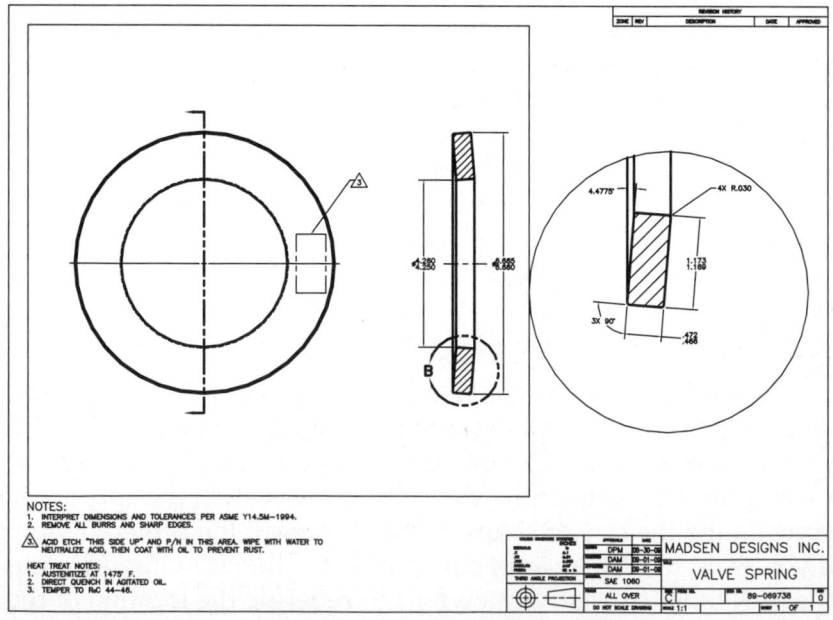

affects objects drawn in model space and paper space. For example, if you change the color of a layer in model space and lock the layer, all objects drawn on that layer in paper space also change color and become locked.

AutoCAD provides the option to freeze layers in a floating viewport and apply *layer property overrides.* These features expand the function of the layer system and improve your ability to reuse drawing content.

Use the **LAYER** tool and the corresponding **Layer Properties Manager** palette to control layer display in floating viewports. See **Figure 30-14.** This is the same palette used to manage layers throughout this textbook. The **NEW VP Freeze, VP Freeze, VP Color, VP Linetype, VP Lineweight,** and **VP Plot Style** columns control layer display options for floating viewports. Except for the **NEW VP Freeze** column, these columns appear only in layout mode. You probably need to use the scroll bar at the bottom of the palette to see the columns. The options can apply to layout content, such as the viewport boundary. However, layer settings typically apply to an active floating viewport. Be sure the floating viewport to which you want to apply layer control settings is active as you work through the following sections.

layer property overrides: Color, linetype, lineweight, and plot style properties applied to specific viewports in paper space.

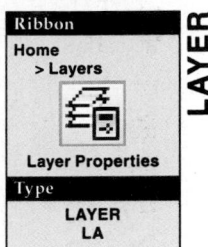

NOTE

The **VPLAYER** tool is a text-based tool that also controls layer display in floating viewports. The **Layer Properties Manager** palette is faster and easier to use than the **VPLAYER** tool.

Freezing and Thawing

Freeze layers in the active viewport to create unique views using a single drawing. For example, **Figure 30-15A** shows the model space display of a floor plan with electrical plan content added directly to the floor plan using electrical plan layers. **Figure 30-15B** shows two layouts from the same drawing file. One layout creates a floor plan with no electrical information, and the other layout creates an electrical plan without specific floor plan content.

In this example, you draw many objects, such as doors, walls, and windows on layers that maintain the global **Thaw** setting. As a result, these objects appear in model space and both floating viewports. Freeze layers in specific viewports (**VP Freeze**) to create two different drawings. This example shows viewport layer freezing in two different viewports, each viewport in a different layout, but you can apply the same concept to multiple viewports in the same layout.

Figure 30-14.
Use the **Layer Properties Manager** to control the display of layers in floating viewports.

Controls freezing in new viewports   Controls freezing in the active viewport   Layer property overrides

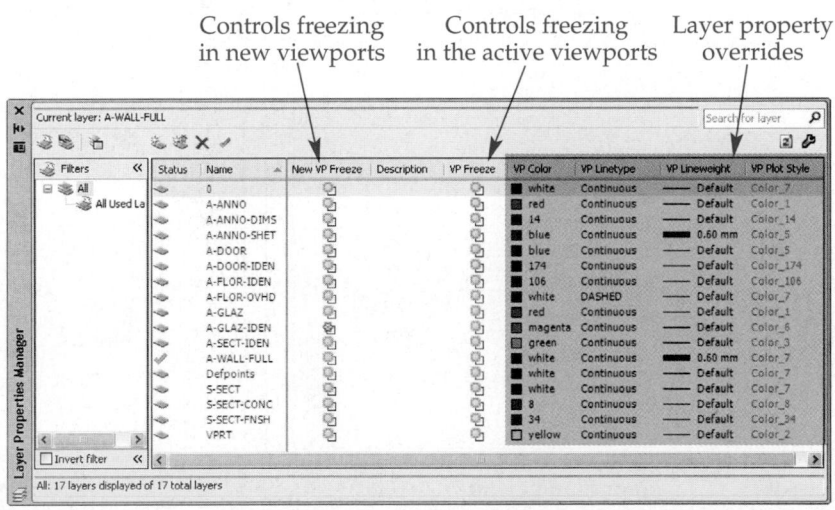

VP   **VP**
Freeze **Thaw**

The **VP Freeze** column of the **Layer Properties Manager** palette controls freezing and thawing layers in the current viewport. Pick the **VP Thaw** icon or the **VP Freeze** icon to toggle freezing and thawing in the current viewport. Using the **VP Freeze** icon freezes layers only in the selected floating viewport, while the **Freeze** icon freezes layers globally in all floating viewports. You can freeze or thaw a layer in all layout viewports, including those created before picking the **VP Freeze** icon or **VP Thaw** icon, by right-clicking and picking **VP Freeze Layer in All Viewports** or **VP Thaw Layer in All Viewports**.

PROFESSIONAL TIP

The **VP Freeze** function is also available in the **Layer Control** drop-down list in the **Layers** panel on the **Home** ribbon tab. This provides a quick way to freeze and thaw layers in a viewport without accessing the **Layer Properties Manager** palette.

Figure 30-15.
A—An example of "overlapping" layers in model space. B—Layers frozen in separate layouts to create unique drawing views.

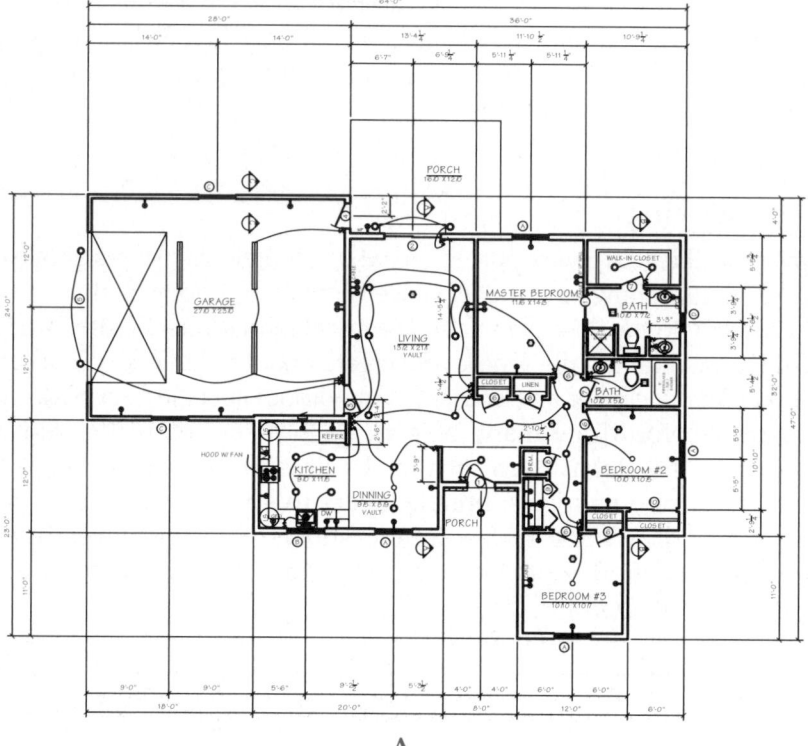

A

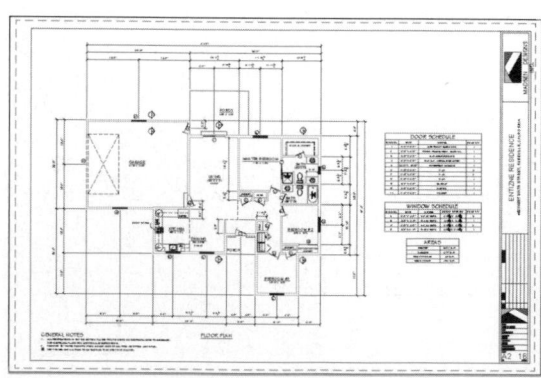

Floor Plan Layout
Created by freezing electrical layers
in the floating viewport

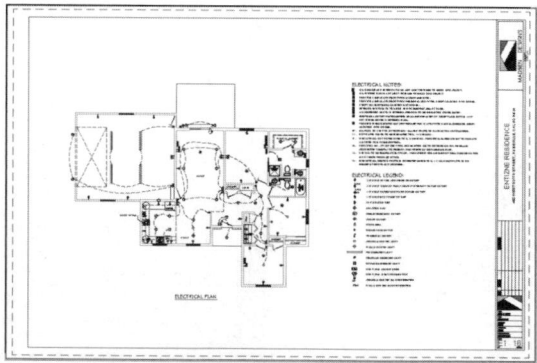

Electrical Plan Layout
Created by freezing floor plan
layers in the floating viewport

B

The **New VP Freeze** column of the **Layer Properties Manager** palette controls freezing and thawing of layers in *newly created* floating viewports. Pick the **VP Thaw** icon or the **VP Freeze** icon to toggle freezing and thawing in any new floating viewport. This feature has no effect on the active viewport. Use the **New VP Freeze** option to freeze specific layers in any new floating viewports.

New VP New VP
Freeze Thaw

NOTE

Right-click on a layer in the **Layer Properties Manager** palette and select **New Layer VP Frozen in All Viewports** to create a new layer preset with the **VP Freeze** and **New VP Freeze** icons selected.

Exercise 30-6

Access the Student Web site (www.g-wlearning.com/CAD) and complete Exercise 30-6.

Layer Property Overrides

Use layer property overrides to create unique views without changing individual object properties, creating separate drawing files, or readjusting global layer properties. For example, **Figure 30-16A** shows the model space display of a hopper and conveyer system with unique layers assigned to the hopper and conveyer. **Figure 30-16B** shows a layout with two floating viewports. The viewport on the left shows the hopper and

Figure 30-16.
A—A hopper and conveyer drawn in model space. B—Using property overrides to create a unique layout view.

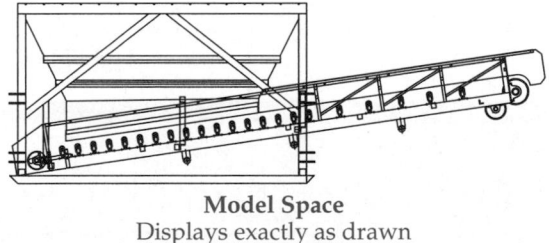

Model Space
Displays exactly as drawn

A

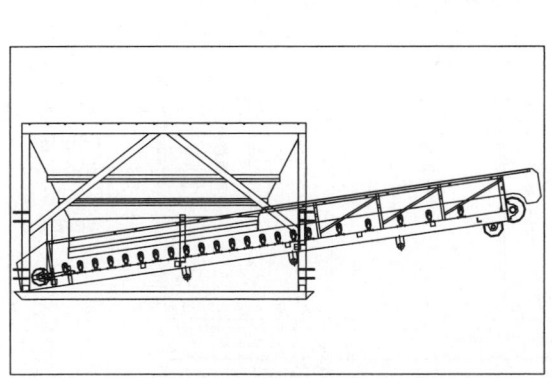

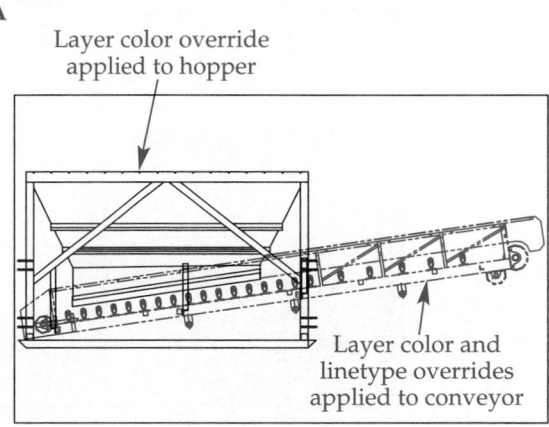

Layer color override
applied to hopper

Layer color and
linetype overrides
applied to conveyor

Left Viewport
Displays model space
objects exactly as drawn

Right Viewport
Layer property
overrides applied

B

conveyer with global layer settings applied, as in model space. The viewport on the right shows the hopper with a layer color override and the conveyer with a layer color and linetype override. In this example, layer property overrides create a view that clearly shows the two separate components. Phantom lines highlight the conveyer as the mechanism.

The **VP Color, VP Linetype, VP Lineweight,** and **VP Plot Style** columns in the **Layer Properties Manager** palette control the property overrides assigned to layers. The **VP Plot Style** column appears only when a named plot style is in use. Layer property overrides apply only to floating viewports in paper space. Layers that contain layer property overrides are not uniquely identified in model space.

The process of overriding a layer property is just like that for changing a global value. For example, to override the color assigned to a layer, pick the color swatch and choose a color from the **Select Color** dialog box. The difference is that layer property overrides apply only to specific layers in an active floating viewport. Object properties do not change from **Bylayer,** and the model space display does not change.

When viewed in paper space, the **Properties** palette, **Layer Properties Manager** palette, and **Layer Control** drop-down list on the ribbon indicate which layers include layer property overrides. See **Figure 30-17.** The **Layer Properties Manager** palette identifies layers that contain layer property overrides with a sheet and viewport icon in the status column. The layer names, global properties affected by the overrides, and the property overrides are highlighted. Use the **Viewport Overrides** filter to quickly display and manage only those layers that include layer property overrides. You can also save layer property overrides in a layer state.

The **Properties** palette identifies layers that contain layer property overrides with a highlighted layer name. Properties affected by the override also appear highlighted and are defined as **Bylayer (VP).** Layers that include layer property overrides also appear highlighted in the **Layer Control** drop-down list on the ribbon.

Figure 30-17.
Layers with property overrides are highlighted in the **Properties Manager** palette, **Layer Properties Manager** palette, and **Layer Control** drop-down list on the ribbon.

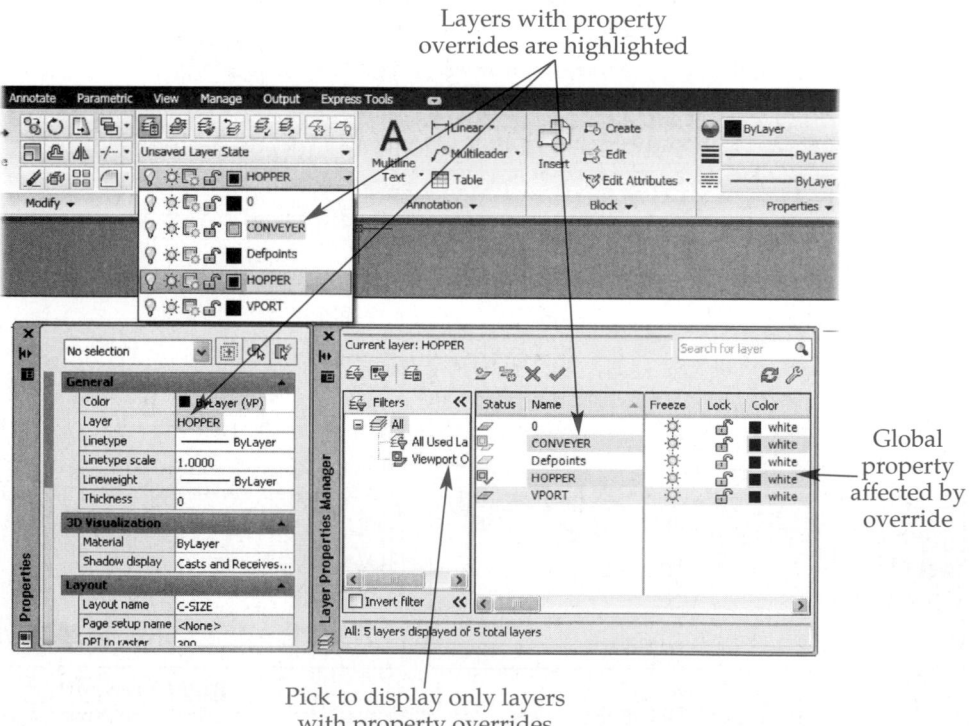

Layers with property overrides are highlighted

Global property affected by override

Pick to display only layers with property overrides

If layer property overrides are no longer necessary, you must remove the overrides from the layer. Changing a property back to the original, or global, value does not remove the override. Right-click on a layer that contains layer property overrides in the **Layer Properties Manager** palette and pick **Remove Viewport Layer Overrides for** to access a cascading submenu of options for removing layer property overrides. Pick **Selected Layers** and then **In Current Viewport Only** or **In All Viewports** to remove layer property overrides from the current viewport or from all viewports that include overrides.

Exercise 30-7

Access the Student Web site (www.g-wlearning.com/CAD) and complete Exercise 30-7.

Turning Off Floating Viewport Objects

By default, objects appear in floating viewports, allowing you to view model space through the viewports You can hide objects in the floating viewport without removing the viewport, which is convenient if, for example, you want to plot a certain view, but still have access to the viewport. One option to toggle the display of objects in the viewport on and off is to select a viewport in paper space and right-click. From the **Display Viewport Objects** cascading menu, select **No** to hide objects or select **Yes** to display objects.

Another option is to select a viewport in paper space and access the **Properties** palette. From the **On** drop-down list, select **Yes** to show objects or select **No** to hide objects. You can also use the **ON** and **OFF** options of the **MVIEW** tool. Follow the prompts to select the viewport(s) to display or hide objects.

Maximizing Floating Viewports

When you activate a floating viewport, you are working in model space from within the paper space display. The primary function of activating a floating viewport is to adjust the display of model space to prepare a final drawing. Typically, you should avoid working inside an active viewport to make changes to model space objects.

One alternative to activating a floating viewport is to maximize it by picking the **Maximize Viewport** button on the status bar or by selecting a viewport, right-clicking, and choosing **Maximize Viewport**. When you maximize a viewport, you fill the entire drawing window with the selected floating viewport. See **Figure 30-18.** This allows you to work more effectively than when the layout content covers much of the window. In addition, a maximized viewport displays objects exactly as they appear in the floating viewport, including frozen layers and layer overrides. Typically, you should maximize a floating viewport to use view tools such as **ZOOM** and **PAN** and make changes to objects in model space, while remaining in paper space.

Figure 30-18.
Maximize a floating viewport to work in a model-space-like environment, but with layout characteristics, such as layers frozen in the viewport and layer property overrides.

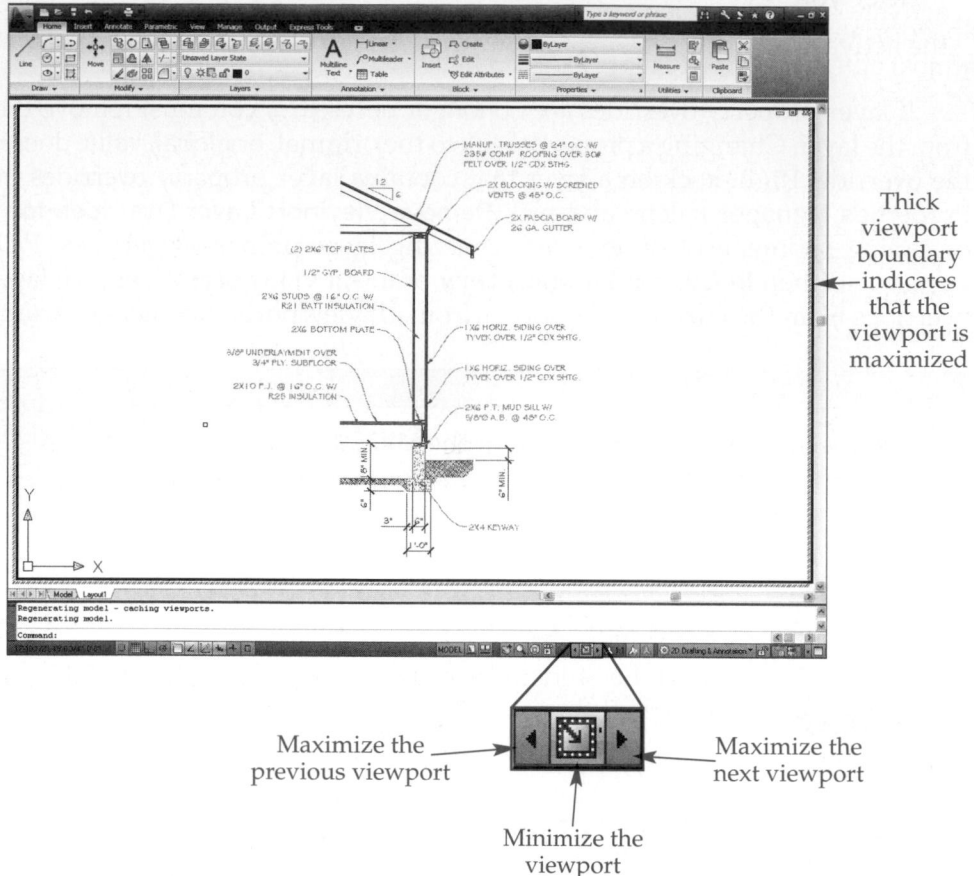

Thick
viewport
boundary
indicates
that the
viewport is
maximized

Maximize the
previous viewport

Maximize the
next viewport

Minimize the
viewport

If the drawing includes multiple viewports, use the **Maximize Previous Viewport** and **Maximize Next Viewport** buttons on the status bar to change to other floating viewports in a maximized display. To redisplay the entire layout, pick the **Minimize Viewport** button on the status bar, right-click and choose **Minimize Viewport**, or type **VPMIN**.

NOTE

You can maximize a floating viewport even if the viewport is not active.

PROFESSIONAL TIP

If you do not want to see floating viewport boundaries on the plotted sheet, remember to freeze or turn off the layer assigned to the floating viewport before plotting.

Exercise 30-8

Access the Student Web site (www.g-wlearning.com/CAD) and complete Exercise 30-8.

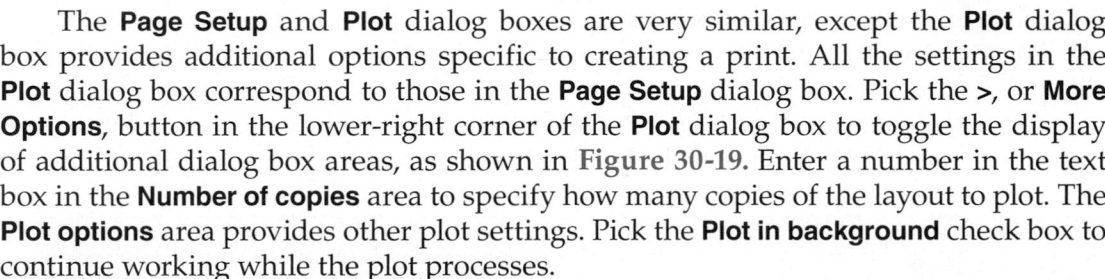

Quick Access

Plot

Ribbon

Output
> Plot
File
> Print

Plot

Type

PLOT
[CTRL]+[P]

PLOT

Plotting

After you prepare a layout for plotting, you are ready to plot. If you develop an appropriate page setup and layout, the process of creating the actual print should be almost automatic. Select the layout to plot and access the **PLOT** tool. The **Plot** dialog box appears with the name of the layout displayed on the title bar. See **Figure 30-19.**

NOTE

You can also access the **Plot** dialog box by selecting from the shortcut menu available from the model or layout tab, or by picking the **Plot** button in a **Model** or a layout thumbnail image in the **Quick View Layouts** or **Quick View Drawings** tool display.

The **Page Setup** and **Plot** dialog boxes are very similar, except the **Plot** dialog box provides additional options specific to creating a print. All the settings in the **Plot** dialog box correspond to those in the **Page Setup** dialog box. Pick the **>**, or **More Options**, button in the lower-right corner of the **Plot** dialog box to toggle the display of additional dialog box areas, as shown in **Figure 30-19.** Enter a number in the text box in the **Number of copies** area to specify how many copies of the layout to plot. The **Plot options** area provides other plot settings. Pick the **Plot in background** check box to continue working while the plot processes.

Most of the **Plot** dialog box settings are the same as those found in the **Page Setup** dialog box. Changing plot settings in the **Plot** dialog box is an effective way to override the page setup for a unique plotting requirement. This is a convenient way to make a plot using slightly modified plot settings without creating a new page setup. For

---

Figure 30-19.
Use the **Plot** dialog box to finalize the layout and send the drawing to the printer, plotter, or file.

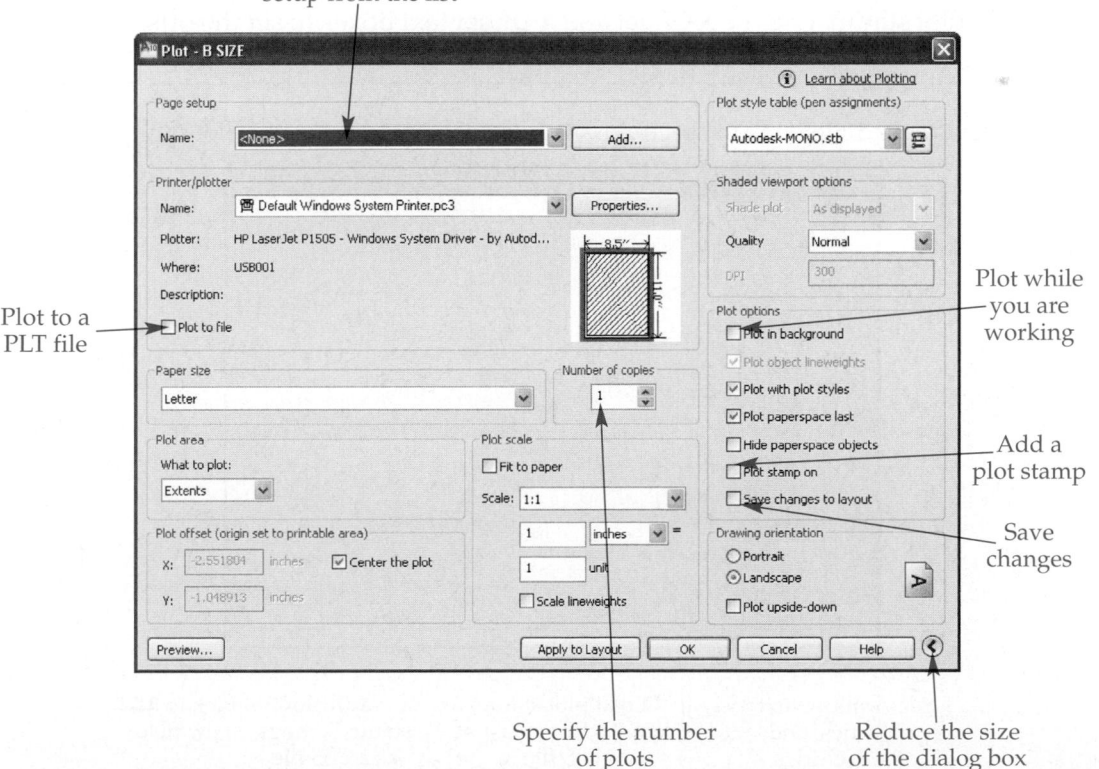

Select a different page setup from the list

Plot to a PLT file

Specify the number of plots

Plot while you are working

Add a plot stamp

Save changes

Reduce the size of the dialog box

example, you can make a "check print" by selecting a printer, using an A- or B-size sheet and scaling the plot to fit the paper. After printing, the settings return to those originally assigned in the page setup, allowing you to plot the final drawing using the appropriate printer, sheet size, and scale (1:1).

Adding a Plot Stamp

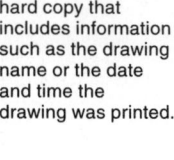

plot stamp: Text added only to the hard copy that includes information such as the drawing name or the date and time the drawing was printed.

Pick the **Plot stamp on** check box in the **Plot options** area of the **Plot** dialog box to add a *plot stamp* to the plot. When you select the check box, the **Plot Stamp Settings...** button appears. Pick the button to display the **Plot Stamp** dialog box. See **Figure 30-20**.

Pick the check boxes located in the **Plot stamp fields** area to identify the information to include in the plot stamp. To create additional plot stamp items, pick the **Add/Edit** button in the **User defined fields** area, and use the **User Defined Fields** dialog box to add, edit, and delete custom fields. For example, add a field for the client name, project name, or contractor who uses the drawing. Select the fields from the drop-down lists in the **User defined fields** area.

The **Preview** area provides a preview of the location and orientation of the plot stamp. The preview does not show the actual plot stamp text. Plot stamp settings are saved in a plot stamp parameter (PSS) file. Pick the **Save As** button to save the current settings as a new PSS file, or pick the **Load** button to access and use an existing PSS file.

> **NOTE**
>
> The log file settings are independent of the plot stamp settings. You can produce a log file without creating a plot stamp or have a plot stamp without producing a log file.

Pick the **Advanced** button to display the **Advanced Options** dialog box shown in **Figure 30-21**. The **Location and offset** area includes options to define the position of the plot stamp. Use the **Location** drop-down list to select the corner where the plot stamp begins. To print the plot stamp upside-down, pick the **Stamp upside-down** check box. Pick **Horizontal** or **Vertical** from the **Orientation** drop-down list to specify the orientation of the plot stamp. Use the **X Offset** and **Y Offset** text boxes to set the offset distances

Figure 30-20.
Use the **Plot Stamp** dialog box to specify the information included in the plot stamp.

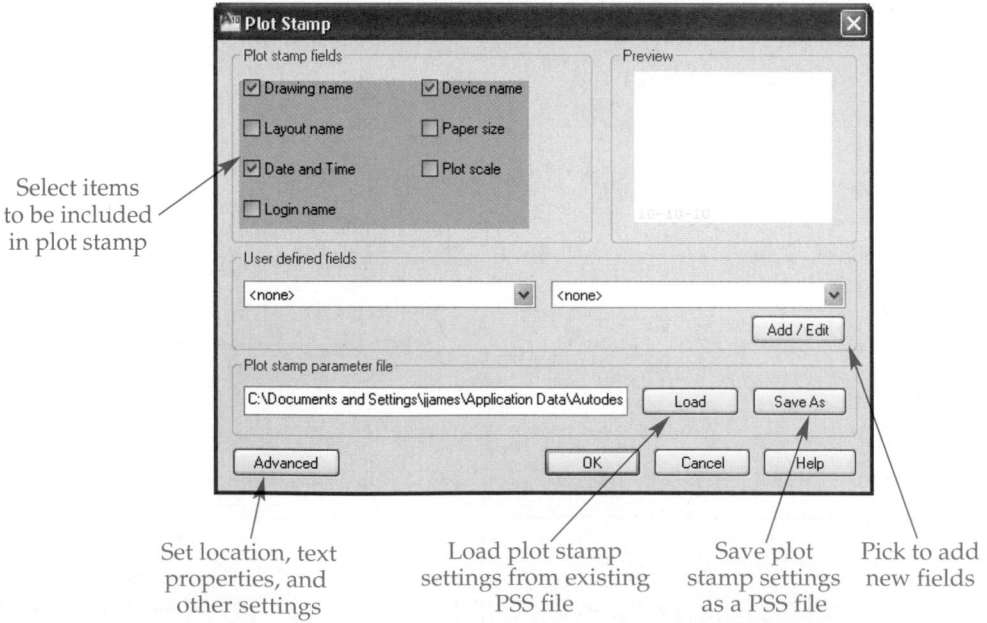

Select items to be included in plot stamp

Set location, text properties, and other settings

Load plot stamp settings from existing PSS file

Save plot stamp settings as a PSS file

Pick to add new fields

AutoCAD and Its Applications—Basics

Figure 30-21.
The **Advanced Options** dialog box allows you to define the plot stamp location, orientation, text font and size, and units.

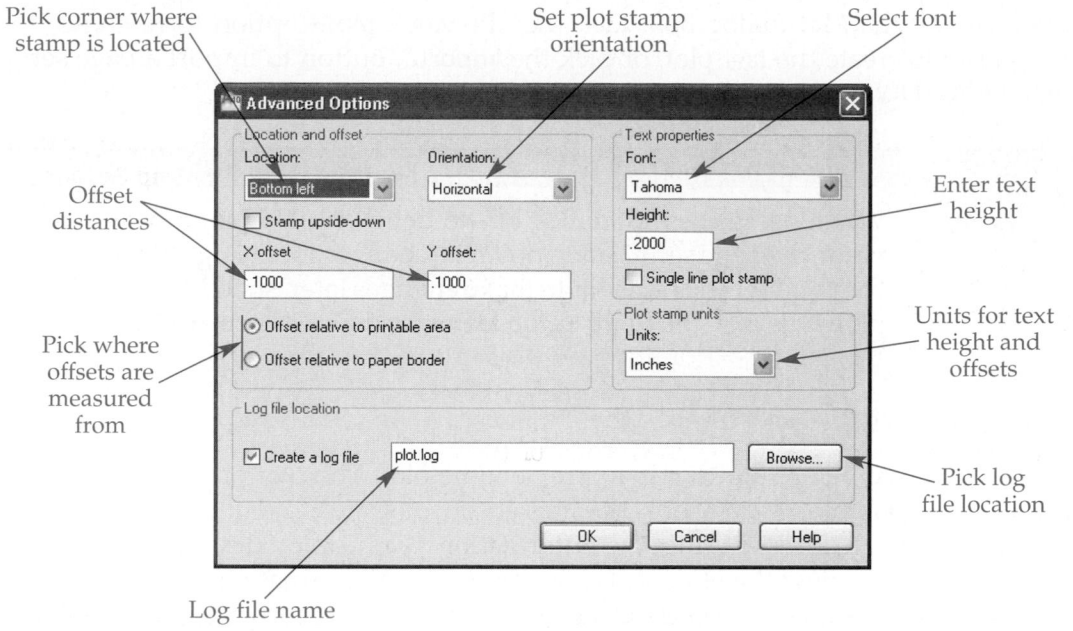

Pick corner where stamp is located

Set plot stamp orientation

Select font

Offset distances

Enter text height

Pick where offsets are measured from

Units for text height and offsets

Pick log file location

Log file name

---

for the plot stamp and pick whether the distances measure from the edge of the printable area or the paper border.

The **Text properties** area provides options for controlling plot stamp text characteristics. Use the **Font** drop-down list to select a font and the **Height** text box to specify the text height. Pick the **Single line plot stamp** check box to contain the plot stamp to a single line. If left unchecked, the plot stamp prints on two lines.

Use the **Units** drop-down list to select the units for the plot stamp offset and text height. The plot stamp units can be different from the drawing units. Select the **Log file location** check box to create a log file of plotted items. Specify the name of the log file in the text box. Pick the **Browse...** button to specify the location of the log file.

> **NOTE**
>
> You can also configure plot stamp settings by picking the **Plot Stamp Settings...** button on the **Plot and Publish** tab of the **Options** dialog box.

Saving Changes to the Layout

If you make changes in the **Plot** dialog box and want to save changes to the layout page setup for future plots, pick the **Save changes to layout** check box in the **Plot options** area. You can also save changes by picking the **Apply to Layout** button. If you do not select the **Save changes to layout** check box or pick the **Apply to Layout** button, changes made in the **Plot** dialog box are discarded, and the original page setup is used the next time you open the **Plot** dialog box.

Page Setup Options

The **Plot** dialog box provides an alternate means of creating a page setup. To apply this technique, access the **Plot** dialog box and make changes to plot settings, just as you would in the **Page Setup** dialog box. Then select the **Add...** button in the **Page setup** area to display the **Add Page Setup** dialog box. Enter a name for the page setup in the **New page setup name:** text box. All current settings in the **Plot** dialog box are saved with the new page setup. Select a page setup from the **Name:** drop-down list to restore the settings in the **Plot** dialog box. Pick the **<Previous plot>** option to reference the setting used to create the last plot, or pick the **Import...** button to import a page setup from a DWG, DWT, or DXF file.

> **NOTE**
>
> When using the **Plot** dialog box to define settings for a page setup, name the page setup *after* you make changes to settings. If you name the page setup and want to make changes later, such as changes to a plot style, use the **Page Setup Manager** dialog box instead.

Previewing the Plot

The final step before plotting is to preview the plot. The plot preview shows you exactly what your plot *should* look like, based on plot and layout settings. Always preview the plot before sending the information to the plot device to check the drawing for errors, view the effects of plot settings, and eliminate unnecessary plots. To preview the plot, pick the **Preview** button in the lower-left corner of the **Plot** dialog box to enter preview mode. See **Figure 30-22.** What you see on-screen is exactly what will plot, assuming you use a color plotter to make color prints and load the correct sheet size in the plot device.

The **Realtime Zoom** tool is activated automatically. Additional view tools are available from the toolbar near the top of the window or from a shortcut menu. Use these tools to help confirm that the plot settings are correct. When you finish previewing the plot and are ready to plot, pick the **Plot** button on the toolbar, or right-click and select **Plot**. To exit the preview without plotting, and return to the **Plot** dialog box, pick the **Close** button on the toolbar, press [Esc] or [Enter], or right-click and select **Exit**. Pick the **OK** button to send the plot to the plot device and close the **Plot** dialog box.

Figure 30-22.
Previewing a plot is an excellent way to confirm that the plot will be correct before sending the information to the plot device.

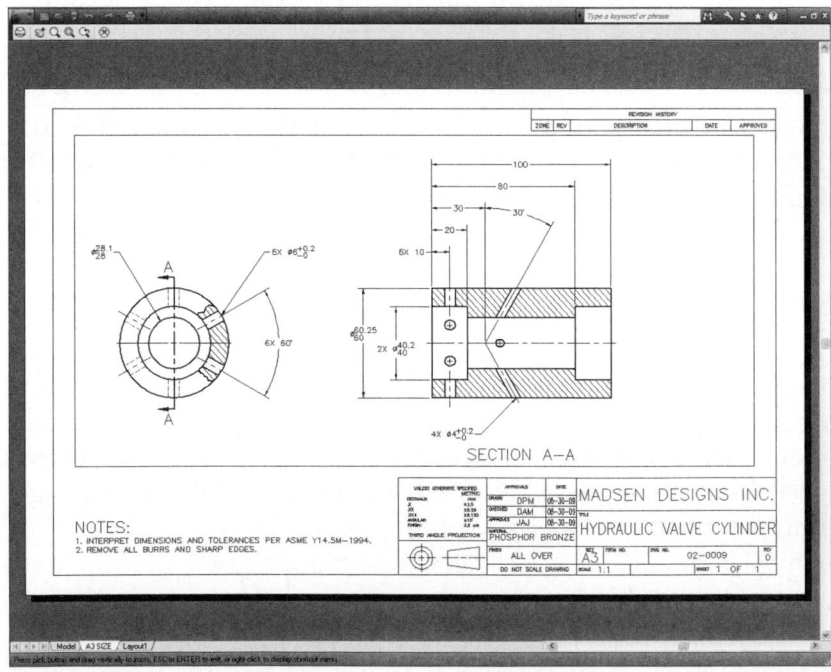

Exercise 30-9

Access the Student Web site (www.g-wlearning.com/CAD) and complete Exercise 30-9.

Plotting to a File

If a plot device is not available, but you are ready to plot, an alternative is to plot to a file. A plot file saves with a PLT extension. The file stores all the drawing geometry, plot styles, and plot settings assigned to the drawing. In offices or schools with only one printer or plotter, a *plot spooler* can be attached to the printer or plotter to plot a PLT file. This device usually allows you to take a PLT file from a storage disk and copy it to the plot spooler, which in turn plots the drawing.

plot spooler: A disk drive with memory that allows you to plot files.

To plot to a file, open the **Plot** dialog box, select the plot device from the **Name:** drop-down list, and check the **Plot to file** check box. The location in which the plot file is saved is set in the **Plot and Publish** tab of the **Options** dialog box. To specify the path, pick the ellipsis (**...**) button for the **Select default location for all plot-to-file operations** dialog box.

Supplemental Material

Additional Plotting Options

For information about several additional plot settings in the **Plot and Publish** tab of the **Options** dialog box, go to the Student Web site (www.g-wlearning.com/CAD), select this chapter, and select **Additional Plotting Options**.

Template Development

Chapter 30

For detailed instructions on adding layouts to each drawing template, go to the Student Web site (www.g-wlearning.com/CAD), select this chapter, and select **Template Development**.

Chapter Test

Answer the following questions. Write your answers on a separate sheet of paper or go to the Student Web site (www.g-wlearning.com/CAD) and complete the electronic chapter test.

1. Name the two types of content that are brought together to create a complete drawing.
2. What tools can you use to modify the boundary of a floating viewport?
3. What **MVIEW** option can form a floating viewport outline using a polyline?
4. What **MVIEW** option can you use to convert any closed object drawn in paper space into a floating viewport?
5. How do you activate a floating viewport?
6. How can you tell that a viewport is activated in paper space?
7. How do you reactivate paper space after activating a floating viewport for editing?
8. How does the scale you assign to a floating viewport compare with the drawing scale?
9. To what value should the **CELTSCALE**, **PSLTSCALE**, and **MSLTSCALE** system variables be set so that the **LTSCALE** value applies correctly in model space and paper space?
10. Viewport edges may "cut off" the drawing when the viewport is correctly scaled. List three things you can do if you want to display the entire view.
11. Why should you lock a viewport after you have adjusted the drawing in the viewport to reflect the proper scale and view?
12. Give an example of why you would want to hide objects in the floating viewport without removing the viewport.
13. What is a plot stamp?
14. If you make changes to the page setup using the **Plot** dialog box, how can you save these changes to the page setup so that the changes apply to future plots?
15. Give at least two reasons why you should always preview a plot before sending the information to the plot device.

Drawing Problems

Note: Some of the following problems refer to drawings or templates created in previous chapters. If you have not yet created those drawings, you will need to do so before working these problems.

▼ Basic

1. Follow the instructions in the Template Development portion of the Student Web site to add and set up layouts for the Mechanical-Inch template file.

2. Follow the instructions in the Template Development portion of the Student Web site to add and set up layouts for the Mechanical-Metric template file.

3. Follow the instructions in the Template Development portion of the Student Web site to add and set up layouts for the Architectural-US template file.

4. Follow the instructions in the Template Development portion of the Student Web site to add and set up layouts for the Architectural-METRIC template file.

5. Follow the instructions in the Template Development portion of the Student Web site to add and set up layouts for the Civil-US template file.

6. Follow the instructions in the Template Development portion of the Student Web site to add and set up layouts for the Civil-METRIC template file.

7. Open P29-4 and save as P30-7. The P30-7 file should be active. Make the **B-SIZE** layout current. Create a new layer named **VPORT**. Delete the default floating viewport and create a single floating viewport .5″ in from the edges of the sheet on the **VPORT** layer. Scale model space in the viewport to 1:1. Plot the layout, leaving the **VPORT** layer on and thawed. Resave the problem.

8. Open P29-5 and save as P30-8. The P30-8 file should be active. Activate the **A2-SIZE** layout. Create a new layer named **VPORT**. Delete the default floating viewport and create a single floating viewport 10 mm from the edges of the sheet on the **VPORT** layer. Scale model space in the viewport to 1:1. Plot the layout, leaving the **VPORT** layer on and thawed. Resave the problem.

9. Open P29-6 and save as P30-9. The P30-9 file should be active. Activate the **B-SIZE** layout. Create a new layer named **VPORT**. Delete the default floating viewport and create a single floating viewport .5″ from the edges of the sheet on the **VPORT** layer. Scale model space in the viewport to 1:1. Plot the layout, leaving the **VPORT** layer on and thawed. Resave the problem.

▼ Intermediate

10. Open P29-7 and save as P30-10. The P30-10 file should be active. Create a floating viewport and scale model space in the viewport using an appropriate scale. Plot the layout, leaving the **VPORT** layer on and thawed. Resave the problem.

11. Open P8-1 and save as P30-11. The P30-11 file should be active. Delete the default **Layout2**. Create a new A-size sheet layout according to the following steps:
    A. Rename the default **Layout1** to **A-SIZE**.
    B. Select the **A-SIZE** layout and access the **Page Setup Manager**.
    C. Modify the **A-SIZE** page setup according to the following settings:
    - **Printer/Plotter:** Select a printer or plotter that can plot an A-size sheet
    - **Paper size:** Select the appropriate A-size sheet (varies with printer or plotter)
    - **Plot area:** Layout
    - **Plot offset:** 0,0
    - **Plot scale:** 1:1 (1 in. = 1 unit)
    - **Plot style table:** monochrome.ctb
    - **Plot with plot styles**
    - **Plot paper space last**
    - Do not check **Hide paper space objects**
    - **Drawing orientation:** Select the appropriate orientation (varies with printer or plotter)

    D. Create a new layer named **VPORT**.
    E. Delete the default floating viewport and create a single floating viewport .5″ from the edges of the sheet on the **VPORT** layer.
    F. Scale model space in the viewport to 1:2. Plot the layout, leaving the **VPORT** layer on and thawed.
    Resave the problem.

12. Open P8-7 and save as P30-12. The P30-12 file should be active. Create layouts and floating viewports as needed to plot the drawing at an appropriate scale.

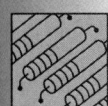

▼ Advanced

13. Open P30-13 from the Student Web site supplied with this textbook. Create a layout, plot style, and page setup so the layout can be plotted as follows: Using color-dependent plot styles, have the equipment (shown in color in the diagram) plot with a lineweight of 0.8 mm and 80% screening on an A-size sheet oriented horizontally. Plotted text height should be 1/8″. Plot in paper space at 1:1. Save the drawing as P30-13.

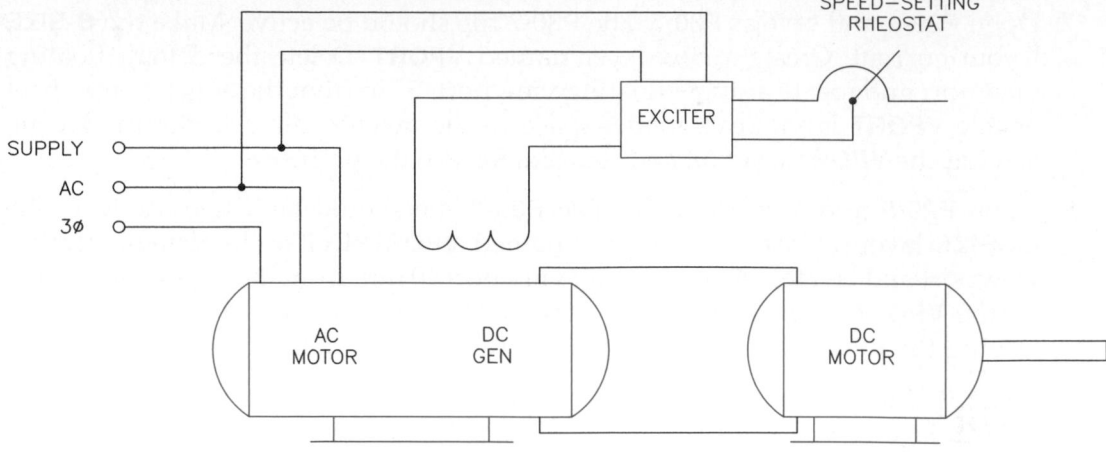

14. Open P30-14 from the Student Web site supplied with this textbook. Create four layouts with the names and displays as follows:
    - The **Entire Schematic** layout plots the entire schematic on a B-size sheet.
    - The **3 Wire Control** layout plots only the 3 Wire Control diagram on an A-size sheet, horizontally oriented.
    - The **Motor** layout plots the motor symbol and connections in the lower center of the schematic on an A-size sheet, oriented vertically.
    - The **Schematic** layout plots schematic without the 3 Wire Control and motor components on an A-size sheet, oriented horizontally.

    Set up the layouts so they can plot with a text height of 1/8″. Plot in paper space at a scale of 1:1. Save the drawing as P30-14.

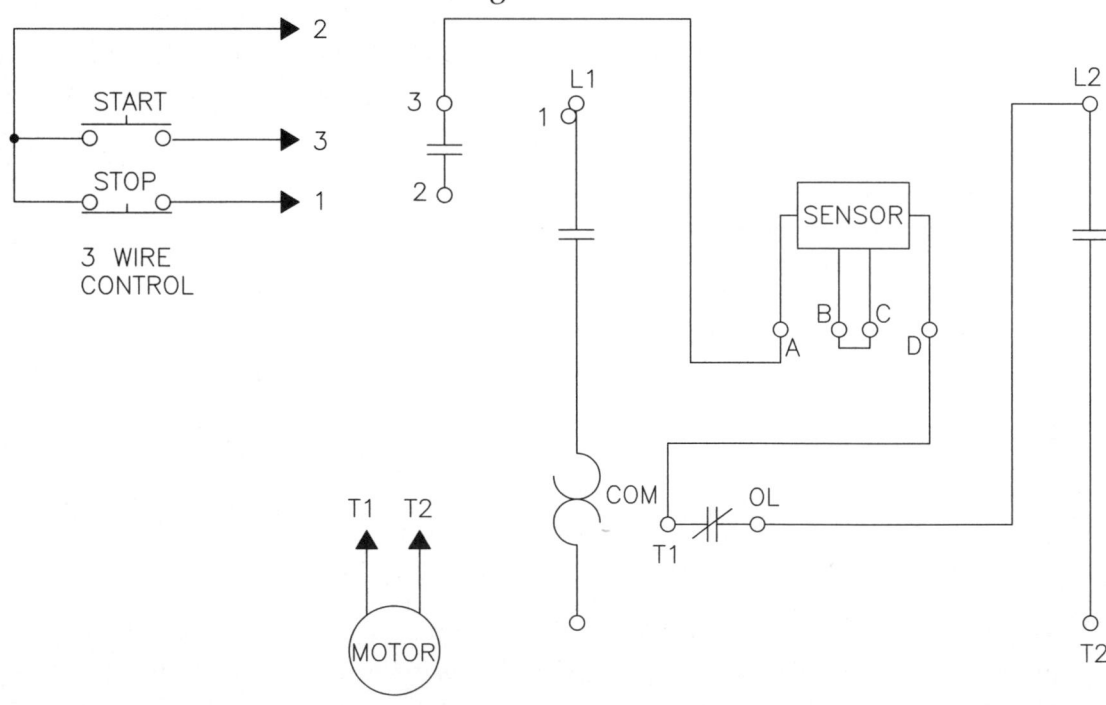

Drawing Problems - Chapter 30

CHAPTER 31

Annotative Objects

Learning Objectives

After completing this chapter, you will be able to do the following:
- ✓ Explain the differences between manual and annotative object scaling.
- ✓ Specify objects as annotative.
- ✓ Create and use annotative objects in model space.
- ✓ Display annotative objects in scaled layout viewports.
- ✓ Adjust the scale of annotations according to a new drawing scale.
- ✓ Use annotative objects to help prepare multiview drawings.

You must scale *annotations* and related items, such as dimension objects and hatch patterns, so that information appears on-screen and plots correctly relative to scaled objects. AutoCAD provides annotative tools to automate the process of scaling *annotative objects*. Annotative tools also provide additional flexibility for working with layouts to create multiview drawings.

annotations: Letters, numbers, words, and notes used to describe information on a drawing.

annotative objects: AutoCAD objects that can be made to adapt automatically to the current drawing scale.

Introduction to Annotative Objects

As explained in previous chapters, you should always draw objects at their actual size, or full scale, in model space, regardless of the size of the objects. For example, if you draw a small machine part and the length of a line in the drawing is 2 mm, draw the line 2 mm long in model space. If you draw a building and the length of a line in the drawing is 80′, draw the line 80′ long in model space. These examples describe drawing objects that are too small or too large for layout and printing purposes. You must *scale* the objects to fit properly on a sheet, according to a specific drawing scale.

When you scale a drawing, you increase or decrease the *displayed* size of model space objects. A properly scaled floating viewport in a layout allows for this process. Scaling a drawing greatly affects the display of items added to objects in model space, such as annotations, because these items should be the same size on a plotted sheet, regardless of the displayed size, or scale, of the rest of the drawing. See **Figure 31-1.**

Traditionally, annotations, hatches, and other objects are scaled manually, which means you determine the scale factor of the drawing scale and then multiply the scale factor by the plotted size of the objects. In contrast, annotative objects are scaled

scale: The ratio between the actual size of drawing objects and the size at which objects plot on a sheet of paper. Also the process of enlarging or reducing objects to fit properly on a sheet of paper.

Figure 31-1.
The large drawing features in this example require scaling in order to fit on a standard size sheet. Annotations are scaled according to the plotted size of the drawing; otherwise, they would be too small to see.

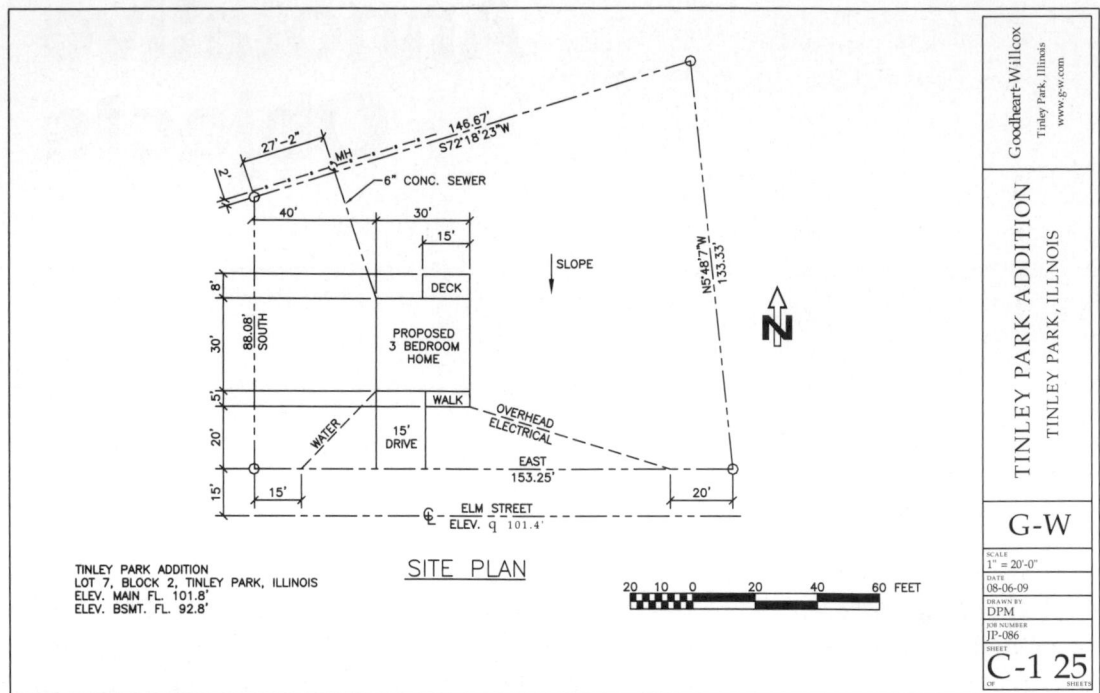

automatically according to the selected annotation scale, which is the same as the drawing scale. This eliminates the need for you to calculate the scale factor and manually adjust the size of objects according to the drawing scale.

PROFESSIONAL TIP

You should use annotative objects instead of traditional manual scaling even if you do not anticipate using a drawing scale other than 1:1.

Defining Annotative Objects

Annotative objects include single-line and multiline text, dimensions, leaders and multileaders, GD&T symbols created using the **TOLERANCE** tool, hatch patterns, blocks, and attributes. The method used to define objects as annotative varies depending on the object type. You can make objects annotative when you first draw them or convert non-annotative objects to annotative status.

Creating New Annotative Objects

Single-line and multiline text is annotative when drawn using an annotative text style. To make a text style annotative, pick the **Annotative** check box in the **Size** area of the **Text Style** dialog box. See **Figure 31-2.** A drawing may include a combination of annotative and non-annotative text, dimension, and multileader styles. An example of text that is typically *not* annotative is text added directly to a layout printed at a scale of 1:1.

AutoCAD and Its Applications—Basics

Figure 31-2.
Single-line and multiline text objects are annotative if they are drawn using an annotative text style.

Pick to make the text style annotative

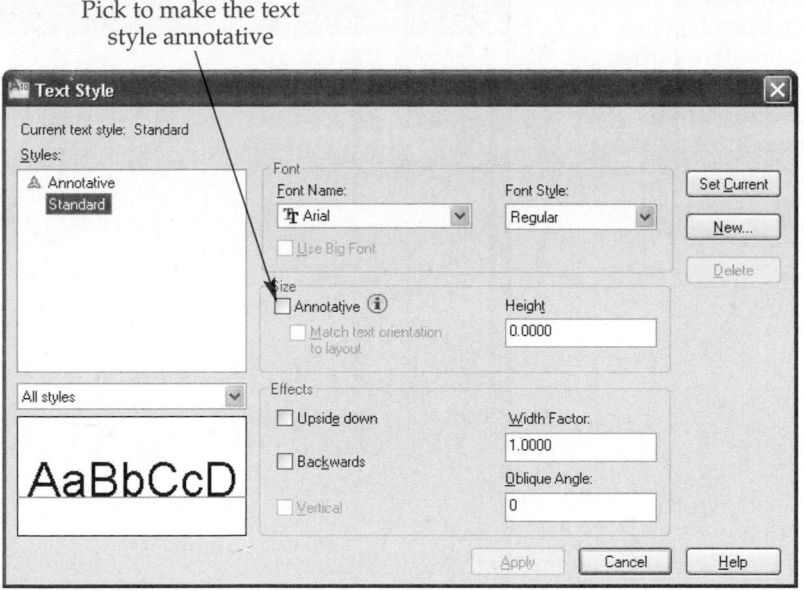

Dimensions, standard leaders, and GD&T symbols created using the **TOLERANCE** tool are annotative when drawn using an annotative dimension style. To make a dimension style annotative, pick the **Annotative** check box in the **Fit** tab of the **New** (or **Modify**) **Dimension Style** dialog box. See **Figure 31-3.**

Multileaders are annotative when drawn using an annotative multileader style. To make a multileader style annotative, pick the **Annotative** check box in the **Leader Structure** tab of the **Modify Multileader Style** dialog box. See **Figure 31-4.**

Figure 31-3.
Dimensions, leaders, and GD&T symbols created using the **TOLERANCE** tool are annotative if they are drawn using an annotative dimension style.

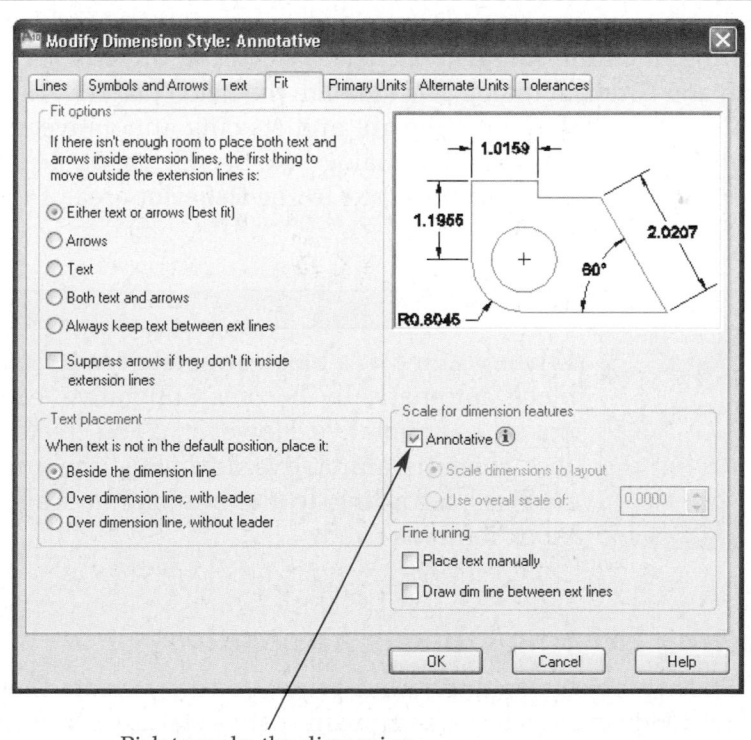

Pick to make the dimension style annotative

Figure 31-4.
Multileaders are
annotative if they
are drawn using
an annotative
multileader style.

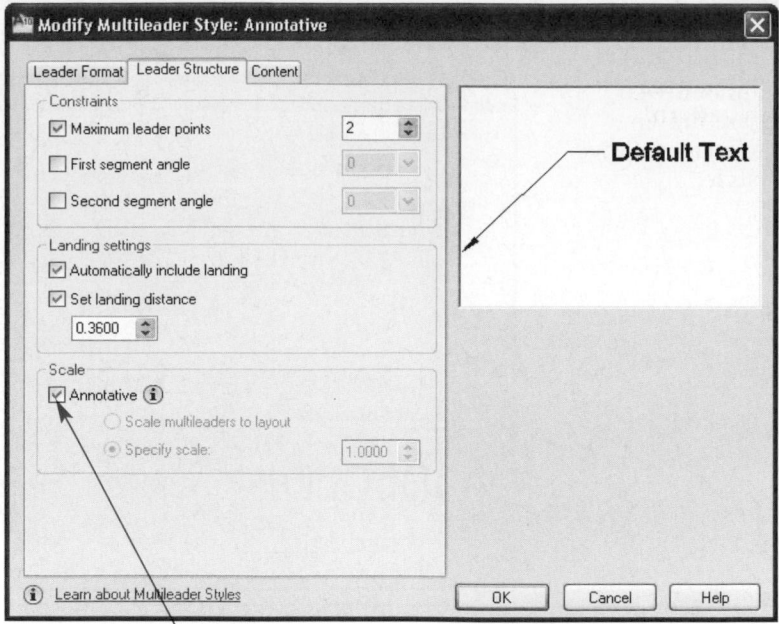

Pick to make the multileader
style annotative

> **NOTE**
>
> When you create an annotative multileader using the block multi-leader type, the block automatically becomes annotative, even if the block is not set as annotative.

Hatch patterns are annotative when you set the hatch scale as annotative during hatching. Pick the **Annotative** check box in the **Options** area on the **Hatch** tab of the **Hatch and Gradient** dialog box to make the hatch pattern annotative. See **Figure 31-5**.

To make attribute text height and spacing annotative, pick the **Annotative** check box in the **Attribute Definition** dialog box. See **Figure 31-6A.** To make a block annotative, pick the **Annotative** check box in the **Behavior** area of the **Block Definition** dialog box. See **Figure 31-6B.**

> **NOTE**
>
> When you make a block annotative, any attributes included in the block automatically become annotative, even if the attributes are not set as annotative. However, if you create a non-annotative block that contains annotative attributes, the annotative attribute scale changes according to the annotation scale, while the size of the block remains fixed.

Making Existing Objects Annotative

Specify objects as annotative when you first create them in model space when possible. However, you can assign annotative status to any objects originally drawn as non-annotative. The appropriate style controls the annotative status of single-line and multiline text, dimensions, standard leaders and multileaders, and GD&T symbols created using the **TOLERANCE** tool. Change the style assigned to the object to an annotative style to make these objects annotative. You must edit or recreate existing hatch patterns, blocks, and attributes in order to make the objects annotative.

AutoCAD and Its Applications—Basics

Figure 31-5.
Set the hatch pattern
scale to annotative
when you create the
hatch pattern.

Pick to make the hatch
scale annotative

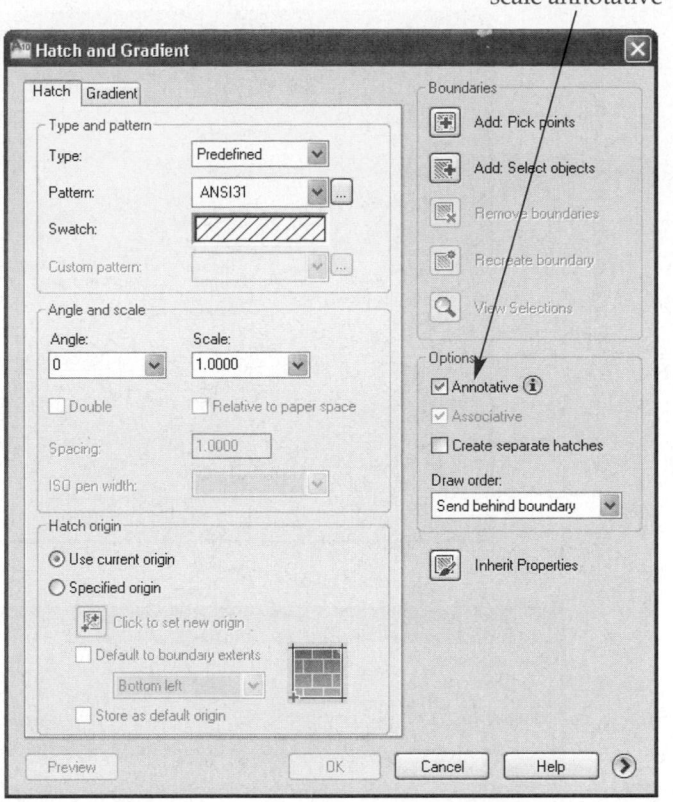

One method you can use to make existing objects annotative is to override the
non-annotative status using the **Properties** palette. This technique is most effective to
make a limited number of objects annotative. The location of the annotative properties
in the **Properties** palette varies depending on the selected object. The **Annotative** and
Annotative scale properties are common to all annotative objects. Select **Yes** from the
Annotative drop-down list to make non-annotative objects annotative, or choose **No** to
make annotative objects non-annotative.

CAUTION

Use caution when overriding an object to annotative status. Assign
annotative objects an appropriate annotative style or status instead
of overriding specific objects when possible.

NOTE

The **MATCHPROP** tool allows you to select the properties of annotative objects and apply those properties to existing objects, making
the objects annotative.

Exercise 31-1

Access the Student Web site (www.g-wlearning.com/CAD) and
complete Exercise 31-1.

Figure 31-6.
Set attributes (A) and blocks (B) as annotative during definition.

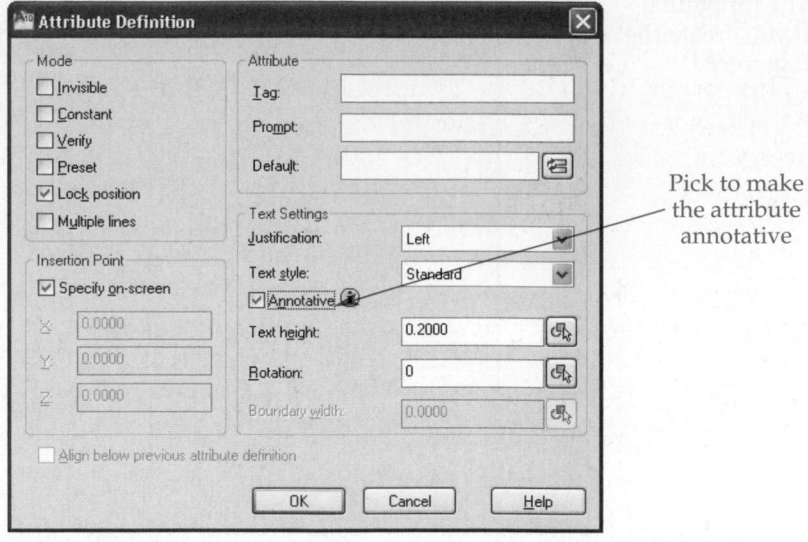

A

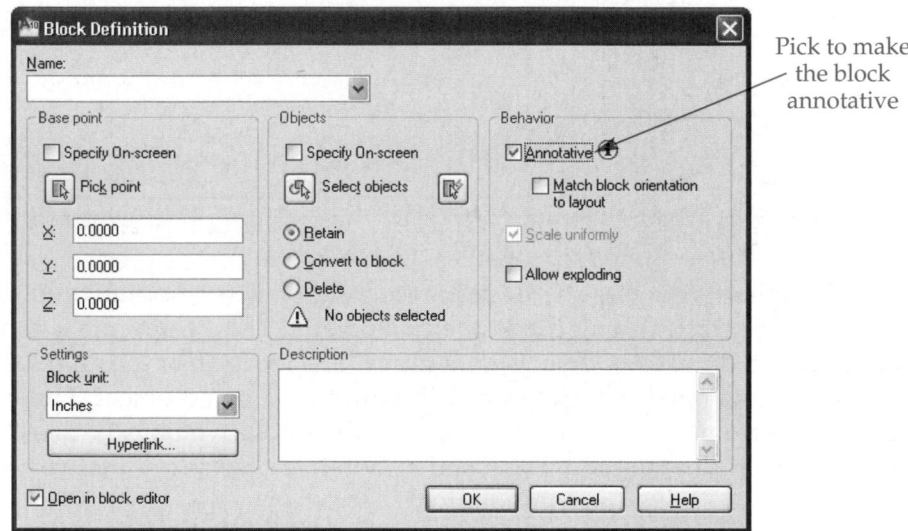

B

Drawing Annotative Objects

Using annotative objects reduces the need to determine the drawing scale factor. However, you must still identify the appropriate drawing scale, which is the same as the *annotation scale*. Ideally, determine drawing scale during template development and incorporate the scale into the settings in your template files. If you do not apply drawing scale to settings in your templates, identify the scale before beginning a drawing, or at least before you begin placing annotations.

annotation scale:
The scale AutoCAD uses to calculate the scale factor applied to annotative objects.

Setting Annotation Scale

You should set the annotation scale before you begin adding annotations so that annotations scale automatically. However, it may be necessary to adjust the annotation scale throughout the drawing process, especially if the drawing scale changes or when preparing multiple drawings with different scales on one sheet. Approach scaling annotations in model space by first selecting an annotation scale and then placing annotative objects. To draw differently scaled annotations, select the new annotation scale before placing the annotative objects.

When you add an annotative object, the **Select Annotation Scale** dialog box may appear. This is a very convenient way to set annotation scale while creating the object. The other primary means of specifying the annotation scale is to choose a scale from the **Annotation Scale** flyout on the status bar. See **Figure 31-7**. Remember that the annotation scale is typically the same as the drawing scale. You can also set the annotation scale in the **Properties** palette by selecting the annotation scale from the **Annotation Scale** option in the **Misc** category. This option is available when no objects are selected.

If a certain scale is not available, or if you want to change existing scales, pick the **Annotation Scale** flyout on the status bar and choose **Custom...** to access the **Edit Scale List** dialog box. The **Edit Scale List** dialog box is also available by picking the **Edit Scale List...** button in the **User Preferences** tab of the **Options** dialog box. The **Edit Scale List** dialog box is the same dialog box used to edit floating viewport scales, as explained in Chapter 30.

NOTE

Annotation scale sets the drawing scale in model space for controlling annotative objects. Viewport scale sets the drawing scale in a layout floating viewport to define the drawing scale. Both scales should be the same and should match the drawing scale.

Figure 31-7.
The status bar includes several annotation scale options. If you display the drawing status bar, the **Annotation Scale** button moves from the application status bar to the drawing status bar.

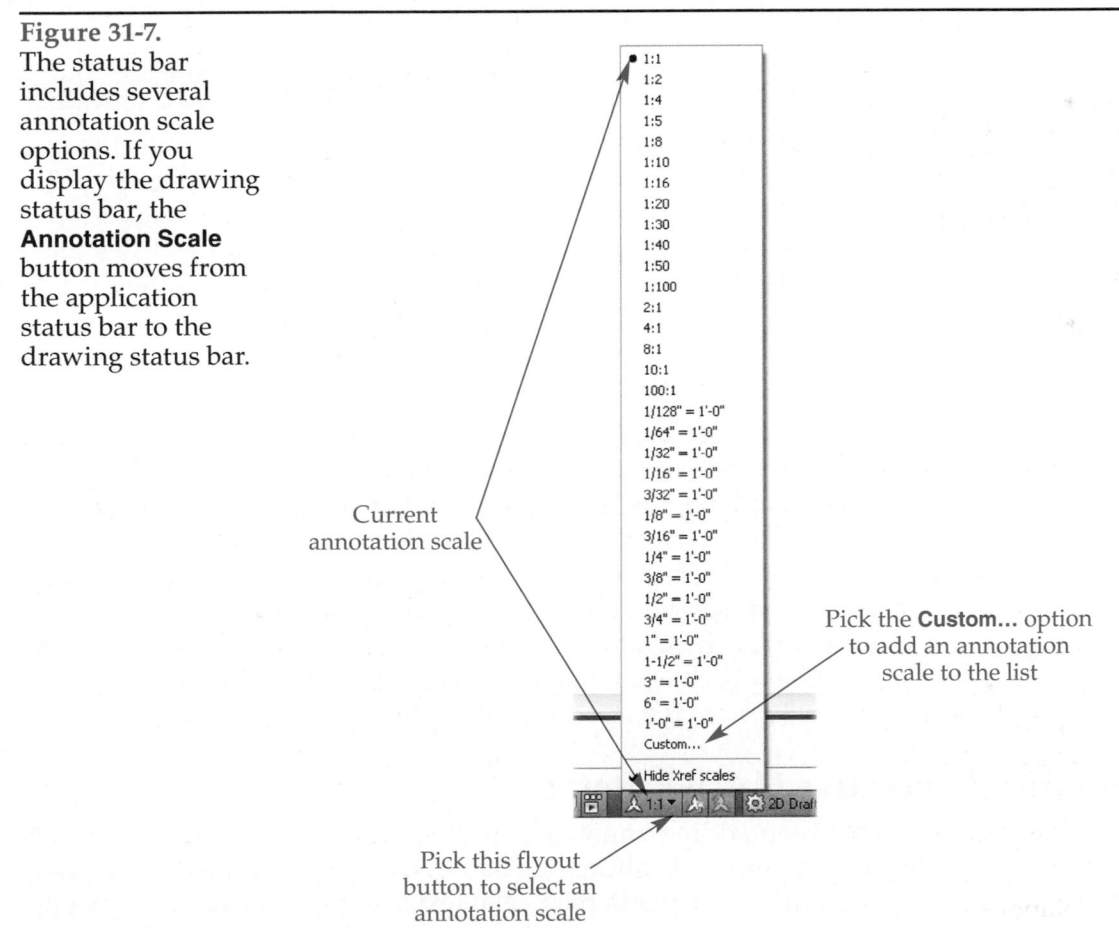

Controlling Model Space Linetype Scale

The **CELTSCALE**, **PSLTSCALE**, and **MSLTSCALE** system variables control how the **LTSCALE** system variable applies to linetypes in model space and paper space. Leave the **CELTSCALE**, **PSLTSCALE**, and **MSLTSCALE** system variables at their default setting of 1 to apply the **LTSCALE** value correctly according to the current annotation scale. However, when you change the annotation scale, you must remember to use the **REGEN** tool to regenerate the display. Otherwise, the linetype scale will not update according to the new scale.

> **PROFESSIONAL TIP**
>
> When you open a drawing in AutoCAD 2010 that was created in an AutoCAD version earlier than AutoCAD 2008, the **MSLTSCALE** system variable is set to 0. Change the value to 1 to take advantage of annotative linetype scaling.

Drawing Annotative Text

Draw annotative text using the same tools as non-annotative text. The difference is the value you enter for text height. To create annotative multiline text, select the **Annotative** button and enter the paper text height, such as 1/4″, in the **Size** text box. See **Figure 31-8.** The text scale, which includes spacing, width, and paragraph settings, automatically adjusts according to the current annotation scale. To create annotative single-line text, after you pick the start point, specify the paper height. The text scale automatically adjusts according to the current annotation scale.

> **NOTE**
>
>
> The **Properties** palette contains specific annotative text properties in addition to those displayed for all annotative objects. For example, use the **Paper text height** property to enter a paper text height. The **Model text height** property is a reference value that identifies the height of the text after the scale factor is automatically applied.

Drawing Annotative Dimensions

Draw annotative dimensions, leaders, GD&T symbols created using the **TOLERANCE** tool, and multileaders using the same tools as non-annotative dimensions. Once you activate an annotative dimension or multileader style and select the appropriate annotation scale, the process of placing correctly scaled dimensions is automatic.

However, you must still determine the correct dimension and text location and spacing from objects when you add dimensions and text to scaled drawings. This involves multiplying the scale factor by the plotted spacing. For example, if the first dimension line should be 3/4″ from an object when plotted, using a 1/4″ = 1′-0″ scale, the correct spacing in model space is 36″ from the object (a scale factor of 48 × 3/4″ = 36″).

Adding Annotative Hatch Patterns

The difference between adding annotative and non-annotative hatch patterns is the way in which the drawing scale affects the hatch scale. When you create annotative hatch patterns, the scale you enter in the **Scale:** text box produces the same results regardless of the specified annotation scale. For example, if you enter a value in the **Scale:** text box that is appropriate for an annotation scale of 1/4″ = 1′-0″, and then

Figure 31-8.
Creating annotative multiline text. Multiline text can contain annotative and non-annotative text.

Select the **Annotative** button   Select the plotted (paper) text height

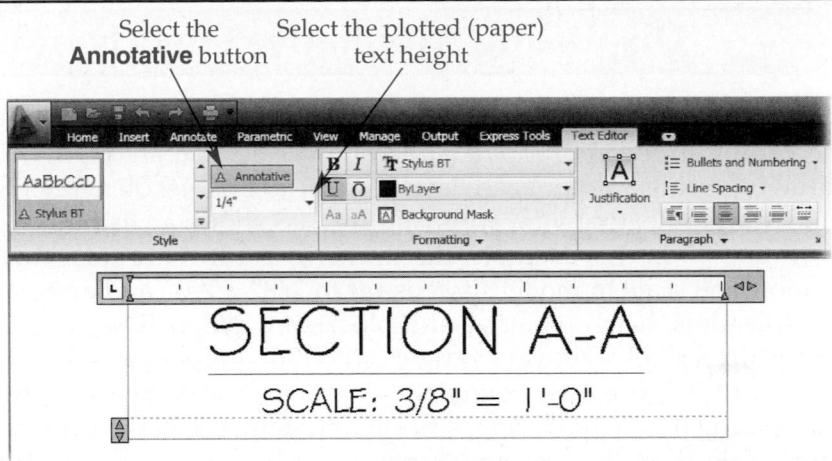

SECTION A-A
SCALE: 3/8" = 1'-0"

change the annotation scale to 1" = 1'-0", the hatch pattern scale does not change relative to the drawing display. It looks the same on the 1/4" = 1'-0" scaled drawing as on the 1" = 1'-0" scaled drawing.

In contrast, when you create non-annotative hatch patterns, if you enter a value in the **Scale:** text box that is appropriate for a drawing scaled to 1/4" = 1'-0" and then change the drawing scale to 1" = 1'-0", the displayed scale of the hatch pattern increases. It looks four times as large on the 1" = 1'-0"scaled drawing as on the 1/4" = 1'-0" scaled drawing.

Placing Annotative Blocks and Attributes

Annotative blocks, often classified as *schematic blocks*, are commonly used for annotation purposes. When you insert an annotative schematic block, AutoCAD determines the block scale based on the current annotation scale, eliminating the need for you to enter a scale factor. For most applications, insert annotative blocks at a scale of 1 to apply the annotation scale correctly. Entering a scale other than 1 adjusts the scale of the block by multiplying the block scale by the annotation scale.

schematic block:
A block originally drawn at a 1:1 scale.

PROFESSIONAL TIP

When you create unit and schematic blocks that contain text and attributes, you should usually not make the text and attributes annotative. The height you specify is set according to the full-scale size of the block, not necessarily the paper height. Any non-annotative text and attributes selected when you make a block annotative also automatically become annotative.

Exercise 31-2

Access the Student Web site (www.g-wlearning.com/CAD) and complete Exercise 31-2.

Displaying Annotative Objects in Layouts

Once you create drawing features and symbols and add annotative objects according to the appropriate annotation scale, you are ready to display and plot the drawing using a paper space layout. Refer to Chapter 30 to review the process of using and scaling floating viewports. **Figure 31-9** shows an example of a drawing scaled to 3/8" = 1'-0". In this example, the drawing features are full scale in model space. The annotation scale in model space is set to 3/8" = 1'-0", and annotative text, dimensions, multileaders, hatch patterns, and blocks are added. The annotative objects are automatically scaled according to the 3/8" = 1'-0" annotation scale.

In the **Figure 31-9** example, the viewport scale and the annotation scale are the same, which is typical when scaling annotative objects. If you select a different viewport scale from the **Viewport Scale** flyout button, the annotation scale automatically adjusts according to the viewport scale. However, if you adjust the viewport scale by zooming, for example, the annotation scale does not change. The viewport scale and the annotation scale must match in order for your drawing and annotative objects to be scaled correctly. You can pick the button to the right of the **Viewport Scale** flyout button, identified in **Figure 31-9,** to synchronize the viewport and annotation scales.

The **Properties** palette also allows you to control viewport and annotation scale. In order to use this method, you must be in paper space to access the viewport properties. Then choose a viewport scale from the **Standard scale** drop-down list. Adjust the annotation scale using the **Annotation scale** option. See **Figure 31-10**.

PROFESSIONAL TIP

Lock the viewport display to avoid zooming and disassociating the viewport scale from the annotation scale. Refer to Chapter 30 for more information on locking and unlocking floating viewports.

Figure 31-9.
Scale a drawing in a floating paper space viewport. Picking the **Viewport Scale** flyout button is one of the easiest ways to set the viewport scale. A button is also available to synchronize the viewport and annotation scale if they do not match.

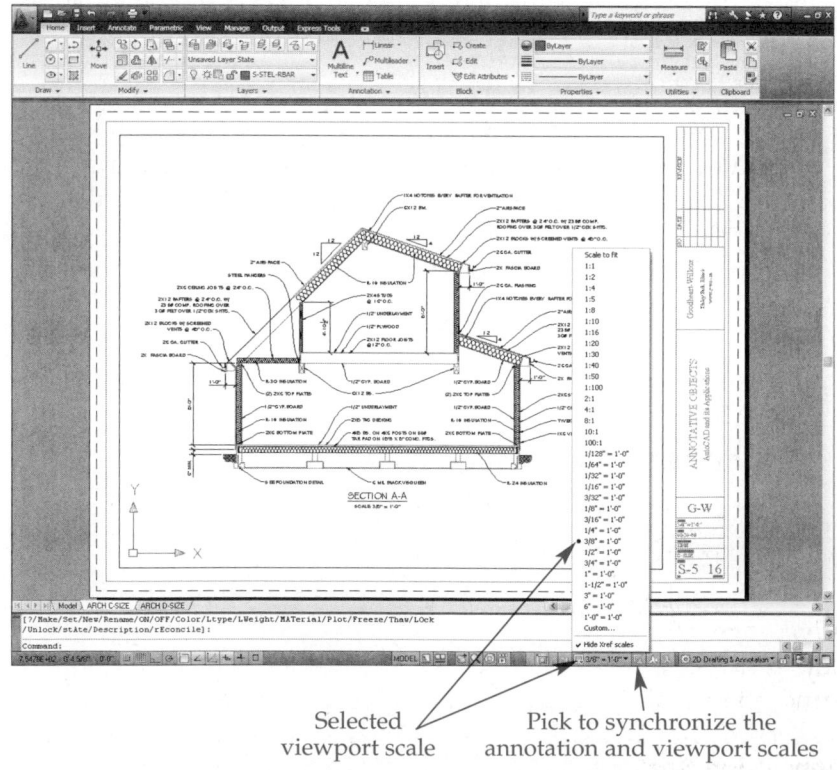

Selected viewport scale

Pick to synchronize the annotation and viewport scales

Figure 31-10.
The **Properties** palette also allows you to set the viewport and annotation scale.

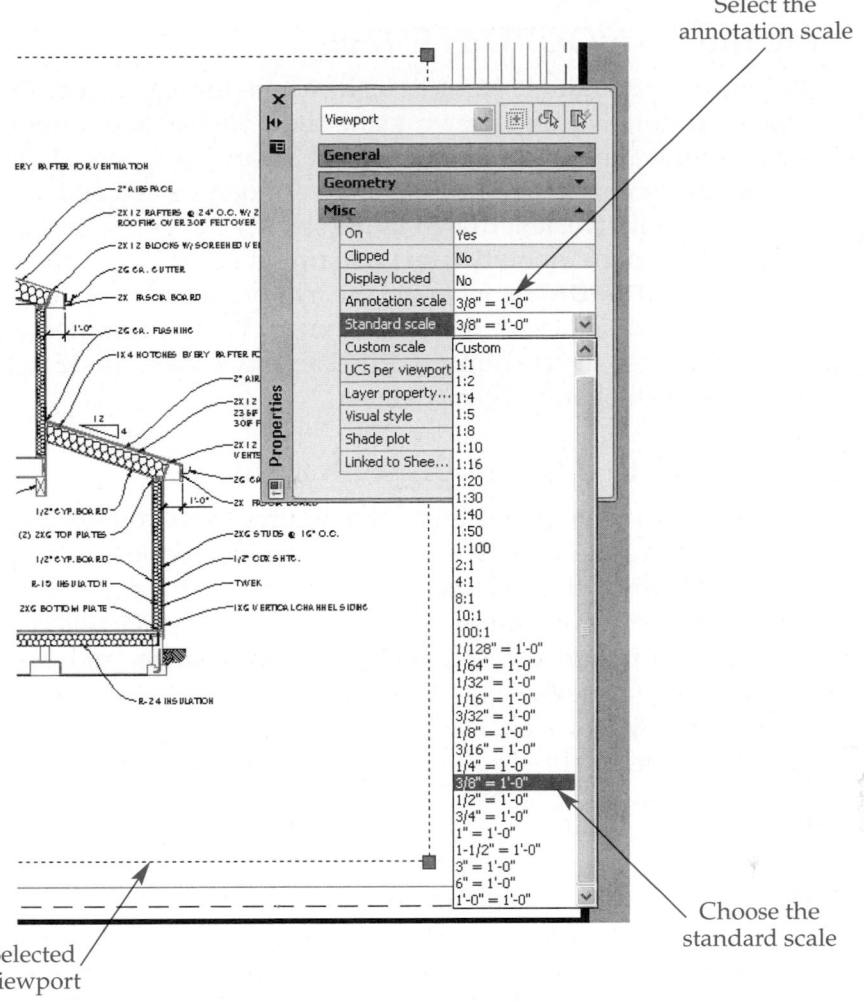

Select the annotation scale

Choose the standard scale

Selected viewport

Exercise 31-3

Access the Student Web site (www.g-wlearning.com/CAD) and complete Exercise 31-3.

Changing Drawing Scale

No matter how much you plan a drawing, drawing scale can change throughout the drawing process. Drawing scale may be reduced if it is necessary to use a smaller sheet. Drawing scale may increase if drawing features are redesigned and become larger, or if additional drawing detail is required.

Changing the drawing scale affects the size and position of annotations. A major advantage of using annotative objects is the ease with which the annotation scale adjusts to different drawing scales. When changing drawing scale, remember that the annotation scale is the same as the drawing scale.

To change annotation scale in model space, select a new annotation scale from the **Annotation Scale** flyout button. To change the annotation scale in an active viewport in a layout, adjust the viewport scale by selecting the drawing scale from the **Viewport**

Scale flyout button. Again, the viewport and annotation scale should be set to the same scale for most applications.

Using the ANNOUPDATE Tool

When you create single-line text using a non-annotative text style, and then change the style to annotative, text drawn using the style becomes annotative. However, the properties of the annotative text remain set according to the non-annotative text style. When you create annotative text using an annotative text style, and then change the style to non-annotative, text drawn in the style becomes non-annotative. However, the properties of the non-annotative text remain set according to the annotative style.

Type
ANNOUPDATE

Use the **ANNOUPDATE** tool to update text properties to reflect the current properties of the text style in which the text is drawn. When prompted to select objects, pick the text to update to the current, modified text style. After making the selections, press [Enter] to exit the tool and update the text.

Introduction to Scale Representations

The previous content of this chapter assumes that you develop a drawing using a single annotation scale. In order for annotative object scale to change when the drawing scale changes, annotative objects must support the new scale. This involves assigning new annotation scales to annotative objects. If annotative objects do not support the new scale, the annotative object scale does not change, and can actually cause the objects to become invisible.

Figure 31-11A shows an example of a drawing prepared at a 3/8″ = 1′-0″ scale and placed on an architectural C-size sheet. The annotation scale in this example is set to 3/8″ = 1′-0″, to automatically scale annotative objects according to a 3/8″ = 1′-0″ drawing scale. In order to change the scale of the drawing to 1/2″ = 1′-0″ to display additional detail, you must ensure that the annotative objects support a 1/2″ = 1′-0″ scale.

Once you add the 1/2″ = 1′-0″ annotation scale to the annotative objects, you can change the annotation scale or the viewport scale to 1/2″ = 1′-0″ to correctly scale annotative objects. See **Figure 31-11B.** The annotative objects in this example support two annotation scales: 3/8″ = 1′-0″ and 1/2″ = 1′-0″. As a result, two *annotative object representations* are available.

annotative object representation:
Display of an annotative object at an annotation scale that the object supports.

> **NOTE**
>
> Annotative objects display an icon when you hover the crosshairs over the objects. Objects that support a single annotation scale display the annotative icon shown in **Figure 31-12A**. Annotative objects that support more than one annotation scale display the annotative icon shown in **Figure 31-12B**. These icons appear by default according to selection preview settings in the **Selection** tab of the **Options** dialog box.

Understanding Annotation Visibility

Before changing the current annotation scale, you should understand how the annotation scale effects annotative object visibility. The annotative object scale does not change if annotative objects do not support an annotation scale. In addition, annotative objects disappear when an annotation scale that the objects do not support is current. For example, if annotative objects only support an annotation scale of 3/8″ = 1′-0″, and an annotation scale of 1/2″ = 1′-0″ is set current, the annotative object scale remains set at 3/8″ = 1′-0″, and the objects become invisible.

Figure 31-11.
A—A drawing created using an annotation scale of 3/8″ = 1′-0″ on an architectural C-size sheet. Annotative objects automatically appear at the correct scale. B—The same drawing shown in A, modified to an annotation scale of 1/2″ = 1′-0″ and placed on an architectural D-size sheet. An annotation scale of 1/2″ = 1′-0″ is added to all of the annotative objects, allowing the objects to adapt to the new scale automatically.

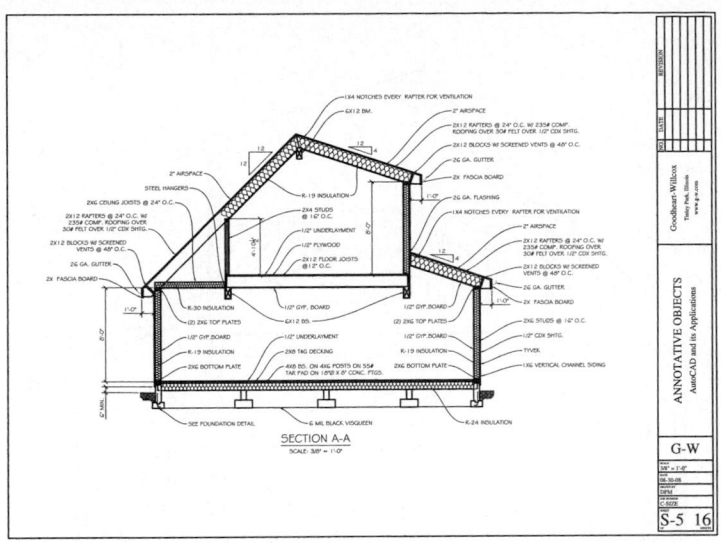

A

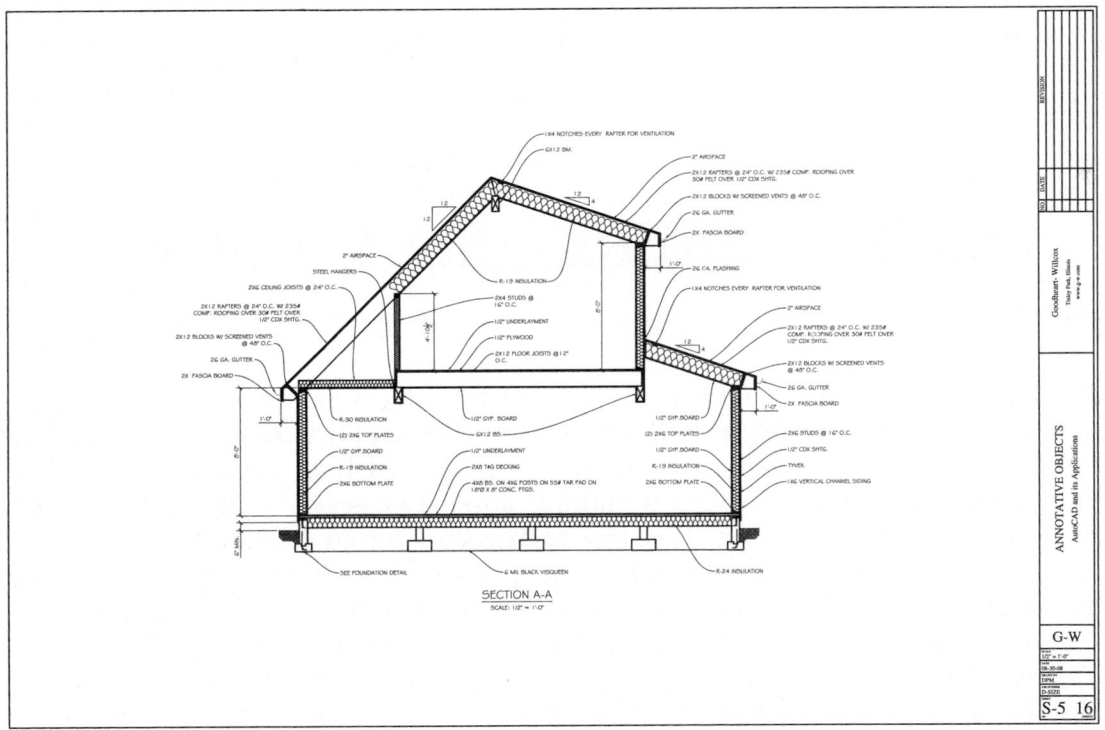

B

Figure 31-12.
Examples of annotative objects that support single and multiple annotation scales.

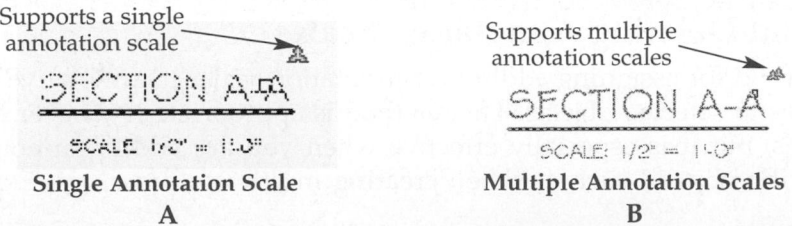

Single Annotation Scale	Multiple Annotation Scales
A	**B**

Figure 31-13.
A—The annotative objects in this example support only a 3/8″ = 1′-0″ annotation scale. However, with annotation visibility turned on, all annotative objects appear, even with the annotation scale set to 1/2″ = 1′-0″. B—The **Annotation Visibility** button on the status bar controls this feature. If you display the drawing status bar, the **Annotation Visibility** button moves from the application status bar to the drawing status bar.

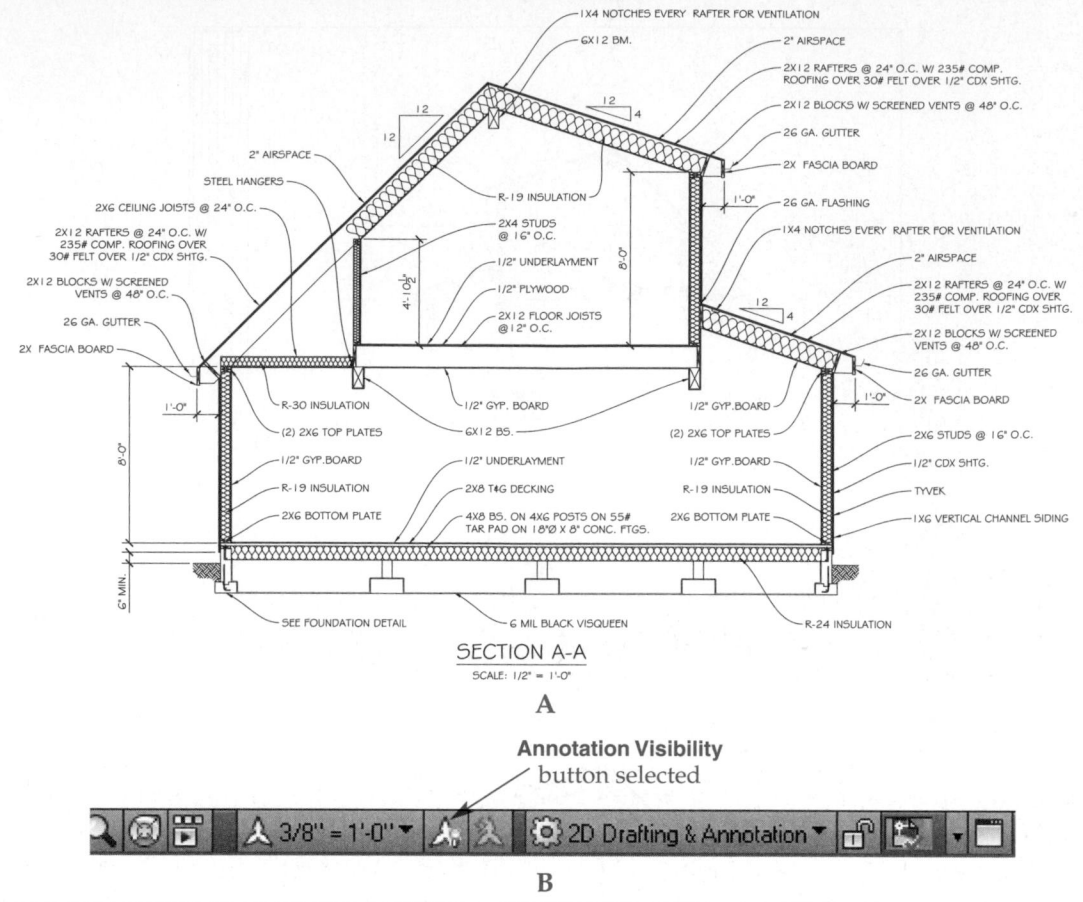

The easiest way to turn on and off annotative object visibility according to the current annotation scale is to pick the **Annotation Visibility** button on the status bar. See **Figure 31-13.** This is most effective when adding and deleting annotation scales to or from annotative objects. If you add multiple annotation scales to annotative objects, the annotative object representation appears based on the current scale.

Deselect the **Annotation Visibility** button to display only the annotative objects that support the current annotation scale. Any annotative objects unsupported by the current annotation scale become invisible. See **Figure 31-14.** This process is most effective when you want to annotate a drawing, or a portion of a drawing, using a different annotation scale without showing annotative object representations specific to a different annotation scale. Turning off the visibility of annotative objects that do not support the current annotation scale is also extremely effective for preparing multiview drawings because it eliminates the need to create separate layers for objects displayed at different scales. This practice is described later in this chapter.

Adding and Deleting Annotation Scales

One method for assigning additional annotation scales to annotative objects is to add the scales to selected objects. This method is appropriate whenever the drawing scale changes, but it is especially effective when you are adding annotation scales only to specific objects, such as when creating multiview drawings. Examples that

AutoCAD and Its Applications—Basics

Figure 31-14.
Deselect the **Annotation Visibility** button to display only those annotative objects that support the current annotation scale. The annotative objects in this example do not appear because they support only a 3/8″ = 1′-0″ annotation scale, and the current annotation scale is 1/2″ = 1′-0″.

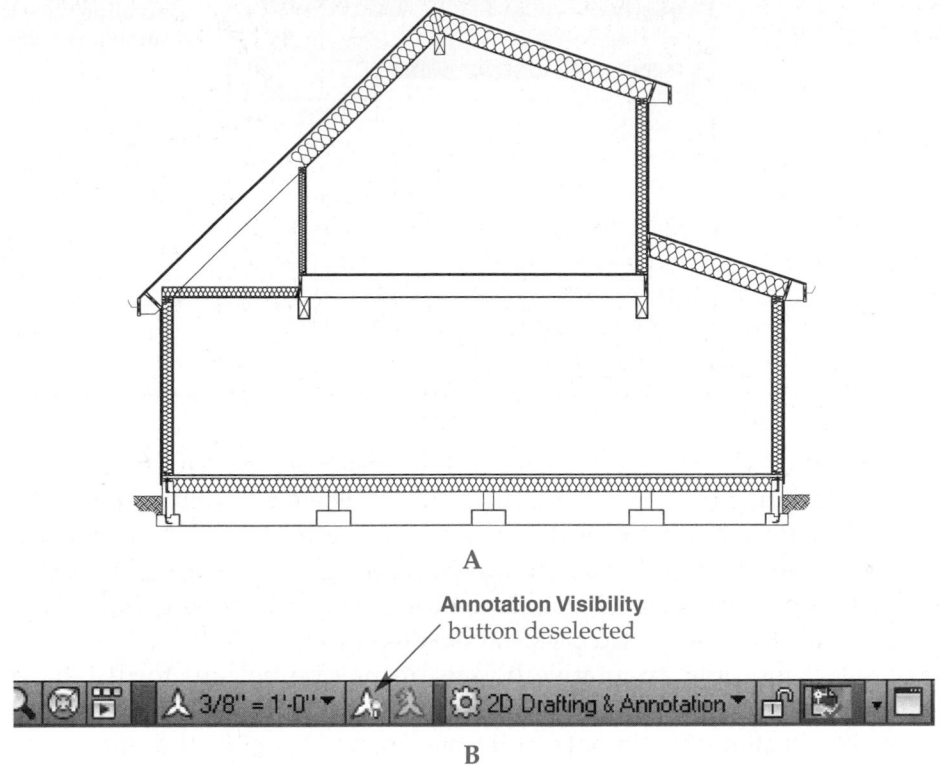

Annotation Visibility
button deselected

B

demonstrate this practice are described later in this chapter. You can add annotation scales to selected objects using annotation scaling tools or the **Properties** palette.

You can delete an annotation scale from annotative objects if the annotation scale is no longer in use, should not display in a specific view, or makes it difficult to work with annotative objects. When you delete an annotation scale from annotative objects, the scale no longer applies. You can delete annotation scales from selected objects using annotation scaling tools or the **Properties** palette.

Using the OBJECTSCALE Tool

The **OBJECTSCALE** tool provides one method of adding and deleting annotation scales supported by annotative objects. A quick way to access the **OBJECTSCALE** tool is to select an annotative object and then right-click and pick **Add/Delete Scale…** from the **Annotative Objects Scales** cascading submenu. If you activate the **OBJECTSCALE** tool by right-clicking on objects, the **Annotation Object Scale** dialog box appears, allowing you to add or remove annotation scales from the selected objects. See **Figure 31-15.** If you access the tool before selecting objects, all annotative objects display, even those objects that do not support the current annotation scale. Select the annotative objects to modify and press [Enter] to display the **Annotation Object Scale** dialog box.

The **Object Scale List** shows all of the annotation scales associated with the selected annotative object. A scale must appear in the list in order for the scale to apply to the annotative object. If you select a different annotation scale, and that scale does not display in the **Object Scale List**, annotative objects do not adapt to the new annotation scale, and you have the option of turning off the annotative objects' visibility. Using the previous example, 1/2″ = 1′-0″ must appear in the **Object Scale List** in order for the annotative objects to adapt to the new annotation scale of 1/2″ = 1′-0″.

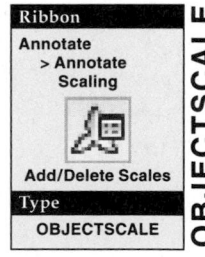

Ribbon
Annotate
> Annotate
Scaling

Add/Delete Scales
Type
OBJECTSCALE

OBJECTSCALE

Figure 31-15.
The **Annotation Object Scale** dialog box allows you to add and delete annotation scales to and from annotative objects.

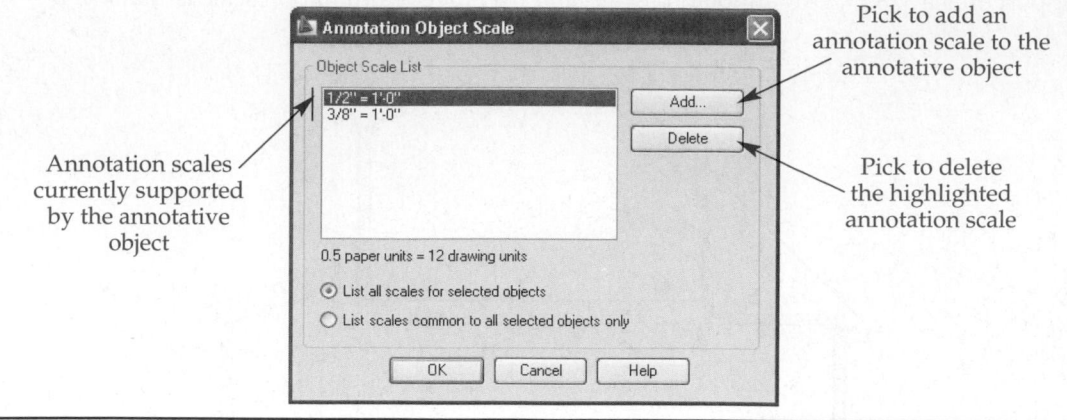

Annotation scales currently supported by the annotative object

Pick to add an annotation scale to the annotative object

Pick to delete the highlighted annotation scale

Pick the **Add...** button to add a scale to the **Object Scale List**. This opens the **Add Scales to Object** dialog box. Highlight scales in the **Scale List** and pick the **OK** button to add the scales to the **Object Scale List**. Once you add a scale to the **Object Scale List**, picking an annotation scale that corresponds to any of the listed scales automatically scales the selected annotative object. To remove a scale from the **Object Scale List**, highlight the scale to remove and pick the **Delete** button.

If you select multiple annotative objects, it may be helpful to display only the annotative scales that are common to the selected objects by picking the **List scales common to all selected objects only** radio button. To show all the annotation scales associated with any of the selected objects, even if some of the objects do not support the listed scales, pick the **List all scales for selected objects** radio button. Picking this option is helpful to delete a listed scale that applies only to certain objects.

> **NOTE**
>
> If a desired scale is not available in the **Add Scales to Object** dialog box, you must close the **Annotation Object Scale** dialog box and access the **Edit Scale List** dialog box to add a new scale to the list of available scales.

Using the Properties Palette

The **Properties** palette also allows you to add annotation scales to selected annotative objects. See **Figure 31-16.** The location of the annotative properties in the **Properties** palette varies depending on the selected object. The **Annotative scale** property displays the annotation scale currently applied to the selected annotative object and contains an ellipsis button (**...**) that opens the **Annotation Object Scale** dialog box when selected.

Automatically Adding Annotation Scales

Another technique for assigning additional annotation scales to annotative objects is to add a selected annotation scale automatically to all annotative objects in the drawing. This eliminates the need to add annotation scales to individual annotative objects and quickly produces newly scaled drawings.

The **ANNOAUTOSCALE** system variable controls the ability to add an annotation scale to all existing annotative objects. Enter 1, –1, 2, –2, 3, –3, 4, or –4, depending on the desired effect. **Figure 31-17A** describes each option. Once you enter the initial value, the easiest way to toggle this system variable on and off is to pick the button on the status bar shown in **Figure 31-17B.**

Figure 31-16.
The **Annotation scale** property in the **Properties** palette is another way to access the **Annotation Object Scale** dialog box.

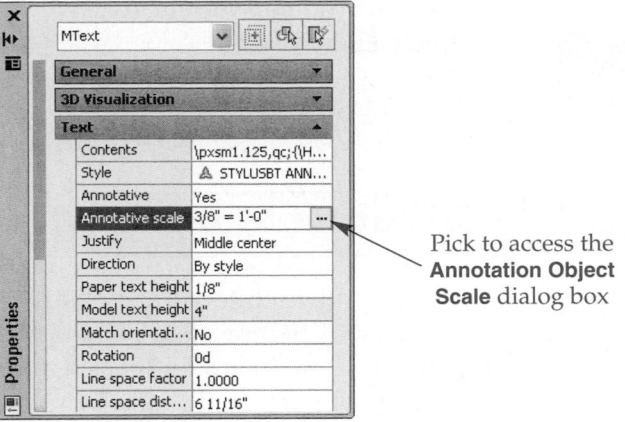

Pick to access the **Annotation Object Scale** dialog box

Figure 31-17.
A—**ANNOAUTOSCALE** system variable options. B—Once you enter the initial **ANNOAUTOSCALE** system variable, use the button on the status bar to toggle **ANNOAUTOSCALE** on and off. If you display the drawing status bar, the **ANNOAUTOSCALE** button moves from the application status bar to the drawing status bar.

Value	Mode	Description
1	On	Adds the selected annotation scale to annotative objects, not including those drawn on a layer that is turned off, frozen, locked, or frozen in a viewport.
−1	Off	1 behavior is used when **ANNOAUTOSCALE** is turned back on.
2	On	Adds the selected annotation scale to annotative objects, not including those drawn on a layer that is turned off, frozen, or frozen in a viewport.
−2	Off	2 behavior is used when **ANNOAUTOSCALE** is turned back on.
3	On	Adds the selected annotation scale to annotative objects, not including those drawn on a layer that is locked.
−3	Off	3 behavior is used when **ANNOAUTOSCALE** is turned back on.
4	On	Adds the selected annotation scale to all annotative objects regardless of the status of the layer on which the annotative object is drawn. 4 is the AutoCAD default setting when toggled on.
−4	Off	4 behavior is used when **ANNOAUTOSCALE** is turned back on. −4 is the AutoCAD default setting when toggled off.

A

Pick to toggle the **ANNOAUTOSCALE** system variable on or off

A 3/8" = 1'-0" ▼

B

CAUTION

Use caution when adding annotation scales automatically. Due to the effectiveness and transparency of the tool, annotation scales are often added to annotative objects unintentionally. Though you can later delete scales, this causes additional work and confusion.

Exercise 31-4

Access the Student Web site (www.g-wlearning.com/CAD) and complete Exercise 31-4.

Preparing Multiview Drawings

Mechanical drawings and architectural construction drawings often contain sections and details drawn at different scales. Using annotative objects offers several advantages, especially when views in model space appear at different scales in layouts. Use scaled viewports to display multiple views using a single file. You can assign a different annotation scale to each drawing view that contains annotative objects, reducing the need to calculate multiple drawing scale factors, while maintaining the appropriate scale of previously drawn annotative objects. Additionally, by adjusting annotative scale representation visibility and position, you can prepare differently scaled multiview drawings, while eliminating the need to use separate, scale-specific layers and annotations.

Creating Differently Scaled Drawings

Figure 31-18A shows an example of two different drawing views, both drawn at full scale in model space. The full section in **Figure 31-18A** uses a 3/8″ = 1′-0″ scale. To prepare this view, set the annotation scale in model space to 3/8″ = 1′-0″, and then add annotative objects. The annotative objects automatically scale according to the 3/8″ = 1′-0″ annotation scale. The stair section in **Figure 31-18A** uses a 1/2″ = 1′-0″ scale. To prepare this view, change the annotation scale in model space from 3/8″ = 1′-0″ to 1/2″ = 1′-0″, and then add annotative objects. The annotative objects automatically scale according to the 1/2″ = 1′-0″ annotation scale. If you look closely, you can see the different scales applied to the drawing views.

With annotation visibility on, as shown in **Figure 31-18A,** you can see all annotative objects, and observe the effects of using different scales. With annotation visibility off, as shown in **Figure 31-18B,** only annotative objects that support the current annotation scale, which is 1/2″ = 1′-0″ in this example, appear.

The next step is to display and plot the drawing using multiple paper space viewports. **Figure 31-19** shows an architectural D-size sheet layout with two floating viewports. One viewport displays the full section at a viewport scale of 3/8″ = 1′-0″. The other viewport displays the stair section at a viewport scale of 1/2″ = 1′-0″. Notice how the annotative objects are the same size in both views.

Exercise 31-5

Access the Student Web site (www.g-wlearning.com/CAD) and complete Exercise 31-5.

Figure 31-18.
Two different drawing views drawn at full scale in model space. The full section uses an annotation scale of 3/8″ = 1′-0″, and the stair section uses an annotation scale of 1/2″ = 1′-0″. A—Annotation visibility is on. B—Annotation visibility is off with the current annotation scale set to 1/2″ = 1′-0″ (stair section view scale).

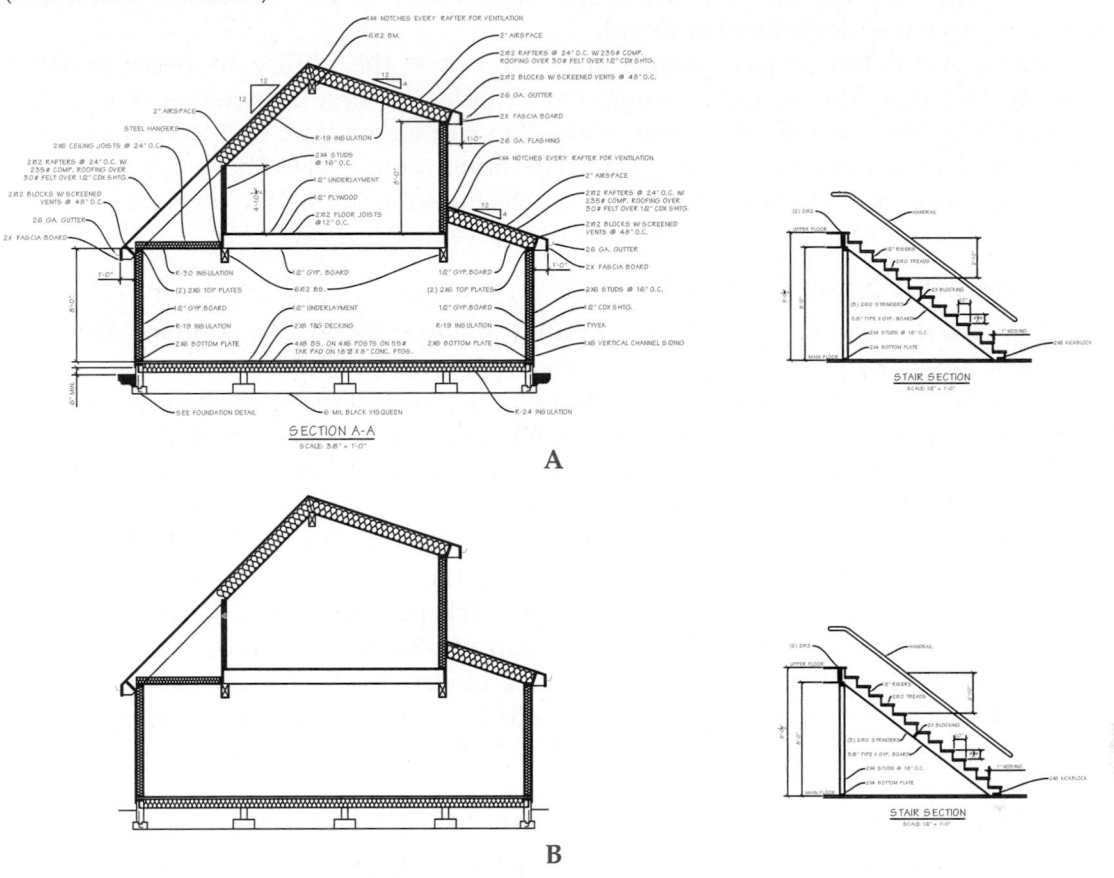

Figure 31-19.
Using viewports with different scales to create a multiview drawing. Notice that the annotative objects are the same size in both views.

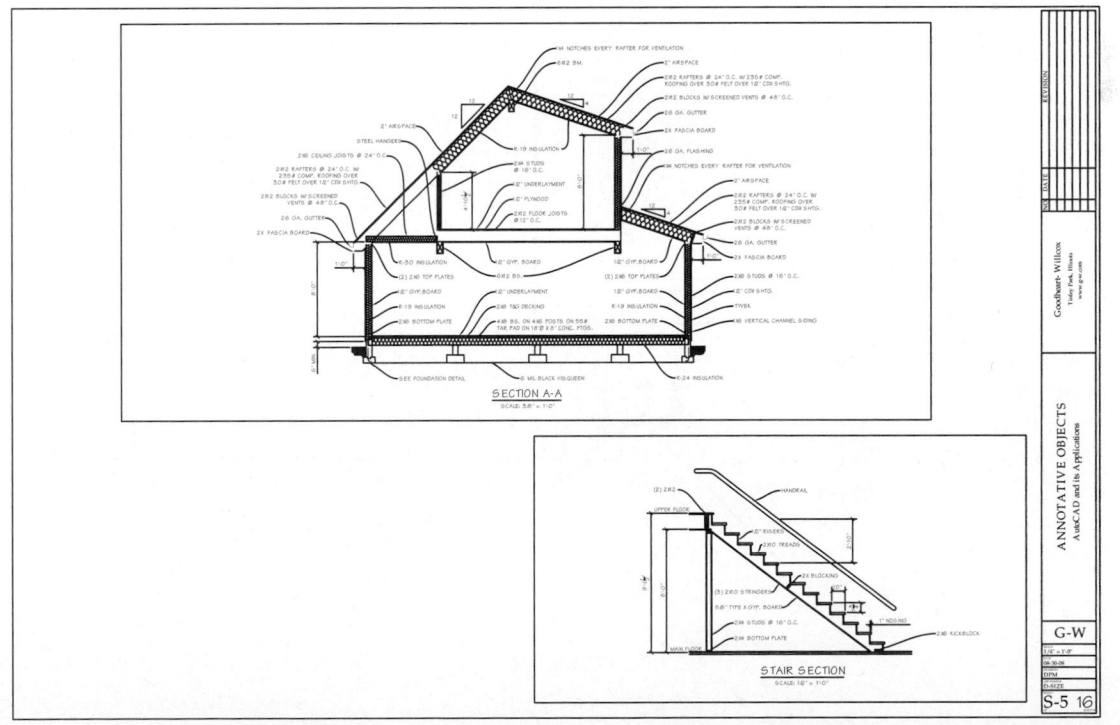

Reusing Annotative Objects

Often the same drawing features appear in different views at different scales. For example, you may want to plot a drawing on a large sheet using a large scale, and plot the same drawing on a smaller sheet using a smaller scale. Another example is preparing a view enlargement or detail.

Using annotative objects significantly improves the ability to reuse existing drawing features. You can use annotation visibility to hide annotative objects not supported by the current annotation scale. You can also adjust the position of scale representations according to the appropriate annotation scale. These options allow you to include differently scaled annotative objects on the same drawing sheet without creating copies of the objects and without using scale-specific layers.

Using Invisible Scale Representations

If annotative objects do not support an annotation scale, the annotative objects disappear when the annotation scale that the objects do not support is current. This is a valuable technique for displaying certain items at a specific scale. Pick the **Annotation Visibility** button on the status bar to turn on and off annotative object visibility.

The following example shows how adjusting the visibility of annotative objects that only support the current annotation scale allows you to create an additional view from existing drawing features. This example uses an annotation scale of 3/4″ = 1′-0″ to create a foundation detail. To begin constructing the foundation detail, add the 3/4″ = 1′-0″ annotation scale to the existing earth hatch pattern so it will appear on the full section and the foundation detail. See **Figure 31-20.** Next, with the current annotation scale set to 3/4″ = 1′-0″, add annotative objects to the foundation detail. See **Figure 31-21.** These objects only support the 3/4″ = 1′-0″ annotation scale, hiding the objects on the full section, which uses a 3/8″ = 1′-0″ scale.

Figure 31-20.
Reuse the earth hatch pattern by adding the 3/4″ = 1′-0″ foundation detail scale to the annotative hatch pattern.

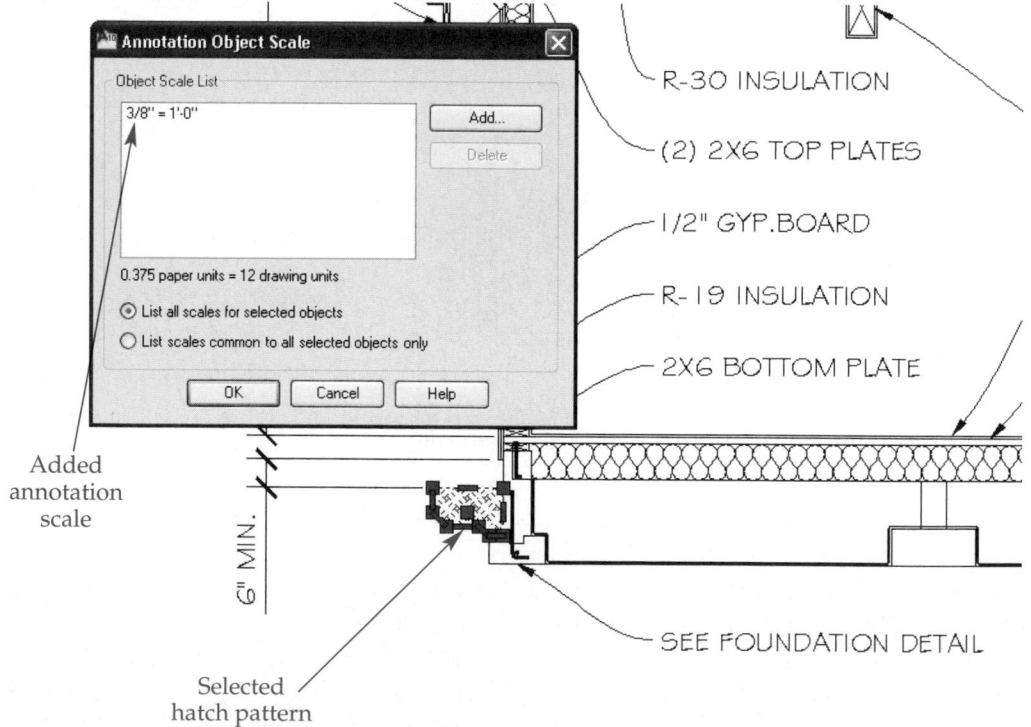

AutoCAD and Its Applications—Basics

Figure 31-21.
Adding annotative text, dimensions, multileaders, and hatch patterns using a 3/4″ = 1′-0″ annotation scale.

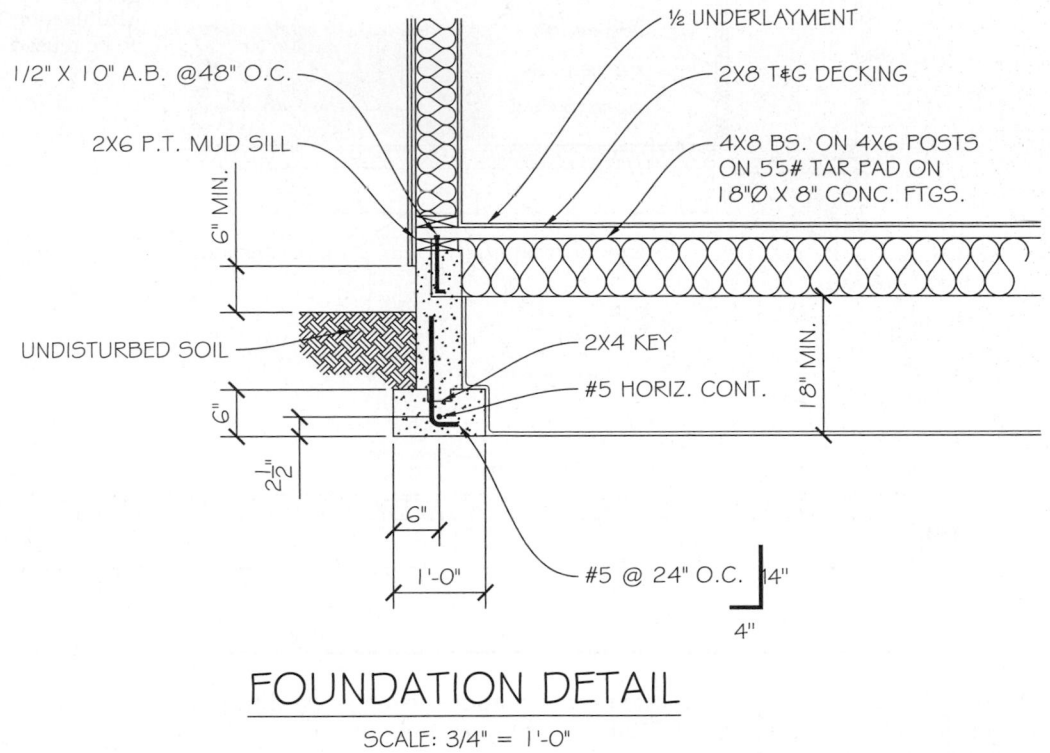

FOUNDATION DETAIL

SCALE: 3/4″ = 1′-0″

PROFESSIONAL TIP

If objects already support an annotation scale, but you do not want to display those annotations at the current scale, delete the annotation scale from the objects.

Adjusting Scale Representation Position

A major benefit of using annotative objects is the ability to reuse objects for differently scaled drawing views. The previous example of adding a 3/4″ = 1′-0″ annotation scale to the earth hatch pattern highlights this concept. When you reuse annotative objects, the location and spacing of annotative objects on one scale are often not appropriate for another scale. You can reposition each scale representation to overcome this issue.

In the foundation detail example, some of the existing 3/8″ = 1′-0″ scaled full section dimensions and multileaders are reused in the foundation detail. The first step is to add a 3/4″ = 1′-0″ annotation scale to the objects. Next, with **Annotation Visibility** turned off, as shown in **Figure 31-22,** you can see the resulting position of the selected objects, which is initially the same location as the 3/8″ = 1′-0″ objects. The only difference is that now the 3/8″ = 1′-0″ objects also support a 3/4″ = 1′-0″ scale.

Adjust the position of annotation scale representations using grip editing methods. When you select annotative objects that support more than one annotation scale, all scale representations appear by default. See **Figure 31-23.** An annotative object is a single object, but it can contain several scale representations. Grips attach to the scale representation that corresponds to the current annotation scale. Using grips to edit scale representations is similar to editing the object used to create the scale representation. The difference when editing a scale representation is that you are adjusting a scaled "copy" of the object. **Figure 31-24** shows the effects of editing the position of dimension and multileader scale representations on the foundation detail. The

Figure 31-22.
A—Reusing some of the existing 3/8″ = 1′-0″ scaled objects to create another drawing view. B—Adding a 3/4″ = 1′-0″ annotation scale to existing objects and setting the annotation scale to 3/4″ = 1′-0″.

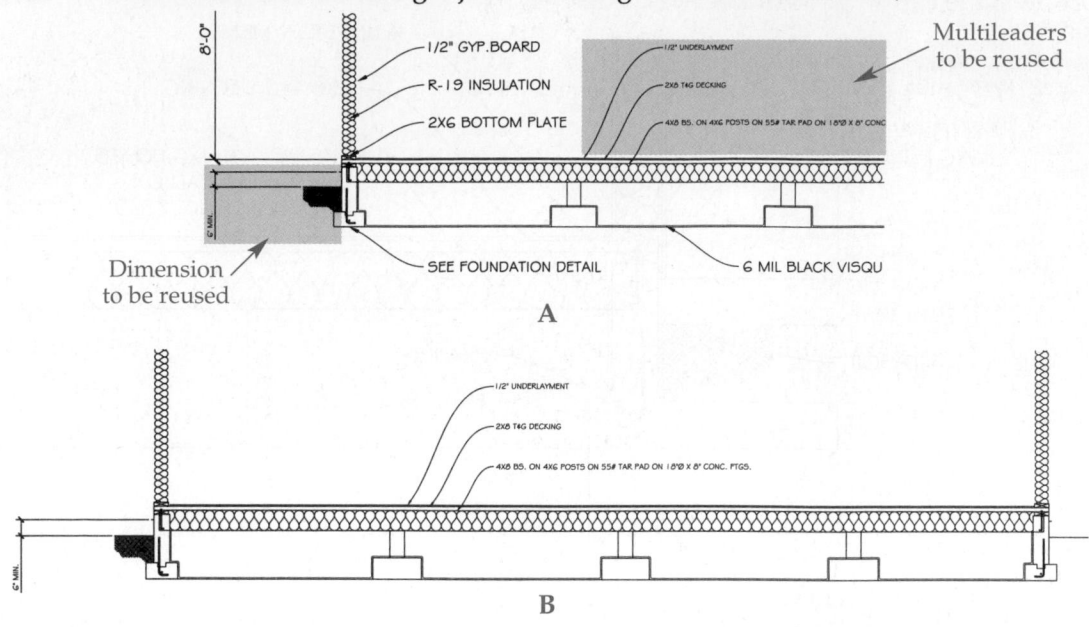

Figure 31-23.
Adjust the position of annotation scale representations using grip editing techniques. When you select annotative objects that support more than one annotation scale, all scale representations appear by default.

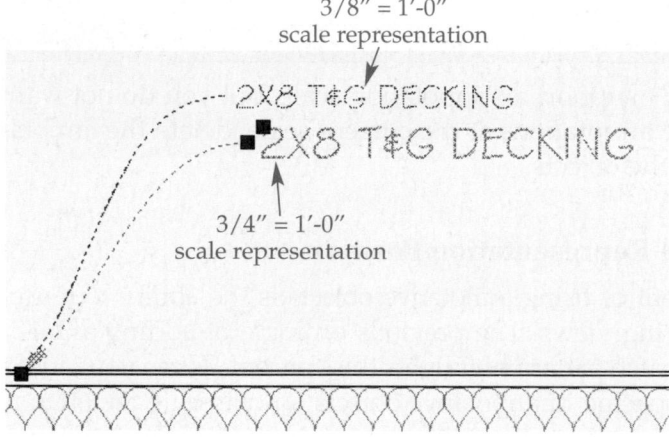

Figure 31-24.
Editing the position of scale representations is much like creating scaled copies of existing annotations.

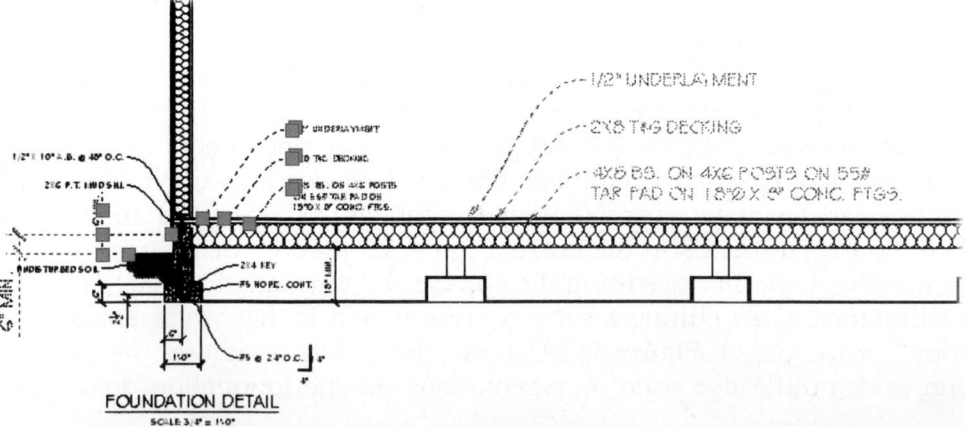

representations are selected to help demonstrate the effects of editing scale representation position. Notice that you can edit all elements of the scale representation to produce the desired annotations at the appropriate locations.

PROFESSIONAL TIP

Use the **DIMSPACE** and **MLEADERALIGN** tools to adjust dimension spacing and multileader alignment after the drawing scale changes.

The **SELECTIONANNODISPLAY** system variable controls the display of selected scale representations and is set to 1 by default. As a result, all scale representations display and appear dimmed when you pick an annotative object that supports multiple annotation scales. Refer again to **Figure 31-23.** The display can be confusing if the selected object supports several annotation scales. Set the **SELECTIONANNODISPLAY** system variable to 0 to display only the scale representation that corresponds to the current annotation scale.

NOTE

You can only edit scale representations individually using grip editing techniques. When you use modify tools to edit an annotative object, all scale representations are edited at once.

Resetting Scale Representation Position

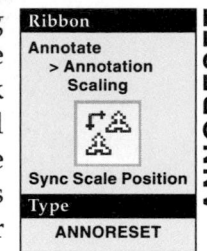

The **ANNORESET** tool removes all unique scale representation positions, allowing you to change the position of all selected scale representations to the position of the scale representation that is set according to the current annotation scale. A quick way to access the **ANNORESET** tool is to select annotative objects, right-click and pick the option from the **Annotative Object Scale** cascading submenu. If you activate the **ANNORESET** tool by right-clicking on objects, the position of the selected objects resets. If you access the tool before selecting objects, pick the annotative objects. After making your selections, press [Enter] to exit the tool and reset the scale representation positions.

Ribbon
Annotate > Annotation Scaling
Sync Scale Position
Type
ANNORESET

ANNORESET

Completing a Multiview Drawing

The last step in creating a multiview drawing is to display and plot the drawing using multiple paper space viewports. **Figure 31-25** shows an architectural D-size sheet layout with three floating viewports. One viewport displays the full section at a 3/8″ = 1′-0″ viewport scale. A second viewport displays the stair section at a 1/2″ = 1′-0″ viewport scale. A third viewport displays the foundation detail at a 3/4″ = 1′-0″ viewport scale.

PROFESSIONAL TIP

When you save drawings using annotative objects to earlier versions of AutoCAD that do not support annotative objects, scale representations may convert to non-annotative objects, but automatically become assigned to unique layers. To use this function, select the **Maintain visual fidelity for annotative objects** check box in the **Open and Save** tab of the **Options...** dialog box.

Figure 31-25.
A complete multiview drawing created using annotative objects.

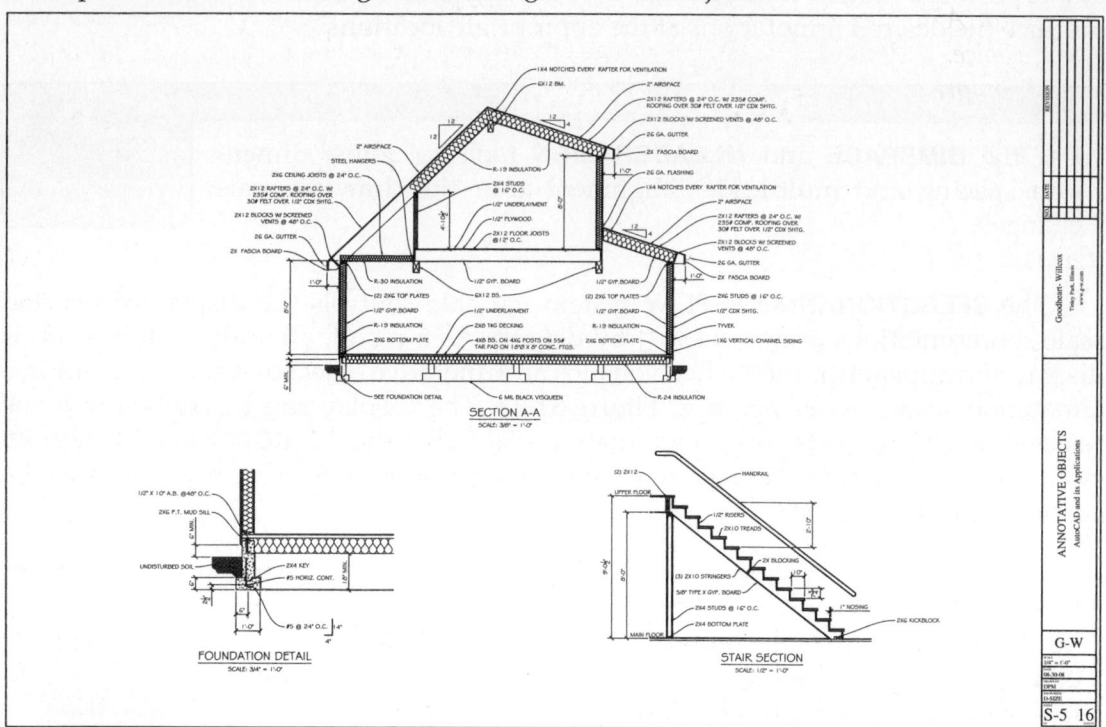

Exercise 31-6

Access the Student Web site (www.g-wlearning.com/CAD) and complete Exercise 31-6.

Chapter Test

Answer the following questions. Write your answers on a separate sheet of paper or go to the Student Web site (www.g-wlearning.com/CAD) and complete the electronic chapter test.

1. What are annotative objects?
2. Explain the practical differences between manual and annotative object scaling.
3. Identify at least four types of objects that can be made annotative.
4. How do you set the text scale, including spacing, width, and paragraph settings, to adjust automatically according to the current annotation scale?
5. Identify an important relationship between the viewport scale and the annotation scale.
6. Which **MSLTSCALE** system variable setting should you use so you do not have to calculate the drawing scale factor when entering an **LTSCALE** value?
7. Name the tool used to update text properties according to the current properties of the text style on which the text is drawn.
8. What is an annotative object representation?
9. Briefly describe the result of setting the **ANNOAUTOSCALE** system variable to a value of 4.
10. Briefly explain the effect of turning annotation visibility on and off.

AutoCAD and Its Applications—Basics

Drawing Problems

Start AutoCAD if it is not already started. Start a new drawing using an appropriate template of your choice. The template should include layers, text styles, dimension styles, and multileader styles appropriate for drawing the given objects. Add layers, text styles, dimension styles, and multileader styles as needed. Draw all objects using appropriate layers, text styles, dimension styles, multileader styles, justification, and format. Follow the specific instructions for each problem. Use your own judgment and approximate dimensions when necessary.

▼ Basic

1. Open P23-9 and save as P31-1. The P31-1 file should be active. Convert all the non-annotative objects to annotative objects. Resave the drawing.

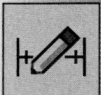

2. Open P25-10 and save as P31-2. The P31-2 file should be active. Convert all of the non-annotative objects to annotative objects. Resave the drawing.

▼ Intermediate

3. Create the section view and side view shown. Use annotative objects to prepare a full-scale drawing of the part. Change the annotation scale to 2:1 and adjust the scale representations as needed according to the new scale. Save the drawing as P31-3.

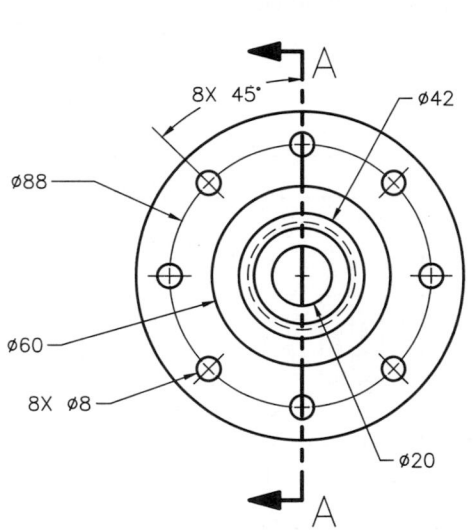

Name: Hub
Material: Cast Iron

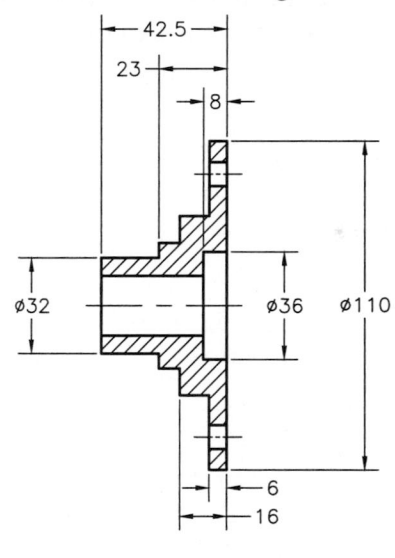

SECTION A—A

4. Create the section view and side views shown. Use annotative objects to prepare a full-scale drawing of the part. Change the annotation scale to 2:1 and adjust the scale representations as needed according to the new scale. Save the drawing as P31-4.

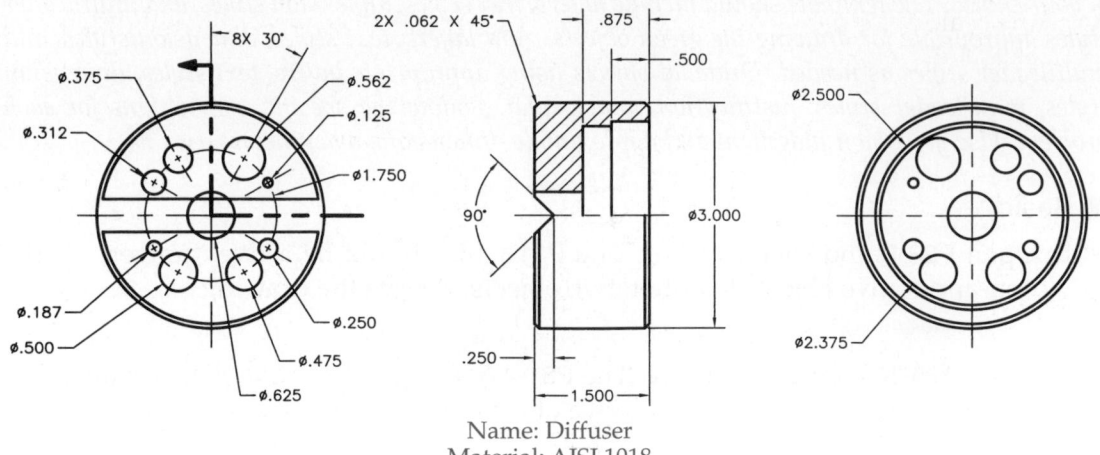

Name: Diffuser
Material: AISI 1018

5. Draw the fan shown at full scale in model space. Use annotative objects to prepare a full-scale view of the fan as shown and a view enlargement of the motor. You should not have to create a copy of the motor or develop scale specific layers. Save the drawing as P31-5.

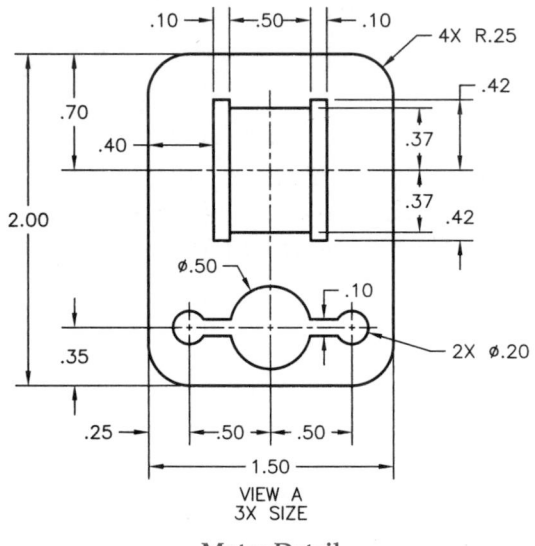

VIEW A
3X SIZE

Motor Detail

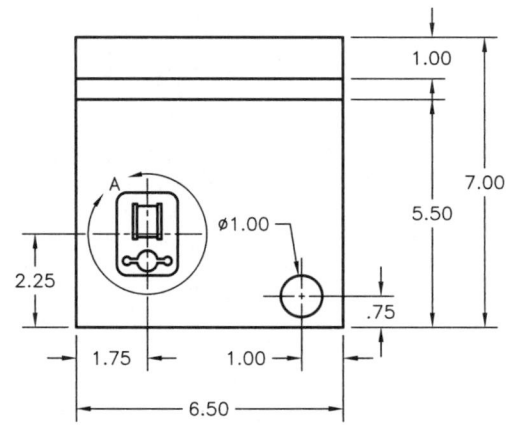

Fan

▼ Advanced

6. Draw the floor plan shown at full scale in model space. Use annotative objects to prepare a 1/4″ = 1′-0″ view. Change the annotation scale to 1/8″ = 1′-0″ and adjust the scale representations as needed according to the new scale. Save the drawing as P31-6.

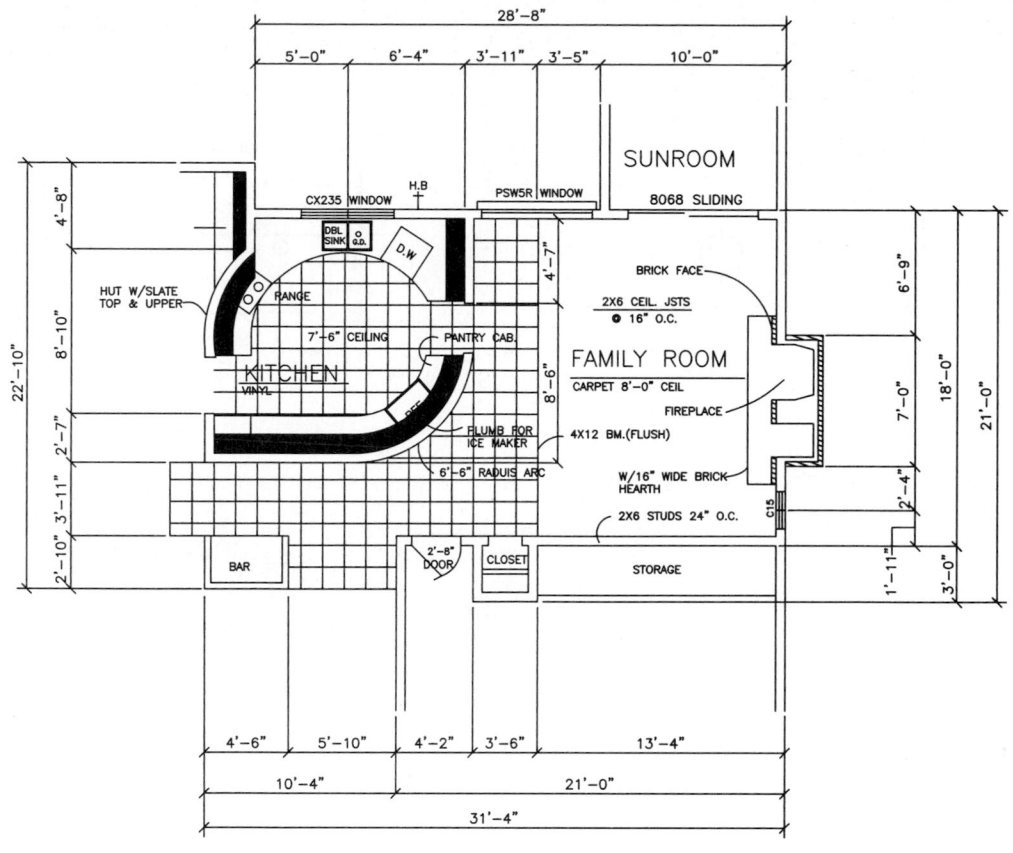

7. Draw the part shown at full scale in model space. Use annotative objects to prepare a full-scale view and a view enlargement of the part as shown. You should not have to create a copy of the part or develop scale specific layers. Save the drawing as P31-7.

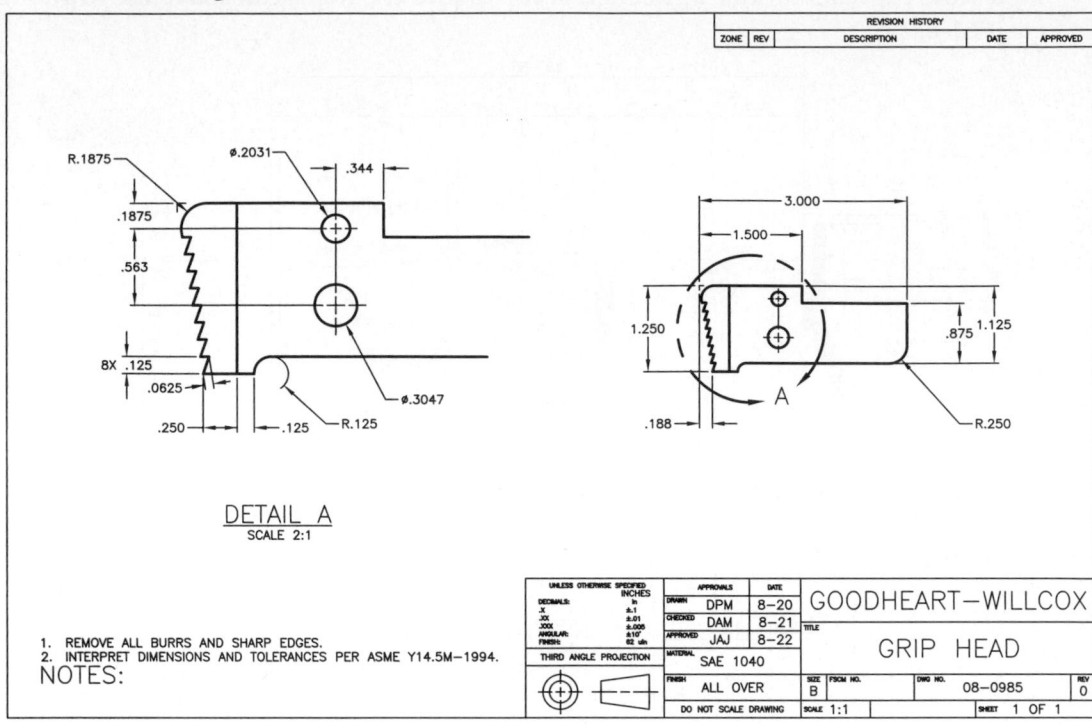

Learning Objectives

After completing this chapter, you will be able to do the following:
- ✓ Explain the function of external references.
- ✓ Attach an existing drawing to the current drawing.
- ✓ Use **DesignCenter** and tool palettes to attach external references.
- ✓ Bind external references and selected dependent objects to a drawing.
- ✓ Edit external references in the current drawing.

External references (xrefs) expand on the concept of reusing existing content, which is a major benefit of designing and drafting with AutoCAD. Xrefs are excellent for applications in which existing base drawings, complex symbols, images, and details are used often, are required to develop new drawings, or are shared by several users.

> **external reference (xref):** A DWG, DWF, raster image, DNG, or PDF file incorporated into a drawing for reference only.

Introduction to Xrefs

You can reference drawing (DWG), design web format (DWF and DWFx), raster image, digital negative (DNG), and portable document format (PDF) files into the *host drawing*, also known as the *master drawing*. Using an xref is similar to inserting an entire drawing as a block. However, unlike a block, which is actually located in the file in which you insert it, xref file geometry is not added to the host drawing. File data appears on-screen for reference only. The result is usable information, but a much smaller host file size than would occur if you inserted a block or copied and pasted objects.

> **host (master) drawing:** The drawing into which xrefs are incorporated.

In addition to reducing file size, one of the greatest benefits of using xrefs is that changes made to original xref files update in the host drawing to display the most recent reference content. AutoCAD reloads each xref whenever the host drawing loads. This allows an individual or a group of people to work on a multi-file project, with the assurance that any revisions to xrefs are displayed in any drawing in which the xrefs are used.

Types of Xref Files

Files that you can reference into a current drawing include existing DWG, DWF, DWFx, raster image, DNG, and PDF files. DWF and DWFx files are AutoCAD drawing files or other application files compressed for publication and viewing on the Web. DWF and DWFx file references are commonly used for sharing information from the Web or from an application other than AutoCAD.

Raster image and DNG file reference is appropriate whenever you need to add an image to a drawing, such as for a company logo in a title block. PDF file reference allows you to reuse PDF file content. Externally referencing an image or PDF file into a drawing is an excellent technique, because the large file sizes often associated with images and PDF files are not reproduced in the current drawing.

Xref Applications

DWG files are the most common externally referenced files and are the focus of this chapter. The term *xref* often applies specifically to referenced DWG files. In general, use external references to reuse existing drawing information to develop another drawing. There are countless applications for xref drawings in every drafting field. The following sections provide typical xref drawing applications. As you work with AutoCAD, you will discover a variety of uses for xref drawings.

Reference Existing Geometry

One of the most common applications for xref drawings is to reference existing geometry into the current drawing to use as a pattern or source of needed information. For example, a floor plan includes size and shape information required to prepare additional plans, elevations, sections, and details. **Figure 32-1** shows an example of referencing a floor plan file into a new drawing to use as an outline for creating a roof plan file. In this example, the floor plan xref attaches to the roof plan file to serve as an outline for creating the roof plan.

Figure 32-2 shows an example of the roof plan file created in **Figure 32-1** attached to an elevation file as an xref and then used to project an elevation. The roof plan xref includes a *nested* floor plan xref. In this example, the elevation file references the roof plan. The roof plan in turn references the floor plan.

nested xrefs: Xrefs contained within other xrefs.

Create a Multiview Drawing

Another xref application is to create commonly used drawings, such as sections and details, as separate drawing files and then attach each drawing as an xref to a host drawing. This method is similar to using blocks. Use floating viewports and layer viewport freezing to create a multiview layout. You can prepare a multiview drawing entirely from existing xref drawings or from a combination of objects created "in place" and attached xrefs.

Figure 32-1.
An example of using a floor plan xref drawing as a pattern, or outline, to draw a roof plan.

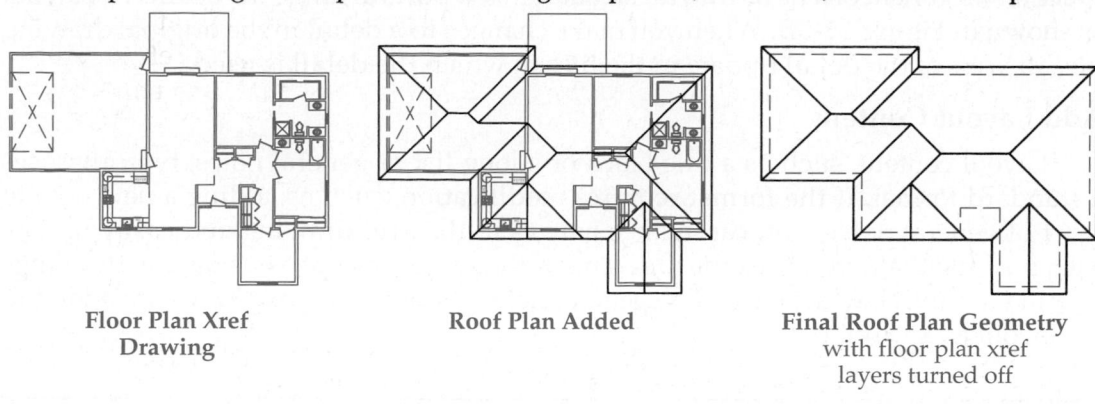

Floor Plan Xref Drawing

Roof Plan Added

Final Roof Plan Geometry
with floor plan xref
layers turned off

Figure 32-2.
Using a roof plan xref drawing that contains a nested floor plan xref drawing as a pattern for projecting geometry needed to create an elevation. Projection lines, drawn using the **XLINE** tool, are for reference.

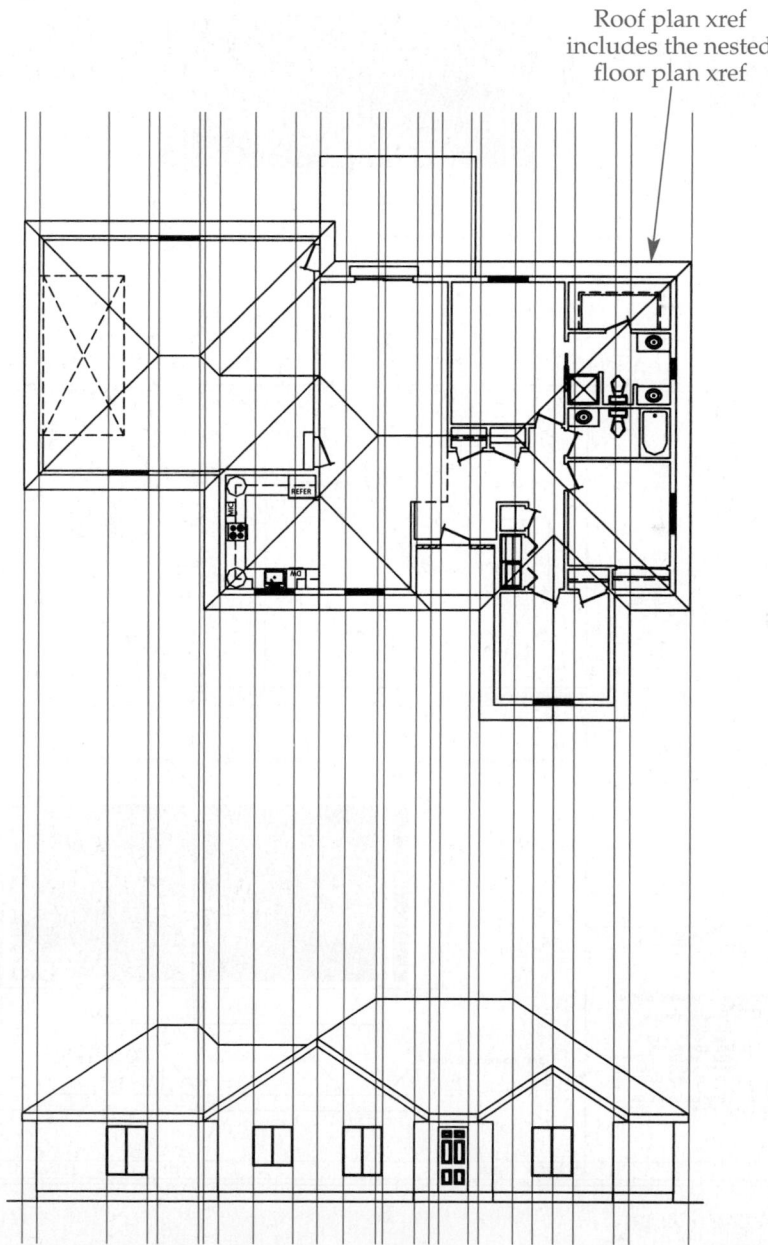

Roof plan xref
includes the nested
floor plan xref

Figure 32-3A shows an example of several stock details referenced into the model space environment of a new drawing. Floating viewports arrange the details in a layout as shown in **Figure 32-3B.** When you make changes to a detail in the original drawing file, all xrefs of the detail update in the files in which the detail is used.

Add Layout Content

Layout content, such as a title block or a long list of general notes, typically uses a standard format. If the format requires modification, such as adding a new note to a list of general notes, you can make changes to the xref drawing and easily update each host file that references the xref. This is the same concept as using xref drawings to build a multiview drawing. The general notes shown in **Figure 32-3B** are added to the layout as an xref.

Figure 32-3.
A—Xref frequently used drawing views into model space, reducing the size of the file and providing the ability to change instances of the view used in multiple drawings. B—Arrange referenced views in floating viewports like other model space objects.

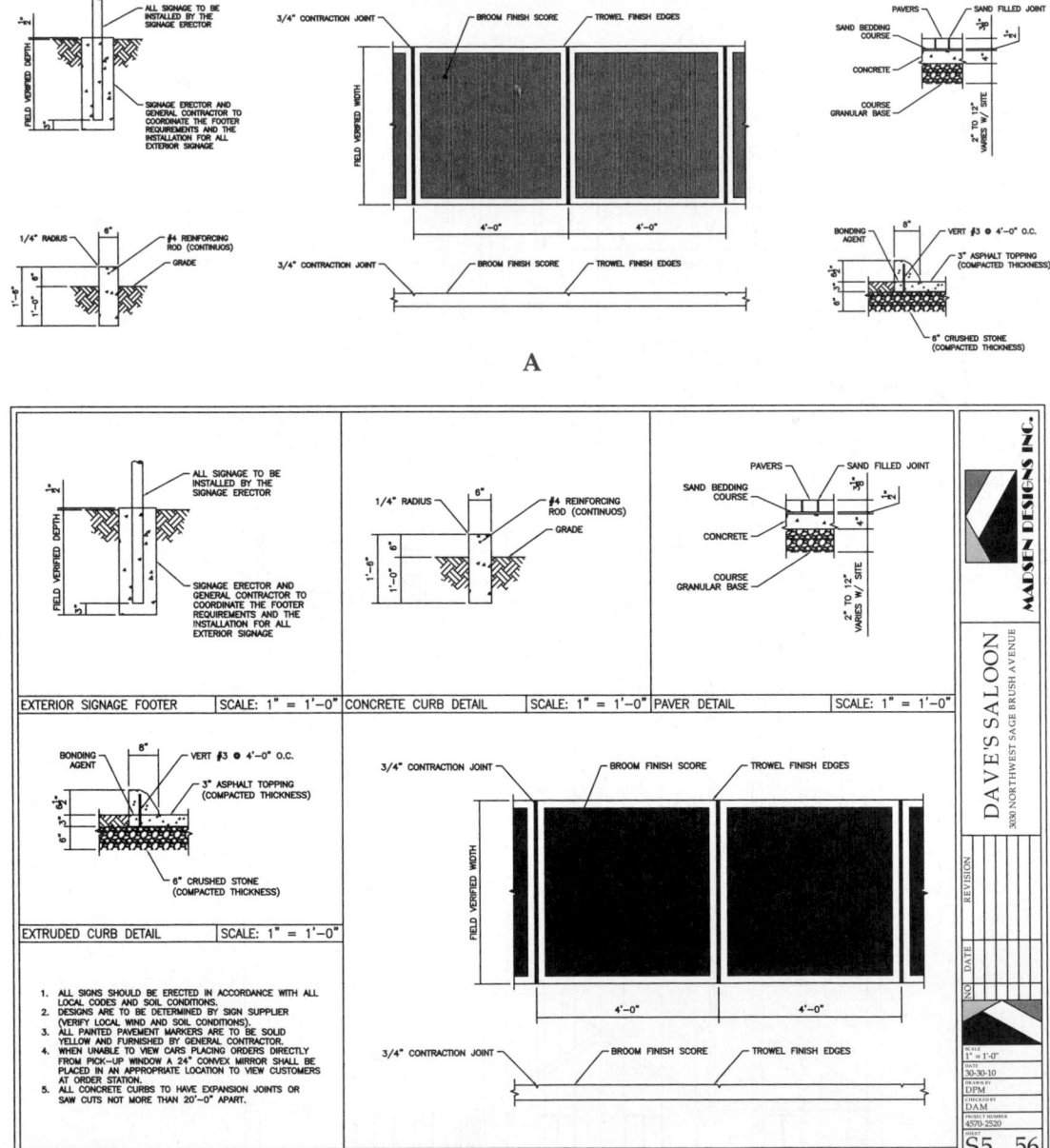

Arrange Sheet Views

You can use external references to arrange sheet views in layouts when you are working with sheet sets. Sheet sets are described in Chapter 33.

Preparing Xref and Host Drawings

Before you begin placing xref drawings, you should prepare the xref and host drawing files for xref insertion. When you place an xref drawing, everything you see in model space is inserted into the host file as a single item. Layout content is not xrefed. The default insertion base point for an xref file is the model space origin, or 0,0,0 point. This is the point attached to the crosshairs or located at the specified insertion point when you insert the xref into the host drawing. If it is critical that xref objects coincide with the 0,0,0 point for insertion, move all objects in model space as needed.

An alternative to moving objects to the origin is to use the **BASE** tool to change the insertion base point of the drawing. Access the **BASE** tool, and then select a new insertion base point. Save the drawing before using it as an xref.

If you use an appropriate template, little preparation should be necessary to prepare the host file to accept an xref. The host file should include a unique layer (named XREF or A-ANNO-REFR, for example) on which to place xrefs. As you will learn, layers in a referenced drawing file remain intact when you add the xref to a host drawing. Therefore, properties and states that you assign to the XREF layer have no effect on xref objects. The main purpose of the XREF layer is to contain the xref on a specific layer. Set the XREF layer current and proceed to place the xref drawing.

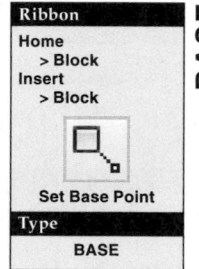

Ribbon

Home
> Block
Insert
> Block

Set Base Point

Type

BASE

BASE

Placing External Reference Drawings

To place an xref, access the **ATTACH** tool to display the **Select Reference File** dialog box. The dialog box is set to display all file types by default. Pick **Drawing (*.dwg)** from the **Files of type:** drop-down list show and reference only drawings. Use the **Select Reference File** dialog box to locate the drawing file to add to the host file as an xref. Then pick the **Open** button to display the **Attach External Reference** dialog box. See **Figure 32-4.**

The **Attach External Reference** dialog box includes options for specifying how and where to place the selected file in the host drawing as an xref. If an external reference already exists in the current drawing, you can place another copy by choosing the file from the **Name:** drop-down. To place a different xref drawing, pick the **Browse...** button and select the new file in the **Select Reference File** dialog box.

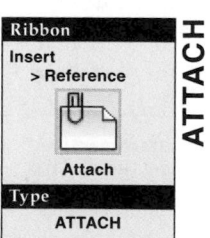

Ribbon

Insert
> Reference

Attach

Type

ATTACH

ATTACH

Figure 32-4.
The **Attach External Reference** dialog box allows you to specify how to place an xref in the host drawing. Pick the **Show Details** button to display additional file details, as shown.

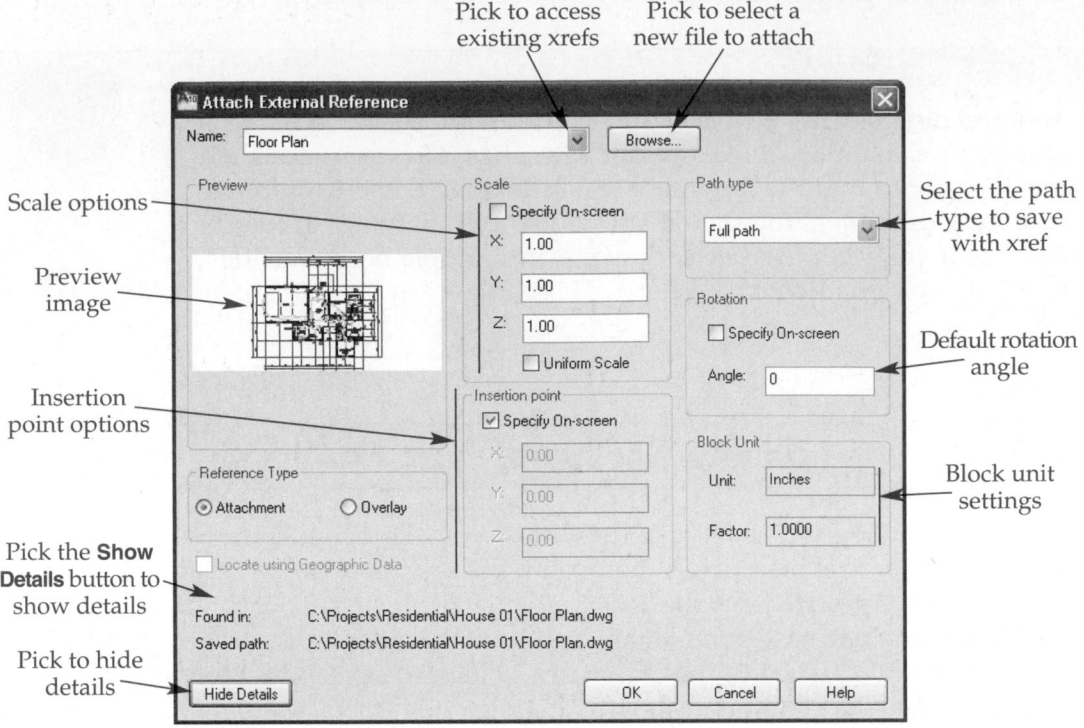

You can also place an xref using the **External References** palette shown in **Figure 32-5.** The **External References** palette is a complete external reference management tool. To place an xref drawing using the **External References** palette, pick the **Attach DWG** button from the **Attach** flyout, or right-click on the **File References** pane and select **Attach DWG...**. The **Select Reference File** dialog box appears, displaying only drawing files. Locate and select a file to add as an xref and pick the **Open** button to display the **Attach External Reference** dialog box.

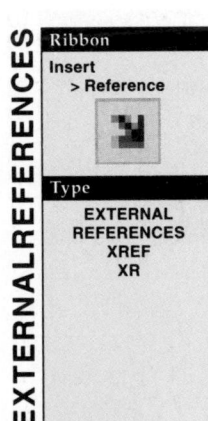

EXTERNALREFERENCES

Ribbon
Insert
> Reference

Type
EXTERNAL
REFERENCES
XREF
XR

XATTACH

Type
XATTACH
XA

> **NOTE**
>
> The **XATTACH** tool is identical to the **ATTACH** tool, but it initially displays only drawing files in the **Select Reference File** dialog box.

Attachment vs. Overlay

You can specify to insert an xref drawing as an *attachment* or an *overlay* by selecting the **Attachment** or **Overlay** radio button in the **Reference Type** area. Attach xrefs for most applications. Most often, an xref overlay allows you to share content with others in a design drafting team, typically while working in a networked environment. In this situation, you can overlay drawings without referencing nested xrefs.

Nesting occurs when an xref file contains, or references, another xref file. An attached xref with nested xrefs is the *parent xref*. When you attach an xref, the host drawing receives any nested xrefs that the xref contains. This does not happen when you overlay an xref. Furthermore, if you overlay an xref in a host drawing and then attach the host drawing to another drawing, the overlaid xref does not appear, even though the host drawing is attached.

For example, you should typically attach a floor plan xref to a host file to create a foundation plan. You can then attach the foundation plan xref to a host file to draw a

attachment: An xref linked with or referenced into the current drawing.

overlay: An xref displayed as an xref without being attached to the current drawing.

parent xref: An xref that contains one or more other xrefs.

Figure 32-5.
The **External References** palette provides access to all options for externally referenced files.

Pick to add an xref
drawing to the host file

Host file

Information
about the
highlighted file

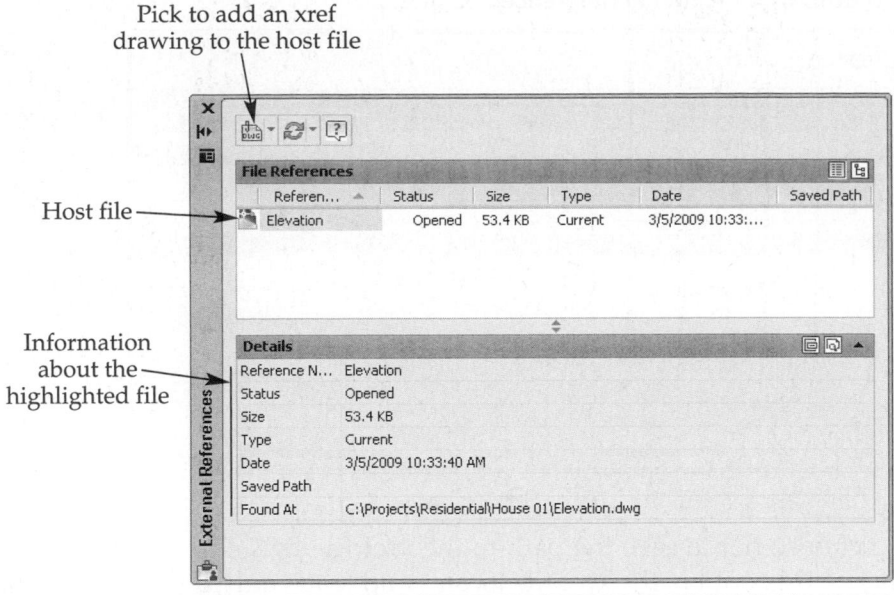

section. Attaching the foundation plan brings the foundation and floor plan geometry into the section file for reference. If a member of your design team uses your section, or is working on a drawing that already has the floor and/or foundation plan attached, she or he can overlay the section xref into a drawing without bringing in the floor plan and foundation plan.

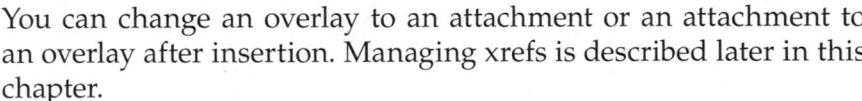 

NOTE

You can change an overlay to an attachment or an attachment to an overlay after insertion. Managing xrefs is described later in this chapter.

Selecting the Path Type

Use the **Path type** drop-down list in the **Path Type** area to set how AutoCAD stores the path to the xref file. The path locates the xref file when you open the host file. The path appears in the **Attach External Reference** dialog box when you pick the **Show Details** button, and later appears in the **External References** palette. See **Figure 32-6.**

The **Full path** option saves an *absolute path* and is active by default. When using this option, you must locate xref drawings in the same drive and folder specified in the saved path. You can move the host drawing to any location, but the xref drawings must remain in the saved path. This option is acceptable if it is unlikely that you will move or copy the host and xref drawings to another computer, drive, or folder.

absolute path: A path to a file defined by the location of the file on the computer system.

If you share your drawings with a client or eventually archive the drawings, the **Relative path** option is often more appropriate. This option saves a *relative path*, which you cannot use if the xref file is on a local or network drive other than the drive that stores the host file. If the host drawing and xref files are located in a single folder and subfolders, you can copy the folder to any location without losing the connection between files. For example, you can copy the folder from the C: drive of one computer to the D: drive of another computer, to a folder on a CD, or to an archive server. If you perform these types of transfers with the **Full path** option, you need to open the host drawing after copying and redefine the saved paths for all xref files.

relative path: A path to a file defined according to its location relative to the host drawing.

Figure 32-6.
You can reference a file using a full path, a relative path, or no path. The path type displays in the **Save Path** column in the **External References** palette.

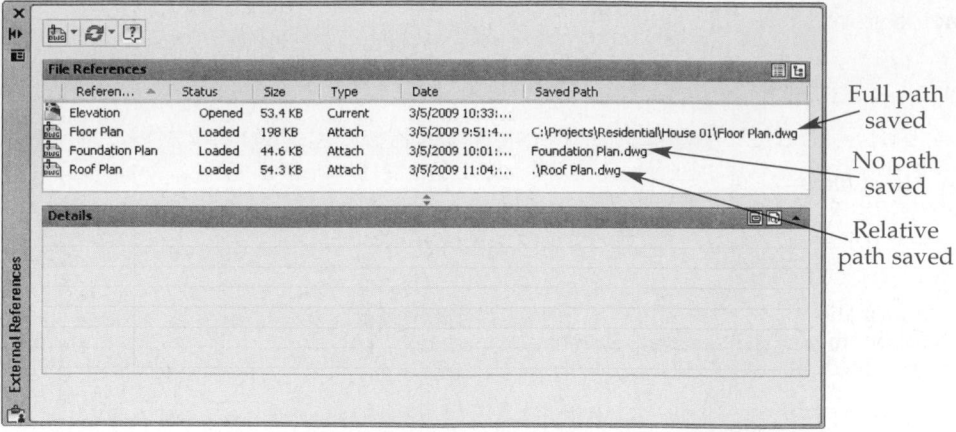

Full path saved

No path saved

Relative path saved

You can also choose not to save the path to the xref file by selecting the **No Path** option. If you choose this option, the xref file locates and loads only if you include the path to the file in one of the Support File Search Path locations or if the xref file is in the same folder as the host file. Specify the Support File Search Path locations in the **Files** tab of the **Options** dialog box.

> **NOTE**
>
> AutoCAD also searches for xref files in all paths of the current project name. These paths are listed under the Project Files Search Path in the **Files** tab of the **Options** dialog box. You can create a new project as follows:
>
> 1. Pick Project Files Search Path to highlight it, and then pick the **Add...** button.
> 2. Enter a project name if desired.
> 3. Pick the plus sign icon (+), and then pick the word Empty.
> 4. Pick the **Browse...** button and locate the folder that is to become part of the project search path. Then pick **OK**.
> 5. Complete the project search path definition by entering the **PROJECTNAME** system variable and specifying the same name used in the **Options** dialog box.

Additional Xref Placement Options

The remaining items in the **Attach External Reference** dialog box allow you to control or identify xref insertion location, scaling, rotation angle, and block unit settings. Use the text boxes in the **Insertion point** area to enter 2D or 3D coordinates for insertion of the xref if the **Specify On-screen** check box is deselected. Activate the **Specify On-screen** check box to specify the insertion location on-screen. The **Locate using Geographic Data** check box is active if the xref and host drawings include *geographic data*. Pick the check box to position the xref using geographic data.

Use the **Scale** area to set xref scale factor. By default, the X, Y, and Z scale factors are set to 1. Enter different values in the corresponding text boxes or activate the **Specify On-screen** check box to display scaling prompts when you insert the xref. Select the **Uniform Scale** check box to apply the X scale factor to the Y and Z scale factors.

geographic data: Information added to a drawing to describe specific locations and directions on Earth.

AutoCAD and Its Applications—Basics

The rotation angle for the inserted xref is 0 by default. Specify a different rotation angle in the **Angle:** text box, or select the **Specify On-screen** check box to display a rotation prompt for the rotation angle. The **Block Unit** area displays the unit and scale factor stored with the selected drawing file.

Inserting the Xref

After adjusting xref specifications in the **Attach External Reference** dialog box, pick the **OK** button to insert the xref into the host drawing. If you chose the **Specify On-screen** check box in the **Insertion point** area, the xref attaches to the crosshairs and a prompt asks for the insertion point. Specify an appropriate insertion point for the xref.

The options for attaching an xref are essentially the same as those used to insert a block. However, remember that xrefs are not added to the database of the host drawing file, as are inserted blocks. Therefore, using external references helps keep your drawing file size to a minimum.

Exercise 32-1

Access the Student Web site (www.g-wlearning.com/CAD) and complete Exercise 32-1.

Placing Xrefs with DesignCenter and Tool Palettes

To place an xref into the current drawing using **DesignCenter**, first use the **Tree View** area to locate the folder containing the drawing to attach. Then display the drawing files located in the selected folder in the **Content** area. Right-click on the drawing file in the **Content** area and select **Attach as Xref...**. Another method is to drag and drop the drawing into the current drawing area using the *right* mouse button. When you release the button, select the **Attach as Xref...** option. The **Attach External Reference** dialog box appears. Enter the appropriate values and pick the **OK** button to place the xref.

You must add an xref or drawing file to a tool palette in order to use the **Tool Palettes** palette to place the file as an xref. To add an xref to a tool palette, drag an existing xref from the current drawing or an xref from the **Content** area of **DesignCenter** into the **Tool Palettes** palette. Use drag and drop to attach the xref to the current drawing from the palette.

Xref files in tool palettes receive an external reference icon. A drawing file, not an xref, added to a tool palette from the current drawing or using **DesignCenter**, is designated as a block tool. To convert the block tool to an xref tool, right-click on the image in the **Tool Palettes** palette and select **Properties...** to display the **Tool Properties** dialog box. Then change the **Insert as** field status from Block to Xref using the **Insert as** drop-down list. The **Reference type** row controls whether the xref is inserted as an attachment or an overlay.

Working with Xref Objects

An xref is inserted as a single object. You can use editing tools such as **MOVE** and **COPY** to modify the xref as needed. However, there are some significant differences between xrefs and other objects. For example, if you erase an xref, the xref definition remains in the file, similar to an erased block. You must *detach* an xref to remove the xref from the file completely.

Xref drawings appear faded by default to help differentiate the xref from the host drawing. Xref fading is an on-screen display function only. Xrefs do not plot faded. The **XDWGFADECTL** system variable controls fading of xref drawings on-screen. The easiest way to adjust fading is to use the options in the expanded **Reference** panel of the **Insert** ribbon tab. See **Figure 32-7A**. Pick the **Xref Fading** button to activate or deactivate xref fading, and use the slider or text box to increase or decrease fading. The default value of 70% creates significant fading. See **Figure 32-7B**.

> **NOTE**
>
> In order to select an xref, you must pick an object displayed on-screen that is part of the xref.

Dependent Objects

dependent objects: Objects displayed in the host drawing, but defined in the xref drawing.

When you place an xref in a drawing, the host file receives all named objects in the xref file, such as layers and blocks, as *dependent objects*, even if the xref file does not use the objects. Dependent objects display in the host drawing for reference only. The xref drawing stores the actual object definitions.

Dependent objects are renamed when you attach an xref so that the xref file name precedes the actual object name, with a vertical bar symbol (|) separating the names. For example, a layer named A-DOOR within an xref drawing file named Floor Plan comes into the host drawing as Floor plan|A-DOOR. See **Figure 32-8**. The unique name distinguishes xref-dependent layers from layers that may have the same name in the host drawing. This also makes it easier to manage layers with several xrefs attached to the host drawing, because file names prefix the layers from each reference file. You cannot rename xref-dependent objects.

When you attach an xref, dependent objects such as layers are added to the host drawing only in order to support the display of the objects in the xref file. You cannot set xref layers current, and as a result, you cannot draw on xref layers. However, you can turn them on and off, thaw and freeze them, and lock or unlock them as needed. You can also change the colors and linetypes of xref layers.

Figure 32-7.
A—Use options in the ribbon to control xref fading.
B—Default fading applied to xref objects.

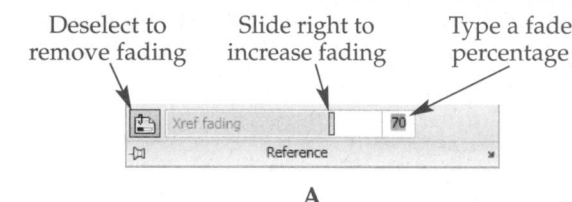

A

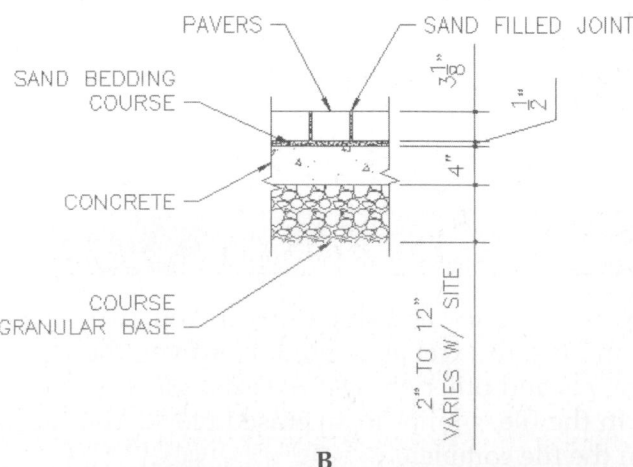

B

Figure 32-8.
The xref drawing name and a vertical bar symbol (|) precede xref-dependent layer names in the host drawing.

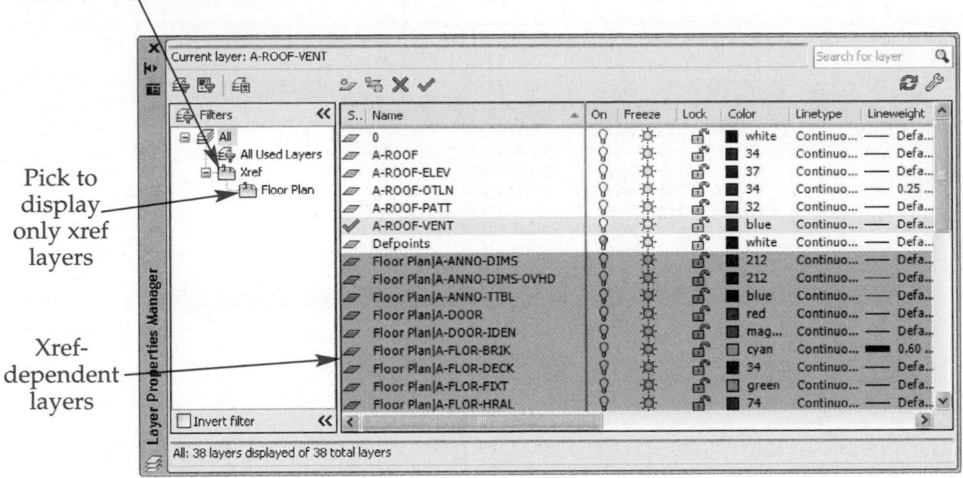

Filters available by default
when you place an xref

Pick to display only xref layers

Xref-dependent layers

NOTE

Use the xref filter in the **Layer Properties Manager** palette to display and manage dependent layers. You can also save dependent layers in a layer state.

PROFESSIONAL TIP

When you attach a drawing as an xref, the reference file comes into the host drawing with the same layer colors and linetypes used in the original file. If you reference a drawing to check the relationship of objects between two drawings, it is a good idea to change the xref layer colors to make it easier to differentiate between the content of the host drawing and the xref drawing. Changing xref layer colors affects only the display in the current drawing and does not alter the original reference file.

Exercise 32-2

Access the Student Web site (www.g-wlearning.com/CAD) and complete Exercise 32-2.

Managing Xrefs

The **External References** palette is the primary tool for managing and accessing current information about xrefs found in a host drawing. The **External References** palette displays an upper **File References** pane and a lower **Details** pane. See **Figure 32-9.** You can display the **File References** pane in list view or tree view and with details or a preview.

Figure 32-9.
The **External References** palette allows you to view and manage referenced files. The **File References** pane appears in **List View** mode and the **Details** pane displays in **Details** mode.

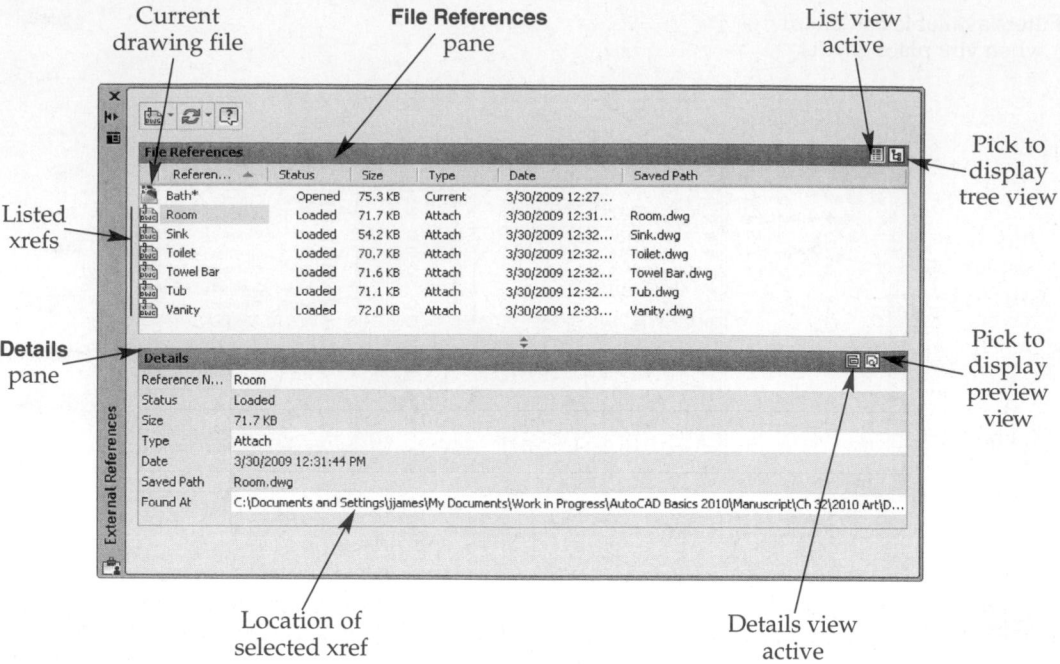

Current drawing file

File References pane

List view active

Listed xrefs

Pick to display tree view

Details pane

Pick to display preview view

Location of selected xref

Details view active

List View Display

The list view display shown in **Figure 32-9** is active by default. Pick the **List View** button or press the [F3] key to activate list view mode while in tree view mode. The labeled columns displayed in list view provide information about and management options for xrefs.

The **Reference Name** column displays the current drawing file name followed by the names of all existing xrefs in alphabetical or chronological order. The standard AutoCAD drawing file icon identifies the host drawing, and a sheet of paper with a paper clip icon identifies xref drawings. Each xref type displays a unique icon. The **Status** column describes the status of each xref, which can be:

- **Loaded.** The xref is attached to the drawing.
- **Unloaded.** The xref is attached but does not display or regenerate.
- **Unreferenced.** The xref has nested xrefs that are not found or are unresolved. An unreferenced xref does not display.
- **Not Found.** The xref file is not found in the specified search paths.
- **Unresolved.** The xref file is missing or cannot be found.
- **Orphaned.** The parent of the nested xref cannot be found.

The **Size** column lists the file size for each xref. The **Type** column indicates whether the xref is attached or referenced as an overlay. The **Date** column indicates the date the xref was last modified.

The **Saved Path** column lists the path name saved with the xref. If only a file name appears, the path is not saved. Prefixes describe the relative paths to xref files. In **Figure 32-6,** the characters .\ precede the Roof Plan reference file. The period (.) represents the folder containing the host drawing. From that folder, AutoCAD looks in the House 01 folder that contains the Roof Plan drawing. The Elevation reference file in **Figure 32-10** uses a similar specification. In this example, the same folder contains the Elevation xref and the host drawing. The characters ..\ precede the specification for the Wall xref. The double period instructs AutoCAD to move up one folder level from the current location. The double period repeats to move up multiple folder levels. For example, AutoCAD locates the Panel xref in **Figure 32-10** by moving up two folder levels from the folder of the host drawing and opening the Symbols folder.

Figure 32-10.
Relationship between the symbols in the **Saved Path** list and file locations within the folder structure.

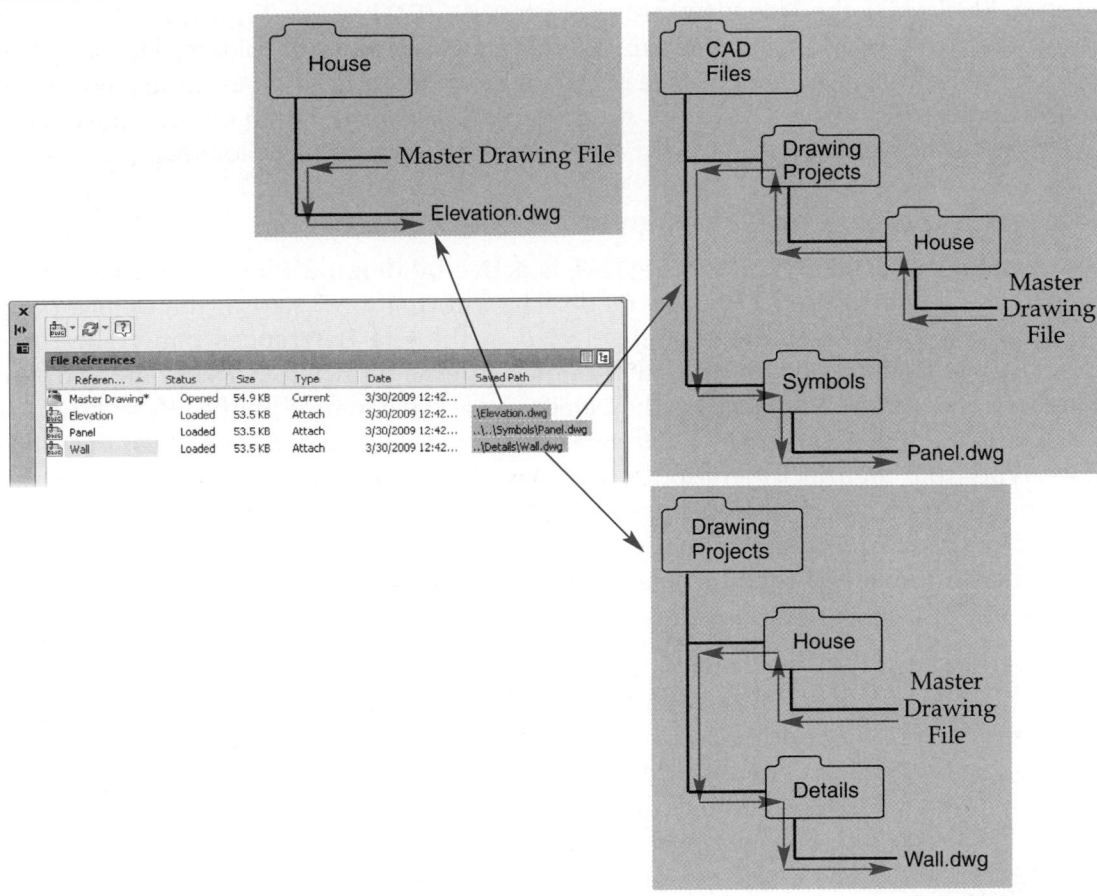

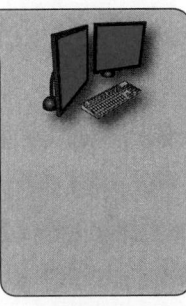

Chapter 32   External References

Tree View Display

To see a list of externally referenced files in the **File References** pane, and show nesting levels, pick the **Tree View** button or press the [F4] key. See **Figure 32-11.** Nesting levels are displayed in a format similar to the arrangement of folders. The status of the xref determines the appearance of the xref icon. An xref with an unloaded or not found status has a grayed-out icon. An upward arrow shown with the icon means the xref was reloaded, and a downward arrow means the xref was unloaded.

Viewing Details or a Preview

Details mode, shown in **Figure 32-9,** is active by default. Pick the **Details** button to display details while in **Preview** mode. The information listed in the **Details** pane corresponds to the host file or xref selected in the **File References** pane. The rows displayed in **Detail** mode are the same as the columns found in **List View** mode of the **File References** pane. However, in the **Details** pane, you can modify the reference name by entering a new name in the **Reference Name** text box. You can also adjust the reference type from an attachment to an overlay or from an overlay to an attachment by picking the appropriate option from the **Type** drop-down list. In addition, the **Details** pane contains a **Found At** row that you can use to update the location of an xref path. To display an image of the xref selected in the **File References** pane, pick the **Preview** button on the **Details** pane. Refer again to **Figure 32-11.**

Detaching, Reloading, and Unloading Xrefs

detach: Remove an xref from a host drawing.

Each time you open a host drawing containing an attached xref, the xref loads and appears on-screen. This association remains permanent until you *detach* the xref. Erasing an xref does not remove the xref from the host drawing. To detach an xref, right-click on the reference name in the **File References** pane of the **External References** palette and pick **Detach**. All instances of the xref and all nested xrefs are removed from the current drawing, along with all referenced data.

reload: Update an xref in the host drawing.

In some situations, you may need to update, or *reload,* an xref file in the host drawing. For example, if you edit an xref while the host drawing is open, the updated version may be different from the version you see. To update the xref, right-click the

Figure 32-11.
The **File References** pane in **Tree View** mode shows nested xref levels. The **Details** pane in **Preview** mode shows a thumbnail preview of the selected xref.

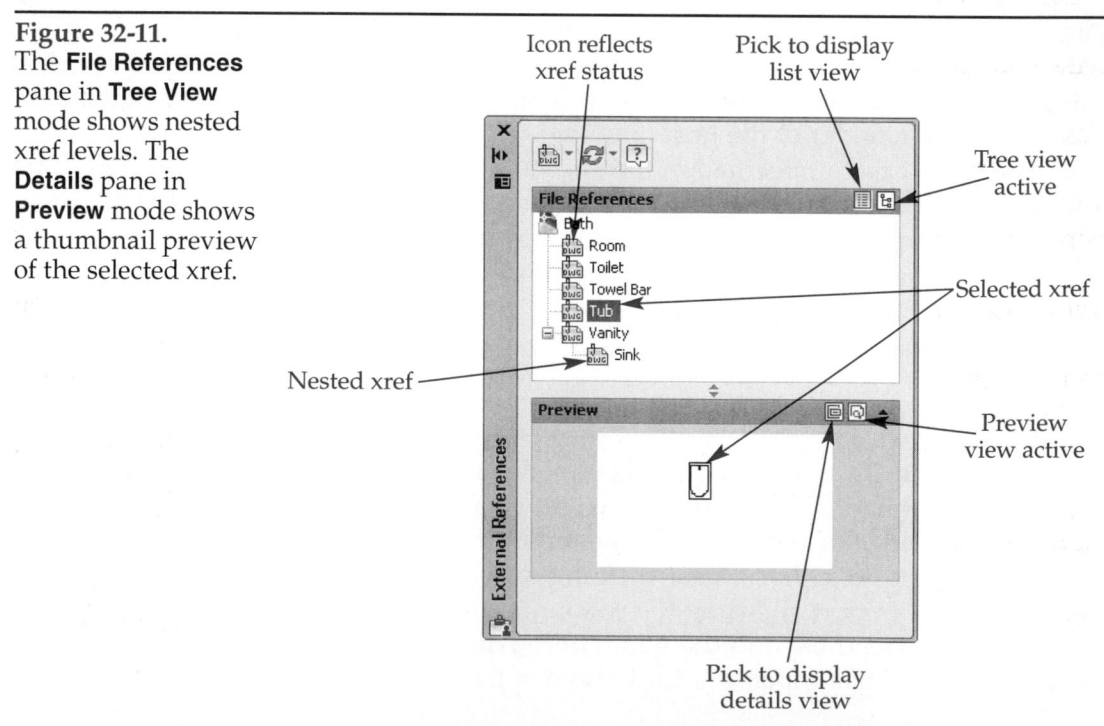

AutoCAD and Its Applications—Basics

reference name in the **File References** pane of the **External References** palette and pick **Reload**, or pick the **Reload All References** button from the flyout to reload all unloaded xrefs. Reloading xrefs forces AutoCAD to read and display the most recently saved version of each xref.

To *unload* an xref, right-click on the reference name in the **File References** pane of the **External References** palette and pick **Unload**. An unloaded xref does not display or regenerate, increasing performance. Reload the xref to redisplay it.

unload: Suppress the display of an xref without removing it from the host drawing.

> **NOTE**
>
> If AutoCAD cannot find an xref, an alert appears when you open the host drawing. Choose the appropriate option to ignore the problem or fix the problem using the **External References** palette.

Updating the Xref Path

A file path saved with an xref is displayed in the **Saved Path** column of the **File References** pane and the **Saved Path** row of the **Details** pane in the **External References** palette. If the **Saved Path** location does not include an xref file, when you open the host drawing, AutoCAD searches along the *library path*. A link to the xref forms if a file with a matching name is found. In such a case, the **Saved Path** location differs from where the file was actually found.

library path: The path AutoCAD searches by default to find an xref file, including the current folder and locations set in the **Options** dialog box.

Check for matching paths in the **External References** palette by comparing the path listed in the **Saved Path** column of the **File References** pane and **Saved Path** row of the **Details** pane with the listing in the **Found At** row of the **Details** pane. When you move an xref and the new location is not in the library path, the xref status is Not Found. To update or find the **Saved Path** location, select the path in the **Found At** edit box and pick the **Browse...** button to the right of the edit box to access the **Select new path** dialog box. Use this dialog box to locate the new folder and select the desired file. Then pick the **Open** button to update the path.

The Manage Xrefs Icon

By default, when you edit, save, and close an xref, and then open the host drawing, changes made to the xref automatically appear without any notification. When you make changes to an xref while the host drawing is open, a notification appears in the status bar tray. Changes indicate by the appearance of the **Manage Xrefs** icon, a balloon message, or both.

The **Tray Settings** dialog box controls notifications in the status bar tray for xref changes and other system updates. Select **Tray Settings...** from the status bar shortcut menu to access the **Tray Settings** dialog box. Select the **Display icons from services** check box to display the **Manage Xrefs** icon in the status bar tray when you attach an xref to the current drawing. If you modified an xref in the current file since opening the file, the **Manage Xrefs** icon appears with an exclamation sign. Pick the **Manage Xrefs** icon or right-click on the **Manage Xrefs** icon and select **External References...** to open the **External References** palette to reload the xref.

Select the **Display notifications from services** check box in the **Tray Settings** dialog box to display a balloon message notification with the name of the modified xref file. See **Figure 32-12A**. You can then pick the xref name in the balloon message to reload the file. The example in **Figure 32-12** shows adding a Towel Bar xref to the Room parent xref drawing. The xref is then reloaded in the host drawing named Bath. See **Figure 32-12B**. You can also reload xrefs by right-clicking on the **Manage Xrefs** icon and selecting **Reload DWG Xrefs**.

Figure 32-12.
The **Manage Xrefs** icon in the AutoCAD status bar tray provides a notification when you modify and save an xref file. A—A balloon message appears with an exclamation point over the icon. B—Reloading the xref file updates the current drawing and changes the appearance of the icon.

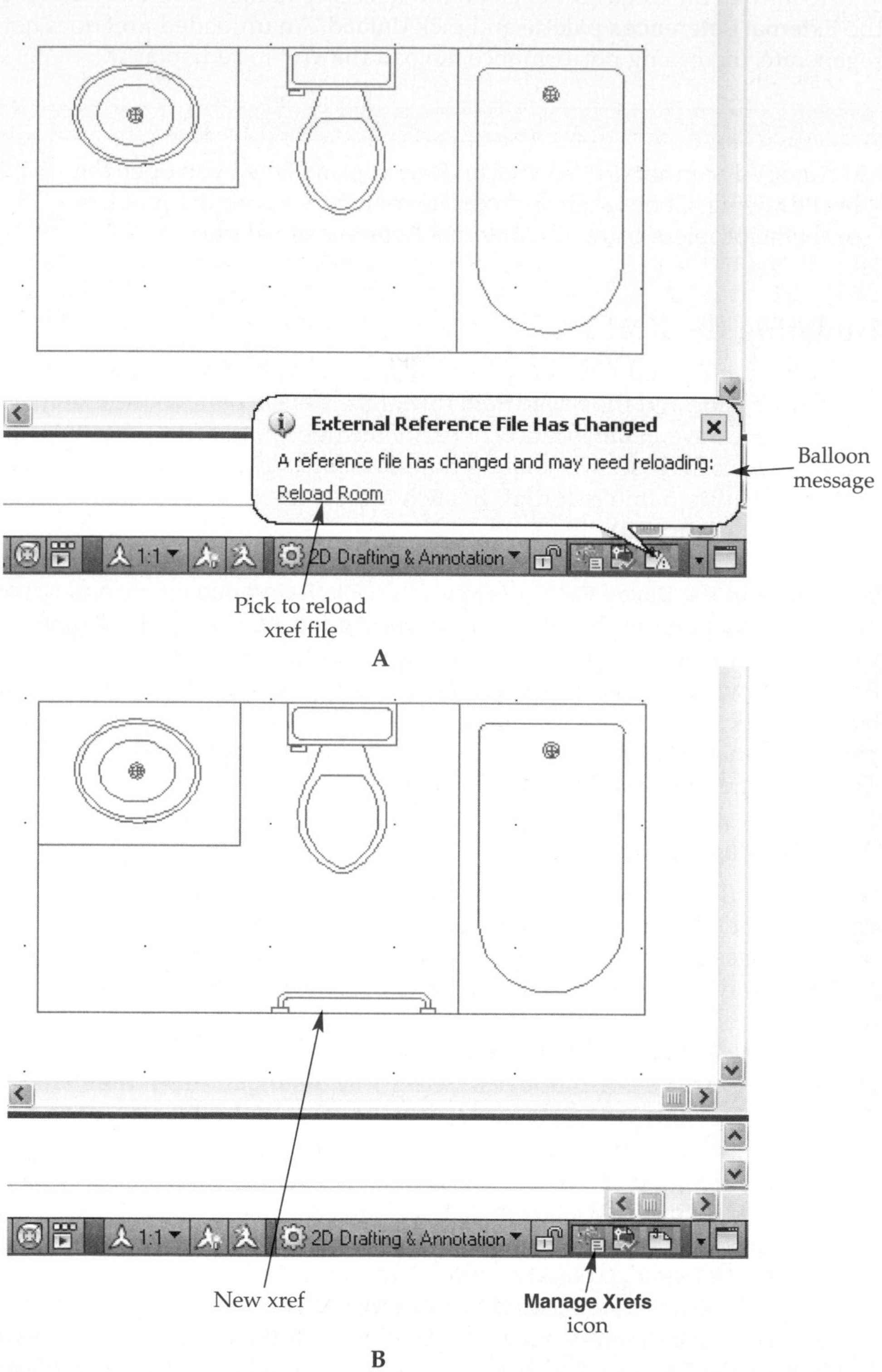

External Reference File Has Changed

A reference file has changed and may need reloading:

Reload Room

Balloon message

Pick to reload xref file

A

New xref

Manage Xrefs icon

B

Exercise 32-3

Access the Student Web site (www.g-wlearning.com/CAD) and complete Exercise 32-3.

Clipping Xrefs

To display only a specific portion of an xref, AutoCAD allows you to create a boundary that displays an xref *subregion*. All geometry that falls outside the boundary is invisible, and objects that are partially within the subregion appear trimmed at the boundary. Although clipped objects appear trimmed, the xref file does not change. Clipping applies to a selected instance of an xref, not to the actual xref definition.

Use the **XCLIP** tool to create and modify clipping boundaries. A quick way to access the **XCLIP** tool is to select an object that is part of the xref file, then right-click and select **Clip Xref**. If you access the **XCLIP** tool while the xref is not selected, pick an object associated with the xref to clip. Then press [Enter] to accept the default **New boundary** option and select the clipping boundary.

When you select the **New boundary** option, a prompt asks you to specify the clipping boundary. Use the default **Rectangular** option to create a rectangular boundary. Then pick the corners of the rectangular boundary. **Figure 32-13** shows an example of using the **XCLIP** tool and a rectangular boundary. Note that the geometry outside the clipping boundary is no longer displayed after the clip is completed. The **New boundary** option includes additional options for specifying the clip boundary and area to clip, as briefly described in **Figure 32-14.**

> **subregion:** The displayed portion of a clipped xref.

Ribbon

Insert
> Reference

Clip

Type

XCLIP
XC

XCLIP

Figure 32-13.
A clipping boundary allows you to clip selected areas of an xref. A—Using the **Rectangular** boundary selection option. B—The clipped xref.

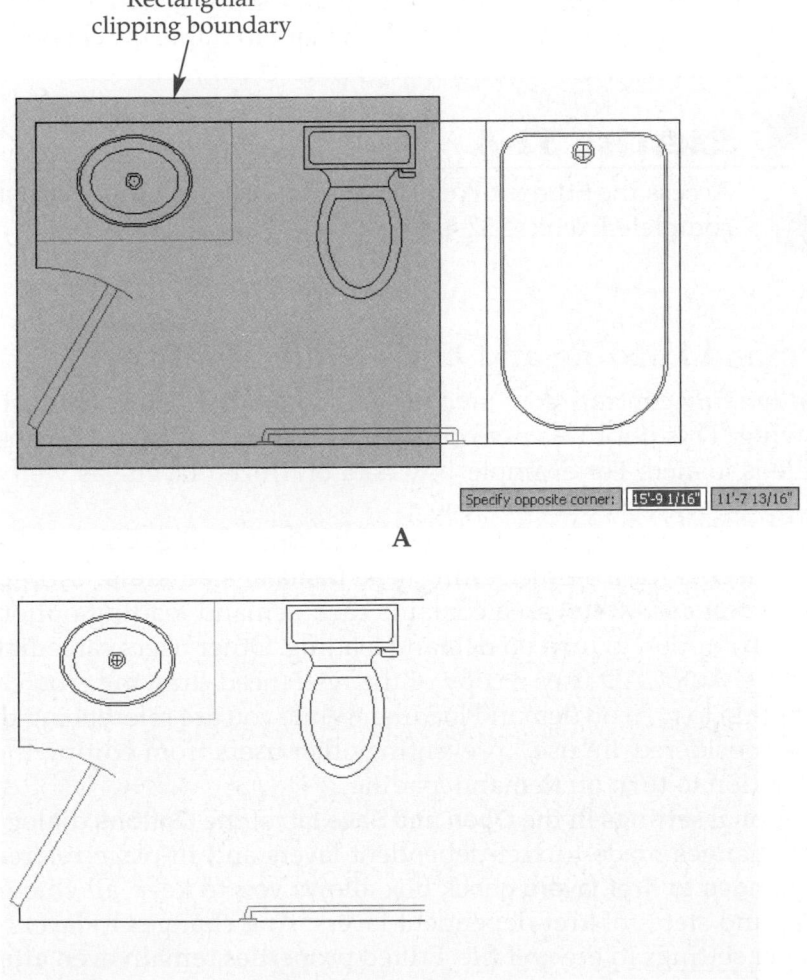

Rectangular clipping boundary

Specify opposite corner: 15'-9 1/16" 11'-7 13/16"

A

B

Figure 32-14.
Additional options available when using the **New boundary** function of the **XCLIP** tool.

Option	Description
Select Polyline	Select an existing polyline object as the clip boundary. If the polyline does not close, the start and endpoints of the boundary connect.
Polygonal	Draw an irregular polygon as a boundary.
Invert clip	Inverts the selection so that the portion of the xref that lies outside of the clipping boundary is clipped. Only the portion of the xref outside of the boundary is displayed.

You can edit a clipped xref as you would an unclipped xref. The clipping boundary moves with the xref. Note that nested xrefs are clipped according to the clipping boundary for the parent xref.

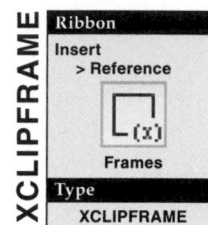
The clipping boundary, or frame, is invisible by default. Use the **XCLIPFRAME** system variable to toggle the display of the clipping boundary frame. Set the value of **XCLIPFRAME** to 1 to turn on the frame.

The other options of the **XCLIP** tool function after a clip boundary has been defined. The **ON** and **OFF** options turn the clipping feature on or off as needed. The **Clipdepth** option allows you to define front and back clipping planes to control the portion of a 3D drawing that displays. Clipping 3D models is described in *AutoCAD and Its Applications—Advanced*. Use the **Delete** option to remove an existing clipping boundary, returning the xref to its unclipped display. Use the **generate Polyline** option to create and display a polyline object at the clip boundary to frame the clipped portion.

Exercise 32-4

Access the Student Web site (www.g-wlearning.com/CAD) and complete Exercise 32-4.

Using Demand Loading and Xref Editing Controls

demand loading:
Loading only the part of an xref file necessary to regenerate the host drawing.

Demand loading controls how much of an xref loads when you attach the xref to the host drawing. This improves performance and saves disk space because only part of the xref file is loaded. For example, any data on frozen layers, as well as any data outside of clipping regions, does not load.

Demand loading occurs by default. Use the **Open and Save** tab of the **Options** dialog box to check or change the setting. The **Demand load Xrefs:** drop-down list in the **External References (Xrefs)** area contains each demand loading option. Select the **Enabled with copy** option to turn on demand loading. Other users can edit the original drawing because AutoCAD uses a copy of the referenced drawing. You can also pick the **Enabled** option to turn on demand loading. While you are referencing the drawing, the xref file is considered "in use," preventing other users from editing the file. Select the **Disabled** option to turn off demand loading.

Two additional settings in the **Open and Save** tab of the **Options** dialog box control the effects of changes made to xref-dependent layers and in-place reference editing. The **Retain changes to Xref layers** check box allows you to keep all changes made to the properties and states of xref-dependent layers. Any changes to layers take precedence over layer settings in the xref file. Edited properties remain even after the xref is reloaded. The **Allow other users to Refedit current drawing** check box controls whether the current drawing can be edited in place by others while it is open and when it is referenced by another file. Both check boxes are selected by default.

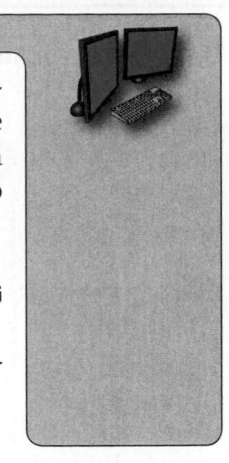

If you plan to use a drawing as an external reference, it is good practice to save the file with *spatial indexes* and *layer indexes*. These lists help improve performance when you reference drawings with frozen layers and clipping boundaries. Use the following procedure to create spatial and layer indexes:

1. Access the **Save Drawing As** dialog box.
2. Pick **Options...** from the **Tools** flyout button and select the **DWG Options** tab of the **Saveas Options** dialog box.
3. Select the type of index required from the **Index type:** drop-down list.
4. Pick the **OK** button and save the drawing.

spatial index: A list of objects ordered according to their locations in 3D space.

layer index: A list of objects ordered according to the layers on which they reside.

Binding an Xref

You can make an xref a permanent part of the host drawing as if you were inserting the file using the **INSERT** tool. This is called *binding* an xref. Binding is useful when you need to send the full drawing file to another location or user, such as a plotting service or client. To bind an xref using the **External References** palette, right-click on the reference name and pick **Bind...**. This displays the **Bind Xrefs** dialog box, which contains **Bind** and **Insert** radio buttons.

binding: Converting an xref to a permanently inserted block in the host drawing.

Using the Insert and Bind Options

The **Insert** option converts the xref into a normal block, as if you had used the **INSERT** tool to place the file. In addition, the drawing is entered into the block definition table, and all named objects, such as layers, blocks, and styles, are incorporated into the host drawing as named in the xref. For example, if you bind an xref file named PLATE that contains a layer named OBJECT, the xref-dependent layer PLATE|OBJECT becomes the locally defined layer OBJECT. All other xref-dependent objects lose the xref name and assume the properties of the locally defined objects with the same name. The **Insert** binding option provides the best results for most purposes.

The **Bind** option also converts the xref into a normal block. However, the xref name remains with all dependent objects. Two dollar signs with a number in between replace the vertical line in each name. For example, an xref layer named Title|Notes becomes Title0Notes. The number inside the dollar signs automatically increments if a local object definition with the same name exists. For example, if Title0Notes already exists in the drawing, the layer is renamed to Title1Notes. In this manner, all xref-dependent object definitions that are bound receive unique names. Rename named objects using the **RENAME** tool.

Binding Specific Dependent Objects

Binding an xref allows you to make all dependent objects in an xref file a permanent part of the host drawing. Dependent objects include named items such as blocks, dimension styles, layers, linetypes, and text styles. Before binding, you cannot directly use any dependent objects from a referenced drawing in the host drawing. For example, you cannot make an xref layer or text style current in the host drawing.

In some cases, you may only need to incorporate one or more specific named objects, such as a layer or block, from an xref into the host drawing, instead of binding the entire xref. If you only need selected items, it can be counterproductive to bind an entire drawing. Instead, use the **XBIND** tool and corresponding **Xbind** dialog box, shown in **Figure 32-15,** to select specific named objects to bind.

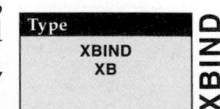

Type
XBIND
XB

XBIND

Reference drawing

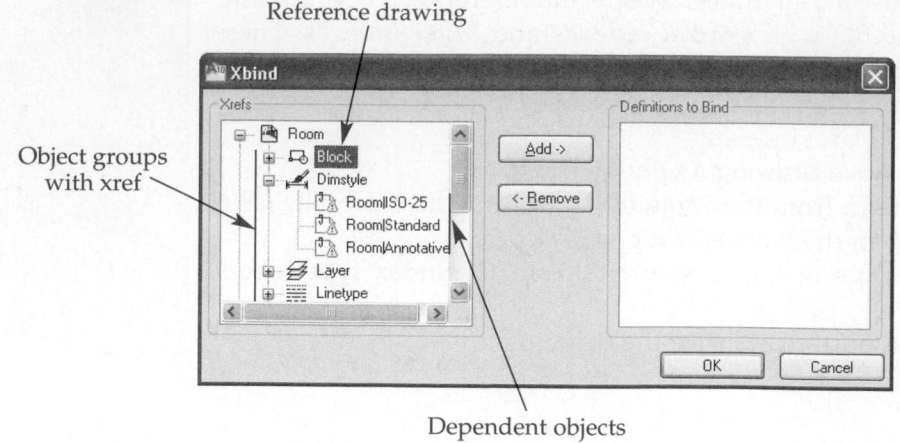

Object groups
with xref

Dependent objects

Xrefs are displayed using AutoCAD drawing file icons. Expand a group to select an individually named object from the corresponding group. To select an object for binding, highlight it and pick the **Add** button. The names of all objects selected and added display in the **Definitions to Bind** list. Pick the **OK** button to complete the operation. A message displayed on the command line indicates how many objects of each type are bound.

Individual objects bound using the **XBIND** tool are renamed in the same manner as objects bound using the **Bind** option in the **Bind Xrefs** dialog box. An automatic linetype bind performs so that a layer that includes a linetype not loaded in the host drawing can reference the required linetype definition. The linetype includes a new linetype name, such as xref1$0$hidden. In a similar manner, a previously undefined block may automatically bind to the host drawing because of binding nested blocks. Rename bound objects using the **RENAME** tool.

Exercise 32-5

Access the Student Web site (www.g-wlearning.com/CAD) and complete Exercise 32-5.

Editing Xref Drawings

reference editing:
Editing reference
drawings from within
the host file.

One option for editing an xref drawing is to use in-place, or *reference editing,* within the host drawing. You can save any changes made to the xref to the original drawing from within the host drawing. Alternatively, you can edit the xref in a separate drawing window as you would any other drawing file.

Reference Editing

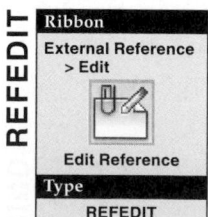

REFEDIT

Ribbon
**External Reference
> Edit**

Edit Reference

Type
REFEDIT

The **REFEDIT** tool allows you to edit xref drawings in place. A quick way to initiate reference editing is to double-click on an xref, or select an xref and then right-click and select **Edit Xref In-place**. If you access the **REFEDIT** tool without first selecting an xref, you must then pick the xref to edit. The **Reference Edit** dialog box opens with the **Identify Reference** tab active. See **Figure 32-16.** The example shows the Room reference

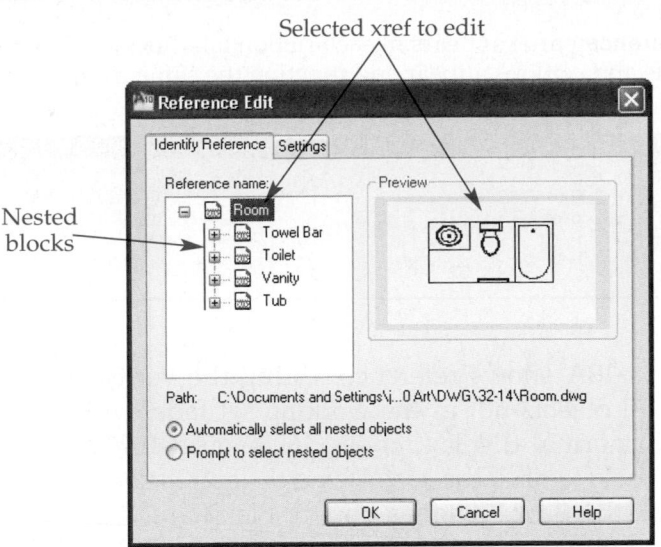

Figure 32-16.
The **Reference Edit** dialog box lists the name of the selected reference drawing and displays an image preview.

Selected xref to edit

Nested blocks

drawing selected for editing. Notice that nested blocks, like the Bath Tub 26 x 60 in. block found in the Tub reference, are listed under the parent xref.

The **Automatically select all nested objects** radio button in the **Path:** area is selected by default. Use this option to make all xref objects available for editing. To edit specific xref objects, pick the **Prompt to select nested objects** radio button. The Select nested objects: prompt displays after you pick the **OK** button, allowing you to pick objects that belong to the previously selected xref. Pick all geometry to edit and press [Enter]. The nested objects you select make up the *working set*. If multiple instances of the same xref appear, be sure to pick objects from the original xref you select.

Additional options for reference editing are available in the **Settings** tab of the **Reference Edit** dialog box. The **Create unique layer, style, and block names** option controls the naming of selected layers and *extracted* objects. Check the box to assign the prefix n, with *n* representing an incremental number, to object names. This is similar to the renaming method used when you bind an xref.

The **Display attribute definitions for editing** option is available if you select a block object in the **Identify Reference** tab of the **Reference Edit** dialog box. Check the box to edit any attribute definitions included in the reference. To prevent accidental changes to objects that do not belong to the working set, check the **Lock objects not in working set** option. This makes all objects outside of the working set unavailable for selection in reference editing mode.

If the selected xref file contains other references, the **Reference name:** area lists all nested xrefs and blocks in tree view. In the example given, Toilet, Tub, Vanity, and Towel Bar are nested xrefs in the Room xref. If you pick the drawing file icon next to Vanity in the tree view, for example, an image preview appears and the selected xref becomes highlighted in the drawing window.

When you finish adjusting settings, pick the **OK** button to begin editing the xref. The primary difference between the drawing and reference editing environments is the **Edit Reference** panel that appears in each ribbon tab. See **Figure 32-17.** Use the tools in the **Edit Reference** panel to add objects to the working set, remove objects from the working set, and save or discard changes to the original xref file.

Any object drawn during the in-place edit is automatically added to the working set. Use the **Add to Working Set** button to add existing objects to the working set. When you add an object to the working set, it is extracted, or removed, from the host drawing. The **Remove from Working Set** button allows you to remove selected objects from the working set. Removing a previously extracted object adds the object back to the host drawing.

working set: Nested objects selected for editing during a **REFEDIT** operation.

extracted: Temporarily removed from the drawing for editing purposes.

Figure 32-17.
The **Edit Reference** panel appears in each ribbon tab during reference editing. Most other drawing tools and options appear and function the same in the drawing and reference editing environments.

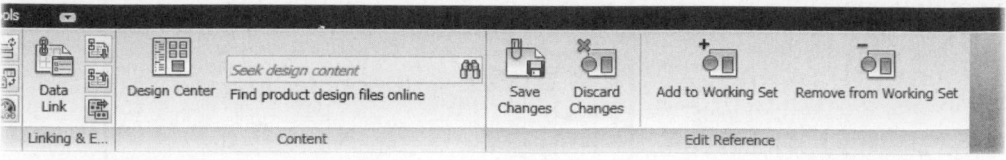

Figure 32-18A shows reference-editing the Vanity xref nested in the Room xref. Notice that all objects not in the working set fade, while the objects in the working set appear in normal display mode. Once you define the working set, use drawing and editing tools to alter the xref. When the edit is complete and you want to save the changes, pick the **Save Changes** button. Pick the **OK** button when AutoCAD asks if you want to continue with the save and redefine the xref. All instances of the xref are updated. **Figure 32-18B** shows the xref after editing to redesign the sink and add a faucet. To exit reference editing without saving changes, pick the **Discard Changes** button.

Figure 32-18.
Reference editing. A—Objects in the drawing that are not a part of the working set are grayed out during the reference-editing session. B—All instances of the xref immediately update after reference editing.

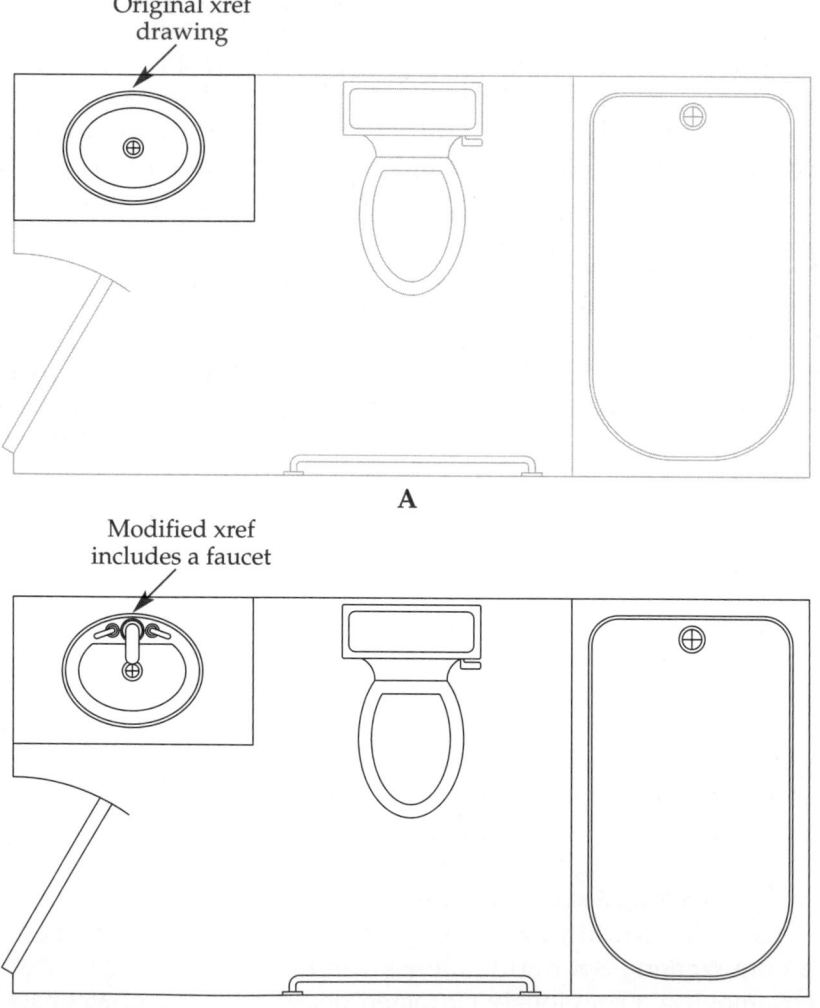

Original xref drawing

A

Modified xref includes a faucet

B

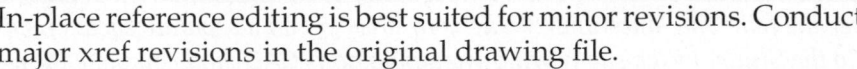 

In-place reference editing is best suited for minor revisions. Conduct major xref revisions in the original drawing file.

All reference edits made using reference editing are saved back to the original drawing file and affect any host drawing that references the file. For this reason, it is critically important that you edit external references only with the permission of your instructor or supervisor.

Opening an Xref File

The **XOPEN** tool allows you to open an xref in a separate drawing window from within the host drawing. A quick way to access the **XOPEN** tool is to select an xref, and then right-click and select **Open Xref**. This is essentially the same procedure as using the **OPEN** tool, but faster. If you access the **XOPEN** tool without first selecting an xref, you must then pick the xref to open.

Type
XOPEN

The xref drawing file opens in a separate drawing window. After you make changes and save the xref file, you must reload the xref file in the host file. Use the **External References** palette or the **Manage Xrefs** icon in the status bar to reload the modified xref file. This ensures that the host file is up-to-date.

You can also open an xref in the **External References** palette by right-clicking the xref name and selecting **Open**.

Select an xref to display the **External References** ribbon tab. This is a convenient location to access tools for reference editing, opening, and clipping an xref. An option is also available for accessing the **External References** palette.

Exercise 32-6

Access the Student Web site (www.g-wlearning.com/CAD) and complete Exercise 32-6.

Chapter Test

1. What five types of files can you reference into an AutoCAD drawing?
2. What effect does the use of referenced drawings have on drawing file size?
3. What is a nested xref?
4. List at least three common applications for xrefs.
5. On what layer should you insert xrefs into a drawing?
6. Which tool allows you to attach an xref drawing to the current file?
7. What is the difference between an overlaid xref and an attached xref?
8. What is the difference between an absolute path and a relative path?
9. Describe the process of placing an xref using **DesignCenter**.
10. What must you do before you can use a tool palette to place an xref?
11. If you attach an xref file named FPLAN to the current drawing, and FPLAN contains a layer called ELECTRICAL, what name will appear for this layer in the **Layer Properties Manager**?
12. What is the purpose of the **Detach** option in the **External References** palette?
13. When do xrefs update in the host drawing?
14. What could you do to suppress an xref temporarily without detaching it from the master drawing?
15. Which tool allows you to display only a specific portion of an externally referenced drawing?
16. What are spatial and layer indexes, and what function do they perform?
17. Why would you want to bind a dependent object to a master drawing?
18. What does the layer name WALL0NOTES mean?
19. What tool allows you to edit external references in place?
20. What tool allows you to open a parent xref drawing into a new AutoCAD drawing window by selecting the xref in the master drawing?

Drawing Problems

Start AutoCAD if it is not already started. Start a new drawing using an appropriate template of your choice. The template should include layers, text styles, dimension styles, and multileader styles appropriate for drawing the given objects. Add layers, text styles, dimension styles, and multileader styles as needed. Draw all objects using appropriate layers, text styles, dimension styles, multileader styles, justification, and format. Follow the specific instructions for each problem. Use your own judgment and approximate dimensions when necessary.

▼ Basic

1. Attach a dimensioned problem from Chapter 18 into a new drawing as an xref. Save the drawing as P32-1.

2. Attach a dimensioned problem from Chapter 19 into a new drawing as an xref. Save the drawing as P32-2.

3. Attach a dimensioned problem from Chapter 20 into a new drawing as an xref. Save the drawing as P32-3.

▼ Intermediate

4. Attach the EX30-9.dwg file used in Exercise 30-9 into a new drawing as an xref. Copy the xref three times. Use the **XCLIP** tool to create a clipping boundary on each view. Apply an inverted rectangular clip to the original xref, a polyline boundary on the first copy, and a polygonal boundary on the second copy. Save the drawing as P32-4.

5. Attach the EX30-9.dwg file used in Exercise 30-9 into a new drawing as an xref. Bind the xref to the new drawing. Rename the layers to the names assigned to the original EX30-9 (xref) file. Explode the block created by binding the xref. Save the drawing as P32-5.

6. Create the multi-detail drawing shown according to the following information:
   - Use the MECHANICAL-INCH.dwt drawing template file available on the Student Web site.
   - Set drawing units to fractional.
   - Xref the following files into model space: Detail-Item 1.dwg, Detail-Item 2.dwg, Detail-Item 3.dwg, Detail-Item 4.dwg, Detail-Item 5.dwg, and Detail-Item 6.dwg. These files are available on the Student Web site.
   - Use six floating viewports on the **C-SIZE** layout to arrange and scale the details. Use a 1:2 scale.
   - Plot the drawing.
   Save the drawing as P32-6.

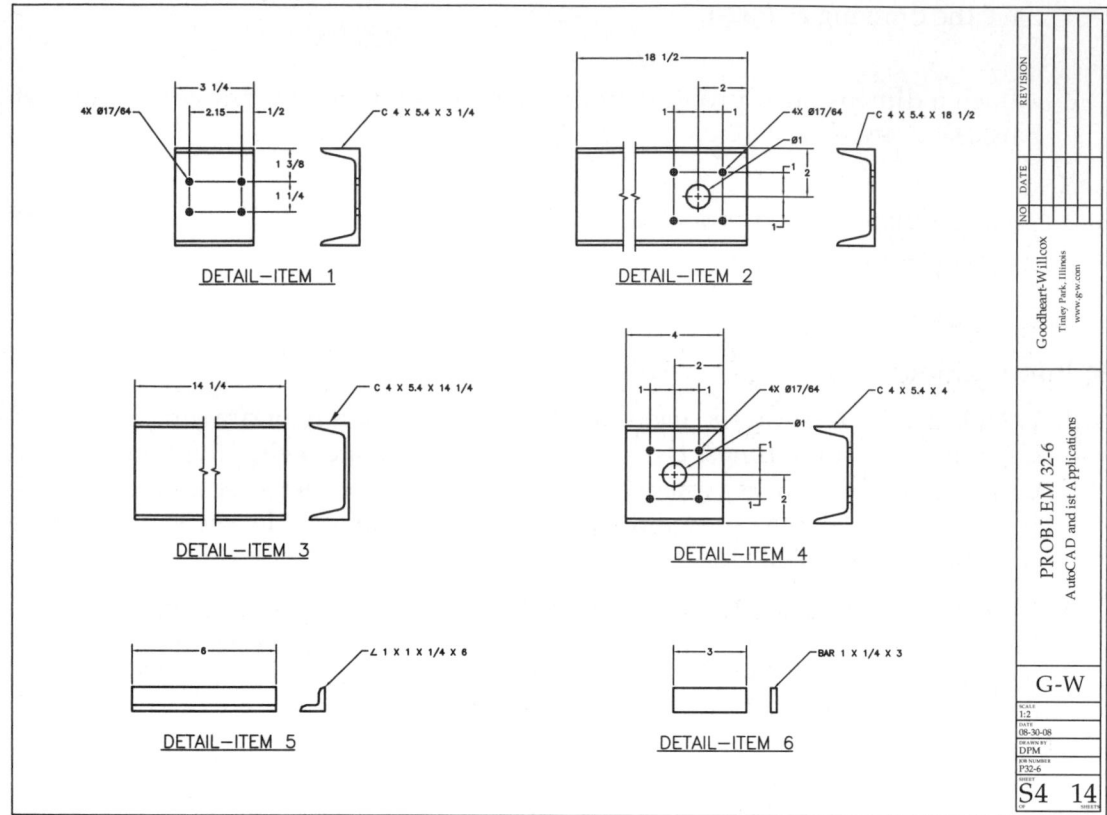

▼ Advanced

7. Design and draw a basic residential floor plan using an appropriate template. Save the file as P32-7FLOOR. Xref the P32-7FLOOR file into a new file as an attachment. Use the xref to help draw a roof plan. Save the roof plan file as P32-7ROOF.

8. Xref the P32-7ROOF file into a new file as an attachment. Use the xref to help draw front and rear elevations. Save the elevation file as P32-8.

9. Use a word processor to write a report of approximately 250 words explaining the purpose of external references. Include a brief description of the types of files that you can reference. Cite at least three examples from actual industry applications of using external references to help prepare drawings. Use at least four sketches to illustrate your report.

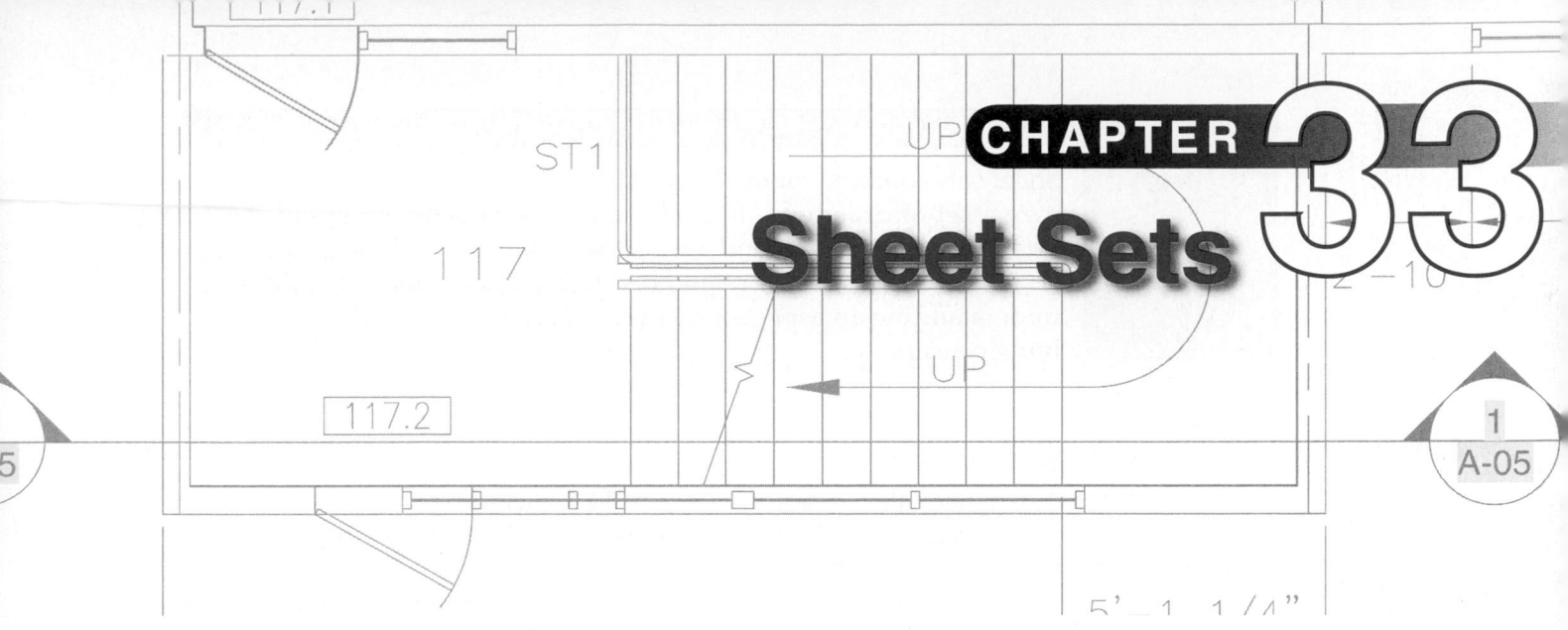

Learning Objectives

After completing this chapter, you will be able to do the following:

✓ Identify and describe the functions of the **Sheet Set Manager**.
✓ Create sheet sets and subsets.
✓ Add sheets and sheet views to a sheet set.
✓ Plot or publish a set of sheets.
✓ Insert callout blocks and view labels into sheet views.
✓ Set up custom properties for a sheet set.
✓ Create a sheet list table.
✓ Archive a set of electronic files for a sheet set.

A design project typically requires a set of drawings and documents that completely specify design requirements. Often a number of sources, such as clients and vendors, share project information. Effectively organizing and distributing drawings during the course of a project is critical to delivering accurate drawings in an orderly and timely manner. *Sheet sets* help simplify the management of a project that contains multiple drawings and views. This chapter describes how to use sheet sets to structure different drawing layouts into groups of files for reviewing, plotting, and publishing.

sheet set: A collection of drawing sheets for a project.

Introduction to Sheet Sets

AutoCAD *sheets* created in drawing file layouts can have additional project-specific properties. All sheets in a sheet set can use a single template. The template can include a title block and other layout content with attributes containing *fields*. Fields allow values such as project name and sheet number to change during the course of the project. The field modifies in the template, automatically applying changes to all sheets within the set. When you insert a new sheet into a sheet set, the sheet numbers and all sheet references update. This automation saves time and improves drawing set accuracy.

sheet: A printed drawing or electronic layout produced for a project.

fields: Special text objects that display values that update automatically.

Once a sheet set is complete, you can print, *publish*, and archive the entire set in a single operation. This is very efficient. For example, it is far easier to plot a sheet set containing twenty sheets than to open and plot twenty separate drawings.

publish: Create electronic files for distribution or plotting.

The Sheet Set Manager

SHEETSET

Quick Access

Sheet Set Manager

Ribbon

View
> Palettes

Sheet Set Manager

Type

**SHEETSET
SSM**

The **Sheet Set Manager** palette, shown in **Figure 33-1,** includes three tabs that allow you to create, organize, and access sheet sets. Use the **Sheet Set Control** drop-down list at the top of the **Sheet Set Manager** to open and create sheet sets. The buttons next to the drop-down list control and manage the items listed in the **Sheet Set Manager**. The buttons vary depending on the current tab. Like other palettes, you can resize, dock, and set the **Sheet Set Manager** to auto-hide. The **Sheet Set Manager** also makes use of extensive tooltips that describe and allow you to preview items in the palette, as well as shortcut menus for accessing tools and options.

Figure 33-1.
The **Sheet Set Manager** palette contains the **Sheet List**, **Sheet Views**, and **Model Views** tabs. The **Sheet Set Control** drop-down list contains options for creating and opening a sheet set.

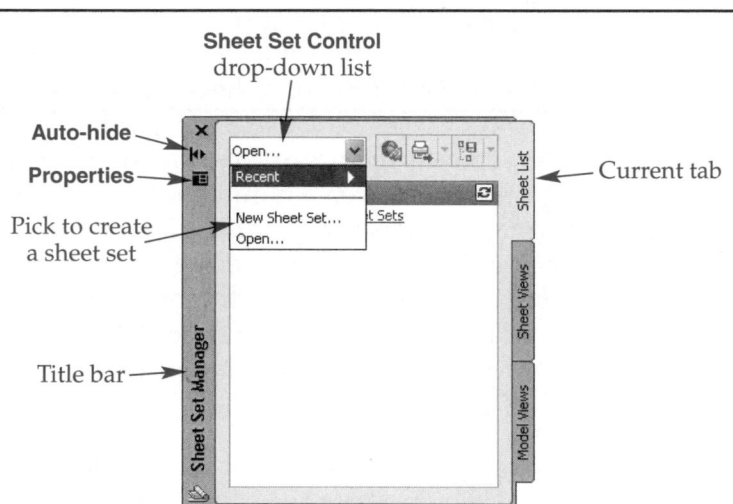

Sheet Set Control drop-down list

Auto-hide

Properties

Pick to create a sheet set

Title bar

Current tab

Open...

Recent

New Sheet Set...
Open...

Sheet List

Sheet Views

Model Views

Sheet Set Manager

Creating Sheet Sets

Create sheet sets from within an open file using the **Create Sheet Set** wizard. See **Figure 33-2.** Access the wizard from within the **Sheet Set Manager** by picking **New Sheet Set...** from the **Sheet Set Control** drop-down list. You can create sheet sets from an example sheet set or by collecting existing drawing files.

Figure 33-2.
Select the **An example
sheet set** radio button
on the **Begin** page
to use an AutoCAD
sheet set or another
existing sheet set as
a model. Pick the
Existing drawings
radio button to create
a new sheet set from
an existing drawing
project.

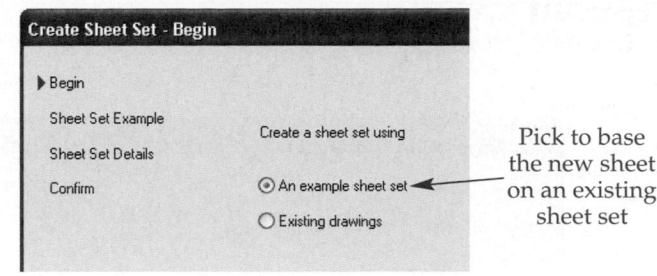

Pick to base
the new sheet
on an existing
sheet set

Using an Example Sheet Set

An example sheet set uses an existing sheet set as a model, similar to a template, for developing a new sheet set. You can modify the example sheet as needed to fit the needs of the new sheet set. AutoCAD provides several example sheet sets based on different drafting disciplines. You are not limited to the examples provided—you can use any existing sheet set as an example sheet set. To create a new sheet set from an example sheet set, access the **Create Sheet Set** wizard, and at the **Begin** page, pick the **An example sheet set** radio button.

Sheet Set Example Page

Pick the **Next** button to display the **Sheet Set Example** page. See **Figure 33-3.** Sheet set information is saved in a sheet set data file (DST). When you create a sheet set from an example sheet set, you start from an existing DST file. The **Select a sheet set to use as an example** radio button is active by default, and a list box displays all DST files in the default Template folder. Pick a sheet set from the list box, or select the **Browse to another sheet set to use as an example** radio button and then select the ellipsis (...) button and use the **Browse for sheet set** dialog box to locate a DST file in another folder. After selecting the DST file for the example sheet set, pick the **Next** button to display the **Sheet Set Details** page. See **Figure 33-4.**

Sheet Set Details Page

The **Sheet Set Details** page allows you to modify the existing sheet set data and create settings for the new project. Enter the name, or title, of the sheet set in the **Name of new sheet set** text box. The name is typically the project number or a short description of the project. Type a description for the sheet set in the **Description (optional)**

Figure 33-3.
Use the **Sheet Set
Example** page to
select an example
sheet set.

List of sheet sets
in **Template** folder

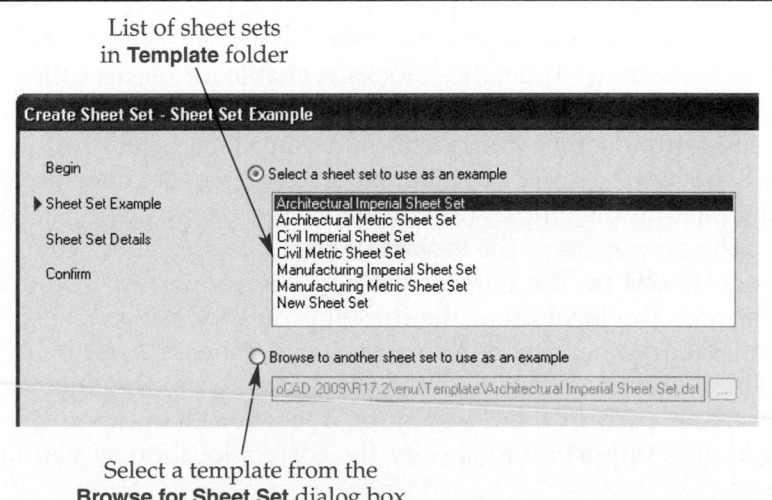

Select a template from the
Browse for Sheet Set dialog box

Figure 33-4.
Enter a name, description, and file path location for the new sheet set on the **Sheet Set Details** page.

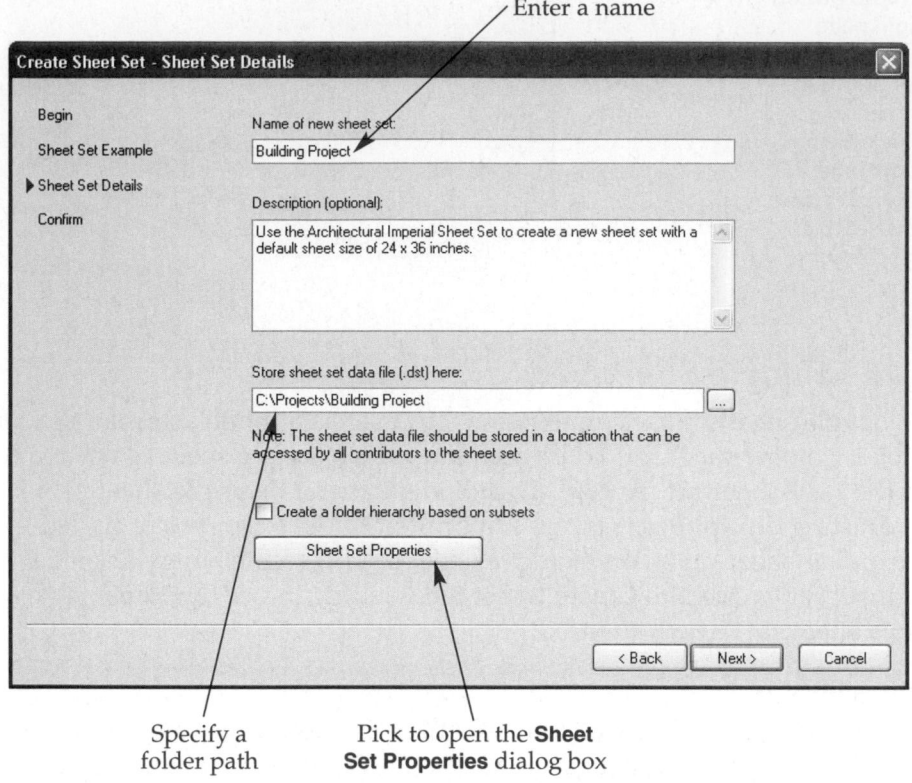

Enter a name

Specify a folder path

Pick to open the **Sheet Set Properties** dialog box

area if desired. The **Store sheet set data file (.dst) here** text box determines where the sheet set file is saved. Pick the ellipsis (**…**) button and select a folder in the **Browse for sheet set folder** dialog box to redefine the default sheet set file location. Pick the **Create a folder hierarchy based on subsets** check box to organize a sheet set so that folders group into *subsets*. Layouts in each folder form under each subset.

subsets: Groups of layouts based on folder hierarchy.

Pick the **Sheet Set Properties** button to open the **Sheet Set Properties** dialog box. See **Figure 33-5**. The **Sheet Set** category includes basic sheet set properties. The **Name** field contains the name entered in the **Name of new sheet set** text box of the **Sheet Set Details** page. The **Sheet set data file** field shows the location of the DST file, specified in the **Store sheet set data file (.dst) here** text box of the **Sheet Set Details** page. The **Description** field contains the description entered in the **Description** area of the **Sheet Set Details** page.

The **Model view** field specifies the folder(s) containing the drawing files used for the sheet set. The **Label block for views** field specifies the block used to label views. The **Callout blocks** field specifies blocks available for use as callout blocks. The **Page setup overrides file** field specifies the location of a drawing template (DWT) file containing a page setup used to override the existing sheet layout settings. You can change all of the **Sheet Set** category settings, except the **Sheet set data file** option, using the text box or by picking the ellipsis (**…**) button to navigate to a specific location.

The properties in the **Project Control** category allow you to store and update information based on the current project. The properties in the **Sheet Creation** category determine the location of the drawing files for new sheets and the template used to create them. When you add a new sheet to a sheet set, AutoCAD creates a new drawing file based on the template and layout specified in the **Sheet creation template** setting. The folder path in the **Sheet storage location** field determines where the new file is saved. It is important to specify the correct location so you know where the files are saved.

Figure 33-5.
The **Sheet Set Properties** dialog box stores the main properties of a sheet set.

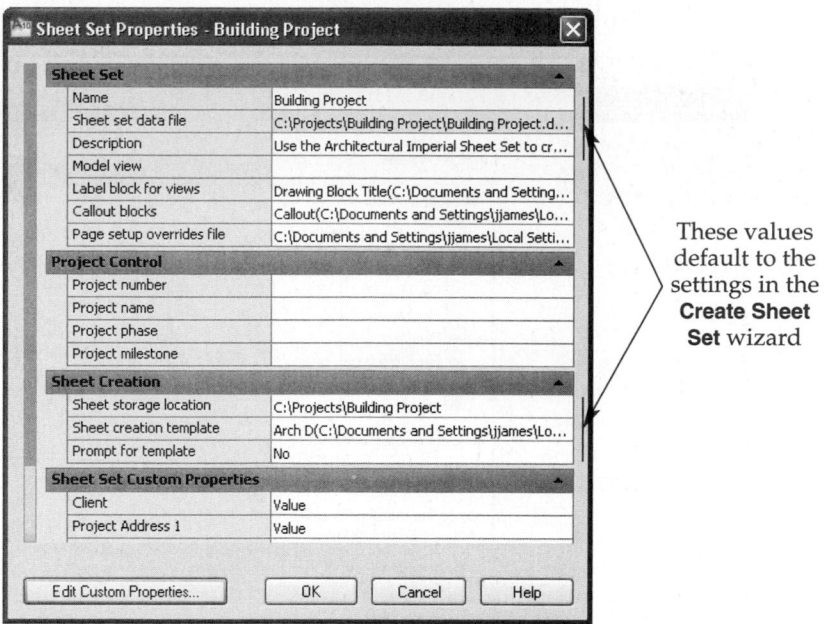

These values default to the settings in the **Create Sheet Set** wizard

When selecting the **Sheet creation template** value, you must specify a template file and a layout. To modify this setting, pick in the text box and then pick the ellipsis (**...**) button to display the **Select Layout as Sheet Template** dialog box. See **Figure 33-6.** All layouts in the selected template appear in the list box. Select the layout and then pick the **OK** button.

If the value in the **Prompt for template** property is set to the default **No**, AutoCAD automatically uses the template layout specified in the **Sheet creation template** field when creating a new sheet. Select **Yes** to choose a different layout when creating a new sheet. Information specific to the project is set up in the **Sheet Set Custom Properties** section, as described later in this chapter. Once all the values in the **Sheet Set Properties** dialog box are set, pick the **OK** button to return to the **Sheet Set Details** page. Pick the **Next** button to continue creating the new sheet set.

Figure 33-6.
Selecting an existing layout as a template for new sheets in a sheet set.

Pick to select a different template file

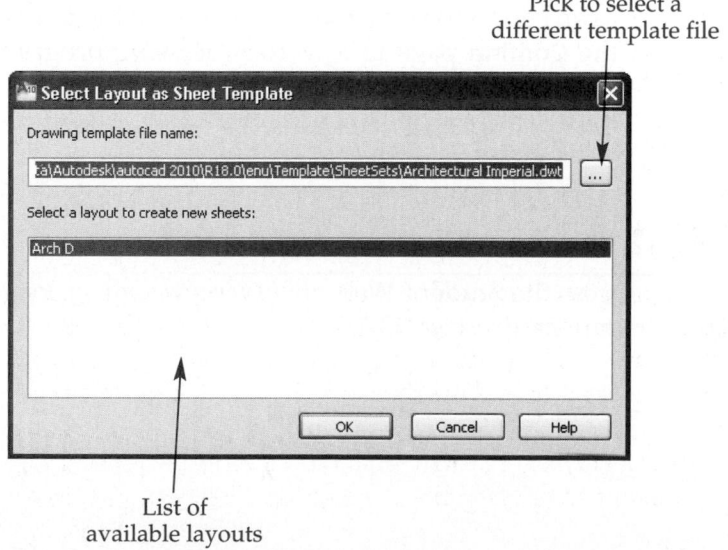

List of available layouts

Figure 33-7.
Use the **Confirm** page to preview settings before creating the sheet set.

Subsets in new sheet set

Scroll down to preview more information

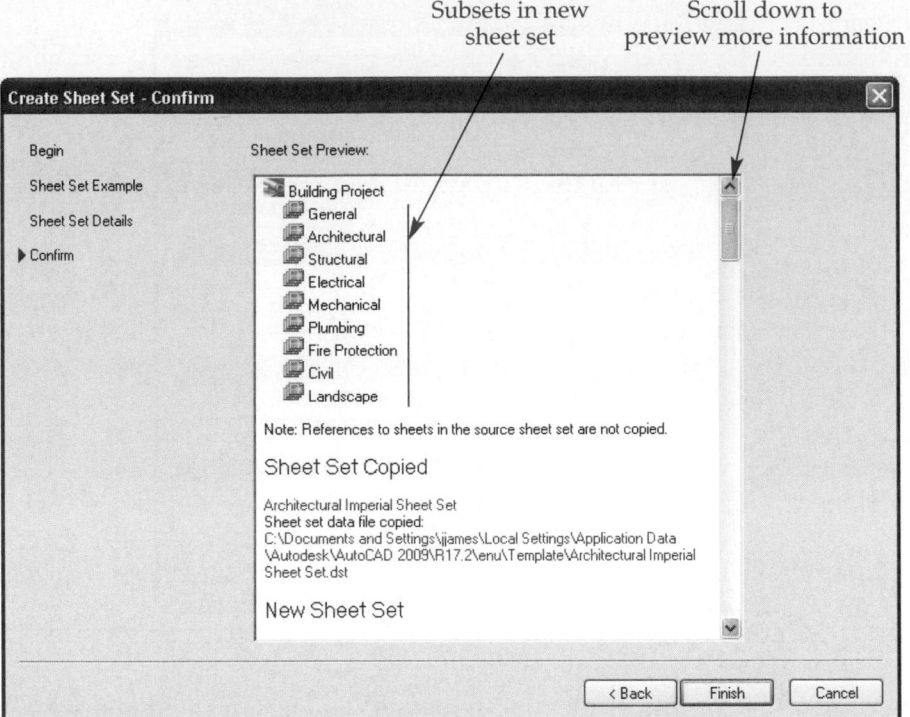

Confirm Page

The **Sheet Set Preview** area of the **Confirm** page displays all information associated with the sheet set. See **Figure 33-7.** The example shown includes a sheet set named Building Project. This sheet set contains a number of subsets related to the project, such as General and Architectural. After creating the sheet set, you can add sheets to each subset. Use the **Back** button to return to a previous page to make changes. After reviewing the information on the **Confirm** page, pick the **Finish** button to create the sheet set.

The sheet set data file is saved to the specified location. You can then open the sheet set in the **Sheet Set Manager**. Since a sheet set is not associated with a particular drawing file, you can open any sheet set, regardless of the active drawing file.

PROFESSIONAL TIP

You can copy and paste the information in the **Sheet Set Preview** area of the **Confirm** page to a word processing program for saving and printing.

Exercise 33-1

Access the Student Web site (www.g-wlearning.com/CAD) and complete Exercise 33-1.

Collecting Existing Drawing Files

You can also import existing layouts from drawing files to create sheets. Each layout in each drawing becomes a separate sheet. When you create a sheet set in this manner, organize all the files used in the project in a structured hierarchy of folders. You should place one layout in each drawing file to simplify access to different layout tabs. In addition, prepare a sheet creation template and specify the template in the **Sheet Set Properties** dialog box to ensure that all sheets use the same layout settings.

To create a new sheet set from an existing drawing project, access the **Create Sheet Set** wizard. At the **Begin** page, select the **Existing drawings** radio button and pick the **Next** button to display the **Sheet Set Details** page. Specify a name and description for the sheet set and the location to save the data file. Pick the **Sheet Set Properties** button to specify additional sheet set properties. When you are finished, pick the **Next** button to display the **Choose Layouts** page shown in **Figure 33-8**.

The **Choose Layouts** page allows you to specify the drawings and layouts added to the sheet set. Pick the **Browse...** button to display the **Browse for Folder** dialog box, and select the folder containing the drawing files with the desired layouts. The selected folder, the drawing files it contains, and all layouts within those drawings are displayed in the list box. When you first open a folder in the **Choose Layouts** page, all drawing files with layouts in the selected folder appear for selection.

The items you check are added to the new sheet set. Uncheck a layout box to remove the layout from the new sheet set. Uncheck a drawing file to uncheck all related layouts. Uncheck a folder to uncheck all related drawing files and layouts. In **Figure 33-8**, only the Furniture.dwg and associated layouts are not included in the sheet set. Use the **Browse for Folder** dialog box as needed to add more folders to the **Choose Layouts** page.

Figure 33-8.
Use the **Choose Layouts** page to import existing layouts to a new sheet set.

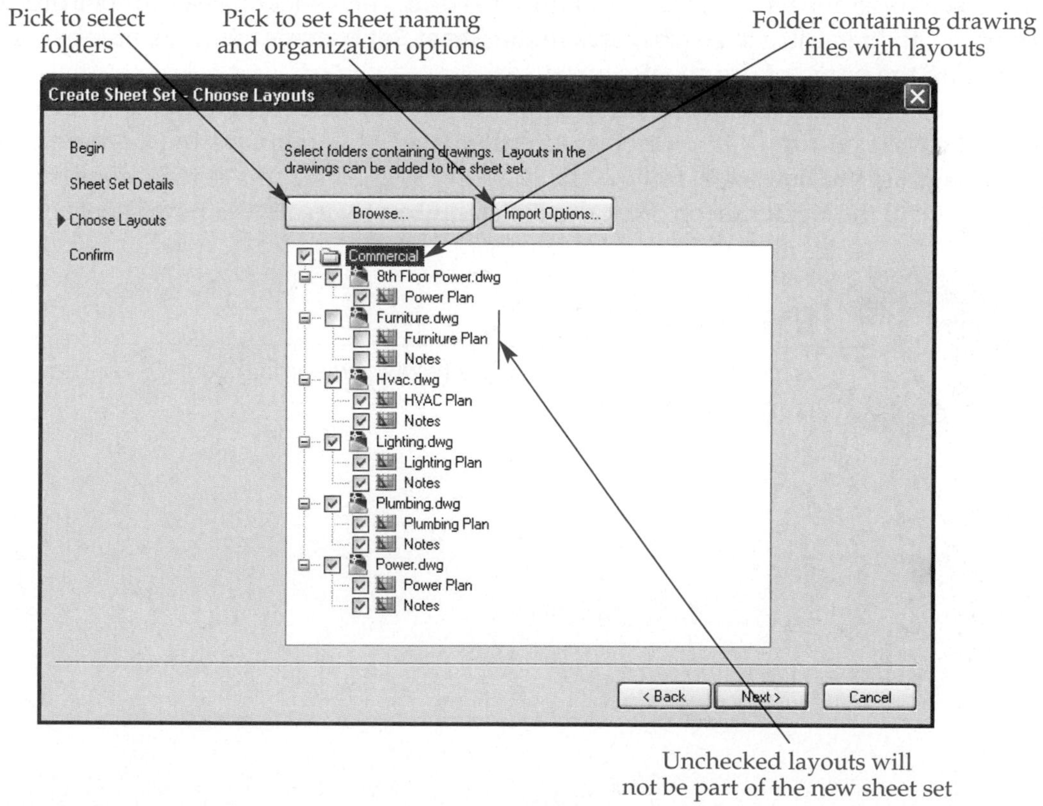

Pick to select folders

Pick to set sheet naming and organization options

Folder containing drawing files with layouts

Unchecked layouts will not be part of the new sheet set

Figure 33-9.
The **Import Options** dialog box allows you to specify naming conventions for sheets and folder structuring options.

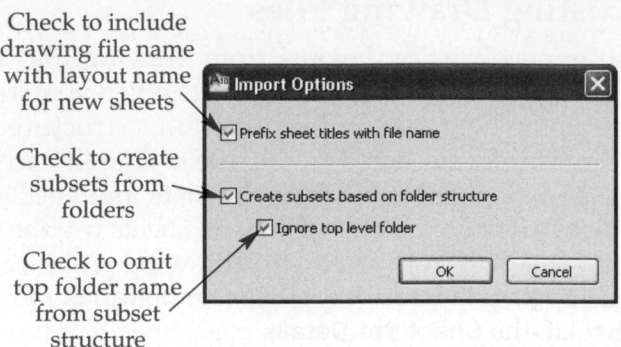

Check to include drawing file name with layout name for new sheets

Check to create subsets from folders

Check to omit top folder name from subset structure

To adjust sheet naming and subset options, pick the **Import Options...** button to display the **Import Options** dialog box. See **Figure 33-9.** When you create a sheet set using existing layouts, the name for a new sheet can be the same as the layout name, or it can be the drawing file name combined with the layout name. If you select the **Prefix sheet titles with file name** check box, the name of the layouts that become sheets includes the drawing file name and the name of the layout. For example, the sheet name Electrical Plan – First Floor Electrical forms from a layout named First Floor Electrical imported from the drawing file Electrical Plan.dwg. To have sheets take on only the layout name, deselect the **Prefix sheet titles with file name** check box.

Select the **Create subsets based on folder structure** check box to organize a sheet set so that folders are grouped into subsets. Layouts in each folder form under each subset. The **Ignore top level folder** option determines whether the folder name at the top level creates a subset. **Figure 33-10A** shows the **Choose Layouts** page with layouts imported from the Residential folder for the Residential Project sheet set. This sheet set uses the **Create subsets based on folder structure** option. The **Sheet Set Manager** in **Figure 33-10B** shows the result of the configuration. Creating subsets for sheet sets helps organize sheets. As shown in **Figure 33-10B,** each sheet has a number preceding its name. By default, a sheet displays in the **Sheet Set Manager** with its number, a dash, and then the name of the sheet.

Once you select all folders and layouts for the new sheet set and adjust all settings, pick the **Next** button on the **Choose Layouts** page. This displays the **Confirm** page. In the **Sheet Set Preview** area, review the sheet set properties. If a setting requires modification, use the **Back** button. Pick the **Finish** button to create the new sheet set.

Exercise 33-2

Access the Student Web site (www.g-wlearning.com/CAD) and complete Exercise 33-2.

Figure 33-10.
Creating a sheet set named Residential Project with subsets. A—Layouts imported from the Architectural and Structural subfolders in the Residential folder. Subfolders are designated as subsets for the new sheet set. B—Subsets shown in the **Sheet Set Manager**.

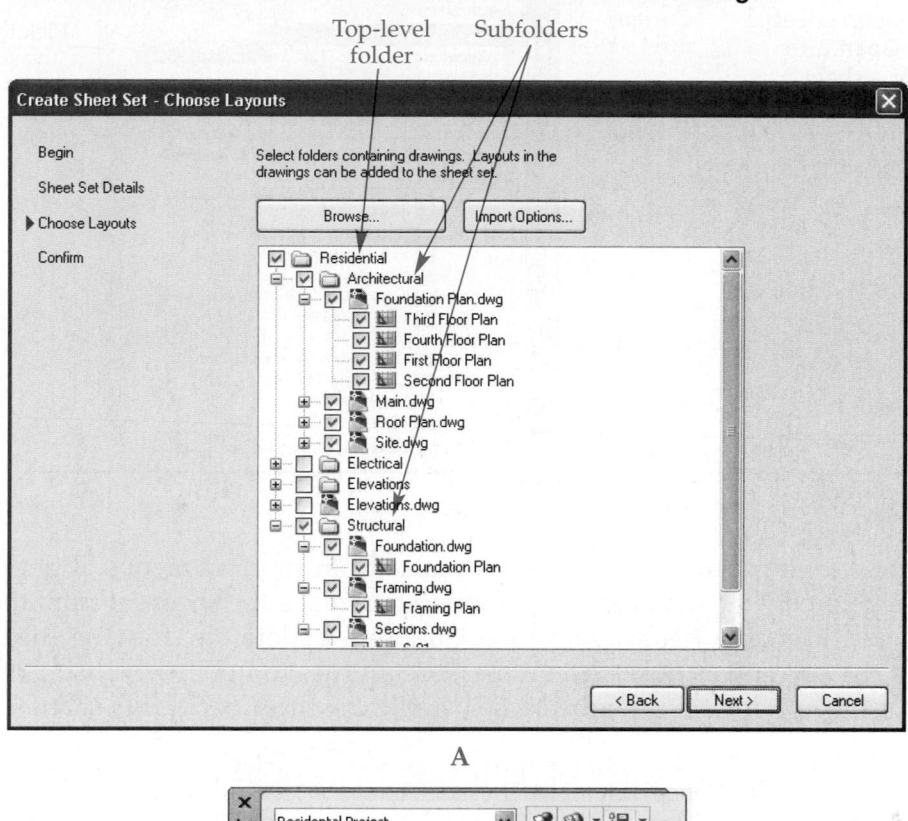

A

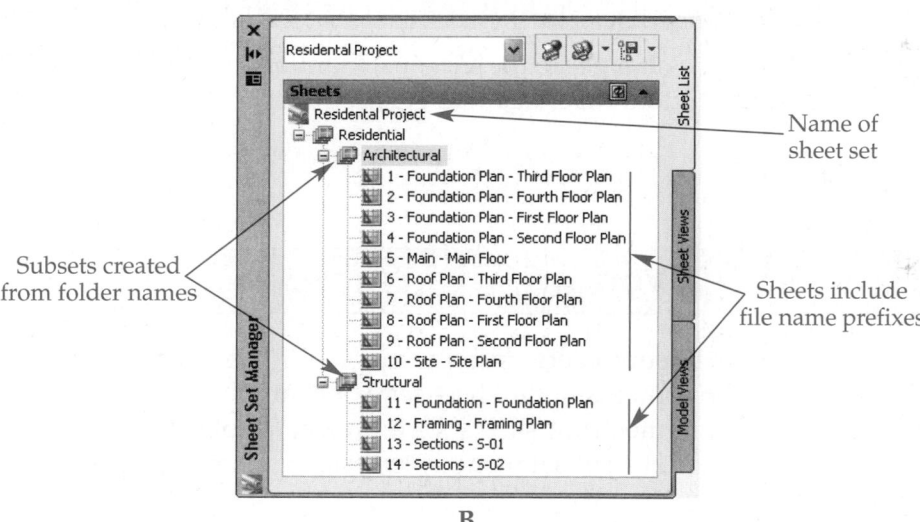

B

Working with Sheet Sets

Open an existing sheet set from within the **Sheet Set Manager** by picking a sheet set from the **Sheet Set Control** drop-down list. See Figure 33-11. The top portion of the drop-down list displays sheet sets opened in the current AutoCAD session. The area clears when you close AutoCAD. Select **Recent** to display a list of the most recently opened sheet sets. Select **Open...** to display the **Open Sheet Set** dialog box. Then navigate to a sheet set data (DST) file and open the file in the **Sheet Set Manager**.

Use the **Sheet List** tab to manage sheets, the **Sheet Views** tab to manage sheet views, and the **Model Views** tab to manage drawing files with layouts. Almost all of the

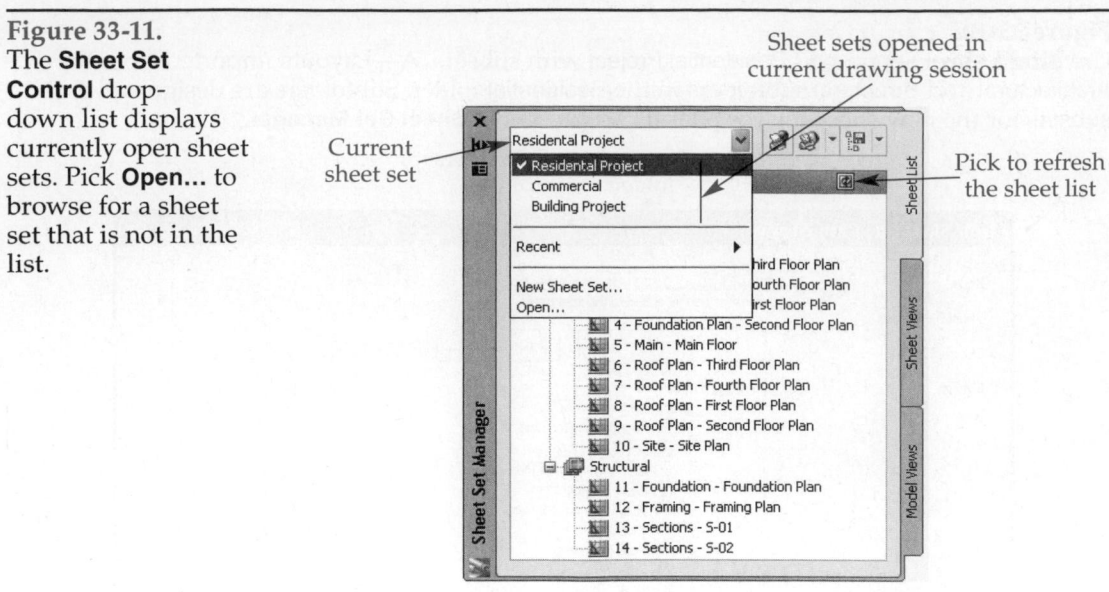

Figure 33-11.
The **Sheet Set Control** drop-down list displays currently open sheet sets. Pick **Open...** to browse for a sheet set that is not in the list.

Current sheet set

Sheet sets opened in current drawing session

Pick to refresh the sheet list

options for working with sheet sets are available from shortcut menus. Right-click on a sheet set to access the **Close Sheet Set** option to remove the sheet set from the **Sheet Set Manager**. Select the **Resave All Sheets** option to update the drawing files associated with the current sheet set. You must close all files to update the list. Additional shortcut menu options are described when applicable throughout this chapter.

NOTE

To update changes to the sheet list manually, pick the **Refresh Sheet Status** button. Refer again to **Figure 33-11.**

Working with Subsets

Subsets are similar to subfolders under a top-level folder in Windows Explorer. Subsets help manage the contents of the sheet set. For example, create a subset named Architectural to store all architectural sheets for a project, a subset named Electrical to store all electrical sheets, and a subset named Plumbing to store all plumbing sheets.

Creating a New Subset

To create a new subset, right-click on the sheet set name or an existing subset in the **Sheet Set Manager** and select **New Subset...** to open the **Subset Properties** dialog box. See **Figure 33-12.** Type the name of the new subset in the **Subset name** text box. For example, if a subset will contain all of the electrical sheets in a sheet set, name the subset Electrical. When you add a new sheet to the subset using a template, the sheet is saved as a drawing file. To create a new folder for the subset, select **Yes** from the **Create Folder Hierarchy** drop-down list. The folder structure mimics the subset structure.

The **Publish Sheets in Subset** option determines whether sheet sets in the subset are published. The **Do Not Publish Sheets** setting does not publish sheets in the subset. An icon identifies subsets set not to publish. The **New Sheet Location** setting determines the path to which new sheets are saved. The default value is the location specified when you created the sheet set.

       AutoCAD and Its Applications—Basics

Figure 33-12.
The **Subset Properties** dialog box allows you to define a new subset.

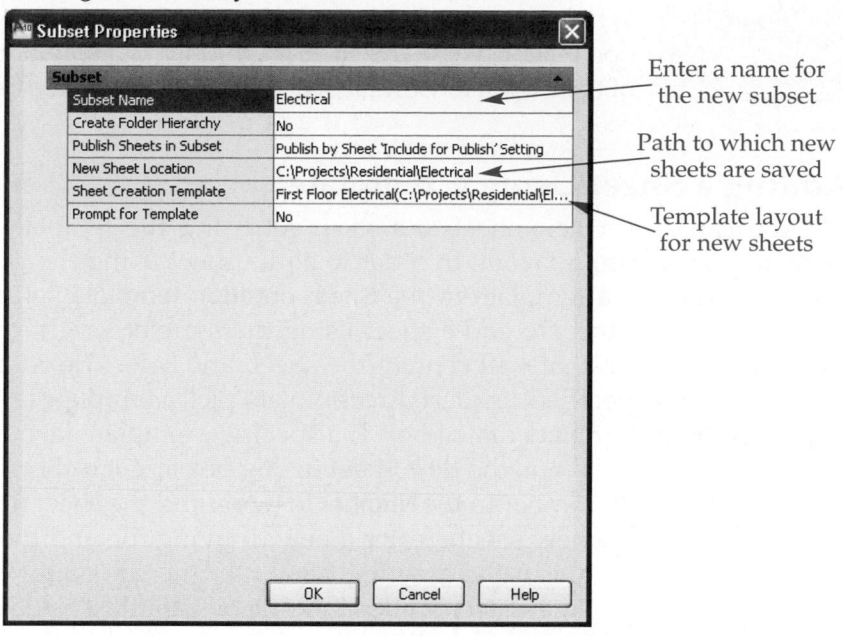

Enter a name for the new subset

Path to which new sheets are saved

Template layout for new sheets

Each subset can also have its own template and layout for new sheets, as specified in the **Sheet Creation Template** setting. For example, if electrical sheets use their own title block and notes, assign a template sheet with the appropriate settings. The procedure for specifying the template and layout for a subset is identical to the procedure used to select the sheet set properties. Use the **Prompt for Template** drop-down list to indicate if a prompt should ask for a sheet template instead of using the specified sheet creation template.

Modifying a Subset

To make changes to an existing subset, right-click on the subset and select **Properties...** to redisplay the **Subset Properties** dialog box. The **Rename Subset...** shortcut menu option also opens the **Subset Properties** dialog box. To delete a subset, right-click on the subset and select **Remove Subset**. This option is unavailable if the subset contains sheets.

Exercise 33-3

Access the Student Web site (www.g-wlearning.com/CAD) and complete Exercise 33-3.

Working with Sheets

One of the most useful features of the **Sheet Set Manager** is the ability to open a sheet quickly for review or modification. To open a sheet, double-click on the sheet or right-click on the sheet and select **Open**. The drawing file that contains the referenced layout tab opens and displays the layout corresponding to the sheet.

> **NOTE**
>
> When you open files from the **Sheet Set Manager**, they are added to the open files list. Use **Quick View Drawings** or the window control features of the ribbon to view all open files. Save and close unused files.

Adding a Sheet Using a Template

You can add a new sheet to a sheet set using the template layout sheet or by importing an existing layout. In order to add a sheet using the template layout sheet, you must specify a template in the **Sheet creation template** setting of the **Sheet Set** properties dialog box. To add a sheet using the template, right-click on the sheet set name or the subset that will contain the sheet, and select **New Sheet…**. If there is no template layout specified, an alert directs you to pick a template layout using the **Select Layout as Sheet Template** dialog box. If a specified template layout exists, or after you select the template layout, the **New Sheet** dialog box appears. See **Figure 33-13.**

Type the sheet number in the **Number** text box and the sheet name in the **Sheet title** text box. Creating a new sheet creates a new drawing file and the sheet title becomes the name of the layout in the drawing file. Enter the file name in the **File name** text box. The file name is the sheet number and title by default. The **Folder path** display box shows where the drawing file will be saved, as specified in the **Subset Properties** or **Sheet Set Properties** dialog box.

Adding an Existing Layout As a Sheet

You can add an existing drawing layout to a sheet set using the **Sheet Set Manager** or directly from an open drawing. To add an existing layout to a sheet set using the **Sheet Set Manager**, right-click on the sheet set name or the subset that will contain the sheet, and select **Import Layout as Sheet…**. This displays the **Import Layouts as Sheets** dialog box. See **Figure 33-14.** Pick the **Browse for Drawings** button to select a drawing file. The layouts from the drawing file appear in the list box. The **Status** field indicates whether the layout can be imported into the sheet set. You cannot import a layout that is already part of a sheet set. By default, all layouts in the drawing are checked for import. Uncheck a box to exclude a layout from import. Select the **Prefix sheet titles with file name** check box to include the name of the file in the sheet title. To import the sheets, pick the **Import Checked** button.

Use a layout tab or the **Quick View Layouts** or **Quick View Drawings** tool to add an existing layout to a sheet set directly from an open drawing. Right-click on the layout tab or thumbnail image to import and select **Import Layout as Sheet…**. The **Import Layouts as Sheets** dialog box appears with the selected layout automatically listed. You must save the drawing and set up the layout to make the **Import Layout as Sheet…** shortcut menu available.

Figure 33-13.
Use the **New Sheet** dialog box to define a new sheet from a template.

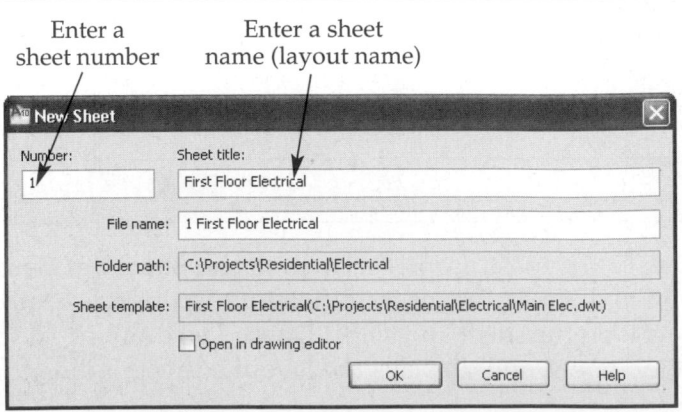

Enter a sheet number

Enter a sheet name (layout name)

Figure 33-14.
Add existing layouts as sheets to a sheet set from the **Import Layouts as Sheets** dialog box.

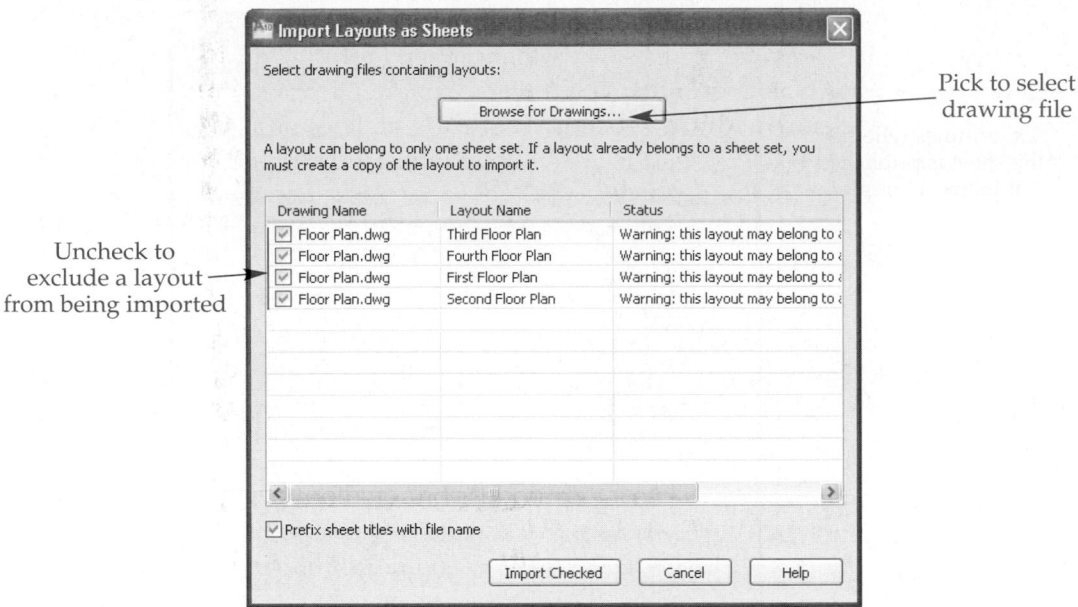

Pick to select drawing file

Uncheck to exclude a layout from being imported

Exercise 33-4

Access the Student Web site (www.g-wlearning.com/CAD) and complete Exercise 33-4.

Modifying Sheet Properties

To modify sheet properties such as the name, number, and description, right-click on the sheet name in the **Sheet Set Manager** to display a shortcut menu. Select **Rename & Renumber...** to change the sheet name and number using the **Rename & Renumber Sheet** dialog box, which is similar to the **New Sheet** dialog box. If the sheet is one of several in a subset, pick the **Next** and **Previous** buttons to access different sheets in the subset.

The **Sheet Properties** dialog box, shown in **Figure 33-15,** also allows you to change the sheet name and number, along with the description and the publish option. To open the **Sheet Properties** dialog box, right-click on the sheet name and select **Properties....** Type a description of the sheet in the **Description** text box. The **Include for publish** option determines whether the sheet is published or plotted with the sheet set. The default value is **Yes**.

The **Expected layout** and **Found layout** text boxes display the file path where the sheet was originally saved and the file path where the sheet was found. If the paths are different, update the **Expected layout** setting by picking the ellipsis (**...**) button and using the **Import layout as sheet** dialog box. To delete a sheet from a sheet set, select the **Remove Sheet** option. This does not delete the drawing file; it only removes the sheet from the sheet set.

NOTE

If you modify the location of a drawing file and the drawing file has layouts associated with a sheet set, the association is broken. You must re-import the layouts into the sheet set or update the specified path to the drawing file in the **Sheet Properties** dialog box.

Figure 33-15.
The **Sheet Properties** dialog box allows you to modify sheet properties.

Determines whether the sheet is published or included in plot

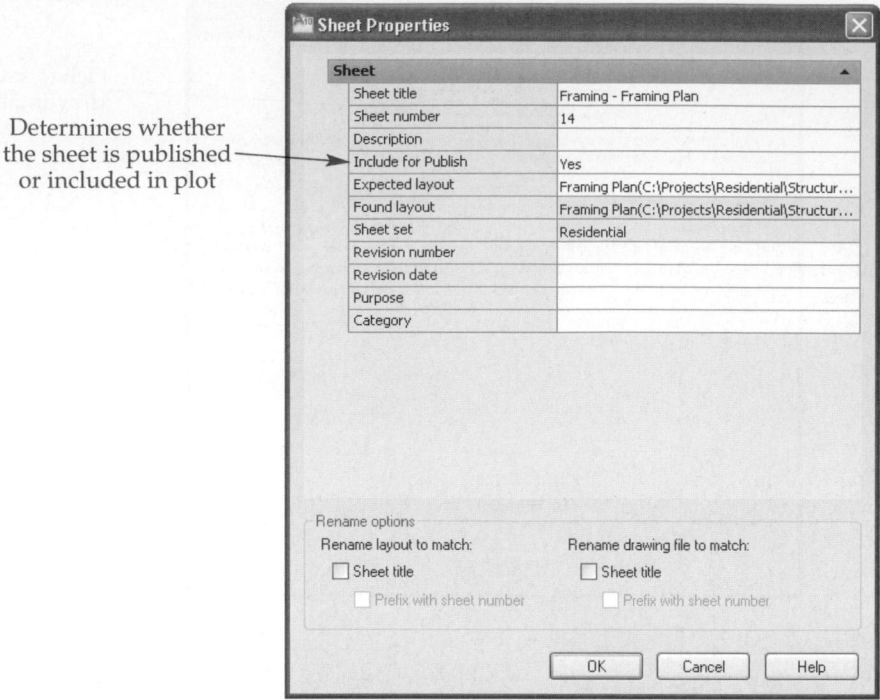

Publishing a Sheet Set

You can publish sheet sets by creating drawing web format (DWF) files, DWFx files supported by Windows Vista, or portable document format (PDF) files. DWF and DWFx files are compressed, vector-based files viewable with the Autodesk Design Review software. You can also publish a sheet set by sending the sheet set to a plotter. For more information about outputting DWF, DWFx, and PDF files, refer to *AutoCAD and Its Applications—Advanced.*

Using the Publish Shortcut Menu

A **Publish** shortcut menu is available in the **Sheet Set Manager** for publishing or plotting an entire sheet set. To access publish options, pick the **Publish** flyout on the **Sheet Set Manager** toolbar, as shown in **Figure 33-16,** or select the **Publish** cascading submenu option available when you right-click in the **Sheets** list.

NOTE

You can select a sheet set, subset, or individual sheets for publishing. Select the appropriate items by pressing [Shift] and [Ctrl] in the **Sheet Set Manager**.

Pick the appropriate **Publish to DWF, Publish to DWFx**, or **Publish to PDF** option to create a DWF, DWFx, or PDF file from the sheet set or the selected sheets. A dialog box appears, allowing you to specify a name and location for the file. The file creates multiple pages with each sheet on a separate page. Use the **Publish to Plotter** option to plot the sheet set or selected sheets to the default plotter or printer using the plot settings from each layout. Select an override from the **Publish using Page Setup Override** cascading submenu to force the sheet to use the selected page setup settings

Figure 33-16.
The **Publish** flyout provides options for preparing a sheet set for publishing or plotting.

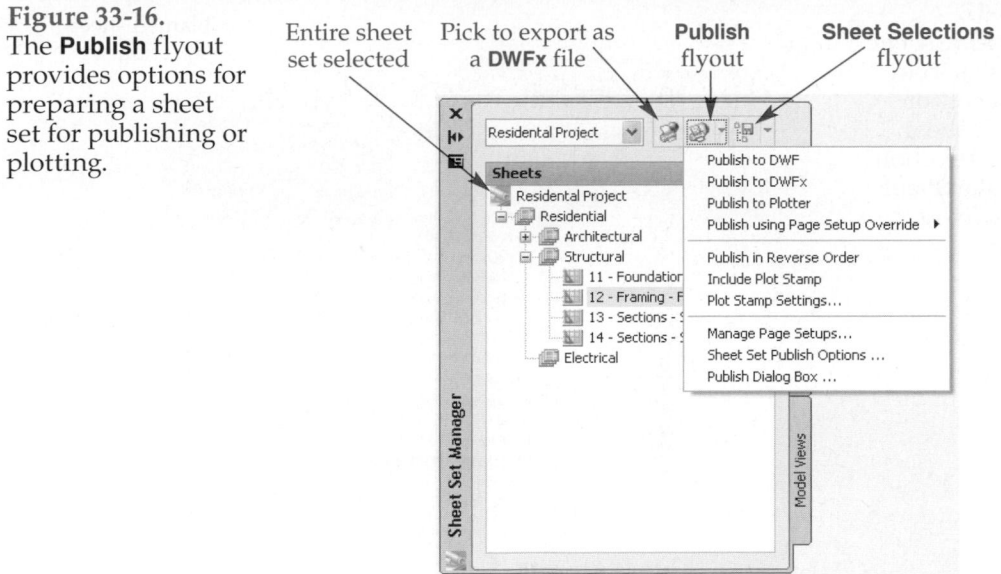

Entire sheet set selected — Pick to export as a **DWFx** file — **Publish** flyout — **Sheet Selections** flyout

instead of the plot settings saved with the layout. This option is unavailable if you have not specified a page setup override for the sheet set or subset. Use the **Include for Publish** option to identify whether the selected items will publish. Pick the **Edit Subset and Sheet Publish Settings...** option to display the **Publish Sheets** dialog box. Check specific sheets to include for publishing.

The **Publish in Reverse Order** option publishes sheets in the opposite order from the order displayed in the **Sheet Set Manager**. When using certain printers, publishing a sheet set in reverse order is helpful so that the last sheet is on the bottom of the stack, at the end of the entire set. Select the **Include Plot Stamp** option to place the plot stamp information assigned to the layout on the sheet during plotting. Select the **Plot Stamp Settings** option to open the **Plot Stamp** dialog box to specify the plot stamp settings. The **Manage Page Setups** option opens the **Page Setup Manager**, allowing you to create a new page setup or modify an existing one. The **Sheet Set Publish Options** selection displays the **Sheet Set Publish Options** dialog box, which displays the available settings for creating a DWF, DWFx, or PDF file. The **Publish Dialog Box** option opens the **Publish** dialog box, which lists the sheets in the current sheet set or the sheet selection.

Creating Sheet Selection Sets

During the course of a project, you may need to publish the same set of sheets often. Save sheets as a selection set to accommodate this need. To save a sheet selection set, select the sheets or subsets to include in the set. Then pick the **Sheet Selections** flyout on the **Sheet Set Manager** toolbar and select **Create...** to access the **New Sheet Selection** dialog box. Enter a name for the selection set and pick the **OK** button. The new selection set appears when you pick the **Sheet Selections** button. **Figure 33-17** shows three different sheet selection sets suitable for common publishing or plotting requirements. When you select a selection set, the sheets become highlighted in the **Sheet Set Manager**.

To rename or delete a sheet selection set, pick **Manage...** from the **Sheet Selections** flyout to display the **Sheet Selections** dialog box. Select the sheet selection set and then pick the **Rename** or **Delete** button.

Figure 33-17.
You can create sheet selection sets from selected sheets or subsets in a sheet set. Access selection sets from the **Sheet Selections** flyout.

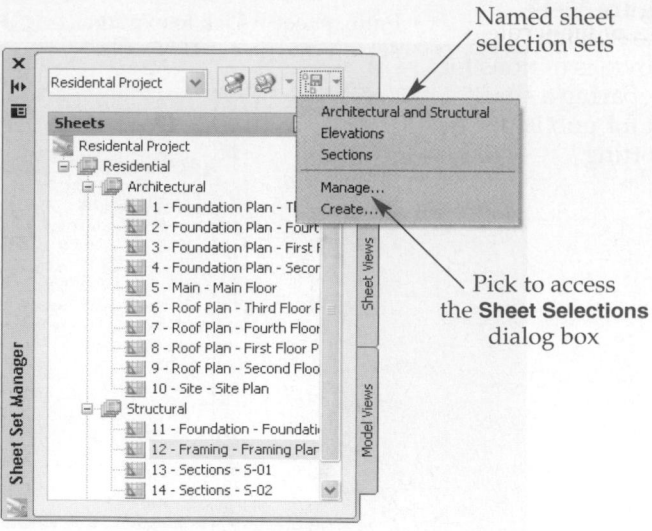

Named sheet selection sets

Pick to access the **Sheet Selections** dialog box

Using Sheet Views

sheet view: A referenced portion of a drawing set, such as an elevation, a section, or a detail.

Sheets can contain *sheet views* that you can use to provide links between sheets and drawing content in a sheet set. Sheet views provide automated labeling and referencing using blocks with attributes containing fields. Sheet view field values update automatically to reflect changes in sheet numbering and organization. Use the **Sheet Views** tab of the **Sheet Set Manager** to group views by category and open views for viewing and editing. Special tools in the **Sheet Set Manager** allow you to identify views with numbers, labels, and callout blocks.

Adding a View Category

View categories organize views in the **Sheet Views** tab, and function similar to the subsets created in the **Sheet List** tab. To create a new view category, make the **Sheet Views** tab current and select the **View by category** button. See **Figure 33-18.** Pick the **New View Category** button or right-click on the sheet set name and select **New View Category...** to open the **View Category** dialog box. See **Figure 33-19.** Enter a name for the category in the **Category name** text box. For example, specify a view category named Elevations and add four elevation views to the new category.

The **View Category** dialog box lists all available callout blocks for the current view category. Check the box next to the callout block to make the block available for all views added to the category. If a block is not in the list, use the **Add Blocks...** button to select the block from a drawing file. After you select the necessary callout blocks, pick the **OK** button to create the new category. Callout blocks are described later in this chapter.

Modifying a View Category

To modify view category properties, right-click on the category name in the **Sheet Set Manager** and select **Rename...** or **Properties...** to reopen the **View Category** dialog box. Change the category name and add different callout blocks to the category as needed. To delete a category, right-click on the category name and select **Remove Category**. The **Remove Category** option is unavailable if there are views in the category. You must remove all of the views before deleting the category.

Figure 33-18.
Create view categories in the **Sheet Views** tab of the **Sheet Set Manager**.

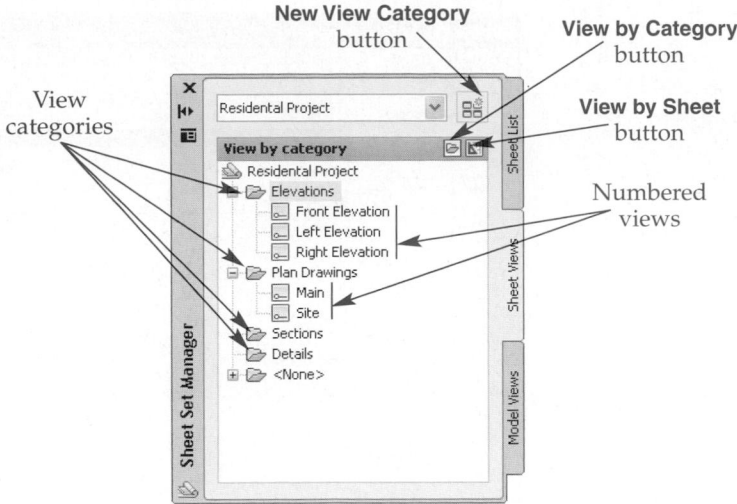

Figure 33-19.
The **View Category** dialog box allows you to name the category and select callout blocks for use with views.

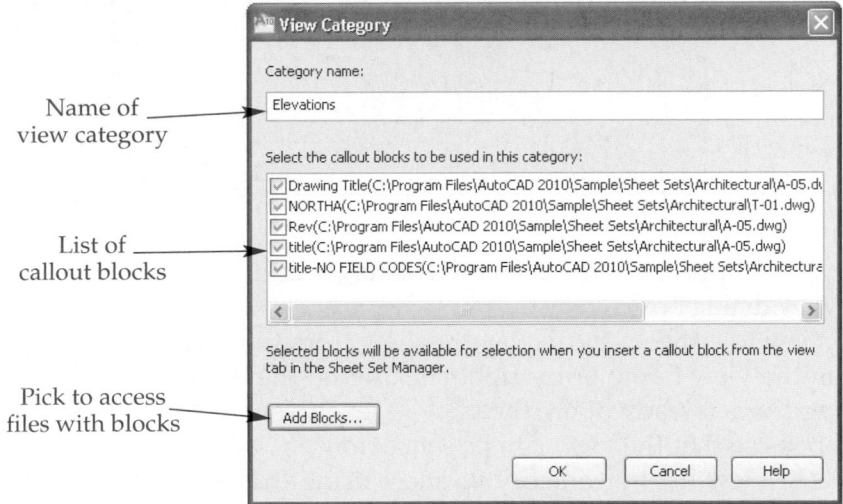

Creating Sheet Views in an Existing Sheet

The **Sheet Set Manager** allows you to add new views to sheets and organize views within sheet sets. Use the following procedure to add a view to an existing sheet set:

1. Open the desired sheet set and add a category for the view if a suitable category does not exist.
2. A sheet must be a part of the sheet set to include a view. If the sheet is not included with the sheet set, add the sheet now.
3. Open the drawing file and set the layout tab in which you will create the new view current.
4. Use display tools to orient the view as needed and then use the **VIEW** tool to access the **View Manager**.
5. In the **View Manager**, pick the **New...** button to open the **New View** dialog box.
6. Select the category to associate with the view from the **View category** drop-down list. See **Figure 33-20**.
7. Specify the remaining view settings and pick the **OK** button to save the view.

Figure 33-20.
The **View Category**
drop-down list
displays the
available view
categories for the
view definition.

Name of
new view

Sheet set
view
categories
available
in current
layout

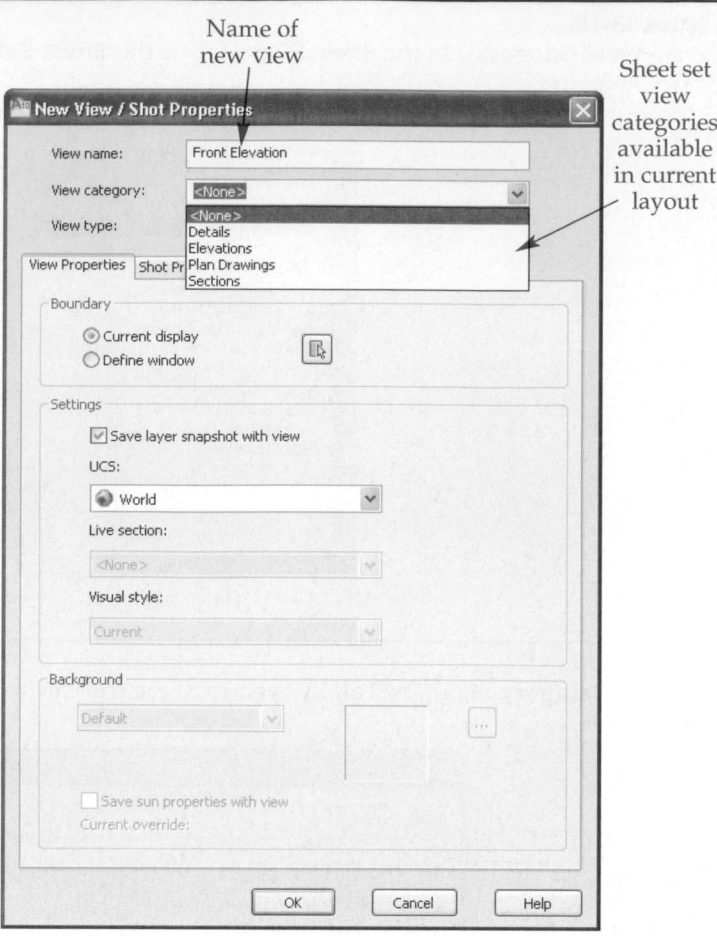

The newly saved view now appears in the **Sheet Set Manager** under the view category selected in the **New View** dialog box.

Once you add a view to a sheet set, display the view from the **Sheet Set Manager** by double-clicking on the view name or by right-clicking on the name and selecting **Display**. If the drawing file is already open, the view is set current. If the drawing file is not open, the file opens so that the view can be set current.

You can display the view list by category or sheet using the appropriate button shown in **Figure 33-18**. Pick the **View by category** button to display all categories, and access views by expanding the category and the sheet. Pick the **View by sheet** button to display the sheet name. Expand the sheet name to display saved views within the sheet.

Creating Sheet Views from Resource Drawings

resource drawings: Drawing files that contain model space views referenced for use as sheet views.

You can also create sheet views from *resource drawings* listed in the **Model Views** tab. The sheet view can be the entire model space drawing or a model space view. When you insert a model space view or drawing into a sheet, the resource drawing becomes an external reference of the sheet drawing.

Folders containing reference drawings appear in the **Model Views** tab, as shown in **Figure 33-21**. To add a new folder, double-click on the Add New Location entry or pick the **Add New Location** button and select a folder. You cannot select specific drawing files—you must select the folder containing the drawing. You can only insert drawings listed in the **Model Views** tab into a sheet to create a new sheet view. If a required drawing does not appear, you must add the folder containing the drawing to the resource drawing list.

Figure 33-21.
You can create sheet views by inserting model space views and drawings from the **Model Views** tab. To see a thumbnail and details, hover the cursor over the drawing or view.

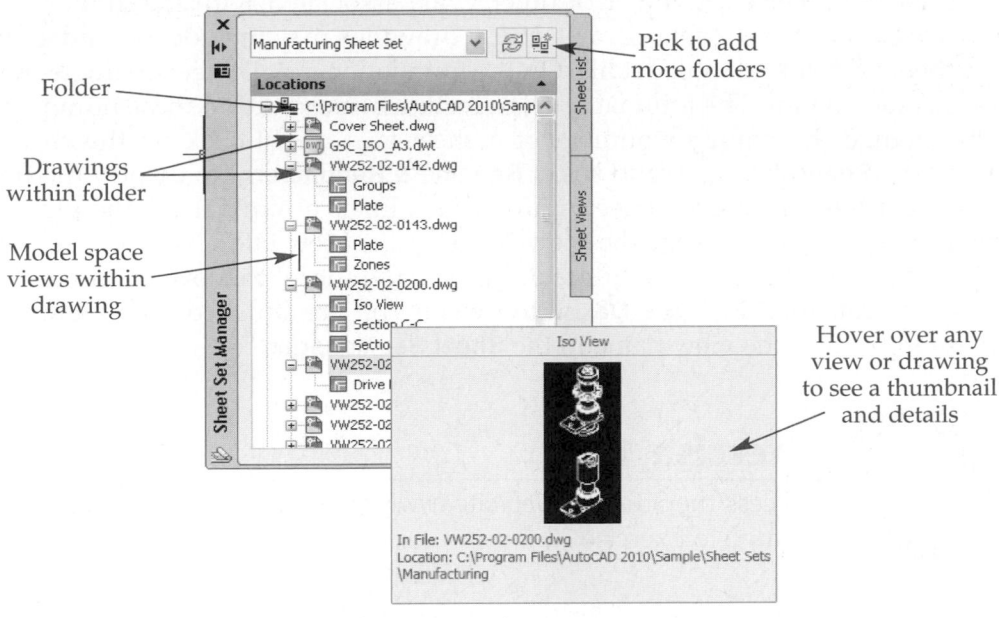

The folder and all of the drawing files found in the folder now appear in the **Locations** list area. The model space views saved in the drawing are listed under the drawing file. The options available for a drawing file are located in the drawing file shortcut menu. To display the menu, right-click on a drawing file.

Pick the **Open** option to open the drawing file and set model space current. Double-clicking on the drawing file also opens the file. Select the **Open read-only** option to open the drawing file as read-only so that file changes are not possible. The **Place on Sheet** option inserts the file into the current sheet as a sheet view. AutoCAD prompts you to specify an insertion point and creates a viewport automatically in the sheet. The **See Model Space Views** option expands the list of model space views in the drawing. This is the same as picking the + sign next to the drawing file. Select the **eTransmit** option to open the **Create Transmittal** dialog box to package the selected file and its associated files together.

If you have not saved a model space view in the drawing, it is listed under the drawing file name. You can insert the model space view as a sheet view in a sheet. To do so, right-click on the model space view name, select **Place on Sheet**, and pick an insertion point in the sheet.

When you insert a model space view or drawing into a sheet, AutoCAD creates a viewport and an xref to the selected drawing. AutoCAD assigns a scale for the viewport, or you can right-click before selecting the insertion point and select the scale for the sheet view. The scale is stored as the **ViewportScale** property of the **SheetView** field and is often displayed in the view label block.

When you create sheet views from resource drawings, an entry is added to the **Sheet Views** tab. If you insert a model space view, the view name is added to the **Sheet Views** tab. If you insert a drawing, the drawing name is added to the **Sheet Views** tab. To delete a location from a sheet set, right-click on the location and select **Remove Location**.

Naming and Numbering Sheet Views

Most projects contain several components, elevations, sections, or details. These items typically receive a number or other value associated with the drawing set for reference. For example, a set of architectural drawings may include a foundation plan and a sheet with foundation details. On the foundation detail sheet, a unique number identifies each detail. The foundation plan includes references to these numbers.

To change the name or number of a sheet view, right-click on the sheet view name in the **Sheet Views** tab and select **Rename & Renumber...** to display the **Rename & Renumber View** dialog box. See **Figure 33-22.** Enter a number for the view in the **Number** text box, and modify the view name in the **View title** text box. Use the **Next** and **Previous** buttons to access different sheets in the subset to move to different views in the view category. Pick the **OK** button when you are finished. The view number displays in front of the view name in the **Sheet Set Manager**.

Exercise 33-5

Access the Student Web site (www.g-wlearning.com/CAD) and complete Exercise 33-5.

Working with Sheet View Blocks

The components, elevations, sections, details, and other drawings shown in sheet views are often located on a specific sheet and referenced on different sheets. When using sheet views, you can insert blocks to identify the sheet view name, number, and scale on the sheet with the sheet view and the sheet that refers to the sheet views. Typically, callout blocks and view label blocks are required for these applications. Using sheet view blocks can help automate the process of adding drawing titles, labels, and similar annotations.

Using Callout Blocks

callout block: A block inserted to indicate a reference to another sheet.

A *callout block* refers to a sheet view. For example, a section line drawn through a building receives a callout block at the end of the section line. The callout block indicates the sheet or the location of the section view and provides information about the viewing direction. You can also use a callout block on a foundation plan, for example, to identify an area addressed by a detail drawing. The callout block is typically located on a different sheet from the sheet view it references.

Several styles of callout blocks are available from AutoCAD to use for different types of sheet views. See **Figure 33-23.** The upper value in a callout block is typically the sheet view number, and the lower value is the drawing on which the sheet view appears. In the default callout blocks, the upper value is an attribute containing the **ViewNumber** property of the **SheetView** field. See **Figure 33-24.** This lists the sheet view

---

Figure 33-22.
Use the **Rename & Renumber View** dialog box to renumber a view.

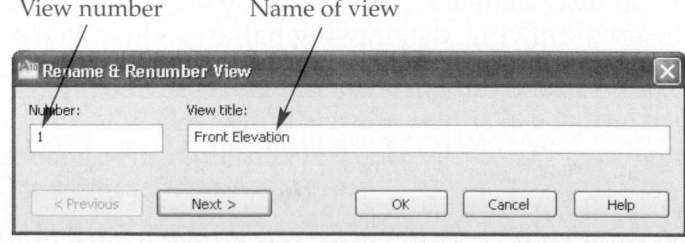

Figure 33-23.
Callout blocks provide reference information for views and sheets. A—Elements of an elevation symbol. B—Examples of commonly used callout blocks.

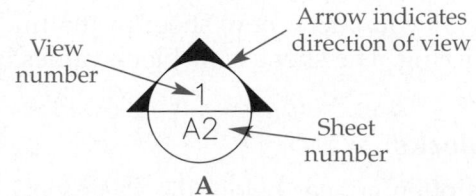

A

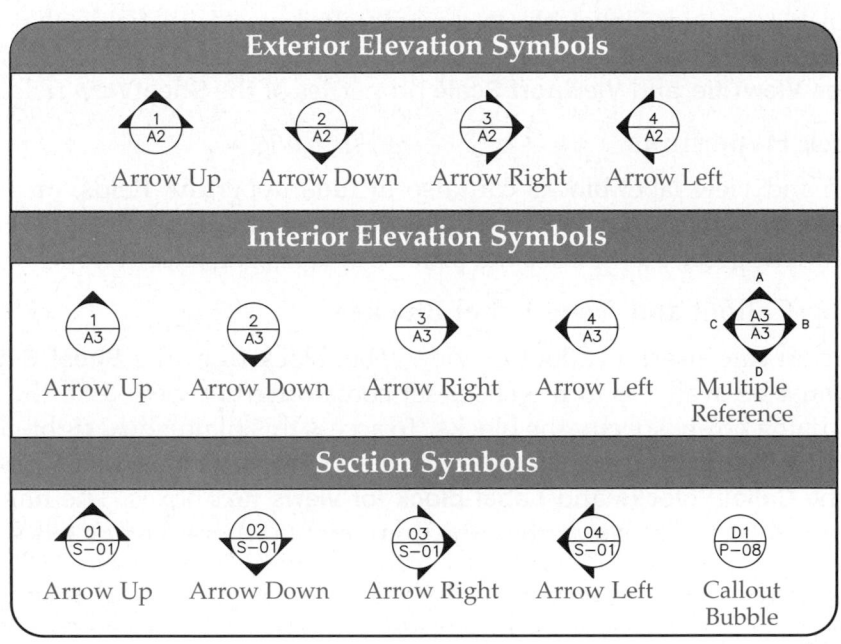

B

Figure 33-24.
The **ViewNumber** property displays the sheet view number. Callout blocks use this field property.

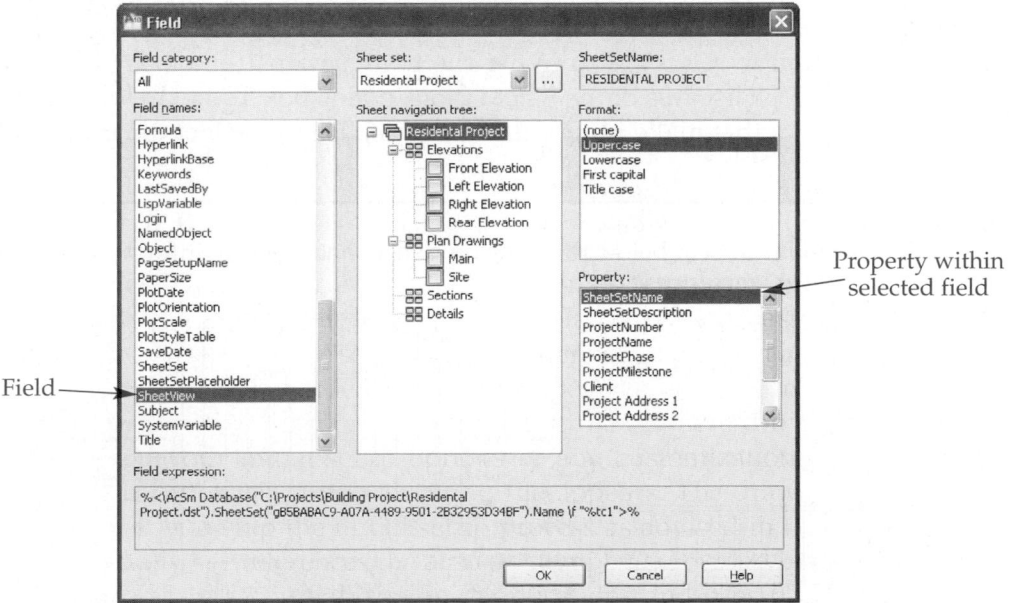

number specified for the sheet view. The lower value is the **SheetNumber** property of the **SheetSet** value. This lists the sheet number of the sheet containing the sheet view.

Sheet view blocks use fields to update attribute values if changes occur to the sheet set. For example, if you add a new sheet in the middle of a sheet set, all later sheets require renumbering. The sheet view block values update automatically as the sheet numbers change.

Using View Label Blocks

View label blocks often appear below the sheet view, as shown in **Figure 33-25**. Like callout blocks, view label blocks include attributes with fields that update to reflect changes to the sheet set or sheet views. View label blocks typically include the **ViewNumber**, **ViewTitle**, and **Viewport Scale** properties of the **SheetView** fields.

Using Block Hyperlinks

Callout and view label blocks can also include hyperlink fields, or *hyperlinks*. You can pick the hyperlink on a callout block to access the referenced detail, section, or elevation. This greatly simplifies the process of accessing sheet views.

Associating Callout and View Label Blocks

Before you can insert a callout or view label block from the **Sheet Set Manager**, the block must be available to the sheet set containing the view. Use the **Sheet Set Properties** dialog box to specify the blocks. To access this dialog box, right-click on the sheet set name in the **Sheet Set Manager** and select **Properties...**. Specify the available blocks in the **Callout blocks** and **Label block for views** text boxes. The name of each block appears, followed by the path to the drawing file where the block is saved.

PROFESSIONAL TIP

A sheet set or view category can have multiple callout blocks available, but only one view label block.

To add a callout block to a sheet set, pick in the **Callout blocks** text box and then pick the ellipsis (...) button to open the **List of Blocks** dialog box. See **Figure 33-26**. Pick the **Add...** button to display the **Select Block** dialog box. In this dialog box, pick the ellipsis (...) button to select the drawing file that contains the block. Then select the block from the block list area of the **Select Block** dialog box. If the drawing file consists of only the objects that make up the drawing file, use the **Select the drawing file as a**

Figure 33-25.
View labels normally appear below the view on a sheet and indicate information such as the view name, number, and scale.

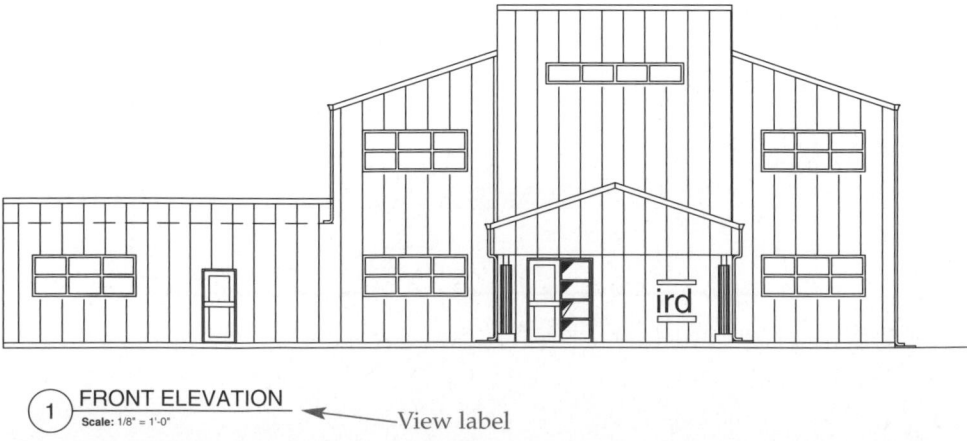

FRONT ELEVATION
Scale: 1/8" = 1'-0"
View label

Figure 33-26.
All callout blocks available to a sheet set appear in the **List of Blocks** dialog box.

Pick to access **Select Block** dialog box

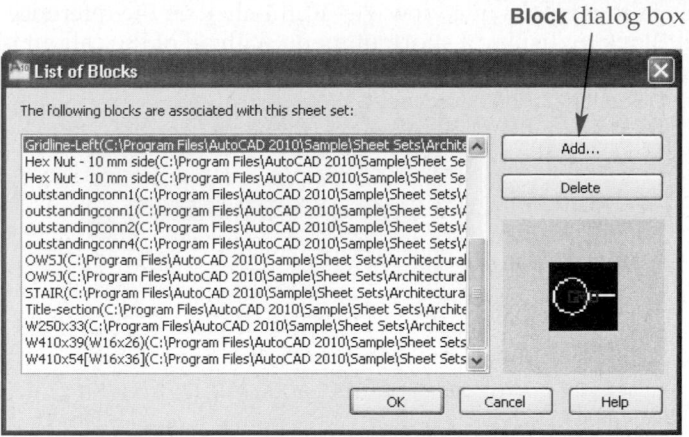

block option. To delete a block from the block list, select it in the **List of Blocks** dialog box and pick the **Delete** button.

Specifying a view title block is similar to specifying a callout block. However, only one view title block can be specified for the sheet set, so the **List of Blocks** dialog box does not display. You can assign each view category its own callout blocks. This way, only the blocks needed for the views in a category are available. For example, a category named Section may only need a section callout bubble, while a category named Elevation may need ten different types of elevation symbols. To modify the callout blocks available for a view category, right-click on the category name and select **Properties...** to open the **View Category** dialog box.

By default, the callout blocks and view title block assigned to a sheet set appear in the block list area. To make a block available to the view category, check the box next to the block. Refer again to **Figure 33-19**. This makes the block available to all of the views in the view category. To add new blocks to the view category, pick the **Add Blocks...** button to access the **Select Block** dialog box.

Inserting Callout and View Label Blocks

To insert a callout block, open the sheet on which the reference is to appear. In the **Sheet Views** tab of the **Sheet Set Manager**, right-click on the sheet view name and select the block from the **Place Callout Block** cascading submenu. See **Figure 33-27A**. Then specify an insertion point for the block. Follow the prompts to scale and rotate the block as needed. When you insert the block, AutoCAD gives the block the same sheet view number and sheet number as the reference view and sheet. See **Figure 33-27B**. AutoCAD automatically renumbers the block if the reference information changes.

The process of inserting a view label block is similar to that for inserting a callout block. In the **Sheet Set Manager**, right-click on the sheet view name and select **Place View Label Block**. Follow the prompts to specify an insertion point, scale, and rotation angle. After you insert the block, the label appears with the view name and number. If the view name or number changes in the sheet set, AutoCAD automatically updates the information.

Exercise 33-6

Access the Student Web site (www.g-wlearning.com/CAD) and complete Exercise 33-6.

Figure 33-27.
Placing callout blocks in a view. A—Right-click on the reference view name and select **Place Callout Block** to display a shortcut menu with all of the callout blocks available. B—Callout blocks placed in the 1-Main Floor Plan view in the A-01 sheet to reference the section view named 1-Section in the A-05 sheet.

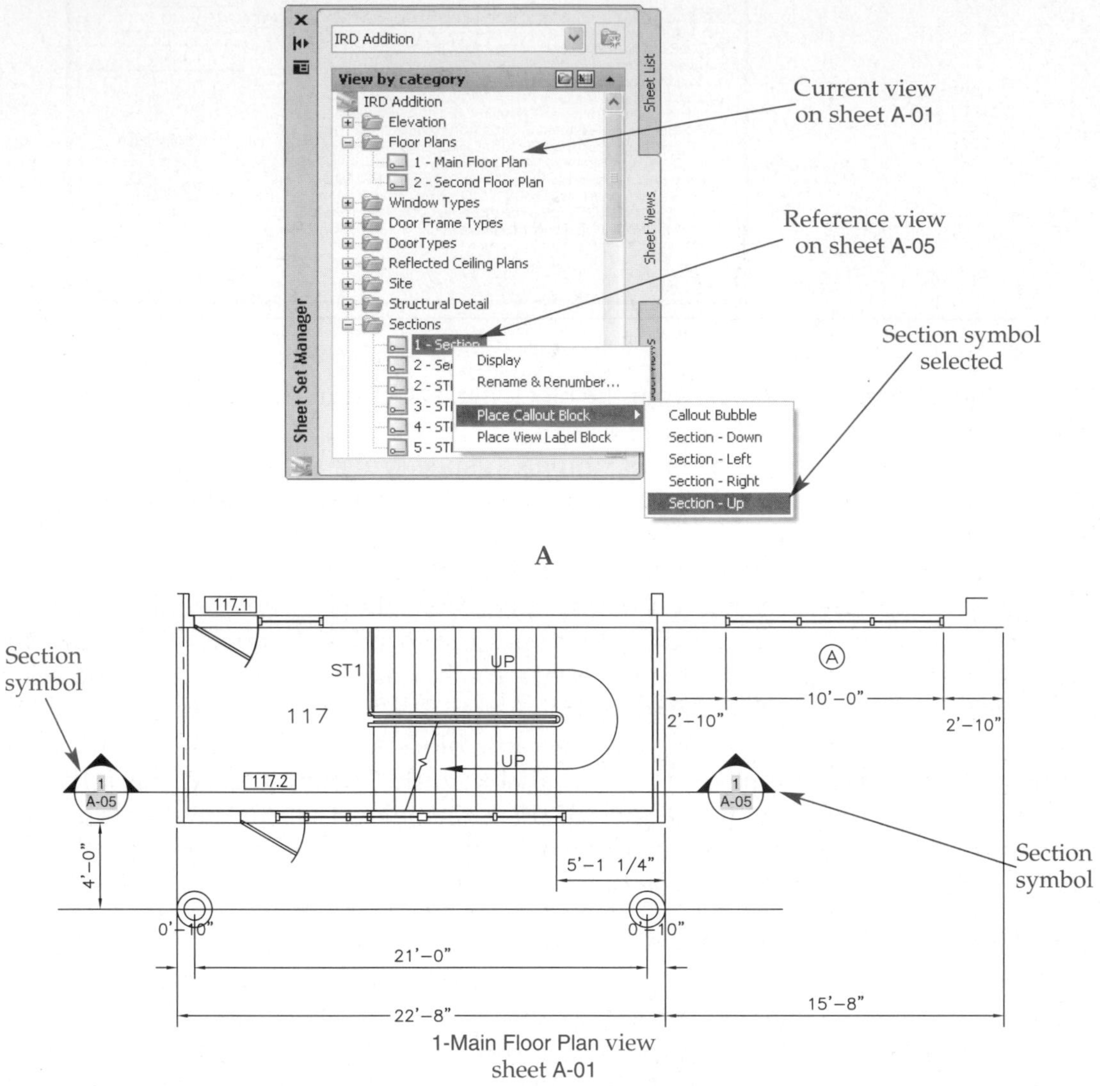

A

B

1-Main Floor Plan view
sheet A-01

Sheet Set Fields

Fields are valuable features for sheet sets. As the project develops, you can set field text on sheets to display up-to-date information when changes occur. For example, create title block items such as the number of sheets, drawing number, project number, and the date the sheet was plotted as fields to automate the process of making changes to the values. To create a field, access the **Field** dialog box from within the **MTEXT** or **TEXT** tool, from the ribbon, or by typing FIELD.

Specifying Fields for Sheet Sets

AutoCAD provides specific field types for use with sheet sets. In the **Field** dialog box, pick **SheetSet** from the **Field category:** drop-down list to display a list of predefined field types in the **Field names:** list box. See **Figure 33-28.** Use these fields to display

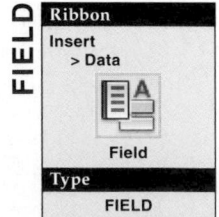

FIELD

Ribbon
Insert
> Data

Field

Type

FIELD

Figure 33-28.
Select **SheetSet** in the **Field category:** list in the **Field** dialog box to display the many fields related to sheet sets.

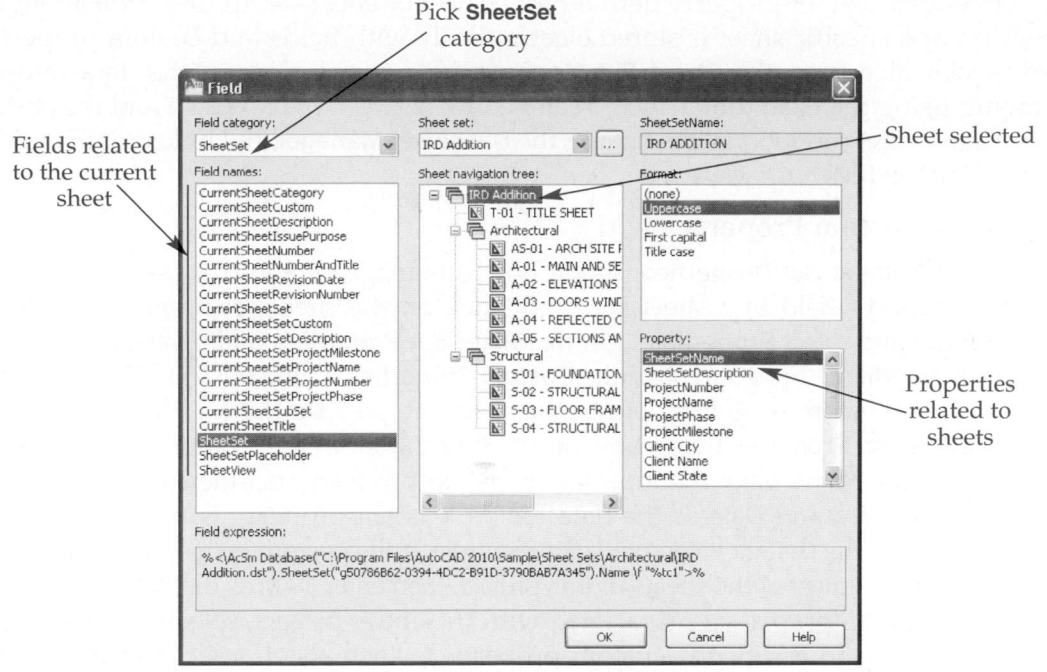

values defined in the sheet, sheet view, or sheet set, such as the sheet title, number, and description. Some of the fields also have several properties. Select one of the field types or properties to display the related value in the **Field** dialog box. For example, select the **CurrentSheetNumber** field to insert a field that displays the sheet number of the current sheet. If the sheet renumbers at a later date, the field changes to display the most current information.

Select the **SheetSet** field to display the **Sheet navigation tree**, which provides options for inserting several values. If you select the sheet set at the top of the tree, a set of properties related to the entire set appears in the **Property** list box. These properties include settings that can apply to all sheets in the set, such as project information and client information. **SheetSet** field properties display the same values on all sheets.

Pick a sheet in the **Sheet navigation tree** to display properties related to the individual sheet, including **SheetTitle**, **SheetNumber**, **Drawn By**, and **Checked By** values. When you include these fields in the sheet set template title block, each sheet can display a unique value. If you add a new sheet to a sheet set, the fields automatically update.

Select the **SheetSetPlaceholder** field to insert a field that contains a *placeholder*. Choose a placeholder from the **Placeholder type:** list box to assign a temporary value to the associated field, such as SheetNumber. Placeholders insert temporary field values in user-defined callout blocks and view labels. When defined with attributes in a callout block, placeholders update to display the correct values when you insert the block onto a sheet from the **Sheet Set Manager**.

placeholder: A temporary value for a field.

Select the **SheetView** field to display the **Sheet navigation tree** with a view list for the sheet set. Sheet set properties appear when you pick the sheet set name in the **Sheet navigation tree**. These properties are identical to those displayed with the **SheetSet** field. Sheet view properties display when you pick a sheet view name in the **Sheet navigation tree**. These properties are specific to a sheet view and include **ViewTitle**, **ViewNumber**, and **ViewScale**. Sheet view fields are common in callout and view label blocks.

Using Custom Properties

Select the **CurrentSheetCustom** or **CurrentSheetSetCustom** field to insert a field that links to a custom property defined for a sheet or sheet set. Information about the sheet set or a specific sheet is stored electronically with fields and custom properties and is viewable from the **Sheet Set Manager**. You can also insert the data into the drawing using the **Field** tool, which creates a link between the text data and the custom field data. When you modify the data in the **Sheet Set Manager**, the linked data updates in the drawing files.

Adding a Custom Property Field

Use the **Sheet Set Properties** dialog box to manage custom properties. To add a custom property field to a sheet set, right-click on the sheet set name in the **Sheet Set Manager** and select **Properties…**. In the **Sheet Set Properties** dialog box, pick the **Edit Custom Properties…** button to open the **Custom Properties** dialog box. See **Figure 33-29**.

To add a custom property field to the sheet set, pick the **Add…** button to display the **Add Custom Property** dialog box. See **Figure 33-30**. Enter a name for the custom property in the **Name** field. Examples of a custom property include Job Number, Client Name, Checked By, and Date. If the data for the custom property is usually the same value, enter it in the **Default value** field. For example, if the custom property is Checked by, and DAM checks most of the sheets in this project, then enter DAM as the default value.

If the custom property is associated with the entire project, select **Sheet Set** from the **Owner** area. To assign the custom property to each sheet, select **Sheet** from the **Owner** area. When you select **Sheet**, the custom property is available in the **Sheet Properties** dialog box and the data attaches to each sheet. Pick the **OK** button to add the custom property to the sheet set.

Entering Custom Property Data

To modify or enter information into a custom property field for a sheet set, open the **Sheet Set Properties** dialog box and modify the value. If a sheet includes custom properties, individual sheets display the custom fields. To modify or enter information into a sheet custom property field, right-click on the sheet and select **Properties…**. Custom properties are listed under the **Sheet Custom Properties** heading of the **Sheet Properties** dialog box. See **Figure 33-31**.

Deleting a Custom Property

To delete a custom property field that you no longer need, right-click on the sheet set and select **Properties…** to open the **Sheet Set Properties** dialog box. Then pick the **Edit Custom Properties…** button. In the **Custom Properties** dialog box, select the custom property to remove and pick the **Delete** button.

Figure 33-29.
Attach information to a sheet set in the **Custom Properties** dialog box.

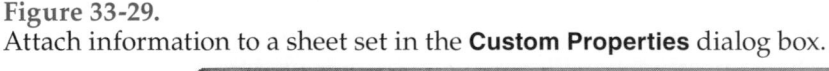
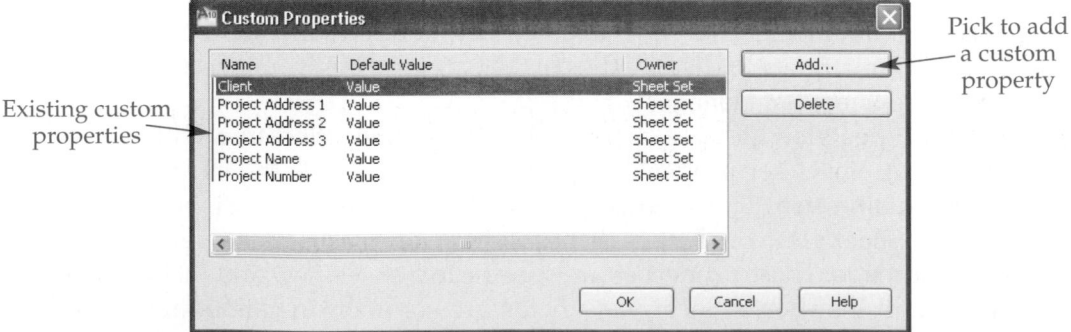

Figure 33-30.
Enter the information for the custom property in the **Add Custom Property** dialog box.

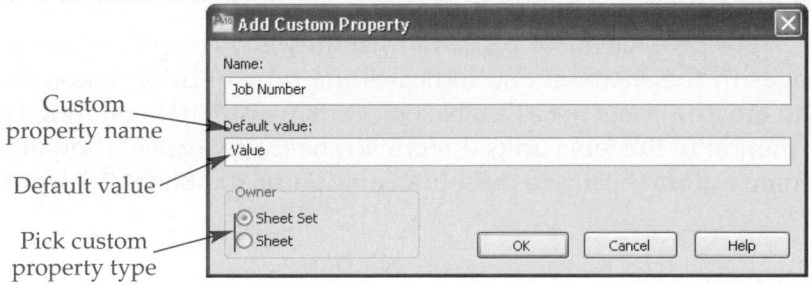

Custom property name

Default value

Pick custom property type

Figure 33-31.
After you add custom properties to individual sheets, the properties are available in the **Sheet Properties** dialog box.

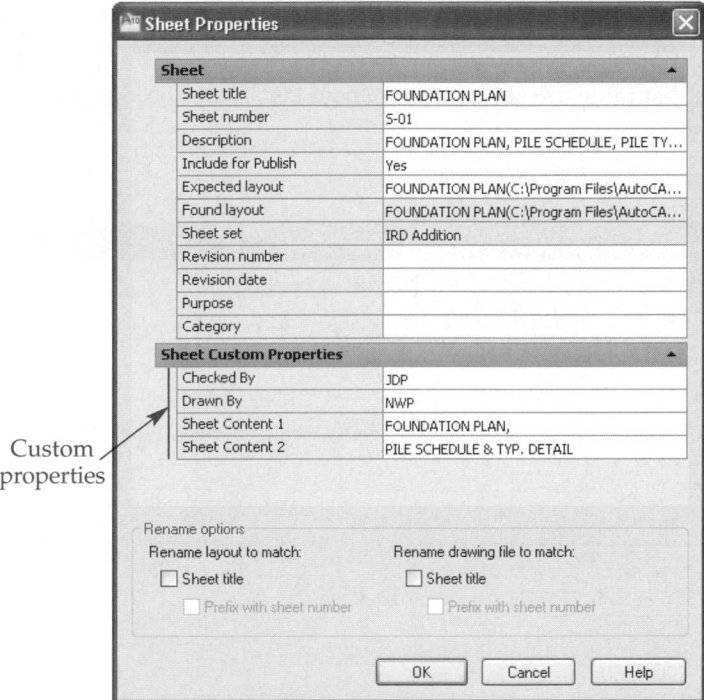

Custom properties

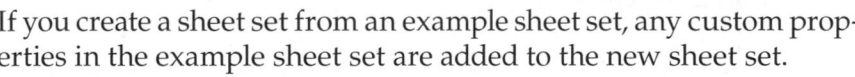

NOTE

If you create a sheet set from an example sheet set, any custom properties in the example sheet set are added to the new sheet set.

Exercise 33-7

Access the Student Web site (www.g-wlearning.com/CAD) and complete Exercise 33-7.

Creating a Sheet List Table

sheet list: A list of all the pages in a sheet set and the type of information found on each sheet.

One of the first pages of a sheet set typically includes a *sheet list*. A sheet list displays all of the pages in the sheet set and indicates the type of information found on each sheet. You can create a sheet list as a table object using information from sheet properties. The information in the table links directly to sheet properties, allowing the sheet list table to update automatically if sheet information in the **Sheet Set Manager** changes.

Inserting a Sheet List Table

To insert a sheet list table, access the **Sheet Set Manager** and open the sheet that will receive the table. Once you open a sheet, right-click on the sheet set name in the **Sheet Set Manager** and select **Insert Sheet List Table...** to open the **Sheet List Table** dialog box. See **Figure 33-32**.

Figure 33-32.
Properties for the sheet list table are set up in the **Sheet List Table** dialog box.

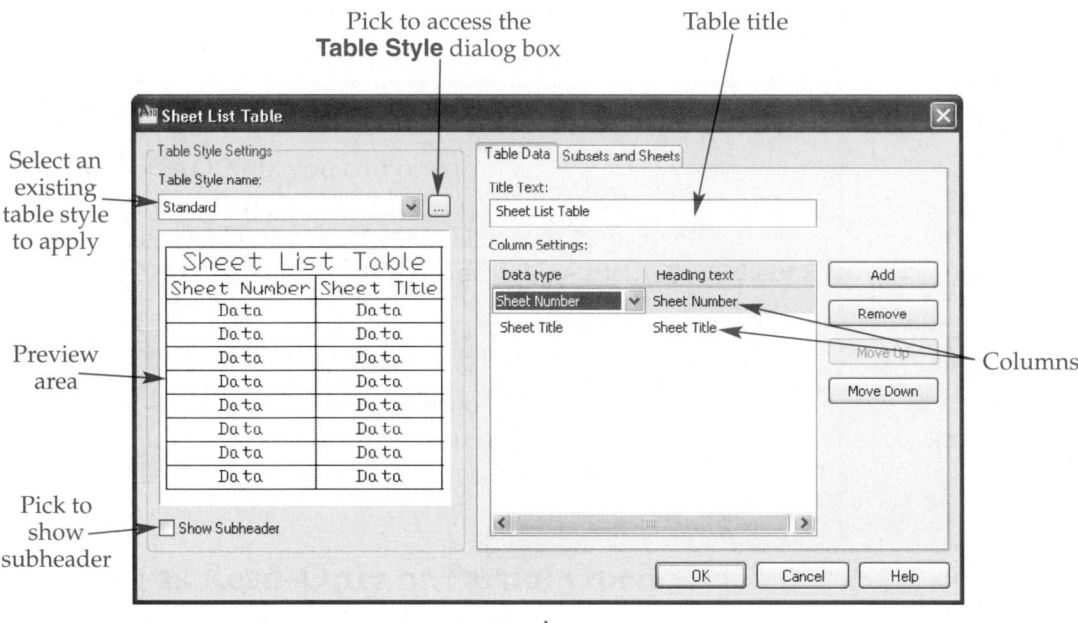

A

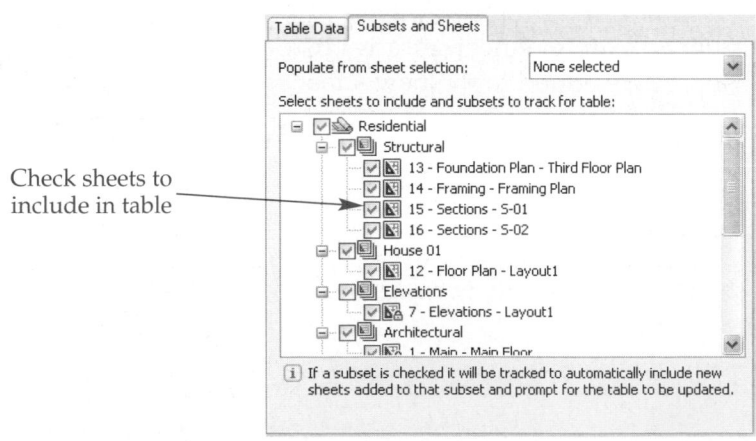

B

AutoCAD and Its Applications—Basics

Select a preset table style for the sheet list from the **Table Style name** drop-down list, or pick the ellipsis (...) button to access the **Table Style** dialog box. A representation of the table is displayed in the preview area. The **Show Subheader** check box determines whether the table includes a subheader row. **Figure 33-33** shows a table that includes Architectural and Structural subheaders created by referencing subset names.

Specifying the Title and Columns

Use the **Table Data** tab, shown in **Figure 33-32A,** to specify the table title, columns, and column organization. Enter the title for the sheet list in the **Title Text** text box. A sheet list table can include various types of information from the drawing file and the sheet set. The default table includes the Sheet Number and Sheet Title fields. To specify the data type, pick the name in the **Data type** column to activate a drop-down list and then select a property from the list. Select the heading text in the **Heading text** column to enter a new value for the column heading.

The data types that are available in the drop-down list come from sheet set properties and drawing properties. To add a different data type to the list, you need to add a custom property to the sheet set. Use the **Add, Remove, Move Up,** and **Move Down** buttons as necessary to add, remove, and organize columns. A new column initially displays under the last column in the list. The column at the top of the list appears on the far right side of the table as the first column in the table.

Specifying Rows

The **Subsets and Sheets** tab, shown in **Figure 33-32B,** allows you to specify table rows by selecting check boxes corresponding to sheets to include in the table. Each row is a sheet reference. If you right-click on a sheet set to access the **Sheet List Table** dialog box, all subsets and sheets in the sheet set are automatically checked for addition to the table. Right-clicking on a subset or sheet initially limits the sheets in the table, although you can add and remove sheets from the table using the appropriate check boxes.

Checked sheets are the only items that appear in the table. You do not need to check subsets to display subheaders. However, you must check sheet sets and subsets to include sheet sets and subsets during updates. If you make changes to the sheet set, a prompt appears to update only the subsets you check in the **Subsets and Sheets** tab. If a sheet selection is available from the **Populate from sheet selection** drop-down list, choosing the selection checks only the sheets associated with the saved selection.

After you specify the table content, pick the **OK** button and select an insertion point to insert the table. **Figure 33-33** shows a sheet list table that uses the sheet number and sheet description fields.

Figure 33-33.
A sheet list table displays information about selected sheets in the sheet set.

SHEET INDEX	
Sheet Number	Sheet Description
T-01	SHEET INDEX, VICINITY MAP, BUILDING CODE ANALYSIS
Architectural	
AS-01	ARCHITECTURAL SITE PLAN, NOTES
A-01	MAIN FLOOR PLAN, SECOND FLOOR PLAN, WALL TYPE NOTES
A-02	EXTERIOR ELEVATIONS
A-03	DOOR & FRAME SCHEDULE, ROOM FINISH SCHEDULE, DOOR, DOOR FRAME & WINDOW TYPES
A-04	MAIN & SECOND FLOOR REFLECTED CEILING PLANS
A-05	STAIR SECTIONS AND DETAILS
Structural	
S-01	FOUNDATION PLAN, PILE SCHEDULE, PILE TYPICAL DETAIL
S-02	STRUCTURAL SECTIONS AND DETAILS
S-03	FLOOR FRAMING PLAN AND SECTIONS
S-04	STRUCTURAL SECTIONS

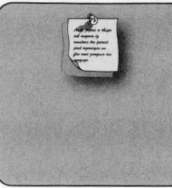

Editing a Sheet List Table

The information in the sheet list table links directly to the data source field. For example, if you modify the sheet numbers in the **Sheet Set Manager**, the sheet list table can update to reflect those changes. To update, select the sheet list table in the drawing, right-click, and select **Update Sheet List Table**. The option is also available from the **Sheet List Table** cascading submenu that appears when you select inside a table header or data cell and then right-click.

To modify the properties for the table, select inside a table header or data cell and then right-click. Choose the **Edit Sheet List Table Settings…** option from the **Sheet List Table** cascading submenu to reopen the **Sheet List Table** dialog box. After making the changes, pick the **OK** button to update the sheet list table.

Using Sheet List Table Hyperlinks

If you include the **Sheet Number** or **Sheet Title** columns in the sheet list table, hyperlinks are assigned to the data automatically. To open a sheet using a hyperlink, hover the crosshairs over a sheet number or sheet title until the hyperlink icon and tooltip appear. The tooltip displays the message CTRL + click to follow link. Hold down the [Ctrl] key and pick the hyperlink to open the selected sheet. This is another way to open a sheet quickly.

Exercise 33-8

Access the Student Web site (www.g-wlearning.com/CAD) and complete Exercise 33-8.

Archiving a Sheet Set

archiving:
Gathering and storing all of the electronic drawing files related to a project.

At different periods throughout a project, you may want to *archive* the drawing set. For example, when you present a set of drawings in a project to the client for the first time, the client may want to make some changes. It may be wise to archive the files for future reference before making modifications. Archive copies of all drawing and related files to a single location. Related files include external references, font files, plot style table files, and template files.

Figure 33-34.
A—The **Archive a Sheet Set** dialog box allows you to specify files to archive and archive settings. B—Use the **Files Tree** tab to archive documents that relate to a project along with the AutoCAD files.

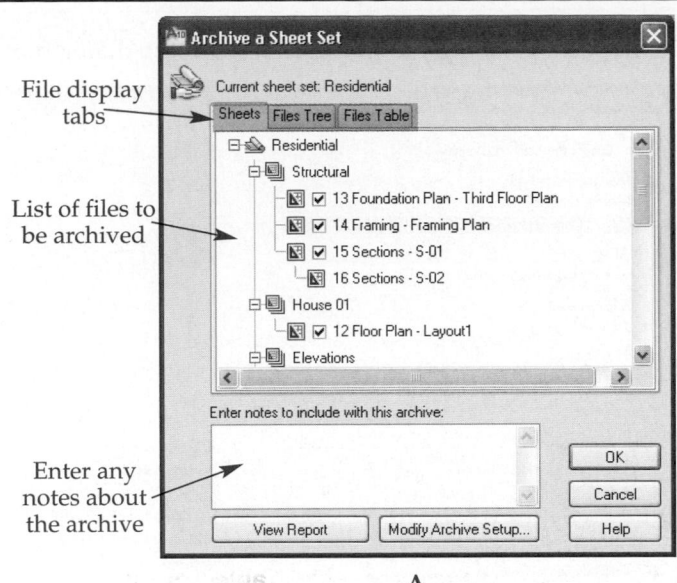

File display tabs

List of files to be archived

Enter any notes about the archive

A

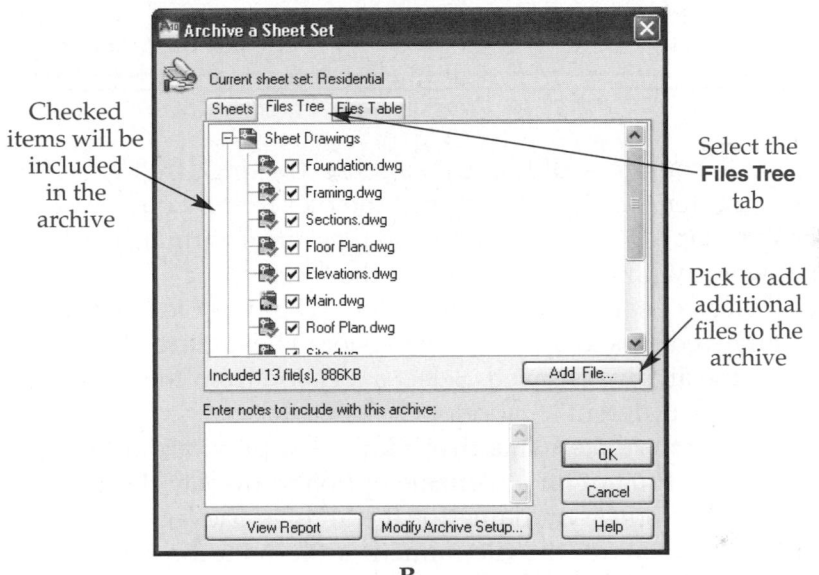

Checked items will be included in the archive

Select the **Files Tree** tab

Pick to add additional files to the archive

B

To archive a sheet set, right-click on the sheet set name in the **Sheet Set Manager** and select **Archive...**, or type ARCHIVE. The **Archive a Sheet Set** dialog box opens. See **Figure 33-34.** The **Sheets** tab, shown in **Figure 33-34A,** displays all of the subsets and sheets in the sheet set. Check the sheets to archive. Drawing files and related files are listed in the **Files Tree** tab, shown in **Figure 33-34B**. Pick the + sign next to a file to display related files.

You can include a file that is not part of the sheet set in the archive by picking the **Add a File** button on the **Files Tree** tab. This opens the **Add File to Archive** dialog box. You can include any type of file with the archive—the archive is not limited to AutoCAD files. To include a file, type the file in the **Enter notes to include with this archive** text box. The **View Report** button lists all of the files included in the archive. Pick the **Save As...** button to save the information to a text file.

The **Modify Archive Setup** dialog box, shown in **Figure 33-35** allows you to specify the location to which the archive is saved, the type of archive, and additional settings. To open the dialog box, pick the **Modify Archive Setup...** button. Use the **Archive package type** drop-down list to specify the archive format. Choose the **Folder (set of files)** option to copy all archived files into a single folder. Select the **Self-extracting executable (*.exe)**

Figure 33-35.
Specify archive file settings using the **Modify Archive Setup** dialog box.

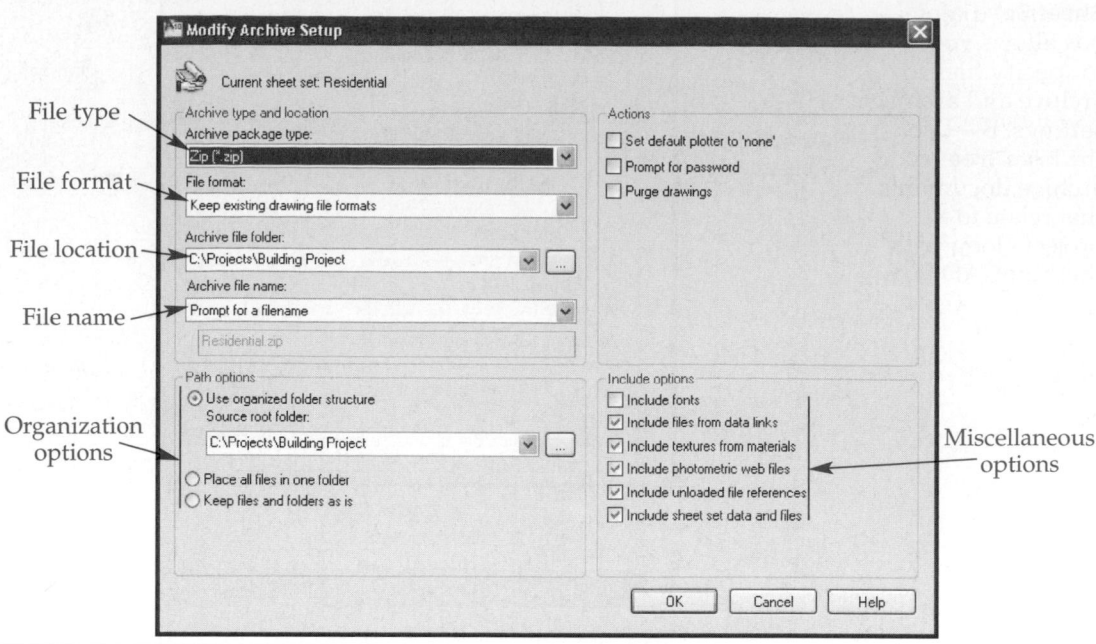

option to compress all files into a self-extracting *zip file*. Self-extracting zip files have an EXE extension. You can extract the files later by double-clicking on the file. Choose the **Zip (*.zip)** option to compress all files into a normal zip file. You must use a program that works with zip files to extract the files.

Select an earlier version of AutoCAD in the **File Format** drop-down list to convert the archived files to the selected version. The **Archive file folder** drop-down list defines where the archive is saved. Select a location from the list or pick the **Browse...** button to choose a different location.

The **Archive file name** drop-down list provides options for naming the archive. Choose the **Prompt for a filename** option to display the **Specify Zip File** dialog box so that you can specify a name for the archive package. Pick the **Overwrite if necessary** option to overwrite the file name if a file with the same name already exists. Select the **Increment file name if necessary** option to create a new file with an incremental number added to the file name if a file with the same name already exists. You can save multiple versions of the archive package using this option.

Use the **Archive Options** area to specify additional settings for the archive package. The first option determines how the folder structure is saved. If you select the **Use organized folder structure** radio button, the archive file duplicates the folder structure for the files, and the **Source root folder** setting determines the root folder for files that use relative paths, such as xrefs. To archive all files into a single folder, pick the **Place all files in one folder** radio button. Select the **Keep files and folders as is** radio button to use the same folder structure for all the files in the sheet set. Pick the **Include fonts** check box to include all fonts used in the drawings in the archive. The **Set default plotter to 'none'** check box disassociates the plotter name from the drawing files. This is useful if you send files to someone using a different plotter. Select the **Prompt for password** check box to set a password for the archive. The password is then required to open the archive package. Pick the **Include sheet set data and files** check box to include the sheet set data file with the archive package.

Chapter Test

Answer the following questions. Write your answers on a separate sheet of paper or go to the Student Web site (www.g-wlearning.com/CAD) and complete the electronic chapter test.

1. What is a sheet set?
2. What does the term *sheet* refer to in relation to a sheet set and a drawing file?
3. Briefly explain two methods for creating a new sheet set.
4. What file extension applies to sheet sets?
5. What are subsets in relation to a sheet set?
6. What is the purpose of the **Create subsets based on folder structure** option in the **Import Options** dialog box?
7. Explain how to create a new subset in a sheet set and specify a template file and layout for creating new sheets in the subset.
8. List two ways to open a sheet from the **Sheet Set Manager**.
9. What is the purpose of the **Import Layouts as Sheets** dialog box?
10. How do you modify a sheet name or number?
11. Briefly explain how to publish a sheet set to a DWF, DWFx ,or PDF file. How are the sheets organized in the resulting file?
12. How do you create a sheet selection set?
13. What are sheet views and how can they be referenced to each other within a sheet set?
14. What tab in the **Sheet Set Manager** allows you to manage sheet views?
15. Briefly explain how to create a view category for a sheet set and associate callout blocks to the category.
16. Explain how to add a drawing file to a sheet set to place views in the drawing on a sheet.
17. What information do the upper and lower values in a callout block typically provide?
18. Explain why AutoCAD callout blocks and view labels automatically update when you make changes to the related sheet set.
19. Explain how to insert a callout block into a drawing.
20. List three examples of fields that are often used in a sheet set.
21. What is the purpose of custom sheet set properties?
22. How do you add a custom property to a sheet set?
23. What is a sheet list?
24. Explain how to add a column heading to a sheet list table.
25. How can you update a sheet list table to reflect changes made in the **Sheet Set Manager**?
26. What happens to edits made manually to a sheet list table when the **Update Sheet List Table** tool is used?
27. What is the purpose of archiving a sheet set?
28. What is a zip file?
29. List the three packaging types available for archiving a sheet set.
30. How can you password-protect an archive?

Drawing Problems

▼ Basic

1. Create a new sheet set using the **Create Sheet Set** wizard and an example sheet set. Use the Civil Imperial Sheet Set example sheet set. Name the new sheet set Civil Sheet Set. Finish creating the sheet set.

2. Create a new sheet set using the **Create Sheet Set** wizard and an example sheet set. Use the New Sheet Set example sheet set. Name the new sheet set My Sheet Set. Finish creating the sheet set.

▼ Intermediate

3. Create a new sheet set using the **Create Sheet Set** wizard and the **Existing drawings** option. Name the new sheet set Schematic Drawings. On the **Choose Layouts** page, pick the **Browse...** button and browse to the folder where the P30-14.dwg file from Chapter 30 is saved. Import all of the layouts from the file into the new sheet set. Continue creating the sheet set as follows:
   A. In the **Sheet Set Properties** dialog box, assign the layout named ISO A1 Layout from the Tutorial-mMfg.dwt template file in the AutoCAD 2010 Template folder as the sheet creation template.
   B. Open a new drawing file using the template of your choice and create a block for a view label. Save the drawing file and then assign the block to the sheet set using the **Label block for views** setting in the **Sheet Set Properties** dialog box.
   C. Create a new view category and name it Schematics.
   D. Open the 3 Wire Control layout, create a new view, and add it to the Schematics view category. Double-click on the new view name in the **Sheet Views** tab and insert the view label block you previously created. Renumber the view and save the drawing.
   E. Add a custom property to the sheet set named Checked by and set the **Owner** type to **Sheet**. Add another custom property named Client and set the **Owner** type to **Sheet Set**.

4. Create a new sheet set using the **Create Sheet Set** wizard and an example sheet set. Use the Architectural Imperial Sheet Set example sheet set. Name the new sheet set Floor Plan Drawings. Finish creating the sheet set. Under the Architectural subset, create a new sheet named Floor Plan. Number the sheet A1. In the **Model Views** tab, add a new location by browsing to the folder where the P18-16.dwg file from Chapter 18 is saved. Open the P18-16.dwg file and continue as follows:
   A. Create three model space views named Kitchen, Living Room, and Dining Room. Orient each display as needed to describe the area of the floor plan. Save and close the drawing.
   B. Open the A1-Floor Plan sheet. Create a new layer named Viewport and set it current.
   C. In the **Model Views** tab, expand the listing under the P18-16.dwg file. Right-click on each view name and select **Place on Sheet**. Insert each view into the layout. Delete the default view labels inserted with the views. Double-click inside each viewport and set the viewport scale as desired.
   D. In the **Sheet Views** tab, renumber the views. Insert a new view label block under each view.
   E. Save and close the drawing.

5. Open the Floor Plan Drawings sheet set created in Problem 33-4. Create an archive of the sheet set using the self-extracting zip executable (EXE) file format.

Drawing Problems - Chapter 33

▼ Advanced

6. Use a word processor to write a report of approximately 250 words explaining the purpose of sheet sets. Cite at least three examples from actual industry applications of using sheet sets to help manage a set of drawings. Use at least two sketches to illustrate your report.

7. Plan a new shopping center for your area. Determine how many stores will be included. If possible, obtain a copy of a survey for vacant land in your area suitable for building the shopping center. Determine the components of a complete set of plans for the shopping center, including a site plan, floor plans, foundation plans, roof plans, elevations, and any needed sections. Establish the components for a new sheet set to organize the drawings and layouts.

8. Plan a new residence with approximately 3500–4000 square feet, four bedrooms, three baths, a den/office, kitchen, dining room, nook, family room, and three-car garage. Create a complete set of plans for the residence, including a site plan, floor plans, foundation plans, roof plans, elevations, and any needed sections. Prepare layouts for each drawing. Then create a new sheet set to organize the drawings and layouts. Archive the final sheet set.

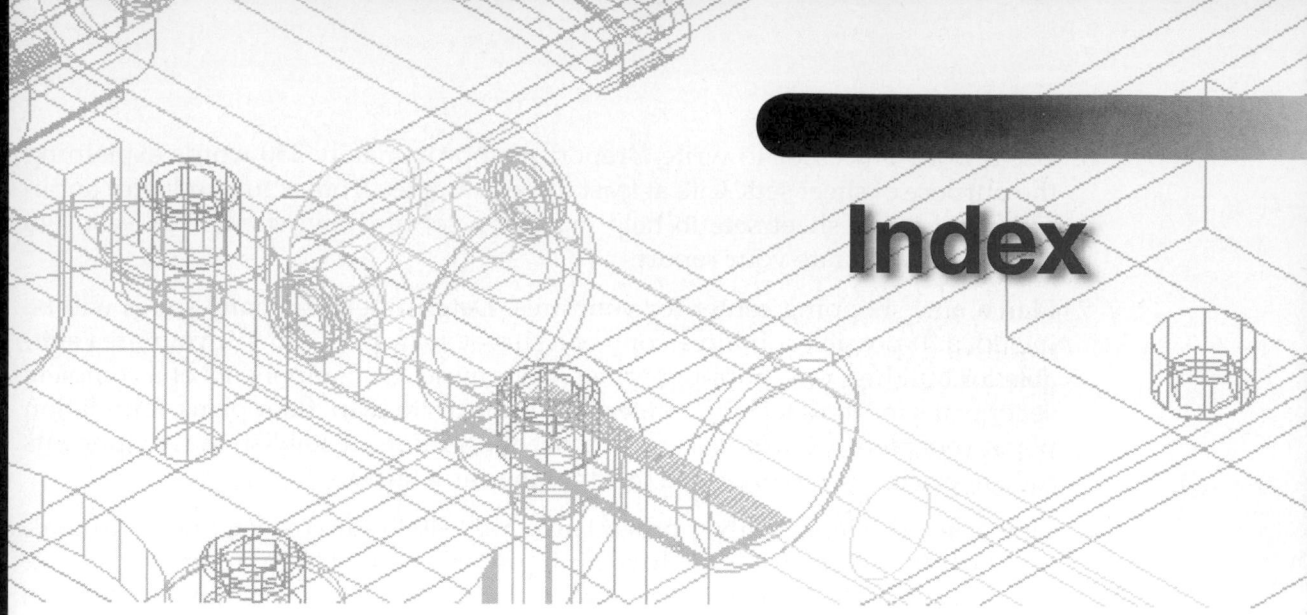

Index

B

AutoCAD and Its Applications—Basics

Rotation option, 115–116
Enclosed Text with Object, 290
Endpoint object snap, 197, 199, 232
Enhanced Attribute Editor, 705–707
    Attribute tab, 705–706
    Properties tab, 706–707
    Text Options tab, 706–707
equal constraints, 591
ERASE tool, 77, 92, 93, 94, 96, 155, 295, 557, 600, 806, 807, 811
escape key, 45
eTransmit feature, 917
EXIT tool, 27
EXPLODE tool, 345, 409, 559, 681–682
exploding objects, 345
EXPORTLAYOUT tool, 783
expressions, 319–320
    entering with **QuickCalc**, 431–432
Express Tools, 290, 412, 639
Extended Intersection object snap override, 202
extending objects, 337–338
EXTEND tool, 335, 337–338
extension line settings, 457
extension lines, oblique, 563–564
Extension object snap, 201
extension path, 201
External References palette, 864–865, 869–873, 881
external references, 859–881
    adding layout content, 862
    additional placement options, 866
    attachment vs. overlay, 864–865
    binding, 877–878
    clipping, 875–876
    creating multiview drawings, 860, 862
    demand loading and xref editing controls, 876
    detaching, reloading, and unloading, 872
    editing, 878–881
    inserting, 867
    list view display, 870–871
    Manage Xrefs icon, 873–874
    managing, 869–877
    opening xref files, 881
    placing with **DesignCenter** and **Tool Palettes**, 867
    placing xref drawings, 863–867
    preparing xref and host drawings, 863
    referencing existing geometry, 860
    selecting path type, 865–866
    tree view display, 872

    updating xref path, 873
    viewing details or a preview, 872
    working with xref objects, 867–869
    xref applications, 860, 862–863
    xref file types, 860
EXTERNALREFERENCES tool, 864
extract, 699
extracted objects, 879

F

Fast Select, 412
features, 447
Federal Supply Code for Manufacturers (FSCM), 711
fence selection, 95
Field dialog box, 281–283, 322, 429, 714, 908–909
fields, 281, 429, 885
    displaying information with, 429
    editing, 283
    inserting, 281–282
    inserting in table cells, 315
    referencing attributes, 714
    sheet set, 908–911
    updating, 282
FIELD tool, 281, 429, 714, 908
Field tool, 322
files
    closing, 57–58
    finding, 60–61
    opening old, 62
    opening saved, 58–62
    recovering damaged, 57
fillet radius, 328–329
FILLET tool, 327–329
    Multiple option, 329
    Polyline option, 328
    Radius option, 327–328
    Trim option, 328–329
fillets, 122, 327–329
    dimensioning, 502
    multiple, 329
FILL mode, 124
filtering layers, 157–160
Find and Replace dialog box, 288–289, 706
Find and Replace tool, 288
Find dialog box, 60
FIND tool, 288–289, 706
fit curve, 356
fit data, editing, 361–362
fit format, 464

fit options, 464–465
fit points, 361–362
fix constraint, 591
flip actions, 738
flip parameters, 737–738
float, 31
floating viewports, 186, 776, 806–822, 862
    activating and deactivating, 813
    adjusting boundaries, 811–813
    adjusting views, 815–816
    controlling layer display, 816–821
    converting objects into, 810
    creating, 807–810
    locking and unlocking, 816
    maximizing, 821–822
    polygonal, 809
    scaling, 814
    turning off objects, 821
flyout, 31
font, 251
fonts, samples of, 253
font style, setting, 251–252
foreshortened, 230
foreshortening, 652
formulas, 319–322
    other options, 322
    sum, average, and count, 320–322
fractions, 78
freezing and thawing layers, 155, 817–818
From snap, 207
front view, 229
Full Navigation Wheel, 179
full sections, 618
fully constrained, 582
function, 435
function keys, 45–46

G

gap tolerance, 631
Gb, 25
general notes, 448
geographic data, 674, 866
GEOMCONSTRAINT tool, 588–591
geometric constraints, 581
    adding, 587–595
    coincident, 588
    collinear and tangent, 590
    concentric and equal, 591
    constraint bars, 594
    horizontal and vertical, 589
    in **Block Editor**, 760–761

    managing, 594–595
    parallel and perpendicular, 589
    symmetric, fix, and smooth, 591
    using, 760–761
geometric dimensioning and tolerancing
    (GD&T), 546
Get Selection Set, 412
global attribute editing, 706–707
global layer settings, 816
global linetype scale, 152, 403
glossaries, 22
grab bars, 31
gradient fills, 631–633
graphic patterns, 615
graphical user interface (GUI), 28
graphics window, 28
grid, 80
grid and snap considerations, 82
GRID tool, 80
grip tools, 395
    Copy option, 399
grips, 310, 316, 317, 361
    adjusting dimension lines, 567
    adjusting leaders, 567
    copying objects, 399
    editing nonassociative hatch patterns,
        638–639
    mirroring objects, 398
    moving objects, 397
    rotating objects, 397
    scaling objects, 398
    stretching objects, 395–396
    using grip tools, 395
    using, 393–399
    viewing basic geometry dimensions,
        419
grips shortcut menu, 395
group filter, 158
grouped balloons, 576
GUI, 28
Guo Biao, 25
gutter, 271

H

half sections, 619
hard copy, 166
Hatch and Gradient dialog box, 620,
    624–627, 629, 633
    Gradient tab, 631–632
    Hatch tab, 620
hatch boundary

Mid Between 2 Points snap, 207
Midpoint object snap, 199, 206, 208–209
minor axis, 113
MINSERT tool, 675
mirroring objects, 376–377, 398
mirror line, 376
MIRROR tool, 376, 660
   using grips, 398
MIRRTEXT system variable, 377
MLEADER tool, 500, 503, 513–514
MLEADERALIGN tool, 506, 573–575, 853
MLEADERCOLLECT tool, 506, 576
MLEADEREDIT tool, 572
MLEADERSTYLE tool, 504–513
model, 70, 775
modeless dialog boxes, 38
model space, 70, 775
   dimension scale, 466
   plotting in, 166
   rotating content, 812–813
   working in, 70–71
Modify Archive Setup dialog box, 915–916
Modify Multileader Style dialog box, 505–506
   Content tab, 508–512
   Leader Format tab, 506
   Leader Structure tab, 507–508
Modify Table Style dialog box, 303
move action, 724, 732–733
MOVE tool, 371–372, 557, 638, 737, 806, 811
   Displacement option, 371
   using grips, 397
moving objects, 371–372
   using grips, 397
MSLTSCALE system variable, 815, 838
MSPACE tool, 778
MTEDIT tool, 287
MTEXT tool, 245, 256, 278, 714, 908
multileaders
   adding and removing leader lines, 572–573
   aligning, 573–575
   annotative, 833–834
   editing, 572–576
   grouping, 576
   inserting, 513–516
Multileader Style Manager, 504–505, 512
multileader styles, 504–513
   content settings, 508–512
   creating new, 505
   importing with DesignCenter, 513
   format settings, 506
   structure settings, 507–508
   modifying, 512

   renaming and deleting, 512
   setting current, 512
   working with, 504
multilines, 120
multiline text
   background mask, 262
   character formatting, 261
   columns, 268–271
   creating, 256–271
   importing text, 271
   lists, 266–268
   paragraph formatting, 262–266
   stacking, 258–259
   style settings, 261
   symbols, 259, 261
   using Text Editor, 256–258
multiple documents, 63–67
   additional window control tools, 66–67
   controlling windows, 63
   copy and paste between, 410
   Quick View Drawings tool, 63–64
multiview drawings, 228–231
   auxiliary views, 230–231
   choosing additional views, 229
   completing, 853
   constructing, 232, 234
   creating different scaled drawings, 848
   creating with xrefs, 860, 862
   front view, 229
   preparing, 848–853
   reusing annotative objects, 850–851, 853
MVIEW tool, 807, 808–809
   Lock option, 816
   Object option, 810
   ON and OFF options, 821
   Polygonal option, 809
MVSETUP tool, 813

N

named objects, purging, 687–688
named plot style table, 792–793
named views, creating, 184–186
naming drawings, 53–54
navigation wheels, 179
NAVSWHEEL tool, 179
Nearest object snap, 206, 737
nested xrefs, 860
nesting, 671
New (or Modify) Dimension Style dialog box, 455–456, 471

parameter options, 723
plot styles, 796
polar parameters, 732
rotation parameters, 734
scale action, 731
stretch action, 728, 729
Table and **Table Breaks** categories, 406–407
Text category, 404–405
turning off floating viewport objects, 821
Value Set category, 741, 764
PROPERTIES tool, 402
property filter, 158
Property Lookup Table dialog box, 756–758
Property Settings dialog box, 407
PSLTSCALE system variable, 815, 838
publish, 885
Publish shortcut menu, 898–899
purge, 678–688
Purge dialog box, 687–688
PURGE tool, 687–688

Q

QDIM tool, 487–488
editing dimensions, 564
QLEADER tool, 504
QNEW tool, 33, 52, 53
QSAVE tool, 55
QSELECT tool, 410
quadrant, 200
Quadrant object snap, 200
quadratic curve 356
Quick Access toolbar, 34–35
QuickCalc, 431–438
additional options, 438
clearing input and history, 433
converting units, 434
drawing values, 436
entering expressions, 431–432
palette, 431, 437
scientific calculations, 433
toolbar, 433, 436
using with **Properties** palette, 438
using with tools, 436–437
variables, 435–436
Quick Properties panel, 400–401, 403, 404, 412
assigning different dimension styles, 561
dimension style overrides, 565

editing blocks, 678
overriding specific multileader properties, 572
Quick Select dialog box, 410–412
Quick View Drawings tool, 58, 63–65, 780
Quick View Drawings toolbar, 64, 65
Quick View Layouts tool, 778–780
QVDRAWING tool, 780
QVLAYOUT tool, 778–780

R

radio button, 71–72
rays, 226–228
RAY tool, 228, 338
read-only, 62
real block, 672
realtime panning, 178
realtime zooming, 176, 826
recovering a damaged file, 57
rectangles, drawing, 121–124
RECTANGLE tool, 43, 118, 121–124
Area option, 123
Chamfer option, 121–122
Dimensions option, 123
Elevation and **Thickness** options, 124
Fillet option, 121–122
Rotation option, 123
Width option, 122–123
rectangular array, 378–380
rectangular coordinate dimensioning without dimension lines, 516
rectangular coordinates, 79
rectangular coordinate system, 448
rectangular pattern, 378–380
redlining, 520–522
REDO tool, 35, 98
redrawing, 190
REFEDIT tool, 681, 710, 878
reference dimension, 582
Reference Edit dialog box, 878–879
reference editing, 878–881
REGENAUTO tool, 190
regenerating, 190
region, 359, 629
regular polygons, 120
dimensioning, 450
relative coordinate entry, 88
relative coordinates, 84
relative path, 865
reload, 872
removed sections, 618

Index–Basics

Index–Basics

Index–Basics

AutoCAD
and Its Applications

ADVANCED

2010

by

Terence M. Shumaker
Faculty Emeritus
Former Chairperson
Drafting Technology
Autodesk Premier Training Center
Clackamas Community College, Oregon City, Oregon

David A. Madsen
President
Madsen Designs, Inc.

Faculty Emeritus
Former Chairperson
Drafting Technology
Autodesk Premier Training Center
Clackamas Community College, Oregon City, Oregon

Director Emeritus
American Design Drafting Association

Jeffrey A. Laurich
Instructor, Mechanical Design Technology
Fox Valley Technical College
Appleton, WI

Publisher
The Goodheart-Willcox Company, Inc.
Tinley Park, Illinois
www.g-w.com

Library of Congress Catalog Card Number 2009004717

ISBN 978-1-60525-162-2

1 2 3 4 5 6 7 8 9 – 10 – 14 13 12 11 10 09

The Goodheart-Willcox Company, Inc. Brand Disclaimer: Brand names, company names, company names, and illustrations for products and services included in this text are provided for educational purposes only and do not represent or imply endorsement or recommendation by the author or the publisher.

The Goodheart-Willcox Company, Inc., Safety Notice: The reader is expressly advised to carefully read, understand, and apply all safety precautions and warnings described in this book or that might also be indicated in undertaking the activities and exercises described herein to minimize risk of personal injury or injury to others. Common sense and good judgment should also be exercised and applied to help avoid all potential hazards. The reader should always refer to the appropriate manufacturer's technical information, directions, and recommendations; then proceed with care to follow specific equipment operating instructions. The reader should understand these notices and cautions are not exhaustive.

The publisher makes no warranty or representation whatsoever, either expressed or implied, including, but not limited to, equipment, procedures, and applications described or referred to herein, their quality, performance, merchantability, or fitness for a particular purpose. The publisher assumes no responsibility for any changes, errors, or omissions in this book. The publisher specifically disclaims any liability whatsoever, including any direct, indirect, incidental, consequential, special, or exemplary damages resulting, in whole or in part, from the reader's use or reliance upon the information, instructions, procedures, warnings, cautions, applications or other matter contained in this book. The publisher assumes no responsibility for the activities of the reader.

Cover source: © Marco Cristifori/zefa/Corbis

Library of Congress Cataloging-in-Publication Data

Shumaker, Terence M.
  AutoCAD and its applications: advanced, 2010/ by Terence
M. Shumaker, David A. Madsen—17th edition
      p. cm.

  Includes index
  ISBN: 978-1-60525-162-2
    1. Computer graphics. 2. AutoCAD. I. Madsen, David A.
II. Title.

T385.S46123 2010
  620'.00420285536—dc22   2009004717

Introduction

AutoCAD and Its Applications—Advanced provides complete instruction in mastering three-dimensional design and modeling using AutoCAD. This text also provides complete instruction in customizing AutoCAD and introduces programming AutoCAD. These topics are covered in an easy-to-understand sequence and progress in a way that allows you to become comfortable with the commands as your knowledge builds from one chapter to the next. In addition, *AutoCAD and Its Applications—Advanced* offers:

- Examples and discussions of industrial practices and standards.
- Professional tips explaining how to effectively and efficiently use AutoCAD.
- Exercises to reinforce the chapter topics. These exercises should be completed where indicated in the text as they build on previously learned material.
- Chapter tests for review of commands and key AutoCAD concepts.
- A large selection of modeling and customizing problems supplement each chapter. Problems are presented as 3D illustrations, actual plotted drawings, and engineering sketches.

Fonts Used in This Text

Different typefaces are used throughout each chapter to define terms and identify AutoCAD commands. Important terms appear in **bold-italic face, serif** type. AutoCAD menus, commands, variables, dialog box names, and toolbar button names are printed in **bold-face, sans serif** type. File names, folder names, and paths appear in the body of the text in Roman, sans serif type. Keyboard keys are shown inside of square brackets [ ] and appear in Roman, sans serif type. For example, [Enter] means to press the enter (return) key.

Other Text References

This text focuses on advanced AutoCAD applications. Basic AutoCAD applications are covered in *AutoCAD and Its Applications—Basics,* which is available from Goodheart-Willcox Publisher. *AutoCAD and Its Applications* texts are also available for previous releases of AutoCAD. For advanced AutoCAD programming applications, refer to the texts *Visual LISP Programming* and *VBA for AutoCAD,* which are both available from Goodheart-Willcox Publisher.

Introducing the AutoCAD Commands

There are several ways to select AutoCAD drawing and editing commands. Selecting commands from the ribbon is slightly different than entering them from the keyboard. When a command is introduced, the command-entry methods are illustrated in the margin next to the text reference.

The example in the margin next to this paragraph illustrates the various methods of initiating the **CONE** command to draw a solid cone primitive while the 3D environment is active. The 3D environment consists of a drawing file based on the acad3D.dwt template and the 3D Modeling workspace current, as described in Chapter 1. This book assumes the 3D environment is current for all procedures and discussions.

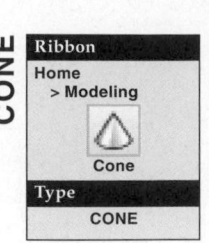

CONE

Ribbon
Home
> Modeling
Cone
Type
CONE

Flexibility in Design

Flexibility is the keyword when using *AutoCAD and Its Applications—Advanced*. This text is an excellent training aid for both individual and classroom instruction. It is also an invaluable resource for any professional using AutoCAD. *AutoCAD and Its Applications—Advanced* teaches you how to apply AutoCAD to common modeling and customizing tasks.

When working through the text, you will see a variety of notices. These include Professional Tips, Notes, and Cautions that help you develop your AutoCAD skills.

PROFESSIONAL TIP

These ideas and suggestions are aimed at increasing your productivity and enhancing your use of AutoCAD commands and techniques.

NOTE

A note alerts you to important aspects of a command, function, or activity that is being discussed. These aspects should be kept in mind while you are working through the text.

CAUTION

A caution alerts you to potential problems if instructions or commands are incorrectly used or if an action can corrupt or alter files, folders, or storage media. If you are in doubt after reading a caution, always consult your instructor or supervisor.

AutoCAD and Its Applications—Advanced provides several ways for you to evaluate your performance. Included are:

- **Exercises.** The student website contains exercises for each chapter. These exercises allow you to perform tasks that reinforce the material just presented. You can work through the exercises at your own pace. However, the exercises are intended to be completed when called out in the text.

- **Chapter test.** Each chapter includes a written test at the end of the chapter. Questions require you to give the proper definition, command, option, or response to perform a certain task. You may also be asked to explain a topic or list appropriate procedures. An electronic version of the test is available on the student website.

- **Drawing problems.** There are a variety of drawing, design, and customizing problems at the end of each chapter. These are presented as real-world CAD drawings, 3D illustrations, and engineering sketches. The problems are designed to make you think, solve problems, use design techniques, research and use proper drawing standards, and correct errors in the drawings or engineering sketches. Graphics are used to represent the discipline to which a drawing problem applies.

 These problems address mechanical drafting and design applications, such as manufactured part designs.

 These problems address architectural and structural drafting and design applications, such as floor plans, furniture, and presentation drawings.

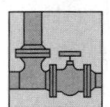 These problems address piping drafting and design applications, such as tank drawings and pipe layout.

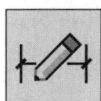 These problems address a variety of general drafting, design, and customization applications. These problems should be attempted by everyone learning advanced AutoCAD techniques for the first time.

NOTE

Some problems presented in this text are given as engineering sketches. These sketches are intended to represent the kind of material from which a drafter is expected to work in a real-world situation. As such, engineering sketches often contain errors or slight inaccuracies and are most often not drawn according to proper drafting conventions and applicable standards. Additionally, other drawings may contain errors or inaccuracies. Errors in these problems are *intentional* to encourage you to apply appropriate techniques and standards in order to solve the problem. As in real-world applications, sketches should be considered preliminary layouts. Always question inaccuracies in sketches and designs and consult the applicable standards or other resources.

Student Website

The student website is located at www.g-wlearning.com/CAD. On this home page, select the entry for *AutoCAD and Its Applications—Advanced* to access the material for this book. The student website contains the exercises and chapter test for each chapter. Additionally, the programming chapters, Chapters 27 through 30, are provided on the student website, not in the printed book. The appendix material is also presented on the student website. The icon shown in the margin here appears throughout the text to indicate a reference to the student website.

As you work through each chapter, exercises on the student website are referenced. The exercises are intended to be completed as the references are encountered in the text. The solid modeling tutorial in Appendix A should be completed after Chapter 14. The remaining appendix material is intended as reference material.

Also included on the student website is the student software supplement. This contains a variety of student activities that are intended to supplement the exercises on the student website. These activities are referenced within the appropriate exercises and can be completed as additional practice. For each chapter that includes associated student software supplemental activities, the student website offers a ZIP file containing the DWG files for the activities. Download the ZIP file and extract the DWG files to the folder of your choice. Then, open the appropriate DWG file and complete the activity.

About the Authors

Terence M. Shumaker is Faculty Emeritus, the former Chairperson of the Drafting Technology Department and former Director of the Autodesk Premier Training Center at Clackamas Community College. Terence taught at the community college level for over 28 years. He has professional experience in surveying, civil drafting, industrial piping, and technical illustration. He is the author of Goodheart-Willcox's *Process Pipe Drafting* and coauthor of the *AutoCAD and Its Applications* series (Releases 10 through 2009 editions) and *AutoCAD Essentials*.

David A. Madsen is the president of Madsen Designs, Inc. (www.madsendesigns. com). David is Faculty Emeritus, the former Chairperson of Drafting Technology and the Autodesk Premier Training Center at Clackamas Community College and former member of the American Design and Drafting Association (ADDA) Board of Directors. David was honored by the ADDA with Director Emeritus status at the annual conference in 2005. David was an instructor and a department chair at Clackamas Community College for nearly 30 years. In addition to community college experience, David was a Drafting Technology instructor at Centennial High School in Gresham, Oregon. David also has extensive experience in mechanical drafting, architectural design and drafting, and construction practices. He is the author of Goodheart-Willcox's *Geometric Dimensioning and Tolerancing* and coauthor of the *AutoCAD and Its Applications* series (Releases 10 through 2009 editions), *Architectural Drafting Using AutoCAD, AutoCAD Architecture and Its Applications, Architectural Desktop and Its Applications, Architectural AutoCAD*, and *AutoCAD Essentials*.

Jeffrey A. Laurich has been an instructor in Mechanical Design Technology at Fox Valley Technical College in Appleton, WI, since 1991. He has also taught business and industry professionals in the Autodesk Premier Training Center at FVTC. Jeff teaches drafting, AutoCAD, Autodesk MAP, 3ds max, and other programs. He created a certificate program at FVTC entitled Computer Rendering and Animation that integrates the 3D capabilities of AutoCAD and 3ds max. Jeff has professional experience in furniture design, surveying, and cartography and holds a degree in Natural Resources Technology. In his consulting business, Jeff uses 3ds max to create renderings and animations for manufacturers and architects. He was a contributing author on *AutoCAD and Its Applications—Advanced* for releases 2007, 2008, and 2009.

Acknowledgments

The authors and publisher would like to thank the following individuals and companies for their assistance and contributions.

Contributing Authors

The authors wish to acknowledge the following contributors for their professional expertise in providing in-depth research and testing, technical assistance, reviews, and development of new materials.

J.C. Malitzke for Chapter 3. J.C. is the department chair of Computer Integrated Technologies and professor of Mechanical Design and Drafting/CAD at Moraine Valley Community College in Palos Hills, IL. He also manages and teaches for the Authorized Autodesk Training Center at Moraine Valley. J.C. has been teaching for over 30 years and actively using and teaching Autodesk products for over 22 years. He is a founding member and past chair of the Autodesk Training Center Executive Committee and currently is chair of the Autodesk Advisory Board. J.C. has been the coauthor and principal investigator on two National Science Foundation grants. He has won numerous awards including: Educator of the Year by the Illinois Drafting Educators Association; the Instructor Quality Award by Autodesk; Autodesk University Instructor Award; Professor of the Year, Co-Innovator of the Year, and Co-Master Teacher awards at Moraine Valley Community College; and the Illinois Outstanding Faculty Member of the Year awarded by the Illinois Community College Trustees Association. J.C. holds a Bachelor's degree in Education and a Master's degree in Industrial Technology from Illinois State University.

Craig Black for Chapters 21, 23, 24, and 27 through 30. Craig is an instructor of Mechanical Design and the former manager of the Autodesk Premier Training Center at Fox Valley Technical College in Appleton, WI. He has been teaching at Fox Valley Technical College since 1990. Craig has served two terms on the Autodesk Training Center Executive Committee (now known as the Advisory Board) and chaired the committee in 2001. He has presented various topics at a number of Autodesk University annual training sessions and has been contracted to teach training sessions on Autodesk products across the United States. Craig not only teaches, but also does AutoCAD customization and AutoLISP and DCL programming for area businesses and industries. Prior to his current position, Craig worked in the civil, architectural, electrical, and mechanical drafting and design disciplines.

Adam Ferris for Chapters 22, 25, and 26. Adam is the Director of Digital Technology at Wilson Architects in Boston, MA. Wilson Architects is a medium-size architecture firm specializing in the science and technology field. Adam is an award-winning Autodesk Certified Instructor with over 14 years of experience on training Autodesk products in the Boston area. He has trained over 4000 students in New England. Prior to joining Wilson Architects, Adam owned an Autodesk products/CAD consulting firm aiding businesses in efficient AutoCAD configurations and management, network management, and 3D design with Autodesk 3ds max and Autodesk VIZ.

Contribution of Materials

Autodesk, Inc.
Bill Fane
CADENCE magazine
EPCM Services Ltd.
Fitzgerald, Hagan, & Hackathorn
Kunz Associates

Brief Contents

Three-Dimensional Design and Modeling

Model Visualization and Presentation

Customizing AutoCAD

Programming AutoCAD

Expanded Table of Contents

Three-Dimensional Design and Modeling

Model Visualization and Presentation

Customizing AutoCAD

Programming AutoCAD

www.g-wlearning.com/CAD

www.g-wlearning.com/CAD

www.g-wlearning.com/CAD

Chapter 30

Student Website Contents

Using the Student Website

Chapters 27 through 30

Chapter Exercises

Chapter Tests

Student Software Supplement

Appendices

Appendix A Solid Modeling Tutorial
Appendix B Common File Extensions
Appendix C AutoCAD Command Aliases
Appendix D Advanced Application Commands
Appendix E Advanced Application System Variables
Appendix F Basic AutoLISP Commands
Appendix G Toolbar Flyouts and Marking Menus

Reference Materials

Drawing Sheet Sizes, Settings, and Scale Parameters
Standards and Related Documents
Drafting Symbols
Standard Tables
Project and Drawing Problem Planning Sheet

Related Web Links

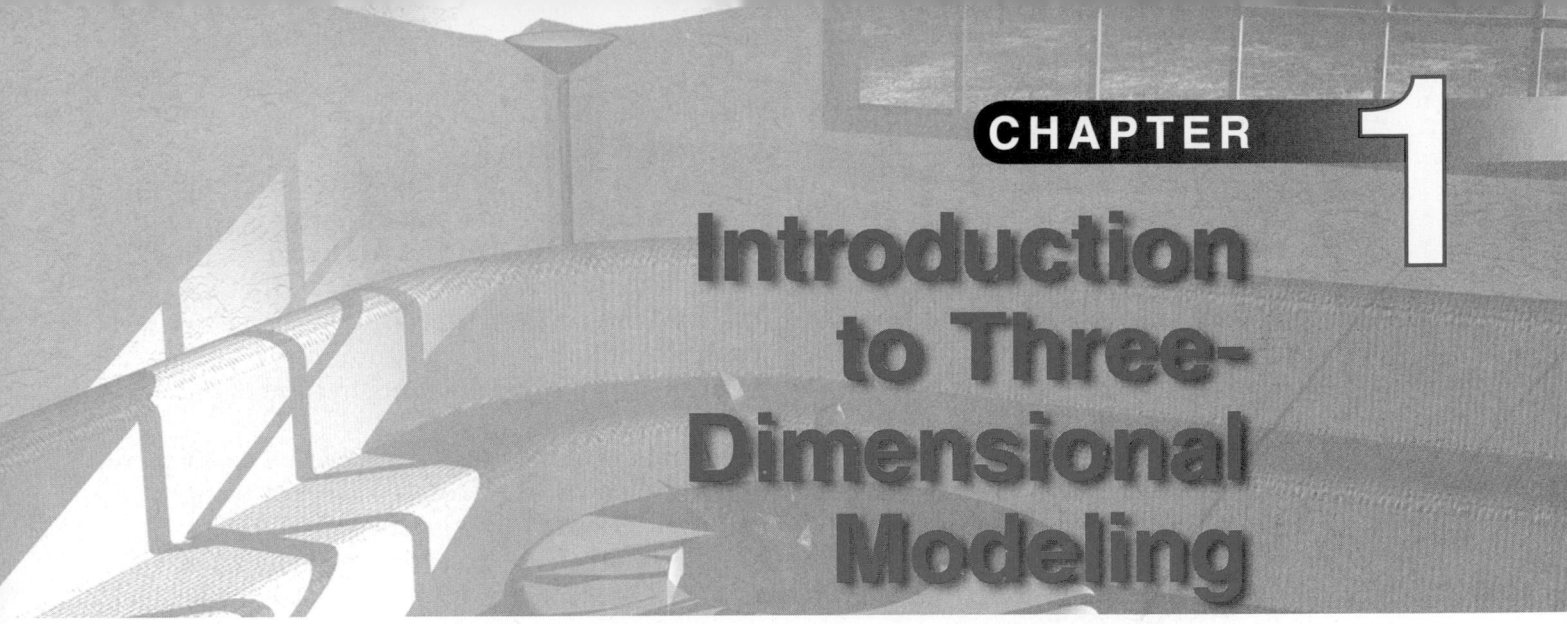

CHAPTER 1

Introduction to Three-Dimensional Modeling

Learning Objectives

After completing this chapter, you will be able to:

✓ Describe how to locate points in 3D space.
✓ Describe the right-hand rule of 3D visualization.
✓ Explain the function of the ribbon.
✓ Display 3D objects from preset isometric viewpoints.
✓ Display 3D objects from any desired viewpoint.
✓ Set a visual style current.

Three-dimensional (3D) design and modeling is a powerful tool for use in design, visualization, testing, analysis, manufacturing, assembly, and marketing. Three-dimensional models also form the basis of computer animations, architectural walk-throughs, and virtual worlds used with virtual reality systems and gaming platforms. Drafters who can design objects, buildings, and "worlds" in 3D are in demand for a wide variety of positions, both inside and outside of the traditional drafting and design disciplines.

The first twelve chapters of this book present a variety of techniques for drawing and designing in 3D. The skills you learn will provide you with the ability to construct any object in 3D and prepare you for entry into an exciting aspect of graphic communication.

To be effective in creating and using 3D objects, you must first have good 3D visualization skills. These include the ability to see an object in three dimensions and to visualize it rotating in space. These skills can be obtained by using 3D techniques to construct objects and by trying to see two-dimensional sketches and drawings as 3D models. This chapter provides an introduction to several aspects of 3D drawing and visualization. Subsequent chapters expand on these aspects and provide a detailed examination of 3D drawing, editing, visualization, and display techniques.

Using Rectangular 3D Coordinates

In two-dimensional drawing, you see one plane defined by two dimensions. These dimensions are usually located on the X and Y axes and what you see is the XY plane. However, in 3D drawing, another coordinate axis—the Z axis—is added. This results in two additional planes—the XZ plane and the YZ plane. If you are looking at a standard AutoCAD screen after AutoCAD is launched using the acad.dwt template, the positive Z axis comes directly out of the screen toward you. AutoCAD can only draw lines in 3D if it knows the X, Y, and Z coordinate values of each point on the object. For 2D drawing, only two of the three coordinates (X and Y) are needed.

Compare the 2D and 3D coordinate systems shown in **Figure 1-1**. Notice that the positive values of Z in the 3D coordinate system come up from the XY plane. In a new drawing based on the acad.dwt template, consider the surface of your computer screen as the XY plane. Anything behind the screen is negative Z and anything in front of the screen is positive Z.

The object in **Figure 1-2A** is a 2D drawing showing the top view of an object. The XY coordinate values of the origin and each point are shown. Think of the object as being drawn directly on the surface of your computer screen. However, this is actually a 3D object. When displayed in a pictorial view, the Z coordinates can be seen. Notice in **Figure 1-2B** that the first two values of each coordinate match the X and Y values of the 2D view. Three-dimensional coordinates are always expressed as (X,Y,Z). The 3D object was drawn using positive Z coordinates. Therefore, the object comes out of your computer screen when it is viewed from directly above. The object can also be drawn using negative Z coordinates. In this case, the object would extend behind, or into, the screen.

Study the nature of the rectangular 3D coordinate system. Be sure you understand Z values before you begin constructing 3D objects. It is especially important that you carefully visualize and plan your design when working with 3D constructions.

Figure 1-1.
A comparison of 2D and 3D coordinate systems.

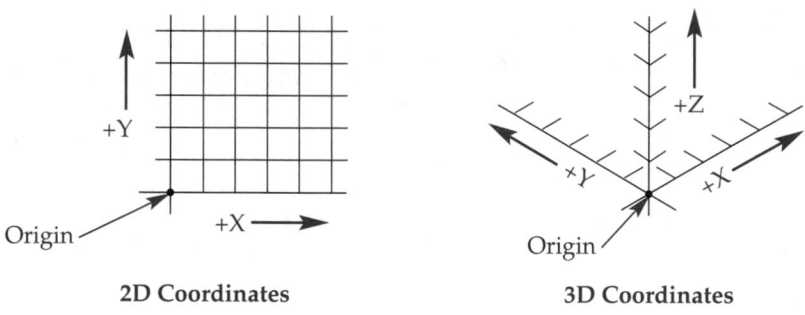

2D Coordinates   3D Coordinates

Figure 1-2.
A—The points making up a 2D object require only two coordinates. B—Each point of a 3D object must have an X, Y, and Z value. Notice that the first two coordinates (X and Y) are the same for each endpoint of a vertical line.

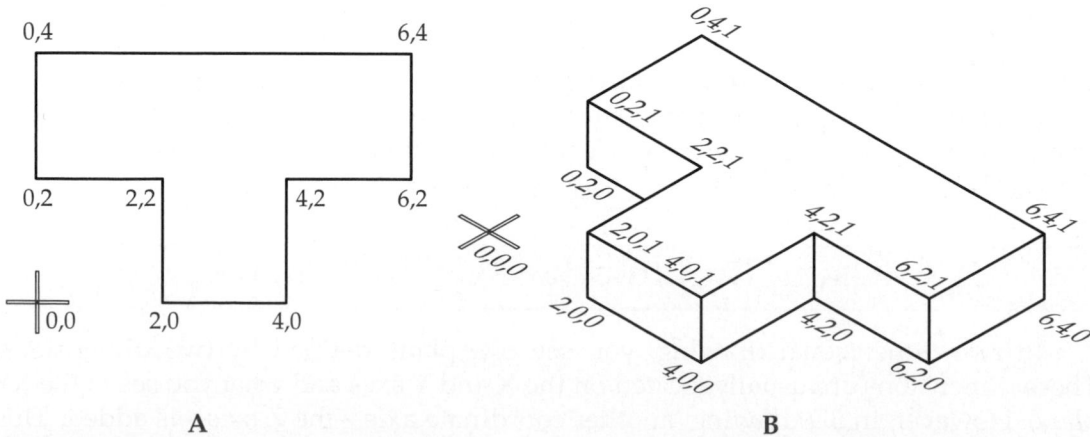

A   B

All points in three-dimensional space can be drawn using one of three coordinate entry methods—rectangular, spherical, or cylindrical. This chapter uses the rectangular coordinate entry method. Complete discussions on the spherical and cylindrical coordinate entry methods are provided in Chapter 6.

Exercise 1-1

Complete the exercise on the student website.
www.g-wlearning.com/CAD

Right-Hand Rule of 3D Drawing

In order to effectively draw in 3D, you must be able to visualize objects in 3D space. The *right-hand rule* is a simple method for visualizing the 3D coordinate system. It is a representation of the positive coordinate values in the three axis directions. Auto-CAD's world coordinate system (WCS) and various user coordinate systems (UCS) are based on this concept of visualization.

To use the right-hand rule, position the thumb, index finger, and middle finger of your right hand as shown in **Figure 1-3.** Although this may seem a bit unusual, it can do wonders for your understanding of the three axes. Imagine that your thumb is the X axis, your index finger is the Y axis, and your middle finger is the Z axis. Hold your hand in front of you so that your middle finger is pointing directly at you, as shown in **Figure 1-3.** This is the plan view of the XY plane. The positive X axis is pointing to the right and the positive Y axis is pointing up. The positive Z axis comes toward you and the origin of this system is the palm of your hand.

Figure 1-3.
Positioning your hand to use the right-hand rule to understand the relationship of the X, Y, and Z axes.

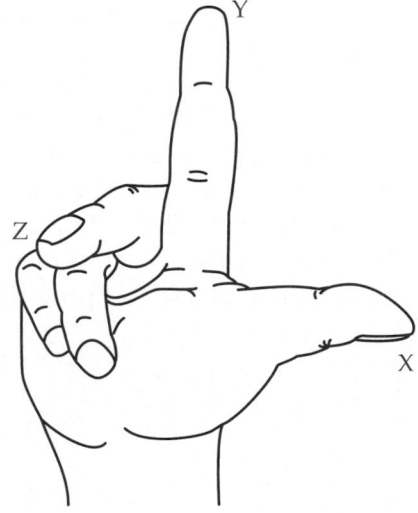

Figure 1-4.
A comparison of the
UCS icon and the
right-hand rule.

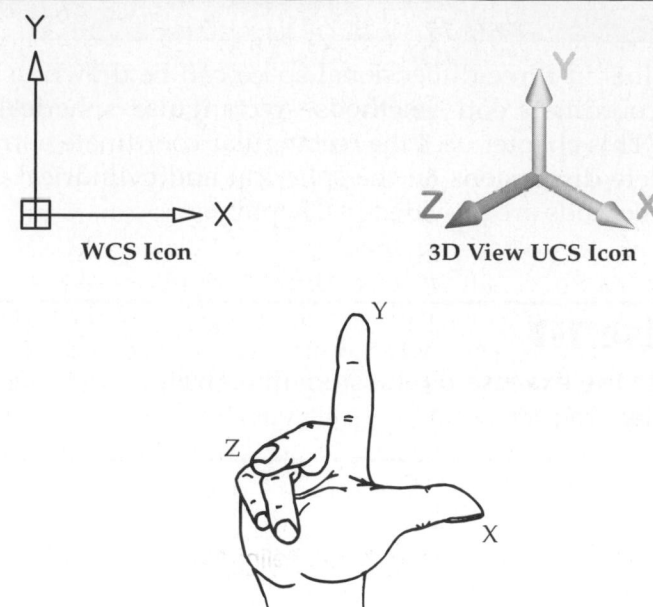

WCS Icon

3D View UCS Icon

Right-Hand Rule

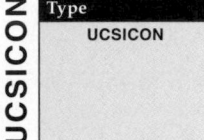
The concept behind the right-hand rule can be visualized even better if you are sitting at a computer and the AutoCAD graphics window is displayed. Make sure the current drawing is based on the acad.dwt template. If the UCS icon is not displayed in the lower-left corner of the screen, turn it on by using the **UCSICON** command. Now, orient your right hand as shown in **Figure 1-3** and position it next to the UCS (or WCS) icon. Your index finger and thumb should point in the same directions as the Y and X axes, respectively, on the UCS icon. Your middle finger will be pointing out of the screen directly at you, which is the Z axis. See **Figure 1-4.** Notice the illustration on the right in the figure. This is the shaded UCS icon. It is displayed when the visual style is not **2D Wireframe**. Visual styles are introduced later in this chapter and discussed in detail in Chapter 4.

The right-hand rule can be used to eliminate confusion when rotating the UCS. The UCS can rotate on any of the three axes, just like a wheel rotates on an axle. Therefore, if you want to visualize how to rotate about the X axis, keep your thumb stationary and turn your hand either toward or away from you. If you wish to rotate about the Y axis, keep your index finger stationary and turn your hand to the left or right. When rotating about the Z axis, you must keep your middle finger stationary and rotate your entire arm.

If your 3D visualization skills are weak or you are having trouble visualizing different orientations of the UCS, use the right-hand rule. It is a useful technique for improving your 3D visualization skills. Rotating the UCS around one or more of the axes can become confusing if proper techniques are not used to visualize the rotation angles. A complete discussion of UCSs is provided in Chapter 6.

Basic Overview of the Interface

AutoCAD provides three working environments tailored to either 2D or 3D drawing or annotating a drawing. These environments are called *workspaces* and can be quickly restored. The workspace for 2D development based on the traditional AutoCAD screen layout is called AutoCAD Classic. The 2D Drafting & Annotation workspace is designed for drawing in 2D and annotating a drawing. It is similar to a streamlined version of the AutoCAD Classic layout with the ribbon displayed in place of toolbars and pull-down menus. The workspace for 3D development is called 3D Modeling. A fourth workspace, Initial Setup, is displayed when AutoCAD is executed for the first time. It remains in effect until another workspace is selected. Once a different workspace is selected, the Initial Setup workspace is no longer available.

Workspaces can be created, customized, and saved to allow a variety of graphical user interface configurations. The 3D Modeling workspace provides quick access to all of the tools required to construct, edit, view, and visualize 3D models. This section provides an overview of the 3D Modeling workspace and the layout of the ribbon and its panels.

NOTE

In order to use the default 3D "environment," you must start a new drawing file based on the acad3D.dwt template and set the 3D Modeling workspace current. All discussions in the remainder of this book assume that AutoCAD is in this default 3D environment.

Workspaces

A *workspace* is a drawing environment in which menus, toolbars, palettes, and ribbon panels are displayed for a specific task. A workspace stores not only which of these tools are visible, but also their on-screen locations. You can quickly change workspaces using the **Workspace Switching** tool on the status bar, **WSCURRENT** command, or **Workspaces** toolbar, as shown in **Figure 1-5.** By default, toolbars are only displayed in the AutoCAD Classic workspace. The default 3D Modeling workspace is shown in **Figure 1-6.**

Figure 1-5.
Switching workspaces.
A—Using the **Workspaces** toolbar (which is not displayed by default in the 3D Modeling workspace).
B—Using the **Workspaces Switching** tool.

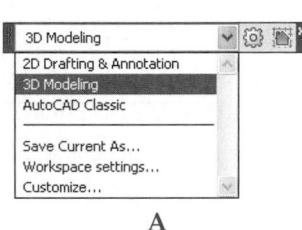

A

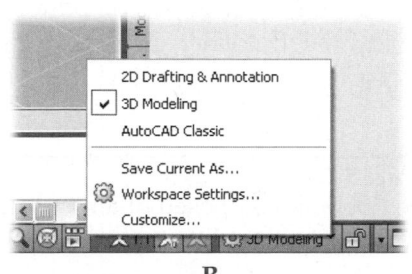

B

Figure 1-6.
The 3D Modeling workspace with a drawing file based on the acad3D.dwt template.

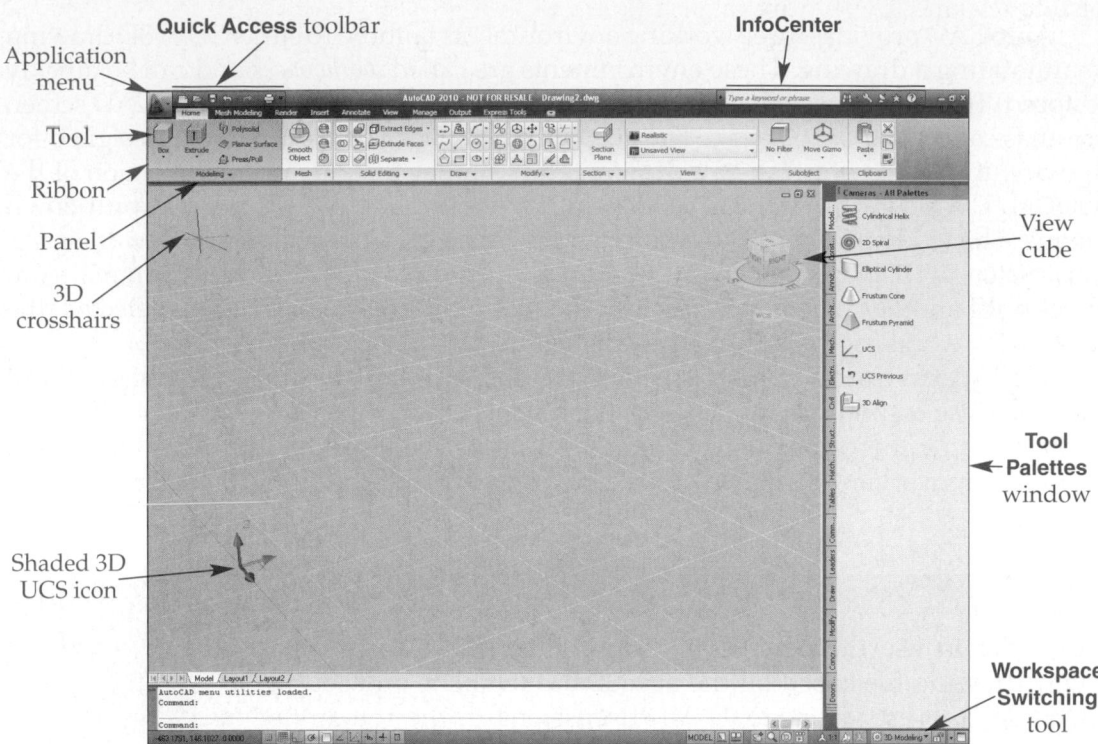

Ribbon Panels

Within the 3D Modeling workspace, the **Home** tab on the ribbon displays nine panels. The tools in these panels provide all of the functions needed to design and view your 3D model. See **Figure 1-7.** The title appears at the bottom of the panel. The button at the right-hand end of the tab names controls the display of the ribbon tabs, panels and panel titles. Picking it cycles through the display of the full ribbon, tab and panel titles only, or tab titles only.

Many panels contain more tools than displayed in the default view. These panels can be expanded by picking the flyout arrow to the right of the panel title. See **Figure 1-8A.** The flyout portion is displayed as long as the cursor remains over the panel. It retracts when the cursor moves off of the panel. To retain the flyout display, pick the pushpin icon in the lower-left corner of the expanded panel. See **Figure 1-8B.** The panel continues to be displayed until you pick the pushpin icon again to release it.

Figure 1-7.
The ribbon is composed of tabs and panels that provide access to 3D modeling and viewing commands without using toolbars and menus.

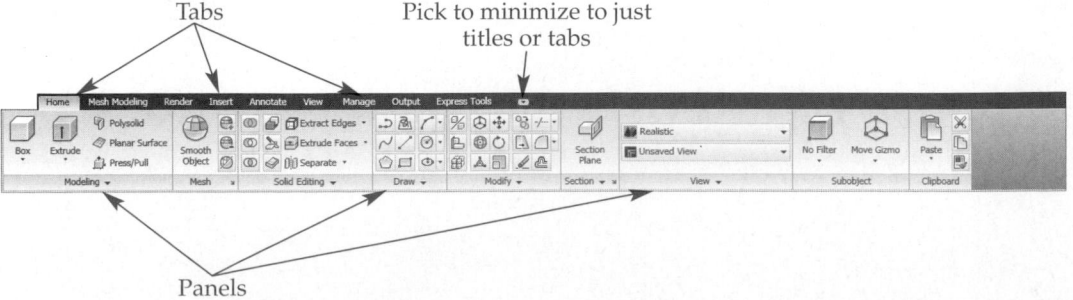

Multiple panel flyouts can be displayed using the pushpin, but may overlap in the process. This may obscure menu tools in adjacent panels. Simply pick on the panel you wish to use and its tools are displayed in full.

In some cases, a dialog box or palette may be related to the tools in a panel. This is indicated by an arrow in the lower-right corner of the panel title. See **Figure 1-8A**. Picking on the arrow displays the related dialog box.

You can display only those control panels that you need. Right-click anywhere on the ribbon tab to display the shortcut menu. Select **Panels** to display a cascading menu that contains a list of the control panels available for that tab. See **Figure 1-9.** The panels that are currently displayed in the ribbon have a check mark next to their name. Select any of the checked control panels that you wish to remove from the current display. Unchecked panels can be displayed by selecting their name. Each tab has its own set of available control panels.

Each panel contains command tools. These are discussed in detail throughout the Modeling and Presentation sections of this book.

Figure 1-8.
A—Picking the flyout arrow expands the panel. The flyout is displayed as long as the cursor remains over the panel. It retracts when the cursor moves off of the panel. B—Multiple panel flyouts can be displayed using the pushpin, but may overlap in the process. Pick the panel you wish to use to display its tools in full.

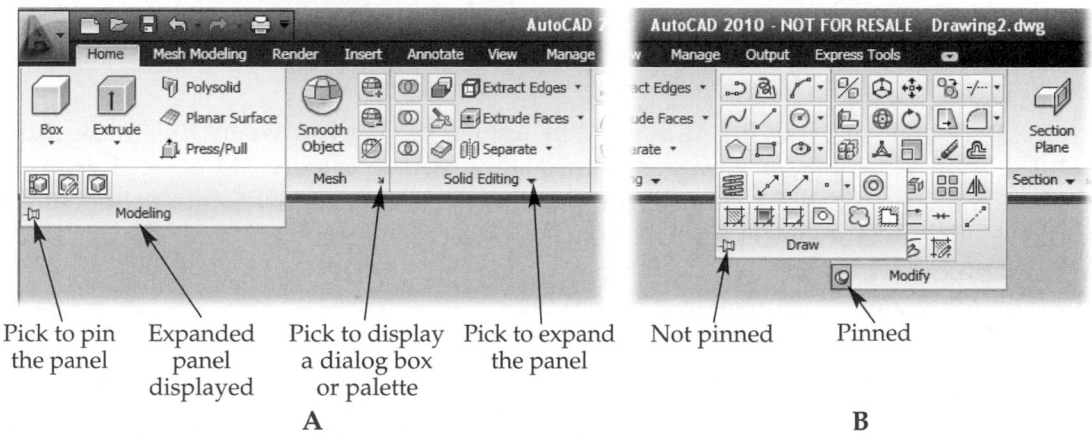

Pick to pin the panel Expanded panel displayed Pick to display a dialog box or palette Pick to expand the panel Not pinned Pinned

A   B

Figure 1-9.
Panels in a tab can be displayed or hidden using the shortcut menu.

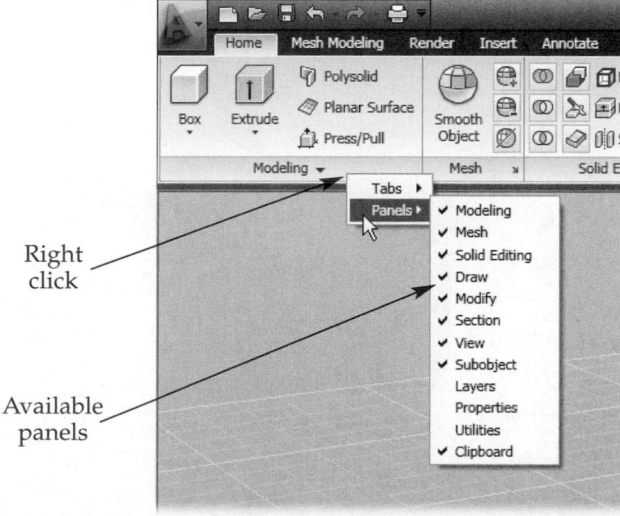

Right click

Available panels

Displaying 3D Views

AutoCAD provides several methods of changing your viewpoint to produce different pictorial views. The default view in the 2D environment based on the acad.dwt template is a plan, or top, view of the XY plane. The default view in the 3D environment based on the acad3D.dwt template is a pictorial, or 3D, view. The *viewpoint* is the location in space from which the object is viewed. The methods for changing your viewpoint include preset isometric and orthographic viewpoints, the view cube, and camera lens settings. Camera settings are discussed in detail in Chapter 19.

Isometric and Orthographic Viewpoint Presets

A 2D isometric drawing is based on angles of 120° between the three axes. AutoCAD provides preset viewpoints that allow you to view a 3D object from one of four isometric locations. See **Figure 1-10.** Each of these viewpoints produces an isometric view of the object. In addition, AutoCAD has presets for the six standard orthographic views of an object. The isometric and orthographic viewpoint presets are based on the world coordinate system (WCS).

Figure 1-10.
There are four preset isometric viewpoints in AutoCAD. This illustration shows the direction from which the cube will be viewed for each of the presets. The grid represents the XY plane of the WCS.

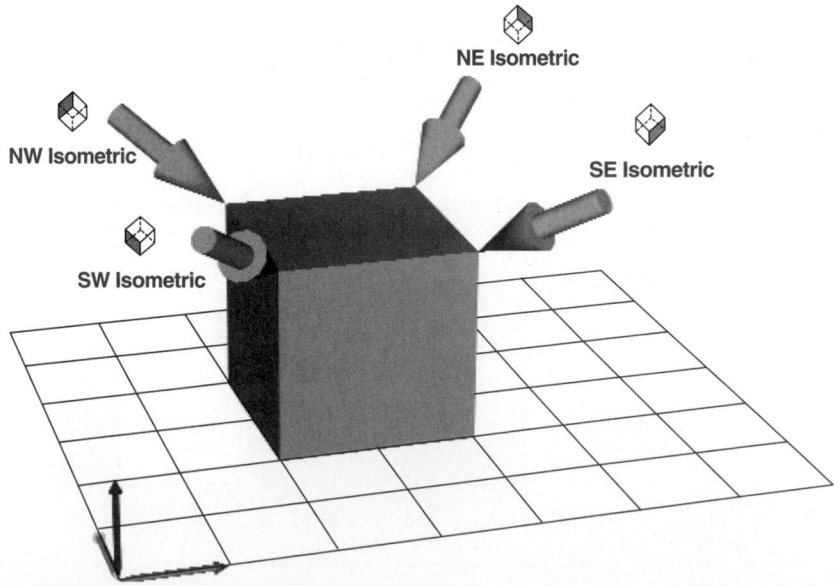

Figure 1-11.
Selecting preset views using the **Views** panel in the **View** tab of the ribbon.

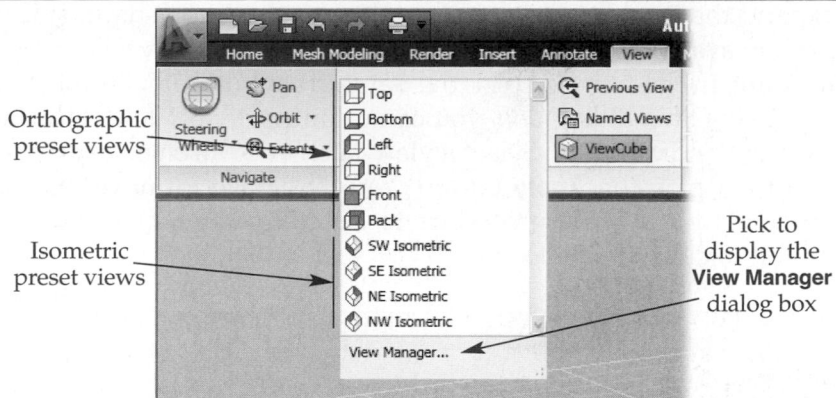

Orthographic preset views

Isometric preset views

Pick to display the **View Manager** dialog box

The four preset isometric views are southwest, southeast, northeast, and northwest. The six orthographic presets are top, bottom, left, right, front, and back. To switch your viewpoint to one of these presets, pick the drop-down list in the **Views** panel in the **View** tab of the ribbon and select the view name. See **Figure 1-11.**

Once you select a view, the viewpoint in the current viewport is automatically changed to display an appropriate isometric or orthographic view. Since these presets are based on the WCS, selecting a preset produces the same view of the object regardless of the current user coordinate system (UCS).

A view that looks straight down on the current drawing plane is called a *plan view.* An important aspect of the orthographic presets is that selecting one not only changes the viewpoint, but, by default, it also changes the UCS to be plan to the orthographic view. All new objects are created on that UCS instead of the WCS (or previous UCS). Working with UCSs is explained in detail in Chapter 6. However, to change the UCS to the WCS type UCS to access the **UCS** command and then type W for the **World** option.

When an isometric or other 3D view is displayed, you can easily switch to a plan view of the current UCS by typing the **PLAN** command and selecting the **Current** option. The **PLAN** command is discussed in more detail in Chapter 4.

View Manager Dialog Box

The **View Manager** dialog box allows you to work with any named view, orthographic preset, or isometric preset. See **Figure 1-12.** To select a preset viewpoint, first

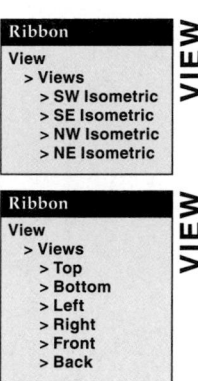

Ribbon
View
> Views
> SW Isometric
> SE Isometric
> NW Isometric
> NE Isometric

Ribbon
View
> Views
> Top
> Bottom
> Left
> Right
> Front
> Back

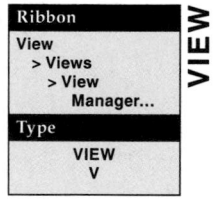

Ribbon
View
> Views
> View
Manager...

Type
VIEW
V

Figure 1-12.
The **View Manager** dialog box allows you to work with any named view and orthographic and isometric preset views.

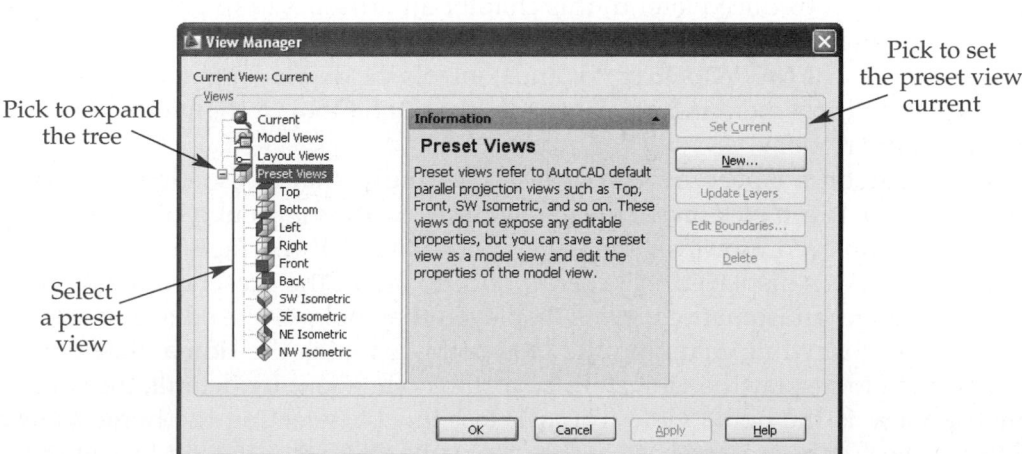

Pick to expand the tree

Select a preset view

Pick to set the preset view current

expand the Preset Views branch in the tree on the left-hand side of the dialog box. The presets available here are the same as described above. To set a preset current, select its name in the tree and pick the **Set Current** button.

Using this dialog box, you can examine a view to determine if you like it before closing the dialog box. Select a view, such as SW Isometric, pick the **Set Current** button, and then pick the **Apply** button. You may have to move the dialog box to view the model. Use the same procedure to examine different views before you pick the **OK** button to close the dialog box.

NOTE

Selecting an orthographic view of a model using one of the methods described above produces a plan view, but it may not achieve the results you desire. Three-dimensional models can be displayed in AutoCAD using either parallel or perspective projection. Displaying a plan view in either projection is possible. However, a true plan view, as used in 2D orthographic projections, can only be created when the model is displayed as a parallel projection. You can quickly change the display from perspective to parallel, or vice versa, by right-clicking on the view cube, then picking either **Parallel** or **Perspective** in the shortcut menu. The current projection is checked in the shortcut menu.

Exercise 1-2

Complete the exercise on the student website.
www.g-wlearning.com/CAD

Introduction to the View Cube

You are not limited to the preset isometric viewpoints. In fact, you can view a 3D object from an unlimited number of viewpoints. The *view cube* allows you to display all of the preset isometric and orthographic views without making panel or menu selections. It also provides quick access to additional pictorial views and easy dynamic manipulation of all orthographic views. It allows you to dynamically rotate the view of the objects to create a new viewpoint.

The view cube is on by default in any 3D shaded or wireframe visual style. It is not displayed in the 2D Wireframe visual style, nor can it be displayed in this visual style. Visual styles are introduced later in this chapter and discussed in detail in Chapter 4.

When displayed, the view cube appears in the upper-right corner of the screen. See **Figure 1-13.** The **NAVVCUBE** command controls the display of the view cube. It can also be quickly turned on and off using the **View Cube** tool in the **Views** panel of the **View** tab on the ribbon.

As the cursor is moved over the view cube, individual faces, edges, and corners are highlighted. If you pick one of the named faces on the view cube, that orthographic plan view is displayed. However, the UCS is not changed. If you pick one of the corners, an isometric view is displayed that corresponds to one of the preset isometric views. If you pick an edge, an isometric view is displayed that looks at the edge you selected.

If you get lost while changing the viewpoint, just pick the **Home View** button on the view cube to display the view defined as the home view. By default, the southwest isometric view is the home view. You can test this by selecting the home view and looking at the compass at the base of the view cube. Note that the left face of the cube

NAVVCUBE

Ribbon
View
> Views

View Cube
Type
NAVVCUBE

Figure 1-13.
Using the view cube to change the viewpoint.

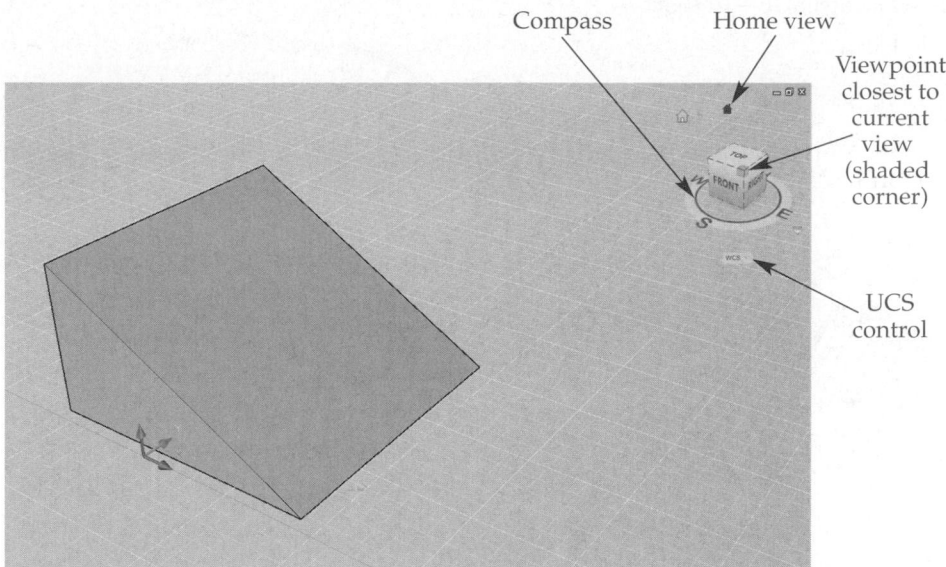

Compass   Home view

Viewpoint closest to current view (shaded corner)

UCS control

aligns with the west point on the compass. The front surface aligns with the south point on the compass. Therefore, the top-front corner points to the southwest.

One edge, corner, or face of the view cube is always shaded or highlighted. Refer to **Figure 1-13.** This shading indicates the viewpoint on the cube that is closest to the current view.

The **NAVVCUBE** command has many options. This discussion is merely an introduction to the command. The command options and view cube features are covered in detail in Chapter 4.

PROFESSIONAL TIP

The **UNDO** command reverses the effects of the **NAVVCUBE** command.

Exercise 1-3

Complete the exercise on the student website.
www.g-wlearning.com/CAD

Introduction to Visual Styles

A 3D model can be displayed in a variety of visual styles. A *visual style* controls the display of edges and shading in a viewport. There are five basic visual styles—2D wireframe, 3D wireframe, 3D hidden, conceptual, and realistic. A *wireframe display* shows all lines on the object, including those representing back or internal features. A *hidden display* suppresses the display of lines that would normally be hidden.

Examples of the basic visual styles are shown in **Figure 1-14.** To change styles, you can use the **VSCURRENT** command or the drop-down list in the **View** panel on the **Home** tab or **Render** tab in the ribbon. See **Figure 1-15.**

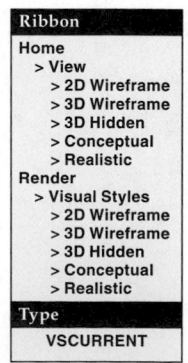

Ribbon
Home
> View
  > 2D Wireframe
  > 3D Wireframe
  > 3D Hidden
  > Conceptual
  > Realistic
Render
> Visual Styles
  > 2D Wireframe
  > 3D Wireframe
  > 3D Hidden
  > Conceptual
  > Realistic
Type
VSCURRENT

VSCURRENT

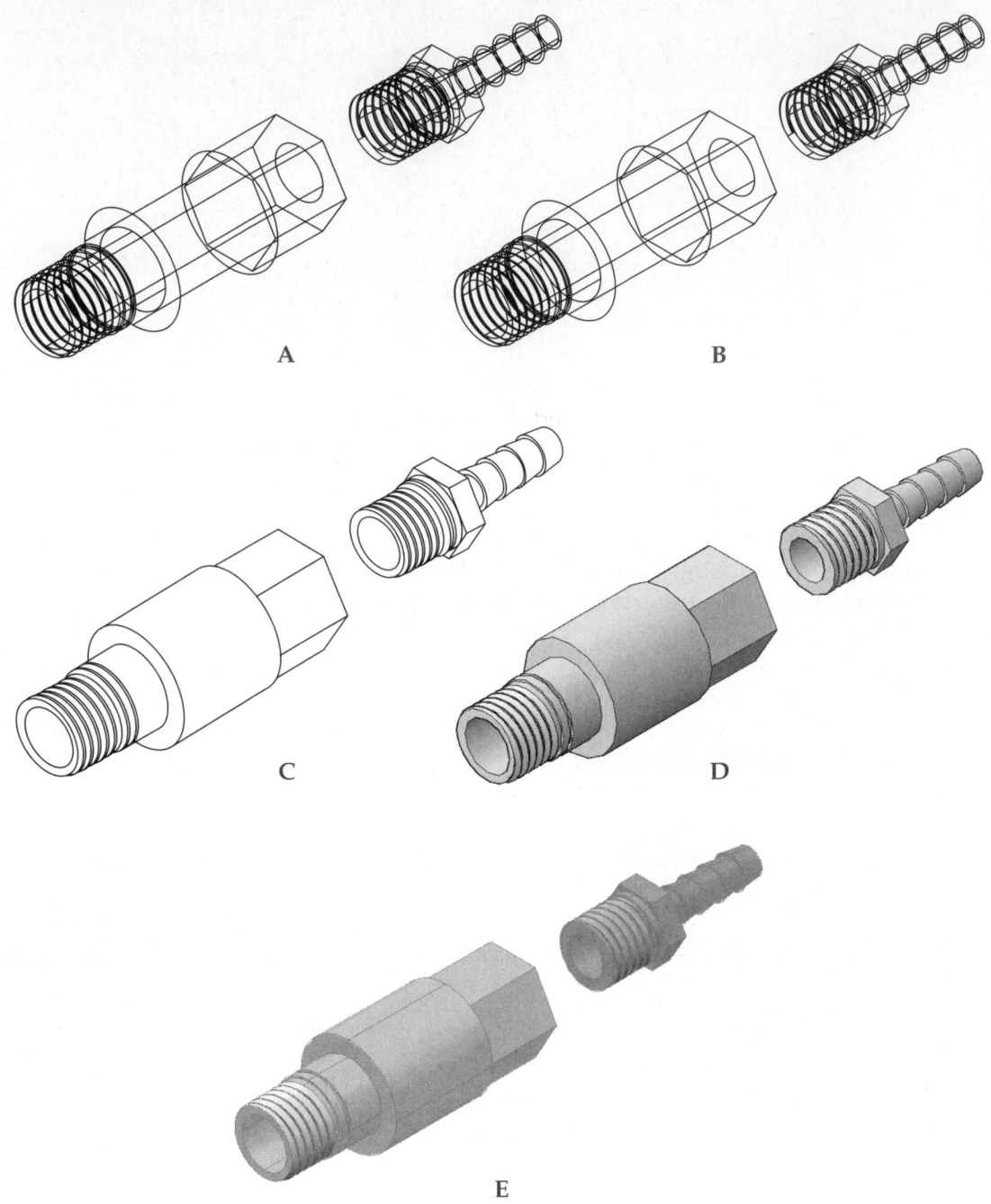

Figure 1-15.
A visual style can be set current using the drop-down list in the **View** panel in the **Home** tab or the **Visual Styles** panel in the **Render** tab of the ribbon.

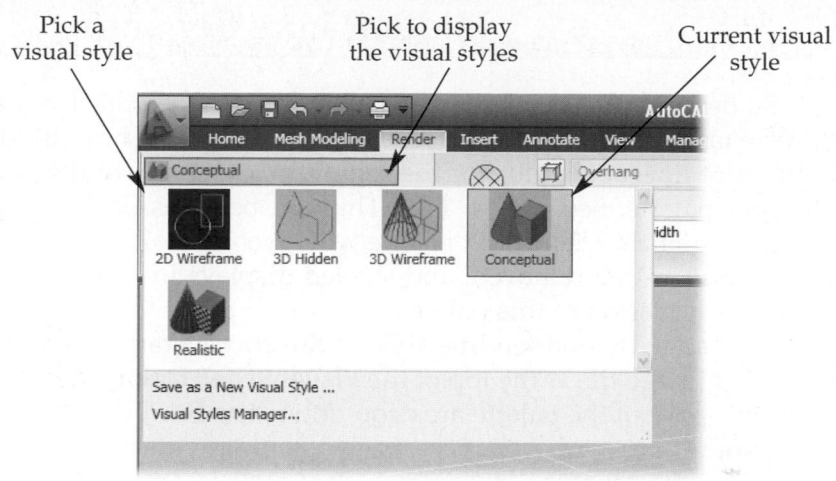

Pick a visual style   Pick to display the visual styles   Current visual style

In the default 3D environment based on the acad3D.dwt template, the default display mode, or visual style, is Realistic. In this visual style, all objects appear as solids and are displayed in their assigned layer colors. Other display options are available. These options are discussed in detail in Chapter 4, but are given here as an introduction.

- **2D Wireframe.** Displays all lines of the model using assigned linetypes and lineweights. The 2D UCS icon and 2D grid are displayed, if turned on. If the **HIDE** command is used to display a hidden-line view, use the **REGEN** command to redisplay the wireframe view.
- **3D Wireframe.** Displays all lines of the model. The 3D grid and the 3D UCS icon are displayed, if turned on.
- **3D Hidden.** Displays all visible lines of the model from the current viewpoint and hides all lines not visible. Objects are not shaded or colored.
- **Conceptual.** The object is smoothed and shaded with transitional colors to help highlight details.
- **Realistic.** Displays the shaded and smoothed model using assigned layer colors and materials (see Chapter 15).

NOTE

When the visual style is 2D Wireframe, you can quickly view the model with hidden lines removed by typing HIDE. The **HIDE** command can be used at any time to remove hidden lines from a wireframe display. If **HIDE** is used when the current visual style is 3D Wireframe, Conceptual, or Realistic, the 3D Hidden visual style is set current.

Exercise 1-4

Complete the exercise on the student website.
www.g-wlearning.com/CAD

Hidden Line Settings

VISUALSTYLES

Ribbon
Home
> Visual Style
Manager...
Render
> Visual Styles
> Visual Style
Manager...
(options button)
View
> 3D Palettes

Visual Styles

Type
VISUALSTYLES

By default, the **HIDE** command removes hidden lines from the display when the 2D Wireframe visual style is current. However, you can have hidden lines displayed in a different linetype and color instead of removed. To set this, open the **Visual Styles Manager** palette. See **Figure 1-16.** This can be accessed by typing the command or using the ribbon. Using this palette, you can control all available settings for wireframe, hidden-line removed, and shaded displays in AutoCAD. See Chapter 15 for a detailed discussion of this palette.

To change the hidden line style in the 2D Wireframe visual style, select the corresponding image tile at the top of the **Visual Styles Manager.** In the **2D Hide - Obscured Lines** category of the palette are drop-down lists from which you can select a linetype and color.

When the **Linetype** drop-down list is set to Off, the display of hidden lines is suppressed by the **HIDE** command. This is the default setting. When a linetype is selected from the drop-down list, hidden lines are displayed in that linetype after the **HIDE** command is used. The linetypes available in the drop-down list are not the same as the linetypes loaded into your drawing.

You can also change the display color of the hidden lines. Simply pick a color in the **Color** drop-down list. To have the hidden lines displayed in the same color as the object, select ByEntity, which is the default. The color setting has no effect when the **Linetype** drop-down list is set to Off.

NOTE

Making changes in the **Visual Styles Manager** redefines the selected visual style.

Figure 1-16.
The manner in which hidden lines appear in hidden displays when the 2D Wireframe visual style is current is controlled in the **Visual Styles Manager** palette.

Pick

Set the color

Set the linetype

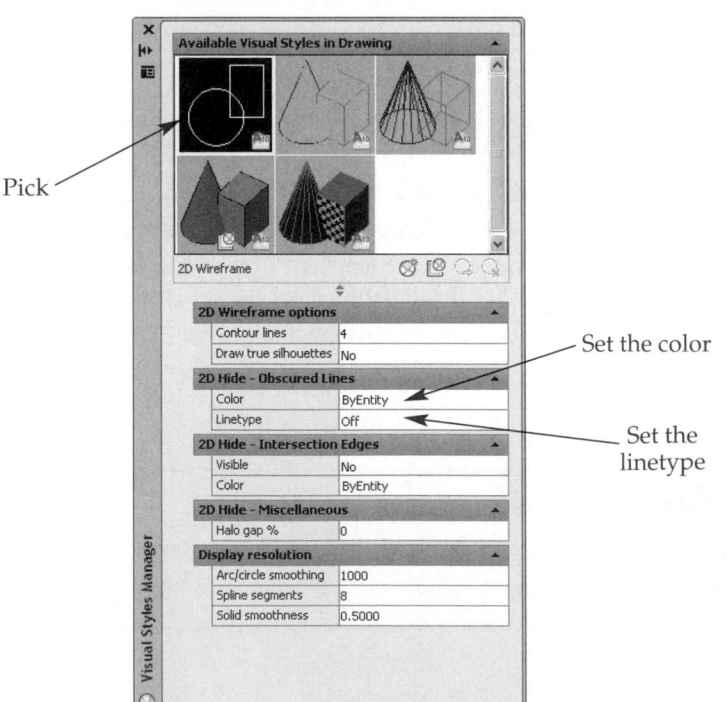

Before constructing a 3D model, you should determine the purpose of your design. What will the model be used for—manufacturing, analysis, or presentation? This helps you determine which tools you should use to construct and display the model. Three-dimensional objects can be drawn as solids, meshes, or surfaces and displayed in wireframe, hidden-line removed, and shaded views.

A *wireframe object,* or model, is an object constructed of lines in 3D space. Wireframe models are hard to visualize because it is difficult to determine the angle of view and the nature of the surfaces represented by the lines. The **HIDE** command has no effect on a wireframe model because there is nothing to hide. All lines are always visible because there are no surfaces or faces between the lines. Wireframe models have very limited applications.

Surface modeling represents solid objects by creating a skin in the shape of the object. However, there is nothing inside of the object. Think of a surface model as a balloon filled with air. A surface model looks more like the real object than a wireframe and can be used for rendering. True surface models are constructed in specialized software used for applications such as civil engineering terrain modeling, automobile body design, sheet metal design and fabrication, and animation.

Like surface modeling, *solid modeling* represents the shape of objects, but also provides data related to the physical properties of the objects. Solid models can be analyzed to determine mass, volume, moments of inertia, and centroid. A solid model is not just a skin, it represents a solid object. Some third-party programs allow you to perform finite element analysis on the model. In addition, solid models can be rendered. Most 3D objects are created as solid models.

In AutoCAD, solid models can be created from primitives. *Primitives* are basic shapes used as the foundation to create complex shapes. Some of these basic shapes include boxes, cylinders, spheres, and cones. Detailed shapes and primitives can be created using 3D mesh primitives and mesh modeling techniques. A 3D *mesh object,* which is a type of surface model, can have a free-flowing shape because the size of the mesh can be adjusted to achieve various levels of smoothness. Mesh objects can be converted to solids for use in model construction. Solid primitives also can be modified to create a finished product. See **Figure 1-17.** See Chapter 3 for a detailed discussion of 3D mesh modeling.

Figure 1-17.
A—These two cylinders and the box are solid primitives. B—With a couple of quick modifications, the large cylinder becomes a shaft with a machined keyway.

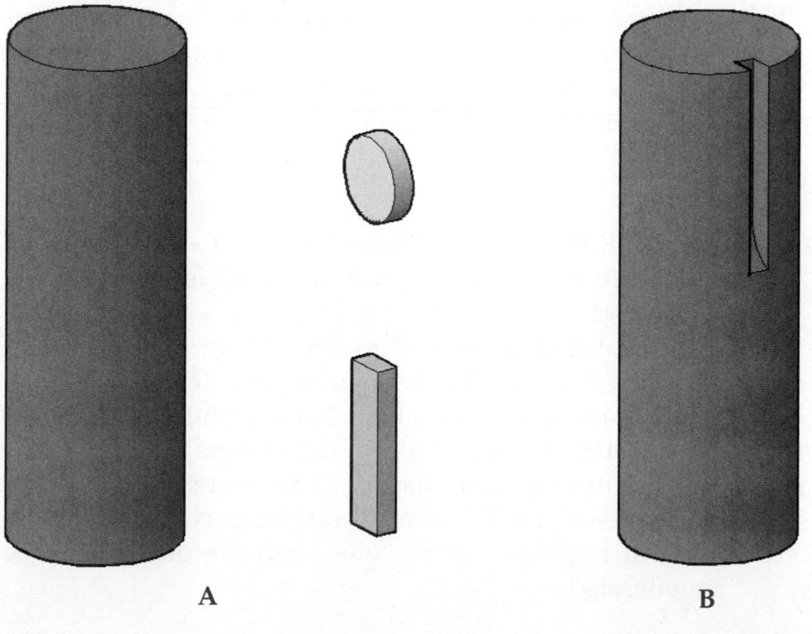

A   B

Surface and solid models can be exported from AutoCAD for use in animation and rendering software, such as Autodesk 3ds max®. Rendered models can be used in any number of presentation formats, including slide shows, black and white or color prints, and animation recorded to video files. Surface and solid models can also be used to create virtual worlds for virtual reality and gaming applications.

Guidelines for Working with 3D Drawings

Working in 3D, like working with 2D drawings, requires careful planning to efficiently produce the desired results. The following guidelines can be used when working in 3D.

Planning

- Determine the type of final drawing you need and the manner in which it will be displayed. Then, choose the method of 3D construction that best suits your needs—wireframe, surface, mesh, or solid.
- For an object requiring only one pictorial view, it actually may be quicker to draw an object in 3D rather than in AutoCAD's isometric mode. AutoCAD's 3D solid modeling tools enable you to quickly create an accurate model, then display it in the required isometric format using preset views. The **FLATSHOT** command can then be used to create a 2D drawing of the model.
- It is best to use AutoCAD's 3D commands to construct objects and layouts that need to be viewed from different angles for design purposes.
- Construct only the details needed for the function of the drawing. This saves space and time and makes visualization much easier.
- Use object snap modes in a pictorial view in conjunction with a dynamic UCS to save having to create new UCSs.
- Keep in mind that when the grid is displayed, the pattern appears at the current elevation and parallel to the XY plane of the current UCS.
- Create layers having different colors for different drawing objects. Turn them on and off as needed or freeze those not being used.

Editing

- Use the **Properties** palette to change the color, layer, or linetype of 3D objects.
- Use grips to edit a solid-modeled object (see Chapter 12).
- Do as much editing as possible from a 3D viewpoint. It is quicker and the results are immediately seen.

Displaying

- Use the **HIDE** command and visual styles to help visualize complex drawings.
- To quickly change views, use the preset isometric views, view cube, and **PLAN** command.
- Use the **VIEW** command to create and save 3D views for quicker pictorial displays. This avoids having to repeatedly use the **NAVVCUBE** command.
- Freeze unwanted layers before displaying objects in 3D and especially before using **HIDE**. AutoCAD regenerates layers that are turned off, which may cause an inaccurate hidden display to be created. Frozen layers are not regenerated.
- Before using **HIDE**, zoom in on the part of a drawing to display. This saves time in regenerating the view because only the objects that are visible are regenerated.

- You may have to slightly move objects that touch or intersect if the display removes a line you need to see or plot. However, be sure to move the objects back to maintain accuracy in the model.

Chapter Test

Answer the following questions. Write your answers on a separate sheet of paper or complete the electronic chapter test on the student website.
www.g-wlearning.com/CAD

1. What are the three coordinates needed to locate any point in 3D space?
2. In a 2D drawing, what is the value for the Z coordinate?
3. What purpose does the right-hand rule serve?
4. Which three fingers are used in the right-hand rule?
5. What is the definition of a *viewpoint?*
6. What is the function of the *ribbon* and its panels?
7. How do you turn the display of individual panels on or off in the ribbon?
8. How can you quickly change the display from perspective projection to parallel projection, or vice versa?
9. How many preset isometric viewpoints does AutoCAD have? List them.
10. How does changing the UCS impact using one of the preset isometric viewpoints?
11. List the six preset orthographic viewpoints.
12. When selecting a preset orthographic viewpoint, what happens to the UCS?
13. Which command allows you to dynamically change your viewpoint using an on-screen cube icon?
14. Define *wireframe display.*
15. Define *hidden display.*
16. Define *wireframe object.*
17. Define *surface model.*
18. Define *solid model.*
19. Define *primitive.*
20. Define *mesh object.*

This table was created from five solid primitives. An additional primitive was used to notch the feet. All of the solids were combined into a single solid using a Boolean operation. The solid object is now ready for materials, lighting, and rendering.

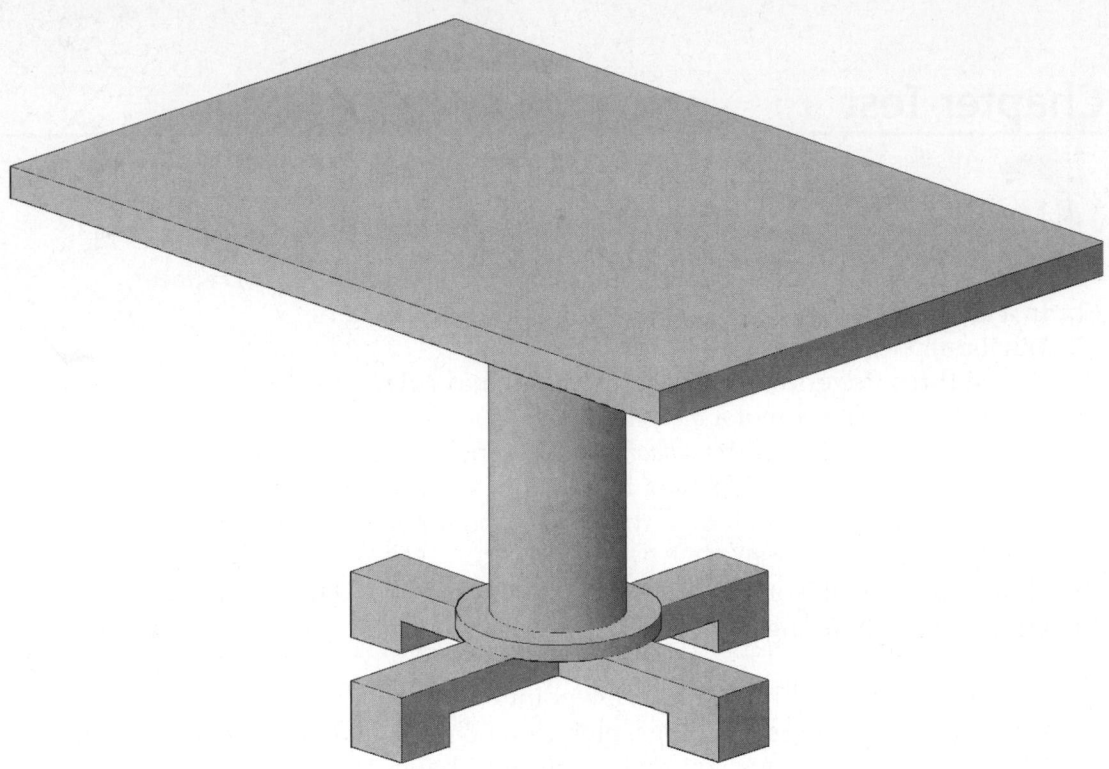

AutoCAD and Its Applications—Advanced

Creating Primitives and Composites

Learning Objectives

After completing this chapter, you will be able to:

✓ Construct 3D solid primitives.
✓ Explain the dynamic feedback presented when constructing solid primitives.
✓ Create complex solids using the **UNION** command.
✓ Remove portions of a solid using the **SUBTRACT** command.
✓ Create a new solid from the interference volume between two solids.
✓ Create regions.

Overview of Solid Modeling

In Chapter 1 you were introduced to the three basic forms of 3D modeling—wireframe objects, solid models, and surface models. Solid models are probably the most useful and, hence, most common type of 3D modeling. A solid model accurately and realistically represents the shape and form of a final object. In addition, a solid model contains data related to the object's volume, mass, and centroid.

Solid modeling is very flexible. A model can start with solid primitives, such as a box, cone, or cylinder, and a variety of editing functions can then be performed. Think of creating a solid model as working with modeling clay. Starting with a basic block of clay, you can add more clay, remove clay, cut holes, round edges, etc., until you have arrived at the final shape and form of the object.

PROFESSIONAL TIP

Snaps can be used on solid objects. For example, you can snap to the center of a solid sphere using the **Center** object snap. The **Endpoint** object snap can be used to select the corners of a box, apex of a cone, corners of a wedge, etc.

Constructing Solid Primitives

As you learned in Chapter 1, a primitive is a basic building block. The eight *solid primitives* in AutoCAD are a box, cone, cylinder, polysolid, pyramid, sphere, torus, and wedge. These primitives can also be used as building blocks for complex solid models. This section provides detailed information on drawing all of the solid primitives. All of the 3D modeling primitive commands can be accessed using the **Modeling** panel in **Home** tab of the ribbon, the **Modeling** toolbar or by typing the name of the 3D modeling primitive. See **Figure 2-1.**

The information required to construct a solid primitive depends on the type of primitive being drawn. For example, to draw a solid cylinder you must provide a center point for the base, a radius or diameter of the base, and the height of the cylinder. A variety of command options are available when creating primitives, but each primitive is constructed using just a few basic dimensions. These are shown in **Figure 2-2.**

Certain familiar editing commands can be used on solid primitives. For example, you can fillet or chamfer the edges of a solid primitive. In addition, there are other editing commands that are specifically for use on solids. You can also perform Boolean operations on solids. These operations allow you to add one solid to another, subtract one solid from another, or create a new solid based on how two solids overlap.

PROFESSIONAL TIP

Solid objects of a more free-form nature can be created by first constructing a mesh primitive, editing it, and then converting it to a solid. This process is covered in detail in Chapter 3.

Using Dynamic Input and Dynamic Feedback

Dynamic input enables you to construct models in a "heads up" fashion with minimal eye movement around the screen. When a command is initiated, the command prompts are then displayed in the dynamic input area, which is at the lower-right corner of the crosshairs. As the pointer is moved, the dynamic input area follows it. The dynamic input area displays values of the cursor location, dimensions, command prompts, and command options (in a drop-down list). Coordinates and dimensions

Figure 2-1.
The **Modeling** panel in the **Home** tab of the ribbon.

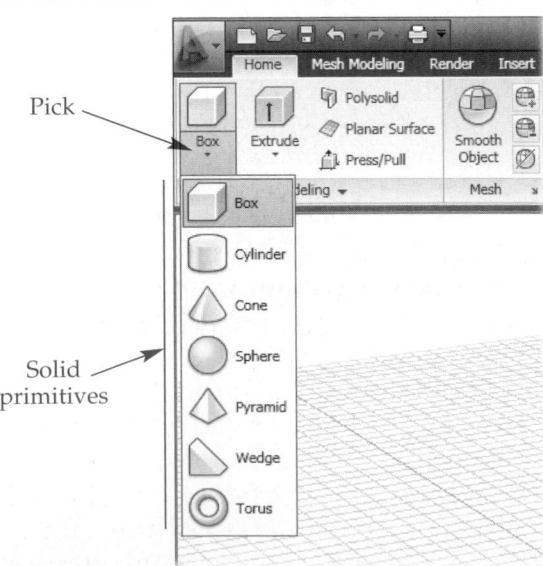

Figure 2-2.
An overview of AutoCAD's solid primitives and the dimensions required to draw them.

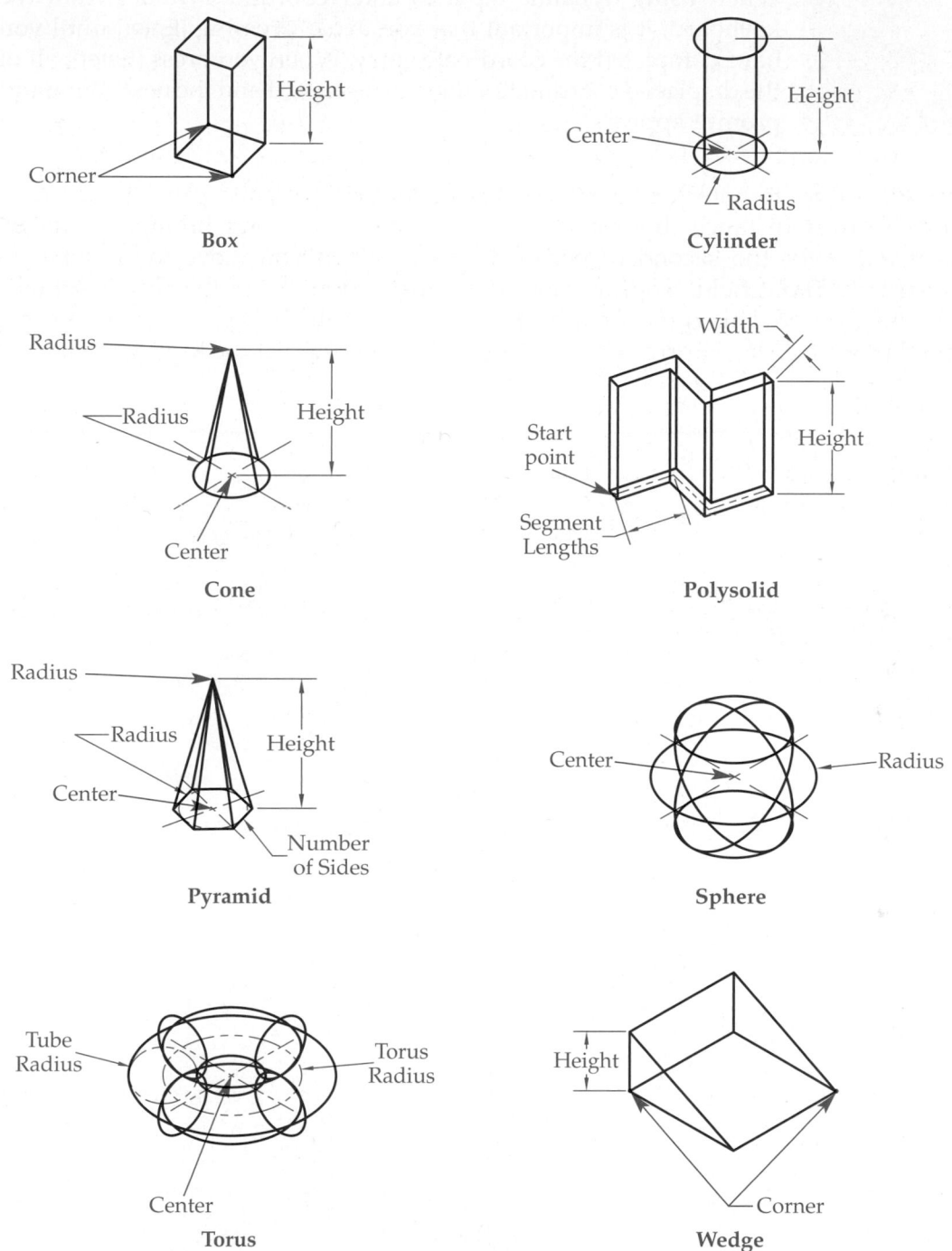

are displayed in boxes called *input fields*. When command options are available, a drop-down list arrow appears. Press the down arrow key on the keyboard to display the list. You can use your pointer to select the option or press the down arrow key until a dot appears by the desired option and press [Enter].

For example, after selecting a modeling command such as **BOX**, the first item that appears in the dynamic input area is the prompt to specify the first corner and a display of the X and Y coordinate values of the crosshairs. At this point you can use the pointer to specify the first corner or type coordinate values. Type the X value and then a comma or the [Tab] key to move to the Y value input box. This locks the typed value and any movement of the pointer will not change it.

When using dynamic input to enter coordinate values from the keyboard, it is important that you avoid pressing [Enter] until you have completed the coordinate entry. When you press [Enter], all of the displayed coordinate values are accepted and the next command prompt appears.

In addition to entering coordinate values for sizes of solid primitives, you can provide direct distance dimensions. For example, the second prompt of the **BOX** command is for the second corner of the base. When you move the pointer, two dimensional input fields appear. Also, notice that a preview of the base is shown in the drawing area. This is the *dynamic feedback* that AutoCAD provides as you create a solid primitive. See **Figure 2-3A**. If you enter a dimension at the keyboard and press

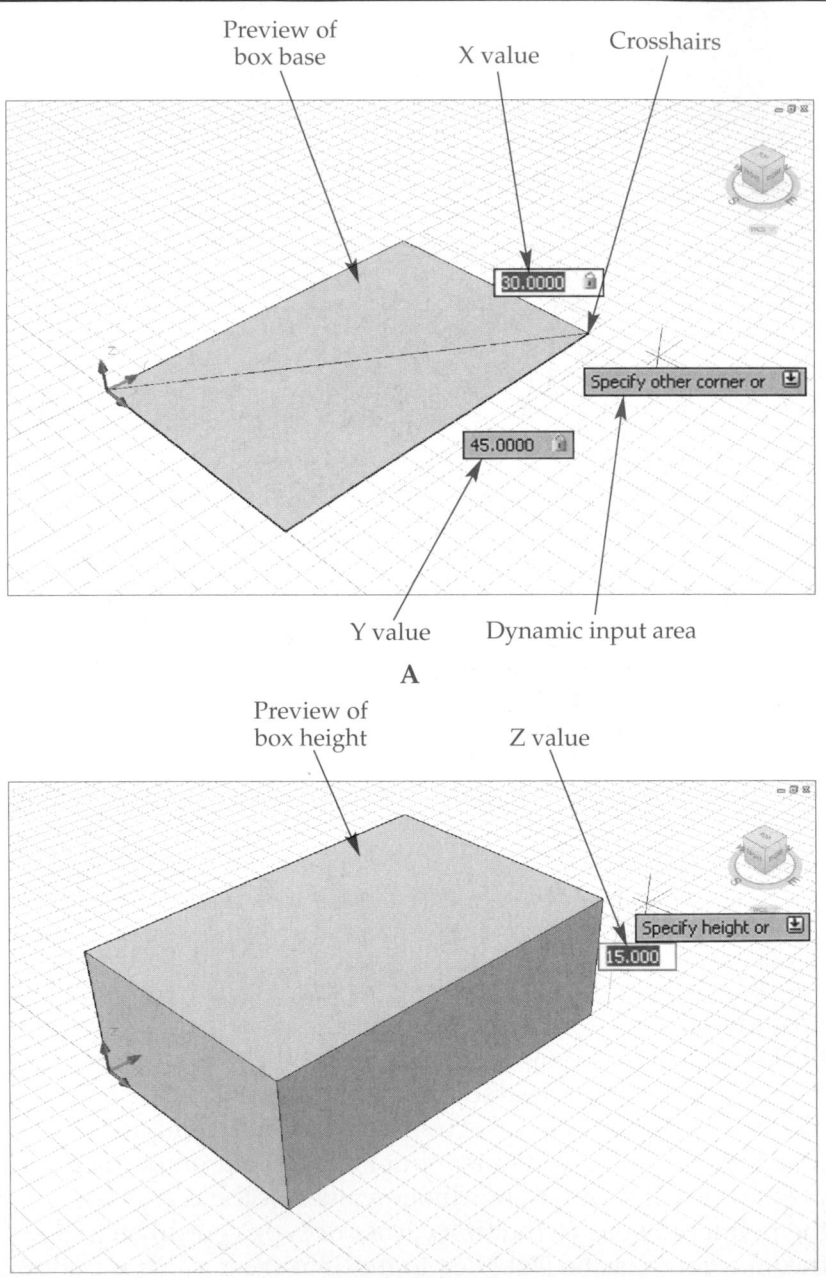

Figure 2-3.
A—Specifying the base of a box with dynamic input on. Notice the preview of the base.
B—Setting the height of a box with dynamic input on. Notice the preview of the height.

the [Tab] key, the value is the length of the side. Then, press the left mouse button to set the base. But, if you enter a value followed by a comma, the dynamic input area changes to display X and Y coordinate boxes. In this case, the values entered are the X and Y coordinates of the opposite corner of the box base. If X, Y, *and* Z coordinates are entered, the point is the opposite corner of the box.

After establishing the location and size of the box base, the next prompt asks you to specify the height. Again, you can either enter a direct dimension value and press [Enter] or select the height with the pointer. See **Figure 2-3B.** AutoCAD provides dynamic feedback on the height of the box as the pointer is moved.

> **NOTE**
>
>
>
> The techniques described above can be used with any form of dynamic input. The current input field is always highlighted. You can always enter a value and use the [Tab] key to lock the input and move to the next field.

Box

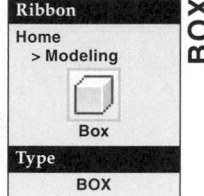

A *box* has six flat sides forming square corners. It can be constructed starting from an initial corner or the center. See **Figure 2-4.** A cube can be constructed, as well as a box with unequal sides.

When the command is initiated, you are prompted to select the first corner or enter the **Center** option. The first corner is one corner on the base of the box. The center is the geometric center of the box, as shown in **Figure 2-4.** If you select the **Center** option, you are next prompted to select the center point.

After selecting the first corner or center, you are prompted to select the other corner or enter the **Cube** or **Length** option. The "other" corner is the opposite corner of the box base if you enter an XY coordinate or the opposite corner of the box if you enter an XYZ coordinate. If the **Length** option is entered, you are first prompted for the length of one side. If dynamic input is on, you can also specify a rotation angle. After entering the length, you are prompted for the width of the box base. If the **Cube** option is selected, the length value is applied to all sides of the box.

Once the length and width of the base are established, you are prompted for the height, unless the **Cube** option was selected. Either enter the height or select the **2point** option. This option allows you to pick two points on screen to set the height. The box is created.

Figure 2-4.
A—A box created using the **Cube** option. B—A box created by selecting the center point.

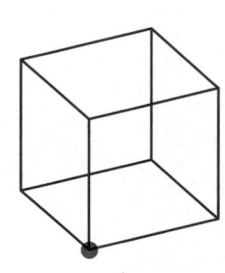 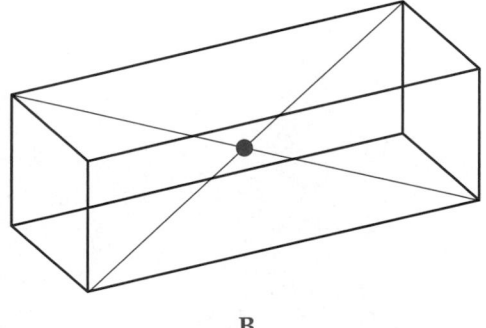

A   B

Exercise 2-1

Complete the exercise on the student website.
www.g-wlearning.com/CAD

Cone

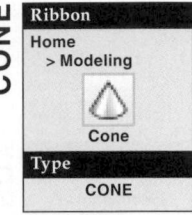

CONE

Ribbon
Home
> Modeling
Cone
Type
CONE

A *cone* has a circular or elliptical base with edges that converge at a single point. The cone may be *truncated* so the top is flat and the cone does not have an apex. See **Figure 2-5.** When the command is initiated, you are prompted for the center point of the cone base or to enter an option. If you pick the center, you must then set the radius of the base. To specify a diameter, enter the **Diameter** option after specifying the center.

The **3P**, **2P**, and **Ttr** options are used to define a circular base using either three points on the circle, two points on the circle, or two points of tangency on the circle and a radius. The **Elliptical** option is used to create an elliptical base.

If the **Elliptical** option is entered, you are prompted to pick both endpoints of one axis and then one endpoint of the other axis of an ellipse that defines the base. If the **Center** option is entered after the **Ellipse** option, you are asked to select the center of the ellipse and then pick an endpoint on each of the axes.

After the base is defined, you are asked to specify a height. You can enter a height or enter the **2point**, **Axis endpoint**, or **Top radius** option. The **2point** option is used to set the height by picking two points on screen. The distance between the points is the height. The height is always applied perpendicular to the base.

Figure 2-5.
A—A circular cone. B—A frustum cone. C—An elliptical cone.

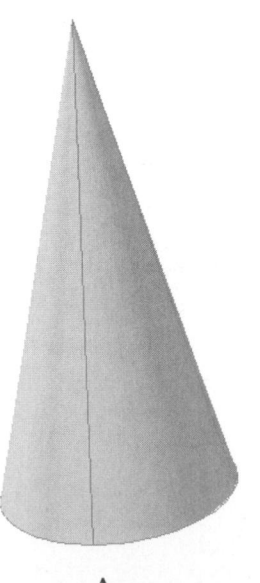

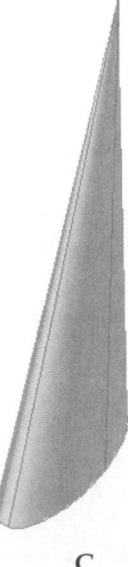

A   B   C

Figure 2-6.
A—Cones can be positioned relative to other objects using the **Axis endpoint** option.
B—The cone is subtracted from the box.

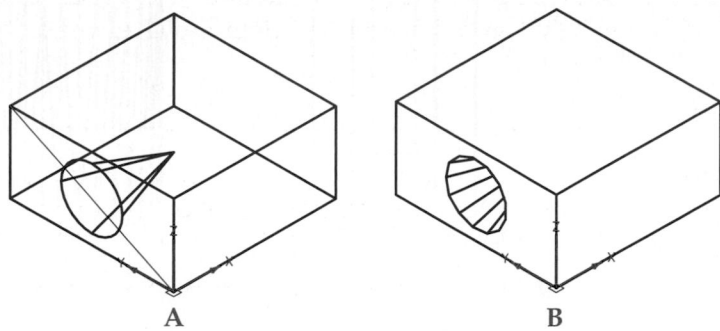

A   B

The **Axis endpoint** option allows you to orient the cone at any angle, regardless of the current UCS. For example, to place a tapered cutout in the end of a block, first create a construction line. Refer to **Figure 2-6.** Then, locate the cone base and give a coordinate location of the apex, or axis endpoint. You can then use editing commands to subtract the cone from the box to create the tapered hole. See Chapters 11 and 12 for model editing details.

The **Top radius** option allows you to specify the radius of the top of the cone. If this option is not used, the radius is zero, which creates a pointed cone. Setting the radius to a value other than zero produces a *frustum cone,* or a cone where the top is truncated and does not come to a point.

Cylinder

A *cylinder* has a circular or elliptical base and edges that extend perpendicular to the base. See **Figure 2-7.** When the command is initiated, you are prompted for the center point of the cylinder base or to enter an option. If you pick the center, you must then set the radius of the base. To specify a diameter, enter the **Diameter** option after specifying the center.

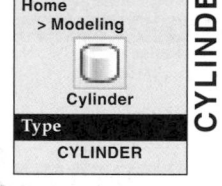

The **3P**, **2P**, and **Ttr** options are used to define a circular base using either three points on the circle, two points on the circle, or two points of tangency on the circle and a radius. The **Elliptical** option is used to create an elliptical base.

If the **Elliptical** option is entered, you are prompted to pick both endpoints of one axis and then one endpoint of the other axis of an ellipse defining the base. If the **Center** option is entered, you are asked to select the center of the ellipse and then pick an endpoint on each of the axes.

Figure 2-7.
A—A circular cylinder. B—An elliptical cylinder.

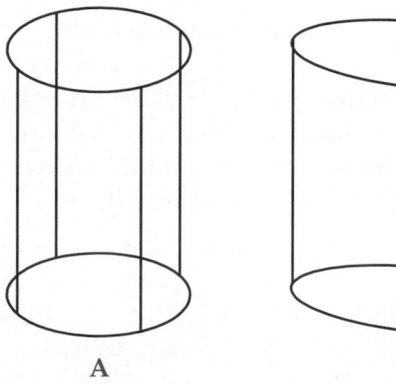

A   B

Figure 2-8.
A—A cylinder is drawn inside of another cylinder using the **Axis endpoint** option. B—The large cylinder has a hole after **SUBTRACT** is used to remove the small cylinder.

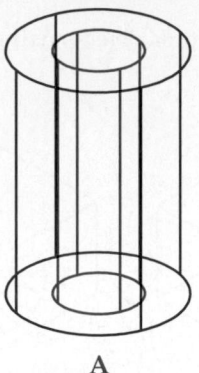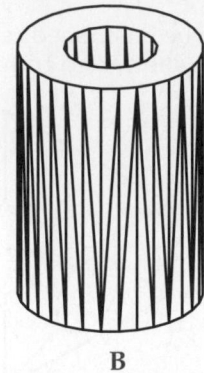

A   B

After the base is defined, you are asked to specify a height or to enter the **2point** or **Axis endpoint** option. The **2point** option is used to set the height by picking two points on screen. The distance between the points is the height. The **Axis endpoint** option allows you to orient the cylinder at any angle, regardless of the current UCS, just as with a cone.

The **Axis endpoint** option is useful for placing a cylinder inside of another object to create a hole. The cylinder can then be subtracted from the other object to create a hole. Refer to **Figure 2-8.** If the axis endpoint does not have the same X and Y coordinates as the center of the base, the cylinder is tilted from the XY plane.

If polar tracking is on when using the **Axis endpoint** option, you can rotate the cylinder axis 90° from the current UCS Z axis, and then turn the cylinder to any preset polar increment. See **Figure 2-9A.** If the polar tracking vector is parallel to the Z axis of the current UCS, the tooltip displays a positive or negative Z value. See **Figure 2-9B.**

Polysolid

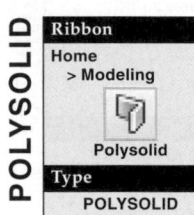

POLYSOLID

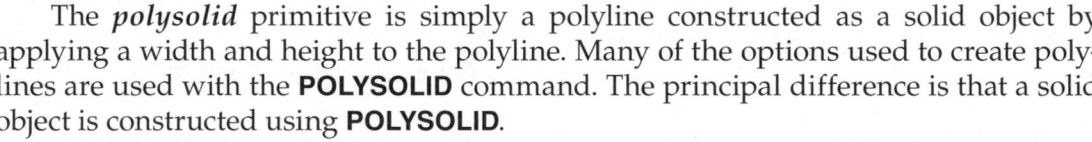

Ribbon
Home
> Modeling

Polysolid

Type
POLYSOLID

The *polysolid* primitive is simply a polyline constructed as a solid object by applying a width and height to the polyline. Many of the options used to create polylines are used with the **POLYSOLID** command. The principal difference is that a solid object is constructed using **POLYSOLID**.

When the command is initiated, you are prompted to select the first point or enter an option. By default, the width of the polysolid is equally applied to each side of the line you draw. This is center justification. Using the **Justify** option, you can set the justification to center, left, or right. The justification applies to all segments created in this command session. See **Figure 2-10.** If you select the wrong justification option, you must exit the command and begin again.

The default width is .25 units and height is four units. These values can be changed using the **Height** and **Width** options of the command. The height value is saved in the **PSOLHEIGHT** system variable. The width value is saved in the **PSOLWIDTH** system variable. Using these system variables, the default width and height can be set outside of the command.

The **Object** option allows you to convert an existing 2D object into a polysolid. AutoCAD entities such as lines, circles, arcs, polylines, polygons, and rectangles can be converted. The 2D object cannot be self intersecting. Some objects, such as 3D polylines and revision clouds, cannot be converted.

Figure 2-9.
A—If polar tracking is on, you can rotate the cylinder axis 90° from the current UCS Z axis and then move the cylinder to any angle in the XY plane. B—If the polar tracking vector is moved parallel to the current Z axis of the UCS, the tooltip displays a positive or negative Z dimension.

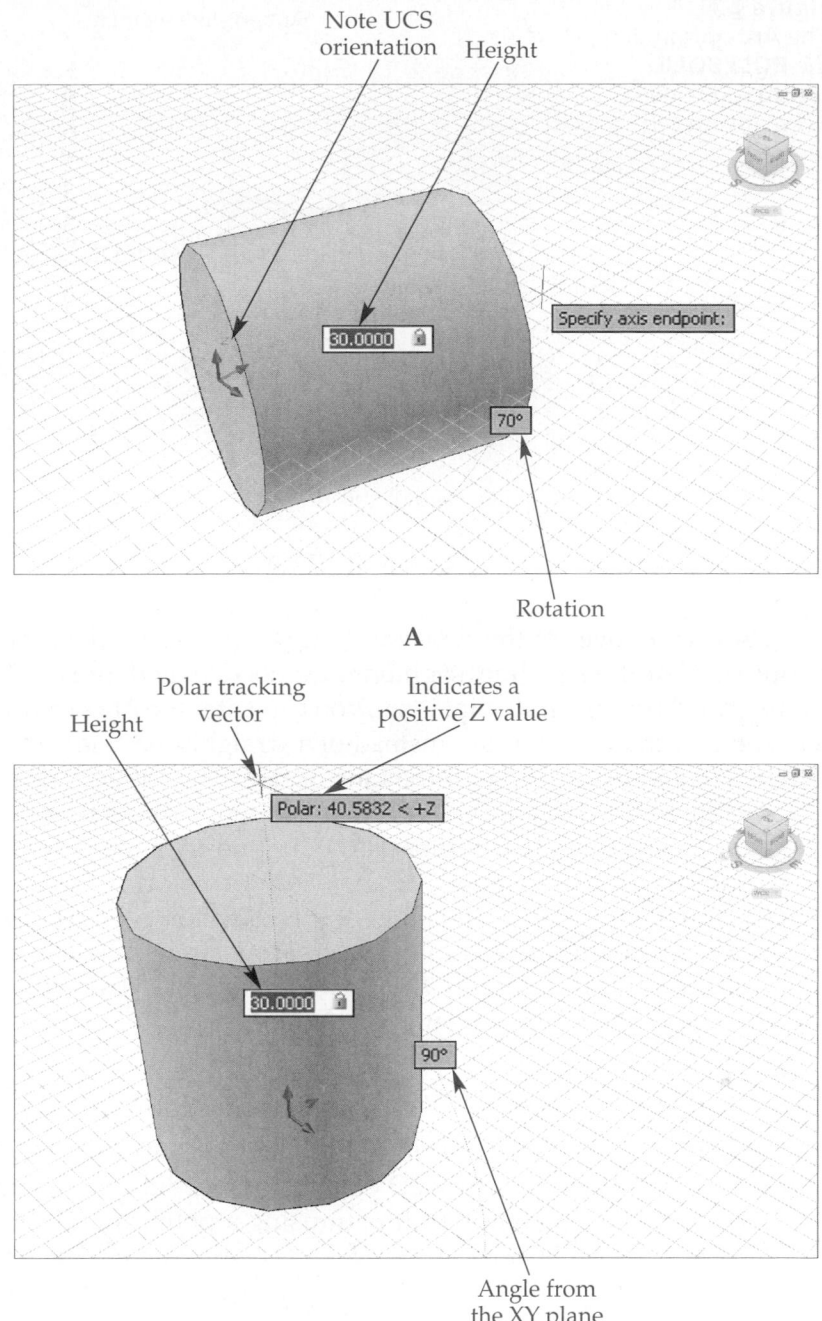

A

B

Figure 2-10.
When you begin the **POLYSOLID** command, use the **Justify** option to select the alignment.

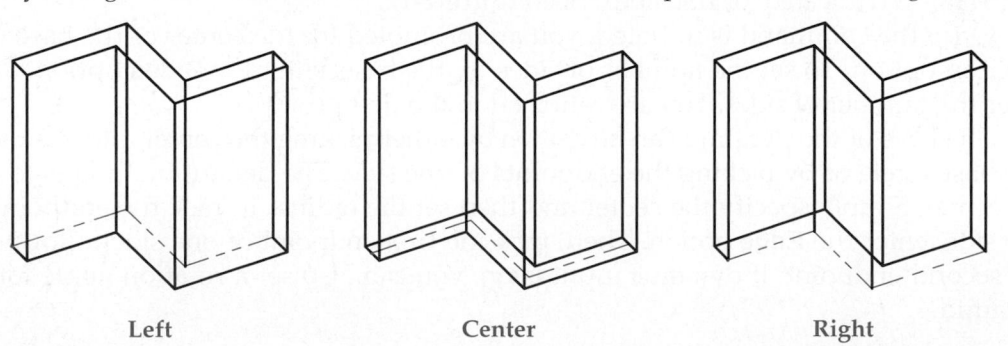

Left   Center   Right

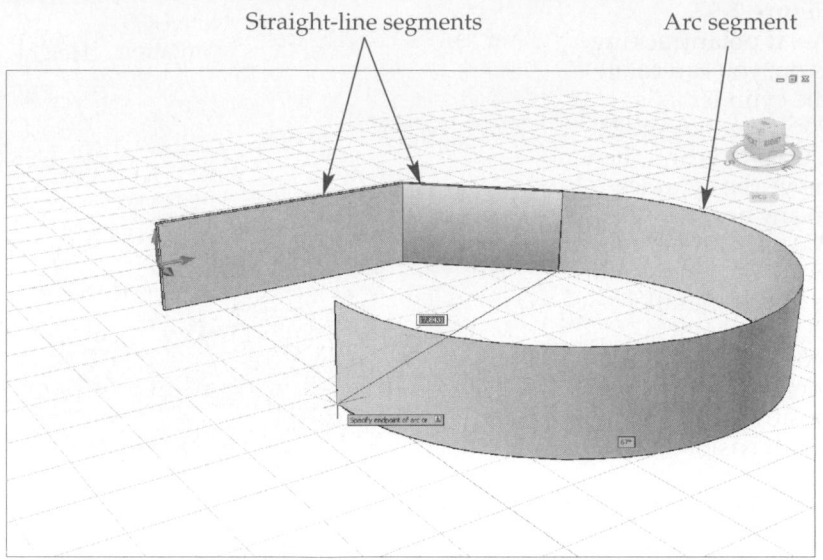

Figure 2-11.
The **Arc** option of the **POLYSOLID** command is used to create curved segments.

Straight-line segments   Arc segment

Once you have set the first point on the polysolid, pick the endpoint of the first segment. Continue adding segments as needed and press [Enter] to complete the command. After the first point is set, you can enter the **Arc** option. The current segment will then be created as an arc instead of a straight line. See **Figure 2-11.** Arc segments will be created until you enter the **Line** option. The suboptions for the **Arc** option are:

- **Close.** If there are two or more segments, this option creates an arc segment between the active point and the first point of the polysolid. You can also close straight-line segments.
- **Direction.** Specifies the tangent direction for the start of the arc.
- **Line.** Returns the command to creating straight-line segments.
- **Second point.** Locates the second point of a two-point arc. This is not the endpoint of the segment.

PROFESSIONAL TIP

The **Object** option of the **POLYSOLID** command is a powerful tool for converting 2D objects to 3D solids. For example, you can create a single-line wall plan using a polyline and then quickly convert it to a 3D model.

Pyramid

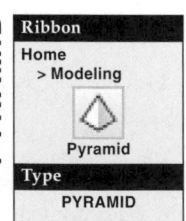

Ribbon
Home
> Modeling

Pyramid

Type
PYRAMID

A *pyramid* has a base composed of straight-line segments and edges that converge at a single point. The pyramid base can be composed of three to 32 sides, much like a 2D polygon. A pyramid may be drawn with a pointed apex or as a *frustum pyramid,* which has a truncated, or flat, apex. See **Figure 2-12.**

Once the command is initiated, you are prompted for the center of the base or to enter an option. To set the number of sides on the base, enter the **Sides** option. Then, enter the number of sides. You are returned to the first prompt.

The base of the pyramid can be drawn by either picking the center and the radius of a base circle or by picking the endpoints of one side. The default method is to pick the center. Simply specify the center and then set the radius. To pick the endpoints of one side, enter the **Edge** option. Then, pick the first endpoint of one side followed by the second endpoint. If dynamic input is on, you can also set a rotation angle for the pyramid.

PYRAMID

Figure 2-12.
A sampling of pyramids that can be constructed with the **PYRAMID** command.

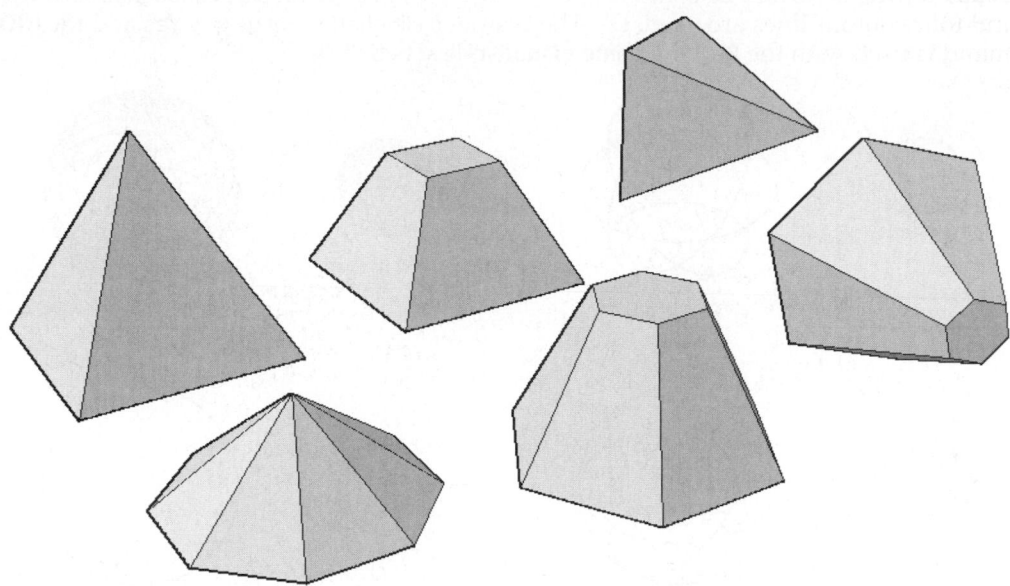

If drawing the base from the center point, the polygon is circumscribed about the base circle by default. To inscribe the polygon on the base circle, enter the **Inscribed** option before setting the radius. To change back to a circumscribed polygon, enter the **Circumscribed** option before setting the radius.

After locating and sizing the base, you are prompted for the height. To create a frustum pyramid, enter the **Top radius** option. Then, set the radius of the top circle. The top will be either inscribed or circumscribed based on the base circle. You are then returned to the height prompt.

The height value can be set by entering a direct distance. You can also use the **2point** option to set the height. With this option, pick two points on screen. The distance between the two points is the height value. The **Axis endpoint** option can also be used to specify the center of the top in the same manner as for a cone or cylinder.

PROFESSIONAL TIP

Six-sided frustum pyramids can be used for bolt heads and as "blanks" for creating nuts.

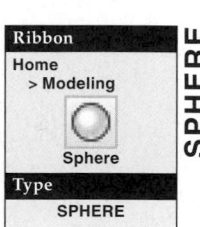

Sphere

A *sphere* is a round, smooth object like a baseball or globe. Once the command is initiated, you are prompted for the center of the sphere or to enter an option. If you pick the center, you must then set the radius of the sphere. To specify a diameter, enter the **Diameter** option after specifying the center. The **3P**, **2P**, and **Ttr** options are used to define the sphere using either three points on the surface of the sphere, two points on the surface of the sphere, or two points of tangency on the surface of the sphere and a radius.

Ribbon
Home
> Modeling
Sphere
Type
SPHERE

SPHERE

Spheres and other curved objects can be displayed in a number of different ways. The manner in which you choose to display these objects should be governed by the display requirements of your work. Notice in **Figure 2-13A** the lines that define the shape of the spheres in a wireframe display. These lines are called *contour lines*, also known as *tessellation lines*. The **Visual Styles Manager** can be used to set the display

Figure 2-13.
A—The Draw true silhouettes setting is No and four contour lines are used. B—The Draw true silhouettes setting is No and 20 contour lines are used. C—The Draw true silhouettes setting is Yes and four contour lines are used. D—The Draw true silhouettes setting is Yes and the **HIDE** command is used with the **2D Wireframe** visual style set current.

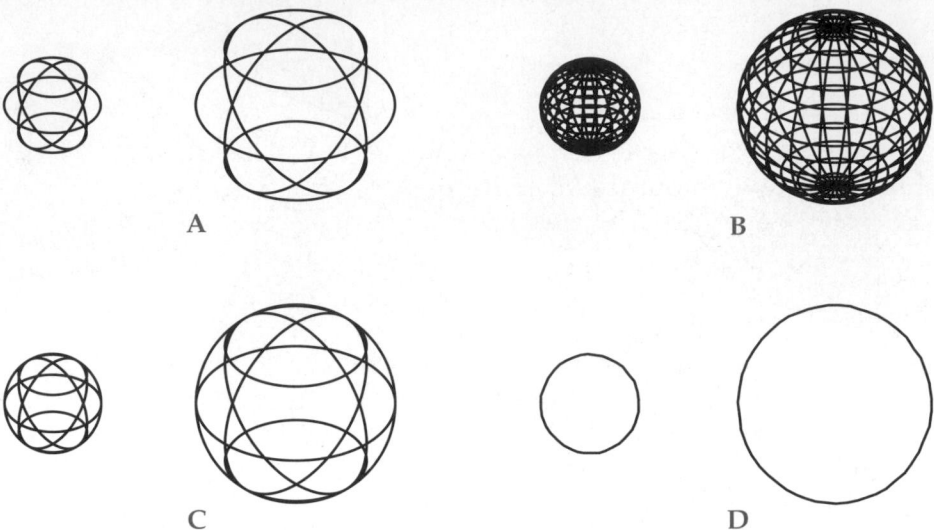

of contour lines and silhouettes on spheres and other curved 3D surfaces for a given visual style. See **Figure 2-14.**

With the **Visual Styles Manager** displayed, select the 2D Wireframe image tile. The Contour lines setting in the **2D Wireframe options** area establishes the number of lines used to show the shape of curved objects. A similar setting appears in the 3D Wireframe, 3D Hidden, Conceptual, and Realistic visual styles if their Edge mode entry is set to Isolines. The default value is four, but can be set to a value from zero to 2047. **Figure 2-13B** displays spheres with 20 contour lines. It is best to use a lower number during construction and preliminary displays of the model and, if needed, higher settings for more

Figure 2-14.
The **2D Wireframe options** area of the **Visual Styles Manager** is used to control the display of contour lines and silhouettes on spheres and other curved 3D surfaces in a given visual style.

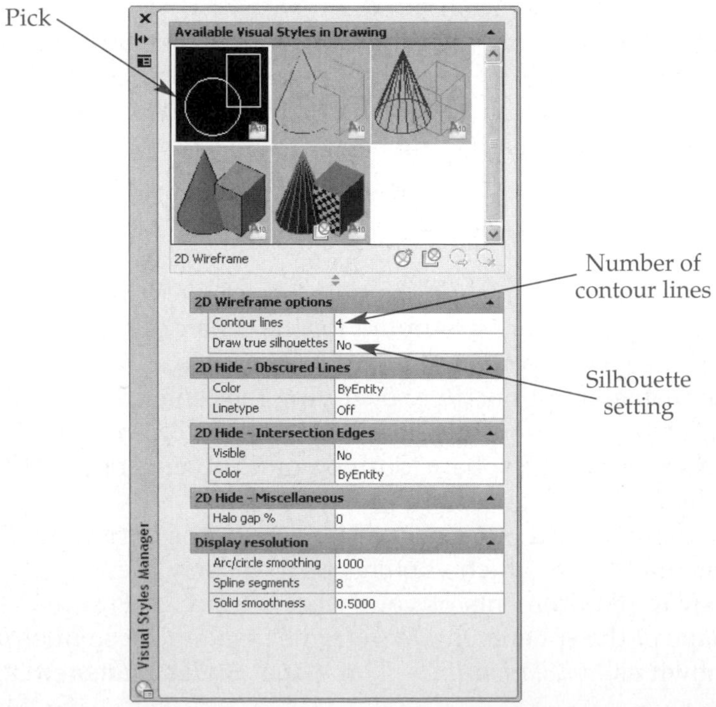

realistic visualization. The contour lines setting is also available in the **Display** tab of the **Options** dialog box or by typing ISOLINES.

The Draw true silhouettes setting in the **2D Wireframe options** area controls the display of silhouettes on 3D solid curved surfaces. The setting is either Yes or No. Notice the sphere silhouette in **Figures 2-13C** and **2-13D**. The Draw true silhouettes setting is stored in the **DISPSILH** system variable.

Torus

Ribbon
Home
> Modeling

Torus

Type
TORUS

TORUS

A basic *torus* is a cylinder bent into a circle, similar to a doughnut or inner tube. There are three types of tori. See **Figure 2-15.** A torus with a tube diameter that touches itself is called *self intersecting* and has no center hole. To create a self-intersecting torus, the tube radius must be greater than the torus radius. The third type of torus looks like a football. It is drawn by entering a negative torus radius and a positive tube radius of greater absolute value, i.e. –1 and 1.1.

Once the command is initiated, you are prompted for the center of the torus or to enter an option. If you pick the center, you must then set the radius of the torus. To specify a diameter, enter the **Diameter** option after specifying the center. This defines a base circle that is the centerline of the tube. The **3P**, **2P**, and **Ttr** options are used to define the base circle of the torus using either three points, two points, or two points of tangency and a radius.

Once the base circle of the torus is defined, you are prompted for the tube radius or to enter an option. The tube radius defines the cross-sectional circle of the tube. To specify a diameter of the cross-sectional circle, enter the **Diameter** option. You can also use the **2point** option to pick two points on screen that define the diameter of the cross-sectional circle.

Figure 2-15.
The three types of tori are shown as wireframes and with hidden lines removed.

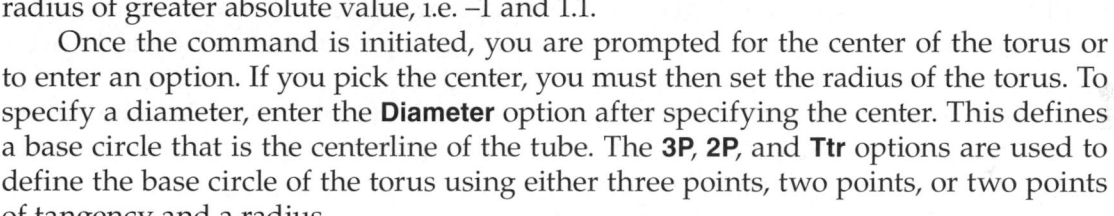

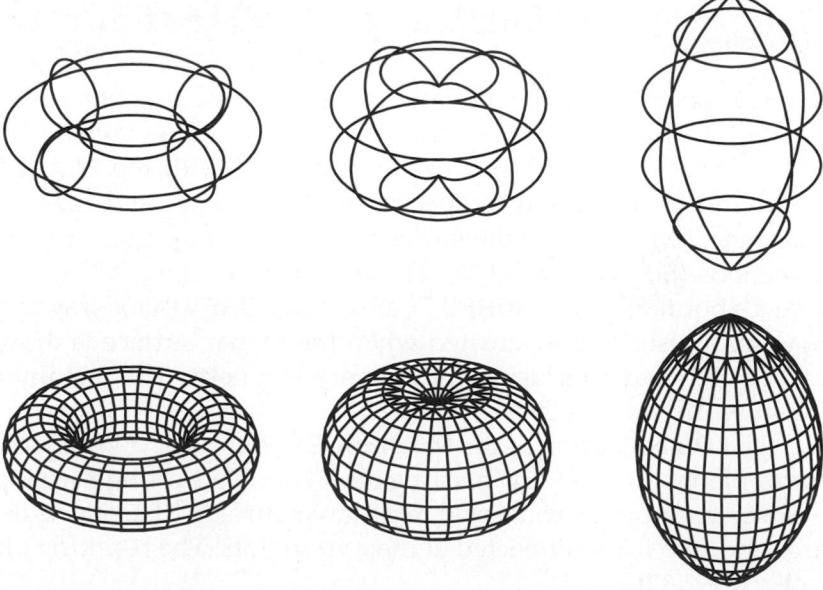

Figure 2-16.
A—A wedge drawn by picking corners and specifying a height. B—A wedge drawn using the **Center** option. Notice the location of the center.

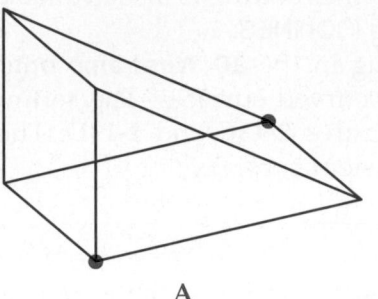

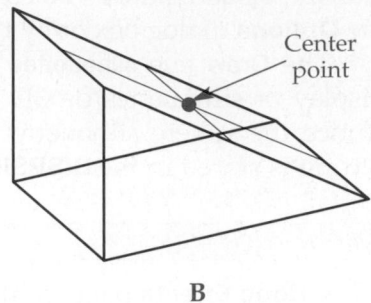

Center point

A

B

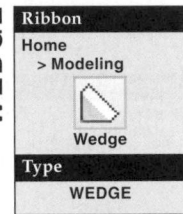
WEDGE

Ribbon
Home
> Modeling
Wedge
Type
WEDGE

Wedge

A *wedge* has five sides, four of which are at right angles and the fifth at an angle other than 90°. See **Figure 2-16.** Once the command is initiated, you are prompted to select the first corner of the base or to enter an option. By default, a wedge is constructed by picking diagonal corners of the base and setting a height. To pick the center point, enter the **Center** option. The center point of a wedge is the middle of the angled surface. You must then pick a point to set the width and length before entering a height.

After specifying the first corner or the center, you can enter the length, width, and height instead of picking a second corner. When prompted for the second corner, enter the **Length** option and specify the length. You are then prompted for the width. After the width is entered, you are prompted for the height.

To create a wedge with equal length, width, and height, enter the **Cube** option when prompted for the second corner. Then, enter a length. The same value is automatically used for the width and height.

Exercise 2-2

Complete the exercise on the student website.
www.g-wlearning.com/CAD

Constructing a Planar Surface

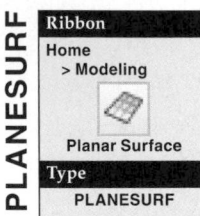
PLANESURF

Ribbon
Home
> Modeling
Planar Surface
Type
PLANESURF

A *planar surface* primitive is an object consisting of a single plane and is created parallel to the current XY plane. The surface that is created has zero thickness and is composed of a mesh of lines. It is created with the **PLANESURF** command. The command prompts you to specify the first corner and then the second corner of a rectangle. Once drawn, the surface is displayed as a mesh with lines in the X and Y directions. See **Figure 2-17A.** These lines are called *isolines* and do not include the object's boundary. The **SURFU** (Y axis) and **SURFV** (X axis) system variables determine how many isolines are created when the planar surface is drawn. The isoline values can be changed later using the **Properties** palette. The maximum number of isolines in either direction is 200.

The **Object** option of the **PLANESURF** command allows you to convert a 2D object into a planar surface. Any existing object or objects lying in a single plane and forming a closed area can be converted to a planar surface. The objects in **Figure 2-17B** are two arcs and two lines connected at their endpoints. The resulting planar surface is shown in **Figure 2-17C.**

Although a planar surface is not a solid, it can be converted into a solid in a single step. For example, the object in **Figure 2-17C** is converted into a solid using the **THICKEN** command. See **Figure 2-17D**. The object that started as two arcs and two lines is now a solid model and can be manipulated and edited like any other solid. This capability allows you to create intricate planar shapes and quickly convert them to a solid for use in advanced modeling applications. Model editing procedures are discussed in detail in Chapters 11 through 13.

Figure 2-17.
A—A rectangular planar surface with four isolines in the Y direction and eight isolines in the X direction. B—These two arcs and two lines form a closed area and lie on a single plane. C—The arcs and curves are converted into a planar surface. D—The planar surface is converted into a solid.

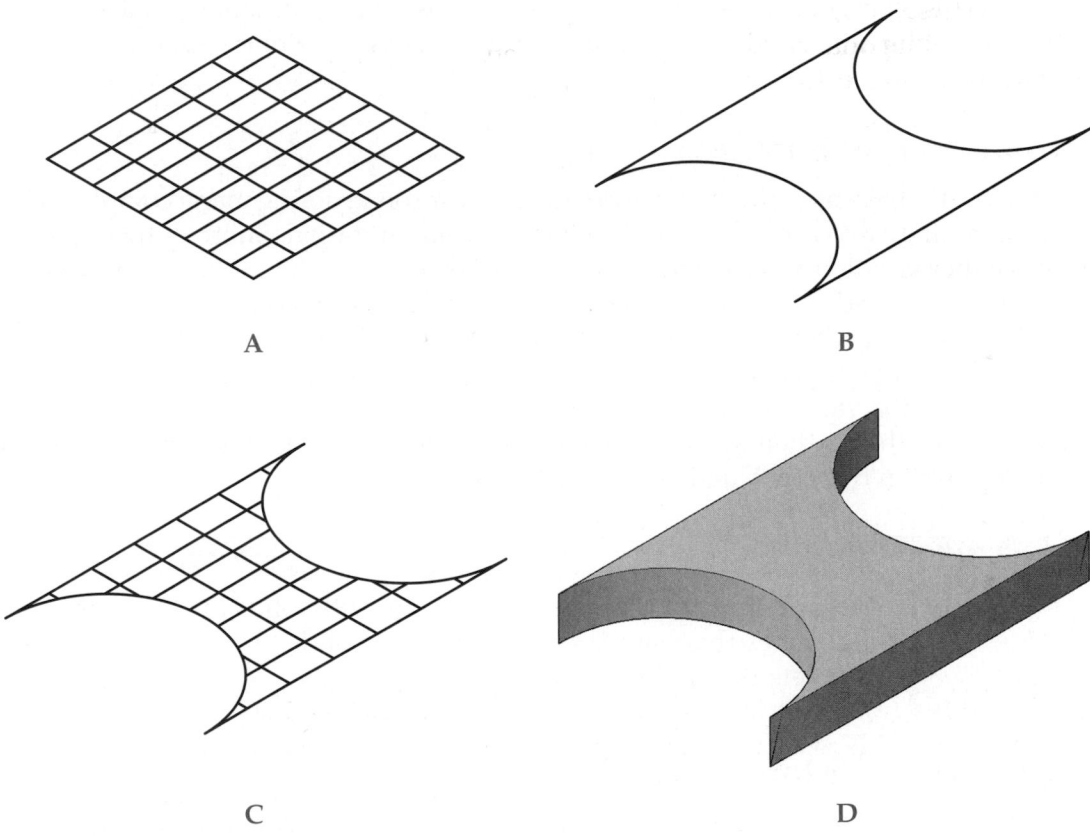

A

B

C

D

Figure 2-18.
Selecting a Boolean
command in the
Solid Editing panel
in the **Home** tab of
the ribbon.

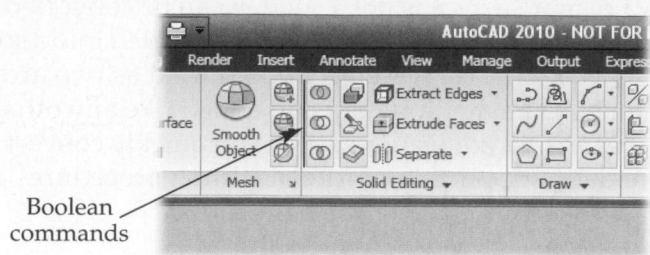

Boolean
commands

Creating Composite Solids

A *composite solid* is a solid model constructed of two or more solids, often primitives. Solids can be subtracted from each other, joined to form a new solid, or overlapped to create an intersection or interference. The commands used to create composite solids are found in the **Solid Editing** panel of the **Home** tab in the ribbon. See **Figure 2-18.**

Introduction to Booleans

There are three operations that form the basis of constructing many complex solid models. Joining two or more solids is called a *union* operation. Subtracting one solid from another is called a *subtraction* operation. Forming a solid based on the volume of overlapping solids is called an *intersection* operation. Unions, subtractions, and intersections as a group are called *Boolean operations.* George Boole (1815–1864) was an English mathematician who developed a system of mathematical logic where all variables have the value of either one or zero. Boole's two-value logic, or *binary algebra,* is the basis for the mathematical calculations used by computers and, specifically, for those required in the construction of composite solids.

> **NOTE**
>
> Boolean operations used to create composite solids can also be used on meshes that have been converted to solids. See Chapter 3 for a complete discussion of meshes.

Joining Two or More Solid Objects

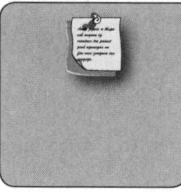

The **UNION** command is used to combine solid objects, **Figure 2-19.** The solids do not need to touch or intersect to form a union. Therefore, accurately locate the primitives when drawing them. After selecting the objects to join, just press [Enter] and the action is completed.

In the examples shown in **Figure 2-19B,** notice that lines, or edges, are shown at the new intersection points of the joined objects. This is an indication that the features are one object, not separate objects.

Subtracting Solids

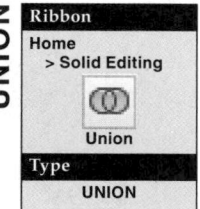

The **SUBTRACT** command allows you to remove the volume of one or more solids from another solid. Several examples are shown in **Figure 2-20.** The first object selected in the subtraction operation is the object *from* which volume is to be subtracted. The next object is the object to be subtracted from the first. The completed object will be a new solid. If the result is the opposite of what you intended, you may have selected the objects in the wrong order. Just undo the operation and try again.

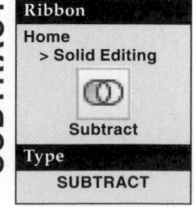

UNION

Ribbon
Home
> Solid Editing
Union
Type
UNION

SUBTRACT

Ribbon
Home
> Solid Editing
Subtract
Type
SUBTRACT

Figure 2-19.
A—The solid primitives shown here have areas of intersection and overlap. B—Composite solids after using the **UNION** command.

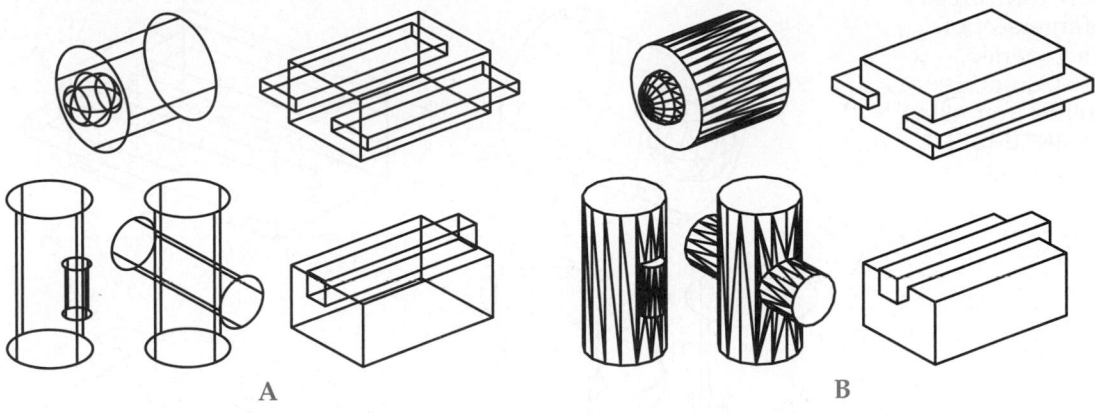

A   B

Figure 2-20.
A—The solid primitives shown here have areas of intersection and overlap. B—Composite solids after using the **SUBTRACT** command.

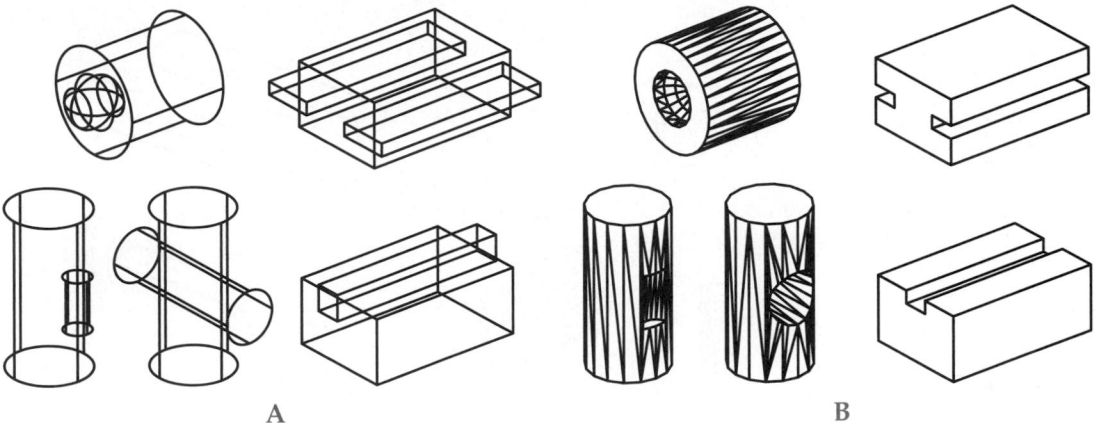

A   B

Creating New Solids from the Intersection of Solids

When solid objects intersect, the overlap forms a common volume, a space that both objects share. This shared space is called an *intersection*. An intersection (common volume) can be made into a composite solid using the **INTERSECT** command. **Figure 2-21** shows several examples. A solid is formed from the common volume. The original objects are removed.

The **INTERSECT** command is also useful in 2D drawing. For example, if you need to create a complex shape that must later be used for inquiry calculations or hatching, draw the main object first. Then, draw all intersecting or overlapping objects. Next, create regions of the shapes. Finally, use **INTERSECT** to create the final shape. The resulting shape is a region and has solid properties. Regions are discussed later in this chapter.

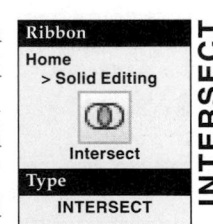

Ribbon
Home
> Solid Editing

Intersect

Type
INTERSECT

INTERSECT

Exercise 2-3

Complete the exercise on the student website.
www.g-wlearning.com/CAD

Figure 2-21.
A—The solid primitives shown here have areas of intersection and overlap. B—Composite solids after using the **INTERSECT** command.

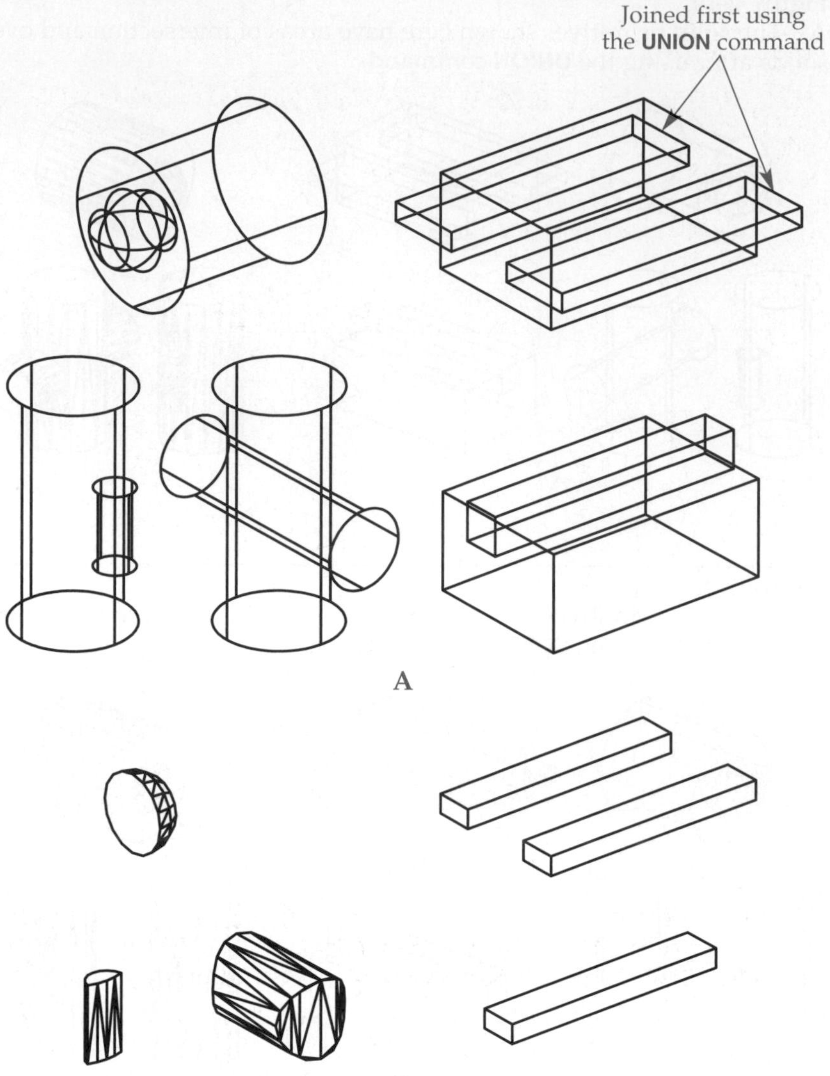

Joined first using the **UNION** command

A

B

Creating New Solids Using the Interfere Command

INTERFERE

Ribbon
Home
> Solid Editing

Interfere

Type
INTERFERE

When you use the **SUBTRACT**, **UNION**, and **INTERSECT** commands, the original solids are deleted. They are replaced by the new composite solid. The **INTERFERE** command does not do this. A new solid is created from the interference (common volume) as if the **INTERSECT** command is used, but the original objects can be either deleted or retained.

Once the command is initiated, you are prompted to select the first set of solids or to enter an option. The **Settings** option opens the **Interference Settings** dialog box, which is used to change the visual style and color of the interference solid and the visual style of the viewport. The **Nested selection** option allows you to check the interference of separate solid objects within a nested block. A *nested block* is one that is composed of other blocks. When any needed options are set, select the first set of solids and press [Enter].

You are prompted to select the second set of solids or to enter an option. Entering the **Check first set** option tells AutoCAD to check the objects in the first set for interference. There is no second set when this option is used. Otherwise, select the second set of solids and press [Enter].

Figure 2-22.
The **Interference Checking** dialog box is used to check for interference between solids. To retain the interference solid, uncheck the **Delete interference objects created on Close** check box.

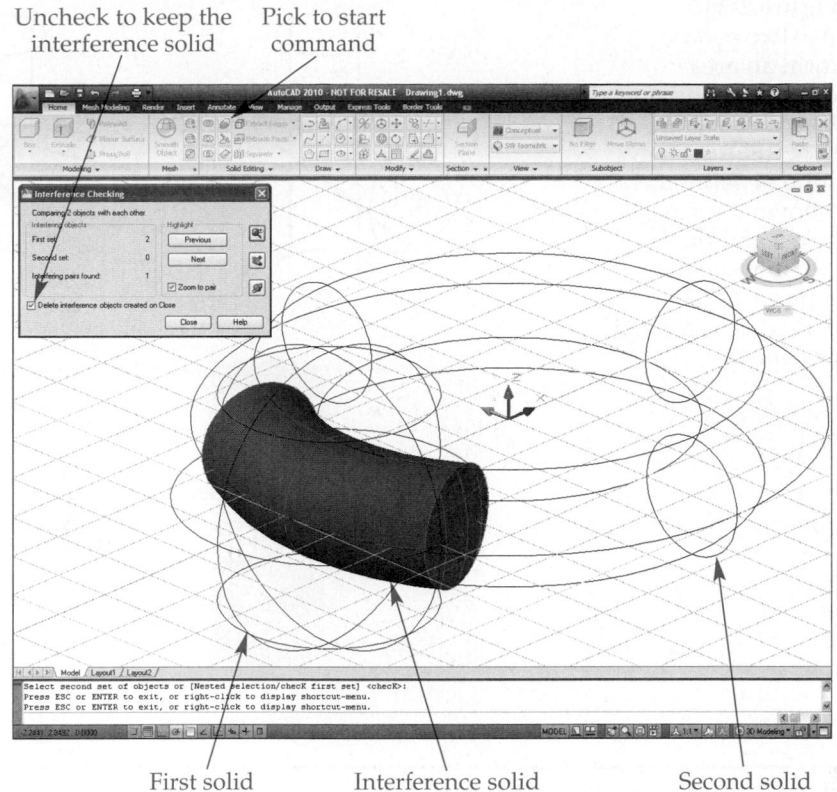

Uncheck to keep the interference solid   Pick to start command

First solid   Interference solid   Second solid

AutoCAD zooms in on the highlighted interference solid and displays the **Interference Checking** dialog box. See **Figure 2-22.** The visual style is set to a wireframe display by default and the interference solid is shaded in a color, which is red by default.

In the **Interfering objects** area of the **Interference Checking** dialog box, the number of objects selected in the first and second sets is displayed. The number of interfering pairs found in the selected objects is also displayed.

The buttons in the **Highlight** area of the dialog box are used to highlight the previous or next interference object. If the **Zoom to pair** check box is checked, AutoCAD zooms to the interference objects when the **Previous** or **Next** button is selected.

To the right of the **Highlight** area are three navigation buttons—**Zoom Realtime**, **Pan Realtime**, and **3D Orbit**. Selecting one of these display options temporarily hides the dialog box and activates the selected command. This allows you to navigate in the viewport. When the command is ended, the dialog box is redisplayed.

By default, the **Delete interference objects created on Close** check box is checked. This means that the object(s) created by interference is deleted. In order to retain the new solid(s), uncheck this box.

An example of interference checking and the result is shown in **Figure 2-23.** Notice that the original solids are intact, but new lines indicate the new solid. The new solid is retained as a separate object because the **Delete interference objects created on Close** check box was unchecked. The new solid can be moved, copied, and manipulated just like any other object. **Figure 2-23C** shows the new object after it has been moved and a conceptual display generated.

When the **INTERFERE** command is used, AutoCAD compares the first set of solids to the second set. Any solids that are selected for both the first and second sets are automatically included as part of the first selection set and eliminated from the second. If you do not select a second set of objects or the **Check first set** option is used, AutoCAD calculates the interference between the objects in the first selection set.

Figure 2-23.
A—Two solids form an area of intersection. B—After using **INTERFERE**, a new solid is defined (shown here in color) and the original solids remain. C—The new solid can be moved or copied.

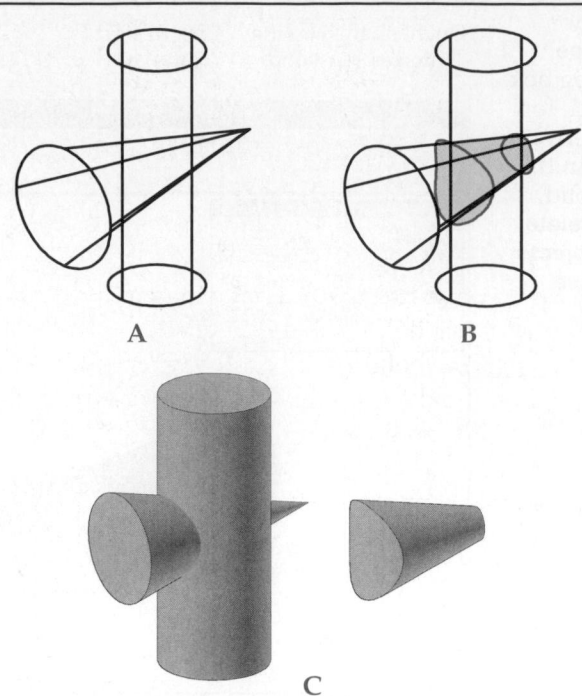

A   B

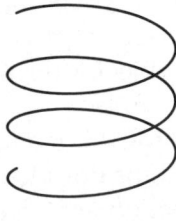 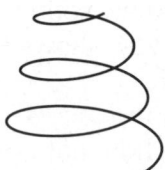

C

Exercise 2-4

Complete the exercise on the student website.
www.g-wlearning.com/CAD

Creating a Helix

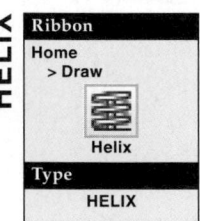

A *helix* is a spline in the form of a spiral and can be created as a 2D or 3D object. See **Figure 2-24.** It is not a solid object. However, it can be used as the path or framework for creating solid objects such as springs and spiral staircases.

When the command is initiated, you are prompted for the center of the helix base. After picking the center, you are prompted to enter the radius of the base. If you want to specify the diameter, enter the **Diameter** option. After the base is defined, you are prompted for the radius of the top. You can use the **Diameter** option to enter a diameter. The top and bottom can be different sizes. Entering different sizes creates a tapered helix, if the helix is 3D. A 2D helix should have different sizes for the top and bottom.

Figure 2-24.
Three types of helices. From left to right, equal top and bottom diameters, unequal top and bottom diameters, and unequal top and bottom diameters with the height set to zero.

After the top and bottom sizes are set, you are prompted to set the height or enter an option. To specify the number of turns in the helix, enter the **Turns** option. Then, enter the number of turns. The maximum is 500 and you can enter values less than one, but greater than zero.

By default, the helix turns in a counterclockwise manner. To change the direction in which the helix turns, enter the **Twist** option. Then, enter CW for clockwise or CCW for counterclockwise.

The height of the helix can be set in one of three ways. First, you can enter a direct distance. To do this, type the height value or pick with the mouse to set the height. To create a 2D helix, enter a height of zero.

You can also set the height for one turn of the helix using the **Turn height** option. In this case, the total height is the number of turns multiplied by the turn height. If you provide a value for the turn height and then specify the helix height, the number of turns is automatically calculated and the helix is drawn. Conversely, if you provide values for both the turn height and number of turns, the helix height is calculated by AutoCAD.

Finally, you can pick a location for the axis endpoint using the **Axis endpoint** option. This is the same option available with a cone, cylinder, or pyramid.

As an example, a solid model of a spring can be created by constructing a helix and a circle and then using the **SWEEP** command to sweep the circle along the helix path. See **Figure 2-25.** The **SWEEP** command is discussed in detail in Chapter 10.

First, determine the diameter of the spring wire and then draw a circle using that value. For this example, you will create two springs each with a wire diameter of .125 units, so draw two circles of that diameter, **Figure 2-26.** Their locations are not important. Next, determine the diameter of the spring and draw a corresponding helix. For this example, draw a helix anywhere on screen with a bottom diameter of one unit and a top diameter of one unit. Set the number of turns to eight and specify a height of two units. Draw another helix with the same settings, except make the top diameter .5 units.

Initiate the **SWEEP** command. You are first prompted to select the objects to sweep; pick one circle and press [Enter]. Next, you are prompted to select the sweep path. Select one of the helices. The first sweep, or spring, is completed. Repeat the procedure for the other circle and helix. The drawing is now composed of the two original, single-line helices and the two new swept solids. The circles are consumed by the **SWEEP** command.

Ribbon
Home
> Modeling
Sweep
Type
SWEEP

SWEEP

Exercise 2-5

Complete the exercise on the student website.
www.g-wlearning.com/CAD

Figure 2-25.
A helix can be used as a path to create a spring.

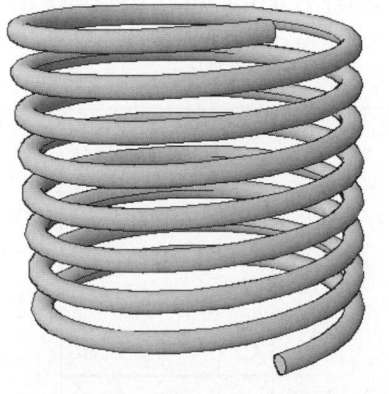
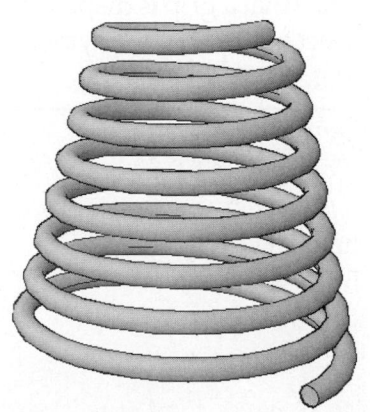

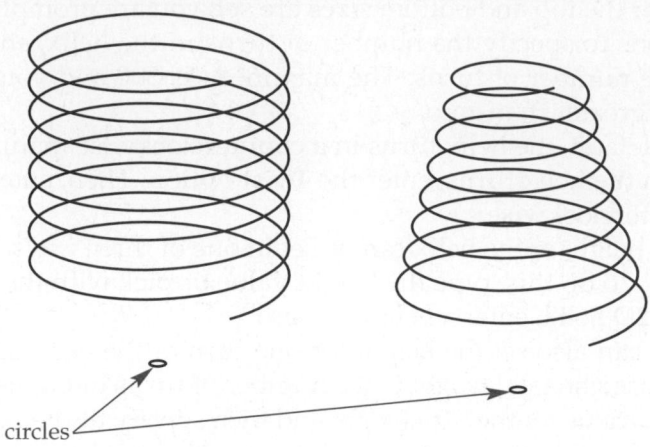

Figure 2-26.
To create a spring, first draw a circle the same diameter as the spring wire. Then, draw the helix and sweep the circle along the helix. Shown here are the two helices used to create the springs in Figure 2-25.

Ø.125 circles

Working with Regions

A *region* is a closed, two-dimensional solid. It is a solid model without thickness (Z value). A region can be analyzed for its mass properties. Therefore, regions are useful for 2D applications where area and boundary calculations must be quickly obtained from a drawing.

Boolean operations can be performed on regions. When regions are unioned, subtracted, or intersected, a *composite region* is created. A composite region is also called a *region model.*

A region can be quickly and easily given a thickness, or *extruded* into a 3D solid object. This means that you can convert a 2D shape into a 3D solid model in just a few steps. An application is drawing a 2D section view, converting it into a region, and extruding the region into a 3D solid model. Extruding is covered in Chapter 9.

Constructing a 2D Region Model

The following example creates, as a region, the plan view of a base for a support bracket. In Chapter 9, you will learn how to extrude the region into a solid. First, start a new drawing. Next, create the profile geometry in **Figure 2-27** using the **RECTANGLE** and **CIRCLE** commands. These commands create 2D objects that can be converted into regions. The **PLINE** and **LINE** commands can also be used to create closed 2D objects.

The **REGION** command allows you to convert closed, two-dimensional objects into regions. When the command is initiated, you are prompted to select objects. Select the rectangle and four circles and then press [Enter]. The rectangle and each circle are now separate regions and the original objects are deleted. You may need to switch to a wireframe visual style in order to see the circles. You can individually pick the regions. If you pick a circle, notice that a grip is displayed in the center, but not at the four quadrants. This is because the object is not a circle anymore. However, you can still snap to the quadrants.

REGION

Ribbon
Home
> Draw
Region
Type
REGION
REG

Figure 2-27.
These 2D shapes can be made into a region. The region can then be made into a 3D solid.

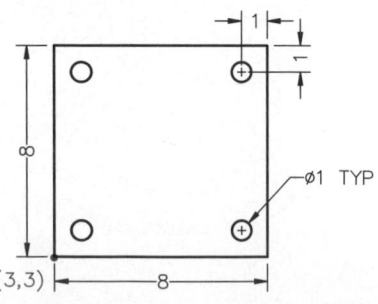

AutoCAD and Its Applications—Advanced

Figure 2-28.
Once the circular regions are subtracted from the rectangular region, they appear as holes. This is clear when the Conceptual or Realistic visual style is set current.

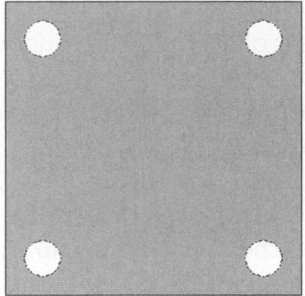

In order to create the proper solid, the circular regions must be subtracted from the rectangular region. Using the **SUBTRACT** command, select the rectangle as the object to be subtracted *from* and then all of the circles as the objects to subtract. Now, if you select the rectangle or any of the circles, you can see that a single region has been created from the five separate regions. If you set the Conceptual or Realistic visual style current, you can see that the circles are now holes in the region. See **Figure 2-28.**

Using the Boundary Command to Create a Region

The **BOUNDARY** command is often used to create a polyline for hatching or an inquiry. In addition, this command can be used to create a region. When the command is initiated, the **Boundary Creation** dialog box is displayed. See **Figure 2-29.**

Next, select **Region** from the **Object type:** drop-down list in the **Boundary retention** area of the dialog box. Also, you can refine the boundary selection method by turning island detection on or off. When the **Island detection** check box above the **Boundary retention** area is checked, island detection is on.

- **On.** When an internal point is selected in the object, AutoCAD creates separate regions from any islands that reside within the object.
- **Off.** When an internal point is selected in the object, AutoCAD ignores islands that reside within the object when creating the region.

Finally, select the **Pick Points** button. The dialog box is closed and you are prompted to select an internal point. Pick a point inside of the object that you wish to convert to a region. Press [Enter] when you are finished and the region is created. You can always check to see if an object is a polyline or region by using the **LIST** command and selecting the object.

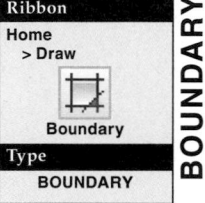

Ribbon
Home
> Draw

Boundary

Type
BOUNDARY

BOUNDARY

Figure 2-29.
Regions can be created using the **Boundary Creation** dialog box.

Pick to select a point inside the boundary

Turn island detection on and off

Select the type of object to be created

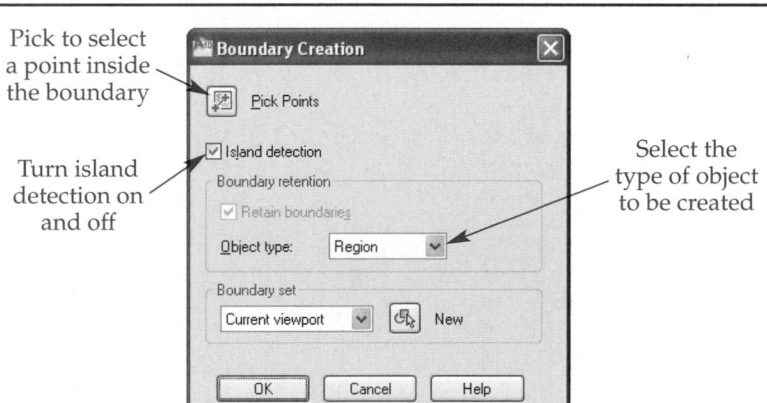

Calculating the Area of a Region

A region is not a polyline. It is an enclosed area called a *loop.* Certain properties of the region, such as area, are stored as a value of the region. The **AREA** command can be used to determine the length of all sides and the area of the loop. This can be a useful advantage of a region.

For example, suppose a parking lot is being repaved. You need to calculate the surface area of the parking lot to determine the amount of material needed. This total surface area excludes the space taken up by planting dividers, sidewalks, and lampposts because you will not be paving under these items. If the parking lot and all objects inside of it are drawn as a region, the **AREA** command can give you this figure in one step using the **Object** option. If a polyline is used to draw the parking lot, all internal features must be subtracted each time the **AREA** command is used.

PROFESSIONAL TIP

Regions can prove valuable when working with many items:
- Roof areas excluding chimneys, vents, and fans.
- Bodies of water, such as lakes, excluding islands.
- Lawns and areas of grass excluding flower beds, trees, and shrubs.
- Landscaping areas excluding lawns, sidewalks, and parking lots.
- Concrete surfaces, such as sidewalks, excluding openings for landscaping, drains, and utility covers.

You can find many other applications for regions that can help in your daily tasks.

Exercise 2-6

Complete the exercise on the student website.
www.g-wlearning.com/CAD

Chapter Test

1. What is a *solid primitive?*
2. How is a solid cube created?
3. How is an elliptical cylinder created?
4. Where is the center of a wedge located?
5. What is a *frustum pyramid?*
6. What is a *polysolid?*
7. Name at least four AutoCAD 2D entities that can be converted to a polysolid.
8. What type of entity does the **HELIX** command create and how can it be converted into a solid model?
9. What is a *composite solid?*
10. Which type of mathematical calculations are used in the construction of solid models?
11. How are two or more solids combined to make a composite solid?
12. What is the function of the **INTERSECT** command?
13. How does the **INTERFERE** command differ from **INTERSECT** and **UNION**?
14. What is a *region?*
15. How can a 2D section view be converted to a 3D solid model?

Drawing Problems

Draw the objects in the following problems using the appropriate solid primitive commands and Boolean operations. Use your own measurements for objects shown without dimensions. Do not add dimensions to the models. Save the drawings as P2-(problem number). Display and plot the problems as indicated by your instructor.

1.

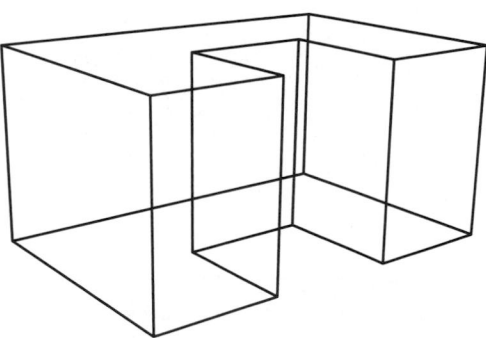

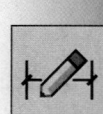

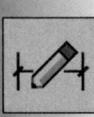

2.

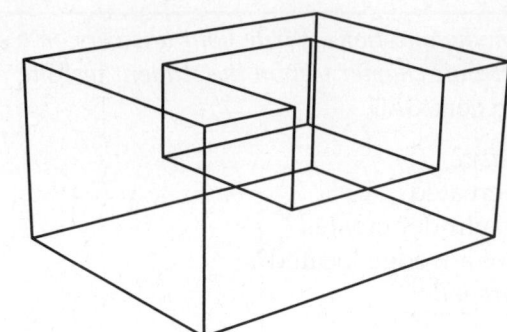

3.

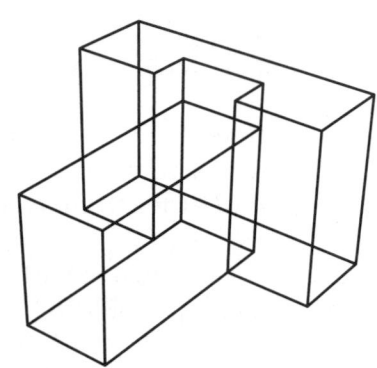

4.

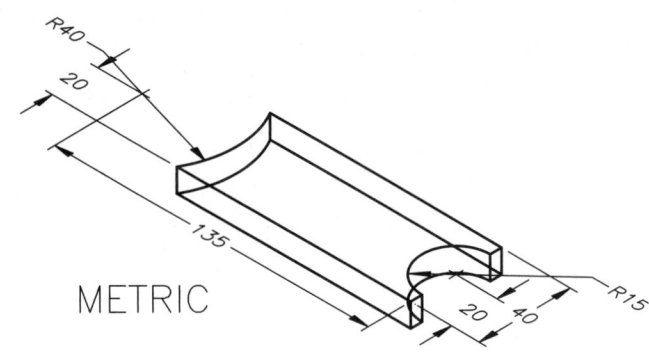

METRIC

5.

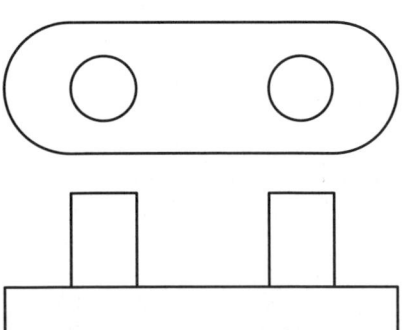

6.

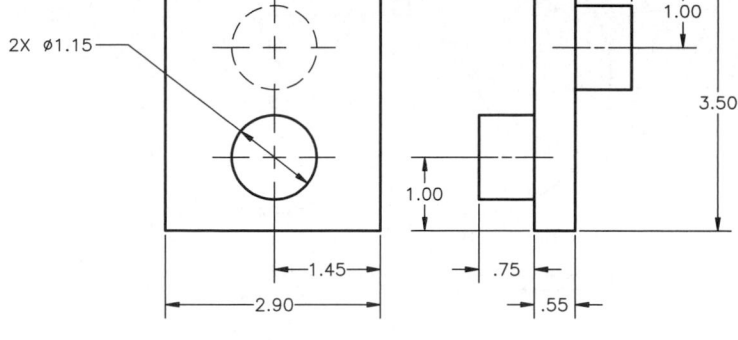

7.

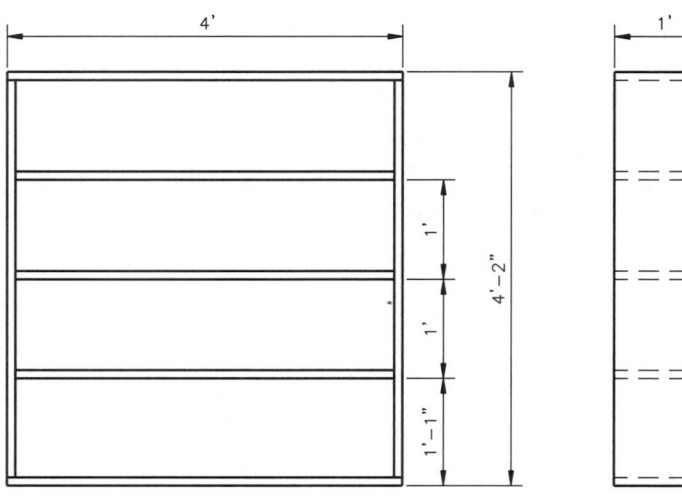

8.

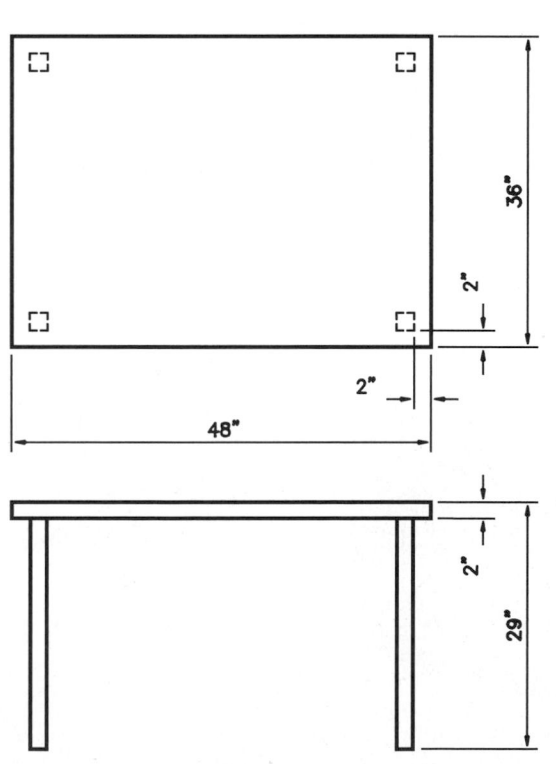

TABLE

9.

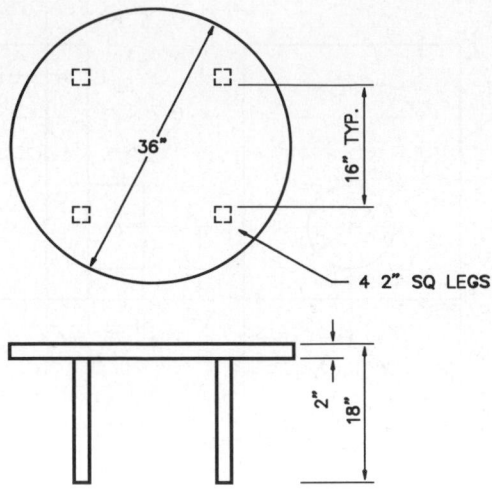

10.

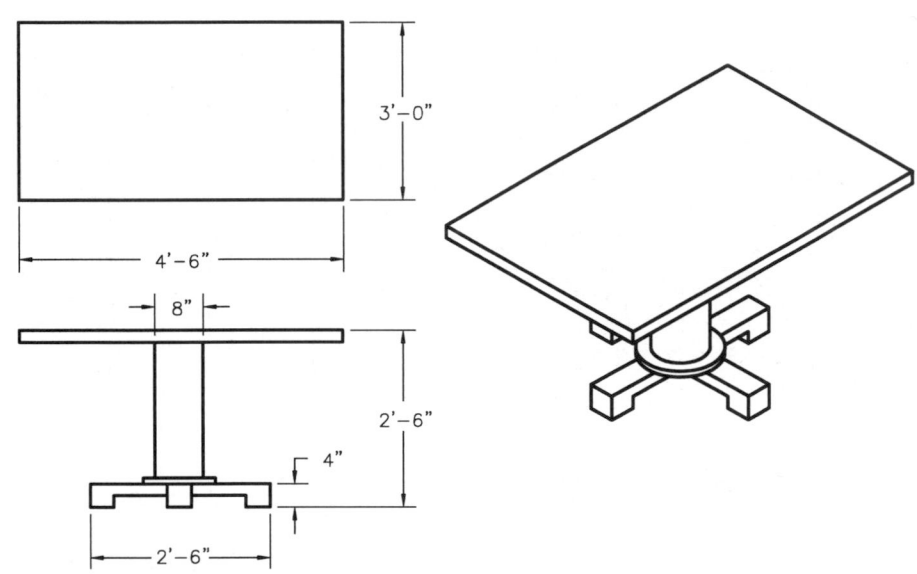

11. Draw the seat for a kitchen chair shown below. In later chapters, you will complete this chair. The seat is 1″ thick.

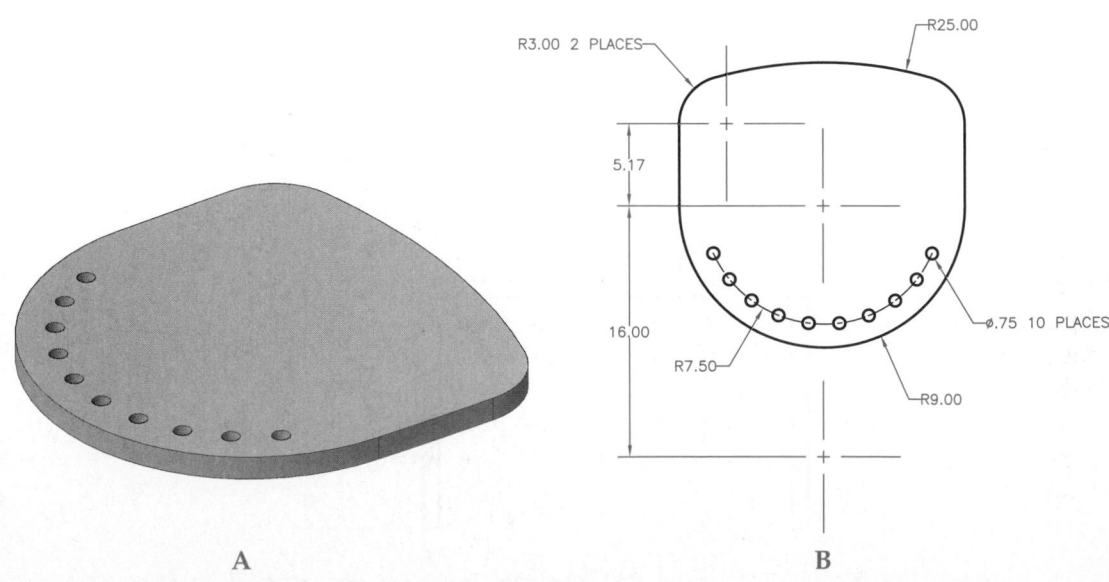

A   B

CHAPTER 3

Mesh Modeling

Learning Objectives

After completing this chapter, you will be able to:

✓ Explain tessellation division and values.
✓ Create mesh primitives.
✓ Create a smoothed mesh object.
✓ Create a refined mesh object.
✓ Construct mesh forms.
✓ Generate a mesh by converting a solid.
✓ Generate a mesh by converting a surface.
✓ Generate a surface by converting a mesh.
✓ Generate a solid by converting a mesh.
✓ Execute editing on mesh objects.
✓ Create a split face on a mesh.
✓ Produce an extruded mesh face.
✓ Apply a crease to mesh subobjects.

Overview of Mesh Modeling

Mesh primitives and mesh forms can be used to create freeform designs. See **Figure 3-1.** The tools for creating and editing meshes extend the capability of AutoCAD's 3D modeling tools. There are two key workflows that the designer considers:

- The creation of 3D models, which can be solids, surfaces, or meshes.
- Editing the 3D models to create unique shapes.

Mesh models can be created as mesh primitives, mesh forms, or freeform mesh shapes. A *mesh model* consists of vertices, edges, and faces. You can modify or refine a mesh by adding smoothness, creases, extrusions, and splits. You can also distort a mesh to create unique freeform shapes.

Mesh models are a type of surface model. *Subdivision surfaces* is another term for mesh models. They do *not* have volume or mass. Rather, mesh models only define the shape of the design.

Figure 3-1
Constructing an ergonomic mouse as a mesh model. A—The basic mesh primitive. B—Using editing tools, the mesh is reformed. C—The completed mesh model.

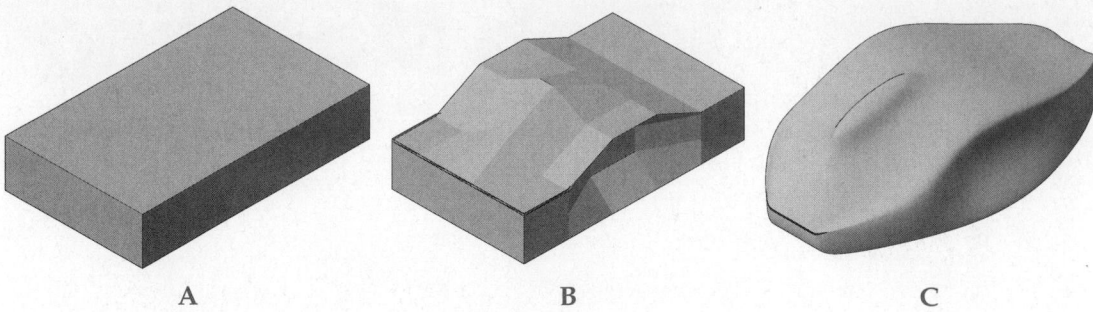

Mesh objects can be created using one of these methods:
- Construct mesh primitives (**MESH** command).
- Construct mesh forms that are ruled, revolved, tabulated, or edge-defined objects (**RULESURF**, **REVSURF**, **TABSURF** and **EDGESURF** commands).
- Convert an existing solid or surface into a mesh object (**MESHSMOOTH** command).
- Convert legacy surface objects into mesh objects using commands such as **3DFACE**, **3DMESH** and **PFACE** commands.

The tools used to create and modify meshes are found on the **Mesh Modeling** tab in the ribbon. The commands used to create mesh primitives are similar to those used to create solid primitives, which are discussed in Chapter 2. However, mesh primitives have face mesh objects that are divided into smaller faces. These divisions are based on tessellation division values (smoothness), as discussed in this chapter.

Tessellation Division Values

Tessellation divisions are the basic foundation for the smoothness of a mesh object. Tessellation divisions on a mesh object consist of planar shapes that fit together to form the surface. See **Figure 3-2**. These divisions display the edges of a mesh face that can then be edited.

When creating mesh primitives, set the mesh tessellation divisions *before* creating a mesh primitive shape. Setting the proper mesh tessellation division value ensures the model has enough faces, edges, and vertices for editing. The default tessellation divisions are listed in **Mesh Primitive Options** dialog box, which is discussed in the next section.

Figure 3-2.
Tessellation divisions are key to mesh modeling as they define the smoothness of the mesh model.

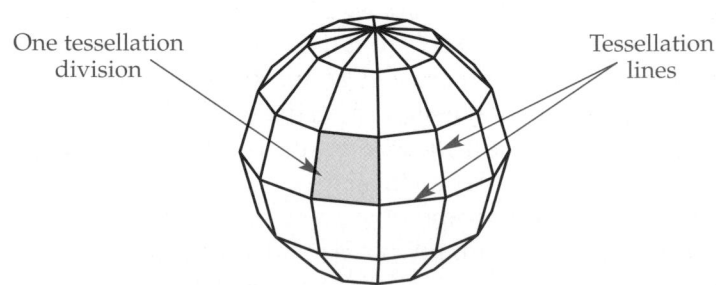

One tessellation division

Tessellation lines

Figure 3-3.
This box mesh primitive is drawn with the default settings. A—Displayed with the 2D Wireframe visual style current. B—After the **HIDE** command is used.

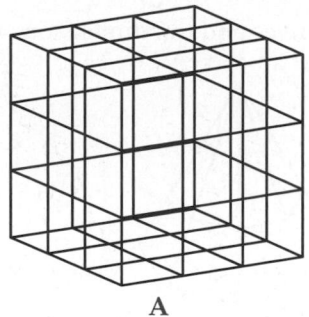

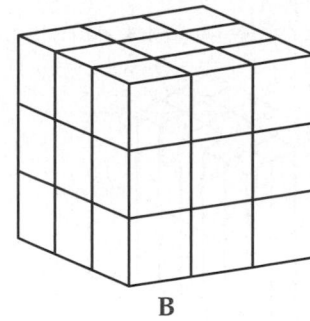

A   B

You can change the default smoothness in the **Mesh Primitive Options** dialog box or, if the primitive is being drawing using the command line, by entering the **Settings** option of the **MESH** command. **Figure 3-3** shows an example of a box mesh primitive created using the default tessellation divisions. The default settings create a box with no smoothness, length divisions of three, width divisions of three, and height divisions of three.

Drawing Mesh Primitives

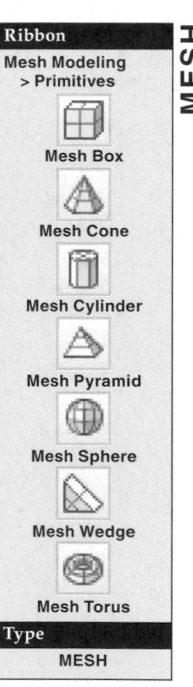

A primitive is a basic building block. Just as there are solid primitives in AutoCAD, there are mesh primitives. The seven *mesh primitives* are mesh box, mesh cone, mesh cylinder, mesh pyramid, mesh sphere, mesh wedge, and mesh torus. See **Figure 3-4.** These primitives can be used as the starting point for creating complex freeform mesh models.

The **Mesh Primitive Options** dialog box is used to set the number of tessellation subdivisions for each mesh primitive object created. See **Figure 3-5.** This dialog box is displayed by picking the dialog box launcher button at the lower-right corner of the **Primitives** panel in the **Mesh Modeling** tab of the ribbon or by using the **MESHPRIMITIVEOPTIONS** command. Set the number of tessellation subdivisions in the **Mesh Primitive Options** dialog box *before* creating a mesh primitive. There is no way to change the number of subdivisions after the primitive is created.

The tessellation subdivisions for each primitive are based on the dimensions required to create the primitive. For example, a box has length, width, and height subdivisions. On the other hand, a mesh cylinder has axis, height, and base subdivisions. To set the subdivisions, select the primitive in the tree on the left-hand side of the **Mesh Primitive Options** dialog box. The subdivision properties are then displayed below the tree. Enter the number of subdivisions as required and then close the dialog box. All new primitives of that type will have this number of subdivisions until the setting is changed in the dialog box. Existing primitives are not affected.

The following is an example of how to creating a mesh box primitive. The mesh box is the mesh primitive used for most base shapes. Refer to **Figure 3-4** for the information required to draw mesh primitives.

1. Open the **Mesh Primitive Options** dialog box.
2. Select the box primitive in the tree.
3. Enter 3 in each of the Length, Width, and Height property boxes to set the subdivisions.
4. Close the dialog box.
5. Select the command for drawing a mesh box.
6. Specify the first corner of the base of the mesh box.
7. Specify the opposite corner of the base of the mesh box.
6. Specify the height of the mesh box.

Figure 3-4.
An overview of AutoCAD's mesh primitives and the dimensions required to draw them.

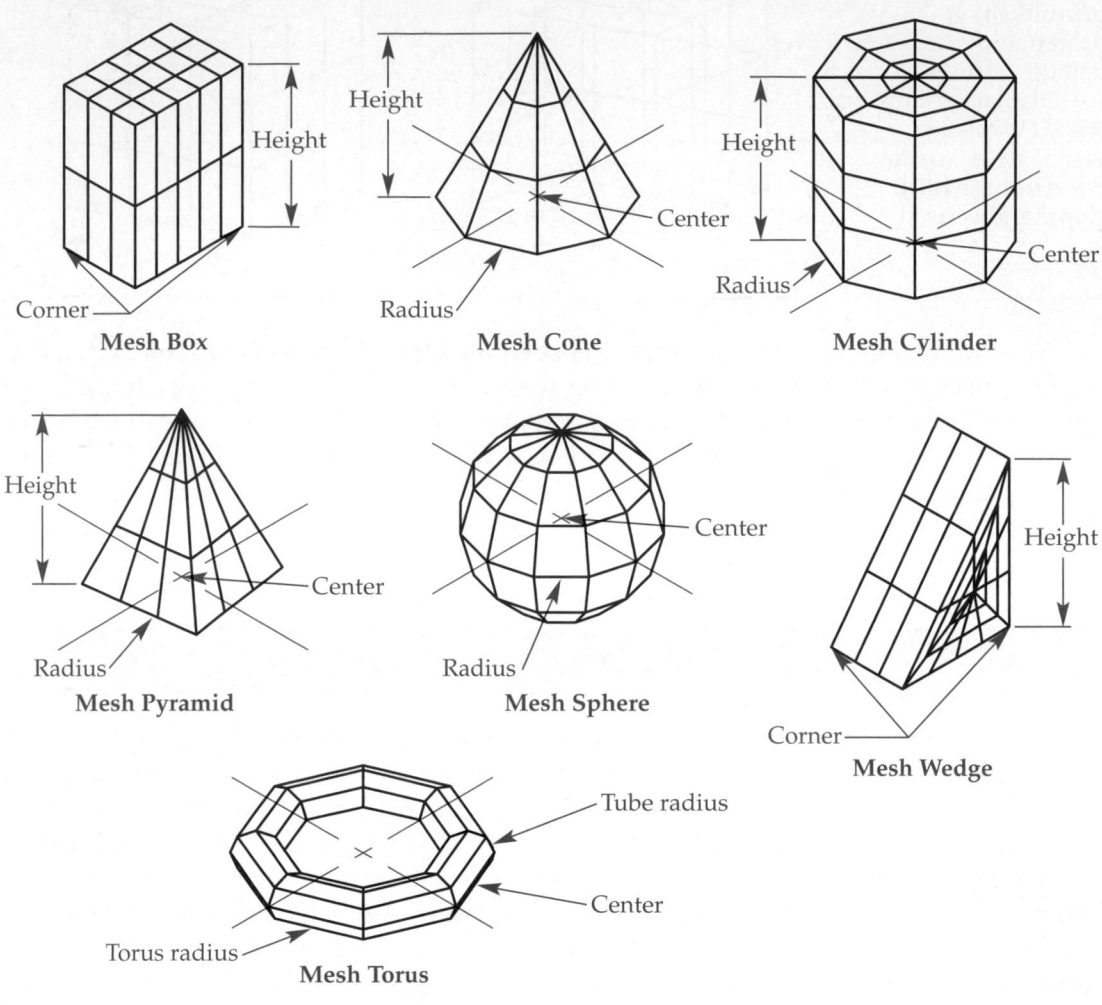

Mesh Box

Mesh Cone

Mesh Cylinder

Mesh Pyramid

Mesh Sphere

Mesh Wedge

Mesh Torus

Figure 3-5.
The **Mesh Primitive Options** dialog box.

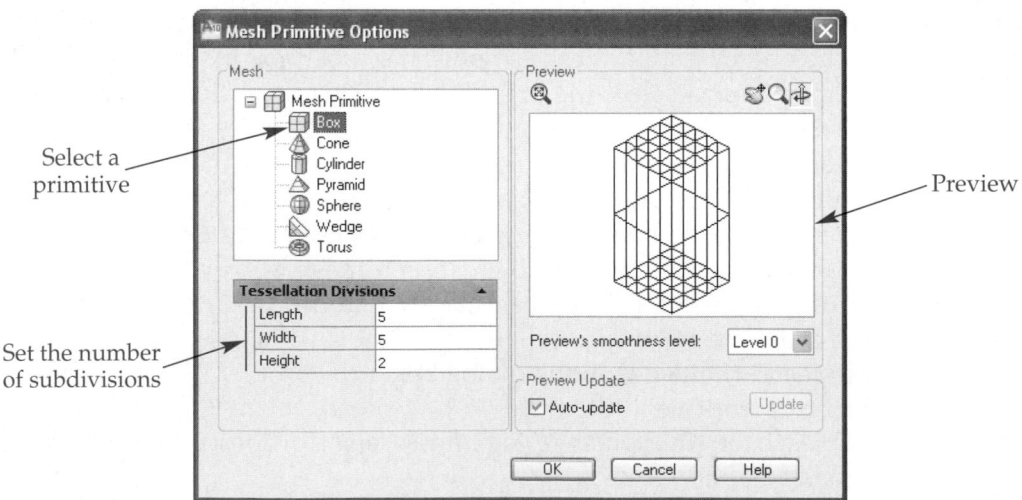

Select a primitive

Set the number of subdivisions

Preview

Exercise 3-1

Complete the exercise on the student website.
www.g-wlearning.com/CAD

Drawing Mesh Forms

You are not limited to the mesh primitives discussed in the previous section. You can also create *mesh forms* based on arcs, lines, and polylines. Mesh forms can be created by four basic methods:

- Using the **RULESURF** command to add a mesh between two objects such as arcs, lines or polylines. This is called a ruled-surface mesh.
- Revolving an open shape with the **REVSURF** command. This is called a revolved-surface mesh.
- Using the **TABSURF** command to extrude an open shape along a directional path. This is called a tabulated-surface mesh.
- Using the **EDGESURF** command to create the mesh based on four adjoining edges, such as lines, arcs, polylines or splines. This is called an edged-surface mesh.

These four commands can create either meshes or legacy surfaces. The **MESHTYPE** system variable controls which type of object is created. The default setting of 1 results in the commands creating mesh objects. A setting of 0 results in the commands creating legacy surfaces (polygon or polyface objects).

NOTE

It is strongly recommended the **MESHTYPE** system variable be set to 1 so the **RULESURF**, **REVSURF**, **TABSURF**, and **EDGESURF** commands will create mesh objects. Legacy surfaces are not widely used.

Ruled-Surface Mesh

A mesh can be constructed between two objects using the **RULESURF** command, **Figure 3-6.** This is called a *ruled-surface mesh.* The two objects can be points, lines, arcs, circles, polylines, splines, or enclosed objects. A ruled-surface mesh can be created between a point and any of the objects listed. However, AutoCAD cannot generate a ruled-surface mesh between an open and a closed object, such as an arc and a circle.

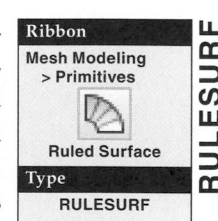

Ribbon
Mesh Modeling
> Primitives

Ruled Surface
Type
RULESURF

RULESURF

The number of tessellation subdivisions that compose the ruled-surface mesh is determined by the **SURFTAB1** system variable. The greater the number of subdivisions, the smoother the surface appears.

When using **RULESURF** to create a mesh between two objects, it is important to select both objects near the same end. If you pick near opposite ends of each object, the resulting mesh may not be what you want. The mesh may be twisted or reversed, somewhat like a butterfly shape. Refer to **Figure 3-6.**

Figure 3-6.
Lines, arcs, polylines, or splines can be used as the basis for the mesh. A—The original objects. B—If you pick incorrectly, the mesh is twisted. C—The mesh created by correctly picking the objects.

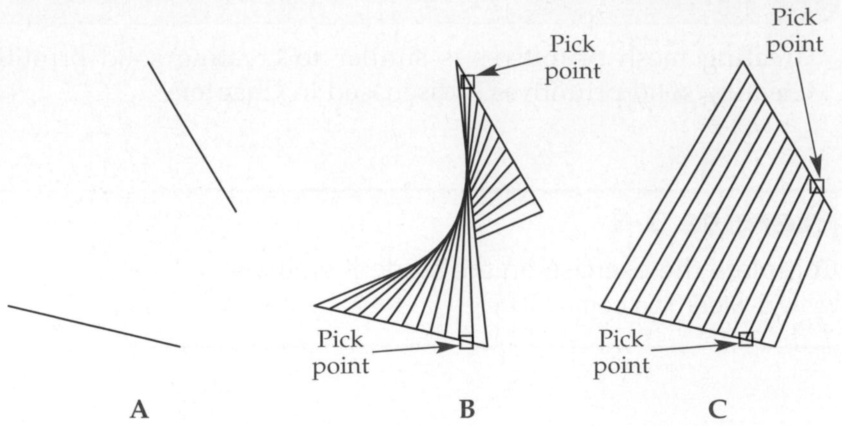

Pick point

Pick point

Pick point

Pick point

A   B   C

Revolved-Surface Mesh

REVSURF

Ribbon
Mesh Modeling
> Primitives

Revolved Surface
Type
REVSURF

With the **REVSURF** command, you can draw a profile and then rotate that profile around an axis to create a symmetrical object. This is called a *revolved-surface mesh*. The **REVSURF** command is a powerful tool and will greatly assist anyone who needs to draw a symmetrical mesh. The profile, or path curve, can be drawn using lines, arcs, circles, ellipses, elliptical arcs, polylines, or donuts. The rotation axis can be a line or an open polyline. Notice the initial layout path curves shown in **Figure 3-7** and the resulting meshes. The **REVSURF** command is powerful because it can create a symmetrical surface from any profile.

After selecting the path curve and the axis, you are prompted to specify the start angle. This allows you to specify an offset angle at which to start the surface revolution. This is useful when the profile is not revolved through 360°. The next prompt requests the included angle. This lets you draw the object through 360° of rotation or just a portion of that, as shown in **Figure 3-7.**

The **SURFTAB1** and **SURFTAB2** system variables control the number of tessellation subdivisions on a revolved-surface mesh. The **SURFTAB1** value determines the number of divisions in the direction of rotation around the axis. The **SURFTAB2** value divides the path curve into the specified number of subdivisions of equal size.

Figure 3-7.
Creating revolved-surface mesh objects. The profile shape and axis can be polylines or splines.

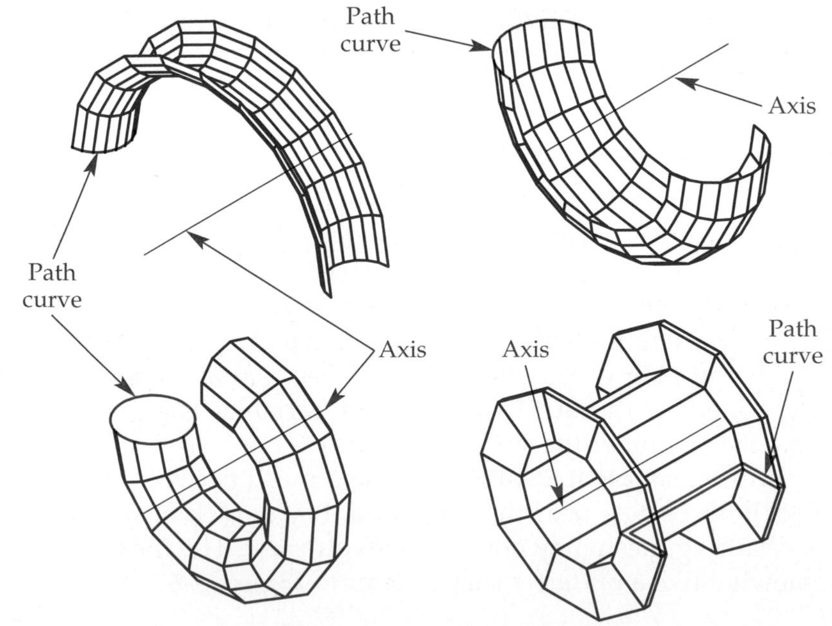

Path curve

Path curve

Axis

Path curve

Axis

Axis

Path curve

AutoCAD and Its Applications—Advanced

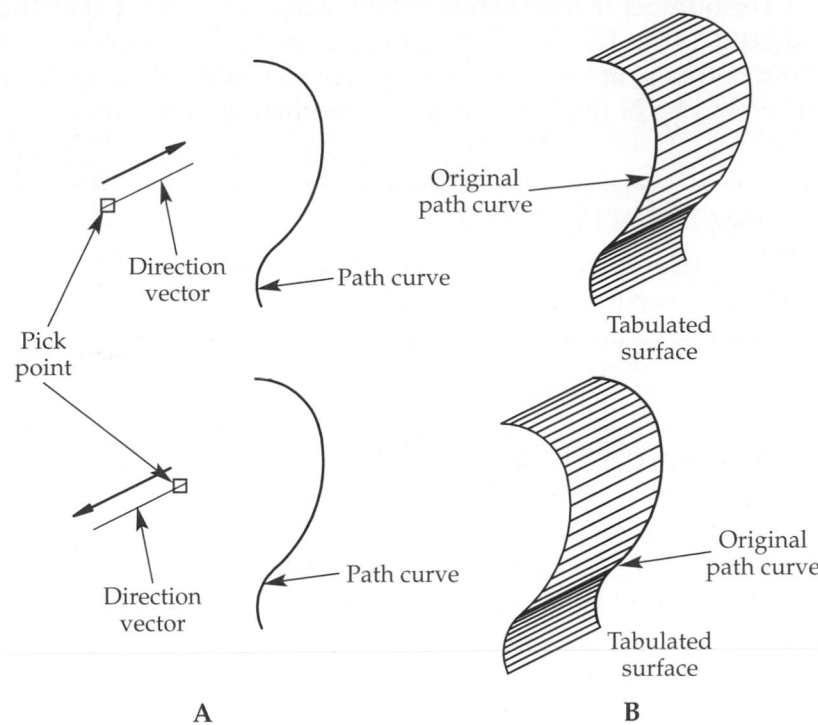

Figure 3-8.
Creating a tabulated-surface mesh object. Lines, arcs, polylines, or splines can be used as the shape and vector. A—The objects on which the mesh will be based. B—The completed tabulated-surface mesh.

Pick point

Direction vector

Path curve

Original path curve

Tabulated surface

Direction vector

Path curve

Original path curve

Tabulated surface

A

B

Tabulated-Surface Mesh

A *tabulated-surface mesh* is similar to a ruled-surface mesh. However, the shape is based on only one entity, **Figure 3-8.** This entity is called the path curve. Lines, arcs, circles, ellipses, 2D polylines, and 3D polylines can all be used as the path curve. A line called the direction vector is also required. This line indicates the direction and length of the tabulated-surface mesh. AutoCAD finds the endpoint of the direction vector closest to your pick point. It sets the direction toward the opposite end of the vector line. The mesh follows the direction and length of the direction vector. The **SURFTAB1** system variable controls the number of "steps" that are constructed. **Figure 3-8** shows the difference the location of the pick point makes when selecting the vector.

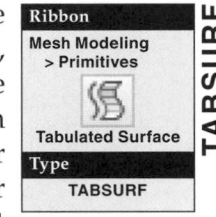

Ribbon
Mesh Modeling
> Primitives

Tabulated Surface

Type
TABSURF

TABSURF

Edged-Surface Mesh

The **EDGESURF** command allows you to construct a 3D mesh between four edges. **Figure 3-9.** The edges can be lines, polylines, splines, or arcs. The endpoints of the objects must precisely meet. However, a closed polyline cannot be used. The four objects can be selected in any order. The resulting mesh is smooth and called an *edged-surface mesh*.

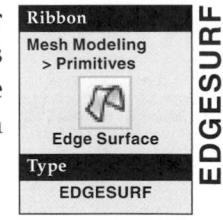

Ribbon
Mesh Modeling
> Primitives

Edge Surface

Type
EDGESURF

EDGESURF

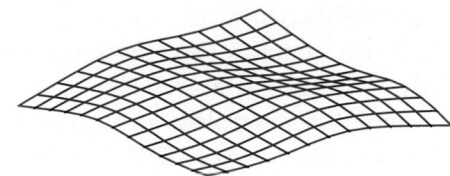

Figure 3-9.
Creating an edged-surface mesh object. Four lines, arcs, polylines, or splines can be used as the four edges. A—The objects on which the mesh will be based. B—The completed edged-surface mesh.

The number of tessellation subdivisions is determined by the variables **SURFTAB1** and **SURFTAB2**. The default value for each of these variables is 6. Higher values increase the smoothness of the mesh. The current values are always displayed on the command line when one of the mesh surfacing commands is executed.

Exercise 3-2

Complete the exercise on the student website.
www.g-wlearning.com/CAD

Converting between Mesh and Surface or Solid Objects

AutoCAD offers the flexibility of converting between solid, surface, and mesh objects. This allows you to select the type of modeling that offers the best tools for the task at hand, then convert the model into a form appropriate for the end result. The next sections discuss converting between solid, surface, and mesh objects and the settings that control the conversion.

PROFESSIONAL TIP

The **DELOBJ** system variable plays an important role when working with meshes, solids, and surfaces. Set the **DELOBJ** system variable to 0 to retain the geometry used to create the mesh, solid, or surface. When the variable is set to 1, the geometry used to create the mesh, solid, or surface is deleted.

Mesh Tessellation Options

The **Mesh Tessellation Options** dialog box contains many options, **Figure 3-10.** This dialog box is displayed by picking the dialog box launcher button at the lower-right corner of the **Mesh** panel in the **Mesh Modeling** tab of the ribbon or by using the **MESHOPTIONS** command.

When converting to mesh objects, the resulting mesh is one of three different mesh types. The **FACETERMESHTYPE** system variable controls which type of mesh is created. Selecting Smooth Mesh Optimized in the **Mesh type:** drop-down list converts objects to the optimized mesh type (**FACETERMESHTYPE** = 0). This is the default and recommended setting. Selecting Mostly Quads in the **Mesh type:** drop-down list creates faces that are mostly quadrilateral (**FACETERMESHTYPE** = 1). Selecting Triangle in the **Mesh type:** drop-down list creates faces that are mostly triangular (**FACETERMESHTYPE** = 2).

The **Mesh distance from original faces** setting is the maximum deviation of the mesh faces (**FACETERDEVSURFACE**). Simply put, this setting determines how closely a converted mesh shape matches the original solid or surface shape. No smoothness values are set for this system variable.

The **Maximum angle between new faces** setting is the maximum angle of a surface normal of two adjoining faces (**FACETERDEVNORMAL** system variable). The higher the value, the more faces are created in very curved areas and the less faces are created in flat areas. Increasing the value is good for objects that have curved areas, such as fillets, rounds, holes, or other tightly curved areas. No smoothness values are set for this system variable.

The **Maximum aspect ratio for new faces** setting is the upper limit for the ratio of height to width for new faces (**FACETERGRIDRATIO** system variable). By adjusting this value, long faces can be avoided, such as would be created from a cylindrical object

MESHOPTIONS

Ribbon

Mesh Modeling
> Mesh
  > Mesh
    Tessellation
    Options
  (options button)

Type

MESHOPTIONS

Figure 3-10.
The **Mesh Tessellation Options** dialog box.

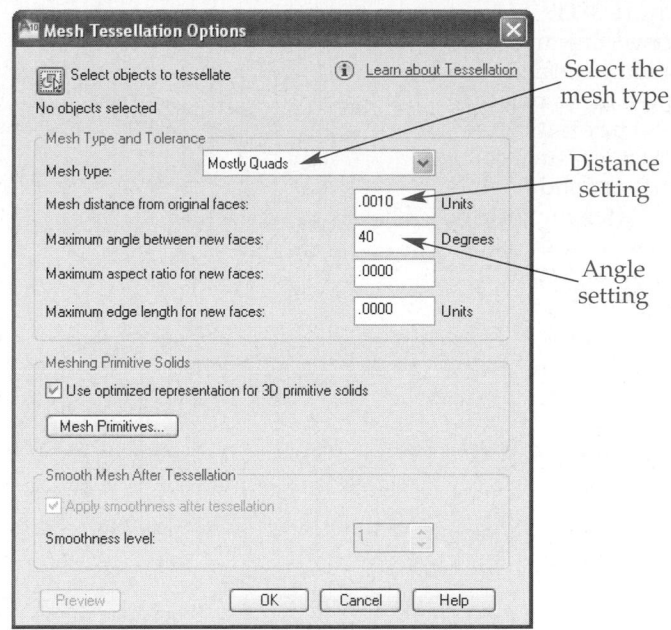

Select the mesh type

Distance setting

Angle setting

during the conversion process. A low value setting value will create a cleaner look in the formed faces. No smoothness values are set for this system variable.

- 0 = no limitation is applied.
- 1 = height must be equal to width.
- >1 = height may exceed width.
- <1 = width may exceed height.

The **Maximum edge length for new faces** setting is the maximum length any edge can be (**FACETERMAXEDGELENGTH** system variable). The default setting is 0, which allows the size of the mesh to be determined by the size of the 3D model. A setting of 1 or higher results in a reduced number of and less accurate faces compared to the original model. No smoothness values are set for this system variable.

NOTE

Converting swept solids and surfaces, regions, closed polylines, 3D face objects, and legacy polygon and polyface mesh objects may produce unexpected results. If this happens, undo the operation and try making setting adjustments in the **Mesh Tessellation Options** dialog box for better results.

Converting from a Solid or Surface to a Mesh

The **MESHSMOOTH** command is used to convert a solid or surface object into a mesh object. See **Figure 3-11.** The command is easy to use. First, enter the command. Then, select the solids or surfaces to be converted. Finally, press [Enter] and the objects are converted into a mesh. If you select an object that is not a primitive, you may receive a message indicating that the command works best on primitives. If you receive this message, choose to create the mesh.

Ribbon
Home
> Mesh

Smooth Object
Mesh Modeling
> Mesh

Smooth Object
Type
MESHSMOOTH

MESHSMOOTH

PROFESSIONAL TIP

With a good understanding of solid modeling, create a solid model first. Then, convert it to a mesh object using the **MESHSMOOTH** command and edit the mesh to create a freeform design.

Figure 3-11.
Converting an existing
solid or surface
into a mesh object.
A—The existing
object is a surface
(left) or solid (right).
B—After converting
it into a mesh using
the **MESHSMOOTH**
command.

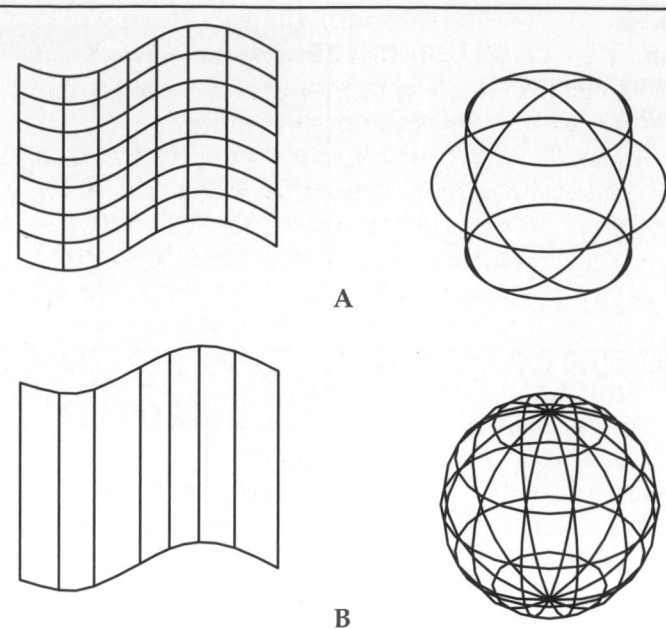

A

B

Exercise 3-3

Complete the exercise on the student website.
www.g-wlearning.com/CAD

Converting from a Mesh to a Solid or Surface

Mesh objects can be converted into solids or surfaces using the **CONVTOSOLID** and **CONVTOSURFACE** commands. The faces on the resulting solid or surface can be smoothed or faceted and optimized or not. This is controlled by the **SMOOTHMESHCONVERT** system variable. Select one of the four possible settings *before* converting a mesh to a solid or surface.

- **Smoothed and optimized.** Coplanar faces are merged into a single face. The overall shape of some faces can change. Edges of faces that are not coplanar are rounded. (**SMOOTHMESHCONVERT** = 0)
- **Smoothed and not optimized.** Each original mesh face is retained in the converted object. Edges of faces that are not coplanar are rounded. (**SMOOTHMESHCONVERT** = 1)
- **Faceted and optimized.** Coplanar faces are merged into a single, flat face. The overall shape of some faces can change. Edges of faces that are not coplanar are creased or angular. (**SMOOTHMESHCONVERT** = 2)
- **Faceted and not optimized.** Each original mesh face is converted to a flat face. Edges of faces that are not coplanar are creased or angular. (**SMOOTHMESHCONVERT** = 3)

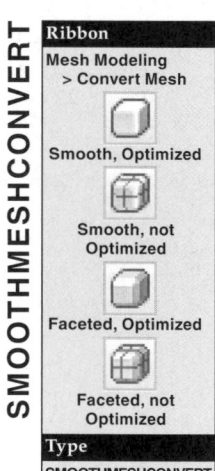
To convert a mesh to a solid, first set the smoothing option, as described above. Then, select the **CONVTOSOLID** command. Next, select the mesh objects to convert and press the [Enter] key. The objects are converted from mesh objects to solid objects based on the selected smoothing option.

To convert a mesh to a suface, first set the smoothing option, as described above. Then, select the **CONVTOSURFACE** command. Next, select the mesh objects to convert and press the [Enter] key. The objects are converted from mesh objects to surface objects based on the selected smoothing option.

The examples shown in **Figure 3-12A** are simple mesh objects. In **Figure 3-12B,** the mesh objects have been converted into solid objects with the **Faceted, Optimized** button selected in the **Convert Mesh** panel on the **Mesh Modeling** tab in the ribbon. In **Figure 3-12C,** the mesh objects have been converted into surface objects with the **Smooth, Optimized** button selected.

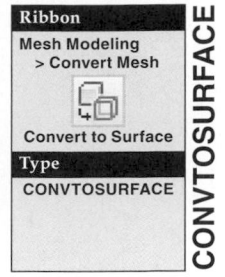

CONVTOSURFACE

PROFESSIONAL TIP

There are some mesh shapes that cannot be converted to a 3D solid. If using grips to edit the mesh, gaps or holes between the faces may be created. Smooth the mesh object to close the gaps or holes. Also, during the mesh editing process, mesh faces may be created that intersect each other and cannot be converted to a 3D solid. Converting this type of a mesh into a 3D solid will result in the following error message:

> Mesh not converted because it is not closed or it self-intersects. Object cannot be converted.

In some cases, there may be a mesh shape that cannot be converted to a solid object, but can be converted to a surface.

Figure 3-12.
A—Three basic mesh objects.
B—The mesh objects converted into faceted solids.
C—The mesh objects converted into smoothed surfaces.

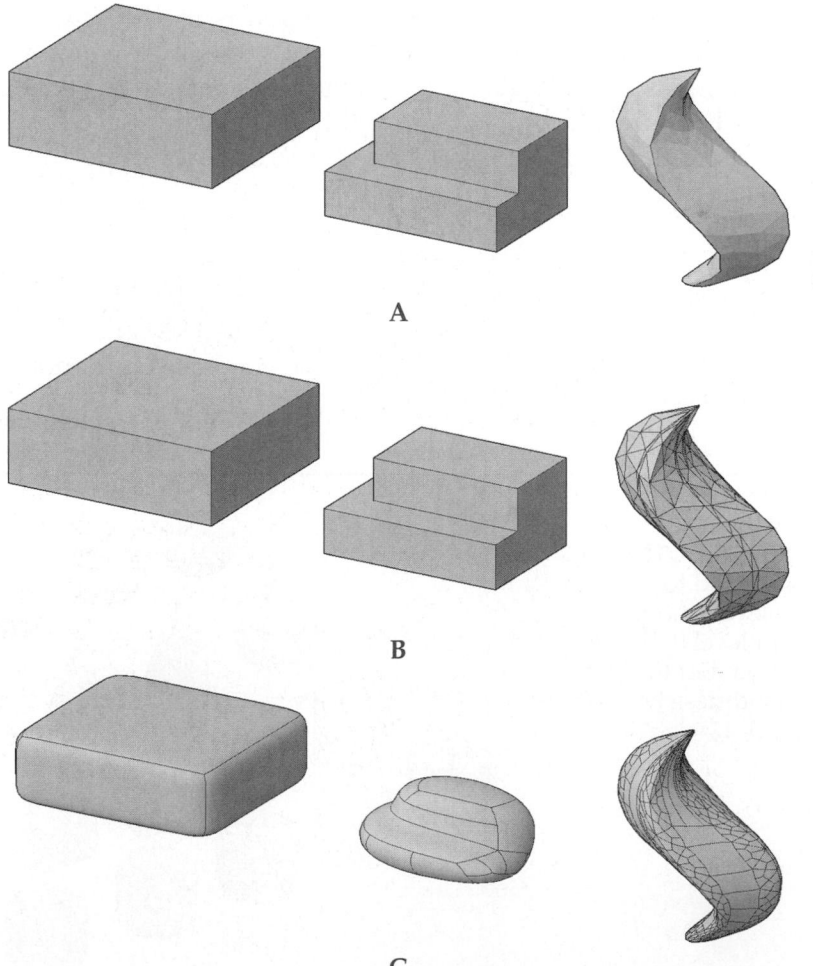

A

B

C

Exercise 3-4

Complete the exercise on the student website.
www.g-wlearning.com/CAD

Smoothing and Refining a Mesh Object

The roundness of a mesh object is increased by increasing the smoothness level. The smoothness can be set before creating the mesh, as described earlier in this chapter. The mesh object can also be refined or have its smoothness increased after it is created. This is described in the following sections. When smoothing or refining a mesh object, the number of tessellation subdivisions increase. The lowest smoothness level is 0 and the highest smoothness level is 4. The default smoothness level is 0.

Adjusting the Smoothness of a Mesh

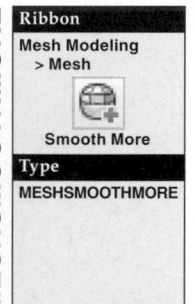

MESHSMOOTHMORE

Ribbon
Mesh Modeling
> Mesh

Smooth More

Type
MESHSMOOTHMORE

When a mesh is smoothed, it changes the form to more closely represent a rounded shape. The **MESHSMOOTHMORE** command is used to increase the level smoothness on a mesh object. There are five levels of smoothness, ranging from level 0 to level 4. Once the command is selected, pick the mesh objects for which to increase the smoothness. Then, press the [Enter] key and the level is increased by one. See **Figure 3-13.** If you select an object that is not a mesh, you have the opportunity to either filter out the non-mesh objects or convert them to meshes.

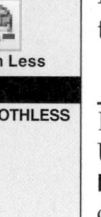

MESHSMOOTHLESS

Ribbon
Mesh Modeling
> Mesh

Smooth Less

Type
MESHSMOOTHLESS

When a mesh is desmoothed, it changes the form to more closely represent a boxed shape. The **MESHSMOOTHLESS** command is used to decrease the level smoothness on a mesh object. Once the command is selected, pick the mesh objects for which to decrease the smoothness. Then, press the [Enter] key and the level is decreased by one. If you select an object that is not a mesh, you have the opportunity to either filter out the non-mesh objects or convert them to meshes.

Figure 3-13.
Use the
MESHSMOOTHMORE
command to increase
the level smoothness
from level 0 to level
4. From left to right,
smoothness levels 0,
1, 2, 3, 4.

Exercise 3-5

Complete the exercise on the student website.
www.g-wlearning.com/CAD

Refining a Mesh

Refining a mesh increases the number of subdivisions in a mesh object. Refining adds detail to the mesh to make it more realistic with smooth flowing lines. This also gives a greater selection for editing the mesh faces, vertices, or edges. The **MESHREFINE** command is used to refine a mesh. When the command is used, the number of subdivisions is quadrupled. See **Figure 3-14.** The smoothness level on the original object must be level 1 or higher.

Ribbon
Mesh Modeling > Mesh
Refine Mesh
Type
MESHREFINE

MESHREFINE

The entire mesh can be refined (all faces at once) or a selected face can be refined. When you refine a mesh *model*, the increased amount of face subdivisions becomes a new smoothness level 0. However, if you refine an individual *face*, the level of smoothness is not reset.

Be careful not to create a too dense of a mesh. Creating a mesh that is too dense may result in subobjects (the mesh shapes) that are very small. This may make it difficult to select subobjects and edit the mesh.

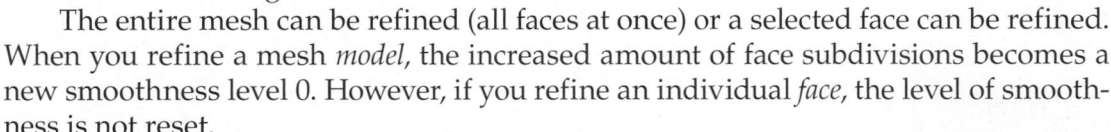

NOTE

As the mesh density increases, the computer's performance will be slower.

Exercise 3-6

Complete the exercise on the student website.
www.g-wlearning.com/CAD

Figure 3-14.
Refining a mesh.
A—The original mesh with a smoothness level of 1. B—The refined mesh, which now has more faces and a smoothness level of 0.

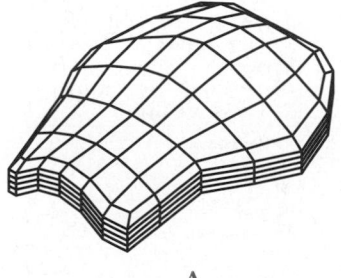
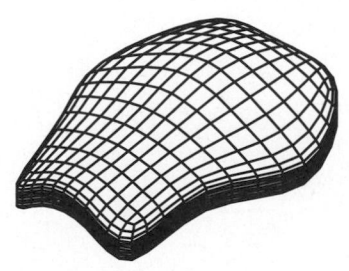

A

B

Editing Meshes

As discussed earlier, the second key workflow of mesh modeling is the ability to edit the mesh. The tools for editing a mesh are found in the **Mesh Edit** and **Subobject** panels on the **Mesh Modeling** tab of the ribbon. See **Figure 3-15.** The face, edge, and vertex subobjects can be edited to change the shape of the mesh. These subobjects can be moved, rotated, or scaled. A face on the mesh can be split or extruded.

Subobject filters are used to assist in the selection of a mesh face, edge, or vertex *before* it is moved, rotated, or scaled. This is especially true for a very dense mesh. First, right-click in the drawing window and select **Subobject Selection Filter** to display the cascading menu, **Figure 3-16.** Then, select the filter you wish to use. You can also use the filter in the **Subobject** panel of the **Mesh Modeling** tab in the ribbon.

Subobject editing is discussed in detail in Chapter 12. The same procedures discussed in that chapter for solids can be applied to mesh models.

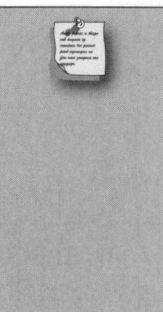

NOTE

Use the [Ctrl] key and left mouse button to select the subobjects or use the **SUBOBJSELECTIONMODE** system variable for subobject filtering. These settings apply:
- 0 = off
- 1 = vertices
- 2 = edges
- 3 = faces
- 4 = solid history

Figure 3-15.
The tools for editing a mesh are found in the **Mesh Edit** and **Subobject** panels of the **Mesh Modeling** tab in the ribbon.

Mesh-editing tools

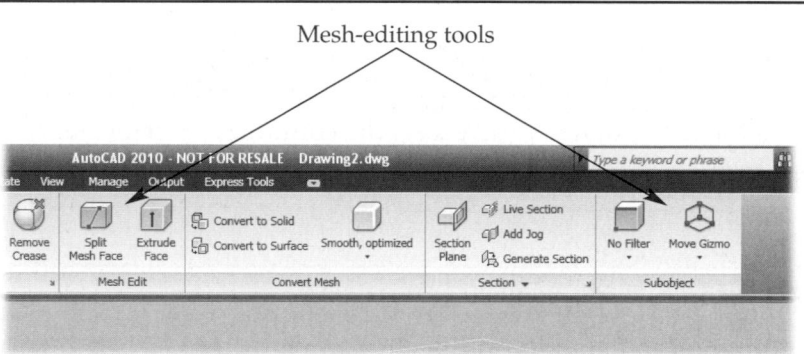

Figure 3-16.
Selecting a subobject filter.

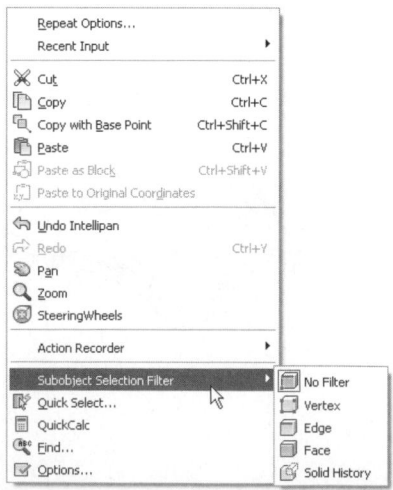

Gizmos

A *gizmo,* also called a grip tool, appear when a subobject is selected. This tool is used to specify how the transformation (movement, rotation, or scaling) is applied. A visual style other than 2D Wireframe must be current in order for the gizmo to appear. Gizmos (grip tools) are discussed in detail in Chapter 11.

The move gizmo allows movement of a subobject along the X, Y, or Z axis or on the XY, XZ, or YZ plane. The rotate gizmo allows rotation about the X, Y, or Z axis. The scale gizmo allows scaling along the X, Y, or Z axis or XY, XZ, or YZ plane.

You can switch between the three gizmos by picking the button in the drop-down list in the **Subobject** panel on the **Mesh Modeling** tab of the ribbon. You can also cycle between the gizmos by pressing the [Enter] key.

For example, to use the move gizmo, pick any axis to move along that axis. See **Figure 3-17.** To move along a plane, pick the rectangular area at the intersection of the axes. To use the rotate gizmo, pick the circle with the center about which you wish to rotate the selection. To use the scale gizmo, pick an axis to nonuniformly scale along that axis. Pick the triangular area at the intersection of the axes to uniformly scale the selection. Pick the area between the inner and outer triangular areas to nonuniformly scale along that plane.

Extrude a Face

You can select a face subobject on a mesh and extrude it. Extruding a mesh face adds new features to the mesh. This creates new faces that can be edited. The **EXTRUDE** command is used to extrude mesh faces. The command sequence to extrude a mesh face is similar to extruding to create a solid shape. This command is covered in detail in Chapter 9.

To extrude a face, enter the command. Then, hold down the [Ctrl] key and pick the face subobject to extrude. Finally, enter an extrusion height. In **Figure 3-18,** a camera is being developed as a mesh model. The **EXTRUDE** command is used to extrude an individual face as the first step in creating the lens tube.

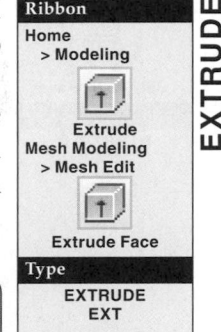

Ribbon
Home > Modeling
Extrude Mesh Modeling > Mesh Edit
Extrude Face
Type
EXTRUDE EXT

EXTRUDE

PROFESSIONAL TIP

Extrude a face instead of moving a face. This will give greater editing control over an individual face.

Figure 2-17.
Editing a mesh using one of the three gizmos is an important part of mesh modeling.
A—Move gizmo. B—Rotate gizmo. C—Scale gizmo.

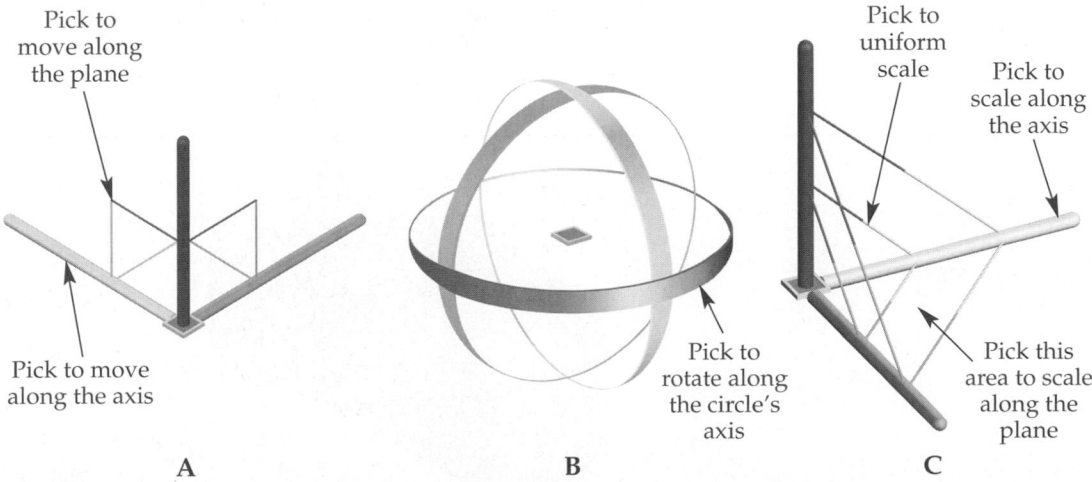

Pick to move along the plane

Pick to move along the axis

Pick to rotate along the circle's axis

Pick to uniform scale

Pick to scale along the axis

Pick this area to scale along the plane

A   B   C

Figure 3-18.
Extruding a mesh face. A—Select the face to extrude. B—The process adds faces to the model.

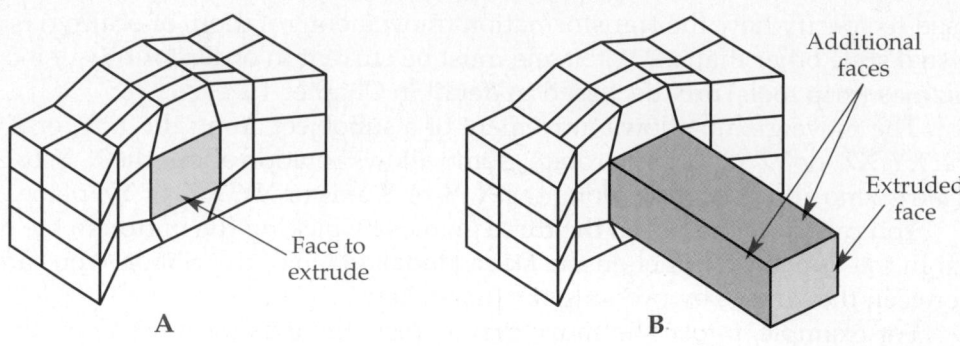

Additional faces

Extruded face

Face to extrude

A

B

Split a Face

Splitting a mesh face is used to increase the number of faces on the model without refining a mesh. Splitting one face creates two faces. This is easier than refining the mesh. Think of these new split faces as subdivisions of an existing face.

The **MESHSPLIT** command is used to split a face. Once the command is entered, select the face to split. Then, specify the starting point and the ending point of a line defining the split. You can specify any two points on the mesh face. **Figure 3-19** shows a mesh model with three faces that have been split.

Once the faces are split, mesh editing options, such as extruding, moving, rotating, or scaling, can be used on the new faces. In **Figure 3-20**, the split faces from **Figure 3-19** have been extruded and then the model smoothed to level 3.

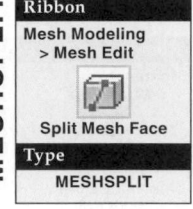

MESHSPLIT

Ribbon
Mesh Modeling
> Mesh Edit

Split Mesh Face
Type
MESHSPLIT

Figure 3-19.
Three faces have been split on this mesh model. Each face has been split into two faces.

Split lines

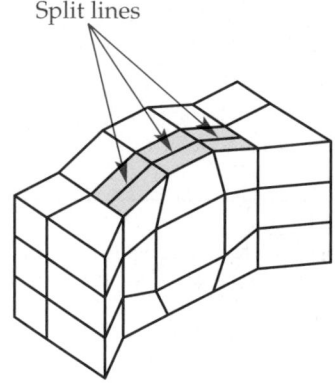

Figure 3-20.
Smoothing is applied to this model. Notice how its edges are rounded.

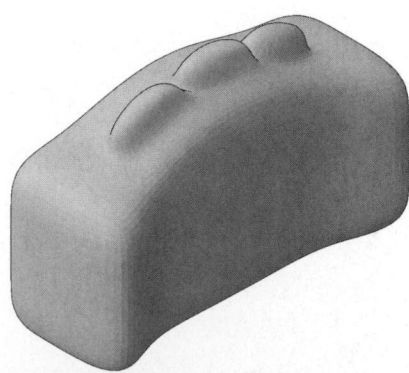

Figure 3-21.
Adding creases to a
mesh. A—Before the
creases are applied,
the bottom of the
mouse is rounded.
This is obvious if
the object is rotated
(right). B—After the
creases are added,
the bottom is flat.

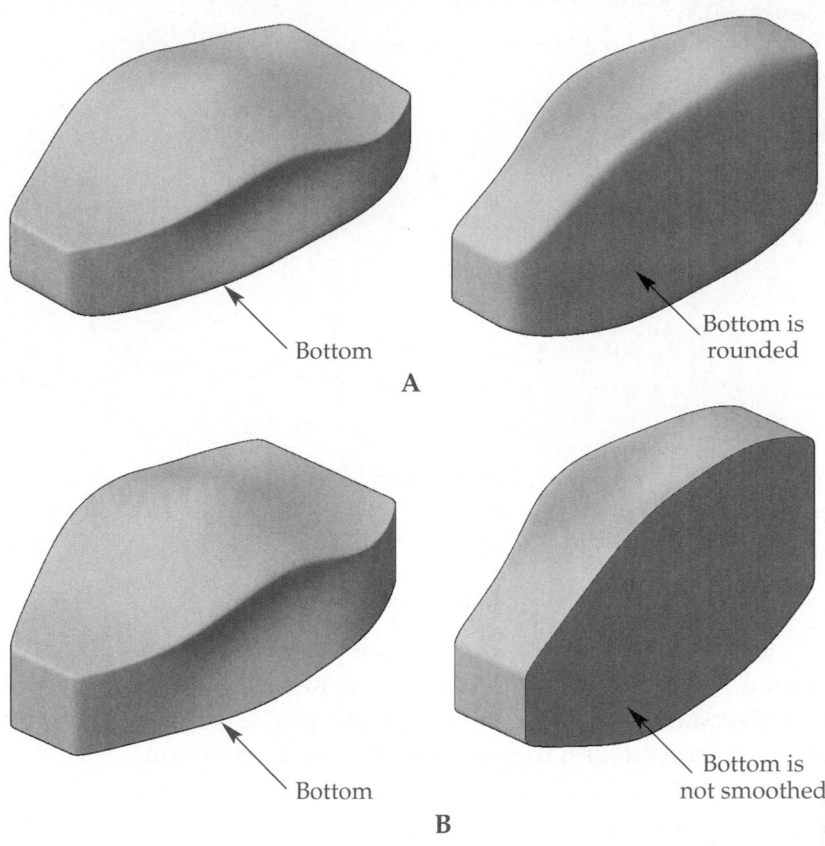

Bottom

Bottom is
rounded

A

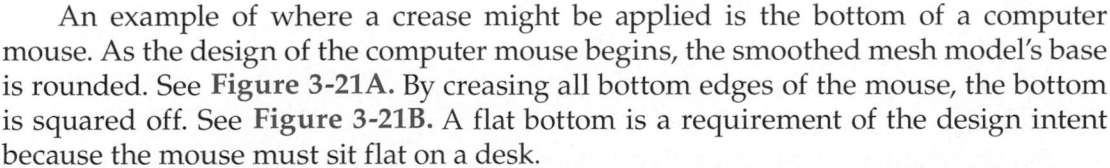

Bottom

Bottom is
not smoothed

B

Exercise 3-7

Complete the exercise on the student website.
www.g-wlearning.com/CAD

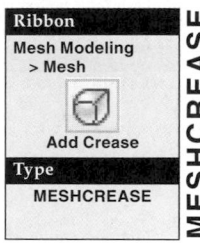

Applying a Crease to a Mesh Model

A *crease* is a sharpening of a mesh subobject, much like a crease in folded paper.
A crease sharpens or squares off an edge or flattens a face. This prevents the subobject
from being smoothed. Creases can be applied to faces, edges, and vertices. A smoothness level of 1 or higher must be assigned to the mesh object for the creases to have
an effect, but they can be applied at any smoothness level. Once a crease is applied,
any existing smoothing is removed from the subobject. If the mesh smoothness is
increased, any creased subobjects are not smoothed. Also, creasing an edge *before* an
object is smoothed will limit the mesh editing capabilities.

An example of where a crease might be applied is the bottom of a computer
mouse. As the design of the computer mouse begins, the smoothed mesh model's base
is rounded. See **Figure 3-21A.** By creasing all bottom edges of the mouse, the bottom
is squared off. See **Figure 3-21B.** A flat bottom is a requirement of the design intent
because the mouse must sit flat on a desk.

In another example, the trigger pads on the game controller shown in **Figure 3-22**
have been creased to create a flat surface. When the conceptual design is finished,
it can be converted to a solid. This will give the designer a unique shape to which
buttons and joysticks can be added. This unique shape would be difficult to create
from scratch as a solid model.

Ribbon

Mesh Modeling
> Mesh

Add Crease

Type

MESHCREASE

MESHCREASE

Figure 3-22
In this example,
creases have been
added to the mesh
to create flat areas
on the game pad for
buttons.

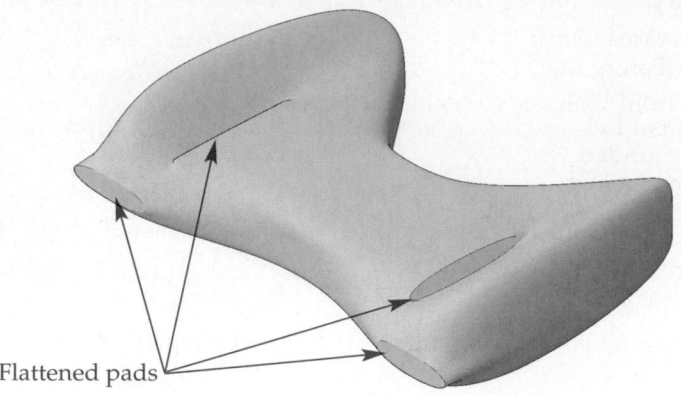

Flattened pads

Once the command is entered, select the subobjects to crease. You do not need to press the [Ctrl] key to select subobjects, but filters should be used to make selection easier. Once the subobjects are selected, press the [Enter] key. This prompt appears:

Specify crease value or [Always] <Always>:

If you enter a crease value, this is the highest smoothing level for which the crease is retained. If the smoothing level is set higher than the crease value, the creased subobject is smoothed. The Always option forces the crease to be retained for all smoothness levels. In most cases, this is the recommended option.

Exercise 3-8

Complete the exercise on the student website.
www.g-wlearning.com/CAD

Removing a Crease

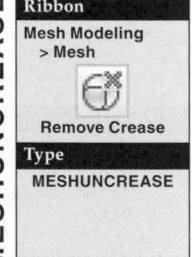

Removing a crease is a simple process. The **MESHUNCREASE** command is used to do this. Enter the command and then select the subobjects from which a crease is to be removed. Then, press the [Enter] key. The crease is removed and smoothing is applied as appropriate.

Ribbon
Mesh Modeling
> Mesh

Remove Crease

Type
MESHUNCREASE

MESHUNCREASE

Chapter Test

Answer the following questions. Write your answers on a separate sheet of paper or complete the electronic chapter test on the student website.
www.g-wlearning.com/CAD

1. Of what does a mesh model consist?
2. What is another term for *mesh modeling*?
3. What are *tessellations divisions*?
4. When creating a mesh primitive, when should mesh tessellation divisions be set?
5. For what is the **Mesh Primitive Options** dialog box used?
6. How is a mesh box created?
7. How is a mesh sphere created?
8. How is a mesh torus created?
9. List the four commands used to create mesh forms.
10. What is the purpose of the **DELOBJ** system variable?

11. Which command converts a mesh object to a surface object?
12. Which command converts a mesh object to a solid object?
13. Describe the smoothness of a mesh object.
14. How many smoothness levels are there?
15. Which command is used to convert an existing solid or surface to a mesh object?
16. List two ways to decrease the smoothness of a mesh.
17. What happens to the mesh when you refine it?
18. How many types of subobjects does a mesh have? List them.
19. Which keyboard key is used to select subobjects for editing?
20. Name the three operations that can be performed with a gizmo.
21. How do you cycle through the three different gizmos?
22. Briefly describe the process for extruding a mesh face.
23. What is the process for splitting a mesh face?
24. Why would you crease a mesh model?
25. Which command is used to remove a crease?

Drawing Problems

1. In this problem, you will create an ergonomic computer mouse as a mesh shape. Change visual styles as needed throughout your work.
   A. Set the tessellation divisions for the box primitive to length = 5, width = 3, and height = 3.
   B. Create a 5 × 3 × 1.04 mesh box primitive.
   C. Move the middle faces inward .25 units on both sides of the mesh box, as shown below in A.
   D. Move edges on the mesh model to form the mouse shape shown below in B.
   E. Smooth the model.
   F. Continue editing faces, edges, and vertices to create a mouse shape similar to the one shown below in C.
   G. Crease the model to create a flat bottom.
   H. Save the model.

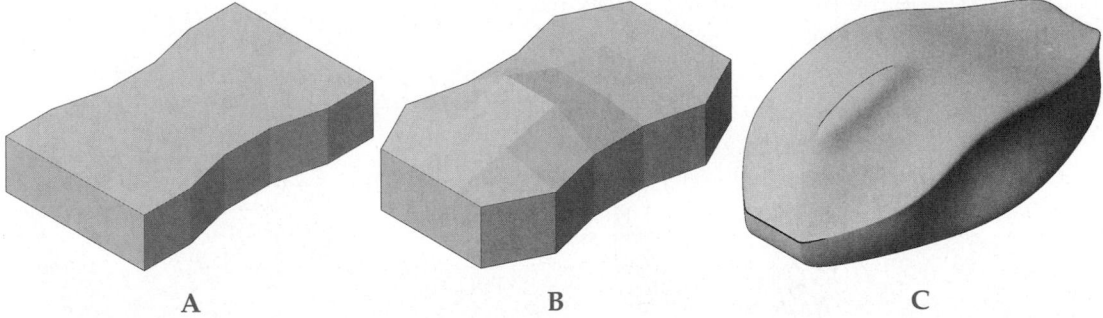

A   B   C

2. In this problem, you will create a wireless phone charger cradle as a mesh shape. Change visual styles as needed throughout your work.
   A. Set the tessellation divisions for the box primitive to length = 5, width = 3, and height = 2.
   B. Create a 3 × 2 × 1.5 mesh box primitive.
   C. Move the middle-back edges up 1 unit and the front edges down .75 units, as shown on the next page in A.
   D. Crease the middle face, then move it down 1 unit.
   E. Crease all of the faces on the bottom.
   F. Smooth the mesh to a level 3 mesh.

Drawing Problems - Chapter 3

G. On the top of the model, nonuniformly scale the middle face (which is now round) to create an elliptical shape. Then, move the face toward the back of the model.

H. Move the middle three edges in the front of the model .5 units toward the back of the model, as shown below in B.

I. Save the model.

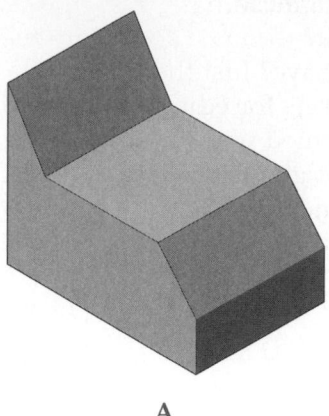 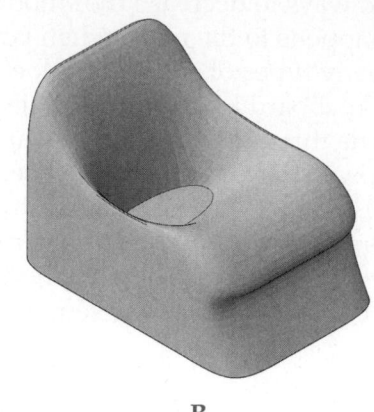

A   B

3. In this problem, you will create a game pad controller as a mesh shape. Change visual styles as needed throughout your work.

A. Set the tessellation divisions for the box primitive to length = 5, width = 3, and height = 2.

B. Create a 6 × 4 × 1 mesh box primitive.

C. Move the middle faces on the back side (long side) inward 1.25 units.

D. Move top edges on the right and left side up 1 unit, as shown below in A.

E. Move the front middle face inward .5 units.

F. Crease the front two corner faces. The top corner of these faces is formed by the edges moved up earlier.

G. Smooth the model to level 3.

H. Move the two creased areas outward .25 units.

I. Split the top face to the left of the indention into three unique faces, as shown below in B.

J. Extrude the three split faces upward .375 units.

K. Save the model.

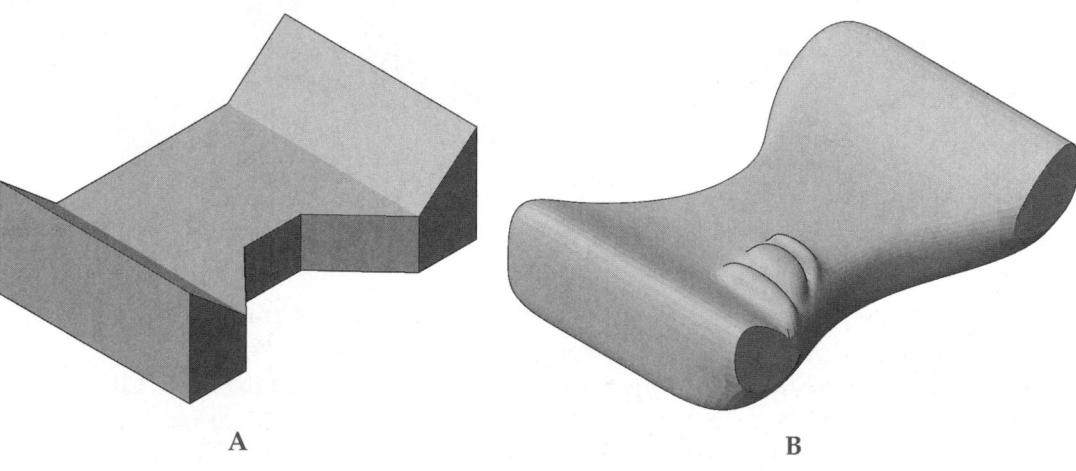

A   B

AutoCAD and Its Applications—Advanced

4. In this problem, you will create a starship for a gaming application as a mesh shape. Change visual styles as needed throughout your work.
   A. Set the tessellation divisions for the sphere primitive to axis = 12 and height = 6.
   B. Create a Ø4 mesh sphere primitive.
   C. Move the vertex on the top of the sphere up 4 units.
   D. Move the vertex on the right and left sides outward 2 units, as shown below in A.
   E. Move the vertex on the bottom of the sphere into the sphere by 1 unit.
   F. Smooth the model to level 2.
   G. On the bottom of the ship, as indicated in A, crease the four middle faces, then move them into the sphere by 1 unit.
   H. Extrude the four middle faces a distance of .75 with a taper of 30°. This forms the landing gear.
   I. Select two faces on the front of the ship and crease them to create windows, as shown below in B.
   J. Increase the smoothness to level 4.
   K. Save the model.

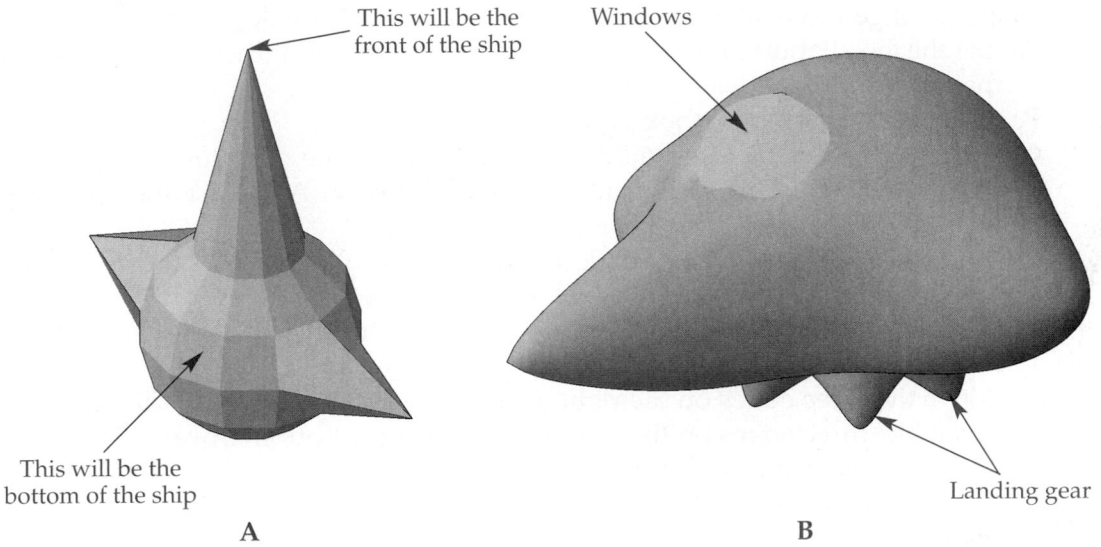

This will be the front of the ship

Windows

This will be the bottom of the ship

Landing gear

A

B

5. In this problem, you will create a home theater chair as a mesh shape. Change visual styles as needed throughout your work.
   A. Set the tessellation divisions for the box primitive to length = 5, width = 3, and height = 2.
   B. Create a 3 × 2 × 1.5 mesh box primitive.
   C. Move the top-back edges (short side) up 1 unit and the top-front edges down .75 units.
   D. Crease the one face in the middle of the top, then move it down 1 unit, as shown on the next page in A.
   F. Crease all faces on the bottom.
   G. Smooth the model to level 3.
   H. Nonuniformly scale the middle face on the top, which is now circular, to create an elliptical shape. Then, move the face .5 units toward the chairback.
   I. Move the three top edges on the front .5 units toward the chairback.
   J. Move the middle face in the front top down 1 unit, as shown on the next page in B.
   K. Rotate the top three back edges 45° to create a sloped chairback.
   L. Move the single edge in the middle of the back inward .5 units, as shown on the next page in C.
   M. Increase the smoothness to level 4.
   N. Save the model.

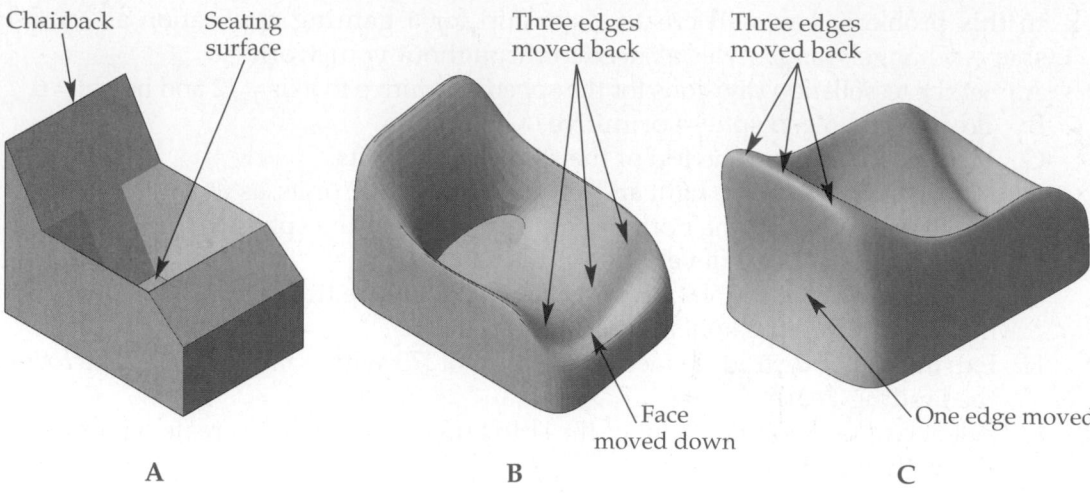

Chairback Seating surface Three edges moved back Three edges moved back

Face moved down One edge moved

A B C

6. In this problem, you will create a conceptual design for a pocket camera as a mesh shape. Change visual styles as needed throughout your work.
   A. Set the tessellation divisions for the box primitive to length = 5, width = 3, and height = 5.
   B. Create a 3 × 1 × 3 mesh box primitive.
   C. Extrude the top face second from the edge to create a button. Extrude to a height of .25 with a taper of 10°, as shown below in A. Crease the top of the extruded face.
   D. Nonuniformly scale the middle face on the front by a factor of two. Then, crease the face.
   E. Move the five edges on the left side of the front outward .25 units, as shown below in B.
   F. Move the three edges on the right-hand side of the top inward .375 units.
   G. Move the three edges on the right-hand side of the bottom inward .375 units.
   H. Crease all faces on the bottom.
   I. Smooth the model to level 4.
   J. Extrude the middle face on the front (which was earlier creased) inward .125 units, as shown below in C.
   K. Save the model.

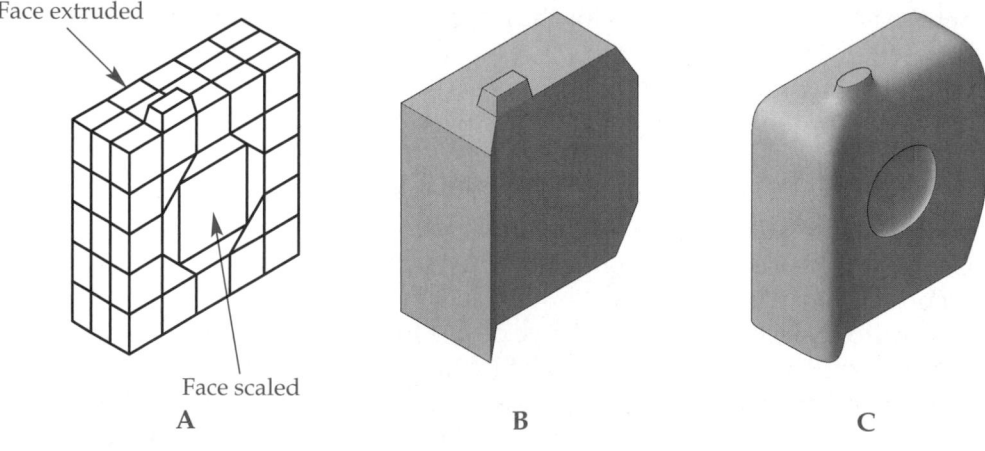

Face extruded

Face scaled
A B C

CHAPTER 4

Viewing and Displaying Three-Dimensional Models

Learning Objectives

After completing this chapter, you will be able to:

✓ Use the view cube to dynamically rotate the view of the model in 3D space.
✓ Use the view cube to display orthographic plan views of all sides on the model.
✓ Use steering wheels to display a 3D model from any angle.
✓ Use the visual style options to create face and edge style display variations.
✓ Render a 3D model.

AutoCAD provides several tools with which you can display and present 3D models in pictorial and orthographic views:

- Preset isometric viewpoints, discussed in Chapter 1.
- Dynamic model display using the view cube. This on-screen tool provides access to preset and dynamic display options.
- Complete 3D model display using steering wheels.
- The **3DORBIT**, **3DFORBIT**, and **3DCORBIT** commands provide dynamic display and continuous orbiting tools for demonstrations and presentations.

Once a viewpoint has been selected, you can enhance the display by applying visual styles. The **View** panel in the **Home** tab of the ribbon provides a variety of ways to display a model, including wireframe, hidden line removal, and simple rendering. An introduction to visual styles is provided in Chapter 1 and additional details are discussed later in this chapter.

A more advanced rendering can be created with the **RENDER** command. It produces the most realistic image with highlights, shading, and materials, if applied. **Figure 4-1** shows a 3D model of a cast iron plumbing cleanout after using **HIDE**, setting the Conceptual visual style current, and using **RENDER**. Notice the difference in the three displays.

PROFESSIONAL TIP

AutoCAD has three powerful display tools: the view cube, steering wheels, and show motion. The power of these three tools has virtually eliminated the need for legacy display functions such as **DVIEW**, **VPOINT**, and **3DORBIT**. However, **3DORBIT**, **3DFORBIT**, and **3DCORBIT** still possess some usefulness and are discussed later in this chapter.

Figure 4-1.
A—Hidden display (hidden lines removed). B—The Conceptual visual style set current.
C—Rendered with lights and materials.

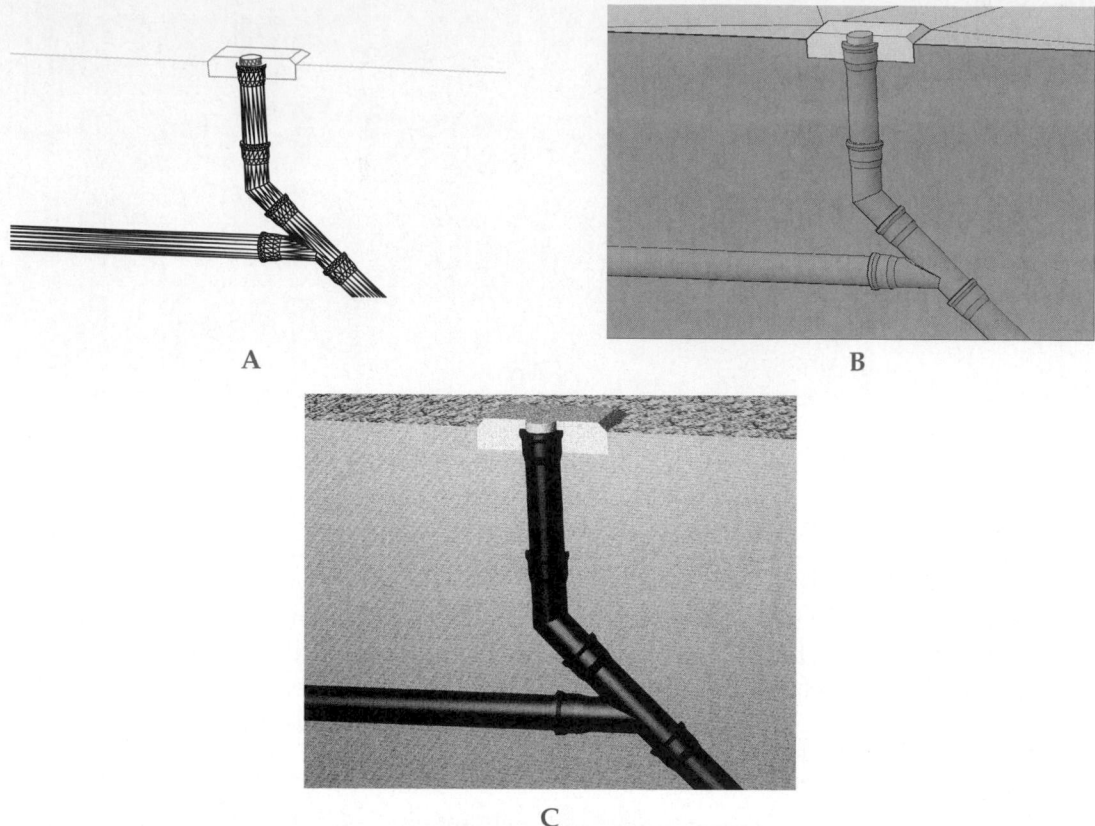

A

B

C

Dynamically Displaying Models with the View Cube

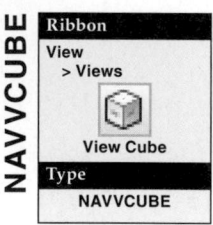

NAVVCUBE

Ribbon
View
  > Views

View Cube

Type
  NAVVCUBE

The *view cube* navigation tool allows you to quickly change the current view of the model to a preset pictorial or orthographic view or any number of dynamically user-defined 3D views. The tool is displayed, by default, in the upper-right corner of the drawing area when a 3D visual style is set current. If the view cube is not displayed, pick the **View Cube** button from the **Views** panel of the **View** tab of the ribbon. The view cube cannot be displayed if the 2D Wireframe visual style is current.

The **View Cube Settings** dialog box offers many settings for expanding and enhancing the view cube. This dialog box is discussed in detail later in this chapter.

Understanding the View Cube

The view cube tool is composed of a cube labeled with the names of all six orthographic faces. See **Figure 4-2.** A compass rests at the base of the cube and is labeled with the four compass points (N, S, E, and W). The compass can be used to change the view. It also provides a visual cue to the orientation of the model in relation to the current user coordinate system (UCS). The *user coordinate system* describes the orientation of the X, Y, and Z axes. A complete discussion of user coordinate systems is given in Chapter 6.

Below the cube and compass is a button labeled WCS, which displays a shortcut menu. The label on this button displays the name of the current user coordinate system, which is, by default, the world coordinate system (WCS). This shortcut menu gives you the ability to switch from one user coordinate system to another. A user coordinate system

Figure 4-2.
The cube in the view cube tool is labeled with the names of all six orthographic faces. The view cube also contains a **Home** icon, compass, and UCS shortcut menu.

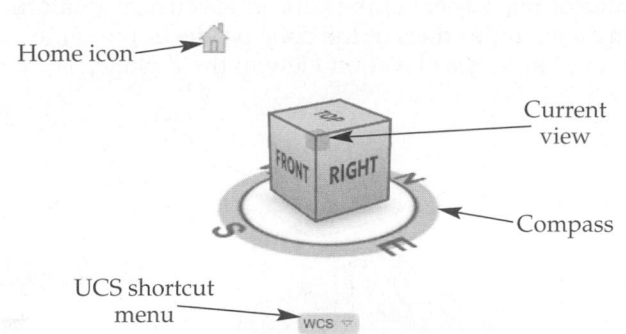

(UCS) is any coordinate system that is not the world coordinate system. The benefit of creating new user coordinate systems is the increase in efficiency and productivity when working on 3D models.

When the cursor is not over the view cube, the tool is displayed in its dimmed or *inactive state.* When the cursor is moved onto the view cube, the tool is displayed in its *active state* and a little house appears to the upper-left of the cube. This house is the **Home** icon. Picking this icon always restores the same view, called the *home view.* Later in this chapter, you will learn how to change the home view.

Notice that a corner, edge, or face on the cube is highlighted. In the default view of a drawing based on the acad3D.dwt template, the corner between the top, right, and front faces is shaded. The shaded part of the cube represents the standard view that is closest to the current view. Move the cursor over the cube and notice that edges, corners, and faces are highlighted as you move the pointer. If you pick a highlighted edge, corner, or face, a view is displayed that looks directly at the selected feature. By selecting standard views on the view cube, you can quickly move between pictorial and orthographic views.

PROFESSIONAL TIP

Picking the top face of the view cube displays a plan view of the current UCS XY plane. This is often quicker than using the **PLAN** command, which is discussed later in this chapter.

Dynamic displays

The easiest way to change the current view using the view cube is to pick and drag the cube. Simply move the cursor over the cube, press and hold the left mouse button, and drag the mouse to change the view. The view cube and model view dynamically change with the mouse movement. When you have found the view you want, release the mouse button. This is similar to the **Orbit** tool in steering wheels, which are discussed later in the chapter.

When dragging the view cube, you are not restricted to the current XY plane, as you are when using the compass. The compass is discussed later.

Pictorial displays

The view cube provides immediate access to 20 different preset pictorial views. When one of the corners of the cube is selected, a standard isometric pictorial view is displayed. See **Figure 4-3A.** Eight corners can be selected. Picking one of the four corners on the top face of the cube restores one of the preset isometric views. You can select a preset isometric view from the drop-down list in the **Views** panel on the **View** tab of the ribbon. However, using the view cube may be a quicker method.

Figure 4-3.
A—When one of the corners of the cube is selected, a standard isometric view is displayed.
B—Selecting one of the edges of the cube produces the same rotation in the XY plane as an
isometric view, but a zero elevation view in the Z plane.

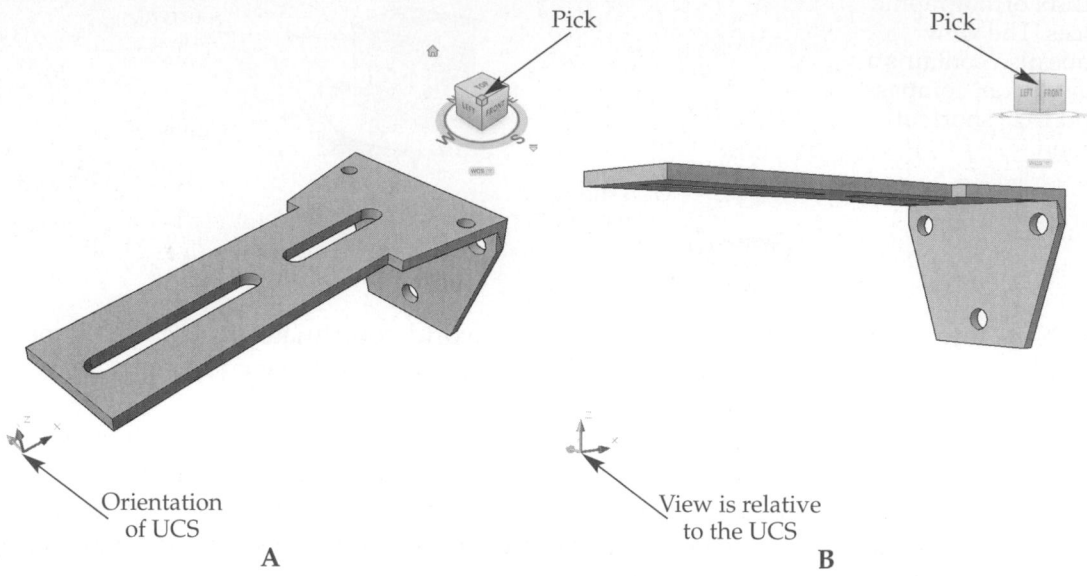

Pick   Pick

Orientation
of UCS
A

View is relative
to the UCS
B

Selecting one of the edges of the cube sets the view perpendicular to that edge.
See **Figure 4-3B.** There are 12 edges that can be selected. These standard views are not
available on the ribbon.

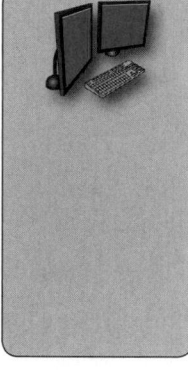

PROFESSIONAL TIP

The quickest method of dynamically rotating a 3D model is achieved
by using the mouse wheel button. Simply press and hold the [Shift]
key while pressing down and holding the mouse wheel. Now move
the mouse in any direction and the view rotates accordingly. This
is a transparent function that executes the **3DORBIT** command and
can be used at any time. Since it is transparent, it can even be used
while you are in the middle of a command. This is an excellent technique to use because it does not require selecting another tool or
executing a command.

Projection

The view displayed in the graphics window can be in one of two projections.
The *projection* refers to how lines are applied to the viewing plane. In a pictorial
view, lines in a *perspective projection* appear to converge as they recede into the
background. The points at which the lines converge are called *vanishing points.* In 2D
drafting, it is common to represent an object in pictorial as a one- or two-point perspective, especially in architectural drafting. In a *parallel projection,* lines remain parallel
as they recede. This is how an orthographic or axonometric (isometric, dimetric, and
trimetric) view is created.

To change the projection, right-click on the view cube to display the shortcut menu.
Three display options are given:
- **Parallel.** Displays the model as a parallel projection. This creates an ortho-
  graphic or axonometric view.
- **Perspective.** Displays the model in the more realistic, perspective projection.
  Lines recede into the background toward invisible vanishing points.

Figure 4-4.
In a parallel projection, parallel lines remain parallel. In a perspective projection, parallel lines converge to a vanishing point. Notice the three receding lines on the boxes. If these lines are extended, they will intersect.

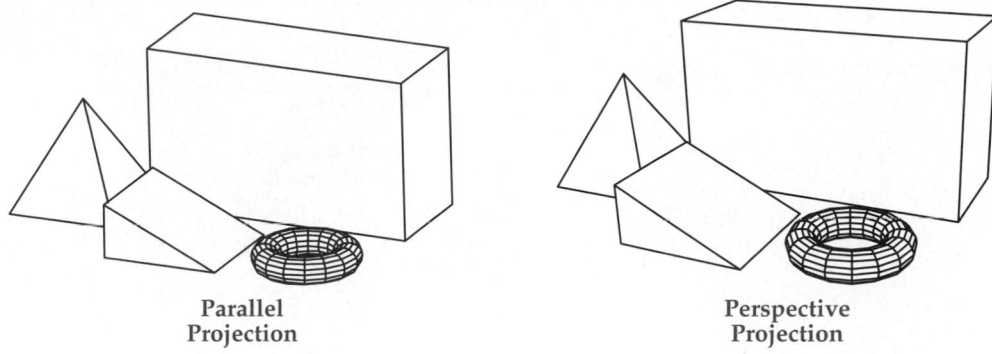

Parallel
Projection

Perspective
Projection

- **Perspective with Ortho Faces.** Displays the model in perspective projection when a pictorial view is displayed and parallel projection when an orthographic view is displayed. The parallel projection is only set current if a face on the view cube is selected to display the orthographic view. It is not set current if the **PLAN** command is used.

Figure 4-4 illustrates the difference between parallel and perspective projection.

Exercise 4-1

Complete the exercise on the student website.
www.g-wlearning.com/CAD

Orthographic displays

The view cube faces are labeled with orthographic view names, such as Top, Front, Left, and so on. Picking on a view cube face produces an orthographic display of that face. Keep in mind, if the current projection is perspective, the view will not be a true orthographic view. See **Figure 4-5.** If you plan to work in perspective projection, but also want to view proper orthographic faces, turn on **Perspective with Ortho Faces** in the view cube shortcut menu. Another way to achieve a proper orthographic view is to turn on parallel projection.

When a face is selected on the view cube, the cube rotates to orthographically display the named face. The view rotates accordingly. In addition, notice that a series of triangles point to the four sides of the cube (when the cursor is over the tool). See **Figure 4-6A.** Picking one of these triangles displays the orthographic view corresponding to the face to which the triangle is pointing, **Figure 4-6B.** This is a quick and efficient method to precisely rotate the display between orthographic views.

When an orthographic view is displayed, two *roll arrows* appear on the view cube. Picking either of these arrows rotates the current view 90° in the selected direction and about an axis perpendicular to the view. See **Figure 4-6C.** Using the triangles and roll arrows on the view cube provides the greatest flexibility in manipulating the model between orthographic views.

Figure 4-5.
A—To properly view orthographic faces using the view cube, turn on **Perspective with Ortho Faces** in the shortcut menu. B—When an orthographic view is set current with perspective projection on, the view is not a true orthographic view. Notice how you can see the receding surfaces.

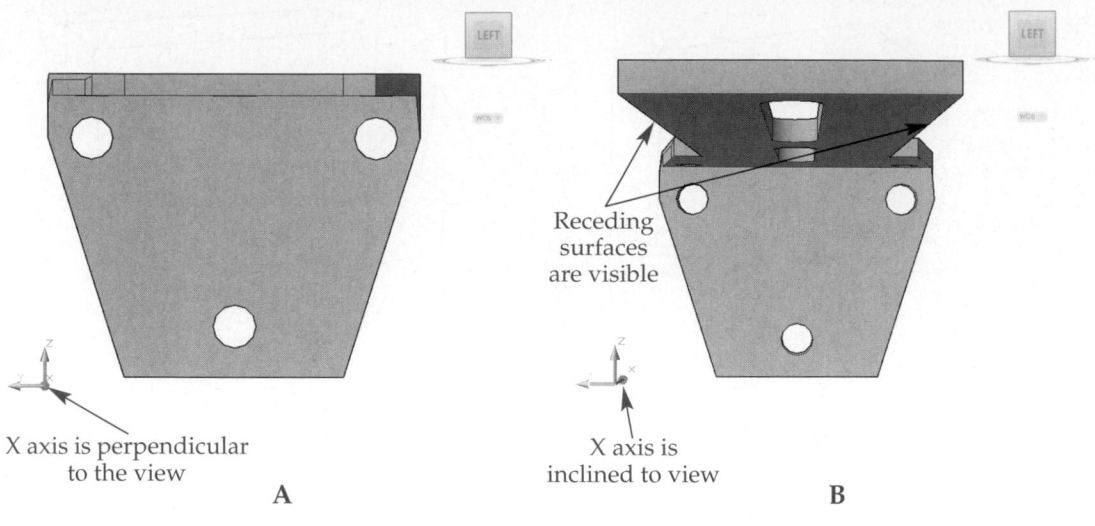

X axis is perpendicular to the view

A

Receding surfaces are visible

X axis is inclined to view

B

Figure 4-6.
A—The selected orthographic face is surrounded by triangles. Pick one of the triangles to display that orthographic face. B—The orthographic face corresponding to the picked triangle is displayed. Picking a roll arrow rotates the current view 90° in the selected direction. C—The rotated view.

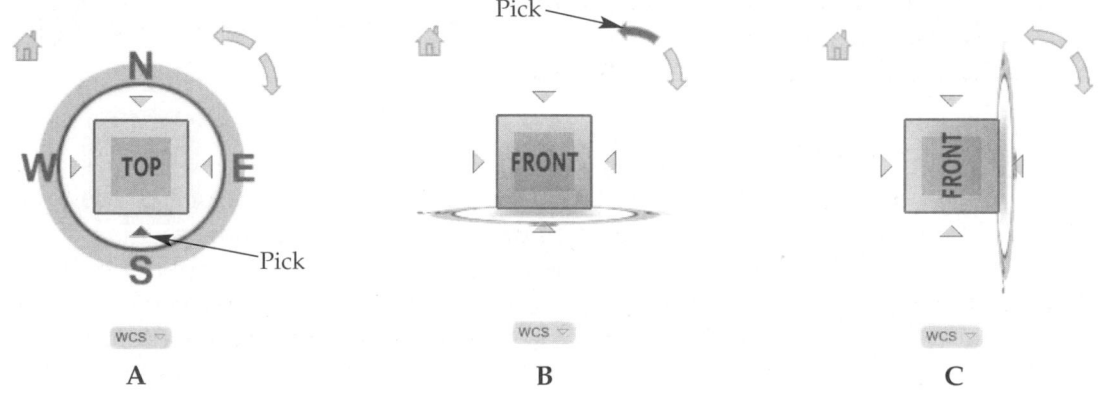

A

B

C

Setting views with the compass

The compass allows you to dynamically rotate the model in the XY plane. To do so, pick and hold on one of the four labels (N, S, E, or W). Then, drag the mouse to rotate the view. The view pivots about the Z axis of the current UCS. Try this a few times and notice that you can completely rotate the model by continuously moving the cursor off of the screen. For example, pick the letter W and move the pointer either right or left (or up or down). Notice as you continue to move the mouse in one direction the model continues to rotate in that direction.

You can use the compass to display the model in a view plan to the right, left, front, or back face of the cube. When the pointer is moved to one of the four compass directions, the letter is highlighted, **Figure 4-7A.** Simply single pick on the compass label that is next to the face you wish to view. See **Figure 4-7B.**

Figure 4-7.
A—When the pointer is moved over one of the four compass directions, the letter is highlighted.
B—If you pick the letter, the orthographic view from that compass direction is displayed.

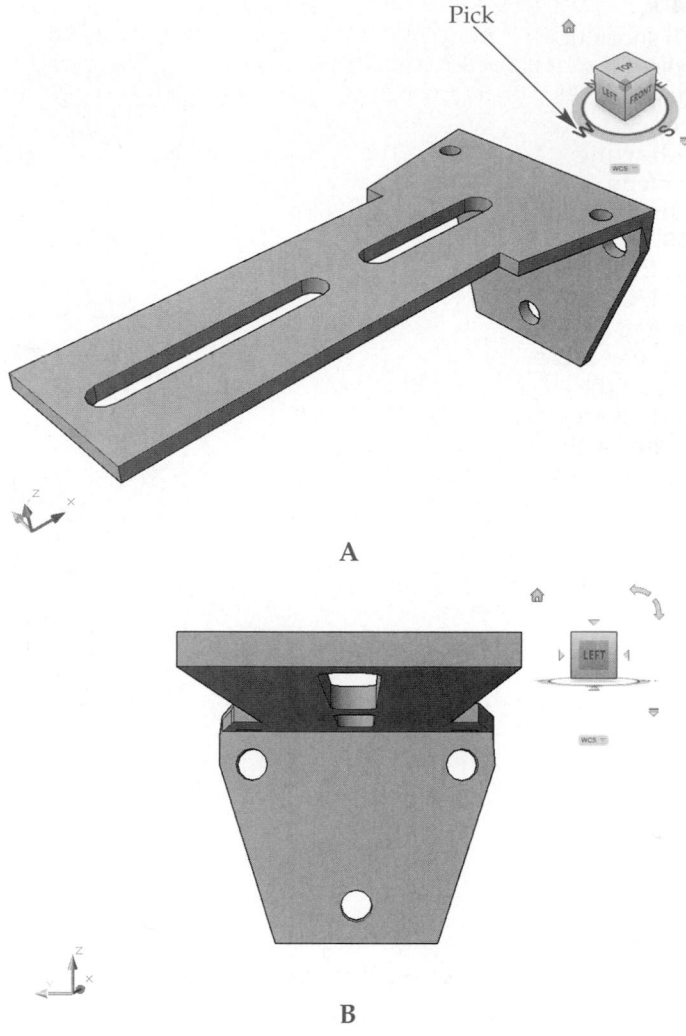

A

B

Home view

The default home view is the southwest isometric view of the WCS. Picking the **Home** icon in the view cube always displays the view defined as the home view, regardless of the current UCS. You can easily set the home view to any display you wish. First, using any navigation method, display the model as required. Next, right-click on the view cube and pick **Set Current View as Home** from the shortcut menu. Now, when you pick the **Home** icon in the view cube, this view is set current. Remember, the **Home** icon does not appear until the cursor is over the view cube.

PROFESSIONAL TIP

Should you become disoriented after repeated use of the view cube, it is far more efficient to pick the **Home** icon than it is to use **UNDO** or try to select an appropriate location on the view cube.

UCS settings

The UCS shortcut menu in the view cube lists the WCS and the names of all named UCSs in the current drawing. If there are no named UCSs, the listing is **WCS** and **New UCS**. See **Figure 4-8A.** To create a new UCS, pick **New UCS** from the shortcut menu. Next, use the appropriate UCS command options to create the new UCS. See Chapter 6 for complete coverage of the **UCS** command.

Figure 4-8.
The UCS shortcut menu below the view cube lists all named UCSs in the current drawing. A—The menu entries are **WCS** and **New UCS**. In this example, there are no saved UCSs in the drawing. B—The names of new UCSs are added to the UCS shortcut menu. The current UCS is indicated with a checkmark.

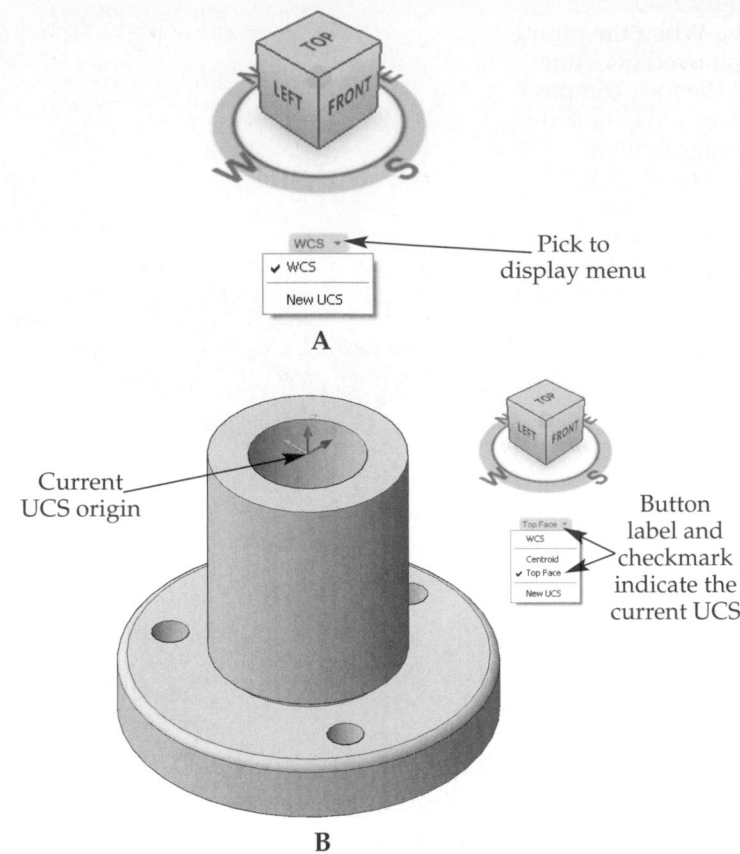

Pick to display menu

A

Current UCS origin

Button label and checkmark indicate the current UCS

B

As new UCSs are created and saved, their names are added to the UCS shortcut menu. See **Figure 4-8B.** Now, if you wish to work on the model using a specific UCS, simply select it from the list. The UCS is then restored.

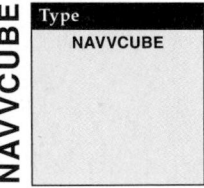

Exercise 4-2

Complete the exercise on the student website.
www.g-wlearning.com/CAD

View Cube Settings Dialog Box

Type
NAVVCUBE

NAVVCUBE

The appearance and function of the view cube can be changed using the **View Cube Settings** dialog box, **Figure 4-9.** This dialog box is displayed by selecting **View Cube Settings...** from the view cube shortcut menu or by using the **Settings** option of the **NAVVCUBE** command. The next sections discuss the options found in the dialog box.

Display

Options in the **Display** area of the **View Cube Settings** dialog box control the appearance of the view cube tool. The thumbnail dynamically previews any changes made to the display options. There are four options in this area of the dialog box.

On-Screen Position. The view cube can be placed in one of four locations in the drawing area: top-right, bottom-right, top-left, or bottom-left corner. By default, it is located in the top-right corner of the screen. To change the location, select it in the **On-screen position:** drop-down list. The **NAVVCUBELOCATION** system variable controls this setting.

AutoCAD and Its Applications—Advanced

Figure 4-9.
The **View Cube**
Settings dialog box.

Set the size of the view cube

Set the opacity of the view cube

Check to display the UCS shortcut menu in the view cube

Select a location for the view cube

Preview of view cube

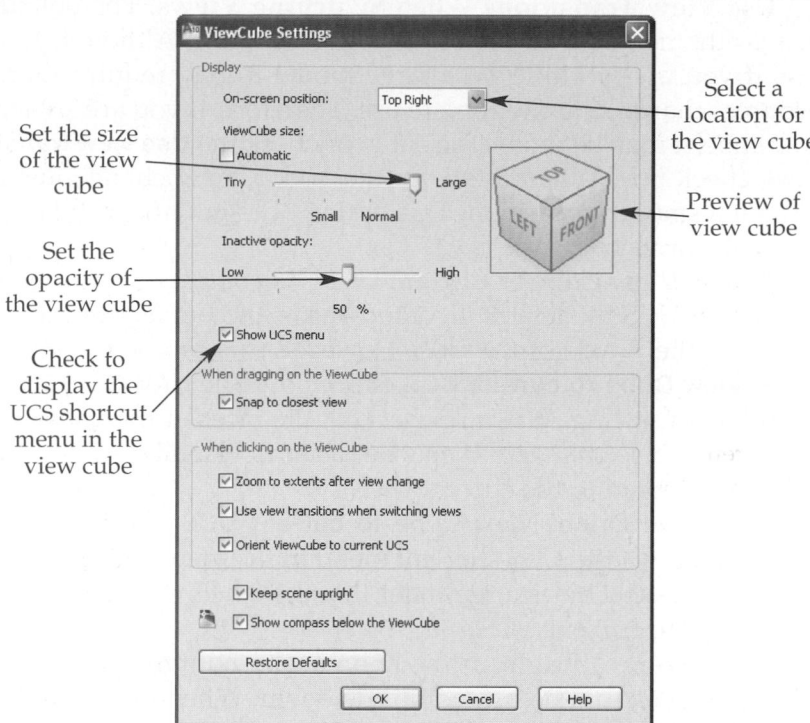

View Cube Size. The view cube can be displayed in one of four sizes. Use the **View Cube Size:** slider to set the size to either tiny, small, normal, or large. The slider is unavailable if the **Automatic** check box is checked, in which case AutoCAD sets the size based on the available screen area. The **NAVVCUBESIZE** system variable controls this setting.

Inactive Opacity. When the view cube is inactive, it is displayed in a semitransparent state. Remember, the view cube is in the inactive state whenever the pointer is not over it. When the view cube is in the active state, it is displayed at 100% opacity. Use the **Inactive opacity:** slider to set the level of opacity (transparency). The value can be from 0% to 100% with a default value of 50%. The percentage is displayed below the slider. A value of zero results in the view cube being hidden until the cursor is moved over it. If opacity is set to 100%, there is no difference between the inactive and active states. The **NAVVCUBEOPACITY** system variable controls this setting.

Show UCS menu. By default, the UCS shortcut menu is displayed in the view cube. This menu is displayed by picking the UCS button below the cube. If you wish to remove the UCS menu from the view cube display, uncheck the **Show UCS menu** check box.

When dragging on the view cube

By default, when you drag the view cube, the view "snaps" to the closest standard view that can be displayed by the view cube. This is because the **Snap to closest view** check box is checked by default. Uncheck this check box if you want the view to freely rotate without snapping to a preset standard view as you drag the view cube.

When clicking on the view cube

Options in the **When clicking on the View Cube** area control how the final view is displayed and how the labels on the view cube can be related to the UCS. There are three options in this area.

Zoom to Extents After View Change. If the **Zoom to extents after change** check box is checked, the model is zoomed to the extents of the drawing whenever the view cube is used to change the view. Uncheck this if you want to use the view cube to change the display without fitting the model to the current viewport.

Use View Transitions When Switching Views. The default transition from one view to the next is a smooth rotation of the view. Although this transition may look nice, if you are working on a large model it may require more time and computer resources than you are willing to use. Therefore, if you are switching views a lot using view cube, it may be more efficient to uncheck the **Use view transitions when switching views** check box. When this is unchecked, a view change just cuts to the new view without a smooth transition. This option does not affect the view when dragging the view cube or its compass.

Orient View Cube to Current UCS. As you have seen, the view cube is aligned to the current UCS by default. In other words, the top face of the cube is always perpendicular to the Z axis of the UCS. However, this can be turned off by unchecking the **Orient View Cube to current UCS** check box. The **NAVVCUBEORIENT** system variable controls this setting. When unchecked, the faces of the view cube are not reoriented when the UCS is changed. It may be easier to visualize view changes if the view cube faces are oriented to the current UCS.

When the **Orient View Cube to current UCS** check box is unchecked, WCS is displayed above the UCS shortcut menu in the view cube (unless the WCS is current). See **Figure 4-10.** This is a reminder that the labels on the view cube relate to the WCS and not to the current UCS.

Keep Scene Upright. If the **Keep scene upright** check box is checked, the view of the model cannot be turned upside down. When unchecked, you may accidentally rotate the view so it is upside down. This can be confusing, so it is best to leave this check box checked.

Show Compass below the View Cube. The compass is displayed by default in the view cube. However, if you do not find the compass useful, you can turn it off. To hide the compass, uncheck the **Show compass below the View Cube** check box.

Restore Defaults. After making several changes in the **View Cube Settings** dialog box you may get confused about the affect of different option settings on the model display. In this case, it is best to pick the **Restore Defaults** button to return all settings to their original values. Then change one setting at a time and test it to be sure the view cube functions as you intended.

Exercise 4-3
Complete the exercise on the student website.
www.g-wlearning.com/CAD

Figure 4-10.
When the **Orient View Cube to current UCS** option is off in the **View Cube Settings** dialog box, the UCS shortcut has WCS displayed above it.

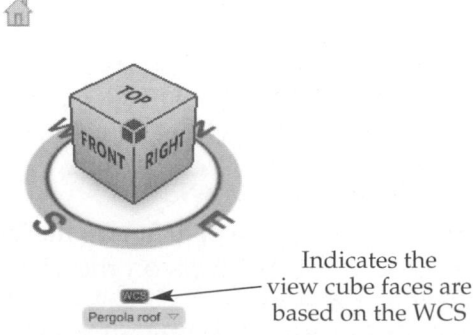

Indicates the view cube faces are based on the WCS

Figure 4-11.
A—Selecting the **3DCORBIT** command. B—This is the continuous orbit cursor in the **3DCORBIT** command (or **Continuous** option of the **3DORBIT** command). Pick and hold the left mouse button. Then, move the cursor in the direction in which you want the view to rotate and release the mouse button.

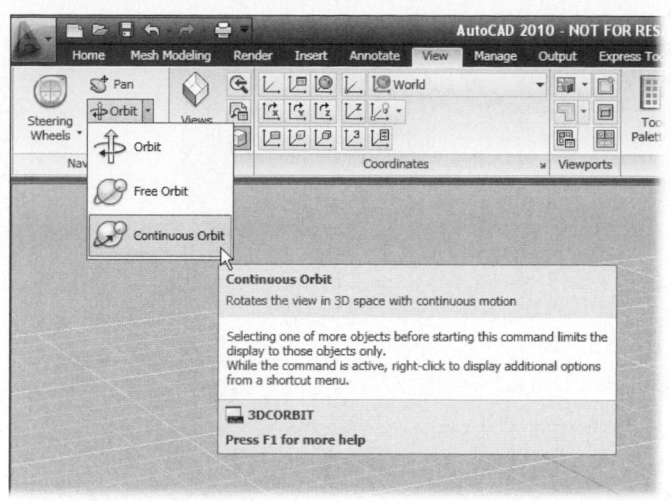

A   B

Creating a Continuous 3D Orbit

The **3DCORBIT** command provides the ability to create a continuous orbit of a model. By moving your pointing device, you can set the model in motion in any direction and at any speed, depending on the power of your computer. An impressive display can be achieved using this command. The command is located in the **Orbit** drop-down list in the **Navigate** panel of the **View** tab on the ribbon. See **Figure 4-11A.**

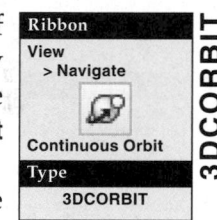

Once the command is initiated, the continuous orbit cursor is displayed. See **Figure 4-11B.** Press and hold the pick button and move the pointer in the direction that you want the model to rotate and at the desired speed of rotation. Release the button when the pointer is moving at the appropriate speed. The model will continue to rotate until you pick the left mouse button, press [Enter] or [Esc], or right-click and pick **Exit** or another option. At any time while the model is orbiting, you can left-click and adjust the rotation angle and speed by repeating the process for starting a continuous orbit.

Plan Command Options

The **PLAN** command, introduced in Chapter 1, allows you to create a plan view of any user coordinate system (UCS) or the world coordinate system (WCS). The **PLAN** command automatically performs a **ZOOM Extents**. This fills the graphics window with the plan view. The command options are:

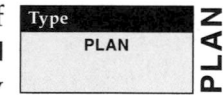

- **Current UCS.** This creates a view of the object that is plan to the current UCS.
- **UCS.** This displays a view plan to a named UCS. The preset UCSs are not considered named UCSs.
- **World.** This creates a view of the object that is plan to the WCS. If the WCS is the current UCS, this option and the **Current UCS** option produce the same results.

This command may have limited usefulness when working with 3D models. The dynamic capabilities of the view cube are much more intuitive and may be quicker to use. However, you may find instances where the **PLAN** command is easier to use, such as when the view cube is not currently displayed.

Figure 4-12.
A—Full Navigation
wheel. B—View Object
wheel. C—Tour Building
wheel.

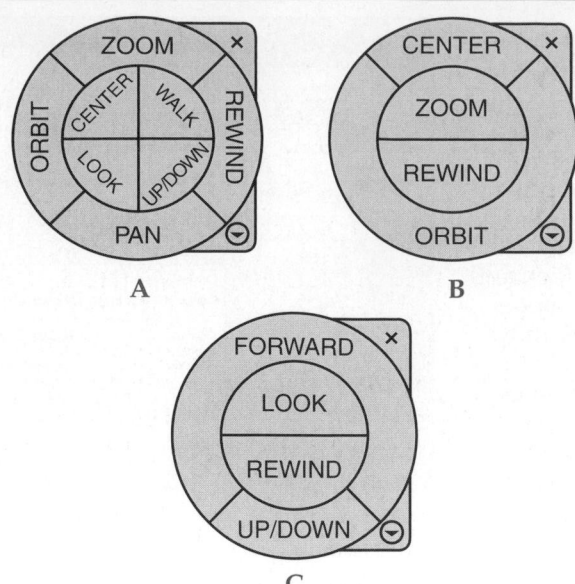

Displaying Models with Steering Wheels

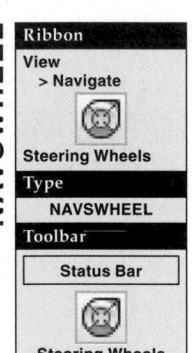

NAVSWHEEL

Ribbon
View
> Navigate

Steering Wheels

Type

NAVSWHEEL

Toolbar

Status Bar

Steering Wheels

Steering wheels, or *wheels,* are dynamic menus that provide quick access to view-navigation tools. A steering wheel follows the cursor as it is moved around the drawing. Each wheel is divided into wedges and each wedge contains a tool. See **Figure 4-12.** The **NAVSWHEEL** command is used to display a steering wheel.

AutoCAD provides two basic types of wheels: View Object and Tour Building. The options contained in these two wheels are combined to form the Full Navigation wheel. Each of these three types can be used in a full-wheel display or a minimized format. This section discusses the Full Navigation wheel and the two basic wheels in their full and mini formats. There is also a 2D Navigation wheel that is displayed in paper (layout) space or when the **NAVSWHEELMODE** system variable is set to 3. The functions of this wheel are covered in detail in *AutoCAD and Its Applications—Basics.*

Using a Steering Wheel

When displayed the very first time, the steering wheel is pinned. This is discussed later in the Steering Wheel Settings section. Once a steering wheel is displayed, move the cursor around the screen. Notice as you move the cursor, the wheel follows it. When you stop the cursor, the wheel stops. If you move the cursor anywhere inside of the wheel, the wheel remains stationary. Note also that as you move the cursor inside of the wheel, a wedge (tool) is highlighted. If you pause the cursor over a tool, a tooltip is displayed that describes the tool.

To use a specific tool, simply pick and hold on the highlighted wedge, then move the cursor as needed to change the view of the model. Once you release the pick button, the tool ends and the wheel is redisplayed. Some options display a *center point* about which the display will move. The **Center** tool is used to set the center point. These features are all discussed in the next sections.

To change between wheels, right-click to display the shortcut menu. Then, select **Full Navigation Wheel** from the menu to display that wheel. Or, select **Basic Wheels** to display a cascading menu. Select either **View Object Wheel** or **Tour Building Wheel** from the cascading menu to display that wheel. Refer to **Figure 4-13.**

Figure 4-13.
The shortcut menu
allows you to switch
between steering
wheels.

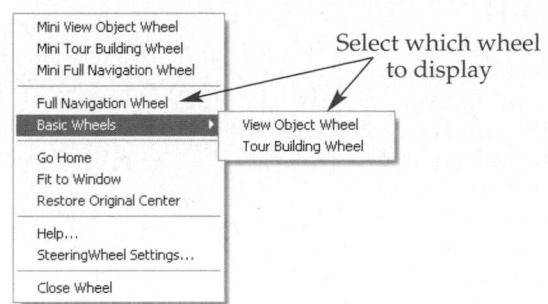

Select which wheel
to display

The view cube is a valuable visualization tool, especially when used in conjunction with a steering wheel. As you use the wheel options to manipulate the model, watch the view cube move. This provides you with a dynamic "map" of where your viewpoint is at all times in relation to the model, the UCS, and the view cube compass.

Exercise 4-4

Complete the exercise on the student website.
www.g-wlearning.com/CAD

Full Navigation Wheel

The Full Navigation wheel contains all of the tools available in the View Object and Tour Building wheels. Refer to **Figure 4-12A.** These tools are discussed in the following sections. It also contains the **Pan** and **Walk** tools, which are not available in either of the other two wheels. These tools are discussed next. The tools shared with the View Object and Tour Building tools are discussed in the sections corresponding to those tools.

In addition, a number of settings are available to change the appearance of the steering wheel and the manner in which some of the tools function. Refer to the Steering Wheel Settings section later in this chapter for a complete discussion of these settings.

Pan

The **Pan** tool allows you to move the model in the direction that you drag the cursor. This tool functions exactly the same as the AutoCAD **RTPAN** command. When you pick and hold on the tool, the cursor changes to four arrows with the label Pan Tool below the cursor.

If you use the **Pan** tool with the perspective projection current, it may appear that the model is slowly rotating about a point. This is not the case. What you are seeing is merely the effect of the vanishing points. As you pan, the relationship between the viewpoint and the vanishing points changes. You can quickly test this by closing the wheel, right-clicking on the view cube, and picking **Parallel** from the shortcut menu. Now display the wheel again and use the **Pan** tool. Notice the difference. The model pans without appearing to rotate.

Walk

The **Walk** tool is used to simulate walking toward, through, or away from the model. When you pick and hold on the tool, the center circle icon is displayed at the bottom-center of the drawing area. See **Figure 4-14.** The cursor changes to an arrow

Figure 4-14.
The **Walk** tool
displays the center
circle icon. As you
move the cursor
around the icon, one
of the arrows shown
here is displayed
to indicate the
direction in which
the view is being
moved.

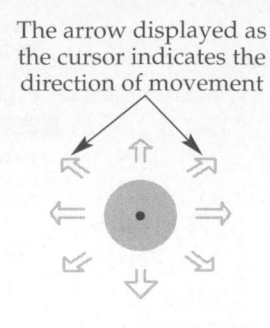

The arrow displayed as
the cursor indicates the
direction of movement

Press Up/Down arrows to adjust height, '+' key to speedup

pointing away from the center of the circle. The arrow indicates the direction in which the view will move as you move the mouse. This gives the illusion of walking in that direction in relation to the model.

If you hold down the [Shift] key while clicking the **Walk** tool, the **Up/Down** slider is displayed. This allows you to change the Y axis orientation of the view. Releasing the [Shift] key returns you to the standard walk mode. The up and down arrow keys can also be used to change the "height" of the view. The speed of walking can be increased with the minus key (–).

The **Up/Down** slider is also used with the **Up/Down** tool on the Tour Building wheel. It is explained in that section.

View Object Wheel

If you think of your model as a building, the tools in the View Object wheel are used to view the outside of the building. This wheel contains four navigation tools: **Center**, **Zoom**, **Rewind**, and **Orbit**. Refer to **Figure 4-12B.**

Center

The **Center** tool is used to set the center point for the current view. Many tools, such as **Zoom** and **Orbit**, are applied in relation to the center point. Pick and hold the **Center** tool, then move the cursor to a point on the model and release. The display immediately changes to center the model on that point. See **Figure 4-15.** The selected point must be on an object, but it does not need to be on a solid. Notice that the center point icon resembles a globe with three orbital axes. These axes relate to the three axes of the model shown on the UCS icon.

Figure 4-15.
The **Center** tool
allows you to select
a new center point
for the current view.
This becomes the
point about which
many steering wheel
tools operate.

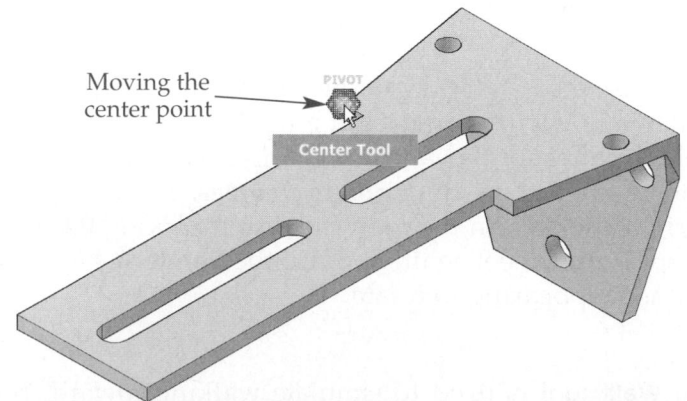

Moving the
center point

Zoom

The **Zoom** tool is used to dynamically zoom the view in and out, just as with the AutoCAD **RTZOOM** command. The tool uses the center point set with the **Center** tool. When you pick and hold the **Zoom** tool in the wheel, the center point is displayed at its current location. The cursor changes to a magnifying glass with the label Zoom Tool displayed below it. See **Figure 4-16.** There are three different ways to use the **Zoom** tool, as discussed in the next sections.

When the **Zoom** tool is accessed in the Full Navigation wheel, the center point is relocated to the position of the steering wheel. If you wish to zoom on the existing pivot point when using the Full Navigation wheel, first press the [Ctrl] key and then access the **Zoom** tool. This prevents the tool from relocating the center point. You can also move the center point using the **Center** tool, then switch to the View Object wheel and access the **Zoom** tool from that wheel. In either case, the zoom is relative to the location of the center point.

> **NOTE**
>
> Once you close the steering wheel, the center point is reset. The next time a steering wheel is displayed, the center point will be in the middle of the current view.

Pick and drag. To dynamically zoom, similar to realtime zoom, pick and drag the cursor. As the cursor is moved up or to the right, the viewpoint moves closer (zoom in). Move the pointer to the left or down and the viewpoint moves farther away (zoom out). The zooming is based on the current center point. When you have achieved the appropriate zoom location, release the pointer button.

Single click. If you select the **Zoom** tool with a single click, the view of the model zooms in by an incremental percentage. Each time you single click on the tool, the view is zoomed by 25%. The zoom is in relation to the center point.

In order for this function to work when the **Zoom** tool is accessed from the Full Navigation wheel, you must check the **Enable single click incremental zoom** check box in the **Steering Wheel Settings** dialog box. This is discussed in detail later.

Shift and click. If you press and hold the [Shift] key and then pick the **Zoom** tool, the view is zoomed out by 25%. As with the single-click method, the **Enable single click incremental zoom** check box must be checked in order to use this with the Full Navigation wheel.

Figure 4-16.
The **Zoom** tool operates in relation to the center point. If the tool is selected from the Full Navigation wheel, the center point is automatically relocated to the cursor location.

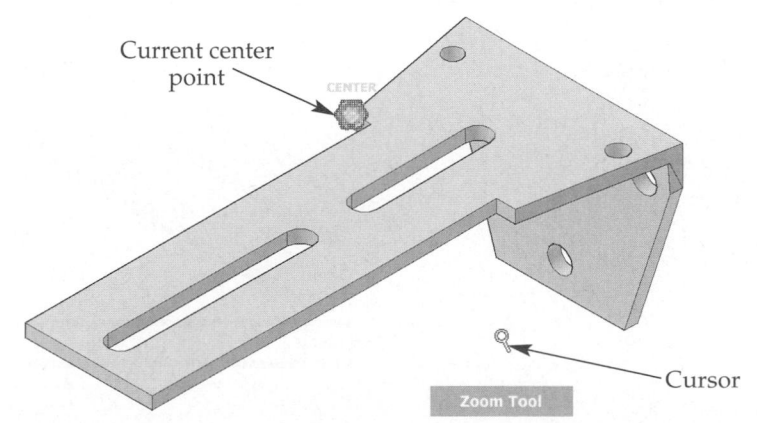

Current center point

Cursor

Rewind

The **Rewind** tool allows you to step back through previous views. A single pick on this tool displays the previous view. If you pick and hold the tool, a "slide show" of previous views is displayed as thumbnail images. See **Figure 4-17**. The most recent view is displayed on the right-hand side. The oldest view is displayed on the left-hand side. The slide representing the current view is highlighted with an orange frame and a set of brackets. While holding the pick button, move the cursor to the left. Notice that the set of brackets moves with the cursor. As a slide is highlighted, the corresponding view is restored in the viewport. Release the pick button when you find the view you want and it is set current.

The navigation history is maintained in the drawing file and is different for each open drawing. However, it is not saved when a drawing is closed. The **Steering Wheel Settings** dialog box allows you to control when thumbnail images are created and saved in the navigation history. This dialog box is discussed later in the chapter.

Orbit

The **Orbit** tool allows you to completely rotate your point of view around the model in any direction. The view pivots about the center point set with the **Center** tool. When using the Full Navigation wheel, the center point can be quickly set by pressing and holding the [Ctrl] key, then picking and holding the **Orbit** tool. Next, drag the center point to the desired pivot point on the model and release. Now you can use the **Orbit** tool.

To use the **Orbit** tool, pick and hold on the tool in the steering wheel. The current center point is displayed with the label Pivot. Also, the cursor changes to a point surrounded by two circular arrows. See **Figure 4-18A**. Move the cursor around the screen and the view of the model pivots about the center point. If this is not the result you wanted, just reset the pivot point.

When the **Orbit** tool is selected in the Full Navigation wheel, an option of the tool allows you to *roll* the view of the model, as opposed to orbiting (freely rotating). When rolling the view, the view axis parallel to your line of sight (perpendicular to the screen) is not altered. To enable this option, first open the **Steering Wheels Settings** dialog box by right-clicking on the wheel and picking **Steering Wheel Settings...** from the shortcut menu. Uncheck the **Maintain Up direction for Orbit tool** check box in the lower-left corner of the dialog box and pick the **OK** button to exit the dialog.

Figure 4-17.
A single pick on the **Rewind** tool displays the previous view. A "slide show" of previous views is displayed as thumbnail images if you press and hold the pick button.

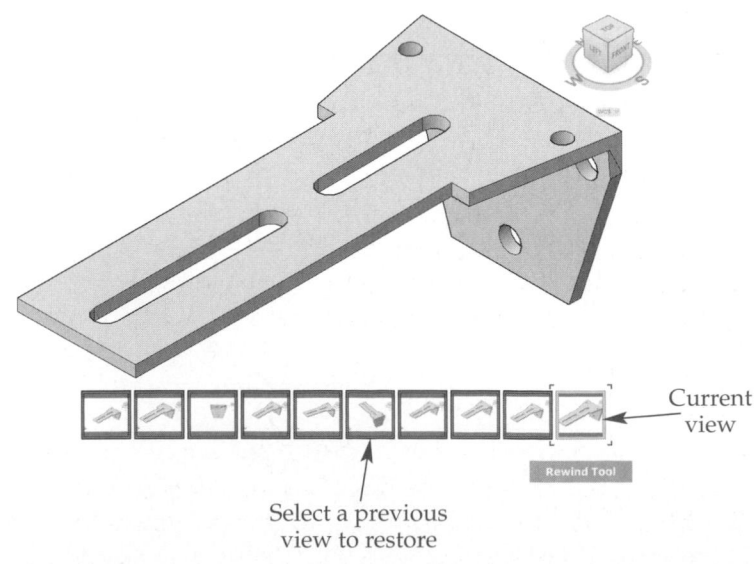

Current view

Rewind Tool

Select a previous view to restore

Figure 4-18.
A—The **Orbit** tool allows you to rotate your point of view completely around the model in any direction. This is the cursor displayed for the tool. B—Hold down the [Shift] key to roll the view when using the **Orbit** tool. The view will only rotate in a circular or "rolling" manner. This is the cursor displayed when rolling the view.

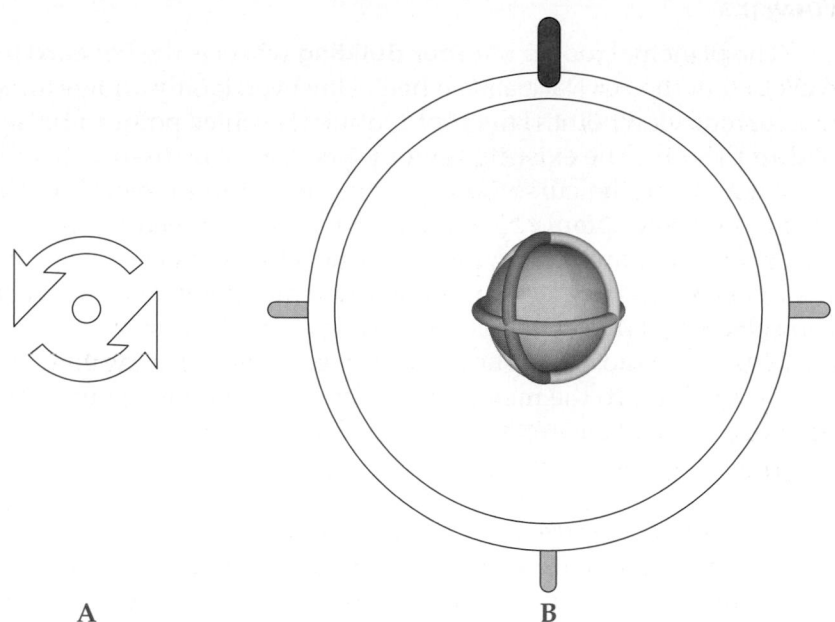

A   B

Now, pick and hold the **Orbit** tool in the Full Navigation wheel. You can orbit the view as described above. However, notice the message at the bottom of the view. Hold down the [Shift] key and the roll icon is displayed on the center point. See **Figure 4-18B**. As you move the cursor, the view only rotates in a circular or "rolling" manner about an axis perpendicular to the screen. Release the [Shift] key and you are returned to normal orbit mode. Remember, this option is *not* available when the **Orbit** tool is selected in the View Object wheel.

PROFESSIONAL TIP

Since the **Orbit** tool is most often used to quickly move your view-point to another side of the model, it is more intuitive to locate the pivot point somewhere on the model. First use the **Center** tool to establish the pivot point. Then, when you use the **Orbit** tool, the view pivots about that part of the model. If the center point is not set, it defaults to the center of the screen.

CAUTION

When the **Zoom** tool is selected from the Full Navigation wheel, the center point is changed to the steering wheel location. Therefore, that center point is used as the pivot point for the **Orbit** tool. When the **Zoom** tool is selected from the View Object wheel, the center point is not changed. Therefore, it is best to switch to the View Object wheel to use the **Zoom** tool or to press the [Ctrl] key before accessing the tool in the Full Navigation wheel.

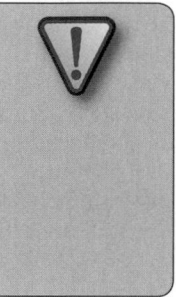

Tour Building Wheel

Where the tools in the View Object wheel are used to view the outside of the "building," the tools in the Tour Building wheel are used to move around inside of the "building." This wheel contains four navigation tools. The **Forward**, **Look**, and **Up/Down** tools are discussed here. The **Rewind** tool is discussed earlier.

Forward

The principal tool in the **Tour Building** wheel is the **Forward** tool. It is similar to the **Walk** tool in the Full Navigation wheel. However, it only allows forward movement from the current viewpoint. This tool requires a center point to be set on the model from within the tool. The existing center point cannot be used.

First, move the cursor and steering wheel to the point on the model that will be the target (center point). Next, pick and hold the **Forward** tool. The pick point becomes the center point and a drag distance indicator is displayed. See **Figure 4-19.** This indicator shows the starting viewpoint, the center point, and the surface of the model that you selected. Hold the mouse button down while moving the pointer up. The orange location slider moves to show the current viewpoint relative to the center point. As you move closer to the model, the green center point icon increases in size, which also provides a visual cue to the zoom level.

Look

The **Look** tool is used to rotate the view about the center of the view. When the tool is activated, the cursor appears as a half circle with arrows. See **Figure 4-20.** As you move the cursor down, the model moves up in the view as if you are actually tilting your head down to see the top of the model. Similarly, as you "look" away from the model to the right or left, the model appears to move away from your line of sight. The distance between you and the model remains the same and the orientation of the model does not change. Therefore, you would not want to use this tool if you wanted to see another side of the model.

Up/Down

As the name indicates, the **Up/Down** tool moves the view up or down along the Y axis of the screen, regardless of the orientation of the current UCS. Pick and hold on the tool and the vertical distance indicator appears. See **Figure 4-21.** Two marks on this indicator show the upper and lower limits within which the view can be moved. The orange slider shows the position of the view as you move the cursor. When the tool is first activated, the view is at the top position. The **Up/Down** tool has limited value. The **Pan** tool is far more versatile.

Figure 4-19.
The drag distance indicator is displayed when using the **Forward** tool. This indicator shows the start point of the view and the selected surface of the model. The slider indicates the current view position relative to the starting point.

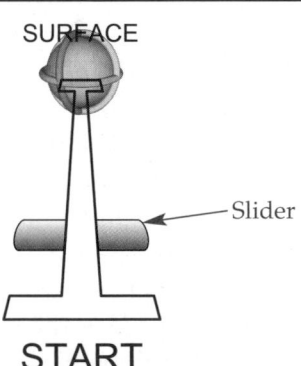

SURFACE

Slider

START

Figure 4-20.
The **Look** tool cursor appears as a half circle with arrows.

Figure 4-21.
The indicator displayed when using the **Up/Down** tool shows the upper and lower limits within which the view can be moved. The top position is the location of the view when the tool is selected.

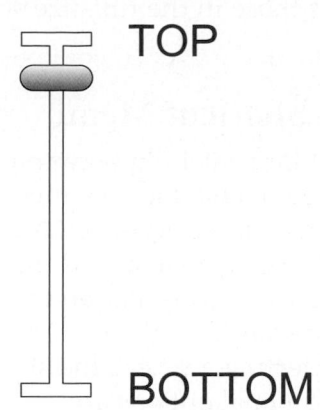

TOP

BOTTOM

Using steering wheel tools such as **Forward**, **Look**, **Orbit**, and **Walk**, you can manipulate your view to fully explore the model. The **Rewind** tool can then be used to replay all of the previous views saved in the navigation history. This process allows you to find views of the model that you may want to save as named views for later use in model construction or for shots created with show motion (discussed in Chapter 5). Remember, views created using steering wheels are not saved with the drawing file. Therefore, it may be a time-saver to create named views in this manner if there is a possibility they will be needed later.

Exercise 4-5

Complete the exercise on the student website.
www.g-wlearning.com/CAD

Mini Wheels

The three wheels discussed above were presented in their *full wheel* formats. As you gain familiarity with the use and function of each wheel and its tools, you may wish to begin using the abbreviated formats. The abbreviated formats are called *mini wheels.* See **Figure 4-22.** The mini wheels can be selected in the steering wheel shortcut menu. When you select a mini wheel, it replaces the cursor pointer.

As you move the mouse, the mini wheel follows. Slowly move the mouse in a small circle and notice that each wedge of the wheel is highlighted. The name of the currently highlighted tool appears below the mini wheel. When the tool you need is highlighted, simply click and hold to activate the tool. All tools in the mini wheels

Figure 4-22.
In addition to full-size steering wheels, mini wheels can be used. A—Mini Full Navigation wheel. B—Mini View Object wheel. C—Mini Tour Building wheel.

Current tool

Name of tool

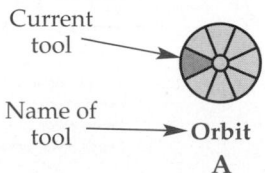
Orbit
A

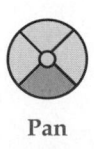
Pan
B

Walk
C

function the same as those in the full-size wheels. The only difference is the appearance of the wheels.

Steering Wheel Shortcut Menu

A quick method for switching between the different wheel formats is to use the steering wheel shortcut menu. Pick the menu arrow at the lower-right corner of a full wheel to display the menu. See **Figure 4-23.** Additionally, you can right-click when any full or mini wheel is displayed to access the menu. To select a wheel or change wheel formats, simply select the appropriate entry in the menu. Note that a check mark is *not* placed by the current wheel.

In addition to selecting a wheel, the shortcut menu provides options for viewing the model. These additional options are:

- **Go Home.** Returns the display to the home view. This is the same as picking the **Home** icon in the view cube.
- **Fit to Window.** Resizes the current view to fit all objects in the drawing inside of the window. This is essentially a zoom extents operation.
- **Restore Original Center.** Restores the original center point of the drawing using the current drawing extents. This does not change the current zoom factor. Therefore, if you are zoomed close into an object and pick this option, the object may disappear from view.
- **Level Camera.** The camera (your viewpoint) is rotated to be level with the XY ground plane.
- **Increase Walk Speed.** The speed used by the **Walk** tool is increased by 100%.
- **Decrease Walk Speed.** The speed used by the **Walk** tool is decreased by 50%.
- **Help.** Displays the online documentation (help file) for steering wheels.
- **Steering Wheel Settings.** Displays the **Steering Wheel Settings** dialog box, which is discussed in the next section.
- **Close Wheel.** Closes the steering wheel. This is the same as pressing [Esc] to close the wheel.

Exercise 4-6

Complete the exercise on the student website.
www.g-wlearning.com/CAD

Figure 4-23.
Select the down arrow to display the steering wheel shortcut menu. Both full-size wheels and mini wheels can be displayed using this menu.

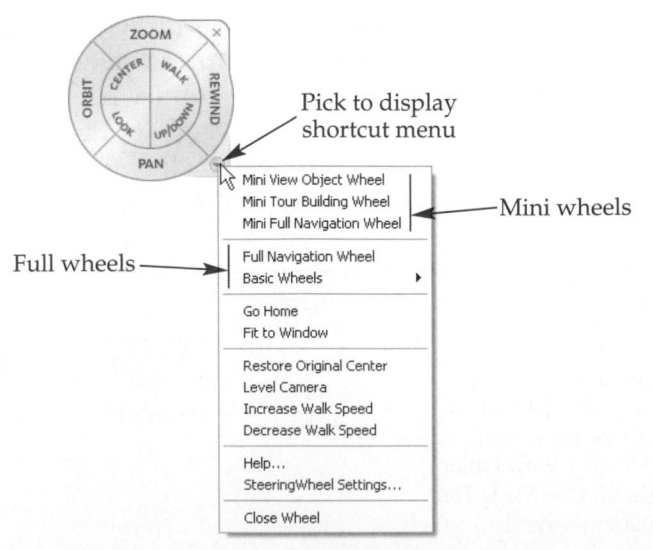

AutoCAD and Its Applications—Advanced

Steering Wheel Settings

The **Steering Wheels Settings** dialog box provides options for wheel appearance. See **Figure 4-24.** It also contains settings for the display and operation of several tools. It is displayed by picking **Steering Wheels Settings** in the shortcut menu.

Changing wheel appearance

The two areas at the top of the **Steering Wheels Settings** dialog box allow you to change the size and opacity of all wheels. The settings in the **Big Wheels** area are for the full-size wheels. The settings in the **Mini Wheels** area are for the mini wheels. The **Wheel size:** slider in each area is used to display the wheels in small, normal, or large size. The mini wheel has a fourth, extra large size. See **Figure 4-25.** These sliders set the **NAVSWHEELSIZEBIG** and **NAVSWHEELSIZEMINI** system variables.

Figure 4-24.
The **Steering Wheels Settings** dialog box provides options for wheel appearance and the display and operation of several tools.

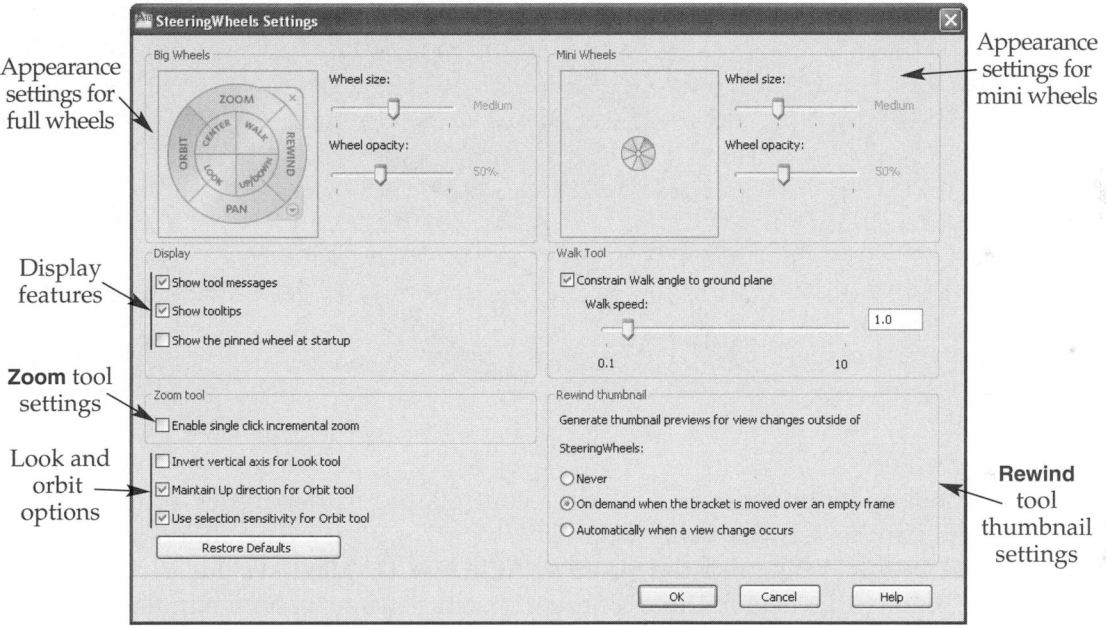

Figure 4-25
 The size of a steering wheel can be set in the **Steering Wheels Settings** dialog box or by using a system variable. A—The **NAVSWHEELSIZEBIG** system variable allows you to display full-size wheels in small, normal, or large size. B—The **NAVSWHEELSIZEMINI** system variable allows you to display mini wheels in small, normal, large, or extra large size.

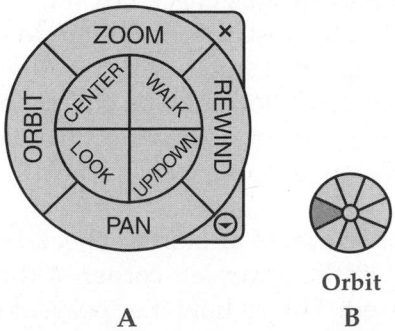

Orbit

A   B

Figure 4-26.
The opacity of a steering wheel can be changed. A—Opacity of 25%. B—Opacity of 50%. C—Opacity of 90%.

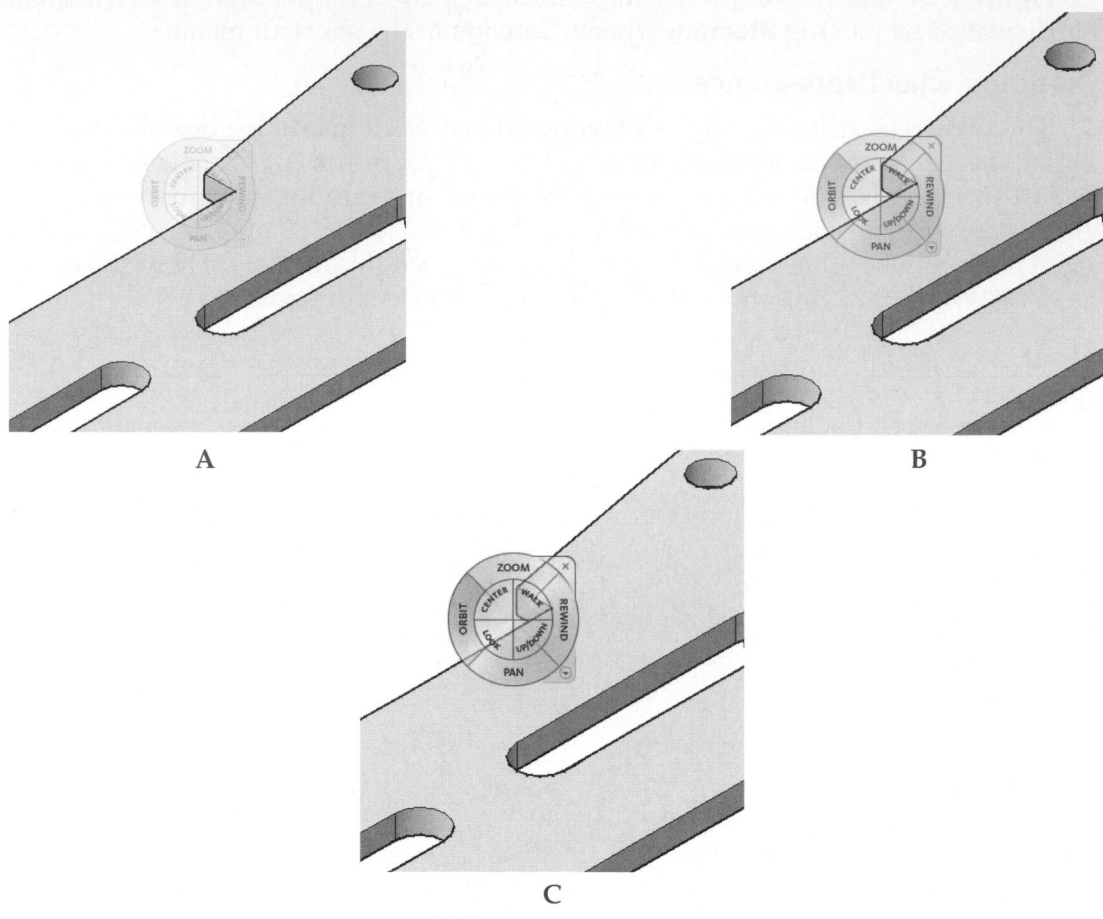

A

B

C

The **Wheel opacity:** slider in each area controls the transparency of the wheels. These sliders can be set to a value from 25% to 90% opacity. The sliders control the **NAVSWHEELOPACITYBIG** and the **NAVSWHEELOPACITYMINI** system variables. The appearance of the full wheel in three different opacity settings is shown in **Figure 4-26.**

Display features

The **Display** area of the **Steering Wheels Settings** dialog box controls three features of the wheel display. These settings determine whether messages and tooltips are displayed. There is also a setting for pinning the wheel at startup.

When a tool is selected, its name is displayed below the cursor. In addition, some tools have features or restrictions that can be indicated in a tool message. To see these messages, the **Show tool messages** check box must be checked. Otherwise, the messages are not displayed, but the restrictions remain in effect.

A tooltip is a short message that appears below the wheel when the cursor is hovered over the wheel. When the **Show tooltips** check box is checked, you can hold the cursor stationary over a tool for approximately three seconds and the tooltip will appear. Then, as you move the cursor over tools in the wheel, the appropriate tool tip is immediately displayed.

When the **Show the pinned wheel at startup** check box is checked, the steering wheel is displayed "pinned" in the lower-left corner of the graphics screen when the **NAVSWHEEL** command is used. This is how it appears the very first time a steering wheel is used when AutoCAD is installed. When the pointer is moved over the wheel, the "first contact" balloon is displayed. See **Figure 4-27.** You can select the **New to 3D** tab

Figure 4-27.
The steering wheel
"first contact"
balloon. A—The
Familiar with 3D tab.
B—The **New to 3D**
tab.

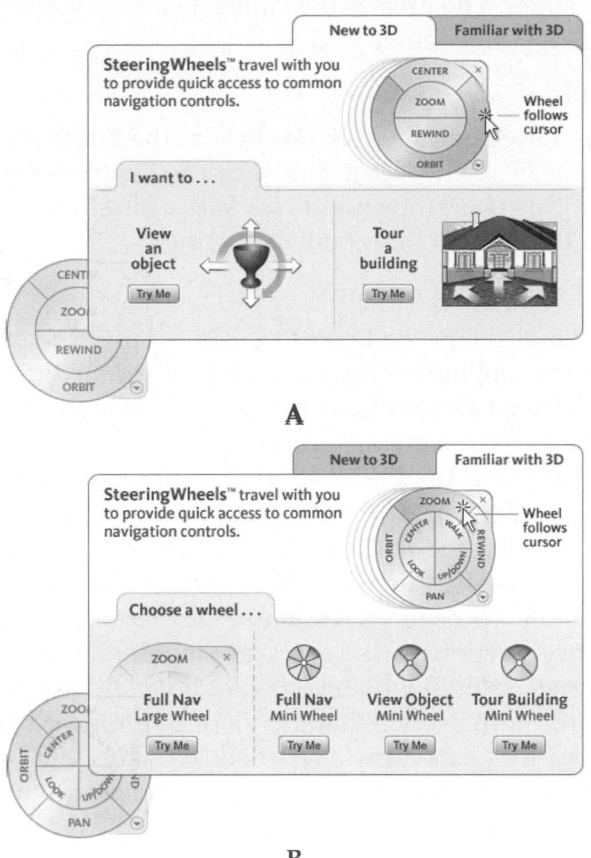

A

B

or **Familiar with 3D** tab. Each tab displays basic information on using steering wheels. This balloon is useful should you need reminders on the function and appearance of each type of wheel. To use the steering wheel, select an entry in the "first contact" balloon.

If you check the **Show the pinned wheel at startup** check box and exit the **Steering Wheels Settings** dialog box, the wheel is not pinned. This option will take affect the next time the **NAVSWHEEL** command is used. The steering wheel is pinned each time thereafter until the setting is changed.

PROFESSIONAL TIP

As soon as you become familiar with steering wheels, turn the **Show the pinned wheel at startup** option off. The "first contact" balloon is designed for a new user and quickly becomes a nuisance.

Walk tool

By default, when the **Walk** tool is used, you move parallel to the ground plane. This is because the **Constrain walk angle to ground plane** check box is checked. Test this by selecting the **Walk** tool and then move the cursor toward the top of the screen. You appear to be "walking" over the top of the model. Now, open the **Steering Wheels Settings** dialog box and uncheck the **Constrain walk angle to ground plane** box. This allows you to "fly" in the direction the cursor is moved when using the **Walk** tool. Exit the dialog box and again select the **Walk** tool. Move the cursor toward the top of the screen. This time it appears that you are flying directly toward or into the model.

The speed at which you walk through or around the model is controlled by the **Walk speed:** slider. The value can also be changed by typing in the text box at the right-hand end of the slider. The greater the value, the faster you will move as the cursor is moved away from the center circle icon.

Zoom tool

As discussed earlier, a single click on the **Zoom** tool zooms in on the current view by a factor of 25%. This is controlled by the **Enable single click incremental zoom** check box. When this check box is not checked, a single click on the tool has no effect. Some users find the single-click zoom confusing.

Look and orbit tool options

By default, when the **Look** tool is used and the cursor is moved downward, the view of the model moves up, just as if you were moving your eyes down. If you check the **Invert vertical axis for Look tool** option, this movement is reversed. In this case, the model moves in the same direction as the cursor.

If the **Maintain Up direction for Orbit tool** check box is not checked, it is possible to turn the model upside down while using the **Orbit** tool. This may not be desirable because it can be disorienting. To prevent this, be sure to leave the option checked. Uncheck this option only when you want to use **Orbit** in a "free-floating" mode.

By default, you can select objects in the model prior to using the wheel and the center of the selection is used as the pivot point for the **Orbit** tool. See **Figure 4-28.** First, select all objects about which you want the view to pivot. Next, display the wheel and pick the **Orbit** tool. Notice that the pivot point is located in the middle of the selected objects. Now, rotate the view as needed. This behavior is controlled by the **Use selection sensitivity for Orbit tool** check box in the **Steering Wheels Settings** dialog box. When unchecked, the selection set has no effect on the center point.

NOTE

If you select objects, display the wheel, and use the **Center** tool to change the center point, that point is used for the **Orbit** tool, not the center of the selection.

Figure 4-28
A—When no objects are selected, the default center point is used as the pivot point.
B—Select the objects about which you want the **Orbit** tool to pivot the view. Then, pick the **Orbit** tool at any point on the screen and notice that the pivot point is located in the middle of the selected objects.

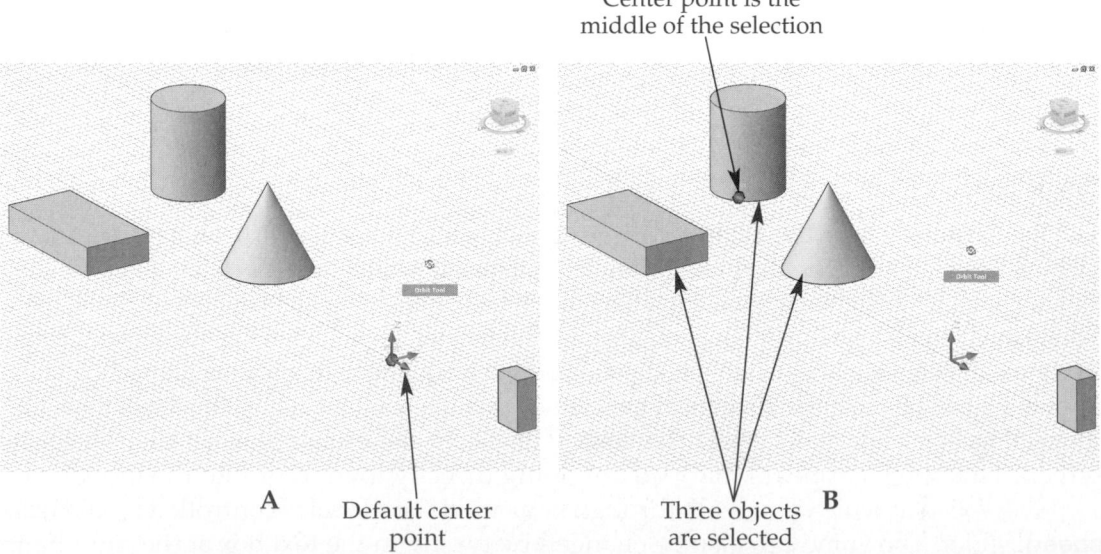

A   Default center point

Center point is the middle of the selection

Three objects are selected   B

Rewind thumbnail

Options in the **Rewind thumbnail** section of the **Steering Wheels Settings** dialog box control when and how thumbnail images are generated for use with the **Rewind** tool when view changes are made without using a wheel. There are three options in this area. Only one option can be on.

When the **Never** radio button is on, thumbnail images are never generated for view changes made outside of a wheel. Thumbnail images are created for view changes made with a wheel, which is true for all three options.

When the **On demand when the bracket is moved over an empty frame** radio button is on, thumbnail images are not automatically generated for view changes made outside of a wheel. The frames for these views display a double arrow icon when the **Rewind** tool in a wheel is used. However, as the brackets are moved over these frames, thumbnail images are generated. This is the default setting. See **Figure 4-29.**

When the **Automatically when a view change occurs** option is selected, a thumbnail image is generated any time a view change is made outside of a wheel. When the **Rewind** tool in a wheel is used, the frames for these views automatically display thumbnail images.

Restore defaults

Picking the **Restore Defaults** button in the **Steering Wheels Settings** dialog box returns all of the settings in the dialog box to their default values. Select this when at any time you are not sure how the settings are affecting the appearance and function of the steering wheel tools. Then, make changes one at a time as needed.

PROFESSIONAL TIP

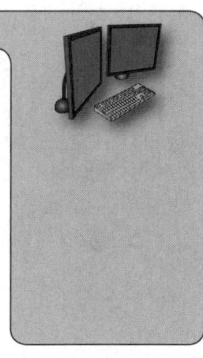

The view cube and steering wheels provide you with quick, dynamic methods of easily changing the viewpoint of the model. Prior versions of AutoCAD provided the **3DORBIT** command as the best dynamic method of changing your viewpoint. The functionality of this command has been replaced with the **Orbit** tool in the steering wheels. Although the **3DORBIT** command is still available for use, you may find that the view cube and steering wheels offer the most productive options for changing views.

Figure 4-29.
A—By default, when the **Rewind** tool is selected, frames representing view changes made outside of a steering wheel display double-arrow icons. B—As the brackets are moved over the blank frames, thumbnail images are generated of the views created outside of a wheel.

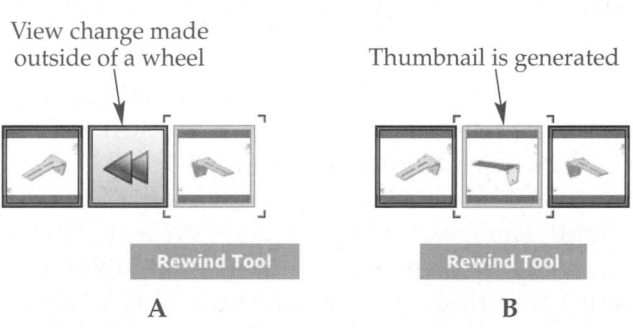

Exercise 4-7

Complete the exercise on the student website.
www.g-wlearning.com/CAD

Displaying a 3D Model in Different Visual Styles

The *display* of a 3D model is how the model is presented. This does not refer to the viewing angle, but rather colors, edge display, and shading or rendering. An object can be shaded from any viewpoint. A shaded model can be edited while still keeping the object shaded. This can make it easier to see how the model is developing without having to reshade the drawing. However, when editing a shaded object, it may also be more difficult to select features.

There are four basic ways in which a model can be displayed. The first is called a *wireframe display.* This is a display in which all lines are shown. The simplest "shaded" display technique is to remove hidden lines using the **HIDE** command or the 3D Hidden visual style to create a *hidden display.* However, this is not really a "shaded" display. A *shaded display* of the model can be created by setting either the Conceptual or Realistic visual style current. The Realistic visual style is considered the most realistic *shaded* display. A more detailed shaded model, a *rendered display* of the model, can be created with the **RENDER** command. A rendering is the most realistic presentation.

Visual styles are introduced in Chapter 1. The following sections discuss AutoCAD's visual styles and introduce rendering in AutoCAD. Detailed discussions on rendering, materials, lights, and animations appear in Chapter 15 through Chapter 18.

Exercise 4-8

Complete the exercise on the student website.
www.g-wlearning.com/CAD

Using the Visual Styles and Edge Effects Panels

A *visual style* controls the manner in which the edges and shading of a model are displayed in a viewport. The **Visual Styles** and **Edge Effects** panels in the **Render** tab of the ribbon provide quick and dynamic access to a variety of settings that create instant changes to the model display. See **Figure 4-30.** This section presents all of the settings available for visual styles that do not rely on the use of lights and materials. The application of lights, cameras, and materials is presented in Chapters 15 through 18.

You learned in Chapter 1 that the **Visual Styles Manager** palette allows access to the full range of settings available to create a visual style. On the other hand, the **Visual Styles** and **Edge Effects** panels in the ribbon display a group of intuitive controls that enable you to quickly alter the display of the model on the screen without redefining the visual style. It may be easier and quicker to first use these panels to change settings when working with variations of model display. These changes provide instant visual feedback, not only while constructing a model, but also when displaying it for evaluation or presentation purposes. Then, should you wish to make detailed changes to the visual style using specific settings and values, use the **Visual Styles Manager** tool palette. A complete discussion of the **Visual Styles Manager** is provided in Chapter 15.

Figure 4-30.
The **Visual Styles** and **Edge Effects** panels in the **Render** tab of the ribbon provide access to options for setting a visual style current and modifying the properties set by a visual style.

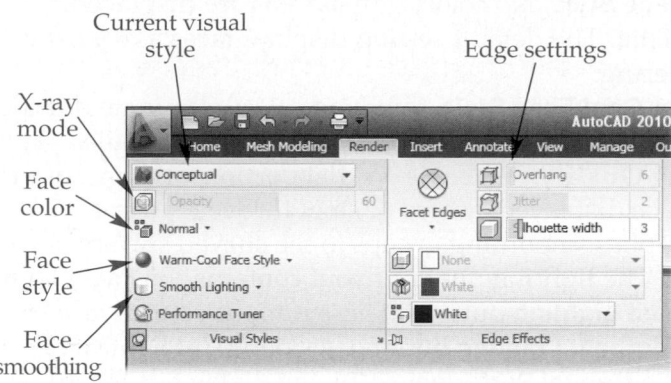

Current visual style

Edge settings

X-ray mode

Face color

Face style

Face smoothing

Visual style face settings

Four buttons in the **Visual Styles** panel give you the ability to change the transparency, face colors, face style (shading type), and lighting quality (face smoothing) on the model when the 3D Hidden, Conceptual, or Realistic visual style is current. Two of the settings, face style and lighting quality, are located in the expanded **Visual Styles** panel, as shown in **Figure 4-30.**

Face Style. The face style flyout contains three buttons that determine the style in which the model faces are displayed. This flyout is located in the expanded **Visual Styles** panel. The use of these options is shown in **Figure 4-31.**

Figure 4-31.
A—The Conceptual visual style is set current, then the **No Face Style** button is selected in the **Visual Style** panel. No colors or materials are displayed. B—The **Warm-Cool Face Style** button is selected. Subdued colors eliminate darkness and highlights that might otherwise obscure or hide faces and details. This is the default setting for the Conceptual visual style. C—The **Realistic** button is selected. The faces are displayed as realistically as possible without rendering.

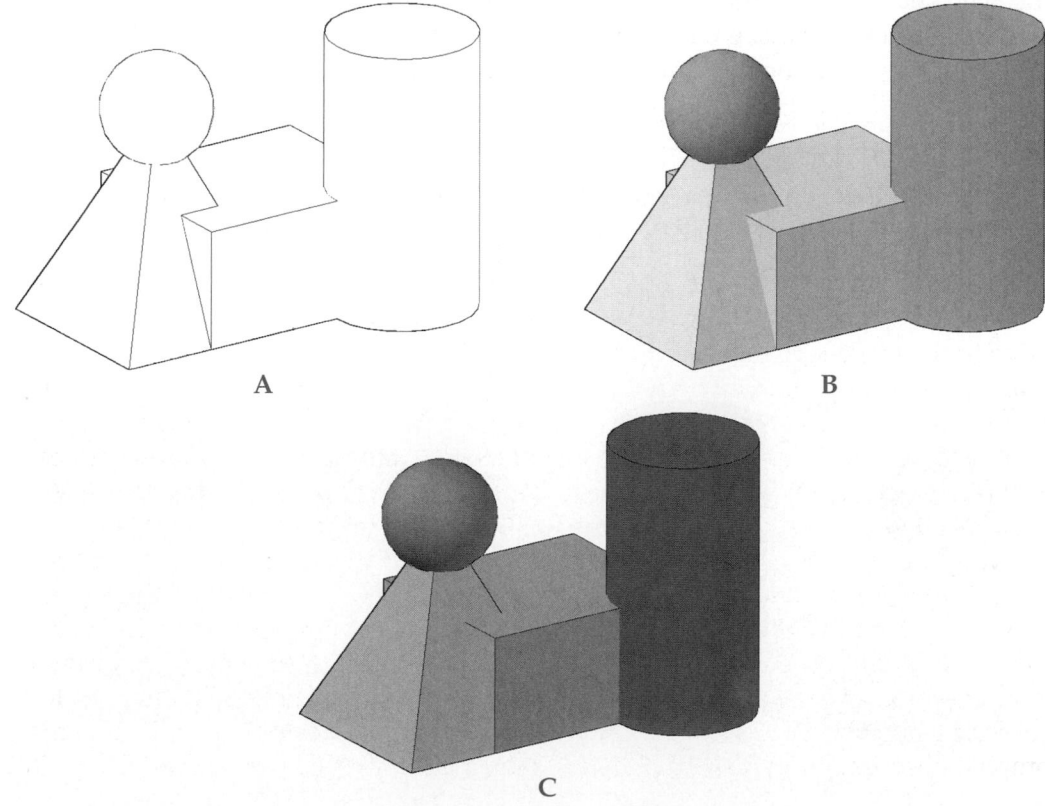

A

B

C

- **No Face Style.** No colors or materials are displayed.
- **Realistic.** The default setting displays faces as realistically as possible without rendering.
- **Warm-Cool Face Style.** Subdued colors eliminate darkness and highlights that might otherwise obscure or hide faces and details.

Lighting Quality. A *facet* is one flat portion, or plane, of a curved surface. Curved surfaces can be displayed faceted or smooth. See **Figure 4-32.** When displayed smooth, AutoCAD applies smoothing groups to the curved surfaces, which can increase regeneration time. The lighting quality flyout contains the **Facet Lighting**, **Smooth Lighting**, and **Smoothest Lighting** buttons. These options control face smoothing.

The **Smoothest Lighting** button uses per-pixel lighting. Per-pixel lighting must be enabled in the **Manual Performance Tuning** dialog box for this setting to have an effect. Otherwise, the **Smooth Lighting** setting is used when **Smoothest Lighting** is selected.

Face Colors. The face colors flyout contains four buttons that determine the manner in which the colors of the model faces are displayed. The color display is based on settings in the **Visual Styles Manager** palette, which is discussed in Chapter 15.

When the **Normal** button is selected in the flyout, face color options are not used. The faces are displayed in their assigned color. This may be ByLayer or an explicit color.

Faces are displayed in shades of a specified color when the **Monochrome** button is selected in the flyout. The default color is white, resulting in shades of gray. The color is controlled by the **VSMONOCOLOR** system variable.

When the **Tint** button is selected in the flyout, a tint is applied to the colors assigned to faces. The hue and saturation of the assigned colors is altered by applying a selected color. The color is controlled by the **VSMONOCOLOR** system variable.

When the **Desaturate** button is selected in the flyout, the faces are displayed in their assigned colors. However, the colors are softened, or desaturated, and appear lighter.

Opacity. When the **X-Ray Effect** button is on, all faces in the viewport are transparent. This is a toggle button that is off by default. When on, the button is highlighted in blue. The slider bar for the **X-Ray Effect** button controls the opacity (transparency) of the model faces. Slide the bar to the left for greater transparency and to the right to make the faces more opaque. You can enter a number in the text box from 1 to 100.

Figure 4-32.
A—The **Facet Lighting** button in the **Visual Styles** panel turns on the display of facets.
B—The **Smooth Lighting** button in the **Visual Styles** panel turns the facet display off so that curved surfaces appear smooth.

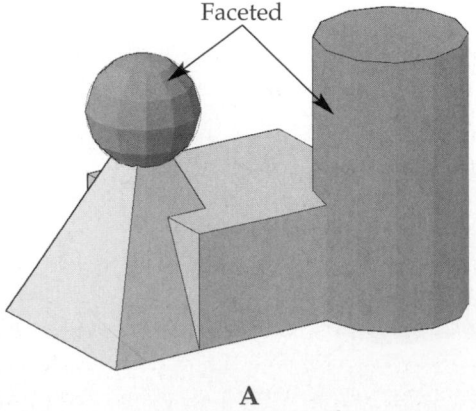
Faceted

A

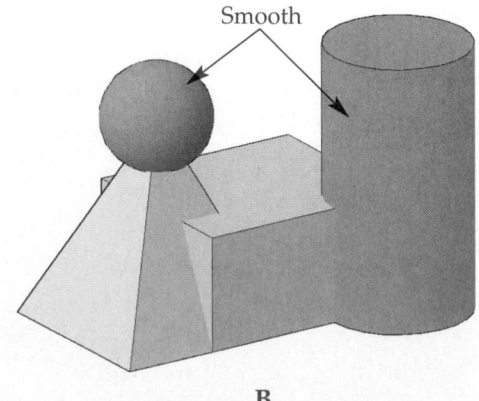
Smooth

B

Figure 4-33.
A—The Conceptual visual style is set current. X-ray effect is off. B—X-ray effect is turned on.
C—The 3D Hidden visual style is set current and X-ray effect is off. D—X-ray effect is turned on.

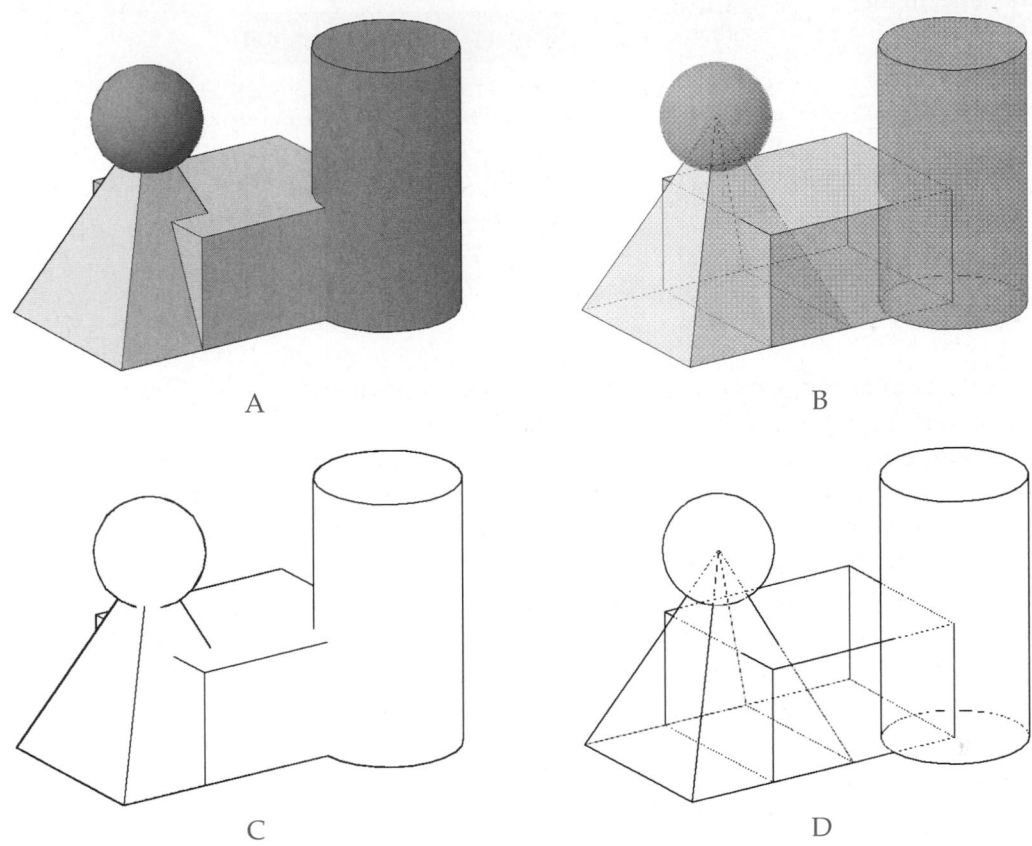

A

B

C

D

Both the numbers 0 and 100 produce an opaque model. This setting is controlled by the **VFACEOPACITY** system variable. The results of turning X-ray effect on and off are shown in **Figure 4-33.**

NOTE

Whenever you make a change to the face or edge settings, the name of the visual style in the visual styles drop-down list in the **Visual Styles** panel changes to *Current*. If you select a named visual style from the drop-down list, the settings of that visual style override the changes you made.

Visual style edge settings

There are seven settings in the **Edge Effects** panel that control the appearance of both visible and obscured edges. Refer to **Figure 4-30.** Obscured edges are those that would normally be hidden by the object or other objects.

Edge Mode. The edge mode flyout at the left side of the **Edge Effects** panel in the **Render** tab of the ribbon controls the lines displayed to define solids. See **Figure 4-34.** The edge mode flyout contains three options:

- **No Edges.** Object edges are not shown. The edge options at the bottom of the panel are disabled. This is the same as setting the **VSEDGES** system variable to 0.
- **Isolines.** Displays isolines based on the current **ISOLINES** system variable setting. The isolines are shown in the color set in the color drop-down list at the bottom of the expanded **Edge Effects** panel. The **Obscured Edge** and

Figure 4-34.
The options in the expanded **Edge Effects** panel in the **Render** tab of the ribbon.

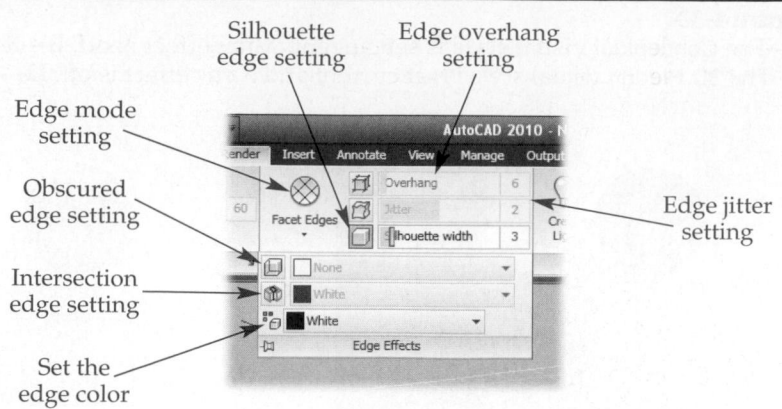

Silhouette edge setting

Edge overhang setting

Edge mode setting

Obscured edge setting

Intersection edge setting

Edge jitter setting

Set the edge color

Intersection Edge options in the expanded panel are disabled. This is the same as setting the **VSEDGES** system variable to 1.

- **Facet Edges.** Edges of 3D faces are displayed in the color selected in the color drop-down list at the bottom of the expanded panel. The **Obscured Edge** and **Intersection Edge** options in the expanded panel are enabled. This is the same as setting the **VSEDGES** system variable to 2.

Edge Overhang. This setting determines if straight edges extend or "overhang" beyond corners. See **Figure 4-35A.** Extended edges are used to create the look of an architectural sketch. The **Edge Overhang** button is either on or off. When on, the button is highlighted in the **Edge Effects** panel. Use the **Overhang** slider bar next to the button to increase or decrease the amount of overhang. As you drag the slider, the overhang dynamically changes in the viewport. This setting is controlled by the **VSEDGEOVER-HANG** system variable and can have a value from 1 to 100.

Edge Jitter. Edge jitter creates a sketch look along the entire length of edges by drawing multiple lines for the edges. See **Figure 4-35B.** The **Edge Jitter** button is either on or off. When on, the button is highlighted in the **Edge Effects** panel. The number of lines drawn for edges can be from zero to three. This can be set using the **Jitter** slider bar next to the button or by changing the **VSEDGEJITTER** system variable.

Silhouette Edges. The **Silhouette Edges** button determines whether or not a highlighting line is applied to the silhouettes of objects. See **Figure 4-35C.** The button is either on or off. When on, the button is highlighted and the silhouette line width is applied. The intersecting edges of shapes are not silhouetted. Use the **Silhouette width** slider bar next to the button to set the thickness of the highlighting line. This is controlled by the **VSSILHEDGES** system variable and is independent of the **DISPSILH** system variable.

Obscured Edges. The **Obscured Edges** button in the expanded **Edge Effects** panel determines whether or not edges that are normally hidden in the current view are displayed. The button is either on or off. When on, the button is highlighted and hidden edges are displayed. See **Figure 4-35D.** This setting is controlled by the **VSOB-SCUREDEDGES** system variable. Additionally, when the button is on, the edge color can be changed using the drop-down list next to the button.

Intersection Edges. The **Intersection Edges** button in the expanded **Edge Effects** panel determines whether or not lines are drawn where solids overlap. The button is either on or off. When on, the button is highlighted and a line is drawn at the intersection. This setting is controlled by the **VSINTERSECTIONEDGES** system variable. Additionally, when the button is on, the color of the intersecting edges can be changed using the drop-down list next to the button.

Edge Color. The **Edge Color** drop-down list in the expanded **Edge Effects** panel controls the color of visible object edges. The standard colors and color palette can be selected here. This setting is controlled by the **VSEDGECOLOR** system variable

Figure 4-35.
A—The model is displayed with extended edges. B—The model is displayed with edge jitter. C—The model is displayed with the silhouette edges highlighted. D—The model is displayed with obscured edges visible.

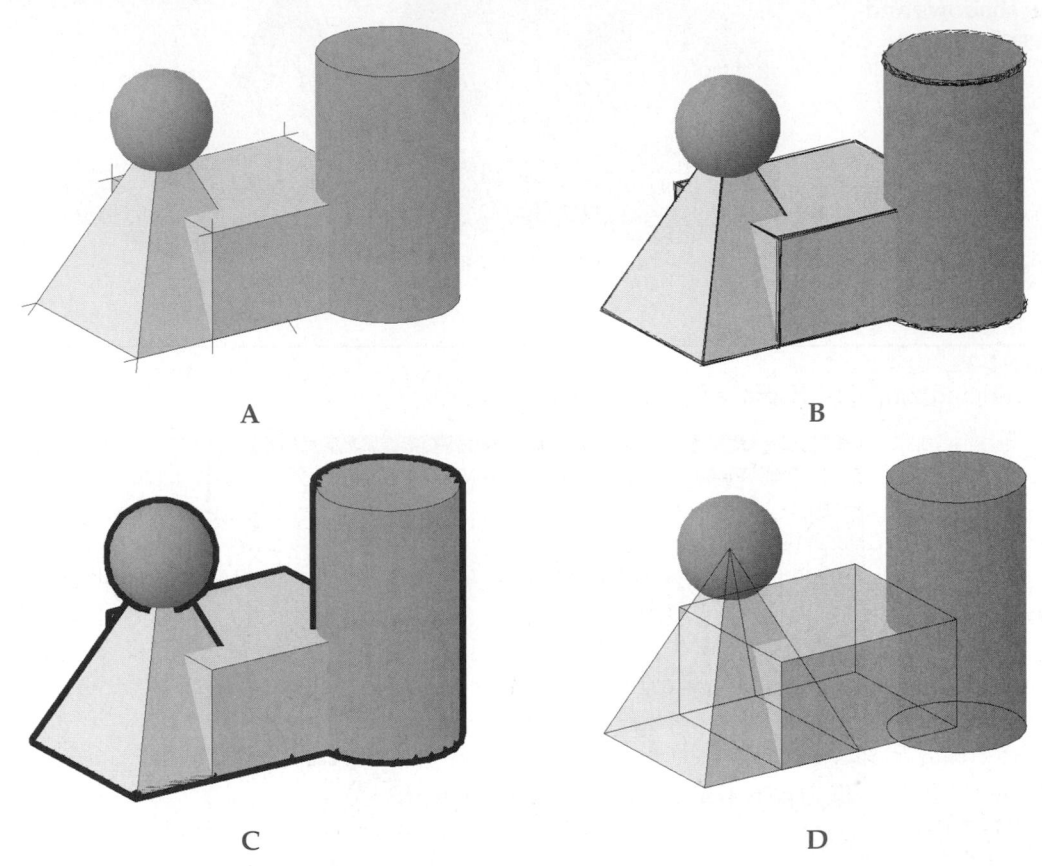

A

B

C

D

Exercise 4-9

Complete the exercise on the student website.
www.g-wlearning.com/CAD

Rendering a Model

The **RENDER** command creates a realistic image of a model, **Figure 4-36.** However, rendering an image takes longer than shading an image. There are a variety of settings that you can change with the **RENDER** command that allow you to fine-tune renderings. These include lights, materials, backgrounds, fog, and preferences. Render settings are discussed in detail in Chapter 15 through Chapter 18.

When the command is initiated, the render window is displayed and the image is rendered. See **Figure 4-37.** The rendering that is produced is based on a variety of advanced render settings that are discussed in Chapter 15 through Chapter 18. The default render settings create an image using a single light source located behind the viewer. The light intensity is set to 1 and, if no materials are applied, the objects are rendered with a matte material that is the same color as the object display color.

Ribbon
Render > Render
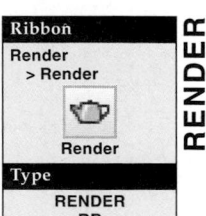
Render
Type
RENDER
RR

RENDER

Figure 4-36.
Rendering produces
the most realistic
display and can
show shadows and
materials.

Figure 4-37.
The rendered model is displayed in the **Render** window.

Rendered
image

Information about
the rendering

Renderings
completed in
this drawing
session

NOTE

If the image is rendered in the viewport, clean the screen using the
ZOOM, **PAN**, **REGEN**, or **REDRAW** command. Setting the rendering
destination as the viewport is discussed in Chapter 15.

Exercise 4-10

Complete the exercise on the student website.
www.g-wlearning.com/CAD

Chapter Test

Answer the following questions. Write your answers on a separate sheet of paper or complete the electronic chapter test on the student website.
www.g-wlearning.com/CAD

1. How do you select a standard isometric preset view using the view cube?
2. How is a standard orthographic view displayed using the view cube?
3. What is the difference between *parallel projection* and *perspective projection?*
4. What happens when one of the four view cube compass letters is picked?
5. Which command generates a continuous 3D orbit?
6. Which command can be used to produce a view that is parallel to the XY plane of the current UCS?
7. What is a *steering wheel?*
8. Briefly describe how to use a steering wheel.
9. The principal tool in the Tour Building wheel is the **Forward** tool. What is the purpose of this tool?
10. Which visual style is considered the highest level of shading?
11. What is the most realistic presentation?
12. Which visual style face setting produces a transparent image?
13. Which visual style edge setting produces the look of an architectural sketch?
14. How do you change the color of obscured edges in a visual style?
15. What is the function of the **RENDER** command?

Drawing Problems

1. Open one of your 3D drawings from Chapter 2 or 3. Do the following.
    A. Use the view cube to create a pictorial view of the drawing.
    B. Set the Conceptual visual style current.
    C. Using the **Render** tab in the ribbon, display the object so that faces are shown in object colors and edges are highlighted in the color of your choice. Change the edges to different colors.
    D. Save the drawing as P04_01.

2. Open one of your 3D drawings from Chapter 2 or 3. Do the following.
    A. Display the objects in warm-cool face style.
    B. Toggle the projection from parallel to perspective and turn off the view cube compass.
    C. Pick an edge on the view cube. Set the parallel projection current. Create a named view based on this display.
    D. Create three additional views based on view cube edges. Alternate between parallel and perspective projection.
    E. Put the model into a continuous orbit.
    F. Save the drawing as P04_02.

3. Open one of your 3D drawings from Chapter 2 that was created with solid primitives. Do the following.
    A. Create three named views, each having a different viewpoint.
    B. Use a different visual style face color option in each view.
    C. Save the drawing as P04_03.

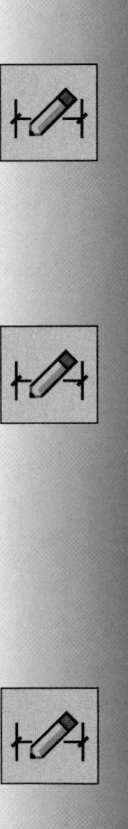

Drawing Problems - Chapter 4

4. Open drawing P04_03 and do the following.
   A. Display one view and set obscured edges to blue.
   B. Change the edge display to isolines. What happened?
   C. Display a different view and turn facet edges on.
   D. Display intersection edges as black.
   E. Create a rendering of the object in one of the views.
   F. Save the drawing as P04_04.

AutoCAD and Its Applications—Advanced

CHAPTER 5

Using Show Motion to View a Model

Learning Objectives

After completing this chapter, you will be able to:

✓ Explain the show motion tool.
✓ Create still shots of 3D models.
✓ Create walk shots of 3D models.
✓ Create cinematic shots of 3D models.
✓ Replay single shots and a sequence of shots.
✓ Change the properties of a shot.

AutoCAD's *show motion* tool is a powerful function that allows you to create named views, animated shots, and basic walkthroughs. It can quickly display a variety of named shots. It is also used to create basic animated presentations and displays. This capability is especially useful for animating 3D models that do not require the complexity and detail of fully textured, rendered animations and walkthroughs. Advanced rendering and walkthroughs are presented later in this book.

A *view* is a single-frame display of a model or drawing from any viewpoint. A *shot* is the manner in which the model is put in motion and the way the camera *moves* to that view. Therefore a single, named view can be modified to create several different shots using camera motion and movement techniques. Using show motion, you can create shots in one of three different formats:

- Still.
- Walk.
- Cinematic.

A *still shot* is exactly the same as a named view, but show motion allows you to add transition effects to display it. A *walk shot* requires that you use the **Walk** tool to define a camera motion path to create an animated shot. A *cinematic shot* is a single view to which you can add camera motion and movement effects to display the view. These shots are all created using the **New View/Shot Properties** dialog box. This is the same dialog box used to create named views.

Understanding Show Motion

Show motion is simply a means for creating, manipulating, and displaying named views. The **NAVSMOTION** command is used for show motion. The process involves the creation of shots using the **ShowMotion** toolbar and the **New View/Shot Properties** dialog box. Once a shot is created, you can give it properties that enable it to be displayed in many different ways. If a saved shot does not display in the manner you desire, it is easily modified.

A *view category* is a heading under which different views are filed. It is not necessary to create categories, especially if you will be making just a few views. On the other hand, if you are working on a complex model and need to create a number of views with a variety of cinematic and motion characteristics, it may be wise to create view categories.

The process of using show motion to create, modify, and display shots begins by first using the **ShowMotion** toolbar. All of your work with shots will be performed using the **New View/Shot Properties** dialog box. Options in this dialog box change based on the type of shot that is selected. These options are discussed in the sections that follow relating to each kind of shot.

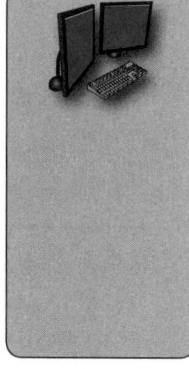

PROFESSIONAL TIP

As you first start working with show motion, you may want to create categories named Still, Cinematic, and Walk. As you create shots, file each under its appropriate category name. This will assist you in seeing how the different shot-creation techniques work. Later, you can apply the view categories to specific components of a complex model. In that case, for example, a subassembly of a 3D model may have a view category that contains all three types of shots for use in different types of modeling work or presentations. Creating and using view categories is discussed in detail at the end of this chapter.

Show Motion Toolbar

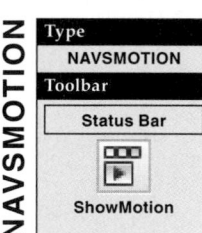

Type	
NAVSMOTION	
Toolbar	
Status Bar	
ShowMotion	

The **ShowMotion** toolbar is displayed at the bottom of the screen when the **NAVSMOTION** command is entered. See **Figure 5-1.** The quickest way to display the **NAVSMOTION** command is to pick the **ShowMotion** button on the status bar. The **ShowMotion** toolbar provides controls for creating and manipulating views:

- **Unpin ShowMotion/Pin ShowMotion**
- **Play all**
- **Stop**
- **Turn on Looping**
- **New Shot...**
- **Close ShowMotion**

When the toolbar is pinned, it remains displayed if you execute other commands, minimize the drawing, change ribbon panels, or switch to another software application. The **Unpin ShowMotion** button is used to unpin the toolbar. The **Pin ShowMotion** button is then displayed in its place. If you unpin the toolbar you must execute the **NAVSMOTION** command each time you wish to use show motion.

Press the **Play all** button to play all of the views and categories displayed as thumbnails above the toolbar. The playback of these views will loop (repeat) if the **Turn on Looping** button is selected. If looping is turned on, a view, category, or all categories are

Figure 5-1.
The **ShowMotion** toolbar is displayed at the bottom of the screen and provides controls for creating and manipulating shots.

ShowMotion toolbar

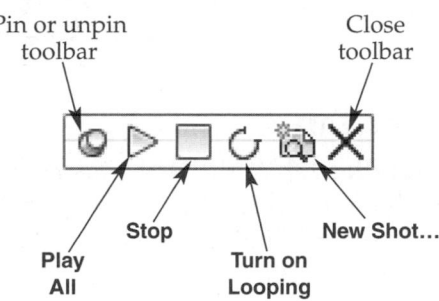

Pin or unpin toolbar

Close toolbar

Play All

Stop

Turn on Looping

New Shot...

displayed in a loop whenever the **Play all** button is selected. The **Turn on Looping** button is a toggle. The image changes to indicate whether or not looping is turned on.

The **New Shot...** button opens the **New View/Shot Properties** dialog box. This dialog box is discussed in the next section. Picking the **Close ShowMotion** button closes the **ShowMotion** toolbar.

New View/Shot Properties Dialog Box

All shot creation takes place inside of the **New View/Shot Properties** dialog box. It is accessed by using the **ShowMotion** toolbar described above or using the **NEWSHOT** command. The dialog box can also be displayed from within the **View Manager** dialog box by picking the **New...** button. The **New View/Shot Properties** dialog box is shown in **Figure 5-2**.

You must supply a shot name and a shot type. A view category is not required, but can help organize shots. This feature is discussed later in the chapter. The dialog box provides two tabs containing options that enable you to define the overall view and then specify the types of movement and motion desired in the shot. These tabs are discussed in the next sections.

Toolbar

ShowMotion

New Shot...

NEWSHOT

Figure 5-2.
The **New View/Shot Properties** dialog box is used to create the three types of shots: still, walk, and cinematic.

Enter a name

Select a category

Select a view type

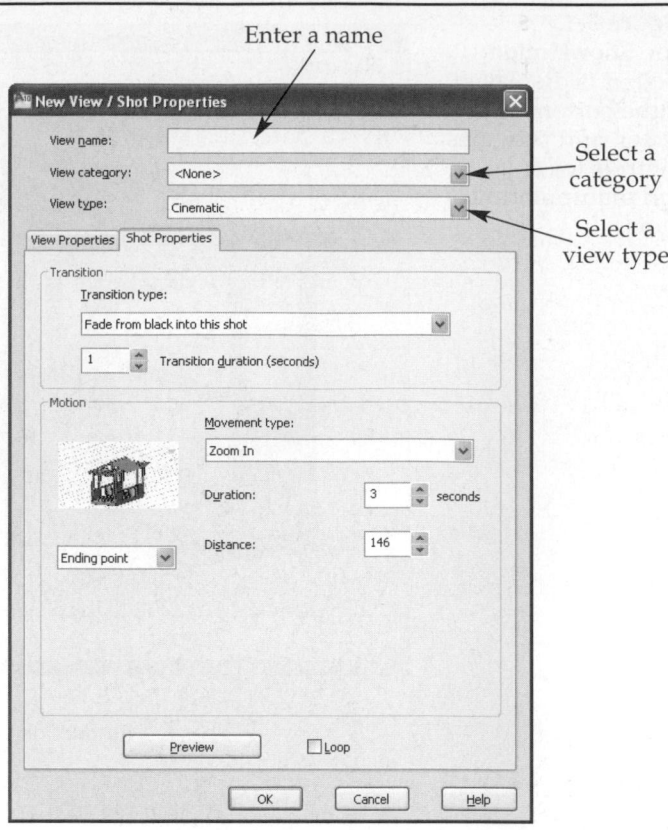

View Properties Tab

The **View Properties** tab of the **New View/Shot Properties** dialog box is composed of three areas in which you can specify the overall presentation of the model view. See **Figure 5-3.** The overall presentation includes the view boundary, visual settings, and background.

Figure 5-3.
The **View Properties** tab of the **New View/ Shot Properties** dialog box.

Boundary setting

Visual settings

Background settings

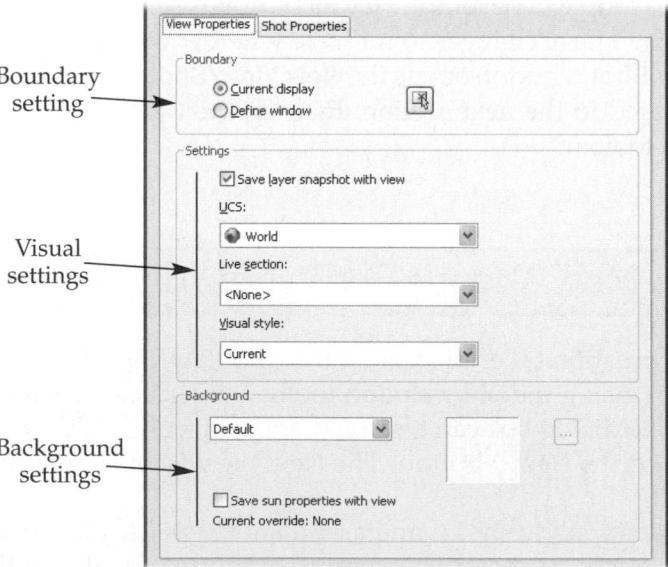

AutoCAD and Its Applications—Advanced

Boundary settings

The setting in the **Boundary** area of the **View Properties** tab determines what is displayed in the shot. The **Current display** button is on by default. This means the shot will be composed of what is currently shown in the drawing area.

You can adjust the view by picking the **Define window** radio button. This temporarily closes the dialog box so you can draw a rectangular window to define the view. After picking the second corner, you can adjust the view by picking the first and second corners of the window again. Press [Enter] to accept the window and return to the dialog box.

Visual settings

The **Settings** area of the **View Properties** tab contains options that apply to the overall display of the model in the shot. When the **Save layer snapshot with view** check box is checked, all of the current layer visibility settings are saved with the new shot. This is checked by default.

Any UCS currently defined in the drawing can be selected for use with the new shot. Use the **UCS:** drop-down list to select the UCS. When the shot is restored, that UCS is restored too. If you select <None> in the drop-down list, there is no UCS associated with the shot.

Live sectioning is a tool that enables you to view the internal features of 3D solids that are cut by a section plane object. This feature is covered in detail in Chapter 14. When a section plane object is created, it is given a name. Therefore, if a model contains one or more section plane objects, their names appear in the **Live section:** drop-down list. If a section plane object is selected in this list, the new shot shows the live sectioning for that plane.

The **Visual style:** drop-down list contains all visual styles in the drawing plus the options of Current and <None>. Selecting a visual style from the list results in that style being set current when the shot is played. Selecting Current or <None> will cause the shot to be displayed in the visual style currently displayed on the screen.

> **NOTE**
>
> A shot created with live sectioning on will be displayed in that manner even though live sectioning may be currently turned off in the drawing. However, subsequent displays of the model that were created with live sectioning off are shown with it on. Keep this in mind as you develop shots for show motion.

Background settings

The **Background** area of the **View Properties** tab provides options for changing the background of the new shot. The drop-down list in this area allows you to select a solid, gradient, image, or sun and sky background. You can also choose to retain the default background. If you pick Solid, Gradient, or Image, the **Background** dialog box appears. See **Figure 5-4.** If you select Sun & Sky, the **Adjust Sun & Sky Background** dialog box is displayed. See **Figure 5-5.** These dialog boxes are used to set the background. They are discussed in detail in Chapter 17.

The **Save sun properties with view** setting is used to have the sunlight data with the view. If you are displaying an architectural model using sunlight, you will likely want this check box checked for show motion. Sunlight and geographic location are discussed in detail in Chapter 17.

Figure 5-4.
The **Background** dialog box is used to add a background to the shot. Here, a gradient background is being created.

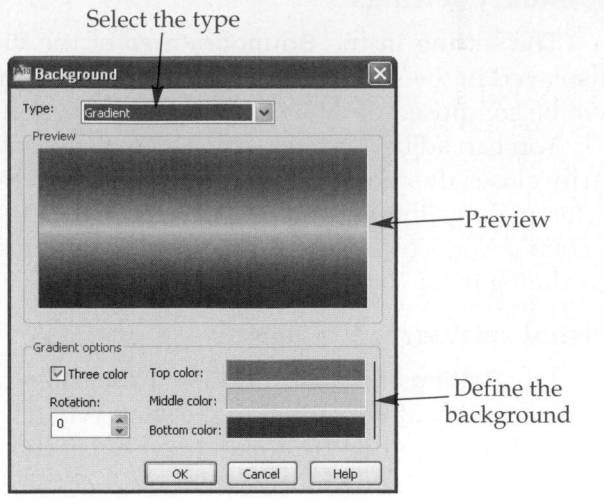

Select the type

Preview

Define the background

Figure 5-5.
When the background is set to Sun & Sky, the **Adjust Sun & Sky Background** dialog box is used to change settings for the background.

Preview

Properties

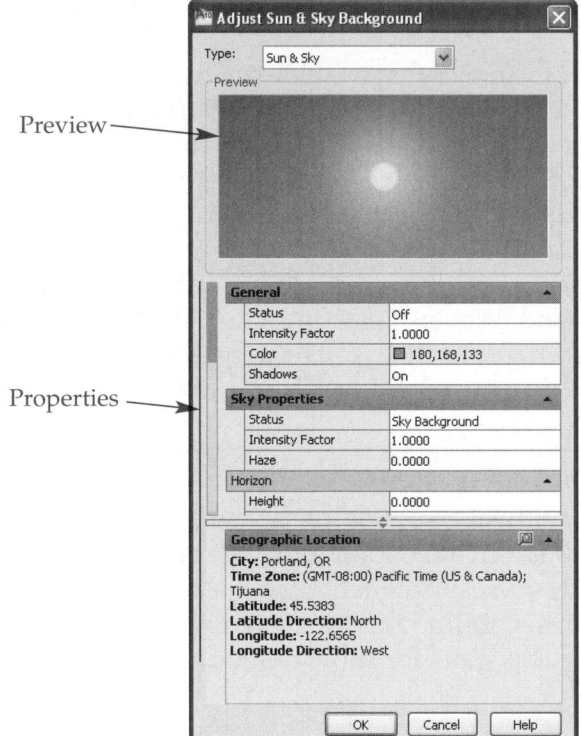

Shot Properties Tab

Settings in the **Shot Properties** tab of the **New View/Shot Properties** dialog box provide options for controlling the transition and motion of shots. There are numerous movement and motion options in this tab. The specific options available are based on the type of shot selected: still, walk, or cinematic. In addition, the cinematic shot contains a variety of motions that can be applied to the shot and each type of motion contains a number of variables. The options in the **Shot Properties** tab are discussed later in this chapter as they apply to different shots, movements, and motions.

Creating a Still Shot

A still shot is the same as a named view, but with a transition. Open the **New View/Shot Properties** dialog box and enter a name in the **View name:** text box. Next, pick in the text box for the **View category**: drop-down list and type a category name or select an existing category. Remember, it is not necessary to create or select a category at this time. Now, select Still from the **View type:** drop-down list.

In the **Shot Properties** tab, select a transition from the **Transition type:** drop-down list. You can select from one of three transitions:

- **Fade from black into this shot.** The screen begins totally black and fades into the current background color.
- **Fade from white into this shot.** The screen begins totally white and fades into the current background color.
- **Cut to shot.** The view is immediately displayed without a transition and the shot movements are applied.

The fade transitions will not function unless hardware acceleration is enabled. Turn on hardware acceleration in the following manner. Refer to **Figure 5-6.**

1. Enter the **3DCONFIG** command to display the **Adaptive Degradation and Performance Tuning** dialog box.
2. Pick the **Manual Tune** button to display the **Manual Performance Tuning** dialog box.
3. Check the **Enable hardware acceleration** check box.
4. Close both dialog boxes.

Figure 5-6.
Hardware acceleration must be activated for the fade transitions to function.

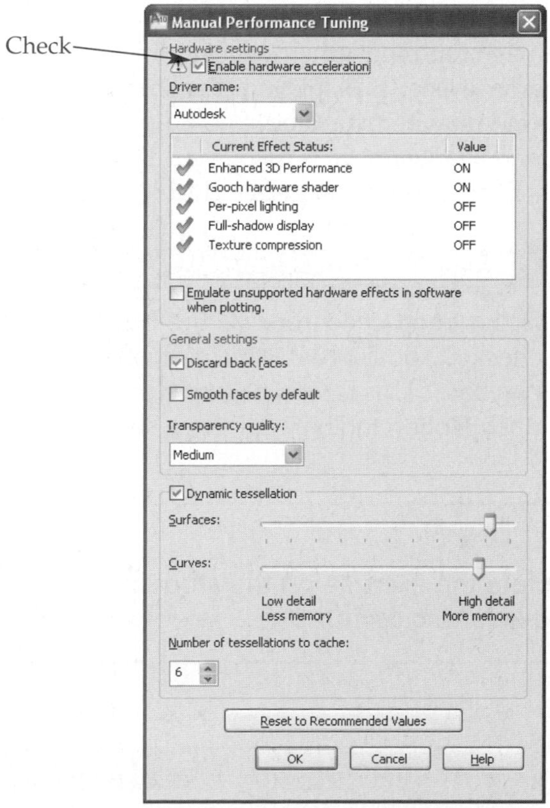

Check—

Figure 5-7.
A—The large thumbnail image on the bottom is for the category. The small thumbnail image is for the shot just created. Its name may be truncated. B—Move your cursor into the shot thumbnail image and it converts to a large image and the category thumbnail image is reduced to a small image.

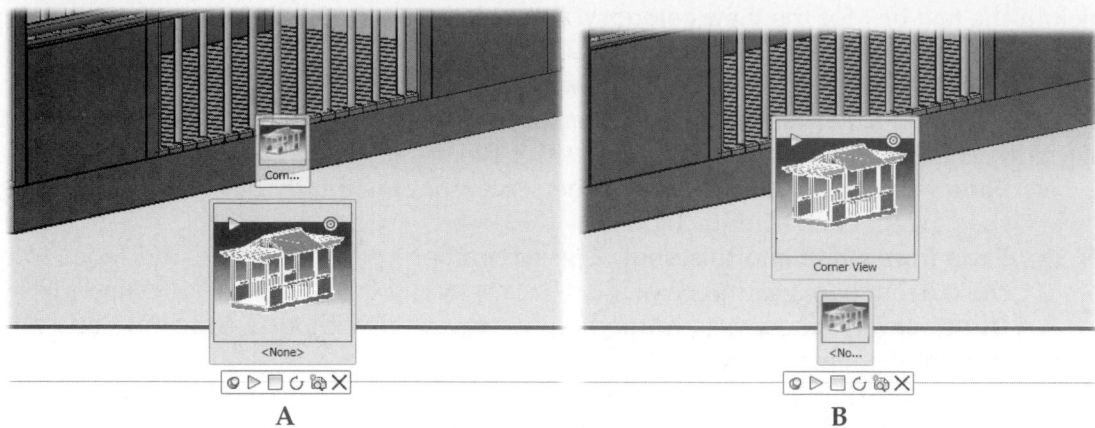

A   B

After hardware acceleration has been enabled, display the **New View/Shot Properties** dialog box, enter a name, and select a transition. Next, in the **Transition duration (seconds)** text box, enter a length of time over which the transition will occur. If you want a fade transition to be complete, be sure to enter a value in the **Duration:** text box that is equal to or greater than the transition duration. Pick the **Preview** button to view the shot, then edit the transition and motion values as needed.

When finished, pick the **OK** button to close the **New View/Shot Properties** dialog box. The thumbnail image for the new shot is displayed above the **ShowMotion** toolbar. The large thumbnail image represents the category. Since there was no view category selected for the view, the name <None> is displayed. The small thumbnail image represents the shot just created. Its name may be truncated. See **Figure 5-7A.** Move the cursor into the shot thumbnail image and a large image is displayed. The name of the shot should appear in its entirety. The category thumbnail image is reduced to a small image. See **Figure 5-7B.**

NOTE

If the **Loop** check box at the bottom of the **Shot Properties** tab is checked, the shot will continuously loop through the transition during playback. This is similar to picking the **Turn on Looping** button in the **ShowMotion** toolbar. All three shot types have this option.

Exercise 5-1

Complete the exercise on the student website.
www.g-wlearning.com/CAD

Creating a Walk Shot

To create a walk shot, open the **New View/Shot Properties** dialog box, enter a name, and select a category, if needed. Next, select Recorded Walk from the **View type:** drop-down list. In the **View Properties** tab, set up the view as described for a still shot.

In the **Shot Properties** tab, set up the transition as described for a still shot. The only option in the **Motion** area of the tab is the **Start recording** button. See **Figure 5-8.** The **Duration:** text box is grayed out because the value is based on how long you record the walk. The camera drop-down list below the preview image is also grayed out because the walk begins at the current display and ends at the point where you terminate it.

This type of shot uses the same **Walk** tool found in the steering wheels and discussed in Chapter 4. After picking the **Start recording** button, the **New View/Shot Properties** dialog box is hidden and the **Walk** tool message is displayed. Pick and drag to active the **Walk** tool. As soon as you pick, the center circle icon is displayed. If needed, you can hold down the [Shift] key to move the view up or down. When the [Shift] key is released (with the mouse button still held down), you can resume walking. The mouse button must be depressed the entire time to record all movements. As soon as you release the button, the recording ends and the dialog box is redisplayed.

Finally, preview the shot. If you need to re-record it, pick the **Start recording** button and begin again. You cannot add to the shot; you must start over. Pick the **OK** button to save the shot and a new thumbnail is displayed in the **ShowMotion** toolbar.

NOTE

The walk shot capability of show motion is limited in its ability to produce a true "walkthrough." The path of your "walk" is inexact and it may take several times to create the effect you need. Should you wish to create a genuine walkthrough of an architectural or structural model, it is better to use tools such as **3DWALK** (walk-through), **3DFLY** (flyby), or a **ANIPATH** (motion path animation). These powerful commands are used to create professional walk-throughs and animations and are covered in Chapter 19.

Figure 5-8.
Creating a walk shot.

Transition settings

Pick to record the walk

Exercise 5-2
Complete the exercise on the student website.
www.g-wlearning.com/CAD

Creating a Cinematic Shot

To create a cinematic shot, open the **New View/Shot Properties** dialog box, enter a name, and select a category, if needed. Next, select Cinematic from the **View type:** drop-down list. In the **View Properties** tab, set up the view as described for a still shot. In the **Shot Properties** tab, pick a transition type and duration as described for a still shot.

The specific options in the **Motion** area of the **Shot Properties** tab are discussed in the next sections, **Figure 5-9.** Pick a type of movement in the **Movement type:** drop-down list. The movement options available will change depending on the type of movement you select. Refer to the chart in **Figure 5-10** as a quick reference for movement options when creating a cinematic shot. Finally, pick the current position of the camera from the camera drop-down list below the preview image.

Pick the **Preview** button to view the shot. Make any changes required before picking the **OK** button to save the shot. Once the **OK** button is picked, the new view is displayed as a small thumbnail image above the large category thumbnail image. See **Figure 5-11.**

Figure 5-9.
Creating a cinematic shot.

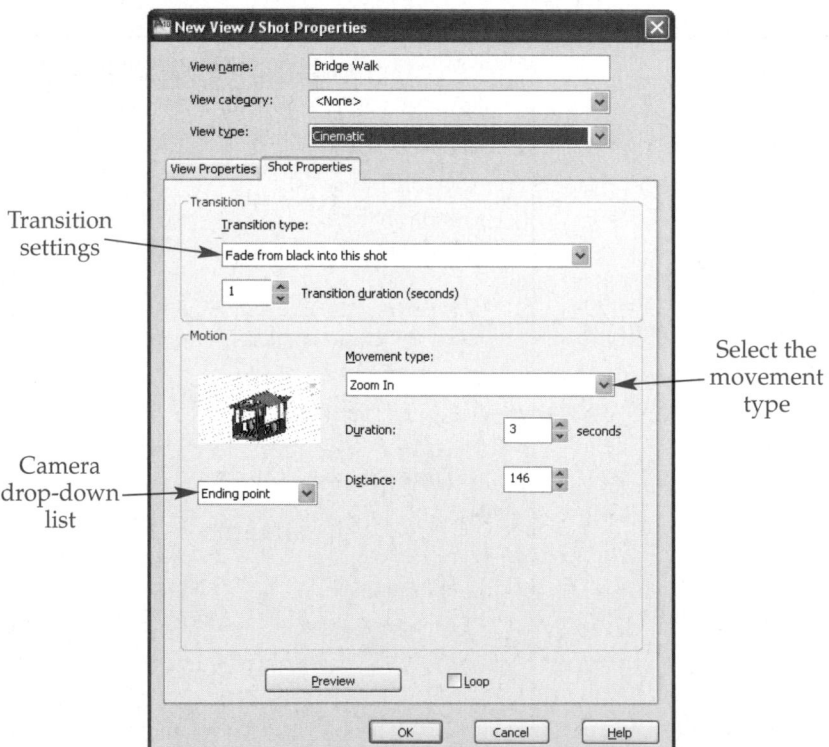

AutoCAD and Its Applications—Advanced

Figure 5-10.
This chart shows the different movement options based on the movement type selected in the **Movement type:** drop-down list in the **New View/Shot Properties** dialog box.

	Duration	Distance	Look at Camera Point	Distance Up	Distance Back	Distance Down
Zoom in	X	X				
Zoom out	X	X				
Track left	X	X	X			
Track right	X	X	X			
Crane up	X		X	X	X	
Crane down	X		X			X
Look	X					
Orbit	X					

Figure 5-11.
The thumbnail image for the new cinematic shot is displayed above the category thumbnail image.

Cinematic Basics

The motion and movement options available for a cinematic shot allow you to create a final display that appears to move into position as if the camera is traveling in a path toward the object. The **Motion** area of the **Shot Properties** tab contains a variety of options for creating an array of cinematic shots. Refer to **Figure 5-9.** The options can be confusing unless you understand some basics about essential components of a cinematic shot. The most important aspect is the preview of the current view. All of the motion actions revolve around this view. This image tile is the current position of the camera and is also referred to as the *key position* of a shot. This is the position that is displayed when you pick the **Go** button on a thumbnail image above the **ShowMotion** toolbar.

The two elements of a cinematic shot are motion and movement. *Motion* relates to the behavior of the object and how it appears to be in motion during the cinematic shot. In addition, it refers to the position of the model at a specified point in the animation. *Movement* in a cinematic shot is the manner in which the camera moves in relation to the object.

Camera Drop-Down List

The camera drop-down list is located below the preview image in the **Motion** area of the **Shot Properties** tab. The preview image represents the position of the camera based on the option selected in the camera drop-down list. The following three options are available in the drop-down list, **Figure 5-12.**

- **Ending point.** The view in the preview image is the display that will be shown at the end of the cinematic shot. All movement options take the shot to this point.
- **Starting point.** The view in the preview image is the display that will be shown at the start of the cinematic shot. All movement options begin at this point.
- **Half-way point.** The view in the preview image is the display that will be shown at the half-way point of the cinematic shot.

These options, in part, determine how the cinematic shot is created. If Starting point is selected, the cinematic shot begins with the current view. During the animation, the model may move off of the screen. If this happens, you may want to select Ending point so the model appears to move into position and stay there.

Exercise 5-3

Complete the exercise on the student website.
www.g-wlearning.com/CAD

Figure 5-12.
Selecting what the key position represents.

Select a camera position

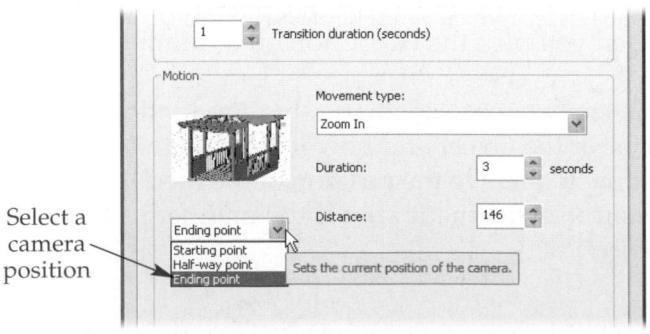

Movement Type

The **Movement type:** drop-down list in the **Shot Properties** tab of the **New View/ Shot Properties** dialog box is used to select the motion for the cinematic shot. There are eight types of camera movement that can be used with a cinematic shot:

- Zoom in.
- Zoom out.
- Track left.
- Track right.
- Crane up.
- Crane down.
- Look.
- Orbit.

The options available in the **Motion** area of the tab are based on which movement type is selected. The movement types and their options are discussed in the next sections.

Zoom in

When Zoom in is selected in the **Movement type:** drop-down list, the camera appears to zoom into the model in the shot. This movement type has two options, **Figure 5-13.** The value in the **Duration:** text box is the length of time over which the animation is recorded. The value in the **Distance:** text box is the distance the camera travels during the animation. The camera zooms in to cover the distance in the specified duration of time. When Zoom in is selected, the camera drop-down list is automatically set to Ending point. Keep in mind that with this option the current position of the camera represents the final display after the cinematic shot is complete.

Zoom out

When Zoom out is selected in the **Movement type:** drop-down list, the camera appears to zoom away from the model in the shot. This movement type has **Duration:** and **Distance:** text options, as described for the Zoom in movement type.

When Zoom out is selected, the camera drop-down list is automatically set to Starting point. With this setting, the camera will zoom out the specified distance and the final image in the shot will be smaller than the preview image. If this is not the effect you want, it may be better to pick Ending point for the current position of the camera. Then, the model will appear to move from behind your view to stop at the current view.

Track left

When Track left is selected in the **Movement type:** drop-down list, the camera will move from right to left. This results in the view moving from left to right. This movement occurs over the specified distance and duration. The Track left movement type has **Duration:** and **Distance:** text boxes, as described above, and the **Always look at camera pivot point** check box, **Figure 5-14.**

Figure 5-13.
The settings for the
Zoom in movement
type.

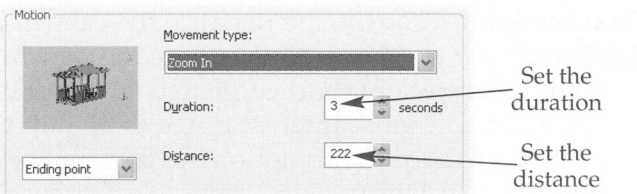

Figure 5-14.
The settings for the
Track left movement
type.

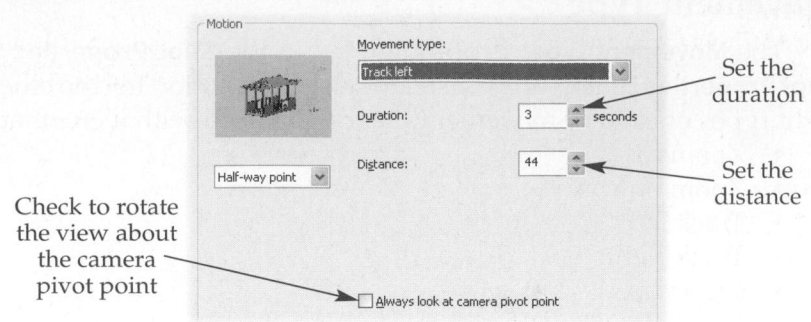

Set the
duration

Set the
distance

Check to rotate
the view about
the camera
pivot point

If the **Always look at camera pivot point** check box is checked, the center of the view remains stationary. As a result, the view in the shot appears to rotate about this point instead of sliding across the screen. The best way to visualize this motion is to use the **Preview** button with the option checked and then unchecked. This option is available for all "track" and "crane" movement types.

When Track left is selected, the camera drop-down list is automatically set to Half-way point. This means the preview image is the middle point of the animation. It will be displayed at the midpoint of the **Duration:** value.

PROFESSIONAL TIP

If the **Distance:** value is large, the screen may be blank for a few moments until the camera moves enough to bring the model into view. If this is not what you want, decrease the **Distance:** value or check the **Always look at camera pivot point** check box.

Track right

When Track right is selected in the **Movement type:** drop-down list, the camera will move from left to right. This results in the view moving from right to left. This movement type has **Distance:** and **Duration:** text boxes and the **Always look at camera pivot point** check box. These options function the same as described for Track left. When Track right is selected, the camera drop-down list is automatically set to Half-way point.

Crane up

Where the "track" movement types move the view left and right, the "crane" movement types move the view up and down. When Crane up is selected in the **Movement type:** drop-down list, the camera will move from bottom to top and then backward. This results in the view moving from top to bottom and zooming out. This movement type has **Distance:** and **Duration:** text boxes and the **Always look at camera pivot point** check box described above, which is checked by default.

This movement type has three additional options, **Figure 5-15.** The value in the **Distance Up:** text box is the distance the camera is moved upward. The value in the **Distance Back:** text box is the distance the camera is moved backward. The backward movement is typically short compared to the upward movement.

You can add more interest to the motion in the shot by shifting the view left or right. First, enable the shift option by checking the check box below the **Distance Back:** setting. Refer to **Figure 5-15.** Then, select either Shift left or Shift right from the drop-down list. Finally, enter a distance in the text box next to the drop-down list. The camera will be shifted left or right for the distance specified in this text box, resulting in the view shifting in the opposite direction. This shifting option is available for both "crane" motion types.

Figure 5-15.
The settings for the
Crane up movement
type.

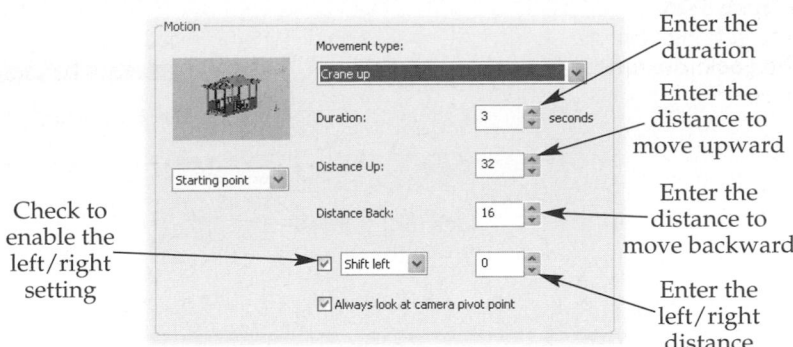

Check to
enable the
left/right
setting

Enter the
duration

Enter the
distance to
move upward

Enter the
distance to
move backward

Enter the
left/right
distance

When Crane up is selected, the camera drop-down list is automatically set to Starting point. This means the preview image shows the beginning of the shot before the movement is applied. If the **Always look at camera pivot point** check box is not checked, and depending on the movement settings, the model may move off of the screen in the shot.

Crane down

When Crane down is selected in the **Movement type:** drop-down list, the camera will move from top to bottom and then forward. This results in the view moving from bottom to top and zooming in. This movement type has **Distance:** and **Duration:** text boxes and the **Always look at camera pivot point** check box described above, which is checked by default. It also has the left/right shifting option described in the previous section.

Instead of **Distance Up:** and **Distance Back:** settings, this movement type has **Distance Down:** and **Distance Forward:** settings. The value in the **Distance Down:** text box is the distance the camera cranes down in the shot. The value in the **Distance Forward:** text box is the distance the camera is moved forward in the shot.

When Crane down is selected, the camera drop-down list is automatically set to Ending point. This means the preview image shows the end of the shot before the movement is applied. If the **Always look at camera pivot point** check box is not checked, and depending on the movement settings, the model may start off of the screen in the shot.

NOTE

All distance values related to the Crane up and Crane down movement types represent how far away from the view the camera must begin before the cinematic shot is started. For example, a large **Distance Back:** value means that the camera may have to begin at a point beyond or even around the view in order to travel the distance back to display the current view (when Ending point is selected in the camera drop-down list). It is always good practice to make a single change to the movements, then preview the shot. This is especially true for the Crane up and Crane down movement types.

Figure 5-16.
The settings for
the Look movement
type.

Pick left
or right

Pick up
or down

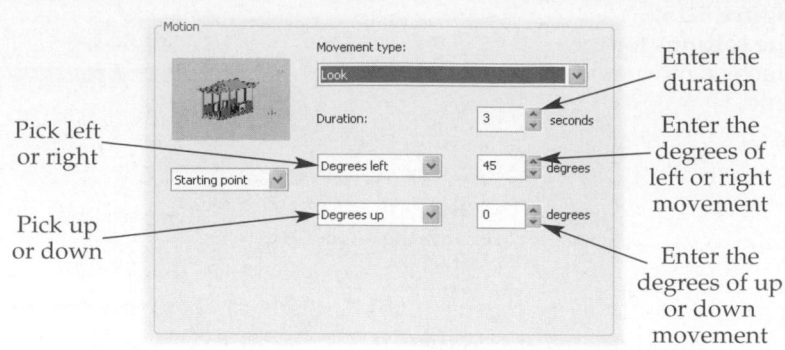

Enter the
duration

Enter the
degrees of
left or right
movement

Enter the
degrees of up
or down
movement

Look

When Look is selected in the **Movement type:** drop-down list, the camera pans based on the values for left/right and up/down to display the view, **Figure 5-16.** For example, if the movement is set to look up 45° and Ending point is selected in the camera drop-down list, then the camera begins at a 45° angle below the view and looks up to display it.

The value in the **Duration:** text box is the length of time over which the animation is recorded. This option is the same as described for the other movement types.

The first drop-down list below the **Duration:** text box is used to specify either left or right movement. Select Degrees left in the drop-down list to have the camera move from right to left, resulting in the view moving from left to right. Select Degrees right to have the camera move from left to right and the view right to left. Next, specify the angular value for this movement in the degrees text box to the right of the drop-down list. For example, suppose you select Degrees left and enter an angle of 15°. In this case, the camera will start 15° to the *right* of the model and rotate to the left in the shot.

The second drop-down list below the **Duration:** text box is used to specify either up or down movement. Select Degrees up in the drop-down list to have the camera move from bottom to top, resulting in the view moving from top to bottom. Select Degrees down to have the camera and view move in the opposite direction. Specify the angular value for this movement in the degrees text box to the right of the drop-down list.

When Look is selected in the **Movement type:** drop-down list, the camera drop-down list is automatically set to Starting point. This means the shot will start with the current view and then apply the movement settings.

Orbit

When Orbit is selected in the **Movement type:** drop-down list, the camera rotates in place based on the values set for left/right and up/down movements. The Orbit movement type has the same options as the Look movement type, as described in the previous section.

When Orbit is selected in the **Movement type:** drop-down list, the camera drop-down list is automatically set to Starting point. This means the shot will start with the current view and then apply the movement settings. Unlike the Look movement type, the view center remains stationary in the shot. As a result, you do not need to worry about the model moving off of the screen.

Displaying or Replaying a Shot

As each shot is created, its thumbnail image is placed above the **ShowMotion** toolbar. The shot name is displayed below the thumbnail image. By default, the category thumbnail image is large and each shot thumbnail image is small. If the cursor is moved over one of the shot thumbnail images, all of the shot thumbnail images are enlarged and the category thumbnail image is reduced in size.

Each thumbnail image is composed of an image of the view in the shot, the shot name, and viewing controls. See **Figure 5-17**. The viewing controls are only displayed when the cursor is moved over the thumbnail image. The three viewing controls are:

- **Play.** Plays the shot. If looping is enabled, the shot repeats. Otherwise, the shot is played once. When this button is picked it changes to **Pause**. You can also play the shot by picking anywhere on its thumbnail image. The **VIEWPLAY** command can be used to replay a shot.
- **Pause.** Pauses the shot. When this button is picked it changes to **Play**. Picking anywhere inside the image or on the **Play** button restarts playing of the shot.
- **Go.** Displays the key position of the view without playing the shot. The **VIEWGO** command also restores a named view.

You can play all of the shots in sequence by picking the **Play** button in the category thumbnail image or by picking anywhere inside of the category thumbnail image. Additionally, you can move to the key position of the first shot in a view category by picking the **Go** button on the view category thumbnail image.

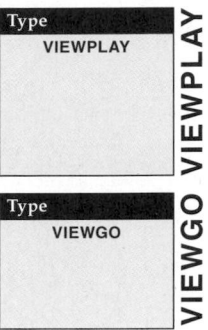

Exercise 5-4

Complete the exercise on the student website.
www.g-wlearning.com/CAD

Figure 5-17.
Each thumbnail image is composed of the shot image, the shot name, and the viewing controls.

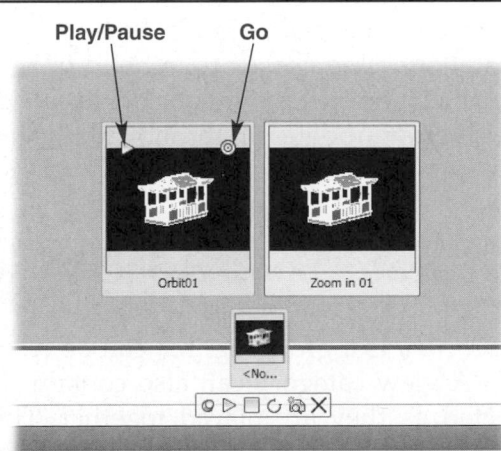

Figure 5-18.
Right-click on a shot
or view category
thumbnail image
in the **ShowMotion**
toolbar to display
this shortcut menu.

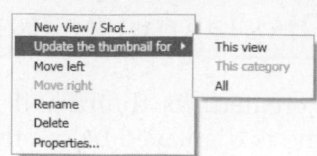

Thumbnail Shortcut Menu

Right-clicking on a shot or view category thumbnail image displays the shortcut menu shown in **Figure 5-18.** This menu provides quick access for modifying and manipulating shots and their thumbnail images.

Picking the **New View/Shot...** entry displays the **New View/Shot Properties** dialog box. Picking the **Properties...** entry displays the **View/Shot Properties** dialog box. This dialog box is the same as the **New View/Shot Properties** dialog box. However, all of the settings of the shot are displayed and can be changed. This option is only available to change the properties of a shot, not a category.

To rename a shot, pick the **Rename** entry in the shortcut menu. The name is highlighted below the thumbnail image. Type the new name and press [Enter]. To delete a shot, select **Delete** from the shortcut menu. There is no warning; the shot is simply deleted. The **UNDO** command can reverse this action.

If there is more than one shot in a category, you can rearrange the order. Right-click on a thumbnail image and pick either **Move left** or **Move right** in the shortcut menu. This moves the shot one step in the selected direction.

When a change is made to the model, it is not automatically reflected in the thumbnail images. If you do not update thumbnail images, they will remain in their original format, regardless of how many changes are made to the model. Selecting **Update the thumbnail for** in the shortcut menu displays a cascading menu with options for updating the thumbnail image:

- **This view.** Updates only the thumbnail image for the shot on which you right-clicked.
- **This category.** Updates all thumbnail images in the category. This option is only enabled when you right-click on a category thumbnail image.
- **All.** This option updates all thumbnail images in all categories and shots.

Creating and Using View Categories

A *view category* is a grouping that can be created in order to separate different types of shots. A view category can also contain shots arranged in a sequence that appear connected as they are played together. This is an optional feature, but an efficient method of separating different types of shots or grouping shots to use for a specific purpose. When a new category is created, it is represented by a category thumbnail image displayed above the **ShowMotion** toolbar.

To create a new view category, simply pick in the **View category:** drop-down list text box in the **New View/Shot Properties** dialog box. Then, enter a name. See **Figure 5-19.** The name is added to the drop-down list. If you wish to organize shots by categories, be sure to select the view category from the drop-down list before picking the **OK** button to exit the dialog box and create the shot.

Figure 5-19.
To create a new view category, type its name in the **View category:** drop-down list text box.

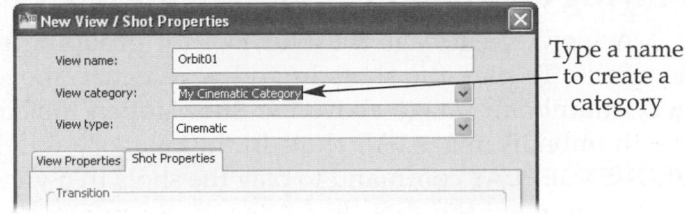

Type a name to create a category

You must create a shot to create a view category. If you delete the only shot in a view category, you also delete the category. The view category will no longer be available in the **New View/Shot Properties** dialog box.

View Category Basics

After a new view category is created, it is represented by a thumbnail image above the **ShowMotion** toolbar. The category name is shown below the thumbnail image. As you create more shots within a category, the thumbnail images for the new shots are placed to the right of existing shots above the category thumbnail image. The thumbnail image for the first shot in a view category is displayed as the view category thumbnail image.

An entire view category and all of the views included in it can be quickly deleted. Simply right-click on the view category thumbnail image and pick **Delete** in the shortcut menu. An alert box appears asking you to confirm the deletion, **Figure 5-20.** The shots in a deleted view category cannot be recovered, so be sure there are no shots in the category you wish to save before deleting. However, you can use the **UNDO** command to reverse the deletion.

PROFESSIONAL TIP

Remember, you can change the order of the shots in a view category using the shortcut menu. Picking **Move left** or **Move right** moves the shot one step. Continue moving shots until the order is appropriate.

Figure 5-20.
To remove a view category, right-click on its thumbnail image in the **ShowMotion** toolbar and pick **Delete** in the shortcut menu. This alert box appears to confirm the deletion.

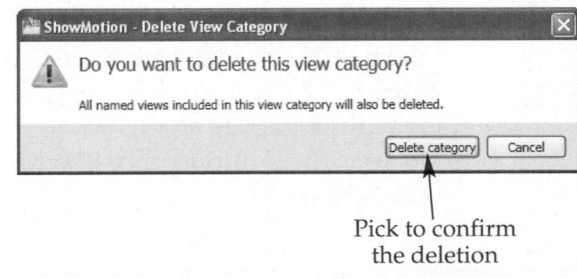

Pick to confirm the deletion

Playing and Looping Shots in a View Category

A view category is a useful tool for grouping shots to be played together in a sequence. To play the shots in a view category, move the cursor over the view category thumbnail image above the **ShowMotion** toolbar. Then, pick the **Play** button in the thumbnail image. All shots in the category will be played. You can also use the **SEQUENCEPLAY** command to play the shots in a view category.

If you want the shots in a category to run on a continuous loop, pick the **Turn on Looping** button in the **ShowMotion** toolbar. Then, when the **Play** button is picked, the shots in the view category will display until the **Pause** button is picked either on the **ShowMotion** toolbar or in the view category thumbnail image. Pressing the [Esc] key also stops playback. To return to single-play mode, pick the **Turn off Looping** button on the **ShowMotion** toolbar. This button replaces the **Turn on Looping** button.

Changing a Shot's View Category and Properties

If you put a shot in the wrong view category, it is simple to move the shot to a different category. Right-click on the shot thumbnail image and select **Properties...** in the shortcut menu. This displays the **New View/Shot Properties** dialog box. Select the proper category name in the **View category:** drop-down list and pick the **OK** button. The shot thumbnail image will move into position above its new view category thumbnail image.

To change shot properties, first right-click on the shot thumbnail image and pick **Properties...** in the shortcut menu. The **New View/Shot Properties** dialog box is displayed. All properties of the shot can be changed. Change the settings as needed. Always preview the shot before you pick the **OK** button to save it.

PROFESSIONAL TIP

Try this procedure for creating a series of shots in a view category that are to be played in sequence.
1. Play each shot to determine where they should be located in the sequence.
2. Move shots left or right in the category as needed to create the proper sequence.
3. Play the category to determine if the shots properly transition from one to the next.
4. If a subsequent shot does not begin where the previous shot ended, right-click on that shot, pick **Properties...**, and make the necessary adjustments in the **View/Shot Properties** dialog box.
5. Play the category again.
6. Repeat the editing process with each shot until the category plays smoothly.

Exercise 5-5

Complete the exercise on the student website.
www.g-wlearning.com/CAD

Chapter Test

Answer the following questions. Write your answers on a separate sheet of paper or complete the electronic chapter test on the student website.
www.g-wlearning.com/CAD

1. For what is the *show motion tool* used?
2. Define *view* as it relates to the show motion tool.
3. Define *shot* as it relates to the show motion tool.
4. List the formats in which a shot can be created.
5. List the six buttons on the **ShowMotion** toolbar.
6. What is *live sectioning* and how can it be included in a shot?
7. Which type of shot is the same as a named view, but with a transition?
8. Which type of shot requires you to navigate through the view as you record the motion?
9. What is the *key position* of a cinematic shot?
10. Define *motion* and *movement* as they relate to a cinematic shot.
11. What is the purpose of the camera drop-down list in the **Shot Properties** tab of the **New View/Shot Properties** or **View/Shot Properties** dialog box?
12. List the eight types of camera movement for a cinematic shot.
13. Briefly describe two ways to play a single shot.
14. What is a *view category?*
15. How is an entire category of shots replayed?

Drawing Problems

1. Open one of your 3D drawings from Chapter 2 or 3. Do the following.
   A. Display a pictorial view of the drawing.
   B. Create a still shot. Use settings of your choice. Create a view category and place the shot in it.
   C. Display a different view of the drawing and create another still shot. Place the shot in the view category you created.
   D. Display a third view of the drawing and create a third still shot. Place the shot in the view category you created.
   E. Play each shot. If necessary, rearrange the shots. Then, play all shots in the category.
   F. Save the drawing as P05_01.

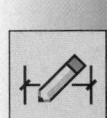

2. Open one of your 3D drawings from Chapter 2 or 3. Do the following.
   A. Display the objects in the Conceptual or Realistic visual style.
   B. Toggle the projection from parallel to perspective.
   C. Create two still shots and four cinematic shots of the model. Use a different motion type with each of the cinematic shots.
   D. Create two new categories and place one still and two cinematic shots in each category.
   E. Edit the shots in each category to create a smooth motion sequence.
   F. Save the drawing as P05_02.

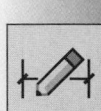

Drawing Problems - Chapter 5

3. Open one of your 3D drawings from Chapter 2 that was created with solid primitives. Do the following.
   A. Display a pictorial view.
   B. Create a walk shot. Place it in a view category named Walk Shots.
   C. Play the shot. How does this compare to a cinematic shot as far as ease of creation?
   D. Save the drawing as P05_03.

4. Open drawing P05_03 and do the following.
   A. Create a new cinematic shot using the Orbit motion type. Place it in a view category named Cinematic Shots.
   B. Create another cinematic shot using the Track left or Track right movement type. Check the **Always look at camera pivot** check box. Place the view in the Cinematic Shots category.
   C. Create a third cinematic shot using the Crane up or Crane down movement type. Make sure the **Always look at camera pivot** check box is checked. Place the view in the Cinematic Shots category.
   D. Play the Cinematic Shots category. Edit the shots as needed to create a smooth display.
   E. Create three still shots. Place them in a view category named Still Shots. Try creating three different gradient backgrounds to simulate dawn, noon, and dusk.
   F. Play the Still Shots category. Edit the shots as needed to create a smooth display.
   F. Save the drawing as P05_04.

Understanding Three-Dimensional Coordinates and User Coordinate Systems

Learning Objectives

After completing this chapter, you will be able to:

✓ Describe rectangular, spherical, and cylindrical methods of coordinate entry.
✓ Draw 3D polylines.
✓ Describe the function of the world and user coordinate systems.
✓ Move the user coordinate system to any surface.
✓ Rotate the user coordinate system to any angle.
✓ Change the user coordinate system to match the plane of a geometric object.
✓ Use a dynamic UCS.
✓ Save and manage user coordinate systems.
✓ Restore and use named user coordinate systems.
✓ Control user coordinate system icon visibility in viewports.

As you learned in Chapter 1, any point in space can be located using X, Y, and Z coordinates. This type of coordinate entry is called *rectangular coordinates.* Rectangular coordinates are most commonly used for coordinate entry. However, there are actually three ways in which to locate a point in space. The other two methods of coordinate entry are spherical coordinates and cylindrical coordinates. These two coordinate entry methods are discussed in the following sections. In addition, this chapter introduces working with user coordinate systems (UCSs).

Introduction to Spherical Coordinates

Locating a point in 3D space with *spherical coordinates* is similar to locating a point on Earth using longitudinal and latitudinal values, with the center of Earth representing the origin. Lines of longitude connect the North and South Poles and provide an east-west measurement on Earth's surface. Lines of latitude horizontally extend around Earth and provide a north-south measurement. The origin (Earth's center) can be that of the default world coordinate system (WCS) or the current user coordinate system (UCS). See **Figure 6-1A.**

When entering spherical coordinates, the longitude measurement is expressed as the angle *in* the XY plane and the latitude measurement is expressed as the angle *from* the XY plane. See **Figure 6-1B.** A distance from the origin is also provided. The coordinates

Figure 6-1.
A—Lines of longitude, representing the highlighted latitudinal segments in the illustration, run from north to south. Lines of latitude, representing the highlighted longitudinal segments, run from east to west. B—Spherical coordinates require a distance, an angle in the XY plane, and an angle from the XY plane.

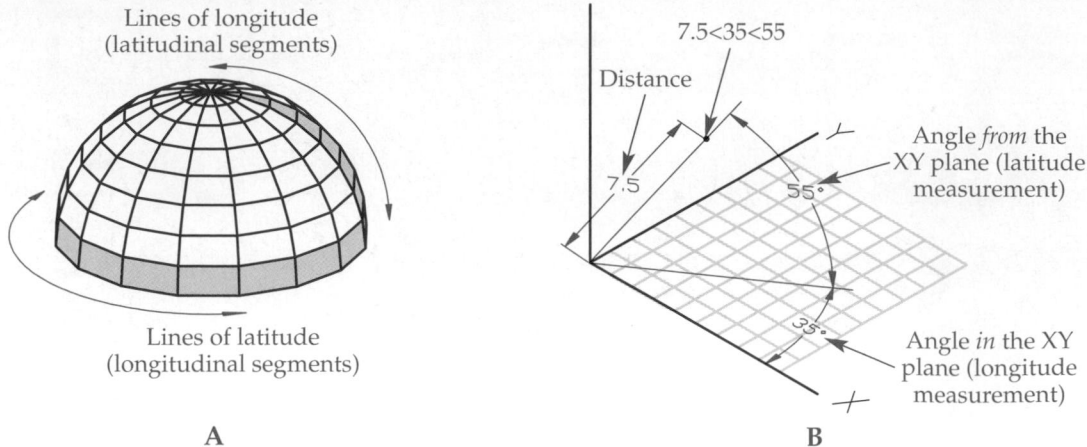

A

B

represent a measurement from the equator toward either the North or South Pole on Earth's surface. The following spherical coordinate entry is shown in **Figure 6-1B.**

7.5<35<55

This coordinate represents an absolute spherical coordinate, which is measured from the origin of the current UCS. Spherical coordinates can also be entered as relative coordinates. For example, a point drawn with the relative spherical coordinate @2<35<45 is located two units from the last point, at an angle of 35° *in* the XY plane, and at a 45° angle *from* the XY plane.

PROFESSIONAL TIP

Spherical coordinates are useful for locating features on a spherical surface. For example, they can be used to specify the location of a hole drilled into a sphere or a feature located from a specific point on a sphere. If you are working on such a spherical object, you might consider locating a UCS at the center of the sphere, then creating several different user coordinate systems rotated at different angles on the surface of the sphere. Any time a location is required, spherical coordinates can be used. Working with UCSs is introduced later in this chapter.

Using Spherical Coordinates

Spherical coordinates are well-suited for locating points on the surface of a sphere. In this section, you will draw a solid sphere and then locate a second solid sphere with its center on the surface of the first sphere.

To draw the first sphere, select the **SPHERE** command. Specify the center point as 7,5 and a radius of 1.5 units. Display a southeast isometric pictorial view of the sphere. Alternately, you can use the view cube to create a different pictorial view. Also, set the 3D Wireframe visual style current and switch to a parallel projection. Your drawing should look similar to **Figure 6-2A.**

Figure 6-2.
A—A three-unit diameter sphere shown from the southeast isometric viewpoint.
B—A .8-unit diameter sphere is drawn with its center located on the surface of the original sphere. Also, lines have been drawn between the poles of the spheres. Notice how the polar axes are parallel. C—The objects after the Conceptual visual style is set current.

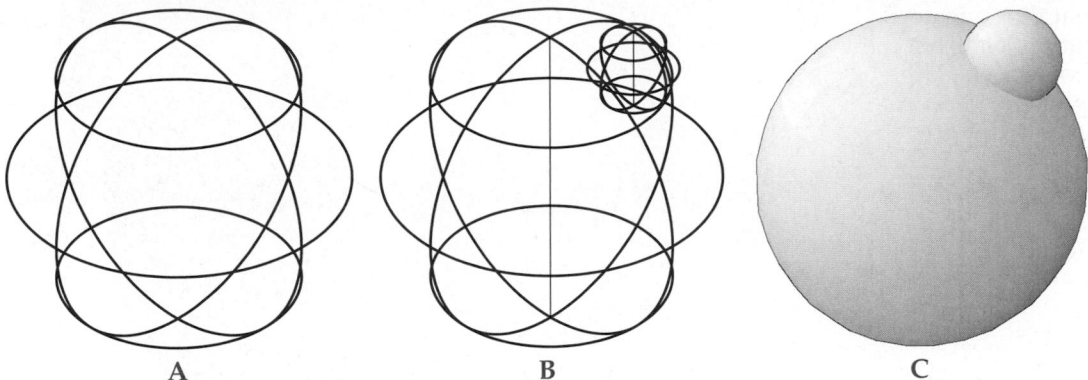

A   B   C

Since you know the radius of the sphere, but the center of the sphere is not at the origin of the current UCS (the WCS), a relative spherical coordinate will be used to draw the second sphere. The sphere you drew is a solid and, as such, you can snap to its center using object snap. Set **Center** as a running object snap and then enter the **SPHERE** command again to draw the second sphere:

> Specify center point or [3P/2P/Ttr]: **FROM**↵
> Base point: *(use the* **Center** *object snap to select the center of the existing sphere)*
> <Offset>: **@1.5<30<60**↵ *(1.5 is the radius of the first sphere)*
> Specify radius or [Diameter]: **.4**↵

The objects should now appear as shown in **Figure 6-2B.** The center of the new sphere is located on the surface of the original sphere. This is clear after setting the Conceptual visual style current, **Figure 6-2C.** If you want the surfaces of the spheres to be tangent, add the radius value of each sphere (1.5 + .4) and enter this value when prompted for the offset from the center of the first sphere:

> <Offset>: **@1.9<30<60**↵

Notice in **Figure 6-2B** that the polar axes of the two spheres are parallel. This is because both objects were drawn using the same UCS, which can be misleading unless you understand how objects are constructed based on the current UCS. Test this by locating a cone on the surface of the large sphere, just below the small sphere. First, display a 3D wireframe view of the objects. Then, select the **CONE** command and continue as follows.

> Specify center point of base or [3P/2P/Ttr/Elliptical]: **FROM**↵
> Base point: **CEN**↵
> of *(pick the large sphere)*
> <Offset>: **@1.5<30<30**↵
> Specify base radius or [Diameter]: **.25**↵
> Specify height or [2Point/Axis endpoint/Top radius]: **1**↵

The result of this construction with the Conceptual visual style set current is shown in **Figure 6-3.** Notice how the axis of the cone is parallel to the polar axis of the sphere. To draw the cone so that its axis projects from the center of the sphere, you will need to change the UCS. This is discussed later in the chapter.

Figure 6-3.
The axis lines of
objects drawn
in the same user
coordinate system
are parallel. Notice
that the cone does
not project from the
center of the large
sphere.

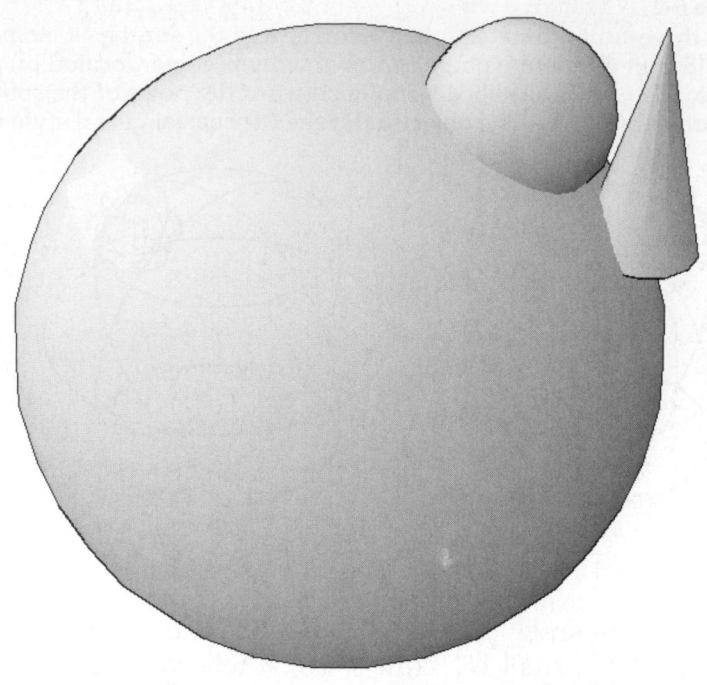

Introduction to Cylindrical Coordinates

Locating a point in space with *cylindrical coordinates* is similar to locating a point on an imaginary cylinder. Cylindrical coordinates have three values. The first value represents the horizontal distance from the origin, which can be thought of as the radius of a cylinder. The second value represents the angle in the XY plane, or the rotation of the cylinder. The third value represents a vertical dimension measured up from the polar coordinate in the XY plane, or the height of the cylinder. See **Figure 6-4.** The absolute cylindrical coordinate shown in the figure is:

7.5<35,6

Like spherical coordinates, cylindrical coordinates can also be entered as relative coordinates. For example, a point drawn with the relative cylindrical coordinate @1.5<30,4 is located 1.5 units from the last point, at an angle of 30° in the XY plane of the previous point, and at a distance of four units up from the XY plane of the previous point.

Figure 6-4.
Cylindrical
coordinates require
a horizontal distance
from the origin,
an angle in the
XY plane, and a Z
dimension.

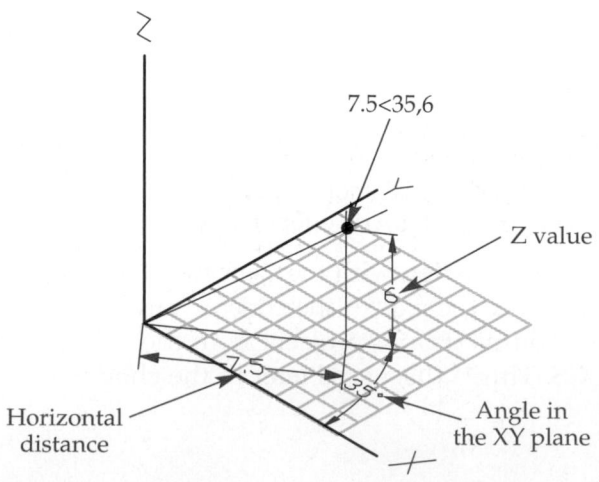

Using Cylindrical Coordinates

Cylindrical coordinates work well for attaching new objects to a cylindrical shape. An example of this is specifying coordinates for a pipe that must be attached to another pipe, tank, or vessel. In **Figure 6-5**, a pipe must be attached to a 12′ diameter tank at a 30° angle from horizontal and 2′-6″ above the floor. In order to properly draw the pipe as a cylinder, you will have to change the UCS, which you will learn how to do later in this chapter. An attachment point for the pipe can be drawn using the **POINT** command and cylindrical coordinates. First, set the **PDMODE** system variable to 3. Then, enter the **POINT** command and continue:

> Current point modes: PDMODE=3 PDSIZE=0.0000
> Specify a point: **FROM**↵
> Base point: **CEN**↵
> of *(pick the base of the cylinder)*
> <Offset>: **@6′<30,2′6″**↵ *(The radius of the tank is 6′.)*

The point can now be used as the center of the pipe (cylinder), **Figure 6-5B**. However, if you draw the pipe now, it will be parallel to the tank (large cylinder). By changing the UCS, as shown in **Figure 6-5C**, the pipe can be correctly drawn. Working with the UCS is introduced later in this chapter.

Exercise 6-1

Complete the exercise on the student website.
www.g-wlearning.com/CAD

Figure 6-5.
A—A plan view of a tank shows the angle of the pipe attachment. B—A 3D view from the southeast quadrant shows the pipe attachment point located with cylindrical coordinates. C—By creating a new UCS, the pipe can be drawn as a cylinder and correctly located without editing.

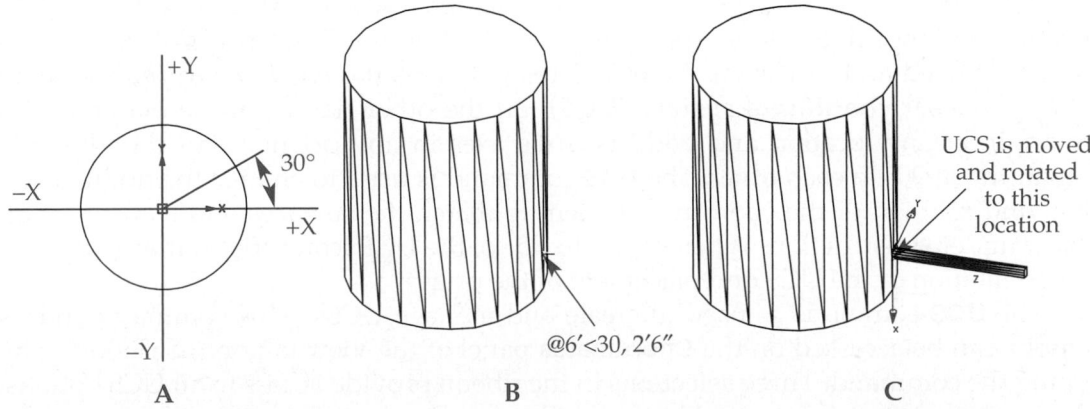

Figure 6-6.
A regular 3D polyline and the B-spline curve version after using the **PEDIT** command.

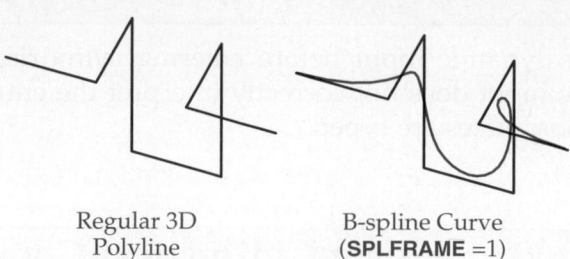

Regular 3D
Polyline

B-spline Curve
(**SPLFRAME** =1)

3D Polylines

A polyline drawn with the **PLINE** command is a 2D object. All segments of the polyline must be drawn parallel to the XY plane of the current UCS. A *3D polyline,* on the other hand, can be drawn in 3D space. The Z coordinate value can vary from point to point in the polyline.

The **3DPOLY** command is used to draw 3D polylines. Any form of coordinate entry is valid for drawing 3D polylines. If polar tracking is on when using the **3DPOLY** command, you can pick points in the Z direction if the polar tracking alignment path is parallel to the Z axis.

The **Close** option can be used to draw the final segment and create a closed shape. There must be at least two segments in the polyline to use the **Close** option. The **Undo** option removes the last segment without canceling the command.

The **PEDIT** command can be used to edit 3D polylines. The **PEDIT Spline** option is used to turn the 3D polyline into a B-spline curve based on the vertices of the polyline. A regular 3D polyline and the same polyline turned into a B-spline curve are shown in **Figure 6-6.** The **SPLFRAME** system variable controls the display of the original polyline frame and is either turned on (1) or off (0).

Exercise 6-2

Complete the exercise on the student website.
www.g-wlearning.com/CAD

Introduction to Working with User Coordinate Systems

All points in a drawing or on an object are defined with XYZ coordinate values (rectangular coordinates) measured from the 0,0,0 origin. Since this system of coordinates is fixed and universal, AutoCAD refers to it as the *world coordinate system (WCS).* A *user coordinate system (UCS),* on the other hand, can be defined with its origin at any location and with its three axes in any orientation desired, while remaining at 90° to each other. The **UCS** command is used to change the origin, position, and rotation of the coordinate system to match the surfaces and features of an object under construction. When set up to do so, the UCS icon reflects the changes in the orientation of the UCS and placement of the origin.

The **UCS** command is used to create and manage UCSs. This command and its options can be accessed on the **Coordinates** panel of the **View** tab on the ribbon or by typing the command. Three selections in the ribbon provide access to all UCS options. These selections are introduced here and discussed in detail later in this chapter.

UCS

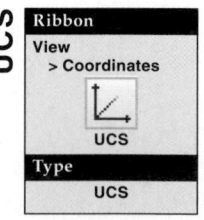

Ribbon
View
> Coordinates

UCS

Type

UCS

- **UCS.** This button executes the **UCS** command which enables you to create a named UCS, and provides access to all of the command options. The **UCS** command can also be selected by picking on the UCS drop-down list below the view cube. All of these options are covered later in this chapter.
- **Named UCS Combo Control.** This drop-down list contains the six orthographic UCS options, which are covered later in this chapter.
- **Named.** This button displays the **UCS** dialog box. The three tabs in the dialog box contain a variety of UCS and UCS icon options and settings. These options and settings are described as you progress through the chapter.

Earlier in this chapter, you used spherical coordinates to locate a small sphere on the surface of a larger sphere. You also drew a cone with the center of its base on the surface of the large sphere. However, the axis of the cone, which is a line from the center of the base to the tip of the cone, is not pointing to the center of the sphere. Refer to **Figure 6-3.** This is because the Z axes of the large sphere and cone are parallel to the world coordinate system (WCS) Z axis. The WCS is the default coordinate system of AutoCAD.

In order for the axis of the cone to project from the sphere's center point, the UCS must be changed using the **UCS** command. Working with different UCSs is discussed in the next section. However, the following is a quick overview and describes how to draw a cone with its axis projecting from the center of the sphere.

First, draw a three-unit diameter sphere with its center at 7,5. Display the drawing from the southeast isometric preset. To help see how the UCS is changing, make sure the UCS icon is displayed at the origin of the current UCS. The UCS icon drop-down list is found in the **Coordinates** panel of the **View** tab of the ribbon and provides access to three options of the **UCSICON** command. See **Figure 6-7.** Pick the **Show UCS Icon at Origin** button to ensure the icon is displayed at the origin. Also, set the 3D Wireframe visual style current.

Now, the sphere is drawn and the UCS icon is displayed at the origin of the current UCS (or at the lower-left corner of the screen, depending on the zoom level). However, the WCS is still the current user coordinate system. You are ready to start changing the UCS to meet your needs.

Begin by moving the UCS origin to the center of the sphere using the **Origin** option of the **UCS** command and the center object snap. Notice that the UCS icon is now displayed at the center of the sphere, **Figure 6-8.** Also, if the grid is displayed, the red X and green Y axes intersect at the center of the sphere.

Study **Figure 6-9,** type the **UCS** command, and continue as follows. Keep in mind that the point you are locating—the center of the cone on the sphere's surface—is 30° from the X axis and 30° from the XY plane.

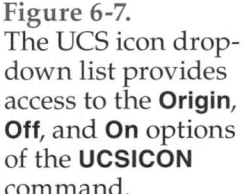

Ribbon	UCS
View > Coordinates	
Origin	
Type	
UCS	

Ribbon	UCS
View > Coordinates	
Z	
Type	
UCS	

Figure 6-7.
The UCS icon drop-down list provides access to the **Origin**, **Off**, and **On** options of the **UCSICON** command.

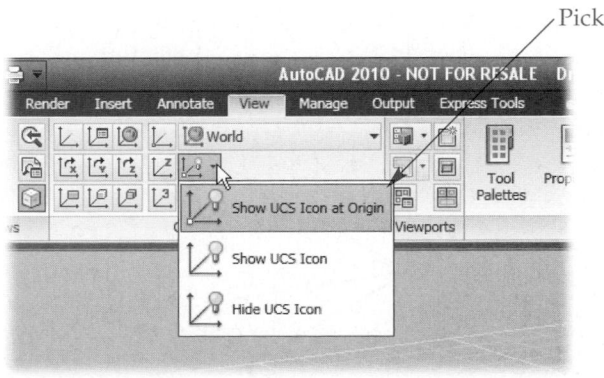

Figure 6-8.
The UCS origin is
moved to the center
of the sphere.

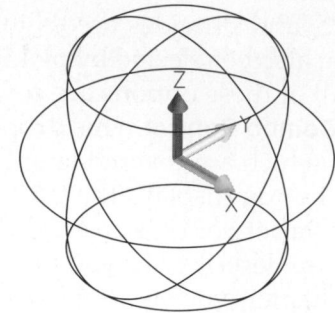

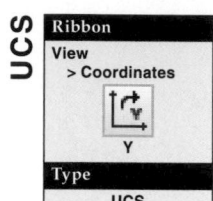

Current ucs name: *NO NAME*
Specify origin of UCS or [Face/NAmed/OBject/Previous/View/World/X/Y/Z/ZAxis]
   <World>: **Z**↵
Specify rotation angle about Z axis <90>: **30.**↵ *(See Figure 6-9B.)*
Command: *(press [Enter] or the spacebar to reissue the **UCS** command)*
Current ucs name: *NO NAME*
Specify origin of UCS or [Face/NAmed/OBject/Previous/View/World/X/Y/Z/ZAxis]
   <World>: **Y.**↵
Specify rotation angle about Y axis <90>: **60.**↵ *(See Figure 6-9D.)*

If the view cube is set to be oriented to the current UCS, then it rotates when the UCS is changed. Additionally, the grid rotates to match the new UCS. If the grid is not on, turn it on by pressing [Ctrl]+[G]. Remember, the grid is displayed on the XY plane of the current UCS.

Figure 6-9.
A—The world
coordinate system.
B—The new UCS is
rotated 30° in the XY
plane about the Z axis.
C—A line rotated up
30° from the XY plane
represents the axis of
the cone. D—The UCS
is rotated 60° about the
Y axis. The centerline
of the cone coincides
with the Z axis of this
UCS.

Figure 6-10.
A—A new UCS is created with the Z axis projecting from the center of the sphere.
B—A cone is drawn using the new UCS. The axis of the cone projects from the center of the sphere.
C—The objects after the Conceptual visual style is set current.

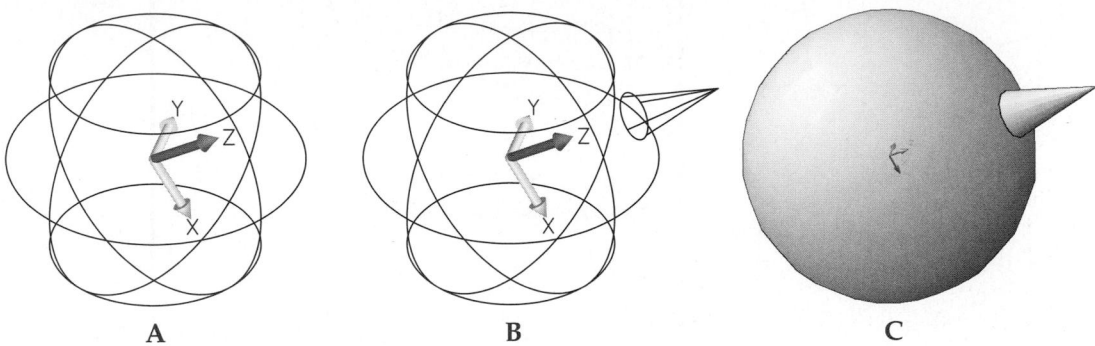

A   B   C

The new UCS can be used to construct a cone with its axis projecting from the center of the sphere. **Figure 6-10A** shows the new UCS located at the center of the sphere. With the UCS rotated, rectangular coordinates can be used to draw the cone. Enter the **CONE** command and specify the center as 0,0,1.5 (the radius of the sphere is 1.5 units). Enter a radius of .25 and a height of 1. The completed cone is shown in **Figure 6-10B.** You can see that the axis projects from the center of the sphere. **Figure 6-10C** shows the objects after setting the Conceptual visual style current.

This same basic procedure can be used in the tank and pipe example presented earlier in this chapter. To correctly locate the pipe (cylinder), first rotate the UCS 30° about the Z axis. Then, rotate the UCS 90° about the Y axis. The Z axis of this new UCS aligns with the long axis of the pipe. Finally, use rectangular coordinates to draw the cylinder with its center at the point drawn in **Figure 6-5B.**

Once you have changed to a new UCS, you can quickly return to the WCS by picking WCS in the view cube drop-down list or by using the **World** option of the **UCS** command. The WCS provides a common "starting place" for creating new UCSs.

Exercise 6-3

Complete the exercise on the student website.
www.g-wlearning.com/CAD

Working with User Coordinate Systems

Once you understand a few of the basic options of user coordinate systems, creating 3D models becomes an easy and quick process. The following sections show how to display the UCS icon, change the UCS in order to work on different surfaces of a model, and name and save a UCS. As you saw in the previous section, working with UCSs is easy.

Displaying the UCS Icon

The symbol that identifies the orientation of the coordinate system is called the *UCS icon.* When AutoCAD is first launched based on the acad3D.dwt template, the UCS icon is located at the WCS origin in the middle of the viewport. The display of this symbol is controlled by the **UCSICON** command. If your drawing does not require

Figure 6-11.
Setting UCS and
UCS icon options in
the **UCS** dialog box.

UCS icon
options

UCS
options

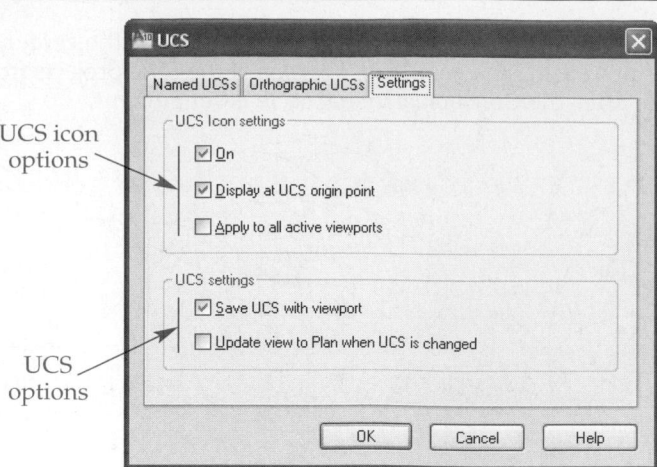

viewports and altered coordinate systems, you may want to turn the icon off using the **Off** option of the command. The icon disappears until you turn it on again using the **On** option of the command.

You can also turn the icon on or off and set the icon to display at the origin using the options in the **Settings** tab of the **UCS** dialog box. Refer to **Figure 6-11.** This dialog box is displayed by picking the **Named** button on the **Coordinates** panel in the **View** tab of the ribbon or typing the **UCSMAN** command.

PROFESSIONAL TIP

It is recommended that you have the UCS icon turned on at all times when working in 3D drawings. It provides a quick indication of the current UCS.

Modifying the UCS Icon

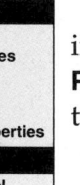

UCSICON

Ribbon
View
> Coordinates

UCS Icon Properties
Type
UCSICON

The appearance of the wireframe UCS icon can be changed using the settings in the **UCS Icon** dialog box. See **Figure 6-12.** This dialog box is accessed using the **Properties** option of the **UCSICON** command. You can modify three characteristics of the UCS icon.

Figure 6-12.
The **UCS Icon**
dialog box allows
you to change the
appearance of the
wireframe UCS icon.

Set style
for icon

Set size

Set icon
color

Set line
thickness

Preview
of settings

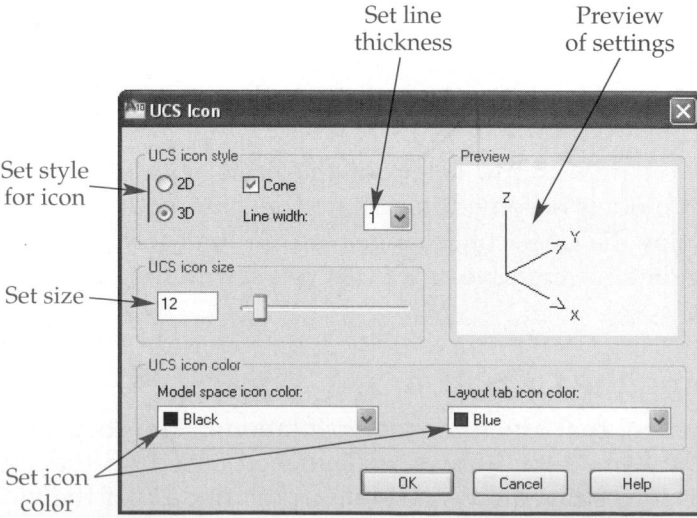

- **Style.** Select either a 2D or 3D icon in the **UCS icon style** area. The 2D style was used by earlier releases of AutoCAD. If the 3D style is selected, which is the default, the **Cone** and **Line width:** options are available. The line width can be one, two, or three pixels.
- **Size.** The **UCS icon size** area contains a text box and a slider. The value in the text box is the size of the UCS icon expressed as a percentage of the viewport size. Enter a new value in the text box or adjust the slider.
- **Color.** Use the drop-down lists in the **UCS icon color** area to set the color of the UCS icon. Notice that different colors can be set for model space and paper space (layout tab).

NOTE

The settings in the **UCS Icon** dialog box have no effect on the shaded 3D UCS icon. This icon is displayed when the 3D Wireframe, 3D Hidden, Conceptual, or Realistic visual style is set current. Also, if a perspective projection is set current when the dialog box is displayed, the preview is shown as dashed lines without arrowheads.

Changing the Coordinate System

To construct a three-dimensional object, you must draw shapes at many different angles. Different planes are needed to draw features on angled surfaces. To construct these features, it is easiest to rotate the UCS to match any surface on an object. The following example illustrates this process.

The object in **Figure 6-13** has a cylinder on the angled surface. A solid modeling command called **EXTRUDE**, which is discussed in Chapter 9, is used to create the base of the object. The cylinder is then drawn on the angled feature. In Chapter 8, you will learn how to dimension the object as shown in **Figure 6-13**.

The first step in creating this model is to draw the side view of the base as a wireframe. You could determine the X, Y, and Z coordinates of each point on the side view and enter the coordinates. However, a lot of typing can be saved if all points share a Z value of 0. By rotating the UCS, you can draw the side view entering only X and Y coordinates. Start a new drawing and display the southeast isometric view. If the UCS icon is off, turn it on and display it at the origin.

Now, rotate the UCS 90° about the X axis. The new UCS is parallel to the side of the object. The UCS icon is displayed at the origin of the UCS. If needed, pan the view so the UCS icon is near the center.

Next, use the **PLINE** command to draw the outline of the side view. Refer to the coordinates shown in **Figure 6-14**. When entering coordinates, you may want to turn off dynamic input so direct-distance entry is disabled. The **PLINE** command is used

---

Figure 6-13.
This object can be constructed by changing the orientation of the coordinate system.

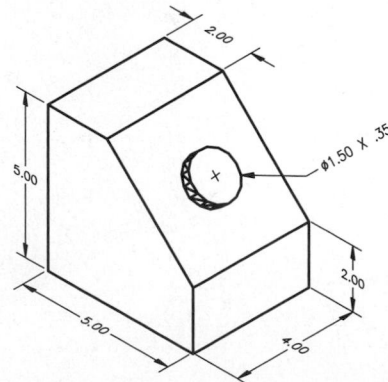

Figure 6-14.
A wireframe of one side of the base is created. Notice the orientation of the UCS.

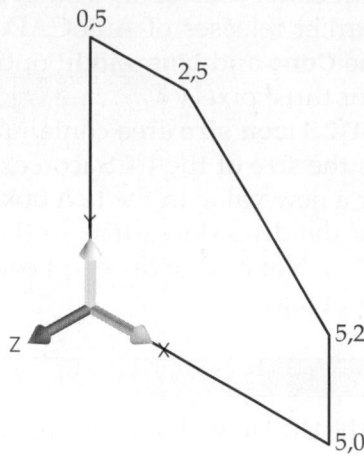

instead of the **LINE** command because a closed polyline can be extruded into a solid. Be sure to use the **Close** option to draw the final segment. A wireframe of one side of the object is created. Notice the orientation of the UCS icon.

Now, the **EXTRUDE** command is used to create the base as a solid. This command is covered in detail in Chapter 9. On the same UCS used to create the wireframe side, enter the command:

Command: **EXTRUDE**↵
Current wire frame density: ISOLINES = *current*
Select objects to extrude: *(pick the polyline)*
Select objects to extrude: ↵
Specify height of extrusion or [Direction/Path/Taper angle]: **−4**↵
Command:

By entering a negative value for the height of the extrusion, the resulting object extends behind (negative Z) the XY plane of the current UCS. You can also move the cursor so the preview extends below the UCS XY plane and enter positive 4. The base is created as a solid. See **Figure 6-15.** You may want to switch to a parallel projection, as shown in the figure.

Figure 6-15.
The wireframe is extruded to create the base as a solid.

AutoCAD and Its Applications—Advanced

Figure 6-16.
Saving a new UCS.
Select **Rename** to
enter a name and
save the Unnamed
UCS.

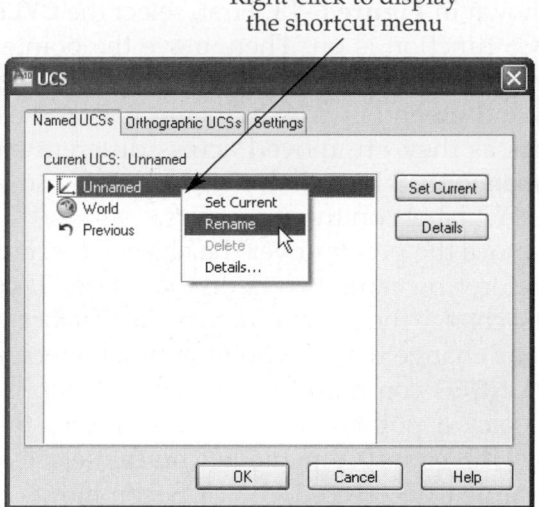

Right-click to display
the shortcut menu

Saving a Named UCS

Once you have created a new UCS that may be used again, it is best to save it for future use. For example, you just created a UCS used to draw the wireframe of one side of the object. You can save this UCS using the **Save** option of the **UCS** command or the **UCS** dialog box.

If using the dialog box, right-click on the entry Unnamed and pick **Rename** in the shortcut menu. See **Figure 6-16.** You can also pick once or double-click on the highlighted name. Then, type the new name in place of Unnamed and press [Enter]. A name can have up to 255 characters. Numbers, letters, spaces, dollar signs ($), hyphens (–), and underscores (_) are valid. Use this method to save a new UCS or to rename an existing one. Now, the coordinate system is saved and can be easily recalled for future use.

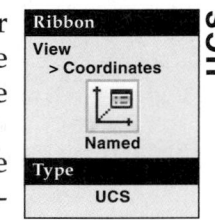

UCS

Ribbon
View
> Coordinates

Named

Type
UCS

PROFESSIONAL TIP

Most drawings can be created by rotating the UCS as needed without saving it. If the drawing is complex with several planes, each containing a large amount of detail, you may wish to save a UCS for each detailed face. Then, restore the proper UCS as needed. For example, when working with architectural drawings, you may wish to establish a different UCS for each floor plan and elevation view and for roofs and walls that require detail work.

Dynamic UCS

A powerful tool for 3D modeling is the *dynamic UCS function*. A dynamic UCS is a UCS temporarily located on any existing face of a 3D model. The function is activated by picking the **Allow/Disallow Dynamic UCS** button in the status bar, pressing the [Ctrl]+[D] key combination, or setting the **UCSDETECT** system variable to 1. When the pointer is moved over a model surface, the XY plane of the UCS is aligned with that surface. This is especially useful when adding primitives or shapes to model surfaces. In addition, dynamic UCSs are useful when inserting blocks and xrefs, locating text, editing 3D geometry, editing with grips, and area calculations.

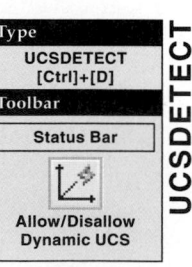

UCSDETECT

Type
UCSDETECT
[Ctrl]+[D]

Toolbar
Status Bar

Allow/Disallow
Dynamic UCS

An example of using a dynamic UCS is to draw the cylinder on the angled face of the object shown in **Figure 6-13.** First, select the **CYLINDER** command. Make sure the dynamic UCS function is on. Then, move the pointer over one of the surfaces of the object. Notice that the 3D crosshairs change when they are moved over a new surface. The red (X) and green (Y) crosshairs are flat on the face. For ease of visualizing the 3D crosshairs as they are moved across different surfaces, right-click on the **Allow/ Disallow Dynamic UCS** button in the status bar and select **Display crosshair labels** to turn on the XYZ labels on the crosshairs.

As you move the pointer over the object faces, note that hidden faces are not highlighted, therefore you cannot work on those faces. If you wish to work on a hidden face you must first change the viewpoint to make that face visible. The view cube can be used to dynamically change the viewpoint without interrupting the current command.

The **CYLINDER** command is currently prompting to select a center point of the base. If you pick a point, this sets the center of the cylinder base *and* temporarily relocates the UCS so its XY plane lies on the selected face. Once the point is selected and the dynamic UCS created, the UCS icon moves to the temporary UCS. When the command is ended, the UCS and UCS icon revert to their previous locations. To locate a 1.5″ diameter cylinder in the center of the angled face, use the following procedure.

1. At the "specify center point of base" prompt, [Shift] + right-click in the drawing area and pick **Mid Between 2 Points** from the shortcut menu. See **Figure 6-17A.**
2. Use the **Endpoint** snap to pick two opposite corners of the angled face. See **Figure 6-17B.** You can also use the **Midpoint** snap and pick the midpoint of the two sides or top and bottom edges. The UCS is temporarily moved to the angled face at the pick point, which is the center of the face.
3. Specify the 1.5 unit diameter for the base.
4. Specify a cylinder height of .35 units. See **Figure 6-17C.**
5. The cylinder is properly located on the angled face and the UCS automatically returns to the previous location. See **Figure 6-17D.**

When setting a dynamic UCS, experiment with the behavior of the crosshairs as they are moved over different surfaces. The orientation of the crosshairs is related to the edge of the face that they are moved over. Can you determine the pattern by which the crosshairs are turned? The X axis of the crosshairs is always aligned with the edge that is crossed.

PROFESSIONAL TIP

If you want to temporarily turn off the dynamic UCS function while working in a command, press and hold the [Shift]+[Z] key combination while moving the pointer over a face. As soon as you release the keys, dynamic UCS function is reinstated.

Exercise 6-4

Complete the exercise on the student website.
www.g-wlearning.com/CAD

AutoCAD and Its Applications—Advanced

Figure 6-17.
Using a dynamic UCS allows you to draw a cylinder on the angled face of the object shown here without creating a new UCS. A—To set the center point of the base and select the angled face for the dynamic UCS, use the **Mid Between 2 Points** snap and select corners or midpoints of the face. B—Set the radius or diameter of the base. C—Set the height of the cylinder. D—When the command is ended, the previous UCS is restored.

Pick

Pick

Specify the radius or diameter

UCS icon is moved to the origin of the temporary UCS

Pick

A

B

Previous UCS is restored

C

D

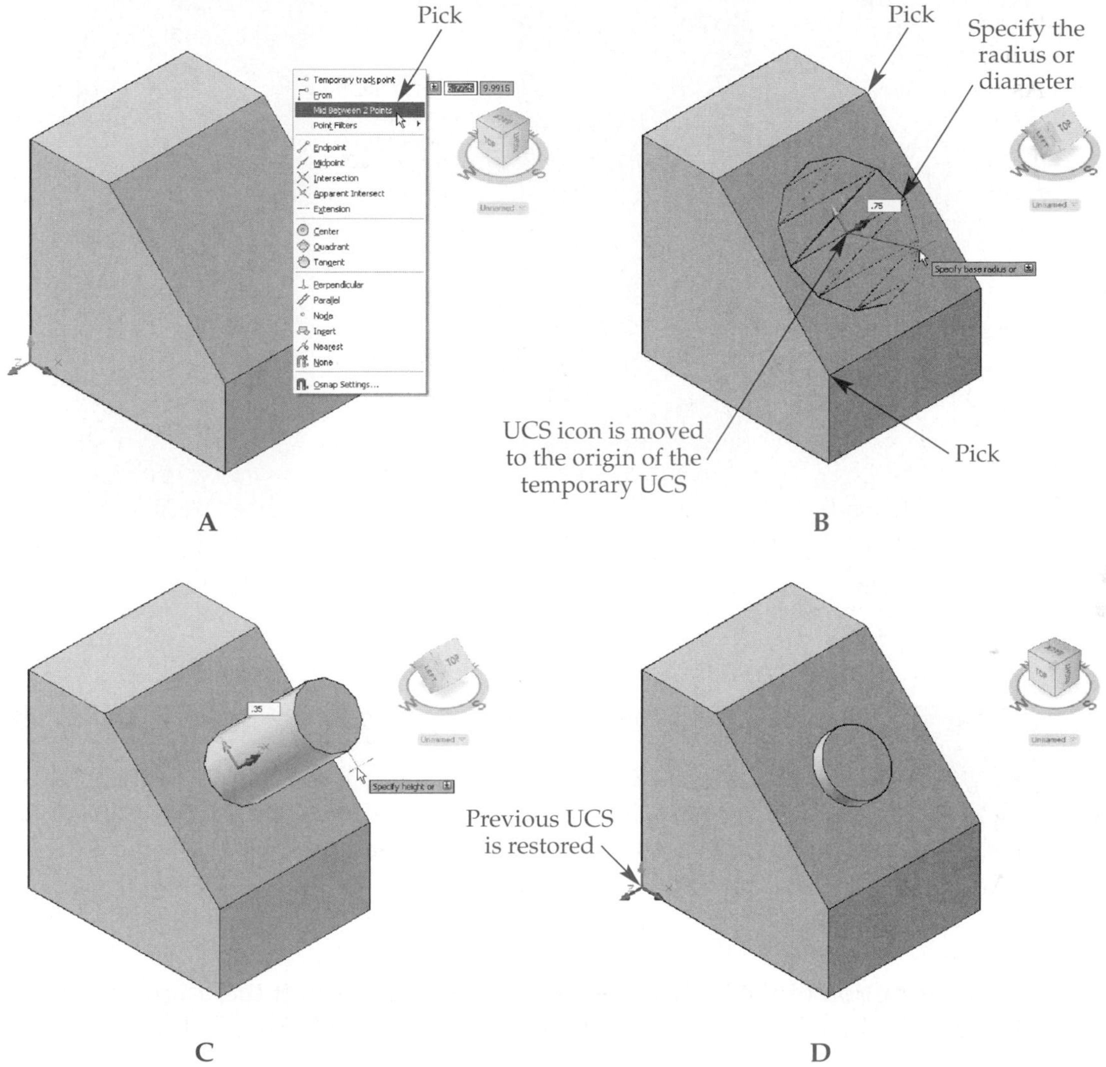

Additional Ways to Change the UCS

There are other ways to change the UCS. These options include picking three points, selecting a new Z axis, and setting the UCS to an existing object. The next sections cover these options.

Selecting Three Points to Create a New UCS

The **3 Point** option of the **UCS** command can be used to change the UCS to any flat surface. This option requires that you first locate a new origin, then a point on the positive X axis, and finally a point on the XY plane that has a positive Y value. Refer to **Figure 6-18.** Use object snaps to select points that are not on the current XY plane. After you pick the third point—the point on the XY plane—the UCS icon changes its orientation to align with the plane defined by the three points.

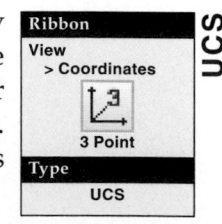

Ribbon

View
> Coordinates

3 Point

Type

UCS

UCS

Figure 6-18.
A—A new UCS can be established by picking three points. P1 is the origin, P2 is on the positive X axis, and P3 is on the XY plane and has a positive Y value. B—The new UCS is created.

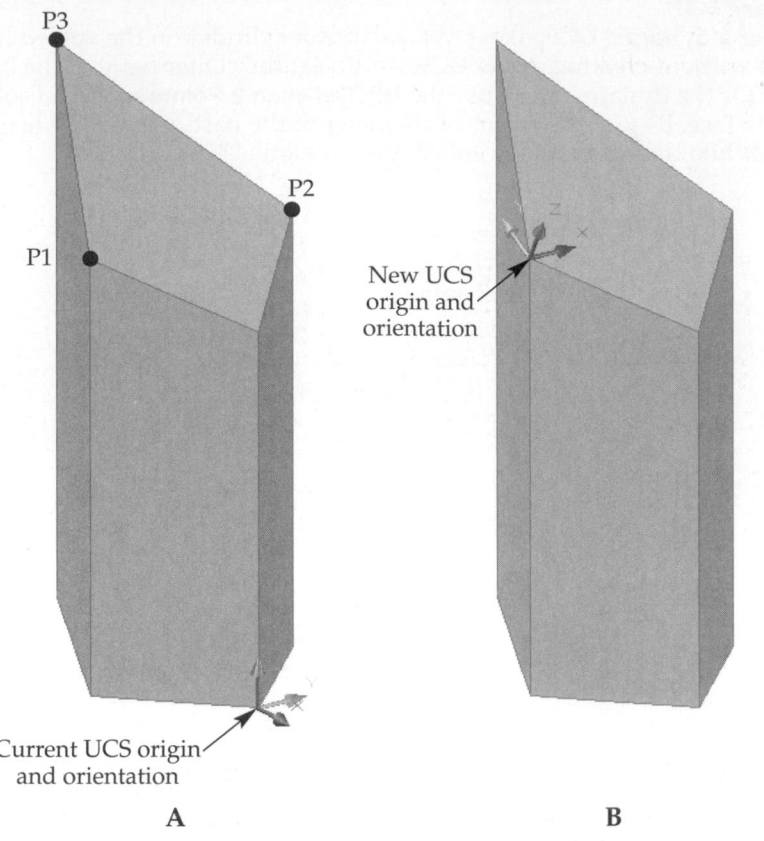

Current UCS origin and orientation

New UCS origin and orientation

A

B

PROFESSIONAL TIP

When typing the **UCS** command, enter 3 at the Specify origin of UCS or [Face/NAmed/OBject/Previous/View/World/X/Y/Z/ZAxis] <World>: prompt. Notice that the option is not listed in the prompt.

Selecting a New Z Axis

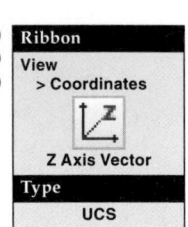

The **ZAxis** option of the **UCS** command allows you to select the origin point and a point on the positive Z axis. Once the new Z axis is defined, AutoCAD sets the new X and Y axes.

You will now add a cylinder to the lower face of the base created earlier. The cylinder extends into the base. Refer to the location of the UCS in **Figure 6-17D.** This is the UCS after adding the cylinder to the angled face with a dynamic UCS. The Z axis does not project perpendicular to the lower face. Therefore, a new UCS must be created on the lower-right face. Change the UCS after entering the **ZAxis** option as follows.

1. Pick the origin of the new UCS. See **Figure 6-19A.** You may have to use an object snap to select the origin.
2. Pick a point on the positive portion of the new Z axis.
3. The new UCS is established and it can be saved if necessary.

Now, use auto-tracking or object snaps to draw a ∅.5" cylinder centered on the lower face and extending 3" into the base. Then, subtract the cylinder from the base part to create the hole, as shown in **Figure 6-19B.**

Figure 6-19.
A—Using the **ZAxis** option to establish a new UCS. B—The new UCS is used to create a cylinder, which is then subtracted from the base to create a hole.

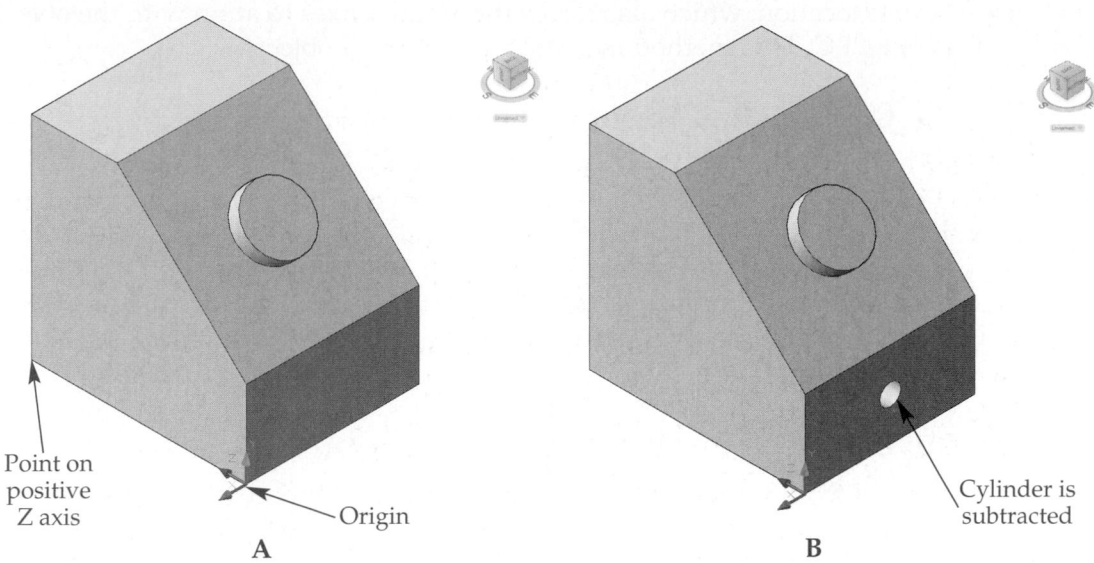

Point on positive Z axis — Origin

A

Cylinder is subtracted

B

Setting the UCS to an Existing Object

The **Object** option of the **UCS** command can be used to define a new UCS on an object. However, there are some objects on which this option cannot be used: 3D polylines, 3D meshes, and xlines. There are also certain rules that control the orientation of the UCS. For example, if you select a circle, the center point becomes the origin of the new UCS. The pick point on the circle determines the direction of the X axis. The Y axis is relative to X and the UCS Z axis is the same as the Z axis of the selected object.

Look at **Figure 6-20A**. The circle is rotated an unknown number of degrees from the XY plane of the WCS. However, you need to create a UCS in which the circle is lying on the XY plane. Select the **Object** option of the **UCS** command and then pick the circle. The UCS icon may look like the one shown in **Figure 6-20B**. Notice how the X and Y axes are not aligned with the quadrants of the circle, as indicated by the grip locations. This may not be what you expected. The X axis orientation is determined by the pick point on the circle. Notice how the X axis is pointing at the pick point.

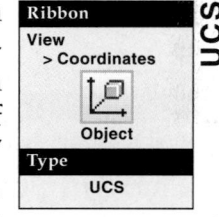

Ribbon	
View	
> Coordinates	
Object	
Type	
UCS	

UCS

Figure 6-20.
A—This circle is rotated off of the WCS XY plane by an unknown number of degrees. It will be used to establish a new UCS. B—The circle is on the XY plane of the new UCS. However, the X and Y axes do not align with the circle's quadrants. C—The **ZAxis** option of the **UCS** command is used to align the UCS with the quadrants of the circle.

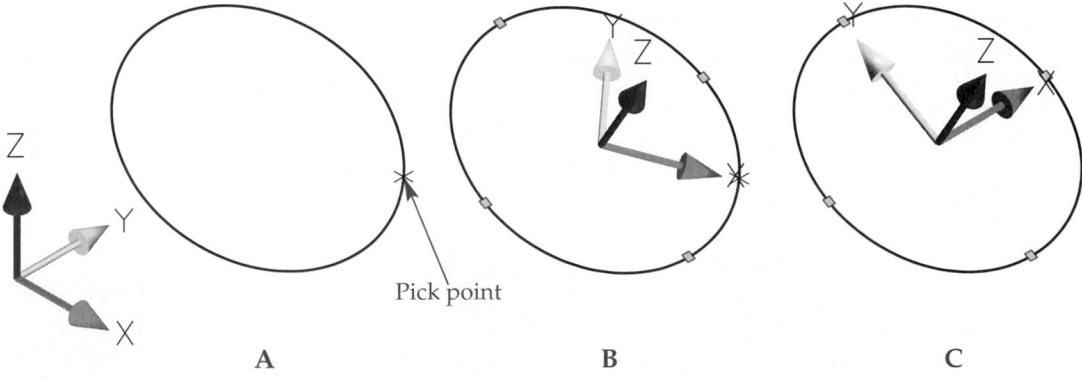

Pick point

A   **B**   **C**

To rotate the UCS in the current plane so the X and Y axes of the UCS are aligned with the quadrants of the circle, use the **ZAxis** option of the **UCS** command. Select the center of the circle as the origin and then enter the absolute coordinate 0,0,1. This uses the current Z axis location, which also forces the X and Y axes to align with the object. Refer to **Figure 6-20C**. This method may not work with all objects.

Setting the UCS to the Face of a 3D Solid

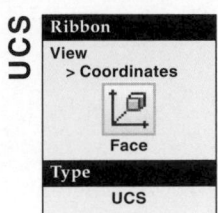

The **Face** option of the **UCS** command allows you to orient the UCS to any face on a 3D solid object. This option does not work on surface objects. Select the command and then pick a face on the solid. After you have selected a face on a 3D solid, you have the options of moving the UCS to the adjacent face or flipping the UCS 180° on the X, Y, or both axes. Use the **Next**, **Xflip**, or **Yflip** options to move or rotate the UCS as needed. Once you achieve the UCS orientation you want, press [Enter] to accept. Notice in **Figure 6-21** how many different UCS orientations can be selected for a single face.

Setting the UCS Perpendicular to the Current View

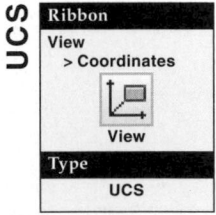

You may need to add notes or labels to a 3D drawing that are plan to the current view, such as that shown in **Figure 6-22**. The **View** option of the **UCS** command makes this easy to do. Immediately after selecting the **View** option, the UCS rotates to a position so the new XY plane is perpendicular to the current line of sight. Now, anything added to the drawing is plan to the current view. The command works on the current viewport only; other viewports are unaffected.

Applying the Current UCS to a Viewport

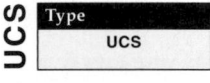

The **Apply** option of the **UCS** command allows you to apply the UCS in the current viewport to any or all model space or paper space viewports. Using the **Apply** option, you can have a different UCS displayed in every viewport or you can apply one UCS to all viewports. With the viewport that contains the UCS to apply active, enter the **Apply** option. This option is only available on the command line. However, the option does not appear in the command prompt. Enter either A or APPLY to select the option. Then, pick a viewport to which the current UCS will be applied and press [Enter]. To apply the current UCS to all viewports, enter the **All** option.

Preset UCS Orientations

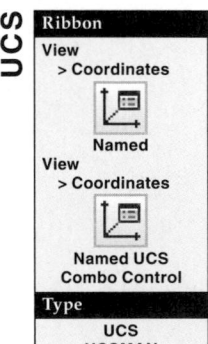

AutoCAD has six preset orthographic UCSs that match the six standard orthographic views. With the current UCS as the top view (plan), all other views are arranged as shown in **Figure 6-23**. These orientations can be selected by using the **Named UCS Combo Control**

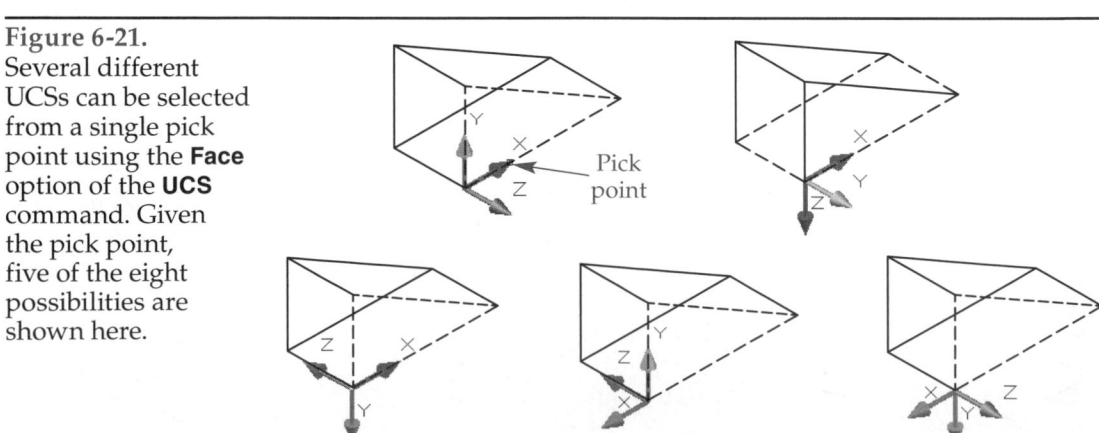

Figure 6-21.
Several different UCSs can be selected from a single pick point using the **Face** option of the **UCS** command. Given the pick point, five of the eight possibilities are shown here.

Figure 6-22.
The **View** option of the **UCS** command **allows** you to place text plan to the current view.

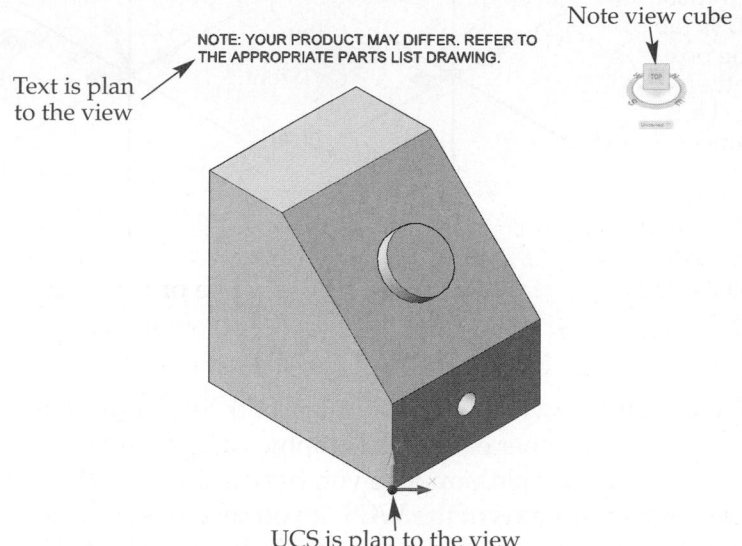

Note view cube

NOTE: YOUR PRODUCT MAY DIFFER. REFER TO
THE APPROPRIATE PARTS LIST DRAWING.

Text is plan
to the view

UCS is plan to the view

drop-down list in the **Coordinates** tab of the **View** panel in the ribbon, entering the command on the command line, or using the **Orthographic UCSs** tab of the **UCS** dialog box. When using the command line, type the name of the UCS (FRONT, BACK, RIGHT, etc.) at the first prompt:

> Specify origin of UCS or [Face/NAmed/OBject/Previous/View/World/X/Y/Z/ZAxis]
> <World>: **FRONT** *or* **FR**↵

Note that there is no orthographic option listed for the command.

Figure 6-23.
The standard orthographic UCSs coincide with the six basic orthographic views.

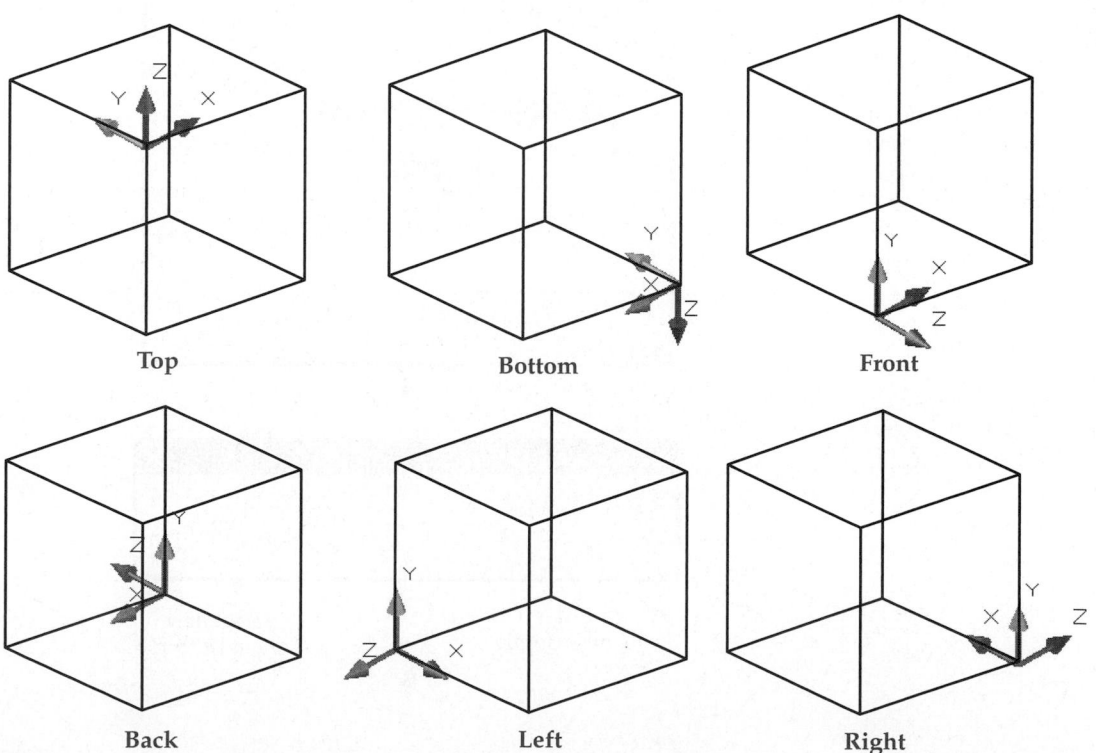

Top

Bottom

Front

Back

Left

Right

Figure 6-24.
The **Relative to:** drop-down list entry in the **Orthographic UCSs** tab of the **UCS** dialog box determines whether the orthographic UCS is based on a named UCS or the WCS. The UCS icon here represents the named UCS. A—Relative to the WCS. B—Relative to the named UCS.

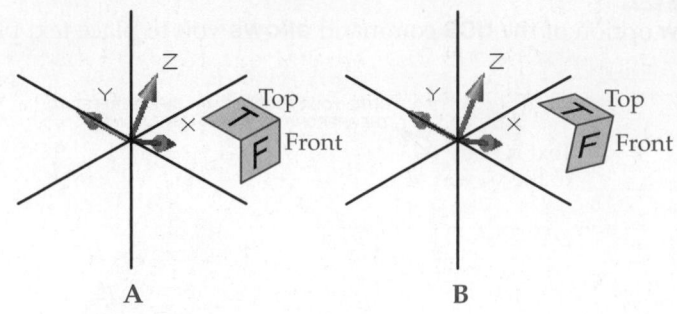

A   B

The **Relative to:** drop-down list at the bottom of the **Orthographic UCSs** tab of the **UCS** dialog box specifies whether each orthographic UCS is relative to a named UCS or absolute to the WCS. For example, suppose you have a saved UCS named Front Corner that is rotated 30° about the Y axis of the WCS. If you set current the top UCS relative to the WCS, the new UCS is perpendicular to the WCS, **Figure 6-24A.** However, if the top UCS is set current relative to the named UCS Front Corner, the new UCS is also rotated from the WCS, **Figure 6-24B.**

The Z value, or depth, of a preset UCS can be changed in the **Orthographic UCSs** tab of the **UCS** dialog box. First, right-click on the name of the UCS you wish to change. Then, pick **Depth** from the shortcut menu, **Figure 6-25A.** This displays the **Orthographic UCS depth** dialog box. See **Figure 6-26B.** You can either enter a new depth value or specify the new location on screen by picking the **Select new origin** button. Once the new depth has been selected, it is reflected in the preset UCS list.

Figure 6-25.
A—The Z value, or depth, of a preset UCS can be changed by right-clicking on its name and selecting **Depth**. B—Enter a new depth value or pick the **Select new origin** button to pick a new location on screen.

Right-click to display the shortcut menu

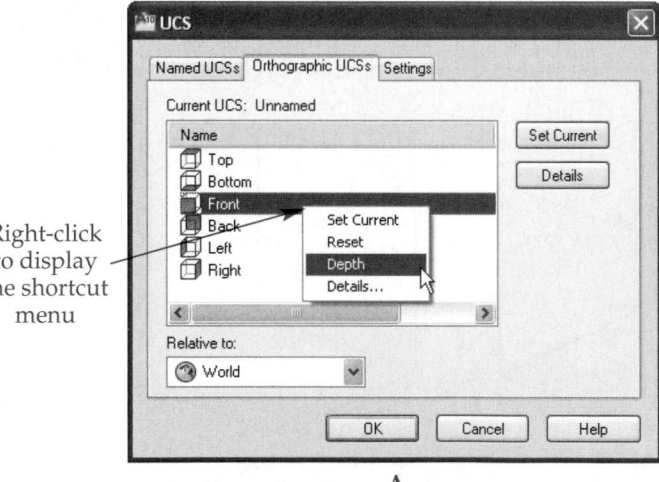

A

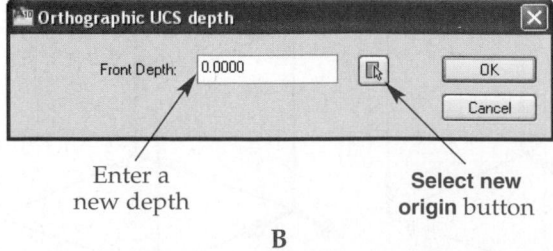

Enter a new depth

Select new origin button

B

PROFESSIONAL TIP

Changing the Relative to: setting affects *all* preset UCSs and *all* preset viewpoints! Therefore, leave this set to World unless absolutely necessary to change it.

Managing User Coordinate Systems and Displays

You can create, name, and use as many user coordinate systems as needed to construct your model or drawing. As you saw earlier, AutoCAD allows you to name (save) coordinate systems for future use. User coordinate systems can be created, renamed, set current, and deleted using the **Named UCSs** tab of the **UCS** dialog box, **Figure 6-26.**

The **Named UCSs** tab contains the **Current UCS:** list box. This list box contains the names of all saved coordinate systems plus World. If other coordinate systems have been used in the current drawing session, Previous appears in the list. Unnamed appears if the current coordinate system has not been named. The current UCS is indicated by a small triangle to the left of its name in the list. To make any of the listed coordinate systems active, highlight the name and pick the **Set Current** button.

A list of coordinate and axis values of the highlighted UCS can be displayed by picking the **Details** button. This displays the **UCS Details** dialog box shown in **Figure 6-27.**

If you right-click on the name of a UCS in the list in the **Named UCSs** tab, a shortcut menu is displayed. Using this menu, you can rename the UCS. Saving a UCS is discussed earlier in this chapter. You can also set the UCS current or delete it using the shortcut menu. The Unnamed UCS cannot be deleted, nor can World be deleted.

Figure 6-26.
The **UCS** dialog box allows you to rename, list, delete, and set current an existing UCS.

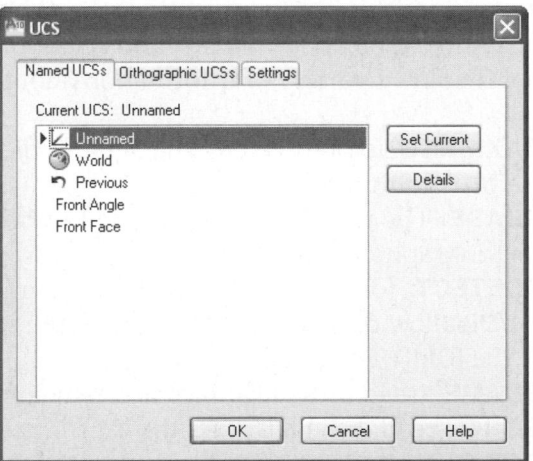

Figure 6-27.
The **UCS Details** dialog box displays the coordinate values of the selected UCS.

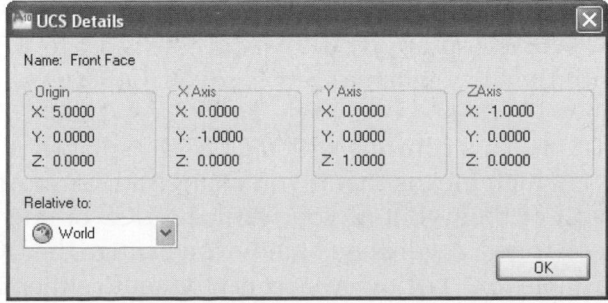

Chapter 6   Understanding Three-Dimensional Coordinates and User Coordinate Systems   **161**

Setting an Automatic Plan Display

After changing the UCS, a plan view is often needed to give you a better feel for the XYZ directions. While you should try to draw in a pictorial view when possible as you construct a 3D object, some constructions may be much easier in a plan view. AutoCAD can be set to automatically make your view of the drawing plan to the current UCS. This is especially useful if you will be changing the UCS often, but want to work in a plan view.

The **UCSFOLLOW** system variable is used to automatically display a plan view of the current UCS. When it is set to 1, a plan view is automatically created in the current viewport when the UCS is changed. Viewports are discussed in Chapter 7. The default setting of **UCSFOLLOW** is 0 (off). After setting the variable to 1, a plan view will be automatically generated the next time the UCS is changed. The **UCSFOLLOW** variable generates the plan view only after the UCS is changed, not immediately after the variable is changed. However, if you select a different viewport, the previous viewport is set plan to the UCS if **UCSFOLLOW** has been set to 1 in that viewport. The **UCSFOLLOW** variable can be individually set for each viewport.

UCS Settings and Variables

As discussed in the previous section, the **UCSFOLLOW** system variable allows you to change how an object is displayed in relation to the UCS. There are also system variables that display a variety of information about the current UCS. These variables include:

- **UCSAXISANG.** (stored value) The default rotation angle for the **X**, **Y**, or **Z** option of the **UCS** command.
- **UCSBASE.** The name of the UCS used to define the origin and orientation of the orthographic UCS settings. It can be any named UCS.
- **UCSDETECT.** (on or off) Turns the dynamic UCS function on and off. The **Allow/Disallow Dynamic UCS** button on the status bar controls this variable, as does the [Ctrl]+[D] key combination.
- **UCSNAME.** (read only) Displays the name of the current UCS.
- **UCSORG.** (read only) Displays the XYZ origin value of the current UCS.
- **UCSORTHO.** (on or off) If set to 1 (on), the related orthographic UCS setting is automatically restored when an orthographic view is restored. If turned off, the current UCS is retained when an orthographic view is restored. Depending on your modeling preferences, you may wish to set this variable to 0.
- **UCSVIEW.** (on or off) If this variable is set to 1 (on), the current UCS is saved with the view when a view is saved. Otherwise, the UCS is not saved with the view.
- **UCSVP.** This controls which UCS is displayed in viewports. The default value is 1, which means that if you change a UCS in one viewport, the UCS changes in all of them. But, if you want the UCS of one or more viewports to remain unchanged, regardless of how you change the UCS in other viewports, set this variable to 0. Each viewport can be set to either 0 or 1.

- **UCSXDIR.** (read only) Displays the XYZ value of the X axis direction of the current UCS.
- **UCSYDIR.** (read only) Displays the XYZ value of the Y axis direction of the current UCS.

UCS options and variables can also be managed in the **Settings** tab of the **UCS** dialog box. See **Figure 6-11.** The settings in this tab are:
- **Save UCS with viewport.** If checked, the current UCS settings are saved with the viewport and the **UCSVP** system variable is set to 1. This variable can be set for each viewport in the drawing. Viewports in which this setting is turned off, or unchecked, will always display the UCS settings of the current active viewport.
- **Update view to Plan when UCS is changed.** This setting controls the **UCSFOLLOW** system variable. When checked, the variable is set to 1. When unchecked, the variable is set to 0.

Chapter Test

Answer the following questions. Write your answers on a separate sheet of paper or complete the electronic chapter test on the student website.
www.g-wlearning.com/CAD

1. Explain *spherical coordinate entry.*
2. Explain *cylindrical coordinate entry.*
3. A new point is to be drawn 4.5″ from the last point. It is to be located at a 63° angle in the XY plane, and at a 35° angle from the XY plane. Write the proper spherical coordinate notation.
4. Write the proper cylindrical coordinate notation for locating a point 4.5″ in the horizontal direction from the origin, 3.6″ along the Z axis, and at a 63° angle in the XY plane.
5. Name the command that is used to draw 3D polylines.
6. Why is the command in question 5 needed?
7. Which command is used to change a 3D polyline into a B-spline curve?
8. How does the **SPLFRAME** system variable affect the B-spline curve created with the command in question 7?
9. What is the *WCS?*
10. What is a *user coordinate system (UCS)?*
11. What effect does the **Origin** option of the **UCSICON** command have on the UCS icon display?
12. Describe how to rotate the UCS so that the Z axis is tilted 30° toward the WCS X axis.
13. How do you return to the WCS from any UCS?
14. Which command controls the display of the user coordinate system icon?
15. What is a *dynamic UCS* and how is one activated?
16. What is the function of the **3 Point** option of the **UCS** command?
17. How do you automatically create a display that is plan to a new UCS?
18. What do you do so that the UCS icon is displayed at the origin of the current user coordinate system?
19. How do you move the UCS along the current Z axis?
20. What is the function of the **Object** option of the **UCS** command?
21. The **Face** option of the **UCS** command can be used on which types of objects?
22. What is the function of the **Apply** option of the **UCS** command?
23. In which dialog box is the **Orthographic UCSs** tab located?
24. Which command displays the **UCS** dialog box?
25. What appears in the **Named UCSs** tab of the **UCS** dialog box if the current UCS has not been saved?

Drawing Problems

For Problems 1–4, draw each object using solid primitives and Boolean commands to create composite solids. Measure the objects directly to obtain the necessary dimensions. Plot the drawings at a 3:1 scale using display methods specified by your instructor. Save the drawings as *P06_(problem number)*.

1.

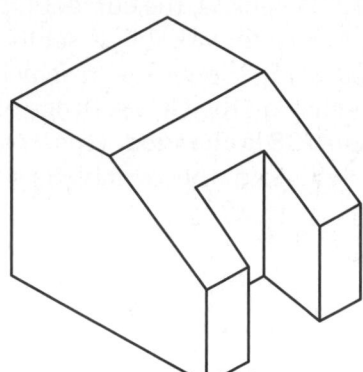

2.

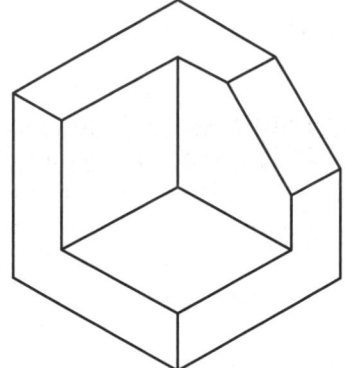

3.

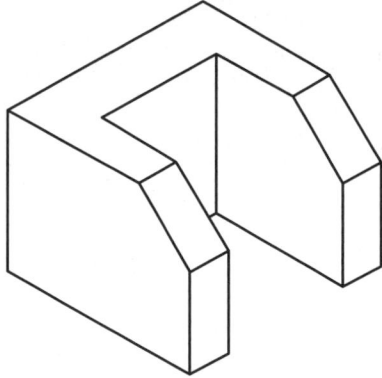

4.

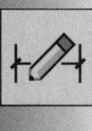

AutoCAD and Its Applications—Advanced

5. Create the mounting bracket shown below. Save the file as P06_05.

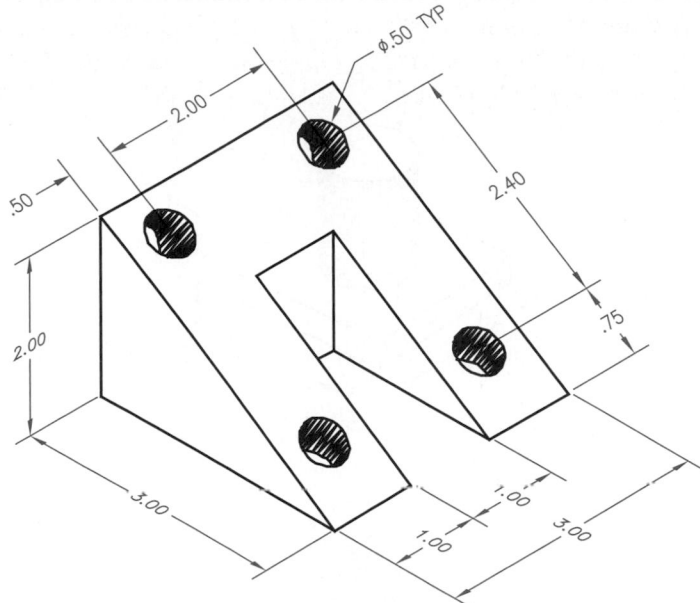

6. Create the computer speaker as shown below. The large-radius, arched surface is created by drawing a three-point arc. The second point of the arc passes through the point located by the .26 and 2.30 dimensions. Save the file as P06_06.

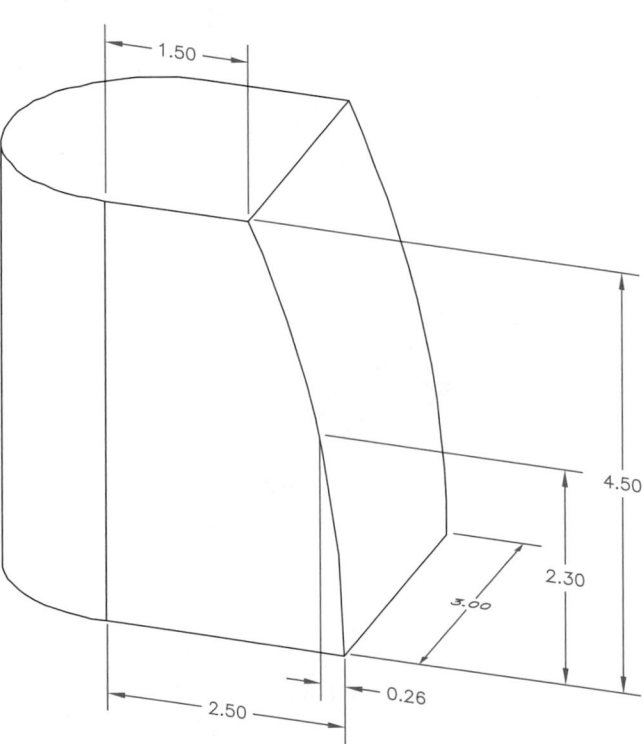

For Problems 7–9, draw each object using solid primitives and Boolean commands to create composite solids. Use the dimensions provided. Save the drawings as *P06_(problem number)*.

7.

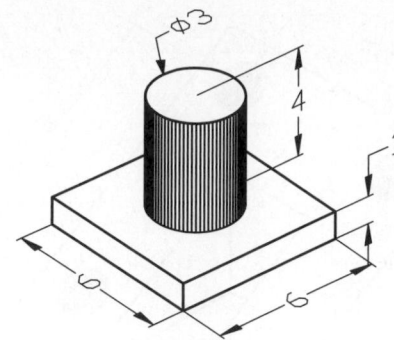

Pedestal #1

8.

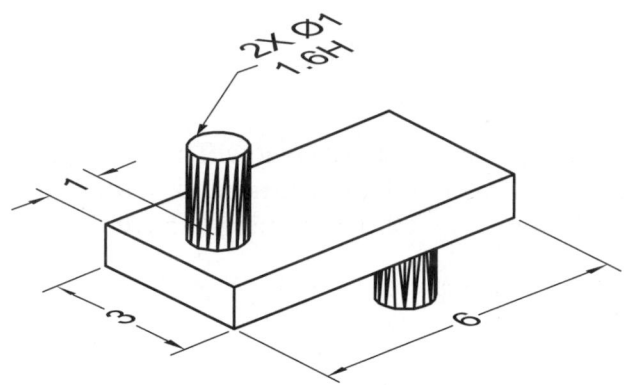

Locking Plate

9.

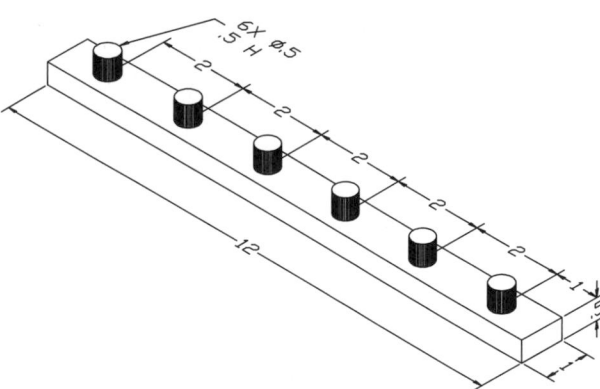

Pin Bar

10. Draw the Ø8″ pedestal shown. It is .5″ thick. The four feet are centered on a Ø7″ circle and are .5″ high. Save the drawing as P06_10.

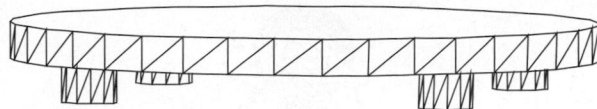

Pedestal #2

11. Four legs (cones), each 3″ high with a Ø1″ base, support this Ø10″ globe. Each leg tilts at an angle of 15° from vertical. The base is Ø12″ and .5″ thick. The bottom surface of the base is 8″ below the center of the globe. Save the drawing as P06_11.

Globe

12. The table legs (A) are 2″ square and 17″ tall. They are 2″ in from each edge. The tabletop (B) is 24″ × 36″ × 1″. Save the drawing as P06_12.

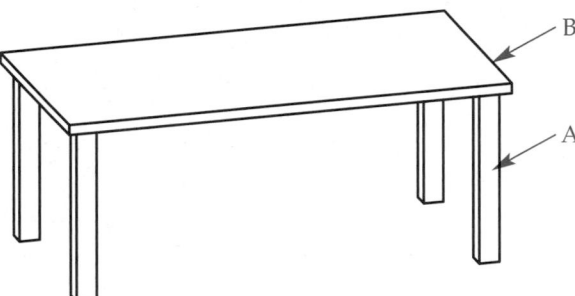

Table #1

13. The table legs (A) for the large table are Ø2″ and 17″ tall. The tabletop (B) is 24″ × 36″ × 1″. The table legs (C) for the small table are Ø2″ and 11″ tall. The tabletop (D) is 24″ × 14″ × 1″. All legs are 1″ in from the edges of the table. Save the drawing as P06_13.

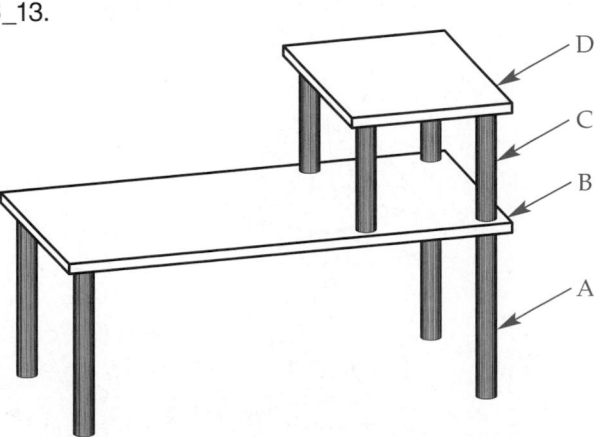

Table #2

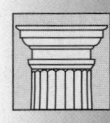

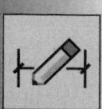

14. The spherical objects (A) are ⌀4″. Object B is 6″ long and ⌀1.5″. Save the drawing as P06_14.

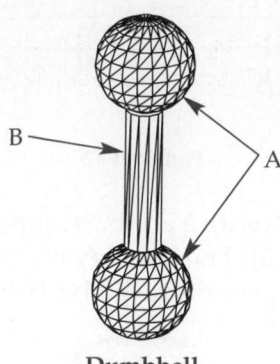

Dumbbell

15. Create the model of the globe using the dimensions shown. Save the file as P06_15.

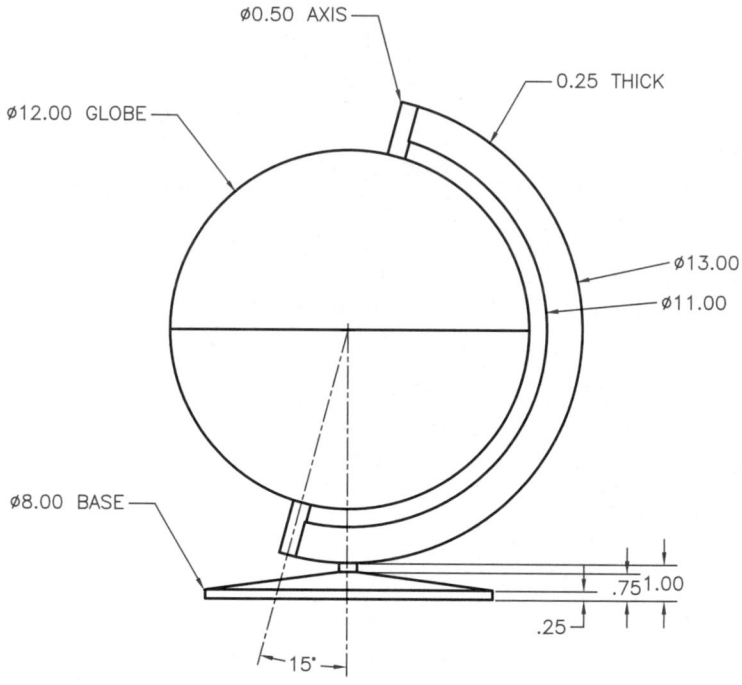

16. Object A is a ⌀8″ cylinder that is 1″ tall. Object B is a ⌀5″ cylinder that is 7″ tall. Object C is a ⌀2″ cylinder that is 6″ tall. Object D is a .5″ × 8″ × .125″ box, and there are four pieces. The top surface of each piece is flush with the top surface of Object C. Object E is a ⌀18″ cone that is 12″ tall. Create a smaller cone and hollow out Object E. Save the drawing as P06_16.

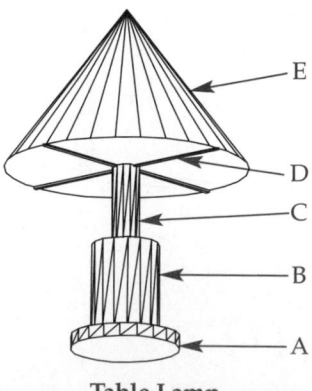

Table Lamp

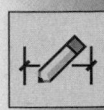

17. Objects A and B are brick walls that are 5' high. The walls are two bricks thick. Research the dimensions of standard brick and draw accordingly. Wall B is 7' long and Wall A is 5' long. Lamps are placed at each end of the walls. Object C is Ø2" and 8" tall. The center is offset from the end of the wall by a distance equal to the width of one brick. Object D is Ø10". Save the drawing as P06_17.

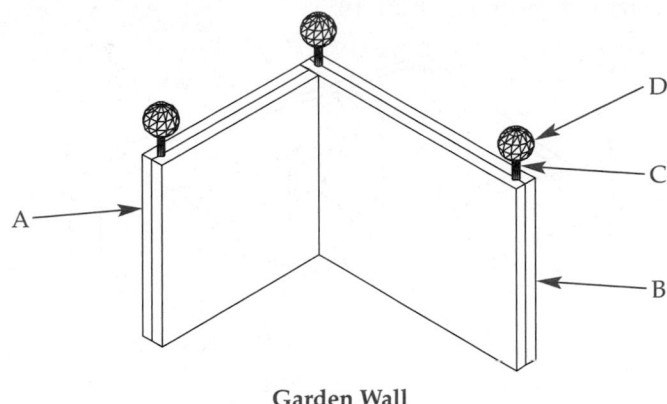

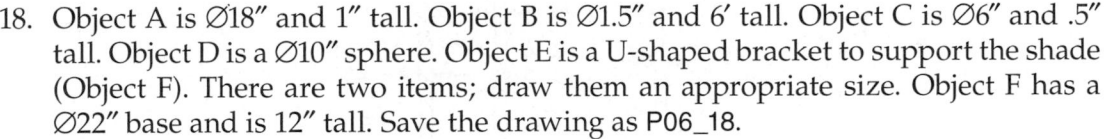

Garden Wall

18. Object A is Ø18" and 1" tall. Object B is Ø1.5" and 6' tall. Object C is Ø6" and .5" tall. Object D is a Ø10" sphere. Object E is a U-shaped bracket to support the shade (Object F). There are two items; draw them an appropriate size. Object F has a Ø22" base and is 12" tall. Save the drawing as P06_18.

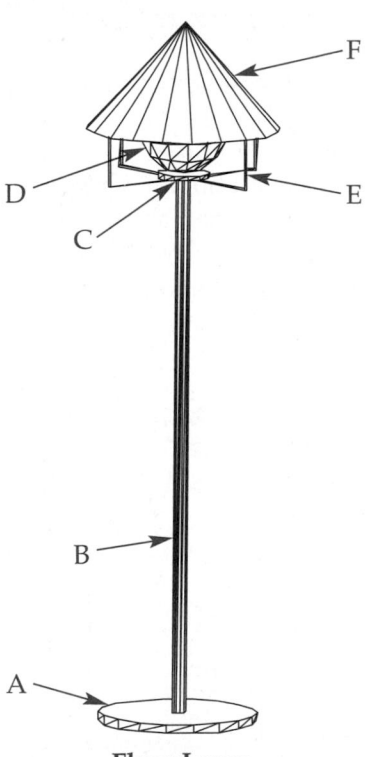

Floor Lamp

19. This is a concept sketch of a desk organizer. Create a solid model using the dimensions given. Use a dynamic UCS when appropriate or create and save new UCSs as needed. Inside dimensions of compartments can vary, but the thickness between compartments should be consistent. Do not add dimensions to the drawing. Plot your drawing on a B-size sheet of paper in a visual style specified by your instructor. Save the drawing as P06_19.

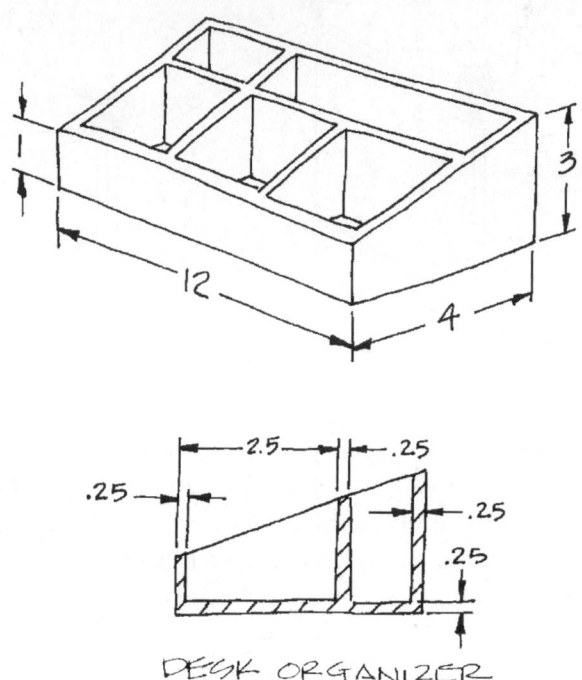

DESK ORGANIZER

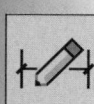

20. This is a concept sketch of a pencil holder. Create a solid model using the dimensions given. Use a dynamic UCS when appropriate or create and save new UCSs as needed. Do not add dimensions to the drawing. Plot your drawing on a B-size sheet of paper in a visual style specified by your instructor. Save the drawing as P06_20.

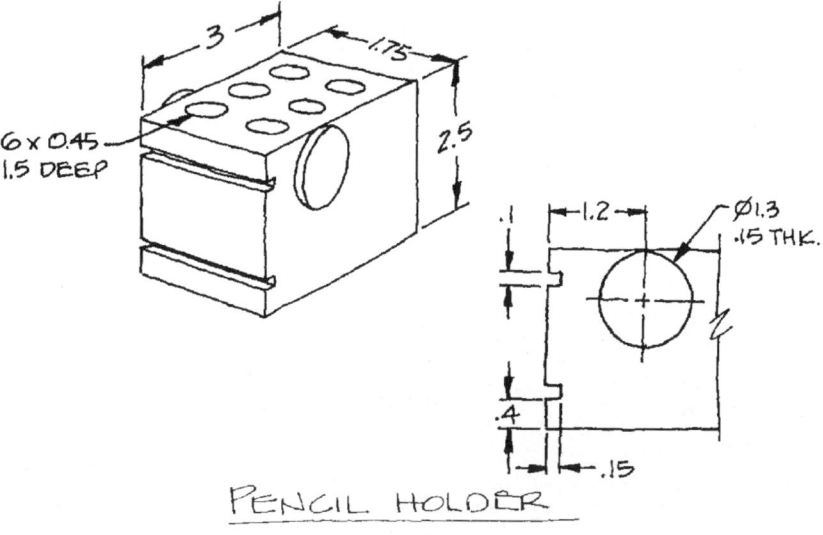

PENCIL HOLDER

21. This is an engineering sketch of a window blind mounting bracket. Create a solid model using the dimensions given. Use a dynamic UCS when appropriate or create and save new UCSs as needed. Do not add dimensions to the drawing. Create two plots, each of a different view, on B-size paper in the visual styles specified by your instructor. Save the drawing as P06_21.

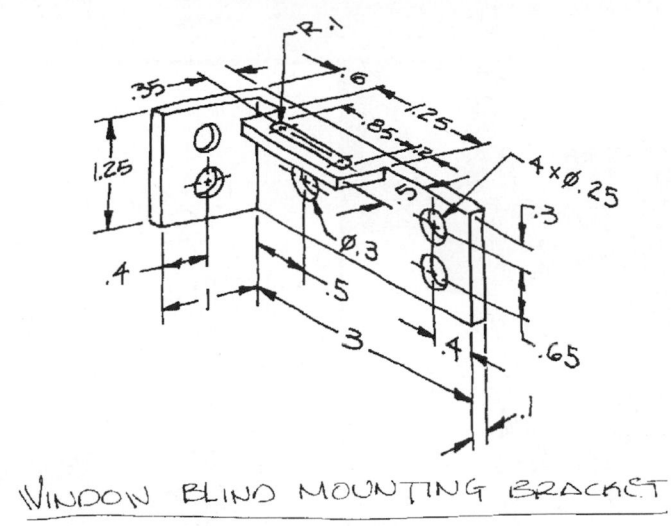

WINDOW BLIND MOUNTING BRACKET

Viewports can be used to increase your modeling efficiency. However, too many viewports can actually decrease your efficiency. Generally, the maximum number of viewports you should use is three or four.

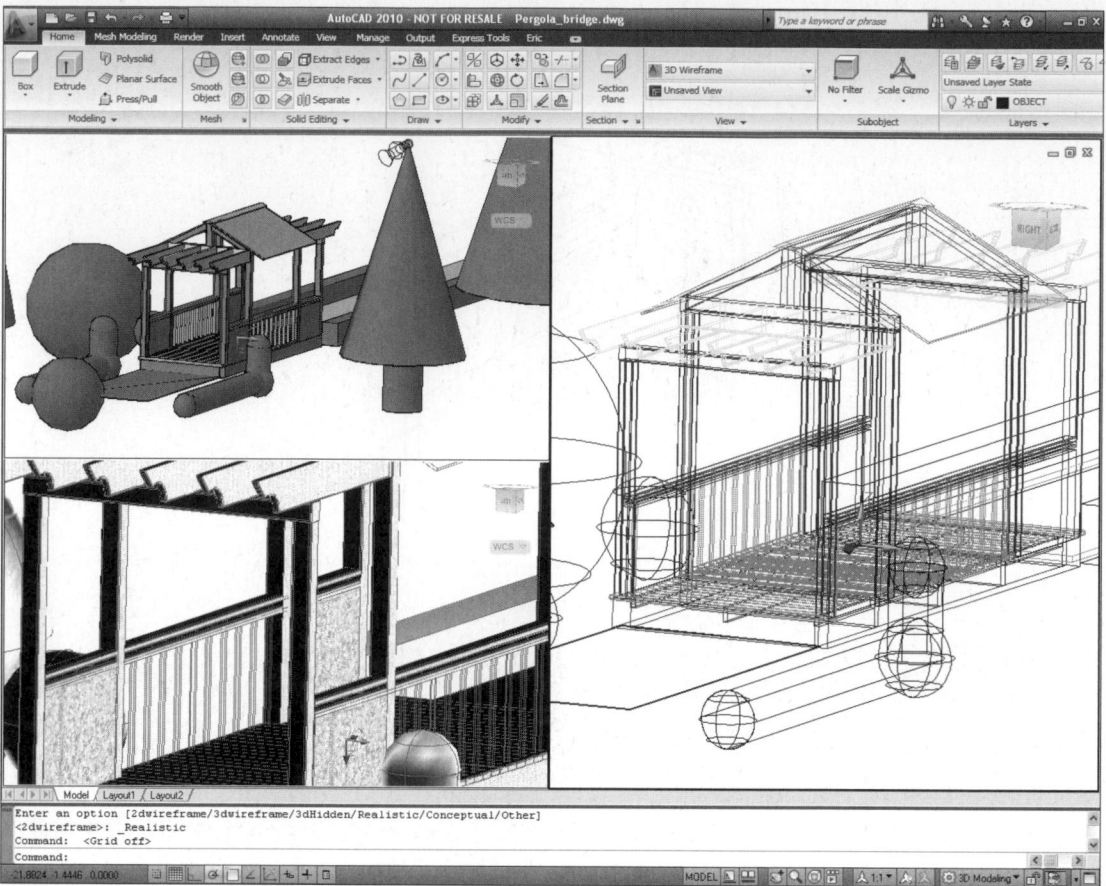

AutoCAD and Its Applications—Advanced

CHAPTER 7

Using Model Space Viewports

Learning Objectives

After completing this chapter, you will be able to:

✓ Describe the function of model space viewports.
✓ Create and save viewport configurations.
✓ Alter the current viewport configuration.
✓ Use multiple viewports to construct a drawing.

A variety of views can be displayed in a drawing at one time using model space viewports. This is especially useful when constructing 3D models. Using the **VPORTS** command, you can divide the drawing area into two or more smaller areas. These areas are called *viewports.* Each viewport can be configured to display a different 2D or 3D view of the model.

The *active viewport* is the viewport in which a command will be applied. Any viewport can be made active, but only one can be active at a time. As objects are added or edited, the results are shown in all viewports. A variety of viewport configurations can be saved and recalled as needed. This chapter discusses the use of viewports and shows how they can be used for 3D constructions.

Understanding Viewports

The AutoCAD drawing area can be divided into a maximum of 64 viewports. However, this is impractical due to the small size of each viewport. Four viewports are usually the maximum number practical to display at one time. The number of viewports you need depends on the model you are drawing. Each viewport can show a different view of an object. This makes it easier to construct 3D objects.

There are two types of viewports used in AutoCAD. The type of viewport created depends on whether it is defined in model space or paper space. *Model space* is the space, or mode, where the drawing is constructed. *Paper space,* or layout space, is the space where a drawing is laid out to be plotted. Viewports created in model space are called *tiled viewports.* Viewports created in paper space are called *floating viewports.*

Model space is active by default when you start AutoCAD. Model space viewports are created with the **VPORTS** command. These viewport configurations cannot be plotted because they are for display purposes only. If you plot from model space, the content of the active viewport is plotted. Tiled viewports are not AutoCAD objects. They are referred to as *tiled* because the edges of each viewport are placed side to side, as with floor tile, and they cannot overlap.

Floating (paper space) viewports are used to lay out the views of a drawing before plotting. They are described as *floating* because they can be moved around and over-lapped. Paper space viewports are objects and they can be edited. These viewports can be thought of as "windows" cut into a sheet of paper to "see into" model space. You can then insert, or *reference,* different scaled drawings (views) into these windows. For example, architectural details or sections and details of complex mechanical parts may be referenced. Detailed discussions of paper space viewports are provided in *AutoCAD and Its Applications—Basics.*

The **VPORTS** command can be used to create viewports in a paper space layout. The process is very similar to that used to create model space viewports, which is discussed next. You can also use the **MVIEW** command to create paper space viewports.

Creating Viewports

Creating model space viewports is similar to working with a multiview layout in manual drafting. In a manual multiview layout, several views are drawn on the same sheet. You can switch from one view to another simply by moving your pencil. With model space viewports, you simply pick with your pointing device in the viewport in which you wish to work. The picked viewport becomes active. Using viewports is a good way to construct 3D models because all views are updated as you draw. However, viewports are also useful when creating 2D drawings.

The project on which you are working determines the number of viewports needed. Keep in mind that the more viewports you display on your screen, the smaller the view in each viewport. Small viewports may not be useful to you. Four different viewport configurations are shown in **Figure 7-1.** As you can see, when 16 viewports are displayed, the viewports are very small. Normally, two to four viewports are used.

Figure 7-1.
A—Two vertical viewports. B—Two horizontal viewports. C—Three viewports, with the largest viewport positioned at the right. D—Sixteen viewports.

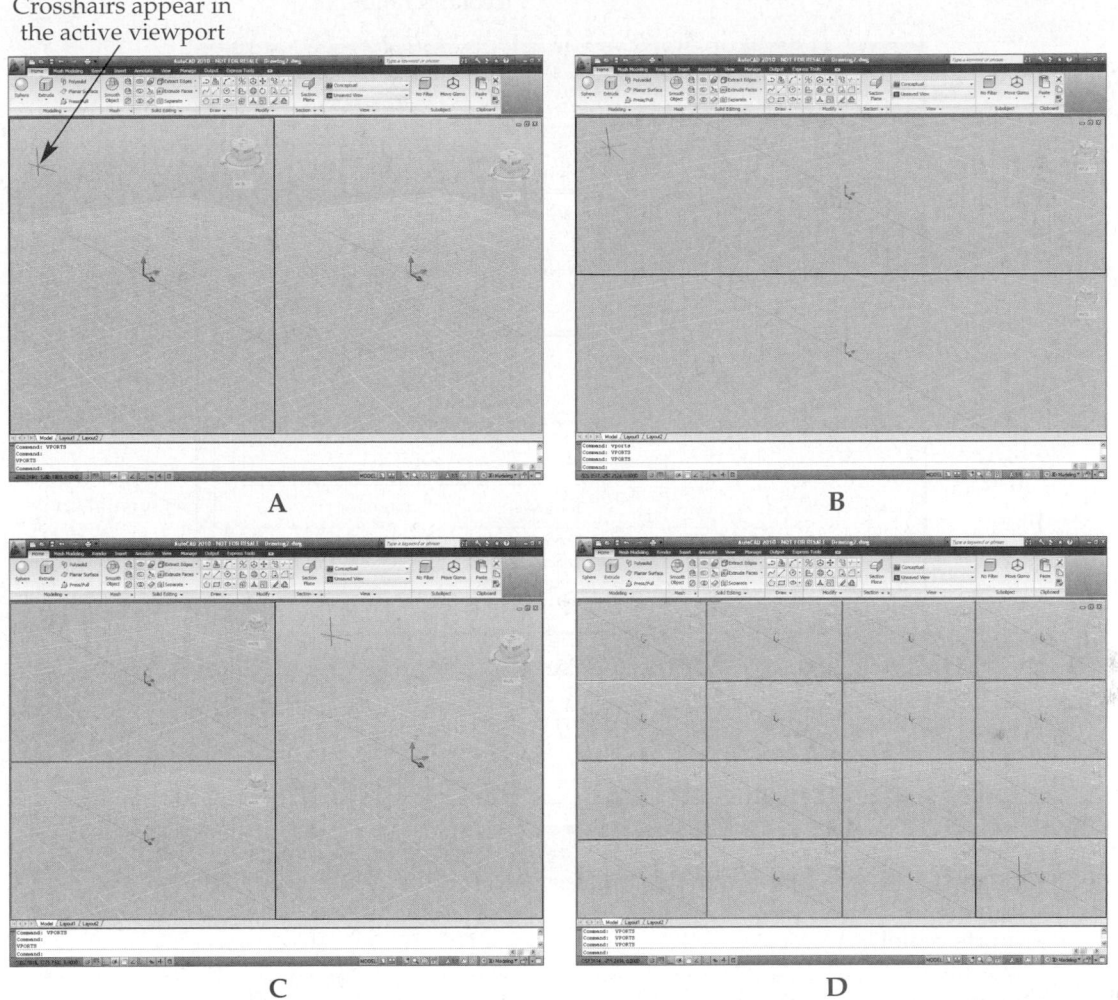

Crosshairs appear in the active viewport

A

B

C

D

A layout of viewport configurations can be quickly created by using the **New Viewports** tab of the **Viewports** dialog box, which is displayed with the **VPORTS** command, **Figure 7-2.** You can also use the **Viewport Configurations** drop-down list in the **Viewports** panel on the **View** tab of the ribbon. See **Figure 7-3.**

There are 12 preset viewport configurations from which to choose in the **New Viewports** tab, including six different options for three-viewport configurations. See **Figure 7-4.** When you pick the name of a configuration in the **Standard viewports:** list, the viewport arrangement is displayed in the **Preview** area. After you have made a selection, you can save the configuration by entering a name in the **New name:** text box and then picking **OK** to close the dialog box. When the **Viewports** dialog box closes, the configuration is displayed on screen.

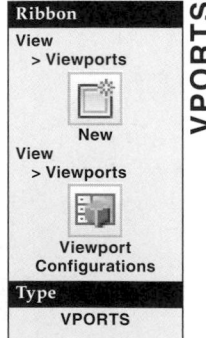

Ribbon
View
> Viewports

New

View
> Viewports

Viewport Configurations

Type
VPORTS

VPORTS

Figure 7-2.
Viewports are created using the **New Viewports** tab of the **Viewports** dialog box.

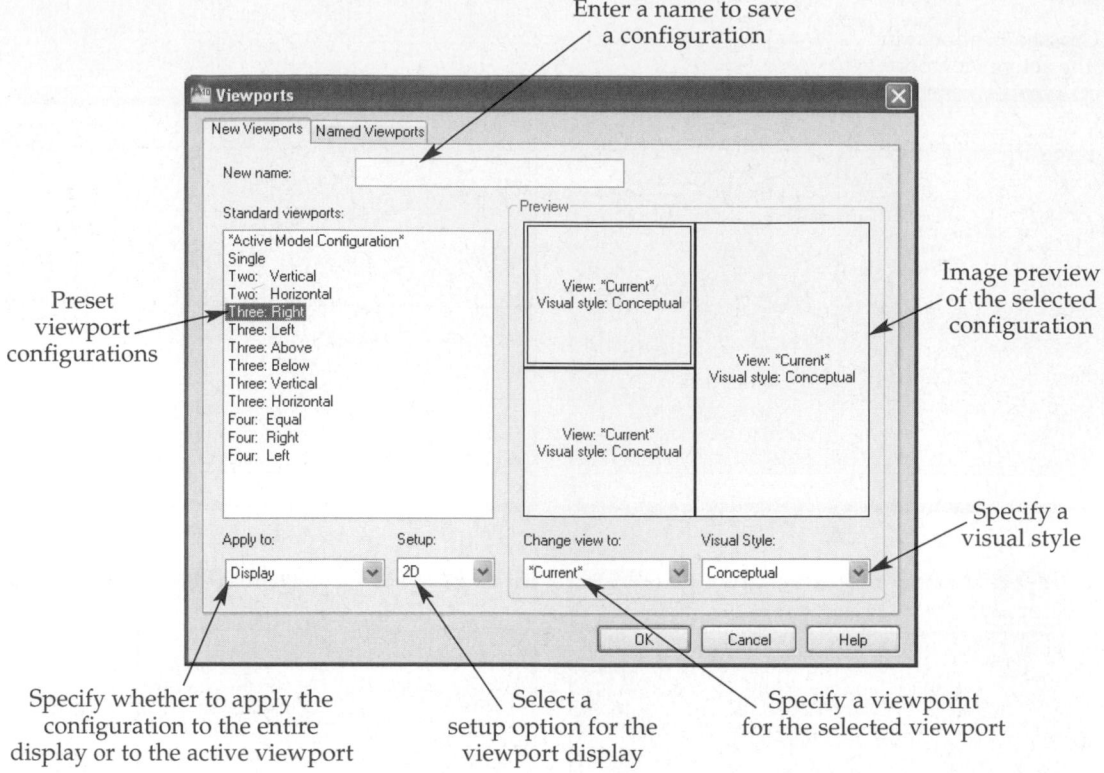

Enter a name to save a configuration

Preset viewport configurations

Image preview of the selected configuration

Specify a visual style

Specify whether to apply the configuration to the entire display or to the active viewport

Select a setup option for the viewport display

Specify a viewpoint for the selected viewport

Figure 7-3.
The **Viewport Configurations** drop-down list in the **Viewports** panel on the **View** tab of the ribbon offers a quick way to recall a preset viewport configuration.

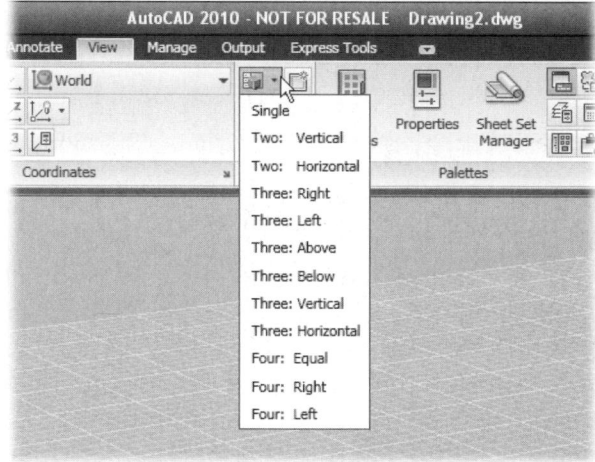

Notice in **Figure 7-1** that the UCS icon and view cube are displayed in all viewports. This is an easy way to tell that several separate screens are displayed, rather than different views of the drawing.

AutoCAD and Its Applications—Advanced

Figure 7-4.
Twelve preset tiled viewport configurations are provided in the **Viewports** dialog box.

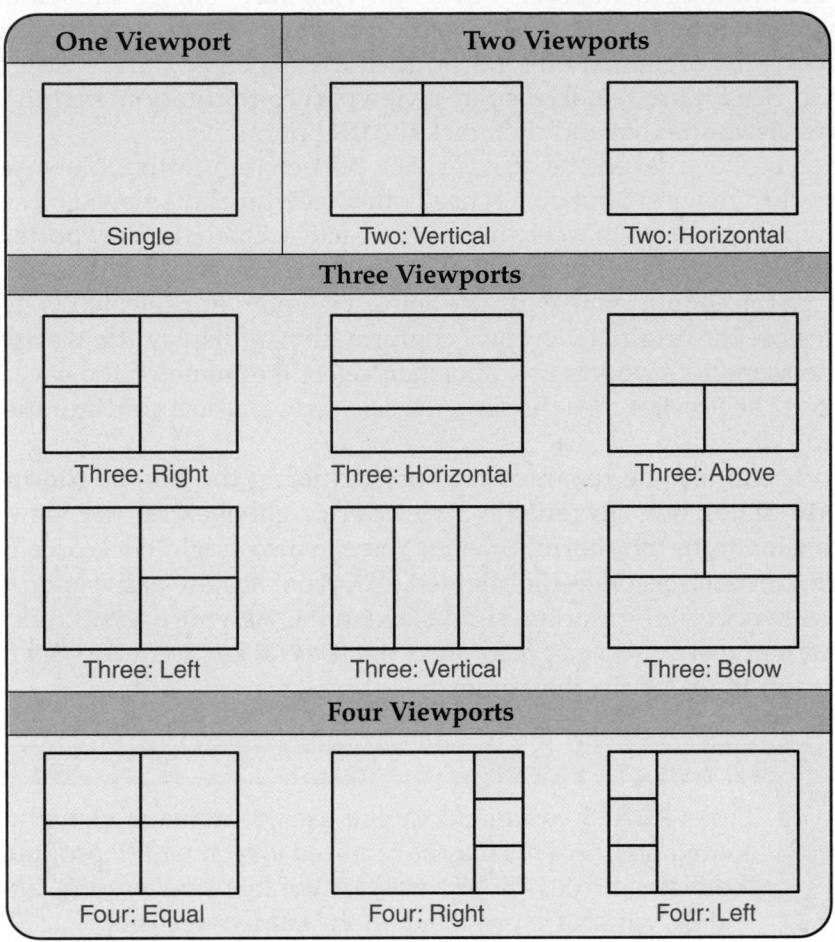

Making a Viewport Active

After a viewport configuration has been created, a thick line surrounds the active viewport. When the screen cursor is moved inside of the active viewport, it appears as crosshairs. When moved into an inactive viewport, the standard Windows cursor appears.

Any viewport can be made active by moving the cursor into the desired viewport and pressing the pick button. You can also press the [Ctrl]+[R] key combination to switch viewports, or use the **CVPORT** (current viewport) system variable. Only one viewport can be active at a time.

    Command: **CVPORT**⏎
    Enter new value for CVPORT <*current*>: **3**⏎

The current value given is the ID number of the active viewport. The ID number is automatically assigned by AutoCAD, starting with 2. To change viewports with the **CVPORT** system variable, simply enter a different ID number. Using the **CVPORT** system variable is also a good way to determine the ID number of a viewport. The number 1 is not a valid viewport ID number.

PROFESSIONAL TIP

Each viewport can have its own view, viewpoint, UCS, zoom scale, limits, grid spacing, and snap setting. Specify the drawing aids in all viewports before saving the configuration. When a viewport is restored, all settings are restored as well.

Ribbon

View
> Viewports

Named Viewports...

Type
VPORTS

Managing Defined Viewports

If you are working with several different viewport configurations, it is easy to restore, rename, or delete existing viewports. You can do so using the **Viewports** dialog box. To access a list of named viewports, open the dialog box and select the **Named Viewports** tab. See **Figure 7-5**. To display a viewport configuration, highlight its name in the **Named viewports:** list and then pick the **OK** button.

Assume you have saved the current viewport configuration. Now, you want to work in a specific viewport, but do not need other viewports displayed on screen. First, pick the viewport you wish to work in to make it active. Open the **Viewports** dialog box, pick the **New Viewports** tab, and then pick **Single**. The **Preview** area displays the single viewport. Pick the **OK** button to exit. The active viewport you selected is displayed on screen. To restore the original viewport configuration, redisplay the **Viewports** dialog box, pick the **Named Viewports** tab, and then select the name of the saved viewport configuration. The **Preview** area displays the selected viewport configuration. Pick the **OK** button to exit.

Viewports can also be renamed and deleted using the **Named Viewports** tab of the **Viewports** dialog box. To rename a viewport, right-click on the viewport name and pick **Rename** from the shortcut menu. You can also single-click on a highlighted name. When the name becomes highlighted text, type the new name and press [Enter]. To delete a viewport configuration, right-click on the viewport name and pick **Delete** from the shortcut menu. You can also press the [Delete] key to delete the highlighted viewport. Press **OK** to exit the dialog box.

PROFESSIONAL TIP

The **-VPORTS** command can be used to manage viewports on the command line. This may be required for some LISP programs where the dialog box cannot be used. AutoLISP is covered in Chapters 27 and 28, which are provided on the student website.

www.g-wlearning.com/CAD

Figure 7-5.
The **Named Viewports** tab of the **Viewport** dialog box lists all named viewports and displays the selected configuration in the **Preview** area.

Select a named configuration

Preview of selected configuration

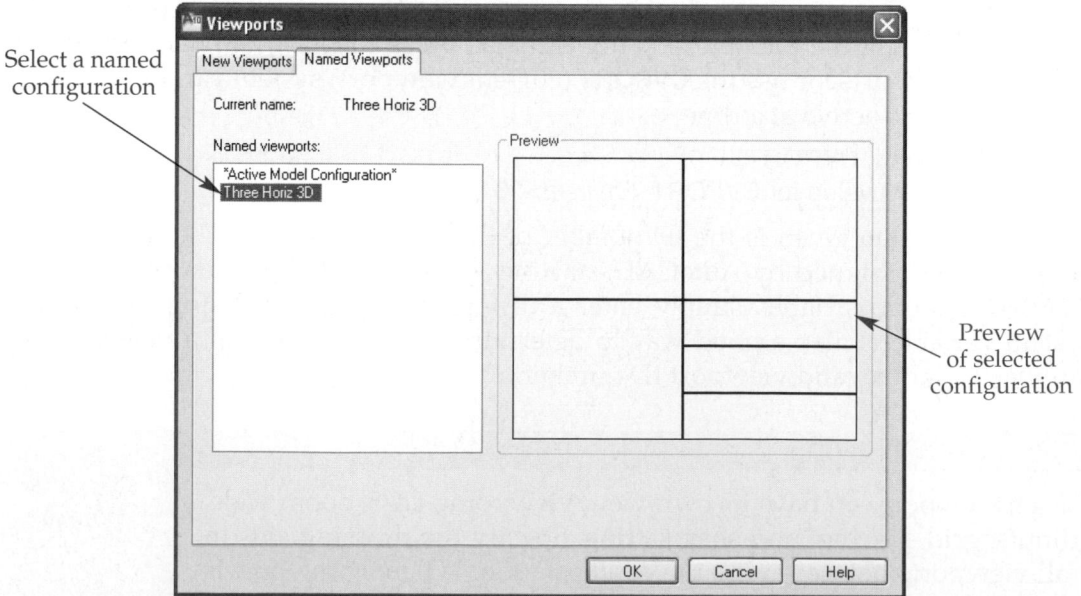

Figure 7-6.
The **Viewports** panel in the **View** tab of the ribbon.

Select a preset configuration

Displays the **New Viewports** tab of the **Viewports** dialog box

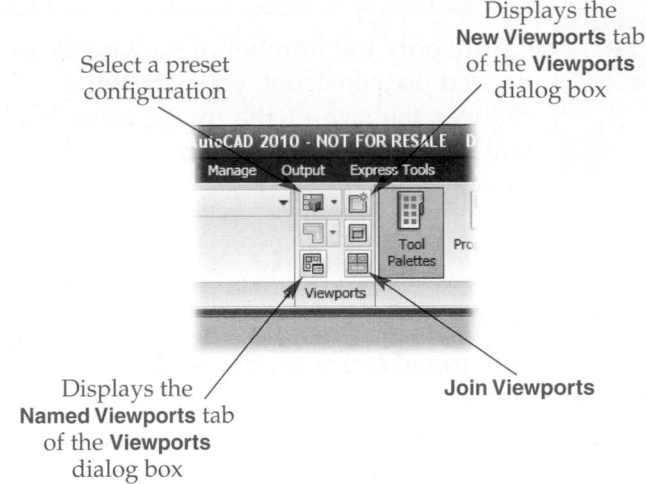

Displays the **Named Viewports** tab of the **Viewports** dialog box

Join Viewports

Using the Viewports Panel

The **Viewports** panel in the **View** tab of the ribbon is shown in **Figure 7-6.** It is used with both model space and paper space viewports. The **Named Viewports...** and **New** buttons on the panel display the **Viewports** dialog box. The **Join** button is used to combine two viewports into a single viewport, as described in the next section. The **Viewport Configurations** drop-down list contains the twelve preset viewport configurations. Selecting a configuration in the list sets it current. The remaining buttons apply to paper space viewports in a layout. A complete discussion of paper space viewports is given in *AutoCAD and Its Applications—Basics.*

Joining Two Viewports

You can join two adjacent viewports in an existing configuration to form a single viewport. This process is often quicker than creating an entirely new configuration. However, the two viewports must form a rectangle when joined, **Figure 7-7.**

When you enter the **Join** option, AutoCAD first prompts you for the *dominant viewport.* All aspects of the dominant viewport are used in the new (joined) viewport. These aspects include the limits, grid, UCS, and snap settings.

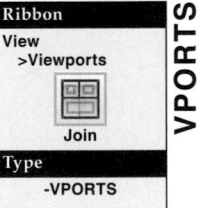

Select dominant viewport <current viewport>: *(select the viewport by picking in it or press [Enter] to set the current viewport as the dominant viewport)*
Select viewport to join: *(select the other viewport)*

The two selected viewports are joined into a single viewport. If you select two viewports that do not form a rectangle, AutoCAD returns the message:

The selected viewports do not form a rectangle.

Figure 7-7.
Two viewports can be joined if they will form a rectangle. If the two viewports will not form a rectangle, they cannot be joined.

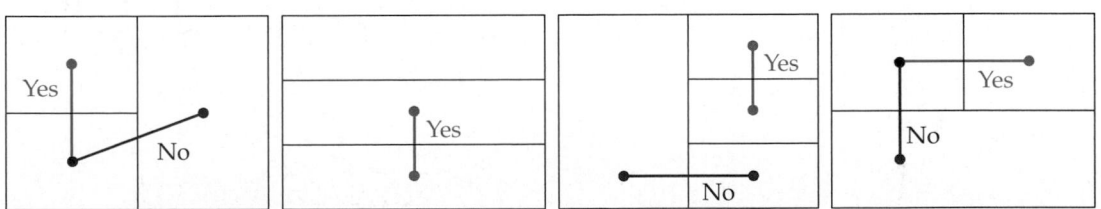

Create only the number of viewports and viewport configurations needed to construct your drawing. Using too many viewports reduces the size of the image in each viewport and may confuse you. Also, it helps to zoom each view so that the objects fill the viewport.

Exercise 7-1

Complete the exercise on the student website.
www.g-wlearning.com/CAD

Applying Viewports to Existing Configurations and Displaying Different Views

You have total control over what is displayed in model space viewports. In addition to displaying various viewport configurations, you can divide an existing viewport into additional viewports or assign a different viewpoint to each viewport. The options for these functions are provided in the **New Viewports** tab of the **Viewports** dialog box. These options are located along the bottom of the **Viewports** dialog box, as shown in **Figure 7-2**:

- **Apply to**.
- **Setup**.
- **Change view to**.
- **Visual style**.

Apply to

When a preset viewport configuration is selected from the **Standard viewports:** list, it can be applied to either the entire display or the current viewport. The previous examples have shown how to create viewports that replace the entire display. Applying a configuration to the active viewport rather than the entire display can be useful when you need to display additional viewports.

For example, first create a configuration of three viewports using the Three: Right configuration option. Then, with the right (large) viewport active, open the **Viewports** dialog box again. Notice that the drop-down list under **Apply to:** is grayed out. Now, pick one of the standard configurations. This enables the **Apply to:** drop-down list. The default option is Display, which means the selected viewport configuration will replace the current display. Pick the drop-down list arrow to reveal the second option, Current Viewport. Pick this option and then pick the **OK** button. Notice that the selected viewport configuration has been applied to only the active (right) viewport. See **Figure 7-8**.

Figure 7-8.
The selected viewport configuration has been applied to the active viewport within the original configuration.

New configuration is applied to the viewport

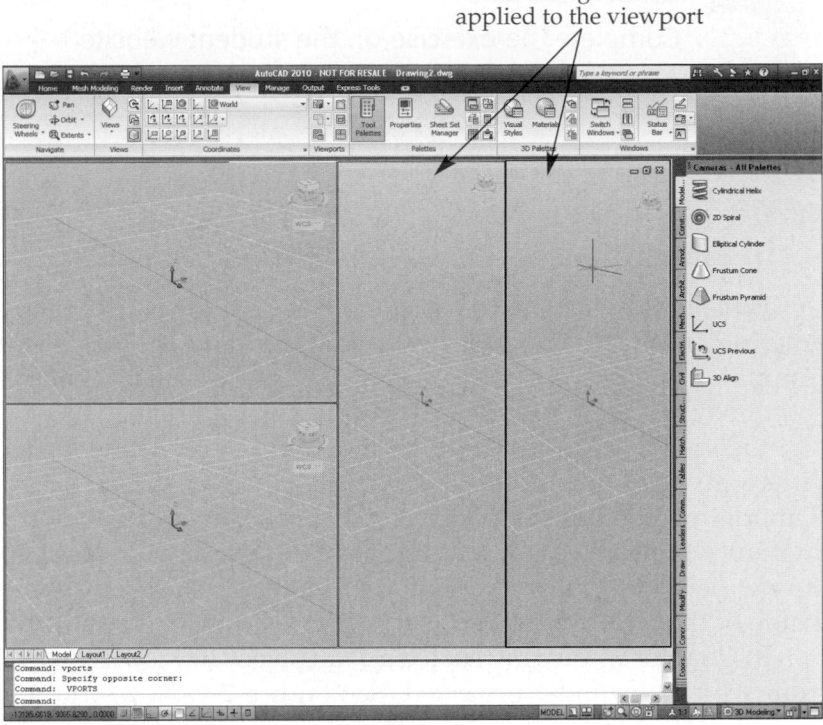

Setup

Viewports can be set up to display views in 2D or 3D. The 2D and 3D options are provided in the **Setup:** drop-down list. Displaying different views while working on a drawing allows you to see the results of your work on each view, since changes are reflected in each viewport as you draw. The selected viewport **Setup:** option controls the types of views available in the **Change view to:** drop-down list.

Change view to

The views that can be displayed in a selected viewport are listed in the **Change view to:** drop-down list. If the **Setup:** drop-down list is set to 2D, the views available to be displayed are limited to the current view and any named views. If 3D is active, the options include all of the standard orthographic and isometric views along with named views. When an orthographic or isometric view is selected for a viewport, the resulting orientation is shown in the **Preview** area. To assign a different viewpoint to a viewport, simply pick within a viewport in the **Preview** area to make it active and then pick a viewpoint from the **Change view to:** drop-down list. Important: if you set a viewport to one of the orthographic preset views, the UCS is also (by default) changed in that viewport to the corresponding preset.

Visual style

A visual style can be specified for a viewport. Pick within a viewport in the **Preview** area to make it active and then select a visual style from the **Visual Style:** drop-down list. All preset and saved visual styles are available in the drop-down list.

PROFESSIONAL TIP

Displaying saved viewport configurations can be automated by using custom menus. Custom menus are easy to create. If a standard naming convention is used, these named configurations can be saved with template drawings. This creates a consistent platform that all students or employees can use.

Exercise 7-2
Complete the exercise on the student website.
www.g-wlearning.com/CAD

Drawing in Multiple Viewports

When used with 2D drawings, viewports allow you to display a view of the entire drawing, plus views showing portions of the drawing. This is similar to using the **VIEW** command, except you can have several views on screen at once. You can also adjust the zoom magnification in each viewport to suit different areas of the drawing.

Viewports are also a powerful aid when constructing 3D models. You can specify different viewpoints in each viewport and see the model take shape as you draw. A model can be quickly constructed because you can switch from one viewport to another while drawing and editing. For example, you can draw a line from a point in one viewport to a point in another viewport simply by changing viewports while inside of the **LINE** command. The result is shown in each viewport.

In Chapter 5, you constructed a solid object. It was a base that had an angled surface from which a cylinder projected. See **Figure 7-9.** Now, you will construct the object using two viewports. First, create a vertical configuration of two viewports. In the **Viewports** dialog box, set the right-hand viewport to display the southeast isometric. Also, select the Conceptual visual style. Set the left-hand viewport to display the front view. Remember, this will also set the UCS to the front preset orthographic UCS in that viewport. Also, select the 3D Wireframe visual style for the left-hand viewport. Close the dialog box and make the left-hand viewport active. You may also want to set the parallel projection current in both viewports.

Next, draw a polyline using the coordinates shown in **Figure 7-10.** You may want to turn off dynamic input to disable direct-distance entry. Be sure to use the **Close** option for the last segment. As you construct the side view, you can clearly see its true size and shape in the left-hand viewport. At the same time, you can see the construction in 3D in the right-hand viewport. Notice that each viewport has a different UCS, as indicated by the UCS icon.

VPORTS

Ribbon
View
> Viewports
New
Type
VPORTS

Figure 7-9.
You will construct the object from Chapter 6 using multiple viewports.

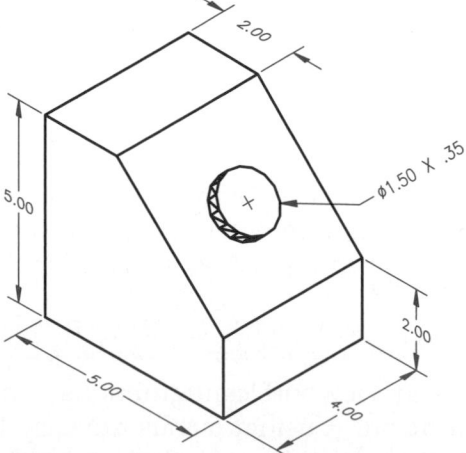

Figure 7-10.
The screen is divided into two viewports. A side view of the object appears in the left-hand viewport and a 3D view appears in the right-hand viewport. Notice the UCS icons.

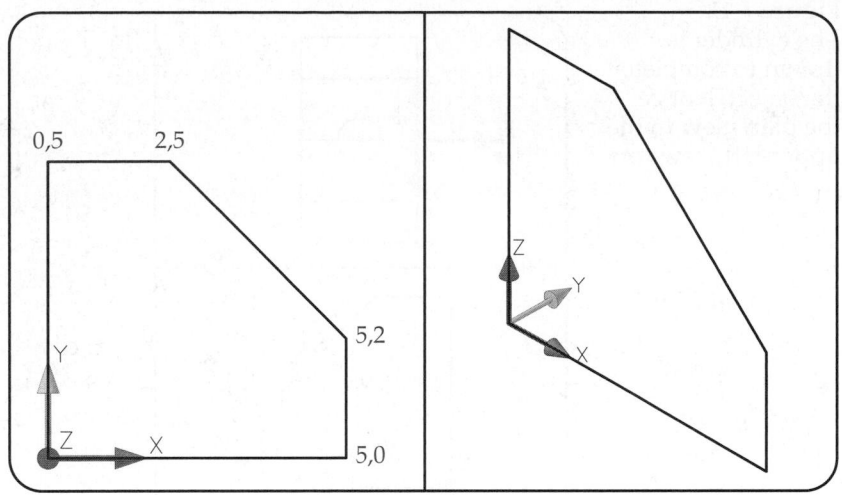

The next step is to extrude the shape to create the base. The **EXTRUDE** command is used to do so, as was the case in Chapter 6. In the left-hand viewport, select the **EXTRUDE** command, pick the polyline, and enter an extrusion height of −4 units. The front face of the object is now complete, **Figure 7-11.**

Now, the cylinder needs to be created on the angled face. First, split the left-hand viewport into two horizontal viewports (top and bottom) using the **New Viewports** tab of the **Viewports** dialog box. Set both of the new viewports to display the current view. Pick the **OK** button to close the dialog box. Then, make the upper-left viewport current and set it up to always display a plan view of the current UCS by setting the **UCSFOLLOW** system variable to 1.

Next, create a new UCS on the angled face. Use the **3 Point** option of the **UCS** command, which is described in Chapter 6. The pick points are shown in **Figure 7-11;** pick them in the right-hand viewport. Notice how the view in the upper-left viewport automatically changes to a plan view of the new current UCS.

Now, set the **Midpoint** object snap and turn on object snap tracking to draw the cylinder. Work in the right-hand viewport. When specifying the center of the base, acquire the midpoints of sides on the angled surface that are perpendicular to each other. Then, enter a diameter of 1.5 units and a height of .35 units. If you are having problems acquiring points, try switching to a parallel display instead of a perspective display.

Figure 7-11.
The base of the object is now complete. A new UCS will be created based on the pick points shown here. Alternately, a dynamic UCS can be used, as shown in Chapter 6.

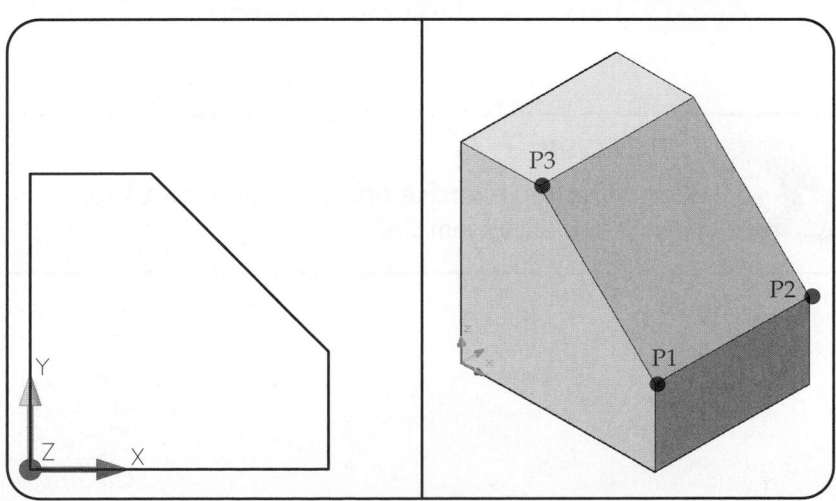

Figure 7-12.
The cylinder is drawn to complete the object. Notice the plan view in the upper-left viewport.

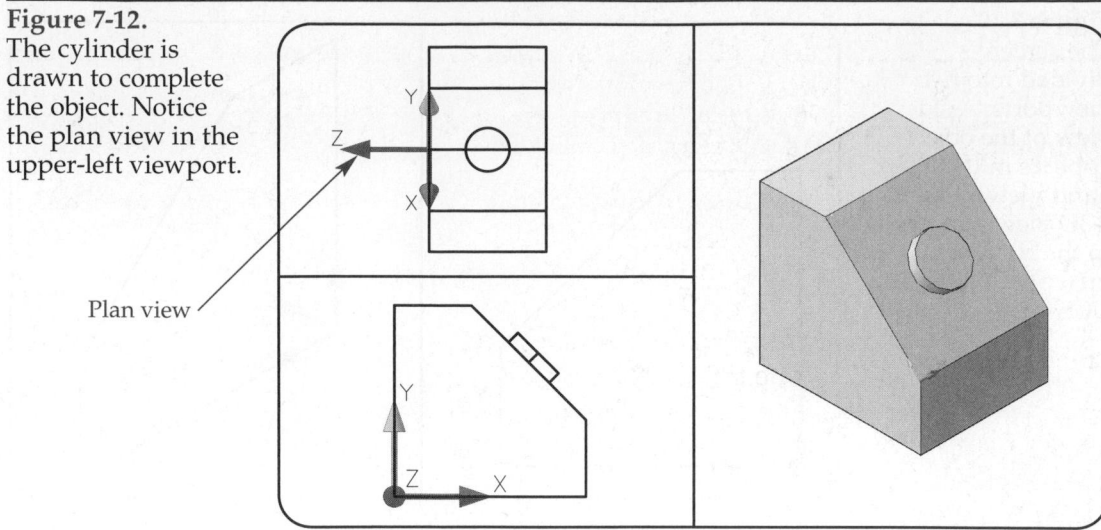

Plan view

The object is now complete, **Figure 7-12.** Notice how the lower-left and right-hand viewports have different UCSs. Each viewport can have its own UCS. The view in the upper-left viewport is the plan view of the current UCS. If the lower-left viewport is made active, the plan view will be of the UCS in that viewport. By default, the UCS orientation in one viewport is not affected by a change to the UCS in another viewport unless the **UCSFOLLOW** system variable is set to 1 in a viewport. If a viewport arrangement is saved with several different UCS configurations, every named UCS remains intact and is displayed when the viewport configuration is restored.

The **REGEN** command affects only the current viewport. To regenerate all viewports at the same time, use the **REGENALL** command. This command can be entered by typing REGENALL.

The Quick Text mode is controlled by the **REGEN** command. Therefore, if you are working with text displayed with the Quick Text mode in viewports, be sure to use the **REGENALL** command in order for the text to be regenerated in all viewports.

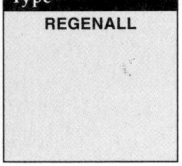

REGENALL

Type
REGENALL

NOTE

The UCS configuration in each viewport is controlled by the **UCSVP** system variable. When **UCSVP** is set to 1 in a viewport, the UCS is independent from all other UCSs, which is the default. If **UCSVP** is set to 0 in a viewport, its UCS will change to reflect any changes to the UCS in the current viewport.

Exercise 7-3

Complete the exercise on the student website.
www.g-wlearning.com/CAD

Chapter Test

Answer the following questions. Write your answers on a separate sheet of paper or complete the electronic chapter test on the student website.
www.g-wlearning.com/CAD

1. What is the purpose of *viewports?*
2. How do you name a configuration of viewports?
3. What is the purpose of saving a configuration of viewports?
4. Explain the difference between *tiled* and *floating* viewports.
5. Name the system variable controlling the maximum number of viewports that can be displayed at one time.
6. How can a named viewport configuration be redisplayed on screen?
7. How can a list of named viewport configurations be displayed?
8. What relationship must two viewports have before they can be joined?
9. What is the significance of the dominant viewport when two viewports are joined?
10. When creating a new viewport configuration, how can you set a visual style in a viewport?

Drawing Problems

1. Construct seven template drawings, each with a preset viewport configuration. Use the following configurations and names. Save the templates under the same name as the viewport configuration.

Number of Viewports	Configuration	Name
2	Horizontal	TWO-H
2	Vertical	TWO-V
3	Right	THREE-R
3	Left	THREE-L
3	Above	THREE-A
3	Below	THREE-B
3	Vertical	THREE-V

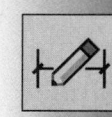

2. Construct one of the problems from Chapter 4 using viewports. Use one of your template drawings from problem 7-1. Save the drawing as P07_02.

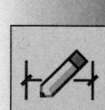

Three-dimensional models are not typically dimensioned, but there are cases where dimensions may be applied to assist workers on the shop floor. Whenever a 3D model is dimensioned, follow accepted drafting practices when possible, but you may need to violate certain rules in order to clearly describe the part.

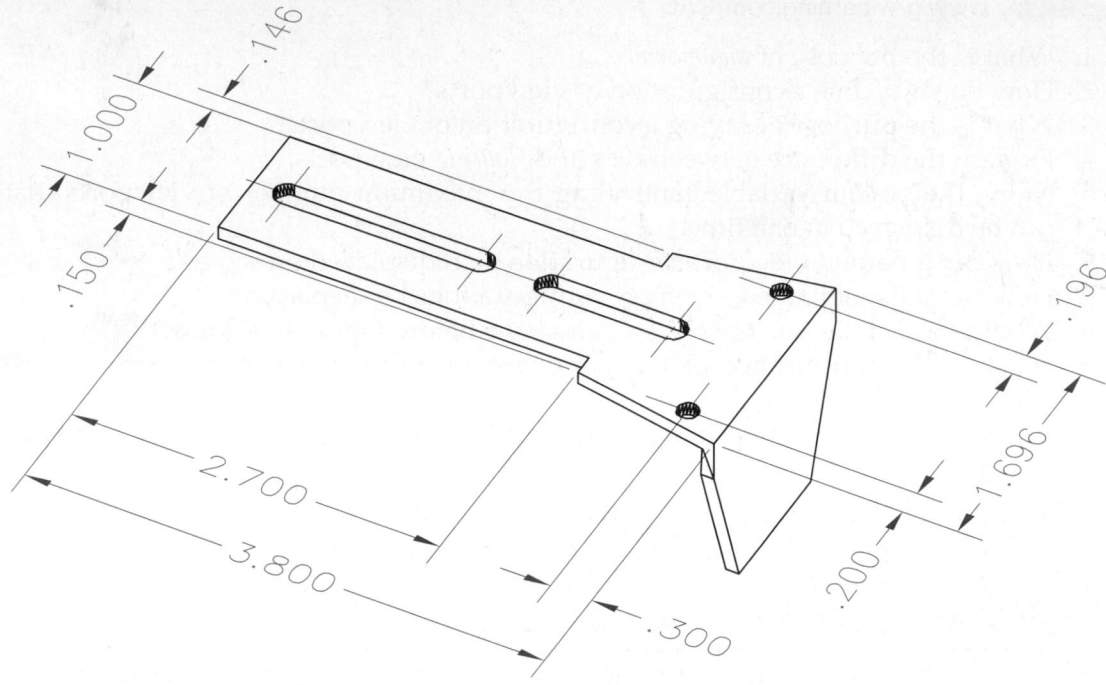

Text and Dimensions in 3D

Learning Objectives

After completing this chapter, you will be able to:

✓ Create text with a thickness.
✓ Draw text that is plan to the current view.
✓ Dimension a 3D drawing.

Creating Text with Thickness

A thickness can be applied to text after it is created. This is done using the **Properties** palette. The thickness setting is located in the **General** section. Once a thickness is applied, the hidden lines can be removed using the **HIDE** command. **Figure 8-1** shows six different fonts as they appear after being given a thickness and with the Conceptual visual style set current.

Ribbon
View
  > Palettes

Properties Palette
Type
PROPERTIES
PR
[Ctrl]+[1]

PROPERTIES

Only text created using the **TEXT** and **DTEXT** commands (text object) can be assigned thickness. Text created with the **MTEXT** command (mtext object) cannot have thickness assigned to it. In addition, only AutoCAD SHX fonts can be given thickness. AutoCAD SHP shape fonts can be compiled into SHX fonts with the **COMPILE** command. The compiled fonts can then be used to create text with thickness. Windows TrueType fonts *cannot* be used to create text with thickness.

Figure 8-1.
Six different fonts
with thickness after
the Conceptual visual
style is set current.

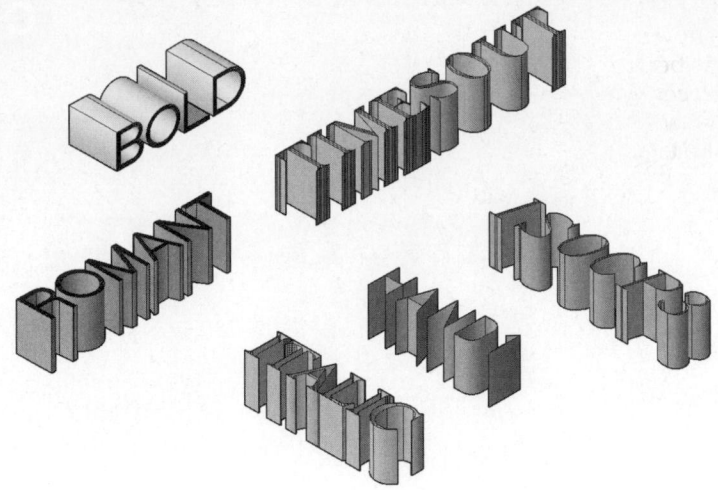

Text and the UCS

Text is created parallel to the XY plane of the UCS in which it is drawn. Therefore, if you wish to show text appearing on a specific plane, establish a new UCS on that plane before placing the text. You can make use of a dynamic UCS. **Figure 8-2** shows several examples of text on different UCS XY planes.

Changing the Orientation of a Text Object

If text is improperly placed or created using the wrong UCS, it can be edited using grips or editing commands. Editing commands and grips are relative to the current UCS. For example, if text is drawn with the WCS current, you can use the **ROTATE** command to change the orientation of the text in the XY plane of the WCS. However, to rotate the text so it tilts up from the XY plane of the WCS, you will need to change the UCS. Rotate the UCS as needed so the Z axis of the new UCS aligns with the axis about which you want to rotate. Then, the **ROTATE** command can be used to rotate the text. The **3DROTATE** command can also be used to avoid rotating the UCS. This command is discussed in Chapter 11.

Figure 8-2.
Text located using
three different
UCSs.

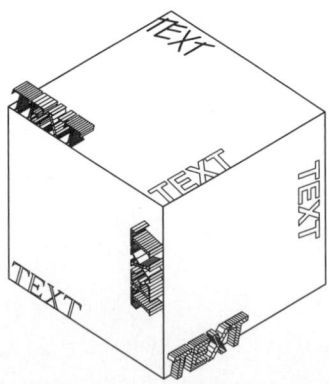

Figure 8-3.
This title (shown
in color) has been
correctly placed
using the **View**
option of the **UCS**
command.

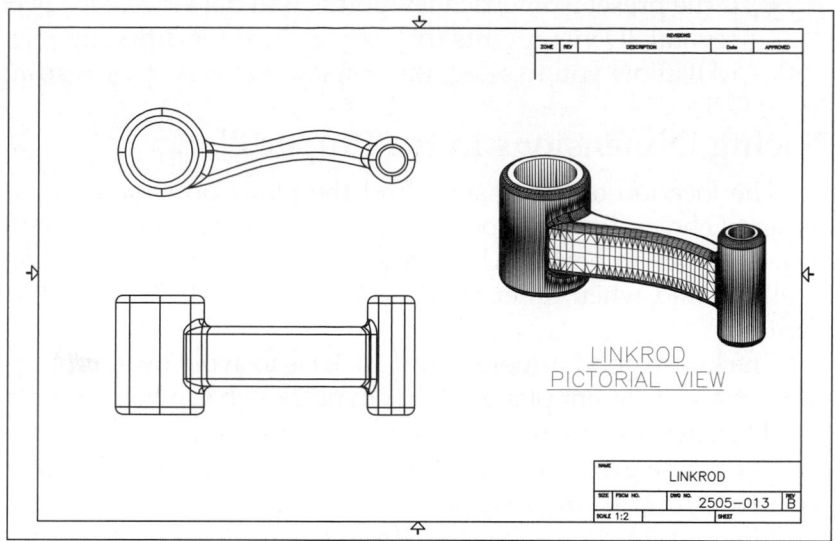

Using the UCS View Option to Create a Title

It is often necessary to create a pictorial view of an object, but with a note or title that is plan to your point of view. For example, you may need to insert the title of a 3D view. See **Figure 8-3.** This is done with the **View** option of the **UCS** command, which is discussed in Chapter 6. With this option, a new UCS is created perpendicular to your viewpoint. However, the view remains unchanged. Inserted text will be horizontal (or vertical) in the current view. Name and save the UCS if you will use it again.

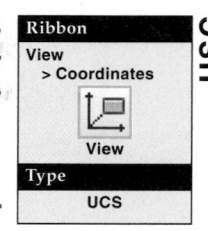

Exercise 8-1

Complete the exercise on the student website.
www.g-wlearning.com/CAD

Dimensioning in 3D

Three-dimensional objects are seldom dimensioned for manufacturing, but may be used for assembly. Dimensioned 3D drawings are most often used for some sort of presentation, such as displays, illustrations, parts manuals, or training manuals. All dimensions, including those shown in 3D, must be clear and easy to read. The most important aspect of applying dimensions to a 3D object is planning. That means following a few basic guidelines.

Creating a 3D Dimensioning Template Drawing

If you often create dimensioned 3D drawings, make a template drawing containing a few 3D settings. Starting a drawing based on one of these templates will speed up the dimensioning process because the settings will already be made for you.

- Create named dimension styles with appropriate text heights. See *AutoCAD and Its Applications—Basics* for detailed information on dimensioning and dimension styles.
- Establish several named user coordinate systems that match the planes on which dimensions will be placed.

- If the preset isometric viewpoints will not serve your needs, establish and save several 3D viewpoints that can be used for different objects. These viewpoints will allow you to select the display that is best for reading dimensions.

Placing Dimensions in the Proper Plane

The location of dimensions and the plane on which they are placed are often a matter of choice. For example, **Figure 8-4** shows several options for placing a thickness dimension on an object. All of these are correct. However, several of the options can be eliminated when other dimensions are added. This illustrates the importance of planning.

The key to good dimensioning in 3D is to avoid overlapping dimension and extension lines in different planes. A freehand sketch can help you plan this. As you lay out the 3D sketch, try to group information items together. Dimensions, notes, and item tags should be grouped so that they are easy to read and understand. This technique is called *information grouping.*

Figure 8-5A shows the object from **Figure 8-4** fully dimensioned using the aligned technique. Notice that the location dimension for the hole is placed on the top surface. This avoids dimensioning to hidden points. **Figure 8-5B** shows the same object dimensioned using the unilateral technique.

To create dimensions that properly display, it may be necessary to modify the dimension text rotation. The dimension shown in **Figure 8-6A** is inverted because the

Figure 8-4.
A thickness dimension can be located in many different places. All locations shown here are acceptable.

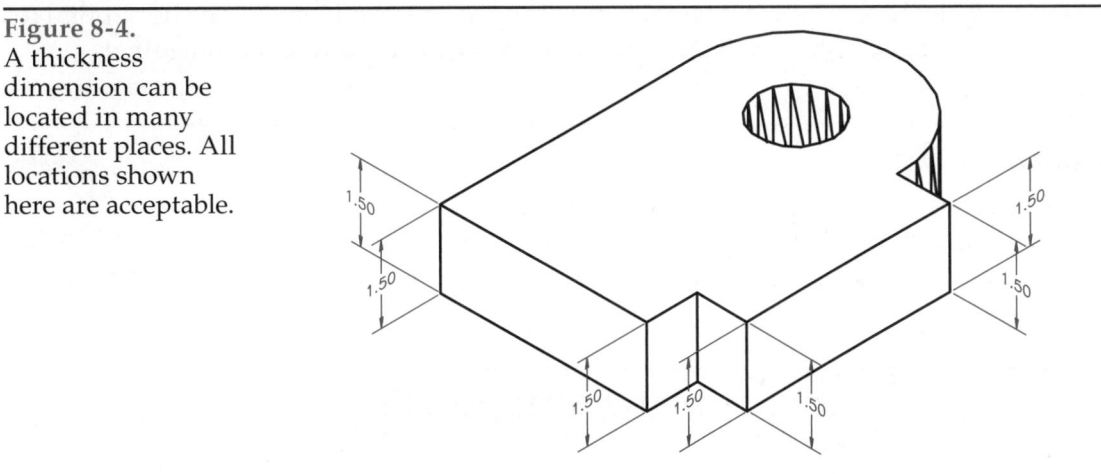

Figure 8-5.
A—An example of a 3D object dimensioned using the aligned technique. B—The object dimensioned with unilateral dimensions.

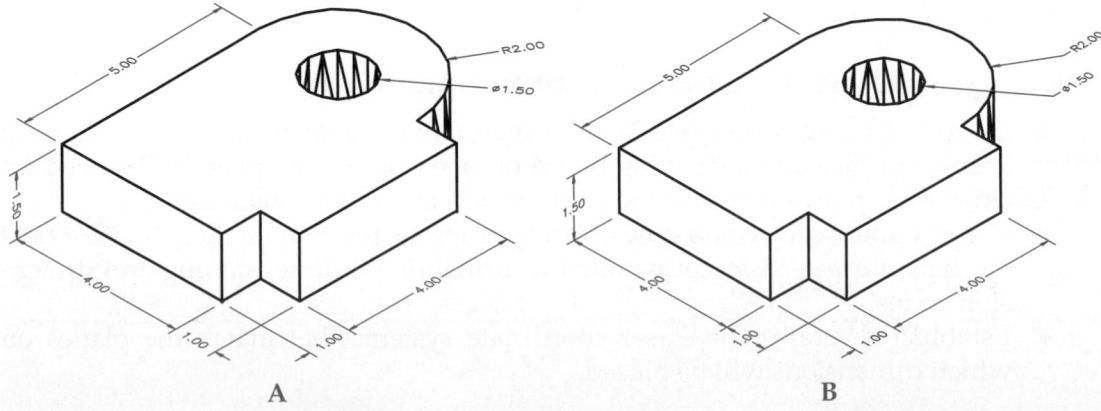

A   B

Figure 8-6.
A—This dimension text is inverted. B—The rotation value of the text is changed and the text reads correctly.

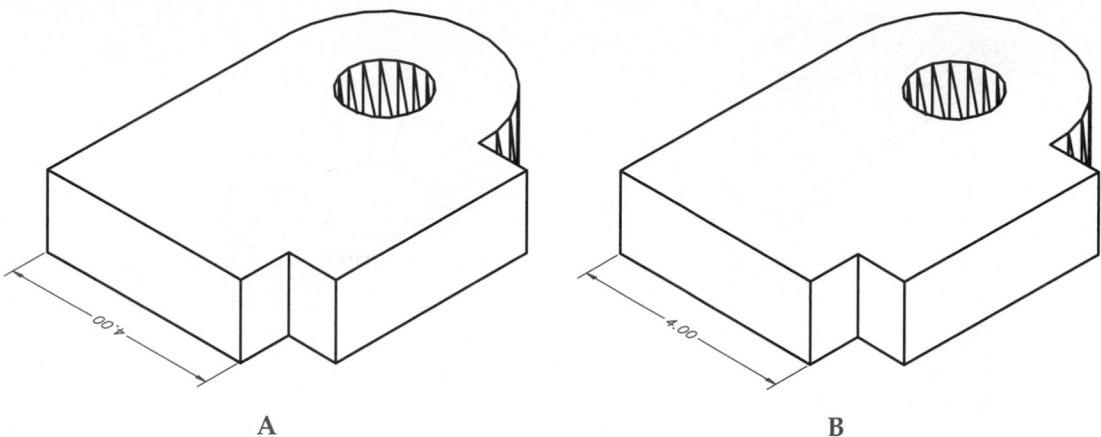

A   B

positive X and Y axes are incorrectly oriented. Using the **Properties** palette, change the text rotation value to 180. The dimension text is then properly displayed, **Figure 8-6B.** Alternately, you can rotate the UCS before drawing the dimension, but this may be more time-consuming.

PROFESSIONAL TIP

Prior to placing dimensions on a 3D drawing, you should determine the purpose of the drawing. For what will it be used? Just as dimensioning a drawing for manufacturing purposes is based on the function of the part, 3D dimensioning is based on the function of the drawing. This determines whether you use chain, datum, arrowless, architectural, or some other style of dimensioning. It also determines how completely the object is dimensioned.

Placing Leaders and Radial Dimensions in 3D

Although standards such as ASME Y14.5M should be followed when possible, the nature of 3D drawing and the requirements of the project may determine how dimensions and leaders are placed. Remember, the most important aspect of dimensioning a 3D drawing is its presentation. Is it easy to read and interpret?

Leaders and radial dimensions can be placed on or perpendicular to the plane of the feature. **Figure 8-7A** shows the placement of leaders on the plane of the top surface. **Figure 8-7B** illustrates the placement of leaders and radial dimensions on two planes that are perpendicular to the top surface of the object. Remember that text, dimensions, and leaders are always created on the XY plane of the current UCS. Therefore, to create the layout in **Figure 8-7B** you must use more than one UCS.

Exercise 8-2

Complete the exercise on the student website.
www.g-wlearning.com/CAD

Figure 8-7.
A—Leaders placed in the plane of the top surface. B—Leaders placed using two UCSs that are perpendicular to the top face.

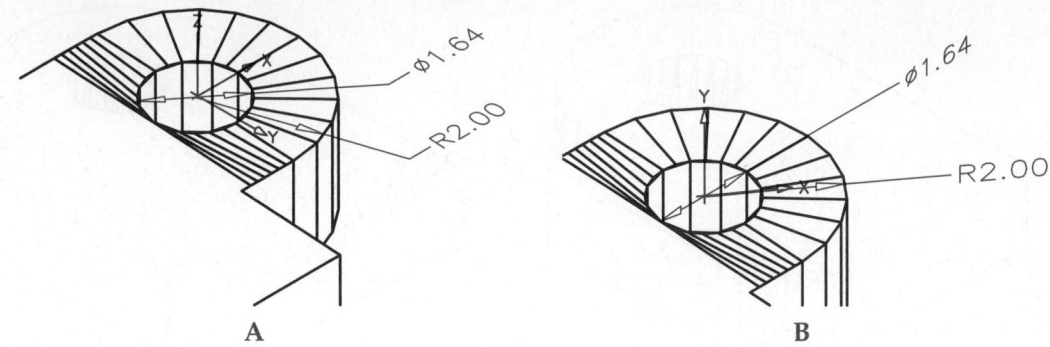

A   B

Chapter Test

Answer the following questions. Write your answers on a separate sheet of paper or complete the electronic chapter test on the student website.
www.g-wlearning.com/CAD

1. How can you create 3D text with thickness?
2. If text is placed using the wrong UCS, how can it be edited to appear on the correct one?
3. How can text be horizontally placed based on your viewpoint if the object is displayed in 3D?
4. Name three items that should be a part of a 3D dimensioning template drawing.
5. What is *information grouping?*

Drawing Problems

1. This is a two-view orthographic drawing of a window valance mounting bracket. Create it as a solid model. Use solid primitives and Boolean commands as needed. Use the dimensions given. Similar holes have the same offset dimensions. Create new UCSs as needed. Display an appropriate pictorial view of the drawing. Then, add dimensions. Finally, add the material note so it is plan to the 3D view. Plot the drawing to scale on a C-size sheet of paper. Save the drawing as P08_01.

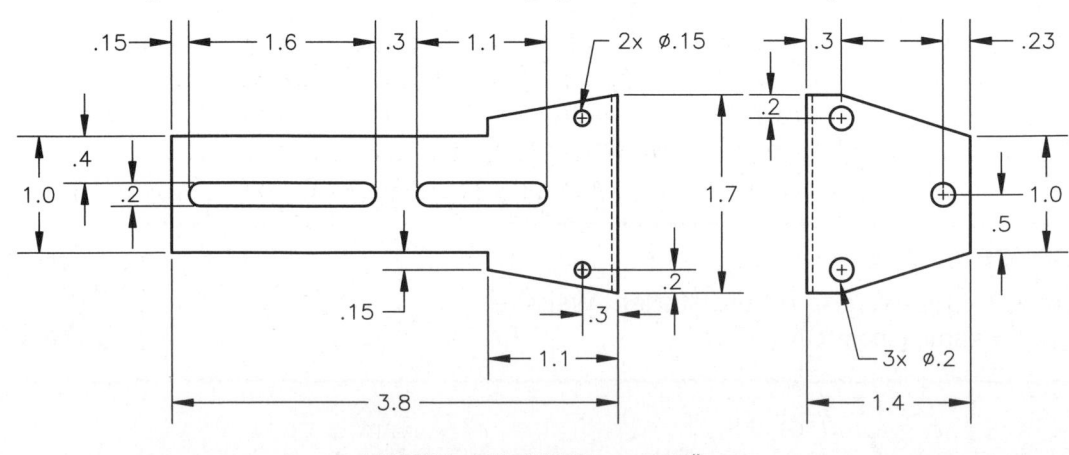

MATERIAL THICKNESS = .125"

2. This is an orthographic drawing of a light fixture bracket. Create it as a solid model. Use solid primitives and Boolean commands as needed. Use the dimensions given. Similar holes have the same offset dimensions. Create new UCSs as needed. Display an appropriate pictorial view of the drawing. Then, add dimensions. Plot the drawing to scale on a C-size sheet of paper. Save the drawing as P08_02.

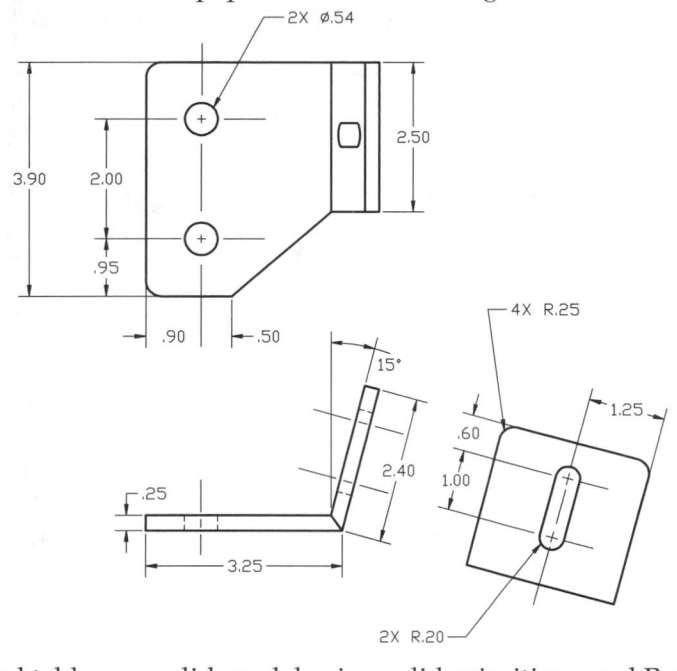

3. Create the end table as a solid model using solid primitives and Boolean commands as needed. The end result should be a single object. As a test of your object editing skills, try drawing the entire model by starting with only a single rectangle. You can copy, resize, extrude, and move objects as you create them from the single rectangle. Use the dimensions given and the following information to construct the model.
   A. Table height is 24".
   B. Top of bottom shelf is 5" off of the floor.
   C. Table legs must be located no less than 1/2" from the tabletop edge.
   D. Shelf must be no closer than .75" from the outside of table legs.
   E. Dimension the table as shown.
   F. Save the drawing as P08_03.

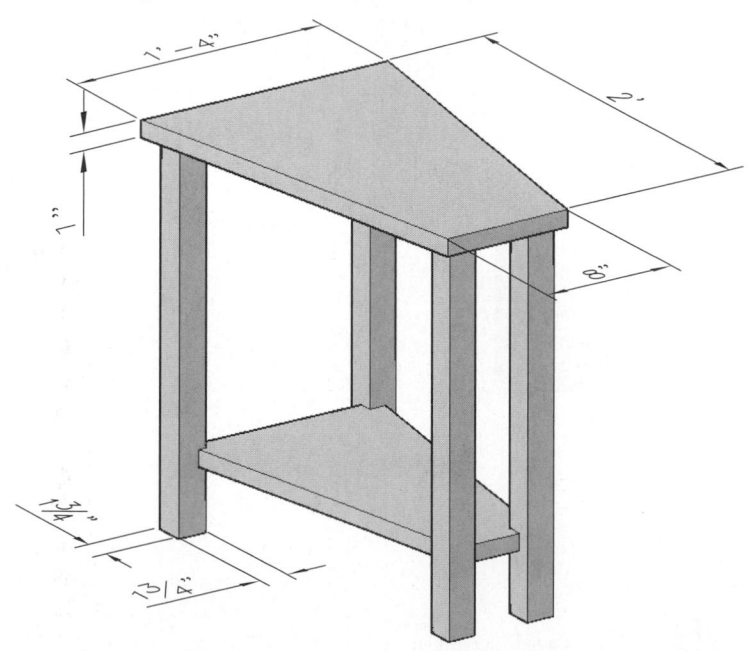

4. Shown below are the profiles of a roof gutter (for the collection of rainwater) and a gutter downspout. Draw the profiles in 3D using the dimensions shown. Use the following additional information to construct a 3D model like the one shown in the shaded view.

   A. Offset the gutter profile to create a material thickness of .025". Be sure to close the ends to create a closed polyline so a 3D solid is created when it is extruded.
   B. Extrude the gutter profile 12" to create a one-foot section.
   C. Relocate the downspout profile on the underside of the gutter.
   D. Construct an extrusion path for the downspout. Refer to the shaded view shown below, but use your own design.
   E. Extrude the downspout profile along the path.
   F. Dimension the end of the gutter profile in 3D.
   G. Save the drawing as P08_04.

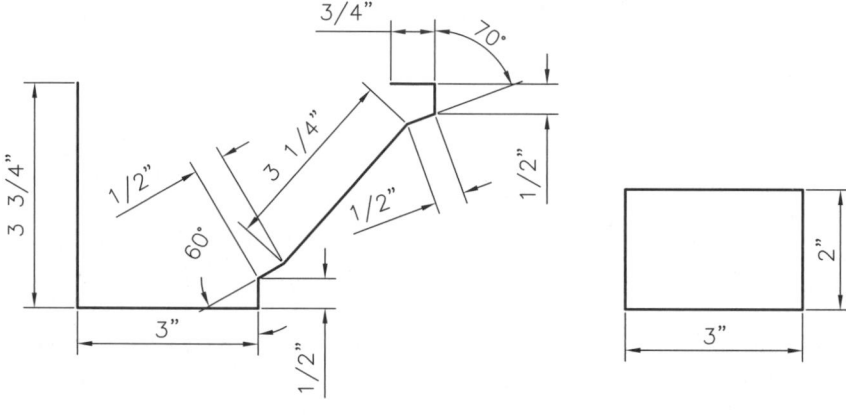

Gutter Profile Downspout Profile

A

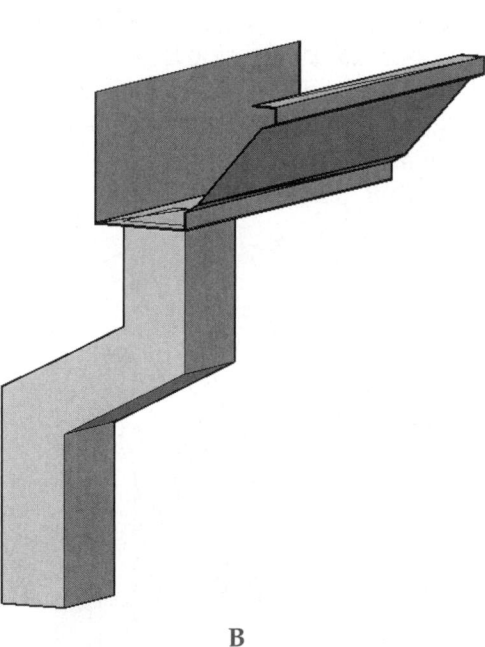

B

*Problems 5–7. These problems are mechanical parts. Create a solid model of each part. Dimension each model. Place the title of each model so it is plan to the pictorial view. Plot the finished drawings on B-size paper. Save each drawing as **P08_**(problem number).*

5.

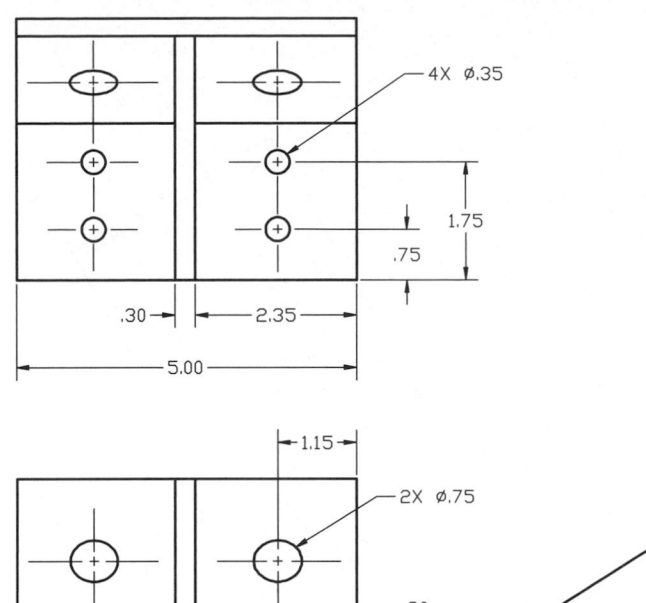

Angle Bracket

6.

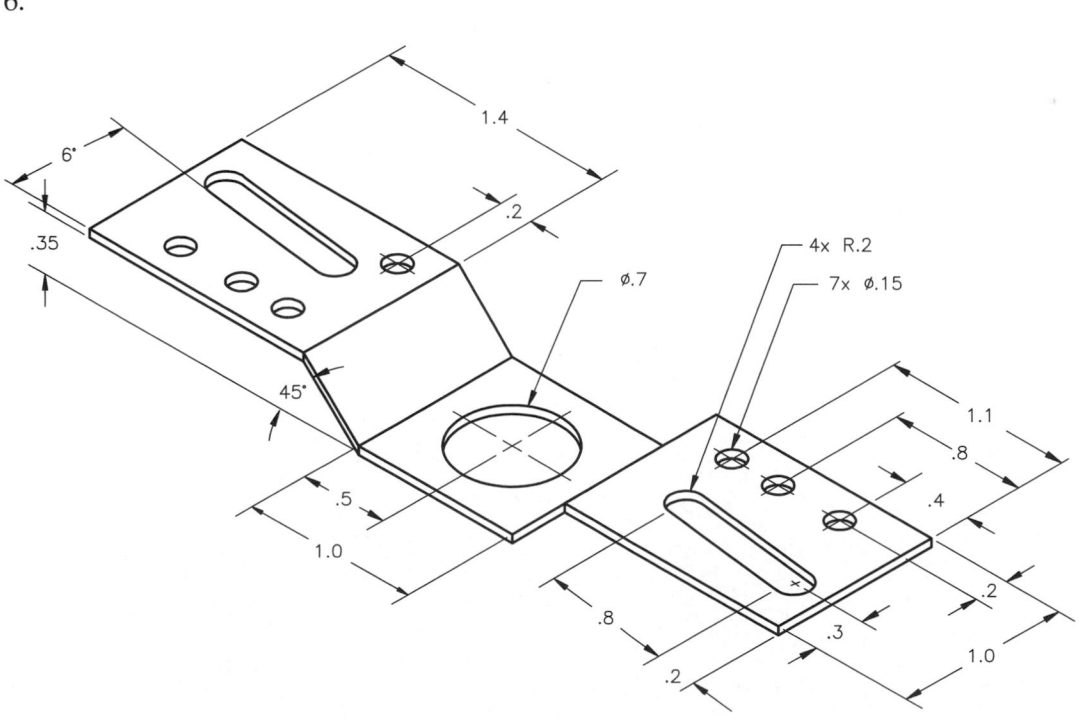

Guide Bracket

7.

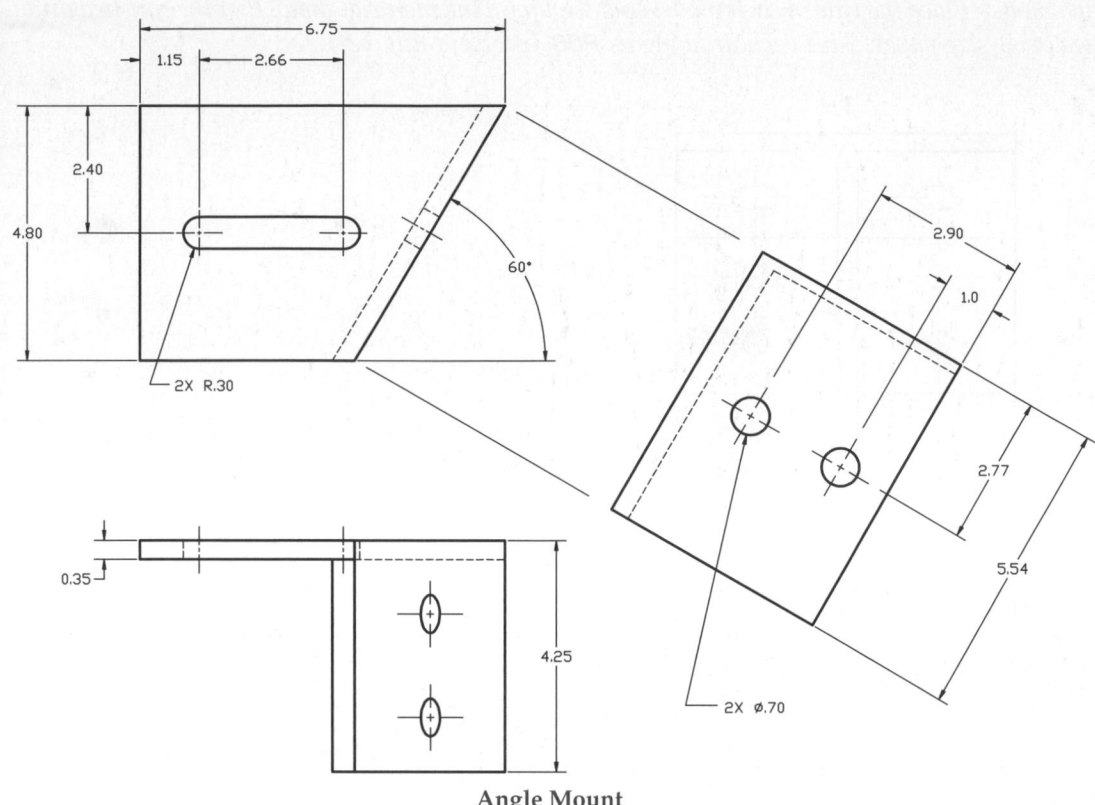

6.75

1.15

2.66

2.40

4.80

60°

2X R.30

2.90

1.0

2.77

5.54

2X ⌀.70

0.35

4.25

Angle Mount

CHAPTER 9

Solid Model Extrusions and Revolutions

Learning Objectives

After completing this chapter, you will be able to:

✓ Create solids and surfaces by extruding 2D profiles.
✓ Extrude planar surfaces.
✓ Create symmetrical 3D solids and surfaces by revolving 2D profiles.
✓ Revolve planar surfaces.
✓ Use solid extrusions and revolutions as construction tools.

Complex shapes can be created by applying a thickness to a two-dimensional profile. This is called *extruding* the shape. You have been introduced to the operation in previous chapters. Two or more profiles can be extruded to intersect. The resulting union can form a new shape by performing a Boolean operation. Symmetrical objects can be created by revolving a 2D profile about an axis to create a new solid.

Creating Solid Model Extrusions

A *solid extrusion* is a closed, two-dimensional shape that has been given thickness. The **EXTRUDE** command allows you to create extrusions from lines, arcs, elliptical arcs, 2D polylines, 2D splines, circles, ellipses, 2D solids, regions, planar surfaces, and donuts. Objects in a block cannot be extruded. Closed objects, such as circles, polygons, closed polylines, and donuts, are converted to solids when they are extruded. Open-ended objects, such as lines, arcs, polylines, elliptical arcs, and splines, are converted to a *surface extrusion* when they are extruded. Surface extrusions have no mass properties.

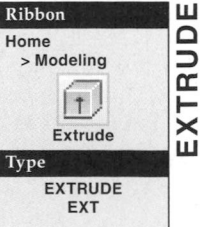

Extrusions can be created along a straight line or along a path curve. A taper angle can also be applied as you extrude an object. **Figure 9-1** illustrates a polygon extruded into a solid.

When the **EXTRUDE** command is selected, you are prompted to select the objects to extrude. Select the objects and press [Enter]. You are then prompted for the extrusion height. The height is always applied along the Z axis of the object, not the current UCS. A positive value extrudes above the XY plane of the object. A negative height value extrudes below the XY plane. If a pictorial view is displayed, you can drag the mouse to set the extrusion above or below the XY plane and then enter the height value.

Figure 9-1.
The **EXTRUDE** command creates a solid or surface by adding thickness to a 2D profile.
A—The initial, closed 2D profile. B—The extruded solid object shown with hidden lines removed.

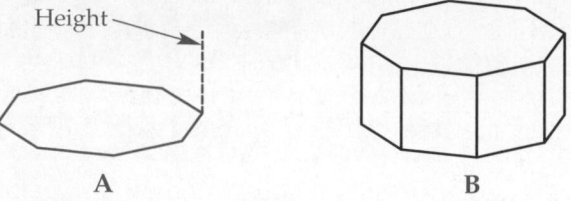

A   B

Figure 9-2.
A—A positive
angle tapers to the
inside of the object
from the base.
B—A negative angle
tapers to the outside
of the object.

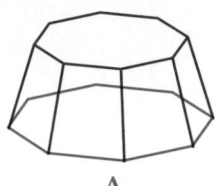 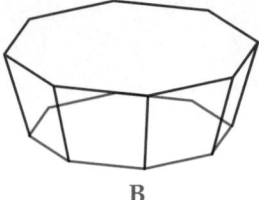

A   B

Before entering a height, you can specify a taper angle. The taper angle can be any value *between* +90° and –90°. A positive angle tapers to the inside of the object from the base. A negative angle tapers to the outside of the object from the base. See **Figure 9-2**. However, the taper angle cannot result in edges that "fold into" the extruded object.

PROFESSIONAL TIP

Objects such as polylines, lines, and arcs that have a thickness can be converted to surfaces using the **CONVTOSURFACE** command. Circles and closed polylines with a thickness can be converted to solids using the **CONVTOSOLID** command.

Extrusions along a Path

A 2D shape can be extruded along a path to create a 3D solid or surface. The path can be a line, circle, arc, ellipse, polygon, polyline, or spline. Line segments and other objects can be first joined to form a polyline path. The corners of angled segments on the extruded object are mitered, while curved segments are smooth. See **Figure 9-3**.

When open objects, such as lines, arcs, polylines, elliptical arcs, and splines, are used as the profile, they are converted to a swept surface when extruded along a path. A *sweep* is a solid or surface that is created when an open or closed curve is pulled,

Figure 9-3.
A—Angled
segments are
mitered when
extruded.
B—Curves are
smoothed when
extruded.

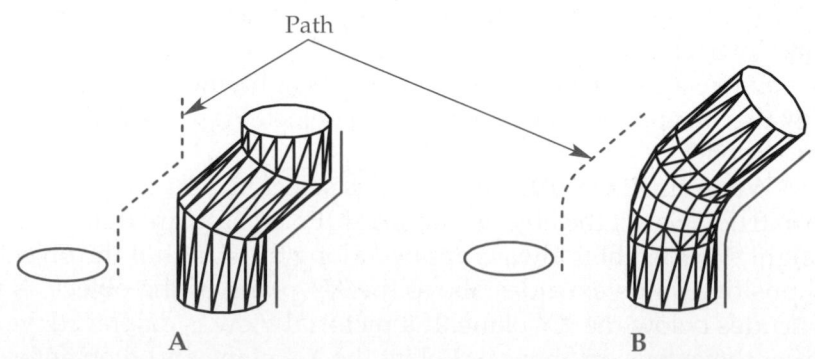

Path

A   B

or swept, along a 2D or 3D path. An extrusion is really a form of a sweep. Sweeps are discussed in detail in Chapter 10.

To extrude along a path, enter the **EXTRUDE** command and select the objects to extrude. When prompted for the height of the extrusion, enter the **Path** option. If needed, first enter a taper angle. Then, pick the object to be used as the extrusion path.

Objects can also be extruded along a line at an angle to the base object, **Figure 9-4.** Notice that the plane at the end of the extruded object is parallel to the original object. Also notice that the length of the extrusion is the same as that of the path. The path does not need to be perpendicular to the object.

If the path begins perpendicular to the profile, the cross section of the resulting extrusion is perpendicular to the path, regardless if the path is a straight line, curve, or spline. See **Figure 9-5.** If the path is a spline or curve that does not begin perpendicular to the profile, the profile may not remain perpendicular to the path as it is extruded.

If one of the endpoints of the path is not on the plane of the object to be extruded, the path is temporarily moved to the center of the profile. The extrusion is then created as if the path were connected to the original object, as shown in **Figure 9-4.**

NOTE

The **DELOBJ** system variable allows you to delete or retain the original extruded objects and path definitions. The settings are:

0   All original geometry and path definitions are retained.

1   Objects used for extrusion (profile curves) are deleted. This is the default.

2   All geometry used to define the extrusion, including path definitions, is deleted.

–1   You are prompted to delete objects used for the extrusion (profile curves).

–2   You are prompted to delete all geometry used to define the extrusion, including path definitions.

The **DELOBJ** system variable also affects the **REVOLVE**, **SWEEP**, and **LOFT** commands.

Extruding Regions

In Chapter 2, you learned how to create 2D regions. As an example, you created the top view of the base shown in **Figure 9-6A** as a region. Regions can be extruded to create 3D solids. The base you created in Chapter 2 can be extruded to create the final solid shown in **Figure 9-6B.** Any features of the region, such as holes, are extruded the

---

Figure 9-4.
A—An object extruded along a path. B—The end of an object extruded along an angled path is parallel to the original object.

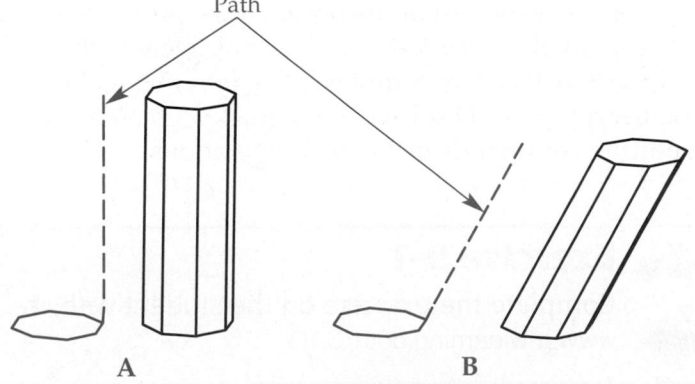

Figure 9-5.
A—Splines can be used as extrusion paths. Notice that the profile on the right is not perpendicular to the start of the path. B—The resulting extrusions.

A

B

Figure 9-6.
A—The 2D region that will be extruded. B—The solid object created by extruding the region, shown in a 3D wireframe display.

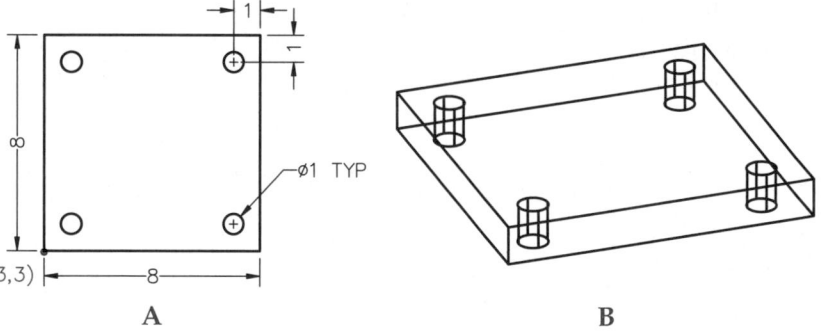

A

B

same thickness as the rest of the object. If the profile was created as polylines, the holes must be separately extruded and then subtracted from the solid. Using this method, you can construct a fairly complex 2D region that includes curved profiles, holes, slots, etc. Then, a complex 3D solid can be quickly created. Additional details can be added using editing commands or Boolean operations.

Exercise 9-1

Complete the exercise on the student website.
www.g-wlearning.com/CAD

Extruding a Planar Surface

A planar surface can be extruded into a solid object in the same manner as a region. Nonplanar (curved) surfaces cannot be extruded. Whereas both surfaces and regions have no thickness, the surface is an object composed of a mesh and the region is actually a solid that possesses mass properties. A surface can be quickly converted to a solid using the **EXTRUDE** command. Simply select the surface when prompted to select objects. The surface can be extruded in a specific direction, along a path, or at a taper angle.

Any closed object, such as a circle, rectangle, polygon, or polyline, can be converted into a surface with the **Object** option of the **PLANESURF** command. This surface can then be extruded into a 3D solid.

PROFESSIONAL TIP

You can also extrude a face on an existing solid into a new solid. When prompted to select objects, press the [Ctrl] key and pick the face to extrude. A face is a *subobject* of a solid. Subobject editing is covered in detail in Chapter 12.

Creating Solid Model Revolutions

The **REVOLVE** command allows you to create solids and surfaces by revolving a shape about an axis. Shapes that can be revolved include lines, arcs, circles, ellipses, polygons, polylines, closed splines, regions, planar surfaces, and donuts. The selected object can be revolved at any angle up to 360°. A *solid revolution* is created when a closed shape is revolved about an axis. A *surface revolution* is created when an open shape is revolved about an axis. Surface revolutions have no mass properties.

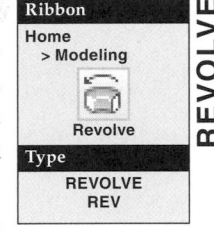

Ribbon
Home
> Modeling
Revolve
Type
REVOLVE
REV

REVOLVE

When the command is selected, you are prompted to pick the objects to revolve. Then, you must define the axis of revolution. The default option is to pick the two endpoints of an axis of revolution. This is shown in **Figure 9-7.** You can also revolve about an object or the X, Y, or Z axis of the current UCS. Once the axis is defined, you are prompted to enter the angle through which the profile will be revolved. When the angle is specified, the revolution is created.

PROFESSIONAL TIP

When creating solid models, keep in mind that the final part will most likely need to be manufactured. Be aware of manufacturing processes and methods as you design parts. It is easy to create a part in AutoCAD with internal features that may be impossible to manufacture, especially when revolving a profile.

Figure 9-7.
Points P1 and P2 are selected as the axis of revolution for the profile.

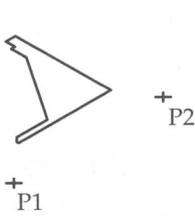

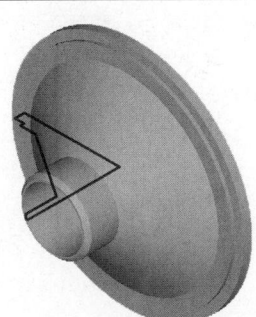

Figure 9-8.
An axis of revolution can be selected using the **Object** option of the **REVOLVE** command. Here, the line is selected as the axis.

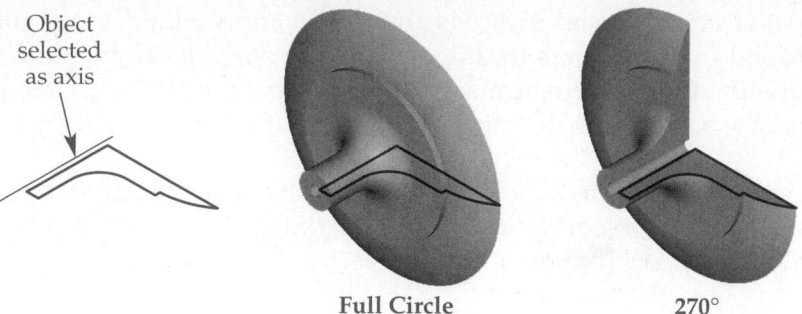

Full Circle   270°

Revolving about an Axis-Line Object

You can select an object, such as a line, as the axis of revolution. **Figure 9-8** shows a solid created using the **Object** option of the **REVOLVE** command. Both a full-circle (360°) revolution and a 270° revolution are shown. Enter the **Object** option when prompted for the axis of revolution. Then, pick the axis object and enter the angle through which the profile will be rotated. You can use the **Start Angle** option before entering an angle of revolution. This allows you to specify the point at which the revolution starts and then the angle of revolution.

Revolving about the X, Y, or Z Axis

The X axis of the current UCS can be used as the axis of revolution by selecting the **X** option of the **REVOLVE** command. The origin of the current UCS is used as one end of the X axis line. Notice in **Figure 9-9** that two different shapes can be created from the same 2D profile by changing the UCS origin. No hole appears in the object in **Figure 9-9B** because the profile was revolved about an edge that coincides with the X axis. The Y or Z axis can also be used as the axis of revolution. See **Figure 9-10.**

Exercise 9-2

Complete the exercise on the student website.
www.g-wlearning.com/CAD

Figure 9-9.
A—A solid is created using the X axis as the axis of revolution. B—A different object is created with the same profile by changing the UCS origin.

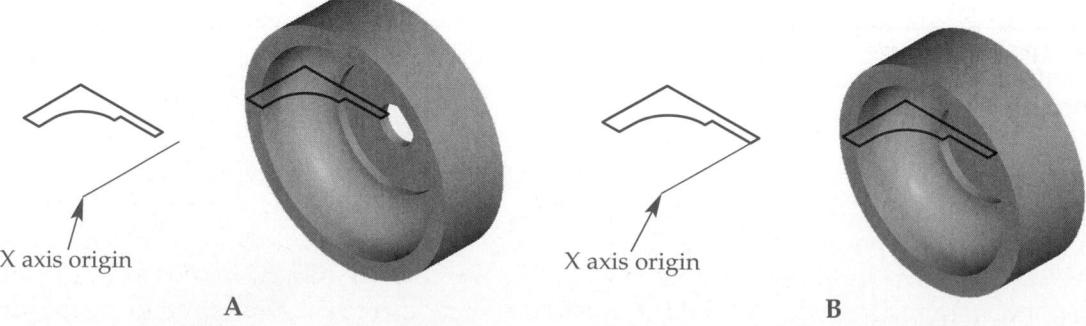

X axis origin   X axis origin

A   B

Figure 9-10.
A—A solid is created using the Y axis as the axis of revolution. B—A different object is created by changing the UCS origin.

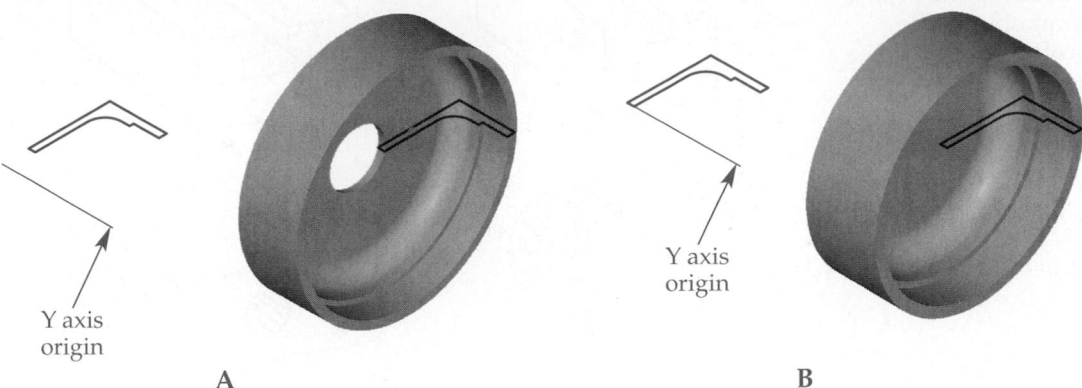

A

B

Revolving Regions

Earlier in this chapter, you learned that regions can be extruded. In this manner, holes, slots, keyways, etc., can be created. Regions can also be revolved. A complex 2D shape can be created using Boolean operations on regions. Then, the region can be revolved. One advantage of this method is it may be easier to create a region than trying to create a complex 2D profile as a single, closed polyline.

Revolving a Planar Surface

Just as a planar surface can be extruded into a solid object, it can also be revolved into a solid object. Nonplanar (curved) surfaces cannot be revolved. When the **REVOLVE** command is selected, simply pick the surface when prompted to select objects. The surface can be revolved about an axis defined by two pick points, an object, or the X, Y, or Z axis of the current UCS.

PROFESSIONAL TIP

You can also revolve a face on an existing solid into a new solid. When prompted to select objects, press the [Ctrl] key and pick the face to revolve. Subobject editing is covered in detail in Chapter 12.

Using Extrude and Revolve as Construction Tools

It is unlikely that an extrusion or revolution will result in a finished object. Rather, these operations will be used with other solid model construction methods, such as Boolean operations, to create the final object. The next sections discuss how to use **EXTRUDE** and **REVOLVE** with other construction methods to create a finished object.

Creating Features with Extrude

You can create a wide variety of features with the **EXTRUDE** command. Study the shapes shown in **Figure 9-11.** These detailed solid objects were created by drawing a profile and then using the **EXTRUDE** command. The objects in **Figures 9-11C** and **9-11D** must first be constructed as regions before they are extruded. For example, the five holes (circles) in **Figure 9-11D** must be removed from the base region using the **SUBTRACT** command.

Figure 9-11.
Detailed solids can be created by extruding the profile of an object. The profiles are shown here in color.

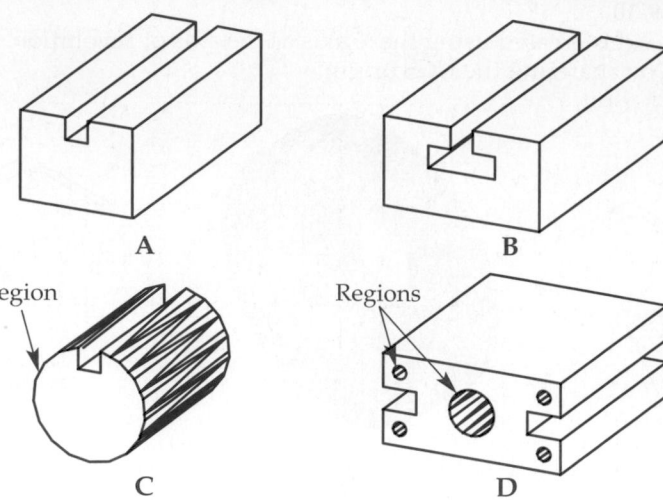

A

B

Region

Regions

C

D

Figure 9-12.
Most of this object can be created by extruding a profile. However, the holes must be added after the extruded solid is created.

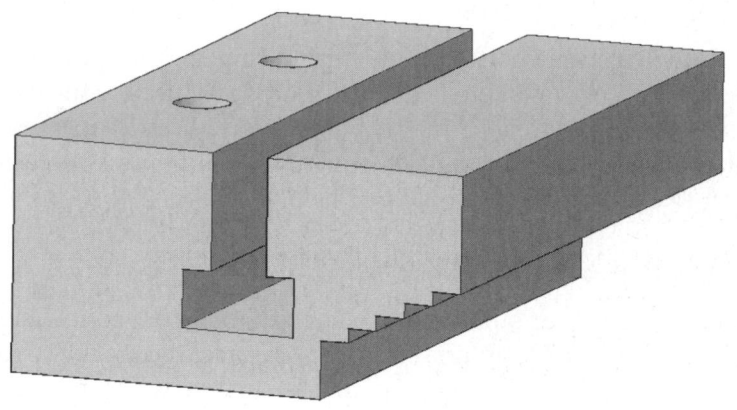

Look at **Figure 9-12.** This is part of a clamping device used to hold parts on a mill table. There is a T-slot milled through the block to receive a T-bolt and one side is stair-stepped, under which parts are clamped. If you look closely at the end of the object, most of the detail can be drawn as a 2D region and then extruded. However, there are also two holes in the top of the block to allow for bolting the clamp to the mill table. These features must be added to the extruded solid.

First, change the UCS to the front preset orthographic UCS. Display a plan view of the UCS. Then, draw the profile shown in **Figure 9-13** using the **PLINE** command. You can draw it in stages, if you like, and then use the **PEDIT** command to join all segments into a single polyline.

Next, use the **EXTRUDE** command to create the 3D solid. Extrude the profile a distance of –6 units with a 0° taper. This will extrude the object away from you. Display the object from the southeast isometric preset viewpoint or use the view cube to display a pictorial view. The object should look similar to **Figure 9-12** without the holes in the top. Set the Conceptual visual style current, if you like.

The two holes are ⌀.5 units and evenly spaced on the surface through which they pass. Change to the WCS and draw a construction line from midpoint to midpoint, as shown in **Figure 9-14.** Then, set **PDMODE** to an appropriate value, such as 3, and use the **DIVIDE** command to divide the construction line into three parts. The two points created by the **DIVIDE** command are equally spaced on the surface and can be used to locate the two holes.

There are two ways to create a hole. You can draw a circle and extrude it to create a cylinder or you can draw a solid cylinder. Either way, you need to subtract the cylinder

EXTRUDE

Ribbon
Home
> Modeling

Extrude

Type
EXTRUDE
EXT

AutoCAD and Its Applications—Advanced

Figure 9-13.
This is the profile that will be extruded for the clamping block.

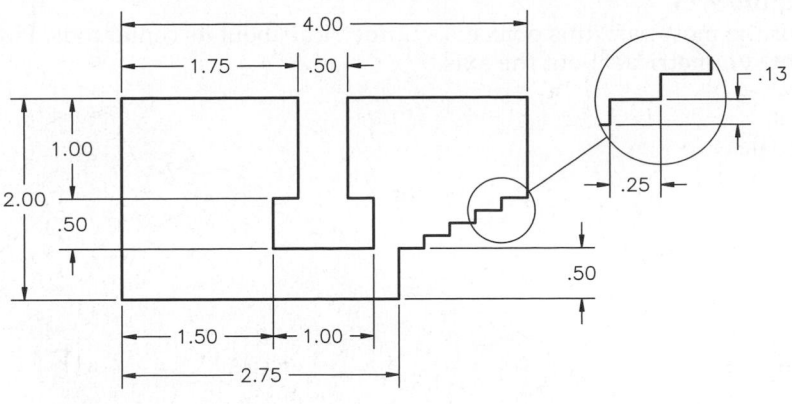

Figure 9-14.
Draw a construction line (shown here in color) and divide it into three parts.

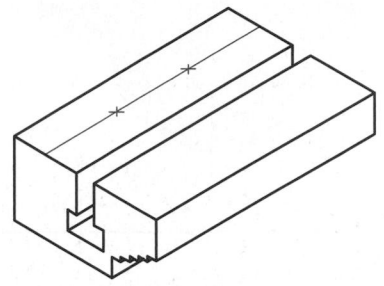

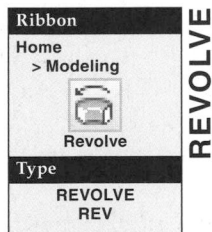

to create the hole. Drawing a solid cylinder is probably easiest. When prompted for a center, use the **Node** object snap to select the point. Then, enter the diameter. Finally, enter a negative height so that the cylinder extends into the solid or drag the cylinder down in the 3D view so it extends all of the way through the block. The actual height is not critical, as long as it extends through the block.

You can either copy the first cylinder to the second point or draw another cylinder. When both cylinders are located, use the **SUBTRACT** command to remove them from the solid. The object is now complete and should look like **Figure 9-12.**

Creating Features with Revolve

The **REVOLVE** command is very useful for creating symmetrical, round objects. However, many times the object you are creating is not completely symmetrical. For example, look at the camshaft in **Figure 9-15.** For the most part, this is a symmetrical, round object. However, the cam lobes are not symmetrical in relation to the shaft and bearings. The **REVOLVE** command can be used to create the shaft and bearings. Then, the cam lobes can be created and added.

Start a new drawing and make sure the WCS is the current UCS. Using the **PLINE** command, draw the profile shown in **Figure 9-16A.** This profile will be revolved through 360°, so you only need to draw half of the true plan view of the cam profile. The profile represents the shaft and three bearings.

Next, display the drawing from the southwest isometric preset viewpoint. Then, use the **REVOLVE** command to create the base camshaft. Pick the endpoints shown in **Figure 9-16A** as the axis of revolution. Revolve the profile through 360°. Zoom extents and set the Conceptual visual style current to clearly see the object.

Now, you need to create one cam lobe. Change the UCS to the left orthographic preset. Then, draw a construction point in the center of the left end of the camshaft. Use the **Center** object snap and an appropriate **PDMODE** setting. Next, draw the profile shown in **Figure 9-16B.** Use the construction point as the center of the large radius. You may want to create a new layer and turn off the display of the base camshaft.

Ribbon
Home
> Modeling

Revolve

Type
REVOLVE
REV

REVOLVE

Figure 9-15.
For the most part, this object is symmetrical about its center axis. However, the cam lobes are not symmetrical about the axis.

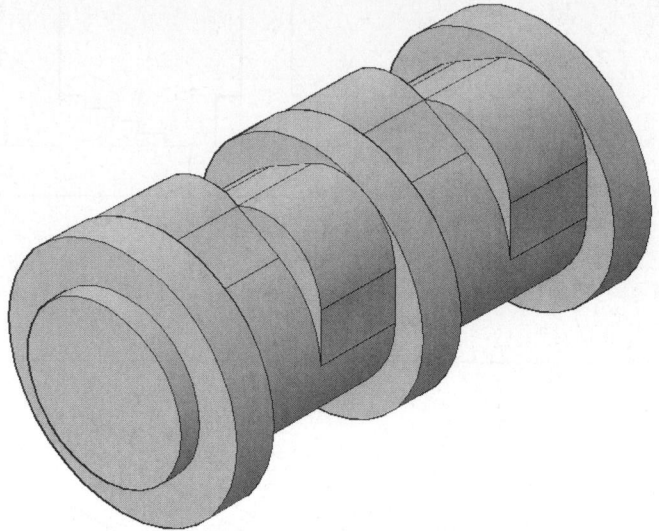

Figure 9-16.
A—This profile will be revolved to create the shaft and bearings. B—This is the profile of one cam lobe, which will be extruded.

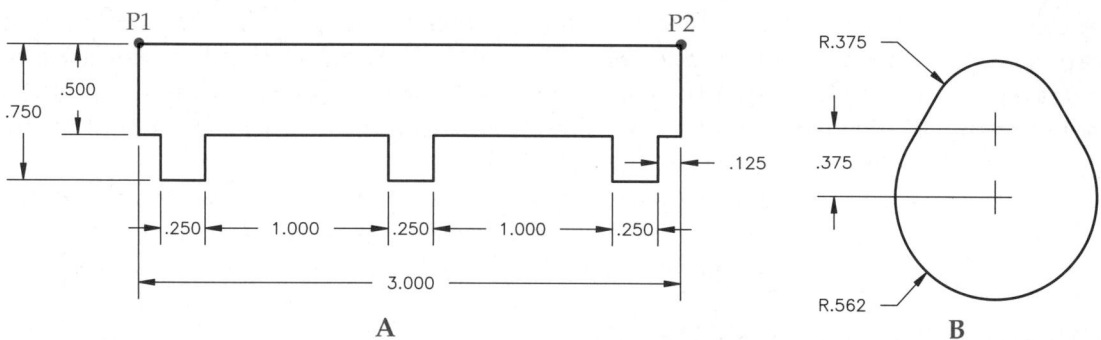

Once the cam lobe profile is created, use the **REGION** command to create a region. Then, use the **EXTRUDE** command to extrude the region a height of –.5 units (into the camshaft). The extrusion should have a 0° taper. If you turned off the display of the base camshaft, turn it back on now.

One cam lobe is created, but it is not in the proper position. With the left UCS current, move the cam lobe –.375 units on the Z axis. If a different UCS is current, the axis of movement will be different. This places the front surface of the cam lobe on the back surface of the first bearing. Now, make a copy of the lobe that is located –.5 units on the Z axis. Finally, copy the first two cam lobes –1.25 units on the Z axis.

You now need to rotate the four cam lobes to their correct orientations. Make sure the left UCS is still current. Then, rotate the first and third cam lobes 30°. If a different UCS is current , you can use the **3DROTATE** command. The center of rotation should be the center of the shaft. There are many points on the shaft to which the **Center** object snap can snap; they are all acceptable. You can also use the construction point as the center of rotation. Rotate the second and fourth cam lobes –30° about the same center.

Finally, use the **UNION** command to join all objects. The final object should appear as shown in **Figure 9-15.** Use the view cube to see all sides of the object. You can also create a rotating display using the **3DORBIT** command.

AutoCAD and Its Applications—Advanced

Multiple Intersecting Extrusions

Many solid objects have complex curves and profiles. These can often be constructed from the intersection of two or more extrusions. The resulting solid is a combination of only the intersecting volumes of the extrusions. The following example shows the construction of a coat hook.

1. Construct the first profile, **Figure 9-17A.**
2. Construct the second profile located on a common point with the first, **Figure 9-17B.**
3. Construct the third profile located on the common point, **Figure 9-17C.**
4. Extrude each profile the required dimension into the same area. Be careful to specify positive or negative heights for each extrusion, **Figures 9-17D** and **9-17E.**
5. Use the **INTERSECT** command to create a composite solid from the volume shared by the three extrusions, **Figure 9-17F.**

Figure 9-17.
Constructing a coat hook. A—Draw the first profile. B—Draw the second profile. C—Draw the third profile. All three profiles should have a common origin. D—Extrude each profile so that the extruded objects intersect. E—The extruded objects after the Conceptual visual style is set current. F—Use the **INTERSECT** command to create the composite solid. The final solid is shown here with the Conceptual visual style set current.

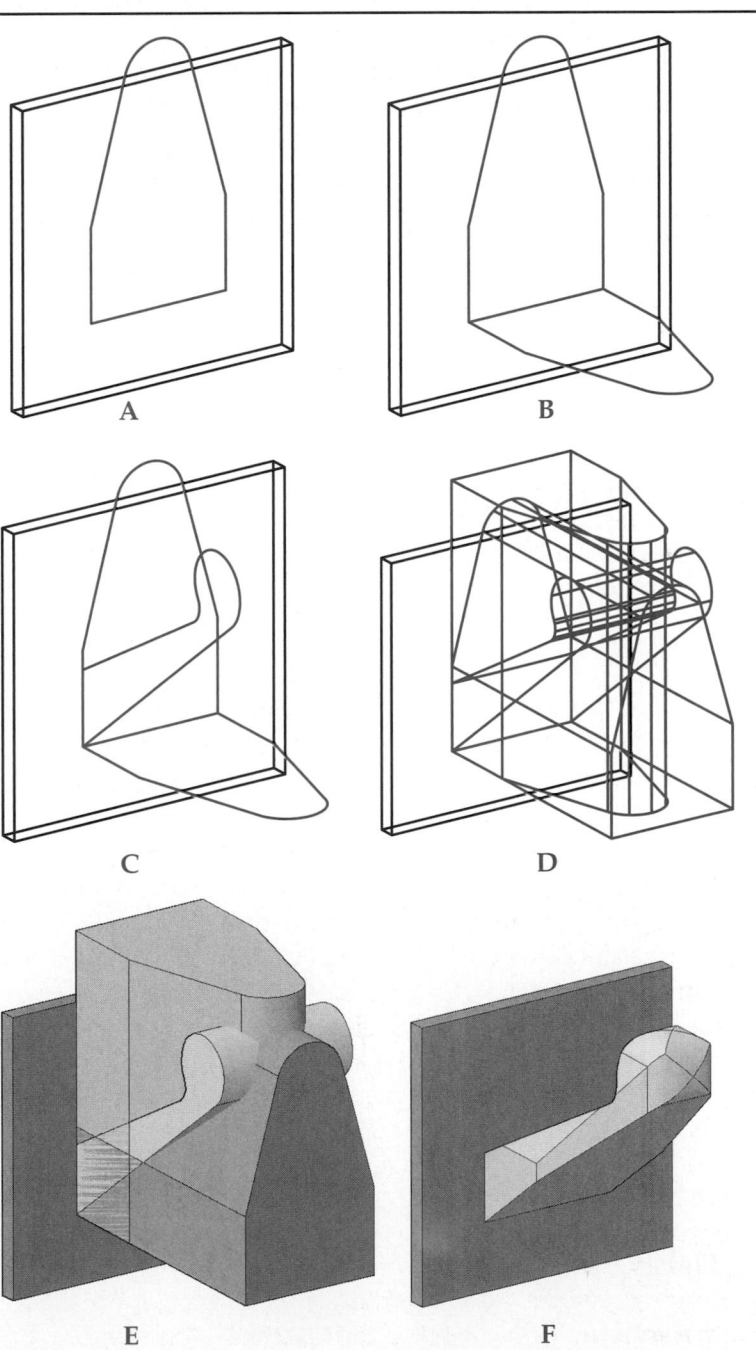

Chapter Test

1. What is an *extrusion?*
2. How do you create a surface extrusion?
3. Briefly describe how to create a solid extrusion.
4. Which command can be used to convert circles and closed polylines with a thickness to solids?
5. How can an extrusion be constructed to extend below the XY plane of the current UCS?
6. What is the range in which a taper angle can vary?
7. How can a curved extrusion be constructed?
8. Which system variable allows you to delete or retain the original extruded objects and path definitions?
9. How is the height of an extrusion applied in relation to the original object?
10. Which type(s) of surface(s) can be extruded?
11. What is a *surface revolution?*
12. How do you create a solid revolution?
13. What are the five different options for selecting the axis of revolution for a revolved solid?
14. How can a given profile be revolved twice (or more) about the same axis and create different shaped solids?
15. What is one advantage of revolving a region over revolving a polyline?

Drawing Problems

1. Construct a 12′ long section of wide flange structural steel with the cross section shown below. Use the dimensions given. Save the drawing as P09_01.

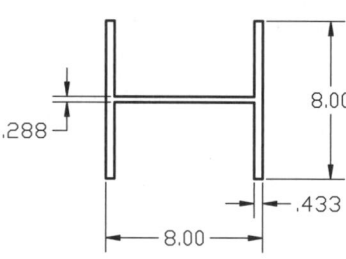

Drawing Problems - Chapter 9

Problems 2–7. These problems require you to use a variety of solid modeling methods to construct the objects. Use **EXTRUDE**, **REVOLVE**, *solid primitives, new UCSs, and Boolean commands to assist in construction. Do not create section views. Save each as* P09_*(problem number).*

2.

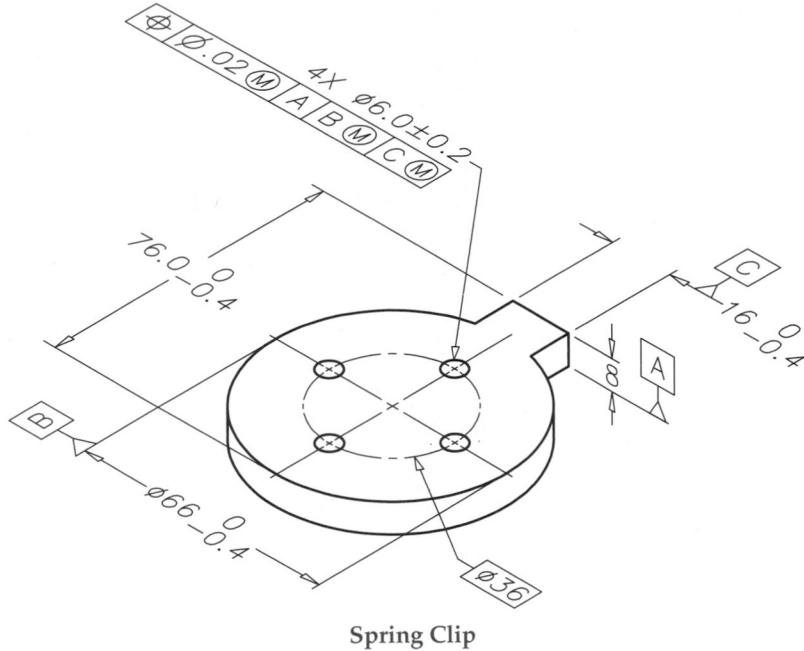

Spring Clip

3.

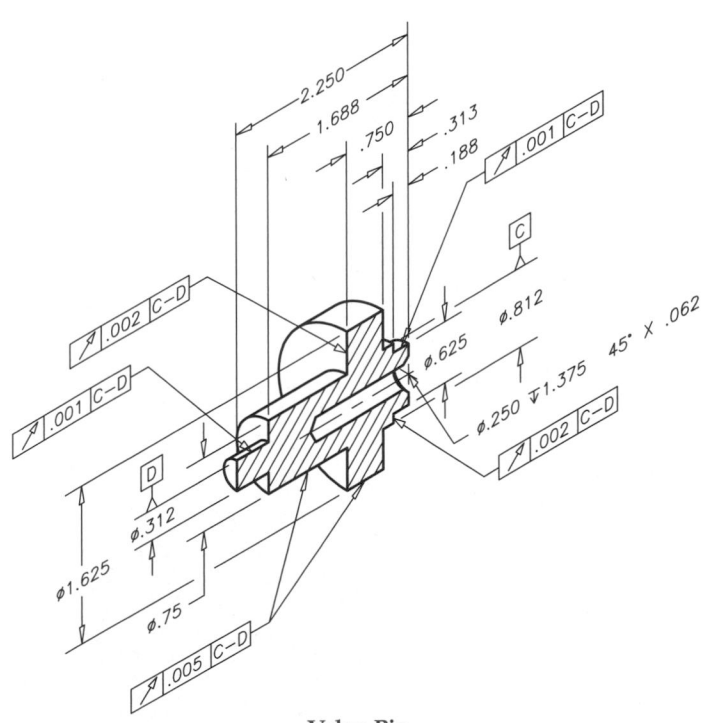

Valve Pin

4.

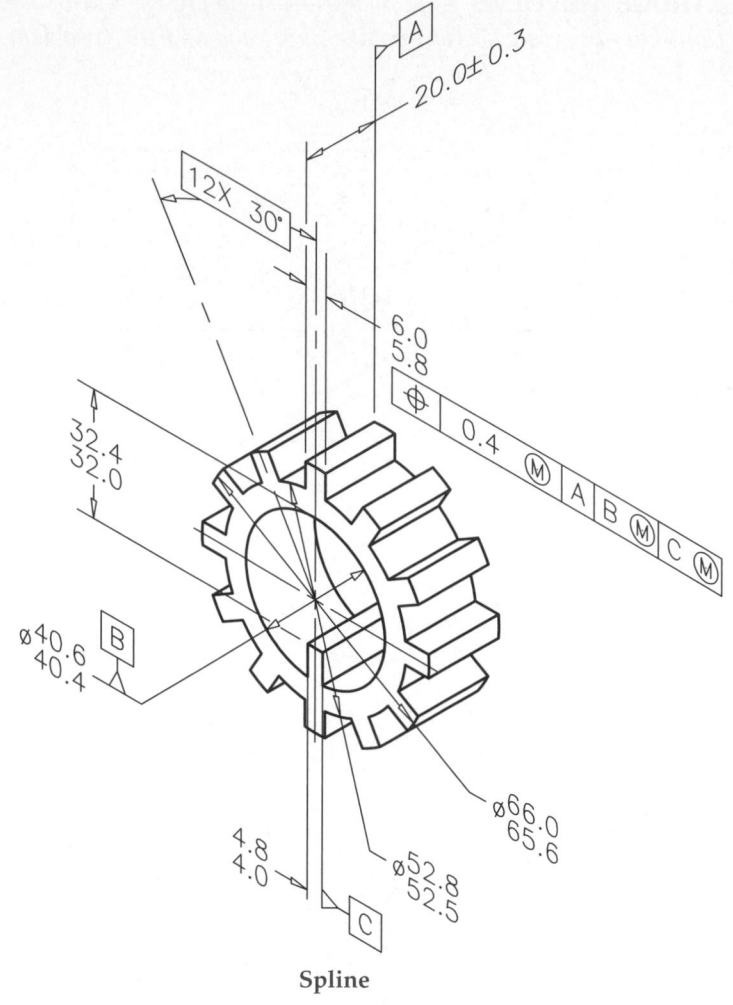

Spline

5.

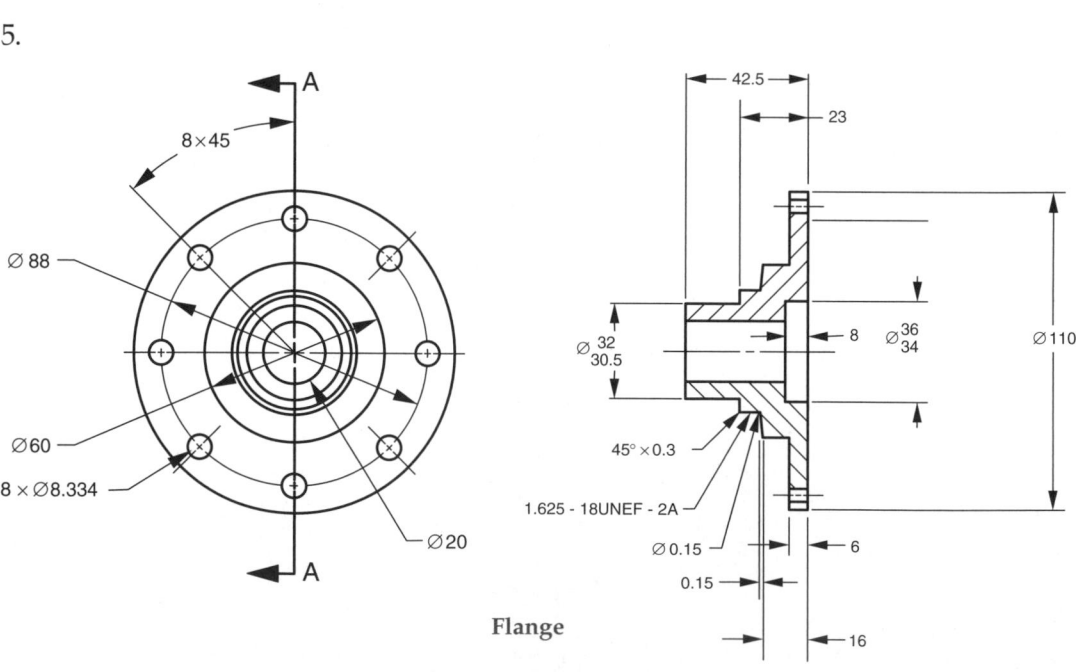

Flange

6.

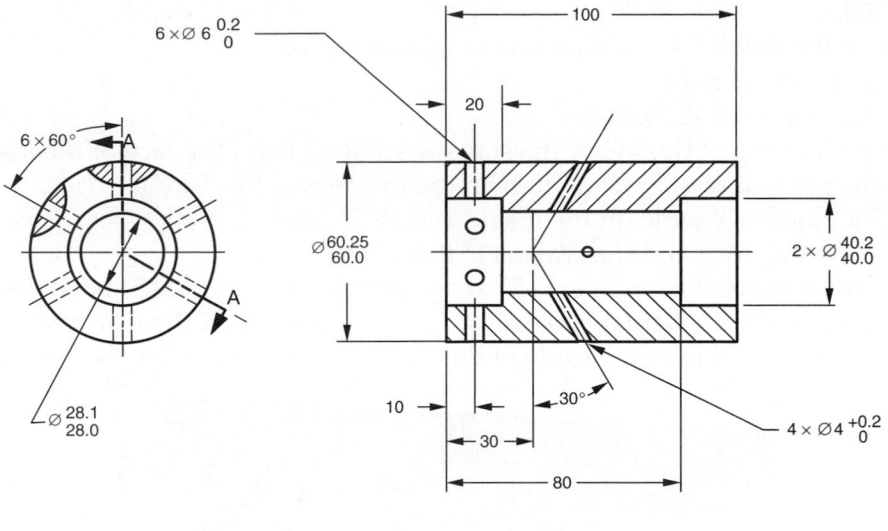

SECTION A-A

Nozzle

7.

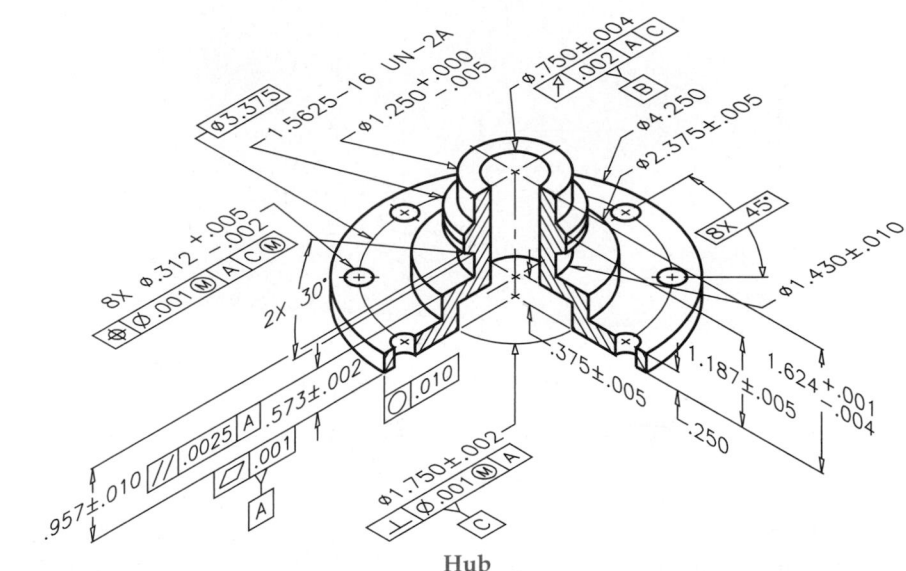

Hub

8. Create the stairway shown below using the following parameters. Save the file as P09_08.
   A. Use the detail for the riser and tread dimensions.
   B. There are 13 risers.
   C. The stairs are 42″ wide.
   D. The landing at the top of the stairs is 48″ long from the face of the last riser.
   E. The vertical wall is 15′-1″ high on the inside and 15′-7″ long.
   F. The floor is 8′ wide on the inside and 15′-7″ long.
   G. Draw the floor and the wall as 1″ thick.
   H. The center of the banister is 3″ away from the wall and 34″ above the steps.
   I. The ends of the banister are directly above the face of the first and last riser.
   J. Use the detail for the profile of the banister. Use **EXTRUDE** as needed.

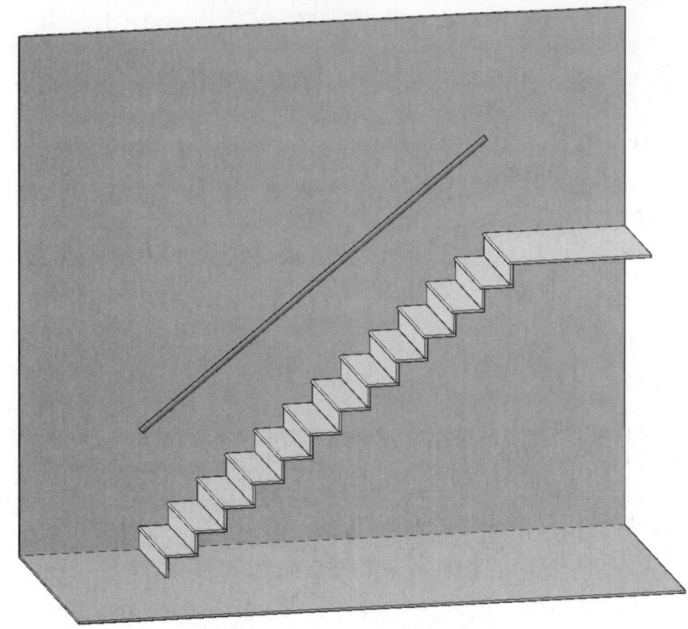

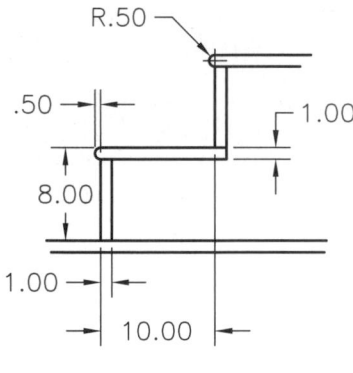

STAIR DETAIL

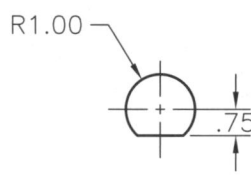

BANISTER DETAIL

AutoCAD and Its Applications—Advanced

9. Construct picture frame moldings using the profiles shown below.
   A. Draw each of the closed profiles shown. Use your own dimensions for the details of the moldings.
   B. The length and width of A and B should be no larger than 1.5″ × 1″.
   C. The length and width of C and D should be no larger than 3″ × 1.5″.
   D. Construct an 8″ × 12″ picture frame using moldings A and B.
   E. Construct a 12″ × 24″ picture frame using moldings C and D.
   F. Save the drawing as P09_09.

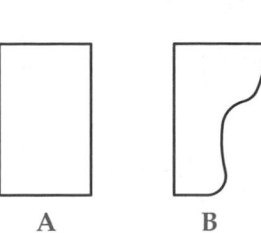 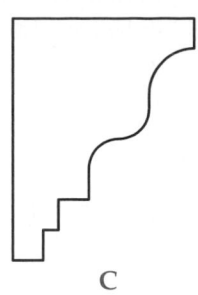 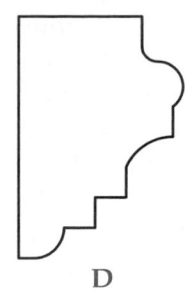

A   B   C   D

10. In this problem, you will refine the seat of the kitchen chair that you started in Chapter 2. You will use an extrusion following a path to create a curved, receding edge under the seat.
   A. Open P02_11 from Chapter 2. If you have not yet completed this model, do so now.
   B. Create a path for the extrusion by drawing a polyline that exactly matches either the upper or lower edge of the seat. Refer to the drawing shown below.
   C. Change the view and UCS as needed to display a plan view of the edge of the seat. Draw the profile shown below.
   D. Extrude the profile along the path and then subtract the extrusion from the seat.
   E. Save the drawing as P09_10.

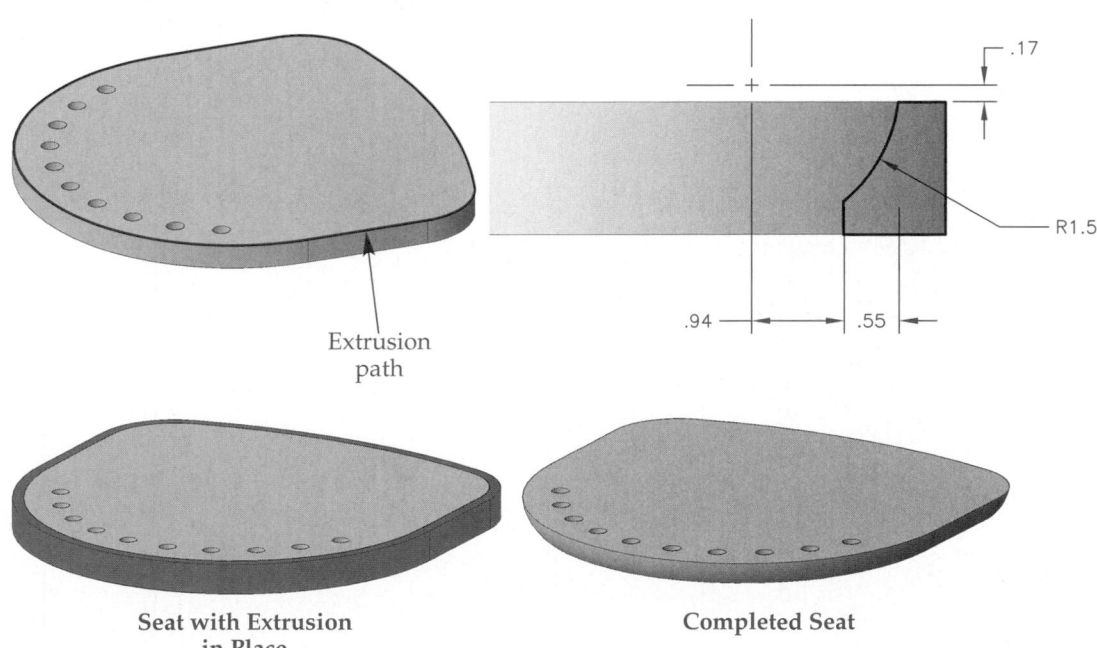

Extrusion path

Seat with Extrusion in Place

Completed Seat

Drawing Problems - Chapter 9

11. In this problem, you will be taking a manually drawn layout from an archive. You are to create a 3D model of the garage to update the archive.

  A. Review the manually drawn layout. Make note of the construction details shown.

  B. Research any additional details needed to construct the model. For example, the thickness of the doors is not listed as these are purchased items. However, you need to know these dimensions to draw the 3D model.

  C. Using what you have learned, create the garage as a solid model. Be sure to create all components, including the studs in the walls, the footings, and the anchor bolts.

  D. Create layers as needed. For example, you may wish to place the wall sheathing on a layer so it can be hidden to show the studs.

  E. Save the drawing as P09_11.

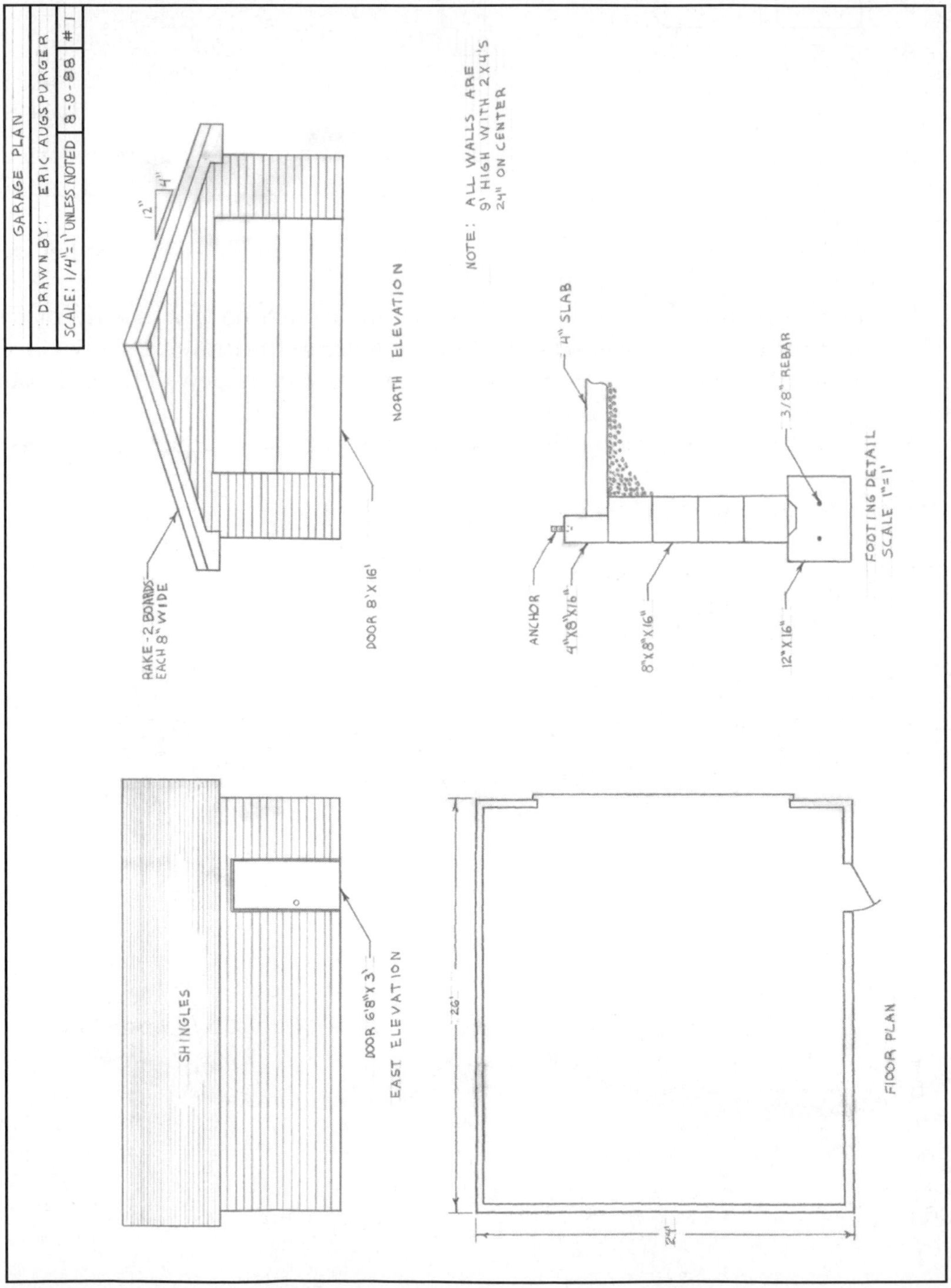

GARAGE PLAN

DRAWN BY: ERIC AUGSPURGER

8-9-88 #1

SCALE: 1/4" = 1' UNLESS NOTED

NORTH ELEVATION

RAKE - 2 BOARDS EACH 8" WIDE

DOOR 8' X 16'

NOTE: ALL WALLS ARE 9' HIGH WITH 2X4'S 24" ON CENTER

4" SLAB

ANCHOR

4"X8"X16"

8"X8"X16"

12"X16"

3/8" REBAR

FOOTING DETAIL SCALE 1" = 1'

SHINGLES

DOOR 6'8" X 3'

EAST ELEVATION

26'

24'

FLOOR PLAN

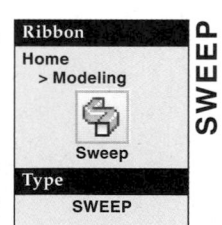

CHAPTER **10**

Sweeps and Lofts

Learning Objectives

After completing this chapter, you will be able to:

✓ Sweep 2D shapes along a 2D or 3D path to create a solid or surface object.
✓ Create 3D solid or surface objects by lofting a series of cross sections.

In the previous chapter, you learned about extruded solids and surfaces. Sweeps and lofts are similar to extrusions. In fact, an extrusion is really just a type of sweep. A *sweep* is an object created by extruding a single 2D profile along a path object. Sweeping an open shape along the path results in a surface object. If a closed shape is swept, a solid object is created. A *loft* is an object created by extruding between two or more 2D profiles. The shape of the loft object blends from one cross-sectional profile to the next. The profiles can control the loft or it can be controlled by one path or multiple guide curves. As with a sweep, open shapes result in surfaces and closed shapes give you solids. Open and closed shapes cannot be used together in the same loft.

Creating Swept Surfaces and Solids

The **SWEEP** command is used to create swept surfaces and solids. The command requires at least two objects:

- 2D shape to be swept.
- 2D or 3D shape to be used as the sweep path.

The profile can be aligned with the path, you can specify the base point, a scale factor can be applied, and the profile can be twisted as it is swept. The command procedure and options are the same for both swept solids and surfaces.

Sweeping an open shape creates a surface. See **Figure 10-1**. The objects that can be swept to create surfaces include lines, arcs, elliptical arcs, 2D polylines, 2D splines, and traces. Sweeping a closed shape creates a solid. See **Figure 10-2**. The objects that can be swept to create a solid include circles, ellipses, closed 2D polylines, closed 2D splines, regions, planar surfaces, and planar faces of solids. The sweep path for either a surface or a solid can be a line, arc, circle, ellipse, elliptical arc, 2D polyline, 2D spline, 3D polyline, 3D spline, helix, or the edge of a surface or solid.

Ribbon
Home
> Modeling
Sweep
Type
SWEEP

SWEEP

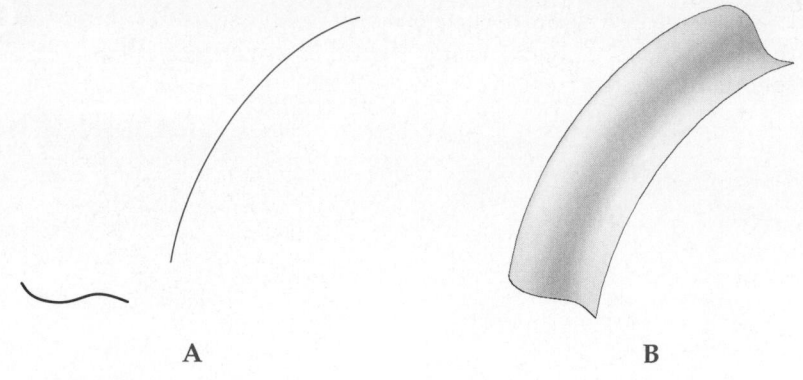

Figure 10-1.
A—This open shape will be swept along the path (shown in color). B—The resulting surface.

A   B

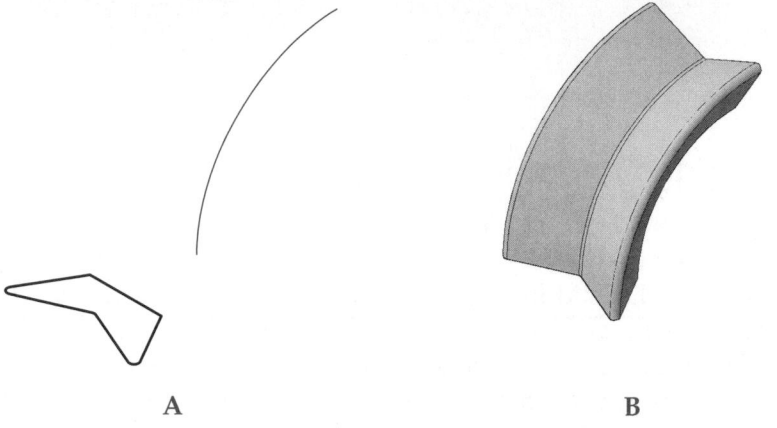

Figure 10-2.
A—This closed shape will be swept along the path (shown in color). B—The resulting solid.

A   B

When the command is initiated, you are prompted to select the objects to sweep. Select the profile(s) and press [Enter]. Planar faces of solids may be selected by holding the [Ctrl] key as you select. Multiple profiles can be selected. They are swept along the same path, but separate objects are created.

Next, you are prompted to select the path. The path and profile can lie on the same plane. Select the object to be used as the sweep path and press [Enter]. To select the edge of a surface or solid as the path, press the [Ctrl] key and then select the edge. The profile is then moved to be perpendicular to the path and extruded along the path. The sweep starts at the endpoint of the path nearest to where you selected it.

Exercise 10-1

Complete the exercise on the student website.
www.g-wlearning.com/CAD

Changing the Alignment of the Profile

By default, the profile is aligned perpendicular to the sweep path. However, you can create a sweep where the profile is not perpendicular to the path. See **Figure 10-3.** After the **SWEEP** command is initiated, select the profile and press [Enter]. Before selecting the path, enter the **Alignment** option. The default setting of **Yes** means that profile will be moved so it is perpendicular to the path. If you select **No**, the profile is kept in the same position relative to the path as it is swept. The position of the 2D shape determines the alignment.

Figure 10-3.
A—The profile and path for the sweep. B—By default, the profile is aligned perpendicular to the path when swept. C—Using the **Alignment** option, the profile can be swept so it is not perpendicular to the path.

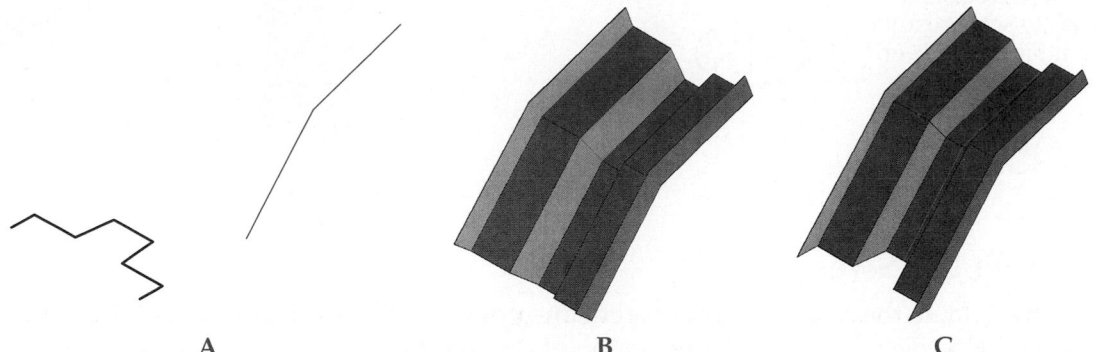

A   B   C

Changing the Base Point

The base point is the location on the shape that will be moved along the path to create the sweep. By default, if the 2D shape intersects the path, the profile is swept along the path at the point of intersection. If the 2D shape does not intersect the path, the default base point depends on the type of object being swept. When lines and arcs are swept, the default base point is their midpoint. Open polylines have a default base point at the midpoint of their total length.

The base point can be any point on the 2D shape or anywhere in the drawing. See **Figure 10-4.** To change the base point, use the **Base point** option of the **SWEEP** command. When the command is initiated, select the profile and press [Enter]. Before selecting the path, enter the **Base point** option. Next, pick the new base point. It does not have to be on an existing object. Once the new base point is selected, pick the path to create the sweep.

PROFESSIONAL TIP

If the location of the base point in relationship to the path is important, line up the shape with the path before starting the **SWEEP** command. Turn off the **Alignment** option in this situation.

Figure 10-4.
A—The profile and path for the sweep. B—The sweep is created with the default base point. C—The end of the path is selected as the base point. Notice the difference in this sweep and the one shown in B.

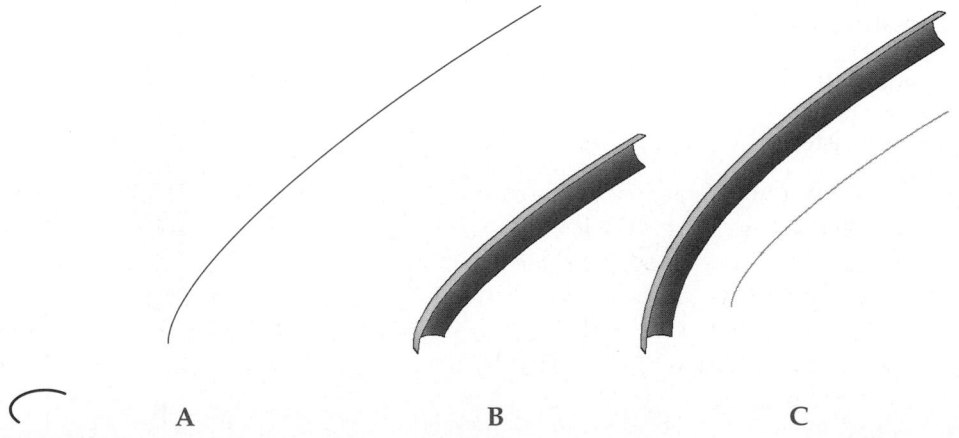

A   B   C

Chapter 10 Sweeps and Lofts

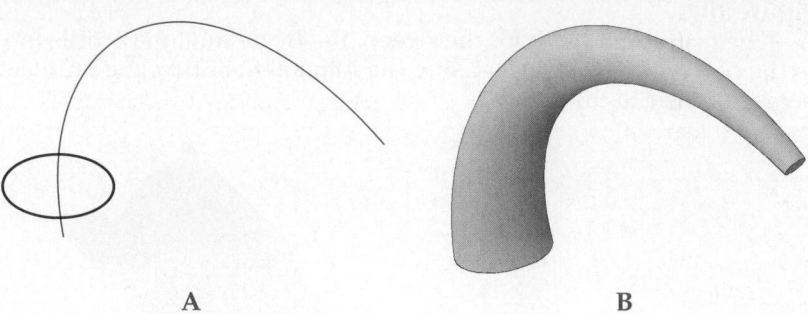

Figure 10-5.
A—The profile and path for the sweep. B—The resulting sweep. Notice how the .25 scale results in a tapered sweep.

A   B

Scaling the Sweep Profile

By default, the size of the profile remains uniform from the beginning of the path to the end. However, using the **Scale** option of the **SWEEP** command, you can change the scale of the profile at the end of the path. This, in effect, tapers the sweep. **Figure 10-5** shows a .25 scale applied to a sweep object. A 2D polyline must be edited using the **Fit** or **Spline** option in order to be used as a path. Sharp corners will not work with the **Scale** option. A 3D polyline path must be a spline.

Once the **SWEEP** command is initiated, select the profile and press [Enter]. Before selecting the path, enter the **Scale** option. You are prompted for the scale. Enter the scale value and press [Enter]. The scale value must be greater than zero. You can also enter the **Reference** option. With this option, pick two points for the first reference line and then two points for the second reference line. The difference in scale between the two distances is the scale value. Once the scale is set, pick the path to create the sweep.

Twisting the Sweep

The profile can be rotated as it is swept along the length of the path by using the **Twist** option of the **SWEEP** command. The angle that you enter indicates the rotation of the shape along the path of the sweep. The higher the number, the more twists in the sweep. **Figure 10-6** shows how a simple, closed profile and a straight line can be used to create a milling tool. The profile was swept with a 270° twist.

Once the **SWEEP** command is initiated, select the profile and press [Enter]. Then, before selecting the path, enter the **Twist** option. You are prompted for the twist angle or to enter the **Bank** option.

Banking is the natural rotation of the profile on a 3D sweep path, similar to a banked curve on a racetrack. See **Figure 10-7**. The path must be 3D (nonplanar) to set

Figure 10-6
A—The profile and path for creating the end mill. B—The resulting end mill model. Notice how the profile is twisted (rotated) as it is swept.

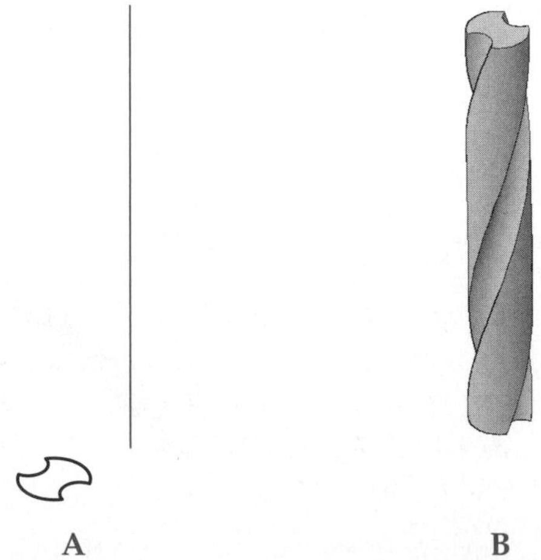

A   B

Figure 10-7.
A—The profile and path for the sweep are shown in red. B—Banking is off for this sweep. When viewed from the side, you can see that the profile does not bank through the curve. Look at the upper-right corner. C—Banking is on for this sweep. Notice how the profile banks, or leans, through the curve. Compare this to B.

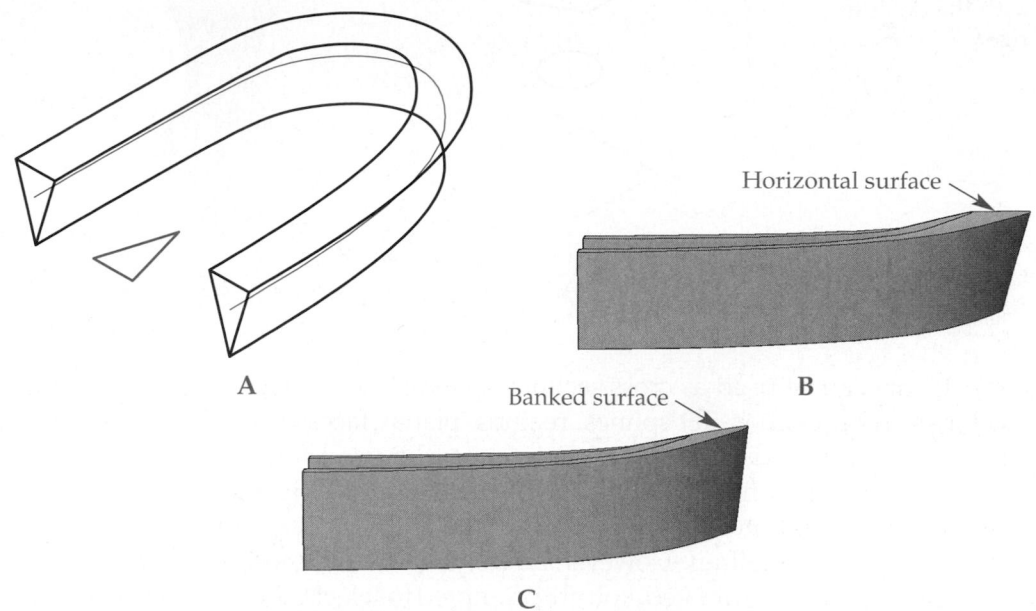

A

Horizontal surface

B

Banked surface

C

banking. The banking option is disabled for a 2D path, although you can go through the motions of turning it on when creating the sweep. Once you use the **Bank** option to turn banking on, it is on by default the next time the **SWEEP** command is used. To turn it off, enter a twist angle of zero (or the twist angle you wish to use).

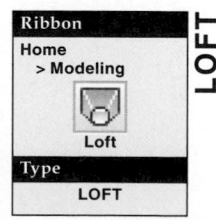

PROFESSIONAL TIP

The sweep options can be changed after the sweep is created using the **Properties** palette. In the **Geometry** section, you will find Profile rotation (alignment), Bank (banking), Twist along path (twist angle), and Scale along path (scale) settings.

Exercise 10-2

Complete the exercise on the student website.
www.g-wlearning.com/CAD

Creating Lofted Objects

The **LOFT** command is used to create lofted surfaces and solids based on a series of cross-sectional profiles. **Figure 10-8** shows an example of a loft formed from a rectangle, circle, and polygon. The loft may be guided by only the cross sections, as shown in the figure, by a path, or by guide curves. Lofting open shapes results in a surface object, while lofting closed shapes creates a solid. Open and closed shapes cannot be combined in the same loft.

Ribbon
Home
> Modeling

Loft
Type
LOFT

LOFT

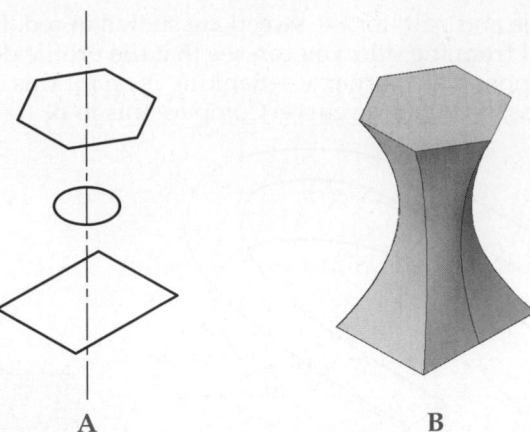

Figure 10-8.
A—The three profiles will be lofted to create a solid.
B—The resulting loft with the default settings.

A   B

Objects that can be used as cross sections include lines, circles, arcs, points, ellipses, elliptical arcs, 2D polylines, 2D splines, regions, planar faces of solids, planar surfaces, planar 3D faces, 2D solids, and traces. Points may be used for the first and last cross sections only. The loft path may be a line, circle, arc, ellipse, elliptical arc, spline, helix, or 2D or 3D polyline. Guide curves may be composed of lines, arcs, elliptical arcs, 2D or 3D splines, and 2D or 3D polylines. However, 2D polylines are limited to only one segment.

Once the command is initiated, you are prompted to select the cross-sectional profiles. Pick each profile in the order in which it should appear in the loft and press [Enter]. Be sure to individually select the cross sections in the order of the loft creation. You may not get the desired loft if you randomly select them or use a window selection.

Next, you are prompted to select how the loft is to be controlled. As mentioned earlier, you can control the loft by the cross sections, a path, or guide curves. These options are discussed in the next sections.

Controlling the Loft with Cross Sections

The **Cross-sections only** option of the **LOFT** command is useful when the 2D cross sections are drawn in their proper locations in space. The command determines the transition from one cross section to the next. The cross sections are not moved by the command.

When the **Cross-sections only** option is selected, the **Loft Settings** dialog box appears, **Figure 10-9.** The settings in this dialog box control the transition or contour

Figure 10-9
When the loft is controlled by cross sections only, the **Loft Settings** dialog box is used to control the transition between profiles.

Select a contour setting

Check to connect the first and last cross sections

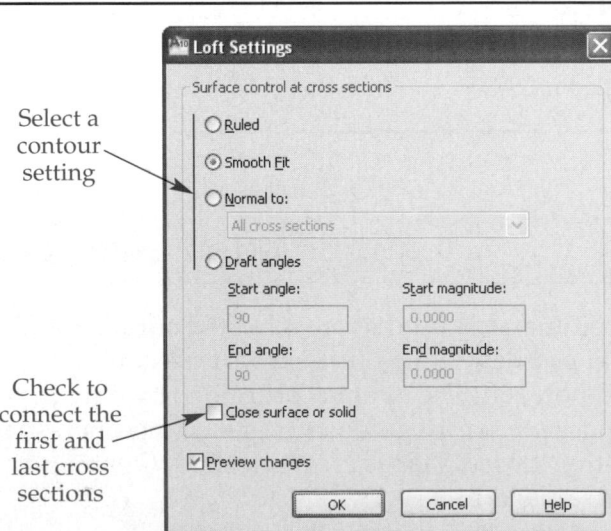

between cross sections. If the **Preview changes** check box at the bottom of the dialog box is checked, the current settings are previewed in the drawing area. As settings are changed, the preview is updated. When all settings have been made, pick the **OK** button to close the dialog box and create the loft.

When the **Ruled** option is selected in the dialog box, the loft has straight transitions between the cross sections. Sharp edges are created at each cross section. **Figure 10-10** shows the same cross sections in **Figure 10-8A** lofted with the **Ruled** option on. Compare this to **Figure 10-8B.**

The **Smooth Fit** option creates a smooth transition between the cross sections. Sharp edges are only created at the first and last cross sections. This is the default setting and the one used to create the loft shown in **Figure 10-8B.**

When the **Normal to:** option is selected in the dialog box, you can choose how the normal of the transition is treated at the cross sections. A *normal* is a vector extending perpendicular to the cross section. When the transition is normal to a cross section, it is perpendicular to the cross section. You can set the transition normal to the first cross section, last cross section, both first and last cross sections, or all cross sections. See **Figure 10-11.** Select the normal setting in the drop-down list. You will have to experiment with these settings to get the desired loft shape.

In manufacturing, plastic or metal parts are sometimes formed in a two-part mold. A slight angle is designed into the parts on the inside and outside surfaces to make removing the part from the mold easier. This taper is called a *draft angle.* The **Draft angles** option allows you to add a taper to the beginning and end of the loft.

When setting the draft angle, you can set the angle and magnitude. See **Figure 10-12.** The default draft angle is 90°, which means the transition is perpendicular to the cross section. The magnitude represents the relative distance from the cross section, in the same direction as the draft angle, before the transition starts to curve toward the next cross section. Magnitude settings depend on the size of the cross sections, the draft angle values, and the distance between the cross sections. You may have to experiment with different magnitude and angle settings to get the desired loft shape.

The **Close surface or solid** option is used to connect the last cross section to the first cross section. See **Figure 10-13.** This option "closes" the loft, similar to the **Close** option of the **LINE** or **PLINE** command. The shapes from **Figure 10-8** are shown in **Figure 10-14** with the **Close** option on. This option is only available when the **Ruled** or **Smooth Fit** radio button is selected.

PROFESSIONAL TIP

The settings in the **Loft Settings** dialog box are retained as the default, so get in the habit of checking them each time you create a loft. The **LOFTNORMALS** system variable controls which surface control radio button is current.

Figure 10-10.
The profiles in Figure 10-8A are lofted with the **Ruled** option selected in the **Loft Settings** dialog box. Compare this to Figure 10-8B.

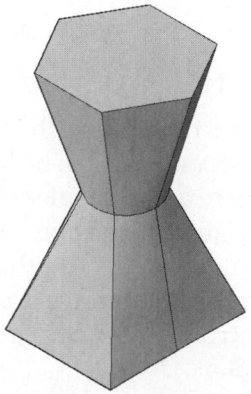

Figure 10-11.
The profiles in Figure 10-8A are lofted with the **Normal to:** option on in the **Loft Settings** dialog box. Cross sections were selected from bottom to top. Compare these results with Figure 10-8B and Figure 10-10. A—**Start cross section**. B—**End cross section**. C—**Start and End cross sections**. D—**All cross sections**.

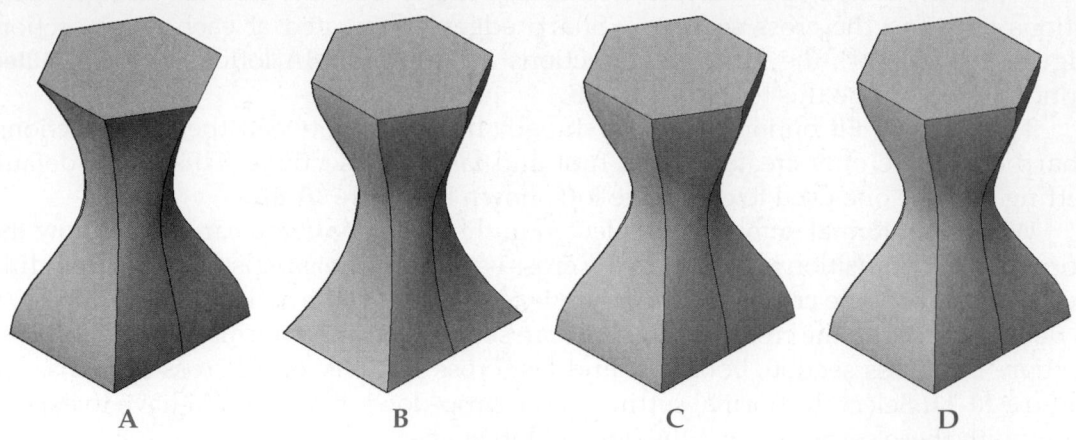

A   B   C   D

Figure 10-12.
When setting the draft angle, you can set the angle and the magnitude. A—Draft angle of 90° and a magnitude of zero. B—Draft angle of 30° and a magnitude of 180. C—Draft angle of 60° and a magnitude of 180.

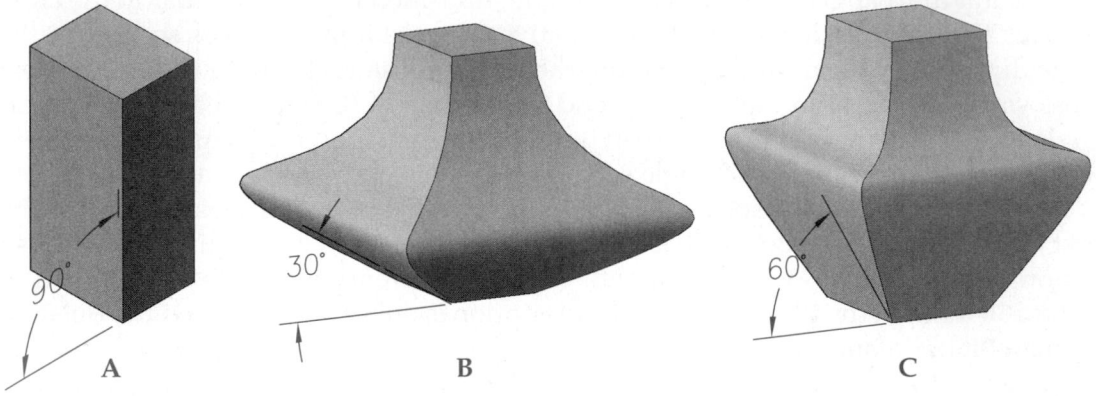

A   B   C

Exercise 10-3

Complete the exercise on the student website.
www.g-wlearning.com/CAD

Controlling the Loft with Guide Curves

Guide curves are lines that control the shape of the transition between cross sections. They do not have to be *curves*. They can be lines, arcs, elliptical arcs, splines (2D or 3D), or polylines (2D or 3D). There are four rules to follow when using guide curves:
* The guide curve should start on the first cross section.
* The guide curve should end on the last cross section.
* The guide curve should intersect all other cross sections.
* The surface control in the **Loft Settings** dialog box must be set to **Smooth Fit** (**LOFTNORMALS** = 1).

Figure 10-13.
A—These profiles will be used to create a sealing ring. They should be selected in a counterclockwise direction starting with the first cross section. B—The resulting loft with the default settings. Notice the gap between the first and last cross sections. C—By checking the **Close surface or solid** check box in the **Loft Settings** dialog box, the loft continues from the last cross section to the first cross section.

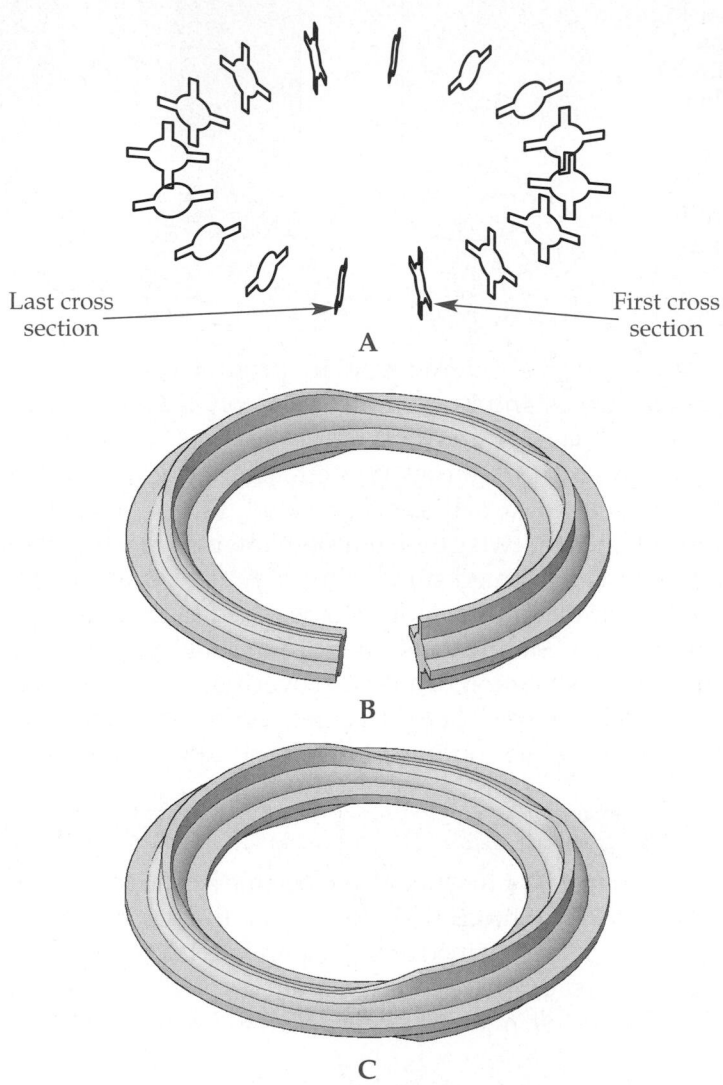

Last cross section

First cross section

A

B

C

Figure 10-14.
The same cross sectional shapes from Figure 10-8 are used in this loft, however the **Close** option has been applied. Notice how the loft is inside-out.

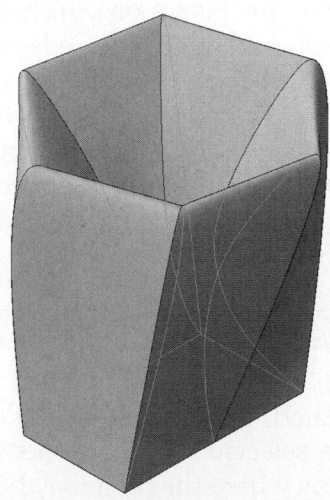

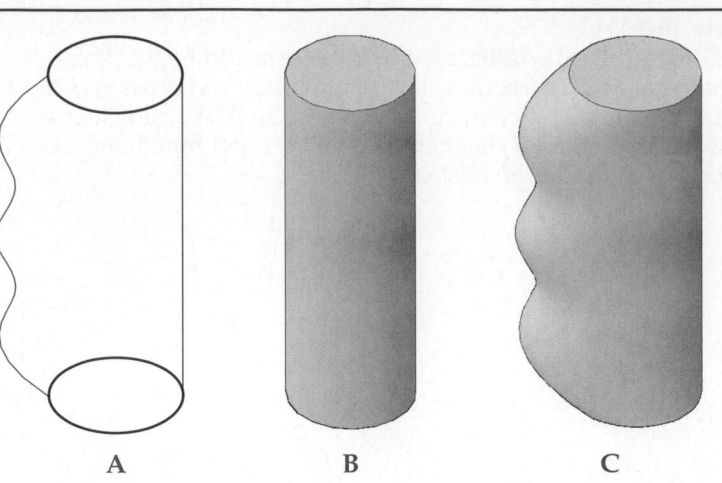

Figure 10-15.
A—These two circles will be lofted. The lines shown in color will be used as guide curves. B—When the circles are lofted using the **Cross-sections only** option, a cylinder is created. C—When the **Guides** option is used and the guide curves shown in A are selected, the resulting loft is shaped like a handle or grip.

A   B   C

When the **Guides** option is entered, you are prompted to select the guide curves. Select all of the guide curves and press [Enter]. The loft is created. The order in which guide curves are selected is not important.

For example, **Figure 10-15A** shows two circles that will be lofted. If the **Cross-sections only** option is used, a cylinder is created, **Figure 10-15B.** However, if the **Guides** option is used and the two guide curves shown in **Figure 10-15A** are selected, one side of the cylinder is deformed similar to a handle or grip. See **Figure 10-15C.**

Lofting is used to create open-contour shapes such as fenders, automobile interior parts, fabrics, and other ergonomic consumer products. **Figure 10-16** shows the use of open 2D splines in the construction of a fabric covering. Notice how each cross section is intersected by the guide curve. There is a cross section at the beginning of the guide curve and one at the end. These conditions fulfill the rules outlined earlier.

CAUTION

Guide curves only work well when the surface control is set to **Smooth Fit** (**LOFTNORMALS** = 1). Remember, the settings in the **Loft Settings** dialog box are retained after the previous **LOFT** command. If you get an error message when using guide curves or the curves are not reflected in the end result, make sure **LOFTNORMALS** is set to 1 and try it again.

Controlling the Loft with a Path

The **Path** option of the **LOFT** command places the cross sections along a single path. The path must intersect the planes on which each of the cross sections lie. However, the path does *not* have to physically touch the edge of each cross section, as is required of guide curves. When the **Path** option is entered, you are prompted to select the path. Once the path is picked, the loft is created. The cross sections remain in their original positions.

Figure 10-17 shows how 2D shapes can be positioned at various points on a path to create a loft. The rectangular shape does not cross the path. However, as long as the path intersects the plane of the rectangle, which it does, the shape will be included in the loft definition. The last shape at the top of the helix is a point object, causing the loft to taper.

PROFESSIONAL TIP

The **LOFT** command does not allow self-intersecting lofts to be created. Unfortunately, the error message you receive only states: The selected entities are not valid. If you see this message, look for areas where the path may be closing in on itself.

Figure 10-16.
A—The open profiles shown in black and the guide curve shown in color will be used to create a fabric covering for the three solid objects. B—The resulting fabric covering. This is a surface because the profiles were open.

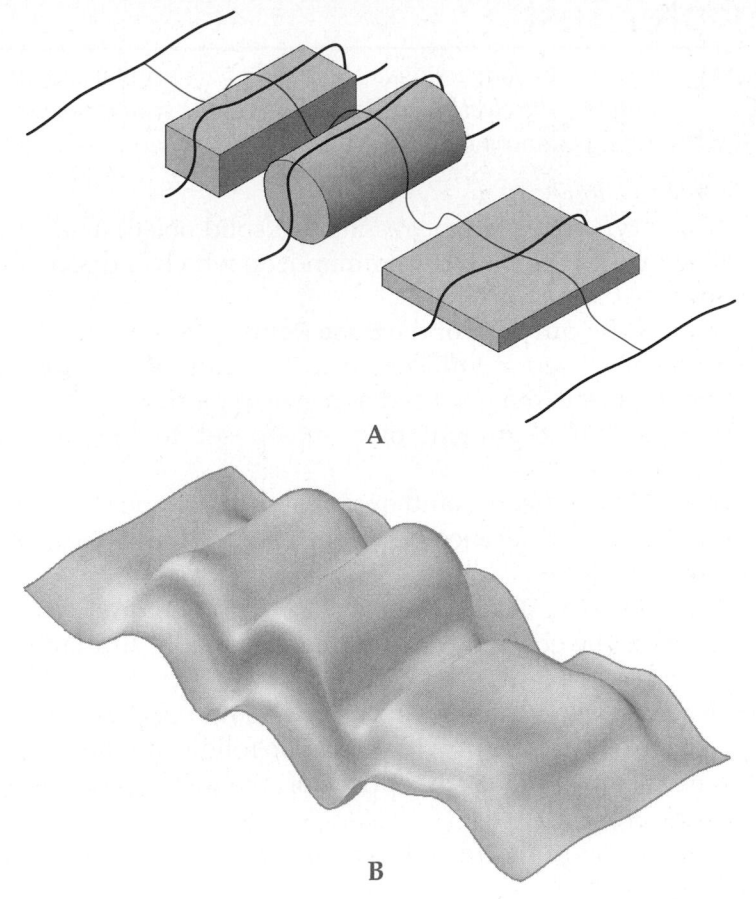

A

B

Figure 10-17.
A—The profiles shown in black will be lofted along the path shown in color. Notice how the rectangular profile is not intersected by the path, but the path does intersect the plane on which the rectangle lies. B—The resulting loft.

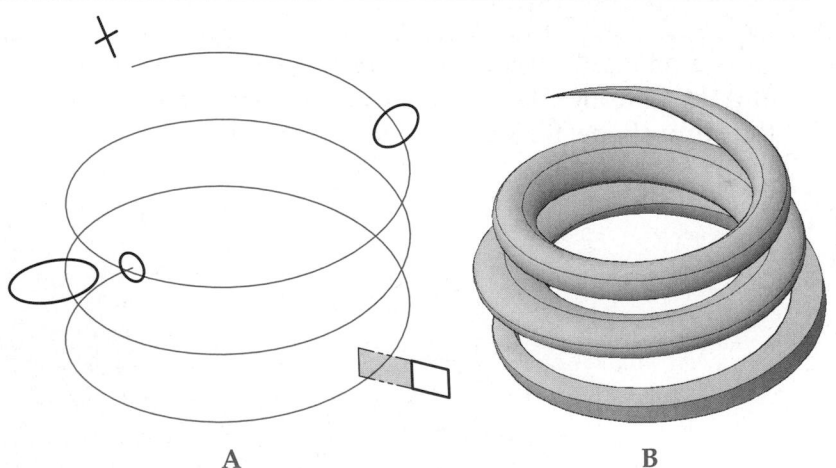

A

B

Exercise 10-4

Complete the exercise on the student website.
www.g-wlearning.com/CAD

Chapter Test

Answer the following questions. Write your answers on a separate sheet of paper or complete the electronic chapter test on the student website.
www.g-wlearning.com/CAD

1. What is a *loft*?
2. Which type of 2D shape results in a solid object when swept or lofted?
3. When using the **SWEEP** command, on which endpoint of the path does the sweep start?
4. What is the purpose of the **Base Point** option of the **SWEEP** command?
5. After the sweep or loft is created, how can the creation options be changed?
6. Which objects may be used as a sweep path?
7. How is the alignment of a sweep set to be perpendicular to the start of the path?
8. Which **SWEEP** command option is used to taper the sweep?
9. What is the difference between the **Ruled** and **Smooth Fit** options in the **LOFT** command?
10. What does the **Bank** option of the **LOFT** command do?
11. Where is the check box that will close the loft, similar to a polyline, and what is its name?
12. List five objects that may be used as guide curves in a loft.
13. What are the four rules that must be followed when using guide curves?
14. When using the **Path** option of the **LOFT** command, what must the path intersect?
15. How can a loft be created so it tapers to a point at its end?

Drawing Problems

1. Create the lamp shade shown below. Create two separate loft objects for the top and the bottom. Then, union the two pieces. Finally, scale a copy and hollow out the lamp shade. Save the drawing as P10_01.

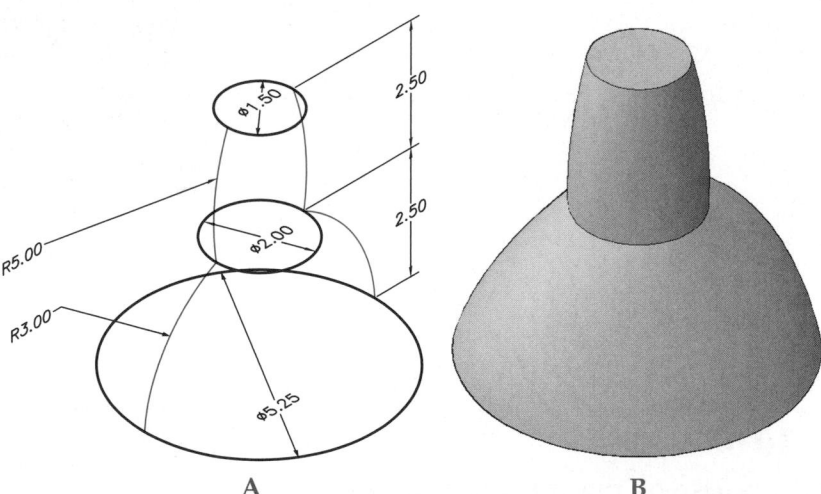

A   B

2. Create the two shampoo bottles shown below. One design uses cross sections only and the other uses a guide curve. Each bottle is made up of two loft objects. Join the pieces so each bottle is one solid. Save the drawing as P10_02.

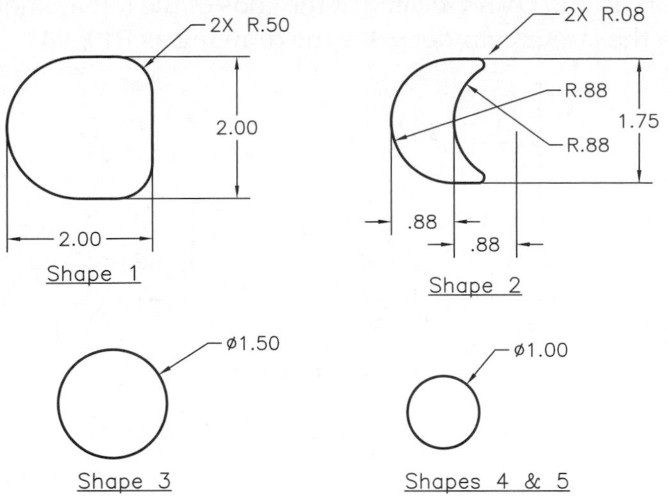

2X R.50

2.00

2.00

Shape 1

2X R.08

R.88

1.75

R.88

.88

.88

Shape 2

Ø1.50

Shape 3

Ø1.00

Shapes 4 & 5

A

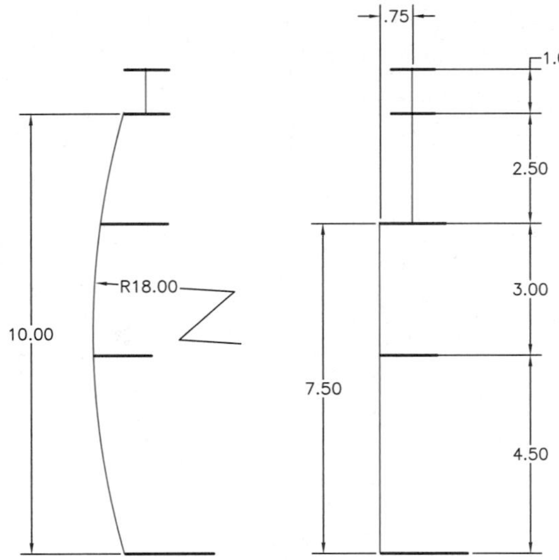

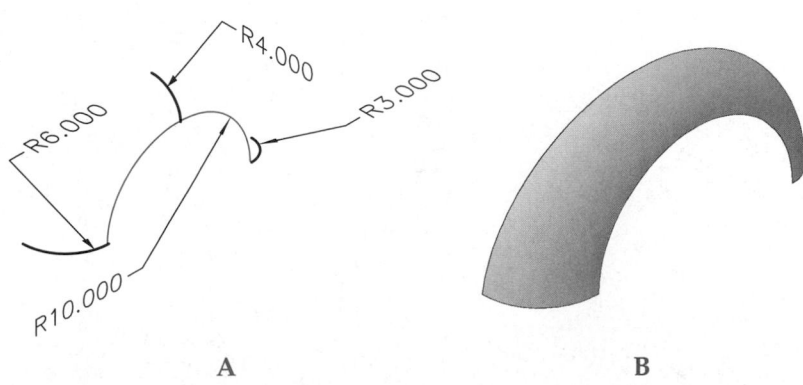

.75

1.00

2.50

3.00

4.50

R18.00

10.00

7.50

B

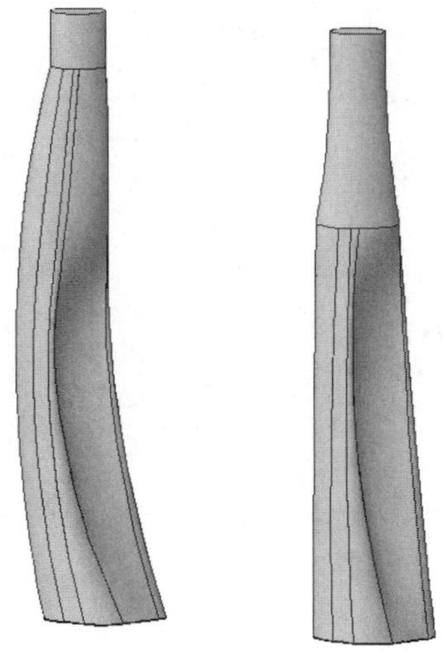

C

3. Create as a loft the automobile fender shown below. Use either the **Guide** or the **Path** option and the line shown in color. Save the drawing as P10_03.

R4.000

R6.000

R3.000

R10.000

A

B

4. Draw as a loft the C-clamp shown below. Use the shapes (A, B, C, and D) as the cross sections and the polyline (in color) as the guide curve. Add Ø1 unit cylinders to the ends. Make one cylinder .125H and the other 1.125H. The cylinders should be centered on profile D and located at the ends of the loft as shown. Make a Ø.625 hole through the larger cylinder. Save the drawing as P10_04.

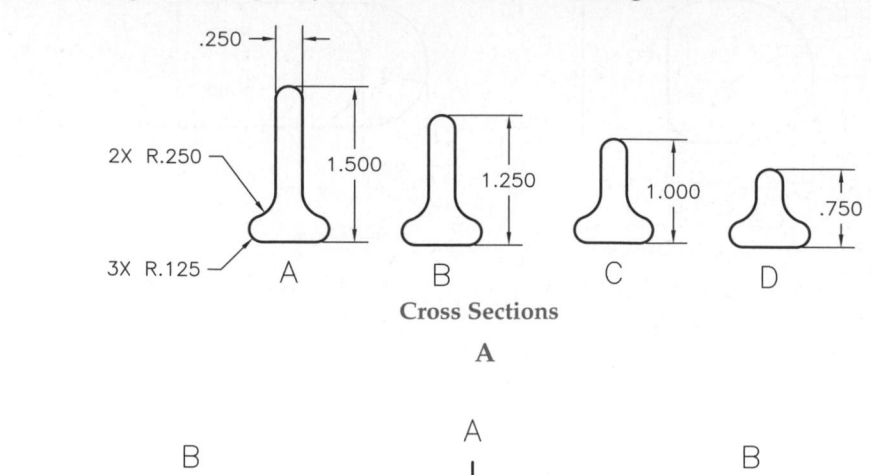

Cross Sections

A

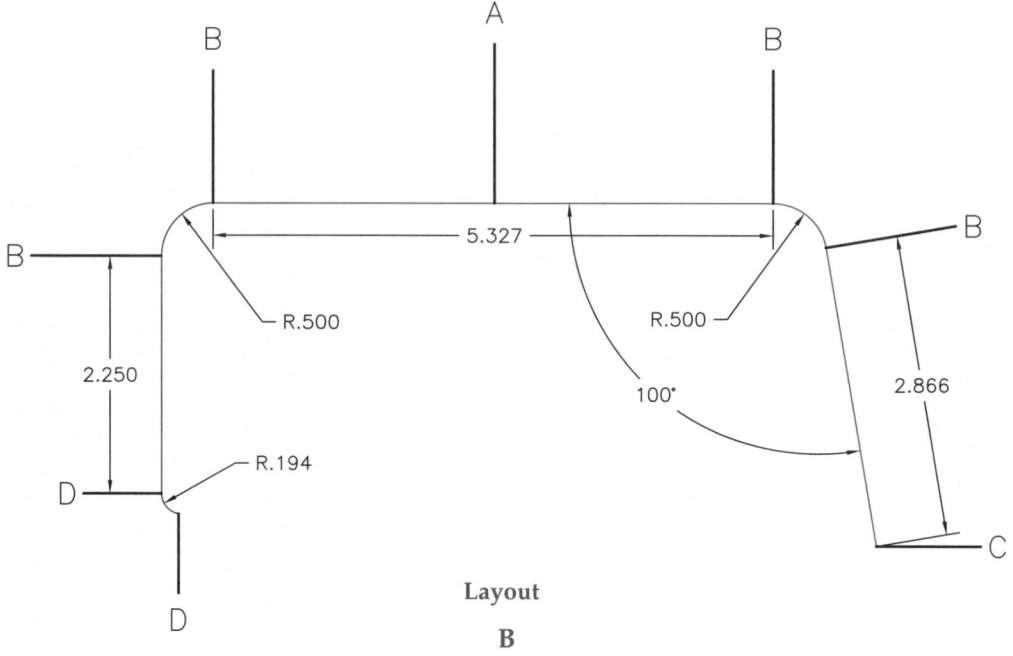

Layout

B

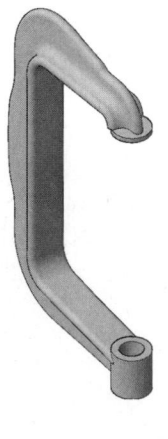

C

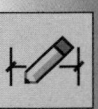

5. In this problem, you will draw a racetrack for toy cars by sweeping a 2D shape along a polyline path.
   A. Draw the polyline path shown with the coordinates given. Turn it into a spline.
   B. Draw the 2D profile shown using the dimensions given. Turn it into a region or a polyline.
   C. Use the **SWEEP** command to create the racetrack, as shown in the shaded view.
   D. You may have to use the **Properties** palette to adjust the sweep after it is drawn.
   E. Save the drawing as P10_05.

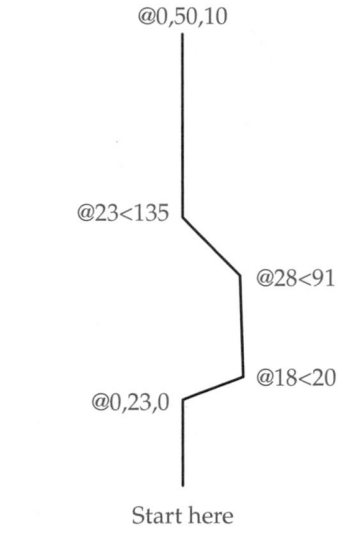

@0,50,10

@23<135

@28<91

@18<20

@0,23,0

Start here

Polyline Path

A

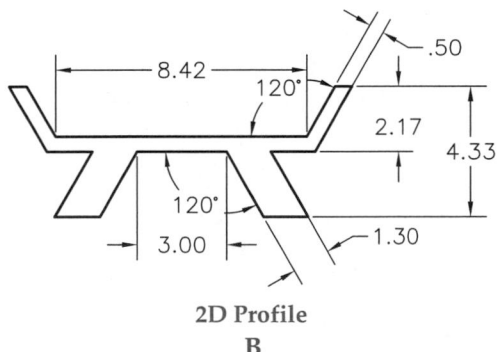

2D Profile

B

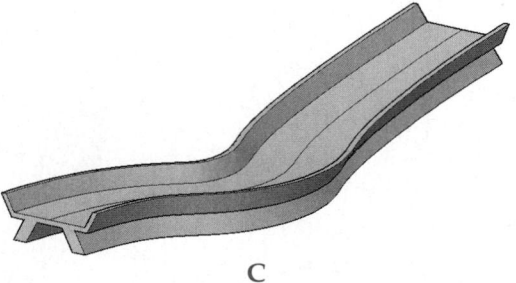

C

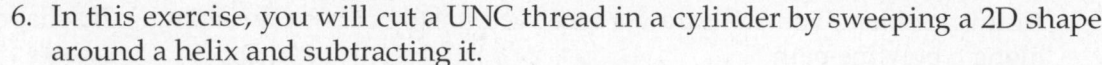

6. In this exercise, you will cut a UNC thread in a cylinder by sweeping a 2D shape around a helix and subtracting it.
   A. Draw a ∅.25 cylinder that is 1.00 in height.
   B. Draw the thread cutter profile shown below. The long edge of the cutter should be aligned with the vertical edge of the cylinder.
   C. Draw a helix centered on the cylinder with base and top radii of .125, turn height of .050, and a total height of 1.000.
   D. Sweep the 2D shape along the helix. Then, subtract the resulting solid from the cylinder. Refer to the shaded view shown below.
   E. If time allows, create another cutter profile to cut a .0313 × 45° chamfer on the end of the thread. Use a circle as a sweep path or revolve the profile about the center of the cylinder.
   F. Save the drawing as P10_06.

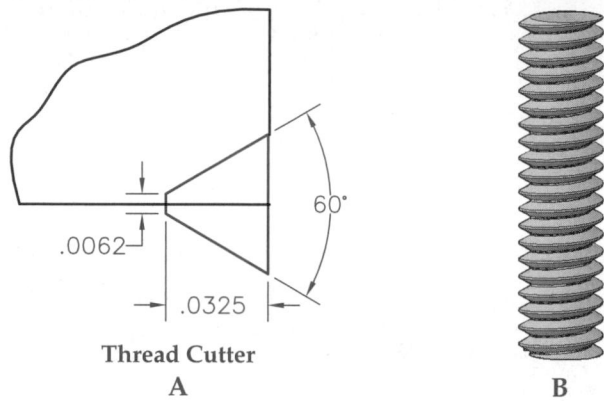

Thread Cutter

A   B

7. In this problem, you will add a seatback to the kitchen chair you started modeling in Chapter 2. In Chapter 9, you refined the seat.

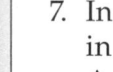

   A. Open P09_10 from Chapter 9.
   B. Draw an arc for the top of the bow. Using a 14.25″ length of the arc, divide it into seven equal segments.
   C. Position eight ∅.50 circles at the division points.
   D. Draw circles at the top of each hole in the seat.
   E. Create a loft between each lower circle and each upper circle.
   F. Using the information in the drawings, create the outer bow for the seatback.
   G. Save the drawing as P10_07.

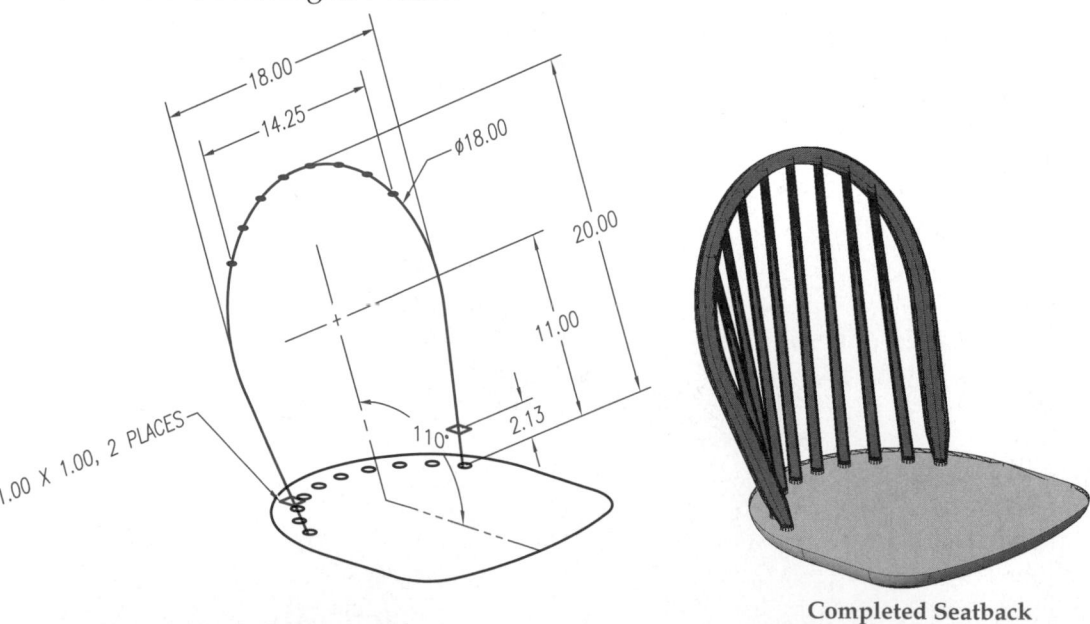

Completed Seatback

Creating and Working with Solid Model Details

Learning Objectives

After completing this chapter, you will be able to:

✓ Change properties on solids.
✓ Align objects.
✓ Rotate objects in three dimensions.
✓ Mirror objects in three dimensions.
✓ Create 3D arrays.
✓ Fillet solid objects.
✓ Chamfer solid objects.
✓ Thicken a surface into a solid.
✓ Convert planar objects into surfaces.
✓ Slice a solid using various methods.
✓ Construct details on solid models.
✓ Remove features from solid models.

Changing Properties

Properties of 3D objects can be modified using the **Properties** palette, which is thoroughly discussed in *AutoCAD and Its Applications—Basics*. This palette is displayed using the **PROPERTIES** command. You can also double-click on a solid object or select the solid, right-click, and pick **Properties** from the shortcut menu.

The **Properties** palette lists the properties of the currently selected object. For example, **Figure 11-1** lists the properties of a selected solid sphere. AutoCAD offers some parametric solid modeling options. A *parametric solid modeling program* allows you to change the parameters, such as a sphere's diameter, in the **Properties** palette. You can also change the sphere's position, linetype, linetype scale, color, layer, lineweight, and visual settings. The categories and properties available in the **Properties** palette depend on the selected object.

To modify an object property, select the property. Then, enter a new value in the right-hand column. The drawing is updated to reflect the changes. You can leave the **Properties** palette open as you continue with your work.

Ribbon
View
> Palettes

Properties Palette

Type
PROPERTIES
PR
[Ctrl]+[1]

PROPERTIES

Figure 11-1.
The **Properties** palette can be used to change many of the properties of a solid.

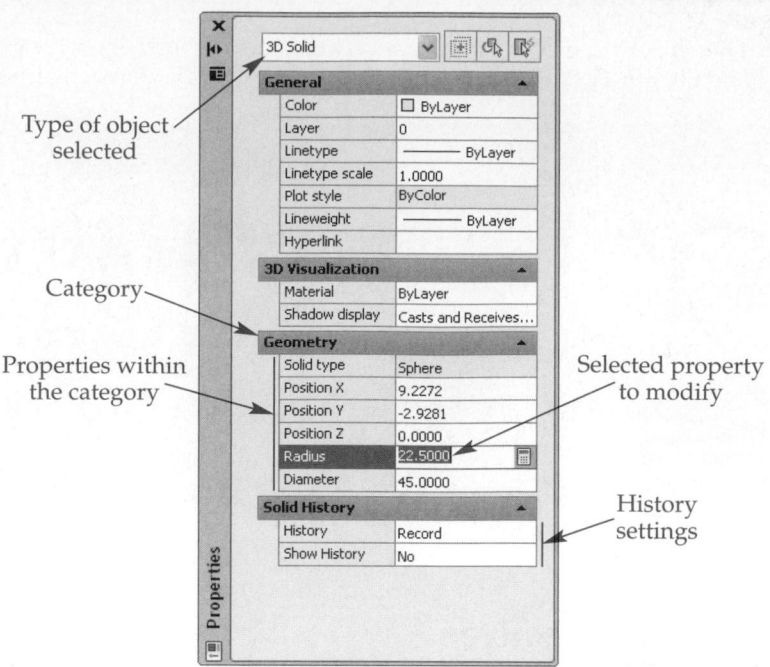

Type of object selected

Category

Properties within the category

Selected property to modify

History settings

Solid Model History

AutoCAD can automatically record a history of a composite solid model's construction. The control of the history setting is found in the **Solid History** category of the **Properties** palette. By default, the History property in the **Properties** palette is set to Record, which means that the history will be saved. See **Figure 11-1.** It is generally a good idea to have the history recorded. Then, at any time, you can graphically display all of the geometry that was used to create the model.

If the **SOLIDHIST** system variable is set to a value of 1, all new solids have their History property set to Record. This is the default. If the system variable is set to 0, all new solids have their History property set to None (no recording). With either setting of the system variable, the **Properties** palette can be used to change the setting for individual solids.

To view the graphic history of the composite solid, set the Show History property in the **Properties** palette to Yes. All of the geometry used to construct the model is displayed. If the **SHOWHIST** system variable is set to 0, the Show History property is set to No for all solids and cannot be changed. If this system variable is set to 2, the Show History property is set to Yes for all solids and cannot be changed. A **SHOWHIST** setting of 1 allows the Show History property to be individually set for each solid.

An example of showing the history on a composite solid is provided in **Figure 11-2.** The object appears in its current edited format in **Figure 11-2A.** The Conceptual visual style is set current and the Show History property is set to No. In **Figure 11-2B**, the Show History property is set to Yes. Isolines have also been turned on. You can see the geometry that was used in the Boolean subtraction operations. Using subobject editing techniques, the individual geometry can be selected and edited. Subobject editing is discussed in detail in Chapter 12.

> **NOTE**
>
> If the Solid History property is set to Yes to display the components of the composite solid, as seen in **Figure 11-2**, the components will appear when the drawing is plotted. Be sure to set the Show History property to No before you print or plot.

Figure 11-2.
A—The object appears in its current edited format with Show History property turned off.
B—The Show History property is set to Yes and the display of isolines has been turned on.

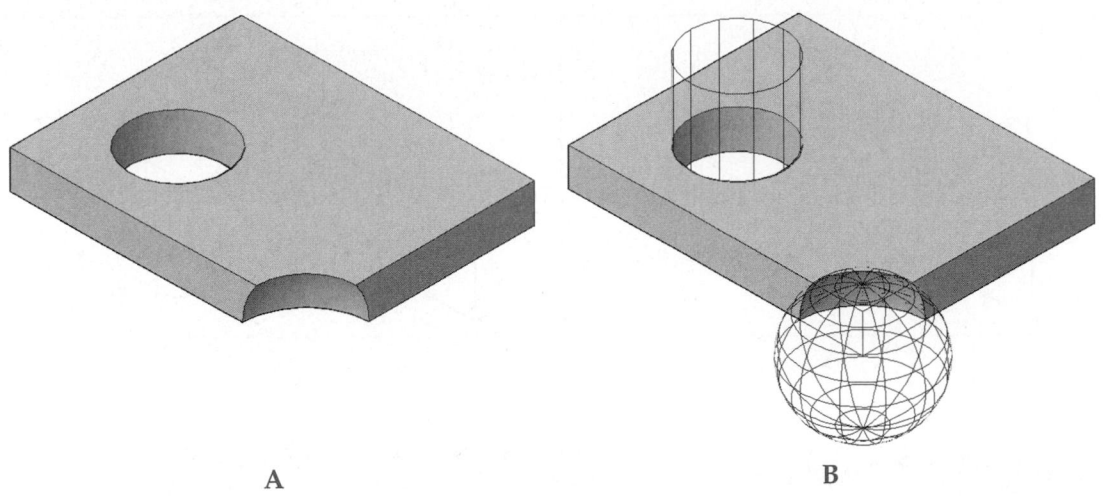

A   B

Aligning Objects in 3D

AutoCAD provides two different methods with which to move and rotate objects in a single command. This is called *aligning* objects. The simplest method is to align 3D objects by picking source points on the first object and then picking destination points on the object to which the first one is to be aligned. This is accomplished with the **3DALIGN** command, which allows you to both relocate and rotate the object. The second and much more versatile method allows you to not only move and rotate an object, but also to scale the object being aligned. This is possible with the **ALIGN** command.

Move and Rotate Objects in 3D Space

The basic function of moving and rotating an object relative to a second object or set of points is done with the **3DALIGN** command. It allows you to reorient an object in 3D space. Using this command, you can correct errors of 3D construction and quickly manipulate 3D objects. The **3DALIGN** command requires existing points (source) and the new location of those existing points (destination).

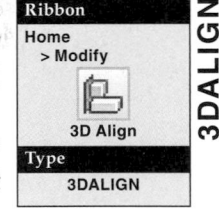

For example, refer to **Figure 11-3**. The wedge in **Figure 11-3A** is aligned in its new position in **Figure 11-3B** as follows. Set the **Intersection** or **Endpoint** running object snap to make point selection easier. Refer to the figure for the pick points.

```
Select objects: (pick the wedge)
1 found
Select objects: ↵
    Specify source plane and orientation…
Specify base point or [Copy]: (pick P1)
Specify second point or [Continue] <C>: (pick P2)
Specify third point or [Continue] <C>: (pick P3)
    Specify destination plane and orientation…
Specify first destination point: (pick P4)
Specify second destination point or [eXit] <X>: (pick P5)
Specify third destination point or [eXit] <X>: ↵
```

Figure 11-3.
The **3DALIGN** command can be used to properly orient 3D objects. A—Before aligning. Note the pick points. B—After aligning.

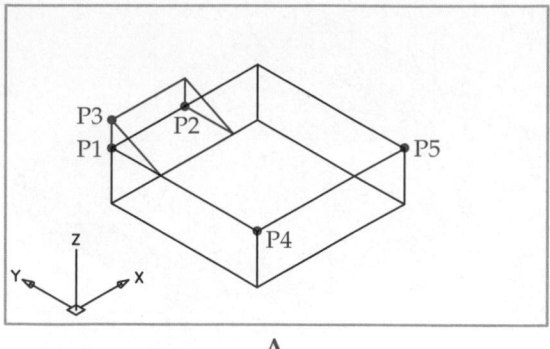

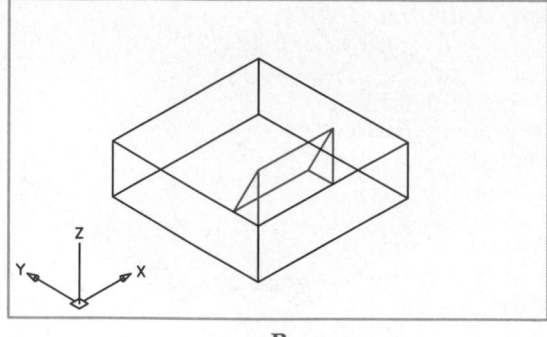

A

B

Exercise 11-1

Complete the exercise on the student website.
www.g-wlearning.com/CAD

Move, Rotate, and Scale Objects in 3D Space

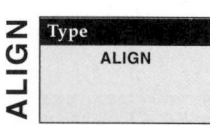
The **ALIGN** command has the same functions of the **3DALIGN** command, but adds the ability to scale an object. Refer to **Figure 11-4.** The 90° bend must be rotated and scaled to fit onto the end of the HVAC assembly. Two source points and two destination points are required, **Figure 11-4A.** Then, you can choose to scale the object.

> Select objects: (*pick the 90° bend*)
> 1 found
> Select objects: ↵
> Specify first source point: (*pick P1*)
> Specify first destination point: (*pick P2; a line is drawn between the two points*)
> Specify second source point: (*pick P3*)
> Specify second destination point: (*pick P4; a line is drawn between the two points*)
> Specify third source point or <continue>: ↵
> Scale objects based on alignment points? [Yes/No] <N>: **Y**↵

The 90° bend is aligned and uniformly scaled to meet the existing ductwork object. See **Figure 11-4B.** You can also align using three source and three destination points. However, when doing so, you cannot scale the object.

Figure 11-4.
Using the **ALIGN** command. A—Two source points and two destination points are required. Notice how the bend is not at the proper scale. B—You can choose to scale the object during the operation. Notice how the aligned bend is also properly scaled.

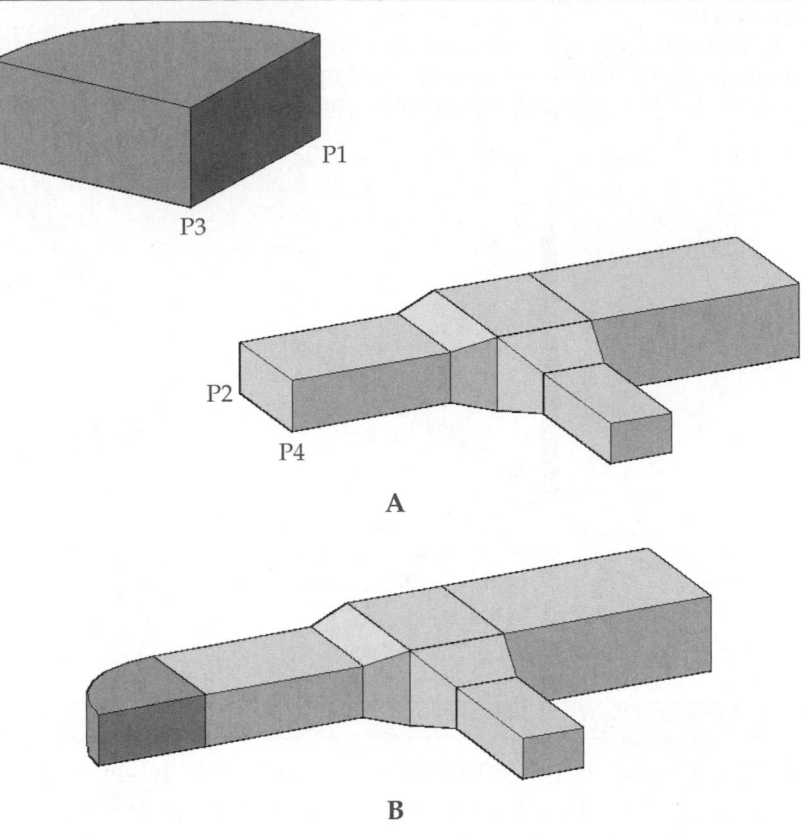

A

B

Exercise 11-2

Complete the exercise on the student website.
www.g-wlearning.com/CAD

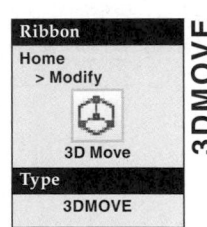

3D Moving

The **3DMOVE** command allows you to quickly move an object along any axis or plane of the current UCS. When the command is initiated, you are prompted to select the objects to move. After selecting the objects, press [Enter]. The *move grip tool* is displayed in the center of the selection set. By default, the move grip tool is also displayed when a solid is selected with no command active.

The move grip tool is a tripod that appears similar to the shaded UCS icon. See **Figure 11-5A.** You can relocate the tool by right-clicking on the tool and selecting **Relocate Gizmo** from the shortcut menu. Then, move the tool to a new location and pick. You can also realign the tool using the shortcut menu.

If you move the pointer over the X, Y, or Z axis of the grip tool, the axis changes to yellow. To restrict movement along that axis, pick the axis. If you move the pointer over one of the right angles at the origin of the tool, the corresponding two axes turn yellow. Pick to restrict the movement to that plane. You can complete the movement by either picking a new point or by direct distance entry.

Ribbon
Home
> Modify
3D Move
Type
3DMOVE

3DMOVE

Figure 11-5.
A—The move grip tool is a tripod that appears similar to the shaded UCS icon. B—This is the rotate grip tool. The three axes of rotation are represented by the circles. The origin of the rotation is where you place the center grip.

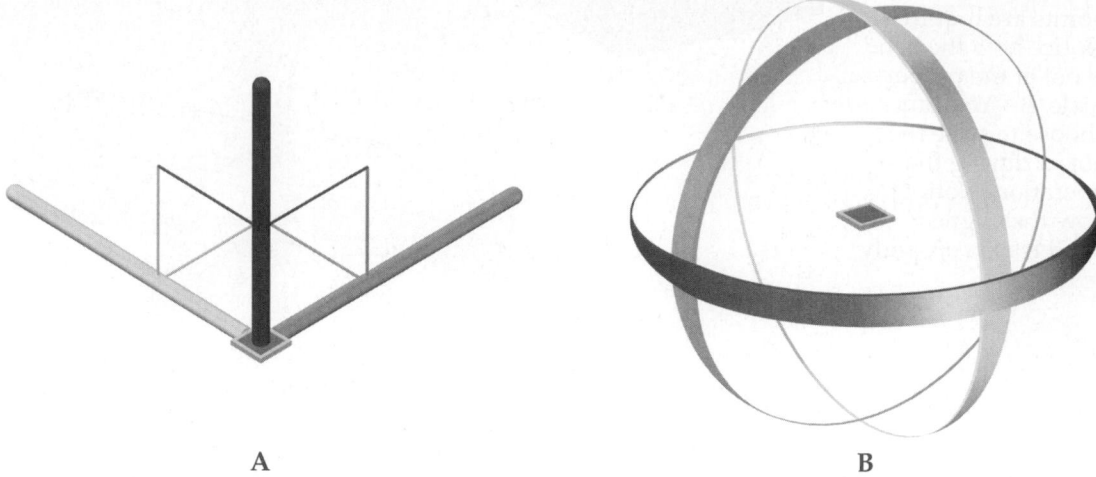

A   B

NOTE

If the **GTAUTO** system variable is set to 1, the move grip tool is displayed when a solid is selected with no command active. If the **GTLOCATION** system variable is set to 0, the grip tool is placed on the UCS icon (not necessarily the UCS *origin*). Both variables are set to 1 by default.

3D Rotating

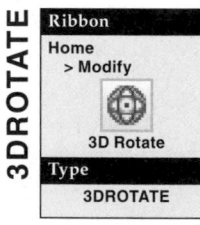

3DROTATE

Ribbon
Home
> Modify
3D Rotate

Type
3DROTATE

As you have seen in earlier chapters, the **ROTATE** command can be used to rotate 3D objects. However, the command can only rotate objects in the XY plane of the current UCS. This is why you had to change UCSs to properly rotate objects. The **3DROTATE** command, on the other hand, can rotate objects on any axis regardless of the current UCS. This is an extremely powerful editing and design tool.

When the command is initiated, you are prompted to select the objects to rotate. After selecting the objects, press [Enter]. The *rotate grip tool* is displayed in the center of the selection set. See **Figure 11-5B**. If the 2D Wireframe visual style is current, the visual style is temporarily changed to the 3D Wireframe because the grip tool is not displayed in 2D mode. The grip tool provides you with a dynamic, graphic representation of the three axes of rotation. After selecting the objects, you must specify a location for the grip tool, which is the base point for rotation.

Now, you can use the grip tool to rotate the objects about the tool's local X, Y, or Z axis. As you hover the cursor over one of the three circles in the grip tool, a vector is displayed that represents the axis of rotation. To rotate about the tool's X axis, pick the red circle on the grip tool. To rotate about the Y axis, pick the green circle. To rotate about the Z axis, pick the blue circle. Once you select a circle, it turns yellow and you are prompted for the start point of the rotation angle. You can enter a direct angle at this prompt or pick the first of two points defining the angle of rotation. When the rotation angle is defined, the object is rotated about the selected axis.

The following example rotates the bend in the HVAC assembly shown in **Figure 11-6A.** Set the **Midpoint** object snap and turn on object tracking. Then, select the command and continue:

    Current positive angle in UCS: ANGDIR=*(current)* ANGBASE=*(current)*
    Select objects: *(pick the bend)*
    1 found
    Select objects: ↵
    Specify base point: *(acquire the midpoint of the vertical and horizontal edges, then pick to place the grip tool in the middle of the rectangular face)*
    Pick a rotation axis: *(pick the green circle)*
    Specify angle start point or type an angle: **180.**↵

Note that the rotate grip tool remains visible through the base point and the angle of rotation selections. The rotated object is shown in **Figure 11-6B.**

If you need to rotate an object on an axis that is not parallel to the current X, Y, or Z axes, use a dynamic UCS with the **3DROTATE** command. Chapter 5 discussed the benefits of using a dynamic UCS when creating objects that need to be parallel to a surface other than the XY plane. With the object selected for rotation and the dynamic UCS option active (pick the **Allow/Disallow Dynamic UCS** button on the status bar), move the rotate grip tool over a face of the object. The grip tool aligns itself with the surface so that the Z axis is perpendicular to the face. Carefully place the grip tool over the point of rotation using object snaps. Make sure that the tool is correctly positioned before picking to locate it. Then, enter an angle or use polar tracking to rotate the object about the appropriate axis on the tool.

Figure 11-6.
A—Use object tracking or object snaps to place the grip tool in the middle of the rectangular face. Then, select the axis of rotation.
B—The completed rotation.

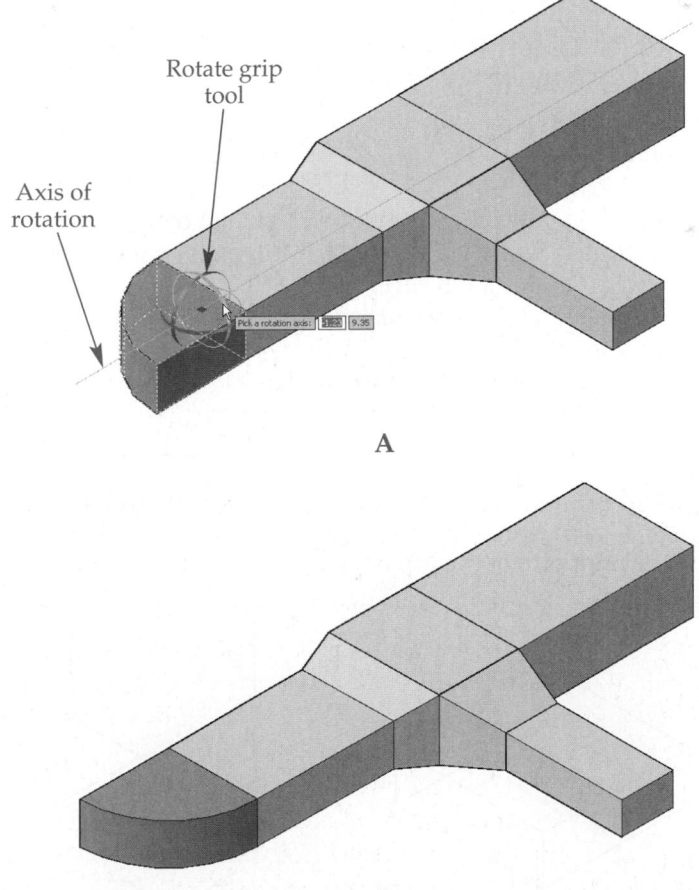

A

B

Exercise 11-3

Complete the exercise on the student website.
www.g-wlearning.com/CAD

3D Mirroring

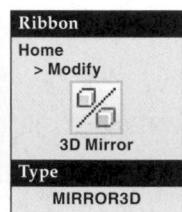

MIRROR3D

Ribbon
Home
> Modify

3D Mirror

Type
MIRROR3D

The **MIRROR** command can be used to rotate 3D objects. However, like the **ROTATE** command, the **MIRROR** command can only work in the XY plane of the current UCS. Often, to properly mirror objects with this command, you have to change UCSs. The **MIRROR3D** command, on the other hand, allows you to mirror objects about any plane regardless of the current UCS.

The default option of the command is to define a mirror plane by picking three points on that plane, **Figure 11-7A.** Object snap modes should be used to accurately define the mirror plane. To mirror the wedge in **Figure 11-7A,** set the **Midpoint** running object snap, select the command, and use the following sequence. The resulting drawing is shown in **Figure 11-7B.**

Select objects: *(pick the wedge)*
1 found
Select objects: ↵
Specify first point of mirror plane (3 points) or
[Object/Last/Zaxis/View/XY/YZ/ZX/3points] <3points>: *(pick P1, which is the mid-point of the box's top edge)*
Specify second point on mirror plane: *(pick P2)*
Specify third point on mirror plane: *(pick P3)*
Delete source objects? [Yes/No] <N>: ↵

Figure 11-7.
The **MIRROR3D** command allows you to mirror objects about any plane regardless of the current UCS. A—The mirror plane defined by the three pick points is shown here in color. Point P1 is the midpoint of the top edge of the base. B—A copy of the original is mirrored.

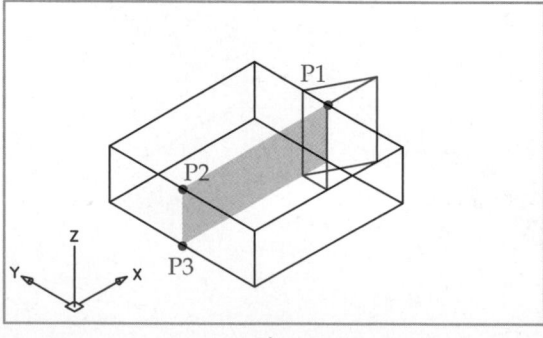

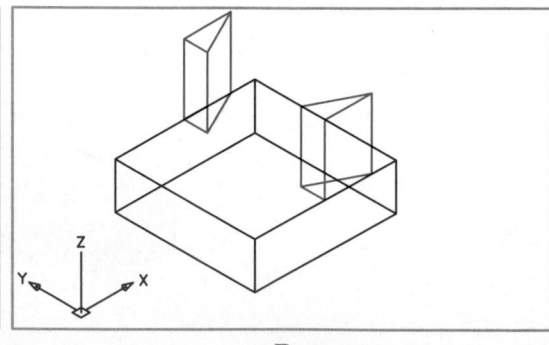

A

B

There are several different ways to define a mirror plane with the **MIRROR3D** command. These are:

- **Object.** The plane of the selected circle, arc, or 2D polyline segment is used as the mirror plane.
- **Last.** Uses the last mirror plane defined.
- **Zaxis.** Defines the plane with a pick point on the mirror plane and a point on the Z axis of the mirror plane.
- **View.** The viewing direction of the current viewpoint is aligned with a selected point to define the plane.
- **XY, YZ, ZX.** The mirror plane is placed parallel to one of the three basic planes of the current UCS and passes through a selected point.
- **3points.** Allows you to pick three points to define the mirror plane, as shown in the above example.

Exercise 11-4

Complete the exercise on the student website.
www.g-wlearning.com/CAD

Creating 3D Arrays

The **ARRAY** command can be used to create either a rectangular or polar array of a 3D object on the XY plane of the current UCS. You probably used this command to complete some of the problems in previous chapters. The **3DARRAY** command allows you to array an object in 3D space. There are two types of 3D arrays—rectangular and polar.

Rectangular 3D Arrays

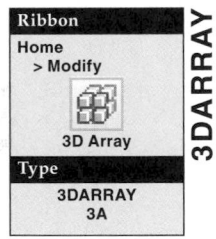

In a *rectangular 3D array,* as with a rectangular 2D array, you must enter the number of rows and columns. However, you must also specify the number of *levels,* which represents the third (Z) dimension. The command sequence is similar to that used with the 2D array command, with two additional prompts.

An example of where a rectangular 3D array may be created is the layout of structural steel columns on multiple floors of a commercial building. In **Figure 11-8A,** you can see two concrete floor slabs of a building and a single steel column. It is now a simple matter of arraying the steel column in rows, columns, and levels.

To draw a rectangular 3D array, select the **3DARRAY** command. Pick the object to array and press [Enter]. Then, specify the **Rectangular** option:

```
Enter the type of array [Rectangular/Polar] <R>: R↵
Enter the number of rows (- - -) <1>: 3↵
Enter the number of columns (¦¦¦) <1>: 5↵
Enter the number of levels (...) <1>: 2↵
Specify the distance between rows (- - -): 10'↵
Specify the distance between columns (¦¦¦): 10'↵
Specify the distance between levels (...): 12'8↵
```

The result is shown in **Figure 11-8B.** Constructions like this can be quickly assembled for multiple levels using the **3DARRAY** command only once.

Figure 11-8.
A—Two floors and one steel column are drawn. B—A rectangular 3D array is used to place all of the required steel columns on both floors at the same time.

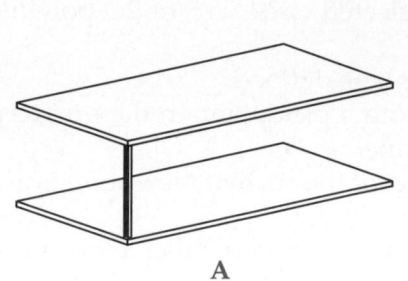
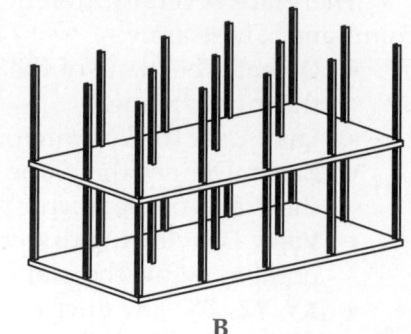

A   B

Polar 3D Arrays

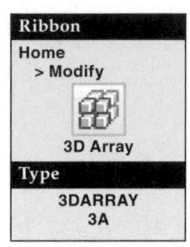

Ribbon
Home
> Modify

3D Array

Type
3DARRAY
3A

3DARRAY

A *polar 3D array* is similar to a polar 2D array. However, the axis of rotation in a 2D polar array is parallel to the Z axis of the current UCS. In a 3D polar array, you can define a centerline axis of rotation that is not parallel to the Z axis of the current UCS. In other words, you can array an object in a UCS different from the current one. Unlike a rectangular 3D array, a polar 3D array does not allow you to create levels of the object. The object is arrayed in a plane defined by the object and the selected centerline (Z) axis.

To draw a polar 3D array, select the **3DARRAY** command. Pick the object to array and press [Enter]. Then, specify the **Polar** option:

Enter the type of array [Rectangular/Polar] <R>: **P**↵

For example, the four mounting flanges on the lower part of the duct in **Figure 11-9A** must be placed on the opposite end. However, notice the orientation of the UCS. First, copy one flange and rotate it to the proper orientation. Then, use the **3DARRAY** command as follows. Make sure polar tracking is on.

Select objects: *(select the copied flange)*
1 found
Select objects: ↵
Enter the type of array [Rectangular/Polar] <R>: **P**↵
Enter the number of items in the array: **4**↵
Specify the angle to fill (+=ccw, −=cw) <360>: ↵
Rotate arrayed objects? [Yes/No] <Y>: ↵
Specify center point of array: **CEN**↵
of: *(pick the center of the upper duct opening)*
Specify second point on axis of rotation: *(move the cursor so the ortho line projects out of the center of the duct opening and pick)*

The completed 3D polar array is shown in **Figure 11-9B.** If additional levels of a polar array are needed, they can be created by copying the array just created.

Exercise 11-5

Complete the exercise on the student website.
www.g-wlearning.com/CAD

Figure 11-9.
A—A ductwork elbow with four flanges in place. Copies of these flanges need to be located on the opposite end. Start by creating one copy as shown. B—By creating a 3D polar array, the flanges are properly oriented without changing the UCS.

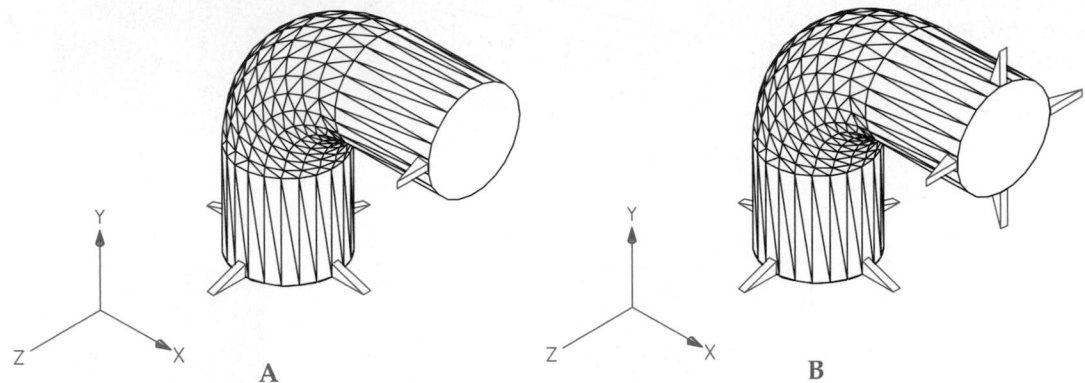

Filleting Solid Objects

A *fillet* is a rounded interior edge on an object, such as a box. A *round* is a rounded exterior edge. The **FILLET** command is used to create both fillets and rounds. Before a fillet or round is created at an intersection, the solid objects that intersect need to be joined using the **UNION** command. Then, use the **FILLET** command. See **Figure 11-10.** Since the object being filleted is actually a single solid and not two objects, only one edge is selected. In the following sequence, the fillet radius is set at .25, then the fillet is created. First, select the **FILLET** command and then continue as follows.

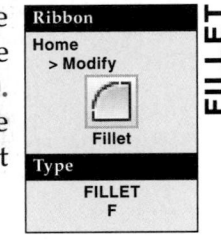

Ribbon

Home
> Modify

Fillet

Type

FILLET
F

FILLET

> Current settings: Mode = *current*, Radius = *current*
> Select first object or [Undo/Polyline/Radius/Trim/Multiple]: **R.↵**
> Specify fillet radius <*current*>: **.25↵**
> Select first object or [Undo/Polyline/Radius/Trim/Multiple]: *(pick edge to be filleted or rounded)*
> Enter fillet radius <0.2500>: ↵
> Select an edge or [Chain/Radius]: ↵ *(this fillets the selected edge, but you can also select other edges at this point)*
> 1 edge(s) selected for fillet.

Examples of fillets and rounds are shown in **Figure 11-11.**

Figure 11-10.
A—Pick the edge where two unioned solids intersect to create a fillet. B—The fillet after rendering.

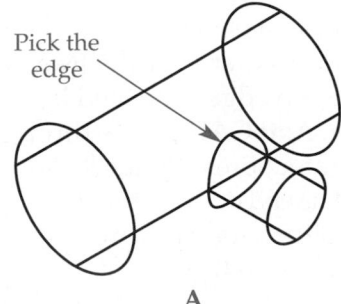

Pick the edge

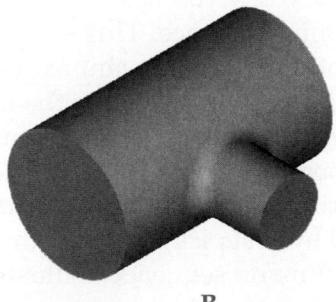

A

B

Figure 11-11.
Examples of fillets
and rounds. The
wireframe displays
show the objects
before the **FILLET**
command is used.

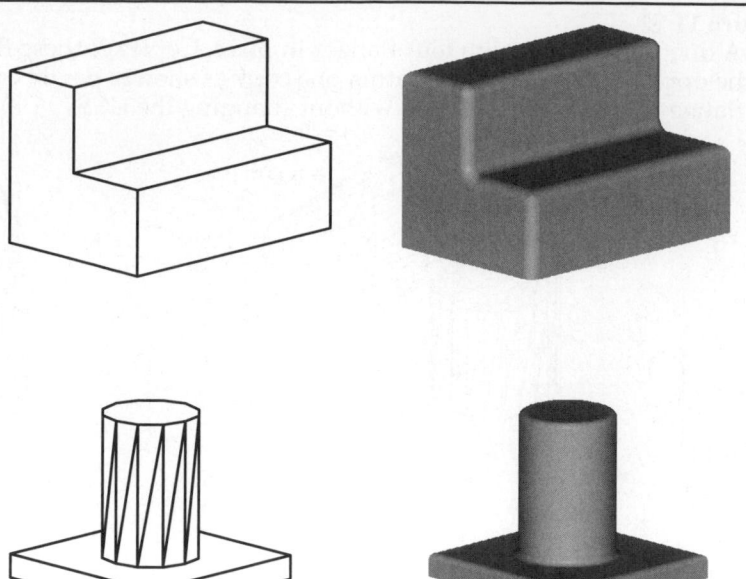

PROFESSIONAL TIP

You can construct and edit solid models while the object is displayed in a shaded view. If your computer has sufficient speed and power, it is often much easier to visualize the model in a 3D view with the Conceptual or Realistic visual style set current. This allows you to realistically view the model. If an edit or construction does not look right, just undo and try again.

Chamfering Solid Objects

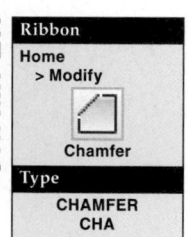

CHAMFER	
Ribbon	
Home	
> Modify	
	Chamfer
Type	
CHAMFER	
CHA	

A *chamfer* is a small square edge on the edges of an object. To create a chamfer on a 3D solid, use the **CHAMFER** command. Just as when chamfering a 2D line, there are two chamfer distances. Therefore, you must specify which surfaces correspond to the first and second distances. The detail to which the chamfer is applied must be constructed before chamfering. For example, if you are chamfering a hole, the object (cylinder) must first be subtracted to create the hole. If you are chamfering an intersection, the two objects must first be unioned.

After you enter the command, you must pick the edge you want to chamfer. The edge is actually the intersection of two surfaces of the solid. One of the two surfaces is highlighted when you select the edge. The highlighted surface is associated with the first chamfer distance. This surface is called the *base surface.* If the highlighted surface is not the one you want as the base surface, enter N at the [Next/OK] prompt and press [Enter]. This highlights the next surface. An edge is created by two surfaces. Therefore, when you enter N for the next surface, AutoCAD cycles through only two surfaces. When the proper base surface is highlighted, press [Enter].

Chamfering a hole is shown in **Figure 11-12A.** The end of the cylinder in **Figure 11-12B** is chamfered by first picking one of the vertical isolines, then picking the top edge. The following command sequence is illustrated in **Figure 11-12A.**

Figure 11-12.
A—A hole is chamfered by picking the top surface, then the edge of the hole. B—The end of a cylinder is chamfered by first picking the side, then the end. Both ends can be chamfered at the same time, as shown here.

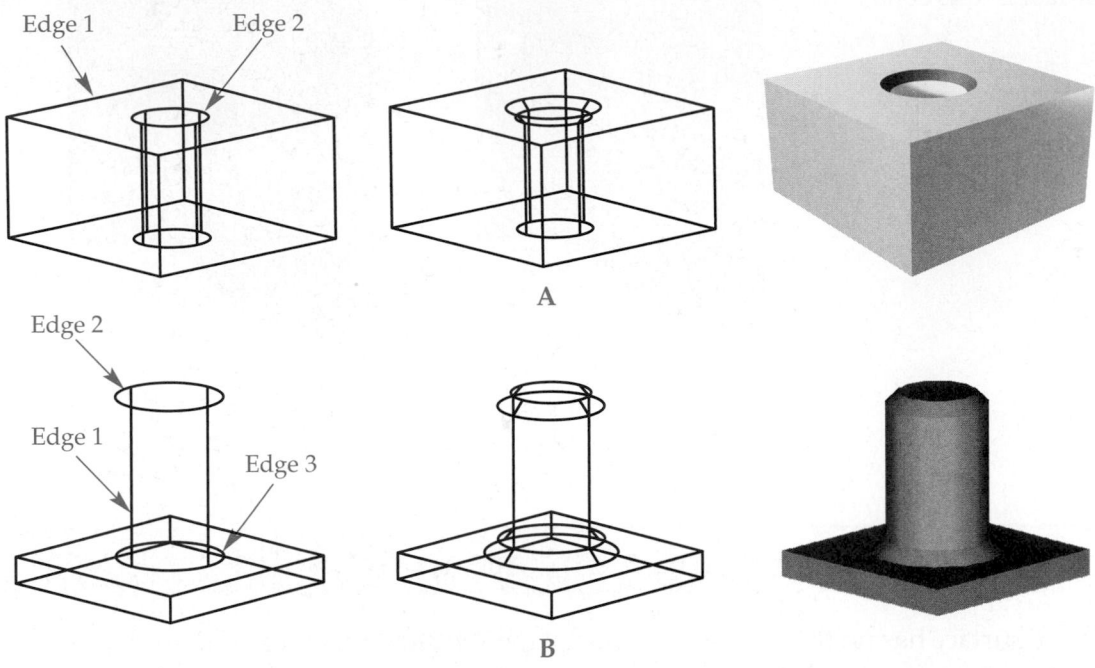

(TRIM mode) Current chamfer Dist1 = *current*, Dist2 = *current*
Select first line or [Undo/Polyline/Distance/Angle/Trim/mEthod/Multiple]: *(pick edge 1)*
Base surface selection…
(if the side surface is highlighted, change to the top surface as follows)
Enter surface selection option [Next/OK (current)] <OK>: **N↵**
(the top surface should be highlighted)
Enter surface selection option [Next/OK (current)] <OK>: ↵
Specify base surface chamfer distance <*current*>: **.125↵**
Specify other surface chamfer distance <*current*>: **.125↵**
Select an edge or [Loop]: *(pick edge 2, the edge of the hole)*
Select an edge or [Loop]: ↵

PROFESSIONAL TIP

If you improperly create a fillet or chamfer, it is best to undo and try again as opposed to trying to fix it with editing methods. Faces and edges can be edited using the **SOLIDEDIT** command. Grips can also be used to edit solids. This procedure is discussed in Chapter 12; the **SOLIDEDIT** command is discussed in Chapter 13.

Exercise 11-6

Complete the exercise on the student website.
www.g-wlearning.com/CAD

Figure 11-13.
A—This surface will be thickened into a solid.
B—The thickened surface is a 3D solid.

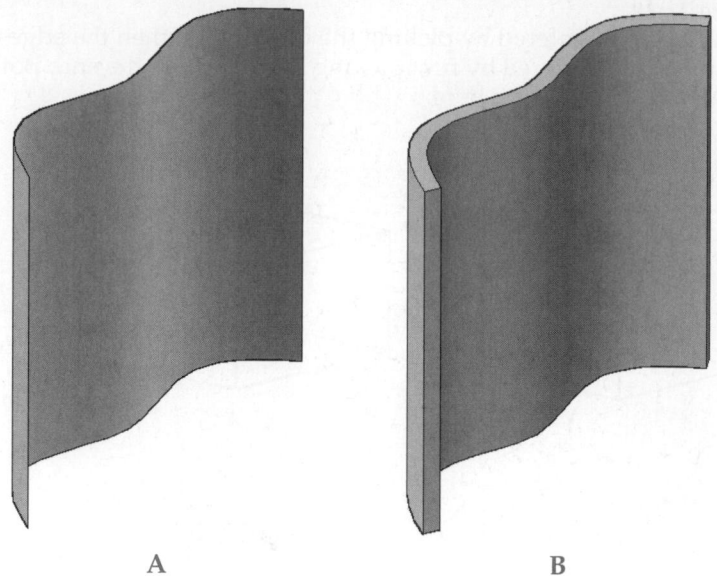

A   B

Thickening a Surface into a Solid

Ribbon
Home
> Solid Editing
Thicken
Type
THICKEN

A surface has no thickness. The value of the **THICKNESS** command does not affect the thickness of a planar surface, unlike for entities such as lines, polylines, polygons, and circles. But, a surface can be quickly converted to a 3D solid using the **THICKEN** command.

To add thickness to a surface, enter the command. Then, pick the surface(s) to thicken and press [Enter]. You are then prompted for the thickness. Enter a thickness value or pick two points on screen to specify the thickness. See **Figure 11-13.**

By default, the original surface object is deleted when the 3D solid is created with **THICKEN**. This is controlled by the **DELOBJ** system variable. To preserve the original surface, change the **DELOBJ** value to 0.

Converting to Surfaces

AutoCAD provides a great deal of flexibility in converting and transforming objects. For example, a simple line can be quickly turned into a 3D solid in just a few steps. Refer to **Figure 11-14.**

CONVTOSURFACE

Ribbon
Home
> Solid Editing
Convert to Surface
Type
CONVTOSURFACE

1. Use the **Properties** palette to give the line a thickness. Notice that the object is still a line object, as indicated in the drop-down list at the top of the **Properties** palette.
2. Select the **CONVTOSURFACE** command.
3. Pick the line. Its property type is now listed in the **Properties** palette as a surface extrusion.
4. Use the **THICKEN** command to give the surface a thickness. Its property type is now a 3D solid.

In this process, the **CONVTOSURFACE** and **THICKEN** commands were instrumental in creating a 3D solid from a line. Other objects that can be converted to surfaces using the **CONVTOSURFACE** command are 2D solids, arcs with thickness, open polylines with a thickness and no width, regions, and planar 3D faces.

Figure 11-14.
The stages of converting a line into a solid. First, draw the line. Next, give the line a thickness using the **Properties** palette. Then, convert the line to a surface using the **CONVTOSURFACE** command. Finally, use the **THICKEN** command to give the surface a thickness.

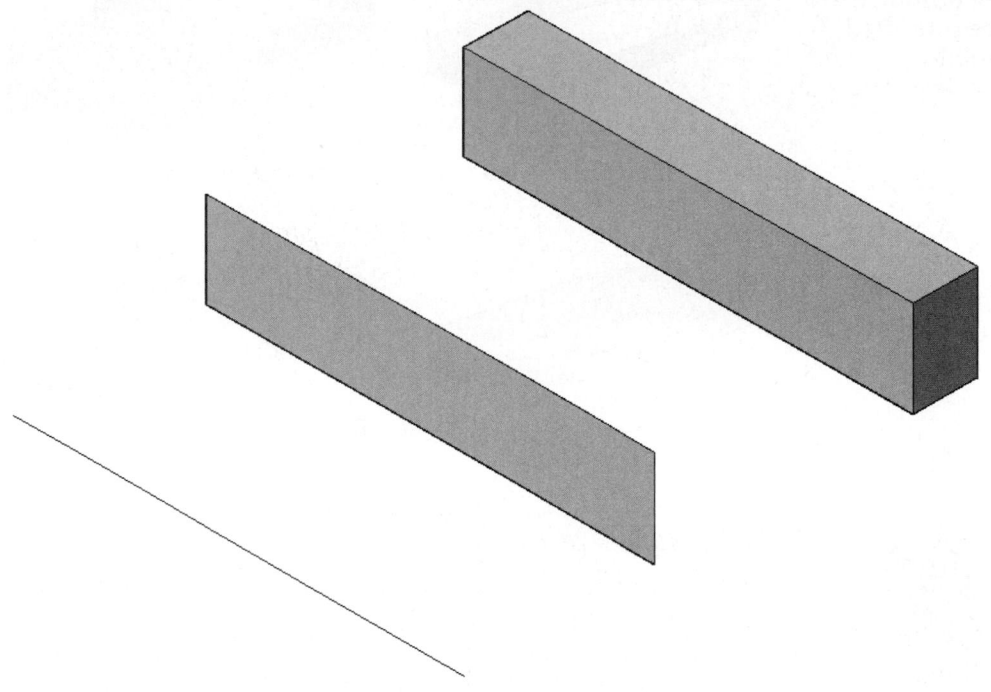

Exercise 11-7

Complete the exercise on the student website.
www.g-wlearning.com/CAD

Converting to Solids

Additional flexibility in creating solids is provided by the **CONVTOSOLID** command. This command allows you to directly convert certain closed objects into solids. You can convert:

- Circles with thickness.
- Wide, uniform-width polylines with thickness. This includes polygons and rectangles.
- Closed, zero-width polylines with thickness. This includes polygons, rectangles, and closed revision clouds.
- Mesh primitives. Keep in mind that the smoothness level applied to the mesh primitives appears on the object when it is converted to a solid

First, select the command. Then, select the objects to convert and press [Enter]. The objects are instantly converted with no additional input required. **Figure 11-15** shows the three different objects before and after conversion to a solid.

If an object that appears to be a closed polyline with a thickness does not convert to a solid and the command line displays the message Cannot convert an open curve, the polyline was not closed using the **Close** option of the **PLINE** command. Use the **PEDIT** or **PROPERTIES** command to close the polyline and use the **CONVTOSOLID** command again.

Ribbon
Home
> Solid Editing
Convert to Solid
Type
CONVTOSOLID

CONVTOSOLID

Figure 11-15.
A—From left to right, two polylines and an edited mesh sphere primitive that will be converted into solids.
B—The resulting solids.

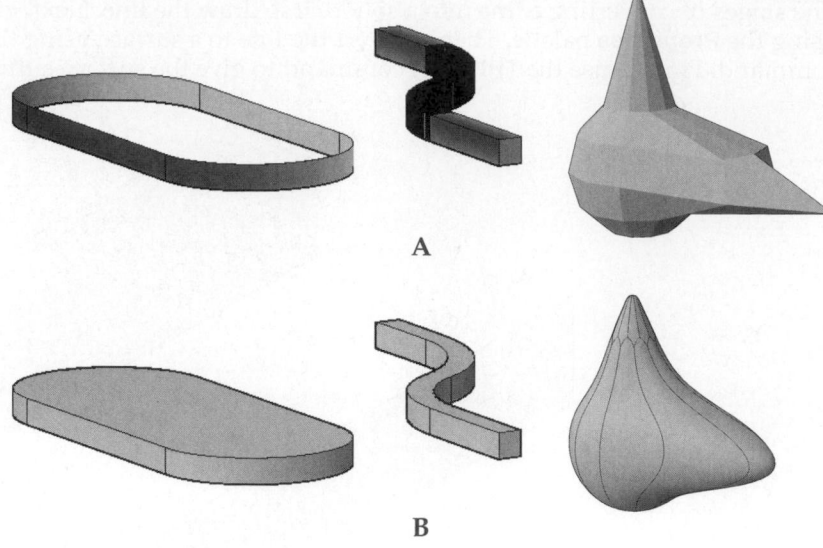

A

B

To quickly create a straight section of pipe, draw a donut with the correct ID and OD of the pipe. Then, use the **Properties** palette to give the donut a thickness equal to the length of the section you are creating. Finally, use the **CONVTOSOLID** command to turn the donut into a solid.

Slicing a Solid

A 3D solid can be sliced at any location by using existing objects such as circles, arcs, ellipses, 2D polylines, 2D splines, or surfaces. Additionally, you can specify a slicing line by picking two points or specify a slicing plane by picking three points. After slicing the solid, you can choose to retain either or both sides of the model. The slices can then be used for model construction or display and presentation purposes.

The **SLICE** command is used to slice solids. When the command is initiated, you are asked to select the solids to be sliced. Select the objects and press [Enter]. Next, you must define the slicing path. The default method of defining a path requires you to specify two points on a slicing plane. The plane passes through the two points and is perpendicular to the XY plane of the current UCS. Refer to **Figure 11-16** as you follow this sequence:

1. Select the command and pick the object to be sliced.
2. Pick the start point of the slicing plane. See **Figure 11-16A.**
3. Pick the second point on the slicing plane.
4. You are prompted to specify a point on the desired side to keep. Select anywhere on the back half of the object. The point does not have to be *on* the object. It must simply be on the side of the cutting plane that you want to keep.
5. The object is sliced and the front half is deleted. See **Figure 11-16B.**

SLICE

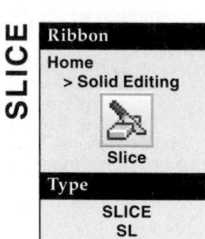

Ribbon
Home
> Solid Editing

Slice

Type
SLICE
SL

Figure 11-16.
Slicing a solid by picking two points. A—Select two points on the cutting plane. The plane passes through these points and is perpendicular to the XY plane of the current UCS. B—The sliced solid.

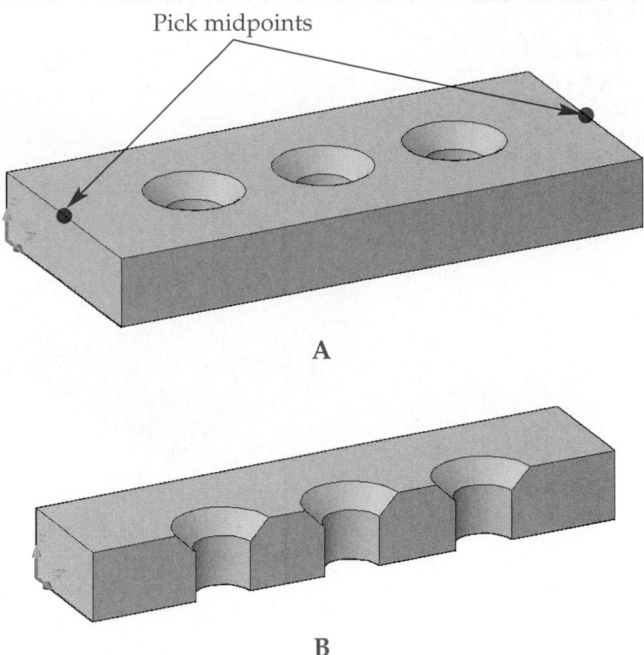

Pick midpoints

A

B

When prompted to select the side to keep, you can press [Enter] to keep both sides. If both sides are retained, two separate 3D solids are created. Each solid can then be manipulated for construction, design, presentation, or animation purposes.

There are several additional options for specifying a slicing path. These options are listed here and described in the following sections.

- **Planar Object**
- **Surface**
- **Zaxis**
- **View**
- **XY**
- **YZ**
- **ZX**
- **3points**

NOTE

Once the **SLICE** command has been used, the history of the solid to that point is removed. If a history of the work is important, then save a copy of the file or place a copy of the object on a frozen layer prior to performing the slice.

Planar Object

A second method to create a slice through a 3D solid is to use an existing planar object. Planar objects include circles, arcs, ellipses, 2D polylines, and 2D splines. See **Figure 11-17A.** The plane on which the planar object lies must intersect the object to be sliced. The current UCS has no effect on this option.

Be sure that the planar object has been moved to the location of the slice. Then, select the **SLICE** command, pick the object to slice, and press [Enter]. Next, select the **Planar Object** option and select the slicing path object (the circle, in this case). Finally, specify which side is to be retained. See **Figure 11-17B.** Again, if both sides are kept, they are separate objects and can be individually manipulated.

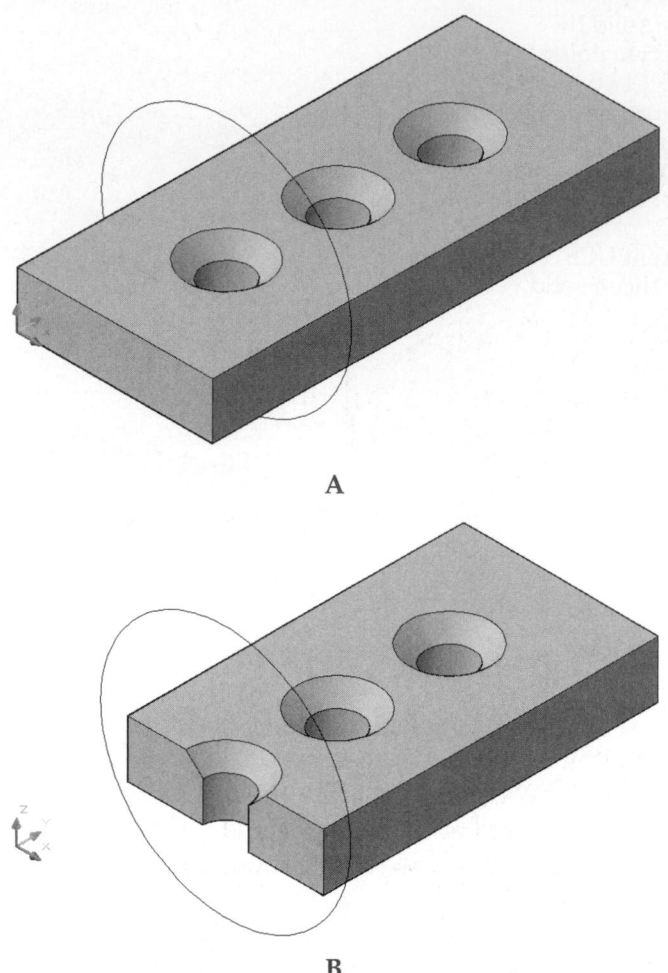

Figure 11-17.
Slicing a solid with a planar object. A—The circle is drawn at the proper orientation and in the correct location. B—The completed slice.

A

B

Surface

A surface object can be used as the slicing path. The surface can be planar or nonplanar (curved). This method can be used to quickly create a mating die. For example, refer to **Figure 11-18.** First, draw the required surface. The surface should exactly match the stamped part that will be manufactured, **Figure 11-18A.** Then, draw a box that encompasses the surface. Next, select the **SLICE** command, pick the box, and press [Enter]. Then, enter the **Surface** option and select the surface. You may need to do this in a wireframe display. Finally, when prompted to select the side to keep, press [Enter] to keep both sides. The two halves of the die can now be moved and rotated as needed, **Figure 11-18B.**

Z Axis

You can specify one point on the cutting plane and one point on the Z axis of the plane. See **Figure 11-19.** This allows you to have a cutting plane that is not parallel to the current UCS XY plane. First, select the **SLICE** command, pick the object to slice, and press [Enter]. Next, enter the **Zaxis** option. Then, pick a point on the XY plane of the cutting plane followed by a point on the Z axis of the cutting plane. Finally, pick the side of the object to keep.

AutoCAD and Its Applications—Advanced

Figure 11-18.
Slicing a solid with
a surface. A—Draw
the surface and
locate it within
the solid to be
sliced. The solid is
represented here
by the wireframe.
B—The completed
slice with both sides
retained. The top
can now be moved
and rotated as
shown here.

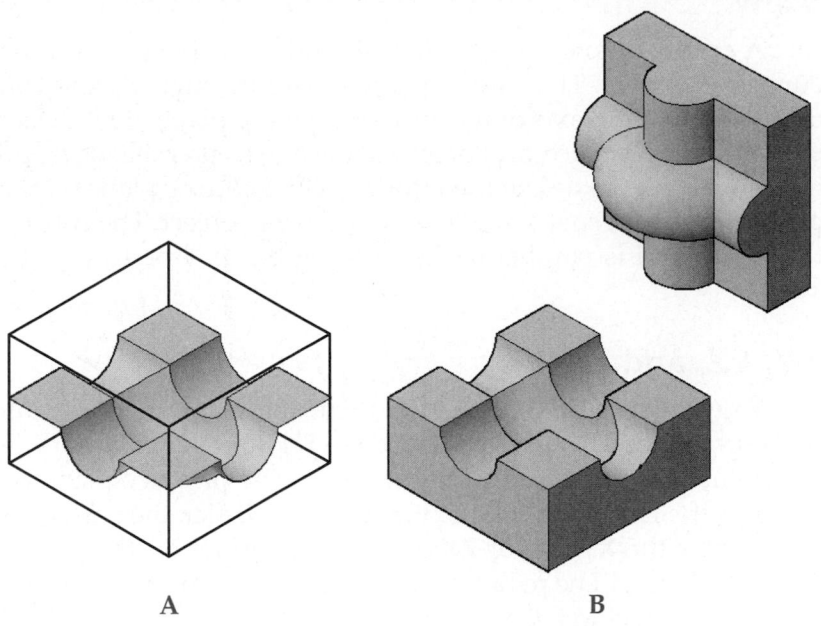

A

B

Figure 11-19.
Slicing a solid using
the **Zaxis** option.
A—Pick one point
on the cutting plane
and a second point
on the Z axis of the
cutting plane.
B—The resulting
slice.

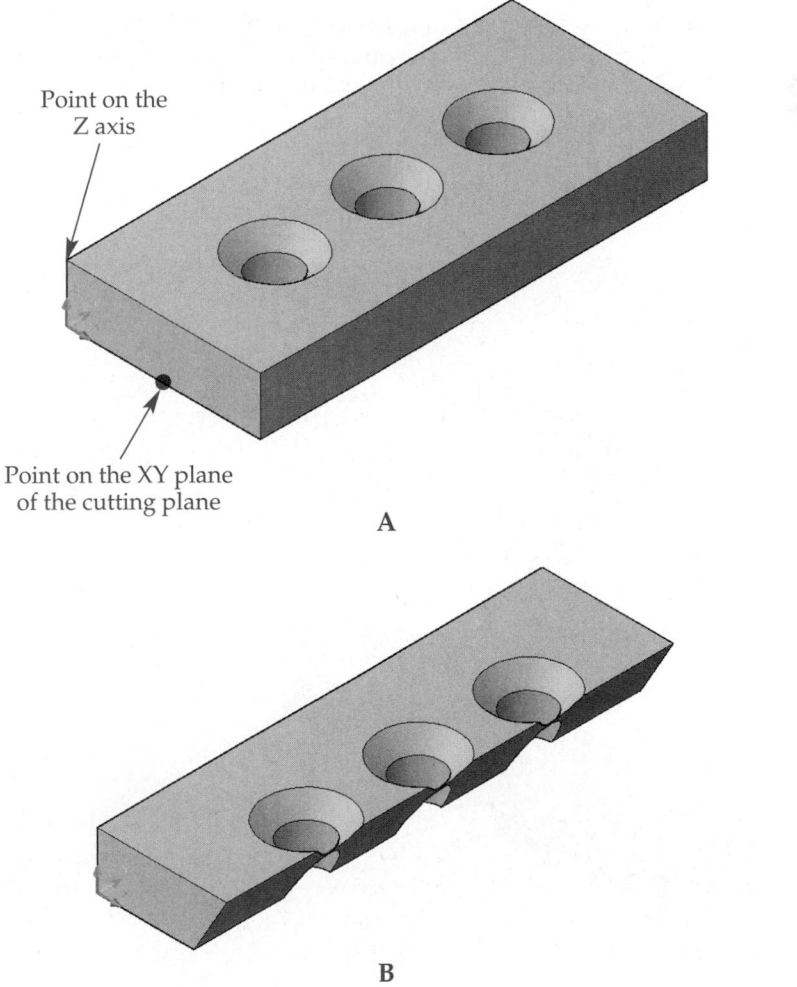

Point on the
Z axis

Point on the XY plane
of the cutting plane

A

B

View

A cutting plane can be established that is aligned with the viewing plane of the current viewport. The cutting plane passes through a point you select, which sets the depth along the Z axis of the current viewing plane. First, select the **SLICE** command, pick the object to slice, and press [Enter]. Next, enter the **View** option. Then, pick a point in the viewport to define the location of the cutting plane on the Z axis of the viewing plane. Use object snaps to select a point on an object. The cutting plane passes through this point and is parallel to the viewing plane. Finally, pick the side of the object to keep.

XY, YZ, and ZX

You can slice an object using a cutting plane that is parallel to any of the three primary planes of the current UCS. See **Figure 11-20.** The cutting plane passes through the point you select and is aligned with the primary plane of the current UCS that you specify. First, select the **SLICE** command, pick the object to slice, and press [Enter]. Next, enter the **XY**, **YZ**, or **ZX** option, depending on the primary plane to which the cutting plane will be parallel. Then, pick a point on the cutting plane. Finally, pick the side of the object to keep.

Figure 11-20.
Slicing a solid using the **XY**, **YZ**, and **ZX** options. A—The object before slicing. The UCS origin is in the center of the first hole and at the midpoint of the height. B—The resulting slice using the **XY** option. C—The resulting slice using the **YZ** option. D—The resulting slice using the **ZX** option.

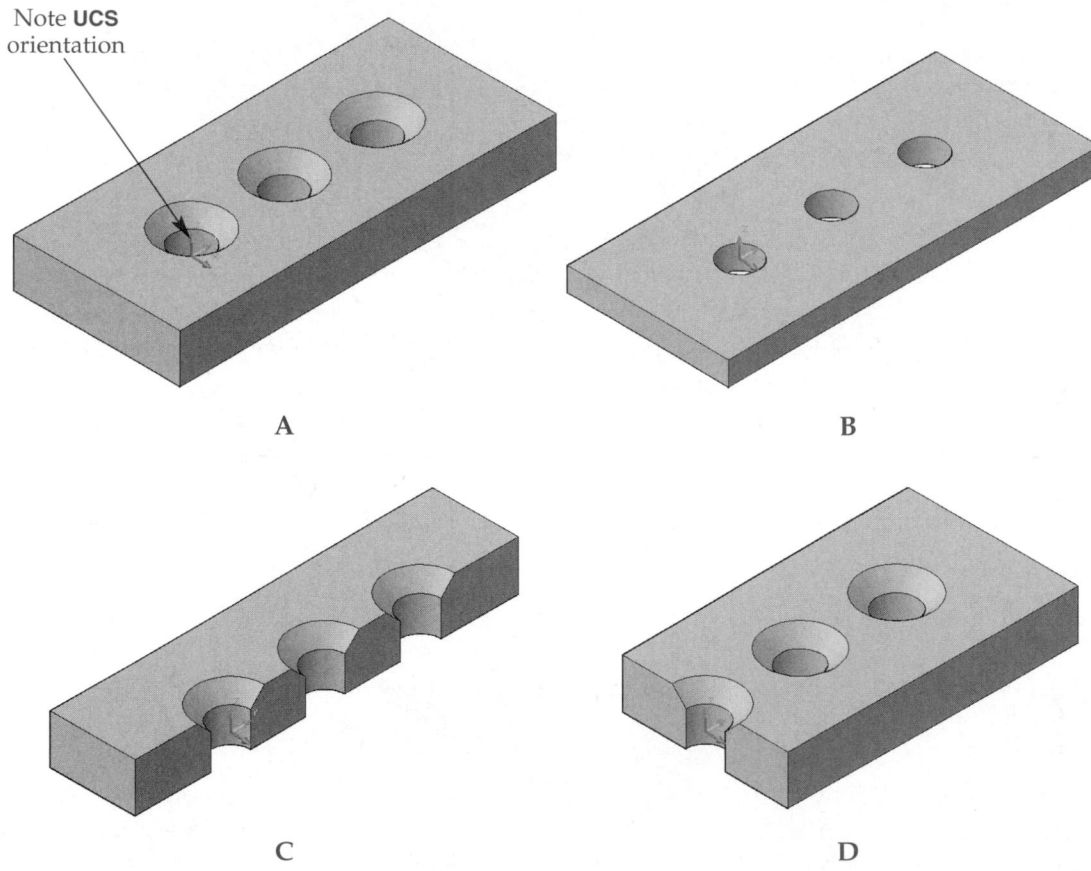

Figure 11-21.
Slicing a solid using the **3points** option. A—Specify three points to define the cutting plane. B—The resulting slice.

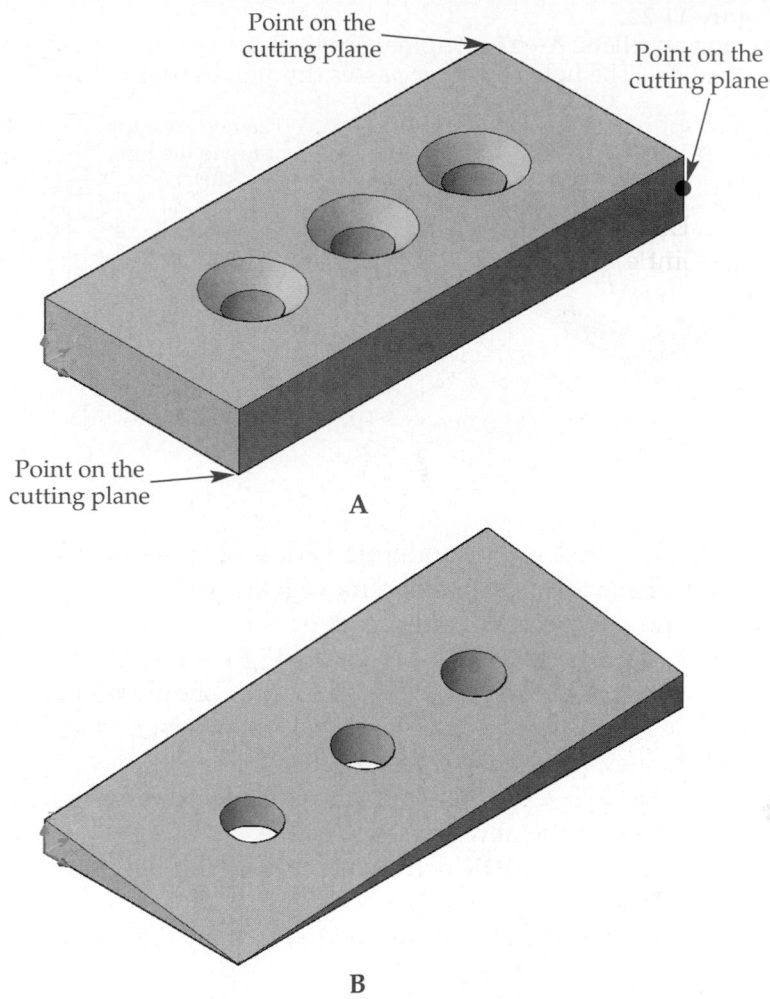

Point on the cutting plane

Point on the cutting plane

Point on the cutting plane

A

Point on the cutting plane

B

Three Points

Three points can be used to define the cutting plane. This allows the cutting plane to be aligned at any angle, similar to the **Zaxis** option. See **Figure 11-21.** First, select the **SLICE** command, pick the object to be sliced, and press [Enter]. Then, enter the **3points** option. Pick three points on the cutting plane and then select the side of the object to keep.

Exercise 11-8

Complete the exercise on the student website.
www.g-wlearning.com/CAD

Removing Details and Features

Sometimes, it may be necessary to remove a detail that has been constructed. For example, suppose you placed a R.5 fillet on an object based on an engineering sketch. Then, the design is changed to a R.25 fillet. The **UNDO** command can only be used in the current drawing session. Also, even if the command can be used, you may have to step back through several other commands to undo the fillet. In another example, suppose an object has a bolt hole that is no longer needed. You will need to remove this feature.

Figure 11-22.
Removing fillets. A—The original object. B—A new base is added and the new fillets are created. C—The hole no longer passes through the object. D—The corrected object.

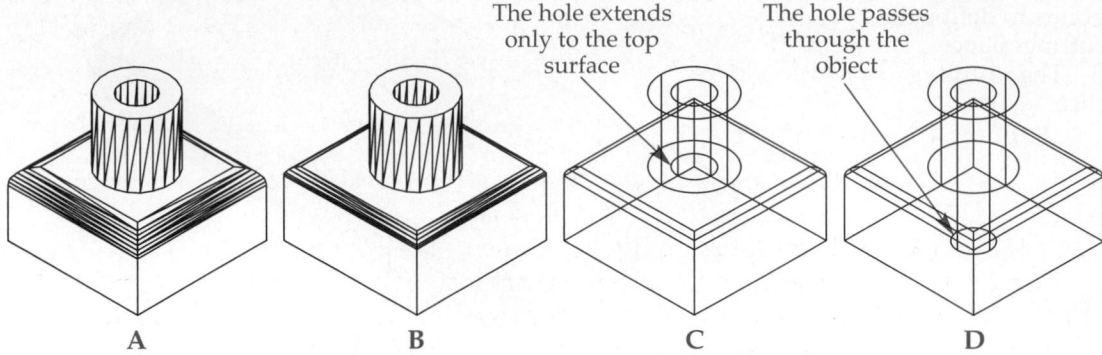

The hole extends only to the top surface

The hole passes through the object

A   B   C   D

In Chapter 2, solid modeling is described as working with modeling clay. If you think in these terms, you can remove features by adding "clay" to the object. Then, the new "clay" can be molded as needed.

For example, look at the object in **Figure 11-22A.** There are R.5 rounds (fillets) on the top surface of the base. However, these should be R.25 rounds. You cannot simply place the new fillets on the object. You must first add material to create a square edge. Then, the new fillets can be added.

1. Draw a solid box with the same dimensions as the base without the rounds. Center the new box on the base.
2. Use the **UNION** command to add the new box to the object. This, in effect, removes the rounds.
3. Use the **FILLET** command to place the R.25 rounds on the top edge of the base, **Figure 11-22B.**

The rounds have now, in effect, been changed from R.5 to R.25. However, there is an unseen problem. Display the object in wireframe. Notice how the hole no longer passes through the object, **Figure 11-22C.** To correct this problem, draw a solid cylinder of the same dimensions as the hole and centered in the hole. Then, subtract the cylinder from the object. The hole now passes through the object, **Figure 11-22D.**

This technique of adding material can be used to remove any internal feature and some external features, such as fillets (rounds). Other external features, such as a boss, can be removed by drawing a solid over the top of the feature. The feature to be removed must be completely enclosed by the new solid. Then, subtract the new solid from the original object. Be sure to "redrill" holes and other internal features as needed.

PROFESSIONAL TIP

There are several other methods for editing solids. These are covered in detail in Chapters 12 and 13. The above procedure can be simplified with these editing methods.

Constructing Details and Features on Solid Models

A variety of machining, structural, and architectural details can be created using some basic solid modeling techniques. The features discussed in the next sections are just a few of the possibilities.

Counterbore and Spotface

A *counterbore* is a recess machined into a part, centered on a hole, that allows the head of a fastener to rest below the surface. Create a counterbore as follows.

1. Draw a cylinder representing the diameter of the hole, **Figure 11-23A.**
2. Draw a second cylinder that is the diameter of the counterbore and center it at the top of the first cylinder. Move the second cylinder so it extends below the surface of the object to the depth of the counterbore, **Figure 11-23B.**
3. Subtract the two cylinders from the base object, **Figure 11-23C.**

A *spotface* is similar to a counterbore, but is not as deep. See **Figure 11-24.** It provides a flat surface for full contact of a washer or underside of a bolt head. Construct it in the same way as a counterbore.

Figure 11-23.
Constructing a counterbore. A—Draw a cylinder to represent a hole. B—Draw a second cylinder to represent the counterbore. C—Subtract the two cylinders from the base object.

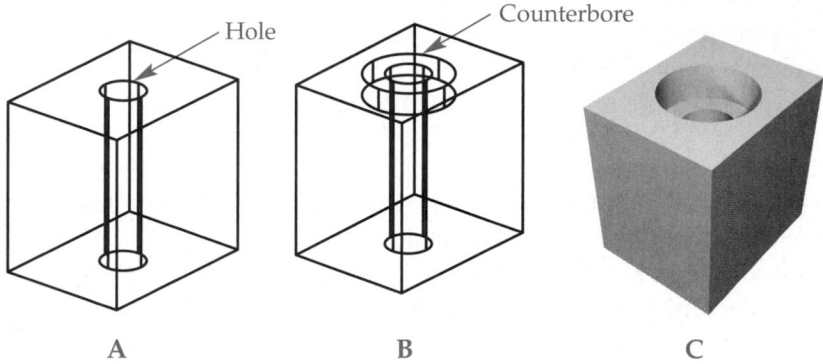

Figure 11-24.
Constructing a spotface. A—The bottom of the second, larger-diameter cylinder should be located at the exact depth of the spotface. However, the height may extend above the surface of the base. Then, subtract the two cylinders from the base. B—The finished solid.

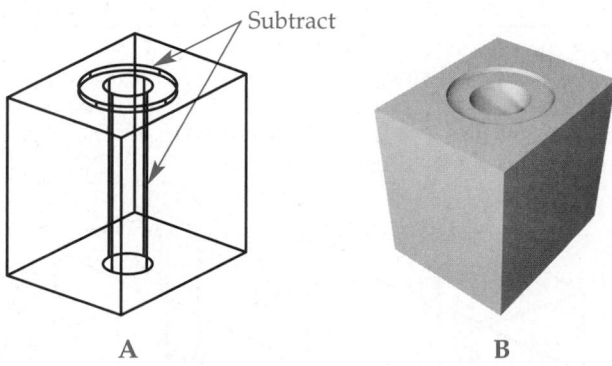

Figure 11-25.
Constructing a countersink. A—Subtract the cylinder from the base to create the hole. B—Chamfer the top of the hole to create a countersink.

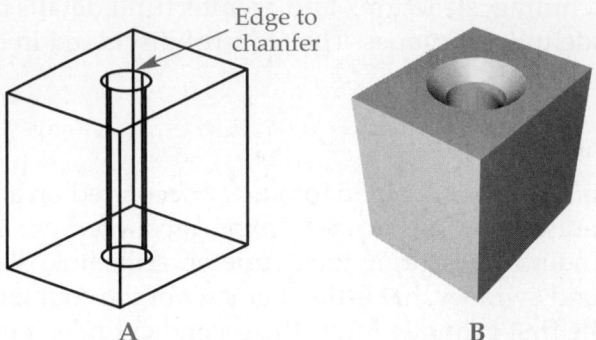

Countersink

A *countersink* is like a counterbore with angled sides. The sides allow a flat-head machine screw or wood screw to sit flush with the surface of an object. A countersink can be drawn in one of two ways. You can draw an inverted cone centered on a hole and subtract it from the base or you can chamfer the top edge of a hole. Chamfering is the quickest method.

1. Draw a cylinder representing the diameter of the hole, **Figure 11-25A**.
2. Subtract the cylinder from the base object.
3. Select the **CHAMFER** command.
4. Select the top edge of the base object.
5. Enter the chamfer distance(s).
6. Pick the top edge of the hole, **Figure 11-25B**.

Boss

A *boss* serves the same function as a spotface. However, it is an area raised above the surface of an object. Draw a boss as follows.

1. Draw a cylinder representing the diameter of the hole. Extend it above the base object higher than the boss is to be, **Figure 11-26A**.
2. Draw a second cylinder the diameter of the boss. Place the base of this cylinder above the top surface of the base object a distance equal to the height of the boss. Give the cylinder a negative height value so that it extends inside of the base object, **Figure 11-26B**.
3. Union the base object and the second cylinder (boss). Subtract the hole from the unioned object, **Figure 11-26C**.
4. Fillet the intersection of the boss with the base object, **Figure 11-26D**.

Figure 11-26.
Constructing a boss. A—Draw a cylinder for the hole so it extends above the surface of the object. B—Draw a cylinder the height of the boss on the top surface of the object. C—Union the large cylinder to the base. Then, subtract the small cylinder (hole) from the unioned objects. D—Fillet the edge to form the boss.

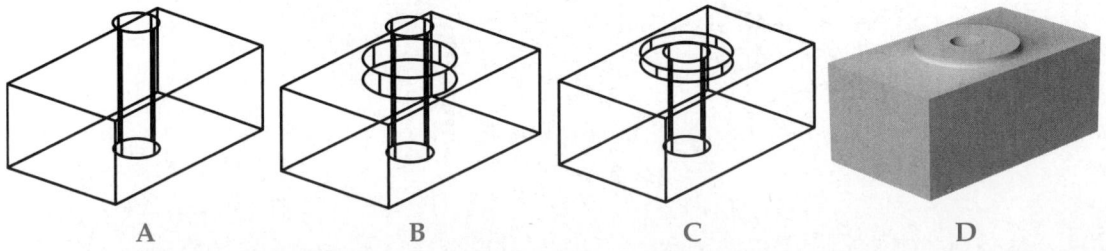

O-Ring Groove

An *O-ring* is a circular seal that resembles a torus. It sits inside of a groove constructed so that part of the O-ring is above the surface. An *O-ring groove* can be constructed by placing the center of a circle on the outside surface of a cylinder. Then, revolve the circle around the cylinder. Finally, subtract the revolved solid from the cylinder.

1. Construct the cylinder to the required dimensions, **Figure 11-27A.**
2. Rotate the UCS on the X axis (or appropriate axis).
3. Draw a circle with a center point on the surface of the cylinder, **Figure 11-27B.**
4. Revolve the circle 360° about the center of the cylinder, **Figure 11-27C.**
5. Subtract the revolved object from the cylinder, **Figure 11-27D.**

Architectural Molding

Architectural molding details can be quickly constructed using extrusions. First, construct the profile of the molding as a closed shape, **Figure 11-28A.** Then, extrude the profile the desired length, **Figure 11-28B.**

Corner intersections of molding can be quickly created by extruding the same shape in two different directions and then joining the two objects. First, draw the molding profile. Then, copy and rotate the profile to orient the local Z axis in the desired direction, **Figure 11-29A.** Next, extrude the two profiles the desired lengths, **Figure 11-29B.** Finally, union the two extrusions to create the mitered corner molding, **Figure 11-29C.**

Figure 11-27.
Constructing an O-ring groove. A—Construct a cylinder; this one has a round placed on one end. B—Draw a circle centered on the surface of the cylinder. C—Revolve the circle 360° about the center of the cylinder. D—Subtract the revolved object from the cylinder. E—The completed O-ring groove.

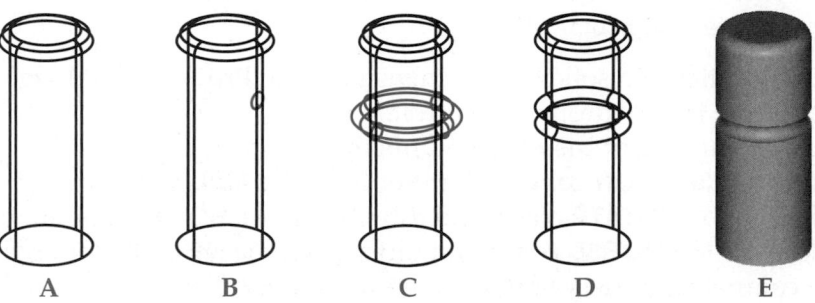

A   B   C   D   E

Figure 11-28.
A—The molding profile. B—The profile extruded to the desired length.

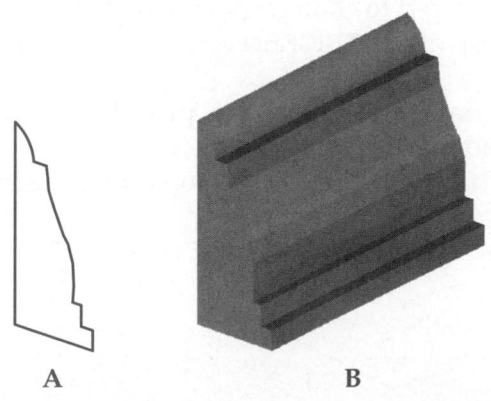

A   B

Figure 11-29.
Constructing corner molding. A—Copy and rotate the molding profile. B—Extrude the profiles to the desired lengths. C—Union the two extrusions to create the mitered corner. Note: The view has been rotated. D—The completed corner.

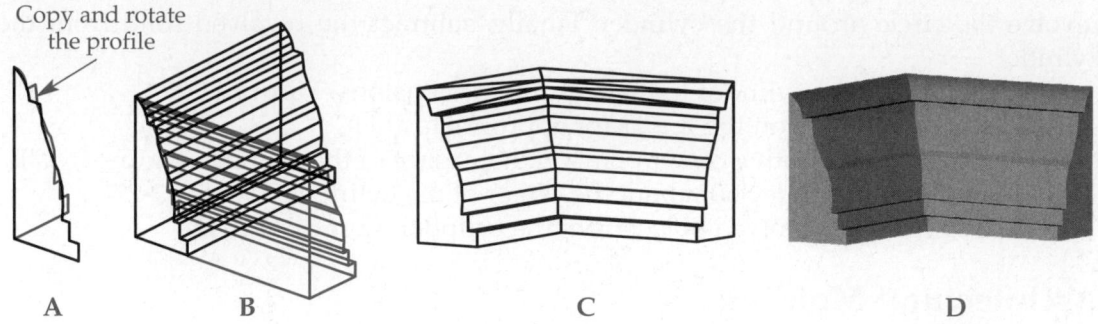

Copy and rotate the profile

A   B   C   D

Exercise 11-9

Complete the exercise on the student website.
www.g-wlearning.com/CAD

Chapter Test

Answer the following questions. Write your answers on a separate sheet of paper or complete the electronic chapter test on the student website.
www.g-wlearning.com/CAD

1. Which properties of a solid can be changed in the **Properties** palette?
2. What does the History property control?
3. What is the purpose of the **ALIGN** command?
4. How does the **3DALIGN** command differ from the **ALIGN** command?
5. How does the **3DROTATE** command differ from the **ROTATE** command?
6. How does the **MIRROR3D** command differ from the **MIRROR** command?
7. Which command allows you to create a rectangular array by defining rows, columns, and levels?
8. How does a 3D polar array differ from a 2D polar array?
9. How many levels can a 3D polar array have?
10. Which command is used to fillet a solid object?
11. Which command is used to chamfer a solid object?
12. What is the purpose of the **THICKEN** command and which type of object does it create?
13. Which system variable allows you to preserve the original object when the **THICKEN** command is used?
14. List four objects that can be converted to surfaces using the **CONVTOSURFACE** command.
15. Briefly describe the function of the **SLICE** command.

Drawing Problems

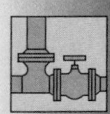

1. Construct an 8″ diameter tee pipe fitting using the dimensions shown below. Hint: Extrude and union two solid cylinders before subtracting the cylinders for the inside diameters.
   A. Use **EXTRUDE** to create two sections of pipe at 90° to each other, then **UNION** the two pieces together.
   B. Use **FILLET** and **CHAMFER** to finish the object. The chamfer distance is .25″ × .25″.
   C. The outside diameter of all three openings is 8.63″ and the pipe wall thickness is .322″.
   D. Save the drawing as P11_01.

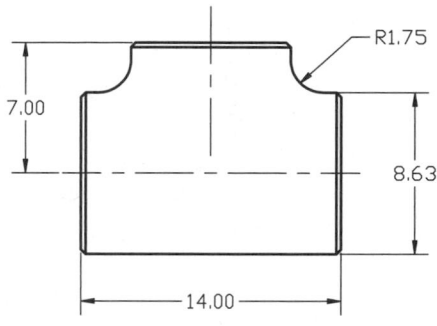

2. Construct an 8″ diameter, 90° elbow pipe fitting using the dimensions shown below.

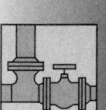

   A. Use **EXTRUDE** or **SWEEP** to create the elbow.
   B. Chamfer the object. The chamfer distance is .25″ × .25″. Note: You cannot use the **CHAMFER** command.
   C. The outside diameter is 8.63″ and the pipe wall thickness is .322″.
   D. Save the drawing as P11_02.

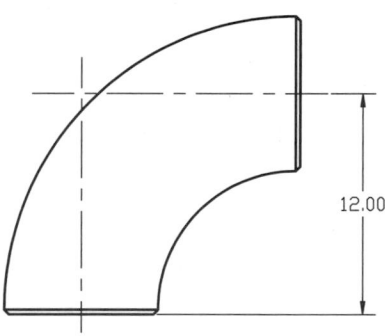

Chapter 11   Creating and Working with Solid Model Details

Problems 3–6. These problems require you to use a variety of solid modeling functions to construct the objects. Use all of the solid modeling and editing commands you have learned so far to assist in construction. Use a dynamic UCS when practical and create new UCSs as needed. Use **SOLIDHIST** and **SHOWHIST** to record and view the steps used to create the solid models. Create copies of the completed models and split them as required to show the internal features visible in the section views. Save each drawing as P11_(problem number).

3.

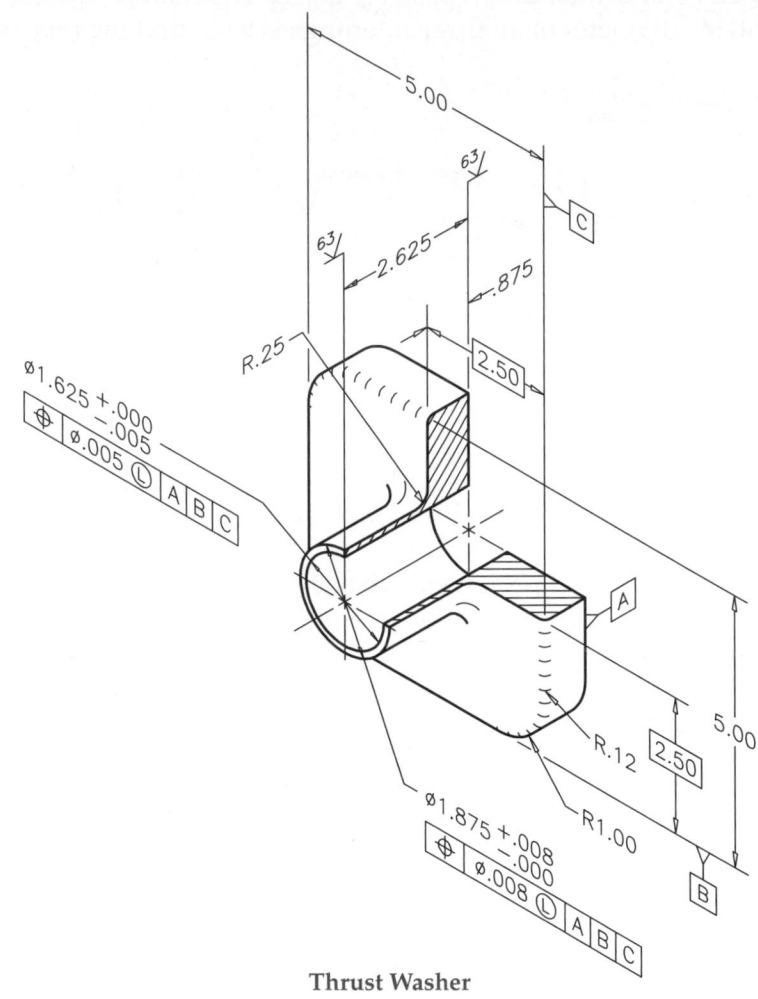

Thrust Washer

4.

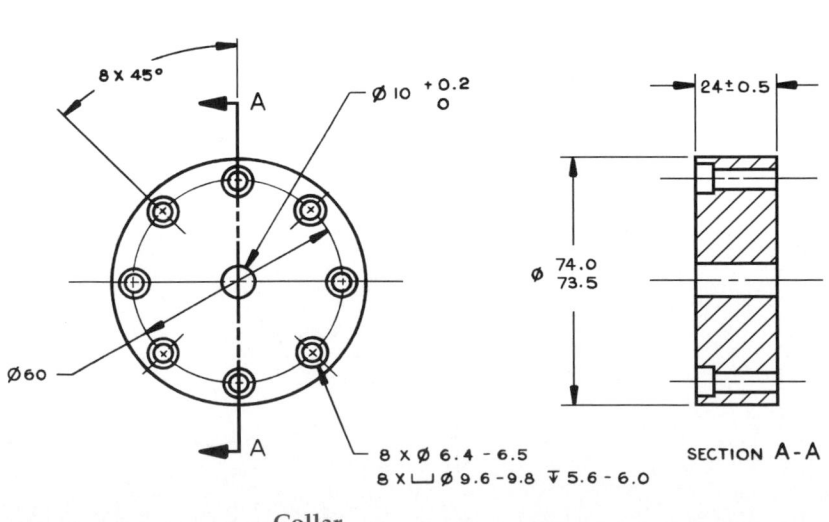

Collar

5.

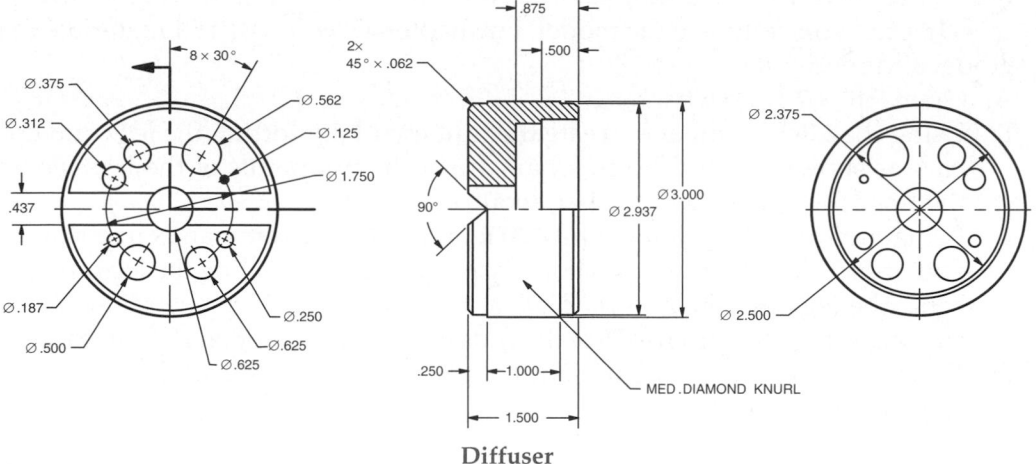

Diffuser

6.

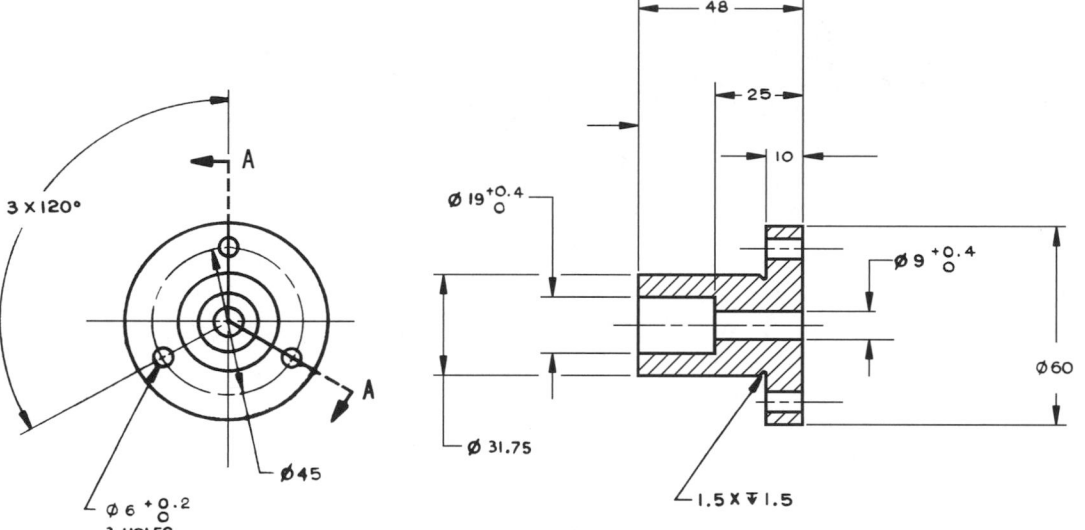

SECTION A - A

Bushing

7. In this problem, you will add legs to the kitchen chair you started in Chapter 2. In Chapter 9 you refined the seat and in Chapter 10 you added the seatback. In this chapter, you complete the model. In Chapter 17, you will add materials to the model and render it.

   A. Open P10_07 from Chapter 10.

   B. Using the **LINE** command, create the framework for lofting the legs and cross-bars as shown below. The crossbars are at the midpoints of the legs. Position the lines for the double crossbar about 4.5″ apart.

   C. Using a combination of the **3DROTATE** and **3DMOVE** commands in conjunction with new UCSs, draw and position circles as shapes for lofting the legs and cross-bars. The legs transition from ⌀1.00 at the ends to ⌀1.25 at the midpoints. The crossbars transition from ⌀.75 at the ends to ⌀1.00 at a position 2″ from each end.

   D. Create the legs and crossbars. The completed model is shown below.

   F. Save the drawing as P11_07.

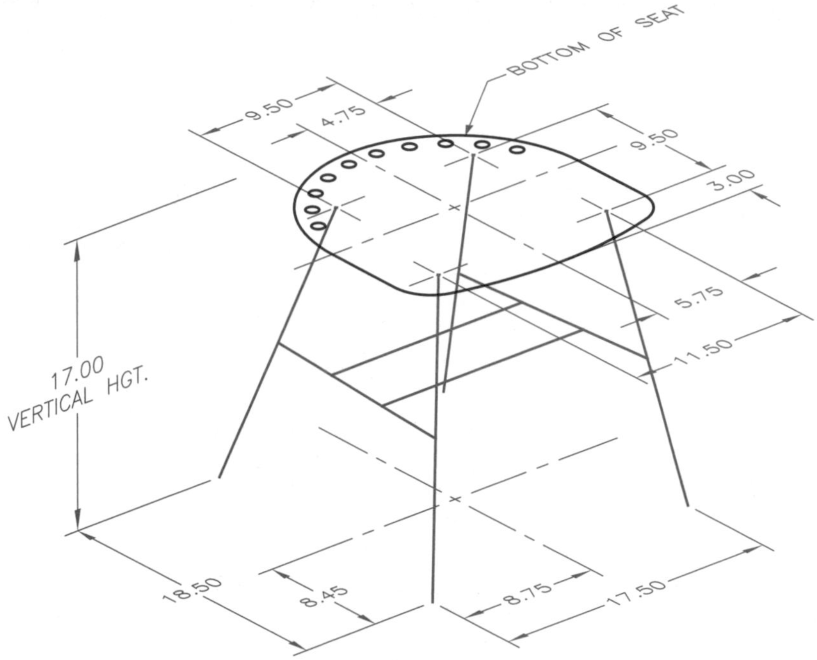

Completed Model

CHAPTER 12

Subobject Editing

Learning Objectives

After completing this chapter, you will be able to:

✓ Select subobjects (faces, edges, and vertices).
✓ Edit solids using grips.
✓ Edit face subobjects.
✓ Edit edge subobjects.
✓ Edit vertex subobjects.
✓ Extrude a closed boundary using the **PRESSPULL** command.
✓ Extract a wireframe from a 3D solid using the **XEDGES** command.

Grip Editing

There are three basic types of 3D solids in AutoCAD. The commands **BOX**, **WEDGE**, **PYRAMID**, **CYLINDER**, **CONE**, **SPHERE**, and **TORUS** create 3D solid *primitives. Sweeps* are 2D profiles given thickness by the **EXTRUDE**, **REVOLVE**, **SWEEP**, and **LOFT** commands to create a 3D solid. Finally, 3D solid *composites* are created by a Boolean operation or by using the **SOLIDEDIT** command. The **SOLIDEDIT** command is discussed in Chapter 13. Smooth-edged solid primitives are achieved by converting mesh objects to solids. These are also composite solids.

There are two types of grips—base and parameter—that may be associated with a solid object. These grips provide an intuitive means of modifying solids. Base grips are square and parameter grips are typically arrows. The editing that can be performed with these grips are discussed in the next sections.

Primitives

The 3D solid primitives all have basically the same grips. However, not all grips are available on all primitives. All primitives have a base grip at the centroid of the base. This grip functions like a standard grip in 2D work. It can be used to stretch, move, rotate, scale, or mirror the solid.

Boxes, wedges, and pyramids have square base grips at the corners that allow the size of the base to be changed. See **Figure 12-1.** The object dynamically changes in the viewport as you select and move the grip or after you type the new coordinate location

Figure 12-1.
Boxes, wedges, and pyramids have square base grips at the corners and parameter grips on the sides of the base and center of the top face, edge, or vertex.

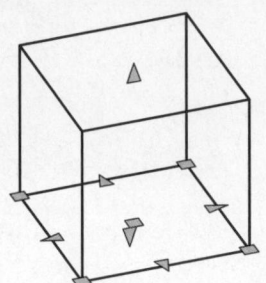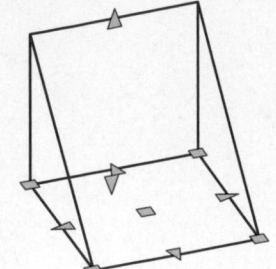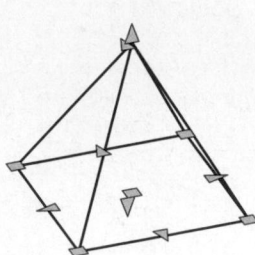

for the grip and press [Enter]. If ortho is off, the length and width can be changed at the same time by dragging the grip, except in the case of a pyramid. The triangular parameter grips on the base allow the length or width to be changed. The height can be changed using the top parameter grip(s). Each object has one parameter grip for changing the height of the apex and one for changing the height of the plane on which the base sits. A pyramid also has a parameter grip at the apex for changing the radius of the top.

Cylinders, cones, and spheres have four parameter grips for changing the radius of the base, or the cross section in the case of a sphere. See **Figure 12-2.** Cylinders and cones also have parameter grips for changing the height of the apex and the height of the plane on which the base sits. Additionally, a cone has a parameter grip at the apex for changing the radius of the top.

A torus has a parameter grip located at the center of the tube. See **Figure 12-3.** This grip is used to change the radius of the torus. Parameter grips at each quadrant of the tube are used to change the radius of the tube.

A polysolid does not have parameter grips. Instead, a base grip appears at each corner of the starting face of the solid. See **Figure 12-4.** Use these grips to change the cross-sectional shape of the polysolid. The corners do not need to remain square. Base grips also appear at the endpoint of each segment centerline. Use these to change the location of each segment's endpoints.

Figure 12-2.
Cylinders, cones, and spheres have four parameter grips for changing their radius. Cylinders and cones have parameter grips for changing their height. Cones also have a parameter grip for changing the radius of the top.

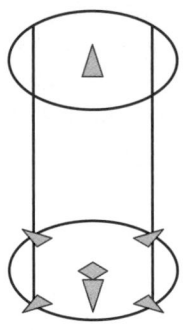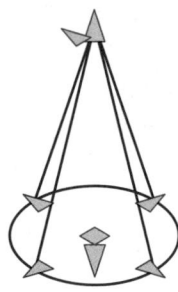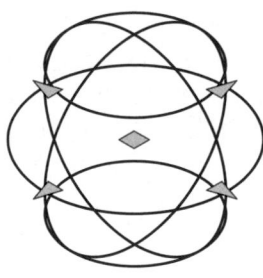

Figure 12-3.
A torus has a parameter grip located at the center of the tube for changing the radius of the torus. There are also parameter grips for changing the radius of the tube.

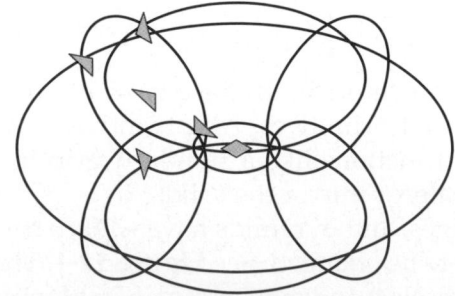

AutoCAD and Its Applications—Advanced

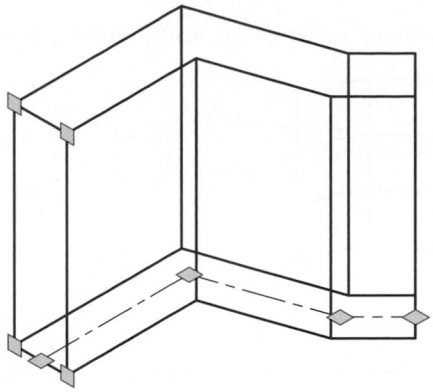

Figure 12-4.
A polysolid has a base grip at each corner of the starting face of the solid and one at the endpoint of each segment.

Swept Solids

Extrusions, revolutions, sweeps, and lofts are considered swept solids and typically have base grips located at the vertices of the 2D profiles. These can be used to change the size of the profile and, thus, the solid. Other grips that appear include:

- A parameter grip appears on the upper face of extrusions for changing the height.
- A base grip appears on the axis of revolved solids for changing the location of the axis in relation to the profile.
- Base grips appear on the vertices of the path of sweeps for changing the shape of the path.

Composite Solids

The Boolean commands (**UNION**, **SUBTRACT**, and **INTERSECT**) create composite solids. Solids that have been modified using any of the options of the **SOLIDEDIT** command also become composite solids, as do meshes converted into solids. The solid may still look like a primitive, sweep, loft, etc., but it is a composite. The grips available with the previous objects are no longer available, unless performing subobject editing on a composite created with a Boolean command (discussed later in this chapter). Composite solids have a base grip located at the centroid of the base surface. This grip can be used to stretch, move, rotate, scale, or mirror the solid. Composite solids converted from meshes have several grips, but these all act as base grips.

NOTE

When performing grip editing on a solid, AutoCAD must be able to "solve" the end result. If it cannot, the edit is not applied.

Using Grips with Surfaces

Surfaces can be edited using grips in the same manner as discussed with solids. A planar surface created with **PLANESURF** can be moved, rotated, scaled, and mirrored, but not stretched. Base grips are located at each corner.

As you learned in Chapter 9, a variety of AutoCAD objects can be extruded to create a surface. Three of these objects—arc, line, and polyline—are shown extruded into surfaces in **Figure 12-5.** Notice the location and type of grips on the surface extrusions. Base grips are located on the original profile that was extruded to make the surface. These grips enable you to alter the shape of the surface. A parameter grip located on the top of the surface is used to change the height of the extrusion.

Figure 12-5.
Surfaces extruded from an arc, line, and polyline. Notice the grips.

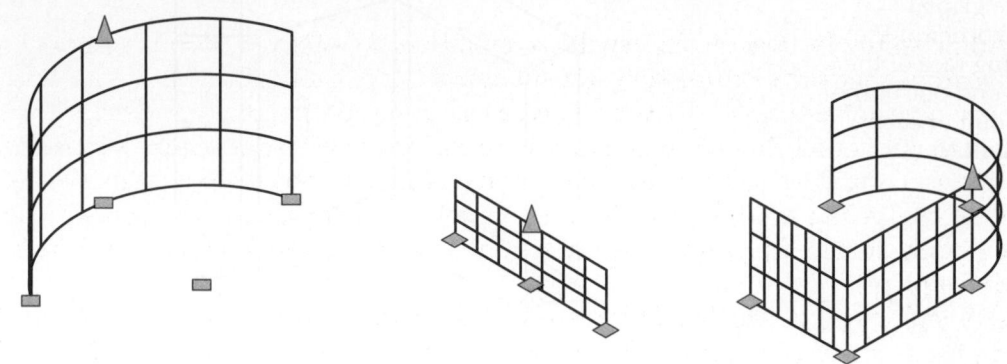

Figure 12-6.
A—Grips can be used to modify the path on this swept surface. B—The swept surface after grip editing.

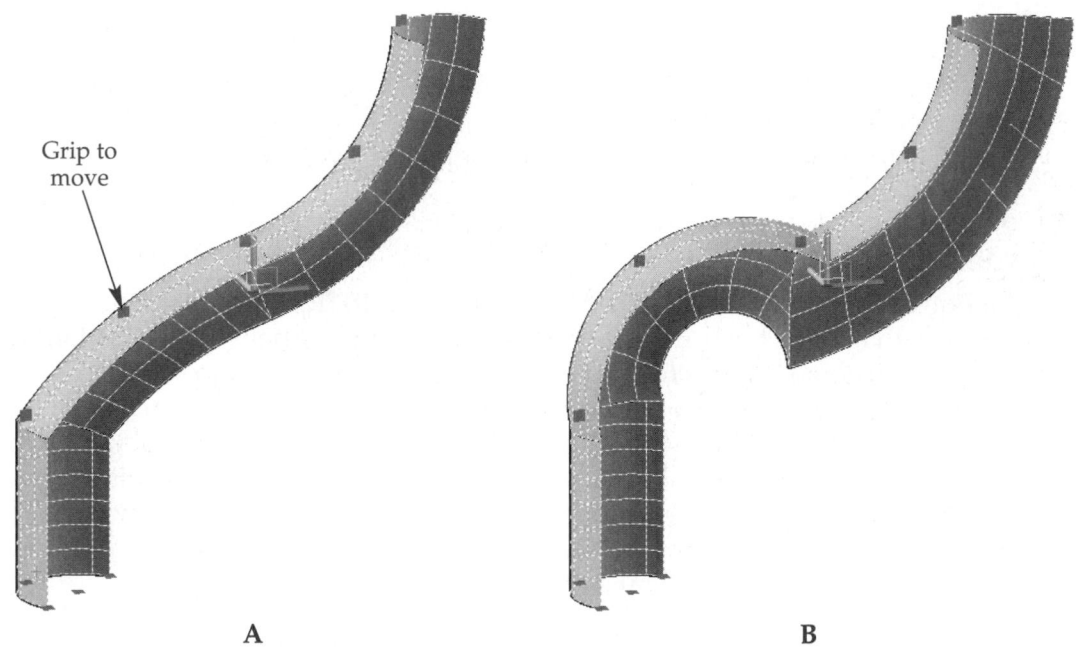

Grip to move

A   B

Surfaces that have been extruded, or swept, along a path can be edited with grips. Also, the grips located on the path allow you to change the shape of the surface extrusion. See **Figure 12-6.**

Exercise 12-1

Complete the exercise on the student website.
www.g-wlearning.com/CAD

Overview of Subobject Editing

AutoCAD solid primitives, such as cylinders, wedges, and boxes, and mesh objects are composed of three types of subobjects: faces, edges, and vertices. In addition, the objects that are used with Boolean commands to create a composite solid are considered subobjects, if the history is recorded. The primitive subobjects can be edited. See **Figure 12-7.** Once selected, the primitive subobjects can even be deleted from the composite solid. **Figure 12-8** illustrates the difference between a composite solid model, the solid primitives used to construct it, and an individual subobject of one of the primitives.

Subobjects can be easily edited using grips, which provide an intuitive and flexible method of solid model design. For example, suppose you need to rotate a face subobject in the current XY plane. You can select the subobject, pick its base grip, and then cycle through the editing functions to **ROTATE**. You can also use the **ROTATE** command on the selected subobject.

Figure 12-7.
A—Selecting a subobject solid primitive within a composite solid displays its grips. B—The grips on the primitive can be used to edit the primitive.

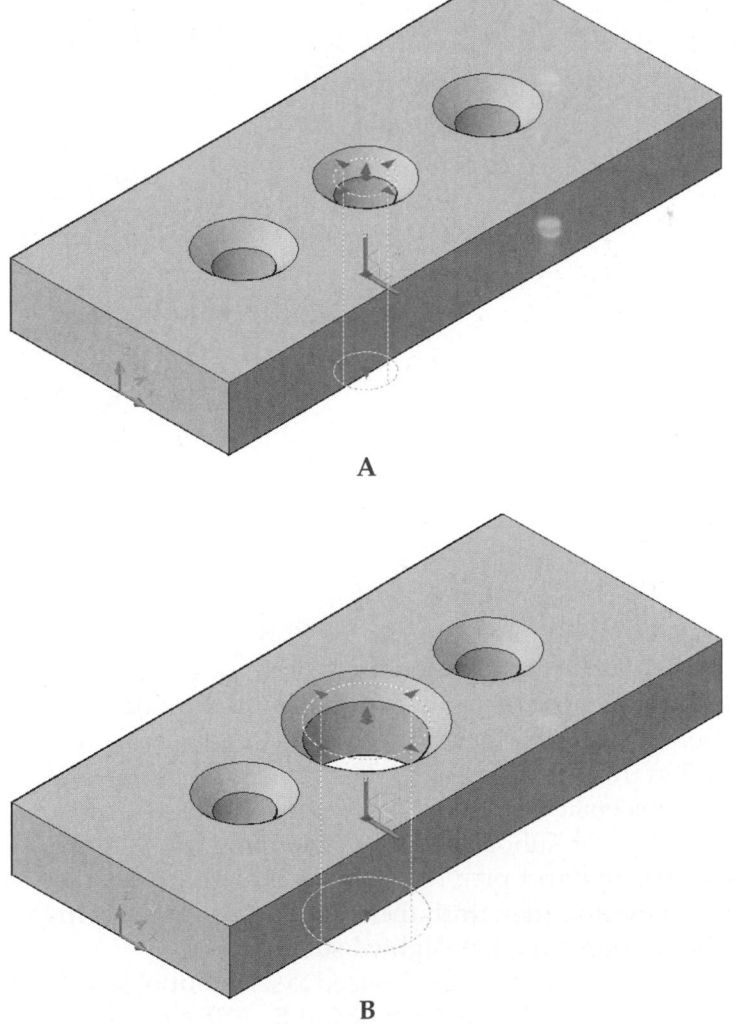

A

B

Figure 12-8.
A—The composite solid model is selected. Notice the single base grip. B—The wedge primitive subobject has been selected. Notice the grips associated with the primitive. C—An edge subobject within the primitive subobject is selected for editing.

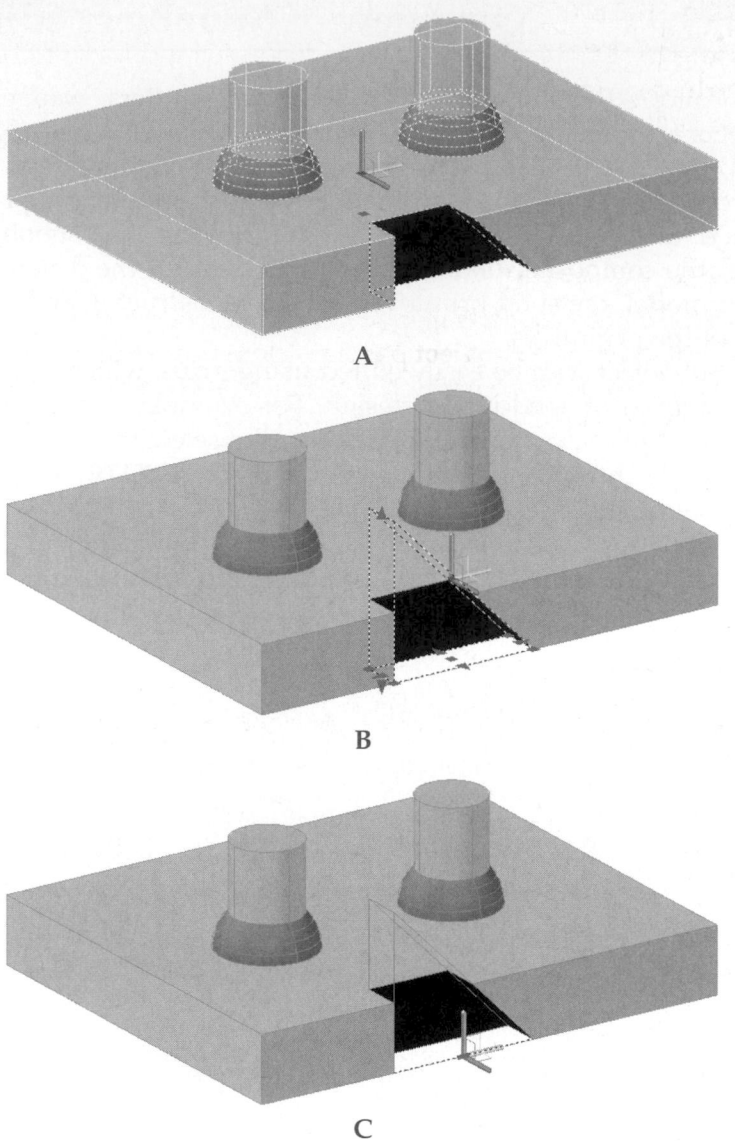

A

B

C

Selecting Subobjects

To select a subobject, press the [Ctrl] key and pick the subobject. You can select multiple subobjects and subobjects on multiple objects. To select a subobject that is hidden in the current view, first display the model as a wireframe. After creating a selection set, select a grip and edit the subobject as needed. Multiple objects can be selected in this manner. To deselect objects, press the [Shift]+[Ctrl] key combination and pick the objects to be removed from the selection set.

If objects or subobjects are overlapping, press the [Ctrl] key and the spacebar to turn on cycling and pick the subobject. Then, release the spacebar, continue holding the [Ctrl] key, and pick until the subobject you need is highlighted. Press [Enter] or the spacebar to select the highlighted subobject.

The [Ctrl] key method can be used to select subobjects for use with editing commands such as **MOVE**, **COPY**, **ROTATE**, **SCALE**, **ARRAY**, and **ERASE**. Some commands, like **ARRAY**, **STRETCH**, and **MIRROR**, are applied to the entire solid. Other operations may not be applied at all, depending on which type of subobject is selected. You can also use the **Properties** palette to change the color of edge and face subobjects or the material assigned to a face. The color of a subobject primitive can also be changed, but not the color of its subobjects.

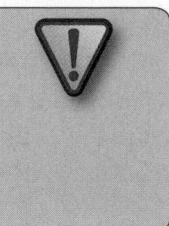
Using the **SOLIDEDIT** command (discussed in Chapter 13) removes the history from a composite solid. Therefore, the original objects—the subobjects—are no longer available for subobject editing. However, you may still be able to perform some subobject edits, such as moving the original objects.

Subobject Selection Buttons

The subobject filters help to quickly select the specific type subobject. The filters are located in the **Subobject** panel in the **Home** and **Mesh Modeling** tabs of the ribbon. See **Figure 12-9.** The filter buttons set the **SUBOBJSELECTIONMODE** system variable. The default setting is 0, which is the same as picking the **No Filter** button in the subobject filter drop-down list.

For example, if you want to work with faces only, pick the **Face** button. Then, when you press the [Ctrl] key to make a selection, notice the face-selection icon near the cursor. A different icon is displayed depending on the subobject filter that is selected. See **Figure 12-10.**

Figure 12-9.
Select a specific subobject filter in the filter drop-down list.

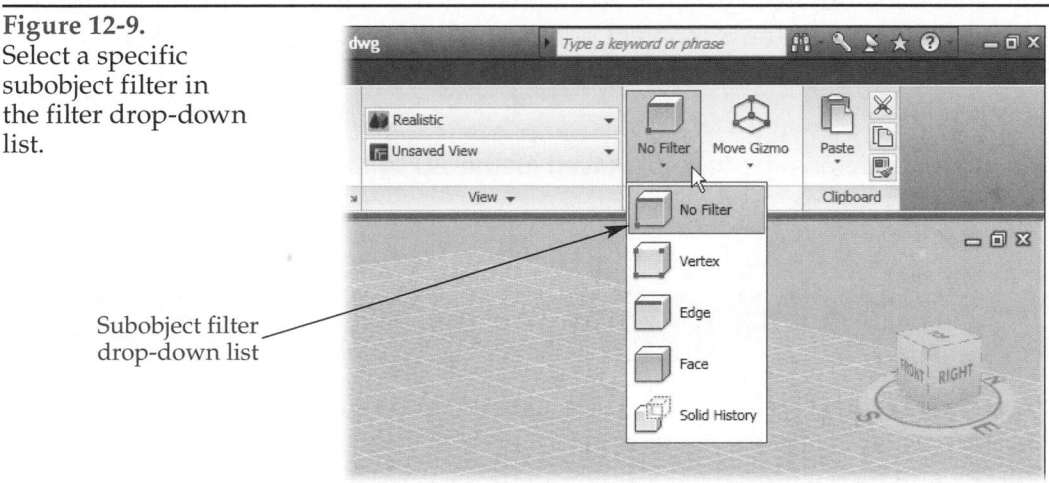

Subobject filter drop-down list

Figure 12-10.
SUBOBJSELECTIONMODE icons are displayed at the cursor to indicate the current selection filter.

Icon	System Variable Setting	Result
	SUBOJSELECTIONMODE = 1	Only vertices can be selected.
	SUBOJSELECTIONMODE = 2	Only edges can be selected.
	SUBOJSELECTIONMODE = 3	Only faces can be selected.
	SUBOJSELECTIONMODE = 4	Only compound subobjects (history objects) within compound objects can be selected.
	SUBOJSELECTIONMODE = 0	Subobject filtering is turned off and all subobjects can be selected.

DEFAULTGIZMO

Ribbon

Home
> Subobject

Move Gizmo

Rotate Gizmo

Scale Gizmo

No Gizmo

Mesh Modeling
> Subobject

Move Gizmo

Rotate Gizmo

Scale Gizmo

No Gizmo

Type

DEFAULTGIZMO

Subobject Editing Gizmos

Each of the following sections discusses methods for editing subobject faces, edges, and vertices. When using the **3DMOVE**, **3DROTATE**, and **3DSCALE** commands, a specialized *gizmo,* also called a *grip tool,* is displayed on the subobject selected for editing. Rather than first selecting one of these 3D editing commands, you can pick the required gizmo from the **Subobject** panel in the **Home** tab of the ribbon in combination with one of the subobject filters. See **Figure 12-11.** Then, when you select the subobject to edit, the appropriate gizmo is displayed. This process eliminates the need to cycle through entity selection or to first execute a specific editing command.

For example, if you wanted to rotate a face, first pick the **Face** and **Rotate Gizmo** buttons in the **Subobject** panel. Next, press the [Ctrl] key and pick the face. The face is immediately highlighted, the rotate gizmo is displayed, and you can proceed with the edit.

The **DEFAULTGIZMO** system variable controls the display of gizmos. Changing the system variable value is the same as picking a button in the gizmo drop-down list in the **Subobject** panel. The values are:

- 0 = **Move Gizmo** button
- 1 = **Rotate Gizmo** button
- 2 = **Scale Gizmo** button
- 3 = **No Gizmo** button

If these values for **DEFAULTGIZMO** are entered on the command line, the button is immediately changed in the **Subobject** panel.

> **NOTE**
>
> Gizmos are *not* displayed if the 2D Wireframe visual style is set current.

Face Subobject Editing

Faces of 3D solids can be modified using commands such as **MOVE**, **ROTATE**, and **SCALE** or by using grips and gizmos (grip tools). To select a face on a 3D solid, press the [Ctrl] key and pick within the boundary of the face. Do not pick the edge of the face. The face subobject filter can also be used to limit the selection to a face.

Figure 12-11.
Gizmos, or grip tools, can be selected from the **Subobject** panel of the **Home** tab.

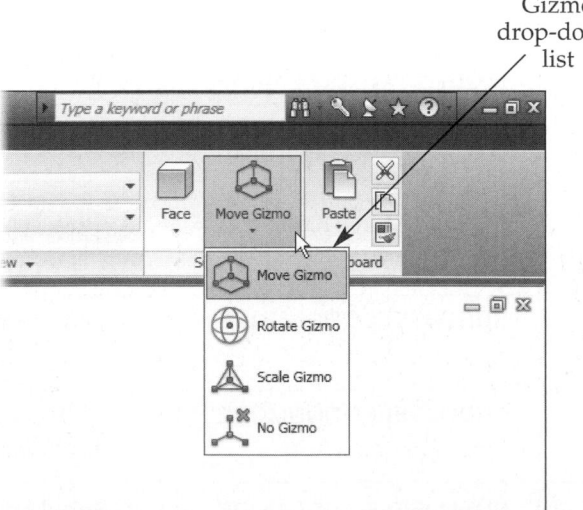

Gizmo
drop-down
list

Face grips are circular and located in the center of the face, as shown in **Figure 12-12**. In the case of a sphere, the grip is located in the center of the sphere since there is only one face. The same is true of the curved face on a cylinder or cone.

By default, the history is recorded for all solid primitives and composites. If you select a primitive or a primitive subobject within a composite solid, all of the grips associated with that primitive are displayed. See **Figure 12-13A**. If you edit a 3D solid primitive face, the history of the primitive is deleted and the object becomes a composite solid. Then, when the object is selected, a single base grip is displayed. See **Figure 12-13B**.

While pressing the [Ctrl] key and selecting a face, it may be difficult to select the face you want or to deselect faces you do not need. Use the view cube to transparently change the viewpoint or use [Shift]+mouse wheel button to activate the transparent **3DORBIT** command and change your viewpoint.

NOTE

The recorded history of a solid composite can be displayed by selecting the solid, opening the **Properties** palette, and changing the Show History property in the **Solid History** category to Yes.

Figure 12-12.
Face grips are located in the center of face subobjects.

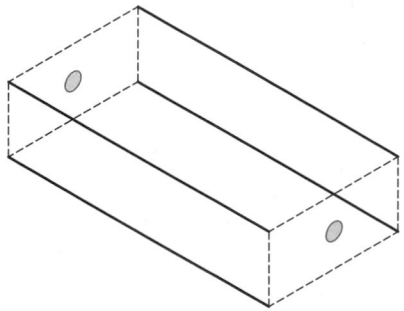 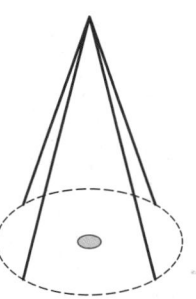

Figure 12-13.
A—This primitive is selected for editing. Notice the grips associated with the primitive.
B—If the primitive is edited, the history of that primitive is deleted and a single base grip is displayed.

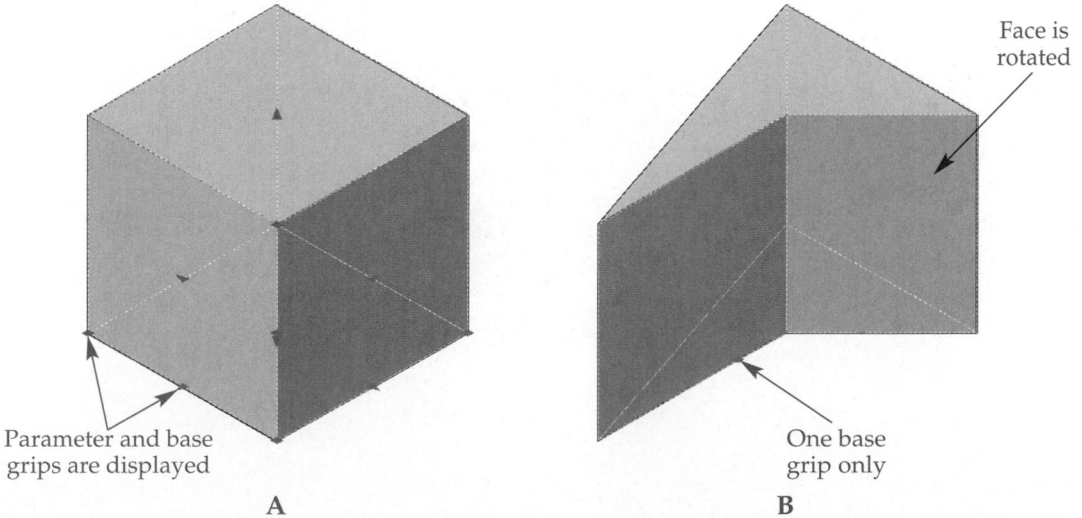

Face is rotated

Parameter and base grips are displayed

One base grip only

A   B

Moving Faces

When a face of a 3D solid is moved, all adjacent faces are dragged and stretched with it. The shape of the original 3D primitive or solid determines the manner in which the face can be moved and how adjacent faces react. A face can be moved using the **MOVE** command, **3DMOVE** command, move gizmo, or by dragging the face's base grip. When moving a face, use the gizmo, polar tracking, or direct distance entry. Otherwise, the results may appear correct in the view in which the edit is made, but, when the view is changed, the actual result may not be what you wanted. See **Figure 12-14.**

The move gizmo, discussed in Chapter 11, is displayed by default (**GTAUTO** = 1) when the face is selected. To use this gizmo, move the pointer over the X, Y, or Z axis of the gizmo; the axis changes to yellow. To restrict movement along that axis, pick the axis. If you move the pointer over one of the right angles at the origin of the gizmo, the corresponding two axes turn yellow. Pick to restrict the movement to that plane. You can complete the movement by either picking a new point or by direct distance entry.

The [Ctrl] key is used to access options when dynamically moving a face. First, select the face. See **Figure 12-15A.** Then, pick the face grip or gizmo and press and release the [Ctrl] key to cycle through the options.

If the [Ctrl] key is not pressed, the moved face maintains its size, shape, and orientation. The shape and plane of adjacent faces are changed. See **Figure 12-15B.** Pressing the [Ctrl] key three times resets the function, as if the [Ctrl] key had not been pressed.

If the [Ctrl] key is pressed once, the moved face maintains its shape and orientation. However, its size is modified because the planes of adjacent faces are maintained. See **Figure 12-15C.** Adjacent faces are not subdivided.

Figure 12-14.
A—The original solid primitives. B—The box and wedge are dynamically edited without using exact coordinates or distances. C—When the viewpoint is changed, you can see that dynamic editing has produced unexpected results.

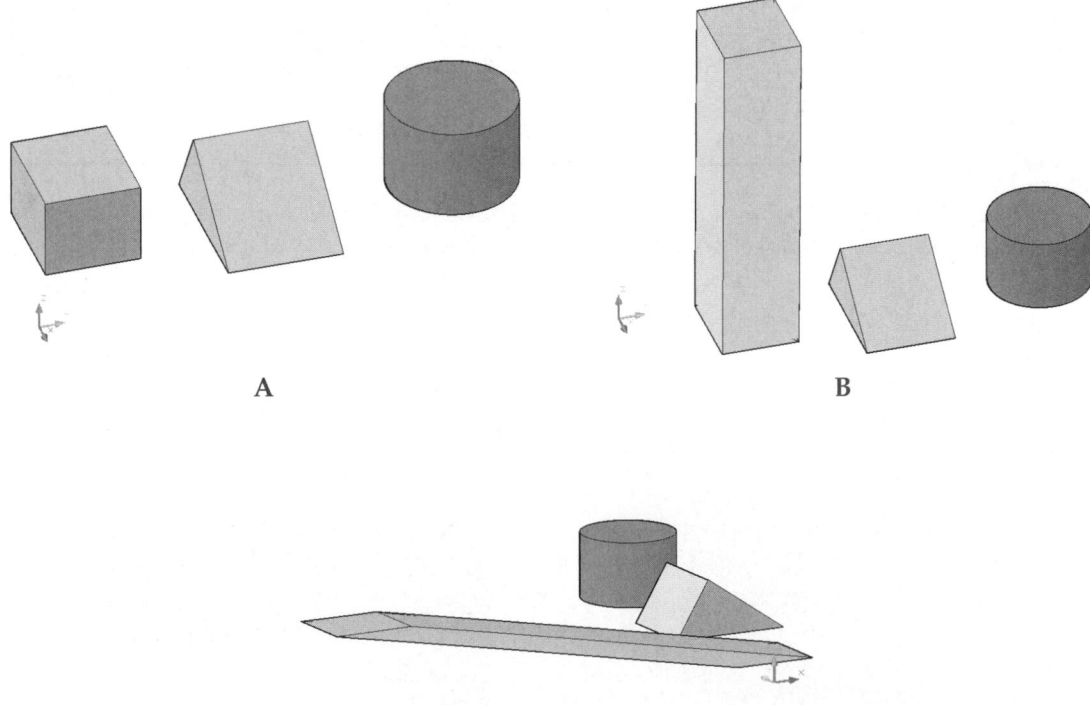

A

B

C

Figure 12-15.
A—The original solid primitive. B—Without pressing the [Ctrl] key, the face maintains its shape and orientation. C—Pressing the [Ctrl] key once keeps the adjacent faces in their original planes, but alters the modified face.

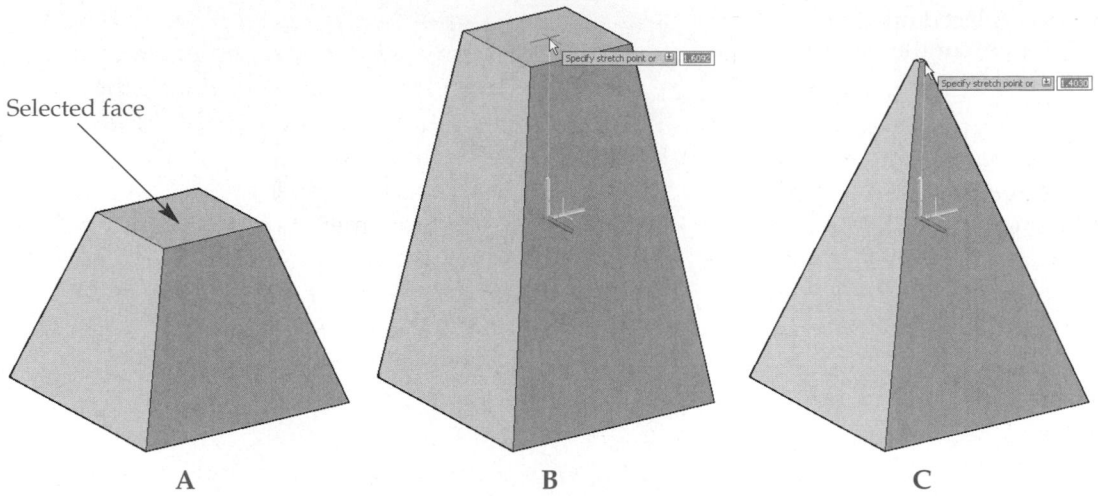

Selected face

A   B   C

If the [Ctrl] key is pressed twice, the moved face maintains its size, shape, and orientation. However, adjacent faces are subdivided into triangular faces, if needed.

PROFESSIONAL TIP

It is always important to keep the design intent of your solid model in mind. If you are creating a conceptual design, you may be able to use subobject grip editing and gizmos without entering precise coordinates. But, if you are working on a design for manufacturing or production, it is usually critical to use tools such as direct distance entry, gizmos with exact values, and polar tracking for greater accuracy.

Rotating Faces

Before rotating any primitive or subobject you must know in which plane the rotation is to occur. The **ROTATE** command permits a rotation in the current XY plane. But, you can get around this limitation by using the **3DROTATE** command. This command allows you to select a rotation plane by means of the rotate gizmo discussed in Chapter 11. When a subobject is selected, you cannot use the spacebar to switch to the rotate gizmo from the move gizmo. The rotate gizmo must be placed using the **3DROTATE** command.

The rotate gizmo provides a dynamic, graphic representation of the three axes of rotation. To rotate about the tool's X axis, pick the red circle on the gizmo. To rotate about the Y axis, pick the green circle. To rotate about the Z axis, pick the blue circle. Once you select a circle, it turns yellow and you are prompted for the start point of the rotation angle. You can enter a direct angle at this prompt or pick a point to define the angle of rotation. When the rotation angle is defined, the face is rotated about the selected axis.

For example, in **Figure 12-16A,** the top face is selected and the rotate gizmo is placed on a corner of the face. After picking the rotation axis on the gizmo, specify the angle start point and then the angle end point. Notice in **Figure 12-16B** that dynamic input can be used to enter an exact angle value. The result is shown in **Figure 12-16C.**

Figure 12-16.
Using the rotate gizmo to rotate a face. A—The top face is selected and the gizmo is placed on a base point of the face. B—The axis of rotation and a starting point for the angle are selected. C—The completed rotation.

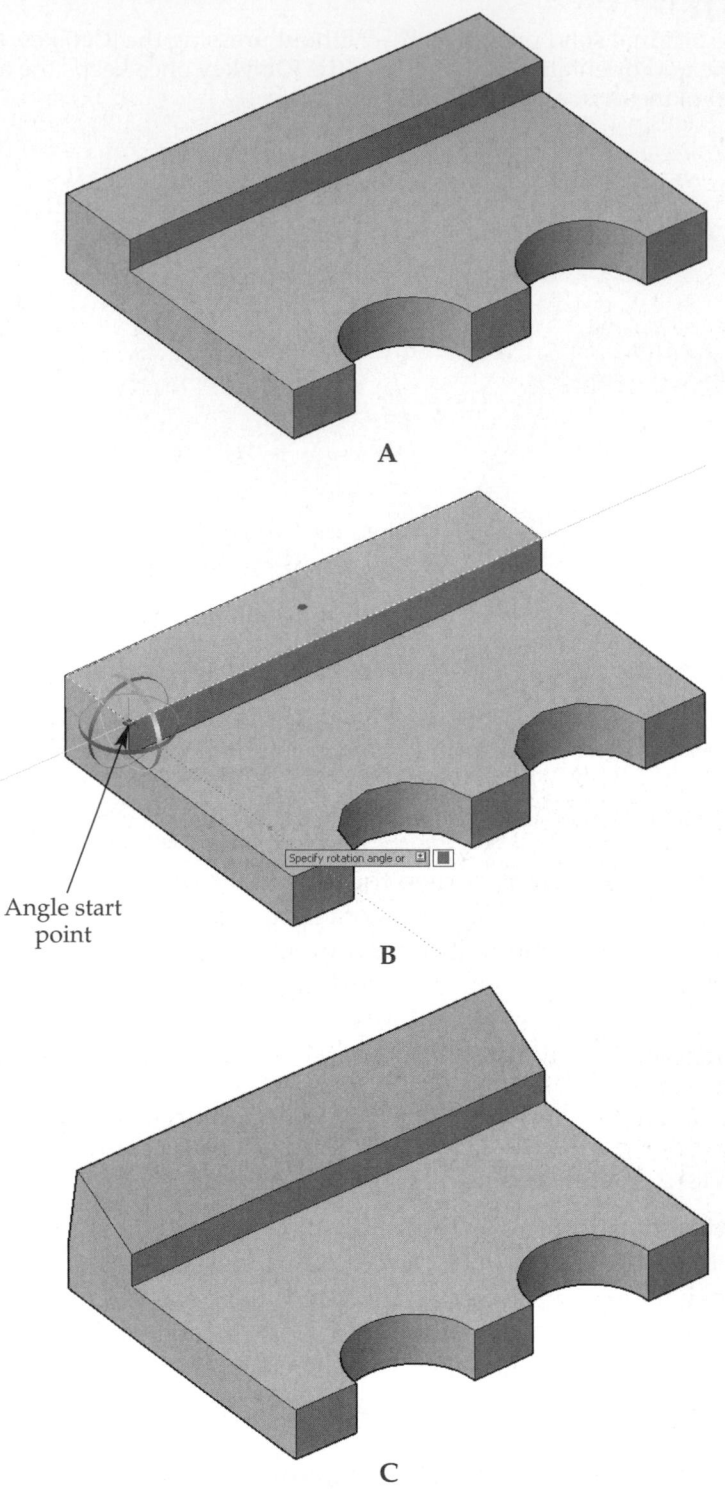

Angle start point

A

B

C

The [Ctrl] key is used to access options when dynamically rotating a face. Pressing the [Ctrl] key while rotating a face affects adjacent faces in the same manner as discussed for moving faces. **Figure 12-17A** shows a rotation without pressing [Ctrl]. The shape and size of the face being rotated is maintained, while the adjacent faces change. **Figure 12-17B** shows a rotation after pressing [Ctrl] once. The shape and size of the face being rotated changes, while the plane and shape of adjacent faces are maintained. Pressing the [Ctrl] key a second time maintains the shape and orientation of the selected face, but triangular faces may be created on adjacent faces. Pressing the [Ctrl] key a third time resets the function.

AutoCAD and Its Applications—Advanced

Figure 12-17.
A—Rotating a face without pressing the [Ctrl] key. The large, top face on the object shown in Figure 10-16A has been selected for rotation.
B—Pressing the [Ctrl] key once keeps the adjacent faces in their original planes. The shape and size of the face being rotated changes.

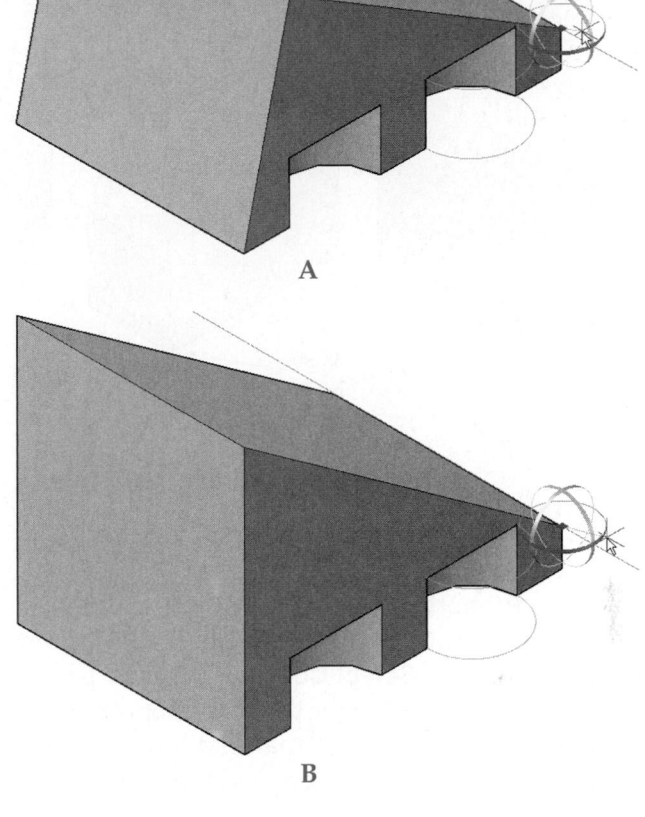

Face being rotated

A

B

Scaling Faces

Scaling a face is a simple procedure. First, select the face to be scaled. Then, select a base point and dynamically pick to change the scale or use a scale factor. See **Figure 12-18.** Pressing the [Ctrl] key has no effect on the scaling process, except to turn it off or on, if the base point is on the same plane as the face. However, if the base point is not on the same plane as the selected face, then pressing the [Ctrl] key has the same effect as for the other face-editing operations.

Figure 12-18.
Scaling a face. A—The original solid. B—The dark face is scaled down. C—The dark face is scaled up.

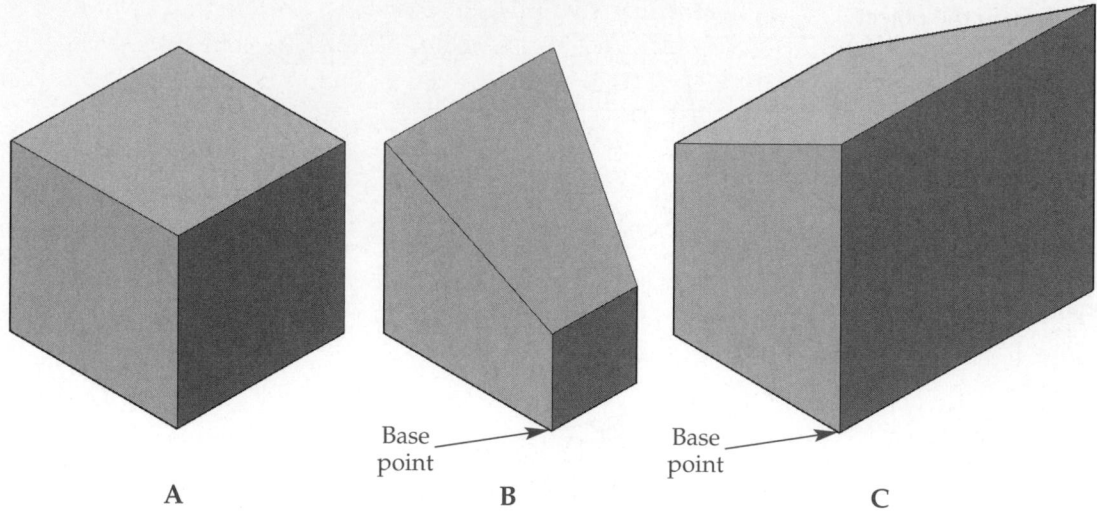

Base point
Base point

A   B   C

Coloring Faces

To change the color of a face, use the [Ctrl] key selection method to select the face. Next, open the **Properties** palette. See **Figure 12-19.** In the **General** category, pick the drop-down list for the Color property. Select the desired color or pick Select Color... and choose a color from the **Select Color** dialog box. To change the material applied to the face, pick the drop-down list for the Material property in the **3D Visualization** category. Select a material from the list. A material must be loaded into the drawing to be available in this drop-down list. Materials are discussed in detail in Chapter 16.

Figure 12-19.
Changing the color of a face or the material assigned to it.

A face is selected

Pick to select a color

Select a material

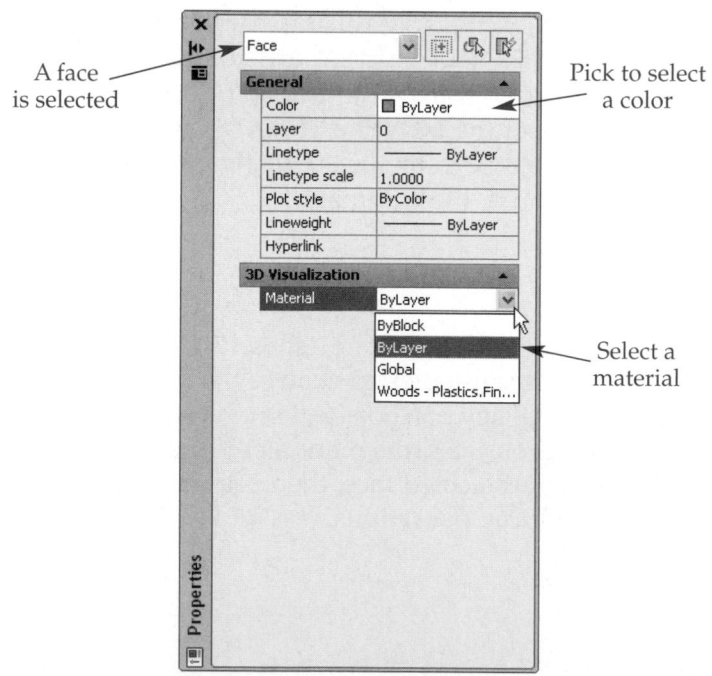

Extruding a Solid Face

Planar faces on 3D solids can be extruded into new solids. Refer to the HVAC duct assembly shown in **Figure 12-20A.** A new, reduced trunk needs to be created on the left end of the assembly. This requires two pieces: a reducer and the trunk.

First, select the **EXTRUDE** command. At the "select objects" prompt, press the [Ctrl] key and pick the face subobject to be extruded. Next, since this is a reduced trunk, specify a taper angle. Enter the **Taper angle** option and specify the angle. In this case, a 15° angle is used. Finally, specify the extrusion height. The height of the reducer is 12". See **Figure 12-20B.**

Now, the new trunk needs to be created. Select the **EXTRUDE** command. Press the [Ctrl] key and pick the face to extrude. Since this piece is not tapered, enter the extrusion height, which in this case is 44". See **Figure 12-20C.** The two new pieces are separate solid objects. If the assembly is to be one solid, use the **UNION** command and join the two new solids to the assembly.

Figure 12-20.
A—A new, reduced trunk needs to be created on the left end of the HVAC assembly. The face shown in color will be extruded. B—The **Taper angle** option of the **EXTRUDE** command is used to create the reducer. The face shown in color will be extruded to create the extension. C—The **EXTRUDE** command is used to create an extension from the reducer.

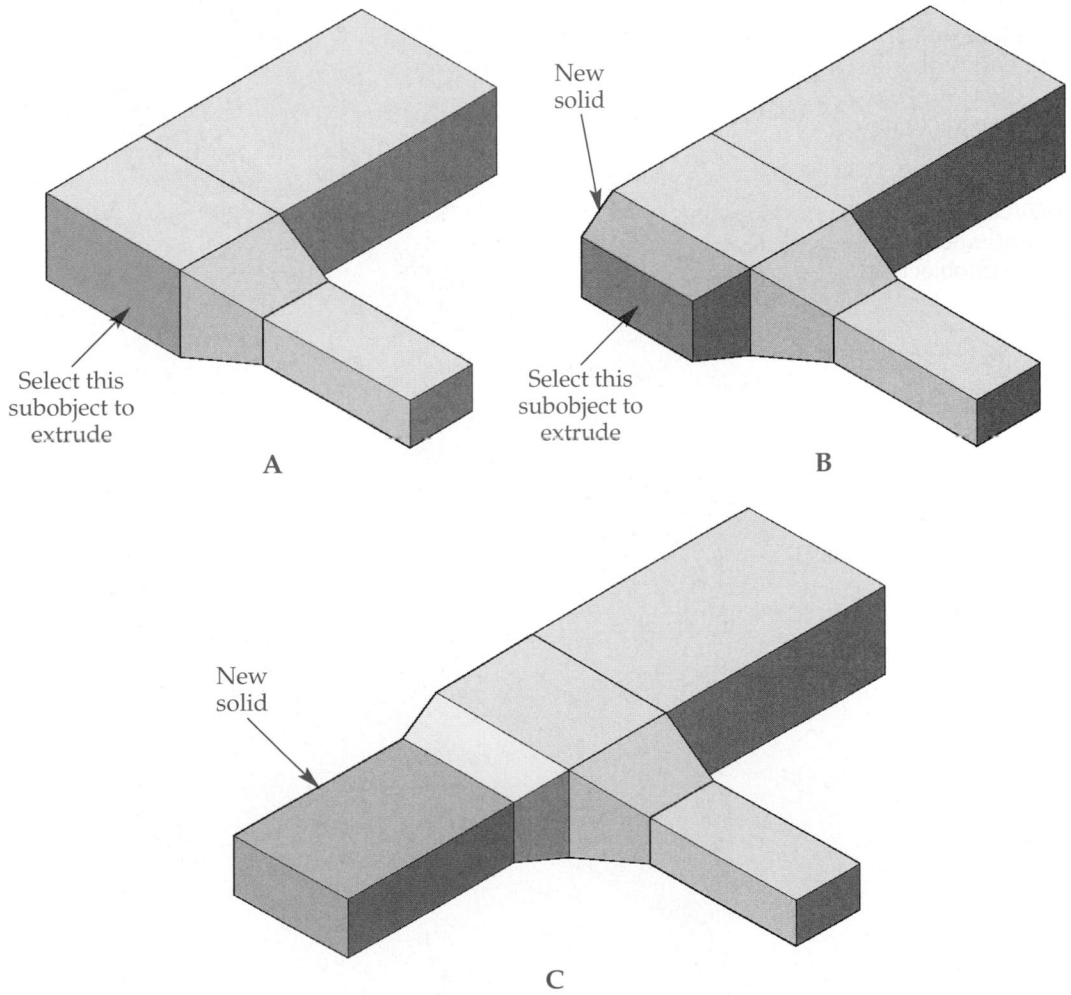

Revolving a Solid Face

Planar faces on 3D solids can be revolved in the same manner as other AutoCAD objects to create new solids. Refer to **Figure 12-21A.** The face on the left end of the HVAC duct created in the last section needs to be revolved to create a 90° bend. First, select the **REVOLVE** command. At the "select objects" prompt, press the [Ctrl] key and pick the face subobject to be revolved.

Next, the axis of revolution needs to be specified. You can pick the two endpoints of the vertical edge, but you can also pick the edge subobject. Enter the **Object** option of the command, press the [Ctrl] key, and select the edge subobject.

Finally, the 90° angle of revolution needs to be specified. **Figure 12-21B** shows the face revolved into a new solid. The bend is a new, separate solid. If necessary, use the **UNION** command to join the bend to the assembly.

Exercise 12-2

Complete the exercise on the student website.
www.g-wlearning.com/CAD

Figure 12-21.
A—The face on the left end of the HVAC duct (shown in color) needs to be revolved to create a 90° bend. B—Use the **REVOLVE** command and pick the face subobject to be revolved.

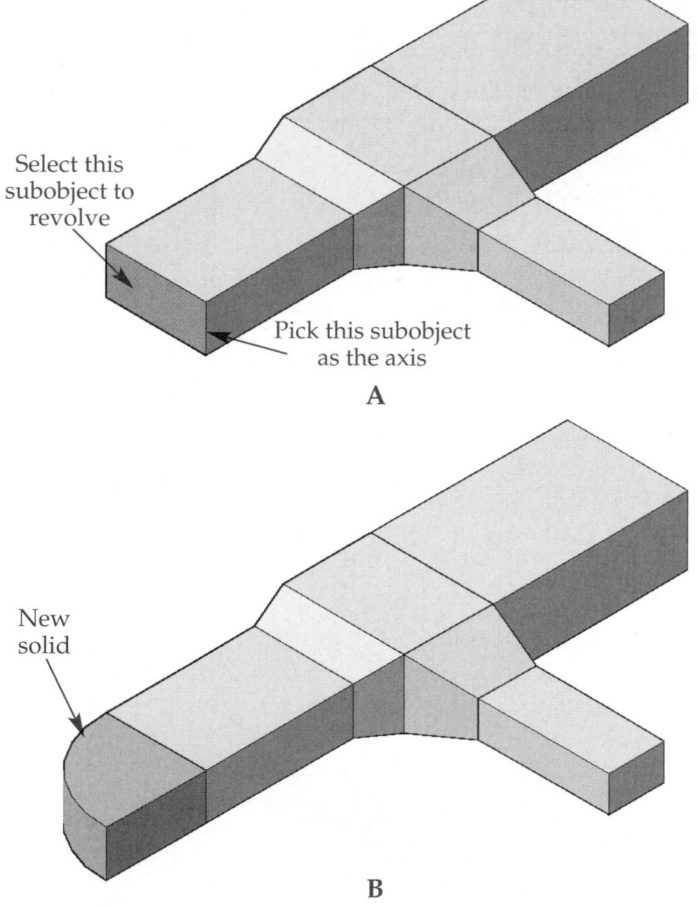

Select this subobject to revolve

Pick this subobject as the axis

A

New solid

B

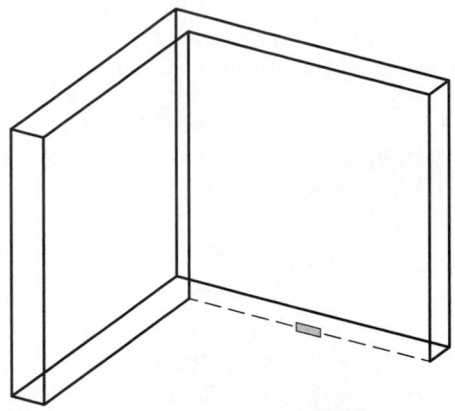

Figure 12-22.
Edge grips are rectangular and displayed in the middle of the edge.

Edge Subobject Editing

Individual edges of a solid can be edited using grips and gizmos in the same manner as faces. To select an edge subobject, press the [Ctrl] key and pick the edge. The edge subobject filter can also be used to limit the selection to an edge.

Grips on linear edges are rectangular and appear in the middle of the edge, **Figure 12-22.** In addition to solid edges, the edges of regions can be altered using **MOVE**, **ROTATE**, and **SCALE**, but grips are not displayed on regions as they are on solid subobjects.

Remember, editing subobjects of a primitive removes the primitive's history. This should always be a consideration if it is important to preserve the solid primitives that were used to construct a 3D solid model. Instead of editing the primitive subobjects at their subobject level, it may be better to add or remove material with a Boolean operation, thus preserving the solid's history.

NOTE

There are several ways to perform a 3D edit. Keep the following options in mind when working with subobject editing.

- Selecting the **MOVE**, **ROTATE**, or **SCALE** commands and picking a subobject will not display a gizmo, regardless of the current gizmo button displayed in the **Subobject** panel.
- Selecting the **3DMOVE**, **3DROTATE**, or **3DSCALE** commands and picking a subobject will display the appropriate gizmo, regardless of the current gizmo button displayed in the **Subobject** panel.
- Picking a subobject without having selected a command first will display the gizmo corresponding to the current gizmo button in the **Subobject** panel.

Moving Edges

To move an edge, select it using the [Ctrl] key, as previously discussed. See **Figure 12-23A.** By default, the gizmo corresponding to the gizmo button in the **Subobject** panel appears. If the move gizmo is not current, select the **Move Gizmo** button in the **Subobject** panel. Select the appropriate axis handle and dynamically move the edge or use direct distance entry, **Figure 12-23B.** The move gizmo remains active until the [Esc] key is pressed.

Figure 12-23.
A—Select the edge subobject to be moved. B—The edge is moved. Notice how the size of the primitive used to subtract the cutout is not affected.

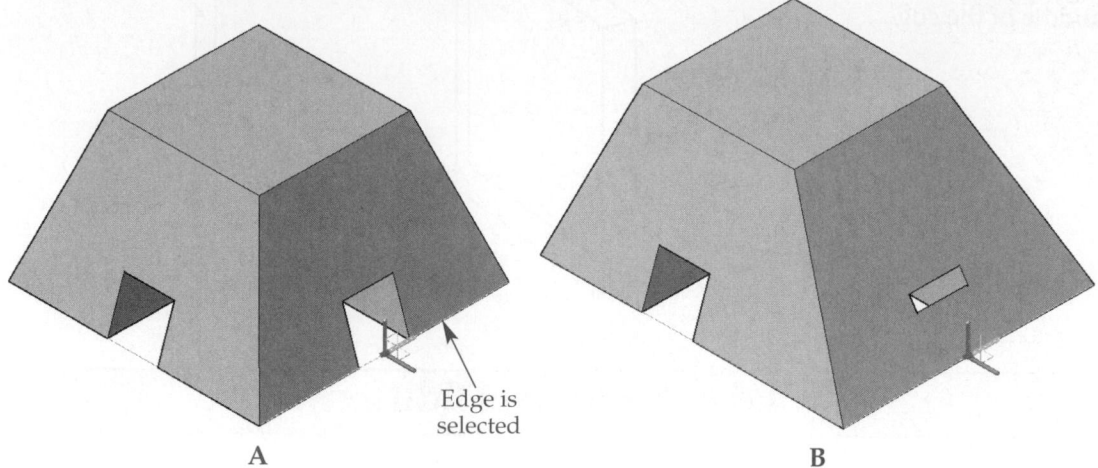

Edge is
selected

A   B

If you pick the edge grip to turn it hot, the gizmo is bypassed. This places you in the standard grip editing mode. You can stretch, move, rotate, scale, and mirror the edge. In this case, the **STRETCH** function works the same as the move gizmo, but less reliably. You must be careful to use either ortho, polar tracking, or direct distance entry, but the possibility for error still exists.

There are a few options to achieve different results when dynamically moving an edge. The [Ctrl] key is used to access these options. First, select the edge. Then, pick the edge grip or gizmo and press and release the [Ctrl] key to cycle through the options.

If the [Ctrl] key is not pressed, the moved edge maintains its length and orientation. However, the shape and planes of adjacent faces are changed. See **Figure 12-24A.** If the [Ctrl] key is pressed three times, the function is reset, as if the [Ctrl] key had not been pressed.

If the [Ctrl] key is pressed once, the moved edge maintains its orientation, but its length is modified. This is because the planes and orientation of adjacent faces are maintained. See **Figure 12-24B.**

If the [Ctrl] key is pressed twice, the moved edge maintains its length and orientation. But, if the move alters the planes of adjacent faces, those faces may become *nonplanar.* In other words, the face may now be located on two or more planes. If this happens, adjacent faces are divided into triangles, **Figure 12-24C.** This is visible when the object in **Figure 12-24C** is displayed in two orthographic views. See **Figure 12-25.**

PROFESSIONAL TIP

You can quickly enter the **MOVE**, **ROTATE**, or **SCALE** command for a subobject by selecting the subobject, right-clicking, and picking the command from the shortcut menu.

Rotating Edges

Before you select an edge to rotate, do a little planning. Since there are a wide variety of edge rotation options, it will save time if you first decide on the location of the base point about which the edge will rotate. Next, determine the direction and angle of rotation. Based on these criteria, choose the option that will accomplish the task the quickest.

To rotate an edge, enter the **ROTATE** or **3DROTATE** command. Then, pick the edge using the [Ctrl] key. Select a base point and then enter the rotation. You can also select the edge, pick the edge grip, and cycle to the **ROTATE** mode.

Figure 12-24.
Moving an edge. A—If the [Ctrl] key is not pressed, the edge maintains its length and orientation, but the shape and planes of adjacent faces are changed. B—If the [Ctrl] key is pressed once, the moved edge maintains its orientation, but its length is modified because the planes of adjacent faces are maintained. C—If the [Ctrl] key is pressed twice, the adjacent faces may be triangulated.

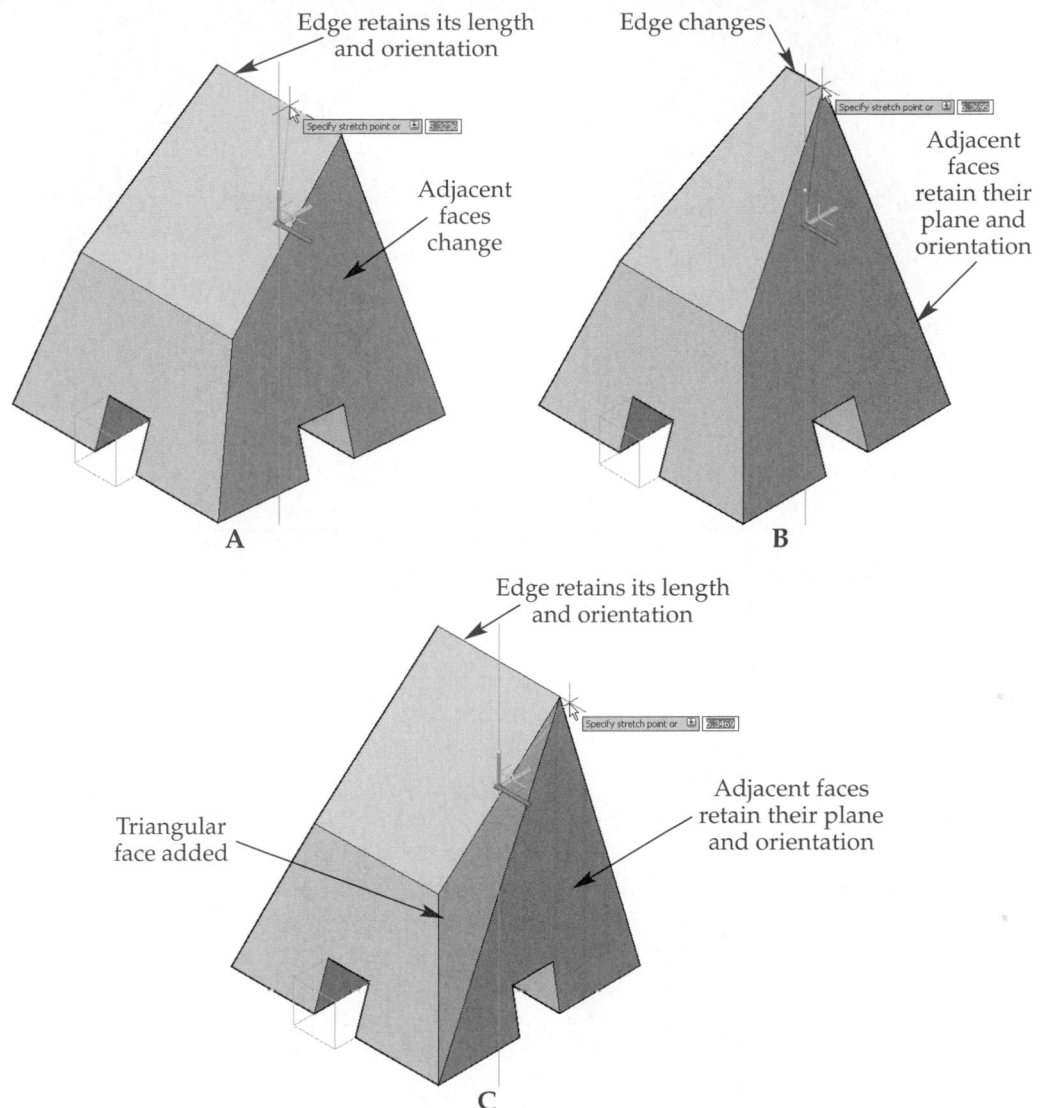

Edge retains its length and orientation

Adjacent faces change

Specify stretch point or

A

Edge changes

Adjacent faces retain their plane and orientation

Specify stretch point or

B

Edge retains its length and orientation

Triangular face added

Adjacent faces retain their plane and orientation

Specify stretch point or

C

Edges are best rotated using the rotate gizmo. It provides a graphic visualization of the axis of rotation. If you select a dynamic UCS while using the **3DROTATE** command, you have a variety of rotation axes to use because the gizmo can be located on all planes adjacent to the selected edge.

There are a few options to achieve different results when dynamically rotating an edge. The [Ctrl] key is used to access these options. First, select the edge. Then, pick the edge grip or gizmo and press and release the [Ctrl] key to cycle through the options.

Pressing the [Ctrl] key while rotating a face affects adjacent faces in the same manner as discussed in moving faces. If the [Ctrl] key is not pressed, the rotated edge maintains its length, but the shape and planes of adjacent faces are changed. See **Figure 12-26A.** If the [Ctrl] key is pressed once, the length of the rotated edge is modified because the planes of adjacent faces are maintained. See **Figure 12-26B.** If the [Ctrl] key is pressed twice, the rotated edge maintains its length, but if the rotation causes faces to become nonplanar, the adjacent faces may be triangulated. See **Figure 12-26C.** Pressing the [Ctrl] key a third time resets the function.

Figure 12-25.
Triangulated faces are clear in plan views. A—Front plan view. B—Side plan view.

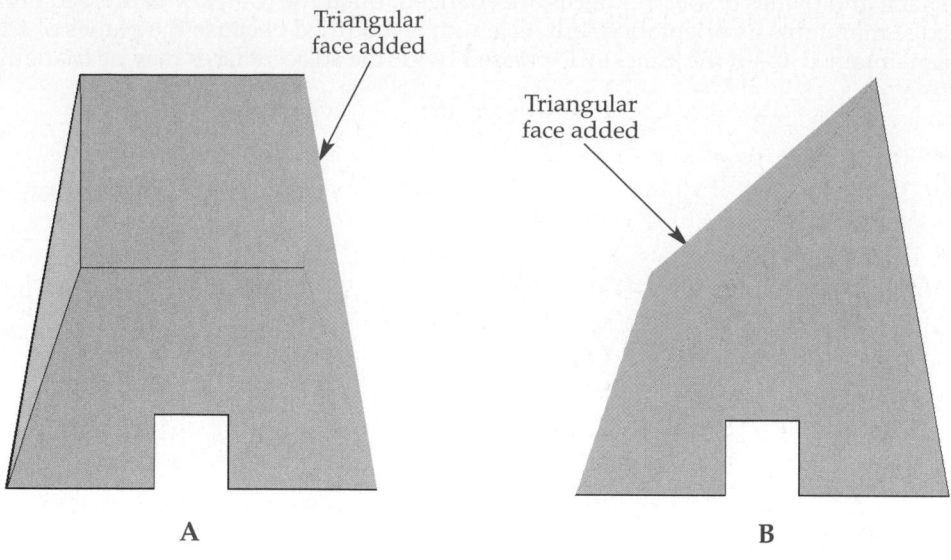

Figure 12-26.
Rotating an edge. A—The [Ctrl] key is not pressed. B—The [Ctrl] key is pressed once. Notice the top edge of the dark face. C—The [Ctrl] key is pressed twice.

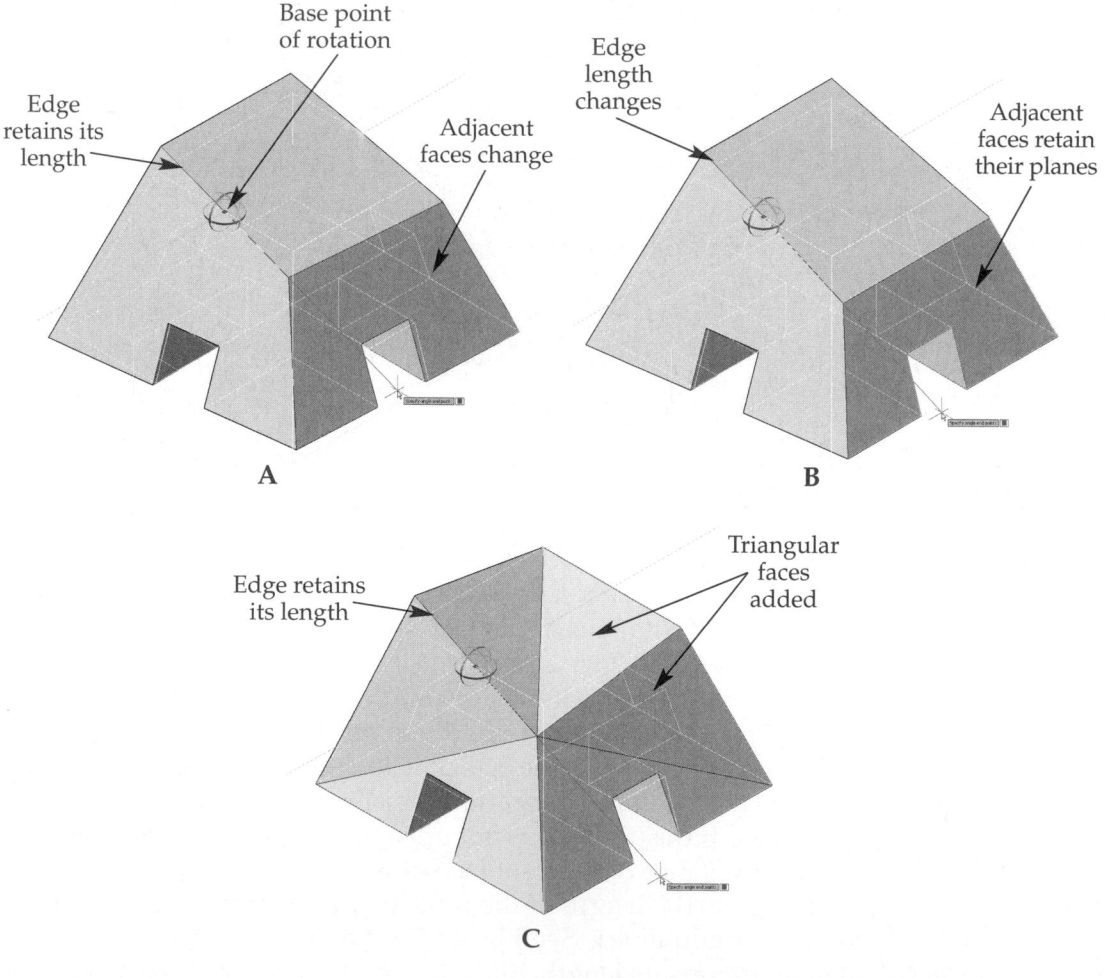

Scaling Edges

Only linear (straight-line) edges can be scaled. Circular edges, such as the ends of cylinders, can be modified using grips or the **SOLIDEDIT** command. These tools can be used to change the diameter or establish taper angles. See Chapter 13 for a complete discussion of the **SOLIDEDIT** command.

To scale a linear edge, enter the **SCALE** command. Select the edge using the [Ctrl] key. Pick a base point for the operation and enter a scale factor. You can also select the edge, pick the edge grip, and cycle to the **SCALE** mode. However, using the scale gizmo may be the best option.

The direction of the scaled edge is related to the base point you select. The base point remains stationary, while the vertices in either direction are scaled. If you enter the **SCALE** command, you are prompted for the base point. If you select the edge grip, the grip becomes the base point. The differences in opposite end and midpoint scaling of an edge are shown in **Figure 12-27.**

There are a few options to achieve different results when dynamically scaling an edge. The [Ctrl] key is used to access these options. First, select the edge. Then, pick the edge grip or enter the **SCALE** command and press and release the [Ctrl] key to cycle through the options.

If the [Ctrl] key is not pressed, the edge is scaled. The shape and planes of adjacent faces are changed to match the scaled edge. See **Figure 12-28A.**

If the [Ctrl] key is pressed once, the edge is, in effect, not scaled. This is because the planes of adjacent faces are maintained.

If the [Ctrl] key is pressed twice, the edge is scaled, as are edges attached to the modified edge. However, if the scaling causes faces to become nonplanar, they may be triangulated. See **Figure 12-28B.**

Figure 12-27.
The differences in opposite end and midpoint scaling of an edge. A—The original object. B—The edge is scaled down with a base point on the left corner. C—The edge is scaled down to the same scale factor, but the base point is on the right corner. D—The edge is scaled down to the same scaled factor with the base point at the middle of the edge.

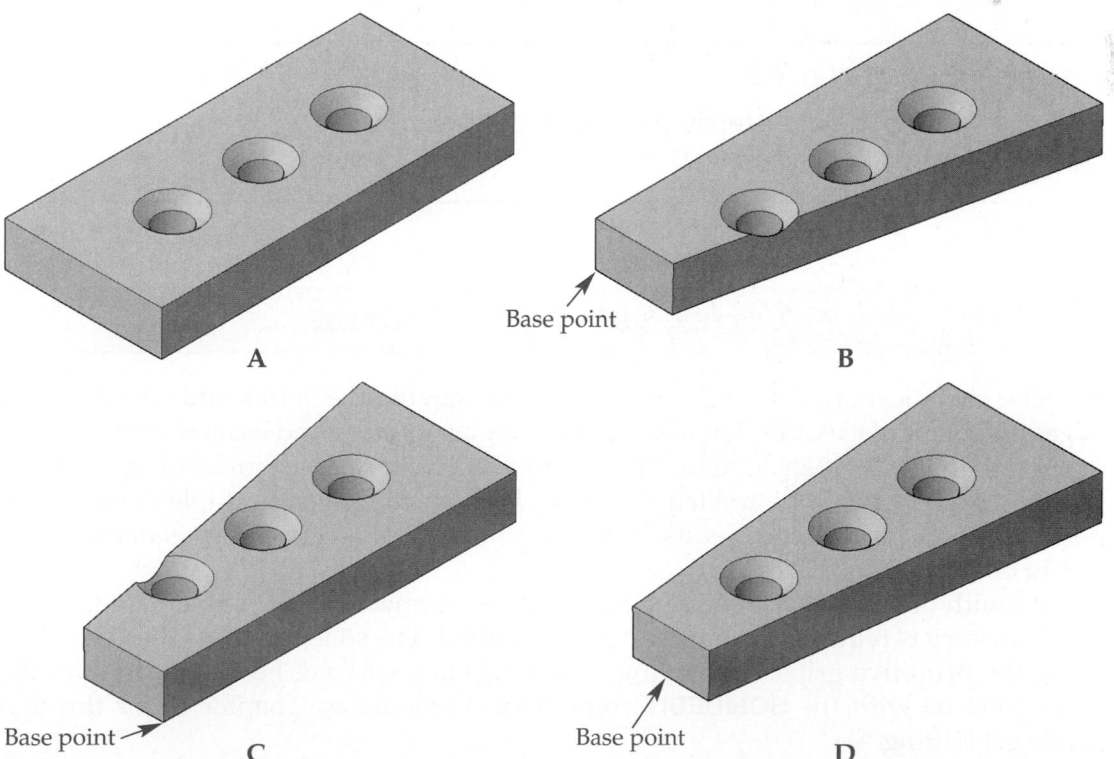

Figure 12-28.
Scaling the front edge with the base point in the middle of the edge. The original object is shown in Figure 10-24A. A—If the [Ctrl] key is not pressed, the edge is scaled and the shape and planes of adjacent faces are changed. B—If the [Ctrl] key is pressed twice, the edge is scaled, as are edges attached to it. If the scaling causes faces to become nonplanar, the adjacent faces may be triangulated.

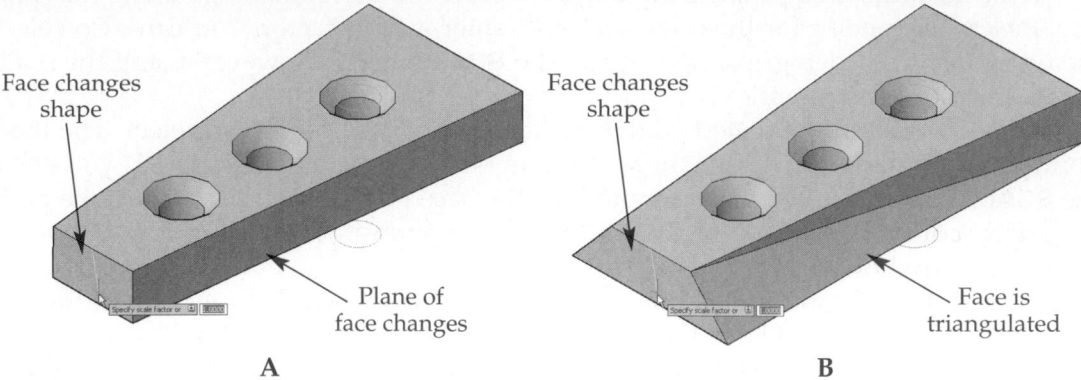

Face changes shape

Plane of face changes

A

Face changes shape

Face is triangulated

B

Coloring Edges

To change the color of an edge, use the [Ctrl] key selection method to select the edge. Next, open the **Properties** palette. In the **General** category, pick the drop-down list for the Color property. Select the desired color or pick Select Color... and choose a color from the **Select Color** dialog box. Edges cannot have materials assigned to them.

Deleting Edges

Edges can be deleted in certain situations. In order for an edge to be deleted, it must completely divide two faces that lie on the same plane. If this condition is met, the **ERASE** command or the [Delete] key can be used to remove the edge. The two faces become a single face.

Exercise 12-3

Complete the exercise on the student website.
www.g-wlearning.com/CAD

Vertex Subobject Editing

The modification of a single vertex involves moving the vertex and stretching all edges and planar faces attached to it. Vertex grips are circular and located on the vertex, as shown in **Figure 12-29.** Use the vertex subobject filter to assist in selecting vertices. A single vertex cannot be rotated or scaled, but you can select multiple vertices and perform rotating and scaling edits. When editing multiple vertices in this manner you are, in effect, editing edges.

As with other subobject editing functions performed on a 3D solid primitive, the solid's history is removed when a vertex is modified. The solid can no longer be edited using the primitive grips; only a single base grip is displayed. Further editing of the solid must be with the **SOLIDEDIT** command, discussed in Chapter 13, or through subobject editing.

AutoCAD and Its Applications—Advanced

Figure 12-29.
Vertex grips are circular and placed on the vertex.

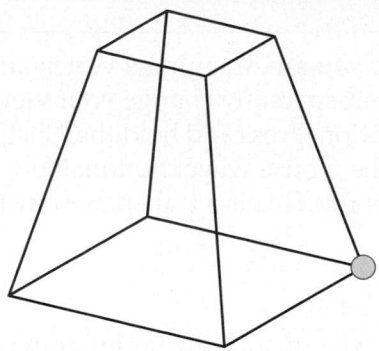

Moving Vertices

To move a vertex, select it using the [Ctrl] key method. If the move gizmo is not displayed, select it from the gizmo drop-down list in the **Subobject** panel. You can use the move gizmo, the **MOVE** command, or standard grip editing modes to move the vertex. If the [Ctrl] key is not pressed while dynamically moving a vertex, adjacent faces are triangulated by the move. Pressing the [Ctrl] key once allows the vertex to be moved without triangulating adjacent faces, but the faces may change shape. In some cases, AutoCAD may deem it necessary to triangulate faces. See **Figure 12-30.**

Figure 12-30.
Moving a vertex. A—The original object. B—Without pressing the [Ctrl] key, adjacent faces are triangulated. C—Pressing the [Ctrl] key once moves the vertex and changes some of the adjacent faces.

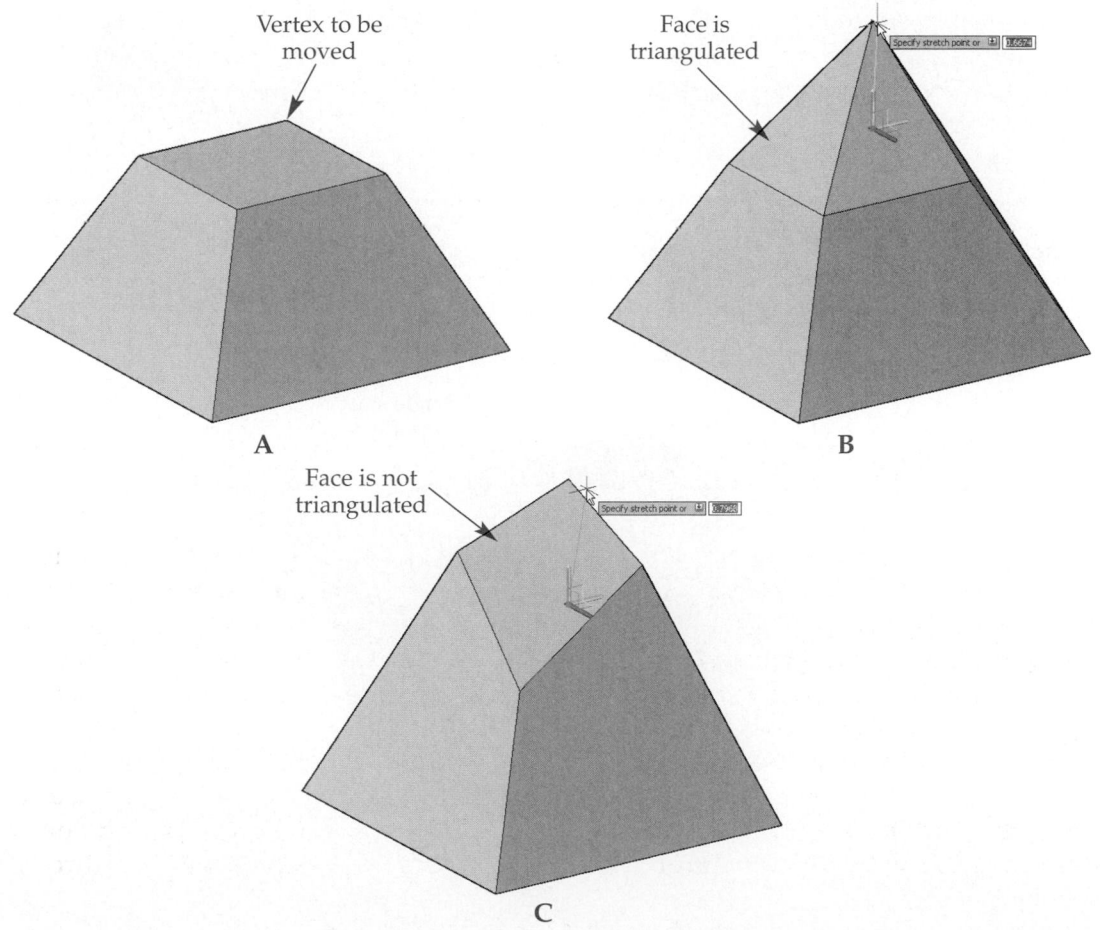

Rotating Vertices

As previously stated, a single vertex cannot be rotated or scaled, but two or more vertices can be. Since two vertices define a line, or edge, any edit is an edge modification. However, the process is slightly different than the edge modifications described earlier in this chapter.

To rotate an edge by selecting its endpoints, press the [Ctrl] key and select each vertex. See **Figure 12-31A.** You may need to use the [Ctrl]+space bar option to turn on cycling. Notice that grips appear at each selected vertex, but the edges between the vertices are not highlighted.

Figure 12-31.
To rotate or scale vertices, multiple vertices must be selected. In effect, the edges are modified. A—Vertices are selected to be rotated. B—The rotate gizmo is placed at the base of rotation and the rotation axis is selected. C—The vertices are rotated.

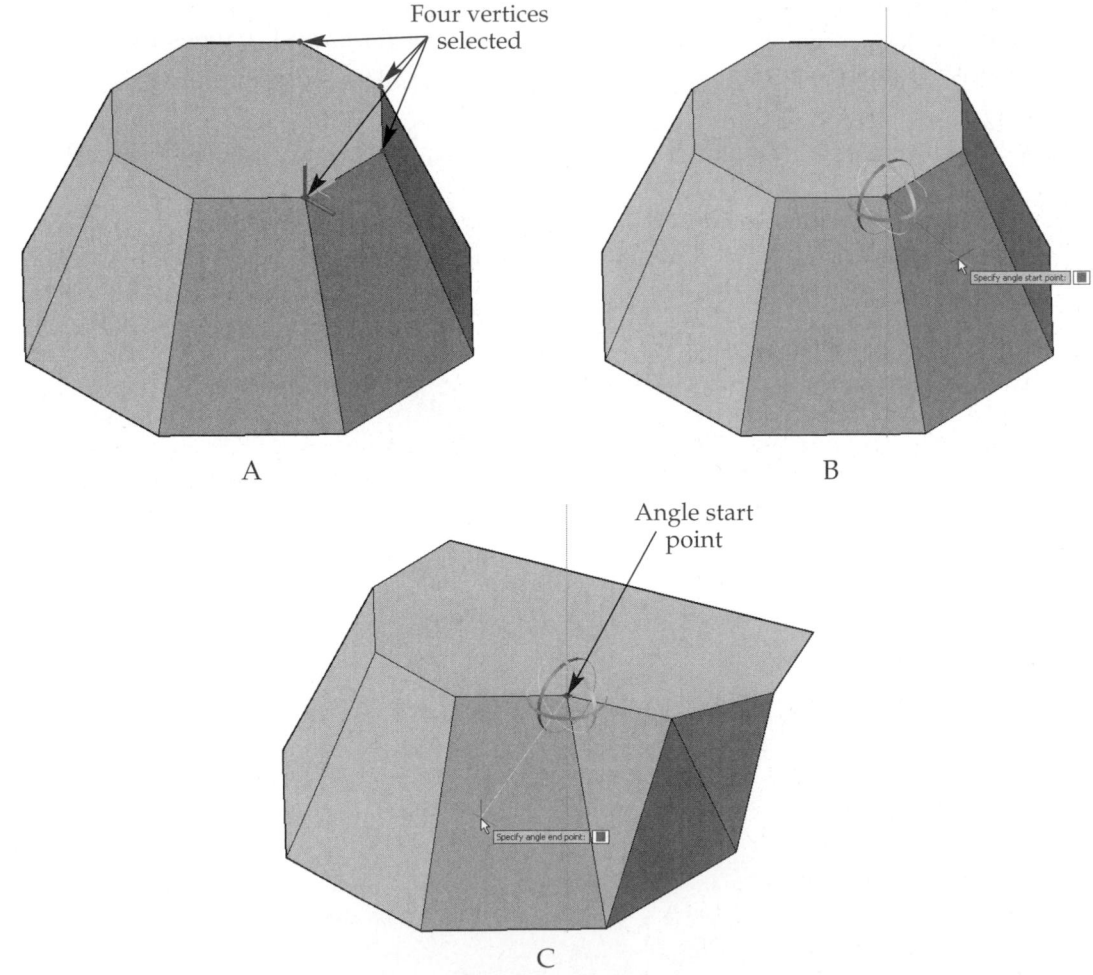

The **ROTATE** command can now be used to rotate the vertices (if **PICKFIRST** is set to 1). However, a more efficient method for rotating vertices is to use the **3DROTATE** command. The combination of the rotate gizmo and the UCS icon enable you to graphically view the rotation plane. Once the command is initiated, move the base point of the rotate gizmo if needed. See **Figure 12-31B**. Then, select the axis of revolution. Finally, pick the angle start point and enter the rotation. See **Figure 12-31C**.

If the [Ctrl] key is not pressed while dynamically rotating the vertices, the area of the selected vertices does not change and adjacent faces are triangulated. This is because the edges of the adjacent faces are attached to the selected vertices, so their edge length changes as the selected edge is rotated. If the [Ctrl] key is pressed once, the adjacent faces are not triangulated unless necessary, but the faces may change shape.

NOTE

If the selected edge does not dynamically rotate at the "angle end point" prompt, then the desired rotation is not possible.

Scaling Vertices

As mentioned earlier, it is not possible to scale a single vertex. However, two or more vertices can be selected for scaling. This, in effect, scales edges. The selection methods are the same as discussed for rotating vertices and the use of the [Ctrl] key while dragging produces the same effects. As the pointer is dragged, the dynamic display of scaled edges may be difficult to visualize. Therefore, it is best to use a scale factor or the **Reference** option to achieve properly scaled edges.

Exercise 12-4

Complete the exercise on the student website.
www.g-wlearning.com/CAD

Using Subobject Editing as a Construction Tool

This section provides an example of how subobject editing can be used to not only make changes to existing solids and composites, but as a powerful construction tool. Some of the procedures of subobject editing, such as editing faces, edges, and vertices, are used to construct an HVAC assembly. The entire model is constructed from a single, solid cube. This is the only primitive you will draw.

Editing Faces

1. Begin by setting the units to Architectural and drawing a 24″ cube. Display the model from the southeast isometric viewpoint.
2. Select the left-hand face and move it 60″ to the left. Also, select the front face and move it out 12″. This forms the first duct. See **Figure 12-32**.
3. Using the **EXTRUDE** command, select the left-hand face and extrude it 36″ to create a new solid. This is a tee junction from which two branches will extend.
4. Select the front face of the new solid and extrude it 28″ with a taper angle of 10° to create a new solid that is a reducer.
5. Select the left-hand face of the tee junction and extrude it 20″ with a taper angle of 10° to create a new solid that is a second reducer. See **Figure 12-33**.

Figure 12-32.
The left-hand face of
the cube is moved
60″. The front face is
then moved 12″.

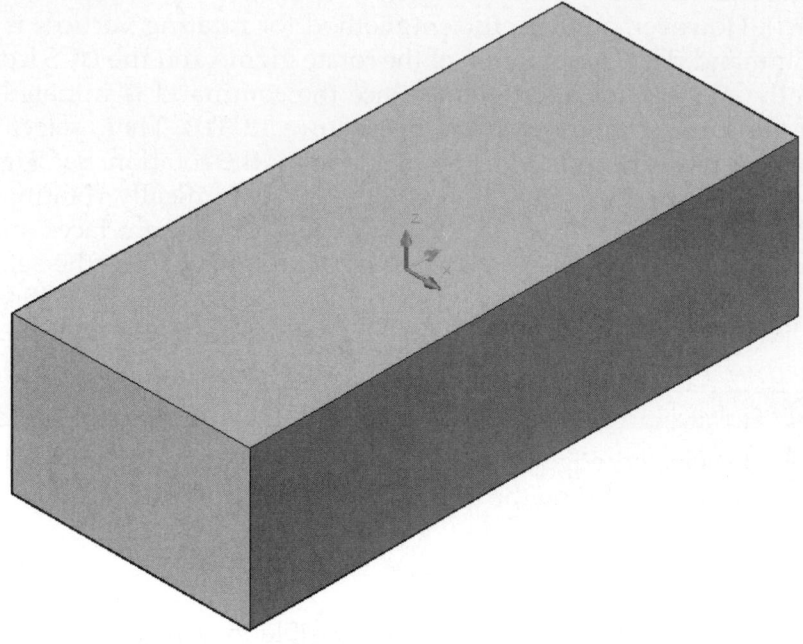

Figure 12-33.
Two reducers are
created by extruding
faces from the tee
junction.

6. Select the left-hand face of the 20″ reducer and move it 3 17/32″ along the positive Z axis. This places the top surface of the reducer level with the trunk of the duct. Next, extrude the left-hand face of this reducer 60″ into a new solid.

7. Use the **REVOLVE** command to turn the left-hand face of the 60″ extension into a new solid that is a 90° bend. Your drawing should now look like **Figure 12-34**.

Editing Edges and Vertices

1. The bottom surface of the 28″ reducer must be level with the bottom of the tee junction and main trunk. Select the bottom edge of the reducer's front face and move it down 4 15/16″.

2. Select the two top vertices on the 28″ reducer's front face and move them down (negative Z) 6″.

Figure 12-34.
The left end of
the 60″ extrusion
is revolved 90° to
create an elbow.

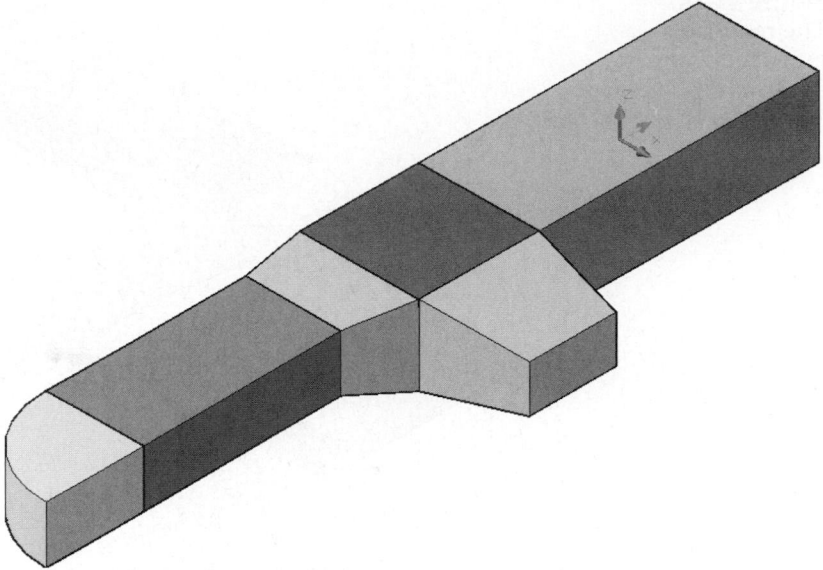

3. Select the front, rectangular face of the 90° bend and extrude it 72″ into a new solid.
4. Select the front face of the 28″ reducer and extrude it 108″ into a new solid.
5. Select the front face of the new solid created in step 4 and extrude it 26″ to create a new solid that will be a tee junction.
6. Extrude the left-hand face of the tee junction 20″. See **Figure 12-35.**
7. Move the top edge of the left-hand face on the 20″ extrusion created in step 6 down 4″.
8. Move each vertical edge of the 20″ extrusion 6″ toward the center of the duct.
9. Mirror a copy of the 20″ extrusion to the opposite side of the tee junction. The completed drawing should look like **Figure 12-36.**

Figure 12-35.
The face of the
revolved elbow is
extruded 72″. The
face of the right
branch is extruded
108″. The right duct
is then extruded
26″ and the left face
of that extrusion is
extruded by 20″.

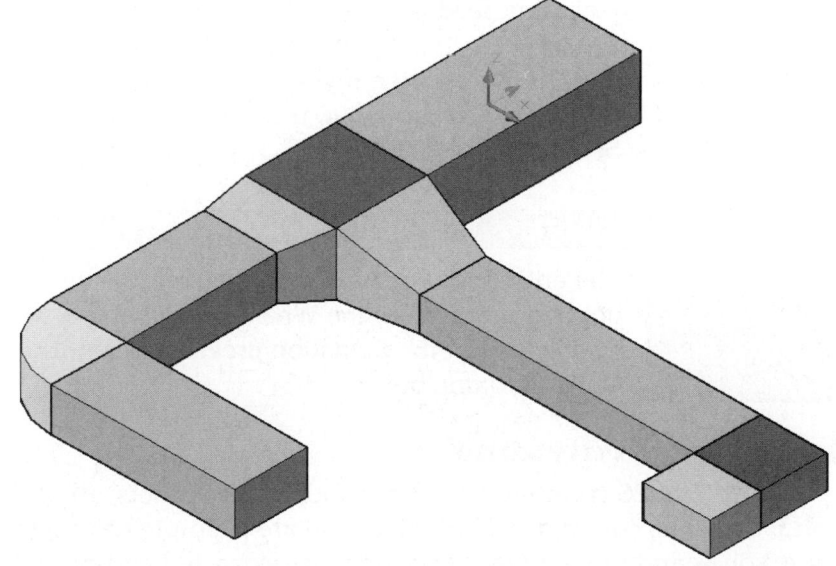

Figure 12-36.
The reducer is
mirrored to create
the final assembly.

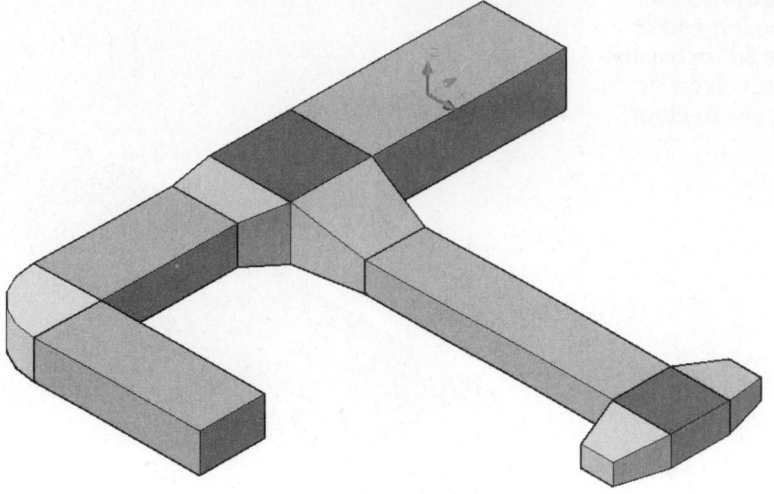

Other Solid Editing Tools

There are other tools that can be used in solid model editing. As you will learn in Chapter 13, the **SOLIDEDIT** tool can be used to edit faces, edges, and vertices much like subobject editing. In addition, you can extrude a closed boundary with the **PRESSPULL** command, extract a wireframe from a solid using the **XEDGES** command, and explode a solid. The **PRESSPULL** and **XEDGES** commands and exploding a solid are discussed in the next sections.

Presspull

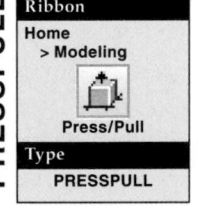

PRESSPULL

Ribbon
Home
> Modeling

Press/Pull

Type
PRESSPULL

The **PRESSPULL** command allows any closed boundary to be extruded. The boundary can be a flat surface, closed polyline, circle, or region. The extrusion is always applied perpendicular to the plane of the boundary, but can be in the positive or negative direction. When applied to the face of a solid, it is very similar to the **Extrude Face** option of the **SOLIDEDIT** command, though dynamic feedback is provided for the extrusion with **PRESSPULL**.

Once the command is initiated, you are prompted to pick inside of the bounded areas to extrude. Move the pointer inside of a boundary and pick. Then, drag the boundary to a new location and pick or, if dynamic input is on, enter the distance to extrude the face. See **Figure 12-37.**

> **NOTE**
>
> The entire boundary must be visible on the screen or the loop will not be found. Also, the **When a command is active** check box must be checked in the **Selection preview** area of the **Selection** tab in the **Options** dialog box.

Extracting a Wireframe

Ribbon
Home
> Solid Editing

Extract Edges

Type
XEDGES

The **XEDGES** command creates copies of, or extracts, all of the edges on a selected solid. Once the command is initiated, you are prompted to select objects. Select one or more solids and press [Enter]. The edges are extracted and placed on top of the existing edges. See **Figure 12-38.** The new objects are created on the current layer.

Straight edges and the curved edges where cylindrical surfaces intersect with flat or other cylindrical surfaces are the only edges extracted. Spheres and tori have no edges that can be extracted. The round bases of cylinders and cones are the only edges of those objects that will be extracted.

AutoCAD and Its Applications—Advanced

Figure 12-37.
Using the **PRESSPULL** command. A—Pick inside of a boundary (shown in color) and drag the boundary to a new location. B—The completed operation.

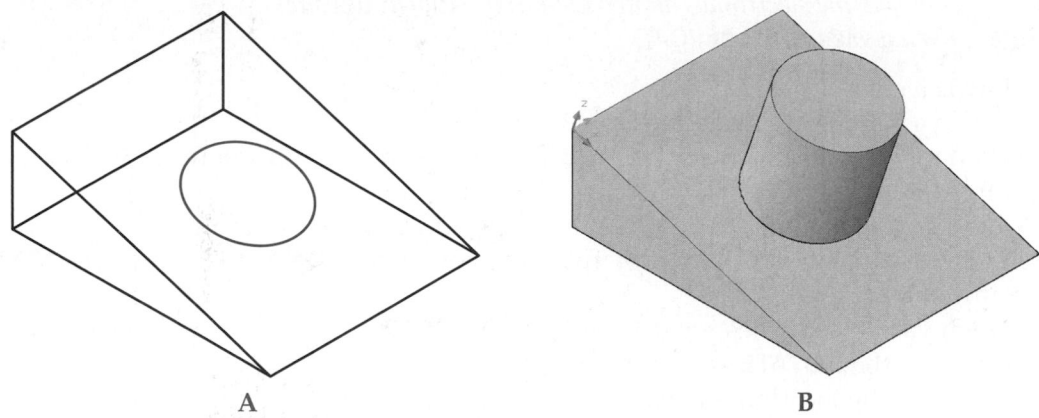

A

B

Figure 12-38.
Extracting edges with the **XEDGES** command. A—The original object. B—The extracted wireframe (edges).

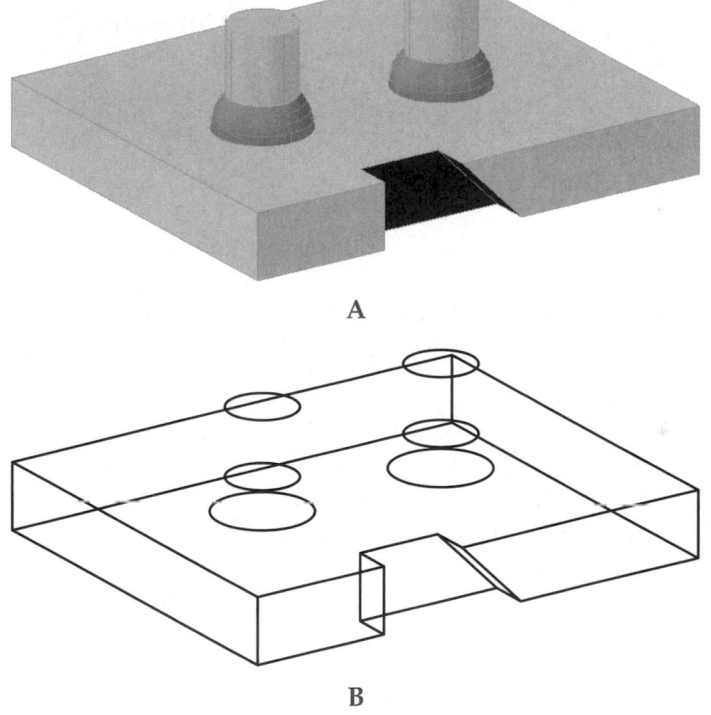

A

B

Exploding a Solid

A solid can be exploded. This turns the solid into surfaces and/or regions. Flat surfaces on the solid are turned into regions. Curved surfaces on the solid are turned into surfaces. To explode a solid, select the **EXPLODE** command. Then, pick the solid(s) to explode and press [Enter].

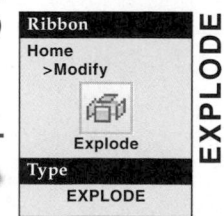

Exercise 12-5

Complete the exercise on the student website.
www.g-wlearning.com/CAD

Chapter Test

Answer the following questions. Write your answers on a separate sheet of paper or complete the electronic chapter test on the student website.
www.g-wlearning.com/CAD

1. How is a subobject selected?
2. How is a subobject deselected?
3. When moving a face on a solid primitive, how can you accurately control the axis of movement?
4. How can you change the results of moving a face while dragging it?
5. Describe a major difference of function between the **ROTATE** and **3DROTATE** commands.
6. Which variable enables you to use the **3DROTATE** command in a 3D view even if you select the **ROTATE** command?
7. How does the location and shape of an edge grip differ from a face grip?
8. What is the most efficient tool to use when rotating an edge and how is it displayed?
9. What is the only type of edge that can be scaled?
10. What is the only editing function that can be done to a single vertex?
11. How are two or more vertices selected for editing?
12. What is the function of the **PRESSPULL** command?
13. On which objects can the **PRESSPULL** command be used?
14. What is the purpose of the **XEDGES** command?
15. When a solid object is exploded, which type of object is created?

Drawing Problems

1. Draw the bookcase shown below using the dimensions given. The final result should be a single solid object. Then, use grip and subobject editing procedures to edit the object as follows.
   A. Change the width of the bookcase to 3′.
   B. Change the height of the bookcase by eliminating the top section. The resulting height should be 3′-2″.
   C. Save the drawing as P12_01.

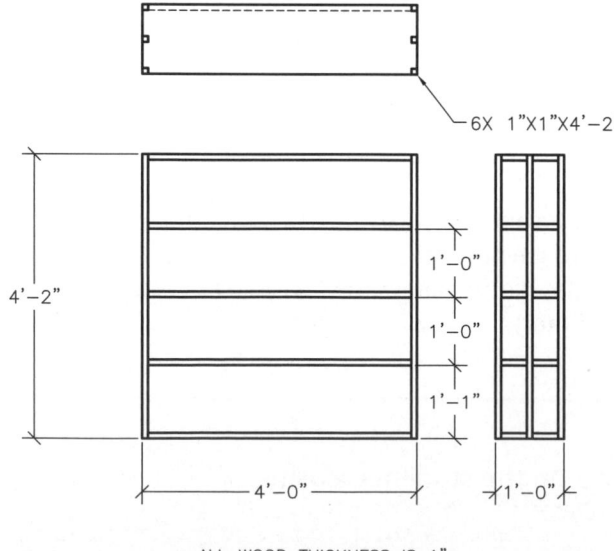

ALL WOOD THICKNESS IS 1″

2. Open problem P12_01. Save it as P12_02. Use primitive and subobject editing procedures to create the following edits.
   A. Change the depth of the top of the bookcase to 6-1/2″.
   B. Change the depth of the bottom of the bookcase to 24″.
   C. Reduce the height of the front uprights so they are flush with the top surface of the next lower shelf.
   D. Extend the front of the second lowest shelf to match the front of the bottom. Add two uprights at the front corners between the bottom and this shelf.
   E. Save the drawing.

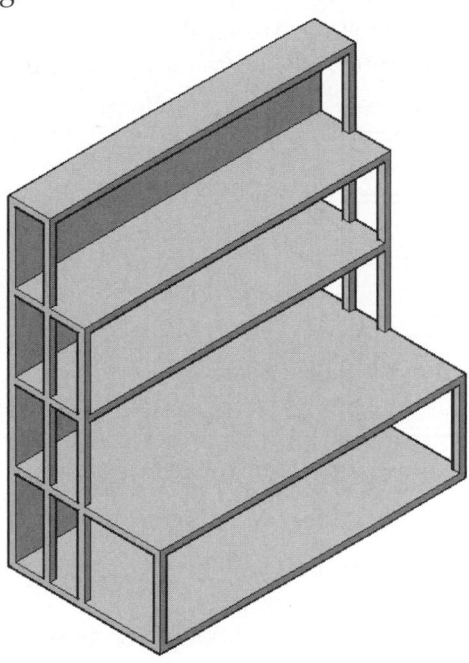

3. Draw the mounting bracket shown below. Then, use primitive and subobject editing procedures to create the following edits.
   A. Change the 3.00″ dimension to 3.50″.
   B. Change the 2.50″ dimension in the front view to 2.75″.
   C. Change the location of the slot in the auxiliary view from .60″ to .70″ and change the length of the slot to 1.15″.
   D. Change the width of each foot in the top view from 2.00″ to 1.50″. The overall dimension (5.00″) should not change.
   E. Change the angle of the bend from 15° to 45°.
   F. Save the drawing as P12_03.

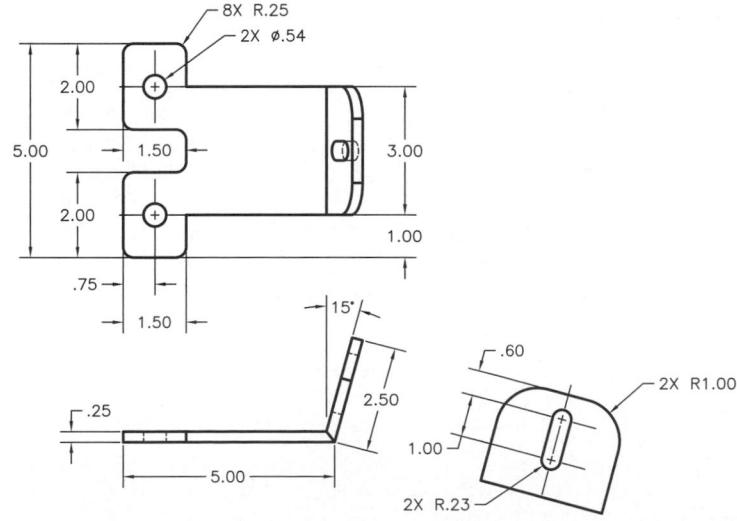

4. Draw as a single composite solid the desk organizer shown in the orthographic views below. Then, use primitive and subobject editing procedures to create the following edits. The final object should look like the shaded view below.
   A. Change the 3″ height to 3.25″.
   B. Change the 2″ height to 1.85″.
   C. Increase the thickness of the long compartment divider to .5″. The increase in thickness should be evenly applied along the centerline of the divider. Locate three evenly spaced, ∅5/16″ × 1.5″ holes in this divider.
   D. Angle the top face of the rear compartments by 30°. The height of the rear of the organizer should be approximately 4.5″ and all corners on the bottom of the organizer should remain square.
   E. Save the drawing as P12_04.

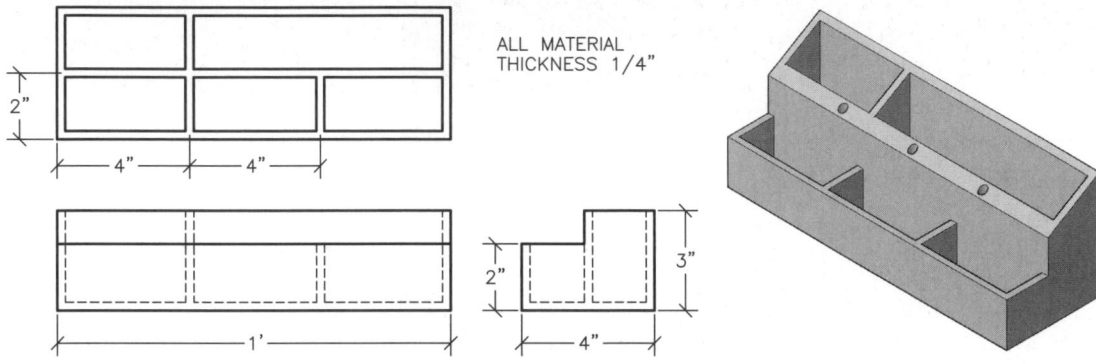

5. Draw the pencil holder shown below. Then, use primitive and subobject editing procedures to create the following edits.
   A. Change the depth of the base to 4.000″. The base should be rectangular, not square, and the grooves should become shorter.
   B. Change the height of the top groove from .250″ to .125″.
   C. Change the diameter of two holes from ∅.450″ to ∅.625″.
   D. Change the diameter of the other two holes from ∅.450″ to ∅1.000″.
   E. Rotate the top face 15° away from the side with the grooves. The planes of the adjoining faces should not change. Refer to the shaded view shown below.
   F. Save the drawing as P12_05.

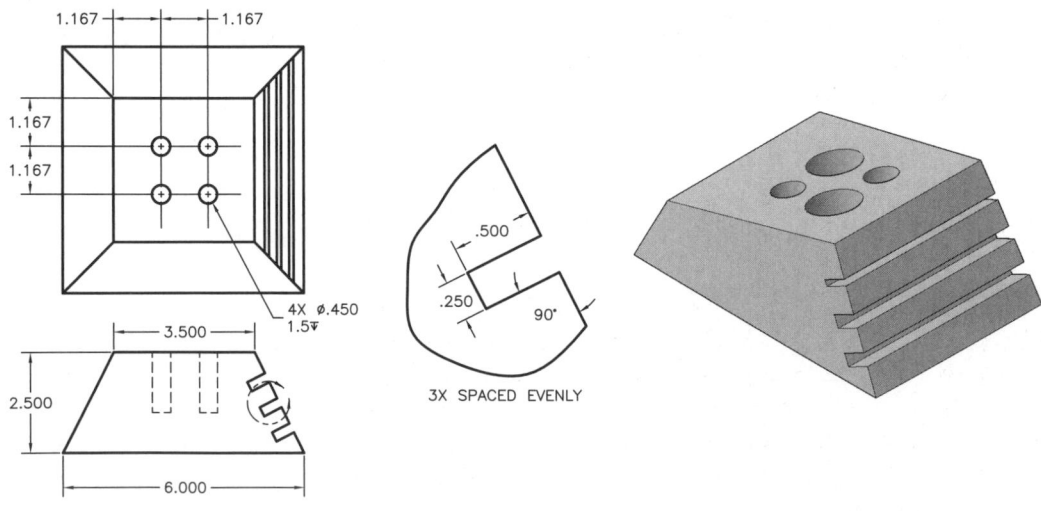

CHAPTER 13

Solid Model Editing

Learning Objectives

After completing this chapter, you will be able to:

✓ Change the shape and configuration of solid object faces.
✓ Copy and change the color of solid object edges and faces.
✓ Break apart a composite solid composed of physically separate entities.
✓ Use the **SOLIDEDIT** command to construct and edit a solid model.

AutoCAD provides expanded capabilities for editing solid models. As you saw in the previous chapter, grips can be used to edit a solid model. Also, the subobjects that make up a solid, such as faces, edges, and endpoints, can be edited. Additionally, a single command, **SOLIDEDIT**, enables you to edit faces, edges, or the entire body of the solid.

NOTE

Mesh objects cannot be modified using the **SOLIDEDIT** command. The mesh object must be converted to a solid first. If you select a mesh for editing with the **SOLIDEDIT** command, you are given the option of converting it to a solid, as long as the display of the dialog box has not been turned off.

Overview of the SOLIDEDIT Command

The **SOLIDEDIT** command allows you to edit the faces, edges, and body of a solid. Many of the subobject editing functions can also be performed with the **SOLIDEDIT** command. The features of the **SOLIDEDIT** command can be accessed in the **Solid Editing** panel on the **Home** tab of the ribbon or by typing SOLIDEDIT. See **Figure 13-1**. The quickest method of entering the command is by using the ribbon. For example, directly select a face-editing option from the drop-down list in the **Solid Editing** panel, as shown in **Figure 13-1**.

Figure 13-1.
Accessing the
SOLIDEDIT
command options.

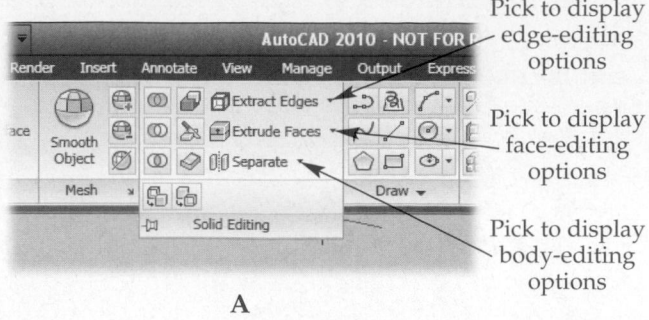

Pick to display
edge-editing
options

Pick to display
face-editing
options

Pick to display
body-editing
options

A

Select the
face-editing option

Pick

Expanded
help text

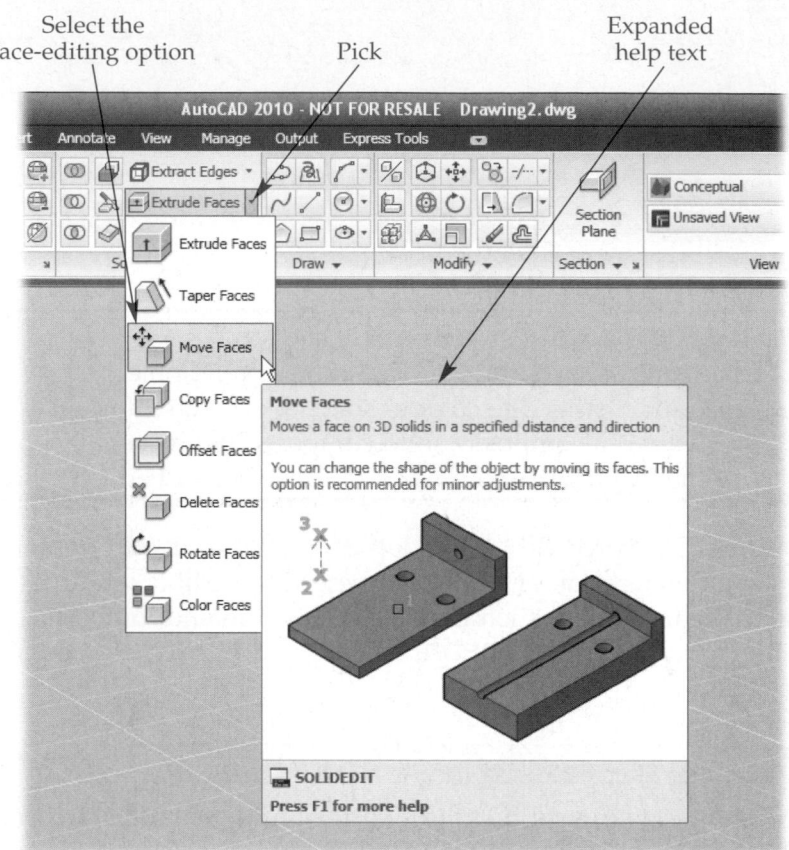

B

When the **SOLIDEDIT** command is typed, you are first asked to select the component of the solid with which you wish to work. Specify either **Face**, **Edge**, or **Body**. The editing options for the selected component are then displayed and are the same as those seen in **Figure 13-1.** The editing function is directly entered when the option is selected from the ribbon. This is why using the ribbon is the most efficient method of entering the **SOLIDEDIT** command options.

The following sections provide an overview of the solid model editing features of the **SOLIDEDIT** command. Each option is explained and the results of each are shown. A tutorial later in the chapter illustrates how these options can be used to construct a model.

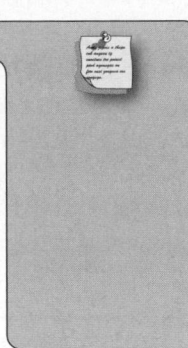

NOTE

AutoCAD displays a variety of error messages when illegal solid editing operations are attempted. Rather than trying to interpret the wording of these messages, just realize that what you tried to do will not work. Actions that may cause errors include trying to rotate a face into other faces or extruding and tapering an object at too great of an angle. When an error occurs, try the operation again with different parameters or determine a different approach to solving the problem in order to maintain the design intent.

Face Editing

The basic components of a solid are its faces and the greatest number of **SOLIDEDIT** options are for editing faces. All eight face editing options ask you to select faces. It is important to make sure you select the correct part of the model for editing. Remember the following three steps when using any of the face editing options.

1. First, select a face to edit. If you pick an edge, AutoCAD selects the two faces that share the edge. If this happens, use the **Remove** option to deselect the unwanted face. A more intuitive approach is to select the open space of the face as if you were touching the side of a part. AutoCAD highlights only that face.

2. Adjust the selection set at the Select faces or [Undo/Remove/ALL]: prompt. The following options are available.
   - **Undo.** Removes the previous selected face(s) from the selection set.
   - **Remove.** Allows you to select faces to remove from the selection set. This is only available when **Add** is current.
   - **All.** Adds all faces on the model to the selection set. This is only available after selecting at least one face. It can also be used to remove all faces if **Remove** is current.
   - **Add.** Allows you to add faces to the selection set. This is only available when **Remove** is current.

3. Press [Enter] to continue with face editing.

Extruding Faces

An extruded face is moved, or stretched, in a selected direction. The extrusion can be straight or have a taper. To extrude a face, select the command and pick the **Face>Extrude** option. Remember, the option is directly entered when picking the button in the ribbon. You are then prompted to select the face(s) to extrude. Nonplanar (curved) faces cannot be extruded. As you pick faces, the prompt verifies the number of faces selected. For example, when an edge is selected, the prompt reads 2 faces found. When done selecting faces, press [Enter] to continue.

Next, the height of the extrusion needs to be specified. A positive value adds material to the solid, while a negative value subtracts material from the solid. A taper can also be given.

> Specify height of extrusion or [Path]: *(enter height)*
> Specify angle of taper for extrusion <0>: *(enter an angle or accept the default)*
> Solid validation started.
> Solid validation completed.
> Enter a face editing option
> [Extrude/Move/Rotate/Offset/Taper/Delete/Copy/coLor/mAterial/Undo/eXit] <eXit>: **X⏎**
> Solids editing automatic checking: SOLIDCHECK=1
> Enter a solids editing option [Face/Edge/Body/Undo/eXit] <eXit>: **X⏎**

Figure 13-2 shows an original solid object and the result of extruding the top face with a 0° taper angle and a 30° taper angle. It also shows the original solid object with two adjacent faces extruded with 15° taper angles.

In addition to extruding a face perpendicular to itself, the extruded face can follow a path. Select the **Path** option at the Specify height of extrusion or [Path]: prompt. The path of extrusion can be a line, circle, arc, ellipse, elliptical arc, polyline, or spline. The extrusion height is the exact length of the path. See **Figure 13-3.**

Exercise 13-1

Complete the exercise on the student website.
www.g-wlearning.com/CAD

Figure 13-2.
Extruding faces on an object. A—The original object. B—The top face is extruded with a 0° taper angle. C—The top face of the original is extruded with a 30° taper angle. D—The top and right-hand faces of the original are extruded with 15° taper angles.

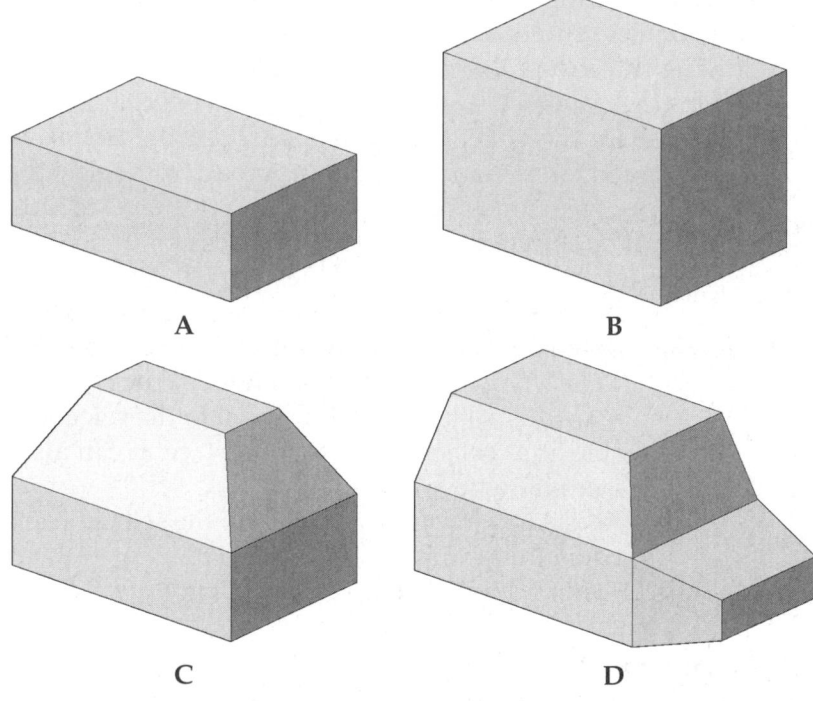

A

B

C

D

Figure 13-3.
The path of extrusion can be a line, circle, arc, ellipse, elliptical arc, polyline, or spline. Here, the paths are shown in color.

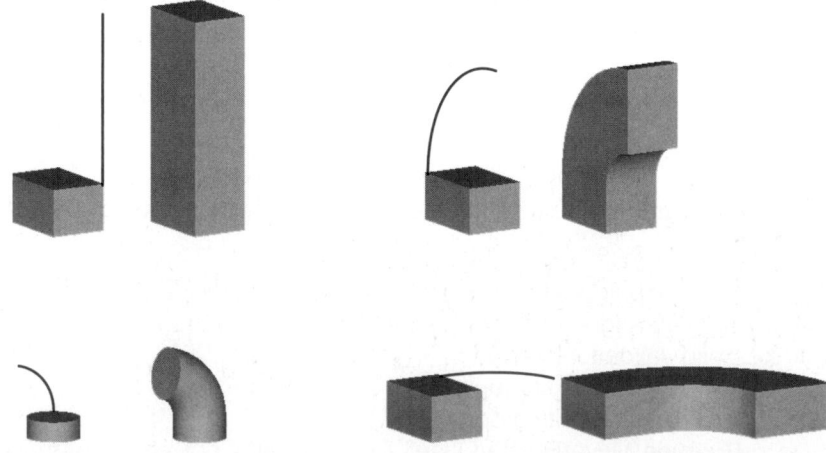

AutoCAD and Its Applications—Advanced

Moving Faces

The **Move Faces** option moves a face in the specified direction and lengthens or shortens the solid object. In another application, a solid model feature (such as a hole) that has been subtracted from an object to create a composite solid can be moved with this option. Object snaps may interfere with the operation of this option, so they may need to be toggled off during the operation.

To move a face, select the command and pick the **Face>Move** option. If the button is picked in the ribbon, the option is directly entered. You are then prompted to select the face(s) to move. When done selecting faces, press [Enter] to continue. Next, you are prompted to select a base point of the operation:

> Specify a base point or displacement: *(pick a base point)*
> Specify a second point of displacement: *(pick a second point or enter coordinates)*
> Solid validation started.
> Solid validation completed.
> Enter a face editing option
> [Extrude/Move/Rotate/Offset/Taper/Delete/Copy/coLor/mAterial/Undo/eXit] <eXit>: **X↵**
> Solids editing automatic checking: SOLIDCHECK=1
> Enter a solids editing option [Face/Edge/Body/Undo/eXit] <eXit>: **X↵**

When adjacent faces are perpendicular, the edited face is moved in a direction so the new position keeps the face parallel to the original. See **Figures 12-4A** and **12-4B**. Faces that are normal to the current UCS can be moved by picking a new location or entering a direct distance. If you are moving a face that is not normal to the current

Figure 13-4.
A—The hole will be moved using the **Move** option of the **SOLIDEDIT** command. B—The hole is moved. C—When the angled face is moved, a portion of it is altered to be coplanar with the vertical face. D—If the angled face is moved more, it becomes completely coplanar to the vertical face. This is a new, single face.

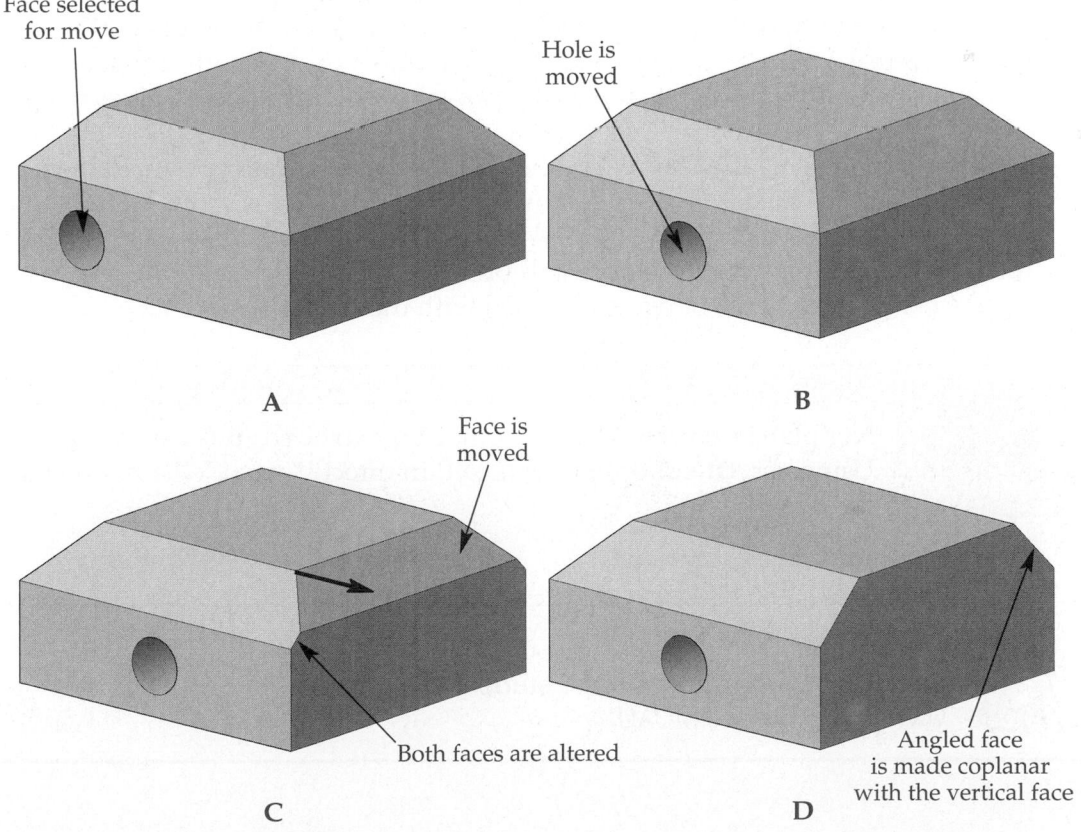

Face selected for move

Hole is moved

A

B

Face is moved

Both faces are altered

Angled face is made coplanar with the vertical face

C

D

Ribbon
Home
> Solid Editing

Move Faces
Type
SOLIDEDIT

SOLIDEDIT

UCS, you can enter coordinates for the second point of displacement, but it may be easier to first use the **Face** option of the **UCS** command to align the UCS with the face to be moved.

When adjacent faces join at angles other than 90°, the moved face will be relocated as stated above, but only if the movement is less than the dimensional offset of the two faces. For example, in **Figure 13-4B** the top edge of the angled face is in .5″ from the vertical face. If the angled face is moved outward a distance of less than .5″, it is altered as shown in **Figure 13-4C.** A portion of the angled face becomes coplanar with the vertical face. If the angled face is moved outward a distance greater than .5″, it is altered so that it forms a single plane with the adjacent face. What has actually happened is that the angled face is moved beyond the adjacent face, while remaining parallel to its original position. Thus, in effect, it has disappeared because the adjacent, vertical face cannot be altered. See **Figure 13-4D.** In this example, the angled face was moved .75″. The new vertical face that is created can now be moved.

Exercise 13-2

Complete the exercise on the student website.
www.g-wlearning.com/CAD

Offsetting Faces

SOLIDEDIT

Ribbon
Home
> Solid Editing

Offset Faces
Type
SOLIDEDIT

The **Offset** option may seem the same as the **Extrude** option because it moves faces by a specified distance or through a specified point. Unlike the **OFFSET** command in AutoCAD, this option moves all selected faces a specified distance. It is most useful when you wish to change the size of features such as slots, holes, grooves, and notches in solid parts. A positive offset distance increases the size or volume of the solid (adds material), a negative distance decreases the size or volume of the solid (removes material). Therefore, if you wish to make the width of a slot wider, provide a negative offset distance to decrease the size of the solid. Picking points to set the offset distance and direct distance entry are always taken as a positive value, so negative values must be entered using the keyboard.

To offset a face, select the command and pick the **Face>Offset** option. Remember, the option is directly entered when picking the button in the ribbon. You are then prompted to select the face(s) to offset. When done selecting faces, press [Enter] to continue. Next, enter the offset distance and press [Enter] and then exit the command. See **Figure 13-5** for examples of features edited with the **Offset** option.

PROFESSIONAL TIP

Nonplanar (curved) faces cannot be extruded, but can be offset. Using the **Offset** option, you can, in effect, "extrude" a nonplanar face.

Exercise 13-3

Complete the exercise on the student website.
www.g-wlearning.com/CAD

Figure 13-5.
Offsetting faces. A—The original objects. The hole is selected to offset. The interior of the L is also selected to offset. B—A positive offset distance increases the size or volume of the solid. C—A negative offset distance decreases the size or volume of the solid.

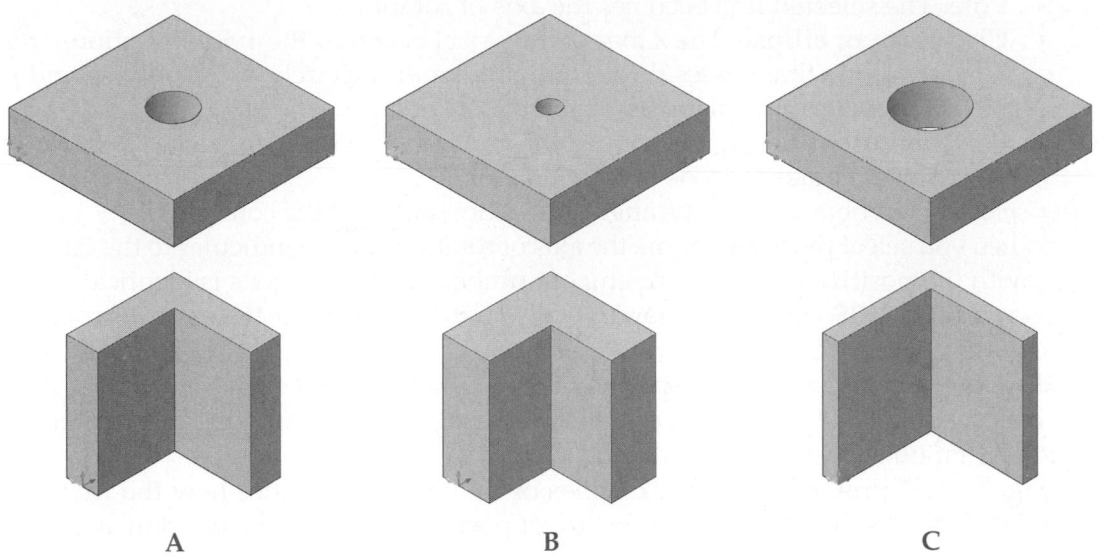

A   B   C

Deleting Faces

The **Delete** option deletes selected faces. This is a quick way to remove features such as chamfers, fillets, holes, and slots. To delete a solid face, select the command and pick the **Face**>**Delete** option. If the button is picked in the ribbon, the option is directly entered. You are then prompted to select the face(s) to delete. When done selecting faces, press [Enter] to continue and then exit the command. When a face is deleted, existing faces extend to fill the gap. No additional faces are created. For instance, the inclined surface of a wedge cannot be deleted as there are no existing faces that can be extended to fill the gap. When deleting the face that is a chamfered or filleted edge, the adjacent edges are extended to fill the gap. See **Figure 13-6.**

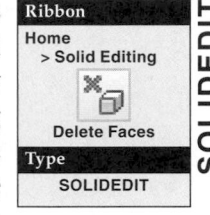

Rotating Faces

The **Rotate** option rotates a face about a selected axis. To rotate a solid face, select the command and pick the **Face**>**Rotate** option. Remember, the option is directly entered when picking the button in the ribbon. You are then prompted to select the face(s) to rotate. When done selecting faces, press [Enter] to continue. There are several methods by which a face can be rotated.

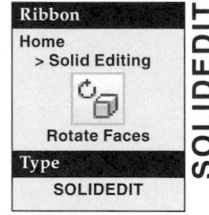

Figure 13-6.
Deleting faces.
A—The original
objects with three
rounds. B—The faces
of two rounds have
been deleted.

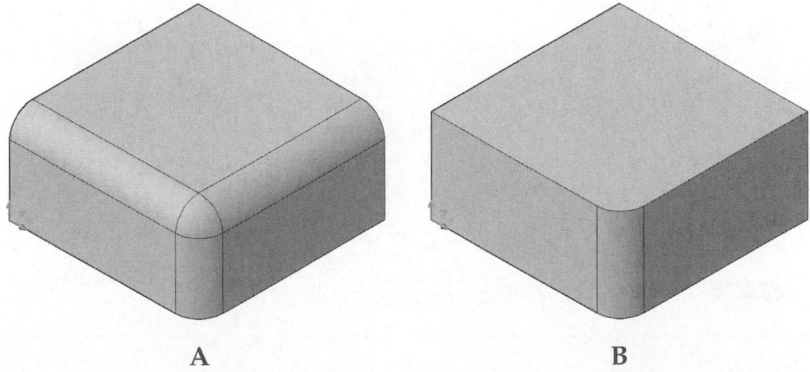

A   B

The **2points** option is the default. Pick two points to define the "hinge" about which the face will rotate. Then, provide the rotation angle and exit the command.

The **Axis by object** option allows you to use an existing object to define the axis of rotation. You can select the following objects.

- **Line.** The selected line becomes the axis of rotation.
- **Circle, arc, or ellipse.** The Z axis of the object becomes the axis of rotation. This Z axis is a line that passes through the center of the circle, arc, or ellipse and is perpendicular to the plane on which the 2D object lies.
- **Polyline or spline.** A line connecting the polyline or spline's start point and endpoint becomes the axis of rotation.

After selecting an object, enter the angle of rotation and exit the command.

When you select the **View** option, the axis of rotation is perpendicular to the current view, with the positive direction coming out of the screen. This axis is identical to the Z axis when the **UCS** command **View** option is used. Next, enter the angle of rotation and exit the command.

The **Xaxis**, **Yaxis**, and **Zaxis** options prompt you to select a point. Either the X, Y, or Z axis that passes through that point is used as the axis of rotation. Then, enter the angle of rotation and exit the command.

Figure 13-7 provides several examples of rotated faces. Notice how the first and second pick points determine the direction of positive and negative rotation angles.

NOTE

A positive rotation angle moves the face in a clockwise direction looking from the first pick point to the second. Conversely, a negative angle rotates the face counterclockwise. If the rotated face will intersect or otherwise interfere with other faces, an error message indicates that the operation failed or that no solution was calculated. In this case, you may wish to try a negative angle if you previously entered a positive one. In addition, you can try selecting the opposite edge of the face as the axis of rotation.

Figure 13-7.
When rotating faces, the first and second pick points determine the direction of positive and negative rotation angles.

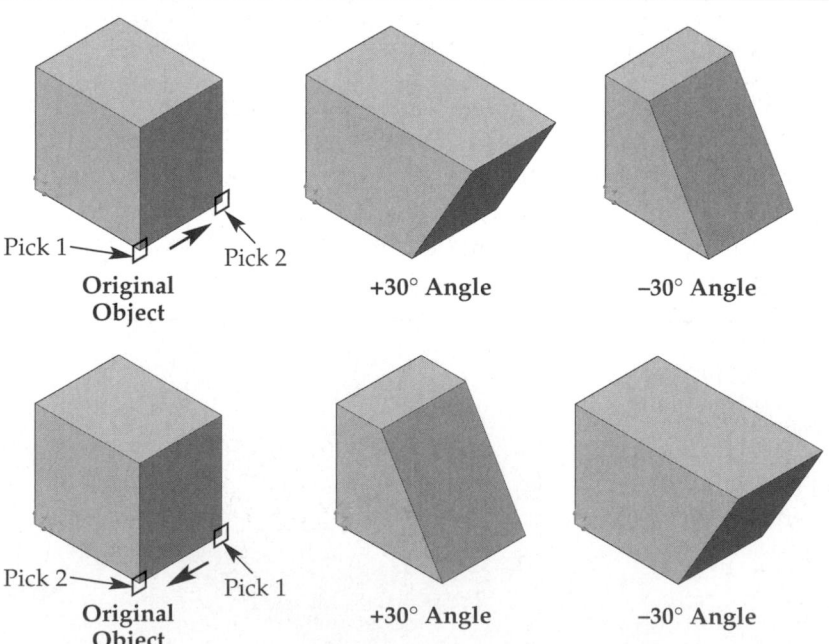

Pick 1 ——► Pick 2
Original Object

+30° Angle

−30° Angle

Pick 2 ——► Pick 1
Original Object

+30° Angle

−30° Angle

AutoCAD and Its Applications—Advanced

Exercise 13-4

Complete the exercise on the student website.
www.g-wlearning.com/CAD

Tapering Faces

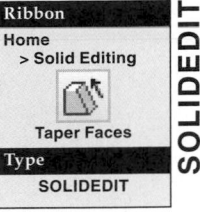
Ribbon
Home
> Solid Editing

Taper Faces
Type
SOLIDEDIT

SOLIDEDIT

The **Taper** option tapers a face at the specified angle, from the first pick point to the second. To taper a solid face, select the command and pick the **Face>Taper** option. If the button is picked in the ribbon, the option is directly entered. You are then prompted to select the face(s) to taper. When done selecting faces, press [Enter] to continue:

> Specify the base point: *(pick the base point)*
> Specify another point along the axis of tapering: *(pick a point along the taper axis)*
> Specify the taper angle: *(enter a taper value)*

Tapers work differently depending on whether the faces being tapered describe the outer boundaries of the solid, a cavity, or a removed portion of the solid. A positive taper angle always removes material. A negative taper angle always adds material. For example, if a positive taper angle is entered for a solid cylinder, the selected object is tapered in on itself from the base point along the axis of tapering, thus removing material. A negative angle tapers the object out away from itself to increase its size along the axis of tapering, thus adding material. See **Figure 13-8**.

Figure 13-8.
Tapering faces.
A—The original objects. The dark face of the box and the circumference of the cylinder are selected. B—Positive taper angle. C—Negative taper angle.

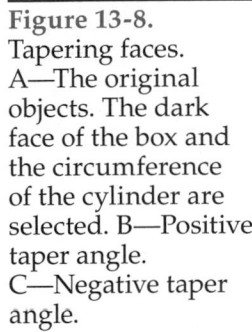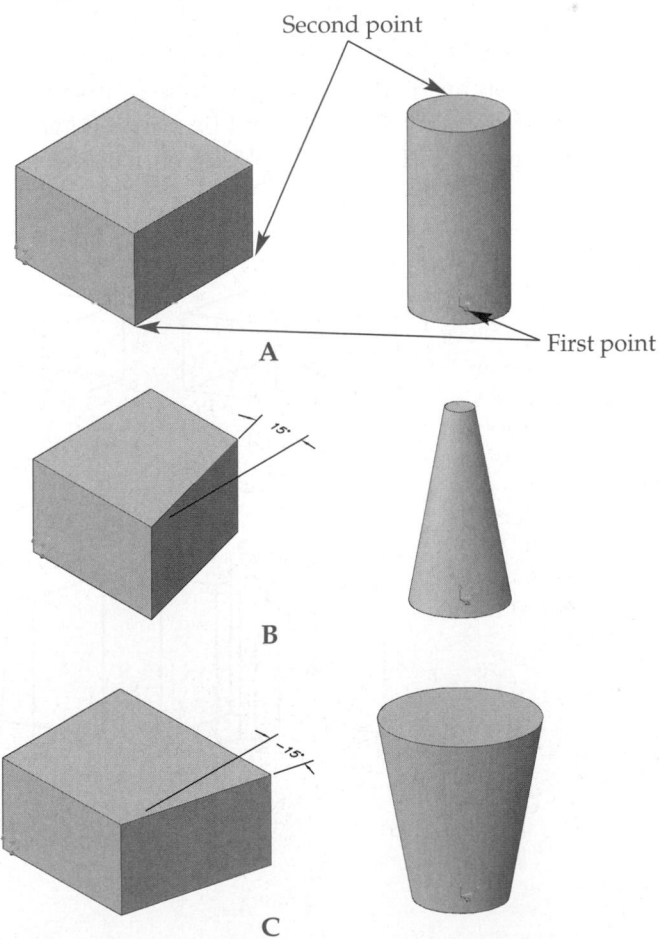

On the other hand, if the faces of a feature such as a slot or hole are tapered, a positive taper angle increases the size of the feature along the axis of tapering. For example, if a round hole is tapered using a positive taper angle, its diameter increases from the base point along the axis of tapering, thus removing material from the solid. Conversely, if the same round hole is tapered using a negative taper angle, its diameter decreases from the base point along the axis of tapering, thus adding material to the solid. **Figure 13-9** shows some examples of this operation.

Exercise 13-5

Complete the exercise on the student website.
www.g-wlearning.com/CAD

Copying Faces

Ribbon
Home
> Solid Editing

Copy Faces

Type
SOLIDEDIT

SOLIDEDIT

The **Copy** option copies a face to the location or coordinates given. The copied face is *not* part of the original solid model. It is actually a region, which can later be extruded, revolved, swept, etc., into a solid. This may be useful when you wish to construct a mating part in an assembly that has the same features on the mating faces or the same outline. This option is quick to use because you can pick a base point on the face, then enter a single direct distance value for the displacement. Be sure an appropriate UCS is set if you wish to use direct distance entry.

To copy a solid face, select the command and pick the **Face>Copy** option. Remember, the option is directly entered when picking the button in the ribbon. You are then prompted to select the face(s) to copy. When done selecting faces, press [Enter] to continue. You are prompted for a base point for the copy. Pick this point and then pick a second point of displacement or press [Enter] to use the first point as a displacement. See **Figure 13-10** for examples of copied faces.

Figure 13-9.
If a hole or slot is tapered using a positive taper angle, its diameter or width increases from the base point along the axis of tapering, thus removing material from the solid. A negative taper angle increases the volume of the solid.

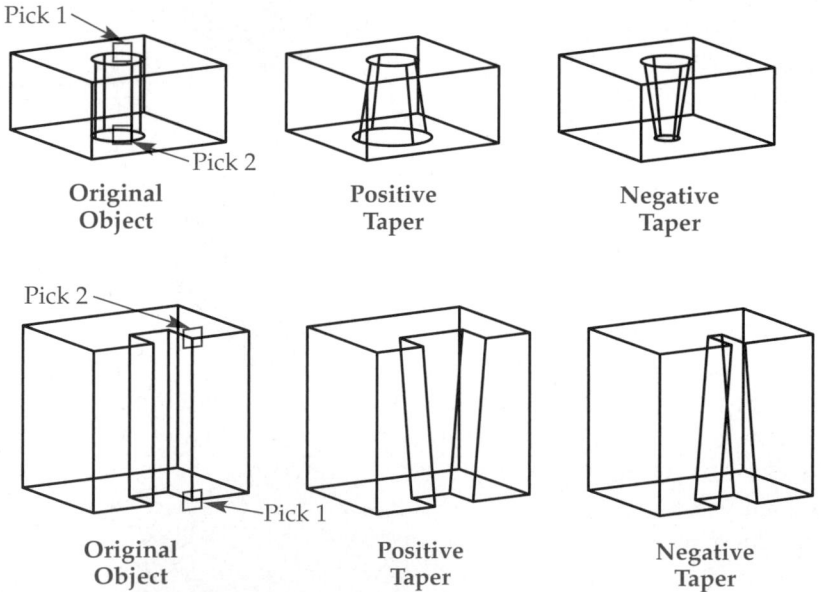

Figure 13-10.
A face can be quickly copied by picking a base point on the face and then entering a direct distance value for the displacement.

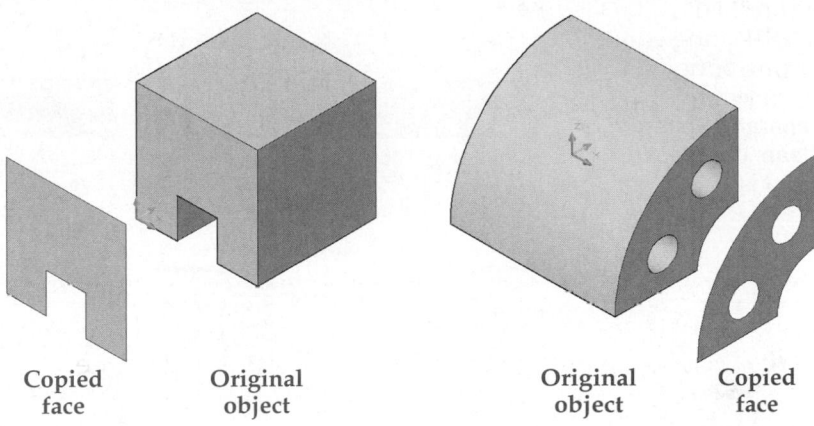

Copied face   Original object   Original object   Copied face

PROFESSIONAL TIP

Copied faces can also be useful for creating additional views. For example, you can copy a face to create a separate plan view with dimensions and notes. A copied face can also be enlarged to show details and to provide additional notation for design or assembly.

Coloring Faces

You can quickly change a selected face to a different color using the **Color** option. Select the command and pick the **Face>Color** option. If the button is picked in the ribbon, the option is directly entered. You are then prompted to select the face(s) to color. When done selecting faces, press [Enter] to continue. Next, choose the desired color from the **Select Color** dialog box that is displayed. Remember, the color of the object (or face) determines the shaded color.

Ribbon
Home
> Solid Editing
Color Faces
Type
SOLIDEDIT
SOLIDEDIT

Exercise 13-6

Complete the exercise on the student website.
www.g-wlearning.com/CAD

Edge Editing

Edges can be edited in only two ways. They can be copied from the solid. Also, the color of an edge can be changed.

Copying an edge is similar to copying a face. To copy a solid edge, select the command and pick the **Edge>Copy** option. Remember, the option is directly entered when picking the button in the ribbon. See **Figure 13-11A.** You are then prompted to select the edge(s) to copy. When done selecting edges, press [Enter] to continue. You are prompted for a base point for the copy. Pick this point and then pick a second point of displacement or press [Enter] to use the first point as a displacement. The edge is copied as a line, arc, circle, ellipse, or spline.

To color a solid edge, select the command and pick the **Edge>Color** option. You are then prompted to select the edge(s) to color. When done selecting edges, press [Enter] to continue. Next, choose the desired color from the **Select Color** dialog box that is displayed and pick the **OK** button. The edges are now displayed with the new color. You may need to set a wireframe or hidden visual style current to see the change.

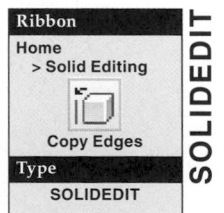

Ribbon
Home
> Solid Editing
Copy Edges
Type
SOLIDEDIT
SOLIDEDIT

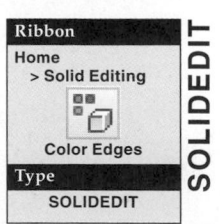

Ribbon
Home
> Solid Editing
Color Edges
Type
SOLIDEDIT
SOLIDEDIT

Figure 13-11.
A—Selecting edge-editing options.
B—Selecting the **Separate**, **Separate**, **Clean**, or **Check** options.

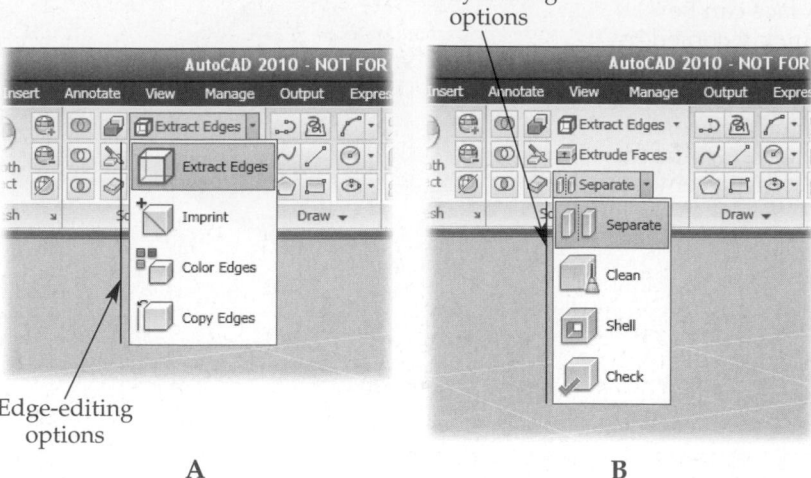

Edge-editing options

Body-editing options

A   B

Body Editing

The body editing options of the **SOLIDEDIT** command perform editing operations on the entire body of the solid model. The body options are **Imprint**, **Separate**, **Shell**, **Clean**, and **Check**. The next sections cover these body editing options.

The **Imprint** option is a body-editing function. However, since it modifies a body by adding edges, the option is located in the edges drop-down list in the **Solid Editing** panel. The **Separate**, **Shell**, **Clean**, and **Check** options are found in body drop-down list. See **Figure 13-11.**

Imprint

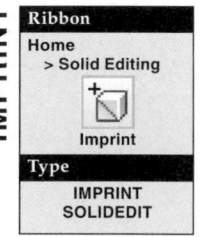

IMPRINT

Ribbon
Home
> Solid Editing

Imprint

Type
IMPRINT
SOLIDEDIT

Arcs, circles, lines, 2D and 3D polylines, ellipses, splines, regions, bodies, and 3D solids can be imprinted onto a solid, if the object intersects the solid. The imprint becomes a face on the surface based on the overlap between the two intersecting objects. Once the imprint has been made, the new face can be modified.

To imprint an object on a solid, select the command and pick the **Body>Imprint** option. If **IMPRINT** is typed or the button is picked on the ribbon, the option is directly entered. Once the option is activated, you are prompted to select the solid. This is the object on which the other objects will be imprinted. Then, select the objects to be imprinted. You have the option of deleting the source objects.

The imprinted face can be modified using face-editing options. **Figure 13-12** illustrates objects imprinted onto a solid model. Two of these are then extruded into the solid to create holes. The **PRESSPULL** command is used on the third object to create a cylindrical feature.

> **NOTE**
>
> Remember that objects are drawn on the XY plane of the current UCS unless you enter a specific Z value. Therefore, before you draw an object to be imprinted onto a solid model, be sure you have set an appropriate UCS for proper placement of the object by using a dynamic UCS or the **UCS** command. Alternately, you can draw the object on the XY plane and then move the object onto the solid object.

Figure 13-12.
Imprinted objects form new faces that can be extruded into the solid. A—A solid box with three objects on the plane of the top face. B—The objects are imprinted, then two of the new faces are extruded through the solid. The **PRESSPULL** command is used on the third new face to create the cylindrical feature.

Objects are on the
plane of the top face

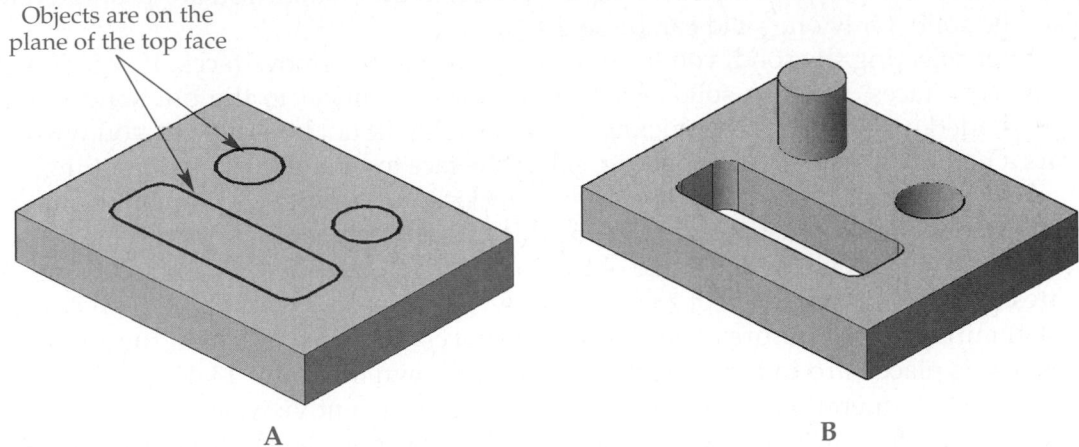

A   B

Separate

The **Separate** option separates two objects that are both a part of a single solid composite, but appear as separate physical entities. This can happen when modifying solids using the Boolean commands. The **Separate** option may be seldom used, but it has a specific purpose. If you select a solid model and an object physically separate from the model is highlighted, the two objects are parts of the same composite solid. If you wish to work with them as individual solids, they must first be separated.

To separate a solid body, select the command and pick the **Body>Separate** option. If the button is picked in the ribbon, the option is directly entered. You are then prompted to select a 3D solid. After you pick the solid, it is automatically separated. No other actions are required and you can exit the command. However, if you select a solid in which the parts are physically joined, AutoCAD indicates this by prompting The selected solid does not have multiple lumps. A "lump" is a physically separate solid entity. In order to separate a solid, it must be composed of multiple lumps. See **Figure 13-13.**

Figure 13-13.
A—After the cylinder is subtracted from the box, the remaining solid is considered one solid. B—Use the **Separate** option to turn this single solid into two solids.

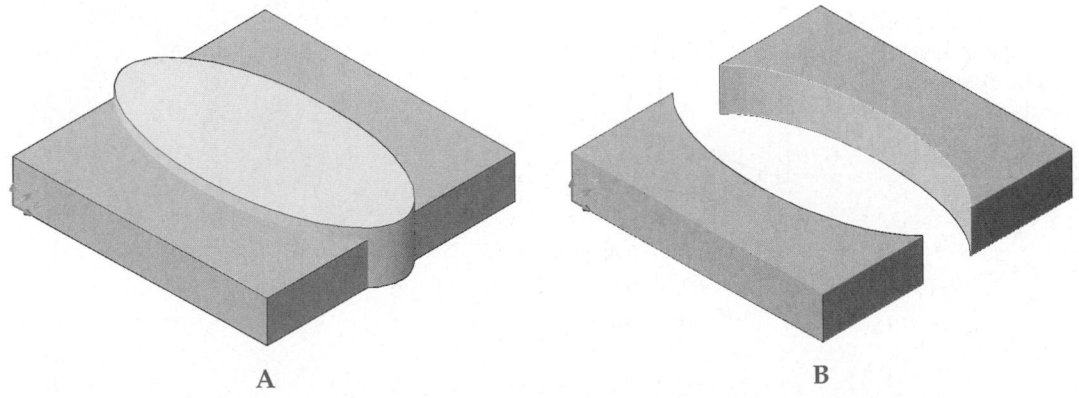

A   B

SOLIDEDIT

Ribbon
Home
> Solid Editing
Shell
Type
SOLIDEDIT

Shell

A *shell* is a solid that has been "hollowed out." The **Shell** option creates a shell of the selected object using a specified offset distance, or thickness. To create a shell of a solid body, select the command and pick the **Body>Shell** option. Remember, the option is directly entered when picking the button in the ribbon. You are prompted to select the solid. Only one solid can be selected.

After selecting the solid, you have the opportunity to remove faces. If you do not remove any faces, the new solid object will appear identical to the old solid object when shaded or rendered. The thickness of the shell will not be visible. If you wish to create a hollow object with an opening, select the face to be removed (the opening).

After selecting the object and specifying any faces to be removed, you are prompted to enter the shell offset distance. This is the thickness of the shell. A positive shell offset distance creates a shell on the inside of the solid body. A negative shell offset distance creates a shell on the outside of the solid body. See **Figure 13-14.** If you shell a solid that contains internal features, such as holes, grooves, and slots, a shell of the specified thickness is placed around those features. This is shown in **Figure 13-15.**

If the shell operation is not readily visible in the current view, you can rotate the view by using the [Shift] + the mouse wheel button to enter the transparent **3DORBIT** command or by using the view cube. You can also see the results by picking the **X-Ray Effect** button in the **Visual Styles** panel on the **Render** tab of the ribbon.

PROFESSIONAL TIP

The **Shell** option of the **SOLIDEDIT** command is very useful in applications such as solid modeling of metal castings or injection-molded plastic parts.

Exercise 13-7

Complete the exercise on the student website.
www.g-wlearning.com/CAD

Figure 13-14.
A—The right-front, bottom, and left-back faces (marked here by gray lines) are removed from the shell operation. B—The resulting object after the shell operation.

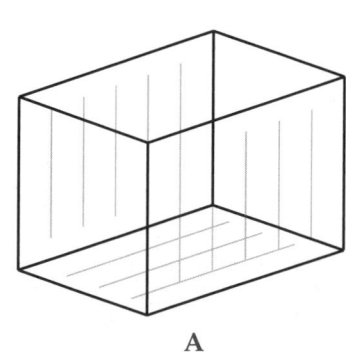

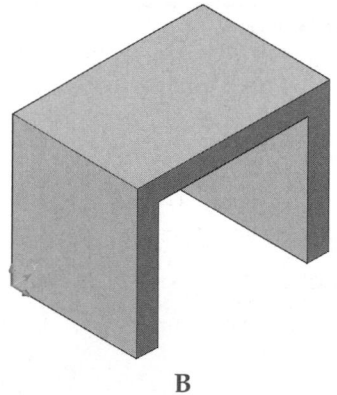

A

B

Figure 13-15.
If you shell a solid that contains internal features, such as holes, grooves, and slots, a shell of the specified thickness is also placed around those features.
A—Solid object with holes subtracted.
B—Wireframe display after shelling with a negative offset.
C—The Conceptual visual style is set current.

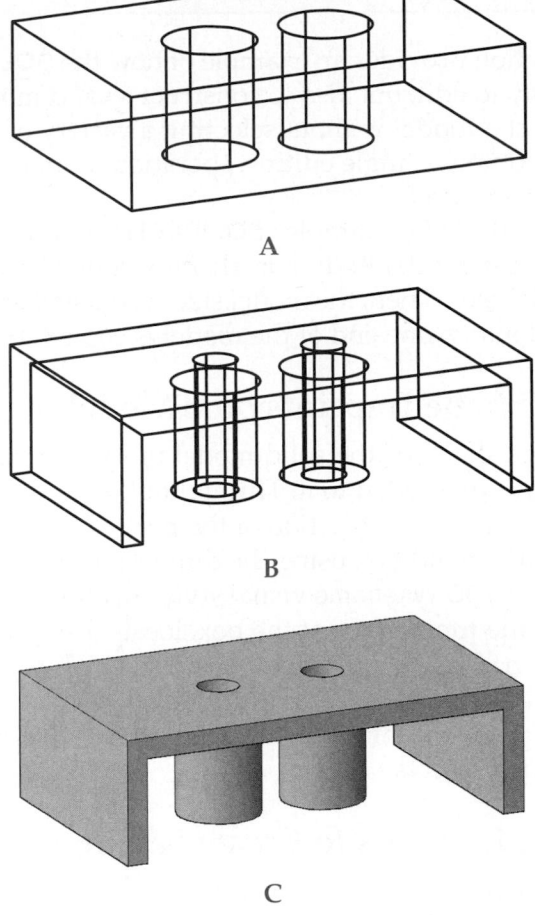

Clean

The **Clean** option removes all unused objects and shared surfaces. Imprinted objects are not removed. Select the command and pick the **Body>Clean** option. If the button is picked in the ribbon, the option is directly entered. Then, pick the solid to be cleaned. No further input is required. You can exit the command.

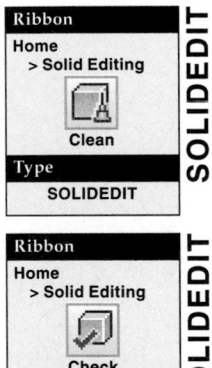

Check

The **Check** option simply determines if the selected object is a valid 3D solid. If a true 3D solid is selected, AutoCAD displays the prompt This object is a valid Shape-Manager solid. and you can exit the command. If the object selected is not a 3D solid, the prompt reads A 3D solid must be selected. and you are prompted to select a 3D solid. To access the **Check** option, select the command and pick the **Body>Check** option. Remember, the option is directly entered when picking the button in the ribbon. Then, select the object to check.

Ribbon
Home
> Solid Editing
Clean
Type
SOLIDEDIT

Ribbon
Home
> Solid Editing
Check
Type
SOLIDEDIT

SOLIDEDIT

SOLIDEDIT

Using SOLIDEDIT as a Construction Tool

This section provides an example of how the **SOLIDEDIT** command options can be used not only to edit, but also to construct a solid model. This makes it easy to design and construct a model without selecting a variety of commands. It also gives you the option of undoing a single editing operation or an entire editing session without ever exiting the command.

In the following example, **SOLIDEDIT** command options are used to imprint shapes onto the model body and then extrude those shapes into the body to create countersunk holes. Then, the model size is adjusted and an angle and taper are applied to one end. Finally, one end of the model is copied to construct a mating part.

Creating Shape Imprints on a Model

The basic shape of the solid model in this tutorial is drawn as a solid box, then shape imprints are added to it. Throughout this exercise, you may wish to change the UCS to assist in the construction of the part.

1. Draw a solid box using the dimensions shown in **Figure 13-16**.
2. Set the 3D Wireframe visual style current.
3. On the top surface of the box, locate a single Ø.4 circle using the dimensions given. Then, copy or array the circle to the other three corners as shown in the figure.
4. Use the **Imprint** option to imprint the circles onto the solid box. Delete the source objects.

Extruding Imprints to Create Features

The imprinted 2D shapes can now be extruded to create new 3D solid features on the model. Use the **Extrude Faces** option to extrude all four imprinted circles.

1. When you select the edge of the first circle, all features on that face are highlighted, but only the circle you picked and the top face have actually been selected. If you pick inside the circle, only the circle is selected and highlighted. In either case, be sure to also pick the remaining three circles.
2. Remove the top face of the box from the selection set, if needed.
3. The depth of the extrusion is .16 units. Remember to enter –.16 for the extrusion height since the holes remove material. The angle of taper for extrusion should be 35°. Your model should look like **Figure 13-17A**.

Figure 13-16.
The initial setup for
the tutorial model.

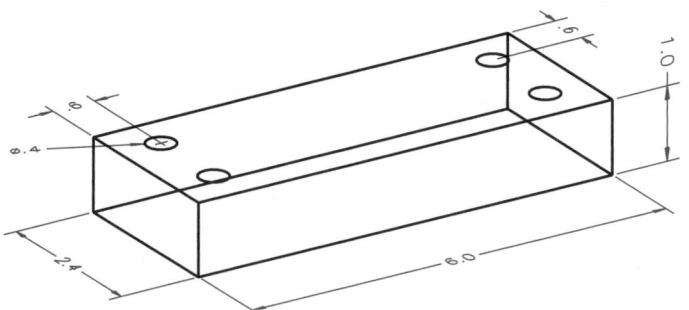

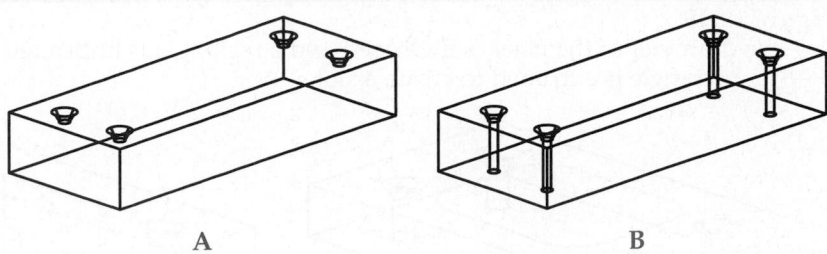

Figure 13-17.
A—The imprinted circles are extruded with a taper angle of 35°. B—Holes are created by further extrusion with a taper angle of 0°.

A

B

4. Extrude the small diameter of the four tapered holes so they intersect the bottom of the solid body. Select the holes by picking the small diameter circles. Instead of calculating the distance from the bottom of the chamfer to the bottom surface, you can simply enter a value that is greater than this distance, such as the original thickness of the object. Again, since the goal is to remove material, use a negative value for the height of the extrusion. There is no taper angle. Your model should now look like **Figure 13-17B.**

Moving Faces to Change Model Size

The next step is to use the **Move Faces** option to decrease the length and thickness of the solid body.
1. Select either end face and the two holes nearest to it. Be sure to select the holes *and* the countersinks. Move the two holes and end face two units toward the other end, thus changing the object length to four units.
2. Select the bottom face and move it .5 units up toward the top face, thus changing the thickness to .5 units. See **Figure 13-18.**

Offsetting a Feature to Change Its Size

Now, the **Offset Faces** option is used to increase the diameter of the four holes and to adjust a rectangular slot that will be added to the solid.
1. Using **Offset Faces**, select the four small hole diameters. Be sure to remove from the selection set any other faces that may be selected.
2. Enter an offset distance of –.05. This increases the hole diameter and decreases the solid volume. Exit the **SOLIDEDIT** command.
3. Select the **RECTANG** command. Set the fillet radius to .4 and draw a 2 × 1.6 rectangle centered on the top face of the solid. See **Figure 13-19A.**
4. Imprint the rectangle on the solid. Delete the source object.
5. Extrude the rectangle completely through the solid (.5 units). Remember to remove from the selection set any other faces that may be selected.
6. Offset the rectangular feature using an offset distance of .2 units. You will need to select all faces of the feature. This decreases the size of the rectangular opening and increases the solid volume. Your drawing should appear as shown in **Figure 13-19B.**

Figure 13-18.
The length of the object is shortened and the height is reduced.

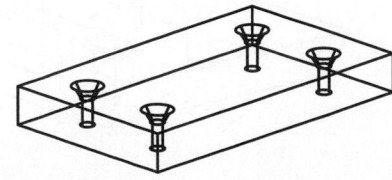

Figure 13-19.
A—The diameter of the holes is increased and a rectangle is imprinted on the top surface.
B—The rectangle is extruded to create a slot.

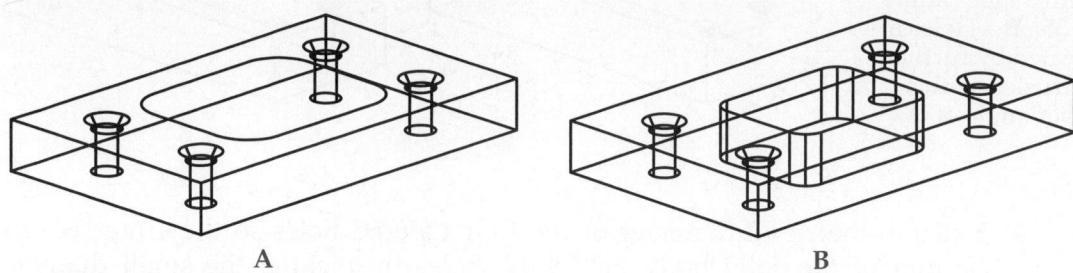

A   B

Tapering Faces

One side of the part is to be angled. The **Taper Faces** option is used to taper the left end of the solid.

1. Using **Taper Faces**, pick the face at the left end of the solid.
2. Pick point 1 in **Figure 13-20** as the base point and point 2 as the second point along the axis of tapering.
3. Enter a value of –10 for the taper angle. This moves the upper-left end away from the solid, creating a tapered end.

Rotating Faces

Next, use the **Rotate Faces** option to rotate the tapered end of the object. The top edge of the face will be rotated away from the holes, adding volume to the solid.

1. Using **Rotate Faces**, pick the face at the left end of the solid.
2. Pick point 1 in **Figure 13-20** as the first axis point and point 2 as the second point.
3. Enter a value of –30 for the rotation angle. This rotates the top edge of the tapered end away from the solid. See **Figure 13-21**.

Figure 13-20.
The left end of the object is tapered. Notice the pick points. These points are also used when selecting an axis of rotation.

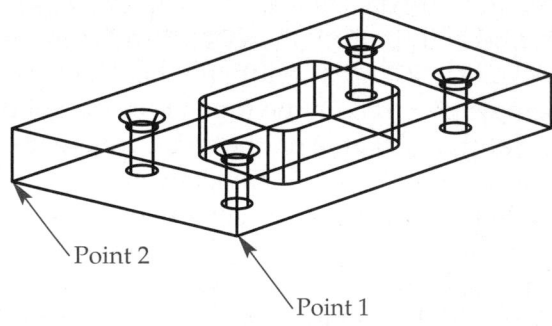

Point 2

Point 1

Figure 13-21.
The tapered end of the object is modified by rotating the face.

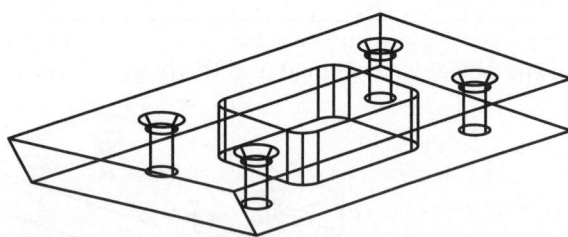

Figure 13-22.
Creating a mating part. A—The angled face is copied. B—The copied face is extruded into a solid.

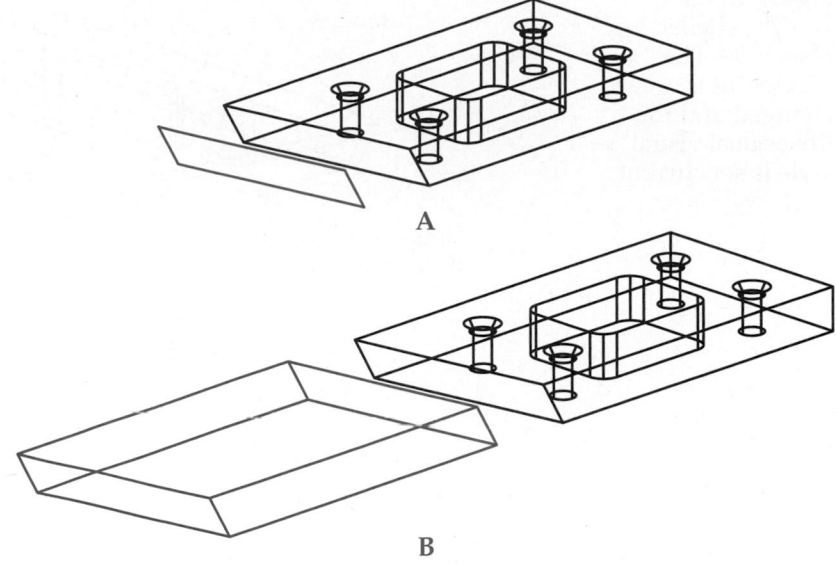

A

B

Copying Faces

A mating part will now be created. This is done by first copying the face on the tapered end of the part.

1. Using the **Copy Faces** option, pick the angled face on the left end of the solid.
2. Pick one of the corners as a base point and copy the face one unit to the left. This face can now be used to create a new solid. See **Figure 13-22A.**
3. Draw a line four units in length on the negative X axis from the lower-right corner of the copied face. Use the **EXTRUDE** command on the copied face to create a new solid. Select the **Path** option and use the line as the extrusion path. See **Figure 13-22B.** If you do not use the **Path** option, the extrusion is projected perpendicular to the face.

NOTE

The **Extrude Faces** option of the **SOLIDEDIT** command cannot be used to turn a copied face into a solid body.

Creating a Shell

The bottom surface of the original solid will now be shelled out. Keep in mind that features such as the four holes and the rectangular slot will not be cut off by the shell. Instead, a shell will be placed around these features. This becomes clear when the operation is performed.

1. Select the **Shell** option and pick the original solid.
2. Remove the lower-left and lower-right edges of the solid. See **Figure 13-23A.** This removes the two side faces and the bottom face.
3. Enter a shell offset distance of .15 units. The shell is created and should appear similar to **Figure 13-23A.**
4. Use the view cube or **3DORBIT** command to view the solid from the bottom. Also, set the Conceptual visual style current. Your model should look like the one shown in **Figure 13-23B.**

Figure 13-23.
A—The shelled
object. B—The
viewpoint is
changed and the
Conceptual visual
style is set current.

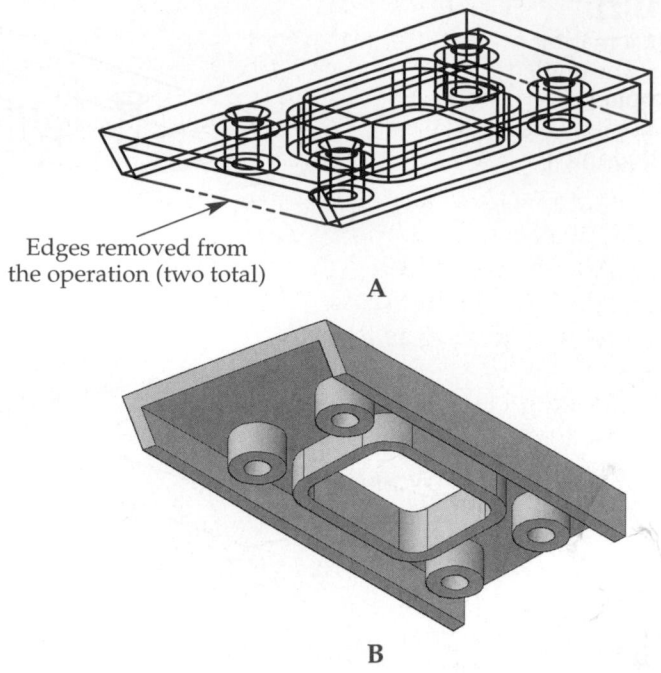

Edges removed from
the operation (two total)

A

B

Chapter Test

Answer the following questions. Write your answers on a separate sheet of paper or complete the electronic chapter test on the student website.
www.g-wlearning.com/CAD

1. What are the three components of a solid model?
2. When using the **SOLIDEDIT** command, how many faces are highlighted if you pick an edge?
3. How do you deselect a face that is part of the selection set?
4. How can you select a single face?
5. Which two operations can the **Extrude Faces** option perform?
6. How does the shape and length of an object selected as the path of an extrusion affect the final extrusion?
7. What is one of the most useful aspects of the **Offset Faces** option?
8. How do positive and negative offset distance values affect the volume of the solid?
9. How is a single object, such as a cylinder, affected by entering a positive taper angle when using the **Taper Faces** option?
10. When a shape is imprinted onto a solid body, which component of the solid does the imprinted object become and how can it be used?
11. In which situation would you use the **Separate** option?
12. How does the **Shell** option affect a solid that contains internal features such as holes, grooves, and slots?
13. How can you determine if an object is a valid 3D solid?
14. Describe two ways to change the view of your model while you are inside of a command.
15. How can you extrude a face in a straight line, but not perpendicular to the face?

Drawing Problems

1. Complete the tutorial presented in this chapter. Then, perform the following additional edits to the original solid.
   A. Lengthen the right end of the solid by .5 units.
   B. Taper the right end of the solid with the same taper angle used on the left end, but taper it in the opposite direction.
   C. Fillet the two long, top edges of the solid using a fillet radius of .2 units.
   D. Rotate the face at the right end of the solid with the same rotation angle used on the left end, but rotate it in the opposite direction.
   E. Save the drawing as P13_01.

2. Construct the solid part shown below using as many **SOLIDEDIT** options as possible. After completing the object, make the following modifications.
   A. Lengthen the 1.250" diameter feature by .250".
   B. Change the .750" diameter hole to .625" diameter.
   C. Change the thickness of the .250" thick flange to .375" (toward the bottom).
   D. Extrude the end of the 1.250" diameter feature .250" with a 15° taper inward.
   E. Save the drawing as P13_02.

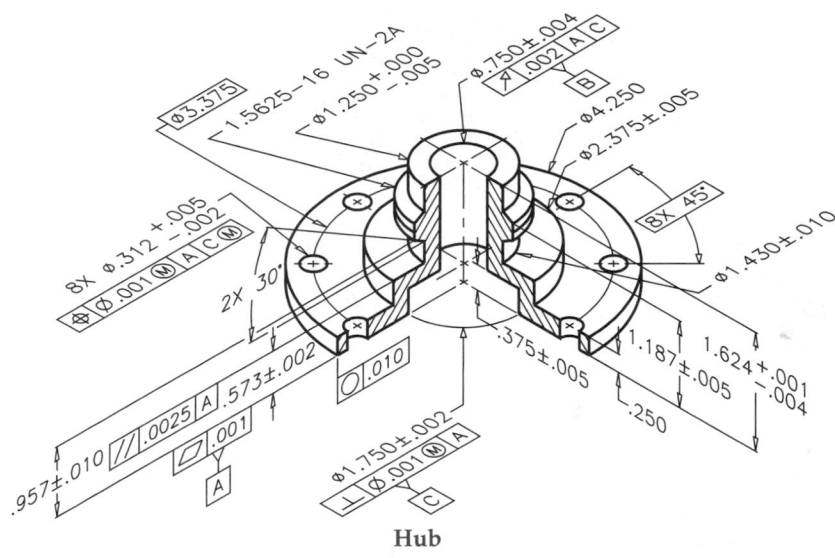

Hub

3. Construct the solid part shown below. Then, perform the following edits on the solid using the **SOLIDEDIT** command.
   A. Change the diameter of the hole to 35.6/35.4.
   B. Add a 5° taper to each inner side of each tooth (the bottom of each tooth should be wider while the top remains the same).
   C. Change the width of the 4.8/4.0 key to 5.8/5.0.
   D. Save the drawing as P13_03.

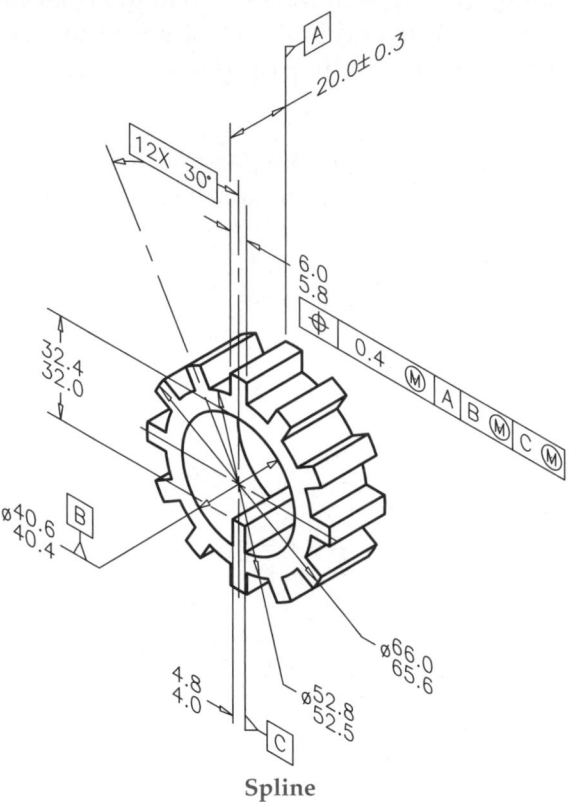

Spline

4. Construct the solid part shown below using as many **SOLIDEDIT** options as possible. Then, perform the following edits on the solid.
   A. Change the depth of the counterbore to 10 mm.
   B. Change the color of all internal surfaces to red.
   C. Save the drawing as P13_04.

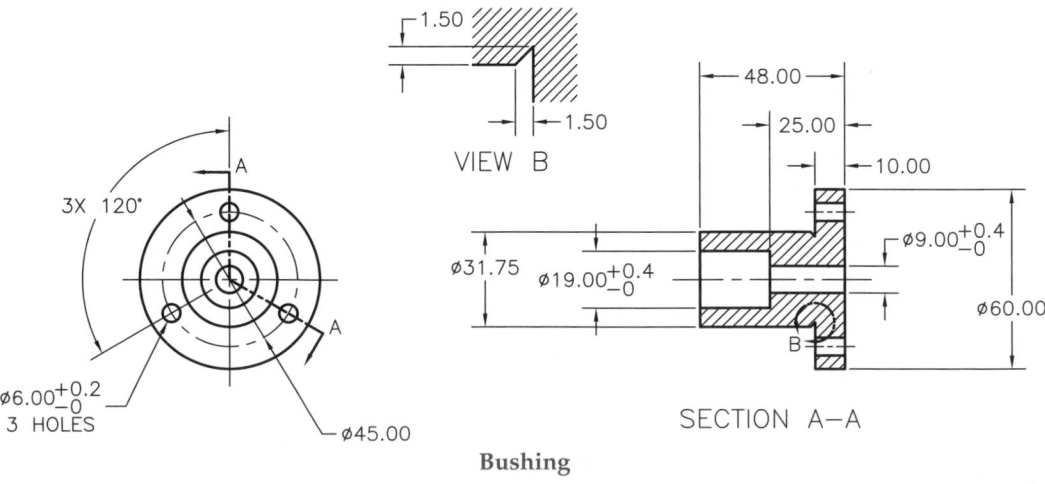

Bushing

5. Construct the solid part shown below using as many **SOLIDEDIT** options as possible. Then, perform the following edits on the solid.
   A. Change the 2.625″ height to 2.325″.
   B. Change the 1.625″ internal diameter to 1.425″.
   C. Taper the outside faces of the .875″ high base at a 5° angle away from the part. Hint: The base cannot be directly tapered.
   D. Save the drawing as P13_05.

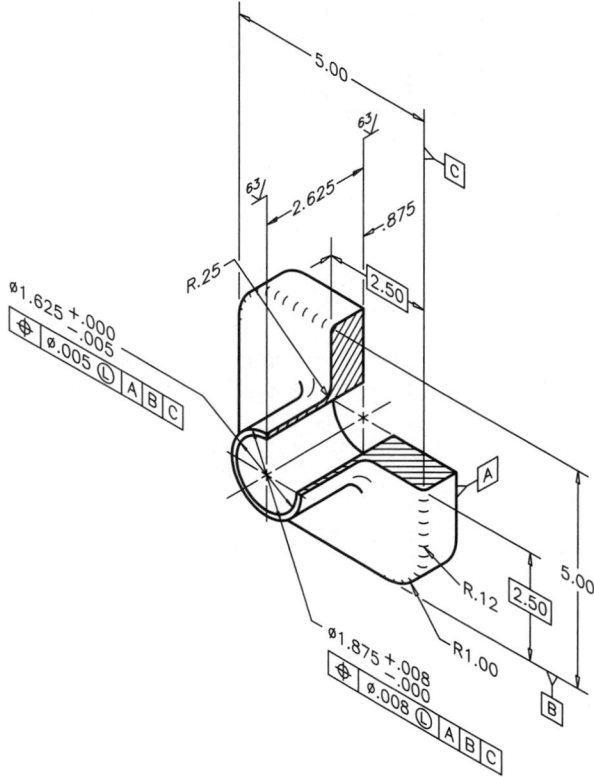

Thrust Washer

6. Construct the solid part shown below using as many **SOLIDEDIT** options as possible. Then, change the dimensions on the model as follows. Save the drawing as P13_06.

Existing	New
100	106
80	82
Ø60	Ø94
Ø40	Ø42
30°	35°

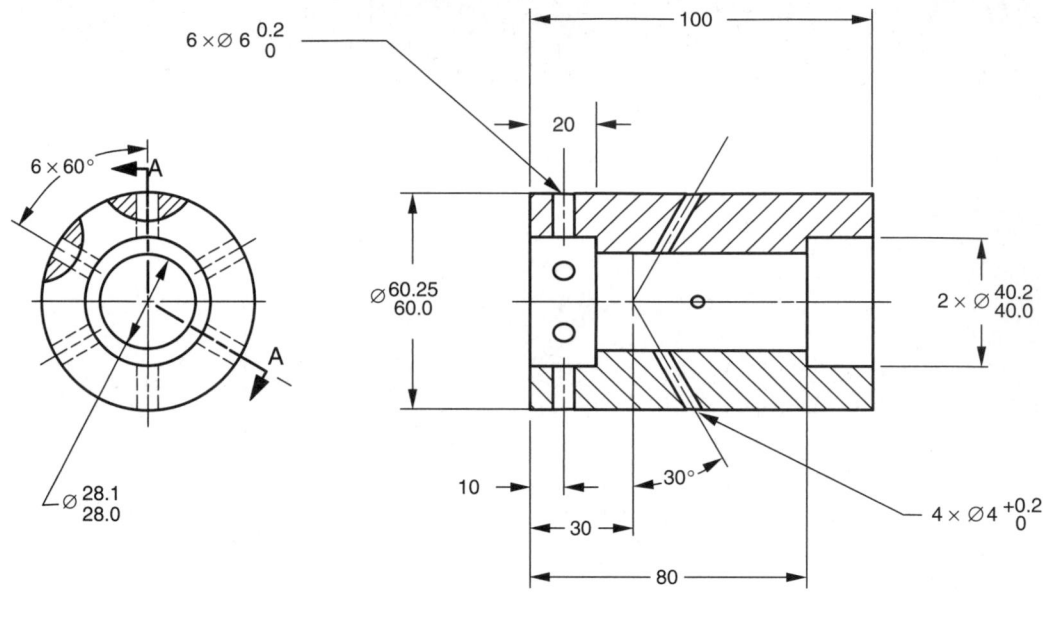

SECTION A-A

Nozzle

CHAPTER 14

Solid Model Display and Analysis

Learning Objectives

After completing this chapter, you will be able to:

✓ Control the display of solid models.
✓ Construct a 3D section plane through a solid model.
✓ Adjust the size and location of section planes.
✓ Create a dynamic section of a 3D solid model.
✓ Construct 2D and 3D section blocks.
✓ Create a flat, 2D projection of a 3D solid model.
✓ Create a multiview layout of a solid model using **SOLVIEW** and **SOLDRAW**.
✓ Construct a profile of a solid using **SOLPROF**.
✓ Perform an analysis of a solid model.
✓ Export and import solid model data.

Certain aspects of a solid model's appearance are controlled by the **ISOLINES**, **DISPSILH**, and **FACETRES** system variables. The **ISOLINES** system variable, introduced in Chapter 4, controls the number of lines used to define solids in wireframe displays. The **FACETRES** system variable controls the number of lines used to define solids in hidden and shaded displays. The **DISPSILH** system variable is used to display a silhouette.

Internal features of the model can be shown using the **SECTIONPLANE** command. This command can create 2D and 3D section views on an object. The **FLATSHOT** command creates a 2D projection of the current view. This chapter also looks at how sections can be combined with 2D projections created with the **SOLVIEW** and **SOLDRAW** commands to create a drawing layout for plotting. In addition, this chapter covers how a profile of a solid can be created using the **SOLPROF** command.

Controlling Solid Model Display

AutoCAD solid models can be displayed as wireframes, with hidden lines removed, shaded, or rendered. A wireframe is the default display when a drawing is started based on the acad.dwt template and is the quickest to display. The hidden, shaded, and rendered displays require a longer regeneration time. When a drawing is started based on the acad3D.dwt or acadiso3D.dwt template, the default display is the Realistic visual style, which is a shaded display.

Isolines

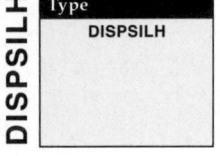
The appearance of a solid model in a wireframe display is controlled by the **ISOLINES** system variable. *Isolines* represent the edges and curved surfaces of a solid model. This setting does *not* affect the final shaded or rendered object. However, if the Edge mode property for the visual style is set to Isolines in the **Visual Styles Manager** palette, isolines are displayed when the visual style is set current. The default **ISOLINES** value is four. It can have a value from zero to 2047. All solid objects in the drawing are affected by changes to the **ISOLINES** value, as are all visual styles with their Edge mode property set to Isolines. **Figure 14-1** illustrates the difference between **ISOLINES** settings of four and 12.

The setting of the **ISOLINES** system variable can be changed in the **Visual Styles Manager** palette. For the 2D Wireframe visual style, change the Contour lines property in the **2D Wireframe Options** category. For all other visual styles, change the Number of lines property in the **Edge Settings** category. See **Figure 14-2A**. The Edge mode property must be set to Isolines to display the Number of lines property. The **ISOLINES** setting can also be changed in the **Contour lines per surface** text box found in the **Display resolution** area of the **Display** tab in the **Options** dialog box, or by typing ISOLINES and then entering a new value. See **Figure 14-2B**.

Creating a Display Silhouette

When the Edge mode property of a 3D visual style is set to Facet Edges, objects are defined by *tessellation lines.* The **Facet Lighting** and **Smooth Lighting** buttons in the **Visual Styles** panel on the **Render** tab of the ribbon also control the Edge mode setting. When the **HIDE** command is used with the 2D Wireframe visual style current, tessellation lines are also displayed. The number of tessellation lines is controlled by the **FACETRES** system variable, which is discussed in the next section.

In the 2D Wireframe visual style, a model can also appear smooth with only a silhouette displayed, similar to the 3D Hidden visual style. This is controlled by the **DISPSILH** (display silhouette) system variable. The **DISPSILH** system variable has two values, 0 (off) and 1 (on). **Figure 14-3** shows solids with **DISPSILH** set to 0 and 1 after setting the 2D Wireframe visual style current and then using **HIDE**.

Figure 14-1.
Isolines define
curved surfaces.
A—**ISOLINES** = 4.
B—**ISOLINES** = 12.

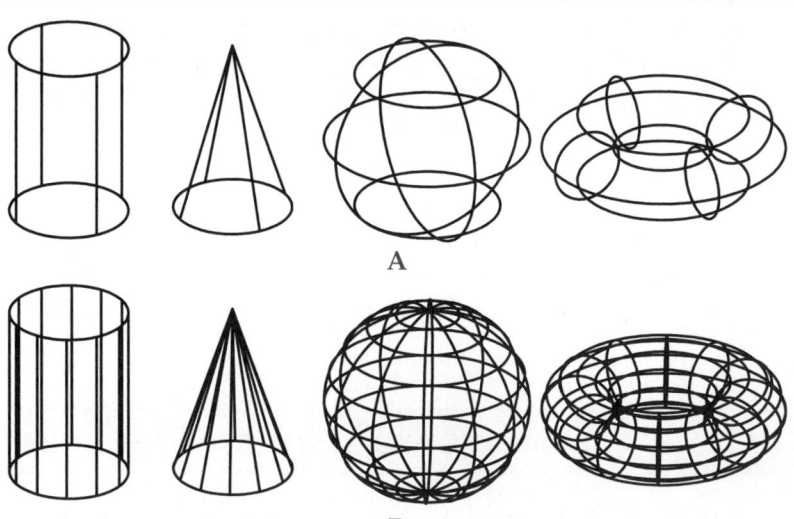

A

B

Figure 14-2.
A—The **ISOLINES** and **DISPSILH** values can be set in the **Visual Styles Manager**. B—The **ISOLINES**, **FACETRES**, and **DISPSILH** values can be set in the **Options** dialog box.

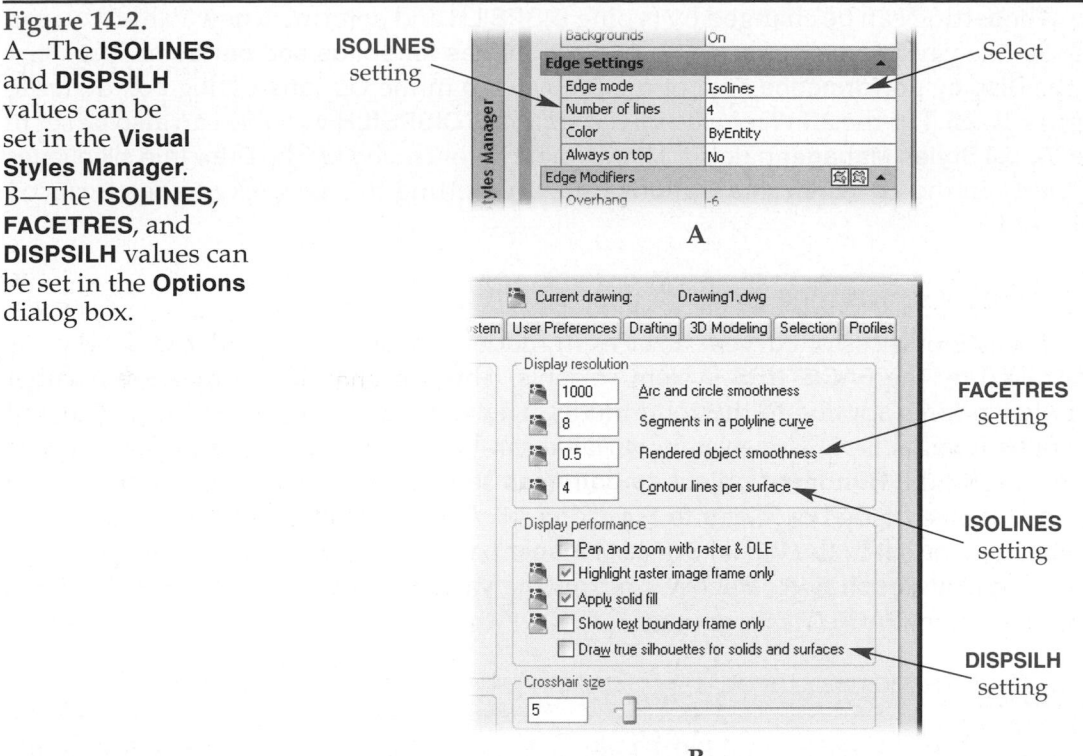

Figure 14-3.
A—The **HIDE** command used when **DISPSILH** is set to 0. Objects are displayed faceted. B—The **HIDE** command used when **DISPSILH** is set to 1. Facets are eliminated and only the silhouette is displayed.

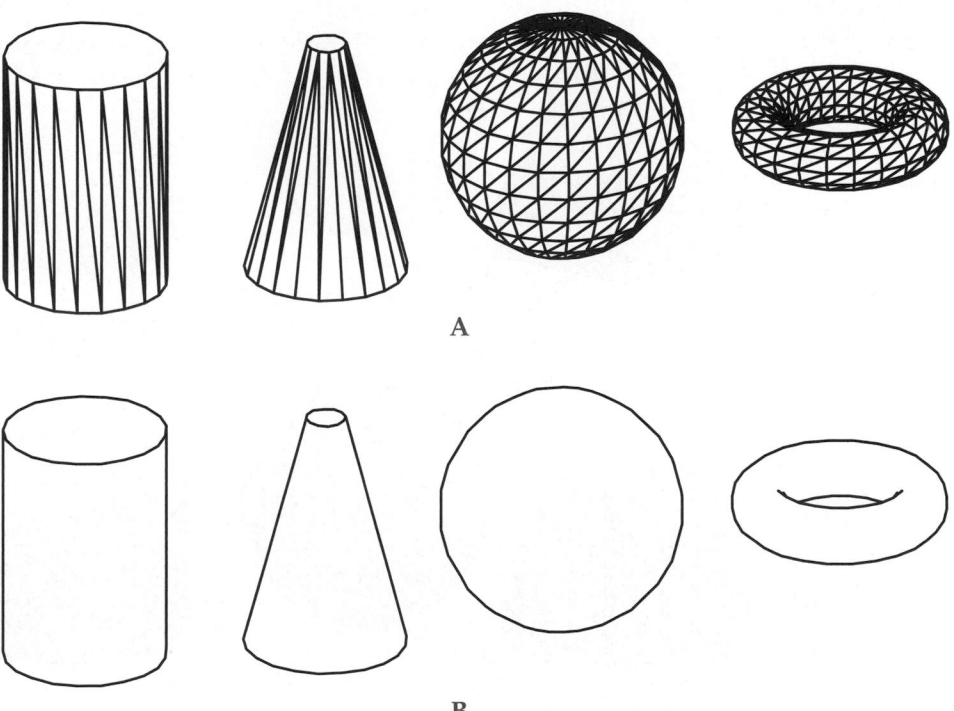

The setting can be changed by typing DISPSILH and entering a new value. You can also set the variable using the **Draw true silhouettes for solids and surfaces** check box in the **Display performance** area of the **Display** tab in the **Options** dialog box. Refer to **Figure 14-2B.** For the 2D Wireframe visual style, the **DISPSILH** variable can also be set in the **Visual Styles Manager** palette. The variable is controlled by the Draw true silhouettes property in the **2D Wireframe Options** category. Setting this property to Yes turns on silhouettes.

Controlling Surface Smoothness

The smoothness of curved surfaces in hidden, shaded, and rendered displays is controlled by the **FACETRES** system variable. This variable determines the number of polygon faces applied to the solid model. The value can range from .01 to 10.0 and the default value is .5. This system variable can be changed by typing FACETRES or by changing the **Rendered object smoothness** setting in the **Display resolution** area in the **Options** dialog box. Refer to **Figure 14-2B.** For the 2D Wireframe visual style, the variable can be set in the **Visual Styles Manager** palette. The Solid Smoothness property in the **Display Resolution** category controls the variable. **Figure 14-4** shows the effect of two different **FACETRES** settings.

> **CAUTION**
>
> Avoid setting **FACETRES** any higher than necessary. The additional edges and faces added by a higher setting will slow down system performance. Always use the lowest setting that will produce the results required by the project.

Figure 14-4.
A—The **FACETRES** setting is .5 and the Realistic visual style set current. B—The **FACETRES** setting is 5.0 and the Realistic visual style is set current.

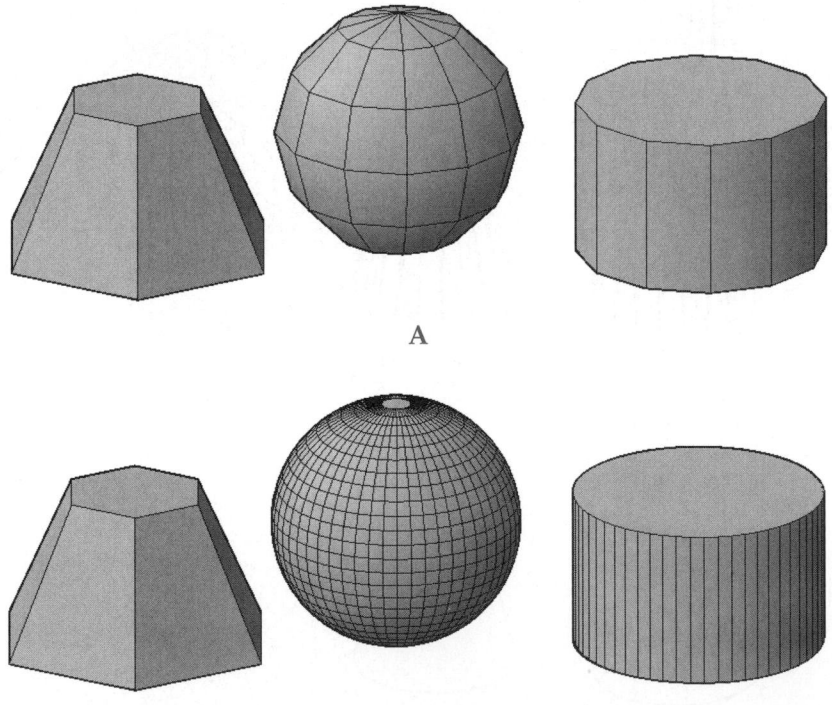

A

B

AutoCAD and Its Applications—Advanced

Exercise 14-1

Complete the exercise on the student website.
www.g-wlearning.com/CAD

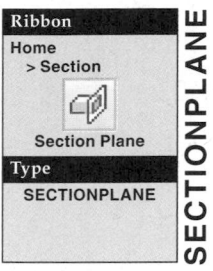
Creating Section Planes

The **SECTIONPLANE** command offers a powerful visualization and display tool. It enables you to construct a section plane, known as an AutoCAD *section object*, that can then be used as a plane to cut through a 3D model. Once the section object is drawn, it can be moved to any location, jogs can be added to it, and it can be rendered "live" so that internal features and sectioned material are dynamically visible as the cutting plane is moved. A variety of section settings allow you to customize the appearance of section features. Additionally, you can generate 2D sections/elevations or 3D sections that can be inserted into the drawing as a block. Once the command is initiated, you are prompted to select a face, the first point on the section object, or to enter an option. Sectioning options are located in the **Section** panel of the **Home** tab on the ribbon.

Pick a Face to Construct a Section Plane

The simplest way to create a section plane is to pick a flat face on the 3D object. Once the command is initiated, move the pointer until the face you wish to select is highlighted, then pick it. A transparent section object is placed on the face you selected and the model is cut at the plane. See **Figure 14-5.** The section plane can now be moved to create a section anywhere along the 3D model.

Pick Two Points to Construct a Section Plane

A second method for defining a section plane is to pick two points through which the section object passes. The section object is perpendicular to the XY plane of the current UCS. When the command is initiated, pick the first point, which cannot be on a face. See P1 in **Figure 14-6.** It may be best to turn off dynamic UCSs or you could end up picking a face as the first point instead of a point. After picking the first point, move the pointer and notice that the section plane rotates about the first point. Next, pick the second point (P2) to define a line that cuts through the model. After the second point is picked, the section object is created. The section plane extends just beyond the edges of the model.

Figure 14-5.
Creating a section object on a face. A—The object before the face is selected. B—The section object is created.

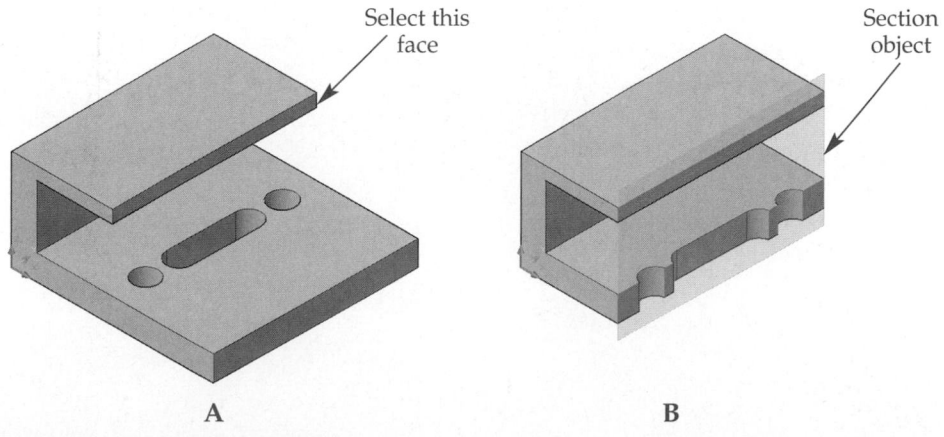

Select this face

Section object

A   B

Figure 14-6.
Creating a section
object by selecting
two points.

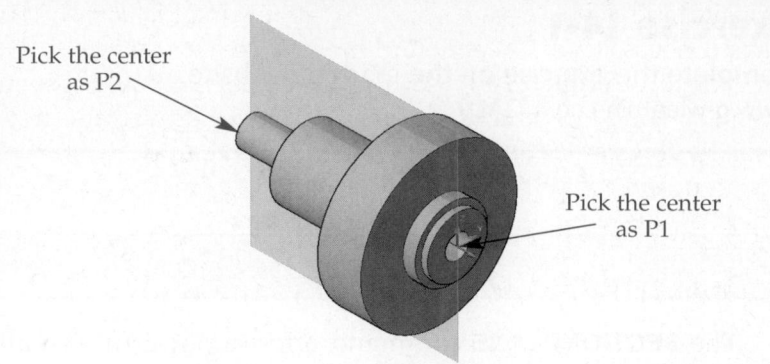

Pick the center
as P2

Pick the center
as P1

NOTE

When the section object is created by picking two points, notice that the model is not automatically cut as it is when a face is selected. This is because *live sectioning* is not turned on when picking two points, but is turned on when picking a face. To turn live sectioning on or off, select the section object, right-click, and select **Activate live sectioning** from the shortcut menu. Live sectioning is discussed later in this chapter.

Pick Multiple Points to Construct a Section Plane

The previous method accepts only two points to construct a single section plane. Using the **Draw section** option, you can specify multiple points in order to create section plane *jogs*. In engineering drawing terminology, a section object drawn in this manner can represent an *offset* or *aligned* section plane.

Once the command is initiated, select the **Draw section** option. Pick the start point, using object snaps if necessary. See **Figure 14-7**. Continue picking points as needed. After picking the last point to define the section plane, press [Enter]. You are then prompted to specify a point in the direction of the section view. This point is on the

Figure 14-7.
Using the **Draw** option of the **SECTIONPLANE** command to create a section object with multiple segments.

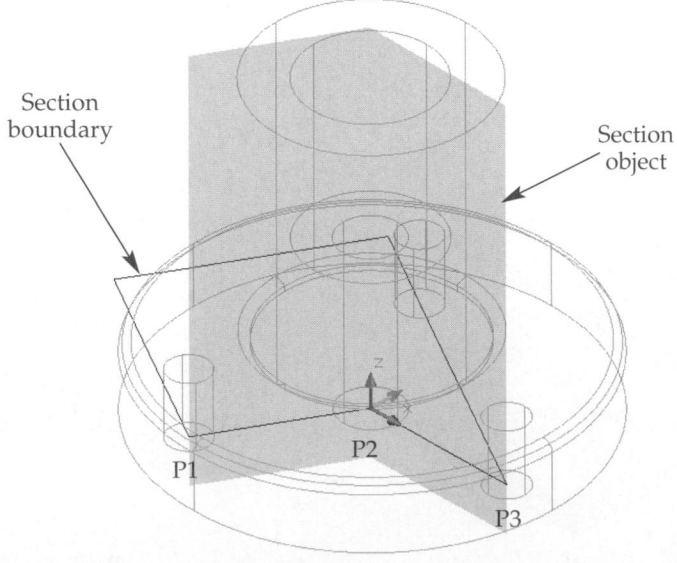

Section boundary

Section object

P1

P2

P3

AutoCAD and Its Applications—Advanced

opposite side of the section object as the viewer. Pick a point on the model using object snaps if necessary. The section plane is created.

Notice in **Figure 14-7** that the pick points created a section object that does not extend beyond the boundary of the model because the hole centers were selected. Also, the command "squares up" the section boundary to create a closed profile. Using section object grips, the section plane can be easily edited to include the entire solid. This is discussed in detail later in the chapter.

Create Orthographic Section Planes

The **Orthographic** option enables you to quickly place a section plane through the front, back, top, bottom, left, or right side of the object. See **Figure 14-8.** The origin is the center point of all objects in the model. Once the command is initiated, select the **Orthographic** option. Then, specify which orthographic plane you want to use as the section plane. The section object is created and all objects in the drawing are affected by it.

You may encounter a situation in which there is more than one solid on the screen and you want to use the **Orthographic** option to create a section object based on just one object. In this case, create a new layer, move objects you do not want to section to this layer, and then freeze the layer. The section object will be created based on the object that is visible. However, if the section plane passes through the objects on the frozen layers, those objects will be sectioned when the layers are thawed. A section plane affects all visible objects.

Figure 14-8.
Examples of orthographic **SECTIONPLANE** options. A—Top. B—Bottom. C—Left. D—Right.

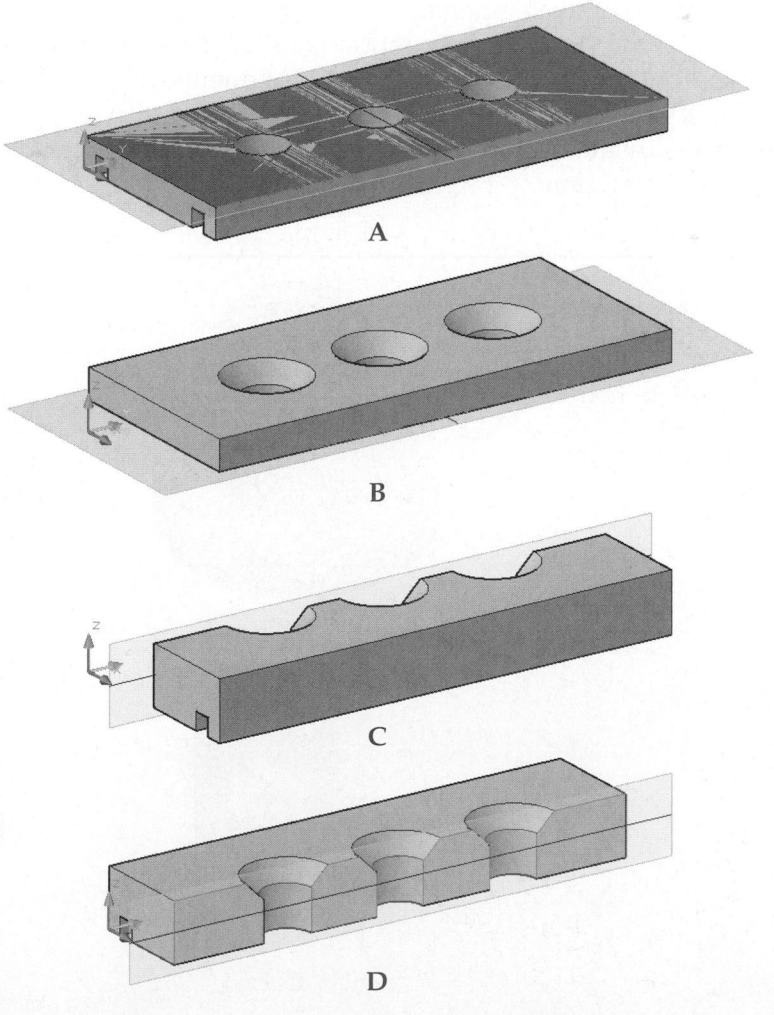

A

B

C

D

Exercise 14-2

Complete the exercise on the student website.
www.g-wlearning.com/CAD

Editing and Using Section Planes

A wide range of section object editing and display options are available. However, there is ribbon or command line access to these procedures. Instead, you must first select the section object, then right-click to display the shortcut menu. From this menu, you can access all of the display and editing functions that apply to the section object.

Section Object States

There are three possible states for the section object created by the **SECTIONPLANE** command—section plane, section boundary, and section volume. See **Figure 14-9.** The section object can be changed from one state to another. Depending on which state is active, the section object will produce different results on the solid(s).

When the section object is created by picking a face, picking two points, or using the **Orthographic** option, the object is in the *section plane state.* A transparent plane is displayed on each segment of the section object and a line connects the pick points (or the edges of the section object). See **Figure 14-9B.** The section plane extends infinitely in the section object's Z direction and along the direction of the object segment (unless connected to other segments).

When the **Draw section** option is used, the *section boundary state* is applied. A transparent plane is displayed on each segment of the section object. A 2D box extends to the XY-plane boundaries of the section object. See **Figure 14-9C.** The sectioned object fits inside of this footprint. The section plane extends infinitely in the section object's Z direction.

Figure 14-9.
Section object states.
A—The original object. B—Section plane. C—Section boundary.
D—Section volume.

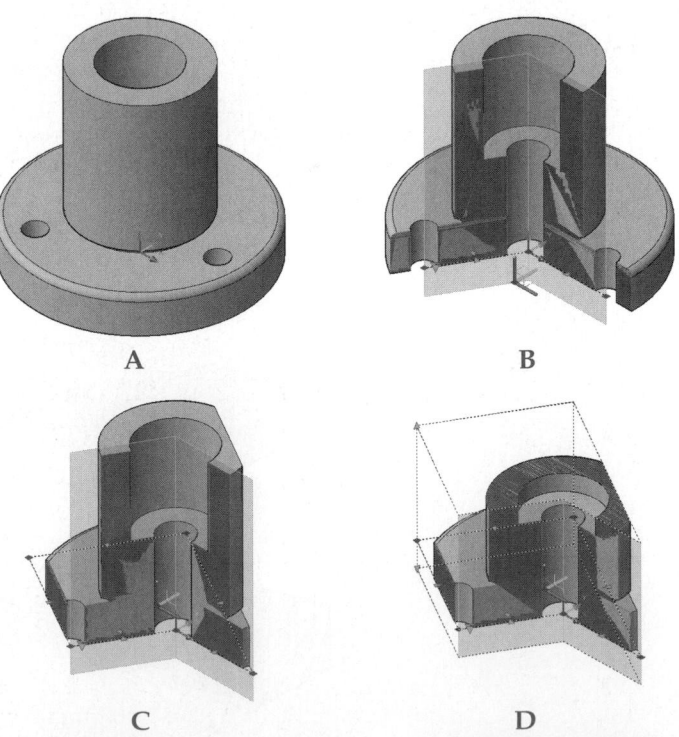

A

B

C

D

AutoCAD and Its Applications—Advanced

The *section volume state* is not applied when the section object is created. The section object must be switched to this state once the object is created, as described in the next section. A transparent plane is displayed on each segment of the section object. In addition, a 3D box extends to the XYZ boundaries of the section object. The sectioned object fits inside of this box. See **Figure 14-9D.**

Section Object Properties

Once created, the properties of the section object can be changed. The **Section Object** category in the **Properties** palette contains properties specific to the section object. See **Figure 14-10.** These properties are described in the next sections.

Name

The default name of the first section object is Section Plane(1). Subsequent section planes are sequentially numbered, such as Section Plane(2), Section Plane(3), and so on. It may be beneficial to rename section objects so the name is representative of the section. For example, Front Half Section is much more descriptive than Section Plane(1). To rename a section object, select the Name property. Then, type a new name in the text box.

Type

As discussed earlier, the section object is in one of three states. The three states are section plane, section boundary, and section volume. To change the state of the section object, select the Type property. Then, pick the state in the drop-down list. The state can also be changed using the menu grip on the section object.

Live section

Live sectioning is a tool that enables you to dynamically view the internal features of a solid, surface, or region as the section object is moved. This tool is discussed later in the chapter. To turn live sectioning on or off, select the Live Section property. Then, pick either Yes (on) or No (off) in the drop-down list. This is the same as turning live sectioning on or off using the shortcut menu or the ribbon.

Plane transparency

The Plane Transparency property determines the opacity of the plane for the section object. The property value can range from 1 to 100. The lower the value, the more opaque the section plane object. See **Figure 14-11.**

Figure 14-10.
The properties of a section object can be changed in the **Properties** palette.

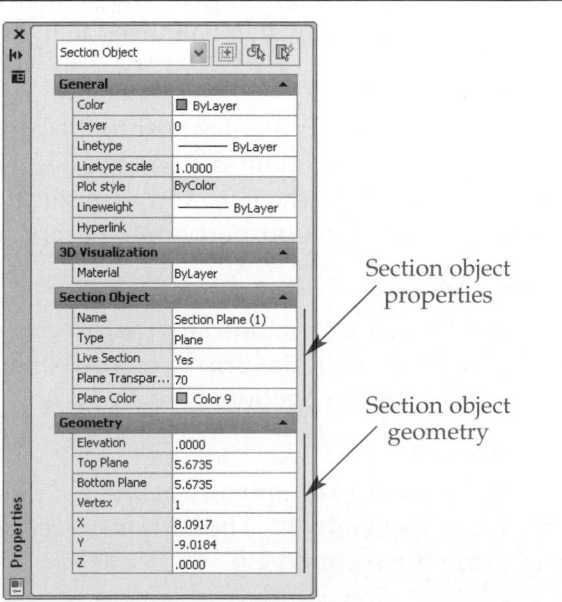

Section object properties

Section object geometry

Figure 14-11.
A—The Plane
Transparency
property of the
section plane object
is set to 1 (or 1%
transparent).
B—The Plane
Transparency
property of the
section plane object
is set to 85 (or 85%
transparent).

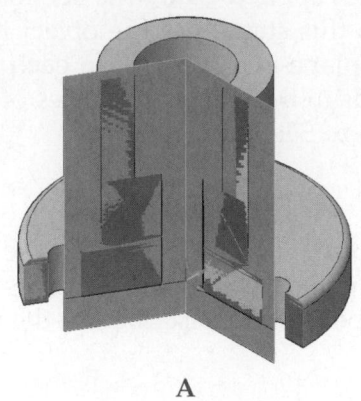
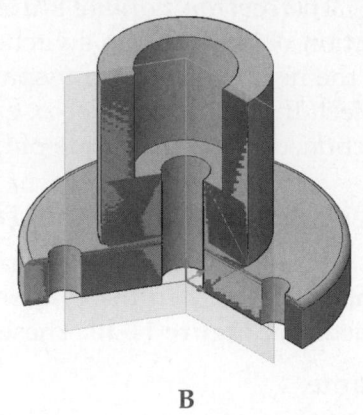

A   B

Plane color

The plane of the section object can be set to any color available in the **Select Color** dialog box. To change the color, select the Plane Color property and then select a color from the drop-down list. To choose a color in the **Select Color** dialog box, pick the Select Color... entry in the drop-down list. This property only affects the plane of the section object, not the lines defining the boundary, volume, or section line. The color of these lines is controlled by the Color property in the **General** category.

Editing the Section Object

When the translucent planes of a section object or the lines representing the section object state are picked, grips are displayed. The specific grips displayed are related to the current section object state. Refer to the grips shown in **Figure 14-9.** The types of grips are:
- Base grip.
- Menu grip.
- Direction grip.
- Second grip.
- Arrow grips.
- Segment end grips.

Base grip

The *base grip* appears at the first point picked when defining the section object. See **Figure 14-12.** It is the grip about which the section object can be rotated and scaled. The section object can also be moved using this grip.

Menu grip

The *menu grip* is always next to the base grip. Refer to **Figure 14-12.** Picking this grip displays the section state menu. See **Figure 14-13.** To switch the section object between states, pick the grip and then select the state from the menu. This is the same as changing the Type property in the **Properties** palette.

Direction grip

The *direction grip* indicates the direction in which the section will be viewed. Refer to **Figure 14-12.** Pick the grip to rotate the view 180°. The direction grip also shows the direction of the live section. Live sectioning is discussed later in this chapter.

Second grip

The *second grip* appears at the second point picked when defining the section object. See **Figure 14-12.** The section object can be rotated and stretched about the base grip using the second grip.

Figure 14-12
The types of grips displayed on a section object.

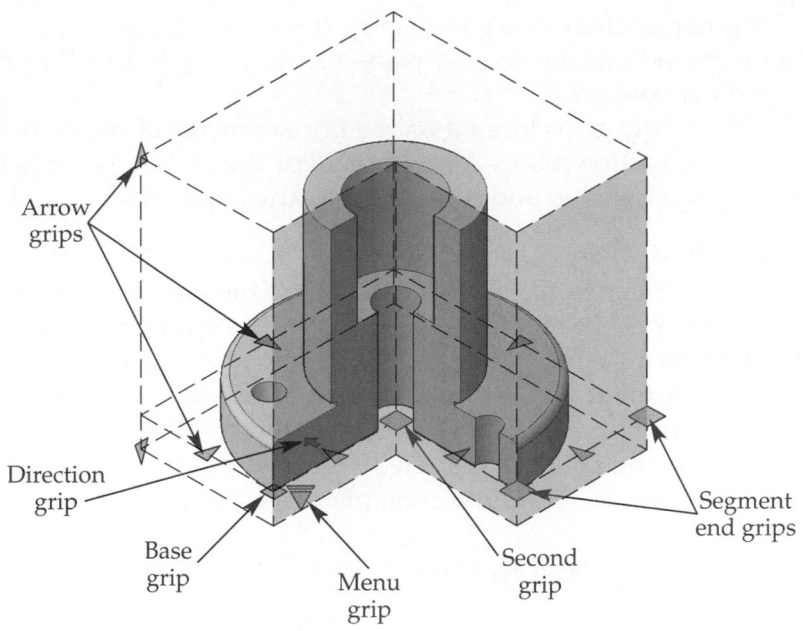

Arrow grips

Direction grip

Base grip

Menu grip

Second grip

Segment end grips

Figure 14-13.
Changing section object states.

Section Plane
✓ Section Boundary
Section Volume

Arrow grips

Arrow grips are located on all of the lines that represent the section plane, boundary, and volume. Refer to **Figure 14-12.** These grips are used to lengthen or shorten the section plane object segments or adjust the height of the section volume. The arrow grips at the top and bottom of the boundary box are used to change the height. Regardless of where the pointer is moved, the section object only extends in the segment's current plane. Changing the length of one segment of the section plane does not affect other segments.

In **Figure 14-7,** you saw an example of using the **Draw section** option to create a section object. The way in which the section object was created resulted in the section plane not extending beyond the solid object. This can quickly be corrected using the arrow grips. Notice in **Figure 14-14** that the arrow grip is being used to extend the

---

Figure 14-14.
A—The arrow grip is being used to extend the left side of the section plane past the boundary of the solid model. B—The edited section object. The right side can be corrected in the same manner.

Drag the grip beyond the edge of the solid object

Section extends beyond the solid object

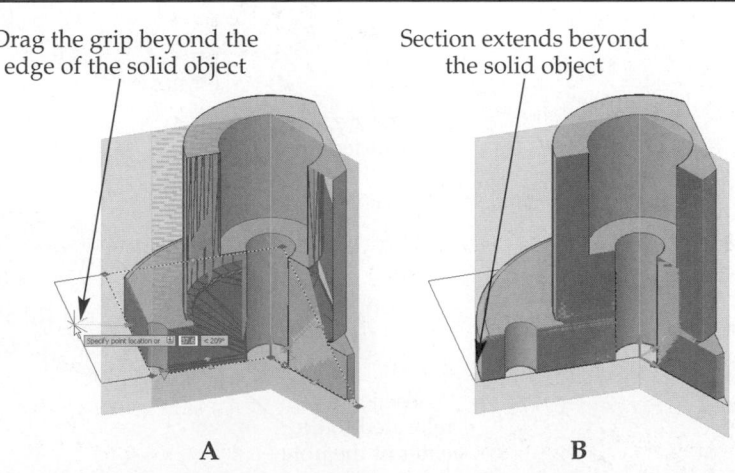

A

B

left side of the section plane past the boundary of the solid model. This allows any subsequent section views to display the entire object rather than just a portion of it. The right side of the section plane can be extended in the same manner using the opposite arrow grip.

The arrow grips located on the line segments of the section plane move the position of the section plane. As a segment of the section plane is moved, it maintains its angular relationship and connection to any adjacent section plane segment.

Segment end grips

The *segment end grips* are located at the end of each line segment defining the section object state. Refer to **Figure 14-12.** The number of displayed segment end grips depends on whether the section object is in the section plane, section boundary, or section volume state. These grips provide access to the standard grip editing options of stretch, move, copy, rotate, scale, and mirror. If the rotate option is used, the section plane is rotated about the selected segment end grip. Moving a segment end grip can change the angle between section plane segments.

Adding Jogs to a Section

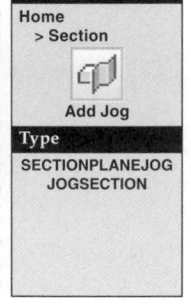

SECTIONPLANEJOG

Ribbon
Home
> Section

Add Jog

Type
SECTIONPLANEJOG
JOGSECTION

You can quickly add a jog, or offset, to an existing section object. First, select the section object. Then, right-click to display the shortcut menu and select **Add jog to section**. You can also enter the **SECTIONPLANEJOG** command. You are then prompted:

Specify a point on the section line to add jog:

Select a point directly on the section line. If any object snap is active, the **Nearest** object snap is temporarily turned on to ensure you pick the line. Once you pick, the jog is automatically added perpendicular to the line segment. See **Figure 14-15A.**

Figure 14-15.
Adding a jog to a section object. A—Pick a point on the section line to add a jog. B—The jog is added, but it is not in the proper location. C—Using the arrow grip, the jog is moved to the proper location.

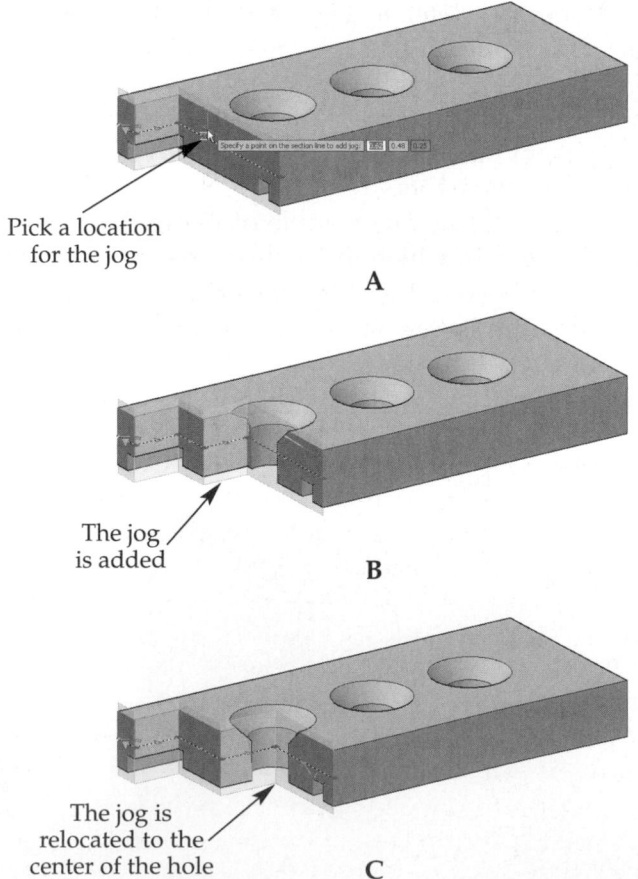

Pick a location for the jog

A

The jog is added

B

The jog is relocated to the center of the hole

C

AutoCAD and Its Applications—Advanced

It is not critical that you pick the exact location on the line where you want the jog to occur. Remember, grips allow you to easily adjust the section plane location. Notice in **Figure 14-15B** that the second jog barely cuts through the first hole. The intention is to run the section plane through the middle of the hole. To fix this, drag the arrow grip so the section plane segment is in the desired location, **Figure 14-15C.**

PROFESSIONAL TIP

If the section plane is not properly located, you can quickly change it. Simply pick the section object and right-click to display the shortcut menu. Then, select **Move**, **Scale**, or **Rotate** from the shortcut menu. Finally, adjust the section object location as needed.

Exercise 14-3

Complete the exercise on the student website.
www.g-wlearning.com/CAD

Live Sectioning

Live sectioning is a tool that enables you to view the internal features of 3D solids, surfaces, and regions that are cut by the section plane of the section object. The view is dynamically updated as the section object is moved. This tool is used to visualize internal features and for establishing section locations from which 2D and 3D section views can be created. Live sectioning is either on or off.

As you have seen, if the section plane is created by selecting a face, live sectioning is automatically turned on. However, when picking two points or using the **Draw** option of the **SECTIONPLANE** command, live sectioning is off. Live sectioning can be turned on and off for individual section objects, but only one section object can be "live" at any given time.

To turn live sectioning on or off, select the section object. Then, right-click to display the shortcut menu and pick **Activate live sectioning**. See **Figure 14-16.** A check mark appears next to the menu item when live sectioning is on. You can also use the ribbon or enter the **LIVESECTION** command and select the section object to toggle the on/off setting. When live sectioning is turned on, the material behind the viewing direction of the section plane is removed. The cross section of the 3D object is shown in gray and the internal shape of the 3D object is visible.

A wide variety of options allow you to change the appearance of not only the live sectioning display, but also of 2D and 3D section blocks that can be created from the sectioned display. These settings are found in the **Section Settings** dialog box. See **Figure 14-17.** To open this dialog box, select the section object, right-click, and pick **Live section settings...** in the shortcut menu. You can also pick the dialog box launcher button in the lower-right corner of the **Section** panel in the **Home** tab of the ribbon.

To change the settings for live sectioning, pick the **Live section settings** radio button at the top of the **Section Settings** dialog box. The categories displayed in the dialog box contain properties related to live sectioning. Settings for 2D and 3D sections and elevations are discussed later in this chapter.

The three categories for live sectioning are **Intersection Boundary**, **Intersection Fill**, and **Cut-Away Geometry**. To display a brief description of any property, hover the cursor over the property in the **Section Settings** dialog box. The description is displayed in a tooltip. A check box at the bottom of the **Section Settings** dialog box allows you to apply the properties to all section objects or to just the selected section object.

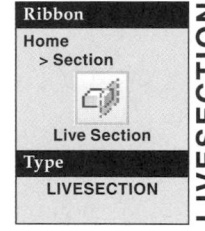

Ribbon
Home
> Section

Live Section

Type
LIVESECTION

LIVESECTION

Figure 14-16.
Live sectioning can be turned on for any section state by selecting the section object, right-clicking to display the shortcut menu, and picking **Activate live sectioning**.

	Repeat CUI	
	Recent Input	▶
	Annotative Object Scale	▶
✂	Cut	Ctrl+X
	Copy	Ctrl+C
	Copy with Base Point	Ctrl+Shift+C
	Paste	Ctrl+V
	Paste as Block	Ctrl+Shift+V
	Paste to Original Coordinates	
	Erase	
	Move	
	Copy Selection	
	Scale	
	Rotate	
	Draw Order	▶
✔	Activate live sectioning	
	Show cut-away geometry	
	Live section settings...	
	Generate 2D/3D section...	
	Add jog to section	
	Deselect All	
	Action Recorder	▶
	Subobject Selection Filter	▶
	Quick Select...	
	QuickCalc	
	Find...	
	Properties	
	Quick Properties	

Turn live sectioning on and off

Figure 14-17.
Section settings. A—For a 2D block. B—For a 3D block. C—For live sectioning.

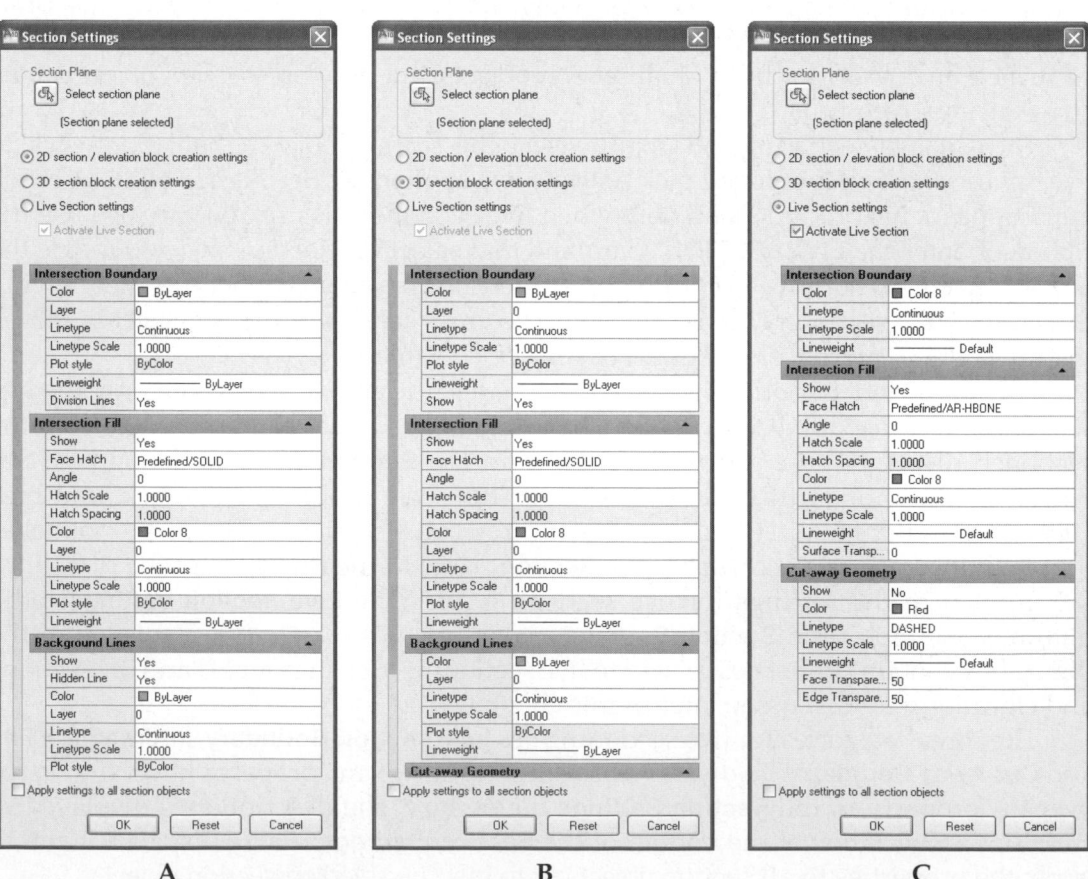

A   B   C

Live sectioning can be quickly turned on or off by double-clicking on the section plane object.

Intersection boundary

The intersection boundary is where the model is intersected by the section object. It is represented by line segments. You can set the color, linetype, linetype scale, and lineweight of the intersection boundary lines. Any linetype loaded into AutoCAD can be used.

Intersection fill

The intersection fill is the material visible on the model surface where the section object cuts. It is displayed as a solid fill, by default. Any hatch pattern available in AutoCAD can be used as the intersection fill. The angle, hatch scale, and hatch spacing, and color can be set. In addition, the linetype, linetype scale, and lineweight can be changed. The fill pattern can even be set to be transparent.

Cutaway geometry

The cutaway geometry is the part of the model removed by the live sectioning. By default, this geometry is not displayed. Changing the Show property to Yes displays the geometry. See **Figure 14-18.** You can set the color, linetype, linetype scale, and lineweight of the lines representing the cutaway geometry. In addition, the Face Transparency and Edge Transparency properties allow you to create a see-through effect, as seen in **Figure 14-19.** Each of these two properties is set to 50 by default.

Figure 14-18.
A—The intersection fill can be displayed as a hatch pattern in any specified color. B—The cutaway geometry is displayed.

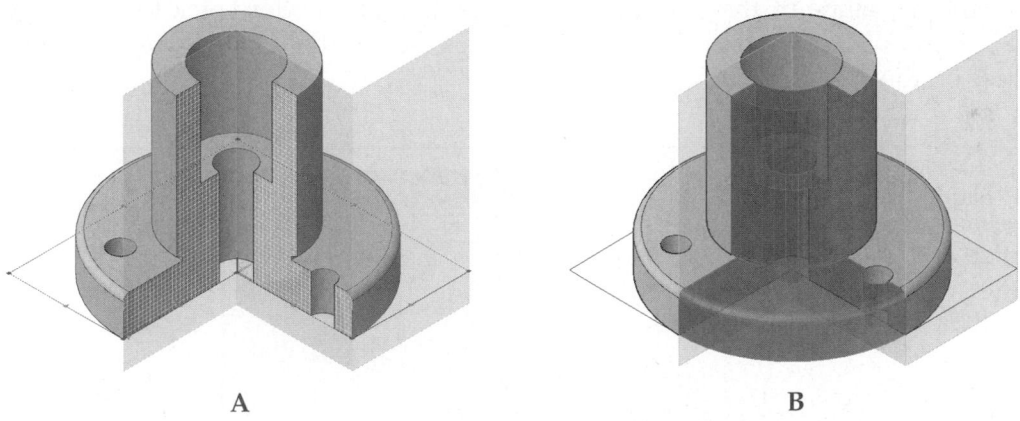

A   B

Figure 14-19.
The cutaway geometry is displayed with 100% transparent faces and solid black lines.

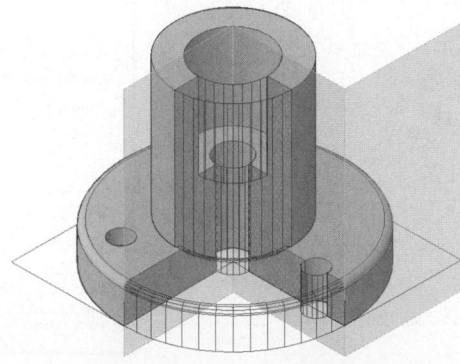

Exercise 14-4

Complete the exercise on the student website.
www.g-wlearning.com/CAD

Generating 2D and 3D Sections and Elevations

The **SECTIONPLANE** command provides a fast and efficient method of creating sections. The sections can be either 2D or 3D. Not only can the sections be displayed on the current drawing, they can also be exported as a file that can then be used in any other drawing or document for display, technical drawing, or manufacturing purposes.

Creating sections

SECTIONPLANETOBLOCK

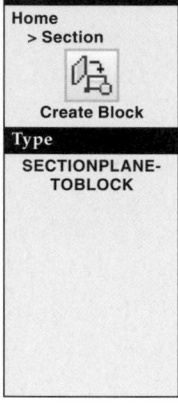

Ribbon
Home
> Section

Create Block

Type
**SECTIONPLANE-
TOBLOCK**

To create a section, enter the **SECTIONPLANETOBLOCK** command. You can also select the section object, right-click, and pick **Generate 2D/3D section...** from the shortcut menu. The **Generate Section/Elevation** dialog box is displayed. See **Figure 14-20.** In this dialog box, you can specify whether the section will be 2D or 3D, select what is included in the section, and specify a destination for the section. To expand the dialog box, pick the **Show details** button, which looks like a down arrow.

To create a 2D section, pick the **2D Section/Elevation** radio button in the **2D/3D** area of the dialog box. A 2D section is projected onto the section plane, but is placed flat on the XY plane of the current UCS. To create a 3D section, pick the **3D Section** radio button. A 3D section is placed so its surfaces are parallel to the corresponding cut surfaces on the 3D object.

Figure 14-20.
The expanded
**Generate Section/
Elevation** dialog box.

Pick 2D or 3D

Select objects
to include

Select where
the section will
be placed

Pick to expand
or collapse the
dialog box

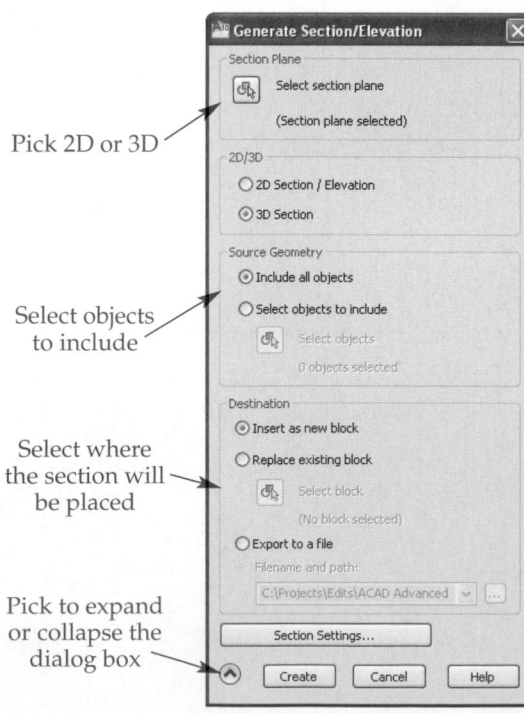

In the **Source Geometry** area of the dialog box, you can specify which geometry is included in the section. Picking the **Include all objects** radio button includes all 3D solids, surfaces, and regions in the section. To limit the section to certain objects, pick the **Select objects to include** radio button. Then, pick the **Select objects** button, select the objects on-screen, and press [Enter]. The number of selected objects is then displayed in the dialog box.

The **Destination** area of the dialog box is where you specify how the section will be placed. To place the section into the current drawing, pick the **Insert as new block** radio button. To update an existing section block, pick the **Replace existing block** radio button. Then, pick the **Select block** button, select the block on-screen, and press [Enter]. You will need to do this if the section object is changed. To save the section to a file for use in other drawings, pick the **Export to a file** radio button. Then, enter a path and file name in the text box.

Once all settings have been made, pick the **Create** button. The section is attached to the cursor and can be placed like a regular block. See **Figure 14-21**. Additionally, the options available are the same as if a regular block is being inserted. Once the block is inserted it can be moved, rotated, and scaled as needed.

Section settings

The **Section Settings...** button at the bottom of the **Generate Section/Elevation** dialog box opens the **Section Settings** dialog box discussed earlier. Using this dialog box, you can adjust all of the properties associated with the type of section being created. Depending on whether the **2D Section** or **3D Section** radio button is selected in the **Generate Section/Elevation** dialog box, the appropriate categories and properties are displayed in the **Section Settings** dialog box. Refer to **Figure 14-17**.

The categories discussed earlier related to the **Live Section Settings** radio button are available, although not all of the properties are displayed. Also, two additional categories are displayed for 2D and 3D sections:

- **Background Lines.** Available for 2D and 3D sections.
- **Curved Tangency Lines.** Only available for 2D sections.

Figure 14-21.
Inserting a 2D section block.

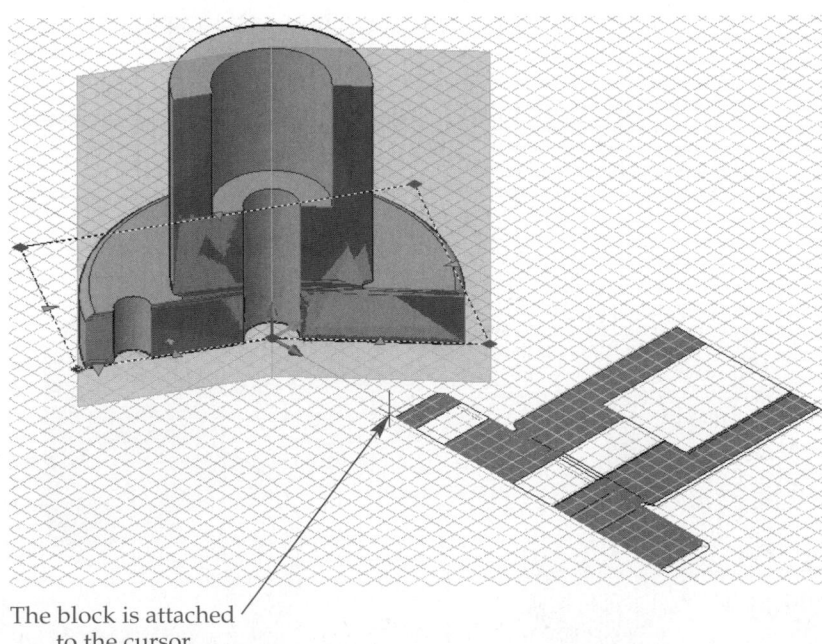

The block is attached
to the cursor

Examples of 2D and 3D sections inserted as blocks in the drawing are shown in **Figure 14-22.** Notice how properties can be set to show cutaway geometry in a different color and to change the section pattern, color, and linetype scale.

Background Lines. The properties in the **Background Lines** category provide control over the appearance of all lines that are not on the section plane. You can choose to have visible background lines, hidden background lines, or both displayed. They can be emphasized with color, linetype, or lineweight. The layer, linetype scale, and plot style can also be changed. These settings are applied to both visible and hidden background lines.

Curved Tangency Lines. The properties in the **Curved Tangency Lines** category apply to lines of tangency behind the section plane. For example, the object shown in **Figure 14-22** has a round on the top of the base. This results in a line of tangency behind the section plane where the round meets the vertical edge. You can have these lines displayed or suppressed. In general, lines of tangency are not shown in a section view. If you choose to display these lines, you can set the color, layer, linetype, linetype scale, plot style, and lineweight of the lines.

NOTE

When a 3D section is created, you must turn off live sectioning to see the complete sectioned object in the block. With live sectioning on, only the cut surfaces appear in the block.

Figure 14-22.
A—The section object is created. B—A 2D section block is inserted into the drawing and the view is made plan to the block. C—A 3D section block is inserted into the drawing. Notice how the hatch pattern is displayed. D—The 3D section block is updated and now the cutaway geometry is displayed.

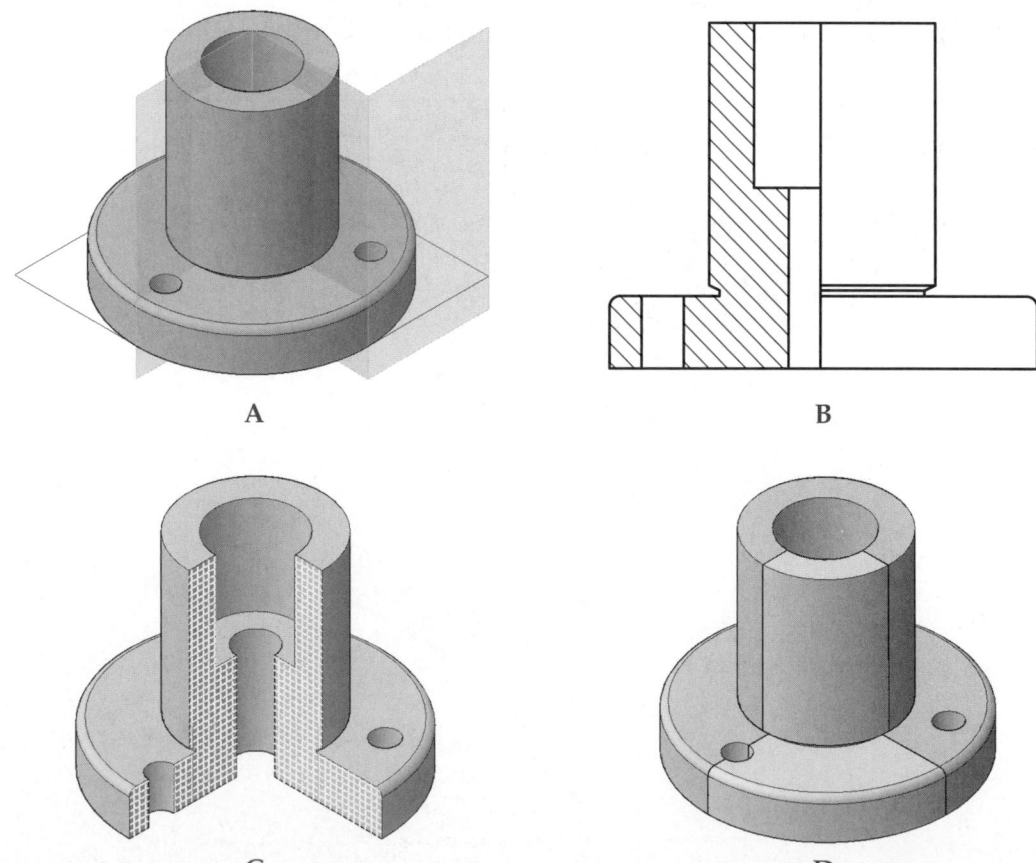

A

B

C

D

Updating the section view

Once the section view is created, it is not automatically updated if the section object is changed. To update the section view, select the section object (not the block), right-click, and pick **Generate 2D/3D section...** from the shortcut menu. Then, in the **Destination** area of the **Generate Section/Elevation** dialog box, pick the **Replace existing block** radio button. If necessary, pick the **Select block** button and select the section block in the drawing. If you want to change the appearance of the section view, pick the **Section Settings...** button and adjust the properties as needed. Finally, pick the **Create** button in the **Generate Section/Elevation** dialog box to update the section block.

Exercise 14-5

Complete the exercise on the student website.
www.g-wlearning.com/CAD

Creating a Flat Display of a 3D Model

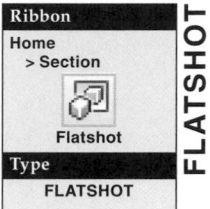

Ribbon
Home
> Section

Flatshot

Type
FLATSHOT

FLATSHOT

The **FLATSHOT** command creates a flat projection of the 3D objects in the drawing from the current viewpoint. The view that is created is composed of 2D geometry and is projected onto the XY plane of the current UCS. This capability is useful for creating technical documents in which pictorial views of 3D objects are required.

Once the command is initiated, the **Flatshot** dialog box is displayed. See **Figure 14-23.** The options in this dialog box are similar to those found in the **Section Settings** dialog box. However, the display properties of foreground and obscured lines are limited to color and linetype. You can choose whether or not obscured lines are displayed in the flat view. You can also choose whether or not tangential edges are included.

Figure 14-23.
The **Flatshot** dialog box.

Select where the flat view will be placed

Settings for foreground lines

Settings for obscured lines

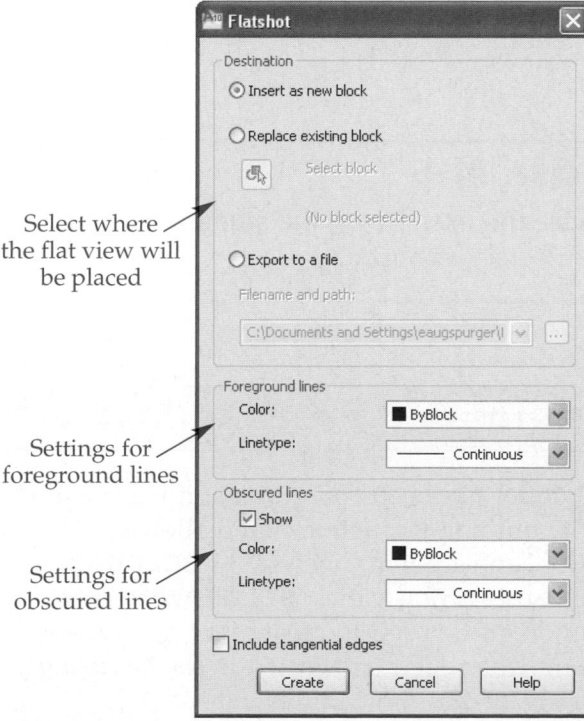

Figure 14-24.
A—The 3D view from which a flat view will be generated. B—The inserted flat view. The viewpoint is plan to the block.

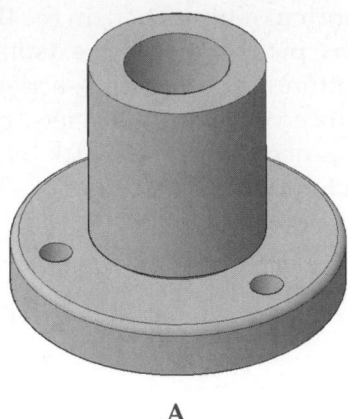

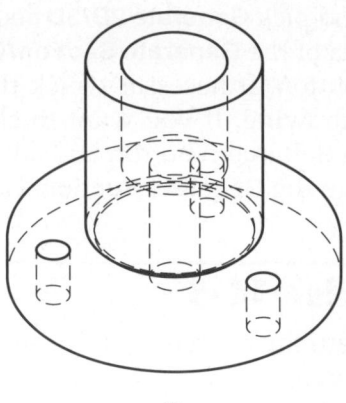

A

B

Select a destination for the flat view. Then, change the foreground and obscured lines settings as needed. Finally, pick the **Create** button. The flat view is inserted into the drawing as a block. Therefore, all of the ensuing prompts are those of a block insertion.

Use the **PLAN** command to obtain a plan view of the current UCS. Since the flat view is a block, it can be edited using the **BEDIT** command. **Figure 14-24** shows a pictorial view of a 3D object and a plan view of the resulting flat view. To update the flat view to one from a different viewpoint, repeat the command and select the **Replace existing block** radio button in the **Destination** area of the dialog box.

PROFESSIONAL TIP

If the intention is to create a block to be used for a technical document, it may be best to export the flat view block to a file. It can then be inserted into a new AutoCAD drawing or copied into a document file.

Exercise 14-6

Complete the exercise on the student website.
www.g-wlearning.com/CAD

Creating and Using Multiview Layouts

Once a solid model has been constructed, it is easy to create a multiview layout using the **SOLVIEW** command. This command allows you to create a layout containing orthographic, section, and auxiliary views. The **SOLDRAW** command can then be used to complete profile and section views. **SOLDRAW** must be used after **SOLVIEW**. The **SOLPROF** command can be used to create a profile of the solid in the current view. These commands may be typed or selected in the **Modeling** panel of the **Home** tab on the ribbon. See **Figure 14-25.**

AutoCAD and Its Applications—Advanced

Figure 14-25.
The **SOLVIEW**, **SOLDRAW**, and **SOLPROF** commands are found in the **Modeling** panel of the **Home** tab on the ribbon.

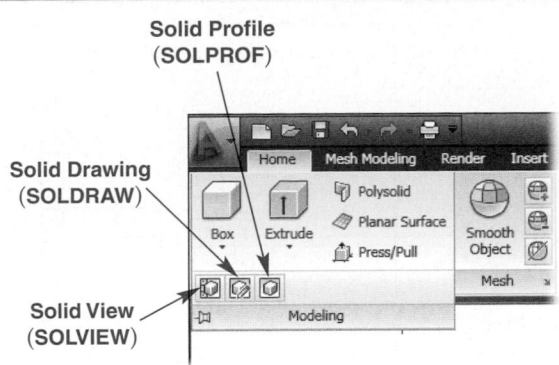

Solid Profile (SOLPROF)

Solid Drawing (SOLDRAW)

Solid View (SOLVIEW)

Creating Views with SOLVIEW

The **SOLVIEW** command is used to create new floating viewports and to establish the display within those viewports. Therefore, you may want to delete the default viewport in the layout (paper space) tab before using the **SOLVIEW** command.

First restore the WCS. This will help avoid any confusion. Then, display a plan view. See **Figure 14-26.** It helps to have additional user coordinate systems created prior to using **SOLVIEW**. This allows you to construct orthographic views based on a specific named UCS.

Before using the **SOLVIEW** command, visualize which view is going to be the top view (or plan view) and how you would like the model rotated in relationship to the layout. With this in mind, look at the current UCS icon and make sure that the X axis is pointing to the "right" and the Y axis is pointing "up" in your imagined layout. If this is not the case, then you must restore the WCS, rotate the current UCS, or restore a saved UCS to correctly align the axes. Then, when you enter the **SOLVIEW** command in the layout, you can simply select the current UCS and you will be creating the top or plan view of your model.

When you enter the **SOLVIEW** command while in model space, AutoCAD automatically switches to layout space (paper space). Next, create an initial view from which other views can project. This is normally the top or front. In the following example, the top view is constructed first by using the plan view of a UCS named Leftside.

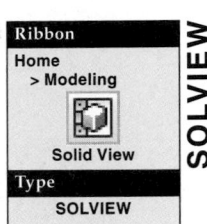

Ribbon
Home
> Modeling

Solid View

Type
SOLVIEW

SOLVIEW

> Enter an option [Ucs/Ortho/Auxiliary/Section]: **U**↵
> Enter an option [Named/World/?/Current] <Current>: **N**↵
> Enter name of UCS to restore: **LEFTSIDE**↵
> Enter view scale <1.0>: **.5**↵
> Specify view center: *(pick a location in the layout for the center of the view)*
> Specify view center <specify viewport>: ↵
> Specify first corner of viewport: *(pick the first corner of a paper space viewport outside of the object)*
> Specify opposite corner of viewport: *(pick the opposite corner of the viewport)*
> Enter view name: **TOPVIEW**↵ *(the left of the object in AutoCAD is the top of the part)*
> Enter an option [Ucs/Ortho/Auxiliary/Section]: *(leave the command active at this time)*

You must provide a name for the view. The result is shown in **Figure 14-27.**

The **SOLVIEW** command remains active until you press the [Enter] or [Esc] key. If you exit **SOLVIEW**, you can still return to the drawing and create additional orthographic viewports. With the command active, continue and create a section view to the right of the top view:

Figure 14-26.
Before using
SOLVIEW, display
a plan view of the
WCS.

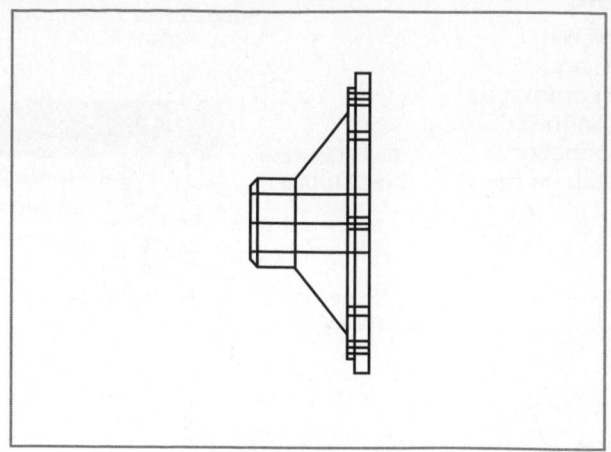

Figure 14-27.
The initial view
created with the **Ucs**
option of **SOLVIEW**.

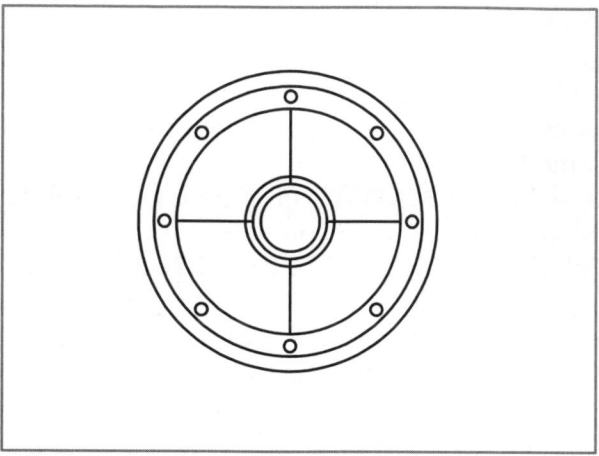

Enter an option [Ucs/Ortho/Auxiliary/Section]: **S↵**
Specify first point of cutting plane: *(pick the quadrant at point 1 in Figure 14-28)*
Specify second point of cutting plane: *(pick the quadrant at point 2)*
Specify side to view from: *(pick point 3)*
Enter view scale <0.5>: ↵
Specify view center: *(pick the center of the new section view)*
Specify view center <specify viewport>: ↵ *(this prompt remains active until [Enter] is pressed to allow you to adjust the view location if necessary)*
Specify first corner of viewport: *(pick one corner of the viewport)*
Specify opposite corner of viewport: *(pick the opposite corner of the viewport)*
Enter view name: **SECTION↵**
Enter an option [Ucs/Ortho/Auxiliary/Section]: ↵

Figure 14-28.
The section view
created with
SOLVIEW (shown
on the right) does
not show projection
lines. The pick
points are shown on
the left.

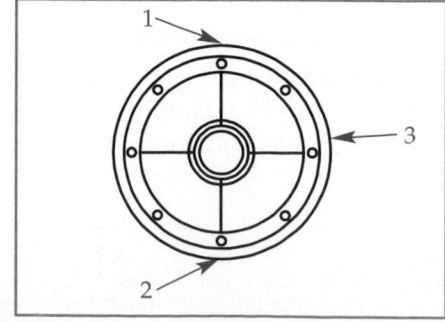
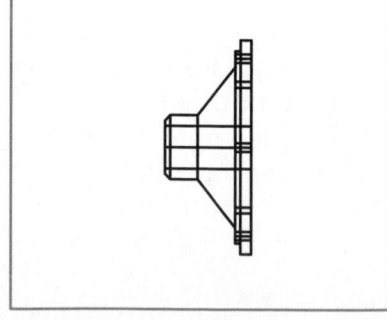

Figure 14-29.
An orthographic front view is created with the **Ortho** option of **SOLVIEW**. This is the view shown at the lower left.

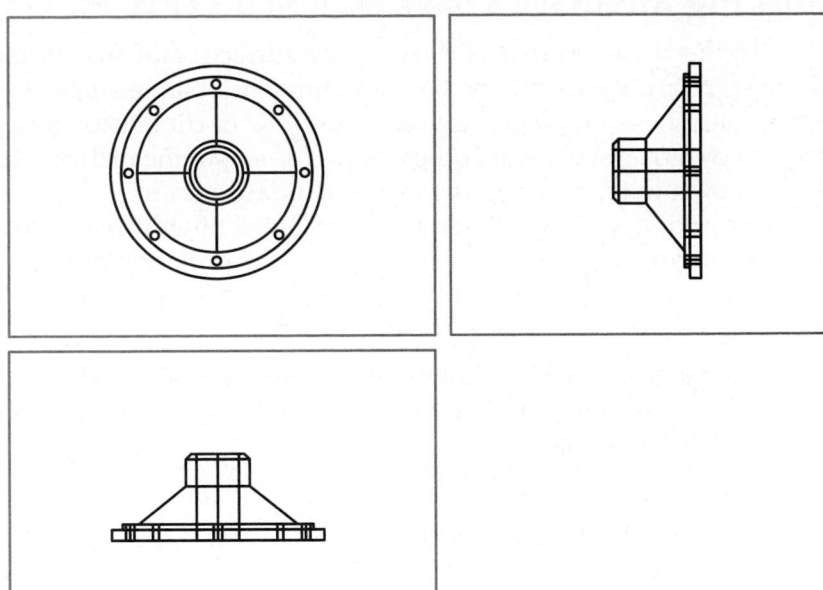

Notice in **Figure 14-28** that the new view is shown in the current visual style and not as a section. This is normal. **SOLVIEW** is used to create the views. The **SOLDRAW** command draws the section lines. **SOLDRAW** is discussed later in this chapter.

A standard orthographic view can be created using the **Ortho** option of **SOLVIEW**. This is illustrated in the following example. The new orthographic view is shown in **Figure 14-29.**

    Enter an option [Ucs/Ortho/Auxiliary/Section]: **O**↵
    Specify side of viewport to project: *(pick the bottom edge of the left viewport)*
    Specify view center: *(pick the center of the new view)*
    Specify view center <specify viewport>: ↵
    Specify first corner of viewport: *(pick one corner of the viewport)*
    Specify opposite corner of viewport: *(pick the opposite corner of the viewport)*
    Enter view name: **FRONTVIEW**↵

The **SOLVIEW** command creates new layers that are used by **SOLDRAW** when profiles and sections are created. The layers are used for the placement of visible, hidden, dimension, and section lines. Each layer is named as the name of the view with a three letter tag, as shown in the following table. The use of these layers is discussed later in this chapter.

Layer Name	Object
View name-VIS	Visible lines
View name-HID	Hidden lines
View name-DIM	Dimension lines
View name-HAT	Hatch patterns (sections)

Exercise 14-7

Complete the exercise on the student website.
www.g-wlearning.com/CAD

Creating Auxiliary Views with SOLVIEW

Auxiliary views are used to display a surface of an object that is not parallel to any of the standard views. It may be an inclined or oblique surface. Refer to **Figure 14-30**. Sometimes these views are necessary to show or dimension a feature that is not being displayed in true size in any other view. The slot in the inclined surface in **Figure 14-30** is not shown in true size in any of the standard views.

The auxiliary view is taken from one of the other views where the inclined surface is shown as an edge. The auxiliary view will be projected perpendicular to this surface. The auxiliary view is created by picking two points on the surface in the front view and another point to indicate the line of sight.

> Enter an option [Ucs/Ortho/Auxiliary/Section]: **A**↵
> Specify first point of inclined plane: *(using object snaps, pick a point on one end of the inclined surface)*
> Specify second point of inclined plane: *(pick a point on the other end of the inclined surface)*
> Specify side to view from: *(pick a point on the side of the surface from which you want to view it)*
> Specify the view center: *(pick the center)*
> Specify the view center: <specify viewport>↵
> Specify first corner of viewport: *(pick one corner of the viewport)*
> Specify opposite corner of viewport: *(pick the opposite corner of the viewport)*
> Enter view name: **AUXILIARYVIEW**↵

Auxiliary views are often incomplete views, so it is acceptable to cut off portions of the view that are not necessary when you specify the corners of the viewport.

PROFESSIONAL TIP

When creating an auxiliary view, you may want to move other viewports that may be in the way to make room for the view.

Figure 14-30.
An auxiliary view (shown in color) is created from the inclined plane in the front view.

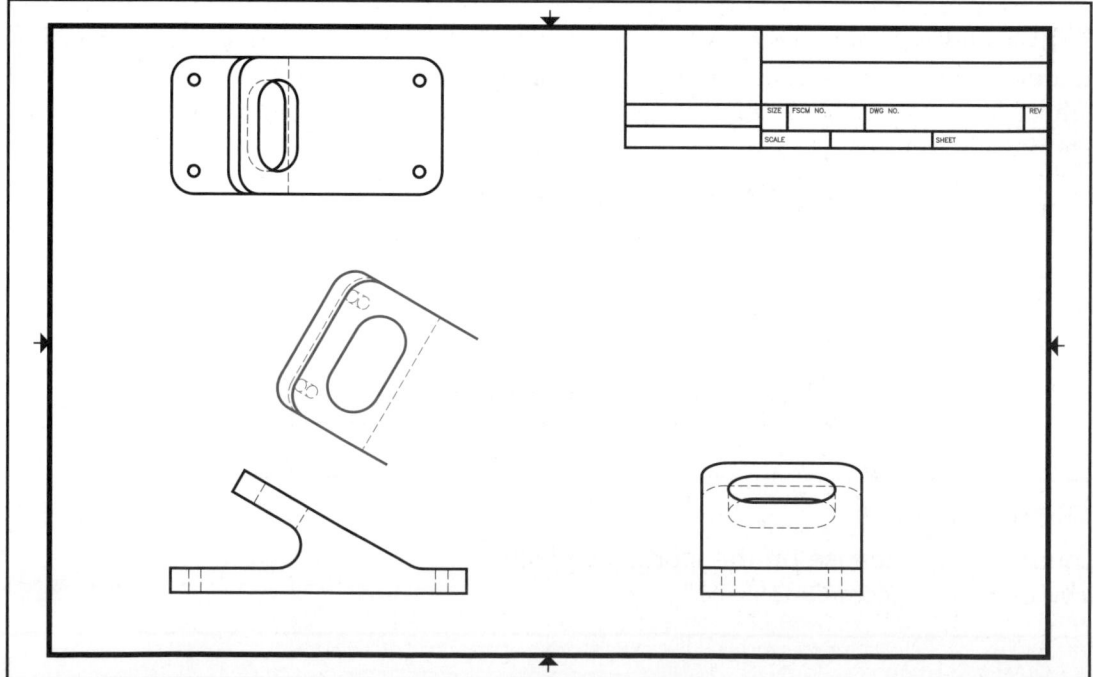

Creating Finished Views with SOLDRAW

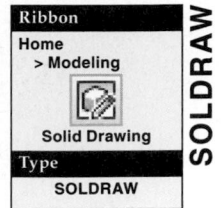

Ribbon
Home
> Modeling

Solid Drawing
Type
SOLDRAW

SOLDRAW

The **SOLVIEW** command saves information specific to each viewport when a new view is created. This information is used by the **SOLDRAW** command to construct a finished profile or section view. **SOLDRAW** first deletes any information currently on the *view name*-VIS, *view name*-HID, and *view name*-HAT layers for the selected view. Visible, hidden, and section lines are automatically placed on the appropriate layer. Therefore, you should avoid placing objects on any layer other than the *view name*-DIM layer.

The **SOLDRAW** command automatically creates a profile or section in the selected viewport. If you select a viewport that was created using the **Section** option of **SOLVIEW**, the **SOLDRAW** command uses the current values of the **HPNAME**, **HPSCALE**, and **HPANG** system variables to construct the section. These three variables control the angle, scale factor, and name of the hatch pattern.

If a view is selected that was not created as a section in **SOLVIEW**, the **SOLDRAW** command constructs a profile view. All new visible and hidden lines are placed on the *view name*-VIS or *view name*-HID layer. All existing objects on those layers are deleted.

Once the command is initiated, you are prompted to select objects. Pick the border of the viewport(s) for which you want the profile or section generated. When all viewports are selected, press [Enter] and the profiles and sections are created.

After the profile construction is completed, lines that should be a hidden linetype are still visible (solid). This is because the linetype set for the *view name*-HID layer is Continuous. Change the linetype for the layer to Hidden and the drawing should appear as shown in Figure 14-31, depending on the current visual style and hatch settings. You may also want to change other layer properties such as color, lineweight, and plot style.

Revising the 3D Model

If changes are needed after these views are created, the best practice is to modify the original 3D solid. However, the views created with **SOLVIEW** and **SOLDRAW** will not immediately reflect changes. To update the views, simply start the **SOLDRAW** command, select the viewports, and press [Enter]. The views are then updated with the changes.

Figure 14-31.
The new front profile view shows hidden lines after the linetype is set to Hidden for the FRONTVIEW-HID layer.

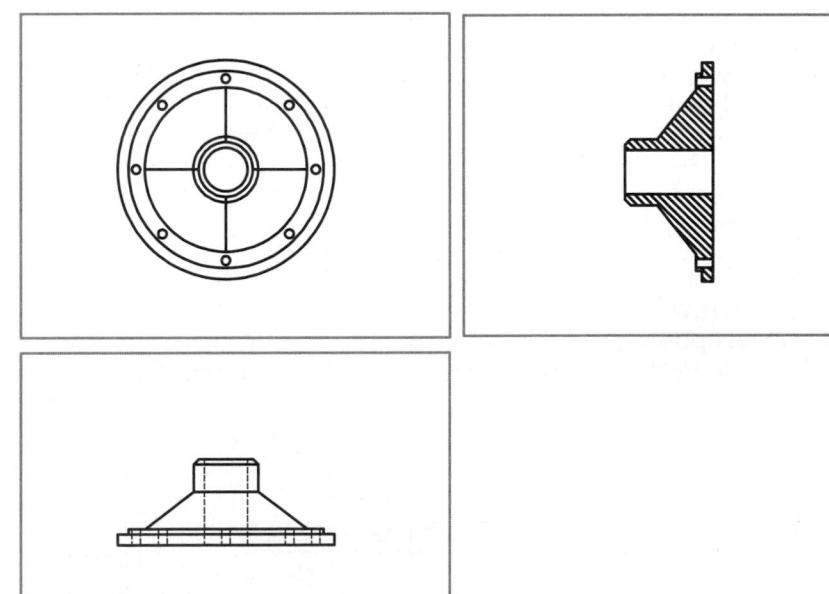

When you go to model space to edit the solid, you may find it difficult to work on the original model. The 2D views created with **SOLDRAW** are projected on the top, bottom, left, and right, and sometimes within the model itself. It may be a good idea to set up a layer filter to temporarily freeze these layers while making changes. Remember to thaw the layers before updating the viewports with **SOLDRAW**.

Adding a 3D View in Paper Space to the Drawing Layout

If you want to add a paper space viewport that contains a 3D (pictorial) view of the solid, use the **MVIEW** command. Create a single viewport by picking the corners. The object will appear in the viewport. Next, activate the viewport and use any of the orbit commands or a preset isometric viewpoint to achieve the desired 3D view. Pan and zoom as necessary. Change to the parallel or perspective projection if needed. You can also use the **Visual Styles** panel in the ribbon to adjust the display of the 3D viewport. The visual style set current for this viewport does not affect the displays in the other viewports. See **Figure 14-32**.

In order to have the hidden display correctly plotted, use the **MVIEW Shadeplot** option on the 3D viewport. Enter the command and select the **Shadeplot** option. Then, set the option to **Hidden**. If you have a hidden display shown in the viewport, you can also select **As displayed**. Then, pick the viewport when prompted to select objects.

Alternately, you can select the viewport and use the **Properties** palette to set the Shade plot property to As Displayed or Hidden. Any visual style display of the viewport can also be plotted in this manner by setting **MVIEW Shadeplot** to **As Displayed** (when the view is shaded) or **Rendered**.

Figure 14-32.
Create a 3D viewport with the **MVIEW** command. You can hide the lines in the viewport, as shown at the lower right. To plot the viewport as a hidden display, use the **MVIEW Shadeplot** option and set it to Hidden.

Tips

Remember the following points when working with **SOLVIEW** and **SOLDRAW**.

- Use **SOLVIEW** first and then **SOLDRAW**.
- Do not draw on the *view name*-HID and *view name*-VIS layers.
- Place model space dimensions for each view on the *view name*-DIM layer for that specific view or simply dimension in paper space on a layer not created by **SOLVIEW**.
- After using **SOLVIEW**, use **SOLDRAW** on all viewports in order to create hidden lines or section views.
- Change the linetype on the *view name*-HID layer to Hidden and adjust other layer properties as needed.
- Create 3D viewports with the **MVIEW** or **VPORTS** command and an orbit command or a preset isometric view. Remove hidden lines when plotting with the **MVIEW Shadeplot** option set to **Hidden**.
- Plot the drawing in layout (paper) space at the scale of 1:1.

Exercise 14-8

Complete the exercise on the student website.
www.g-wlearning.com/CAD

Creating a Profile with SOLPROF

The **SOLPROF** command creates a profile view from a 3D solid model. This is similar to the **Profile** option of the **SOLVIEW** command. However, **SOLPROF** is limited to creating a profile view of the solid for the current view only.

Ribbon
Home
> Modeling
Solid Profile

Type
SOLPROF

SOLPROF creates a block of all lines forming the profile of the objcct. It also creates a block of the hidden lines of the object. The original 3D object is retained. Each of these blocks is placed on a new layer with the name of PH-*view handle* and PV-*view handle*. A *view handle* is a name composed of numbers and letters that is automatically given to a viewport by AutoCAD. For example, if the view handle for the current viewport is 2C9, the **SOLPROF** command creates the layers PH-2C9 and PV-2C9.

You must be in layout (paper) space and have a model space viewport active to use the command. Once the command is initiated, you are prompted to select objects:

> Select objects: *(pick the solid)*
> 1 found
> Select objects: ↵
> Display hidden profile lines on separate layer? [Yes/No] <Y>: ↵
> Project profile lines onto a plane? [Yes/No] <Y>: ↵

If you answer yes to this prompt, the 3D profile lines are projected to a 2D plane and converted to 2D objects. This produces a cleaner profile.

> Delete tangential edges? [Yes/No] <Y>: ↵

Answering yes to this prompt produces a proper 2D view by eliminating lines that would normally appear at tangent points of arcs and lines. If you wish to display the profile lines in a 3D view, do not delete the tangential edges. Once the profile is created, freeze the layer of the original object in the viewport. The original object and the profile created with **SOLPROF** are shown in **Figure 14-33**.

NOTE

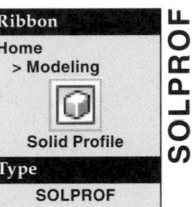

When plotting views created with **SOLPROF**, hidden lines may not be displayed unless you freeze the layer that contains the original 3D object.

SOLPROF

Figure 14-33.
A—The original solid. B—A profile created with **SOLPROF**.

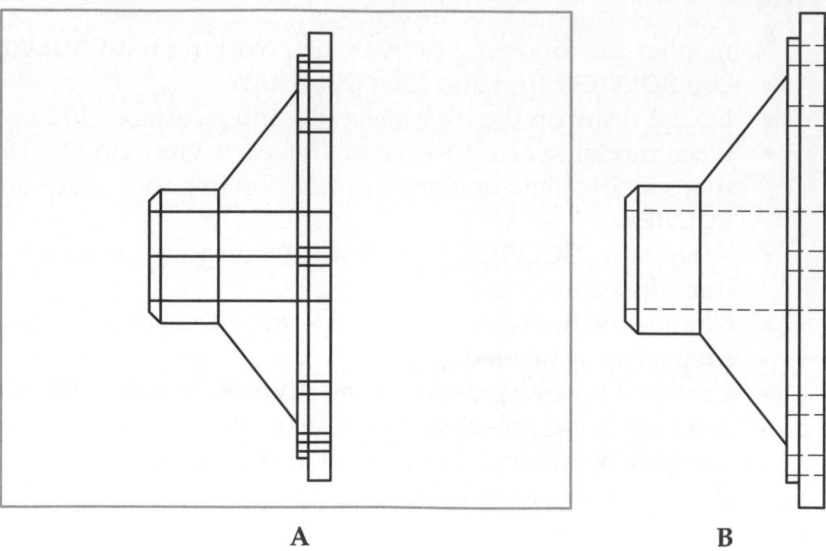

A   B

Solid Model Analysis

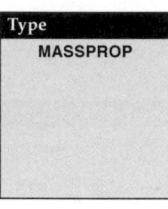
The **MASSPROP** command allows you to analyze a solid model for its physical properties. The data obtained from **MASSPROP** can be retained for reference by saving the data to a file. The default file name is the drawing name. The file is an ASCII text file with a .mpr (mass properties) extension. The analysis can be used for third party applications to produce finite element analysis, material lists, or other testing studies.

Once the command is initiated, you are prompted to select objects. Pick the objects for which you want the mass properties displayed and press [Enter]. AutoCAD analyzes the model and displays the results in the AutoCAD text window. See **Figure 14-34.** The following properties are listed.

- **Mass.** A measure of the inertia of a solid. In other words, the more mass an object has, the more inertia it has. Note: Mass is *not* a unit of measurement of inertia.
- **Volume.** The amount of 3D space the solid occupies.

Figure 14-34.
The **MASSPROP** command displays a list of solid properties in the AutoCAD text window.

```
AutoCAD Text Window - Drawing5.dwg
Edit
Select objects: 1 found

Select objects:
---------------    SOLIDS    ---------------

Mass:                   13.1986
Volume:                 13.1986
Bounding box:       X: -0.5000  --  5.5000
                    Y: 0.0000   --  2.5000
                    Z: -2.5000  --  0.5000
Centroid:           X: 2.5420
                    Y: 0.6029
                    Z: -1.0000
Moments of inertia: X: 32.7882
                    Y: 137.9144
                    Z: 124.0954
Products of inertia: XY: 18.5713
                     YZ: -7.9569
                     ZX: -33.5515
Radii of gyration:   X: 1.5761
                     Y: 3.2325
                     Z: 3.0663
Principal moments and X-Y-Z directions about centroid:

Press ENTER to continue:
```

- **Bounding box.** The dimensions of a 3D box that fully encloses the solid.
- **Centroid.** A point in 3D space that represents the geometric center of the mass.
- **Moments of inertia.** A solid's resistance when rotating about a given axis.
- **Products of inertia.** A solid's resistance when rotating about two axes at a time.
- **Radii of gyration.** Similar to moments of inertia. Specified as a radius about an axis.
- **Principal moments and X-Y-Z directions about a centroid.** The axes about which the moments of inertia are the highest and lowest.

PROFESSIONAL TIP

Advanced applications of solid model design and analysis are possible with Autodesk Inventor® software. This product allows you to create parametric designs and assign a wide variety of physical materials to the solid model.

Solid Model File Exchange

AutoCAD drawing files can be converted to files that can be used for testing and analysis. Use the **ACISOUT** command or **Export Data** dialog box to create a file with a .sat extension. These files can be imported into AutoCAD with the **ACISIN** or **IMPORT** command.

Solids can also be exported for use with stereolithography software. These files have a .stl extension. Use the **STLOUT** command or the **Export Data** dialog box to create STL files.

Importing and Exporting Solid Model Files

A solid model is frequently used with analysis and testing software or in the manufacture of a part. The **ACISOUT** and **EXPORT** commands allow you to create a type of file that can be used for these purposes. Once the **ACISOUT** command is initiated, you are prompted to select objects. After selecting objects and pressing [Enter], a standard save dialog box is displayed. See **Figure 14-35.** When using the **EXPORT** command, the standard save dialog box appears first. After entering a file name and selecting a file type (SAT), you are then prompted to select objects.

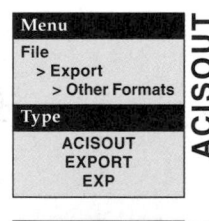

An SAT file can be imported into AutoCAD and automatically converted into a drawing file using the **ACISIN** and **IMPORT** commands. Once either command is initiated, a standard open dialog box appears. Change the file type to SAT, locate the file, and pick the **Open** button.

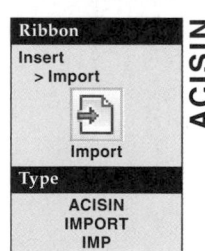

Stereolithography Files

Stereolithography is a technology that creates plastic, prototype 3D models using a computer-generated solid model, a laser, and a vat of liquid polymer. This technology is also called *rapid prototyping* or *3D printing.* A prototype 3D model can be designed and formed in a short amount of time without using standard manufacturing processes. The **FACETRES** setting affects the "resolution" of the solid in an exported STL file and, thus, the final stereolithograph.

Most software used to create a stereolithograph can read STL files. AutoCAD can export a drawing file to the STL format, but *cannot* import STL files. Also, the solid model must be positioned in the current UCS in such a way so the entire object has positive XYZ coordinates.

Figure 14-35.
Exporting an ACIS
file.

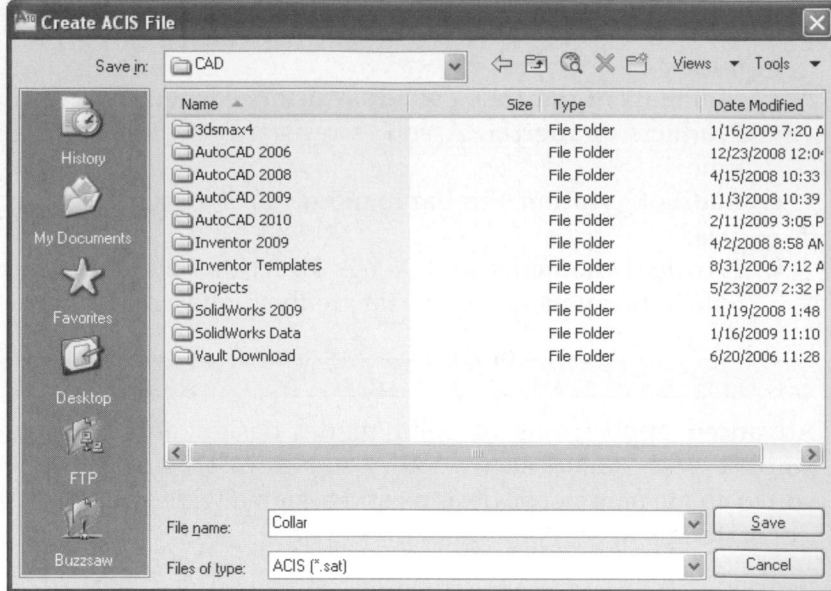

STLOUT

Menu
File
  > Export
    > Other Formats
Type
  STLOUT
  EXPORT
  EXP

The **STLOUT** and **EXPORT** commands can be used to create an STL file. Once the **STLOUT** command is initiated, you are prompted to select solids or watertight meshes. A *watertight mesh* is completely closed and contains no openings. You are then asked if you want to create a binary STL file. If you answer no to the prompt, an ASCII file is created. Keep in mind that a binary STL file may be as much as five times smaller than the same file in ASCII format. Use the file type required by your stereolithography machine or service provider. After you choose the type of file, a standard save dialog box is displayed. Type the file name in the **File name:** edit box and pick **Save** or press [Enter].

Once the **EXPORT** command is initiated, the standard save dialog box appears. Name the file and select the STL file type. Once the dialog box is closed, you are prompted to select an object. After you select the object and press [Enter], the file is created. You are not given the option of selecting a binary or ASCII format for the file. A binary file is automatically created.

PROFESSIONAL TIP

The **3DPRINT** command can be used to streamline the process of sending a solid model to a stereolithography service provider.

Exercise 14-9

Complete the exercise on the student website.
www.g-wlearning.com/CAD

Chapter Test

Answer the following questions. Write your answers on a separate sheet of paper or complete the electronic chapter test on the student website.
www.g-wlearning.com/CAD

1. What does the **SECTIONPLANE** command create?
2. How is the **Face** option of the **SECTIONPLANE** command used?
3. Which option of the **SECTIONPLANE** command is used to create sections with jogs?
4. When a section object is created by picking a face or two points or using the **Orthographic** option of the **SECTIONPLANE** command, which section object state is established?
5. Which section object grips are used to accomplish the following tasks?
   A. Change the section object state.
   B. Lengthen or shorten the section object segment.
   C. Rotate the section view 180°.
6. How is live sectioning turned on or off?
7. Which category in the **Section Settings** dialog box provides control over the material that is removed by the section object?
8. What are the two types of section view blocks that can be created from a section object?
9. Which command is used to create a flat view of the objects projected from the current viewpoint?
10. Which command should be used first, **SOLDRAW** or **SOLVIEW**?
11. Which option of the **SOLVIEW** command is used to create an orthographic view?
12. Name the layer(s) that the **SOLVIEW** command automatically create(s).
13. Which layer(s) in question 12 should you avoid drawing on?
14. Which command can automatically complete a section view using the current settings of **HPNAME**, **HPSCALE**, and **HPANG**?
15. Which command creates a profile view from a 3D model?
16. What is the function of the **MASSPROP** command?
17. What is the extension of the ASCII file that can be created by **MASSPROP**?
18. What is a *centroid?*
19. Which commands export and import solid models?
20. Which type of file has an .stl extension?

Drawing Problems

1. Open one of your solid model problems from a previous chapter and do the following.
   A. Use the **Face** option of the **SECTIONPLANE** command to create a section object.
   B. Alter the section so that the section plane object cuts through features of the model.
   C. Change the section settings to display an ANSI hatch pattern.
   D. Save the drawing as P14_01.

2. Open one of your solid model problems from a previous chapter and do the following.
   A. Construct a section through the model using the **Draw** option of the **SECTIONPLANE** command. Cut through as many features as possible.
   B. Display cutaway geometry with a 50% transparency.
   C. Display section lines using an appropriate hatch pattern.
   D. Generate a 3D section block that displays the cutaway geometry in a color of your choice.
   E. Create a layout with a viewport for the 3D block displayed at half the size of the original model.
   F. Save the drawing as P14_02.

3. Open one of your solid model problems from a previous chapter and do the following.
   A. Create a multiview layout of the model. One of the views should be a section view. Use a total of three 2D views.
   B. Use **SOLVIEW** and **SOLDRAW** to create the views. Be sure that section lines and hidden lines are properly displayed.
   C. Create a fourth viewport that contains a 3D view of the solid. Place the label PICTORIAL VIEW within the viewport.
   D. Plot the drawing so the 3D view is displayed with hidden lines removed.
   E. Save the drawing as P14_03.

4. Open one of your solid model problems from a previous chapter and do the following.
   A. Display the model in a plan view.
   B. Use **SOLPROF** to create a profile view. Save the profile view as a block in a file named P14_04PLN.
   C. Display the original model in a 3D view.
   D. Use **SECTIONPLANE** to construct a 2D front-view section of the model.
   E. Display the section as a plan view.
   F. Insert the block P14_04PLN above the section view. Adjust the views so they are properly aligned.
   G. Save the drawing as P14_04.

5. Choose five solid model problems from previous chapters and copy them to a new folder. Then, do the following.
   A. Open the first drawing. Export it as an SAT file.
   B. Do the same for the remaining four files.
   C. Compare the sizes of the SAT files with the DWG files. Compare the combined sizes of both types of files.
   D. Begin a new drawing and import one of the SAT files.

AutoCAD and Its Applications—Advanced

6. Draw the object shown below as a solid model. Do not dimension the object. Then, do the following.
   A. Construct a section object that creates a full section along the centerline of the hole.
   B. Generate a 2D section and display it on the drawing at half the size of the original. Specify section settings as desired.
   C. Generate a 3D section and display it on the drawing at half the size of the original. Do not display cutaway geometry. Specify section settings as desired.
   D. Activate live sectioning. Do not display the cutaway geometry.
   E. On the original solid model, display the intersection fill as an ANSI hatch pattern.
   F. Save the drawing as P14_06.

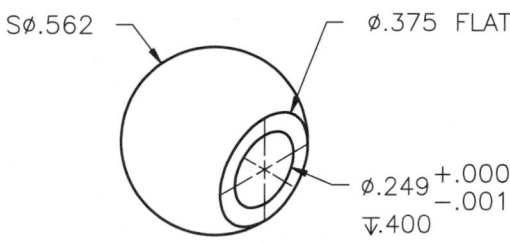

7. Draw the object shown below as a solid model. Only half of the object is shown; draw the complete object. Do not dimension the object. Then, do the following.
   A. Construct a section plane that creates a full section, as shown.
   B. Display the intersection fill as an ANSI hatch pattern.
   C. Activate live sectioning and view the cutaway geometry with a high level of transparency.
   D. Save the drawing as P14_07.

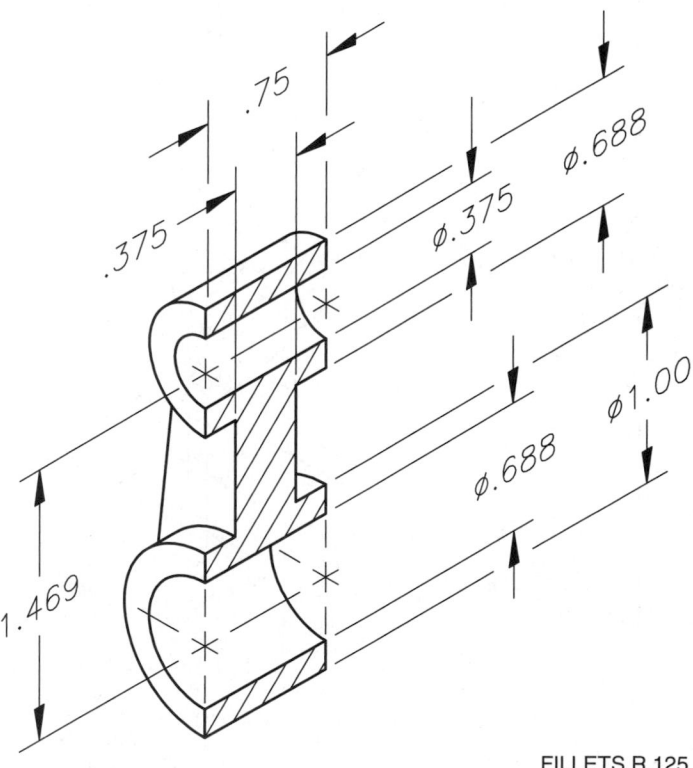

FILLETS R.125

Drawing Problems - Chapter 14

8. Draw the object shown below as a solid model. Only half of the object is shown; draw the complete object. Do not dimension the object. Then, do the following.
   A. Construct a section plane that creates a half section.
   B. Display the intersection fill as an ANSI hatch pattern.
   C. Activate live sectioning and view the cutaway geometry with a low level of transparency.
   D. Generate a 3D section and save it as a block.
   E. Use **SOLVIEW** to create a two-view orthographic layout. Use an appropriate scale to plot on a B-size sheet.
   F. Create a third floating viewport and insert the 3D section block scaled to half the size of the drawing.
   G. Save the drawing as P14_08.

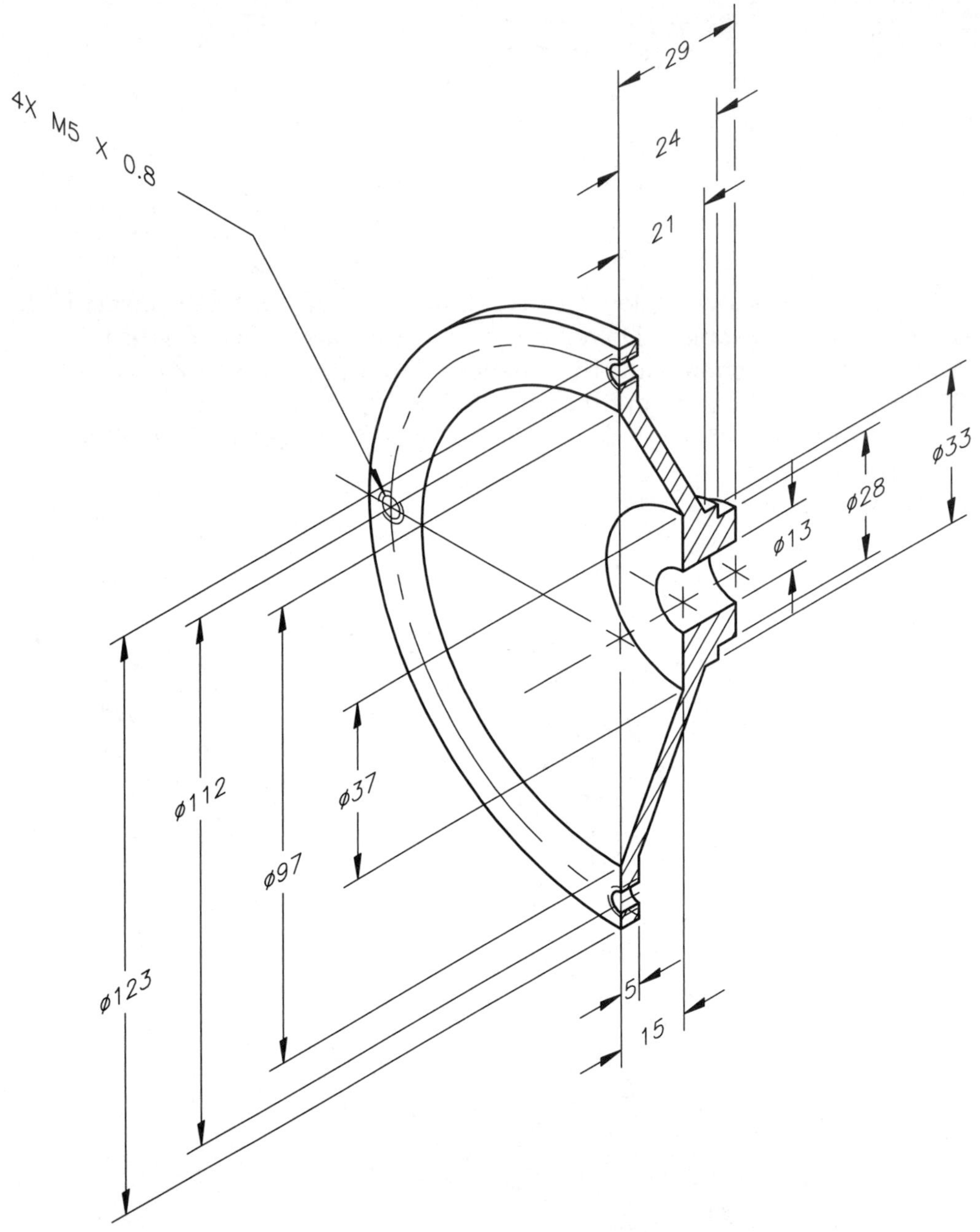

4X M5 X 0.8

29

24

21

ø33

ø28

ø13

ø112

ø37

ø97

ø123

5

15

9. Draw the object shown below as a solid model. Use your own dimensions. Then, do the following.
   A. Construct a section plane that creates an offset section. The section should pass through the center of two holes in the base and through the large central hole.
   B. Display the intersection fill as an ANSI hatch pattern.
   C. Activate live sectioning and view the cutaway geometry with a low level of transparency in the color red.
   D. Generate a 3D section of the sectioned solid model and save it as a block.
   E. Use **SOLVIEW** to create a two-view orthographic layout. One view should be a half section. Use an appropriate scale to plot on a B-size sheet.
   F. Create a third floating viewport and insert the 3D section block scaled to half the size of the drawing.
   G. Save the drawing as P14_09.

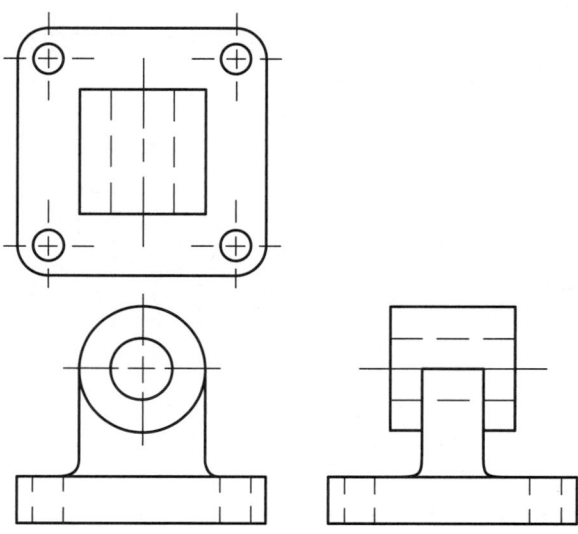

10. Draw the object shown below as a solid model. Do not dimension the object. Then, do the following.
    A. Display a 3D view of the model, generate a flat view, and save it as a block named P14_10_ FLATSHOT.
    B. Construct a section plane that creates a half section.
    C. Display the intersection fill as an ANSI hatch pattern.
    D. Activate live sectioning and view the cutaway geometry with a low level of transparency in the color red.
    E. Alter the section plane to create the section shown below.
    F. Generate a 3D section and save it as a block.
    G. Use **SOLVIEW** to create a two-view orthographic layout. One view should be a full section. Use an appropriate scale to plot on an A-size sheet.
    H. Create a third floating viewport and insert the 3D section block scaled to half the size of the drawing.
    I. Create a fourth viewport and insert the P14_10_FLATSHOT block scaled to half the size of the drawing.
    J. Save the drawing as P14_10.

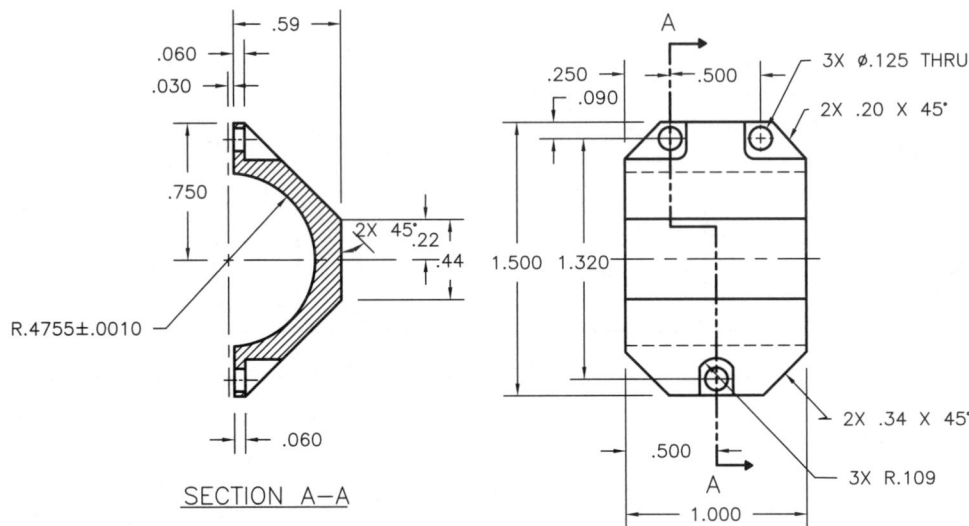

SECTION A—A

AutoCAD and Its Applications—Advanced

CHAPTER 15

Visual Style Settings and Basic Rendering

Learning Objectives

After completing this chapter, you will be able to:

✓ Describe the **Visual Style Manager** palette.
✓ Change the settings for visual styles.
✓ Create custom visual styles.
✓ Export visual styles to a tool palette.
✓ Render a scene using sunlight.
✓ Save a rendered image from the **Render** window.

In Chapter 1, you were introduced to the default visual styles. In Chapter 4, you learned how to use the **Render** tab on the ribbon to adjust several settings related to how the visual style represents objects. This is a way to quickly and easily change the appearance of the scene. In this chapter, you will learn about other visual style settings and how to redefine the visual style. You will also learn how to create your own visual style. Finally, this chapter provides an introduction to lights and rendering.

Overview of the Visual Styles Manager

The **Visual Styles Manager** palette provides access to all of the visual style settings. This palette is a floating window similar to the **Properties** palette. See **Figure 15-1.** Changes made in the **Visual Styles Manager** redefine the visual style.

At the top of the **Visual Styles Manager** are image tiles for the defined visual styles. See **Figure 15-2.** The default visual styles are 2D Wireframe, 3D Hidden, 3D Wireframe, Conceptual, and Realistic. User-defined visual styles also appear as image tiles. The image on the tile is a preview of the visual style settings. Selecting an image tile provides access to the properties of the visual style in the palette below. The name of the currently selected visual style appears below the image tiles and the corresponding image tile is surrounded by a yellow border.

Ribbon
View
> 3D Palettes

Visual Styles
Type
VISUALSTYLES

VISUALSTYLES

Figure 15-1.
The **Visual Styles Manager** palette.

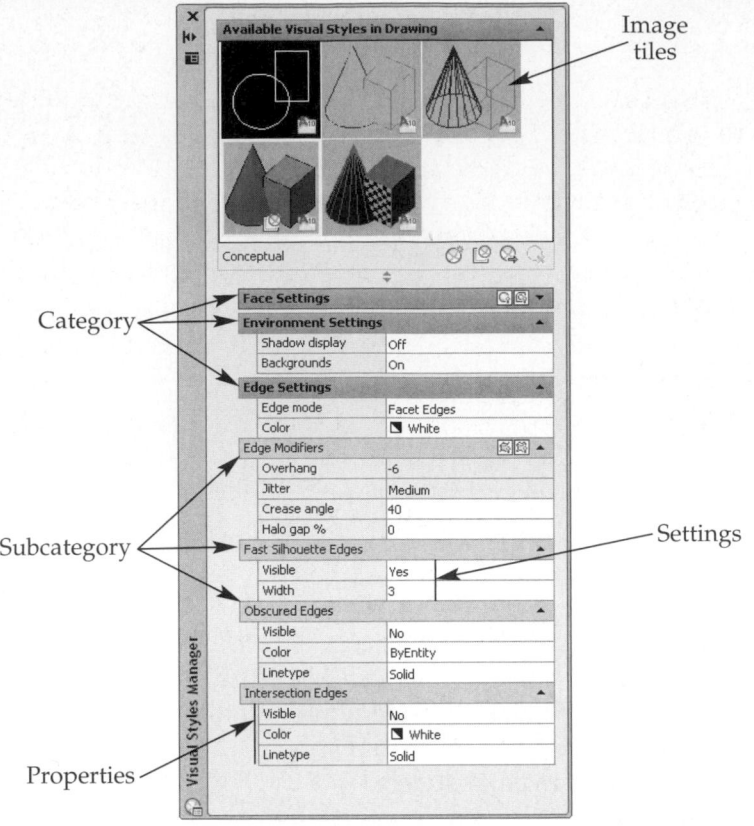

Image tiles

Category

Subcategory

Settings

Properties

Figure 15-2.
The image tiles correspond to the visual styles. The image on the tile is a preview of the visual style's settings.

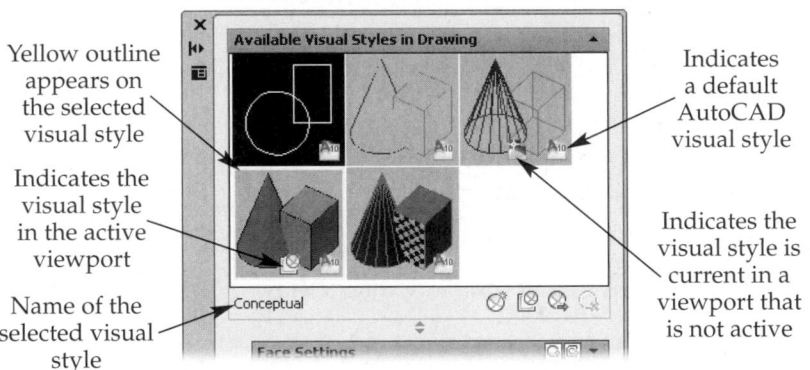

Yellow outline appears on the selected visual style

Indicates the visual style in the active viewport

Name of the selected visual style

Indicates a default AutoCAD visual style

Indicates the visual style is current in a viewport that is not active

To set a different visual style current using the **Visual Styles Manager**, double-click on the image tile. You can also select the image tile and pick the **Apply Selected Visual Style to Current Viewport** button immediately below the image tiles. An icon containing a white star is displayed in the image tile of the visual style that is current in the active viewport, as shown in **Figure 15-2.** A drawing icon appears in the image tile if the visual style is current in a viewport that is not active. The AutoCAD icon appears in the image tiles of the default visual styles.

Exercise 15-1

Complete the exercise on the student website.
www.g-wlearning.com/CAD

As you saw in Chapter 3, the **Visual Styles** and **Edge Effects** panels on the **Render** tab of the ribbon provide several settings for altering the visual style. These settings are also available in the **Visual Styles Manager**. In addition, there are settings in the **Visual Styles Manager** that are not available on the ribbon. The next sections discuss settings available in the **Visual Styles Manager** for the default visual styles. Remember, changing any setting in the **Visual Styles Manager** redefines the visual style. Changes made using the ribbon are temporary.

2D Wireframe

When the 2D Wireframe visual style is set current, lines and curves are used to show the edges of 3D objects. Assigned linetypes and lineweights are displayed. All edges are visible as if the object is constructed of pieces of wire soldered together at the intersections (thus, the name *wireframe*). Either the 2D or 3D wireframe UCS icon is displayed and the 2D grid is displayed, if it is turned on. OLE objects will display normally. In addition, the drawing window display changes to the 2D Model Space context and parallel projection. For the 2D Wireframe visual style, the **Visual Styles Manager** displays the following categories. See **Figure 15-3**.

- **2D Wireframe Options**
- **2D Hide—Obscured Lines**
- **2D Hide—Intersection Edges**
- **2D Hide—Miscellaneous**
- **Display Resolution**

2D wireframe options

The Contour lines property controls the **ISOLINES** system variable. Isolines are the lines used to define curved surfaces on solid objects when displayed in a wireframe view. The setting is 4 by default and can range from 0 to 2047. Isolines are suppressed when the **HIDE** command is used with the 2D Wireframe visual style set current.

The Draw true silhouettes property controls the **DISPSILH** system variable. This determines whether or not silhouette edges are shown on curved surfaces. It is set to Off by default, which is equivalent to a **DISPSILH** setting of 0. This property is different from the **Silhouette Edges** setting in the **Edge Effects** panel in **Render** tab of the ribbon.

Figure 15-3.
The categories and properties available for the 2D Wireframe visual style.

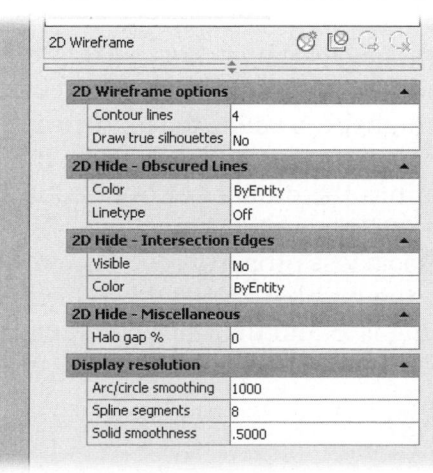

2D hide—obscured lines

The Color property in this category controls the **OBSCUREDCOLOR** system variable. This property determines the color of obscured lines when the **HIDE** command is used. The default setting is ByEntity. This means that, when displayed, obscured lines are shown in the same color as the object.

The Linetype property controls the **OBSCUREDLTYPE** system variable. This property determines whether or not obscured lines are displayed and in which linetype they are displayed. The default setting is Off, which means that obscured lines are not displayed when the **HIDE** command is used. The available linetypes are: Solid, Dashed, Dotted, Short Dash, Medium Dash, Long Dash, Double Short Dash, Double Medium Dash, Double Long Dash, Medium Long Dash, and Sparse Dot.

NOTE

The above linetypes are not the same as the linetypes loaded into the **Linetype Manager** dialog box. They are independent of zoom levels, which means the dash size will stay the same when zooming in and out.

2D hide—intersection edges

This category is used to toggle the display of polylines at the intersection of 3D surfaces and set the color of the lines. The Visible property controls the **INTERSECTIONDISPLAY** system variable. This property determines whether or not polylines are displayed at the intersection of non-unioned 3D surfaces. The default setting is Off, which means that polylines are not displayed when the **HIDE** command is used.

The Color property in this category controls the **INTERSECTIONCOLOR** system variable. This property determines the color of the polylines displayed at intersection edges. By default, the setting is ByEntity. This means that, when displayed, the polylines at intersection edges are shown in the same color as the object.

2D hide—miscellaneous

The Halo Gap % property controls the **HALOGAP** system variable. This property determines the gap that is displayed where one object partially obscures another (between the foreground edge and where the background edge starts to show). The default setting is 0 and the value can range from 0 to 100. The value refers to a percentage of one unit. The gap is only displayed when the **HIDE** command is used. It is not affected by the zoom level.

Display resolution

The Arc/circle smoothing property controls the zoom percentage set by the **VIEWRES** command. This determines the resolution of circles and arcs. The value can range from 1 to 20,000. The higher the value, the higher the resolution of circles and arcs.

The Spline segments property controls the **SPLINESEGS** system variable. This property determines the number of line segments in a spline-fit polyline. The value can range from –32,768 to 32,767.

The Solid smoothness property controls the **FACETRES** system variable. This property determines the number of polygonal faces applied to curved surfaces on solids. The default setting is .5 and the value can range from .01 to 10.0. A higher **FACETRES** value will create a smoother finish when 3D printing.

Exercise 15-2

Complete the exercise on the student website.
www.g-wlearning.com/CAD

3D Hidden

The 3D Hidden visual style removes obscured lines from your view and makes 3D objects appear solid. The previous projection is retained and that context is set current. The benefit of using 3D Hidden is that you get sufficient 3D display, but it does not push the graphics system too hard. Objects are not shaded or colored. This is very useful when working on complex drawings and/or using a slow computer.

3D Wireframe

The 3D Wireframe visual style is similar to the 2D Wireframe visual style. All edges are visible and the shaded UCS icon is displayed. The previous projection is retained and that context is set current. When working in 3D, a wireframe view is sometimes necessary to select objects normally hidden from your view. While a 3D view can be displayed with the 2D Wireframe visual style, setting the 3D Wireframe current automatically displays grid lines and the shaded UCS icon (if they are turned on).

Conceptual

When the Conceptual visual style is set current, objects are smoothed and shaded. The shading is a transition from cool to warm colors. The transitional colors help highlight details. The previous projection is retained and that context is set current.

Realistic

As with the Conceptual visual style, the objects have smoothing and shading applied to them when the Realistic visual style is set current. In addition, if materials are applied to the objects, the materials are displayed. The previous projection is retained and that context is set current. This visual style is good for a final look at the scene before rendering.

Settings for 3D Wireframe, 3D Hidden, Conceptual, and Realistic Visual Styles

The 3D Wireframe, 3D Hidden, Conceptual, and Realistic visual styles share similar categories and settings in the **Visual Styles Manager**. The visual styles have **Face Settings**, **Environmental Settings**, and **Edge Settings** categories. See **Figure 15-4**. These categories and the properties available in them are discussed in the next sections.

Face settings

VSFACESTYLE

Ribbon
Render
>Visual Styles
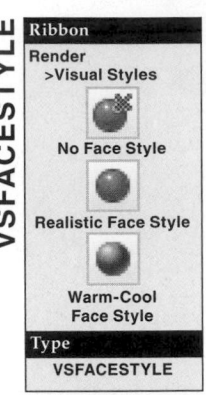
No Face Style
Realistic Face Style
Warm-Cool Face Style
Type
VSFACESTYLE

The Face style property controls the **VSFACESTYLE** system variable. This is also the same as selecting the **No Face Style**, **Realistic Face Style**, or **Warm-Cool Face Style** button in the **Visual Styles** panel on the **Render** tab of the ribbon. These settings are discussed in Chapter 4. The default setting for the 3D Wireframe and 3D Hidden visual styles is None, for the Conceptual visual style is Gooch, and for the Realistic visual style is Real.

VSLIGHTINGQUALITY

Ribbon
Render
>Visual Styles
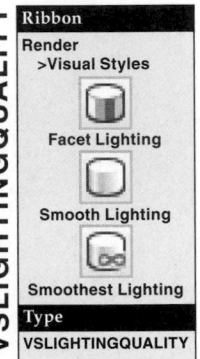
Facet Lighting
Smooth Lighting
Smoothest Lighting
Type
VSLIGHTINGQUALITY

The Lighting quality property controls the **VSLIGHTINGQUALITY** system variable. This property determines whether curved surfaces are displayed smooth or as a series of flat faces. This is also the same as selecting the **Facet Lighting**, **Smooth Lighting**, or **Smoothest Lighting** button in the **Visual Styles** panel on the **Render** tab of the ribbon, as discussed in Chapter 4. No effect is produced if the Face style property is set to None, nor can the property be changed. The default setting for the 3D Wireframe, 3D Hidden, Conceptual, and Realistic visual styles is Smooth.

The Highlight intensity property controls the **VSFACEHIGHLIGHT** system variable. This property determines the size of the highlight on faces to which no material is assigned. A small highlight on an object makes it look smooth and hard. A large highlight on an object makes it look rough or soft. The initial value for the 3D Wireframe, 3D Hidden, Conceptual, and Realistic visual styles is –30, the value can range from –100 to 100. The higher the setting is above 0, the larger the highlight. Settings below 0 set the value, but turn off the effect. To quickly turn the effect on or off, pick the **Highlight intensity** button on the **Face Settings** category title bar. See **Figure 15-5**. This changes the value from negative to positive or vice versa. This property cannot be changed if the Face style property is set to None.

Figure 15-4.
The categories and properties available for the 3D Wireframe, 3D Hidden, Conceptual, and Realistic visual styles.

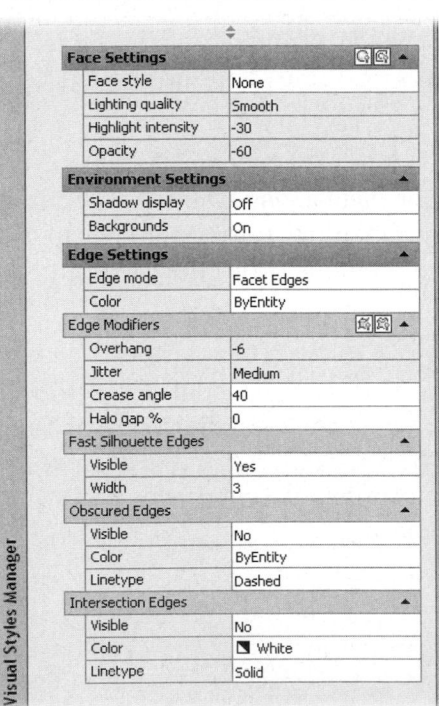

Figure 15-5.
The **Face Settings** category for the 3D Wireframe, 3D Hidden, Conceptual, and Realistic visual styles.

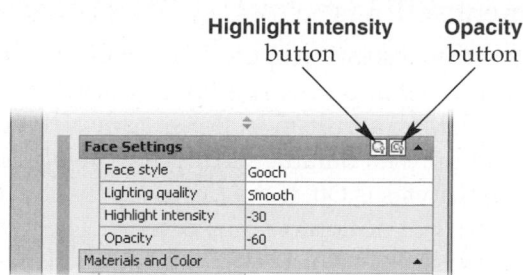

Highlight intensity button   Opacity button

The Opacity property controls the **VSFACEOPACITY** system variable. This property determines how transparent or opaque faces are in the viewport. The default setting for the 3D Wireframe, 3D Hidden, Conceptual, and Realistic visual styles is −60. The value can range from −100 to 100. When the setting is 0, the faces are completely transparent. When the setting is 100, the faces are completely opaque. Settings below 0 set the value, but turn off the effect. To quickly turn the effect on or off, pick the **Opacity** button on the **Face Settings** category title bar. See **Figure 15-5.** This changes the value from negative to positive, or vice versa. This property cannot be changed if the Face style property is set to None.

Opacity may also be controlled by picking the **X-Ray Effect** button in the **Visual Styles** panel on the **Render** tab of the ribbon. The adjacent **Opacity** slider sets the value. However, remember, picking a button or making settings on the ribbon is only temporary. The change is not saved to the visual style.

The **Materials and Color** subcategory is only displayed when the Face style property is set to Real or Gooch. There are three properties in this subcategory—Material display, Face color mode, and Monochrome color or Tint color, depending on which face color mode is current.

The Material display property controls the **VSMATERIALMODE** system variable. The default setting for the 3D Wireframe, 3D Hidden, Conceptual, and Realistic visual styles is Off. This means that objects display in their assigned color. When the setting is changed to Materials, the objects display the color of the material, but not the textures. When the setting is changed to Materials and textures, full materials are displayed.

The Face color mode property controls the **VSFACECOLORMODE** system variable. This property determines how color is applied to the faces of an object. It is the same as picking a button in the face colors flyout in the **Visual Styles** panel on the **Render** tab of the ribbon. The choices are:

- Normal. The object color is applied to faces.
- Monochrome. One color is applied to all faces. This also displays and enables the Monochrome color property.
- Tint. A combination of the object color and a specified color is applied to faces. This also displays and enables the Tint color property. The Tint property only works when the Material display property is set to Materials.
- Desaturate. The object color is applied to faces, but the saturation of the color is reduced by 30%.

The Monochrome color and Tint color properties control the **VSMONOCOLOR** system variable. This system variable determines the color that is applied when the Face color mode property is set to Monochrome or Tint.

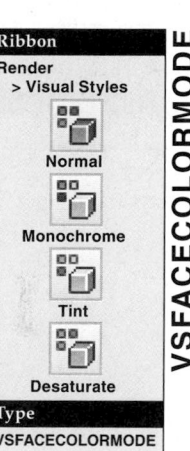

Ribbon
Render
> Visual Styles

Normal

Monochrome

Tint

Desaturate

Type
VSFACECOLORMODE

VSFACECOLORMODE

CAUTION

Displaying materials and textures on 3D objects in a complex drawing will slow system performance. Set the Face color mode property to Materials and textures only when it is absolutely necessary.

Environment settings

Ribbon
Render
> Lights

No Shadows

Ground Shadows

Full Shadows

Type
VSSHADOWS

The Shadow display property controls the **VSSHADOWS** system variable. This property controls if and how shadows are cast when the visual style is set current. It is the same as picking a button in the shadows flyout in the **Lights** panel on the **Render** tab of the ribbon. The default setting for the 3D Wireframe, 3D Hidden, Conceptual, and Realistic visual styles is Off. If the property is set to Ground shadow, objects cast shadows on the ground, but not onto other objects. The "ground" is the XY plane of the WCS. The Full shadows setting only works if lights have been placed in the scene and hardware acceleration is enabled.

The Backgrounds property controls the **VSBACKGROUNDS** system variable. This property determines whether or not the preselected background is displayed in the viewport. The default setting for the 3D Wireframe, 3D Hidden, Conceptual, and Realistic visual styles is On. Backgrounds can only be assigned to a view when a named view is created or edited. After the view is created, restore the view to display the background.

NOTE

Use the **3DCONFIG** command to enable hardware acceleration.

Edge settings

VSEDGES

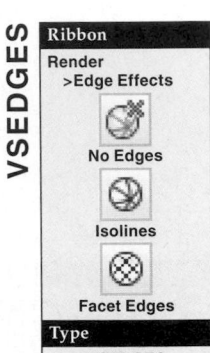

Ribbon
Render
>Edge Effects

No Edges

Isolines

Facet Edges

Type
VSEDGES

The Edge mode property controls the **VSEDGES** system variable. This property determines how edges on solid objects are represented when the visual style is set current. This is the same as picking a button in the edge flyout in the **Edge Effects** panel on the **Render** tab of the ribbon. The default for the 3D Wireframe and Realistic visual styles is Isolines. This means that isolines are displayed. The default for the 3D Hidden and Conceptual visual styles is Facet Edges. This means that faceted edges are displayed. Setting this property to None turns off isolines and facets and displays no edges. If the Face style property is set to None, this property cannot be set to None.

The Color property controls the **VSEDGECOLOR** system variable. This property determines the color of all edges on objects in the drawing. It is disabled when the Edge mode property is set to None.

The Number of lines and Always on top properties are displayed when the Edge mode property is set to Isolines. The Number of lines property controls the **ISOLINES** system variable. The Always on top property controls the **VSISOONTOP** system variable. This property determines if isolines are displayed when the model is shaded or hidden. The default for the 3D Wireframe, 3D Hidden, Conceptual, and Realistic visual styles is No. When set to Yes, edges are always displayed.

Edge Modifiers. This subcategory is not displayed if the Edge mode property is set to None. The Overhang property controls the **VSEDGEOVERHANG** system variable. This property can be used to create a hand-sketched appearance by extending the ends of edges. See **Figure 15-6A.** In order to make changes to this property, the **Overhanging edges** button must be on in the **Edge Modifiers** subcategory title bar. See **Figure 15-7.** The **Edge Overhang** button in the **Edge Effects** panel on the **Render** tab of the ribbon can also be turned on. The value for Overhang property can range from –100 to 100, which is the number of pixels. The higher the setting, the longer the overhang. A negative value sets the overhang length, but turns off the property. Picking either button makes the value positive and applies the effect (or makes the value negative and turns off the effect).

VSEDGEOVERHANG

Ribbon
Render
>Edge Effects

Edge Overhang

Type
VSEDGEOVERHANG

To adjust the overhang setting using the **Edge Effects** panel, first turn on the effect by picking the button. Then, move the cursor over the **Overhang** slider. The cursor changes to left and right arrows. Pick and drag the slider to the left to decrease the overhang or to the right to increase the overhang. The value is displayed on the right-hand side of the slider. Remember, however, that settings made in the ribbon are not saved to the visual style.

The Jitter property controls the **VSEDGEJITTER** system variable. Jitter makes edges of objects look as if they were sketched with a pencil. See **Figure 15-6B.** In order to make changes to this property, the **Jitter edges** button must be on in the **Edge Modifiers** subcategory title bar. See **Figure 15-7.** The **Edge Jitter** button in the **Edge Effects** panel on the **Render** tab of the ribbon can also be turned on. There are four settings from which to choose: Off, Low, Medium, and High. The number of sketched lines increases at each higher setting.

To adjust the jitter setting using the **Edge Effects** panel, first turn on the effect by picking the button. Then, move the cursor over the slider. The cursor changes to left and right arrows. Pick and drag the slider to the left to decrease the jitter or to the right to increase the jitter. The value is displayed on the right-hand side of the slider. Remember, there are only four possible settings for jitter.

When the Edge mode property is set to Facet Edges, the Crease angle and Halo gap % properties are displayed in the **Edge Modifiers** subcategory. The Crease angle property controls the **VSEDGESMOOTH** system variable. This property determines how facet edges within a face are displayed based on the angle between adjacent faces. It does not affect edges between faces. See **Figure 15-8.** The value can range from 0 to 180. This is

Ribbon
Render
>Edge Effects
Edge Jitter
Type
VSEDGEJITTER

VSEDGEJITTER

Figure 15-6.
A—Overhanging edges have been turned on for this visual style.
B—Edge jitter has been turned on for this visual style.

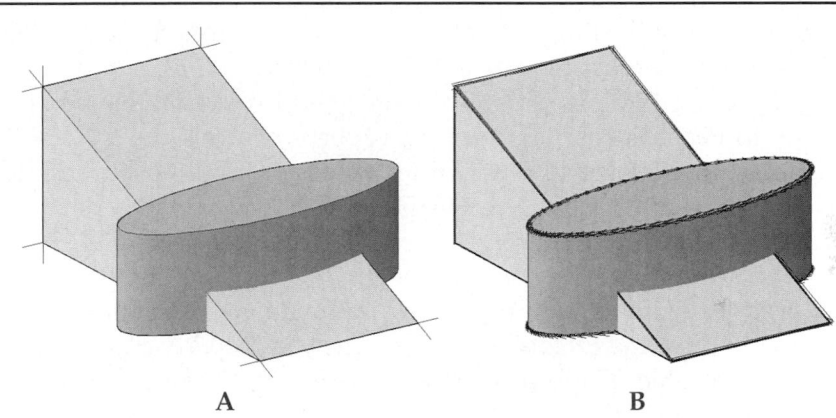

A   B

Figure 15-7.
The **Edge Modifiers** subcategory for the 3D Wireframe, 3D Hidden, Conceptual, and Realistic visual styles.

Overhanging edges button   **Jitter edges** button

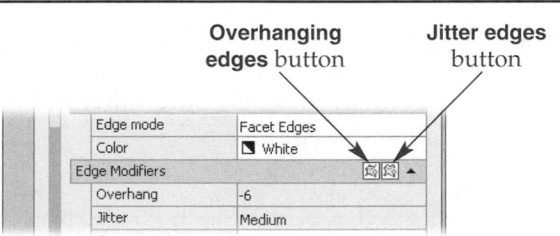

Figure 15-8.
A—The Crease angle property is set to 0. Notice the edges between facets within each face.
B—The Crease angle property is set to 10. The edges are no longer displayed.

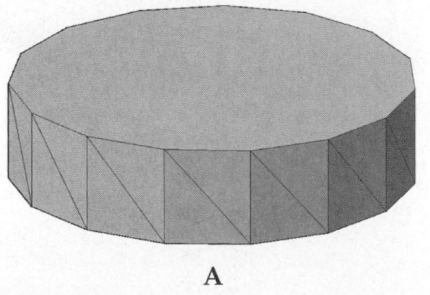
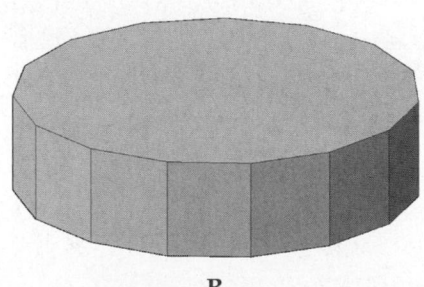

A   B

the number of degrees between edges below which a line is displayed. The Halo gap % property is similar to the setting discussed earlier in the 2D Hide—Miscellaneous section; however, this property controls the **VSHALOGAP** system variable.

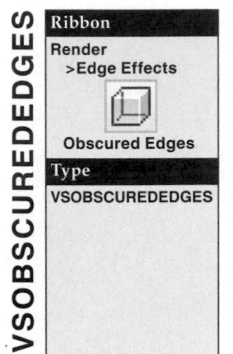

VSSILHEDGES

Ribbon
Render
>Edge Effects

Silhouette Edges

Type
VSSILHEDGES

Fast Silhouette Edges. This subcategory is available for all of the Edge mode settings. The Visible property controls the **VSSILHEDGES** system variable. It determines whether or not silhouette edges are displayed around the outside edges of all objects. The default for the 3D Wireframe and Realistic visual styles is No. The default for the 3D Hidden and Conceptual visual styles is Yes. The Yes setting is the same as turning on the **Silhouette Edges** button in the **Edge Effects** panel on the **Render** tab of the ribbon.

The Width property controls the **VSSILHWIDTH** system variable. This property determines the width of silhouette lines. It is measured in pixels and the value can range from 1 to 25. Changing this property is the same as adjusting the **Silhouette width** slider on the **Edge Effects** panel. To adjust this slider, the **Silhouette Edges** button must be on (or the Visible property set to Yes).

Obscured Edges. This subcategory is only available when the Edge mode property is set to Facet Edges. The Visible property controls the **VSOBSCUREDEDGES** system variable. This property determines whether or not obscured edges are displayed in a hidden or shaded view. See **Figure 15-9.** Setting this property is the same as picking the **Obscured Edges** button in the **Edge Effects** panel on the **Render** tab of the ribbon.

The Color property controls the **VSOBSCUREDCOLOR** system variable. Setting this color is the same as picking a color in the drop-down list next to the **Obscured Edges** button in the **Edge Effects** panel. The Linetype property controls the **VSOBSCUREDLTYPE** system variable. These properties function the same as those discussed earlier in this chapter in the 2D Hide—Obscured Lines section.

VSOBSCUREDEDGES

Ribbon
Render
>Edge Effects

Obscured Edges

Type
VSOBSCUREDEDGES

VSINTERSECTIONEDGES

Ribbon
Render
>Edge Effects

Intersection Edges

Type
VSINTERSECTION-
EDGES

Intersection Edges. This subcategory is only available when the Edge mode property is set to Facet Edges. The Visible property controls the **VSINTERSECTIONEDGES** system variable. This property determines whether or not lines are displayed where one 3D object intersects another 3D object. See **Figure 15-10.** Setting this property to Yes is the same as turning on the **Intersection Edges** button in the **Edge Effects** panel on the **Render** tab of the ribbon.

Figure 15-9.
A—Obscured lines are not shown. B—The Visible property is set to Yes and obscured lines are shown.

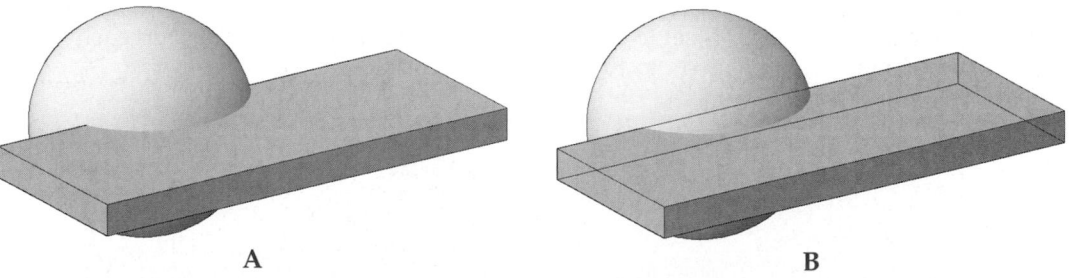

A   B

AutoCAD and Its Applications—Advanced

Figure 15-10.
A—A line does not appear where these two objects intersect. B—The Visible property is set to Yes and a line appears at the intersection.

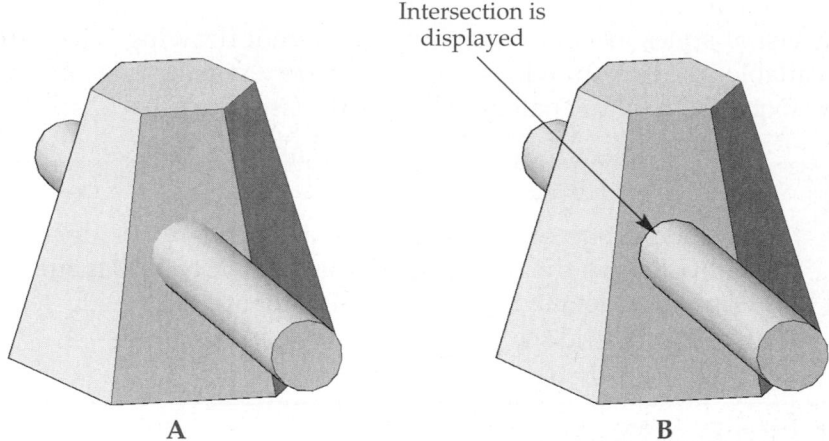

Intersection is displayed

A   B

The Color property controls the **VSINTERSECTIONCOLOR** system variable. Setting this variable is the same as picking a color in the drop-down list next to the **Intersection Edges** button in the **Edge Effects** panel. The Linetype property controls the **VSINTERSECTIONLTYPE** system variable. These properties function the same as those discussed earlier in this chapter in the 2D Hide—Intersection Edges section.

PROFESSIONAL TIP

Setting the intersection edges Color property to a color that contrasts with the objects in your model is a good way to quickly check for interference between 3D objects.

Creating Your Own Visual Style

As you saw in the previous sections, you can customize the default AutoCAD visual styles. However, you may also want to create a number of different visual styles to quickly change the display of the scene. Custom visual styles are easy to create.

To create a custom visual style, open the **Visual Style Manager**. Then, pick the **Create New Visual Style** button below the image tiles. You can also right-click in the image tile area and select **Create New Visual Style...** from the shortcut menu. In the **Create New Visual Style** dialog box that appears, type a name for the new style and give it a description. See **Figure 15-11**. Then, pick the **OK** button to create the new visual style.

Figure 15-11.
Creating a new visual style.

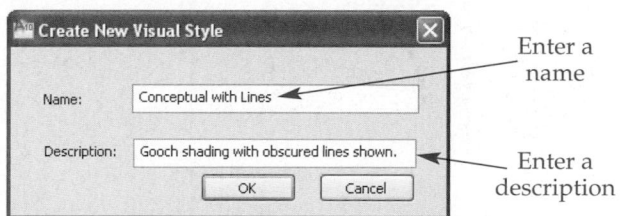

Enter a name

Enter a description

An image tile is created for the new visual style. The name and description of the visual style appear as help text when the cursor is over the image tile. Select the image tile to display the default properties for the new visual style. Then, change the settings as needed to meet your requirements.

Custom visual styles are only saved in the current drawing. They are not automatically available in other drawings. To use the new visual styles in any drawing, they must be exported to a tool palette. This is discussed in the next section.

PROFESSIONAL TIP

To return one of AutoCAD's visual styles to its default settings, right-click on the image tile in the **Visual Styles Manager** and select **Restore to default** from the shortcut menu.

Exercise 15-3
Complete the exercise on the student website.
www.g-wlearning.com/CAD

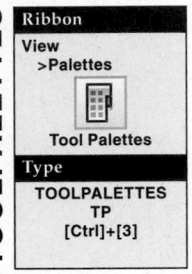

TOOLPALETTES

Ribbon
View
>Palettes

[icon]

Tool Palettes

Type
TOOLPALETTES
TP
[Ctrl]+[3]

Steps for Exporting Visual Styles to a Tool Palette

To have custom visual styles available in other drawings, export them to a tool palette. Use the following procedure.

1. Create and customize a visual style as described in the previous section.
2. Open the **Tool Palettes** window.
3. Right-click on the **Tool Palettes** title bar and pick **New Palette** from the shortcut menu.
4. Type the name of the new palette, such as My Visual Styles, in the text box that appears. See **Figure 15-12A**.

Figure 15-12.
A—Creating a new tool palette on which to place visual style tools.
B— A visual style has been copied to the tool palette as a tool.

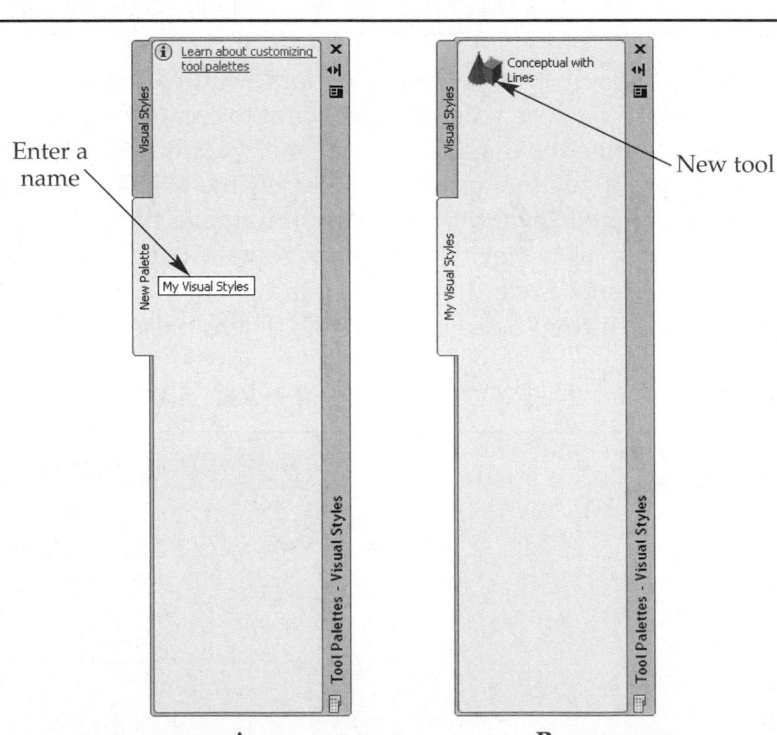

Enter a name

New tool

A   B

5. The new palette is added and active. You are ready to export your custom visual styles into it.

6. Select the image tile of the visual style in the **Visual Styles Manager**. Remember, a yellow border appears around the selected image tile.

7. Pick the **Export the Selected Visual Style to the Tool Palette** button below the image tiles in the **Visual Styles Manager**. You can also right-click on the image tile and select **Export to Active Tool Palette** from the shortcut menu.

A new tool now appears in the palette with the same image, name, and description as the visual style in the **Visual Styles Manager**. See **Figure 15-12B**. Selecting the tool applies the visual style to the current viewport. You can also right-click on the tool to display a shortcut menu. Using this menu, you can apply the visual style to the current viewport, all viewports, or add the visual style to the current drawing. The shortcut menu also allows you to rename the tool, access the properties of the visual style, and delete the visual style from the palette.

PROFESSIONAL TIP

A visual style can be added as a tool on a tool palette by dragging its image tile from the **Visual Styles Manager** and dropping it onto the tool palette.

Exercise 15-4

Complete the exercise on the student website.
www.g-wlearning.com/CAD

Deleting Visual Styles from the Visual Styles Manager

Custom visual styles can be deleted from the **Visual Styles Manager**. Pick the image tile of the visual style you want to delete. Then, pick the **Delete the Selected Visual Style** button below the image tiles. You can also right-click on the image tile and select **Delete** from the shortcut menu. You are *not* warned about the deletion. The default AutoCAD visual styles cannot be deleted, nor can a visual style that is currently in use.

Plotting Visual Styles

A visual style not only affects the on-screen display, it also affects plots. To plot objects with a specific visual style, use the following guidelines.

Plotting a Visual Style from Model Space

There are two basic methods for plotting from model space. The method you use simply depends on your preference.

Method 1. Open the **Plot** dialog box and expand it by picking the **More Options** (>) button. Then, select the desired display from the **Shade plot** drop-down list in the **Shaded viewport options** area. Finally, plot the drawing. With this method, the current visual style is irrelevant.

Method 2. Set the desired visual style current. Then, open the **Plot** dialog box. Select As displayed from the **Shade plot** drop-down list in the **Shaded viewport options** area. Finally, plot the drawing.

Plotting a Visual Style from Layout (Paper) Space

When plotting from layout (paper) space, the shade plot properties of the viewports govern how the viewport is plotted. The viewports can be set to plot visual styles in three different ways.

Method 1. Select the viewport in layout space and right-click to display the shortcut menu. Pick **Shade plot** to display the cascading menu. Then, select the appropriate visual style.

Method 2. Use the **Properties** palette to set the Shade plot property of the viewport. To do this, select the viewport in layout space and open the **Properties** palette. Pick the Shade plot property in the **Misc** category and change the setting to the desired option. The Shade plot property is also available in the **Quick Properties** palette.

Method 3. Use the **Visual Styles** suboption of the **Shadeplot** option of the **MVIEW** command. When prompted to select objects, pick the border of the viewport. Do not pick the objects in the viewport.

> **NOTE**
>
> The visual style of the viewport may also be selected when you create a viewport configuration in the **Viewports** dialog box (**VPORTS** command). Select the viewport in the **Preview** area of the dialog box. Then, pick the visual style desired from the **Visual Style:** drop-down list at the bottom of the dialog box. The **VPORTS** command can be used in model space or layout (paper) space.

Introduction to Rendering

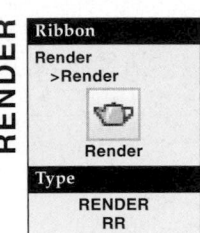

Visual styles provide a way to plot your 3D scene to paper or a file, but control over the appearance is limited to the visual style settings. In Chapter 4, you were briefly introduced to the **RENDER** command. The **RENDER** command offers complete control over the scene and, with its features, you can create photorealistic images. In this chapter, you will be introduced to AutoCAD's rendering and lighting tools. Materials are discussed in later chapters along with more advanced rendering and lighting features.

When you render a scene, you are making a realistic image of your design that can be printed, displayed on a web page, or used in a presentation. To create an attractive rendering, you have to figure out what view you want to display, where the lights should be placed, what types of materials need to be applied to the 3D objects, and the kind of output that is needed. This section shows you how to create a quick rendering of your scene.

Introduction to Lights

Lights provide the illumination to a scene and are essential for rendering. There are three types of lighting in AutoCAD—default lighting, sunlight, and user-created lighting. AutoCAD automatically creates two default light sources in every scene. These lights ensure that all surfaces on the model are illuminated and visible. The types of lighting are discussed in more detail in Chapter 17.

A scene can be rendered with the default lights, but the results are usually not adequate to produce a photorealistic image. See **Figure 15-13.** The appearance is very artificial and no shadows are created. Shadows anchor objects to the scene and make them look real. See **Figure 15-14.** Without shadows, objects appear to float in space. Because the default lights do not cast shadows, other lights must be added to the scene and set to cast shadows. When a light is added to a scene, the default lights must be turned off. The first time you add a light, you receive a warning to this effect (unless the warning has been disabled).

Figure 15-13.
A scene rendered with the default AutoCAD lighting.

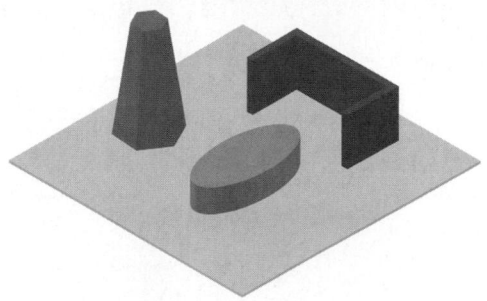

Figure 15-14.
A light has been added and set to cast shadows. Compare this rendering with Figure 15-13.

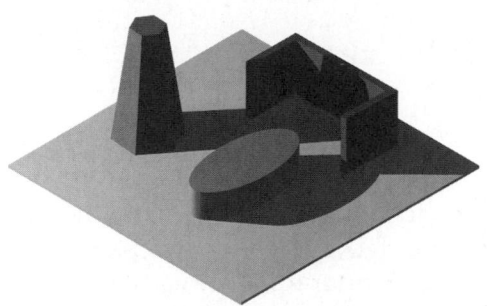

In this section, you will learn how to add sunlight to the scene. Chapter 17 provides detailed information on lighting. Sunlight is produced by an automated distant light. Sunlight can be turned on by picking the **Sun Status** button in the **Sun & Location** panel on the **Render** tab of the ribbon. The button background is blue when sunlight is on. See **Figure 15-15.**

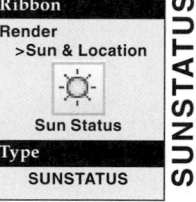

If the **Default Lighting** button is on in the **Lights** panel when the **Sun Status** button is turned on, a warning dialog box is displayed. This dialog box gives you choices to either turn off default lighting or keep it on. You cannot see the effects of sunlight with default lighting turned on, so it is recommended to turn it off.

If the current visual style is set to display full shadows, you should now see shadows in the scene, provided there are areas to receive shadows. Remember, hardware acceleration must be enabled to display full shadows.

The **Date** and **Time** sliders in the **Sun & Location** panel on the **Render** tab of the ribbon are active when sunlight is turned on. You can drag the sliders to adjust the date and time. The current date and time are displayed on the right-hand end of the sliders. As you drag the sliders, the shadows in the scene change to reflect the settings.

Figure 15-15.
The **Lights** and **Sun & Location** panels on the **Render** tab of the ribbon.

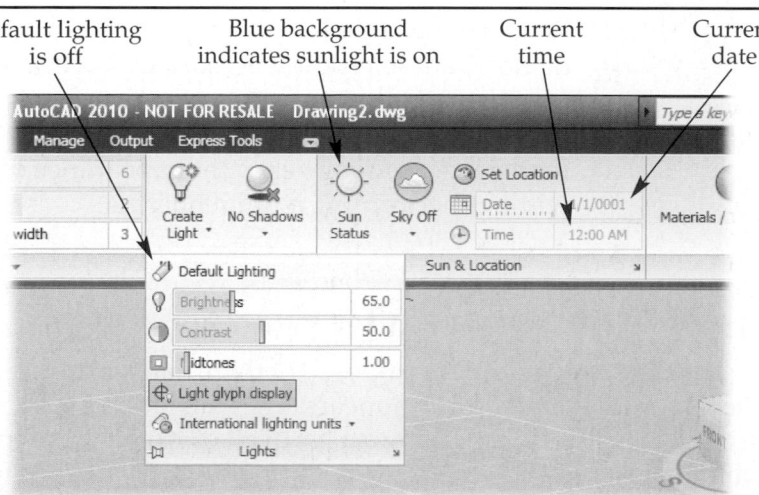

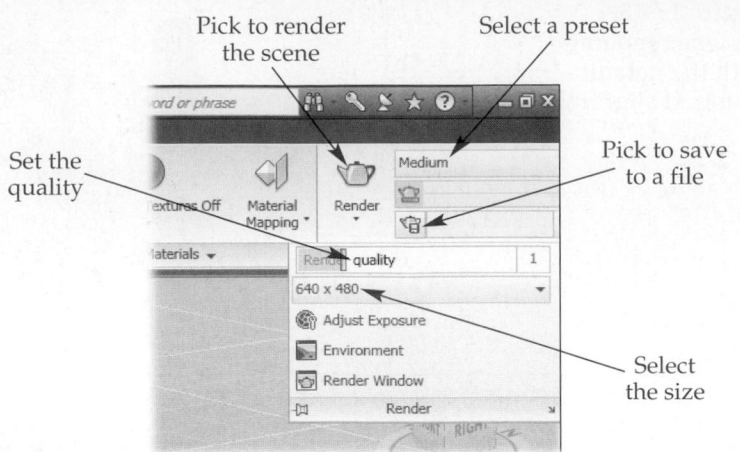

Figure 15-16.
The **Render** panel on the **Render** tab of the ribbon.

Pick to render the scene

Select a preset

Set the quality

Pick to save to a file

Select the size

Rendering the Scene

The **Render** panel on the **Render** tab of the ribbon is shown in **Figure 15-16.** There are two buttons located in this control panel to initiate a rendering. If you pick the **Render** button, the **Render** window appears (by default) and AutoCAD immediately starts rendering the viewport. You will see the rendered tiles appear in the image pane as they are calculated. The **Render** window is explained more in the next section.

If you pick the **Render Region** button, which is located in the flyout, you are prompted to pick two points in the viewport, similar to performing a window selection. The selected area is rendered in the viewport. Rendering a cropped area is often used to test areas of the scene for possible problems before performing the final rendering.

Also in the **Render** panel you will find the render preset drop-down list. This is located in the upper-right corner of the panel. This list gives you a selection of rendering presets based on image quality. The choices are:

- Draft
- Low
- Medium
- High
- Presentation

The Draft entry produces the lowest-quality rendering. The Presentation entry produces the highest-quality rendering. The better the quality, the longer it takes to complete the rendering process.

PROFESSIONAL TIP

Rendering a complex drawing may take a very long time and you do not want to repeat it because of some small error. It is important to make sure that everything in the scene is perfect before the final rendering. By rendering a cropped region and using lower-quality renderings, you can verify the appearance of any questionable areas without performing a full rendering.

Introduction to the Render Window

The **Render** window is composed of three main areas. See **Figure 15-17.** The image pane is where the rendering appears. The statistics pane shows the current rendering settings. The history pane shows a list of all of the images rendered from the drawing, with the most recent at the top.

AutoCAD and Its Applications—Advanced

Figure 15-17.
The **Render** window.

Image pane

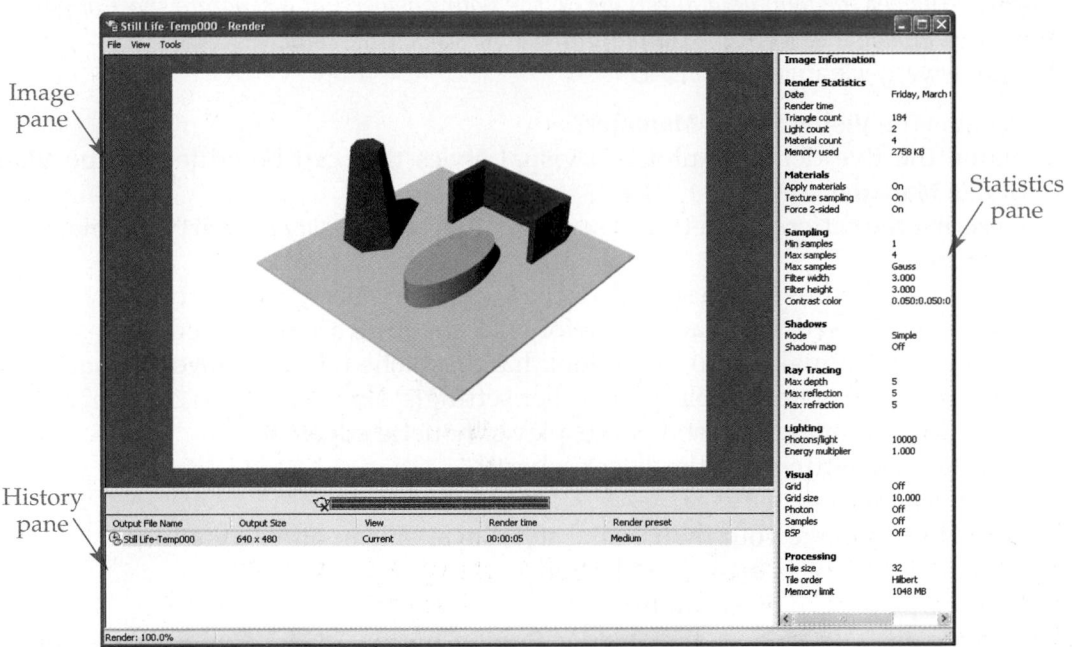

Statistics pane

History pane

You can zoom into the image in the image pane for detailed inspection. Use the mouse scroll wheel or the **Zoom +** and **Zoom –** entries in the **Tools** pull-down menu of the **Render** window. In the **File** pull-down menu of the **Render** window, select **Save...** to save the image selected in the history pane to an image file. The symbol in front of the image in the history pane changes to a folder with a green check mark on it. The **Save Copy...** option in the **File** pull-down menu of the **Render** window creates a copy of the image without modifying the original in the history pane.

NOTE

Advanced rendering is discussed in Chapter 18.

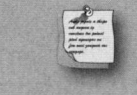

Exercise 15-5

Complete the exercise on the student website.
www.g-wlearning.com/CAD

Chapter Test

Answer the following questions. Write your answers on a separate sheet of paper or complete the electronic chapter test on the student website.
www.g-wlearning.com/CAD

1. What is the **Visual Styles Manager**?
2. Name the five default AutoCAD visual styles that can be edited in the **Visual Styles Manager**.
3. Describe the difference between setting the Lighting quality property to Smooth and Faceted.
4. What does the Desaturate setting of the Face color mode property do?
5. What has to be added to a scene before full shadows are displayed?
6. If you want to make your scene look hand sketched, but the Overhang and Jitter properties are not available, what other setting(s) do you have to change?
7. How do you set a visual style to display silhouette edges?
8. List the four settings for the Jitter property.
9. What is an *intersection edge*?
10. How do you make your own visual styles available in other drawings?
11. Which visual styles cannot be deleted?
12. How can you turn on sunlight?
13. Explain the function of the **Render Region** button in the **Render** panel on the ribbon.
14. Name the three main areas of the **Render** window.
15. How can you save a rendered image in the **Render** window?

Drawing Problems

1. In this problem, you will construct a living room scene using some simple shapes and blocks available through **DesignCenter**.
    A. Draw a 12′ × 12′ × 1″ box.
    B. Draw two boxes to represent walls, 12′ × 4″ × 8′. Position them as shown on the next page.
    C. Open **DesignCenter** and select the **Autodesk Seek Design Content** button (you must be connected to the Internet).
    D. Search for 3D drawing files for the following objects, drag and drop them into the scene, and position them as shown: sofa, table, end table, lamp, plant, entertainment center, and chair. The blocks do not have to be exactly the same as shown on the next page and may need to be scaled up or down.
    E. Apply each of the five default visual styles to the viewport and plot each. Use the As Displayed option in the **Plot** dialog box. Note the differences in each one.
    F. Save the drawing as P15_01.

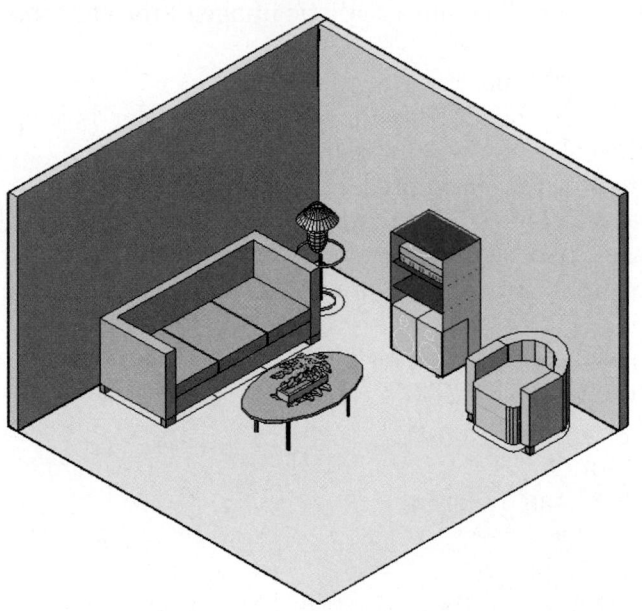

2. In this problem, you will create a new visual style to display the scene as if it is hand sketched.
   A. Open drawing P15_01.
   B. Create a new visual style named Hand Sketched with a description of Displays objects as sketched.
   C. Change the overhang and jitter settings to make the scene look as shown below.
   D. Change any other settings you like.
   E. Plot the scene. Select the new visual style in the **Shade plot** drop-down list.
   F. Export the new visual style to a tool palette so that it can be used in other drawings.
   G. Save the drawing as P15_02.

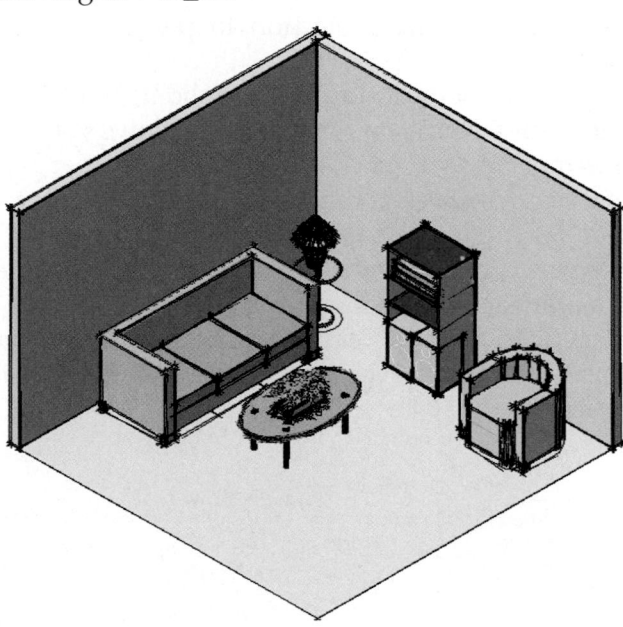

3. In this problem, you will create a realistic image of the car fender that you created in Chapter 10.
   A. Open drawing P10_03.
   B. Freeze any layers needed so that only the fender is displayed. Display the fender in the color you want it to be.
   C. Draw a planar surface to represent the ground.
   D. Set the Realistic visual style current. Then, turn on the highlight intensity and full shadows. Also, set the Edge Mode property to None.
   E. Turn on sunlight. Adjust the **Date** and **Time** sliders to make the shadows look as shown.
   F. Render the scene and save it as a JPEG image. Name the file P15_03.jpg.
   G. Save the drawing as P15_03.

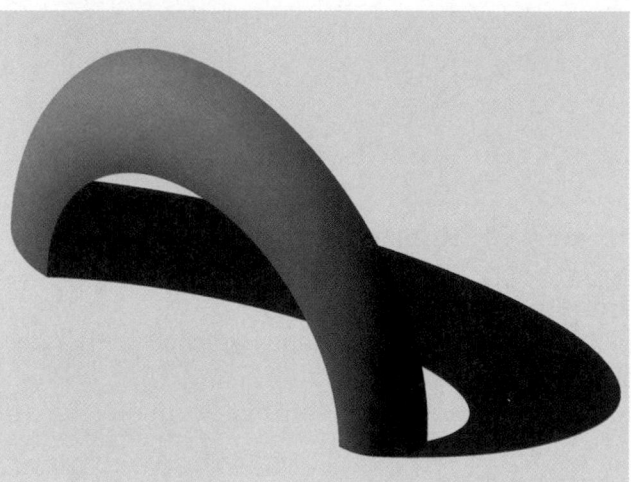

4. This problem demonstrates the differences in rendering time and image quality of the five different rendering presets.
   A. Open any 3D drawing from a previous chapter and display an appropriate isometric view. Change the projection to perspective, if it is not already current.
   B. Turn on sunlight and set the **Date** and **Time** sliders to place the shadows where you want them. Tip: Turning on full shadows allows you to locate the shadows without rendering.
   C. Render the scene once for each rendering preset: Draft, Low, Medium, High, and Presentation.
   D. In the history pane of the **Render** window, note the differences between the rendering time for each rendering.
   E. Save each image with a corresponding name: P15_04_Draft.jpg, P15_04_Low.jpg, P15_04_Medium.jpg, P15_04_High.jpg, and P15_04_Presentation.jpg.
   F. Save the drawing as P15_04.

Materials in AutoCAD

Learning Objectives

After completing this chapter, you will be able to:

✓ Attach materials to the objects in a drawing.
✓ Change the properties of existing materials.
✓ Create new materials.

A *material* is simply an image stretched over an object to make it appear as though the object is made out of wood, marble, glass, brick, or various other materials. AutoCAD provides an assortment of materials that can be used in your drawings to create a realistic scene. The materials are grouped into categories to make them easier to find.

Materials are easy to attach. They can be dragged and dropped onto the objects, attached to all selected objects, and even attached based on the object's layer. Once the material is attached, you can adjust how the material is *mapped* to the object. If the current visual style is set to display materials in the viewport, you can immediately see the effects on the object. The properties of a material can also be changed to make it look shinier, softer, smoother, rougher, and so on. When you finally render the scene, you will see the full effect of the materials.

PROFESSIONAL TIP

The "typical" installation of AutoCAD includes approximately 75 sample materials. However, approximately 425 materials are available if you choose to install the materials library. If you did not choose that option during installation of AutoCAD, you can use the installation DVD to add the full materials library. For information describing how to do this, pick the **Help** button in the **InfoCenter** or press [F1] to display the online documentation. Then, in the **Contents** tab, select the **Installation and Licensing Guides** topic. Browse the topic for information on installing additional features such as the materials library.

Materials Library

TOOLPALETTES

Ribbon
View
> Palettes

Tool Palettes

Type
TOOLPALETTES
TP
[Ctrl]+[3]

The *materials library* is the location where all materials are stored. AutoCAD uses tool palettes as the materials library. The **Materials** tool palette group contains eight sample palettes:

- **Concrete—Materials Sample**
- **Doors and Windows—Materials Sample**
- **Fabric—Materials Sample**
- **Finishes—Materials Sample**
- **Flooring—Materials Sample**
- **Masonry—Materials Sample**
- **Metals—Materials Sample**
- **Woods and Plastics—Materials Sample**

To display only the palettes in the **Materials** group, right-click on the title bar of the **Tool Palettes** window and select **Materials** from the shortcut menu, **Figure 16-1.** Notice the other tool palette groups in the shortcut menu.

Each palette contains several material tools. See **Figure 16-2.** Each material is displayed in the palette as a sphere on a checkered background. The background is used to make the transparent materials, such as glass, more visible. The material name is, by default, shown to the right of the sphere. You may want to increase the width of the **Tool Palettes** window to see the complete name of each material. Tool palette options, such as display options, are covered in detail in Chapter 24.

The properties of the individual materials can be accessed by right-clicking on the tool in the palette and picking **Properties...** from the shortcut menu. In the **Tool Properties** dialog box that is displayed, you can change the name, description of the material, and material properties. See **Figure 16-3.** The **Tool Properties** dialog box is

Figure 16-1.
Displaying only the tool palettes in the **Materials** group. Notice that the materials library has also been installed.

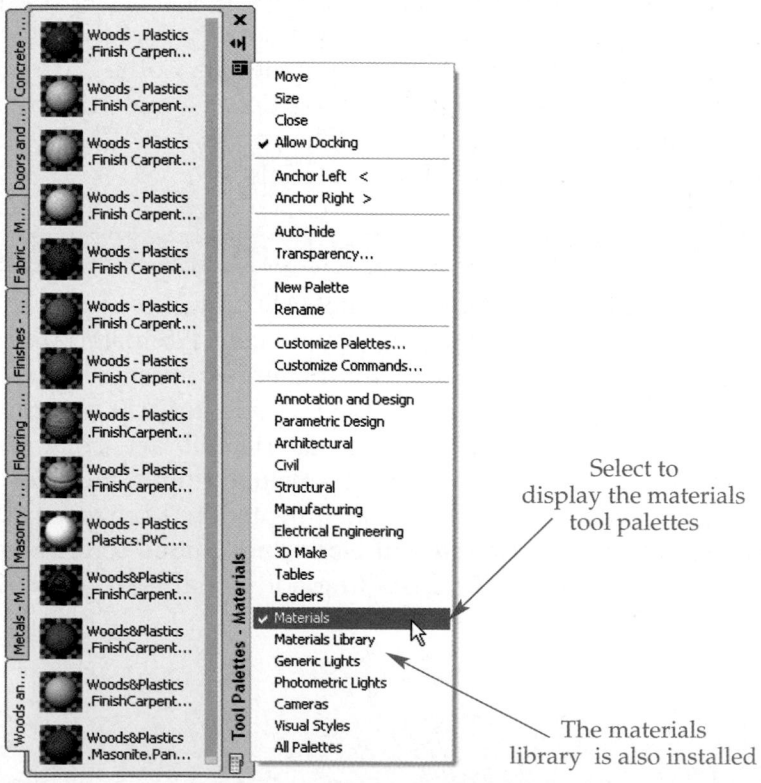

AutoCAD and Its Applications—Advanced

Figure 16-2.
Selected tool palettes in the **Materials** group. A—**Doors and Windows**. B—**Flooring**.
C—**Masonry**. D—**Woods and Plastics**.

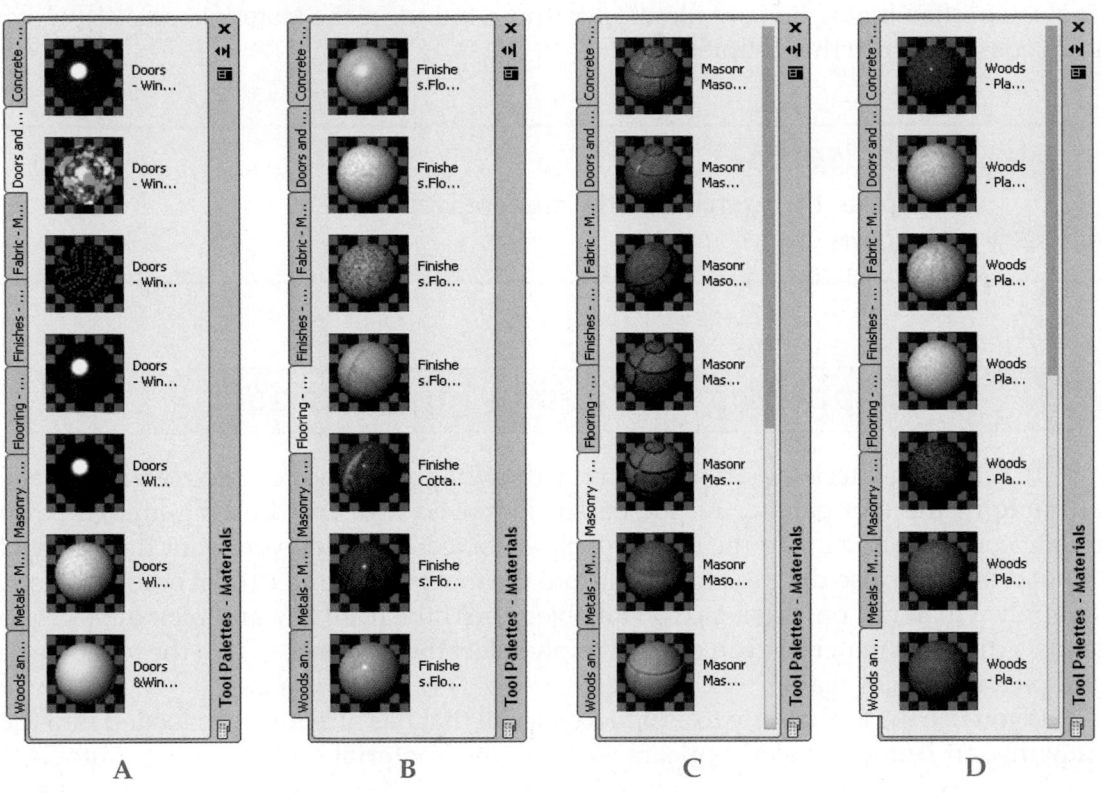

Figure 16-3.
The properties of a
material tool.

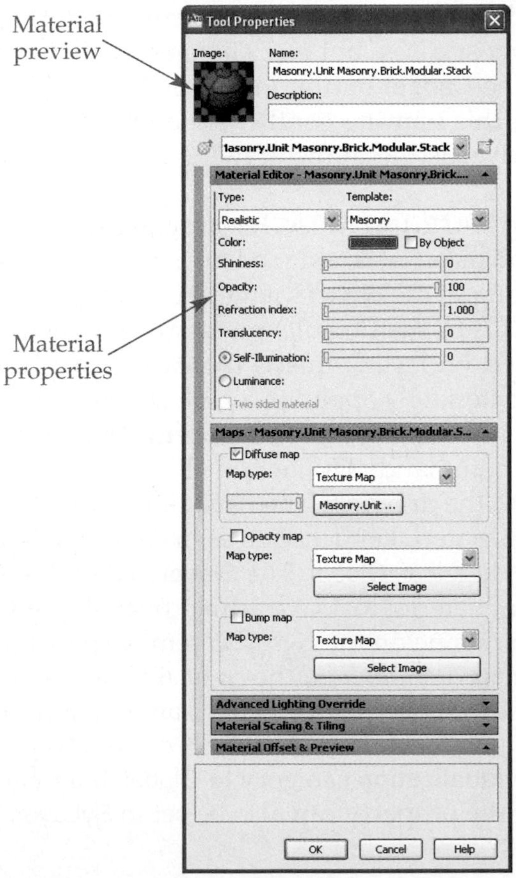

Material
preview

Material
properties

discussed in more detail in Chapter 24. Changes made to the material in this dialog box only affect the material tool. To apply these changes to objects in the current scene, the material must be reattached to the objects. The features in the material editor in the **Tool Properties** dialog box are identical to those found in the **Materials** palette, which is discussed later in this chapter.

Exercise 16-1

Complete the exercise on the student website.
www.g-wlearning.com/CAD

Applying and Removing Materials

To attach a material to an object in the drawing, pick once on the material image or name in the tool palette. As the cursor is moved into the drawing area, a paint brush icon appears next to the cursor. Pick the object to which you want the material applied. You can also drag the material from the tool palette and drop it onto an object. To apply a material only to a face on an object, hold the [Ctrl] key and pick the face. To apply a different material to an object, simply select the new material in the tool palette and pick the object again.

If you use the tool palette to assign a material that has already been loaded into the drawing, an AutoCAD alert appears warning of a material name conflict. AutoCAD needs to know how to handle the duplicate material. If you pick **Save this Material as a Copy**, the material is added to the drawing as Copy of *material name* (or Copy 1 of *material name*, Copy 2 of *material name*, and so on). Picking **Overwrite the Material** replaces the material in the drawing with the one you are attempting to attach. This is probably the best option in most cases, but any changes made to the existing material in the drawing are lost. The **Cancel** button in the alert dialog box allows you to cancel the operation. To avoid this naming-conflict situation, use the **Materials** palette to apply any materials already existing in the drawing to other objects. The **Materials** palette is discussed later.

You can use the **MATERIALATTACH** command to assign materials to the layers in your drawing. Once a material is assigned to a layer, any object on that layer is displayed in the material, as long as the object's material property is set to ByLayer. When objects are created in AutoCAD, the default "material" assigned to them is ByLayer. If your objects are organized on layers, this is the easiest way to attach materials. You can override the layer material by applying a material to individual objects.

Figure 16-4 shows the **Material Attachment Options** dialog box displayed by the **MATERIALATTACH** command. The list on the left side of the dialog box shows the materials loaded into the drawing. The right side of the dialog box shows the layers in the drawing and the material attached to each layer. When no material is attached to a layer, the material is listed as Global. The Global material is a "blank" material in every drawing. To attach a material to a layer, drag the material from the list on the left and drop it onto the layer name on the right. To remove a material from a layer, pick the **X** button next to the material name on the right side of the dialog box.

The easiest way to remove a material from an individual object is with the **Properties** palette. To remove a material from an object or subobject, simply change its Material property in the **3D Visualization** category to Global. If a material has not been assigned to the object's layer, the property can also be set to ByLayer. See **Figure 16-5.**

MATERIALATTACH

Ribbon

Render
> Materials

Attach By Layer

Type

MATERIALATTACH

Figure 16-4.
Attaching materials to layers.

Layers in
drawing

Materials in
the drawing

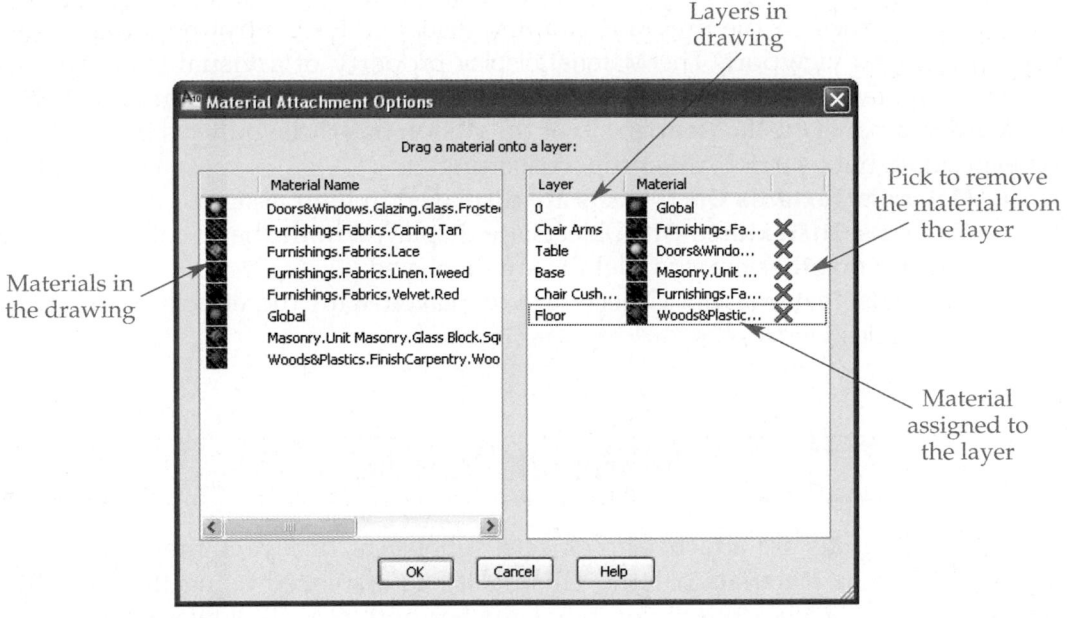

Pick to remove
the material from
the layer

Material
assigned to
the layer

Figure 16-5.
Removing a material from an object. A—The material is assigned. B—The material is removed.

Material
assigned

Material
removed

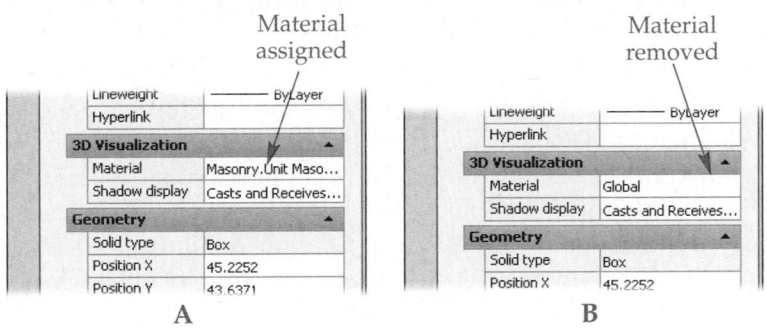

A   B

PROFESSIONAL TIP

A material can be applied to an object by dragging the material from the tool palette and dropping it onto the object. Also, a material can be loaded into the drawing without attaching it to an object by picking the material tool once in the tool palette and pressing [Enter] or by dragging and dropping it into a blank area of the drawing. This makes the material available in the drawing.

Exercise 16-2

Complete the exercise on the student website.
www.g-wlearning.com/CAD

Material Display Options

VSMATERIALMODE

Ribbon

Render
> Materials

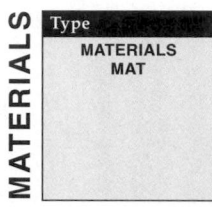

Materials/Textures
Off

Materials On/
Textures Off

Materials/Textures
On

Type
VSMATERIALMODE

As you learned in the previous chapter, visual styles control how materials are displayed in the viewport. The Material display property of a visual style can be set to display materials and textures, materials only, or neither materials nor textures. The **Materials** panel on the **Render** tab of the ribbon has three buttons in a flyout that correspond to, but override, this property setting:

- **Materials/Textures Off.** Objects are displayed in their assigned colors.
- **Materials On/Textures Off.** Objects are displayed in the basic color of the material, but no other material details are displayed.
- **Materials/Textures On.** Objects are displayed with the effects of all material properties visible.

Materials Palette

When materials are attached to objects, subobjects, or layers, they are automatically added to the **Materials** palette. This palette is displayed using the **MATERIALS** command or by picking the dialog box launcher button at the lower-right corner of the **Materials** panel in the **Render** tab on the ribbon. The **Materials** palette contains all of the materials loaded into the current drawing. It also provides a material editor for modifying the materials. See **Figure 16-6**.

At the top of the window is the **Available Materials in Drawing** pane. Samples (swatches) are displayed in this pane representing the materials that have been loaded into the drawing. A drawing icon in the lower-right corner of a swatch indicates that the material is currently in use in the drawing. In addition to the loaded materials, the default AutoCAD material Global appears in the list of swatches. The swatch outlined in yellow is the currently selected material. Its properties are displayed in the material editor. The material editor consists of several panes, including the **Material Editor**, **Maps**, **Advanced Lighting Override**, **Material Scaling & Tiling**, and **Material Offset & Preview** panes. Depending on the selected material, some of these panels may not be displayed.

Swatch Options

Above the material swatches, at the right-hand end of the **Available Materials in Drawing** title bar, there is a square button. This is the **Toggle Display Mode** button, which toggles the swatch area between the display of multiple materials to a single material. See **Figure 16-7**. The single-swatch display mode provides a much better view of the details of the selected material. Arrow buttons on either side of the single-material swatch allow you to select the next or previous material swatch. Immediately below the swatches are several buttons:

- **Swatch Geometry**
- **Checkered Underlay**
- **Preview Swatch Lighting Model**
- **Create New Material**
- **Purge from Drawing**
- **Indicate Materials in Use**
- **Apply Material to Objects**
- **Remove Materials from Selected Objects**

The sample geometry in the swatch can be displayed as a sphere, box, or cylinder. Pick the **Swatch Geometry** flyout and select the geometry to display in the swatch.

Figure 16-6.
The **Materials** palette provides swatches of the materials in the drawing and a material editor for modifying material properties.

Selected material

Select the sample geometry

Icon indicates the material is attached

Pick to turn the background on or off

Properties of the selected material

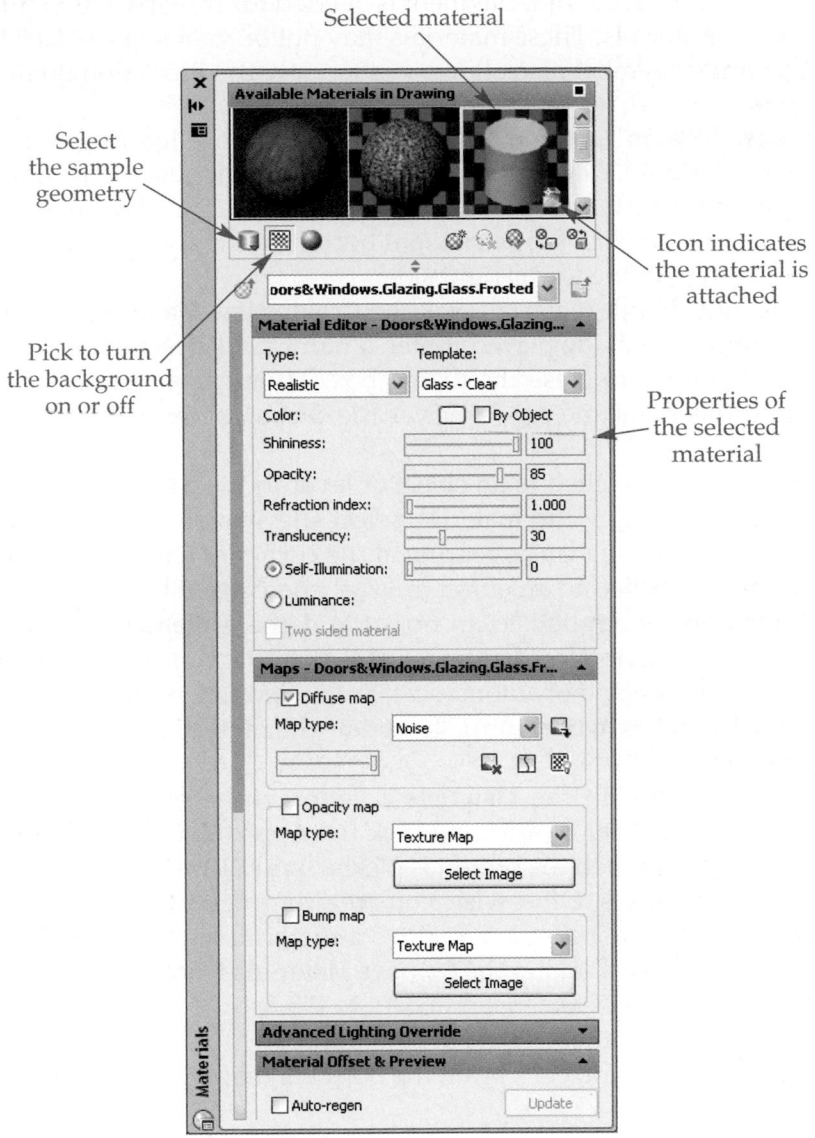

Figure 16-7.
The material swatches in the **Materials** palette can be displayed in different sizes. A—The medium setting. B—The full setting.

Pick to toggle the display mode

Pick to display the previous material

Pick to display the next material

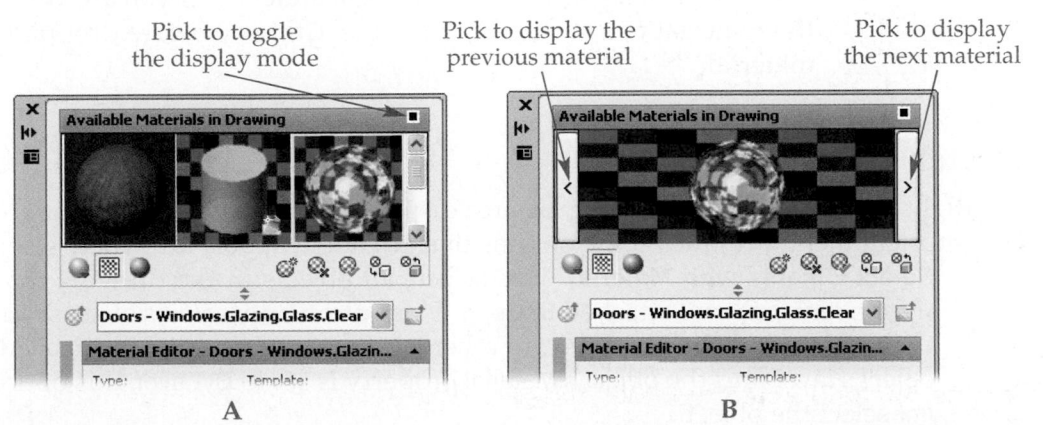

A   B

This allows you to preview the material on an object of a similar shape to the object on which the material will be used.

The **Checkered Underlay** button toggles the background in the swatch from checkered to black. The checkered underlayment is needed for transparent, semitransparent, and dark-colored materials. These materials may not be visible on a black background. The checkered underlayment also allows you to view the refraction quality of a transparent material.

The **Preview Swatch Lighting Model** determines how lighting is applied to the geometry in the sample. You can choose between single and dual lighting sources. Pick the **Single Light** button in the flyout to have one light illuminate the geometry from the upper-left side. Pick the **Back Light** button in the flyout to add a second light behind the geometry on the lower-right side.

To create a new, blank material, pick the **Create New Material** button. The **Create New Material** dialog box is displayed. Enter a name and description for the material and pick the **OK** button to close the dialog box. The new material is displayed as a new swatch and has the same properties as the Global material. Changing properties is discussed later.

If a material is not attached to an object or layer in the drawing, it can be removed from the drawing. To purge the material, select the swatch and pick the **Purge from Drawing** button. A drawing icon appearing at the corner of a material swatch indicates that the material is attached to an object or layer and cannot be purged.

When a material is applied to an object and the **Materials** palette is open, the drawing icon indicating an attached material is not automatically added to the material swatch. This icon is also not automatically removed when the material is detached. Pick the **Indicate Materials in Use** button to update the material swatches to show which materials are attached.

Any material shown in the **Materials** palette can be attached to objects in the drawing. Select the material swatch and pick the **Apply Material to Objects** button. If any objects are selected when the button is picked, and **PICKFIRST** is set to 1, the material is applied to the objects. Otherwise, you are prompted to select objects.

You learned earlier that you can remove a material from an object by setting its Material property to Global. Picking the **Remove Materials from Selected Objects** button allows you to set an object's Material property to ByLayer. If no material is assigned to the object's layer, this, in effect, removes the material from the object. However, if a material is assigned to the object's layer, the object is displayed in that material.

PROFESSIONAL TIP

A material can be assigned to an object by dragging and dropping from the **Materials** palette. Select the material swatch in the **Materials** palette, drag the swatch into the drawing, and drop it onto the object to which you want the material attached. If you are attaching the material to a subobject, press the [Ctrl] key before dropping the material.

Swatch Shortcut Menu

Right-clicking in the swatch display area displays a shortcut menu. See **Figure 16-8.** This shortcut menu provides some options that are not available anywhere else:

- **Select Objects with Material.** This selects all objects in the drawing that have the current material attached to them. This option only selects objects that have the material attached *explicitly.* In other words, if the material is attached to an object's layer and the object's Material property is set to ByLayer, this option will *not* select the object.

Figure 16-8.
The shortcut menu displayed by right-clicking in the swatch display area of the **Materials** palette. If an object is currently selected, the **Apply Material...** entry is replaced with **Apply Material To Selection**.

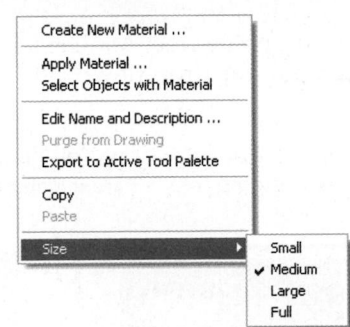

- **Edit Name and Description.** Displays a dialog box in which the name and description of the material can be changed.
- **Export to Active Tool Palette.** Exports the selected material to the current tool palette, as long as the palette is not read-only. Tool palettes are discussed in detail in Chapter 24.
- **Copy and Paste options.** These two options allow you to copy the selected material and paste it back into the swatch area as a new material with the same properties. These options are very useful when creating a group of similar materials.
- **Size.** Displays a cascading menu with options for the display size of the material swatches—**Small**, **Medium**, **Large**, and **Full**.

Exercise 16-3

Complete the exercise on the student website.
www.g-wlearning.com/CAD

Creating and Modifying Materials

Before creating and modifying materials, it is important to know your way around the **Materials** palette. There are six basic panes that may be displayed in the **Materials** palette:

- **Available Materials in Drawing**
- **Material Editor**
- **Maps**
- **Advanced Lighting Override**
- **Material Scaling & Tiling**
- **Material Offset & Preview**

The **Available Materials in Drawing** pane, discussed in the previous section, contains the material-preview swatches and related buttons. Depending on the material, some or all of the other five panes will be displayed or additional panes may be displayed.

Below the **Available Materials in Drawing** pane and above the **Material Editor** pane are three controls for working with nested maps. See **Figure 16-9A.** Maps are discussed later in this chapter. The drop-down list contains the *mapping tree.* The name in bold at the top of the list is the name of the current level of navigation. See **Figure 16-9B.** Picking one of the other levels in the tree navigates to that level. The name of that level is moved to the top of the drop-down list and displayed in bold. The panes displayed in the **Materials** palette are only those containing property settings for the currently selected level.

Figure 16-9.
Navigating the material mapping tree. A—The top level. Note the navigation tools. B—The mapping tree is displayed in the drop-down list. The name of the current level is displayed in bold.

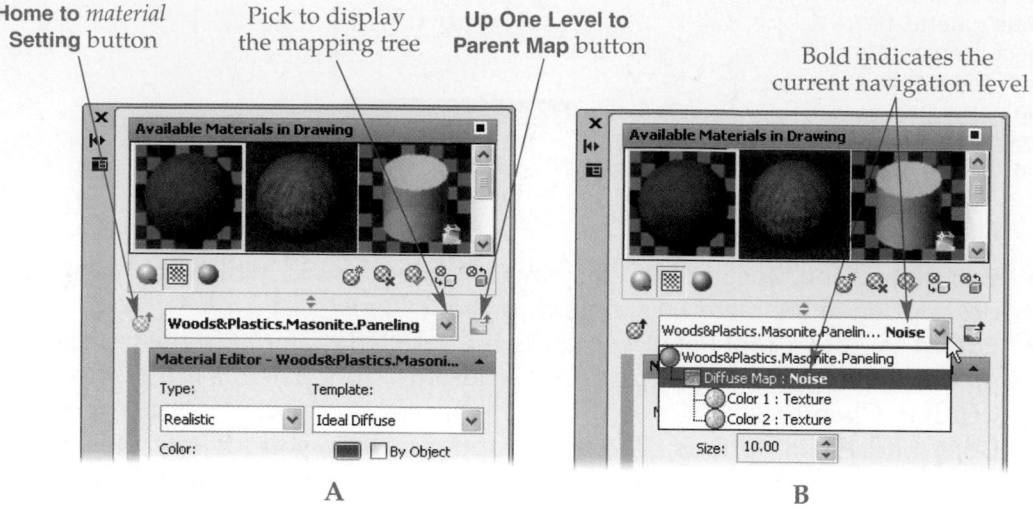

A

B

The button to the right of the drop-down list is **Up One Level to Parent Map**. Picking this button navigates up one step in the mapping tree. On the left side of the drop-down list is the **Home to** *material* **Setting** button. Picking this button navigates to the top level of the mapping tree, or the top of the material definition, no matter where you are in the mapping tree.

Creating New Materials

The basic properties of the material selected in the **Materials** palette are displayed in the **Material Editor** pane. Additional properties are displayed in the other panes in the **Materials** palette. Every material created in AutoCAD is based on a template. The specific properties available in the various panes are determined by the template on which the material is based. Templates are discussed in the next section.

To create a new material, open the **Materials** palette. Then, pick the **Create New Material** button below the material swatches or right-click in the swatch display area and select **Create New Material...** from the shortcut menu. In the **Create New Material** dialog box that appears, name the material and provide a description, **Figure 16-10.** When you pick the **OK** button, the material is automatically selected in the **Materials** palette, based on the Global material, and ready to be modified. To modify an existing material, simply select the material swatch and it is ready to be modified.

Types and Templates

A material can be one of four material types—realistic, realistic metal, advanced, or advanced metal. The *material type* determines the basic properties available for the material. It is set in the **Type:** drop-down list at the upper-left corner of the **Material Editor** pane. See **Figure 16-11A.**

In addition to a material type, you can select a material template. A *material template* provides you with a starting point for creating your own materials. It has settings already established that can be easily modified to give you the appearance that you are looking for. To select a template, pick it in the **Template:** drop-down list at the upper-right corner of the **Material Editor** pane, **Figure 16-11B.** Material templates are only available for the realistic and realistic metal material types. Each has its own set of templates. The advanced and advanced metal material types do not display the **Template:** drop-down list.

Figure 16-10.
Creating a new material.

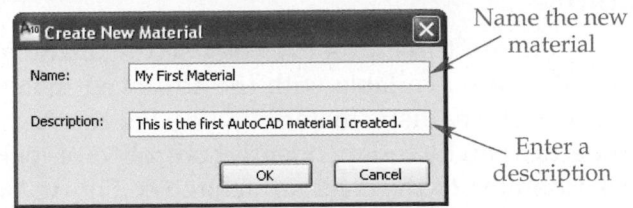

Name the new material

Enter a description

Figure 16-11.
Selecting a material type and a template on which to base the new material. A—Selecting the material type. B—Selecting a template.

Pick to select a material template

Pick to select a material type

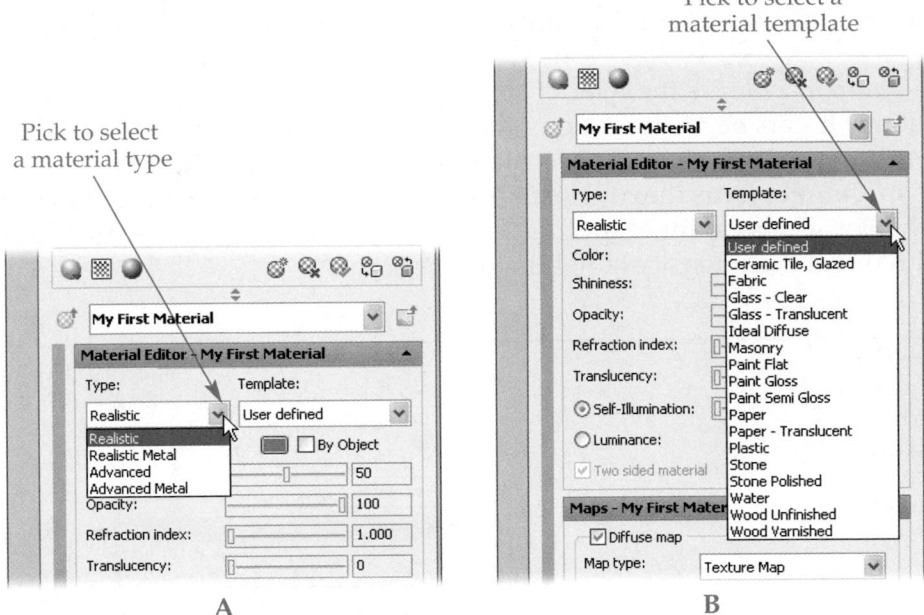

A

B

The realistic or realistic metal material type is a good starting point if you are new to material creation. Each provides basic material properties. Realistic materials are based on the physical qualities of the material: color, shininess, opacity, refraction index, translucency, self illumination, and luminance. If a template other than User defined is selected, some of these properties have preset values applied to them. This gives you a starting point to create your own fabric, glass, metal, and various other materials.

When you get comfortable with creating basic materials, the two advanced material types—advanced and advanced metal—offer more material properties to provide additional control over the material appearance. The main difference between a realistic material type and an advanced material type is the addition of ambient, diffuse, and specular color settings and a reflection property.

PROFESSIONAL TIP

AutoCAD provides fantastic-looking materials to dress up the scene and make it look real. However, after you get comfortable with creating materials, start a library of your own materials. If your project is presented to a customer along with projects from competitors, and your competitors are using standard AutoCAD materials, your project will stand out from the crowd.

Color

There are three possible color settings: ambient, diffuse, and specular. All three properties are available with the advanced material types, but only diffuse is available with the realistic material types. For the advanced material types, the three color properties can be independently controlled or locked together. To lock colors, pick the lock icon next to the color swatches. See **Figure 16-12.** When locked, the diffuse color is always the dominant color. Checking the **By Object** check box turns off the color swatch. The color reverts to the object color (ByLayer, for example). To set the color, pick the color swatch to display the **Select Color** dialog box, **Figure 16-13.** Then, select a color and pick the **OK** button to close the dialog box.

The *diffuse color* is the color of the object in lighted areas, or the perceived color of the material. See **Figure 16-14.** It is the predominant color you see when you look at the object. Set this color first. The other two colors are typically based on the diffuse color.

The *ambient color* is the color of the object where light does not directly provide illumination. It can be thought of as the color of an object in shadows. In nature, shadows cast by an object typically contain some of the ambient color.

The *specular color* is the color of the highlight (the shiny spot). It is typically white or a light color. The amount of specular color shown is determined by the shininess of the material and the intensity of lighting in the scene.

Figure 16-12.
The ambient and diffuse colors are locked. The diffuse and specular colors are not locked.

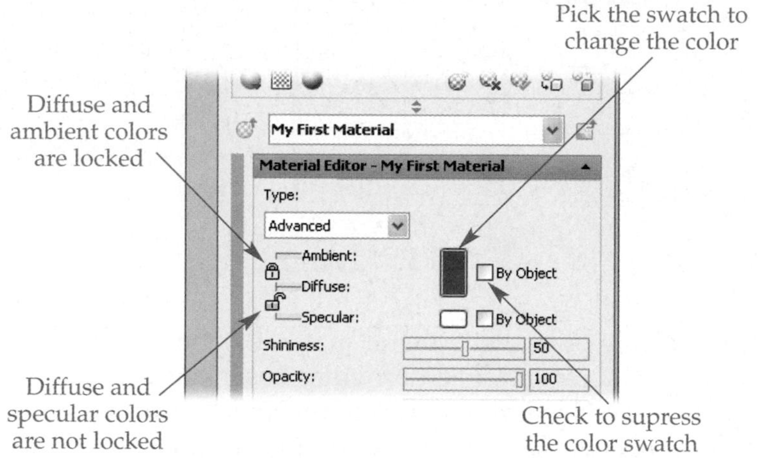

Figure 16-13.
Setting a color for a material property.

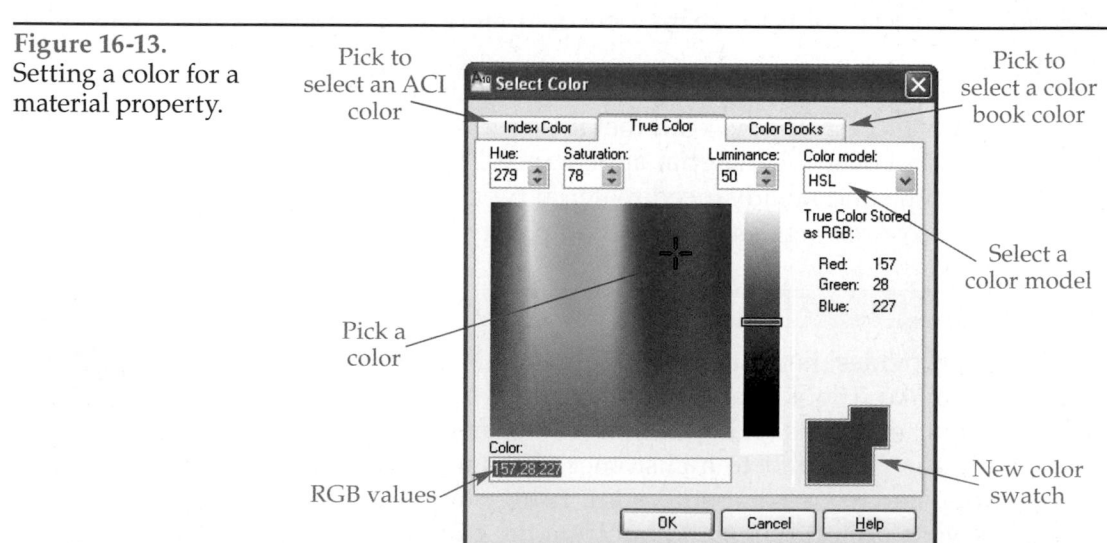

AutoCAD and Its Applications—Advanced

Figure 16-14.
The three colors
of a material are
indicated here.

Diffuse color

Specular color

Ambient color

PROFESSIONAL TIP

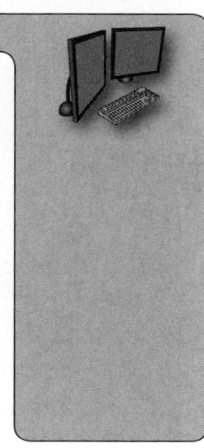

Specular highlights can be seen everywhere. Look around you right now at edges and inclined surfaces. The diffuse color of the surface typically has little to do with the color of the highlight. The color of the light source usually determines the predominant highlight color. The majority of highlights are white or near white because most light sources are white or nearly white. However, highlights in the interior of a home may have a yellow cast to them because incandescent lightbulbs generally cast yellow light. Compact fluorescent lights cast a slightly different color. Outside with a clear sky and bright sun, highlights may have a slight blue cast. These small details are what make a scene realistic.

Shininess

Shininess is a measure of the surface roughness. Smooth surfaces are very shiny and have a small, sharp highlight. These surfaces reflect in one direction most of the light that hits the surface. Rough surfaces tend to diffuse, or break up, light as it is reflected. Therefore, these surfaces do not appear very shiny and have a large, soft highlight. See **Figure 16-15.** To set the shininess, drag the **Shininess:** slider left to decrease the value or right to increase the value. You can also enter a value in the text box at the right-hand end of the slider.

Opacity

Opacity is a measure of a material's transparency, or how "see through" the material is. **Figure 16-16** shows an example of using transparent materials to show the internal workings of a mechanical assembly. To change the opacity value, drag the **Opacity:** slider to the left to make the material more transparent or right to make it more opaque. You can also enter a value in the text box at the right-hand end of the slider. A value of 100 creates an opaque material. Lower values create semitransparent materials. Realistic and advanced material types have an opacity property.

Chapter 16 Materials in AutoCAD

Figure 16-15.
Three different shininess settings are illustrated here.

Shininess = 25   Shininess = 50   Shininess = 75
      A   B   C

Figure 16-16.
The material used for the housing on this mechanism has an opacity setting of five.

Reflection

The *reflection* is a mirror image of the other objects in the scene. See **Figure 16-17.** Only the advanced and advanced metal material types have a reflection property. To make a material reflective, drag the **Reflection:** slider. Dragging the slider to the right increases the reflectivity of the material and to the left decreases reflectivity. You can also enter a value in the text box at the right-hand end of the slider.

Refraction Index

The *refraction index,* also known as the index of refraction (IOR), is a measure of how much light is bent (refracted) as it passes through transparent or semitransparent materials. Refraction is what causes objects to appear distorted when viewed through a bottle or glass of water. See **Figure 16-18.** The higher the refraction index value, the more light is bent as it passes through the material. To set the value, drag the **Refraction index:** slider to the right to increase the value or left to decrease the value. You can also enter a value in the text box at the right-hand end of the slider. A value of 1.000 is the refraction index of a vacuum. The value of water is 1.3333 and glass is around 1.500. Generally, the refraction index is not set much above 1.700 and is usually somewhere between 1.000 and 1.500. Realistic and advanced material types have a refraction index property.

Figure 16-17.
The effect of increasing reflectivity. A—The reflection value of the material on the box is zero. B—The reflection value is increased to 100.

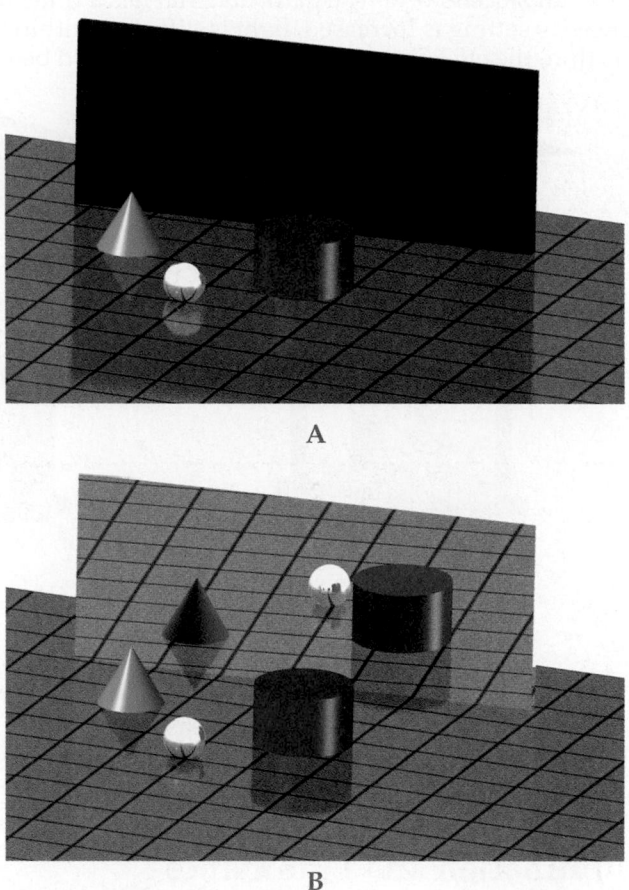

A

B

Figure 16-18.
The effect of refraction. A—The transparent material on the sphere has a refraction index of zero. B—When the refraction index is increased, the cylinder behind the sphere is distorted as light is refracted by the material.

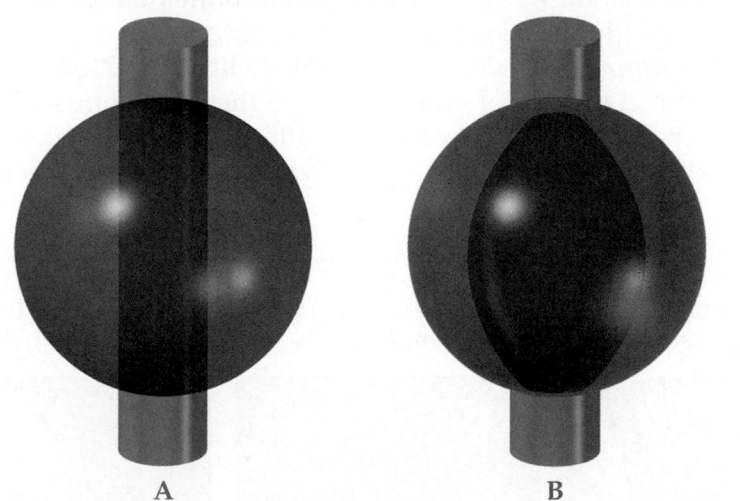

A   B

Translucency

Translucency is a quality of transparent and semitransparent materials that causes light to be diffused (scattered) as it passes through the material. See **Figure 16-19.** This makes any object with the material applied to it appear as if it is being illuminated from within, or glowing. The thicker the material, the more pronounced the effect. In AutoCAD, the translucency setting affects transparent, semitransparent, and opaque materials. With a higher setting, light appears to travel through an object, lighting the

Figure 16-19.
The effect of translucency. A—The glass material has a translucency setting of zero. B—When the translucency setting is increased, light is diffused within the material. In this case, since the glass is thin, the effect is not as "glowing" as it would be for a thicker material.

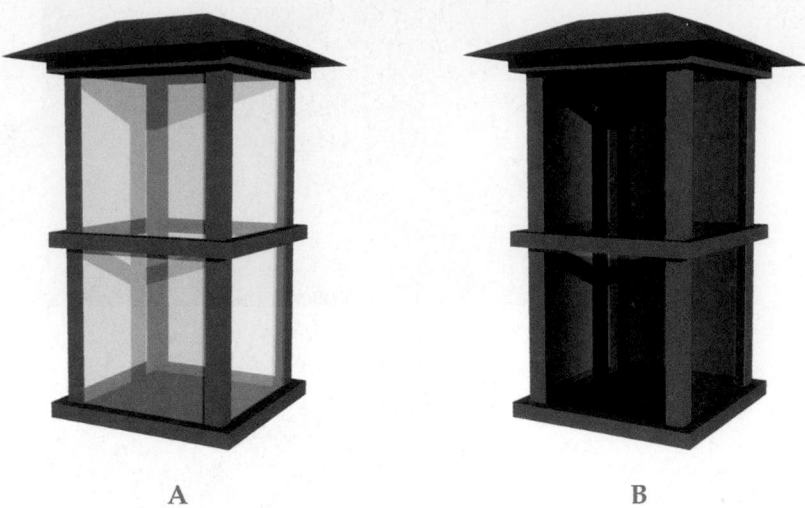

A   B

opposite side. To change the translucency value, drag the **Translucency:** slider to the right to increase the value or left to decrease the value. You can also enter a value in the text box at the right-hand end of the slider. Realistic and advanced material types have a translucency property.

Self Illumination and Luminance

Self illumination is an effect of a material producing illumination. See **Figure 16-20.** For example, the surface of a neon tube glows. However, in AutoCAD, a material with self illumination will not actually add illumination to a scene. All four material types have a self illumination property.

Luminance is defined as the value of light reflected off a surface. For realistic and realistic metal material types, you have the choice of using either self illumination or luminance. They both have a similar affect on the material. To use self illumination, pick the **Self-Illumination:** radio button. To use luminance, pick the **Luminance:** radio button.

Figure 16-20.
The effect of self illumination/luminance. A—The globe of this lightbulb does not have any self illumination. B—Self illumination is applied to the globe material.

A   B

To set the self illumination value, drag the **Self-Illumination:** slider to the right to increase the value or left to decrease the value. You can also enter a value in the text box at the right-hand end of the slider. Valid values for self illumination are from 1 to 100.

Luminance is expressed in candelas per square meter (cd/m^2). For example, $1\ cd/m^2$ is the equivalent of 1 candela of light radiating from a surface that is 1 square meter. To set luminance, enter a value in the text box that is displayed when the **Luminance:** radio button is picked. You may want to use luminance if the maximum self illumination setting of 100 is not making the material bright enough for you. A luminance setting of $1500\ cd/m^2$ is about the same as a self illumination setting of 100. Luminance can be set as high as 100 million cd/m^2.

Material Maps

Below the **Material Editor** pane in the **Materials** palette is the **Maps** pane. A *texture map* is simply an image applied to a material property. This type of map is known as a *2D map* because it is applied to the surface of an object and does not extend into it. On the other hand, a *procedural map* is mathematically generated based on the colors and values you select. This type of map is known as a *3D map* because it extends through the object. A material that has a map applied to at least one of its properties is called a *mapped material.*

Maps

There are nine types of maps that can be applied to material properties: texture, checker, gradient ramp, marble, noise, speckle, tiles, waves, and wood. Each map has unique settings. A material can have separate diffuse, reflection, opacity, and bump maps, as discussed later in this chapter.

Texture Map. A texture map is an image file, such as a digital photograph, that is applied to one of the material's properties. **Figure 16-21** shows an object with an image of a forest applied to the diffuse color of the material attached to it. To specify a texture map, first select Texture Map in the drop-down list in the appropriate area of the **Maps** pane. Then, pick the **Select Image** button in the same area. The **Select Image File** dialog box is displayed, which is a standard open dialog box. Browse to the folder where the image file is saved, select the file, and pick the **Open** button.

The "select image" button is now labeled with the file name of the image file. Also, a slider is displayed to the left of the button. See **Figure 16-22**. The slider is used to set the percentage of the image file that is applied to the property. At 100% (fully right),

Figure 16-21.
A material with a texture map. A—The background object's material does not have any maps applied. B—A texture map has been applied to the diffuse color component of the background object's material.

A   B

Figure 16-22.
Use the slider to adjust the percentage of the texture map that is contributed to the property. All properties that can be mapped have this slider once a map is applied.

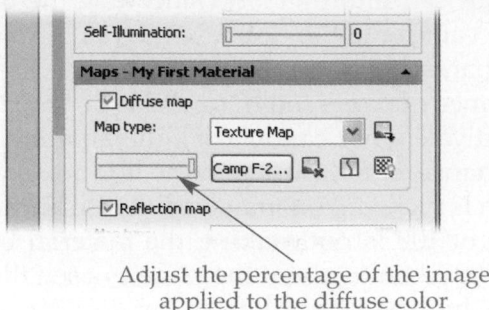

Adjust the percentage of the image
applied to the diffuse color

the entire image is applied to the property. At 50% (in the middle), the image appears to be 50% transparent, which is applied to the property. All of the map types have this slider for adjusting the amount of the map that is applied to the property.

The image settings are controlled by picking the **Click for Texture Map settings** button to the right of the **Map type:** drop-down list. This navigates to the map level of the material tree. Three additional buttons are also displayed to the right of the "select image" button. These are used to adjust and delete the map from the property. These buttons are available for all map types and are discussed later in the section Adjusting Material Maps.

Checker. A *checker map* creates a two-color checkerboard pattern. By default, the colors are black and white, but different colors or images can be used as well. This map type can be used for checkerboard pattern floor materials. However, by changing various properties, you can simulate many different effects and use the map for other materials.

To specify a checker map, first select Checker in the drop-down list in the appropriate area of the **Maps** pane. Then, pick the **Click for Checker settings** button next to the **Map type:** drop-down list to navigate to the checker map level of the mapping tree. The **Checker** pane is now displayed in the **Materials** palette. See **Figure 16-23.** In this pane, the two colors that make up the checker pattern are defined. By default, they are both solid colors. Pick on the color swatch to change the color. To swap the color definitions, pick the **Swaps the Map Types** button in the middle of the pane. You are not limited to solid colors. All of the map types are listed in the **Map type:** drop-down list, along with Solid Color. For example, you can add a texture map to the color 1 definition and a noise map to the color 2 definition. The possibilities are endless.

The **Soften:** setting is used to blur the edges between the checkers. To change the setting, enter a value in the text box or use the up and down arrows. A value of 0.00 creates sharp edges between the checkers. The maximum setting is 5.00 and produces edges that are very blurred.

Figure 16-23.
The map-level properties of a checker map.

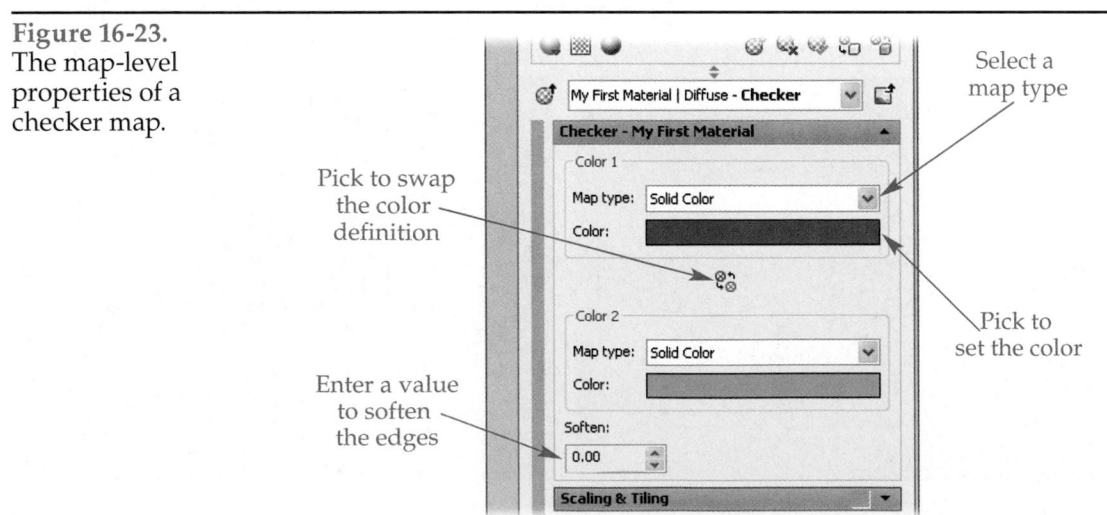

Select a map type

Pick to swap the color definition

Pick to set the color

Enter a value to soften the edges

Gradient Ramp. The gradient ramp map is a texture map that allows you to create a material blending colors and textures in different patterns. It is similar to the **HATCH** command's **Gradient color** option. Select Gradient Ramp in the **Map type:** drop-down list. Then, pick the **Click for Gradient Ramp settings** button to navigate to the map level of the material tree.

The **Gradient Ramp** panel is now displayed in the **Materials** palette. See **Figure 16-24.** The gradient ramp is represented with three nodes at the bottom edge of the ramp. Each node represents a different color in the ramp. By default, the left node (node 1) is black, the middle node (node 3) is gray, and the right node (node 2) is white. The middle node can be moved left or right to change where the color transitions from black to gray to white.

There must be at least three nodes, but you are not limited to three nodes. Picking anywhere in the ramp creates a new node. It can then be moved and its properties changed. Right-clicking on a node opens a shortcut menu. This menu contains options to copy, paste, or delete the node.

Selecting a node by picking on it changes the panel directly below the ramp to the properties for that node. The **Map type:** drop-down list contains all of the map types. This means you are not limited to creating gradients of colors. You can blend patterns and images to create many different variations of gradients. See **Figure 16-25.**

Figure 16-24.
The map-level properties of a gradient map.

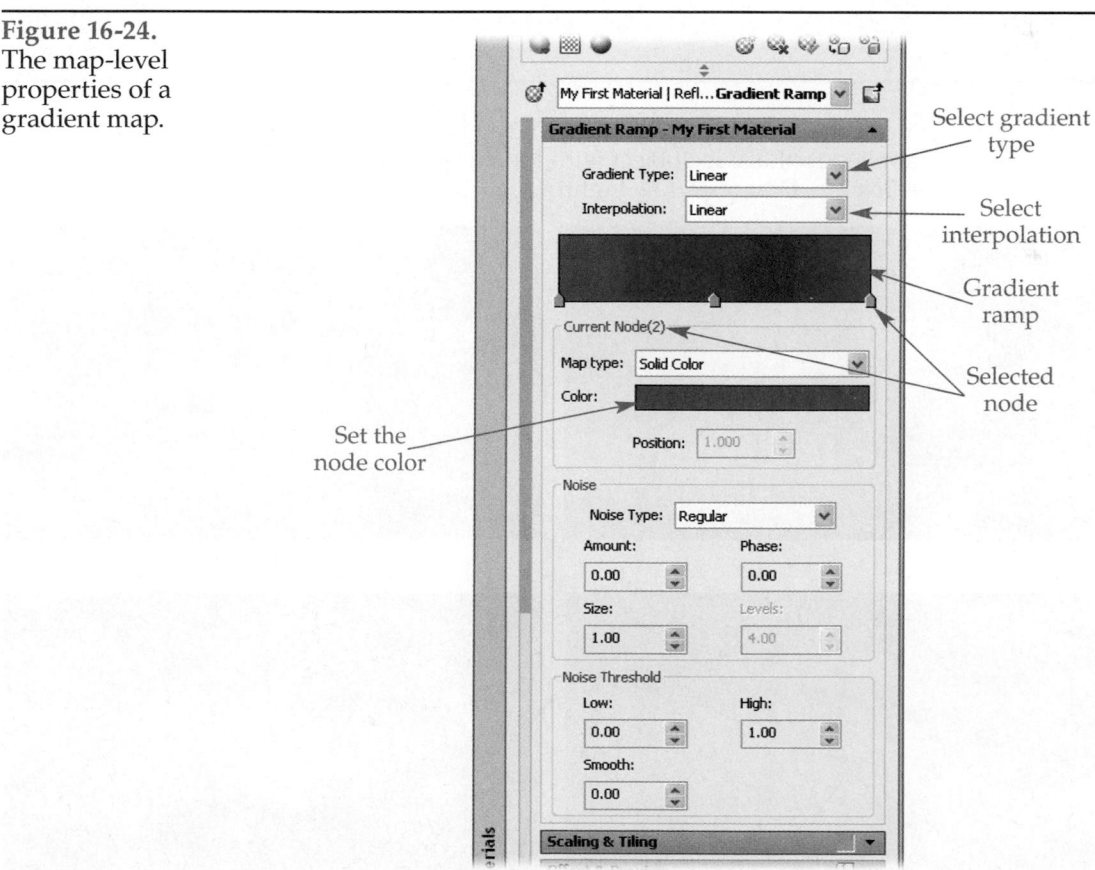

- Select gradient type
- Select interpolation
- Gradient ramp
- Selected node
- Set the node color

Above the ramp are the **Gradient Type:** and **Interpolation:** drop-down lists. The setting in the **Gradient Type:** drop-down list controls the pattern of the gradient ramp. See **Figure 16-26.** The default pattern is linear. This results in a typical pattern similar to the ramp display. The other options are:

- **4 Corner.** Creates a linear transition that is asymmetrical.
- **Box.** The transition of colors or maps is in the shape of a square.
- **Diagonal.** This transition is linear, but rotated on the surface.

Figure 16-25.
This floor tile material is created using a gradient map. A texture map is alternated with solid white stripes. The box type and solid interpolation are used.

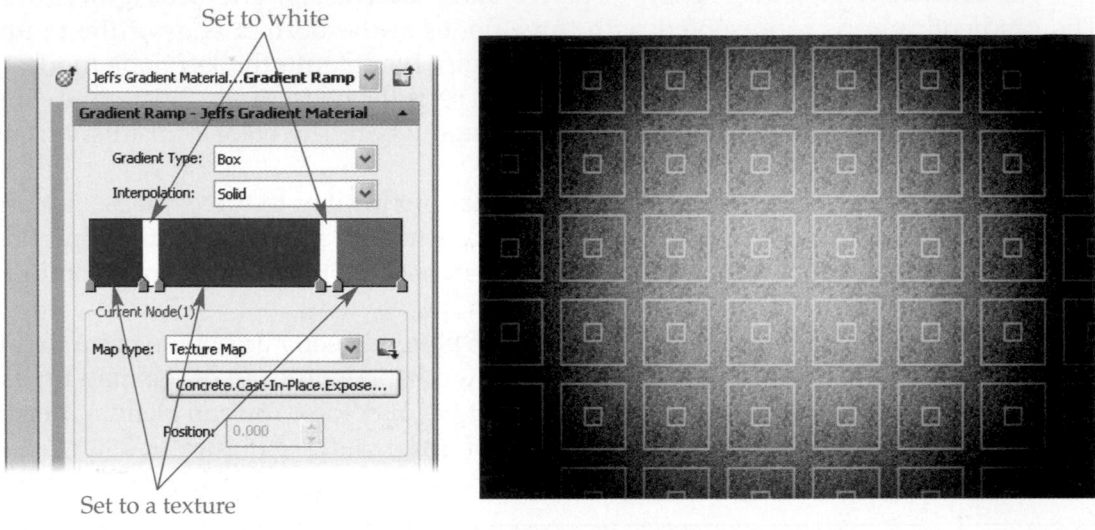

Figure 16-26.
There are 12 different gradient types available for use in a gradient ramp map. The four shown here are applied to the same object with the same lighting and mapping coordinates. A—4 Corner. B—Box. C—Diagonal. D—Lighting.

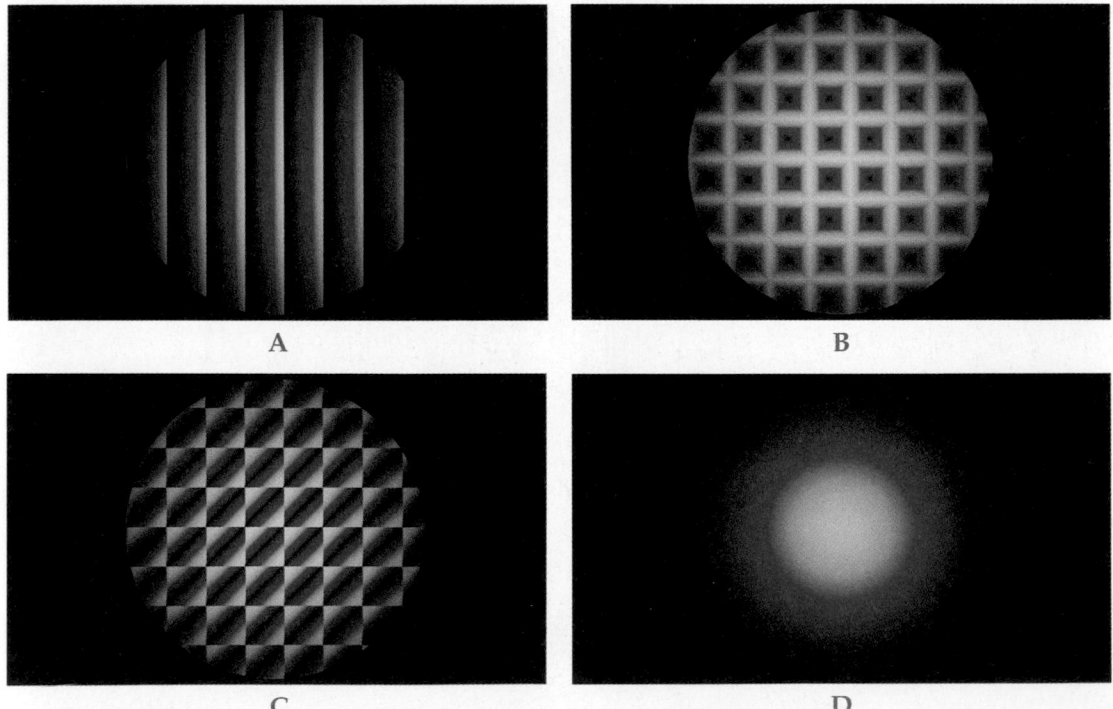

- **Lighting.** The intensity of the light source determines where the transition takes place. The right side of the ramp corresponds to the highest intensity of light and the left side of the ramp is equal to no light.
- **Linear.** This is the default. It is a smooth transition from one node to the next.
- **Mapped.** With this type you can assign one of the maps to determine the gradient. When selected, the **Gradient Map** area appears just below the **Gradient Type:** drop-down list. Use this area to select the map and change its properties.
- **Normal.** The angle between the camera direction and the surface normal controls how the pattern is displayed. The left side of the ramp is 0° between the normal and the camera viewpoint. The right side is 90° between the normal and the camera viewpoint.
- **Pong.** This gradient is a rotated linear transition, similar to the diagonal type, but it pivots about the corner of a box and reverses in the middle of the pattern.
- **Radial.** Colors and maps are arranged in a circular pattern similar to a target.
- **Spiral.** The gradient sweeps about a central point similar to the movement on a radar screen.
- **Sweep.** Similar to the spiral type, but the center of the sweep is at a corner instead of in the center. Also, this does not repeat like the pong type.
- **Tartan.** This resembles a plaid pattern. It is very similar to the box type.

The **Interpolation:** drop-down list controls the transition of colors and maps from one node to the next. Transitions are applied to the nodes from left to right, regardless of the node number. The options are:

- **Custom.** When this is selected, an **Interpolation** drop-down list is displayed in the **Current Node** panel in the **Gradient Ramp** pane. This is used to select a specific transition at each node. Each node can have a different setting in this drop-down list.
- **Ease In.** Shifts the transition closer to the node on the right.
- **Ease In Out.** Shifts the transition toward the node, but it remains more or less centered on the node.
- **Ease Out.** Shifts the transition closer to the node on the left.
- **Linear.** This is the default. The transition is constant from one node to the next.
- **Solid.** No transition between nodes. There is an abrupt change at each node.

Below the **Current Node** area in the **Gradient Ramp** panel are the **Noise** and **Noise Threshold** areas. Noise may be added to the gradient map to create an uneven appearance. The settings are similar to those for the noise map. The noise map is described later in this section.

Marble. A *marble map* is a procedural map based on the colors and values you set. To specify a marble map, first select Marble in the drop-down list in the appropriate area of the **Maps** pane. To set the marble properties, pick the **Click for Marble settings** button next to the **Map type:** drop-down list to navigate to the marble map level of the mapping tree. The **Marble** pane is now displayed in the **Materials** palette. See **Figure 16-27.**

A marble map is based on two colors—stone and vein. The two color swatches in the **Marble** pane are used to specify these colors. You can swap the vein and stone colors by picking the **Swaps the Colors** button to the right of the swatches.

The **Vein spacing:** setting determines the relative distance between each vein in the marble. The **Vein width:** setting determines the relative width of each vein. Each of these settings can range from 0.00 to 100.00 and has a default of 1.00.

Figure 16-27.
The map-level
properties of a
marble map.

Pick to set
the color

Set the vein
spacing

Pick to swap
the colors

Set the
vein width

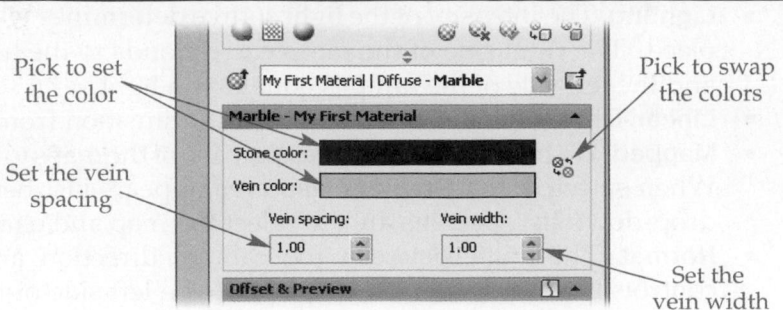

> **NOTE**
>
> The mathematical calculations that create the marble map are based on the world coordinate system. If you move or rotate the object, a different vein result is produced.

Noise. A *noise map* is a random pattern of two colors used to create an uneven appearance on the material. It is most often used to simulate materials such as concrete, soil, asphalt, grass, and so on. To specify a noise map, first select Noise in the drop-down list in the appropriate area of the **Maps** pane. To set the noise properties, pick the **Click for Noise settings** button next to the **Map type:** drop-down list to navigate to the noise map level of the mapping tree. The **Noise** pane is now displayed in the **Materials** palette. See **Figure 16-28.**

First, you need to select the type of noise. The options in the **Noise Type:** drop-down list are:

- **Regular.** This is "plain" noise and useful for most applications.
- **Fractal.** This creates the noise pattern using a fractal algorithm. When this is selected, the **Level:** control in the **Noise Threshold** area of the pane is enabled.
- **Turbulence.** This is similar to fractal, except that it creates fault lines.

The **Size:** setting below the **Noise Type:** drop-down list controls the size scale of the noise. The larger the value, the larger the size of the noise. The default value is 1.00 and the value can range from 0.00 to 1 billion.

Figure 16-28.
The map-level
properties of a noise
map.

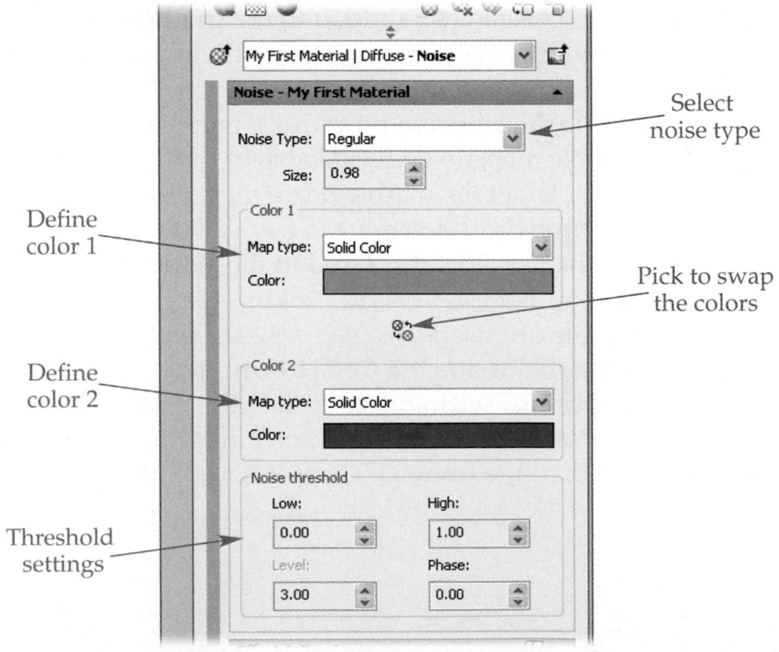

Select
noise type

Define
color 1

Pick to swap
the colors

Define
color 2

Threshold
settings

The **Color 1** and **Color 2** areas of the pane control the color of the pattern of noise. To change the color, pick the color swatch. You can also select a map for the color definition using the **Map type:** drop-down list. To swap the color definitions, pick the **Swaps the Map Types** button.

The settings in the **Noise Threshold** area of the **Noise** pane are used to fine-tune the noise effect. The settings in this area are:

- **Low.** The closer this setting is to 1.00, the more dominate color 1 is. The default setting is 0.00 and it can range from 0.00 to 1.00.
- **High.** The closer this setting is to 0.00, the more dominate color 2 is. The default setting is 1.00 and it can range from 0.00 to 1.00.
- **Level.** Sets the energy amount for fractal and turbulence. Lower values make the fractal noise appear blurry and the turbulence lines more defined. The default setting is 3.00 and it can range from 0.00 upward.
- **Phase.** Randomly changes the noise pattern with each value. This allows you to have materials with the same noise map settings look slightly different. You should have different patterns on different materials. This adds a level of realism to your scene.

Speckle. A *speckle map* is a random pattern of dots based on two colors. This map is great for textured walls, sand, granite, and so on. To specify a speckle map, first select Speckle in the drop-down list in the appropriate area of the **Maps** pane. To set the speckle map properties, pick the **Click for Speckle settings** button next to the **Map type:** drop-down list to navigate to the speckle map level of the mapping tree. The **Speckle** pane is now displayed in the **Materials** palette. See **Figure 16-29.** The settings for a speckle map are very simple. Pick colors for the color 1 and color 2 definitions. You cannot use maps, only colors. The **Size:** setting controls the size of the speckles.

Tiles. A *tile map* is a pattern of rectangular, colored blocks surrounded by colored grout lines. This may be the most versatile map in the whole collection. Tiles are used to simulate tile floors, ceiling grids, hardwood floors, and many different types of brick walls. To specify a tile map, first select Tiles in the drop-down list in the appropriate area of the **Maps** pane. To set the tile map properties, pick the **Click for Tile settings** button next to the **Map type:** drop-down list to navigate to the tile map level of the mapping tree. The **Tiles** pane is now displayed in the **Materials** palette. See **Figure 16-30.**

In the **Tiles** pane, first you need to select the pattern for the map. In the **Pattern type:** drop-down list, select one of the seven predefined tile patterns or Custom Pattern to create your own. The names of the predefined patterns bring to mind brick walls. For example, a mason may use a stack bond to build a brick wall. However, remember these are only *patterns*. You can also use a brick pattern to create tile floors and acoustic ceiling panels. Four of the tile patterns are shown in **Figure 16-31.**

The **Random seed:** setting below the **Pattern type:** drop-down list is used to create a random color variation in the tiles. Ceramic tile floors, for example, look more realistic if each tile is slightly different in color. This variation is automatically applied, but entering a different random seed changes the pattern. Each material that uses a tile pattern should have a different random seed.

Figure 16-29.
The map-level properties of a speckle map.

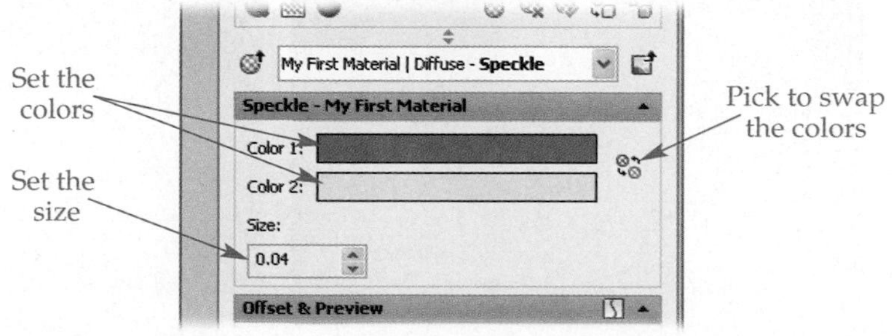

Set the colors

Set the size

Pick to swap the colors

Figure 16-30.
The map-level
properties of a tile
map.

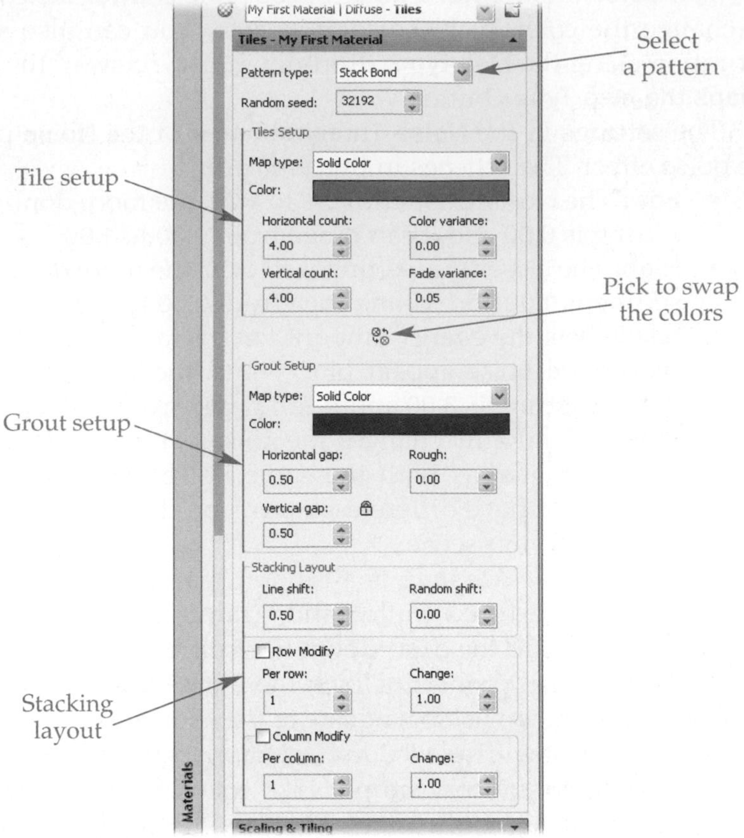

Tile setup

Grout setup

Stacking
layout

Materials

Select
a pattern

Pick to swap
the colors

Figure 16-31.
A tile map can have a custom pattern or predefined pattern. Four of the predefined patterns are shown here. A—Running bond. B—English bond. C—Stack bond. D—Fine running bond.

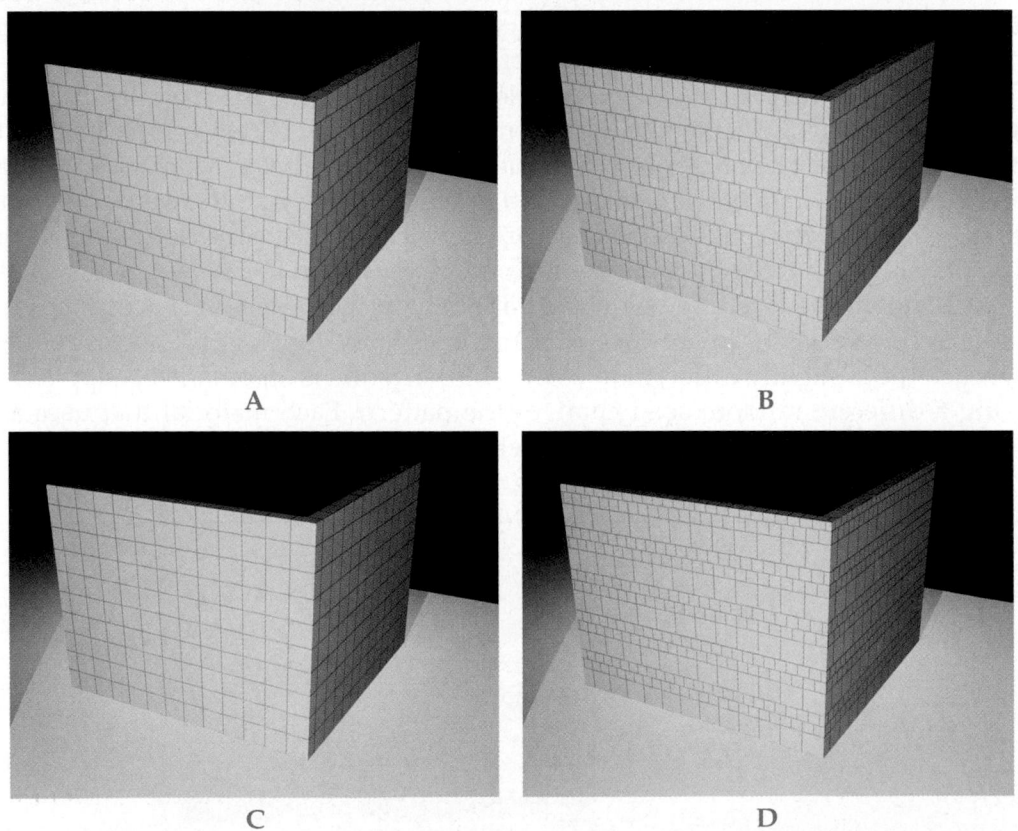

A

B

C

D

AutoCAD and Its Applications—Advanced

A tile pattern is really made up of tiles and grout. The **Tiles** pane contains **Tiles Setup** and **Grout Setup** areas in which these elements are defined. Each area has a drop-down list for selecting a map type and, if the map type is Solid Color, a color swatch for setting the color. The number and size of the tiles is controlled by the **Horizontal count:** and **Vertical count:** settings in the **Tiles Setup** area. The **Color variance:** setting in the **Tiles Setup** area can be used to slightly alter the color of each tile to create a more realistic appearance. The **Fade variance:** setting in the **Tiles Setup** area is used to slightly fade the color of each tile. You will have to experiment with the color variance and fading to create the look you need. The grout setup is mainly just controlling size of the grout with the **Horizontal gap:** and **Vertical gap:** settings. Most of the time, these values will be the same and they can be locked together with the lock button. In some cases, such as for a hardwood floor material, you will have to scale the pattern differently on the horizontal and vertical axes to make the gap thicker in one direction.

In the **Stacking Layout** area, the value in the **Line shift:** text box changes the location of the vertical grout lines in every other row to create an alternate pattern of tiles. The default value is 0.50 and the range is from 0.00 to 100.00. The value in the **Random shift:** text box randomly moves the same lines. This works nicely for hardwood floor materials. Default value is 0.00 and the range is 0.00 to 100.00. The **Stacking Layout** area is only available with the Custom Pattern type.

The settings in the **Row Modify** and **Column Modify** areas are available with all tile pattern types, but may be disabled by default. To enable the settings, check the check box by the area name. The settings in these areas allow you to change the number of grout lines in the horizontal and vertical directions to create your own pattern. The **Per row:** and **Per column:** settings determine which rows and columns will be changed. When set to 0, no changes take place in the row or column. When set to 1, every row or column will be changed. When set to 2, every other row or column will be changed, and so on. The value must be a whole number. The setting in the **Change:** text box controls the size of the tiles in the row or column. A setting of 1 means that the tiles remain their original size. A setting of 0.50 makes the tiles one-half of their original size, a setting of 2 makes the tiles twice their original size, and so on. A setting of 0.00, in effect, completely turns off the row or column and the underlying color (usually black) shows through.

Waves. A *wave map* creates a pattern of concentric circles. Imagine dropping two or three stones into a pool of water and watching the ripples intersect with each other. A number of wave centers are randomly generated and a pattern created by the overlapping waves is the result. As the name implies, the wave map is usually used to simulate water. To specify a wave map, first select Waves in the drop-down list in the appropriate area of the **Maps** pane. To set the wave map properties, pick the **Click for Waves settings** button next to the **Map type:** drop-down list to navigate to the wave map level of the mapping tree. The **Waves** pane is now displayed in the **Materials** palette. See **Figure 16-32.**

In the **Waves** pane, first specify the two colors that will be used in the pattern. Maps cannot be used for the color definitions. To set a color, pick on the swatch and choose a color in the **Select Color** dialog box.

Below the color swatches are two radio buttons next to the **Distribution:** label. The radio button that is selected determines how the wave centers are distributed on the object. Picking the **3D** radio button means that the wave centers are randomly distributed over the surface of an imaginary sphere. This distribution affects all sides of an object. On the other hand, picking the **2D** radio button means that the wave centers are distributed on the XY plane. This is much better for nearly flat surfaces, such as the surface of a pond or lake.

The remaining settings in the **Waves** pane define the pattern of waves. The value in the **Number of waves:** text box is the number of wave centers that are generating the waves. The **Wave radius:** value is the radius of the circle or sphere from which

Figure 16-32.
The map-level properties for a wave map.

Specify colors

Pick to swap colors

Set 2D or 3D

Adjust the effect

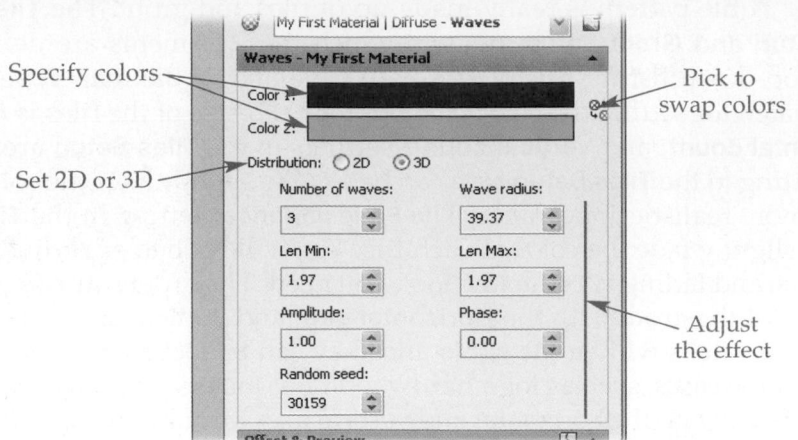

the waves originate. The **Len Min:** and **Len Max:** settings define the minimum and maximum interval for each wave. The value in the **Amplitude:** text box can be thought of as the "power" of the wave. The default is 1.00, but the value can range from 0.00 to 10000.00. A value less than 1.00 makes color 1 more dominant. For a value greater than 1.00, color 2 is more dominant. The **Phase:** text box is used to shift the pattern and the **Random seed:** text box is used to redistribute the wave centers.

PROFESSIONAL TIP

To see how your changes affect the map, check the **Auto-regen** check box in the **Offset & Preview** pane. This will automatically update the preview in that pane when you change a setting. You can also change the material swatch geometry to a cube, sphere, or cylinder. Some maps, like a wave map, are easier to understand when displayed on a cube.

Wood. A *wood map* is a procedural map that generates a wood grain based on the colors and values you select. See **Figure 16-33.** To specify a wood map, first select Wood in the drop-down list in the appropriate area of the **Maps** pane. To set the wood map properties, pick the **Click for Wood settings** button next to the **Map type:** drop-down list to navigate to the wave map level of the mapping tree. The **Wood** pane is now displayed in the **Materials** palette. See **Figure 16-34.**

Figure 16-33.
A wood map is a procedural, or 3D, map. This type of map passes through the entire object to which it is applied.

Figure 16-34.
The map-level properties of a wood map.

Set the colors

Adjust the pattern

Pick to swap the colors

A wood map is based on two colors. The two color swatches in the **Wood** pane are used to specify these colors, usually one dark and one light color. You can swap the two colors by picking the **Swaps the Colors** button to the right of the swatches. The **Radial noise:** setting determines the waviness of the wood's rings. The rings are found by cutting a tree crosswise. The **Axial noise:** setting determines the waviness of the length of the tree trunk. The **Grain thickness:** setting determines the relative width of the grain.

NOTE

The mathematical calculations that create the wood map are based on the world coordinate system. If you move or rotate the object, a different grain pattern is produced.

Diffuse map

A *diffuse map* is applied to the diffuse color property of a material. See **Figure 16-35.** This is assigned in the **Maps** pane, which is displayed at the top of the material mapping tree. The **Diffuse map** check box in the pane toggles the specified diffuse map off and on. When unchecked, the object color defined in the **Material Editor** pane controls the color of the object. If no map is specified, the toggle has no affect.

Figure 16-35.
A—The diffuse color property of the material applied to this box has no map. B—A texture map is applied to the diffuse color property of the material.

A

B

To assign a diffuse map, select the type of map to apply using the **Map type:** drop-down list in the **Diffuse map** area of the **Maps** pane. Then, define the map as described earlier. Once the map is defined, return to the top of the material mapping tree. In the **Diffuse map** area of the **Maps** pane, use the slider to adjust how much of the map is applied to the diffuse color.

Exercise 16-4

Complete the exercise on the student website.
www.g-wlearning.com/CAD

Reflection map

The advanced and advanced metal material types have a reflection property. A *reflection map* is applied to this property. This is often done for shiny materials in outdoor scenes. An image of clouds or blue sky is applied as a reflection map because the sky is not modeled. A reflection map is assigned in the **Maps** pane, which is displayed at the top of the material mapping tree. The **Reflection map** check box in the pane toggles the specified reflection map off and on.

To assign a reflection map, select the type of map to apply using the **Map type:** drop-down list in the **Reflection map** area of the **Maps** pane. Then, define the map as described earlier. Once the map is defined, return to the top of the material mapping tree. In the **Reflection map** area of the **Maps** pane, use the slider to adjust how much of the map is applied to the reflection property. Often, only a small percentage of the reflection map is applied to the material.

Opacity map

An *opacity map* is applied to the opacity property of a material to make an object appear transparent in different areas. Without an opacity map, the opacity property is equally applied to the object. Black areas of the map are transparent, white areas are opaque, and gray areas are semitransparent. See **Figure 16-36.** If a color appears in the map, the grayscale value of the color is used to calculate transparency. An opacity map is assigned in the **Maps** pane, which is displayed at the top of the material mapping tree. The **Opacity map** check box in the pane toggles the specified opacity map off and on.

Figure 16-36.
The effect of an opacity map. A—This black and white image will be used as the opacity map. B—The material on the plane is completely opaque. C—When the opacity map is applied to the material, the dark areas of the map produce transparent areas on the object.

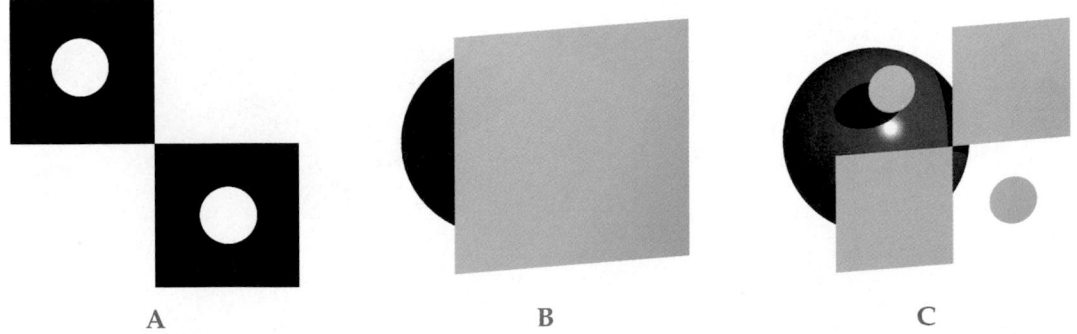

A   B   C

To assign an opacity map, select the type of map to apply using the **Map type:** drop-down list in the **Opacity map** area of the **Maps** pane. Then, define the map as described earlier. Once the map is defined, return to the top of the material mapping tree. In the **Opacity map** area of the **Maps** pane, use the slider to adjust how much of the map is applied to the opacity property.

Bump map

A *bump map* is a map applied to the material to make some areas of the material appear raised and other areas depressed. The black, white, and grayscale values of the map are used to determine raised and depressed areas. Dark areas of the map make the material surface appear depressed and light areas make the material surface appear unchanged. For example, to show the texture of a brick wall, you could physically model the grooves into the wall. This would take a lot of time to model and would immensely increase the rendering time because of the increased complexity of the geometry. Using a bump map is an easier and more efficient way to accomplish the same task. **Figure 16-37** shows a bump map used to represent an embossed stamp on a metal case. A bump map is assigned in the **Maps** pane, which is displayed at the top of the material mapping tree. The **Bump map** check box in the pane toggles the specified bump map off and on.

To assign a bump map, select the type of map to apply using the **Map type:** drop-down list in the **Bump map** area of the **Maps** pane. Then, define the map as described earlier. Once the map is defined, return to the top of the material mapping tree. In the **Bump map** area of the **Maps** pane, use the slider to adjust how much of the map is applied to the bump property. Typically, the bump map is set to a low percentage.

Exercise 16-5

Complete the exercise on the student website.
www.g-wlearning.com/CAD

Figure 16-37.
The effect of a bump map. A—This image will be used as the bump map.
B—When applied to the material, the bump map simulates an embossed stamp in the metal case.

A

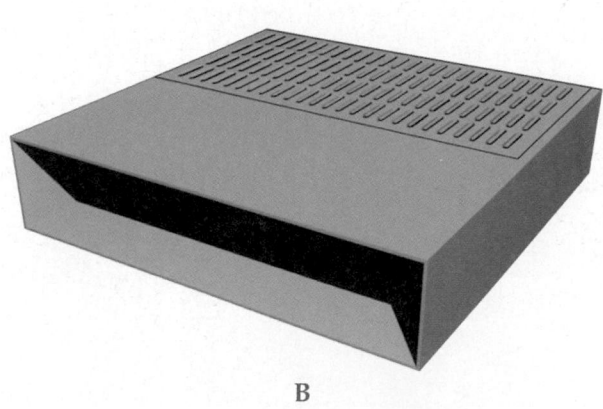

B

Simply applying a map to a material rarely results in a realistic scene when the scene is rendered. The maps often need to be adjusted to produce the desired results. Maps can be adjusted at the material level or the object level. A combination of these two adjustments is usually required to produce a photorealistic rendering.

Material-Level Adjustments

Whenever a map is applied to the diffuse, reflection, opacity, or bump property, three buttons are displayed in the corresponding area in the **Maps** pane below the **Map type:** drop-down list. See **Figure 16-38.** To remove the map from the material property, simply pick the **Delete map information from material** button. A warning may appear indicating that subtextures will also be deleted. Once the map is removed, the three buttons and the slider are removed from the area. If a texture map was removed, the "select image" button is again labeled **Select Image**, which indicates that there is no map attached to the property.

The middle button below the **Map type:** drop-down list is used to synchronize or unsynchronize the map channels within the same material definition. For example, to make a realistic tile floor, a tile map is applied as a diffuse map. Another tile map is also used as a bump map to make the grout look recessed. If the scaling is changed for the bump map, but not synchronized, the grout colors and indentations will not match. See **Figure 16-39.** The synchronize button ensures that all of these settings are the same.

Figure 16-38.
The controls for a map. Once applied to a property, all maps display these controls.

Figure 16-39.
A—This material has diffuse color and bump maps. B—If the bump map is not synchronized with the material, it will not align with the grout lines if scaled.

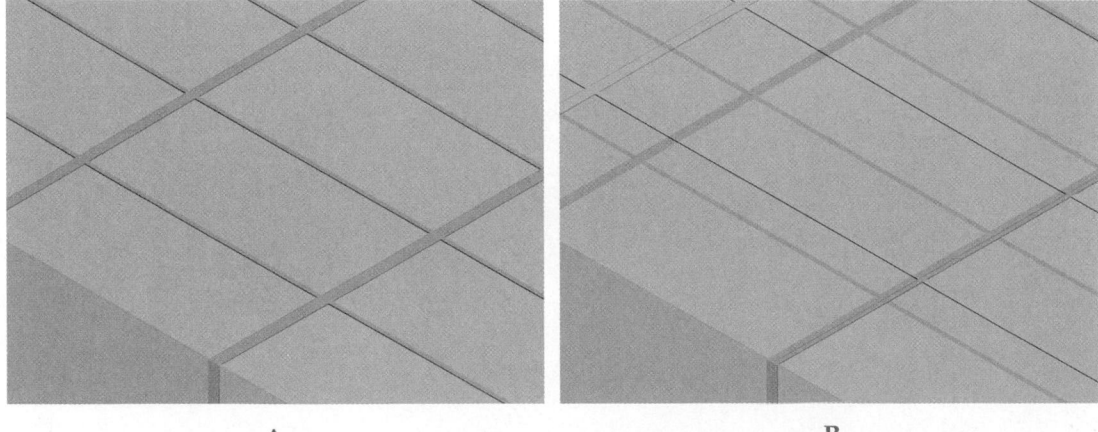

A   B

Maps are synchronized by default. When synchronized, the button appears closed. When unsynchronized, the button appears open. When the button is picked to turn off synchronization, a warning message is displayed indicating that the other properties will be unaffected. It is recommended that you leave the synchronize feature on.

The right-hand button below the **Map type:** drop-down list opens the **Map Preview** dialog box. This dialog box displays the map channel in a larger, 2D view. This dialog box remains open until you close it. Checking the **Auto-update** check box forces the image to automatically update when a change is made to the map. If it is unchecked, the image is not updated until you manually update it by picking the **Update** button in the same dialog box.

The settings in the **Advanced Lighting Override** pane in the **Materials** palette are used to adjust the appearance of materials illuminated by indirect lighting when global illumination or final gathering is used to render the scene. These topics are discussed in Chapter 18. The **Advanced Lighting Override** pane is only displayed for realistic and realistic metal material types.

The **Material Scaling & Tiling** and **Material Offset & Preview** panes in the **Materials** palette allow adjustments to be made that affect how the map image fits on the material. See **Figure 16-40.** Settings made in these panes impact all objects in the drawing that have this material applied to them. Each property map has similar panes at the map level in the mapping tree. These are named **Scaling & Tiling** and **Offset & Preview** and contain the same controls. Notice the title bars of the panes at the map level have synchronize buttons, which are the same buttons found in the **Maps** pane at the top of the material mapping tree. If unsynchronized, the settings in the **Scaling & Tiling** and **Offset & Preview** panes at the map level affect only the map. The settings in the **Material Scaling & Tiling** and **Material Offset & Preview** panes at the top of the material mapping tree are applied to all maps in the material definition.

Figure 16-40.
Making map adjustments to the material. A—The **Material Scaling & Tiling** pane (or map-level **Scaling & Tiling** pane) is available for texture (2D) maps. B—The **Material Offset & Preview** pane (or map-level **Offset & Preview** pane) is available for all maps.

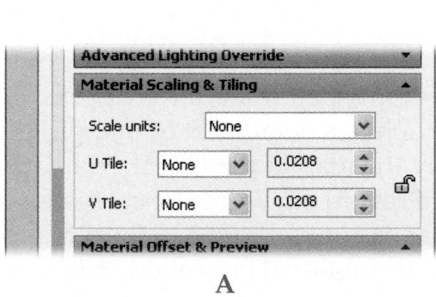

A

B

Scaling and tiling

The map-level **Scaling & Tiling** pane is only available for texture maps (2D maps): texture map, checker, gradient ramp, and tiles. By the same token, the **Material Scaling & Tiling** pane is only displayed if one of the material properties has a texture map.

At the top of the **Material Scaling & Tiling** pane (or map-level **Scaling & Tiling** pane), select the units that will be used to apply the scaling. In the **Scale units:** drop-down list, you can select None, Fit to Gizmo, or a standard unit of measure. The gizmo is used for object-level material adjustments, as described in the next section.

After selecting the units for the scale, select the type of tiling for the U and V axes (or width and height, if real-world units are specified) using the left-hand drop-down lists. The U and V axes are similar to the X and Y axes in a 2D drawing, but are relative to the map image. You can select None, Tile, or Mirror. None means the map will not be repeated. Mirror means the map is repeated, but each tile is a mirror image of its neighbor. Finally, enter the scale on the U and V axes using the right-hand text boxes. The value is the number of times the map fits within a one-unit square (defined by the selected units). The aspect ratio can be locked using the lock button to the right of the text boxes.

Offset and preview

At the top of the **Material Offset & Preview** pane (or map-level **Offset & Preview** pane) there is the **Auto-regen** check box. With this checked, the image automatically updates when changes are made. If it is not checked, you must pick the **Update** button to see the changes.

The value in the **Preview size:** text box is a zoom factor for the preview image. If the image is not properly displaying in the preview window, use this to zoom in or out. You can also use the zoom in (+) and zoom out (–) buttons to zoom in and out.

The **V Tile** and **U Tile** sliders change the scale of the map. As the slider is moved, the corresponding text box value in the **Material Scaling & Tiling** pane changes. These sliders are not displayed if a real-world unit is selected in the **Scale units:** drop-down list in the **Material Scaling & Tiling** pane.

The **U Offset:** and **V Offset:** text boxes at the bottom of the **Material Offset & Preview** pane set the location of the map image within the material. The offsets can also be changed by picking and dragging the image in the preview pane.

The value in the **Rotation:** text box determines the rotation of the map about the W axis. The W axis is similar to the Z axis, but is local to the image. When spherical or cylindrical mapping is applied to an object, this setting has no effect. Object-level mapping is discussed in the next section.

Object-Level Adjustments

Material mapping refers to specifying how a mapped material is applied to an object. When a mapped material is attached to an object, a default set of mapping coordinates, or simply *default mapping*, is used to apply the map to the object. Many times, the **Material Scaling & Tiling** and **Material Offset & Preview** panes (or the corresponding map-level panes) can be used to alter how the map is applied to the default mapping. But, since these changes affect all objects to which the material is applied, this may not be acceptable.

Fortunately, AutoCAD allows you to adjust mapping at the object level for texture-mapped (2D-mapped) materials. The **MATERIALMAP** command applies a grip tool, or *gizmo,* based on one of four mapping types: planar, box, spherical, or cylindrical. See **Figure 16-41.** The colored edge represents the start and end of the map. For best results, select the mapping type based on the general shape of the object to which mapping is applied. Do not be afraid to experiment with other mapping types, however. Any mapping type can be used on any object, regardless of the object's shape. However, only one mapping type can be applied to an object at any given time.

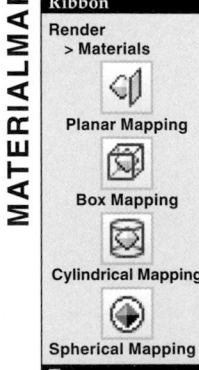

MATERIALMAP

Ribbon
Render
> Materials

Planar Mapping

Box Mapping

Cylindrical Mapping

Spherical Mapping

Type
MATERIALMAP

Figure 16-41.
These are the four material map gizmos. From left to right: planar, box, spherical, and cylindrical. The colored edge represents the start and end of the map.

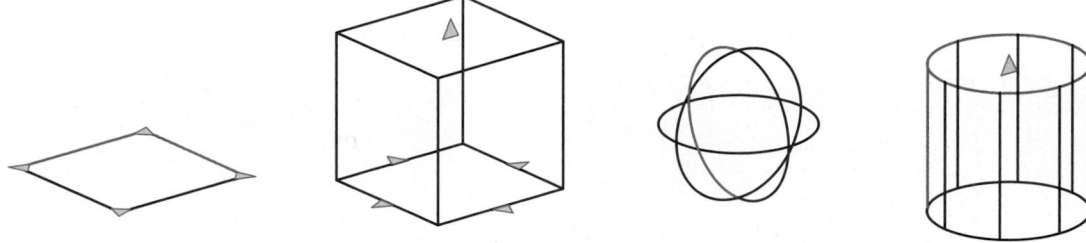

After one of the mapping types is selected, you are prompted to select the faces or objects. You can select multiple objects or faces. After making a selection, the gizmo is placed on the selection set. The command remains active for you to adjust the mapping or enter an option.

Drag the grips on the gizmo to stretch or scale the material. The effects of editing a diffuse color map are dynamically displayed if the current visual style is set to display materials and textures. Otherwise, exit the command and render the scene to see the effect of the edit. To readjust the mapping, select the same mapping type and pick the object again. The gizmo is displayed in the same location as before.

The **Move** and **Rotate** options of the command toggle between the move and rotate grip tools. Using the grip tools, you can move and rotate the map on the object. The **Reset** option of the command restores the default mapping to the object. The **Switch mapping mode** option allows you to change between the four types of mapping.

If the command is typed, there is an additional option. The **Copy mapping to** option is a quick and easy way to apply the changes made to the current object to other objects in the scene. Enter this option, select the face or object to copy from, and then select the faces or objects to copy to. This option is also available if the **Switch mapping mode** option is entered.

For example, look at **Figure 16-42A.** The grain on the stair risers is running vertically when it should run horizontally. First, apply a planar map to the bottom riser. Next, rotate the mapping 90°, **Figure 16-42B.** Finally, use the **Copy mapping to** option to copy the mapping to the other risers, **Figure 16-42C.**

PROFESSIONAL TIP

If you are using a bump, reflection, or opacity map and need to adjust it at the object level, apply the same map as a diffuse color map. Also, set the visual style to display materials and textures. Then, adjust the object mapping as needed. The edits are dynamically displayed in the viewport. When the image is in the correct location, remove the diffuse color map from the material.

Exercise 16-6

Complete the exercise on the student website.
www.g-wlearning.com/CAD

Figure 16-42.
Correcting material mapping. A—The grain on the risers runs vertically instead of horizontally. B—Rotating the map with the rotate gizmo. C—The corrected rendering. (Model courtesy of Arcways, Inc., Neenah, WI)

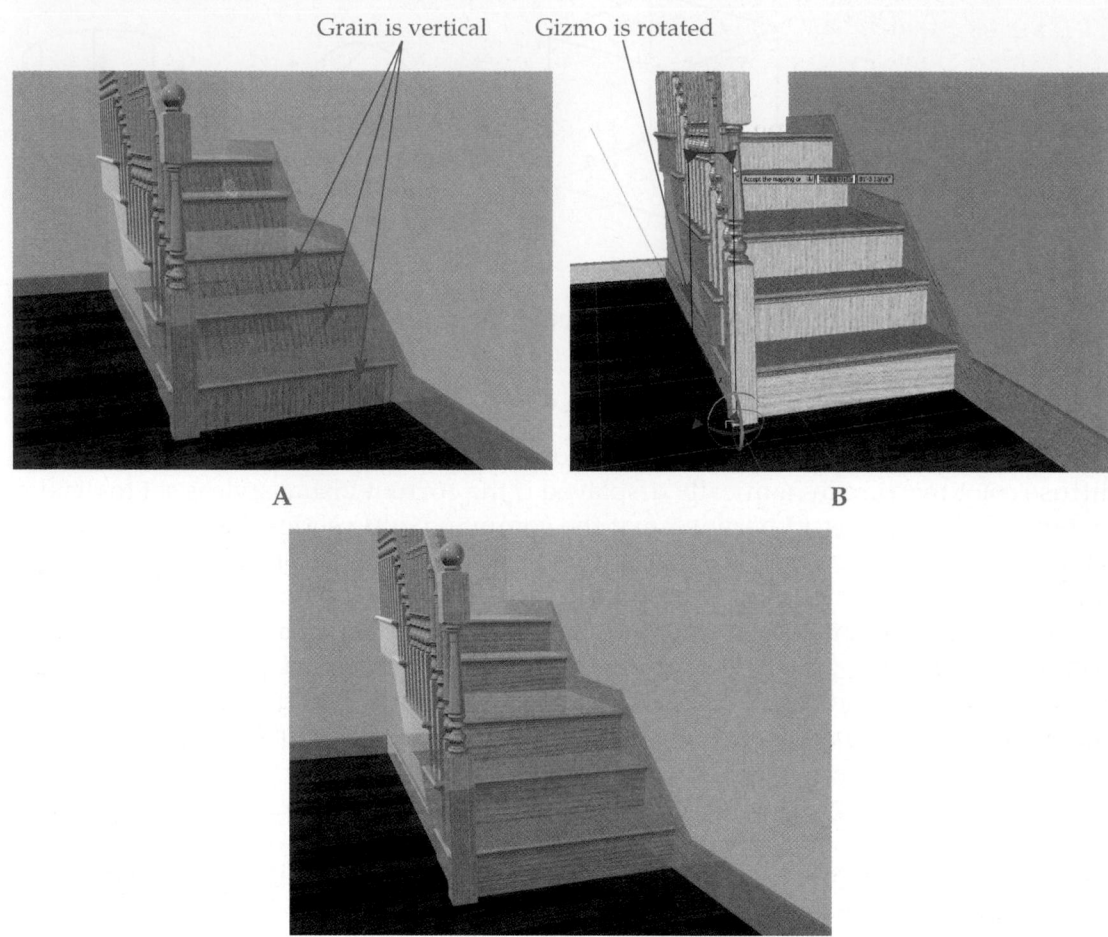

Grain is vertical   Gizmo is rotated

A

B

C

Chapter Test

Answer the following questions. Write your answers on a separate sheet of paper or complete the electronic chapter test on the student website.
www.g-wlearning.com/CAD

1. Define *material*.
2. Define *materials library*.
3. Why do the materials in the tool palettes and **Materials** palette have a checkered background?
4. Describe how to attach a material using a tool palette.
5. How can materials be attached to layers?
6. By default, which material is attached to newly created objects?
7. Which material is used as the base material for creating new materials?
8. Name the three material display options for a visual style that can also be set using the **Materials** panel on the ribbon.
9. How do you know if a material in the **Materials** palette is being used in the drawing?
10. How can the name and description of an existing material be changed?
11. Name the four basic material types.
12. What are the templates available for the realistic metal material type?

13. To create a reflective material, which material type(s) can be used?
14. What are the three material color settings? Explain each.
15. Describe the difference between a transparent material and a translucent material.
16. How much illumination does a self-illuminated material add to a scene?
17. How is a marble material created?
18. Explain how black and white areas of an opacity map affect the transparency of a material.
19. Which objects do changes made in the **Material Offset & Preview** pane affect?
20. Name the four types of mapping available for adjusting texture maps at the object level.

Drawing Problems

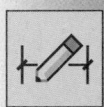

1. In this problem, you will create a scene with basic 3D objects, attach materials to the objects, and adjust the settings of the materials.
   A. Start a new drawing and set the units to architectural.
   B. Draw a 15′ × 15′ planar surface to represent the floor.
   C. Draw two boxes to represent two walls. Make the boxes 15′×4″×9′. Position them to form a 90° corner. Alternately, you can draw a polysolid of the same dimensions.
   D. Draw a R2′ × 5′H cone in the center of the room.
   E. Open the **Tool Palettes** window, display the **Materials** palette group, and display the **Flooring** tool palette. Attach the material Finishes.Flooring.Tile.Square.Terra Cotta to the floor.
   F. Using the **Finishes-Paint, Plaster, Walls, Wax** material tool palette, attach the material Finishes.Plaster.Stucco.Troweled.White to the wall.
   G. Using the same material tool palette, attach the material Finishes.Wall Covering. Stripes.Vertical.Blue-Grey to the cone.
   H. Turn on the sun and adjust the time to create good shadows. Refer to Chapter 15 for an introduction to sun settings.
   I. Render the scene. Save the rendering as P16_01.jpg.
   J. Save the drawing as P16_01.

2. In this problem, you will attach materials to the objects in an existing drawing and render the scene.
    A. Open the drawing P15_01 from Chapter 15. If you did not complete this problem, do so now. Save the drawing as P16_02.
    B. Open the **Tool Palettes** window and display the **Materials** palette group.
    C. Attach the materials of your choice to the objects in the scene. Do not be restricted by the names of the materials. For example, a concrete material may be suitable for foliage or even carpet with a simple color change. Be creative.
    D. If the items in the scene were inserted using the **Autodesk Seek Design Content** button in **DesignCenter**, they may be blocks with nested layers. Instead of exploding the blocks, use the **MATERIALATTACH** command and attach materials to the layers on which the nested objects reside.
    E. Turn on the sun and adjust the time to create good shadows. Refer to Chapter 15 for an introduction to sun settings.
    F. Render the scene. Save the rendering as P16_02.jpg.
    G. Save the drawing.

3. In this problem, you will create custom wood and marble materials.
   A. Start a new drawing and save it as P16_03.
   B. Draw two 5 × 5 × 5 boxes and position them near each other. Using other primitives, cut notches and holes in the boxes. The boxes will be used to test the custom materials.
   C. In the **Materials** palette, create two new materials. Name one Wood-*your initials* and the other Marble-*your initials.*
   D. Attach the wood material to one of the boxes and the marble material to the other box.
   E. Render the scene and make note of the wood grain and marble veins.
   F. Use the **Wood** and **Marble** panes in the material editor to change the properties of the materials.
   G. Render the scene again and make note of the changes. Using the **Render Region** button on the **Render** panel in the **Render** tab of the ribbon can save time when testing material changes.
   H. When you are satisfied with the materials, save the rendering as P16_03.
   I. Save the drawing.

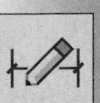

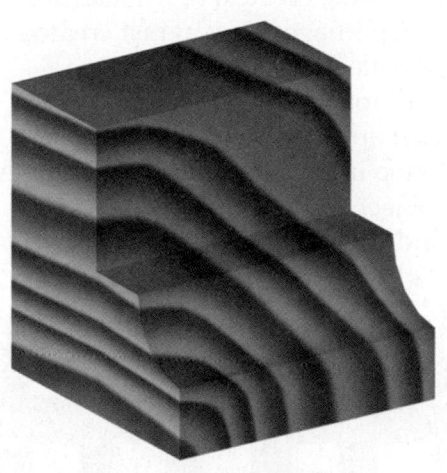

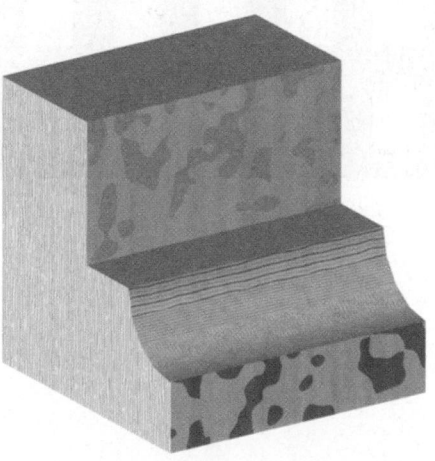

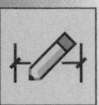

4. In this problem, you will create a bitmap and use it as opacity and bump maps.
    A. Draw a rectangle with an array of smaller rectangles inside of it, as shown below. Sizes are not important and the pattern can be varied if you like.
    B. Display a plan view of the rectangles. Then, copy all of the objects to the Windows clipboard by pressing [Ctrl]+[C] and selecting the objects.
    C. Launch Windows Paint. Then, paste the objects into the blank file. Notice how the AutoCAD background outside of the large rectangle is also included.
    D. Use the select tool (rectangle) in Paint to draw a window around the large rectangle created in AutoCAD and the smaller rectangles within it. Copy this to the Windows clipboard by pressing [Ctrl]+[C].
    E. Start a new Paint file without saving the current one and paste the image from the clipboard into the new blank file. Now, the unwanted AutoCAD background is no longer displayed. If needed, change the small rectangles to black and the lattice to white using the tools in Paint. The colors should be the reverse of what is shown below. Then, save the image file as P16_04.bmp and close Paint.
    F. In AutoCAD, draw a solid box of any size.
    G. Using the **Materials** palette, create a new material.
    H. Assign the P16_04.bmp image file you just created as an opacity map. Adjust the map so that it is scaled to fit to the object.
    I. Render the scene and note the effect.
    J. Turn off the opacity map.
    K. Assign the P16_04.bmp image file as a bump map. Adjust the map so that it is scaled to fit to the object.
    L. Render the scene and note the effect.
    M. Save the drawing as P16_04.

CHAPTER **17**

Lighting

Learning Objectives

After completing this chapter, you will be able to:
- ✓ Describe the types of lighting in AutoCAD.
- ✓ List the user-created lights available in AutoCAD.
- ✓ Change the properties of lights.
- ✓ Generate and modify shadows.
- ✓ Add a background to your scene and control its appearance.

In Chapter 15, you were introduced to lighting. You learned how to adjust lighting by turning off the default lights and adding sunlight. In this chapter, you will learn all about the lights available in AutoCAD. You will learn lighting tips and tricks to help make the scene look its best.

Types of Lights

Ambient light is like natural light just before sunrise. It is the same intensity everywhere. All faces of the object receive the same amount of ambient light. Ambient light cannot create highlights, nor can it be concentrated in one area. AutoCAD does not have an ambient light setting. Instead, it relies on indirect illumination, which is discussed in Chapter 18.

A *point light* is like a lightbulb. Light rays from a point light shine out in all directions. A point light can create highlights. The intensity of a point light falls off, or weakens, over distance. Other programs, such as Autodesk 3ds max®, may call these lights *omni lights.* A *target point light* is the same as a standard point light except that a target is specified. The illumination of the target point light is directed toward the target.

A *distant light* is a directed light source with parallel light rays. This acts much like the Sun. Rays from a distant light strike all objects in your model on the same side and with the same intensity. The direction and intensity of a distant light can be changed.

A *spotlight* is like a distant light, but it projects in a cone shape. Its light rays are not parallel. A spotlight is placed closer to the object than a distant light. Spotlights have a hotspot and a falloff. The light from a standard spotlight is directed toward a target. A *free spotlight* is the same as a standard spotlight, but without a target.

A *weblight* is a directed light that represents real-world distribution of light. The illumination is based on photometric data that can be entered for each light. The light from a standard weblight is directed toward a target. A *free weblight* is the same as a standard weblight, but without a target point.

Properties of Lights

There are several factors that affect how a light illuminates an object. These include the angle of incidence, reflectivity of the object's surface, and the distance that the light is from the object. In addition, the ability to cast shadows is a property of light. Shadows are discussed in detail later in this chapter.

Angle of Incidence

AutoCAD renders the faces of a model based on the angle at which light rays strike the faces. This angle is called the *angle of incidence.* See **Figure 17-1.** A face that is perpendicular to light rays receives the most light. As the angle of incidence decreases, the amount of light striking the face also decreases.

Reflectivity

The angle at which light rays are reflected off of a surface is called the *angle of reflection.* The angle of reflection is always equal to the angle of incidence. Refer to **Figure 17-1.**

The "brightness" of light reflected from an object is actually the number of light rays that reach your eyes. A surface that reflects a bright light, such as a mirror, is reflecting most of the light rays that strike it. The amount of reflection you see is called the *highlight.* The highlight is determined by the angle from the viewpoint relative to the angle of incidence. Refer to **Figure 17-1.**

The surface quality of the object affects how light is reflected. A smooth surface has a high specular factor. The *specular factor* indicates the number of light rays that have the same angle of reflection. Surfaces that are not smooth have a low specular factor. These surfaces are called *matte.* Matte surfaces *diffuse,* or "spread out," the light as it strikes the surface. This means that few of the light rays have the same angle of reflection. **Figure 17-2** illustrates the difference between matte and high specular

Figure 17-1.
The amount of reflection, or highlight, you see depends on the angle from which you view the object.

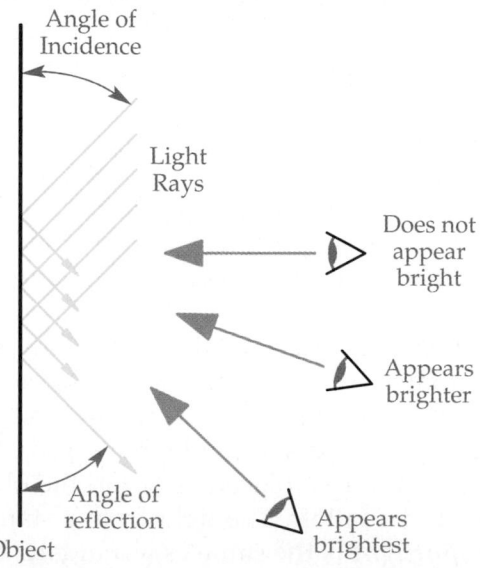

AutoCAD and Its Applications—Advanced

Figure 17-2.
Matte surfaces produce diffuse light. This is also referred to as having a low specular factor. Shiny surfaces evenly reflect light and have a high specular factor.

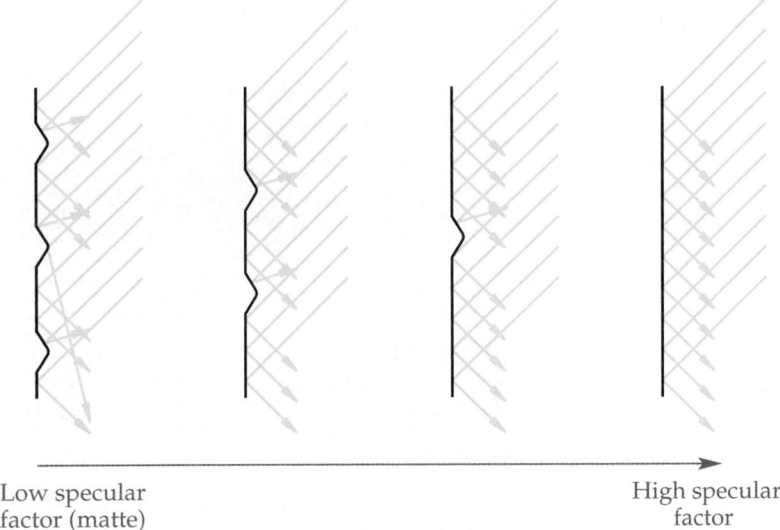

Low specular factor (matte)

High specular factor

finishes. Surfaces can also vary in *roughness.* Roughness is a measure of the polish on a surface. This also affects how diffused the reflected light is.

Hotspot and Falloff

A spotlight produces a cone of light. The *hotspot* is the central portion of the cone, where the light is brightest. See **Figure 17-3.** The *falloff* is the outer portion of the cone, where the light begins to blend to shadow. The hotspot and falloff of a spotlight are not affected by the distance the light is from an object. Spotlights are the only lights with hotspot and falloff properties.

Attenuation

The farther an object is from a point light or spotlight, the less light will reach the object. See **Figure 17-4.** The intensity of light decreases over distance. This decrease is called *attenuation.* All lights in AutoCAD, except distant lights, have some kind of attenuation. Often, attenuation is called *falloff* or *decay.* However, do not confuse this

Figure 17-3.
The hotspot of a spotlight is the area that receives the most light. The smaller cone is the hotspot. The falloff receives light, but less than the hotspot. The larger cone is the falloff.

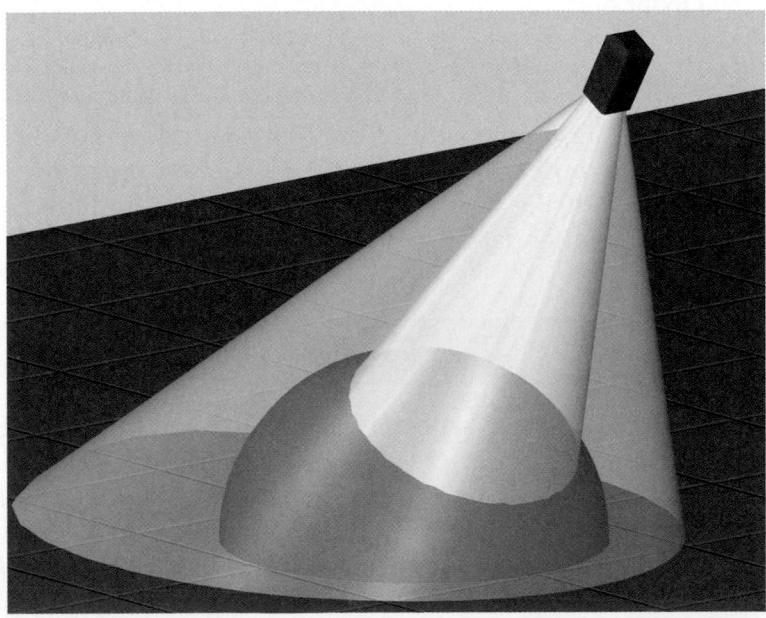

Figure 17-4.
Attenuation is the intensity of light decreasing over distance. Attenuation has been turned on in this scene.

Less illumination

More illumination

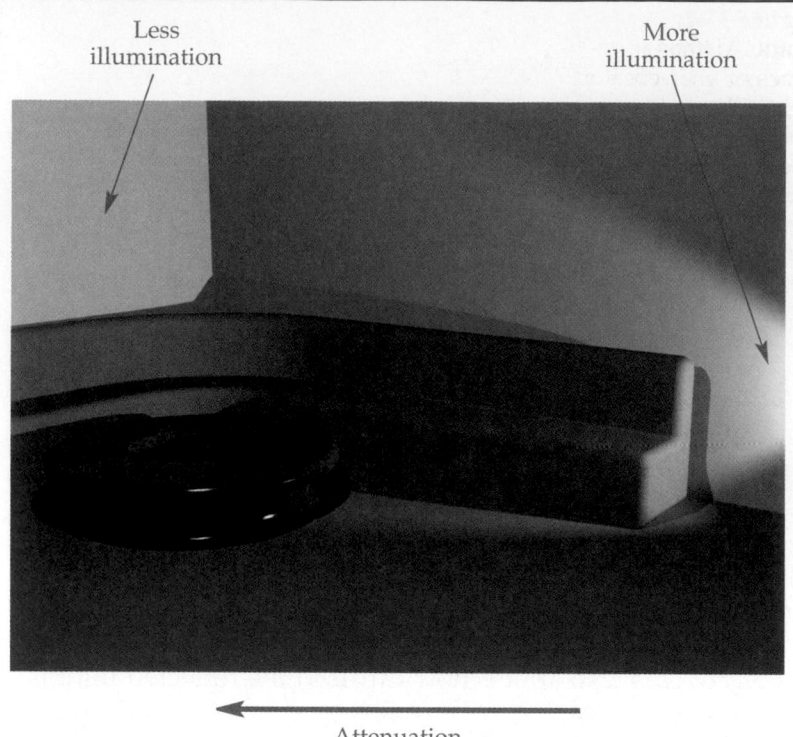

Attenuation

with the falloff of a spotlight, which is the outer edge of the cone of illumination. The following attenuation settings are available in AutoCAD.

- **None.** Applies the same light intensity regardless of distance. In other words, no attenuation is calculated.
- **Inverse Linear.** The illumination of an object decreases in inverse proportion to the distance. For example, if an object is two units from the light, it receives 1/2 of the full light. If the object is four units away, it receives 1/4 of the full light.
- **Inverse Squared.** The illumination of an object decreases in inverse proportion to the square of the distance. For example, if an object is two units from the light, it receives $(1/2)^2$, or 1/4, of the full light. If the object is four units away, it receives $(1/4)^2$, or 1/16, of the full light. As you can see, attenuation is greater for each unit of distance with the **Inverse Squared** option than with the **Inverse Linear** option.

PROFESSIONAL TIP

The intensity of the Sun's rays does not diminish from one point on Earth to another. They are weakened by the angle at which they strike Earth. Therefore, since distant lights are similar to the Sun, attenuation is not a factor with distant lights.

AutoCAD Lights

AutoCAD has three types of lighting: default lighting, sunlight with or without sky illumination, and user-created lighting. *Default lighting* is the lighting automatically available in the scene. It is composed of two light sources that evenly illuminate all surfaces. As the viewpoint is changed, the light sources follow to maintain an even illumination of the scene. There is no control over default lighting and it must be shut off whenever one of the other types of lighting is used.

Figure 17-5.
AutoCAD has four types of user-created lights: distant, point, weblights, and spotlights. A weblight is really a targeted point light. It projects in all directions, but may be predominant in one direction.

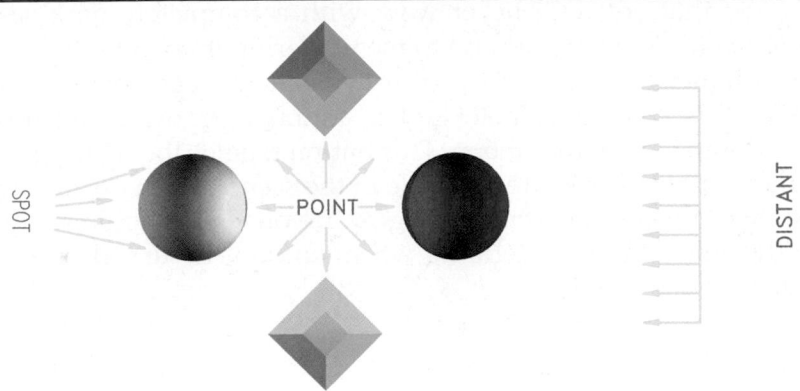

As you saw in Chapter 15, *sunlight* may be added to any scene. AutoCAD uses a distant light to simulate the parallel rays of the Sun. The date and time of day can be adjusted to create different sunlight illumination. *Sky illumination* may also be added with sunlight to simulate light bouncing off of objects in the scene and particles in the atmosphere. This helps create a more-natural feel.

User-created lighting results when you add AutoCAD light objects to the drawing. There are four types of user-created lights: distant light, weblight, point light, and spotlight. See **Figure 17-5**. A distant light is a directed light source with parallel light rays. A weblight is a directional point light containing light intensity (photometric) data. A point light is like a lightbulb with light rays shining out in all directions. A spotlight is like a distant light, but it projects light in a cone shape instead of having parallel light rays.

When created, point lights, weblights, and spotlights are represented by *light glyphs,* or icons, in the drawing. To suppress the display of light glyphs, pick the **Light Glyph Display** button in the expanded area of the **Lights** panel on the **Render** tab of the ribbon. The button is blue when light glyphs are displayed. The default lights, sun, and distant lights are not represented by glyphs.

In this section, you will learn how to add lights. You will also learn how to adjust the various properties of sunlight and AutoCAD light objects. The tools for working with lights can be accessed using the command line, **Generic Lights** tool palette, the tool palettes in the **Photometric Lights** tool palette group, and **Lights** panel on the **Render** tab of the ribbon. See **Figure 17-6**.

Ribbon

Render
> Lights

Light Glyph Display

Type

LIGHTGLYPHDISPLAY

LIGHTGLYPHDISPLAY

Figure 17-6.
The tools for adding and controlling lights.

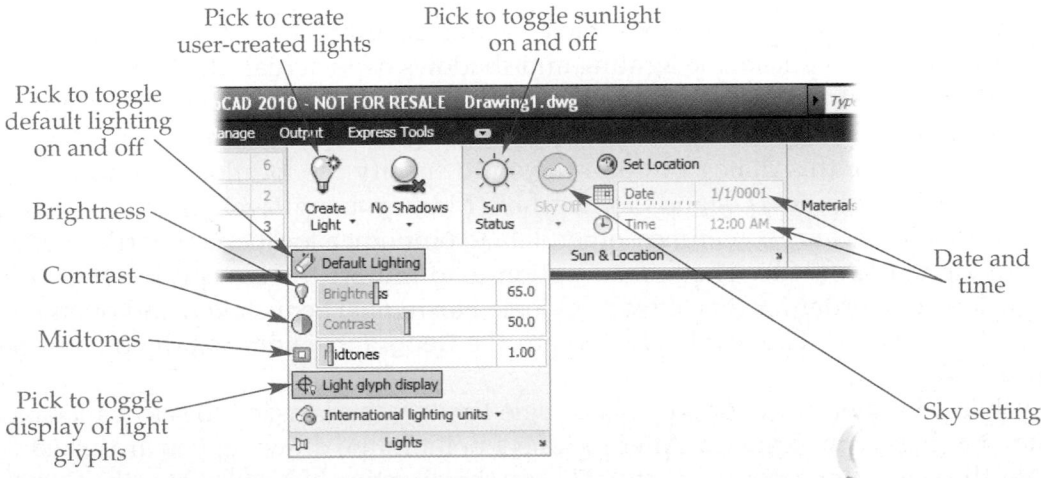

So that you will never work with a completely dark scene, default lighting is applied in the viewport and to the rendering if no other lights are added. In order for your lights to be applied, you must switch between default lighting and user lighting. To do this, pick the **Default Lighting** button in the expanded area of the **Lights** panel on the **Render** tab of the ribbon. This button toggles the lighting between default lighting and whatever lights are available in the scene. When default lighting is on, the button is blue. When off, the button is grey. If you elected to do so, AutoCAD will automatically shut off default lighting when sunlight is turned on or a user-created light is added to the scene.

Lighting Units

There are three types of *lighting units* available in AutoCAD: standard (generic), international (SI), and US customary (American). Generic lighting is the type of lighting that was used in AutoCAD prior to AutoCAD 2008. This lighting provides very nice results, but the settings are not based on any real measurements. International or US customary lighting is called photometric lighting. *Photometric lighting* is physically correct and attenuates at the square of the distance from the source. For more accuracy, photometric data files can be imported from lighting manufacturers.

The **LIGHTINGUNITS** system variable sets which type of lighting is used. A setting of 0 means that standard (generic) lighting is used. However, for more realistic lighting, it is recommended that photometric lighting be used. Enter a value of 1 and US customary (American) lighting units are used. A setting of 2 turns on international lighting units. This is the default setting. A setting of 1 or 2 results in photometric lighting. The only difference between a setting of 1 and 2 is that US customary (American) units are displayed as *candelas* and international units are displayed as *lumens*.

Sunlight

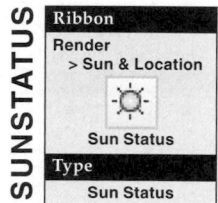

To turn sunlight on or off, pick the **Sun Status** button in the **Sun & Location** panel on the **Render** tab of the ribbon. This button is blue when sunlight is on. Sunlight can also be turned on or off in the **Sun Properties** palette, which is discussed later. Sunlight is not represented by a light glyph. The date, time, and geographic location can also be set in the **Sun & Location** panel on the **Render** tab of the ribbon. These properties determine how the scene is illuminated by the sun.

To change the current date, drag the **Date** slider left or right. Sunlight must be on for the slider to be enabled. As you drag the slider, the date is displayed on the right-hand end of the slider. The time is changed in the same manner as the date. Drag the **Time** slider left or right. As you drag the slider, the time is displayed on the right-hand end of the slider.

The location of the scene can be set to an actual geographic location. This is important if you want to replicate the lighting and shadows of an actual site. Picking the **Set Location** button on the **Sun & Location** panel in the **Render** tab of the ribbon opens a dialog box titled **Geographic Location – Define Geographic Location**. See **Figure 17-7.** The first option in this dialog box allows you to specify the location by importing a KML or a KMZ file. *KML* stands for Keyhole Markup Language. This file contains latitude, longitude, and, sometimes, other data to pinpoint a location on Earth. A KMZ file is a zipped KML file. The second option is to import the current location from Google Earth. In order for this to work, Google Earth must be installed and open with the location selected. The third option opens the **Geographic Location** dialog box. See **Figure 17-8.**

When a location is imported from Google Earth, you are asked to select a location in the drawing for the location. After picking a point in the drawing, you are asked for the North vector. Pick a point that should be in the direction of North. Once this is set, a

Figure 17-7.
This dialog box provides the options for defining the geographic location of the model.

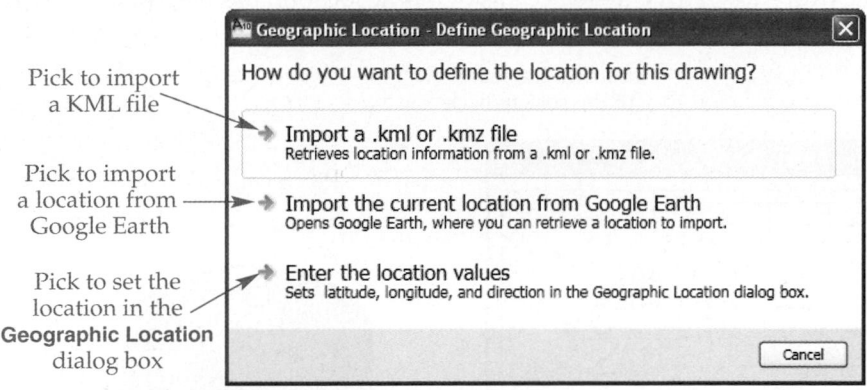

Pick to import a KML file

Pick to import a location from Google Earth

Pick to set the location in the **Geographic Location** dialog box

Figure 17-8.
The **Geographic Location** dialog box is used to input geographic location data.

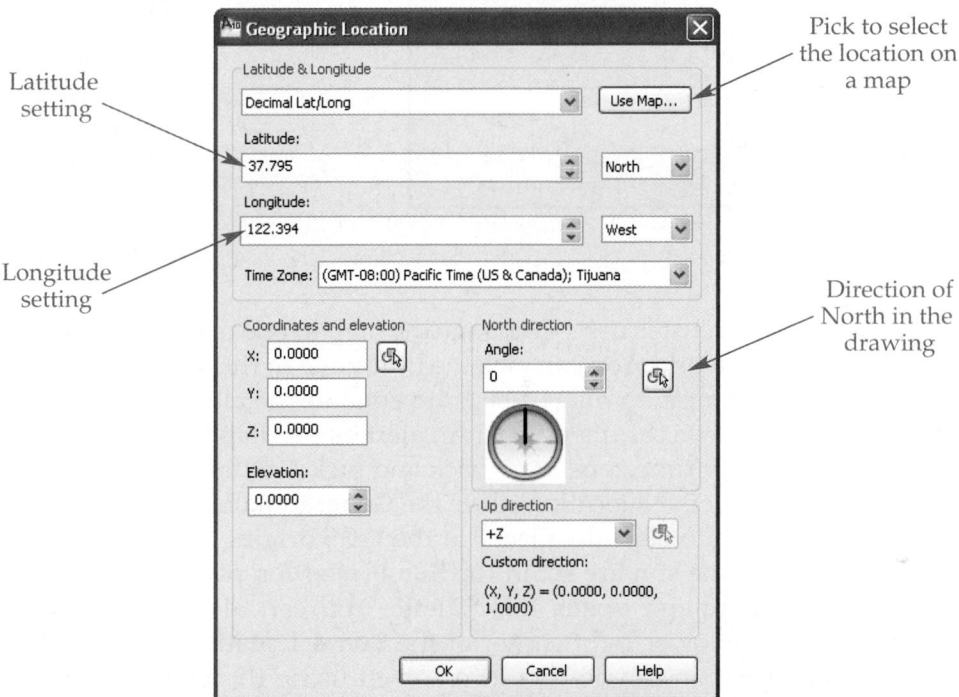

Latitude setting

Longitude setting

Pick to select the location on a map

Direction of North in the drawing

geographic marker is added at the point selected for the location. This marker looks like a red and white thumbtack. Its visibility is controlled with the **GEOMARKERVISIBILITY** system variable. See **Figure 17-9.**

To set the location in the **Geographic Location** dialog box, enter the latitude and longitude. You can select to use either decimal or degrees/minutes/seconds values for these settings. Also, pick a time zone in the **Time Zone:** drop-down list and set XYZ coordinates and elevation for AutoCAD's world coordinate system. You can specify the angle for North in the **Angle:** text box or by picking points in the drawing window. The "up direction" is normally +Z, but you can change it to –Z. You can also set it to anything you want if you select Custom Direction in the drop-down list in the **Up direction** area.

The quickest way to set the location is to select it on a map. Pick the **Use Map...** button to open the **Location Picker** dialog box. See **Figure 17-9.** The crosshairs on the map indicate the current location. By default, picking a point on the map selects the nearest big city. Many other cities are available in the **Nearest City:** drop-down list.

Figure 17-9.
A—The **Location Picker** dialog box is a quick way to set the location. Pick a location on the map or select it in the drop-down lists. B—Once a geographic location has been set, the marker is added to the WCS origin.

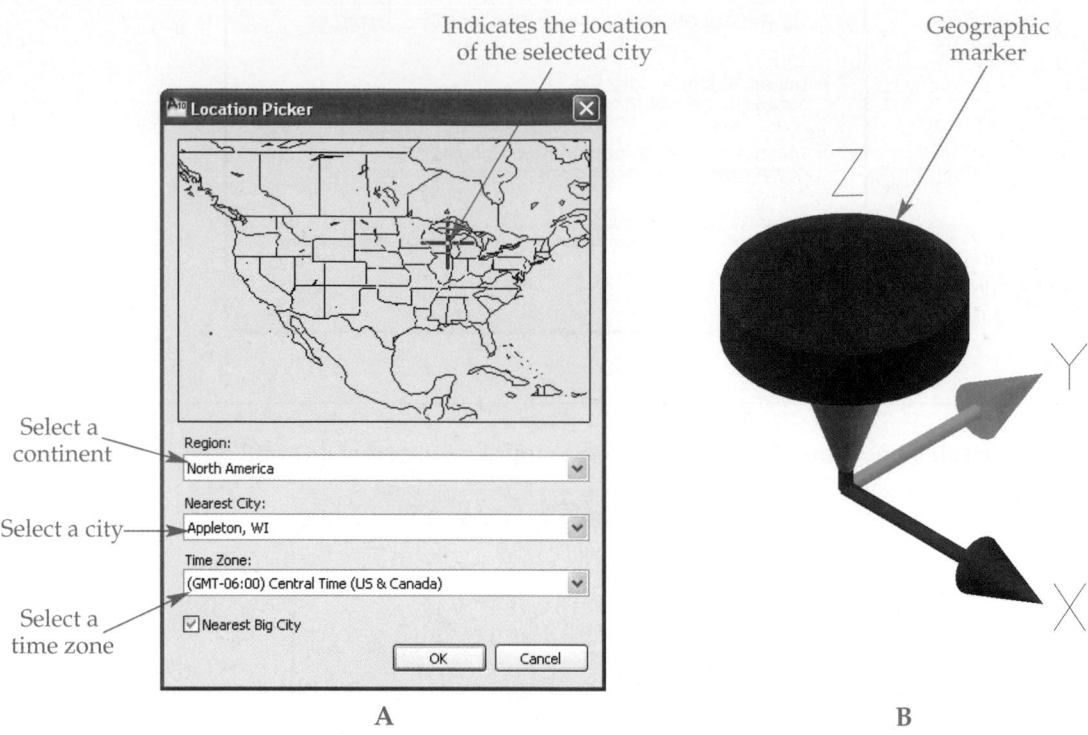

Indicates the location of the selected city

Geographic marker

Select a continent

Select a city

Select a time zone

A

B

The **Region:** drop-down list is used to change the map to one of eight different areas of the world or the entire world. You can pick the time zone in the **Time Zone:** drop-down list, but AutoCAD attempts to match the time zone to the city you select when you pick the **OK** button to close the dialog box. An alert box is displayed that gives you the option to accept the new time zone or go back and pick a different one.

Once you have selected a location, close the **Geographic Location** dialog box. The geographic marker is automatically placed at the WCS origin.

The properties of the sun are set in the **Sun Properties** palette, Figure 17-10. The **SUNPROPERTIES** command opens this palette. You can also pick the dialog box launcher button at the lower-right corner of the **Sun & Location** panel in the **Render** tab of the ribbon. The sun can be turned on or off using the **Sun Properties** palette. The date, time, and time zone can also be changed in the palette. These settings are the same as previously discussed. There are other properties of the sun that are only available in the **Sun Properties** palette. This palette contains several categories, which are discussed in the next sections.

Type
SUNPROPERTIES

General

The Intensity Factor property in the **General** category determines the brightness of the sun. Setting this property to zero, in effect, turns off sunlight. Increasing the property makes the sunlight brighter. The maximum value for the property is determined by the capabilities of your computer.

The Color property in the **General** category is used to set the color of the sun, if photometric lighting is off. By default, sunlight is white (true color 255, 255, 255). To change the color, pick the drop-down list and choose a new color. If you pick the Select Color... entry, the **Select Color** dialog box is displayed for selecting a color. Sunlight can be changed to any color, but be aware that changing the color of the light may drastically alter the appearance of a scene. This is especially true if materials are attached

SUNPROPERTIES

Figure 17-10.
The **Sun Properties** palette.

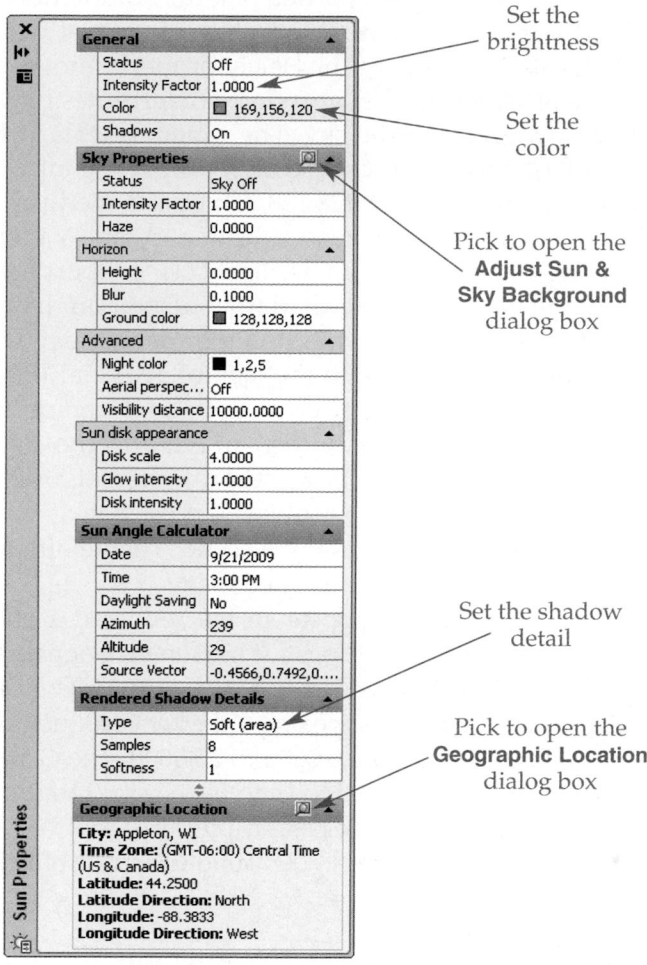

Set the brightness

Set the color

Pick to open the **Adjust Sun & Sky Background** dialog box

Set the shadow detail

Pick to open the **Geographic Location** dialog box

to objects in the drawing. The color of sunlight is often set to a very light blue for an outdoor scene to help convey a bright blue sky.

The Shadows property in the **General** category determines whether or not the sun casts shadows. The property is either on or off. Shadows are discussed in detail later in this chapter.

NOTE

Photometric lighting must be off (**LIGHTINGUNITS** = 0) in order to set the sun color. If photometric lighting is on, the Color property is disabled and set to a preselected color based on the geographic location, date, and time.

Sky properties

The settings in the **Sky Properties** category control the sky. The sky is used in conjunction with sunlight to generate more-realistic lighting in the scene. This category is only displayed if photometric lighting is on. The Status property determines if the sky effect is on or off. The value in the Intensity property is a multiplier for the illumination provided by the sky. The Haze property controls how the sky illumination is diffused. The value can range from 0.0000 to 15.0000. The preset sun color is affected by this value.

Notice the button in the title bar of the **Sky Properties** category. Picking this button opens the **Adjust Sun & Sky Background** dialog box. This dialog box contains the same settings found in the **Sun Properties** palette, but includes a preview of the sun disk.

The settings in the **Horizon** subcategory control what the horizon looks like and where it is located. Changing the Height property moves the horizon up or down. The default value is 0.0000. The Blur property determines how much the horizon is blurred between the ground and the sky. The range for this value is from 0.0000 to 10.0000 with a default of 0.1000. The Ground color property controls the color of the ground. The default color is true color 128,128,128, which is a medium gray.

The settings in the **Advanced** subcategory allow you to control some of the more artistic settings of your scene. The Night color property sets the color of the night sky. This is only visible if sky illumination is turned on. In a city, the night sky may have an orange tint to it because of the streetlamps in the city. However, in the country, the night sky is nearly black due to the lack of artificial illumination. The Aerial perspective property determines whether or not aerial perspective is applied. This is a way of simulating distance between the camera and the sky/background. The setting is either on or off. The Visibility distance property sets the distance from the camera at which haze obscures 10% of the objects in the background. This is a very useful tool for creating the illusion of depth in a scene. The default value is 10000.0000, but it can range from 0.0000 to whatever is needed.

Normally, lights do not appear in the rendered scene at all, only the illumination provided by the lights. The settings in the **Sun disk appearance** subcategory control what the sun looks like in the sky, or on the background. The Disk scale property sets the size of the sun, or solar disk, as it appears on the background. The default value is 4.0000 with the range of values being from 0.0000 to 25.0000. The Glow intensity value determines the size of the glowing halo around the sun in the sky. Default value is 1.0000 and can range from 0.0000 to 25.0000. The Disk intensity property controls the brightness of the sun on the background. The default value is 1.0000 and the range of values is from 0.0000 to 25.0000.

Sun angle calculator

The settings in the **Sun Angle Calculator** category determine the angle of the sun in relationship to the XY plane. The Date and Time properties are discussed earlier and can be controlled from the **Sun & Location** panel in the **Render** tab of the ribbon. This category in the **Sun Properties** palette also includes the Daylight Saving property. This property is used to turn daylight saving on or off. The Azimuth, Altitude, and Source Vector properties display the current settings, but are read only in the **Sun Properties** palette. These values are automatically calculated by the settings in the **Geographic Location** dialog box.

Rendered shadow details

The Type property in the **Rendered Shadow Details** category determines the type of shadow cast by the sun, if shadows are cast. When the property is set to Sharp, raytraced shadows are cast. These shadows have sharp edges. Raytracing produces accurate shadows, but rendering may take longer. When the property is set to either Soft (mapped) or Soft (area), shadow-mapped shadows are cast. This type of shadow has soft edges. Shadow-mapped shadows may be calculated quicker than raytraced shadows, but the resulting shadows are less precise. In addition, soft shadows do not work with transparent surfaces like windows. When the Type property is set to Soft (mapped), two additional settings are available in the category:

- Map Size. This property determines the number of subdivisions, or samples, used to create the shadow. By default, shadow maps are 256 × 256 pixels in size. If shadows look grainy, increasing this setting will make them look better.
- Softness. This property determines the sharpness of the shadow's edge. The value ranges from 1 to 10. The higher the value, the softer (less sharp) the edge of the shadow.

When the Type property is set to Soft (area), the two additional settings are:

- Samples. This property sets the number of samples used on the solar disk. The value can be from 0.0000 to 1000.0000.
- Softness. This property determines the sharpness of the shadow's edge, as described above.

When the Type property is set to Sharp, the other settings are read only and not applied.

NOTE

With photometric lighting on (**LIGHTINGUNITS** = 1 or 2), only the Soft (area) selection is available in the Type property drop-down list.

Geographic location

The **Geographic Location** category at the bottom of the **Sun Properties** palette displays the current geographic location settings. Changes cannot be made here, but picking the **Launch Geographic Location** button in the category's title bar opens the **Geographic Location** dialog box, which is described earlier.

Exercise 17-1

Complete the exercise on the student website.
www.g-wlearning.com/CAD

Distant Lights

Distant lights are user-created lights that have parallel light rays. See **Figure 17-11**. When a distant light is created, the location from where the light is originating must be specified along with the direction of the light rays. Distant lights are not represented in the drawing by light glyphs. Distant lights are often used to create even, uniform, overhead illumination, such as you would encounter in an office situation. The distance of the objects in the scene to the distant light has no effect on the intensity of the illumination. Distant lights do not attenuate. A distant light can also be used

Figure 17-11.
This example shows the use of a distant light to simulate sunlight shining through a window. Notice how the edges of the shadows of the grill are parallel.

Ribbon
Render
> Lights

Distant

Type
DISTANTLIGHT
LIGHT

to simulate sunlight without having to set up a time and location. However, the light must be manually moved to change the illuminating effect.

The **DISTANTLIGHT** or **LIGHT** command is used to create a distant light. You are first prompted to specify the direction from which the light is originating or to enter the **Vector** option. If you pick a point, it is the location of the light. Next, you are prompted to specify the point to which the light is pointing. This is simply the location where the light is aimed.

If you enter the **Vector** option instead of picking a "from" point, you must type the endpoint coordinates (in WCS units) of the direction vector. The light will point from the WCS origin to the entered endpoint.

After the light location and direction are determined, several other options are available. You can name the light, set the intensity, turn the light on or off, determine if and how shadows are cast, and set the light color.

AutoCAD provides a default name for new lights based on the type of light and a sequential number, such as Distantlight1, Distantlight2, and so on. It is a good idea to provide a meaningful name for a light. This is especially true if there are other lights in the scene. To rename a light, enter the **Name** option. Then, type the name of the light and press [Enter].

To set the brightness of the light, enter the **Intensity** option. Then, type a value and press [Enter]. The default value is 1.00. Setting the value to 0.00, in effect, turns off the light. The maximum value depends on the capabilities of your computer. This option is called **Intensity Factor** if photometric lighting is enabled.

When a light is created, it is on. To turn the light off, enter the **Status** option. Then, change the setting to **Off**.

The **Photometry** option is available when photometric lighting is active. It controls the luminous qualities of visible light sources. Once you enter the **Photometry** options, you can select one of three settings:

- **Intensity**. This is the power of the light source and can be entered as candelas (cd), luminous flux (lx), or foot-candles (fc), depending on the current lighting units.
- **Color**. This is the color of the light source and can be changed by typing in a name (to get a list of color names, enter ?) or by the Kelvin temperature value (k).
- **Exit**. Exits the command option.

The **Shadow** option is used to determine if and how shadows are cast by the distant light. To turn off shadow casting, enter the option and select the **Off** setting. The **Sharp** setting creates raytraced shadows. The **Softmapped** setting casts shadow-mapped shadows. Shadows are discussed later in this chapter.

By default, new distant lights cast white light (true color 255, 255, 255). To change the color of the light, enter the **Color** option. This option is called **Filter Color** if photometric lighting is enabled. To specify a new true color, simply specify the RGB values and press [Enter]. To specify a color based on hue, saturation, and luminance (HSL), enter the **Hsl** option and specify the values. To enter an AutoCAD color index (ACI) number, enter the **Index color** option and specify the ACI number. To specify a color book color, enter the **Color Book** option and then specify the name of the color book followed by the name of the color.

Once all settings for the distant light have been made, use the **Exit** option to end the command and create the light. Do not press [Esc] to end the command. Doing so actually cancels the command and the light is not created.

Exercise 17-2

Complete the exercise on the student website.
www.g-wlearning.com/CAD

Point Lights

Point lights are user-created lights that have light rays projecting in all directions. See **Figure 17-12.** When a point light is created, its location must be specified. Since point lights illuminate in all directions, there is no "to" location for a point light. A light glyph represents point lights in the drawing. See **Figure 17-13.** Point lights can be set to attenuate. In this case, the distance of the objects in the scene to the point light affects the intensity of the illumination.

The **POINTLIGHT** or **LIGHT** command is used to create a point light. You are first prompted to specify the location of the point light. Once the location is established, several options for the light are available. You can name the light, set the intensity, turn the light on or off, adjust the photometry settings, determine if and how shadows are cast, set the attenuation, and set the light color. The **Name**, **Intensity** (or **Intensity**

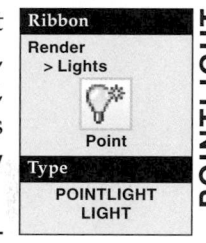

Ribbon
Render
> Lights
Point

Type
POINTLIGHT
LIGHT

POINTLIGHT

Figure 17-12.
A—A point light is placed inside of the lamp fixture. Notice how the light projects in all directions. B—When shadow casting is turned on for the light, the lampshade blocks the light from illuminating objects below the shade.

A   B

Figure 17-13.
This is the light glyph for a point light.

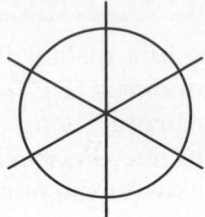

Factor), **Status**, **Photometry**, **Shadow**, and **Color** (or **Filter Color**) options work the same as the corresponding options for a distant light.

The **Attenuation** option is used to set attenuation for the point light. When this option is selected, five more options are available:

- **Attenuation Type**
- **Use Limits**
- **Attenuation Start Limit**
- **Attenuation End Limit**
- **Exit**

The **Exit** option returns you to the previous prompt.

The **Attenuation Type** option is used to turn attenuation on and off and to set the type of attenuation. To turn attenuation off, select the option and then enter **None**. To turn attenuation on, select the option and then enter either **Inverse Linear** or **Inverse Squared**. Attenuation is discussed in detail earlier in this chapter.

The **Use Limits** option determines if the attenuation of the light has a beginning and an end. When this option is set to **Off**, attenuation starts at the light and ends when the illumination reaches zero. When set to **On**, attenuation begins at the starting limit and ends at the ending limit.

To set the starting point for attenuation, enter the **Attenuation Start Limit** option. Then, specify the distance from the point light where attenuation will begin. The full intensity of the light provides illumination up to this point. From this point to the attenuation end limit, the light falls off.

To set the point where the illumination attenuates to zero, enter the **Attenuation End Limit** option. Then, specify the distance from the point light where the illumination is zero. Beyond this point, AutoCAD does not calculate the effect of the light.

Target Point Lights

The target point light is like a regular point light except that the command starts by asking for a *source* location and a *target* location. The rest of the options are the same. You can type TARGETPOINT to access this command or pick the **Targetpoint** option in the **LIGHT** command.

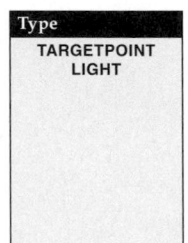

Type

TARGETPOINT

TARGETPOINT
LIGHT

> **CAUTION**
>
> It is important to set attenuation limits. If there is no end limit for the light, AutoCAD may calculate the illumination beyond the boundary of the scene, even if there is nothing there to see. To speed processing time, tell AutoCAD where illumination stops for point and spotlights.

A point light may be used as an incandescent lightbulb, such as in a table lamp. Most of these lightbulbs cast a yellow light. In these cases, you may want to change the color of the light to a light yellow. Compact fluorescent lightbulbs may cast white, light blue, or light yellow light. LED-based lights cast a color based on the color of the LED. Other colors can be used to give the impression of heat or colored lights.

Exercise 17-3

Complete the exercise on the student website.
www.g-wlearning.com/CAD

Spotlights

Spotlights are user-created lights that have light rays projecting in a cone shape in one direction. See **Figure 17-14.** When a spotlight is created, the location from where the light is originating must be specified along with the direction in which the light rays travel. A light glyph represents spotlights in the drawing. See **Figure 17-15.** Spotlights can be set to attenuate. In this case, the distance of the objects in the scene to the spotlight affects the intensity of the illumination.

Figure 17-14.
Three spotlights are used to simulate recessed ceiling lights. Notice how the light from each spotlight projects in a cone.

Figure 17-15.
This is the light glyph for a spotlight.

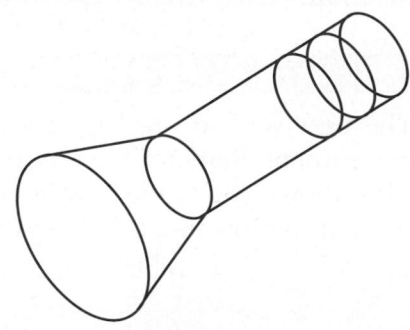

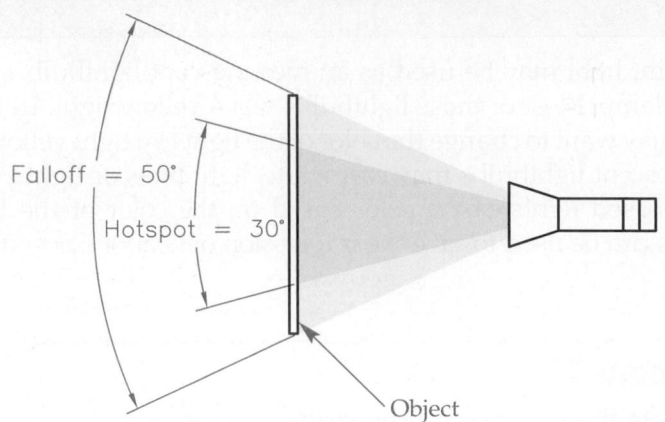

Figure 17-16.
Hotspot and falloff for a spotlight are angular measurements.

Falloff = 50°

Hotspot = 30°

Object

SPOTLIGHT

Ribbon

Render
> Lights

Spot

Type
SPOTLIGHT
LIGHT

The **SPOTLIGHT** or **LIGHT** command is used to create a spotlight. You are first prompted to specify the location of the light. This is from where the light rays will originate. Next, you are prompted for the target location. This is simply the location where the light is aimed.

Once the location and target are established, several options for the light are available. The **Name**, **Intensity** (or **Intensity Factor**), **Status**, **Photometry**, **Shadow**, and **Color** (or **Filter Color**) options work the same as the corresponding options for a point light. However, a spotlight also has hotspot and falloff settings.

The *hotspot* is the inner cone of illumination for a spotlight. Refer to **Figure 17-16**. This is measured in degrees. To set the hotspot, enter the **Hotspot** option and then specify the number of degrees for the hotspot.

The *falloff*, not to be confused with attenuation, is the outer cone of illumination for a spotlight. Like the hotspot, it is measured in degrees. The falloff value must be greater than or equal to the hotspot value. It cannot be less than the hotspot value. In practice, the falloff value is often much greater than the hotspot value. To set the falloff, enter the **Falloff** option and then specify the number of degrees for the falloff.

Once the light is created and you select it in the viewport, grips are displayed. If you hover the cursor over a grip, a tooltip is displayed indicating what the grip will modify. You can use grips to change the location of the spotlight and its target, the hotspot, and the falloff. If you hover over a falloff or hotspot grip, the current angle is displayed in the wireframe cone (if dynamic input is on).

Free Spotlight

A free spotlight is like a standard spotlight except that you do not specify a target, only the light location. The rest of the options are the same. You can type FREESPOT to access this command or pick the **Freespot** option in the **LIGHT** command. When created, a free spotlight points down the Z axis (from positive to negative) of the current UCS. A free spotlight may be easier to control than a standard spotlight because you do not have to worry about the target point. If you want to change the angle or position of the light, use the **3DMOVE** and **ROTATE3D** commands.

PROFESSIONAL TIP

The best way to see how colored lights affect your model is to experiment. Remember, you can render selected areas of the scene. This allows you to see how light intensity and color change objects without performing a full render.

Weblight

A photometric weblight is really just a targeted point light. The difference is that a weblight provides a more precise representation of the light. Real-world lights appear to evenly illuminate from their source, but, in reality, the shape of the light, the material used in its manufacture, and other factors make all lights distribute their energy in different ways. These data are provided by light manufacturers in the form of light distribution data. Light distribution data can be loaded into the **Photometric Web** subcategory of the **General** category in the **Properties** palette when the light is selected. Select the Web file property, then pick the browse button (...) and select an IES file. IES stands for Illuminating Engineering Society. AutoCAD's online documentation has additional information on IES files.

Think of the web of a weblight as a spherical cage surrounding the light source. If the light is evenly distributed from its source, the cage is a true sphere. In actuality, a light may emit more light energy in the X direction than in the Z direction. In this case, the cage bulges out further in the X direction. The position of this bulge may be important to the illumination of the scene and you may need to rotate the web to apply more or less light in one direction or another.

To add a weblight to the scene, you can type WEBLIGHT or pick the **Web** option in the **LIGHT** command. You are prompted for source and target locations. The **Name**, **Intensity Factor**, **Status**, **Photometry**, **Shadow**, and **Filter Color** options work the same as the corresponding options for the previously discussed lights. However, weblights have an additional **Web** option. When this option is activated, these options are presented:

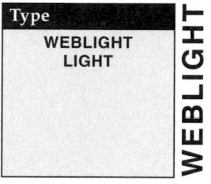

- **File.** Allows you to select an IES file.
- **X.** Rotates the web around the X axis.
- **Y.** Rotates the web around the Y axis.
- **Z.** Rotates the web around the Z axis.
- **Exit.** Exits the **Web** option.

Point and spotlights can be converted to weblights, and vice versa, using the **Properties** palette. Simply select an existing light and open the **Properties** palette. In the **General** category, the Type property determines whether the light is a point light, spotlight, or weblight. Select the type in the drop-down list. Using the **Properties** palette with lights is discussed in detail later in this chapter.

Free Weblight

A free weblight is the same as a standard weblight except there is no target. Only the source location is specified when placing the light. To change the location and direction of the light, use the **ROTATE3D** and **3DMOVE** commands.

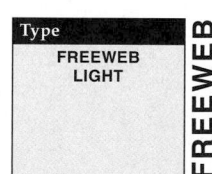

NOTE

To create either standard or free weblights, photometric lighting must be enabled (**LIGHTINGUNITS** = 1 or 2).

Photometric Lights Tool Palette Group

Photometric lights may be easily added to the drawing using the tool palettes in the **Photometric Lights** tool palette group. This palette group contains four palettes: **Fluorescent**, **High Intensity Discharge**, **Incandescent**, and **Low Pressure Sodium**. Lights created with these tools have preset properties for **Intensity Factor**, **Shadow**, and **Filter Color**. The glyph for the long fluorescent lights has a yellow line passing through its center indicating the direction of the light. The high intensity discharge, low-pressure sodium, and regular incandescent lights are point lights. The incandescent halogen lights are free spotlights. The recessed incandescent lights are a special weblight designed for recessed light fixtures.

Lights in Model and Properties Palettes

The **Lights in Model** palette is extremely useful for controlling the lights in your scene, **Figure 17-17.** Used in conjunction with the **Properties** palette, you can manage and edit all of the lights in a scene.

Light list

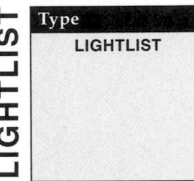

The **LIGHTLIST** command displays the **Lights in Model** palette. This can also be displayed by picking the dialog box launcher button at the lower-right corner of the **Lights** panel in the **Render** tab of the ribbon. All user-created lights in the scene are displayed in the list. To modify the properties of a light, either double-click on the light name or right-click on it and select **Properties** from the shortcut menu. This opens the **Properties** palette. See **Figure 17-18.** If the **Properties** palette is already open, you can simply select a light in the **Lights in Model** palette. You can select more than one light by pressing the [Ctrl] key and selecting the names in the **Lights in Model** palette, which allows you to change all of their settings at the same time. This is an excellent way to make the lighting in your scene uniform or to control a series of lights with a single edit.

A light can be deleted from the scene using the **Lights in Model** palette. To do so, simply right-click on the name of the light and select **Delete Light** from the shortcut menu. The light is removed from the drawing. Using the **UNDO** command restores the light.

Figure 17-17.
All user-created lights are listed in the **Lights in Model** palette.

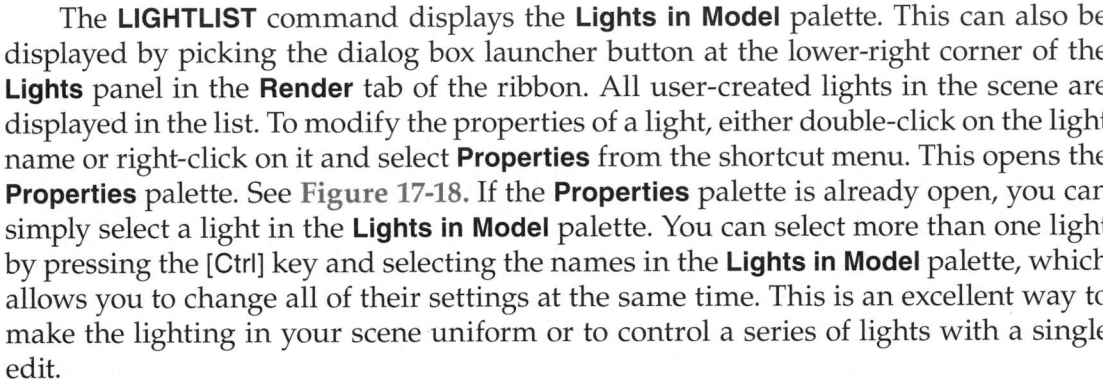

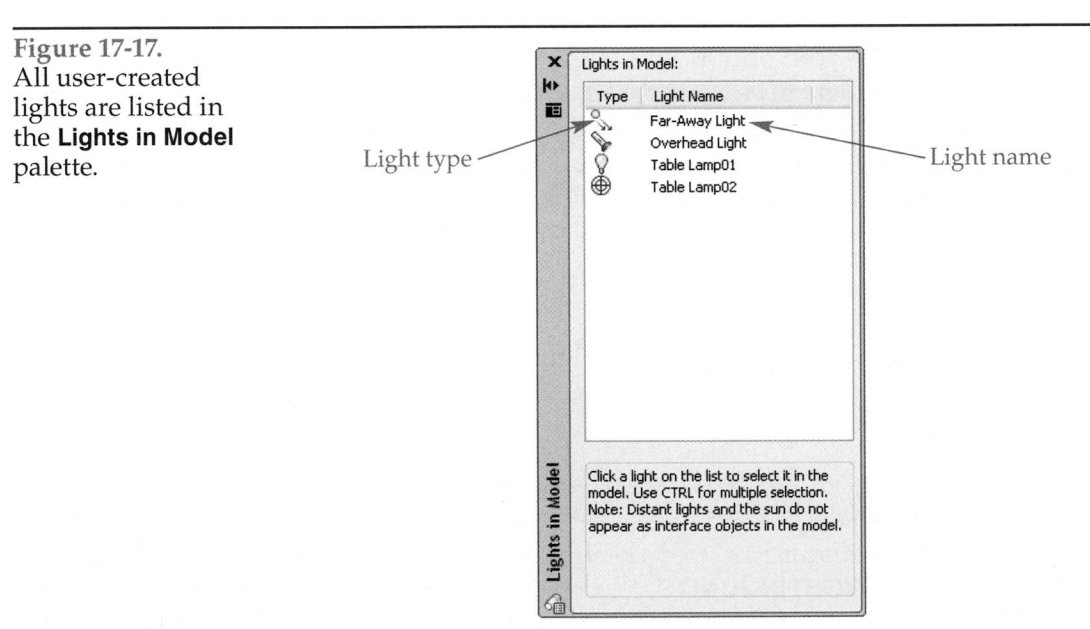

Figure 17-18
The **Properties**
palette with a
weblight selected.

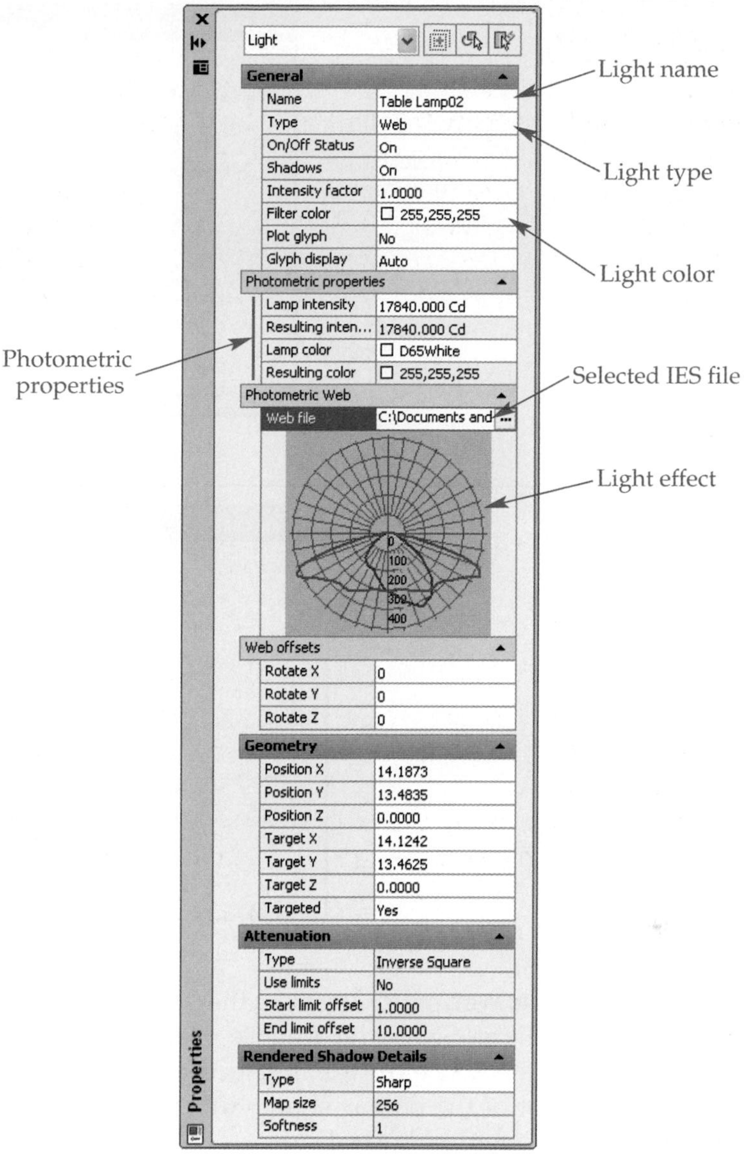

Photometric
properties

Light name

Light type

Light color

Selected IES file

Light effect

Properties palette

The **Photometric Properties** subcategory of the **General** category in the **Properties** palette has special settings for photometric lights. The Lamp intensity property determines the brightness of the light. The value may be expressed in candelas (cd), lumens (lm), or illuminance (lux) values. When you select the Lamp intensity property, a button is displayed to the right of the value. Picking this button opens the **Lamp Intensity** dialog box, **Figure 17-19.** In this dialog box, you can change the illumination units and set the intensity (Lamp intensity property). You can also set an intensity scale factor. This is multiplied by the Lamp intensity property to obtain the actual illumination supplied by the light. The read-only Resulting intensity property in the **Properties** palette displays the result.

The Lamp color property in the **Photometric Properties** subcategory in the **General** category controls the color of the light. If you select the property, a button is displayed to the right of the value. Picking this button opens the **Lamp Color** dialog box. See **Figure 17-20.** This dialog box gives you the option to control the color of the light by either standard spectra colors or Kelvin colors. The color selected in the **Filter color:** drop-down list is applied to the color of the light. The **Resulting color:** swatch displays the color cast by the light once the filter color is applied. If the filter color is white (255,255,255), then the light color is the color cast by the light.

Figure 17-19.
The **Lamp Intensity** dialog box is used to set the intensity for a light.

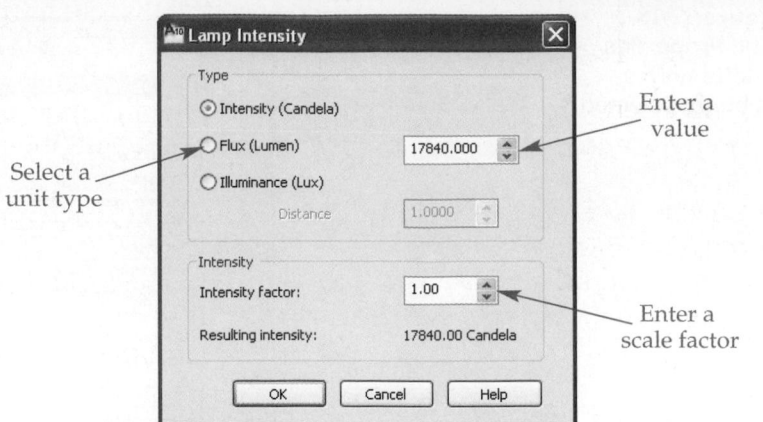

Select a unit type

Enter a value

Enter a scale factor

Figure 17-20.
The **Lamp Color** dialog box is used to set the color for the light and a filter color, if needed.

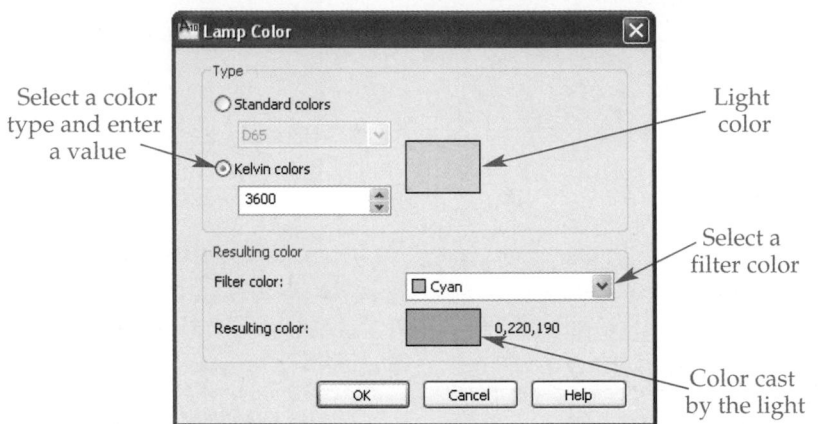

Select a color type and enter a value

Light color

Select a filter color

Color cast by the light

The **Photometric Web** subcategory in the **General** category is where you can specify an IES file for the light. Select the Web file property, then pick the browse button (**...**) and select the IES file. Once the file is selected, its location is displayed in the Web file property. The effect of the data is shown in a graph at the bottom of the **Photometric Web** subcategory. Refer to **Figure 17-18.**

The **Web offsets** subcategory in the **General** category allows you to rotate the web around the X, Y, and Z axes. This is discussed earlier in the Weblight section.

In the **Geometry** category, you can change the X, Y, and Z coordinates of the light. You can also change the X, Y, and Z coordinates of the light's target. If the light is not targeted, the Target X, Target Y, and Target Z properties are not displayed. To change the light from targeted to free, and vice versa, select Yes or No in the Targeted property drop-down list.

The properties in the **Attenuation** category are the same as those discussed earlier in this chapter in the Point Lights section. In order to change these properties in the **Properties** palette, photometric lighting must be off (**LIGHTINGUNITS** = 0).

The last category in the **Properties** palette is **Rendered Shadow Details**. The properties in this category are used to control shadows. Shadows are discussed later in this chapter.

PROFESSIONAL TIP

It is important to give your lights names that make them easy to identify in a list. If you accept the default names for lights, they will be called Pointlight1, Spotlight5, Distantlight7, Weblight2, etc., making them difficult to identify. Use the **Name** option when creating the light or, after the light is created, the **Properties** palette to change the name of the light.

AutoCAD and Its Applications—Advanced

Determining Proper Light Intensity

As a general rule, the object nearest to a point light or spotlight should receive the full illumination, or full intensity, of the light. Full intensity of any light that has an attenuation property is a value of one. Remember, attenuation is calculated using either the inverse linear or inverse square method. Therefore, you must calculate the appropriate intensity.

For example, suppose you have drawn an object and placed a point light and a spotlight. The point light is 55 units from the object. The spotlight is 43 units from the object. Use the following calculations to determine the correct intensity settings for the lights.

- **Inverse linear.** If the point light is 55 units from the object, the object receives 1/55 of the light. Therefore, set the intensity of the point light to 55 so the light intensity striking the object has a value of 1 (55/55 = 1). Since the spotlight is 43 units from the object, set its light intensity to 43 (43/43 = 1).
- **Inverse square.** If the point light is 55 units from the object, the object receives $(1/55)^2$, or 1/3025 ($55^2 = 3025$), of the light. Therefore, set the intensity of the point light to 3025 (3025/3025 = 1). The object receives $(1/43)^2$, or 1/1849 ($43^2 = 1849$), of the spotlight's illumination. Therefore, set the intensity of the spotlight to 1849 (1849/1849 = 1).

However, it should be noted that these settings are merely a starting point. You will likely spend some time adjusting lighting to produce the desired results. In some cases, it may take longer to light the scene than it did to model it.

PROFESSIONAL TIP

If you render a scene and the image appears black, all of the lights may have been turned off or have their intensity set to zero. A scene with no lights placed in it will be rendered with default lighting.

Shadows

Shadows are critical to the realism of a rendered 3D model. A model without shadows appears obviously fake. On the other hand, a model with realistic materials and shadows may be hard to recognize as computer generated. In AutoCAD, the sun, distant lights, point lights, spotlights, and weblights all can cast shadows. AutoCAD's default lighting does not cast shadows. There are two types of shadows that AutoCAD can create: shadow mapped and raytrace. The **Advanced Rendering Settings** palette provides settings for controlling the creation of shadows when rendering. This palette and its options are discussed in the next chapter.

The options for creating shadows are the same for all lights. The options can be set when the light is created or adjusted later using the **Properties** palette. In the case of sunlight, the **Sun Properties** palette is used to set the options.

Shadow-Mapped Shadow Settings

A *shadow-mapped shadow* is a bitmap generated by AutoCAD. A shadow map has soft edges that can be adjusted. Creating shadow-mapped shadows is the only way to produce a soft-edge shadow. However, shadow maps do not transmit object color from transparent objects onto the surfaces behind the object. **Figure 17-21** shows the difference between shadow-mapped shadows and raytraced shadows.

Figure 17-21.
The shadow from
the object in the
foreground is a
shadow-mapped
shadow. The shadow
from the object in
the background is a
raytraced shadow.

To specify shadow-mapped shadows and set the quality, or resolution, of the shadow, select the light and open the **Properties** palette (or **Sun Properties** palette). Use the **Lights in Model** palette to select a distant light and open the **Properties** palette for it. At the bottom of the **Properties** palette is the **Rendered Shadow Details** category, **Figure 17-22.** To specify shadow-mapped shadows, set the Type property to Soft (shadow map).

The Map size property determines the quality of the shadow. The value is the number of samples used to create the shadow. The higher the setting, the better quality of the generated shadow. However, the higher the setting, the longer it will take to render.

The value of the Softness property determines how soft the edge of the shadow is. The higher the value, the softer or blurrier the edge of the shadow. A low value can produce a very hard edge. The value can range from 1 to 5.

A variation of shadow-mapped shadows is created when the Type property is set to Soft (sampled). This type of shadow map must be used with photometric lighting (**LIGHTINGUNITS** = 1 or 2). In this case, different properties are displayed. The Samples property determines the number of "rays" used to generate the shadows. However, this is not considered raytracing. The Visible in rendering property determines whether the shape of the light is rendered. The Shape property sets the shape of the light. For spotlights, the shape can be either rectangular or circular (disk). For point and weblights, the shape can be linear, rectangular, circular (disk) cylindrical, or spherical. The remaining properties are based on the selected shape and are used to define the size of the shape.

Figure 17-22.
The **Rendered
Shadow Details**
category in the
Properties palette.

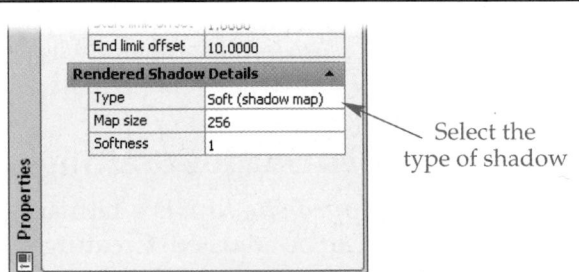

Select the
type of shadow

AutoCAD and Its Applications—Advanced

Raytrace Shadow Settings

A *raytrace shadow* is created by beams, or rays, from the light source. These rays trace the path of light as they strike objects to create a shadow. In addition, rays can pass through transparent objects, such as green glass, and project color onto surfaces behind the object. Raytrace shadows have a well-defined edge. They cannot be adjusted to produce a soft edge. Raytrace shadows can be used with standard and photometric lighting.

All lights set to cast shadows, except those set for shadow-mapped shadows, cast raytraced shadows. To switch from shadow-mapped shadows to raytrace shadows, select the light object and open the **Properties** palette. In the **Rendered Shadow Details** category, set the Type property to Sharp. The other properties are disabled because they only apply to shadow-mapped shadows.

Adding a Background

A *background* is the backdrop for your 3D model. The background can be a solid color, a gradient of colors, a bitmap file, the sun and sky, or the current AutoCAD drawing background color. By default, the background is the drawing background

To change the background for your drawing, you must first create a named view with the **VIEW** command. In the **View Manager** dialog box, pick the **New...** button to display the **New View/Shot Properties** dialog box. See **Figure 17-23**. The view name, category, and type are specified at the top of the dialog box. Near the bottom of the **View Properties** tab is the **Background** area. The drop-down list in this area is used to specify the type of background. The choices are: Default, Solid, Gradient, Image, and Sun & Sky. The Default setting uses the current AutoCAD viewport color. Photometric lighting must be on (**LIGHTINGUNITS** = 1 or 2) for Sun & Sky to appear in the drop-down list.

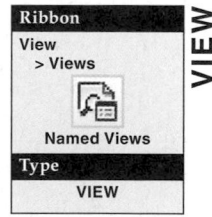

Ribbon
View
  > Views

Named Views
Type
  VIEW

VIEW

Figure 17-23.
The **View Properties** tab of the **New View/ Shot Properties** dialog box. A named view must be created before you can use a background in your scene.

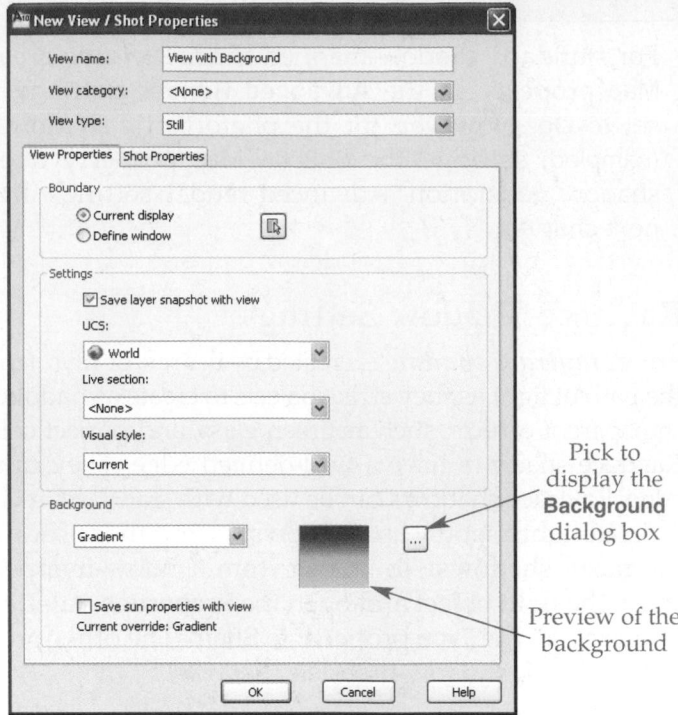

Pick to display the **Background** dialog box

Preview of the background

Solid Backgrounds

If you select Solid in the drop-down list, the **Background** dialog box is displayed. The Type: drop-down list in this dialog box is automatically set to Solid and the default color is displayed in the **Preview** area. See **Figure 17-24.** In the **Solid options** area of the dialog box, pick the horizontal **Color:** bar to open the **Select Color** dialog box. Then, select the background color that you desire. When the **Select Color** dialog box is closed, the color you picked is displayed in the **Preview** area of the **Background** dialog box. Close the **Background** dialog box, save the view, set the new view current, and close the **View Manager** dialog box.

Figure 17-24.
Creating a solid background.

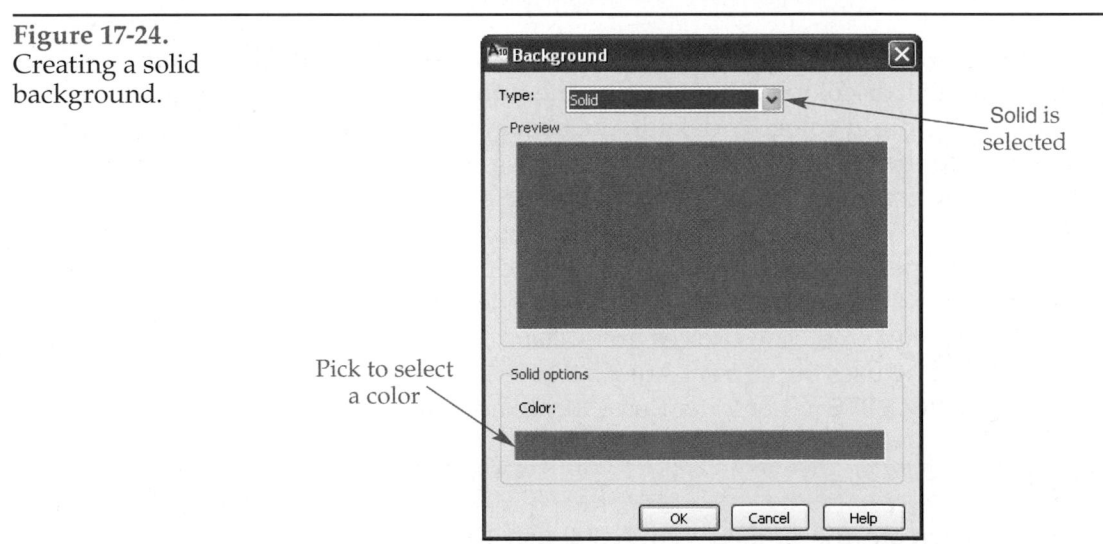

Solid is selected

Pick to select a color

Make sure to do a test rendering with the background color you selected. The final result may look quite different than your expectations.

Gradient Backgrounds

A gradient background can be composed of two or three colors. If you select Gradient in the drop-down list in the **New View/Shot Properties** dialog box, the **Background** dialog box is displayed with Gradient selected and the default gradient colors displayed in the **Preview** area. See **Figure 17-25.** The **Top color:**, **Middle color:**, and **Bottom color:** swatches are displayed on the right-hand side of the **Gradient options** area. Selecting a swatch opens the **Select Color** dialog box for changing the color. To create a two-color gradient composed of the top and bottom colors, uncheck the **Three Color** check box. The **Rotation:** text box provides the option of rotating the gradient. Close the **Background** dialog box, save the view, set the view current, and close the **View Manager** dialog box.

Convincing, clear blue skies can be simulated using the **Gradient** option. Initially, set the **Top**, **Middle**, and **Bottom** color values the same. Then, change the lightness (luminance) in the **True Color** tab in the **Select Colors** dialog box. Preview the background and make adjustments as needed.

Using an Image as a Background

An image can be used as a background. This technique can be used to produce realistic or imaginative settings for your models. If you select Image in the drop-down list in the **New View/Shot Properties** dialog box, the **Background** dialog box is displayed with Image selected and a blank image displayed in the **Preview** area. See **Figure 17-26.** To locate the image file, pick the **Browse...** button to display a standard open dialog box. These image file types may be used for the background: TGA, BMP, PNG, JFIF (JPEG), TIFF, GIF, and PCX.

Once the image file is selected, it must be adjusted. The **Preview** area of the **Background** dialog box shows the image with a preview of a drawing sheet. This drawing sheet indicates how the image is going to be positioned in the view. Pick the **Adjust Image...** button to open the **Adjust Background Image** dialog box. See **Figure 17-27.**

In the **Image position:** drop-down list, pick how the image is applied to the viewport. The Center option centers the image in the view without changing its aspect ratio or scale. The Stretch option centers the image and stretches or shrinks it to fill

Figure 17-25.
Creating a gradient background.

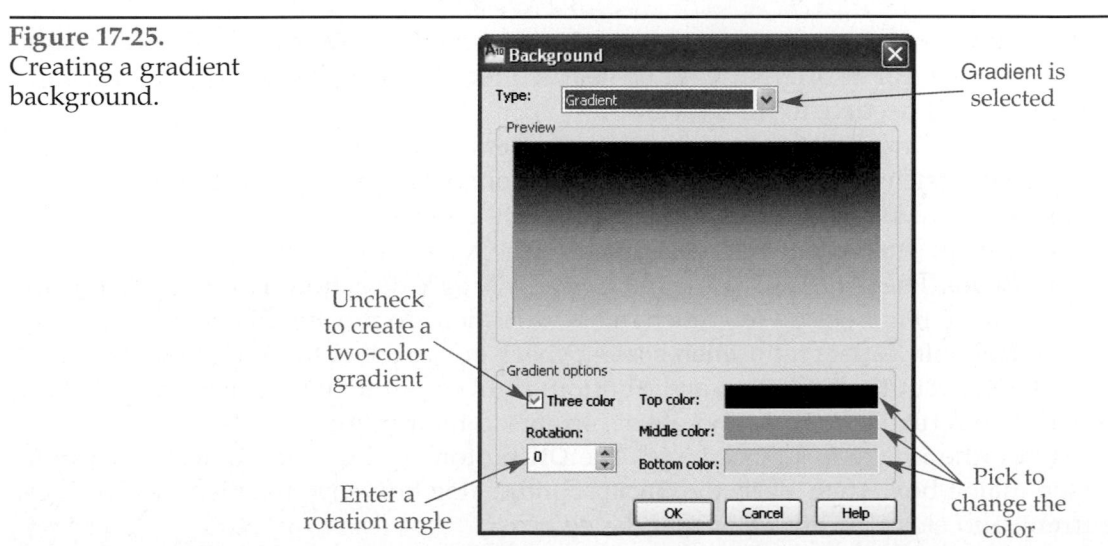

Gradient is selected

Uncheck to create a two-color gradient

Enter a rotation angle

Pick to change the color

Figure 17-26.
Setting an image as
the background.

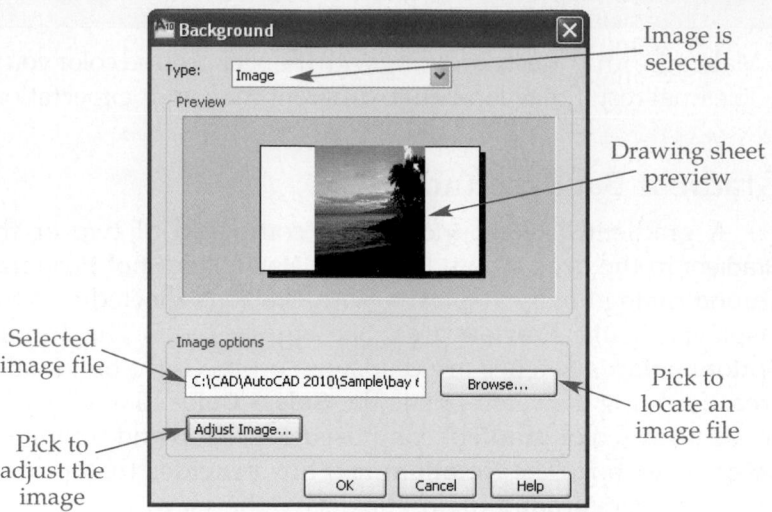

Image is
selected

Drawing sheet
preview

Selected
image file

Pick to
locate an
image file

Pick to
adjust the
image

Figure 17-27.
Adjusting the
background image.

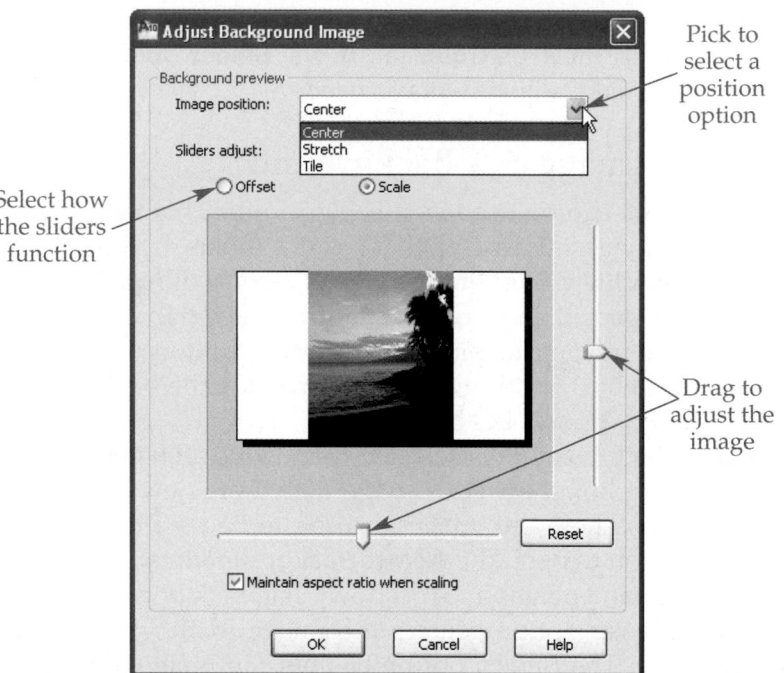

Pick to
select a
position
option

Select how
the sliders
function

Drag to
adjust the
image

the entire view. This is one way to plot an image file from AutoCAD. The Tile option keeps the image at its original size and shape, but moves it to the upper-left corner and duplicates it, if needed, to fill the view.

After the image is positioned, use the sliders to adjust it further. The sliders are disabled if Stretch is selected in the **Image position:** drop-down list. The slider function is based on which radio button is picked above the image:

- **Offset.** The sliders move the image in the X or Y direction.
- **Scale.** The sliders scale the image in the X or Y direction. This may distort the image if it is scaled too much in one direction. To prevent distortion, check the **Maintain aspect ratio when scaling** check box at the bottom of the dialog box.

The **Reset** button is located at the bottom-right corner of the preview pane. Picking this button returns the scale and offset settings to their original values.

Once the image is adjusted, pick the **OK** button to close the **Adjust Background Image** dialog box. Then, close the **Background** dialog box, save the view, set the view current, and close the **View Manager** dialog box.

Figure 17-28.
The **Adjust Sun &
Sky Background**
dialog box contains
settings for the
sun and sky
illumination.

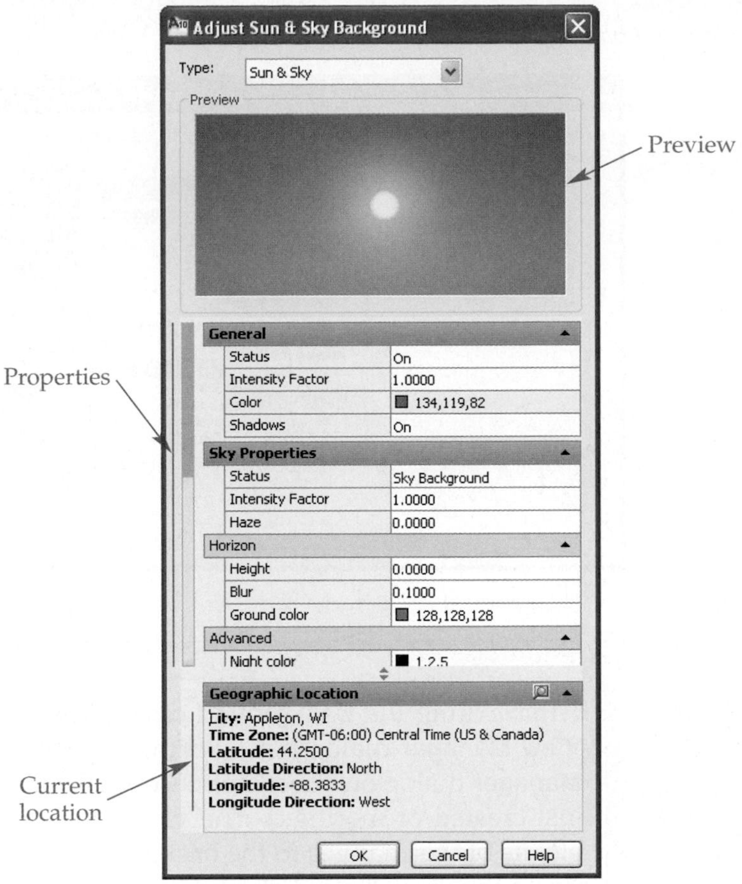

Preview

Properties

Current
location

Sun and Sky

If you select Sun & Sky in the drop-down list in the **New View** dialog box, the
Adjust Sun & Sky Background dialog box is displayed. See **Figure 17-28.** AutoCAD
uses the settings in this dialog box to simulate the sun in the sky. This dialog box
has a preview tile at the top and the **General**, **Sky Properties**, **Sun Angle Calculator**,
Rendered Shadow Details, and **Geographic Location** categories. The settings in these
categories are discussed earlier in this chapter in the Sky Properties section.

Once the sky is set, pick the **OK** button to close the **Adjust Sun & Sky Background**
dialog box. Then, save the view, set the view current, and close the **View Manager**
dialog box.

Changing the Background on an Existing View

To change the background of existing named views, open the **View Manager** dialog
box. Select the view name in the **Views** tree on the left-hand side of the dialog box.
Then, in the **General** category in the middle of the dialog box, select the Background
override property. See **Figure 17-29.** Next, pick the drop-down list for the property and
select the type of background you want applied: None, Solid, Gradient, Image, Sun &
Sky, and Edit. Picking None sets the background to the AutoCAD default background.
Setting the property to Solid, Gradient, or Image opens the **Background** dialog where
you can make settings for that type. Selecting Sun & Sky opens the **Adjust Sun & Sky
Background** dialog box. Picking Edit opens the **Background** or the **Adjust Sun & Sky
Background** dialog box with the settings of the current background. Once the back-
ground type has been changed or the existing background edited, pick the **OK** button
to save the view and close the **View Manager** dialog box.

Figure 17-29.
Changing the background of an existing, named view.

Select the view

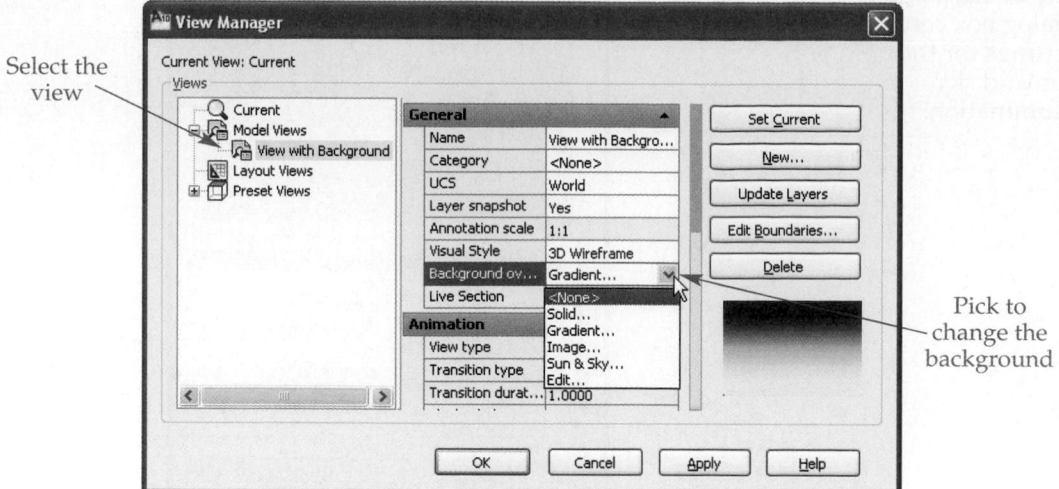

Pick to change the background

NOTE

After exiting the **Background** dialog box, you are returned to the **View Manager** dialog box. Picking the **OK** button to exit the **View Manager** dialog box does not necessarily activate the view that you just created or modified. The view must be set current to see the effects of the changes to the background. A view can be set current using the drop-down list on the **Viewports** panel in the **View** tab of the ribbon or it can be set current in the **View Manager** dialog box.

Exercise 17-5

Complete the exercise on the student website.
www.g-wlearning.com/CAD

Chapter Test

Answer the following questions. Write your answers on a separate sheet of paper or complete the electronic chapter test on the student website.
www.g-wlearning.com/CAD

1. Compare and contrast *ambient light, distant lights, point lights, spotlights,* and *weblights.*
2. Define *angle of incidence.*
3. Define *angle of reflection.*
4. A smooth surface has a(n) _____ specular factor.
5. Describe *hotspot* and *falloff.* Which lights have these properties?
6. What is *attenuation?*
7. What are the three types of lighting in AutoCAD?
8. What are the four types of light objects in AutoCAD?
9. What are light glyphs and which lights have them?

10. List the types of shadows that can be created in AutoCAD. Which type(s) can have soft edges?
11. Which type of shadow must be created for light to pass through transparent objects?
12. What must be created before a background can be added to a scene?
13. What are the four types of backgrounds in AutoCAD, other than the default background?
14. Why would you draw a line between the "from" point and "to" point of a distant light?
15. Describe how a gradient background can be used to represent a clear blue sky.

Drawing Problems

1. In this problem, you will draw some basic 3D shapes to create a building similar to an ancient structure, place lights in the drawing, and render the scene with shadows.
    A. Begin a new drawing and set the units to architectural. Save the drawing as P17_01.
    B. Draw a 32′ × 22′ planar surface to represent the floor. Using the tool palettes in the **Materials Library** palette group (or **Materials** group if the materials library is not installed), attach a material of your choice to the floor.
    C. Draw cylinders to represent pillars. Make each ∅2′ × 15′ tall. There are ten pillars per side. Attach a suitable material to the pillars.
    D. The roof is 32′ × 22′ and 5′ tall at the ridge. Attach an appropriate material.
    E. Create a perspective viewpoint looking into the building.
    F. Turn on sunlight and turn off the default lighting. Set the geographic location to Athens, Greece. Change the date and time to whatever you wish. Make sure the sun is set to cast shadows.
    G. Render the scene.
    H. Save the drawing.

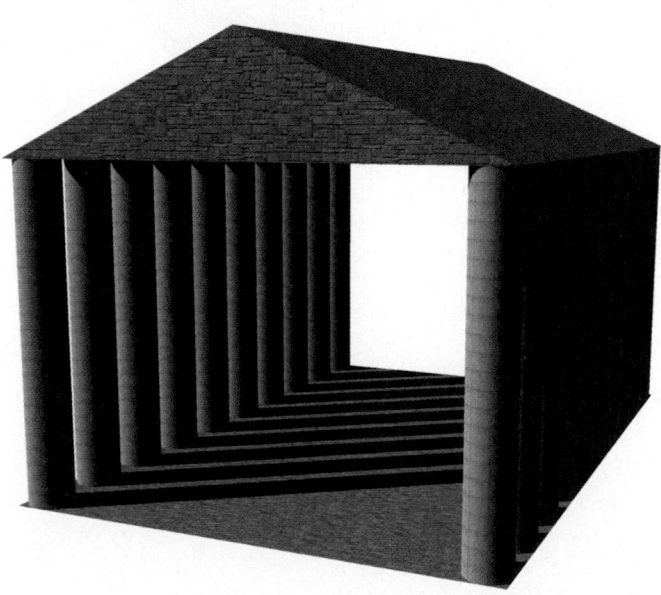

2. Using the drawing from problem 1, you will experiment with different lighting types.
   A. Open drawing P17_01 and save it as P17_02.
   B. Turn off the sun.
   C. Place three point lights inside of the building. Evenly space the lights along the centerline of the ceiling. Adjust the light intensity so that the interior is not washed out. Set the color of the middle light to white. Set the color of the outside lights to red or blue. Render the scene.
   D. Turn off the point lights.
   E. Place two spotlights, one pointing from the front corner to the rear corner and the other pointing between the pillars on the left side of the building. Target them at the floor. Render the scene. Adjust the intensity, hotspot, and falloff as needed.
   F. Turn the point lights back on and render the scene with all six lights active. Adjust the light intensities again if the rendering is too washed out with light.
   G. Save the drawing.

3. Using a previously created mechanical model, you will apply materials and lights to make it ready for presentation.
   A. Open drawing P09_05 created in Chapter 9. Save it as P17_03.
   B. Draw a planar surface below the flange to represent a tabletop.
   C. Using the tool palettes, attach an appropriate material, such as a wood or tile material, to the surface.
   D. Create a new material based on the Advanced material type. Attach a texture map from the "all users" AutoCAD folder to the diffuse color property. There are several metal texture maps located in the \Textures folder. Apply a low reflection value to the material.
   E. Place two spotlights in the drawing and target them at the flange from different angles. Adjust their hotspot, falloff, and intensity to get the proper lighting.
   F. Create a perspective view of the scene. Then, render the scene.
   G. Save the drawing.

4. The building shown below will be used to study passive solar heating at different times of the year. Model the building using the overall dimensions given. Use your own dimensions for everything else. The side with the windows should be facing South (–Y in AutoCAD).

A. Set the geographical location to a city in the northern hemisphere.
B. Set the date to midsummer and the time to noon.
C. Turn on sunlight and the default lighting off.
D. Render the scene and note the location of the shadows inside the building.
E. Change the date to late winter, render the scene again, and note the new location of the shadows. You can easily switch between the rendered images in the **Render** window by selecting each rendering in the **History** pane (this is discussed more in the next chapter).
F. Observing the changes in the shadow locations, what design changes can be made to maximize Sun exposure in the cold winter months? What design changes can be made to minimize Sun exposure in the heat of summer?
G. Change the geographical location to somewhere closer to the equator. Then, render the scene in summer and follow. How do the shadows compare to those in the previous location?
H. Save the drawing as P17_04.

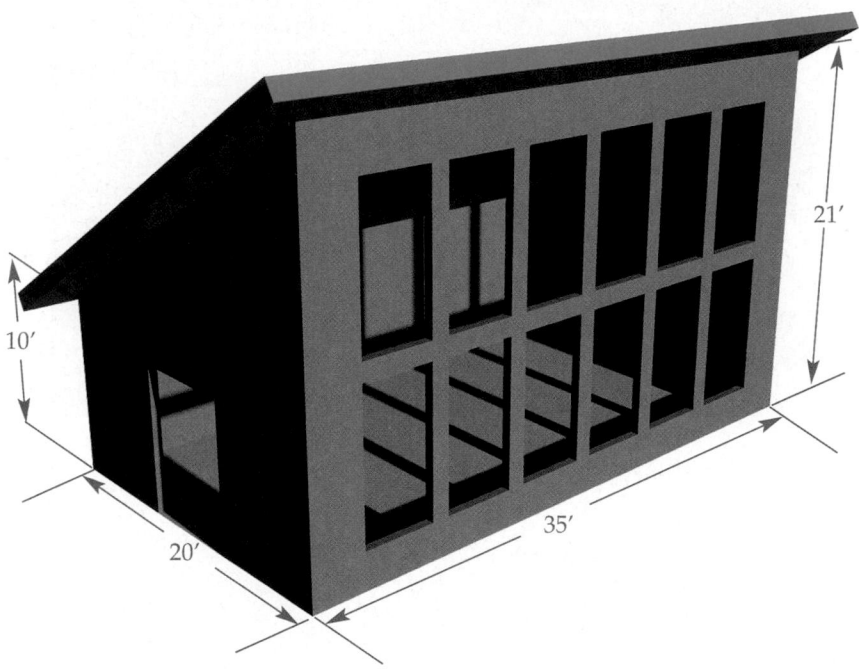

5. In this problem, you will be adding lights to a model and controlling their properties to create a pleasing scene. Model the courtyard shown below. The overall dimensions are 15' × 16' × 4' (wall height). Use your own dimensions for everything else. Add lights as follows.

A. Turn on the sun and turn off the default lighting. Set the time to late in the day so that the sun is close to the horizon.

B. In the **Sky Properties** category of the **Sun Properties** palette, select Sky Background and Illumination for the Status property.

C. Add a point light at the center of each sphere.

D. Attach the Doors & Windows.Glazing.Glass.Frosted material to the spheres.

E. Create a fill light above to illuminate the scene. This can be a point or spotlight.

F. Render the scene using the low preset to see the lighting effects.

G. Adjust the sun properties to create the look that you want.

H. Adjust the properties of the point lights and any other lights in the scene. You may have to increase the intensity of the lights quite a bit to properly illuminate the scene.

I. When the scene is illuminated the way you want it, render the scene using medium or high preset.

J. Save the drawing as P17_05.

6. In Chapter 11, you completed the kitchen chair model that you started in Chapter 2. In this problem, you will be adding lights to the model and attaching materials to the various components in the chair.
   A. Open P11_07 from Chapter 11.
   B. Create a layer for each of the chair components: seat, legs/crossbars, seatback bow, and seatback spindles.
   C. Draw a planar surface to represent the floor. Place this on its own layer.
   D. Assign materials to each of the layers. You can use materials from the materials library or create your own materials.
   E. Set the Realistic visual style current.
   F. Adjust material mapping as needed.
   G. Add lighting to the scene.
   H. Render the scene. If you experience problems with the materials, try attaching by object instead of by layer.
   I. Save the drawing as P17_06.

CHAPTER 18

Advanced Rendering

Learning Objectives

After completing this chapter, you will be able to:

✓ Make advanced rendering settings.
✓ Set the resolution for a rendering.
✓ Save a rendering to an image file.
✓ Add fog/depth cueing to a scene.

In Chapter 15, you learned how to create a view of your scene that is more realistic than a visual style. In that chapter, you used AutoCAD's sunlight feature to create a simple rendering with mostly default settings. In this chapter, you will discover how to make a rendering look truly realistic, or photorealistic. Some of the advanced rendering features add significantly to the rendering time and you must learn how to balance the quality of the rendering with an acceptable time frame to get the job done.

Render Window

By default, a drawing is rendered in the **Render** window, unless you are rendering a cropped area. This window allows you to inspect the rendering, save it to a file, compare it with previous renderings, and take note of the statistics. See **Figure 18-1**. There are three main areas of the **Render** window—the image, history, and statistics panes.

Image Pane

As AutoCAD processes the scene, the image begins to appear in the image pane in its final form. There may be as many as four phases that the rendering goes through as it is being processed:

- **Translation.** Processes the drawing information and determines light intensity, shadow placement, colors, and so on. This phase is always completed.
- **Photon emission.** *Photon emission* is a technique for calculating indirect illumination that traces photons emitted by the light source until they come to rest on a diffuse surface. It determines which areas will be illuminated by indirect, or bounced, light. The photon emission phase may or may not be processed, depending on settings in the **Advanced Render Settings** palette.

Figure 18-1.
The **Render** window.

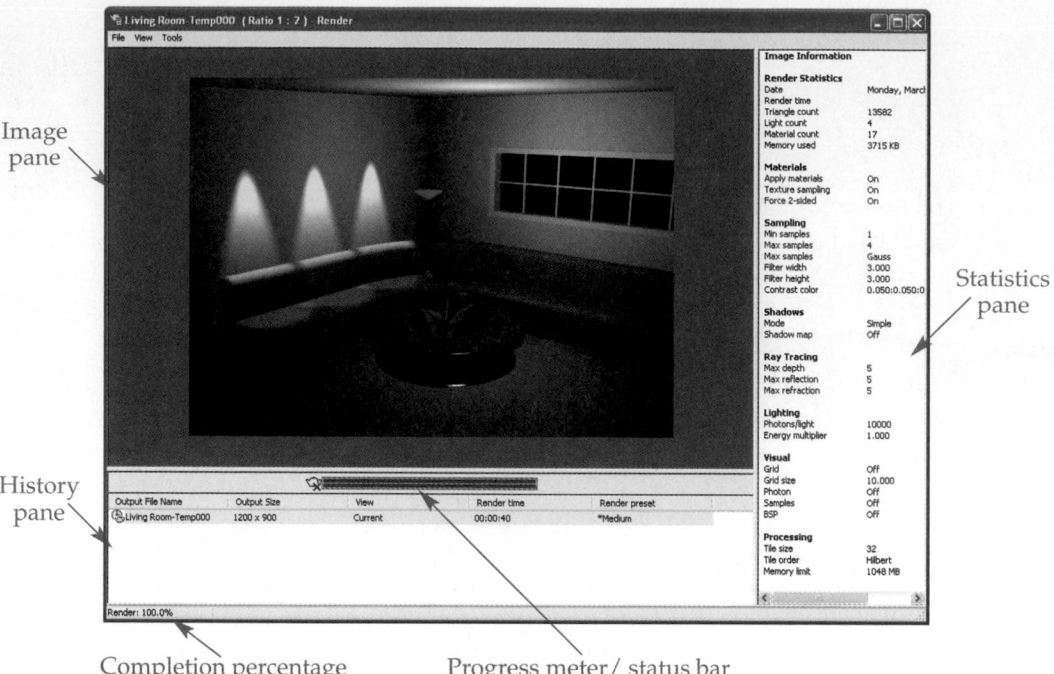

Image pane

Statistics pane

History pane

Completion percentage   Progress meter/ status bar

- **Final gather.** *Final gather* increases the number of rays used to calculate global illumination (GI). This phase will be processed if it is turned on in the **Advanced Render Settings** palette.
- **Render.** Converts the data into an image. This phase is always completed.

Immediately below the image pane is the progress meter/status bar. The top bar displays the progress of the current phase and the bottom bar indicates the progress of the entire rendering. Also, at the very bottom of the **Render** window, below the history pane, the status of the phase is shown with its percentage complete. The rendering can be cancelled at any time by pressing the [Esc] key or picking the **X** button to the left of the progress meter.

As discussed in Chapter 15, you can zoom the rendering in and out to inspect it. You can also save it to an image file using the **File** pull-down menu in the **Render** window.

History Pane

The history pane contains a list of all of the renderings that were created in this drawing since it was created, not just in this drawing session. The items in this list are called *history entries.* There are two types of history entries, which are indicated by icons. See **Figure 18-2.**

Figure 18-2.
The icon in front of the name indicates if the entry is normal or temporary.

Temporary entry

Normal entry

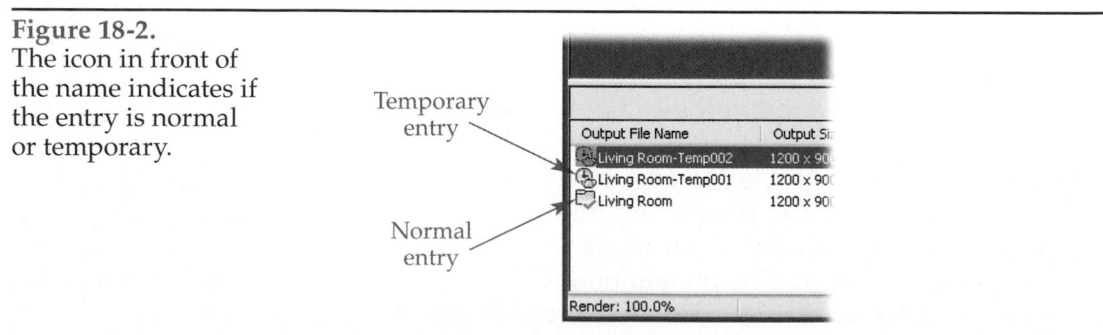

- **Normal.** The entry is saved to file. A link is maintained to that file. If the drawing is saved, closed, and reopened, you can pick the entry to view the rendering in the image pane.
- **Temporary.** The entry is available in the current drawing session, but is not saved to a file. If the drawing is closed, the image is lost. The name of the entry in the Output File Name column ends with -Temp*x*.

Right-clicking on an entry in the history pane displays a shortcut menu. The options in this menu can be used to save the image, render the image again, and manage the entry. The options in the shortcut menu are:

- **Render Again.** Renders the scene again using the same settings. A new entry is not added to the history pane.
- **Save.** Saves the rendered image to a file using a standard save dialog box. This changes the entry from a temporary entry into a normal entry.
- **Save Copy.** Saves the rendered image to a new file without changing the current entry.
- **Make Render Settings Current.** Makes all of the rendering settings of the entry the current rendering settings in the drawing. This allows you to render the current scene using the settings of the entry.
- **Remove From the List.** Deletes the entry from the history pane, but any image files saved from the entry remain.
- **Delete Output File.** Deletes the image file created by saving the entry. The entry remains in the history pane and any image files that were created as copies are retained.

Statistics Pane

The statistics pane shows the details of the rendering that is selected in the history pane. By selecting renderings in the history pane, you can see in the image pane which version provides the best result. Then, you can use the statistics pane to view the settings. The information under the Render Statistics heading (date, render time, etc.) is added when the rendering is completed. The rest of the information reflects the settings in the **Advanced Render Settings** palette and the **Render Presets Manager** dialog box at the time the rendering was created.

Exercise 18-1

Complete the exercise on the student website.
www.g-wlearning.com/CAD

Advanced Render Settings

The quickest and easiest way to control the quality of a rendering is with render presets. AutoCAD provides five standard render presets: Draft, Low, Medium, High, and Presentation. The Draft preset provides the lowest-quality rendering. Each preset above Draft changes the advanced render settings to gradually improve the rendering quality, peaking with the Presentation preset. However, as the quality is improved, the rendering time increases. The presets can be selected in the **Render** panel on the **Render** tab of the ribbon or from the drop-down list at the top of the **Advanced Render Settings** palette. Creating and using your own render presets is covered later in this chapter.

The **Advanced Render Settings** palette provides settings that give you complete control over how a rendering is created. The **RPREF** command opens the palette. The palette can also be displayed by picking the dialog box launcher button at the lower-right corner of the **Render** panel in the **Render** tab on the ribbon. There are five main categories in this palette: **General**, **Ray Tracing**, **Indirect Illumination**, **Diagnostic**, and **Processing**. These categories are explained in the next sections.

General

The **General** category provides properties for controlling the rendering destination, materials, sampling, and shadows, **Figure 18-3.** It contains four subcategories: **Render Context**, **Materials**, **Sampling**, and **Shadows**.

Render context

The **Render Context** subcategory contains general properties that control the rendering. The Procedure property determines what will be rendered. The settings are View, Crop, and Selected. View is the default and renders whatever you see in the drawing window. Crop allows you to specify an area of the scene to render. This is very useful when you want to do a test rendering, but do not want to wait for the whole scene to render. The Selected setting allows you to pick which objects to render.

The Destination property determines where the rendered scene will be displayed. You can choose to have the rendering placed in the viewport or **Render** window.

If the **Determines if File is Written** button in the subcategory title bar is picked (depressed), the Output File Name property is enabled. This property sets the name and location of the file to which the rendering will automatically be saved.

The Output Size property sets the resolution, measured in pixels × pixels, for the rendered image. You can select standard resolutions or pick Specify Output Size... for a custom resolution. In the **Output Size** dialog box that is displayed when Specify Output Size... is selected, you can enter any resolution that you want. See **Figure 18-4.** If you want to prevent the image from stretching, make sure the **Lock image aspect** button is selected in the dialog box so that the height and width remain proportional. When you change the resolution, it is stored in the drawing.

Figure 18-3.
The **General** category of the **Advanced Render Settings** palette.

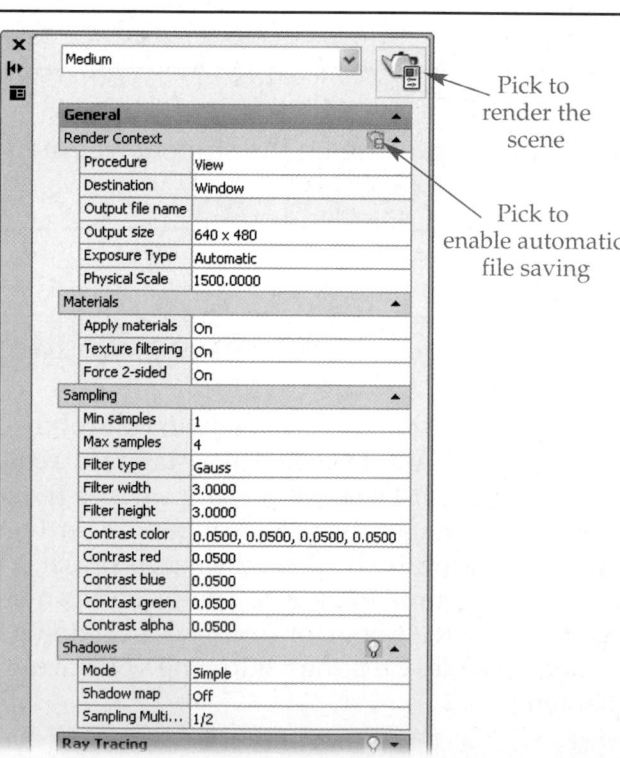

Pick to render the scene

Pick to enable automatic file saving

Figure 18-4.
Setting a custom
resolution.

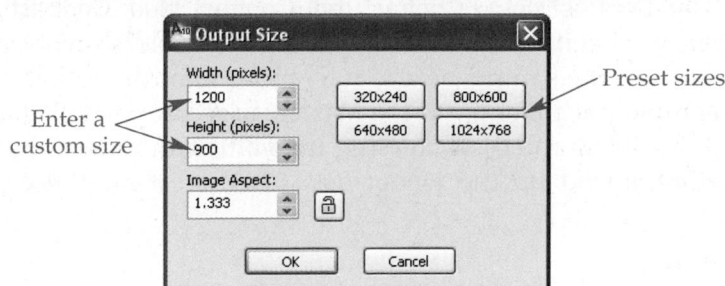

The Exposure Type property can be set to Automatic or Logarithmic. When set to Automatic, the entire image is sampled and some of the dim lighting effects are enhanced to make them more visible. When set to Logarithmic, the brightness and contrast are used to map physical values to RGB values. This is better for scenes with high dynamic ranges.

Exposure control needs a scale to work with and, if you are using non-physical lights (**LIGHTINGUNITS** set to 0), the Physical Scale property provides a scale. Standard lights have an Intensity Factor property that is multiplied by the Physical Scale value to determine the actual brightness of the light. The default value is 1500. In other words, a point light with an Intensity Factor value of 2 has an actual lamp intensity of 3000 candelas when the Physical Scale property is set to 1500.

Materials

The properties in the **Materials** subcategory determine how materials are handled in the rendering. The Apply Materials property controls whether or not materials attached to objects are rendered. The property can be set to On or Off. If set to Off, objects are rendered in their own colors. The Texture Filtering property determines whether or not antialiasing is applied to texture maps when rendered. Antialiasing is a way of reducing "jaggies" in the rendered image. The Force 2-sided property determines if AutoCAD renders both sides of all faces. This can fix problems where objects disappear in a rendering, but will increase rendering time.

Sampling

Sampling is a technique that tests the scene color at each pixel and then determines what the final color should be. This is most important in transition areas, such as edges of objects or shadows. Increasing the sampling will smooth out the jagged edges and incorrect coloring, but increase rendering time. You may also notice thicker lines in the final rendering.

The Min samples and Max samples properties set the minimum and maximum number of samples computed per pixel. A value of 1 means one sample per pixel. A value of 1/4 means one sample for every four pixels. The Filter type property determines how the samples are brought together to determine the pixel value:

- Box. Quickest method; evenly combines samples and gives them equal weight.
- Triangle. Weights the samples based on a pyramid with samples in the center of the filter area receiving the most weight.
- Gauss. Weights the samples based on a bell curve with samples in the center of the filter area receiving the most weight.
- Mitchell. Most accurate. Weights samples based on a curve centered on the filter area, like Gauss; however, this curve is steeper.
- Lanczos. Weights samples based on a curve centered on the filter area, like Mitchell, but it diminishes the weight of samples at the edge of the filter area.

The Filter width and Filter height properties determine the size of the filter area. A larger filter area softens the image, but increases rendering time.

The Contrast color, Contrast red, Contrast blue, Contrast green, and Contrast alpha properties specify the threshold value of the colors involved in sampling. If a sample differs from the sample next to it by more than this color, AutoCAD takes more than one sample per pixel up to the Max samples property. Values can be from 0.0 (black) to 1.0 (fully saturated). Increasing the value can reduce the amount of sampling and, therefore, speed up the rendering. However, it may also reduce the quality of the image.

Shadows

The properties in the **Shadows** subcategory control how the renderer handles shadows generated by the lights in the scene. For shadows to be applied, the button in the subcategory title bar must be on (yellow).

The Mode property controls a shader function that calculates light effects. There are three modes that determine how shading is calculated:
- Simple. Shaders are randomly created.
- Sorted. Shaders are called in order from the object to the light.
- Segment. Shaders are called in order from the volume shaders to the segments of the light rays between the object and the light.

The Shadow Map property determines whether shadow-mapped or raytraced shadows are created. When this property is set to On, shadow-mapped shadows are generated. When it is set to Off, raytraced shadows are created.

The Sampling Multiplier property limits shadow sampling for area lights. The values are preset for the rendering presets: Draft = 0, Low = 1/4, Medium = 1/2, High = 1, and Presentation = 1. However, these values can be changed. This is the same principle that is described in the Sampling section, but instead of sampling pixels for object color, it is sampling for shadows.

Raytracing

The **Ray Tracing** category provides properties for controlling how the rendered image is shaded, **Figure 18-5.** *Raytracing* is a method of calculating reflections, refractions, and shadows by tracing the path of the light rays from the light sources. This is more accurate at producing shadows than shadow mapping, but it takes more time and the shadow edge is always sharp. To enable raytracing, pick the button in the category's title bar. If this is off (not yellow), there will be no raytracing and the properties are disabled.

The Max reflections property is the maximum number of times that a ray can be reflected. The Max refractions property is the maximum number of times that a ray can be refracted. The Max depth property is the maximum number of reflections and refractions. For example, if this property is set to 5 and the Max reflections property is set to 3, then no more than two refractions will occur. A good way to figure out the required maximum depth is to imagine a light ray traveling through transparent objects or bouncing off of reflective objects in your scene. Count how many surfaces the object must contact and that is the maximum depth.

Figure 18-5.
The **Ray Tracing** category of the **Advanced Render Settings** palette.

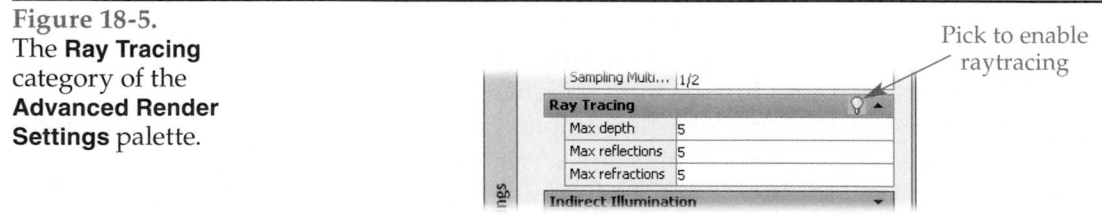

Pick to enable raytracing

Indirect Illumination

Indirect illumination is a method in AutoCAD that simulates natural, bounced light. If indirect illumination is turned off and light does not directly strike an object, the object is black. Without indirect illumination enabled, other lights must be added to the scene to simulate indirect illumination. The properties in the **Indirect Illumination** category allow you to create a natural-looking scene. There are three subcategories in the **Indirect Illumination** category: **Global Illumination**, **Final Gather**, and **Light Properties**. See **Figure 18-6.**

Global illumination

Global illumination (GI) is indirect illumination. Bounced light is simulated by generating photon maps on surfaces in the scene. These maps are created by tracing photons from the light source. Photons bounce around the scene from one object to the next until they finally strike a diffuse surface. When a photon strikes a surface, it is stored in the photon map. To enable global illumination, pick the button in the subcategory's title bar. If this button is off (not yellow), there will be no indirect illumination.

The Photons/sample property sets the number of photons used to generate the photon map. The higher the value, the less noise global illumination produces. However, rendering time is longer and the image is blurrier.

The Use radius property determines whether the photons are a default radius or a user-specified radius. When the property is set to On, the Radius property sets the size of the photon. When set to Off, the radius of each photon is 1/10th of the scene's radius.

The Max reflections property is the maximum number of times that a photon can be reflected. The Max refractions property is the maximum number of times that a photon can be refracted. The Max depth property is the maximum number of reflections and refractions. These properties function as explained in the Ray Tracing section.

Final gathering

The settings in the **Global Illumination** subcategory may result in dark and light areas in the scene. *Final gathering* increases the number of rays in the rendering and cleans up these artifacts. It will also greatly increase rendering time. Final gathering works the best with scenes that contain overall diffuse lighting. See **Figure 18-7.** The Mode property for final gathering can be set to:

- On. Turns on global illumination for final gathering.
- Off. Turns off global illumination for final gathering.
- Auto. Global illumination is turned on or off based on the sky light status. This is the default setting.

Figure 18-6.
The **Indirect Illumination** category of the **Advanced Render Settings** palette.

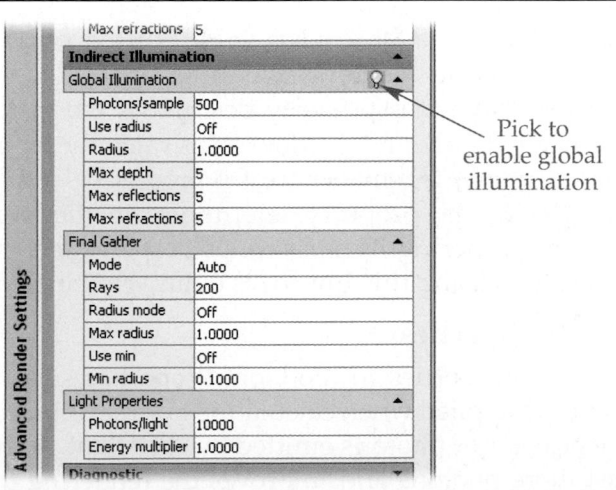

Pick to enable global illumination

Figure 18-7.
A—This scene has a single point light. B—Global illumination is turned on. Notice the unevenness of the lighting. This can be seen especially on the sofa and in the corner of the walls. C—Final gathering cleans up artifacts and provides a more even illumination.

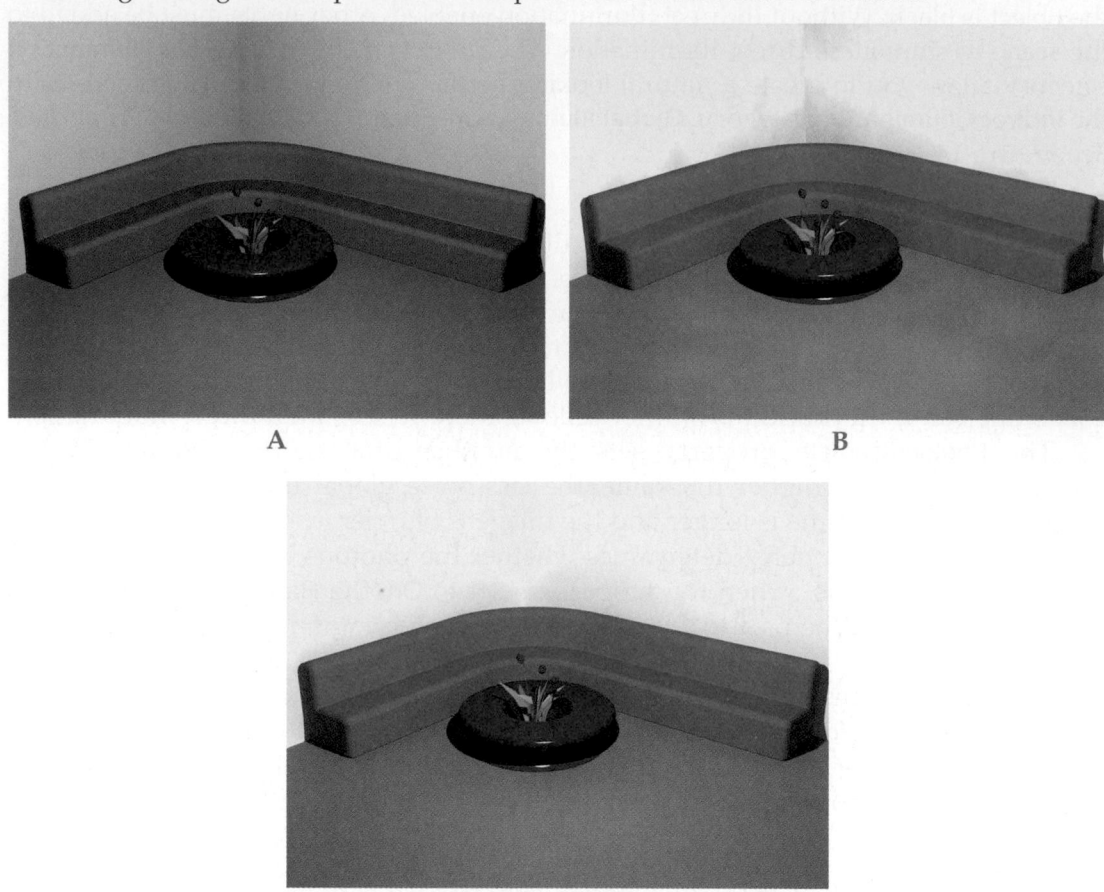

A   B

C

The Rays property sets the number of rays used to calculate indirect illumination. The higher the value, the better the result, but the longer it takes to render the scene.

The Radius mode property determines how the Max radius property is applied during final gathering. There are three possible settings:

- On. The Max radius value is used for final gathering and it is measured in world units.
- Off. The radius of each area processed by final gathering is 10% of the maximum model radius.
- View. The Max radius value is used for final gathering, but it is measured in pixels instead of world units.

The Max radius property determines the maximum radius of each area processed during final gathering. The lower this value, the higher the quality of the rendering because a larger number of smaller areas is processed. However, rendering time is higher.

The Use min property determines whether or not the Min radius property is applied for final gathering. The Min radius property sets the minimum radius of the processed areas. Increasing this improves quality, but increases rendering time.

Light properties

The properties in the **Light Properties** subcategory control how the lights in the scene are applied when calculating indirect illumination. The Photons/light property sets the number of photons emitted by each light. Increasing this number makes each light cast more photons and improves the rendering quality. The Energy multiplier property determines how much light energy is used in global illumination. The default value of 1.0000 does not increase or decrease the light energy. Values less than the default decrease the light energy. Values greater than the default increase the light energy.

PROFESSIONAL TIP

If your scene looks washed out (flooded with light) with indirect illumination enabled, experiment with reducing the energy multiplier. This can have a dramatic effect on the scene.

Diagnostic

The properties in the **Diagnostic** category control tools to help you understand why the rendering produced the results it did, **Figure 18-8.** The scene can be rendered with photon maps, grids, and irradiance shown. These tools can help you diagnose and correct problems.

The Grid property determines if a coordinate grid is shown in the rendered image. The Grid size property sets the size of the grid. When the Grid property is set to Off, which is the default, the grid is not shown. There are three other settings:

- Object. A colored grid displays local coordinates (UVW). Each object has its own set of local coordinates.
- World. World coordinates (XYZ) are displayed in a colored grid, **Figure 18-9A.**
- Camera. Coordinates of a UCS corresponding to the camera or current view are displayed in a colored grid, **Figure 18-9B.**

The Photon property controls whether or not the effect of a photon map is shown in the rendering. When the property is set to Density or Irradiance and global illumination is on, the scene is rendered and overlaid with an image representing the photon map. See **Figure 18-10.**

Figure 18-8.
The **Diagnostic** category of the **Advanced Render Settings** palette.

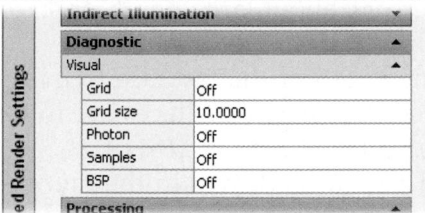

Figure 18-9.
Applying a grid to the rendering. A—The Grid property is set to World. B—The Grid property is set to Camera.

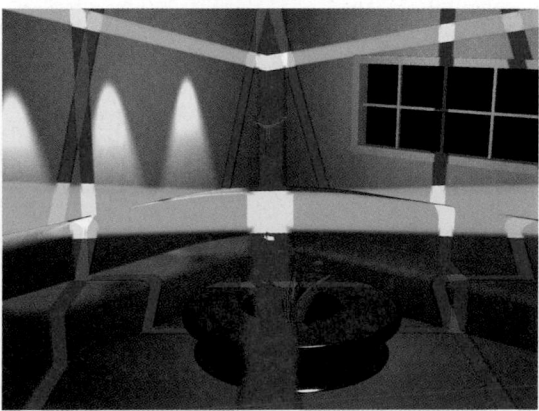

A   B

Figure 18-10.
A photon map
is applied to the
rendering
(Photon property
set to Density).

- Density. Shows the photon map projected onto the scene. Higher-density areas are red and lower-density areas are the cooler colors.
- Irradiance. Similar to density, but the photons are shaded based on their irradiance value. Maximum irradiance is red and lower irradiance values are shown in the cooler colors.

The Samples property can be set to On or Off. When set to On, a grid is rendered plan to the view and varying shades of gray and white are displayed in the scene. This tool is another way to evaluate the lighting in the scene.

The BSP property determines whether or not the effects of *binary space partitioning (BSP)* are shown. BSP is a raytrace-acceleration method. When rendering, if you receive a message about large depth or size values or the rendering is very slow, this tool may help you locate the problem.

- Depth. The depth of the raytrace tree is displayed. Top faces are displayed in bright red. The deeper the faces are in the tree, the cooler the colors in which they are displayed, **Figure 18-11A.**
- Size. The size of the leaves in the raytrace tree are displayed. Different colors are used to identify different leaf sizes, **Figure 18-11B.**

Figure 18-11.
Showing the effects of binary space partitioning. A—The BSP property is set to Depth.
B—The BSP property is set to Size.

A

B

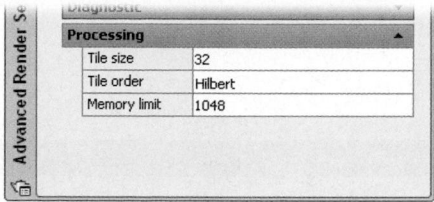

Figure 18-12.
The **Processing** category of the **Advanced Render Settings** palette.

Processing

The properties in the **Processing** category control how the final render processing takes place, **Figure 18-12.** The Tile size property controls the size of the tiles into which the total image is subdivided. The larger the tile size, the fewer tiles that have to be rendered and the fewer times the image has to update. Larger tiles usually mean a shorter rendering time. The Tile order property controls the order in which the tiles are rendered:

- Hilbert. The "cost" of switching to the next tile determines which tile is rendered next.
- Spiral. The rendering begins with the tiles in the center of the image and then spirals outward.
- Left to Right. The tiles are rendered from bottom to top and left to right in columns.
- Right to Left. The tiles are rendered from bottom to top and right to left in columns.
- Top to Bottom. The tiles are rendered from right to left and top to bottom in rows.
- Bottom to Top. The tiles are rendered from right to left and bottom to top in rows.

The Memory limit property specifies the maximum memory allocated for the rendering process. When this limit is reached, some objects may be removed from rendering.

Exercise 18-2

Complete the exercise on the student website.
www.g-wlearning.com/CAD

Render Presets

Once settings have been established that create a rendering with the desired results, the settings can be saved to a custom render preset. The **Render Presets** dialog box is used to create custom render presets, **Figure 18-13.** The **RENDERPRESETS** command opens this dialog box.

The left side of the dialog box displays a tree that contains the standard render presets and any custom render presets. In the middle of the dialog box are all of the properties for the selected render preset. These are the same properties available in the **Advanced Render Settings** palette. On the right side of the dialog box are three buttons that allow you to make a preset current, make a copy of a preset, or delete a preset. You cannot delete one of the default presets.

The easiest way to create a custom preset is to start with a standard render preset and modify the properties until the desired result is produced. This preset will be indicated as the current preset in the **Render Presets** dialog box, but there will be an asterisk (*) in front of its name. The asterisk indicates that the preset has been changed from its original settings. Next, pick the **Create Copy** button in the **Render Presets** dialog box to make a copy. In the **Copy Render Preset** dialog box that appears, name the new render preset, provide a description, and pick the **OK** button. The new preset is saved in the Custom Render Presets branch of the tree.

Ribbon
Render > Render Manager Render Presets...
Type
RENDERPRESETS RP

RENDERPRESETS

Figure 18-13.
The **Render Presets Manager** dialog box.

Saved custom presets

Pick to create a copy with the current settings

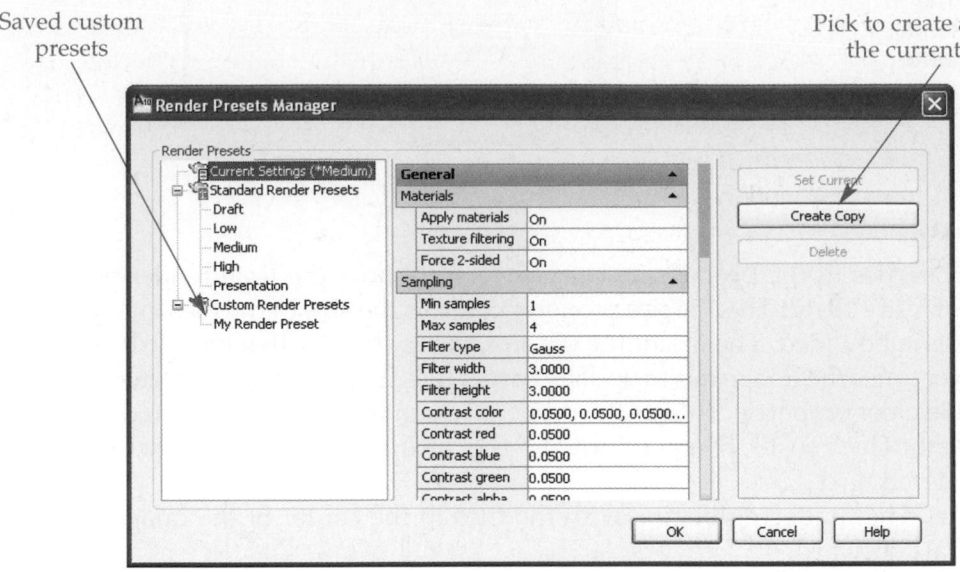

Render Exposure

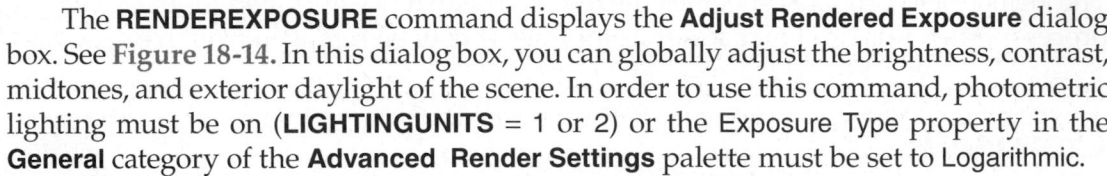

RENDEREXPOSURE

Ribbon

Render
> Render

Adjust Exposure

Type

RENDEREXPOSURE

The **RENDEREXPOSURE** command displays the **Adjust Rendered Exposure** dialog box. See **Figure 18-14.** In this dialog box, you can globally adjust the brightness, contrast, midtones, and exterior daylight of the scene. In order to use this command, photometric lighting must be on (**LIGHTINGUNITS** = 1 or 2) or the Exposure Type property in the **General** category of the **Advanced Render Settings** palette must be set to Logarithmic.

The **Preview** area in the **Adjust Rendered Exposure** dialog box displays the rendered scene with the changes you make in the dialog box so you can see how the scene will be altered. This saves the step of re-rendering the scene. Simply change the settings until the preview looks correct and then close the dialog box. The properties in this dialog box are:

- Brightness. Controls the brightness of the colors. The default value is 65.0000 and it can range from 0.0000 to 200.0000. Increasing the value increases how light the colors in the scene appear.

Figure 18-14.
Using the
RENDEREXPOSURE
command to adjust
the rendering.

Preview

Settings

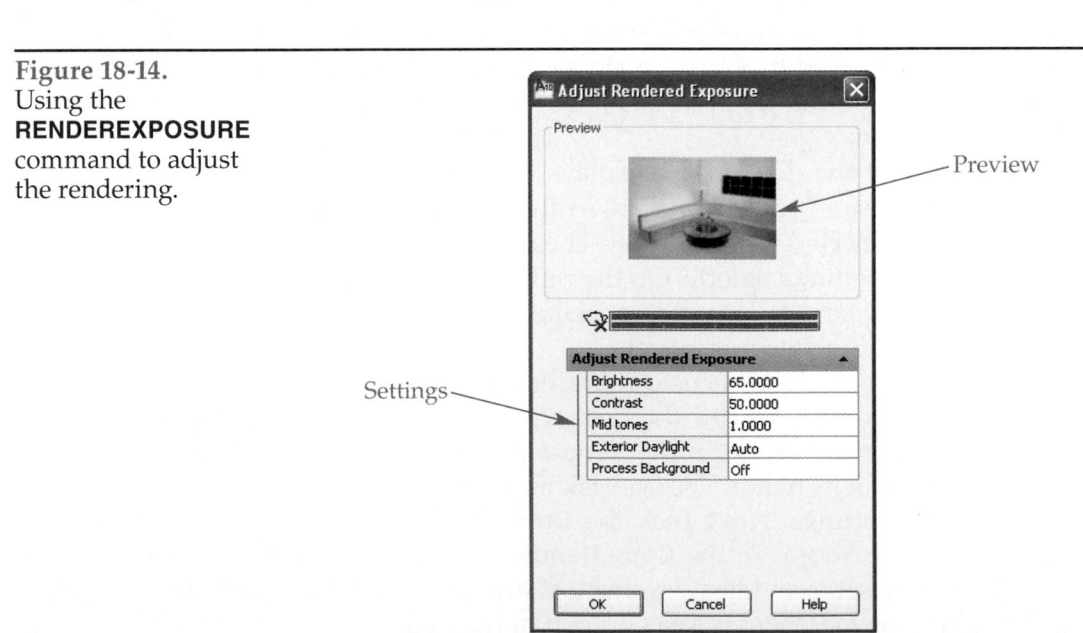

AutoCAD and Its Applications—Advanced

- Contrast. Controls the contrast of the colors in the scene. The default value is 50.0000 and it can range from 0.0000 to 100.0000. Increasing the value increases the difference between similar colors, in effect increasing the brightness of the scene.
- Mid tones. Controls the midtone values of the colors. The midtone colors are neither light nor dark. The default value is 1.0000 and it can range from 0.0000 to 20.0000.
- Exterior Daylight. Sets the exposure for scenes illuminated with sunlight. It is either on, off, or automatic. The default setting is Auto.
- Process Background. Specifies whether or not the background is processed by exposure control when the scene is rendered. It is either on or off.

To force the preview to update, pick the button to the left of the rendering progress bars below the preview that looks like a teapot and an X. The preview is updated with the current settings.

Render Environment

The render environment allows for the addition of fog or depth cueing to the scene. *Fog* and *depth cueing* in AutoCAD are actually ways of using color to visually represent the distance between the camera (viewer) and objects in the model. See **Figure 18-15.** This is similar to looking at an object from a distance and seeing that the object is a little obscured from haze in the air. The only difference between fog and

Figure 18-15.
A—This scene has no fog/depth cueing applied. B—The scene has white fog applied (including the background). C—The scene has black depth cueing applied (including the background).

A

B

C

Figure 18-16.
The **Render Environment** dialog box is used to add fog/depth cueing to the scene.

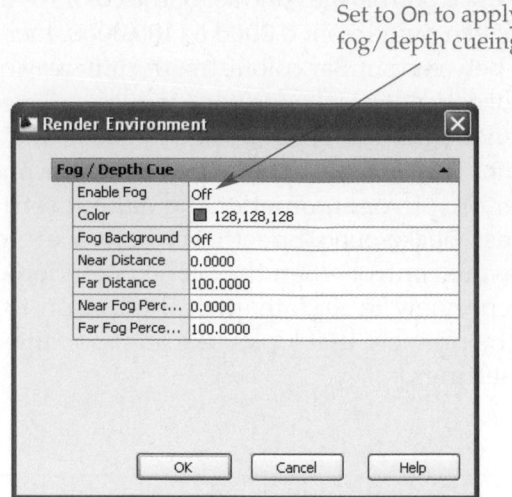

Set to On to apply fog/depth cueing

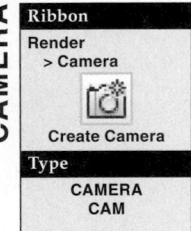
depth cueing is the color. Fog is displayed as white or another light color and depth cueing is generally displayed as black. The **Render Environment** dialog box is used to add fog/depth cueing, **Figure 18-16.**

Creating a Camera

Before adding fog/depth cueing, a camera must be created that shows the view you want. Then, start the **3DCLIP** command and adjust the back clipping plane to where you want the effect to end. Only the back clipping plane needs to be active. The fog/depth cueing references this plane and the camera location.

Creating a camera and adjusting clipping planes is discussed in detail in Chapter 19. However, to create a camera and turn on the clipping plane(s), first select the command. Note: the **Camera** pane is not displayed by default in the **Render** tab of the ribbon. Then, pick a location for the camera followed by the location for its target. Next, enter the **Clipping** option. Turn on the front clipping plane, if desired, and enter the offset distance. Then, turn on the back clipping plane and enter the offset distance. Finally, end the command (do not press [Esc]).

> **PROFESSIONAL TIP**
>
> A camera is automatically created when a view is saved as a named view. This is another good reason to save your views.

Adding Fog

Once a camera is created, open the **Render Environment** dialog box. To turn on fog/depth cueing, set the Enable Fog property to On. To set the color of the effect, select the Color property. Then, choose a color in the drop-down list. The Select Color... entry displays the **Select Color** dialog box. The Fog Background property determines whether or not the background is affected by the fog/depth cueing just like everything else.

The Near Distance property sets where the fog/depth cueing begins. This is a distance from the camera. The value can be from 0.0000 to 100.0000, which is a percentage of the total distance between the camera and the back clipping plane. The back clipping plane is where the target is located. The Far Distance property sets where the fog ends. This is also a distance from the camera. The value is also a percentage of the total distance from the camera to the back clipping plane and can be from 0.0000 to 100.0000. In other words, 100% ends at the back clipping plane.

The Near Fog Percentage property determines the opacity of the fog at its starting location. A value of 100 means the fog is 100% opaque. The near percentage is usually set to 0, or 0% opaque. The Far Fog Percentage property determines the opacity of the fog at its ending location. The fog/depth cueing will increase in opacity from the near distance to the far distance starting with the near fog percentage and ending with the far fog percentage.

Exercise 18-3

Complete the exercise on the student website.
www.g-wlearning.com/CAD

Chapter Test

Answer the following questions. Write your answers on a separate sheet of paper or complete the electronic chapter test on the student website.
www.g-wlearning.com/CAD

1. Describe the three panes of the **Render** window.
2. What are the three possible destinations for render output?
3. Once a rendering is completed and displayed in the **Render** window, how can it be saved to a file?
4. List the render presets AutoCAD provides.
5. What is *sampling* and what do the properties in the **Sampling** subcategory in the **Advanced Render Settings** palette control?
6. Raytracing calculates shadows, _____, and _____.
7. How does global illumination simulate bounced light?
8. What is the benefit of final gathering?
9. For what is the Energy multiplier property in the **List Properties** subcategory in the **Advanced Render Settings** palette used?
10. For what are the properties in the **Diagnostic** category of the **Advanced Render Settings** palette used?
11. Describe how to create a custom render preset.
12. List the properties that can be changed in the **Adjust Render Exposure** dialog box.
13. What is *fog/depth cueing?*
14. Which color is normally used to display depth cueing?
15. What must be created before fog or depth cueing is added to the scene?

Drawing Problems

1. Using the drawing from problem 2 in Chapter 17, you will experiment with advanced render settings.
    A. Open drawing P17_02 and save it as P18_01.
    B. Make sure all of the lights are active and render the scene to the **Render** window using the Medium or High render preset.
    C. In the **Advanced Render Settings** palette, enable global illumination. Then, render the scene again.
    D. Enable final gathering and render the scene again. This time it will probably take much longer to render.
    E. Which rendering has the best quality?
    F. Which setting impacted render time the most?
    G. Save the last image as P18_01.jpg.
    H. Save the drawing.

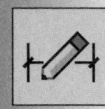

2. In this problem, you will set up fog/depth cueing.
    A. Start a new drawing and save it as P18_02.
    B. Draw a planar surface that is 50 units × 20 units.
    C. Randomly place various objects (cones, boxes, spheres, etc.) on the plane. Assign a different color or material to each object.
    D. Create a viewpoint that is almost at ground level looking down the length of the plane. Try to get as many objects in the view as possible. Save this as a named view and add a background of some type.
    E. Add a distant light source. Position it and adjust its intensity so that interesting shadows are created in the scene, but the objects are sufficiently illuminated.
    F. With the **3DCLIP** command, set up clipping planes with the back clipping plane at the far end of the plane. Make sure the back clipping plane is on.
    G. In the **Render Environment** dialog box, turn on fog and set the color to black. The far distance should be 100 and the percentage should be around 75.
    H. Render the scene.
    I. Change the fog color to white and render the scene again.
    J. Set the fog to affect the background and render the scene again.
    K. Save the image as P18_02.jpg.
    L. Save the drawing.

3. In this problem, you will experiment with the **RENDEREXPOSURE** command and final gathering. Open P17_05 created in Chapter 17 and save it as P18_03. If you did not complete this problem, do so now.

A. Open the **Advanced Render Settings** palette. In the **General** category, set the Exposure Type property to Logarithmic.
B. In the **Indirect Illumination** category, change the Mode property in the **Final Gather** subcategory to Off.
C. Render the scene and note the appearance.
D. Use the **RENDEREXPOSURE** command to display the **Adjust Rendered Exposure** dialog box. Note the appearance of the preview image.
E. Change the brightness setting to 80 and note how the preview changes.
F. Pick the **OK** button to close the **Adjust Rendered Exposure** dialog box and render the scene again. Does the rendered scene match the preview in the **Adjust Rendered Exposure** dialog box?
G. Turn on final gathering (Mode property = On) and open the **Adjust Rendered Exposure** dialog box. How does the preview look different? Why?
H. Adjust the brightness setting to get the exposure that you want in the preview. Then, close the dialog box and render the scene again.
I. Experiment with the other settings in the **Adjust Rendered Exposure** dialog box until you get the scene the way you want it.
J. Render the scene one last time and save the image as a file called P18_03.jpg.
K. Save the drawing.

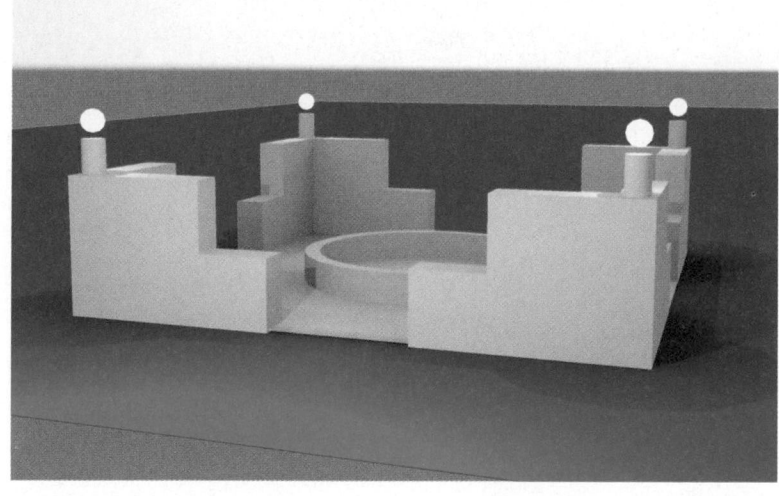

4. Using the same drawing from problem 18-3, you will perform diagnostics to determine the effects of the lights on the final rendering.
   A. Open drawing P18_03 and save it as P18_04.
   B. In the **Advanced Render Settings** palette, select the Medium rendering preset. In the **Indirect Illumination** category, turn off final gathering (Mode property = Off).
   C. In the **Diagnostic** category of the **Advanced Render Settings** palette, set the Grid property to Object. Render the scene.
   D. In the **Diagnostic** category of the **Advanced Render Settings** palette, set the Grid property to World. Render the scene.
   E. In the **Diagnostic** category of the **Advanced Render Settings** palette, set the Grid property to Camera. Render the scene.
   F. Describe the differences and explain why this is helpful in analyzing a scene.
   G. In the **Indirect Illumination** category of the **Advanced Render Settings** palette, turn on global illumination. In the **Diagnostic** category, turn off the grid and set the Photon property to Density.
   H. Render the scene. Describe the effect and what can be learned from it.
   I. Set the Photon property to Irradiance and render the scene. What does this effect tell about the lighting in the scene?
   J. Which diagnostic worked the best and why?
   K. Save the drawing.

CHAPTER 19

Cameras, Walkthroughs, and Flybys

Learning Objectives

After completing this chapter, you will be able to:
- ✓ Create a camera to define a static 3D view.
- ✓ Activate and adjust front and back clipping planes.
- ✓ Record a walkthrough of a 3D model to a movie file.
- ✓ Record a flyby of a 3D model to a movie file.
- ✓ Create walkthroughs and flybys by following a path.
- ✓ Control the viewpoint, speed, and quality of the animation.

Once you have a 3D design complete, or even while still in the conceptual phase of design, you may want to take a stroll through the model and have a look around. You may also want to strap on some wings and fly over and around the model to see it from above. A *walkthrough animation* shows a scene as a person would view it walking through the scene. Walkthroughs are typically used to show the interior of a building, but can be created for exterior scenes as well. A *flyby animation* is similar to a walkthrough, except that the person is not bound by gravity. In other words, the scene is viewed as a bird flying above would see it. Flybys often show the exterior of a building.

The **3DWALK** command is used to create a walkthrough by recording views as a camera "walks" through the scene. The **3DFLY** command is very similar, but the movement of the camera is not limited to a single Z value. A path can also be drawn and the camera linked to the path. This chapter discusses these commands and other methods needed to create the animation you need. In addition, creating and using cameras are discussed.

Creating Cameras

Cameras are used in AutoCAD to store a viewpoint and easily recall it later when needed for viewing or rendering the scene. After the camera is established, you can zoom, pan, and orbit as needed and then come back to the camera view. It is not necessary to create a camera before using the **3DWALK**, **3DFLY**, and **ANIPATH** commands (discussed later) because these commands create their own cameras.

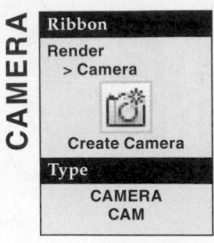

CAMERA

Ribbon
Render
> Camera

Create Camera

Type

CAMERA
CAM

The **CAMERA** command allows you to add a camera to the scene. Note: the **Camera** panel is not displayed by default in the **Render** tab of the ribbon. Cameras are normally placed in the plan view of the scene to make it easy for you to pick where you want to "stand" and where you want to "look." Once the command is selected, you are first prompted to specify the camera location. A camera glyph is placed in the scene at the camera location, **Figure 19-1.** Next, you must specify the target location. As you move the cursor before picking the target location, a pyramid-shaped field of view indicates what will be seen in the view (if a 3D visual style is current). Once you select the target location, the command remains active for you to select an option:

Enter an option [?/Name/LOcation/Height/Target/LEns/Clipping/View/eXit]<eXit>:

The list, or **?**, option allows you to list the cameras in the drawing. Select this option and type an asterisk (*) to show all of the cameras in the drawing. You can also enter a name or part of a name and an asterisk. For example, entering HOUSE* will list all of the cameras whose name begins with HOUSE, such as HOUSE_SW, HOUSE_SE, and HOUSE_PLAN.

The **Name** option allows you to change the name of the camera as you create it. If you do not rename the camera, it is given a default, sequential name, such as Camera1, Camera2, Camera3, and so on. It is always a good idea to provide meaningful names for cameras. Names such as Living Room_SW, Corner, or Hallway_Looking East leave no doubt as to what the camera shows. If you choose not to rename the camera at this point, it can be renamed later using the **Properties** palette.

The **Location** option allows you to change the placement of the camera. Enter the option and then specify the new location. You can enter coordinates or pick a location in the drawing.

The **Height** option allows you to change the vertical location of the camera. Enter the option and then enter the height of the camera. The value you enter is the number of units from the current XY plane. If you are placing the camera in a plan view, this option is used to tilt the view up or down from the current XY plane.

The **Target** option allows you to change the placement of the camera target. Enter the option and then specify the new location. You can enter coordinates or pick a location in the drawing.

The **Lens** option allows you to change the focal length of the camera lens. If you change the lens focal length, you are really changing the field of view, or the area of the drawing that the camera covers. The lower the lens focal length, the wider the field of view angle. The focal length is measured in millimeters.

The **Clipping** option is used to turn the front and back clipping planes on or off. These planes are used to limit what is shown in the camera view. Clipping planes are discussed later in this chapter.

Figure 19-1.
A camera is represented by a glyph. When the camera is selected, the field of view (shown in color) and grips are displayed.

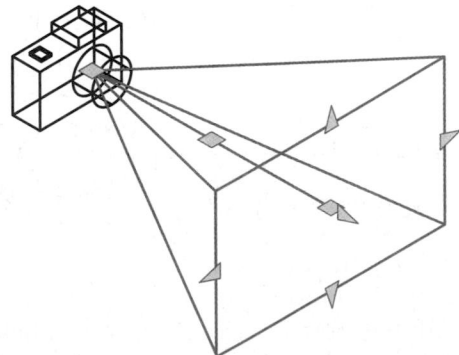

The **View** option is used to change the current view to that shown by the camera. This option has two choices—**Yes** or **No**. If you select **Yes**, the active viewport switches to the camera view and the **CAMERA** command ends. If you select **No**, the previous prompt returns.

Once you have made all settings, press [Enter] or select the **Exit** option to end the command. The view (camera) is listed with the other saved views in the drop-down list in the **Views** panel on the **View** tab of the ribbon. It is also listed under the Model Views branch in the **View Manager** dialog box. Selecting the view makes it the current view in the active viewport.

PROFESSIONAL TIP

In addition to the camera name, many other camera properties can be changed in the **Properties** palette. The camera and target locations can be changed, the lens focal length and field of view can be adjusted, and the clipping planes can be modified. Also, you can change the roll angle, which is the rotation about a line from the camera to the target, and set the camera glyph to plot.

Camera System Variables

The **CAMERADISPLAY** system variable controls the visibility of camera glyphs. When set to 1, which is the default, camera glyphs are displayed. When set to 0, camera glyphs are not displayed. Creating a camera automatically sets the variable to 1. The **Show Cameras** button in the **Camera** panel on the **Render** tab of the ribbon toggles the display of camera glyphs off and on. Remember, this panel is not displayed by default.

When creating a camera, if you pick the camera and target locations without using object snaps, you may assume that the camera and target are located on the XY plane (Z coordinate of 0) of the current UCS. This may or may not be true. The **CAMERAHEIGHT** system variable determines the default height of the camera if a Z coordinate is not provided. It is a good idea to set this variable to a typical eye height before placing cameras. There is no corresponding system variable for the target because the target is usually placed by snapping to an object of interest. If X and Y coordinates are entered for the target location, but a Z coordinate is not provided, the Z value is automatically 0. If a camera was previously created in the drawing session and the **Height** option was used, that height value becomes the default camera height.

Camera Tool Palette

The **Camera** tool palette provides a quick way to add a camera, but the default tools do not allow for the options described earlier. The **Normal Camera** tool creates a camera with a 50 mm focal length. This camera simulates normal human vision. The **Wide-angle Camera** tool creates a camera with a 35 mm focal length. This type of view is commonly used for scenery or interior views where it is important to show as much as possible with minimal distortion. The **Extreme Wide-angle Camera** tool creates a camera with a 6 mm focal length. This camera produces a fish-eye view, which is very distorted and mainly useful for special effects.

Changing the Camera View

Once the camera is placed, it is easy to manipulate. If you select a camera, the **Camera Preview** window is displayed by default. This window shows the view through the camera, **Figure 19-2.** The view in the window can be displayed in the 3D Hidden, 3D Wireframe, Conceptual, Realistic, or any other named visual style. Select the visual style

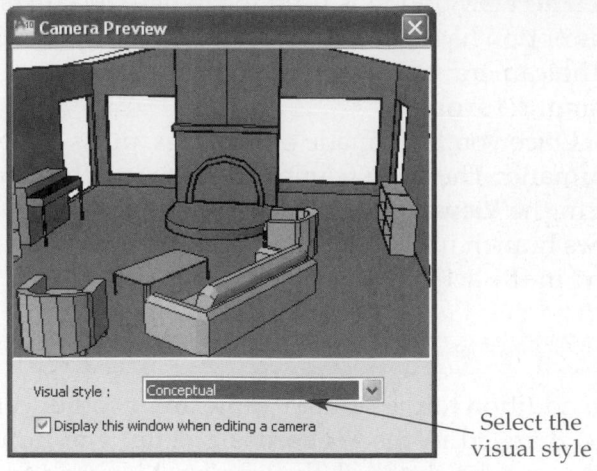

Select the visual style

in the drop-down list in the window. If the **Display this window when editing a camera** check box at the bottom of the window is unchecked, the window is not displayed the next time a camera is selected. The next time the drawing is opened, this setting is restored (checked).

When a camera is selected, grips are displayed. Refer to **Figure 19-1.** If you hover the cursor over a grip, a tooltip appears indicating what the grip will alter. Picking the base grip on the camera allows you to reposition the camera in the scene. If the **Camera Preview** window is open, watch the preview as you move the camera to help guide you. Selecting the grip on the target allows the target to be repositioned. Again, use the preview in the **Camera Preview** window as a guide. The grip at the midpoint between the camera and target can be used to reposition the camera and target at the same time. If you pick and move one of the arrow grips on the end of the field of view, the lens focal length and field of view are changed.

Camera Clipping Planes

Clipping planes allow you to suppress objects in the foreground or background of your scene. Picture these clipping planes as flat, 2D objects perpendicular to the line of sight that can be moved closer to or farther from the viewer. Only the objects between the front and back clipping planes and within the field of view are seen in the camera view. This is helpful for eliminating walls, roofs, or any other clutter that may take away from the focus of the scene. Also, as mentioned in Chapter 18, the back clipping plane should be enabled when applying fog/depth cueing using the **Render Environment** dialog box. Clipping planes can be set while creating the camera or later using the **Properties** palette and adjusted using grips.

To set the clipping planes while creating the camera, enter the **Clipping** option. You are prompted:

Enable the front clipping plane? [Yes/No] <No>:

To enable the front clipping plane, enter YES. You are then asked to specify the offset from the target plane. This is described next. Once you enter the offset, or if you answer **No**, you are prompted:

Enable the back clipping plane? [Yes/No] <No>:

To enable the back clipping plane, enter YES and then specify the offset from the target plane.

The *target plane* is the 2D plane that is perpendicular to the line of sight and passing through the camera's target point. Offsets for both front and back clipping planes are from this plane. Positive values place the clipping planes between the camera and the target plane. Negative values place the planes on the opposite side of the target plane from the camera. You can place the clipping planes anywhere in the scene from the camera location to infinity. You cannot, however, place the back clipping plane between the front clipping plane and the camera.

The best way to adjust clipping planes is using the **Properties** palette or grips. Create the camera and then display a plan view of the camera and target (an approximate plan view is okay). Select the camera and open the **Properties** palette. In the **Clipping** category, select the Clipping property. In the property drop-down list, select Front on, Back on, or Front and back on to turn on the appropriate clipping plane(s). Notice that the clipping planes are visible in the viewport, **Figure 19-3.** Next, enter offset values for the Front plane and Back plane properties as appropriate or use grips to set the locations of the clipping planes. By displaying a plan view of the camera and target, you can see where the clipping planes are located and visualize their effect on the scene. If the **Camera Preview** window is open, the clipping is reflected in the preview.

Exercise 19-1

Complete the exercise on the student website.
www.g-wlearning.com/CAD

Figure 19-3.
Adjusting the clipping planes for a camera.

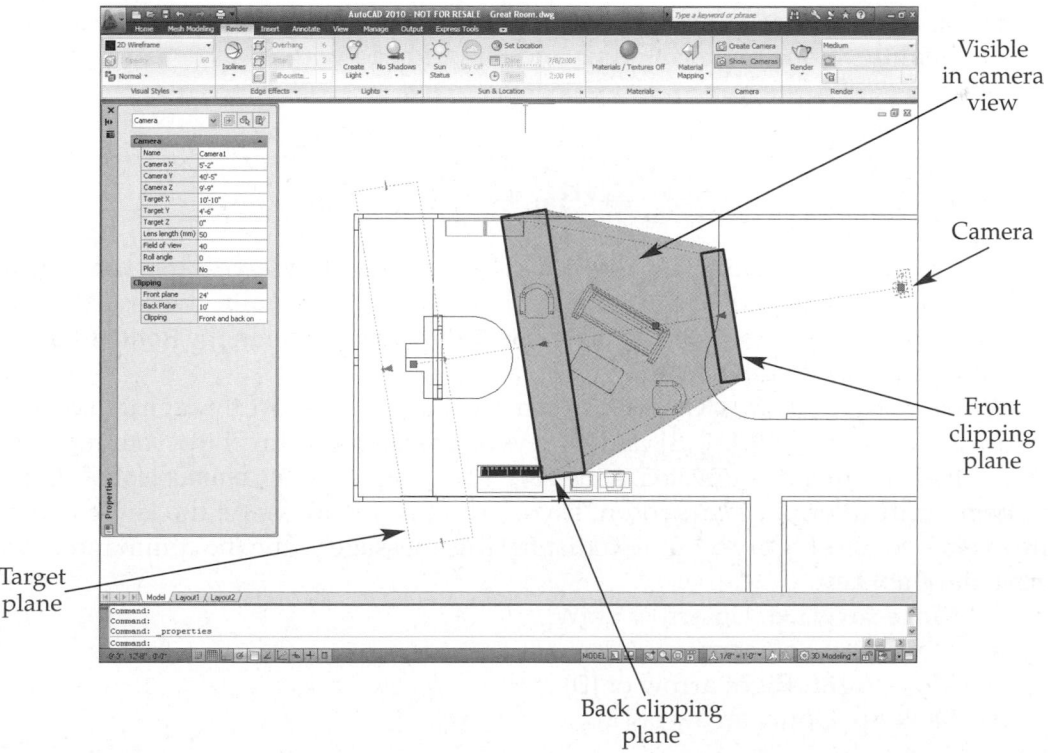

Animation Preparation

The tools presented in this chapter make it easy to lay out a path, plan camera angles, and record the movement of the camera. The resulting animation can be directly output to a number of movie file types that can be shared with others. However, there are some decisions to make first.

It is important to plan out exactly what you want to see in the animation. Think like a movie director and plan the "shots." Ask these questions:

- What will be visible from each camera angle?
- Is there a background in place?
- Is the lighting appropriate?
- Will a simple walkthrough suffice or will a flyby be necessary?
- How close is the viewer (camera) going to be to the objects in the scene?

The answers to these questions will help determine the modeling detail required. Do not model anything that will not be seen. Also, do not place detailed materials on objects that are not the focus of the animation. Processing the animation may take a long time. Unnecessary detail may bog down the computer. In addition, walkthroughs and flybys must be created in perspective, not parallel, views.

The "visual quality" of the scene has the biggest impact on the time involved in rendering the animation. An animation can be rendered in any visual style or using any render preset that is available in the drawing. It is a natural tendency to render at the highest level to make the animation look the best. However, a computer animation has a playback rate of 30 frames per second (fps). If a single frame (view) takes three minutes to render using the Presentation render preset, how long will it take to render a 30 second animation? An animation 30 seconds in length has 900 frames (30 fps × 30 seconds). If each frame takes three minutes to render, the entire animation will take 2700 minutes, or 45 hours, to render.

Are you willing to wait two or three days for a 30 second movie? How about your boss or your client? There are trade-offs and concessions to be made. Perform test renderings on static views and note the rendering time. Then, decide on the acceptable level of quality versus rendering time and move ahead with it.

Walking and Flying

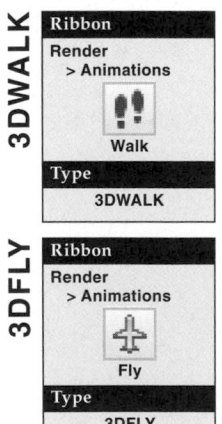

The process for creating a walkthrough or a flyby is the same. First, the command is initiated. Then, the movement is defined and recorded. Finally, the recorded movement is saved to an animation file. Note: the **Animations** panel in the **Render** tab on the ribbon is not displayed by default.

When using the **3DWALK** and **3DFLY** commands, you can move through the scene using the arrow keys or the [W], [A], [S], and [D] keys on the keyboard to control your movements. Once either command is initiated, a message appears from the **Communication Center**, if balloon notifications are turned on. If you expand this message, the key movements are explained. See **Figure 19-4.** To redisplay this message while the command is active, press the [Tab] key.

- **Move forward.** Up arrow or [W].
- **Move left.** Left arrow or [A].
- **Move right.** Right arrow or [D].
- **Back up.** Down arrow or [S].

Figure 19-4.
This message from the **Communication Center** shows the keys that can be used to navigate through an animation.

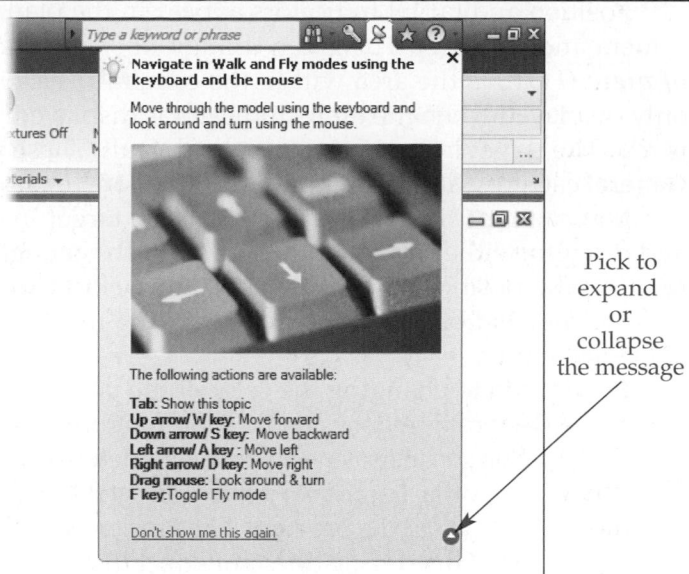

Pick to expand or collapse the message

You can also navigate through the scene using the mouse. Press and hold the left mouse button and then drag the mouse in the active viewport to "steer" through the scene. With the **3DWALK** command, the camera remains at the same Z value. With the **3DFLY** command, the Z position of the camera can change. The steps for creating a walkthrough or flyby are provided at the end of this section.

Position Locator

When the **3DWALK** or **3DFLY** command is initiated, the **Position Locator** palette appears. Sec **Figure 19-5.** The preview in this palette shows a plan view of the scene. The purpose of this window is to provide an overview of the scene, in plan, while you develop the animation. It does not need to be displayed to create an animation and can be closed if it takes up too much space or slows down the rendering.

Figure 19-5.
The **Position Locator** palette.

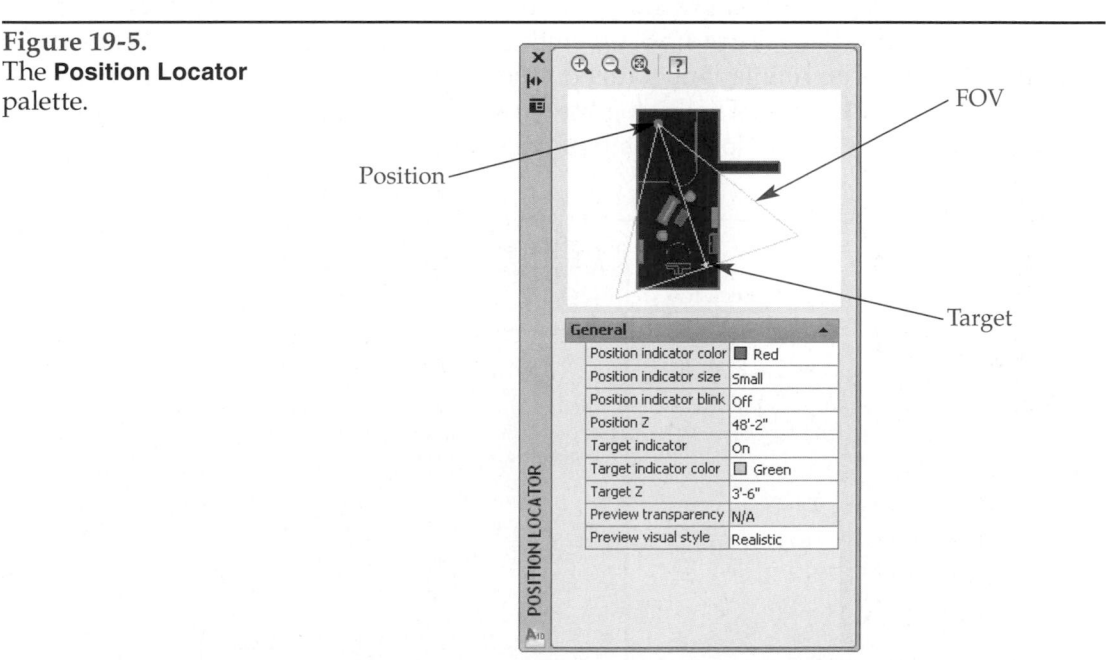

Position

FOV

Target

Position and target indicators appear in the plan view to show the location of the camera and its target. The green triangular shape displays the field of view. The *field of view (FOV)* is the area within the camera's "vision." The field of view indicator is only displayed when the target indicator is displayed. By default, the position indicator is red. The target indicator is green by default. These properties can be changed in the **General** category at the bottom of the **Position Locator** palette.

You can reposition the camera and the target in the plan view simply by picking and dragging either indicator. The effect of the change is visible in the active viewport. Moving the position and target indicators closer together reduces the field of view. Picking the field of view lines and dragging moves the position and target indicators at the same time.

In addition to changing the color of the position and target indicators, the properties in the **General** category can be used to modify the display in the **Position Locator** palette. The Position indicator size property determines if both indicators are displayed small, medium, or large. If the Position indicator blink property is set to On, both indicators flash on and off in the preview. The Preview visual style property sets the visual style for the preview. This setting does not affect the current viewport or the animation. The Preview transparency property is set to 50% by default, but can be changed to whatever you want. If the view in the **Position Locator** palette is obscured by something (a roof, perhaps), you may want to set the Preview visual style property to 3D Hidden and the Preview transparency property to 80% or 90%. This will make the objects under the roof visible. If hardware acceleration is on (**3DCONFIG** command), then the Preview transparency property is disabled.

Exercise 19-2

Complete the exercise on the student website.
www.g-wlearning.com/CAD

Walk and Fly Settings

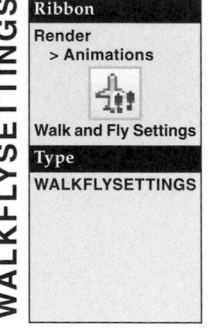
General settings for walkthroughs and flybys are made in the **Walk and Fly Settings** dialog box. See **Figure 19-6.** Open this dialog box by picking the **Walk and Fly Settings** button in the **Animations** panel on the ribbon (in the **Walk** flyout). This panel is not displayed by default. The dialog box can also be displayed by picking the **Walk and Fly...** button in the **3D Modeling** tab of the **Options** dialog box.

Figure 19-6.
General settings for the walkthrough or flyby are made in the **Walk and Fly Settings** dialog box.

Check to automatically display the **Position Locator** palette

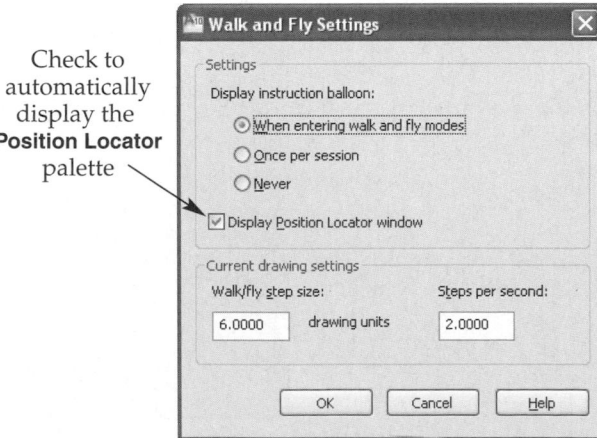

AutoCAD and Its Applications—Advanced

The three radio buttons at the top of the dialog box are used to determine when the message shown in **Figure 19-4** is displayed. The check box determines if the **Position Locator** palette is automatically displayed when the **3DWALK** or **3DFLY** command is entered.

The text boxes in **Current drawing settings** area determine the size of each step and the number of steps per second. The **Walk/fly step size:** setting controls the **STEPSIZE** system variable. This is the number of units that the camera moves in one step. The **Steps per second:** setting controls the **STEPSPERSEC** system variable. This is the number of steps the camera takes each second. Together, these two settings determine how fast the camera moves in the animation.

PROFESSIONAL TIP

You will have to experiment with step size and steps per second values to make an animation that is easy to watch. Start with low numbers (for slow movement) and work your way up. Fast movement is disorienting and makes the viewer feel like they are missing something. The viewer should be able to take their time and get a good look at your design.

To get a feel for the proper speed for a walkthrough, pay attention to the next movie or TV show that you watch. When the director wants you to get a good look at the setting for the scene, the camera very slowly pans around the room. To emphasize distance, the camera slowly zooms in to a target object or person.

Camera tools

The expanded **View** panel on the **Home** tab of the ribbon contains some tools for quickly adjusting the camera before starting the animation. See **Figure 19-7.** The **Lens length** slider controls how much of the scene is seen by the camera. The *lens length* refers to the focal length of the camera lens. The higher the number, the closer you are to the subject. The range is from about 1 to 100000; 50 is a good starting point. There are stops on the slider for standard lens lengths. You can enter a specific value for the lens length or field of view by selecting the text in the slider, typing the value, and pressing [Enter]. The current view must be a camera view for the slider to be enabled.

Below the **Lens length** slider are text boxes for the camera and target positions. These can be used to change the X, Y, and Z coordinates of the camera or target. It is usually easier to change the X and Y locations in the **Position Locator** palette, but the Z coordinate cannot be set there. The Z coordinate determines eye level.

Figure 19-7. Camera tools are located in the expanded **View** panel on the **Home** tab of the ribbon.

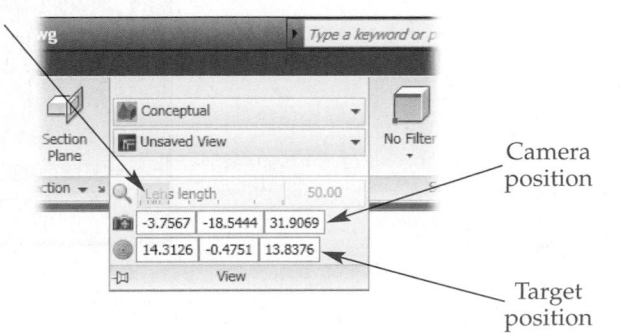

Animation tools

The **Animations** panel on the **Render** tab of the ribbon contains the tools for controlling the recording and playback of the animation. See **Figure 19-8.** Remember, this panel is not displayed by default. The **Animation Record** button is used to initiate recording of camera movement. After the **3DWALK** or **3DFLY** command is activated, pick the button to start recording. Make sure that you are ready to start moving when you pick the button because recording starts as soon as it is picked.

Picking the **Animation Pause** button temporarily stops recording. This allows you to make adjustments to the view without recording the adjustment. When you are ready to begin recording again, pick the record button to resume.

Picking the **Animation Play** button stops the recording and opens the **Animation Preview** dialog box in which the animation is played, **Figure 19-9.** The controls in this dialog box can be used to rewind, pause, and play the animation. The slider can be dragged to preview part of the animation or move to a specific frame. The visual style can also be set using the drop-down list. If the animation is created using a render preset, the file must be played in Windows Media Player or another media player to view the rendered detail.

Picking the **Save** button in the **Animation** panel stops recording and opens the **Save As** dialog box. Name the animation file, navigate to a location, and pick the **Save** button.

Figure 19-8.
The **Animations** panel on the **Render** tab of the ribbon is where you can record and play back the walkthrough or flyby animation. This panel is not displayed by default.

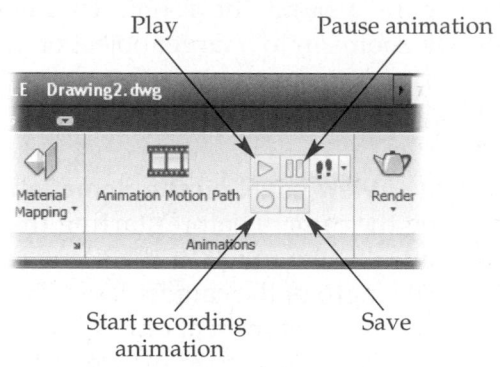

Figure 19-9.
The animation is played in the **Animation Preview** dialog box.

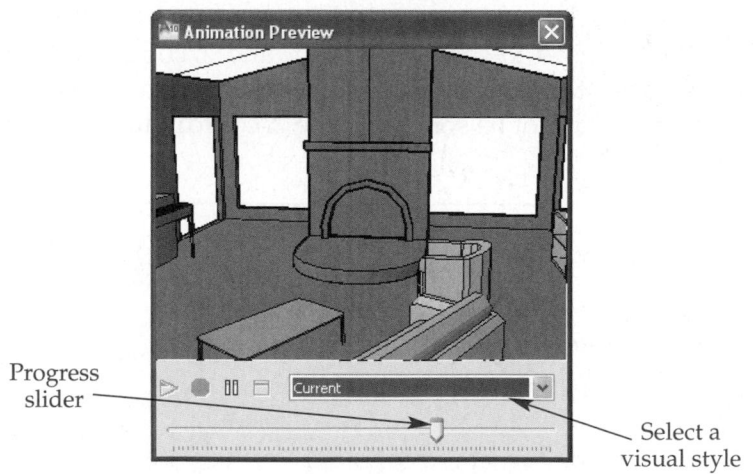

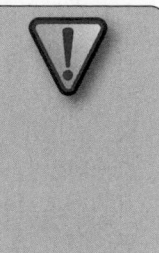

While the **3DWALK** or **3DFLY** command is active and the record button is on, you are creating an animation. If you move the camera in the **Position Locator** palette and start re-recording the animation to correct a problem, but do not first exit the current **3DWALK** or **3DFLY** command session, you are adding another segment to the animation you just previewed. To start over, first exit the current command session.

Exercise 19-3

Complete the exercise on the student website.
www.g-wlearning.com/CAD

Animation settings

The **Animation Settings** dialog box may contain the most important settings pertaining to walkthroughs and flybys. See **Figure 19-10.** These animation settings determine how good the animation looks, how long it is going to take to complete, and how big the file will be. The dialog box is displayed by picking the **Animation settings...** button in the **Save As** dialog box displayed when saving an animation.

The **Visual style:** drop-down list is used to set the shading level in the animation. The name of this drop-down list is a little misleading because visual styles and render presets are available. The higher the shading or rendering level selected in this drop-down list, the longer the rendering will take to process and the bigger the file will be. If you have numerous lights casting shadows, detailed materials, and global illumination and final gathering enabled, settle in for a long wait. A simple, straight-ahead walkthrough of 10 or 15 feet can easily result in 300 frames of animation. If each frame takes about five seconds to render, that equals 1500 seconds, or 25 minutes, to create an animation file that is only 10 seconds long.

The **Frame rate (FPS):** text box sets the number of frames per second for the playback. In other words, this sets the speed of the animation playback. The default is 30 fps, which is a common playback rate for computers.

The **Resolution:** drop-down list offers standard choices of resolution, from 160×120 to 1024×768. These are measured in pixels × pixels. Remember, higher resolutions mean longer processing times and larger file sizes.

The **Format:** drop-down list is used to select the output file type. The file type must be set in this dialog box. It cannot be changed in the **Save As** dialog box. The choices of output file type are:

- **WMV.** The standard movie file format for Windows Media Player.
- **AVI.** Audio-Video Interleaved is the Windows standard for movie files.

Figure 19-10.
The **Animation Settings** dialog box contains important settings pertaining to walkthroughs and flybys.

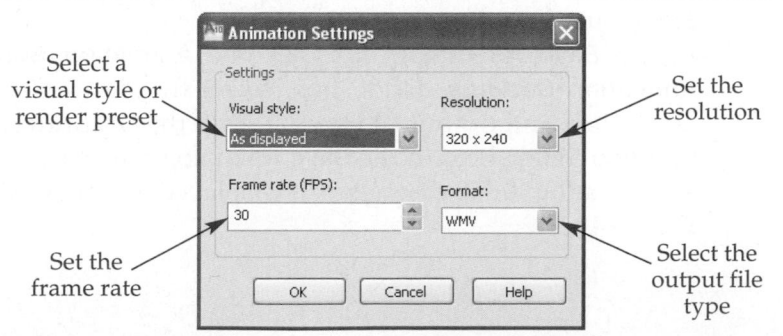

- **MOV.** QuickTime® Movie is the standard file format for Apple® movie files.
- **MPG.** Moving Picture Experts Group (MPEG) is another very common movie file format.

Depending on the configuration of your computer, you may not have all of these file type options or you may have additional options not listed here.

PROFESSIONAL TIP

Other AutoCAD navigation modes may be used to create a walk-through or flyby. Anytime that you enter constrained, free, or continuous orbit, you can pick the **Start Recording Animation** button to record the movements. After you activate the **3DWALK** or **3DFLY** command, right-click and experiment with some of the other options as you record your animation. You can even combine some of these navigation modes with walking or flying. For example, use **3DFLY** to zoom into a scene, pause the animation, switch to constrained orbit, restart the recording, and slowly circle around your model. Animation settings are also available in this shortcut menu.

Exercise 19-4

Complete the exercise on the student website.
www.g-wlearning.com/CAD

Steps to Create a Walkthrough or Flyby

1. Plan your animation: where you are moving from and to, what you are going to be looking at, and what will be the focal point of the scene.
2. Set up a multiple-viewport configuration of three or four viewports.
3. In one of the viewports, create or restore a named view with the appropriate starting viewpoint. Make sure a background is set up, if desired.
4. Start the **3DWALK** or **3DFLY** command and note in the **Position Locator** palette the location of the camera and target and the field of view. Adjust these in the expanded **View** panel on the ribbon, if necessary.
5. Position your fingers over the navigation keys on the keyboard.
6. Pick the **Animation Record** button.
7. Practice navigating through the view and then pick the **Animation Play** button to preview your animation. When you are done practicing, make sure to cancel the command and reposition the camera at the starting point.
8. Pick the **Animation Record** button to start over.
9. Start navigating through the view. Try to keep the movements as smooth as possible. Any jerks and shakes will be visible in the animation.
10. When you are done, stop moving forward and then pick the stop (**Animation Save**) button.
11. In the **Save As** dialog box pick the **Animation Settings...** button. In the **Animation Settings** dialog box, select the desired visual style, frame rate, resolution, and output file format. Pick the **OK** button to close the **Animation Settings** dialog box. In the **Save As** dialog box, name and save the file.
12. The **Creating Video** dialog box is displayed as AutoCAD processes the frames, **Figure 19-11.**

Figure 19-11.
The **Creating Video** dialog box is displayed as AutoCAD generates the animation.

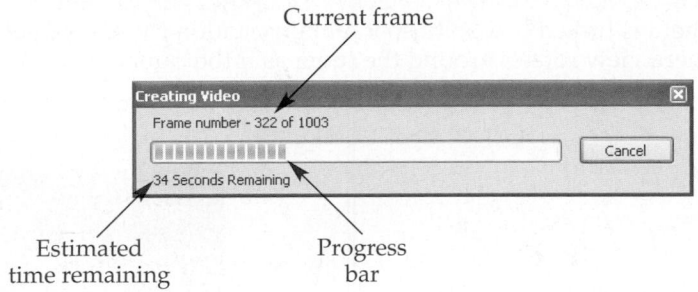

Current frame

Estimated time remaining

Progress bar

13. When the **Creating Video** dialog box is automatically closed, the animation file is saved and you can take a look at it. Pick the **Play** button and watch the animation in the **Animation Preview** window. You can also locate the file using Windows Explorer. Then, double-click on the file to play the animation in Media Player (or whichever program is associated with the file type).

14. Exit the command. If you are not satisfied with the results and want to try it again, make sure to exit the command before you make another attempt at the walkthrough or flyby.

Motion Path Animation

You may have found it hard to create smooth motion using the keyboard and mouse. Fortunately, AutoCAD provides an easy way to create a nice, smooth animated walkthrough or flyby. This is done through the use of a motion path. A *motion path* is simply a line along which the camera, target, or both travel during the animation.

One method of using a motion path is to link the camera and target to a single path. The camera and its line of sight then follow the path much like a train follows tracks. See **Figure 19-12**.

Another option when using a motion path is to link the camera to a single point in the scene and the target to a path. For example, the target can be set to follow a circle or arc. The camera swivels on the point and "looks at" the path as if it is being rotated on a tripod. See **Figure 19-13**.

Figure 19-12.
A—The camera and target are linked to the same path (shown in color). B—The camera looks straight ahead as it moves along the path.

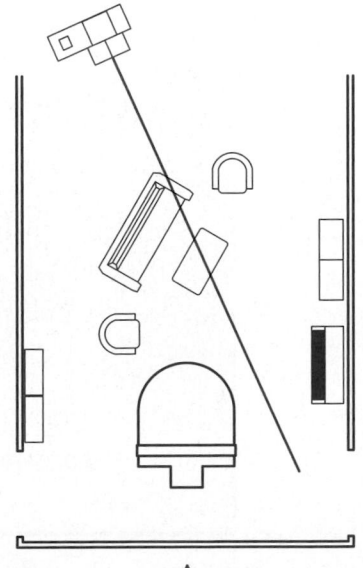

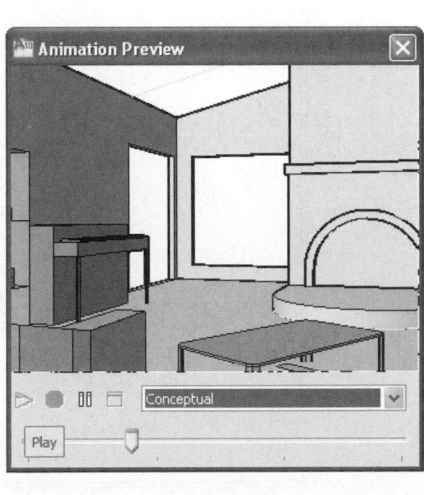

A

B

Figure 19-13.
A—The camera is linked to a point so it remains stationary. The target is linked to the circle.
B—The camera view rotates around the room as if the camera is on a swivel tripod.

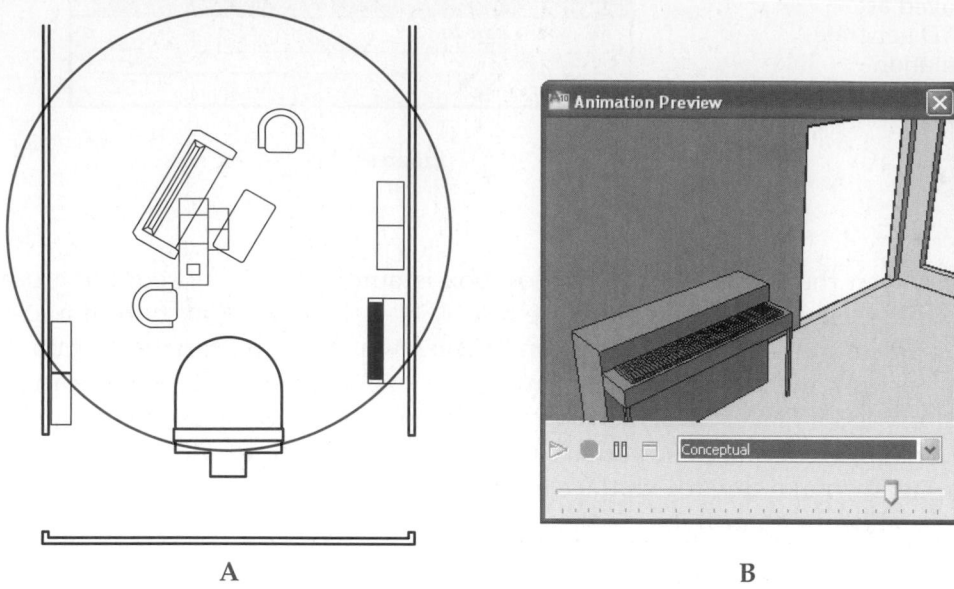

A third way to use a motion path is to have the camera follow a path, but have the target locked onto a stationary point. This is similar to riding in a vehicle and watching an object of interest on the side of the road. As the vehicle moves, your gaze remains fixed on the object. See **Figure 19-14.**

The fourth method of using a motion path is to have both the camera and target follow separate paths. Picture yourself walking into an unfamiliar room. As you walk into the center of the room, your gaze sweeps left and right across the room. In this case, the camera (you) follows a straight line path and the target (your gaze) follows an arc from one side of the room to the other.

Figure 19-14.
A—The camera is linked to the spline path and the target is linked to the point (shown in color). B—As the camera moves along the path, it always looks at the point.

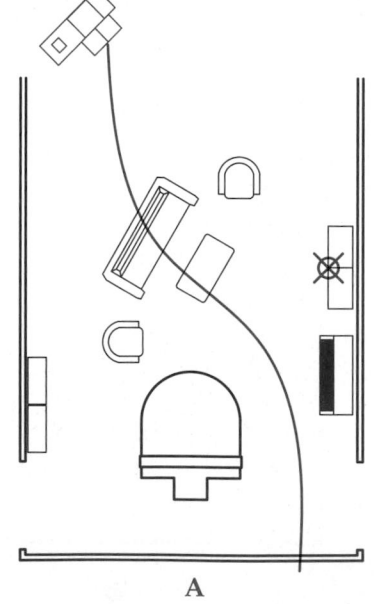
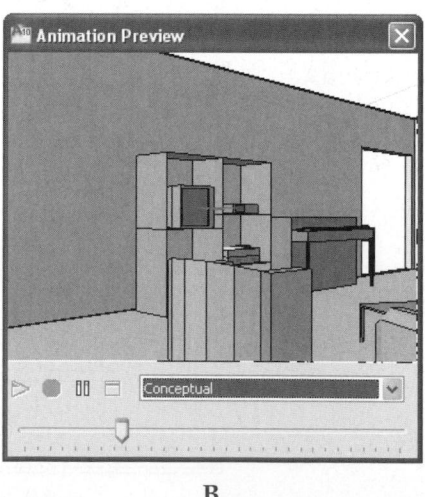

Figure 19-15.
The **Motion Path Animation** dialog box is used to create an animation that follows a path.

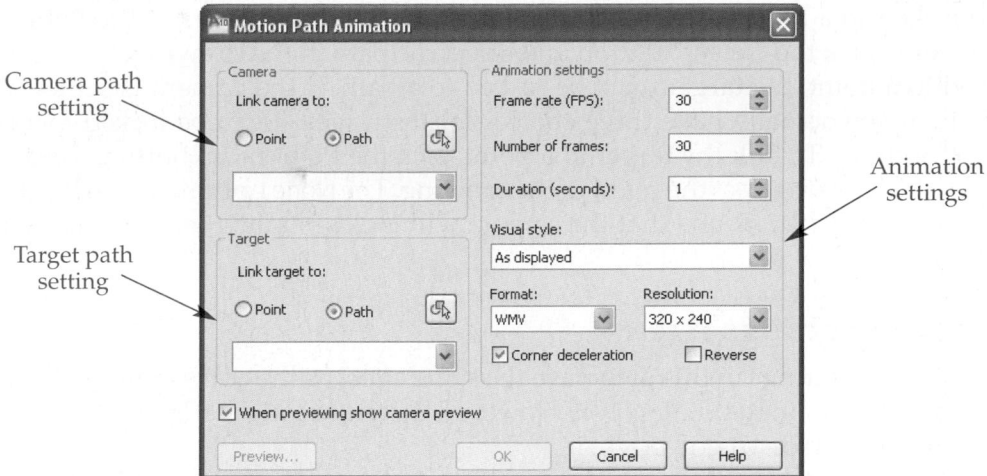

Camera path setting

Target path setting

Animation settings

The **ANIPATH** command is used to assign motion paths. The command opens the **Motion Path Animation** dialog box. See **Figure 19-15.** This dialog box has three main areas: **Camera**, **Target**, and **Animation settings**. These areas are described in detail in the next sections. The steps for creating a motion path animation are provided at the end of this section.

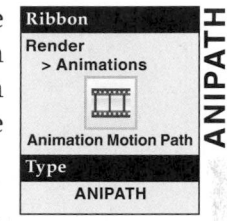

Ribbon
Render
> Animations
Animation Motion Path

Type
ANIPATH

ANIPATH

NOTE

Selecting a motion path automatically creates a camera. You cannot add a motion path to an existing camera.

Camera Area

The camera can be linked to a path or a point. To select a path, pick the **Path** radio button and then pick the "select" button next to the radio buttons. The dialog box is temporarily closed for you to select the path in the drawing. The path may be a line, arc, circle, ellipse, elliptical arc, polyline, 3D polyline, spline, or helix, but it must be drawn before the **ANIPATH** command is used. Splines are nice for motion paths because they are smooth and have gradual curves. The camera moves from the first point on the path to the last point on the path, so create paths with this in mind.

To select a stationary point, pick the **Point** radio button. Then, pick the "select" button next to the radio buttons. When the dialog box is hidden, specify the location in the drawing. You can use object snaps or enter coordinates. It may be a good idea to have a point drawn and use object snaps to select the point.

The camera must be linked to either a path or a point. If neither is selected, the command cannot be completed. If you want the camera to remain stationary as the target moves, select the **Point** radio button and then pick the stationary point in the drawing.

Once a point or path has been selected, it is added to the drop-down list. All named motion paths and selected motion points in the drawing appear in this list. Instead of using the "select" button, you can select the path or point in this drop-down list.

Target Area

The target is the location where the camera points. Like the camera, the target can be linked to a point or a path. To link the target to a path, select the **Path** radio button. Then, pick the "select" button and select the path in the drawing. If the camera is linked to a point, the target must be linked to a path. If the camera is set to follow a path, then you actually have three choices for the target. It can be linked to a path, point, or nothing. To link the target to a point, pick the **Point** radio button. Then, pick the "select" button to select the point in the drawing. The None option, which is selected in the drop-down list, means that the camera will look straight ahead down the path as it moves.

Animation Settings Area

Most of the settings in this area have the same effect as the corresponding settings in the **Animation Settings** dialog box. However, there are four settings unique to the **Motion Path Animation** dialog box.

The **Number of frames:** text box is used to set the total number of frames in the animation. Remember, a computer has a playback rate of 30 fps. Therefore, if the frame rate is set to 30, set the number of frames to 450 to create an animation that is 15 seconds long ($30 \times 15 = 450$).

The value in the **Duration (seconds):** text box is the total time of the animation. This value is automatically calculated based on the frame rate and number of frames. However, you can enter a duration value. Doing so will automatically change the number of frames based on the frame rate.

By default, the **Corner deceleration** check box is checked. This slows down the movement of the camera and target as they reach corners and curves on the path. If this is unchecked, the camera and target move at the same speed along the entire path, creating very jerky motion on curves and at corners. It is natural to decelerate on curves.

The **Reverse** check box simply switches the starting and ending points of the animation. If the camera (or target) travels from the first endpoint to the second endpoint, checking this check box makes the camera (or target) travel from the second endpoint to the first.

Previewing and Completing the Animation

To preview the animation, pick the **Preview...** button at the bottom of the **Motion Path Animation** dialog box. The camera glyph moves along the path in all viewports. If the **When previewing show camera preview** check box is checked, the **Animation Preview** window is also displayed and shows the animation.

To finish the animation, pick the **OK** button in the **Motion Path Animation** dialog box. The **Save As** dialog box is displayed. Name the file and specify the location. If you need to change the file type, pick the **Animation settings...** button to open the **Animation Settings** dialog box. Change the file type, close the dialog box, and continue with the save.

Steps to Create a Motion Path Animation

1. Plan your animation: where you are moving from and to, what you are going to be looking at, and what will be the focal point of the scene.
2. Draw the paths and points to which the camera and target will be linked. Draw the path in the direction the camera should travel (first point to last point). Do not draw any sharp corners on the paths and make sure that the Z value (height) is correct.
3. Start the **ANIPATH** command.

4. Pick the camera path or point.
5. Pick the target path or point (or None).
6. Adjust the frames per second, number of frames, and duration to set the length and speed of the animation.
7. Select a visual style, the file format, and the resolution.
8. Preview the animation. Adjust settings, if needed.
9. Save the animation to a file.

Exercise 19-5

Complete the exercise on the student website.
www.g-wlearning.com/CAD

Chapter Test

Answer the following questions. Write your answers on a separate sheet of paper or complete the electronic chapter test on the student website.
www.g-wlearning.com/CAD

1. Which system variable controls the display of camera glyphs?
2. Name the three camera tools available on the **Camera** tool palette and explain the differences between them.
3. When is the **Camera Preview** window displayed, by default?
4. From where is the offset distance for the camera clipping planes measured?
5. What is the difference between the **3DWALK** and **3DFLY** commands?
6. How do you "steer" your movement when creating a walkthrough or flyby animation?
7. What is the *field of view?*
8. What is the purpose of the **Position Indicator** palette?
9. In the **Walk and Fly Settings** dialog box, which settings combine to control the speed of the animation?
10. How do you start recording a walkthrough or flyby?
11. What must be done before correcting a motion error in a walkthrough or flyby?
12. Motion path animation involves linking a camera or target to _____ or _____.
13. Which types of objects may be used as a motion path?
14. If None is selected as the target "path," what does the camera do in the animation?
15. Explain *corner deceleration.*

Drawing Problems

1. In this problem, you will create and manipulate a camera in a drawing from a previous chapter.
   A. Open drawing P18_02 from Chapter 18 and save it as P19_01.
   B. Create at least two viewports and display a plan view in one of them.
   C. Use the **CAMERA** command to create a camera looking at the objects from the southwest quadrant. Change the camera settings as needed to display a pleasing view of the scene.
   D. Name the camera SW View.
   E. Turn on both the front and back clipping planes. Adjust them to eliminate one object in the front and one object in the back.
   F. Open the **Camera** tool palette and, using the tools in the palette, create three more cameras looking at the scene from various locations. Change their names to Normal, Wide-angle, and Fish-eye to match the type of camera.
   G. Save the drawing.

2. In this problem, you will draw some basic 3D shapes to represent equipment in a small workshop. Then, you will create an animated walkthrough.
   A. Start a new drawing and set the units to architectural. Save it as P19_02.
   B. Draw a planar surface that is 15′ × 30′.
   C. Draw three 9′ tall walls enclosing the two long sides and one short side.
   D. Use boxes and a cylinder to represent equipment. Refer to the illustration shown below. Use your own dimensions.
   E. Use the **3DWALK** command to create an animation of walking into the workshop. Turn and look at the shelves at the end of the animation.
   F. Set the visual style to Conceptual and the resolution to 640 × 480.
   G. Save the animation to a file named P19_02.avi (or the format of your choice).
   H. Save the drawing.

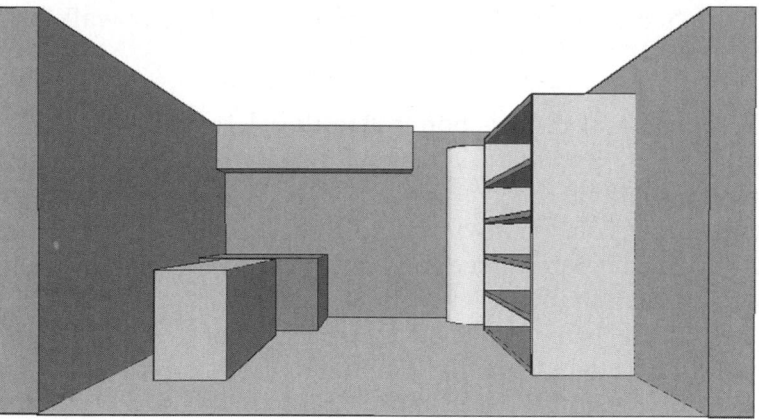

3. In this problem, you will create a motion path animation for the workshop drawn in problem 19-2.
   A. Open drawing P19_02 and save it as P19_03.
   B. Draw a line and an arc similar to those shown in color below. The dimensions are not important.
   C. Move both objects so they are 4′ off of the floor.
   D. Using the **ANIPATH** command, link the camera to the line and the target to the arc. Set the resolution to 640 × 480.
   E. Preview the animation. Adjust the animation settings as necessary. You may need to slow down the animation quite a bit. How do you do this?
   F. Save the animation as a file named P19_03.wmv (or the file format of your choice).
   G. Save the drawing.

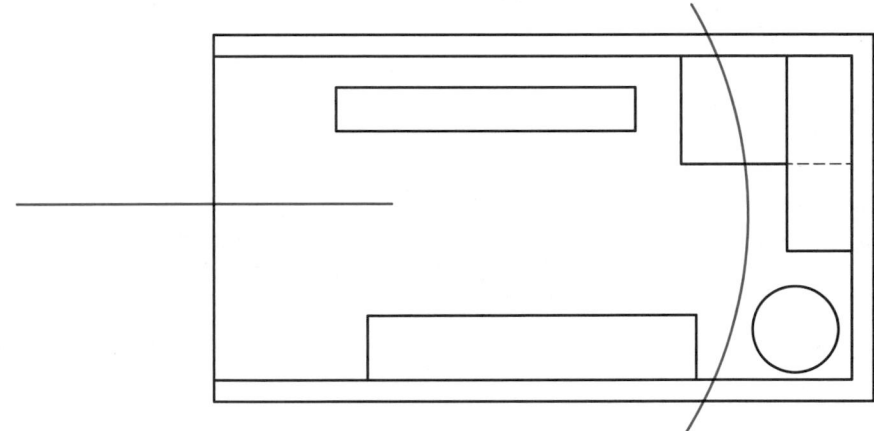

4. In this problem, you will create a motion path animation for presentation of a mechanical drawing from a previous chapter.
   A. Open drawing P17_03 and save it as P19_04.
   B. Draw a circle centered on the flange with a radius of 300.
   C. Move the circle 200 units in the Z direction.
   D. Using the **ANIPATH** command, link the camera to the circle and the target to the center of the flange.
   E. Preview the animation and adjust the animation settings as necessary. Due to the materials in the scene, the preview may play slowly, depending on the capabilities of your computer.
   F. Select a visual style that your computer can handle. Set the resolution to 320 × 240.
   G. Save the animation to a file named P19_04.avi (or the file format of your choice).
   H. Save the drawing.

5. In this problem, you will create a flyby of the building that you created in Chapter 17.
   A. Open drawing P17_01 and save it as P19_05.
   B. Create a perspective view of the scene that shows the building from slightly above it. Save the view.
   C. Start the **3DFLY** command. Practice with the movement keys to make sure you know how to fly around the building. Then, cancel the command.
   D. Restore the starting view and select the **3DFLY** command.
   E. Record the flyby and save the animation as P19_05.wmv (or the file format of your choice).
   F. Save the drawing.

20

Using Raster, Vector, and Web Graphics

Learning Objectives

After completing this chapter, you will be able to:

✓ Compare raster and vector files.
✓ Import and export raster files using AutoCAD.
✓ Import and export vector files using AutoCAD.
✓ Set image commands to manipulate inserted raster files.
✓ Create DWF, DWFx, and PDF files.

One of the important aspects of drawing in AutoCAD is the ability to share information. Generally, this means sharing drawing data and geometry between CAD software, either other AutoCAD workstations or workstations using different software. AutoCAD creates drawing data files in a format known as a *vector* file. However, you can also share your work, as images, with photo editing and desktop publishing software. In Chapter 15 through Chapter 19, you learned how to create realistic scenes and render them to files. A scene rendered to a file is a *raster* image. However, raster images used in AutoCAD do not have to be created in AutoCAD. They may also come from digital photographs, scanned images, or Internet sources. This chapter introduces using AutoCAD to work with raster and vector graphics files. This includes importing, exporting, and setting various parameters.

Introduction to Raster and Vector Graphics

In the world of electronic imaging, there are two basic types of files—raster and vector. AutoCAD drawings are called vector graphics. A *vector* is an object defined by XYZ coordinates. In other words, AutoCAD stores the mathematical definition of an object. *Pixels* (picture elements) are the "dots" or "bits" in the monitor that make up the display screen. When drawing vector objects in AutoCAD, your monitor uses pixels to create a representation of the object on the monitor. However, there is no relationship between the physical pixels in your monitor and a vector object. Pixels simply show the object at the current zoom percentage. Some common vector files are DWG, DXF, AI, and EPS.

Many illustrations created with drawing, painting, and presentation software are saved as raster files. A *raster file* creates a picture or image file using the location and color of the screen pixels. In other words, a raster file is made up of "dots." Raster files are usually called *bitmaps.* There are several types of raster files used for presentation graphics and desktop publishing. Some common raster file types include TIFF, JPEG, and GIF.

Working with Raster Files

Raster images inserted into AutoCAD drawings are treated much like externally referenced drawings (xrefs). Therefore, they are managed in the **External References** palette. See **Figure 20-1.** This can be displayed with the **EXTERNALREFERENCES** command or by picking the dialog box launcher button at the lower-right corner of the **Reference** pane on the **Insert** tab of the ribbon. Raster images are not added to the drawing database, but are attached and referenced by a path to the file's location. Any changes to the image content must be made to the original file. Settings and commands in AutoCAD can, however, control the portion of the image shown and its appearance. Images can be inserted, removed, and modified using commands found in the **Reference** panel in the **Insert** tab of the ribbon, **Figure 20-2.** These functions are discussed in detail in this section.

At the top of the **External References** palette is a drop-down list containing buttons for attaching drawings (DWG), image files, DWF files, DGN files, and PDF files. The **File References** area lists all files currently attached to the drawing, whether they are draw-

EXTERNALREFERENCES

Ribbon

View
>Palettes

External
References Palette

Type

EXTERNAL-
REFERENCES
XREF
IMAGE
IM

Figure 20-1.
The **External References** palette is used to manage attached images. The tooltip can be configured to display only the file name, a preview, details, or both a preview and details, as shown here.

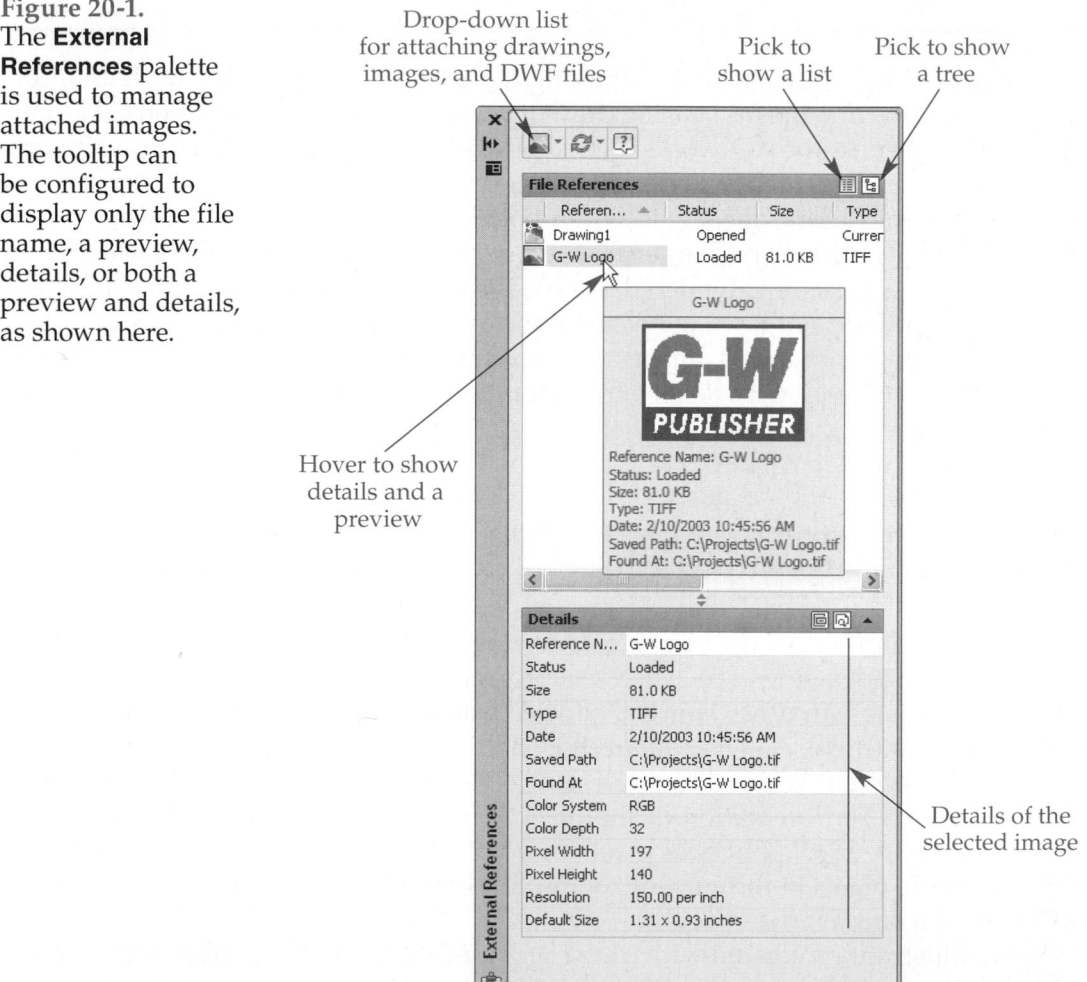

484   AutoCAD and Its Applications—Advanced

Figure 20-2.
Image commands
located on the
Reference panel in
the ribbon.

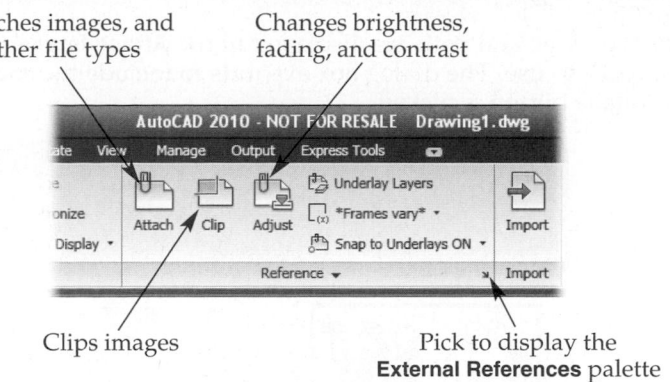

Attaches images, and
other file types

Changes brightness,
fading, and contrast

Clips images

Pick to display the
External References palette

ings, images, DWF, DGN, or PDF files. Right-clicking on an entry displays a shortcut menu that allows you to open, attach, unload, reload, and detach the files. If you hover the cursor over an entry in the palette, a preview window is displayed as help text. By default, this preview provides details related to the file and a preview of the file.

There are many different types of raster files. Some raster files used in industry today are:

- **Tagged Image File Format (TIFF).** A file format developed by Aldus Corporation and Microsoft Corporation. This is one of the most commonly used image file types.
- **Joint Photographic Experts Group (JPEG).** A highly compressed graphics image file. This type of file is very common on websites. Also known as a JPG file.
- **Graphics Interchange Format (GIF).** A file format developed to allow the exchange of graphic images over an online computer service, such as the Internet. This type of file is sometimes found on websites, often animated.
- **Portable Network Graphics (PNG).** Developed in the mid 1990s as a replacement for the GIF format. This file type is extensively used for electronic transmission, such as via e-mail or as website graphics.

Other raster file types can also be imported into AutoCAD. If you have a raster image that cannot be directly imported, you will need to first import the file into a paint or draw program. Then, export the image in a format that AutoCAD can read.

Inserting Raster Images

The **IMAGEATTACH** command is used to attach an image file to a drawing. When the command is selected, the **Select Reference File** dialog box is displayed. Next, navigate to the folder containing the raster file, select the file, and pick **Open**. This displays the **Attach Image** dialog box, **Figure 20-3**. When the **OK** button is picked and the image placed, it is displayed in the drawing area. See **Figure 20-4**.

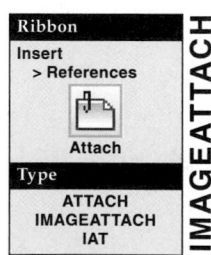

Ribbon

Insert
> References

Attach

Type

ATTACH
IMAGEATTACH
IAT

IMAGEATTACH

The image name is displayed at the top of the dialog box. You can choose to save the full path, a relative path, or no path. The type of path is selected in the **Path type** drop-down list.

A *full path* specifies the complete location of the image file, such as c:\images\building.tif. If the image file is moved from this location, AutoCAD cannot find it.

A *relative path* specifies the location of the image file based on the location of the drawing file. For example, the path .\images tells AutoCAD that the image file is located in a subfolder (named images) of the folder where the drawing is located. The entry ..\images tells AutoCAD to look for the file by moving up one folder from where the drawing is stored and then in the subfolder \images. The entry ..\..\images tells AutoCAD to move up two folders and then look in the subfolder \images. The current drawing must be saved in order to specify a relative path.

Figure 20-3.
The image name and path are displayed in the **Attach Image** dialog box. Be sure to select the type of path to use. The dialog box expands to include the **Image Information** area when the **Show Details** button is picked.

Select the type of path

Actual path

Path that will be saved

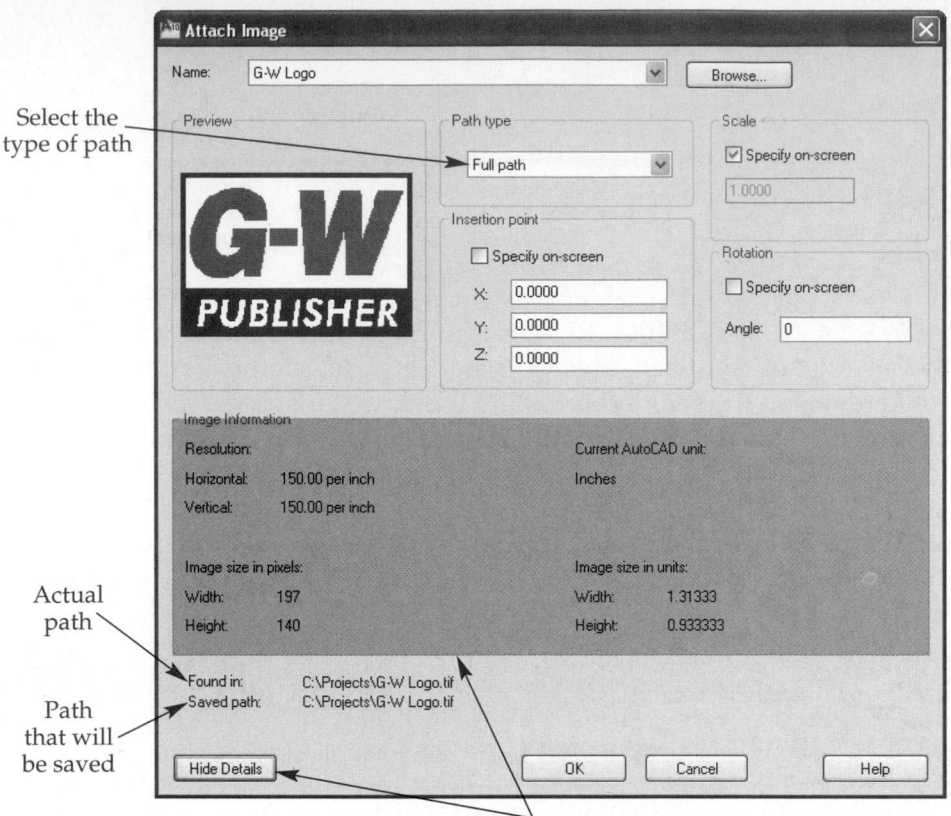

Picking this button toggles the display of the lower portion

Figure 20-4.
The raster image attached to an AutoCAD drawing.

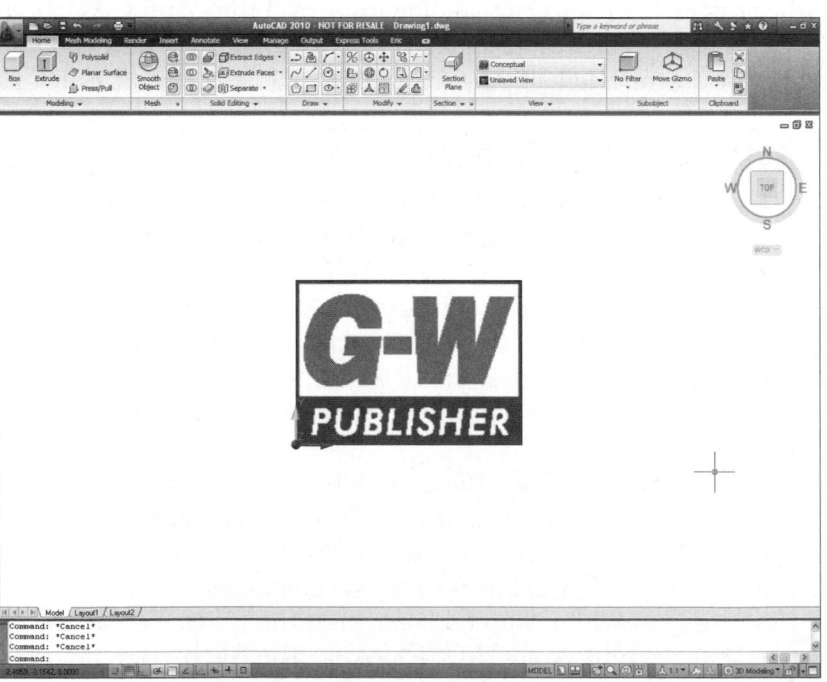

AutoCAD and Its Applications—Advanced

The *no path* option tells AutoCAD that the image is located in the same folder as the drawing. If the image file is not found in that folder, AutoCAD looks in the path specified by the **PROJECTNAME** system variable, then in the support files search path defined in the **Files** tab of the **Options** dialog box.

The path to the image file is displayed at the bottom of the **Image Information** area. Also displayed is the path that will be saved in the drawing. The **Image Information** area is displayed by picking the **Show Details** button at the bottom of the dialog box.

You can preset image parameters (insertion point, scale, and rotation) or choose to specify them on-screen. Image resolution information is displayed in the **Image Information** area of the expanded dialog box. See **Figure 20-3.**

PROFESSIONAL TIP

If you are working on a project that uses xrefs and attached images, adding a "project subfolder" below the folder where the drawings are stored may be beneficial. Then, use relative paths when inserting images or xrefs. This allows all related files for a project to be found by AutoCAD, even if the folder structure is moved to a different drive or "root" folder.

Managing Attached Images

As stated earlier, the **External References** palette is used to control the raster images inserted into a drawing. When you hover over the file name in the **File References** area, the preview window displays the image name, its status (loaded or unloaded), file size, type, date the image was last saved, and the saved path. Refer to **Figure 20-1.**

Right-clicking on the image name displays a shortcut menu containing options to help you manage the image. The five options are:
- **Open.** This opens the image in the program associated with the file type of the image. For example, if Microsoft Photo Editor is associated with the TIFF file type, the TIFF image is displayed in this program.
- **Attach.** This opens the **Attach Image** dialog box, discussed in the previous section, for attaching an additional image to the drawing.
- **Unload.** Unloads the selected image, but retains its path information. The Status column displays Unloaded if this option is selected. Display the list view to see the columns. An unloaded image is displayed as a frame until reloaded.
- **Reload.** Reloads the selected image file.
- **Detach.** Removes, or detaches, the selected image file from the drawing.

Right-clicking in **File References** area, but not on a file name, displays a different shortcut menu. See **Figure 20-5.** This shortcut menu contains ten options:
- **Reload All References.** Reloads any files attached to the current drawing.
- **Select All.** Selects all of the files listed in the **File References** area.
- **Attach DWG.** Allows you to attach other drawings as xrefs.
- **Attach Image.** Allows you to attach additional images.
- **Attach DWF.** Allows you to attach a DWF file as an xref.
- **Attach DGN.** Allows you to attach a DGN file (Microstation drawing) as an xref.
- **Attach PDF.** Allows you to attach a PDF (Adobe Acrobat) file.
- **Tooltip Style.** The tooltip (help text) that appears when you hover over one of the items is customizable. The tooltip can display the file name only, a preview only in the size you select, file details (including the name), or details and a preview.
- **Preview/Details Pane.** Turns on or off the pane at the bottom of the **External References** palette. If the tooltip is set to display details and a preview, you probably will not need to display the preview/details pane.
- **Close.** Closes the **External References** palette.

Figure 20-5.
The tooltip in the **External References** palette can be configured to display various information.

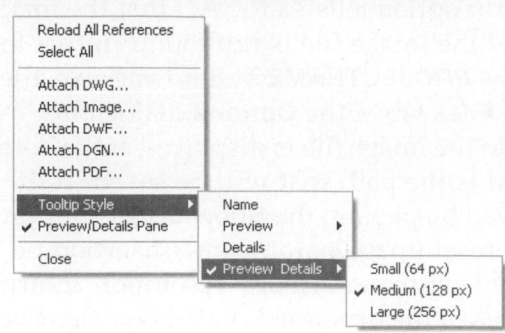

If drawing-regeneration time is becoming long, unload attached image files that are not needed for the current drawing session.

Controlling Image File Displays

Once an image is attached to the current drawing, its display can be adjusted if needed. The **IMAGECLIP**, **IMAGEADJUST**, **IMAGEQUALITY**, **TRANSPARENCY**, and **IMAGEFRAME** commands are used to do so. These commands are discussed in this section. All of these commands (except **IMAGEQUALITY**) are available in the **Image** contextual tab. This tab is displayed in the ribbon when the border of the image is selected. See **Figure 20-6.**

Clipping an image

Ribbon
Image
> Clipping

Create Clipping
Boundary

Type
IMAGECLIP
CLIP
ICL

The **IMAGECLIP** command allows you to trim away a portion of the image that does not need to be seen. The clipping frame, also called a *clipping path*, can be rectangular or polygonal. By default, the **IMAGECLIP** command removes portions of the image outside of the boundary. You can remove portions inside of the boundary with the **Invert Clip** option.

Once the command is selected, you are prompted to pick the image to clip, unless the image is selected when the command is selected. To create a rectangular clipping frame, continue:

```
Outside mode – Objects outside boundary will be hidden.
Specify clipping boundary or select invert option:
[Select polyline/Polygonal/Rectangular/Invert clip] <Rectangular>: R↵
Specify first corner point: (pick the first corner of the clipping boundary)
Specify opposite corner point: (pick the second corner)
```

Figure 20-6.
The **Image** contextual tab is displayed in the ribbon when an image is selected.

Settings for adjusting the image

Remove a clipping path

Change the image transparency

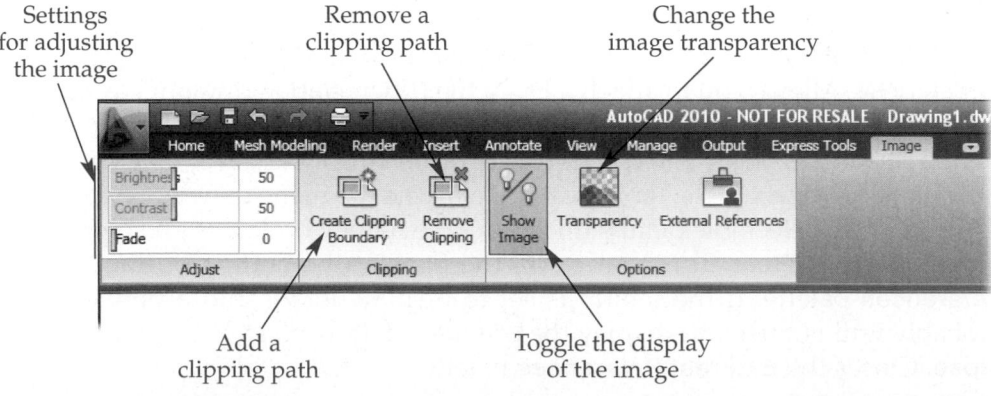

Add a clipping path

Toggle the display of the image

AutoCAD and Its Applications—Advanced

The image outside of the rectangular frame is hidden. The image is selected and grips are displayed at the corners with an arrow grip on the left-hand edge. Pick and drag the corner grips to change the size of the clip boundary. Picking the arrow grip inverts what is hidden. This is the same as using the **Invert Clip** option.

The **Polygonal** option allows you to construct a clipping frame composed of three or more points. Select the command, pick the image to clip, and continue:

```
Enter image clipping option [ON/OFF/Delete/New boundary] <New>: N↵
Outside mode – Objects outside boundary will be hidden.
Specify clipping boundary or select invert option:
[Select polyline/Polygonal/Rectangular/Invert clip] <Rectangular>: P↵
Specify first point: (pick first point to be used for the clipping boundary)
Specify next point or [Undo]: (pick second point)
Specify next point or [Undo]: (pick third point)
Specify next point or [Close/Undo]: (pick additional points as needed)
Specify next point or [Close/Undo]: ↵
```

The image outside of the polygonal frame is hidden. The grips function the same as they do with a rectangular boundary. **Figure 20-7** shows the results of using the **Rectangular** and **Polygonal** options of the **IMAGECLIP** command on a raster image. Three additional options of **IMAGECLIP** allow you to work with the display of the clipped image.

- **ON.** Turns the clipping frame on to display only the clipped area.
- **OFF.** Turns off the clipping frame to display the entire original image and frame.
- **Delete.** Deletes the clipping frame and displays the entire original image.

NOTE

You can pick an unclipped image frame to display the grips for editing. The grips are attached to the image itself. If one grip is stretched, it affects the entire image by proportionally enlarging or reducing it. On the other hand, if you select a clipped image for grip editing, the grips are attached to the clipping frame. Stretching the clipping frame does not change the size or shape of the image, but alters the frame and retains the size of the image.

Figure 20-7.
A—A rectangular image clip. B—A polygonal image clip. The path is shown here in color for illustration.

Polygonal path

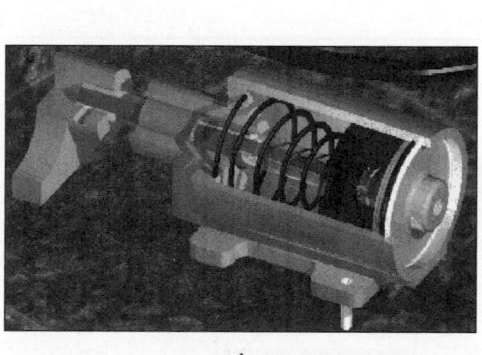
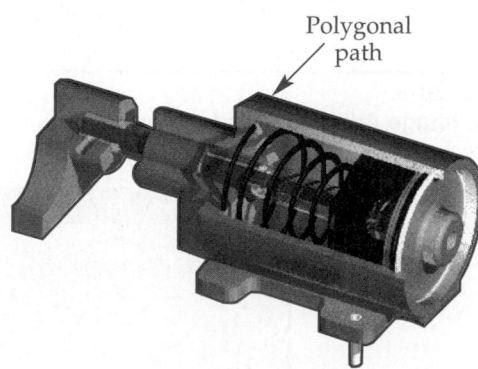

A   B

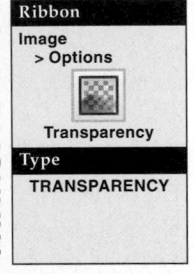

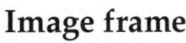

Adjusting an image

The **IMAGEADJUST** command provides control over the brightness, contrast, and fade of the image. These adjustments are made in the **Adjust** panel on the **Image** contextual tab in the ribbon, **Figure 20-6**. The adjustments are applied to all selected images.

If the command is accessed by typing, the **Image Adjust** dialog box is displayed, **Figure 20-8**. Once the command is selected, you are prompted to pick an image, unless the image is selected before the command is accessed. If you want the same settings applied to multiple images, you can pick them all at the same time. When done picking objects, press [Enter] to display the dialog box.

Values can be changed by typing in the text boxes or by using the slider bars in either the ribbon or dialog box. The preview tile in the dialog box dynamically changes as the sliders are moved. Picking the **Reset** button in the dialog box returns all values to their defaults.

- **Brightness.** Controls pixel whiteness and indirectly affects the contrast. Values can range from 0 to 100, with 50 as the default value. Higher values increase the brightness.
- **Contrast.** Controls the contrast of the image, or how close each pixel is moved toward its primary or secondary color. Values can range from 0 to 100, with 50 as the default value. Higher values increase the contrast.
- **Fade.** Controls the fading of the image, or how close the image is to the background color. Values can range from 0 to 100, with 0 as the default value. Higher values increase the fading.

The **IMAGEQUALITY** command provides two options: **High** and **Draft**. The high quality setting produces the best image display, but requires more time to regenerate. If you are working with several images in a drawing, it is best to set the **Draft** option current. The image displayed is lower quality, but requires less time to display. The setting applies to all images in the drawing.

Transparency

Some raster images have transparent background pixels. The **TRANSPARENCY** command controls the display of these pixels. If **TRANSPARENCY** is on, the drawing will show through the image background. Images are inserted with this feature turned off. The setting applies to individual images. Multiple images can be selected at the same time. Remember, only images containing transparent pixels are affected.

Image frame

The **IMAGEFRAME** command controls the appearance of frames around all images in the current drawing. When attaching (inserting) images, AutoCAD places a frame around the image in the current layer color and linetype. There are three settings for

---

Figure 20-8.
In the **Image Adjust** dialog box, brightness, contrast, and fade values can be numerically entered or set using the sliders.

the **IMAGEFRAME** command. The default setting of 1 turns on the display of the frame and allows the frame to be plotted. A setting of 2 turns on the display of the frame, but the frame is not plotted. A setting of 0 turns off the display of the frame. It is not plotted and cannot be selected for editing. The setting applies to all images in the drawing.

Uses of Raster Files in AutoCAD

One use of raster images is as a background for sketching or tracing. For example, you may need a line drawing of an image that is only available as a continuous tone (print) photograph. The photo can be scanned, which produces a raster image. After importing the raster image with the **IMAGEATTACH** command, use the appropriate drawing commands to sketch or trace the image. After the object is sketched, the original raster image can be deleted, frozen, or unloaded, leaving the tracing. You can then add other elements to the tracing to create a full drawing. See **Figure 20-9.**

Raster files can be combined with AutoCAD drawing and modeling features in many ways to complete or complement the design. For example, company watermarks or logos can be easily added to title blocks, style sheets, and company drawing standards. Drawings that require designs, labels, and a variety of text fonts can be created using raster files in conjunction with the wide variety of TrueType fonts available with AutoCAD. Archived manual drawings can also be scanned, brought into AutoCAD, and then traced to create a CAD drawing.

Figure 20-9.
Using a raster image as a model for a drawing. A—The imported raster image. B—Use AutoCAD commands to trace the image. Then, either delete the image or freeze its layer. C—The completed drawing plotted on a title block.

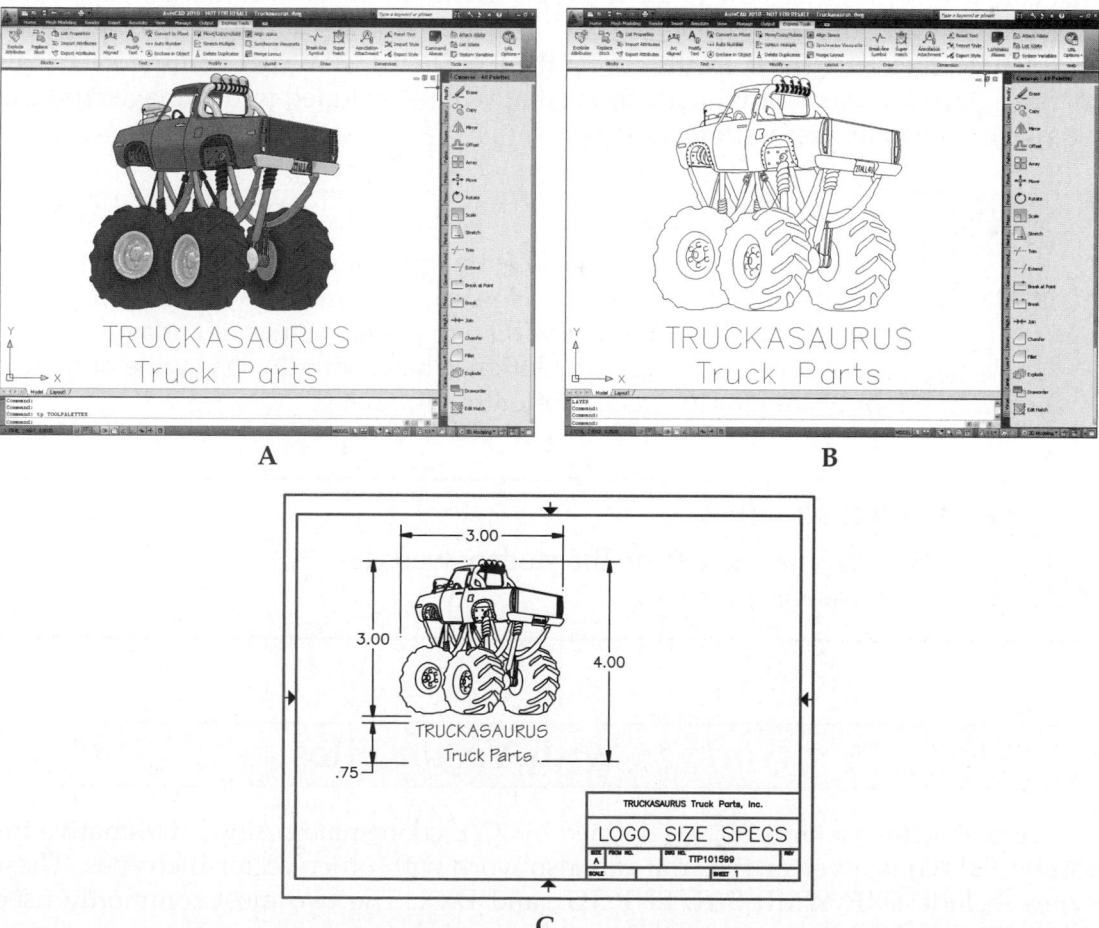

You can add features to complement raster files. For example, you can import a raster file, dimension or annotate it, and even add special shapes to it. Then, export it as the same type of file. Now, you can use the revised file in the original software in which it was created. As with any creative process, let your imagination and the job requirements determine how you use this capability of AutoCAD.

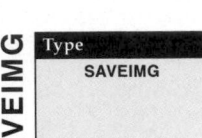

SAVEIMG

Type	
	SAVEIMG

Exercise 20-1

Complete the exercise on the student website.
www.g-wlearning.com/CAD

Exporting a Drawing to a Raster File

You can save a rendering to a raster file. This is discussed in Chapter 18. However, 2D objects are not rendered and, therefore, do not appear in the file. If you want what is displayed in the current viewport, including 2D objects, saved as a raster file, you must use the **SAVEIMG** command. This command saves the current AutoCAD viewport as an image file. What you see in the viewport is what you will get in the file, including the effect of the current visual style.

Once the command is selected, the **Render Output File** dialog box is displayed. This is a standard save dialog box. The **Files of type:** drop-down list displays the file types to which the image can be saved. Select the file type based on the type required for a particular process, application, job, or client. The best thing to do is ask whomever you are creating the file for which type of file will work best. After selecting the file type, give the file a name and pick the **Save** button. Another dialog box is displayed that contains settings specific to the file type. Make settings as needed and close this dialog box to save the file.

A BMP file can also be created using the **EXPORT** or **BMPOUT** command. In this manner, you can select individual objects that will be included in the image. You can also save shaded images with this method.

Exercise 20-2

Complete the exercise on the student website.
www.g-wlearning.com/CAD

Working with Vector Files

A vector file contains objects defined by XYZ coordinates. AutoCAD's native file format (DWG) is a vector file. You can also work with other vector file types. These types include DXF, WMF, SAT, EPS, STL, and DXX. The two most commonly used types, DXF and WMF, are covered in the next sections.

Exporting and Importing DXF Files

DXF is a generic file type that defines AutoCAD geometry in an ASCII text file. Other programs that recognize the DXF format can then "read" this file. The DXF file format retains the mathematical definitions of AutoCAD objects in vector form. The DXF objects imported into other vector-based programs, or opened in AutoCAD, can be edited as needed.

Exporting DXF files

The **DXFOUT** command is used to save a DXF file. Once the command is selected, the **Save Drawing As** dialog box is displayed. See **Figure 20-10.** Select the DXF file type from the **Files of type:** drop-down list. Name the file and specify a location where you want to save it. Notice that you can select different versions of DXF. This is to ensure that the file you save is "backward compatible." For example, if you are sharing the file with somebody using AutoCAD 2000, save the DXF as that version to ensure AutoCAD 2000 can read the file.

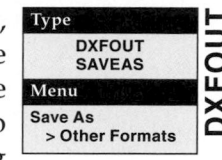

Since this is a "save as" operation, the current drawing is saved as a DXF file. If you continue to work on the drawing, you are working on the DXF version, *not* the DWG version. In order to work on the original drawing, you must open the DWG file. However, if you continue to work on the drawing in DXF form and attempt to save or close the drawing, the **Save Drawing As** dialog box is displayed. You can save the drawing as a DWG or replace the previously saved DXF file. If you save the drawing as a DXF file, you are also informed that the drawing is not a DWG and given the opportunity to save it in that format.

When a DXF file is saved, all geometry in the drawing is saved, regardless of the current zoom percentage or selected objects. The DXF file format saves any surfaced or solid 3D objects as 3D geometry. When a DXF file containing 3D geometry is opened, the surfaced or solid 3D geometry remains intact. In addition, the current visual style is saved in the DXF file.

> **NOTE**
>
> Not all programs that can import DXF files are capable of correctly "reading" 3D objects or the visual style.

Figure 20-10.
The **Save Drawing As** dialog box is used to save a DXF file.

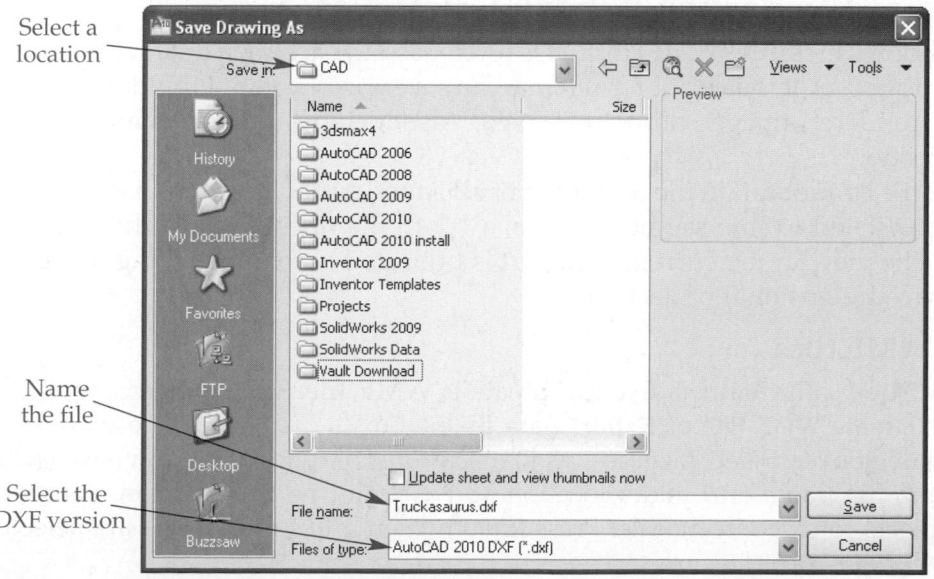

Figure 20-11.
The **Select File** dialog box is used to import a DXF file.

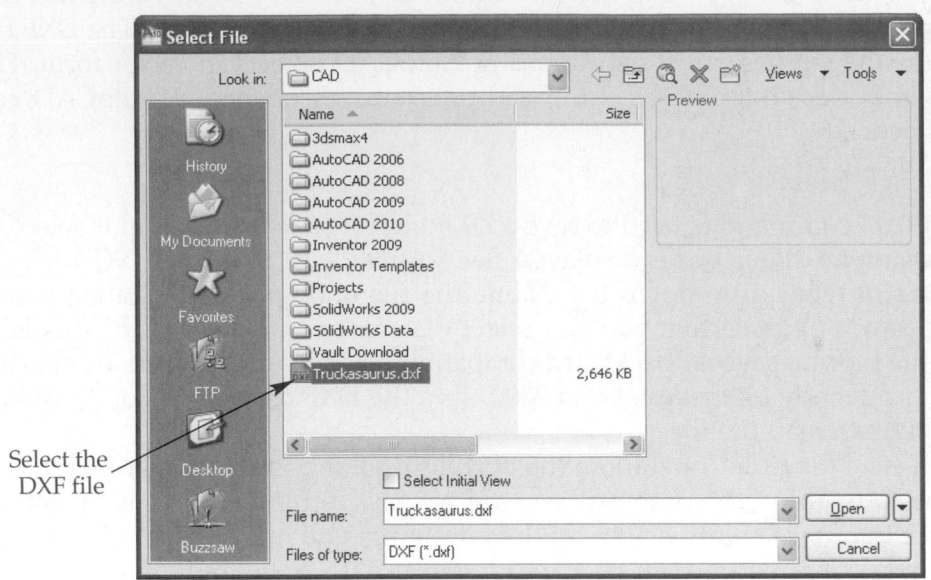

Select the
DXF file

Importing DXF files

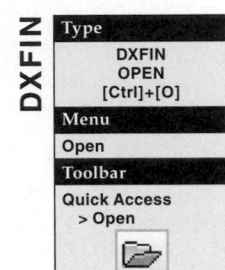
To open a DXF file, use the **DXFIN** command. Once the command is selected, the **Select File** dialog box is displayed, **Figure 20-11.** Select DXF (*.dxf) from the **Files of type:** drop-down list. Then, select the DXF file you want to open. Notice that there is no preview when the file is selected. AutoCAD does not support previews for the DXF file type. Finally, pick the **Open** button.

The DXF file is opened in a new document window. To place a DXF file into the *current* drawing, insert it as a block. If you do not want it inserted as a block, open the file (**DXFIN**), copy it to the clipboard ([Ctrl]+[C]), and paste ([Ctrl]+[V]) it into the current drawing.

> **NOTE**
>
> If you open a DXF file and try to save it, the **Save Drawing As** dialog box appears. You can save it as DXF, overwriting the existing file, or under a new name or as another file type.

Exporting and Importing Windows Metafiles

The Windows metafile (WMF) format is often used to exchange data with desktop publishing programs. It is a vector format that can save wireframe and hidden displays. Shaded and rendered images cannot be saved. Also, perspective views are saved in parallel projection.

A WMF file cannot retain the definition of all AutoCAD object types. For example, circles are translated to line segments. Also, a WMF file does *not* save three-dimensional data. The view in the current viewport is projected onto the viewing plane and saved as a two-dimensional projection.

Exporting WMF files

The **WMFOUT** command is used to create a WMF file. When the command is selected, the **Create WMF File** or **Export Data** dialog box is displayed. These are standard save dialog boxes. Select Metafile (*.wmf) in the **Files of type:** drop-down list. After specifying the file name and folder location and picking the **Save** button, you must select the objects to place in the file. Press [Enter] when all of the objects are selected and the WMF file is saved.

Only the portions of selected objects that are visible on-screen are written into the file. If part of a selected object is not visible on screen, that part is "clipped." Also, the current view resolution affects the appearance of a Windows metafile. For example, when **VIEWRES** is set low, circles in your AutoCAD drawing may look like polygons in the WMF file. When saved to a Windows metafile, curved objects are composed of line segments rather than defined as circles or arcs.

Importing WMF files

Use the **WMFIN** command to import a Windows metafile into a drawing. When the command is selected, the **Import WMF** or **Import File** dialog box is displayed. Select Metafile (*.wmf) in the **Files of type:** drop-down list, then select a file.

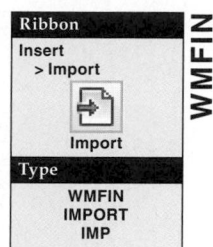

A Windows metafile is imported as a block consisting of all of the objects in the file. You can explode the block if you need to edit the objects within it. If an object is not filled, it is created as a polyline when brought into AutoCAD. This includes arcs and circles. Objects composed of several closed polylines to represent fills are created from solid fill objects, as if created using the **SOLID** command with the **FILL** system variable off.

There are two settings used to control the appearance of Windows metafiles imported into AutoCAD. Type WMFOPTS to display the **WMF In Options** dialog box, **Figure 20-12.** You can also pick the **Options...** button in the **Tools** drop-down menu in the "import" dialog box. The **WMF In Options** dialog box contains the following two check boxes.

- **Wire Frame (No Fills).** When checked, filled areas are imported only as outlines. Otherwise, filled areas are imported as filled objects (when **FILL** is on).
- **Wide Lines.** When this option is checked, the relative line widths of lines and borders from the WMF file are maintained. Otherwise, they are imported using a zero width.

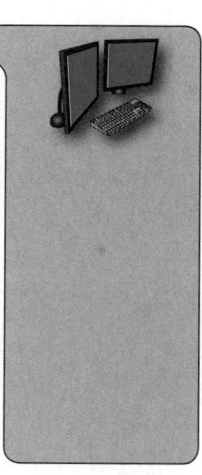

PROFESSIONAL TIP

PostScript is a copyrighted page description language developed by Adobe Systems. PostScript files are widely used in desktop publishing. AutoCAD can export PostScript files with the **PSOUT** or **EXPORT** commands. In addition, AutoCAD has several PostScript patterns that can be used as fills for closed polylines. The **PSFILL** command is used to add these patterns. However, professionals rarely use AutoCAD's PostScript functions. AutoCAD cannot import, view, or print PostScript files. In addition, any drawing that contains PostScript patterns (fills) must be sent to a printer or plotter that is PostScript compatible. These printers are not common outside of the graphics industry.

Exercise 20-3

Complete the exercise on the student website.
www.g-wlearning.com/CAD

Figure 20-12.
Setting options for imported WMF files.

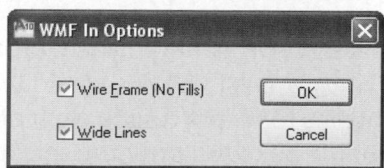

DWF and PDF Files

Often it is necessary to share AutoCAD drawing data with other "non-AutoCAD" systems. Drawings saved as DWF and PDF files are easily attached to an e-mail or saved to a company server where others can access them. With the associated viewing software, these files can be opened, displayed, and printed as desired without the use of AutoCAD.

You can save an AutoCAD drawing as a *Design Web Format (DWF)* file. A DWF file is a highly compressed, vector file that can be viewed using the Autodesk Design Review or Autodesk DWF Viewer. In addition, when one of these programs is installed in conjunction with a supported Internet browser, you can view DWF files on the web. Autodesk Design Review is installed when AutoCAD is installed or it can be downloaded for free from the Autodesk website.

As of AutoCAD 2009, you can also save a drawing in DWFx format. A DWFx file serves the same purpose as a "standard" DWF file, but it is based on Microsoft's XML Paper Specification (XPS) format. DWFx files can be viewed with the Autodesk Design Review or with the Microsoft XPS Viewer which is supplied with the Microsoft Vista operating system. The process for creating and using DWFx files is the same as for DWF files. When a DWF or DWFx file is opened in Autodesk Design Review, the view cube and orbit commands are accessible to view the 3D object from different angles.

You can export an AutoCAD drawing as a *Portable Document Format (PDF)* file. A PDF file is a vector-based file, like DWFx and DWF. Anyone can view them using Adobe Reader. This is a free utility that can be downloaded from the Adobe website.

Drawings are saved as DWF, DWFx, or PDF files using either the **3DDWF**, **EXPORT**, or **PUBLISH** command. If you are creating a single-sheet file, use the **3DDWF** or **EXPORT** command. If you are creating a multi-sheet file, use the **PUBLISH** command.

PROFESSIONAL TIP

The **ETRANSMIT** command prepares a transmittal for e-mail that contains font files, plot styles, table files, and xrefs associated with the drawing. This can be a time-saving feature if you share drawings that contain xrefs.

Exporting to DWF and DWFx

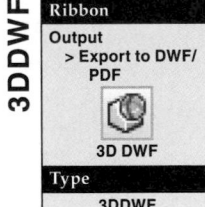

3DDWF

Ribbon
Output
> Export to DWF/PDF

3D DWF

Type
3DDWF

The **3DDWF** command must be used to retain the 3D data in the DWF or DWFx file. When the **3DDWF** command is launched, the **Export 3D DWF** dialog box is displayed. See **Figure 20-13A.** This is a standard save dialog box. You can select either the DWFx or DWF format from the **Files of type:** drop-down list. Then, enter a file name and select a location.

If you save the file at this point, all of the objects in model space will be exported. Selecting **Options...** in the **Tools** pull-down menu in this dialog box displays the **3D DWF Publish** dialog box, **Figure 20-13B.** In this dialog box, you can select which objects you wish to export and also whether or not to include materials.

To create the file, pick the **Save** button in the **Export 3D DWF** dialog box. Once the file is saved, a message appears indicating this fact and offering you the opportunity to view the file.

Settings for Exporting to DWF, DWFx, and PDF Files

The **Export to DWF/PDF** panel in the **Output** tab on the ribbon provides different options for exporting to DWF, DWFx, or PDF files, **Figure 20-14.** The export flyout button contains buttons for exporting to each of these file formats. Each button opens a standard save as dialog box with only the selected file type available. All other options are the same. The **Save As DWFx** dialog box is shown in **Figure 20-15.**

Figure 20-13.
A—The **3DDWF** command displays a standard save dialog box. You can set options using the **Tools** pull-down menu. B—Setting options for 3D DWF files.

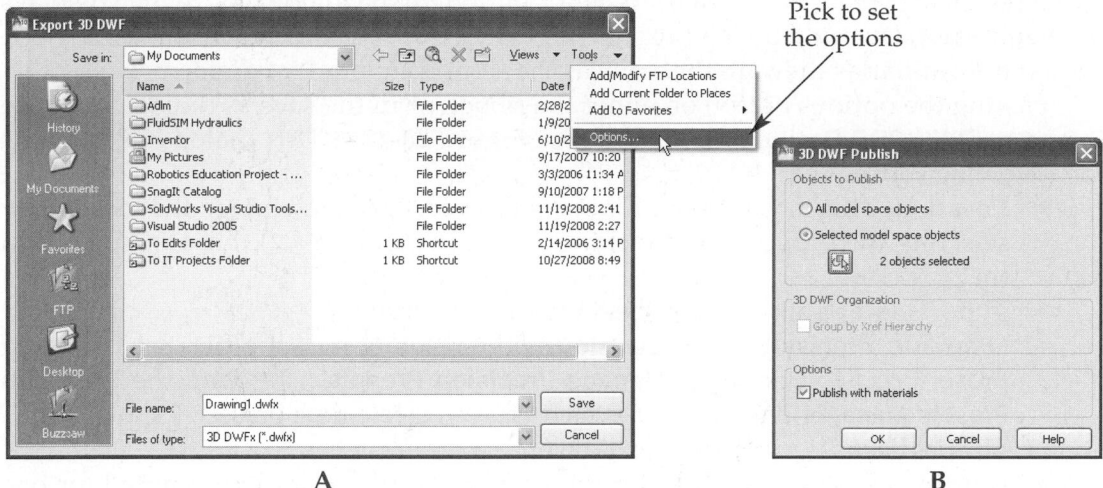

Pick to set the options

A

B

Figure 20-14.
The **Export to DWF/PDF** panel provides tools for exporting the AutoCAD drawing to DWF, DWFx, and PDF formats.

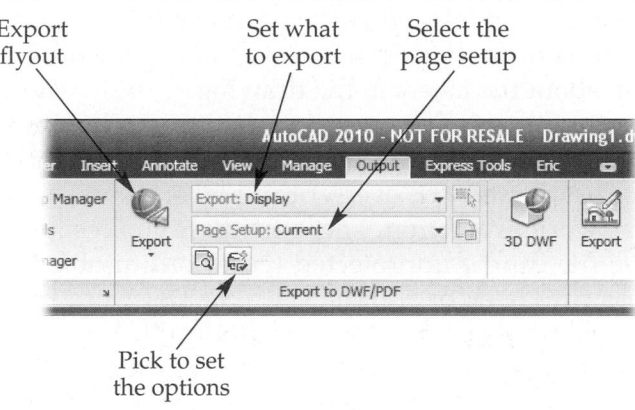

Export flyout

Set what to export

Select the page setup

Pick to set the options

Figure 20-15.
The **Save As DWFx** dialog box is used to save a drawing as a DWFx file.

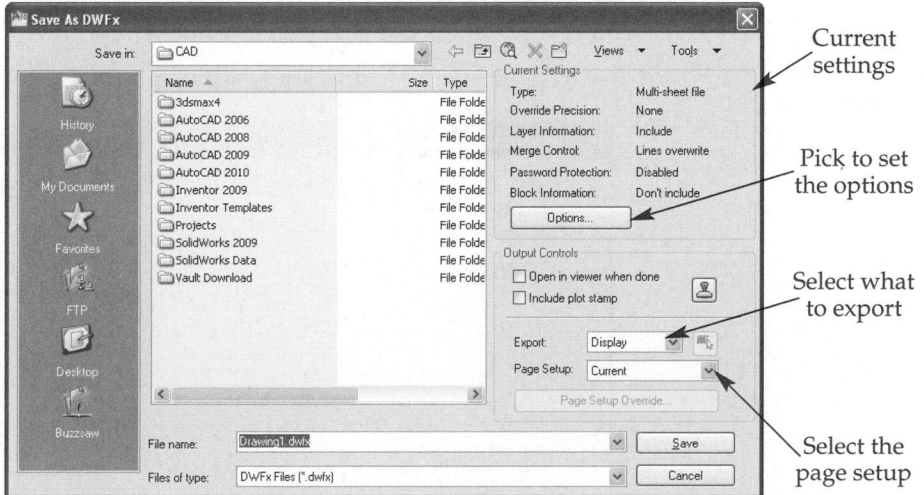

Current settings

Pick to set the options

Select what to export

Select the page setup

You can select what to export in either the panel or the dialog box. Your options are display, extents, or window. If you select Current for the page setup, AutoCAD creates the exported file with whatever page setup is current in the drawing. Selecting Override enables up the Page Setup Override… button (in the dialog box), which displays the **Page Setup Override** dialog box. See **Figure 20-16.** The settings in this dialog box should be familiar as they are the same as the settings in the **Plot** dialog box.

Picking the options button on either the ribbon or in the save as dialog opens the **Export to DWF/PDF Options** dialog box. See **Figure 20-17A.** This dialog box contains the following settings.

- **Location.** This is where the file will be saved. You can change the location by selecting the property and picking the browse button (...) that is displayed in the property.
- **Type.** Select either a single sheet or multi-sheet file.
- **Override Precision.** The precision is the resolution of the file. Select from a preset precision or select **Manage Precision Presets…** to open the **Precision Presets Manager** dialog box in which you can create new presets, **Figure 20-17B.** This is discussed in more detail below.
- **Naming.** Select either to enter a name for the file or to be prompted for one later.
- **Name.** This property is read-only (N/A) unless you have selected to specify a name. In this case, enter the file name here.
- **Layer information.** You can elect to include or not include in the file information about the layers in the drawing.
- **Merge control.** This property determines how the colors of overlapping lines are handled. You can choose to merge (blend) the colors or to have the color of the last plotted line overwrite all other colors.
- **Password protection.** There are three choices for password-protecting the file. The file can be unprotected (password disabled), you can be prompted for a password later, or you can specify a password in this dialog box.
- **Password.** This property is read-only (N/A) unless you have selected to specify a password. In this case, enter the password here.
- **Block information.** You can choose to include or not include in the file attribute information for blocks in the drawing.
- **Block template file.** This property is read-only (N/A) unless you have selected to include block information. If block attribute information is to be included,

Figure 20-16.
The **Page Setup Override** dialog box has options similar to the **Plot** dialog box.

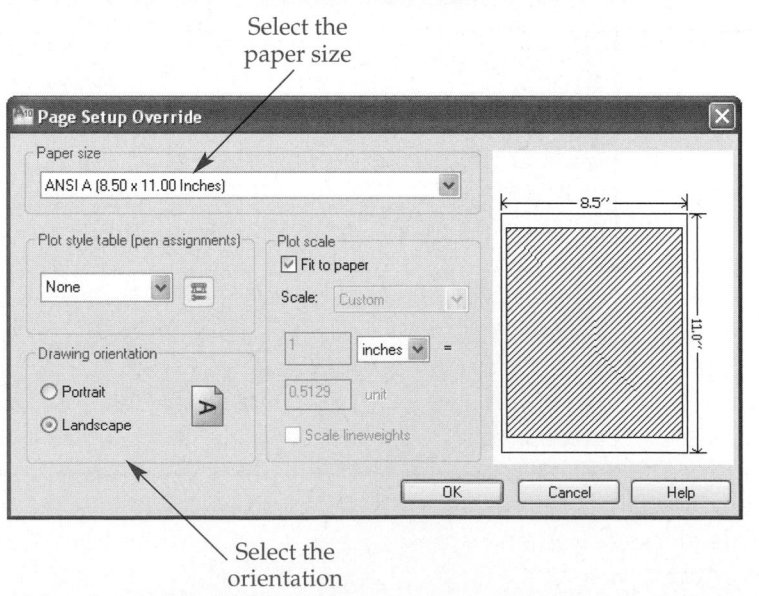

Select the paper size

Select the orientation

Figure 20-17.
A—Setting options in the **Export to DWF/PDF Options** dialog box. B—The **Precision Presets Manager** is used to create new presets.

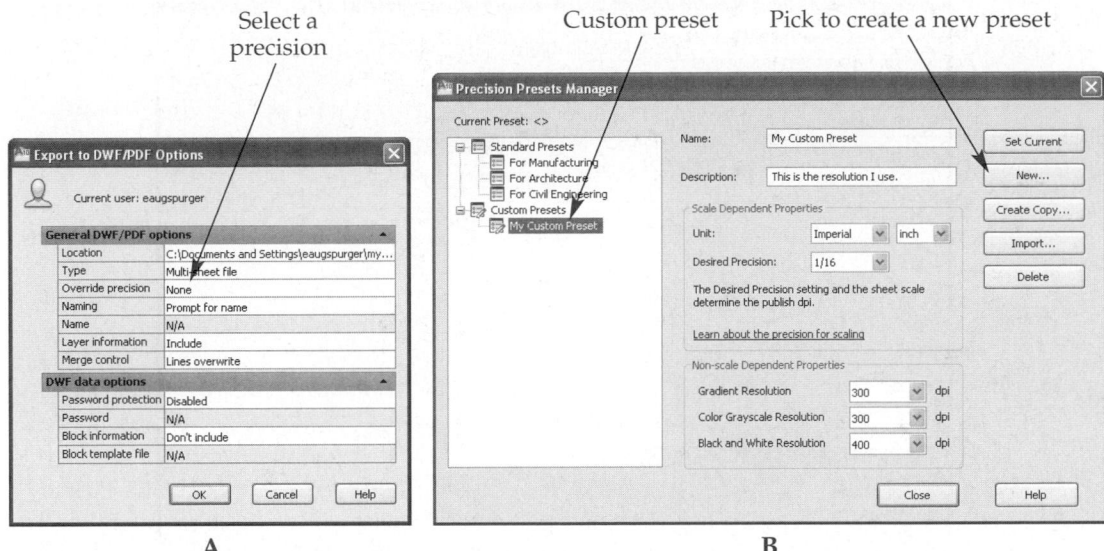

A

B

this property displays a drop-down list and a browse button (...) so that you can either select, create, or edit a block template (DXE) file.

The Override precision property contains standard precision presets for manufacturing, architectural, or civil drafting disciplines. This helps to create scaled prints when the DWF or PDF files are printed. The Manage Precision Presets... option opens the **Precision Presets Manager** dialog box, shown in **Figure 20-17B**. In this dialog box, you can view the discipline-specific presets or create new custom presets with your own settings. The default presets are read-only and cannot be changed. The properties that do not depend on scale—gradient, color grayscale, and black and white resolutions—determine the quality of the resulting DWF or PDF files.

Exercise 20-4

Complete the exercise on the student website.
www.g-wlearning.com/CAD

Publishing DWF, DWFx, and PDF Files

To *publish* a DWF, DWFx, or PDF file, first enter the **PUBLISH** command. The **Publish** dialog box is displayed, **Figure 20-18**. Select the file format in the **Publish to:** drop-down list in the upper-left corner of the dialog box. To change settings for the file that will be published, pick the **Publish Options...** button. This displays the **Publish Options** dialog box. The options in this dialog box are the same as those described above for the **Export to DWF/PDF Options** dialog box. Change any settings as necessary and pick the **OK** button to return to the **Publish** dialog box.

When you are ready to create the file, pick the **Publish** button in the **Publish** dialog box. A standard save dialog box appears. Enter a name for the file, select a location, and pick the **Select** button. You cannot change the file type in this dialog box. Next, you are asked if you want to save the current list of sheets. After you make a selection in this message box, the drawing is published to the file. This may take a few seconds to complete.

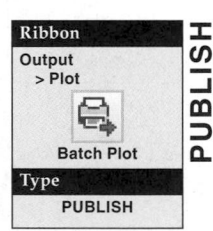

Ribbon
Output
> Plot

Batch Plot

Type
PUBLISH

PUBLISH

Figure 20-18.
Using the **PUBLISH** command to create a DWF, DWFx, or PDF file.

Choose
DWF, DWFx,
or PDF

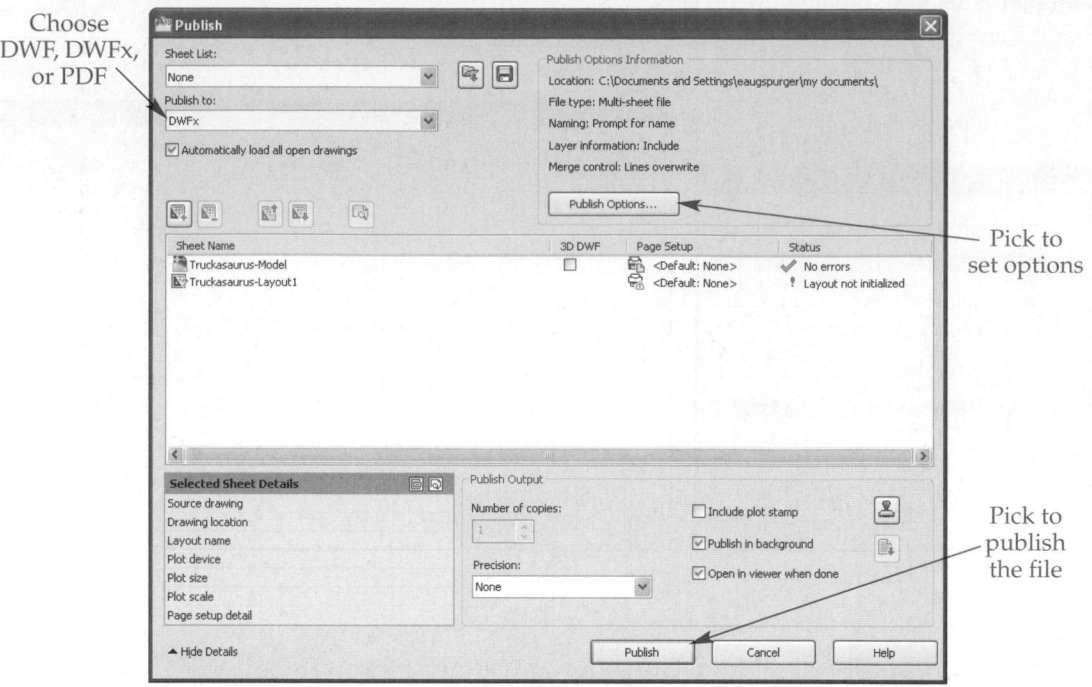

Pick to
set options

Pick to
publish
the file

Commands that control the on-screen display of geometry, such as **VIEWRES**, **FACETRES**, and **DISPSILH**, and the current visual style affect the resulting DWFx or DWF file.

Plotting DWF, DWFx, and PDF Files

Instead of publishing DWF, DWFx, and PDF files, you can *plot* them using the **PLOT** command. Use the command to open the **Plot** dialog box. In the **Name:** drop-down list in the **Printer/plotter** area, select the DWF6 ePlot.pc3 (for DWF files), the DWFx ePlot.pc3 (for DWFx files), or DWG To PDF.pc3 (for PDF files).

As discussed in the previous sections, there are settings that control the final output for these files. After the PC3 file is selected, pick the **Properties...** button in the **Printer/plotter** area. In the **Plotter Configuration Editor** dialog box that is displayed, select Custom Properties in the tree on the **Device and Document Settings** tab. Then, pick the **Custom Properties...** button to display the **DWF6 ePlot Properties**, the **DWFx ePlot Properties**, or the **DWG to PDF Properties** dialog box. The latter is shown in **Figure 20-19**.

The two "resolution" areas in the properties dialog box have settings that control the accuracy of the resulting file. A medium resolution is best in most cases. A file created with high resolution may be too large for practical electronic transmission. A lower resolution will create a smaller file. Small files make for easy electronic transmission. However, the resulting file may not display as accurately as one created at a higher resolution. You can separately set a maximum resolution for vector graphics, gradients, color/grayscale images, and black and white images. See **Figure 20-20** for a comparison of resolution settings in a plotted PDF file. The file size increases as the resolution increases.

Once you have made all settings as needed, pick the **OK** button. Then, pick **OK** in the **Plotter Configuration Editor**. If changes were made to the settings, a dialog box

Figure 20-19.
Setting the properties for a PDF file in the **DWG To PDF Properties** dialog box.

Resolution settings

DWG To PDF Properties

Vector and Gradient Resolution (dpi)
Vector resolution:
600 dpi
Custom vector resolution:
40000 dpi

Gradient resolution:
400 dpi
Custom gradient resolution:
200 dpi

Raster Image Resolution (dpi)
Color and grayscale resolution:
400 dpi
Custom color resolution:
200 dpi

Black and white resolution:
400 dpi
Custom black and white resolution:
400 dpi

Font Handling
○ Capture none   ⊙ Capture some   ○ Capture all
Edit Font List...   ☐ As geometry

Additional Output Setting
☑ Include layer information
☑ Open in PDF viewer when done

OK   Cancel   Help

Figure 20-20.
A comparison of low-resolution and high-resolution PDF files. A—The lines in the low-resolution file have jaggies. This is especially apparent on the windows. B—The lines in the high-resolution file are cleaner.

A

B

appears asking if you want to apply the changes on a one-time basis or save the configuration to a PC3 plotter configuration file.

Use all of the other settings in the **Plot** dialog box just as you would when plotting a hard copy. Refer to *AutoCAD and Its Applications—Basics* for detailed information on plotting. When you pick the **OK** button to "plot", the **Browse for Plot File** dialog box is displayed. This is a standard save dialog box with only the selected file type available. The default filename is the drawing name and current space name separated by a hyphen. Use that name or enter a new name, navigate to the location where you want to save the file, and pick the **Save** button.

PROFESSIONAL TIP

Using the **PLOT** command to produce DWF, DWFx, or PDF files is a better option than exporting when you want to include information in your layout such as borders, title blocks, notes, dimensions, etc. Also, remember that the floating viewports in a layout may be set to any visual style or rendering preset. This gives you the capability of creating a fully rendered DWF, DWFx, or PDF file from your 3D model.

Exercise 20-5

Complete the exercise on the student website.
www.g-wlearning.com/CAD

Chapter Test

Answer the following questions. Write your answers on a separate sheet of paper or complete the electronic chapter test on the student website.
www.g-wlearning.com/CAD

1. Name four common formats of raster images that can be imported into AutoCAD.
2. Which command allows you to attach a raster file to the current AutoCAD drawing?
3. What is the display status in the drawing of an inserted image that has been unloaded?
4. Which two shapes can be used to clip a raster image?
5. What is the function of the **IMAGEADJUST** command?
6. Name two commands that allow you to export bitmap files.
7. Give the name and file type of the vector file that can be exchanged between object-based programs (object definitions are retained), other than DWG.
8. Name the commands that allow you to import and export the file type in question 7.
9. How are three-dimensional objects treated when exported to a WMF file?
10. How can a DXF file be inserted into the *current* drawing?
11. The DWF, _____, and _____ are three file types commonly used to share AutoCAD drawing information with people who do not have AutoCAD.
12. Give two examples of software products that may be used to view DWFx files.
13. Why would you use the **3DDWF** command to export to a DWF or DWFx file?
14. If you are using the **PLOT** command to create a PDF file, what must be selected in the **Name:** drop-down list in the **Printer/plotter** area of the **Plot** dialog box?
15. When plotting to a DWF, DWFx, or PDF file, what is a disadvantage to using high resolution settings? What is a disadvantage of using low resolution settings?

Drawing Problems

1. Locate some sample raster files with .jpg, .png, or .tif file extensions. These files are often included as samples with software. They can also be downloaded as freeware from the Internet. With the permission of your instructor or supervisor, create a folder on your hard drive and copy the raster files there. Create a new drawing and attach each of the raster files. Place text below each image indicating the source of the image. Save the drawing as P20_01.

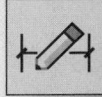

2. Choose a small raster file and attach it to a new AutoCAD drawing.
   A. Insert the image so it fills the entire screen.
   B. Undo and insert the image again using a scale factor that fills half the screen with the image.
   C. Stretch the original object using grips, then experiment with different clipping boundaries. Stretch the image after it has been clipped and observe the result.
   D. Create a layer named Raster. Create a second layer named Object. Give each layer the color of your choice. Set the current layer to Raster.
   E. Import the same image next to the previous one at the same scale factor.
   F. Set the current layer to Object and use AutoCAD drawing commands to trace the outline of the second raster image.
   G. Unload the raster image or freeze the Raster layer.
   H. Save the drawing as P20_02.

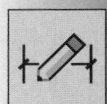

3. For this problem, you will import several raster files into AutoCAD. Then, you will trace the object in each file and save it as a block or wblock to be used on other drawings.
   A. Find several raster files that contain simple objects, shapes, or figures that you might use in other drawings.
   B. Create a template drawing containing Object and Raster layers.
   C. Import each raster file into AutoCAD on the Raster layer using the appropriate command. Set the Object layer current and trace the shape or objects using AutoCAD drawing commands.
   D. Detach the raster information, keeping only the traced lines of the object.
   E. Save the object as a block or wblock using an appropriate file-naming system.
   F. After all blocks have been created, insert each one into a single drawing and label each with its name. Include a path if necessary.
   G. Save the drawing as P20_03.
   H. Print or plot the final drawing.

4. In this problem, you will create a memo outlining your progress on a flange.
   A. Open drawing P17_03 from Chapter 17 and save it as P20_04.
   B. Set the 3D Hidden visual style current.
   C. Set the background color to white.
   D. Use the **SAVEIMG** command and save the scene as a monochrome BMP file.
   E. Render the scene using the Presentation render preset and a resolution of 320 × 480. Save the rendering as a BMP file.
   F. Open a word processor capable of importing BMP files, such as Microsoft Word or Wordpad.
   G. Write a memo related to the project. A sample appears below. The memo should discuss how you created the drawing, the BMP file, and the rendered file. Insert the BMP files as appropriate.
   F. Save the document as P20_04. Print the document.

MEMO

To: Otto Desque
From: Ima Drafter
Date: Thursday, March 14
Subject: Project Progress

Dear Otto,

I have completed the initial drawing. As you can see from the drawing shown here, the project is complying with design parameters. The drawing is ready for transfer to the engineering department for approval.

I have also included a rendered image of the project. The material spec'ed by the engineering department is represented in the rendering. This may help in evaluation of the design.

Respectfully,

Ima

---

5. Begin a new drawing.
   A. Insert the blocks you created in problem 3. Arrange them in any order.
   B. Add any notes you need to identify this drawing as a sheet of library shapes. Be sure each shape is identified with its file name and location (path).
   C. Create a PDF file of the drawing.
   D. Save the drawing as P20_05.
   E. Open the PDF in Adobe Reader and print it. If Adobe Reader is not installed, obtain permission from your instructor or supervisor to download and install it.

6. Add a raster image to one of your title block template drawings as a design element or a company logo. A sample is shown below. Import an existing raster image or create your own using a program such as Windows Paint. Save the template drawing.

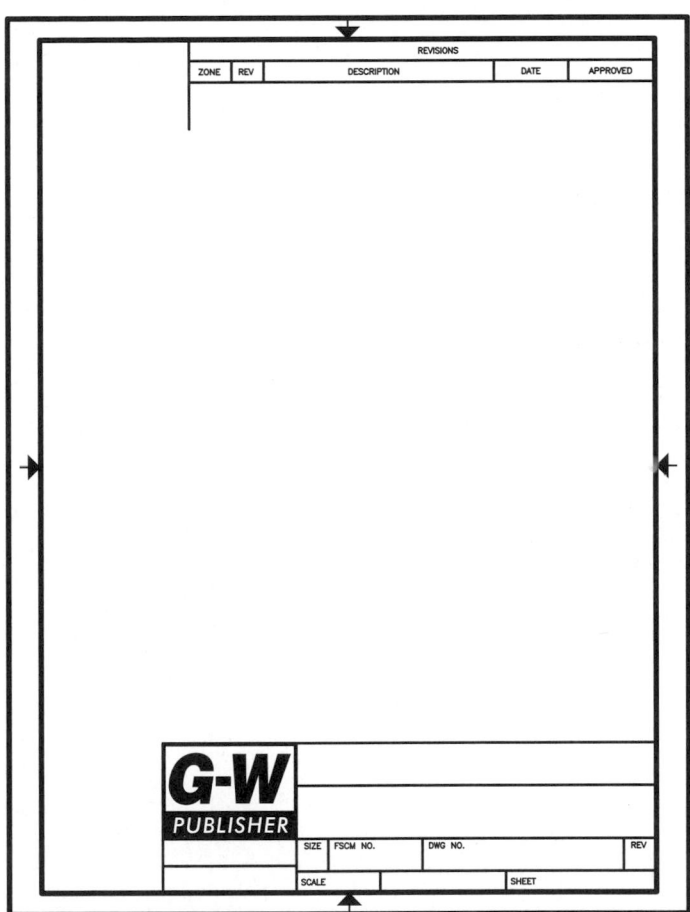

7. Open a 3D drawing from a previous chapter and save it as P20_07.
   A. Export the model to a DWF file. Save it as P20_07.dwf.
   B. Use the 3DDWF command to export the model as a 3D DWF file. Name this file P20_07.dwf.
   C. Use Autodesk Design Review to open the DWF files and compare them.
   D. Which viewing commands are only available in the 3D file?
   E. Note any other differences between the two files.

8. Open a 3D drawing from a previous chapter and save it as P20_08.
    A. Use the **PLOT** command to create a DWF plot file. Name it P20_08.dwf.
    B. Use the **PLOT** command to create a DWFx plot file. Name it P20_08.dwfx.
    C. Use the **PLOT** command to create a PDF plot file. Name it P20_08.pdf.
    D. Open these three files with either the Autodesk Design Review or Adobe Acrobat Reader.
    E. What are the differences between these files?
    F. Why would you use the **PLOT** command to create these files rather than exporting them as was done in problem 7?
    G. What capabilities are different between Autodesk Design Review and Adobe Acrobat Reader?

AutoCAD and Its Applications—Advanced

Customizing the AutoCAD Environment

Learning Objectives

After completing this chapter, you will be able to:

✓ Set system variables.
✓ Assign colors and fonts to the text and graphics windows.
✓ Control general AutoCAD system variables.
✓ Set options that control display quality and AutoCAD performance.
✓ Control shortcut menus.

AutoCAD provides a variety of options for customizing the user interface and working environment. These options permit users to configure the software to suit personal preferences. You can define colors for the individual window elements, assign preferred fonts to the command line window, control shortcut menus, and assign properties to program icons.

The options for customizing the AutoCAD user interface and working environment are found in the **Options** dialog box, **Figure 21-1.** This dialog box is commonly accessed by right-clicking in the drawing area or command line with nothing selected and no command active and picking **Options...** from the shortcut menu. It can also be displayed by picking the **Options** button at the bottom of the application menu (menu browser).

Changes made in the **Options** dialog box do not take effect until either the **Apply** or **OK** button is picked. If you pick the **Cancel** button or the close button (**X**) before picking **Apply**, all changes are discarded. Each time you change the options settings, the system registry is updated and the changes are used in this and subsequent drawing sessions. Settings that are stored within the drawing file have the AutoCAD drawing icon next to them. These settings do not apply to other drawings. Settings without the icon affect all AutoCAD drawing sessions.

Figure 21-1.
The **Options** dialog box is used to customize the AutoCAD working environment. Each tab contains a variety of options and settings.

Select the appropriate tab

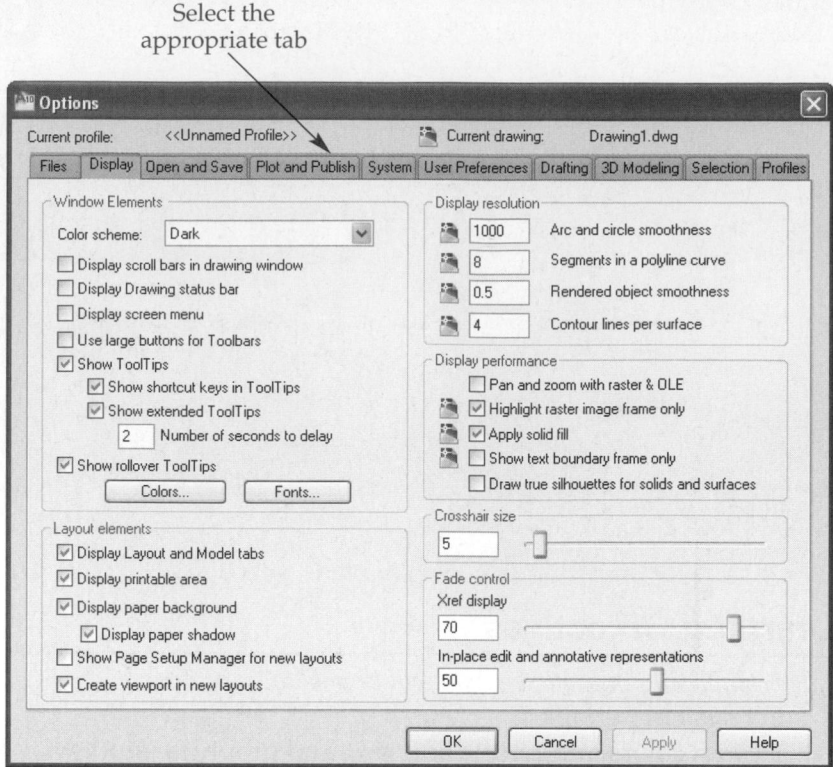

Setting AutoCAD System Variables

There are numerous settings that control the manner in which AutoCAD behaves in the Windows environment. These settings are made through the use of system variables. *System variables* are used to specify such items as which folders to search for driver and menu files and the location of temporary and support files. The default settings created during installation are usually adequate, but changing the settings may result in better performance. While several different options exist for setting many of the system variables, the simplest method is to use the **Options** dialog box.

File Locations

When AutoCAD is used in a network environment, some files pertaining to AutoCAD may reside on a network drive so all users can access them. Other files may reside in folders specifically created for a particular AutoCAD user. These files may include external reference files, custom menu files, and drawings containing blocks.

The **Files** tab of the **Options** dialog box is used to specify the path AutoCAD searches to find support files and driver files. It also contains the paths where certain types of files are saved and where AutoCAD looks for specific types of files. Support files include text fonts, menus, AutoLISP files, ObjectARX files, blocks, linetypes, and hatch patterns.

The folder names shown under the Support File Search Path heading in the **Search paths, file names, and file locations:** list are automatically created by AutoCAD during the installation. For example, **Figure 21-2** shows that the support files are stored in six different folders. Folders are searched in the order in which they are listed under Support File Search Path. As previously mentioned, some of these paths are created for a specific user. The first path listed is long and ultimately ends with the \Support folder. In the example shown, this path starts on the C: drive in a folder called \Documents and Settings. The folder listed immediately after the \Documents and Settings folder is the user-specific folder. The next folders to be searched are, in order, \Support, \Fonts, \Help,

Figure 21-2.
Folder paths can be customized in the **Files** tab of the **Options** dialog box.

Folder for
specific user

Add a new folder
to the selected path

Folders in
support
search
path

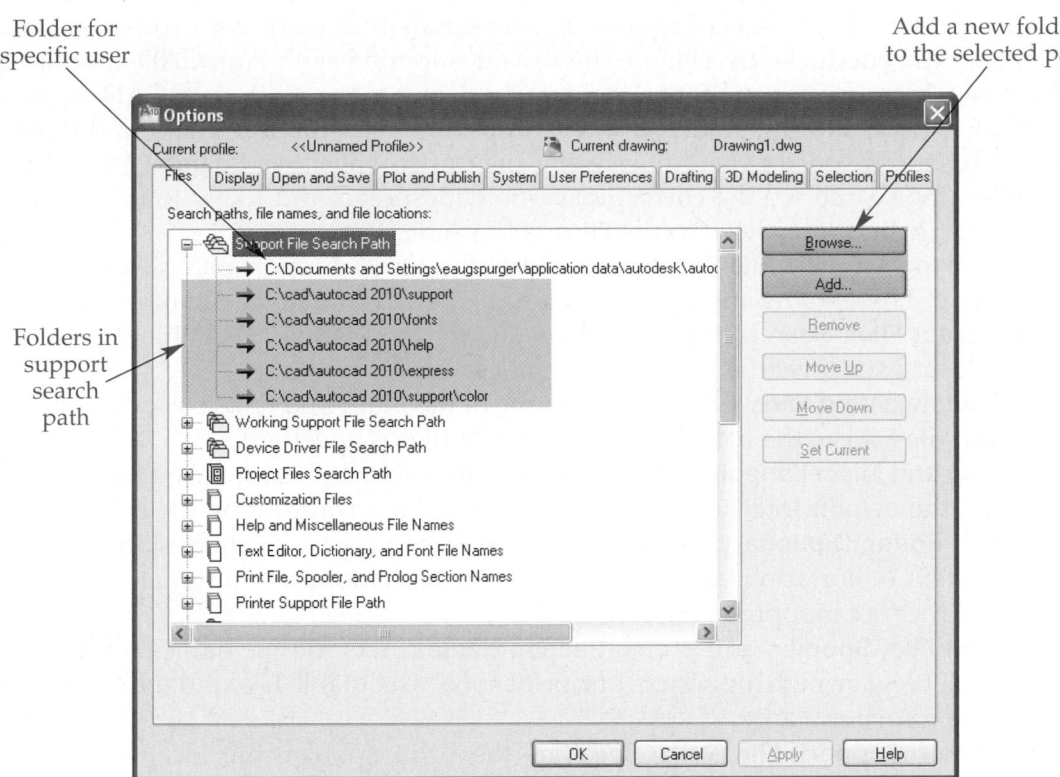

\Express, and \Support\Color. These paths are not user specific; they are located in the AutoCAD installation path.

You can add the path of any new folders you create that contain support files. As an example, suppose all of the blocks you typically use are stored in a separate folder named \Blocks. By placing this folder name in the support files search path, it is not necessary to include the entire path to the blocks in any command macros you create to automatically insert your blocks (this is discussed in Chapter 22). The path can also be omitted when using the Command: prompt with the file dialog box interface turned off.

A folder can be added to the existing search path in two ways. The first method is to highlight the Support File Search Path heading and pick the **Add...** button. This places a new, empty listing under the heading. You can now type C:\Blocks to complete the entry. Alternately, instead of typing the path name, after picking **Add...** you can pick the **Browse...** button to display the **Browse for Folder** dialog box. You can then use this dialog box to select the desired folder. The new setting takes effect as soon as you pick **Apply** or the **OK** button and close the **Options** dialog box.

PROFESSIONAL TIP

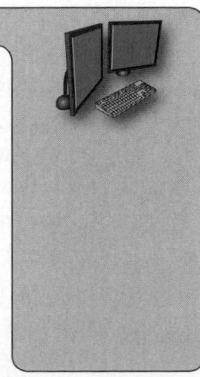

In addition to the paths listed under Support File Search Path, AutoCAD will search two other folders. The folder that contains the AutoCAD executable file, acad.exe, (typically, C:\Program Files\ AutoCAD 2010) is searched, as is the folder that contains the current drawing file. These folders are searched only if the desired file name is not found in any of the listed folders. It is not advisable to store files such as block files in the AutoCAD folder. However, if you store block files in the same folder as the current drawing, you do not need to add that folder to the search path.

Other File Settings

Another setting that can be specified in the **Files** tab is the location of device driver files. *Device drivers* are specifications for peripherals that work with AutoCAD and other Autodesk products. By default, the drivers supplied with AutoCAD are placed in the \Drv folder. If you purchase a third-party driver to use with AutoCAD, be sure to load the driver into this folder. If the third-party driver must reside in a different folder, you should specify that folder using the Device Driver File Search Path setting. Otherwise, the search for the correct driver is widespread and likely to take longer. Some other file locations listed in the **Files** tab include:

- **Working Support File Search Path.** Lists the active support paths AutoCAD is using. These paths are only for reference; they cannot be added or amended.
- **Project Files Search Path.** Sets the value for the **PROJECTNAME** system variable and specifies the project path names.
- **Customization Files.** Specifies the name of the main and enterprise customization files. Also, the location of custom icon files is specified.
- **Help and Miscellaneous File Names.** Specifies which files are used for the help file, the default Internet location, and where the configuration file is located.
- **Text Editor, Dictionary, and Font File Names.** Specifies which files are used for the text editor application, main and custom dictionaries, alternate font files, and the font mapping file.
- **Print File, Spooler, and Prolog Section Names.** Sets the file names for the plot file for legacy plotting scripts, the print spool executable file, and the PostScript prolog section name.
- **Printer Support File Path.** Specifies the print spooler file location, printer configuration search path, printer description file search path, and plot style table search path.
- **Automatic Save File Location.** Sets the path where the autosave (.sv$) file is stored. An autosave file is only created if the **Automatic save** option is checked in the **Open and Save** tab of the **Options** dialog box.
- **Color Book Locations.** Specifies the path for color book files that can be used when specifying colors in the **Select Color** dialog box.
- **Data Sources Location.** Specifies the path for database source files (.udl).
- **Template Settings.** Specifies the default location for drawing and sheet set template files and the file name for the defaults.
- **Tool Palettes File Locations.** Specifies the path for tool palette support files.
- **Authoring Palette File Locations.** Specifies the location of authoring palette files.
- **Log File Location.** Specifies the path for the AutoCAD log file. A log file is only created if the **Maintain a log file** option is checked in the **Open and Save** tab of the **Options** dialog box.
- **Action Recorder Settings.** Specifies search and storage paths for the action recorder macro files (.actm).
- **Plot and Publish Log File Location.** Specifies the path for the log file for "plot and publish" operations. A log file is only created if the **Automatically save plot and publish log** check box in the **Plot and Publish** tab of the **Options** dialog box is checked.
- **Temporary Drawing File Location.** Sets the folder where AutoCAD stores temporary drawing files.
- **Temporary External Reference File Location.** Indicates where temporary external reference files are placed.
- **Texture Maps Search Path.** Location of texture map files for rendering.
- **Web File Search Path.** Specifies the folders to search for files associated with photometric weblight lighting.

- **i-drop Associated File Location.** Specifies the folder used by default to store downloaded i-drop content.
- **DGN Mapping Settings.** Specifies the location of the mapping files for translation to and from Microstation drawings (.dgn). This folder must have read/write file permissions enabled in order for the Microstation commands to properly function.

Customizing the Graphics Window

Numerous options are available to customize the graphics window to your personal liking. Select the **Display** tab in the **Options** dialog box to view the display-control options, **Figure 21-3**.

The **Window Elements** area has a setting for the color scheme. The color scheme controls the outline color of the ribbon and status bar. Using the drop-down list, you can select between light and dark color schemes.

There are also check boxes for turning the scroll bars, status bar, and screen menu on or off; using large buttons for toolbars; showing tooltips and whether or not to show shortcut keys and extended commands in tooltips; and turning rollover tooltips on or off. The element colors and font settings can be changed using the buttons at the bottom of the **Window Elements** area.

On the right-hand side of the tab near the bottom, the **Crosshair size** setting is a percentage of the drawing screen area. The higher the value, the further the crosshairs extend. The settings in the Face control area determine the display intensity of the unselected objects in reference edit mode and in-place edit and annotative representations. A higher value means the unselected objects are less visible. Other options are discussed in the next sections.

Figure 21-3.
Use the **Display** tab to set up many of the visual elements of the AutoCAD environment.

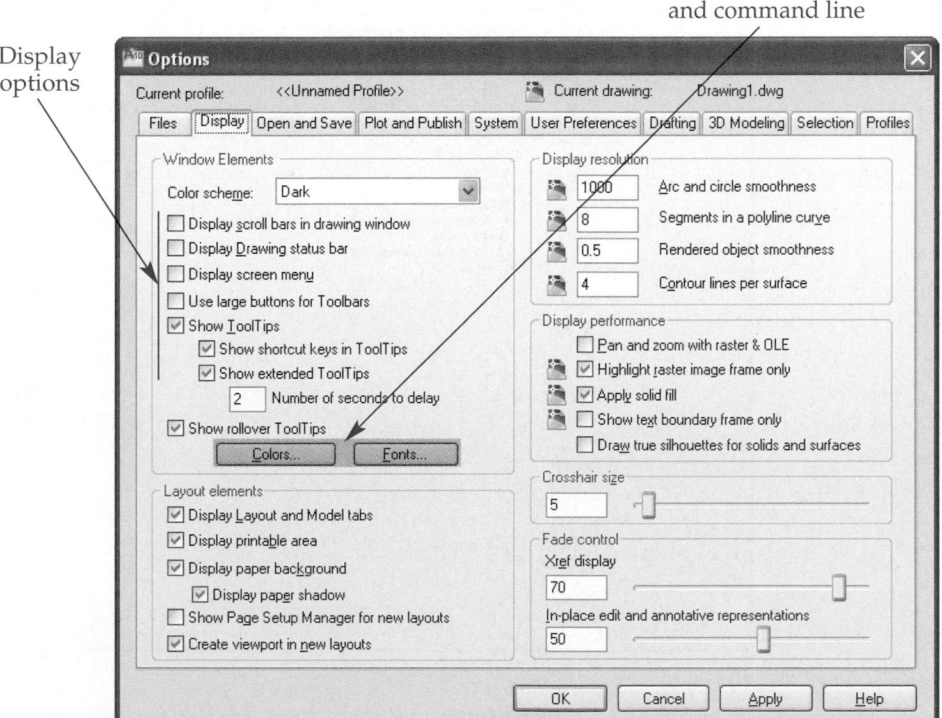

Changing Colors

By customizing colors, you can add your personal touch and make AutoCAD stand out among other active Windows applications. AutoCAD provides this capability with the **Drawing Window Colors** dialog box, **Figure 21-4.** This dialog box is accessed by picking the **Colors...** button in the **Window Elements** area of the **Display** tab in the **Options** dialog box.

To change a color, first select a context. A *context* is one of the environments, or modes, in AutoCAD, such as the 3D perspective projection mode that is set current when a new drawing is started based on the acad3D.dwt template. The **Context:** list box contains the names of all contexts. The context that was current when the **Options** dialog box is opened is initially selected. A preview of the context and its settings is displayed in the **Preview:** area at the bottom of the dialog box.

Each context contains several interface elements. An *interface element* is an item that is visible, or can be made visible, in a given context, such as the grid axis, autosnap marker, or light glyphs. Once a context is selected, pick the element to change in the **Interface element:** list.

With a context and element selected, the color of the element can be changed. Use the **Color:** drop-down list to change the color. If you pick the Select Color... entry, the **Select Color** dialog box is displayed. Below the **Color:** drop-down list is the **Tint for X, Y, Z** check box. This check box is available when certain elements are selected. When checked, a tint is applied along the X, Y, and Z axes. The elements to which a tint can be applied are: crosshairs, autotrack vector, drafting tooltip background, grid major lines, grid minor lines, and grid axis lines.

Along the right side of the dialog box are buttons for restoring the default settings. Picking the **Restore current element** button resets the currently selected element to its default color. Picking the **Restore current context** button resets *all* of the elements of the currently selected context to their default colors. Picking the **Restore all contexts**

Figure 21-4.
Change AutoCAD color settings using the **Drawing Window Colors** dialog box.

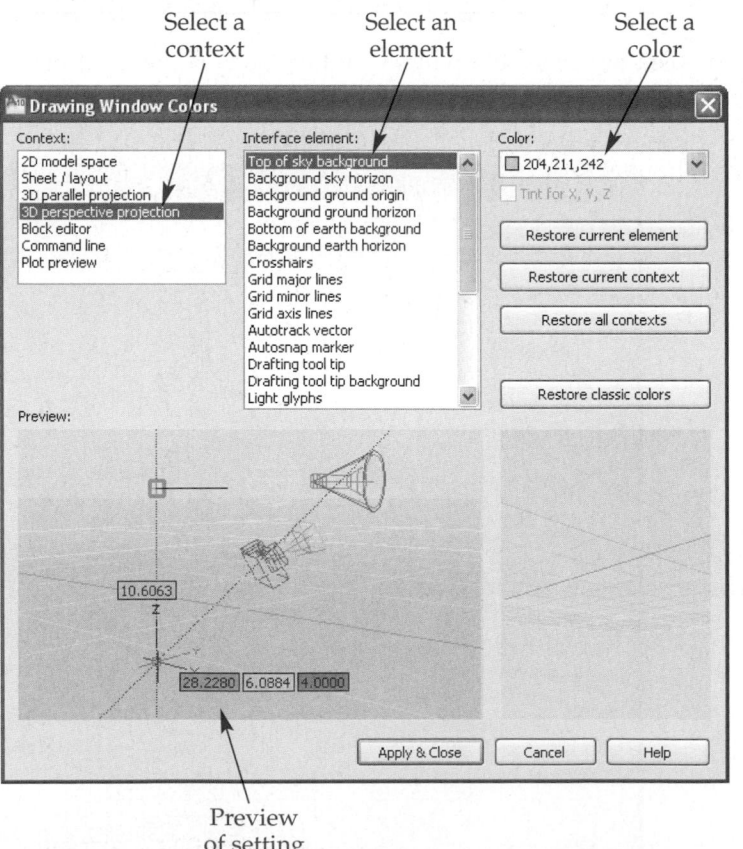

Select a context

Select an element

Select a color

Preview of setting

button resets *all* of the elements in *all* of the contexts to their default colors. Lastly, the **Restore classic colors** button sets the AutoCAD graphics screen to the traditional black background in model space and the grip and glyph colors to appropriate colors for the black background. These are the settings found in older releases of AutoCAD.

Once the colors are changed as needed, pick the **Apply & Close** button. Then, pick the **OK** button in the **Options** dialog box. The graphics window regenerates and displays the color changes you made.

PROFESSIONAL TIP

When setting the background color to black, first use the **Restore classic colors** button. Then, adjust the other color settings from that point forward.

Changing Fonts

You can change the fonts used in the **Command Line** window. The font you select has no effect on the text in your drawings, nor is the font used in the AutoCAD dialog boxes, pull-down menus, or screen menus.

To change the font used in the **Command Line** window, pick the **Fonts...** button in the **Display** tab of the **Options** dialog box. The **Command Line Window Font** dialog box appears, **Figure 21-5**.

The default font used by AutoCAD for the graphics window is Courier New. The font style for Courier New is regular (not bold or italic) and the default size is 10 points. Select a new font from the **Font:** list. This list displays the system fonts available for use. Also, set a style and size. The **Sample Command Line Font** area displays a sample of the selected font. Once you have selected the desired font, font style, and font size for the **Command Line** window, pick the **Apply & Close** button to assign the new font.

NOTE

The Windows system "menu" font controls the font style for the text displayed in the screen menu.

PROFESSIONAL TIP

The **UNDO** command does not affect changes made to your system using the **Options** dialog box. If you have made changes you do not want to save, pick **Cancel** to dismiss the **Options** dialog box. However, picking **Cancel** does *not* dismiss changes that have been applied using the **Apply** button.

Figure 21-5.
The **Command Line** window can be changed to suit your preference.

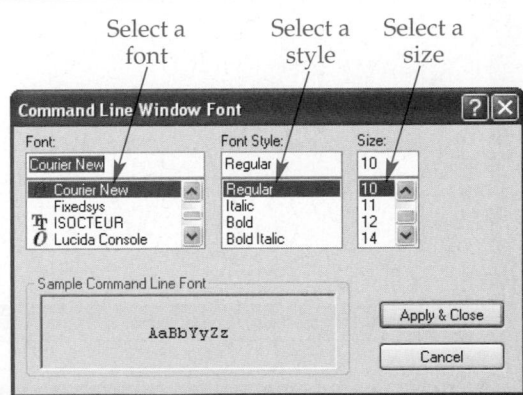

Exercise 21-1

Complete the exercise on the student website.
www.g-wlearning.com/CAD

Layout Display Settings

The appearance of a layout (paper space) tab is different than the appearance of the **Model** tab. The theory behind the default layout tab settings is to provide a picture of what the drawing will look like when plotted. You can see if the objects will fit on the paper or if some of the objects are outside of the margins. The following options, which are found in the **Layout elements** area of the **Display** tab in the **Options** dialog box, are illustrated in **Figure 21-6**.

- **Display Layout and Model tabs.** Displays the **Model** and layout tabs at the bottom of the drawing screen area. This is checked by default.
- **Display printable area.** The margins of the printable area are shown as dashed lines on the layout paper. Any portion of an object outside of the margins is not plotted.
- **Display paper background.** Displays the paper size specified in the page setup.
- **Display paper shadow.** Displays a shadow to the right and bottom of the paper. This option is only available if **Display paper background** is checked.
- **Show Page Setup Manager for new layouts.** Determines if the **Page Setup** dialog box is displayed when a new layout is selected or created. By default, this is unchecked.
- **Create viewport in new layouts.** Determines whether a viewport is automatically created when a new layout is selected or created. Many users uncheck this option since they will be creating their own floating viewports.

Figure 21-6.
Customizing the display of layouts. These options are set in the **Options** dialog box.

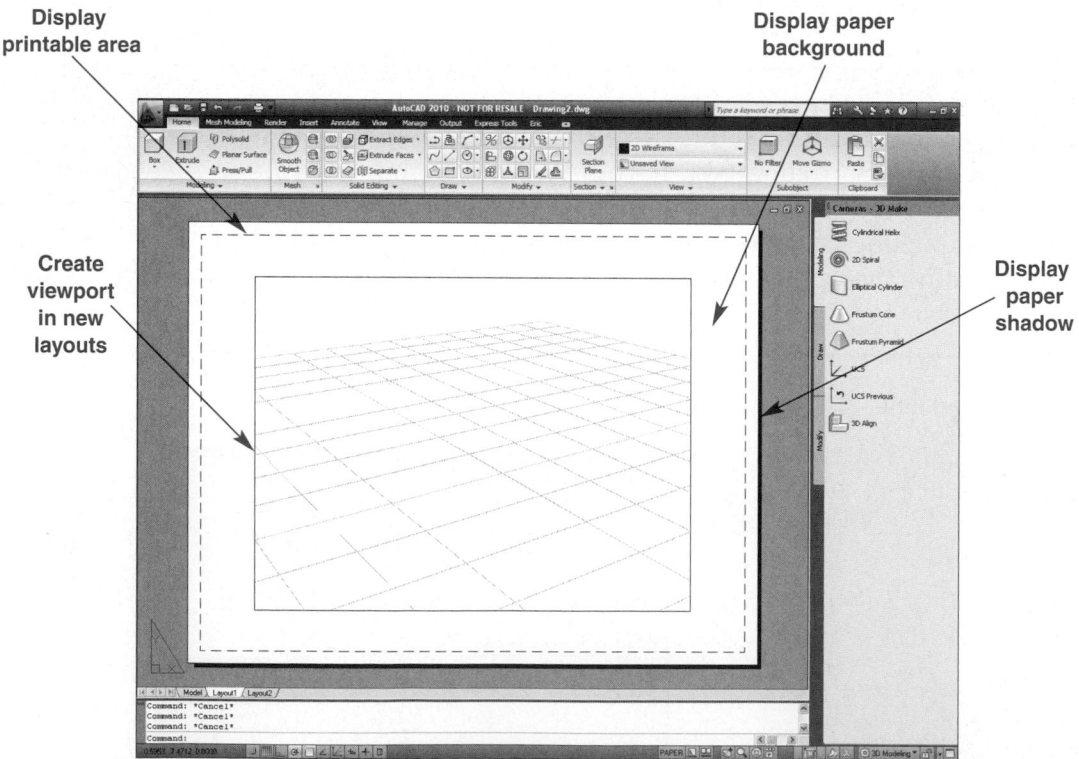

AutoCAD and Its Applications—Advanced

Display Resolution and Performance Settings

The settings in the **Display resolution** and **Display performance** areas of the **Display** tab in the **Options** dialog box affect the performance of AutoCAD. The settings can affect regeneration time and realtime panning and zooming. The following options are available in the **Display resolution** area. If the AutoCAD drawing icon is shown next to the setting, the value is saved in the current drawing, not in the AutoCAD system registry.

- **Arc and circle smoothness.** This setting controls the smoothness of circles, arcs, and ellipses. The default value is 1000; the range is from 1 to 20000. The system variable equivalent is **VIEWRES**.
- **Segments in a polyline curve.** This value determines how many line segments will be generated for each polyline curve. The default value is 8; the range is a nonzero value from –32768 to 32767. The system variable equivalent is **SPLINESEGS**.
- **Rendered object smoothness.** This setting controls the smoothness of curved solids when they are hidden, shaded, or rendered. This value is multiplied by the **Arc and circle smoothness** value. The default value is 0.5; the range is from 0.01 to 10. The system variable equivalent is **FACETRES**.
- **Contour lines per surface.** This value controls the number of contour lines per surface on solid objects. The default value is 4; the range is from 0 to 2047. The system variable equivalent is **ISOLINES**.

The following options are available in the **Display performance** area.

- **Pan and zoom with raster & OLE.** If this is checked, raster images are displayed when panning and zooming. If it is unchecked, only the frame is displayed during the operation. The system variable equivalent is **RTDISPLAY**.
- **Highlight raster image frame only.** If this is checked, only the frame around a raster image is highlighted when the image is selected. If this option is unchecked, the image displays a diagonal checkered pattern to indicate selection. The system variable equivalent is **IMAGEHLT**.
- **Apply solid fill.** Controls the display of solid fills in objects. Affected objects include hatches, wide polylines, solids, multilines, and traces. The system variable equivalent is **FILLMODE**.
- **Show text boundary frame only.** This setting controls the Quick Text mode. When checked, text is replaced by a rectangular frame. The system variable equivalent is **QTEXTMODE**.
- **Draw true silhouettes for solids and surfaces.** Controls whether or not the silhouette curves are displayed for solid objects. The system variable equivalent is **DISPSILH**.

NOTE

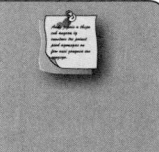

After changing display settings, use the **REGEN** or **REGENALL** command to make the settings take effect on the objects in the drawing.

PROFESSIONAL TIP

If you notice performance slowing down, you may want to adjust display settings. For example, if there is a lot of text in the drawing, you can activate Quick Text mode to improve performance. When the drawing is ready for plotting, deactivate Quick Text mode.

File Saving Options

The settings specified in the **Open and Save** tab of the **Options** dialog box deal with how drawing files are saved, safety precautions for files, how file names display in the application menu (menu browser), the behavior of xrefs, and the loading of ObjectARX applications and proxy objects. This tab is shown in **Figure 21-7**. The options in this tab are discussed in the next sections.

Default Settings for Saving Files

The settings in the **File Save** area determine the defaults for saving files. The setting in the **Save as:** drop-down list determines the default file type. You may want to change this setting if you are saving drawing files as a previous release of AutoCAD or saving drawings as DXF files.

The **Maintain visual fidelity for annotative objects** check box controls how annotative objects are displayed when the drawing is opened in AutoCAD 2007 or earlier versions. If you work primarily in model space, this can be left unchecked. If you use layouts and expect the drawing files to be saved for an older version of AutoCAD, this should be checked. When checked and the drawing is saved and then opened in an older version of AutoCAD, the scaled representations of annotative objects are divided into separate objects. These objects are stored in anonymous blocks saved on separate layers. The block names are based on the layer's original name appended with a number. When the drawing is opened once again in AutoCAD 2008 or later, the annotative objects are restored to normal. The system variable equivalent for this toggle is **SAVEFIDELITY**. Checking the check box sets this variable to 1 (on). Unchecking it sets the variable to 0 (off).

The **Maintain drawing size compatibility** check box determines how drawings with individual objects greater than 256 MB are handled. Files created with versions previous

---

Figure 21-7.
The **Open and Save** tab settings control default save options, file safety features, how file names are displayed in the application menu, xref options, and ObjectARX application options.

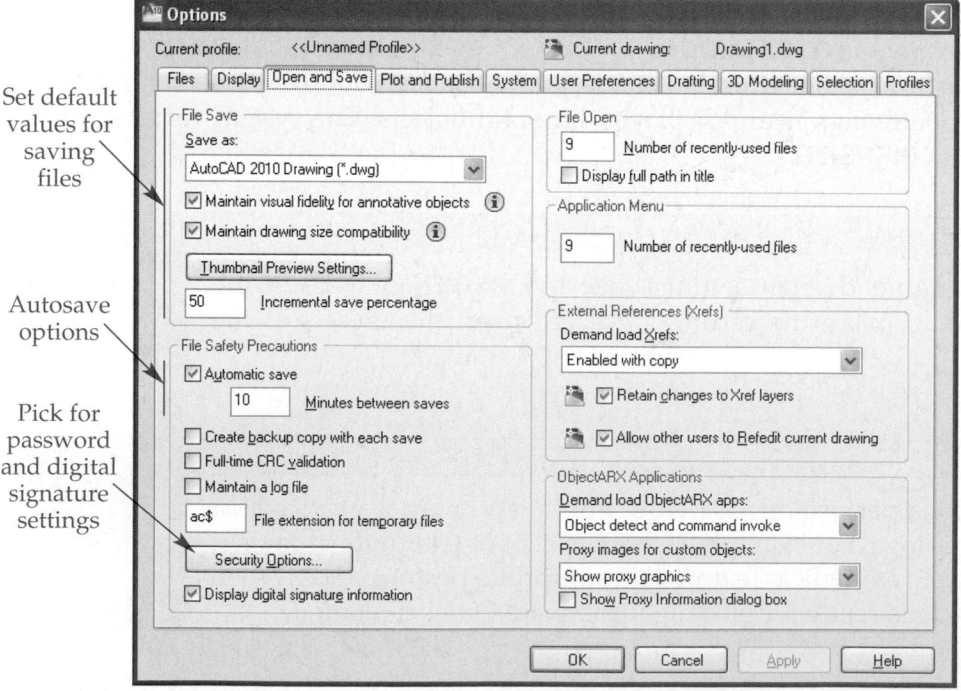

Set default values for saving files

Autosave options

Pick for password and digital signature settings

to AutoCAD 2010 are not allowed to have objects larger than 256 MB. When saving a drawing to a legacy format (2009 or earlier), problems may be encountered when attempting to open that drawing due to compatibility issues with these large objects. Checking this option maintains compatibility with earlier versions of AutoCAD by telling AutoCAD to check for objects larger than 256 MB when attempting to save the drawing. An alert box will be displayed noting that the issue needs to be resolved before the file can be saved. If you do not plan on working on your drawings in earlier versions of AutoCAD, this option box can be unchecked. The system variable equivalent for this toggle is **LARGEOBJECTSUPPORT**. Unchecking the **Maintain drawing size compatibility** check box sets the system variable to 1 (on) and allows the ability to create large objects. When the check box is checked, the system variable is set to 0 (off).

The **Incremental save percentage** value determines how much of the drawing is saved when a **SAVE** or **QSAVE** is performed. If the quantity of new data in a drawing file reaches the specified percentage, a full save is performed. To force a full save to be performed, set the value to 0.

If you pick the **Thumbnail Preview Settings...** button, the **Thumbnail Preview Settings** dialog box is displayed, **Figure 21-8.** If the **Save a thumbnail preview image** check box is checked, a preview image of the drawing will be displayed in the **Select File** dialog box when the drawing is selected for opening. The system variable equivalent for this setting is **RASTERPREVIEW**; 1 creates a preview.

The two radio buttons below the **Save a thumbnail preview image** check box set what is used as the basis for the thumbnail. The **Use view when drawing last saved** radio button bases the thumbnail image on the last zoom location of the drawing. The **Use Home View** radio button bases the thumbnail on the view defined as the home view.

The home view can be set to the current view by picking the **Set current View as Home** button in the **Home view** area. This can also be done in the drawing area by using the view cube. See chapter 4 for a detailed explanation of the view cube. To change the home view to the default setting, pick the **Reset Home to default** button. A preview of the current home view is shown to the left of the buttons.

Figure 21-8.
The **Thumbnail Preview Settings** dialog box.

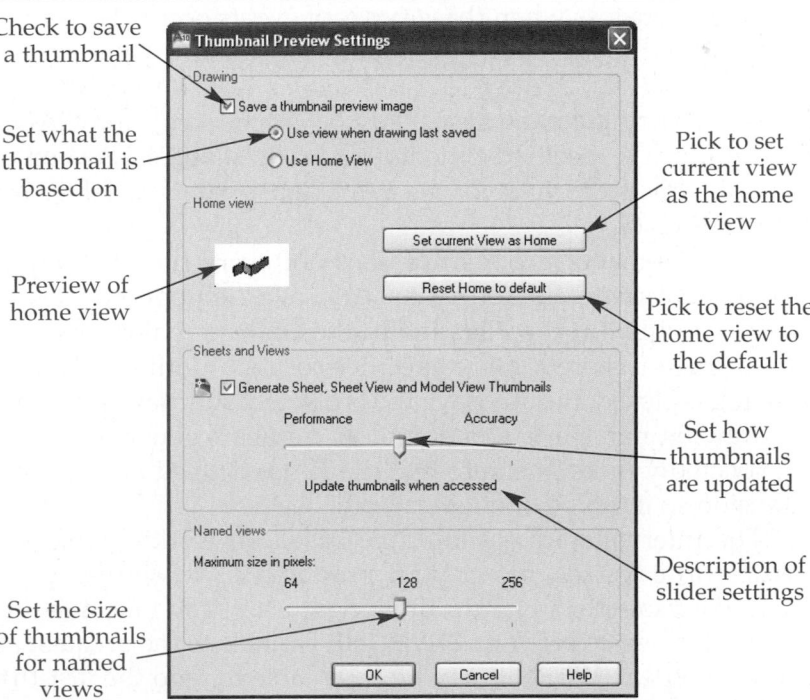

Check to save a thumbnail

Set what the thumbnail is based on

Preview of home view

Pick to set current view as the home view

Pick to reset the home view to the default

Set how thumbnails are updated

Description of slider settings

Set the size of thumbnails for named views

When the **Generate Sheet, Sheet View, and Model View Thumbnails** check box is checked in the **Thumbnail Preview Settings** dialog box, the thumbnails in the **Sheet Set Manager** are updated based on the position of the slider below this check box. The slider can be set to one of three positions. A description of the current setting appears below the slider. When the slider is in the middle position (default), thumbnails are updated when they are accessed. When the slider is in the left-hand position, thumbnails must be manually updated. When the slider is in the right-hand position, the thumbnails are updated when the drawing is saved. The system variable equivalent is **UPDATETHUMBNAIL**. The settings are:

- **0.** The **Generate Sheet, Sheet View, and Model View Thumbnails** check box is unchecked.
- **7.** The check box is checked and the slider is in the left-hand position.
- **15.** The check box is checked and the slider is in the middle position.
- **23.** The check box is checked and the slider is in the right-hand position.

The **Maximum size in pixels:** slider in the **Thumbnail Preview Settings** dialog box controls size of thumbnails for named views. The slider can be set to 64, 128, or 256 pixels. This is the square size of the thumbnail. The system variable equivalent is **THUMBSIZE**, where a setting of 0 is 64 pixels, 1 is 128 pixels, and 2 is 256 pixels.

CAUTION

To maintain forward compatibility of drawings, annotative objects should not be edited in older versions of AutoCAD. Doing so may compromise the annotative properties. For example, exploding an annotative block in an older version of AutoCAD then opening that drawing in AutoCAD 2010 results in each of the scaled representations becoming a separate annotative object.

Autosave Settings

When working in AutoCAD, data loss can occur due to a sudden power outage or an unforeseen system error. AutoCAD provides several safety precautions to help minimize data loss when these types of events occur. The settings for the precautions are found in the **File Safety Precautions** area in the **Open and Save** tab of the **Options** dialog box.

When the **Automatic save** check box is checked, AutoCAD automatically creates a backup file at a specified time interval. The **Minutes between saves** edit box sets this interval. This is the value of the **SAVETIME** system variable. Removing the check sets **SAVETIME** to 0.

The automatic save feature does not overwrite the source drawing file with its incremental saves. Rather, AutoCAD saves temporary files. The path for autosave files is specified in the **Files** tab in the **Options** dialog box, as discussed earlier. The autosave file is stored in the specified location until the drawing is closed. When the drawing is closed, the autosave file is deleted. Autosave files have a .sv$ extension with the drawing name and some random numbers generated by AutoCAD. If AutoCAD unexpectedly quits, the autosave file is not deleted and can be renamed with a .dwg extension so it can be opened in AutoCAD.

The interval setting should be based on working conditions and file size. It is possible to adversely affect your productivity by setting your **SAVETIME** value too small. For example, in larger drawings, a save can take a significant amount of time. Ideally, it is best to set your **SAVETIME** variable to the greatest amount of time you can afford to repeat. While it may be acceptable to redo the last fifteen minutes or less of work, it is unlikely that you would feel the same about having to redo the last hour of work.

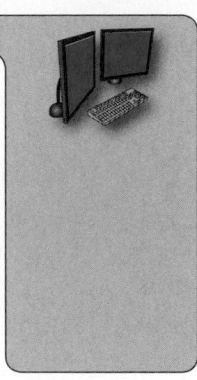

Setting and resetting the **SAVETIME** variable according to any given situation is often the best approach. The factors that should influence the current setting include not only file size, but also the working conditions. If your computer system is experiencing frequent lockups or crashes, your automatic saves should occur often. Weather can also be a factor. Wind or electrical storms should be an immediate cue to reduce the value of the **SAVETIME** variable. In addition to automatic saves, most veteran users have developed the positive habit of "save early, save often."

Backup Files

AutoCAD can create a backup of the current drawing file whenever the current drawing is saved. The backup file uses the same name as the drawing, but has a .bak file extension. The backup is not overwritten when a different drawing is opened or saved. When the **Create backup copy with each save** check box in the **Open and Save** tab of the **Options** dialog box is checked, the backup file feature is enabled. If not checked, the file is not backed up when you save. Unless you prefer to take unnecessary risks, it is usually best to have this feature enabled.

CRC Validation

A *cyclic redundancy check,* or CRC, verifies that the number of data bits sent is the same as the number received. **Full-time CRC validation** is a feature you can use when drawing files are being corrupted and you suspect a hardware or software problem. When using full-time CRC validation, the CRC check is done every time data are read into the drawing. This ensures that all data are correctly received.

Log Files

The log file can serve a variety of purposes. The source of drawing errors can be determined by reviewing the commands that produced the incorrect results. Additionally, log files can be reviewed by a CAD manager to determine the need for staff training or customization of the system.

When the **Maintain a log file** check box is activated in the **Open and Save** tab of the **Options** dialog box, AutoCAD creates a file named with the drawing name, a code, and the .log file extension. The name and location of the log file can be specified using the Log File Location listing in the **Files** tab of the **Options** dialog box. When activated, all prompts, messages, and responses that appear in the **Command Line** window are saved to this file. The log file status can also be set using the **LOGFILEON** and **LOGFILEOFF** commands.

Toggle the log file open before listing any saved layers, blocks, views, or user coordinate systems. You can then print the log file contents and keep a hard copy at your workstation as a handy reference.

File Opening Settings

The **File Open** area of the **Open and Save** tab in the **Options** dialog box contains two settings. The value in the **Number of recently-used files to list** text box controls the number of drawing files listed in the **File** pull-down menu. The pull-down menus are not displayed, by default, unless the AutoCAD Classic workspace is set current. The

value can be from 0 to 9. The **Display full path in title** check box controls whether the entire drawing file path (when checked) or just the file name (when unchecked) is displayed in the title bar of the AutoCAD window.

Application Menu Settings

The **Application Menu** area of the **Open and Save** tab in the **Options** dialog box controls the number of drawing files displayed in the **Recent Documents** entry in the application menu (menu browser). The setting in the **Number of recently used files** text box is the number of files displayed and can range from 0 to 50.

External Reference Settings

The external reference options in the **Open and Save** tab of the **Options** dialog box are important if you are working with xrefs. These options are found in the **External Reference (Xrefs)** area of the tab. The **Demand load Xrefs:** setting can affect system performance and the ability for another user to edit a drawing currently referenced into another drawing. You can select Enabled, Disabled, or Enabled with copy from the drop-down list. This setting is also controlled by the **XLOADCTL** system variable.

If the **Retain changes to Xref layers** option is checked, xref layer settings are saved with the drawing file. The **VISRETAIN** system variable also controls this setting.

The **Allow other users to Refedit current drawing** setting controls whether or not the drawing can be edited in-place when it is referenced by another drawing. This setting is also controlled by the **XEDIT** system variable.

ObjectARX Options

The **ObjectARX Applications** area of the **Open and Save** tab of the **Options** dialog box controls the loading of ObjectARX applications and the displaying of proxy objects. The **Demand load ObjectARX apps:** setting specifies if and when AutoCAD loads third-party applications associated with objects in the drawing. The **Proxy images for custom objects:** setting controls how objects created by a third-party application are displayed. When a drawing with proxy objects is opened, the **Proxy Information** dialog box is displayed. To disable the dialog box, uncheck the **Show Proxy Information dialog box** option. This is unchecked by default.

System Settings

Options for the pointing device, graphic settings, general system options, and dbConnect can be found in the **System** tab of the **Options** dialog box, **Figure 21-9.** These settings affect the interaction between AutoCAD and your operating system.

In the **3D Performance** area is the **Performance Settings** button. Selecting this button displays the **Adaptive Degradation and Performance Tuning** dialog box. The options available in this dialog box are discussed in the next section.

The **Current Pointing Device** area determines the pointing device used with AutoCAD. The default is the current system pointing device (usually your mouse). If you have a digitizer tablet, you will want to select the Wintab Compatible Digitizer option. You must configure your tablet before it can be used.

The **Layout Regen Options** setting determines what is regenerated and when it is regenerated when working with layout tabs. The following options are available in the **dbConnect Options** area.

Figure 21-9.
General AutoCAD system options and hardware settings can be controlled in the **System** tab.

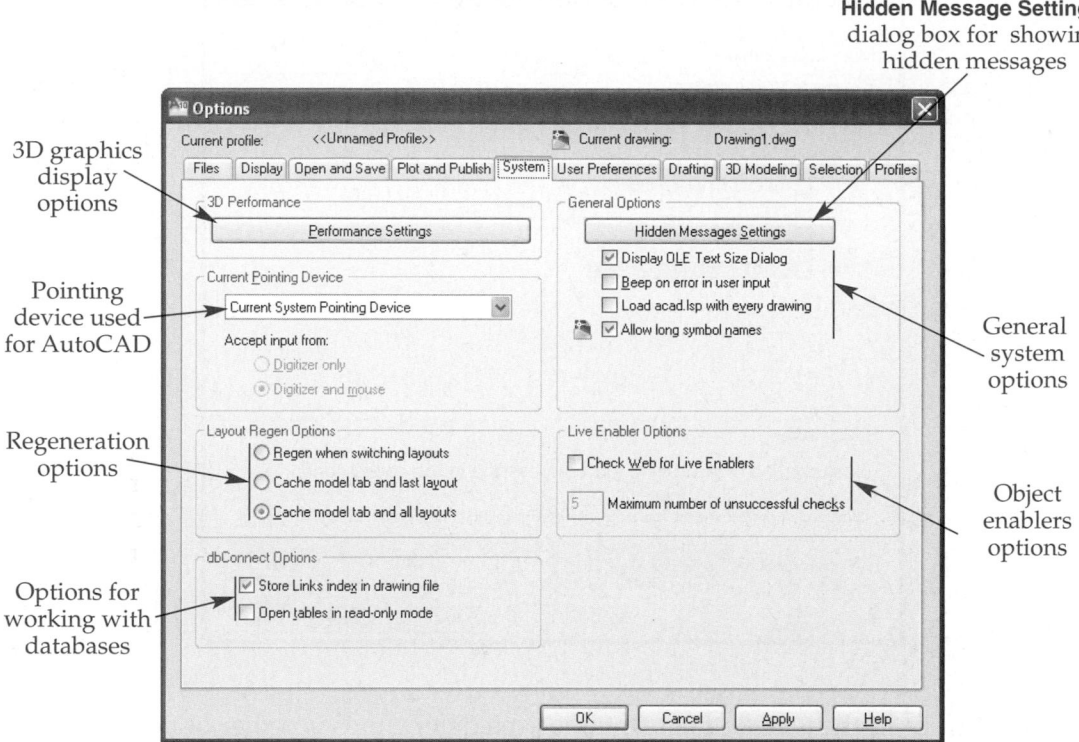

Pick to display
Hidden Message Settings
dialog box for showing
hidden messages

3D graphics
display
options

Pointing
device used
for AutoCAD

Regeneration
options

Options for
working with
databases

General
system
options

Object
enablers
options

- **Store Links index in drawing file.** When this option is checked, the database index is saved within the drawing file. This makes the link selection operation quicker, but increases the drawing file size.
- **Open tables in read-only mode.** Determines whether tables are opened in read-only mode.

The settings in the **General Options** area control general system functions. When the **Hidden Messages Settings** button is picked, the **Hidden Message Settings** dialog box is displayed, **Figure 21-10.** All messages hidden by the user picking the "do not show again" option in a message box are available in this dialog box. For example, if you select a solid object, a message is displayed indicating to use the [Ctrl] key to select a subobject. If you choose to always hide this message, it appears in the **Hidden Message Settings** dialog box. Check the entry for this message in the dialog box to have it once again displayed when you select a solid. The following additional options are available in the **General Options** area of the **System** tab.

- **Display OLE Text Size Dialog.** When inserting an OLE object, the **OLE Text Size** dialog box is displayed if this option is checked.
- **Beep on error in user input.** Specifies whether AutoCAD alerts you of incorrect user input with an audible beep. By default, this is off.
- **Load acad.lsp with every drawing.** This setting turns the persistent AutoLISP feature on or off. By default, it is off.
- **Allow long symbol names.** Determines if long symbol names can be used in AutoCAD. If this option is checked, up to 255 characters can be used for layers, dimension styles, blocks, linetypes, text styles, layouts, UCS names, views, and viewport configurations. If unchecked, symbol names are limited to 31 characters. By default, this option is checked. Also, it is saved in the drawing, not the AutoCAD system registry. The system variable is **EXTNAMES**.

Figure 21-10.
Turning on the display of previously hidden messages.

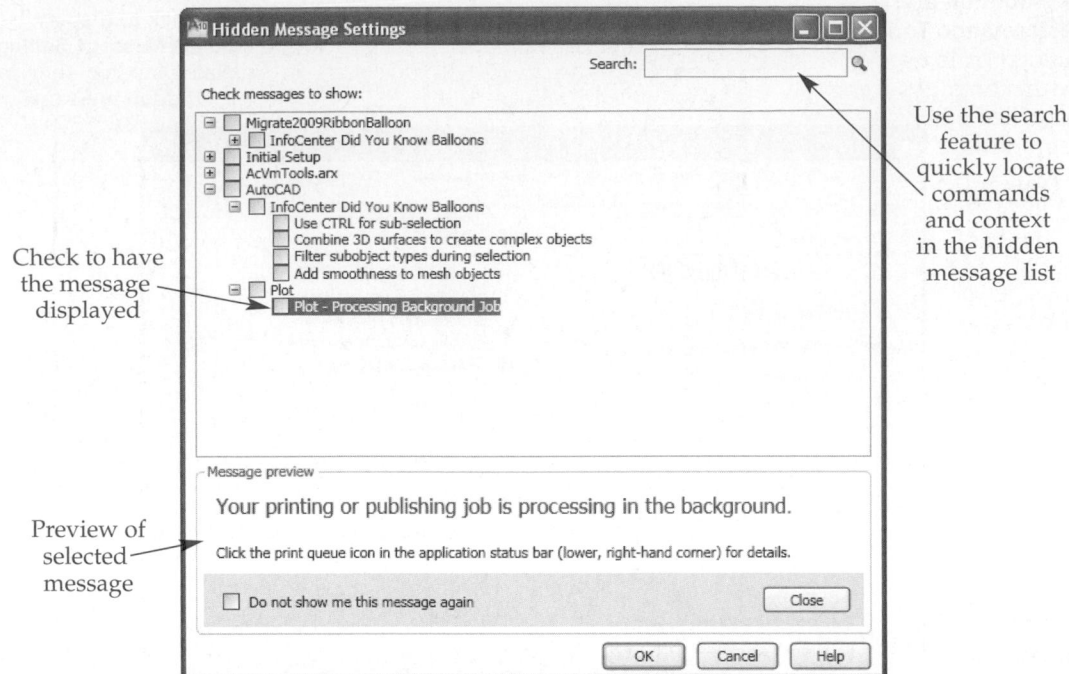

Check to have the message displayed

Preview of selected message

Use the search feature to quickly locate commands and context in the hidden message list

The setting in the **Live Enabler Options** area determines whether or not AutoCAD searches the Autodesk website for object enablers. An object enabler allows your version of AutoCAD to open and manipulate drawings created in applications like AutoCAD Architecture without the display of proxy object errors. It also allows the use of the custom-made objects created in those applications. Object enablers are provided for free by Autodesk. The system variable that controls this setting is **PROXYWEB-SEARCH**. The **Maximum number of unsuccessful checks** edit box indicates how many times AutoCAD will search the website.

3D Performance Settings

With all of the powerful 3D and solid modeling features that are built into AutoCAD, there are many display-related tasks being handled by AutoCAD, the graphics card, and the computer itself. Materials, lights, shadows, shading, and rendering require a lot of computing power in order to project a quality representation of the model onto the monitor screen. Often there is no reduction in quality to any of the desired effects if the materials are not too complicated, few lights are used, or if you have shadows turned off. Sometimes, in order to make one effect look good, fewer resources have to be assigned to other effects. The software and hardware, working together, usually do an adequate job assigning these resources. However, it may be necessary for you to assist in this decision-making process. The settings for this process are made in the **Adaptive Degradation and Performance Tuning** dialog box, **Figure 21-11.** To display this dialog box, enter the **3DCONFIG** command or pick the **Performance Settings** button in the **Systems** tab of the **Options** dialog box.

The left-hand side of the **Adaptive Degradation and Performance Tuning** dialog box contains settings for controlling adaptive degradation. *Adaptive degradation* controls system performance by turning off features or preventing them from using resources. The check box at the top of the area controls whether or not adaptive degradation

Figure 21-11.
The **Adaptive Degradation and Performance Tuning** dialog box is used to turn on and prioritize adaptive degradation.

Frame rate

Pick to manually tune performance

Effects that will degrade, listed in order of priority

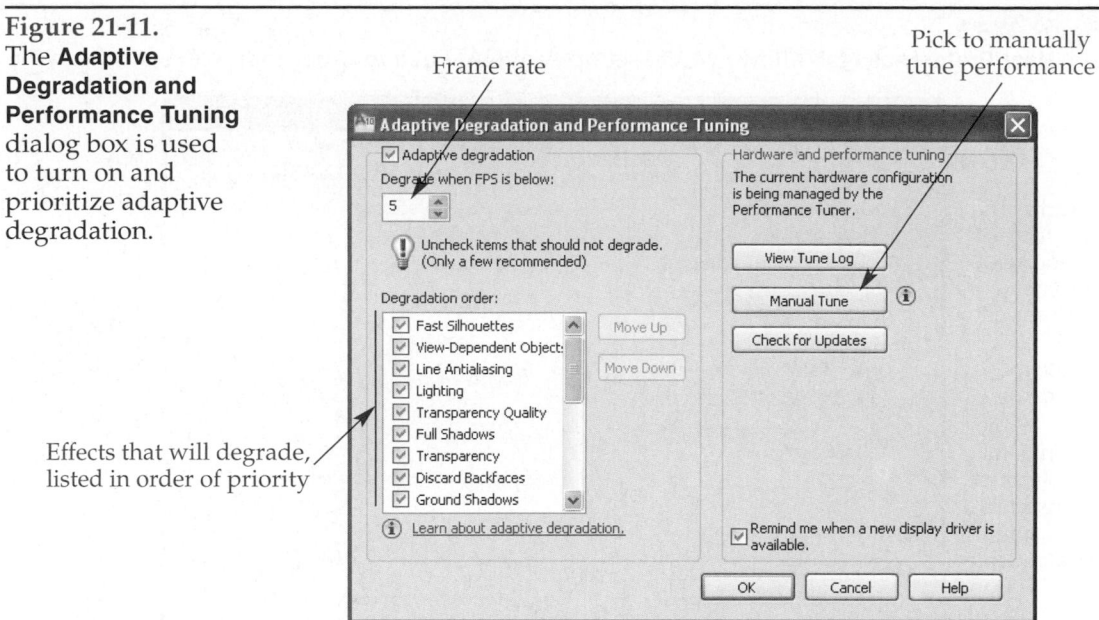

is being used. When unchecked, adaptive degradation is turned off and all effects are using all resources. This may result in graphics lagging or becoming slow and "choppy" as you zoom and pan around your drawing. The orbiting commands are even more affected by this being turned off. By checking the check box, adaptive degradation is activated.

AutoCAD tracks its graphics performance in terms of frames per second (fps). Just below the **Adaptive degradation** check box is a text box for setting this value. You may enter a new value in the text box or use the arrows to increase or decrease the value. The higher the number, the sooner resources start being reassigned. If performance dips below this level, resources are taken away from the various effects that create the displayed graphics.

The effects that can be controlled while adaptive degradation is turned on are shown in the **Degradation order:** list box. Certain effects that you deem important can be unchecked so they are not degraded and operate using maximum resources. The top-to-bottom order in which the effects are listed determines the priority in which resources are removed. This order can be changed by selecting an effect and picking the **Move Up** or **Move Down** button on the right side of the list.

On the right-hand side of the **Adaptive Degradation and Performance Tuning** dialog box is the **Hardware and performance tuning** area. Picking the **View Tune Log** button displays a log of any features or effects that have been turned off. Information regarding your computer, amount of RAM, and 3D graphics card are also shown. The log can be saved as a file. The **Manual Tune** button displays the **Manual Performance Tuning** dialog box. This dialog box allows control over hardware settings (*hardware acceleration,* graphics card driver name, and the effects the graphics card is capable of), general settings (discard back faces and quality of transparency), and dynamic tessellation settings (surface and curve tessellation settings and number of tessellations to cache). At the bottom of the **Adaptive Degradation and Performance Tuning** dialog box is the **Remind me when a new driver is available** check box. When checked, AutoCAD displays a pop-up message in the graphics area whenever a new graphics card driver is available.

When the settings have been adjusted as desired in the **Adaptive Degradation and Performance Tuning** dialog box, pick the **OK** button to return to the **Options** dialog box. Then, close the **Options** dialog box.

Figure 21-12.
The **User Preferences** tab allows you to set up AutoCAD in a manner that works best for you.

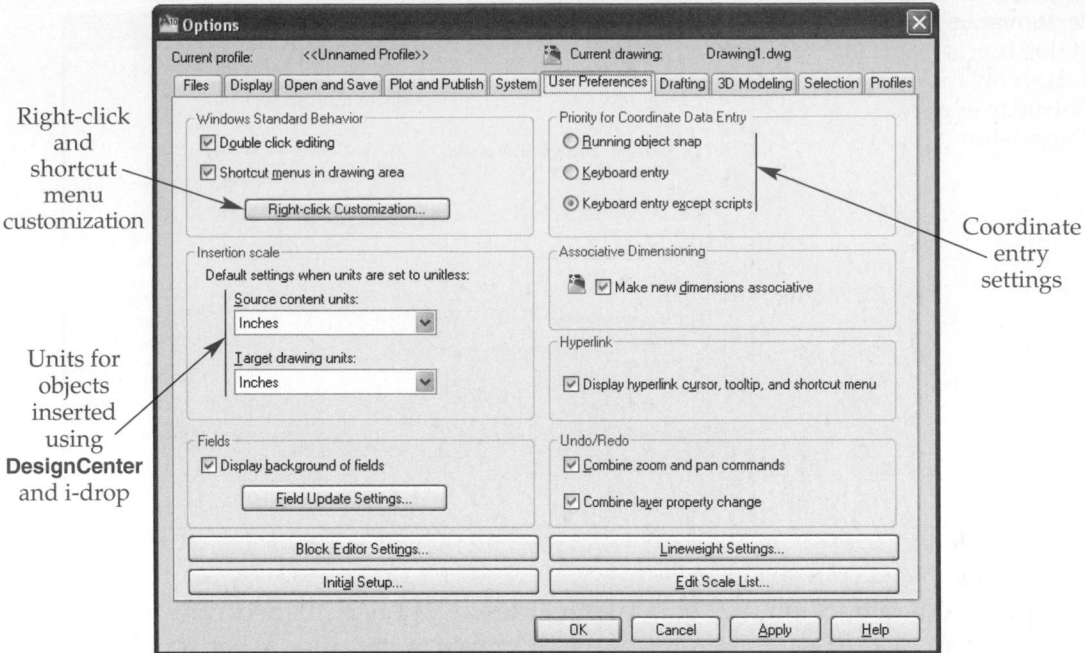

Right-click and shortcut menu customization

Units for objects inserted using **DesignCenter** and i-drop

Coordinate entry settings

User Preferences

A variety of settings are found in the **User Preferences** tab of the **Options** dialog box. See **Figure 21-12.** AutoCAD allows users to optimize the way they work in AutoCAD by providing options for double-click editing, shortcut menu functions, **DesignCenter** units, working with fields, coordinate data entry, associative dimensions, hyperlink icon display, undo/redo control, default lineweight settings, and scale list settings. All of these are controlled in this tab.

Shortcut Menus and Double-Click Editing

AutoCAD has tools that provide easy access to commonly used editing commands and options. Two of these tools are double-click editing and shortcut menus. Double-clicking on an object calls the most appropriate editing tool for that object type, often the **Properties** palette. Shortcut menus are displayed by right-clicking and are *context sensitive*. This means the options available in the shortcut menu are determined by the active command, cursor location, or selected object.

To enable double-click editing, check the **Double click editing** check box in the **Windows Standard Behavior** area in the **User Preferences** tab of the **Options** dialog box. To enable shortcut menus, check the **Shortcut menus in drawing area** check box in the same area. Disabling the shortcut menus makes a right mouse click the equivalent of pressing the [Enter] key. In general, this is not recommended.

You can also customize the setting for the right mouse button. Pick the **Right-click Customization...** button to access the **Right-Click Customization** dialog box. See **Figure 21-13.** The **Turn on time-sensitive right-click:** check box controls the right-click behavior. A quick click is the same as pressing [Enter]. A longer click displays a shortcut menu. You can set the duration of the longer click in milliseconds. If the check box is checked, the **Default Mode** and **Command Mode** areas of the dialog box are disabled.

Different settings can be used for the three different shortcut menu modes. Each of the three menu modes has a separate area in the **Right-Click Customization** dialog box.

Figure 21-13.
Use this dialog box
to customize the
right mouse button.

Controls
right-click
timing

Select
behavior for
each mode

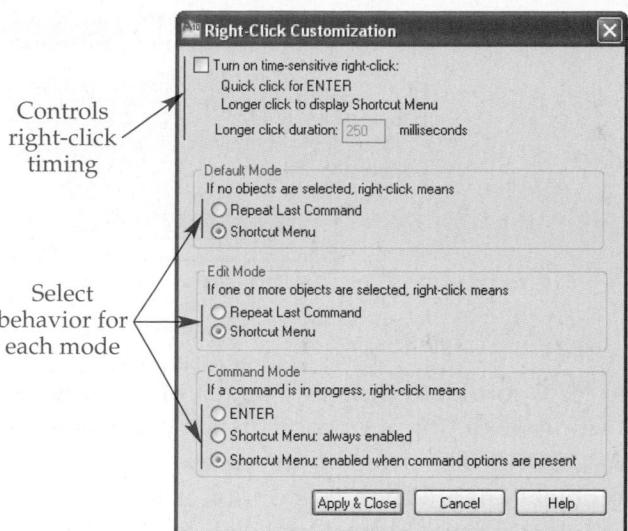

- **Default Mode.** In this mode, no objects are selected and no command is active. The **Repeat Last Command** option activates the last command issued. The **Shortcut Menu** option displays the shortcut menu.
- **Edit Mode.** In this mode, an object is selected, but no command is active. The **Repeat Last Command** option activates the last command issued. The **Shortcut Menu** option displays the shortcut menu.
- **Command Mode.** In this mode, a command is active. The **ENTER** option makes a right-click the same as pressing [Enter]. The **Shortcut Menu: always enabled** option means that the shortcut menu is always displayed in command mode. The **Shortcut Menu: enabled when command options are present** option means the shortcut menu is only displayed when command options are available on the command line. When there are no command options, a right-click is the same as [Enter]. This is the default option.

Insertion Scale

In the **Insertion scale** area of the **User Preferences** tab, unit values can be set for objects when they are inserted into a drawing. This applies to "unitless" objects dragged from **DesignCenter** or inserted using the i-drop method. The **Source content units:** setting specifies the units for objects being inserted into the current drawing. The **Target drawing units:** setting determines the units in the current drawing. These settings are used when there are no units set with the **INSUNITS** system variable.

Fields

A *field* is a special type of text object that displays a specific property value, setting, or characteristic. Fields can display information related to a specific object, general drawing properties, or information related to the current user or computer system. The text displayed in the field can change if the value being displayed changes. Refer to *AutoCAD and Its Applications—Basics* for more information on using fields.

In the **Fields** area of the **User Preferences** tab, you can set whether or not a field is displayed with a nonplotting background. When the **Display background of fields** check box is checked, the field background is displayed in light gray.

Picking the **Field Update Settings...** button in the **Fields** area opens the **Field Update Settings** dialog box, **Figure 21-14**. In this dialog box, you can set when fields are automatically updated. The five options are **Open**, **Save**, **Plot**, **eTransmit**, and **Regen**. Check as many of the options as appropriate.

Figure 21-14.
Setting when fields
are automatically
updated.

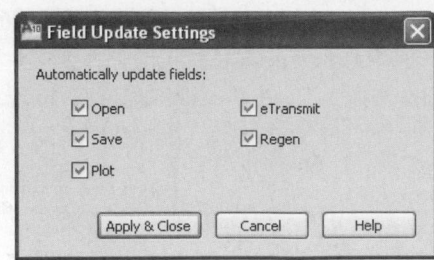

Coordinate Data Priority

The **Priority for Coordinate Data Entry** area of the **User Preferences** tab controls how AutoCAD responds to input of coordinate data. The system variable equivalent for this is **OSNAPCOORD**. The three options are:

- **Running object snap.** When this option is selected, object snaps always override coordinate entry. This is equivalent to an **OSNAPCOORD** setting of 0.
- **Keyboard entry.** When this option is selected, coordinate entry always overrides object snaps. This is equivalent to an **OSNAPCOORD** setting of 1.
- **Keyboard entry except scripts.** When this option is selected, coordinate entry will override object snaps except those object snaps contained within scripts. This is the default and equivalent to an **OSNAPCOORD** setting of 2.

Associative Dimensions

By default, all new dimensions are associative. This means that the dimension value automatically changes when a dimension's defpoints are moved. However, you can turn this option off in the **User Preferences** tab of the **Options** dialog box. When the **Make new dimensions associative** check box in the **Associative Dimensioning** area is unchecked, any dimensions drawn do *not* have associativity. This is equivalent to a **DIMASSOC** setting of 1.

Hyperlinks

In the **Hyperlink** area of the **User Preferences** tab, you can set whether or not the hyperlink cursor and tooltip are displayed when the cursor is over a hyperlink. If **Display hyperlink cursor, tooltip, and shortcut menu** is checked, the hyperlink icon appears next to the crosshairs when the cursor is over an object containing a hyperlink. The tooltip is also displayed. Additional hyperlink options are available from the shortcut menu when an object with a hyperlink is selected.

Undo/Redo

The **Undo/Redo** area of the **User Preferences** tab allows you to control how multiple, consecutive zooms and pans are handled within the **UNDO** and **REDO** commands. By checking the **Combine zoom and pan commands** check box, back-to-back zooms and pans are considered a single operation for undo and redo purposes. In other words, performing an undo or redo undoes or redoes the entire zoom/pan sequence. Unchecking the check box allows each zoom or pan to be considered a separate operation.

When the **Combine layer property change** check box is checked, all changes made in the **Layer Properties Manager** palette are considered one operation. If this is unchecked, each change is considered a separate operation when using **UNDO** and **REDO**.

Figure 21-15.
Use the **Block Editor Settings** dialog box to adjust the appearance of objects when working in the **Block Editor** window.

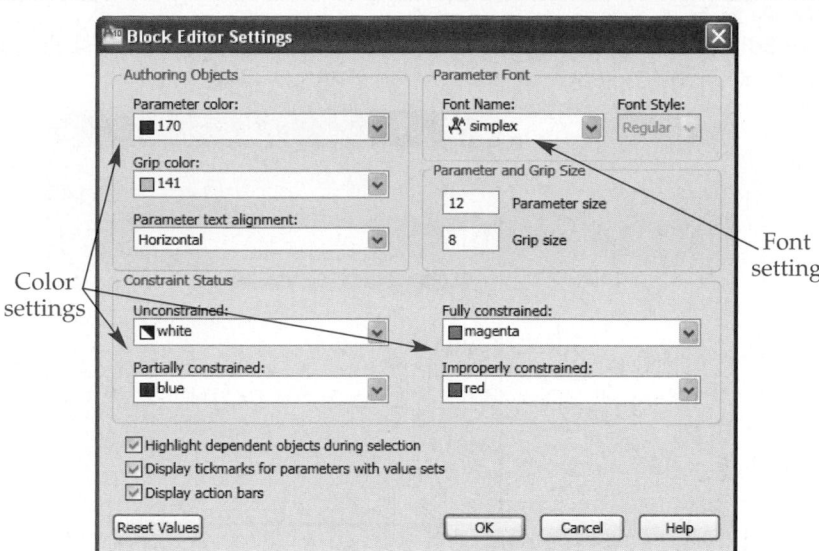

Color settings

Font setting

Block Editor Settings, Initial Setup, Lineweight Settings, and Edit Scale List

At the bottom of the **User Preferences** tab are the **Block Editor Settings...**, **Initial Setup...**, **Lineweight Settings...**, and **Edit Scale List...** buttons. Picking one of these buttons opens a dialog box with corresponding settings.

Picking the **Block Editor Settings...** button opens a dialog box that contains settings for the colors, fonts, and sizes of the various elements that appear when working in the **Block Editor** window, **Figure 21-15**. The **Block Editor** window is discussed in detail in *AutoCAD and Its Applications—Basics.* At the bottom of the dialog box there are three check boxes that control the highlight dependent objects during selection, display of tick marks for parameters with value sets, and display of action bars. These check boxes control the **BDEPENDENCYHIGHLIGHT**, **BTMARKDISPLAY**, and **BACTIONBAR-MODE** system variables.

Picking the **Initial Setup...** button opens the **Initial Setup** dialog box that was presented when installing the software. This dialog box can be used to adjust any of the settings chosen at installation.

The **Lineweight Settings...** button opens the **Lineweight Settings** dialog box in which you can change default lineweight settings. This is discussed in detail in *AutoCAD and Its Applications—Basics.*

A default list of scales appears in various dialog boxes related to viewports, page setups, and plot scaling. You can add custom scales to or remove scales from this list so that it is more appropriate for your application. Picking the **Edit Scale List...** button at the bottom of the **User Preferences** tab displays the **Edit Scale List** dialog box. See **Figure 21-16**. The dialog box displays the current list of scales. The buttons on the right side of the dialog box allow you to add a new scale, edit an existing scale, move a scale up or down within the list, delete a scale, or reset the list to the default set of scales.

To add a scale, pick the **Add...** button. In the **Add Scale** dialog box that appears, enter a name for the scale in the **Name appearing in scale list:** text box. Then, in the **Scale Properties** area of the dialog box, enter values to indicate how many paper space units equal how many drawing units. Finally, pick the **OK** button to return to the **Edit Scale List** dialog box. The new scale appears in the list and is available wherever the scale list is displayed.

Figure 21-16.
The **Edit Scale List** dialog box allows you to change the scale list that appears when using various viewport, page setup, and plot scaling dialog boxes.

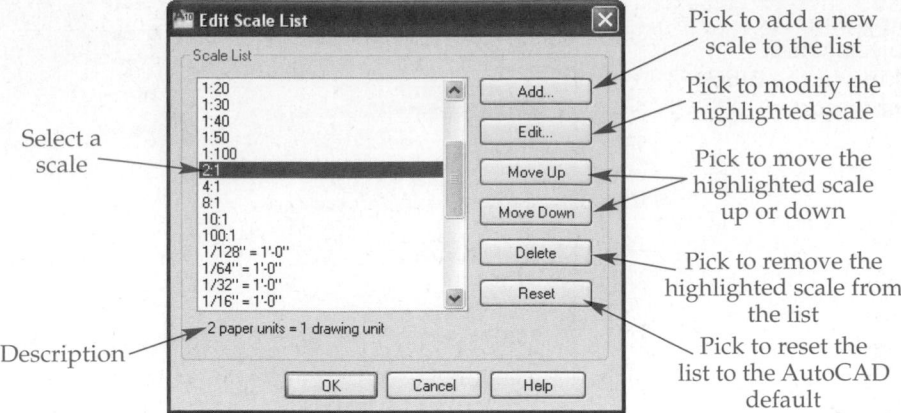

Select a scale

Description

Pick to add a new scale to the list

Pick to modify the highlighted scale

Pick to move the highlighted scale up or down

Pick to remove the highlighted scale from the list

Pick to reset the list to the AutoCAD default

3D Display Properties

There are many ways to customize your system specifically for working in a 3D environment. The **3D Modeling** tab of the **Options** dialog box allows you to control the various settings having to do with working in 3D, **Figure 21-17.**

The **3D Crosshairs** area of the tab contains check boxes for displaying the Z axis on the crosshairs, labeling the axes of standard crosshairs, and labeling the axes of the dynamic UCS icon. There are three labeling possibilities from which to choose:

- X, Y, and Z.
- N (north), E (east), and z.
- Or you can specify custom labels for each axes.

The check boxes in the **Display View Cube or UCS Icon** area determine if the UCS icon is displayed in 2D model space. For a model space display in a visual style other

Figure 21-17.
Use the **3D Modeling** tab to customize settings when working in 3D.

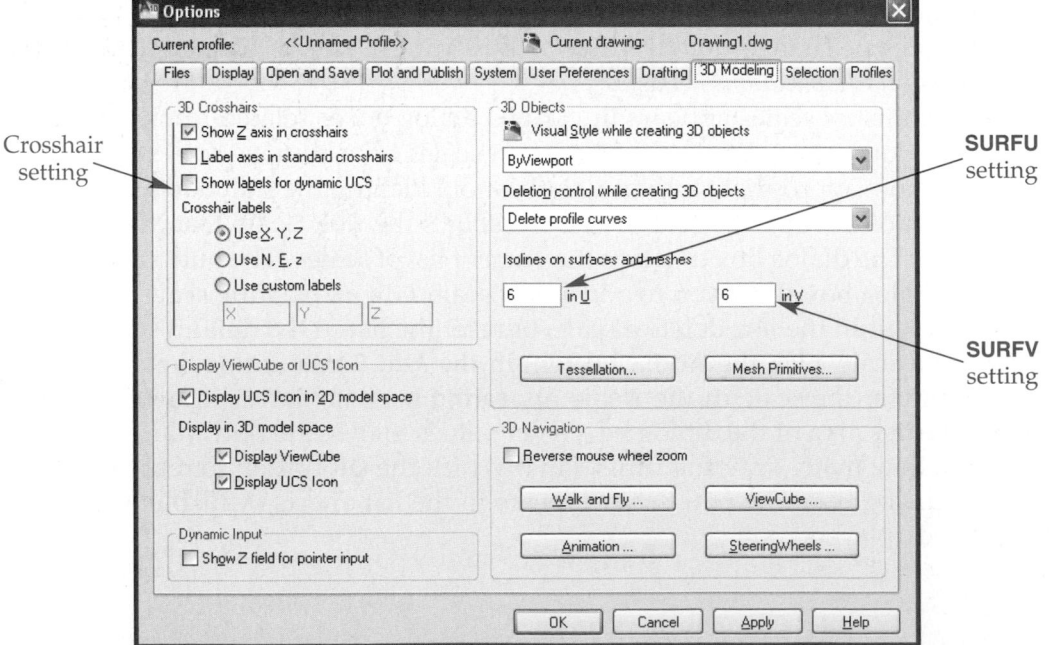

Crosshair setting

SURFU setting

SURFV setting

AutoCAD and Its Applications—Advanced

than 2D Wireframe, you have control of both the UCS icon and the view cube. The **Dynamic Input** area has a check box for showing the Z field for dynamic input.

The setting in the **Visual Style while creating 3D objects** drop-down list in the **3D Objects** area determines which visual style is set current when objects are created. The **Deletion control while creating 3D objects** drop-down list determines how geometry is handled when creating 3D objects. For example, when **Delete profile curves** is selected, the profile and path curves are deleted after a sweep is created. The two edit boxes in the **3D Objects** area set the **SURFU** and **SURFV** system variables for mesh objects and old-style surfaces. The **Tessellation...** and **Mesh Primitives...** buttons open dialog boxes that contain settings for controlling the options associated with mesh tessellation and mesh primitives. These settings are discussed thoroughly in Chapter 3.

The **3D Navigation** area has a check box for reversing the zoom direction of the mouse wheel. There are also four buttons in this area that allow access to settings for walkthroughs/flybys, animations, the view cube, and steering wheels. Selecting the **Walk and Fly settings...** button opens the **Walk and Fly Settings** dialog box. This dialog box contains settings used when creating walkthroughs and flybys. Selecting the **Animation settings...** button opens the **Animation Settings** dialog box. This dialog box contains settings that control the actual animation of a walkthrough or flyby. The **Walk and Fly Settings** and **Animation Settings** dialog boxes are discussed in detail in Chapter 19. Selecting the **View Cube** button opens the **View Cube Settings** dialog box. Selecting the **Steering Wheels** button opens the **Steering Wheels Settings** dialog box. These dialog boxes are discussed in Chapter 4.

Chapter Test

 *Answer the following questions. Write your answers on a separate sheet of paper or complete the electronic chapter test on the student website.*
www.g-wlearning.com/CAD

1. List two methods used to open the **Options** dialog box.
2. List the tabs found in the **Options** dialog box.
3. AutoCAD resides in the C:\Program Files\AutoCAD 2010 folder on your workstation. You have created two folders under \AutoCAD 2010 named \Projects and \Symbols. You want to store your drawings in the \Projects folder and your blocks in the \Symbols folder. What should you enter in the Support File Search Path area so these folders are added to the search path?
4. How do you open the **Drawing Window Colors** dialog box to change the color of AutoCAD screen elements?
5. For which AutoCAD features can you customize the font (not within a drawing)?
6. In which tab of the **Options** dialog box can you change settings for layout tabs?
7. Briefly describe how to turn on the automatic save feature and specify the save interval.
8. How do you select the folder in which the autosave file is saved?
9. What are the advantages of toggling the log file open?
10. Name the two commands that toggle the log file on and off.
11. How would you set the right mouse button to perform an [Enter], rather than displaying shortcut menus?
12. How do you open the **Edit Scale List** dialog box from within the **Options** dialog box?
13. On which tab of the **Options** dialog box is the **Block Editor Settings...** button located?
14. How do you change the number of seconds that AutoCAD waits to display an extended tooltip?
15. How do you access the view cube setting from within the **Options** dialog box?

Drawing Problems

1. Using the methods described in this chapter, change the AutoCAD screen colors to your liking. Customize the 3D parallel and 3D perspective contexts.

2. Set up AutoCAD so that profile curves are retained when creating 3D objects. Also, increase the isolines for meshes. To determine the best default setting, draw various meshes, change the setting, and draw additional meshes.

CHAPTER 22

Customizing Tools and Tool Locations

Learning Objectives

After completing this chapter, you will be able to:

✓ Explain the features of the **Customize User Interface** dialog box.
✓ Describe partial CUIx files.
✓ Create custom commands.
✓ Create new toolbars, ribbon tabs, and ribbon panels.
✓ Customize ribbon tabs and panels with submenus and drop-down lists.
✓ Explain how to customize menus.

One of the easiest ways to alter the AutoCAD environment is by customizing interface elements. *Interface elements* are graphic command-entry components of AutoCAD, such as the ribbon, toolbars, and menus. Existing ribbon panels, toolbars, and menus can be quickly modified by removing and adding commands. New commands can also be created and assigned to an existing ribbon panel, toolbar, or menu or to new interface elements. The most powerful aspect of customizing ribbon panels, toolbars, and menus is the ability to quickly create entirely new functions to help you in your work. The key to good customization can be broken down into four simple rules:

- Always make a backup of the original files, such as the acad.cuix file, *before* customizing.
- Do not over customize your work. Plan your customization in steps to minimize confusion and maximize productivity. Anticipate workflow and where needs exist.
- It is best to locate customized files in folders other than the default AutoCAD folders. This makes it easier to upgrade AutoCAD in the future.
- Thoroughly test your customizations before implementing them. This will save many headaches for you and the end user.

Customize User Interface Dialog Box

The **Customize User Interface** dialog box funnels all major graphical user interface elements of AutoCAD into one central area where they can be tailored for productivity, **Figure 22-1.** All AutoCAD commands are linked for customization. The interface

Figure 22-1.
The **Customize User Interface** dialog box is used to edit existing interface elements, create new interface elements, create custom commands, and create command icons.

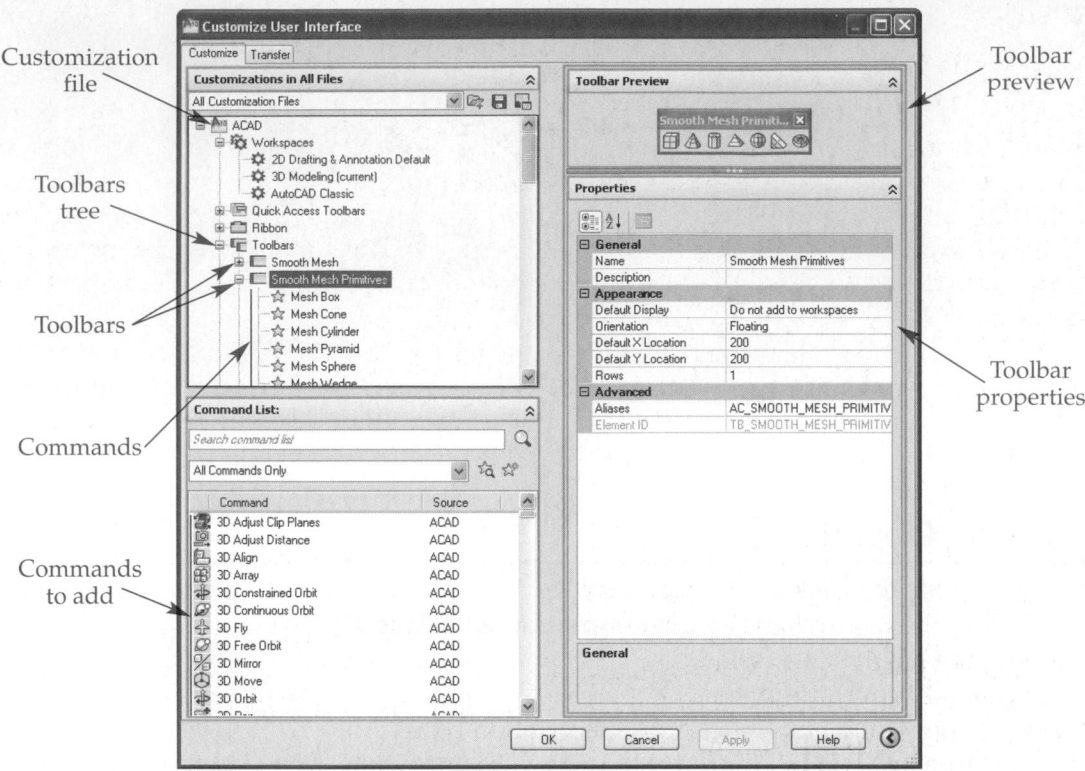

Customization file

Toolbars tree

Toolbars

Commands

Commands to add

Toolbar preview

Toolbar properties

elements that will be discussed in this chapter are the ribbon tabs and panels, toolbars, and menus located in the menu bar.

The **CUI** command displays the **Customize User Interface** dialog box. This can also be displayed by right-clicking on a toolbar and picking **Customize...** from the shortcut menu. The changes made in the **Customize User Interface** dialog box are saved in a customization (CUIx) file. By default, this is the acad.cuix file.

The upper-left pane of the **Customize User Interface** dialog box is initially labeled **Customizations in All Files**. The drop-down list located below the pane name contains the name of the main CUIx file and any other currently loaded partial CUIx files. Partial CUIx files are discussed later. By default, the main file acad.cuix and partial CUIx files custom.cuix, acimpression.cuix, autodeskseek.cuix, and acetmain.cuix files are installed. If you select a different entry from the drop-down list, the name of the pane changes to reflect the selection, either **All Customization Files** or **Main CUI File**.

By default, the customization file acad.cuix is the main CUIx file. In the box located below the drop-down list, the selected CUIx file is displayed in a tree. The top level of the tree is the ACAD branch, which is the name of the selected customization file, and the AutoCAD logo icon is shown next to it. The tree under the ACAD branch lists the various customizable interface elements. For example, to see the list of available toolbars, expand the Toolbars branch (node), by picking the plus sign located just to its left. Any other partially loaded CUIx files that have toolbars in them will be listed under the Partial Customization Files branch in the tree and can have their toolbar list similarly expanded.

The shortcut menus displayed in the **Customize User Interface** dialog box provide editing options based on the branch or item selected. Options are available for creating new items; renaming, deleting, copying/pasting, and duplicating items; adding menus or submenus; and inserting separators. For example, to delete a ribbon panel from the

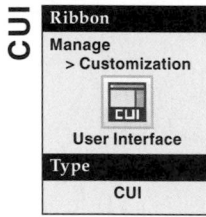

CUI

Ribbon
Manage
> Customization

CUI
User Interface

Type
CUI

interface, select it in the Ribbon Panels branch, right-click, and pick **Delete** from the shortcut menu. When prompted, pick the **Yes** button to delete the item. Then, pick the **Apply** or **OK** button in the **Customize User Interface** dialog box to make the deletion permanent. However, a better method is to remove a ribbon panel from the workspace. This way, the ribbon panel is still available to other workspaces. Refer to Chapter 25 for complete details on workspaces.

To rename content, such as a toolbar, select the toolbar in the Toolbars branch, right-click, and pick **Rename** from the shortcut menu. The existing name of the toolbar in the tree turns into an edit box with the current name highlighted. Type a new name in the edit box and press [Enter]. The new toolbar name is displayed in the tree. You can also rename content by editing the Name property in the **Properties** pane on the right-hand side of the dialog box. Pick the **Apply** or **OK** button in the **Customize User Interface** dialog box to make the change permanent.

You can modify all existing content by deleting and adding commands. The **Command List:** pane of the **Customize User Interface** dialog box contains all commands, including those that are not by default available on an interface element. Custom commands can be assigned to all interface elements, such as ribbons, toolbars, and menus.

The right-hand side of the **Customize User Interface** dialog box displays specific information of the highlighted content. Panes that will appear are **Information, Preview,** and **Properties**, depending on the selected content. It is easy to read and navigate the **Customize User Interface** dialog box by remembering the general information is stored in the upper-left pane, proceeding down to the **Command List:** pane and over to the panes on the right-hand side of the dialog box for more specific information.

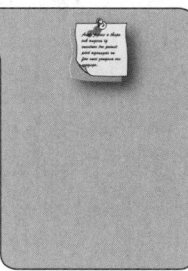

NOTE

All changes made in the **Customize User Interface** dialog box are saved in the CUIx file, including workspace, ribbon, toolbar, menu, and shortcut key customizations. Customizing the ribbon, toolbars, and menus is discussed later in this chapter. Customizing shortcut keys is discussed in Chapter 23. Customizing workspaces and the **Quick Access** toolbar is covered in Chapter 25.

Adding a Command to an Interface Element

All commands are available in the **Command List:** pane of the **Customize User Interface** dialog box. Included are many commands not found on the default ribbon, toolbars, or menus. Any custom commands you have created are also available in this pane.

To add a command to any interface element, first expand the tree for the element in the **Customizations in All Files** pane. For example, to add a command to a toolbar, expand the Toolbars branch so the branch is visible for toolbar to which you want the command added. Then, select a command from the **Command List:** pane. The list is alphabetized. If you hover the cursor over a command, the macro or command is displayed in a tooltip. See **Figure 22-2.**

You can search the command list by picking the **Find command or text** button at the top of the **Command List:** pane. The drop-down list at the top of the pane can be used to filter the list so that only commands in a certain category appear in the list. You can also filter the list by typing in the text box at the top of the pane. Only those commands containing the characters in this text box are displayed in the list.

Figure 22-2.
The **Commands List:** pane of the **Customize User Interface** dialog box displays all predefined and custom commands. These commands can be added to interface elements.

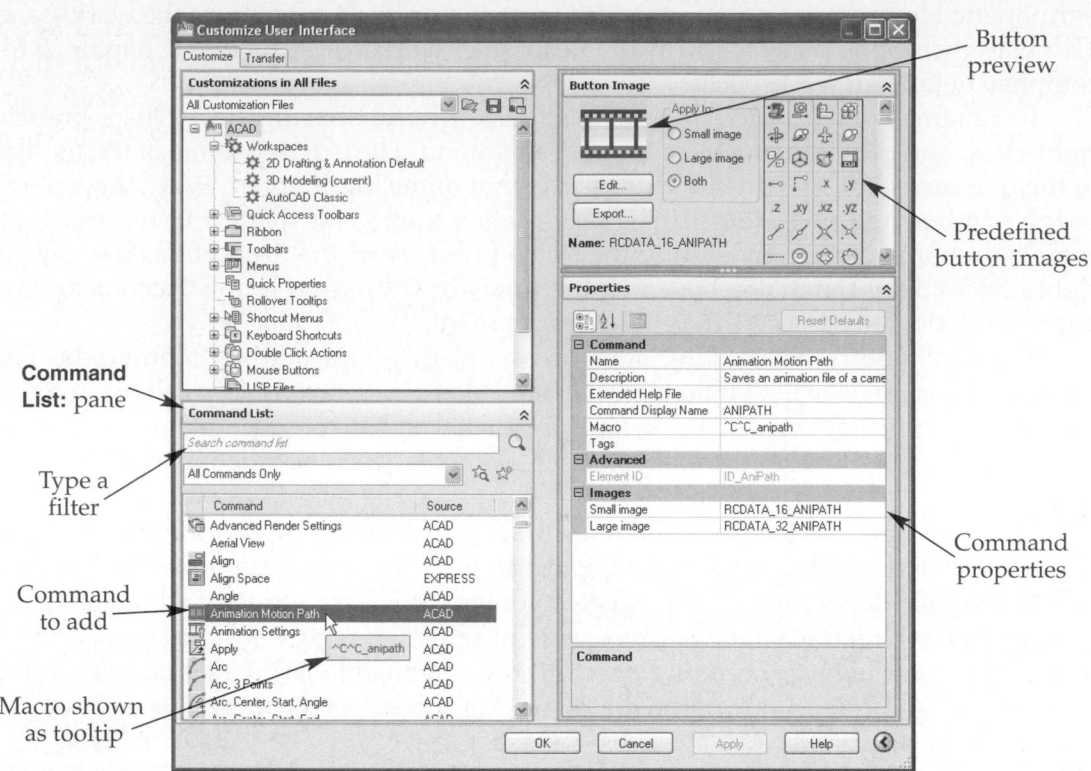

Once the command is located, pick and hold on the command in the **Command List:** pane and drag it into the **Customizations in All Files** pane. A horizontal "I-bar" appears in the pane as you drag the command. This represents the location where the command will be inserted. Position the new command between the commands where you would like it to appear and release the left mouse button. The new command is added to the branch for the interface element. Pick the **Apply** or **OK** button in the **Customize User Interface** dialog box to make the addition permanent.

> **NOTE**
>
> Selecting the **Cancel** button in the **Customize User Interface** dialog box after selecting the **Apply** button does *not* cancel the changes made before the **Apply** button was selected. Also, picking the Windows close button (the X) before picking the **Apply** button *cancels* the changes.

Deleting a Command from an Interface Element

To delete a command from an interface element, expand the tree for the element in the **Customizations in All Files** pane. You may need to expand more than one level to see the command you wish to delete. All of the commands currently on the interface element are displayed as branches below the element name.

Select the command you wish to delete, right-click, and pick **Remove** from the shortcut menu. You can also select the command and press the [Delete] key. Pick the **Apply** or **OK** button in the **Customize User Interface** dialog box to make the deletion permanent. The command is, however, still available in the **Command List:** pane of the **Customize User Interface** dialog box. It can be referenced to another interface element.

Moving and Copying Commands

You can move and copy commands between any interface element. First, in the upper-left pane of the **Customize User Interface** dialog box, expand the tree for both elements that you wish to edit. To move a command from one element to another, pick and hold on the command and drag it to the other interface element. The horizontal "I-bar" cursor appears as you drag. Position the cursor between the commands where you want the new command to appear and release the left mouse button. The command is moved from the first element to the second.

Use this same process to copy a command between elements, but hold the [Ctrl] key before you release the left mouse button. The command remains on the first element and a copy is placed on the second element.

You can also drag commands from the **Customize User Interface** dialog box and drop them onto toolbars and tool palettes that are currently displayed in the AutoCAD window. However, this method cannot be used to add a command tool to the ribbon.

A command can be removed from a displayed toolbar while the **Customize User Interface** dialog box is open by dragging it from the toolbar into the drawing area and releasing. A message appears asking if you want to remove the button. Pick **OK** to remove the button.

Buttons can also be rearranged on displayed toolbars while the **Customize User Interface** dialog box is open by simply dragging a button to a new position. However, it is recommended that you use the **Customize User Interface** dialog box to make edits to toolbars until you are completely comfortable with the drag-and-drop method.

PROFESSIONAL TIP

When dragging a command to a tool palette, if the desired palette is not current (on top), simply pause the cursor over the palette name until the palette is made current. Tool palette customization is discussed in detail in Chapter 24.

Adding a Separator to an Interface Element

A *separator* is a vertical or horizontal line that can be used in toolbars and menus to create visual groupings of related commands. On ribbon panels, a separator is a gap between tools. For example, look at the **Draw** panel in the **Home** tab of the ribbon with the 3D Modeling workspace set current. Between the **3D Polyline** button and the **Arc** drop-down list there is a gap between the two indicating a separator. There is also a separator between the **Line** button and the **Circle** drop-down list and between the **Rectangle** button and **Ellipse** drop-down list. These separators are vertical gaps because the ribbon is docked along the top edge. If the ribbon is docked along the left or right side of the screen, the separators will be horizontal gaps. Likewise, if a toolbar is horizontal, the separator is a vertical line. If a toolbar is vertical, the separator is a horizontal line.

To add separators to an interface element, first open the **Customize User Interface** dialog box. Then, in the **Customizations in All Files** pane, expand the branch for the element to which you want separators added.

For a toolbar or menu, right-click on the command in the tree below which you want the separator added. Select **Insert Separator** from the shortcut menu. A separator, represented by two dashes, appears in the tree below the selected command.

For a ribbon panel, right-click on the row in the panel to which you want a separator added. Then, select **Add Separator** from the shortcut menu. A separator is added to the bottom of the row's branch.

Once a separator is added, it can be dragged to a new location in the tree. When done adding and moving separators, pick the **OK** button to close the **Customize User Interface** dialog box.

A ribbon panel can also have a panel separator, called a *slideout*. The commands in the tree below the slideout appear in the expanded panel. For example, the **Draw** panel in the **Home** tab of the ribbon (with the 3D Modeling workspace current) has a slideout below the row of buttons containing the **Polygon**, **Rectangle** and **Ellipse** buttons. The slideout is automatically added to all panels.

<table><tr><td>NOTE</td></tr></table>

A separator can be removed from an interface element in the same manner as removing a command.

Partial CUIx Files

A *partial CUIx file* is any CUIx file that is not the main CUIx file (acad.cuix). To load a partial CUIx file, pick the Open… entry in the drop-down list in the **Customizations in All Files** pane. Remember, the name of this pane may be different, depending on what is currently selected in the drop-down list. You can also pick the **Load partial customization file** button to the right of the drop-down list. Next, in the **Open** dialog box that is displayed, navigate to the folder where the CUIx file is located, select the file, and pick the **Open** button. If the partial CUIx file that has been opened contains any workspaces, the AutoCAD alert shown in **Figure 22-3** is displayed. Any workspace information contained in the CUIx file is not automatically available. Workspaces are covered in Chapter 25.

Once you open the CUIx file, it is automatically selected in the drop-down list. The name of the pane changes to **Customizations in Main CUI**. Now, you can manage the items contained within the partial CUIx.

If you select either the main CUIx or All Customization Files in the drop-down list, the Partial Customization Files branch appears in the tree. Expanding this branch, you can see the partial CUIx files that are loaded. Expanding the branch for a partial CUIx file, you can see the items contained within the CUIx file. These items can be copied from the partial CUI file to the main CUI file as needed.

To unload a partial CUIx file, select All Customization Files in the drop-down list in the "customizations" pane. Then, expand the Partial Customization Files branch, right-click on the name of the CUIx file, and select **Unload** *CUIx_file_name* from the shortcut menu.

<table><tr><td>PROFESSIONAL TIP</td></tr></table>

You can also unload a partial CUIx file by typing MENULOAD or MENUUNLOAD at the Command: prompt. Then, in the **Load/Unload Customizations** dialog box, select the CUIx file to unload and pick the **Unload** button.

Figure 22-3.
This warning appears when loading a partial CUIx file that contains workspaces. Workspaces are covered in Chapter 25.

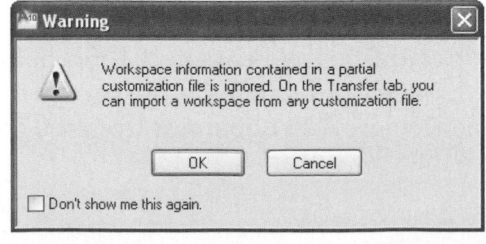

Creating New Commands

You are not limited to AutoCAD's predefined commands. *Custom commands* can be created and then added to ribbon panels, menus, tool palettes, and toolbars. First, however, you must create the new command. To create a custom command, first pick the **Create a new command** button in the **Command List:** pane of the **Customize User Interface** dialog box. This button is to the right of the drop-down list. A new command is added to the list in the **Command List:** pane. Also, the **Button Image** and **Properties** panes are displayed for the new command. See **Figure 22-4.**

General Command Properties

By default, the new command name is **Command***n*, where *n* is a sequential number. To give the command a descriptive name, highlight the command in the **Command List:** pane. Next, pick in the Name property edit box in the **Command** category of the **Properties** pane. Then, type the new name and press [Enter]. This property is displayed as the command name on the status bar and in the tooltip. The name should be logical and short, such as **Draw Box**. The entry in the Command Display Name property is what appears in the command-line section of the tooltip.

The text that appears in the Description property text box in the **Command** category of the **Properties** pane appears on the AutoCAD status line when the cursor is over the button. This text, called the *help string,* should also be logical, but can be longer and more descriptive than the command name.

The Extended Help File property is used to specify an Extensible Application Markup Language (XAML) file to use as extended help. The *extended help* is displayed in the tooltip when the cursor is paused over a tool for a longer period of time. By

---

Figure 22-4.
The first step in adding a custom command to an interface element is to create the custom command.

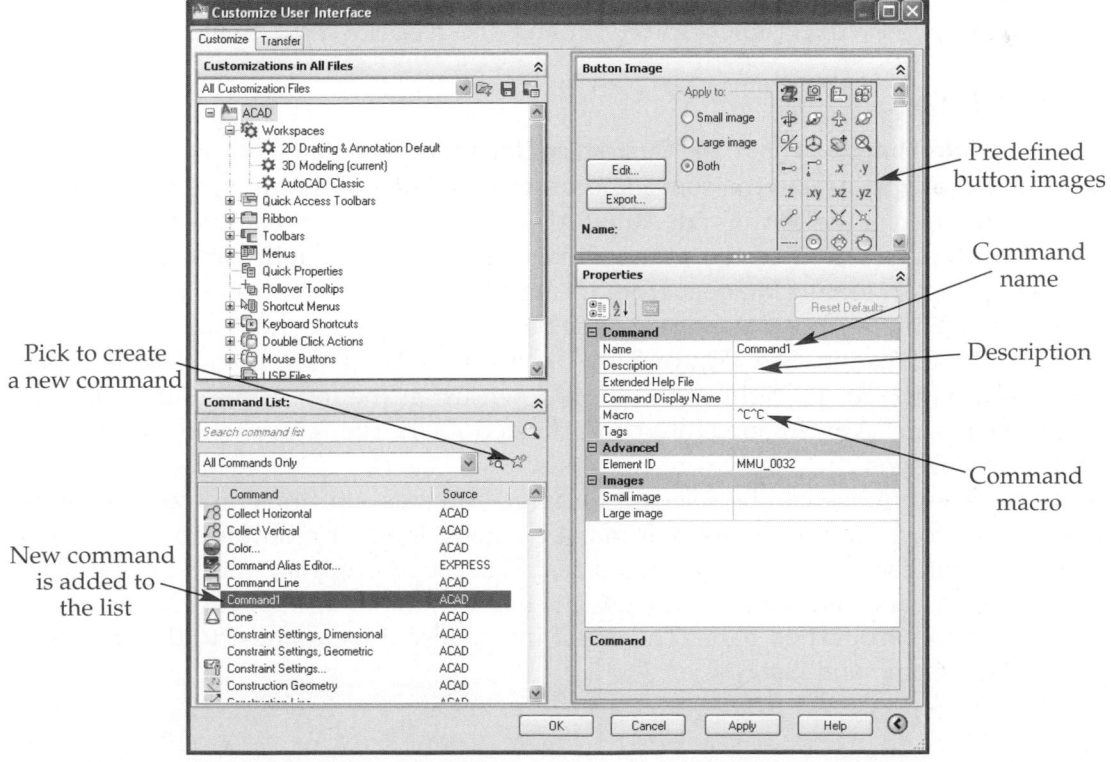

Figure 22-5.
The custom command is named and a help string and tag are assigned to it.

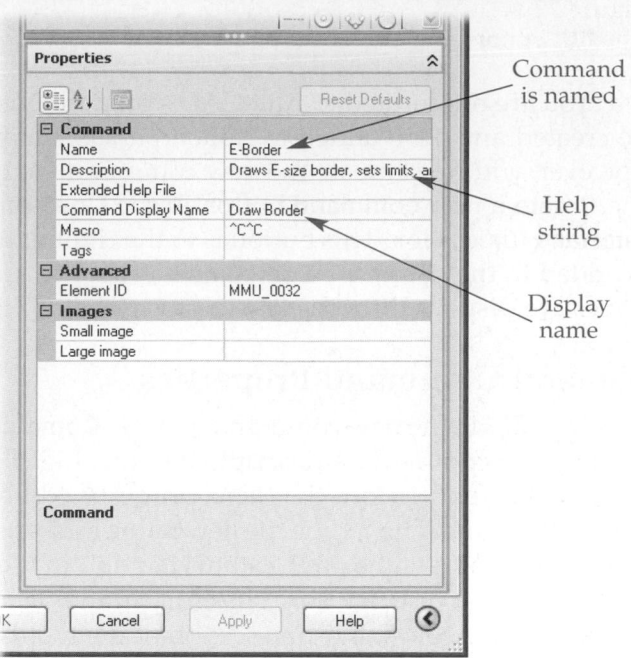

Command is named

Help string

Display name

default, if you pause the cursor for two seconds, the extended help is displayed (if the tool contains extended help). To assign an XAML file, select the property and pick the ellipsis button (...) on the right-hand side of the text box. Then, in the standard open dialog box that appears, locate and open the file. For information on creating XAML files, search the Internet for resources. Many resources can be found on the Microsoft website.

As an example, you will create a command that draws a rectangular border for an E-size sheet (44″ × 34″) using a wide polyline, sets the drawing limits, and finishes with **ZOOM Extents**. To start, create a new command and enter **E-Border** as the name. Also, enter Draws E-size border, sets limits, and zooms extents. for the Description property. In the Command Display Name property, enter Draw Border. See **Figure 22-5.** In the next sections, you will complete the command and its associated image.

Button Image

The **Button Image** pane in the **Customize User Interface** dialog box is used to define the image that appears on the command button. The image should graphically represent the function of the command. AutoCAD provides several predefined images. One of these can be selected as the button image. You can also right-click on the list of images and select **Import Image...** from the shortcut menu to import an image.

The **CUI Editor—Image Manager** dialog box can be used to control and store custom images in a CUIx file. See **Figure 22-6.** To display this dialog box, pick the **Image Manager...** button to the right of the drop-down list at the top of the **Customizations in All Files** pane. Any image stored in a loaded CUIx file is available in the list of predefined images.

A different image can be selected for large and small buttons or you can use the same image for both button sizes. It may be a good idea for a button to have a separate image for each of the two button sizes. Pick the appropriate radio button in the **Button Image** pane and select an image. The name of the image appears in the **Images** category in the **Properties** pane. The small image also appears next to the command name in the **Command List:** pane.

Figure 22-6.
The **CUI Editor—Image Manager** dialog box is used to store button images in a CUIx file.

Selected
CUIx file

Image stored
in CUIx file

Preview
of image

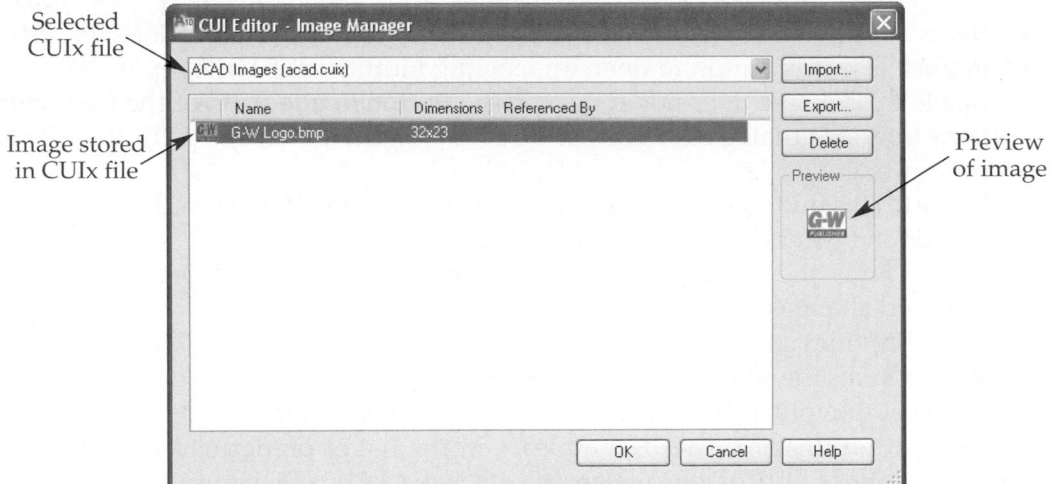

However, confusion may arise if your custom command has the same button image as an existing AutoCAD command. It is best to create custom button images for use with your custom commands. The **CUI Editor—Image Manager** dialog box makes it easy to manage the button images. You can either modify an existing button image or create a new image from scratch. In either case, a predefined image must be selected from the list of existing images. Then, pick the **Edit...** button in the **Button Image** pane to open the **Button Editor** dialog box. This is described in the next section.

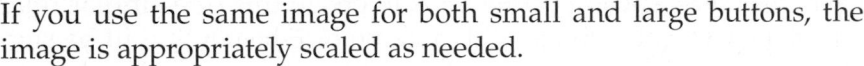

NOTE

If you use the same image for both small and large buttons, the image is appropriately scaled as needed.

Creating a custom button image

The **Button Editor** dialog box has basic "pixel-painting" tools and several features to simplify the editing process. The four tools are shown as buttons at the top of the dialog box. The pencil paints individual pixels. The line tool allows you to draw a line between two points. The circle tool allows you to draw center/radius style ellipses and circles. The erase tool clears the color from individual pixels. The current color is selected from the color palette on the left-hand side of the dialog box and indicated by a depressed color button. Anything you draw appears in the current color. A preview of the button image appears to the right of the tools.

Drawing a button image is usually much easier with the grid turned on. The grid provides outlines for each pixel in the graphic. Each square represents one pixel. Picking the **Grid** check box toggles the state of the grid.

When the toolbar buttons are set to their default, small size, the button editor provides a drawing area of 16 pixels × 16 pixels. If **Use large buttons for Toolbars** is turned on in the **Display** tab of the **Options** dialog box, then the button image drawing area is 32 pixels × 32 pixels. The images in the **Customize User Interface** dialog box are displayed at the current size setting (small or large).

There are several other tools available in the **Button Editor** dialog box. These include the following.

- **Clear.** If you want to erase everything and start over, pick the **Clear** button to clear the drawing area. This is the button you will use to clear the existing image and start a button image from scratch.

- **Undo.** You can undo the last operation by picking this button. Only the last operation can be undone. An operation that has been undone cannot be redone
- **Save.** Names the current button image and saves it to the current CUIx file.
- **Import.** Use this button to open an existing bitmap (BMP) file, up to 380 × 380 pixels in size, that does not appear in the **Button Image** pane of the **Customize User Interface** dialog box. The image is automatically resized to fit the current button size.
- **Export.** This button saves a file using a standard save dialog box. Use this when you do not want to alter the original button image.
- **Close.** Ends the **Button Editor** session. A message is displayed if you have unsaved changes.
- **Help.** Provides context-sensitive help.
- **More.** Opens the standard **Select Color** dialog box. This allows you to use colors in the button other than those in the default color palette.

Once a button image is saved, it appears in the list of predefined images in the **Button Image** pane of the **Customize User Interface** dialog box. All images saved for use as button images must be stored where AutoCAD will find them. AutoCAD provides the \Icons folder within the user's support file search path. This is the default folder when using the **Export...** button in the **Button Editor** dialog box. If you choose to use a different folder, it must be added to the support file search path, which is specified in the **Files** tab of the **Options** dialog box.

Rather than using an existing button image for the **E-Border** command, an entirely new button image will be created. With **E-Border** highlighted in the **Command List:** pane, select any one of the images in the **Button Image** pane and pick the **Edit...** button. The **Button Editor** dialog box is displayed. Now, select the **Clear** button to completely remove the existing image.

Figure 22-7A shows a 16 × 16 pixel image created for the **E-Border** button with the **Grid** option activated. Use the pencil and line tools to create this or a similar image. Using the **Save...** button, save your image with a name of E-border; it will be stored in the current CUIx. Pick the **Close** button to return to the **Customize User Interface** dialog box. Your newly created image now appears in the list of existing images in the **Button Image** pane, as shown in **Figure 22-7B.** It is automatically associated with the command.

Figure 22-7.
A—A custom button image is created in the **Button Editor** dialog box. B—The new button image has been saved and appears in the list.

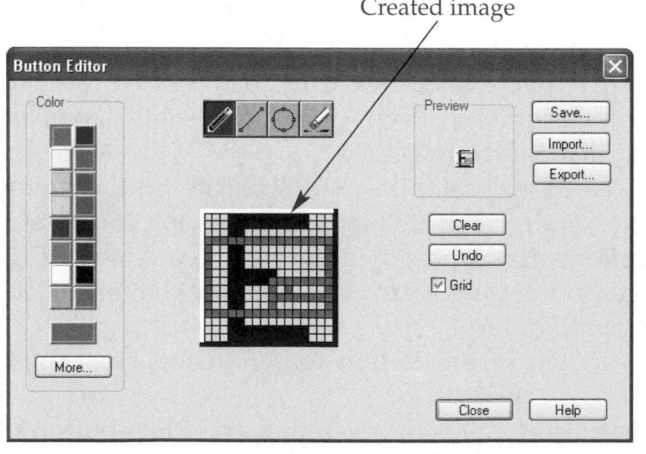
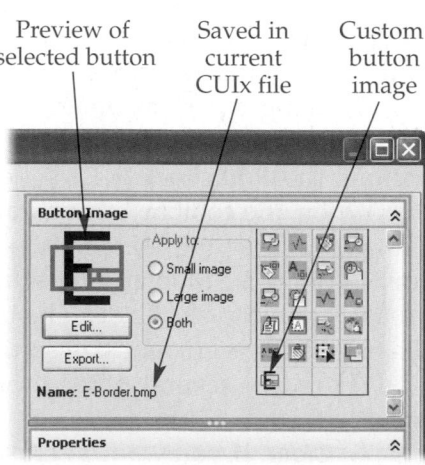

A   B

Associating a custom image with a command

There are two ways to associate a new, custom button image with a command. You can use the **Button Image** pane or the **Properties** pane in the **Customize User Interface** dialog box. Once a button image is associated with a command, the image is used for that command on *all* ribbon panels, menus, and toolbars where the command is inserted.

To use the **Button Image** pane to assign an image to a command, first make sure the command is selected in the **Command List:** pane. Then, select the **Large**, **Small**, or **Both** radio button in the **Button Image** pane to determine for which size of button the image will be used. Next, pick the button image in the list of predefined button images. Finally, pick the **Apply** button at the bottom of the **Customize User Interface** dialog box to assign the image to the button.

You can also use the **Properties** pane to associate a saved button image file with the command. Make sure the command is selected in the **Command List:** pane. Then, in the **Properties** pane, expand the **Images** category to display the Small image and Large image properties. If there is an image currently associated with the property, the path to the image is displayed in the text box, **Figure 22-8.** If there is no path displayed,

Figure 22-8.
The custom button image has been assigned to the custom command.

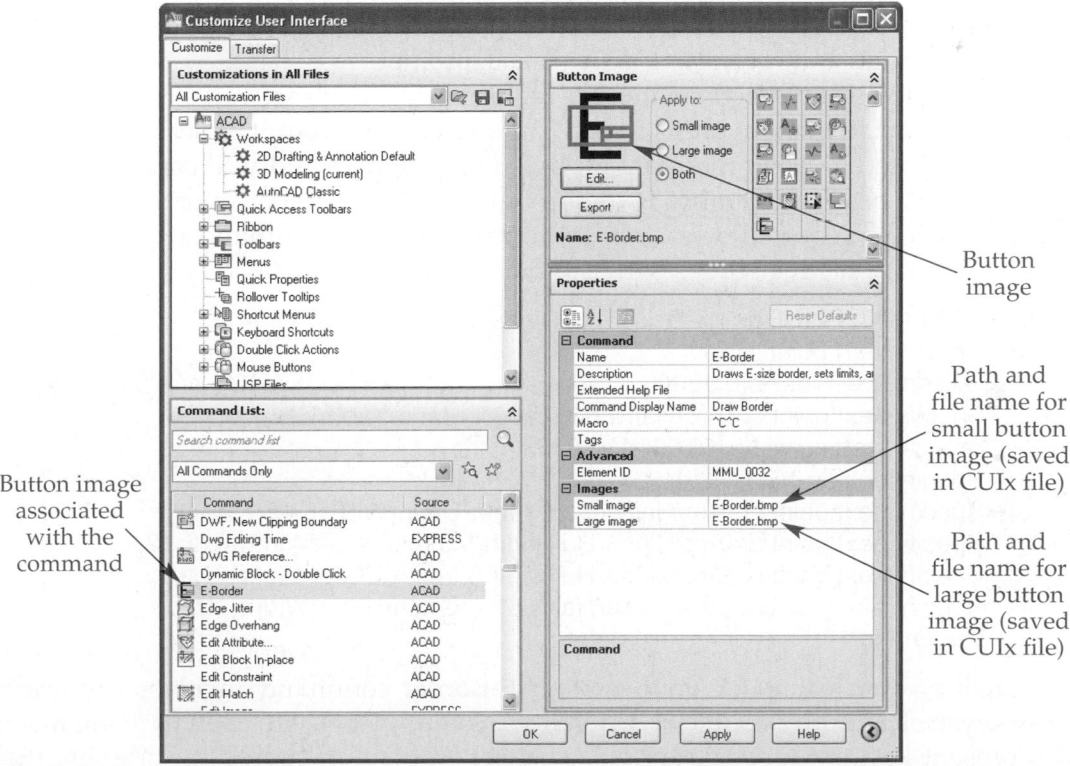

the image is saved in a CUIx file. Pick in each property text box and type the path and file name of the saved image files. Alternately, you can pick the ellipsis button (…) to display a standard open dialog box and locate the file. This button appears when the property is selected. Finally, pick the **Apply** button at the bottom of the **Customize User Interface** dialog box to assign the image(s) to the button.

If you only designate an image file for small buttons, the button for the command will be blank when you switch to large buttons. This is because no image has been designated for that size. Be sure to specify an image for both small and large buttons.

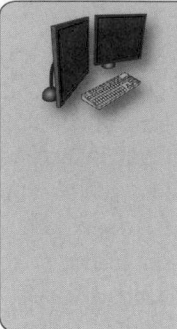

PROFESSIONAL TIP

To open and edit a button image that is not shown in the button image list, select any button image and then pick the **Edit…** button to display the **Button Editor**. Then, use the **Import…** button to open the button image you want to edit. To assign an image that does not appear in the image list, use the **Properties** pane as described above. Additionally, you may use the **CUI Editor—Image Manager** dialog box to import images into a loaded CUIx file, which are then available in the image list.

Defining a Custom Command

Now, you need to define the action that the custom command will perform. A text string called a *macro* defines the action performed by the command. This text string appears in the Macro property text box in the **Command** category in the **Properties** pane of the **Customize User Interface** dialog box. In many cases, this "command" is actually a macro that invokes more than one command. By default, the text ^C^C appears in the text box. The text ^C is a cancel command. This is the same as pressing the [Esc] key. The default text, then, represents two cancels.

Two cancels are required to be sure you begin at the Command: prompt. One cancel may not completely exit some commands. In this case, a second cancel is required to fully exit the command. Whenever a command is not required to operate transparently, it is best to begin the macro with two cancel keystrokes (^C^C) to fully exit any current command and return to the Command: prompt.

The macro must perfectly match the requirements of the activated commands. For example, if the **LINE** command is issued, the subsequent prompt expects a coordinate point to be entered. Any other data input is inappropriate and will cause an error in the macro. It is best to manually "walk through" the desired macro, writing down each step and the data required by each prompt. The following command sequence creates the rectangular polyline border with a .015 line width.

```
Command: PLINE↵
Specify start point: 1,1↵
Current line-width is 0.0000
Specify next point or [Arc/Halfwidth/Length/Undo/Width]: W↵
Specify starting width <0.0000>: .015↵
Specify ending width <0.0150>: ↵
Specify next point or [Arc/Halfwidth/Length/Undo/Width]: 42,1↵
Specify next point or [Arc/Close/Halfwidth/Length/Undo/Width]: 42,32↵
Specify next point or [Arc/Close/Halfwidth/Length/Undo/Width]: 1,32↵
Specify next point or [Arc/Close/Halfwidth/Length/Undo/Width]: C↵
Command:
```

Creating the macro for your custom **E-Border** command involves duplicating these keystrokes, with a couple of differences. Some symbols are used in menu macros to represent keystrokes. For example, a cancel (^C) is not entered by pressing [Esc].

Instead, the [Shift]+[6] key combination is used to place the *caret* symbol, which is used to represent the [Ctrl] key in combination with the subsequent character (a C in this case). Another keystroke represented by a symbol is the [Enter] key. An [Enter] is placed in a macro as a semicolon (;). A space can also be used to designate [Enter]. However, the semicolon is more commonly used because it is very easy to count to make sure that the correct number of "enters" are supplied.

AutoCAD system variables and control characters can be used in menus. They can be included to increase the speed and usefulness of your menu commands. Become familiar with these variables so you can make use of them in your menus.

- **^B.** Snap mode toggle.
- **^C.** Cancel.
- **^D.** Dynamic UCS toggle.
- **^E.** Isoplane crosshair toggle.
- **^G.** Grid mode toggle.
- **^H.** Issues a backspace.
- **^I.** Issues a tab.
- **^M.** Issues a return.
- **^O.** Ortho mode toggle.
- **^P. MENUECHO** system variable toggle.
- **^Q.** Toggles echoing of prompts, status listings, and input to the printer.
- **^T.** Tablet toggle.
- **^V.** Switches current viewport.
- **^Z.** Suppresses the addition of the automatic [Enter] at the end of a command macro.
- **\.** Pauses for user input

Keeping the above guidelines in mind, the following macro draws the polyline border.

> ^C^CPLINE;1,1;W;.015;;42,1;42,32;1,32;C;

Compare this with the command line entry example to identify each part of the macro.

The next steps that the command will perform are to set the limits and zoom to display the entire border. To do this at the command line requires the following entries.

> Command: **LIMITS**⏎
> Reset Model space limits:
> Specify lower left corner or [ON/OFF] <0.0000,0.0000>: **0,0**⏎
> Specify upper right corner <12.0000,9.0000>: **44,34**⏎
> Command: **ZOOM**⏎
> Specify corner of window, enter a scale factor (nX or nXP), or
> [All/Center/Dynamic/Extents/Previous/Scale/Window/Object] <real time>: **E**⏎ *(this prompt will differ if the current view is perspective, but the entry is the same)*
> Command:

Continue to develop the macro by adding the following text string (shown in color) immediately after the previous one.

> ^C^CPLINE;1,1;W;.015;;42,1;42,32;1,32;C;**LIMITS;0,0;44,34;ZOOM;E**

An "enter" is automatically issued at the end of the macro, so it is not necessary to place a semicolon at the end. The macro for the custom command is now complete.

To assign the macro to your custom **E-Border** command, first make sure the command is selected in the **Command List:** pane of the **Customize User Interface** dialog box. Then, pick in the Macro property text box in the **Properties** pane and enter the complete macro shown above. For a long macro such as this one, you can pick the ellipsis button (...) at the end of the text box to display the **Long String Editor**. See **Figure 22-9.** Enter the macro in this dialog box and pick the **OK** button to return to

Figure 22-9.
The **Long String Editor** dialog box can be used to write longer macros. The text string will automatically "wrap" in this dialog box, which does not affect the macro.

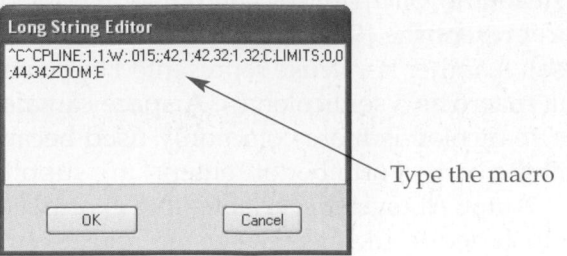

Long String Editor

^C^CPLINE;1,1;W;.015;;42,1;42,32;1,32;C;LIMITS;0,0
;44,34;ZOOM;E

Type the macro

OK   Cancel

the **Customize User Interface** dialog box. Finally, pick the **Apply** button to associate the macro with the custom command.

Exercise 22-1

Complete the exercise on the student website.
www.g-wlearning.com/CAD

Placing a Custom Command on an Interface Element

The custom command is now fully defined. The macro has been written and associated with the command. A custom button image has also been created and associated with the command. Now, you can add the custom command to a ribbon panel, menu, or toolbar just as you would one of the predefined AutoCAD commands. This is introduced earlier and covered in detail later in this chapter in the specific sections on the ribbon, toolbars, and menus.

After adding the custom command to an interface element, it should be fully functional when you exit the **Customize User Interface** dialog box. Once you close the dialog box, test the command to make sure it works. If it does not, edit the macro in the **Customize User Interface** dialog box as needed.

Overview of the Ribbon

By default, the ribbon is docked at the top of the graphics area. The ribbon can be floating or docked to the left or right as well. The ribbon contains commands and tools on *panels*. The panels are grouped on *tabs* that can be individually displayed. Think of the tabs as the containers that hold the ribbon panels. Together, the panels and tabs make up the ribbon.

To the right of the last tab name is an arrow icon. This button is used to change the appearance of the docked ribbon. The three options are to show the full ribbon, minimize to panel tiles, or minimize to tabs. When minimized to panel tiles, hover the cursor over a title and the corresponding panel is displayed. When minimized to tabs, pick the tab name and the tab is displayed. See **Figure 22-10.**

A ribbon panel may contain rows of command buttons, drop-down lists, or sliders. You can choose which panels are visible by right-clicking on the ribbon to display a shortcut menu. See **Figure 22-11.** Select either **Tabs** or **Panels** and choose which content to display. The currently displayed items are checked in the submenus. The items displayed in the **Panels** submenu are based on which tab is current (on top). There are separate panels for each tab.

Figure 22-10.
The three display states of the ribbon when it is docked. A—Full. B—Minimized to panel titles. C—Minimized to tabs.

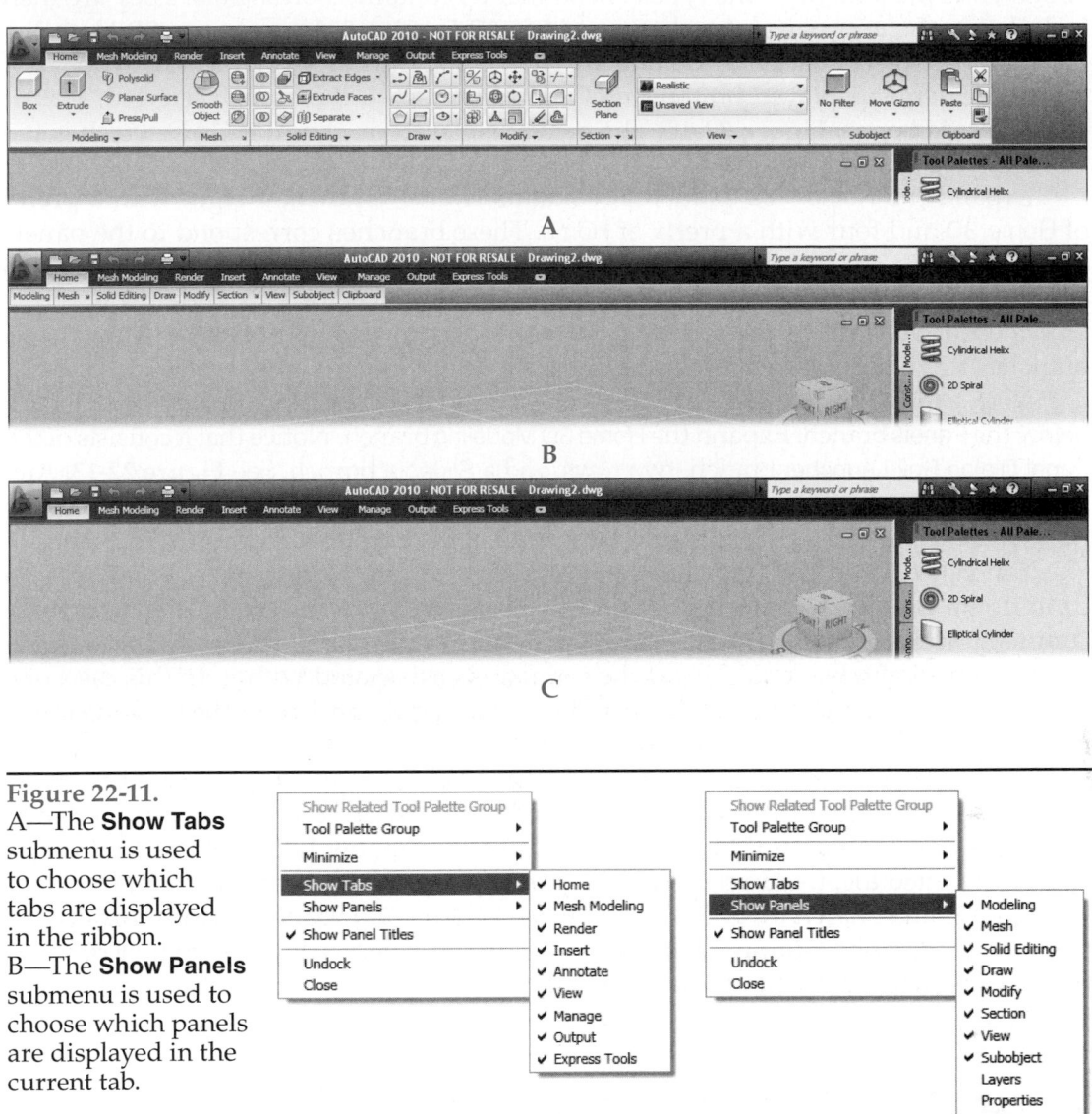

A

B

C

Figure 22-11.
A—The **Show Tabs** submenu is used to choose which tabs are displayed in the ribbon.
B—The **Show Panels** submenu is used to choose which panels are displayed in the current tab.

A

B

Workspaces are typically used to set which ribbon components are displayed. Chapter 25 discusses customizing workspaces. Also, the ribbon can dynamically change when commands are accessed. *Contextual tabs* may be displayed on the ribbon when a command is active and then hidden when the command is finished. Contextual tabs are discussed further in the next section.

Tabs and Panels

The ribbon has three customization branches in the **Customize User Interface** dialog box: the Tabs branch, Panels branch, and Contextual Tab States branch. These branches are located in the Ribbon branch. Since the tabs contain the panels, they are displayed first in the tree. In this section, you will examine each area to gain a better understanding of the composition of the panels and how panels relate to tabs.

In the **Customizations in All Files** pane, expand the Tabs branch. There are 12 default tabs associated with the main user interface. These appear at the top of the branch. The remaining tabs have "contextual" in their name. These are used to reference the contextual tab states. Some of the main ribbon tab names have the suffix 2D or 3D. These help identify the types of commands contained on the tab. They are then included in the appropriate workspace, either 2D Drafting & Annotation or 3D Modeling. For example, select the Home - 3D branch. Notice that the name of this branch does not match the name displayed on the AutoCAD screen. With the branch selected, look at the **Properties** pane, **Figure 22-12.** The name displayed on the AutoCAD screen is the value in Display Text property.

Expand the Home - 3D branch. It contains twelve branches: eight with a prefix of Home 3D and four with a prefix of Home. These branches correspond to the panels associated with the **Home** tab in the 3D Modeling workspace. Notice that there are no branches below these. The Tabs branch contains branches for tabs and, below those, branches for panels. The Panels branch contains branches for panels and, below those, branches for the commands on each panel.

Now, expand the Panels branch. All of the available panels are shown as branches below the Panels branch. Expand the Home 3D Modeling branch. Notice that it consists of the Panel Dialog Box Launcher branch, two rows, and a Slideout branch. See **Figure 22-13.** Any row listed below the Slideout branch is only visible when the ribbon panel is expanded. For the **Modeling** panel in the ribbon, row 2 is located in the expanded portion of the panel.

You can expand the branches for the rows in a tab. Notice row 1 contains two drop-down lists and a subpanel. The icon for a drop-down list looks similar to a command icon, but the star has a small arrow at the bottom. If you expand the branch for the drop-down list, you can see the commands associated with it. In this case, one drop-down list contains the solid primitive commands and the other one contains commands such as **LOFT** and **EXTRUDE**.

Next, expand the branch for the subpanel (Sub-Panel 1). This subpanel contains three rows. Expanding the branch for each row displays the commands contained in it. Notice how the drop-down list and subpanel contained in row 1 of the Home 3D Modeling branch are fitted together in the **Modeling** panel in the ribbon. See **Figure 22-14.**

Expand the branch for row 2 in the Home 3D Modeling branch. Notice this branch contains commands, but no drop-down lists or subpanels. Also, notice that row 2 is below the Slideout branch, which is the panel separator. This means it is displayed in the expanded panel. Refer to **Figure 22-14.**

Figure 22-12.
The properties of the Home 3D tab.

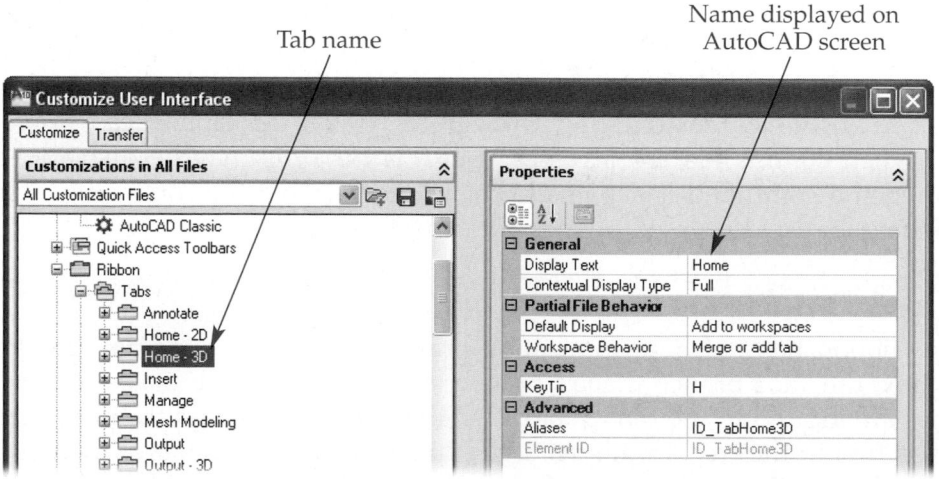

Figure 22-13.
Notice how a ribbon panel is composed in the **Customize User Interface** dialog box.

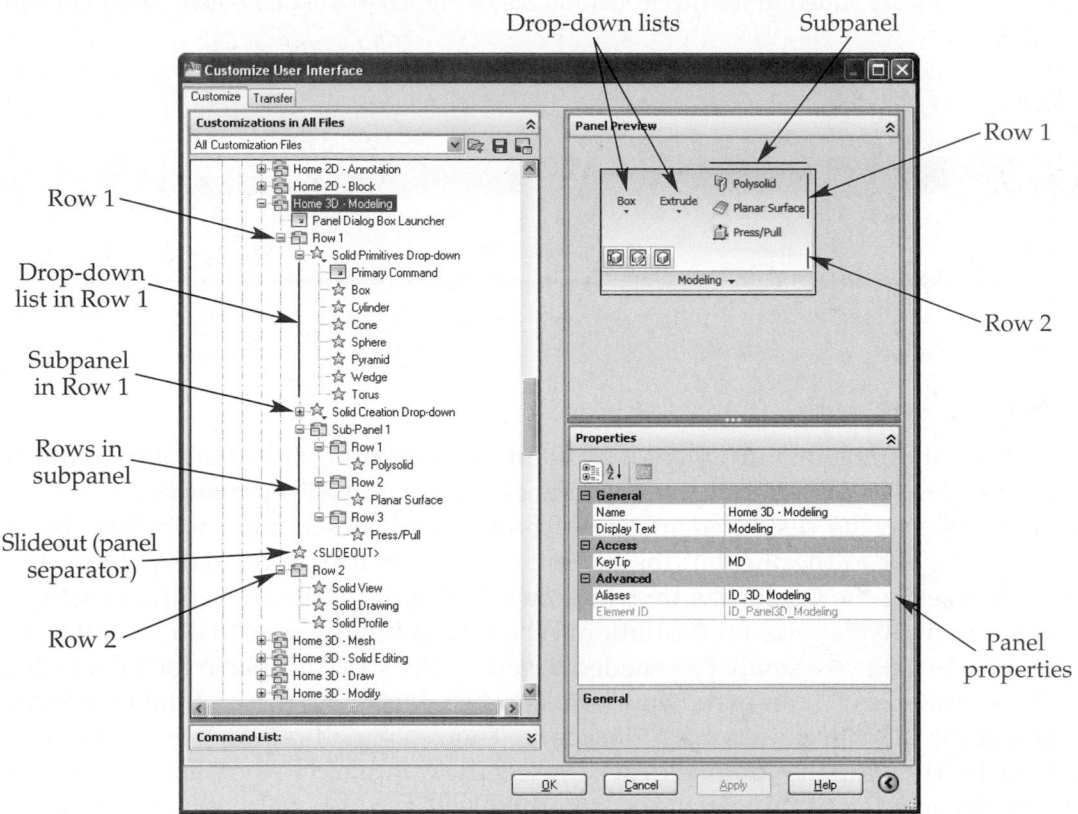

Drop-down lists · Subpanel · Row 1 · Row 2

Row 1

Drop-down list in Row 1

Subpanel in Row 1

Rows in subpanel

Slideout (panel separator)

Row 2

Panel properties

Figure 22-14.
Notice how a ribbon panel is composed on the ribbon.

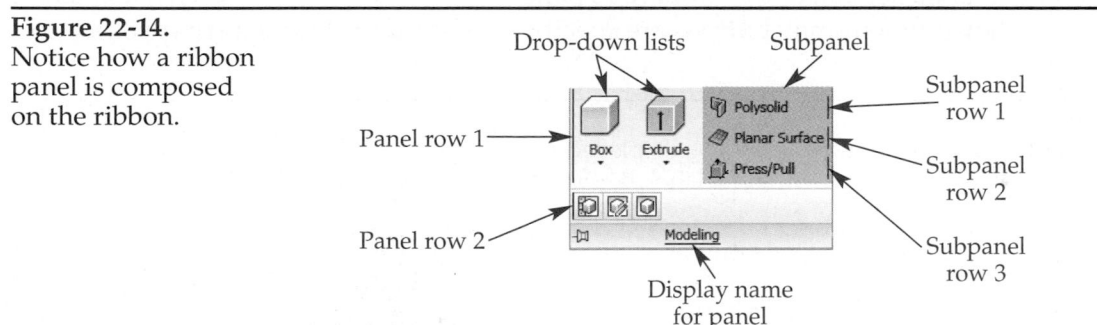

Drop-down lists · Subpanel

Panel row 1 · Panel row 2

Subpanel row 1 · Subpanel row 2 · Subpanel row 3

Display name for panel

The final branch in the Ribbon branch is Contextual Tab States. Expand this branch and several branches of AutoCAD commands and features are displayed. Some branches have contextual tab panels assigned to them and others do not. For example, expand the Text Editor in progress branch. Below it is the Text Editor Contextual Tab branch. This indicates the **Text Editor** contextual tab will be displayed when text is being created in the drawing.

For example, use the **MTEXT** command to create sample multiline text. While the command is active, the **Text Editor** tab is displayed in the ribbon and made active. See **Figure 22-15.** When the command is complete, the tab is automatically removed from the ribbon.

You can associate tabs for any command or feature listed in the Contextual Tab States branch in the **Customize User Interface** dialog box. To do so, drag the tab from the Tab branch and drop it in the desired branch in Contextual Tab States branch.

Figure 22-15.
When the **MTEXT** command is active, the **Text Editor** contextual tab is displayed in the ribbon. Contextual tabs are added in the Contextual Tab States branch in the **Customize User Interface** dialog box.

Contextual tab for
MTEXT command

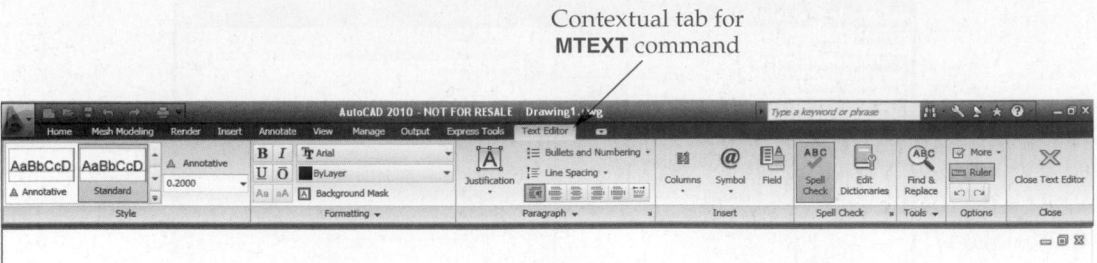

Button Properties

In the ribbon, command buttons have additional properties from the basic command properties. These are located in the **Appearance** section of the **Properties** pane.

Buttons may be displayed in one of four default sizes: large with text below (vertical) or next to (horizontal) the button, small with text, and small without text. The size is set in the **Customize User Interface** dialog box. The size setting is actually the maximum display size for the button. When the subpanel is set to do so, AutoCAD adjusts the button size smaller as needed based on the space available for the ribbon.

To set the size of a command button, expand the Ribbon branch and then the Panels branch in the **Customizations in All Files** pane. Then, expand the branches for the panel and row that contain the command. Next, select the command in the **Customizations in All Files** pane. If the panel branch is currently selected, you can also pick the button in the **Panel Preview** pane to select the command. Finally, set the Button Style property in the **Appearance** section of the **Properties** pane. See **Figure 22-16**. Generally, text labels are not shown when small buttons are specified. When large buttons are specified, the

Figure 22-16.
A—Buttons in a ribbon panel can be displayed in one of four sizes. The small size can be displayed either with or without a label. B—Setting the appearance of a button in a panel.

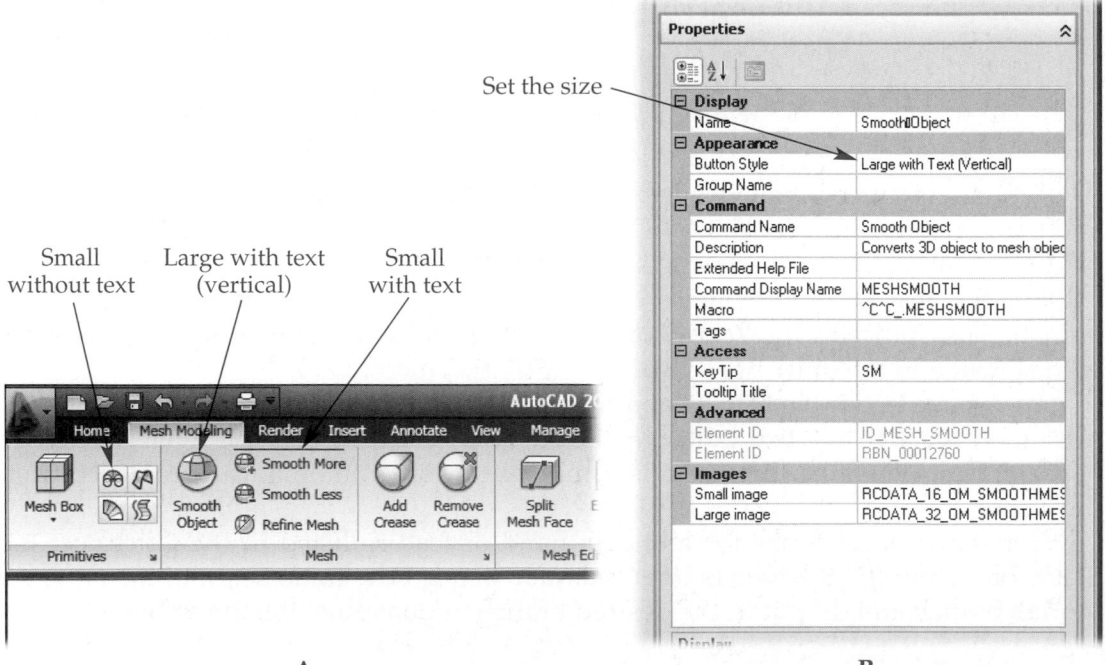

Small without text

Large with text (vertical)

Small with text

Set the size

A

B

vertical orientation is usually selected so the text is below the button. This helps reduce the width of the panel to preserve space on the ribbon.

Customizing a Panel

To add a command to a panel, open the **Customize User Interface** dialog box. In the **Customizations in All Files** pane, expand the Ribbon branch and then the Panels branch. Next, expand the branches for the row to which the command will be added. In the **Commands List:** pane, locate the command to add to the panel. Drag the command from the **Commands List:** pane and drop it into position in the tree in the **Customizations in All Files** pane.

To remove a command from a panel, right-click on the command in the panel's branch in the **Customizations in All Files** pane and select **Remove** from the shortcut menu. You can also select the command and press the [Delete] key.

Row 1 is the top of the panel. In addition, the top of a row branch is the left-hand side of the panel. Commands are displayed in this order on the panel. The commands in a panel can be rearranged. In the **Customizations in All Files** pane, select the command to move and drag it to a new location, either within its current row or in a different row. Rows can also be rearranged by dragging them within the tree in the **Customization in All Files** pane. After you drag a row to a new location, all rows are automatically renumbered. The row at the top of the panel branch is always row 1 and all other rows are sequentially numbered. The panel separator (Slideout branch) can also be dragged to a new location. Remember, rows listed after the panel separator are not displayed until the panel is expanded.

A new row can be added to a panel. In the **Customizations in All Files** pane, right-click on the row *after* which you would like the new row added. To add a new first row, right-click on the panel branch name. Then, select **New Row** from the shortcut menu. The new row is added and all other rows are renumbered. Once a row is added, commands can be added to it.

NOTE

There are certain conditions in which rows in a panel cannot be rearranged. If you attempt to drag a row to a different location and you cannot, just realize you have encountered one of these situations.

Adding a Drop-Down List to a Ribbon Panel

A drop-down list is added to a row in a ribbon panel using the **Customize User Interface** dialog box. To add a drop-down list, right-click on the row branch in the **Customizations in All Files** pane. Then, select **New Drop-down** from the shortcut menu. A branch for the new drop-down list is added to the bottom of the row's branch. Now you can drag commands from the **Command List:** pane into the drop-down list branch.

Notice the Primary Command branch below the drop-down list branch. The command directly below this is displayed as the button for the drop-down list on the ribbon. For example, the **Modeling** panel has the **Box** button displayed for the primitives drop-down list. In the **Customize User Interface** dialog box, the **Box** command is listed directly below the Primary Command branch.

With the branch for the drop-down list selected in the **Customizations in All Files** pane, look at the **Properties** pane. See **Figure 22-17**. Drop-down lists have appearance properties in addition to Button Style: Behavior and Split Button List Style. The Button Style properties are the same as discussed earlier. These settings determine how the drop-down list on the ribbon is controlled. The Behavior property sets whether the top button in the drop-down list executes a command or displays the Name property. The

Figure 22-17.
Adding a drop-down list to a ribbon panel.

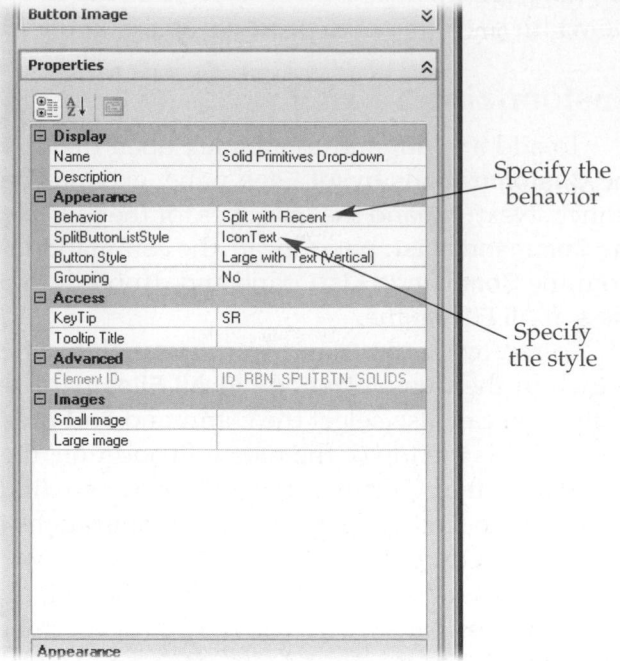

Specify the behavior

Specify the style

Split Button List Style property determines how the buttons and names appear when the list drops down from the panel.

The default Behavior setting is Split with Recent and the default Split Button List Style setting is Icon Text. These settings are typically most desirable for standard AutoCAD workflow. This behavior means that the drop-down list on the ribbon will display in two parts (split). The upper part is the most recent command and the lower part, displayed when the button is pushed, shows additional command icons in the drop-down list.

Creating a New Tab or Panel

To create a new panel, open the **Customize User Interface** dialog box. Then, in the **Customizations in All Files** pane, right-click on the Panels branch and select **New Panel** from the shortcut menu. A new panel with the default name of Panel*x* is added to the bottom of the Panels branch. The name appears in a text box in the tree. Enter a name for the new panel, either in the tree or in the **Properties** pane. Expand the branch for the new panel and notice that the Dialog Box Launcher, Row 1, and Slideout branches are automatically added when the panel is created. Add commands and rows to the new panel as needed. A new tab is similarly created by right-clicking on the Tabs branch.

You will now create a new ribbon panel containing the custom command you created earlier. First, create a new panel and name it My Panel. Then, in the **Command List:** pane, locate the **E-Border** command. Drag the command into the tree in the **Customizations in All Files** panel. The I-bar cursor appears as you drag through the tree. When the I-bar is below the Row 1 branch, drop the command. See **Figure 22-18.**

AutoCAD and Its Applications—Advanced

Figure 22-18
Adding a command to a custom ribbon panel.

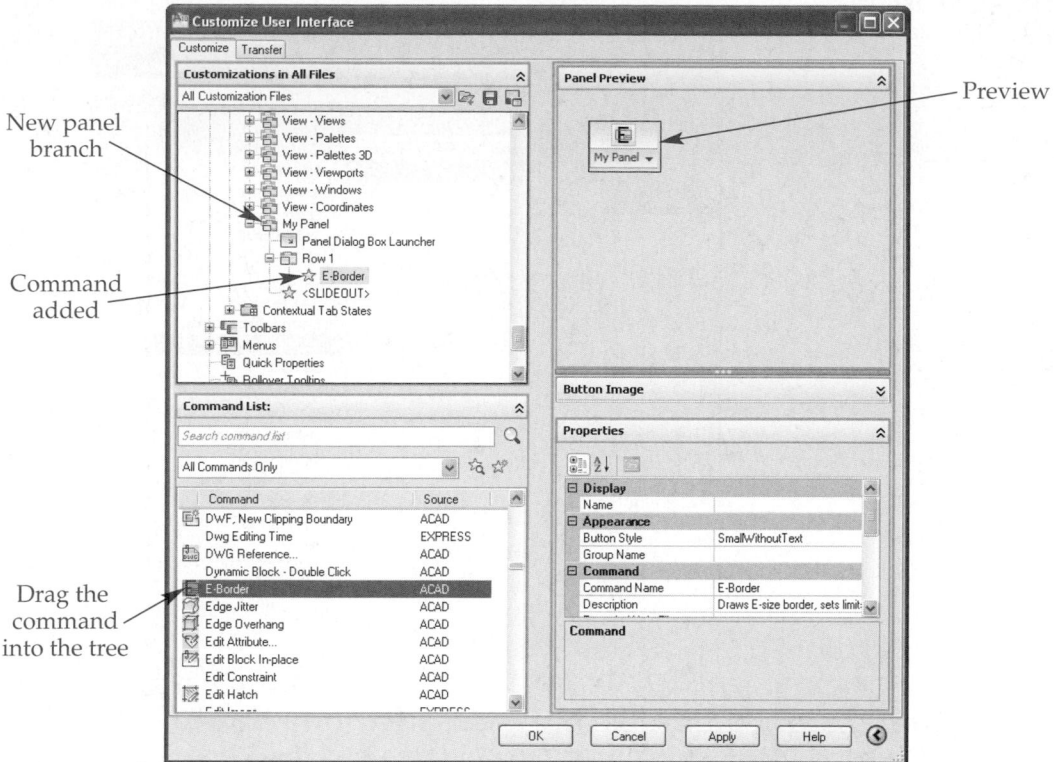

Now, create a new tab to hold the new panel. Name the tab My Stuff. In the **Properties** pane for the new tab, change the Display Text property to Border Tools. Next, locate the My Panel branch and drag it into the My Stuff branch. As you drag, the same I-bar appears in the tree. When the I-bar is below the My Stuff branch, release the mouse button.

For the tab to be displayed, it must be added to the current workspace. Workspaces are covered in detail in Chapter 25. To add the tab, select the workspace in the **Customizations in All Files** pane. Then, drag the tab from the **Customizations in All Files** pane and drop it into the Ribbon Tabs branch in the **Workspace Contents** pane.

Close the **Customize User Interface** dialog box. After the menu compiles, the new tab is displayed in the ribbon, **Figure 22-19.** It is located on the right-hand side of the ribbon. Test the **E-Border** button on the panel to make sure the command properly functions.

Figure 22-19.
A custom tab, panel, and command have been added to the ribbon. Notice the display name for the tab matches the Display Text property for the tab.

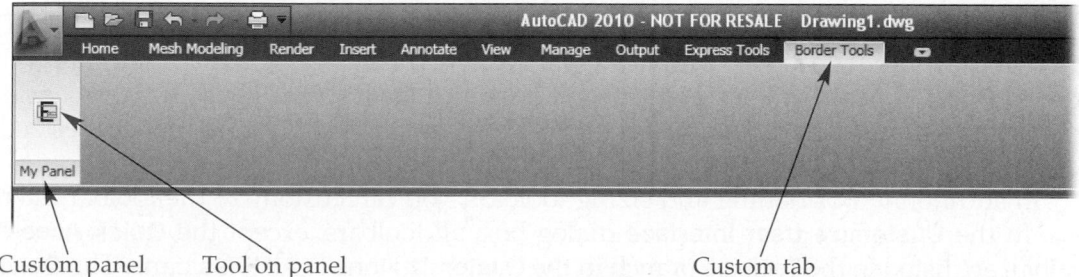

Figure 22-20.
Associating a tool palette group with a ribbon tab.

Right-click on panel

Select tool palette group

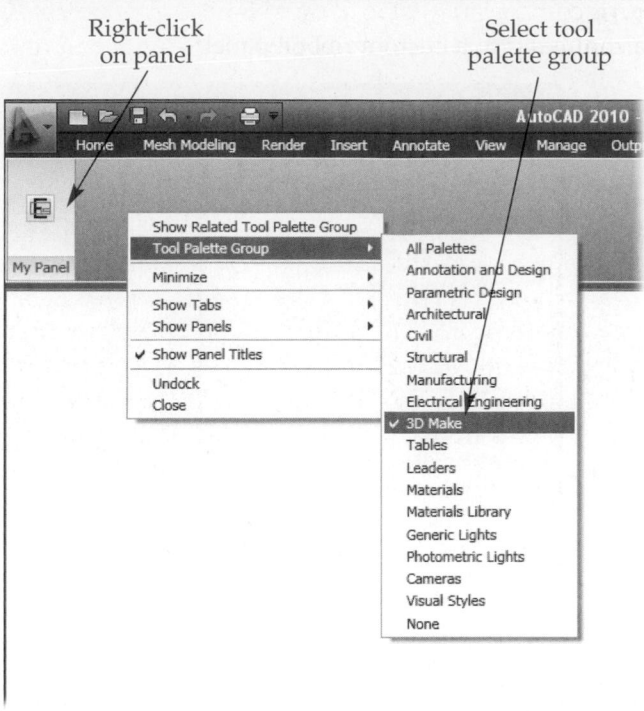

Associating a Tool Palette Group with a Ribbon Tab

A tool palette group can be associated with a tab in the ribbon. Then, the associated tool palette group is displayed in the **Tool Palettes** window when you right-click on the tab and select **Show Related Tool Palette Group** from the shortcut menu.

To associate a tool palette group with a tab, right-click on the tab in the ribbon. This is done with the **Customize User Interface** dialog box closed. Next, select **Tool Palette Group** in the shortcut menu and then the name of the group in the submenu. See Figure 22-20. You can now right-click on the tab and select **Show Related Tool Palette Group** from the shortcut menu. The tool palette group you associated with the tab is displayed in the **Tool Palettes** window.

Exercise 22-2

Complete the exercise on the student website.
www.g-wlearning.com/CAD

Overview of Toolbars

Although the ribbon is the primary graphic interface for accessing the main AutoCAD commands, *toolbars* can also provide quick access to many AutoCAD commands with one or two quick "picks." This interface provides additional flexibility, especially considering toolbars are a fraction of the size of the full ribbon. Toolbars are moved, resized, docked, and floated in the same way as in all Windows-compatible software.

In addition to positioning and sizing toolbars, you can customize the toolbar interface. In the **Customize User Interface** dialog box, all toolbars, except the **Quick Access** toolbar, are listed in the Toolbars branch in the **Customizations in All Files** pane. The **Quick Access** toolbar is customized via workspaces, which are discussed in Chapter 25.

AutoCAD and Its Applications—Advanced

When a command is placed on a toolbar, it is represented by a button. You can add new command buttons or reposition existing command buttons for quicker access. Infrequently used commands can be removed from the toolbar or repositioned to a less prominent location. Entirely new toolbars can be created and filled with predefined or custom commands.

Toolbar Visibility

By default, toolbars other than the **Quick Access** toolbar are not displayed except in the AutoCAD Classic workspace. Use the **-TOOLBAR** command to display toolbars. You are first prompted for the toolbar name. The complete toolbar name consists of the menu group and toolbar name, separated by a period. For example, the toolbar name for the **Draw** toolbar is ACAD.DRAW. The menu group name can be omitted when only one menu is currently loaded or if the toolbar name is not duplicated in another menu group. After specifying the toolbar name (or selecting **ALL** for all toolbars), you can select an option.

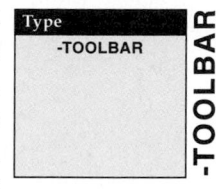

> Command: **-TOOLBAR**↵
> Enter toolbar name or [ALL]: **ACAD.DRAW**↵
> Enter an option [Show/Hide/Left/Right/Top/Bottom/Float] <Show>:

These options are used to hide, show, or specify a location for the toolbar.
- **Show.** Makes the toolbar visible.
- **Hide.** Causes the toolbar to be invisible.
- **Left.** Places the toolbar in a docked position at the left side of the AutoCAD window.
- **Right.** Places the toolbar in a docked position at the right side of the AutoCAD window.
- **Top.** Places the toolbar in a docked position at the top of the AutoCAD window.
- **Bottom.** Places the toolbar in a docked position at the bottom of the AutoCAD window.
- **Float.** Places the toolbar as a floating toolbar.

For example, to dock the **Zoom** toolbar on the left side of the AutoCAD window, use the following command sequence.

> Command: **-TOOLBAR**↵
> Enter toolbar name or [ALL]: **ACAD.ZOOM**↵
> Enter an option [Show/Hide/Left/Right/Top/Bottom/Float] <Show>: **LEFT**↵
> Enter new position (horizontal, vertical) <0,0>: ↵
> Command:

Another way to hide a floating toolbar is to pick its menu control button. This is the X in the corner of the toolbar. If you wish to hide a docked toolbar, you can first move it away from the edge to make it a floating toolbar. Then, pick the menu control button. When you hide a previously docked toolbar in this manner, it will appear in the floating position when you again make it visible.

When using floating toolbars, it is also possible to overlap the toolbars to save screen space. To bring a toolbar to the front, simply pick on it. Be sure to leave part of each toolbar showing.

Toolbar Display Options

Located in the **Window Elements** area of the **Display** tab of the **Options** dialog box are four check boxes and a text box relating to toolbars. See **Figure 22-21.**

When the **Use large buttons for Toolbars** check box is checked, the size of toolbar buttons is increased from 16 × 16 pixels to 32 × 32 pixels. At higher screen resolutions, such as 1280 × 1024, the small buttons may be difficult to see. At lower screen resolutions, such as 800 × 600, the large buttons take up too much of the display area.

Figure 22-21.
The **Display** tab of
the **Options** dialog
box contains settings
for toolbars.

Toolbar
options

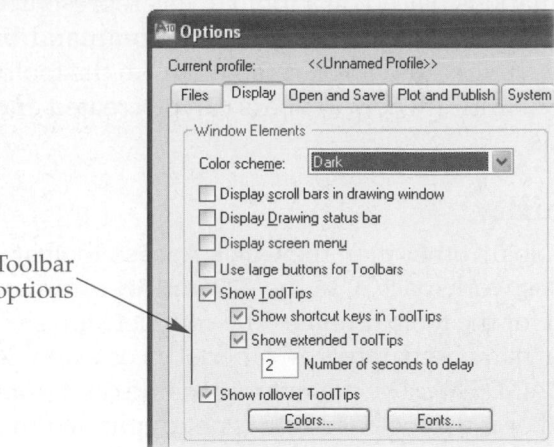

When the **Show ToolTips** check box is checked, the name of the button to which you are pointing is displayed next to the cursor. Below this check box is the **Show shortcut keys in ToolTips** check box. When this option is checked, the shortcut key combination for the command is displayed in the tooltip. The **Show extended ToolTips** check box determines whether extended tooltips are displayed. When checked, extended tooltips are displayed when the cursor is hovered over a button for the number of seconds entered in the **Number of seconds to delay** text box. When tooltips are turned off, these two check boxes and the text box are grayed out.

Creating a New Toolbar

To create a new toolbar, open the **Customize User Interface** dialog box. Then, right-click on the Toolbars branch in the upper-left pane and pick **New Toolbar** in the shortcut menu. A new toolbar is added at the bottom of the Toolbars branch. An edit box is displayed in place of the toolbar name with a default name highlighted. Type a descriptive name for the toolbar and press [Enter].

After the new toolbar is named, it is highlighted in the Toolbars branch. The properties for the toolbar are displayed in the **Properties** pane of the **Customize User Interface** dialog box. See **Figure 22-22.** A preview of the toolbar also appears in the **Toolbar Preview** pane, but since the new toolbar is empty, there is not currently a preview. You can change the name of the toolbar and add a description in the **General** category of the **Properties** pane. The description appears on the AutoCAD status bar when the cursor is over the docked toolbar. In the **Appearance** category, you can specify the default settings for the toolbar, including whether it is included in the current workspace, floating or docked, the location of the toolbar's upper-left corner, and the number of rows for the toolbar. The settings in the **Advanced** category are used for programming applications.

Adding a Command to a Toolbar

To add a command to a toolbar, first expand the Toolbars branch in the **Customizations in All Files** pane in the **Customize User Interface** dialog box. Next, expand the branch for the toolbar to which the command will be added. Then, select a command from the **Command List:** pane and drag it into position in the tree in the **Customizations in All Files** pane. As you drag the command in the tree, an I-bar cursor is displayed. When the I-bar is below the command where you want the new command placed, release the mouse button.

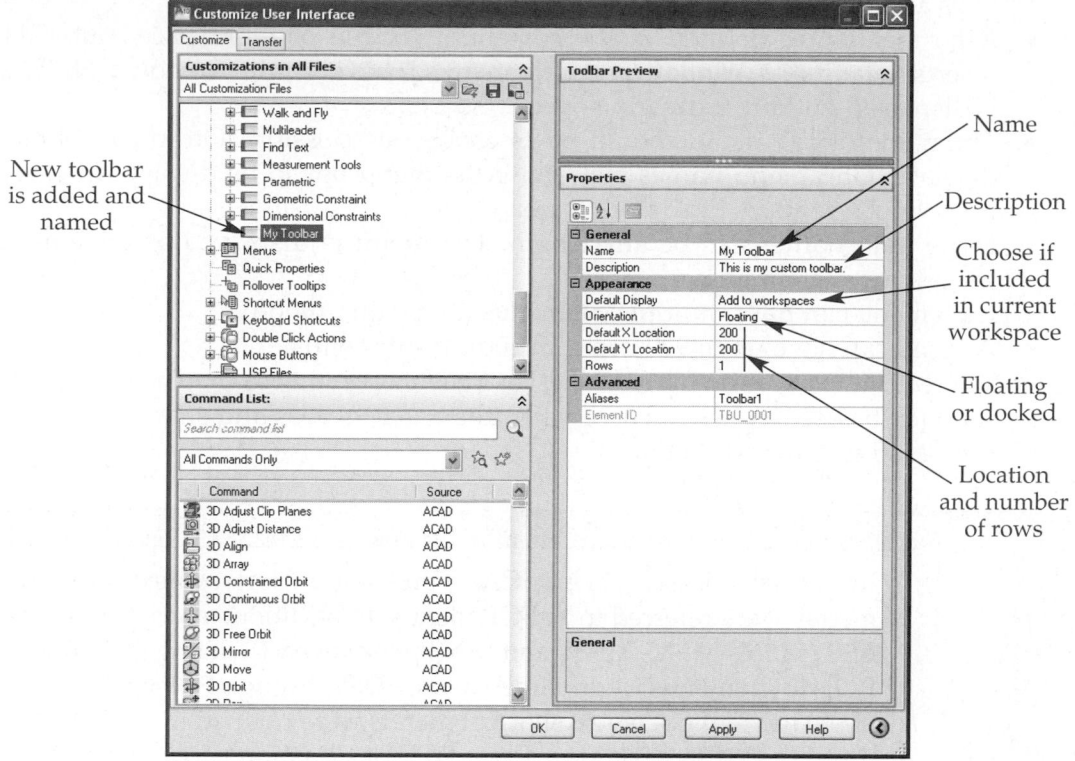

New toolbar is added and named

Name

Description

Choose if included in current workspace

Floating or docked

Location and number of rows

PROFESSIONAL TIP

If the toolbar branch is selected, you can also add a command by dragging and dropping the command into the **Panel Preview** pane, rather than into the tree in the **Customizations in All Files** pane. However, the toolbar must contain at least one command in order for the preview to appear.

Exercise 22-3

Complete the exercise on the student website.
www.g-wlearning.com/CAD

Overview of Menus

When the menu bar is displayed, the names of the standard (classic) *menus* appear at the top of the AutoCAD graphics window. The menu bar is displayed in the AutoCAD Classic workspace. To display the menu bar in other workspaces, pick the arrow icon at the right-hand end of the **Quick Access** toolbar and select **Show Menu Bar** in the shortcut menu or type the **MENUBAR** command. Menus are selected by placing the cursor over the menu name and picking. You can also use the access (mnemonic) keys to select menus.

Once you understand how menus are designed, you can customize existing menus and create your own. Some basic information about menus includes:

- By default, AutoCAD has 13 menus displayed on the menu bar. If the Express tools are not installed, there are 12 menus.
- If no menus are defined in the current CUIx file or workspace, AutoCAD inserts default **File**, **Window**, and **Help** menus. This is similar to how AutoCAD is displayed without a drawing open.
- The name of the menu should be as concise as possible. On low-resolution displays, long menu names may cause the menu bar to be displayed on two lines, which reduces the drawing area.
- Menu item names can be any length. The menu is displayed as wide as its longest menu item name.
- Each menu can have multiple submenus (cascading menu).
- A menu can have up to 999 items, including submenus.
- To create an access (mnemonic) key for a menu or menu item, place an ampersand (&) before the desired access key character. Access and shortcut keys are discussed in the next section.

NOTE

In earlier releases of AutoCAD, pull-down and context shortcut menus were referred to as POP menus. In addition, a series of menu files (MNU, MNC, MNR, and MNS) were used to define the menus. Pull-down menus were defined in the POP1 through POP499 sections of the menu file. Context shortcut menus were defined in the POP500 through POP999 sections. These POP designations are still used by AutoCAD for the sake of compatibility with older menus being used in the current version of AutoCAD. In the **Customize User Interface** dialog box, the POP names appear as aliases for these menus.

Shortcut and Access Keys

Before getting started with menu customization, it is important to understand the difference between shortcut keys and access keys. *Shortcut keys,* also called *accelerator keys,* are key combinations used to initiate a command. For example, [Ctrl]+[1] displays or closes the **Properties** palette. Custom shortcut keys can be created to initiate specific AutoCAD commands or macros. Creating custom shortcut keys is covered in Chapter 23.

Access keys, also called *mnemonic keys,* are keys used to access a menu or menu item via the keyboard. Pressing the [Alt] key activates the access keys for the menus. The access keys are shown as underlined (underscored) letters. Most access keys (underscores) are not displayed in the menu bar until the [Alt] key is depressed. For example, notice that the letter M is underlined in the **Modify** menu name. Pressing the [M] key accesses the **Modify** menu.

Any letter in the menu or menu item name can be defined as the access key, but an access key must be unique for a menu or submenu. Notice within the **Modify** pull-down menu that the M is used for **Match Properties**, so **Mirror** and **Move** use the i and v, respectively. The letter T can be used for both **Trim** and **Text** because **Text** is in the **Object** submenu, while **Trim** is in the "main" **Modify** menu. When creating custom menus, you can add custom access keys to the menu.

PROFESSIONAL TIP

Once the access (mnemonic) keys are activated, you can use the arrow keys to navigate through the pull-down menu structure.

AutoCAD and Its Applications—Advanced

Creating a New Menu

A new menu is created within the **Customize User Interface** dialog box. First, a menu is added to the Menus branch. Then, commands are added to the new menu. The process is basically the same as creating a new ribbon panel or toolbar, as described earlier in this chapter. The basic procedure is:

1. Open the **Customize User Interface** dialog box.
2. In the **Customizations in All Files** pane, expand the Menus branch. All of the existing menus are displayed.
3. Right-click on the Menus branch to display the shortcut menu. Pick **New Menu** from the shortcut menu. A new menu is added to the bottom of the list of existing menus. See **Figure 22-23A**. The name is highlighted in an edit box so

Figure 22-23.
A—Adding a new menu. B—Commands have been added to the new menu.

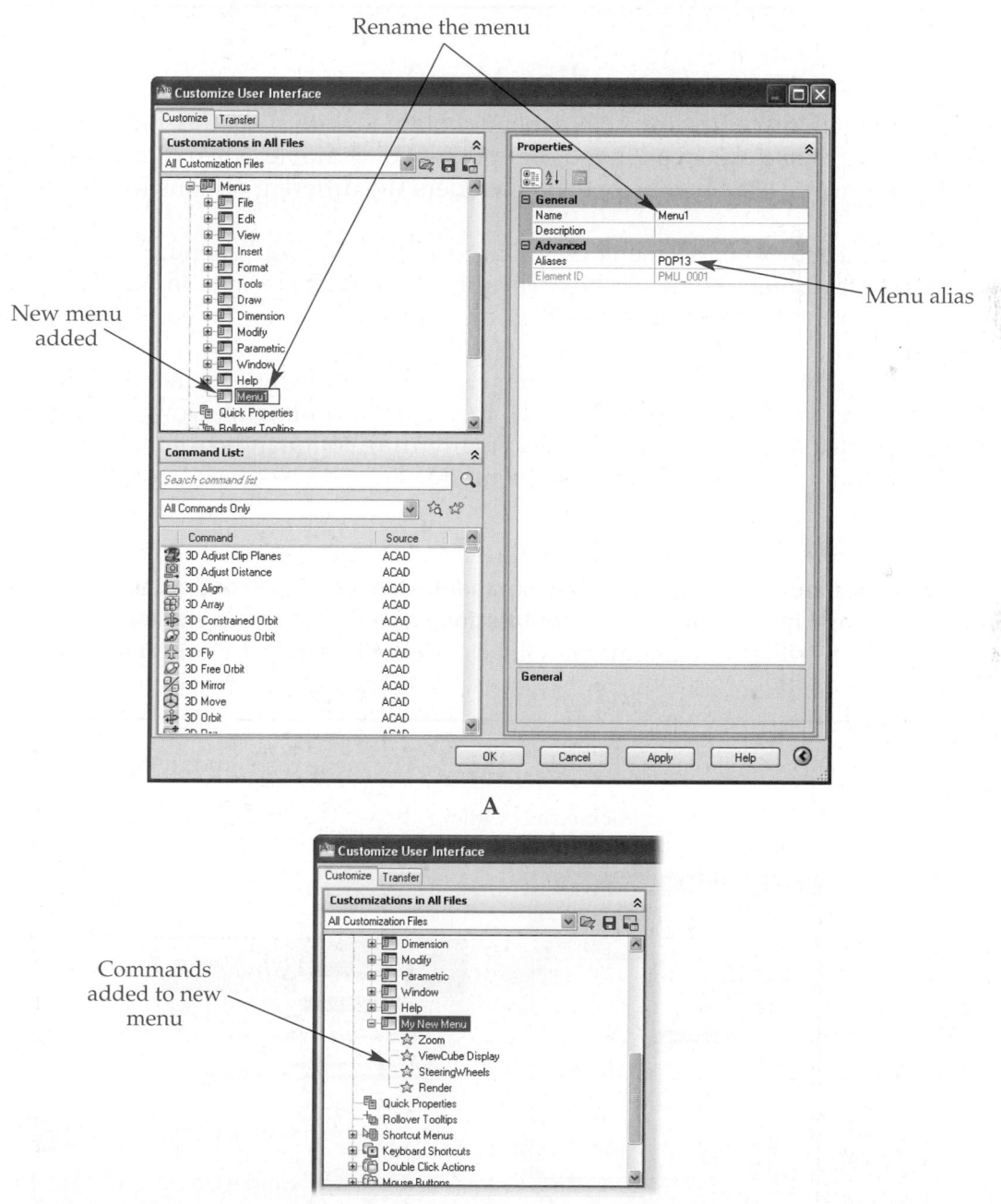

A

B

the default name can be changed.

4. Give the menu an appropriate name.

5. Drag the desired commands from the **Commands List:** pane and drop them into the new menu. See **Figure 22-26B.**

6. To add a separator, right-click on the command below which it should be inserted and select **Insert Separator** from the shortcut menu.

When adding a new menu, it is automatically assigned an alias of POP*n*, where *n* is the next available integer. Also, the menu is automatically available in all workspaces. Workspaces are covered in detail in Chapter 25.

Exercise 22-4

Complete the exercise on the student website.
www.g-wlearning.com/CAD

Adding a Submenu (Cascading Menu)

A *submenu* (cascading menu) is a menu contained within another menu. It can be used to help group similar commands or options. For example, when **Circle** is selected in the **Draw** menu, a submenu appears that offers the different options for drawing a circle.

Adding a submenu to a menu is similar to adding a "main" menu. First, open the **Customize User Interface** dialog box. Then, in the **Customizations in All Files** pane, expand the branch for the menu to which the submenu is to be added. Right-click on the command after which the submenu should appear. In the shortcut menu that is displayed, pick **New Sub-menu**. A new menu is added within the first menu. Notice that the icon in the tree indicates this item is a menu, not a command. Now, the submenu can be renamed to an appropriate name. Finally, drag commands from the **Command List:** pane and drop them into the new menu. See **Figure 22-24.**

Adding a Command to a Menu

To add a command to a menu, first expand the Menus branch in the **Customizations in All Files** pane in the **Customize User Interface** dialog box. Next, expand the branch for the menu to which the command will be added. Then, select a command from the

Figure 22-24.
A—A submenu has been added to the new menu. B—The menu displayed in the menu bar.

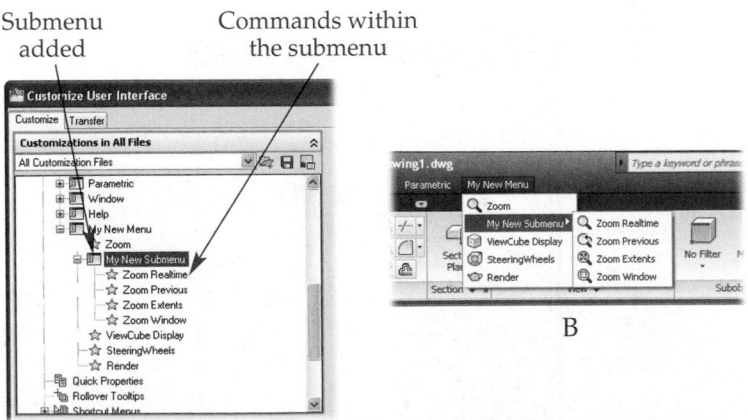

Command List: pane and drag it into position in the tree in the **Customizations in All Files** pane. As you drag the command in the tree, an I-bar cursor is displayed. When the I-bar is below the command where you want the new command placed, release the mouse button.

Removing a Menu

If you want to permanently remove a menu, open the **Customize User Interface** dialog box. Then, expand the tree in the **Customizations in All Files** pane to display the menu to be deleted. Right-click on the menu and pick **Delete** from the shortcut menu. You can also highlight the menu in the tree and press the [Delete] key.

The above procedure is not recommended because the menu is permanently removed. To "restore" the menu in the future, it must be rebuilt. A better way to remove any unwanted menus is by deleting them from the workspace. Managing workspaces is covered in detail in Chapter 25.

Some Notes about Menus

Here are a few more things to keep in mind when developing menus.
- Menus are disabled during **DTEXT** after the rotation angle is entered and during **SKETCH** after the record increment is set.
- A menu label can be as long as needed, but should be as brief as possible for easy reading. The menu width is automatically created to fit the width of the longest item.
- Menus that are longer than the screen display are truncated to fit on the screen.

Sample Custom Commands

The following examples show how AutoCAD commands and options can be used to create commands. These custom commands can be placed on ribbon panels, toolbars, or menus. Remember, an ampersand (&) preceding a character in a menu or item name defines the keyboard access (mnemonic) key used to enable it. The examples are listed using the following three-step process.
- Step 1. A description of the macro.
- Step 2. The key strokes required for the macro.
- Step 3. The name and macro for the new command as entered in the **Properties** pane of the **Customize User Interface** dialog box.

Example 1

1. This **HEXAGON** command will start the **POLYGON** command and draw a six-sided polygon inscribed in a circle.
2. **POLYGON**⏎
   6⏎
   (select center)
   I⏎
3. Name: &Hexagon
   Macro: *^C^Cpolygon;6;\i

The asterisk in front of the ^C^C repeats the command until it is canceled. The \ in front of i indicates that the macro will wait for user input, in this case the center of the polygon, before continuing.

Example 2

1. This **DOT** command draws a solid dot that is .1 unit in diameter. Use the **DONUT** command. The inside diameter is 0 (zero) and the outside diameter is .1.
2. **DONUT.⏎**
   0.⏎
   .1.⏎
3. Name: &Dot
   Macro: ^C^Cdonut;0;.1

Example 3

1. This **X-POINT** command sets the **PDMODE** system variable to 3 and draws an X at the pick point. The command should repeat.
2. **PDMODE.⏎**
   3.⏎
   POINT.⏎
   (pick the point)
3. Name: &X-Point
   Macro: *^C^Cpdmode;3;point

Example 4

1. This command, named **NOTATION**, could be used by a drawing checker or instructor. It allows them to circle features on a drawing and then add a leader and text. It first sets the color to red, then draws a circle, snaps a leader to the nearest point that is picked on the circle, and prompts for the text. User input for text is provided, then a cancel [Esc] returns the Command: prompt and the color is set to ByLayer.
2. **-COLOR.⏎**
   RED.⏎
   CIRCLE.⏎
   (pick center point)
   (pick radius)
   LEADER.⏎
   NEA.⏎
   (pick a point on the circle)
   (pick end of leader)
   (press [Enter] for automatic shoulder)
   (enter text) ⏎
   (press [Enter] to cancel)
   -COLOR.⏎
   BYLAYER.⏎
3. Name: &Notation
   Macro: ^C^C-color;red;circle;\\leader;nea;\\;\;-color;bylayer

Example 5

1. This is a repeating command named **MULTISQUARE** that draws one-unit squares oriented at a 0° horizontal angle until the command is canceled.
2. **RECTANG.⏎**
   (pick lower-left corner)
   @1,1.⏎
3. Name: &Multisquare
   Macro: *^C^Crectang;\@1,1

AutoCAD and Its Applications—Advanced

PROFESSIONAL TIP

Some commands, such as the **COLOR** command, display a dialog box. Menu macros can provide input to the command line, but cannot control dialog boxes. To access the command-line version of a command, prefix the command name with a hyphen (-), as shown in example 4. However, not all commands that display a dialog box have a command-line equivalent.

Exercise 22-5

Complete the exercise on the student website.
www.g-wlearning.com/CAD

Chapter Test

Answer the following questions. Write your answers on a separate sheet of paper or complete the electronic chapter test on the student website.
www.g-wlearning.com/CAD

1. What is an *interface element* in AutoCAD?
2. Which command is used to access the **Customize User Interface** dialog box?
3. In which pane of the **Customize User Interface** dialog box can you find all predefined commands?
4. How do you add a command to an interface element?
5. How do you remove a command from an interface element?
6. How can you copy a command to a new location on a different interface element?
7. What is a *partial CUI file?*
8. Briefly describe how to create a custom command.
9. What is an *extended help file?*
10. Name the four drawing tools that are provided in the **Button Editor** dialog box.
11. What is the default, small size (in pixels) of the button editor drawing area?
12. Where is the **Use large buttons for Toolbars** check box located? What function does this check box perform?
13. How should you develop and test a new macro before entering it into a custom command definition?
14. Name two ways to specify an [Enter] in a macro. Which of the two methods is recommended?
15. Briefly describe the composition of the ribbon.
16. What determines which commands in a ribbon panel appear in the expanded panel?
17. How do you create a new ribbon tab and add a new panel to it?
18. Briefly describe how to customize a toolbar.
19. How do you create a new toolbar?
20. Explain how to display a toolbar.
21. Briefly describe a contextual tab.
22. What is a *drop-down list?*
23. How do you add a drop-down list to a ribbon panel?
24. How wide is a menu?
25. Interpret the following menu item.

    ^C^Crectang;\@1,1

Drawing Problems

Before customizing any toolbars, menus, or the ribbon, check with your instructor or supervisor for specific instructions or guidelines.

1. Create a new toolbar using the following information.
   A. Name the toolbar **Draw/Modify**.
   B. Copy at least three, but no more than six, commonly used drawing commands onto the new toolbar. Use only existing commands; do not create new ones.
   C. Copy at least three, but no more than six, commonly used editing commands onto the new toolbar. Use only existing commands; do not create new ones.
   D. Dock the new **Draw/Modify** toolbar at the upper-left side of the screen.

2. Create a new toolbar using the following information.
   A. Name the toolbar **My 3D Tools**.
   B. Copy the following solid primitive commands onto the new toolbar.

Box	**Pyramid**
Cone	**Sphere**
Cylinder	**Torus**

   C. Copy the following view commands onto the new toolbar.

Top	**Bottom**	**Left**
Right	**Front**	**Back**

   D. Copy the following UCS commands onto the new toolbar.

3 Point	**Object**	**World**
Face UCS	**Origin**	**UCS Previous**

   E. Dock the toolbar below the toolbar created in problem 1.

3. Create a new ribbon panel using the following information.
   A. The displayed name of the panel should be **Paper Space Viewports**.
   B. The panel should contain eight custom commands that use the **MVIEW** command to create paper space viewports:
      - **1 Viewport**—allow user to pick location
      - **1 Viewport (Fit)**
      - **2 Viewports (Horizontal)**—allow user to pick location
      - **2 Viewports (Vertical)**
      - **3 Viewports**—allow user to pick orientation and location
      - **3 Viewports (Right)**
      - **4 Viewports**—allow user to pick location
      - **4 Viewports (Fit)**
   C. Construct button graphics for the custom commands. Save the images in the default \Icons folder or create a new folder (be sure to add it to the AutoCAD support environment).
   D. Create a custom command that will switch from one viewport to another.
   E. Place a button on the panel that executes the **PLOT** command.

Drawing Problems - Chapter 22

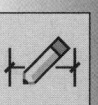

4. Create a new ribbon panel for inserting title block drawings. Name the panel **Title Blocks**.
   A. The panel should contain six custom commands that do the following.
      - Insert the ANSI A title block drawing (plot style of your choice)
      - Insert the ANSI B title block drawing (plot style of your choice)
      - Insert the ANSI C title block drawing (plot style of your choice)
      - Insert the ANSI D title block drawing (plot style of your choice)
      - Insert the ANSI E title block drawing (plot style of your choice)
      - Insert the Architectural title block drawing (plot style of your choice)
   B. Create button graphics for each of the custom commands. Save the images in a new folder and add the folder to the AutoCAD support environment.

5. Add a drop-down list to the **Paper Space Viewports** panel created in problem 3. Add the six custom commands created in problem 4 to this flyout.

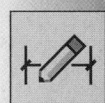

6. Create a new dimensioning menu. Place as many dimensioning commands as you need in the menu. Use submenus if necessary. One or more of the submenus should be dimensioning variables. Include menu access (mnemonic) keys.

7. Create a menu for 3D objects. Include menu access (mnemonic) keys. The menu should include the following items.
   - At least three 3D solid objects
   - **HIDE** command
   - At least three visual style commands
   - **VPORTS** command
   - **NAVSWHEEL** command

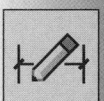

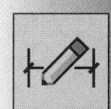

8. Create a new menu named **Special**. The menu should include the following drawing and editing commands.

LINE	MOVE
ARC	COPY
CIRCLE	STRETCH
POLYLINE	TRIM
POLYGON	EXTEND
RECTANGLE	CHAMFER
DTEXT	FILLET
ERASE	

Use submenus, if necessary. Include a separator line between the drawing and editing commands and specify appropriate menu access (mnemonic) keys.

9. Create a menu to insert a variety of blocks or symbols. These symbols can be for any drawing discipline that suits your needs. Use submenus and menu access (mnemonic) keys, if necessary.

10. Create a new ribbon panel named **My 3D Tools**. Evaluate which tools you use most often for creating and rendering 3D models. Place these commands on the new panel, even if they are already contained on another panel. The purpose of this new panel is to streamline your modeling and rendering work. Use drop-down lists as necessary. Arrange the commands on the panel so the panel does not need to be expanded to access the most frequently used commands.

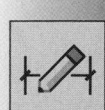

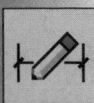

11. Create a ribbon tab with the display name **My Tools**. Set the tab name to your initials. Add the panels created in problems 3 and 10 to this tab. Review the tabs and tools to be sure they properly function.

CHAPTER 23

Customizing Key and Click Actions

Learning Objectives

After completing this chapter, you will be able to:
✓ Assign shortcut keys to commands.
✓ Explain how shortcut menus function.
✓ Edit existing shortcut menus.
✓ Create custom shortcut menus.
✓ Customize an object's quick properties.
✓ Create custom rollover tooltips.
✓ Describe double-click actions.
✓ Edit double-click actions.
✓ Create custom double-click actions.

AutoCAD has many tools that can be used in "heads-up design." Heads-up design is a concept of working in which your eyes remain focused on the drawing area. For example, when dynamic input is on, you do not need to look at the command line to see the options for the current command. The options are displayed near the cursor in the drawing area. AutoCAD's shortcut menus and double-click actions also contribute to heads-up design. Shortcut menus are displayed by right-clicking. Double-click actions are initiated when an object is double-clicked. Like much of the graphic content in AutoCAD (the ribbon, toolbars, etc.), shortcut menus and double-click actions can be customized.

In order to use the shortcut menu and double-click action customization techniques discussed in this chapter, shortcut menus and double-click editing need to be enabled. To do this, open the **Options** dialog box and select the **User Preferences** tab. Then, check the **Double click editing** and **Shortcut menus in drawing area** check boxes, as shown in **Figure 23-1.**

The use of shortcut menus can be further refined by picking the **Right-click Customization...** button that appears below the check boxes. This displays the **Right-Click Customization** dialog box. See **Figure 23-2.** The settings in this dialog box allow you to define what a right-click does when in default mode, edit mode, or command mode. For this chapter, pick the **Shortcut menu** radio buttons in the **Default Mode** and **Edit Mode** areas. Also, pick the **Shortcut Menu: always enabled** radio button in the **Command Mode** area. Then, close the **Right-Click Customization** and **Options** dialog boxes.

Figure 23-1.
The settings for enabling shortcut menus and double-click editing are found in the **Options** dialog box.

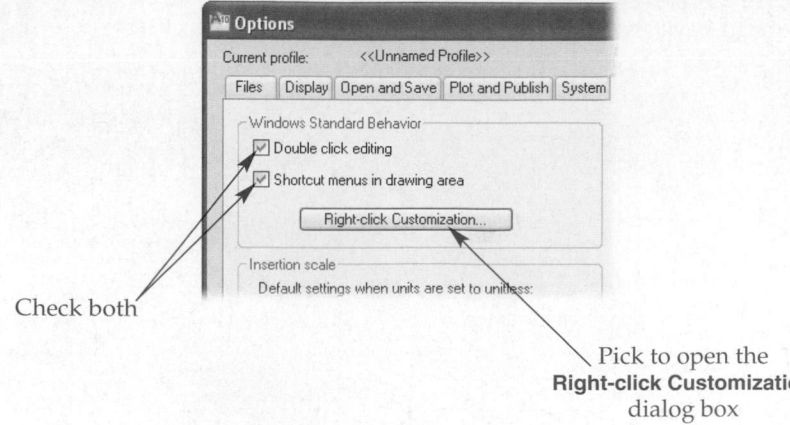

Check both

Pick to open the **Right-click Customization** dialog box

Figure 23-2.
The settings in the **Right-Click Customization** dialog box allow you to define what a right-click does when in default mode, edit mode, or command mode.

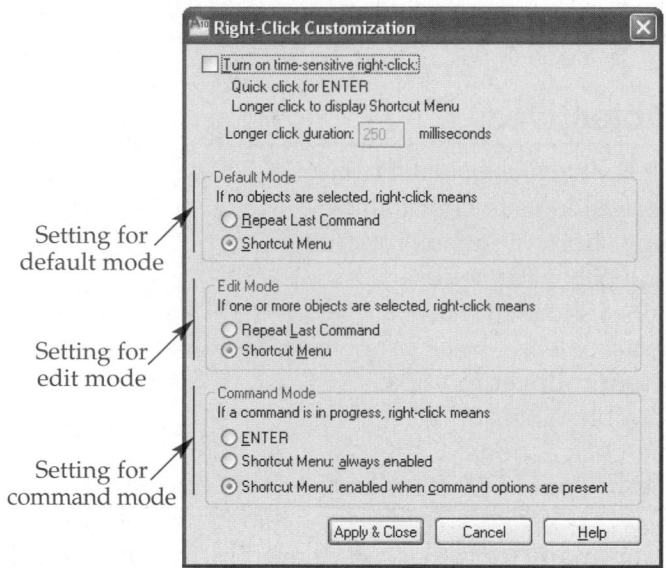

Setting for default mode

Setting for edit mode

Setting for command mode

Customizing Shortcut Keys

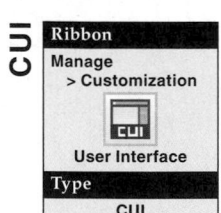

CUI

Ribbon
Manage
> Customization

CUI

User Interface

Type
CUI

You can define your own custom shortcut keys (accelerator keys) for AutoCAD commands and custom macros. The **Customize User Interface** dialog box is used to define shortcut keys. To see the commands to which shortcut keys are assigned, expand the Keyboard Shortcuts branch in the **Customizations in All Files** pane. Then, expand the Shortcut Keys branch. All commands that have a shortcut key assigned to them appear in this branch. See **Figure 23-3.**

When the Shortcut Keys branch is selected, the **Shortcuts** pane is displayed in the upper-right corner of the **Customize User Interface** dialog box. A command that has a shortcut key assigned to it can be selected in this pane to display its properties in the **Information** pane in the lower-right corner of the dialog box.

Assigning a Shortcut Key

To assign a shortcut key to a command, first locate the command in the **Command List:** pane of the **Customize User Interface** dialog box. Next, drag the command into the Shortcut Keys branch in the **Customizations in All Files** pane. The command is added to the list of shortcut keys (although it may not be immediately visible) and the **Properties** pane is displayed for the command. See **Figure 23-4.** In the **Access** category of the

AutoCAD and Its Applications—Advanced

Figure 23-3.
Shortcut keys are added to commands in the **Customize User Interface** dialog box.

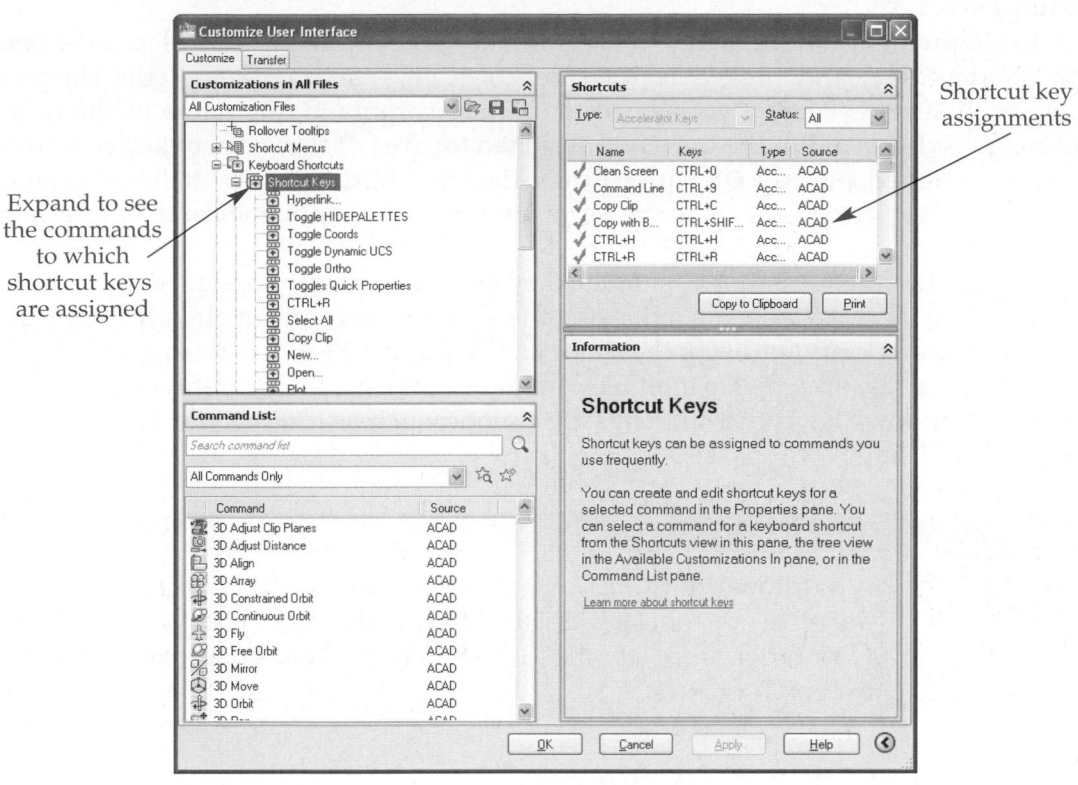

Expand to see the commands to which shortcut keys are assigned

Shortcut key assignments

Figure 23-4.
Adding a shortcut key for the **RENDER** command. Drag the **RENDER** command up from the **Command List:** pane to the Shortcut Keys branch.

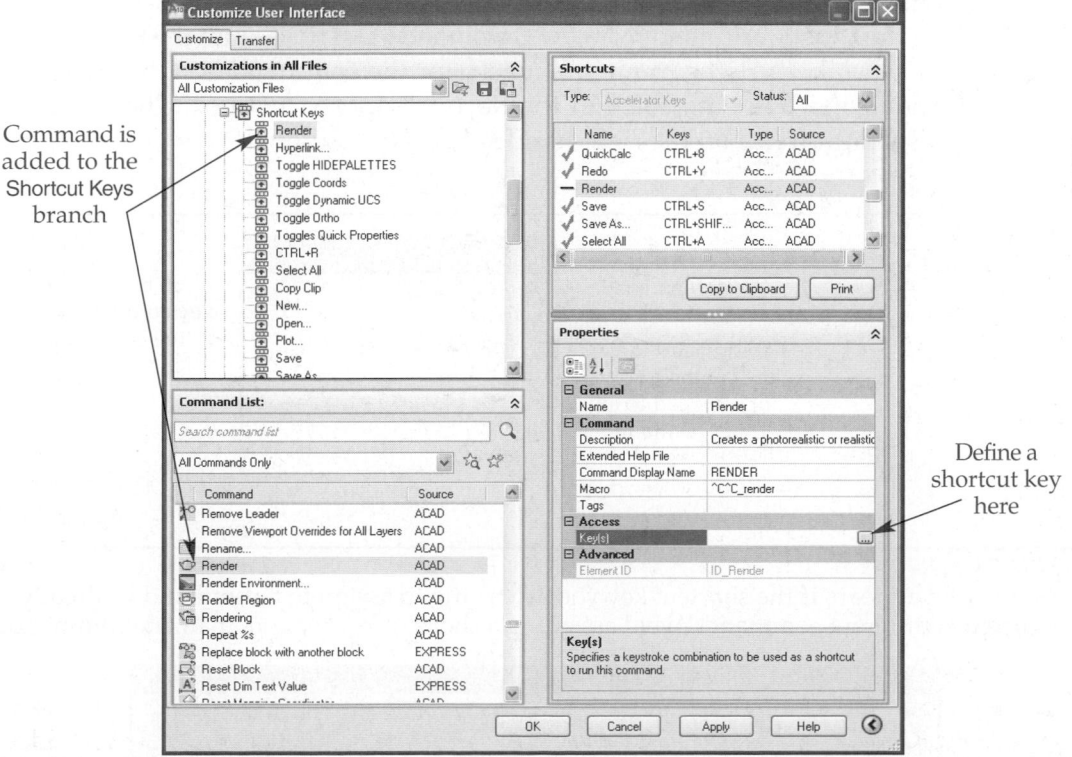

Command is added to the Shortcut Keys branch

Define a shortcut key here

Properties pane, pick in the Key(s) property text box. Next, pick the ellipsis button (**...**) on the right-hand end of the text box to display the **Shortcut Keys** dialog box. See **Figure 23-5.**

To assign a new shortcut key to the command, pick in the text box labeled **Press new shortcut key:** and press a combination of [Ctrl] + another key. If the shortcut key combination is currently assigned to another command, the name of the other command is displayed in the **Currently assigned to:** area. If the shortcut key combination is unassigned, pick the **OK** button to associate the shortcut key with the command. The shortcut key then appears in the Key(s) property in the **Customize User Interface** dialog box.

If you attempt to assign a shortcut key that is currently assigned to another command, an alert box appears indicating the shortcut assignment already exists and explaining the priority for using the shortcut. See **Figure 23-6.** It is not a good idea to have a shortcut key assigned to multiple commands. Be especially careful to ensure the standard Windows keyboard shortcuts are unique, such as [Ctrl]+[X] for cut, [Ctrl]+[C] for copy, and [Ctrl]+[V] for paste.

PROFESSIONAL TIP

In addition to [Ctrl]+*key*, a shortcut can be [Ctrl]+[Shift]+*key*, [Ctrl]+[Alt]+*key*, or [Ctrl]+[Shift]+[Alt]+*key*. The [Caps Lock] key must be off in order to specify the [Shift] key in the **Press new shortcut key:** text box.

Example Shortcut Key Assignment

To provide an example of customizing shortcut keys, this section shows how to assign the shortcut key [Ctrl]+[Alt]+[C] to the **CLOSE** command. Do the following:
1. Open the **Customize User Interface** dialog box.
2. Expand the Keyboard Shortcuts branch in the **Customizations in All Files** pane.
3. Expand the Shortcut Keys branch.
4. Drag the **Close** command from the **Commands List:** pane into the Shortcut Keys branch. Make sure the command macro for the command is ^C^C_close.
5. In the **Properties** pane, pick in the Key(s) property text box. Then, pick the ellipsis button (**...**) on the right-hand side of the text box.

Figure 23-5.
The **Shortcut Keys** dialog box is where a shortcut key is specified for the command.

Selected key combination

Indicates the key combination is not assigned to any other command

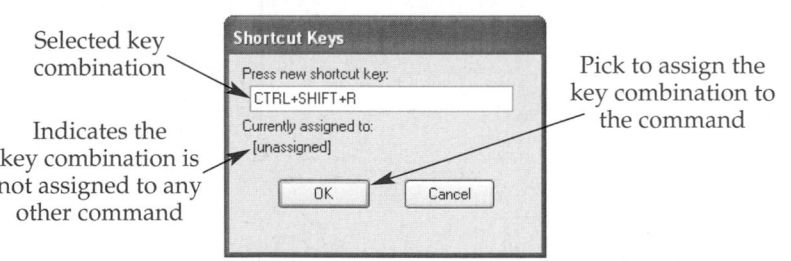

Pick to assign the key combination to the command

Figure 23-6.
This warning appears if the shortcut key you are trying to assign to a command is already assigned to a different command. Avoid assigning a shortcut key to more than one command.

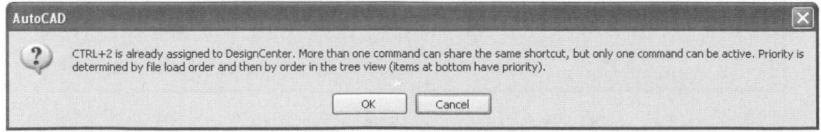

6. In the **Shortcut Keys** dialog box, pick in the **Press new shortcut key:** text box.
7. Press the [Ctrl] key, [Alt] key, and [C] key at the same time. The message at the bottom of the dialog box should indicate that this shortcut key is unassigned.
8. Pick the **OK** button to close the **Shortcut Keys** dialog box.
9. Pick the **OK** button to close the **Customize User Interface** dialog box and apply the change.
10. Test the [Ctrl]+[Alt]+[C] shortcut key. When the shortcut key is used, either the current drawing should close or you should be prompted to save the changes to the drawing before closing.

PROFESSIONAL TIP

Shortcut keys (accelerator keys) have some specific limitations. For example, a shortcut cannot pause for user input or use repeating commands. Be aware of this when assigning shortcut keys to custom commands.

Exercise 23-1

Complete the exercise on the student website.
www.g-wlearning.com/CAD

Examining Existing Shortcut Menus

Shortcut menus are context-sensitive menus that appear at the cursor location when using the right-hand button on the mouse (right-clicking). *Context sensitive* means that the displayed shortcut menu is dependent on what is occurring at the time of the right-click. For example, if no command is active, there is no object selection, and you right-click in the drawing area, the shortcut menu shown in **Figure 23-7A** is displayed. If no command is active and you right-click in **Command Line** window, the shortcut menu shown in **Figure 23-7B** is displayed. If the **CIRCLE** command is active

---

Figure 23-7.
A—Displayed when no command is active and no object is selected. B—Displayed when no command is active and you right-click in the **Command Line** window. C—Displayed when the **CIRCLE** command is active and before any point is selected.

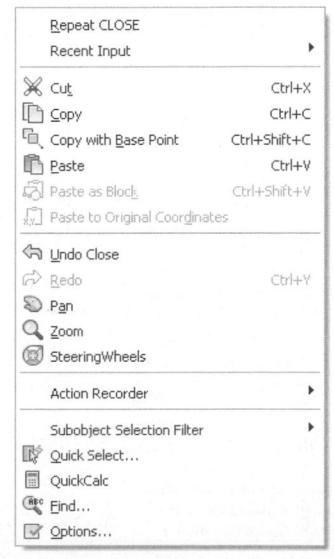

and you right-click in the drawing area before any point is selected, the shortcut menu shown in **Figure 23-7C** is displayed. Other menus appear when right-clicking in other situations, too, such as when grips are being used or when an object is selected in the drawing window. In the case of a selected object, the shortcut menu is based on the type of object that is selected.

Before learning how to customize shortcut menus, examine the existing shortcut menus. Open the **Customize User Interface** dialog box and look at the **Customizations in All Files** pane. The name of this pane is based on what is selected in the drop-down list. Expand the Shortcut Menus branch in the tree. All of the existing shortcut menu names are displayed as branches. See **Figure 23-8**. There are command-specific, object-specific, and generic shortcut menus. The generic shortcut menus are:

- **Command Menu.** This menu appears when right-clicking in the drawing window while a command is active. Any command options for the active command are inserted into this menu. See **Figure 23-9**.
- **Default Menu.** This menu appears when right-clicking in the drawing window while no command is active and no objects are selected. See **Figure 23-10**.
- **Edit Menu.** This menu appears when right-clicking in the drawing window when no command is active and an object is selected. In order for this menu to be displayed, the **PICKFIRST** system variable must be set to 1. If an object menu is available for the type of object selected, it is inserted into this menu. See **Figure 23-11**.
- **Grips Cursor Menu.** This menu appears when grips are being used. See **Figure 23-12**. An object must be selected and at least one grip must be hot.
- **Object Snap Cursor Menu.** This menu appears when holding the down [Shift] key and right-clicking. See **Figure 23-13**. It also appears as the Snap Overrides branch in the Command Menu branch, meaning it is displayed as a cascading menu.

Figure 23-8.
Existing shortcut menus are displayed as branches in the Shortcut Menus branch in the **Customize User Interface** dialog box.

Expand the Shortcut Menus branch

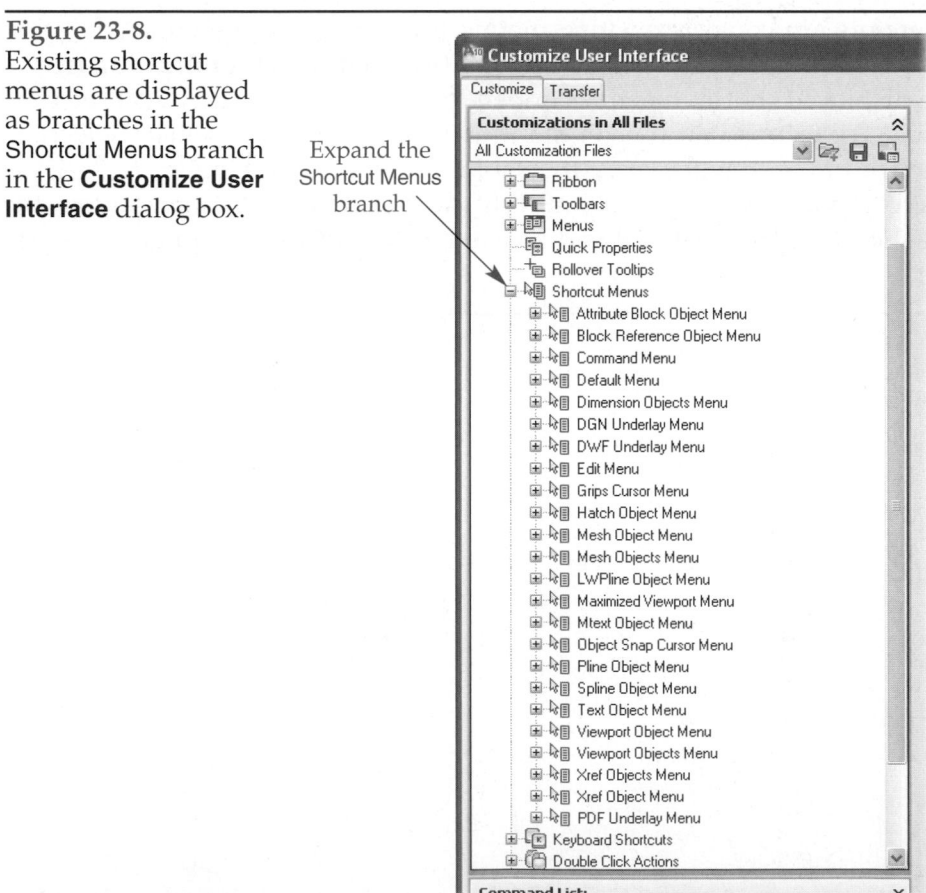

Figure 23-9.
A—The Command
Menu branch in the
**Customize User
Interface** dialog box.
B—The command
shortcut menu.

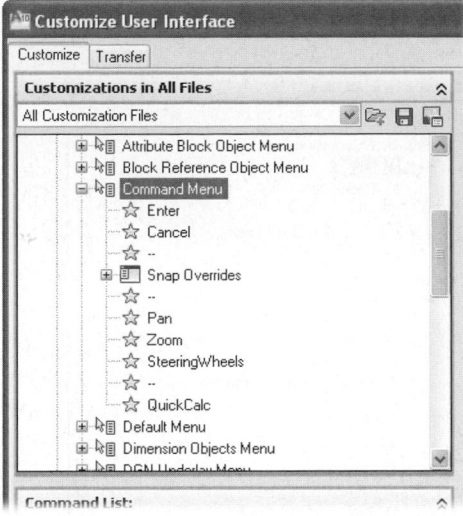

B

A

Figure 23-10.
A—The Default
Menu branch in the
**Customize User
Interface** dialog box.
B—The default
shortcut menu.

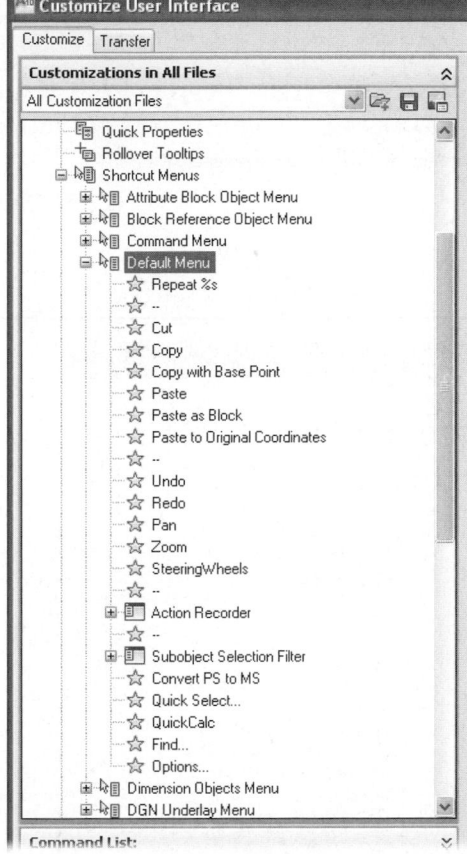

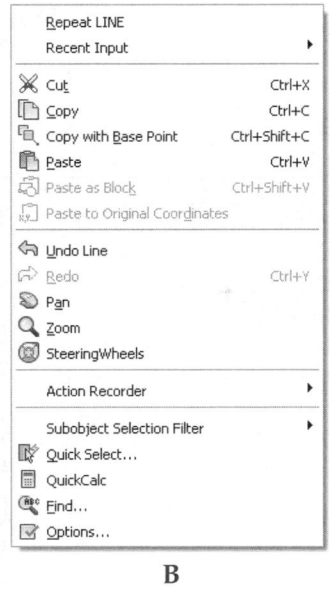

B

A

Figure 23-11.
A—The Edit Menu
branch in the
**Customize User
Interface** dialog box.
B—The edit shortcut
menu.

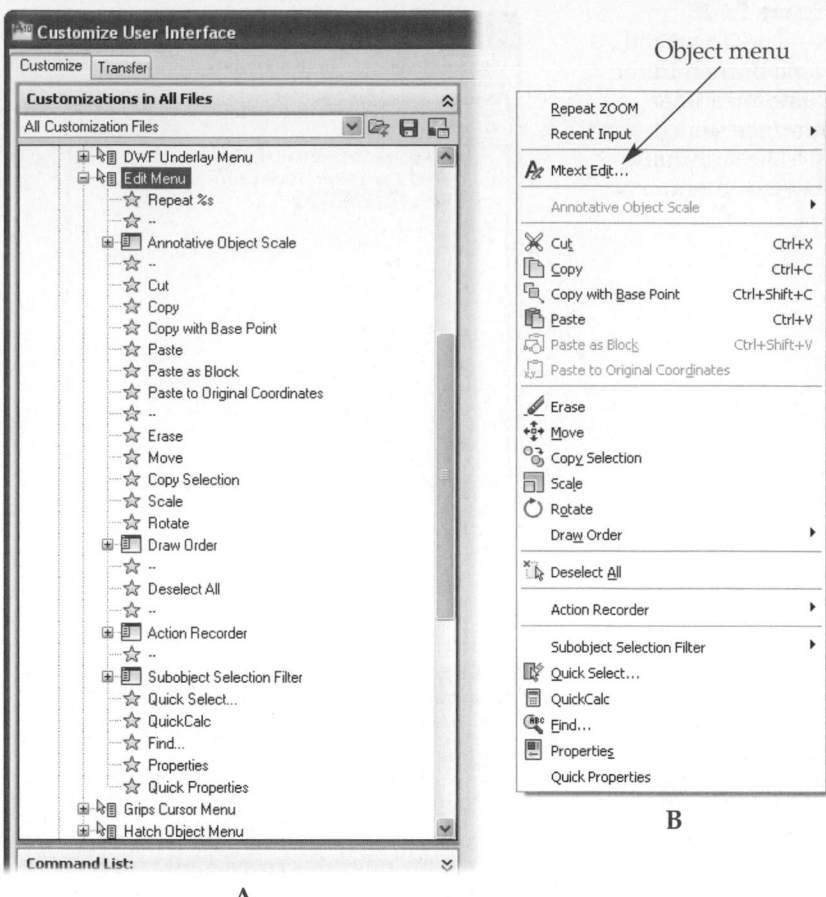

A

B

Figure 23-12.
A—The Grips Cursor
Menu branch in the
**Customize User
Interface** dialog
box. B—The grips
shortcut menu.

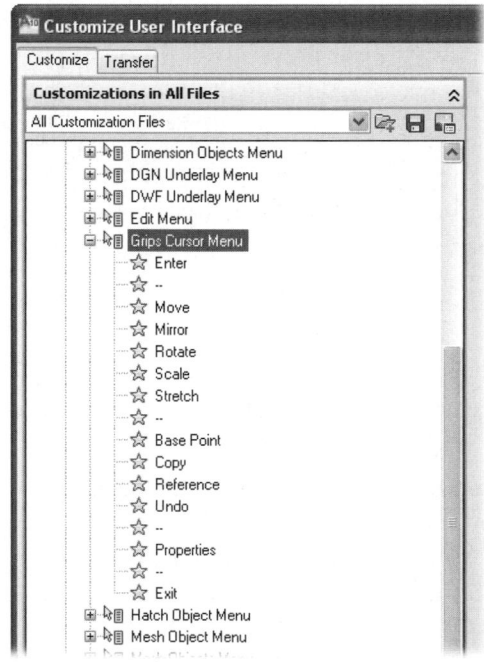

A

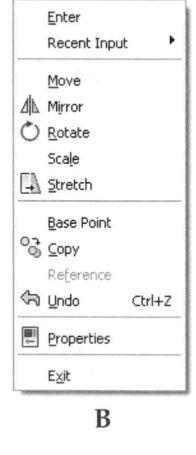

B

Figure 23-13.
A—The Object Snap
Cursor Menu branch
in the **Customize
User Interface** dialog
box. B—The object
snap shortcut menu.

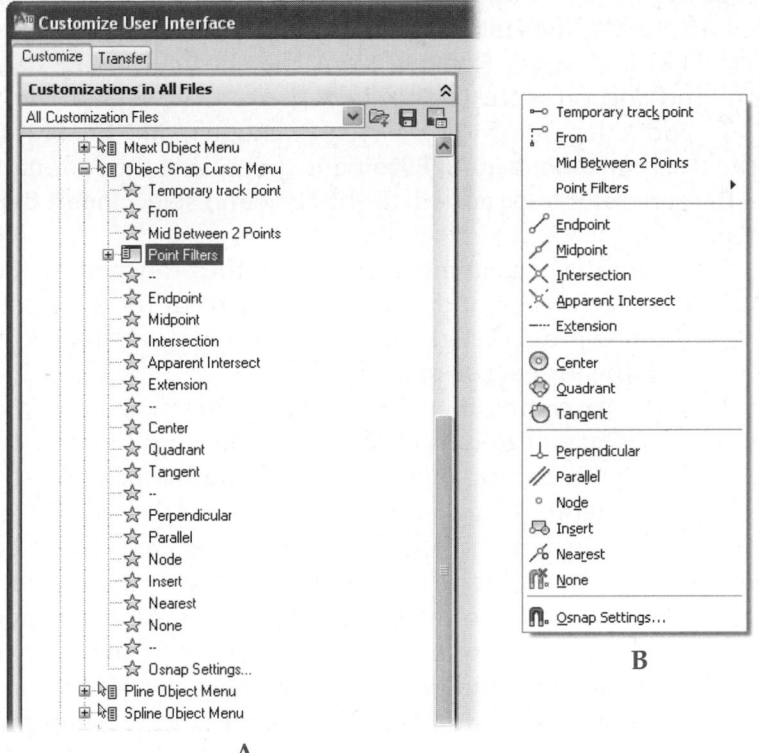

A

B

The remaining menus are object-specific menus that appear when right-clicking while a certain type of object is selected. Notice that there are menu branches named Attribute Block Objects Menu, Block Reference Objects Menu, Dimension Objects Menu, Hatch Object Menu, and others. These menus contain items that can be used on the type of object selected. For example, in the Shortcut Menus branch, the menu branch Dimension Objects Menu contains commands for editing the dimension text position, dimension text precision, and dimension style.

Exercise 23-2

Complete the exercise on the student website.
www.g-wlearning.com/CAD

Customizing Shortcut Menus

The existing shortcut menus can be customized by adding or removing commands. Shortcut menus can also be customized by visually grouping commands using separators. Cascading menus can be added to shortcut menus.

To add a command to a shortcut menu, open the **Customize User Interface** dialog box and, in the **Customizations in All Files** pane, expand the branch for the shortcut menu you would like to customize. Next, locate the command you wish to add in the **Command List:** pane. Then, drag the command into the desired position within the shortcut menu in the **Customizations in All Files** pane and drop it when the bar appears.

To remove a command from a shortcut menu, expand the branch for the shortcut menu in the **Customizations in All Files** pane. Highlight the command to be removed, right-click, and select **Remove** from the shortcut menu. You can also highlight the command and press the [Delete] key.

To add a separator to a shortcut menu, expand the branch for the shortcut menu in the **Customizations in All Files** pane. Highlight the command *after* which you would like the separator to be added. Right-click and select **Insert Separator** from the shortcut menu.

To rename a shortcut menu, highlight the branch in the **Customizations in All Files** pane. Then, right-click and select **Rename** from the shortcut menu. Finally, type the new name and press [Enter]. The shortcut menu can also be renamed using the Name property in the **Properties** pane.

To add a cascading menu to a shortcut menu, expand the branch for the shortcut menu in the **Customizations in All Files** pane. Highlight the command *after* which you would like the cascading menu to appear. Right-click and select **New Sub-menu** from the shortcut menu. A new shortcut menu branch with the default name of Menu*x* is added to the current shortcut menu. The new menu can be renamed. Now, in the **Command List:** pane, locate the commands you wish to add to the new shortcut menu. Drag the commands into the **Customizations in All Files** pane and drop them next to the name of the new shortcut menu. When the arrow appears next to the new menu name, drop the command to add it to the new shortcut menu.

Creating a new, custom shortcut menu is a two-step process. First, make a new shortcut menu and then drag commands into it. Follow these steps to make a new shortcut menu:

1. Open the **Customize User Interface** dialog box.
2. In the **Customizations in All Files** pane, right-click on the Shortcut Menus branch and select **New Shortcut Menu** in the shortcut menu that is displayed.
3. Enter a name for the shortcut menu.
4. In the **Properties** pane, add a description for the shortcut menu in the **General** category.
5. In the **Advanced** category of the **Properties** pane, add an alias. This alias is in addition to the automatic, sequential POP5*xx* alias that AutoCAD creates. Select the property, pick the ellipsis button (**...**) at the right-hand end of the text box, and type the alias in the **Aliases** dialog box that appears. Each alias must be on its own line in this dialog box. Close the **Aliases** dialog box.
6. Drag commands from the **Command List:** pane into the new shortcut menu.
7. Pick the **Apply** or **OK** button to apply the changes.

There are two types of custom shortcut menus: object specific and command oriented. The next sections describe the two types of custom shortcut menus in detail.

Creating Object-Specific Shortcut Menus

When creating an object-specific shortcut menu, there can actually be two menus available. One menu is displayed for instances when just a single object of a given type is selected. The other menu is displayed when more than one object is selected.

The name assigned to the object menu should follow the same syntax used for naming AutoCAD's default object-specific menus: *object_type* **Object Menu** or *object_type* **Objects Menu** (with an S). In this way, when looking at the shortcut menus in the **Customizations in All Files** pane in the **Customize User Interface** dialog box, you will easily recognize which object type that menu applies to and whether it is for multiple selected objects or a single selected object. The use of this syntax is optional. Menus can be named using whatever naming scheme you wish. However, it is recommended to follow the naming syntax described here.

The alias for the shortcut menu has a syntax that *must* be followed. It is this alias that AutoCAD uses in determining to which object or objects the menu applies. The syntax for the alias must take on the form of OBJECT_*type* or OBJECTS_*type* and must be exactly followed in order for AutoCAD to properly display the shortcut menu.

As an example, the following procedure creates a shortcut menu that allows access to the **LENGTHEN** and **BREAK** commands when a single line is selected.

1. Open the **Customize User Interface** dialog box.
2. Right-click on the Shortcut Menus branch in the **Customizations in All Files** pane and select **New Shortcut Menu**.
3. Name the shortcut menu **Line Object Menu**.
4. In the **Properties** pane, select the Alias property in the **Advanced** category. Then, pick the ellipsis button to open the **Aliases** dialog box. On the second line, enter the alias OBJECT_LINE and then close the **Aliases** dialog box. Since the **LENGTHEN** and **BREAK** commands can only be applied to a single object, be sure to use the OBJECT_*type* syntax (without the S).
5. Drag the **LENGTHEN** and **BREAK** commands from the **Command List:** pane into the Line Object Menu branch in **Customizations in All Files** pane. See **Figure 23-14A**.
6. Pick the **OK** button to close the **Customize User Interface** dialog box and apply the changes.

Now, draw a line, select it, and right-click. Notice that **Lengthen** and **Break** entries appear in the shortcut menu. See **Figure 23-14B**. Selecting either entry executes the command. If the command accepts a preselected object, it is executed on the selected line. Neither **LENGTHEN** nor **BREAK** accepts preselected objects; you must reselect the line. Having the entries in the shortcut menu provides for quicker access to the command.

Figure 23-14.
A—The new object-specific shortcut menu is created.
B—The **BREAK** and **LENGTHEN** commands are now available in the shortcut menu.

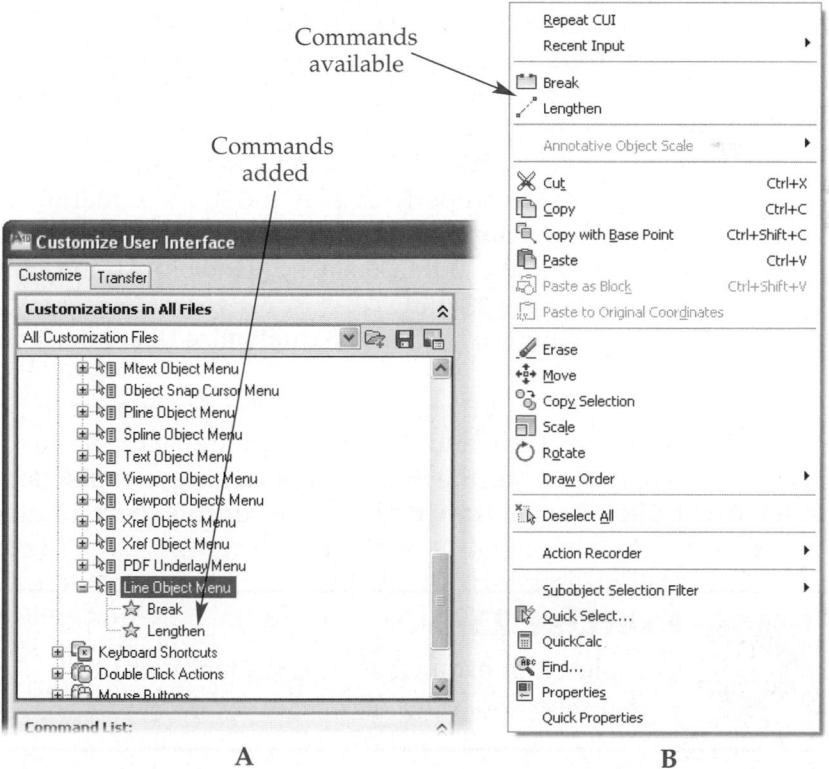

A   B

Creating Command-Oriented Shortcut Menus

When a command is being executed, any command options appear in the shortcut menu. For example, when the **CIRCLE** command prompts for a radius, you can right-click and select **Diameter** from the shortcut menu. Custom shortcut menus can be created for use when certain commands are active. This allows you to add options to the shortcut menu that is displayed when a command is active. Quicker access to object snaps and object selection methods are just a couple of applications that custom, command-oriented shortcut menus could allow for within commonly used commands.

A command-oriented shortcut menu is created in the same way as an object-oriented shortcut menu, as discussed in the previous section. However, the syntax for the alias is slightly different. The alias must be in the form of COMMAND_*command_name*, where *command_name* is the name of the command with which you want the shortcut menu associated.

Also, if the command step does not have any default options, such as a Select objects: prompt, right-clicking is, by default, interpreted as the [Enter] key. Therefore, in the **Right-Click Customization** dialog box, the **Shortcut Menu: always enabled** radio button must be selected in the **Command Mode** area, as described earlier.

As an example, the following procedure creates a custom shortcut menu that displays **Previous**, **Last**, and **Fence** selection options at the Select objects: prompt for the **MOVE** command.

1. Open the **Customize User Interface** dialog box.
2. Make custom commands for the three selection options. Name the commands **Previous**, **Last**, and **Fence**. For the macros, remove the ^C^C that is automatically placed in the macro and type the selection option; for example, PREVIOUS for the **Previous** command.
3. Right-click on the Shortcut Menus branch in the **Customizations in All Files** pane and select **New Shortcut Menu**.
4. Name the shortcut menu **Move Command Menu**.
5. In the **Properties** pane, select the Alias property in the **Advanced** category. Then, pick the ellipsis button to open the **Aliases** dialog box. On the second line, enter the alias COMMAND_MOVE and then close the **Aliases** dialog box. The syntax of COMMAND_*command_name* must be exactly followed in order for AutoCAD to properly display the shortcut menu.
6. Drag the **Previous**, **Last**, and **Fence** custom commands from the **Command List:** pane into the Move Command Menu branch in **Customizations in All Files** pane. See **Figure 23-15A.**
7. Pick the **OK** button to close the **Customize User Interface** dialog box and apply the changes.

Now, initiate the **MOVE** command. At the Select objects: prompt, right-click and notice that **Previous**, **Last**, and **Fence** entries are available in the shortcut menu. See **Figure 23-15B.** Remember, the **Shortcut Menu: always enabled** radio button must be on in the **Right-Click Customization** dialog box for this shortcut menu to appear.

Exercise 23-3

Complete the exercise on the student website.
www.g-wlearning.com/CAD

Figure 23-15.
A—The new command-specific shortcut menu is created.
B—The **Previous**, **Last**, and **Fence** "command options" (actually, custom commands) are available in the shortcut menu.

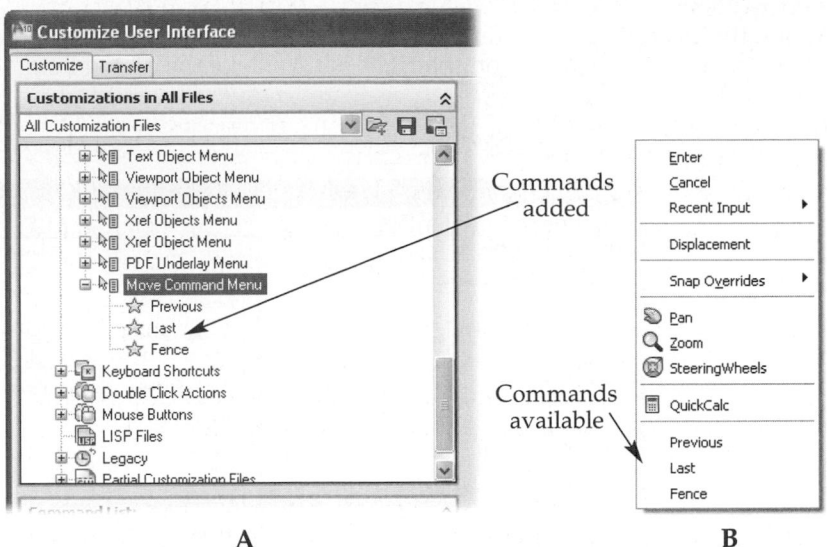

Commands added

Commands available

A   B

Customizing Quick Properties and Rollover Tooltips

The **Quick Properties** palette appears when you select objects in the graphics area. This is a streamlined version of the **Properties** palette that displays *quick properties,* which provide certain information about the object. In order for the **Quick Properties** palette to be displayed, the **Quick Properties** button on the status bar must be on. The palette can also be displayed by right-clicking with an object selected and selecting **Quick Properties** in the shortcut menu so it is checked.

Rollover tooltips provide certain information about an object, also called quick properties, in a graphic tooltip. This tooltip is displayed as the cursor is hovered over the object. A rollover tooltip serves a similar purpose as the **Quick Properties** palette, but it only provides information. Properties cannot be changed in the tooltip.

The quick properties displayed in the **Quick Properties** palette and a rollover tooltip can be customized. Each object type can have different quick properties displayed. Additionally, different quick properties can be displayed in the **Quick Properties** palette and rollover tooltip for a given object type. The steps for customizing quick properties are the same for the **Quick Properties** palette and rollover tooltips, as discussed in this section.

Open the **Customize User Interface** dialog box. In the **Customizations in All Files** pane, select the Quick Properties branch. Two columns are displayed on the right-hand side of the dialog box. See **Figure 23-16.** When you select Rollover Tooltips in the **Customizations in All Files** pane, the same two columns are displayed. However, the settings may be different between the Quick Properties branch and the Rollover Tooltips branch.

The left-hand column displays the *object type list,* which is a list of AutoCAD object types. You may add or remove object types by picking the **Edit Object Type List** button at the top of the column. This displays the **Edit Object Type List** dialog box, **Figure 23-17.** All available AutoCAD object types are listed in this dialog box. Those that are checked appear in the object type list in the **Customize User Interface** dialog box.

The object type list controls which objects display quick properties. The right-hand column displays a list of quick properties that can be displayed. The properties that are checked appear in the **Quick Properties** palette or rollover tooltip. Remember, the palette and tooltip can display different properties.

Figure 23-16.
When the Quick Properties branch is selected in the **Customize User Interface** dialog box, the object type list is displayed on the right-hand side of the dialog box.

Object type list

Pick to edit the object type list

Check the properties to display

Pick to list only general properites

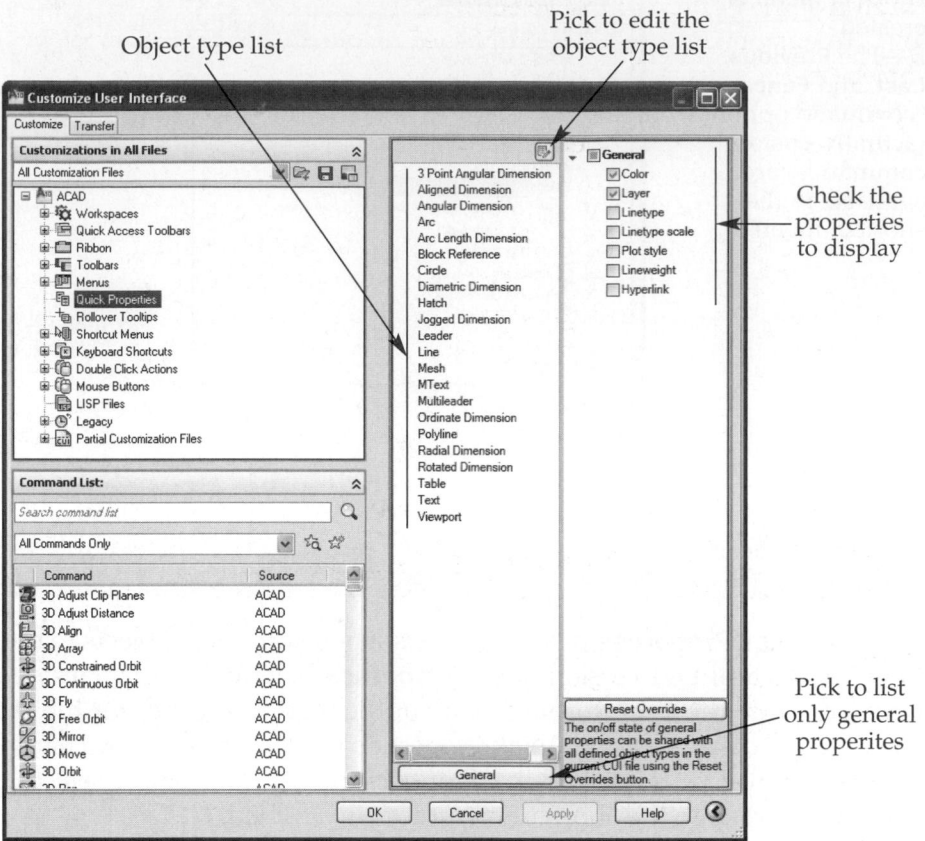

Figure 23-17.
Determining which object types appear in the object type list in the **Customize User Interface** dialog box.

Check the object types to display

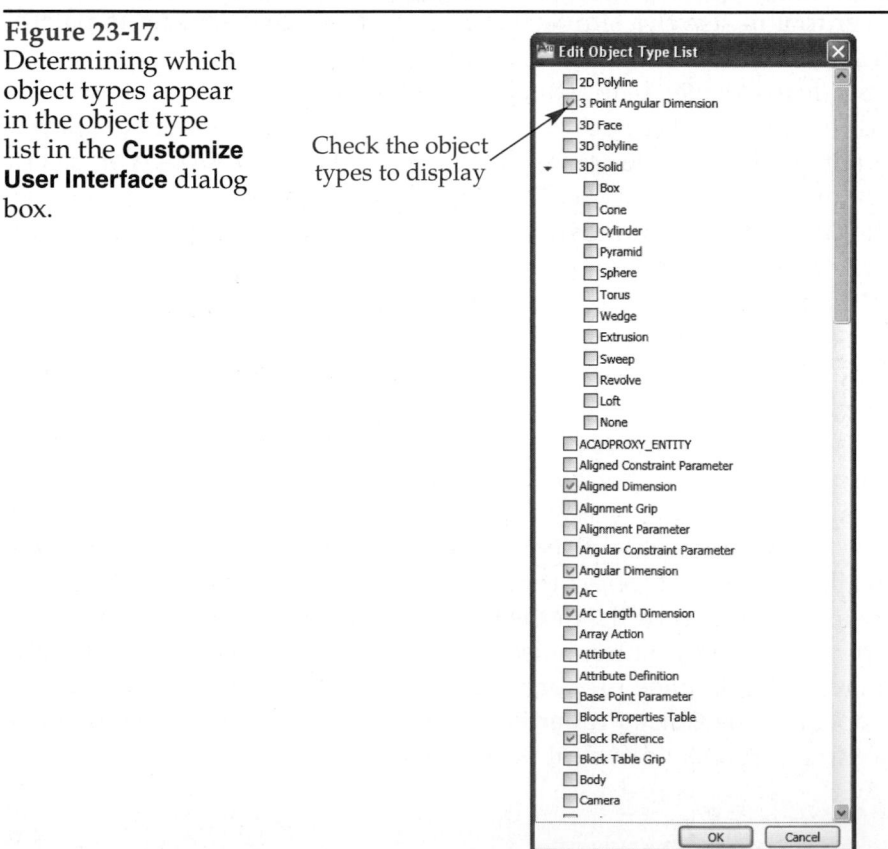

All objects have a General category containing similar quick properties. By default, the color and layer properties are turned on for all objects in the list. To display the General category for objects that do not have defined quick properties, pick the **General** button at the bottom of the object type list.

To customize the quick properties displayed for a specific object type, select the object in the object type list. The category list in the right-hand column displays all categories and quick properties available for that object. Check the properties that you want displayed and uncheck the properties you do not want displayed.

Figure 23-18A shows the line object type selected in the object type list. Notice that this object has 3D Visualization and Geometry categories, as well as the General category. The Material and Angle properties are set to display in addition to the default settings. Once you close the **Customize User Interface** dialog box to apply the changes, the new quick properties are displayed for a selected line. See **Figure 23-18B.** You may need to pause the cursor over the **Quick Properties** palette to expand the palette.

PROFESSIONAL TIP

By customizing quick properties, you may improve your AutoCAD workflow by having quick access to properties throughout the design process. For example, you may wish to have the Show History property displayed in the **Quick Properties** palette for solid primitives, sweeps, lofts, and revolutions. This will allow you to quickly show or hide the history of the solid. Additionally, you may wish to have the Material property displayed in the rollover tooltip for these objects. This will provide a quick indication of which material is assigned to the object.

Figure 23-18.
Customizing the quick properties for a line. A—Two properties in addition to the default properties are set to display. B—The **Quick Properties** palette contains the additional properties.

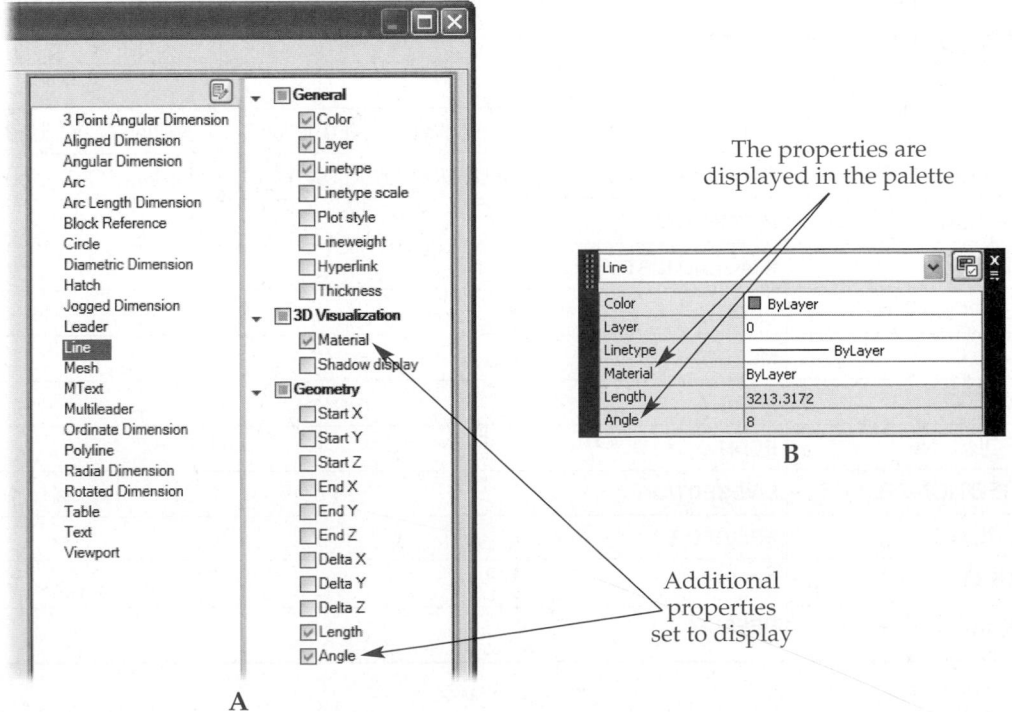

Customizing Double-Click Actions

By double-clicking on certain objects, an appropriate editing command is automatically executed. Which command is initiated is determined by the *double-click action* associated with the object type. Some AutoCAD objects have very specific editing tools available. For example, multiline text (mtext) objects are edited with the in-place text editor and hatches are edited in the **Hatch Edit** dialog box.

A list of AutoCAD objects that have default double-click actions associated with them, other than the **PROPERTIES** command, is shown in **Figure 23-19.** If you double-click on one of the object types listed in the table, the command or macro listed in the Associated Double-Click Action column is executed. If the object type is not listed in the table, it is likely the **Properties** palette is displayed, by default, when the object is double-clicked. This is the double-click action associated with most objects.

Assigning Double-Click Actions

Double-click actions are assigned to specific object types in the **Customize User Interface** dialog box. In the **Customizations in All Files** pane, expand the Double Click Actions branch. All of the AutoCAD object types are listed. See **Figure 23-20.** Expand each branch and notice that many double-click actions call the **Properties** palette, while the objects listed in the table in **Figure 23-19** have double-click actions that call the object-specific editing command.

Figure 23-19.
AutoCAD objects to which a default double-click action other than **PROPERTIES** is assigned.

AutoCAD Object Type	Associated Double-Click Action
ATTDEF	DDEDIT
ATTBLOCKREF	EATTEDIT
ATTDYNBLOCKREF	EATTEDIT
ATTRIB	ATTIPEDIT
BLOCKREF	$M=$(if,$(and,$(>,$(getvar,blockeditlock),0)),^C^C_properties,^C^C_bedit)
DYNBLOCKREF	$M=$(if,$(and,$(>,$(getvar,blockeditlock),0)),^C^C_properties,^C^C_bedit)
HATCH	HATCHEDIT
IMAGE	IMAGEADJUST
LWPOLYLINE	PEDIT
MLINE	MLEDIT
MTEXT	MTEDIT
POLYLINE	PEDIT
SECTIONOBJECT	LIVESECTION
SPLINE	SPLINEDIT
TEXT	DDEDIT
XREF	REFEDIT

Figure 23-20.
All of the AutoCAD object types are displayed in the Double Click Actions branch in the **Customize User Interface** dialog box.

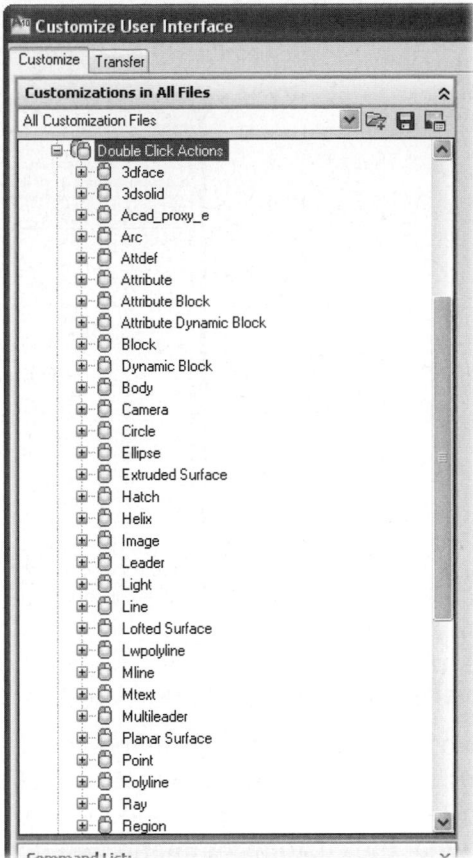

Use the following procedure to change the double-click editing action associated with an object type. For this example, the **DDPTYPE** command will be associated with the point object type so the **Point Style** dialog box appears when a point object is double-clicked.

1. Open the **Customize User Interface** dialog box.
2. In the **Customizations in All Files** pane, expand the Double Click Actions branch.
3. Expand the Point branch under the Double Click Actions branch. Notice that the **PROPERTIES** command is associated with the point object type.
4. In the **Command List:** pane, select the **Point Style...** command. This is the **DDPTYPE** command, as indicated in the **Properties** pane when the command is selected.
5. Drag the **Point Style...** command from the **Command List:** pane and drop it into the Point branch in the **Customizations in All Files** pane. See **Figure 23-21.** The command replaces the existing command as there can only be one double-click action.
6. Pick the **OK** button to close the **Customize User Interface** dialog box and apply the change.

Now, draw a point using the **POINT** command. Double-click on the point and the **Point Style** dialog box appears. Select a new point style in the dialog box and pick the **OK** button. All existing points in the drawing should update to the new style. If not, use the **REGEN** command to update the display.

Figure 23-21.
The double-click action associated with the point object type is changed.

New double-click action assigned

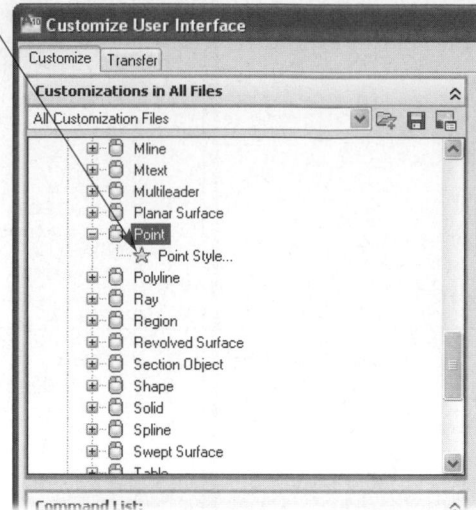

Custom Double-Click Action

You can also create a custom command and assign it to an object type as a double-click action. In this section, you will create a custom command for editing the radius of a circle to one-half of its current value when the circle double-clicked. Here is a breakdown of what the custom command will do:

• Cancel any commands in progress. (^C^C)
• Execute the **SCALE** command with the circle that was double-clicked being the object to be scaled. (scale;)
• Set the center of the circle, which is the point last created, as the base point. (@;)
• Scale the circle to half of its original size. (0.5)

This can be a handy tool should you accidentally enter the intended diameter in response to the radius prompt for the **CIRCLE** command. On noticing you just made that mistake, double-click the circle and the problem is fixed.

Follow these steps to create a custom command and assign it as a double-click action for the dimension object type:

1. Open the **Customize User Interface** dialog box.
2. Pick the **Create a new command** button in the **Command List:** pane.
3. Name the custom command **CirRadToDia**.
4. In the **Properties** pane, select the Macro property. Then, enter the macro ^C^Cscale;@;0.5 in the text box.
5. In the **Customizations in All Files** pane, expand the Double Click Actions branch and locate the Circle branch below it. Notice that the **PROPERTIES** command is currently associated with the circle object type.
6. In the **Command List:** pane, select the new **CirRadToDia** command and drag it into the Circle branch in the **Customizations in All Files** pane.
7. Pick the **OK** button to close the **Customize User Interface** dialog box and apply the change.

Draw a circle using the **CIRCLE** command. Now, double-click on the circle. The circle becomes one-half of its original size. In other words, the previous radius value is the new diameter value.

Chapter Test

Answer the following questions. Write your answers on a separate sheet of paper or complete the electronic chapter test on the student website.
www.g-wlearning.com/CAD

1. What is the key combination called that allows you to press the [Ctrl] key and an additional key to execute a command?
2. How can you disable all shortcut menus and all double-click actions?
3. In which dialog box can you define what a right-click does when in default mode, edit mode, and command mode?
4. Why are shortcut menus *context sensitive?*
5. When does the command shortcut menu appear?
6. What must the **PICKFIRST** setting be in order for the edit shortcut menu to appear?
7. Briefly describe how to create a new shortcut menu.
8. Describe the syntax for the name of an object-specific shortcut menu.
9. Why is CIRCLE_OBJECT *not* a valid alias for a shortcut menu?
10. Describe the difference between an OBJECT_*type* shortcut menu and an OBJECTS_*type* shortcut menu.
11. What is the syntax for the alias for a command-specific shortcut menu?
12. List the steps to add an alias to a shortcut menu.
13. How do you turn on the display of the **Quick Properties** palette?
14. What is a *quick property?*
15. How does a rollover tooltip differ from the **Quick Properties** palette?
16. In which dialog box do you select the objects that appear in the object type list in the **Customize User Interface** dialog box?
17. What is a *double-click action?*
18. What is the most common double-click action?
19. In which dialog box is a double-click action assigned?
20. List the basic steps for modifying the double-click action associated with an object.

Drawing Problems

Before customizing AutoCAD, check with your instructor or supervisor for specific instructions or guidelines.

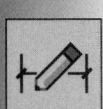

1. Create a shortcut key for each of the drawing and editing commands listed below. Be sure not to use any existing shortcut keys.

LINE	**MOVE**
ARC	**COPY**
CIRCLE	**STRETCH**
POLYLINE	**TRIM**
POLYGON	**EXTEND**
RECTANGLE	**CHAMFER**
DTEXT	**FILLET**
ERASE	

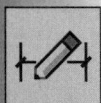

2. In this problem, create an object-specific shortcut menu. The shortcut menu should be displayed when an arc object is selected. The shortcut menu should contain the **LENGTHEN** and **BREAK** commands.

3. In this problem, create a command-specific shortcut menu. The shortcut menu should be displayed when the **CIRCLE** command is active. Add the **Circle, Tan, Tan, Tan** command available in the **Customize User Interface** dialog box to the shortcut menu. Add two selection options, such as **Last** or **Window**, as described in this chapter.

4. By default, double-clicking on a circle displays the **Properties** palette Take the steps necessary so that a **REGEN** is performed instead.

5. Create a custom command that changes the color of an object to blue. Then, assign this command as the double-click action for the hatch object type.

CHAPTER 24

Tool Palette Customization

Learning Objectives

After completing this chapter, you will be able to:

✓ Modify the appearance of the **Tool Palettes** window.
✓ Compare and contrast block insertion, hatch insertion, and command tools.
✓ Create new tool palettes from scratch and using **DesignCenter**.
✓ Add tools to existing tool palettes.
✓ Explain how tool palettes are formatted.
✓ Adjust the properties of tools.
✓ Create a flyout tool.
✓ Organize tool palette tabs into groups.
✓ Export and import tools and tool palettes.

Tool palettes are a user-interface method for the easy insertion of blocks and hatch patterns, for command entry, and for attaching materials. Blocks and hatch patterns can be simply dragged and dropped from a tool palette directly into a drawing. In addition, tool palettes serve as a materials library in AutoCAD. Tool palettes are contained within the **Tool Palettes** window.

Tool Palette Overview

The **TOOLPALETTES** command is used to display the **Tool Palettes** window. Notice that the **Tool Palettes** window has a number of tabs on its edge. Each of these tabs corresponds to a tool palette. To make a tool palette active, pick its tab. If there are more tabs than can be displayed, pick on the "stack" at the bottom of the tabs to display a shortcut menu in which you can select the tool palette to display. To use a tool on any tool palette, drag it from the tool palette and drop it into the drawing.

By default, the **Tool Palettes** window contains the **Modeling**, **Constraints**, **Annotation**, **Architectural**, **Mechanical**, **Electrical**, **Civil**, **Structural**, **Hatches and Fills**, **Tables**, **Command Tool Samples**, **Leaders**, **Draw**, **Modify**, **Cameras**, **Visual Styles**, eight material palettes (more if the materials library is installed), and five light palettes. Each of these is described below.

- **Modeling palette.** Contains tools for creating specific variations of some solid primitives. Also contains two UCS tools and a 3D align tool.

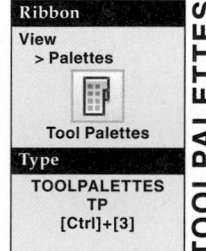

Ribbon

View
> Palettes

Tool Palettes

Type

TOOLPALETTES
TP
[Ctrl]+[3]

TOOLPALETTES

- **Constraints palette.** Contains tools for inserting geometric and dimensional constraints.
- **Annotation palette.** Contains blocks that are typically inserted in paper space.
- **Architectural, Mechanical, Electrical, Civil, and Structural palettes.** Can be used to insert blocks that are meant to be used within the discipline for which the palette is named. Certain preset properties are already attached to the symbols, such as scale and rotation.
- **Hatches and Fills palette.** Contains some commonly used hatches and sample gradient fills that can be quickly inserted into a drawing.
- **Tables palette.** Allows you to insert sample tables in US customary (Imperial) and metric formats.
- **Command Tool Samples palette.** Allows you to execute certain commands by picking the tool.
- **Leaders palette.** Allows you to place leaders with or without text or with various types of callout balloons in both US customary (Imperial) and metric scales.
- **Draw palette.** Contains many of the same tools found in the **Draw** panel on the **Home** tab of the ribbon and some tools for inserting blocks, attaching images, and attaching xrefs.
- **Modify palette.** Contains many of the same tools found in the **Modify** panel on the **Home** tab of the ribbon.
- **Material palettes.** The tools in these palettes allow you to quickly drag and drop various materials into the drawing. The material palettes are AutoCAD's materials library. If the materials library is not installed, eight palettes containing sample materials are available.
- **Light palettes.** Contain tools for inserting many types of lights.
- **Cameras palette.** Contains tools for inserting three different cameras into the drawing. The cameras have differing lens lengths and fields of view.
- **Visual Styles palette.** Contains tools for three variations of the default visual styles.

For detailed instruction on how tool palettes can be used to insert blocks and hatch patterns, see *AutoCAD and Its Applications—Basics.* This chapter describes how to customize existing tool palettes and create your own tool palettes.

Tool Palette Appearance

There are a number of methods for altering the appearance of the **Tool Palettes** window, either for productivity or personal preference. For example, the tabs can be renamed. Right-click on the title of the tab and select **Rename Palette** from the shortcut menu. An edit box is displayed near the current name with the name highlighted. Type a new name and press [Enter] to rename the palettes. Other methods of changing the appearance of the **Tool Palettes** window are covered in this section.

Docking

By default, the **Tool Palettes** window is floating on the right side of the screen in the AutoCAD Classic workspace, docked on the right side of the screen in the 3D Modeling workspace, and not displayed in the 2D Drafting & Annotation workspace. If docked, it can be moved to a new, floating location by picking and holding on the title bar, dragging the window to the desired location, and releasing the pick button. Moving the window to the far side of the drawing window forces the title bar to flip to the other side of the **Tool Palettes** window so the bar is toward the outer edge of the drawing window.

By default, the **Tool Palettes** window can be docked, just like a toolbar. Moving the window outside of the drawing window to the left or right docks it. The **Tool Palettes** window cannot be docked at the top or bottom. To prevent docking, right-click on the **Tool Palettes** window title bar to display the shortcut menu. Select **Allow Docking** to remove the check mark. When a check mark appears next to **Allow Docking**, the window can be docked.

The **Tool Palettes** window can be resized just like standard windows. While floating, the top and bottom edges, the vertical area just below the tabs, and the corners can be used to resize the window. While docked, only the right or left edge can be used for resizing. As with standard windows, move the cursor to one of the edges of the window until a double arrow appears. Then, press and hold the pick button, drag the edge until the window reaches the desired size, and release the pick button.

Transparency

Using the **Tool Palettes** window while it is floating may cause occasional visibility problems because it covers up part of the drawing window. However, the window can be made partially transparent so that the part of the drawing under the window can be seen. Transparency will not be active when the window is in a docked position. Hardware acceleration may need to be turned off to set transparency, depending on your system configuration.

Right-click on the title bar of the **Tool Palettes** window and pick **Transparency...** from the shortcut menu. The window must be floating for this option to appear. The **Transparency** dialog box is then displayed, **Figure 24-1**. To set the level of transparency, move the slider in the **General** area to a lower value. The slider controls the level of transparency applied to the window. The further to the left that the slider is placed, the more transparent the **Tool Palettes** window, **Figure 24-2**. Placing the slider all of the way to the right makes the **Tool Palettes** window opaque. The slider in the **Rollover** area sets the transparency level for when the cursor is over the window. This setting must be equal to or greater than the slider setting in the **General** area. To see the transparency level when the cursor is over the window, pick the **Preview** button.

If you find a particular level of transparency that you like, but wish to make the **Tool Palettes** window opaque for a short time while you perform an operation or two, you can just use the toggle to turn off transparency. Open the **Transparency** dialog box

Figure 24-1.
The **Transparency** dialog box is used to control the transparency of the **Tool Palettes** window.

Check to apply the settings to all palettes

Uncheck to enable transparency for all windows

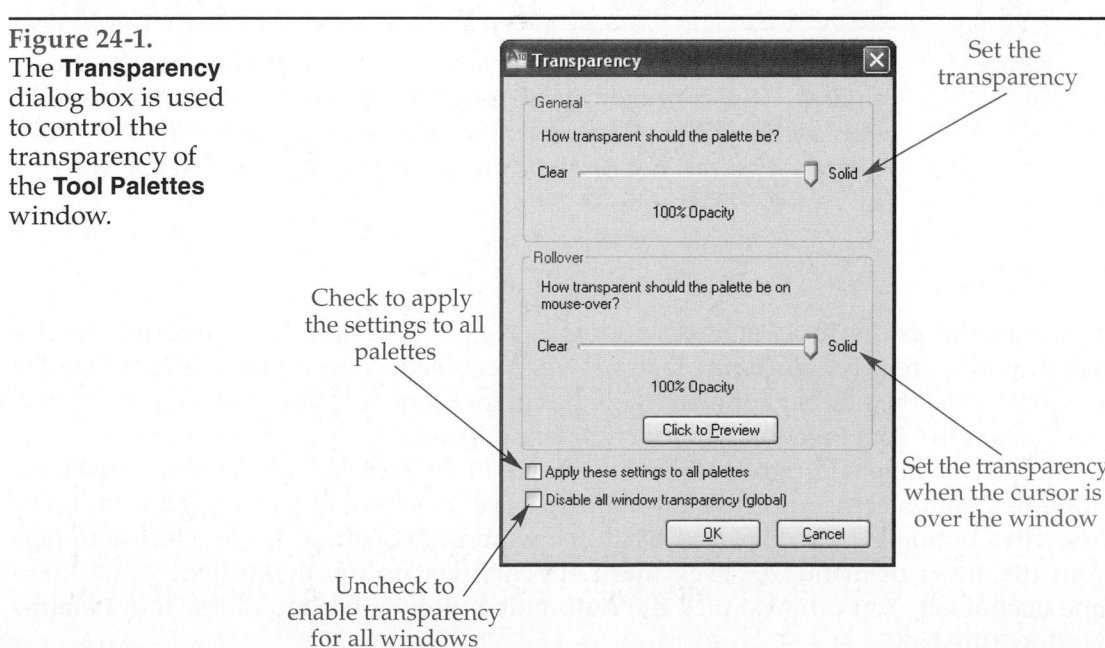

Set the transparency

Set the transparency when the cursor is over the window

Figure 24-2.
A—The **Tool Palettes** window has a low transparency setting. B—The **Tool Palettes** window has a high transparency setting.

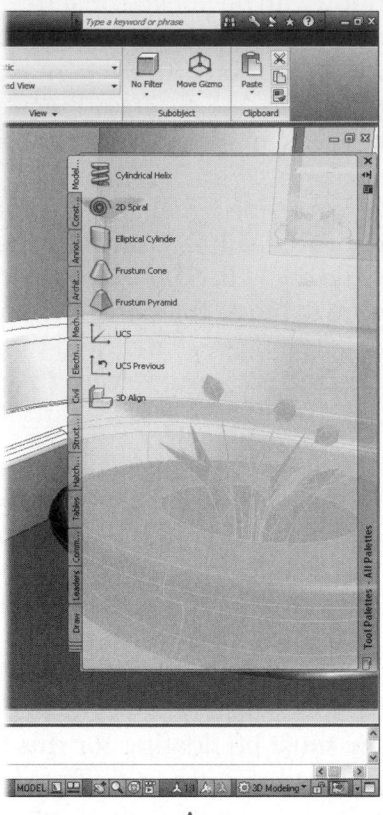
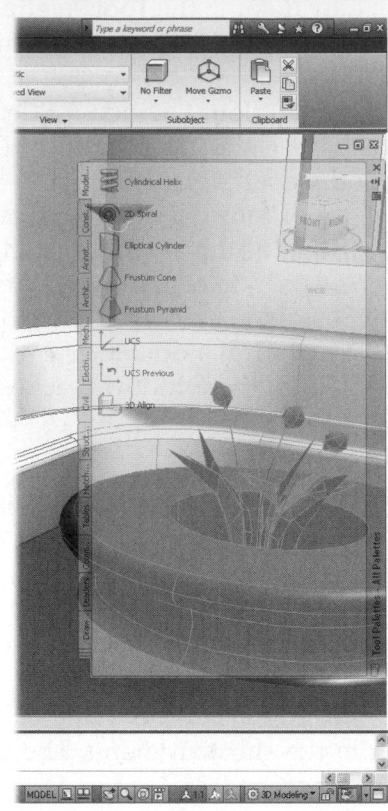

A   B

and check the **Disable all window transparency (global)** check box. Then, when you want to go back to your previous level of transparency, simply open the **Transparency** dialog box again and uncheck the check box. You will not have to adjust the slider; just toggle the transparency back on.

Also notice in the **Transparency** dialog box the **Apply these settings to all palettes** check box. This allows you to apply the changes made in the dialog box to all AutoCAD palettes. Often, a user will prefer the same level of transparency for all palettes.

PROFESSIONAL TIP

Although transparency allows you to *see* through the **Tool Palettes** window, you cannot *work* through it. You cannot access points behind the window because the cursor is actually on the **Tool Palettes** window, not on the drawing underneath it. This is true of any AutoCAD palette.

Autohide

As useful as the **Tool Palettes** window is, it does take up a large amount of valuable drawing area. The *autohide* feature, when enabled, compresses the **Tool Palettes** window so just the title bar appears when the cursor is not over the window, **Figure 24-3.** This allows the **Tool Palettes** window to take up less room when not being used.

To turn on autohide, right-click on the title bar of the **Tool Palettes** window or pick the **Properties** button at the top of the title bar to display the shortcut menu, **Figure 24-4.** The **Properties** button is only displayed when the window is floating. Then, select **Auto-hide** from the shortcut menu. A check mark appears next to the menu item when autohide is enabled. You can also pick the **Auto-hide** button at the top of the **Tool Palettes** window title bar.

Figure 24-3.
When the autohide feature is enabled, the **Tool Palettes** window appears as only the title bar when the cursor is not over it.

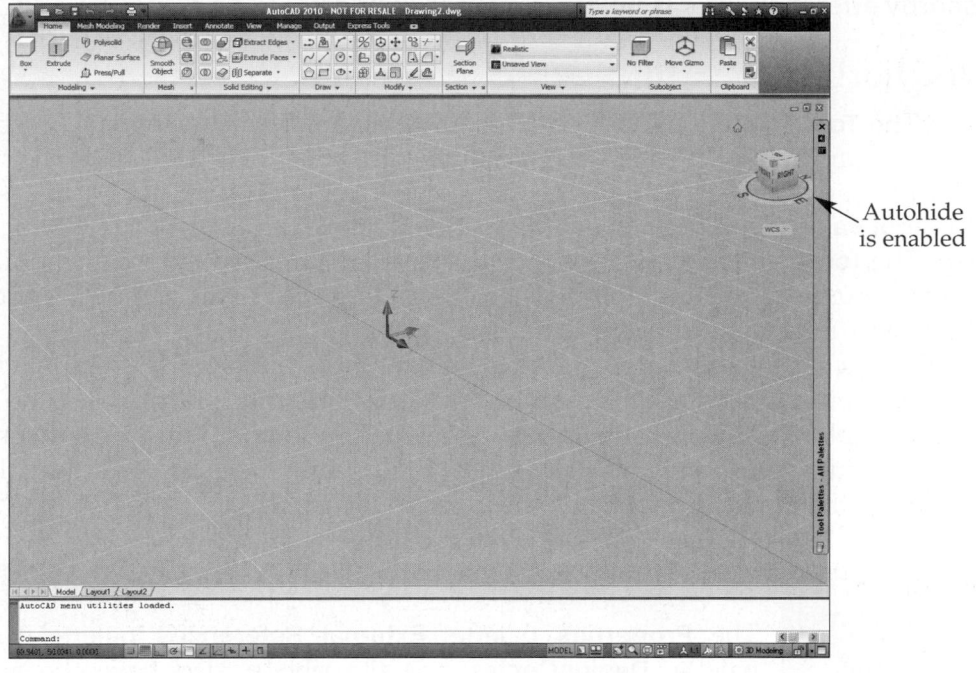

Autohide
is enabled

Figure 24-4.
Using the shortcut menu to enable autohide.

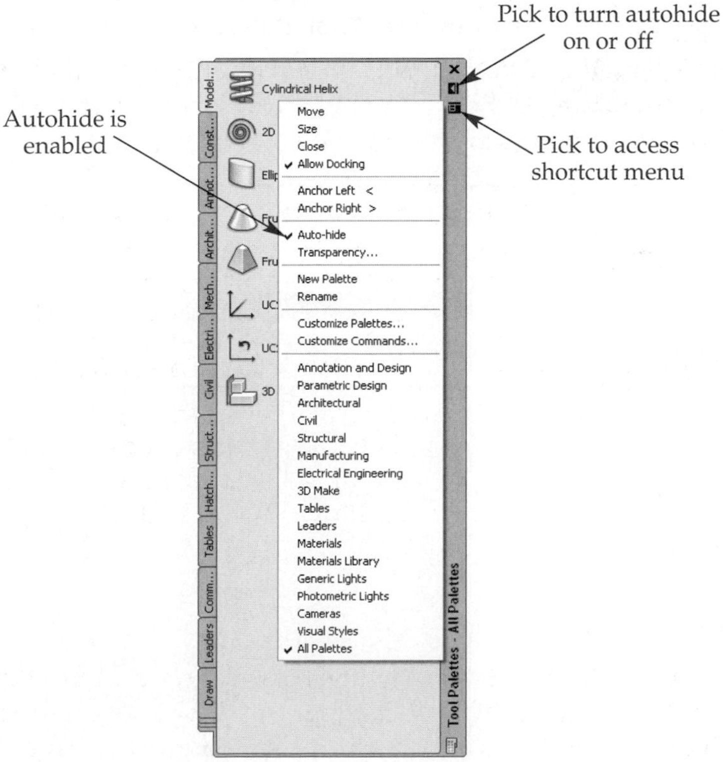

Pick to turn autohide
on or off

Autohide is
enabled

Pick to access
shortcut menu

Cylindrical Helix

Move
Size
Close
✓ Allow Docking

Anchor Left <
Anchor Right >

✓ Auto-hide
Transparency...

New Palette
Rename

Customize Palettes...
Customize Commands...

Annotation and Design
Parametric Design
Architectural
Civil
Structural
Manufacturing
Electrical Engineering
3D Make
Tables
Leaders
Materials
Materials Library
Generic Lights
Photometric Lights
Cameras
Visual Styles
✓ All Palettes

To use the **Tool Palettes** window when autohide is enabled, move the cursor over the title bar and the window expands to its normal size. The tool palettes can then be used in the standard way. After using the **Tool Palettes** window, it is again hidden shortly after the cursor is no longer over the window.

Anchoring

The **Tool Palettes** window can also be *anchored.* Anchoring is a combination of docking and autohide. When a tool palette is anchored, it is docked on the right or left side of the drawing area, but it is compressed to just a title bar. See **Figure 24-5.** To use the **Tool Palettes** window when it is anchored, move the cursor over the anchored title bar. The **Tool Palettes** window is then displayed floating next to the anchored window. It can be used just as if autohide is enabled. Once the cursor is moved off of the **Tool Palettes** window, the window is hidden.

To anchor the tool palette, docking must first be enabled. Then, right-click on the title bar of the **Tool Palettes** window or pick the **Properties** button at the top of the title bar to display the shortcut menu. Next, select either **Anchor Left <** or **Anchor Right >** in the shortcut menu. To return the window to floating mode, pick and drag the title bar back into the drawing area while the window is displayed.

PROFESSIONAL TIP

The **Properties** palette, **External References** palette, **QuickCalc** palette, **DesignCenter**, and the ribbon also have the anchoring feature. You can create a very productive drawing-window arrangement by anchoring these items along with the **Tool Palettes** window. Anchor the **Tool Palettes** window on one side of the drawing window and the **Properties** palette and **DesignCenter** on the other side. By default, the ribbon is docked at the top of the screen.

Figure 24-5.
When the **Tool Palettes** window is anchored, it is docked, but compressed to just its title bar.

The window is anchored

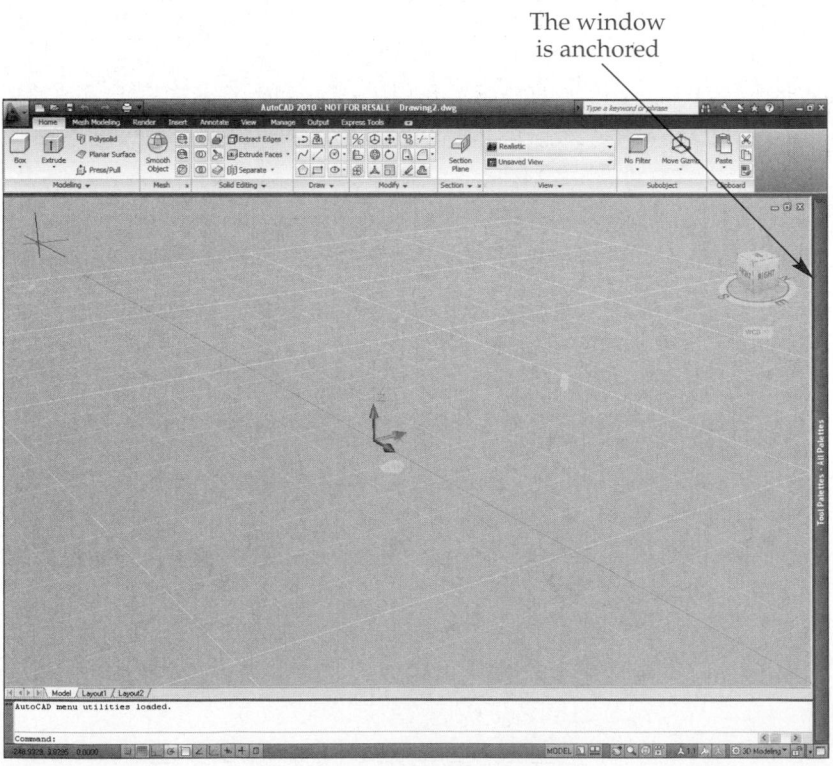

AutoCAD and Its Applications—Advanced

Tool Appearance

The way in which the tools are shown in the palettes can be customized. To do so, right-click in a blank area of the current tool palette (not on the title bar). Then, select **View Options...** from the shortcut menu. The **View Options** dialog box is displayed, **Figure 24-6.**

The **Image Size:** area of the dialog box is used to set the size of the tool icons in the palette. Drag the slider to the left or right to change the size. Dragging the slider to the left decreases the size of the icon. Dragging the slider to the right increases the size of the icon. To the left of the slider is a preview that represents the size of the icon.

The **View Style:** area controls how the tools on the tool palettes are displayed. When the **Icon only** radio button is selected, the tools are represented as an image only, **Figure 24-7A.** Tooltips will be displayed if you hold your cursor over a tool. Selecting

Figure 24-6.
The **View Options** dialog box is used to set how the tools appear in tool palettes.

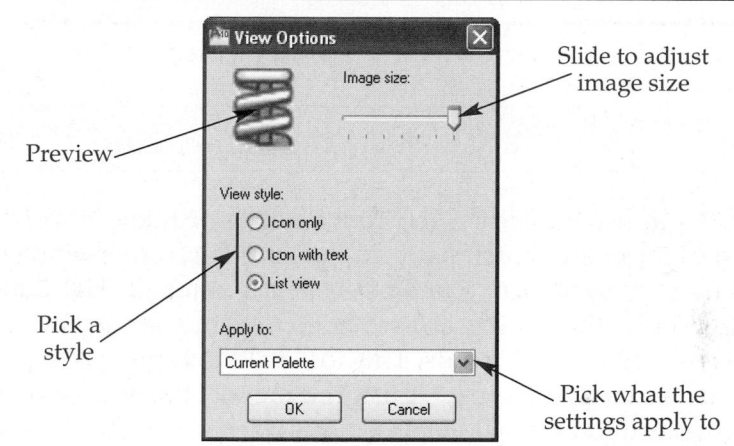

Figure 24-7.
The various ways in which tools can appear in tool palettes. A—Icons only. B—Icons and text. C—As a list.

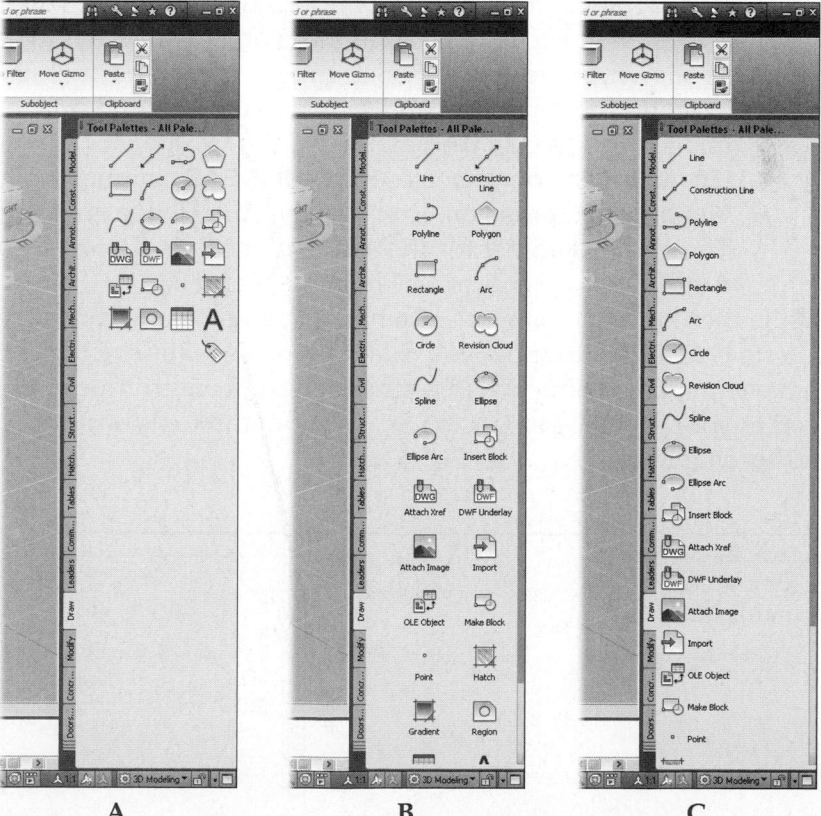

the **Icon with text** radio button represents the tools with an image and the tool name below the image, **Figure 24-7B.** When the **List view** radio button is selected, the tools are represented with an image and the tool name to the side of the image, **Figure 24-7C.** This is the default view style.

The **Apply to:** drop-down list at the bottom of the dialog box determines where the settings are applied. To have the settings applied to the current tool palette, select Current Palette from the drop-down list. To have the settings applied to all tool palettes, select All Palettes from the drop-down list. When finished making settings, pick the **OK** button to close the **View** options dialog box.

Exercise 24-1

Complete the exercise on the student website.
www.g-wlearning.com/CAD

Commands in a Tool Palette

As indicated earlier, the **Tool Palettes** window not only offers a means to easily insert blocks and hatch patterns, it can be used to execute commands. Make the **Tool Palettes** window active and select the **Command Tool Samples** palette, **Figure 24-8.** The tools on this palette are provided to demonstrate how tool palettes can be customized by adding commands. This tool palette is provided with the intention that it will be customized by the user. Customizing tool palettes is discussed later in this chapter. The default tools provided are:

- **Line.** Executes the **LINE** command. Notice that this tool has a small triangle, or arrow, to the right of the icon. The arrow indicates that the tool acts as a *flyout tool,* similar to flyout buttons on the ribbon. Picking the arrow displays a graphic shortcut menu containing other command tools attached to this tool, as shown in **Figure 24-8.** Selecting a command tool from the shortcut menu executes the command and causes that command image to become the default image displayed in the palette.
- **Linear Dimension.** Executes the **DIMLINEAR** command. This tool is also a flyout.
- **VisualLisp Expression.** Executes the AutoLISP expression (entget (car (entsel))). The entity data list for the selected entity is displayed on the command line. AutoLISP is discussed in Chapters 27 and 28.

While these three tools on the **Command Tools** palette can increase productivity, this palette is included to show you examples of ways in which tools can be customized to make you more productive in your own design environment. It is meant to be customized to your own needs. Customizing command tools on a tool palette is covered later in this chapter.

Figure 24-8.
A flyout tool contains other tools.

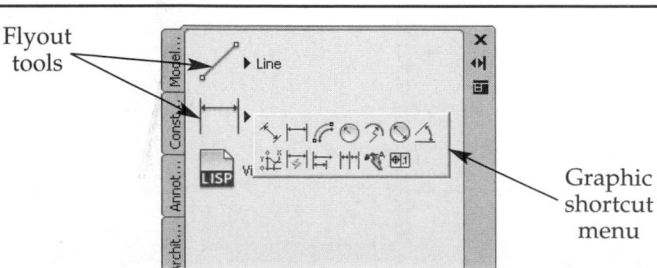

Flyout tools

Graphic shortcut menu

Adding Tool Palettes

As discussed later in this chapter, you can add new tools to tool palettes. You can also create new tool palettes and then add tools to them. To add a new tool palette, right-click on a blank area of an existing tool palette or on the title bar of the **Tool Palettes** window. Then, select **New Palette** from the shortcut menu. A new, blank palette is added and a text box appears next to the name. Type the desired name for the new palette and press [Enter].

Notice the help link at the top of the new tool palette, **Figure 24-9**. This link offers assistance in customizing tool palettes. It will disappear once you add a tool to the tool palette.

The new tool palette can be reordered within the tabs by right-clicking on the tab name to display the shortcut menu. Then, select **Move Up** or **Move Down** to reorder the palettes. You may need to do this several times in order to get the palette in the position you want. The other palettes can be reordered in this same manner. If you need to reorder multiple palettes, you can use the **Customize** dialog box. This dialog box is discussed in detail later in this chapter.

DesignCenter can be used to create a new palette fully populated with all of the blocks contained in a drawing. In the **Folders** tab of the **DesignCenter** palette, navigate to the drawing from which you are making the tool palette, right-click on the drawing name, and select **Create Tool Palette** from the shortcut menu. A new tool palette is added to the **Tool Palettes** window with the same name as the drawing file.

For example, open **DesignCenter** and navigate to the Fasteners-US drawing located in the \Sample\DesignCenter folder. Right-click on the file name and select **Create Tool Palette** from the shortcut menu, **Figure 24-10A**. A tool palette named **Fasteners-US** is added to the **Tool Palettes** window. All of the blocks contained in the Fasteners-US drawing are available on the tool palette, **Figure 24-10B**.

Ribbon	
View	
> Palettes	
DesignCenter	**ADCENTER**
Type	
ADCENTER DC ADC [Ctrl]+[2]	

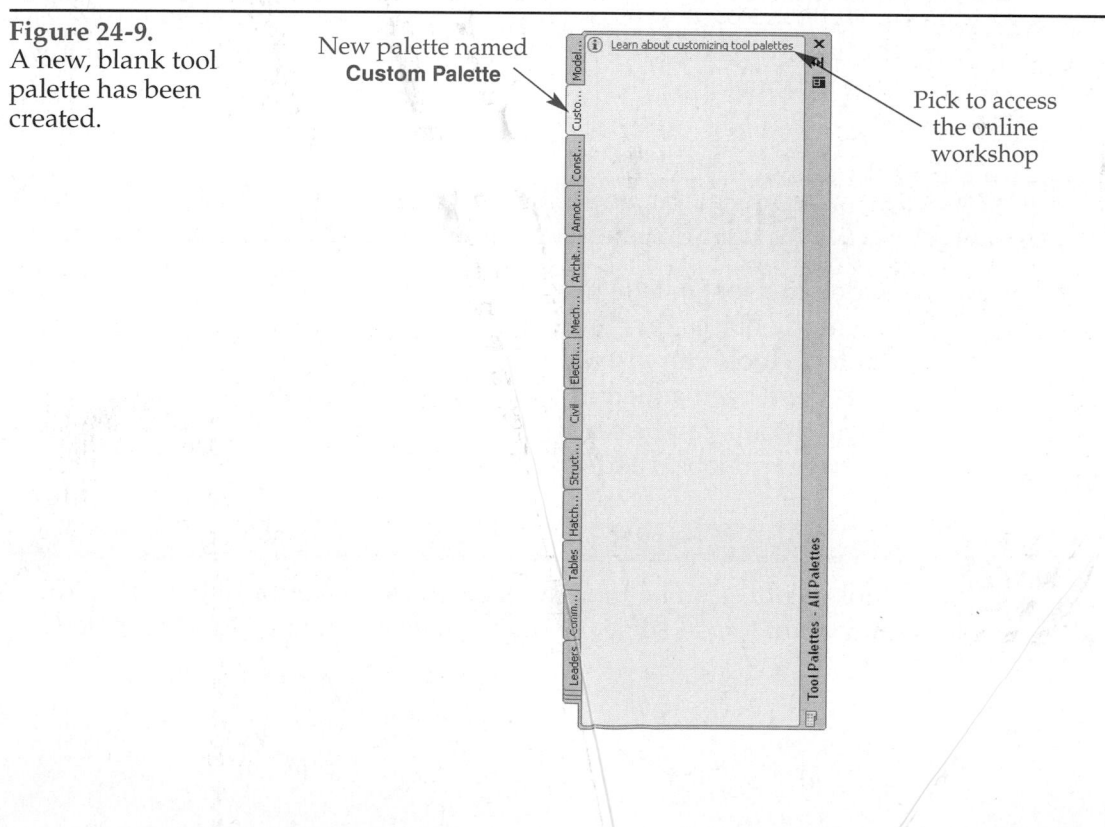

Figure 24-9.
A new, blank tool palette has been created.

New palette named **Custom Palette**

Pick to access the online workshop

Figure 24-10.
A—Creating a tool palette from the blocks contained within a drawing. B—The new tool palette is added.

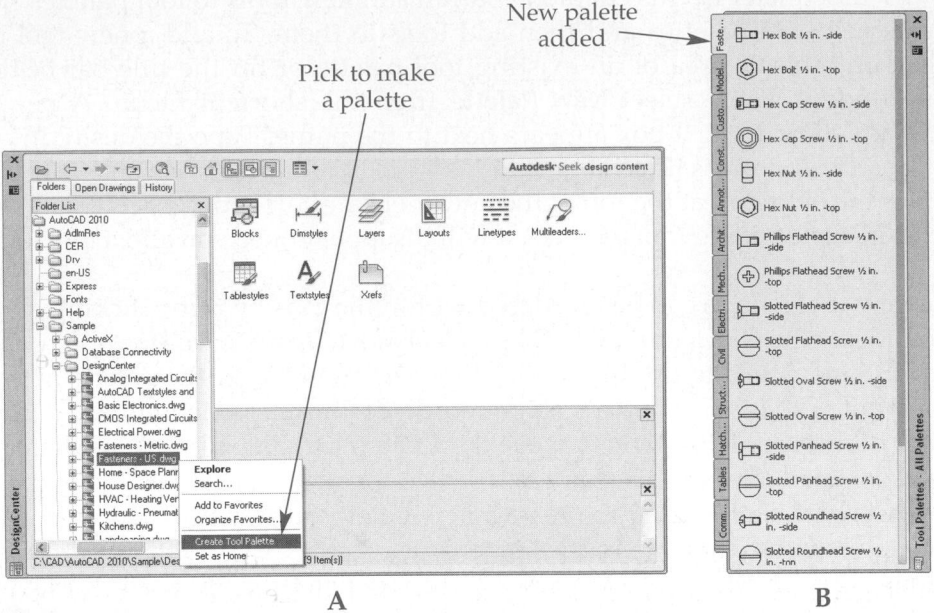

A

B

CAUTION

When using any "block insertion" tool from a tool palette, the block is actually being imported from the source drawing—the drawing from which the block tool on the palette was created. An error occurs if the source file has been moved or deleted. The source file must be restored to its original location to allow the tool to work or the tool must be recreated from the source drawing in the drawing's new location.

Adding Tools to a Tool Palette

Tools can be added to a tool palette in a variety of ways. Tools can be created from toolbar buttons, geometric objects in the current drawing, and objects in other drawings (via **DesignCenter**). Tools can also be copied to and pasted from the Windows clipboard. Once tools have been added to a tool palette, they can be arranged to suit your preference. Related tools can be separated into distinct areas on the tool palette and those areas can have text labels added to them.

PROFESSIONAL TIP

Tool palettes can be thought of as an extension of ribbon tabs. You may want to add some of your favorite commands to a tool palette.

Creating a Tool from a Toolbar

Toolbar buttons can be directly dragged and dropped onto a tool palette. To do this, the **Customize** dialog box for tool palettes must be open. This is *not* the **Customize User Interface** dialog box that is used to customize the ribbon, as described in previous chapters. To display the **Customize** dialog box, use the **CUSTOMIZE** command or right-click on a tool palette and select **Customize Palettes...** from the shortcut menu.

You do not actually use the **Customize** dialog box to copy a toolbar button to a tool palette, but the dialog box must be open. The **Customize** dialog box is discussed in detail later in this chapter.

Make sure the palette you want the button added to is current (on top). With the **Customize** dialog box open, move your cursor to the desired toolbar button. Flyout buttons cannot be added to a tool palette. Pick and hold on the toolbar button and drag it to the desired location in the tool palette, **Figure 24-11**. A horizontal "I-bar" appears in the tool palette to indicate where the new tool will be inserted. Drop the toolbar button when it is in the desired position. The tool is inserted in the tool palette. Close the **Customize** dialog box.

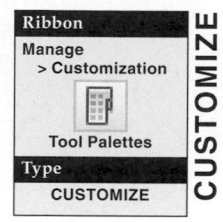

Ribbon
Manage
> Customization

Tool Palettes

Type
CUSTOMIZE

CUSTOMIZE

NOTE

Tools cannot be dragged from the ribbon to a tool palette, only from toolbars. Toolbars are not displayed by default except in the AutoCAD Classic workspace.

Creating a Tool from an Object in the Current Drawing

Another way to add a drawing command to a tool palette is to drag an object in the current drawing, such as a line, hatch, block, dimension, camera, or light, and drop it onto the tool palette. The appropriate command to create that object is added to the tool palette. Not all objects support this method. For example, solid primitives cannot be dragged onto a tool palette to create a tool.

Figure 24-11.
Creating a tool on a tool palette from a toolbar button.

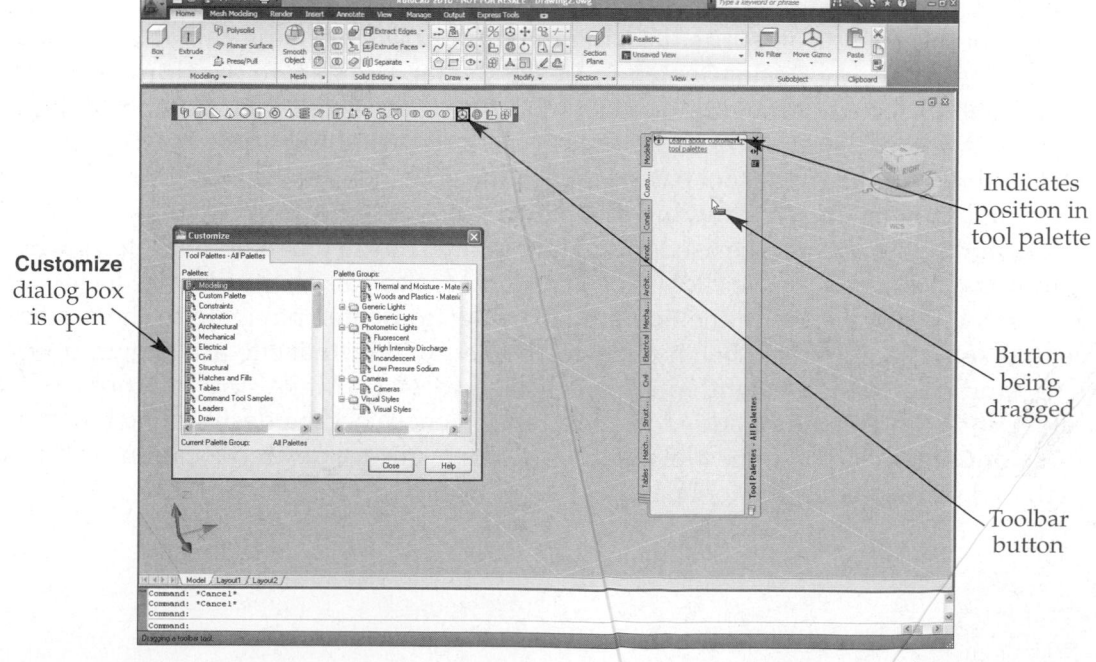

Customize dialog box is open

Indicates position in tool palette

Button being dragged

Toolbar button

First, ensure that the **PICKFIRST** system variable is set to 1. Also, make sure the tool palette to which you want the tool added to is current (on top). Next, with no command active, select the desired object. Move the cursor directly onto the selected object (not a grip). Then, press and hold down either mouse button. Finally, drag the object to the desired position on the tool palette and drop it. The appropriate drawing command is inserted into the tool palette.

Using the Windows Clipboard to Create a Tool

The copy-and-paste feature of the Windows operating system is another way to transfer objects in the drawing to a tool palette. First, ensure that the **PICKFIRST** system variable is set to 1. Then, with no command active, right-click on the object and select **Copy** from the shortcut menu. Next, make current the tool palette to which you want the tool added. Finally, right-click on a blank area in the tool palette and select **Paste** from the shortcut menu. The appropriate command is added as the last tool on the tool palette.

The copy-and-paste technique can also be used to transfer tools from one tool palette to another. First, make current the tool palette that contains the tool to be transferred. Right-click on the tool to transfer and select **Copy** in the shortcut menu. If you want to *move* the tool from the first tool palette to the second, select **Cut** in the shortcut menu. Next, make current the tool palette to which you want the tool transferred. Right-click and select **Paste** from the shortcut menu. The tool is added to the second tool palette as the last tool on the palette.

Adding Block and Hatch Tools from DesignCenter

Earlier, you saw how to create a tool palette consisting of all of the blocks in a single drawing by using **DesignCenter**. It is also possible to add individual blocks from a drawing to a tool palette using **DesignCenter**. To do this, open **DesignCenter**. In the **Folders** tab, navigate to the drawing that contains the desired block. Expand that drawing's branch to see the named objects within the drawing. Select the Blocks branch. The blocks that are defined in the drawing appear on the right-hand side of the **DesignCenter** palette. Make sure the tool palette you want the tool added to is current. Then, select the block in **DesignCenter** and drag it to the tool palette, **Figure 24-12.** Move the cursor to the desired position on the tool palette and drop the block. The new block tool is inserted in the tool palette.

A tool that inserts an entire drawing into the current drawing can also be added to a tool palette using **DesignCenter**. In **DesignCenter**, navigate to the drawing in the **Folders** tab. Select the drawing on the right-hand side of the **DesignCenter** palette, drag it to the tool palette, and drop it in the desired position, **Figure 24-13.** The new block tool is inserted in the tool palette. When the new tool is used, the entire drawing is inserted into the current drawing as a block.

DesignCenter can also be used to add hatch patterns to a tool palette. Hatch pattern definitions are stored in two files—acad.pat and acadiso.pat. These files are located in the user's \Support folder. In the **Folders** tab of **DesignCenter**, navigate to the acad.pat file and select it. All of the hatch patterns defined within that file are shown on the right-hand side of the **DesignCenter** palette, **Figure 24-14.** Make sure the tool palette you want the hatch pattern added to is current. Then, select the desired hatch pattern in **DesignCenter**, drag it to the tool palette, and drop it in the desired location. The new hatch tool is inserted in the tool palette.

Figure 24-12.
Adding an individual block contained within a drawing as a tool on a tool palette.

Selected block

Indicates the position

Drag the block

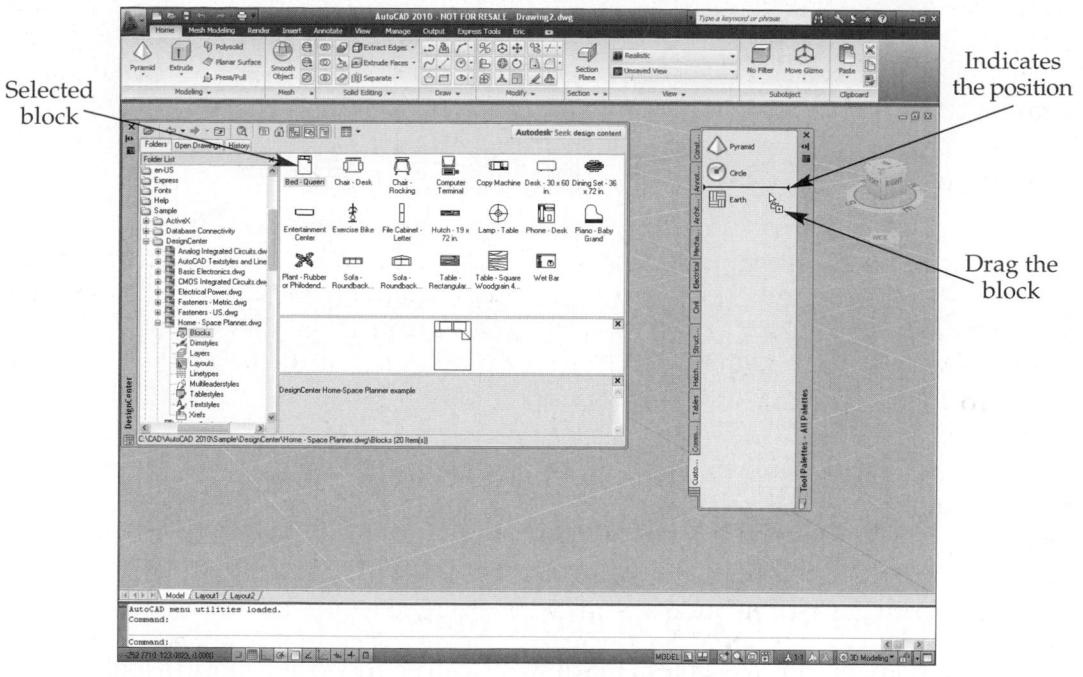

Figure 24-13.
Adding an entire drawing as a tool on a tool palette.

Selected drawing

Drag the drawing

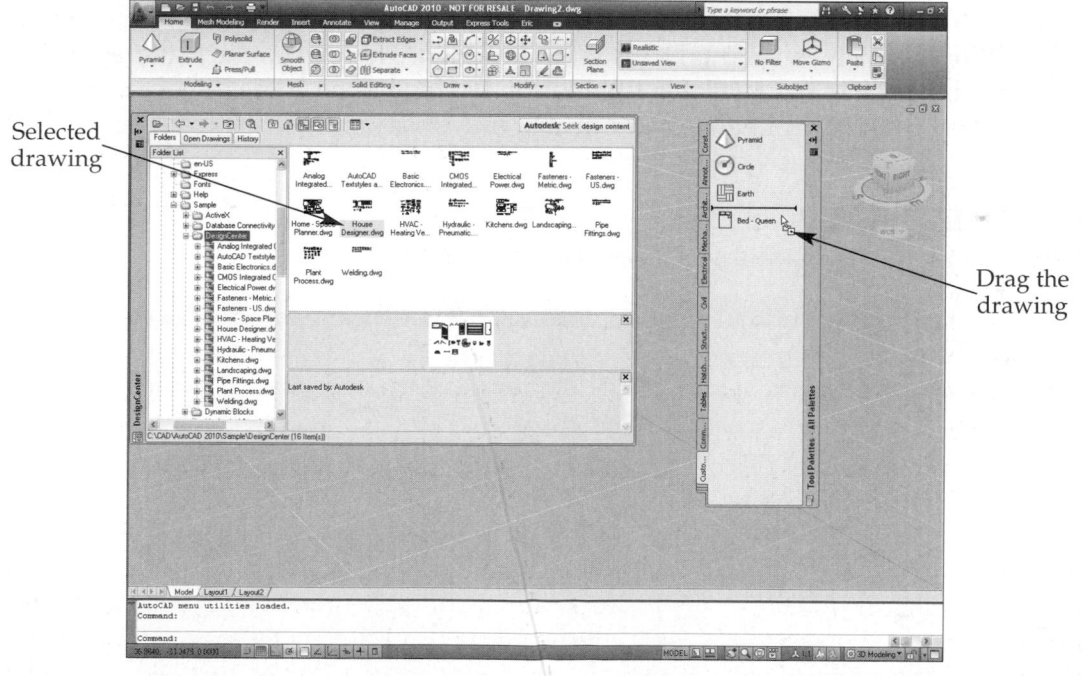

Figure 24-14.
Hatch patterns can be selected in **DesignCenter** and dragged to a tool palette to create a new tool.

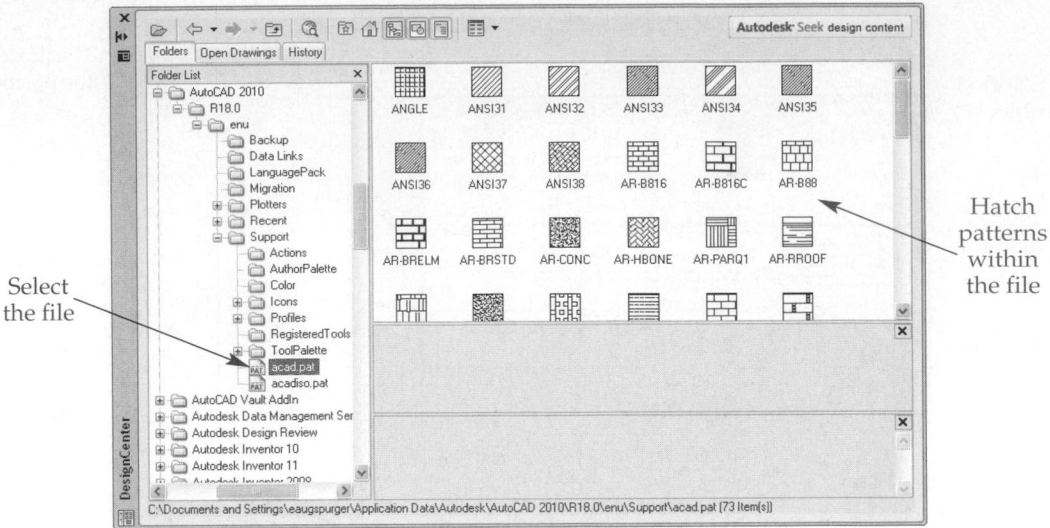

Select the file

Hatch patterns within the file

Adding Visual Style and Material Tools

Visual styles that are available in the drawing can be added as a tool to a tool palette. To do this, first make current the tool palette to which you want the visual style added. Then, open the **Visual Styles Manager**. Select the icon for the visual style at the top of the **Visual Style Manager**, drag it to the tool palette, and drop it into position. See **Figure 24-15.**

Materials that are available in the current drawing can also be added to a tool palette. This is how you create and manage a materials library. First, make current the tool palette to which you want the visual style added. Then, open the **Materials** palette. Select the material in the **Available Materials in Drawing** pane at the top of the **Materials** palette, drag it to the tool palette, and drop it into position. See **Figure 24-16.**

Figure 24-15.
Adding a visual style as a tool on a tool palette.

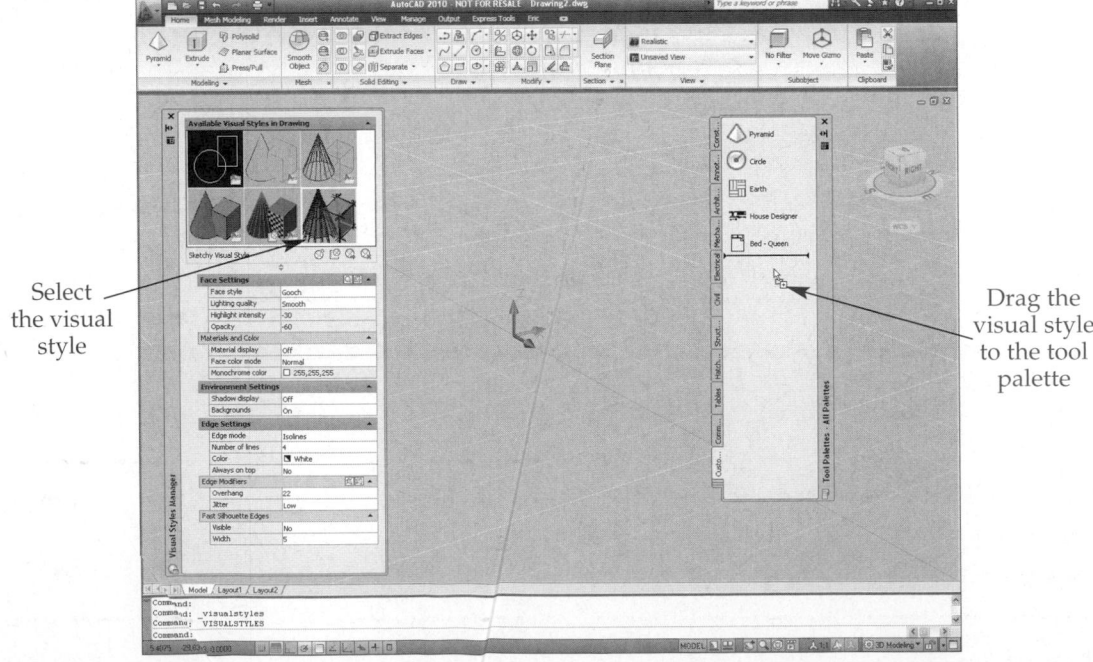

Select the visual style

Drag the visual style to the tool palette

Figure 24-16.
Adding a material as a tool on a tool palette.

Select the material

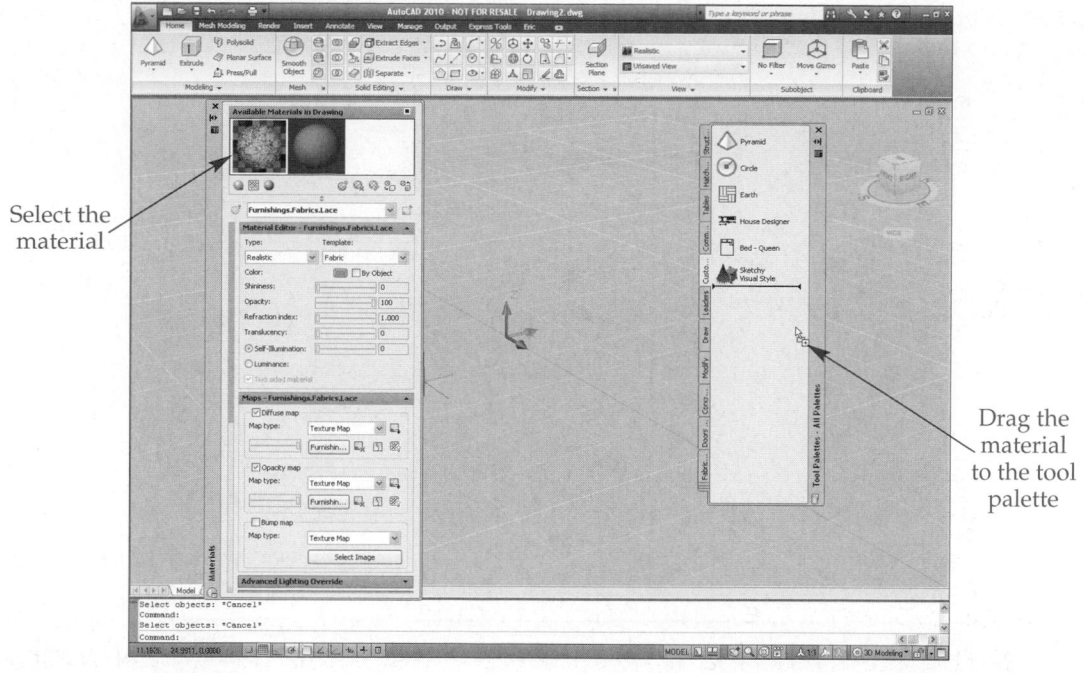

Drag the material to the tool palette

PROFESSIONAL TIP

You can also drag and drop blocks and hatch patterns from the current drawing onto a tool palette. To drag a block from the drawing to a tool palette, the drawing must be saved (cannot be unnamed) because the tool references the source file of the block.

Rearranging Tools on a Tool Palette

The tools on a tool palette can be rearranged into a more productive order. To move a tool within a tool palette, simply select the tool and drag it to a new location. Remember, the horizontal I-bar indicates where the tool will be moved. In this way, you can place your drawing tools together, your block tools together, and so on.

You can further separate the tools within a tool palette by adding separator bars and text labels. Move the cursor so that it is between the two tools where you would like to add the separator. Then, right-click and select **Add Separator** from the shortcut menu. A horizontal bar is added to the tool palette. To add text to a tool palette, right-click between the two items where the label should be and select **Add Text** from the shortcut menu. A text box is displayed with the default text highlighted. Type the text that you want for the label and press [Enter]. The text label is added to the tool palette. Separators and text labels can be used together to make the visual grouping of tools even more apparent, **Figure 24-17.** You can move separators and text labels to different locations within the palette just as you can tools.

Exercise 24-2

Complete the exercise on the student website.
www.g-wlearning.com/CAD

Figure 24-17.
Separators and text
labels can be added
to tool palettes to
help group tools.

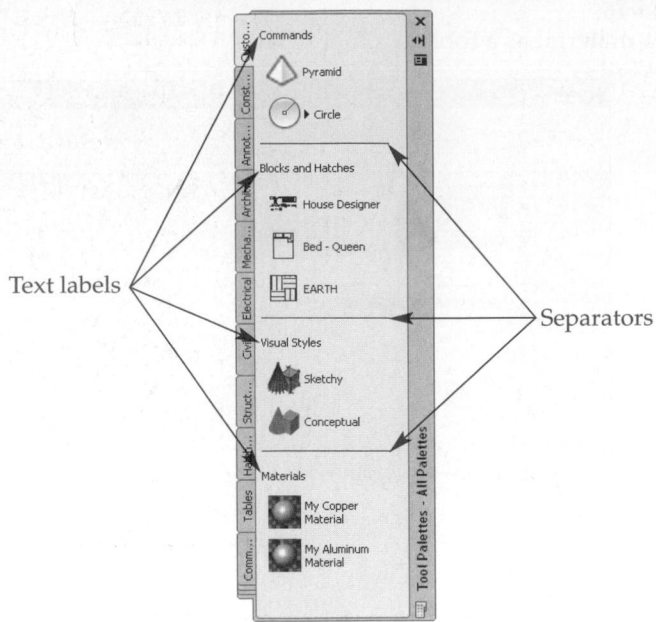

Text labels

Separators

Modifying the Properties of a Tool

A tool on a tool palette can basically do one of these operations:
- Insert a block.
- Insert a hatch pattern.
- Insert a gradient fill.
- Initiate a command.
- Insert a light.
- Insert a camera.
- Apply a visual style.
- Apply a material.

Another type of tool called a flyout is discussed later in this chapter. Each of these tools has general properties assigned to it, such as color, layer, or linetype. Each of these tools also has some tool-specific properties assigned to it, depending on the operation associated with the tool. Most of these assigned properties can be customized to your own needs.

For example, select the **Hatches and Fills** palette in the **Tool Palettes** window. Right-click on the **Curved** gradient tool and select **Properties...** in the shortcut menu. The **Tool Properties** dialog box is displayed, **Figure 24-18.** Notice that the lower half of the dialog box is divided into two sections. The **General** category contains settings for properties such as Color, Layer, and Linetype. The **Pattern** category contains settings for properties specific to this tool's operation—inserting a gradient fill. Notice that properties such as Color 1, Color 2, and Gradient angle are shown. Other types of tools will have a different category in place of the **Pattern** category. Select the **Cancel** button to close the dialog box.

Now, select the **Command Tool Samples** palette in the **Tool Palettes** window, right-click on the **VisualLisp Expression** tool in the tool palette, and select **Properties...** from the shortcut menu. The **Tool Properties** dialog box is displayed, **Figure 24-19.** This is the same dialog box displayed for the **Curved** gradient tool. However, in place of the **Pattern** category is the **Command** category. The settings in the **Command** category are specific to the **VisualLisp Expression** tool. Notice that the **General** category contains the same properties as for the **Curved** gradient tool. There are some other general properties listed that are not applicable to a gradient (Linetype scale, Text style, and Dimension style). Select the **Cancel** button to close the dialog box.

     AutoCAD and Its Applications—Advanced

Figure 24-18.
Modifying the
properties of a
gradient fill tool.

Icon

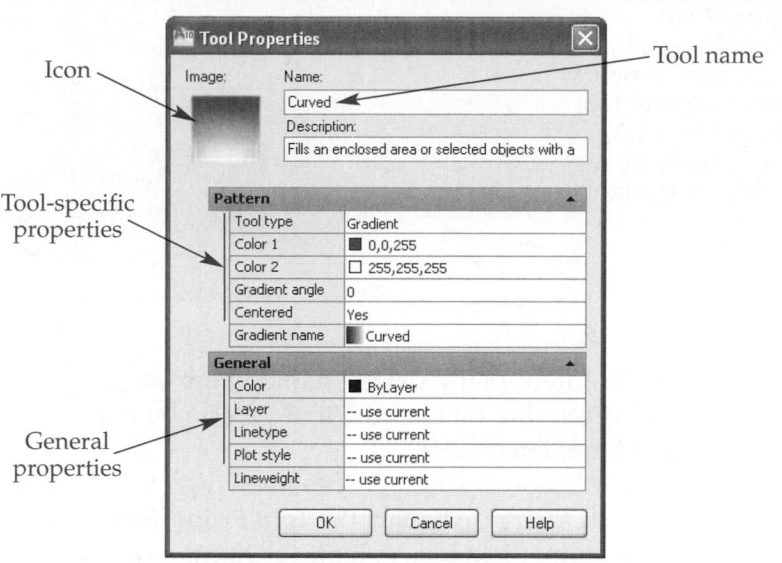

Tool name

Tool-specific
properties

General
properties

Figure 24-19.
Modifying the
properties of a
command tool.

Tool-specific
properties

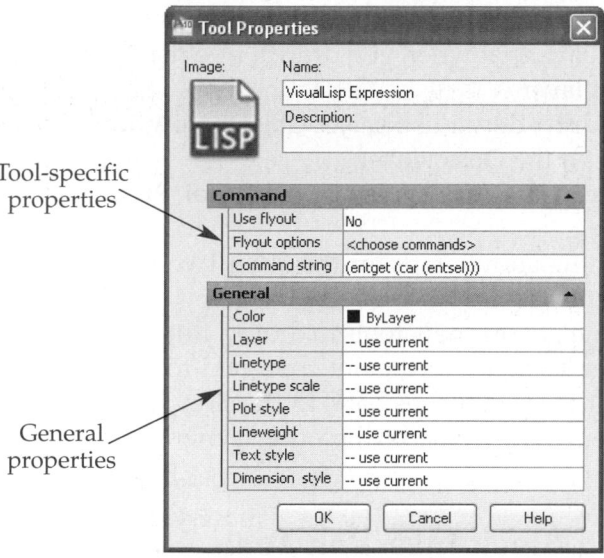

General
properties

Customizing General Properties

The three items at the top of the **Tool Properties** dialog box are available for all types of tools. The **Image:** area shows the image that is assigned to the tool. This can be modified within the dialog box for command, camera, light, and visual style tools. The **Name:** text box displays the name of the tool. You can enter a new name for the tool. The **Description:** text box displays the current description of the tool. You can change the existing description or enter a new description. The name and description appear in the tooltip that is displayed when the cursor is held over the tool, **Figure 24-20.**

The **General** category in the **Tool Properties** dialog box can be used to assign specific values to the general properties that are a part of nearly all AutoCAD objects. The properties that can be customized are Color, Layer, Linetype, Linetype scale, Plot style, Lineweight, Text style, and Dimension style. Depending on the type of tool, some of these properties may not be available.

The values assigned in the **Tool Properties** dialog box override the current property settings in the drawing when the tool is used. For instance, create three layers called Object, Hidden, and Center in the current drawing. A separate line tool can now be created for each of these layers:

Figure 24-20.
The Name: and
Description: property
settings in the **Tool
Properties** dialog
box are used as the
tooltip for the tool.

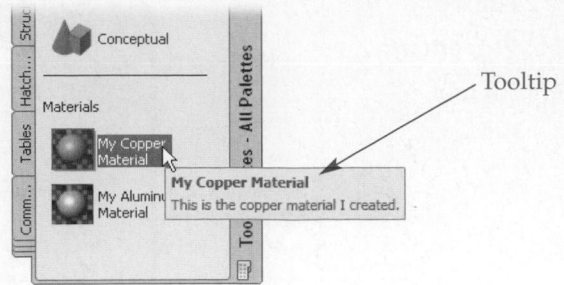

1. Create a new, blank palette named **Line Tools**.
2. Add three line tools to the tool palette by copying them from the **Command Tool Samples** palette.
3. Right-click on the first of these new line tools and select **Properties** from the shortcut menu to display the **Tool Properties** dialog box.
4. In the **Name:** text box, enter Line-Object as the name.
5. In the **Description:** text box, enter Draws a line on the Object layer. as the description.
6. In the **General** category, pick the Layer property, which is set to —use current. This means that when you draw a line using the tool, the line is drawn on the current layer.
7. In the drop-down list, select Object. Now, any lines drawn with the tool are placed on the Object layer.
8. Pick the **OK** button to close the **Tool Properties** dialog box and save the changes.
9. Repeat the above steps for the other two line tools setting them to the Hidden and Center layers.

Now, use one of your new tools to draw a line. Notice that when you select the tool the current layer switches to the one assigned to the tool. When you finish using the tool, the current layer switches back to the previous layer. Try each of the other new tools. Experiment with some of the other general properties, such as Color, Linetype, and Lineweight.

Customizing Block Insertion Tools

When a block insertion tool is modified in the **Tool Properties** dialog box, properties specific to block insertion are displayed in a category labeled **Insert**, **Figure 24-21.** These properties are described as follows.

Name

The Name property contains the name of the block to be inserted. It is not usually modified.

Source file

The Source file property lists the name of and path to the source drawing containing the block. If the drawing file has been moved to a new location, specify the correct location in the text box for this property. When you pick in the text box, an ellipses button (...) appears at the right-hand side of the box. You can pick this button to browse for the drawing file.

Scale

The Scale property value is the scale that will be applied to the block when it is inserted into the drawing. The scale is applied equally in the X, Y, and Z directions. The default value is 1.000.

AutoCAD and Its Applications—Advanced

Figure 24-21.
The **Tool Properties**
dialog box for a
block insertion tool.

Properties
specific
to block
insertion

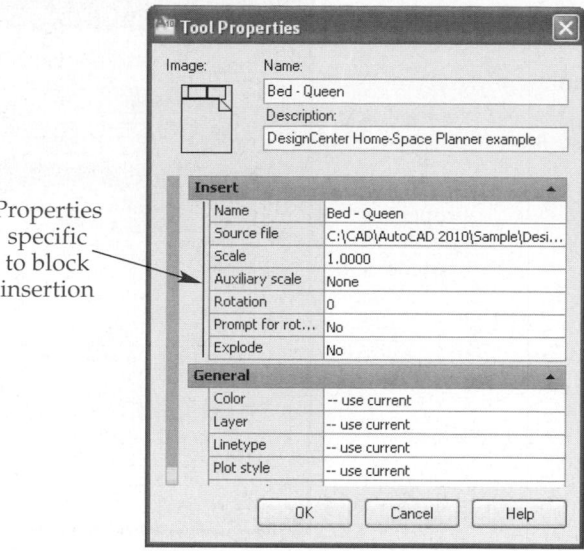

Auxiliary scale

The block is inserted at a scale calculated by multiplying the Auxiliary scale property value by the Scale property value. This drop-down list allows you to apply either the dimension scale or the plot scale to the insertion scale. The default value is None.

Rotation

The rotation angle for the block when it is inserted is set by the Rotation property. The default value is 0.

Prompt for rotation

The Prompt for rotation property determines whether or not the user is prompted for a rotation angle when the block is inserted. The default value in the drop-down list is No.

Explode

The Explode property determines whether or not the block is inserted as a block or as its component objects (exploded). The default value in the drop-down list is No, which means the block is inserted unexploded.

Customizing Hatch Pattern Tools

When a hatch pattern insertion tool is modified in the **Tool Properties** dialog box, properties specific to hatch patterns are displayed in the **Pattern** category, **Figure 24-22.** These properties are described as follows. Which properties are disabled or enabled is determined by the pattern type.

Tool type

The Tool type property determines if the hatch pattern is a standard hatch or a gradient fill. Choosing Gradient in the drop-down list changes the rest of the properties found in this area of the dialog box. Gradient fill properties are discussed later.

Type

The Type property determines the type of hatch pattern. To change the type, select the property and then pick the ellipses button (...) at the right-hand end of the entry. The **Hatch Pattern Type** dialog box is displayed, **Figure 24-23.** In the **Pattern type:** drop-down list of this dialog box, select User-defined, Predefined, or Custom.

Figure 24-22.
The **Tool Properties** dialog box for a hatch insertion tool.

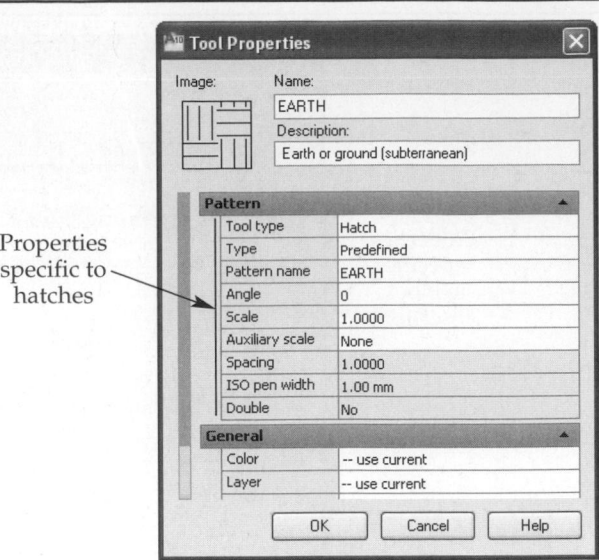

Properties specific to hatches

Figure 24-23.
The **Hatch Pattern Type** dialog box is used to determine the type of hatch. A predefined hatch pattern can also be selected in this dialog box.

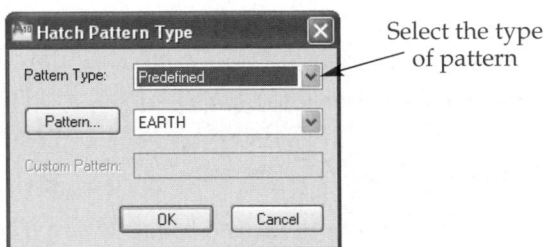

Select the type of pattern

If Predefined is selected, the **Pattern...** button and drop-down list are enabled. Select a pattern from the drop-down list or pick the button to select a pattern in the **Hatch Pattern Palette** dialog box. This is the same dialog box used with the **BHATCH** command.

If User-defined is selected, the rest of the items in the dialog box are disabled. The properties for user-defined hatch patterns are set in the **Tool Properties** dialog box, as discussed next.

Selecting Custom disables the **Pattern...** button and drop-down list and enables the **Custom Pattern:** text box. In this text box, enter the name of the custom pattern to use.

Pattern name

If the pattern type is set to Predefined, you can change the pattern using the Pattern name property. Select the property and pick the ellipses button (**...**) to open the **Hatch Pattern Palette** dialog box, **Figure 24-24.** If you selected the pattern in the **Hatch Pattern Type** dialog box, you will not need to select it using this property. This property is read-only when the type is set to User-defined or Custom.

Angle

The Angle property allows you to rotate the hatch pattern. This is the same as entering a rotation when using the **BHATCH** command. The default value is 0.

Scale

The Scale property determines the scale to be applied to the hatch pattern. Pick in the text box and enter a scale factor for the pattern. The default value is 1.000.

Figure 24-24.
Selecting a predefined hatch pattern.

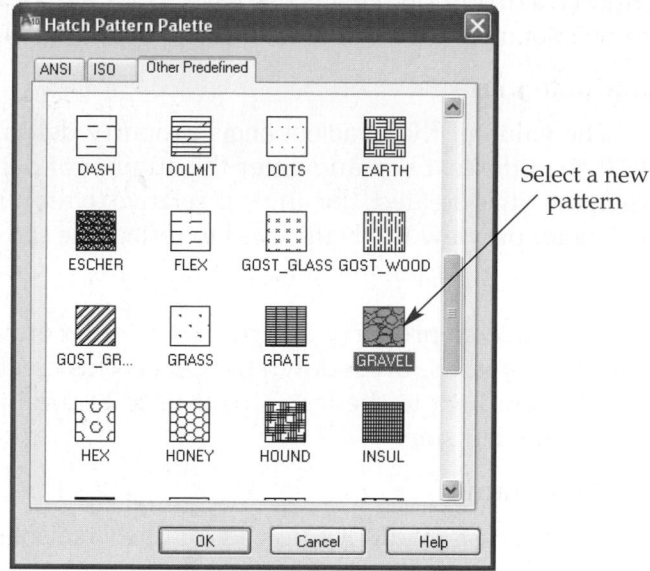

Select a new pattern

Auxiliary scale

The hatch is inserted at a scale calculated by multiplying the Auxiliary scale property value by the Scale property value. The Auxiliary scale drop-down list allows you to apply either the dimension scale or the plot scale to the insertion scale. The default value is None.

Spacing

The Spacing property determines the spacing of lines in a user-defined hatch pattern. For other hatch pattern types, this property is read-only. To change the spacing, pick in the text box and enter the relative distance between lines. The default value is 1.000.

ISO pen width

The ISO pen width property allows you to set the pen width for ISO hatch patterns. It is disabled for other hatch patterns. The default value in the drop-down list is 1.00 mm.

Double

The Double property determines whether or not the user-defined hatch is a crosshatch pattern. For other hatch pattern types, this property is read-only. The default value in the drop-down list is No. To make a crosshatch pattern, select Yes.

Customizing Gradient Fill Tools

When a gradient fill hatch pattern insertion tool is modified in the **Tool Properties** dialog box, properties specific to gradient fills are displayed in the **Pattern** category. See **Figure 24-18**. These properties are described as follows.

Tool type

The Tool type property determines if the hatch pattern is a standard hatch or a gradient fill. Choosing Hatch in the drop-down list changes the rest of the properties found in this area of the dialog box, as described earlier.

Color 1 and Color 2

The Color 1 property is used to specify the first color of the gradient fill. The Color 2 property is used to specify the second color of the gradient fill. When you pick the property, a drop-down list is displayed. Pick a color from the drop-down list or choose Select Color... to pick a color in the **Color Selector** dialog box.

Gradient angle

The value of the Gradient angle property determines the rotation of the gradient fill. Pick in the text box and enter the number of degrees for the angle of the gradient. When the fill is created, the angle is relative to the current UCS. When you press [Enter] the **Image:** preview tile is updated to reflect the setting.

Centered

The Centered property determines whether or not the gradient fill is centered. The default value in the drop-down list is Yes. This creates a symmetrical fill. Selecting No shifts the gradient to the left. This is used to simulate a light source illuminating the fill from the left side.

Gradient name

The type of gradient fill is set by the Gradient name property. AutoCAD's preset gradient fills appear in the drop-down list. Select the type of gradient fill to use.

Customizing Command Tools

When a command tool is modified in the **Tool Properties** dialog box, properties specific to commands are displayed in the **Command** category, **Figure 24-25.** These properties are described as follows.

- **Use flyout.** This property determines whether or not the tool is a flyout. Flyouts are discussed later in this chapter.
- **Flyout options.** This property allows you to pick which commands are associated with the flyout, as discussed later.
- **Command string.** The command macro for the tool is entered in this text box, if the tool is not a flyout. Creating custom commands is discussed in Chapter 22. If the tool is a flyout, this text box is disabled.

For example, suppose you need a tool that will draw three concentric circles with diameters of .50, 1.00, and 1.50. This tool is shown in **Figure 24-25.** Use the following procedure.

1. In the current drawing, create a circle of any diameter at any location.
2. Drag and drop the circle onto a tool palette to create a new tool.

Figure 24-25.
The **Tool Properties** dialog box for a command tool.

Properties specific to commands

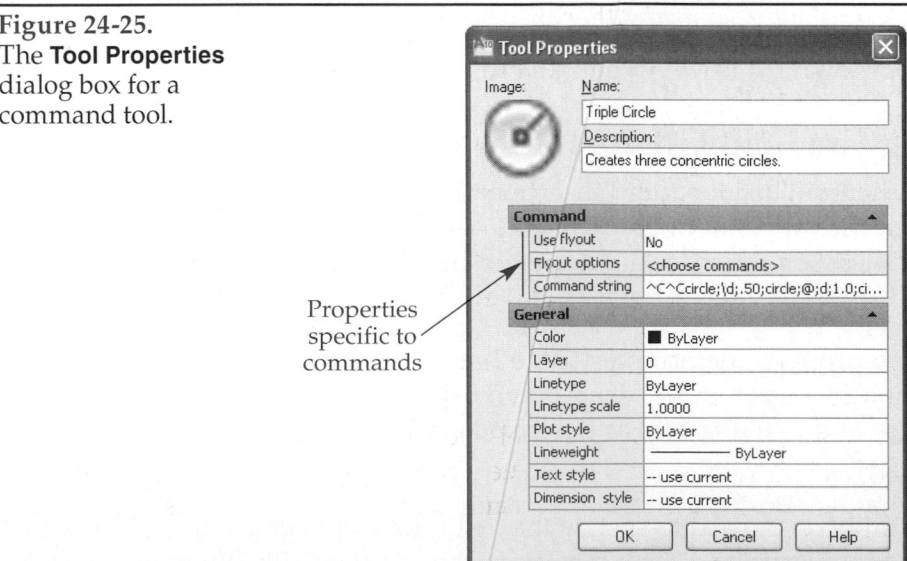

AutoCAD and Its Applications—Advanced

3. Right-click on the new **Circle** tool and select **Properties...** from the shortcut menu.
4. In the **Tool Properties** dialog box, change the following properties. The Use flyout property must be set to No to enable the Command string property.
   Name: Triple Circle
   Description: Creates three concentric circles.
   Use Flyout: No
   Command string: ^C^Ccircle;\d;.50;circle;@;d;1.0;circle;@;d;1.5
5. Pick the **OK** button to close the **Tool Properties** dialog box.
6. Test the new tool.

Look closely at the command macro for the **Triple Circle** tool you just created. Can you identify each component of the macro? If not, use the tool and then display the **AutoCAD Text Window** by pressing [F2]. Using the text window, determine what function each component of the macro performs.

NOTE

The **VisualLisp Expression** tool included on the **Command Tool Samples** palette is not actually a "command" tool. It is merely there to let you know that you can use a command tool to execute Visual LISP expressions. Visual LISP functions can be added in the Command string property in the **Tool Properties** dialog box. For more information on AutoLISP and Visual LISP, refer to Chapters 27 and 28 of this text and to the text *Visual LISP Programming* from Goodheart-Willcox Publisher.

Exercise 24-3

Complete the exercise on the student website.
www.g-wlearning.com/CAD

Customizing Camera and Light Tools

Cameras and lights are stored in the drawing file. By placing camera and light tools on a tool palette, you can store your favorite settings for use in all drawings. Productivity is increased since you do not have to use **DesignCenter** to browse for your favorite cameras and lights within other drawings. In fact, AutoCAD provides several tool palettes with different light tools. These tools palettes are provided for lights:

- **Generic Lights**
- **Fluorescent**
- **High Intensity Discharge**
- **Incandescent**
- **Low Pressure Sodium**

AutoCAD also has a **Cameras** tool palette containing three camera tools. These tools are provided on the **Cameras** tool palette:

- **Normal Camera**
- **Wide-Angle Camera**
- **Extreme Wide-Angle Camera**

When customizing a camera or light tool, the **Tool Properties** dialog box provides all of the properties that are required to create a camera or light. See **Figure 24-26**. For detailed information about setting up lights and cameras, see Chapters 17 and 19.

Figure 24-26.
A—The **Tool Properties** dialog box for a light tool. B—The **Tool Properties** dialog box for a camera tool.

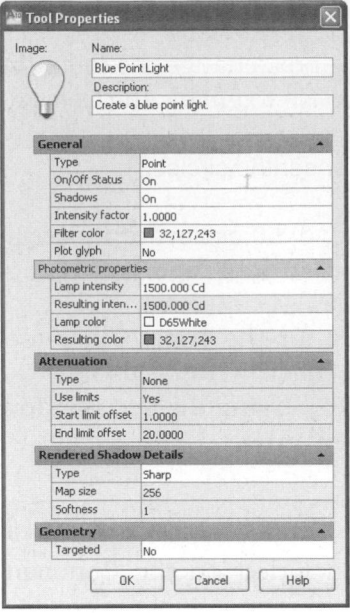

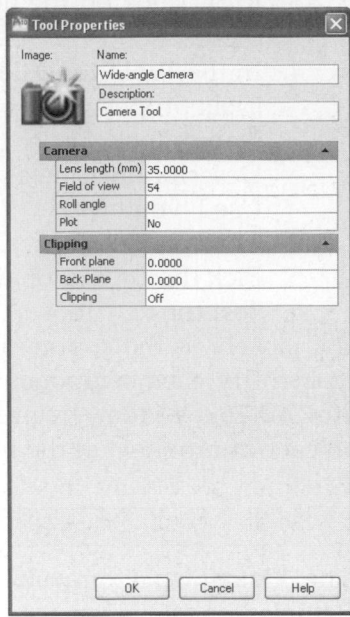

A

B

Customizing Visual Style and Material Tools

When either a visual style or material tool is modified in the **Tool Properties** dialog box, properties specific to the type of tool appear in the dialog box. The properties available are identical to those available in the **Visual Style Manager** or **Materials** palette. See **Figure 24-27.** Refer to Chapter 15 for information on creating and modifying visual styles and Chapter 16 for information on creating and modifying materials.

NOTE

Modifying a material tool does not alter the material in the drawing if it has been attached to any objects.

Figure 24-27.
A—The **Tool Properties** dialog box for a visual style tool. B—The **Tool Properties** dialog box for a material tool.

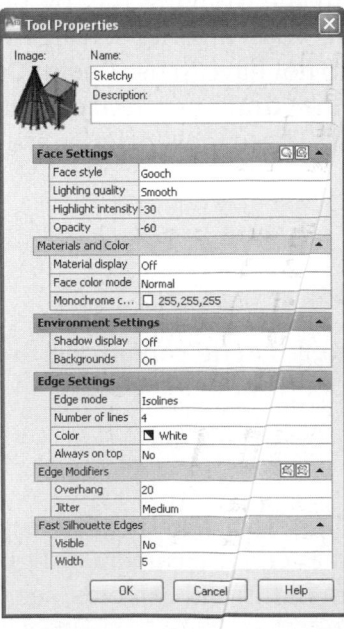

A

B

Changing a Tool Icon

You can change the icon associated with a tool. This can be done from within the **Tool Properties** dialog box or directly on the tool palette. Right-click on the icon, either in the dialog box or on the tool palette, and select **Specify Image...** from the shortcut menu. The **Select Image File** dialog box is displayed. This is a standard "open" dialog box. Navigate to the image file, select it, and pick the **Open** button. The icon displays the new image.

To change an icon back to the original image, right-click on the icon (either in the dialog box or on the tool palette) and select **Remove specified image** from the shortcut menu. The icon returns to its original image. Only certain tools allow the image to be removed.

NOTE

Some tools, such as material tools, have an icon image based on the settings of the tool. You should not specify an image for these tools. If the material is redefined, right-click on the tool icon in the tool palette and select **Update tool image** from the shortcut menu.

Working with Flyouts

A command tool in a tool palette can be set to function as a flyout. A flyout tool is similar to a flyout button found on the ribbon or a toolbar. The tool icon for a flyout displays a small arrow to the right of the tool. When the arrow is picked, a graphic shortcut menu is displayed that contains the command tools in the flyout.

To set a command tool as a flyout, open the **Tool Properties** dialog box for the tool. Then, in the **Command** category, set the Use flyout property to Yes. The tool is now a flyout. Notice that the **Command** string text box in the dialog box is disabled. You must now select the commands that will be displayed in the flyout.

The Flyout options property is used to specify which commands will be displayed in the flyout for a tool. This property is disabled unless the Use flyout property is set to Yes. To choose the commands to appear in the flyout, select the property and pick the ellipses button (...) at the right-hand side. The **Flyout Options** dialog box appears, **Figure 24-28.**

The commands that will appear in the flyout tool are indicated with a check in the **Flyout Options** dialog box. To prevent a command from being displayed in the flyout, remove the check mark next to its name. Then, close the **Flyout Options** dialog box to return to the **Tool Properties** dialog box. Finally, close the **Tool Properties** dialog box and test the tool.

The commands displayed in the **Flyout Options** dialog box depend on which command tool is being modified. **Figure 24-28A** shows the commands available when a dimensioning command tool is modified. When a drawing command tool is modified, the commands shown in **Figure 24-28B** are available. These also act as the default tools for other commands, such as modify commands or custom commands.

PROFESSIONAL TIP

When a drawing or dimensioning command is added to a tool palette, the tool is automatically a flyout. It can then be modified to customize or remove its flyout capabilities.

Figure 24-28.
A—The commands available for a dimensioning command flyout tool. B—The commands available for a drawing command flyout tool.

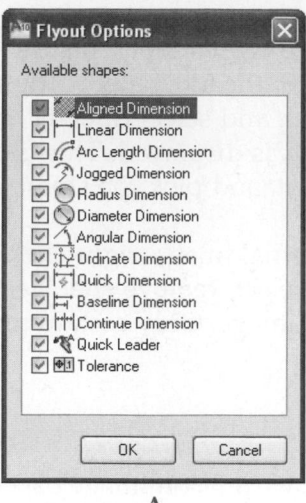

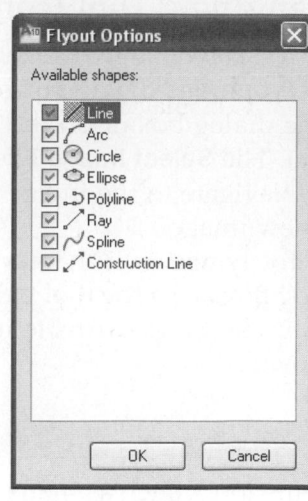

A

B

Organizing Tool Palettes into Groups

Tool palettes are very useful, productive tools. You may find it beneficial to create many tool palettes to meet your design needs. However, having too many palettes visible at once can be counterproductive. The tool palette names become abbreviated in the **Tool Palettes** window or the tabs may be stacked on top of each other. This is the case if all of the default AutoCAD tool palettes are displayed. Fortunately, tool palettes can be divided into named groups and then a single group can be displayed.

By default, all of the tool palettes are shown in the **Tool Palettes** window, but the existing tool palettes are divided into named groups. Right-click on the title bar of the **Tool Palettes** window and look at the shortcut menu. At the bottom of the shortcut menu, notice the entries below the last separator, **Figure 24-29**. These are the palette groups. The check mark next to **All Palettes** indicates that all of the defined tool palettes are being displayed. Select **Annotation and Design** from the shortcut menu and notice that only the block-related tool palettes are visible. This is indicated in the title bar. Right-click on the title bar again and select **Materials Library** from the shortcut menu. Notice that only the tool palettes related to materials are displayed. These tool palettes represent AutoCAD's materials library.

The **Customize** dialog box is used to make new palette groups or customize existing groups. To open this dialog box, right-click on the **Tool Palettes** window title bar and select **Customize Palettes...** from the shortcut menu. You can also use the **CUSTOMIZE** command. All of the currently defined tool palettes are listed in the **Palettes:** area on the left side of the dialog box, **Figure 24-30**. The currently defined palette groups are listed in the **Palette Groups:** area on the right side of the dialog box. Palette groups are shown as folders. The tool palettes contained within the group are shown in the tree below the folder. The current palette group is indicated at the bottom of the dialog box and by the bold folder name in the **Palette Groups:** area.

You can customize one of the existing palette groups by adding tool palettes to or removing tool palettes from the group. To remove a tool palette from a group, select the palette name under the group name in the **Palette Groups:** area. Then, press the [Delete] key or right-click and select **Remove** from the shortcut menu. To delete a palette group, select it in the **Palette Groups:** area and press the [Delete] key or right-click and select **Delete** from the shortcut menu. To add a palette to a group, simply select the palette in the **Palettes:** area and drag it into the group in the **Palette Groups:** area. A tool palette can be a member of more than one palette group.

CUSTOMIZE

Ribbon
Manage
> Customization
Tool Palettes
Type
CUSTOMIZE

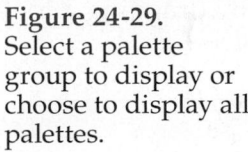
Figure 24-29.
Select a palette group to display or choose to display all palettes.

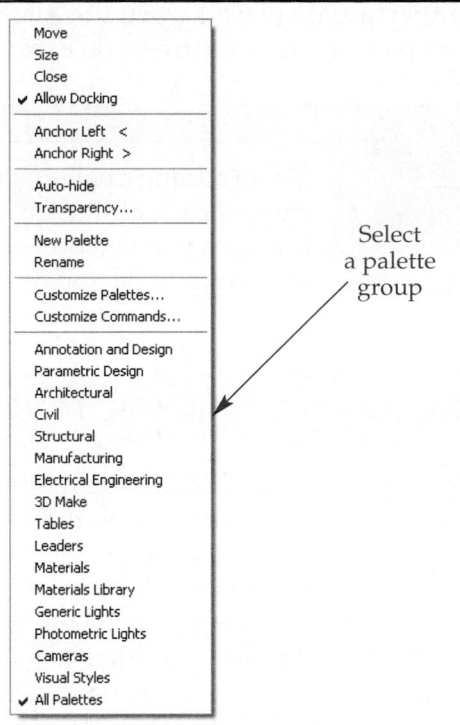

Select a palette group

Figure 24-30.
The **Customize** dialog box is used to create and manage palette groups.

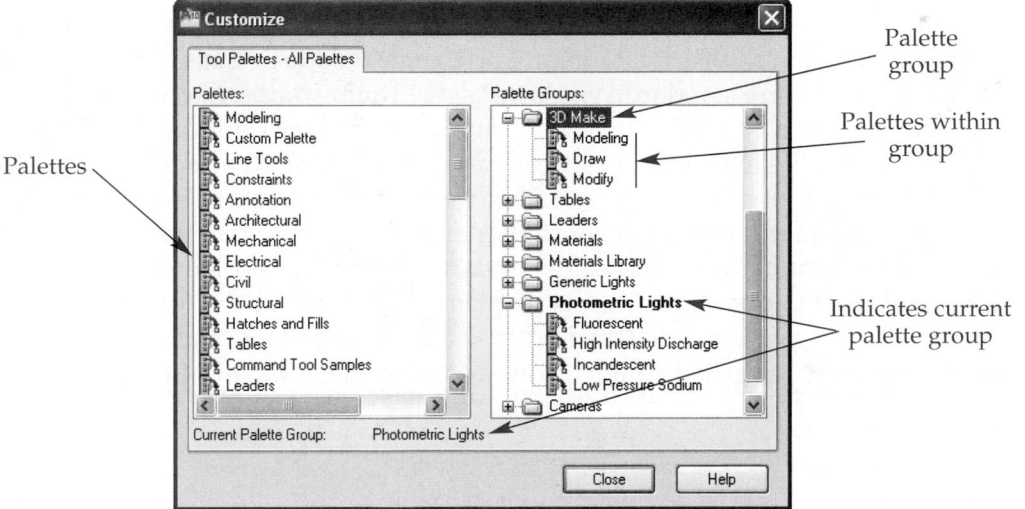

Palettes

Palette group

Palettes within group

Indicates current palette group

To create a new palette group, right-click in a blank area of the **Palette Groups:** area and select **New Group** from the shortcut menu. A new folder appears in the tree with the default name highlighted in a text box. Type a name for the new palette group and press [Enter]. Then, drag tool palettes from the **Palettes:** area and drop them into the new palette group.

The order in which tool palettes appear in the **Tool Palettes** window is determined by their positions in the tree. To reorder the tool palettes, simply drag them to different positions within the palette group. To reorder the palette groups, drag the folders to different positions within the tree in the **Palette Groups:** area. The order in which the palette groups appear in the **Palette Groups:** area determines the order in which they appear in the shortcut menu displayed by right-clicking on the **Tool Palettes** window title bar. The order in which tool palettes appear in the **Palettes:** area is the order in which they are displayed when the **All Palettes** option is selected. By dragging a tool palette up or down in the tree, you can reorder the list.

Saving and Sharing Tool Palettes and Palette Groups

Tool palettes and palette groups can be exported from and imported into AutoCAD. Both operations are performed in the **Customize** dialog box. Some precautions about tool palette files:

- Tool palette files can only be imported into the same version of AutoCAD as the version from which the file was exported.
- Tool palette files exported from AutoCAD and imported into AutoCAD LT may have tools that will not work or behave the same. For example, color property tools using a color other than an AutoCAD Color Index (ACI) color are converted to ByLayer in AutoCAD LT. Also, gradient fill tools convert to hatch tools in AutoCAD LT. Raster image tools do not work in AutoCAD LT.

To export a tool palette, right-click on the palette to be exported in the **Palettes:** area of the **Customize** dialog box and select **Export...** from the shortcut menu. To export a palette group, right-click on the name of the group in the **Palette Groups:** area and select **Export...** from the shortcut menu. The **Export Palette** or **Export Group** dialog box is displayed. These are standard Windows "save" dialog boxes. Name the file, navigate to the folder where you want to save it, and pick the **Save** button. Tool palettes are saved with a .xtp file extension. Palette groups are saved with a .xgp file extension.

To import a tool palette, right-click in the **Palettes:** area of the **Customize** dialog box and select **Import...** from the shortcut menu. To import a palette group, right-click in the **Palette Groups:** area and select **Import...** from the shortcut menu. The **Import Palette** or **Import Group** dialog box is displayed. Navigate to the folder where the file is saved, select it, and pick the **Open** button. The imported tool palette is added to the **Palettes:** area. An imported palette group is added to the **Palette Groups:** area.

Chapter Test

1. Name three ways to open the **Tool Palettes** window.
2. How do you dock the **Tool Palettes** window?
3. When enabled, what does the autohide feature do to the **Tool Palettes** window? How is the **Tool Palettes** window accessed when autohide is enabled?
4. Name two ways to toggle the autohide feature on the **Tool Palettes** window.
5. Describe how anchoring the **Tool Palettes** window differs from using the autohide feature or docking it.
6. How do you activate the transparency feature for the **Tool Palettes** window?
7. Name the three view styles in which tools can be displayed in the **Tool Palettes** window.
8. How do you create a new, blank tool palette?
9. How do you add a tool palette that contains all of the blocks in a particular drawing?
10. Which dialog box must be open to create a tool on a tool palette from a toolbar button? How is it opened?
11. What are the names of the two files that store hatch pattern definitions?
12. How do you rearrange tools on a tool palette?
13. Name four of the general properties that can be customized on individual tools on a tool palette.
14. What is the purpose of the auxiliary scale property for block and hatch insertion tools?
15. What are the two types of patterns that can be inserted using a tool in a tool palette?
16. If you are creating a command tool that performs a custom function, where is the command macro entered?
17. Which two types of commands can be included in a flyout tool?
18. What is the purpose of creating tool palette groups?
19. What is the extension given to an exported tool palette file?
20. How do you import a tool palette file?

Drawing Problems

Before customizing or creating any tool palettes, check with your instructor or supervisor for specific instructions or guidelines.

1. Design a complete tool palette system for your chosen discipline. Incorporate:
    *   Multiple tool palettes
    *   Block tools
    *   Hatch tools
    *   Drawing command tools (with and without flyouts)
    *   Modification command tools
    *   Inquiry command tools
    *   Custom macro tools
    *   Dimensioning command tools (with and without flyouts)
    A. On the tool palettes that use multiple types of command tools, use separators and text labels to group the types of tools.
    B. Create groups for the multiple tool palettes.

2. Export the tool palette groups created in problem 1. Copy the files to removable media or an archive drive.

CHAPTER 25

User Profiles and Workspaces

Learning Objectives

After completing this chapter, you will be able to:

✓ Describe user profiles.
✓ Create user profiles.
✓ Restore a user profile.
✓ Describe workspaces.
✓ Explain the **Quick Access** toolbar.
✓ Create workspaces.
✓ Customize a workspace.
✓ Restore a workspace.
✓ Customize the **Quick Access** toolbar.

In a school or company, there is often more than one person who will use the same AutoCAD workstation. Each drafter has a unique style for creating a drawing. While there are often general rules to follow, many times the method used to arrive at the end result is not important. As you learned in previous chapters, there are many ways to customize AutoCAD. You can set screen colors and other features of the AutoCAD environment, create custom menus, and customize the ribbon. Many of these settings can be saved in a user profile or workspace. User profiles and workspaces allow you to quickly and easily restore a group of custom settings.

User Profiles

A *user profile* is a group of settings for devices and AutoCAD functions. Some of the settings and values a profile can contain are:

- Temporary drawing file location.
- Template drawing file location.
- Text display format.
- Startup dialog box display.
- Minutes between automatic saves.
- File extension for temporary files.
- AutoCAD screen or pull-down menu display (on/off).
- Color and font settings for AutoCAD's text and graphics screens.

- Type of pointer and length of crosshairs.
- Default locations of printers, plotters (PC3 files), and plot styles (CTB/STB files).

Multiple profiles can be saved by a single user for different applications or several users can create individual profiles for their own use. A user profile should not be confused with settings found in a drawing. Template files are used to save settings relating to a drawing session, such as units, limits, object snap settings, drafting settings, grip settings, arc and circle smoothness, dimension styles, and text styles. A user profile, on the other hand, saves settings related to the performance and appearance of the software and hardware.

Creating a User Profile

A user profile is basically a collection of all things you have customized in AutoCAD *except* ribbon tabs and panels, application and pull-down menus, tool palettes, and toolbars. These customizations are usually done to make AutoCAD easier for you to use. For example, as you gain experience in AutoCAD, you realize that you may:

- Prefer the crosshairs extending to the edges of the graphics window.
- Need to have the plotters and plot styles shared on a network.
- Prefer the graphics window background color to be gray.

Through the course of several drawing sessions, you have customized AutoCAD to reflect these preferences. Now, so you do not lose your preferred settings, you should create a user profile.

First, open the **Options** dialog box and pick the **Profiles** tab, **Figure 25-1.** Pick the **Add to List...** button on the right side of the tab. The **Add Profile** dialog box is opened, **Figure 25-2.** Enter a name and description. Then, pick the **Apply & Close** button to close the **Add Profile** dialog box. The current settings are saved to the user profile and the new user profile is now listed in the **Profiles** tab of the **Options** dialog box. The user profile is saved and will be available in the current and future AutoCAD drawing sessions.

Figure 25-1.
Settings can be saved in a profile.

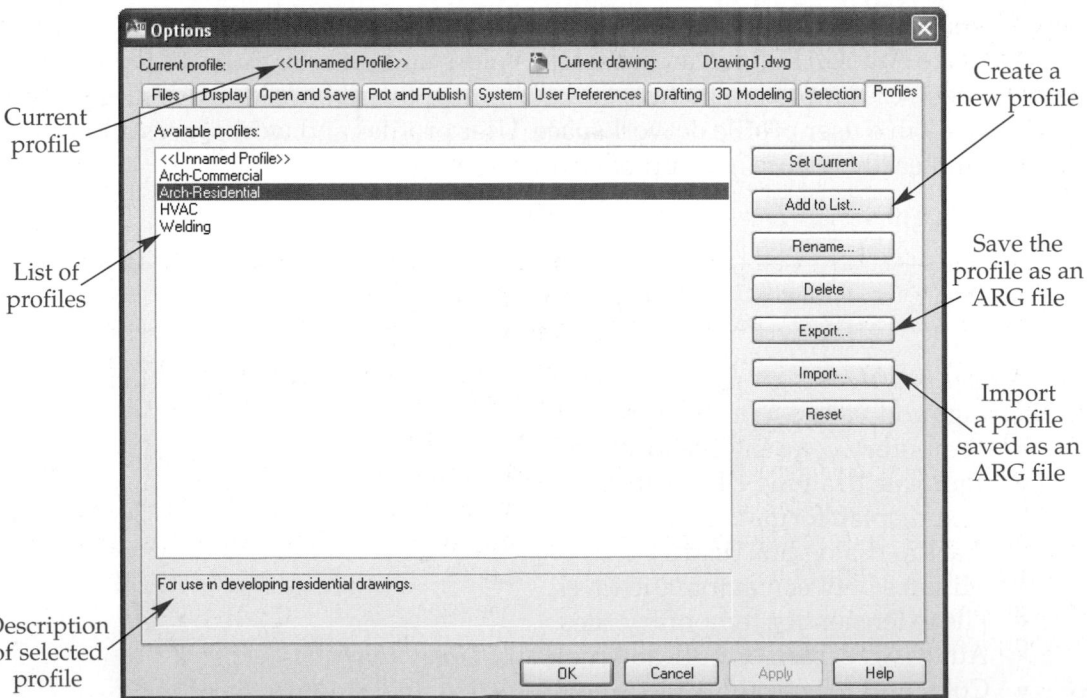

Figure 25-2.
The **Add Profile**
dialog box is used to
create a new profile.

Enter name for new profile

Enter description for new profile

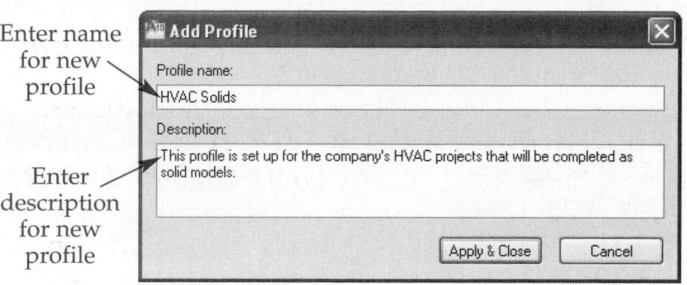

To change the name of a user profile, pick the **Rename...** button in the **Profiles** tab. In the **Change Profile** dialog box, enter a new name and description. To delete a user profile, highlight it in the **Profiles** tab and pick the **Delete** button. You cannot delete the current user profile. If you pick the **Reset** button, the highlighted user profile has all of its settings restored to AutoCAD defaults.

PROFESSIONAL TIP

Changes made to the environment (screen color, toolbar display, etc.) are automatically saved to the current profile.

Restoring a User Profile

Once a user profile is saved, it is available in the current and future AutoCAD drawing sessions. To set any saved user profile as the current user profile, first open the **Options** dialog box. The current profile is indicated at the top of the **Options** dialog box. Then, pick the **Profiles** tab. Highlight the name of the profile to restore in the **Available profiles:** list. Then, pick the **Set Current** button. You can also double-click on the name in the **Available profiles:** list to restore a profile. All of the settings in the user profile are applied while the **Options** dialog box is still open. Close the dialog box to return to the drawing editor.

Exercise 25-1

Complete the exercise on the student website.
www.g-wlearning.com/CAD

Importing and Exporting User Profiles

A user profile can be exported and imported. This allows you to take your user profile to a different AutoCAD workstation. A user profile is saved as an ARG file.

To export a user profile, open the **Options** dialog box and pick the **Profiles** tab. Highlight the profile to export and pick the **Export...** button. The **Export Profiles** dialog box is opened, **Figure 25-3.** Then, select a folder and name the file. When you pick the **Save** button, the user profile is saved with the .arg file extension.

To import a user profile, pick the **Import...** button in the **Profiles** tab. The **Import Profile** dialog box is displayed. This is a standard "open" dialog box. Then, navigate to the appropriate folder, select the proper ARG file, and pick the **Open** button. A second dialog box named **Import Profile** is displayed. See **Figure 25-4.** You can rename the user profile, change the description, and choose to include the file path. Pick the **Apply & Close** button to complete the process. The user profile is then available in the **Profiles** tab.

Figure 25-3.
The **Export Profile** dialog box is used to save a user profile as an ARG file, which can be transferred to another AutoCAD workstation.

Select a folder

Name the user profile

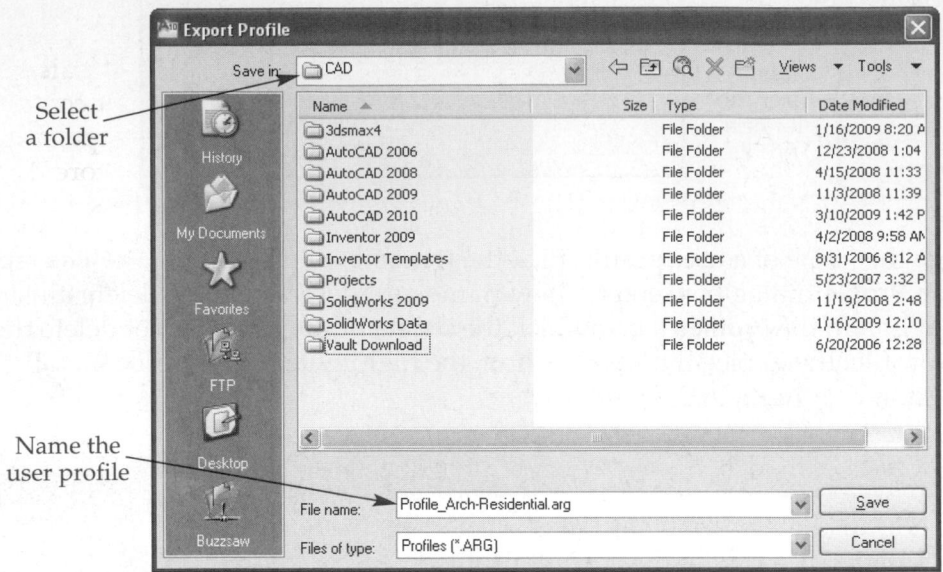

Figure 25-4.
Importing a user profile.

Profile name

Description

Check to include the path

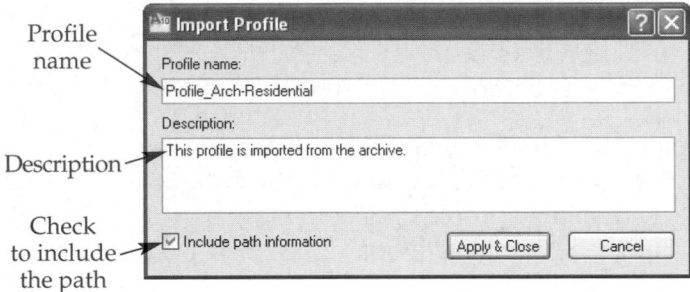

PROFESSIONAL TIP

A variety of settings can be changed in the **Options** dialog box. Settings that have the AutoCAD drawing icon located next to them can be stored in a template file (or drawing). Using the appropriate template file to begin a drawing automatically resets these settings. Settings within the **Options** dialog box that do not have the AutoCAD drawing icon next to them can typically be restored through the use of user profiles.

Exercise 25-2

Complete the exercise on the student website.
www.g-wlearning.com/CAD

Workspaces

As you have seen, a user profile allows you to set and restore settings for screen colors, drafting settings, and file locations. On the other hand, a workspace allows you to set and restore settings for the **Quick Access** toolbar, ribbon, menus, and palettes (**DesignCenter**, **Properties** palette, **Tool Palettes** window, etc.), but it does not contain environmental settings. A *workspace* is a collection of displayed toolbars, palettes, ribbon tabs and panels, and menus and their configurations. A workspace stores not only which of these graphic tools are visible, but also their on-screen locations.

PROFESSIONAL TIP

A user profile also stores which toolbars are displayed and their location. However, the configurations of the menus (application and pull-down menus) and ribbon and display status of palettes are not saved in a user profile. Therefore, if a workspace is restored, the settings for toolbar display and location override the current user profile settings.

Creating a Workspace

AutoCAD has three default workspaces—2D Drafting and Annotation, 3D Modeling, and AutoCAD Classic. When AutoCAD is first installed, it also includes the Initial Setup workspace, but this workspace is available only until another workspace is selected. A workspace can be set current using the **Workspace Switching** button on the right-hand side of the status bar. See **Figure 25-5**. Picking this button displays a shortcut menu containing the names of all available workspaces. It also contains options for working with workspaces. When AutoCAD is launched, the workspace that was last active is restored. AutoCAD can be set up so that any changes made to the ribbon, toolbars, menus, and palettes are saved to this workspace. However, it is best to create your own workspaces.

The first step in setting up and storing your own workspace is to arrange the ribbon, toolbars, tool palettes, and menus to your liking. Refer to Chapter 22. Next, use the **WSSAVE** command to open the **Save Workspace** dialog box, **Figure 25-6**. You can easily access the **WSSAVE** command by selecting **Save Current As...** in the shortcut menu displayed by picking the **Workspace Switching** button. In the **Save Workspace** dialog box, enter a name for the workspace, such as Normal Design or Standard Arrangement, and then pick the **Save** button.

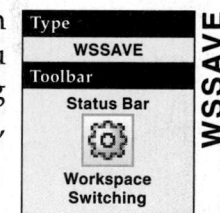

Type
WSSAVE
Toolbar
Status Bar
Workspace
Switching

WSSAVE

Figure 25-5.
Switching
workspaces.

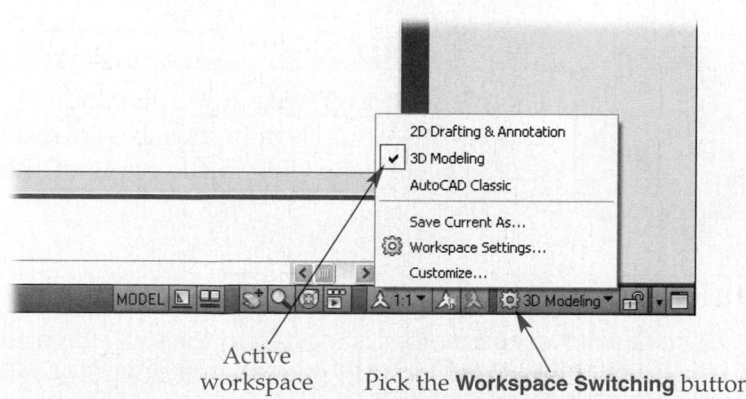

Active
workspace   Pick the **Workspace Switching** button

Figure 25-6.
Creating a new
workspace.

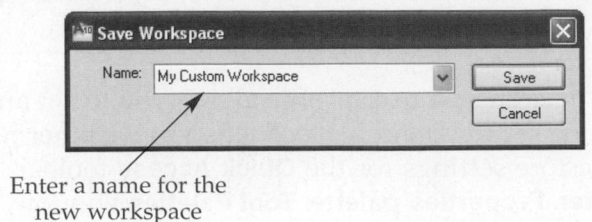

Enter a name for the
new workspace

The current settings for toolbars, pull-down menus, and tool palettes are now stored in the new workspace, which is also made current. The workspace name also appears in the shortcut menu displayed by picking the **Workspace Switching** button. The current workspace is indicated in this menu by a check mark.

PROFESSIONAL TIP

Workspaces are saved in a CUIx file. By default, they are saved to the main CUIx file (acad.cuix).

Restoring a Workspace

Once a workspace has been saved, it can easily be restored. A list of available workspaces appears in the pop-up list displayed by picking the **Workspace Switching** button. To select a different workspace, simply pick the name of the workspace in the pop-up list. A workspace can also be restored using the command line:

Command: **WORKSPACE**↵
Enter workspace option [setCurrent/SAveas/Edit/Rename/Delete/SEttings/?]
    <setCurrent>: **C**↵
Enter name of workspace to make current [?] <*current*>: **MY CUSTOM**
    WORKSPACE↵
Command:

Notice that you can manage workspaces using this command.

You can also use the **Customize User Interface** dialog box to restore a workspace. First, open the dialog box. Then, expand the Workspaces branch in the **Customizations in All Files** pane. All of the workspaces defined in the default CUIx file and any open CUIx files are displayed in this branch. The label (current) follows the name of the current workspace. To restore a workspace, right-click on its name in the Workspaces branch and select **Set Current** from the shortcut menu. Its name is now followed by (current). Pick the **OK** button to close the **Customize User Interface** dialog box and make the workspace current.

PROFESSIONAL TIP

The **WSCURRENT** system variable indicates the current workspace. You can use this system variable to restore a workspace. Simply set the system variable to the name of the workspace you want to restore.

Quick Access Toolbar

The **Quick Access** toolbar is located in the top-left corner of the screen to the right of the application menu icon, **Figure 25-7**. It contains essential tools for AutoCAD. By default, it displays the **New**, **Open**, **Save**, **Undo**, **Redo**, and **Plot** tools. In addition, you can display other tools on the toolbar by picking the down arrow button on the

AutoCAD and Its Applications—Advanced

Figure 25-7.
The **Quick Access**
toolbar is customized
as part of the
workspace.

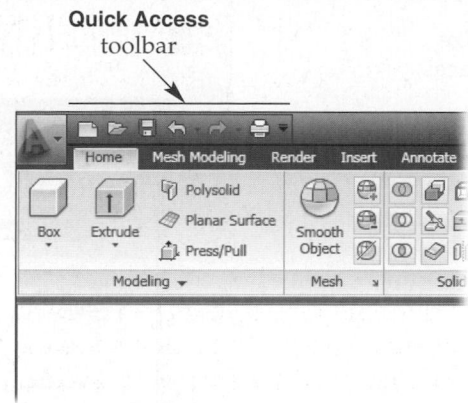

Quick Access
toolbar

Quick Access toolbar to display a shortcut menu. In this menu, select the tools to show and hide. The additional available tools are **Match Properties**, **Batch Plot**, **Plot Preview**, **Properties**, **Sheet Set Manager**, and **Render**.

Unlike the toolbars discussed in Chapter 23, the **Quick Access** toolbar is customized inside of the workspace. You may create more than one **Quick Access** toolbar, but only one may be active in a workspace. When creating a new **Quick Access** toolbar, each one is automatically created with the commands outlined above. Customizing them is discussed in the next section.

PROFESSIONAL TIP

Any tool on the ribbon can be quickly added to the **Quick Access** toolbar. Right-click on the tool icon and select **Add to Quick Access Toolbar** from the shortcut menu.

Customizing a Workspace

An existing workspace can be customized using the **Customize User Interface** dialog box. First, select the workspace to be customized in the Workspaces branch of the **Customizations in All Files** pane. Remember, the name of this pane will change based on the selection in the drop-down list. The **Workspace Contents** pane at the upper-right corner of the dialog box displays the contents of the selected workspace. There are Quick Access Toolbar, Toolbars, Menus, Palettes, and Ribbon Tabs branches. Refer to **Figure 25-8.** Expand a branch to see which components the workspace contains.

To customize the workspace, pick the **Customize Workspace** button at the top of the **Workspace Contents** pane. The tree in the pane turns blue to indicate you are in customization mode and the button changes to the **Done** button. Also, notice that the tree in the **Customizations in All Files** pane has changed. Several branches have disappeared and the Menus branch has a green check mark next to them. If you expand the branches in the **Customizations in All Files** pane, you will see that a green check mark also appears next to the components that currently are in the workspace. See **Figure 25-9.**

To add a component to the workspace, pick the blank box in front of its name to place a check mark in the box. The component also appears in the **Workspace Contents** pane. To remove a component from the workspace, pick the check mark in front of its name to clear the box. The component is also removed from the **Workspace Contents** pane.

The components in the Quick Access Toolbar branch are commands. To add a command to the Quick Access Toolbar branch, locate it in the **Command List:** pane. Then, drag and drop it into the branch in the **Workspace Contents** pane. To remove a command from the Quick Access Toolbar branch, select it in the **Workspace Contents** pane. Then, press the [Delete] key.

Figure 25-8.
You can customize existing workspaces.

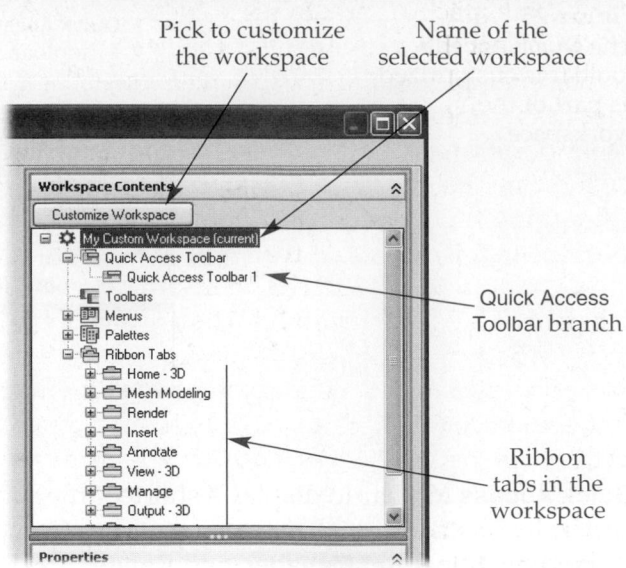

Pick to customize the workspace

Name of the selected workspace

Quick Access Toolbar branch

Ribbon tabs in the workspace

Figure 25-9.
Specifying which toolbars, menus, and ribbon tabs are included in the workspace.

Check mark indicates the component is in the workspace

No check mark indicates it is not in the current workspace

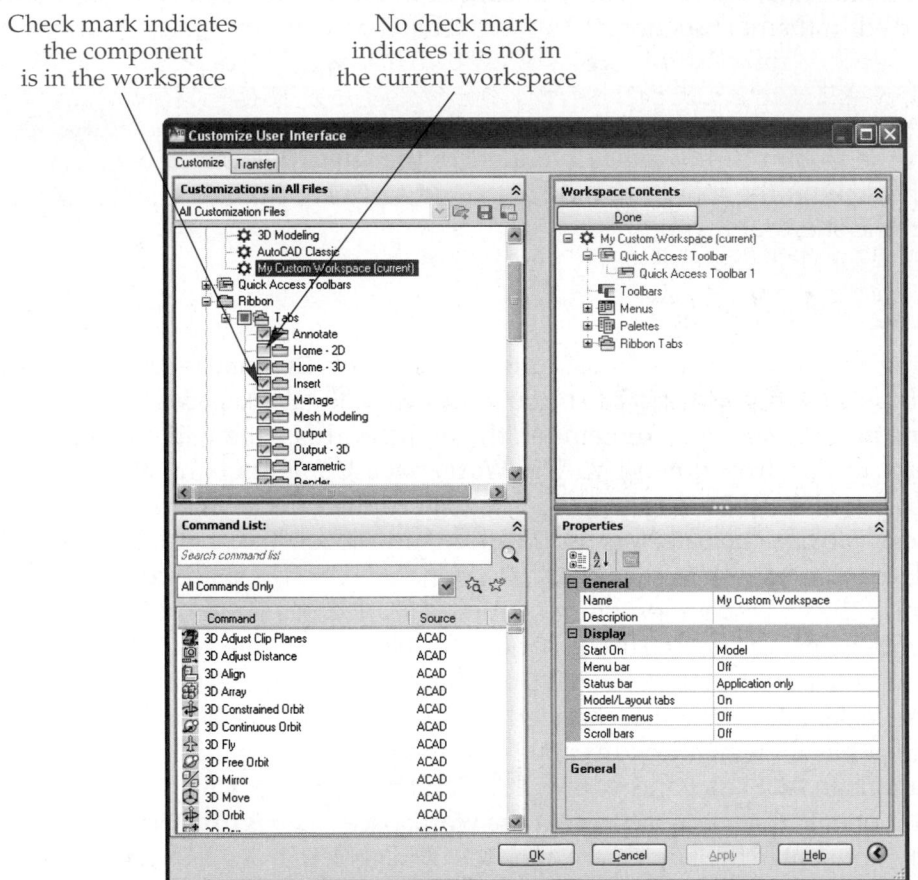

When in customization mode for workspaces, you may choose which **Quick Access** toolbar is displayed in the workspace. However, you cannot be in customization mode to customize a **Quick Access** toolbar.

When done customizing the workspace, pick the **Done** button in the **Workspace Contents** pane. The tree is no longer displayed in blue. Also, all branches are once again displayed in the **Customizations in All Files** pane. You can now expand a branch

AutoCAD and Its Applications—Advanced

in the **Workspace Contents** pane, select a component name, and use the **Properties** pane to adjust the component's properties.

For example, suppose you have added the **Layers** toolbar to a workspace. In the **Properties** pane, you can set the orientation to floating, specify the location of its anchor point, and set the number of rows for the toolbar. You can also change the order in which menus appear in the application menu (menu browser) and on the menu bar by dragging them to a new location in the tree in the **Workspace Contents** pane. The top of the tree is the top of the application menu or the left-hand side of the menu bar.

You may have noticed that there is not a Palettes branch in the **Customizations in All Files** pane. All palettes are automatically available in all workspaces. You can, however, specify whether a palette is displayed or hidden in a workspace. You can also change other properties of a palette, such as floating/docked status, its size, and whether or not the autohide feature is enabled. To change the properties of a palette, first select it in the Palettes branch in the **Workspace Contents** pane. Then, in the **Properties** pane, adjust the properties as needed. When you pick **OK** to close the **Customize User Interface** dialog box, the new default properties of the palette are set for that workspace.

For example, the **Tool Palette** window in the 3D Modeling workspace is, by default, docked on the right-hand side of the screen. You can hide the **Tool Palette** window when in the drawing editor by simply picking the **Close** button (**X**). You can also drag the palette into the drawing area to float it. However, these actions do not change the default setting for the workspace. If you restore the workspace in the future, the **Tool Palette** window will again be docked on the right-hand side of the screen. You must alter the default settings for the **Tool Palette** window in the workspace. First, select Tool Palette in the Palettes branch in the **Workspace Contents** pane. Then, in the **Properties** pane, change the Show property to No. See **Figure 25-10**. Now, when you restore the

Figure 25-10.
Changing the default properties of a palette for a given workspace.

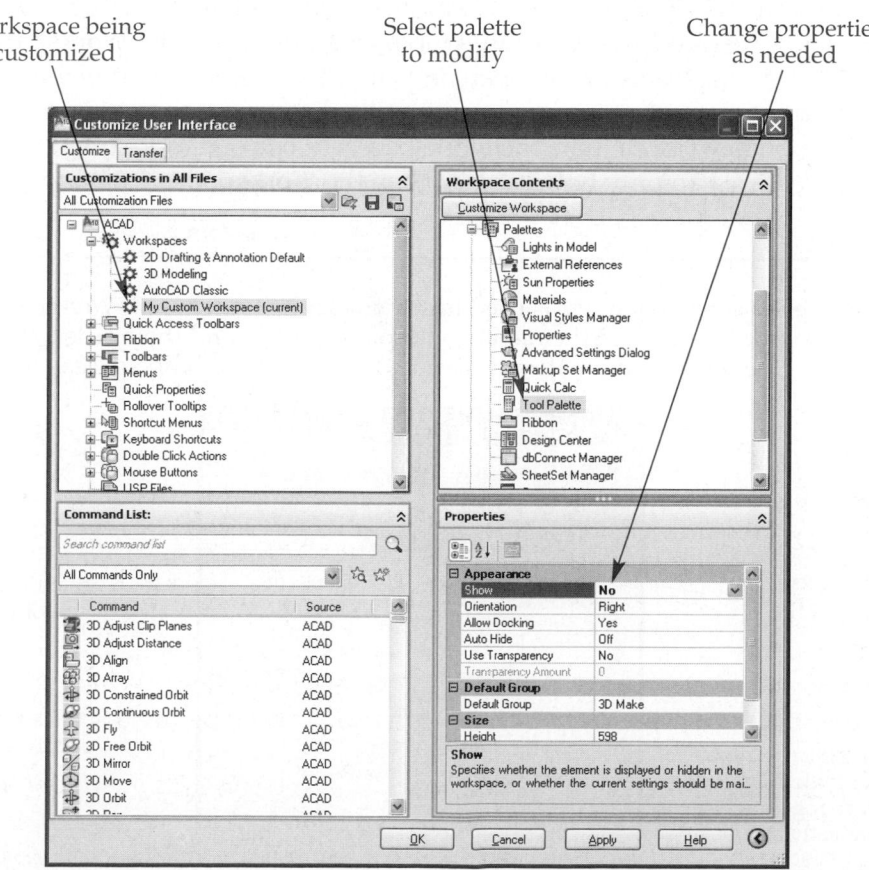

Workspace being customized

Select palette to modify

Change properties as needed

workspace, the **Tool Palette** window will not be displayed. It can, of course, be manually displayed as needed.

You can also set up a workspace so that it displays model space or layout (paper) space when restored. By default, a workspace displays model space when it is set as current. To change this, highlight the workspace name in either the **Customizations in All Files** pane or the **Workspace Contents** pane. Then, in the **Properties** pane, change the Start On property to Layout or Do not change. If Model is specified for the Start On property, model space is displayed when the workspace is restored. If Layout is specified for the Start On property, the most recently active layout tab is displayed when the workspace is restored. If Do not change is specified for the Start On property, the current tab remains active when the workspace is restored.

PROFESSIONAL TIP

If the **Customize User Interface** dialog box is open, but you are not in workspace customization mode, you can also remove a component from the workspace by right-clicking on it in the tree in the **Workspace Contents** pane and selecting **Remove from Workspace** in the shortcut menu. This is true for all component types except palettes, as all palettes are always available to all workspaces.

Workspace Settings

There are various settings related to workspaces. These are set in the **Workspace Settings** dialog box. See **Figure 25-11**. The **WSSETTINGS** command opens this dialog box. It can also be displayed by selecting **Workspace Settings...** from the shortcut menu displayed by picking the **Workspace Switching** button on the status bar.

At the top of the **Workspace Settings** dialog box is the **My Workspace =** drop-down list. All saved workspaces in the CUIx file appear in this list. The workspace that is selected in the list is defined as My Workspace. The workspace that is designated as My Workspace is restored when the **My Workspace** button on the **Workspaces** toolbar is picked. This can be useful if one person primarily uses a machine, but others may temporarily use the machine with their own workspace settings. You may also find this useful if you have more than one workspace, but use one more often than all of the others. This feature can only be accessed when the **Workspaces** toolbar is displayed.

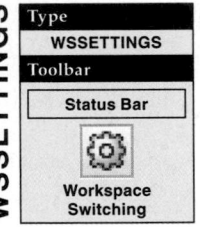

WSSETTINGS

Type
WSSETTINGS
Toolbar
Status Bar
⚙
Workspace Switching

Figure 25-11.
The **Workspace Settings** dialog box is used to set which workspace is My Workspace, specify which workspaces are shown in the menu and drop-down list and their order, and set whether or not changes are automatically saved when a different workspace is restored.

The **Menu Display and Order** area of the **Workspace Settings** dialog box contains a list of all workspaces saved in the current CUIx file. The order of this list determines the order of the list that appears in the **Workspace Switching** shortcut menu and in the drop-down list on the **Workspaces** toolbar. The order of the list can be modified by highlighting one of the workspaces and using the **Move Up** and **Move Down** buttons. The **Add Separator** button is used to add a horizontal line, or menu separator, to the list. A separator is used to logically group workspace names within the list. The separator can be relocated within the list just like a workspace name.

You can prevent a workspace name or separator from being displayed in the menu or drop-down list by removing the check box next to its name. The check box next to the current workspace and the workspace designated as My Workspace can be cleared. However, these workspaces will always be displayed in the menu and drop-down list.

At the bottom of the **Workspace Settings** dialog box is the **When Switching Workspaces** area. The radio buttons in this area determine whether or not changes you have made since you last saved the workspace, such as the visibility of toolbars, are saved when you switch to a different workspace. To retain the settings as you last saved them, pick the **Do not save changes to workspace** radio button. Changes made since the workspace was last saved are discarded when a different workspace is restored. If the **Automatically save workspace changes** radio button is on, any "as you work" changes are automatically saved to the workspace when a different workspace is restored.

At any time, you can manually save the settings to the current workspace. Pick the Save Current As… selection in the **Workspace Switching** shortcut menu. When the **Save Workspace** dialog box appears, select the current workspace from the drop-down list. Then, pick the **Save** button. An alert is displayed stating that a workspace with that name already exists and asking if you would like to replace it. Pick the **Yes** button to save the changes to the current workspace.

PROFESSIONAL TIP

Unlike a user profile, changes to the environment (toolbar display, menu configuration, etc.) are not necessarily automatically saved. To ensure the changes are only saved when you decide to save them, be sure the **Do not save workspace changes** radio button is on in the **Workspace Settings** dialog box.

Exercise 25-3

Complete the exercise on the student website.
www.g-wlearning.com/CAD

Controlling the Display of Ribbon Tabs and Panels

Right-click anywhere on the ribbon to display the shortcut menu. Notice the **Tabs** and **Panels** cascading menus. See **Figure 25-12.** The default ribbon tabs in the 3D Modeling workspace are **Home, Mesh Modeling, Render, Insert, Annotate, View, Manage, Output,** and **Express Tools** (if installed). The **Panels** cascading menu in the shortcut menu displays the panels of the current tab. For example, the default panels of the **Home** tab in the 3D Modeling workspace are **Modeling, Mesh, Solid Editing, Draw, Modify, Section, View, Subobject, Layers, Properties, Utilities,** and **Clipboard.** In the **Tabs** or **Panels** cascading menu, the currently displayed tabs and panels have a check mark next to their name. Selecting a name toggles the visibility of the tab or panel. For example, notice that for the **Home** tab the **Properties**, **Utilities**, and **Clipboard** panels are not displayed by default and their names are unchecked in the shortcut menu, as shown in **Figure 25-12B.**

While the visibility of the ribbon tabs and panels can be adjusted "on the fly" as described above, visibility can also be controlled by customizing the workspace. This is a more efficient method.

1. Open the **Customize User Interface** dialog box and expand the Workspaces branch in the **Customizations in All Files** pane.

2. Select the workspace for which you wish to adjust panel visibility.

3. In the **Workspace Contents** pane, pick the **Customize Workspace** button.

4. Expand the Ribbon Tabs branch in the **Workspace Contents** pane. The tabs visible in the workspace appear in the tree. The panels visible in the workspace appear as branches below the tabs.

5. In the **Customizations in All Files** pane, expand the Ribbon branch and then the Tabs branch. All available tabs appear in the branch. The currently displayed tabs have a check mark next to their name. See **Figure 25-13.**

6. Check the tabs to display and uncheck the tabs to hide.

7. To control which panels are displayed for a tab, expand the tab's branch in the **Workspace Contents** pane. Then, select the panel to hide and, in the **Properties** pane, change its Show property to No. To display a panel, change the property to Yes.

8. In the **Workspace Contents** pane, pick the **Done** button.

9. Exit the **Customize User Interface** dialog box.

Tabs and panels appear in the ribbon in the order they are displayed in the tree of the **Workspace Contents** pane. Top to bottom equals left to right when the ribbon is horizontal. As you add panels to the workspace, they are added to the bottom of the Ribbon Tabs branch. Refer to Chapter 22 for information on creating custom ribbon tabs and panels. To rearrange the tabs or panels, pick and drag them within the **Workspace Contents** pane. You do not need to be in customization mode. You can also remove a tab from the workspace by right-clicking on its name in the **Workspace Contents** pane and picking **Remove from Workspace** from the shortcut menu.

Figure 25-12.
A shortcut menu is displayed when you right-click on the ribbon. A—The **Tabs** cascading menu. B—The **Panels** cascading menu. This is for the **Home** tab. Notice there are three panels hidden by default.

A

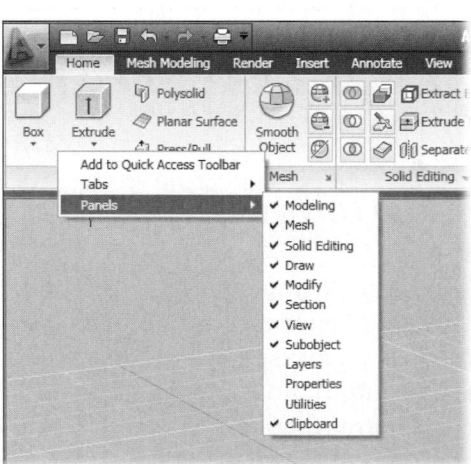

B

AutoCAD and Its Applications—Advanced

Figure 25-13.
Setting which ribbon tabs and panels are displayed for a workspace.

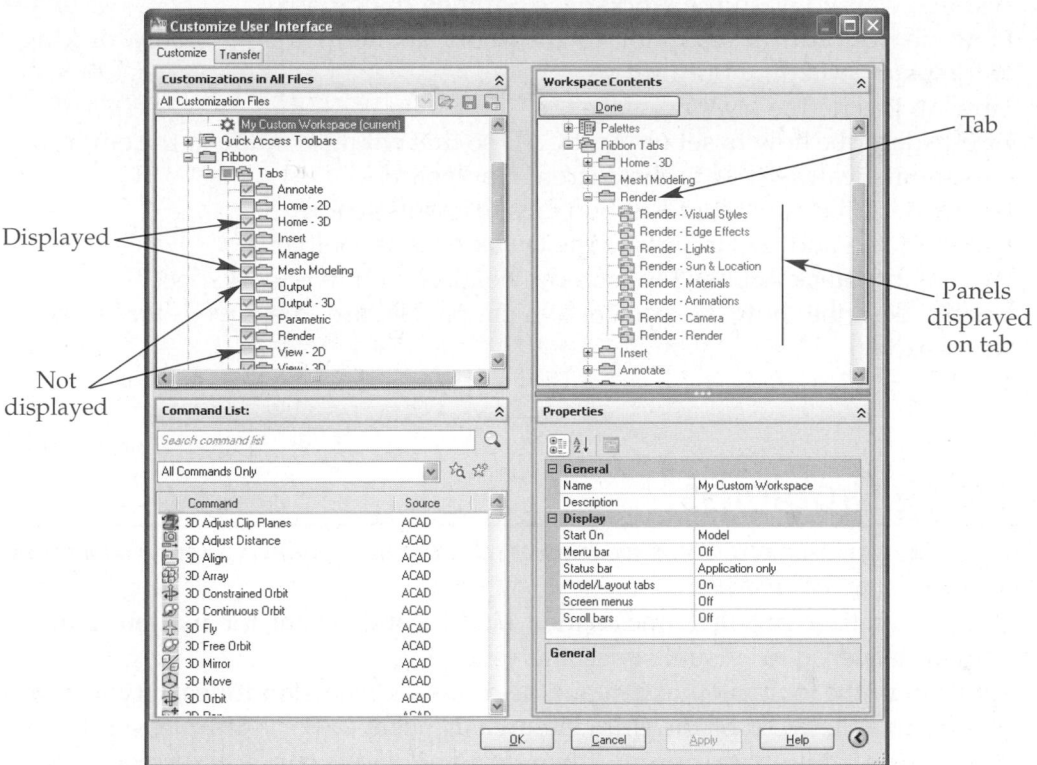

Displayed

Not displayed

Tab

Panels displayed on tab

PROFESSIONAL TIP

You do not need to be in customization mode to change a panel's Show property. Also, when a panel's Show property is set to No, the panel is still available in the **Panels** cascading menu in the shortcut menu displayed by right-clicking on the ribbon.

Chapter Test

Answer the following questions. Write your answers on a separate sheet of paper or complete the electronic chapter test on the student website.
www.g-wlearning.com/CAD

1. What is a *user profile?*
2. How and why are profiles used?
3. What is the file extension used for a profile when it is exported?
4. In which dialog box is a user profile created?
5. How do you restore a user profile?
6. Why would you export a user profile?
7. Briefly describe how to import a user profile.
8. Define *workspace* as related to AutoCAD.
9. List three ways to open the **Save Workspace** dialog box.
10. How do you restore a workspace? List three methods.
11. When customizing a workspace, how do you determine which menus, ribbon tabs, and toolbars are displayed?

12. Briefly describe how to change the default settings for a palette for a given workspace.
13. List two ways to open the **Workspace Settings** dialog box.
14. How do you add a separator to the shortcut menu displayed by picking the **Workspace Switching** button?
15. How do you define My Workspace?
16. Briefly describe how to set up AutoCAD so that changes made to the environment are automatically saved to the current workspace.
17. Briefly describe how to add ribbon tabs to a workspace.
18. How can you add commands to the **Quick Access** toolbar?
19. How many **Quick Access** toolbars can be displayed in a workspace?
20. Briefly describe how to control which panels are displayed for a tab in a workspace.

Drawing Problems

Before creating any user profiles or workspaces, check with your instructor or supervisor for specific instructions or guidelines.

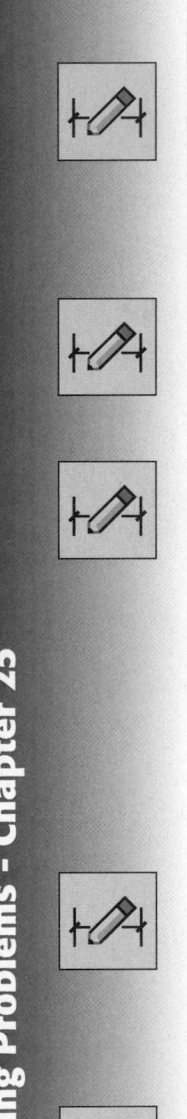

1. Create two user profiles, one named Model Development for modeling and one named Rendering for visual styles and rendering.
   A. Display the toolbars that contain the commands needed for each type of work.
   B. Change the color of the drawing area as needed. For example, some drafters prefer a white background when drawing. However, a black background is often desired when shading and rendering the model.

2. Export the user profiles created in problem 1 to ARG files. Then, delete each profile from AutoCAD. Restart AutoCAD and verify that the profiles are no longer available. Next, import each profile from file. Restore each profile to verify the settings.

3. Create two workspaces, one named Design Development and one named Dimensioning.
   A. For the Design Development workspace, display the ribbon tabs and panels related to drawing and editing. Also, rearrange the menus to group the drawing and editing/modifying menus together. Remove any menus that are not needed.
   B. For the Dimensioning workspace, display the ribbon tabs and panels related to dimensioning the drawing. You may consider removing editing/modifying menus.
   C. Create custom ribbon tabs and panels as needed. Include flyouts when advantageous.

4. Customize the two workspaces created in problem 3.
   A. Change default on/off status of any palettes to suit the purpose of the workspace. For example, you may want the **DesignCenter**, **Tool Palette**, and **Properties** palettes displayed for the Design Development workspace.
   B. Enable the autohide status of all displayed palettes.
   C. Change the Dimensioning workspace so that it starts in a layout.

5. Set up the workspaces from problem 4 so that "on the fly" changes are automatically saved when a different workspace is made current.

CHAPTER 26

Recording and Using Actions

Learning Objectives

After completing this chapter, you will be able to:
✓ Describe the action recorder.
✓ Create and store an action.
✓ Run an action.
✓ Modify an action.
✓ Explain the limitations of the action recorder.

In schools and workplaces, certain design elements and customizing features can be enhanced by automation. The AutoCAD *action recorder* is a tool that allows the user to automate a task or function by saving the steps needed to complete it. These steps are called an *action*. Creating a library of recorded actions allows a user to increase productivity by more efficiently working in AutoCAD.

Action Recorder Overview

The **Action Recorder** panel is located in the **Manage** tab of the ribbon. It contains seven tools: **Record/Stop**, **Insert Message**, **Insert Base Point**, **Pause for User Input**, **Play**, **Preference**, and **Manage Action Macros**. In addition, a drop-down list displays all recorded actions. See **Figure 26-1.** The expanded **Action Recorder** panel displays the action tree. The *action tree* shows the steps used to create the current action that is selected in the drop-down list. The steps are listed in order from top to bottom. The first step is located at the top of the tree.

The action recorder saves all steps in an action to an ACTM file. By saving the action to a file, it can repeatedly be used. The ACTM file name cannot contain spaces or special characters. Also, the name cannot be the same as an existing AutoCAD command name.

The folder where actions are stored is defined in the **Files** tab of the **Options** dialog box. To change the location where actions are stored, expand the Action Recorder Setting branch in the **Files** tab, **Figure 26-2.** Then, pick the Actions Recording File Locations branch, pick the **Browse...** button, and locate the new folder.

Figure 26-1.
The **Action Recorder** panel on the **Manage** tab of the ribbon with the action tree displayed.

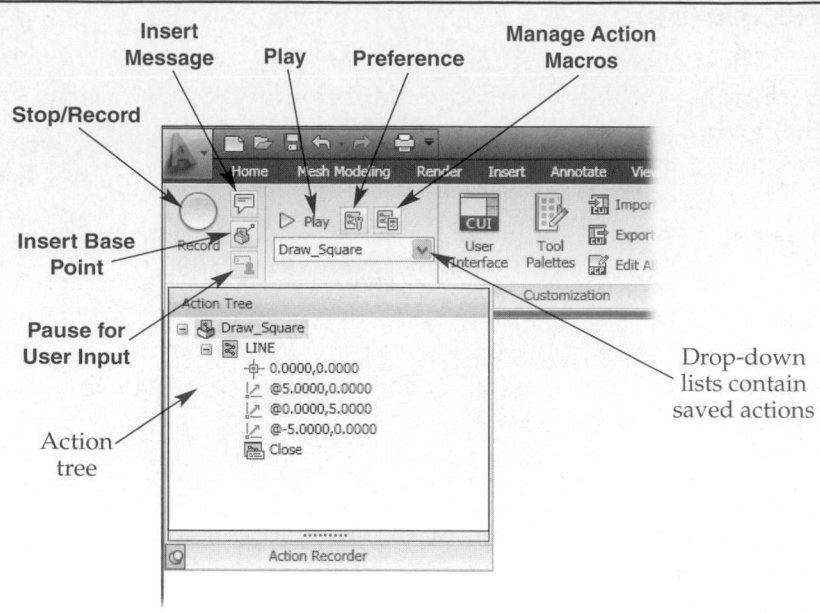

Insert Message

Play

Preference

Manage Action Macros

Stop/Record

Insert Base Point

Pause for User Input

Action tree

Drop-down lists contain saved actions

Figure 26-2.
The location where actions are stored is set in the **Files** tab of the **Options** dialog box.

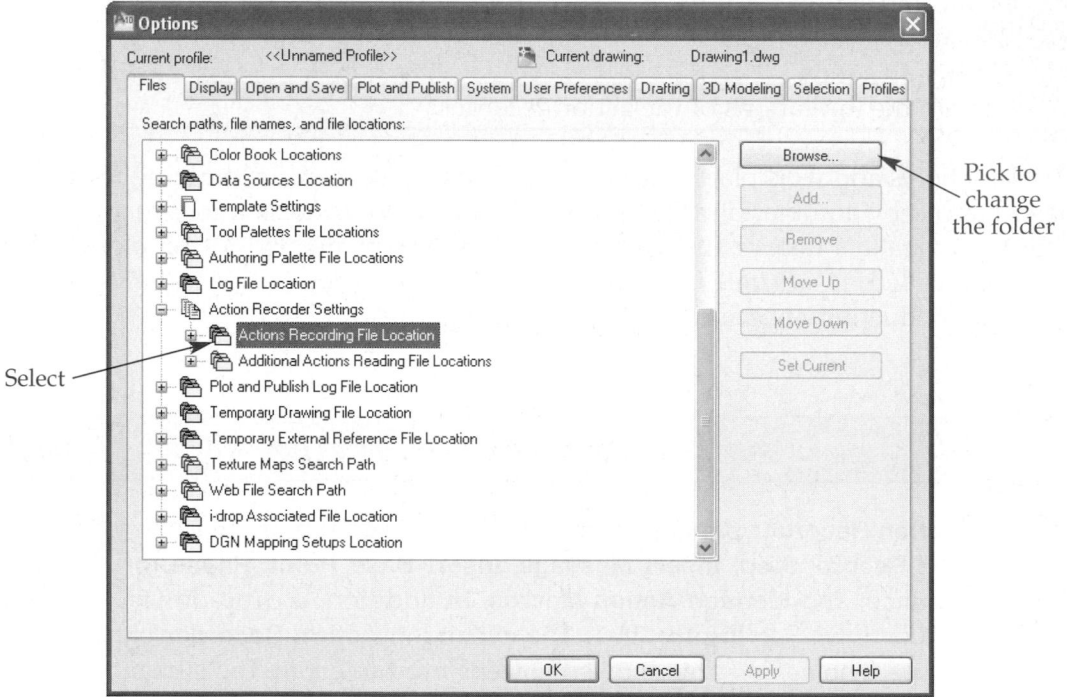

Select

Pick to change the folder

Record/Stop

The **Record** tool launches AutoCAD into record mode. Then, you can begin performing the steps needed to define your action. Once the **Record** tool is activated, the button changes to the **Stop** button. This is used when the steps are completed and you have finished recording your action.

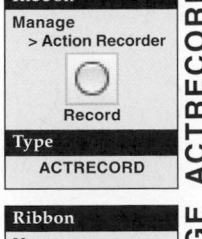

Insert Message

When in record mode or while editing a macro, the **Insert Message** tool is enabled. This tool allows you to insert a message dialog box that will be displayed for the user during action playback. A message is especially useful for directing the user's attention to what is required. One can also be used at the beginning of the action to introduce the user to the action and how it will perform.

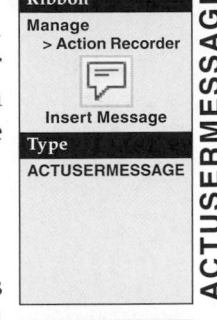

Insert Base Point

When in record mode or while editing a macro, the **Insert Base Point** tool is enabled. This tool allows you to choose a starting location for a particular command or series of commands. This is used to override the default absolute coordinates that are automatically entered while creating an object. It can be used during recording, but only when applicable. The tool is disabled when not available. Additionally, when editing a macro, you may add this option in front of a particular command.

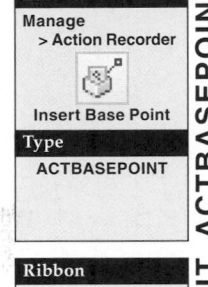

Pause for User Input

When in record mode or while editing a macro, the **Pause for User Input** tool is enabled. Inserting a request for user input temporarily halts the action at that point during playback. The user is prompted for input, such as selecting objects or picking locations on the screen.

Play

The **Play** tool plays the current macro. The current macro is displayed at the top of the drop-down list in the **Action Recorder** panel. This drop-down list is discussed later.

Preferences

Picking the **Preference** tool displays the **Action Recorder Preferences** dialog box, **Figure 26-3.** This dialog box contains three check boxes for setting how the **Action Recorder** panel functions. When the **Expand on playback** check box is checked, **Action Recorder** panel is automatically expanded to show the action tree during action playback. This is unchecked by default. When the **Expand on recording** check box is checked, the **Action Recorder** panel automatically expands to display the action tree during action recording. This is checked by default. When the **Prompt for action macro name** check box is checked, the **Action Macro** dialog box is displayed after recording of the action is complete (the **Stop** button is picked). This dialog box prompts for the action name, which is the name of the ACTM file. See **Figure 26-4.**

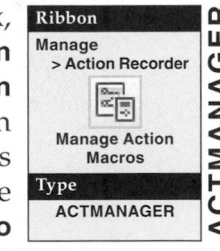

Figure 26-3.
The **Action Recorder Preferences** dialog box is used to set certain behavior of the action recorder.

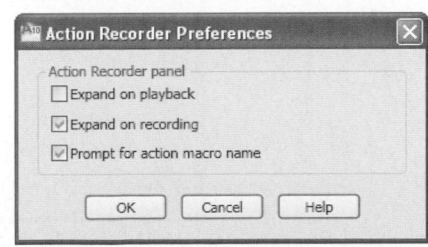

Figure 26-4.
The **Action Macro** dialog box is used to save an action.

Name the action

File name matches the action name

Enter a description for the action

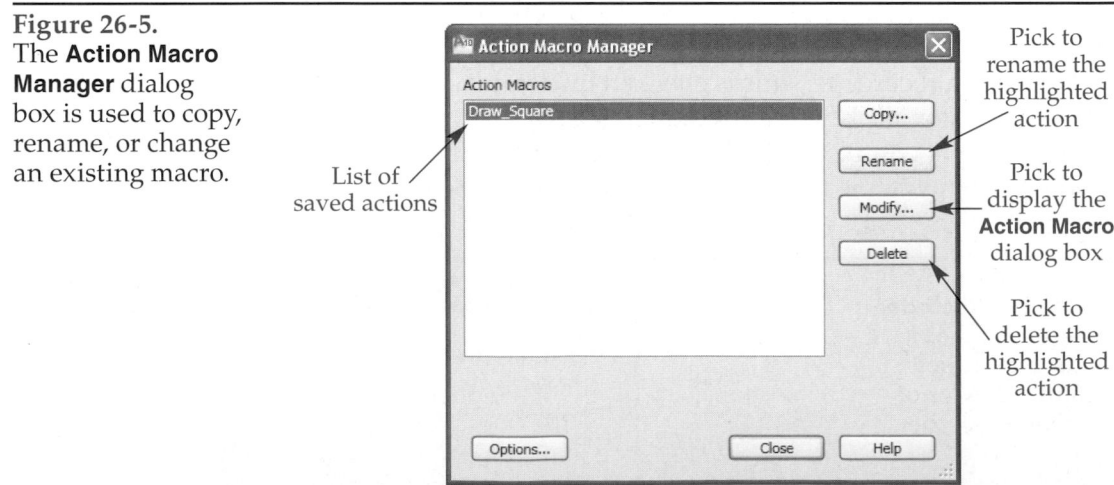

Action Macro

Action Macro Command Name:
DrawTriangle

File Name:
DrawTriangle.actm

Folder Path:
C:\Documents and Settings\eaugspurger\application data\autodesk\autocad :

Description:
This action drawn an equilateral traqingle with legs of one unit and the lower-left corner on 0,0.

Restore pre-playback view
☑ When pausing for user input
☐ Once playback finishes

☑ Check for inconsistencies when playback begins

OK   Cancel   Help

Manage Action Macros

The **Action Macro Manager** dialog box allows you to rename saved actions or change their descriptions. Picking the **Manage Action Macros** button in the **Action Recorder** panel displays this dialog box. See **Figure 26-5.** The dialog box contains a list of actions saved in the action macro folder specified in the **Options** dialog box.

Picking the **Copy** button displays the **Action Macro** dialog box, which is also displayed by default when recording of an action is complete. Enter a name and description for the new action and set the other properties. When you pick the **OK** button, the steps in the action highlighted in the **Manage Action Macros** dialog box are saved under the new name. The original action is not altered.

Picking the **Rename** button replaces the name of the highlighted macro with an edit box. This allows you to enter a new name for the action. Renaming the action also changes the name of the ACTM file.

Picking the **Modify** button also displays the **Action Macro** dialog box. However, when you make changes to the name, description, or other properties, the action highlighted in the **Manage Action Macros** dialog box is altered.

Figure 26-5.
The **Action Macro Manager** dialog box is used to copy, rename, or change an existing macro.

List of saved actions

Action Macro Manager

Action Macros
Draw_Square

Copy...
Rename
Modify...
Delete

Options...   Close   Help

Pick to rename the highlighted action

Pick to display the **Action Macro** dialog box

Pick to delete the highlighted action

Picking the **Delete** button deletes the action and associated ACTM file. Picking the **Options...** button displays the **Options** dialog box with the Action Recorder Settings branch highlighted in the **Files** tab.

Action Tree and Available Actions

The **Available Action Macro** drop-down list displays all action files (ACTM) located in the action folder. As explained earlier, the action folder is set in the **Files** tab of the **Options** dialog box. The current action is displayed at the top of the list. To see all of the available actions, pick the drop-down list. To set a different action current, simply pick it in the drop-down list.

At the bottom of the expanded **Action Recorder** panel is the action tree. The action tree contains all of the steps in the action in graphic form. See **Figure 26-6.** The name of the action is the top of the tree. Each step is a branch in the tree. In addition, the command branches can be expanded to see the values entered into the action.

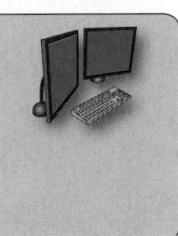

PROFESSIONAL TIP

Inserting messages is especially useful to direct the user about the steps of an action. Inserting requests for user input in an action pauses the action and allows the user to select objects or on-screen points. These features make actions more dynamic and help ensure proper usage.

Exercise 26-1

Complete the exercise on the student website.
www.g-wlearning.com/CAD

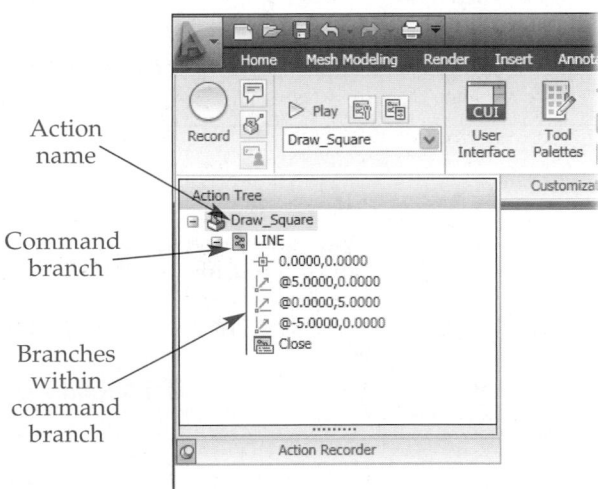

Figure 26-6.
The action tree contains all of the steps in the current action.

Action name

Command branch

Branches within command branch

Creating and Storing an Action

Now that you have investigated the features of the action recorder, it is time to create an action. This example action will create a layer named A-Anno-Iden, set the layer color to orange (ACI = 30), and make the layer current. The action will also set the annotation scale to 1/4″ = 1′-0″. In this section, you will create the basic action. Later, you will modify the action to make it more interactive. Follow these steps to create the basic action named AnnoQtr.

1. Start a new drawing.
2. Pick the **Record** button in the **Action Recorder** panel on the **Manage** tab of the ribbon. Notice that the panel is shaded in red and a red dot appears next to the crosshairs. These are indications that you are in record mode.
3. Display the **Layer Properties Manager** palette.
4. Pick the **New Layer** button in the palette's toolbar.
5. In the edit box, name the new layer A-Anno-Iden.
6. Pick the color swatch for the layer and, in the **Select Color** dialog box, pick color 30 on the **Index Color** tab. Then, pick the **OK** button to close the **Select Color** dialog box.
7. Right-click on the newly created layer and pick **Set Current** from the shortcut menu.
8. Close the **Layer Properties Manager** palette.
9. Using the **Annotation Scale** button on the status bar, set the 1/4″ = 1′-0″ scale current. See **Figure 26-7.** If the drawing is based on the acad3D.dwt template, you must type the **CANNOSCALE** command since the status bar button is unavailable.
10. Pick the **Stop** button on the **Action Recorder** panel in the **Tools** tab of the ribbon. This ends recording of the action.
11. The **Action Macro** dialog box appears when you pick the **Stop** button, unless you have turned off the "prompt for name" option.
12. In the **Action Macro** dialog box, type AnnoQtr in the **Action Macro Command Name:** text box.
13. In the **Description:** text box, type Creates a new layer with color 30 and sets the 1/4″ = 1′-0″ annotation scale current or a similar description.
14. Pick the **OK** button in the **Action Macro** dialog box to save the action.

Figure 26-7.
Setting the annotation scale for the action being recorded.

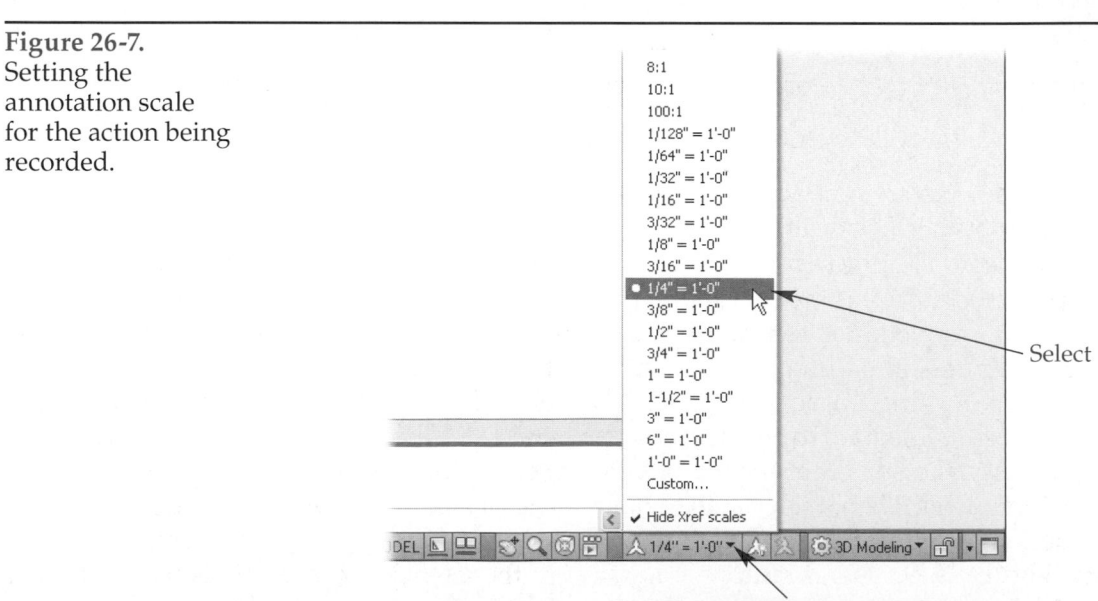

Pick the **Annotation Scale** button

Figure 26-8.
The new action is automatically set current. Its action tree is displayed in the expanded panel.

New action is set current

Action tree for new action

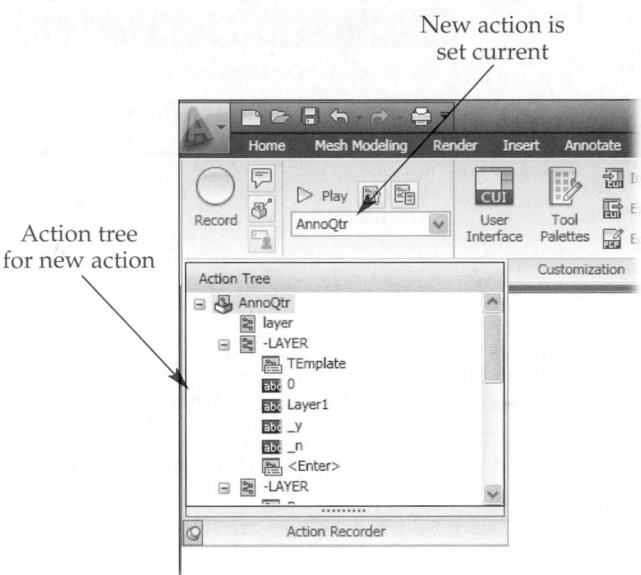

The AnnoQtr action is now available to run. It is automatically selected as the current action in the **Available Action Macro** drop-down list in the **Action Recorder** panel. Also, the action tree for the action is displayed in the expanded panel. See **Figure 26-8.**

PROFESSIONAL TIP

In a design environment, saving actions in a shared folder on a network can streamline productivity. This creates an environment in which AutoCAD users can share their knowledge by creating actions drawn from their own design experiences.

Exercise 26-2

Complete the exercise on the student website.
www.g-wlearning.com/CAD

Running an Action

To run the current action, pick the **Play** button in the **Action Recorder** panel on the **Manage** tab in the ribbon. The current action is shown at the top of the **Available Action Macro** drop-down list in the **Action Recorder** panel. To run a different action, simply select it in the drop-down list and then pick the **Play** button. Another method of running an action is to type its name on the command line. This is why the action name cannot be the same as an existing AutoCAD command.

Before testing an action, be sure to delete all graphics from the screen unless needed for the action. To test the action recorded in the previous section, start a new drawing. Make sure the AnnoQtr action is set current in the **Available Action Macro** drop-down list. Then, pick the **Play** button and watch the action run. By default, when the action is complete, AutoCAD displays the **Action Macro—Playback Complete** dialog box, **Figure 26-9.** This is simply a message to the user that the action is finished. Pick the **OK** button to close the dialog box.

Figure 26-9.
This dialog box is displayed when an action is complete. If you check the **Do not show this message again** check box, this dialog box will not be displayed at the end of action playback.

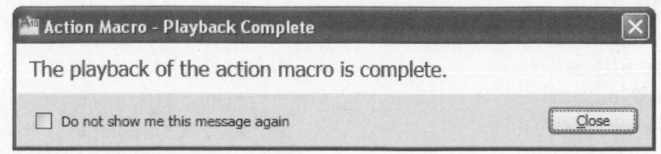

Actions run just like regular AutoCAD commands. Since they can be executed on the command line, actions can be made into custom AutoCAD commands. They can be placed in a customized ribbon panel, toolbar, menu, or palette. When creating a custom command from an action, use the syntax ^C^C_action_name. Refer to Chapter 22 for information on creating custom commands.

PROFESSIONAL TIP

Actions can streamline the customization process by creating setup procedures to minimize macro writing when creating custom commands. For example, suppose you have an annotation block that you want inserted at a set scale and on a specific layer. You can create an action similar to the example above (AnnoQtr). Then, you can insert the action in the command string prior to the block insertion. This will streamline your customization process.

Modifying an Action

The action tree can be used to edit an action. Examine the action tree for the AnnoQtr example, as shown in **Figure 26-10.** At the top of the action tree is the action name. It is always the top branch of the tree. The icon to the left of the name indicates that it is an action. The branches below the action name contain the commands and steps in the action.

Notice the dash in front of the **LAYER** command. This dash indicates the command is a command-line command. Even though you used the **Layer Properties Manager** palette when recording the action, the action recorder translated your picks into the command-line version of the steps.

Branches below the command names contain the specific information used to create each item. Each branch is automatically converted to the command-line entry. The icon next to each branch indicates the type of action or input represented by the branch. These branches are also called *nodes*. For example, reviewing the **LAYER** command branches, you can see a text icon next to the name A-Anno-Iden. This icon indicates that the branch, or node, is a text entry. Additionally, the icon for the branch where the layer color is set to 30 is a color wheel. This indicates the branch is a change in color value.

An icon with a head-and-shoulders image in the lower-right corner indicates the action will pause at this point to allow the user to enter information. This icon is called a user icon. The AnnoQtr action does not currently contain any user icons.

Now, you will modify the action recorded earlier. You will change some of the steps to require user input. For example, you will have the user type the annotation scale instead of automatically setting it to 1/4″ = 1′-0″ scale. So the user knows what is

AutoCAD and Its Applications—Advanced

Figure 26-10.
The action tree for the recorded action prior to editing the action.

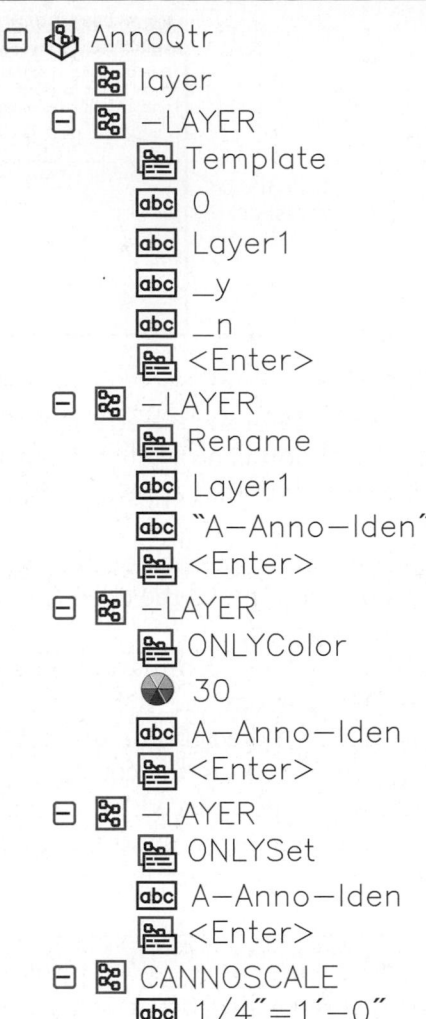

required, you will add a message explaining the step. Since the scale is not automatically set, the action also needs to be renamed to remove the Qtr (quarter inch scale) designation. Complete the following steps.

1. Expand the **Action Recorder** panel to see the action tree. You may want to pin the panel in the expanded state.
2. Right-click on the action name AnnoQtr and select **Rename...** from the shortcut menu. To retain the original action instead of renaming it, select **Copy...** from the shortcut menu. The **Action Macro** dialog box is displayed.
3. In **Action Macro Command Name:** text box, type AnnoUserScale and then pick the **OK** button to rename the action. Note: the edits in steps 2 and 3 may also be accomplished using the **Action Macro Manager** dialog box.
4. In the action tree, locate the CANNOSCALE branch. Right-click on the 1/4″ = 1′-0″ branch below it and select **Pause for User Input** from the shortcut menu. The value in the branch (1/4″ = 1′-0″) is grayed out and the icon for the branch now contains the user icon.
5. To add a message describing how to type the correct scale, right-click on the 1/4″ = 1′-0″ branch and select **Insert User Message...** from the shortcut menu. The **Insert User Message** dialog box appears, **Figure 26-11.**
6. In the **Insert User Message** dialog box, type the message:

> Enter the desired annotation scale on the command line. Please use proper syntax. For example, type 1/8″ = 1′-0″ for the 1/8″ Architectural scale. Include spaces on each side of the equals symbol.

Figure 26-11.
Adding a user
message to the
action.

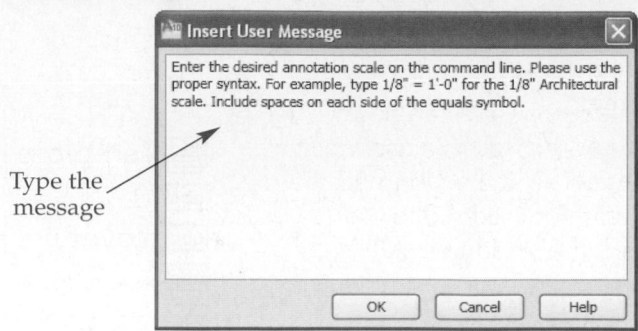

Type the
message

Notice that a User Message branch is added to the action tree above the branch that was right clicked. The action is now altered. Start a new drawing and run the altered action. Notice the dialog box that appears to guide the user through the inputs. See **Figure 26-12.**

To remove user input, right-click on the branch and select **Pause for User Input** from the shortcut menu to remove the check mark. The action then uses the value set when the action was recorded.

If you would like to delete a command from the action, right-click on the command branch and select **Delete** from the shortcut menu. Once deleted, the command is permanently removed from the action. This cannot be undone.

To edit a value, right-click on the corresponding branch and select **Edit** from the shortcut menu. The value is replaced by a text box. Type the new value and press [Enter]. Careful; if you move the cursor off of the panel while the text box is displayed, whatever is displayed in the text box is entered as the value.

NOTE

When recording the action, use the buttons on the **Action Recorder** panel to insert messages, base points, and user-input requests.

Exercise 26-3

Complete the exercise on the student website.
www.g-wlearning.com/CAD

Figure 26-12.
The message
displayed to the user
during playback of
the example action.

Message to
the user

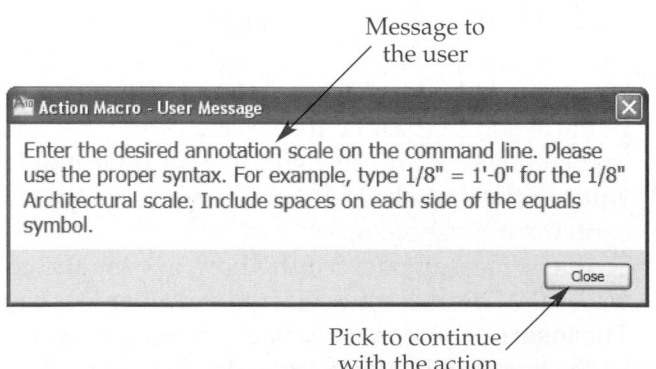

Pick to continue
with the action

This example creates a rectangle and fills it with a hatch pattern. First, you will record an action that draws a rectangle at a user-specified location and then uses the **ZOOM** command to center the rectangle. The action will draw a brick hatch pattern inside of the rectangle. Then, the hatch pattern is edited to change/reset its origin. In this example, commands will be entered on the command line. Although it is more time-consuming to create an action in this way, it allows easy access to, and editing of, the action steps.

1. Start a new drawing.
2. Begin recording a new action.
3. Select the **Insert Base Point** button and choose any location on the screen. This allows the user to place the rectangle anywhere on the screen during future uses of the macro.
4. Type the **RECTANG** command. When prompted to specify the first corner point, pick anywhere on the screen. When prompted for the other corner point, type @120,60.
5. Type the **ZOOM** command and enter the extents option. Enter the command again and type .7x to center the rectangle on the screen.
6. Type the **-HATCH** command. Use the **Select object** option to select the rectangle.
7. Use the **Properties** option and select the hatch AR-B816. This is the architectural brick pattern. Enter a scale of 1.000 and 0° rotation angle. Be sure to enter the values; do not press [Enter] to accept the defaults. Finish the **-HATCH** command.
8. Type the **-HATCHEDIT** command and select the brick pattern.
9. Select the **Origin** option, then select the **Set new origin** option. Pick the lower, right-hand endpoint of the rectangle. When prompted, store this location as the default origin.
10. Stop recording of the action and name it BrickHatchEdit.

Next, edit the action to allow user input and display messages. This will aid the user in selecting points.

11. For the <Select Objects> node of the **-HATCHEDIT** command, change it to pause for user input. This allows the user to pick the hatch to edit.
12. Insert a user message for the **-HATCHEDIT** command just above the <Select Objects> node. The message should read Select the brick hatch object. This gives instruction to the user.
13. For the point location under the **Set** option of the **-HATCHEDIT** command, change it to pause for user input. In this way, the user selects the origin point for where the hatch pattern begins.
14. Insert a user message for the point location under the **Set** option of the **-HATCHEDIT** command. The message should read Edit the hatch and select a new origin point when prompted. This gives instruction to the user.
15. Test the action. When prompted for the new origin, select the midpoint of the lower line of the rectangle. Notice the result is different from the pattern created when recording the action.

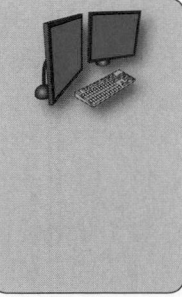

The action recorder works best with command-line versions of commands. Many commands can be run on the command line only by placing a hyphen (-) in front of the command name. For example, the **LAYER** command displays the **Layer Properties Managers** palette. However, the **-LAYER** command displays all options and settings on the command line without displaying a palette or dialog box. Not all commands have a command-line version.

Limitations of the Action Recorder

As a customizing tool, the action recorder is very powerful. However, as in all customization, there is syntax to follow and limitations to address. For example, the AutoCAD commands shown in **Figure 26-13** cannot be recorded in an action.

The following chapters discuss AutoLISP and other programming features of AutoCAD. Actions are able to call and run AutoLISP routines during the recording process. However, the specific AutoLISP routines, Object ARX files, VBA macros, .NET assemblies, and associated files must all be loaded into AutoCAD in order for the programming feature to function during playback of the action. See Chapters 27 through 30 for further details on programming AutoCAD. These chapters are located on the student website.

When creating an action for use in a professional or school environment, you must always consider the source files. For example, if you create an action that reads from a template file or loads particular blocks, those external files must always be either loaded into AutoCAD or accessible to the user.

Finally, you cannot append an existing action. If you need to add steps to the action, you must re-record the existing steps along with the new steps. For this reason, it is very important to plan an action before recording it.

PROFESSIONAL TIP

You can create actions with other Autodesk applications, such as AutoCAD Architecture. If you create an action using AEC objects, for example, the action can only properly function if the action is played back in AutoCAD Architecture.

Figure 26-13.
These commands cannot be used in an action.

ACTSTOP
ACTUSERINPUT
ACTUSERMESSAGE
-ACTUSERMESSAGE
DXFIN
FILEOPEN
NEW
OPEN
PARTIALOPEN
PRESSPULL
QNEW
RECOVER
TABLEDIT
VBAIDE
VBALOAD
-VBALOAD
VBAMAN
VBANEW
VBAPREF
VBARUN
-VBARUN
VBASTMT
VBAUNLOAD
XOPEN

AutoCAD and Its Applications—Advanced

Chapter Test

Answer the following questions. Write your answers on a separate sheet of paper or complete the electronic chapter test on the student website.
www.g-wlearning.com/CAD

1. What is an *action?*
2. What is used to record an action?
3. In which file format is an action saved?
4. By default, where are action files stored?
5. What is the *action tree?*
6. How do you set an action current?
7. Name two ways to play an action.
8. How do you pause an action for user input?
9. How can you display a custom message in an action?
10. Briefly describe how to change the value associated with a branch in an action.

Drawing Problems

1. In this problem, you will create an action that sets up a drawing for architectural work. The action will create layers and set the units. The action should:
   A. Display a message to the user indicating what the action will do.
   B. Create a layer named A-Wall and set its color to ACI 113 (or the color of your choice).
   C. Create a layer named A-Door and set its color to ACI 31 (or the color of your choice).
   D. Create a layer named A-Demo and set its color to ACI 1 (or the color of your choice).
   E. Set the units to architectural with a precision of 1/4".
   F. Save the action as ArchSetup.actm. If the file is saved in a folder other than the default, make sure that folder is added to the AutoCAD search path.

2. In this problem, you will create an action that allows the user to select objects and then places those objects at an elevation of 12 units. The action should:
   A. Display a message to the user indicating what the action will do.
   B. Prompt the user to select objects.
   C. Change the elevation property of all objects to 12 units.
   D. Save the action as 1FtElevChange.actm. If the file is saved in a folder other than the default, make sure that folder is added to the AutoCAD search path.

3. In this problem, you will modify the action created in problem 2. Edit the action to allow the user to enter the new elevation. Be sure to display a message to the user indicating what is being asked of them. This is a good tool for flattening linework or resetting solids. Save the action as ElevChange.actm. If the file is saved in a folder other than the default, make sure that folder is added to the AutoCAD search path.

Drawing Problems - Chapter 26

 *The following chapters are available on the student website. These chapters discuss various aspects of programming AutoCAD, including working with AutoLISP, creating dialog boxes using DCL, and working with Visual Basic for Applications (VBA). For detailed information on programming AutoCAD, refer to* Visual LISP Programming *and* VBA for AutoCAD *from Goodheart-Willcox Publisher.* www.g-wlearning.com/CAD

Programming AutoCAD

Chapter 27
Introduction to AutoLISP

What Is AutoLISP?
Basic AutoLISP Functions
AutoLISP Program Files
Introduction to the Visual LISP Editor
Defining New AutoCAD Commands
Creating Your First AutoLISP Program
Creating Specialized Functions
Basic AutoLISP Review

Chapter 28
Beyond AutoLISP Basics

Providing for Additional User Input
Using the Values of System Variables
Working with Lists
Using Polar Coordinates and Angles
Locating AutoCAD's AutoLISP Files
Sample AutoLISP Programs

Chapter 29
Introduction to Dialog Control Language (DCL)

DCL File Formats
AutoLISP and DCL
DCL Tiles
Associating Functions with Tiles
Working with DCL

Chapter 30
Introduction to Visual Basic for Applications (VBA)

Object-Oriented Programming
Methods, Properties, and Events
Understanding How a VBA Program Works
Using the VBA Manager
Using the Visual Basic Editor
Dealing with Data Types
Using Variables
Allowing for User Interaction
Creating AutoCAD Entities
Editing AutoCAD Entities
Creating a Dialog Box Using VBA
Running a VBA Macro from the Keyboard

Index

The index entries shown in black refer to chapters located in this printed book. The index entries shown in color or page number listings in color refer to chapters that are located on the student website.

 www.g-wlearning.com/CAD

The index entries shown in color refer to chapters that are located on the student website.

Index–Advanced

The index entries shown in color refer to chapters that are located on the student website.

The index entries shown in color refer to chapters that are located on the student website.

The index entries shown in color refer to chapters that are located on the student website.

visual style and material
tools, 608
workspaces, 621–624
CUSTOMIZE command, 595, 610
Customize dialog box, 595,
610–612
Customize User Interface dialog
box, 531–536
adding commands to
interface elements,
533–534
button images, 538–542
creating new commands,
537–544
customizing workspaces,
621–624
deleting commands from
interface elements,
534
display of ribbon tabs and
panels, 626
double-click actions, 580–582
menus, 557–559
moving and copying
commands, 535
partial CUIx files, 536
quick properties and
rollover tooltips,
577, 579
restoring workspaces, 620
ribbon, 545–552
shortcut keys, 566–569
shortcut menus, 570–576
toolbars, 552, 554–555
cutaway geometry, 331–332
CVPORT system variable, 177
cyclic redundancy check, 519
cylinder, 41–42, 262
CYLINDER command, 41–42, 154
cylindrical coordinates, 144–145

D

data types, 696
DEFAULTGIZMO system
variable, 268
default lighting, 414, 416
defining command, 542–544
DELOBJ system variable, 70,
199, 244
depth cueing, 459
Design Web Format (DWF) files.
See DWF files.
DesignCenter, 593–594
adding block and hatch
tools from, 596
details,
constructing, 253–255

removing, 251–252
device drivers, 510
Dialog Control Language
(DCL), 676–686
and AutoLISP, 677–678
associating functions with
tiles, 679–680
file formats, 676–677
tiles, 676, 679–680
working with, 683–686
diffuse color, 384
diffuse map, 399–400
dimensioning, 189–191
creating template drawings,
189–190
placing in proper plane,
190–191
placing leaders and radial
dimensions, 191
DIMLINEAR command, 592
direction grip, 326
display, 110
display resolution and format
settings, 515
DISPSILH system variable, 114,
317–318, 320, 355, 515
distance function, 657, 659
distant light, 411, 421–423
DISTANTLIGHT command, 422
DIVIDE command, 204
docking, 586–587
dominant viewport, 179
double-click actions, 580–583
assigning, 580–581
custom, 582–583
double-click editing, 524–525
draft angle, 221
Drawing Window Colors dialog
box, 512
drop-down lists, adding to
panels, 549–550
DTEXT command, 187, 559
DWF files, 496, 498–499
exporting to, 496, 498–499
plotting, 500, 502
publishing, 499
DWFx files, 496
exporting to, 496, 498–499
plotting, 500, 502
publishing, 499
DWG To PDF Properties dialog
box, 500–501
DXF files, exporting and
importing, 493–494
DXFIN command, 494
DXFOUT command, 493
dynamic displays, 87
dynamic feedback, 38

dynamic input, 36–39
and cylindrical coordinates,
145
dynamic UCS, 40, 153–154, 237

E

edged-surface mesh, 69
edge editing, 303
Edge Effects panel, 113–114
EDGEOVERHANG system
variable, 360
edge settings, 113–114
edge subobject editing, 277–282
coloring edges, 282
deleting edges, 282
moving edges, 277–278
rotating edges, 278–279
scaling edges, 281
EDGESURF command, 64, 67, 69
edit scale list, 527
Edit Scale List dialog box, 527–528
embedded, 691
ERASE command, 266, 282
ETRANSMIT command, 496
events, 690
EXPLODE command, 289
exploding solids, 289
Export 3D DWF dialog box, 496
EXPORT command, 345–346,
492, 495–496
exporting solid model files, 345
Export Profile dialog box, 617–618
Export to DWF/PDF Options
dialog box, 498–499
extended help, 537–538
EXTERNALREFERENCES
command, 484
External References Palette,
484–485, 487
EXTNAMES system variable, 521
EXTRUDE command, 75, 151–152,
183, 197, 199, 201, 206,
275, 285
creating features with,
203–204
extruded, 56
extruding faces, 77
extrusions, 197–201
along paths, 198–199
multiple intersecting, 207

F

face editing, 295–303
coloring, 303
copying, 302–303

The index entries shown in color refer to chapters that are located on the student website.

The index entries shown in color refer to chapters that are located on the student website.

AutoCAD and Its Applications—Advanced

The index entries shown in color refer to chapters that are located on the student website.

Index–Advanced

The index entries shown in color refer to chapters that are located on the student website.

Index-Advanced

The index entries shown in color refer to chapters that are located on the student website.

The index entries shown in color refer to chapters that are located on the student website.

The index entries shown in color refer to chapters that are located on the student website.

The index entries shown in color refer to chapters that are located on the student website.

The index entries shown in color refer to chapters that are located on the student website.

The index entries shown in color refer to chapters that are located on the student website.